Rational (Reciprocal) Function

$f(x) = \dfrac{1}{x}$

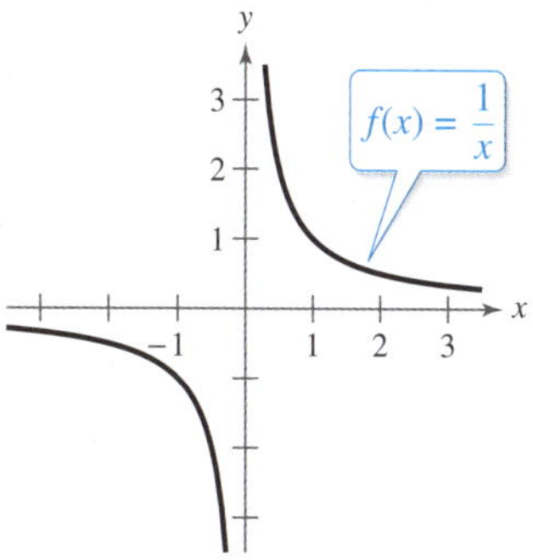

Domain: $(-\infty, 0) \cup (0, \infty)$
Range: $(-\infty, 0) \cup (0, \infty)$
No intercepts
Decreasing on $(-\infty, 0)$ and $(0, \infty)$
Odd function
Origin symmetry
Vertical asymptote: y-axis
Horizontal asymptote: x-axis

Exponential Function

$f(x) = a^x,\ a > 1$

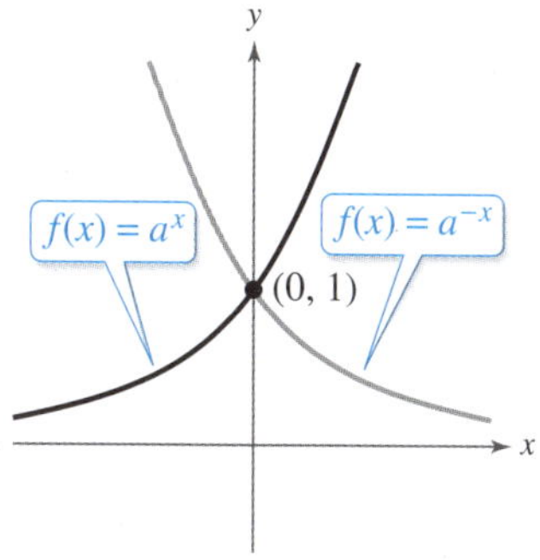

Domain: $(-\infty, \infty)$
Range: $(0, \infty)$
Intercept: $(0, 1)$
Increasing on $(-\infty, \infty)$ for $f(x) = a^x$
Decreasing on $(-\infty, \infty)$ for $f(x) = a^{-x}$
Horizontal asymptote: x-axis
Continuous

Logarithmic Function

$f(x) = \log_a x,\ a > 1$

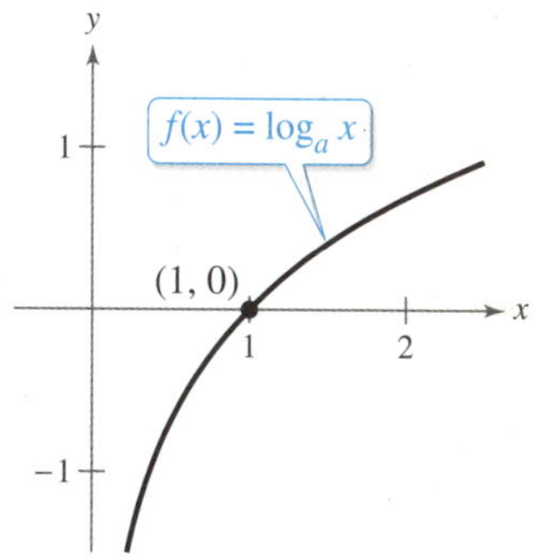

Domain: $(0, \infty)$
Range: $(-\infty, \infty)$
Intercept: $(1, 0)$
Increasing on $(0, \infty)$
Vertical asymptote: y-axis
Continuous
Reflection of graph of $f(x) = a^x$ in the line $y = x$

Sine Function

$f(x) = \sin x$

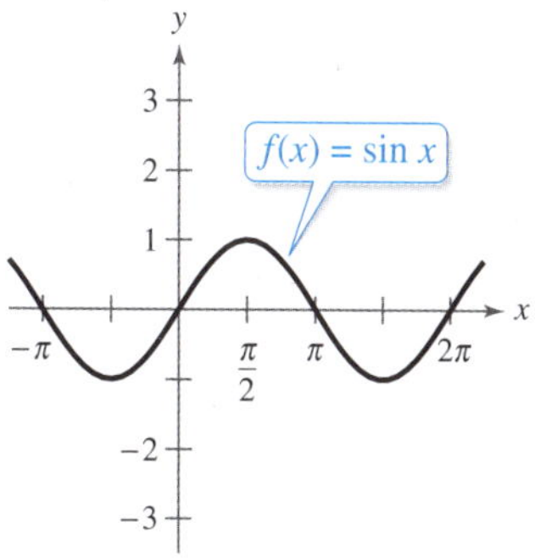

Domain: $(-\infty, \infty)$
Range: $[-1, 1]$
Period: 2π
x-intercepts: $(n\pi, 0)$
y-intercept: $(0, 0)$
Odd function
Origin symmetry

Cosine Function

$f(x) = \cos x$

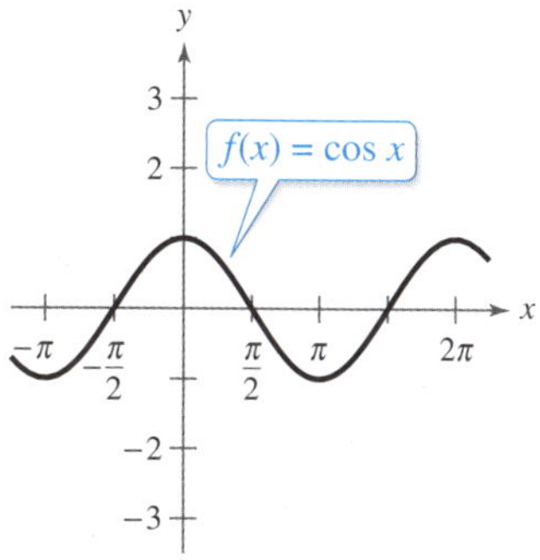

Domain: $(-\infty, \infty)$
Range: $[-1, 1]$
Period: 2π
x-intercepts: $\left(\dfrac{\pi}{2} + n\pi, 0\right)$
y-intercept: $(0, 1)$
Even function
y-axis symmetry

Tangent Function

$f(x) = \tan x$

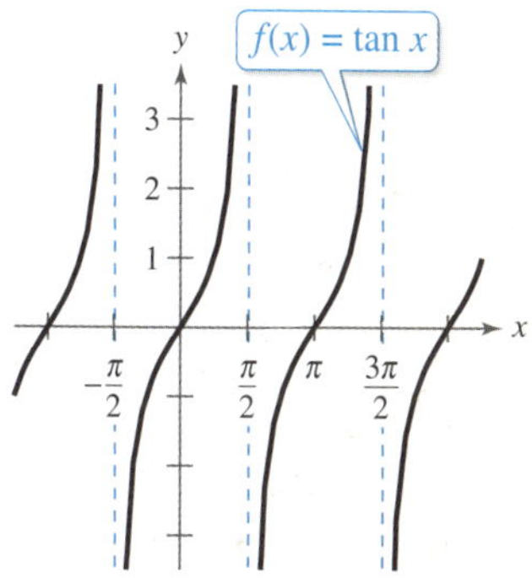

Domain: all $x \neq \dfrac{\pi}{2} + n\pi$
Range: $(-\infty, \infty)$
Period: π
x-intercepts: $(n\pi, 0)$
y-intercept: $(0, 0)$
Vertical asymptotes:

$$x = \frac{\pi}{2} + n\pi$$

Odd function
Origin symmetry

AF727414

Cosecant Function

$f(x) = \csc x$

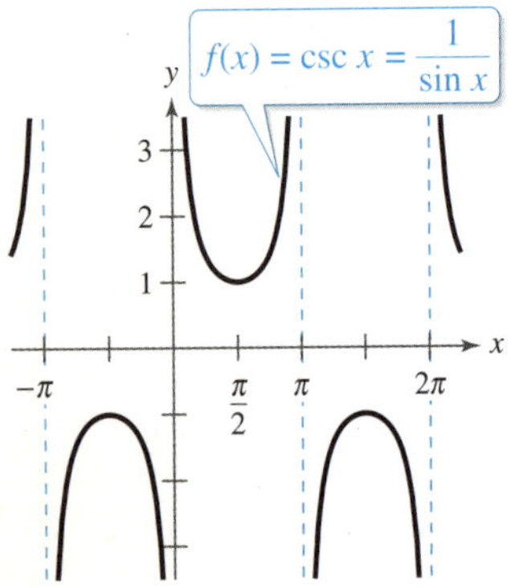

Domain: all $x \neq n\pi$
Range: $(-\infty, -1] \cup [1, \infty)$
Period: 2π
No intercepts
Vertical asymptotes: $x = n\pi$
Odd function
Origin symmetry

Secant Function

$f(x) = \sec x$

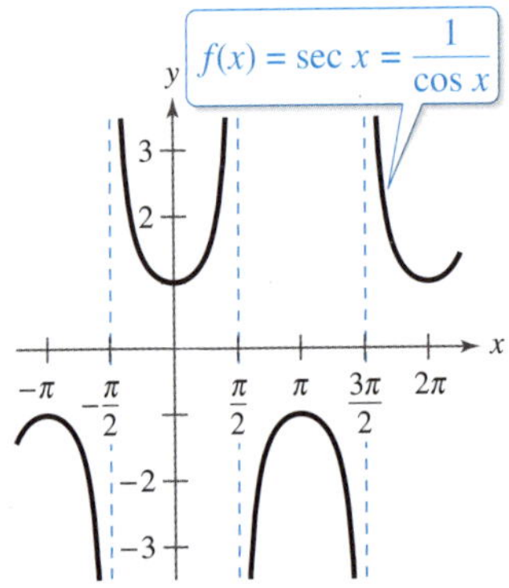

Domain: all $x \neq \frac{\pi}{2} + n\pi$

Range: $(-\infty, -1] \cup [1, \infty)$
Period: 2π
y-intercept: $(0, 1)$
Vertical asymptotes:

$$x = \frac{\pi}{2} + n\pi$$

Even function
y-axis symmetry

Cotangent Function

$f(x) = \cot x$

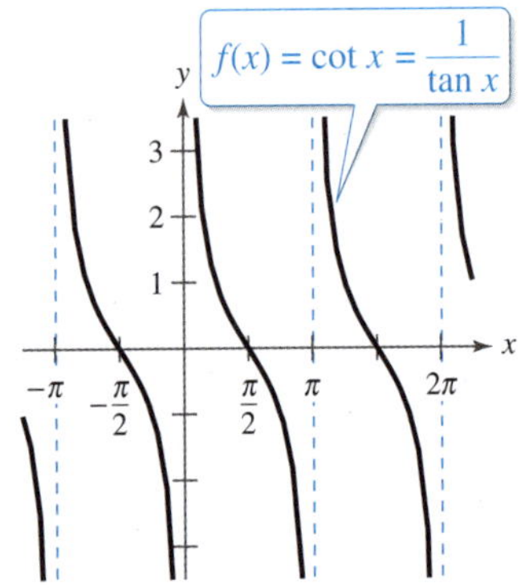

Domain: all $x \neq n\pi$
Range: $(-\infty, \infty)$
Period: π

x-intercepts: $\left(\frac{\pi}{2} + n\pi, 0\right)$

Vertical asymptotes: $x = n\pi$
Odd function
Origin symmetry

Inverse Sine Function

$f(x) = \arcsin x$

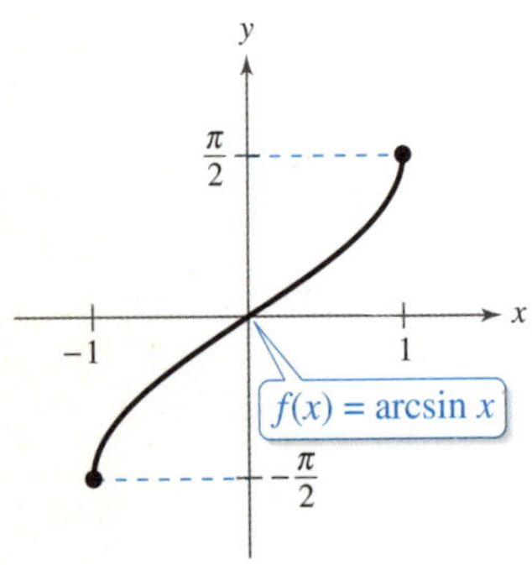

Domain: $[-1, 1]$

Range: $\left[-\frac{\pi}{2}, \frac{\pi}{2}\right]$

Intercept: $(0, 0)$
Odd function
Origin symmetry

Inverse Cosine Function

$f(x) = \arccos x$

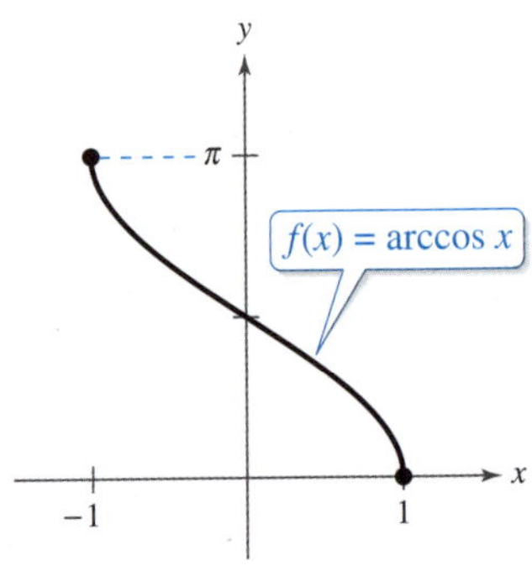

Domain: $[-1, 1]$
Range: $[0, \pi]$

y-intercept: $\left(0, \frac{\pi}{2}\right)$

Inverse Tangent Function

$f(x) = \arctan x$

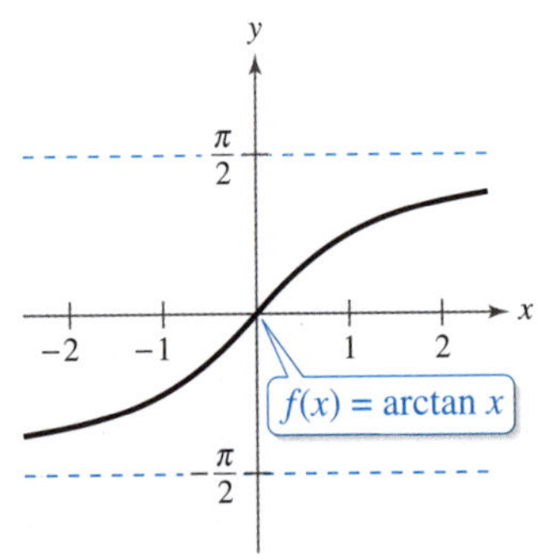

Domain: $(-\infty, \infty)$

Range: $\left(-\frac{\pi}{2}, \frac{\pi}{2}\right)$

Intercept: $(0, 0)$
Horizontal asymptotes:

$$y = \pm\frac{\pi}{2}$$

Odd function
Origin symmetry

Precalculus

EWA Student Start Guide

Ron Larson

CENGAGE Learning®

Australia • Brazil • Japan • Korea • Mexico • Singapore • Spain • United Kingdom • United States

CENGAGE Learning

Precalculus
EWA Student Start Guide

Senior Project Development Manager:
Linda deStefano

Market Development Manager:
Heather Kramer

Senior Production/Manufacturing Manager:
Donna M. Brown

Production Editorial Manager:
Kim Fry

Sr. Rights Acquisition Account Manager:
Todd Osborne

Printed in the United States of America

Precalculus
Ron Larson

© 2014, 2011, 2007 Cengage Learning. All rights reserved.

Library of Congress Control Number: 2012948319

ALL RIGHTS RESERVED. No part of this work covered by the copyright herein may be reproduced, transmitted, stored or used in any form or by any means graphic, electronic, or mechanical, including but not limited to photocopying, recording, scanning, digitizing, taping, Web distribution, information networks, or information storage and retrieval systems, except as permitted under Section 107 or 108 of the 1976 United States Copyright Act, without the prior written permission of the publisher.

For product information and technology assistance, contact us at
Cengage Learning Customer & Sales Support, 1-800-354-9706
For permission to use material from this text or product,
submit all requests online at **cengage.com/permissions**
Further permissions questions can be emailed to
permissionrequest@cengage.com

This book contains select works from existing Cengage Learning resources and was produced by Cengage Learning Custom Solutions for collegiate use. As such, those adopting and/or contributing to this work are responsible for editorial content accuracy, continuity and completeness.

Compilation © 2013 Cengage Learning

ISBN-13: 978-1-285-89548-2

ISBN-10: 1-285-89548-7

Cengage Learning
5191 Natorp Boulevard
Mason, Ohio 45040
USA
Cengage Learning is a leading provider of customized learning solutions with office locations around the globe, including Singapore, the United Kingdom, Australia, Mexico, Brazil, and Japan. Locate your local office at:
international.cengage.com/region.

Cengage Learning products are represented in Canada by Nelson Education, Ltd.
For your lifelong learning solutions, visit **www.cengage.com/custom.**
Visit our corporate website at **www.cengage.com.**

WebAssign

WebAssign Student Quick Start Guide

You can use WebAssign to access your homework, quizzes, and tests — whatever your instructor chooses — at any time of day or night, from any computer with a connection to the Internet and a Web browser. Your instructor creates your assignments, schedules them, and decides how many submissions you get. Your instructor also determines if you can have an extension, if you can save your work without submitting it at the time, and how much feedback you get after you submit an assignment.

The WebAssign support staff cannot change your username or password, give extensions, change your score, give you extra submissions, or help you with the content of your assignments.

Logging In

You can log in to WebAssign using any Web browser connected to the Internet. There are two different ways to log in to WebAssign. Each requires information from your teacher. If you are unsure about how to log in, please check with your teacher or another student in your class.

Go to the login page at http://webassign.net/login.html or the web address provided by your teacher. The way you log in depends on how your instructor set up the class:

- If your teacher created a WebAssign account for you, they will provide you with a **Username**, an **Institution** code and a **Password**. Simply enter this information in the boxes provided and click the **Log In** button.

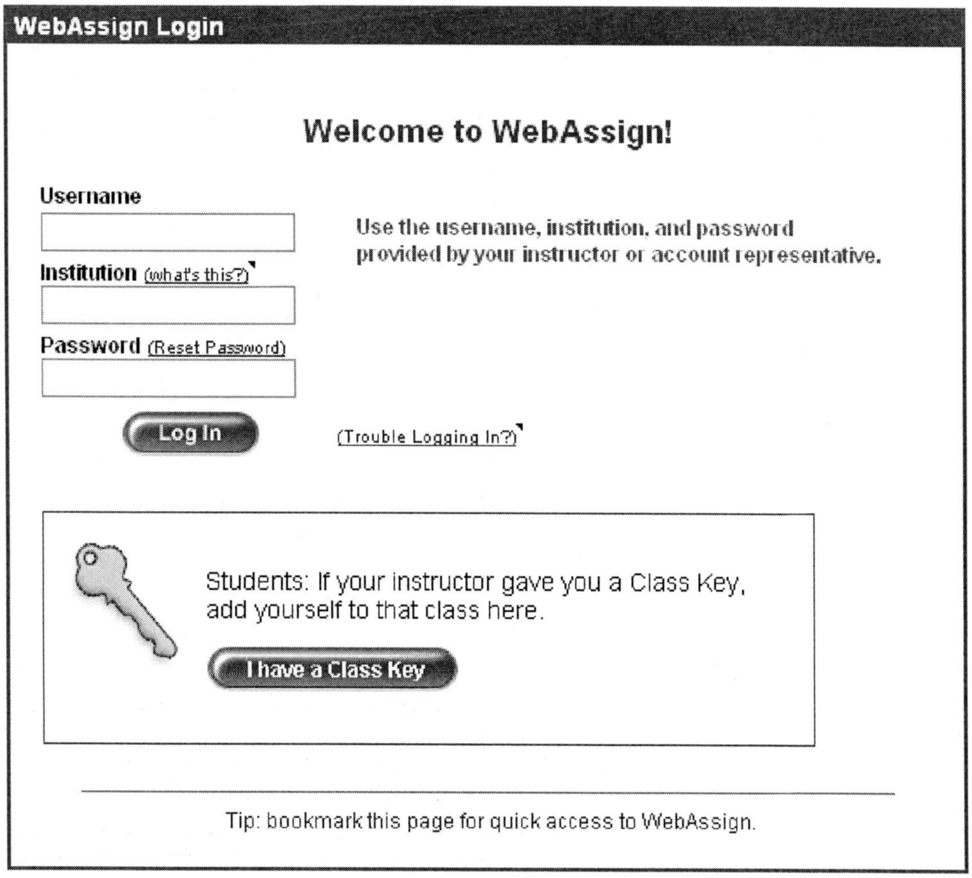

WebAssign

- If your teacher wants you to **Self-Enroll** in the WebAssign course they will provide you with a **Class Key**. You will create your own username and password. It is important that you remember this information so you can log in for the remainder of the class. In this case, just click the **I have a Class Key** button. You don't need to enter any other information on this page.

  Then, enter the **Class Key** your instructor provided and click **Submit.** Verify you are enrolling in the correct class on the next page.

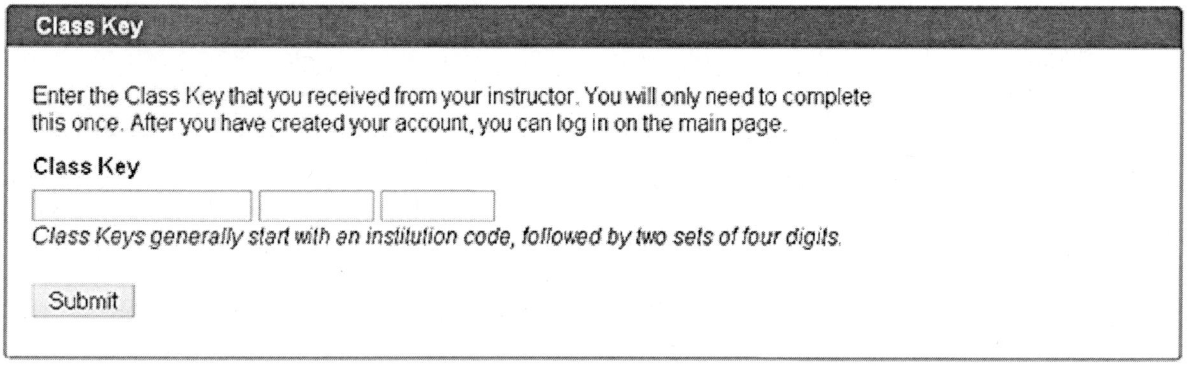

Class Key

Enter the Class Key that you received from your instructor. You will only need to complete this once. After you have created your account, you can log in on the main page.

Class Key

Class Keys generally start with an institution code, followed by two sets of four digits.

Submit

- Enter your preferred Login and Student information.
- Click the **Create My Account** button to complete the enrollment process.
- A review screen will display, showing your username, institution code, and password. **Retain a copy of this information.** You will need it to log into WebAssign.

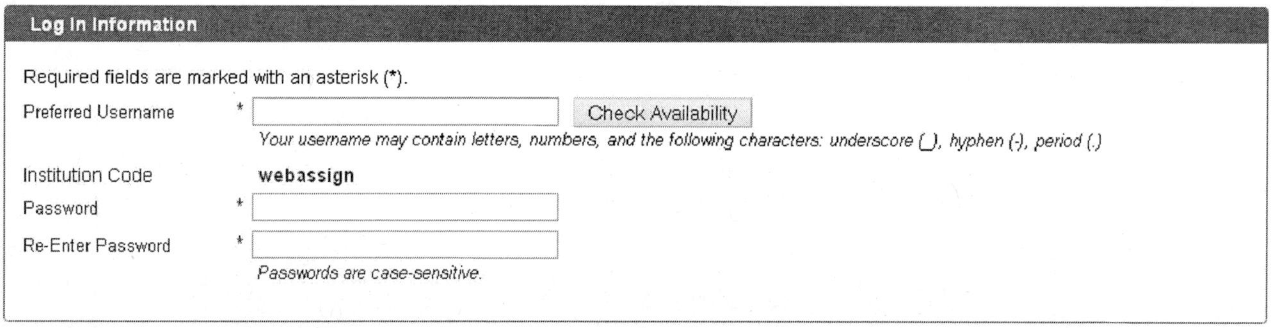

Log In Information

Required fields are marked with an asterisk (*).

Preferred Username * Check Availability

Your username may contain letters, numbers, and the following characters: underscore (_), hyphen (-), period (.)

Institution Code **webassign**

Password *

Re-Enter Password *

Passwords are case-sensitive.

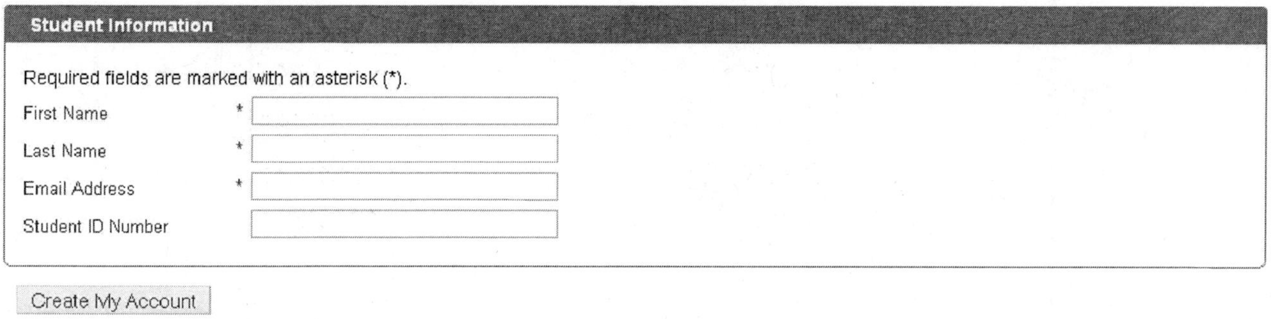

Student Information

Required fields are marked with an asterisk (*).

First Name *

Last Name *

Email Address *

Student ID Number

Create My Account

WebAssign

Access Codes

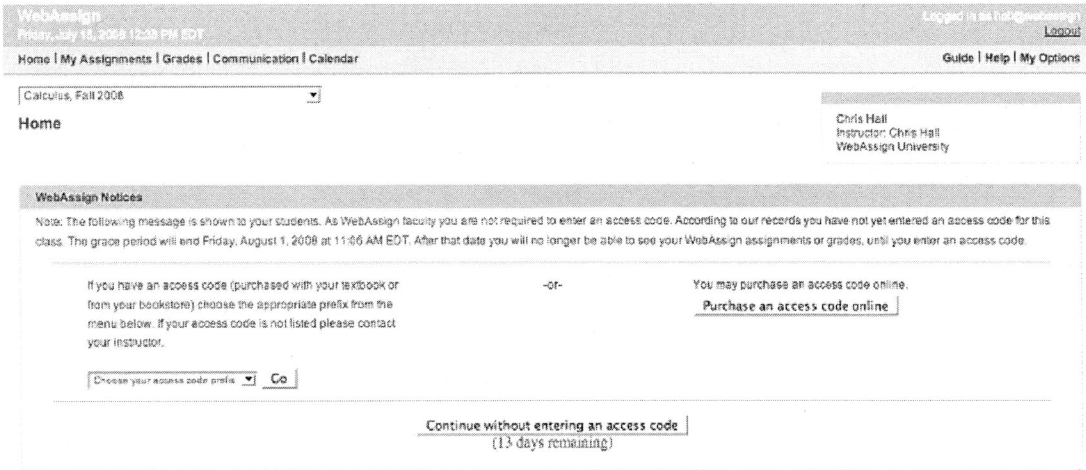

Once you log in, you may see a WebAssign Notice about entering an access code for your class. You can get an Access Code from any of the following places if you need to use one:

- A new textbook you purchased for the class.
- Your bookstore, which may sell Access Code cards.
- Online, where you can purchase an access code with a credit card.

You have a 14 day grace period to use WebAssign, starting with the WebAssign class start date. During this time you can work on and view your WebAssign assignments without registering a code.

After the grace period is over you will only see the code registration message until you submit or purchase a code.

There are two types of WebAssign access code cards. The small card requires you to scratch off the silver surface in order to reveal the complete access code.

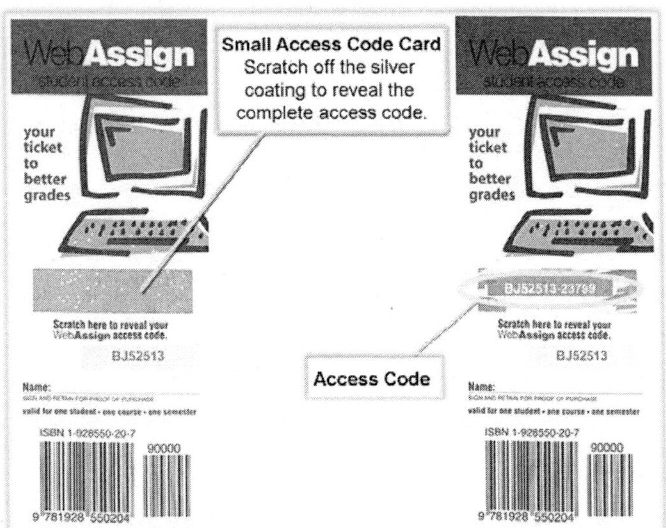

The larger security envelope card requires you to open the card to reveal the access code number.

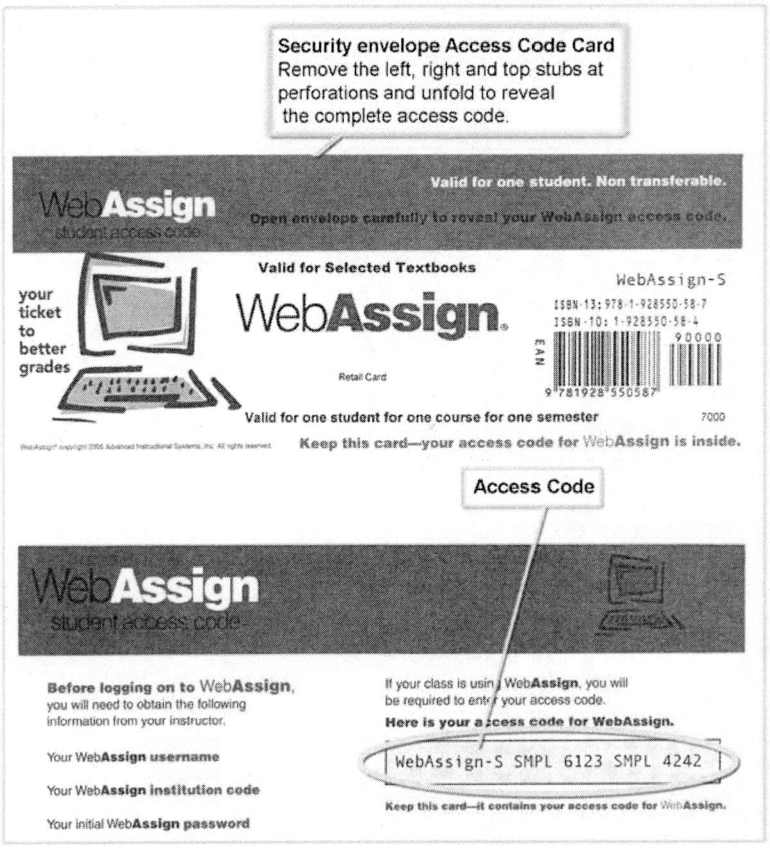

If you would like to purchase an access code directly from WebAssign online, you may do so with a credit card. Your code will be automatically registered to your WebAssign account as soon as the transaction is complete. You will receive an email confirmation. Please keep a copy for your records.

WebAssign.

Your WebAssign Home Page

Once you have successfully logged in you will see your WebAssign homepage. If you are taking more than one WebAssign class, you will need to select which class you wish to view first.

The upper right corner features links to a complete Student **Guide**, as well as a link to WebAssign Technical Support under **Help**. If you want to change your password or add or update your email address, simply click **My Options** in the upper right hand corner.

You will see your assignments and due dates listed, as well as any Communications, Grades, and Announcements posted by your teacher.

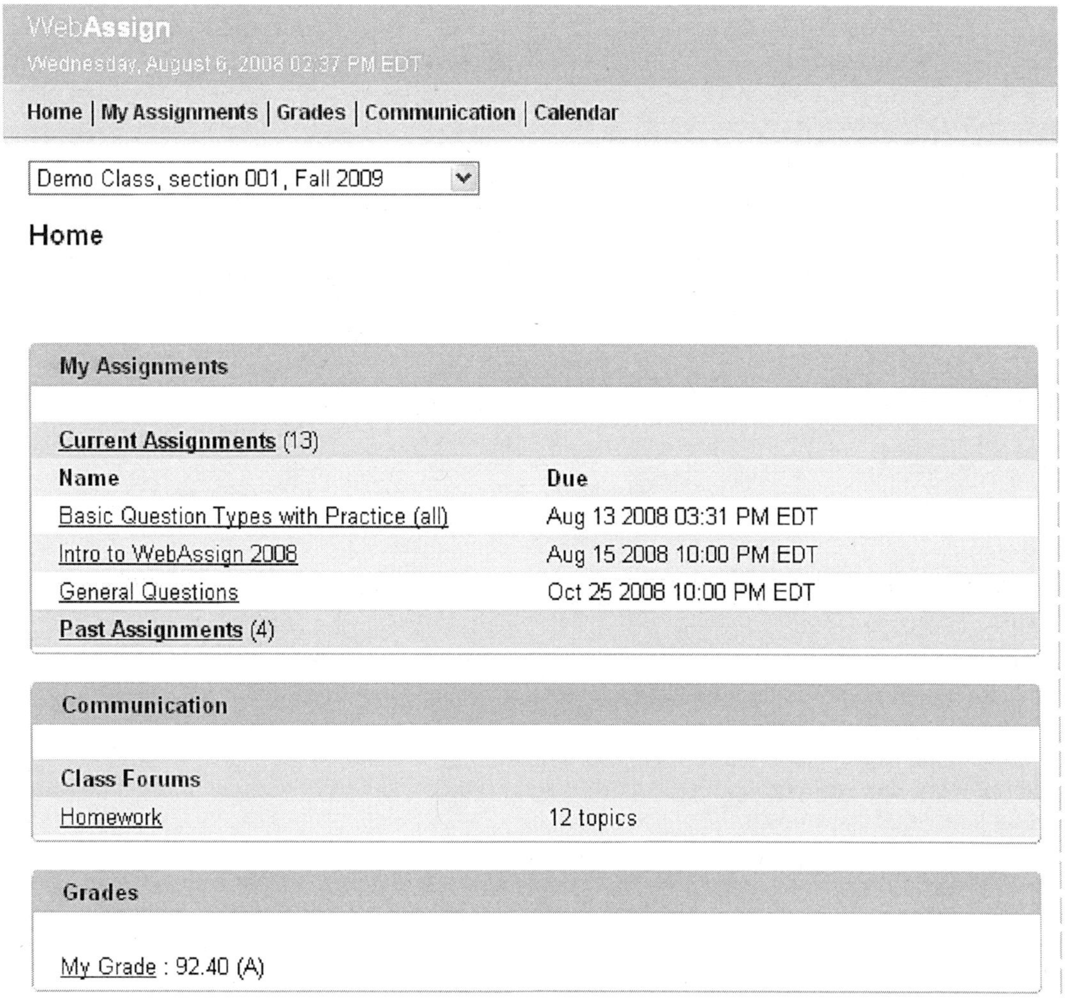

WebAssign

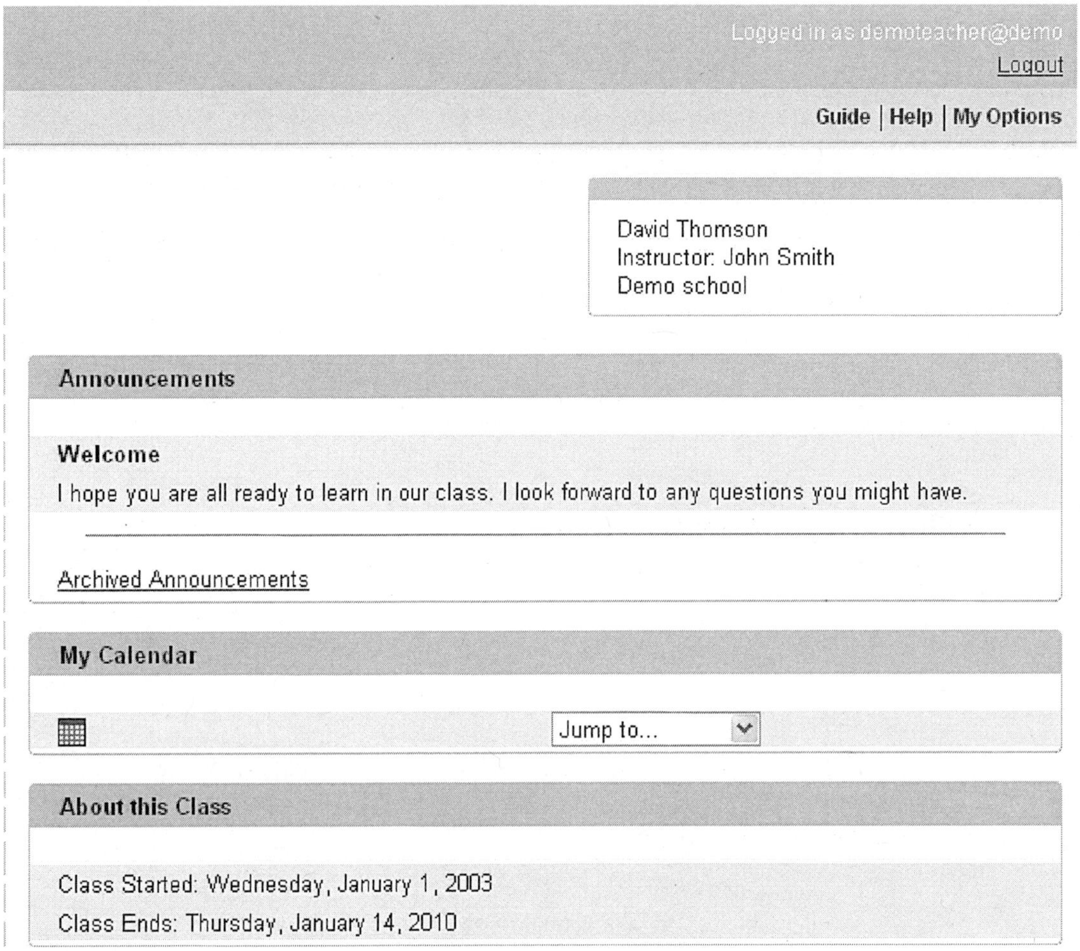

Answering Questions

WebAssign has a variety of different question types, ranging from multiple choice to fill-in-the-blank to symbolic questions. Here are some things to keep in mind as you work through your assignments:

- Some questions may include numbers or words that appear in red. This signifies that the number or word has been randomized, so that you receive a different version of the same basic question from your classmates.

- Some WebAssign questions check the number of significant figures in your answer. If you enter the correct value with the wrong number of significant figures, you will not receive credit, but you will receive a hint that your number does not have the correct number of significant figures.

- Some questions require entering symbolic notation. Answer symbolic questions by using calculator notation. You must use the exact variables specified in the questions. The order is not important as long as it is mathematically correct. Clicking on the eye button previews the expression you enter in proper mathematical notation. Clicking on the symbolic formatting help button provides tips for using the correct keystrokes.

WebAssign

- When you click on some WebAssign chemistry or math questions an input palette will open. These palettes, called chemPad and mathPad, will help you enter your answer in proper notation.

- Some questions may require the use of an Active Figure simulation. Active Figures require the free Macromedia Flash Player plug-in, downloadable from www.macromedia.com.

- If your instructor allows it, you can save your work without grading by selecting the Save Work button at the end of the question. After you save your work, it will be available to you the next time you click the assignment.

- Please note that WebAssign will **not** automatically submit your answers for scoring if you only **Save** your work. Your teacher will not be able to see your submissions. Please be sure to **Submit** prior to the due date and time.

- If your instructor allows it, you can submit answers by question part or for the entire assignment. To submit an individual question answer for grading, click the **Submit New Answers to Question __** button at the bottom of each question. To submit the entire assignment for grading, click the **Submit All New Answers** button at the end of the assignment.

Technical Support

If you are having difficulty logging in, please be sure to check with your teacher and verify whether an account has been created for you or whether you need to self-enroll. In either case your teacher needs to provide the appropriate information (username, institution code and password OR Class Key).

To email WebAssign Support go to http://www.webassign.net/info/support/report.html. This page also lists answers to **Common Problems**, and provides links to the **Student Guide**.

August 7, 2008

Contents

Appendices

*Available at the text-specific website *www.cengagebrain.com*

Preface

Welcome to *Precalculus*, Ninth Edition. I am proud to present to you this new edition. As with all editions, I have been able to incorporate many useful comments from you, our user. And while much has changed in this revision, you will still find what you expect—a pedagogically sound, mathematically precise, and comprehensive textbook. Additionally, I am pleased and excited to offer you something brand new—a companion website at **LarsonPrecalculus.com.**

My goal for every edition of this textbook is to provide students with the tools that they need to master precalculus. I hope you find that the changes in this edition, together with **LarsonPrecalculus.com,** will help accomplish just that.

New To This Edition

NEW LarsonPrecalculus.com

This companion website offers multiple tools and resources to supplement your learning. Access to these features is free. View and listen to worked-out solutions of Checkpoint problems in English or Spanish, download data sets, work on chapter projects, watch lesson videos, and much more.

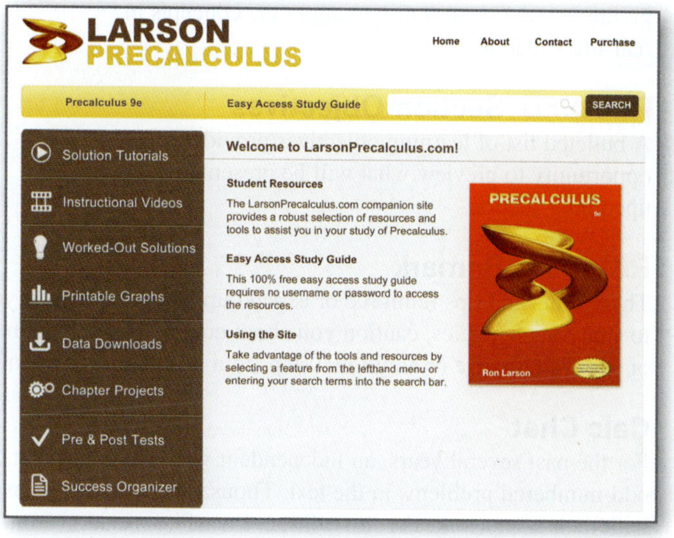

NEW Chapter Opener

Each Chapter Opener highlights real-life applications used in the examples and exercises.

96. HOW DO YOU SEE IT? The graph represents the height h of a projectile after t seconds.

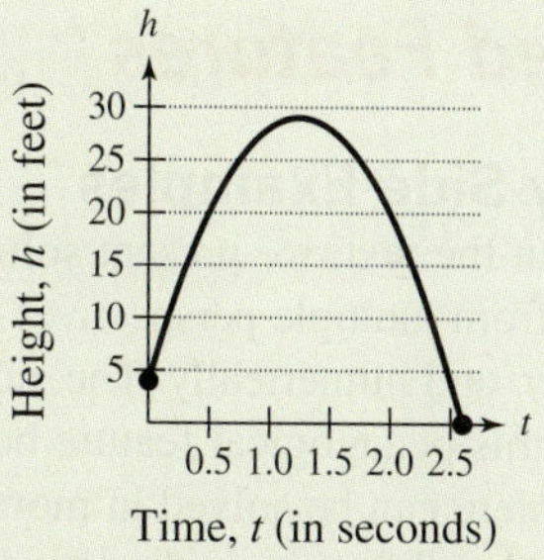

(a) Explain why h is a function of t.

(b) Approximate the height of the projectile after 0.5 second and after 1.25 seconds.

(c) Approximate the domain of h.

(d) Is t a function of h? Explain.

NEW Summarize

The Summarize feature at the end of each section helps you organize the lesson's key concepts into a concise summary, providing you with a valuable study tool.

NEW How Do You See It?

The How Do You See It? feature in each section presents a real-life exercise that you will solve by visual inspection using the concepts learned in the lesson. This exercise is excellent for classroom discussion or test preparation.

NEW Checkpoints

Accompanying every example, the Checkpoint problems encourage immediate practice and check your understanding of the concepts presented in the example. View and listen to worked-out solutions of the Checkpoint problems in English or Spanish at LarsonPrecalculus.com.

NEW Data Spreadsheets

Download these editable spreadsheets from LarsonPrecalculus.com, and use the data to solve exercises.

DATA

Year	Number of Tax Returns Made Through E-File
2003	52.9
2004	61.5
2005	68.5
2006	73.3
2007	80.0
2008	89.9
2009	95.0
2010	98.7

Spreadsheet at LarsonPrecalculus.com

REVISED Exercise Sets

The exercise sets have been carefully and extensively examined to ensure they are rigorous and relevant and to include all topics our users have suggested. The exercises have been **reorganized and titled** so you can better see the connections between examples and exercises. Multi-step, real-life exercises reinforce problem-solving skills and mastery of concepts by giving you the opportunity to apply the concepts in real-life situations.

REVISED Section Objectives

A bulleted list of learning objectives provides you the opportunity to preview what will be presented in the upcoming section.

REVISED Remark

These hints and tips reinforce or expand upon concepts, help you learn how to study mathematics, caution you about common errors, address special cases, or show alternative or additional steps to a solution of an example.

Calc Chat

For the past several years, an independent website—CalcChat.com—has provided free solutions to all odd-numbered problems in the text. Thousands of students have visited the site for practice and help with their homework. For this edition, I used information from CalcChat.com, including which solutions students accessed most often, to help guide the revision of the exercises.

Trusted Features

Side-By-Side Examples

Throughout the text, we present solutions to many examples from multiple perspectives—algebraically, graphically, and numerically. The side-by-side format of this pedagogical feature helps you to see that a problem can be solved in more than one way and to see that different methods yield the same result. The side-by-side format also addresses many different learning styles.

Algebra Help

Algebra Help directs you to sections of the textbook where you can review algebra skills needed to master the current topic.

Technology

The technology feature gives suggestions for effectively using tools such as calculators, graphing calculators, and spreadsheet programs to help deepen your understanding of concepts, ease lengthy calculations, and provide alternate solution methods for verifying answers obtained by hand.

Historical Notes

These notes provide helpful information regarding famous mathematicians and their work.

Algebra of Calculus

Throughout the text, special emphasis is given to the algebraic techniques used in calculus. Algebra of Calculus examples and exercises are integrated throughout the text and are identified by the symbol f.

Vocabulary Exercises

The vocabulary exercises appear at the beginning of the exercise set for each section. These problems help you review previously learned vocabulary terms that you will use in solving the section exercises.

TECHNOLOGY You can use a graphing utility to check that a solution is reasonable. One way to do this is to graph the left side of the equation, then graph the right side of the equation, and determine the point of intersection. For instance, in Example 2, if you graph the equations

$y_1 = 6(x - 1) + 4$ The left side

$y_2 = 3(7x + 1)$ The right side

in the same viewing window, they should intersect at $x = -\frac{1}{3}$, as shown in the graph below.

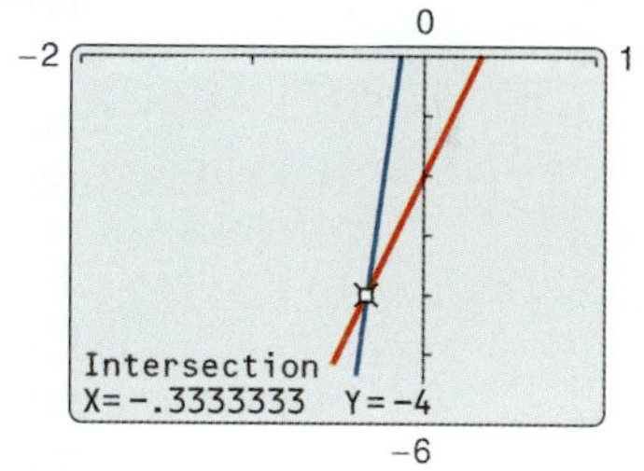

Project: Department of Defense The table shows the total numbers of military personnel P (in thousands) on active duty from 1980 through 2010. *(Source: U.S. Department of Defense)*

Year	Personnel, P	Year	Personnel, P
1980	2051	1995	1518
1981	2083	1996	1472
1982	2109	1997	1439
1983	2123	1998	1407
1984	2138	1999	1386
1985	2151	2000	1384
1986	2169	2001	1385
1987	2174	2002	1414
1988	2138	2003	1434
1989	2130	2004	1427
1990	2044	2005	1389
1991	1986	2006	1385
1992	1807	2007	1380
1993	1705	2008	1402
1994	1610	2009	1419
		2010	1431

(a) Use a graphing utility to plot the data. Let t represent the year, with $t = 0$ corresponding to 1980.

(b) A model that approximates the data is given by

$$P = \frac{9.6518t^2 - 244.743t + 2044.77}{0.0059t^2 - 0.131t + 1}$$

where P is the total number of personnel (in thousands) and t is the year, with $t = 0$ corresponding to 1980. Construct a table showing the actual values of P and the values of P obtained using the model.

Project

The projects at the end of selected sections involve in-depth applied exercises in which you will work with large, real-life data sets, often creating or analyzing models. These projects are offered online at LarsonPrecalculus.com.

Chapter Summaries

The Chapter Summary now includes explanations and examples of the objectives taught in each chapter.

Enhanced WebAssign

Enhanced WebAssign combines exceptional Precalculus content that you know and love with the most powerful online homework solution, WebAssign. Enhanced WebAssign engages you with immediate feedback, rich tutorial content and interactive, fully customizable eBooks (YouBook) helping you to develop a deeper conceptual understanding of the subject matter.

Instructor Resources

Print

Annotated Instructor's Edition
ISBN-13: 978-1-133-94902-2

This AIE is the complete student text plus point-of-use annotations for you, including extra projects, classroom activities, teaching strategies, and additional examples. Answers to even-numbered text exercises, Vocabulary Checks, and Explorations are also provided.

Complete Solutions Manual
ISBN-13: 978-1-133-95442-2

This manual contains solutions to all exercises from the text, including Chapter Review Exercises, and Chapter Tests.

Media

PowerLecture with ExamView™
ISBN-13: 978-1-133-95440-8

The DVD provides you with dynamic media tools for teaching Precalculus while using an interactive white board. PowerPoint® lecture slides and art slides of the figures from the text, together with electronic files for the test bank and a link to the Solution Builder, are available. The algorithmic ExamView allows you to create, deliver, and customize tests (both print and online) in minutes with this easy-to-use assessment system. The DVD also provides you with a tutorial on integrating our instructor materials into your interactive whiteboard platform. Enhance how your students interact with you, your lecture, and each other.

Solution Builder
(*www.cengage.com/solutionbuilder*)
This online instructor database offers complete worked-out solutions to all exercises in the text, allowing you to create customized, secure solutions printouts (in PDF format) matched exactly to the problems you assign in class.

ENHANCED WebAssign
www.webassign.net
Printed Access Card: 978-0-538-73810-1
Online Access Code: 978-1-285-18181-3

Exclusively from Cengage Learning, Enhanced WebAssign combines the exceptional mathematics content that you know and love with the most powerful online homework solution, WebAssign. Enhanced WebAssign engages students with immediate feedback, rich tutorial content, and interactive, fully customizable eBooks (YouBook), helping students to develop a deeper conceptual understanding of their subject matter. Online assignments can be built by selecting from thousands of text-specific problems or supplemented with problems from any Cengage Learning textbook.

Student Resources

Print

Student Study and Solutions Manual
ISBN-13: 978-1-133-95441-5

This guide offers step-by-step solutions for all odd-numbered text exercises, Chapter and Cumulative Tests, and Practice Tests with solutions.

Text-Specific DVD
ISBN-13: 978-1-133-96287-8

Keyed to the text by section, these DVDs provide comprehensive coverage of the course—along with additional explanations of concepts, sample problems, and application—to help you review essential topics.

Note Taking Guide
ISBN-13: 978-1-133-94904-6

This innovative study aid, in the form of a notebook organizer, helps you develop a section-by-section summary of key concepts.

Media

ENHANCED WebAssign

www.webassign.net
Printed Access Card: 978-0-538-73810-1
Online Access Code: 978-1-285-18181-3

Enhanced WebAssign (assigned by the instructor) provides you with instant feedback on homework assignments. This online homework system is easy to use and includes helpful links to textbook sections, video examples, and problem-specific tutorials.

CengageBrain.com

Visit *www.cengagebrain.com* to access additional course materials and companion resources. At the CengageBrain.com home page, search for the ISBN of your title (from the back cover of your book) using the search box at the top of the page. This will take you to the product page where free companion resources can be found.

Acknowledgements

I would like to thank the many people who have helped me prepare the text and the supplements package. Their encouragement, criticisms, and suggestions have been invaluable.

Thank you to all of the instructors who took the time to review the changes in this edition and to provide suggestions for improving it. Without your help, this book would not be possible.

Reviewers

Timothy Andrew Brown, *South Georgia College*
Blair E. Caboot, *Keystone College*
Shannon Cornell, *Amarillo College*
Gayla Dance, *Millsaps College*
Paul Finster, *El Paso Community College*
Paul A. Flasch, *Pima Community College West Campus*
Vadas Gintautas, *Chatham University*
Lorraine A. Hughes, *Mississippi State University*
Shu-Jen Huang, *University of Florida*
Renyetta Johnson, *East Mississippi Community College*
George Keihany, *Fort Valley State University*
Mulatu Lemma, *Savannah State University*
William Mays Jr., *Salem Community College*
Marcella Melby, *University of Minnesota*
Jonathan Prewett, *University of Wyoming*
Denise Reid, *Valdosta State University*
David L. Sonnier, *Lyon College*
David H. Tseng, *Miami Dade College – Kendall Campus*
Kimberly Walters, *Mississippi State University*
Richard Weil, *Brown College*
Solomon Willis, *Cleveland Community College*
Bradley R. Young, *Darton College*

My thanks to Robert Hostetler, The Behrend College, The Pennsylvania State University, and David Heyd, The Behrend College, The Pennsylvania State University, for their significant contributions to previous editions of this text.

I would also like to thank the staff at Larson Texts, Inc. who assisted with proofreading the manuscript, preparing and proofreading the art package, and checking and typesetting the supplements.

On a personal level, I am grateful to my spouse, Deanna Gilbert Larson, for her love, patience, and support. Also, a special thanks goes to R. Scott O'Neil. If you have suggestions for improving this text, please feel free to write to me. Over the past two decades I have received many useful comments from both instructors and students, and I value these comments very highly.

Ron Larson, Ph.D.
Professor of Mathematics
Penn State University
www.RonLarson.com

1 Functions and Their Graphs

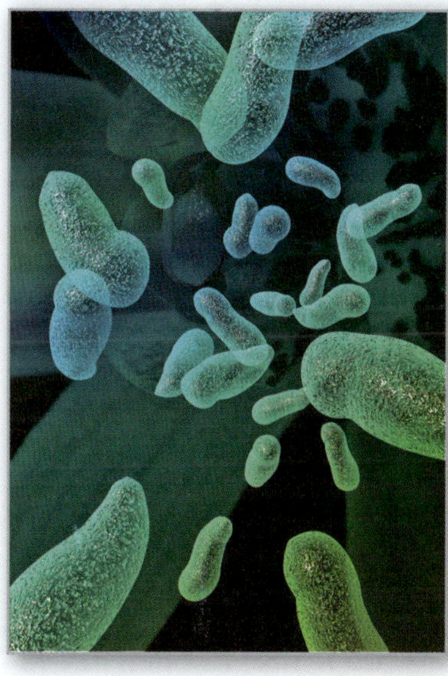

Bacteria *(Example 8, page 80)*

Snowstorm *(Exercise 47, page 66)*

Average Speed *(Example 7, page 54)*

Alternative-Fueled Vehicles *(Example 10, page 42)*

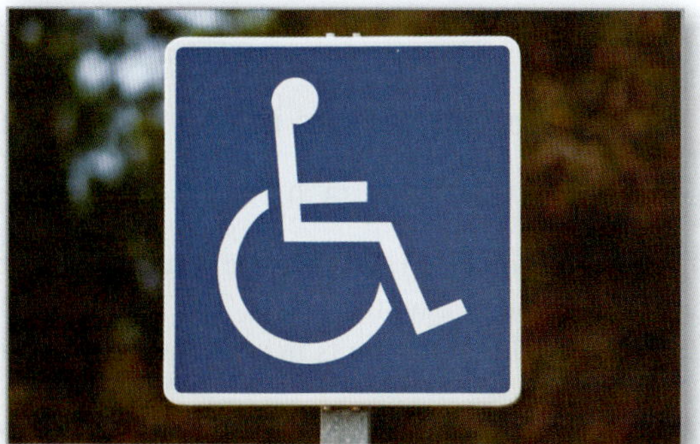

Americans with Disabilities Act *(page 28)*

Clockwise from top left, nulinukas/Shutterstock.com; Fedorov Oleksiy/Shutterstock.com; wellphoto/Shutterstock.com; Jultud/Shutterstock.com; sadwitch/Shutterstock.com

1.1 Rectangular Coordinates

The Cartesian plane can help you visualize relationships between two variables. For instance, in Exercise 37 on page 9, given how far north and west one city is from another, plotting points to represent the cities can help you visualize these distances and determine the flying distance between the cities.

- **Plot points in the Cartesian plane.**
- **Use the Distance Formula to find the distance between two points.**
- **Use the Midpoint Formula to find the midpoint of a line segment.**
- **Use a coordinate plane to model and solve real-life problems.**

The Cartesian Plane

Just as you can represent real numbers by points on a real number line, you can represent ordered pairs of real numbers by points in a plane called the **rectangular coordinate system,** or the **Cartesian plane,** named after the French mathematician René Descartes (1596–1650).

Two real number lines intersecting at right angles form the Cartesian plane, as shown in Figure 1.1. The horizontal real number line is usually called the ***x*-axis,** and the vertical real number line is usually called the ***y*-axis.** The point of intersection of these two axes is the **origin,** and the two axes divide the plane into four parts called **quadrants.**

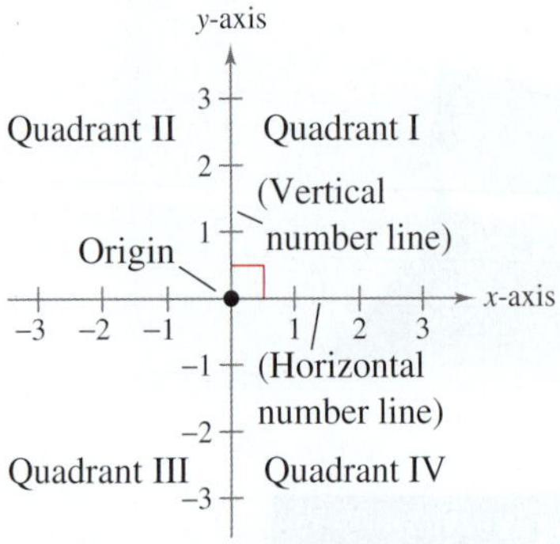

Figure 1.1

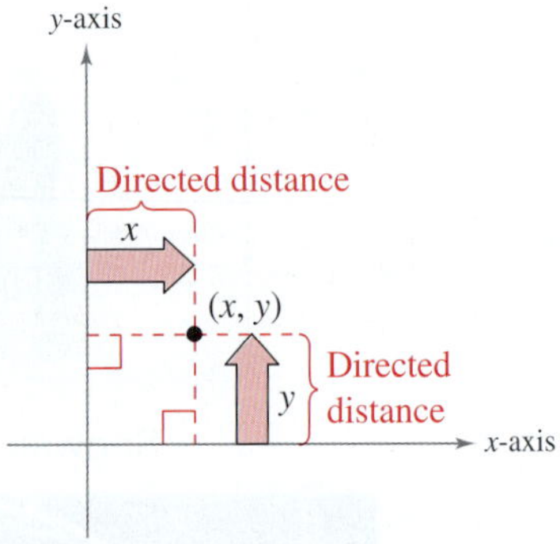

Figure 1.2

Each point in the plane corresponds to an **ordered pair** (x, y) of real numbers x and y, called **coordinates** of the point. The ***x*-coordinate** represents the directed distance from the y-axis to the point, and the ***y*-coordinate** represents the directed distance from the x-axis to the point, as shown in Figure 1.2.

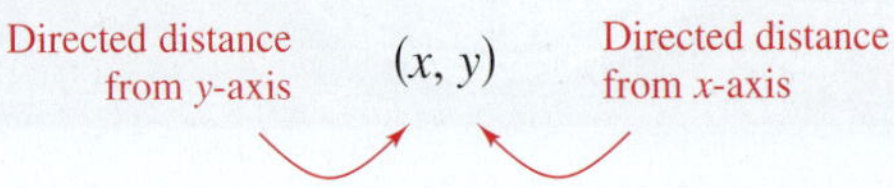

The notation (x, y) denotes both a point in the plane and an open interval on the real number line. The context will tell you which meaning is intended.

EXAMPLE 1 Plotting Points in the Cartesian Plane

Plot the points $(-1, 2)$, $(3, 4)$, $(0, 0)$, $(3, 0)$, and $(-2, -3)$.

Solution To plot the point $(-1, 2)$, imagine a vertical line through -1 on the x-axis and a horizontal line through 2 on the y-axis. The intersection of these two lines is the point $(-1, 2)$. Plot the other four points in a similar way, as shown in Figure 1.3.

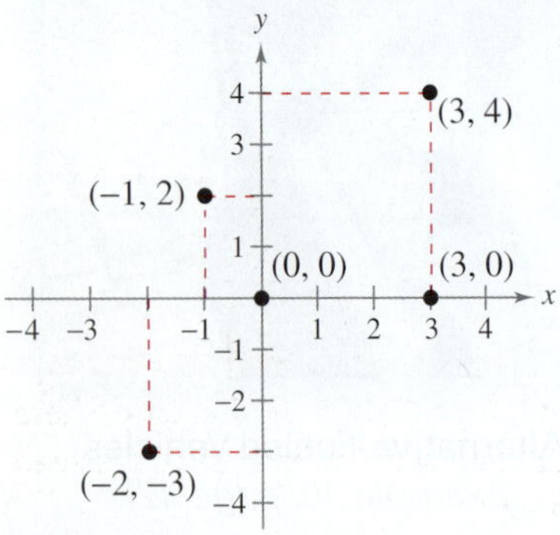

Figure 1.3

✔ ***Checkpoint*** *Audio-video solution in English & Spanish at LarsonPrecalculus.com.*

Plot the points $(-3, 2)$, $(4, -2)$, $(3, 1)$, $(0, -2)$, and $(-1, -2)$.

Fernando Jose Vasconcelos Soares/Shutterstock.com

The beauty of a rectangular coordinate system is that it allows you to *see* relationships between two variables. It would be difficult to overestimate the importance of Descartes's introduction of coordinates in the plane. Today, his ideas are in common use in virtually every scientific and business-related field.

EXAMPLE 2 Sketching a Scatter Plot

Year, t	Subscribers, N
2001	128.4
2002	140.8
2003	158.7
2004	182.1
2005	207.9
2006	233.0
2007	255.4
2008	270.3
2009	290.9
2010	311.0

DATA Spreadsheet at LarsonPrecalculus.com

The table shows the numbers N (in millions) of subscribers to a cellular telecommunication service in the United States from 2001 through 2010, where t represents the year. Sketch a scatter plot of the data. *(Source: CTIA-The Wireless Association)*

Solution To sketch a *scatter plot* of the data shown in the table, represent each pair of values by an ordered pair (t, N) and plot the resulting points, as shown below. For instance, the ordered pair (2001, 128.4) represents the first pair of values. Note that the break in the t-axis indicates omission of the years before 2001.

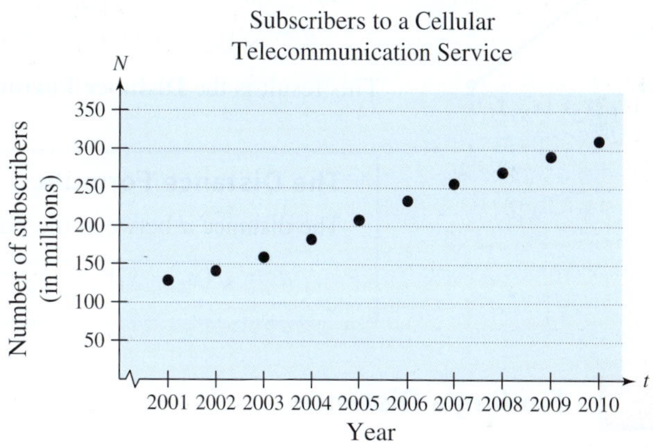

Checkpoint Audio-video solution in English & Spanish at LarsonPrecalculus.com.

The table shows the numbers N (in thousands) of cellular telecommunication service employees in the United States from 2001 through 2010, where t represents the year. Sketch a scatter plot of the data. *(Source: CTIA-The Wireless Association)*

t	N
2001	203.6
2002	192.4
2003	205.6
2004	226.0
2005	233.1
2006	253.8
2007	266.8
2008	268.5
2009	249.2
2010	250.4

DATA Spreadsheet at LarsonPrecalculus.com

In Example 2, you could have let $t = 1$ represent the year 2001. In that case, there would not have been a break in the horizontal axis, and the labels 1 through 10 (instead of 2001 through 2010) would have been on the tick marks.

TECHNOLOGY The scatter plot in Example 2 is only one way to represent the data graphically. You could also represent the data using a bar graph or a line graph. Try using a graphing utility to represent the data given in Example 2 graphically.

The Pythagorean Theorem and the Distance Formula

The following famous theorem is used extensively throughout this course.

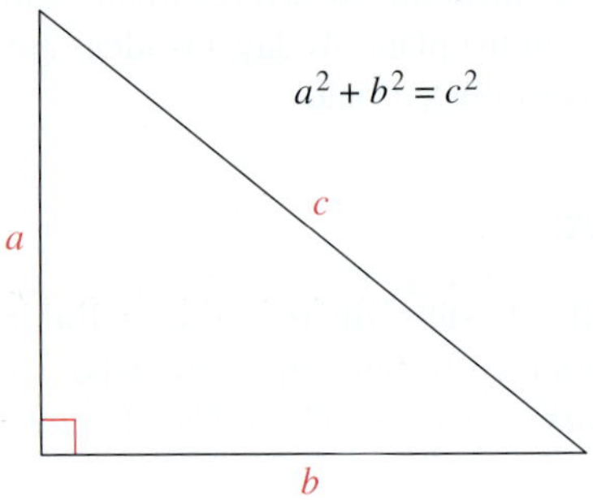

Figure 1.4

> **Pythagorean Theorem**
>
> For a right triangle with hypotenuse of length c and sides of lengths a and b, you have $a^2 + b^2 = c^2$, as shown in Figure 1.4. (The converse is also true. That is, if $a^2 + b^2 = c^2$, then the triangle is a right triangle.)

Suppose you want to determine the distance d between two points (x_1, y_1) and (x_2, y_2) in the plane. These two points can form a right triangle, as shown in Figure 1.5. The length of the vertical side of the triangle is $|y_2 - y_1|$ and the length of the horizontal side is $|x_2 - x_1|$.

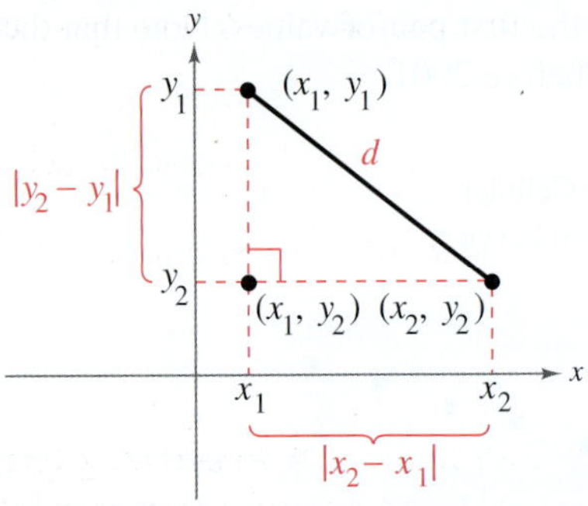

Figure 1.5

By the Pythagorean Theorem,

$$d^2 = |x_2 - x_1|^2 + |y_2 - y_1|^2$$
$$d = \sqrt{|x_2 - x_1|^2 + |y_2 - y_1|^2} = \sqrt{(x_2 - x_1)^2 + (y_2 - y_1)^2}.$$

This result is the **Distance Formula.**

> **The Distance Formula**
>
> The distance d between the points (x_1, y_1) and (x_2, y_2) in the plane is
>
> $$d = \sqrt{(x_2 - x_1)^2 + (y_2 - y_1)^2}.$$

EXAMPLE 3 Finding a Distance

Find the distance between the points $(-2, 1)$ and $(3, 4)$.

Algebraic Solution

Let

$$(x_1, y_1) = (-2, 1) \quad \text{and} \quad (x_2, y_2) = (3, 4).$$

Then apply the Distance Formula.

$d = \sqrt{(x_2 - x_1)^2 + (y_2 - y_1)^2}$	Distance Formula
$= \sqrt{[3 - (-2)]^2 + (4 - 1)^2}$	Substitute for x_1, y_1, x_2, and y_2.
$= \sqrt{(5)^2 + (3)^2}$	Simplify.
$= \sqrt{34}$	Simplify.
≈ 5.83	Use a calculator.

So, the distance between the points is about 5.83 units. Use the Pythagorean Theorem to check that the distance is correct.

$d^2 \stackrel{?}{=} 5^2 + 3^2$	Pythagorean Theorem
$(\sqrt{34})^2 \stackrel{?}{=} 5^2 + 3^2$	Substitute for d.
$34 = 34$	Distance checks. ✓

Graphical Solution

Use centimeter graph paper to plot the points $A(-2, 1)$ and $B(3, 4)$. Carefully sketch the line segment from A to B. Then use a centimeter ruler to measure the length of the segment.

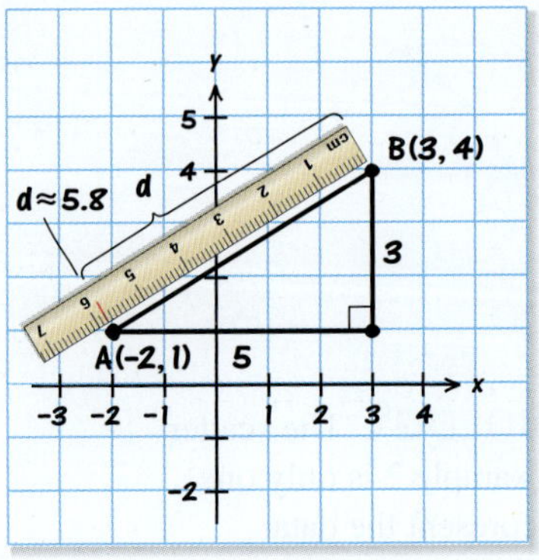

The line segment measures about 5.8 centimeters. So, the distance between the points is about 5.8 units.

✓ **Checkpoint** *Audio-video solution in English & Spanish at LarsonPrecalculus.com.*

Find the distance between the points $(3, 1)$ and $(-3, 0)$.

EXAMPLE 4 **Verifying a Right Triangle**

Show that the points

$(2, 1)$, $(4, 0)$, and $(5, 7)$

are vertices of a right triangle.

Solution The three points are plotted in Figure 1.6. Using the Distance Formula, the lengths of the three sides are as follows.

$$d_1 = \sqrt{(5-2)^2 + (7-1)^2} = \sqrt{9+36} = \sqrt{45}$$

$$d_2 = \sqrt{(4-2)^2 + (0-1)^2} = \sqrt{4+1} = \sqrt{5}$$

$$d_3 = \sqrt{(5-4)^2 + (7-0)^2} = \sqrt{1+49} = \sqrt{50}$$

Because $(d_1)^2 + (d_2)^2 = 45 + 5 = 50 = (d_3)^2$, you can conclude by the Pythagorean Theorem that the triangle must be a right triangle.

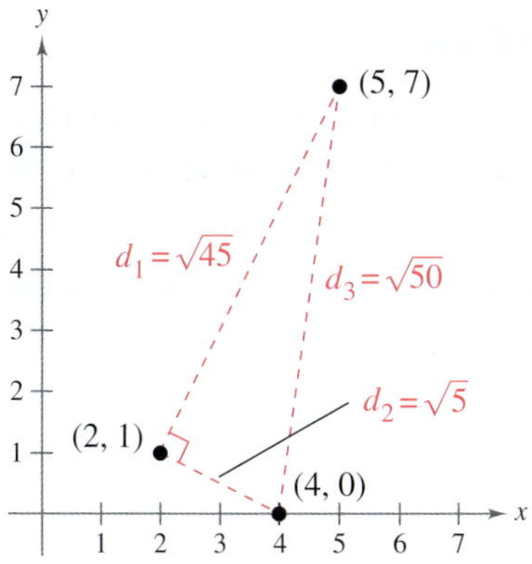

Figure 1.6

✓ **Checkpoint** *Audio-video solution in English & Spanish at LarsonPrecalculus.com.*

Show that the points $(2, -1)$, $(5, 5)$, and $(6, -3)$ are vertices of a right triangle.

▷ **ALGEBRA HELP** You can review the techniques for evaluating a radical in Appendix A.2.

The Midpoint Formula

To find the **midpoint** of the line segment that joins two points in a coordinate plane, you can find the average values of the respective coordinates of the two endpoints using the **Midpoint Formula.**

The Midpoint Formula

The midpoint of the line segment joining the points (x_1, y_1) and (x_2, y_2) is given by the Midpoint Formula

$$\text{Midpoint} = \left(\frac{x_1 + x_2}{2}, \frac{y_1 + y_2}{2}\right).$$

For a proof of the Midpoint Formula, see Proofs in Mathematics on page 110.

EXAMPLE 5 **Finding a Line Segment's Midpoint**

Find the midpoint of the line segment joining the points

$(-5, -3)$ and $(9, 3)$.

Solution Let $(x_1, y_1) = (-5, -3)$ and $(x_2, y_2) = (9, 3)$.

$$\text{Midpoint} = \left(\frac{x_1 + x_2}{2}, \frac{y_1 + y_2}{2}\right) \quad \text{Midpoint Formula}$$

$$= \left(\frac{-5 + 9}{2}, \frac{-3 + 3}{2}\right) \quad \text{Substitute for } x_1, y_1, x_2, \text{ and } y_2.$$

$$= (2, 0) \quad \text{Simplify.}$$

The midpoint of the line segment is $(2, 0)$, as shown in Figure 1.7.

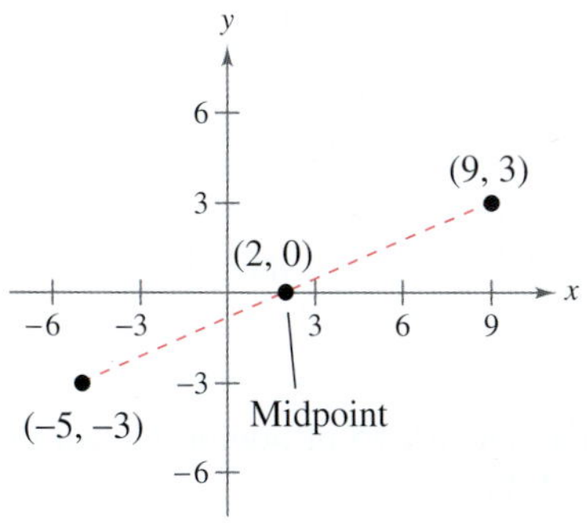

Figure 1.7

✓ **Checkpoint** *Audio-video solution in English & Spanish at LarsonPrecalculus.com.*

Find the midpoint of the line segment joining the points $(-2, 8)$ and $(4, -10)$.

Applications

EXAMPLE 6 Finding the Length of a Pass

A football quarterback throws a pass from the 28-yard line, 40 yards from the sideline. A wide receiver catches the pass on the 5-yard line, 20 yards from the same sideline, as shown in Figure 1.8. How long is the pass?

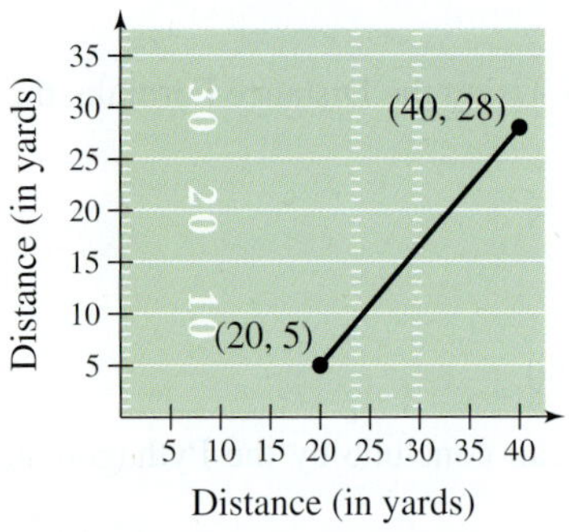

Figure 1.8

Solution You can find the length of the pass by finding the distance between the points (40, 28) and (20, 5).

$$d = \sqrt{(x_2 - x_1)^2 + (y_2 - y_1)^2} \qquad \text{Distance Formula}$$
$$= \sqrt{(40 - 20)^2 + (28 - 5)^2} \qquad \text{Substitute for } x_1, y_1, x_2, \text{ and } y_2.$$
$$= \sqrt{20^2 + 23^2} \qquad \text{Simplify.}$$
$$= \sqrt{400 + 529} \qquad \text{Simplify.}$$
$$= \sqrt{929} \qquad \text{Simplify.}$$
$$\approx 30 \qquad \text{Use a calculator.}$$

So, the pass is about 30 yards long.

✓ ***Checkpoint*** *Audio-video solution in English & Spanish at LarsonPrecalculus.com.*

A football quarterback throws a pass from the 10-yard line, 10 yards from the sideline. A wide receiver catches the pass on the 32-yard line, 25 yards from the same sideline. How long is the pass? ■

In Example 6, the scale along the goal line does not normally appear on a football field. However, when you use coordinate geometry to solve real-life problems, you are free to place the coordinate system in any way that is convenient for the solution of the problem.

EXAMPLE 7 Estimating Annual Sales

Starbucks Corporation had annual sales of approximately \$9.8 billion in 2009 and \$11.7 billion in 2011. Without knowing any additional information, what would you estimate the 2010 sales to have been? *(Source: Starbucks Corporation)*

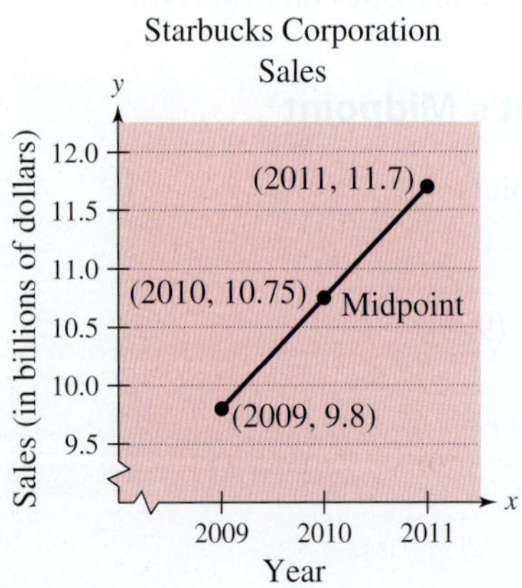

Figure 1.9

Solution One solution to the problem is to assume that sales followed a linear pattern. With this assumption, you can estimate the 2010 sales by finding the midpoint of the line segment connecting the points (2009, 9.8) and (2011, 11.7).

$$\text{Midpoint} = \left(\frac{x_1 + x_2}{2}, \frac{y_1 + y_2}{2}\right) \qquad \text{Midpoint Formula}$$
$$= \left(\frac{2009 + 2011}{2}, \frac{9.8 + 11.7}{2}\right) \qquad \text{Substitute for } x_1, x_2, y_1, \text{ and } y_2.$$
$$= (2010, 10.75) \qquad \text{Simplify.}$$

So, you would estimate the 2010 sales to have been about \$10.75 billion, as shown in Figure 1.9. (The actual 2010 sales were about \$10.71 billion.)

✓ ***Checkpoint*** *Audio-video solution in English & Spanish at LarsonPrecalculus.com.*

Yahoo! Inc. had annual revenues of approximately \$7.2 billon in 2008 and \$6.3 billion in 2010. Without knowing any additional information, what would you estimate the 2009 revenue to have been? *(Source: Yahoo! Inc.)* ■

Much of computer graphics, including this computer-generated tessellation, consists of transformations of points in a coordinate plane. Example 8 illustrates one type of transformation called a translation. Other types include reflections, rotations, and stretches.

EXAMPLE 8 **Translating Points in the Plane**

The triangle in Figure 1.10 has vertices at the points $(-1, 2)$, $(1, -4)$, and $(2, 3)$. Shift the triangle three units to the right and two units up and find the vertices of the shifted triangle, as shown in Figure 1.11.

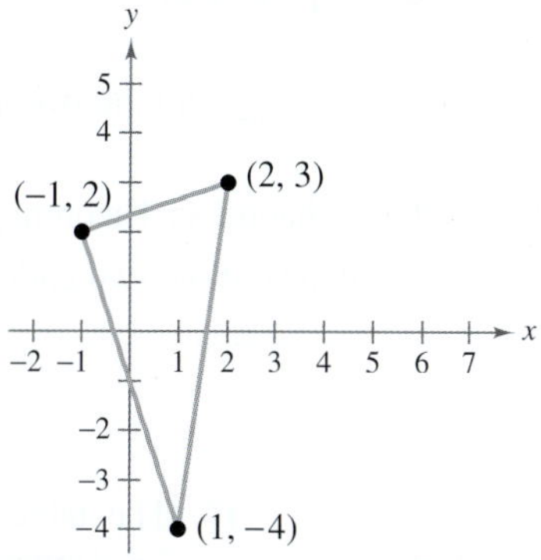

Figure 1.10

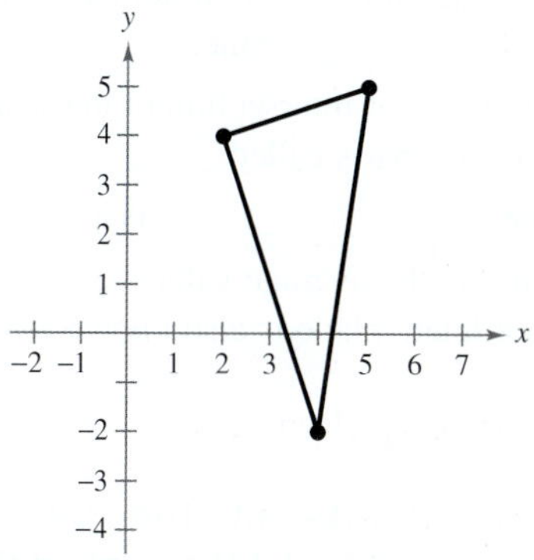

Figure 1.11

Solution To shift the vertices three units to the right, add 3 to each of the x-coordinates. To shift the vertices two units up, add 2 to each of the y-coordinates.

Original Point	Translated Point
$(-1, 2)$	$(-1 + 3, 2 + 2) = (2, 4)$
$(1, -4)$	$(1 + 3, -4 + 2) = (4, -2)$
$(2, 3)$	$(2 + 3, 3 + 2) = (5, 5)$

✓ ***Checkpoint*** *Audio-video solution in English & Spanish at LarsonPrecalculus.com.*

Find the vertices of the parallelogram shown after translating it two units to the left and four units down.

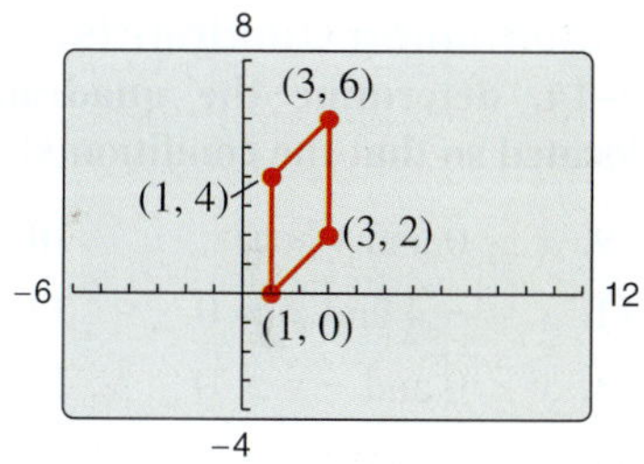

The figures in Example 8 were not really essential to the solution. Nevertheless, it is strongly recommended that you develop the habit of including sketches with your solutions—even when they are not required.

Summarize (Section 1.1)

1. Describe the Cartesian plane *(page 2)*. For an example of plotting points in the Cartesian plane, see Example 1.
2. State the Distance Formula *(page 4)*. For examples of using the Distance Formula to find the distance between two points, see Examples 3 and 4.
3. State the Midpoint Formula *(page 5)*. For an example of using the Midpoint Formula to find the midpoint of a line segment, see Example 5.
4. Describe examples of how to use a coordinate plane to model and solve real-life problems *(pages 6 and 7, Examples 6–8)*.

Matt Antonino/Shutterstock.com

1.1 Exercises

See **CalcChat.com** for tutorial help and worked-out solutions to odd-numbered exercises.

Vocabulary: Fill in the blanks.

1. An ordered pair of real numbers can be represented in a plane called the rectangular coordinate system or the ________ plane.

2. The point of intersection of the x- and y-axes is the ________, and the two axes divide the coordinate plane into four parts called ________.

3. The ________ ________ is a result derived from the Pythagorean Theorem.

4. Finding the average values of the representative coordinates of the two endpoints of a line segment in a coordinate plane is also known as using the ________ ________.

Skills and Applications

Plotting Points in the Cartesian Plane **In Exercises 5 and 6, plot the points in the Cartesian plane.**

5. $(-4, 2), (-3, -6), (0, 5), (1, -4), (0, 0), (3, 1)$

6. $\left(1, -\frac{1}{3}\right), (0.5, -1), \left(\frac{3}{7}, 3\right), \left(-\frac{4}{3}, -\frac{3}{7}\right), (-2, 2.5)$

Finding the Coordinates of a Point **In Exercises 7 and 8, find the coordinates of the point.**

7. The point is located three units to the left of the y-axis and four units above the x-axis.

8. The point is on the x-axis and 12 units to the left of the y-axis.

Determining Quadrant(s) for a Point **In Exercises 9–14, determine the quadrant(s) in which (x, y) is located so that the condition(s) is (are) satisfied.**

9. $x > 0$ and $y < 0$ **10.** $x < 0$ and $y < 0$

11. $x = -4$ and $y > 0$ **12.** $y < -5$

13. $x < 0$ and $-y > 0$ **14.** $xy > 0$

Sketching a Scatter Plot **In Exercises 15 and 16, sketch a scatter plot of the data shown in the table.**

15. The table shows the number y of Wal-Mart stores for each year x from 2003 through 2010. *(Source: Wal-Mart Stores, Inc.)*

DATA

Year, x	Number of Stores, y
2003	4906
2004	5289
2005	6141
2006	6779
2007	7262
2008	7720
2009	8416
2010	8970

Spreadsheet at LarsonPrecalculus.com

16. The table shows the lowest temperature on record y (in degrees Fahrenheit) in Duluth, Minnesota, for each month x, where $x = 1$ represents January. *(Source: NOAA)*

DATA

Month, x	Temperature, y
1	−39
2	−39
3	−29
4	−5
5	17
6	27
7	35
8	32
9	22
10	8
11	−23
12	−34

Spreadsheet at LarsonPrecalculus.com

Finding a Distance **In Exercises 17–22, find the distance between the points.**

17. $(-2, 6), (3, -6)$ **18.** $(8, 5), (0, 20)$

19. $(1, 4), (-5, -1)$ **20.** $(1, 3), (3, -2)$

21. $\left(\frac{1}{2}, \frac{4}{3}\right), (2, -1)$ **22.** $(9.5, -2.6), (-3.9, 8.2)$

Verifying a Right Triangle **In Exercises 23 and 24, (a) find the length of each side of the right triangle, and (b) show that these lengths satisfy the Pythagorean Theorem.**

23.

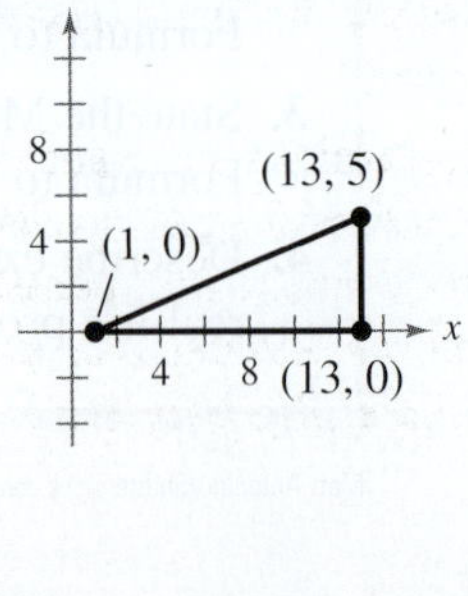

24.

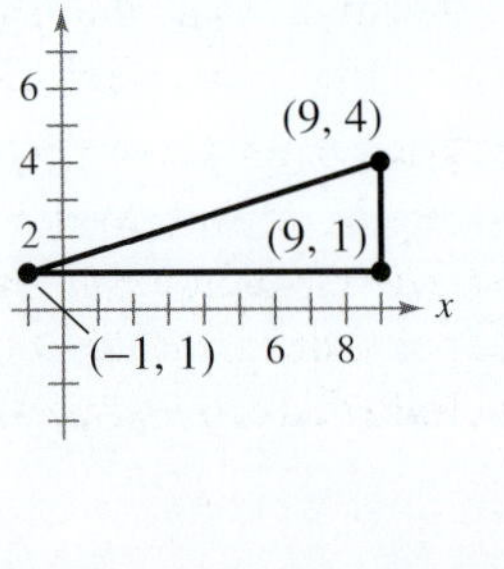

Verifying a Polygon **In Exercises 25–28, show that the points form the vertices of the indicated polygon.**

25. Right triangle: $(4, 0), (2, 1), (-1, -5)$

26. Right triangle: $(-1, 3), (3, 5), (5, 1)$

27. Isosceles triangle: $(1, -3), (3, 2), (-2, 4)$

28. Isosceles triangle: $(2, 3), (4, 9), (-2, 7)$

Plotting, Distance, and Midpoint **In Exercises 29–36, (a) plot the points, (b) find the distance between the points, and (c) find the midpoint of the line segment joining the points.**

29. $(6, -3), (6, 5)$

30. $(1, 4), (8, 4)$

31. $(1, 1), (9, 7)$

32. $(1, 12), (6, 0)$

33. $(-1, 2), (5, 4)$

34. $(2, 10), (10, 2)$

35. $(-16.8, 12.3), (5.6, 4.9)$

36. $\left(\frac{1}{2}, 1\right), \left(-\frac{5}{2}, \frac{4}{3}\right)$

37. Flying Distance An airplane flies from Naples, Italy, in a straight line to Rome, Italy, which is 120 kilometers north and 150 kilometers west of Naples. How far does the plane fly?

38. Sports A soccer player passes the ball from a point that is 18 yards from the endline and 12 yards from the sideline. A teammate who is 42 yards from the same endline and 50 yards from the same sideline receives the pass. (See figure.) How long is the pass?

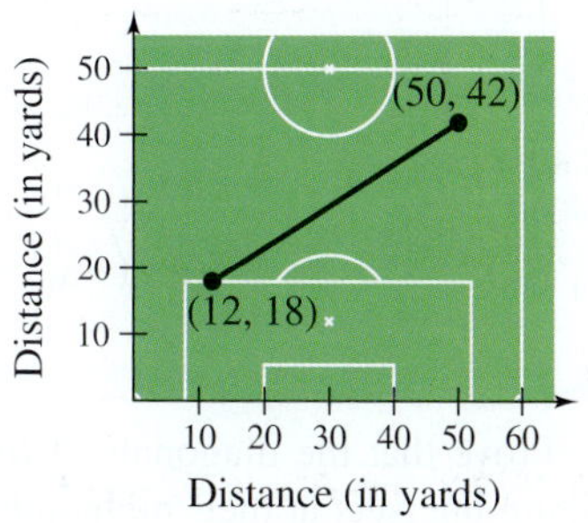

39. Sales The Coca-Cola Company had sales of \$19,564 million in 2002 and \$35,123 million in 2010. Use the Midpoint Formula to estimate the sales in 2006. Assume that the sales followed a linear pattern. *(Source: The Coca-Cola Company)*

40. Earnings per Share The earnings per share for Big Lots, Inc. were \$1.89 in 2008 and \$2.83 in 2010. Use the Midpoint Formula to estimate the earnings per share in 2009. Assume that the earnings per share followed a linear pattern. *(Source: Big Lots, Inc.)*

Translating Points in the Plane **In Exercises 41–44, find the coordinates of the vertices of the polygon after the indicated translation to a new position in the plane.**

41.

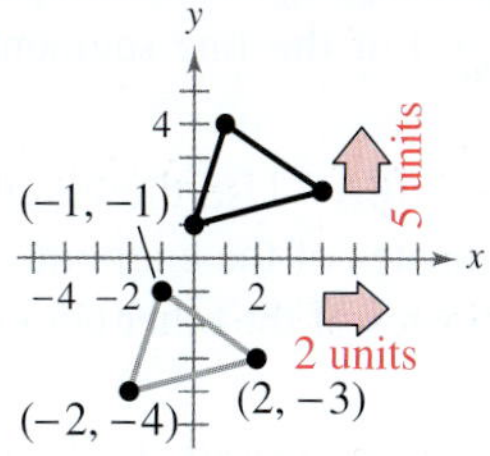

42.

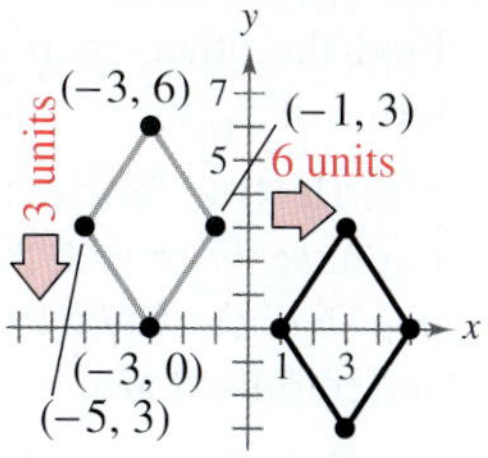

43. Original coordinates of vertices: $(-7, -2), (-2, 2), (-2, -4), (-7, -4)$

Shift: eight units up, four units to the right

44. Original coordinates of vertices: $(5, 8), (3, 6), (7, 6)$

Shift: 6 units down, 10 units to the left

45. Minimum Wage Use the graph below, which shows the minimum wages in the United States (in dollars) from 1950 through 2011. *(Source: U.S. Department of Labor)*

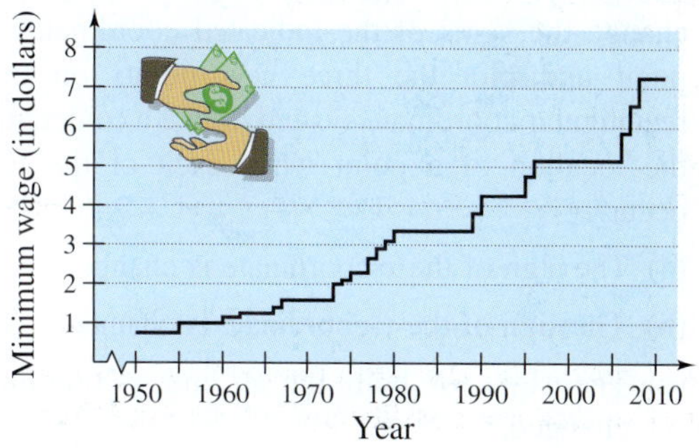

(a) Which decade shows the greatest increase in minimum wage?

(b) Approximate the percent increases in the minimum wage from 1990 to 1995 and from 1995 to 2011.

(c) Use the percent increase from 1995 to 2011 to predict the minimum wage in 2016.

(d) Do you believe that your prediction in part (c) is reasonable? Explain.

46. Data Analysis: Exam Scores The table shows the mathematics entrance test scores x and the final examination scores y in an algebra course for a sample of 10 students.

x	22	29	35	40	44	48	53	58	65	76
y	53	74	57	66	79	90	76	93	83	99

(a) Sketch a scatter plot of the data.

(b) Find the entrance test score of any student with a final exam score in the 80s.

(c) Does a higher entrance test score imply a higher final exam score? Explain.

Fernando Jose Vasconcelos Soares/Shutterstock.com

Exploration

47. Using the Midpoint Formula A line segment has (x_1, y_1) as one endpoint and (x_m, y_m) as its midpoint. Find the other endpoint (x_2, y_2) of the line segment in terms of x_1, y_1, x_m, and y_m.

48. Using the Midpoint Formula Use the result of Exercise 47 to find the coordinates of the endpoint of a line segment when the coordinates of the other endpoint and midpoint are, respectively,

(a) $(1, -2), (4, -1)$ and (b) $(-5, 11), (2, 4)$.

49. Using the Midpoint Formula Use the Midpoint Formula three times to find the three points that divide the line segment joining (x_1, y_1) and (x_2, y_2) into four parts.

50. Using the Midpoint Formula Use the result of Exercise 49 to find the points that divide the line segment joining the given points into four equal parts.

(a) $(1, -2), (4, -1)$ (b) $(-2, -3), (0, 0)$

51. Make a Conjecture Plot the points $(2, 1)$, $(-3, 5)$, and $(7, -3)$ on a rectangular coordinate system. Then change the signs of the indicated coordinates of each point and plot the three new points on the same rectangular coordinate system. Make a conjecture about the location of a point when each of the following occurs.

(a) The sign of the x-coordinate is changed.

(b) The sign of the y-coordinate is changed.

(c) The signs of both the x- and y-coordinates are changed.

52. Collinear Points Three or more points are *collinear* when they all lie on the same line. Use the steps following to determine whether the set of points $\{A(2, 3), B(2, 6), C(6, 3)\}$ and the set of points $\{A(8, 3), B(5, 2), C(2, 1)\}$ are collinear.

(a) For each set of points, use the Distance Formula to find the distances from A to B, from B to C, and from A to C. What relationship exists among these distances for each set of points?

(b) Plot each set of points in the Cartesian plane. Do all the points of either set appear to lie on the same line?

(c) Compare your conclusions from part (a) with the conclusions you made from the graphs in part (b). Make a general statement about how to use the Distance Formula to determine collinearity.

53. Think About It When plotting points on the rectangular coordinate system, is it true that the scales on the x- and y-axes must be the same? Explain.

54. Think About It What is the y-coordinate of any point on the x-axis? What is the x-coordinate of any point on the y-axis?

True or False? **In Exercises 55–57, determine whether the statement is true or false. Justify your answer.**

55. In order to divide a line segment into 16 equal parts, you would have to use the Midpoint Formula 16 times.

56. The points $(-8, 4)$, $(2, 11)$, and $(-5, 1)$ represent the vertices of an isosceles triangle.

57. If four points represent the vertices of a polygon, and the four sides are equal, then the polygon must be a square.

58. HOW DO YOU SEE IT? Use the plot of the point (x_0, y_0) in the figure. Match the transformation of the point with the correct plot. Explain your reasoning. [The plots are labeled (i), (ii), (iii), and (iv).]

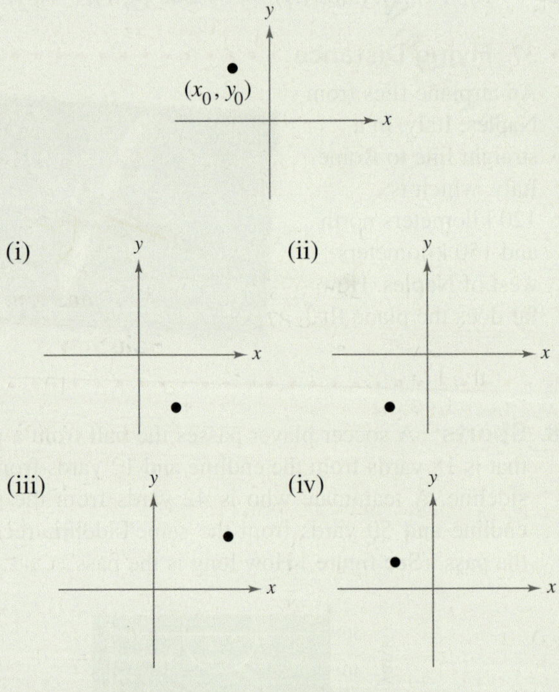

(a) (x_0, y_0) (b) $(-2x_0, y_0)$

(c) $(x_0, \frac{1}{2}y_0)$ (d) $(-x_0, -y_0)$

59. Proof Prove that the diagonals of the parallelogram in the figure intersect at their midpoints.

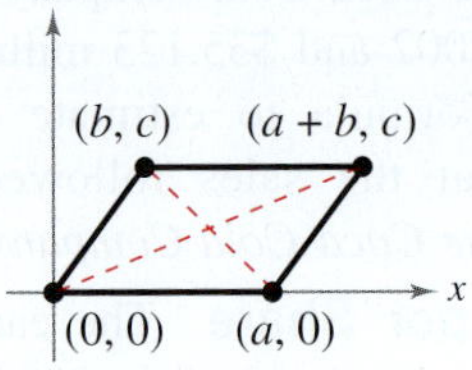

1.2 Graphs of Equations

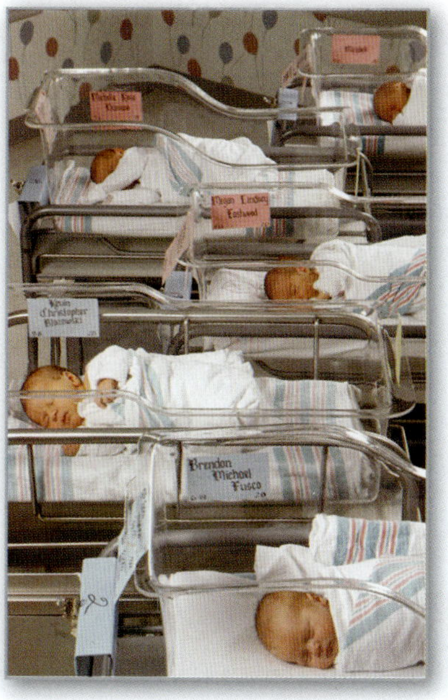

The graph of an equation can help you see relationships between real-life quantities. For example, in Exercise 87 on page 21, you will use a graph to predict the life expectancy of a child born in 2015.

- Sketch graphs of equations.
- Identify x- and y-intercepts of graphs of equations.
- Use symmetry to sketch graphs of equations.
- Write equations of and sketch graphs of circles.
- Use graphs of equations in solving real-life problems.

The Graph of an Equation

In Section 1.1, you used a coordinate system to graphically represent the relationship between two quantities. There, the graphical picture consisted of a collection of points in a coordinate plane.

Frequently, a relationship between two quantities is expressed as an **equation in two variables.** For instance, $y = 7 - 3x$ is an equation in x and y. An ordered pair (a, b) is a **solution** or **solution point** of an equation in x and y when the substitutions $x = a$ and $y = b$ result in a true statement. For instance, $(1, 4)$ is a solution of $y = 7 - 3x$ because $4 = 7 - 3(1)$ is a true statement.

In this section, you will review some basic procedures for sketching the graph of an equation in two variables. The **graph of an equation** is the set of all points that are solutions of the equation.

ALGEBRA HELP When evaluating an expression or an equation, remember to follow the Basic Rules of Algebra. To review these rules, see Appendix A.1.

EXAMPLE 1 Determining Solution Points

Determine whether (a) $(2, 13)$ and (b) $(-1, -3)$ lie on the graph of $y = 10x - 7$.

Solution

a. $y = 10x - 7$ Write original equation.

$13 \stackrel{?}{=} 10(2) - 7$ Substitute 2 for x and 13 for y.

$13 = 13$ $(2, 13)$ is a solution. ✓

The point $(2, 13)$ *does* lie on the graph of $y = 10x - 7$ because it is a solution point of the equation.

b. $y = 10x - 7$ Write original equation.

$-3 \stackrel{?}{=} 10(-1) - 7$ Substitute -1 for x and -3 for y.

$-3 \neq -17$ $(-1, -3)$ is not a solution.

The point $(-1, -3)$ *does not* lie on the graph of $y = 10x - 7$ because it is *not* a solution point of the equation.

✓ ***Checkpoint*** *Audio-video solution in English & Spanish at LarsonPrecalculus.com.*

Determine whether (a) $(3, -5)$ and (b) $(-2, 26)$ lie on the graph of $y = 14 - 6x$.

The basic technique used for sketching the graph of an equation is the **point-plotting method.**

The Point-Plotting Method of Graphing

1. When possible, isolate one of the variables.
2. Construct a table of values showing several solution points.
3. Plot these points in a rectangular coordinate system.
4. Connect the points with a smooth curve or line.

John Griffin/The Image Works

It is important to use negative values, zero, and positive values for x when constructing a table.

EXAMPLE 2 Sketching the Graph of an Equation

Sketch the graph of

$$y = -3x + 7.$$

Solution

Because the equation is already solved for y, construct a table of values that consists of several solution points of the equation. For instance, when $x = -1$,

$$y = -3(-1) + 7$$
$$= 10$$

which implies that

$$(-1, 10)$$

is a solution point of the equation.

x	$y = -3x + 7$	(x, y)
-1	10	$(-1, 10)$
0	7	$(0, 7)$
1	4	$(1, 4)$
2	1	$(2, 1)$
3	-2	$(3, -2)$
4	-5	$(4, -5)$

From the table, it follows that

$$(-1, 10),\ (0, 7),\ (1, 4),\ (2, 1),\ (3, -2),\quad \text{and} \quad (4, -5)$$

are solution points of the equation. After plotting these points and connecting them, you can see that they appear to lie on a line, as shown below.

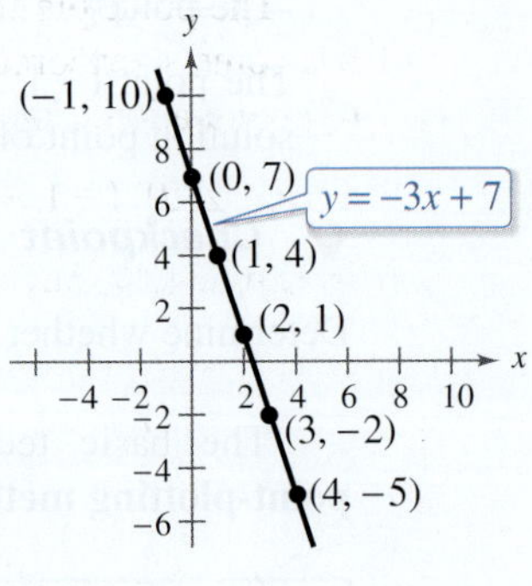

Checkpoint Audio-video solution in English & Spanish at LarsonPrecalculus.com.

Sketch the graph of each equation.

a. $y = -3x + 2$

b. $y = 2x + 1$

EXAMPLE 3 Sketching the Graph of an Equation

Sketch the graph of

$$y = x^2 - 2.$$

Solution

Because the equation is already solved for y, begin by constructing a table of values.

x	-2	-1	0	1	2	3
$y = x^2 - 2$	2	-1	-2	-1	2	7
(x, y)	$(-2, 2)$	$(-1, -1)$	$(0, -2)$	$(1, -1)$	$(2, 2)$	$(3, 7)$

Next, plot the points given in the table, as shown in Figure 1.12. Finally, connect the points with a smooth curve, as shown in Figure 1.13.

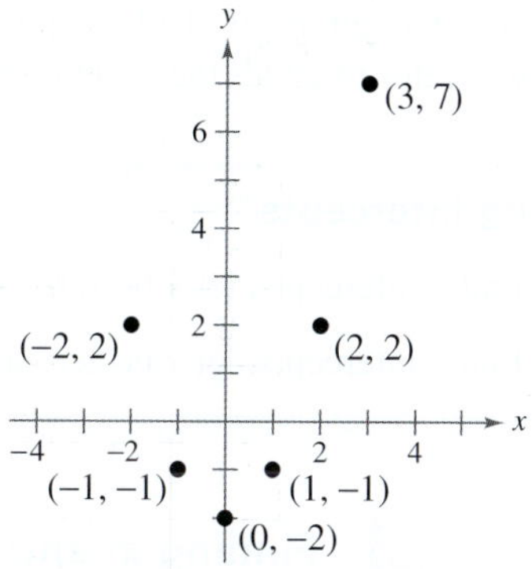

Figure 1.12

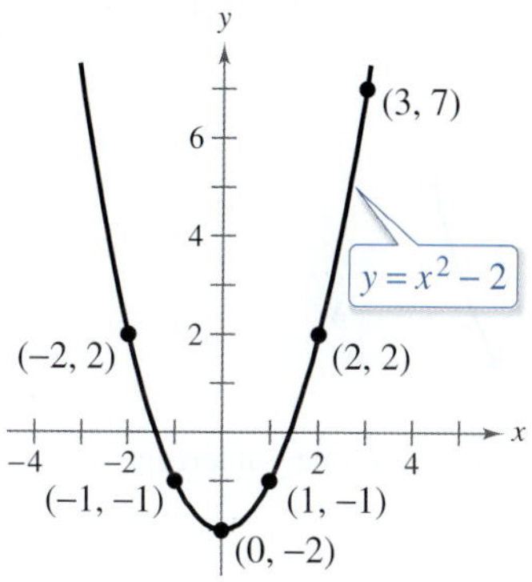

Figure 1.13

REMARK One of your goals in this course is to learn to classify the basic shape of a graph from its equation. For instance, you will learn that the *linear equation* in Example 2 has the form

$$y = mx + b$$

and its graph is a line. Similarly, the *quadratic equation* in Example 3 has the form

$$y = ax^2 + bx + c$$

and its graph is a parabola.

✓ ***Checkpoint*** *Audio-video solution in English & Spanish at LarsonPrecalculus.com.*

Sketch the graph of each equation.

a. $y = x^2 + 3$

b. $y = 1 - x^2$

The point-plotting method demonstrated in Examples 2 and 3 is easy to use, but it has some shortcomings. With too few solution points, you can misrepresent the graph of an equation. For instance, when you only plot the four points

$$(-2, 2), (-1, -1), (1, -1), \quad \text{and} \quad (2, 2)$$

in Figure 1.12, any one of the three graphs below is reasonable.

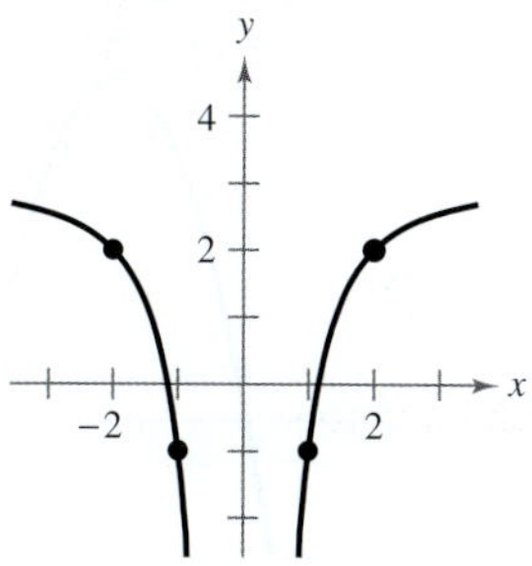

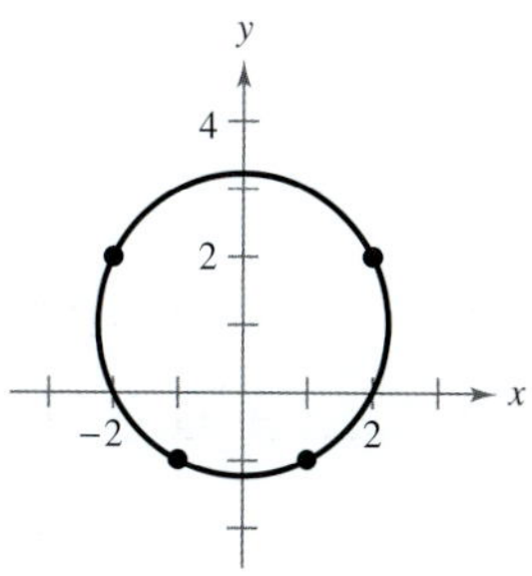

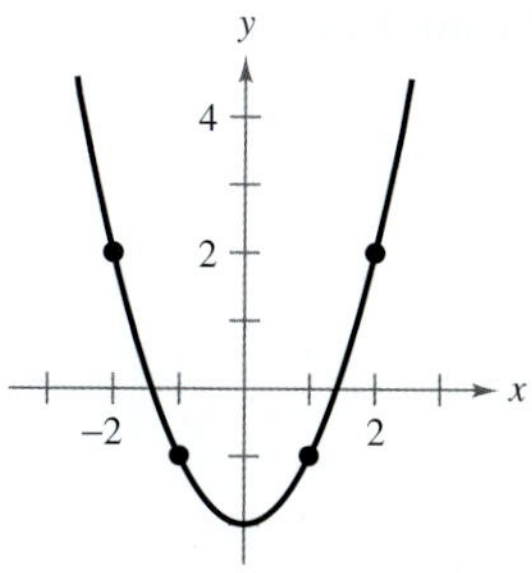

TECHNOLOGY To graph an equation involving x and y on a graphing utility, use the following procedure.

1. Rewrite the equation so that y is isolated on the left side.
2. Enter the equation into the graphing utility.
3. Determine a *viewing window* that shows all important features of the graph.
4. Graph the equation.

Intercepts of a Graph

It is often easy to determine the solution points that have zero as either the x-coordinate or the y-coordinate. These points are called **intercepts** because they are the points at which the graph intersects or touches the x- or y-axis. It is possible for a graph to have no intercepts, one intercept, or several intercepts, as shown in Figure 1.14.

Note that an x-intercept can be written as the ordered pair $(a, 0)$ and a y-intercept can be written as the ordered pair $(0, b)$. Some texts denote the x-intercept as the x-coordinate of the point $(a, 0)$ [and the y-intercept as the y-coordinate of the point $(0, b)$] rather than the point itself. Unless it is necessary to make a distinction, the term *intercept* will refer to either the point or the coordinate.

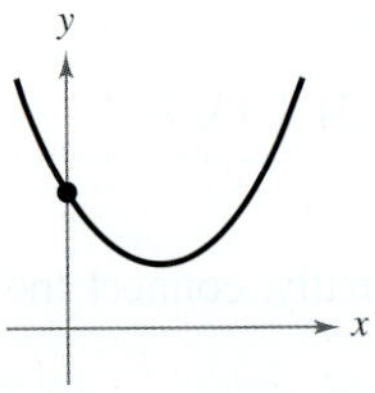

No x-intercepts; one y-intercept

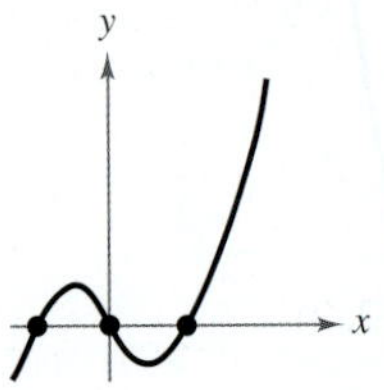

Three x-intercepts; one y-intercept

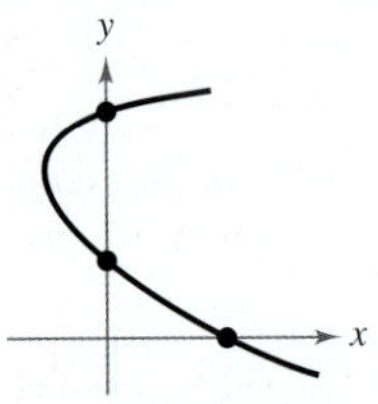

One x-intercept; two y-intercepts

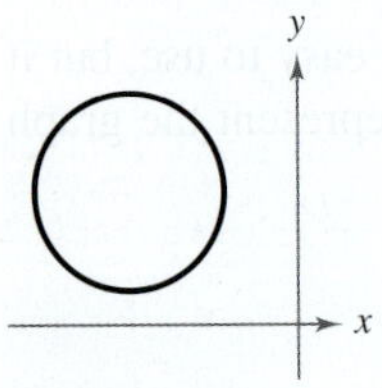

No intercepts

Figure 1.14

Finding Intercepts

1. To find x-intercepts, let y be zero and solve the equation for x.
2. To find y-intercepts, let x be zero and solve the equation for y.

EXAMPLE 4 Finding x- and y-Intercepts

To find the x-intercepts of the graph of $y = x^3 - 4x$, let $y = 0$. Then $0 = x^3 - 4x = x(x^2 - 4)$ has solutions $x = 0$ and $x = \pm 2$.

x-intercepts: $(0, 0), (2, 0), (-2, 0)$ See figure.

To find the y-intercept of the graph of $y = x^3 - 4x$, let $x = 0$. Then $y = (0)^3 - 4(0)$ has one solution, $y = 0$.

y-intercept: $(0, 0)$ See figure.

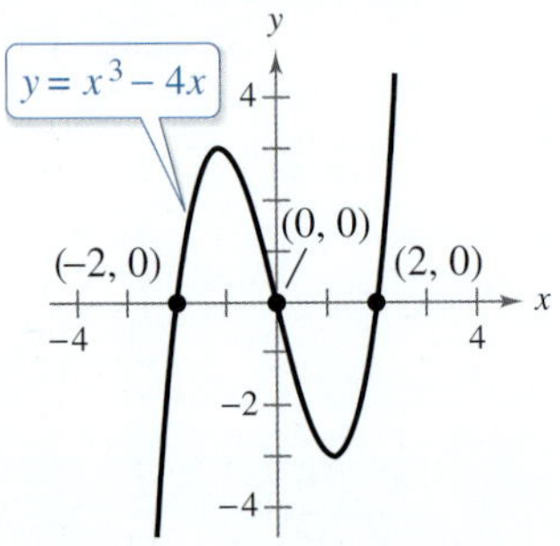

✓ **Checkpoint** *Audio-video solution in English & Spanish at LarsonPrecalculus.com.*

Find the x- and y-intercepts of the graph of $y = -x^2 - 5x$ shown in the figure below.

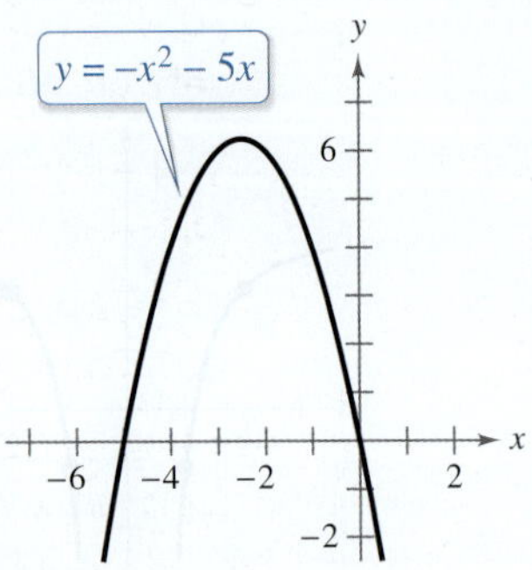

Symmetry

Graphs of equations can have **symmetry** with respect to one of the coordinate axes or with respect to the origin. Symmetry with respect to the x-axis means that when the Cartesian plane is folded along the x-axis, the portion of the graph above the x-axis coincides with the portion below the x-axis. Symmetry with respect to the y-axis or the origin can be described in a similar manner, as shown below.

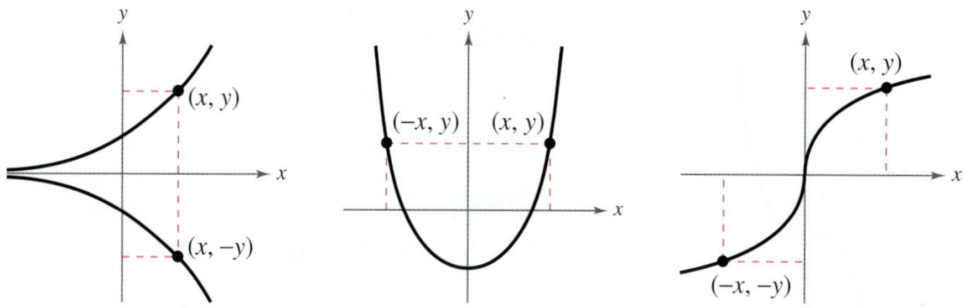

x-Axis symmetry y-Axis symmetry Origin symmetry

Knowing the symmetry of a graph *before* attempting to sketch it is helpful, because then you need only half as many solution points to sketch the graph. There are three basic types of symmetry, described as follows.

Graphical Tests for Symmetry

1. A graph is **symmetric with respect to the x-axis** if, whenever (x, y) is on the graph, $(x, -y)$ is also on the graph.
2. A graph is **symmetric with respect to the y-axis** if, whenever (x, y) is on the graph, $(-x, y)$ is also on the graph.
3. A graph is **symmetric with respect to the origin** if, whenever (x, y) is on the graph, $(-x, -y)$ is also on the graph.

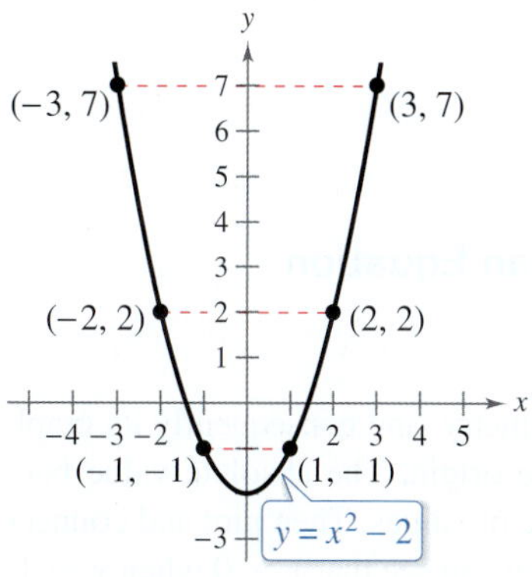

y-Axis symmetry
Figure 1.15

You can conclude that the graph of $y = x^2 - 2$ is symmetric with respect to the y-axis because the point $(-x, y)$ is also on the graph of $y = x^2 - 2$. (See the table below and Figure 1.15.)

x	-3	-2	-1	1	2	3
y	7	2	-1	-1	2	7
(x, y)	$(-3, 7)$	$(-2, 2)$	$(-1, -1)$	$(1, -1)$	$(2, 2)$	$(3, 7)$

Algebraic Tests for Symmetry

1. The graph of an equation is symmetric with respect to the x-axis when replacing y with $-y$ yields an equivalent equation.
2. The graph of an equation is symmetric with respect to the y-axis when replacing x with $-x$ yields an equivalent equation.
3. The graph of an equation is symmetric with respect to the origin when replacing x with $-x$ and y with $-y$ yields an equivalent equation.

EXAMPLE 5 Testing for Symmetry

Test $y = 2x^3$ for symmetry with respect to both axes and the origin.

Solution

x-Axis:	$y = 2x^3$	Write original equation.
	$-y = 2x^3$	Replace y with $-y$. Result is *not* an equivalent equation.
y-Axis:	$y = 2x^3$	Write original equation.
	$y = 2(-x)^3$	Replace x with $-x$.
	$y = -2x^3$	Simplify. Result is *not* an equivalent equation.
Origin:	$y = 2x^3$	Write original equation.
	$-y = 2(-x)^3$	Replace y with $-y$ and x with $-x$.
	$-y = -2x^3$	Simplify.
	$y = 2x^3$	Equivalent equation

Of the three tests for symmetry, the only one that is satisfied is the test for origin symmetry (see Figure 1.16).

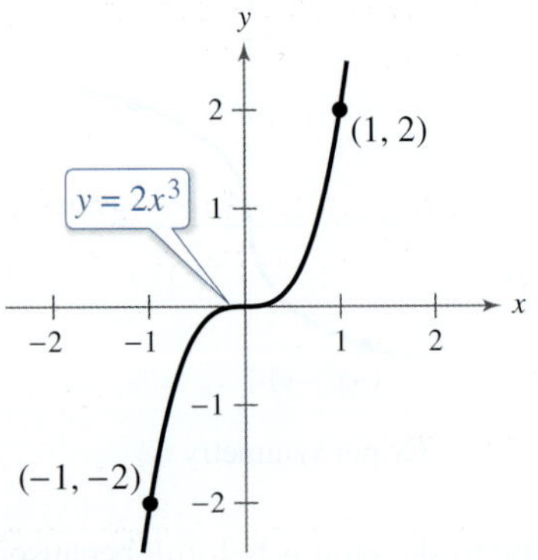

Figure 1.16

✓ ***Checkpoint*** *Audio-video solution in English & Spanish at LarsonPrecalculus.com.*

Test $y^2 = 6 - x$ for symmetry with respect to both axes and the origin.

EXAMPLE 6 Using Symmetry as a Sketching Aid

Use symmetry to sketch the graph of $x - y^2 = 1$.

Solution Of the three tests for symmetry, the only one that is satisfied is the test for x-axis symmetry because $x - (-y)^2 = 1$ is equivalent to $x - y^2 = 1$. So, the graph is symmetric with respect to the x-axis. Using symmetry, you only need to find the solution points above the x-axis and then reflect them to obtain the graph, as shown in Figure 1.17.

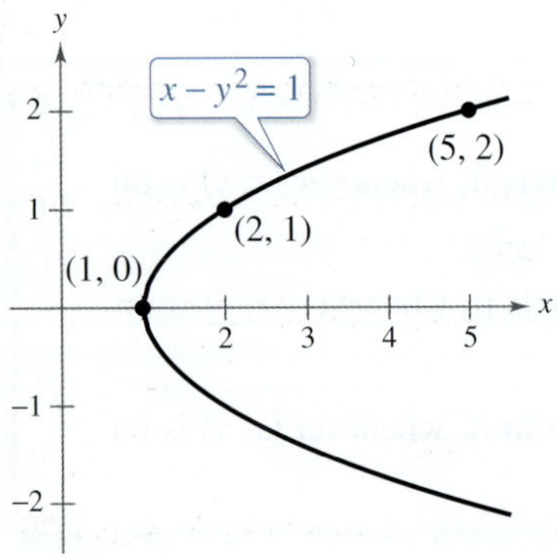

Figure 1.17

✓ ***Checkpoint*** *Audio-video solution in English & Spanish at LarsonPrecalculus.com.*

Use symmetry to sketch the graph of $y = x^2 - 4$.

ALGEBRA HELP In Example 7, $|x - 1|$ is an absolute value expression. You can review the techniques for evaluating an absolute value expression in Appendix A.1.

EXAMPLE 7 Sketching the Graph of an Equation

Sketch the graph of $y = |x - 1|$.

Solution This equation fails all three tests for symmetry, and consequently its graph is not symmetric with respect to either axis or to the origin. The absolute value bars indicate that y is always nonnegative. Construct a table of values. Then plot and connect the points, as shown in Figure 1.18. From the table, you can see that $x = 0$ when $y = 1$. So, the y-intercept is $(0, 1)$. Similarly, $y = 0$ when $x = 1$. So, the x-intercept is $(1, 0)$.

x	-2	-1	0	1	2	3	4
$y = \|x - 1\|$	3	2	1	0	1	2	3
(x, y)	$(-2, 3)$	$(-1, 2)$	$(0, 1)$	$(1, 0)$	$(2, 1)$	$(3, 2)$	$(4, 3)$

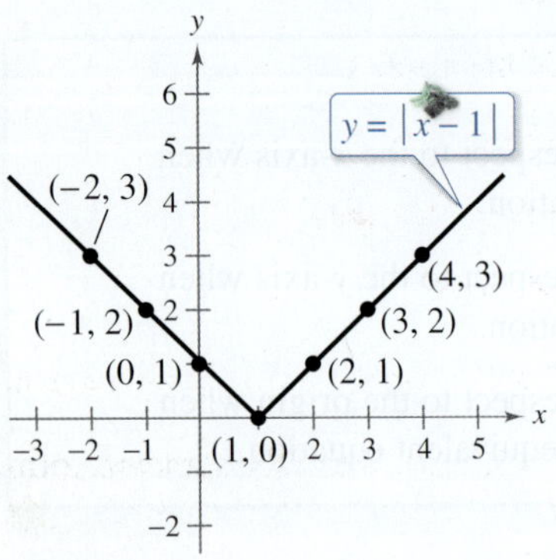

Figure 1.18

✓ ***Checkpoint*** *Audio-video solution in English & Spanish at LarsonPrecalculus.com.*

Sketch the graph of $y = |x - 2|$.

Circles

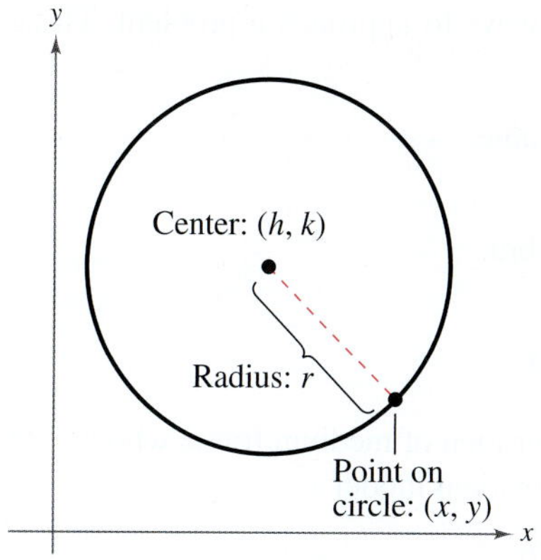

Figure 1.19

Throughout this course, you will learn to recognize several types of graphs from their equations. For instance, you will learn to recognize that the graph of a second-degree equation of the form

$$y = ax^2 + bx + c$$

is a parabola (see Example 3). The graph of a **circle** is also easy to recognize.

Consider the circle shown in Figure 1.19. A point (x, y) lies on the circle if and only if its distance from the center (h, k) is r. By the Distance Formula,

$$\sqrt{(x - h)^2 + (y - k)^2} = r.$$

By squaring each side of this equation, you obtain the **standard form of the equation of a circle.** For example, a circle with its center at $(h, k) = (1, 3)$ and radius $r = 4$ is given by

$$\sqrt{(x - 1)^2 + (y - 3)^2} = 4 \quad \text{Substitute for } h, k, \text{ and } r.$$

$$(x - 1)^2 + (y - 3)^2 = 16. \quad \text{Square each side.}$$

> **Standard Form of the Equation of a Circle**
>
> A point (x, y) lies on the circle of **radius** r and **center** (h, k) if and only if
>
> $$(x - h)^2 + (y - k)^2 = r^2.$$

REMARK Be careful when you are finding h and k from the standard form of the equation of a circle. For instance, to find h and k from the equation of the circle in Example 8, rewrite the quantities $(x + 1)^2$ and $(y - 2)^2$ using subtraction.

$$(x + 1)^2 = [x - (-1)]^2,$$
$$(y - 2)^2 = [y - (2)]^2$$

So, $h = -1$ and $k = 2$.

From this result, you can see that the standard form of the equation of a circle *with its center at the origin,* $(h, k) = (0, 0)$, is simply

$$x^2 + y^2 = r^2. \quad \text{Circle with center at origin}$$

EXAMPLE 8 Writing the Equation of a Circle

The point $(3, 4)$ lies on a circle whose center is at $(-1, 2)$, as shown in Figure 1.20. Write the standard form of the equation of this circle.

Solution

The radius of the circle is the distance between $(-1, 2)$ and $(3, 4)$.

$$r = \sqrt{(x - h)^2 + (y - k)^2} \quad \text{Distance Formula}$$
$$= \sqrt{[3 - (-1)]^2 + (4 - 2)^2} \quad \text{Substitute for } x, y, h, \text{ and } k.$$
$$= \sqrt{4^2 + 2^2} \quad \text{Simplify.}$$
$$= \sqrt{16 + 4} \quad \text{Simplify.}$$
$$= \sqrt{20} \quad \text{Radius}$$

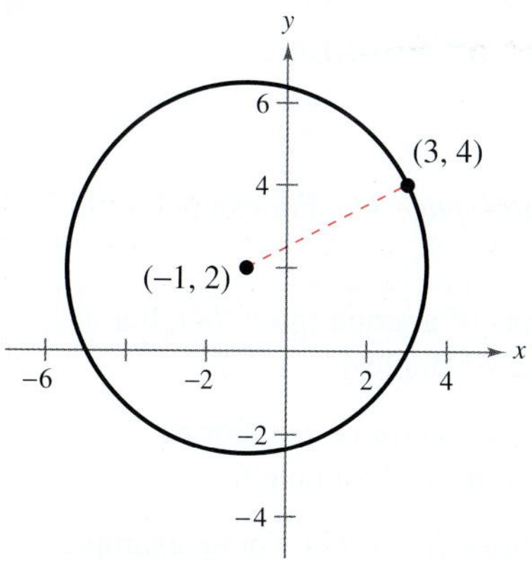

Figure 1.20

Using $(h, k) = (-1, 2)$ and $r = \sqrt{20}$, the equation of the circle is

$$(x - h)^2 + (y - k)^2 = r^2 \quad \text{Equation of circle}$$
$$[x - (-1)]^2 + (y - 2)^2 = (\sqrt{20})^2 \quad \text{Substitute for } h, k, \text{ and } r.$$
$$(x + 1)^2 + (y - 2)^2 = 20. \quad \text{Standard form}$$

Checkpoint *Audio-video solution in English & Spanish at LarsonPrecalculus.com.*

The point $(1, -2)$ lies on a circle whose center is at $(-3, -5)$. Write the standard form of the equation of this circle.

Application

REMARK You should develop the habit of using at least two approaches to solve every problem. This helps build your intuition and helps you check that your answers are reasonable.

In this course, you will learn that there are many ways to approach a problem. Three common approaches are illustrated in Example 9.

A Numerical Approach: Construct and use a table.

A Graphical Approach: Draw and use a graph.

An Algebraic Approach: Use the rules of algebra.

EXAMPLE 9 Recommended Weight

The median recommended weights y (in pounds) for men of medium frame who are 25 to 59 years old can be approximated by the mathematical model

$$y = 0.073x^2 - 6.99x + 289.0, \quad 62 \le x \le 76$$

where x is a man's height (in inches). *(Source: Metropolitan Life Insurance Company)*

a. Construct a table of values that shows the median recommended weights for men with heights of 62, 64, 66, 68, 70, 72, 74, and 76 inches.

b. Use the table of values to sketch a graph of the model. Then use the graph to estimate *graphically* the median recommended weight for a man whose height is 71 inches.

c. Use the model to confirm *algebraically* the estimate you found in part (b).

Solution

DATA

Height, x	Weight, y
62	136.2
64	140.6
66	145.6
68	151.2
70	157.4
72	164.2
74	171.5
76	179.4

Spreadsheet at LarsonPrecalculus.com

a. You can use a calculator to construct the table, as shown on the left.

b. The table of values can be used to sketch the graph of the equation, as shown in Figure 1.21. From the graph, you can estimate that a height of 71 inches corresponds to a weight of about 161 pounds.

Recommended Weight

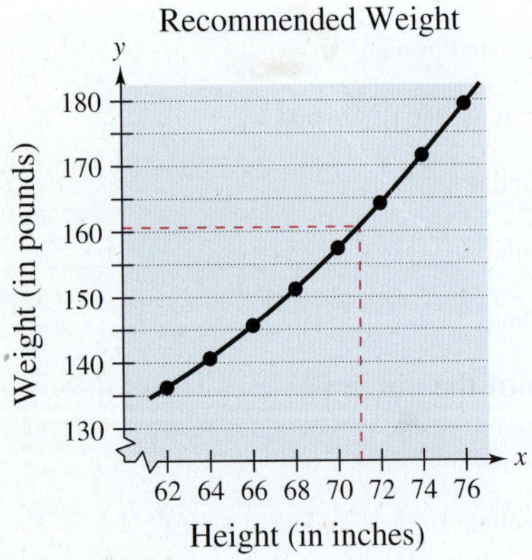

Figure 1.21

c. To confirm algebraically the estimate found in part (b), you can substitute 71 for x in the model.

$$y = 0.073(71)^2 - 6.99(71) + 289.0 \approx 160.70$$

So, the graphical estimate of 161 pounds is fairly good.

✓ **Checkpoint** Audio-video solution in English & Spanish at LarsonPrecalculus.com.

Use Figure 1.21 to estimate *graphically* the median recommended weight for a man whose height is 75 inches. Then confirm the estimate *algebraically*.

Summarize (Section 1.2)

1. Describe how to sketch the graph of an equation *(page 11)*. For examples of graphing equations, see Examples 1–3.
2. Describe how to identify the x- and y-intercepts of a graph *(page 14)*. For an example of identifying x- and y-intercepts, see Example 4.
3. Describe how to use symmetry to graph an equation *(page 15)*. For an example of using symmetry to graph an equation, see Example 6.
4. State the standard form of the equation of a circle *(page 17)*. For an example of writing the standard form of the equation of a circle, see Example 8.
5. Describe how to use the graph of an equation to solve a real-life problem *(page 18, Example 9)*.

1.2 Exercises

See CalcChat.com for tutorial help and worked-out solutions to odd-numbered exercises.

Vocabulary: Fill in the blanks.

1. An ordered pair (a, b) is a ________ of an equation in x and y when the substitutions $x = a$ and $y = b$ result in a true statement.
2. The set of all solution points of an equation is the ________ of the equation.
3. The points at which a graph intersects or touches an axis are called the ________ of the graph.
4. A graph is symmetric with respect to the ________ if, whenever (x, y) is on the graph, $(-x, y)$ is also on the graph.
5. The equation $(x - h)^2 + (y - k)^2 = r^2$ is the standard form of the equation of a ________ with center ________ and radius ________.
6. When you construct and use a table to solve a problem, you are using a ________ approach.

Skills and Applications

Determining Solution Points **In Exercises 7–14, determine whether each point lies on the graph of the equation.**

	Equation	Points	
7.	$y = \sqrt{x + 4}$	(a) $(0, 2)$	(b) $(5, 3)$
8.	$y = \sqrt{5 - x}$	(a) $(1, 2)$	(b) $(5, 0)$
9.	$y = x^2 - 3x + 2$	(a) $(2, 0)$	(b) $(-2, 8)$
10.	$y = 4 - \lvert x - 2 \rvert$	(a) $(1, 5)$	(b) $(6, 0)$
11.	$y = \lvert x - 1 \rvert + 2$	(a) $(2, 3)$	(b) $(-1, 0)$
12.	$2x - y - 3 = 0$	(a) $(1, 2)$	(b) $(1, -1)$
13.	$x^2 + y^2 = 20$	(a) $(3, -2)$	(b) $(-4, 2)$
14.	$y = \frac{1}{3}x^3 - 2x^2$	(a) $\left(2, -\frac{16}{3}\right)$	(b) $(-3, 9)$

Sketching the Graph of an Equation **In Exercises 15–18, complete the table. Use the resulting solution points to sketch the graph of the equation.**

15. $y = -2x + 5$

x	-1	0	1	2	$\frac{5}{2}$
y					
(x, y)					

16. $y = \frac{3}{4}x - 1$

x	-2	0	1	$\frac{4}{3}$	2
y					
(x, y)					

17. $y = x^2 - 3x$

x	-1	0	1	2	3
y					
(x, y)					

18. $y = 5 - x^2$

x	-2	-1	0	1	2
y					
(x, y)					

Identifying x- and y-Intercepts **In Exercises 19–22, identify the x- and y-intercepts of the graph. Verify your results algebraically.**

19. $y = (x - 3)^2$

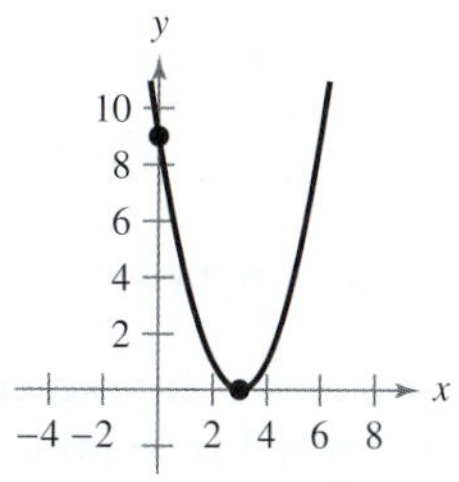

20. $y = 16 - 4x^2$

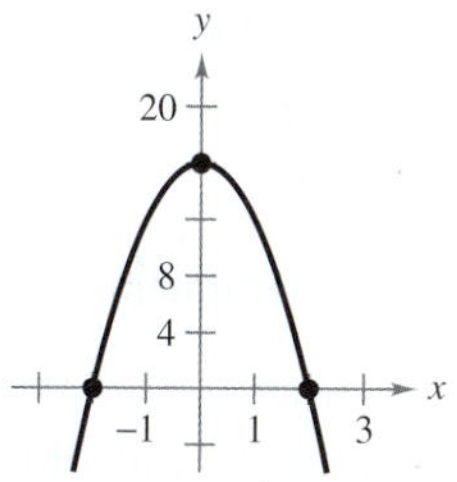

21. $y = \lvert x + 2 \rvert$

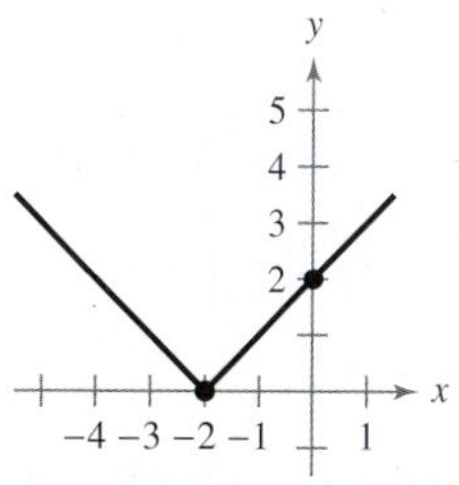

22. $y^2 = 4 - x$

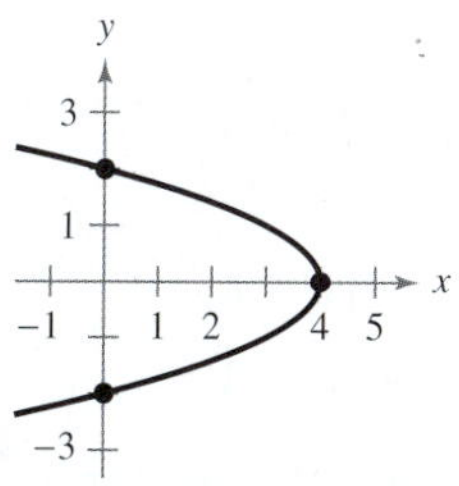

Finding x- and y-Intercepts **In Exercises 23–32, find the x- and y-intercepts of the graph of the equation.**

23. $y = 5x - 6$

24. $y = 8 - 3x$

25. $y = \sqrt{x + 4}$

26. $y = \sqrt{2x - 1}$

27. $y = \lvert 3x - 7 \rvert$

28. $y = -\lvert x + 10 \rvert$

29. $y = 2x^3 - 4x^2$

30. $y = x^4 - 25$

31. $y^2 = 6 - x$

32. $y^2 = x + 1$

Testing for Symmetry **In Exercises 33–40, use the algebraic tests to check for symmetry with respect to both axes and the origin.**

33. $x^2 - y = 0$ **34.** $x - y^2 = 0$

35. $y = x^3$ **36.** $y = x^4 - x^2 + 3$

37. $y = \dfrac{x}{x^2 + 1}$ **38.** $y = \dfrac{1}{x^2 + 1}$

39. $xy^2 + 10 = 0$ **40.** $xy = 4$

Using Symmetry as a Sketching Aid **In Exercises 41–44, assume that the graph has the indicated type of symmetry. Sketch the complete graph of the equation. To print an enlarged copy of the graph, go to *MathGraphs.com*.**

41.

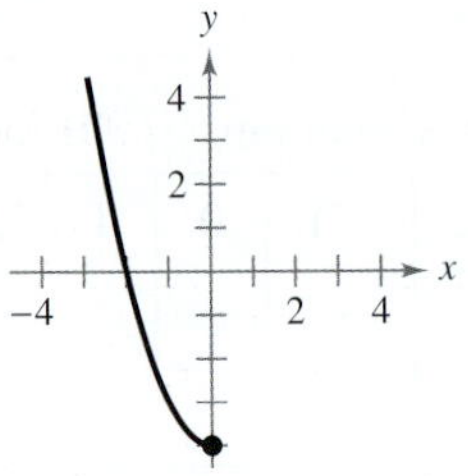

y-Axis symmetry

42.

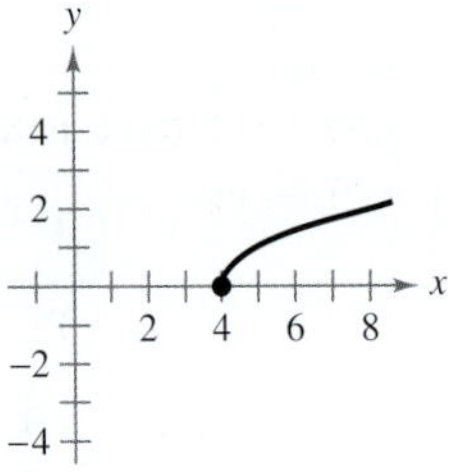

x-Axis symmetry

43.

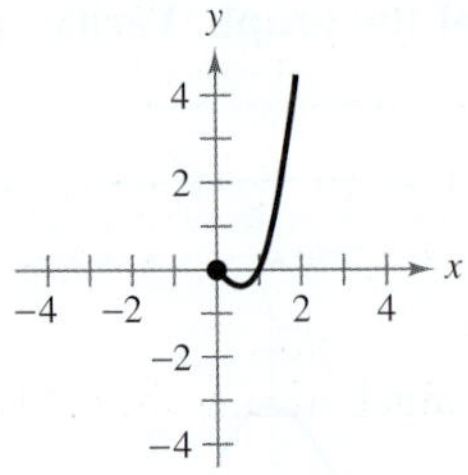

Origin symmetry

44.

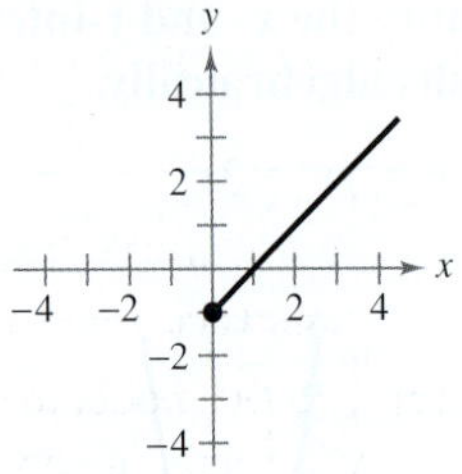

y-Axis symmetry

Sketching the Graph of an Equation **In Exercises 45–56, identify any intercepts and test for symmetry. Then sketch the graph of the equation.**

45. $y = -3x + 1$ **46.** $y = 2x - 3$

47. $y = x^2 - 2x$ **48.** $y = -x^2 - 2x$

49. $y = x^3 + 3$ **50.** $y = x^3 - 1$

51. $y = \sqrt{x - 3}$ **52.** $y = \sqrt{1 - x}$

53. $y = |x - 6|$ **54.** $y = 1 - |x|$

55. $x = y^2 - 1$ **56.** $x = y^2 - 5$

Graphical Analysis **In Exercises 57–68, use a graphing utility to graph the equation. Use a standard setting. Approximate any intercepts.**

57. $y = 3 - \frac{1}{2}x$ **58.** $y = \frac{2}{3}x - 1$

59. $y = x^2 - 4x + 3$ **60.** $y = x^2 + x - 2$

61. $y = \dfrac{2x}{x - 1}$ **62.** $y = \dfrac{4}{x^2 + 1}$

63. $y = \sqrt[3]{x} + 2$ **64.** $y = \sqrt[3]{x + 1}$

65. $y = x\sqrt{x + 6}$ **66.** $y = (6 - x)\sqrt{x}$

67. $y = |x + 3|$ **68.** $y = 2 - |x|$

Writing the Equation of a Circle **In Exercises 69–76, write the standard form of the equation of the circle with the given characteristics.**

69. Center: $(0, 0)$; Radius: 4

70. Center: $(0, 0)$; Radius: 5

71. Center: $(2, -1)$; Radius: 4

72. Center: $(-7, -4)$; Radius: 7

73. Center: $(-1, 2)$; Solution point: $(0, 0)$

74. Center: $(3, -2)$; Solution point: $(-1, 1)$

75. Endpoints of a diameter: $(0, 0)$, $(6, 8)$

76. Endpoints of a diameter: $(-4, -1)$, $(4, 1)$

Sketching the Graph of a Circle **In Exercises 77–82, find the center and radius of the circle. Then sketch the graph of the circle.**

77. $x^2 + y^2 = 25$ **78.** $x^2 + y^2 = 16$

79. $(x - 1)^2 + (y + 3)^2 = 9$ **80.** $x^2 + (y - 1)^2 = 1$

81. $\left(x - \frac{1}{2}\right)^2 + \left(y - \frac{1}{2}\right)^2 = \frac{9}{4}$

82. $(x - 2)^2 + (y + 3)^2 = \frac{16}{9}$

83. Depreciation A hospital purchases a new magnetic resonance imaging (MRI) machine for \$500,000. The depreciated value y (reduced value) after t years is given by $y = 500{,}000 - 40{,}000t$, $0 \le t \le 8$. Sketch the graph of the equation.

84. Consumerism You purchase an all-terrain vehicle (ATV) for \$8000. The depreciated value y after t years is given by $y = 8000 - 900t$, $0 \le t \le 6$. Sketch the graph of the equation.

85. Geometry A regulation NFL playing field (including the end zones) of length x and width y has a perimeter of $346\frac{2}{3}$ or $\frac{1040}{3}$ yards.

(a) Draw a rectangle that gives a visual representation of the problem. Use the specified variables to label the sides of the rectangle.

(b) Show that the width of the rectangle is $y = \frac{520}{3} - x$ and its area is $A = x\left(\frac{520}{3} - x\right)$.

(c) Use a graphing utility to graph the area equation. Be sure to adjust your window settings.

(d) From the graph in part (c), estimate the dimensions of the rectangle that yield a maximum area.

(e) Use your school's library, the Internet, or some other reference source to find the actual dimensions and area of a regulation NFL playing field and compare your findings with the results of part (d).

The symbol indicates an exercise or a part of an exercise in which you are instructed to use a graphing utility.

86. Geometry A soccer playing field of length x and width y has a perimeter of 360 meters.

(a) Draw a rectangle that gives a visual representation of the problem. Use the specified variables to label the sides of the rectangle.

(b) Show that the width of the rectangle is $y = 180 - x$ and its area is $A = x(180 - x)$.

(c) Use a graphing utility to graph the area equation. Be sure to adjust your window settings.

(d) From the graph in part (c), estimate the dimensions of the rectangle that yield a maximum area.

(e) Use your school's library, the Internet, or some other reference source to find the actual dimensions and area of a regulation Major League Soccer field and compare your findings with the results of part (d).

87. Population Statistics The table shows the life expectancies of a child (at birth) in the United States for selected years from 1930 through 2000. *(Source: U.S. National Center for Health Statistics)*

DATA

Spreadsheet at LarsonPrecalculus.com

Year	Life Expectancy, y
1930	59.7
1940	62.9
1950	68.2
1960	69.7
1970	70.8
1980	73.7
1990	75.4
2000	76.8

A model for the life expectancy during this period is

$$y = -0.002t^2 + 0.50t + 46.6, \quad 30 \le t \le 100$$

where y represents the life expectancy and t is the time in years, with $t = 30$ corresponding to 1930.

(a) Use a graphing utility to graph the data from the table and the model in the same viewing window. How well does the model fit the data? Explain.

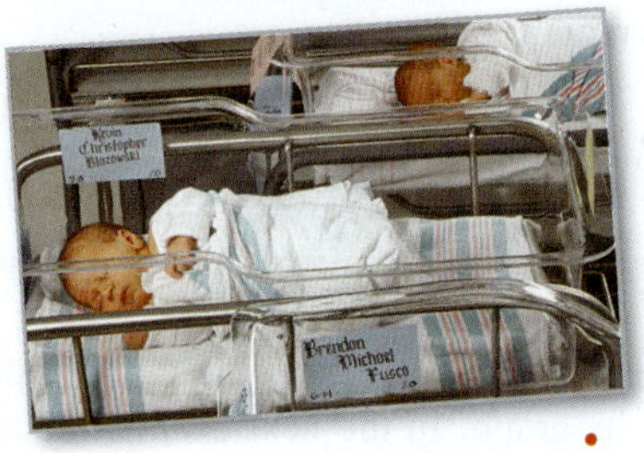

(b) Determine the life expectancy in 1990 both graphically and algebraically.

87. Population Statistics (continued)

(c) Use the graph to determine the year when life expectancy was approximately 76.0. Verify your answer algebraically.

(d) One projection for the life expectancy of a child born in 2015 is 78.9. How does this compare with the projection given by the model?

(e) Do you think this model can be used to predict the life expectancy of a child 50 years from now? Explain.

88. Electronics The resistance y (in ohms) of 1000 feet of solid copper wire at 68 degrees Fahrenheit is

$$y = \frac{10,370}{x^2}$$

where x is the diameter of the wire in mils (0.001 inch).

(a) Complete the table.

x	5	10	20	30	40	50
y						

x	60	70	80	90	100
y					

(b) Use the table of values in part (a) to sketch a graph of the model. Then use your graph to estimate the resistance when $x = 85.5$.

(c) Use the model to confirm algebraically the estimate you found in part (b).

(d) What can you conclude in general about the relationship between the diameter of the copper wire and the resistance?

Exploration

89. Think About It Find a and b if the graph of $y = ax^2 + bx^3$ is symmetric with respect to (a) the y-axis and (b) the origin. (There are many correct answers.)

90. HOW DO YOU SEE IT? The graph of the circle with equation $x^2 + y^2 = 1$ is shown below. Describe the types of symmetry that you observe.

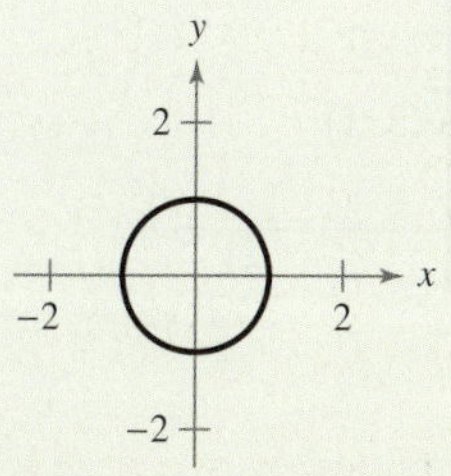

John Griffin/The Image Works

1.3 Linear Equations in Two Variables

Linear equations in two variables can help you model and solve real-life problems. For instance, in Exercise 90 on page 33, you will use a surveyor's measurements to find a linear equation that models a mountain road.

- **Use slope to graph linear equations in two variables.**
- **Find the slope of a line given two points on the line.**
- **Write linear equations in two variables.**
- **Use slope to identify parallel and perpendicular lines.**
- **Use slope and linear equations in two variables to model and solve real-life problems.**

Using Slope

The simplest mathematical model for relating two variables is the **linear equation in two variables** $y = mx + b$. The equation is called *linear* because its graph is a line. (In mathematics, the term *line* means *straight line*.) By letting $x = 0$, you obtain

$$y = m(0) + b = b.$$

So, the line crosses the y-axis at $y = b$, as shown in the figures below. In other words, the y-intercept is $(0, b)$. The steepness or slope of the line is m.

$$y = mx + b$$

Slope (m) — y-Intercept (b)

The **slope** of a nonvertical line is the number of units the line rises (or falls) vertically for each unit of horizontal change from left to right, as shown below.

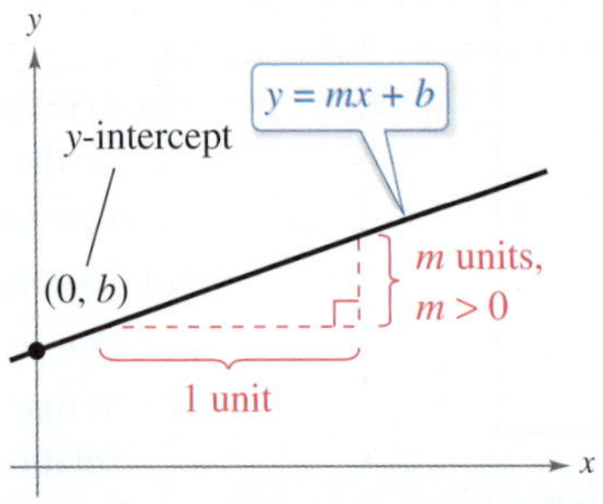

Positive slope, line rises.

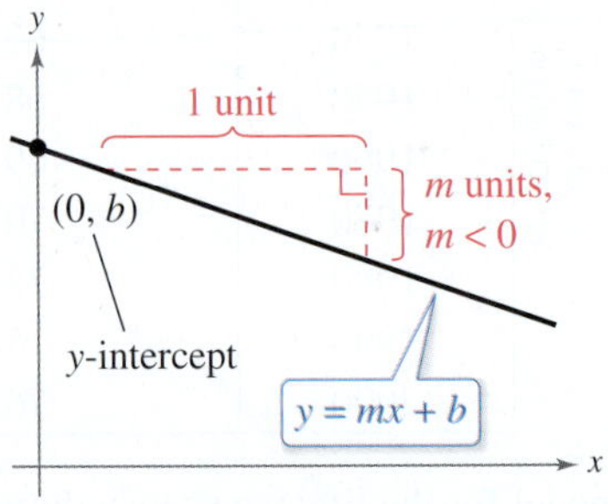

Negative slope, line falls.

A linear equation written in **slope-intercept form** has the form $y = mx + b$.

> **The Slope-Intercept Form of the Equation of a Line**
>
> The graph of the equation
>
> $$y = mx + b$$
>
> is a line whose slope is m and whose y-intercept is $(0, b)$.

Once you have determined the slope and the y-intercept of a line, it is a relatively simple matter to sketch its graph. In the next example, note that none of the lines is vertical. A vertical line has an equation of the form

$$x = a. \quad \text{Vertical line}$$

The equation of a vertical line cannot be written in the form $y = mx + b$ because the slope of a vertical line is undefined, as indicated in Figure 1.22.

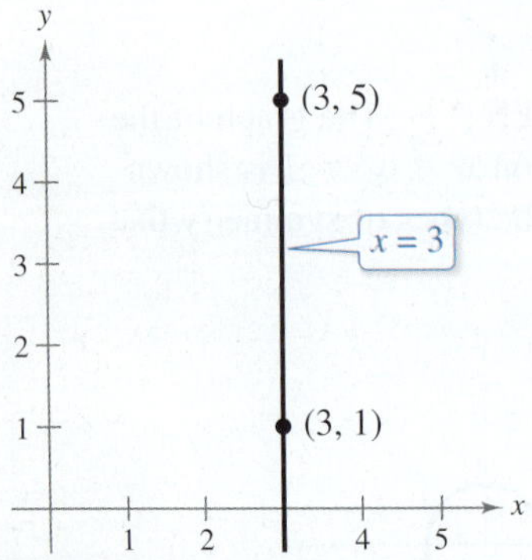

Slope is undefined.
Figure 1.22

Dmitry Kalinovsky/Shutterstock.com

EXAMPLE 1 Graphing a Linear Equation

Sketch the graph of each linear equation.

a. $y = 2x + 1$

b. $y = 2$

c. $x + y = 2$

Solution

a. Because $b = 1$, the y-intercept is $(0, 1)$. Moreover, because the slope is $m = 2$, the line *rises* two units for each unit the line moves to the right.

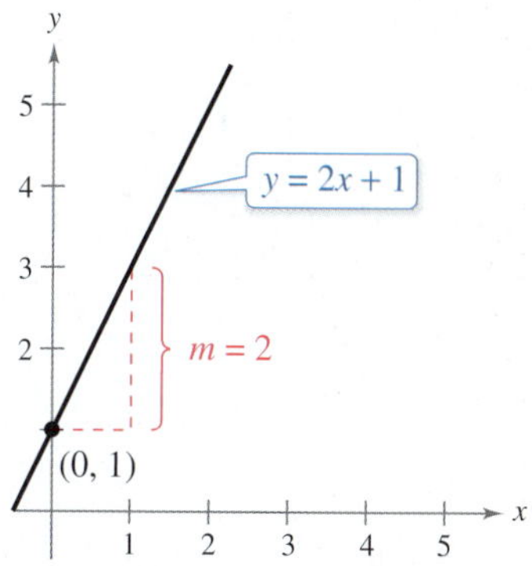

When m is positive, the line rises.

b. By writing this equation in the form $y = (0)x + 2$, you can see that the y-intercept is $(0, 2)$ and the slope is zero. A zero slope implies that the line is horizontal—that is, it does not rise *or* fall.

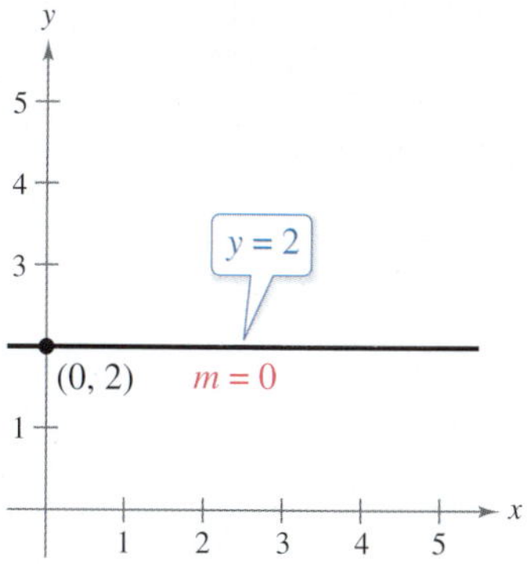

When m is 0, the line is horizontal.

c. By writing this equation in slope-intercept form

$x + y = 2$ — Write original equation.

$y = -x + 2$ — Subtract x from each side.

$y = (-1)x + 2$ — Write in slope-intercept form.

you can see that the y-intercept is $(0, 2)$. Moreover, because the slope is $m = -1$, the line *falls* one unit for each unit the line moves to the right.

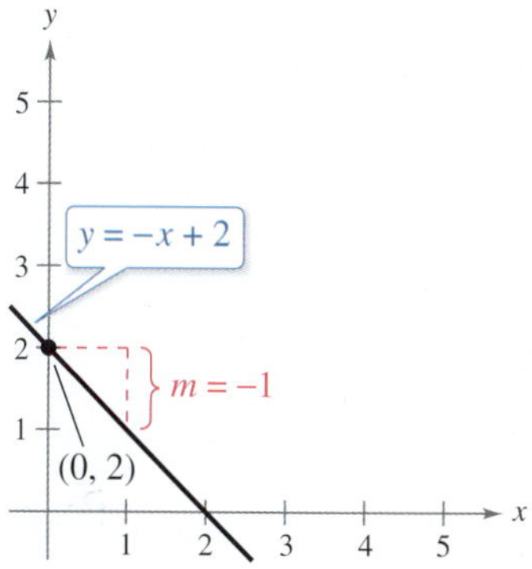

When m is negative, the line falls.

✓ Checkpoint Audio-video solution in English & Spanish at LarsonPrecalculus.com.

Sketch the graph of each linear equation.

a. $y = -3x + 2$

b. $y = -3$

c. $4x + y = 5$

Finding the Slope of a Line

Given an equation of a line, you can find its slope by writing the equation in slope-intercept form. If you are not given an equation, then you can still find the slope of a line. For instance, suppose you want to find the slope of the line passing through the points (x_1, y_1) and (x_2, y_2), as shown below.

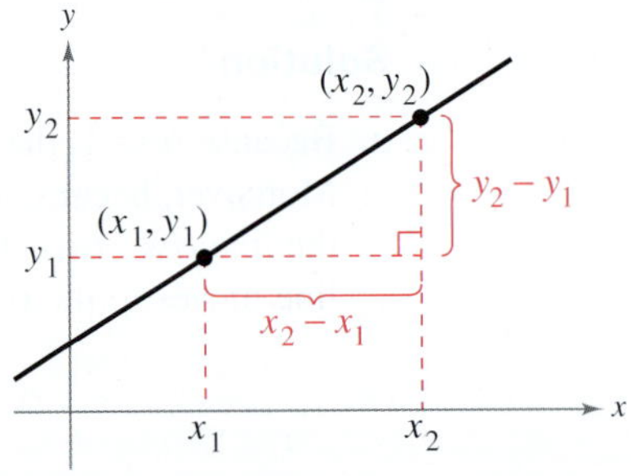

As you move from left to right along this line, a change of $(y_2 - y_1)$ units in the vertical direction corresponds to a change of $(x_2 - x_1)$ units in the horizontal direction.

$$y_2 - y_1 = \text{the change in } y = \text{rise}$$

and

$$x_2 - x_1 = \text{the change in } x = \text{run}$$

The ratio of $(y_2 - y_1)$ to $(x_2 - x_1)$ represents the slope of the line that passes through the points (x_1, y_1) and (x_2, y_2).

$$\text{Slope} = \frac{\text{change in } y}{\text{change in } x} = \frac{\text{rise}}{\text{run}} = \frac{y_2 - y_1}{x_2 - x_1}$$

The Slope of a Line Passing Through Two Points

The **slope** m of the nonvertical line through (x_1, y_1) and (x_2, y_2) is

$$m = \frac{y_2 - y_1}{x_2 - x_1}$$

where $x_1 \neq x_2$.

When using the formula for slope, the *order of subtraction* is important. Given two points on a line, you are free to label either one of them as (x_1, y_1) and the other as (x_2, y_2). However, once you have done this, you must form the numerator and denominator using the same order of subtraction.

$$m = \frac{y_2 - y_1}{x_2 - x_1} \qquad m = \frac{y_1 - y_2}{x_1 - x_2} \qquad m = \frac{y_2 - y_1}{x_1 - x_2}$$

Correct Correct Incorrect

For instance, the slope of the line passing through the points $(3, 4)$ and $(5, 7)$ can be calculated as

$$m = \frac{7 - 4}{5 - 3} = \frac{3}{2}$$

or, reversing the subtraction order in both the numerator and denominator, as

$$m = \frac{4 - 7}{3 - 5} = \frac{-3}{-2} = \frac{3}{2}.$$

EXAMPLE 2 Finding the Slope of a Line Through Two Points

Find the slope of the line passing through each pair of points.

a. $(-2, 0)$ and $(3, 1)$ **b.** $(-1, 2)$ and $(2, 2)$

c. $(0, 4)$ and $(1, -1)$ **d.** $(3, 4)$ and $(3, 1)$

Solution

a. Letting $(x_1, y_1) = (-2, 0)$ and $(x_2, y_2) = (3, 1)$, you obtain a slope of

$$m = \frac{y_2 - y_1}{x_2 - x_1} = \frac{1 - 0}{3 - (-2)} = \frac{1}{5}.$$ See Figure 1.23.

b. The slope of the line passing through $(-1, 2)$ and $(2, 2)$ is

$$m = \frac{2 - 2}{2 - (-1)} = \frac{0}{3} = 0.$$ See Figure 1.24.

c. The slope of the line passing through $(0, 4)$ and $(1, -1)$ is

$$m = \frac{-1 - 4}{1 - 0} = \frac{-5}{1} = -5.$$ See Figure 1.25.

d. The slope of the line passing through $(3, 4)$ and $(3, 1)$ is

$$m = \frac{1 - 4}{3 - 3} = \frac{-3}{0}.$$ See Figure 1.26.

Because division by 0 is undefined, the slope is undefined and the line is vertical.

REMARK In Figures 1.23 through 1.26, note the relationships between slope and the orientation of the line.

a. Positive slope: line rises from left to right

b. Zero slope: line is horizontal

c. Negative slope: line falls from left to right

d. Undefined slope: line is vertical

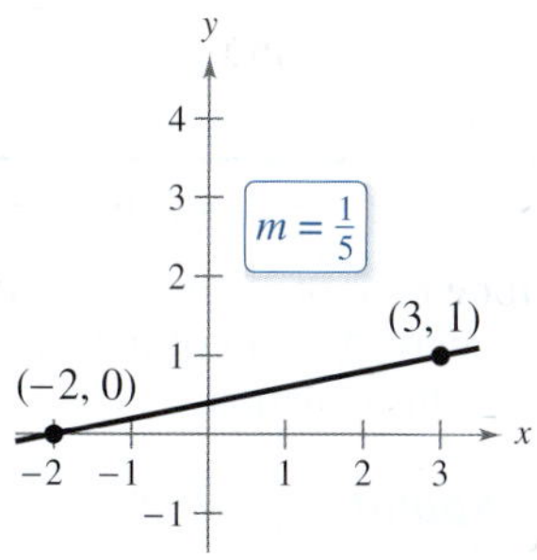

Figure 1.23

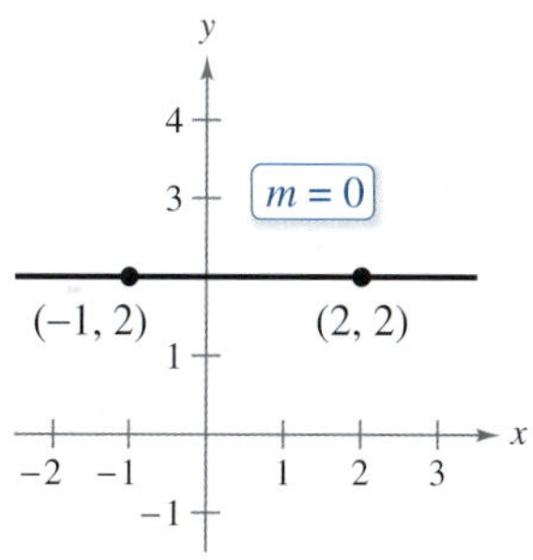

Figure 1.24

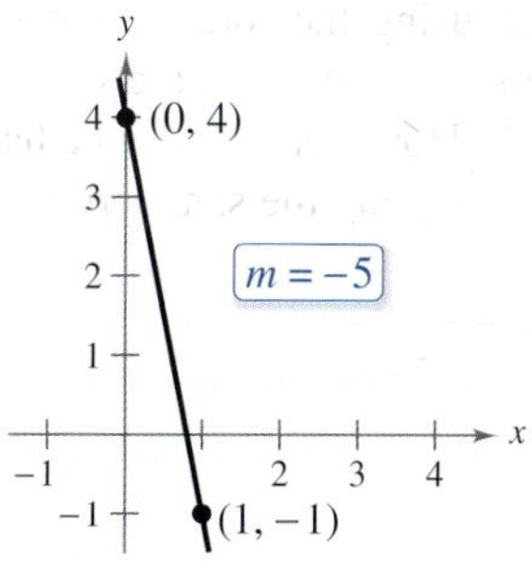

Figure 1.25

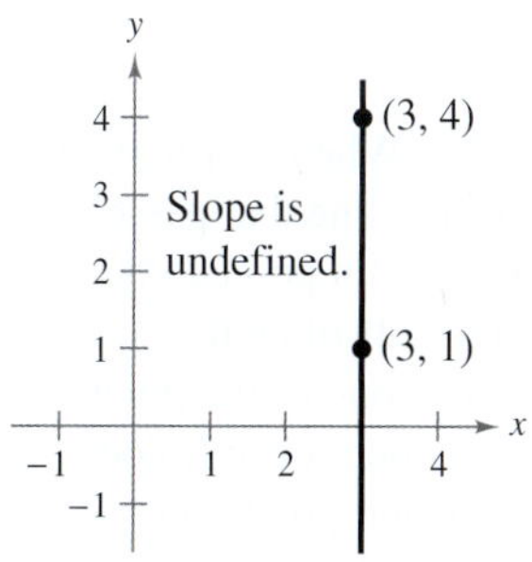

Figure 1.26

✓ Checkpoint Audio-video solution in English & Spanish at LarsonPrecalculus.com.

Find the slope of the line passing through each pair of points.

a. $(-5, -6)$ and $(2, 8)$ **b.** $(4, 2)$ and $(2, 5)$

c. $(0, 0)$ and $(0, -6)$ **d.** $(0, -1)$ and $(3, -1)$

Writing Linear Equations in Two Variables

If (x_1, y_1) is a point on a line of slope m and (x, y) is *any other* point on the line, then

$$\frac{y - y_1}{x - x_1} = m.$$

This equation involving the variables x and y, rewritten in the form

$$y - y_1 = m(x - x_1)$$

is the **point-slope form** of the equation of a line.

> **Point-Slope Form of the Equation of a Line**
>
> The equation of the line with slope m passing through the point (x_1, y_1) is
>
> $$y - y_1 = m(x - x_1).$$

The point-slope form is most useful for *finding* the equation of a line. You should remember this form.

EXAMPLE 3 Using the Point-Slope Form

Find the slope-intercept form of the equation of the line that has a slope of 3 and passes through the point $(1, -2)$.

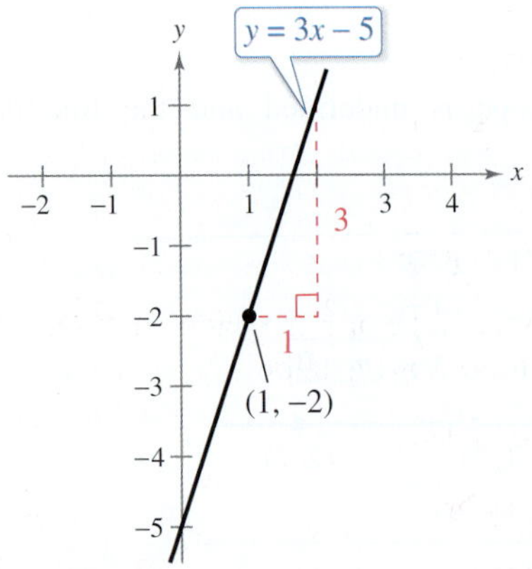

Figure 1.27

Solution Use the point-slope form with $m = 3$ and $(x_1, y_1) = (1, -2)$.

$$y - y_1 = m(x - x_1) \quad \text{Point-slope form}$$

$$y - (-2) = 3(x - 1) \quad \text{Substitute for } m, x_1, \text{ and } y_1.$$

$$y + 2 = 3x - 3 \quad \text{Simplify.}$$

$$y = 3x - 5 \quad \text{Write in slope-intercept form.}$$

The slope-intercept form of the equation of the line is $y = 3x - 5$. Figure 1.27 shows the graph of this equation.

✓ **Checkpoint** *Audio-video solution in English & Spanish at LarsonPrecalculus.com.*

Find the slope-intercept form of the equation of the line that has the given slope and passes through the given point.

a. $m = 2$, $(3, -7)$

b. $m = -\frac{2}{3}$, $(1, 1)$

c. $m = 0$, $(1, 1)$

REMARK When you find an equation of the line that passes through two given points, you only need to substitute the coordinates of one of the points in the point-slope form. It does not matter which point you choose because both points will yield the same result.

The point-slope form can be used to find an equation of the line passing through two points (x_1, y_1) and (x_2, y_2). To do this, first find the slope of the line

$$m = \frac{y_2 - y_1}{x_2 - x_1}, \quad x_1 \neq x_2$$

and then use the point-slope form to obtain the equation

$$y - y_1 = \frac{y_2 - y_1}{x_2 - x_1}(x - x_1). \quad \text{Two-point form}$$

This is sometimes called the **two-point form** of the equation of a line.

Parallel and Perpendicular Lines

Slope can tell you whether two nonvertical lines in a plane are parallel, perpendicular, or neither.

> **Parallel and Perpendicular Lines**
>
> 1. Two distinct nonvertical lines are **parallel** if and only if their slopes are equal. That is,
>
> $$m_1 = m_2.$$
>
> 2. Two nonvertical lines are **perpendicular** if and only if their slopes are negative reciprocals of each other. That is,
>
> $$m_1 = \frac{-1}{m_2}.$$

EXAMPLE 4 Finding Parallel and Perpendicular Lines

Find the slope-intercept form of the equations of the lines that pass through the point $(2, -1)$ and are (a) parallel to and (b) perpendicular to the line $2x - 3y = 5$.

Solution By writing the equation of the given line in slope-intercept form

$2x - 3y = 5$ Write original equation.

$-3y = -2x + 5$ Subtract $2x$ from each side.

$y = \frac{2}{3}x - \frac{5}{3}$ Write in slope-intercept form.

you can see that it has a slope of $m = \frac{2}{3}$, as shown in Figure 1.28.

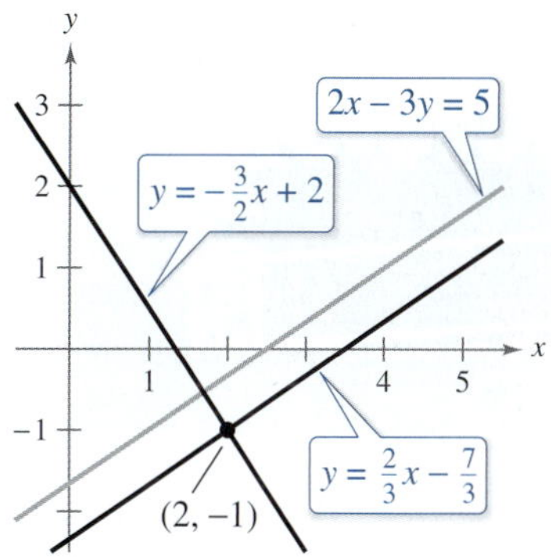

Figure 1.28

a. Any line parallel to the given line must also have a slope of $\frac{2}{3}$. So, the line through $(2, -1)$ that is parallel to the given line has the following equation.

$y - (-1) = \frac{2}{3}(x - 2)$ Write in point-slope form.

$3(y + 1) = 2(x - 2)$ Multiply each side by 3.

$3y + 3 = 2x - 4$ Distributive Property

$y = \frac{2}{3}x - \frac{7}{3}$ Write in slope-intercept form.

b. Any line perpendicular to the given line must have a slope of $-\frac{3}{2}$ (because $-\frac{3}{2}$ is the negative reciprocal of $\frac{2}{3}$). So, the line through $(2, -1)$ that is perpendicular to the given line has the following equation.

$y - (-1) = -\frac{3}{2}(x - 2)$ Write in point-slope form.

$2(y + 1) = -3(x - 2)$ Multiply each side by 2.

$2y + 2 = -3x + 6$ Distributive Property

$y = -\frac{3}{2}x + 2$ Write in slope-intercept form.

✓ Checkpoint *Audio-video solution in English & Spanish at LarsonPrecalculus.com.*

Find the slope-intercept form of the equations of the lines that pass through the point $(-4, 1)$ and are (a) parallel to and (b) perpendicular to the line $5x - 3y = 8$.

Notice in Example 4 how the slope-intercept form is used to obtain information about the graph of a line, whereas the point-slope form is used to write the equation of a line.

TECHNOLOGY On a graphing utility, lines will not appear to have the correct slope unless you use a viewing window that has a square setting. For instance, try graphing the lines in Example 4 using the standard setting $-10 \le x \le 10$ and $-10 \le y \le 10$. Then reset the viewing window with the square setting $-9 \le x \le 9$ and $-6 \le y \le 6$. On which setting do the lines $y = \frac{2}{3}x - \frac{5}{3}$ and $y = -\frac{3}{2}x + 2$ appear to be perpendicular?

Applications

In real-life problems, the slope of a line can be interpreted as either a *ratio* or a *rate*. If the x-axis and y-axis have the same unit of measure, then the slope has no units and is a **ratio.** If the x-axis and y-axis have different units of measure, then the slope is a **rate** or **rate of change.**

EXAMPLE 5 Using Slope as a Ratio

The maximum recommended slope of a wheelchair ramp is $\frac{1}{12}$. A business is installing a wheelchair ramp that rises 22 inches over a horizontal length of 24 feet. Is the ramp steeper than recommended? *(Source: ADA Standards for Accessible Design)*

Solution The horizontal length of the ramp is 24 feet or $12(24) = 288$ inches, as shown below. So, the slope of the ramp is

$$\text{Slope} = \frac{\text{vertical change}}{\text{horizontal change}} = \frac{22 \text{ in.}}{288 \text{ in.}} \approx 0.076.$$

Because $\frac{1}{12} \approx 0.083$, the slope of the ramp is not steeper than recommended.

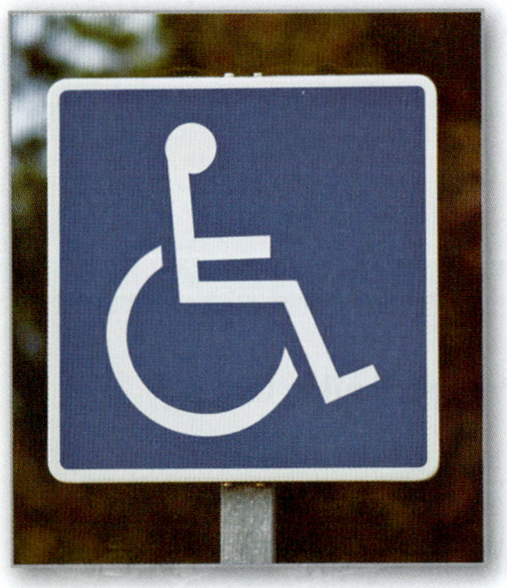

The Americans with Disabilities Act (ADA) became law on July 26, 1990. It is the most comprehensive formulation of rights for persons with disabilities in U.S. (and world) history.

Checkpoint Audio-video solution in English & Spanish at LarsonPrecalculus.com.

The business in Example 5 installs a second ramp that rises 36 inches over a horizontal length of 32 feet. Is the ramp steeper than recommended?

EXAMPLE 6 Using Slope as a Rate of Change

A kitchen appliance manufacturing company determines that the total cost C (in dollars) of producing x units of a blender is

$$C = 25x + 3500. \quad \text{Cost equation}$$

Describe the practical significance of the y-intercept and slope of this line.

Solution The y-intercept $(0, 3500)$ tells you that the cost of producing zero units is \$3500. This is the *fixed cost* of production—it includes costs that must be paid regardless of the number of units produced. The slope of $m = 25$ tells you that the cost of producing each unit is \$25, as shown in Figure 1.29. Economists call the cost per unit the *marginal cost*. If the production increases by one unit, then the "margin," or extra amount of cost, is \$25. So, the cost increases at a rate of \$25 per unit.

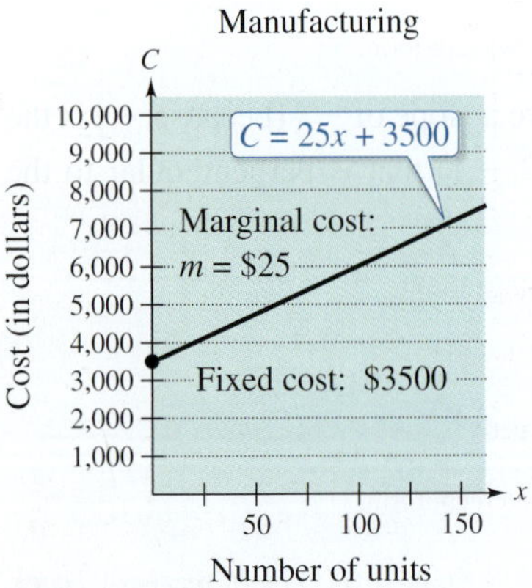

Production cost
Figure 1.29

Checkpoint Audio-video solution in English & Spanish at LarsonPrecalculus.com.

An accounting firm determines that the value V (in dollars) of a copier t years after its purchase is

$$V = -300t + 1500.$$

Describe the practical significance of the y-intercept and slope of this line.

Jultud/Shutterstock.com

Businesses can deduct most of their expenses in the same year they occur. One exception is the cost of property that has a useful life of more than 1 year. Such costs must be *depreciated* (decreased in value) over the useful life of the property. Depreciating the *same amount* each year is called *linear* or *straight-line depreciation*. The *book value* is the difference between the original value and the total amount of depreciation accumulated to date.

EXAMPLE 7 Straight-Line Depreciation

A college purchased exercise equipment worth \$12,000 for the new campus fitness center. The equipment has a useful life of 8 years. The salvage value at the end of 8 years is \$2000. Write a linear equation that describes the book value of the equipment each year.

Solution Let V represent the value of the equipment at the end of year t. Represent the initial value of the equipment by the data point (0, 12,000) and the salvage value of the equipment by the data point (8, 2000). The slope of the line is

$$m = \frac{2000 - 12{,}000}{8 - 0}$$

$$= -\$1250$$

which represents the annual depreciation in *dollars per year*. Using the point-slope form, you can write the equation of the line as follows.

$$V - 12{,}000 = -1250(t - 0) \qquad \text{Write in point-slope form.}$$

$$V = -1250t + 12{,}000 \qquad \text{Write in slope-intercept form.}$$

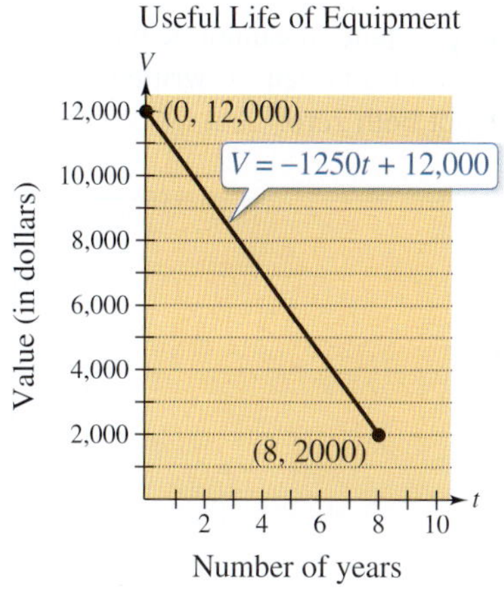

Straight-line depreciation

Figure 1.30

The table shows the book value at the end of each year, and Figure 1.30 shows the graph of the equation.

Year, t	Value, V
0	12,000
1	10,750
2	9500
3	8250
4	7000
5	5750
6	4500
7	3250
8	2000

Checkpoint *Audio-video solution in English & Spanish at LarsonPrecalculus.com.*

A manufacturing firm purchased a machine worth \$24,750. The machine has a useful life of 6 years. After 6 years, the machine will have to be discarded and replaced. That is, it will have no salvage value. Write a linear equation that describes the book value of the machine each year.

In many real-life applications, the two data points that determine the line are often given in a disguised form. Note how the data points are described in Example 7.

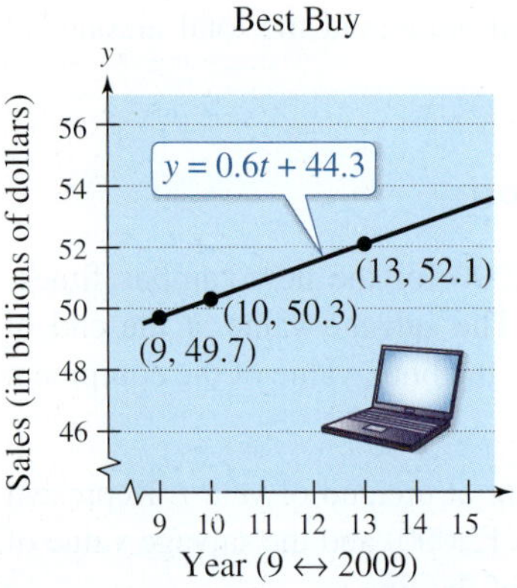

Figure 1.31

EXAMPLE 8 Predicting Sales

The sales for Best Buy were approximately \$49.7 billion in 2009 and \$50.3 billion in 2010. Using only this information, write a linear equation that gives the sales in terms of the year. Then predict the sales in 2013. *(Source: Best Buy Company, Inc.)*

Solution Let $t = 9$ represent 2009. Then the two given values are represented by the data points (9, 49.7) and (10, 50.3). The slope of the line through these points is

$$m = \frac{50.3 - 49.7}{10 - 9} = 0.6.$$

You can find the equation that relates the sales y and the year t to be

$$y - 49.7 = 0.6(t - 9) \quad \text{Write in point-slope form.}$$

$$y = 0.6t + 44.3. \quad \text{Write in slope-intercept form.}$$

According to this equation, the sales in 2013 will be

$$y = 0.6(13) + 44.3 = 7.8 + 44.3 = \$52.1 \text{ billion. (See Figure 1.31.)}$$

Checkpoint *Audio-video solution in English & Spanish at LarsonPrecalculus.com.*

The sales for Nokia were approximately \$58.6 billion in 2009 and \$56.6 billion in 2010. Repeat Example 8 using this information. *(Source: Nokia Corporation)*

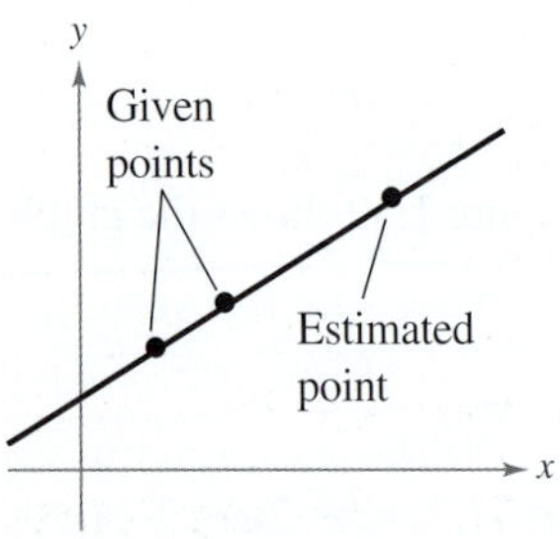

Linear extrapolation
Figure 1.32

The prediction method illustrated in Example 8 is called **linear extrapolation.** Note in Figure 1.32 that an extrapolated point does not lie between the given points. When the estimated point lies between two given points, as shown in Figure 1.33, the procedure is called **linear interpolation.**

Because the slope of a vertical line is not defined, its equation cannot be written in slope-intercept form. However, every line has an equation that can be written in the **general form** $Ax + By + C = 0$, where A and B are not both zero.

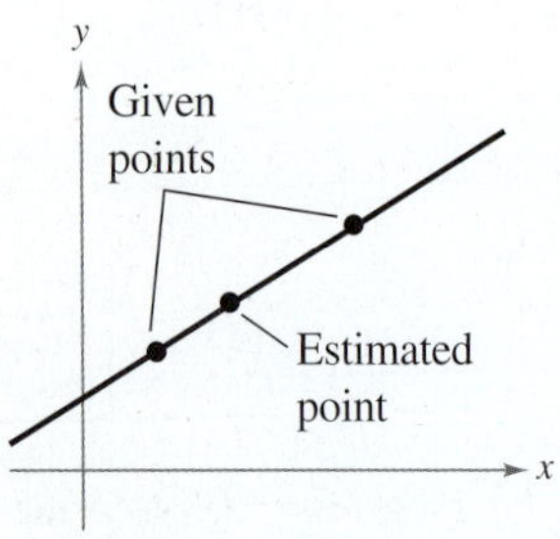

Linear interpolation
Figure 1.33

Summary of Equations of Lines

1. General form: $Ax + By + C = 0$
2. Vertical line: $x = a$
3. Horizontal line: $y = b$
4. Slope-intercept form: $y = mx + b$
5. Point-slope form: $y - y_1 = m(x - x_1)$
6. Two-point form: $y - y_1 = \dfrac{y_2 - y_1}{x_2 - x_1}(x - x_1)$

Summarize (Section 1.3)

1. Explain how to use slope to graph a linear equation in two variables *(page 22)* and how to find the slope of a line passing through two points *(page 24)*. For examples of using and finding slopes, see Examples 1 and 2.
2. State the point-slope form of the equation of a line *(page 26)*. For an example of using point-slope form, see Example 3.
3. Explain how to use slope to identify parallel and perpendicular lines *(page 27)*. For an example of finding parallel and perpendicular lines, see Example 4.
4. Describe examples of how to use slope and linear equations in two variables to model and solve real-life problems *(pages 28–30, Examples 5–8)*.

1.3 Exercises

See CalcChat.com for tutorial help and worked-out solutions to odd-numbered exercises.

Vocabulary: Fill in the blanks.

1. The simplest mathematical model for relating two variables is the ________ equation in two variables $y = mx + b$.
2. For a line, the ratio of the change in y to the change in x is called the ________ of the line.
3. The ________-________ form of the equation of a line with slope m passing through (x_1, y_1) is $y - y_1 = m(x - x_1)$.
4. Two lines are ________ if and only if their slopes are equal.
5. Two lines are ________ if and only if their slopes are negative reciprocals of each other.
6. When the x-axis and y-axis have different units of measure, the slope can be interpreted as a ________.
7. The prediction method ________ ________ is the method used to estimate a point on a line when the point does not lie between the given points.
8. Every line has an equation that can be written in ________ form.

Skills and Applications

Identifying Lines In Exercises 9 and 10, identify the line that has each slope.

9. (a) $m = \frac{2}{3}$
(b) m is undefined.
(c) $m = -2$

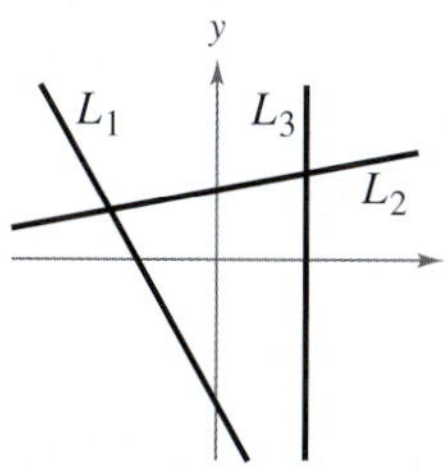

10. (a) $m = 0$
(b) $m = -\frac{3}{4}$
(c) $m = 1$

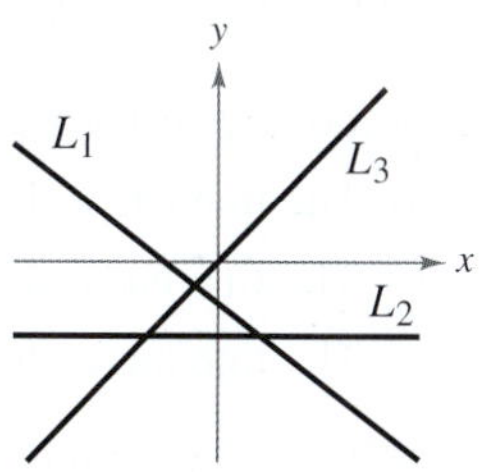

Sketching Lines In Exercises 11 and 12, sketch the lines through the point with the indicated slopes on the same set of coordinate axes.

	Point	Slopes	
11.	(2, 3)	(a) 0	(b) 1
		(c) 2	(d) −3
12.	(−4, 1)	(a) 3	(b) −3
		(c) $\frac{1}{2}$	(d) Undefined

Estimating the Slope of a Line In Exercises 13 and 14, estimate the slope of the line.

13.

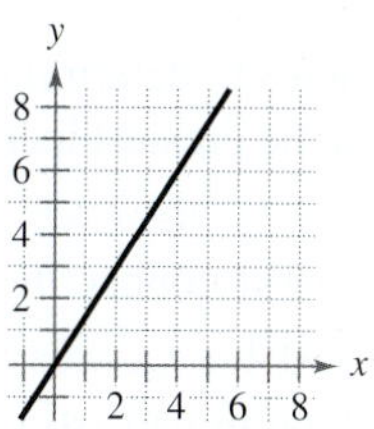

14.

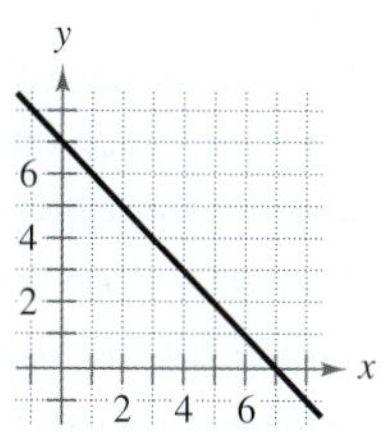

Graphing a Linear Equation In Exercises 15–24, find the slope and y-intercept (if possible) of the equation of the line. Sketch the line.

15. $y = 5x + 3$ **16.** $y = -x - 10$
17. $y = -\frac{1}{2}x + 4$ **18.** $y = \frac{3}{2}x + 6$
19. $y - 3 = 0$ **20.** $x + 5 = 0$
21. $5x - 2 = 0$ **22.** $3y + 5 = 0$
23. $7x - 6y = 30$ **24.** $2x + 3y = 9$

Finding the Slope of a Line Through Two Points In Exercises 25–34, plot the points and find the slope of the line passing through the pair of points.

25. (0, 9), (6, 0) **26.** (12, 0), (0, −8)
27. (−3, −2), (1, 6) **28.** (2, 4), (4, −4)
29. (5, −7), (8, −7) **30.** (−2, 1), (−4, −5)
31. (−6, −1), (−6, 4) **32.** (0, −10), (−4, 0)
33. (4.8, 3.1), (−5.2, 1.6)
34. $\left(\frac{11}{2}, -\frac{4}{3}\right), \left(-\frac{3}{2}, -\frac{1}{3}\right)$

Using a Point and Slope In Exercises 35–42, use the point on the line and the slope m of the line to find three additional points through which the line passes. (There are many correct answers.)

35. (2, 1), $m = 0$ **36.** (3, −2), $m = 0$
37. (−8, 1), m is undefined.
38. (1, 5), m is undefined.
39. (−5, 4), $m = 2$
40. (0, −9), $m = -2$
41. (−1, −6), $m = -\frac{1}{2}$
42. (7, −2), $m = \frac{1}{2}$

Finding an Equation of a Line In Exercises 43–54, find an equation of the line that passes through the given point and has the indicated slope m. Sketch the line.

43. $(0, -2)$, $m = 3$
44. $(0, 10)$, $m = -1$
45. $(-3, 6)$, $m = -2$
46. $(0, 0)$, $m = 4$
47. $(4, 0)$, $m = -\frac{1}{3}$
48. $(8, 2)$, $m = \frac{1}{4}$
49. $(2, -3)$, $m = -\frac{1}{2}$
50. $(-2, -5)$, $m = \frac{3}{4}$
51. $(6, -1)$, m is undefined.
52. $(-10, 4)$, m is undefined.
53. $\left(4, \frac{5}{2}\right)$, $m = 0$
54. $(-5.1, 1.8)$, $m = 5$

Finding an Equation of a Line In Exercises 55–64, find an equation of the line passing through the points. Sketch the line.

55. $(5, -1), (-5, 5)$
56. $(4, 3), (-4, -4)$
57. $(-8, 1), (-8, 7)$
58. $(-1, 4), (6, 4)$
59. $\left(2, \frac{1}{2}\right), \left(\frac{1}{2}, \frac{5}{4}\right)$
60. $(1, 1), \left(6, -\frac{2}{3}\right)$
61. $(1, 0.6), (-2, -0.6)$
62. $(-8, 0.6), (2, -2.4)$
63. $(2, -1), \left(\frac{1}{3}, -1\right)$
64. $\left(\frac{7}{3}, -8\right), \left(\frac{7}{3}, 1\right)$

Parallel and Perpendicular Lines In Exercises 65–68, determine whether the lines are parallel, perpendicular, or neither.

65. L_1: $y = \frac{1}{3}x - 2$
    L_2: $y = \frac{1}{3}x + 3$
66. L_1: $y = 4x - 1$
    L_2: $y = 4x + 7$
67. L_1: $y = \frac{1}{2}x - 3$
    L_2: $y = -\frac{1}{2}x + 1$
68. L_1: $y = -\frac{4}{5}x - 5$
    L_2: $y = \frac{5}{4}x + 1$

Parallel and Perpendicular Lines In Exercises 69–72, determine whether the lines L_1 and L_2 passing through the pairs of points are parallel, perpendicular, or neither.

69. L_1: $(0, -1), (5, 9)$
    L_2: $(0, 3), (4, 1)$
70. L_1: $(-2, -1), (1, 5)$
    L_2: $(1, 3), (5, -5)$
71. L_1: $(3, 6), (-6, 0)$
    L_2: $(0, -1), \left(5, \frac{7}{3}\right)$
72. L_1: $(4, 8), (-4, 2)$
    L_2: $(3, -5), \left(-1, \frac{1}{3}\right)$

Finding Parallel and Perpendicular Lines In Exercises 73–80, write equations of the lines through the given point (a) parallel to and (b) perpendicular to the given line.

73. $4x - 2y = 3$, $(2, 1)$
74. $x + y = 7$, $(-3, 2)$
75. $3x + 4y = 7$, $\left(-\frac{2}{3}, \frac{7}{8}\right)$
76. $5x + 3y = 0$, $\left(\frac{7}{8}, \frac{3}{4}\right)$
77. $y + 3 = 0$, $(-1, 0)$
78. $x - 4 = 0$, $(3, -2)$
79. $x - y = 4$, $(2.5, 6.8)$
80. $6x + 2y = 9$, $(-3.9, -1.4)$

Intercept Form of the Equation of a Line In Exercises 81–86, use the *intercept form* to find the equation of the line with the given intercepts. The intercept form of the equation of a line with intercepts $(a, 0)$ and $(0, b)$ is

$$\frac{x}{a} + \frac{y}{b} = 1, \ a \neq 0, \ b \neq 0.$$

81. x-intercept: $(2, 0)$
    y-intercept: $(0, 3)$
82. x-intercept: $(-3, 0)$
    y-intercept: $(0, 4)$
83. x-intercept: $\left(-\frac{1}{6}, 0\right)$
    y-intercept: $\left(0, -\frac{2}{3}\right)$
84. x-intercept: $\left(\frac{2}{3}, 0\right)$
    y-intercept: $(0, -2)$
85. Point on line: $(1, 2)$
    x-intercept: $(c, 0)$
    y-intercept: $(0, c)$, $c \neq 0$
86. Point on line: $(-3, 4)$
    x-intercept: $(d, 0)$
    y-intercept: $(0, d)$, $d \neq 0$

87. **Sales** The following are the slopes of lines representing annual sales y in terms of time x in years. Use the slopes to interpret any change in annual sales for a one-year increase in time.
    (a) The line has a slope of $m = 135$.
    (b) The line has a slope of $m = 0$.
    (c) The line has a slope of $m = -40$.

88. **Sales** The graph shows the sales (in billions of dollars) for Apple Inc. in the years 2004 through 2010. *(Source: Apple Inc.)*

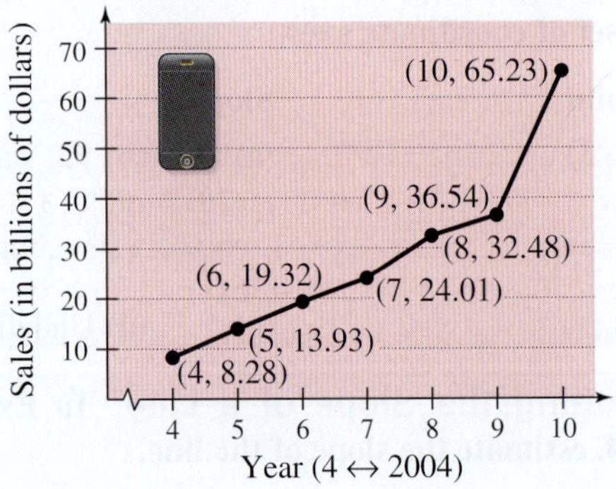

    (a) Use the slopes of the line segments to determine the years in which the sales showed the greatest increase and the least increase.
    (b) Find the slope of the line segment connecting the points for the years 2004 and 2010.
    (c) Interpret the meaning of the slope in part (b) in the context of the problem.

89. Road Grade You are driving on a road that has a 6% uphill grade. This means that the slope of the road is $\frac{6}{100}$. Approximate the amount of vertical change in your position when you drive 200 feet.

90. Road Grade

From the top of a mountain road, a surveyor takes several horizontal measurements x and several vertical measurements y, as shown in the table (x and y are measured in feet).

x	300	600	900	1200
y	-25	-50	-75	-100

x	1500	1800	2100
y	-125	-150	-175

(a) Sketch a scatter plot of the data.

(b) Use a straightedge to sketch the line that you think best fits the data.

(c) Find an equation for the line you sketched in part (b).

(d) Interpret the meaning of the slope of the line in part (c) in the context of the problem.

(e) The surveyor needs to put up a road sign that indicates the steepness of the road. For instance, a surveyor would put up a sign that states "8% grade" on a road with a downhill grade that has a slope of $-\frac{8}{100}$. What should the sign state for the road in this problem?

Rate of Change In Exercises 91 and 92, you are given the dollar value of a product in 2013 and the rate at which the value of the product is expected to change during the next 5 years. Use this information to write a linear equation that gives the dollar value V of the product in terms of the year t. (Let $t = 13$ represent 2013.)

	2013 Value	Rate
91.	\$2540	\$125 decrease per year
92.	\$156	\$4.50 increase per year

93. Cost The cost C of producing n computer laptop bags is given by

$$C = 1.25n + 15{,}750, \quad 0 < n.$$

Explain what the C-intercept and the slope measure.

94. Monthly Salary A pharmaceutical salesperson receives a monthly salary of \$2500 plus a commission of 7% of sales. Write a linear equation for the salesperson's monthly wage W in terms of monthly sales S.

95. Depreciation A sub shop purchases a used pizza oven for \$875. After 5 years, the oven will have to be discarded and replaced. Write a linear equation giving the value V of the equipment during the 5 years it will be in use.

96. Depreciation A school district purchases a high-volume printer, copier, and scanner for \$24,000. After 10 years, the equipment will have to be replaced. Its value at that time is expected to be \$2000. Write a linear equation giving the value V of the equipment during the 10 years it will be in use.

97. Temperature Conversion Write a linear equation that expresses the relationship between the temperature in degrees Celsius C and degrees Fahrenheit F. Use the fact that water freezes at 0°C (32°F) and boils at 100°C (212°F).

98. Brain Weight The average weight of a male child's brain is 970 grams at age 1 and 1270 grams at age 3. *(Source: American Neurological Association)*

(a) Assuming that the relationship between brain weight y and age t is linear, write a linear model for the data.

(b) What is the slope and what does it tell you about brain weight?

(c) Use your model to estimate the average brain weight at age 2.

(d) Use your school's library, the Internet, or some other reference source to find the actual average brain weight at age 2. How close was your estimate?

(e) Do you think your model could be used to determine the average brain weight of an adult? Explain.

99. Cost, Revenue, and Profit A roofing contractor purchases a shingle delivery truck with a shingle elevator for \$42,000. The vehicle requires an average expenditure of \$9.50 per hour for fuel and maintenance, and the operator is paid \$11.50 per hour.

(a) Write a linear equation giving the total cost C of operating this equipment for t hours. (Include the purchase cost of the equipment.)

(b) Assuming that customers are charged \$45 per hour of machine use, write an equation for the revenue R derived from t hours of use.

(c) Use the formula for profit $P = R - C$ to write an equation for the profit derived from t hours of use.

(d) Use the result of part (c) to find the break-even point—that is, the number of hours this equipment must be used to yield a profit of 0 dollars.

Dmitry Kalinovsky/Shutterstock.com

100. Geometry The length and width of a rectangular garden are 15 meters and 10 meters, respectively. A walkway of width x surrounds the garden.

(a) Draw a diagram that gives a visual representation of the problem.

(b) Write the equation for the perimeter y of the walkway in terms of x.

(c) Use a graphing utility to graph the equation for the perimeter.

(d) Determine the slope of the graph in part (c). For each additional one-meter increase in the width of the walkway, determine the increase in its perimeter.

Exploration

True or False? **In Exercises 101 and 102, determine whether the statement is true or false. Justify your answer.**

101. A line with a slope of $-\frac{5}{7}$ is steeper than a line with a slope of $-\frac{6}{7}$.

102. The line through $(-8, 2)$ and $(-1, 4)$ and the line through $(0, -4)$ and $(-7, 7)$ are parallel.

103. Right Triangle Explain how you could use slope to show that the points $A(-1, 5)$, $B(3, 7)$, and $C(5, 3)$ are the vertices of a right triangle.

104. Vertical Line Explain why the slope of a vertical line is said to be undefined.

105. Think About It With the information shown in the graphs, is it possible to determine the slope of each line? Is it possible that the lines could have the same slope? Explain.

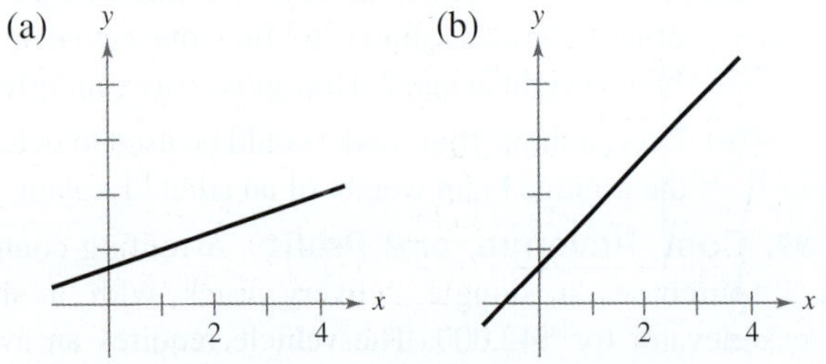

106. Perpendicular Segments Find d_1 and d_2 in terms of m_1 and m_2, respectively (see figure). Then use the Pythagorean Theorem to find a relationship between m_1 and m_2.

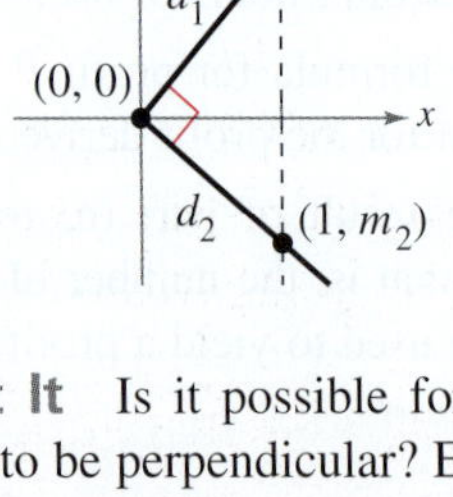

107. Think About It Is it possible for two lines with positive slopes to be perpendicular? Explain.

108. Slope and Steepness The slopes of two lines are -4 and $\frac{5}{2}$. Which is steeper? Explain.

109. Comparing Slopes Use a graphing utility to compare the slopes of the lines $y = mx$, where $m = 0.5$, 1, 2, and 4. Which line rises most quickly? Now, let $m = -0.5$, -1, -2, and -4. Which line falls most quickly? Use a square setting to obtain a true geometric perspective. What can you conclude about the slope and the "rate" at which the line rises or falls?

110. HOW DO YOU SEE IT? Match the description of the situation with its graph. Also determine the slope and y-intercept of each graph and interpret the slope and y-intercept in the context of the situation. [The graphs are labeled (i), (ii), (iii), and (iv).]

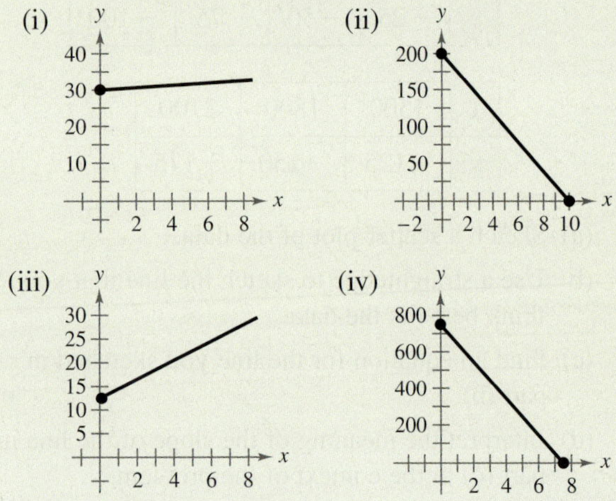

(a) A person is paying \$20 per week to a friend to repay a \$200 loan.

(b) An employee receives \$12.50 per hour plus \$2 for each unit produced per hour.

(c) A sales representative receives \$30 per day for food plus \$0.32 for each mile traveled.

(d) A computer that was purchased for \$750 depreciates \$100 per year.

Finding a Relationship for Equidistance **In Exercises 111–114, find a relationship between x and y such that (x, y) is equidistant (the same distance) from the two points.**

111. $(4, -1), (-2, 3)$

112. $(6, 5), (1, -8)$

113. $\left(3, \frac{5}{2}\right), (-7, 1)$

114. $\left(-\frac{1}{2}, -4\right), \left(\frac{7}{2}, \frac{5}{4}\right)$

Project: Bachelor's Degrees To work an extended application analyzing the numbers of bachelor's degrees earned by women in the United States from 1998 through 2009, visit this text's website at *LarsonPrecalculus.com*. *(Source: National Center for Education Statistics)*

1.4 Functions

Functions can help you model and solve real-life problems. For instance, in Exercise 74 on page 47, you will use a function to model the force of water against the face of a dam.

- **Determine whether relations between two variables are functions, and use function notation.**
- **Find the domains of functions.**
- **Use functions to model and solve real-life problems.**
- **Evaluate difference quotients.**

Introduction to Functions and Function Notation

Many everyday phenomena involve two quantities that are related to each other by some rule of correspondence. The mathematical term for such a rule of correspondence is a **relation.** In mathematics, equations and formulas often represent relations. For instance, the simple interest I earned on \$1000 for 1 year is related to the annual interest rate r by the formula $I = 1000r$.

The formula $I = 1000r$ represents a special kind of relation that matches each item from one set with *exactly one* item from a different set. Such a relation is called a **function.**

> **Definition of Function**
>
> A **function** f from a set A to a set B is a relation that assigns to each element x in the set A exactly one element y in the set B. The set A is the **domain** (or set of inputs) of the function f, and the set B contains the **range** (or set of outputs).

To help understand this definition, look at the function below, which relates the time of day to the temperature.

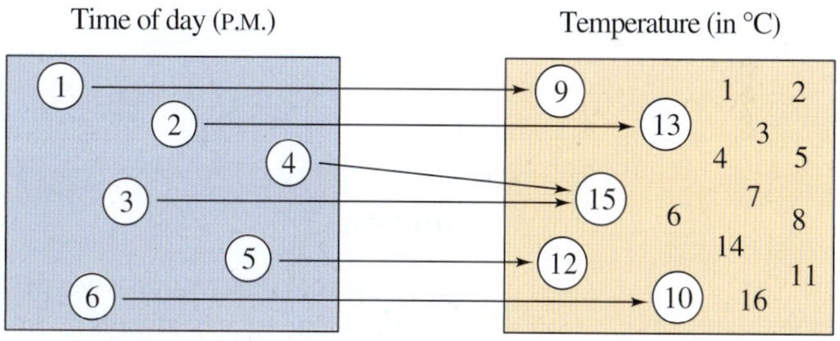

Set A is the domain.
Inputs: 1, 2, 3, 4, 5, 6

Set B contains the range.
Outputs: 9, 10, 12, 13, 15

The following ordered pairs can represent this function. The first coordinate (x-value) is the input and the second coordinate (y-value) is the output.

$$\{(1, 9), (2, 13), (3, 15), (4, 15), (5, 12), (6, 10)\}$$

> **Characteristics of a Function from Set A to Set B**
>
> 1. Each element in A must be matched with an element in B.
> 2. Some elements in B may not be matched with any element in A.
> 3. Two or more elements in A may be matched with the same element in B.
> 4. An element in A (the domain) cannot be matched with two different elements in B.

Lester Lefkowitz/CORBIS

Four common ways to represent functions are as follows.

> **Four Ways to Represent a Function**
>
> 1. *Verbally* by a sentence that describes how the input variable is related to the output variable
> 2. *Numerically* by a table or a list of ordered pairs that matches input values with output values
> 3. *Graphically* by points on a graph in a coordinate plane in which the horizontal axis represents the input values and the vertical axis represents the output values
> 4. *Algebraically* by an equation in two variables

To determine whether a relation is a function, you must decide whether each input value is matched with exactly one output value. When any input value is matched with two or more output values, the relation is not a function.

EXAMPLE 1 Testing for Functions

Determine whether the relation represents y as a function of x.

a. The input value x is the number of representatives from a state, and the output value y is the number of senators.

b.

Input, x	Output, y
2	11
2	10
3	8
4	5
5	1

c.

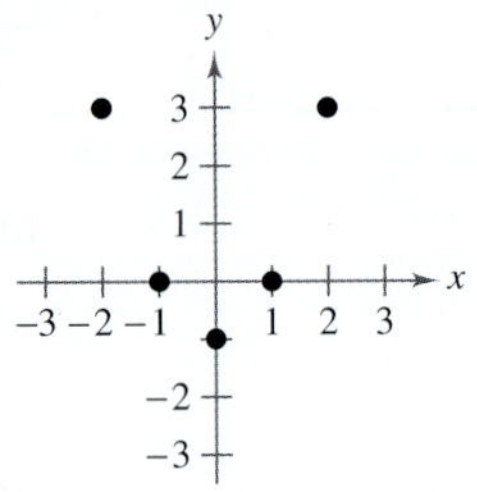

Solution

a. This verbal description *does* describe y as a function of x. Regardless of the value of x, the value of y is always 2. Such functions are called *constant functions*.

b. This table *does not* describe y as a function of x. The input value 2 is matched with two different y-values.

c. The graph *does* describe y as a function of x. Each input value is matched with exactly one output value.

✓ Checkpoint *Audio-video solution in English & Spanish at LarsonPrecalculus.com.*

Determine whether the relation represents y as a function of x.

a. *Domain*, x *Range*, y

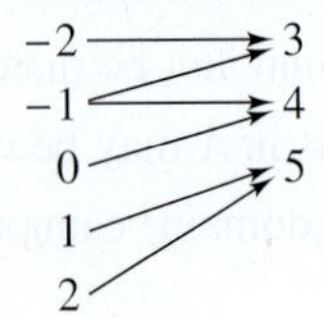

b.

Input, x	0	1	2	3	4
Output, y	−4	−2	0	2	4

HISTORICAL NOTE

Many consider Leonhard Euler (1707–1783), a Swiss mathematician, the most prolific and productive mathematician in history. One of his greatest influences on mathematics was his use of symbols, or notation. Euler introduced the function notation $y = f(x)$.

Representing functions by sets of ordered pairs is common in *discrete mathematics*. In algebra, however, it is more common to represent functions by equations or formulas involving two variables. For instance, the equation

$y = x^2$ $\quad$ *y is a function of x.*

represents the variable y as a function of the variable x. In this equation, x is the **independent variable** and y is the **dependent variable.** The domain of the function is the set of all values taken on by the independent variable x, and the range of the function is the set of all values taken on by the dependent variable y.

EXAMPLE 2 Testing for Functions Represented Algebraically

Which of the equations represent(s) y as a function of x?

a. $x^2 + y = 1$

b. $-x + y^2 = 1$

Solution To determine whether y is a function of x, try to solve for y in terms of x.

a. Solving for y yields

$x^2 + y = 1$ $\quad$ Write original equation.

$y = 1 - x^2.$ $\quad$ Solve for y.

To each value of x there corresponds exactly one value of y. So, y is a function of x.

b. Solving for y yields

$-x + y^2 = 1$ $\quad$ Write original equation.

$y^2 = 1 + x$ $\quad$ Add x to each side.

$y = \pm\sqrt{1 + x}.$ $\quad$ Solve for y.

The $\pm$ indicates that to a given value of x there correspond two values of y. So, y is not a function of x.

✓ ***Checkpoint*** *Audio-video solution in English & Spanish at LarsonPrecalculus.com.*

Which of the equations represent(s) y as a function of x?

a. $x^2 + y^2 = 8$ **b.** $y - 4x^2 = 36$

When using an equation to represent a function, it is convenient to name the function for easy reference. For example, you know that the equation $y = 1 - x^2$ describes y as a function of x. Suppose you give this function the name "f." Then you can use the following **function notation.**

Input	Output	Equation
x	$f(x)$	$f(x) = 1 - x^2$

The symbol $f(x)$ is read as *the value of f at x* or simply *f of x*. The symbol $f(x)$ corresponds to the y-value for a given x. So, you can write $y = f(x)$. Keep in mind that f is the *name* of the function, whereas $f(x)$ is the *value* of the function at x. For instance, the function $f(x) = 3 - 2x$ has *function values* denoted by $f(-1), f(0), f(2)$, and so on. To find these values, substitute the specified input values into the given equation.

For $x = -1$, $\quad f(-1) = 3 - 2(-1) = 3 + 2 = 5.$

For $x = 0$, $\quad f(0) = 3 - 2(0) = 3 - 0 = 3.$

For $x = 2$, $\quad f(2) = 3 - 2(2) = 3 - 4 = -1.$

Bettmann/Corbis

Although f is often used as a convenient function name and x is often used as the independent variable, you can use other letters. For instance,

$$f(x) = x^2 - 4x + 7, \quad f(t) = t^2 - 4t + 7, \quad \text{and} \quad g(s) = s^2 - 4s + 7$$

all define the same function. In fact, the role of the independent variable is that of a "placeholder." Consequently, the function could be described by

$$f(\blacksquare) = (\blacksquare)^2 - 4(\blacksquare) + 7.$$

EXAMPLE 3 Evaluating a Function

Let $g(x) = -x^2 + 4x + 1$. Find each function value.

a. $g(2)$ **b.** $g(t)$ **c.** $g(x + 2)$

Solution

a. Replacing x with 2 in $g(x) = -x^2 + 4x + 1$ yields the following.

$$\begin{aligned} g(2) &= -(2)^2 + 4(2) + 1 \\ &= -4 + 8 + 1 \\ &= 5 \end{aligned}$$

b. Replacing x with t yields the following.

$$\begin{aligned} g(t) &= -(t)^2 + 4(t) + 1 \\ &= -t^2 + 4t + 1 \end{aligned}$$

c. Replacing x with $x + 2$ yields the following.

$$\begin{aligned} g(x + 2) &= -(x + 2)^2 + 4(x + 2) + 1 \\ &= -(x^2 + 4x + 4) + 4x + 8 + 1 \\ &= -x^2 - 4x - 4 + 4x + 8 + 1 \\ &= -x^2 + 5 \end{aligned}$$

REMARK In Example 3(c), note that $g(x + 2)$ is not equal to $g(x) + g(2)$. In general, $g(u + v) \neq g(u) + g(v)$.

✓ ***Checkpoint*** *Audio-video solution in English & Spanish at LarsonPrecalculus.com.*

Let $f(x) = 10 - 3x^2$. Find each function value.

a. $f(2)$ **b.** $f(-4)$ **c.** $f(x - 1)$

A function defined by two or more equations over a specified domain is called a **piecewise-defined function.**

EXAMPLE 4 A Piecewise-Defined Function

Evaluate the function when $x = -1, 0,$ and 1.

$$f(x) = \begin{cases} x^2 + 1, & x < 0 \\ x - 1, & x \geq 0 \end{cases}$$

Solution Because $x = -1$ is less than 0, use $f(x) = x^2 + 1$ to obtain $f(-1) = (-1)^2 + 1 = 2$. For $x = 0$, use $f(x) = x - 1$ to obtain $f(0) = (0) - 1 = -1$. For $x = 1$, use $f(x) = x - 1$ to obtain $f(1) = (1) - 1 = 0$.

✓ ***Checkpoint*** *Audio-video solution in English & Spanish at LarsonPrecalculus.com.*

Evaluate the function given in Example 4 when $x = -2, 2,$ and 3.

EXAMPLE 5 Finding Values for Which $f(x) = 0$

Find all real values of x such that $f(x) = 0$.

a. $f(x) = -2x + 10$ **b.** $f(x) = x^2 - 5x + 6$

Solution For each function, set $f(x) = 0$ and solve for x.

a. $-2x + 10 = 0$ Set $f(x)$ equal to 0.

$-2x = -10$ Subtract 10 from each side.

$x = 5$ Divide each side by -2.

So, $f(x) = 0$ when $x = 5$.

b. $x^2 - 5x + 6 = 0$ Set $f(x)$ equal to 0.

$(x - 2)(x - 3) = 0$ Factor.

$x - 2 = 0 \Rightarrow x = 2$ Set 1st factor equal to 0.

$x - 3 = 0 \Rightarrow x = 3$ Set 2nd factor equal to 0.

So, $f(x) = 0$ when $x = 2$ or $x = 3$.

Checkpoint Audio-video solution in English & Spanish at LarsonPrecalculus.com.

Find all real values of x such that $f(x) = 0$, where $f(x) = x^2 - 16$.

EXAMPLE 6 Finding Values for Which $f(x) = g(x)$

Find the values of x for which $f(x) = g(x)$.

a. $f(x) = x^2 + 1$ and $g(x) = 3x - x^2$

b. $f(x) = x^2 - 1$ and $g(x) = -x^2 + x + 2$

Solution

a. $x^2 + 1 = 3x - x^2$ Set $f(x)$ equal to $g(x)$.

$2x^2 - 3x + 1 = 0$ Write in general form.

$(2x - 1)(x - 1) = 0$ Factor.

$2x - 1 = 0 \Rightarrow x = \frac{1}{2}$ Set 1st factor equal to 0.

$x - 1 = 0 \Rightarrow x = 1$ Set 2nd factor equal to 0.

So, $f(x) = g(x)$ when $x = \dfrac{1}{2}$ or $x = 1$.

b. $x^2 - 1 = -x^2 + x + 2$ Set $f(x)$ equal to $g(x)$.

$2x^2 - x - 3 = 0$ Write in general form.

$(2x - 3)(x + 1) = 0$ Factor.

$2x - 3 = 0 \Rightarrow x = \frac{3}{2}$ Set 1st factor equal to 0.

$x + 1 = 0 \Rightarrow x = -1$ Set 2nd factor equal to 0.

So, $f(x) = g(x)$ when $x = \dfrac{3}{2}$ or $x = -1$.

Checkpoint Audio-video solution in English & Spanish at LarsonPrecalculus.com.

Find the values of x for which $f(x) = g(x)$, where $f(x) = x^2 + 6x - 24$ and $g(x) = 4x - x^2$.

The Domain of a Function

TECHNOLOGY Use a graphing utility to graph the functions $y = \sqrt{4 - x^2}$ and $y = \sqrt{x^2 - 4}$. What is the domain of each function? Do the domains of these two functions overlap? If so, for what values do the domains overlap?

The domain of a function can be described explicitly or it can be *implied* by the expression used to define the function. The **implied domain** is the set of all real numbers for which the expression is defined. For instance, the function

$$f(x) = \frac{1}{x^2 - 4}$$ Domain excludes x-values that result in division by zero.

has an implied domain consisting of all real x other than $x = \pm 2$. These two values are excluded from the domain because division by zero is undefined. Another common type of implied domain is that used to avoid even roots of negative numbers. For example, the function

$$f(x) = \sqrt{x}$$ Domain excludes x-values that result in even roots of negative numbers.

is defined only for $x \geq 0$. So, its implied domain is the interval $[0, \infty)$. In general, the domain of a function *excludes* values that would cause division by zero *or* that would result in the even root of a negative number.

EXAMPLE 7 Finding the Domain of a Function

Find the domain of each function.

a. f: $\{(-3, 0), (-1, 4), (0, 2), (2, 2), (4, -1)\}$ **b.** $g(x) = \dfrac{1}{x + 5}$

c. Volume of a sphere: $V = \frac{4}{3}\pi r^3$ **d.** $h(x) = \sqrt{4 - 3x}$

Solution

a. The domain of f consists of all first coordinates in the set of ordered pairs.

Domain $= \{-3, -1, 0, 2, 4\}$

b. Excluding x-values that yield zero in the denominator, the domain of g is the set of all real numbers x except $x = -5$.

c. Because this function represents the volume of a sphere, the values of the radius r must be positive. So, the domain is the set of all real numbers r such that $r > 0$.

d. This function is defined only for x-values for which

$4 - 3x \geq 0.$

By solving this inequality, you can conclude that $x \leq \frac{4}{3}$. So, the domain is the interval $\left(-\infty, \frac{4}{3}\right]$.

✓ ***Checkpoint*** Audio-video solution in English & Spanish at LarsonPrecalculus.com.

Find the domain of each function.

a. f: $\{(-2, 2), (-1, 1), (0, 3), (1, 1), (2, 2)\}$ **b.** $g(x) = \dfrac{1}{3 - x}$

c. Circumference of a circle: $C = 2\pi r$ **d.** $h(x) = \sqrt{x - 16}$

In Example 7(c), note that the domain of a function may be implied by the physical context. For instance, from the equation

$$V = \frac{4}{3}\pi r^3$$

you would have no reason to restrict r to positive values, but the physical context implies that a sphere cannot have a negative or zero radius.

Applications

EXAMPLE 8 The Dimensions of a Container

You work in the marketing department of a soft-drink company and are experimenting with a new can for iced tea that is slightly narrower and taller than a standard can. For your experimental can, the ratio of the height to the radius is 4.

a. Write the volume of the can as a function of the radius r.

b. Write the volume of the can as a function of the height h.

Solution

a. $V(r) = \pi r^2 h = \pi r^2(4r) = 4\pi r^3$ — Write V as a function of r.

b. $V(h) = \pi r^2 h = \pi\left(\frac{h}{4}\right)^2 h = \frac{\pi h^3}{16}$ — Write V as a function of h.

✓ **Checkpoint** Audio-video solution in English & Spanish at LarsonPrecalculus.com.

For the experimental can described in Example 8, write the *surface area* as a function of (a) the radius r and (b) the height h.

EXAMPLE 9 The Path of a Baseball

A batter hits a baseball at a point 3 feet above ground at a velocity of 100 feet per second and an angle of 45°. The path of the baseball is given by the function

$$f(x) = -0.0032x^2 + x + 3$$

where $f(x)$ is the height of the baseball (in feet) and x is the horizontal distance from home plate (in feet). Will the baseball clear a 10-foot fence located 300 feet from home plate?

Algebraic Solution

When $x = 300$, you can find the height of the baseball as follows.

$f(x) = -0.0032x^2 + x + 3$ — Write original function.

$f(300) = -0.0032(300)^2 + 300 + 3$ — Substitute 300 for x.

$= 15$ — Simplify.

When $x = 300$, the height of the baseball is 15 feet. So, the baseball will clear a 10-foot fence.

Graphical Solution

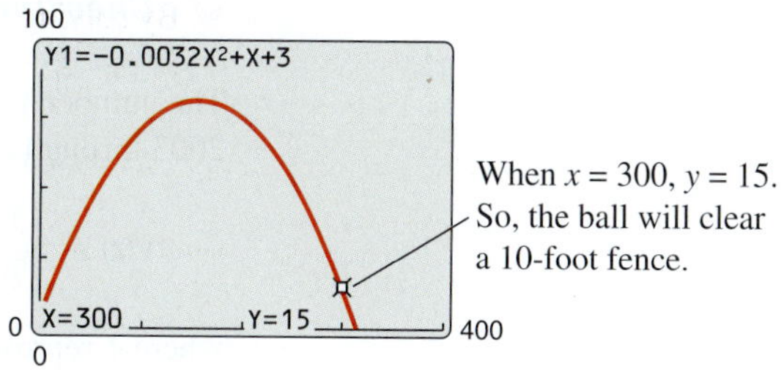

✓ **Checkpoint** Audio-video solution in English & Spanish at LarsonPrecalculus.com.

A second baseman throws a baseball toward the first baseman 60 feet away. The path of the baseball is given by

$$f(x) = -0.004x^2 + 0.3x + 6$$

where $f(x)$ is the height of the baseball (in feet) and x is the horizontal distance from the second baseman (in feet). The first baseman can reach 8 feet high. Can the first baseman catch the baseball without jumping?

EXAMPLE 10 **Alternative-Fueled Vehicles**

Alternative fuels for vehicles include electricity, ethanol, hydrogen, compressed natural gas, liquefied natural gas, and liquefied petroleum gas.

The number V (in thousands) of alternative-fueled vehicles in the United States increased in a linear pattern from 2003 through 2005, and then increased in a different linear pattern from 2006 through 2009, as shown in the bar graph. These two patterns can be approximated by the function

$$V(t) = \begin{cases} 29.05t + 447.7, & 3 \le t \le 5 \\ 65.50t + 241.9, & 6 \le t \le 9 \end{cases}$$

where t represents the year, with $t = 3$ corresponding to 2003. Use this function to approximate the number of alternative-fueled vehicles for each year from 2003 through 2009. *(Source: U.S. Energy Information Administration)*

Number of Alternative-Fueled Vehicles in the U.S.

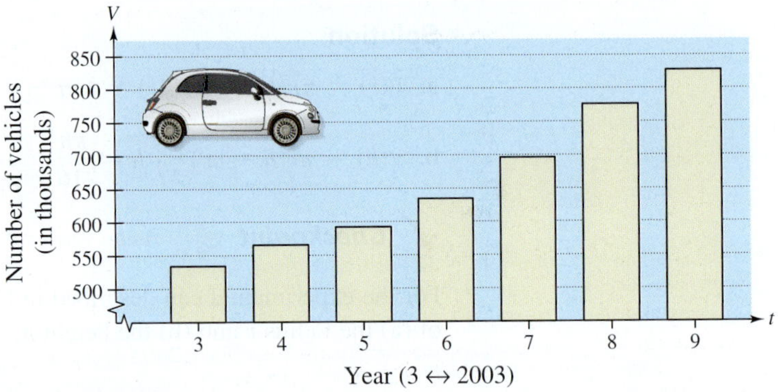

Solution From 2003 through 2005, use $V(t) = 29.05t + 447.7$.

$$\underbrace{534.9}_{2003} \quad \underbrace{563.9}_{2004} \quad \underbrace{593.0}_{2005}$$

From 2006 to 2009, use $V(t) = 65.50t + 241.9$.

$$\underbrace{634.9}_{2006} \quad \underbrace{700.4}_{2007} \quad \underbrace{765.9}_{2008} \quad \underbrace{831.4}_{2009}$$

✓ ***Checkpoint*** *Audio-video solution in English & Spanish at LarsonPrecalculus.com.*

The number V (in thousands) of 85%-ethanol-fueled vehicles in the United States from 2003 through 2009 can be approximated by the function

$$V(t) = \begin{cases} 33.65t + 77.8, & 3 \le t \le 5 \\ 70.75t - 126.6, & 6 \le t \le 9 \end{cases}$$

where t represents the year, with $t = 3$ corresponding to 2003. Use this function to approximate the number of 85%-ethanol-fueled vehicles for each year from 2003 through 2009. *(Source: U.S. Energy Information Administration)*

Difference Quotients

One of the basic definitions in calculus employs the ratio

$$\frac{f(x + h) - f(x)}{h}, \quad h \neq 0.$$

This ratio is called a **difference quotient,** as illustrated in Example 11.

wellphoto/Shutterstock.com

EXAMPLE 11 **Evaluating a Difference Quotient**

For $f(x) = x^2 - 4x + 7$, find $\dfrac{f(x+h) - f(x)}{h}$.

REMARK You may find it easier to calculate the difference quotient in Example 11 by first finding $f(x+h)$, and then substituting the resulting expression into the difference quotient

$$\frac{f(x+h) - f(x)}{h}.$$

Solution

$$\frac{f(x+h) - f(x)}{h} = \frac{[(x+h)^2 - 4(x+h) + 7] - (x^2 - 4x + 7)}{h}$$

$$= \frac{x^2 + 2xh + h^2 - 4x - 4h + 7 - x^2 + 4x - 7}{h}$$

$$= \frac{2xh + h^2 - 4h}{h} = \frac{h(2x + h - 4)}{h} = 2x + h - 4, \quad h \neq 0$$

Checkpoint Audio-video solution in English & Spanish at LarsonPrecalculus.com.

For $f(x) = x^2 + 2x - 3$, find $\dfrac{f(x+h) - f(x)}{h}$.

Summary of Function Terminology

Function: A **function** is a relationship between two variables such that to each value of the independent variable there corresponds exactly one value of the dependent variable.

Function Notation: $y = f(x)$

f is the *name* of the function.

y is the **dependent variable.**

x is the **independent variable.**

$f(x)$ is the *value of the function at x.*

Domain: The **domain** of a function is the set of all values (inputs) of the independent variable for which the function is defined. If x is in the domain of f, then f is said to be *defined* at x. If x is not in the domain of f, then f is said to be *undefined* at x.

Range: The **range** of a function is the set of all values (outputs) assumed by the dependent variable (that is, the set of all function values).

Implied Domain: If f is defined by an algebraic expression and the domain is not specified, then the **implied domain** consists of all real numbers for which the expression is defined.

Summarize (Section 1.4)

1. State the definition of a function and describe function notation *(pages 35–38)*. For examples of determining functions and using function notation, see Examples 1–6.
2. State the definition of the implied domain of a function *(page 40)*. For an example of finding the domains of functions, see Example 7.
3. Describe examples of how functions can model real-life problems *(pages 41 and 42, Examples 8–10)*.
4. State the definition of a difference quotient *(page 42)*. For an example of evaluating a difference quotient, see Example 11.

1.4 Exercises

See **CalcChat.com** for tutorial help and worked-out solutions to odd-numbered exercises.

Vocabulary: Fill in the blanks.

1. A relation that assigns to each element x from a set of inputs, or ________, exactly one element y in a set of outputs, or ________, is called a ________.
2. For an equation that represents y as a function of x, the set of all values taken on by the ________ variable x is the domain, and the set of all values taken on by the ________ variable y is the range.
3. If the domain of the function f is not given, then the set of values of the independent variable for which the expression is defined is called the ________ ________.
4. In calculus, one of the basic definitions is that of a ________ ________, given by $\dfrac{f(x+h)-f(x)}{h}$, $h \neq 0$.

Skills and Applications

Testing for Functions In Exercises 5–8, determine whether the relation represents y as a function of x.

5. *Domain, x* *Range, y*

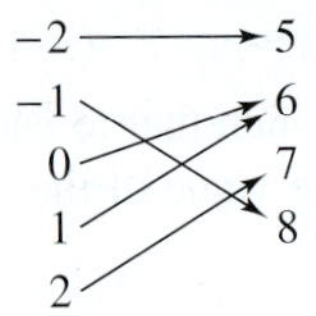

6. *Domain, x* *Range, y*

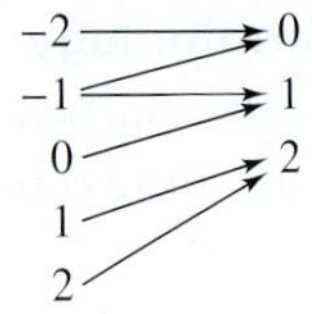

7.

Input, x	10	7	4	7	10
Output, y	3	6	9	12	15

8.

Input, x	0	3	9	12	15
Output, y	3	3	3	3	3

Testing for Functions In Exercises 9 and 10, which sets of ordered pairs represent functions from A to B? Explain.

9. $A = \{0, 1, 2, 3\}$ and $B = \{-2, -1, 0, 1, 2\}$
   (a) $\{(0, 1), (1, -2), (2, 0), (3, 2)\}$
   (b) $\{(0, -1), (2, 2), (1, -2), (3, 0), (1, 1)\}$
   (c) $\{(0, 0), (1, 0), (2, 0), (3, 0)\}$
   (d) $\{(0, 2), (3, 0), (1, 1)\}$
10. $A = \{a, b, c\}$ and $B = \{0, 1, 2, 3\}$
    (a) $\{(a, 1), (c, 2), (c, 3), (b, 3)\}$
    (b) $\{(a, 1), (b, 2), (c, 3)\}$
    (c) $\{(1, a), (0, a), (2, c), (3, b)\}$
    (d) $\{(c, 0), (b, 0), (a, 3)\}$

Testing for Functions Represented Algebraically In Exercises 11–20, determine whether the equation represents y as a function of x.

11. $x^2 + y^2 = 4$
12. $x^2 + y = 4$
13. $2x + 3y = 4$
14. $(x - 2)^2 + y^2 = 4$
15. $y = \sqrt{16 - x^2}$
16. $y = \sqrt{x + 5}$
17. $y = |4 - x|$
18. $|y| = 4 - x$
19. $y = -75$
20. $x - 1 = 0$

Evaluating a Function In Exercises 21–32, evaluate (if possible) the function at each specified value of the independent variable and simplify.

21. $f(x) = 2x - 3$
    (a) $f(1)$ (b) $f(-3)$ (c) $f(x - 1)$
22. $V(r) = \frac{4}{3}\pi r^3$
    (a) $V(3)$ (b) $V\left(\frac{3}{2}\right)$ (c) $V(2r)$
23. $g(t) = 4t^2 - 3t + 5$
    (a) $g(2)$ (b) $g(t - 2)$ (c) $g(t) - g(2)$
24. $h(t) = t^2 - 2t$
    (a) $h(2)$ (b) $h(1.5)$ (c) $h(x + 2)$
25. $f(y) = 3 - \sqrt{y}$
    (a) $f(4)$ (b) $f(0.25)$ (c) $f(4x^2)$
26. $f(x) = \sqrt{x + 8} + 2$
    (a) $f(-8)$ (b) $f(1)$ (c) $f(x - 8)$
27. $q(x) = 1/(x^2 - 9)$
    (a) $q(0)$ (b) $q(3)$ (c) $q(y + 3)$
28. $q(t) = (2t^2 + 3)/t^2$
    (a) $q(2)$ (b) $q(0)$ (c) $q(-x)$
29. $f(x) = |x|/x$
    (a) $f(2)$ (b) $f(-2)$ (c) $f(x - 1)$
30. $f(x) = |x| + 4$
    (a) $f(2)$ (b) $f(-2)$ (c) $f(x^2)$
31. $f(x) = \begin{cases} 2x + 1, & x < 0 \\ 2x + 2, & x \geq 0 \end{cases}$
    (a) $f(-1)$ (b) $f(0)$ (c) $f(2)$
32. $f(x) = \begin{cases} 4 - 5x, & x \leq -2 \\ 0, & -2 < x < 2 \\ x^2 + 1, & x \geq 2 \end{cases}$
    (a) $f(-3)$ (b) $f(4)$ (c) $f(-1)$

Evaluating a Function **In Exercises 33–36, complete the table.**

33. $f(x) = x^2 - 3$

x	-2	-1	0	1	2
$f(x)$					

34. $h(t) = \frac{1}{2}|t + 3|$

t	-5	-4	-3	-2	-1
$h(t)$					

35. $f(x) = \begin{cases} -\frac{1}{2}x + 4, & x \le 0 \\ (x - 2)^2, & x > 0 \end{cases}$

x	-2	-1	0	1	2
$f(x)$					

36. $f(x) = \begin{cases} 9 - x^2, & x < 3 \\ x - 3, & x \ge 3 \end{cases}$

x	1	2	3	4	5
$f(x)$					

Finding Values for Which $f(x) = 0$ **In Exercises 37–44, find all real values of x such that $f(x) = 0$.**

37. $f(x) = 15 - 3x$

38. $f(x) = 5x + 1$

39. $f(x) = \dfrac{3x - 4}{5}$

40. $f(x) = \dfrac{12 - x^2}{5}$

41. $f(x) = x^2 - 9$

42. $f(x) = x^2 - 8x + 15$

43. $f(x) = x^3 - x$

44. $f(x) = x^3 - x^2 - 4x + 4$

Finding Values for Which $f(x) = g(x)$ **In Exercises 45–48, find the value(s) of x for which $f(x) = g(x)$.**

45. $f(x) = x^2, \quad g(x) = x + 2$

46. $f(x) = x^2 + 2x + 1, \quad g(x) = 7x - 5$

47. $f(x) = x^4 - 2x^2, \quad g(x) = 2x^2$

48. $f(x) = \sqrt{x} - 4, \quad g(x) = 2 - x$

Finding the Domain of a Function **In Exercises 49–60, find the domain of the function.**

49. $f(x) = 5x^2 + 2x - 1$

50. $g(x) = 1 - 2x^2$

51. $h(t) = \dfrac{4}{t}$

52. $s(y) = \dfrac{3y}{y + 5}$

53. $g(y) = \sqrt{y - 10}$

54. $f(t) = \sqrt[3]{t + 4}$

55. $g(x) = \dfrac{1}{x} - \dfrac{3}{x + 2}$

56. $h(x) = \dfrac{10}{x^2 - 2x}$

57. $f(s) = \dfrac{\sqrt{s - 1}}{s - 4}$

58. $f(x) = \dfrac{\sqrt{x + 6}}{6 + x}$

59. $f(x) = \dfrac{x - 4}{\sqrt{x}}$

60. $f(x) = \dfrac{x + 2}{\sqrt{x - 10}}$

61. Maximum Volume An open box of maximum volume is to be made from a square piece of material 24 centimeters on a side by cutting equal squares from the corners and turning up the sides (see figure).

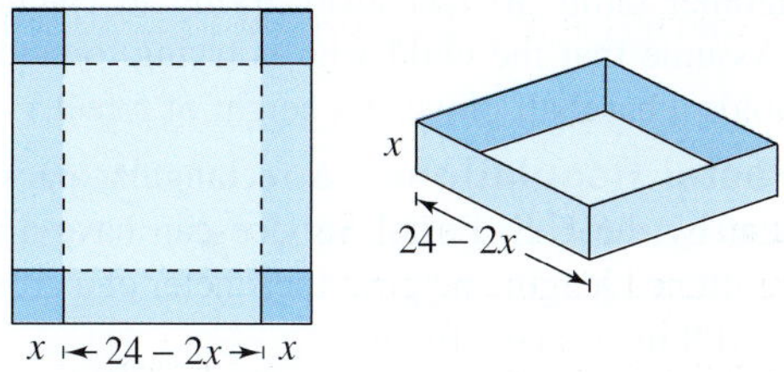

(a) The table shows the volumes V (in cubic centimeters) of the box for various heights x (in centimeters). Use the table to estimate the maximum volume.

Height, x	1	2	3	4	5	6
Volume, V	484	800	972	1024	980	864

(b) Plot the points (x, V) from the table in part (a). Does the relation defined by the ordered pairs represent V as a function of x?

(c) Given that V is a function of x, write the function and determine its domain.

62. Maximum Profit The cost per unit in the production of an MP3 player is \$60. The manufacturer charges \$90 per unit for orders of 100 or less. To encourage large orders, the manufacturer reduces the charge by \$0.15 per MP3 player for each unit ordered in excess of 100 (for example, there would be a charge of \$87 per MP3 player for an order size of 120).

(a) The table shows the profits P (in dollars) for various numbers of units ordered, x. Use the table to estimate the maximum profit.

Units, x	130	140	150	160	170
Profit, P	3315	3360	3375	3360	3315

(b) Plot the points (x, P) from the table in part (a). Does the relation defined by the ordered pairs represent P as a function of x?

(c) Given that P is a function of x, write the function and determine its domain. (*Note:* $P = R - C$, where R is revenue and C is cost.)

63. Geometry Write the area A of a square as a function of its perimeter P.

64. Geometry Write the area A of a circle as a function of its circumference C.

65. Path of a Ball The height y (in feet) of a baseball thrown by a child is

$$y = -\frac{1}{10}x^2 + 3x + 6$$

where x is the horizontal distance (in feet) from where the ball was thrown. Will the ball fly over the head of another child 30 feet away trying to catch the ball? (Assume that the child who is trying to catch the ball holds a baseball glove at a height of 5 feet.)

66. Postal Regulations A rectangular package to be sent by the U.S. Postal Service can have a maximum combined length and girth (perimeter of a cross section) of 108 inches (see figure).

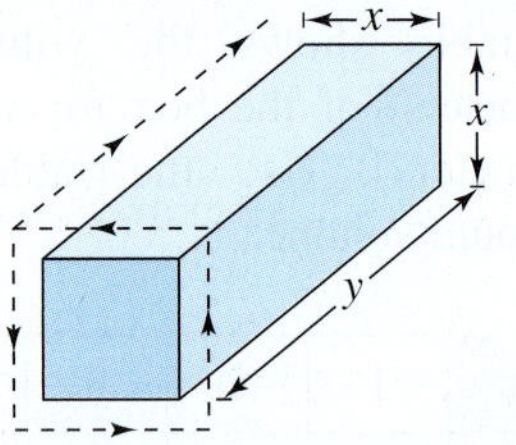

(a) Write the volume V of the package as a function of x. What is the domain of the function?

(b) Use a graphing utility to graph the function. Be sure to use an appropriate window setting.

(c) What dimensions will maximize the volume of the package? Explain your answer.

67. Geometry A right triangle is formed in the first quadrant by the x- and y-axes and a line through the point $(2, 1)$ (see figure). Write the area A of the triangle as a function of x, and determine the domain of the function.

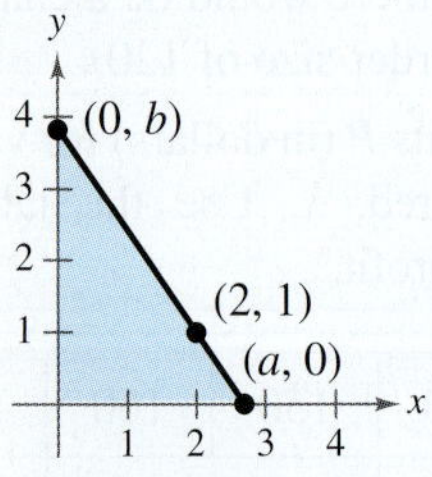

Figure for 67

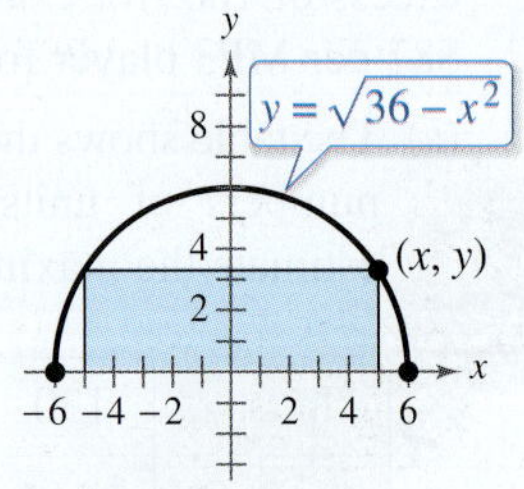

Figure for 68

68. Geometry A rectangle is bounded by the x-axis and the semicircle $y = \sqrt{36 - x^2}$ (see figure). Write the area A of the rectangle as a function of x, and graphically determine the domain of the function.

69. Prescription Drugs The percents p of prescriptions filled with generic drugs in the United States from 2004 through 2010 (see figure) can be approximated by the model

$$p(t) = \begin{cases} 4.57t + 27.3, & 4 \le t \le 7 \\ 3.35t + 37.6, & 8 \le t \le 10 \end{cases}$$

where t represents the year, with $t = 4$ corresponding to 2004. Use this model to find the percent of prescriptions filled with generic drugs in each year from 2004 through 2010. (*Source: National Association of Chain Drug Stores*)

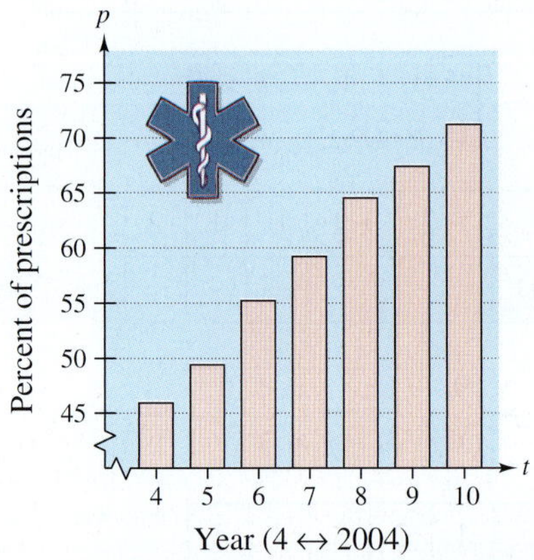

70. Median Sale Price The median sale prices p (in thousands of dollars) of an existing one-family home in the United States from 2000 through 2010 (see figure) can be approximated by the model

$$p(t) = \begin{cases} 0.438t^2 + 10.81t + 145.9, & 0 \le t \le 6 \\ 5.575t^2 - 110.67t + 720.8, & 7 \le t \le 10 \end{cases}$$

where t represents the year, with $t = 0$ corresponding to 2000. Use this model to find the median sale price of an existing one-family home in each year from 2000 through 2010. (*Source: National Association of Realtors*)

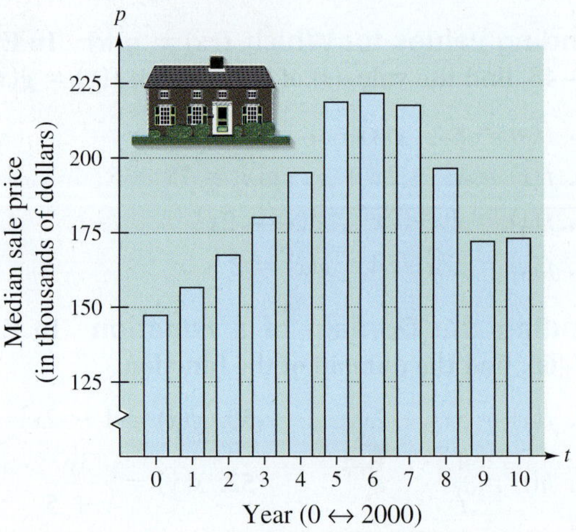

71. Cost, Revenue, and Profit A company produces a product for which the variable cost is \$12.30 per unit and the fixed costs are \$98,000. The product sells for \$17.98. Let x be the number of units produced and sold.

(a) The total cost for a business is the sum of the variable cost and the fixed costs. Write the total cost C as a function of the number of units produced.

(b) Write the revenue R as a function of the number of units sold.

(c) Write the profit P as a function of the number of units sold. (*Note:* $P = R - C$)

72. Average Cost The inventor of a new game believes that the variable cost for producing the game is \$0.95 per unit and the fixed costs are \$6000. The inventor sells each game for \$1.69. Let x be the number of games sold.

(a) The total cost for a business is the sum of the variable cost and the fixed costs. Write the total cost C as a function of the number of games sold.

(b) Write the average cost per unit $\overline{C} = C/x$ as a function of x.

73. Height of a Balloon A balloon carrying a transmitter ascends vertically from a point 3000 feet from the receiving station.

(a) Draw a diagram that gives a visual representation of the problem. Let h represent the height of the balloon and let d represent the distance between the balloon and the receiving station.

(b) Write the height of the balloon as a function of d. What is the domain of the function?

74. Physics

The function $F(y) = 149.76\sqrt{10}y^{5/2}$ estimates the force F (in tons) of water against the face of a dam, where y is the depth of the water (in feet).

(a) Complete the table. What can you conclude from the table?

y	5	10	20	30	40
$F(y)$					

(b) Use the table to approximate the depth at which the force against the dam is 1,000,000 tons.

(c) Find the depth at which the force against the dam is 1,000,000 tons algebraically.

75. Transportation For groups of 80 or more people, a charter bus company determines the rate per person according to the formula

$$\text{Rate} = 8 - 0.05(n - 80), \quad n \ge 80$$

where the rate is given in dollars and n is the number of people.

(a) Write the revenue R for the bus company as a function of n.

(b) Use the function in part (a) to complete the table. What can you conclude?

n	90	100	110	120	130	140	150
$R(n)$							

76. E-Filing The table shows the numbers of tax returns (in millions) made through e-file from 2003 through 2010. Let $f(t)$ represent the number of tax returns made through e-file in the year t. (*Source: Internal Revenue Service*)

DATA

Year	Number of Tax Returns Made Through E-File
2003	52.9
2004	61.5
2005	68.5
2006	73.3
2007	80.0
2008	89.9
2009	95.0
2010	98.7

Spreadsheet at LarsonPrecalculus.com

(a) Find $\dfrac{f(2010) - f(2003)}{2010 - 2003}$ and interpret the result in the context of the problem.

(b) Make a scatter plot of the data.

(c) Find a linear model for the data algebraically. Let N represent the number of tax returns made through e-file and let $t = 3$ correspond to 2003.

(d) Use the model found in part (c) to complete the table.

t	3	4	5	6	7	8	9	10
N								

(e) Compare your results from part (d) with the actual data.

(f) Use a graphing utility to find a linear model for the data. Let $x = 3$ correspond to 2003. How does the model you found in part (c) compare with the model given by the graphing utility?

Lester Lefkowitz/CORBIS

Evaluating a Difference Quotient In Exercises 77–84, find the difference quotient and simplify your answer.

77. $f(x) = x^2 - x + 1, \quad \dfrac{f(2+h) - f(2)}{h}, \quad h \neq 0$

78. $f(x) = 5x - x^2, \quad \dfrac{f(5+h) - f(5)}{h}, \quad h \neq 0$

79. $f(x) = x^3 + 3x, \quad \dfrac{f(x+h) - f(x)}{h}, \quad h \neq 0$

80. $f(x) = 4x^2 - 2x, \quad \dfrac{f(x+h) - f(x)}{h}, \quad h \neq 0$

81. $g(x) = \dfrac{1}{x^2}, \quad \dfrac{g(x) - g(3)}{x - 3}, \quad x \neq 3$

82. $f(t) = \dfrac{1}{t-2}, \quad \dfrac{f(t) - f(1)}{t - 1}, \quad t \neq 1$

83. $f(x) = \sqrt{5x}, \quad \dfrac{f(x) - f(5)}{x - 5}, \quad x \neq 5$

84. $f(x) = x^{2/3} + 1, \quad \dfrac{f(x) - f(8)}{x - 8}, \quad x \neq 8$

Matching and Determining Constants In Exercises 85–88, match the data with one of the following functions

$$f(x) = cx,\ g(x) = cx^2,\ h(x) = c\sqrt{|x|},\ \text{and } r(x) = \frac{c}{x}$$

and determine the value of the constant *c* that will make the function fit the data in the table.

85.

x	-4	-1	0	1	4
y	-32	-2	0	-2	-32

86.

x	-4	-1	0	1	4
y	-1	$-\frac{1}{4}$	0	$\frac{1}{4}$	1

87.

x	-4	-1	0	1	4
y	-8	-32	Undefined	32	8

88.

x	-4	-1	0	1	4
y	6	3	0	3	6

Exploration

True or False? In Exercises 89–92, determine whether the statement is true or false. Justify your answer.

89. Every relation is a function.

90. Every function is a relation.

91. For the function

$$f(x) = x^4 - 1$$

the domain is $(-\infty, \infty)$ and the range is $(0, \infty)$.

92. The set of ordered pairs $\{(-8, -2), (-6, 0), (-4, 0), (-2, 2), (0, 4), (2, -2)\}$ represents a function.

93. Think About It Consider

$$f(x) = \sqrt{x - 1} \quad \text{and} \quad g(x) = \frac{1}{\sqrt{x - 1}}.$$

Why are the domains of f and g different?

94. Think About It Consider

$$f(x) = \sqrt{x - 2} \text{ and } g(x) = \sqrt[3]{x - 2}.$$

Why are the domains of f and g different?

95. Think About It Given

$$f(x) = x^2$$

is f the independent variable? Why or why not?

96. HOW DO YOU SEE IT? The graph represents the height h of a projectile after t seconds.

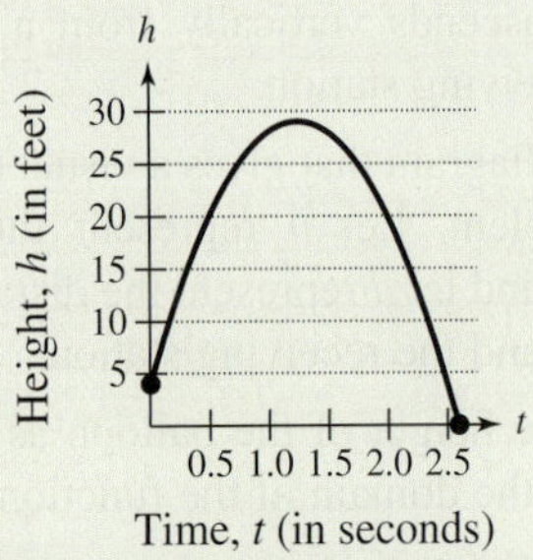

(a) Explain why h is a function of t.

(b) Approximate the height of the projectile after 0.5 second and after 1.25 seconds.

(c) Approximate the domain of h.

(d) Is t a function of h? Explain.

Think About It In Exercises 97 and 98, determine whether the statements use the word *function* in ways that are mathematically correct. Explain your reasoning.

97. (a) The sales tax on a purchased item is a function of the selling price.

(b) Your score on the next algebra exam is a function of the number of hours you study the night before the exam.

98. (a) The amount in your savings account is a function of your salary.

(b) The speed at which a free-falling baseball strikes the ground is a function of the height from which it was dropped.

1.5 Analyzing Graphs of Functions

Graphs of functions can help you visualize relationships between variables in real life. For instance, in Exercise 90 on page 59, you will use the graph of a function to visually represent the temperature of a city over a 24-hour period.

- Use the Vertical Line Test for functions.
- Find the zeros of functions.
- Determine intervals on which functions are increasing or decreasing and determine relative maximum and relative minimum values of functions.
- Determine the average rate of change of a function.
- Identify even and odd functions.

The Graph of a Function

In Section 1.4, you studied functions from an algebraic point of view. In this section, you will study functions from a graphical perspective.

The **graph of a function** f is the collection of ordered pairs $(x, f(x))$ such that x is in the domain of f. As you study this section, remember that

$x =$ the directed distance from the y-axis

$y = f(x) =$ the directed distance from the x-axis

as shown in the figure at the right.

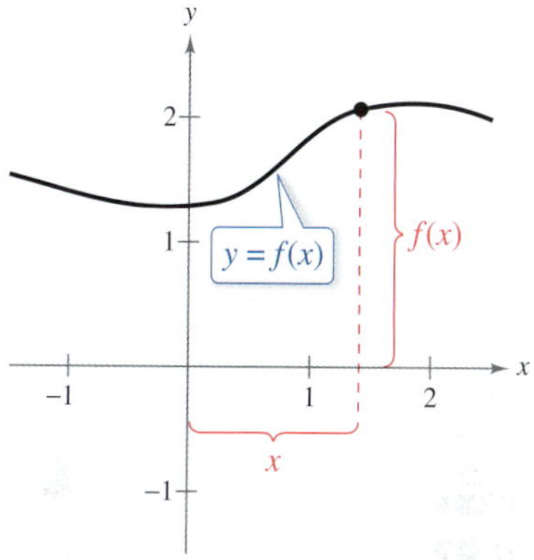

EXAMPLE 1 Finding the Domain and Range of a Function

Use the graph of the function f, shown in Figure 1.34, to find (a) the domain of f, (b) the function values $f(-1)$ and $f(2)$, and (c) the range of f.

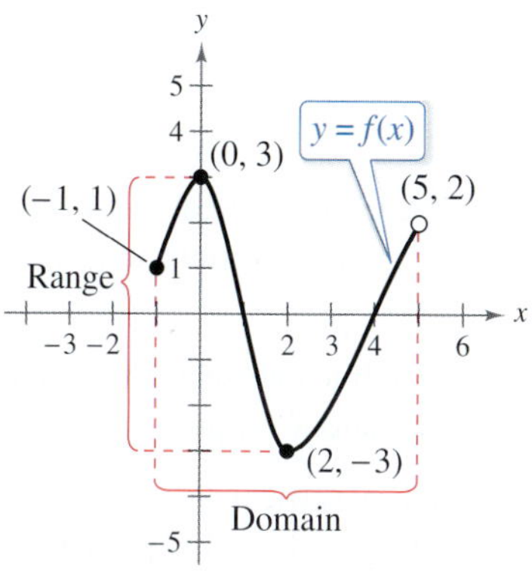

Figure 1.34

Solution

a. The closed dot at $(-1, 1)$ indicates that $x = -1$ is in the domain of f, whereas the open dot at $(5, 2)$ indicates that $x = 5$ is not in the domain. So, the domain of f is all x in the interval $[-1, 5)$.

b. Because $(-1, 1)$ is a point on the graph of f, it follows that $f(-1) = 1$. Similarly, because $(2, -3)$ is a point on the graph of f, it follows that $f(2) = -3$.

c. Because the graph does not extend below $f(2) = -3$ or above $f(0) = 3$, the range of f is the interval $[-3, 3]$.

✓ Checkpoint *Audio-video solution in English & Spanish at LarsonPrecalculus.com.*

Use the graph of the function f to find (a) the domain of f, (b) the function values $f(0)$ and $f(3)$, and (c) the range of f.

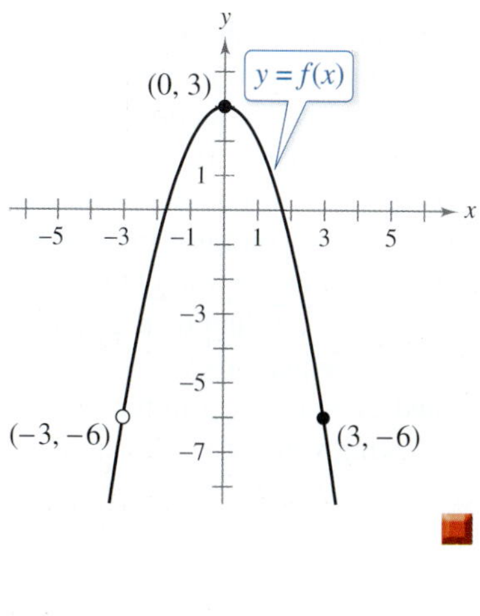

REMARK The use of dots (open or closed) at the extreme left and right points of a graph indicates that the graph does not extend beyond these points. If such dots are not on the graph, then assume that the graph extends beyond these points.

gary718/Shutterstock.com

By the definition of a function, at most one y-value corresponds to a given x-value. This means that the graph of a function cannot have two or more different points with the same x-coordinate, and no two points on the graph of a function can be vertically above or below each other. It follows, then, that a vertical line can intersect the graph of a function at most once. This observation provides a convenient visual test called the **Vertical Line Test** for functions.

Vertical Line Test for Functions

A set of points in a coordinate plane is the graph of y as a function of x if and only if no *vertical* line intersects the graph at more than one point.

EXAMPLE 2 Vertical Line Test for Functions

Use the Vertical Line Test to decide whether each of the following graphs represents y as a function of x.

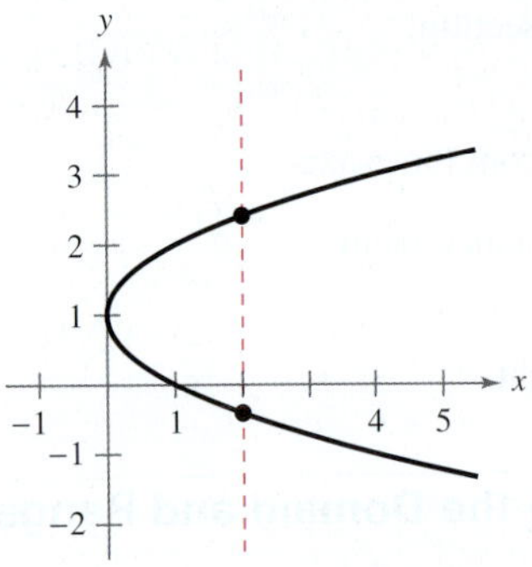

(a)

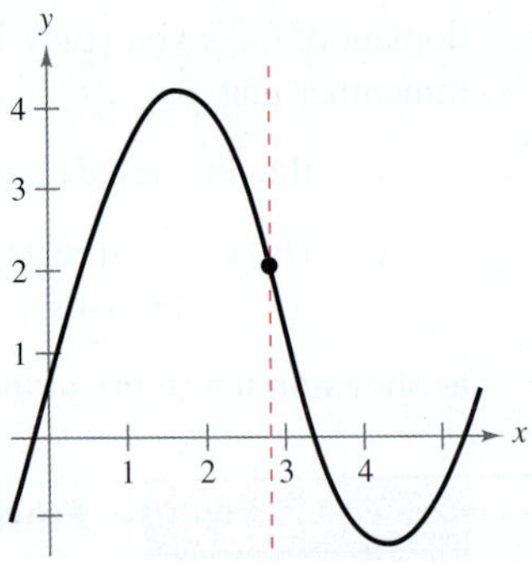

(b)

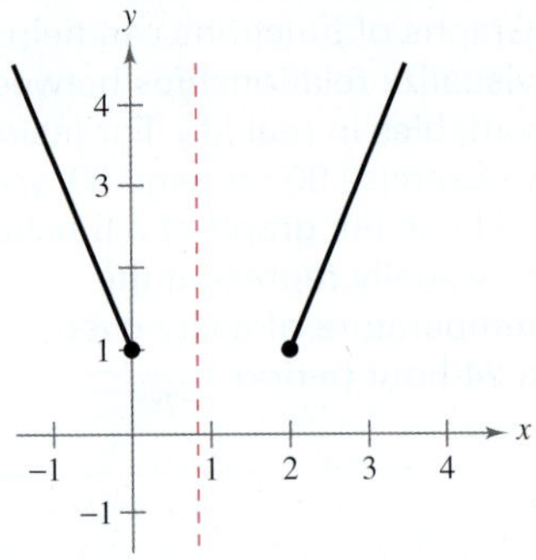

(c)

Solution

a. This *is not* a graph of y as a function of x, because there are vertical lines that intersect the graph twice. That is, for a particular input x, there is more than one output y.

b. This *is* a graph of y as a function of x, because every vertical line intersects the graph at most once. That is, for a particular input x, there is at most one output y.

c. This *is* a graph of y as a function of x. (Note that when a vertical line does not intersect the graph, it simply means that the function is undefined for that particular value of x.) That is, for a particular input x, there is at most one output y.

TECHNOLOGY Most graphing utilities graph functions of x more easily than other types of equations. For instance, the graph shown in (a) above represents the equation $x - (y - 1)^2 = 0$. To use a graphing utility to duplicate this graph, you must first solve the equation for y to obtain $y = 1 \pm \sqrt{x}$, and then graph the two equations $y_1 = 1 + \sqrt{x}$ and $y_2 = 1 - \sqrt{x}$ in the same viewing window.

✓ **Checkpoint** Audio-video solution in English & Spanish at LarsonPrecalculus.com.

Use the Vertical Line Test to decide whether the graph represents y as a function of x.

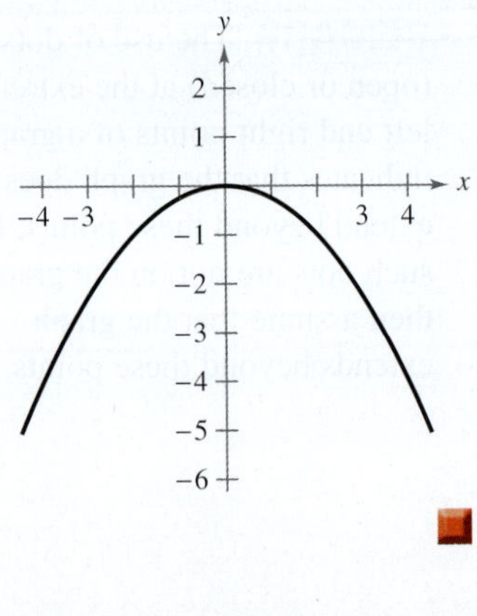

Zeros of a Function

ALGEBRA HELP To do Example 3, you need to be able to solve equations. You can review the techniques for solving equations in Appendix A.5.

If the graph of a function of x has an x-intercept at $(a, 0)$, then a is a **zero** of the function.

> **Zeros of a Function**
>
> The **zeros of a function** f of x are the x-values for which $f(x) = 0$.

EXAMPLE 3 Finding the Zeros of a Function

Find the zeros of each function.

a. $f(x) = 3x^2 + x - 10$

b. $g(x) = \sqrt{10 - x^2}$

c. $h(t) = \dfrac{2t - 3}{t + 5}$

Solution To find the zeros of a function, set the function equal to zero and solve for the independent variable.

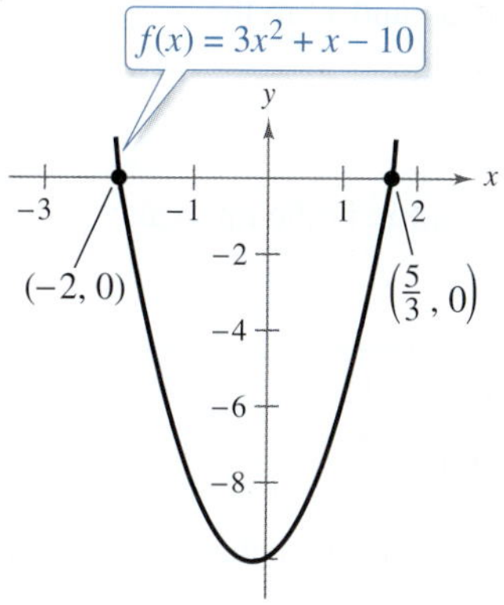

Zeros of f: $x = -2, x = \frac{5}{3}$
Figure 1.35

a.

$3x^2 + x - 10 = 0$ Set $f(x)$ equal to 0.

$(3x - 5)(x + 2) = 0$ Factor.

$3x - 5 = 0 \Rightarrow x = \frac{5}{3}$ Set 1st factor equal to 0.

$x + 2 = 0 \Rightarrow x = -2$ Set 2nd factor equal to 0.

The zeros of f are $x = \frac{5}{3}$ and $x = -2$. In Figure 1.35, note that the graph of f has $\left(\frac{5}{3}, 0\right)$ and $(-2, 0)$ as its x-intercepts.

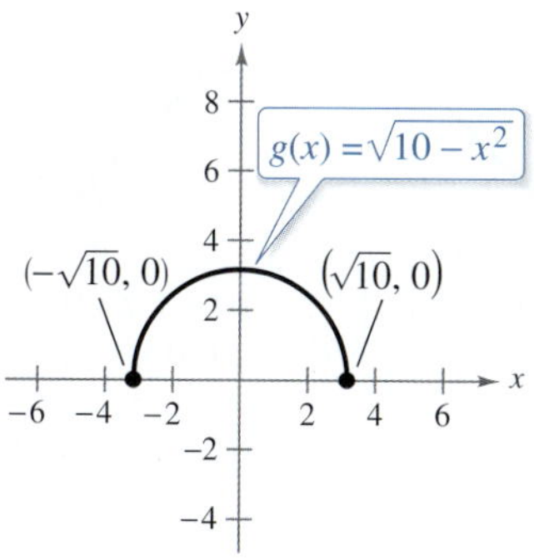

Zeros of g: $x = \pm\sqrt{10}$
Figure 1.36

b.

$\sqrt{10 - x^2} = 0$ Set $g(x)$ equal to 0.

$10 - x^2 = 0$ Square each side.

$10 = x^2$ Add x^2 to each side.

$\pm\sqrt{10} = x$ Extract square roots.

The zeros of g are $x = -\sqrt{10}$ and $x = \sqrt{10}$. In Figure 1.36, note that the graph of g has $\left(-\sqrt{10}, 0\right)$ and $\left(\sqrt{10}, 0\right)$ as its x-intercepts.

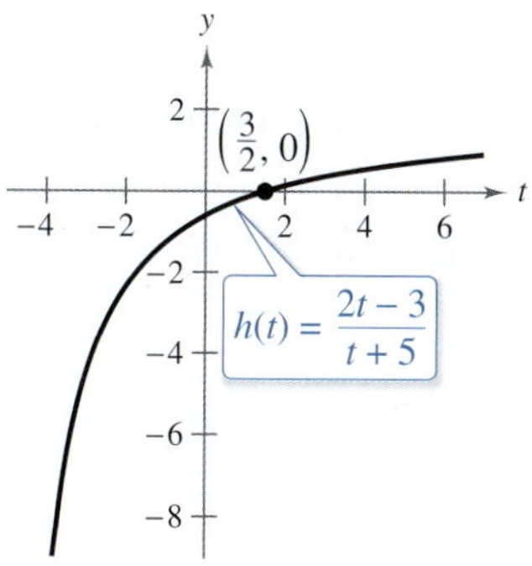

Zero of h: $t = \frac{3}{2}$
Figure 1.37

c.

$\dfrac{2t - 3}{t + 5} = 0$ Set $h(t)$ equal to 0.

$2t - 3 = 0$ Multiply each side by $t + 5$.

$2t = 3$ Add 3 to each side.

$t = \dfrac{3}{2}$ Divide each side by 2.

The zero of h is $t = \frac{3}{2}$. In Figure 1.37, note that the graph of h has $\left(\frac{3}{2}, 0\right)$ as its t-intercept.

Checkpoint Audio-video solution in English & Spanish at LarsonPrecalculus.com.

Find the zeros of each function.

a. $f(x) = 2x^2 + 13x - 24$ **b.** $g(t) = \sqrt{t - 25}$ **c.** $h(x) = \dfrac{x^2 - 2}{x - 1}$

Increasing and Decreasing Functions

The more you know about the graph of a function, the more you know about the function itself. Consider the graph shown in Figure 1.38. As you move from *left to right*, this graph falls from $x = -2$ to $x = 0$, is constant from $x = 0$ to $x = 2$, and rises from $x = 2$ to $x = 4$.

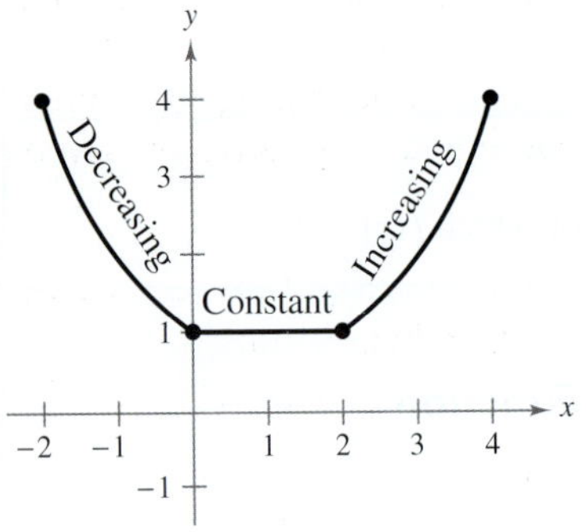

Figure 1.38

> **Increasing, Decreasing, and Constant Functions**
>
> A function f is **increasing** on an interval when, for any x_1 and x_2 in the interval,
>
> $$x_1 < x_2 \quad \text{implies} \quad f(x_1) < f(x_2).$$
>
> A function f is **decreasing** on an interval when, for any x_1 and x_2 in the interval,
>
> $$x_1 < x_2 \quad \text{implies} \quad f(x_1) > f(x_2).$$
>
> A function f is **constant** on an interval when, for any x_1 and x_2 in the interval,
>
> $$f(x_1) = f(x_2).$$

EXAMPLE 4 **Describing Function Behavior**

Use the graphs to describe the increasing, decreasing, or constant behavior of each function.

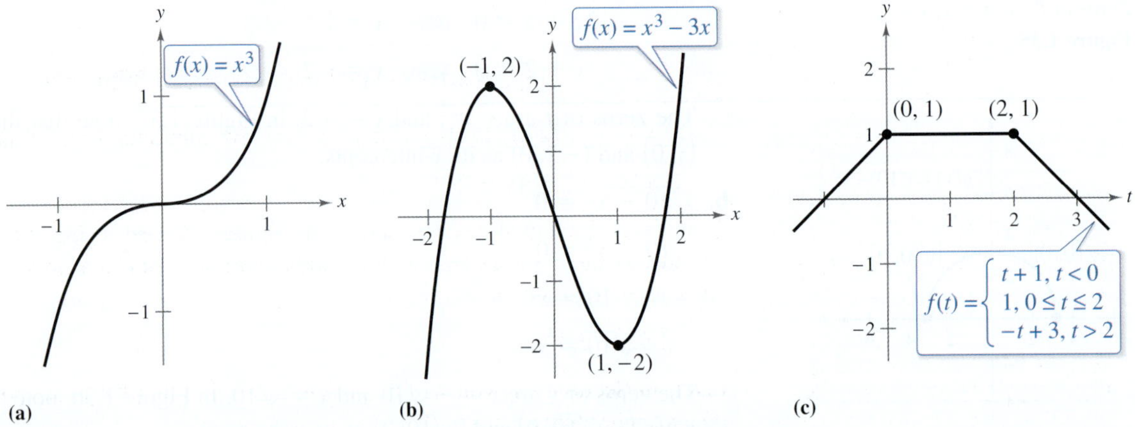

Solution

a. This function is increasing over the entire real line.

b. This function is increasing on the interval $(-\infty, -1)$, decreasing on the interval $(-1, 1)$, and increasing on the interval $(1, \infty)$.

c. This function is increasing on the interval $(-\infty, 0)$, constant on the interval $(0, 2)$, and decreasing on the interval $(2, \infty)$.

✓ ***Checkpoint*** *Audio-video solution in English & Spanish at LarsonPrecalculus.com.*

Graph the function

$$f(x) = x^3 + 3x^2 - 1.$$

Then use the graph to describe the increasing and decreasing behavior of the function.

To help you decide whether a function is increasing, decreasing, or constant on an interval, you can evaluate the function for several values of x. However, you need calculus to determine, for certain, all intervals on which a function is increasing, decreasing, or constant.

The points at which a function changes its increasing, decreasing, or constant behavior are helpful in determining the **relative minimum** or **relative maximum** values of the function.

REMARK A relative minimum or relative maximum is also referred to as a local minimum or local maximum.

Definitions of Relative Minimum and Relative Maximum

A function value $f(a)$ is called a **relative minimum** of f when there exists an interval (x_1, x_2) that contains a such that

$$x_1 < x < x_2 \quad \text{implies} \quad f(a) \le f(x).$$

A function value $f(a)$ is called a **relative maximum** of f when there exists an interval (x_1, x_2) that contains a such that

$$x_1 < x < x_2 \quad \text{implies} \quad f(a) \ge f(x).$$

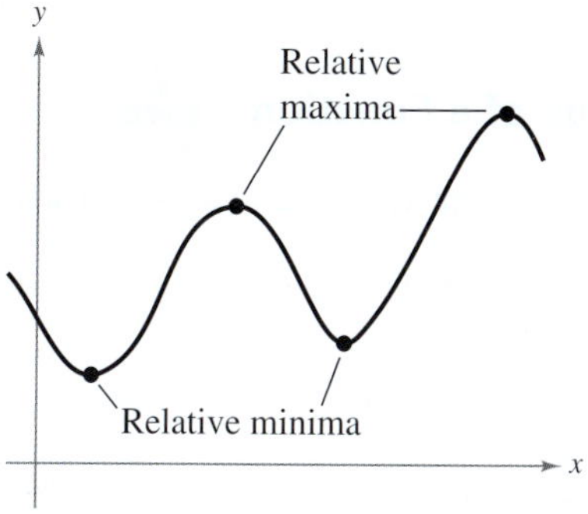

Figure 1.39

Figure 1.39 shows several different examples of relative minima and relative maxima. In Section 2.1, you will study a technique for finding the *exact point* at which a second-degree polynomial function has a relative minimum or relative maximum. For the time being, however, you can use a graphing utility to find reasonable approximations of these points.

EXAMPLE 5 Approximating a Relative Minimum

Use a graphing utility to approximate the relative minimum of the function

$$f(x) = 3x^2 - 4x - 2.$$

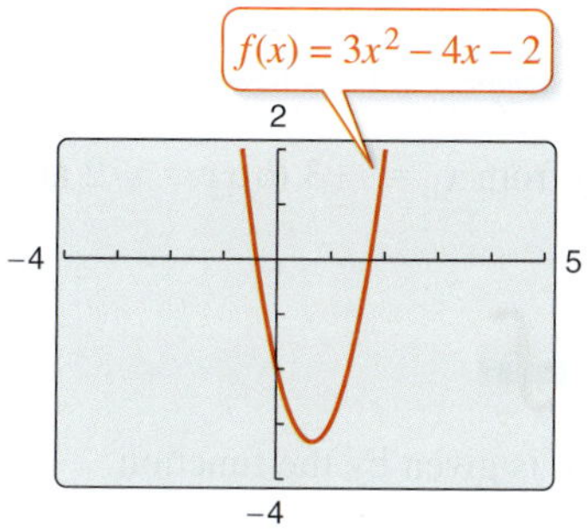

Figure 1.40

Solution The graph of f is shown in Figure 1.40. By using the *zoom* and *trace* features or the *minimum* feature of a graphing utility, you can estimate that the function has a relative minimum at the point

$(0.67, -3.33)$. Relative minimum

Later, in Section 2.1, you will be able to determine that the exact point at which the relative minimum occurs is $\left(\frac{2}{3}, -\frac{10}{3}\right)$.

✔ **Checkpoint** *Audio-video solution in English & Spanish at LarsonPrecalculus.com.*

Use a graphing utility to approximate the relative maximum of the function

$$f(x) = -4x^2 - 7x + 3.$$

You can also use the *table* feature of a graphing utility to numerically approximate the relative minimum of the function in Example 5. Using a table that begins at 0.6 and increments the value of x by 0.01, you can approximate that the minimum of $f(x) = 3x^2 - 4x - 2$ occurs at the point $(0.67, -3.33)$.

TECHNOLOGY When you use a graphing utility to estimate the x- and y-values of a relative minimum or relative maximum, the *zoom* feature will often produce graphs that are nearly flat. To overcome this problem, you can manually change the vertical setting of the viewing window. The graph will stretch vertically when the values of Ymin and Ymax are closer together.

Average Rate of Change

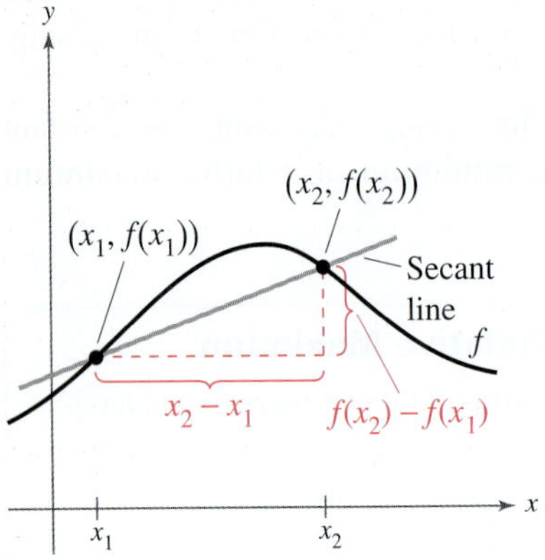

Figure 1.41

In Section 1.3, you learned that the slope of a line can be interpreted as a *rate of change*. For a nonlinear graph whose slope changes at each point, the **average rate of change** between any two points $(x_1, f(x_1))$ and $(x_2, f(x_2))$ is the slope of the line through the two points (see Figure 1.41). The line through the two points is called the **secant line,** and the slope of this line is denoted as m_{sec}.

$$\text{Average rate of change of } f \text{ from } x_1 \text{ to } x_2 = \frac{f(x_2) - f(x_1)}{x_2 - x_1}$$

$$= \frac{\text{change in } y}{\text{change in } x}$$

$$= m_{\text{sec}}$$

EXAMPLE 6 Average Rate of Change of a Function

Find the average rates of change of $f(x) = x^3 - 3x$ (a) from $x_1 = -2$ to $x_2 = -1$ and (b) from $x_1 = 0$ to $x_2 = 1$ (see Figure 1.42).

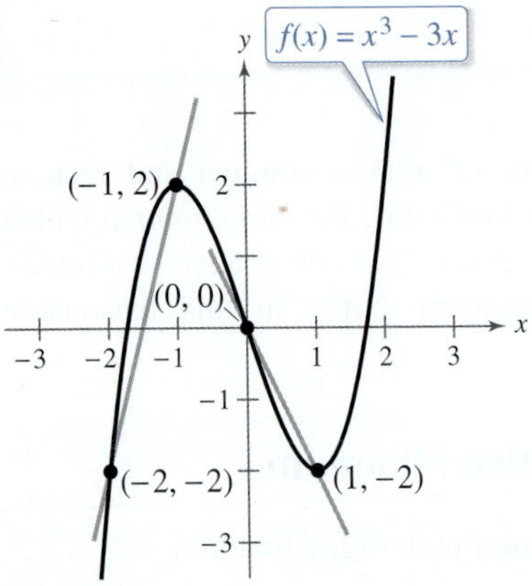

Figure 1.42

Solution

a. The average rate of change of f from $x_1 = -2$ to $x_2 = -1$ is

$$\frac{f(x_2) - f(x_1)}{x_2 - x_1} = \frac{f(-1) - f(-2)}{-1 - (-2)} = \frac{2 - (-2)}{1} = 4.$$

Secant line has positive slope.

b. The average rate of change of f from $x_1 = 0$ to $x_2 = 1$ is

$$\frac{f(x_2) - f(x_1)}{x_2 - x_1} = \frac{f(1) - f(0)}{1 - 0} = \frac{-2 - 0}{1} = -2.$$

Secant line has negative slope.

✓ **Checkpoint** *Audio-video solution in English & Spanish at LarsonPrecalculus.com.*

Find the average rates of change of $f(x) = x^2 + 2x$ (a) from $x_1 = -3$ to $x_2 = -2$ and (b) from $x_1 = -2$ to $x_2 = 0$.

EXAMPLE 7 Finding Average Speed

The distance s (in feet) a moving car is from a stoplight is given by the function

$$s(t) = 20t^{3/2}$$

where t is the time (in seconds). Find the average speed of the car (a) from $t_1 = 0$ to $t_2 = 4$ seconds and (b) from $t_1 = 4$ to $t_2 = 9$ seconds.

Average speed is an average rate of change.

Solution

a. The average speed of the car from $t_1 = 0$ to $t_2 = 4$ seconds is

$$\frac{s(t_2) - s(t_1)}{t_2 - t_1} = \frac{s(4) - s(0)}{4 - 0} = \frac{160 - 0}{4} = 40 \text{ feet per second.}$$

b. The average speed of the car from $t_1 = 4$ to $t_2 = 9$ seconds is

$$\frac{s(t_2) - s(t_1)}{t_2 - t_1} = \frac{s(9) - s(4)}{9 - 4} = \frac{540 - 160}{5} = 76 \text{ feet per second.}$$

✓ **Checkpoint** *Audio-video solution in English & Spanish at LarsonPrecalculus.com.*

In Example 7, find the average speed of the car (a) from $t_1 = 0$ to $t_2 = 1$ second and (b) from $t_1 = 1$ second to $t_2 = 4$ seconds.

sadwitch/Shutterstock.com

Even and Odd Functions

In Section 1.2, you studied different types of symmetry of a graph. In the terminology of functions, a function is said to be **even** when its graph is symmetric with respect to the y-axis and **odd** when its graph is symmetric with respect to the origin. The symmetry tests in Section 1.2 yield the following tests for even and odd functions.

> **Tests for Even and Odd Functions**
>
> A function $y = f(x)$ is **even** when, for each x in the domain of f, $f(-x) = f(x)$.
>
> A function $y = f(x)$ is **odd** when, for each x in the domain of f, $f(-x) = -f(x)$.

EXAMPLE 8 Even and Odd Functions

a. The function $g(x) = x^3 - x$ is odd because $g(-x) = -g(x)$, as follows.

$$\begin{aligned} g(-x) &= (-x)^3 - (-x) && \text{Substitute } -x \text{ for } x. \\ &= -x^3 + x && \text{Simplify.} \\ &= -(x^3 - x) && \text{Distributive Property} \\ &= -g(x) && \text{Test for odd function} \end{aligned}$$

b. The function $h(x) = x^2 + 1$ is even because $h(-x) = h(x)$, as follows.

$$\begin{aligned} h(-x) &= (-x)^2 + 1 && \text{Substitute } -x \text{ for } x. \\ &= x^2 + 1 && \text{Simplify.} \\ &= h(x) && \text{Test for even function} \end{aligned}$$

Figure 1.43 shows the graphs and symmetry of these two functions.

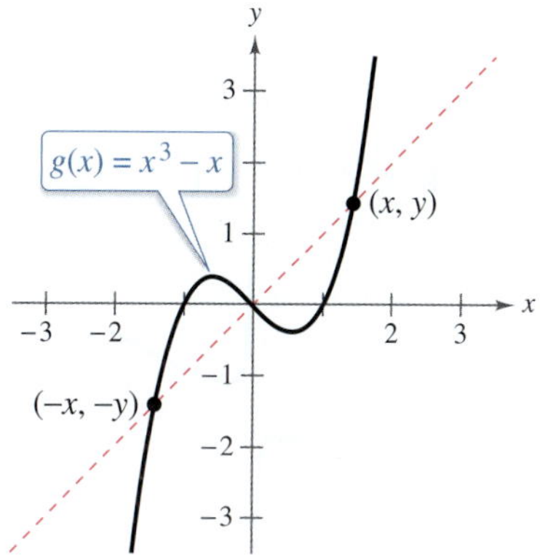

(a) Symmetric to origin: Odd Function

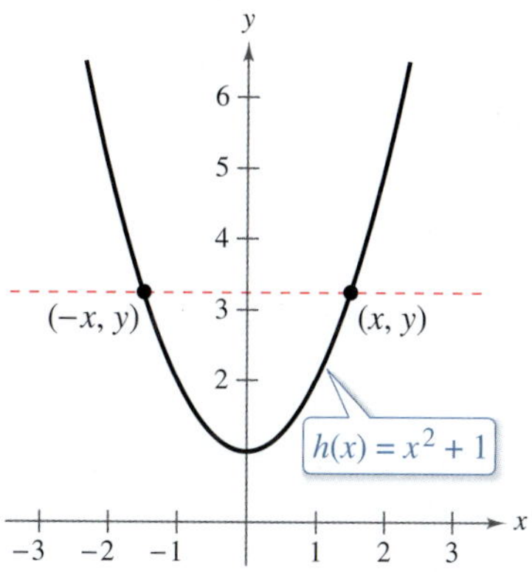

(b) Symmetric to y-axis: Even Function

Figure 1.43

✓ ***Checkpoint*** *Audio-video solution in English & Spanish at LarsonPrecalculus.com.*

Determine whether the function is even, odd, or neither. Then describe the symmetry.

a. $f(x) = 5 - 3x$ **b.** $g(x) = x^4 - x^2 - 1$ **c.** $h(x) = 2x^3 + 3x$

> ## Summarize (Section 1.5)
>
> 1. State the Vertical Line Test for functions *(page 50)*. For an example of using the Vertical Line Test, see Example 2.
> 2. Explain how to find the zeros of a function *(page 51)*. For an example of finding the zeros of functions, see Example 3.
> 3. Explain how to determine intervals on which functions are increasing or decreasing *(page 52)* and how to determine relative maximum and relative minimum values of functions *(page 53)*. For an example of describing function behavior, see Example 4. For an example of approximating a relative minimum, see Example 5.
> 4. Explain how to determine the average rate of change of a function *(page 54)*. For examples of determining average rates of change, see Examples 6 and 7.
> 5. State the definitions of an even function and an odd function *(page 55)*. For an example of identifying even and odd functions, see Example 8.

1.5 Exercises

See CalcChat.com for tutorial help and worked-out solutions to odd-numbered exercises.

Vocabulary: Fill in the blanks.

1. The ________ ________ ________ is used to determine whether the graph of an equation is a function of y in terms of x.
2. The ________ of a function f are the values of x for which $f(x) = 0$.
3. A function f is ________ on an interval when, for any x_1 and x_2 in the interval, $x_1 < x_2$ implies $f(x_1) > f(x_2)$.
4. A function value $f(a)$ is a relative ________ of f when there exists an interval (x_1, x_2) containing a such that $x_1 < x < x_2$ implies $f(a) \geq f(x)$.
5. The ________ ________ ________ ________ between any two points $(x_1, f(x_1))$ and $(x_2, f(x_2))$ is the slope of the line through the two points, and this line is called the ________ line.
6. A function f is ________ when, for each x in the domain of f, $f(-x) = -f(x)$.

Skills and Applications

Domain, Range, and Values of a Function In Exercises 7–10, use the graph of the function to find the domain and range of f and the indicated function values.

7. (a) $f(-2)$ (b) $f(-1)$ (c) $f(\frac{1}{2})$ (d) $f(1)$

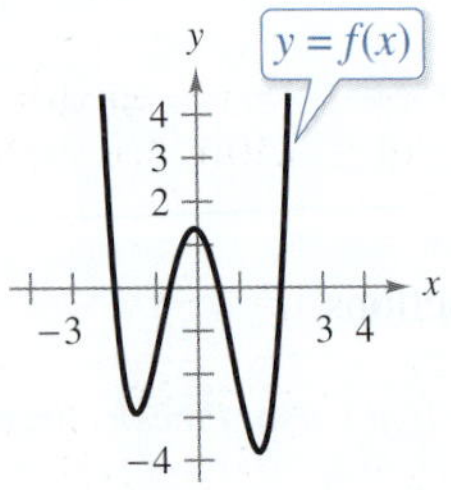

8. (a) $f(-1)$ (b) $f(2)$ (c) $f(0)$ (d) $f(1)$

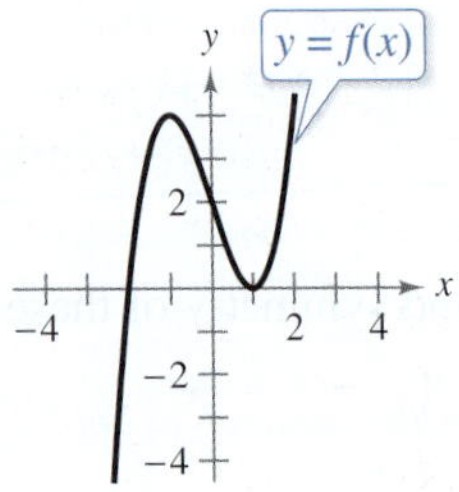

9. (a) $f(2)$ (b) $f(1)$ (c) $f(3)$ (d) $f(-1)$

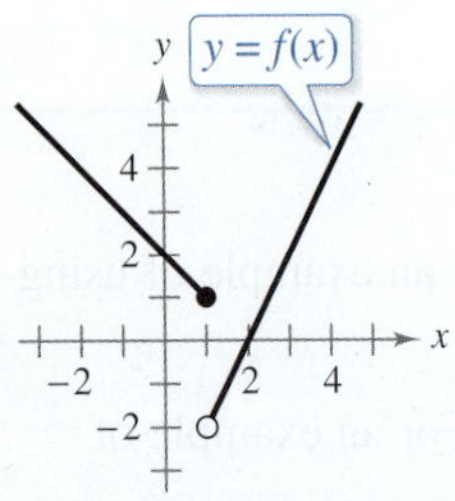

10. (a) $f(-2)$ (b) $f(1)$ (c) $f(0)$ (d) $f(2)$

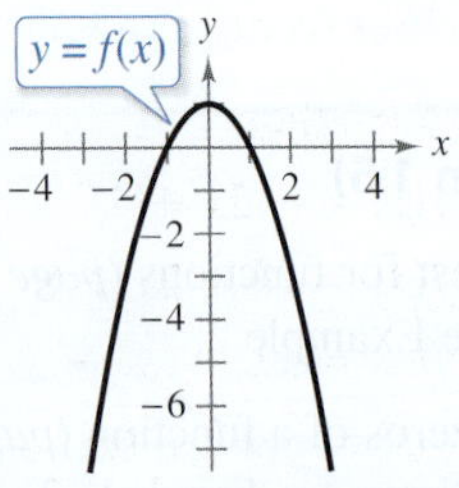

Vertical Line Test for Functions In Exercises 11–14, use the Vertical Line Test to determine whether y is a function of x. To print an enlarged copy of the graph, go to *MathGraphs.com*.

11. $y = \frac{1}{4}x^3$

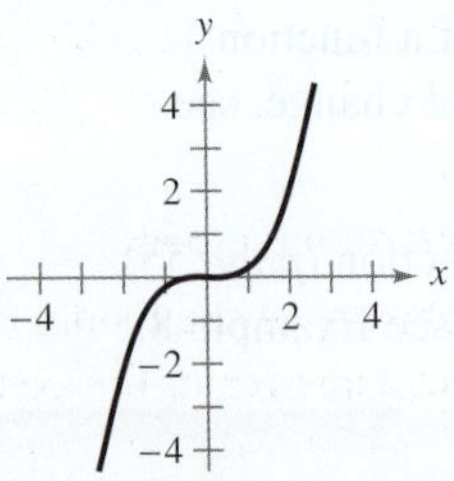

12. $x - y^2 = 1$

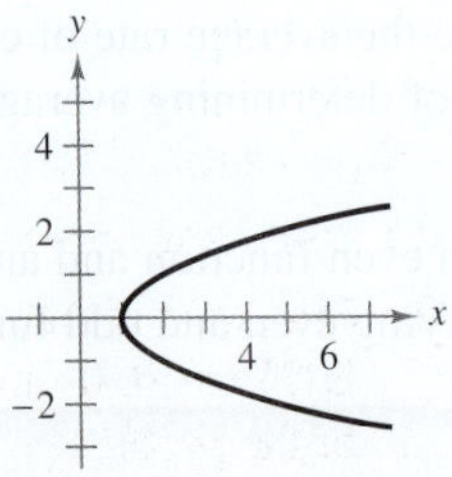

13. $x^2 + y^2 = 25$

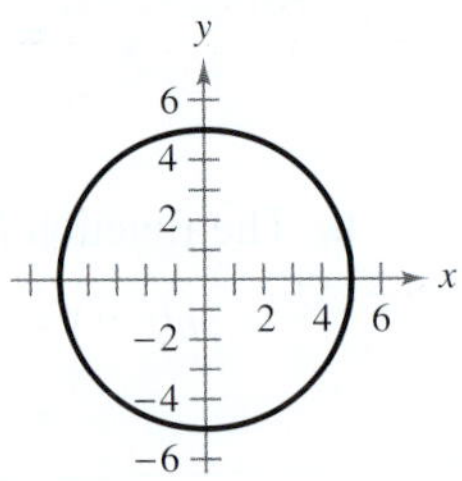

14. $x^2 = 2xy - 1$

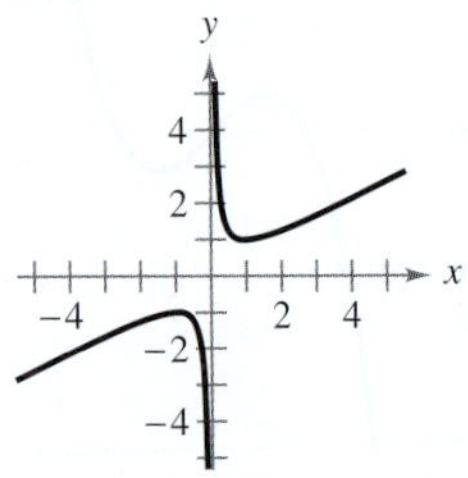

Finding the Zeros of a Function In Exercises 15–24, find the zeros of the function algebraically.

15. $f(x) = 2x^2 - 7x - 30$
16. $f(x) = 3x^2 + 22x - 16$
17. $f(x) = \dfrac{x}{9x^2 - 4}$
18. $f(x) = \dfrac{x^2 - 9x + 14}{4x}$
19. $f(x) = \frac{1}{2}x^3 - x$
20. $f(x) = 9x^4 - 25x^2$
21. $f(x) = x^3 - 4x^2 - 9x + 36$
22. $f(x) = 4x^3 - 24x^2 - x + 6$
23. $f(x) = \sqrt{2x} - 1$
24. $f(x) = \sqrt{3x + 2}$

Graphing and Finding Zeros In Exercises 25–30, (a) use a graphing utility to graph the function and find the zeros of the function and (b) verify your results from part (a) algebraically.

25. $f(x) = 3 + \dfrac{5}{x}$
26. $f(x) = x(x - 7)$
27. $f(x) = \sqrt{2x + 11}$
28. $f(x) = \sqrt{3x - 14} - 8$
29. $f(x) = \dfrac{3x - 1}{x - 6}$
30. $f(x) = \dfrac{2x^2 - 9}{3 - x}$

Describing Function Behavior **In Exercises 31–38, determine the intervals on which the function is increasing, decreasing, or constant.**

31. $f(x) = \frac{3}{2}x$

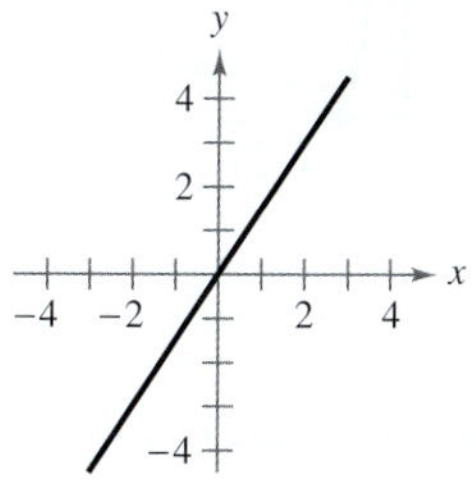

32. $f(x) = x^2 - 4x$

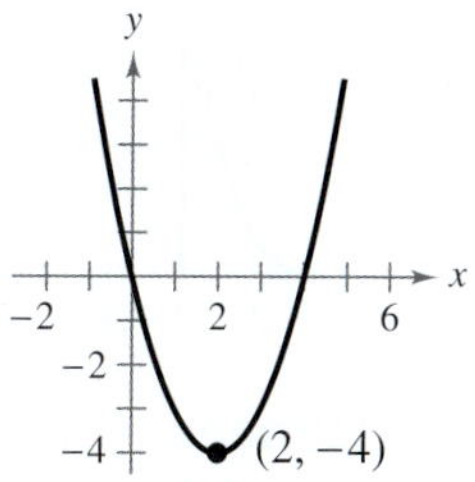

33. $f(x) = x^3 - 3x^2 + 2$

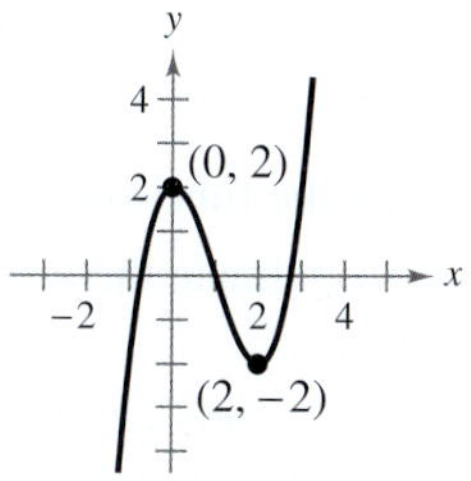

34. $f(x) = \sqrt{x^2 - 1}$

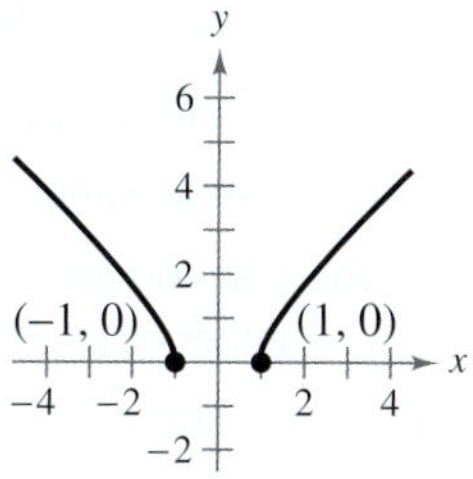

35. $f(x) = |x + 1| + |x - 1|$

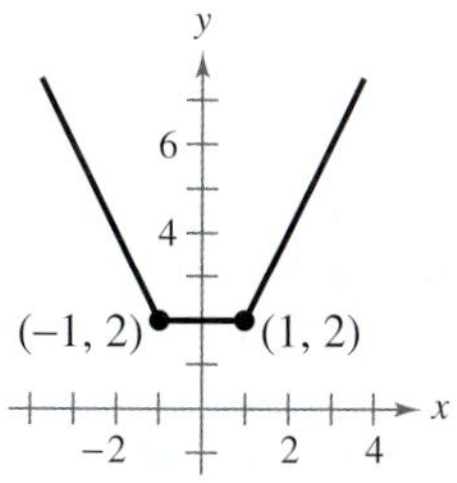

36. $f(x) = \dfrac{x^2 + x + 1}{x + 1}$

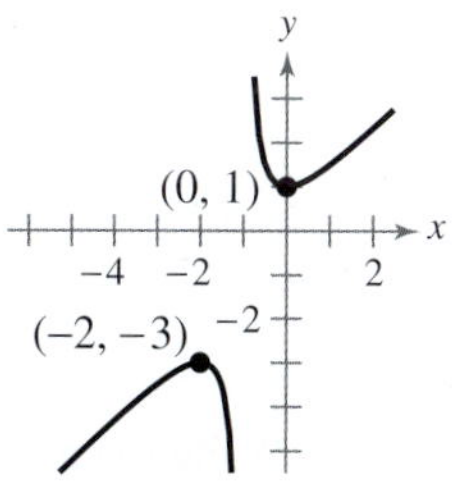

37. $f(x) = \begin{cases} x + 3, & x \le 0 \\ 3, & 0 < x \le 2 \\ 2x + 1, & x > 2 \end{cases}$

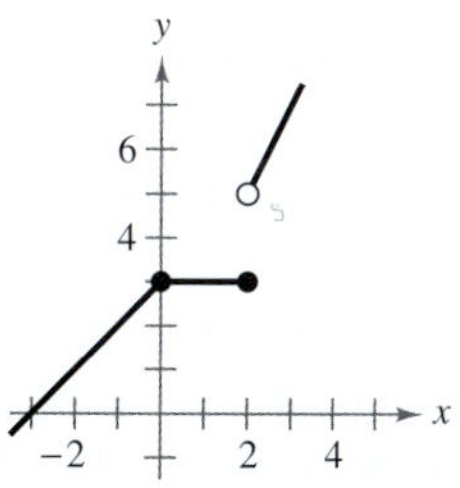

38. $f(x) = \begin{cases} 2x + 1, & x \le -1 \\ x^2 - 2, & x > -1 \end{cases}$

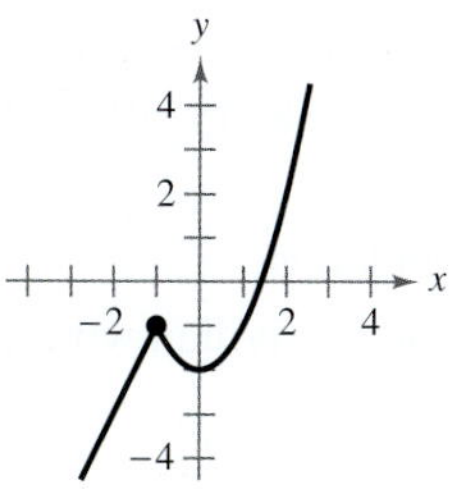

Describing Function Behavior **In Exercises 39–46, (a) use a graphing utility to graph the function and visually determine the intervals on which the function is increasing, decreasing, or constant, and (b) make a table of values to verify whether the function is increasing, decreasing, or constant on the intervals you identified in part (a).**

39. $f(x) = 3$

40. $g(x) = x$

41. $g(s) = \dfrac{s^2}{4}$

42. $f(x) = 3x^4 - 6x^2$

43. $f(x) = \sqrt{1 - x}$

44. $f(x) = x\sqrt{x + 3}$

45. $f(x) = x^{3/2}$

46. $f(x) = x^{2/3}$

Approximating Relative Minima or Maxima **In Exercises 47–54, use a graphing utility to graph the function and approximate (to two decimal places) any relative minima or maxima.**

47. $f(x) = 3x^2 - 2x - 5$

48. $f(x) = -x^2 + 3x - 2$

49. $f(x) = -2x^2 + 9x$

50. $f(x) = x(x - 2)(x + 3)$

51. $f(x) = x^3 - 3x^2 - x + 1$

52. $h(x) = x^3 - 6x^2 + 15$

53. $h(x) = (x - 1)\sqrt{x}$

54. $g(x) = x\sqrt{4 - x}$

Graphical Analysis **In Exercises 55–60, graph the function and determine the interval(s) for which $f(x) \ge 0$.**

55. $f(x) = 4 - x$

56. $f(x) = 4x + 2$

57. $f(x) = 9 - x^2$

58. $f(x) = x^2 - 4x$

59. $f(x) = \sqrt{x - 1}$

60. $f(x) = -(1 + |x|)$

Average Rate of Change of a Function **In Exercises 61–64, find the average rate of change of the function from x_1 to x_2.**

	Function	**x-Values**
61.	$f(x) = -2x + 15$	$x_1 = 0, x_2 = 3$
62.	$f(x) = x^2 - 2x + 8$	$x_1 = 1, x_2 = 5$
63.	$f(x) = x^3 - 3x^2 - x$	$x_1 = 1, x_2 = 3$
64.	$f(x) = -x^3 + 6x^2 + x$	$x_1 = 1, x_2 = 6$

65. Research and Development The amounts y (in millions of dollars) the U.S. Department of Energy spent for research and development from 2005 through 2010 can be approximated by the model

$$y = 56.77t^2 - 366.8t + 8916, \quad 5 \le t \le 10$$

where t represents the year, with $t = 5$ corresponding to 2005. *(Source: American Association for the Advancement of Science)*

(a) Use a graphing utility to graph the model.

(b) Find the average rate of change of the model from 2005 to 2010. Interpret your answer in the context of the problem.

66. **Finding Average Speed** Use the information in Example 7 to find the average speed of the car from $t_1 = 0$ to $t_2 = 9$ seconds. Explain why the result is less than the value obtained in part (b) of Example 7.

Physics In Exercises 67–70, (a) use the position equation $s = -16t^2 + v_0t + s_0$ to write a function that represents the situation, (b) use a graphing utility to graph the function, (c) find the average rate of change of the function from t_1 to t_2, (d) describe the slope of the secant line through t_1 and t_2, (e) find the equation of the secant line through t_1 and t_2, and (f) graph the secant line in the same viewing window as your position function.

67. An object is thrown upward from a height of 6 feet at a velocity of 64 feet per second.

$t_1 = 0, t_2 = 3$

68. An object is thrown upward from a height of 6.5 feet at a velocity of 72 feet per second.

$t_1 = 0, t_2 = 4$

69. An object is thrown upward from ground level at a velocity of 120 feet per second.

$t_1 = 3, t_2 = 5$

70. An object is dropped from a height of 80 feet.

$t_1 = 1, t_2 = 2$

Even, Odd, or Neither? In Exercises 71–76, determine whether the function is even, odd, or neither. Then describe the symmetry.

71. $f(x) = x^6 - 2x^2 + 3$
72. $g(x) = x^3 - 5x$
73. $f(x) = x\sqrt{1 - x^2}$
74. $h(x) = x\sqrt{x + 5}$
75. $f(s) = 4s^{3/2}$
76. $g(s) = 4s^{2/3}$

Even, Odd, or Neither? In Exercises 77–82, sketch a graph of the function and determine whether it is even, odd, or neither. Verify your answer algebraically.

77. $f(x) = -9$
78. $f(x) = 5 - 3x$
79. $f(x) = -|x - 5|$
80. $h(x) = x^2 - 4$
81. $f(x) = \sqrt{1 - x}$
82. $g(t) = \sqrt[3]{t - 1}$

Height of a Rectangle In Exercises 83 and 84, write the height h of the rectangle as a function of x.

83.

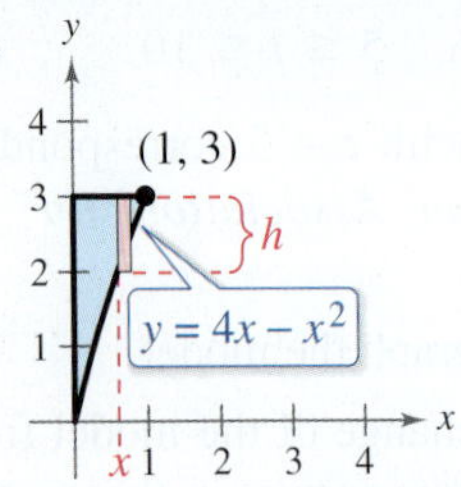

84.

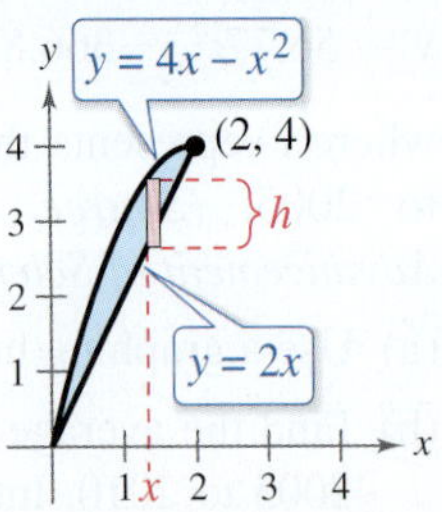

Length of a Rectangle In Exercises 85 and 86, write the length L of the rectangle as a function of y.

85.

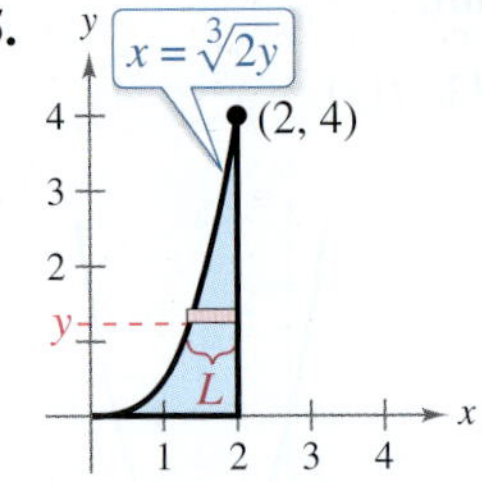

86.

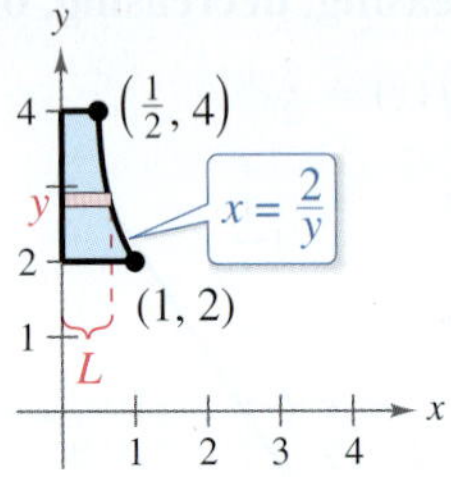

87. **Lumens** The number of lumens (time rate of flow of light) L from a fluorescent lamp can be approximated by the model

$$L = -0.294x^2 + 97.744x - 664.875, \quad 20 \le x \le 90$$

where x is the wattage of the lamp.

(a) Use a graphing utility to graph the function.

(b) Use the graph from part (a) to estimate the wattage necessary to obtain 2000 lumens.

88. **Geometry** Corners of equal size are cut from a square with sides of length 8 meters (see figure).

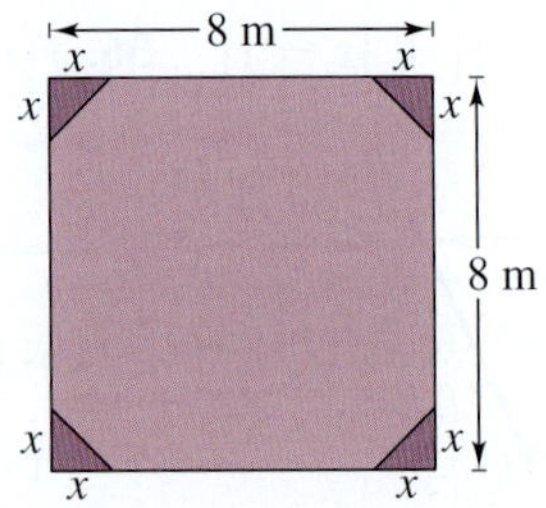

(a) Write the area A of the resulting figure as a function of x. Determine the domain of the function.

(b) Use a graphing utility to graph the area function over its domain. Use the graph to find the range of the function.

(c) Identify the figure that results when x is the maximum value in the domain of the function. What would be the length of each side of the figure?

89. **Coordinate Axis Scale** Each function described below models the specified data for the years 2003 through 2013, with $t = 3$ corresponding to 2003. Estimate a reasonable scale for the vertical axis (e.g., hundreds, thousands, millions, etc.) of the graph and justify your answer. (There are many correct answers.)

(a) $f(t)$ represents the average salary of college professors.

(b) $f(t)$ represents the U.S. population.

(c) $f(t)$ represents the percent of the civilian work force that is unemployed.

90. Data Analysis: Temperature The table shows the temperatures y (in degrees Fahrenheit) in a city over a 24-hour period. Let x represent the time of day, where $x = 0$ corresponds to 6 A.M.

DATA

Time, x	Temperature, y
0	34
2	50
4	60
6	64
8	63
10	59
12	53
14	46
16	40
18	36
20	34
22	37
24	45

Spreadsheet at LarsonPrecalculus.com

A model that represents these data is given by

$$y = 0.026x^3 - 1.03x^2 + 10.2x + 34, \quad 0 \le x \le 24.$$

(a) Use a graphing utility to create a scatter plot of the data. Then graph the model in the same viewing window.

(b) How well does the model fit the data?

(c) Use the graph to approximate the times when the temperature was increasing and decreasing.

(d) Use the graph to approximate the maximum and minimum temperatures during this 24-hour period.

(e) Could this model predict the temperatures in the city during the next 24-hour period? Why or why not?

91. Writing Use a graphing utility to graph each function. Write a paragraph describing any similarities and differences you observe among the graphs.

(a) $y = x$

(b) $y = x^2$

(c) $y = x^3$

(d) $y = x^4$

(e) $y = x^5$

(f) $y = x^6$

92. HOW DO YOU SEE IT? Use the graph of the function to answer (a)–(e).

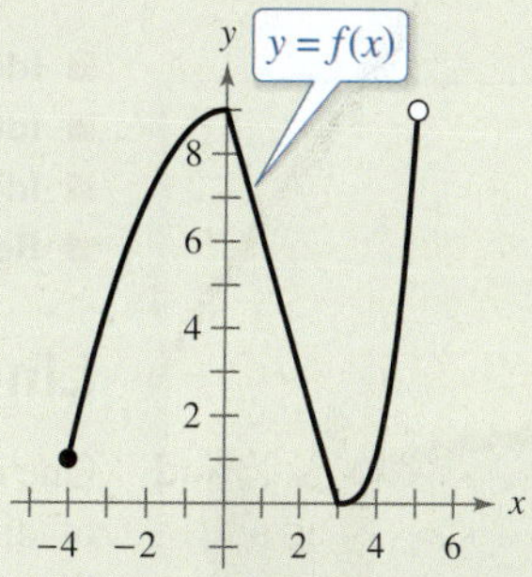

(a) Find the domain and range of f.

(b) Find the zero(s) of f.

(c) Determine the intervals over which f is increasing, decreasing, or constant.

(d) Approximate any relative minimum or relative maximum values of f.

(e) Is f even, odd, or neither?

Exploration

True or False? **In Exercises 93 and 94, determine whether the statement is true or false. Justify your answer.**

93. A function with a square root cannot have a domain that is the set of real numbers.

94. It is possible for an odd function to have the interval $[0, \infty)$ as its domain.

Think About It **In Exercises 95 and 96, find the coordinates of a second point on the graph of a function f when the given point is on the graph and the function is (a) even and (b) odd.**

95. $\left(-\frac{5}{3}, -7\right)$ **96.** $(2a, 2c)$

97. Graphical Reasoning Graph each of the functions with a graphing utility. Determine whether the function is even, odd, or neither.

$f(x) = x^2 - x^4$ $g(x) = 2x^3 + 1$

$h(x) = x^5 - 2x^3 + x$ $j(x) = 2 - x^6 - x^8$

$k(x) = x^5 - 2x^4 + x - 2$ $p(x) = x^9 + 3x^5 - x^3 + x$

What do you notice about the equations of functions that are odd? What do you notice about the equations of functions that are even? Can you describe a way to identify a function as odd or even by inspecting the equation? Can you describe a way to identify a function as neither odd nor even by inspecting the equation?

98. Even, Odd, or Neither? If f is an even function, determine whether g is even, odd, or neither. Explain.

(a) $g(x) = -f(x)$ (b) $g(x) = f(-x)$

(c) $g(x) = f(x) - 2$ (d) $g(x) = f(x - 2)$

gary718/Shutterstock.com

1.6 A Library of Parent Functions

Piecewise-defined functions can help you model real-life situations. For instance, in Exercise 47 on page 66, you will write a piecewise-defined function to model the depth of snow during a snowstorm.

- **Identify and graph linear and squaring functions.**
- **Identify and graph cubic, square root, and reciprocal functions.**
- **Identify and graph step and other piecewise-defined functions.**
- **Recognize graphs of parent functions.**

Linear and Squaring Functions

One of the goals of this text is to enable you to recognize the basic shapes of the graphs of different types of functions. For instance, you know that the graph of the **linear function** $f(x) = ax + b$ is a line with slope $m = a$ and y-intercept at $(0, b)$. The graph of the linear function has the following characteristics.

- The domain of the function is the set of all real numbers.
- When $m \neq 0$, the range of the function is the set of all real numbers.
- The graph has an x-intercept at $(-b/m, 0)$ and a y-intercept at $(0, b)$.
- The graph is increasing when $m > 0$, decreasing when $m < 0$, and constant when $m = 0$.

EXAMPLE 1 **Writing a Linear Function**

Write the linear function f for which $f(1) = 3$ and $f(4) = 0$.

Solution To find the equation of the line that passes through $(x_1, y_1) = (1, 3)$ and $(x_2, y_2) = (4, 0)$, first find the slope of the line.

$$m = \frac{y_2 - y_1}{x_2 - x_1} = \frac{0 - 3}{4 - 1} = \frac{-3}{3} = -1$$

Next, use the point-slope form of the equation of a line.

$$y - y_1 = m(x - x_1) \qquad \text{Point-slope form}$$

$$y - 3 = -1(x - 1) \qquad \text{Substitute for } x_1, y_1, \text{ and } m.$$

$$y = -x + 4 \qquad \text{Simplify.}$$

$$f(x) = -x + 4 \qquad \text{Function notation}$$

The figure below shows the graph of this function.

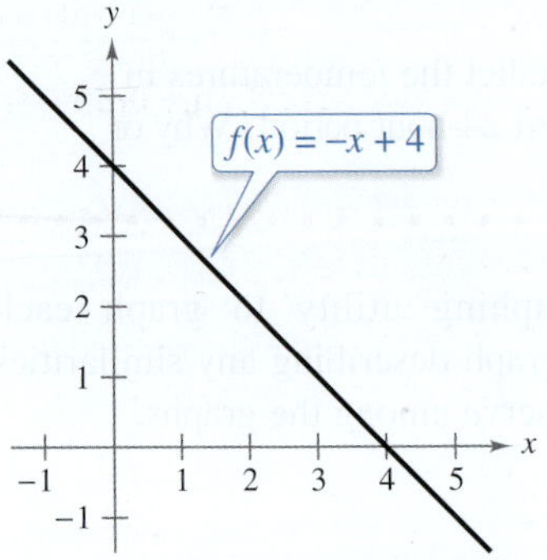

Checkpoint Audio-video solution in English & Spanish at LarsonPrecalculus.com.

Write the linear function f for which $f(-2) = 6$ and $f(4) = -9$.

nulinukas/Shutterstock.com

There are two special types of linear functions, the **constant function** and the **identity function.** A constant function has the form

$$f(x) = c$$

and has the domain of all real numbers with a range consisting of a single real number c. The graph of a constant function is a horizontal line, as shown in Figure 1.44. The identity function has the form

$$f(x) = x.$$

Its domain and range are the set of all real numbers. The identity function has a slope of $m = 1$ and a y-intercept at $(0, 0)$. The graph of the identity function is a line for which each x-coordinate equals the corresponding y-coordinate. The graph is always increasing, as shown in Figure 1.45.

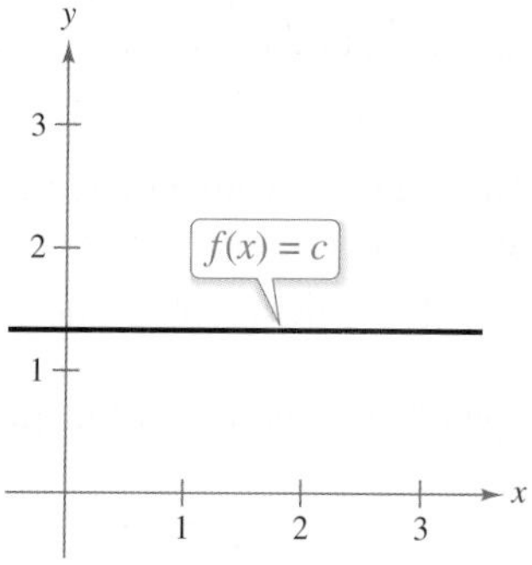

Figure 1.44

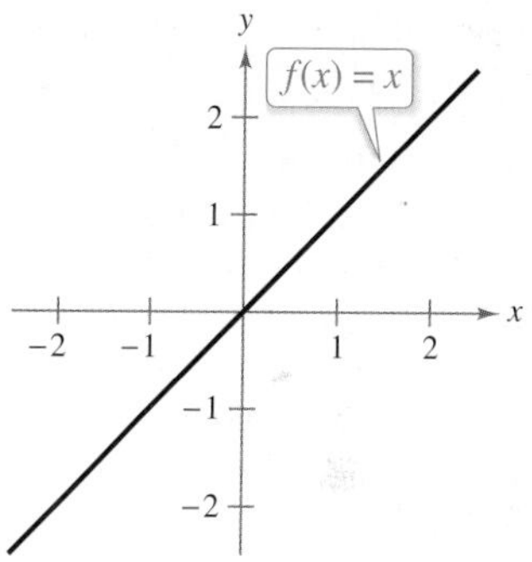

Figure 1.45

The graph of the **squaring function**

$$f(x) = x^2$$

is a U-shaped curve with the following characteristics.

- The domain of the function is the set of all real numbers.
- The range of the function is the set of all nonnegative real numbers.
- The function is even.
- The graph has an intercept at $(0, 0)$.
- The graph is decreasing on the interval $(-\infty, 0)$ and increasing on the interval $(0, \infty)$.
- The graph is symmetric with respect to the y-axis.
- The graph has a relative minimum at $(0, 0)$.

The figure below shows the graph of the squaring function.

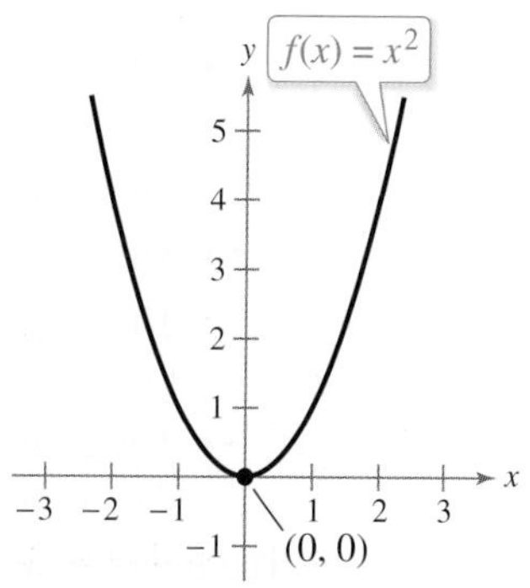

Cubic, Square Root, and Reciprocal Functions

The following summarizes the basic characteristics of the graphs of the **cubic, square root,** and **reciprocal functions.**

1. The graph of the *cubic* function

$$f(x) = x^3$$

has the following characteristics.

- The domain of the function is the set of all real numbers.
- The range of the function is the set of all real numbers.
- The function is odd.
- The graph has an intercept at $(0, 0)$.
- The graph is increasing on the interval $(-\infty, \infty)$.
- The graph is symmetric with respect to the origin.

The figure shows the graph of the cubic function.

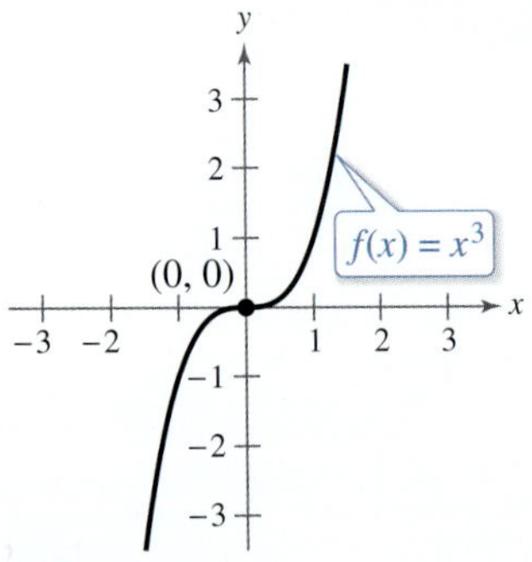

Cubic function

2. The graph of the *square root* function

$$f(x) = \sqrt{x}$$

has the following characteristics.

- The domain of the function is the set of all nonnegative real numbers.
- The range of the function is the set of all nonnegative real numbers.
- The graph has an intercept at $(0, 0)$.
- The graph is increasing on the interval $(0, \infty)$.

The figure shows the graph of the square root function.

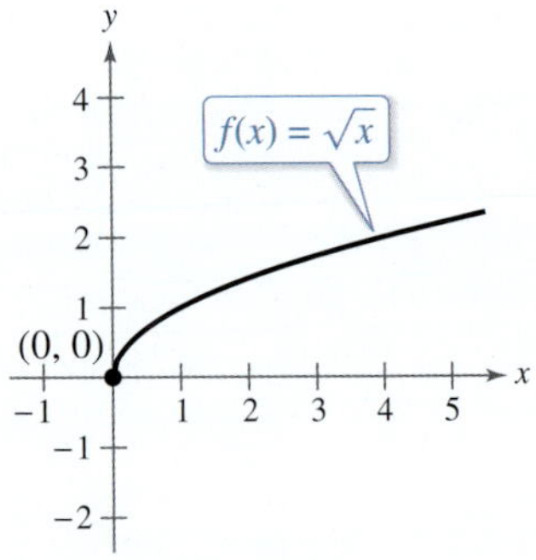

Square root function

3. The graph of the *reciprocal* function

$$f(x) = \frac{1}{x}$$

has the following characteristics.

- The domain of the function is $(-\infty, 0) \cup (0, \infty)$.
- The range of the function is $(-\infty, 0) \cup (0, \infty)$.
- The function is odd.
- The graph does not have any intercepts.
- The graph is decreasing on the intervals $(-\infty, 0)$ and $(0, \infty)$.
- The graph is symmetric with respect to the origin.

The figure shows the graph of the reciprocal function.

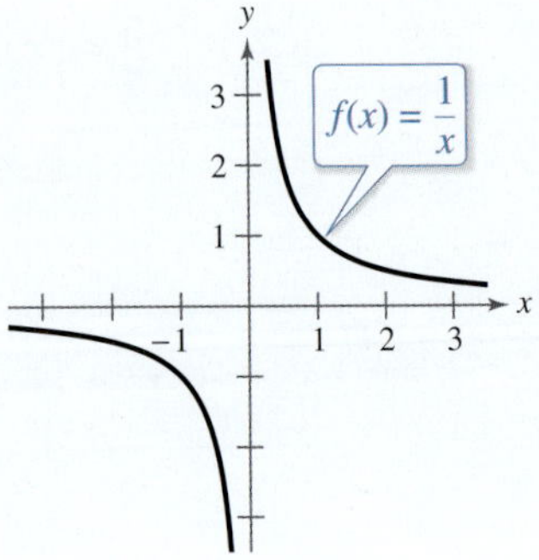

Reciprocal function

Step and Piecewise-Defined Functions

Functions whose graphs resemble sets of stairsteps are known as **step functions.** The most famous of the step functions is the **greatest integer function,** denoted by $[\![x]\!]$ and defined as

$$f(x) = [\![x]\!] = \textit{the greatest integer less than or equal to } x.$$

Some values of the greatest integer function are as follows.

$$[\![-1]\!] = (\text{greatest integer} \le -1) = -1$$

$$[\![-\tfrac{1}{2}]\!] = (\text{greatest integer} \le -\tfrac{1}{2}) = -1$$

$$[\![\tfrac{1}{10}]\!] = (\text{greatest integer} \le \tfrac{1}{10}) = 0$$

$$[\![1.5]\!] = (\text{greatest integer} \le 1.5) = 1$$

$$[\![1.9]\!] = (\text{greatest integer} \le 1.9) = 1$$

The graph of the greatest integer function

$$f(x) = [\![x]\!]$$

has the following characteristics, as shown in Figure 1.46.

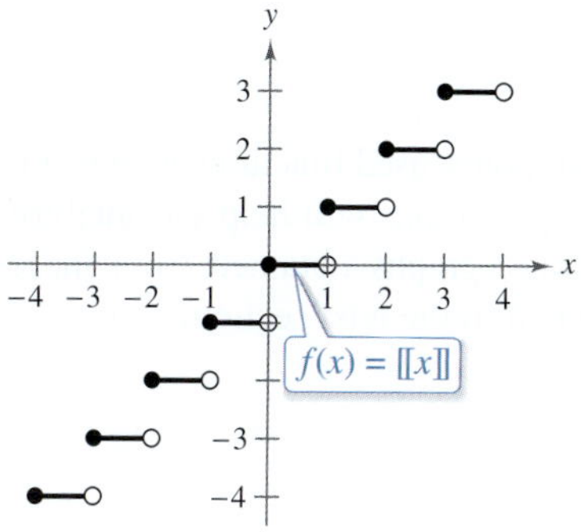

Figure 1.46

- The domain of the function is the set of all real numbers.
- The range of the function is the set of all integers.
- The graph has a y-intercept at $(0, 0)$ and x-intercepts in the interval $[0, 1)$.
- The graph is constant between each pair of consecutive integer values of x.
- The graph jumps vertically one unit at each integer value of x.

TECHNOLOGY When using your graphing utility to graph a step function, you should set your graphing utility to *dot* mode.

EXAMPLE 2 Evaluating a Step Function

Evaluate the function when $x = -1, 2,$ and $\frac{3}{2}$.

$$f(x) = [\![x]\!] + 1$$

Solution For $x = -1$, the greatest integer ≤ -1 is -1, so

$$f(-1) = [\![-1]\!] + 1 = -1 + 1 = 0.$$

For $x = 2$, the greatest integer ≤ 2 is 2, so

$$f(2) = [\![2]\!] + 1 = 2 + 1 = 3.$$

For $x = \frac{3}{2}$, the greatest integer $\le \frac{3}{2}$ is 1, so

$$f\left(\tfrac{3}{2}\right) = [\![\tfrac{3}{2}]\!] + 1 = 1 + 1 = 2.$$

Verify your answers by examining the graph of $f(x) = [\![x]\!] + 1$ shown in Figure 1.47.

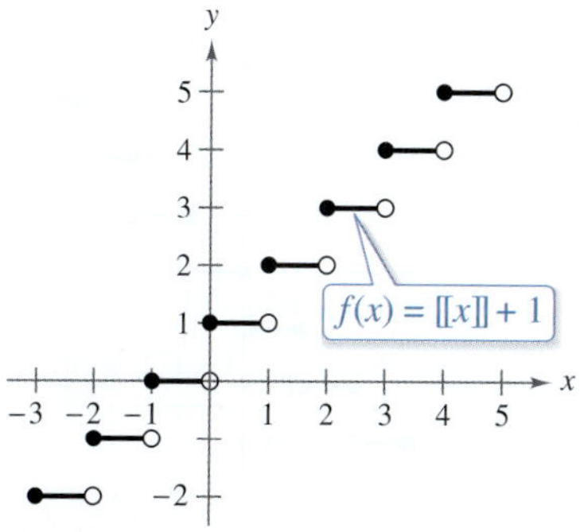

Figure 1.47

Checkpoint Audio-video solution in English & Spanish at LarsonPrecalculus.com.

Evaluate the function when $x = -\frac{3}{2}, 1,$ and $-\frac{5}{2}$.

$$f(x) = [\![x + 2]\!]$$

Recall from Section 1.4 that a piecewise-defined function is defined by two or more equations over a specified domain. To graph a piecewise-defined function, graph each equation separately over the specified domain, as shown in Example 3.

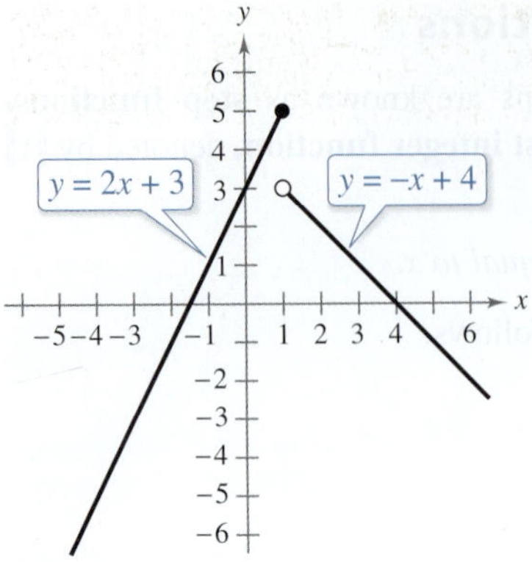

Figure 1.48

EXAMPLE 3 Graphing a Piecewise-Defined Function

Sketch the graph of $f(x) = \begin{cases} 2x + 3, & x \le 1 \\ -x + 4, & x > 1 \end{cases}$.

Solution This piecewise-defined function consists of two linear functions. At $x = 1$ and to the left of $x = 1$, the graph is the line $y = 2x + 3$, and to the right of $x = 1$ the graph is the line $y = -x + 4$, as shown in Figure 1.48. Notice that the point $(1, 5)$ is a solid dot and the point $(1, 3)$ is an open dot. This is because $f(1) = 2(1) + 3 = 5$.

✓ **Checkpoint** *Audio-video solution in English & Spanish at LarsonPrecalculus.com.*

Sketch the graph of $f(x) = \begin{cases} -\frac{1}{2}x - 6, & x \le -4 \\ x + 5, & x > -4 \end{cases}$.

Parent Functions

The eight graphs shown below represent the most commonly used functions in algebra. Familiarity with the basic characteristics of these simple graphs will help you analyze the shapes of more complicated graphs—in particular, graphs obtained from these graphs by the rigid and nonrigid transformations studied in the next section.

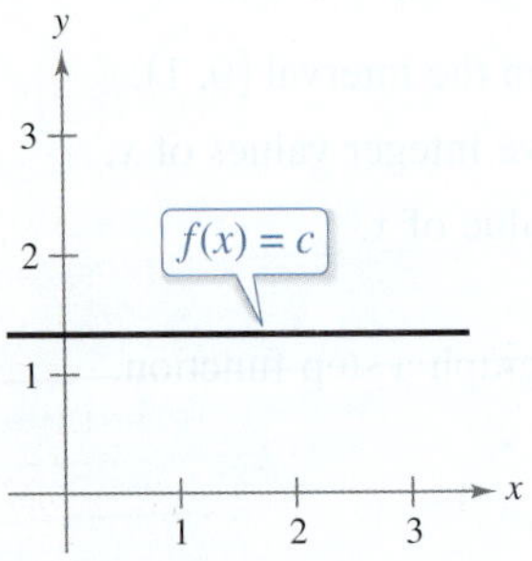

(a) Constant Function

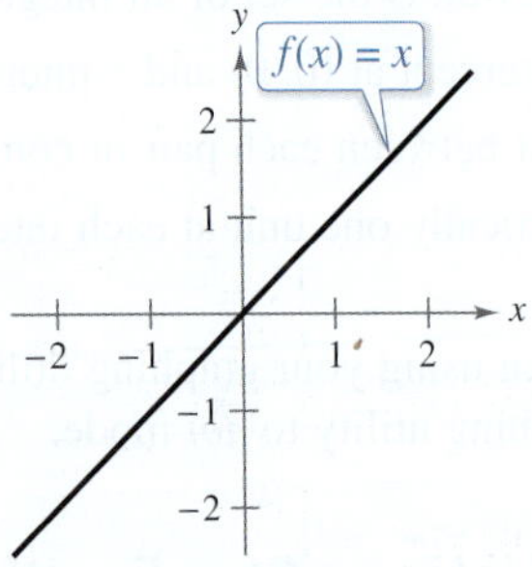

(b) Identity Function

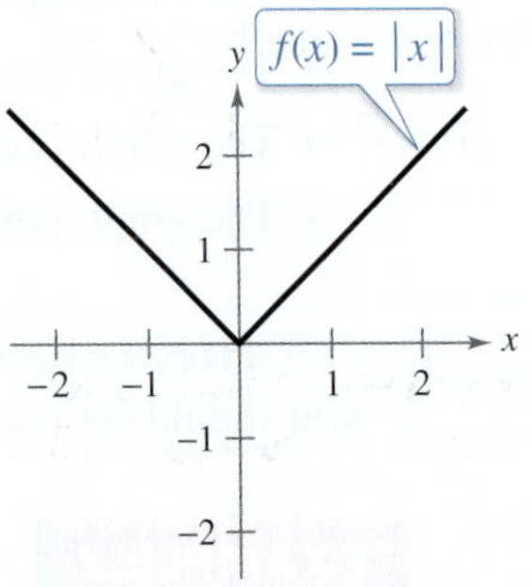

(c) Absolute Value Function

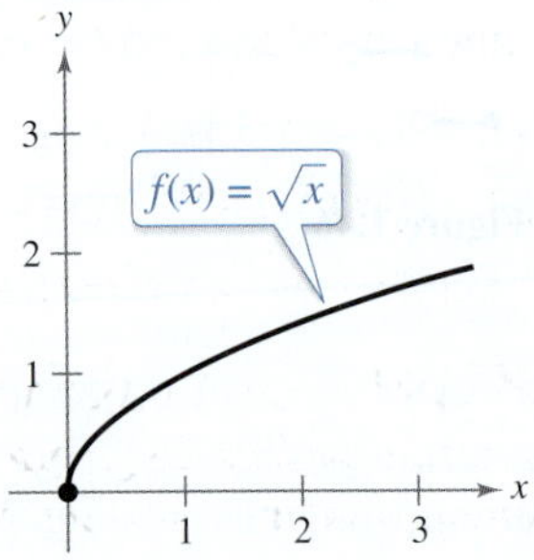

(d) Square Root Function

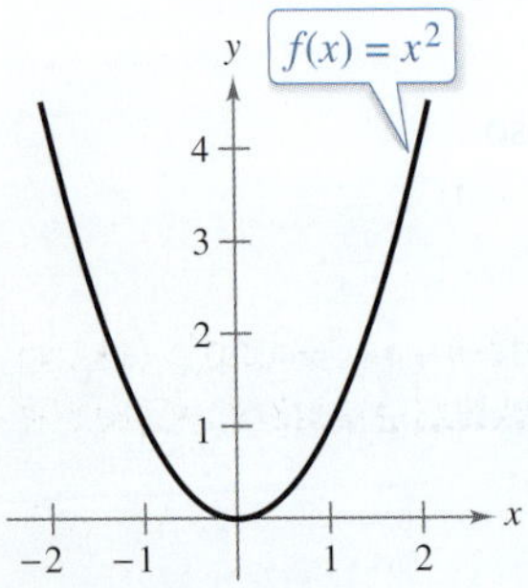

(e) Quadratic Function

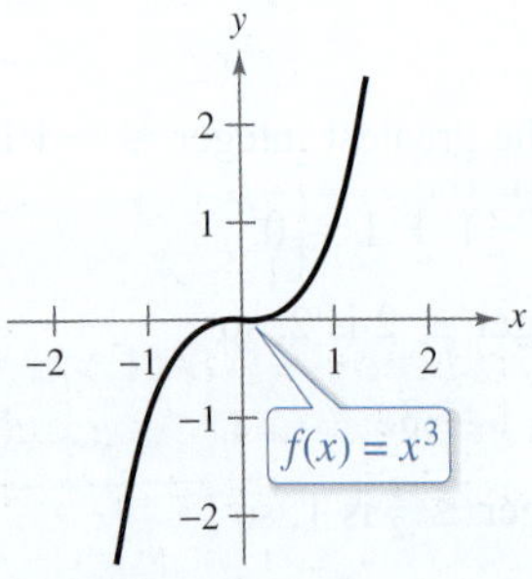

(f) Cubic Function

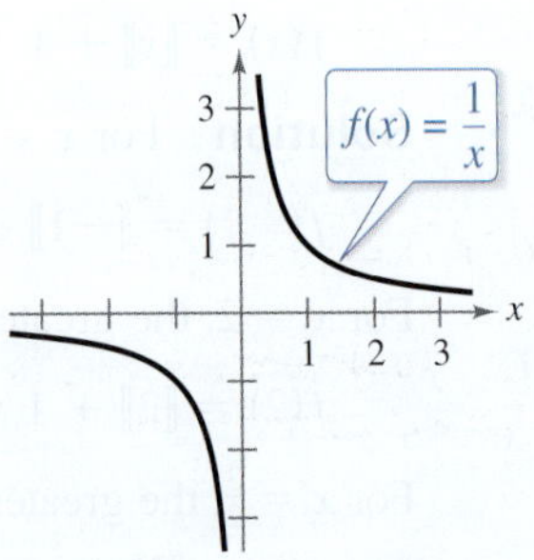

(g) Reciprocal Function

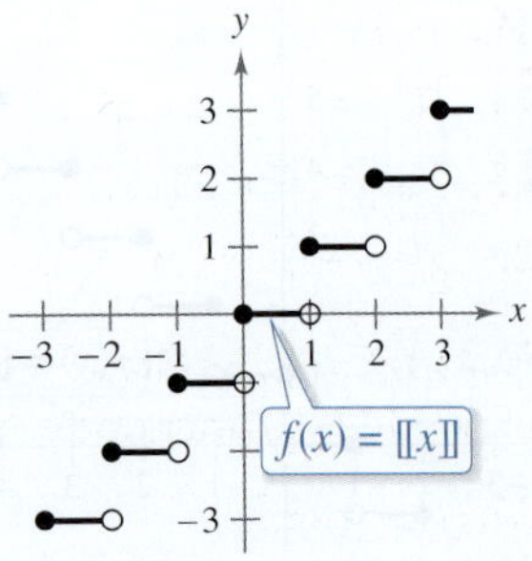

(h) Greatest Integer Function

Summarize (Section 1.6)

1. Explain how to identify and graph linear and squaring functions *(pages 60 and 61)*. For an example involving a linear function, see Example 1.
2. Explain how to identify and graph cubic, square root, and reciprocal functions *(page 62)*.
3. Explain how to identify and graph step and other piecewise-defined functions *(page 63)*. For an example involving a step function, see Example 2. For an example of graphing a piecewise-defined function, see Example 3.
4. State and sketch the graphs of parent functions *(page 64)*.

1.6 Exercises

See CalcChat.com for tutorial help and worked-out solutions to odd-numbered exercises.

Vocabulary

In Exercises 1–9, match each function with its name.

1. $f(x) = [\![x]\!]$
2. $f(x) = x$
3. $f(x) = 1/x$
4. $f(x) = x^2$
5. $f(x) = \sqrt{x}$
6. $f(x) = c$
7. $f(x) = |x|$
8. $f(x) = x^3$
9. $f(x) = ax + b$

(a) squaring function (b) square root function (c) cubic function
(d) linear function (e) constant function (f) absolute value function
(g) greatest integer function (h) reciprocal function (i) identity function

10. Fill in the blank: The constant function and the identity function are two special types of ________ functions.

Skills and Applications

Writing a Linear Function In Exercises 11–14, (a) write the linear function f such that it has the indicated function values and (b) sketch the graph of the function.

11. $f(1) = 4,\quad f(0) = 6$
12. $f(-3) = -8,\quad f(1) = 2$
13. $f(-5) = -1,\quad f(5) = -1$
14. $f(\frac{2}{3}) = -\frac{15}{2},\quad f(-4) = -11$

Graphing a Function In Exercises 15–26, use a graphing utility to graph the function. Be sure to choose an appropriate viewing window.

15. $f(x) = 2.5x - 4.25$
16. $f(x) = \frac{5}{6} - \frac{2}{3}x$
17. $g(x) = -2x^2$
18. $f(x) = 3x^2 - 1.75$
19. $f(x) = x^3 - 1$
20. $f(x) = (x - 1)^3 + 2$
21. $f(x) = 4 - 2\sqrt{x}$
22. $h(x) = \sqrt{x + 2} + 3$
23. $f(x) = 4 + (1/x)$
24. $k(x) = 1/(x - 3)$
25. $g(x) = |x| - 5$
26. $f(x) = |x - 1|$

Evaluating a Step Function In Exercises 27–30, evaluate the function for the indicated values.

27. $f(x) = [\![x]\!]$

(a) $f(2.1)$ (b) $f(2.9)$ (c) $f(-3.1)$ (d) $f(\frac{7}{2})$

28. $h(x) = [\![x + 3]\!]$

(a) $h(-2)$ (b) $h(\frac{1}{2})$ (c) $h(4.2)$ (d) $h(-21.6)$

29. $k(x) = [\![\frac{1}{2}x + 6]\!]$

(a) $k(5)$ (b) $k(-6.1)$ (c) $k(0.1)$ (d) $k(15)$

30. $g(x) = -7[\![x + 4]\!] + 6$

(a) $g(\frac{1}{8})$ (b) $g(9)$ (c) $g(-4)$ (d) $g(\frac{3}{2})$

Graphing a Step Function In Exercises 31–34, sketch the graph of the function.

31. $g(x) = -[\![x]\!]$
32. $g(x) = 4[\![x]\!]$
33. $g(x) = [\![x]\!] - 1$
34. $g(x) = [\![x - 3]\!]$

Graphing a Piecewise-Defined Function In Exercises 35–40, sketch the graph of the function.

35. $g(x) = \begin{cases} x + 6, & x \le -4 \\ \frac{1}{2}x - 4, & x > -4 \end{cases}$

36. $f(x) = \begin{cases} \sqrt{4 + x}, & x < 0 \\ \sqrt{4 - x}, & x \ge 0 \end{cases}$

37. $f(x) = \begin{cases} 1 - (x - 1)^2, & x \le 2 \\ \sqrt{x - 2}, & x > 2 \end{cases}$

38. $f(x) = \begin{cases} x^2 + 5, & x \le 1 \\ -x^2 + 4x + 3, & x > 1 \end{cases}$

39. $h(x) = \begin{cases} 4 - x^2, & x < -2 \\ 3 + x, & -2 \le x < 0 \\ x^2 + 1, & x \ge 0 \end{cases}$

40. $k(x) = \begin{cases} 2x + 1, & x \le -1 \\ 2x^2 - 1, & -1 < x \le 1 \\ 1 - x^2, & x > 1 \end{cases}$

Graphing a Function In Exercises 41 and 42, (a) use a graphing utility to graph the function and (b) state the domain and range of the function.

41. $s(x) = 2(\frac{1}{4}x - [\![\frac{1}{4}x]\!])$
42. $k(x) = 4(\frac{1}{2}x - [\![\frac{1}{2}x]\!])^2$

43. **Wages** A mechanic's pay is \$14.00 per hour for regular time and time-and-a-half for overtime. The weekly wage function is

$$W(h) = \begin{cases} 14h, & 0 < h \le 40 \\ 21(h - 40) + 560, & h > 40 \end{cases}$$

where h is the number of hours worked in a week.

(a) Evaluate $W(30)$, $W(40)$, $W(45)$, and $W(50)$.

(b) The company increased the regular work week to 45 hours. What is the new weekly wage function?

44. Revenue The table shows the monthly revenue y (in thousands of dollars) of a landscaping business for each month of the year 2013, with $x = 1$ representing January.

DATA

Month, x	Revenue, y
1	5.2
2	5.6
3	6.6
4	8.3
5	11.5
6	15.8
7	12.8
8	10.1
9	8.6
10	6.9
11	4.5
12	2.7

Spreadsheet at LarsonPrecalculus.com

A mathematical model that represents these data is

$$f(x) = \begin{cases} -1.97x + 26.3 \\ 0.505x^2 - 1.47x + 6.3. \end{cases}$$

(a) Use a graphing utility to graph the model. What is the domain of each part of the piecewise-defined function? How can you tell? Explain your reasoning.

(b) Find $f(5)$ and $f(11)$, and interpret your results in the context of the problem.

(c) How do the values obtained from the model in part (a) compare with the actual data values?

45. Fluid Flow The intake pipe of a 100-gallon tank has a flow rate of 10 gallons per minute, and two drainpipes have flow rates of 5 gallons per minute each. The figure shows the volume V of fluid in the tank as a function of time t. Determine the combination of the input pipe and drain pipes in which the fluid is flowing in specific subintervals of the 1 hour of time shown on the graph. (There are many correct answers.)

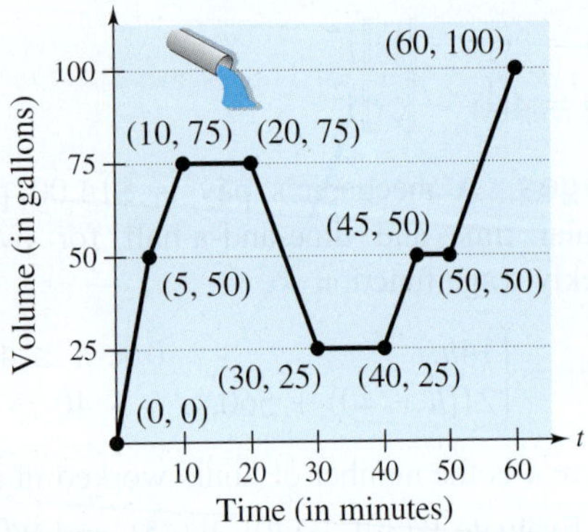

nulinukas/Shutterstock.com

46. Delivery Charges The cost of sending an overnight package from New York to Atlanta is \$26.10 for a package weighing up to, but not including, 1 pound and \$4.35 for each additional pound or portion of a pound.

(a) Use the greatest integer function to create a model for the cost C of overnight delivery of a package weighing x pounds, $x > 0$.

(b) Sketch the graph of the function.

47. Snowstorm

During a nine-hour snowstorm, it snows at a rate of 1 inch per hour for the first 2 hours, at a rate of 2 inches per hour for the next 6 hours, and at a rate of 0.5 inch per hour for the final hour. Write and graph a piecewise-defined function that gives the depth of the snow during the snowstorm. How many inches of snow accumulated from the storm?

Exploration

48. HOW DO YOU SEE IT? For each graph of f shown below, answer (a)–(d).

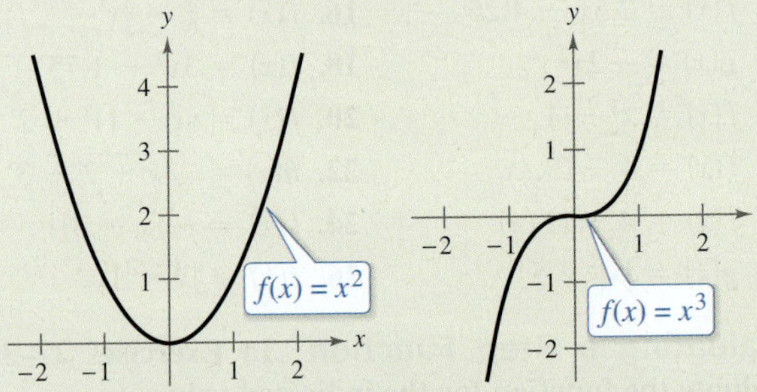

(a) Find the domain and range of f.

(b) Find the x- and y-intercepts of the graph of f.

(c) Determine the intervals on which f is increasing, decreasing, or constant.

(d) Determine whether f is even, odd, or neither. Then describe the symmetry.

True or False? **In Exercises 49 and 50, determine whether the statement is true or false. Justify your answer.**

49. A piecewise-defined function will always have at least one x-intercept or at least one y-intercept.

50. A linear equation will always have an x-intercept and a y-intercept.

1.7 Transformations of Functions

Transformations of functions can help you model real-life applications. For instance, Exercise 69 on page 74 shows how a transformation of a function can model the number of horsepower required to overcome wind drag on an automobile.

- **Use vertical and horizontal shifts to sketch graphs of functions.**
- **Use reflections to sketch graphs of functions.**
- **Use nonrigid transformations to sketch graphs of functions.**

Shifting Graphs

Many functions have graphs that are transformations of the parent graphs summarized in Section 1.6. For example, you can obtain the graph of

$$h(x) = x^2 + 2$$

by shifting the graph of $f(x) = x^2$ *up* two units, as shown in Figure 1.49. In function notation, h and f are related as follows.

$$h(x) = x^2 + 2 = f(x) + 2 \qquad \text{Upward shift of two units}$$

Similarly, you can obtain the graph of

$$g(x) = (x - 2)^2$$

by shifting the graph of $f(x) = x^2$ to the *right* two units, as shown in Figure 1.50. In this case, the functions g and f have the following relationship.

$$g(x) = (x - 2)^2 = f(x - 2) \qquad \text{Right shift of two units}$$

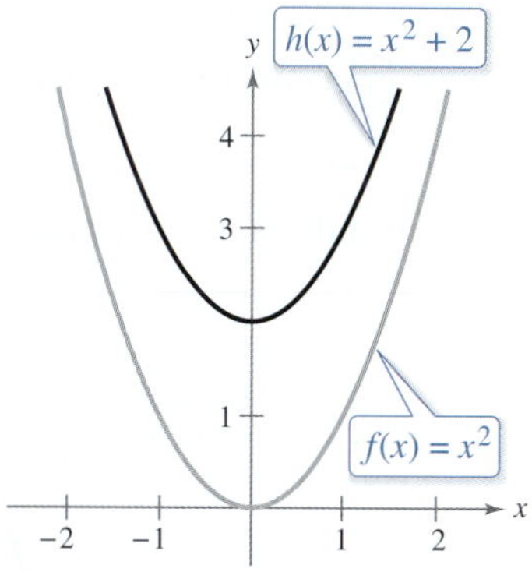

Figure 1.49

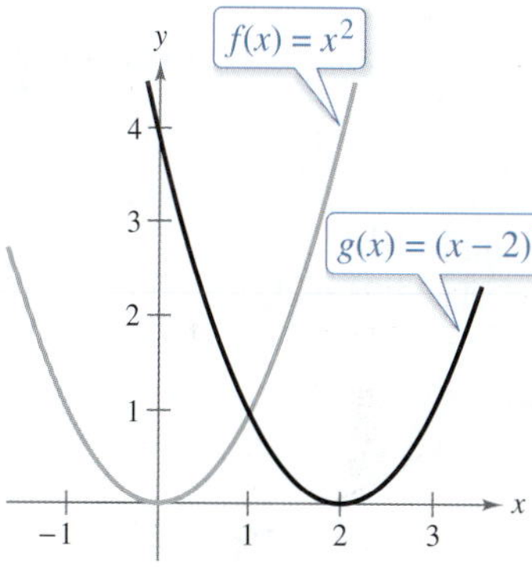

Figure 1.50

The following list summarizes this discussion about horizontal and vertical shifts.

REMARK In items 3 and 4, be sure you see that $h(x) = f(x - c)$ corresponds to a *right* shift and $h(x) = f(x + c)$ corresponds to a *left* shift for $c > 0$.

Vertical and Horizontal Shifts

Let c be a positive real number. **Vertical and horizontal shifts** in the graph of $y = f(x)$ are represented as follows.

1. Vertical shift c units *up*: $h(x) = f(x) + c$
2. Vertical shift c units *down*: $h(x) = f(x) - c$
3. Horizontal shift c units to the *right*: $h(x) = f(x - c)$
4. Horizontal shift c units to the *left*: $h(x) = f(x + c)$

Some graphs can be obtained from combinations of vertical and horizontal shifts, as demonstrated in Example 1(b). Vertical and horizontal shifts generate a *family of functions,* each with the same shape but at a different location in the plane.

Robert Young/Shutterstock.com

EXAMPLE 1 **Shifts in the Graph of a Function**

Use the graph of $f(x) = x^3$ to sketch the graph of each function.

a. $g(x) = x^3 - 1$

b. $h(x) = (x + 2)^3 + 1$

Solution

a. Relative to the graph of $f(x) = x^3$, the graph of

$$g(x) = x^3 - 1$$

is a downward shift of one unit, as shown below.

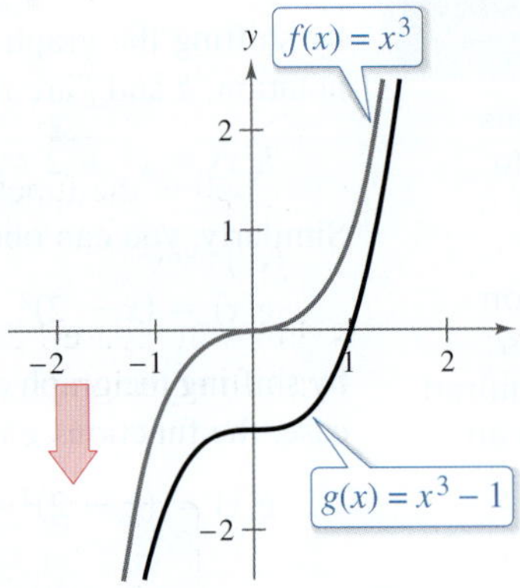

b. Relative to the graph of $f(x) = x^3$, the graph of

$$h(x) = (x + 2)^3 + 1$$

involves a left shift of two units and an upward shift of one unit, as shown below.

REMARK In Example 1(a), note that $g(x) = f(x) - 1$ and in Example 1(b), $h(x) = f(x + 2) + 1$.

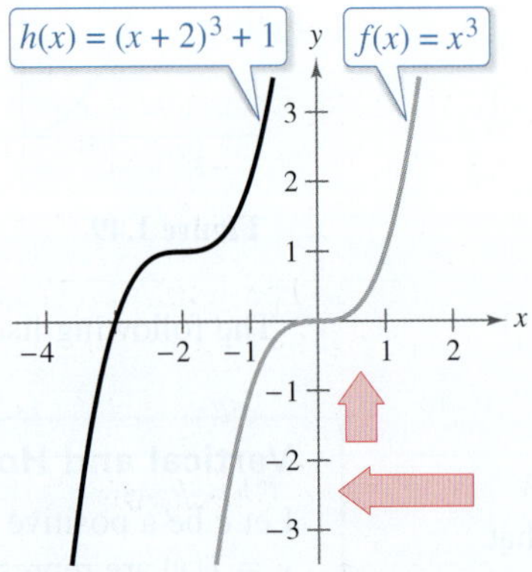

Checkpoint *Audio-video solution in English & Spanish at LarsonPrecalculus.com.*

Use the graph of $f(x) = x^3$ to sketch the graph of each function.

a. $h(x) = x^3 + 5$

b. $g(x) = (x - 3)^3 + 2$

In Example 1(b), you obtain the same result when the vertical shift precedes the horizontal shift *or* when the horizontal shift precedes the vertical shift.

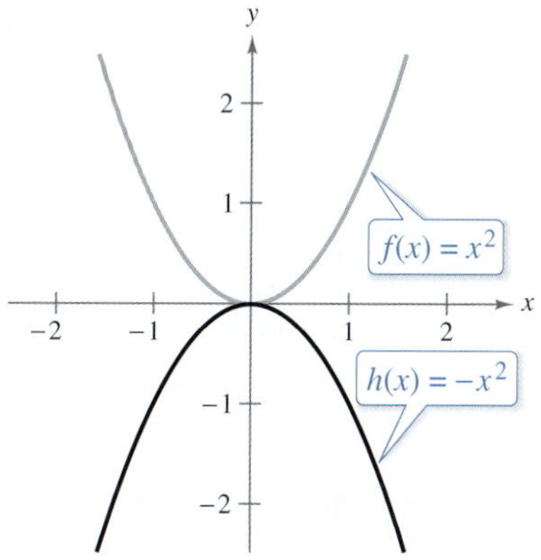

Figure 1.51

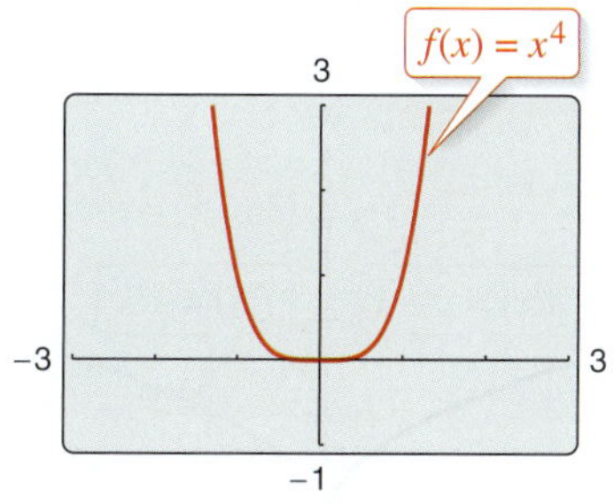

Figure 1.52

Reflecting Graphs

Another common type of transformation is a **reflection.** For instance, if you consider the x-axis to be a mirror, then the graph of $h(x) = -x^2$ is the mirror image (or reflection) of the graph of $f(x) = x^2$, as shown in Figure 1.51.

> **Reflections in the Coordinate Axes**
>
> **Reflections** in the coordinate axes of the graph of $y = f(x)$ are represented as follows.
>
> **1.** Reflection in the x-axis: $h(x) = -f(x)$
>
> **2.** Reflection in the y-axis: $h(x) = f(-x)$

EXAMPLE 2 Writing Equations from Graphs

The graph of the function

$$f(x) = x^4$$

is shown in Figure 1.52. Each of the graphs below is a transformation of the graph of f. Write an equation for each of these functions.

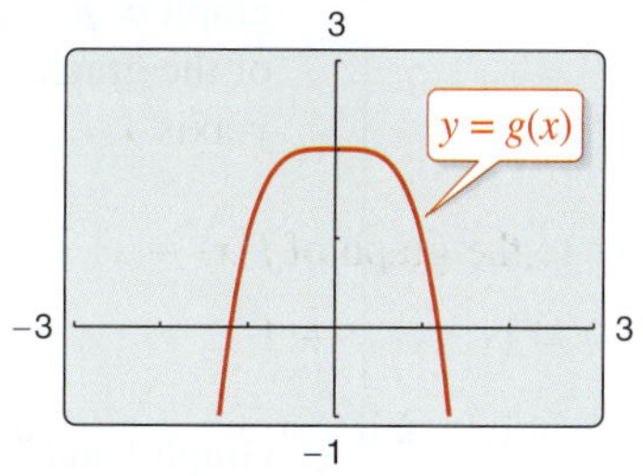

(a)

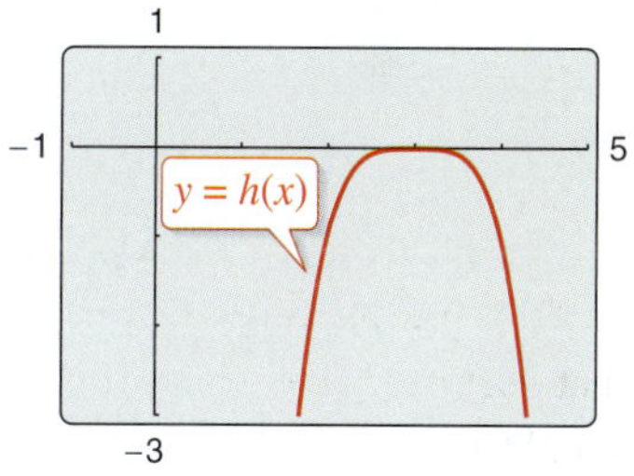

(b)

Solution

a. The graph of g is a reflection in the x-axis *followed by* an upward shift of two units of the graph of $f(x) = x^4$. So, the equation for g is

$$g(x) = -x^4 + 2.$$

b. The graph of h is a horizontal shift of three units to the right *followed by* a reflection in the x-axis of the graph of $f(x) = x^4$. So, the equation for h is

$$h(x) = -(x - 3)^4.$$

Checkpoint *Audio-video solution in English & Spanish at LarsonPrecalculus.com.*

The graph is a transformation of the graph of $f(x) = x^4$. Write an equation for the function.

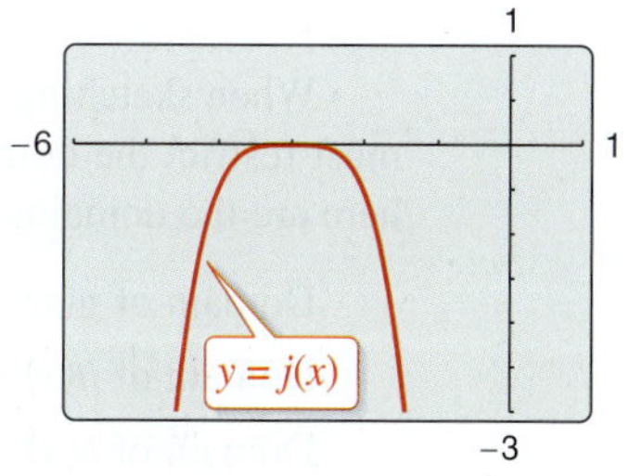

EXAMPLE 3 Reflections and Shifts

Compare the graph of each function with the graph of $f(x) = \sqrt{x}$.

a. $g(x) = -\sqrt{x}$ **b.** $h(x) = \sqrt{-x}$ **c.** $k(x) = -\sqrt{x+2}$

Algebraic Solution

a. The graph of g is a reflection of the graph of f in the x-axis because

$$g(x) = -\sqrt{x}$$
$$= -f(x).$$

b. The graph of h is a reflection of the graph of f in the y-axis because

$$h(x) = \sqrt{-x}$$
$$= f(-x).$$

c. The graph of k is a left shift of two units followed by a reflection in the x-axis because

$$k(x) = -\sqrt{x+2}$$
$$= -f(x+2).$$

Graphical Solution

a. Graph f and g on the same set of coordinate axes. From the graph, you can see that the graph of g is a reflection of the graph of f in the x-axis.

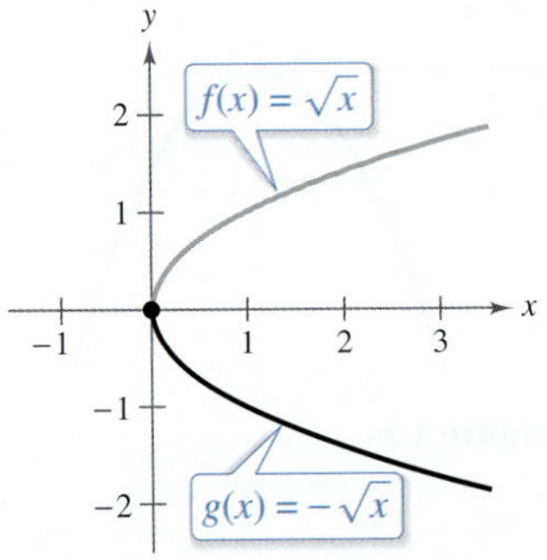

b. Graph f and h on the same set of coordinate axes. From the graph, you can see that the graph of h is a reflection of the graph of f in the y-axis.

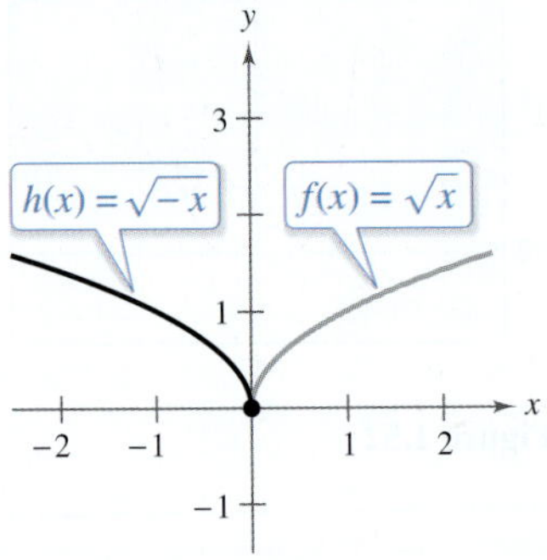

c. Graph f and k on the same set of coordinate axes. From the graph, you can see that the graph of k is a left shift of two units of the graph of f, followed by a reflection in the x-axis.

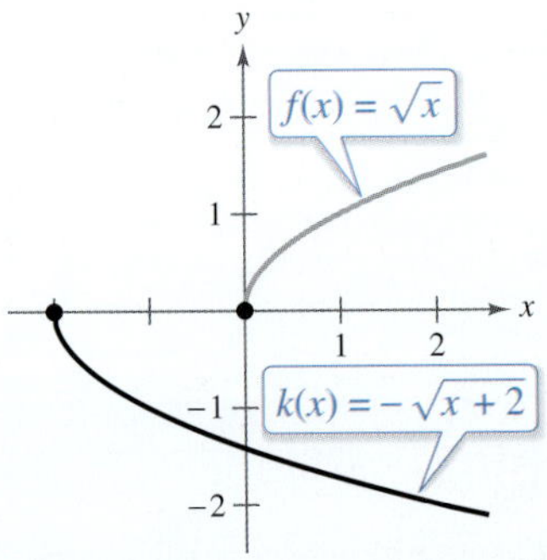

✓ Checkpoint Audio-video solution in English & Spanish at LarsonPrecalculus.com.

Compare the graph of each function with the graph of

$f(x) = \sqrt{x-1}$.

a. $g(x) = -\sqrt{x-1}$ **b.** $h(x) = \sqrt{-x-1}$

When sketching the graphs of functions involving square roots, remember that you must restrict the domain to exclude negative numbers inside the radical. For instance, here are the domains of the functions in Example 3.

Domain of $g(x) = -\sqrt{x}$: $x \geq 0$

Domain of $h(x) = \sqrt{-x}$: $x \leq 0$

Domain of $k(x) = -\sqrt{x+2}$: $x \geq -2$

Nonrigid Transformations

Horizontal shifts, vertical shifts, and reflections are **rigid transformations** because the basic shape of the graph is unchanged. These transformations change only the *position* of the graph in the coordinate plane. **Nonrigid transformations** are those that cause a *distortion*—a change in the shape of the original graph. For instance, a nonrigid transformation of the graph of $y = f(x)$ is represented by $g(x) = cf(x)$, where the transformation is a **vertical stretch** when $c > 1$ and a **vertical shrink** when $0 < c < 1$. Another nonrigid transformation of the graph of $y = f(x)$ is represented by $h(x) = f(cx)$, where the transformation is a **horizontal shrink** when $c > 1$ and a **horizontal stretch** when $0 < c < 1$.

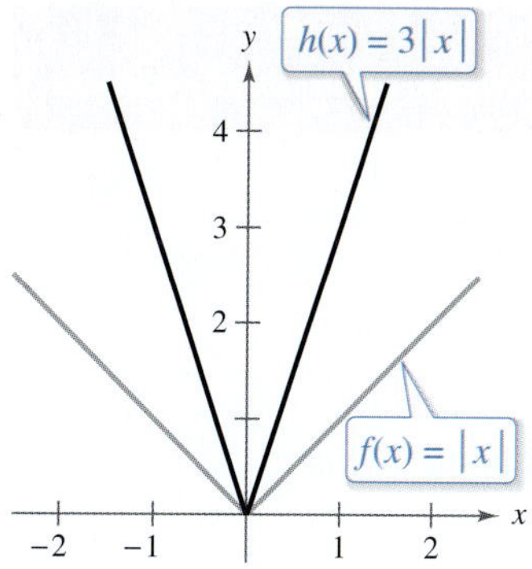

Figure 1.53

EXAMPLE 4 Nonrigid Transformations

Compare the graph of each function with the graph of $f(x) = |x|$.

a. $h(x) = 3|x|$ **b.** $g(x) = \frac{1}{3}|x|$

Solution

a. Relative to the graph of $f(x) = |x|$, the graph of $h(x) = 3|x| = 3f(x)$ is a vertical stretch (each y-value is multiplied by 3) of the graph of f. (See Figure 1.53.)

b. Similarly, the graph of $g(x) = \frac{1}{3}|x| = \frac{1}{3}f(x)$ is a vertical shrink $\left(\text{each } y\text{-value is multiplied by } \frac{1}{3}\right)$ of the graph of f. (See Figure 1.54.)

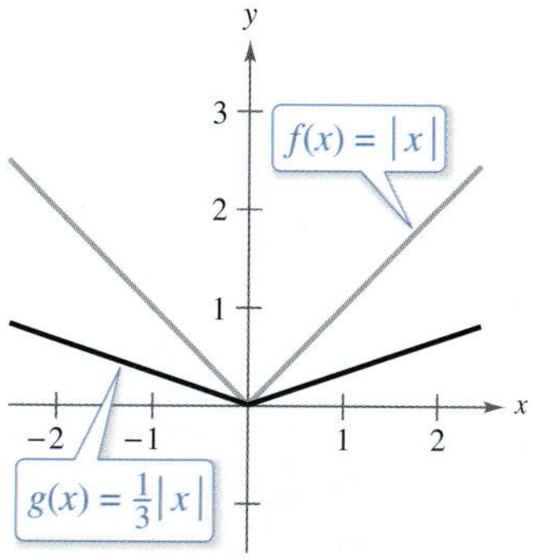

Figure 1.54

✓ **Checkpoint** Audio-video solution in English & Spanish at LarsonPrecalculus.com.

Compare the graph of each function with the graph of $f(x) = x^2$.

a. $g(x) = 4x^2$ **b.** $h(x) = \frac{1}{4}x^2$

EXAMPLE 5 Nonrigid Transformations

Compare the graph of each function with the graph of $f(x) = 2 - x^3$.

a. $g(x) = f(2x)$ **b.** $h(x) = f\left(\frac{1}{2}x\right)$

Solution

a. Relative to the graph of $f(x) = 2 - x^3$, the graph of $g(x) = f(2x) = 2 - (2x)^3 = 2 - 8x^3$ is a horizontal shrink $(c > 1)$ of the graph of f. (See Figure 1.55.)

b. Similarly, the graph of $h(x) = f\left(\frac{1}{2}x\right) = 2 - \left(\frac{1}{2}x\right)^3 = 2 - \frac{1}{8}x^3$ is a horizontal stretch $(0 < c < 1)$ of the graph of f. (See Figure 1.56.)

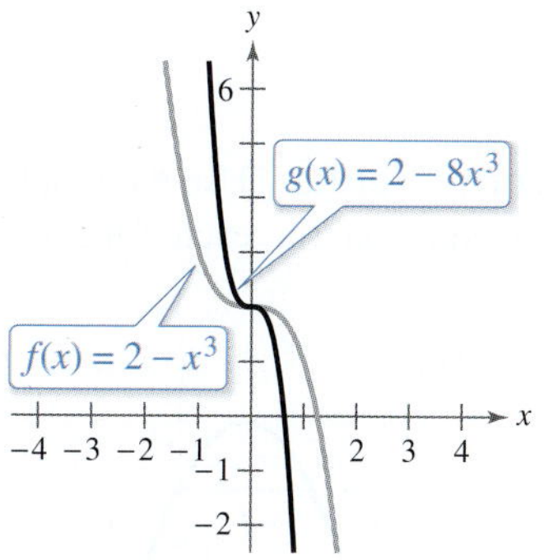

Figure 1.55

✓ **Checkpoint** Audio-video solution in English & Spanish at LarsonPrecalculus.com.

Compare the graph of each function with the graph of $f(x) = x^2 + 3$.

a. $g(x) = f(2x)$ **b.** $h(x) = f\left(\frac{1}{2}x\right)$

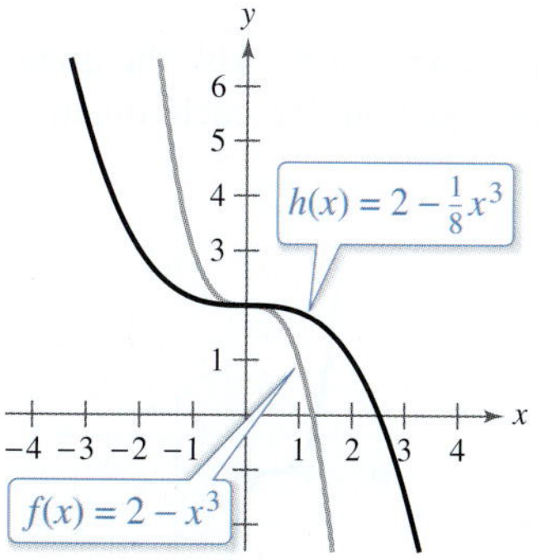

Figure 1.56

Summarize (Section 1.7)

1. Describe how to shift the graph of a function vertically and horizontally *(page 67)*. For an example of shifting the graph of a function, see Example 1.
2. Describe how to reflect the graph of a function in the x-axis and in the y-axis *(page 69)*. For examples of reflecting graphs of functions, see Examples 2 and 3.
3. Describe nonrigid transformations of the graph of a function *(page 71)*. For examples of nonrigid transformations, see Examples 4 and 5.

1.7 Exercises

See CalcChat.com for tutorial help and worked-out solutions to odd-numbered exercises.

Vocabulary

In Exercises 1–3, fill in the blanks.

1. Horizontal shifts, vertical shifts, and reflections are called ________ transformations.
2. A reflection in the x-axis of $y = f(x)$ is represented by $h(x) =$ ________, while a reflection in the y-axis of $y = f(x)$ is represented by $h(x) =$ ________.
3. A nonrigid transformation of $y = f(x)$ represented by $g(x) = cf(x)$ is a ________ ________ when $c > 1$ and a ________ ________ when $0 < c < 1$.

4. Match the rigid transformation of $y = f(x)$ with the correct representation of the graph of h, where $c > 0$.

(a) $h(x) = f(x) + c$	(i) A horizontal shift of f, c units to the right
(b) $h(x) = f(x) - c$	(ii) A vertical shift of f, c units down
(c) $h(x) = f(x + c)$	(iii) A horizontal shift of f, c units to the left
(d) $h(x) = f(x - c)$	(iv) A vertical shift of f, c units up

Skills and Applications

5. **Shifts in the Graph of a Function** For each function, sketch (on the same set of coordinate axes) a graph of each function for $c = -1, 1,$ and 3.
   (a) $f(x) = |x| + c$ (b) $f(x) = |x - c|$
6. **Shifts in the Graph of a Function** For each function, sketch (on the same set of coordinate axes) a graph of each function for $c = -3, -1, 1,$ and 3.
   (a) $f(x) = \sqrt{x} + c$ (b) $f(x) = \sqrt{x - c}$
7. **Shifts in the Graph of a Function** For each function, sketch (on the same set of coordinate axes) a graph of each function for $c = -2, 0,$ and 2.
   (a) $f(x) = [\![x]\!] + c$ (b) $f(x) = [\![x + c]\!]$
8. **Shifts in the Graph of a Function** For each function, sketch (on the same set of coordinate axes) a graph of each function for $c = -3, -1, 1,$ and 3.

$$\text{(a) } f(x) = \begin{cases} x^2 + c, & x < 0 \\ -x^2 + c, & x \ge 0 \end{cases}$$

$$\text{(b) } f(x) = \begin{cases} (x + c)^2, & x < 0 \\ -(x + c)^2, & x \ge 0 \end{cases}$$

Sketching Transformations **In Exercises 9 and 10, use the graph of f to sketch each graph. To print an enlarged copy of the graph, go to *MathGraphs.com*.**

9. (a) $y = f(-x)$
   (b) $y = f(x) + 4$
   (c) $y = 2f(x)$
   (d) $y = -f(x - 4)$
   (e) $y = f(x) - 3$
   (f) $y = -f(x) - 1$
   (g) $y = f(2x)$

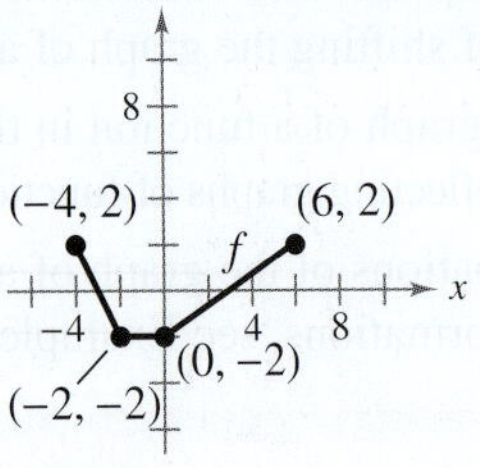

10. (a) $y = f(x - 5)$
    (b) $y = -f(x) + 3$
    (c) $y = \frac{1}{3}f(x)$
    (d) $y = -f(x + 1)$
    (e) $y = f(-x)$
    (f) $y = f(x) - 10$
    (g) $y = f\left(\frac{1}{3}x\right)$

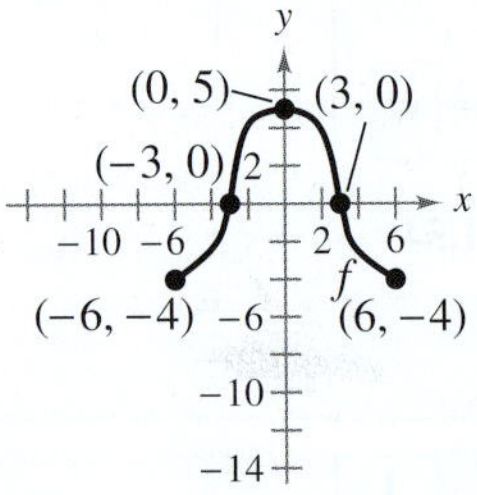

11. **Writing Equations from Graphs** Use the graph of $f(x) = x^2$ to write an equation for each function whose graph is shown.

(a)

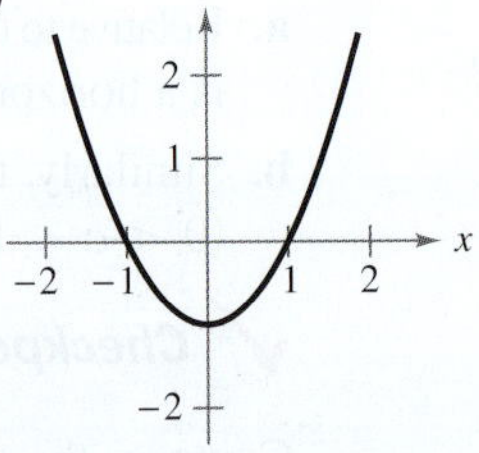

(b)

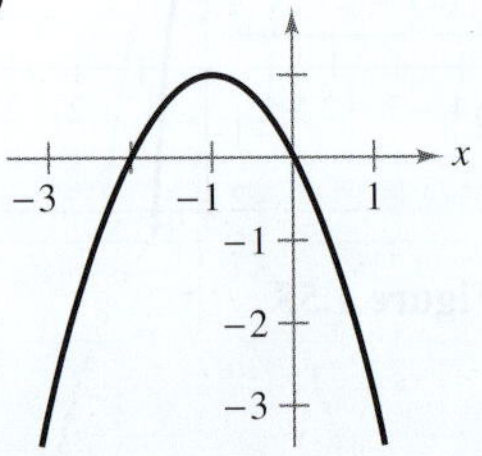

12. **Writing Equations from Graphs** Use the graph of $f(x) = x^3$ to write an equation for each function whose graph is shown.

(a)

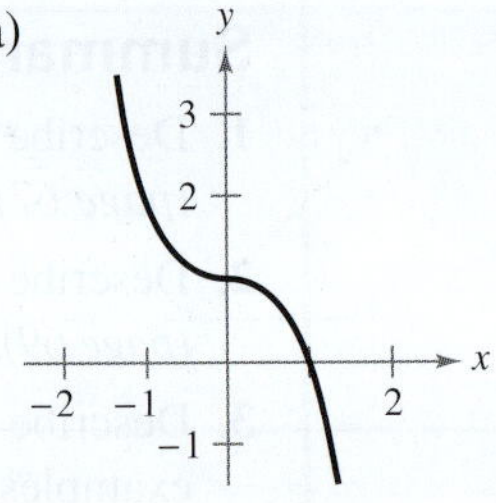

(b)

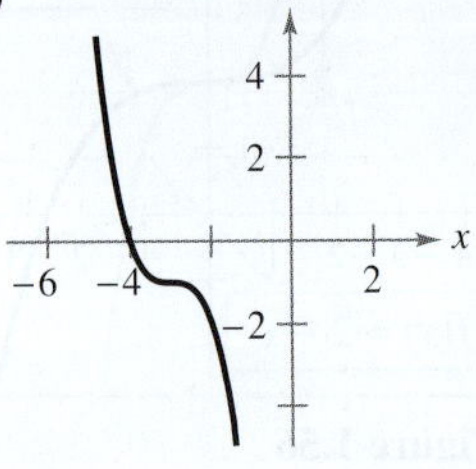

13. Writing Equations from Graphs Use the graph of $f(x) = |x|$ to write an equation for each function whose graph is shown.

(a)

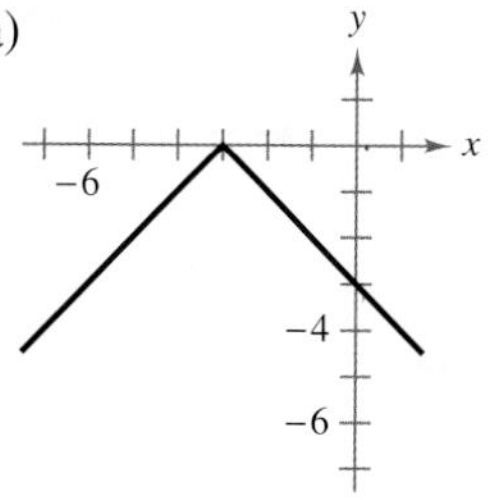

(b)

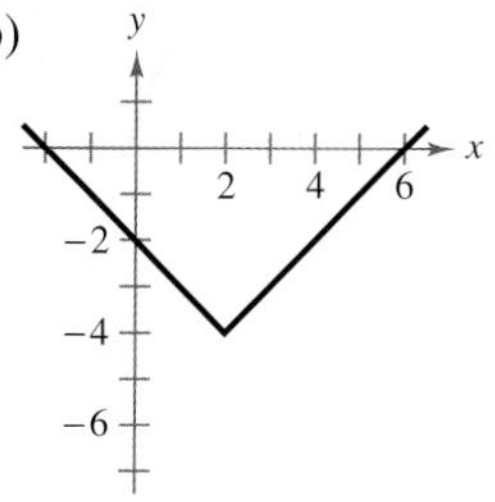

14. Writing Equations from Graphs Use the graph of $f(x) = \sqrt{x}$ to write an equation for each function whose graph is shown.

(a)

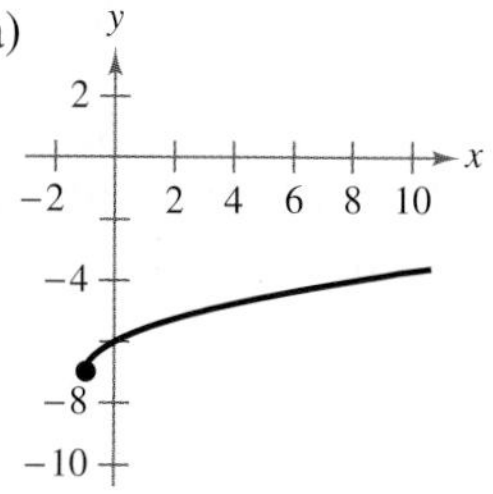

(b)

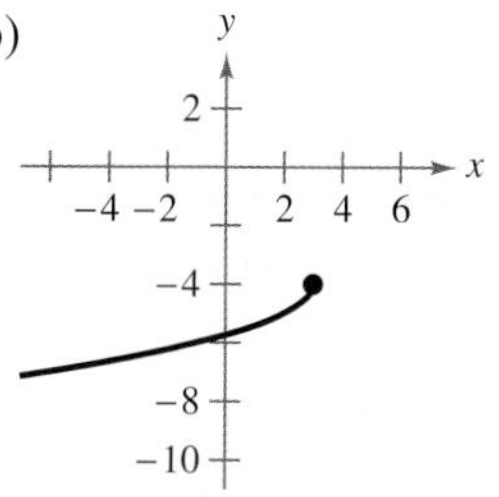

Identifying a Parent Function **In Exercises 15–20, identify the parent function and the transformation shown in the graph. Write an equation for the function shown in the graph.**

15.

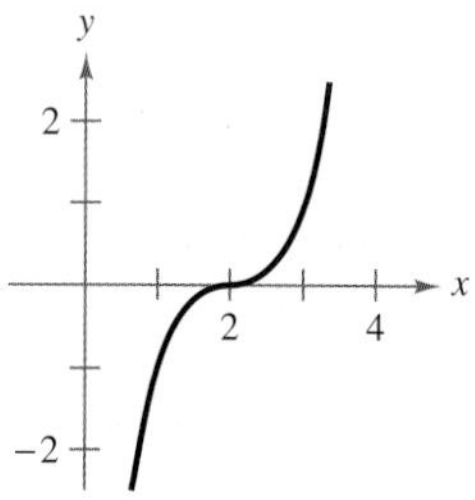

16.

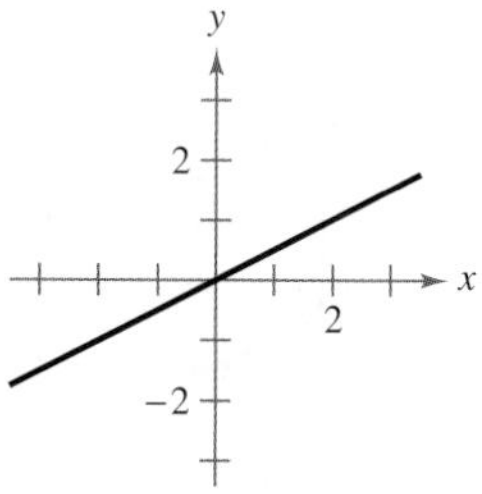

17.

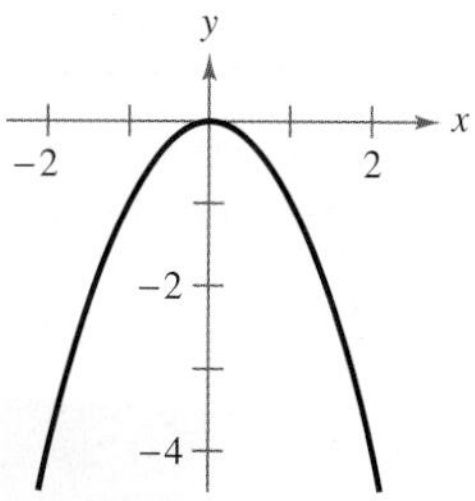

18.

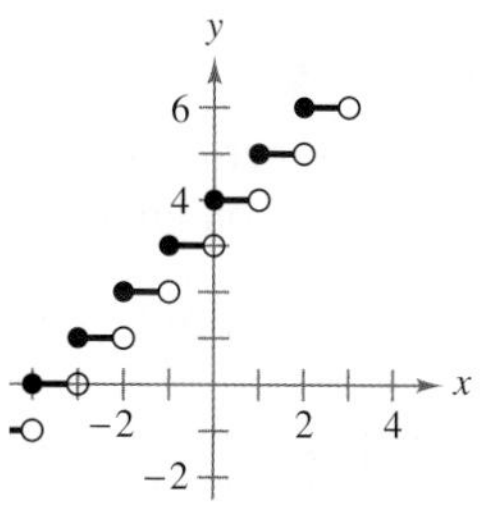

19.

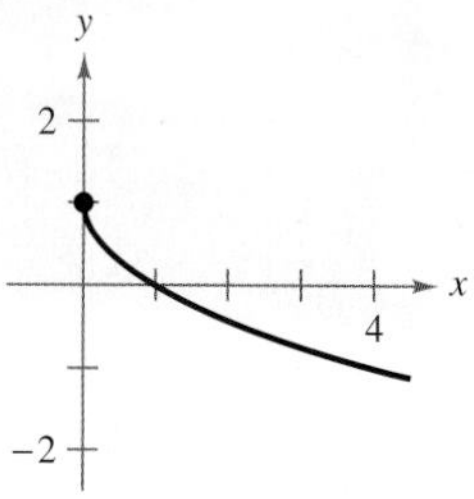

20.

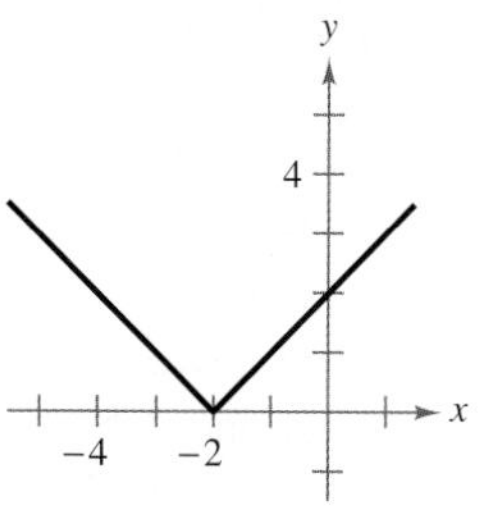

Identifying a Parent Function **In Exercises 21–46, g is related to one of the parent functions described in Section 1.6. (a) Identify the parent function f. (b) Describe the sequence of transformations from f to g. (c) Sketch the graph of g. (d) Use function notation to write g in terms of f.**

21. $g(x) = 12 - x^2$

22. $g(x) = (x - 8)^2$

23. $g(x) = x^3 + 7$

24. $g(x) = -x^3 - 1$

25. $g(x) = \frac{2}{3}x^2 + 4$

26. $g(x) = 2(x - 7)^2$

27. $g(x) = 2 - (x + 5)^2$

28. $g(x) = -\frac{1}{4}(x + 2)^2 - 2$

29. $g(x) = \sqrt{3x}$

30. $g(x) = \sqrt{\frac{1}{4}x}$

31. $g(x) = (x - 1)^3 + 2$

32. $g(x) = (x + 3)^3 - 10$

33. $g(x) = 3(x - 2)^3$

34. $g(x) = -\frac{1}{2}(x + 1)^3$

35. $g(x) = -|x| - 2$

36. $g(x) = 6 - |x + 5|$

37. $g(x) = -|x + 4| + 8$

38. $g(x) = |-x + 3| + 9$

39. $g(x) = -2|x - 1| - 4$

40. $g(x) = \frac{1}{2}|x - 2| - 3$

41. $g(x) = 3 - [\![x]\!]$

42. $g(x) = 2[\![x + 5]\!]$

43. $g(x) = \sqrt{x - 9}$

44. $g(x) = \sqrt{x + 4} + 8$

45. $g(x) = \sqrt{7 - x} - 2$

46. $g(x) = \sqrt{3x} + 1$

Writing an Equation from a Description **In Exercises 47–54, write an equation for the function described by the given characteristics.**

47. The shape of $f(x) = x^2$, but shifted three units to the right and seven units down

48. The shape of $f(x) = x^2$, but shifted two units to the left, nine units up, and then reflected in the x-axis

49. The shape of $f(x) = x^3$, but shifted 13 units to the right

50. The shape of $f(x) = x^3$, but shifted six units to the left, six units down, and then reflected in the y-axis

51. The shape of $f(x) = |x|$, but shifted 12 units up and then reflected in the x-axis

52. The shape of $f(x) = |x|$, but shifted four units to the left and eight units down

53. The shape of $f(x) = \sqrt{x}$, but shifted six units to the left and then reflected in both the x-axis and the y-axis

54. The shape of $f(x) = \sqrt{x}$, but shifted nine units down and then reflected in both the x-axis and the y-axis

55. Writing Equations from Graphs Use the graph of $f(x) = x^2$ to write an equation for each function whose graph is shown.

(a)

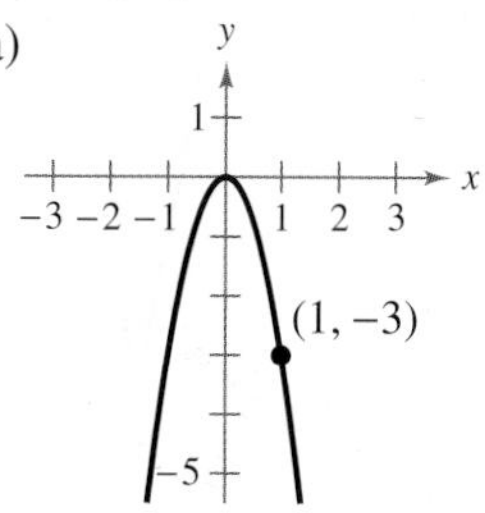

(b)

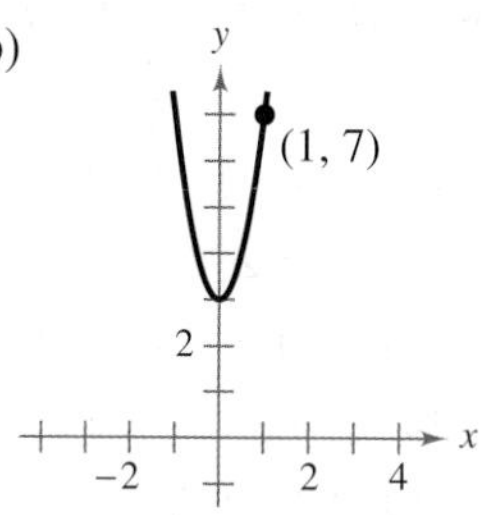

56. Writing Equations from Graphs Use the graph of

$f(x) = x^3$

to write an equation for each function whose graph is shown.

(a)

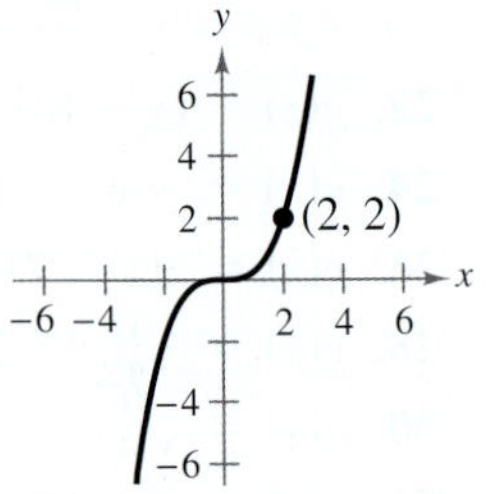

(b)

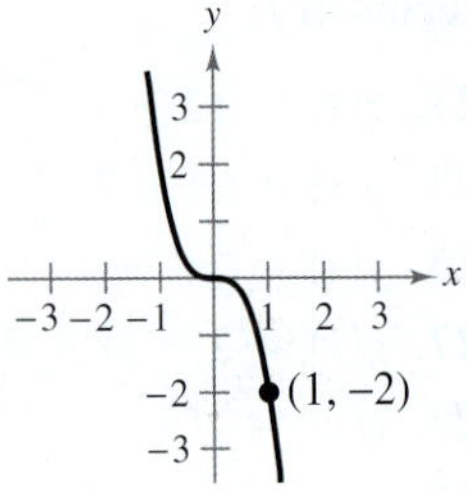

57. Writing Equations from Graphs Use the graph of

$f(x) = |x|$

to write an equation for each function whose graph is shown.

(a)

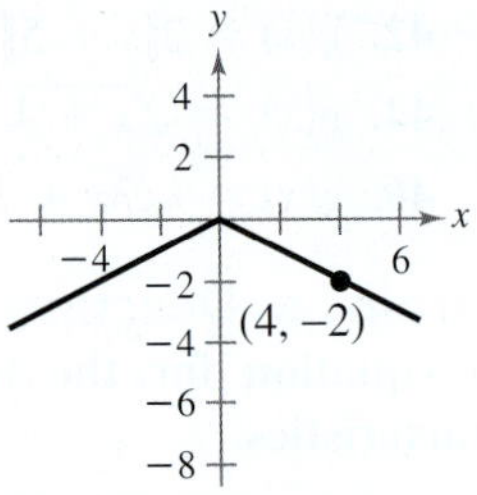

(b)

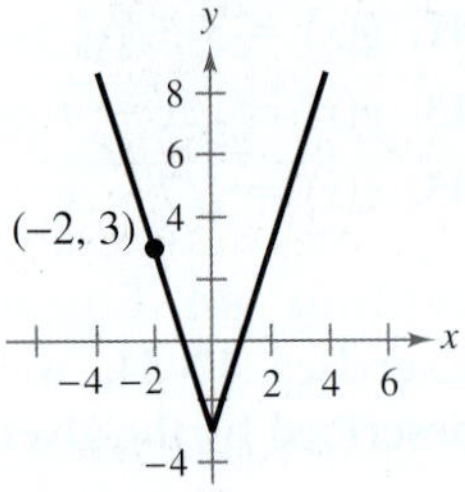

58. Writing Equations from Graphs Use the graph of

$f(x) = \sqrt{x}$

to write an equation for each function whose graph is shown.

(a)

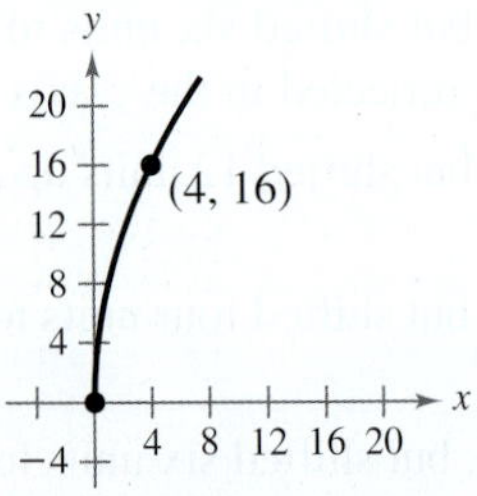

(b)

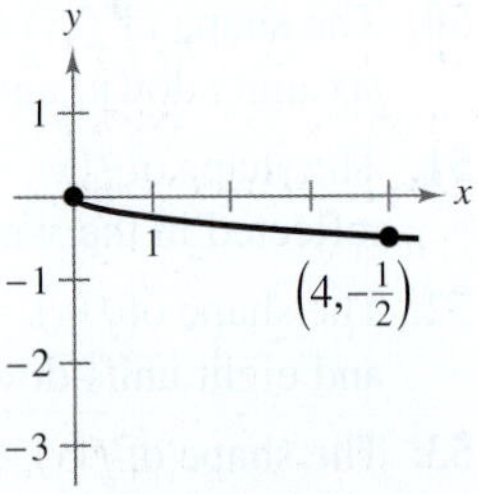

Identifying a Parent Function In Exercises 59–64, identify the parent function and the transformation shown in the graph. Write an equation for the function shown in the graph. Then use a graphing utility to verify your answer.

59.

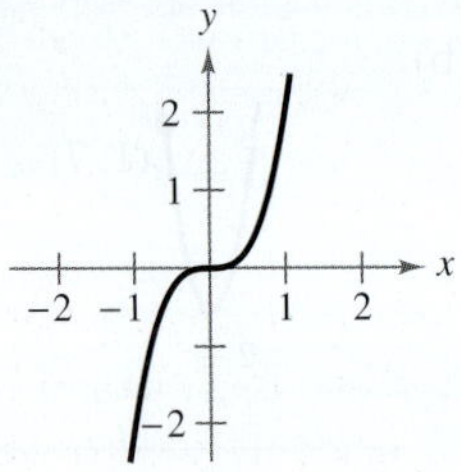

60.

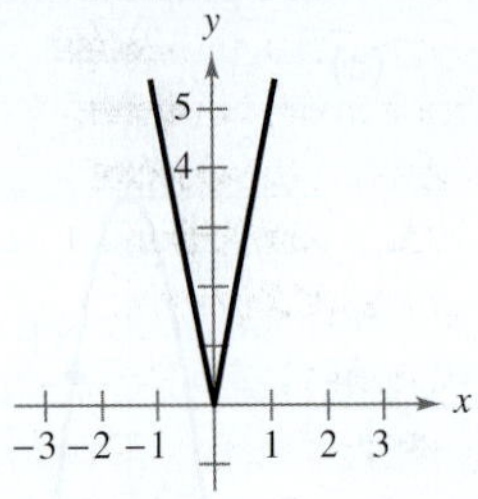

61.

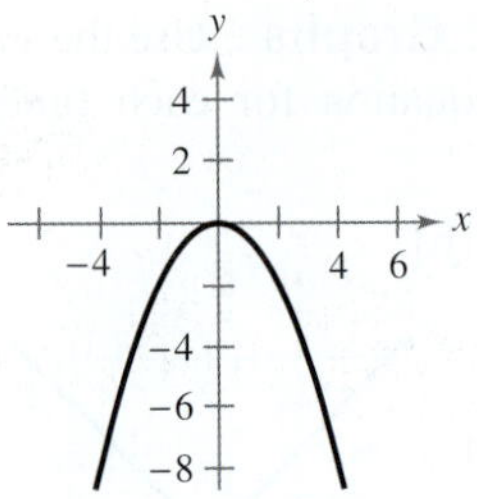

62.

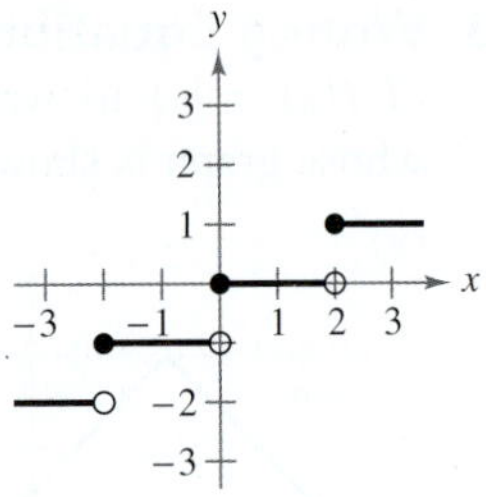

63.

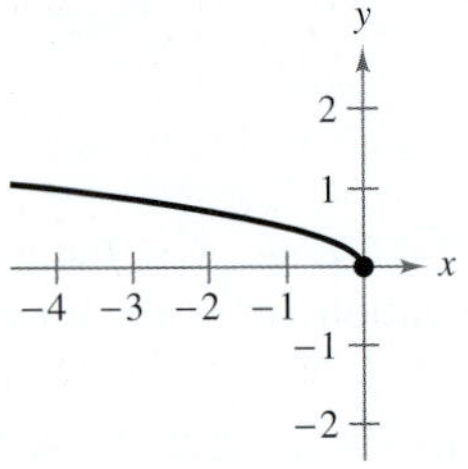

64.

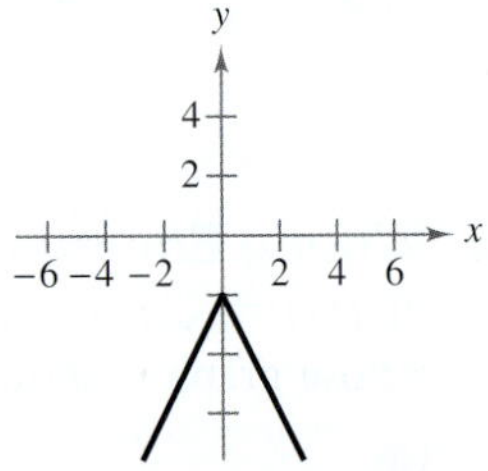

Graphical Analysis In Exercises 65–68, use the viewing window shown to write a possible equation for the transformation of the parent function.

65.

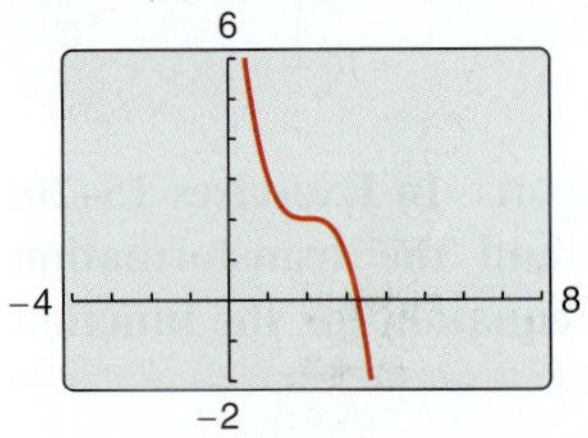

66.

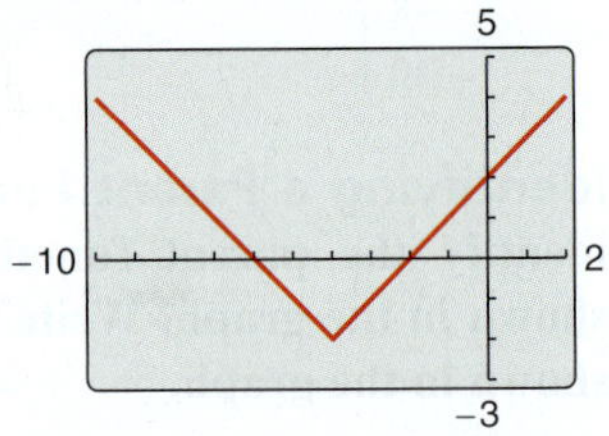

67.

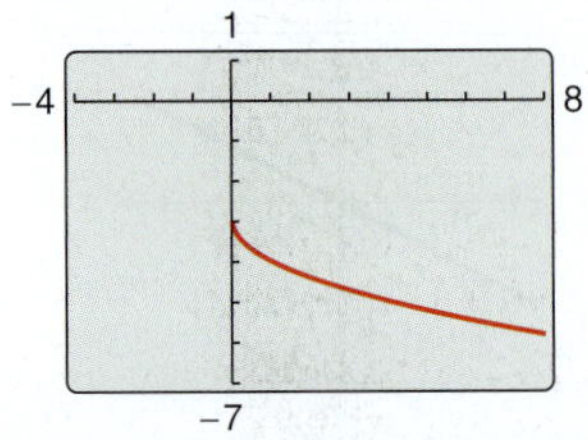

68.

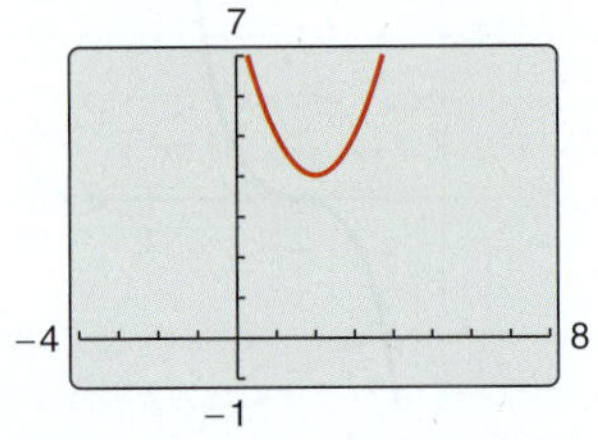

69. Automobile Aerodynamics The number of horsepower H required to overcome wind drag on an automobile is approximated by

$$H(x) = 0.002x^2 + 0.005x - 0.029, \quad 10 \le x \le 100$$

where x is the speed of the car (in miles per hour).

(a) Use a graphing utility to graph the function.

(b) Rewrite the horsepower function so that x represents the speed in kilometers per hour. [Find $H(x/1.6)$.] Identify the type of transformation applied to the graph of the horsepower function.

Robert Young/Shutterstock.com

70. Households The numbers N (in millions) of households in the United States from 2003 through 2010 can be approximated by

$$N = -0.068(x - 13.68)^2 + 119, \quad 3 \le t \le 10$$

where t represents the year, with $t = 3$ corresponding to 2003. *(Source: U.S. Census Bureau)*

(a) Describe the transformation of the parent function $f(x) = x^2$. Then use a graphing utility to graph the function over the specified domain.

(b) Find the average rate of change of the function from 2003 to 2010. Interpret your answer in the context of the problem.

(c) Use the model to predict the number of households in the United States in 2018. Does your answer seem reasonable? Explain.

Exploration

True or False? **In Exercises 71–74, determine whether the statement is true or false. Justify your answer.**

71. The graph of $y = f(-x)$ is a reflection of the graph of $y = f(x)$ in the x-axis.

72. The graph of $y = -f(x)$ is a reflection of the graph of $y = f(x)$ in the y-axis.

73. The graphs of $f(x) = |x| + 6$ and $f(x) = |-x| + 6$ are identical.

74. If the graph of the parent function $f(x) = x^2$ is shifted six units to the right, three units up, and reflected in the x-axis, then the point $(-2, 19)$ will lie on the graph of the transformation.

75. Finding Points on a Graph The graph of $y = f(x)$ passes through the points $(0, 1)$, $(1, 2)$, and $(2, 3)$. Find the corresponding points on the graph of $y = f(x + 2) - 1$.

76. Think About It You can use either of two methods to graph a function: plotting points or translating a parent function as shown in this section. Which method of graphing do you prefer to use for each function? Explain.

(a) $f(x) = 3x^2 - 4x + 1$

(b) $f(x) = 2(x - 1)^2 - 6$

77. Predicting Graphical Relationships Use a graphing utility to graph f, g, and h in the same viewing window. Before looking at the graphs, try to predict how the graphs of g and h relate to the graph of f.

(a) $f(x) = x^2$, $g(x) = (x - 4)^2$, $h(x) = (x - 4)^2 + 3$

(b) $f(x) = x^2$, $g(x) = (x + 1)^2$, $h(x) = (x + 1)^2 - 2$

(c) $f(x) = x^2$, $g(x) = (x + 4)^2$, $h(x) = (x + 4)^2 + 2$

78. HOW DO YOU SEE IT? Use the graph of $y = f(x)$ to find the intervals on which each of the graphs in (a)–(e) is increasing and decreasing. If not possible, then state the reason.

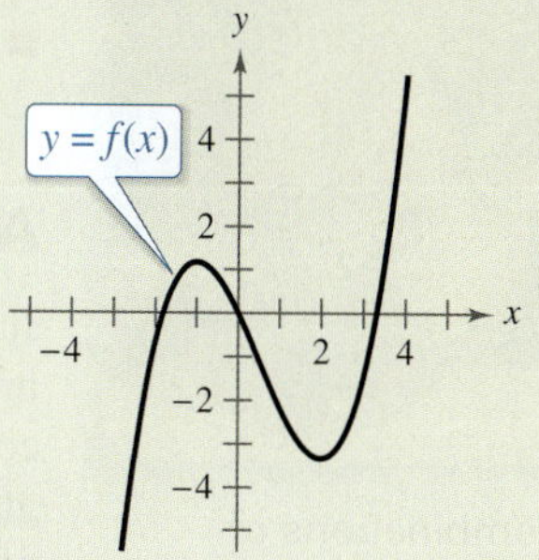

(a) $y = f(-x)$ (b) $y = -f(x)$ (c) $y = \frac{1}{2}f(x)$

(d) $y = -f(x - 1)$ (e) $y = f(x - 2) + 1$

79. Describing Profits Management originally predicted that the profits from the sales of a new product would be approximated by the graph of the function f shown. The actual profits are shown by the function g along with a verbal description. Use the concepts of transformations of graphs to write g in terms of f.

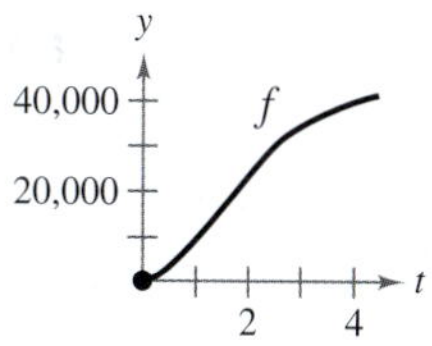

(a) The profits were only three-fourths as large as expected.

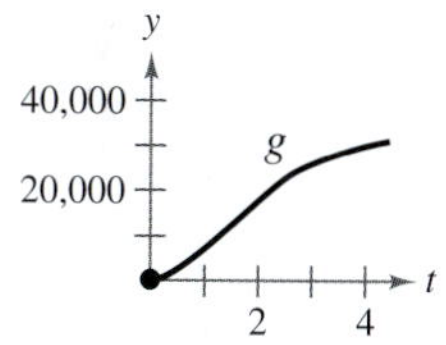

(b) The profits were consistently \$10,000 greater than predicted.

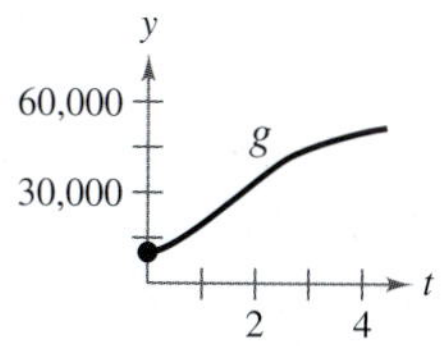

(c) There was a two-year delay in the introduction of the product. After sales began, profits grew as expected.

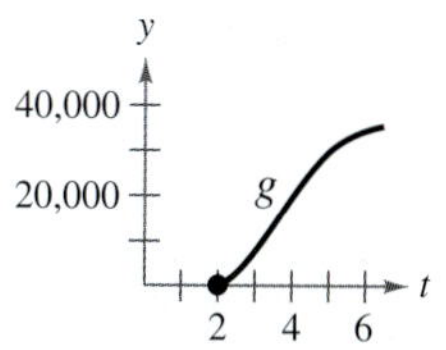

80. Reversing the Order of Transformations Reverse the order of transformations in Example 2(a). Do you obtain the same graph? Do the same for Example 2(b). Do you obtain the same graph? Explain.

1.8 Combinations of Functions: Composite Functions

Arithmetic combinations of functions can help you model and solve real-life problems. For instance, in Exercise 57 on page 82, you will use arithmetic combinations of functions to analyze numbers of pets in the United States.

- **Add, subtract, multiply, and divide functions.**
- **Find the composition of one function with another function.**
- **Use combinations and compositions of functions to model and solve real-life problems.**

Arithmetic Combinations of Functions

Just as two real numbers can be combined by the operations of addition, subtraction, multiplication, and division to form other real numbers, two *functions* can be combined to create new functions. For example, the functions $f(x) = 2x - 3$ and $g(x) = x^2 - 1$ can be combined to form the sum, difference, product, and quotient of f and g.

$$f(x) + g(x) = (2x - 3) + (x^2 - 1) = x^2 + 2x - 4 \qquad \text{Sum}$$

$$f(x) - g(x) = (2x - 3) - (x^2 - 1) = -x^2 + 2x - 2 \qquad \text{Difference}$$

$$f(x)g(x) = (2x - 3)(x^2 - 1) = 2x^3 - 3x^2 - 2x + 3 \qquad \text{Product}$$

$$\frac{f(x)}{g(x)} = \frac{2x - 3}{x^2 - 1}, \quad x \neq \pm 1 \qquad \text{Quotient}$$

The domain of an **arithmetic combination** of functions f and g consists of all real numbers that are common to the domains of f and g. In the case of the quotient $f(x)/g(x)$, there is the further restriction that $g(x) \neq 0$.

Sum, Difference, Product, and Quotient of Functions

Let f and g be two functions with overlapping domains. Then, for all x common to both domains, the *sum, difference, product,* and *quotient* of f and g are defined as follows.

1. Sum: $(f + g)(x) = f(x) + g(x)$
2. Difference: $(f - g)(x) = f(x) - g(x)$
3. Product: $(fg)(x) = f(x) \cdot g(x)$
4. Quotient: $\left(\frac{f}{g}\right)(x) = \frac{f(x)}{g(x)}, \quad g(x) \neq 0$

EXAMPLE 1 Finding the Sum of Two Functions

Given $f(x) = 2x + 1$ and $g(x) = x^2 + 2x - 1$, find $(f + g)(x)$. Then evaluate the sum when $x = 3$.

Solution The sum of f and g is

$$(f + g)(x) = f(x) + g(x) = (2x + 1) + (x^2 + 2x - 1) = x^2 + 4x.$$

When $x = 3$, the value of this sum is

$$(f + g)(3) = 3^2 + 4(3) = 21.$$

✓ **Checkpoint** *Audio-video solution in English & Spanish at LarsonPrecalculus.com.*

Given $f(x) = x^2$ and $g(x) = 1 - x$, find $(f + g)(x)$. Then evaluate the sum when $x = 2$.

Michael Pettigrew/Shutterstock.com

EXAMPLE 2 **Finding the Difference of Two Functions**

Given $f(x) = 2x + 1$ and $g(x) = x^2 + 2x - 1$, find $(f - g)(x)$. Then evaluate the difference when $x = 2$.

Solution The difference of f and g is

$$(f - g)(x) = f(x) - g(x) = (2x + 1) - (x^2 + 2x - 1) = -x^2 + 2.$$

When $x = 2$, the value of this difference is

$$(f - g)(2) = -(2)^2 + 2 = -2.$$

✓ **Checkpoint** Audio-video solution in English & Spanish at LarsonPrecalculus.com.

Given $f(x) = x^2$ and $g(x) = 1 - x$, find $(f - g)(x)$. Then evaluate the difference when $x = 3$.

EXAMPLE 3 **Finding the Product of Two Functions**

Given $f(x) = x^2$ and $g(x) = x - 3$, find $(fg)(x)$. Then evaluate the product when $x = 4$.

Solution The product of f and g is

$$(fg)(x) = f(x)g(x) = (x^2)(x - 3) = x^3 - 3x^2.$$

When $x = 4$, the value of this product is

$$(fg)(4) = 4^3 - 3(4)^2 = 16.$$

✓ **Checkpoint** Audio-video solution in English & Spanish at LarsonPrecalculus.com.

Given $f(x) = x^2$ and $g(x) = 1 - x$, find $(fg)(x)$. Then evaluate the product when $x = 3$.

In Examples 1–3, both f and g have domains that consist of all real numbers. So, the domains of $f + g$, $f - g$, and fg are also the set of all real numbers. Remember to consider any restrictions on the domains of f and g when forming the sum, difference, product, or quotient of f and g.

EXAMPLE 4 **Finding the Quotients of Two Functions**

Find $(f/g)(x)$ and $(g/f)(x)$ for the functions $f(x) = \sqrt{x}$ and $g(x) = \sqrt{4 - x^2}$. Then find the domains of f/g and g/f.

Solution The quotient of f and g is

$$\left(\frac{f}{g}\right)(x) = \frac{f(x)}{g(x)} = \frac{\sqrt{x}}{\sqrt{4 - x^2}}$$

and the quotient of g and f is

$$\left(\frac{g}{f}\right)(x) = \frac{g(x)}{f(x)} = \frac{\sqrt{4 - x^2}}{\sqrt{x}}.$$

The domain of f is $[0, \infty)$ and the domain of g is $[-2, 2]$. The intersection of these domains is $[0, 2]$. So, the domains of f/g and g/f are as follows.

Domain of f/g: $[0, 2)$ Domain of g/f: $(0, 2]$

REMARK Note that the domain of f/g includes $x = 0$, but not $x = 2$, because $x = 2$ yields a zero in the denominator, whereas the domain of g/f includes $x = 2$, but not $x = 0$, because $x = 0$ yields a zero in the denominator.

✓ **Checkpoint** Audio-video solution in English & Spanish at LarsonPrecalculus.com.

Find $(f/g)(x)$ and $(g/f)(x)$ for the functions $f(x) = \sqrt{x - 3}$ and $g(x) = \sqrt{16 - x^2}$. Then find the domains of f/g and g/f.

Composition of Functions

Another way of combining two functions is to form the **composition** of one with the other. For instance, if $f(x) = x^2$ and $g(x) = x + 1$, then the composition of f with g is

$$f(g(x)) = f(x + 1)$$
$$= (x + 1)^2.$$

This composition is denoted as $f \circ g$ and reads as "f composed with g."

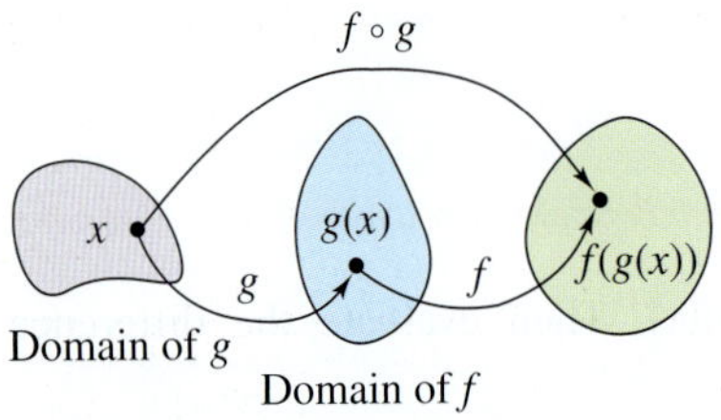

Figure 1.57

> **Definition of Composition of Two Functions**
>
> The **composition** of the function f with the function g is
>
> $$(f \circ g)(x) = f(g(x)).$$
>
> The domain of $f \circ g$ is the set of all x in the domain of g such that $g(x)$ is in the domain of f. (See Figure 1.57.)

EXAMPLE 5 Composition of Functions

Given $f(x) = x + 2$ and $g(x) = 4 - x^2$, find the following.

a. $(f \circ g)(x)$

b. $(g \circ f)(x)$

c. $(g \circ f)(-2)$

Solution

a. The composition of f with g is as follows.

$$(f \circ g)(x) = f(g(x)) \quad \text{Definition of } f \circ g$$
$$= f(4 - x^2) \quad \text{Definition of } g(x)$$
$$= (4 - x^2) + 2 \quad \text{Definition of } f(x)$$
$$= -x^2 + 6 \quad \text{Simplify.}$$

b. The composition of g with f is as follows.

$$(g \circ f)(x) = g(f(x)) \quad \text{Definition of } g \circ f$$
$$= g(x + 2) \quad \text{Definition of } f(x)$$
$$= 4 - (x + 2)^2 \quad \text{Definition of } g(x)$$
$$= 4 - (x^2 + 4x + 4) \quad \text{Expand.}$$
$$= -x^2 - 4x \quad \text{Simplify.}$$

Note that, in this case, $(f \circ g)(x) \neq (g \circ f)(x)$.

c. Using the result of part (b), write the following.

$$(g \circ f)(-2) = -(-2)^2 - 4(-2) \quad \text{Substitute.}$$
$$= -4 + 8 \quad \text{Simplify.}$$
$$= 4 \quad \text{Simplify.}$$

REMARK The following tables of values help illustrate the composition $(f \circ g)(x)$ in Example 5(a).

x	0	1	2	3
$g(x)$	4	3	0	-5

$g(x)$	4	3	0	-5
$f(g(x))$	6	5	2	-3

x	0	1	2	3
$f(g(x))$	6	5	2	-3

Note that the first two tables can be combined (or "composed") to produce the values in the third table.

✓ **Checkpoint** *Audio-video solution in English & Spanish at LarsonPrecalculus.com.*

Given $f(x) = 2x + 5$ and $g(x) = 4x^2 + 1$, find the following.

a. $(f \circ g)(x)$ **b.** $(g \circ f)(x)$ **c.** $(f \circ g)\left(-\frac{1}{2}\right)$

EXAMPLE 6 Finding the Domain of a Composite Function

Find the domain of $(f \circ g)(x)$ for the functions

$$f(x) = x^2 - 9 \quad \text{and} \quad g(x) = \sqrt{9 - x^2}.$$

Algebraic Solution

The composition of the functions is as follows.

$$\begin{aligned}(f \circ g)(x) &= f(g(x)) \\ &= f(\sqrt{9 - x^2}) \\ &= (\sqrt{9 - x^2})^2 - 9 \\ &= 9 - x^2 - 9 \\ &= -x^2\end{aligned}$$

From this, it might appear that the domain of the composition is the set of all real numbers. This, however, is not true. Because the domain of f is the set of all real numbers and the domain of g is $[-3, 3]$, the domain of $f \circ g$ is $[-3, 3]$.

Graphical Solution

The x-coordinates of the points on the graph extend from -3 to 3. So, the domain of $f \circ g$ is $[-3, 3]$.

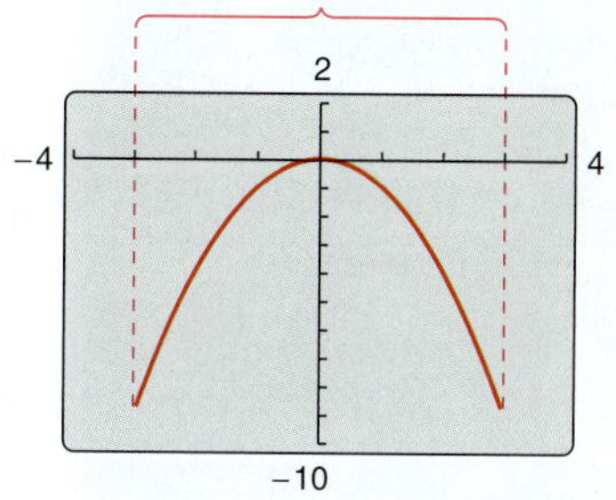

Checkpoint Audio-video solution in English & Spanish at LarsonPrecalculus.com.

Find the domain of $(f \circ g)(x)$ for the functions $f(x) = \sqrt{x}$ and $g(x) = x^2 + 4$.

In Examples 5 and 6, you formed the composition of two given functions. In calculus, it is also important to be able to identify two functions that make up a given composite function. For instance, the function $h(x) = (3x - 5)^3$ is the composition of $f(x) = x^3$ and $g(x) = 3x - 5$. That is,

$$h(x) = (3x - 5)^3 = [g(x)]^3 = f(g(x)).$$

Basically, to "decompose" a composite function, look for an "inner" function and an "outer" function. In the function h above, $g(x) = 3x - 5$ is the inner function and $f(x) = x^3$ is the outer function.

EXAMPLE 7 Decomposing a Composite Function

Write the function $h(x) = \dfrac{1}{(x - 2)^2}$ as a composition of two functions.

Solution One way to write h as a composition of two functions is to take the inner function to be $g(x) = x - 2$ and the outer function to be

$$f(x) = \frac{1}{x^2} = x^{-2}.$$

Then write

$$\begin{aligned}h(x) &= \frac{1}{(x - 2)^2} \\ &= (x - 2)^{-2} \\ &= f(x - 2) \\ &= f(g(x)).\end{aligned}$$

Checkpoint Audio-video solution in English & Spanish at LarsonPrecalculus.com.

Write the function $h(x) = \dfrac{\sqrt[3]{8 - x}}{5}$ as a composition of two functions.

Application

EXAMPLE 8 Bacteria Count

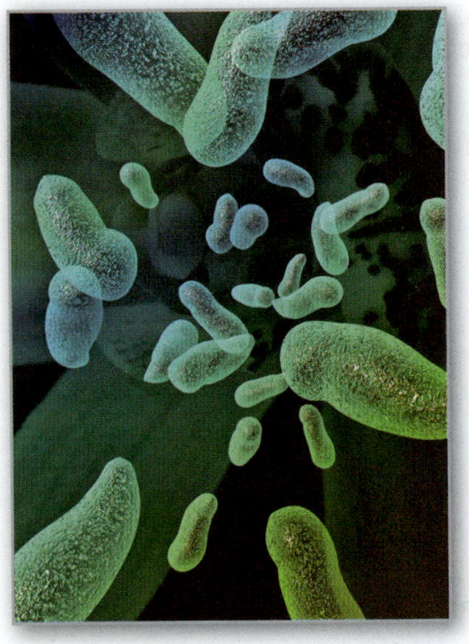

Refrigerated foods can have two types of bacteria: pathogenic bacteria, which can cause foodborne illness, and spoilage bacteria, which give foods an unpleasant look, smell, taste, or texture.

The number N of bacteria in a refrigerated food is given by

$$N(T) = 20T^2 - 80T + 500, \quad 2 \le T \le 14$$

where T is the temperature of the food in degrees Celsius. When the food is removed from refrigeration, the temperature of the food is given by

$$T(t) = 4t + 2, \quad 0 \le t \le 3$$

where t is the time in hours.

a. Find the composition $(N \circ T)(t)$ and interpret its meaning in context.

b. Find the time when the bacteria count reaches 2000.

Solution

a.
$$\begin{aligned}(N \circ T)(t) &= N(T(t)) \\ &= 20(4t + 2)^2 - 80(4t + 2) + 500 \\ &= 20(16t^2 + 16t + 4) - 320t - 160 + 500 \\ &= 320t^2 + 320t + 80 - 320t - 160 + 500 \\ &= 320t^2 + 420\end{aligned}$$

The composite function $(N \circ T)(t)$ represents the number of bacteria in the food as a function of the amount of time the food has been out of refrigeration.

b. The bacteria count will reach 2000 when $320t^2 + 420 = 2000$. Solve this equation to find that the count will reach 2000 when $t \approx 2.2$ hours. Note that when you solve this equation, you reject the negative value because it is not in the domain of the composite function.

✓ **Checkpoint** Audio-video solution in English & Spanish at LarsonPrecalculus.com.

The number N of bacteria in a refrigerated food is given by

$$N(T) = 8T^2 - 14T + 200, \quad 2 \le T \le 12$$

where T is the temperature of the food in degrees Celsius. When the food is removed from refrigeration, the temperature of the food is given by

$$T(t) = 2t + 2, \quad 0 \le t \le 5$$

where t is the time in hours.

a. Find the composition $(N \circ T)(t)$.

b. Find the time when the bacteria count reaches 1000.

Summarize (Section 1.8)

1. Explain how to add, subtract, multiply, and divide functions *(page 76)*. For examples of finding arithmetic combinations of functions, see Examples 1–4.
2. Explain how to find the composition of one function with another function *(page 78)*. For examples that use compositions of functions, see Examples 5–7.
3. Describe a real-life example that uses a composition of functions *(page 80, Example 8)*.

Fedorov Oleksiy/Shutterstock.com

1.8 Exercises

See CalcChat.com for tutorial help and worked-out solutions to odd-numbered exercises.

Vocabulary: **Fill in the blanks.**

1. Two functions f and g can be combined by the arithmetic operations of ________, ________, ________, and ________ to create new functions.
2. The ________ of the function f with g is $(f \circ g)(x) = f(g(x))$.

Skills and Applications

Graphing the Sum of Two Functions **In Exercises 3 and 4, use the graphs of f and g to graph $h(x) = (f + g)(x)$. To print an enlarged copy of the graph, go to *MathGraphs.com*.**

3.

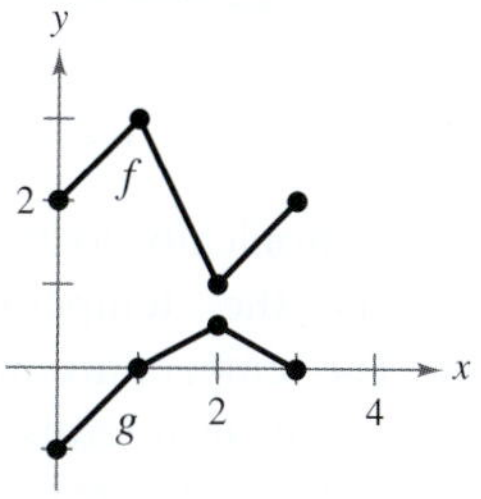

4.

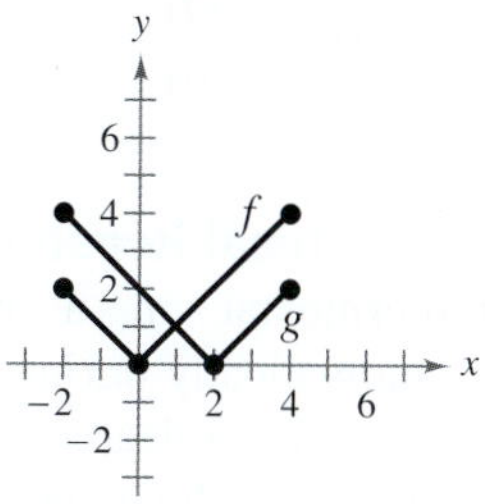

Finding Arithmetic Combinations of Functions **In Exercises 5–12, find (a) $(f + g)(x)$, (b) $(f - g)(x)$, (c) $(fg)(x)$, and (d) $(f/g)(x)$. What is the domain of f/g?**

5. $f(x) = x + 2,\quad g(x) = x - 2$
6. $f(x) = 2x - 5,\quad g(x) = 2 - x$
7. $f(x) = x^2,\quad g(x) = 4x - 5$
8. $f(x) = 3x + 1,\quad g(x) = 5x - 4$
9. $f(x) = x^2 + 6,\quad g(x) = \sqrt{1 - x}$
10. $f(x) = \sqrt{x^2 - 4},\quad g(x) = \dfrac{x^2}{x^2 + 1}$
11. $f(x) = \dfrac{1}{x},\quad g(x) = \dfrac{1}{x^2}$
12. $f(x) = \dfrac{x}{x + 1},\quad g(x) = x^3$

Evaluating an Arithmetic Combination of Functions **In Exercises 13–24, evaluate the indicated function for $f(x) = x^2 + 1$ and $g(x) = x - 4$.**

13. $(f + g)(2)$
14. $(f - g)(-1)$
15. $(f - g)(0)$
16. $(f + g)(1)$
17. $(f - g)(3t)$
18. $(f + g)(t - 2)$
19. $(fg)(6)$
20. $(fg)(-6)$
21. $(f/g)(5)$
22. $(f/g)(0)$
23. $(f/g)(-1) - g(3)$
24. $(fg)(5) + f(4)$

Graphing Two Functions and Their Sum **In Exercises 25 and 26, graph the functions f, g, and $f + g$ on the same set of coordinate axes.**

25. $f(x) = \frac{1}{2}x,\quad g(x) = x - 1$
26. $f(x) = 4 - x^2,\quad g(x) = x$

Graphical Reasoning **In Exercises 27–30, use a graphing utility to graph f, g, and $f + g$ in the same viewing window. Which function contributes most to the magnitude of the sum when $0 \le x \le 2$? Which function contributes most to the magnitude of the sum when $x > 6$?**

27. $f(x) = 3x,\quad g(x) = -\dfrac{x^3}{10}$
28. $f(x) = \dfrac{x}{2},\quad g(x) = \sqrt{x}$
29. $f(x) = 3x + 2,\quad g(x) = -\sqrt{x + 5}$
30. $f(x) = x^2 - \frac{1}{2},\quad g(x) = -3x^2 - 1$

Finding Compositions of Functions **In Exercises 31–34, find (a) $f \circ g$, (b) $g \circ f$, and (c) $g \circ g$.**

31. $f(x) = x^2,\quad g(x) = x - 1$
32. $f(x) = 3x + 5,\quad g(x) = 5 - x$
33. $f(x) = \sqrt[3]{x - 1},\quad g(x) = x^3 + 1$
34. $f(x) = x^3,\quad g(x) = \dfrac{1}{x}$

Finding Domains of Functions and Composite Functions **In Exercises 35–42, find (a) $f \circ g$ and (b) $g \circ f$. Find the domain of each function and each composite function.**

35. $f(x) = \sqrt{x + 4},\quad g(x) = x^2$
36. $f(x) = \sqrt[3]{x - 5},\quad g(x) = x^3 + 1$
37. $f(x) = x^2 + 1,\quad g(x) = \sqrt{x}$
38. $f(x) = x^{2/3},\quad g(x) = x^6$
39. $f(x) = |x|,\quad g(x) = x + 6$
40. $f(x) = |x - 4|,\quad g(x) = 3 - x$
41. $f(x) = \dfrac{1}{x},\quad g(x) = x + 3$
42. $f(x) = \dfrac{3}{x^2 - 1},\quad g(x) = x + 1$

Evaluating Combinations of Functions In Exercises 43–46, use the graphs of f and g to evaluate the functions.

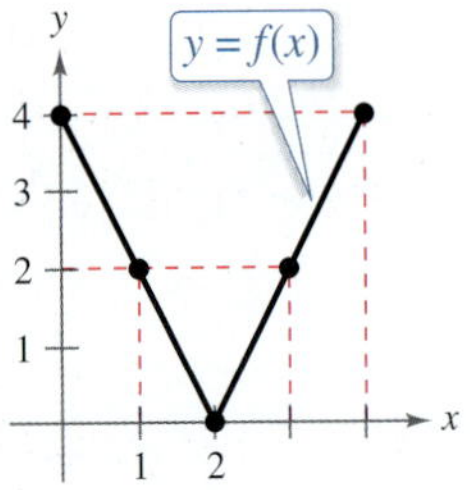

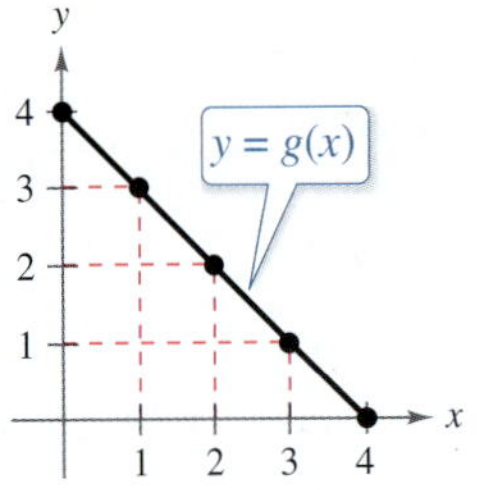

43. (a) $(f + g)(3)$ (b) $(f/g)(2)$

44. (a) $(f - g)(1)$ (b) $(fg)(4)$

45. (a) $(f \circ g)(2)$ (b) $(g \circ f)(2)$

46. (a) $(f \circ g)(1)$ (b) $(g \circ f)(3)$

Decomposing a Composite Function In Exercises 47–54, find two functions f and g such that $(f \circ g)(x) = h(x)$. (There are many correct answers.)

47. $h(x) = (2x + 1)^2$

48. $h(x) = (1 - x)^3$

49. $h(x) = \sqrt[3]{x^2 - 4}$

50. $h(x) = \sqrt{9 - x}$

51. $h(x) = \dfrac{1}{x + 2}$

52. $h(x) = \dfrac{4}{(5x + 2)^2}$

53. $h(x) = \dfrac{-x^2 + 3}{4 - x^2}$

54. $h(x) = \dfrac{27x^3 + 6x}{10 - 27x^3}$

55. Stopping Distance The research and development department of an automobile manufacturer has determined that when a driver is required to stop quickly to avoid an accident, the distance (in feet) the car travels during the driver's reaction time is given by $R(x) = \frac{3}{4}x$, where x is the speed of the car in miles per hour. The distance (in feet) traveled while the driver is braking is given by $B(x) = \frac{1}{15}x^2$.

(a) Find the function that represents the total stopping distance T.

(b) Graph the functions R, B, and T on the same set of coordinate axes for $0 \le x \le 60$.

(c) Which function contributes most to the magnitude of the sum at higher speeds? Explain.

56. Vital Statistics Let $b(t)$ be the number of births in the United States in year t, and let $d(t)$ represent the number of deaths in the United States in year t, where $t = 10$ corresponds to 2010.

(a) If $p(t)$ is the population of the United States in year t, then find the function $c(t)$ that represents the percent change in the population of the United States.

(b) Interpret the value of $c(13)$.

57. Pets

Let $d(t)$ be the number of dogs in the United States in year t, and let $c(t)$ be the number of cats in the United States in year t, where $t = 10$ corresponds to 2010.

(a) Find the function $p(t)$ that represents the total number of dogs and cats in the United States.

(b) Interpret the value of $p(13)$.

(c) Let $n(t)$ represent the population of the United States in year t, where $t = 10$ corresponds to 2010. Find and interpret

$$h(t) = \frac{p(t)}{n(t)}.$$

58. Graphical Reasoning An electronically controlled thermostat in a home lowers the temperature automatically during the night. The temperature in the house T (in degrees Fahrenheit) is given in terms of t, the time in hours on a 24-hour clock (see figure).

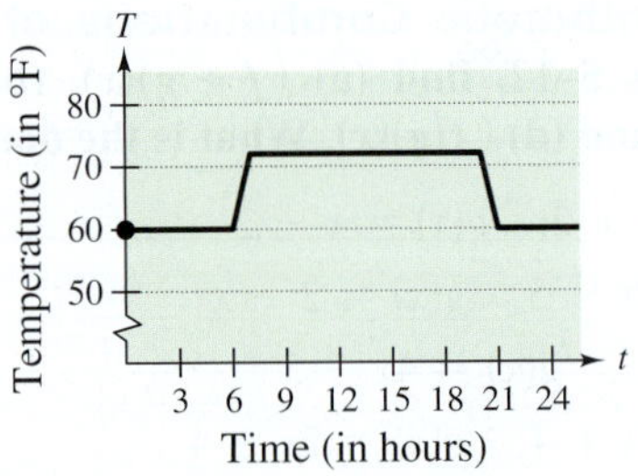

(a) Explain why T is a function of t.

(b) Approximate $T(4)$ and $T(15)$.

(c) The thermostat is reprogrammed to produce a temperature H for which $H(t) = T(t - 1)$. How does this change the temperature?

(d) The thermostat is reprogrammed to produce a temperature H for which $H(t) = T(t) - 1$. How does this change the temperature?

(e) Write a piecewise-defined function that represents the graph.

59. Geometry A square concrete foundation is a base for a cylindrical tank (see figure).

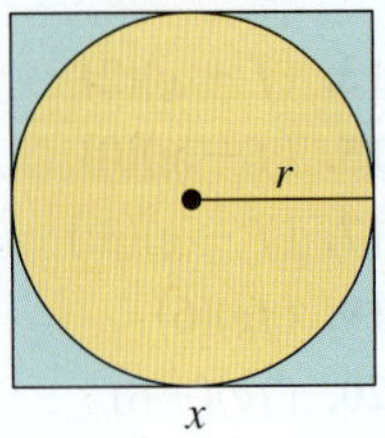

(a) Write the radius r of the tank as a function of the length x of the sides of the square.

(b) Write the area A of the circular base of the tank as a function of the radius r.

(c) Find and interpret $(A \circ r)(x)$.

Michael Pettigrew/Shutterstock.com

60. Bacteria Count The number N of bacteria in a refrigerated food is given by

$$N(T) = 10T^2 - 20T + 600, \quad 2 \le T \le 20$$

where T is the temperature of the food in degrees Celsius. When the food is removed from refrigeration, the temperature of the food is given by

$$T(t) = 3t + 2, \quad 0 \le t \le 6$$

where t is the time in hours.

(a) Find the composition $(N \circ T)(t)$ and interpret its meaning in context.

(b) Find the bacteria count after 0.5 hour.

(c) Find the time when the bacteria count reaches 1500.

61. Salary You are a sales representative for a clothing manufacturer. You are paid an annual salary, plus a bonus of 3% of your sales over \$500,000. Consider the two functions $f(x) = x - 500{,}000$ and $g(x) = 0.03x$. When x is greater than \$500,000, which of the following represents your bonus? Explain your reasoning.

(a) $f(g(x))$

(b) $g(f(x))$

62. Consumer Awareness The suggested retail price of a new hybrid car is p dollars. The dealership advertises a factory rebate of \$2000 and a 10% discount.

(a) Write a function R in terms of p giving the cost of the hybrid car after receiving the rebate from the factory.

(b) Write a function S in terms of p giving the cost of the hybrid car after receiving the dealership discount.

(c) Form the composite functions $(R \circ S)(p)$ and $(S \circ R)(p)$ and interpret each.

(d) Find $(R \circ S)(25{,}795)$ and $(S \circ R)(25{,}795)$. Which yields the lower cost for the hybrid car? Explain.

Exploration

Siblings In Exercises 63 and 64, three siblings are three different ages. The oldest is twice the age of the middle sibling, and the middle sibling is six years older than one-half the age of the youngest.

63. (a) Write a composite function that gives the oldest sibling's age in terms of the youngest. Explain how you arrived at your answer.

(b) If the oldest sibling is 16 years old, then find the ages of the other two siblings.

64. (a) Write a composite function that gives the youngest sibling's age in terms of the oldest. Explain how you arrived at your answer.

(b) If the youngest sibling is two years old, then find the ages of the other two siblings.

True or False? In Exercises 65 and 66, determine whether the statement is true or false. Justify your answer.

65. If $f(x) = x + 1$ and $g(x) = 6x$, then

$$(f \circ g)(x) = (g \circ f)(x).$$

66. When you are given two functions $f(x)$ and $g(x)$, you can calculate $(f \circ g)(x)$ if and only if the range of g is a subset of the domain of f.

67. Proof Prove that the product of two odd functions is an even function, and that the product of two even functions is an even function.

68. HOW DO YOU SEE IT? The graphs labeled L_1, L_2, L_3, and L_4 represent four different pricing discounts, where p is the original price (in dollars) and S is the sale price (in dollars). Match each function with its graph. Describe the situations in parts (c) and (d).

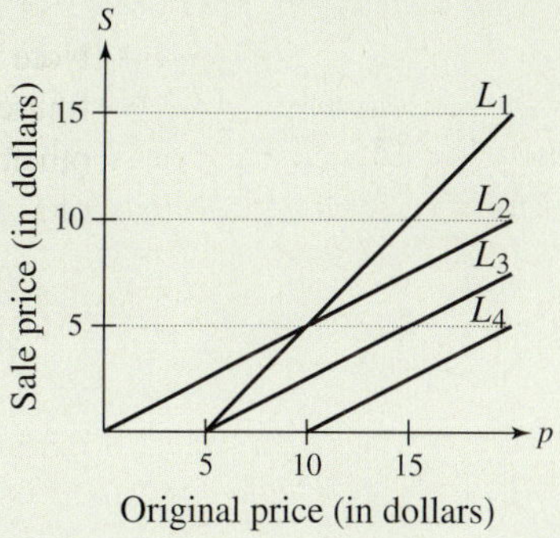

(a) $f(p)$: A 50% discount is applied.

(b) $g(p)$: A \$5 discount is applied.

(c) $(g \circ f)(p)$

(d) $(f \circ g)(p)$

69. Conjecture Use examples to hypothesize whether the product of an odd function and an even function is even or odd. Then prove your hypothesis.

70. Proof

(a) Given a function f, prove that $g(x)$ is even and $h(x)$ is odd, where $g(x) = \frac{1}{2}[f(x) + f(-x)]$ and $h(x) = \frac{1}{2}[f(x) - f(-x)]$.

(b) Use the result of part (a) to prove that any function can be written as a sum of even and odd functions. [*Hint:* Add the two equations in part (a).]

(c) Use the result of part (b) to write each function as a sum of even and odd functions.

$$f(x) = x^2 - 2x + 1, \quad k(x) = \frac{1}{x + 1}$$

1.9 Inverse Functions

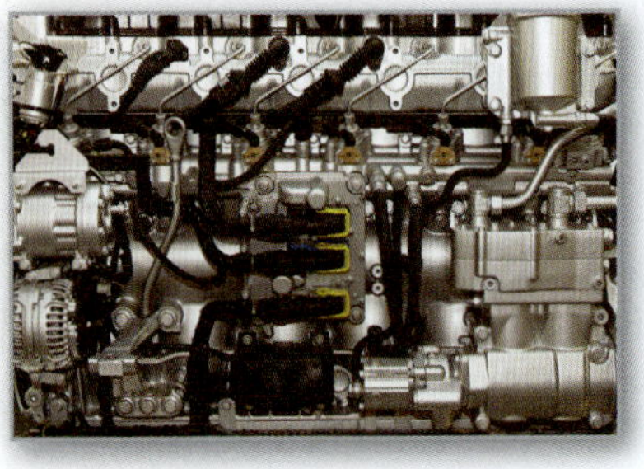

Inverse functions can help you model and solve real-life problems. For instance, in Exercise 94 on page 92, an inverse function can help you determine the percent load interval for a diesel engine.

- **Find inverse functions informally and verify that two functions are inverse functions of each other.**
- **Use graphs of functions to determine whether functions have inverse functions.**
- **Use the Horizontal Line Test to determine whether functions are one-to-one.**
- **Find inverse functions algebraically.**

Inverse Functions

Recall from Section 1.4 that a set of ordered pairs can represent a function. For instance, the function $f(x) = x + 4$ from the set $A = \{1, 2, 3, 4\}$ to the set $B = \{5, 6, 7, 8\}$ can be written as follows.

$$f(x) = x + 4: \ \{(1, 5), (2, 6), (3, 7), (4, 8)\}$$

In this case, by interchanging the first and second coordinates of each of these ordered pairs, you can form the **inverse function** of f, which is denoted by f^{-1}. It is a function from the set B to the set A, and can be written as follows.

$$f^{-1}(x) = x - 4: \ \{(5, 1), (6, 2), (7, 3), (8, 4)\}$$

Note that the domain of f is equal to the range of f^{-1}, and vice versa, as shown in the figure below. Also note that the functions f and f^{-1} have the effect of "undoing" each other. In other words, when you form the composition of f with f^{-1} or the composition of f^{-1} with f, you obtain the identity function.

$$f(f^{-1}(x)) = f(x - 4) = (x - 4) + 4 = x$$

$$f^{-1}(f(x)) = f^{-1}(x + 4) = (x + 4) - 4 = x$$

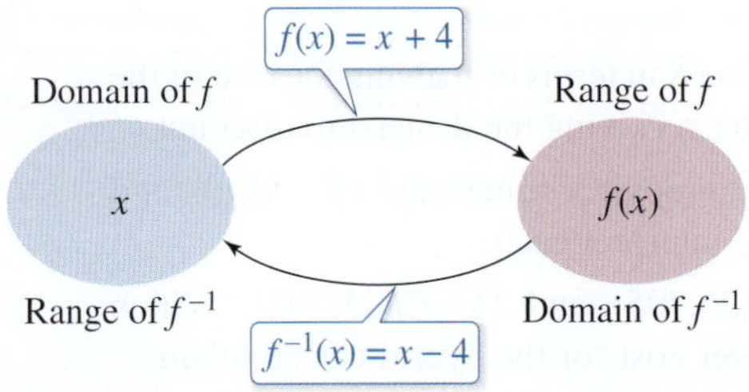

EXAMPLE 1 Finding an Inverse Function Informally

Find the inverse function of $f(x) = 4x$. Then verify that both $f(f^{-1}(x))$ and $f^{-1}(f(x))$ are equal to the identity function.

Solution The function f *multiplies* each input by 4. To "undo" this function, you need to *divide* each input by 4. So, the inverse function of $f(x) = 4x$ is

$$f^{-1}(x) = \frac{x}{4}.$$

Verify that $f(f^{-1}(x)) = x$ and $f^{-1}(f(x)) = x$ as follows.

$$f(f^{-1}(x)) = f\left(\frac{x}{4}\right) = 4\left(\frac{x}{4}\right) = x \qquad f^{-1}(f(x)) = f^{-1}(4x) = \frac{4x}{4} = x$$

✓ **Checkpoint** *Audio-video solution in English & Spanish at LarsonPrecalculus.com.*

Find the inverse function of $f(x) = \frac{1}{5}x$. Then verify that both $f(f^{-1}(x))$ and $f^{-1}(f(x))$ are equal to the identity function.

Baloncici/Shutterstock.com

Definition of Inverse Function

Let f and g be two functions such that

$$f(g(x)) = x \qquad \text{for every } x \text{ in the domain of } g$$

and

$$g(f(x)) = x \qquad \text{for every } x \text{ in the domain of } f.$$

Under these conditions, the function g is the **inverse function** of the function f. The function g is denoted by f^{-1} (read "f-inverse"). So,

$$f(f^{-1}(x)) = x \quad \text{and} \quad f^{-1}(f(x)) = x.$$

The domain of f must be equal to the range of f^{-1}, and the range of f must be equal to the domain of f^{-1}.

Do not be confused by the use of -1 to denote the inverse function f^{-1}. In this text, whenever f^{-1} is written, it *always* refers to the inverse function of the function f and *not* to the reciprocal of $f(x)$.

If the function g is the inverse function of the function f, then it must also be true that the function f is the inverse function of the function g. For this reason, you can say that the functions f and g are *inverse functions of each other.*

EXAMPLE 2 Verifying Inverse Functions

Which of the functions is the inverse function of $f(x) = \dfrac{5}{x-2}$?

$$g(x) = \frac{x-2}{5} \qquad h(x) = \frac{5}{x} + 2$$

Solution By forming the composition of f with g, you have

$$f(g(x)) = f\left(\frac{x-2}{5}\right) = \frac{5}{\left(\dfrac{x-2}{5}\right) - 2} = \frac{25}{x-12} \neq x.$$

Because this composition is not equal to the identity function x, it follows that g *is not* the inverse function of f. By forming the composition of f with h, you have

$$f(h(x)) = f\left(\frac{5}{x} + 2\right) = \frac{5}{\left(\dfrac{5}{x} + 2\right) - 2} = \frac{5}{\left(\dfrac{5}{x}\right)} = x.$$

So, it appears that h *is* the inverse function of f. Confirm this by showing that the composition of h with f is also equal to the identity function, as follows.

$$h(f(x)) = h\left(\frac{5}{x-2}\right) = \frac{5}{\left(\dfrac{5}{x-2}\right)} + 2 = x - 2 + 2 = x$$

✓ Checkpoint *Audio-video solution in English & Spanish at LarsonPrecalculus.com.*

Which of the functions is the inverse function of $f(x) = \dfrac{x-4}{7}$?

$$g(x) = 7x + 4 \qquad h(x) = \frac{7}{x-4}$$

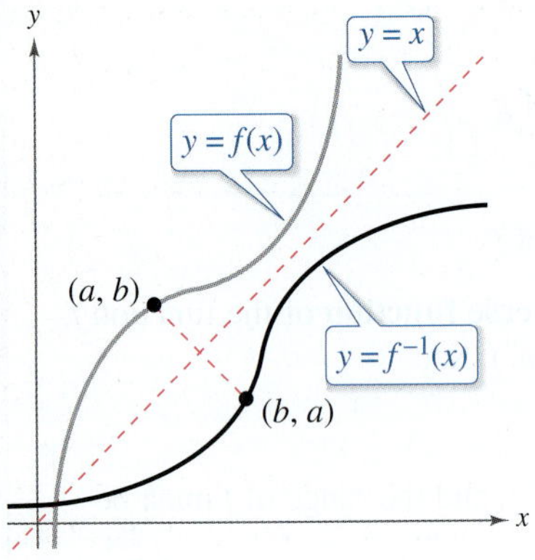

Figure 1.58

The Graph of an Inverse Function

The graphs of a function f and its inverse function f^{-1} are related to each other in the following way. If the point (a, b) lies on the graph of f, then the point (b, a) must lie on the graph of f^{-1}, and vice versa. This means that the graph of f^{-1} is a *reflection* of the graph of f in the line $y = x$, as shown in Figure 1.58.

EXAMPLE 3 Verifying Inverse Functions Graphically

Sketch the graphs of the inverse functions $f(x) = 2x - 3$ and $f^{-1}(x) = \frac{1}{2}(x + 3)$ on the same rectangular coordinate system and show that the graphs are reflections of each other in the line $y = x$.

Solution The graphs of f and f^{-1} are shown in Figure 1.59. It appears that the graphs are reflections of each other in the line $y = x$. You can further verify this reflective property by testing a few points on each graph. Note in the following list that if the point (a, b) is on the graph of f, then the point (b, a) is on the graph of f^{-1}.

Graph of $f(x) = 2x - 3$	Graph of $f^{-1}(x) = \frac{1}{2}(x + 3)$
$(-1, -5)$	$(-5, -1)$
$(0, -3)$	$(-3, 0)$
$(1, -1)$	$(-1, 1)$
$(2, 1)$	$(1, 2)$
$(3, 3)$	$(3, 3)$

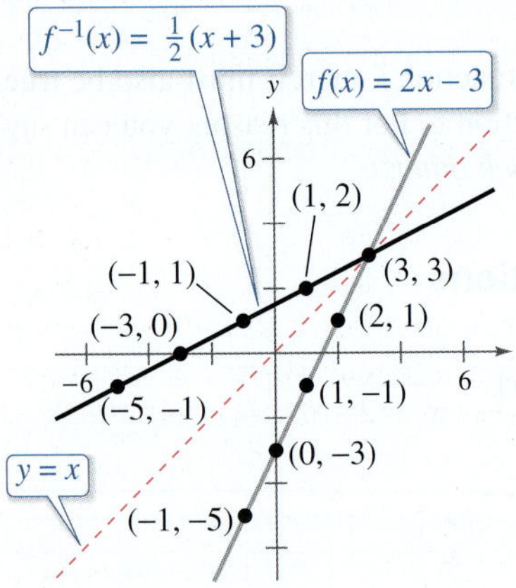

Figure 1.59

✓ **Checkpoint** Audio-video solution in English & Spanish at LarsonPrecalculus.com.

Sketch the graphs of the inverse functions $f(x) = 4x - 1$ and $f^{-1}(x) = \frac{1}{4}(x + 1)$ on the same rectangular coordinate system and show that the graphs are reflections of each other in the line $y = x$.

EXAMPLE 4 Verifying Inverse Functions Graphically

Sketch the graphs of the inverse functions $f(x) = x^2$ $(x \geq 0)$ and $f^{-1}(x) = \sqrt{x}$ on the same rectangular coordinate system and show that the graphs are reflections of each other in the line $y = x$.

Solution The graphs of f and f^{-1} are shown in Figure 1.60. It appears that the graphs are reflections of each other in the line $y = x$. You can further verify this reflective property by testing a few points on each graph. Note in the following list that if the point (a, b) is on the graph of f, then the point (b, a) is on the graph of f^{-1}.

Graph of $f(x) = x^2,\ x \geq 0$	Graph of $f^{-1}(x) = \sqrt{x}$
$(0, 0)$	$(0, 0)$
$(1, 1)$	$(1, 1)$
$(2, 4)$	$(4, 2)$
$(3, 9)$	$(9, 3)$

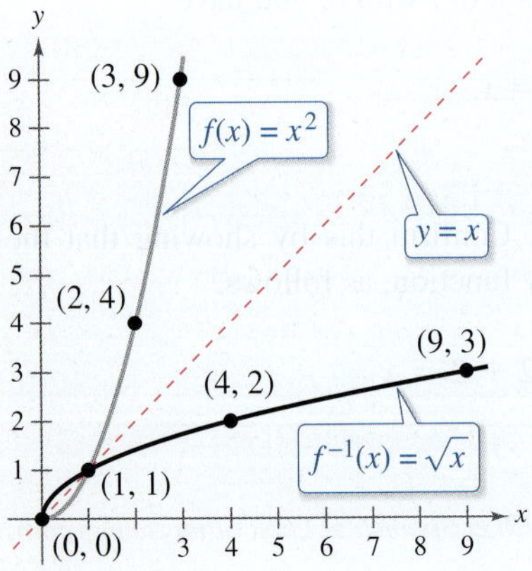

Figure 1.60

Try showing that $f(f^{-1}(x)) = x$ and $f^{-1}(f(x)) = x$.

✓ **Checkpoint** Audio-video solution in English & Spanish at LarsonPrecalculus.com.

Sketch the graphs of the inverse functions $f(x) = x^2 + 1$ $(x \geq 0)$ and $f^{-1}(x) = \sqrt{x - 1}$ on the same rectangular coordinate system and show that the graphs are reflections of each other in the line $y = x$.

One-to-One Functions

The reflective property of the graphs of inverse functions gives you a *geometric* test for determining whether a function has an inverse function. This test is called the **Horizontal Line Test** for inverse functions.

> **Horizontal Line Test for Inverse Functions**
>
> A function f has an inverse function if and only if no *horizontal* line intersects the graph of f at more than one point.

If no horizontal line intersects the graph of f at more than one point, then no y-value matches with more than one x-value. This is the essential characteristic of what are called **one-to-one functions.**

> **One-to-One Functions**
>
> A function f is **one-to-one** when each value of the dependent variable corresponds to exactly one value of the independent variable. A function f has an inverse function if and only if f is one-to-one.

Consider the function $f(x) = x^2$. The table on the left is a table of values for $f(x) = x^2$. The table on the right is the same as the table on the left but with the values in the columns interchanged. The table on the right does not represent a function because the input $x = 4$, for instance, matches with two different outputs: $y = -2$ and $y = 2$. So, $f(x) = x^2$ is not one-to-one and does not have an inverse function.

x	$f(x) = x^2$
-2	4
-1	1
0	0
1	1
2	4
3	9

x	y
4	-2
1	-1
0	0
1	1
4	2
9	3

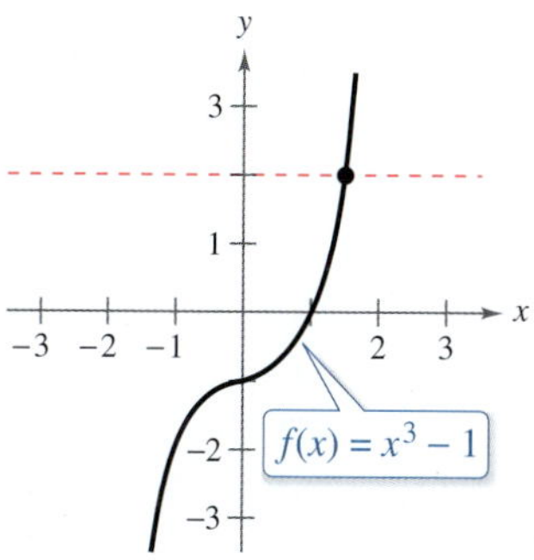

Figure 1.61

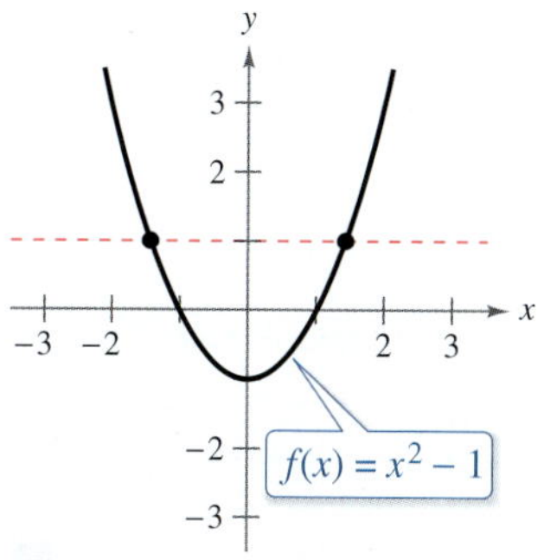

Figure 1.62

EXAMPLE 5 Applying the Horizontal Line Test

a. The graph of the function $f(x) = x^3 - 1$ is shown in Figure 1.61. Because no horizontal line intersects the graph of f at more than one point, f *is* a one-to-one function and *does* have an inverse function.

b. The graph of the function $f(x) = x^2 - 1$ is shown in Figure 1.62. Because it is possible to find a horizontal line that intersects the graph of f at more than one point, f *is not* a one-to-one function and *does not* have an inverse function.

✓ ***Checkpoint*** *Audio-video solution in English & Spanish at LarsonPrecalculus.com.*

Use the graph of f to determine whether the function has an inverse function.

a. $f(x) = \frac{1}{2}(3 - x)$

b. $f(x) = |x|$

Finding Inverse Functions Algebraically

For relatively simple functions (such as the one in Example 1), you can find inverse functions by inspection. For more complicated functions, however, it is best to use the following guidelines. The key step in these guidelines is Step 3—interchanging the roles of x and y. This step corresponds to the fact that inverse functions have ordered pairs with the coordinates reversed.

REMARK Note what happens when you try to find the inverse function of a function that is not one-to-one.

$f(x) = x^2 + 1$ — Original function

$y = x^2 + 1$ — Replace $f(x)$ with y.

$x = y^2 + 1$ — Interchange x and y.

$x - 1 = y^2$ — Isolate y-term.

$y = \pm\sqrt{x - 1}$ — Solve for y.

You obtain two y-values for each x.

Finding an Inverse Function

1. Use the Horizontal Line Test to decide whether f has an inverse function.
2. In the equation for $f(x)$, replace $f(x)$ with y.
3. Interchange the roles of x and y, and solve for y.
4. Replace y with $f^{-1}(x)$ in the new equation.
5. Verify that f and f^{-1} are inverse functions of each other by showing that the domain of f is equal to the range of f^{-1}, the range of f is equal to the domain of f^{-1}, and $f(f^{-1}(x)) = x$ and $f^{-1}(f(x)) = x$.

EXAMPLE 6 **Finding an Inverse Function Algebraically**

Find the inverse function of

$$f(x) = \frac{5 - x}{3x + 2}.$$

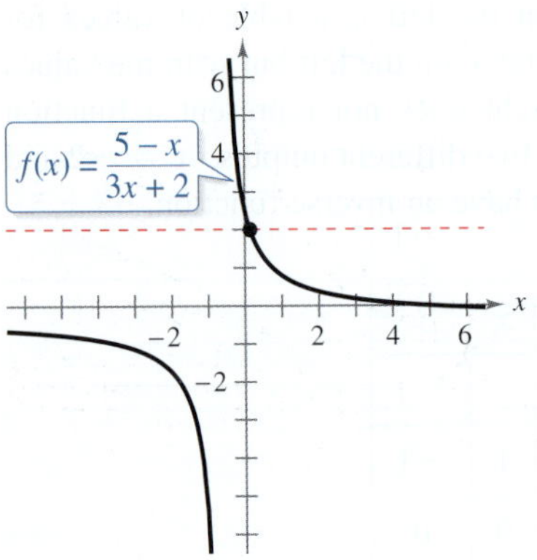

Figure 1.63

Solution The graph of f is shown in Figure 1.63. This graph passes the Horizontal Line Test. So, you know that f is one-to-one and has an inverse function.

$$f(x) = \frac{5 - x}{3x + 2} \quad \text{Write original function.}$$

$$y = \frac{5 - x}{3x + 2} \quad \text{Replace } f(x) \text{ with } y.$$

$$x = \frac{5 - y}{3y + 2} \quad \text{Interchange } x \text{ and } y.$$

$$x(3y + 2) = 5 - y \quad \text{Multiply each side by } 3y + 2.$$

$$3xy + 2x = 5 - y \quad \text{Distributive Property}$$

$$3xy + y = 5 - 2x \quad \text{Collect terms with } y.$$

$$y(3x + 1) = 5 - 2x \quad \text{Factor.}$$

$$y = \frac{5 - 2x}{3x + 1} \quad \text{Solve for } y.$$

$$f^{-1}(x) = \frac{5 - 2x}{3x + 1} \quad \text{Replace } y \text{ with } f^{-1}(x).$$

Check that $f(f^{-1}(x)) = x$ and $f^{-1}(f(x)) = x$.

✓ Checkpoint *Audio-video solution in English & Spanish at LarsonPrecalculus.com.*

Find the inverse function of

$$f(x) = \frac{5 - 3x}{x + 2}.$$

EXAMPLE 7 Finding an Inverse Function Algebraically

Find the inverse function of

$$f(x) = \sqrt{2x - 3}.$$

Solution The graph of f is a curve, as shown in the figure below. Because this graph passes the Horizontal Line Test, you know that f is one-to-one and has an inverse function.

$f(x) = \sqrt{2x - 3}$ Write original function.

$y = \sqrt{2x - 3}$ Replace $f(x)$ with y.

$x = \sqrt{2y - 3}$ Interchange x and y.

$x^2 = 2y - 3$ Square each side.

$2y = x^2 + 3$ Isolate y-term.

$y = \dfrac{x^2 + 3}{2}$ Solve for y.

$f^{-1}(x) = \dfrac{x^2 + 3}{2}, \quad x \geq 0$ Replace y with $f^{-1}(x)$.

The graph of f^{-1} in the figure is the reflection of the graph of f in the line $y = x$. Note that the range of f is the interval $[0, \infty)$, which implies that the domain of f^{-1} is the interval $[0, \infty)$. Moreover, the domain of f is the interval $\left[\frac{3}{2}, \infty\right)$, which implies that the range of f^{-1} is the interval $\left[\frac{3}{2}, \infty\right)$. Verify that $f(f^{-1}(x)) = x$ and $f^{-1}(f(x)) = x$.

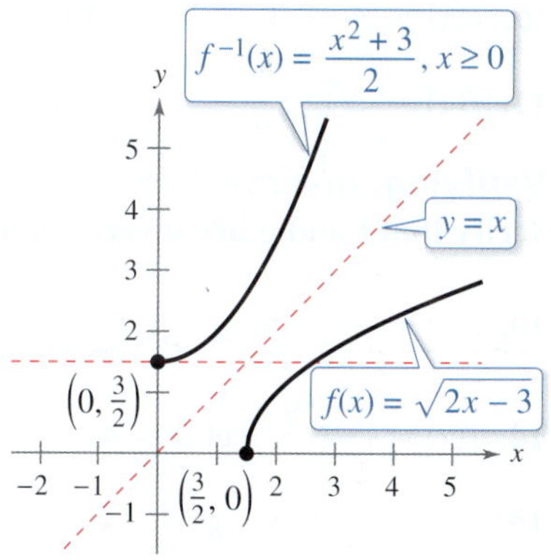

Checkpoint Audio-video solution in English & Spanish at LarsonPrecalculus.com.

Find the inverse function of

$$f(x) = \sqrt[3]{10 + x}.$$

Summarize (Section 1.9)

1. State the definition of an inverse function *(page 85)*. For examples of finding inverse functions informally and verifying inverse functions, see Examples 1 and 2.
2. Explain how to use the graph of a function to determine whether the function has an inverse function *(page 86)*. For examples of verifying inverse functions graphically, see Examples 3 and 4.
3. Explain how to use the Horizontal Line Test to determine whether a function is one-to-one *(page 87)*. For an example of applying the Horizontal Line Test, see Example 5.
4. Explain how to find an inverse function algebraically *(page 88)*. For examples of finding inverse functions algebraically, see Examples 6 and 7.

1.9 Exercises

See CalcChat.com for tutorial help and worked-out solutions to odd-numbered exercises.

Vocabulary: Fill in the blanks.

1. If the composite functions $f(g(x))$ and $g(f(x))$ both equal x, then the function g is the ________ function of f.
2. The inverse function of f is denoted by ________.
3. The domain of f is the ________ of f^{-1}, and the ________ of f^{-1} is the range of f.
4. The graphs of f and f^{-1} are reflections of each other in the line ________.
5. A function f is ________ when each value of the dependent variable corresponds to exactly one value of the independent variable.
6. A graphical test for the existence of an inverse function of f is called the _______ Line Test.

Skills and Applications

Finding an Inverse Function Informally In Exercises 7–12, find the inverse function of f informally. Verify that $f(f^{-1}(x)) = x$ and $f^{-1}(f(x)) = x$.

7. $f(x) = 6x$
8. $f(x) = \frac{1}{3}x$
9. $f(x) = 3x + 1$
10. $f(x) = \dfrac{x-1}{5}$
11. $f(x) = \sqrt[3]{x}$
12. $f(x) = x^5$

Verifying Inverse Functions In Exercises 13–16, verify that f and g are inverse functions.

13. $f(x) = -\dfrac{7}{2}x - 3, \quad g(x) = -\dfrac{2x+6}{7}$
14. $f(x) = \dfrac{x-9}{4}, \quad g(x) = 4x + 9$
15. $f(x) = x^3 + 5, \quad g(x) = \sqrt[3]{x-5}$
16. $f(x) = \dfrac{x^3}{2}, \quad g(x) = \sqrt[3]{2x}$

Sketching the Graph of an Inverse Function In Exercises 17–20, use the graph of the function to sketch the graph of its inverse function $y = f^{-1}(x)$.

17.
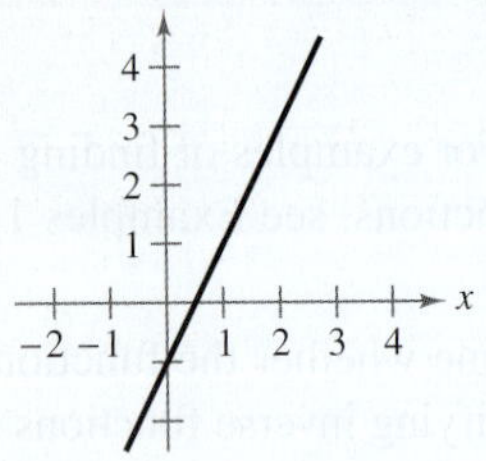

18.
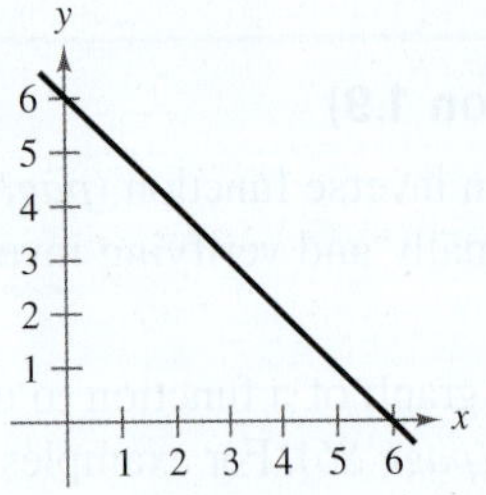

19.
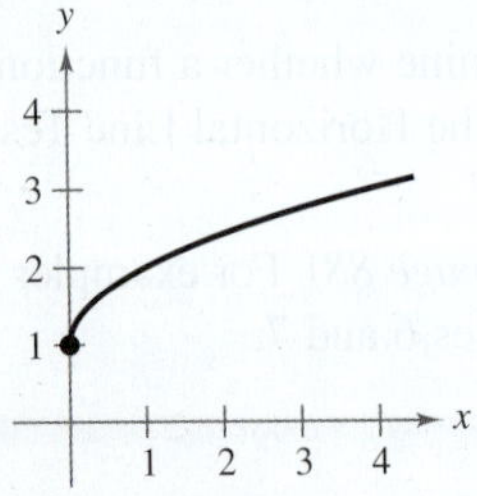

20.
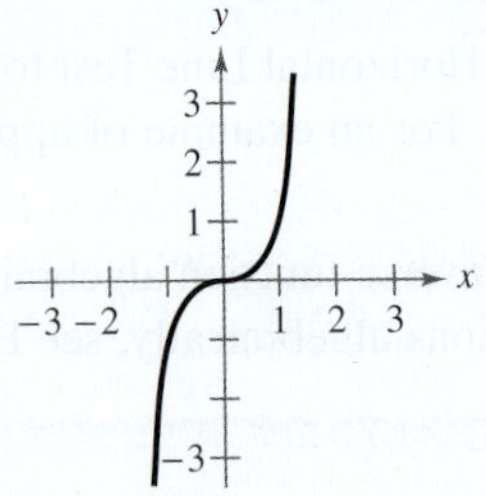

Verifying Inverse Functions In Exercises 21–32, verify that f and g are inverse functions (a) algebraically and (b) graphically.

21. $f(x) = 2x, \quad g(x) = \dfrac{x}{2}$
22. $f(x) = x - 5, \quad g(x) = x + 5$
23. $f(x) = 7x + 1, \quad g(x) = \dfrac{x-1}{7}$
24. $f(x) = 3 - 4x, \quad g(x) = \dfrac{3-x}{4}$
25. $f(x) = \dfrac{x^3}{8}, \quad g(x) = \sqrt[3]{8x}$
26. $f(x) = \dfrac{1}{x}, \quad g(x) = \dfrac{1}{x}$
27. $f(x) = \sqrt{x-4}, \quad g(x) = x^2 + 4, \quad x \geq 0$
28. $f(x) = 1 - x^3, \quad g(x) = \sqrt[3]{1-x}$
29. $f(x) = 9 - x^2, \quad x \geq 0, \quad g(x) = \sqrt{9-x}, \quad x \leq 9$
30. $f(x) = \dfrac{1}{1+x}, \quad x \geq 0, \quad g(x) = \dfrac{1-x}{x}, \quad 0 < x \leq 1$
31. $f(x) = \dfrac{x-1}{x+5}, \quad g(x) = -\dfrac{5x+1}{x-1}$
32. $f(x) = \dfrac{x+3}{x-2}, \quad g(x) = \dfrac{2x+3}{x-1}$

Using a Table to Determine an Inverse Function In Exercises 33 and 34, does the function have an inverse function?

33.

x	-1	0	1	2	3	4
$f(x)$	-2	1	2	1	-2	-6

34.

x	-3	-2	-1	0	2	3
$f(x)$	10	6	4	1	-3	-10

Using a Table to Find an Inverse Function In Exercises 35 and 36, use the table of values for $y = f(x)$ to complete a table for $y = f^{-1}(x)$.

35.

x	-2	-1	0	1	2	3
$f(x)$	-2	0	2	4	6	8

36.

x	-3	-2	-1	0	1	2
$f(x)$	-10	-7	-4	-1	2	5

Applying the Horizontal Line Test In Exercises 37–40, does the function have an inverse function?

37. 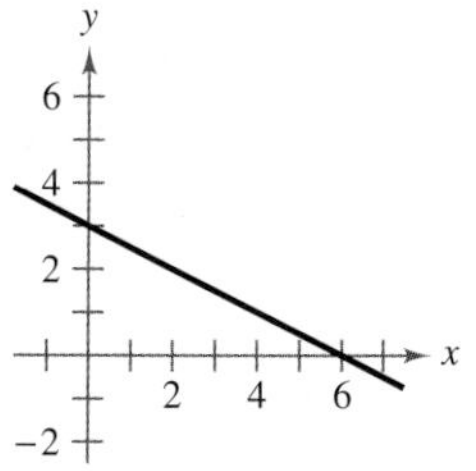

38. 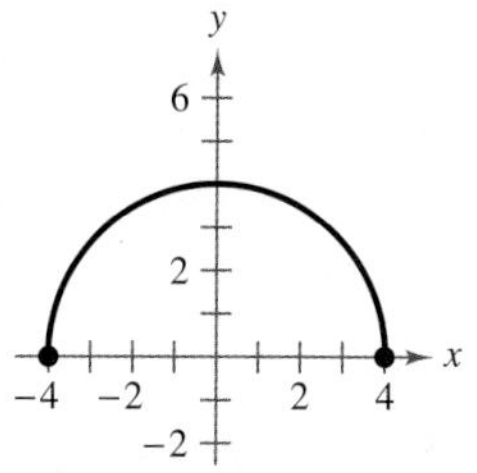

39. 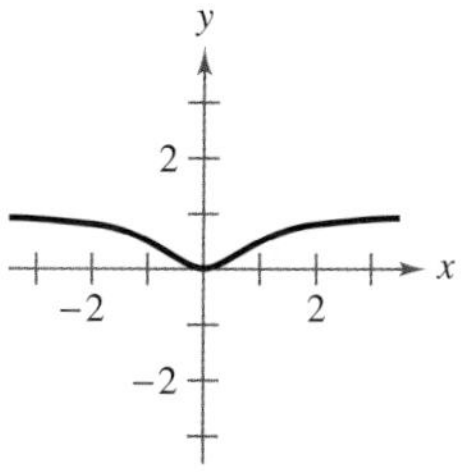

40. 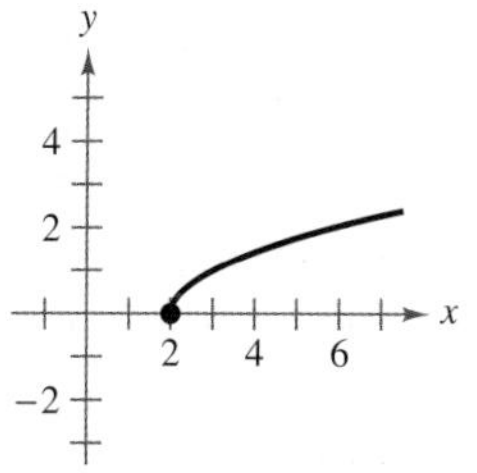

Applying the Horizontal Line Test In Exercises 41–44, use a graphing utility to graph the function, and use the Horizontal Line Test to determine whether the function has an inverse function.

41. $g(x) = (x + 5)^3$

42. $f(x) = \frac{1}{8}(x + 2)^2 - 1$

43. $f(x) = -2x\sqrt{16 - x^2}$

44. $h(x) = |x + 4| - |x - 4|$

Finding and Analyzing Inverse Functions In Exercises 45–56, (a) find the inverse function of f, (b) graph both f and f^{-1} on the same set of coordinate axes, (c) describe the relationship between the graphs of f and f^{-1}, and (d) state the domains and ranges of f and f^{-1}.

45. $f(x) = 2x - 3$

46. $f(x) = 3x + 1$

47. $f(x) = x^5 - 2$

48. $f(x) = x^3 + 1$

49. $f(x) = \sqrt{4 - x^2}, \quad 0 \le x \le 2$

50. $f(x) = x^2 - 2, \quad x \le 0$

51. $f(x) = \dfrac{4}{x}$

52. $f(x) = -\dfrac{2}{x}$

53. $f(x) = \dfrac{x + 1}{x - 2}$

54. $f(x) = \dfrac{x - 3}{x + 2}$

55. $f(x) = \sqrt[3]{x - 1}$

56. $f(x) = x^{3/5}$

Finding an Inverse Function In Exercises 57–72, determine whether the function has an inverse function. If it does, then find the inverse function.

57. $f(x) = x^4$

58. $f(x) = \dfrac{1}{x^2}$

59. $g(x) = \dfrac{x}{8}$

60. $f(x) = 3x + 5$

61. $p(x) = -4$

62. $f(x) = \dfrac{3x + 4}{5}$

63. $f(x) = (x + 3)^2, \quad x \ge -3$

64. $q(x) = (x - 5)^2$

65. $f(x) = \begin{cases} x + 3, & x < 0 \\ 6 - x, & x \ge 0 \end{cases}$

66. $f(x) = \begin{cases} -x, & x \le 0 \\ x^2 - 3x, & x > 0 \end{cases}$

67. $h(x) = -\dfrac{4}{x^2}$

68. $f(x) = |x - 2|, \quad x \le 2$

69. $f(x) = \sqrt{2x + 3}$

70. $f(x) = \sqrt{x - 2}$

71. $f(x) = \dfrac{6x + 4}{4x + 5}$

72. $f(x) = \dfrac{5x - 3}{2x + 5}$

Restricting the Domain In Exercises 73–82, restrict the domain of the function f so that the function is one-to-one and has an inverse function. Then find the inverse function f^{-1}. State the domains and ranges of f and f^{-1}. Explain your results. (There are many correct answers.)

73. $f(x) = (x - 2)^2$

74. $f(x) = 1 - x^4$

75. $f(x) = |x + 2|$

76. $f(x) = |x - 5|$

77. $f(x) = (x + 6)^2$

78. $f(x) = (x - 4)^2$

79. $f(x) = -2x^2 + 5$

80. $f(x) = \frac{1}{2}x^2 - 1$

81. $f(x) = |x - 4| + 1$

82. $f(x) = -|x - 1| - 2$

Composition with Inverses In Exercises 83–88, use the functions $f(x) = \frac{1}{8}x - 3$ and $g(x) = x^3$ to find the indicated value or function.

83. $(f^{-1} \circ g^{-1})(1)$

84. $(g^{-1} \circ f^{-1})(-3)$

85. $(f^{-1} \circ f^{-1})(6)$

86. $(g^{-1} \circ g^{-1})(-4)$

87. $(f \circ g)^{-1}$

88. $g^{-1} \circ f^{-1}$

Composition with Inverses In Exercises 89–92, use the functions $f(x) = x + 4$ and $g(x) = 2x - 5$ to find the specified function.

89. $g^{-1} \circ f^{-1}$

90. $f^{-1} \circ g^{-1}$

91. $(f \circ g)^{-1}$

92. $(g \circ f)^{-1}$

93. Hourly Wage Your wage is \$10.00 per hour plus \$0.75 for each unit produced per hour. So, your hourly wage y in terms of the number of units produced x is $y = 10 + 0.75x$.

(a) Find the inverse function. What does each variable represent in the inverse function?

(b) Determine the number of units produced when your hourly wage is \$24.25.

94. Diesel Mechanics

The function

$$y = 0.03x^2 + 245.50, \quad 0 < x < 100$$

approximates the exhaust temperature y in degrees Fahrenheit, where x is the percent load for a diesel engine.

(a) Find the inverse function. What does each variable represent in the inverse function?

(b) Use a graphing utility to graph the inverse function.

(c) The exhaust temperature of the engine must not exceed 500 degrees Fahrenheit. What is the percent load interval?

Exploration

True or False? **In Exercises 95 and 96, determine whether the statement is true or false. Justify your answer.**

95. If f is an even function, then f^{-1} exists.

96. If the inverse function of f exists and the graph of f has a y-intercept, then the y-intercept of f is an x-intercept of f^{-1}.

Graphical Analysis **In Exercises 97 and 98, use the graph of the function f to create a table of values for the given points. Then create a second table that can be used to find f^{-1}, and sketch the graph of f^{-1} if possible.**

97.

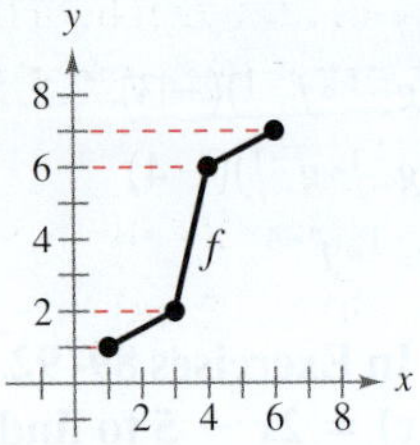

98.

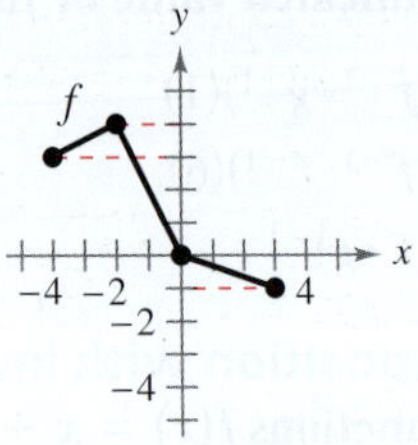

99. Proof Prove that if f and g are one-to-one functions, then $(f \circ g)^{-1}(x) = (g^{-1} \circ f^{-1})(x)$.

100. Proof Prove that if f is a one-to-one odd function, then f^{-1} is an odd function.

101. Think About It The function $f(x) = k(2 - x - x^3)$ has an inverse function, and $f^{-1}(3) = -2$. Find k.

102. Think About It Consider the functions $f(x) = x + 2$ and $f^{-1}(x) = x - 2$. Evaluate $f(f^{-1}(x))$ and $f^{-1}(f(x))$ for the indicated values of x. What can you conclude about the functions?

x	-10	0	7	45
$f(f^{-1}(x))$				
$f^{-1}(f(x))$				

103. Think About It Restrict the domain of $f(x) = x^2 + 1$ to $x \ge 0$. Use a graphing utility to graph the function. Does the restricted function have an inverse function? Explain.

104. HOW DO YOU SEE IT? The cost C for a business to make personalized T-shirts is given by

$$C(x) = 7.50x + 1500$$

where x represents the number of T-shirts.

(a) The graphs of C and C^{-1} are shown below. Match each function with its graph.

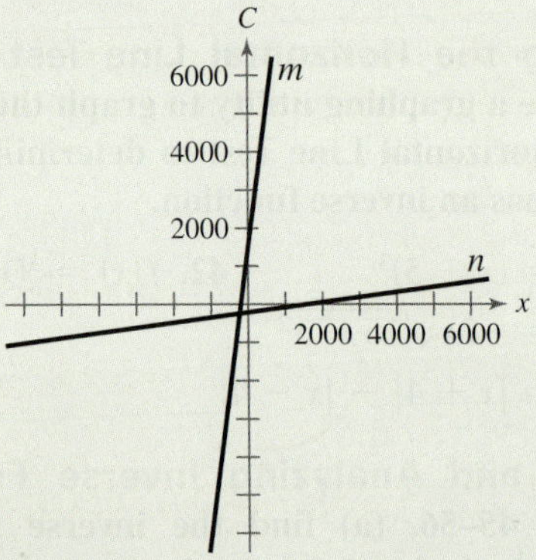

(b) Explain what $C(x)$ and $C^{-1}(x)$ represent in the context of the problem.

One-to-One Function Representation **In Exercises 105 and 106, determine whether the situation could be represented by a one-to-one function. If so, then write a statement that best describes the inverse function.**

105. The number of miles n a marathon runner has completed in terms of the time t in hours

106. The depth of the tide d at a beach in terms of the time t over a 24-hour period

Baloncici/Shutterstock.com

1.10 Mathematical Modeling and Variation

You can use functions as models to represent a wide variety of real-life data sets. For instance, in Exercise 75 on page 103, a variation model can be used to model the water temperatures of the ocean at various depths.

- **Use mathematical models to approximate sets of data points.**
- **Use the *regression* feature of a graphing utility to find the equation of a least squares regression line.**
- **Write mathematical models for direct variation, direct variation as an nth power, inverse variation, combined variation, and joint variation.**

Introduction

In this section, you will study the following techniques for fitting models to data: *least squares regression* and *direct and inverse variation.*

EXAMPLE 1 A Mathematical Model

The populations y (in millions) of the United States from 2003 through 2010 are shown in the table. *(Source: U.S. Census Bureau)*

Year, t	2003	2004	2005	2006	2007	2008	2009	2010
Population, y	290.1	292.8	295.5	298.4	301.2	304.1	306.8	308.7

Spreadsheet at LarsonPrecalculus.com

REMARK Note that the linear model in Example 1 was found using the *regression* feature of a graphing utility and is the line that *best* fits the data. This concept of a "best-fitting" line is discussed on the next page.

A linear model that approximates the data is

$$y = 2.72t + 282.0, \quad 3 \le t \le 10$$

where t represents the year, with $t = 3$ corresponding to 2003. Plot the actual data *and* the model on the same set of coordinate axes. How closely does the model represent the data?

Solution The actual data are plotted in Figure 1.64, along with the graph of the linear model. From the graph, it appears that the model is a "good fit" for the actual data. You can see how well the model fits by comparing the actual values of y with the values of y given by the model. The values given by the model are labeled y^* in the table below.

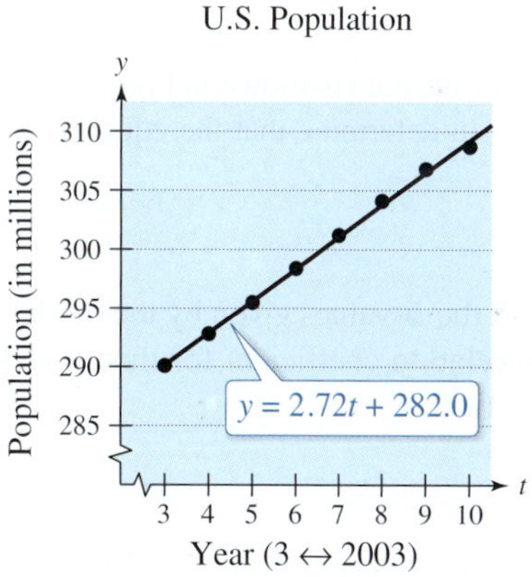

Figure 1.64

t	3	4	5	6	7	8	9	10
y	290.1	292.8	295.5	298.4	301.2	304.1	306.8	308.7
y^*	290.2	292.9	295.6	298.3	301.0	303.8	306.5	309.2

Checkpoint *Audio-video solution in English & Spanish at LarsonPrecalculus.com.*

The median sales prices y (in thousands of dollars) of new homes sold in a neighborhood from 2003 through 2010 are given by the following ordered pairs. *(Spreadsheet at LarsonPrecalculus.com)*

(2003, 179.4) (2005, 191.0) (2007, 202.6) (2009, 214.9)
(2004, 185.4) (2006, 196.7) (2008, 208.7) (2010, 221.4)

A linear model that approximates the data is

$$y = 5.96t + 161.3, \quad 3 \le t \le 10$$

where t represents the year, with $t = 3$ corresponding to 2003. Plot the actual data *and* the model on the same set of coordinate axes. How closely does the model represent the data?

Khoroshunova Olga/Shutterstock.com

Least Squares Regression and Graphing Utilities

So far in this text, you have worked with many different types of mathematical models that approximate real-life data. In some instances the model was given (as in Example 1), whereas in other instances you were asked to find the model using simple algebraic techniques or a graphing utility.

To find a model that approximates a set of data most accurately, statisticians use a measure called the **sum of square differences,** which is the sum of the squares of the differences between actual data values and model values. The "best-fitting" linear model, called the **least squares regression line,** is the one with the least sum of square differences.

Recall that you can approximate this line visually by plotting the data points and drawing the line that appears to best fit the data—or you can enter the data points into a calculator or computer and use the *linear regression* feature of the calculator or computer.

When you use the *regression* feature of a graphing calculator or computer program, you will notice that the program may also output an "r-value." This r-value is the **correlation coefficient** of the data and gives a measure of how well the model fits the data. The closer the value of $|r|$ is to 1, the better the fit.

EXAMPLE 2 Finding a Least Squares Regression Line

The data in the table show the numbers E (in millions) of Medicare enrollees from 2003 through 2010. Construct a scatter plot that represents the data and find the least squares regression line for the data. *(Source: U.S. Centers for Medicare and Medicaid Services)*

DATA

Spreadsheet at LarsonPrecalculus.com

Year	Medicare Enrollees, E
2003	41.2
2004	41.9
2005	42.6
2006	43.4
2007	44.4
2008	45.5
2009	46.6
2010	47.5

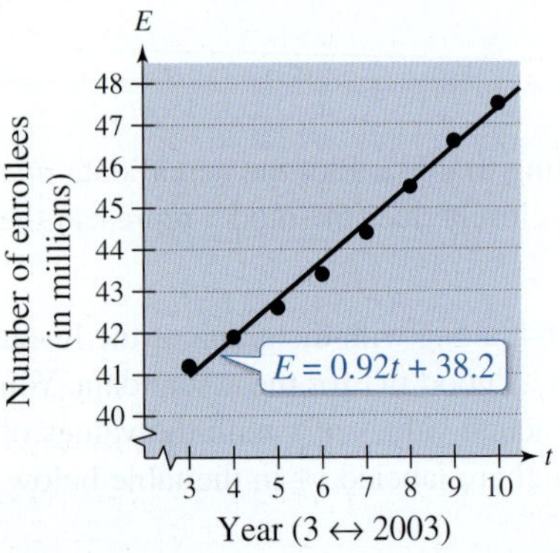

Figure 1.65

Solution Let $t = 3$ represent 2003. The scatter plot for the data is shown in Figure 1.65. Using the *regression* feature of a graphing utility, you can determine that the equation of the least squares regression line is

$$E = 0.92t + 38.2$$

To check this model, compare the actual E-values with the E-values given by the model, which are labeled E^* in the table at the left. The correlation coefficient for this model is $r \approx 0.996$, which implies that the model is a good fit.

t	E	E^*
3	41.2	41.0
4	41.9	41.9
5	42.6	42.8
6	43.4	43.7
7	44.4	44.6
8	45.5	45.6
9	46.6	46.5
10	47.5	47.4

Checkpoint *Audio-video solution in English & Spanish at LarsonPrecalculus.com.*

The total Medicare disbursements (in billions of dollars) are given by the following ordered pairs. *(Spreadsheet at LarsonPrecalculus.com)* Construct a scatter plot that represents the data and find the least squares regression line for the data. *(Source: U.S. Centers for Medicare and Medicaid Services)*

DATA

(2003, 277.8) (2005, 336.9) (2007, 434.8) (2009, 498.2)
(2004, 301.5) (2006, 380.4) (2008, 455.1) (2010, 521.1)

Direct Variation

There are two basic types of linear models. The more general model has a y-intercept that is nonzero.

$$y = mx + b, \quad b \neq 0$$

The simpler model

$$y = kx$$

has a y-intercept that is zero. In the simpler model, y is said to **vary directly** as x, or to be **directly proportional** to x.

> **Direct Variation**
>
> The following statements are equivalent.
>
> **1.** y **varies directly** as x.
>
> **2.** y is **directly proportional** to x.
>
> **3.** $y = kx$ for some nonzero constant k.
>
> k is the **constant of variation** or the **constant of proportionality.**

EXAMPLE 3 Direct Variation

In Pennsylvania, the state income tax is directly proportional to *gross income*. You are working in Pennsylvania and your state income tax deduction is \$46.05 for a gross monthly income of \$1500. Find a mathematical model that gives the Pennsylvania state income tax in terms of gross income.

Solution

Verbal Model: State income tax = k · Gross income

Labels:

State income tax $= y$ (dollars)
Gross income $= x$ (dollars)
Income tax rate $= k$ (percent in decimal form)

Equation: $y = kx$

Pennsylvania Taxes

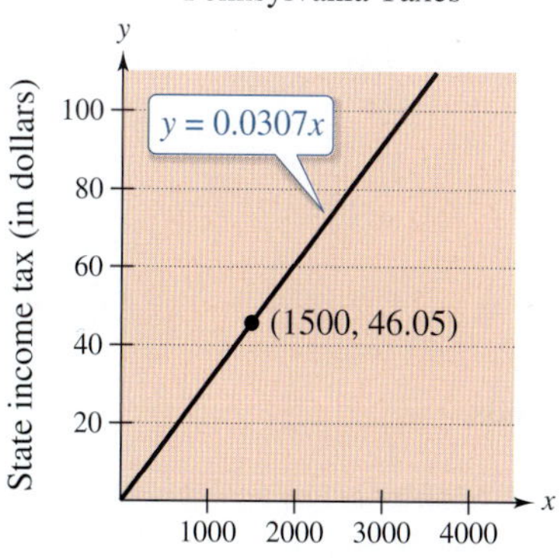

Figure 1.66

To solve for k, substitute the given information into the equation $y = kx$, and then solve for k.

$$y = kx \qquad \text{Write direct variation model.}$$

$$46.05 = k(1500) \qquad \text{Substitute 46.05 for } y \text{ and 1500 for } x.$$

$$0.0307 = k \qquad \text{Simplify.}$$

So, the equation (or model) for state income tax in Pennsylvania is

$$y = 0.0307x.$$

In other words, Pennsylvania has a state income tax rate of 3.07% of gross income. The graph of this equation is shown in Figure 1.66.

✓ ***Checkpoint*** *Audio-video solution in English & Spanish at LarsonPrecalculus.com.*

The simple interest on an investment is directly proportional to the amount of the investment. For instance, an investment of \$2500 will earn \$187.50 after 1 year. Find a mathematical model that gives the interest I after 1 year in terms of the amount invested P.

Direct Variation as an *n*th Power

Another type of direct variation relates one variable to a *power* of another variable. For example, in the formula for the area of a circle

$$A = \pi r^2$$

the area A is directly proportional to the square of the radius r. Note that for this formula, π is the constant of proportionality.

REMARK Note that the direct variation model $y = kx$ is a special case of $y = kx^n$ with $n = 1$.

Direct Variation as an *n*th Power

The following statements are equivalent.

1. y **varies directly as the *n*th power** of x.
2. y is **directly proportional to the *n*th power** of x.
3. $y = kx^n$ for some nonzero constant k.

EXAMPLE 4 Direct Variation as an *n*th Power

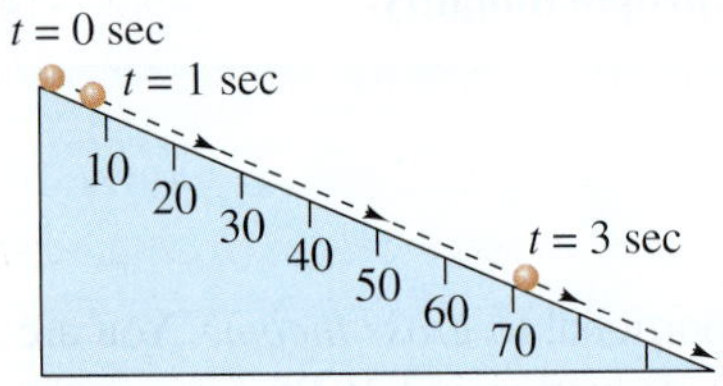

Figure 1.67

The distance a ball rolls down an inclined plane is directly proportional to the square of the time it rolls. During the first second, the ball rolls 8 feet. (See Figure 1.67.)

a. Write an equation relating the distance traveled to the time.

b. How far will the ball roll during the first 3 seconds?

Solution

a. Letting d be the distance (in feet) the ball rolls and letting t be the time (in seconds), you have

$$d = kt^2.$$

Now, because $d = 8$ when $t = 1$, you can see that $k = 8$, as follows.

$d = kt^2$ Write direct variation model.

$8 = k(1)^2$ Substitute 8 for d and 1 for t.

$8 = k$ Simplify.

So, the equation relating distance to time is

$$d = 8t^2.$$

b. When $t = 3$, the distance traveled is

$d = 8(3)^2$ Substitute 3 for t.

$= 8(9)$ Simplify.

$= 72$ feet. Simplify.

So, the ball will roll 72 feet during the first 3 seconds.

Checkpoint *Audio-video solution in English & Spanish at LarsonPrecalculus.com.*

Neglecting air resistance, the distance s an object falls varies directly as the square of the duration t of the fall. An object falls a distance of 144 feet in 3 seconds. How far will it fall in 6 seconds?

In Examples 3 and 4, the direct variations are such that an *increase* in one variable corresponds to an *increase* in the other variable. This is also true in the model $d = \frac{1}{5}F$, $F > 0$, where an increase in F results in an increase in d. You should not, however, assume that this always occurs with direct variation. For example, in the model $y = -3x$, an increase in x results in a *decrease* in y, and yet y is said to vary directly as x.

Inverse Variation

Inverse Variation

The following statements are equivalent.

1. y **varies inversely** as x.
2. y is **inversely proportional** to x.
3. $y = \dfrac{k}{x}$ for some nonzero constant k.

If x and y are related by an equation of the form $y = k/x^n$, then y varies inversely as the nth power of x (or y is inversely proportional to the nth power of x).

EXAMPLE 5 Inverse Variation

According to the National Association of Manufacturers, "manufacturing supports an estimated 17 million jobs in the U.S.—about one in six private sector jobs."

A company has found that the demand for one of its products varies inversely as the price of the product. When the price is \$6.25, the demand is 400 units. Approximate the demand when the price is \$5.75.

Solution

Let p be the price and let x be the demand. Because x varies inversely as p, you have

$$x = \frac{k}{p}.$$

Now, because $x = 400$ when $p = 6.25$, you have

$$x = \frac{k}{p} \qquad \text{Write inverse variation model.}$$

$$400 = \frac{k}{6.25} \qquad \text{Substitute 400 for } x \text{ and 6.25 for } p.$$

$$(400)(6.25) = k \qquad \text{Multiply each side by 6.25.}$$

$$2500 = k. \qquad \text{Simplify.}$$

So, the equation relating price and demand is

$$x = \frac{2500}{p}.$$

When $p = 5.75$, the demand is

$$x = \frac{2500}{p} \qquad \text{Write inverse variation model.}$$

$$= \frac{2500}{5.75} \qquad \text{Substitute 5.75 for } p.$$

$$\approx 435 \text{ units.} \qquad \text{Simplify.}$$

So, the demand for the product is 435 units when the price of the product is \$5.75.

✓ **Checkpoint** *Audio-video solution in English & Spanish at LarsonPrecalculus.com.*

The company in Example 5 has found that the demand for another of its products varies inversely as the price of the product. When the price is \$2.75, the demand is 600 units. Approximate the demand when the price is \$3.25.

Marcin Balcerzak/Shutterstock.com

Combined Variation

Some applications of variation involve problems with *both* direct and inverse variations in the same model. These types of models are said to have **combined variation.**

EXAMPLE 6 Combined Variation

A gas law states that the volume of an enclosed gas varies directly as the temperature *and* inversely as the pressure, as shown in Figure 1.68. The pressure of a gas is 0.75 kilogram per square centimeter when the temperature is 294 K and the volume is 8000 cubic centimeters.

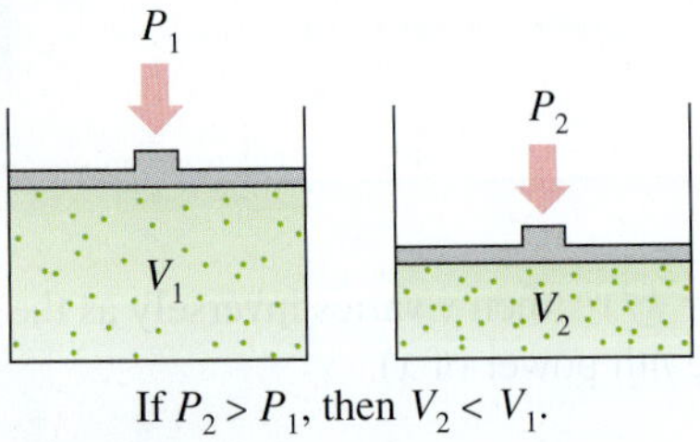

If the temperature is held constant and pressure increases, then the volume decreases.

Figure 1.68

a. Write an equation relating pressure, temperature, and volume.

b. Find the pressure when the temperature is 300 K and the volume is 7000 cubic centimeters.

Solution

a. Let V be volume (in cubic centimeters), let P be pressure (in kilograms per square centimeter), and let T be temperature (in Kelvin). Because V varies directly as T and inversely as P, you have

$$V = \frac{kT}{P}.$$

Now, because $P = 0.75$ when $T = 294$ and $V = 8000$, you have

$$V = \frac{kT}{P} \qquad \text{Write combined variation model.}$$

$$8000 = \frac{k(294)}{0.75} \qquad \text{Substitute 8000 for } V\text{, 294 for } T\text{, and 0.75 for } P.$$

$$\frac{6000}{294} = k \qquad \text{Simplify.}$$

$$\frac{1000}{49} = k. \qquad \text{Simplify.}$$

So, the equation relating pressure, temperature, and volume is

$$V = \frac{1000}{49}\left(\frac{T}{P}\right).$$

b. Isolate P on one side of the equation by multiplying each side by P and dividing each side by V to obtain $P = \frac{1000}{49}\left(\frac{T}{V}\right)$. When $T = 300$ and $V = 7000$, the pressure is

$$P = \frac{1000}{49}\left(\frac{T}{V}\right) \qquad \text{Combined variation model solved for } P.$$

$$= \frac{1000}{49}\left(\frac{300}{7000}\right) \qquad \text{Substitute 300 for } T \text{ and 7000 for } V.$$

$$= \frac{300}{343} \qquad \text{Simplify.}$$

$$\approx 0.87 \text{ kilogram per square centimeter.} \qquad \text{Simplify.}$$

So, the pressure is about 0.87 kilogram per square centimeter when the temperature is 300 K and the volume is 7000 cubic centimeters.

✓ **Checkpoint** *Audio-video solution in English & Spanish at LarsonPrecalculus.com.*

The resistance of a copper wire carrying an electrical current is directly proportional to its length and inversely proportional to its cross-sectional area. Given that #28 copper wire (which has a diameter of 0.0126 inch) has a resistance of 66.17 ohms per thousand feet, what length of #28 copper wire will produce a resistance of 33.5 ohms?

Joint Variation

> **Joint Variation**
>
> The following statements are equivalent.
>
> **1.** z **varies jointly** as x and y. **2.** z is **jointly proportional** to x and y.
>
> **3.** $z = kxy$ for some nonzero constant k.

If x, y, and z are related by an equation of the form $z = kx^n y^m$, then z varies jointly as the nth power of x and the mth power of y.

EXAMPLE 7 Joint Variation

The *simple* interest for a certain savings account is jointly proportional to the time and the principal. After one quarter (3 months), the interest on a principal of \$5000 is \$43.75. (a) Write an equation relating the interest, principal, and time. (b) Find the interest after three quarters.

Solution

a. Let I = interest (in dollars), P = principal (in dollars), and t = time (in years). Because I is jointly proportional to P and t, you have

$$I = kPt.$$

For $I = 43.75$, $P = 5000$, and $t = \frac{1}{4}$, you have

$$43.75 = k(5000)\left(\frac{1}{4}\right)$$

which implies that $k = 4(43.75)/5000 = 0.035$. So, the equation relating interest, principal, and time is

$$I = 0.035Pt$$

which is the familiar equation for simple interest where the constant of proportionality, 0.035, represents an annual interest rate of 3.5%.

b. When $P = \$5000$ and $t = \frac{3}{4}$, the interest is

$$I = (0.035)(5000)\left(\frac{3}{4}\right) = \$131.25.$$

✓ ***Checkpoint*** *Audio-video solution in English & Spanish at LarsonPrecalculus.com.*

The kinetic energy E of an object varies jointly with the object's mass m and the square of the object's velocity v. An object with a mass of 50 kilograms traveling at 16 meters per second has a kinetic energy of 6400 joules. What is the kinetic energy of an object with mass of 70 kilograms traveling at 20 meters per second?

> **Summarize (Section 1.10)**
>
> **1.** Give an example of how you can use a mathematical model to approximate a set of data points *(page 93, Example 1).*
>
> **2.** Describe how you can use the *regression* feature of a graphing utility to find the equation of a least squares regression line *(page 94, Example 2).*
>
> **3.** Describe direct variation, direct variation as an nth power, inverse variation, combined variation, and joint variation *(pages 95–99, Examples 3–7).*

1.10 Exercises

See CalcChat.com for tutorial help and worked-out solutions to odd-numbered exercises.

Vocabulary: Fill in the blanks.

1. Two techniques for fitting models to data are called direct and inverse ________ and least squares ________.
2. Statisticians use a measure called the ________ of ________ ________ to find a model that approximates a set of data most accurately.
3. The linear model with the least sum of square differences is called the ________ ________ ________ line.
4. An r-value of a set of data, also called a ________ ________, gives a measure of how well a model fits a set of data.
5. Direct variation models can be described as "y varies directly as x," or "y is ________ ________ to x."
6. In direct variation models of the form $y = kx$, k is called the ________ of ________.
7. The direct variation model $y = kx^n$ can be described as "y varies directly as the nth power of x," or "y is ________ ________ to the nth power of x."
8. The mathematical model $y = \frac{k}{x}$ is an example of ________ variation.
9. Mathematical models that involve both direct and inverse variation are said to have ________ variation.
10. The joint variation model $z = kxy$ can be described as "z varies jointly as x and y," or "z is ________ ________ to x and y."

Skills and Applications

11. **Labor Force** The total numbers of people 16 years of age and over (in thousands) not in the U.S. civilian labor force from 1998 through 2010 are given by the following ordered pairs. *(Spreadsheet at LarsonPrecalculus.com)*

DATA

(1998, 67,547) (2005, 76,762)
(1999, 68,385) (2006, 77,387)
(2000, 69,994) (2007, 78,743)
(2001, 71,359) (2008, 79,501)
(2002, 72,707) (2009, 81,659)
(2003, 74,658) (2010, 83,941)
(2004, 75,956)

A linear model that approximates the data is $y = 1298.8t + 57{,}094$, $8 \le t \le 20$, where y represents the total number of people 16 years of age and over (in thousands) not in the U.S. civilian labor force and $t = 8$ represents 1998. Plot the actual data and the model on the same set of coordinate axes. How closely does the model represent the data? *(Source: U.S. Bureau of Labor Statistics)*

12. **Sports** The winning times (in minutes) in the women's 400-meter freestyle swimming event in the Olympics from 1948 through 2008 are given by the following ordered pairs. *(Spreadsheet at LarsonPrecalculus.com)*

DATA

(1948, 5.30) (1972, 4.32) (1992, 4.12)
(1952, 5.20) (1976, 4.16) (1996, 4.12)
(1956, 4.91) (1980, 4.15) (2000, 4.10)
(1960, 4.84) (1984, 4.12) (2004, 4.09)
(1964, 4.72) (1988, 4.06) (2008, 4.05)
(1968, 4.53)

A linear model that approximates the data is $y = -0.020t + 5.00$, $-2 \le t \le 58$, where y represents the winning time (in minutes) and $t = -2$ represents 1948. Plot the actual data and the model on the same set of coordinate axes. How closely does the model represent the data? Does it appear that another type of model may be a better fit? Explain. *(Source: International Olympic Committee)*

Sketching a Line **In Exercises 13–16, sketch the line that you think best approximates the data in the scatter plot. Then find an equation of the line. To print an enlarged copy of the graph, go to *MathGraphs.com*.**

13.
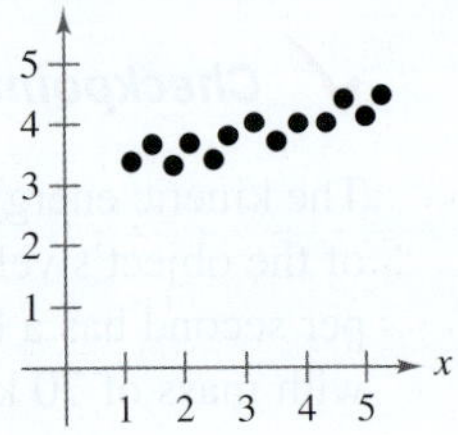

14.
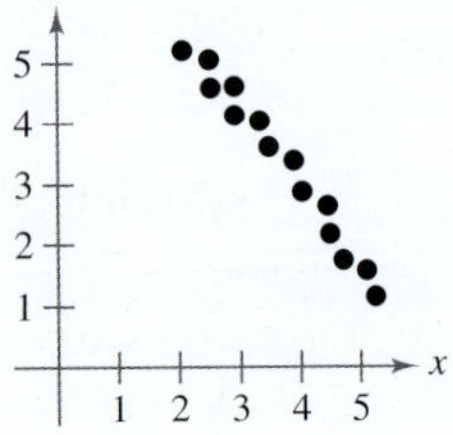

15.
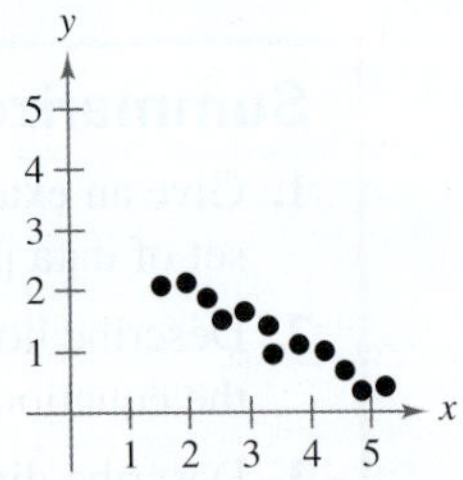

16.
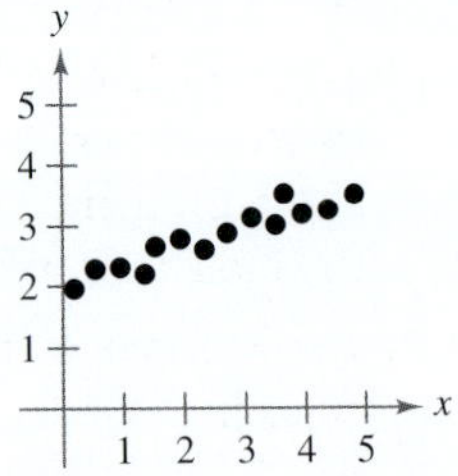

17. Sports The lengths (in feet) of the winning men's discus throws in the Olympics from 1920 through 2008 are given by the following ordered pairs. *(Spreadsheet at LarsonPrecalculus.com)* *(Source: International Olympic Committee)*

DATA

(1920, 146.6)	(1956, 184.9)	(1984, 218.5)
(1924, 151.3)	(1960, 194.2)	(1988, 225.8)
(1928, 155.3)	(1964, 200.1)	(1992, 213.7)
(1932, 162.3)	(1968, 212.5)	(1996, 227.7)
(1936, 165.6)	(1972, 211.3)	(2000, 227.3)
(1948, 173.2)	(1976, 221.5)	(2004, 229.3)
(1952, 180.5)	(1980, 218.7)	(2008, 225.8)

(a) Sketch a scatter plot of the data. Let y represent the length of the winning discus throw (in feet) and let $t = 20$ represent 1920.

(b) Use a straightedge to sketch the best-fitting line through the points and find an equation of the line.

(c) Use the *regression* feature of a graphing utility to find the least squares regression line that fits the data.

(d) Compare the linear model you found in part (b) with the linear model given by the graphing utility in part (c).

18. Data Analysis: Broadway Shows The annual gross ticket sales S (in millions of dollars) for Broadway shows in New York City from 1995 through 2011 are given by the following ordered pairs. *(Spreadsheet at LarsonPrecalculus.com)* *(Source: The Broadway League)*

DATA

(1995, 406)	(2004, 771)
(1996, 436)	(2005, 769)
(1997, 499)	(2006, 862)
(1998, 558)	(2007, 939)
(1999, 588)	(2008, 938)
(2000, 603)	(2009, 943)
(2001, 666)	(2010, 1020)
(2002, 643)	(2011, 1080)
(2003, 721)	

(a) Use a graphing utility to create a scatter plot of the data. Let $t = 5$ represent 1995.

(b) Use the *regression* feature of the graphing utility to find the equation of the least squares regression line that fits the data.

(c) Use the graphing utility to graph the scatter plot you created in part (a) and the model you found in part (b) in the same viewing window. How closely does the model represent the data?

(d) Use the model to predict the annual gross ticket sales in 2017.

(e) Interpret the meaning of the slope of the linear model in the context of the problem.

Direct Variation **In Exercises 19–26, assume that y is directly proportional to x. Use the given x-value and y-value to find a linear model that relates y and x.**

19. $x = 2, y = 14$
20. $x = 5, y = 12$
21. $x = 10, y = 2050$
22. $x = 6, y = 580$
23. $x = 5, y = 1$
24. $x = -24, y = 3$
25. $x = 4, y = 8\pi$
26. $x = \pi, y = -1$

Direct Variation as an *n*th Power **In Exercises 27–30, use the given values of k and n to complete the table for the direct variation model $y = kx^n$. Plot the points in a rectangular coordinate system.**

x	2	4	6	8	10
$y = kx^n$					

27. $k = 1, n = 2$
28. $k = 2, n = 2$
29. $k = \frac{1}{2}, n = 3$
30. $k = \frac{1}{4}, n = 3$

Inverse Variation as an *n*th Power **In Exercises 31–34, use the given values of k and n to complete the table for the inverse variation model $y = k/x^n$. Plot the points in a rectangular coordinate system.**

x	2	4	6	8	10
$y = k/x^n$					

31. $k = 2, n = 1$
32. $k = 5, n = 1$
33. $k = 10, n = 2$
34. $k = 20, n = 2$

Think About It **In Exercises 35 and 36, use the graph to determine whether y varies directly as some power of x or inversely as some power of x. Explain.**

35.

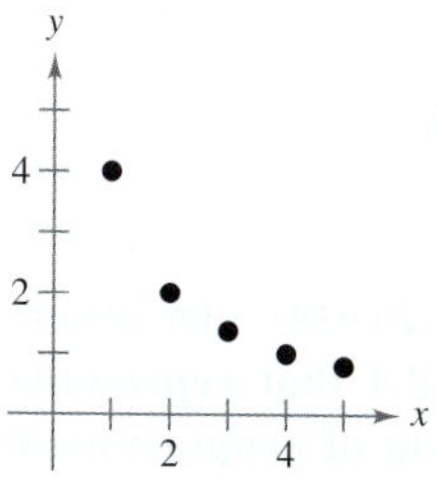

36.

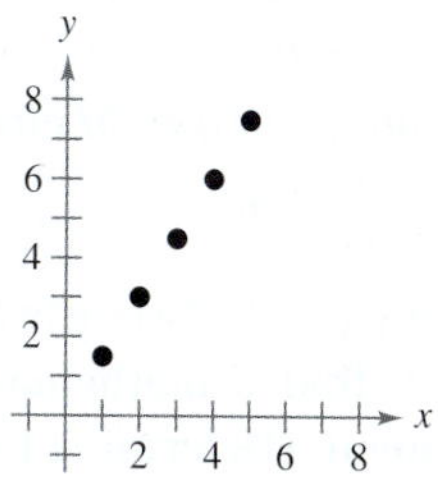

Determining Variation **In Exercises 37–40, determine whether the variation model represented by the ordered pairs (x, y) is of the form $y = kx$ or $y = k/x$, and find k. Then write a model that relates y and x.**

37. $(5, 1), \left(10, \frac{1}{2}\right), \left(15, \frac{1}{3}\right), \left(20, \frac{1}{4}\right), \left(25, \frac{1}{5}\right)$

38. $(5, 2), (10, 4), (15, 6), (20, 8), (25, 10)$

39. $(5, -3.5), (10, -7), (15, -10.5), (20, -14), (25, -17.5)$

40. $(5, 24), (10, 12), (15, 8), (20, 6), \left(25, \frac{24}{5}\right)$

Finding a Mathematical Model **In Exercises 41–50, find a mathematical model for the verbal statement.**

41. A varies directly as the square of r.
42. V varies directly as the cube of e.
43. y varies inversely as the square of x.
44. h varies inversely as the square root of s.
45. F varies directly as g and inversely as r^2.
46. z varies jointly as the square of x and the cube of y.
47. *Newton's Law of Cooling:* The rate of change R of the temperature of an object is directly proportional to the difference between the temperature T of the object and the temperature T_e of the environment in which the object is placed.
48. *Boyle's Law:* For a constant temperature, the pressure P of a gas is inversely proportional to the volume V of the gas.
49. *Logistic Growth:* The rate of growth R of a population is jointly proportional to the size S of the population and the difference between S and the maximum population size L that the environment can support.
50. *Newton's Law of Universal Gravitation:* The gravitational attraction F between two objects of masses m_1 and m_2 is jointly proportional to the masses and inversely proportional to the square of the distance r between the objects.

Describing a Formula **In Exercises 51–54, write a sentence using the variation terminology of this section to describe the formula.**

51. *Surface area of a sphere:*
$S = 4\pi r^2$
52. *Average speed:*
$r = d/t$
53. *Area of a triangle:*
$A = \frac{1}{2}bh$
54. *Volume of a right circular cylinder:*
$V = \pi r^2 h$

Finding a Mathematical Model **In Exercises 55–62, find a mathematical model that represents the statement. (Determine the constant of proportionality.)**

55. A varies directly as r^2. ($A = 9\pi$ when $r = 3$.)
56. y varies inversely as x. ($y = 3$ when $x = 25$.)
57. y is inversely proportional to x. ($y = 7$ when $x = 4$.)
58. z varies jointly as x and y. ($z = 64$ when $x = 4$ and $y = 8$.)
59. F is jointly proportional to r and the third power of s. ($F = 4158$ when $r = 11$ and $s = 3$.)
60. P varies directly as x and inversely as the square of y. $\left(P = \frac{28}{3}\text{ when } x = 42 \text{ and } y = 9.\right)$
61. z varies directly as the square of x and inversely as y. ($z = 6$ when $x = 6$ and $y = 4$.)
62. v varies jointly as p and q and inversely as the square of s. ($v = 1.5$ when $p = 4.1$, $q = 6.3$, and $s = 1.2$.)
63. **Simple Interest** The simple interest on an investment is directly proportional to the amount of the investment. An investment of \$3250 will earn \$113.75 after 1 year. Find a mathematical model that gives the interest I after 1 year in terms of the amount invested P.
64. **Simple Interest** The simple interest on an investment is directly proportional to the amount of the investment. An investment of \$6500 will earn \$211.25 after 1 year. Find a mathematical model that gives the interest I after 1 year in terms of the amount invested P.
65. **Measurement** Use the fact that 13 inches is approximately the same length as 33 centimeters to find a mathematical model that relates centimeters y to inches x. Then use the model to find the numbers of centimeters in 10 inches and 20 inches.
66. **Measurement** Use the fact that 14 gallons is approximately the same amount as 53 liters to find a mathematical model that relates liters y to gallons x. Then use the model to find the numbers of liters in 5 gallons and 25 gallons.

Hooke's Law **In Exercises 67–70, use Hooke's Law for springs, which states that the distance a spring is stretched (or compressed) varies directly as the force on the spring.**

67. A force of 220 newtons stretches a spring 0.12 meter. What force is required to stretch the spring 0.16 meter?
68. A force of 265 newtons stretches a spring 0.15 meter.
    (a) What force is required to stretch the spring 0.1 meter?
    (b) How far will a force of 90 newtons stretch the spring?
69. The coiled spring of a toy supports the weight of a child. The spring is compressed a distance of 1.9 inches by the weight of a 25-pound child. The toy will not work properly if its spring is compressed more than 3 inches. What is the maximum weight for which the toy will work properly?
70. An overhead garage door has two springs, one on each side of the door. A force of 15 pounds is required to stretch each spring 1 foot. Because of a pulley system, the springs stretch only one-half the distance the door travels. The door moves a total of 8 feet, and the springs are at their natural lengths when the door is open. Find the combined lifting force applied to the door by the springs when the door is closed.

71. Ecology The diameter of the largest particle that can be moved by a stream varies approximately directly as the square of the velocity of the stream. A stream with a velocity of $\frac{1}{4}$ mile per hour can move coarse sand particles about 0.02 inch in diameter. Approximate the velocity required to carry particles 0.12 inch in diameter.

72. Music The frequency of vibrations of a piano string varies directly as the square root of the tension on the string and inversely as the length of the string. The middle A string has a frequency of 440 vibrations per second. Find the frequency of a string that has 1.25 times as much tension and is 1.2 times as long.

73. Work The work W done when lifting an object varies jointly with the object's mass m and the height h that the object is lifted. The work done when a 120-kilogram object is lifted 1.8 meters is 2116.8 joules. How much work is done when lifting a 100-kilogram object 1.5 meters?

74. Beam Load The maximum load that can be safely supported by a horizontal beam varies jointly as the width of the beam and the square of its depth and inversely as the length of the beam. Determine the changes in the maximum safe load under the following conditions.

(a) The width and length of the beam are doubled.

(b) The width and depth of the beam are doubled.

75. Data Analysis: Ocean Temperatures An oceanographer took readings of the water temperatures C (in degrees Celsius) at several depths d (in meters). The data collected are shown as ordered pairs (d, C). *(Spreadsheet at LarsonPrecalculus.com)*

DATA

(1000, 4.2) (4000, 1.2)
(2000, 1.9) (5000, 0.9)
(3000, 1.4)

(a) Sketch a scatter plot of the data.

(b) Does it appear that the data can be modeled by the inverse variation model $C = k/d$? If so, find k for each pair of coordinates.

(c) Determine the mean value of k from part (b) to find the inverse variation model $C = k/d$.

(d) Use a graphing utility to plot the data points and the inverse model from part (c).

(e) Use the model to approximate the depth at which the water temperature is 3°C.

76. Data Analysis: Light Intensity A light probe is located x centimeters from a light source, and the intensity y (in microwatts per square centimeter) of the light is measured. The results are shown as ordered pairs (x, y). *(Spreadsheet at LarsonPrecalculus.com)*

DATA

(30, 0.1881) (34, 0.1543) (38, 0.1172)
(42, 0.0998) (46, 0.0775) (50, 0.0645)

A model for the data is $y = 262.76/x^{2.12}$.

(a) Use a graphing utility to plot the data points and the model in the same viewing window.

(b) Use the model to approximate the light intensity 25 centimeters from the light source.

Exploration

True or False? **In Exercises 77 and 78, decide whether the statement is true or false. Justify your answer.**

77. In the equation for the area of a circle, $A = \pi r^2$, the area A varies jointly with π and the square of the radius r.

78. If the correlation coefficient for a least squares regression line is close to -1, then the regression line cannot be used to describe the data.

79. Writing

(a) Given that y varies directly as the square of x and x is doubled, how will y change? Explain.

(b) Given that y varies inversely as the square of x and x is doubled, how will y change? Explain.

80. HOW DO YOU SEE IT? Discuss how well the data shown in each scatter plot can be approximated by a linear model.

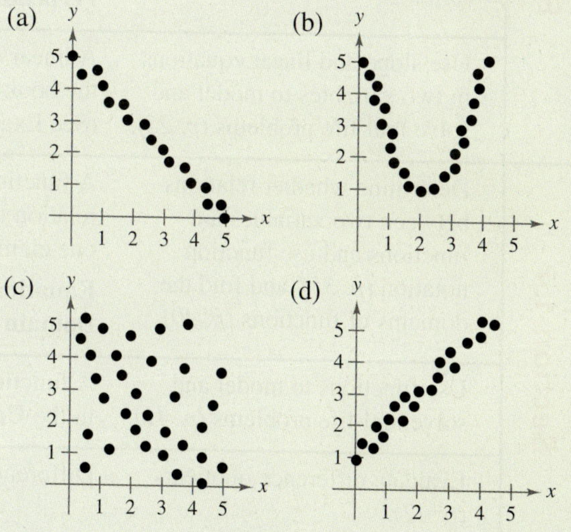

Project: Fraud and Identity Theft To work an extended application analyzing the numbers of fraud complaints and identity theft victims in the United States in 2010, visit this text's website at *LarsonPrecalculus.com*. *(Source: U.S. Federal Trade Commission)*

Khoroshunova Olga/Shutterstock.com

Chapter Summary

	What Did You Learn?	Explanation/Examples	Review Exercises
Section 1.1	Plot points in the Cartesian plane *(p. 2)*, use the Distance Formula *(p. 4)* and the Midpoint Formula *(p. 5)*, and use a coordinate plane to model and solve real-life problems *(p. 6)*.	For an ordered pair (x, y), the x-coordinate is the directed distance from the y-axis to the point, and the y-coordinate is the directed distance from the x-axis to the point. The coordinate plane can help you find the length of a football pass. (See Example 6.)	1–6
Section 1.2	Sketch graphs of equations *(p. 11)*, identify intercepts *(p. 14)*, and use symmetry to sketch graphs of equations *(p. 15)*.	To find x-intercepts, let y be zero and solve for x. To find y-intercepts, let x be zero and solve for y. Graphs can have symmetry with respect to one of the coordinate axes or with respect to the origin.	7–22
	Write equations of and sketch graphs of circles *(p. 17)*.	A point (x, y) lies on the circle of radius r and center (h, k) if and only if $(x - h)^2 + (y - k)^2 = r^2$.	23–27
	Use graphs of equations in solving real-life problems *(p. 18)*.	The graph of an equation can help you estimate the recommended weight for a man. (See Example 9.)	28
Section 1.3	Use slope to graph linear equations in two variables *(p. 22)*.	The graph of the equation $y = mx + b$ is a line whose slope is m and whose y-intercept is $(0, b)$.	29–32
	Find the slope of a line given two points on the line *(p. 24)*.	The slope m of the nonvertical line through (x_1, y_1) and (x_2, y_2) is $m = (y_2 - y_1)/(x_2 - x_1)$, where $x_1 \neq x_2$.	33, 34
	Write linear equations in two variables *(p. 26)*, and identify parallel and perpendicular lines *(p. 27)*.	The equation of the line with slope m passing through the point (x_1, y_1) is $y - y_1 = m(x - x_1)$. **Parallel lines:** $m_1 = m_2$ **Perpendicular lines:** $m_1 = -1/m_2$	35–40
	Use slope and linear equations in two variables to model and solve real-life problems *(p. 28)*.	A linear equation in two variables can help you describe the book value of exercise equipment in a given year. (See Example 7.)	41, 42
Section 1.4	Determine whether relations between two variables are functions and use function notation *(p. 35)*, and find the domains of functions *(p. 40)*.	A function f from a set A (domain) to a set B (range) is a relation that assigns to each element x in the set A exactly one element y in the set B. **Equation:** $f(x) = 5 - x^2$ **$f(2)$:** $f(2) = 5 - 2^2 = 1$ **Domain of $f(x) = 5 - x^2$:** All real numbers	43–50
	Use functions to model and solve real-life problems *(p. 41)*.	A function can model the number of alternative-fueled vehicles in the United States. (See Example 10.)	51, 52
	Evaluate difference quotients *(p. 42)*.	Difference quotient: $[f(x + h) - f(x)]/h, h \neq 0$	53, 54
Section 1.5	Use the Vertical Line Test for functions *(p. 50)*.	A set of points in a coordinate plane is the graph of y as a function of x if and only if no *vertical* line intersects the graph at more than one point.	55, 56
	Find the zeros of functions *(p. 51)*.	**Zeros of $f(x)$:** x-values for which $f(x) = 0$	57–60

	What Did You Learn?	Explanation/Examples	Review Exercises
Section 1.5	Determine intervals on which functions are increasing or decreasing *(p. 52)*, relative minimum and maximum values of functions *(p. 53)*, and the average rates of change of functions *(p. 54)*.	To determine whether a function is increasing, decreasing, or constant on an interval, evaluate the function for several values of x. The points at which the behavior of a function changes can help determine a relative minimum or relative maximum. The average rate of change between any two points is the slope of the line (secant line) through the two points.	61–66
	Identify even and odd functions *(p. 55)*.	**Even:** For each x in the domain of f, $f(-x) = f(x)$. **Odd:** For each x in the domain of f, $f(-x) = -f(x)$.	67, 68
Section 1.6	Identify and graph different types of functions *(p. 60)*, and recognize graphs of parent functions *(p. 64)*.	**Linear:** $f(x) = ax + b$; **Squaring:** $f(x) = x^2$; **Cubic:** $f(x) = x^3$; **Square Root:** $f(x) = \sqrt{x}$; **Reciprocal:** $f(x) = 1/x$ Eight of the most commonly used functions in algebra are shown on page 64.	69–74
Section 1.7	Use vertical and horizontal shifts *(p. 67)*, reflections *(p. 69)*, and nonrigid transformations *(p. 71)* to sketch graphs of functions.	**Vertical shifts:** $h(x) = f(x) + c$ or $h(x) = f(x) - c$ **Horizontal shifts:** $h(x) = f(x - c)$ or $h(x) = f(x + c)$ **Reflection in x-axis:** $h(x) = -f(x)$ **Reflection in y-axis:** $h(x) = f(-x)$ **Nonrigid transformations:** $h(x) = cf(x)$ or $h(x) = f(cx)$	75–84
Section 1.8	Add, subtract, multiply, and divide functions *(p. 76)*, find compositions of functions *(p. 78)*, and use combinations and compositions of functions to model and solve real-life problems *(p. 80)*.	$(f + g)(x) = f(x) + g(x)$ $(f - g)(x) = f(x) - g(x)$ $(fg)(x) = f(x) \cdot g(x)$ $(f/g)(x) = f(x)/g(x), g(x) \neq 0$ The composition of the function f with the function g is $(f \circ g)(x) = f(g(x))$. A composite function can represent the number of bacteria in food as a function of the amount of time the food has been out of refrigeration. (See Example 8.)	85–90
Section 1.9	Find inverse functions informally and verify that two functions are inverse functions of each other *(p. 84)*.	Let f and g be two functions such that $f(g(x)) = x$ for every x in the domain of g and $g(f(x)) = x$ for every x in the domain of f. Under these conditions, the function g is the inverse function of the function f.	91, 92
	Use graphs to find inverse functions *(p. 86)*, use the Horizontal Line Test *(p. 87)*, and find inverse functions algebraically *(p. 88)*.	If the point (a, b) lies on the graph of f, then the point (b, a) must lie on the graph of f^{-1}, and vice versa. In short, the graph of f^{-1} is a reflection of the graph of f in the line $y = x$. To find an inverse function, replace $f(x)$ with y, interchange the roles of x and y, solve for y, and then replace y with $f^{-1}(x)$.	93–98
Section 1.10	Use mathematical models to approximate sets of data points *(p. 93)*, and use the *regression* feature of a graphing utility to find the equation of a least squares regression line *(p. 94)*.	To see how well a model fits a set of data, compare the actual values and model values of y. (See Example 1.) The sum of square differences is the sum of the squares of the differences between actual data values and model values. The least squares regression line is the linear model with the least sum of square differences.	99
	Write mathematical models for direct variation, direct variation as an nth power, inverse variation, combined variation, and joint variation *(pp. 95–99)*.	**Direct variation:** $y = kx$ for some nonzero constant k. **Direct variation as an nth power:** $y = kx^n$ for some nonzero constant k. **Inverse variation:** $y = k/x$ for some nonzero constant k. **Joint variation:** $z = kxy$ for some nonzero constant k.	100, 101

Review Exercises

See CalcChat.com for tutorial help and worked-out solutions to odd-numbered exercises.

1.1 Plotting Points in the Cartesian Plane

In Exercises 1 and 2, plot the points in the Cartesian plane.

1. $(5, 5), (-2, 0), (-3, 6), (-1, -7)$
2. $(0, 6), (8, 1), (5, -4), (-3, -3)$

Determining Quadrant(s) for a Point **In Exercises 3 and 4, determine the quadrant(s) in which (x, y) is located so that the condition(s) is (are) satisfied.**

3. $x > 0$ and $y = -2$
4. $xy = 4$

5. **Plotting, Distance, and Midpoint** Plot the points $(-2, 6)$ and $(4, -3)$. Then find the distance between the points and the midpoint of the line segment joining the points.

6. **Sales** Barnes & Noble had annual sales of \$5.1 billion in 2008 and \$7.0 billion in 2010. Use the Midpoint Formula to estimate the sales in 2009. Assume that the annual sales followed a linear pattern. *(Source: Barnes & Noble, Inc.)*

1.2 Sketching the Graph of an Equation

In Exercises 7–10, construct a table of values. Use the resulting solution points to sketch the graph of the equation.

7. $y = 3x - 5$
8. $y = -\frac{1}{2}x + 2$
9. $y = x^2 - 3x$
10. $y = 2x^2 - x - 9$

Finding x- and y-Intercepts **In Exercises 11–14, find the x- and y-intercepts of the graph of the equation.**

11. $y = 2x + 7$
12. $y = |x + 1| - 3$
13. $y = (x - 3)^2 - 4$
14. $y = x\sqrt{4 - x^2}$

Intercepts, Symmetry, and Graphing **In Exercises 15–22, identify any intercepts and test for symmetry. Then sketch the graph of the equation.**

15. $y = -4x + 1$
16. $y = 5x - 6$
17. $y = 5 - x^2$
18. $y = x^2 - 10$
19. $y = x^3 + 3$
20. $y = -6 - x^3$
21. $y = \sqrt{x + 5}$
22. $y = |x| + 9$

Sketching the Graph of a Circle **In Exercises 23–26, find the center and radius of the circle. Then sketch the graph of the circle.**

23. $x^2 + y^2 = 9$
24. $x^2 + y^2 = 4$
25. $(x + 2)^2 + y^2 = 16$
26. $x^2 + (y - 8)^2 = 81$

27. **Writing the Equation of a Circle** Write the standard form of the equation of the circle for which the endpoints of a diameter are $(0, 0)$ and $(4, -6)$.

28. **Physics** The force F (in pounds) required to stretch a spring x inches from its natural length (see figure) is

$$F = \frac{5}{4}x, \quad 0 \le x \le 20.$$

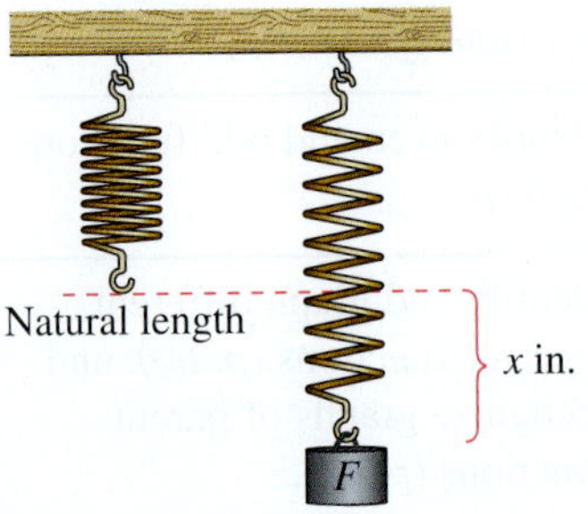

(a) Use the model to complete the table.

x	0	4	8	12	16	20
Force, F						

(b) Sketch a graph of the model.

(c) Use the graph to estimate the force necessary to stretch the spring 10 inches.

1.3 Graphing a Linear Equation

In Exercises 29–32, find the slope and y-intercept (if possible) of the equation of the line. Sketch the line.

29. $y = 3x + 13$
30. $y = -10x + 9$
31. $y = 6$
32. $x = -3$

Finding the Slope of a Line Through Two Points **In Exercises 33 and 34, plot the points and find the slope of the line passing through the pair of points.**

33. $(6, 4), (-3, -4)$
34. $(-3, 2), (8, 2)$

Finding an Equation of a Line **In Exercises 35 and 36, find an equation of the line that passes through the given point and has the indicated slope m. Sketch the line.**

35. $(10, -3), \quad m = -\frac{1}{2}$
36. $(-8, 5), \quad m = 0$

Finding an Equation of a Line **In Exercises 37 and 38, find an equation of the line passing through the points.**

37. $(-1, 0), (6, 2)$
38. $(11, -2), (6, -1)$

Finding Parallel and Perpendicular Lines **In Exercises 39 and 40, write the slope-intercept form of the equations of the lines through the given point (a) parallel to and (b) perpendicular to the given line.**

39. $5x - 4y = 8, \ (3, -2)$
40. $2x + 3y = 5, \ (-8, 3)$

41. **Sales** A discount outlet is offering a 20% discount on all items. Write a linear equation giving the sale price S for an item with a list price L.

42. **Hourly Wage** A microchip manufacturer pays its assembly line workers \$12.25 per hour. In addition, workers receive a piecework rate of \$0.75 per unit produced. Write a linear equation for the hourly wage W in terms of the number of units x produced per hour.

1.4 Testing for Functions Represented Algebraically **In Exercises 43–46, determine whether the equation represents y as a function of x.**

43. $16x - y^4 = 0$

44. $2x - y - 3 = 0$

45. $y = \sqrt{1 - x}$

46. $|y| = x + 2$

Evaluating a Function **In Exercises 47 and 48, evaluate the function at each specified value of the independent variable and simplify.**

47. $f(x) = x^2 + 1$

(a) $f(2)$ (b) $f(-4)$ (c) $f(t^2)$ (d) $f(t + 1)$

48. $h(x) = \begin{cases} 2x + 1, & x \le -1 \\ x^2 + 2, & x > -1 \end{cases}$

(a) $h(-2)$ (b) $h(-1)$ (c) $h(0)$ (d) $h(2)$

Finding the Domain of a Function **In Exercises 49 and 50, find the domain of the function. Verify your result with a graph.**

49. $f(x) = \sqrt{25 - x^2}$

50. $h(x) = \dfrac{x}{x^2 - x - 6}$

Physics **In Exercises 51 and 52, the velocity of a ball projected upward from ground level is given by $v(t) = -32t + 48$, where t is the time in seconds and v is the velocity in feet per second.**

51. Find the velocity when $t = 1$.

52. Find the time when the ball reaches its maximum height. [*Hint:* Find the time when $v(t) = 0$.]

Evaluating a Difference Quotient **In Exercises 53 and 54, find the difference quotient and simplify your answer.**

53. $f(x) = 2x^2 + 3x - 1, \quad \dfrac{f(x + h) - f(x)}{h}, \quad h \neq 0$

54. $f(x) = x^3 - 5x^2 + x, \quad \dfrac{f(x + h) - f(x)}{h}, \quad h \neq 0$

1.5 Vertical Line Test for Functions **In Exercises 55 and 56, use the Vertical Line Test to determine whether y is a function of x. To print an enlarged copy of the graph, go to *MathGraphs.com*.**

55. $y = (x - 3)^2$

56. $x = -|4 - y|$

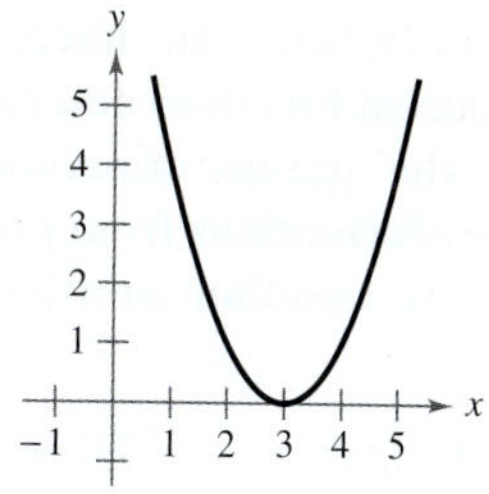

Figure for 55

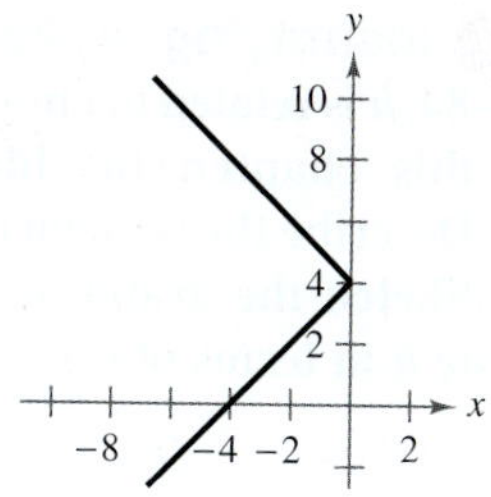

Figure for 56

Finding the Zeros of a Function **In Exercises 57–60, find the zeros of the function algebraically.**

57. $f(x) = 3x^2 - 16x + 21$

58. $f(x) = 5x^2 + 4x - 1$

59. $f(x) = \dfrac{8x + 3}{11 - x}$

60. $f(x) = x^3 - x^2$

Describing Function Behavior **In Exercises 61 and 62, use a graphing utility to graph the function and visually determine the intervals on which the function is increasing, decreasing, or constant.**

61. $f(x) = |x| + |x + 1|$

62. $f(x) = (x^2 - 4)^2$

Approximating Relative Minima or Maxima **In Exercises 63 and 64, use a graphing utility to graph the function and approximate (to two decimal places) any relative minima or maxima.**

63. $f(x) = -x^2 + 2x + 1$

64. $f(x) = x^3 - 4x^2 - 1$

Average Rate of Change of a Function **In Exercises 65 and 66, find the average rate of change of the function from x_1 to x_2.**

65. $f(x) = -x^2 + 8x - 4, \quad x_1 = 0, x_2 = 4$

66. $f(x) = 2 - \sqrt{x + 1}, \quad x_1 = 3, x_2 = 7$

Even, Odd, or Neither? **In Exercises 67–68, determine whether the function is even, odd, or neither. Then describe the symmetry.**

67. $f(x) = x^4 - 20x^2$

68. $f(x) = 2x\sqrt{x^2 + 3}$

1.6 Writing a Linear Function **In Exercises 69 and 70, (a) write the linear function f such that it has the indicated function values, and (b) sketch the graph of the function.**

69. $f(2) = -6, \quad f(-1) = 3$

70. $f(0) = -5, \quad f(4) = -8$

Graphing a Function **In Exercises 71–74, sketch the graph of the function.**

71. $f(x) = 3 - x^2$

72. $f(x) = \sqrt{x + 1}$

73. $g(x) = \dfrac{1}{x + 5}$

74. $f(x) = \begin{cases} 5x - 3, & x \ge -1 \\ -4x + 5, & x < -1 \end{cases}$

1.7 Identifying a Parent Function **In Exercises 75–84, h is related to one of the parent functions described in this chapter. (a) Identify the parent function f. (b) Describe the sequence of transformations from f to h. (c) Sketch the graph of h. (d) Use function notation to write h in terms of f.**

75. $h(x) = x^2 - 9$

76. $h(x) = (x - 2)^3 + 2$

77. $h(x) = -\sqrt{x} + 4$

78. $h(x) = |x + 3| - 5$

79. $h(x) = -(x + 2)^2 + 3$

80. $h(x) = \frac{1}{2}(x - 1)^2 - 2$

81. $h(x) = -[\![x]\!] + 6$

82. $h(x) = -\sqrt{x + 1} + 9$

83. $h(x) = 5[\![x - 9]\!]$

84. $h(x) = -\frac{1}{3}x^3$

1.8 Finding Arithmetic Combinations of Functions **In Exercises 85 and 86, find (a) $(f + g)(x)$, (b) $(f - g)(x)$, (c) $(fg)(x)$ and (d) $(f/g)(x)$. What is the domain of f/g?**

85. $f(x) = x^2 + 3, \quad g(x) = 2x - 1$

86. $f(x) = x^2 - 4, \quad g(x) = \sqrt{3 - x}$

Finding Domains of Functions and Composite Functions **In Exercises 87 and 88, find (a) $f \circ g$ and (b) $g \circ f$. Find the domain of each function and each composite function.**

87. $f(x) = \frac{1}{3}x - 3, \quad g(x) = 3x + 1$

88. $f(x) = x^3 - 4, \quad g(x) = \sqrt[3]{x + 7}$

Bacteria Count **In Exercises 89 and 90, the number N of bacteria in a refrigerated food is given by $N(T) = 25T^2 - 50T + 300$, $1 \le T \le 19$, where T is the temperature of the food in degrees Celsius. When the food is removed from refrigeration, the temperature of the food is given by $T(t) = 2t + 1$, $0 \le t \le 9$, where t is the time in hours.**

89. Find the composition $(N \circ T)(t)$ and interpret its meaning in context.

90. Find the time when the bacteria count reaches 750.

1.9 Finding an Inverse Function Informally **In Exercises 91 and 92, find the inverse function of f informally. Verify that $f(f^{-1}(x)) = x$ and $f^{-1}(f(x)) = x$.**

91. $f(x) = 3x + 8$

92. $f(x) = \dfrac{x - 4}{5}$

Applying the Horizontal Line Test **In Exercises 93 and 94, use a graphing utility to graph the function, and use the Horizontal Line Test to determine whether the function has an inverse function.**

93. $f(x) = (x - 1)^2$

94. $h(t) = \dfrac{2}{t - 3}$

Finding and Analyzing Inverse Functions **In Exercises 95 and 96, (a) find the inverse function of f, (b) graph both f and f^{-1} on the same set of coordinate axes, (c) describe the relationship between the graphs of f and f^{-1}, and (d) state the domains and ranges of f and f^{-1}.**

95. $f(x) = \frac{1}{2}x - 3$

96. $f(x) = \sqrt{x + 1}$

Restricting the Domain **In Exercises 97 and 98, restrict the domain of the function f to an interval on which the function is increasing, and determine f^{-1} on that interval.**

97. $f(x) = 2(x - 4)^2$

98. $f(x) = |x - 2|$

1.10

99. Compact Discs The values V (in billions of dollars) of shipments of compact discs in the United States from 2004 through 2010 are given by the following ordered pairs. *(Spreadsheet at LarsonPrecalculus.com) (Source: Recording Industry Association of America)*

DATA

(2004, 11.45) (2005, 10.52) (2006, 9.37)
(2007, 7.45) (2008, 5.47) (2009, 4.27)
(2010, 3.36)

(a) Use a graphing utility to create a scatter plot of the data. Let t represent the year, with $t = 4$ corresponding to 2004.

(b) Use the *regression* feature of the graphing utility to find the equation of the least squares regression line that fits the data. Then graph the model and the scatter plot you found in part (a) in the same viewing window. How closely does the model represent the data?

100. Travel Time The travel time between two cities is inversely proportional to the average speed. A train travels between the cities in 3 hours at an average speed of 65 miles per hour. How long would it take to travel between the cities at an average speed of 80 miles per hour?

101. Cost The cost of constructing a wooden box with a square base varies jointly as the height of the box and the square of the width of the box. A box of height 16 inches and width 6 inches costs \$28.80. How much would a box of height 14 inches and width 8 inches cost?

Exploration

True or False? **In Exercises 102 and 103, determine whether the statement is true or false. Justify your answer.**

102. Relative to the graph of $f(x) = \sqrt{x}$, the function $h(x) = -\sqrt{x + 9} - 13$ is shifted 9 units to the left and 13 units down, then reflected in the x-axis.

103. If f and g are two inverse functions, then the domain of g is equal to the range of f.

Chapter Test

See CalcChat.com for tutorial help and worked-out solutions to odd-numbered exercises.

Take this test as you would take a test in class. When you are finished, check your work against the answers given in the back of the book.

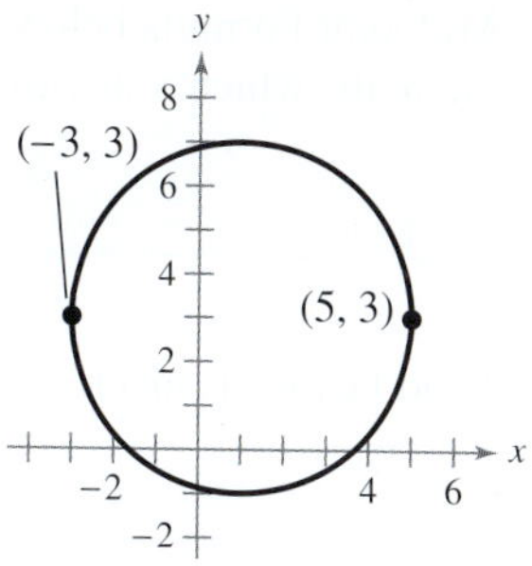

Figure for 6

1. Plot the points $(-2, 5)$ and $(6, 0)$. Then find the distance between the points and the midpoint of the line segment joining the points.
2. A cylindrical can has a radius of 4 centimeters. Write the volume V of the can as a function of the height h.

In Exercises 3–5, use intercepts and symmetry to sketch the graph of the equation.

3. $y = 3 - 5x$ 4. $y = 4 - |x|$ 5. $y = x^2 - 1$

6. Write the standard form of the equation of the circle shown at the left.

In Exercises 7 and 8, find an equation of the line passing through the points.

7. $(2, -3), (-4, 9)$ 8. $(3, 0.8), (7, -6)$

9. Write equations of the lines that pass through the point $(0, 4)$ and are (a) parallel to and (b) perpendicular to the line $5x + 2y = 3$.

10. Evaluate $f(x) = \dfrac{\sqrt{x+9}}{x^2 - 81}$ at each value and simplify: (a) $f(7)$ (b) $f(-5)$ (c) $f(x - 9)$.

11. Find the domain of $f(x) = 10 - \sqrt{3 - x}$.

In Exercises 12–14, (a) find the zeros of the function, (b) use a graphing utility to graph the function, (c) approximate the intervals on which the function is increasing, decreasing, or constant, and (d) determine whether the function is even, odd, or neither.

12. $f(x) = 2x^6 + 5x^4 - x^2$ 13. $f(x) = 4x\sqrt{3 - x}$ 14. $f(x) = |x + 5|$

15. Sketch the graph of $f(x) = \begin{cases} 3x + 7, & x \le -3 \\ 4x^2 - 1, & x > -3 \end{cases}$.

In Exercises 16–18, identify the parent function in the transformation. Then sketch a graph of the function.

16. $h(x) = -[\![x]\!]$ 17. $h(x) = -\sqrt{x + 5} + 8$ 18. $h(x) = -2(x - 5)^3 + 3$

In Exercises 19 and 20, find (a) $(f + g)(x)$, (b) $(f - g)(x)$, (c) $(fg)(x)$, (d) $(f/g)(x)$, (e) $(f \circ g)(x)$, and (f) $(g \circ f)(x)$.

19. $f(x) = 3x^2 - 7$, $g(x) = -x^2 - 4x + 5$ 20. $f(x) = 1/x$, $g(x) = 2\sqrt{x}$

In Exercises 21–23, determine whether the function has an inverse function. If so, find the inverse function.

21. $f(x) = x^3 + 8$ 22. $f(x) = |x^2 - 3| + 6$ 23. $f(x) = 3x\sqrt{x}$

In Exercises 24–26, find a mathematical model that represents the statement. (Determine the constant of proportionality.)

24. v varies directly as the square root of s. ($v = 24$ when $s = 16$.)
25. A varies jointly as x and y. ($A = 500$ when $x = 15$ and $y = 8$.)
26. b varies inversely as a. ($b = 32$ when $a = 1.5$.)

Proofs in Mathematics

What does the word *proof* mean to you? In mathematics, the word *proof* means a valid argument. When you are proving a statement or theorem, you must use facts, definitions, and accepted properties in a logical order. You can also use previously proved theorems in your proof. For instance, the proof of the Midpoint Formula below uses the Distance Formula. There are several different proof methods, which you will see in later chapters.

The Midpoint Formula ***(p. 5)***

The midpoint of the line segment joining the points (x_1, y_1) and (x_2, y_2) is given by the Midpoint Formula

$$\text{Midpoint} = \left(\frac{x_1 + x_2}{2}, \frac{y_1 + y_2}{2}\right).$$

THE CARTESIAN PLANE

The Cartesian plane was named after the French mathematician René Descartes (1596–1650). While Descartes was lying in bed, he noticed a fly buzzing around on the square ceiling tiles. He discovered that he could describe the position of the fly by the ceiling tile upon which the fly landed. This led to the development of the Cartesian plane. Descartes felt that using a coordinate plane could facilitate descriptions of the positions of objects.

Proof

Using the figure, you must show that $d_1 = d_2$ and $d_1 + d_2 = d_3$.

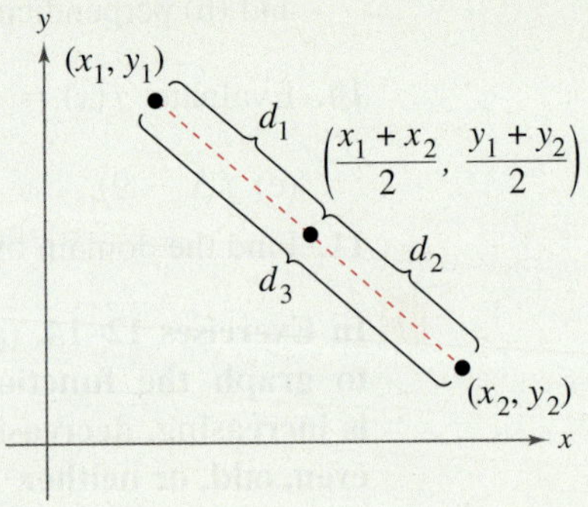

By the Distance Formula, you obtain

$$d_1 = \sqrt{\left(\frac{x_1 + x_2}{2} - x_1\right)^2 + \left(\frac{y_1 + y_2}{2} - y_1\right)^2}$$

$$= \frac{1}{2}\sqrt{(x_2 - x_1)^2 + (y_2 - y_1)^2}$$

$$d_2 = \sqrt{\left(x_2 - \frac{x_1 + x_2}{2}\right)^2 + \left(y_2 - \frac{y_1 + y_2}{2}\right)^2}$$

$$= \frac{1}{2}\sqrt{(x_2 - x_1)^2 + (y_2 - y_1)^2}$$

$$d_3 = \sqrt{(x_2 - x_1)^2 + (y_2 - y_1)^2}$$

So, it follows that $d_1 = d_2$ and $d_1 + d_2 = d_3$. ■

P.S. Problem Solving

1. **Monthly Wages** As a salesperson, you receive a monthly salary of \$2000, plus a commission of 7% of sales. You receive an offer for a new job at \$2300 per month, plus a commission of 5% of sales.
   (a) Write a linear equation for your current monthly wage W_1 in terms of your monthly sales S.
   (b) Write a linear equation for the monthly wage W_2 of your new job offer in terms of the monthly sales S.
   (c) Use a graphing utility to graph both equations in the same viewing window. Find the point of intersection. What does it signify?
   (d) You think you can sell \$20,000 per month. Should you change jobs? Explain.

2. **Telephone Keypad** For the numbers 2 through 9 on a telephone keypad (see figure), create two relations: one mapping numbers onto letters, and the other mapping letters onto numbers. Are both relations functions? Explain.

3. **Sums and Differences of Functions** What can be said about the sum and difference of each of the following?
   (a) Two even functions
   (b) Two odd functions
   (c) An odd function and an even function

4. **Inverse Functions** The two functions

   $f(x) = x$ and $g(x) = -x$

   are their own inverse functions. Graph each function and explain why this is true. Graph other linear functions that are their own inverse functions. Find a general formula for a family of linear functions that are their own inverse functions.

5. **Proof** Prove that a function of the following form is even.

$$y = a_{2n}x^{2n} + a_{2n-2}x^{2n-2} + \cdot \cdot \cdot + a_2x^2 + a_0$$

6. **Miniature Golf** A miniature golf professional is trying to make a hole-in-one on the miniature golf green shown. The golf ball is at the point (2.5, 2) and the hole is at the point (9.5, 2). The professional wants to bank the ball off the side wall of the green at the point (x, y). Find the coordinates of the point (x, y). Then write an equation for the path of the ball.

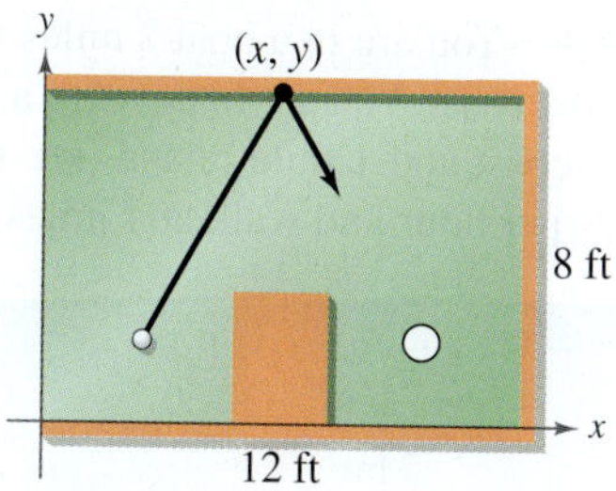

Figure for 6

7. **Titanic** At 2:00 P.M. on April 11, 1912, the *Titanic* left Cobh, Ireland, on her voyage to New York City. At 11:40 P.M. on April 14, the *Titanic* struck an iceberg and sank, having covered only about 2100 miles of the approximately 3400-mile trip.
   (a) What was the total duration of the voyage in hours?
   (b) What was the average speed in miles per hour?
   (c) Write a function relating the distance of the *Titanic* from New York City and the number of hours traveled. Find the domain and range of the function.
   (d) Graph the function from part (c).

8. **Average Rate of Change** Consider the function $f(x) = -x^2 + 4x - 3$. Find the average rate of change of the function from x_1 to x_2.
   (a) $x_1 = 1, x_2 = 2$
   (b) $x_1 = 1, x_2 = 1.5$
   (c) $x_1 = 1, x_2 = 1.25$
   (d) $x_1 = 1, x_2 = 1.125$
   (e) $x_1 = 1, x_2 = 1.0625$
   (f) Does the average rate of change seem to be approaching one value? If so, then state the value.
   (g) Find the equations of the secant lines through the points $(x_1, f(x_1))$ and $(x_2, f(x_2))$ for parts (a)–(e).
   (h) Find the equation of the line through the point $(1, f(1))$ using your answer from part (f) as the slope of the line.

9. **Inverse of a Composition** Consider the functions $f(x) = 4x$ and $g(x) = x + 6$.
   (a) Find $(f \circ g)(x)$.
   (b) Find $(f \circ g)^{-1}(x)$.
   (c) Find $f^{-1}(x)$ and $g^{-1}(x)$.
   (d) Find $(g^{-1} \circ f^{-1})(x)$ and compare the result with that of part (b).
   (e) Repeat parts (a) through (d) for $f(x) = x^3 + 1$ and $g(x) = 2x$.
   (f) Write two one-to-one functions f and g, and repeat parts (a) through (d) for these functions.
   (g) Make a conjecture about $(f \circ g)^{-1}(x)$ and $(g^{-1} \circ f^{-1})(x)$.

10. Trip Time You are in a boat 2 miles from the nearest point on the coast. You are to travel to a point Q, 3 miles down the coast and 1 mile inland (see figure). You row at 2 miles per hour and walk at 4 miles per hour.

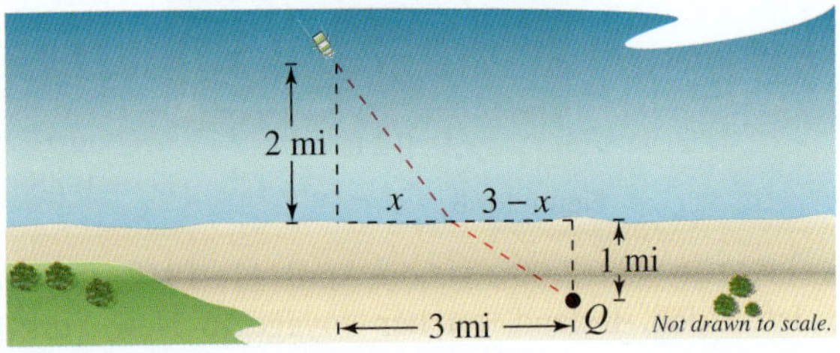

(a) Write the total time T of the trip as a function of x.

(b) Determine the domain of the function.

(c) Use a graphing utility to graph the function. Be sure to choose an appropriate viewing window.

(d) Find the value of x that minimizes T.

(e) Write a brief paragraph interpreting these values.

11. Heaviside Function The **Heaviside function** $H(x)$ is widely used in engineering applications. (See figure.) To print an enlarged copy of the graph, go to *MathGraphs.com*.

$$H(x) = \begin{cases} 1, & x \geq 0 \\ 0, & x < 0 \end{cases}$$

Sketch the graph of each function by hand.

(a) $H(x) - 2$

(b) $H(x - 2)$

(c) $-H(x)$

(d) $H(-x)$

(e) $\frac{1}{2}H(x)$

(f) $-H(x - 2) + 2$

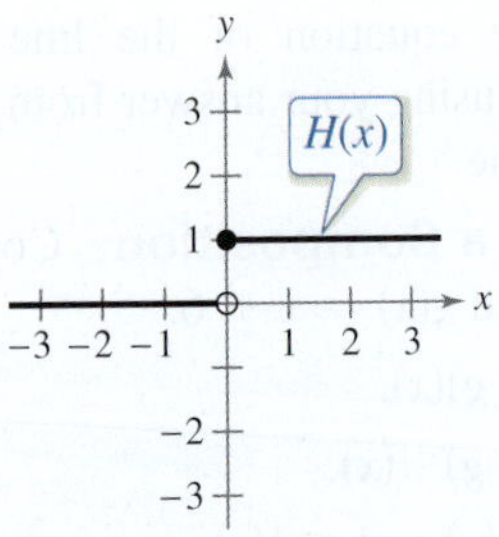

12. Repeated Composition Let $f(x) = \dfrac{1}{1 - x}$.

(a) What are the domain and range of f?

(b) Find $f(f(x))$. What is the domain of this function?

(c) Find $f(f(f(x)))$. Is the graph a line? Why or why not?

13. Associative Property with Compositions Show that the Associative Property holds for compositions of functions—that is,

$$(f \circ (g \circ h))(x) = ((f \circ g) \circ h)(x).$$

14. Graphical Analysis Consider the graph of the function f shown in the figure. Use this graph to sketch the graph of each function. To print an enlarged copy of the graph, go to *MathGraphs.com*.

(a) $f(x + 1)$

(b) $f(x) + 1$

(c) $2f(x)$

(d) $f(-x)$

(e) $-f(x)$

(f) $|f(x)|$

(g) $f(|x|)$

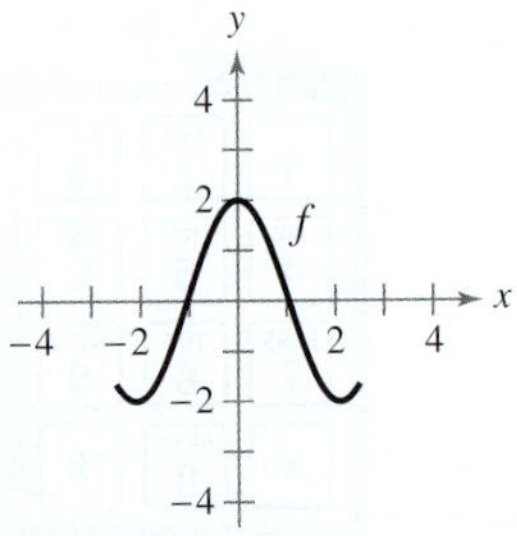

15. Graphical Analysis Use the graphs of f and f^{-1} to complete each table of function values.

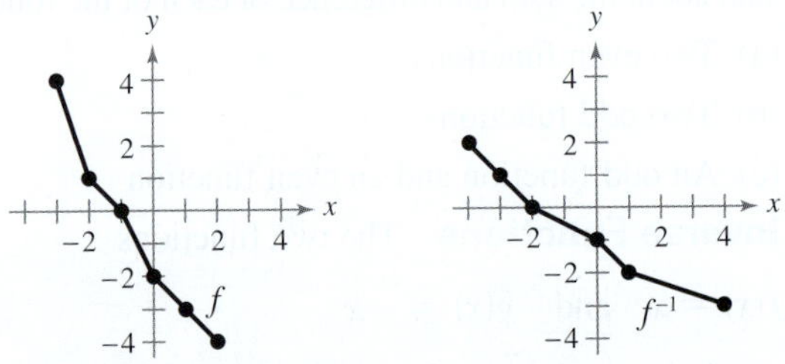

(a)

x	-4	-2	0	4
$(f(f^{-1}(x)))$				

(b)

x	-3	-2	0	1
$(f + f^{-1})(x)$				

(c)

x	-3	-2	0	1
$(f \cdot f^{-1})(x)$				

(d)

x	-4	-3	0	4
$\lvert f^{-1}(x)\rvert$				

2 Polynomial and Rational Functions

Candle Making (*Example 11, page 163*)

Impedance (*Exercise 91, page 153*)

Lyme Disease (*Exercise 86, page 146*)

Tree Growth (*Exercise 104, page 136*)

Path of a Diver (*Exercise 75, page 121*)

Clockwise from top left, hfng/Shutterstock.com; © Richard Megna/Fundamental Photographs; S.Borisov/Shutterstock.com; wellphoto/Shutterstock.com; Dariusz Majgier/Shutterstock.com

2.1 Quadratic Functions and Models

You can use quadratic functions to model the path of an object or person. For instance, in Exercise 75 on page 121, you will use a quadratic function to model the path of a diver.

- **Analyze graphs of quadratic functions.**
- **Write quadratic functions in standard form and use the results to sketch graphs of functions.**
- **Find minimum and maximum values of quadratic functions in real-life applications.**

The Graph of a Quadratic Function

In this and the next section, you will study the graphs of polynomial functions. In Section 1.6, you were introduced to the following basic functions.

$f(x) = ax + b$ Linear function

$f(x) = c$ Constant function

$f(x) = x^2$ Squaring function

These functions are examples of **polynomial functions.**

> **Definition of Polynomial Function**
>
> Let n be a nonnegative integer and let $a_n, a_{n-1}, \ldots, a_2, a_1, a_0$ be real numbers with $a_n \neq 0$. The function
>
> $$f(x) = a_n x^n + a_{n-1}x^{n-1} + \cdots + a_2x^2 + a_1x + a_0$$
>
> is called a **polynomial function of x with degree n.**

Polynomial functions are classified by degree. For instance, a constant function $f(x) = c$ with $c \neq 0$ has degree 0, and a linear function $f(x) = ax + b$ with $a \neq 0$ has degree 1. In this section, you will study second-degree polynomial functions, which are called **quadratic functions.**

For instance, each of the following functions is a quadratic function.

$$f(x) = x^2 + 6x + 2$$

$$g(x) = 2(x + 1)^2 - 3$$

$$h(x) = 9 + \tfrac{1}{4}x^2$$

$$k(x) = -3x^2 + 4$$

$$m(x) = (x - 2)(x + 1)$$

Note that the squaring function is a simple quadratic function that has degree 2.

> **Definition of Quadratic Function**
>
> Let a, b, and c be real numbers with $a \neq 0$. The function
>
> $$f(x) = ax^2 + bx + c \qquad \text{Quadratic function}$$
>
> is called a **quadratic function.**

The graph of a quadratic function is a special type of "U"-shaped curve called a **parabola.** Parabolas occur in many real-life applications—especially those involving reflective properties of satellite dishes and flashlight reflectors. You will study these properties in Section 10.2.

wellphoto/Shutterstock.com

All parabolas are symmetric with respect to a line called the **axis of symmetry,** or simply the **axis** of the parabola. The point where the axis intersects the parabola is the **vertex** of the parabola, as shown below. When the leading coefficient is positive, the graph of

$$f(x) = ax^2 + bx + c$$

is a parabola that opens upward. When the leading coefficient is negative, the graph of

$$f(x) = ax^2 + bx + c$$

is a parabola that opens downward.

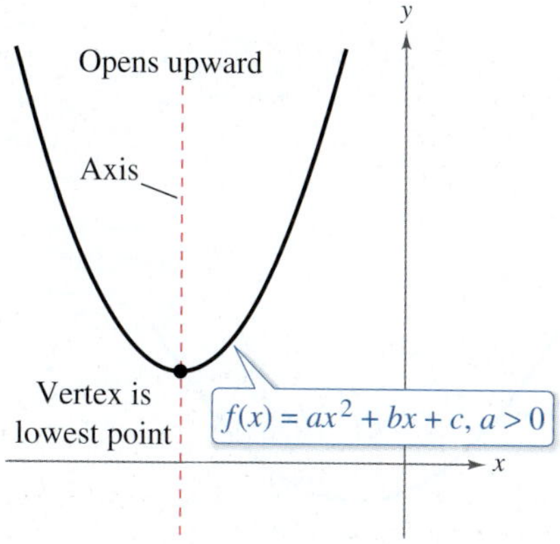

Leading coefficient is positive.

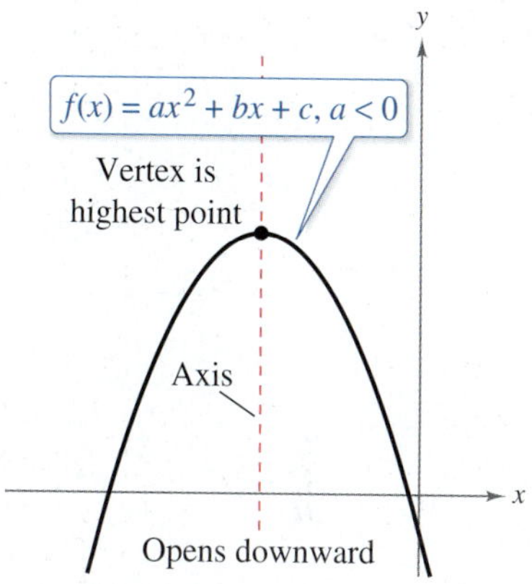

Leading coefficient is negative.

The simplest type of quadratic function is

$$f(x) = ax^2.$$

Its graph is a parabola whose vertex is $(0, 0)$. When $a > 0$, the vertex is the point with the *minimum* y-value on the graph, and when $a < 0$, the vertex is the point with the *maximum* y-value on the graph, as shown below.

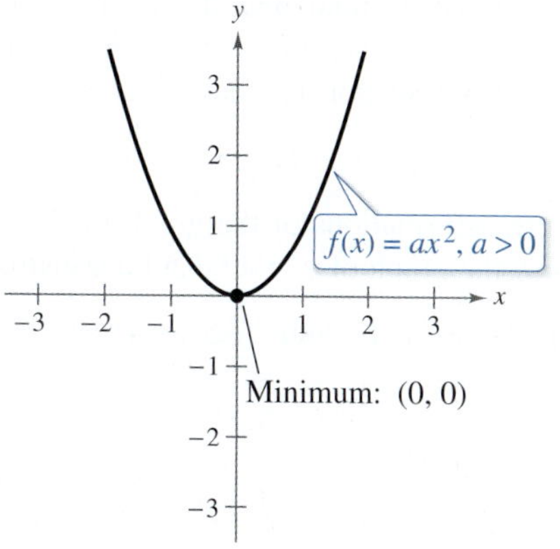

Leading coefficient is positive.

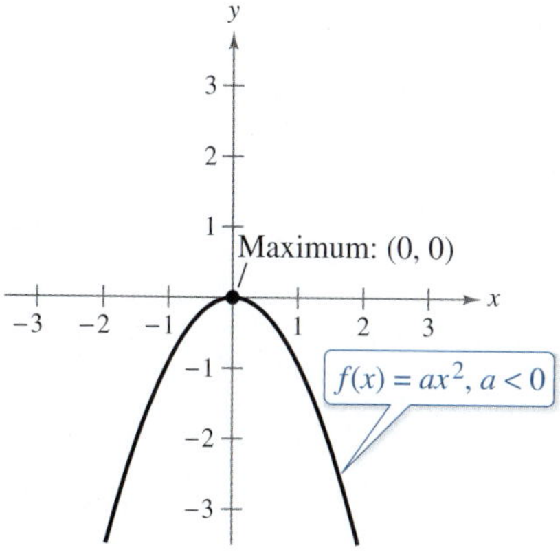

Leading coefficient is negative.

When sketching the graph of

$$f(x) = ax^2$$

it is helpful to use the graph of $y = x^2$ as a reference, as discussed in Section 1.7. There you learned that when $a > 1$, the graph of $y = af(x)$ is a vertical stretch of the graph of $y = f(x)$. When $0 < a < 1$, the graph of $y = af(x)$ is a vertical shrink of the graph of $y = f(x)$. This is demonstrated again in Example 1.

EXAMPLE 1 Sketching Graphs of Quadratic Functions

Sketch the graph of each quadratic function and compare it with the graph of $y = x^2$.

a. $f(x) = \frac{1}{3}x^2$ **b.** $g(x) = 2x^2$

▷ **ALGEBRA HELP** You can review the techniques for shifting, reflecting, and stretching graphs in Section 1.7.

Solution

a. Compared with $y = x^2$, each output of $f(x) = \frac{1}{3}x^2$ "shrinks" by a factor of $\frac{1}{3}$, creating the broader parabola shown in Figure 2.1.

b. Compared with $y = x^2$, each output of $g(x) = 2x^2$ "stretches" by a factor of 2, creating the narrower parabola shown in Figure 2.2.

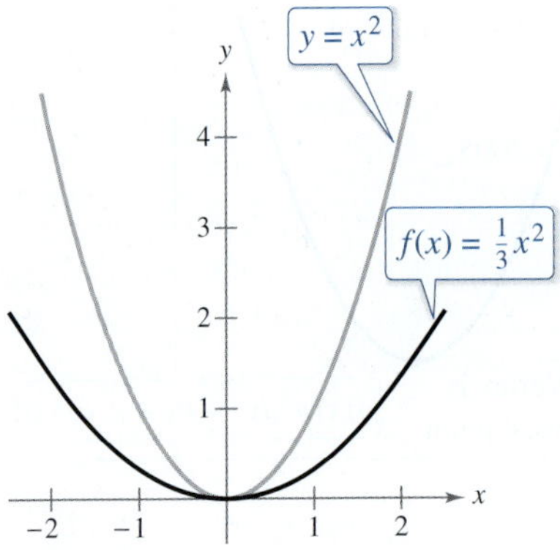

Figure 2.1

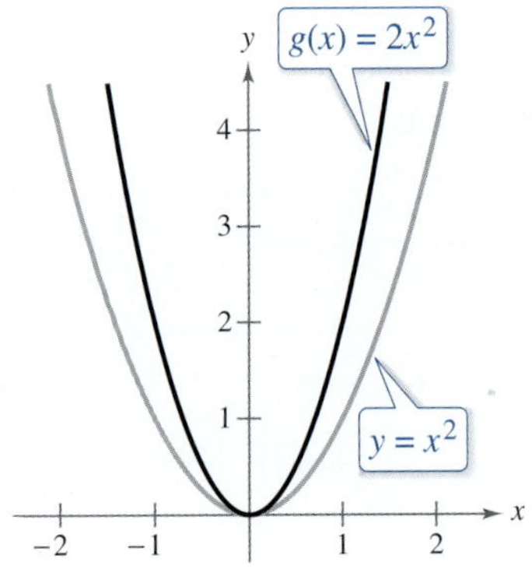

Figure 2.2

✓ ***Checkpoint*** *Audio-video solution in English & Spanish at LarsonPrecalculus.com.*

Sketch the graph of each quadratic function and compare it with the graph of $y = x^2$.

a. $f(x) = \frac{1}{4}x^2$ **b.** $g(x) = -\frac{1}{6}x^2$

c. $h(x) = \frac{5}{2}x^2$ **d.** $k(x) = -4x^2$

In Example 1, note that the coefficient a determines how wide the parabola $f(x) = ax^2$ opens. When $|a|$ is small, the parabola opens wider than when $|a|$ is large. Recall from Section 1.7 that the graphs of

$$y = f(x \pm c), \quad y = f(x) \pm c, \quad y = f(-x), \quad \text{and} \quad y = -f(x)$$

are rigid transformations of the graph of $y = f(x)$. For instance, in the figures below, notice how the graph of $y = x^2$ can be transformed to produce the graphs of

$$f(x) = -x^2 + 1 \quad \text{and} \quad g(x) = (x + 2)^2 - 3.$$

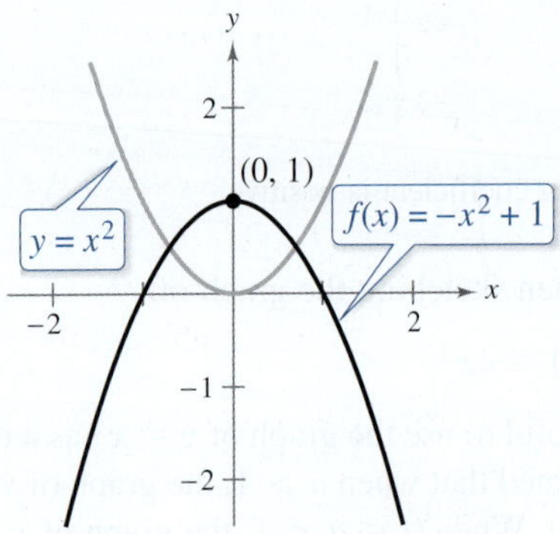

Reflection in x-axis followed by an upward shift of one unit

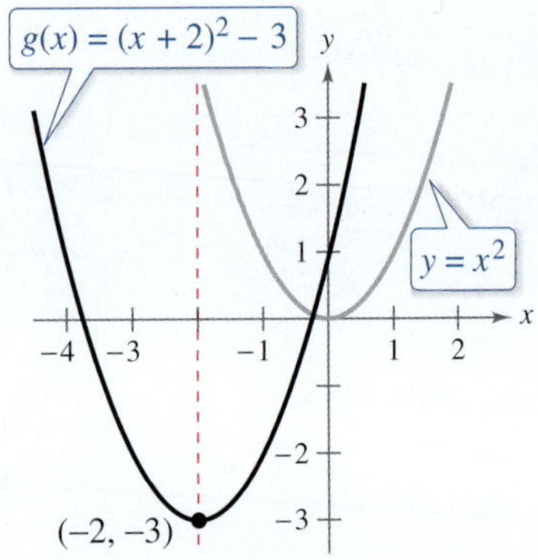

Left shift of two units followed by a downward shift of three units

The Standard Form of a Quadratic Function

REMARK The standard form of a quadratic function identifies four basic transformations of the graph of $y = x^2$.

a. The factor a produces a vertical stretch or shrink.

b. When $a < 0$, the graph is reflected in the x-axis.

c. The factor $(x - h)^2$ represents a horizontal shift of h units.

d. The term k represents a vertical shift of k units.

The **standard form** of a quadratic function is $f(x) = a(x - h)^2 + k$. This form is especially convenient for sketching a parabola because it identifies the vertex of the parabola as (h, k).

> **Standard Form of a Quadratic Function**
>
> The quadratic function
>
> $$f(x) = a(x - h)^2 + k, \quad a \neq 0$$
>
> is in **standard form.** The graph of f is a parabola whose axis is the vertical line $x = h$ and whose vertex is the point (h, k). When $a > 0$, the parabola opens upward, and when $a < 0$, the parabola opens downward.

To graph a parabola, it is helpful to begin by writing the quadratic function in standard form using the process of completing the square, as illustrated in Example 2. In this example, notice that when completing the square, you *add and subtract* the square of half the coefficient of x within the parentheses instead of adding the value to each side of the equation as is done in Appendix A.5.

EXAMPLE 2 Using Standard Form to Graph a Parabola

Sketch the graph of

$$f(x) = 2x^2 + 8x + 7.$$

Identify the vertex and the axis of the parabola.

ALGEBRA HELP You can review the techniques for completing the square in Appendix A.5.

Solution Begin by writing the quadratic function in standard form. Notice that the first step in completing the square is to factor out any coefficient of x^2 that is not 1.

$f(x) = 2x^2 + 8x + 7$ Write original function.

$= 2(x^2 + 4x) + 7$ Factor 2 out of x-terms.

$= 2(x^2 + 4x + 4 - 4) + 7$ Add and subtract 4 within parentheses.

$(4/2)^2$

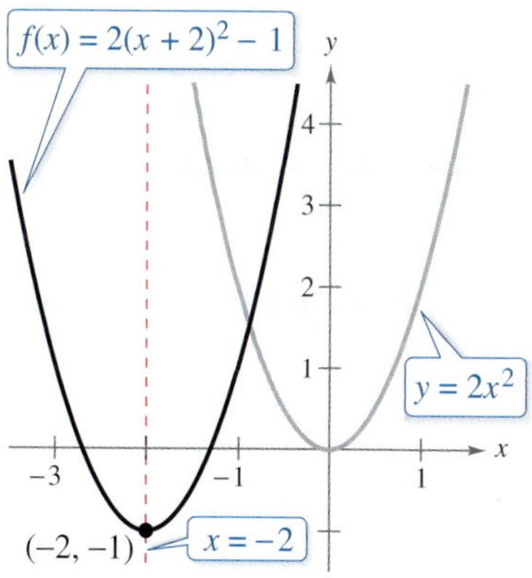

Figure 2.3

After adding and subtracting 4 within the parentheses, you must now regroup the terms to form a perfect square trinomial. The -4 can be removed from inside the parentheses; however, because of the 2 outside of the parentheses, you must multiply -4 by 2, as shown below.

$f(x) = 2(x^2 + 4x + 4) - 2(4) + 7$ Regroup terms.

$= 2(x^2 + 4x + 4) - 8 + 7$ Simplify.

$= 2(x + 2)^2 - 1$ Write in standard form.

From this form, you can see that the graph of f is a parabola that opens upward and has its vertex at $(-2, -1)$. This corresponds to a left shift of two units and a downward shift of one unit relative to the graph of $y = 2x^2$, as shown in Figure 2.3. In the figure, you can see that the axis of the parabola is the vertical line through the vertex, $x = -2$.

Checkpoint Audio-video solution in English & Spanish at LarsonPrecalculus.com.

Sketch the graph of

$$f(x) = 3x^2 - 6x + 4.$$

Identify the vertex and the axis of the parabola.

ALGEBRA HELP You can review the techniques for using the Quadratic Formula in Appendix A.5.

To find the x-intercepts of the graph of $f(x) = ax^2 + bx + c$, you must solve the equation $ax^2 + bx + c = 0$. When $ax^2 + bx + c$ does not factor, you can use the Quadratic Formula to find the x-intercepts. Remember, however, that a parabola may not have x-intercepts.

EXAMPLE 3 Finding the Vertex and x-Intercepts of a Parabola

Sketch the graph of $f(x) = -x^2 + 6x - 8$. Identify the vertex and x-intercepts.

Solution

$$f(x) = -x^2 + 6x - 8 \quad \text{Write original function.}$$
$$= -(x^2 - 6x) - 8 \quad \text{Factor } -1 \text{ out of } x\text{-terms.}$$
$$= -(x^2 - 6x + 9 - 9) - 8 \quad \text{Add and subtract 9 within parentheses.}$$
$$(-6/2)^2$$
$$= -(x^2 - 6x + 9) - (-9) - 8 \quad \text{Regroup terms.}$$
$$= -(x - 3)^2 + 1 \quad \text{Write in standard form.}$$

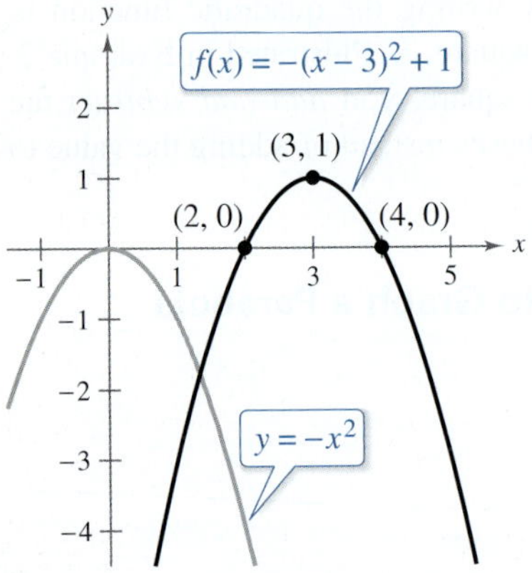

Figure 2.4

From this form, you can see that f is a parabola that opens downward with vertex $(3, 1)$. The x-intercepts of the graph are determined as follows.

$$-(x^2 - 6x + 8) = 0 \quad \text{Factor out } -1.$$
$$-(x - 2)(x - 4) = 0 \quad \text{Factor.}$$
$$x - 2 = 0 \Rightarrow x = 2 \quad \text{Set 1st factor equal to 0.}$$
$$x - 4 = 0 \Rightarrow x = 4 \quad \text{Set 2nd factor equal to 0.}$$

So, the x-intercepts are $(2, 0)$ and $(4, 0)$, as shown in Figure 2.4.

✓ **Checkpoint** *Audio-video solution in English & Spanish at LarsonPrecalculus.com.*

Sketch the graph of $f(x) = x^2 - 4x + 3$. Identify the vertex and x-intercepts.

EXAMPLE 4 Writing the Equation of a Parabola

Write the standard form of the equation of the parabola whose vertex is $(1, 2)$ and that passes through the point $(3, -6)$.

Solution Because the vertex is $(h, k) = (1, 2)$, the equation has the form

$$f(x) = a(x - 1)^2 + 2. \quad \text{Substitute for } h \text{ and } k \text{ in standard form.}$$

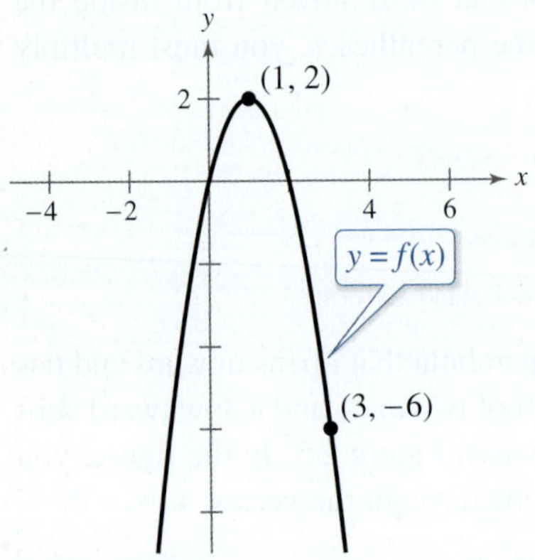

Figure 2.5

Because the parabola passes through the point $(3, -6)$, it follows that $f(3) = -6$. So,

$$f(x) = a(x - 1)^2 + 2 \quad \text{Write in standard form.}$$
$$-6 = a(3 - 1)^2 + 2 \quad \text{Substitute 3 for } x \text{ and } -6 \text{ for } f(x).$$
$$-6 = 4a + 2 \quad \text{Simplify.}$$
$$-8 = 4a \quad \text{Subtract 2 from each side.}$$
$$-2 = a. \quad \text{Divide each side by 4.}$$

The equation in standard form is $f(x) = -2(x - 1)^2 + 2$. The graph of f is shown in Figure 2.5.

✓ **Checkpoint** *Audio-video solution in English & Spanish at LarsonPrecalculus.com.*

Write the standard form of the equation of the parabola whose vertex is $(-4, 11)$ and that passes through the point $(-6, 15)$.

Finding Minimum and Maximum Values

Many applications involve finding the maximum or minimum value of a quadratic function. By completing the square of the quadratic function $f(x) = ax^2 + bx + c$, you can rewrite the function in standard form (see Exercise 93).

$$f(x) = a\left(x + \frac{b}{2a}\right)^2 + \left(c - \frac{b^2}{4a}\right) \qquad \text{Standard form}$$

So, the vertex of the graph of f is $\left(-\frac{b}{2a}, f\left(-\frac{b}{2a}\right)\right)$, which implies the following.

Minimum and Maximum Values of Quadratic Functions

Consider the function $f(x) = ax^2 + bx + c$ with vertex $\left(-\frac{b}{2a}, f\left(-\frac{b}{2a}\right)\right)$.

1. When $a > 0$, f has a *minimum* at $x = -\frac{b}{2a}$. The minimum value is $f\left(-\frac{b}{2a}\right)$.
2. When $a < 0$, f has a *maximum* at $x = -\frac{b}{2a}$. The maximum value is $f\left(-\frac{b}{2a}\right)$.

EXAMPLE 5 The Maximum Height of a Baseball

The path of a baseball after being hit is given by the function $f(x) = -0.0032x^2 + x + 3$, where $f(x)$ is the height of the baseball (in feet) and x is the horizontal distance from home plate (in feet). What is the maximum height of the baseball?

Algebraic Solution

For this quadratic function, you have

$$f(x) = ax^2 + bx + c = -0.0032x^2 + x + 3$$

which implies that $a = -0.0032$ and $b = 1$. Because $a < 0$, the function has a maximum at $x = -b/(2a)$. So, the baseball reaches its maximum height when it is

$$x = -\frac{b}{2a} = -\frac{1}{2(-0.0032)} = 156.25 \text{ feet from home plate.}$$

At this distance, the maximum height is

$$f(156.25) = -0.0032(156.25)^2 + 156.25 + 3 = 81.125 \text{ feet.}$$

Graphical Solution

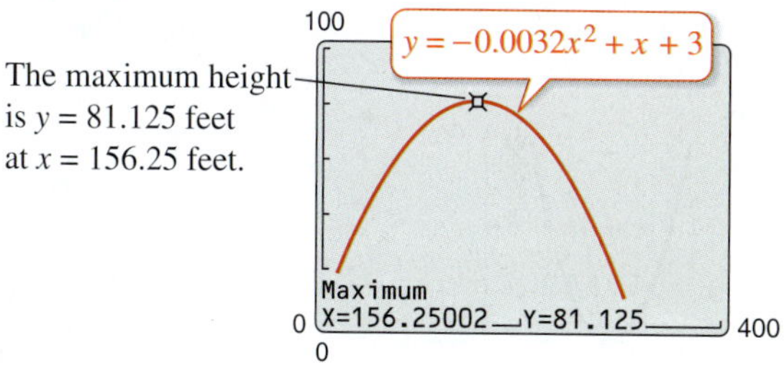

✔ **Checkpoint** *Audio-video solution in English & Spanish at LarsonPrecalculus.com.*

Rework Example 5 when the path of the baseball is given by the function

$$f(x) = -0.007x^2 + x + 4.$$

Summarize (Section 2.1)

1. State the definition of a quadratic function and describe its graph *(pages 114–116)*. For an example of sketching the graphs of quadratic functions, see Example 1.
2. State the standard form of a quadratic function *(page 117)*. For examples that use the standard form of a quadratic function, see Examples 2–4.
3. Describe how to find the minimum or maximum value of a quadratic function *(page 119)*. For a real-life example that involves finding the maximum height of a baseball, see Example 5.

2.1 Exercises

See CalcChat.com for tutorial help and worked-out solutions to odd-numbered exercises.

Vocabulary: Fill in the blanks.

1. Linear, constant, and squaring functions are examples of ________ functions.

2. A polynomial function of x with degree n has the form $f(x) = a_n x^n + a_{n-1}x^{n-1} + \cdots + a_1 x + a_0$ $(a_n \neq 0)$, where n is a ________ ________ and $a_n, a_{n-1}, \ldots, a_1, a_0$ are ________ numbers.

3. A ________ function is a second-degree polynomial function, and its graph is called a ________.

4. The graph of a quadratic function is symmetric about its ________.

5. When the graph of a quadratic function opens upward, its leading coefficient is ________ and the vertex of the graph is a ________.

6. When the graph of a quadratic function opens downward, its leading coefficient is ________ and the vertex of the graph is a ________.

Skills and Applications

Matching **In Exercises 7–12, match the quadratic function with its graph. [The graphs are labeled (a), (b), (c), (d), (e), and (f).]**

(a)

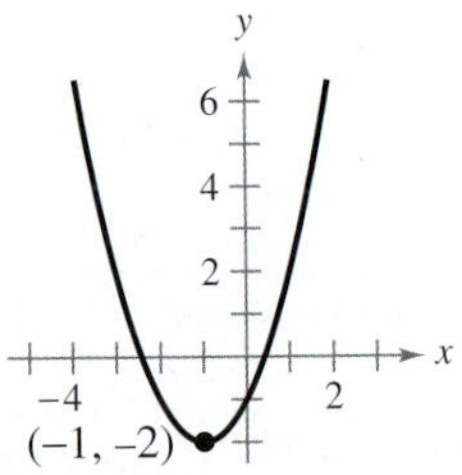

(b)

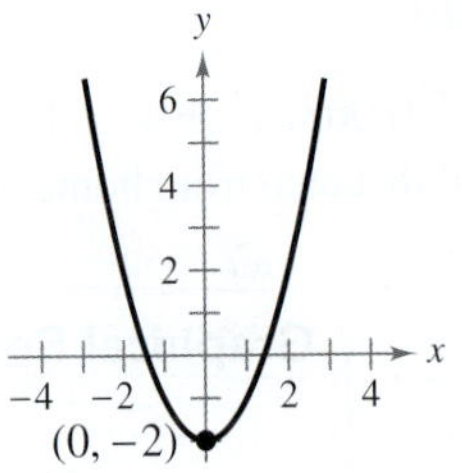

(c)

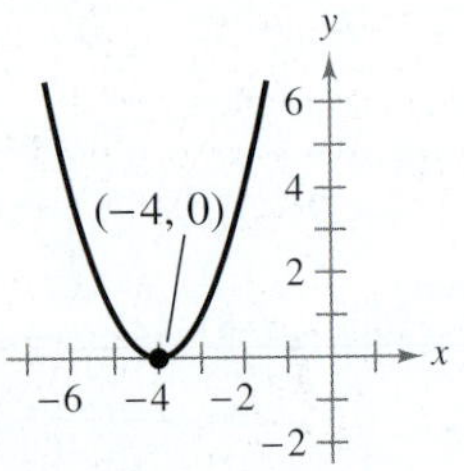

(d)

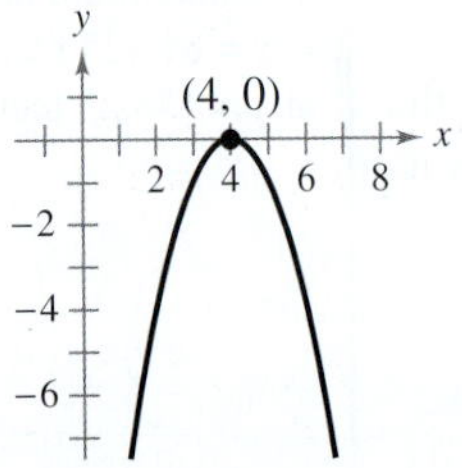

(e)

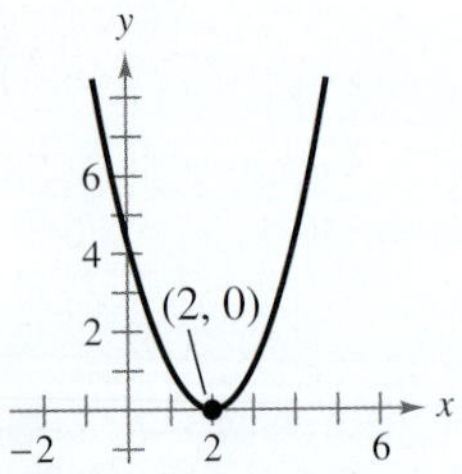

(f)

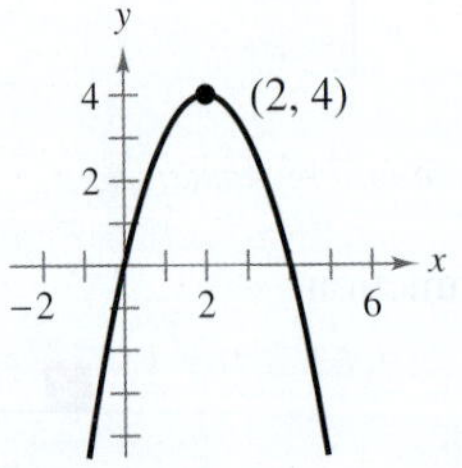

7. $f(x) = (x - 2)^2$ **8.** $f(x) = (x + 4)^2$

9. $f(x) = x^2 - 2$ **10.** $f(x) = (x + 1)^2 - 2$

11. $f(x) = 4 - (x - 2)^2$ **12.** $f(x) = -(x - 4)^2$

Sketching Graphs of Quadratic Functions **In Exercises 13–16, sketch the graph of each quadratic function and compare it with the graph of $y = x^2$.**

13. (a) $f(x) = \frac{1}{2}x^2$ (b) $g(x) = -\frac{1}{8}x^2$
(c) $h(x) = \frac{3}{2}x^2$ (d) $k(x) = -3x^2$

14. (a) $f(x) = x^2 + 1$ (b) $g(x) = x^2 - 1$
(c) $h(x) = x^2 + 3$ (d) $k(x) = x^2 - 3$

15. (a) $f(x) = (x - 1)^2$ (b) $g(x) = (3x)^2 + 1$
(c) $h(x) = \left(\frac{1}{3}x\right)^2 - 3$ (d) $k(x) = (x + 3)^2$

16. (a) $f(x) = -\frac{1}{2}(x - 2)^2 + 1$
(b) $g(x) = \left[\frac{1}{2}(x - 1)\right]^2 - 3$
(c) $h(x) = -\frac{1}{2}(x + 2)^2 - 1$
(d) $k(x) = [2(x + 1)]^2 + 4$

Using Standard Form to Graph a Parabola **In Exercises 17–34, write the quadratic function in standard form and sketch its graph. Identify the vertex, axis of symmetry, and x-intercept(s).**

17. $f(x) = x^2 - 6x$ **18.** $g(x) = x^2 - 8x$

19. $h(x) = x^2 - 8x + 16$ **20.** $g(x) = x^2 + 2x + 1$

21. $f(x) = x^2 + 8x + 13$ **22.** $f(x) = x^2 - 12x + 44$

23. $f(x) = x^2 - 14x + 54$ **24.** $h(x) = x^2 + 16x - 17$

25. $f(x) = x^2 + 34x + 289$ **26.** $f(x) = x^2 - 30x + 225$

27. $f(x) = x^2 - x + \frac{5}{4}$ **28.** $f(x) = x^2 + 3x + \frac{1}{4}$

29. $f(x) = -x^2 + 2x + 5$ **30.** $f(x) = -x^2 - 4x + 1$

31. $h(x) = 4x^2 - 4x + 21$ **32.** $f(x) = 2x^2 - x + 1$

33. $f(x) = \frac{1}{4}x^2 - 2x - 12$ **34.** $f(x) = -\frac{1}{3}x^2 + 3x - 6$

Graphical Analysis **In Exercises 35–42, use a graphing utility to graph the quadratic function. Identify the vertex, axis of symmetry, and x-intercept(s). Then check your results algebraically by writing the quadratic function in standard form.**

35. $f(x) = -(x^2 + 2x - 3)$ **36.** $f(x) = -(x^2 + x - 30)$

37. $g(x) = x^2 + 8x + 11$ **38.** $f(x) = x^2 + 10x + 14$

39. $f(x) = 2x^2 - 16x + 32$

40. $f(x) = -4x^2 + 24x - 41$

41. $g(x) = \frac{1}{2}(x^2 + 4x - 2)$ **42.** $f(x) = \frac{3}{5}(x^2 + 6x - 5)$

Writing the Equation of a Parabola **In Exercises 43–46, write an equation for the parabola in standard form.**

43.

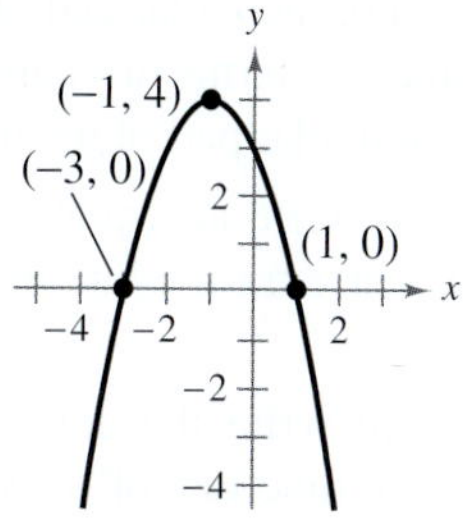

44.

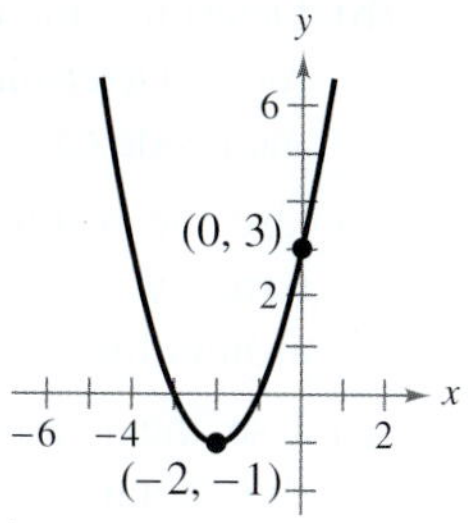

45.

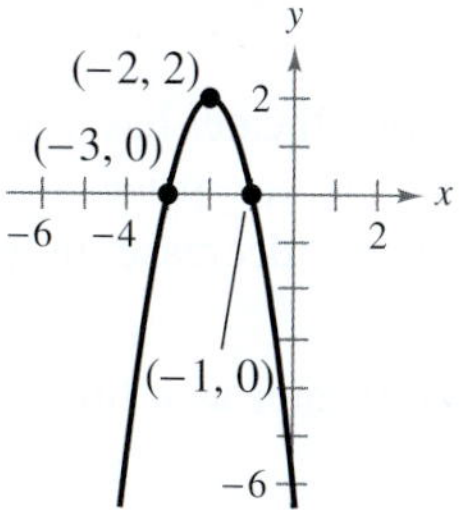

46.

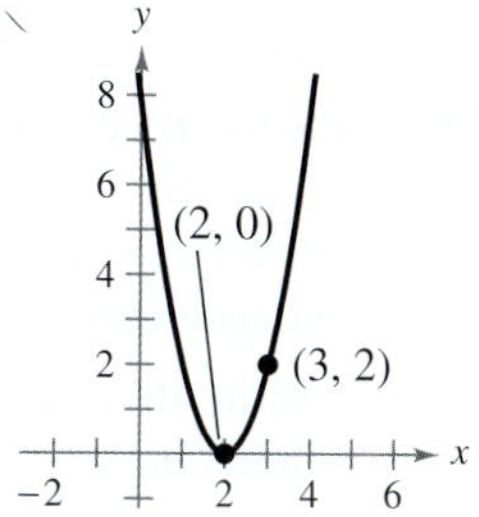

Writing the Equation of a Parabola **In Exercises 47–56, write the standard form of the equation of the parabola that has the indicated vertex and passes through the given point.**

47. Vertex: $(-2, 5)$; point: $(0, 9)$

48. Vertex: $(4, -1)$; point: $(2, 3)$

49. Vertex: $(1, -2)$; point: $(-1, 14)$

50. Vertex: $(2, 3)$; point: $(0, 2)$

51. Vertex: $(5, 12)$; point: $(7, 15)$

52. Vertex: $(-2, -2)$; point: $(-1, 0)$

53. Vertex: $\left(-\frac{1}{4}, \frac{3}{2}\right)$; point: $(-2, 0)$

54. Vertex: $\left(\frac{5}{2}, -\frac{3}{4}\right)$; point: $(-2, 4)$

55. Vertex: $\left(-\frac{5}{2}, 0\right)$; point: $\left(-\frac{7}{2}, -\frac{16}{3}\right)$

56. Vertex: $(6, 6)$; point: $\left(\frac{61}{10}, \frac{3}{2}\right)$

Graphical Reasoning **In Exercises 57 and 58, determine the x-intercept(s) of the graph visually. Then find the x-intercept(s) algebraically to confirm your results.**

57. $y = x^2 - 4x - 5$

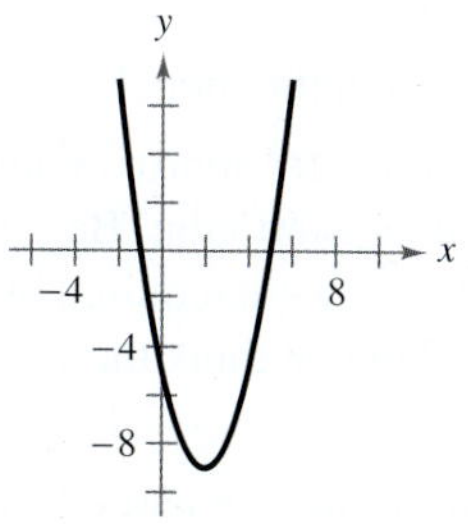

58. $y = 2x^2 + 5x - 3$

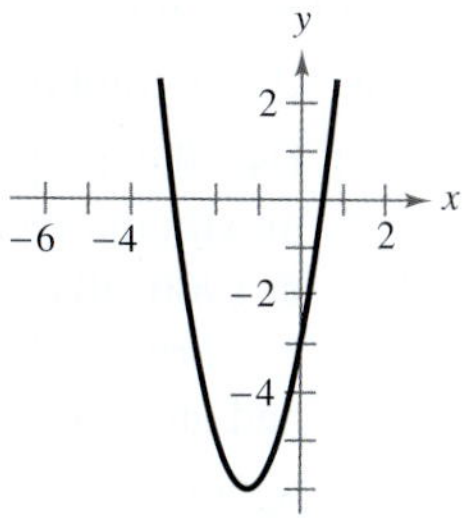

wellphoto/Shutterstock.com

Graphical Analysis **In Exercises 59–64, use a graphing utility to graph the quadratic function. Find the x-intercept(s) of the graph and compare them with the solutions of the corresponding quadratic equation when $f(x) = 0$.**

59. $f(x) = x^2 - 4x$

60. $f(x) = -2x^2 + 10x$

61. $f(x) = x^2 - 9x + 18$

62. $f(x) = x^2 - 8x - 20$

63. $f(x) = 2x^2 - 7x - 30$

64. $f(x) = \frac{7}{10}(x^2 + 12x - 45)$

Finding Quadratic Functions **In Exercises 65–70, find two quadratic functions, one that opens upward and one that opens downward, whose graphs have the given x-intercepts. (There are many correct answers.)**

65. $(-1, 0), (3, 0)$

66. $(-5, 0), (5, 0)$

67. $(0, 0), (10, 0)$

68. $(4, 0), (8, 0)$

69. $(-3, 0), \left(-\frac{1}{2}, 0\right)$

70. $\left(-\frac{5}{2}, 0\right), (2, 0)$

Number Problems **In Exercises 71–74, find two positive real numbers whose product is a maximum.**

71. The sum is 110.

72. The sum is S.

73. The sum of the first and twice the second is 24.

74. The sum of the first and three times the second is 42.

75. Path of a Diver

The path of a diver is given by the function

$$f(x) = -\frac{4}{9}x^2 + \frac{24}{9}x + 12$$

where $f(x)$ is the height (in feet) and x is the horizontal distance from the end of the diving board (in feet). What is the maximum height of the diver?

76. Height of a Ball The path of a punted football is given by the function

$$f(x) = -\frac{16}{2025}x^2 + \frac{9}{5}x + 1.5$$

where $f(x)$ is the height (in feet) and x is the horizontal distance (in feet) from the point at which the ball is punted.

(a) How high is the ball when it is punted?

(b) What is the maximum height of the punt?

(c) How long is the punt?

77. Minimum Cost A manufacturer of lighting fixtures has daily production costs of $C = 800 - 10x + 0.25x^2$, where C is the total cost (in dollars) and x is the number of units produced. How many fixtures should be produced each day to yield a minimum cost?

78. Maximum Profit The profit P (in hundreds of dollars) that a company makes depends on the amount x (in hundreds of dollars) the company spends on advertising according to the model $P = 230 + 20x - 0.5x^2$. What expenditure for advertising will yield a maximum profit?

79. Maximum Revenue The total revenue R earned (in thousands of dollars) from manufacturing handheld video games is given by

$R(p) = -25p^2 + 1200p$

where p is the price per unit (in dollars).

(a) Find the revenues when the prices per unit are \$20, \$25, and \$30.

(b) Find the unit price that will yield a maximum revenue. What is the maximum revenue? Explain your results.

80. Maximum Revenue The total revenue R earned per day (in dollars) from a pet-sitting service is given by $R(p) = -12p^2 + 150p$, where p is the price charged per pet (in dollars).

(a) Find the revenues when the prices per pet are \$4, \$6, and \$8.

(b) Find the unit price that will yield a maximum revenue. What is the maximum revenue? Explain your results.

81. Numerical, Graphical, and Analytical Analysis A rancher has 200 feet of fencing to enclose two adjacent rectangular corrals (see figure).

(a) Write the area A of the corrals as a function of x.

(b) Construct a table showing possible values of x and the corresponding areas of the corral. Use the table to estimate the dimensions that will produce the maximum enclosed area.

(c) Use a graphing utility to graph the area function. Use the graph to approximate the dimensions that will produce the maximum enclosed area.

(d) Write the area function in standard form to find analytically the dimensions that will produce the maximum area.

(e) Compare your results from parts (b), (c), and (d).

82. Geometry An indoor physical fitness room consists of a rectangular region with a semicircle on each end. The perimeter of the room is to be a 200-meter single-lane running track.

(a) Draw a diagram that gives a visual representation of the problem. Let x and y represent the length and width of the rectangular region, respectively.

(b) Determine the radius of each semicircular end of the room. Determine the distance, in terms of y, around the inside edge of each semicircular part of the track.

(c) Use the result of part (b) to write an equation, in terms of x and y, for the distance traveled in one lap around the track. Solve for y.

(d) Use the result of part (c) to write the area A of the rectangular region as a function of x. What dimensions will produce a rectangle of maximum area?

83. Maximum Revenue A small theater has a seating capacity of 2000. When the ticket price is \$20, attendance is 1500. For each \$1 decrease in price, attendance increases by 100.

(a) Write the revenue R of the theater as a function of ticket price x.

(b) What ticket price will yield a maximum revenue? What is the maximum revenue?

84. Maximum Area A Norman window is constructed by adjoining a semicircle to the top of an ordinary rectangular window (see figure). The perimeter of the window is 16 feet.

(a) Write the area A of the window as a function of x.

(b) What dimensions will produce a window of maximum area?

85. Graphical Analysis From 1950 through 2005, the per capita consumption C of cigarettes by Americans (age 18 and older) can be modeled by $C = 3565.0 + 60.30t - 1.783t^2$, $0 \le t \le 55$, where t is the year, with $t = 0$ corresponding to 1950. (*Source: Tobacco Outlook Report*)

(a) Use a graphing utility to graph the model.

(b) Use the graph of the model to approximate the maximum average annual consumption. Beginning in 1966, all cigarette packages were required by law to carry a health warning. Do you think the warning had any effect? Explain.

(c) In 2005, the U.S. population (age 18 and over) was 296,329,000. Of those, about 59,858,458 were smokers. What was the average annual cigarette consumption *per smoker* in 2005? What was the average daily cigarette consumption *per smoker*?

86. Data Analysis: Sales The sales y (in billions of dollars) for Harley-Davidson from 2000 through 2010 are shown in the table. *(Source: U.S. Harley-Davidson, Inc.)*

DATA

Year	Sales, y
2000	2.91
2001	3.36
2002	4.09
2003	4.62
2004	5.02
2005	5.34
2006	5.80
2007	5.73
2008	5.59
2009	4.78
2010	4.86

Spreadsheet at LarsonPrecalculus.com

(a) Use a graphing utility to create a scatter plot of the data. Let x represent the year, with $x = 0$ corresponding to 2000.

(b) Use the *regression* feature of the graphing utility to find a quadratic model for the data.

(c) Use the graphing utility to graph the model in the same viewing window as the scatter plot. How well does the model fit the data?

(d) Use the *trace* feature of the graphing utility to approximate the year in which the sales for Harley-Davidson were the greatest.

(e) Verify your answer to part (d) algebraically.

(f) Use the model to predict the sales for Harley-Davidson in 2013.

Exploration

True or False? **In Exercises 87 and 88, determine whether the statement is true or false. Justify your answer.**

87. The graph of $f(x) = -12x^2 - 1$ has no x-intercepts.

88. The graphs of

$$f(x) = -4x^2 - 10x + 7$$

and

$$g(x) = 12x^2 + 30x + 1$$

have the same axis of symmetry.

Think About It **In Exercises 89–92, find the values of b such that the function has the given maximum or minimum value.**

89. $f(x) = -x^2 + bx - 75$; Maximum value: 25

90. $f(x) = -x^2 + bx - 16$; Maximum value: 48

91. $f(x) = x^2 + bx + 26$; Minimum value: 10

92. $f(x) = x^2 + bx - 25$; Minimum value: -50

93. Verifying the Vertex Write the quadratic function

$$f(x) = ax^2 + bx + c$$

in standard form to verify that the vertex occurs at

$$\left(-\frac{b}{2a}, f\left(-\frac{b}{2a}\right)\right).$$

94. HOW DO YOU SEE IT? The graph shows a quadratic function of the form

$$P(t) = at^2 + bt + c$$

which represents the yearly profits for a company, where $P(t)$ is the profit in year t.

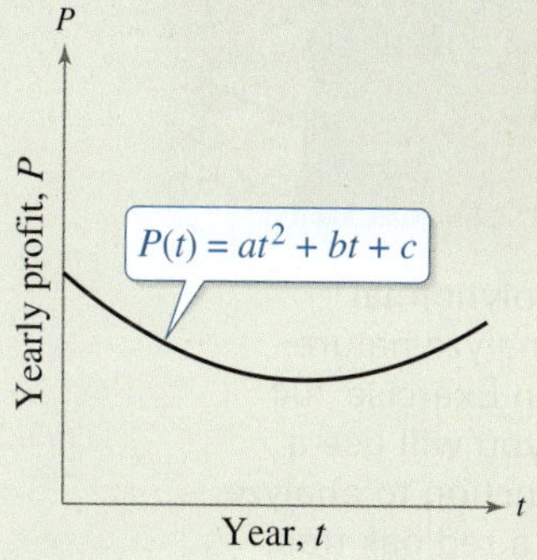

(a) Is the value of a positive, negative, or zero? Explain.

(b) Write an expression in terms of a and b that represents the year t when the company made the least profit.

(c) The company made the same yearly profits in 2004 and 2012. Estimate the year in which the company made the least profit.

(d) Assume that the model is still valid today. Are the yearly profits currently increasing, decreasing, or constant? Explain.

95. Proof Assume that the function

$$f(x) = ax^2 + bx + c, \quad a \neq 0$$

has two real zeros. Prove that the x-coordinate of the vertex of the graph is the average of the zeros of f. (*Hint:* Use the Quadratic Formula.)

Project: Height of a Basketball To work an extended application analyzing the height of a basketball after it has been dropped, visit this text's website at *LarsonPrecalculus.com.*

2.2 Polynomial Functions of Higher Degree

You can use polynomial functions to analyze nature. For instance, in Exercise 104 on page 136, you will use a polynomial function to analyze the growth of a red oak tree.

- **Use transformations to sketch graphs of polynomial functions.**
- **Use the Leading Coefficient Test to determine the end behaviors of graphs of polynomial functions.**
- **Find and use the real zeros of polynomial functions as sketching aids.**
- **Use the Intermediate Value Theorem to help locate the real zeros of polynomial functions.**

Graphs of Polynomial Functions

In this section, you will study basic features of the graphs of polynomial functions. The first feature is that the graph of a polynomial function is **continuous.** Essentially, this means that the graph of a polynomial function has no breaks, holes, or gaps, as shown in Figure 2.6(a). The graph shown in Figure 2.6(b) is an example of a piecewise-defined function that is not continuous.

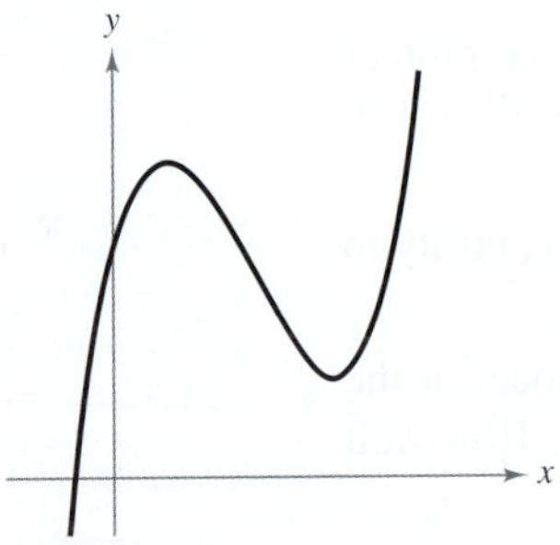

(a) Polynomial functions have continuous graphs.

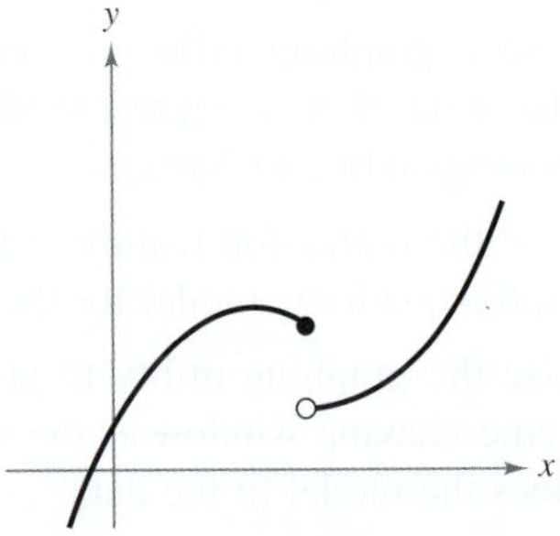

(b) Functions with graphs that are not continuous are not polynomial functions.

Figure 2.6

The second feature is that the graph of a polynomial function has only smooth, rounded turns, as shown in Figure 2.7. A polynomial function cannot have a sharp turn. For instance, the function $f(x) = |x|$, the graph of which has a sharp turn at the point $(0, 0)$, as shown in Figure 2.8, is not a polynomial function.

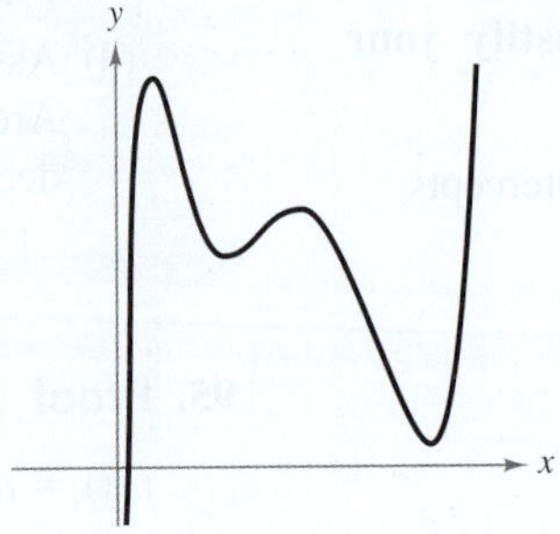

Polynomial functions have graphs with smooth, rounded turns.
Figure 2.7

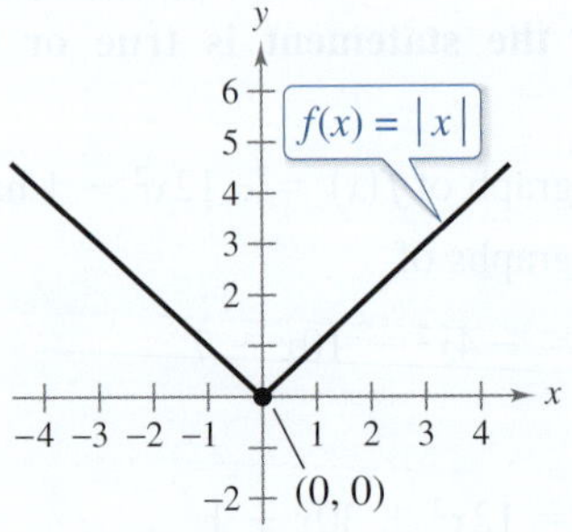

Graphs of polynomial functions cannot have sharp turns.
Figure 2.8

The graphs of polynomial functions of degree greater than 2 are more difficult to analyze than the graphs of polynomial functions of degree 0, 1, or 2. However, using the features presented in this section, coupled with your knowledge of point plotting, intercepts, and symmetry, you should be able to make reasonably accurate sketches *by hand*.

S.Borisov/Shutterstock.com

The polynomial functions that have the simplest graphs are monomial functions of the form $f(x) = x^n$, where n is an integer greater than zero. From Figure 2.9, you can see that when n is *even*, the graph is similar to the graph of $f(x) = x^2$, and when n is *odd*, the graph is similar to the graph of $f(x) = x^3$. Moreover, the greater the value of n, the flatter the graph near the origin.

REMARK For functions given by $f(x) = x^n$, when n is even, the graph of the function is symmetric with respect to the y-axis, and when n is odd, the graph of the function is symmetric with respect to the origin.

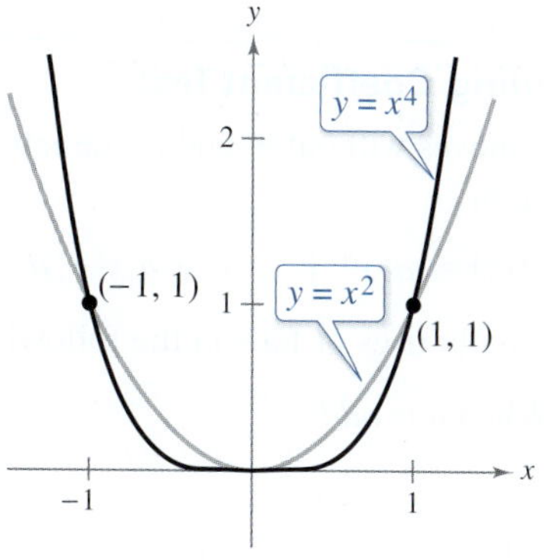

(a) When n is even, the graph of $y = x^n$ touches the axis at the x-intercept.

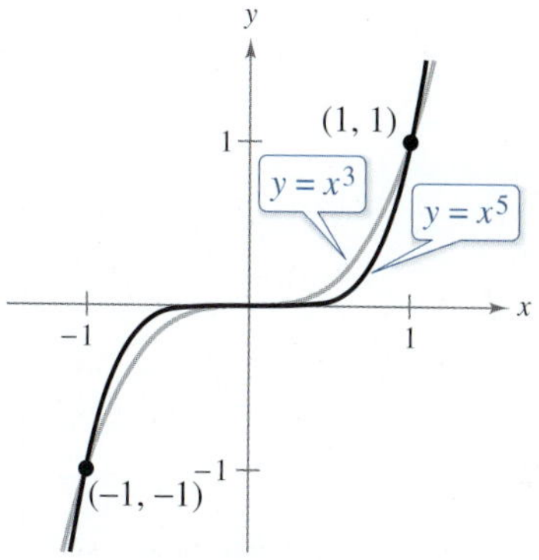

(b) When n is odd, the graph of $y = x^n$ crosses the axis at the x-intercept.

Figure 2.9

EXAMPLE 1 Sketching Transformations of Monomial Functions

Sketch the graph of each function.

a. $f(x) = -x^5$ **b.** $h(x) = (x + 1)^4$

Solution

a. Because the degree of $f(x) = -x^5$ is odd, its graph is similar to the graph of $y = x^3$. In Figure 2.10, note that the negative coefficient has the effect of reflecting the graph in the x-axis.

b. The graph of $h(x) = (x + 1)^4$, as shown in Figure 2.11, is a left shift by one unit of the graph of $y = x^4$.

ALGEBRA HELP You can review the techniques for shifting, reflecting, and stretching graphs in Section 1.7.

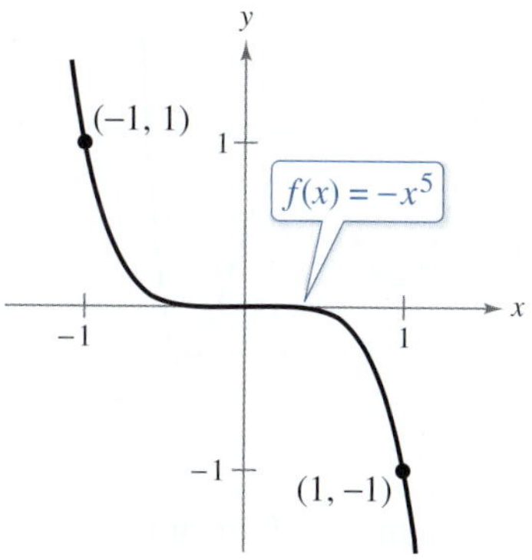

Figure 2.10

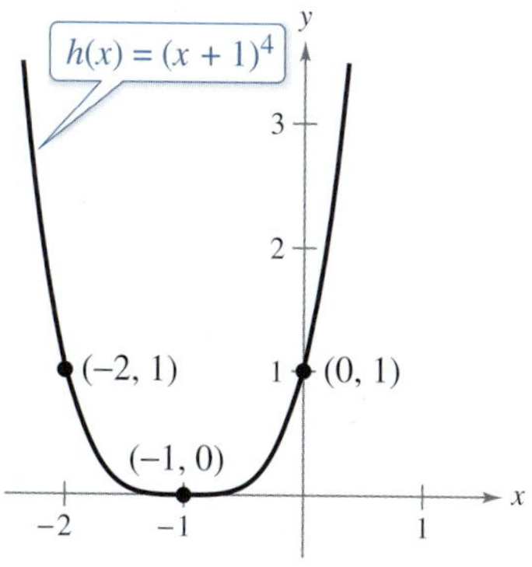

Figure 2.11

Checkpoint Audio-video solution in English & Spanish at LarsonPrecalculus.com.

Sketch the graph of each function.

a. $f(x) = (x + 5)^4$

b. $g(x) = x^4 - 7$

c. $h(x) = 7 - x^4$

d. $k(x) = \frac{1}{4}(x - 3)^4$

The Leading Coefficient Test

In Example 1, note that both graphs eventually rise or fall without bound as x moves to the left or to the right. Whether the graph of a polynomial function eventually rises or falls can be determined by the function's degree (even or odd) and by its leading coefficient (positive or negative), as indicated in the **Leading Coefficient Test.**

REMARK The notation "$f(x) \to -\infty$ as $x \to -\infty$" indicates that the graph falls to the left. The notation "$f(x) \to \infty$ as $x \to \infty$" indicates that the graph rises to the right.

Leading Coefficient Test

As x moves without bound to the left or to the right, the graph of the polynomial function

$$f(x) = a_n x^n + \cdot \cdot \cdot + a_1 x + a_0$$

eventually rises or falls in the following manner.

1. When n is *odd:*

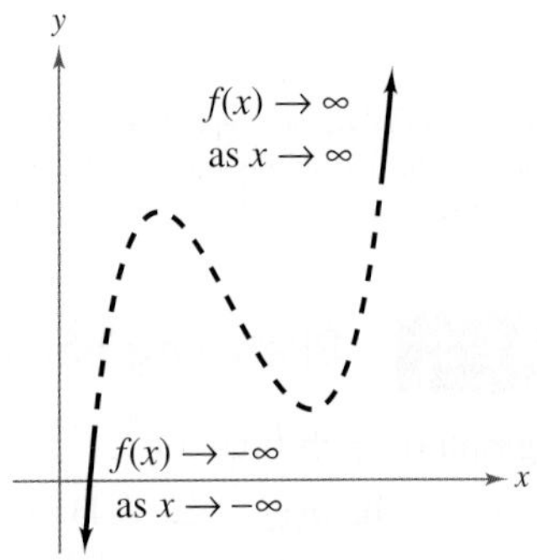

If the leading coefficient is positive ($a_n > 0$), then the graph falls to the left and rises to the right.

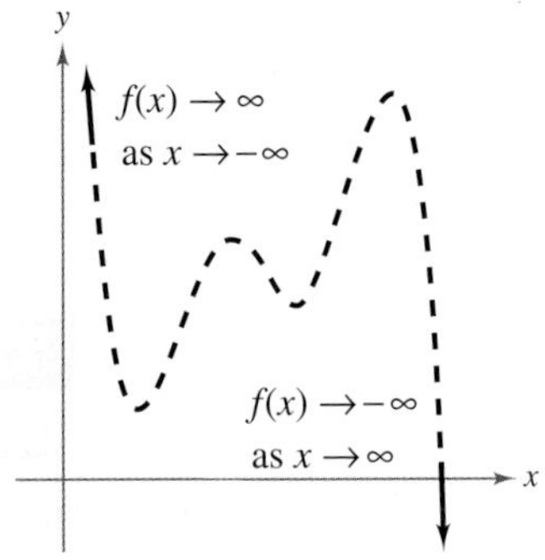

If the leading coefficient is negative ($a_n < 0$), then the graph rises to the left and falls to the right.

2. When n is *even:*

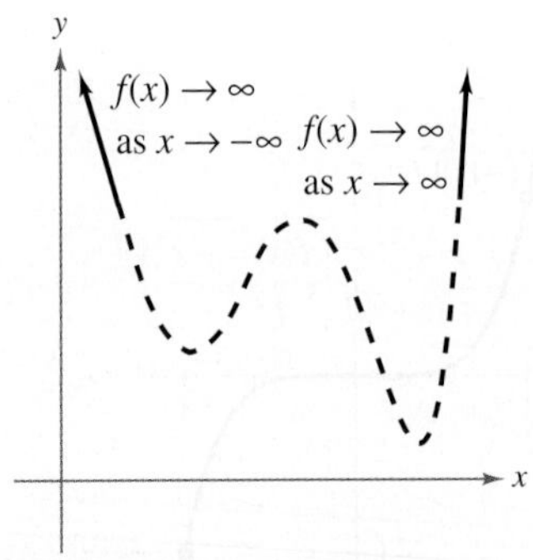

If the leading coefficient is positive ($a_n > 0$), then the graph rises to the left and to the right.

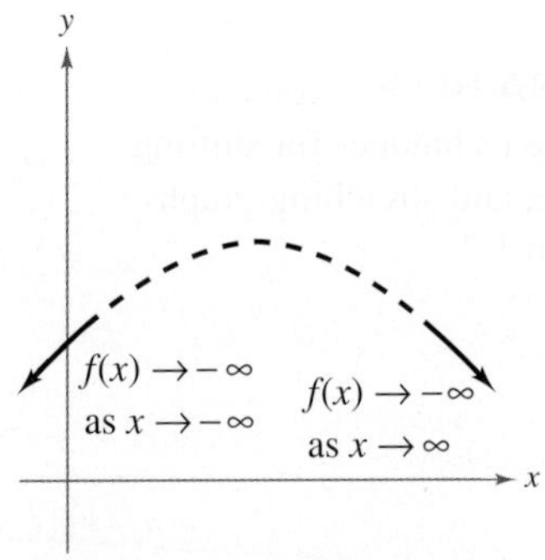

If the leading coefficient is negative ($a_n < 0$), then the graph falls to the left and to the right.

The dashed portions of the graphs indicate that the test determines *only* the right-hand and left-hand behavior of the graph.

As you continue to study polynomial functions and their graphs, you will notice that the degree of a polynomial plays an important role in determining other characteristics of the polynomial function and its graph.

EXAMPLE 2 Applying the Leading Coefficient Test

Describe the right-hand and left-hand behavior of the graph of each function.

a. $f(x) = -x^3 + 4x$ **b.** $f(x) = x^4 - 5x^2 + 4$ **c.** $f(x) = x^5 - x$

Solution

a. Because the degree is odd and the leading coefficient is negative, the graph rises to the left and falls to the right, as shown below.

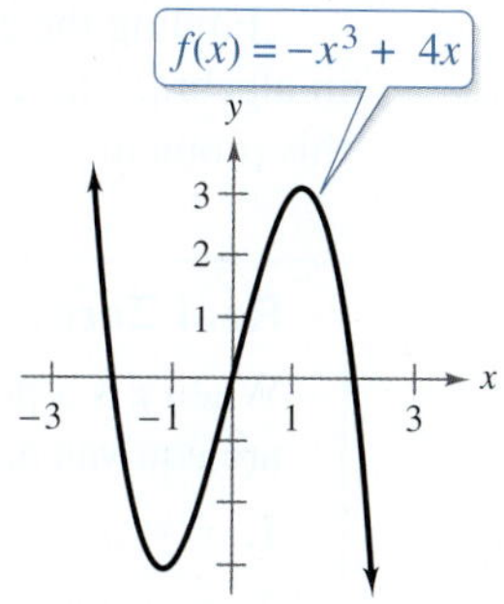

b. Because the degree is even and the leading coefficient is positive, the graph rises to the left and to the right, as shown below.

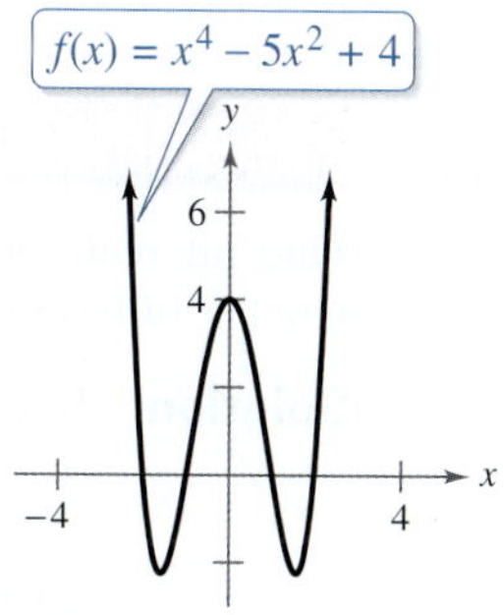

c. Because the degree is odd and the leading coefficient is positive, the graph falls to the left and rises to the right, as shown below.

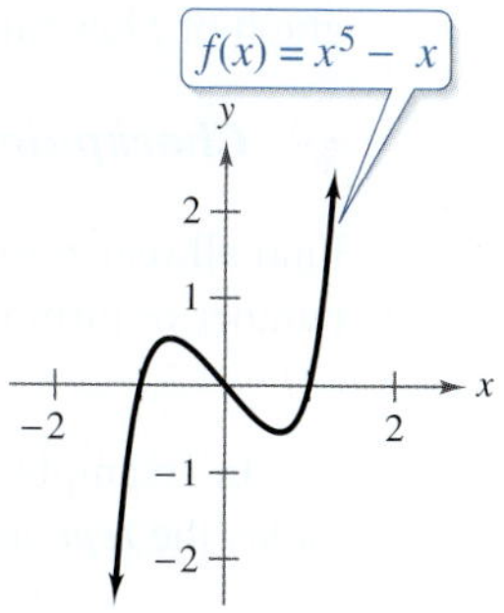

✓ Checkpoint *Audio-video solution in English & Spanish at LarsonPrecalculus.com.*

Describe the right-hand and left-hand behavior of the graph of each function.

a. $f(x) = \frac{1}{4}x^3 - 2x$ **b.** $f(x) = -3.6x^5 + 5x^3 - 1$

In Example 2, note that the Leading Coefficient Test tells you only whether the graph *eventually* rises or falls to the right or to the left. Other characteristics of the graph, such as intercepts and minimum and maximum points, must be determined by other tests.

Real Zeros of Polynomial Functions

It can be shown that for a polynomial function f of degree n, the following statements are true.

1. The function f has, at most, n real zeros. (You will study this result in detail in the discussion of the Fundamental Theorem of Algebra in Section 2.5.)
2. The graph of f has, at most, $n - 1$ turning points. (Turning points, also called relative minima or relative maxima, are points at which the graph changes from increasing to decreasing or vice versa.)

REMARK Remember that the *zeros* of a function of x are the x-values for which the function is zero.

Finding the zeros of a polynomial function is one of the most important problems in algebra. There is a strong interplay between graphical and algebraic approaches to this problem.

Real Zeros of Polynomial Functions

When f is a polynomial function and a is a real number, the following statements are equivalent.

1. $x = a$ is a *zero* of the function f.
2. $x = a$ is a *solution* of the polynomial equation $f(x) = 0$.
3. $(x - a)$ is a *factor* of the polynomial $f(x)$.
4. $(a, 0)$ is an *x-intercept* of the graph of f.

ALGEBRA HELP To do Example 3 algebraically, you need to be able to completely factor polynomials. You can review the techniques for factoring in Appendix A.3.

EXAMPLE 3 Finding Real Zeros of a Polynomial Function

Find all real zeros of $f(x) = -2x^4 + 2x^2$. Then determine the maximum possible number of turning points of the graph of the function.

Solution To find the real zeros of the function, set $f(x)$ equal to zero and solve for x.

$$-2x^4 + 2x^2 = 0 \quad \text{Set } f(x) \text{ equal to 0.}$$

$$-2x^2(x^2 - 1) = 0 \quad \text{Remove common monomial factor.}$$

$$-2x^2(x - 1)(x + 1) = 0 \quad \text{Factor completely.}$$

So, the real zeros are $x = 0$, $x = 1$, and $x = -1$. Because the function is a fourth-degree polynomial, the graph of f can have at most $4 - 1 = 3$ turning points. In this case, the graph of f has three turning points, as shown in Figure 2.12.

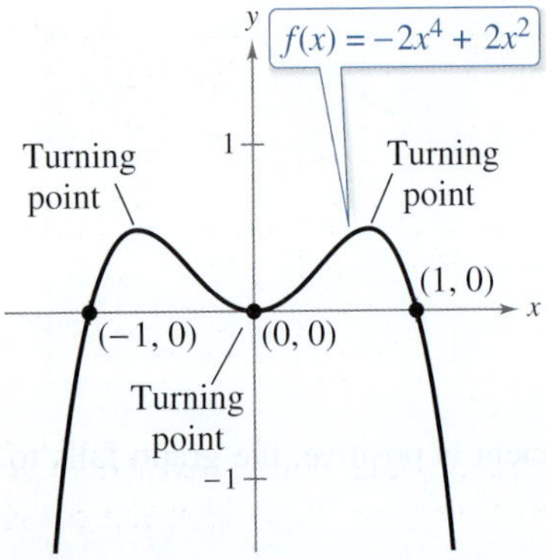

Figure 2.12

✓ **Checkpoint** Audio-video solution in English & Spanish at LarsonPrecalculus.com.

Find all real zeros of $f(x) = x^3 - 12x^2 + 36x$. Then determine the maximum possible number of turning points of the graph of the function.

In Example 3, note that because the exponent is greater than 1, the factor $-2x^2$ yields the *repeated* zero $x = 0$.

Repeated Zeros

A factor $(x - a)^k$, $k > 1$, yields a **repeated zero** $x = a$ of **multiplicity** k.

1. When k is odd, the graph *crosses* the x-axis at $x = a$.
2. When k is even, the graph *touches* the x-axis (but does not cross the x-axis) at $x = a$.

To graph polynomial functions, use the fact that a polynomial function can change signs only at its zeros. Between two consecutive zeros, a polynomial must be entirely positive or entirely negative. (This follows from the Intermediate Value Theorem, which you will study later in this section.) This means that when you put the real zeros of a polynomial function in order, they divide the real number line into intervals in which the function has no sign changes. These resulting intervals are **test intervals** in which you choose a representative x-value to determine whether the value of the polynomial function is positive (the graph lies above the x-axis) or negative (the graph lies below the x-axis).

EXAMPLE 4 Sketching the Graph of a Polynomial Function

Sketch the graph of $f(x) = 3x^4 - 4x^3$.

Solution

1. *Apply the Leading Coefficient Test.* Because the leading coefficient is positive and the degree is even, you know that the graph eventually rises to the left and to the right (see Figure 2.13).
2. *Find the Real Zeros of the Polynomial.* By factoring $f(x) = 3x^4 - 4x^3$ as $f(x) = x^3(3x - 4)$, you can see that the real zeros of f are $x = 0$ and $x = \frac{4}{3}$ (both of odd multiplicity). So, the x-intercepts occur at $(0, 0)$ and $\left(\frac{4}{3}, 0\right)$. Add these points to your graph, as shown in Figure 2.13.
3. *Plot a Few Additional Points.* Use the zeros of the polynomial to find the test intervals. In each test interval, choose a representative x-value and evaluate the polynomial function, as shown in the table.

Test Interval	Representative x-Value	Value of f	Sign	Point on Graph
$(-\infty, 0)$	-1	$f(-1) = 7$	Positive	$(-1, 7)$
$\left(0, \frac{4}{3}\right)$	1	$f(1) = -1$	Negative	$(1, -1)$
$\left(\frac{4}{3}, \infty\right)$	1.5	$f(1.5) = 1.6875$	Positive	$(1.5, 1.6875)$

4. *Draw the Graph.* Draw a continuous curve through the points, as shown in Figure 2.14. Because both zeros are of odd multiplicity, you know that the graph should cross the x-axis at $x = 0$ and $x = \frac{4}{3}$.

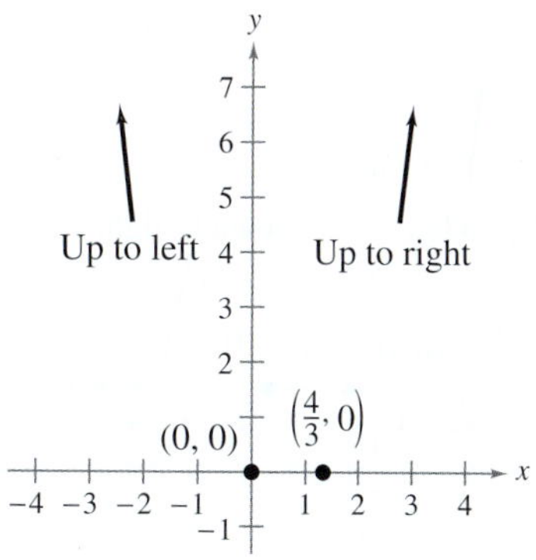

Figure 2.13

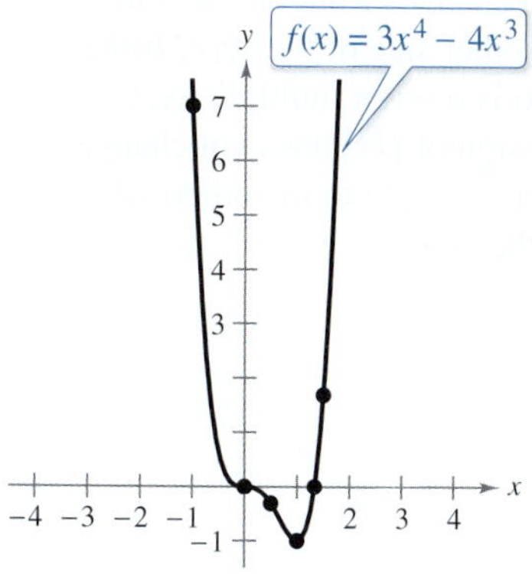

Figure 2.14

✓ **Checkpoint** *Audio-video solution in English & Spanish at LarsonPrecalculus.com.*

Sketch the graph of $f(x) = 2x^3 - 6x^2$.

TECHNOLOGY Example 4 uses an *algebraic approach* to describe the graph of the function. A graphing utility can complement this approach. Remember to find a viewing window that shows all significant features of the graph. For instance, viewing window (a) illustrates all of the significant features of the function in Example 4 while viewing window (b) does not.

(a)

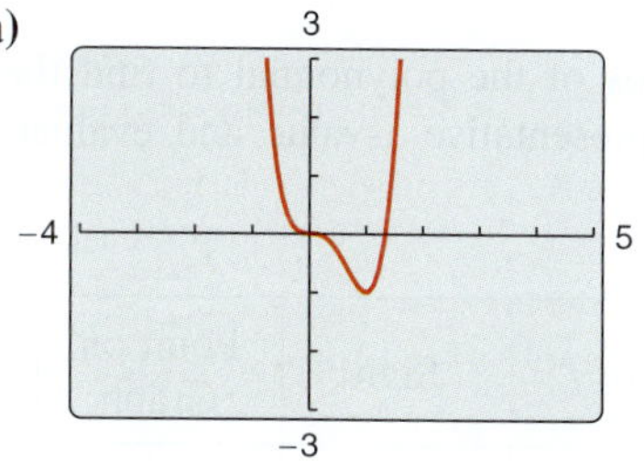

(b)

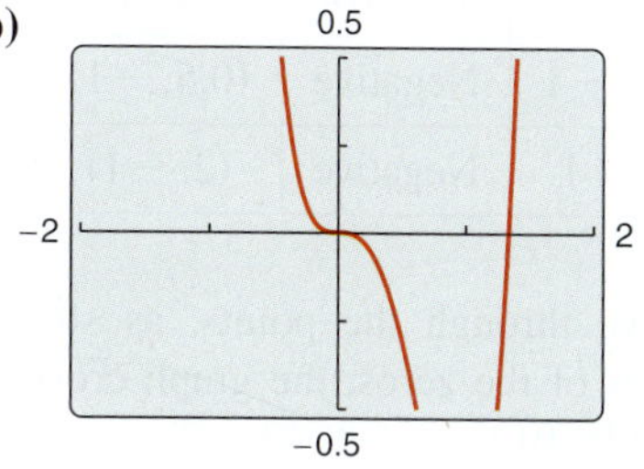

REMARK If you are unsure of the shape of a portion of the graph of a polynomial function, then plot some additional points, such as the point $(0.5, -0.3125)$, as shown in Figure 2.14.

A polynomial function is in **standard form** when its terms are in descending order of exponents from left to right. Before applying the Leading Coefficient Test to a polynomial function, it is a good idea to make sure that the polynomial function is in standard form.

EXAMPLE 5 Sketching the Graph of a Polynomial Function

Sketch the graph of $f(x) = -2x^3 + 6x^2 - \frac{9}{2}x$.

Solution

1. *Apply the Leading Coefficient Test.* Because the leading coefficient is negative and the degree is odd, you know that the graph eventually rises to the left and falls to the right (see Figure 2.15).

2. *Find the Real Zeros of the Polynomial.* By factoring

$$\begin{aligned} f(x) &= -2x^3 + 6x^2 - \tfrac{9}{2}x \\ &= -\tfrac{1}{2}x(4x^2 - 12x + 9) \\ &= -\tfrac{1}{2}x(2x - 3)^2 \end{aligned}$$

you can see that the real zeros of f are $x = 0$ (odd multiplicity) and $x = \frac{3}{2}$ (even multiplicity). So, the x-intercepts occur at $(0, 0)$ and $\left(\frac{3}{2}, 0\right)$. Add these points to your graph, as shown in Figure 2.15.

3. *Plot a Few Additional Points.* Use the zeros of the polynomial to find the test intervals. In each test interval, choose a representative x-value and evaluate the polynomial function, as shown in the table.

Test Interval	Representative x-Value	Value of f	Sign	Point on Graph
$(-\infty, 0)$	-0.5	$f(-0.5) = 4$	Positive	$(-0.5, 4)$
$\left(0, \frac{3}{2}\right)$	0.5	$f(0.5) = -1$	Negative	$(0.5, -1)$
$\left(\frac{3}{2}, \infty\right)$	2	$f(2) = -1$	Negative	$(2, -1)$

REMARK Observe in Example 5 that the sign of $f(x)$ is positive to the left of and negative to the right of the zero $x = 0$. Similarly, the sign of $f(x)$ is negative to the left and to the right of the zero $x = \frac{3}{2}$. This suggests that if the zero of a polynomial function is of *odd* multiplicity, then the sign of $f(x)$ changes from one side to the other side of the zero. If the zero is of *even* multiplicity, then the sign of $f(x)$ does not change from one side to the other side of the zero.

4. *Draw the Graph.* Draw a continuous curve through the points, as shown in Figure 2.16. As indicated by the multiplicities of the zeros, the graph crosses the x-axis at $(0, 0)$ but does not cross the x-axis at $\left(\frac{3}{2}, 0\right)$.

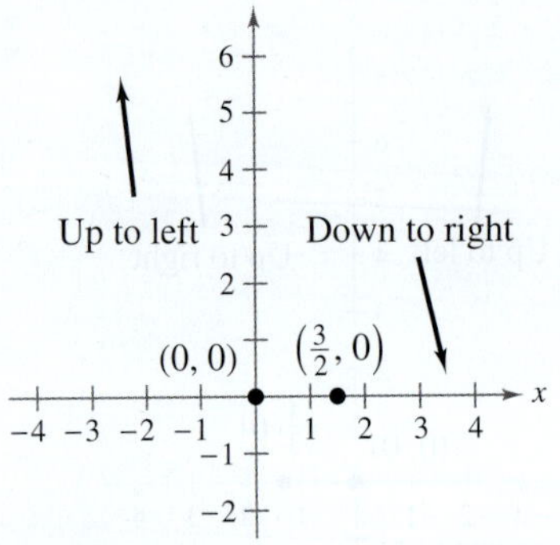

Figure 2.15

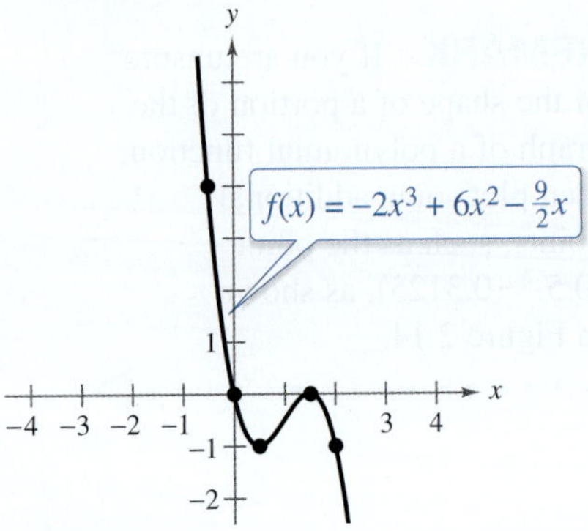

Figure 2.16

✓ **Checkpoint** Audio-video solution in English & Spanish at LarsonPrecalculus.com.

Sketch the graph of $f(x) = -\frac{1}{4}x^4 + \frac{3}{2}x^3 - \frac{9}{4}x^2$.

The Intermediate Value Theorem

The **Intermediate Value Theorem** implies that if

$$(a, f(a)) \quad \text{and} \quad (b, f(b))$$

are two points on the graph of a polynomial function such that $f(a) \neq f(b)$, then for any number d between $f(a)$ and $f(b)$ there must be a number c between a and b such that $f(c) = d$. (See figure.)

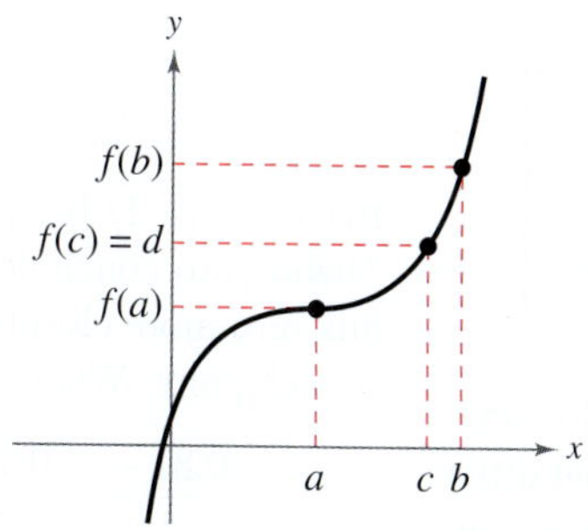

Intermediate Value Theorem

Let a and b be real numbers such that $a < b$. If f is a polynomial function such that $f(a) \neq f(b)$, then, in the interval $[a, b]$, f takes on every value between $f(a)$ and $f(b)$.

The Intermediate Value Theorem helps you locate the real zeros of a polynomial function in the following way. If you can find a value $x = a$ at which a polynomial function is positive, and another value $x = b$ at which it is negative, then you can conclude that the function has at least one real zero between these two values. For example, the function

$$f(x) = x^3 + x^2 + 1$$

is negative when $x = -2$ and positive when $x = -1$. So, it follows from the Intermediate Value Theorem that f must have a real zero somewhere between -2 and -1, as shown below.

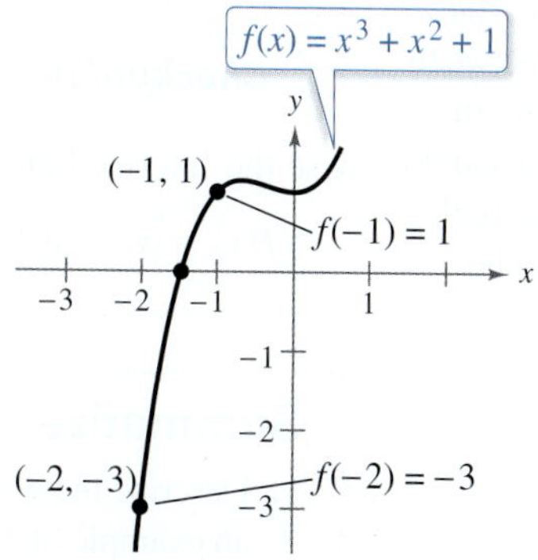

The function f must have a real zero somewhere between -2 and -1.

By continuing this line of reasoning, you can approximate any real zeros of a polynomial function to any desired accuracy. This concept is further demonstrated in Example 6.

EXAMPLE 6 Using the Intermediate Value Theorem

Use the Intermediate Value Theorem to approximate the real zero of

$$f(x) = x^3 - x^2 + 1.$$

Solution Begin by computing a few function values, as follows.

x	-2	-1	0	1
$f(x)$	-11	-1	1	1

Because $f(-1)$ is negative and $f(0)$ is positive, you can apply the Intermediate Value Theorem to conclude that the function has a real zero between -1 and 0. To pinpoint this zero more closely, divide the interval $[-1, 0]$ into tenths and evaluate the function at each point. When you do this, you will find that

$$f(-0.8) = -0.152 \quad \text{and} \quad f(-0.7) = 0.167.$$

So, f must have a real zero between -0.8 and -0.7, as shown below.

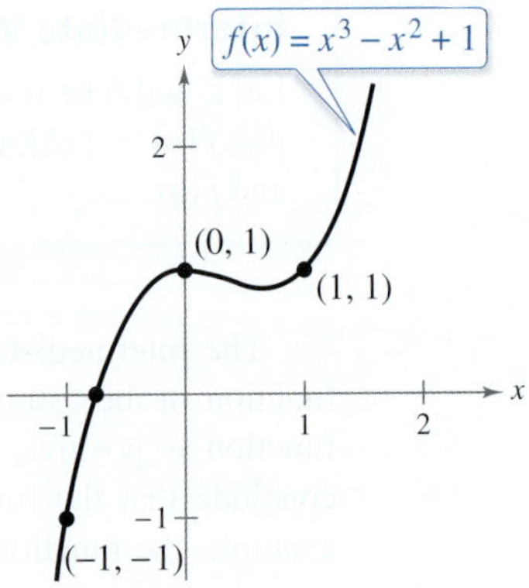

The function f has a real zero between -0.8 and -0.7.

For a more accurate approximation, compute function values between $f(-0.8)$ and $f(-0.7)$ and apply the Intermediate Value Theorem again. By continuing this process, you can approximate this zero to any desired accuracy.

Checkpoint *Audio-video solution in English & Spanish at LarsonPrecalculus.com.*

Use the Intermediate Value Theorem to approximate the real zero of

$$f(x) = x^3 - 3x^2 - 2.$$

Summarize (Section 2.2)

1. Describe the graphs of monomial functions of the form $f(x) = x^n$ *(page 125)*. For an example of sketching transformations of monomial functions, see Example 1.
2. Describe the Leading Coefficient Test *(page 126)*. For an example of applying the Leading Coefficient Test, see Example 2.
3. Describe the real zeros of polynomial functions *(page 128)*. For examples involving the real zeros of polynomial functions, see Examples 3–5.
4. State the Intermediate Value Theorem *(page 131)*. For an example of using the Intermediate Value Theorem, see Example 6.

TECHNOLOGY You can use the *table* feature of a graphing utility to approximate the real zero of the polynomial function in Example 6. Construct a table that shows function values for $-2 \le x \le 4$, as shown below.

X	Y1
-2	-11
-1	-1
0	1
1	1
2	5
3	19
4	49

X=0

Scroll through the table looking for consecutive values that differ in sign. From the table, you can see that $f(-1)$ and $f(0)$ differ in sign. So, you can conclude from the Intermediate Value Theorem that the function has a real zero between -1 and 0. You can adjust your table to show function values for $-1 \le x \le 0$ using increments of 0.1, as shown below.

X	Y1
-1	-1
-.9	-.539
-.8	-.152
-.7	.167
-.6	.424
-.5	.625
-.4	.776

X=-.7

By scrolling through the table, you can see that $f(-0.8)$ and $f(-0.7)$ differ in sign. So, the function has a real zero between -0.8 and -0.7. Repeating this process several times, you should obtain $x \approx -0.755$ as the real zero of the function. Use the *zero* or *root* feature of the graphing utility to confirm this result.

2.2 Exercises

See **CalcChat.com** for tutorial help and worked-out solutions to odd-numbered exercises.

Vocabulary: Fill in the blanks.

1. The graphs of all polynomial functions are ________, which means that the graphs have no breaks, holes, or gaps.
2. The ________ ________ ________ is used to determine the left-hand and right-hand behavior of the graph of a polynomial function.
3. A polynomial function of degree n has at most ________ real zeros and at most ________ turning points.
4. When $x = a$ is a zero of a polynomial function f, the following three statements are true.
   (a) $x = a$ is a ________ of the polynomial equation $f(x) = 0$.
   (b) ________ is a factor of the polynomial $f(x)$.
   (c) $(a, 0)$ is an ________ of the graph of f.
5. When a real zero of a polynomial function is of even multiplicity, the graph of f ________ the x-axis at $x = a$, and when it is of odd multiplicity, the graph of f ________ the x-axis at $x = a$.
6. A factor $(x - a)^k$, $k > 1$, yields a ________ ________ $x = a$ of ________ k.
7. A polynomial function is written in ________ form when its terms are written in descending order of exponents from left to right.
8. The ________ ________ Theorem states that if f is a polynomial function such that $f(a) \neq f(b)$, then, in the interval $[a, b]$, f takes on every value between $f(a)$ and $f(b)$.

Skills and Applications

Matching **In Exercises 9–14, match the polynomial function with its graph. [The graphs are labeled (a), (b), (c), (d), (e), and (f).]**

(a)
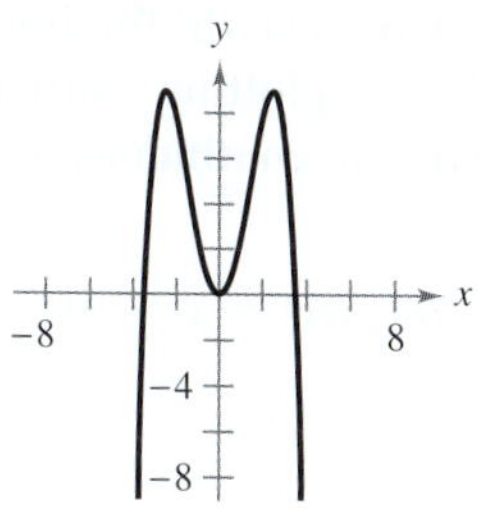

(b)
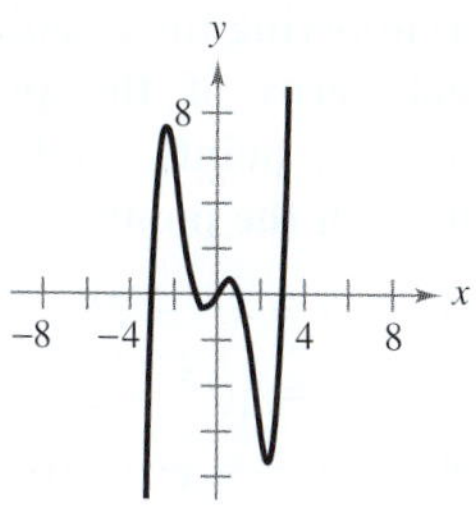

(c)
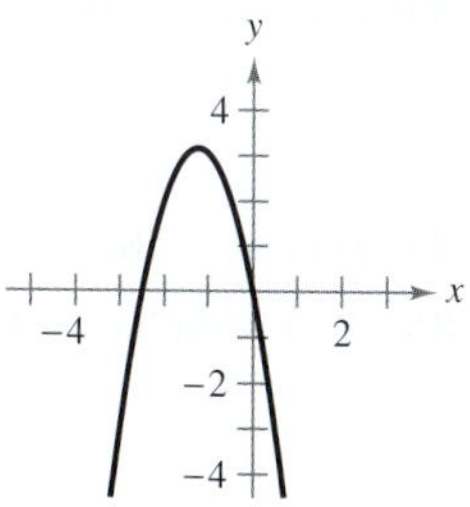

(d)
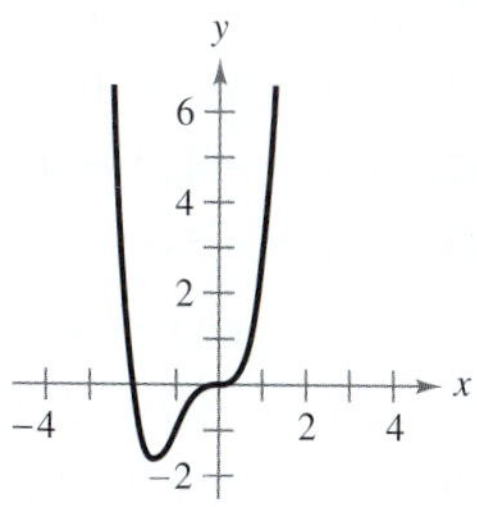

(e)
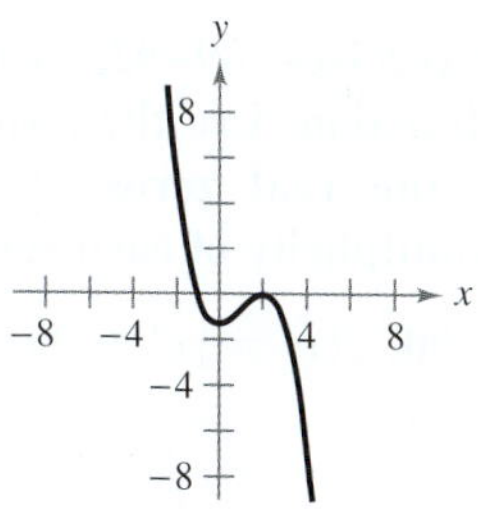

(f)
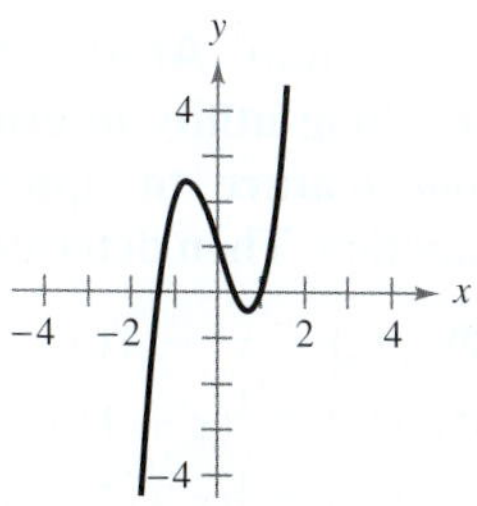

9. $f(x) = -2x^2 - 5x$
10. $f(x) = 2x^3 - 3x + 1$
11. $f(x) = -\frac{1}{4}x^4 + 3x^2$
12. $f(x) = -\frac{1}{3}x^3 + x^2 - \frac{4}{3}$
13. $f(x) = x^4 + 2x^3$
14. $f(x) = \frac{1}{5}x^5 - 2x^3 + \frac{9}{5}x$

Sketching Transformations of Monomial Functions **In Exercises 15–18, sketch the graph of $y = x^n$ and each transformation.**

15. $y = x^3$
   (a) $f(x) = (x - 4)^3$ (b) $f(x) = x^3 - 4$
   (c) $f(x) = -\frac{1}{4}x^3$ (d) $f(x) = (x - 4)^3 - 4$
16. $y = x^5$
   (a) $f(x) = (x + 1)^5$ (b) $f(x) = x^5 + 1$
   (c) $f(x) = 1 - \frac{1}{2}x^5$ (d) $f(x) = -\frac{1}{2}(x + 1)^5$
17. $y = x^4$
   (a) $f(x) = (x + 3)^4$ (b) $f(x) = x^4 - 3$
   (c) $f(x) = 4 - x^4$ (d) $f(x) = \frac{1}{2}(x - 1)^4$
   (e) $f(x) = (2x)^4 + 1$ (f) $f(x) = \left(\frac{1}{2}x\right)^4 - 2$
18. $y = x^6$
   (a) $f(x) = -\frac{1}{8}x^6$ (b) $f(x) = (x + 2)^6 - 4$
   (c) $f(x) = x^6 - 5$ (d) $f(x) = -\frac{1}{4}x^6 + 1$
   (e) $f(x) = \left(\frac{1}{4}x\right)^6 - 2$ (f) $f(x) = (2x)^6 - 1$

Apply the Leading Coefficient Test In Exercises 19–30, describe the right-hand and left-hand behavior of the graph of the polynomial function.

19. $f(x) = \frac{1}{5}x^3 + 4x$
20. $f(x) = 2x^2 - 3x + 1$
21. $g(x) = 5 - \frac{7}{2}x - 3x^2$
22. $h(x) = 1 - x^6$
23. $g(x) = -x^3 + 3x^2$
24. $g(x) = -x^4 + 4x - 6$
25. $f(x) = -2.1x^5 + 4x^3 - 2$
26. $f(x) = 4x^5 - 7x + 6.5$
27. $f(x) = 6 - 2x + 4x^2 - 5x^7$
28. $f(x) = (3x^4 - 2x + 5)/4$
29. $h(t) = -\frac{3}{4}(t^2 - 3t + 6)$
30. $f(s) = -\frac{7}{8}(s^3 + 5s^2 - 7s + 1)$

Graphical Analysis In Exercises 31–34, use a graphing utility to graph the functions f and g in the same viewing window. Zoom out sufficiently far to show that the right-hand and left-hand behaviors of f and g appear identical.

31. $f(x) = 3x^3 - 9x + 1, \quad g(x) = 3x^3$
32. $f(x) = -\frac{1}{3}(x^3 - 3x + 2), \quad g(x) = -\frac{1}{3}x^3$
33. $f(x) = -(x^4 - 4x^3 + 16x), \quad g(x) = -x^4$
34. $f(x) = 3x^4 - 6x^2, \quad g(x) = 3x^4$

Finding Real Zeros of a Polynomial Function In Exercises 35–50, (a) find all real zeros of the polynomial function, (b) determine the multiplicity of each zero, (c) determine the maximum possible number of turning points of the graph of the function, and (d) use a graphing utility to graph the function and verify your answers.

35. $f(x) = x^2 - 36$
36. $f(x) = 81 - x^2$
37. $h(t) = t^2 - 6t + 9$
38. $f(x) = x^2 + 10x + 25$
39. $f(x) = \frac{1}{3}x^2 + \frac{1}{3}x - \frac{2}{3}$
40. $f(x) = \frac{1}{2}x^2 + \frac{5}{2}x - \frac{3}{2}$
41. $f(x) = 3x^3 - 12x^2 + 3x$
42. $g(x) = 5x(x^2 - 2x - 1)$
43. $f(t) = t^3 - 8t^2 + 16t$
44. $f(x) = x^4 - x^3 - 30x^2$
45. $g(t) = t^5 - 6t^3 + 9t$
46. $f(x) = x^5 + x^3 - 6x$
47. $f(x) = 3x^4 + 9x^2 + 6$
48. $f(x) = 2x^4 - 2x^2 - 40$
49. $g(x) = x^3 + 3x^2 - 4x - 12$
50. $f(x) = x^3 - 4x^2 - 25x + 100$

Graphical Analysis In Exercises 51–54, (a) use a graphing utility to graph the function, (b) use the graph to approximate any x-intercepts of the graph, (c) set $y = 0$ and solve the resulting equation, and (d) compare the results of part (c) with any x-intercepts of the graph.

51. $y = 4x^3 - 20x^2 + 25x$
52. $y = 4x^3 + 4x^2 - 8x - 8$
53. $y = x^5 - 5x^3 + 4x$
54. $y = \frac{1}{4}x^3(x^2 - 9)$

Finding a Polynomial Function In Exercises 55–64, find a polynomial function that has the given zeros. (There are many correct answers.)

55. 0, 8
56. 0, -7
57. 2, -6
58. -4, 5
59. 0, -4, -5
60. 0, 1, 10
61. 4, -3, 3, 0
62. -2, -1, 0, 1, 2
63. $1 + \sqrt{3}$, $1 - \sqrt{3}$
64. 2, $4 + \sqrt{5}$, $4 - \sqrt{5}$

Finding a Polynomial Function In Exercises 65–74, find a polynomial of degree n that has the given zero(s). (There are many correct answers.)

	Zero(s)	Degree
65.	$x = -3$	$n = 2$
66.	$x = -12, -6$	$n = 2$
67.	$x = -5, 0, 1$	$n = 3$
68.	$x = -2, 4, 7$	$n = 3$
69.	$x = 0, \sqrt{3}, -\sqrt{3}$	$n = 3$
70.	$x = 9$	$n = 3$
71.	$x = -5, 1, 2$	$n = 4$
72.	$x = -4, -1, 3, 6$	$n = 4$
73.	$x = 0, -4$	$n = 5$
74.	$x = -1, 4, 7, 8$	$n = 5$

Sketching the Graph of a Polynomial Function In Exercises 75–88, sketch the graph of the function by (a) applying the Leading Coefficient Test, (b) finding the real zeros of the polynomial, (c) plotting sufficient solution points, and (d) drawing a continuous curve through the points.

75. $f(x) = x^3 - 25x$
76. $g(x) = x^4 - 9x^2$
77. $f(t) = \frac{1}{4}(t^2 - 2t + 15)$
78. $g(x) = -x^2 + 10x - 16$
79. $f(x) = x^3 - 2x^2$
80. $f(x) = 8 - x^3$
81. $f(x) = 3x^3 - 15x^2 + 18x$
82. $f(x) = -4x^3 + 4x^2 + 15x$
83. $f(x) = -5x^2 - x^3$
84. $f(x) = -48x^2 + 3x^4$
85. $f(x) = x^2(x - 4)$
86. $h(x) = \frac{1}{3}x^3(x - 4)^2$
87. $g(t) = -\frac{1}{4}(t - 2)^2(t + 2)^2$
88. $g(x) = \frac{1}{10}(x + 1)^2(x - 3)^3$

Graphical Analysis In Exercises 89–92, use a graphing utility to graph the function. Use the *zero* or *root* feature to approximate the real zeros of the function. Then determine the multiplicity of each zero.

89. $f(x) = x^3 - 16x$
90. $f(x) = \frac{1}{4}x^4 - 2x^2$
91. $g(x) = \frac{1}{5}(x + 1)^2(x - 3)(2x - 9)$
92. $h(x) = \frac{1}{5}(x + 2)^2(3x - 5)^2$

Using the Intermediate Value Theorem **In Exercises 93–96, (a) use the Intermediate Value Theorem and the *table* feature of a graphing utility to find intervals one unit in length in which the polynomial function is guaranteed to have a zero. (b) Adjust the table to approximate the zeros of the function. Use the *zero* or *root* feature of the graphing utility to verify your results.**

93. $f(x) = x^3 - 3x^2 + 3$

94. $f(x) = 0.11x^3 - 2.07x^2 + 9.81x - 6.88$

95. $g(x) = 3x^4 + 4x^3 - 3$

96. $h(x) = x^4 - 10x^2 + 3$

97. Numerical and Graphical Analysis You plan to construct an open box from a square piece of material, 36 inches on a side, by cutting equal squares with sides of length x from the corners and turning up the sides (see figure).

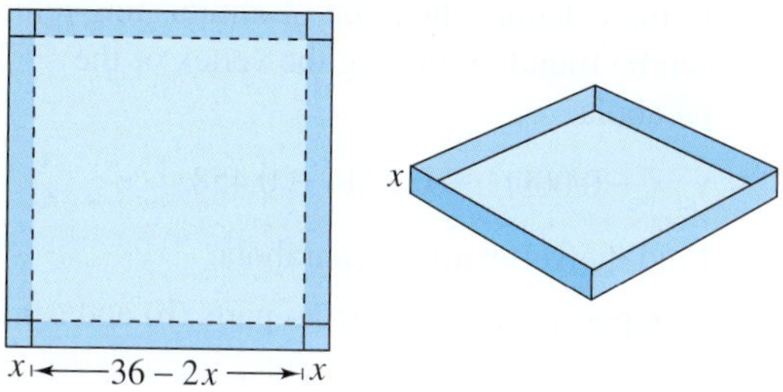

(a) Write a function V that represents the volume of the box.

(b) Determine the domain of the function V.

(c) Use a graphing utility to construct a table that shows the box heights x and the corresponding volumes $V(x)$. Use the table to estimate the dimensions that will produce a maximum volume.

(d) Use the graphing utility to graph V and use the graph to estimate the value of x for which $V(x)$ is a maximum. Compare your result with that of part (c).

98. Maximum Volume You plan to construct an open box with locking tabs from a square piece of material 24 inches on a side by cutting equal squares from the corners and folding along the dashed lines (see figure).

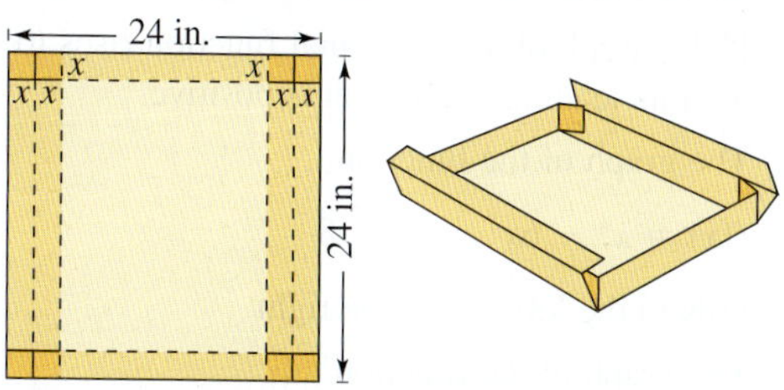

(a) Write a function V that represents the volume of the box.

(b) Determine the domain of the function V.

(c) Sketch a graph of the function and estimate the value of x for which $V(x)$ is a maximum.

99. Construction A roofing contractor is fabricating gutters from 12-inch aluminum sheeting. The contractor plans to use an aluminum siding folding press to create the gutter by creasing equal lengths for the sidewalls (see figure).

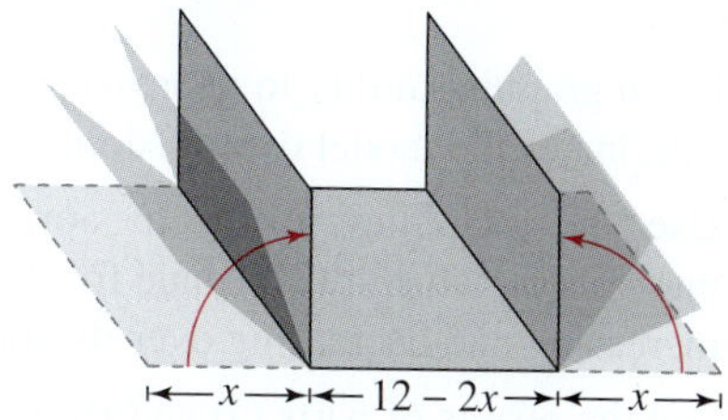

(a) Let x represent the height of the sidewall of the gutter. Write a function A that represents the cross-sectional area of the gutter.

(b) The length of the aluminum sheeting is 16 feet. Write a function V that represents the volume of one run of gutter in terms of x.

(c) Determine the domain of the function V.

(d) Use a graphing utility to construct a table that shows the sidewall heights x and the corresponding volumes $V(x)$. Use the table to estimate the dimensions that will produce a maximum volume.

(e) Use the graphing utility to graph V. Use the graph to estimate the value of x for which $V(x)$ is a maximum. Compare your result with that of part (d).

(f) Would changing the length of the aluminum sheet affect the value of x for which $V(x)$ is a maximum? Explain.

100. Construction An industrial propane tank is formed by adjoining two hemispheres to the ends of a right circular cylinder. The length of the cylindrical portion of the tank is four times the radius of the hemispherical components (see figure).

(a) Write a function V that represents the total volume of the tank in terms of r.

(b) Find the domain of the function V.

(c) Use a graphing utility to graph the function.

(d) Use the graph from part (c) to estimate the radius and length of the cylindrical portion of the tank when the total volume of the tank is 120 cubic feet.

101. Revenue The revenues R (in millions of dollars) for a company from 2003 through 2010 can be modeled by

$$R = 6.212t^3 - 132.87t^2 + 863.2t - 1115, \; 3 \le t \le 10$$

where t represents the year, with $t = 3$ corresponding to 2003.

(a) Use a graphing utility to approximate any relative extrema of the model over its domain.

(b) Use the graphing utility to approximate the intervals on which the revenue for the company is increasing and decreasing over its domain.

(c) Use the results of parts (a) and (b) to describe the company's revenue during this time period.

102. Revenue The revenues R (in millions of dollars) for a company from 2003 through 2010 can be modeled by

$$R = -0.1685t^4 + 4.298t^3 - 39.044t^2 + 149.9t - 185,$$
$$3 \le t \le 10$$

where t represents the year, with $t = 3$ corresponding to 2003.

(a) Use a graphing utility to approximate any relative extrema of the model over its domain.

(b) Use the graphing utility to approximate the intervals on which the revenue for the company is increasing and decreasing over its domain.

(c) Use the results of parts (a) and (b) to describe the company's revenue during this time period.

103. Revenue The revenue R (in millions of dollars) for a company is related to its advertising expense by the function

$$R = \frac{1}{100{,}000}(-x^3 + 600x^2), \quad 0 \le x \le 400$$

where x is the amount spent on advertising (in tens of thousands of dollars). Use the graph of this function to estimate the point on the graph at which the function is increasing most rapidly. This point is called the *point of diminishing returns* because any expense above this amount will yield less return per dollar invested in advertising.

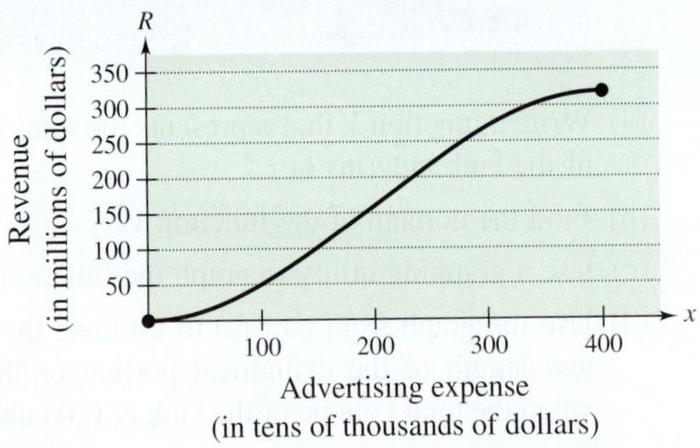

104. Tree Growth The growth of a red oak tree is approximated by the function

$$G = -0.003t^3 + 0.137t^2 + 0.458t - 0.839,$$
$$2 \le t \le 34$$

where G is the height of the tree (in feet) and t is its age (in years).

(a) Use a graphing utility to graph the function.

(b) Estimate the age of the tree when it is growing most rapidly. This point is called the *point of diminishing returns* because the increase in size will be less with each additional year.

(c) Using calculus, the point of diminishing returns can be found by finding the vertex of the parabola

$$y = -0.009t^2 + 0.274t + 0.458.$$

Find the vertex of this parabola.

(d) Compare your results from parts (b) and (c).

Exploration

True or False? **In Exercises 105–111, determine whether the statement is true or false. Justify your answer.**

105. A fifth-degree polynomial function can have five turning points in its graph.

106. If f is a polynomial function of x such that $f(2) = -6$ and $f(6) = 6$, then f has at most one real zero between $x = 2$ and $x = 6$.

107. A polynomial function cannot have more real zeros than it has turning points.

108. If the graph of a polynomial function falls to the right, then its leading coefficient is negative.

109. If the graph of a polynomial function rises to the left, then its leading coefficient is positive.

110. The graph of the function

$$f(x) = x^4 - 6x^3 - 3x - 8$$

falls to the left and to the right.

111. The graph of the function

$$f(x) = 2 + x - x^2 + x^3 - x^4 + x^5 + x^6 - x^7$$

rises to the left and falls to the right.

Artistas/Shutterstock.com

112. HOW DO YOU SEE IT? For each graph, describe a polynomial function that could represent the graph. (Indicate the degree of the function and the sign of its leading coefficient.)

(a)

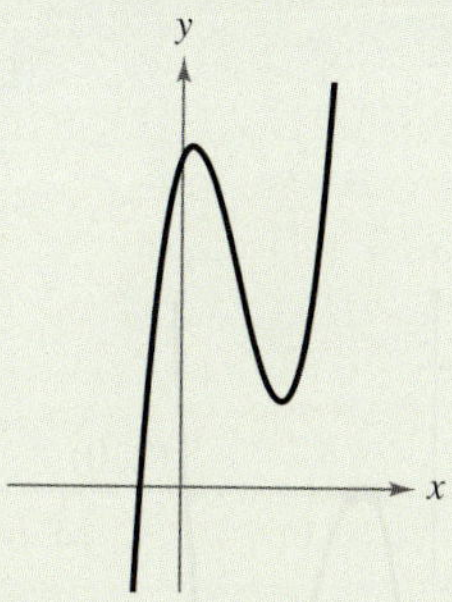

(b)

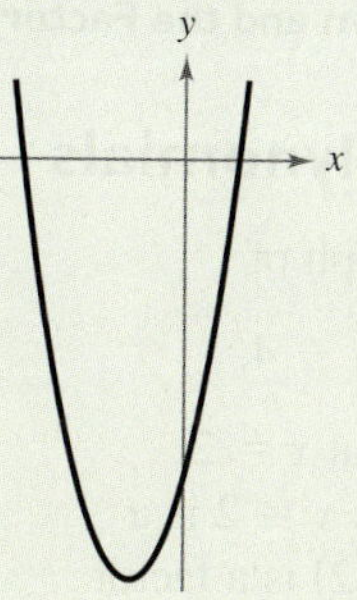

(c)

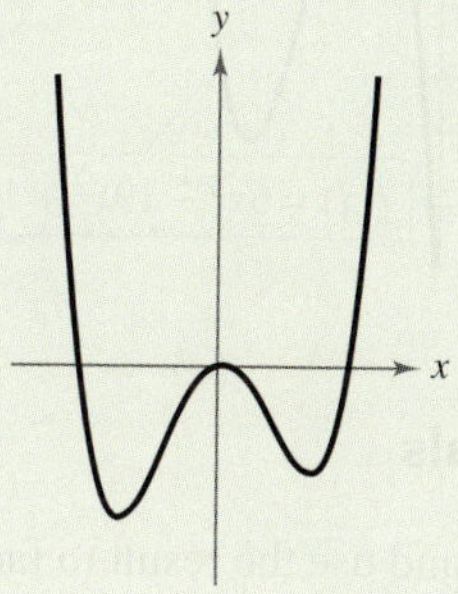

(d)

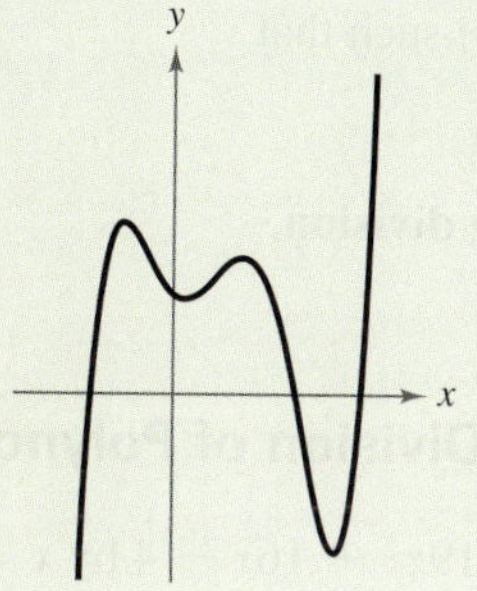

113. Modeling Polynomials Sketch the graph of a polynomial function that is of fourth degree, has a zero of multiplicity 2, and has a negative leading coefficient. Sketch the graph of another polynomial function with the same characteristics except that the leading coefficient is positive.

114. Modeling Polynomials Sketch the graph of a polynomial function that is of fifth degree, has a zero of multiplicity 2, and has a negative leading coefficient. Sketch the graph of another polynomial function with the same characteristics except that the leading coefficient is positive.

115. Graphical Reasoning Sketch a graph of the function $f(x) = x^4$. Explain how the graph of each function g differs (if it does) from the graph of f. Determine whether g is even, odd, or neither.

(a) $g(x) = f(x) + 2$

(b) $g(x) = f(x + 2)$

(c) $g(x) = f(-x)$

(d) $g(x) = -f(x)$

(e) $g(x) = f\left(\frac{1}{2}x\right)$

(f) $g(x) = \frac{1}{2}f(x)$

(g) $g(x) = f(x^{3/4})$

(h) $g(x) = (f \circ f)(x)$

116. Think About It For each function, identify the degree of the function and whether the degree of the function is even or odd. Identify the leading coefficient and whether the leading coefficient is positive or negative. Use a graphing utility to graph each function. Describe the relationship between the degree of the function and the sign of the leading coefficient of the function and the right-hand and left-hand behavior of the graph of the function.

(a) $f(x) = x^3 - 2x^2 - x + 1$

(b) $f(x) = 2x^5 + 2x^2 - 5x + 1$

(c) $f(x) = -2x^5 - x^2 + 5x + 3$

(d) $f(x) = -x^3 + 5x - 2$

(e) $f(x) = 2x^2 + 3x - 4$

(f) $f(x) = x^4 - 3x^2 + 2x - 1$

(g) $f(x) = x^2 + 3x + 2$

117. Think About It Sketch the graph of each polynomial function. Then count the number of real zeros of the function and the numbers of relative minima and relative maxima. Compare these numbers with the degree of the polynomial. What do you observe?

(a) $f(x) = -x^3 + 9x$

(b) $f(x) = x^4 - 10x^2 + 9$

(c) $f(x) = x^5 - 16x$

118. Think About It Explore transformations of the form

$$g(x) = a(x - h)^5 + k.$$

(a) Use a graphing utility to graph the functions

$$y_1 = -\frac{1}{3}(x - 2)^5 + 1$$

and

$$y_2 = \frac{3}{5}(x + 2)^5 - 3.$$

Determine whether the graphs are increasing or decreasing. Explain.

(b) Will the graph of g always be increasing or decreasing? If so, then is this behavior determined by a, h, or k? Explain.

(c) Use the graphing utility to graph the function

$$H(x) = x^5 - 3x^3 + 2x + 1.$$

Use the graph and the result of part (b) to determine whether H can be written in the form

$$H(x) = a(x - h)^5 + k.$$

Explain.

2.3 Polynomial and Synthetic Division

Synthetic division can help you evaluate polynomial functions. For instance, in Exercise 86 on page 146, you will use synthetic division to evaluate a model for the number of confirmed cases of Lyme disease in Maine.

- **Use long division to divide polynomials by other polynomials.**
- **Use synthetic division to divide polynomials by binomials of the form $(x - k)$.**
- **Use the Remainder Theorem and the Factor Theorem.**

Long Division of Polynomials

Suppose you are given the graph of

$$f(x) = 6x^3 - 19x^2 + 16x - 4.$$

Notice that a zero of f occurs at $x = 2$, as shown at the right. Because $x = 2$ is a zero of f, you know that $(x - 2)$ is a factor of $f(x)$. This means that there exists a second-degree polynomial $q(x)$ such that

$$f(x) = (x - 2) \cdot q(x).$$

To find $q(x)$, you can use **long division,** as illustrated in Example 1.

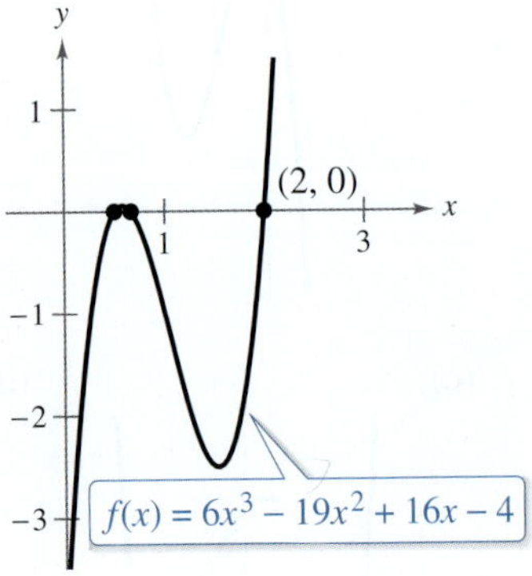

EXAMPLE 1 Long Division of Polynomials

Divide the polynomial $6x^3 - 19x^2 + 16x - 4$ by $x - 2$, and use the result to factor the polynomial completely.

Solution

Think $\frac{6x^3}{x} = 6x^2$.

Think $\frac{-7x^2}{x} = -7x$.

Think $\frac{2x}{x} = 2$.

$$\begin{array}{rll}
 & 6x^2 - 7x + 2 & \\
x - 2 \,\big)\!\! & 6x^3 - 19x^2 + 16x - 4 & \\
 & 6x^3 - 12x^2 & \text{Multiply: } 6x^2(x - 2). \\
 & -7x^2 + 16x & \text{Subtract.} \\
 & -7x^2 + 14x & \text{Multiply: } -7x(x - 2). \\
 & 2x - 4 & \text{Subtract.} \\
 & 2x - 4 & \text{Multiply: } 2(x - 2). \\
 & 0 & \text{Subtract.}
\end{array}$$

From this division, you can conclude that

$$6x^3 - 19x^2 + 16x - 4 = (x - 2)(6x^2 - 7x + 2)$$

and by factoring the quadratic $6x^2 - 7x + 2$, you have

$$6x^3 - 19x^2 + 16x - 4 = (x - 2)(2x - 1)(3x - 2).$$

✓ ***Checkpoint*** *Audio-video solution in English & Spanish at LarsonPrecalculus.com.*

Divide the polynomial $3x^2 + 19x + 28$ by $x + 4$, and use the result to factor the polynomial completely.

Dariusz Majgier/Shutterstock.com

In Example 1, $x - 2$ is a factor of the polynomial

$$6x^3 - 19x^2 + 16x - 4$$

and the long division process produces a remainder of zero. Often, long division will produce a nonzero remainder. For instance, when you divide $x^2 + 3x + 5$ by $x + 1$, you obtain the following.

$$\begin{array}{rl} & x + 2 \quad \longleftarrow \text{Quotient} \\ \text{Divisor} \longrightarrow \; x + 1 & \overline{)\,x^2 + 3x + 5} \quad \longleftarrow \text{Dividend} \\ & \underline{x^2 + x} \\ & 2x + 5 \\ & \underline{2x + 2} \\ & 3 \quad \longleftarrow \text{Remainder} \end{array}$$

In fractional form, you can write this result as follows.

$$\underbrace{\frac{\overbrace{x^2 + 3x + 5}^{\text{Dividend}}}{x + 1}}_{\text{Divisor}} = \overbrace{x + 2}^{\text{Quotient}} + \frac{\overset{\text{Remainder}}{3}}{\underbrace{x + 1}_{\text{Divisor}}}$$

This implies that

$$x^2 + 3x + 5 = (x + 1)(x + 2) + 3 \qquad \text{Multiply each side by } (x + 1).$$

which illustrates the following theorem, called the **Division Algorithm.**

The Division Algorithm

If $f(x)$ and $d(x)$ are polynomials such that $d(x) \neq 0$, and the degree of $d(x)$ is less than or equal to the degree of $f(x)$, then there exist unique polynomials $q(x)$ and $r(x)$ such that

$$\underset{\text{Dividend}}{f(x)} = \underset{\text{Divisor}}{d(x)}\underset{\text{Quotient}}{q(x)} + \underset{\text{Remainder}}{r(x)}$$

where $r(x) = 0$ or the degree of $r(x)$ is less than the degree of $d(x)$. If the remainder $r(x)$ is zero, $d(x)$ *divides evenly* into $f(x)$.

The Division Algorithm can also be written as

$$\frac{f(x)}{d(x)} = q(x) + \frac{r(x)}{d(x)}.$$

In the Division Algorithm, the rational expression $f(x)/d(x)$ is **improper** because the degree of $f(x)$ is greater than or equal to the degree of $d(x)$. On the other hand, the rational expression $r(x)/d(x)$ is **proper** because the degree of $r(x)$ is less than the degree of $d(x)$.

Before you apply the Division Algorithm, follow these steps.

1. Write the dividend and divisor in descending powers of the variable.
2. Insert placeholders with zero coefficients for missing powers of the variable.

Note how these steps are applied in the next example.

EXAMPLE 2 Long Division of Polynomials

Divide $x^3 - 1$ by $x - 1$.

Solution Because there is no x^2-term or x-term in the dividend $x^3 - 1$, you need to rewrite the dividend as $x^3 + 0x^2 + 0x - 1$ before you apply the Division Algorithm.

$$\begin{array}{r l}
x^2 + x + 1 & \\
x - 1 \overline{\big)\, x^3 + 0x^2 + 0x - 1} & \\
\underline{x^3 - x^2} \qquad\qquad & \text{Multiply } x^2 \text{ by } x - 1. \\
x^2 + 0x \qquad & \text{Subtract and bring down } 0x. \\
\underline{x^2 - x} \qquad & \text{Multiply } x \text{ by } x - 1. \\
x - 1 & \text{Subtract and bring down } -1. \\
\underline{x - 1} & \text{Multiply } 1 \text{ by } x - 1. \\
0 & \text{Subtract.}
\end{array}$$

So, $x - 1$ divides evenly into $x^3 - 1$, and you can write

$$\frac{x^3 - 1}{x - 1} = x^2 + x + 1, \quad x \neq 1.$$

You can check this result by multiplying.

$$\begin{aligned}(x - 1)(x^2 + x + 1) &= x^3 + x^2 + x - x^2 - x - 1 \\ &= x^3 - 1\end{aligned}$$

✓ **Checkpoint** *Audio-video solution in English & Spanish at LarsonPrecalculus.com.*

Divide $x^3 - 2x^2 - 9$ by $x - 3$.

Note that the condition $x \neq 1$ in Example 1 is equivalent to the condition in the Division Algorithm that $d(x) \neq 0$. Statement of this condition will be omitted for the remainder of this section, but you should be aware that the results of polynomial division are only valid when the divisor does not equal zero.

EXAMPLE 3 Long Division of Polynomials

Divide $-5x^2 - 2 + 3x + 2x^4 + 4x^3$ by $2x - 3 + x^2$.

Solution Begin by writing the dividend and divisor in descending powers of x.

$$\begin{array}{r l}
2x^2 \qquad\qquad + 1 & \\
x^2 + 2x - 3 \overline{\big)\, 2x^4 + 4x^3 - 5x^2 + 3x - 2} & \\
\underline{2x^4 + 4x^3 - 6x^2} \qquad\qquad & \text{Multiply } 2x^2 \text{ by } x^2 + 2x - 3. \\
x^2 + 3x - 2 & \text{Subtract and bring down } 3x - 2. \\
\underline{x^2 + 2x - 3} & \text{Multiply } 1 \text{ by } x^2 + 2x - 3. \\
x + 1 & \text{Subtract.}
\end{array}$$

Note that the first subtraction eliminated two terms from the dividend. When this happens, the quotient skips a term. You can write the result as

$$\frac{2x^4 + 4x^3 - 5x^2 + 3x - 2}{x^2 + 2x - 3} = 2x^2 + 1 + \frac{x + 1}{x^2 + 2x - 3}.$$

✓ **Checkpoint** *Audio-video solution in English & Spanish at LarsonPrecalculus.com.*

Divide $-x^3 + 9x + 6x^4 - x^2 - 3$ by $1 + 3x$.

Synthetic Division

There is a nice shortcut for long division of polynomials by divisors of the form $x - k$. This shortcut is called **synthetic division.** The pattern for synthetic division of a cubic polynomial is summarized as follows. (The pattern for higher-degree polynomials is similar.)

Synthetic Division (for a Cubic Polynomial)

To divide $ax^3 + bx^2 + cx + d$ by $x - k$, use the following pattern.

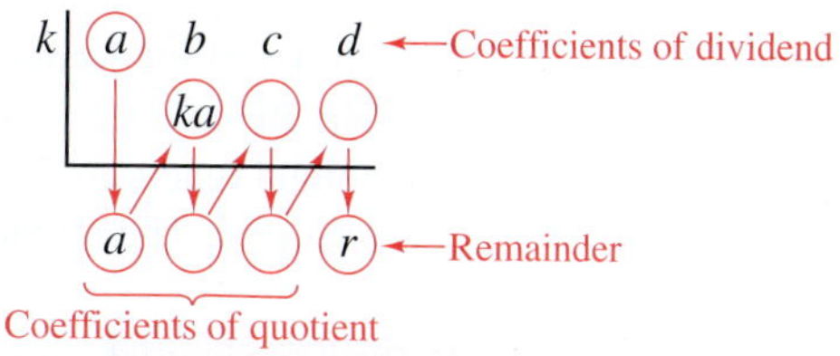

Vertical pattern: Add terms in columns.
Diagonal pattern: Multiply results by k.

This algorithm for synthetic division works only for divisors of the form $x - k$. Remember that $x + k = x - (-k)$.

EXAMPLE 4 Using Synthetic Division

Use synthetic division to divide $x^4 - 10x^2 - 2x + 4$ by $x + 3$.

Solution You should set up the array as follows. Note that a zero is included for the missing x^3-term in the dividend.

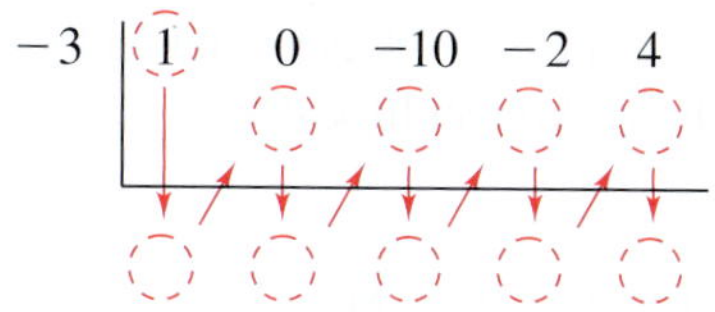

Then, use the synthetic division pattern by adding terms in columns and multiplying the results by -3.

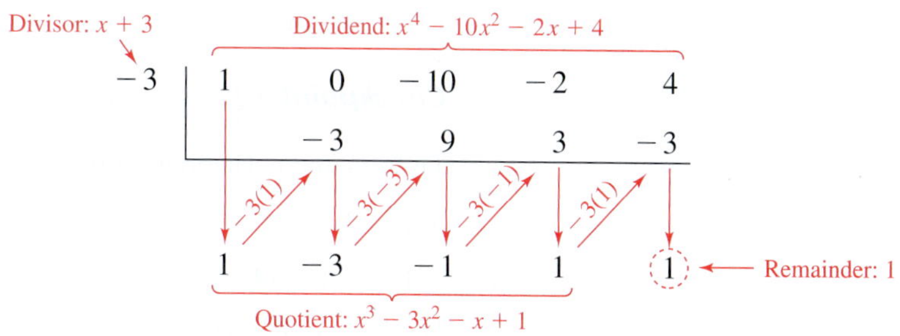

So, you have

$$\frac{x^4 - 10x^2 - 2x + 4}{x + 3} = x^3 - 3x^2 - x + 1 + \frac{1}{x + 3}.$$

✓ **Checkpoint** Audio-video solution in English & Spanish at LarsonPrecalculus.com.

Use synthetic division to divide $5x^3 + 8x^2 - x + 6$ by $x + 2$.

The Remainder and Factor Theorems

The remainder obtained in the synthetic division process has an important interpretation, as described in the **Remainder Theorem.**

> **The Remainder Theorem**
>
> If a polynomial $f(x)$ is divided by $x - k$, then the remainder is
>
> $$r = f(k).$$

For a proof of the Remainder Theorem, see Proofs in Mathematics on page 195.

The Remainder Theorem tells you that synthetic division can be used to evaluate a polynomial function. That is, to evaluate a polynomial function $f(x)$ when $x = k$, divide $f(x)$ by $x - k$. The remainder will be $f(k)$, as illustrated in Example 5.

EXAMPLE 5 Using the Remainder Theorem

Use the Remainder Theorem to evaluate

$$f(x) = 3x^3 + 8x^2 + 5x - 7$$

when $x = -2$.

Solution Using synthetic division, you obtain the following.

$$\begin{array}{r|rrrr} -2 & 3 & 8 & 5 & -7 \\ & & -6 & -4 & -2 \\ \hline & 3 & 2 & 1 & -9 \end{array}$$

Because the remainder is $r = -9$, you can conclude that

$$f(-2) = -9. \qquad r = f(k)$$

This means that $(-2, -9)$ is a point on the graph of f. You can check this by substituting $x = -2$ in the original function.

Check

$$\begin{aligned} f(-2) &= 3(-2)^3 + 8(-2)^2 + 5(-2) - 7 \\ &= 3(-8) + 8(4) - 10 - 7 \\ &= -24 + 32 - 10 - 7 \\ &= -9 \end{aligned}$$

✓ **Checkpoint** Audio-video solution in English & Spanish at LarsonPrecalculus.com.

Use the Remainder Theorem to find each function value given

$$f(x) = 4x^3 + 10x^2 - 3x - 8.$$

a. $f(-1)$ **b.** $f(4)$

c. $f\left(\frac{1}{2}\right)$ **d.** $f(-3)$

▷ **TECHNOLOGY** You can evaluate a function with your graphing utility by entering the function in the equation editor and using the *table* feature in *ask* mode. When the table is set to *ask* mode, you can enter values in the X column and the corresponding function values will be displayed in the function column.

Another important theorem is the **Factor Theorem,** which is stated below.

> **The Factor Theorem**
>
> A polynomial $f(x)$ has a factor $(x - k)$ if and only if $f(k) = 0$.

For a proof of the Factor Theorem, see Proofs in Mathematics on page 195.

Using the Factor Theorem, you can test whether a polynomial has $(x - k)$ as a factor by evaluating the polynomial at $x = k$. If the result is 0, then $(x - k)$ is a factor.

EXAMPLE 6 Factoring a Polynomial: Repeated Division

Show that $(x - 2)$ and $(x + 3)$ are factors of $f(x) = 2x^4 + 7x^3 - 4x^2 - 27x - 18$. Then find the remaining factors of $f(x)$.

Algebraic Solution

Using synthetic division with the factor $(x - 2)$, you obtain the following.

2	2	7	−4	−27	−18	
		4	22	36	18	
	2	11	18	9	0	→ 0 remainder, so $f(2) = 0$ and $(x - 2)$ is a factor.

Take the result of this division and perform synthetic division again using the factor $(x + 3)$.

−3	2	11	18	9	
		−6	−15	−9	
	2	5	3	0	→ 0 remainder, so $f(-3) = 0$ and $(x + 3)$ is a factor.

$2x^2 + 5x + 3$

Because the resulting quadratic expression factors as

$$2x^2 + 5x + 3 = (2x + 3)(x + 1)$$

the complete factorization of $f(x)$ is

$$f(x) = (x - 2)(x + 3)(2x + 3)(x + 1).$$

Graphical Solution

From the graph of $f(x) = 2x^4 + 7x^3 - 4x^2 - 27x - 18$, you can see that there are four x-intercepts (see figure below). These occur at $x = -3$, $x = -\frac{3}{2}$, $x = -1$, and $x = 2$. (Check this algebraically.) This implies that $(x + 3)$, $\left(x + \frac{3}{2}\right)$, $(x + 1)$, and $(x - 2)$ are factors of $f(x)$. $\left[\text{Note that } \left(x + \frac{3}{2}\right) \text{ and } (2x + 3) \text{ are equivalent factors because they both yield the same zero, } x = -\frac{3}{2}.\right]$

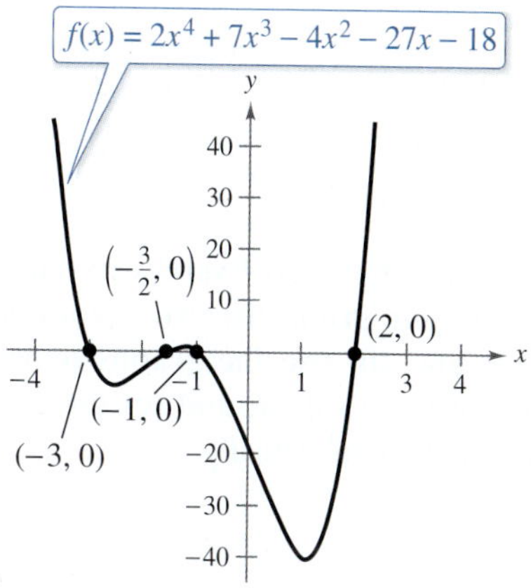

✓ Checkpoint *Audio-video solution in English & Spanish at LarsonPrecalculus.com.*

Show that $(x + 3)$ is a factor of $f(x) = x^3 - 19x - 30$. Then find the remaining factors of $f(x)$.

> **Summarize (Section 2.3)**
>
> **1.** Describe how to divide one polynomial by another polynomial using long division *(pages 138 and 139)*. For examples of long division of polynomials, see Examples 1–3.
>
> **2.** Describe the algorithm for synthetic division *(page 141)*. For an example of using synthetic division, see Example 4.
>
> **3.** State the Remainder Theorem and the Factor Theorem *(pages 142 and 143)*. For an example of using the Remainder Theorem, see Example 5. For an example of using the Factor Theorem, see Example 6.

2.3 Exercises

See CalcChat.com for tutorial help and worked-out solutions to odd-numbered exercises.

Vocabulary

1. Two forms of the Division Algorithm are shown below. Identify and label each term or function.

$$f(x) = d(x)q(x) + r(x) \qquad \frac{f(x)}{d(x)} = q(x) + \frac{r(x)}{d(x)}$$

In Exercises 2–6, fill in the blanks.

2. In the Division Algorithm, the rational expression $r(x)/d(x)$ is ________ because the degree of $r(x)$ is less than the degree of $d(x)$.
3. In the Division Algorithm, the rational expression $f(x)/d(x)$ is ________ because the degree of $f(x)$ is greater than or equal to the degree of $d(x)$.
4. An alternative method to long division of polynomials is called ________ ________, in which the divisor must be of the form $x - k$.
5. The ________ Theorem states that a polynomial $f(x)$ has a factor $(x - k)$ if and only if $f(k) = 0$.
6. The ________ Theorem states that if a polynomial $f(x)$ is divided by $x - k$, then the remainder is $r = f(k)$.

Skills and Applications

Analytical Analysis **In Exercises 7 and 8, use long division to verify that $y_1 = y_2$.**

7. $y_1 = \dfrac{x^2}{x + 2}$, $y_2 = x - 2 + \dfrac{4}{x + 2}$
8. $y_1 = \dfrac{x^4 - 3x^2 - 1}{x^2 + 5}$, $y_2 = x^2 - 8 + \dfrac{39}{x^2 + 5}$

Graphical Analysis **In Exercises 9 and 10, (a) use a graphing utility to graph the two equations in the same viewing window, (b) use the graphs to verify that the expressions are equivalent, and (c) use long division to verify the results algebraically.**

9. $y_1 = \dfrac{x^2 + 2x - 1}{x + 3}$, $y_2 = x - 1 + \dfrac{2}{x + 3}$
10. $y_1 = \dfrac{x^4 + x^2 - 1}{x^2 + 1}$, $y_2 = x^2 - \dfrac{1}{x^2 + 1}$

Long Division of Polynomials **In Exercises 11–26, use long division to divide.**

11. $(2x^2 + 10x + 12) \div (x + 3)$
12. $(5x^2 - 17x - 12) \div (x - 4)$
13. $(4x^3 - 7x^2 - 11x + 5) \div (4x + 5)$
14. $(6x^3 - 16x^2 + 17x - 6) \div (3x - 2)$
15. $(x^4 + 5x^3 + 6x^2 - x - 2) \div (x + 2)$
16. $(x^3 + 4x^2 - 3x - 12) \div (x - 3)$
17. $(x^3 - 27) \div (x - 3)$
18. $(x^3 + 125) \div (x + 5)$
19. $(7x + 3) \div (x + 2)$
20. $(8x - 5) \div (2x + 1)$
21. $(x^3 - 9) \div (x^2 + 1)$
22. $(x^5 + 7) \div (x^3 - 1)$
23. $(3x + 2x^3 - 9 - 8x^2) \div (x^2 + 1)$
24. $(5x^3 - 16 - 20x + x^4) \div (x^2 - x - 3)$
25. $\dfrac{x^4}{(x - 1)^3}$
26. $\dfrac{2x^3 - 4x^2 - 15x + 5}{(x - 1)^2}$

Using Synthetic Division **In Exercises 27–46, use synthetic division to divide.**

27. $(3x^3 - 17x^2 + 15x - 25) \div (x - 5)$
28. $(5x^3 + 18x^2 + 7x - 6) \div (x + 3)$
29. $(6x^3 + 7x^2 - x + 26) \div (x - 3)$
30. $(2x^3 + 14x^2 - 20x + 7) \div (x + 6)$
31. $(4x^3 - 9x + 8x^2 - 18) \div (x + 2)$
32. $(9x^3 - 16x - 18x^2 + 32) \div (x - 2)$
33. $(-x^3 + 75x - 250) \div (x + 10)$
34. $(3x^3 - 16x^2 - 72) \div (x - 6)$
35. $(5x^3 - 6x^2 + 8) \div (x - 4)$
36. $(5x^3 + 6x + 8) \div (x + 2)$
37. $\dfrac{10x^4 - 50x^3 - 800}{x - 6}$
38. $\dfrac{x^5 - 13x^4 - 120x + 80}{x + 3}$
39. $\dfrac{x^3 + 512}{x + 8}$
40. $\dfrac{x^3 - 729}{x - 9}$
41. $\dfrac{-3x^4}{x - 2}$
42. $\dfrac{-3x^4}{x + 2}$
43. $\dfrac{180x - x^4}{x - 6}$
44. $\dfrac{5 - 3x + 2x^2 - x^3}{x + 1}$
45. $\dfrac{4x^3 + 16x^2 - 23x - 15}{x + \frac{1}{2}}$
46. $\dfrac{3x^3 - 4x^2 + 5}{x - \frac{3}{2}}$

Using the Remainder Theorem In Exercises 47–54, write the function in the form $f(x) = (x - k)q(x) + r$ for the given value of k, and demonstrate that $f(k) = r$.

47. $f(x) = x^3 - x^2 - 14x + 11, \quad k = 4$
48. $f(x) = x^3 - 5x^2 - 11x + 8, \quad k = -2$
49. $f(x) = 15x^4 + 10x^3 - 6x^2 + 14, \quad k = -\frac{2}{3}$
50. $f(x) = 10x^3 - 22x^2 - 3x + 4, \quad k = \frac{1}{5}$
51. $f(x) = x^3 + 3x^2 - 2x - 14, \quad k = \sqrt{2}$
52. $f(x) = x^3 + 2x^2 - 5x - 4, \quad k = -\sqrt{5}$
53. $f(x) = -4x^3 + 6x^2 + 12x + 4, \quad k = 1 - \sqrt{3}$
54. $f(x) = -3x^3 + 8x^2 + 10x - 8, \quad k = 2 + \sqrt{2}$

Using the Remainder Theorem In Exercises 55–58, use the Remainder Theorem and synthetic division to find each function value. Verify your answers using another method.

55. $f(x) = 2x^3 - 7x + 3$
(a) $f(1)$ (b) $f(-2)$ (c) $f(\frac{1}{2})$ (d) $f(2)$
56. $g(x) = 2x^6 + 3x^4 - x^2 + 3$
(a) $g(2)$ (b) $g(1)$ (c) $g(3)$ (d) $g(-1)$
57. $h(x) = x^3 - 5x^2 - 7x + 4$
(a) $h(3)$ (b) $h(2)$ (c) $h(-2)$ (d) $h(-5)$
58. $f(x) = 4x^4 - 16x^3 + 7x^2 + 20$
(a) $f(1)$ (b) $f(-2)$ (c) $f(5)$ (d) $f(-10)$

Using the Factor Theorem In Exercises 59–66, use synthetic division to show that x is a solution of the third-degree polynomial equation, and use the result to factor the polynomial completely. List all real solutions of the equation.

59. $x^3 - 7x + 6 = 0, \quad x = 2$
60. $x^3 - 28x - 48 = 0, \quad x = -4$
61. $2x^3 - 15x^2 + 27x - 10 = 0, \quad x = \frac{1}{2}$
62. $48x^3 - 80x^2 + 41x - 6 = 0, \quad x = \frac{2}{3}$
63. $x^3 + 2x^2 - 3x - 6 = 0, \quad x = \sqrt{3}$
64. $x^3 + 2x^2 - 2x - 4 = 0, \quad x = \sqrt{2}$
65. $x^3 - 3x^2 + 2 = 0, \quad x = 1 + \sqrt{3}$
66. $x^3 - x^2 - 13x - 3 = 0, \quad x = 2 - \sqrt{5}$

Factoring a Polynomial In Exercises 67–74, (a) verify the given factors of $f(x)$, (b) find the remaining factor(s) of $f(x)$, (c) use your results to write the complete factorization of $f(x)$, (d) list all real zeros of f, and (e) confirm your results by using a graphing utility to graph the function.

Function	Factors
67. $f(x) = 2x^3 + x^2 - 5x + 2$	$(x + 2), (x - 1)$
68. $f(x) = 3x^3 + 2x^2 - 19x + 6$	$(x + 3), (x - 2)$
69. $f(x) = x^4 - 4x^3 - 15x^2 + 58x - 40$	$(x - 5), (x + 4)$
70. $f(x) = 8x^4 - 14x^3 - 71x^2 - 10x + 24$	$(x + 2), (x - 4)$
71. $f(x) = 6x^3 + 41x^2 - 9x - 14$	$(2x + 1), (3x - 2)$
72. $f(x) = 10x^3 - 11x^2 - 72x + 45$	$(2x + 5), (5x - 3)$
73. $f(x) = 2x^3 - x^2 - 10x + 5$	$(2x - 1), (x + \sqrt{5})$
74. $f(x) = x^3 + 3x^2 - 48x - 144$	$(x + 4\sqrt{3}), (x + 3)$

Graphical Analysis In Exercises 75–80, (a) use the *zero* or *root* feature of a graphing utility to approximate the zeros of the function accurate to three decimal places, (b) determine the exact value of one of the zeros, and (c) use synthetic division to verify your result from part (b), and then factor the polynomial completely.

75. $f(x) = x^3 - 2x^2 - 5x + 10$
76. $g(x) = x^3 - 4x^2 - 2x + 8$
77. $h(t) = t^3 - 2t^2 - 7t + 2$
78. $f(s) = s^3 - 12s^2 + 40s - 24$
79. $h(x) = x^5 - 7x^4 + 10x^3 + 14x^2 - 24x$
80. $g(x) = 6x^4 - 11x^3 - 51x^2 + 99x - 27$

Simplifying Rational Expressions In Exercises 81–84, simplify the rational expression by using long division or synthetic division.

81. $\dfrac{4x^3 - 8x^2 + x + 3}{2x - 3}$
82. $\dfrac{x^3 + x^2 - 64x - 64}{x + 8}$
83. $\dfrac{x^4 + 6x^3 + 11x^2 + 6x}{x^2 + 3x + 2}$
84. $\dfrac{x^4 + 9x^3 - 5x^2 - 36x + 4}{x^2 - 4}$

85. **Profit** A company that produces calculators estimated that the profit P (in dollars) from selling a particular model of calculator was

$$P = -152x^3 + 7545x^2 - 169{,}625, \quad 0 \le x \le 45$$

where x was the advertising expense (in tens of thousands of dollars). For this model of calculator, the advertising expense was \$400,000 ($x = 40$) and the profit was \$2,174,375.

(a) Use a graphing utility to graph the profit function.
(b) Use the graph from part (a) to estimate another amount the company could have spent on advertising that would have produced the same profit.
(c) Use synthetic division to confirm the result of part (b) algebraically.

86. Data Analysis: Lyme Disease The numbers N of confirmed cases of Lyme disease in Maine from 2003 to 2010 are shown in the table, where t represents the year, with $t = 3$ corresponding to 2003. *(Source: Centers for Disease Control and Prevention)*

DATA

Year, t	Number, N
3	175
4	225
5	247
6	338
7	529
8	780
9	791
10	559

Spreadsheet at LarsonPrecalculus.com

(a) Use a graphing utility to create a scatter plot of the data.

(b) Use the *regression* feature of the graphing utility to find a quartic model for the data. Graph the model in the same viewing window as the scatter plot.

(c) Use the model to create a table of estimated values of N. Compare the model with the original data.

(d) Use synthetic division to confirm algebraically your estimated value for the year 2010.

Exploration

True or False? **In Exercises 87–90, determine whether the statement is true or false. Justify your answer.**

87. If $(7x + 4)$ is a factor of some polynomial function $f(x)$, then $\frac{4}{7}$ is a zero of f.

88. $(2x - 1)$ is a factor of the polynomial

$$6x^6 + x^5 - 92x^4 + 45x^3 + 184x^2 + 4x - 48.$$

89. The rational expression

$$\frac{x^3 + 2x^2 - 13x + 10}{x^2 - 4x - 12}$$

is improper.

90. The equation

$$\frac{x^3 - 3x^2 + 4}{x + 1} = x^2 - 4x + 4$$

is true for all values of x.

Think About It **In Exercises 91 and 92, perform the division by assuming that n is a positive integer.**

91. $\dfrac{x^{3n} + 9x^{2n} + 27x^n + 27}{x^n + 3}$

92. $\dfrac{x^{3n} - 3x^{2n} + 5x^n - 6}{x^n - 2}$

93. Writing Briefly explain what it means for a divisor to divide evenly into a dividend.

94. Writing Briefly explain how to check polynomial division, and justify your reasoning. Give an example.

Exploration **In Exercises 95 and 96, find the constant c such that the denominator will divide evenly into the numerator.**

95. $\dfrac{x^3 + 4x^2 - 3x + c}{x - 5}$ **96.** $\dfrac{x^5 - 2x^2 + x + c}{x + 2}$

97. Think About It Find the value of k such that $x - 4$ is a factor of $x^3 - kx^2 + 2kx - 8$.

98. HOW DO YOU SEE IT? The graph below shows a company's estimated profits for different advertising expenses. The company's actual profit was \$936,660 for an advertising expense of \$300,000.

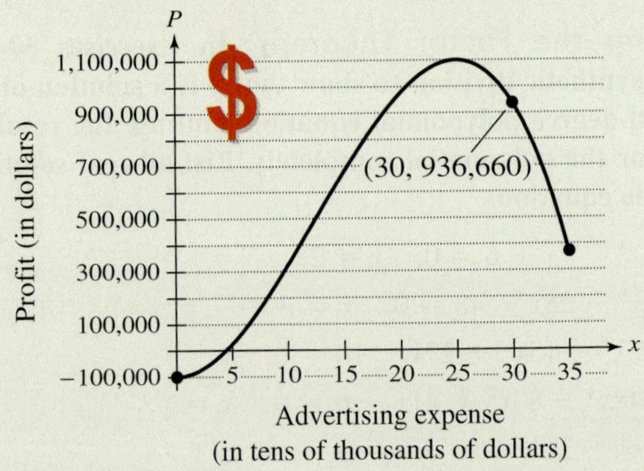

(a) From the graph, it appears that the company could have obtained the same profit for a lesser advertising expense. Use the graph to estimate this expense.

(b) The company's model is

$$P = -140.75x^3 + 5348.3x^2 - 76{,}560, \quad 0 \le x \le 35$$

where P is the profit (in dollars) and x is the advertising expense (in tens of thousands of dollars). Explain how you could verify the lesser expense from part (a) algebraically.

Dariusz Majgier/Shutterstock.com

2.4 Complex Numbers

You can use complex numbers to model and solve real-life problems in electronics. For instance, in Exercise 91 on page 153, you will use complex numbers to find the impedance of an electrical circuit.

- Use the imaginary unit i to write complex numbers.
- Add, subtract, and multiply complex numbers.
- Use complex conjugates to write the quotient of two complex numbers in standard form.
- Find complex solutions of quadratic equations.

The Imaginary Unit i

You have learned that some quadratic equations have no real solutions. For instance, the quadratic equation

$$x^2 + 1 = 0$$

has no real solution because there is no real number x that can be squared to produce -1. To overcome this deficiency, mathematicians created an expanded system of numbers using the **imaginary unit i,** defined as

$$i = \sqrt{-1} \qquad \text{Imaginary unit}$$

where $i^2 = -1$. By adding real numbers to real multiples of this imaginary unit, the set of **complex numbers** is obtained. Each complex number can be written in the **standard form $a + bi$.** For instance, the standard form of the complex number $-5 + \sqrt{-9}$ is $-5 + 3i$ because

$$-5 + \sqrt{-9} = -5 + \sqrt{3^2(-1)} = -5 + 3\sqrt{-1} = -5 + 3i.$$

Definition of a Complex Number

Let a and b be real numbers. The number $a + bi$ is called a **complex number,** and it is said to be written in **standard form.** The real number a is called the **real part** and the real number b is called the **imaginary part** of the complex number.

When $b = 0$, the number $a + bi$ is a real number. When $b \neq 0$, the number $a + bi$ is called an **imaginary number.** A number of the form bi, where $b \neq 0$, is called a **pure imaginary number.**

The set of real numbers is a subset of the set of complex numbers, as shown below. This is true because every real number a can be written as a complex number using $b = 0$. That is, for every real number a, you can write $a = a + 0i$.

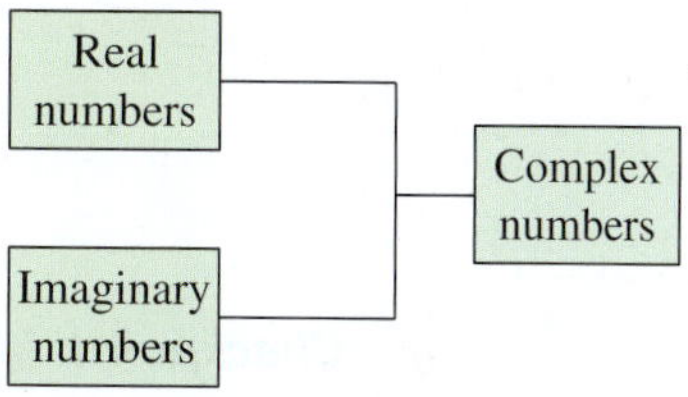

Equality of Complex Numbers

Two complex numbers $a + bi$ and $c + di$, written in standard form, are equal to each other

$$a + bi = c + di \qquad \text{Equality of two complex numbers}$$

if and only if $a = c$ and $b = d$.

© Richard Megna/Fundamental Photographs

Operations with Complex Numbers

To add (or subtract) two complex numbers, you add (or subtract) the real and imaginary parts of the numbers separately.

> **Addition and Subtraction of Complex Numbers**
>
> For two complex numbers $a + bi$ and $c + di$ written in standard form, the sum and difference are defined as follows.
>
> *Sum:* $(a + bi) + (c + di) = (a + c) + (b + d)i$
>
> *Difference:* $(a + bi) - (c + di) = (a - c) + (b - d)i$

The **additive identity** in the complex number system is zero (the same as in the real number system). Furthermore, the **additive inverse** of the complex number $a + bi$ is

$$-(a + bi) = -a - bi. \quad \text{Additive inverse}$$

So, you have

$$(a + bi) + (-a - bi) = 0 + 0i = 0.$$

EXAMPLE 1 Adding and Subtracting Complex Numbers

a. $(4 + 7i) + (1 - 6i) = 4 + 7i + 1 - 6i$ — Remove parentheses.

$= (4 + 1) + (7 - 6)i$ — Group like terms.

$= 5 + i$ — Write in standard form.

b. $(1 + 2i) + (3 - 2i) = 1 + 2i + 3 - 2i$ — Remove parentheses.

$= (1 + 3) + (2 - 2)i$ — Group like terms.

$= 4 + 0i$ — Simplify.

$= 4$ — Write in standard form.

REMARK Note that the sum of two complex numbers can be a real number.

c. $3i - (-2 + 3i) - (2 + 5i) = 3i + 2 - 3i - 2 - 5i$

$= (2 - 2) + (3 - 3 - 5)i$

$= 0 - 5i$

$= -5i$

d. $(3 + 2i) + (4 - i) - (7 + i) = 3 + 2i + 4 - i - 7 - i$

$= (3 + 4 - 7) + (2 - 1 - 1)i$

$= 0 + 0i$

$= 0$

✓ ***Checkpoint*** *Audio-video solution in English & Spanish at LarsonPrecalculus.com.*

Perform each operation and write the result in standard form.

a. $(7 + 3i) + (5 - 4i)$

b. $(3 + 4i) - (5 - 3i)$

c. $2i + (-3 - 4i) - (-3 - 3i)$

d. $(5 - 3i) + (3 + 5i) - (8 + 2i)$

Many of the properties of real numbers are valid for complex numbers as well. Here are some examples.

Associative Properties of Addition and Multiplication

Commutative Properties of Addition and Multiplication

Distributive Property of Multiplication Over Addition

Notice below how these properties are used when two complex numbers are multiplied.

$$(a + bi)(c + di) = a(c + di) + bi(c + di) \quad \text{Distributive Property}$$
$$= ac + (ad)i + (bc)i + (bd)i^2 \quad \text{Distributive Property}$$
$$= ac + (ad)i + (bc)i + (bd)(-1) \quad i^2 = -1$$
$$= ac - bd + (ad)i + (bc)i \quad \text{Commutative Property}$$
$$= (ac - bd) + (ad + bc)i \quad \text{Associative Property}$$

Rather than trying to memorize this multiplication rule, you should simply remember how to use the Distributive Property to multiply two complex numbers.

EXAMPLE 2 Multiplying Complex Numbers

a. $4(-2 + 3i) = 4(-2) + 4(3i)$ Distributive Property

$= -8 + 12i$ Simplify.

b. $(2 - i)(4 + 3i) = 2(4 + 3i) - i(4 + 3i)$ Distributive Property

$= 8 + 6i - 4i - 3i^2$ Distributive Property

$= 8 + 6i - 4i - 3(-1)$ $i^2 = -1$

$= (8 + 3) + (6 - 4)i$ Group like terms.

$= 11 + 2i$ Write in standard form.

c. $(3 + 2i)(3 - 2i) = 3(3 - 2i) + 2i(3 - 2i)$ Distributive Property

$= 9 - 6i + 6i - 4i^2$ Distributive Property

$= 9 - 6i + 6i - 4(-1)$ $i^2 = -1$

$= 9 + 4$ Simplify.

$= 13$ Write in standard form.

d. $(3 + 2i)^2 = (3 + 2i)(3 + 2i)$ Square of a binomial

$= 3(3 + 2i) + 2i(3 + 2i)$ Distributive Property

$= 9 + 6i + 6i + 4i^2$ Distributive Property

$= 9 + 6i + 6i + 4(-1)$ $i^2 = -1$

$= 9 + 12i - 4$ Simplify.

$= 5 + 12i$ Write in standard form.

REMARK The procedure described above is similar to multiplying two binomials and combining like terms, as in the FOIL Method shown in Appendix A.3. For instance, you can use the FOIL Method to multiply the two complex numbers from Example 2(b).

F O I L

$(2 - i)(4 + 3i) = 8 + 6i - 4i - 3i^2$

✓ ***Checkpoint*** *Audio-video solution in English & Spanish at LarsonPrecalculus.com.*

Perform each operation and write the result in standard form.

a. $(2 - 4i)(3 + 3i)$

b. $(4 + 5i)(4 - 5i)$

c. $(4 + 2i)^2$

Complex Conjugates

ALGEBRA HELP You can compare complex conjugates with the method for rationalizing denominators in Appendix A.2.

Notice in Example 2(c) that the product of two complex numbers can be a real number. This occurs with pairs of complex numbers of the form $a + bi$ and $a - bi$, called **complex conjugates.**

$$\begin{aligned}(a + bi)(a - bi) &= a^2 - abi + abi - b^2i^2 \\ &= a^2 - b^2(-1) \\ &= a^2 + b^2\end{aligned}$$

EXAMPLE 3 Multiplying Conjugates

Multiply each complex number by its complex conjugate.

a. $1 + i$ **b.** $4 - 3i$

Solution

a. The complex conjugate of $1 + i$ is $1 - i$.

$$(1 + i)(1 - i) = 1^2 - i^2 = 1 - (-1) = 2$$

b. The complex conjugate of $4 - 3i$ is $4 + 3i$.

$$(4 - 3i)(4 + 3i) = 4^2 - (3i)^2 = 16 - 9i^2 = 16 - 9(-1) = 25$$

✓ ***Checkpoint*** Audio-video solution in English & Spanish at LarsonPrecalculus.com.

Multiply each complex number by its complex conjugate.

a. $3 + 6i$

b. $2 - 5i$

To write the quotient of $a + bi$ and $c + di$ in standard form, where c and d are not both zero, multiply the numerator and denominator by the complex conjugate of the *denominator* to obtain

$$\frac{a + bi}{c + di} = \frac{a + bi}{c + di}\left(\frac{c - di}{c - di}\right) = \frac{(ac + bd) + (bc - ad)i}{c^2 + d^2}.$$

REMARK Note that when you multiply the numerator and denominator of a quotient of complex numbers by

$$\frac{c - di}{c - di}$$

you are actually multiplying the quotient by a form of 1. You are not changing the original expression, you are only creating an expression that is equivalent to the original expression.

EXAMPLE 4 Quotient of Complex Numbers in Standard Form

$$\frac{2 + 3i}{4 - 2i} = \frac{2 + 3i}{4 - 2i}\left(\frac{4 + 2i}{4 + 2i}\right) \quad \text{Multiply numerator and denominator by complex conjugate of denominator.}$$

$$= \frac{8 + 4i + 12i + 6i^2}{16 - 4i^2} \quad \text{Expand.}$$

$$= \frac{8 - 6 + 16i}{16 + 4} \quad i^2 = -1$$

$$= \frac{2 + 16i}{20} \quad \text{Simplify.}$$

$$= \frac{1}{10} + \frac{4}{5}i \quad \text{Write in standard form.}$$

✓ ***Checkpoint*** Audio-video solution in English & Spanish at LarsonPrecalculus.com.

Write $\dfrac{2 + i}{2 - i}$ in standard form.

Complex Solutions of Quadratic Equations

You can write a number such as $\sqrt{-3}$ in standard form by factoring out $i = \sqrt{-1}$.

$$\sqrt{-3} = \sqrt{3(-1)} = \sqrt{3}\sqrt{-1} = \sqrt{3}i$$

The number $\sqrt{3}i$ is called the *principal square root* of -3.

REMARK The definition of principal square root uses the rule

$$\sqrt{ab} = \sqrt{a}\sqrt{b}$$

for $a > 0$ and $b < 0$. This rule is not valid if *both* a and b are negative. For example,

$$\begin{aligned}\sqrt{-5}\sqrt{-5} &= \sqrt{5(-1)}\sqrt{5(-1)}\\ &= \sqrt{5}i\sqrt{5}i\\ &= \sqrt{25}i^2\\ &= 5i^2\\ &= -5\end{aligned}$$

whereas

$$\sqrt{(-5)(-5)} = \sqrt{25} = 5.$$

To avoid problems with square roots of negative numbers, be sure to convert complex numbers to standard form *before* multiplying.

Principal Square Root of a Negative Number

When a is a positive real number, the **principal square root** of $-a$ is defined as

$$\sqrt{-a} = \sqrt{a}i.$$

EXAMPLE 5 Writing Complex Numbers in Standard Form

a. $\sqrt{-3}\sqrt{-12} = \sqrt{3}i\sqrt{12}i = \sqrt{36}i^2 = 6(-1) = -6$

b. $\sqrt{-48} - \sqrt{-27} = \sqrt{48}i - \sqrt{27}i = 4\sqrt{3}i - 3\sqrt{3}i = \sqrt{3}i$

c. $$\begin{aligned}\left(-1 + \sqrt{-3}\right)^2 = \left(-1 + \sqrt{3}i\right)^2 &= (-1)^2 - 2\sqrt{3}i + \left(\sqrt{3}\right)^2(i^2)\\ &= 1 - 2\sqrt{3}i + 3(-1)\\ &= -2 - 2\sqrt{3}i\end{aligned}$$

✓ Checkpoint *Audio-video solution in English & Spanish at LarsonPrecalculus.com.*

Write $\sqrt{-14}\sqrt{-2}$ in standard form.

EXAMPLE 6 Complex Solutions of a Quadratic Equation

Solve $3x^2 - 2x + 5 = 0$.

Solution

$$x = \frac{-(-2) \pm \sqrt{(-2)^2 - 4(3)(5)}}{2(3)} \quad \text{Quadratic Formula}$$

$$= \frac{2 \pm \sqrt{-56}}{6} \quad \text{Simplify.}$$

$$= \frac{2 \pm 2\sqrt{14}i}{6} \quad \text{Write } \sqrt{-56} \text{ in standard form.}$$

$$= \frac{1}{3} \pm \frac{\sqrt{14}}{3}i \quad \text{Write in standard form.}$$

ALGEBRA HELP You can review the techniques for using the Quadratic Formula in Appendix A.5.

✓ Checkpoint *Audio-video solution in English & Spanish at LarsonPrecalculus.com.*

Solve $8x^2 + 14x + 9 = 0$.

Summarize (Section 2.4)

1. Describe how to write complex numbers using the imaginary unit i *(page 147)*.
2. Describe how to add, subtract, and multiply complex numbers *(pages 148 and 149, Examples 1 and 2)*.
3. Describe how to use complex conjugates to write the quotient of two complex numbers in standard form *(page 150, Example 4)*.
4. Describe how to find complex solutions of a quadratic equation *(page 151, Example 6)*.

2.4 Exercises

See CalcChat.com for tutorial help and worked-out solutions to odd-numbered exercises.

Vocabulary: Fill in the blanks.

1. A ________ number has the form $a + bi$, where $a \neq 0$, $b = 0$.
2. An ________ number has the form $a + bi$, where $a \neq 0$, $b \neq 0$.
3. A ________ ________ number has the form $a + bi$, where $a = 0$, $b \neq 0$.
4. The imaginary unit i is defined as $i =$ ________, where $i^2 =$ ________.
5. When a is a positive real number, the ________ ________ root of $-a$ is defined as $\sqrt{-a} = \sqrt{a}\,i$.
6. The numbers $a + bi$ and $a - bi$ are called ________ ________, and their product is a real number $a^2 + b^2$.

Skills and Applications

Equality of Complex Numbers In Exercises 7–10, find real numbers a and b such that the equation is true.

7. $a + bi = -12 + 7i$
8. $a + bi = 13 + 4i$
9. $(a - 1) + (b + 3)i = 5 + 8i$
10. $(a + 6) + 2bi = 6 - 5i$

Writing a Complex Number in Standard Form In Exercises 11–22, write the complex number in standard form.

11. $8 + \sqrt{-25}$
12. $5 + \sqrt{-36}$
13. $2 - \sqrt{-27}$
14. $1 + \sqrt{-8}$
15. $\sqrt{-80}$
16. $\sqrt{-4}$
17. 14
18. 75
19. $-10i + i^2$
20. $-4i^2 + 2i$
21. $\sqrt{-0.09}$
22. $\sqrt{-0.0049}$

Performing Operations with Complex Numbers In Exercises 23–42, perform the operation and write the result in standard form.

23. $(7 + i) + (3 - 4i)$
24. $(13 - 2i) + (-5 + 6i)$
25. $(9 - i) - (8 - i)$
26. $(3 + 2i) - (6 + 13i)$
27. $\left(-2 + \sqrt{-8}\right) + \left(5 - \sqrt{-50}\right)$
28. $\left(8 + \sqrt{-18}\right) - \left(4 + 3\sqrt{2}i\right)$
29. $13i - (14 - 7i)$
30. $25 + (-10 + 11i) + 15i$
31. $-\left(\frac{3}{2} + \frac{5}{2}i\right) + \left(\frac{5}{3} + \frac{11}{3}i\right)$
32. $(1.6 + 3.2i) + (-5.8 + 4.3i)$
33. $(1 + i)(3 - 2i)$
34. $(7 - 2i)(3 - 5i)$
35. $12i(1 - 9i)$
36. $-8i(9 + 4i)$
37. $\left(\sqrt{14} + \sqrt{10}i\right)\left(\sqrt{14} - \sqrt{10}i\right)$
38. $\left(\sqrt{3} + \sqrt{15}i\right)\left(\sqrt{3} - \sqrt{15}i\right)$
39. $(6 + 7i)^2$
40. $(5 - 4i)^2$
41. $(2 + 3i)^2 + (2 - 3i)^2$
42. $(1 - 2i)^2 - (1 + 2i)^2$

Multiplying Conjugates In Exercises 43–50, write the complex conjugate of the complex number. Then multiply the number by its complex conjugate.

43. $9 + 2i$
44. $8 - 10i$
45. $-1 - \sqrt{5}i$
46. $-3 + \sqrt{2}i$
47. $\sqrt{-20}$
48. $\sqrt{-15}$
49. $\sqrt{6}$
50. $1 + \sqrt{8}$

Quotient of Complex Numbers in Standard Form In Exercises 51–60, write the quotient in standard form.

51. $\dfrac{3}{i}$
52. $-\dfrac{14}{2i}$
53. $\dfrac{2}{4 - 5i}$
54. $\dfrac{13}{1 - i}$
55. $\dfrac{5 + i}{5 - i}$
56. $\dfrac{6 - 7i}{1 - 2i}$
57. $\dfrac{9 - 4i}{i}$
58. $\dfrac{8 + 16i}{2i}$
59. $\dfrac{3i}{(4 - 5i)^2}$
60. $\dfrac{5i}{(2 + 3i)^2}$

Performing Operations with Complex Numbers In Exercises 61–64, perform the operation and write the result in standard form.

61. $\dfrac{2}{1 + i} - \dfrac{3}{1 - i}$
62. $\dfrac{2i}{2 + i} + \dfrac{5}{2 - i}$
63. $\dfrac{i}{3 - 2i} + \dfrac{2i}{3 + 8i}$
64. $\dfrac{1 + i}{i} - \dfrac{3}{4 - i}$

Writing a Complex Number in Standard Form In Exercises 65–70, write the complex number in standard form.

65. $\sqrt{-6} \cdot \sqrt{-2}$
66. $\sqrt{-5} \cdot \sqrt{-10}$
67. $\left(\sqrt{-15}\right)^2$
68. $\left(\sqrt{-75}\right)^2$
69. $\left(3 + \sqrt{-5}\right)\left(7 - \sqrt{-10}\right)$
70. $\left(2 - \sqrt{-6}\right)^2$

Complex Solutions of a Quadratic Equation **In Exercises 71–80, use the Quadratic Formula to solve the quadratic equation.**

71. $x^2 - 2x + 2 = 0$ **72.** $x^2 + 6x + 10 = 0$

73. $4x^2 + 16x + 17 = 0$ **74.** $9x^2 - 6x + 37 = 0$

75. $4x^2 + 16x + 15 = 0$ **76.** $16t^2 - 4t + 3 = 0$

77. $\frac{3}{2}x^2 - 6x + 9 = 0$ **78.** $\frac{7}{8}x^2 - \frac{3}{4}x + \frac{5}{16} = 0$

79. $1.4x^2 - 2x - 10 = 0$ **80.** $4.5x^2 - 3x + 12 = 0$

Simplifying a Complex Number **In Exercises 81–90, simplify the complex number and write it in standard form.**

81. $-6i^3 + i^2$ **82.** $4i^2 - 2i^3$

83. $-14i^5$ **84.** $(-i)^3$

85. $(\sqrt{-72})^3$ **86.** $(\sqrt{-2})^6$

87. $\dfrac{1}{i^3}$ **88.** $\dfrac{1}{(2i)^3}$

89. $(3i)^4$ **90.** $(-i)^6$

91. Impedance

The opposition to current in an electrical circuit is called its impedance. The impedance z in a parallel circuit with two pathways satisfies the equation

$$\frac{1}{z} = \frac{1}{z_1} + \frac{1}{z_2}$$

where z_1 is the impedance (in ohms) of pathway 1 and z_2 is the impedance of pathway 2.

(a) The impedance of each pathway in a parallel circuit is found by adding the impedances of all components in the pathway. Use the table to find z_1 and z_2.

(b) Find the impedance z.

	Resistor	Inductor	Capacitor
Symbol	$a\ \Omega$	$b\ \Omega$	$c\ \Omega$
Impedance	a	bi	$-ci$

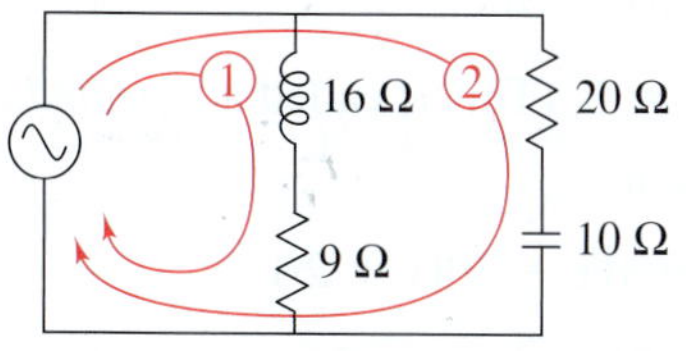

92. Cube of a Complex Number Cube each complex number.

(a) $-1 + \sqrt{3}i$ (b) $-1 - \sqrt{3}i$

Exploration

True or False? **In Exercises 93–96, determine whether the statement is true or false. Justify your answer.**

93. There is no complex number that is equal to its complex conjugate.

94. $-i\sqrt{6}$ is a solution of $x^4 - x^2 + 14 = 56$.

95. $i^{44} + i^{150} - i^{74} - i^{109} + i^{61} = -1$

96. The sum of two complex numbers is always a real number.

97. Pattern Recognition Complete the following.

$i^1 = i$ $i^2 = -1$ $i^3 = -i$ $i^4 = 1$

$i^5 = \square$ $i^6 = \square$ $i^7 = \square$ $i^8 = \square$

$i^9 = \square$ $i^{10} = \square$ $i^{11} = \square$ $i^{12} = \square$

What pattern do you see? Write a brief description of how you would find i raised to any positive integer power.

98. HOW DO YOU SEE IT? The coordinate system shown below is called the *complex plane*. In the complex plane, the point that corresponds to the complex number $a + bi$ is (a, b).

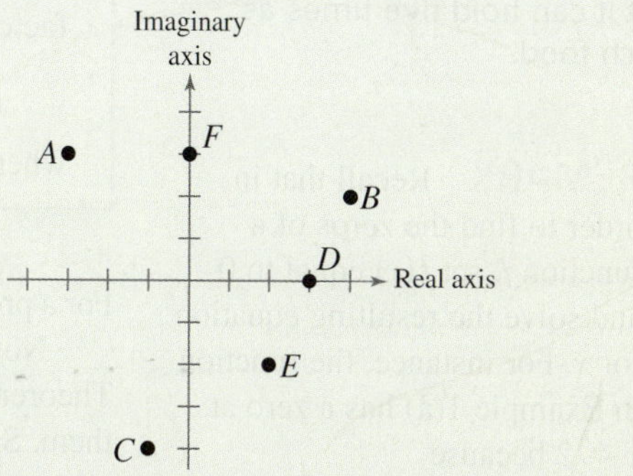

Match each complex number with its corresponding point.

(i) 3 (ii) $3i$ (iii) $4 + 2i$

(iv) $2 - 2i$ (v) $-3 + 3i$ (vi) $-1 - 4i$

99. Error Analysis Describe the error.

$$\sqrt{-6}\sqrt{-6} = \sqrt{(-6)(-6)} = \sqrt{36} = 6$$

100. Proof Prove that the complex conjugate of the product of two complex numbers $a_1 + b_1 i$ and $a_2 + b_2 i$ is the product of their complex conjugates.

101. Proof Prove that the complex conjugate of the sum of two complex numbers $a_1 + b_1 i$ and $a_2 + b_2 i$ is the sum of their complex conjugates.

© Richard Megna/Fundamental Photographs

2.5 Zeros of Polynomial Functions

Finding zeros of polynomial functions is an important part of solving real-life problems. For instance, in Exercise 113 on page 166, the zeros of a polynomial function can help you redesign a storage bin so that it can hold five times as much food.

- **Use the Fundamental Theorem of Algebra to determine the number of zeros of polynomial functions.**
- **Find rational zeros of polynomial functions.**
- **Find conjugate pairs of complex zeros.**
- **Find zeros of polynomials by factoring.**
- **Use Descartes's Rule of Signs and the Upper and Lower Bound Rules to find zeros of polynomials.**
- **Find zeros of polynomials in real-life applications.**

The Fundamental Theorem of Algebra

In the complex number system, every nth-degree polynomial function has *precisely* n zeros. This important result is derived from the **Fundamental Theorem of Algebra,** first proved by the German mathematician Carl Friedrich Gauss (1777–1855).

> **The Fundamental Theorem of Algebra**
>
> If $f(x)$ is a polynomial of degree n, where $n > 0$, then f has at least one zero in the complex number system.

Using the Fundamental Theorem of Algebra and the equivalence of zeros and factors, you obtain the **Linear Factorization Theorem.**

> **Linear Factorization Theorem**
>
> If $f(x)$ is a polynomial of degree n, where $n > 0$, then $f(x)$ has precisely n linear factors
>
> $$f(x) = a_n(x - c_1)(x - c_2) \cdot \cdot \cdot (x - c_n)$$
>
> where $c_1, c_2, \ldots, c_n$ are complex numbers.

For a proof of the Linear Factorization Theorem, see Proofs in Mathematics on page 196.

Note that the Fundamental Theorem of Algebra and the Linear Factorization Theorem tell you only that the zeros or factors of a polynomial exist, not how to find them. Such theorems are called **existence theorems.**

REMARK Recall that in order to find the zeros of a function f, set $f(x)$ equal to 0 and solve the resulting equation for x. For instance, the function in Example 1(a) has a zero at $x = 2$ because

$$x - 2 = 0$$
$$x = 2.$$

EXAMPLE 1 Zeros of Polynomial Functions

a. The first-degree polynomial function $f(x) = x - 2$ has exactly *one* zero: $x = 2$.

b. Counting multiplicity, the second-degree polynomial function

$$f(x) = x^2 - 6x + 9 = (x - 3)(x - 3)$$

has exactly *two* zeros: $x = 3$ and $x = 3$. (This is called a *repeated zero.*)

c. The third-degree polynomial function

$$f(x) = x^3 + 4x = x(x^2 + 4) = x(x - 2i)(x + 2i)$$

has exactly *three* zeros: $x = 0$, $x = 2i$, and $x = -2i$.

ALGEBRA HELP Examples 1(b) and 1(c) involve factoring polynomials. You can review the techniques for factoring polynomials in Appendix A.3.

✓ **Checkpoint** Audio-video solution in English & Spanish at LarsonPrecalculus.com.

Determine the number of zeros of the polynomial function $f(x) = x^4 - 1$.

k45025/Shutterstock.com

The Rational Zero Test

The **Rational Zero Test** relates the possible rational zeros of a polynomial (having integer coefficients) to the leading coefficient and to the constant term of the polynomial.

Although they were not contemporaries, Jean Le Rond d'Alembert (1717–1783) worked independently of Carl Gauss in trying to prove the Fundamental Theorem of Algebra. His efforts were such that, in France, the Fundamental Theorem of Algebra is frequently known as the Theorem of d'Alembert.

> **The Rational Zero Test**
>
> If the polynomial $f(x) = a_n x^n + a_{n-1}x^{n-1} + \cdot \cdot \cdot + a_2 x^2 + a_1 x + a_0$ has *integer* coefficients, then every rational zero of f has the form
>
> $$\text{Rational zero} = \frac{p}{q}$$
>
> where p and q have no common factors other than 1, and
>
> $p =$ a factor of the constant term a_0
>
> $q =$ a factor of the leading coefficient a_n.

To use the Rational Zero Test, you should first list all rational numbers whose numerators are factors of the constant term and whose denominators are factors of the leading coefficient.

$$\text{Possible rational zeros} = \frac{\text{factors of constant term}}{\text{factors of leading coefficient}}$$

Having formed this list of *possible rational zeros,* use a trial-and-error method to determine which, if any, are actual zeros of the polynomial. Note that when the leading coefficient is 1, the possible rational zeros are simply the factors of the constant term.

EXAMPLE 2 Rational Zero Test with Leading Coefficient of 1

Find the rational zeros of

$$f(x) = x^3 + x + 1.$$

Solution Because the leading coefficient is 1, the possible rational zeros are the factors of the constant term.

Possible rational zeros: 1 and -1

By testing these possible zeros, you can see that neither works.

$$f(1) = (1)^3 + 1 + 1$$
$$= 3$$
$$f(-1) = (-1)^3 + (-1) + 1$$
$$= -1$$

So, you can conclude that the given polynomial has *no* rational zeros. Note from the graph of f in Figure 2.17 that f does have one real zero between -1 and 0. However, by the Rational Zero Test, you know that this real zero is *not* a rational number.

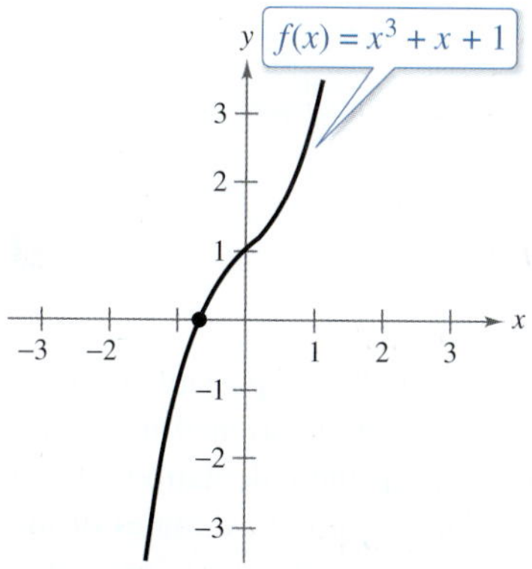

Figure 2.17

✓ Checkpoint Audio-video solution in English & Spanish at LarsonPrecalculus.com.

Find the rational zeros of each function.

a. $f(x) = x^3 - 5x^2 + 2x + 8$

b. $f(x) = x^3 + 2x^2 + 6x - 4$

c. $f(x) = x^3 - 3x^2 + 2x - 6$

Fogg Art Museum/Harvard University

EXAMPLE 3 Rational Zero Test with Leading Coefficient of 1

Find the rational zeros of

$$f(x) = x^4 - x^3 + x^2 - 3x - 6.$$

Solution Because the leading coefficient is 1, the possible rational zeros are the factors of the constant term.

Possible rational zeros: $\pm 1, \pm 2, \pm 3, \pm 6$

REMARK When the list of possible rational zeros is small, as in Example 2, it may be quicker to test the zeros by evaluating the function. When the list of possible rational zeros is large, as in Example 3, it may be quicker to use a different approach to test the zeros, such as using synthetic division or sketching a graph.

By applying synthetic division successively, you can determine that $x = -1$ and $x = 2$ are the only two rational zeros.

$$\begin{array}{r|rrrrr} -1 & 1 & -1 & 1 & -3 & -6 \\ & & -1 & 2 & -3 & 6 \\ \hline & 1 & -2 & 3 & -6 & 0 \end{array}$$

0 remainder, so $x = -1$ is a zero.

$$\begin{array}{r|rrrr} 2 & 1 & -2 & 3 & -6 \\ & & 2 & 0 & 6 \\ \hline & 1 & 0 & 3 & 0 \end{array}$$

0 remainder, so $x = 2$ is a zero.

So, $f(x)$ factors as

$$f(x) = (x + 1)(x - 2)(x^2 + 3).$$

Because the factor $(x^2 + 3)$ produces no real zeros, you can conclude that $x = -1$ and $x = 2$ are the only *real* zeros of f, which is verified in the figure below.

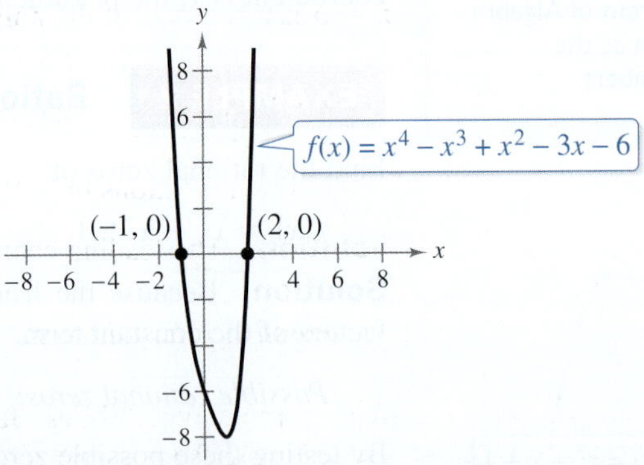

✓ **Checkpoint** Audio-video solution in English & Spanish at LarsonPrecalculus.com.

Find the rational zeros of

$$f(x) = x^3 - 15x^2 + 75x - 125.$$

When the leading coefficient of a polynomial is not 1, the list of possible rational zeros can increase dramatically. In such cases, the search can be shortened in several ways: (1) a programmable calculator can be used to speed up the calculations; (2) a graph, drawn either by hand or with a graphing utility, can give good estimates of the locations of the zeros; (3) the Intermediate Value Theorem along with a table generated by a graphing utility can give approximations of the zeros; and (4) synthetic division can be used to test the possible rational zeros.

Finding the first zero is often the most difficult part. After that, the search is simplified by working with the lower-degree polynomial obtained in synthetic division, as shown in Example 3.

EXAMPLE 4 Using the Rational Zero Test

Find the rational zeros of $f(x) = 2x^3 + 3x^2 - 8x + 3$.

Solution The leading coefficient is 2 and the constant term is 3.

$$\textit{Possible rational zeros: } \frac{\text{Factors of } 3}{\text{Factors of } 2} = \frac{\pm 1, \pm 3}{\pm 1, \pm 2} = \pm 1, \pm 3, \pm\frac{1}{2}, \pm\frac{3}{2}$$

REMARK Remember that when you try to find the rational zeros of a polynomial function with many possible rational zeros, as in Example 4, you must use trial and error. There is no quick algebraic method to determine which of the possibilities is an actual zero; however, sketching a graph may be helpful.

By synthetic division, you can determine that $x = 1$ is a rational zero.

$$\begin{array}{r|rrrr} 1 & 2 & 3 & -8 & 3 \\ & & 2 & 5 & -3 \\ \hline & 2 & 5 & -3 & 0 \end{array}$$

So, $f(x)$ factors as

$$f(x) = (x - 1)(2x^2 + 5x - 3)$$
$$= (x - 1)(2x - 1)(x + 3)$$

and you can conclude that the rational zeros of f are $x = 1$, $x = \frac{1}{2}$, and $x = -3$.

Checkpoint Audio-video solution in English & Spanish at LarsonPrecalculus.com.

Find the rational zeros of

$$f(x) = 2x^4 - 9x^3 - 18x^2 + 71x - 30.$$

Recall from Section 2.2 that if $x = a$ is a zero of the polynomial function f, then $x = a$ is a solution of the polynomial equation $f(x) = 0$.

EXAMPLE 5 Solving a Polynomial Equation

Find all real solutions of $-10x^3 + 15x^2 + 16x - 12 = 0$.

Solution The leading coefficient is -10 and the constant term is -12.

$$\textit{Possible rational solutions: } \frac{\text{Factors of } -12}{\text{Factors of } -10} = \frac{\pm 1, \pm 2, \pm 3, \pm 4, \pm 6, \pm 12}{\pm 1, \pm 2, \pm 5, \pm 10}$$

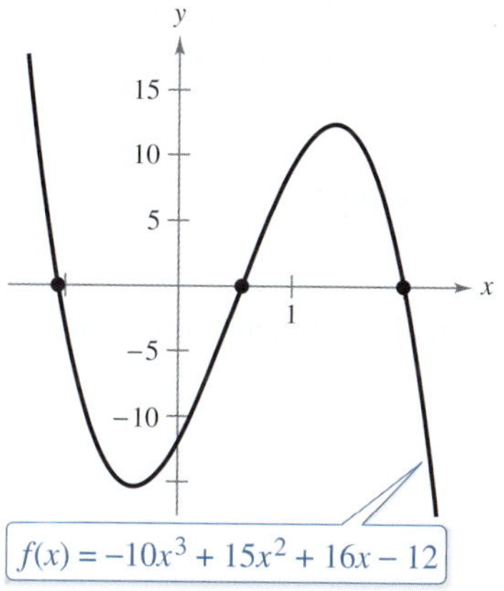

Figure 2.18

With so many possibilities (32, in fact), it is worth your time to stop and sketch a graph. From Figure 2.18, it looks like three reasonable solutions would be $x = -\frac{6}{5}$, $x = \frac{1}{2}$, and $x = 2$. Testing these by synthetic division shows that $x = 2$ is the only rational solution. So, you have

$$(x - 2)(-10x^2 - 5x + 6) = 0.$$

Using the Quadratic Formula for the second factor, you find that the two additional solutions are irrational numbers.

$$x = \frac{-5 - \sqrt{265}}{20} \approx -1.0639$$

and

$$x = \frac{-5 + \sqrt{265}}{20} \approx 0.5639$$

ALGEBRA HELP You can review the techniques for using the Quadratic Formula in Appendix A.5.

Checkpoint Audio-video solution in English & Spanish at LarsonPrecalculus.com.

Find all real solutions of

$$x^3 + 4x^2 - 15x - 18 = 0.$$

Conjugate Pairs

In Example 1(c), note that the pair of complex zeros are conjugates. That is, they are of the form

$$a + bi$$

and

$$a - bi.$$

Complex Zeros Occur in Conjugate Pairs

Let f be a polynomial function that has *real coefficients.* If $a + bi$, where $b \neq 0$, is a zero of the function, then the conjugate $a - bi$ is also a zero of the function.

Be sure you see that this result is true only when the polynomial function has *real coefficients.* For instance, the result applies to the function

$$f(x) = x^2 + 1$$

but not to the function

$$g(x) = x - i.$$

EXAMPLE 6 Finding a Polynomial Function with Given Zeros

Find a fourth-degree polynomial function with real coefficients that has

$$-1, \quad -1, \quad \text{and} \quad 3i$$

as zeros.

Solution Because $3i$ is a zero *and* the polynomial is stated to have real coefficients, you know that the conjugate $-3i$ must also be a zero. So, the four zeros are

$$-1, \quad -1, \quad 3i, \quad \text{and} \quad -3i.$$

Then, using the Linear Factorization Theorem, $f(x)$ can be written as

$$f(x) = a(x + 1)(x + 1)(x - 3i)(x + 3i).$$

For simplicity, let $a = 1$. Then multiply the factors with real coefficients to obtain

$$(x + 1)(x + 1) = x^2 + 2x + 1$$

and multiply the complex conjugates to obtain

$$(x - 3i)(x + 3i) = x^2 + 9.$$

So, you obtain the following fourth-degree polynomial function.

$$\begin{aligned} f(x) &= (x^2 + 2x + 1)(x^2 + 9) \\ &= x^4 + 2x^3 + 10x^2 + 18x + 9 \end{aligned}$$

✓ ***Checkpoint*** *Audio-video solution in English & Spanish at LarsonPrecalculus.com.*

Find a fourth-degree polynomial function with real coefficients that has the given zeros.

a. $2, -2$, and $-7i$

b. $1, 3$, and $4 - i$

c. $-1, 2$, and $3 + i$

Factoring a Polynomial

The Linear Factorization Theorem shows that you can write any nth-degree polynomial as the product of n linear factors.

$$f(x) = a_n(x - c_1)(x - c_2)(x - c_3) \cdot \cdot \cdot (x - c_n)$$

However, this result includes the possibility that some of the values of c_i are complex. The following theorem says that even if you do not want to get involved with "complex factors," you can still write $f(x)$ as the product of linear and/or quadratic factors. For a proof of this theorem, see Proofs in Mathematics on page 196.

Factors of a Polynomial

Every polynomial of degree $n > 0$ with real coefficients can be written as the product of linear and quadratic factors with real coefficients, where the quadratic factors have no real zeros.

A quadratic factor with no real zeros is said to be *prime* or **irreducible over the reals.** Note that this is not the same as being *irreducible over the rationals.* For example, the quadratic $x^2 + 1 = (x - i)(x + i)$ is irreducible over the reals (and therefore over the rationals). On the other hand, the quadratic $x^2 - 2 = (x - \sqrt{2})(x + \sqrt{2})$ is irreducible over the rationals but *reducible* over the reals.

TECHNOLOGY In Example 7, you can use a graphing utility to graph the function, as shown below.

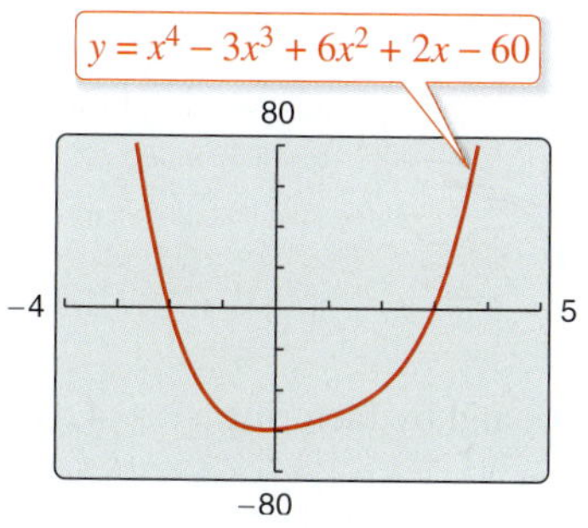

Then you can use the *zero* or *root* feature of the graphing utility to determine that $x = -2$ and $x = 3$ are zeros of the function.

EXAMPLE 7 Finding the Zeros of a Polynomial Function

Find all the zeros of $f(x) = x^4 - 3x^3 + 6x^2 + 2x - 60$ given that $1 + 3i$ is a zero of f.

Solution Because complex zeros occur in conjugate pairs, you know that $1 - 3i$ is also a zero of f. This means that both $[x - (1 + 3i)]$ and $[x - (1 - 3i)]$ are factors of $f(x)$. Multiplying these two factors produces

$$\begin{aligned}[x - (1 + 3i)][x - (1 - 3i)] &= [(x - 1) - 3i][(x - 1) + 3i] \\ &= (x - 1)^2 - 9i^2 \\ &= x^2 - 2x + 10.\end{aligned}$$

Using long division, you can divide $x^2 - 2x + 10$ into $f(x)$ to obtain the following.

$$\begin{array}{r} x^2 - x - 6 \\ x^2 - 2x + 10 \overline{)\, x^4 - 3x^3 + 6x^2 + 2x - 60} \\ \underline{x^4 - 2x^3 + 10x^2 } \\ -x^3 - 4x^2 + 2x \\ \underline{-x^3 + 2x^2 - 10x } \\ -6x^2 + 12x - 60 \\ \underline{-6x^2 + 12x - 60} \\ 0 \end{array}$$

So, you have

$$f(x) = (x^2 - 2x + 10)(x^2 - x - 6) = (x^2 - 2x + 10)(x - 3)(x + 2)$$

and you can conclude that the zeros of f are $x = 1 + 3i$, $x = 1 - 3i$, $x = 3$, and $x = -2$.

ALGEBRA HELP You can review the techniques for polynomial long division in Section 2.3.

✓ ***Checkpoint*** *Audio-video solution in English & Spanish at LarsonPrecalculus.com.*

Find all the zeros of $f(x) = 3x^3 - 2x^2 + 48x - 32$ given that $4i$ is a zero of f.

In Example 7, without being told that $1 + 3i$ is a zero of f, you could still find all the zeros of the function by using synthetic division to find the real zeros -2 and 3. Then you could factor the polynomial as

$$(x + 2)(x - 3)(x^2 - 2x + 10).$$

Finally, by using the Quadratic Formula, you could determine that the zeros are

$$x = -2, \quad x = 3, \quad x = 1 + 3i, \quad \text{and} \quad x = 1 - 3i.$$

Example 8 shows how to find all the zeros of a polynomial function, including complex zeros.

▷ **TECHNOLOGY** In Example 8, the fifth-degree polynomial function has three real zeros. In such cases, you can use the *zoom* and *trace* features or the *zero* or *root* feature of a graphing utility to approximate the real zeros. You can then use these real zeros to determine the complex zeros algebraically.

EXAMPLE 8 Finding the Zeros of a Polynomial Function

Write

$$f(x) = x^5 + x^3 + 2x^2 - 12x + 8$$

as the product of linear factors and list all the zeros of the function.

Solution Because the leading coefficient is 1, the possible rational zeros are the factors of the constant term.

Possible rational zeros: $\pm1, \pm2, \pm4,$ and ±8

Synthetic division produces the following.

$$\begin{array}{r|rrrrrr} 1 & 1 & 0 & 1 & 2 & -12 & 8 \\ & & 1 & 1 & 2 & 4 & -8 \\ \hline & 1 & 1 & 2 & 4 & -8 & 0 \end{array} \quad \longrightarrow \text{1 is a zero.}$$

$$\begin{array}{r|rrrrr} -2 & 1 & 1 & 2 & 4 & -8 \\ & & -2 & 2 & -8 & 8 \\ \hline & 1 & -1 & 4 & -4 & 0 \end{array} \quad \longrightarrow -2 \text{ is a zero.}$$

So, you have

$$f(x) = x^5 + x^3 + 2x^2 - 12x + 8$$
$$= (x - 1)(x + 2)(x^3 - x^2 + 4x - 4).$$

You can factor $x^3 - x^2 + 4x - 4$ as $(x - 1)(x^2 + 4)$, and by factoring $x^2 + 4$ as

$$x^2 - (-4) = \left(x - \sqrt{-4}\right)\left(x + \sqrt{-4}\right)$$
$$= (x - 2i)(x + 2i)$$

you obtain

$$f(x) = (x - 1)(x - 1)(x + 2)(x - 2i)(x + 2i)$$

which gives the following five zeros of f.

$$x = 1, \quad x = 1, \quad x = -2, \quad x = 2i, \quad \text{and} \quad x = -2i$$

From the graph of f shown in Figure 2.19, you can see that the *real* zeros are the only ones that appear as x-intercepts. Note that $x = 1$ is a repeated zero.

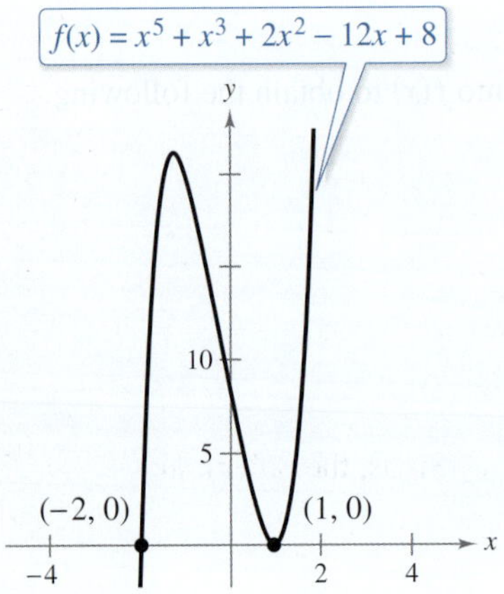

Figure 2.19

✓ ***Checkpoint*** *Audio-video solution in English & Spanish at LarsonPrecalculus.com.*

Write each function as the product of linear factors and list all the zeros of the function.

a. $f(x) = x^4 + 8x^2 - 9$

b. $g(x) = x^3 - 3x^2 + 7x - 5$

c. $h(x) = x^3 - 11x^2 + 41x - 51$

Other Tests for Zeros of Polynomials

You know that an nth-degree polynomial function can have *at most n* real zeros. Of course, many nth-degree polynomials do not have that many real zeros. For instance, $f(x) = x^2 + 1$ has no real zeros, and $f(x) = x^3 + 1$ has only one real zero. The following theorem, called **Descartes's Rule of Signs,** sheds more light on the number of real zeros of a polynomial.

> **Descartes's Rule of Signs**
>
> Let $f(x) = a_n x^n + a_{n-1}x^{n-1} + \cdot \cdot \cdot + a_2x^2 + a_1x + a_0$ be a polynomial with real coefficients and $a_0 \neq 0$.
>
> **1.** The number of *positive real zeros* of f is either equal to the number of variations in sign of $f(x)$ or less than that number by an even integer.
>
> **2.** The number of *negative real zeros* of f is either equal to the number of variations in sign of $f(-x)$ or less than that number by an even integer.

A **variation in sign** means that two consecutive coefficients have opposite signs.

When using Descartes's Rule of Signs, a zero of multiplicity k should be counted as k zeros. For instance, the polynomial $x^3 - 3x + 2$ has two variations in sign, and so it has either two positive or no positive real zeros. Because

$$x^3 - 3x + 2 = (x - 1)(x - 1)(x + 2)$$

you can see that the two positive real zeros are $x = 1$ of multiplicity 2.

EXAMPLE 9 Using Descartes's Rule of Signs

Determine the possible numbers of positive and negative real zeros of

$$f(x) = 3x^3 - 5x^2 + 6x - 4.$$

Solution The original polynomial has *three* variations in sign.

+ to −   + to −

$$f(x) = 3x^3 - 5x^2 + 6x - 4$$

− to +

The polynomial

$$\begin{aligned} f(-x) &= 3(-x)^3 - 5(-x)^2 + 6(-x) - 4 \\ &= -3x^3 - 5x^2 - 6x - 4 \end{aligned}$$

has no variations in sign. So, from Descartes's Rule of Signs, the polynomial

$$f(x) = 3x^3 - 5x^2 + 6x - 4$$

has either three positive real zeros or one positive real zero, and has no negative real zeros. From the graph in Figure 2.20, you can see that the function has only one real zero, $x = 1$.

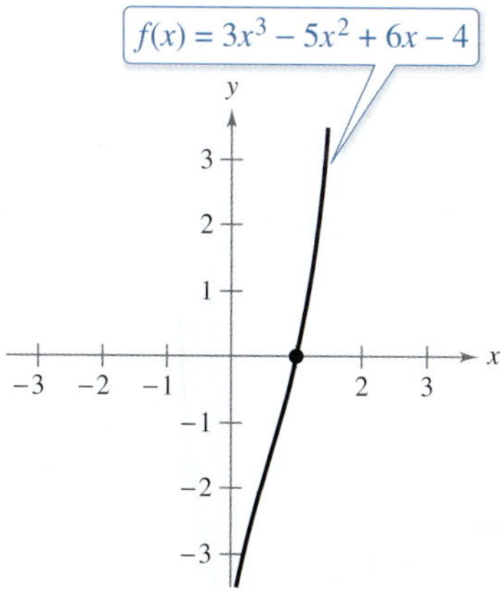

Figure 2.20

✓ **Checkpoint** *Audio-video solution in English & Spanish at LarsonPrecalculus.com.*

Determine the possible numbers of positive and negative real zeros of

$$f(x) = -2x^3 + 5x^2 - x + 8.$$

Another test for zeros of a polynomial function is related to the sign pattern in the last row of the synthetic division array. This test can give you an upper or lower bound for the real zeros of f. A real number c is an **upper bound** for the real zeros of f when no zeros are greater than c. Similarly, c is a **lower bound** when no real zeros of f are less than c.

Upper and Lower Bound Rules

Let $f(x)$ be a polynomial with real coefficients and a positive leading coefficient. Suppose $f(x)$ is divided by $x - c$, using synthetic division.

1. If $c > 0$ and each number in the last row is either positive or zero, then c is an **upper bound** for the real zeros of f.
2. If $c < 0$ and the numbers in the last row are alternately positive and negative (zero entries count as positive or negative), then c is a **lower bound** for the real zeros of f.

EXAMPLE 10 Finding the Zeros of a Polynomial Function

Find all real zeros of

$$f(x) = 6x^3 - 4x^2 + 3x - 2.$$

Solution The possible real zeros are as follows.

$$\frac{\text{Factors of } -2}{\text{Factors of } 6} = \frac{\pm 1, \pm 2}{\pm 1, \pm 2, \pm 3, \pm 6} = \pm 1, \pm\frac{1}{2}, \pm\frac{1}{3}, \pm\frac{1}{6}, \pm\frac{2}{3}, \pm 2$$

The original polynomial $f(x)$ has three variations in sign. The polynomial

$$f(-x) = 6(-x)^3 - 4(-x)^2 + 3(-x) - 2$$
$$= -6x^3 - 4x^2 - 3x - 2$$

has no variations in sign. As a result of these two findings, you can apply Descartes's Rule of Signs to conclude that there are three positive real zeros or one positive real zero, and no negative real zeros. Trying $x = 1$ produces the following.

$$\begin{array}{r|rrrr} 1 & 6 & -4 & 3 & -2 \\ & & 6 & 2 & 5 \\ \hline & 6 & 2 & 5 & 3 \end{array}$$

So, $x = 1$ is not a zero, but because the last row has all positive entries, you know that $x = 1$ is an upper bound for the real zeros. So, you can restrict the search to zeros between 0 and 1. By trial and error, you can determine that $x = \frac{2}{3}$ is a zero. So,

$$f(x) = \left(x - \frac{2}{3}\right)(6x^2 + 3).$$

Because $6x^2 + 3$ has no real zeros, it follows that $x = \frac{2}{3}$ is the only real zero, as shown at the right.

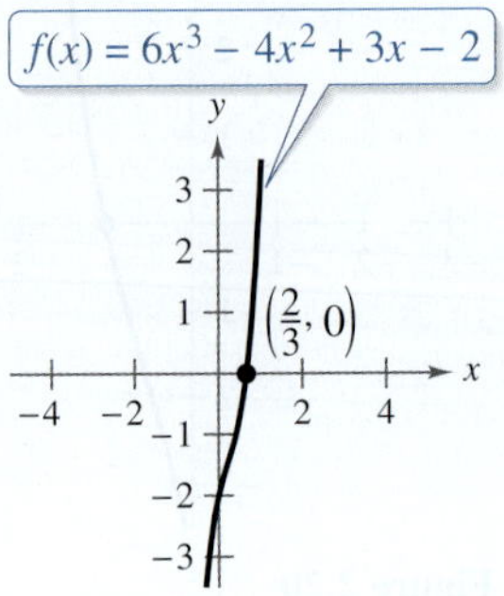

Checkpoint Audio-video solution in English & Spanish at LarsonPrecalculus.com.

Find all real zeros of

$$f(x) = 8x^3 - 4x^2 + 6x - 3.$$

Application

According to the National Candle Association, 7 out of 10 U.S. households use candles.

EXAMPLE 11 Using a Polynomial Model

You are designing candle making kits. Each kit contains 25 cubic inches of candle wax and a mold for making a pyramid-shaped candle. You want the height of the candle to be 2 inches less than the length of each side of the candle's square base. What should the dimensions of your candle mold be?

Solution The volume of a pyramid is $V = \frac{1}{3}Bh$, where B is the area of the base and h is the height. The area of the base is x^2 and the height is $(x - 2)$. So, the volume of the pyramid is $V = \frac{1}{3}x^2(x - 2)$. Substituting 25 for the volume yields the following.

$$25 = \frac{1}{3}x^2(x - 2) \quad \text{Substitute 25 for } V.$$

$$75 = x^3 - 2x^2 \quad \text{Multiply each side by 3, and distribute } x^2.$$

$$0 = x^3 - 2x^2 - 75 \quad \text{Write in general form.}$$

The possible rational solutions are $x = \pm 1, \pm 3, \pm 5, \pm 15, \pm 25, \pm 75$. Use synthetic division to test some of the possible solutions. Note that in this case it makes sense to test only positive x-values. Using synthetic division, you can determine that $x = 5$ is a solution.

$$\begin{array}{r|rrrr} 5 & 1 & -2 & 0 & -75 \\ & & 5 & 15 & 75 \\ \hline & 1 & 3 & 15 & 0 \end{array}$$

The other two solutions that satisfy $x^2 + 3x + 15 = 0$ are imaginary and can be discarded. You can conclude that the base of the candle mold should be 5 inches by 5 inches and the height should be $5 - 2 = 3$ inches.

✓ **Checkpoint** Audio-video solution in English & Spanish at LarsonPrecalculus.com.

Rework Example 11 when each kit contains 147 cubic inches of candle wax and you want the height of the pyramid-shaped candle to be 2 inches more than the length of each side of the candle's square base.

Before concluding this section, here is an additional hint that can help you find the real zeros of a polynomial. When the terms of $f(x)$ have a common monomial factor, it should be factored out before applying the tests in this section. For instance, by writing

$$f(x) = x^4 - 5x^3 + 3x^2 + x = x(x^3 - 5x^2 + 3x + 1)$$

you can see that $x = 0$ is a zero of f and that the remaining zeros can be obtained by analyzing the cubic factor.

Summarize (Section 2.5)

1. State the Fundamental Theorem of Algebra and the Linear Factorization Theorem *(page 154, Example 1)*.
2. Describe the Rational Zero Test *(page 155, Examples 2–5)*.
3. State the definition of conjugates *(page 158, Example 6)*.
4. Describe what it means for a quadratic factor to be irreducible over the reals *(page 159, Examples 7 and 8)*.
5. State Descartes's Rule of Signs and the Upper and Lower Bound Rules *(pages 161 and 162, Examples 9 and 10)*.
6. Describe a real-life situation involving finding the zeros of a polynomial *(page 163, Example 11)*.

hfng/Shutterstock.com

2.5 Exercises

See CalcChat.com for tutorial help and worked-out solutions to odd-numbered exercises.

Vocabulary: Fill in the blanks.

1. The ________ ________ of ________ states that if $f(x)$ is a polynomial of degree n $(n > 0)$, then f has at least one zero in the complex number system.
2. The ________ ________ ________ states that if $f(x)$ is a polynomial of degree n $(n > 0)$, then $f(x)$ has precisely n linear factors, $f(x) = a_n(x - c_1)(x - c_2) \cdot \cdot \cdot (x - c_n)$, where $c_1, c_2, \ldots, c_n$ are complex numbers.
3. The test that gives a list of the possible rational zeros of a polynomial function is the ________ ________ Test.
4. If $a + bi$ is a complex zero of a polynomial with real coefficients, then so is its ________, $a - bi$.
5. Every polynomial of degree $n > 0$ with real coefficients can be written as the product of ________ and ________ factors with real coefficients, where the ________ factors have no real zeros.
6. A quadratic factor that cannot be factored further as a product of linear factors containing real numbers is said to be ________ over the ________.
7. The theorem that can be used to determine the possible numbers of positive real zeros and negative real zeros of a function is called ________ ________ of ________.
8. A real number b is a ________ bound for the real zeros of f when no real zeros are less than b, and is a ________ bound when no real zeros are greater than b.

Skills and Applications

Zeros of Polynomial Functions **In Exercises 9–14, determine the number of zeros of the polynomial function.**

9. $f(x) = x + 3$
10. $f(x) = x^2 + 5x - 6$
11. $g(x) = 1 - x^3$
12. $f(x) = x^6 - x^7$
13. $f(x) = (x + 5)^2$
14. $h(t) = (t - 1)^2 - (t + 1)^2$

Using the Rational Zero Test **In Exercises 15–18, use the Rational Zero Test to list the possible rational zeros of f. Verify that the zeros of f shown on the graph are contained in the list.**

15. $f(x) = x^3 + 2x^2 - x - 2$

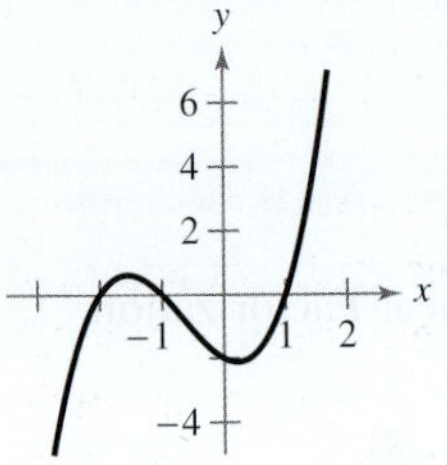

16. $f(x) = x^3 - 4x^2 - 4x + 16$

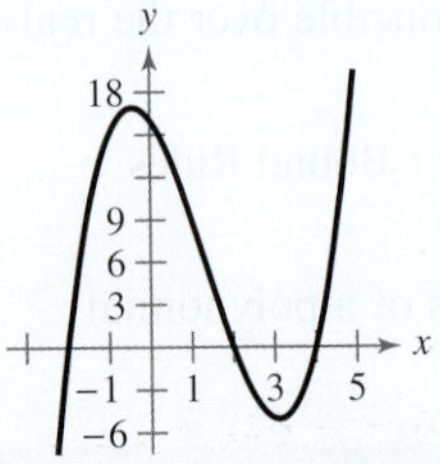

17. $f(x) = 2x^4 - 17x^3 + 35x^2 + 9x - 45$

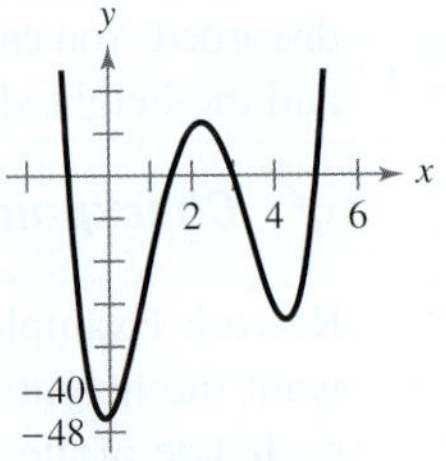

18. $f(x) = 4x^5 - 8x^4 - 5x^3 + 10x^2 + x - 2$

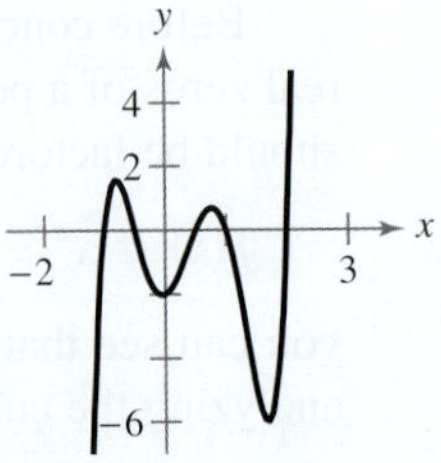

Using the Rational Zero Test **In Exercises 19–28, find the rational zeros of the function.**

19. $f(x) = x^3 - 7x - 6$
20. $f(x) = x^3 - 13x + 12$
21. $g(x) = x^3 - 4x^2 - x + 4$
22. $h(x) = x^3 - 9x^2 + 20x - 12$
23. $h(t) = t^3 + 8t^2 + 13t + 6$
24. $p(x) = x^3 - 9x^2 + 27x - 27$
25. $C(x) = 2x^3 + 3x^2 - 1$
26. $f(x) = 3x^3 - 19x^2 + 33x - 9$
27. $f(x) = 9x^4 - 9x^3 - 58x^2 + 4x + 24$
28. $f(x) = 2x^4 - 15x^3 + 23x^2 + 15x - 25$

Solving a Polynomial Equation In Exercises 29–32, find all real solutions of the polynomial equation.

29. $z^4 + z^3 + z^2 + 3z - 6 = 0$

30. $x^4 - 13x^2 - 12x = 0$

31. $2y^4 + 3y^3 - 16y^2 + 15y - 4 = 0$

32. $x^5 - x^4 - 3x^3 + 5x^2 - 2x = 0$

Using the Rational Zero Test In Exercises 33–36, (a) list the possible rational zeros of f, (b) sketch the graph of f so that some of the possible zeros in part (a) can be disregarded, and then (c) determine all real zeros of f.

33. $f(x) = x^3 + x^2 - 4x - 4$

34. $f(x) = -3x^3 + 20x^2 - 36x + 16$

35. $f(x) = -4x^3 + 15x^2 - 8x - 3$

36. $f(x) = 4x^3 - 12x^2 - x + 15$

Using the Rational Zero Test In Exercises 37–40, (a) list the possible rational zeros of f, (b) use a graphing utility to graph f so that some of the possible zeros in part (a) can be disregarded, and then (c) determine all real zeros of f.

37. $f(x) = -2x^4 + 13x^3 - 21x^2 + 2x + 8$

38. $f(x) = 4x^4 - 17x^2 + 4$

39. $f(x) = 32x^3 - 52x^2 + 17x + 3$

40. $f(x) = 4x^3 + 7x^2 - 11x - 18$

Graphical Analysis In Exercises 41–44, (a) use the *zero* or *root* feature of a graphing utility to approximate the zeros of the function accurate to three decimal places, (b) determine the exact value of one of the zeros (use synthetic division to verify your result), and (c) factor the polynomial completely.

41. $f(x) = x^4 - 3x^2 + 2$

42. $P(t) = t^4 - 7t^2 + 12$

43. $h(x) = x^5 - 7x^4 + 10x^3 + 14x^2 - 24x$

44. $g(x) = 6x^4 - 11x^3 - 51x^2 + 99x - 27$

Finding a Polynomial Function with Given Zeros In Exercises 45–50, find a polynomial function with real coefficients that has the given zeros. (There are many correct answers.)

45. $1, 5i$

46. $4, -3i$

47. $2, 5 + i$

48. $5, 3 - 2i$

49. $\frac{2}{3}, -1, 3 + \sqrt{2}i$

50. $-5, -5, 1 + \sqrt{3}i$

Factoring a Polynomial In Exercises 51–54, write the polynomial (a) as the product of factors that are irreducible over the *rationals*, (b) as the product of linear and quadratic factors that are irreducible over the *reals*, and (c) in completely factored form.

51. $f(x) = x^4 + 6x^2 - 27$

52. $f(x) = x^4 - 2x^3 - 3x^2 + 12x - 18$
(*Hint:* One factor is $x^2 - 6$.)

53. $f(x) = x^4 - 4x^3 + 5x^2 - 2x - 6$
(*Hint:* One factor is $x^2 - 2x - 2$.)

54. $f(x) = x^4 - 3x^3 - x^2 - 12x - 20$
(*Hint:* One factor is $x^2 + 4$.)

Finding the Zeros of a Polynomial Function In Exercises 55–62, use the given zero to find all the zeros of the function.

Function	Zero
55. $f(x) = x^3 - x^2 + 4x - 4$	$2i$
56. $f(x) = 2x^3 + 3x^2 + 18x + 27$	$3i$
57. $f(x) = 2x^4 - x^3 + 49x^2 - 25x - 25$	$5i$
58. $g(x) = x^3 - 7x^2 - x + 87$	$5 + 2i$
59. $g(x) = 4x^3 + 23x^2 + 34x - 10$	$-3 + i$
60. $h(x) = 3x^3 - 4x^2 + 8x + 8$	$1 - \sqrt{3}i$
61. $f(x) = x^4 + 3x^3 - 5x^2 - 21x + 22$	$-3 + \sqrt{2}i$
62. $f(x) = x^3 + 4x^2 + 14x + 20$	$-1 - 3i$

Finding the Zeros of a Polynomial Function In Exercises 63–80, write the polynomial as the product of linear factors and list all the zeros of the function.

63. $f(x) = x^2 + 36$

64. $f(x) = x^2 - x + 56$

65. $h(x) = x^2 - 2x + 17$

66. $g(x) = x^2 + 10x + 17$

67. $f(x) = x^4 - 16$

68. $f(y) = y^4 - 256$

69. $f(z) = z^2 - 2z + 2$

70. $h(x) = x^3 - 3x^2 + 4x - 2$

71. $g(x) = x^3 - 3x^2 + x + 5$

72. $f(x) = x^3 - x^2 + x + 39$

73. $h(x) = x^3 - x + 6$

74. $h(x) = x^3 + 9x^2 + 27x + 35$

75. $f(x) = 5x^3 - 9x^2 + 28x + 6$

76. $g(x) = 2x^3 - x^2 + 8x + 21$

77. $g(x) = x^4 - 4x^3 + 8x^2 - 16x + 16$

78. $h(x) = x^4 + 6x^3 + 10x^2 + 6x + 9$

79. $f(x) = x^4 + 10x^2 + 9$

80. $f(x) = x^4 + 29x^2 + 100$

Finding the Zeros of a Polynomial Function In Exercises 81–86, find all the zeros of the function. When there is an extended list of possible rational zeros, use a graphing utility to graph the function in order to disregard any of the possible rational zeros that are obviously not zeros of the function.

81. $f(x) = x^3 + 24x^2 + 214x + 740$

82. $f(s) = 2s^3 - 5s^2 + 12s - 5$

83. $f(x) = 16x^3 - 20x^2 - 4x + 15$

84. $f(x) = 9x^3 - 15x^2 + 11x - 5$

85. $f(x) = 2x^4 + 5x^3 + 4x^2 + 5x + 2$

86. $g(x) = x^5 - 8x^4 + 28x^3 - 56x^2 + 64x - 32$

Using Descartes's Rule of Signs In Exercises 87–94, use Descartes's Rule of Signs to determine the possible numbers of positive and negative real zeros of the function.

87. $g(x) = 2x^3 - 3x^2 - 3$
88. $h(x) = 4x^2 - 8x + 3$
89. $h(x) = 2x^3 + 3x^2 + 1$
90. $h(x) = 2x^4 - 3x + 2$
91. $g(x) = 5x^5 - 10x$
92. $f(x) = 4x^3 - 3x^2 + 2x - 1$
93. $f(x) = -5x^3 + x^2 - x + 5$
94. $f(x) = 3x^3 + 2x^2 + x + 3$

Verifying Upper and Lower Bounds In Exercises 95–98, use synthetic division to verify the upper and lower bounds of the real zeros of f.

95. $f(x) = x^3 + 3x^2 - 2x + 1$
(a) Upper: $x = 1$ (b) Lower: $x = -4$
96. $f(x) = x^3 - 4x^2 + 1$
(a) Upper: $x = 4$ (b) Lower: $x = -1$
97. $f(x) = x^4 - 4x^3 + 16x - 16$
(a) Upper: $x = 5$ (b) Lower: $x = -3$
98. $f(x) = 2x^4 - 8x + 3$
(a) Upper: $x = 3$ (b) Lower: $x = -4$

Finding the Zeros of a Polynomial Function In Exercises 99–102, find all real zeros of the function.

99. $f(x) = 4x^3 - 3x - 1$
100. $f(z) = 12z^3 - 4z^2 - 27z + 9$
101. $f(y) = 4y^3 + 3y^2 + 8y + 6$
102. $g(x) = 3x^3 - 2x^2 + 15x - 10$

Finding the Rational Zeros of a Polynomial In Exercises 103–106, find the rational zeros of the polynomial function.

103. $P(x) = x^4 - \frac{25}{4}x^2 + 9 = \frac{1}{4}(4x^4 - 25x^2 + 36)$
104. $f(x) = x^3 - \frac{3}{2}x^2 - \frac{23}{2}x + 6 = \frac{1}{2}(2x^3 - 3x^2 - 23x + 12)$
105. $f(x) = x^3 - \frac{1}{4}x^2 - x + \frac{1}{4} = \frac{1}{4}(4x^3 - x^2 - 4x + 1)$
106. $f(z) = z^3 + \frac{11}{6}z^2 - \frac{1}{2}z - \frac{1}{3} = \frac{1}{6}(6z^3 + 11z^2 - 3z - 2)$

Rational and Irrational Zeros In Exercises 107–110, match the cubic function with the numbers of rational and irrational zeros.

(a) Rational zeros: 0; irrational zeros: 1
(b) Rational zeros: 3; irrational zeros: 0
(c) Rational zeros: 1; irrational zeros: 2
(d) Rational zeros: 1; irrational zeros: 0

107. $f(x) = x^3 - 1$
108. $f(x) = x^3 - 2$
109. $f(x) = x^3 - x$
110. $f(x) = x^3 - 2x$

111. **Geometry** You want to make an open box from a rectangular piece of material, 15 centimeters by 9 centimeters, by cutting equal squares from the corners and turning up the sides.
(a) Let x represent the side length of each of the squares removed. Draw a diagram showing the squares removed from the original piece of material and the resulting dimensions of the open box.
(b) Use the diagram to write the volume V of the box as a function of x. Determine the domain of the function.
(c) Sketch the graph of the function and approximate the dimensions of the box that will yield a maximum volume.
(d) Find values of x such that $V = 56$. Which of these values is a physical impossibility in the construction of the box? Explain.

112. **Geometry** A rectangular package to be sent by a delivery service (see figure) can have a maximum combined length and girth (perimeter of a cross section) of 120 inches.

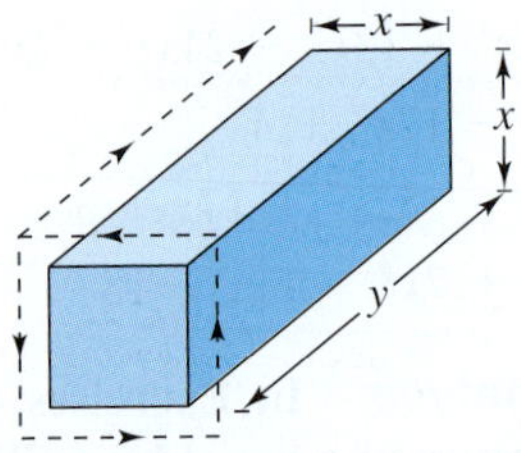

(a) Use the diagram to write the volume V of the package as a function of x.
(b) Use a graphing utility to graph the function and approximate the dimensions of the package that will yield a maximum volume.
(c) Find values of x such that $V = 13{,}500$. Which of these values is a physical impossibility in the construction of the package? Explain.

113. **Geometry**
A bulk food storage bin with dimensions 2 feet by 3 feet by 4 feet needs to be increased in size to hold five times as much food as the current bin. (Assume each dimension is increased by the same amount.)
(a) Write a function that represents the volume V of the new bin.
(b) Find the dimensions of the new bin.

k45025/Shutterstock.com

114. Cost The ordering and transportation cost C (in thousands of dollars) for machine parts is

$$C = 100\left(\frac{200}{x^2} + \frac{x}{x + 30}\right), \quad x \geq 1$$

where x is the order size (in hundreds). In calculus, it can be shown that the cost is a minimum when

$$3x^3 - 40x^2 - 2400x - 36{,}000 = 0.$$

Use a calculator to approximate the optimal order size to the nearest hundred units.

Exploration

True or False? **In Exercises 115 and 116, decide whether the statement is true or false. Justify your answer.**

115. It is possible for a third-degree polynomial function with integer coefficients to have no real zeros.

116. If $x = -i$ is a zero of the function

$$f(x) = x^3 + ix^2 + ix - 1$$

then $x = i$ must also be a zero of f.

Think About It **In Exercises 117–122, determine (if possible) the zeros of the function g when the function f has zeros at $x = r_1$, $x = r_2$, and $x = r_3$.**

117. $g(x) = -f(x)$ **118.** $g(x) = 3f(x)$

119. $g(x) = f(x - 5)$ **120.** $g(x) = f(2x)$

121. $g(x) = 3 + f(x)$ **122.** $g(x) = f(-x)$

123. Think About It A cubic polynomial function f has real zeros -2, $\frac{1}{2}$, and 3, and its leading coefficient is negative. Write an equation for f and sketch its graph. How many different polynomial functions are possible for f?

124. Think About It Sketch the graph of a fifth-degree polynomial function whose leading coefficient is positive and that has a zero at $x = 3$ of multiplicity 2.

Writing an Equation **In Exercises 125 and 126, the graph of a cubic polynomial function $y = f(x)$ is shown. One of the zeros is $1 + i$. Write an equation for f.**

125.

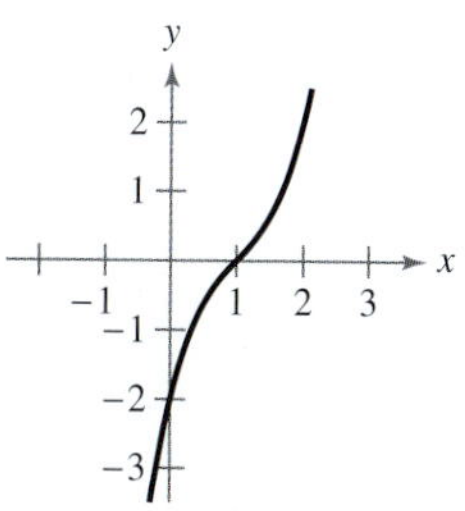

126.

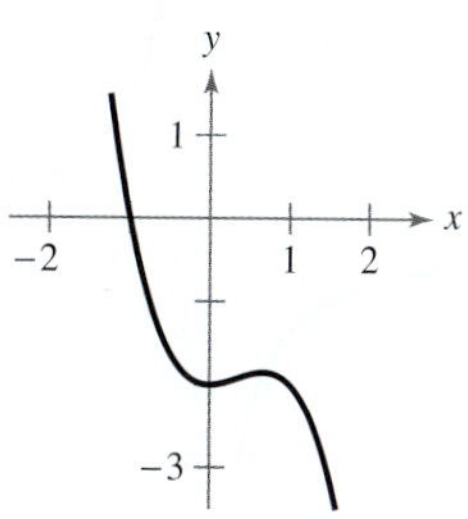

127. Think About It Let $y = f(x)$ be a quartic polynomial with leading coefficient $a = 1$ and $f(i) = f(2i) = 0$. Write an equation for f.

128. Think About It Let $y = f(x)$ be a cubic polynomial with leading coefficient $a = -1$ and $f(2) = f(i) = 0$. Write an equation for f.

129. Writing an Equation Write the equation for a quadratic function f (with integer coefficients) that has the given zeros. Assume that b is a positive integer.

(a) $\pm\sqrt{b}i$

(b) $a \pm bi$

130. HOW DO YOU SEE IT? Use the information in the table to answer each question.

Interval	Value of $f(x)$
$(-\infty, -2)$	Positive
$(-2, 1)$	Negative
$(1, 4)$	Negative
$(4, \infty)$	Positive

(a) What are the three real zeros of the polynomial function f?

(b) What can be said about the behavior of the graph of f at $x = 1$?

(c) What is the least possible degree of f? Explain. Can the degree of f ever be odd? Explain.

(d) Is the leading coefficient of f positive or negative? Explain.

(e) Sketch a graph of a function that exhibits the behavior described in the table.

131. Graphical Reasoning The graph of one of the following functions is shown below. Identify the function shown in the graph. Explain why each of the others is not the correct function. Use a graphing utility to verify your result.

(a) $f(x) = x^2(x + 2)(x - 3.5)$

(b) $g(x) = (x + 2)(x - 3.5)$

(c) $h(x) = (x + 2)(x - 3.5)(x^2 + 1)$

(d) $k(x) = (x + 1)(x + 2)(x - 3.5)$

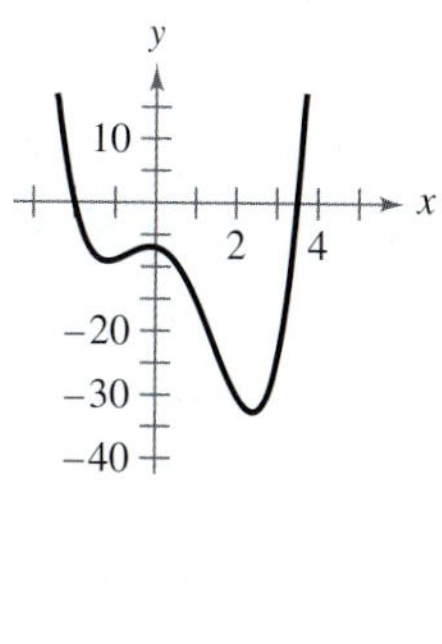

2.6 Rational Functions

Rational functions can help you model and solve real-life problems relating to environmental scenarios. For instance, in Exercise 71 on page 178, you will use a rational function to determine the cost of supplying recycling bins to the population of a rural township.

- Find the domains of rational functions.
- Find the vertical and horizontal asymptotes of the graphs of rational functions.
- Sketch the graphs of rational functions.
- Sketch the graphs of rational functions that have slant asymptotes.
- Use rational functions to model and solve real-life problems.

Introduction

A **rational function** is a quotient of polynomial functions. It can be written in the form

$$f(x) = \frac{N(x)}{D(x)}$$

where $N(x)$ and $D(x)$ are polynomials and $D(x)$ is not the zero polynomial.

In general, the *domain* of a rational function of x includes all real numbers except x-values that make the denominator zero. Much of the discussion of rational functions will focus on their graphical behavior near the x-values excluded from the domain.

EXAMPLE 1 Finding the Domain of a Rational Function

Find the domain of $f(x) = \frac{1}{x}$ and discuss the behavior of f near any excluded x-values.

Solution Because the denominator is zero when $x = 0$, the domain of f is all real numbers except $x = 0$. To determine the behavior of f near this excluded value, evaluate $f(x)$ to the left and right of $x = 0$, as indicated in the following tables.

x	-1	-0.5	-0.1	-0.01	-0.001	$\longrightarrow 0$
$f(x)$	-1	-2	-10	-100	-1000	$\longrightarrow -\infty$

x	$0 \longleftarrow$	0.001	0.01	0.1	0.5	1
$f(x)$	$\infty \longleftarrow$	1000	100	10	2	1

Note that as x approaches 0 *from the left*, $f(x)$ decreases without bound. In contrast, as x approaches 0 *from the right*, $f(x)$ increases without bound. The graph of f is shown below.

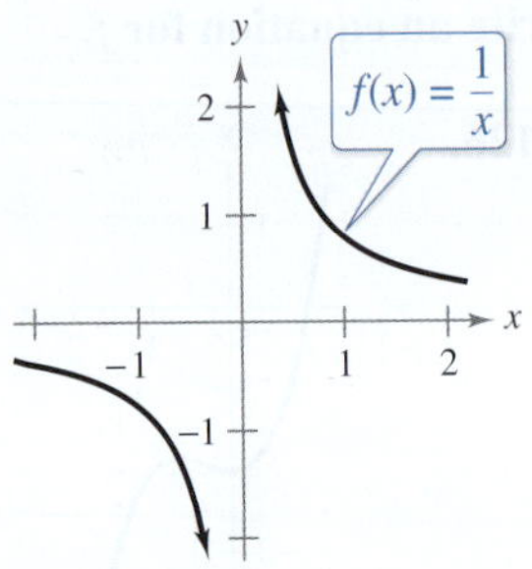

Checkpoint Audio-video solution in English & Spanish at LarsonPrecalculus.com.

Find the domain of the rational function $f(x) = \frac{3x}{x-1}$ and discuss the behavior of f near any excluded x-values.

ZQFotography/Shutterstock.com

Vertical and Horizontal Asymptotes

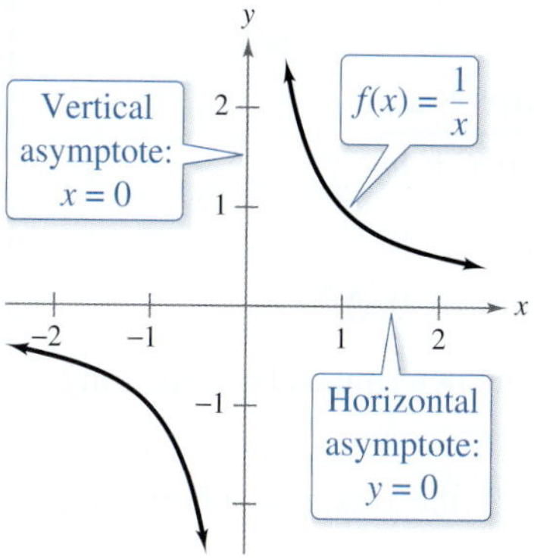

Figure 2.21

In Example 1, the behavior of f near $x = 0$ is denoted as follows.

$$\underbrace{f(x) \longrightarrow -\infty \text{ as } x \longrightarrow 0^-}_{f(x) \text{ decreases without bound as } x \text{ approaches 0 from the left.}} \qquad \underbrace{f(x) \longrightarrow \infty \text{ as } x \longrightarrow 0^+}_{f(x) \text{ increases without bound as } x \text{ approaches 0 from the right.}}$$

The line $x = 0$ is a **vertical asymptote** of the graph of f, as shown in Figure 2.21. From this figure, you can see that the graph of f also has a **horizontal asymptote**—the line $y = 0$. The behavior of f near $y = 0$ is denoted as follows.

$$\underbrace{f(x) \longrightarrow 0 \text{ as } x \longrightarrow -\infty}_{f(x) \text{ approaches 0 as } x \text{ decreases without bound.}} \qquad \underbrace{f(x) \longrightarrow 0 \text{ as } x \longrightarrow \infty}_{f(x) \text{ approaches 0 as } x \text{ increases without bound.}}$$

Definitions of Vertical and Horizontal Asymptotes

1. The line $x = a$ is a **vertical asymptote** of the graph of f when

$$f(x) \longrightarrow \infty \quad \text{or} \quad f(x) \longrightarrow -\infty$$

as $x \longrightarrow a$, either from the right or from the left.

2. The line $y = b$ is a **horizontal asymptote** of the graph of f when

$$f(x) \longrightarrow b$$

as $x \longrightarrow \infty$ or $x \longrightarrow -\infty$.

Eventually (as $x \longrightarrow \infty$ or $x \longrightarrow -\infty$), the distance between the horizontal asymptote and the points on the graph must approach zero. Figure 2.22 shows the vertical and horizontal asymptotes of the graphs of three rational functions.

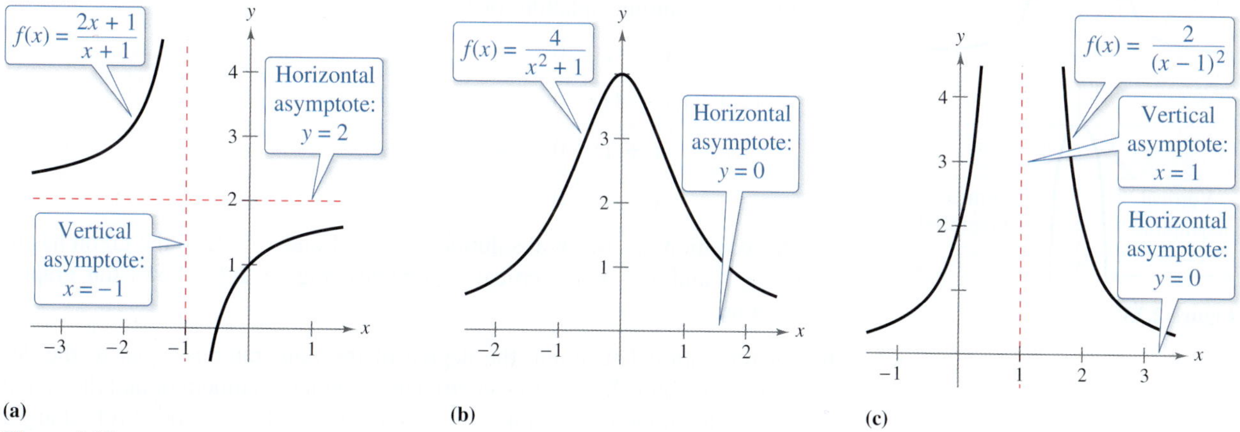

Figure 2.22

The graphs of

$$f(x) = \frac{1}{x}$$

in Figure 2.21 and

$$f(x) = \frac{2x + 1}{x + 1}$$

in Figure 2.22(a) are **hyperbolas.**

Vertical and Horizontal Asymptotes of a Rational Function

Let f be the rational function

$$f(x) = \frac{N(x)}{D(x)} = \frac{a_n x^n + a_{n-1}x^{n-1} + \cdot \cdot \cdot + a_1 x + a_0}{b_m x^m + b_{m-1}x^{m-1} + \cdot \cdot \cdot + b_1 x + b_0}$$

where $N(x)$ and $D(x)$ have no common factors.

1. The graph of f has *vertical* asymptotes at the zeros of $D(x)$.
2. The graph of f has one or no *horizontal* asymptote determined by comparing the degrees of $N(x)$ and $D(x)$.
   a. When $n < m$, the graph of f has the line $y = 0$ (the x-axis) as a horizontal asymptote.
   b. When $n = m$, the graph of f has the line $y = a_n/b_m$ (ratio of the leading coefficients) as a horizontal asymptote.
   c. When $n > m$, the graph of f has no horizontal asymptote.

EXAMPLE 2 **Finding Vertical and Horizontal Asymptotes**

Find all vertical and horizontal asymptotes of the graph of each rational function.

a. $f(x) = \dfrac{2x^2}{x^2 - 1}$ **b.** $f(x) = \dfrac{x^2 + x - 2}{x^2 - x - 6}$

Solution

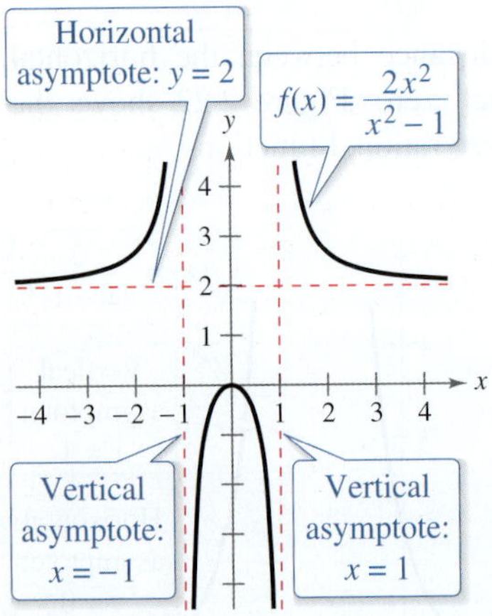

Figure 2.23

a. For this rational function, the degree of the numerator is *equal* to the degree of the denominator. The leading coefficient of the numerator is 2 and the leading coefficient of the denominator is 1, so the graph has the line $y = 2$ as a horizontal asymptote. To find any vertical asymptotes, set the denominator equal to zero and solve the resulting equation for x.

$x^2 - 1 = 0$ Set denominator equal to zero.

$(x + 1)(x - 1) = 0$ Factor.

$x + 1 = 0 \implies x = -1$ Set 1st factor equal to 0.

$x - 1 = 0 \implies x = 1$ Set 2nd factor equal to 0.

This equation has two real solutions, $x = -1$ and $x = 1$, so the graph has the lines $x = -1$ and $x = 1$ as vertical asymptotes. Figure 2.23 shows the graph of this function.

b. For this rational function, the degree of the numerator is equal to the degree of the denominator. The leading coefficient of both the numerator and the denominator is 1, so the graph has the line $y = 1$ as a horizontal asymptote. To find any vertical asymptotes, first factor the numerator and denominator as follows.

$$f(x) = \frac{x^2 + x - 2}{x^2 - x - 6} = \frac{(x - 1)\cancel{(x + 2)}}{\cancel{(x + 2)}(x - 3)} = \frac{x - 1}{x - 3}, \quad x \neq -2$$

By setting the denominator $x - 3$ (of the simplified function) equal to zero, you can determine that the graph has the line $x = 3$ as a vertical asymptote.

▷ **ALGEBRA HELP** You can review the techniques for factoring in Appendix A.3.

✓ **Checkpoint** Audio-video solution in English & Spanish at LarsonPrecalculus.com.

Find all vertical and horizontal asymptotes of the graph of $f(x) = \dfrac{3x^2 + 7x - 6}{x^2 + 4x + 3}$.

2.6 Exercises

See CalcChat.com for tutorial help and worked-out solutions to odd-numbered exercises.

Vocabulary: Fill in the blanks.

1. Functions of the form $f(x) = N(x)/D(x)$, where $N(x)$ and $D(x)$ are polynomials and $D(x)$ is not the zero polynomial, are called ________ ________.
2. When $f(x) \to \pm\infty$ as $x \to a$ from the left or the right, $x = a$ is a ________ ________ of the graph of f.
3. When $f(x) \to b$ as $x \to \pm\infty$, $y = b$ is a ________ ________ of the graph of f.
4. For the rational function $f(x) = N(x)/D(x)$, if the degree of $N(x)$ is exactly one more than the degree of $D(x)$, then the graph of f has a ________ (or oblique) ________.

Skills and Applications

Finding the Domain of a Rational Function In Exercises 5–8, find the domain of the function and discuss the behavior of f near any excluded x-values.

5. $f(x) = \dfrac{1}{x-1}$
6. $f(x) = \dfrac{5x}{x+2}$
7. $f(x) = \dfrac{3x^2}{x^2-1}$
8. $f(x) = \dfrac{2x}{x^2-4}$

Finding Vertical and Horizontal Asymptotes In Exercises 9–16, find all vertical and horizontal asymptotes of the graph of the function.

9. $f(x) = \dfrac{4}{x^2}$
10. $f(x) = \dfrac{1}{(x-2)^3}$
11. $f(x) = \dfrac{5+x}{5-x}$
12. $f(x) = \dfrac{3-7x}{3+2x}$
13. $f(x) = \dfrac{x^3}{x^2-1}$
14. $f(x) = \dfrac{4x^2}{x+2}$
15. $f(x) = \dfrac{3x^2+1}{x^2+x+9}$
16. $f(x) = \dfrac{3x^2+x-5}{x^2+1}$

Sketching the Graph of a Rational Function In Exercises 17–40, (a) state the domain of the function, (b) identify all intercepts, (c) find any vertical or horizontal asymptotes, and (d) plot additional solution points as needed to sketch the graph of the rational function.

17. $f(x) = \dfrac{1}{x+2}$
18. $f(x) = \dfrac{1}{x-3}$
19. $h(x) = \dfrac{-1}{x+4}$
20. $g(x) = \dfrac{1}{6-x}$
21. $C(x) = \dfrac{7+2x}{2+x}$
22. $P(x) = \dfrac{1-3x}{1-x}$
23. $f(x) = \dfrac{x^2}{x^2+9}$
24. $g(s) = \dfrac{4s}{s^2+4}$
25. $f(t) = \dfrac{1-2t}{t}$
26. $f(x) = -\dfrac{1}{(x-2)^2}$
27. $h(x) = \dfrac{x^2-5x+4}{x^2-4}$
28. $g(x) = \dfrac{x^2-2x-8}{x^2-9}$
29. $f(x) = \dfrac{x-4}{x^2-16}$
30. $f(x) = \dfrac{x+1}{x^2-1}$
31. $f(t) = \dfrac{t^2-1}{t-1}$
32. $f(x) = \dfrac{x^2-36}{x+6}$
33. $f(x) = \dfrac{x^2-25}{x^2-4x-5}$
34. $f(x) = \dfrac{x^2-4}{x^2-3x+2}$
35. $f(x) = \dfrac{x^2+3x}{x^2+x-6}$
36. $f(x) = \dfrac{5(x+4)}{x^2+x-12}$
37. $f(x) = \dfrac{x^2-3x-4}{2x^2+x-1}$
38. $f(x) = \dfrac{6x^2-11x+3}{6x^2-7x-3}$
39. $f(x) = \dfrac{2x^2-5x-3}{x^3-2x^2-x+2}$
40. $f(x) = \dfrac{x^2-x-2}{x^3-2x^2-5x+6}$

Matching a Rational Function with Its Graph In Exercises 41–44, match the rational function with its graph. [The graphs are labeled (a), (b), (c), and (d).]

(a)

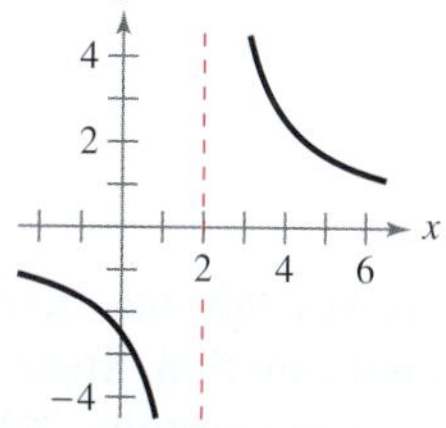

(b)

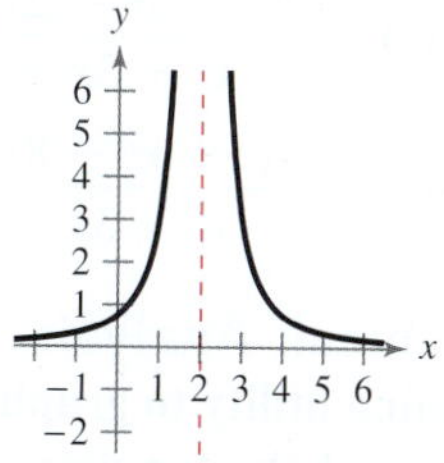

(c)

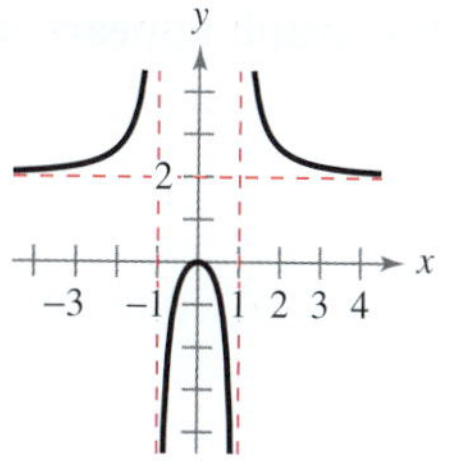

(d)

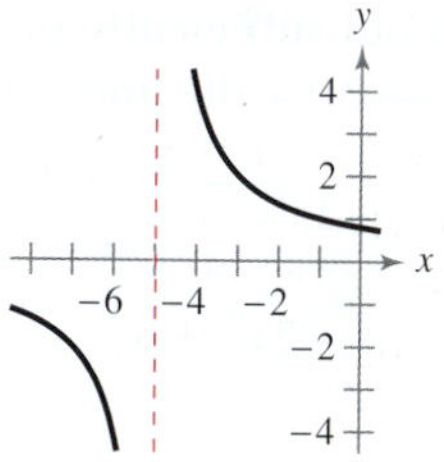

41. $f(x) = \dfrac{4}{x+5}$
42. $f(x) = \dfrac{5}{x-2}$
43. $f(x) = \dfrac{2x^2}{x^2-1}$
44. $f(x) = \dfrac{3}{(x-2)^2}$

Analytical and Graphical Analysis In Exercises 45–48, (a) state the domains of f and g, (b) use a graphing utility to graph f and g in the same viewing window, and (c) explain why the graphing utility may not show the difference in the domains of f and g.

45. $f(x) = \dfrac{x^2 - 1}{x + 1}, \quad g(x) = x - 1$

46. $f(x) = \dfrac{x^2(x - 2)}{x^2 - 2x}, \quad g(x) = x$

47. $f(x) = \dfrac{x - 2}{x^2 - 2x}, \quad g(x) = \dfrac{1}{x}$

48. $f(x) = \dfrac{2x - 6}{x^2 - 7x + 12}, \quad g(x) = \dfrac{2}{x - 4}$

A Rational Function with a Slant Asymptote In Exercises 49–62, (a) state the domain of the function, (b) identify all intercepts, (c) find any vertical or slant asymptotes, and (d) plot additional solution points as needed to sketch the graph of the rational function.

49. $h(x) = \dfrac{x^2 - 9}{x}$

50. $g(x) = \dfrac{x^2 + 5}{x}$

51. $f(x) = \dfrac{2x^2 + 1}{x}$

52. $f(x) = \dfrac{1 - x^2}{x}$

53. $g(x) = \dfrac{x^2 + 1}{x}$

54. $h(x) = \dfrac{x^2}{x - 1}$

55. $f(t) = -\dfrac{t^2 + 1}{t + 5}$

56. $f(x) = \dfrac{x^2}{3x + 1}$

57. $f(x) = \dfrac{x^3}{x^2 - 4}$

58. $g(x) = \dfrac{x^3}{2x^2 - 8}$

59. $f(x) = \dfrac{x^2 - x + 1}{x - 1}$

60. $f(x) = \dfrac{2x^2 - 5x + 5}{x - 2}$

61. $f(x) = \dfrac{2x^3 - x^2 - 2x + 1}{x^2 + 3x + 2}$

62. $f(x) = \dfrac{2x^3 + x^2 - 8x - 4}{x^2 - 3x + 2}$

Graphical Analysis In Exercises 63–66, use a graphing utility to graph the rational function. State the domain of the function and find any asymptotes. Then zoom out sufficiently far so that the graph appears as a line. Identify the line.

63. $f(x) = \dfrac{x^2 + 5x + 8}{x + 3}$

64. $f(x) = \dfrac{2x^2 + x}{x + 1}$

65. $g(x) = \dfrac{1 + 3x^2 - x^3}{x^2}$

66. $h(x) = \dfrac{12 - 2x - x^2}{2(4 + x)}$

Graphical Reasoning In Exercises 67–70, (a) use the graph to determine any x-intercepts of the graph of the rational function and (b) set $y = 0$ and solve the resulting equation to confirm your result in part (a).

67. $y = \dfrac{x + 1}{x - 3}$

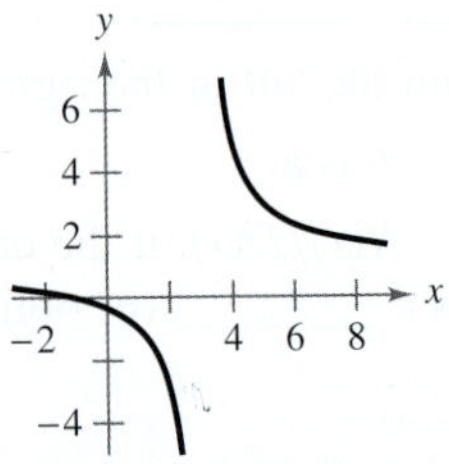

68. $y = \dfrac{2x}{x - 3}$

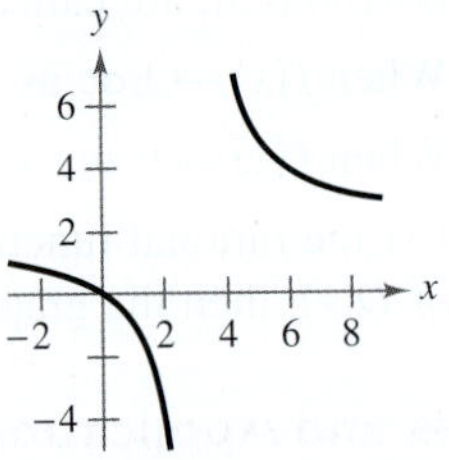

69. $y = \dfrac{1}{x} - x$

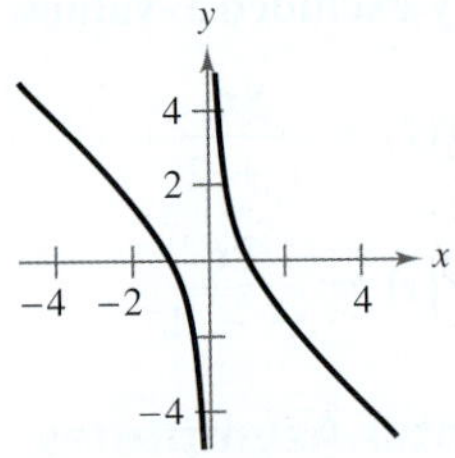

70. $y = x - 3 + \dfrac{2}{x}$

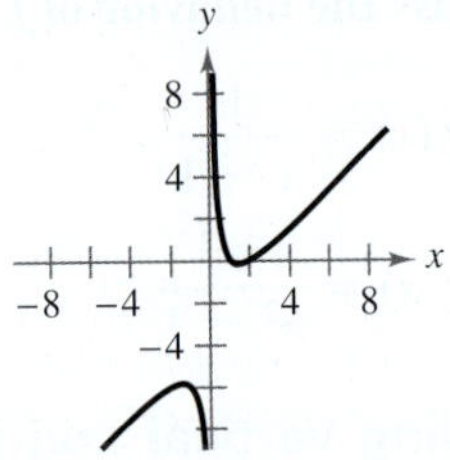

71. Recycling

The cost C (in dollars) of supplying recycling bins to $p\%$ of the population of a rural township is given by

$$C = \frac{25{,}000p}{100 - p}, \quad 0 \le p < 100.$$

(a) Use a graphing utility to graph the cost function.

(b) Find the costs of supplying bins to 15%, 50%, and 90% of the population.

(c) According to the model, would it be possible to supply bins to 100% of the population? Explain.

72. **Population Growth** The game commission introduces 100 deer into newly acquired state game lands. The population N of the herd is modeled by

$$N = \frac{20(5 + 3t)}{1 + 0.04t}, \quad t \ge 0$$

where t is the time in years.

(a) Use a graphing utility to graph the model.

(b) Find the populations when $t = 5$, $t = 10$, and $t = 25$.

(c) What is the limiting size of the herd as time increases?

ZQFotography/Shutterstock.com

73. Page Design A page that is x inches wide and y inches high contains 30 square inches of print. The top and bottom margins are each 1 inch deep, and the margins on each side are 2 inches wide (see figure).

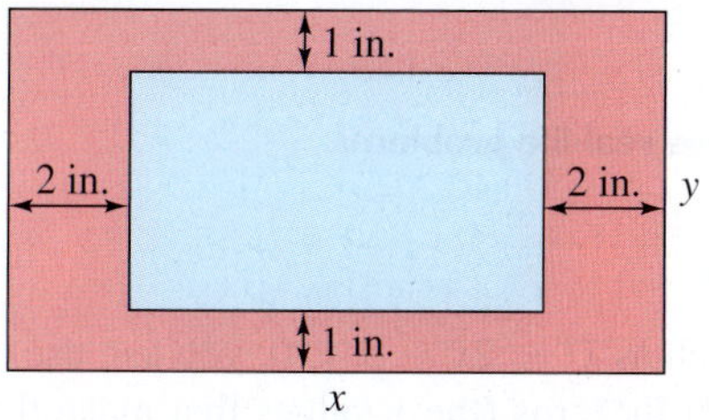

(a) Write a function for the total area A of the page in terms of x.

(b) Determine the domain of the function based on the physical constraints of the problem.

(c) Use a graphing utility to graph the area function and approximate the page size for which the least amount of paper is used.

74. Page Design A rectangular page contains 64 square inches of print. The margins at the top and bottom of the page are each 1 inch deep. The margins on each side are $1\frac{1}{2}$ inches wide. What should the dimensions of the page be so that the least amount of paper is used?

75. Average Speed A driver averaged 50 miles per hour on the round trip between two cities 100 miles apart. The average speeds for going and returning were x and y miles per hour, respectively.

(a) Show that $y = (25x)/(x - 25)$.

(b) Determine the vertical and horizontal asymptotes of the graph of the function.

(c) Use a graphing utility to graph the function.

(d) Complete the table.

x	30	35	40	45	50	55	60
y							

(e) Are the results in the table what you expected? Explain.

(f) Is it possible to average 20 miles per hour in one direction and still average 50 miles per hour on the round trip? Explain.

Exploration

76. Writing Is every rational function a polynomial function? Is every polynomial function a rational function? Explain.

True or False? **In Exercises 77–79, determine whether the statement is true or false. Justify your answer.**

77. The graph of a polynomial function can have infinitely many vertical asymptotes.

78. The graph of a rational function can never cross one of its asymptotes.

79. The graph of a rational function can have a vertical asymptote, a horizontal asymptote, and a slant asymptote.

80. HOW DO YOU SEE IT? The graph of a rational function

$$f(x) = \frac{N(x)}{D(x)}$$

is shown below. Determine which of the statements about the function is false. Justify your answer.

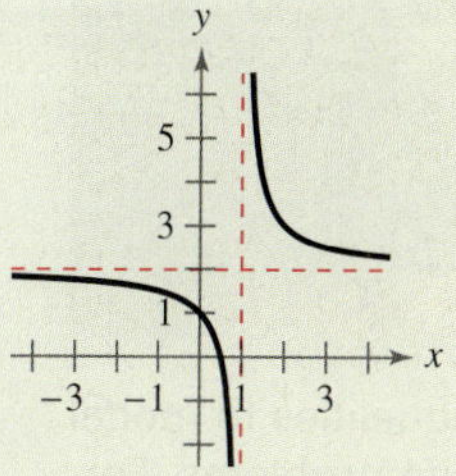

(a) $D(1) = 0$.

(b) The degrees of $N(x)$ and $D(x)$ are equal.

(c) The ratio of the leading coefficients of $N(x)$ and $D(x)$ is 1.

Multiple Choice **In Exercises 81 and 82, identify which rational function the graph represents.**

81.

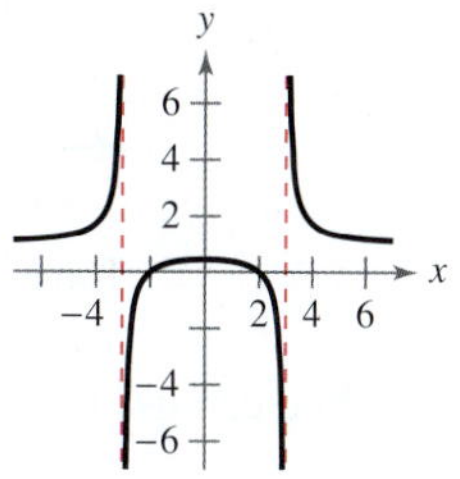

(a) $f(x) = \dfrac{x^2 - 9}{x^2 - 4}$

(b) $f(x) = \dfrac{x^2 - 4}{x^2 - 9}$

(c) $f(x) = \dfrac{x - 4}{x^2 - 9}$

(d) $f(x) = \dfrac{x - 9}{x^2 - 4}$

82.

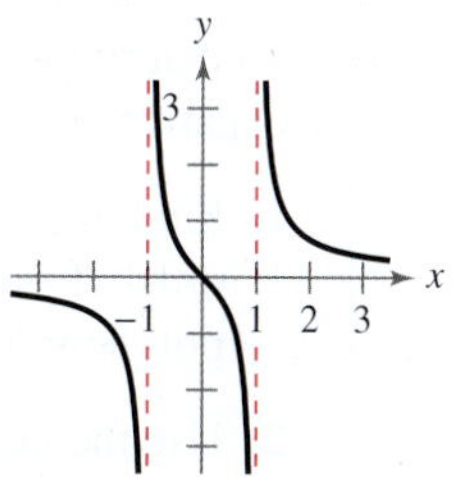

(a) $f(x) = \dfrac{x^2 - 1}{x^2 + 1}$

(b) $f(x) = \dfrac{x^2 + 1}{x^2 - 1}$

(c) $f(x) = \dfrac{x}{x^2 - 1}$

(d) $f(x) = \dfrac{x}{x^2 + 1}$

Project: Department of Defense To work an extended application analyzing the total numbers of military personnel on active duty from 1980 through 2010, visit this text's website at *LarsonPrecalculus.com. (Source: U.S. Department of Defense)*

2.7 Nonlinear Inequalities

You can use inequalities to model and solve real-life problems. For instance, in Exercises 73 and 74 on page 188, you will use a polynomial inequality to model the height of a projectile.

- **Solve polynomial inequalities.**
- **Solve rational inequalities.**
- **Use inequalities to model and solve real-life problems.**

Polynomial Inequalities

To solve a polynomial inequality such as $x^2 - 2x - 3 < 0$, you can use the fact that a polynomial can change signs only at its zeros (the x-values that make the polynomial equal to zero). Between two consecutive zeros, a polynomial must be entirely positive or entirely negative. This means that when the real zeros of a polynomial are put in order, they divide the real number line into intervals in which the polynomial has no sign changes. These zeros are the **key numbers** of the inequality, and the resulting intervals are the *test intervals* for the inequality. For instance, the polynomial above factors as

$$x^2 - 2x - 3 = (x + 1)(x - 3)$$

and has two zeros,

$$x = -1 \quad \text{and} \quad x = 3.$$

These zeros divide the real number line into three test intervals:

$$(-\infty, -1), \quad (-1, 3), \quad \text{and} \quad (3, \infty). \qquad \text{(See figure.)}$$

So, to solve the inequality $x^2 - 2x - 3 < 0$, you need only test one value from each of these test intervals. When a value from a test interval satisfies the original inequality, you can conclude that the interval is a solution of the inequality.

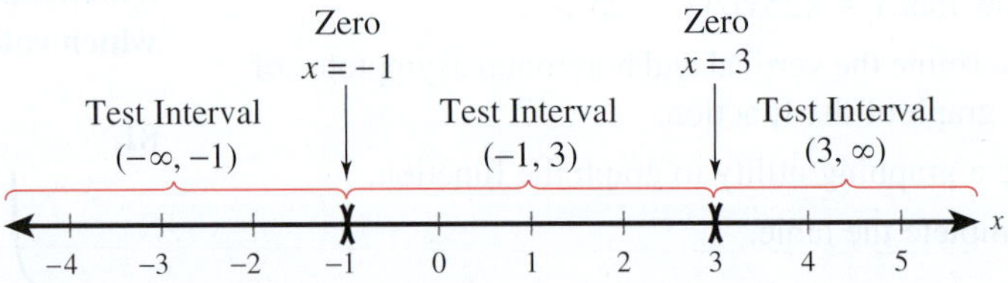

Three test intervals for $x^2 - 2x - 3$

You can use the same basic approach to determine the test intervals for any polynomial.

Finding Test Intervals for a Polynomial

To determine the intervals on which the values of a polynomial are entirely negative or entirely positive, use the following steps.

1. Find all real zeros of the polynomial, and arrange the zeros in increasing order (from least to greatest). These zeros are the key numbers of the polynomial.
2. Use the key numbers of the polynomial to determine its test intervals.
3. Choose one representative x-value in each test interval and evaluate the polynomial at that value. When the value of the polynomial is negative, the polynomial has negative values for every x-value in the interval. When the value of the polynomial is positive, the polynomial has positive values for every x-value in the interval.

Vakhrushev Pavel/Shutterstock.com

EXAMPLE 1 Solving a Polynomial Inequality

Solve $x^2 - x - 6 < 0$. Then graph the solution set.

ALGEBRA HELP You can review the techniques for factoring polynomials in Appendix A.3.

Solution By factoring the polynomial as

$$x^2 - x - 6 = (x + 2)(x - 3)$$

you can see that the key numbers are $x = -2$ and $x = 3$. So, the polynomial's test intervals are

$$(-\infty, -2), \quad (-2, 3), \quad \text{and} \quad (3, \infty). \qquad \text{Test intervals}$$

In each test interval, choose a representative x-value and evaluate the polynomial.

Test Interval	x-Value	Polynomial Value	Conclusion
$(-\infty, -2)$	$x = -3$	$(-3)^2 - (-3) - 6 = 6$	Positive
$(-2, 3)$	$x = 0$	$(0)^2 - (0) - 6 = -6$	Negative
$(3, \infty)$	$x = 4$	$(4)^2 - (4) - 6 = 6$	Positive

From this, you can conclude that the inequality is satisfied for all x-values in $(-2, 3)$. This implies that the solution of the inequality

$$x^2 - x - 6 < 0$$

is the interval $(-2, 3)$, as shown below. Note that the original inequality contains a "less than" symbol. This means that the solution set does not contain the endpoints of the test interval $(-2, 3)$.

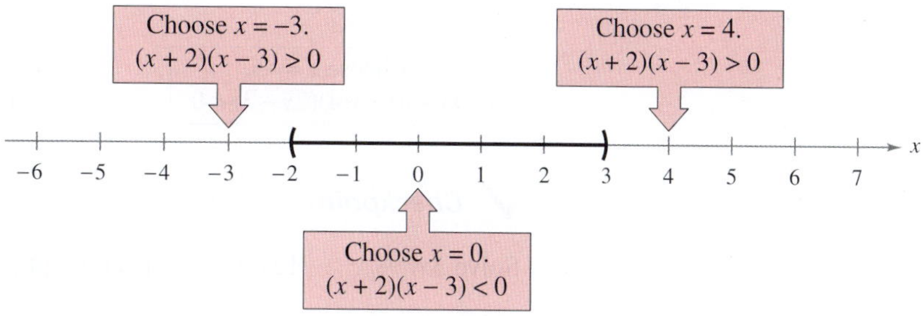

Checkpoint Audio-video solution in English & Spanish at LarsonPrecalculus.com.

Solve $x^2 - x - 20 < 0$. Then graph the solution set.

As with linear inequalities, you can check the reasonableness of a solution by substituting x-values into the original inequality. For instance, to check the solution found in Example 1, try substituting several x-values from the interval $(-2, 3)$ into the inequality

$$x^2 - x - 6 < 0.$$

Regardless of which x-values you choose, the inequality should be satisfied.

You can also use a graph to check the result of Example 1. Sketch the graph of

$$y = x^2 - x - 6$$

as shown in Figure 2.33. Notice that the graph is below the x-axis on the interval $(-2, 3)$.

In Example 1, the polynomial inequality was given in general form (with the polynomial on one side and zero on the other). Whenever this is not the case, you should begin the solution process by writing the inequality in general form.

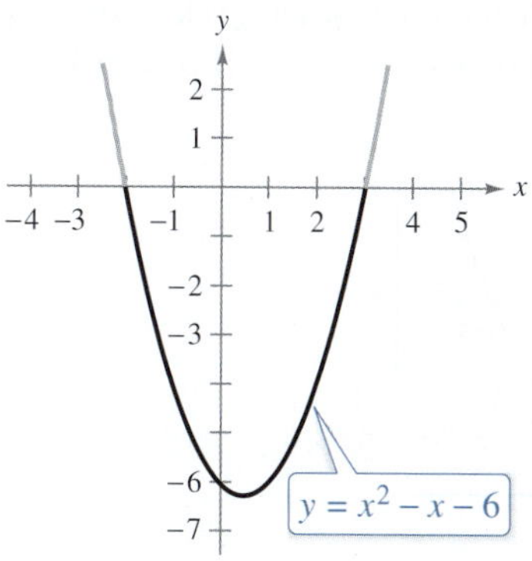

Figure 2.33

EXAMPLE 2 Solving a Polynomial Inequality

Solve $2x^3 - 3x^2 - 32x > -48$. Then graph the solution set.

Solution

$$2x^3 - 3x^2 - 32x + 48 > 0 \quad \text{Write in general form.}$$

$$(x - 4)(x + 4)(2x - 3) > 0 \quad \text{Factor.}$$

The key numbers are $x = -4$, $x = \frac{3}{2}$, and $x = 4$, and the test intervals are $(-\infty, -4)$, $\left(-4, \frac{3}{2}\right)$, $\left(\frac{3}{2}, 4\right)$, and $(4, \infty)$.

Test Interval	*x*-Value	Polynomial Value	Conclusion
$(-\infty, -4)$	$x = -5$	$2(-5)^3 - 3(-5)^2 - 32(-5) + 48 = -117$	Negative
$\left(-4, \frac{3}{2}\right)$	$x = 0$	$2(0)^3 - 3(0)^2 - 32(0) + 48 = 48$	Positive
$\left(\frac{3}{2}, 4\right)$	$x = 2$	$2(2)^3 - 3(2)^2 - 32(2) + 48 = -12$	Negative
$(4, \infty)$	$x = 5$	$2(5)^3 - 3(5)^2 - 32(5) + 48 = 63$	Positive

From this, you can conclude that the inequality is satisfied on the open intervals $\left(-4, \frac{3}{2}\right)$ and $(4, \infty)$. So, the solution set is $\left(-4, \frac{3}{2}\right) \cup (4, \infty)$, as shown below.

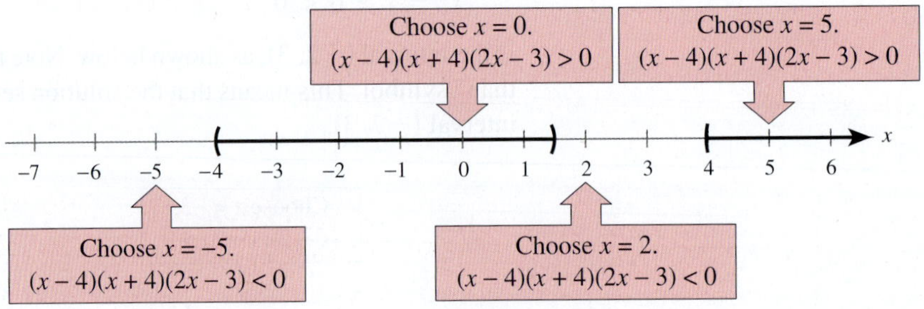

✓ Checkpoint *Audio-video solution in English & Spanish at LarsonPrecalculus.com.*

Solve $3x^3 - x^2 - 12x > -4$. Then graph the solution set.

EXAMPLE 3 Solving a Polynomial Inequality

Solve $4x^2 - 5x > 6$.

Algebraic Solution

$$4x^2 - 5x - 6 > 0 \quad \text{Write in general form.}$$

$$(x - 2)(4x + 3) > 0 \quad \text{Factor.}$$

Key Numbers: $x = -\frac{3}{4}$, $x = 2$

Test Intervals: $\left(-\infty, -\frac{3}{4}\right)$, $\left(-\frac{3}{4}, 2\right)$, $(2, \infty)$

Test: Is $(x - 2)(4x + 3) > 0$?

After testing these intervals, you can see that the polynomial $4x^2 - 5x - 6$ is positive on the open intervals $\left(-\infty, -\frac{3}{4}\right)$ and $(2, \infty)$. So, the solution set of the inequality is $\left(-\infty, -\frac{3}{4}\right) \cup (2, \infty)$.

Graphical Solution

First write the polynomial inequality $4x^2 - 5x > 6$ as $4x^2 - 5x - 6 > 0$. Then use a graphing utility to graph $y = 4x^2 - 5x - 6$. In the figure below, you can see that the graph is *above* the *x*-axis when *x* is less than $-\frac{3}{4}$ *or* when *x* is greater than 2. So, the solution set is $\left(-\infty, -\frac{3}{4}\right) \cup (2, \infty)$.

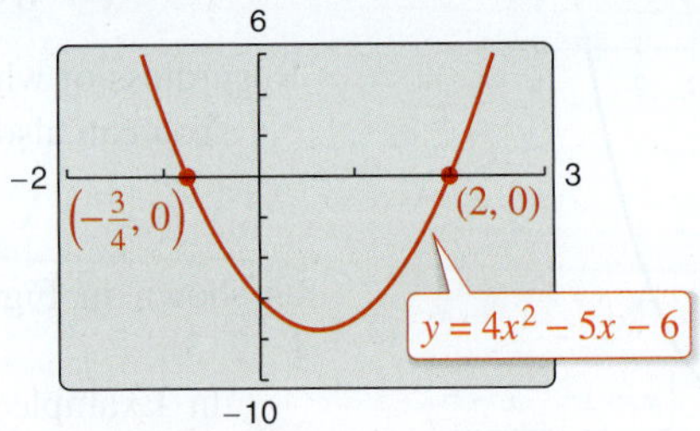

✓ Checkpoint *Audio-video solution in English & Spanish at LarsonPrecalculus.com.*

Solve $2x^2 + 3x < 5$ (a) algebraically and (b) graphically.

You may find it easier to determine the sign of a polynomial from its *factored* form. For instance, in Example 3, when you substitute the test value $x = 1$ into the factored form

$$(x - 2)(4x + 3)$$

you can see that the sign pattern of the factors is

$$(-)(+)$$

which yields a negative result. Try using the factored forms of the polynomials to determine the signs of the polynomials in the test intervals of the other examples in this section.

When solving a polynomial inequality, be sure to account for the particular type of inequality symbol given in the inequality. For instance, in Example 3, note that the original inequality contained a "greater than" symbol and the solution consisted of two open intervals. If the original inequality had been

$$4x^2 - 5x \geq 6$$

then the solution set would have been $\left(-\infty, -\frac{3}{4}\right] \cup [2, \infty)$.

Each of the polynomial inequalities in Examples 1, 2, and 3 has a solution set that consists of a single interval or the union of two intervals. When solving the exercises for this section, watch for unusual solution sets, as illustrated in Example 4.

EXAMPLE 4 Unusual Solution Sets

a. The solution set of

$$x^2 + 2x + 4 > 0$$

consists of the entire set of real numbers, $(-\infty, \infty)$. In other words, the value of the quadratic $x^2 + 2x + 4$ is positive for every real value of x.

b. The solution set of

$$x^2 + 2x + 1 \leq 0$$

consists of the single real number $\{-1\}$, because the quadratic $x^2 + 2x + 1$ has only one key number, $x = -1$, and it is the only value that satisfies the inequality.

c. The solution set of

$$x^2 + 3x + 5 < 0$$

is empty. In other words, the quadratic $x^2 + 3x + 5$ is not less than zero for any value of x.

d. The solution set of

$$x^2 - 4x + 4 > 0$$

consists of all real numbers except $x = 2$. In interval notation, this solution set can be written as $(-\infty, 2) \cup (2, \infty)$.

✓ **Checkpoint** *Audio-video solution in English & Spanish at LarsonPrecalculus.com.*

What is unusual about the solution set of each inequality?

a. $x^2 + 6x + 9 < 0$

b. $x^2 + 4x + 4 \leq 0$

c. $x^2 - 6x + 9 > 0$

d. $x^2 - 2x + 1 \geq 0$

Rational Inequalities

The concepts of key numbers and test intervals can be extended to rational inequalities. Use the fact that the value of a rational expression can change sign only at its *zeros* (the x-values for which its numerator is zero) and its *undefined values* (the x-values for which its denominator is zero). These two types of numbers make up the *key numbers* of a rational inequality. When solving a rational inequality, begin by writing the inequality in general form with the rational expression on the left and zero on the right.

REMARK In Example 5, write 3 as $\frac{3}{1}$. You should be able to see that the LCD (least common denominator) is $(x-5)(1) = x-5$. So, you can rewrite the general form as

$$\frac{2x-7}{x-5} - \frac{3(x-5)}{x-5} \le 0,$$

which simplifies as shown.

EXAMPLE 5 Solving a Rational Inequality

Solve $\dfrac{2x-7}{x-5} \le 3$. Then graph the solution set.

Solution

$$\frac{2x-7}{x-5} \le 3 \qquad \text{Write original inequality.}$$

$$\frac{2x-7}{x-5} - 3 \le 0 \qquad \text{Write in general form.}$$

$$\frac{2x-7-3x+15}{x-5} \le 0 \qquad \text{Find the LCD and subtract fractions.}$$

$$\frac{-x+8}{x-5} \le 0 \qquad \text{Simplify.}$$

Key Numbers: $x = 5, x = 8$ Zeros and undefined values of rational expression

Test Intervals: $(-\infty, 5), (5, 8), (8, \infty)$

Test: Is $\dfrac{-x+8}{x-5} \le 0$?

After testing these intervals, as shown below, you can see that the inequality is satisfied on the open intervals $(-\infty, 5)$ and $(8, \infty)$. Moreover, because

$$\frac{-x+8}{x-5} = 0$$

when $x = 8$, you can conclude that the solution set consists of all real numbers in the intervals $(-\infty, 5) \cup [8, \infty)$. (Be sure to use a bracket to indicate that x can equal 8.)

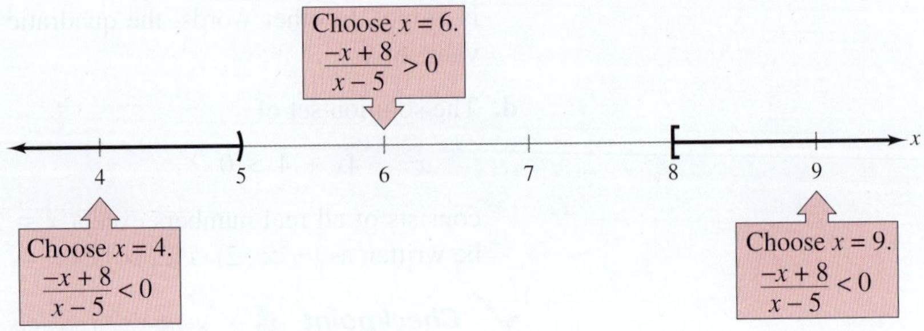

Checkpoint Audio-video solution in English & Spanish at LarsonPrecalculus.com.

Solve each inequality. Then graph the solution set.

a. $\dfrac{x-2}{x-3} \ge -3$

b. $\dfrac{4x-1}{x-6} > 3$

Applications

One common application of inequalities comes from business and involves profit, revenue, and cost. The formula that relates these three quantities is

Profit	=	Revenue	−	Cost

$$P = R - C.$$

EXAMPLE 6 Increasing the Profit for a Product

The marketing department of a calculator manufacturer has determined that the demand for a new model of calculator is

$$p = 100 - 0.00001x, \quad 0 \le x \le 10{,}000{,}000 \qquad \text{Demand equation}$$

where p is the price per calculator (in dollars) and x represents the number of calculators sold. (According to this model, no one would be willing to pay \$100 for the calculator. At the other extreme, the company could not *give* away more than 10 million calculators.) The revenue for selling x calculators is

$$R = xp = x(100 - 0.00001x) \qquad \text{Revenue equation}$$

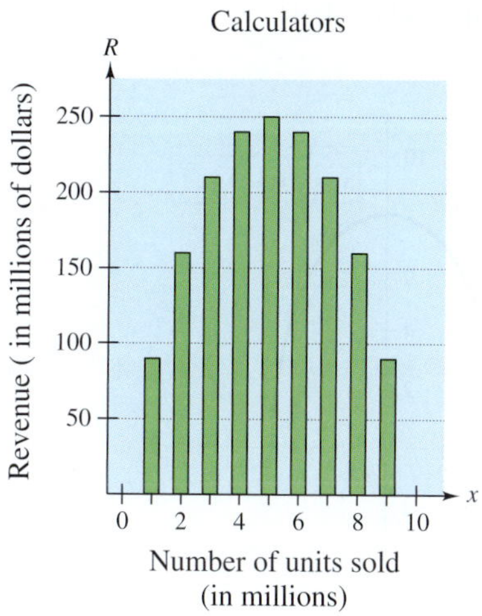

Figure 2.34

as shown in Figure 2.34. The total cost of producing x calculators is \$10 per calculator plus a one-time development cost of \$2,500,000. So, the total cost is

$$C = 10x + 2{,}500{,}000. \qquad \text{Cost equation}$$

What prices can the company charge per calculator to obtain a profit of at least \$190,000,000?

Solution

Verbal Model:

Profit	=	Revenue	−	Cost

Equation: $P = R - C$

$$P = 100x - 0.00001x^2 - (10x + 2{,}500{,}000)$$

$$P = -0.00001x^2 + 90x - 2{,}500{,}000$$

To answer the question, solve the inequality

$$P \ge 190{,}000{,}000$$

$$-0.00001x^2 + 90x - 2{,}500{,}000 \ge 190{,}000{,}000.$$

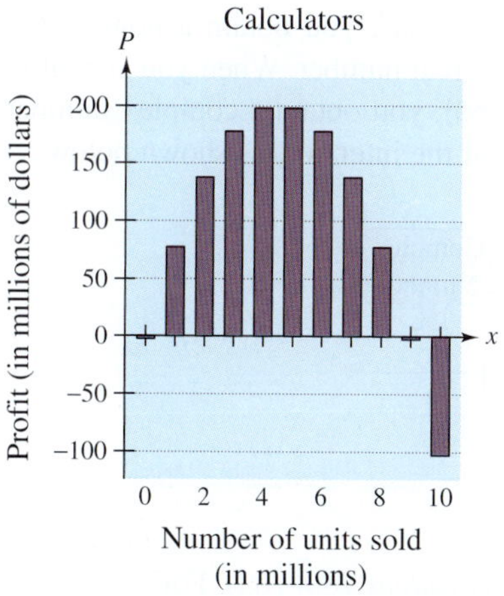

Figure 2.35

When you write the inequality in general form, find the key numbers and the test intervals, and then test a value in each test interval, you can find the solution to be

$$3{,}500{,}000 \le x \le 5{,}500{,}000$$

as shown in Figure 2.35. Substituting the x-values in the original demand equation shows that prices of

$$\$45.00 \le p \le \$65.00$$

will yield a profit of at least \$190,000,000.

✓ **Checkpoint** *Audio-video solution in English & Spanish at LarsonPrecalculus.com.*

The revenue and cost equations for a product are

$$R = x(60 - 0.0001x) \quad \text{and} \quad C = 12x + 1{,}800{,}000$$

where R and C are measured in dollars and x represents the number of units sold. How many units must be sold to obtain a profit of at least \$3,600,000?

Another common application of inequalities is finding the domain of an expression that involves a square root, as shown in Example 7.

EXAMPLE 7 Finding the Domain of an Expression

Find the domain of $\sqrt{64 - 4x^2}$.

Algebraic Solution

Recall that the domain of an expression is the set of all x-values for which the expression is defined. Because $\sqrt{64 - 4x^2}$ is defined (has real values) only if $64 - 4x^2$ is nonnegative, the domain is given by $64 - 4x^2 \geq 0$.

$64 - 4x^2 \geq 0$ Write in general form.

$16 - x^2 \geq 0$ Divide each side by 4.

$(4 - x)(4 + x) \geq 0$ Write in factored form.

So, the inequality has two key numbers: $x = -4$ and $x = 4$. You can use these two numbers to test the inequality, as follows.

Key numbers: $x = -4, x = 4$

Test intervals: $(-\infty, -4), (-4, 4), (4, \infty)$

Test: Is $(4 - x)(4 + x) \geq 0$?

A test shows that the inequality is satisfied in the *closed interval* $[-4, 4]$. So, the domain of the expression $\sqrt{64 - 4x^2}$ is the interval $[-4, 4]$.

Graphical Solution

Begin by sketching the graph of the equation $y = \sqrt{64 - 4x^2}$, as shown below. From the graph, you can determine that the x-values extend from -4 to 4 (including -4 and 4). So, the domain of the expression $\sqrt{64 - 4x^2}$ is the interval $[-4, 4]$.

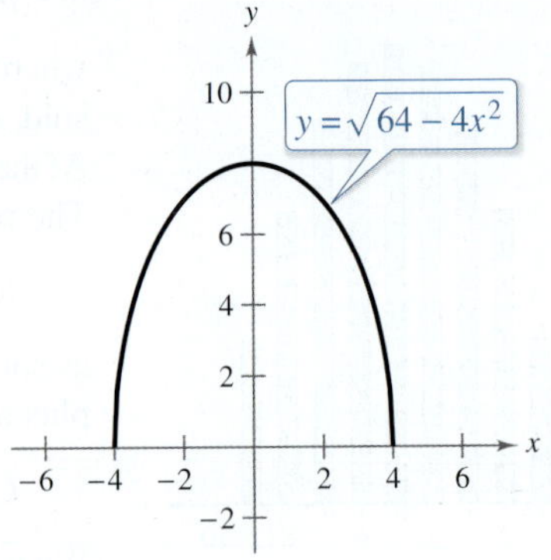

Checkpoint *Audio-video solution in English & Spanish at LarsonPrecalculus.com.*

Find the domain of $\sqrt{x^2 - 7x + 10}$.

To analyze a test interval, choose a representative x-value in the interval and evaluate the expression at that value. For instance, in Example 7, when you substitute any number from the interval $[-4, 4]$ into the expression $\sqrt{64 - 4x^2}$, you obtain a nonnegative number under the radical symbol that simplifies to a real number. When you substitute any number from the intervals $(-\infty, -4)$ and $(4, \infty)$, you obtain a complex number. It might be helpful to draw a visual representation of the intervals, as shown below.

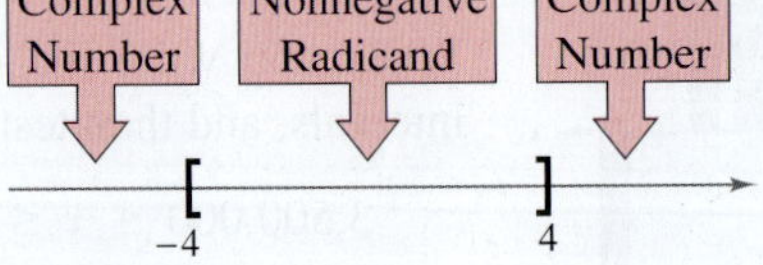

Summarize (Section 2.7)

1. Describe how to solve a polynomial inequality *(pages 180–183)*. For examples of solving polynomial inequalities, see Examples 1–4.
2. Describe how to solve a rational inequality *(page 184)*. For an example of solving a rational inequality, see Example 5.
3. Give an example of an application involving a polynomial inequality *(pages 185 and 186, Examples 6 and 7)*.

2.7 Exercises

See CalcChat.com for tutorial help and worked-out solutions to odd-numbered exercises.

Vocabulary: Fill in the blanks.

1. Between two consecutive zeros, a polynomial must be entirely ________ or entirely ________.
2. To solve a polynomial inequality, find the ________ numbers of the polynomial, and use these numbers to create ________ ________ for the inequality.
3. The key numbers of a rational expression are its ________ and its ________ ________.
4. The formula that relates cost, revenue, and profit is ________.

Skills and Applications

Checking Solutions **In Exercises 5–8, determine whether each value of x is a solution of the inequality.**

	Inequality	Values	
5.	$x^2 - 3 < 0$	(a) $x = 3$	(b) $x = 0$
		(c) $x = \frac{3}{2}$	(d) $x = -5$
6.	$x^2 - x - 12 \geq 0$	(a) $x = 5$	(b) $x = 0$
		(c) $x = -4$	(d) $x = -3$
7.	$\dfrac{x+2}{x-4} \geq 3$	(a) $x = 5$	(b) $x = 4$
		(c) $x = -\frac{9}{2}$	(d) $x = \frac{9}{2}$
8.	$\dfrac{3x^2}{x^2+4} < 1$	(a) $x = -2$	(b) $x = -1$
		(c) $x = 0$	(d) $x = 3$

Finding Key Numbers **In Exercises 9–12, find the key numbers of the expression.**

9. $3x^2 - x - 2$
10. $9x^3 - 25x^2$
11. $\dfrac{1}{x-5} + 1$
12. $\dfrac{x}{x+2} - \dfrac{2}{x-1}$

Solving a Polynomial Inequality **In Exercises 13–34, solve the inequality. Then graph the solution set.**

13. $x^2 < 9$
14. $x^2 \leq 16$
15. $(x + 2)^2 \leq 25$
16. $(x - 3)^2 \geq 1$
17. $x^2 + 4x + 4 \geq 9$
18. $x^2 - 6x + 9 < 16$
19. $x^2 + x < 6$
20. $x^2 + 2x > 3$
21. $x^2 + 2x - 3 < 0$
22. $x^2 > 2x + 8$
23. $3x^2 - 11x > 20$
24. $-2x^2 + 6x \leq -15$
25. $x^2 - 3x - 18 > 0$
26. $x^3 + 2x^2 - 4x \leq 8$
27. $x^3 - 3x^2 - x > -3$
28. $2x^3 + 13x^2 - 8x \geq 52$
29. $4x^3 - 6x^2 < 0$
30. $4x^3 - 12x^2 > 0$
31. $x^3 - 4x \geq 0$
32. $2x^3 - x^4 \leq 0$
33. $(x - 1)^2(x + 2)^3 \geq 0$
34. $x^4(x - 3) \leq 0$

Unusual Solution Sets **In Exercises 35–38, explain what is unusual about the solution set of the inequality.**

35. $4x^2 - 4x + 1 \leq 0$
36. $x^2 + 3x + 8 > 0$
37. $x^2 - 6x + 12 \leq 0$
38. $x^2 - 8x + 16 > 0$

Solving a Rational Inequality **In Exercises 39–52, solve the inequality. Then graph the solution set.**

39. $\dfrac{4x - 1}{x} > 0$
40. $\dfrac{x^2 - 1}{x} < 0$
41. $\dfrac{3x - 5}{x - 5} \geq 0$
42. $\dfrac{5 + 7x}{1 + 2x} \leq 4$
43. $\dfrac{x + 6}{x + 1} - 2 < 0$
44. $\dfrac{x + 12}{x + 2} - 3 \geq 0$
45. $\dfrac{2}{x + 5} > \dfrac{1}{x - 3}$
46. $\dfrac{5}{x - 6} > \dfrac{3}{x + 2}$
47. $\dfrac{1}{x - 3} \leq \dfrac{9}{4x + 3}$
48. $\dfrac{1}{x} \geq \dfrac{1}{x + 3}$
49. $\dfrac{x^2 + 2x}{x^2 - 9} \leq 0$
50. $\dfrac{x^2 + x - 6}{x} \geq 0$
51. $\dfrac{3}{x - 1} + \dfrac{2x}{x + 1} > -1$
52. $\dfrac{3x}{x - 1} \leq \dfrac{x}{x + 4} + 3$

Graphical Analysis **In Exercises 53–60, use a graphing utility to graph the equation. Use the graph to approximate the values of x that satisfy each inequality.**

	Equation	Inequalities	
53.	$y = -x^2 + 2x + 3$	(a) $y \leq 0$	(b) $y \geq 3$
54.	$y = \frac{1}{2}x^2 - 2x + 1$	(a) $y \leq 0$	(b) $y \geq 7$
55.	$y = \frac{1}{8}x^3 - \frac{1}{2}x$	(a) $y \geq 0$	(b) $y \leq 6$
56.	$y = x^3 - x^2 - 16x + 16$	(a) $y \leq 0$	(b) $y \geq 36$
57.	$y = \dfrac{3x}{x - 2}$	(a) $y \leq 0$	(b) $y \geq 6$
58.	$y = \dfrac{2(x - 2)}{x + 1}$	(a) $y \leq 0$	(b) $y \geq 8$
59.	$y = \dfrac{2x^2}{x^2 + 4}$	(a) $y \geq 1$	(b) $y \leq 2$
60.	$y = \dfrac{5x}{x^2 + 4}$	(a) $y \geq 1$	(b) $y \leq 0$

Finding the Domain of an Expression **In Exercises 61–66, find the domain of the expression. Use a graphing utility to verify your result.**

61. $\sqrt{4 - x^2}$ **62.** $\sqrt{x^2 - 4}$

63. $\sqrt{x^2 - 9x + 20}$ **64.** $\sqrt{81 - 4x^2}$

65. $\sqrt{\dfrac{x}{x^2 - 2x - 35}}$ **66.** $\sqrt{\dfrac{x}{x^2 - 9}}$

Solving an Inequality **In Exercises 67–72, solve the inequality. (Round your answers to two decimal places.)**

67. $0.4x^2 + 5.26 < 10.2$

68. $-1.3x^2 + 3.78 > 2.12$

69. $-0.5x^2 + 12.5x + 1.6 > 0$

70. $1.2x^2 + 4.8x + 3.1 < 5.3$

71. $\dfrac{1}{2.3x - 5.2} > 3.4$ **72.** $\dfrac{2}{3.1x - 3.7} > 5.8$

Height of a Projectile

In Exercises 73 and 74, use the position equation

$$s = -16t^2 + v_0t + s_0$$

where s represents the height of an object (in feet), v_0 represents the initial velocity of the object (in feet per second), s_0 represents the initial height of the object (in feet), and t represents the time (in seconds).

73. A projectile is fired straight upward from ground level ($s_0 = 0$) with an initial velocity of 160 feet per second.

(a) At what instant will it be back at ground level?

(b) When will the height exceed 384 feet?

74. A projectile is fired straight upward from ground level ($s_0 = 0$) with an initial velocity of 128 feet per second.

(a) At what instant will it be back at ground level?

(b) When will the height be less than 128 feet?

75. Geometry A rectangular playing field with a perimeter of 100 meters is to have an area of at least 500 square meters. Within what bounds must the length of the rectangle lie?

76. Geometry A rectangular parking lot with a perimeter of 440 feet is to have an area of at least 8000 square feet. Within what bounds must the length of the rectangle lie?

77. Cost, Revenue, and Profit The revenue and cost equations for a product are $R = x(75 - 0.0005x)$ and $C = 30x + 250{,}000$, where R and C are measured in dollars and x represents the number of units sold. How many units must be sold to obtain a profit of at least \$750,000? What is the price per unit?

78. Cost, Revenue, and Profit The revenue and cost equations for a product are $R = x(50 - 0.0002x)$ and $C = 12x + 150{,}000$, where R and C are measured in dollars and x represents the number of units sold. How many units must be sold to obtain a profit of at least \$1,650,000? What is the price per unit?

79. School Enrollment The numbers N (in millions) of students enrolled in schools in the United States from 2000 through 2009 are shown in the table. *(Source: U.S. Census Bureau)*

DATA

Year	Number, N
2000	72.2
2001	73.1
2002	74.0
2003	74.9
2004	75.5
2005	75.8
2006	75.2
2007	76.0
2008	76.3
2009	77.3

Spreadsheet at LarsonPrecalculus.com

(a) Use a graphing utility to create a scatter plot of the data. Let t represent the year, with $t = 0$ corresponding to 2000.

(b) Use the *regression* feature of the graphing utility to find a quartic model for the data. (A quartic model has the form $at^4 + bt^3 + ct^2 + dt + e$, where a, b, c, d, and e are constant and t is variable.)

(c) Graph the model and the scatter plot in the same viewing window. How well does the model fit the data?

(d) According to the model, when did the number of students enrolled in schools exceed 74 million?

(e) Is the model valid for long-term predictions of student enrollment? Explain.

80. Safe Load The maximum safe load uniformly distributed over a one-foot section of a two-inch-wide wooden beam can be approximated by the model

$$\text{Load} = 168.5d^2 - 472.1$$

where d is the depth of the beam.

(a) Evaluate the model for $d = 4$, $d = 6$, $d = 8$, $d = 10$, and $d = 12$. Use the results to create a bar graph.

(b) Determine the minimum depth of the beam that will safely support a load of 2000 pounds.

Vakhrushev Pavel/Shutterstock.com

81. Resistors When two resistors of resistances R_1 and R_2 are connected in parallel (see figure), the total resistance R satisfies the equation

$$\frac{1}{R} = \frac{1}{R_1} + \frac{1}{R_2}.$$

Find R_1 for a parallel circuit in which $R_2 = 2$ ohms and R must be at least 1 ohm.

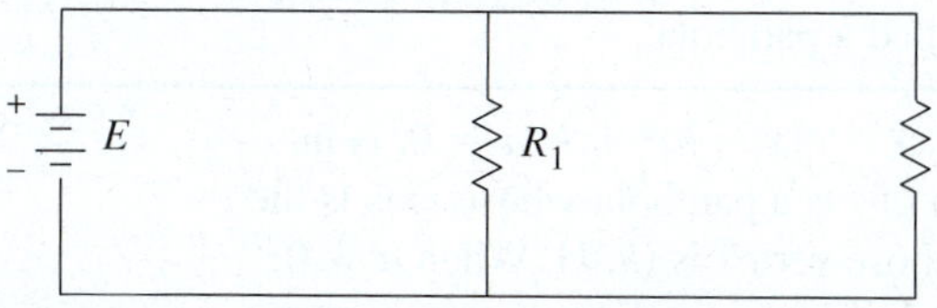

82. Teachers' Salaries The mean salaries S (in thousands of dollars) of public school classroom teachers in the United States from 2000 through 2011 are shown in the table.

DATA

Year	Salary, S
2000	42.2
2001	43.7
2002	43.8
2003	45.0
2004	45.6
2005	45.9
2006	48.2
2007	49.3
2008	51.3
2009	52.9
2010	54.4
2011	54.2

Spreadsheet at LarsonPrecalculus.com

A model that approximates these data is given by

$$S = \frac{42.16 - 0.236t}{1 - 0.026t}, \quad 0 \le t \le 11$$

where t represents the year, with $t = 0$ corresponding to 2000. *(Source: Educational Research Service, Arlington, VA)*

(a) Use a graphing utility to create a scatter plot of the data. Then graph the model in the same viewing window.

(b) How well does the model fit the data? Explain.

(c) Use the model to predict when the salary for classroom teachers will exceed \$60,000.

(d) Is the model valid for long-term predictions of classroom teacher salaries? Explain.

Exploration

True or False? **In Exercises 83 and 84, determine whether the statement is true or false. Justify your answer.**

83. The zeros of the polynomial $x^3 - 2x^2 - 11x + 12 \ge 0$ divide the real number line into four test intervals.

84. The solution set of the inequality $\frac{3}{2}x^2 + 3x + 6 \ge 0$ is the entire set of real numbers.

Conjecture **In Exercises 85–88, (a) find the interval(s) for b such that the equation has at least one real solution and (b) write a conjecture about the interval(s) based on the values of the coefficients.**

85. $x^2 + bx + 4 = 0$

86. $x^2 + bx - 4 = 0$

87. $3x^2 + bx + 10 = 0$

88. $2x^2 + bx + 5 = 0$

89. Graphical Analysis You can use a graphing utility to verify the results in Example 4. For instance, the graph of $y = x^2 + 2x + 4$ is shown below. Notice that the y-values are greater than 0 for all values of x, as stated in Example 4(a). Use the graphing utility to graph $y = x^2 + 2x + 1$, $y = x^2 + 3x + 5$, and $y = x^2 - 4x + 4$. Explain how you can use the graphs to verify the results of parts (b), (c), and (d) of Example 4.

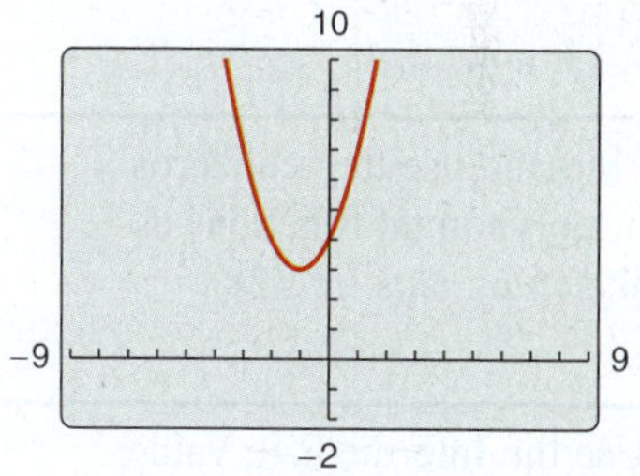

90. HOW DO YOU SEE IT? Consider the polynomial

$(x - a)(x - b)$

and the real number line shown below.

(a) Identify the points on the line at which the polynomial is zero.

(b) In each of the three subintervals of the line, write the sign of each factor and the sign of the product.

(c) At what x-values does the polynomial change signs?

Chapter Summary

Section	What Did You Learn?	Explanation/Examples	Review Exercises
Section 2.1	Analyze graphs of quadratic functions *(p. 114).*	Let a, b, and c be real numbers with $a \neq 0$. The function $f(x) = ax^2 + bx + c$ is called a quadratic function. Its graph is a "U"-shaped curve called a parabola.	1, 2
	Write quadratic functions in standard form and use the results to sketch graphs of functions *(p. 117).*	The quadratic function $f(x) = a(x - h)^2 + k$, $a \neq 0$, is in standard form. The graph of f is a parabola whose axis is the vertical line $x = h$ and whose vertex is (h, k). When $a > 0$, the parabola opens upward. When $a < 0$, the parabola opens downward.	3–8
	Find minimum and maximum values of quadratic functions in real-life applications *(p. 119).*	Consider $f(x) = ax^2 + bx + c$ with vertex $\left(-\frac{b}{2a}, f\left(-\frac{b}{2a}\right)\right)$. When $a > 0$, f has a *minimum* at $x = -b/(2a)$. When $a < 0$, f has a *maximum* at $x = -b/(2a)$.	9, 10
Section 2.2	Use transformations to sketch graphs of polynomial functions *(p. 124).*	The graph of a polynomial function is continuous (no breaks, holes, or gaps) and has only smooth, rounded turns.	11, 12
	Use the Leading Coefficient Test to determine the end behaviors of graphs of polynomial functions *(p. 126).*	Consider the graph of $f(x) = a_nx^n + \cdot \cdot \cdot + a_1x + a_0$. **When n is odd:** If $a_n > 0$, then the graph falls to the left and rises to the right. If $a_n < 0$, then the graph rises to the left and falls to the right. **When n is even:** If $a_n > 0$, then the graph rises to the left and to the right. If $a_n < 0$, then the graph falls to the left and to the right.	13–16
	Find and use the real zeros of polynomial functions as sketching aids *(p. 128).*	When f is a polynomial function and a is a real number, the following are equivalent: (1) $x = a$ is a *zero* of f, (2) $x = a$ is a *solution* of the equation $f(x) = 0$, (3) $(x - a)$ is a *factor* of $f(x)$, and (4) $(a, 0)$ is an *x-intercept* of the graph of f.	17–20
	Use the Intermediate Value Theorem to help locate the real zeros of polynomial functions *(p. 131).*	Let a and b be real numbers such that $a < b$. If f is a polynomial function such that $f(a) \neq f(b)$, then, in $[a, b]$, f takes on every value between $f(a)$ and $f(b)$.	21, 22
Section 2.3	Use long division to divide polynomials by other polynomials *(p. 138).*	Dividend, Quotient, Remainder, Divisor: $\dfrac{x^2 + 3x + 5}{x + 1} = x + 2 + \dfrac{3}{x + 1}$ (Dividend: $x^2 + 3x + 5$; Divisor: $x + 1$; Quotient: $x + 2$; Remainder: 3)	23, 24
	Use synthetic division to divide polynomials by binomials of the form $(x - k)$ *(p. 141).*	Divisor: $x + 3$ Dividend: $x^4 - 10x^2 - 2x + 4$ $\begin{array}{r\|rrrrr} -3 & 1 & 0 & -10 & -2 & 4 \\ & & -3 & 9 & 3 & -3 \\ \hline & 1 & -3 & -1 & 1 & 1 \end{array}$ Remainder: 1 Quotient: $x^3 - 3x^2 - x + 1$	25, 26
	Use the Remainder Theorem and the Factor Theorem *(p. 142).*	**The Remainder Theorem:** If a polynomial $f(x)$ is divided by $x - k$, then the remainder is $r = f(k)$. **The Factor Theorem:** A polynomial $f(x)$ has a factor $(x - k)$ if and only if $f(k) = 0$.	27, 28

	What Did You Learn?	Explanation/Examples	Review Exercises
Section 2.4	Use the imaginary unit i to write complex numbers *(p. 147).*	When a and b are real numbers, $a + bi$ is a complex number. Two complex numbers $a + bi$ and $c + di$, written in standard form, are equal to each other if and only if $a = c$ and $b = d$.	29, 30
	Add, subtract, and multiply complex numbers *(p. 148).*	**Sum:** $(a + bi) + (c + di) = (a + c) + (b + d)i$ **Difference:** $(a + bi) - (c + di) = (a - c) + (b - d)i$	31–34
	Use complex conjugates to write the quotient of two complex numbers in standard form *(p. 150).*	The numbers $a + bi$ and $a - bi$ are complex conjugates. To write $(a + bi)/(c + di)$ in standard form, where c and d are not both zero, multiply the numerator and denominator by $c - di$.	35, 36
	Find complex solutions of quadratic equations *(p. 151).*	When a is a positive real number, the principal square root of $-a$ is defined as $\sqrt{-a} = \sqrt{a}i$.	37, 38
Section 2.5	Use the Fundamental Theorem of Algebra to determine the number of zeros of polynomial functions *(p. 154).*	**The Fundamental Theorem of Algebra** If $f(x)$ is a polynomial of degree n, where $n > 0$, then f has at least one zero in the complex number system.	39, 40
	Find rational zeros of polynomial functions *(p. 155)*, and conjugate pairs of complex zeros *(p. 158).*	The Rational Zero Test relates the possible rational zeros of a polynomial to the leading coefficient and to the constant term of the polynomial. Let f be a polynomial function that has real coefficients. If $a + bi$ $(b \neq 0)$ is a zero of the function, then the conjugate $a - bi$ is also a zero of the function.	41, 42
	Find zeros of polynomials by factoring *(p. 159)*, using Descartes's Rule of Signs *(p. 161)*, and using the Upper and Lower Bound Rules *(p. 162).*	Every polynomial of degree $n > 0$ with real coefficients can be written as the product of linear and quadratic factors with real coefficients, where the quadratic factors have no real zeros.	43–46
Section 2.6	Find the domains *(p. 168)*, and vertical and horizontal asymptotes *(p. 169)*, of the graphs of rational functions.	The domain of a rational function of x includes all real numbers except x-values that make the denominator zero. The line $x = a$ is a vertical asymptote of the graph of f when $f(x) \to \infty$ or $f(x) \to -\infty$ as $x \to a$, either from the right or from the left. The line $y = b$ is a horizontal asymptote of the graph of f when $f(x) \to b$ as $x \to \infty$ or $x \to -\infty$.	47, 48
	Sketch the graphs of rational functions *(p. 171)*, including functions with slant asymptotes *(p. 174).*	Consider a rational function whose denominator is of degree 1 or greater. If the degree of the numerator is exactly *one more* than the degree of the denominator, then the graph of the function has a slant asymptote.	49–56
	Use rational functions to model and solve real-life problems *(p. 175).*	A rational function can help you model the cost of removing a given percent of the smokestack pollutants at a utility company that burns coal. (See Example 8.)	57, 58
Section 2.7	Solve polynomial *(p. 180)* and rational *(p. 184)* inequalities.	Use the concepts of key numbers and test intervals to solve both polynomial and rational inequalities.	59–62
	Use inequalities to model and solve real-life problems *(p. 185).*	A common application of inequalities involves profit P, revenue R, and cost C. (See Example 6.)	63

Review Exercises

See CalcChat.com for tutorial help and worked-out solutions to odd-numbered exercises.

2.1 Sketching Graphs of Quadratic Functions

In Exercises 1 and 2, sketch the graph of each quadratic function and compare it with the graph of $y = x^2$.

1. (a) $g(x) = -2x^2$
   (b) $h(x) = x^2 + 2$
2. (a) $h(x) = (x - 3)^2$
   (b) $k(x) = \frac{1}{2}x^2 - 1$

Using Standard Form to Graph a Parabola **In Exercises 3–8, write the quadratic function in standard form and sketch its graph. Identify the vertex, axis of symmetry, and x-intercept(s).**

3. $g(x) = x^2 - 2x$
4. $f(x) = x^2 + 8x + 10$
5. $h(x) = 3 + 4x - x^2$
6. $f(t) = -2t^2 + 4t + 1$
7. $h(x) = 4x^2 + 4x + 13$
8. $f(x) = \frac{1}{3}(x^2 + 5x - 4)$

9. **Geometry** The perimeter of a rectangle is 1000 meters.
   (a) Write the width y as a function of the length x. Use the result to write the area A as a function of x.
   (b) Of all possible rectangles with perimeters of 1000 meters, find the dimensions of the one with the maximum area.

10. **Minimum Cost** A soft-drink manufacturer has daily production costs of $C = 70{,}000 - 120x + 0.055x^2$, where C is the total cost (in dollars) and x is the number of units produced. How many units should the manufacturer produce each day to yield a minimum cost?

2.2 Sketching a Transformation of a Monomial Function

In Exercises 11 and 12, sketch the graphs of $y = x^n$ and the transformation.

11. $y = x^4,\ f(x) = 6 - x^4$
12. $y = x^5,\ f(x) = \frac{1}{2}x^5 + 3$

Applying the Leading Coefficient Test **In Exercises 13–16, describe the right-hand and left-hand behavior of the graph of the polynomial function.**

13. $f(x) = -2x^2 - 5x + 12$
14. $f(x) = \frac{1}{2}x^3 + 2x$
15. $g(x) = \frac{3}{4}(x^4 + 3x^2 + 2)$
16. $h(x) = -x^7 + 8x^2 - 8x$

Sketching the Graph of a Polynomial Function **In Exercises 17–20, sketch the graph of the function by (a) applying the Leading Coefficient Test, (b) finding the real zeros of the polynomial, (c) plotting sufficient solution points, and (d) drawing a continuous curve through the points.**

17. $g(x) = 2x^3 + 4x^2$
18. $h(x) = 3x^2 - x^4$
19. $f(x) = -x^3 + x^2 - 2$
20. $f(x) = x(x^3 + x^2 - 5x + 3)$

Using the Intermediate Value Theorem **In Exercises 21 and 22, (a) use the Intermediate Value Theorem and the *table* feature of a graphing utility to find intervals one unit in length in which the polynomial function is guaranteed to have a zero. (b) Adjust the table to approximate the zeros of the function. Use the *zero* or *root* feature of the graphing utility to verify your results.**

21. $f(x) = 3x^3 - x^2 + 3$
22. $f(x) = x^4 - 5x - 1$

2.3 Long Division of Polynomials

In Exercises 23 and 24, use long division to divide.

23. $\dfrac{30x^2 - 3x + 8}{5x - 3}$
24. $\dfrac{5x^3 - 21x^2 - 25x - 4}{x^2 - 5x - 1}$

Using the Factor Theorem **In Exercises 25 and 26, use synthetic division to determine whether the given values of x are zeros of the function.**

25. $f(x) = 20x^4 + 9x^3 - 14x^2 - 3x$
    (a) $x = -1$ (b) $x = \frac{3}{4}$ (c) $x = 0$ (d) $x = 1$
26. $f(x) = 3x^3 - 8x^2 - 20x + 16$
    (a) $x = 4$ (b) $x = -4$ (c) $x = \frac{2}{3}$ (d) $x = -1$

Factoring a Polynomial **In Exercises 27 and 28, (a) verify the given factor(s) of $f(x)$, (b) find the remaining factors of $f(x)$, (c) use your results to write the complete factorization of $f(x)$, (d) list all real zeros of f, and (e) confirm your results by using a graphing utility to graph the function.**

	Function	Factor(s)
27.	$f(x) = 2x^3 + 11x^2 - 21x - 90$	$(x + 6)$
28.	$f(x) = x^4 - 4x^3 - 7x^2 + 22x + 24$	$(x + 2), (x - 3)$

2.4 Writing a Complex Number in Standard Form

In Exercises 29 and 30, write the complex number in standard form.

29. $8 + \sqrt{-100}$
30. $-5i + i^2$

Performing Operations with Complex Numbers **In Exercises 31–34, perform the operation and write the result in standard form.**

31. $(7 + 5i) + (-4 + 2i)$
32. $\left(\sqrt{2} - \sqrt{2}i\right) - \left(\sqrt{2} + \sqrt{2}i\right)$
33. $7i(11 - 9i)$
34. $(1 + 6i)(5 - 2i)$

Performing Operations with Complex Numbers **In Exercises 35 and 36, perform the operation and write the result in standard form.**

35. $\dfrac{6 + i}{4 - i}$
36. $\dfrac{4}{2 - 3i} + \dfrac{2}{1 + i}$

Complex Solutions of a Quadratic Equation In Exercises 37 and 38, use the Quadratic Formula to solve the quadratic equation.

37. $x^2 - 2x + 10 = 0$ **38.** $6x^2 + 3x + 27 = 0$

2.5 Zeros of Polynomial Functions In Exercises 39 and 40, determine the number of zeros of the polynomial function.

39. $g(x) = x^2 - 2x - 8$ **40.** $h(t) = t^2 - t^5$

Finding the Zeros of a Polynomial Function In Exercises 41 and 42, find all the zeros of the function.

41. $f(x) = x^3 + 3x^2 - 28x - 60$

42. $f(x) = 4x^4 - 11x^3 + 14x^2 - 6x$

Finding the Zeros of a Polynomial Function In Exercises 43 and 44, write the polynomial as the product of linear factors and list all the zeros of the function.

43. $g(x) = x^3 - 7x^2 + 36$

44. $f(x) = x^4 + 8x^3 + 8x^2 - 72x - 153$

45. Using Descartes's Rule of Signs Use Descartes's Rule of Signs to determine the possible numbers of positive and negative real zeros of $h(x) = -2x^5 + 4x^3 - 2x^2 + 5$.

46. Verifying Upper and Lower Bounds Use synthetic division to verify the upper and lower bounds of the real zeros of $f(x) = 4x^3 - 3x^2 + 4x - 3$.

(a) Upper: $x = 1$ (b) Lower: $x = -\frac{1}{4}$

2.6 Finding Domain and Asymptotes In Exercises 47 and 48, find the domain and the vertical and horizontal asymptotes of the graph of the rational function.

47. $f(x) = \dfrac{3x}{x + 10}$

48. $f(x) = \dfrac{8}{x^2 - 10x + 24}$

Sketching the Graph of a Rational Function In Exercises 49–56, (a) state the domain of the function, (b) identify all intercepts, (c) find any asymptotes, and (d) plot additional solution points as needed to sketch the graph of the rational function.

49. $f(x) = \dfrac{4}{x}$ **50.** $h(x) = \dfrac{x - 4}{x - 7}$

51. $f(x) = \dfrac{x}{x^2 + 1}$ **52.** $f(x) = \dfrac{2x^2}{x^2 - 4}$

53. $f(x) = \dfrac{6x^2 - 11x + 3}{3x^2 - x}$ **54.** $f(x) = \dfrac{6x^2 - 7x + 2}{4x^2 - 1}$

55. $f(x) = \dfrac{2x^3}{x^2 + 1}$ **56.** $f(x) = \dfrac{x^2 + 1}{x + 1}$

57. Seizure of Illegal Drugs The cost C (in millions of dollars) for the federal government to seize $p\%$ of an illegal drug as it enters the country is given by

$$C = \frac{528p}{100 - p}, \quad 0 \le p < 100.$$

(a) Use a graphing utility to graph the cost function.

(b) Find the costs of seizing 25%, 50%, and 75% of the drug.

(c) According to the model, would it be possible to seize 100% of the drug?

58. Page Design A page that is x inches wide and y inches high contains 30 square inches of print. The top and bottom margins are each 2 inches deep, and the margins on each side are 2 inches wide.

(a) Write a function for the total area A of the page in terms of x.

(b) Determine the domain of the function based on the physical constraints of the problem.

(c) Use a graphing utility to graph the area function and approximate the page size for which the least amount of paper is used.

2.7 Solving an Inequality In Exercises 59–62, solve the inequality. Then graph the solution set.

59. $12x^2 + 5x < 2$

60. $x^3 - 16x \ge 0$

61. $\dfrac{2}{x + 1} \ge \dfrac{3}{x - 1}$

62. $\dfrac{x^2 - 9x + 20}{x} < 0$

63. Population of a Species A biologist introduces 200 ladybugs into a crop field. The population P of the ladybugs can be approximated by the model

$$P = \frac{1000(1 + 3t)}{5 + t}$$

where t is the time in days. Find the time required for the population to increase to at least 2000 ladybugs.

Exploration

64. Writing Describe what is meant by an asymptote of a graph.

True or False? In Exercises 65 and 66, determine whether the statement is true or false. Justify your answer.

65. A fourth-degree polynomial with real coefficients can have -5, $-8i$, $4i$, and 5 as its zeros.

66. The domain of a rational function can never be the set of all real numbers.

Chapter Test

See CalcChat.com for tutorial help and worked-out solutions to odd-numbered exercises.

Take this test as you would take a test in class. When you are finished, check your work against the answers given in the back of the book.

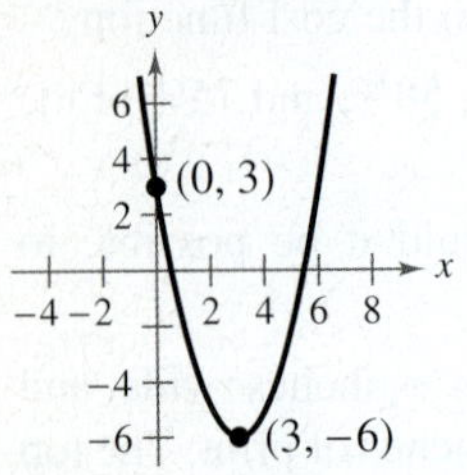

Figure for 2

1. Sketch the graph of each quadratic function and compare it with the graph of $y = x^2$.
   (a) $g(x) = 2 - x^2$
   (b) $g(x) = \left(x - \frac{3}{2}\right)^2$
2. Write the standard form of the equation of the parabola shown in the figure.
3. The path of a ball is given by the function $f(x) = -\frac{1}{20}x^2 + 3x + 5$, where $f(x)$ is the height (in feet) of the ball and x is the horizontal distance (in feet) from where the ball was thrown.
   (a) What is the maximum height of the ball?
   (b) Which number determines the height at which the ball was thrown? Does changing this value change the coordinates of the maximum height of the ball? Explain.
4. Describe the right-hand and left-hand behavior of the graph of the function $h(t) = -\frac{3}{4}t^5 + 2t^2$. Then sketch its graph.
5. Divide using long division.

$$\frac{3x^3 + 4x - 1}{x^2 + 1}$$

6. Divide using synthetic division.

$$\frac{2x^4 - 5x^2 - 3}{x - 2}$$

7. Use synthetic division to show that $x = \frac{5}{2}$ is a zero of the function

$$f(x) = 2x^3 - 5x^2 - 6x + 15.$$

   Use the result to factor the polynomial function completely and list all the zeros of the function.
8. Perform each operation and write the result in standard form.
   (a) $10i - \left(3 + \sqrt{-25}\right)$
   (b) $\left(2 + \sqrt{3}i\right)\left(2 - \sqrt{3}i\right)$
9. Write the quotient in standard form: $\dfrac{5}{2 + i}$.

In Exercises 10 and 11, find a polynomial function with real coefficients that has the given zeros. (There are many correct answers.)

10. $0, 3, 2 + i$
11. $1 - \sqrt{3}i, 2, 2$

In Exercises 12 and 13, find all the zeros of the function.

12. $f(x) = 3x^3 + 14x^2 - 7x - 10$
13. $f(x) = x^4 - 9x^2 - 22x - 24$

In Exercises 14–16, identify any intercepts and asymptotes of the graph of the function. Then sketch a graph of the function.

14. $h(x) = \dfrac{4}{x^2} - 1$
15. $f(x) = \dfrac{2x^2 - 5x - 12}{x^2 - 16}$
16. $g(x) = \dfrac{x^2 + 2}{x - 1}$

In Exercises 17 and 18, solve the inequality. Then graph the solution set.

17. $2x^2 + 5x > 12$
18. $\dfrac{2}{x} \le \dfrac{1}{x + 6}$

Proofs in Mathematics

These two pages contain proofs of four important theorems about polynomial functions. The first two theorems are from Section 2.3, and the second two theorems are from Section 2.5.

The Remainder Theorem ***(p. 142)***

If a polynomial $f(x)$ is divided by $x - k$, then the remainder is

$$r = f(k).$$

Proof

From the Division Algorithm, you have

$$f(x) = (x - k)q(x) + r(x)$$

and because either $r(x) = 0$ or the degree of $r(x)$ is less than the degree of $x - k$, you know that $r(x)$ must be a constant. That is, $r(x) = r$. Now, by evaluating $f(x)$ at $x = k$, you have

$$\begin{aligned} f(k) &= (k - k)q(k) + r \\ &= (0)q(k) + r = r. \end{aligned}$$

To be successful in algebra, it is important that you understand the connection among *factors* of a polynomial, *zeros* of a polynomial function, and *solutions* or *roots* of a polynomial equation. The Factor Theorem is the basis for this connection.

The Factor Theorem ***(p. 143)***

A polynomial $f(x)$ has a factor $(x - k)$ if and only if $f(k) = 0$.

Proof

Using the Division Algorithm with the factor $(x - k)$, you have

$$f(x) = (x - k)q(x) + r(x).$$

By the Remainder Theorem, $r(x) = r = f(k)$, and you have

$$f(x) = (x - k)q(x) + f(k)$$

where $q(x)$ is a polynomial of lesser degree than $f(x)$. If $f(k) = 0$, then

$$f(x) = (x - k)q(x)$$

and you see that $(x - k)$ is a factor of $f(x)$. Conversely, if $(x - k)$ is a factor of $f(x)$, then division of $f(x)$ by $(x - k)$ yields a remainder of 0. So, by the Remainder Theorem, you have $f(k) = 0$.

Linear Factorization Theorem *(p. 154)*

If $f(x)$ is a polynomial of degree n, where $n > 0$, then $f(x)$ has precisely n linear factors

$$f(x) = a_n(x - c_1)(x - c_2) \cdot \cdot \cdot (x - c_n)$$

where $c_1, c_2, \ldots, c_n$ are complex numbers.

THE FUNDAMENTAL THEOREM OF ALGEBRA

The Linear Factorization Theorem is closely related to the Fundamental Theorem of Algebra. The Fundamental Theorem of Algebra has a long and interesting history. In the early work with polynomial equations, the Fundamental Theorem of Algebra was thought to have been not true, because imaginary solutions were not considered. In fact, in the very early work by mathematicians such as Abu al-Khwarizmi (c. 800 A.D.), negative solutions were also not considered.

Once imaginary numbers were accepted, several mathematicians attempted to give a general proof of the Fundamental Theorem of Algebra. These included Gottfried von Leibniz (1702), Jean d'Alembert (1746), Leonhard Euler (1749), Joseph-Louis Lagrange (1772), and Pierre Simon Laplace (1795). The mathematician usually credited with the first correct proof of the Fundamental Theorem of Algebra is Carl Friedrich Gauss, who published the proof in his doctoral thesis in 1799.

Proof

Using the Fundamental Theorem of Algebra, you know that f must have at least one zero, c_1. Consequently, $(x - c_1)$ is a factor of $f(x)$, and you have

$$f(x) = (x - c_1)f_1(x).$$

If the degree of $f_1(x)$ is greater than zero, then you again apply the Fundamental Theorem of Algebra to conclude that f_1 must have a zero c_2, which implies that

$$f(x) = (x - c_1)(x - c_2)f_2(x).$$

It is clear that the degree of $f_1(x)$ is $n - 1$, that the degree of $f_2(x)$ is $n - 2$, and that you can repeatedly apply the Fundamental Theorem of Algebra n times until you obtain

$$f(x) = a_n(x - c_1)(x - c_2) \cdot \cdot \cdot (x - c_n)$$

where a_n is the leading coefficient of the polynomial $f(x)$. ■

Factors of a Polynomial *(p. 159)*

Every polynomial of degree $n > 0$ with real coefficients can be written as the product of linear and quadratic factors with real coefficients, where the quadratic factors have no real zeros.

Proof

To begin, you use the Linear Factorization Theorem to conclude that $f(x)$ can be *completely* factored in the form

$$f(x) = d(x - c_1)(x - c_2)(x - c_3) \cdot \cdot \cdot (x - c_n).$$

If each c_i is real, then there is nothing more to prove. If any c_i is complex ($c_i = a + bi$, $b \neq 0$), then, because the coefficients of $f(x)$ are real, you know that the conjugate $c_j = a - bi$ is also a zero. By multiplying the corresponding factors, you obtain

$$\begin{aligned}(x - c_i)(x - c_j) &= [x - (a + bi)][x - (a - bi)] \\ &= x^2 - 2ax + (a^2 + b^2)\end{aligned}$$

where each coefficient is real. ■

P.S. Problem Solving

1. Verifying the Remainder Theorem Show that if $f(x) = ax^3 + bx^2 + cx + d$, then $f(k) = r$, where $r = ak^3 + bk^2 + ck + d$, using long division. In other words, verify the Remainder Theorem for a third-degree polynomial function.

2. Babylonian Mathematics In 2000 B.C., the Babylonians solved polynomial equations by referring to tables of values. One such table gave the values of $y^3 + y^2$. To be able to use this table, the Babylonians sometimes had to manipulate the equation, as shown below.

$$ax^3 + bx^2 = c \quad \text{Original equation}$$

$$\frac{a^3x^3}{b^3} + \frac{a^2x^2}{b^2} = \frac{a^2c}{b^3} \quad \text{Multiply each side by } \frac{a^2}{b^3}.$$

$$\left(\frac{ax}{b}\right)^3 + \left(\frac{ax}{b}\right)^2 = \frac{a^2c}{b^3} \quad \text{Rewrite.}$$

Then they would find $(a^2c)/b^3$ in the $y^3 + y^2$ column of the table. Because they knew that the corresponding y-value was equal to $(ax)/b$, they could conclude that $x = (by)/a$.

(a) Calculate $y^3 + y^2$ for $y = 1, 2, 3, \ldots, 10$. Record the values in a table.

(b) Use the table from part (a) and the method above to solve each equation.

(i) $x^3 + x^2 = 252$

(ii) $x^3 + 2x^2 = 288$

(iii) $3x^3 + x^2 = 90$

(iv) $2x^3 + 5x^2 = 2500$

(v) $7x^3 + 6x^2 = 1728$

(vi) $10x^3 + 3x^2 = 297$

(c) Using the methods from this chapter, verify your solution of each equation.

3. Finding Dimensions At a glassware factory, molten cobalt glass is poured into molds to make paperweights. Each mold is a rectangular prism whose height is 3 inches greater than the length of each side of the square base. A machine pours 20 cubic inches of liquid glass into each mold. What are the dimensions of the mold?

4. True or False? Determine whether the statement is true or false. If false, provide one or more reasons why the statement is false and correct the statement. Let $f(x) = ax^3 + bx^2 + cx + d$, $a \neq 0$, and let $f(2) = -1$. Then

$$\frac{f(x)}{x+1} = q(x) + \frac{2}{x+1}$$

where $q(x)$ is a second-degree polynomial.

5. Finding the Equation of a Parabola The parabola shown in the figure has an equation of the form $y = ax^2 + bx + c$. Find the equation of this parabola using the following methods.

(a) Find the equation analytically.

(b) Use the regression feature of a graphing utility to find the equation.

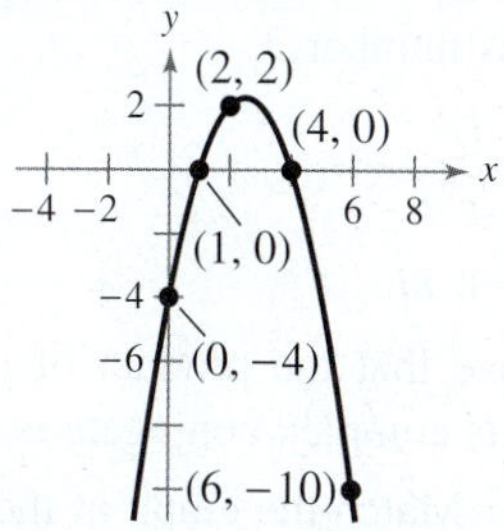

6. Finding the Slope of a Tangent Line One of the fundamental themes of calculus is to find the slope of the tangent line to a curve at a point. To see how this can be done, consider the point (2, 4) on the graph of the quadratic function $f(x) = x^2$, as shown in the figure.

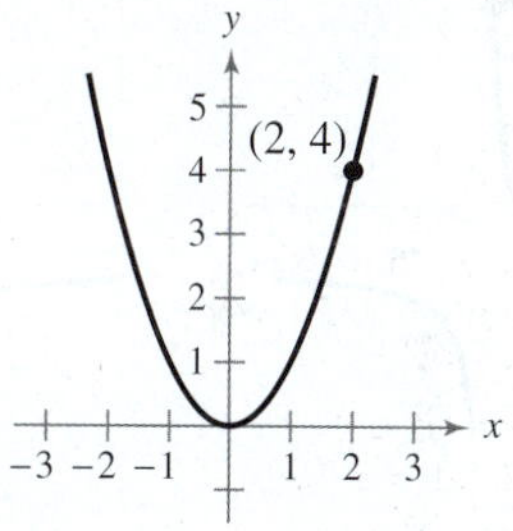

(a) Find the slope m_1 of the line joining (2, 4) and (3, 9). Is the slope of the tangent line at (2, 4) greater than or less than the slope of the line through (2, 4) and (3, 9)?

(b) Find the slope m_2 of the line joining (2, 4) and (1, 1). Is the slope of the tangent line at (2, 4) greater than or less than the slope of the line through (2, 4) and (1, 1)?

(c) Find the slope m_3 of the line joining (2, 4) and (2.1, 4.41). Is the slope of the tangent line at (2, 4) greater than or less than the slope of the line through (2, 4) and (2.1, 4.41)?

(d) Find the slope m_h of the line joining (2, 4) and $(2 + h, f(2 + h))$ in terms of the nonzero number h.

(e) Evaluate the slope formula from part (d) for $h = -1$, 1, and 0.1. Compare these values with those in parts (a)–(c).

(f) What can you conclude the slope $m_{\tan}$ of the tangent line at (2, 4) to be? Explain your answer.

7. **Writing Cubic Functions** Use the form $f(x) = (x - k)q(x) + r$ to write a cubic function that (a) passes through the point $(2, 5)$ and rises to the right and (b) passes through the point $(-3, 1)$ and falls to the right. (There are many correct answers.)

8. **Multiplicative Inverse of a Complex Number** The multiplicative inverse of z is a complex number z_m such that $z \cdot z_m = 1$. Find the multiplicative inverse of each complex number.

   (a) $z = 1 + i$

   (b) $z = 3 - i$

   (c) $z = -2 + 8i$

9. **Proof** Prove that the product of a complex number $a + bi$ and its complex conjugate is a real number.

10. **Matching** Match the graph of the rational function

$$f(x) = \frac{ax + b}{cx + d}$$

with the given conditions.

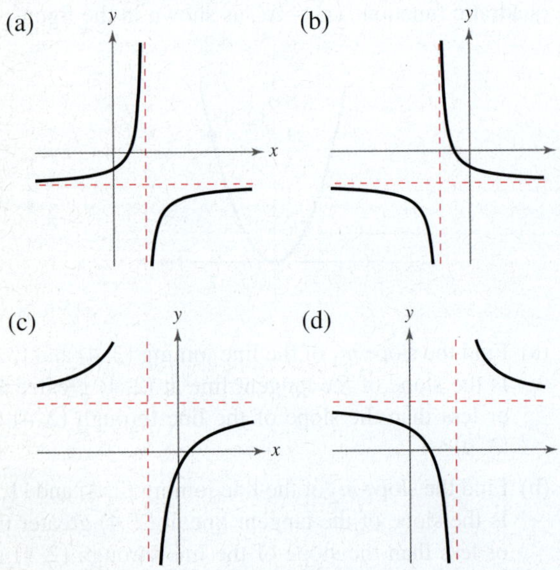

(i)	(ii)	(iii)	(iv)
$a > 0$	$a > 0$	$a < 0$	$a > 0$
$b < 0$	$b > 0$	$b > 0$	$b < 0$
$c > 0$	$c < 0$	$c > 0$	$c > 0$
$d < 0$	$d < 0$	$d < 0$	$d > 0$

11. **Effects of Values on a Graph** Consider the function

$$f(x) = \frac{ax}{(x - b)^2}.$$

   (a) Determine the effect on the graph of f when $b \neq 0$ and a is varied. Consider cases in which a is positive and a is negative.

   (b) Determine the effect on the graph of f when $a \neq 0$ and b is varied.

12. **Distinct Vision** The endpoints of the interval over which distinct vision is possible are called the *near point* and *far point* of the eye (see figure). With increasing age, these points normally change. The table shows the approximate near points y (in inches) for various ages x (in years).

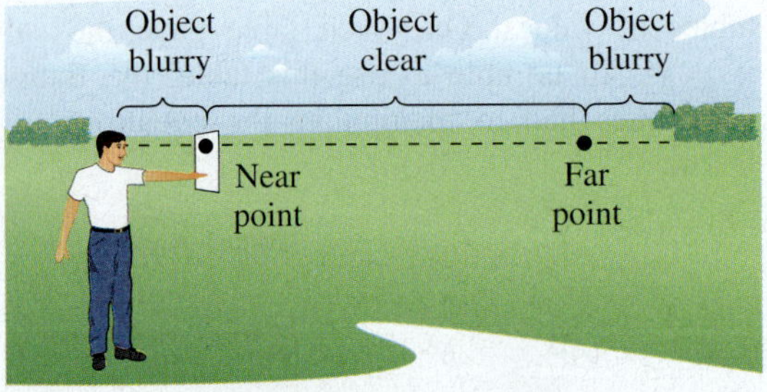

Age, x	Near Point, y
16	3.0
32	4.7
44	9.8
50	19.7
60	39.4

   (a) Use the *regression* feature of a graphing utility to find a quadratic model for the data. Use the graphing utility to plot the data and graph the model in the same viewing window.

   (b) Find a rational model for the data. Take the reciprocals of the near points to generate the points $(x, 1/y)$. Use the *regression* feature of the graphing utility to find a linear model for the data. The resulting line has the form

$$\frac{1}{y} = ax + b.$$

   Solve for y. Use the graphing utility to plot the data and graph the model in the same viewing window.

   (c) Use the *table* feature of the graphing utility to construct a table showing the predicted near point based on each model for each of the ages in the original table. How well do the models fit the original data?

   (d) Use both models to estimate the near point for a person who is 25 years old. Which model is a better fit?

   (e) Do you think either model can be used to predict the near point for a person who is 70 years old? Explain.

13. **Roots of a Cubic Equation** Can a cubic equation with real coefficients have two real zeros and one complex zero? Explain your reasoning.

3 Exponential and Logarithmic Functions

Trees per Acre *(Exercise 83, page 236)*

Earthquakes *(Example 6, page 244)*

Sound Intensity *(Exercises 85–88, page 226)*

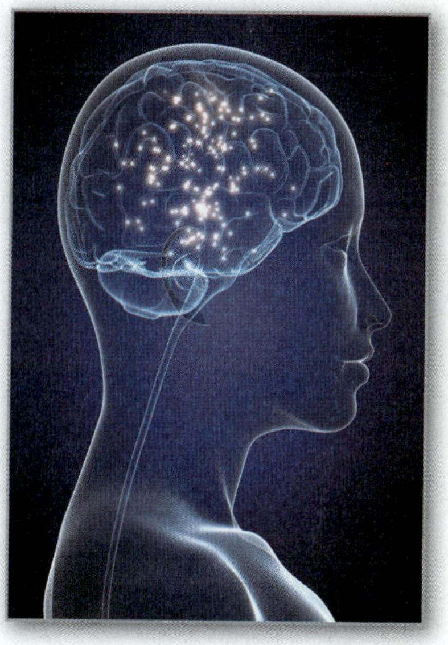

Human Memory Model *(Exercise 81, page 220)*

Nuclear Reactor Accident *(Example 9, page 207)*

Clockwise from top left, James Marshall/CORBIS; Darrenp/Shutterstock.com; Sebastian Kaulitzki/Shutterstock.com; Hellen Sergeyeva/Shutterstock.com; kentoh/Shutterstock.com

3.1 Exponential Functions and Their Graphs

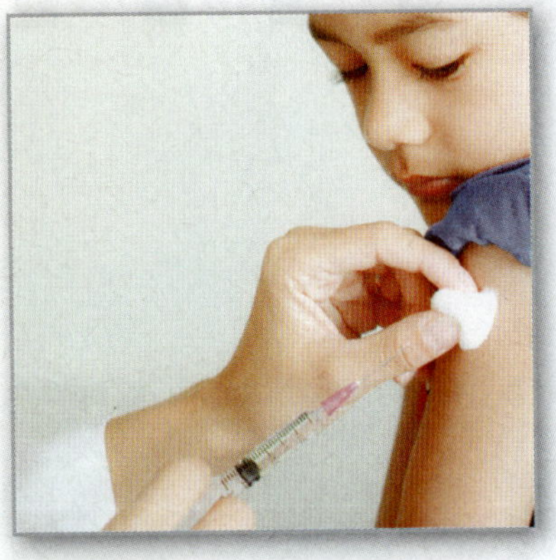

Exponential functions can help you model and solve real-life problems. For instance, Exercise 72 on page 210 uses an exponential function to model the concentration of a drug in the bloodstream.

- **Recognize and evaluate exponential functions with base *a*.**
- **Graph exponential functions and use the One-to-One Property.**
- **Recognize, evaluate, and graph exponential functions with base e.**
- **Use exponential functions to model and solve real-life problems.**

Exponential Functions

So far, this text has dealt mainly with **algebraic functions,** which include polynomial functions and rational functions. In this chapter, you will study two types of nonalgebraic functions—*exponential functions* and *logarithmic functions*. These functions are examples of **transcendental functions.**

> **Definition of Exponential Function**
>
> The **exponential function *f* with base *a*** is denoted by
>
> $$f(x) = a^x$$
>
> where $a > 0$, $a \neq 1$, and x is any real number.

The base $a = 1$ is excluded because it yields $f(x) = 1^x = 1$. This is a constant function, not an exponential function.

You have evaluated a^x for integer and rational values of x. For example, you know that $4^3 = 64$ and $4^{1/2} = 2$. However, to evaluate 4^x for any real number x, you need to interpret forms with *irrational* exponents. For the purposes of this text, it is sufficient to think of $a^{\sqrt{2}}$ (where $\sqrt{2} \approx 1.41421356$) as the number that has the successively closer approximations

$$a^{1.4}, a^{1.41}, a^{1.414}, a^{1.4142}, a^{1.41421}, \ldots .$$

EXAMPLE 1 Evaluating Exponential Functions

Use a calculator to evaluate each function at the indicated value of x.

	Function	Value
a.	$f(x) = 2^x$	$x = -3.1$
b.	$f(x) = 2^{-x}$	$x = \pi$
c.	$f(x) = 0.6^x$	$x = \frac{3}{2}$

Solution

	Function Value	Graphing Calculator Keystrokes	Display
a.	$f(-3.1) = 2^{-3.1}$	2 ^ (−) 3.1 ENTER	0.1166291
b.	$f(\pi) = 2^{-\pi}$	2 ^ (−) π ENTER	0.1133147
c.	$f(\frac{3}{2}) = (0.6)^{3/2}$	.6 ^ (3 ÷ 2) ENTER	0.4647580

✓ Checkpoint *Audio-video solution in English & Spanish at LarsonPrecalculus.com.*

Use a calculator to evaluate $f(x) = 8^{-x}$ at $x = \sqrt{2}$.

When evaluating exponential functions with a calculator, remember to enclose fractional exponents in parentheses. Because the calculator follows the order of operations, parentheses are crucial in order to obtain the correct result.

Sura Nualpradid/Shutterstock.com

Graphs of Exponential Functions

The graphs of all exponential functions have similar characteristics, as shown in Examples 2, 3, and 5.

EXAMPLE 2 **Graphs of $y = a^x$**

▷ **ALGEBRA HELP** You can review the techniques for sketching the graph of an equation in Section 1.2.

In the same coordinate plane, sketch the graph of each function.

a. $f(x) = 2^x$

b. $g(x) = 4^x$

Solution The following table lists some values for each function, and Figure 3.1 shows the graphs of the two functions. Note that both graphs are increasing. Moreover, the graph of $g(x) = 4^x$ is increasing more rapidly than the graph of $f(x) = 2^x$.

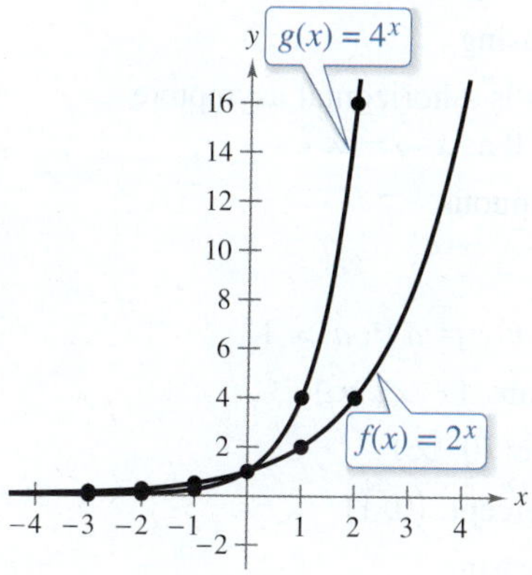

Figure 3.1

x	-3	-2	-1	0	1	2
2^x	$\frac{1}{8}$	$\frac{1}{4}$	$\frac{1}{2}$	1	2	4
4^x	$\frac{1}{64}$	$\frac{1}{16}$	$\frac{1}{4}$	1	4	16

✓ **Checkpoint** Audio-video solution in English & Spanish at LarsonPrecalculus.com.

In the same coordinate plane, sketch the graph of each function.

a. $f(x) = 3^x$ **b.** $g(x) = 9^x$

The table in Example 2 was evaluated by hand. You could, of course, use a graphing utility to construct tables with even more values.

EXAMPLE 3 **Graphs of $y = a^{-x}$**

In the same coordinate plane, sketch the graph of each function.

a. $F(x) = 2^{-x}$

b. $G(x) = 4^{-x}$

Solution The following table lists some values for each function, and Figure 3.2 shows the graphs of the two functions. Note that both graphs are decreasing. Moreover, the graph of $G(x) = 4^{-x}$ is decreasing more rapidly than the graph of $F(x) = 2^{-x}$.

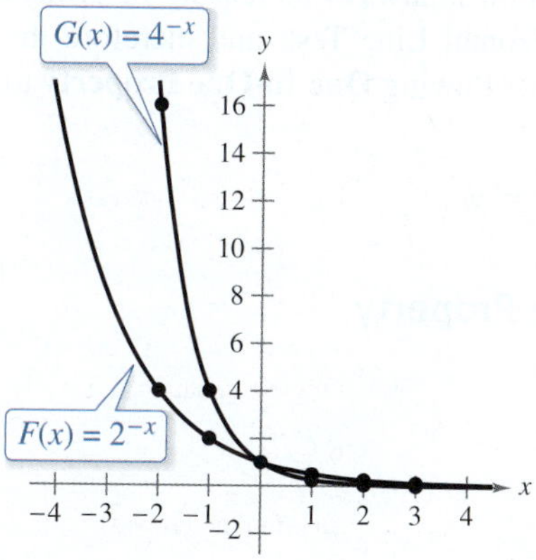

Figure 3.2

x	-2	-1	0	1	2	3
2^{-x}	4	2	1	$\frac{1}{2}$	$\frac{1}{4}$	$\frac{1}{8}$
4^{-x}	16	4	1	$\frac{1}{4}$	$\frac{1}{16}$	$\frac{1}{64}$

✓ **Checkpoint** Audio-video solution in English & Spanish at LarsonPrecalculus.com.

In the same coordinate plane, sketch the graph of each function.

a. $f(x) = 3^{-x}$ **b.** $g(x) = 9^{-x}$

Note that it is possible to use one of the properties of exponents to rewrite the functions in Example 3 with positive exponents, as follows.

$$F(x) = 2^{-x} = \frac{1}{2^x} = \left(\frac{1}{2}\right)^x \quad \text{and} \quad G(x) = 4^{-x} = \frac{1}{4^x} = \left(\frac{1}{4}\right)^x$$

Comparing the functions in Examples 2 and 3, observe that

$$F(x) = 2^{-x} = f(-x) \quad \text{and} \quad G(x) = 4^{-x} = g(-x).$$

Consequently, the graph of F is a reflection (in the y-axis) of the graph of f. The graphs of G and g have the same relationship. The graphs in Figures 3.1 and 3.2 are typical of the exponential functions $y = a^x$ and $y = a^{-x}$. They have one y-intercept and one horizontal asymptote (the x-axis), and they are continuous. The following summarizes the basic characteristics of these exponential functions.

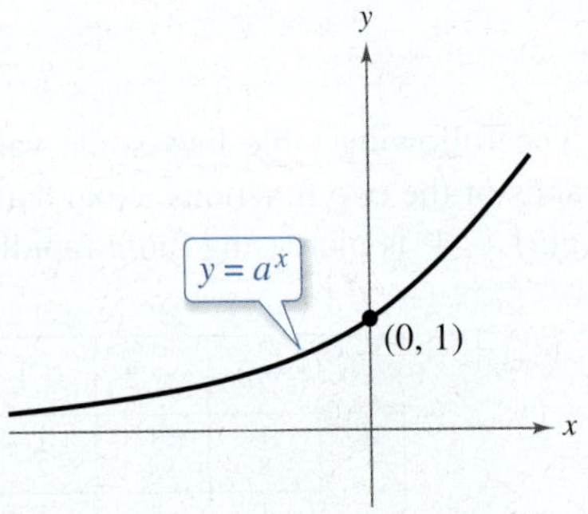

Graph of $y = a^x, a > 1$

- Domain: $(-\infty, \infty)$
- Range: $(0, \infty)$
- y-intercept: $(0, 1)$
- Increasing
- x-axis is a horizontal asymptote ($a^x \to 0$ as $x \to -\infty$).
- Continuous

REMARK Notice that the range of an exponential function is $(0, \infty)$, which means that $a^x > 0$ for all values of x.

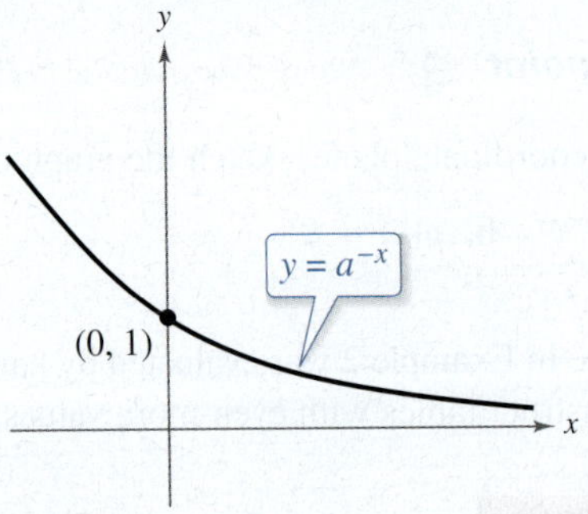

Graph of $y = a^{-x}, a > 1$

- Domain: $(-\infty, \infty)$
- Range: $(0, \infty)$
- y-intercept: $(0, 1)$
- Decreasing
- x-axis is a horizontal asymptote ($a^{-x} \to 0$ as $x \to \infty$).
- Continuous

Notice that the graph of an exponential function is always increasing or always decreasing. As a result, the graphs pass the Horizontal Line Test, and therefore the functions are one-to-one functions. You can use the following **One-to-One Property** to solve simple exponential equations.

For $a > 0$ and $a \neq 1$, $a^x = a^y$ if and only if $x = y$. One-to-One Property

EXAMPLE 4 Using the One-to-One Property

a. $9 = 3^{x+1}$ Original equation

$3^2 = 3^{x+1}$ $9 = 3^2$

$2 = x + 1$ One-to-One Property

$1 = x$ Solve for x.

b. $\left(\frac{1}{2}\right)^x = 8$ Original equation

$2^{-x} = 2^3$ $\left(\frac{1}{2}\right)^x = 2^{-x}, 8 = 2^3$

$x = -3$ One-to-One Property

✓ **Checkpoint** *Audio-video solution in English & Spanish at LarsonPrecalculus.com.*

Use the One-to-One Property to solve the equation for x.

a. $8 = 2^{2x-1}$ **b.** $\left(\frac{1}{3}\right)^{-x} = 27$

In the following example, notice how the graph of $y = a^x$ can be used to sketch the graphs of functions of the form $f(x) = b \pm a^{x+c}$.

▷ **ALGEBRA HELP** You can review the techniques for transforming the graph of a function in Section 1.7.

EXAMPLE 5 Transformations of Graphs of Exponential Functions

Each of the following graphs is a transformation of the graph of $f(x) = 3^x$.

a. Because $g(x) = 3^{x+1} = f(x + 1)$, the graph of g can be obtained by shifting the graph of f one unit to the *left*, as shown in Figure 3.3.

b. Because $h(x) = 3^x - 2 = f(x) - 2$, the graph of h can be obtained by shifting the graph of f *down* two units, as shown in Figure 3.4.

c. Because $k(x) = -3^x = -f(x)$, the graph of k can be obtained by *reflecting* the graph of f in the x-axis, as shown in Figure 3.5.

d. Because $j(x) = 3^{-x} = f(-x)$, the graph of j can be obtained by *reflecting* the graph of f in the y-axis, as shown in Figure 3.6.

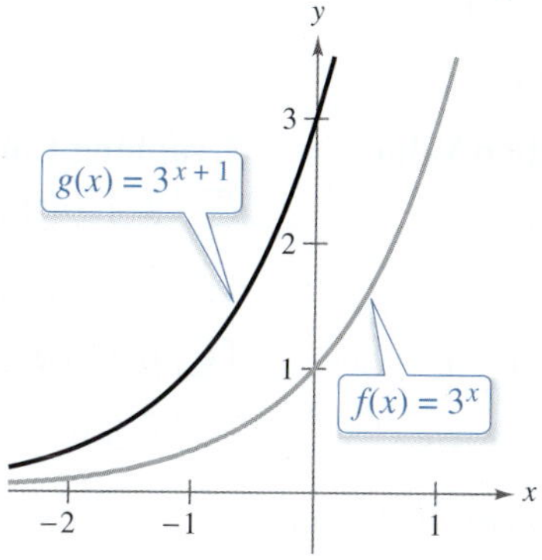

Figure 3.3 Horizontal shift

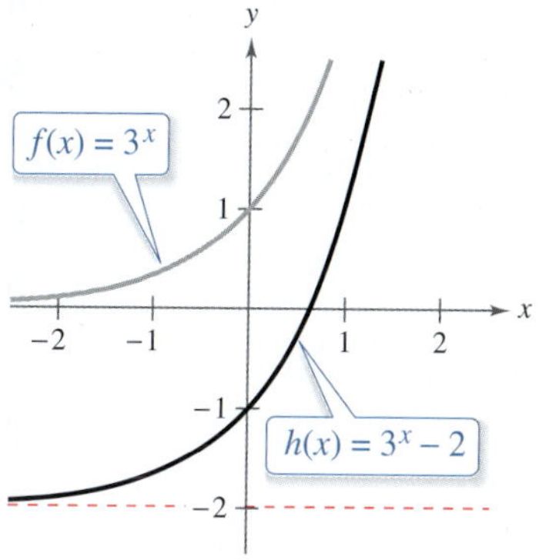

Figure 3.4 Vertical shift

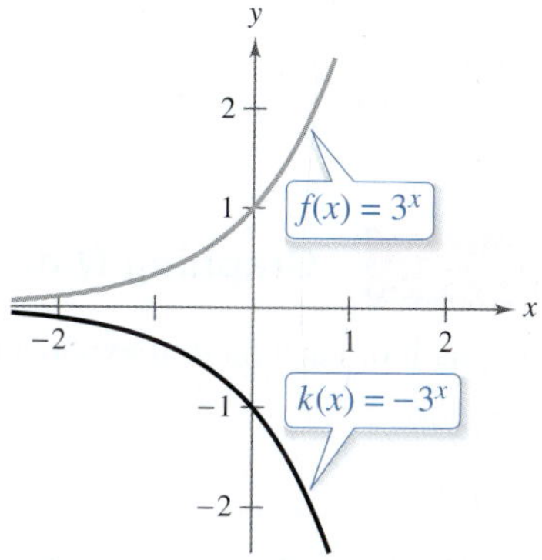

Figure 3.5 Reflection in x-axis

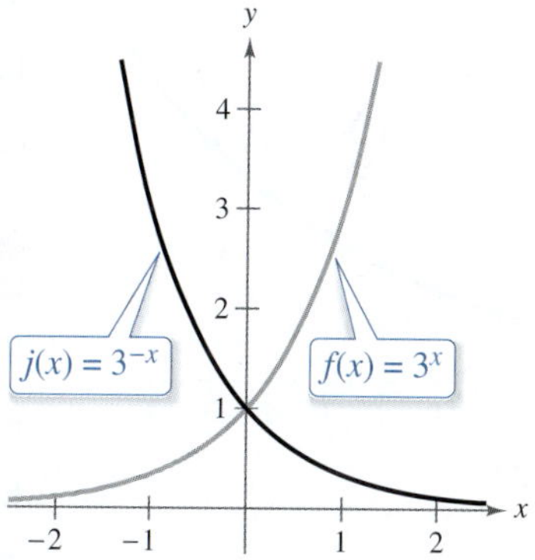

Figure 3.6 Reflection in y-axis

✓ **Checkpoint** Audio-video solution in English & Spanish at LarsonPrecalculus.com.

Use the graph of $f(x) = 4^x$ to describe the transformation that yields the graph of each function.

a. $g(x) = 4^{x-2}$ **b.** $h(x) = 4^x + 3$ **c.** $k(x) = 4^{-x} - 3$

Notice that the transformations in Figures 3.3, 3.5, and 3.6 keep the x-axis as a horizontal asymptote, but the transformation in Figure 3.4 yields a new horizontal asymptote of $y = -2$. Also, be sure to note how each transformation affects the y-intercept.

The Natural Base e

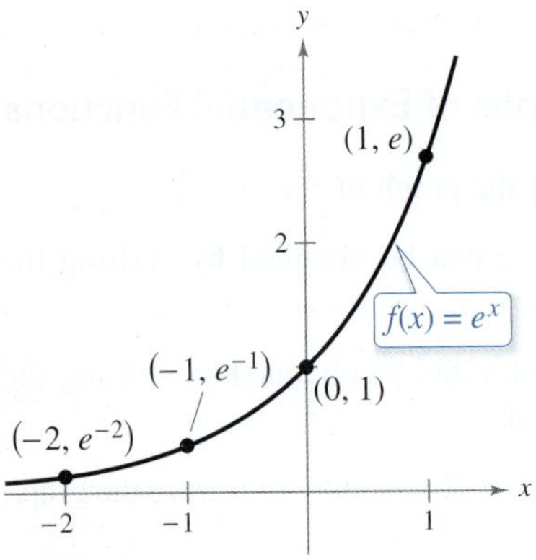

Figure 3.7

In many applications, the most convenient choice for a base is the irrational number

$$e \approx 2.718281828\ldots.$$

This number is called the **natural base.** The function $f(x) = e^x$ is called the **natural exponential function.** Figure 3.7 shows its graph. Be sure you see that for the exponential function $f(x) = e^x$, e is the constant 2.718281828 . . . , whereas x is the variable.

EXAMPLE 6 Evaluating the Natural Exponential Function

Use a calculator to evaluate the function $f(x) = e^x$ at each value of x.

a. $x = -2$

b. $x = -1$

c. $x = 0.25$

d. $x = -0.3$

Solution

	Function Value	Graphing Calculator Keystrokes	Display
a.	$f(-2) = e^{-2}$	e^x (−) 2 ENTER	0.1353353
b.	$f(-1) = e^{-1}$	e^x (−) 1 ENTER	0.3678794
c.	$f(0.25) = e^{0.25}$	e^x 0.25 ENTER	1.2840254
d.	$f(-0.3) = e^{-0.3}$	e^x (−) 0.3 ENTER	0.7408182

✓ **Checkpoint** Audio-video solution in English & Spanish at LarsonPrecalculus.com.

Use a calculator to evaluate the function $f(x) = e^x$ at each value of x.

a. $x = 0.3$

b. $x = -1.2$

c. $x = 6.2$

EXAMPLE 7 Graphing Natural Exponential Functions

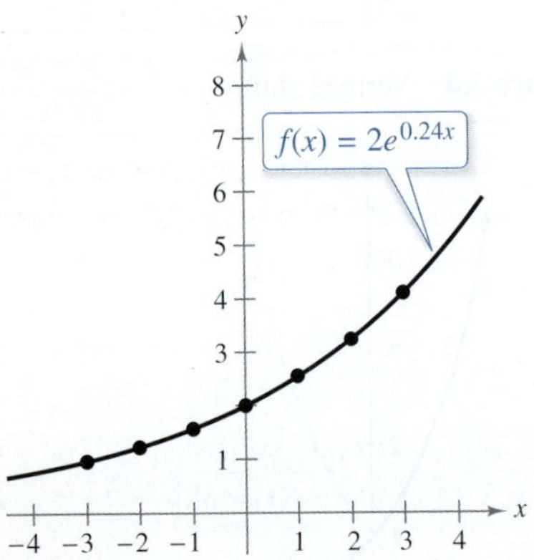

Figure 3.8

Sketch the graph of each natural exponential function.

a. $f(x) = 2e^{0.24x}$

b. $g(x) = \frac{1}{2}e^{-0.58x}$

Solution To sketch these two graphs, use a graphing utility to construct a table of values, as follows. After constructing the table, plot the points and connect them with smooth curves, as shown in Figures 3.8 and 3.9. Note that the graph in Figure 3.8 is increasing, whereas the graph in Figure 3.9 is decreasing.

x	−3	−2	−1	0	1	2	3
$f(x)$	0.974	1.238	1.573	2.000	2.542	3.232	4.109
$g(x)$	2.849	1.595	0.893	0.500	0.280	0.157	0.088

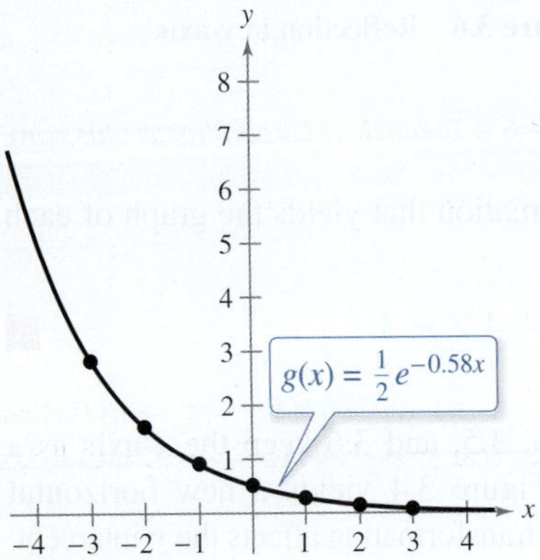

Figure 3.9

✓ **Checkpoint** Audio-video solution in English & Spanish at LarsonPrecalculus.com.

Sketch the graph of $f(x) = 5e^{0.17x}$.

Applications

One of the most familiar examples of exponential growth is an investment earning *continuously compounded interest*. The formula for *interest compounded n times per year* is

$$A = P\left(1 + \frac{r}{n}\right)^{nt}.$$

In this formula, A is the balance in the account, P is the principal (or original deposit), r is the annual interest rate (in decimal form), n is the number of compoundings per year, and t is the time in years. Using exponential functions, you can *develop* this formula and show how it leads to continuous compounding.

Suppose you invest a principal P at an annual interest rate r, compounded once per year. If the interest is added to the principal at the end of the year, then the new balance P_1 is

$$\begin{aligned} P_1 &= P + Pr \\ &= P(1 + r). \end{aligned}$$

This pattern of multiplying the previous principal by $1 + r$ repeats each successive year, as follows.

Year	Balance After Each Compounding
0	$P = P$
1	$P_1 = P(1 + r)$
2	$P_2 = P_1(1 + r) = P(1 + r)(1 + r) = P(1 + r)^2$
3	$P_3 = P_2(1 + r) = P(1 + r)^2(1 + r) = P(1 + r)^3$
$\vdots$	$\vdots$
t	$P_t = P(1 + r)^t$

To accommodate more frequent (quarterly, monthly, or daily) compounding of interest, let n be the number of compoundings per year and let t be the number of years. Then the rate per compounding is r/n and the account balance after t years is

$$A = P\left(1 + \frac{r}{n}\right)^{nt}. \qquad \text{Amount (balance) with } n \text{ compoundings per year}$$

When you let the number of compoundings n increase without bound, the process approaches what is called **continuous compounding.** In the formula for n compoundings per year, let $m = n/r$. This produces

$$\begin{aligned} A &= P\left(1 + \frac{r}{n}\right)^{nt} && \text{Amount with } n \text{ compoundings per year} \\ &= P\left(1 + \frac{r}{mr}\right)^{mrt} && \text{Substitute } mr \text{ for } n. \\ &= P\left(1 + \frac{1}{m}\right)^{mrt} && \text{Simplify.} \\ &= P\left[\left(1 + \frac{1}{m}\right)^{m}\right]^{rt}. && \text{Property of exponents} \end{aligned}$$

m	$\left(1 + \frac{1}{m}\right)^m$
1	2
10	2.59374246
100	2.704813829
1,000	2.716923932
10,000	2.718145927
100,000	2.718268237
1,000,000	2.718280469
10,000,000	2.718281693
↓	↓
∞	e

As m increases without bound (that is, as $m \to \infty$), the table at the left shows that $[1 + (1/m)]^m \to e$. From this, you can conclude that the formula for continuous compounding is

$$A = Pe^{rt}. \qquad \text{Substitute } e \text{ for } (1 + 1/m)^m.$$

REMARK Be sure you see that, when using the formulas for compound interest, you must write the annual interest rate in decimal form. For instance, you must write 6% as 0.06.

Formulas for Compound Interest

After t years, the balance A in an account with principal P and annual interest rate r (in decimal form) is given by the following formulas.

1. For n compoundings per year: $A = P\left(1 + \frac{r}{n}\right)^{nt}$
2. For continuous compounding: $A = Pe^{rt}$

EXAMPLE 8 **Compound Interest**

You invest \$12,000 at an annual rate of 3%. Find the balance after 5 years when the interest is compounded

a. quarterly.

b. monthly.

c. continuously.

Solution

a. For quarterly compounding, you have $n = 4$. So, in 5 years at 3%, the balance is

$$A = P\left(1 + \frac{r}{n}\right)^{nt}$$ Formula for compound interest

$$= 12{,}000\left(1 + \frac{0.03}{4}\right)^{4(5)}$$ Substitute for P, r, n, and t.

$$\approx \$13{,}934.21.$$ Use a calculator.

b. For monthly compounding, you have $n = 12$. So, in 5 years at 3%, the balance is

$$A = P\left(1 + \frac{r}{n}\right)^{nt}$$ Formula for compound interest

$$= 12{,}000\left(1 + \frac{0.03}{12}\right)^{12(5)}$$ Substitute for P, r, n, and t.

$$\approx \$13{,}939.40.$$ Use a calculator.

c. For continuous compounding, the balance is

$$A = Pe^{rt}$$ Formula for continuous compounding

$$= 12{,}000e^{0.03(5)}$$ Substitute for P, r, and t.

$$\approx \$13{,}942.01.$$ Use a calculator.

✓ ***Checkpoint*** *Audio-video solution in English & Spanish at LarsonPrecalculus.com.*

You invest \$6000 at an annual rate of 4%. Find the balance after 7 years when the interest is compounded

a. quarterly.

b. monthly.

c. continuously.

In Example 8, note that continuous compounding yields more than quarterly and monthly compounding. This is typical of the two types of compounding. That is, for a given principal, interest rate, and time, continuous compounding will always yield a larger balance than compounding n times per year.

EXAMPLE 9 Radioactive Decay

The International Atomic Energy Authority ranks nuclear incidents and accidents by severity using a scale from 1 to 7 called the International Nuclear and Radiological Event Scale (INES). A level 7 ranking is the most severe. To date, the Chernobyl accident is the only nuclear accident in history to be given an INES level 7 ranking.

In 1986, a nuclear reactor accident occurred in Chernobyl in what was then the Soviet Union. The explosion spread highly toxic radioactive chemicals, such as plutonium (^{239}Pu), over hundreds of square miles, and the government evacuated the city and the surrounding area. To see why the city is now uninhabited, consider the model

$$P = 10\left(\frac{1}{2}\right)^{t/24,100}$$

which represents the amount of plutonium P that remains (from an initial amount of 10 pounds) after t years. Sketch the graph of this function over the interval from $t = 0$ to $t = 100,000$, where $t = 0$ represents 1986. How much of the 10 pounds will remain in the year 2017? How much of the 10 pounds will remain after 100,000 years?

Solution The graph of this function is shown in the figure at the right. Note from this graph that plutonium has a *half-life* of about 24,100 years. That is, after 24,100 years, *half* of the original amount will remain. After another 24,100 years, one-quarter of the original amount will remain, and so on. In the year 2017 ($t = 31$), there will still be

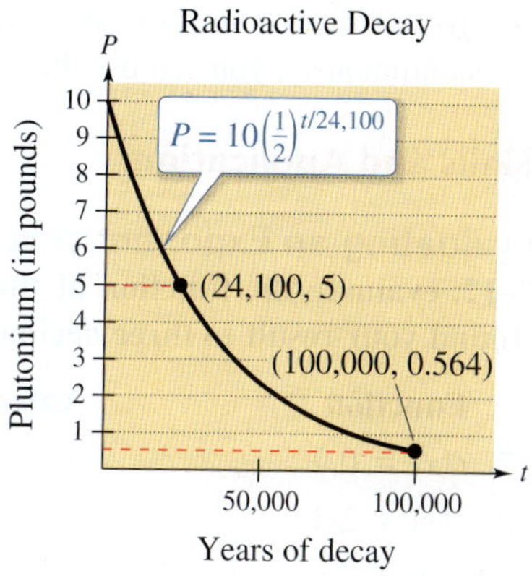

$$P = 10\left(\frac{1}{2}\right)^{31/24,100}$$

$$\approx 10\left(\frac{1}{2}\right)^{0.0012863}$$

$$\approx 9.991 \text{ pounds}$$

of plutonium remaining. After 100,000 years, there will still be

$$P = 10\left(\frac{1}{2}\right)^{100,000/24,100}$$

$$\approx 0.564 \text{ pound}$$

of plutonium remaining.

✓ **Checkpoint** *Audio-video solution in English & Spanish at LarsonPrecalculus.com.*

In Example 9, how much of the 10 pounds will remain in the year 2089? How much of the 10 pounds will remain after 125,000 years?

Summarize (Section 3.1)

1. State the definition of an exponential function f with base a *(page 200)*. For an example of evaluating exponential functions, see Example 1.
2. Describe the basic characteristics of the exponential functions $y = a^x$ and $y = a^{-x}$, $a > 1$ *(page 202)*. For examples of graphing exponential functions, see Examples 2, 3, and 5.
3. State the definitions of the natural base and the natural exponential function *(page 204)*. For examples of evaluating and graphing natural exponential functions, see Examples 6 and 7.
4. Describe examples of how to use exponential functions to model and solve real-life problems *(pages 206 and 207, Examples 8 and 9)*.

Hellen Sergeyeva/Shutterstock.com

3.1 Exercises

See CalcChat.com for tutorial help and worked-out solutions to odd-numbered exercises.

Vocabulary: Fill in the blanks.

1. Polynomial and rational functions are examples of ________ functions.
2. Exponential and logarithmic functions are examples of nonalgebraic functions, also called ________ functions.
3. You can use the ________ Property to solve simple exponential equations.
4. The exponential function $f(x) = e^x$ is called the ________ ________ function, and the base e is called the ________ base.
5. To find the amount A in an account after t years with principal P and an annual interest rate r compounded n times per year, you can use the formula ________.
6. To find the amount A in an account after t years with principal P and an annual interest rate r compounded continuously, you can use the formula ________.

Skills and Applications

Evaluating an Exponential Function In Exercises 7–12, evaluate the function at the indicated value of x. Round your result to three decimal places.

Function	Value
7. $f(x) = 0.9^x$	$x = 1.4$
8. $f(x) = 2.3^x$	$x = \frac{3}{2}$
9. $f(x) = 5^x$	$x = -\pi$
10. $f(x) = \left(\frac{2}{3}\right)^{5x}$	$x = \frac{3}{10}$
11. $g(x) = 5000(2^x)$	$x = -1.5$
12. $f(x) = 200(1.2)^{12x}$	$x = 24$

Matching an Exponential Function with Its Graph In Exercises 13–16, match the exponential function with its graph. [The graphs are labeled (a), (b), (c), and (d).]

(a)

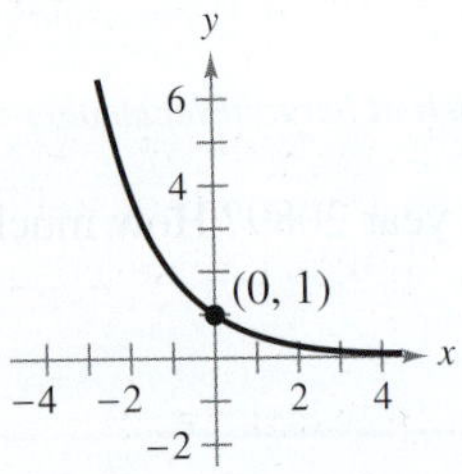

(b)

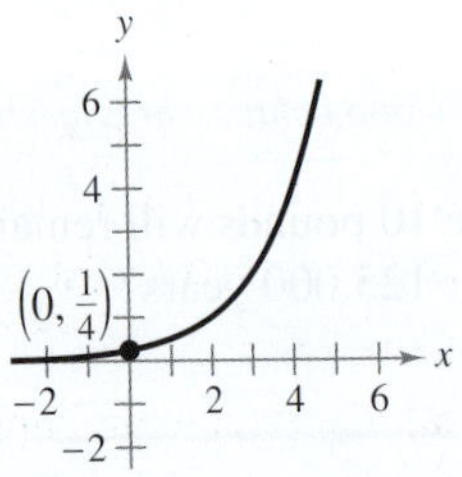

(c)

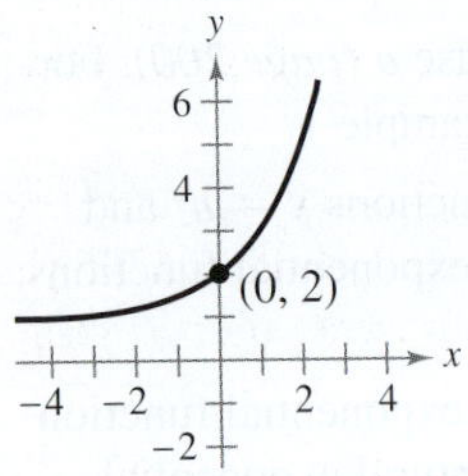

(d)

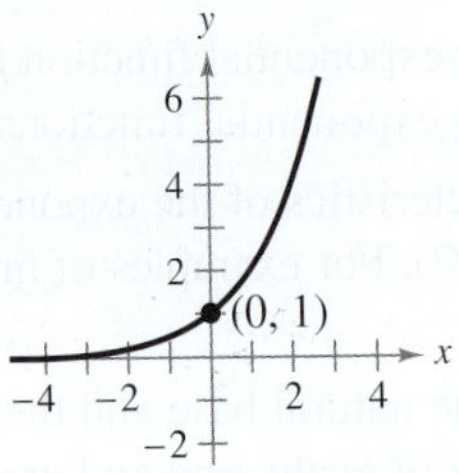

13. $f(x) = 2^x$
14. $f(x) = 2^x + 1$
15. $f(x) = 2^{-x}$
16. $f(x) = 2^{x-2}$

Graphing an Exponential Function In Exercises 17–22, use a graphing utility to construct a table of values for the function. Then sketch the graph of the function.

17. $f(x) = \left(\frac{1}{2}\right)^x$
18. $f(x) = \left(\frac{1}{2}\right)^{-x}$
19. $f(x) = 6^{-x}$
20. $f(x) = 6^x$
21. $f(x) = 2^{x-1}$
22. $f(x) = 4^{x-3} + 3$

Using the One-to-One Property In Exercises 23–26, use the One-to-One Property to solve the equation for x.

23. $3^{x+1} = 27$
24. $2^{x-3} = 16$
25. $\left(\frac{1}{2}\right)^x = 32$
26. $5^{x-2} = \frac{1}{125}$

Transforming the Graph of an Exponential Function In Exercises 27–30, use the graph of f to describe the transformation that yields the graph of g.

27. $f(x) = 3^x, \quad g(x) = 3^x + 1$
28. $f(x) = 10^x, \quad g(x) = 10^{-x+3}$
29. $f(x) = \left(\frac{7}{2}\right)^x, \quad g(x) = -\left(\frac{7}{2}\right)^{-x}$
30. $f(x) = 0.3^x, \quad g(x) = -0.3^x + 5$

Graphing an Exponential Function In Exercises 31–34, use a graphing utility to graph the exponential function.

31. $y = 2^{-x^2}$
32. $y = 3^{-|x|}$
33. $y = 3^{x-2} + 1$
34. $y = 4^{x+1} - 2$

Evaluating a Natural Exponential Function In Exercises 35–38, evaluate the function at the indicated value of x. Round your result to three decimal places.

Function	Value
35. $f(x) = e^x$	$x = 3.2$
36. $f(x) = 1.5e^{x/2}$	$x = 240$
37. $f(x) = 5000e^{0.06x}$	$x = 6$
38. $f(x) = 250e^{0.05x}$	$x = 20$

Graphing a Natural Exponential Function In Exercises 39–44, use a graphing utility to construct a table of values for the function. Then sketch the graph of the function.

39. $f(x) = e^x$
40. $f(x) = e^{-x}$
41. $f(x) = 3e^{x+4}$
42. $f(x) = 2e^{-0.5x}$
43. $f(x) = 2e^{x-2} + 4$
44. $f(x) = 2 + e^{x-5}$

Graphing a Natural Exponential Function In Exercises 45–50, use a graphing utility to graph the exponential function.

45. $y = 1.08e^{-5x}$
46. $y = 1.08e^{5x}$
47. $s(t) = 2e^{0.12t}$
48. $s(t) = 3e^{-0.2t}$
49. $g(x) = 1 + e^{-x}$
50. $h(x) = e^{x-2}$

Using the One-to-One Property In Exercises 51–54, use the One-to-One Property to solve the equation for x.

51. $e^{3x+2} = e^3$
52. $e^{2x-1} = e^4$
53. $e^{x^2-3} = e^{2x}$
54. $e^{x^2+6} = e^{5x}$

Compound Interest In Exercises 55–58, complete the table to determine the balance A for P dollars invested at rate r for t years and compounded n times per year.

n	1	2	4	12	365	Continuous
A						

55. $P = \$1500$, $r = 2\%$, $t = 10$ years
56. $P = \$2500$, $r = 3.5\%$, $t = 10$ years
57. $P = \$2500$, $r = 4\%$, $t = 20$ years
58. $P = \$1000$, $r = 6\%$, $t = 40$ years

Compound Interest In Exercises 59–62, complete the table to determine the balance A for \$12,000 invested at rate r for t years, compounded continuously.

t	10	20	30	40	50
A					

59. $r = 4\%$
60. $r = 6\%$
61. $r = 6.5\%$
62. $r = 3.5\%$

63. Trust Fund On the day of a child's birth, a parent deposits \$30,000 in a trust fund that pays 5% interest, compounded continuously. Determine the balance in this account on the child's 25th birthday.

64. Trust Fund A philanthropist deposits \$5000 in a trust fund that pays 7.5% interest, compounded continuously. The balance will be given to the college from which the philanthropist graduated after the money has earned interest for 50 years. How much will the college receive?

65. Inflation Assuming that the annual rate of inflation averages 4% over the next 10 years, the approximate costs C of goods or services during any year in that decade will be modeled by $C(t) = P(1.04)^t$, where t is the time in years and P is the present cost. The price of an oil change for your car is presently \$23.95. Estimate the price 10 years from now.

66. Computer Virus The number V of computers infected by a virus increases according to the model $V(t) = 100e^{4.6052t}$, where t is the time in hours. Find the number of computers infected after (a) 1 hour, (b) 1.5 hours, and (c) 2 hours.

67. Population Growth The projected populations of the United States for the years 2020 through 2050 can be modeled by $P = 290.323e^{0.0083t}$, where P is the population (in millions) and t is the time (in years), with $t = 20$ corresponding to 2020. *(Source: U.S. Census Bureau)*

(a) Use a graphing utility to graph the function for the years 2020 through 2050.

(b) Use the *table* feature of the graphing utility to create a table of values for the same time period as in part (a).

(c) According to the model, during what year will the population of the United States exceed 400 million?

68. Population The populations P (in millions) of Italy from 2000 through 2012 can be approximated by the model $P = 57.563e^{0.0052t}$, where t represents the year, with $t = 0$ corresponding to 2000. *(Source: U.S. Census Bureau, International Data Base)*

(a) According to the model, is the population of Italy increasing or decreasing? Explain.

(b) Find the populations of Italy in 2000 and 2012.

(c) Use the model to predict the populations of Italy in 2020 and 2025.

69. Radioactive Decay Let Q represent a mass of radioactive plutonium (^{239}Pu) (in grams), whose half-life is 24,100 years. The quantity of plutonium present after t years is $Q = 16\left(\frac{1}{2}\right)^{t/24,100}$.

(a) Determine the initial quantity (when $t = 0$).

(b) Determine the quantity present after 75,000 years.

(c) Use a graphing utility to graph the function over the interval $t = 0$ to $t = 150,000$.

70. Radioactive Decay Let Q represent a mass of carbon 14 (^{14}C) (in grams), whose half-life is 5715 years. The quantity of carbon 14 present after t years is $Q = 10\left(\frac{1}{2}\right)^{t/5715}$.

(a) Determine the initial quantity (when $t = 0$).

(b) Determine the quantity present after 2000 years.

(c) Sketch the graph of this function over the interval $t = 0$ to $t = 10,000$.

71. Depreciation After t years, the value of a wheelchair conversion van that originally cost \$49,810 depreciates so that each year it is worth $\frac{7}{8}$ of its value for the previous year.

(a) Find a model for $V(t)$, the value of the van after t years.

(b) Determine the value of the van 4 years after it was purchased.

72. Drug Concentration

Immediately following an injection, the concentration of a drug in the bloodstream is 300 milligrams per milliliter. After t hours, the concentration is 75% of the level of the previous hour.

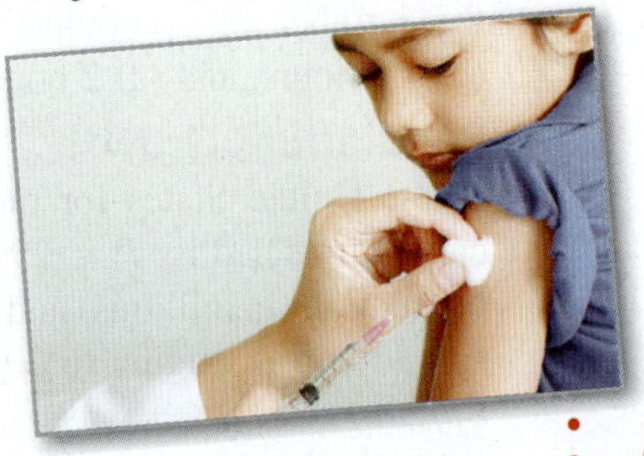

(a) Find a model for $C(t)$, the concentration of the drug after t hours.

(b) Determine the concentration of the drug after 8 hours.

Exploration

True or False? **In Exercises 73 and 74, determine whether the statement is true or false. Justify your answer.**

73. The line $y = -2$ is an asymptote for the graph of $f(x) = 10^x - 2$.

74. $e = \dfrac{271{,}801}{99{,}990}$

Think About It **In Exercises 75–78, use properties of exponents to determine which functions (if any) are the same.**

75. $f(x) = 3^{x-2}$
$g(x) = 3^x - 9$
$h(x) = \frac{1}{9}(3^x)$

76. $f(x) = 4^x + 12$
$g(x) = 2^{2x+6}$
$h(x) = 64(4^x)$

77. $f(x) = 16(4^{-x})$
$g(x) = \left(\frac{1}{4}\right)^{x-2}$
$h(x) = 16(2^{-2x})$

78. $f(x) = e^{-x} + 3$
$g(x) = e^{3-x}$
$h(x) = -e^{x-3}$

79. Solving Inequalities Graph the functions $y = 3^x$ and $y = 4^x$ and use the graphs to solve each inequality.

(a) $4^x < 3^x$ (b) $4^x > 3^x$

80. Graphical Analysis Use a graphing utility to graph each function. Use the graph to find where the function is increasing and decreasing, and approximate any relative maximum or minimum values.

(a) $f(x) = x^2e^{-x}$ (b) $g(x) = x2^{3-x}$

81. Graphical Analysis Use a graphing utility to graph $y_1 = (1 + 1/x)^x$ and $y_2 = e$ in the same viewing window. Using the *trace* feature, explain what happens to the graph of y_1 as x increases.

82. Graphical Analysis Use a graphing utility to graph

$$f(x) = \left(1 + \frac{0.5}{x}\right)^x \quad \text{and} \quad g(x) = e^{0.5}$$

in the same viewing window. What is the relationship between f and g as x increases and decreases without bound?

83. Graphical Analysis Use a graphing utility to graph each pair of functions in the same viewing window. Describe any similarities and differences in the graphs.

(a) $y_1 = 2^x, y_2 = x^2$

(b) $y_1 = 3^x, y_2 = x^3$

84. HOW DO YOU SEE IT? The figure shows the graphs of $y = 2^x$, $y = e^x$, $y = 10^x$, $y = 2^{-x}$, $y = e^{-x}$, and $y = 10^{-x}$. Match each function with its graph. [The graphs are labeled (a) through (f).] Explain your reasoning.

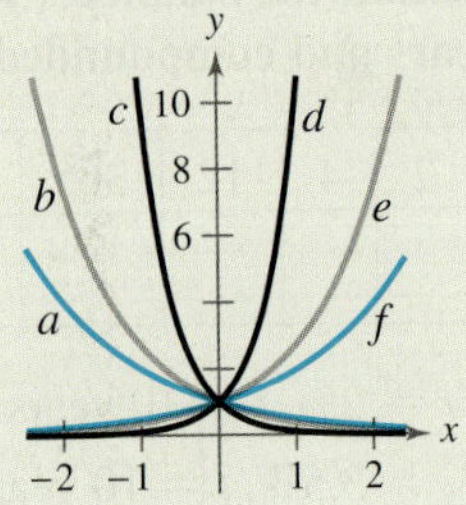

85. Think About It Which functions are exponential?

(a) $3x$

(b) $3x^2$

(c) 3^x

(d) 2^{-x}

86. Compound Interest Use the formula

$$A = P\left(1 + \frac{r}{n}\right)^{nt}$$

to calculate the balance of an investment when $P = \$3000$, $r = 6\%$, and $t = 10$ years, and compounding is done (a) by the day, (b) by the hour, (c) by the minute, and (d) by the second. Does increasing the number of compoundings per year result in unlimited growth of the balance? Explain.

Project: Population per Square Mile To work an extended application analyzing the population per square mile of the United States, visit this text's website at *LarsonPrecalculus.com*. *(Source: U.S. Census Bureau)*

Sura Nualpradid/Shutterstock.com

3.2 Logarithmic Functions and Their Graphs

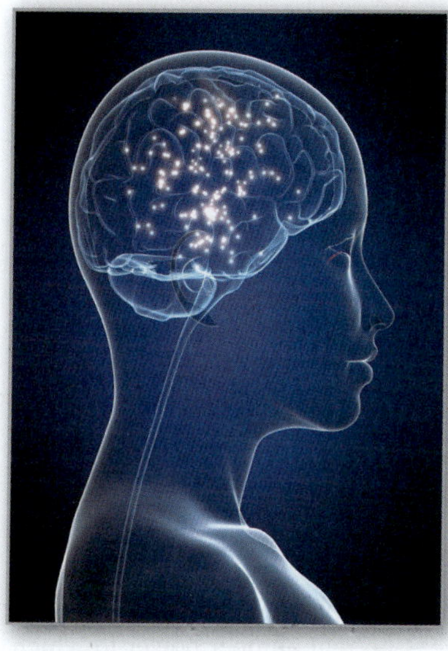

Logarithmic functions can often model scientific observations. For instance, Exercise 81 on page 220 uses a logarithmic function to model human memory.

- **Recognize and evaluate logarithmic functions with base *a*.**
- **Graph logarithmic functions.**
- **Recognize, evaluate, and graph natural logarithmic functions.**
- **Use logarithmic functions to model and solve real-life problems.**

Logarithmic Functions

In Section 1.9, you studied the concept of an inverse function. There, you learned that when a function is one-to-one—that is, when the function has the property that no horizontal line intersects the graph of the function more than once—the function must have an inverse function. By looking back at the graphs of the exponential functions introduced in Section 3.1, you will see that every function of the form $f(x) = a^x$ passes the Horizontal Line Test and therefore must have an inverse function. This inverse function is called the **logarithmic function with base *a*.**

Definition of Logarithmic Function with Base *a*

For $x > 0$, $a > 0$, and $a \neq 1$,

$$y = \log_a x \text{ if and only if } x = a^y.$$

The function

$$f(x) = \log_a x \qquad \text{Read as "log base } a \text{ of } x."$$

is called the **logarithmic function with base *a*.**

The equations $y = \log_a x$ and $x = a^y$ are equivalent. The first equation is in logarithmic form and the second is in exponential form. For example, $2 = \log_3 9$ is equivalent to $9 = 3^2$, and $5^3 = 125$ is equivalent to $\log_5 125 = 3$.

When evaluating logarithms, remember that *a logarithm is an exponent*. This means that $\log_a x$ is the exponent to which a must be raised to obtain x. For instance, $\log_2 8 = 3$ because 2 raised to the third power is 8.

EXAMPLE 1 Evaluating Logarithms

Evaluate each logarithm at the indicated value of x.

a. $f(x) = \log_2 x, \quad x = 32$ **b.** $f(x) = \log_3 x, \quad x = 1$

c. $f(x) = \log_4 x, \quad x = 2$ **d.** $f(x) = \log_{10} x, \quad x = \frac{1}{100}$

Solution

a. $f(32) = \log_2 32 = 5$ because $2^5 = 32$.

b. $f(1) = \log_3 1 = 0$ because $3^0 = 1$.

c. $f(2) = \log_4 2 = \frac{1}{2}$ because $4^{1/2} = \sqrt{4} = 2$.

d. $f\left(\frac{1}{100}\right) = \log_{10} \frac{1}{100} = -2$ because $10^{-2} = \frac{1}{10^2} = \frac{1}{100}$.

✓ ***Checkpoint*** *Audio-video solution in English & Spanish at LarsonPrecalculus.com.*

Evaluate each logarithm at the indicated value of x.

a. $f(x) = \log_6 x, \ x = 1$ **b.** $f(x) = \log_5 x, \ x = \frac{1}{125}$ **c.** $f(x) = \log_{10} x, \ x = 10{,}000$

Sebastian Kaulitzki/Shutterstock.com

The logarithmic function with base 10 is called the **common logarithmic function.** It is denoted by $\log_{10}$ or simply by log. On most calculators, this function is denoted by LOG. Example 2 shows how to use a calculator to evaluate common logarithmic functions. You will learn how to use a calculator to calculate logarithms to any base in the next section.

EXAMPLE 2 Evaluating Common Logarithms on a Calculator

Use a calculator to evaluate the function $f(x) = \log x$ at each value of x.

a. $x = 10$ **b.** $x = \frac{1}{3}$ **c.** $x = 2.5$ **d.** $x = -2$

Solution

Function Value	Graphing Calculator Keystrokes	Display
a. $f(10) = \log 10$	LOG 10 ENTER	1
b. $f\left(\frac{1}{3}\right) = \log \frac{1}{3}$	LOG (1 ÷ 3) ENTER	-0.4771213
c. $f(2.5) = \log 2.5$	LOG 2.5 ENTER	0.3979400
d. $f(-2) = \log(-2)$	LOG (−) 2 ENTER	ERROR

Note that the calculator displays an error message (or a complex number) when you try to evaluate $\log(-2)$. The reason for this is that there is no real number power to which 10 can be raised to obtain -2.

✓ **Checkpoint** *Audio-video solution in English & Spanish at LarsonPrecalculus.com.*

Use a calculator to evaluate the function $f(x) = \log x$ at each value of x.

a. $x = 275$ **b.** $x = 0.275$ **c.** $x = -\frac{1}{2}$ **d.** $x = \frac{1}{2}$

The following properties follow directly from the definition of the logarithmic function with base a.

Properties of Logarithms

1. $\log_a 1 = 0$ because $a^0 = 1$.
2. $\log_a a = 1$ because $a^1 = a$.
3. $\log_a a^x = x$ and $a^{\log_a x} = x$ Inverse Properties
4. If $\log_a x = \log_a y$, then $x = y$. One-to-One Property

EXAMPLE 3 Using Properties of Logarithms

a. Simplify $\log_4 1$. **b.** Simplify $\log_{\sqrt{7}} \sqrt{7}$. **c.** Simplify $6^{\log_6 20}$.

Solution

a. Using Property 1, $\log_4 1 = 0$.

b. Using Property 2, $\log_{\sqrt{7}} \sqrt{7} = 1$.

c. Using the Inverse Property (Property 3), $6^{\log_6 20} = 20$.

✓ **Checkpoint** *Audio-video solution in English & Spanish at LarsonPrecalculus.com.*

a. Simplify $\log_9 9$. **b.** Simplify $20^{\log_{20} 3}$. **c.** Simplify $\log_{\sqrt{3}} 1$.

You can use the One-to-One Property (Property 4) to solve simple logarithmic equations, as shown in Example 4.

EXAMPLE 4 **Using the One-to-One Property**

a. $\log_3 x = \log_3 12$ Original equation

$x = 12$ One-to-One Property

b. $\log(2x + 1) = \log 3x \Rightarrow 2x + 1 = 3x \Rightarrow 1 = x$

c. $\log_4(x^2 - 6) = \log_4 10 \Rightarrow x^2 - 6 = 10 \Rightarrow x^2 = 16 \Rightarrow x = \pm 4$

✓ ***Checkpoint*** *Audio-video solution in English & Spanish at LarsonPrecalculus.com.*

Solve $\log_5(x^2 + 3) = \log_5 12$ for x.

Graphs of Logarithmic Functions

To sketch the graph of $y = \log_a x$, use the fact that the graphs of inverse functions are reflections of each other in the line $y = x$.

EXAMPLE 5 **Graphs of Exponential and Logarithmic Functions**

In the same coordinate plane, sketch the graph of each function.

a. $f(x) = 2^x$ **b.** $g(x) = \log_2 x$

Solution

a. For $f(x) = 2^x$, construct a table of values. By plotting these points and connecting them with a smooth curve, you obtain the graph shown in Figure 3.10.

x	-2	-1	0	1	2	3
$f(x) = 2^x$	$\frac{1}{4}$	$\frac{1}{2}$	1	2	4	8

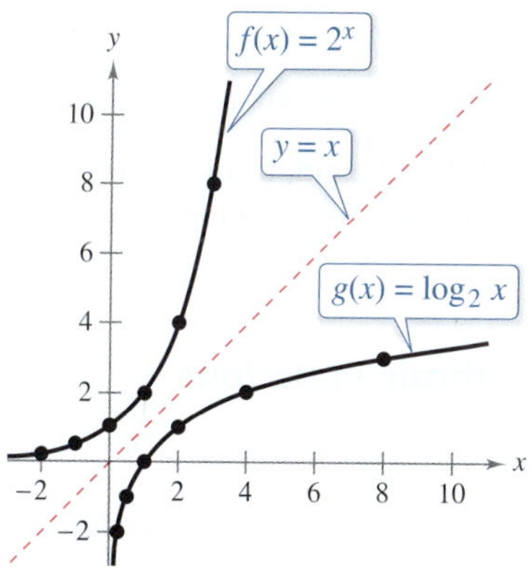

Figure 3.10

b. Because $g(x) = \log_2 x$ is the inverse function of $f(x) = 2^x$, the graph of g is obtained by plotting the points $(f(x), x)$ and connecting them with a smooth curve. The graph of g is a reflection of the graph of f in the line $y = x$, as shown in Figure 3.10.

✓ ***Checkpoint*** *Audio-video solution in English & Spanish at LarsonPrecalculus.com.*

In the same coordinate plane, sketch the graphs of (a) $f(x) = 8^x$ and (b) $g(x) = \log_8 x$.

EXAMPLE 6 **Sketching the Graph of a Logarithmic Function**

Sketch the graph of $f(x) = \log x$. Identify the vertical asymptote.

Solution Begin by constructing a table of values. Note that some of the values can be obtained without a calculator by using the properties of logarithms. Others require a calculator. Next, plot the points and connect them with a smooth curve, as shown in Figure 3.11. The vertical asymptote is $x = 0$ (y-axis).

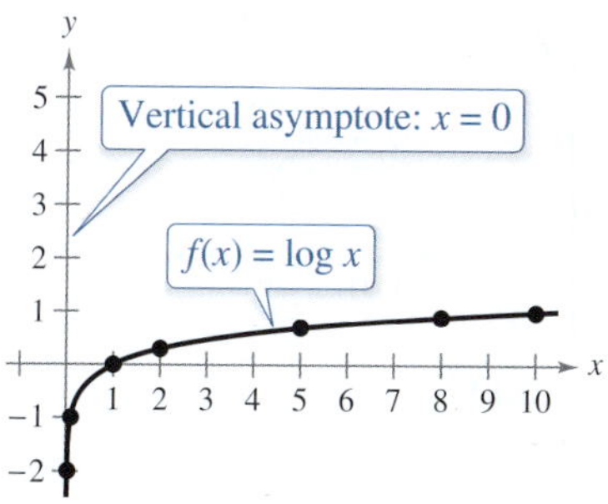

Figure 3.11

	Without calculator				With calculator		
x	$\frac{1}{100}$	$\frac{1}{10}$	1	10	2	5	8
$f(x) = \log x$	-2	-1	0	1	0.301	0.699	0.903

✓ ***Checkpoint*** *Audio-video solution in English & Spanish at LarsonPrecalculus.com.*

Sketch the graph of $f(x) = \log_9 x$. Identify the vertical asymptote.

The nature of the graph in Figure 3.11 is typical of functions of the form $f(x) = \log_a x, a > 1$. They have one x-intercept and one vertical asymptote. Notice how slowly the graph rises for $x > 1$. The following summarizes the basic characteristics of logarithmic graphs.

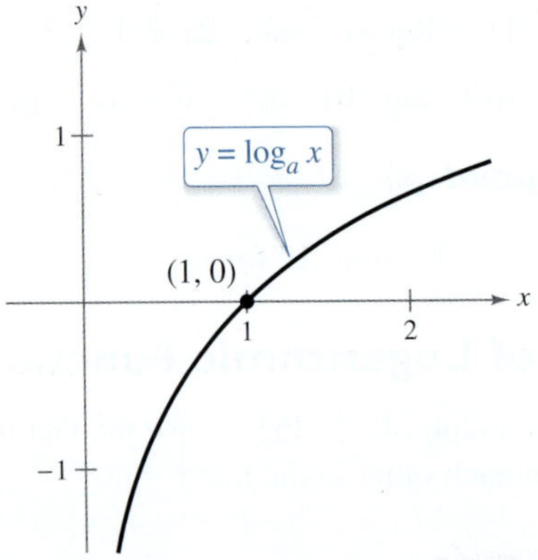

Graph of $y = \log_a x, a > 1$

- Domain: $(0, \infty)$
- Range: $(-\infty, \infty)$
- x-intercept: $(1, 0)$
- Increasing
- One-to-one, therefore has an inverse function
- y-axis is a vertical asymptote ($\log_a x \to -\infty$ as $x \to 0^+$).
- Continuous
- Reflection of graph of $y = a^x$ in the line $y = x$

The basic characteristics of the graph of $f(x) = a^x$ are shown below to illustrate the inverse relation between $f(x) = a^x$ and $g(x) = \log_a x$.

- Domain: $(-\infty, \infty)$
- Range: $(0, \infty)$
- y-intercept: $(0, 1)$
- x-axis is a horizontal asymptote ($a^x \to 0$ as $x \to -\infty$).

The next example uses the graph of $y = \log_a x$ to sketch the graphs of functions of the form $f(x) = b \pm \log_a(x + c)$. Notice how a horizontal shift of the graph results in a horizontal shift of the vertical asymptote.

EXAMPLE 7 Shifting Graphs of Logarithmic Functions

The graph of each of the functions is similar to the graph of $f(x) = \log x$.

a. Because $g(x) = \log(x - 1) = f(x - 1)$, the graph of g can be obtained by shifting the graph of f one unit to the right, as shown in Figure 3.12.

b. Because $h(x) = 2 + \log x = 2 + f(x)$, the graph of h can be obtained by shifting the graph of f two units up, as shown in Figure 3.13.

REMARK Use your understanding of transformations to identify vertical asymptotes of logarithmic functions. For instance, in Example 7(a), the graph of $g(x) = f(x - 1)$ shifts the graph of $f(x)$ one unit to the right. So, the vertical asymptote of the graph of $g(x)$ is $x = 1$, one unit to the right of the vertical asymptote of the graph of $f(x)$.

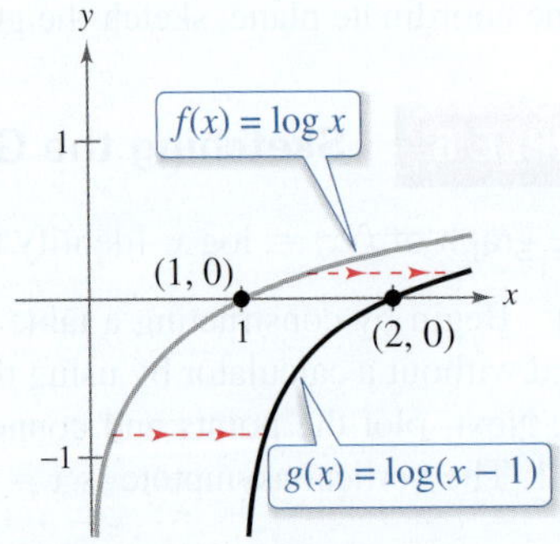

Figure 3.12

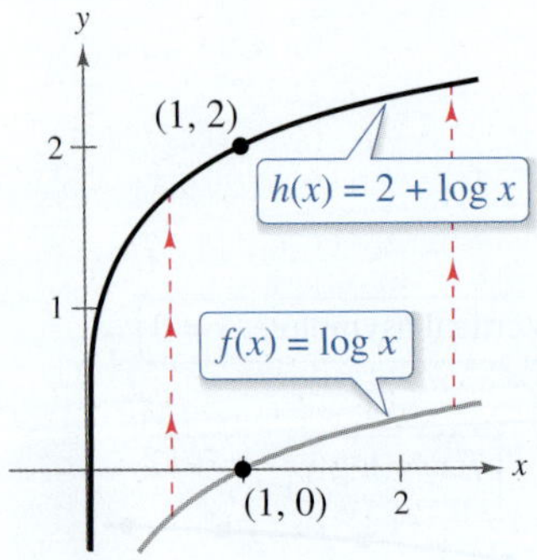

Figure 3.13

Checkpoint Audio-video solution in English & Spanish at LarsonPrecalculus.com.

Use the graph of $f(x) = \log_3 x$ to sketch the graph of each function.

a. $g(x) = -1 + \log_3 x$

b. $h(x) = \log_3(x + 3)$

ALGEBRA HELP You can review the techniques for shifting, reflecting, and stretching graphs in Section 1.7.

The Natural Logarithmic Function

By looking back at the graph of the natural exponential function introduced on page 204 in Section 3.1, you will see that $f(x) = e^x$ is one-to-one and so has an inverse function. This inverse function is called the **natural logarithmic function** and is denoted by the special symbol $\ln x$, read as "the natural log of x" or "el en of x." Note that the natural logarithm is written without a base. The base is understood to be e.

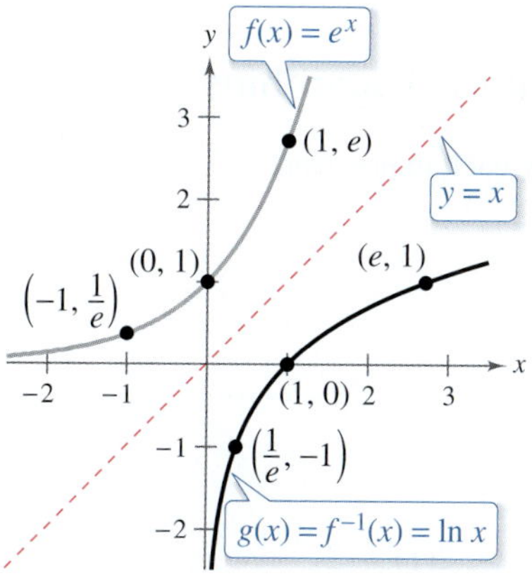

Reflection of graph of $f(x) = e^x$ in the line $y = x$

Figure 3.14

> **The Natural Logarithmic Function**
>
> The function defined by
>
> $$f(x) = \log_e x = \ln x, \quad x > 0$$
>
> is called the **natural logarithmic function.**

The above definition implies that the natural logarithmic function and the natural exponential function are inverse functions of each other. So, every logarithmic equation can be written in an equivalent exponential form, and every exponential equation can be written in an equivalent logarithmic form. That is, $y = \ln x$ and $x = e^y$ are equivalent equations.

Because the functions $f(x) = e^x$ and $g(x) = \ln x$ are inverse functions of each other, their graphs are reflections of each other in the line $y = x$. Figure 3.14 illustrates this reflective property.

On most calculators, [LN] denotes the natural logarithm, as illustrated in Example 8.

EXAMPLE 8 Evaluating the Natural Logarithmic Function

Use a calculator to evaluate the function $f(x) = \ln x$ at each value of x.

a. $x = 2$

b. $x = 0.3$

c. $x = -1$

d. $x = 1 + \sqrt{2}$

Solution

Function Value	Graphing Calculator Keystrokes	Display
a. $f(2) = \ln 2$	[LN] 2 [ENTER]	0.6931472
b. $f(0.3) = \ln 0.3$	[LN] .3 [ENTER]	–1.2039728
c. $f(-1) = \ln(-1)$	[LN] [(–)] 1 [ENTER]	ERROR
d. $f(1 + \sqrt{2}) = \ln(1 + \sqrt{2})$	[LN] [(] 1 [+] [√] 2 [)] [ENTER]	0.8813736

Checkpoint Audio-video solution in English & Spanish at LarsonPrecalculus.com.

Use a calculator to evaluate the function $f(x) = \ln x$ at each value of x.

a. $x = 0.01$ **b.** $x = 4$ **c.** $x = \sqrt{3} + 2$ **d.** $x = \sqrt{3} - 2$

REMARK Notice that as with every other logarithmic function, the domain of the natural logarithmic function is the set of *positive real numbers*—be sure you see that $\ln x$ is not defined for zero or for negative numbers.

In Example 8, be sure you see that $\ln(-1)$ gives an error message on most calculators. (Some calculators may display a complex number.) This occurs because the domain of $\ln x$ is the set of positive real numbers (see Figure 3.14). So, $\ln(-1)$ is undefined.

The four properties of logarithms listed on page 212 are also valid for natural logarithms.

Properties of Natural Logarithms

1. $\ln 1 = 0$ because $e^0 = 1$.
2. $\ln e = 1$ because $e^1 = e$.
3. $\ln e^x = x$ and $e^{\ln x} = x$ — Inverse Properties
4. If $\ln x = \ln y$, then $x = y$. — One-to-One Property

EXAMPLE 9 Using Properties of Natural Logarithms

Use the properties of natural logarithms to simplify each expression.

a. $\ln \frac{1}{e}$ **b.** $e^{\ln 5}$ **c.** $\frac{\ln 1}{3}$ **d.** $2 \ln e$

Solution

a. $\ln \frac{1}{e} = \ln e^{-1} = -1$ Inverse Property

b. $e^{\ln 5} = 5$ Inverse Property

c. $\frac{\ln 1}{3} = \frac{0}{3} = 0$ Property 1

d. $2 \ln e = 2(1) = 2$ Property 2

✓ **Checkpoint** Audio-video solution in English & Spanish at LarsonPrecalculus.com.

Use the properties of natural logarithms to simplify each expression.

a. $\ln e^{1/3}$ **b.** $5 \ln 1$ **c.** $\frac{3}{4} \ln e$ **d.** $e^{\ln 7}$

EXAMPLE 10 Finding the Domains of Logarithmic Functions

Find the domain of each function.

a. $f(x) = \ln(x - 2)$ **b.** $g(x) = \ln(2 - x)$ **c.** $h(x) = \ln x^2$

Solution

a. Because $\ln(x - 2)$ is defined only when $x - 2 > 0$, it follows that the domain of f is $(2, \infty)$. Figure 3.15 shows the graph of f.

b. Because $\ln(2 - x)$ is defined only when $2 - x > 0$, it follows that the domain of g is $(-\infty, 2)$. Figure 3.16 shows the graph of g.

c. Because $\ln x^2$ is defined only when $x^2 > 0$, it follows that the domain of h is all real numbers except $x = 0$. Figure 3.17 shows the graph of h.

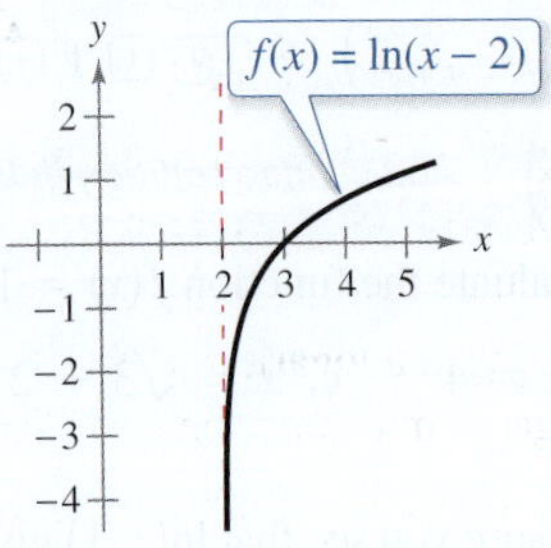

Figure 3.15

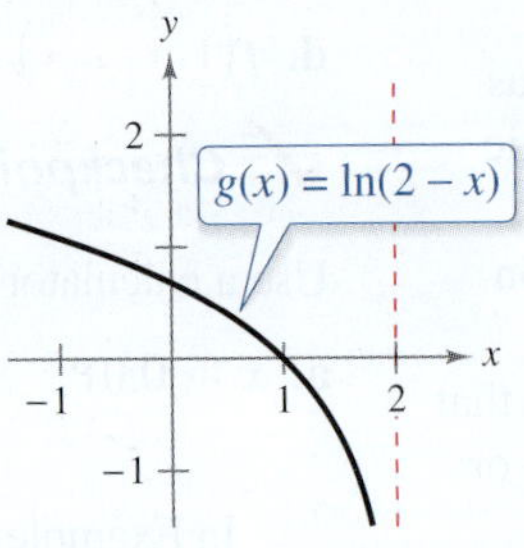

Figure 3.16

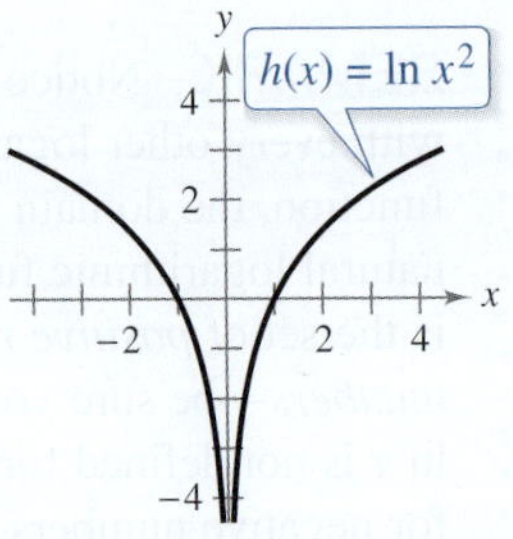

Figure 3.17

✓ **Checkpoint** Audio-video solution in English & Spanish at LarsonPrecalculus.com.

Find the domain of $f(x) = \ln(x + 3)$.

Application

EXAMPLE 11 **Human Memory Model**

Students participating in a psychology experiment attended several lectures on a subject and took an exam. Every month for a year after the exam, the students took a retest to see how much of the material they remembered. The average scores for the group are given by the *human memory model* $f(t) = 75 - 6\ln(t + 1)$, $0 \le t \le 12$, where t is the time in months.

a. What was the average score on the original exam $(t = 0)$?

b. What was the average score at the end of $t = 2$ months?

c. What was the average score at the end of $t = 6$ months?

Algebraic Solution

a. The original average score was

$f(0) = 75 - 6\ln(0 + 1)$	Substitute 0 for t.
$= 75 - 6\ln 1$	Simplify.
$= 75 - 6(0)$	Property of natural logarithms
$= 75.$	Solution

b. After 2 months, the average score was

$f(2) = 75 - 6\ln(2 + 1)$	Substitute 2 for t.
$= 75 - 6\ln 3$	Simplify.
$\approx 75 - 6(1.0986)$	Use a calculator.
$\approx 68.41.$	Solution

c. After 6 months, the average score was

$f(6) = 75 - 6\ln(6 + 1)$	Substitute 6 for t.
$= 75 - 6\ln 7$	Simplify.
$\approx 75 - 6(1.9459)$	Use a calculator.
$\approx 63.32.$	Solution

Graphical Solution

a.

When $t = 0$, $f(0) = 75$. So, the original average score was 75.

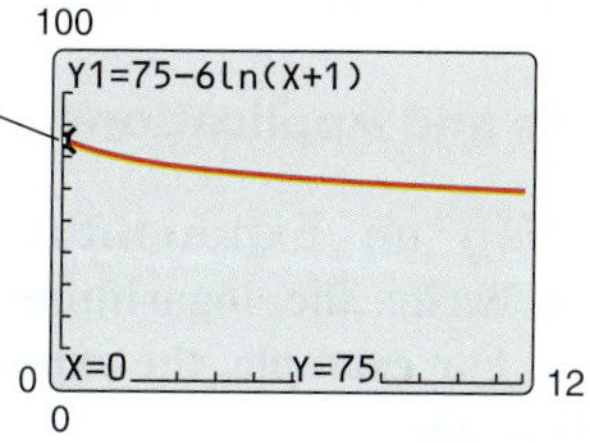

b.

When $t = 2$, $f(2) \approx 68.41$. So, the average score after 2 months was about 68.41.

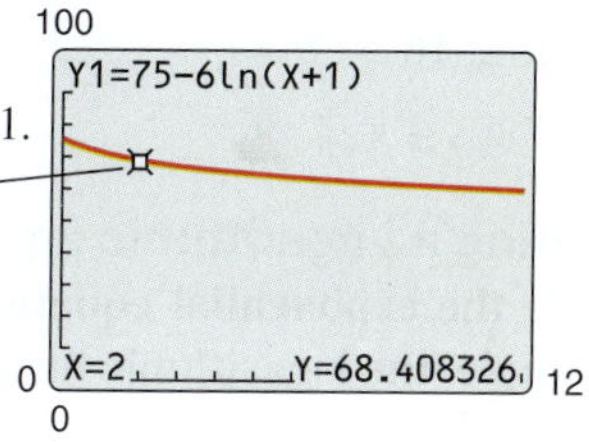

c.

When $t = 6$, $f(6) \approx 63.32$. So, the average score after 6 months was about 63.32.

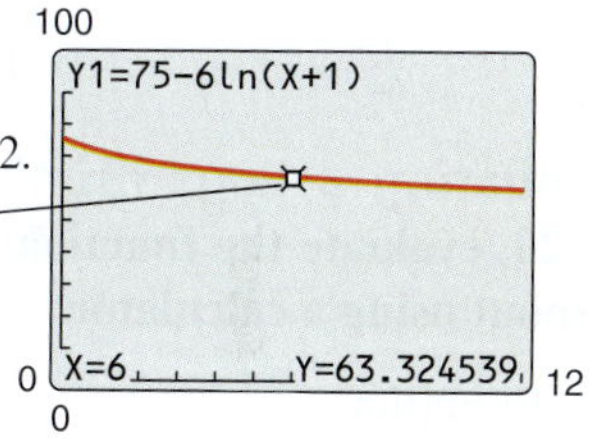

✓ ***Checkpoint*** *Audio-video solution in English & Spanish at LarsonPrecalculus.com.*

In Example 11, find the average score at the end of (a) $t = 1$ month, (b) $t = 9$ months, and (c) $t = 12$ months.

Summarize **(Section 3.2)**

1. State the definition of a logarithmic function with base a *(page 211)* and make a list of the properties of logarithms *(page 212)*. For examples of evaluating logarithmic functions and using the properties of logarithms, see Examples 1–4.
2. Explain how to graph a logarithmic function *(pages 213 and 214)*. For examples of graphing logarithmic functions, see Examples 5–7.
3. State the definition of the natural logarithmic function *(page 215)* and make a list of the properties of natural logarithms *(page 216)*. For examples of evaluating natural logarithmic functions and using the properties of natural logarithms, see Examples 8 and 9.
4. Describe an example of how to use a logarithmic function to model and solve a real-life problem *(page 217, Example 11)*.

3.2 Exercises

See CalcChat.com for tutorial help and worked-out solutions to odd-numbered exercises.

Vocabulary: Fill in the blanks.

1. The inverse function of the exponential function $f(x) = a^x$ is called the ________ function with base a.
2. The common logarithmic function has base ________ .
3. The logarithmic function $f(x) = \ln x$ is called the ________ logarithmic function and has base ________.
4. The Inverse Properties of logarithms state that $\log_a a^x = x$ and ________.
5. The One-to-One Property of natural logarithms states that if $\ln x = \ln y$, then ________.
6. The domain of the natural logarithmic function is the set of ________ ________ ________ .

Skills and Applications

Writing an Exponential Equation **In Exercises 7–10, write the logarithmic equation in exponential form. For example, the exponential form of $\log_5 25 = 2$ is $5^2 = 25$.**

7. $\log_4 16 = 2$
8. $\log_9 \frac{1}{81} = -2$
9. $\log_{32} 4 = \frac{2}{5}$
10. $\log_{64} 8 = \frac{1}{2}$

Writing a Logarithmic Equation **In Exercises 11–14, write the exponential equation in logarithmic form. For example, the logarithmic form of $2^3 = 8$ is $\log_2 8 = 3$.**

11. $5^3 = 125$
12. $9^{3/2} = 27$
13. $4^{-3} = \frac{1}{64}$
14. $24^0 = 1$

Evaluating a Logarithmic Function **In Exercises 15–20, evaluate the function at the indicated value of x without using a calculator.**

	Function	Value
15.	$f(x) = \log_2 x$	$x = 64$
16.	$f(x) = \log_{25} x$	$x = 5$
17.	$f(x) = \log_8 x$	$x = 1$
18.	$f(x) = \log x$	$x = 10$
19.	$g(x) = \log_a x$	$x = a^2$
20.	$g(x) = \log_b x$	$x = b^{-3}$

Evaluating a Common Logarithm on a Calculator **In Exercises 21–24, use a calculator to evaluate $f(x) = \log x$ at the indicated value of x. Round your result to three decimal places.**

21. $x = \frac{7}{8}$
22. $x = \frac{1}{500}$
23. $x = 12.5$
24. $x = 96.75$

Using Properties of Logarithms **In Exercises 25–28, use the properties of logarithms to simplify the expression.**

25. $\log_{11} 11^7$
26. $\log_{3.2} 1$
27. $\log_\pi \pi$
28. $9^{\log_9 15}$

Using the One-to-One Property **In Exercises 29–32, use the One-to-One Property to solve the equation for x.**

29. $\log_5(x + 1) = \log_5 6$
30. $\log_2(x - 3) = \log_2 9$
31. $\log(2x + 1) = \log 15$
32. $\log(5x + 3) = \log 12$

Graphs of Exponential and Logarithmic Functions **In Exercises 33–36, sketch the graphs of f and g in the same coordinate plane.**

33. $f(x) = 7^x$, $g(x) = \log_7 x$
34. $f(x) = 5^x$, $g(x) = \log_5 x$
35. $f(x) = 6^x$, $g(x) = \log_6 x$
36. $f(x) = 10^x$, $g(x) = \log x$

Matching a Logarithmic Function with Its Graph **In Exercises 37–40, use the graph of $g(x) = \log_3 x$ to match the given function with its graph. Then describe the relationship between the graphs of f and g. [The graphs are labeled (a), (b), (c), and (d).]**

(a)

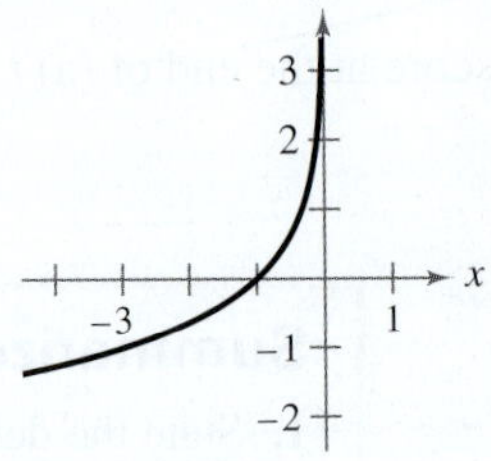

(b)

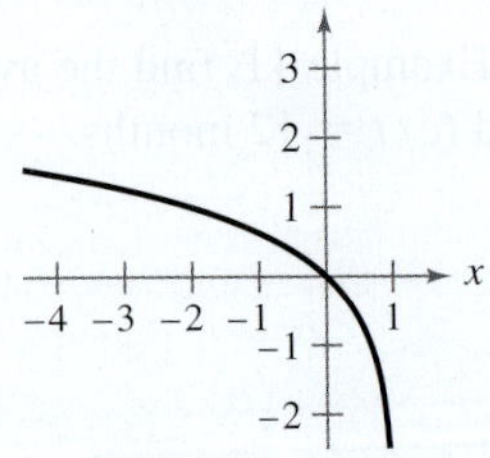

(c)

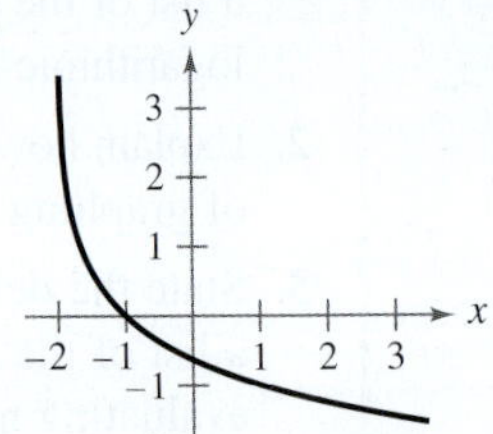

(d)

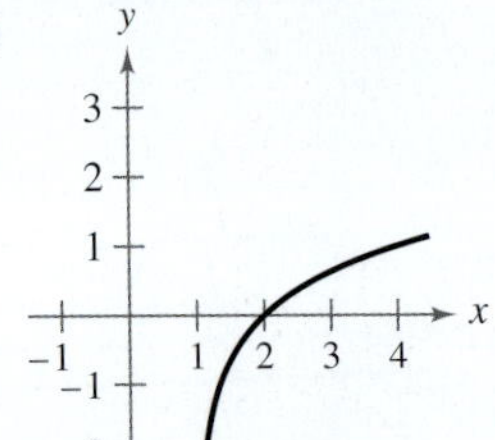

37. $f(x) = -\log_3(x + 2)$
38. $f(x) = \log_3(x - 1)$
39. $f(x) = \log_3(1 - x)$
40. $f(x) = -\log_3(-x)$

Sketching the Graph of a Logarithmic Function **In Exercises 41–48, find the domain, x-intercept, and vertical asymptote of the logarithmic function and sketch its graph.**

41. $f(x) = \log_4 x$
42. $g(x) = \log_6 x$
43. $y = -\log_3 x + 2$
44. $h(x) = \log_4(x - 3)$
45. $f(x) = -\log_6(x + 2)$
46. $y = \log_5(x - 1) + 4$
47. $y = \log\left(\frac{x}{7}\right)$
48. $y = \log(-x)$

Writing a Natural Exponential Equation **In Exercises 49–52, write the logarithmic equation in exponential form.**

49. $\ln \frac{1}{2} = -0.693 \ldots$
50. $\ln 7 = 1.945 \ldots$
51. $\ln 250 = 5.521 \ldots$
52. $\ln 1 = 0$

Writing a Natural Logarithmic Equation **In Exercises 53–56, write the exponential equation in logarithmic form.**

53. $e^2 = 7.3890 \ldots$
54. $e^{1/2} = 1.6487 \ldots$
55. $e^{-0.9} = 0.406 \ldots$
56. $e^{2x} = 3$

Evaluating a Logarithmic Function on a Calculator **In Exercises 57–60, use a calculator to evaluate the function at the indicated value of x. Round your result to three decimal places.**

Function	Value
57. $f(x) = \ln x$	$x = 18.42$
58. $f(x) = 3 \ln x$	$x = 0.74$
59. $g(x) = 8 \ln x$	$x = 0.05$
60. $g(x) = -\ln x$	$x = \frac{1}{2}$

Evaluating a Natural Logarithm **In Exercises 61–64, evaluate $g(x) = \ln x$ at the indicated value of x without using a calculator.**

61. $x = e^5$
62. $x = e^{-4}$
63. $x = e^{-5/6}$
64. $x = e^{-5/2}$

Graphing a Natural Logarithmic Function **In Exercises 65–68, find the domain, x-intercept, and vertical asymptote of the logarithmic function and sketch its graph.**

65. $f(x) = \ln(x - 4)$
66. $h(x) = \ln(x + 5)$
67. $g(x) = \ln(-x)$
68. $f(x) = \ln(3 - x)$

Graphing a Natural Logarithmic Function **In Exercises 69–72, use a graphing utility to graph the function. Be sure to use an appropriate viewing window.**

69. $f(x) = \ln(x - 1)$
70. $f(x) = \ln(x + 2)$
71. $f(x) = \ln x + 8$
72. $f(x) = 3 \ln x - 1$

Using the One-to-One Property **In Exercises 73–76, use the One-to-One Property to solve the equation for x.**

73. $\ln(x + 4) = \ln 12$
74. $\ln(x - 7) = \ln 7$
75. $\ln(x^2 - 2) = \ln 23$
76. $\ln(x^2 - x) = \ln 6$

77. **Monthly Payment** The model

$$t = 16.625 \ln\left(\frac{x}{x - 750}\right), \quad x > 750$$

approximates the length of a home mortgage of \$150,000 at 6% in terms of the monthly payment. In the model, t is the length of the mortgage in years and x is the monthly payment in dollars.

(a) Use the model to approximate the lengths of a \$150,000 mortgage at 6% when the monthly payment is \$897.72 and when the monthly payment is \$1659.24.

(b) Approximate the total amounts paid over the term of the mortgage with a monthly payment of \$897.72 and with a monthly payment of \$1659.24.

(c) Approximate the total interest charges for a monthly payment of \$897.72 and for a monthly payment of \$1659.24.

(d) What is the vertical asymptote for the model? Interpret its meaning in the context of the problem.

78. **Wireless Only** The percents P of households in the United States with wireless-only telephone service from 2005 through 2011 can be approximated by the model

$$P = -4.00 + 1.335t \ln t, \quad 5 \le t \le 11$$

where t represents the year, with $t = 5$ corresponding to 2005. *(Source: National Center for Health Statistics)*

(a) Complete the table.

t	5	6	7	8	9	10	11
P							

(b) Use a graphing utility to graph the function.

(c) Can the model be used to predict the percents of households with wireless-only telephone service beyond 2011? Explain.

79. **Population** The time t (in years) for the world population to double when it is increasing at a continuous rate of r is given by $t = (\ln 2)/r$.

(a) Complete the table and interpret your results.

r	0.005	0.010	0.015	0.020	0.025	0.030
t						

(b) Use a graphing utility to graph the function.

80. Compound Interest A principal P, invested at $5\frac{1}{2}\%$ and compounded continuously, increases to an amount K times the original principal after t years, where $t = (\ln K)/0.055$.

(a) Complete the table and interpret your results.

K	1	2	4	6	8	10	12
t							

(b) Sketch a graph of the function.

81. Human Memory Model

Students in a mathematics class took an exam and then took a retest monthly with an equivalent exam. The average scores for the class are given by the human memory model

$f(t) = 80 - 17 \log(t + 1), \quad 0 \le t \le 12$

where t is the time in months.

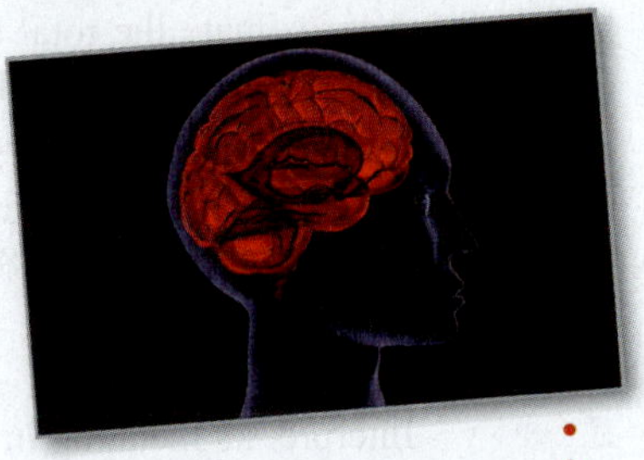

(a) Use a graphing utility to graph the model over the specified domain.

(b) What was the average score on the original exam $(t = 0)$?

(c) What was the average score after 4 months?

(d) What was the average score after 10 months?

82. Sound Intensity The relationship between the number of decibels β and the intensity of a sound I in watts per square meter is

$$\beta = 10 \log\left(\frac{I}{10^{-12}}\right).$$

(a) Determine the number of decibels of a sound with an intensity of 1 watt per square meter.

(b) Determine the number of decibels of a sound with an intensity of 10^{-2} watt per square meter.

(c) The intensity of the sound in part (a) is 100 times as great as that in part (b). Is the number of decibels 100 times as great? Explain.

Exploration

True or False? **In Exercises 83 and 84, determine whether the statement is true or false. Justify your answer.**

83. The graph of $f(x) = \log_6 x$ is a reflection of the graph of $g(x) = 6^x$ in the x-axis.

84. The graph of $f(x) = \log_3 x$ contains the point $(27, 3)$.

85. Graphical Analysis Use a graphing utility to graph f and g in the same viewing window and determine which is increasing at the greater rate as x approaches $+\infty$. What can you conclude about the rate of growth of the natural logarithmic function?

(a) $f(x) = \ln x, \quad g(x) = \sqrt{x}$

(b) $f(x) = \ln x, \quad g(x) = \sqrt[4]{x}$

86. Limit of a Function

(a) Complete the table for the function

$f(x) = (\ln x)/x.$

x	1	5	10	10^2	10^4	10^6
$f(x)$						

(b) Use the table in part (a) to determine what value $f(x)$ approaches as x increases without bound.

(c) Use a graphing utility to confirm the result of part (b).

87. Think About It A student obtained the following table of values by evaluating a function. Determine which of the statements may be true and which must be false.

x	1	2	8
y	0	1	3

(a) y is an exponential function of x.

(b) y is a logarithmic function of x.

(c) x is an exponential function of y.

(d) y is a linear function of x.

88. HOW DO YOU SEE IT? The figure shows the graphs of $f(x) = 3^x$ and $g(x) = \log_3 x$. [The graphs are labeled m and n.]

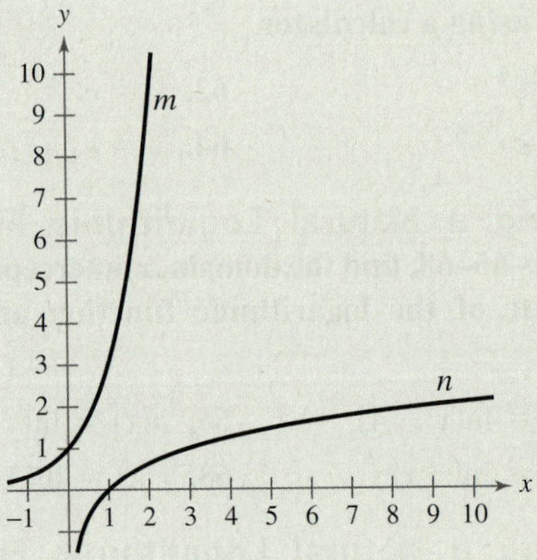

(a) Match each function with its graph.

(b) Given that $f(a) = b$, what is $g(b)$? Explain.

89. Writing Explain why $\log_a x$ is defined only for $0 < a < 1$ and $a > 1$.

YAKOBCHUK VASYL/Shutterstock.com

3.3 Properties of Logarithms

Logarithmic functions can help you model and solve real-life problems. For instance, Exercises 85–88 on page 226 use a logarithmic function to model the relationship between the number of decibels and the intensity of a sound.

- Use the change-of-base formula to rewrite and evaluate logarithmic expressions.
- Use properties of logarithms to evaluate or rewrite logarithmic expressions.
- Use properties of logarithms to expand or condense logarithmic expressions.
- Use logarithmic functions to model and solve real-life problems.

Change of Base

Most calculators have only two types of log keys, one for common logarithms (base 10) and one for natural logarithms (base e). Although common logarithms and natural logarithms are the most frequently used, you may occasionally need to evaluate logarithms with other bases. To do this, use the following **change-of-base formula.**

Change-of-Base Formula

Let a, b, and x be positive real numbers such that $a \neq 1$ and $b \neq 1$. Then $\log_a x$ can be converted to a different base as follows.

Base b	**Base 10**	**Base e**
$\log_a x = \dfrac{\log_b x}{\log_b a}$	$\log_a x = \dfrac{\log x}{\log a}$	$\log_a x = \dfrac{\ln x}{\ln a}$

One way to look at the change-of-base formula is that logarithms with base a are *constant multiples* of logarithms with base b. The constant multiplier is

$$\frac{1}{\log_b a}.$$

EXAMPLE 1 Changing Bases Using Common Logarithms

$$\log_4 25 = \frac{\log 25}{\log 4} \qquad \log_a x = \frac{\log x}{\log a}$$

$$\approx \frac{1.39794}{0.60206} \qquad \text{Use a calculator.}$$

$$\approx 2.3219 \qquad \text{Simplify.}$$

✓ **Checkpoint** *Audio-video solution in English & Spanish at LarsonPrecalculus.com.*

Evaluate $\log_2 12$ using the change-of-base formula and common logarithms.

EXAMPLE 2 Changing Bases Using Natural Logarithms

$$\log_4 25 = \frac{\ln 25}{\ln 4} \qquad \log_a x = \frac{\ln x}{\ln a}$$

$$\approx \frac{3.21888}{1.38629} \qquad \text{Use a calculator.}$$

$$\approx 2.3219 \qquad \text{Simplify.}$$

✓ **Checkpoint** *Audio-video solution in English & Spanish at LarsonPrecalculus.com.*

Evaluate $\log_2 12$ using the change-of-base formula and natural logarithms. ■

kentoh/Shutterstock.com

Properties of Logarithms

You know from the preceding section that the logarithmic function with base a is the *inverse function* of the exponential function with base a. So, it makes sense that the properties of exponents should have corresponding properties involving logarithms. For instance, the exponential property $a^u a^v = a^{u+v}$ has the corresponding logarithmic property $\log_a(uv) = \log_a u + \log_a v$.

REMARK There is no property that can be used to rewrite $\log_a(u \pm v)$. Specifically, $\log_a(u + v)$ is *not* equal to $\log_a u + \log_a v$.

Properties of Logarithms

Let a be a positive number such that $a \neq 1$, and let n be a real number. If u and v are positive real numbers, then the following properties are true.

	Logarithm with Base a	Natural Logarithm
1. Product Property:	$\log_a(uv) = \log_a u + \log_a v$	$\ln(uv) = \ln u + \ln v$
2. Quotient Property:	$\log_a \dfrac{u}{v} = \log_a u - \log_a v$	$\ln \dfrac{u}{v} = \ln u - \ln v$
3. Power Property:	$\log_a u^n = n \log_a u$	$\ln u^n = n \ln u$

For proofs of the properties listed above, see Proofs in Mathematics on page 258.

EXAMPLE 3 Using Properties of Logarithms

Write each logarithm in terms of ln 2 and ln 3.

a. $\ln 6$ **b.** $\ln \dfrac{2}{27}$

Solution

a. $\ln 6 = \ln(2 \cdot 3)$ Rewrite 6 as $2 \cdot 3$.

$= \ln 2 + \ln 3$ Product Property

b. $\ln \dfrac{2}{27} = \ln 2 - \ln 27$ Quotient Property

$= \ln 2 - \ln 3^3$ Rewrite 27 as 3^3.

$= \ln 2 - 3 \ln 3$ Power Property

✓ **Checkpoint** Audio-video solution in English & Spanish at LarsonPrecalculus.com.

Write each logarithm in terms of log 3 and log 5.

a. $\log 75$ **b.** $\log \dfrac{9}{125}$

HISTORICAL NOTE

John Napier, a Scottish mathematician, developed logarithms as a way to simplify tedious calculations. Beginning in 1594, Napier worked about 20 years on the development of logarithms. Napier only partially succeeded in his quest to simplify tedious calculations. Nonetheless, the development of logarithms was a step forward and received immediate recognition.

EXAMPLE 4 Using Properties of Logarithms

Find the exact value of $\log_5 \sqrt[3]{5}$ without using a calculator.

Solution

$$\log_5 \sqrt[3]{5} = \log_5 5^{1/3} = \tfrac{1}{3} \log_5 5 = \tfrac{1}{3}(1) = \tfrac{1}{3}$$

✓ **Checkpoint** Audio-video solution in English & Spanish at LarsonPrecalculus.com.

Find the exact value of $\ln e^6 - \ln e^2$ without using a calculator.

Mary Evans Picture Library

Rewriting Logarithmic Expressions

The properties of logarithms are useful for rewriting logarithmic expressions in forms that simplify the operations of algebra. This is true because these properties convert complicated products, quotients, and exponential forms into simpler sums, differences, and products, respectively.

EXAMPLE 5 Expanding Logarithmic Expressions

Expand each logarithmic expression.

a. $\log_4 5x^3y$ **b.** $\ln \dfrac{\sqrt{3x-5}}{7}$

Solution

a. $\log_4 5x^3y = \log_4 5 + \log_4 x^3 + \log_4 y$ Product Property

$= \log_4 5 + 3\log_4 x + \log_4 y$ Power Property

b. $\ln \dfrac{\sqrt{3x-5}}{7} = \ln \dfrac{(3x-5)^{1/2}}{7}$ Rewrite using rational exponent.

$= \ln(3x-5)^{1/2} - \ln 7$ Quotient Property

$= \dfrac{1}{2}\ln(3x-5) - \ln 7$ Power Property

▷ **ALGEBRA HELP** You can review rewriting radicals and rational exponents in Appendix A.2.

Checkpoint Audio-video solution in English & Spanish at LarsonPrecalculus.com.

Expand the expression $\log_3 \dfrac{4x^2}{\sqrt{y}}$.

Example 5 uses the properties of logarithms to *expand* logarithmic expressions. Example 6 reverses this procedure and uses the properties of logarithms to *condense* logarithmic expressions.

EXAMPLE 6 Condensing Logarithmic Expressions

Condense each logarithmic expression.

a. $\frac{1}{2}\log x + 3\log(x+1)$ **b.** $2\ln(x+2) - \ln x$ **c.** $\frac{1}{3}[\log_2 x + \log_2(x+1)]$

Solution

a. $\frac{1}{2}\log x + 3\log(x+1) = \log x^{1/2} + \log(x+1)^3$ Power Property

$= \log\left[\sqrt{x}(x+1)^3\right]$ Product Property

b. $2\ln(x+2) - \ln x = \ln(x+2)^2 - \ln x$ Power Property

$= \ln \dfrac{(x+2)^2}{x}$ Quotient Property

c. $\frac{1}{3}[\log_2 x + \log_2(x+1)] = \frac{1}{3}\log_2[x(x+1)]$ Product Property

$= \log_2[x(x+1)]^{1/3}$ Power Property

$= \log_2 \sqrt[3]{x(x+1)}$ Rewrite with a radical.

Checkpoint Audio-video solution in English & Spanish at LarsonPrecalculus.com.

Condense the expression $2[\log(x+3) - 2\log(x-2)]$.

Application

One method of determining how the x- and y-values for a set of nonlinear data are related is to take the natural logarithm of each of the x- and y-values. If the points, when graphed, fall on a line, then you can determine that the x- and y-values are related by the equation $\ln y = m \ln x$ where m is the slope of the line.

EXAMPLE 7 Finding a Mathematical Model

The table shows the mean distance from the sun x and the period y (the time it takes a planet to orbit the sun) for each of the six planets that are closest to the sun. In the table, the mean distance is given in terms of astronomical units (where Earth's mean distance is defined as 1.0), and the period is given in years. Find an equation that relates y and x.

DATA

Planet	Mean Distance, x	Period, y
Mercury	0.387	0.241
Venus	0.723	0.615
Earth	1.000	1.000
Mars	1.524	1.881
Jupiter	5.203	11.860
Saturn	9.555	29.420

Spreadsheet at LarsonPrecalculus.com

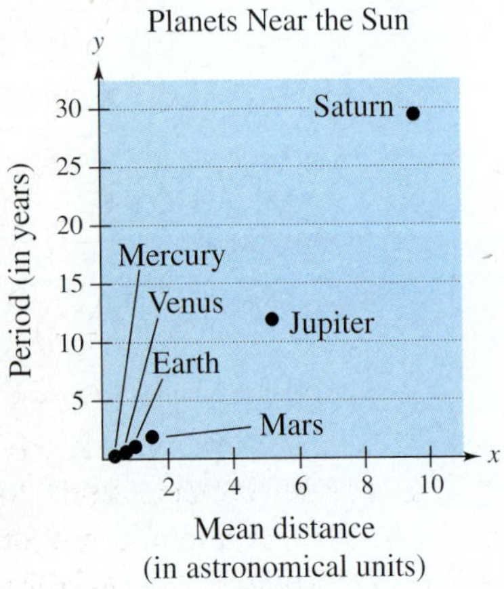

Figure 3.18

Solution Figure 3.18 shows the plots of the points given by the above table. From this figure, it is not clear how to find an equation that relates y and x. To solve this problem, take the natural logarithm of each of the x- and y-values, as shown in the table at the left. Now, by plotting the points in the table at the left, you can see that all six of the points appear to lie in a line (see Figure 3.19). Choose any two points to determine the slope of the line. Using the points (0.421, 0.632) and (0, 0), the slope of the line is

$$m = \frac{0.632 - 0}{0.421 - 0} \approx 1.5 = \frac{3}{2}.$$

By the point-slope form, the equation of the line is $Y = \frac{3}{2}X$, where $Y = \ln y$ and $X = \ln x$. So, $\ln y = \frac{3}{2} \ln x$.

Planet	$\ln x$	$\ln y$
Mercury	−0.949	−1.423
Venus	−0.324	−0.486
Earth	0.000	0.000
Mars	0.421	0.632
Jupiter	1.649	2.473
Saturn	2.257	3.382

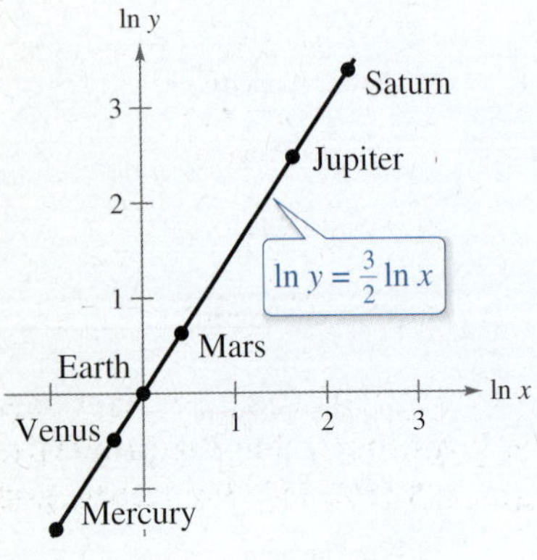

Figure 3.19

✓ ***Checkpoint*** *Audio-video solution in English & Spanish at LarsonPrecalculus.com.*

Find a logarithmic equation that relates y and x for the following ordered pairs.

(0.37, 0.51), (1.00, 1.00), (2.72, 1.95) (7.39, 3.79), (20.09, 7.39)

Summarize (Section 3.3)

1. State the change-of-base formula *(page 221)*. For examples that use the change-of-base formula to rewrite and evaluate logarithmic expressions, see Examples 1 and 2.
2. Make a list of the properties of logarithms *(page 222)*. For examples that use the properties of logarithms to evaluate or rewrite logarithmic expressions, see Examples 3 and 4.
3. Explain how to use the properties of logarithms to expand or condense logarithmic expressions *(page 223)*. For examples of expanding and condensing logarithmic expressions, see Examples 5 and 6.
4. Describe an example of how to use a logarithmic function to model and solve a real-life problem *(page 224, Example 7)*.

3.3 Exercises

See CalcChat.com for tutorial help and worked-out solutions to odd-numbered exercises.

Vocabulary

In Exercises 1–3, fill in the blanks.

1. To evaluate a logarithm to any base, use the ________ formula.
2. The change-of-base formula for base e is given by $\log_a x =$ ________.
3. You can consider $\log_a x$ to be a constant multiple of $\log_b x$; the constant multiplier is ________.

In Exercises 4–6, match the property of logarithms with its name.

4. $\log_a(uv) = \log_a u + \log_a v$ (a) Power Property
5. $\ln u^n = n \ln u$ (b) Quotient Property
6. $\log_a \frac{u}{v} = \log_a u - \log_a v$ (c) Product Property

Skills and Applications

Rewriting a Logarithm **In Exercises 7–10, rewrite the logarithm as a ratio of (a) common logarithms and (b) natural logarithms.**

7. $\log_5 16$
8. $\log_{1/5} x$
9. $\log_x \frac{3}{10}$
10. $\log_{2.6} x$

Using the Change-of-Base Formula **In Exercises 11–14, evaluate the logarithm using the change-of-base formula. Round your result to three decimal places.**

11. $\log_3 7$
12. $\log_{1/2} 4$
13. $\log_9 0.1$
14. $\log_3 0.015$

Using Properties of Logarithms **In Exercises 15–20, use the properties of logarithms to rewrite and simplify the logarithmic expression.**

15. $\log_4 8$
16. $\log_2(4^2 \cdot 3^4)$
17. $\log_5 \frac{1}{250}$
18. $\log \frac{9}{300}$
19. $\ln(5e^6)$
20. $\ln \frac{6}{e^2}$

Using Properties of Logarithms **In Exercises 21–36, find the exact value of the logarithmic expression without using a calculator. (If this is not possible, then state the reason.)**

21. $\log_3 9$
22. $\log_5 \frac{1}{125}$
23. $\log_2 \sqrt[4]{8}$
24. $\log_6 \sqrt[3]{6}$
25. $\log_4 16^2$
26. $\log_3 81^{-3}$
27. $\log_2(-2)$
28. $\log_3(-27)$
29. $\ln e^{4.5}$
30. $3 \ln e^4$
31. $\ln \frac{1}{\sqrt{e}}$
32. $\ln \sqrt[4]{e^3}$
33. $\ln e^2 + \ln e^5$
34. $2 \ln e^6 - \ln e^5$
35. $\log_5 75 - \log_5 3$
36. $\log_4 2 + \log_4 32$

Expanding a Logarithmic Expression **In Exercises 37–58, use the properties of logarithms to expand the expression as a sum, difference, and/or constant multiple of logarithms. (Assume all variables are positive.)**

37. $\ln 4x$
38. $\log_3 10z$
39. $\log_8 x^4$
40. $\log_{10} \frac{y}{2}$
41. $\log_5 \frac{5}{x}$
42. $\log_6 \frac{1}{z^3}$
43. $\ln \sqrt{z}$
44. $\ln \sqrt[3]{t}$
45. $\ln xyz^2$
46. $\log 4x^2y$
47. $\ln z(z-1)^2,\ z > 1$
48. $\ln\left(\frac{x^2-1}{x^3}\right),\ x > 1$
49. $\log_2 \frac{\sqrt{a-1}}{9},\ a > 1$
50. $\ln \frac{6}{\sqrt{x^2+1}}$
51. $\ln \sqrt[3]{\frac{x}{y}}$
52. $\ln \sqrt{\frac{x^2}{y^3}}$
53. $\ln x^2 \sqrt{\frac{y}{z}}$
54. $\log_2 x^4 \sqrt{\frac{y}{z^3}}$
55. $\log_5 \frac{x^2}{y^2z^3}$
56. $\log_{10} \frac{xy^4}{z^5}$
57. $\ln \sqrt[4]{x^3(x^2+3)}$
58. $\ln \sqrt{x^2(x+2)}$

Using Properties of Logarithms **In Exercises 59–66, approximate the logarithm using the properties of logarithms, given $\log_b 2 \approx 0.3562$, $\log_b 3 \approx 0.5646$, and $\log_b 5 \approx 0.8271$.**

59. $\log_b 10$
60. $\log_b \frac{2}{3}$
61. $\log_b 8$
62. $\log_b \sqrt{2}$
63. $\log_b 45$
64. $\log_b(2b)^{-2}$
65. $\log_b(3b^2)$
66. $\log_b \sqrt[3]{3b}$

Condensing a Logarithmic Expression **In Exercises 67–82, condense the expression to the logarithm of a single quantity.**

67. $\ln 2 + \ln x$

68. $\log_5 8 - \log_5 t$

69. $2 \log_2 x + 4 \log_2 y$

70. $\frac{2}{3} \log_7(z - 2)$

71. $\frac{1}{4} \log_3 5x$

72. $-4 \log_6 2x$

73. $\log x - 2 \log(x + 1)$

74. $2 \ln 8 + 5 \ln(z - 4)$

75. $\log x - 2 \log y + 3 \log z$

76. $3 \log_3 x + 4 \log_3 y - 4 \log_3 z$

77. $\ln x - [\ln(x + 1) + \ln(x - 1)]$

78. $4[\ln z + \ln(z + 5)] - 2 \ln(z - 5)$

79. $\frac{1}{3}[2 \ln(x + 3) + \ln x - \ln(x^2 - 1)]$

80. $2[3 \ln x - \ln(x + 1) - \ln(x - 1)]$

81. $\frac{1}{3}[\log_8 y + 2 \log_8(y + 4)] - \log_8(y - 1)$

82. $\frac{1}{2}[\log_4(x + 1) + 2 \log_4(x - 1)] + 6 \log_4 x$

Comparing Logarithmic Quantities **In Exercises 83 and 84, compare the logarithmic quantities. If two are equal, then explain why.**

83. $\dfrac{\log_2 32}{\log_2 4}, \quad \log_2 \dfrac{32}{4}, \quad \log_2 32 - \log_2 4$

84. $\log_7 \sqrt{70}, \quad \log_7 35, \quad \frac{1}{2} + \log_7 \sqrt{10}$

Sound Intensity

In Exercises 85–88, use the following information. The relationship between the number of decibels β and the intensity of a sound I in watts per square meter is given by

$$\beta = 10 \log\left(\frac{I}{10^{-12}}\right).$$

85. Use the properties of logarithms to write the formula in simpler form, and determine the number of decibels of a sound with an intensity of 10^{-6} watt per square meter.

86. Find the difference in loudness between an average office with an intensity of 1.26×10^{-7} watt per square meter and a broadcast studio with an intensity of 3.16×10^{-10} watt per square meter.

87. Find the difference in loudness between a vacuum cleaner with an intensity of 10^{-4} watt per square meter and rustling leaves with an intensity of 10^{-11} watt per square meter.

88. You and your roommate are playing your stereos at the same time and at the same intensity. How much louder is the music when both stereos are playing compared with just one stereo playing?

Curve Fitting **In Exercises 89–92, find a logarithmic equation that relates y and x. Explain the steps used to find the equation.**

89.

x	1	2	3	4	5	6
y	1	1.189	1.316	1.414	1.495	1.565

90.

x	1	2	3	4	5	6
y	1	1.587	2.080	2.520	2.924	3.302

91.

x	1	2	3	4	5	6
y	2.5	2.102	1.9	1.768	1.672	1.597

92.

x	1	2	3	4	5	6
y	0.5	2.828	7.794	16	27.951	44.091

93. Galloping Speeds of Animals Four-legged animals run with two different types of motion: trotting and galloping. An animal that is trotting has at least one foot on the ground at all times, whereas an animal that is galloping has all four feet off the ground at some point in its stride. The number of strides per minute at which an animal breaks from a trot to a gallop depends on the weight of the animal. Use the table to find a logarithmic equation that relates an animal's weight x (in pounds) and its lowest galloping speed y (in strides per minute).

DATA

Weight, x	Galloping Speed, y
25	191.5
35	182.7
50	173.8
75	164.2
500	125.9
1000	114.2

Spreadsheet at LarsonPrecalculus.com

94. Nail Length The approximate lengths and diameters (in inches) of common nails are shown in the table. Find a logarithmic equation that relates the diameter y of a common nail to its length x.

Length, x	Diameter, y
1	0.072
2	0.120
3	0.148
4	0.203
5	0.238

kentoh/Shutterstock.com

95. Comparing Models A cup of water at an initial temperature of 78°C is placed in a room at a constant temperature of 21°C. The temperature of the water is measured every 5 minutes during a half-hour period. The results are recorded as ordered pairs of the form (t, T), where t is the time (in minutes) and T is the temperature (in degrees Celsius).

(0, 78.0°), (5, 66.0°), (10, 57.5°), (15, 51.2°), (20, 46.3°), (25, 42.4°), (30, 39.6°)

(a) The graph of the model for the data should be asymptotic with the graph of the temperature of the room. Subtract the room temperature from each of the temperatures in the ordered pairs. Use a graphing utility to plot the data points (t, T) and $(t, T - 21)$.

(b) An exponential model for the data $(t, T - 21)$ is given by $T - 21 = 54.4(0.964)^t$. Solve for T and graph the model. Compare the result with the plot of the original data.

(c) Take the natural logarithms of the revised temperatures. Use the graphing utility to plot the points $(t, \ln(T - 21))$ and observe that the points appear to be linear. Use the *regression* feature of the graphing utility to fit a line to these data. This resulting line has the form $\ln(T - 21) = at + b$. Solve for T, and verify that the result is equivalent to the model in part (b).

(d) Fit a rational model to the data. Take the reciprocals of the y-coordinates of the revised data points to generate the points

$$\left(t, \frac{1}{T - 21}\right).$$

Use the graphing utility to graph these points and observe that they appear to be linear. Use the *regression* feature of the graphing utility to fit a line to these data. The resulting line has the form

$$\frac{1}{T - 21} = at + b.$$

Solve for T, and use the graphing utility to graph the rational function and the original data points.

(e) Why did taking the logarithms of the temperatures lead to a linear scatter plot? Why did taking the reciprocals of the temperatures lead to a linear scatter plot?

Exploration

96. Graphical Analysis Use a graphing utility to graph the functions $y_1 = \ln x - \ln(x - 3)$ and $y_2 = \ln \dfrac{x}{x - 3}$ in the same viewing window. Does the graphing utility show the functions with the same domain? If so, should it? Explain your reasoning.

True or False? **In Exercises 97–102, determine whether the statement is true or false given that $f(x) = \ln x$. Justify your answer.**

97. $f(0) = 0$

98. $f(ax) = f(a) + f(x), \quad a > 0, \quad x > 0$

99. $f(x - 2) = f(x) - f(2), \quad x > 2$

100. $\sqrt{f(x)} = \frac{1}{2}f(x)$

101. If $f(u) = 2f(v)$, then $v = u^2$.

102. If $f(x) < 0$, then $0 < x < 1$.

Using the Change-of-Base Formula **In Exercises 103–106, use the change-of-base formula to rewrite the logarithm as a ratio of logarithms. Then use a graphing utility to graph the ratio.**

103. $f(x) = \log_2 x$

104. $f(x) = \log_{1/2} x$

105. $f(x) = \log_{1/4} x$

106. $f(x) = \log_{11.8} x$

107. Discussion A classmate claims that the following are true.

(a) $\ln(u + v) = \ln u + \ln v = \ln(uv)$

(b) $\ln(u - v) = \ln u - \ln v = \ln \dfrac{u}{v}$

(c) $(\ln u)^n = n(\ln u) = \ln u^n$

Discuss how you would demonstrate that these claims are not true.

108. HOW DO YOU SEE IT? The figure shows the graphs of $y = \ln x$, $y = \ln x^2$, $y = \ln 2x$, and $y = \ln 2$. Match each function with its graph. (The graphs are labeled A through D.) Explain your reasoning.

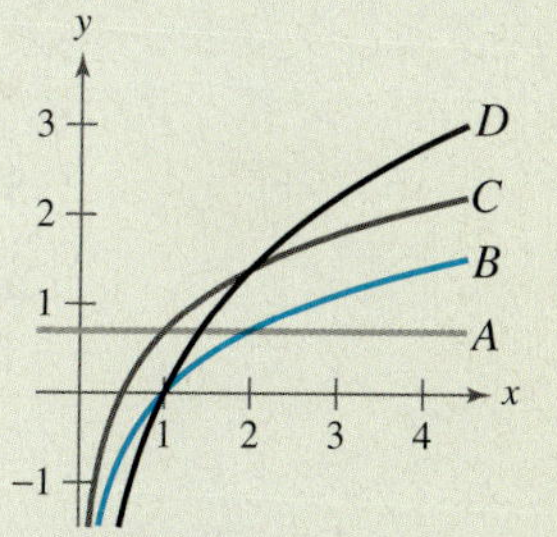

109. Think About It For how many integers between 1 and 20 can you approximate natural logarithms, given the values $\ln 2 \approx 0.6931$, $\ln 3 \approx 1.0986$, and $\ln 5 \approx 1.6094$? Approximate these logarithms (do not use a calculator).

3.4 Exponential and Logarithmic Equations

Exponential and logarithmic equations can help you model and solve life science applications. For instance, Exercise 83 on page 236 uses an exponential function to model the number of trees per acre given the average diameter of the trees.

- Solve simple exponential and logarithmic equations.
- Solve more complicated exponential equations.
- Solve more complicated logarithmic equations.
- Use exponential and logarithmic equations to model and solve real-life problems.

Introduction

So far in this chapter, you have studied the definitions, graphs, and properties of exponential and logarithmic functions. In this section, you will study procedures for *solving equations* involving these exponential and logarithmic functions.

There are two basic strategies for solving exponential or logarithmic equations. The first is based on the One-to-One Properties and was used to solve simple exponential and logarithmic equations in Sections 3.1 and 3.2. The second is based on the Inverse Properties. For $a > 0$ and $a \neq 1$, the following properties are true for all x and y for which $\log_a x$ and $\log_a y$ are defined.

One-to-One Properties	Inverse Properties
$a^x = a^y$ if and only if $x = y$.	$a^{\log_a x} = x$
$\log_a x = \log_a y$ if and only if $x = y$.	$\log_a a^x = x$

EXAMPLE 1 Solving Simple Equations

Original Equation	Rewritten Equation	Solution	Property
a. $2^x = 32$	$2^x = 2^5$	$x = 5$	One-to-One
b. $\ln x - \ln 3 = 0$	$\ln x = \ln 3$	$x = 3$	One-to-One
c. $\left(\frac{1}{3}\right)^x = 9$	$3^{-x} = 3^2$	$x = -2$	One-to-One
d. $e^x = 7$	$\ln e^x = \ln 7$	$x = \ln 7$	Inverse
e. $\ln x = -3$	$e^{\ln x} = e^{-3}$	$x = e^{-3}$	Inverse
f. $\log x = -1$	$10^{\log x} = 10^{-1}$	$x = 10^{-1} = \frac{1}{10}$	Inverse
g. $\log_3 x = 4$	$3^{\log_3 x} = 3^4$	$x = 81$	Inverse

Checkpoint *Audio-video solution in English & Spanish at LarsonPrecalculus.com.*

Solve each equation for x.

a. $2^x = 512$ **b.** $\log_6 x = 3$ **c.** $5 - e^x = 0$ **d.** $9^x = \frac{1}{3}$

The strategies used in Example 1 are summarized as follows.

Strategies for Solving Exponential and Logarithmic Equations

1. Rewrite the original equation in a form that allows the use of the One-to-One Properties of exponential or logarithmic functions.
2. Rewrite an *exponential* equation in logarithmic form and apply the Inverse Property of logarithmic functions.
3. Rewrite a *logarithmic* equation in exponential form and apply the Inverse Property of exponential functions.

goran cakmazovic/Shutterstock.com

Solving Exponential Equations

EXAMPLE 2 Solving Exponential Equations

Solve each equation and approximate the result to three decimal places, if necessary.

a. $e^{-x^2} = e^{-3x-4}$ **b.** $3(2^x) = 42$

Solution

a.

$$e^{-x^2} = e^{-3x-4}$$ Write original equation.

$$-x^2 = -3x - 4$$ One-to-One Property

$$x^2 - 3x - 4 = 0$$ Write in general form.

$$(x+1)(x-4) = 0$$ Factor.

$$(x+1) = 0 \Rightarrow x = -1$$ Set 1st factor equal to 0.

$$(x-4) = 0 \Rightarrow x = 4$$ Set 2nd factor equal to 0.

The solutions are $x = -1$ and $x = 4$. Check these in the original equation.

b.

$$3(2^x) = 42$$ Write original equation.

$$2^x = 14$$ Divide each side by 3.

$$\log_2 2^x = \log_2 14$$ Take log (base 2) of each side.

$$x = \log_2 14$$ Inverse Property

$$x = \frac{\ln 14}{\ln 2} \approx 3.807$$ Change-of-base formula

The solution is $x = \log_2 14 \approx 3.807$. Check this in the original equation.

REMARK Another way to solve Example 2(b) is by taking the natural log of each side and then applying the Power Property, as follows.

$$3(2^x) = 42$$

$$2^x = 14$$

$$\ln 2^x = \ln 14$$

$$x \ln 2 = \ln 14$$

$$x = \frac{\ln 14}{\ln 2} \approx 3.807$$

Notice that you obtain the same result as in Example 2(b).

✓ ***Checkpoint*** *Audio-video solution in English & Spanish at LarsonPrecalculus.com.*

Solve each equation and approximate the result to three decimal places, if necessary.

a. $e^{2x} = e^{x^2-8}$ **b.** $2(5^x) = 32$

In Example 2(b), the exact solution is $x = \log_2 14$, and the approximate solution is $x \approx 3.807$. An exact answer is preferred when the solution is an intermediate step in a larger problem. For a final answer, an approximate solution is easier to comprehend.

EXAMPLE 3 Solving an Exponential Equation

Solve $e^x + 5 = 60$ and approximate the result to three decimal places.

Solution

$$e^x + 5 = 60$$ Write original equation.

$$e^x = 55$$ Subtract 5 from each side.

$$\ln e^x = \ln 55$$ Take natural log of each side.

$$x = \ln 55 \approx 4.007$$ Inverse Property

The solution is $x = \ln 55 \approx 4.007$. Check this in the original equation.

REMARK Remember that the natural logarithmic function has a base of e.

✓ ***Checkpoint*** *Audio-video solution in English & Spanish at LarsonPrecalculus.com.*

Solve $e^x - 7 = 23$ and approximate the result to three decimal places.

EXAMPLE 4 Solving an Exponential Equation

Solve $2(3^{2t-5}) - 4 = 11$ and approximate the result to three decimal places.

Solution

$2(3^{2t-5}) - 4 = 11$ — Write original equation.

$2(3^{2t-5}) = 15$ — Add 4 to each side.

$3^{2t-5} = \dfrac{15}{2}$ — Divide each side by 2.

$\log_3 3^{2t-5} = \log_3 \dfrac{15}{2}$ — Take log (base 3) of each side.

$2t - 5 = \log_3 \dfrac{15}{2}$ — Inverse Property

$2t = 5 + \log_3 7.5$ — Add 5 to each side.

$t = \dfrac{5}{2} + \dfrac{1}{2}\log_3 7.5$ — Divide each side by 2.

$t \approx 3.417$ — Use a calculator.

The solution is $t = \frac{5}{2} + \frac{1}{2}\log_3 7.5 \approx 3.417$. Check this in the original equation.

REMARK Remember that to evaluate a logarithm such as $\log_3 7.5$, you need to use the change-of-base formula.

$$\log_3 7.5 = \frac{\ln 7.5}{\ln 3} \approx 1.834$$

✓ ***Checkpoint*** *Audio-video solution in English & Spanish at LarsonPrecalculus.com.*

Solve $6(2^{t+5}) + 4 = 11$ and approximate the result to three decimal places.

When an equation involves two or more exponential expressions, you can still use a procedure similar to that demonstrated in Examples 2, 3, and 4. However, the algebra is a bit more complicated.

EXAMPLE 5 Solving an Exponential Equation of Quadratic Type

Solve $e^{2x} - 3e^x + 2 = 0$.

Algebraic Solution

$e^{2x} - 3e^x + 2 = 0$ — Write original equation.

$(e^x)^2 - 3e^x + 2 = 0$ — Write in quadratic form.

$(e^x - 2)(e^x - 1) = 0$ — Factor.

$e^x - 2 = 0$ — Set 1st factor equal to 0.

$x = \ln 2$ — Solution

$e^x - 1 = 0$ — Set 2nd factor equal to 0.

$x = 0$ — Solution

The solutions are $x = \ln 2 \approx 0.693$ and $x = 0$. Check these in the original equation.

Graphical Solution

Use a graphing utility to graph $y = e^{2x} - 3e^x + 2$ and then find the zeros.

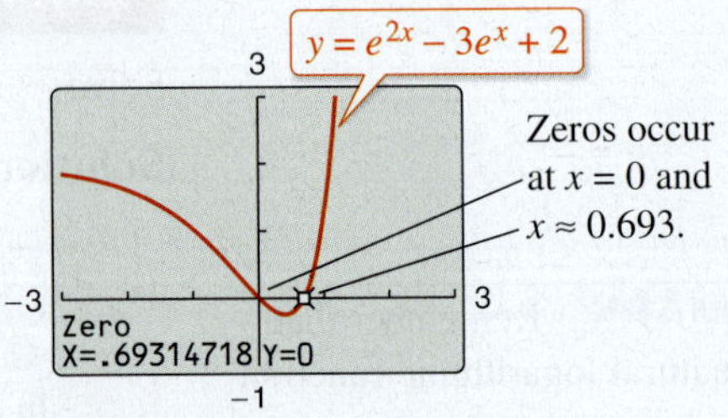

So, you can conclude that the solutions are $x = 0$ and $x \approx 0.693$.

✓ ***Checkpoint*** *Audio-video solution in English & Spanish at LarsonPrecalculus.com.*

Solve $e^{2x} - 7e^x + 12 = 0$.

Solving Logarithmic Equations

To solve a logarithmic equation, you can write it in exponential form.

$\ln x = 3$ Logarithmic form

$e^{\ln x} = e^3$ Exponentiate each side.

$x = e^3$ Exponential form

This procedure is called *exponentiating* each side of an equation.

REMARK Remember to check your solutions in the original equation when solving equations to verify that the answer is correct and to make sure that the answer is in the domain of the original equation.

EXAMPLE 6 Solving Logarithmic Equations

a. $\ln x = 2$ Original equation

$e^{\ln x} = e^2$ Exponentiate each side.

$x = e^2$ Inverse Property

b. $\log_3(5x - 1) = \log_3(x + 7)$ Original equation

$5x - 1 = x + 7$ One-to-One Property

$x = 2$ Solution

c. $\log_6(3x + 14) - \log_6 5 = \log_6 2x$ Original equation

$\log_6\left(\frac{3x + 14}{5}\right) = \log_6 2x$ Quotient Property of Logarithms

$\frac{3x + 14}{5} = 2x$ One-to-One Property

$3x + 14 = 10x$ Multiply each side by 5.

$x = 2$ Solution

✓ **Checkpoint** Audio-video solution in English & Spanish at LarsonPrecalculus.com.

Solve each equation.

a. $\ln x = \frac{2}{3}$ **b.** $\log_2(2x - 3) = \log_2(x + 4)$ **c.** $\log 4x - \log(12 + x) = \log 2$

EXAMPLE 7 Solving a Logarithmic Equation

Solve $5 + 2 \ln x = 4$ and approximate the result to three decimal places.

Algebraic Solution

$5 + 2 \ln x = 4$ Write original equation.

$2 \ln x = -1$ Subtract 5 from each side.

$\ln x = -\frac{1}{2}$ Divide each side by 2.

$e^{\ln x} = e^{-1/2}$ Exponentiate each side.

$x = e^{-1/2}$ Inverse Property

$x \approx 0.607$ Use a calculator.

Graphical Solution

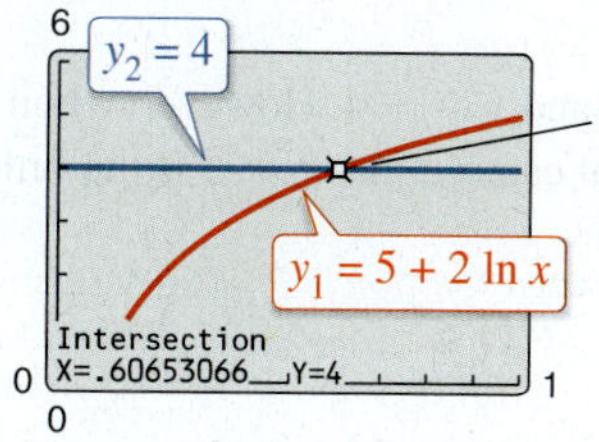

The intersection point is about (0.607, 4). So, the solution is $x \approx 0.607$.

✓ **Checkpoint** Audio-video solution in English & Spanish at LarsonPrecalculus.com.

Solve $7 + 3 \ln x = 5$ and approximate the result to three decimal places.

EXAMPLE 8 Solving a Logarithmic Equation

Solve $2 \log_5 3x = 4$.

Solution

$2 \log_5 3x = 4$ Write original equation.

$\log_5 3x = 2$ Divide each side by 2.

$5^{\log_5 3x} = 5^2$ Exponentiate each side (base 5).

$3x = 25$ Inverse Property

$x = \frac{25}{3}$ Divide each side by 3.

The solution is $x = \frac{25}{3}$. Check this in the original equation.

✓ **Checkpoint** Audio-video solution in English & Spanish at LarsonPrecalculus.com.

Solve $3 \log_4 6x = 9$.

Because the domain of a logarithmic function generally does not include all real numbers, you should be sure to check for extraneous solutions of logarithmic equations.

EXAMPLE 9 Checking for Extraneous Solutions

Solve $\log 5x + \log(x - 1) = 2$.

Algebraic Solution

$\log 5x + \log(x - 1) = 2$ Write original equation.

$\log[5x(x - 1)] = 2$ Product Property of Logarithms

$10^{\log(5x^2 - 5x)} = 10^2$ Exponentiate each side (base 10).

$5x^2 - 5x = 100$ Inverse Property

$x^2 - x - 20 = 0$ Write in general form.

$(x - 5)(x + 4) = 0$ Factor.

$x - 5 = 0$ Set 1st factor equal to 0.

$x = 5$ Solution

$x + 4 = 0$ Set 2nd factor equal to 0.

$x = -4$ Solution

The solutions appear to be $x = 5$ and $x = -4$. However, when you check these in the original equation, you can see that $x = 5$ is the only solution.

Graphical Solution

First, rewrite the original equation as

$$\log 5x + \log(x - 1) - 2 = 0.$$

Then use a graphing utility to graph the equation

$$y = \log 5x + \log(x - 1) - 2$$

and find the zero(s).

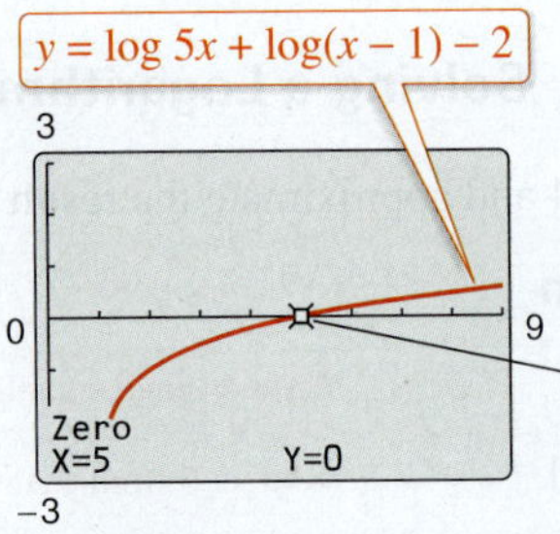

A zero occurs at $x = 5$. So, the solution is $x = 5$.

✓ **Checkpoint** Audio-video solution in English & Spanish at LarsonPrecalculus.com.

Solve $\log x + \log(x - 9) = 1$.

In Example 9, the domain of $\log 5x$ is $x > 0$ and the domain of $\log(x - 1)$ is $x > 1$, so the domain of the original equation is

$$x > 1.$$

Because the domain is all real numbers greater than 1, the solution $x = -4$ is extraneous. The graphical solution verifies this conclusion.

REMARK Notice in Example 9 that the logarithmic part of the equation is condensed into a single logarithm before exponentiating each side of the equation.

Applications

EXAMPLE 10 Doubling an Investment

You invest \$500 at an annual interest rate of 6.75%, compounded continuously. How long will it take your money to double?

Solution Using the formula for continuous compounding, the balance is

$$A = Pe^{rt}$$

$$A = 500e^{0.0675t}.$$

To find the time required for the balance to double, let $A = 1000$ and solve the resulting equation for t.

$$500e^{0.0675t} = 1000 \qquad \text{Let } A = 1000.$$

$$e^{0.0675t} = 2 \qquad \text{Divide each side by 500.}$$

$$\ln e^{0.0675t} = \ln 2 \qquad \text{Take natural log of each side.}$$

$$0.0675t = \ln 2 \qquad \text{Inverse Property}$$

$$t = \frac{\ln 2}{0.0675} \qquad \text{Divide each side by 0.0675.}$$

$$t \approx 10.27 \qquad \text{Use a calculator.}$$

The balance in the account will double after approximately 10.27 years. This result is demonstrated graphically below.

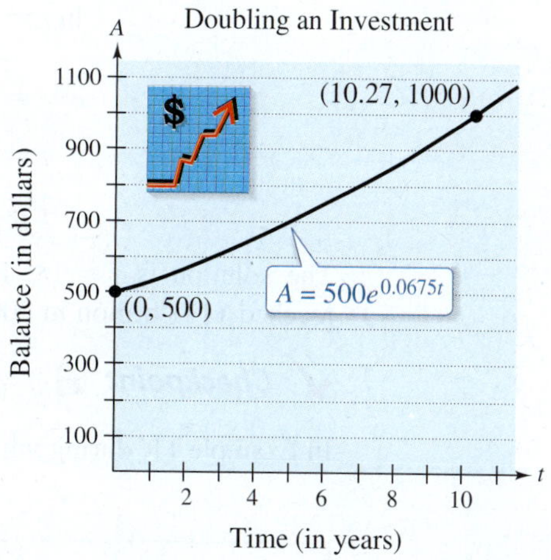

✓ Checkpoint Audio-video solution in English & Spanish at LarsonPrecalculus.com.

You invest \$500 at an annual interest rate of 5.25%, compounded continuously. How long will it take your money to double? Compare your result with that of Example 10.

In Example 10, an approximate answer of 10.27 years is given. Within the context of the problem, the exact solution

$$t = \frac{\ln 2}{0.0675}$$

does not make sense as an answer.

EXAMPLE 11 Retail Sales

The retail sales y (in billions of dollars) of e-commerce companies in the United States from 2002 through 2010 can be modeled by

$$y = -566 + 244.7 \ln t, \quad 12 \le t \le 20$$

where t represents the year, with $t = 12$ corresponding to 2002 (see figure). During which year did the sales reach \$141 billion? *(Source: U.S. Census Bureau)*

Retail Sales of e-Commerce Companies

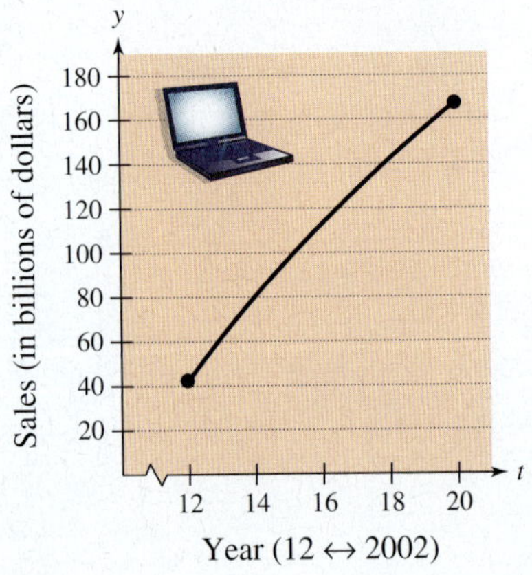

Solution

$$-566 + 244.7 \ln t = y \quad \text{Write original equation.}$$

$$-566 + 244.7 \ln t = 141 \quad \text{Substitute 141 for } y.$$

$$244.7 \ln t = 707 \quad \text{Add 566 to each side.}$$

$$\ln t = \frac{707}{244.7} \quad \text{Divide each side by 244.7.}$$

$$e^{\ln t} = e^{707/244.7} \quad \text{Exponentiate each side.}$$

$$t = e^{707/244.7} \quad \text{Inverse Property}$$

$$t \approx 18 \quad \text{Use a calculator.}$$

The solution is $t \approx 18$. Because $t = 12$ represents 2002, it follows that the sales reached \$141 billion in 2008.

✓ **Checkpoint** *Audio-video solution in English & Spanish at LarsonPrecalculus.com.*

In Example 11, during which year did the sales reach \$80 billion?

Summarize (Section 3.4)

1. State the One-to-One Properties and the Inverse Properties that can help you solve simple exponential and logarithmic equations *(page 228)*. For an example of solving simple exponential and logarithmic equations, see Example 1.
2. Describe strategies for solving exponential equations *(pages 229 and 230)*. For examples of solving exponential equations, see Examples 2–5.
3. Describe strategies for solving logarithmic equations *(pages 231 and 232)*. For examples of solving logarithmic equations, see Examples 6–9.
4. Describe examples of how to use exponential and logarithmic equations to model and solve real-life problems *(pages 233 and 234, Examples 10 and 11)*.

3.4 Exercises

See CalcChat.com for tutorial help and worked-out solutions to odd-numbered exercises.

Vocabulary: Fill in the blanks.

1. To solve exponential and logarithmic equations, you can use the following One-to-One and Inverse Properties.

(a) $a^x = a^y$ if and only if ________.

(b) $\log_a x = \log_a y$ if and only if ________.

(c) $a^{\log_a x} =$ ________

(d) $\log_a a^x =$ ________

2. An ________ solution does not satisfy the original equation.

Skills and Applications

Determining Solutions In Exercises 3–6, determine whether each x-value is a solution (or an approximate solution) of the equation.

3. $4^{2x-7} = 64$
(a) $x = 5$
(b) $x = 2$

4. $4e^{x-1} = 60$
(a) $x = 1 + \ln 15$
(b) $x = \ln 16$

5. $\log_2(x + 3) = 10$
(a) $x = 1021$
(b) $x = 17$
(c) $x = 10^2 - 3$

6. $\ln(2x + 3) = 5.8$
(a) $x = \frac{1}{2}(-3 + \ln 5.8)$
(b) $x = \frac{1}{2}(-3 + e^{5.8})$
(c) $x \approx 163.650$

Solving a Simple Equation In Exercises 7–14, solve for x.

7. $4^x = 16$

8. $\left(\frac{1}{2}\right)^x = 32$

9. $\ln x - \ln 2 = 0$

10. $e^x = 2$

11. $\ln x = -1$

12. $\log x = -2$

13. $\log_4 x = 3$

14. $\log_5 x = \frac{1}{2}$

Approximating a Point of Intersection In Exercises 15 and 16, approximate the point of intersection of the graphs of f and g. Then solve the equation $f(x) = g(x)$ algebraically to verify your approximation.

15. $f(x) = 2^x$
$g(x) = 8$

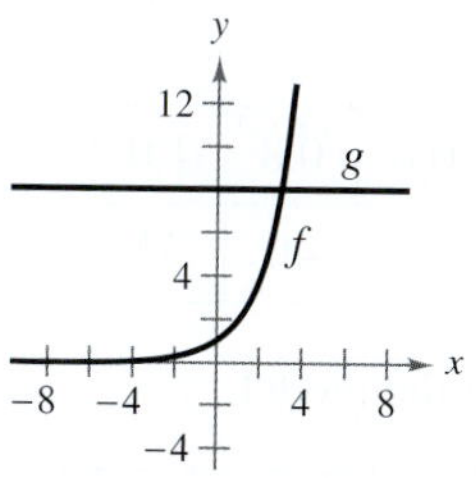

16. $f(x) = \log_3 x$
$g(x) = 2$

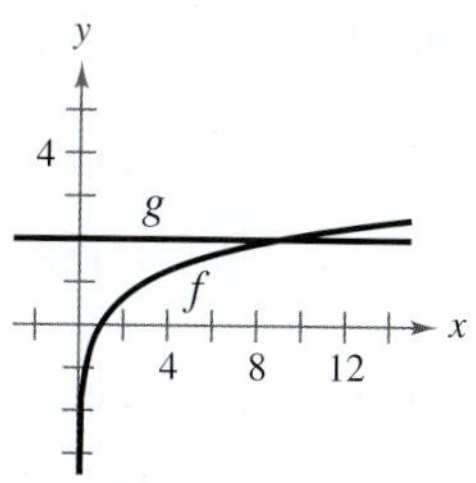

Solving an Exponential Equation In Exercises 17–44, solve the exponential equation algebraically. Approximate the result to three decimal places.

17. $e^x = e^{x^2-2}$

18. $e^{x^2-3} = e^{x-2}$

19. $4(3^x) = 20$

20. $4e^x = 91$

21. $e^x - 9 = 19$

22. $6^x + 10 = 47$

23. $3^{2x} = 80$

24. $4^{-3t} = 0.10$

25. $2^{3-x} = 565$

26. $8^{-2-x} = 431$

27. $8(10^{3x}) = 12$

28. $8(3^{6-x}) = 40$

29. $e^{3x} = 12$

30. $1000e^{-4x} = 75$

31. $7 - 2e^x = 5$

32. $-14 + 3e^x = 11$

33. $6(2^{3x-1}) - 7 = 9$

34. $8(4^{6-2x}) + 13 = 41$

35. $2^x = 3^{x+1}$

36. $2^{x+1} = e^{1-x}$

37. $4^x = 5^{x^2}$

38. $3^{x^2} = 7^{6-x}$

39. $e^{2x} - 4e^x - 5 = 0$

40. $e^{2x} - 5e^x + 6 = 0$

41. $\dfrac{500}{100 - e^{x/2}} = 20$

42. $\dfrac{400}{1 + e^{-x}} = 350$

43. $\left(1 + \dfrac{0.065}{365}\right)^{365t} = 4$

44. $\left(1 + \dfrac{0.10}{12}\right)^{12t} = 2$

Solving a Logarithmic Equation In Exercises 45–62, solve the logarithmic equation algebraically. Approximate the result to three decimal places.

45. $\ln x = -3$

46. $\ln x - 7 = 0$

47. $2.1 = \ln 6x$

48. $\log 3z = 2$

49. $3 \ln 5x = 10$

50. $\ln\sqrt{x - 8} = 5$

51. $2 - 6 \ln x = 10$

52. $2 + 3 \ln x = 12$

53. $6 \log_3(0.5x) = 11$

54. $4 \log(x - 6) = 11$

55. $\ln x - \ln(x + 1) = 2$

56. $\ln x + \ln(x + 1) = 1$

57. $\ln(x + 5) = \ln(x - 1) - \ln(x + 1)$

58. $\ln(x + 1) - \ln(x - 2) = \ln x$

59. $\log(3x + 4) = \log(x - 10)$

60. $\log_2 x + \log_2(x + 2) = \log_2(x + 6)$

61. $\log_4 x - \log_4(x - 1) = \frac{1}{2}$

62. $\log 8x - \log(1 + \sqrt{x}) = 2$

Graphing and Solving an Equation **In Exercises 63–70, use a graphing utility to graph and solve the equation. Approximate the result to three decimal places. Verify your result algebraically.**

63. $5^x = 212$
64. $6e^{1-x} = 25$
65. $8e^{-2x/3} = 11$
66. $e^{0.09t} = 3$
67. $3 - \ln x = 0$
68. $10 - 4\ln(x - 2) = 0$
69. $2\ln(x + 3) = 3$
70. $\ln(x + 1) = 2 - \ln x$

Compound Interest **In Exercises 71 and 72, you invest \$2500 in an account at interest rate r, compounded continuously. Find the time required for the amount to (a) double and (b) triple.**

71. $r = 0.025$
72. $r = 0.0375$

Algebra of Calculus **In Exercises 73–80, solve the equation algebraically. Round your result to three decimal places. Verify your answer using a graphing utility.**

73. $2x^2e^{2x} + 2xe^{2x} = 0$
74. $-x^2e^{-x} + 2xe^{-x} = 0$
75. $-xe^{-x} + e^{-x} = 0$
76. $e^{-2x} - 2xe^{-2x} = 0$
77. $2x\ln x + x = 0$
78. $\dfrac{1 - \ln x}{x^2} = 0$
79. $\dfrac{1 + \ln x}{2} = 0$
80. $2x\ln\left(\dfrac{1}{x}\right) - x = 0$

81. **Average Heights** The percent m of American males between the ages of 20 and 29 who are under x inches tall is modeled by

$$m(x) = \frac{100}{1 + e^{-0.5536(x-69.51)}}, \quad 64 \le x \le 78$$

and the percent f of American females between the ages of 20 and 29 who are under x inches tall is modeled by

$$f(x) = \frac{100}{1 + e^{-0.5834(x-64.49)}}, \quad 60 \le x \le 78.$$

(Source: U.S. National Center for Health Statistics)

(a) Use the graph to determine any horizontal asymptotes of the graphs of the functions. Interpret the meaning in the context of the problem.

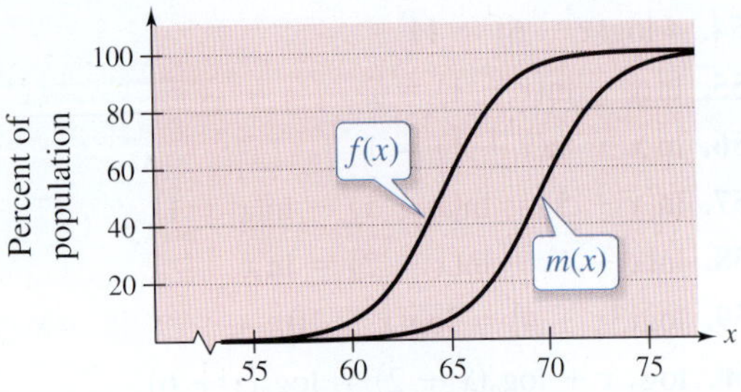

(b) What is the average height of each sex?

82. **U.S. Currency** The values y (in billions of dollars) of U.S. currency in circulation in the years 2000 through 2010 can be modeled by $y = -611 + 507\ln t$, $10 \le t \le 20$, where t represents the year, with $t = 10$ corresponding to 2000. During which year did the value of U.S. currency in circulation exceed \$690 billion? *(Source: Board of Governors of the Federal Reserve System)*

83. **Trees per Acre** The number N of trees of a given species per acre is approximated by the model $N = 68(10^{-0.04x})$, $5 \le x \le 40$, where x is the average diameter of the trees (in inches) 3 feet above the ground. Use the model to approximate the average diameter of the trees in a test plot when $N = 21$.

84. **Demand** The demand equation for a smart phone is

$$p = 5000\left(1 - \frac{4}{4 + e^{-0.002x}}\right).$$

Find the demand x for a price of (a) $p = \$169$ and (b) $p = \$299$.

85. **Automobiles** Engineers design automobiles with crumple zones that help protect their occupants in crashes. The crumple zones allow the occupants to move short distances when the automobiles come to abrupt stops. The greater the distance moved, the fewer g's the crash victims experience. (One g is equal to the acceleration due to gravity.) In crash tests with vehicles moving at 90 kilometers per hour, analysts measured the numbers of g's experienced during deceleration by crash dummies that were permitted to move x meters during impact. The table shows the data. A model for the data is given by $y = -3.00 + 11.88\ln x + (36.94/x)$, where y is the number of g's.

x	0.2	0.4	0.6	0.8	1.0
g's	158	80	53	40	32

(a) Complete the table using the model.

x	0.2	0.4	0.6	0.8	1.0
y					

(b) Use a graphing utility to graph the data points and the model in the same viewing window. How do they compare?

James Marshall/CORBIS

(c) Use the model to estimate the distance traveled during impact, assuming that the passenger deceleration must not exceed 30 g's.

(d) Do you think it is practical to lower the number of g's experienced during impact to fewer than 23? Explain your reasoning.

86. Data Analysis An object at a temperature of 160°C was removed from a furnace and placed in a room at 20°C. The temperature T of the object was measured each hour h and recorded in the table. A model for the data is given by $T = 20[1 + 7(2^{-h})]$. The figure shows the graph of this model.

DATA

Spreadsheet at LarsonPrecalculus.com

Hour, h	Temperature, T
0	160°
1	90°
2	56°
3	38°
4	29°
5	24°

(a) Use the graph to identify the horizontal asymptote of the model and interpret the asymptote in the context of the problem.

(b) Use the model to approximate the time when the temperature of the object was 100°C.

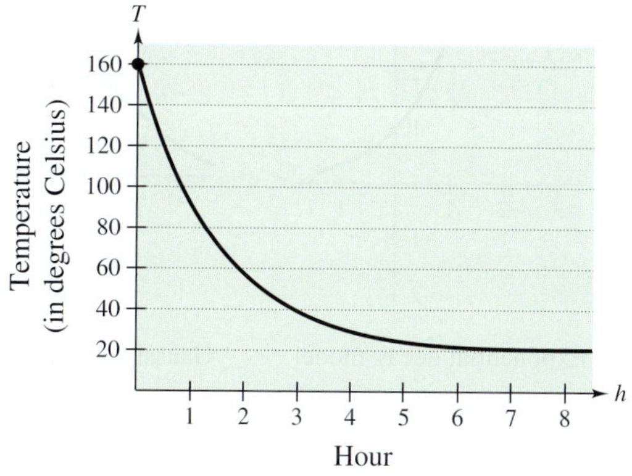

Exploration

True or False? **In Exercises 87–90, rewrite each verbal statement as an equation. Then decide whether the statement is true or false. Justify your answer.**

87. The logarithm of the product of two numbers is equal to the sum of the logarithms of the numbers.

88. The logarithm of the sum of two numbers is equal to the product of the logarithms of the numbers.

89. The logarithm of the difference of two numbers is equal to the difference of the logarithms of the numbers.

90. The logarithm of the quotient of two numbers is equal to the difference of the logarithms of the numbers.

91. Think About It Is it possible for a logarithmic equation to have more than one extraneous solution? Explain.

92. HOW DO YOU SEE IT? Solving $\log_3 x + \log_3(x - 8) = 2$ algebraically, the solutions appear to be $x = 9$ and $x = -1$. Use the graph of $y = \log_3 x + \log_3(x - 8) - 2$ to determine whether each value is an actual solution of the equation. Explain your reasoning.

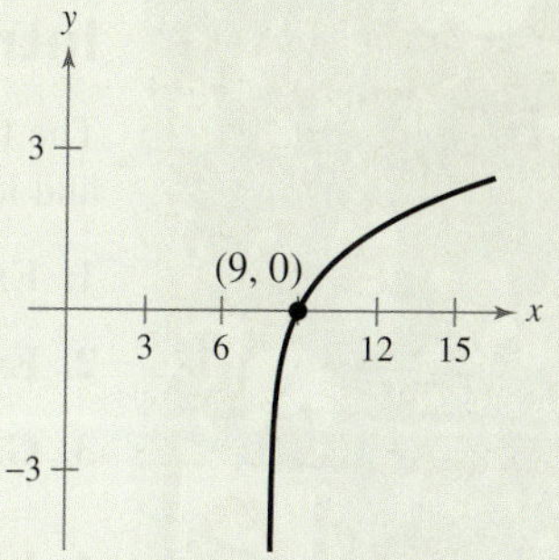

93. Finance You are investing P dollars at an annual interest rate of r, compounded continuously, for t years. Which of the following would result in the highest value of the investment? Explain your reasoning.

(a) Double the amount you invest.

(b) Double your interest rate.

(c) Double the number of years.

94. Think About It Are the times required for the investments in Exercises 71 and 72 to quadruple twice as long as the times for them to double? Give a reason for your answer and verify your answer algebraically.

95. Effective Yield The *effective yield* of an investment plan is the percent increase in the balance after 1 year. Find the effective yield for each investment plan. Which investment plan has the greatest effective yield? Which investment plan will have the highest balance after 5 years?

(a) 7% annual interest rate, compounded annually

(b) 7% annual interest rate, compounded continuously

(c) 7% annual interest rate, compounded quarterly

(d) 7.25% annual interest rate, compounded quarterly

96. Graphical Analysis Let $f(x) = \log_a x$ and $g(x) = a^x$, where $a > 1$.

(a) Let $a = 1.2$ and use a graphing utility to graph the two functions in the same viewing window. What do you observe? Approximate any points of intersection of the two graphs.

(b) Determine the value(s) of a for which the two graphs have one point of intersection.

(c) Determine the value(s) of a for which the two graphs have two points of intersection.

3.5 Exponential and Logarithmic Models

Exponential growth and decay models can often represent the populations of countries. For instance, in Exercise 30 on page 246, you will use exponential growth and decay models to compare the populations of several countries.

- **Recognize the five most common types of models involving exponential and logarithmic functions.**
- **Use exponential growth and decay functions to model and solve real-life problems.**
- **Use Gaussian functions to model and solve real-life problems.**
- **Use logistic growth functions to model and solve real-life problems.**
- **Use logarithmic functions to model and solve real-life problems.**

Introduction

The five most common types of mathematical models involving exponential functions and logarithmic functions are as follows.

1. **Exponential growth model:** $y = ae^{bx}, \quad b > 0$
2. **Exponential decay model:** $y = ae^{-bx}, \quad b > 0$
3. **Gaussian model:** $y = ae^{-(x-b)^2/c}$
4. **Logistic growth model:** $y = \dfrac{a}{1 + be^{-rx}}$
5. **Logarithmic models:** $y = a + b \ln x, \quad y = a + b \log x$

The basic shapes of the graphs of these functions are as follows.

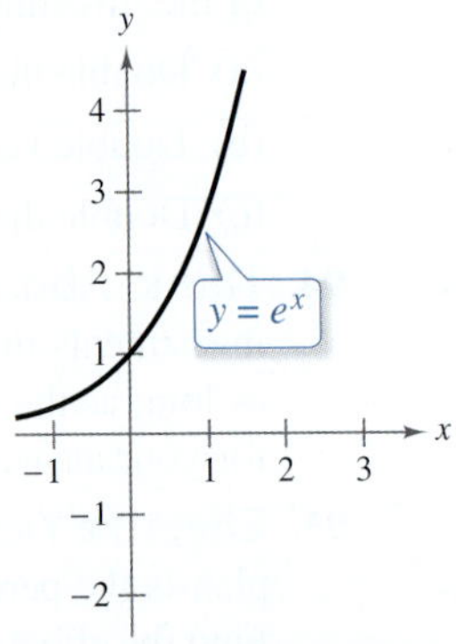

Exponential growth model

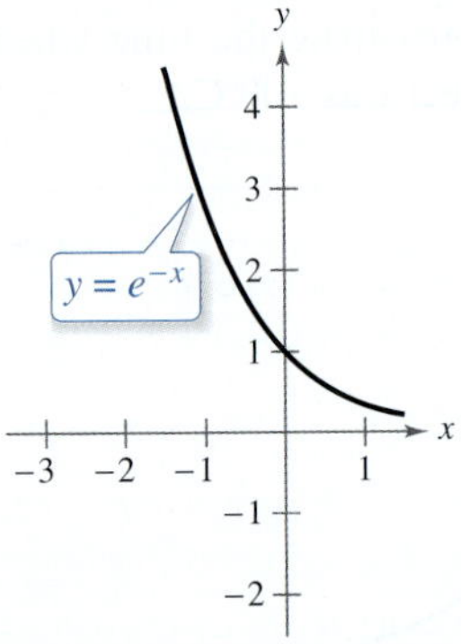

Exponential decay model

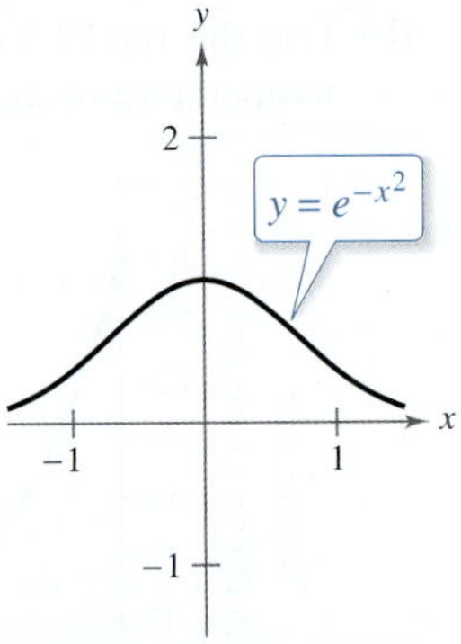

Gaussian model

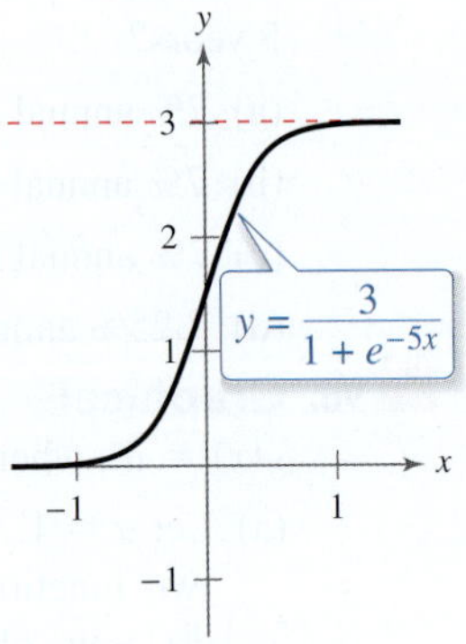

Logistic growth model

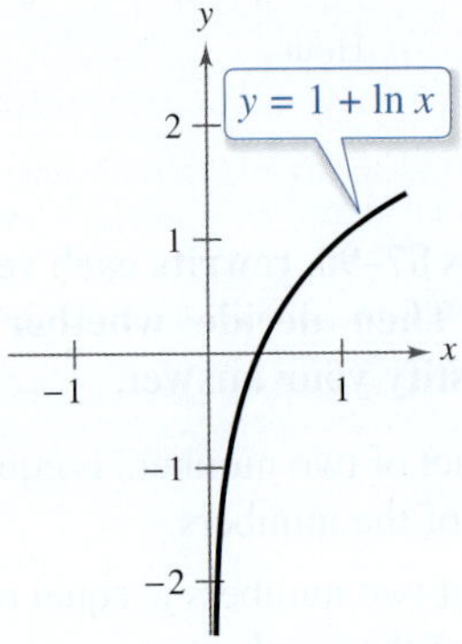

Natural logarithmic model

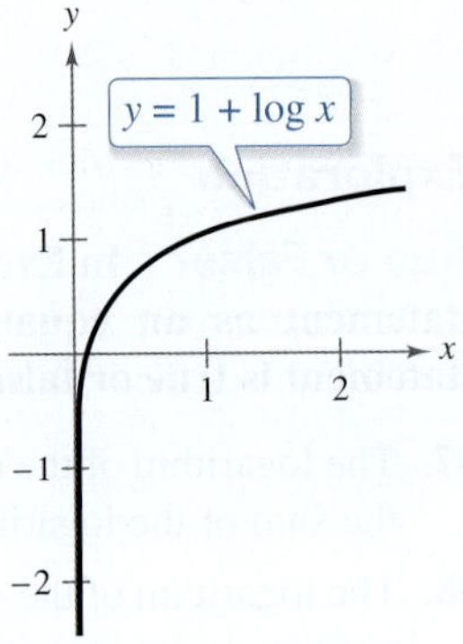

Common logarithmic model

You often gain insight into a situation modeled by an exponential or logarithmic function by identifying and interpreting the asymptotes of the graph of the function. Identify the asymptote(s) of the graph of each function shown above.

Alan Becker/Stone/Getty Images

Exponential Growth and Decay

EXAMPLE 1 Online Advertising

Estimates of the amounts (in billions of dollars) of U.S. online advertising spending from 2011 through 2015 are shown in the table. A scatter plot of the data is shown at the right. *(Source: eMarketer)*

Year	2011	2012	2013	2014	2015
Advertising Spending	31.3	36.8	41.2	45.5	49.5

Online Advertising Spending

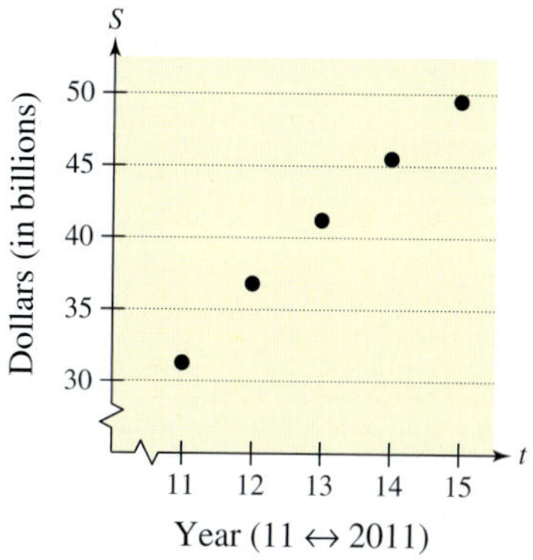

An exponential growth model that approximates the data is given by

$$S = 9.30e^{0.1129t}, \quad 11 \leq t \leq 15$$

where S is the amount of spending (in billions of dollars) and $t = 11$ represents 2011. Compare the values given by the model with the estimates shown in the table. According to this model, when will the amount of U.S. online advertising spending reach \$80 billion?

Algebraic Solution

The following table compares the two sets of advertising spending amounts.

Year	2011	2012	2013	2014	2015
Advertising Spending	31.3	36.8	41.2	45.5	49.5
Model	32.2	36.0	40.4	45.2	50.6

To find when the amount of U.S. online advertising spending will reach \$80 billion, let $S = 80$ in the model and solve for t.

$9.30e^{0.1129t} = S$ — Write original model.

$9.30e^{0.1129t} = 80$ — Substitute 80 for S.

$e^{0.1129t} \approx 8.6022$ — Divide each side by 9.30.

$\ln e^{0.1129t} \approx \ln 8.6022$ — Take natural log of each side.

$0.1129t \approx 2.1520$ — Inverse Property

$t \approx 19.1$ — Divide each side by 0.1129.

According to the model, the amount of U.S. online advertising spending will reach \$80 billion in 2019.

Graphical Solution

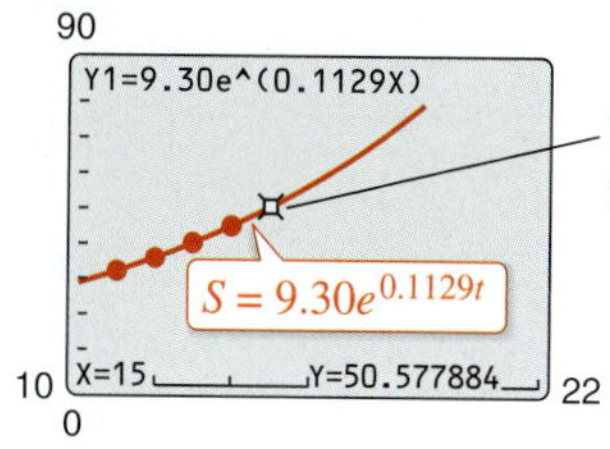

The model appears to fit the data closely.

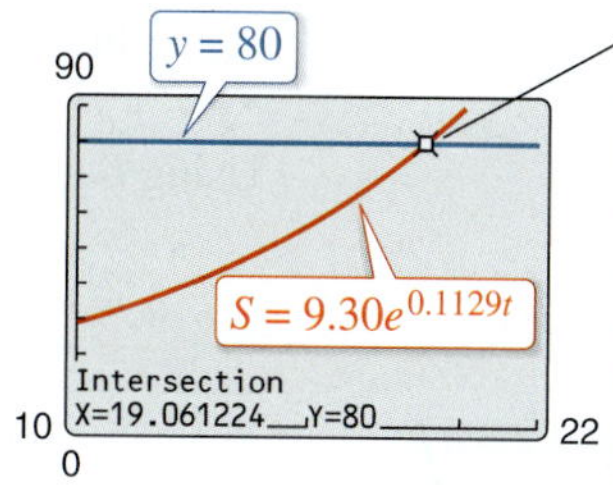

The intersection point of the model and the line $y = 80$ is about (19.1, 80). So, according to the model, the amount of U.S. online advertising spending will reach \$80 billion in 2019.

✓ Checkpoint *Audio-video solution in English & Spanish at LarsonPrecalculus.com.*

In Example 1, when will the amount of U.S. online advertising spending reach \$100 billion?

▷ **TECHNOLOGY** Some graphing utilities have an *exponential regression* feature that can help you find exponential models to represent data. If you have such a graphing utility, try using it to find an exponential model for the data given in Example 1. How does your model compare with the model given in Example 1?

In Example 1, the exponential growth model is given. But if this model were not given, then how would you find such a model? Example 2 demonstrates one technique.

EXAMPLE 2 Modeling Population Growth

In a research experiment, a population of fruit flies is increasing according to the law of exponential growth. After 2 days there are 100 flies, and after 4 days there are 300 flies. How many flies will there be after 5 days?

Solution Let y be the number of flies at time t. From the given information, you know that $y = 100$ when $t = 2$ and $y = 300$ when $t = 4$. Substituting this information into the model $y = ae^{bt}$ produces

$$100 = ae^{2b} \quad \text{and} \quad 300 = ae^{4b}.$$

To solve for b, solve for a in the first equation.

$$100 = ae^{2b} \qquad \text{Write first equation.}$$

$$\frac{100}{e^{2b}} = a \qquad \text{Solve for } a.$$

Then substitute the result into the second equation.

$$300 = ae^{4b} \qquad \text{Write second equation.}$$

$$300 = \left(\frac{100}{e^{2b}}\right)e^{4b} \qquad \text{Substitute } \frac{100}{e^{2b}} \text{ for } a.$$

$$\frac{300}{100} = e^{2b} \qquad \text{Simplify, and divide each side by 100.}$$

$$\ln 3 = 2b \qquad \text{Take natural log of each side.}$$

$$\frac{1}{2}\ln 3 = b \qquad \text{Solve for } b.$$

Using $b = \frac{1}{2}\ln 3$ and the equation you found for a,

$$a = \frac{100}{e^{2[(1/2)\ln 3]}} \qquad \text{Substitute } \tfrac{1}{2}\ln 3 \text{ for } b.$$

$$= \frac{100}{e^{\ln 3}} \qquad \text{Simplify.}$$

$$= \frac{100}{3} \qquad \text{Inverse Property}$$

$$\approx 33.33. \qquad \text{Divide.}$$

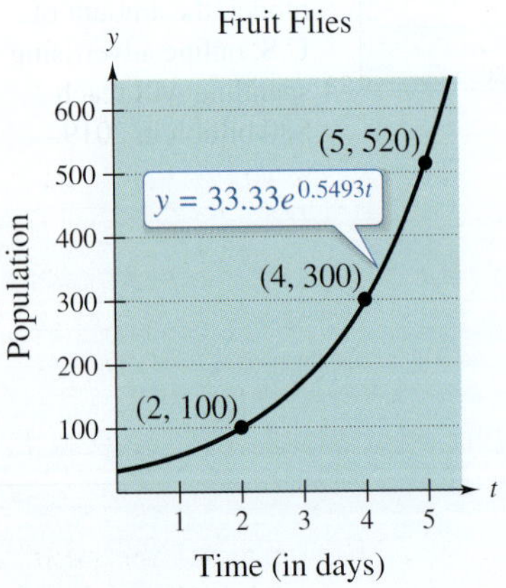

Figure 3.20

So, with $a \approx 33.33$ and $b = \frac{1}{2}\ln 3 \approx 0.5493$, the exponential growth model is

$$y = 33.33e^{0.5493t}$$

as shown in Figure 3.20. After 5 days, the population will be

$$y = 33.33e^{0.5493(5)}$$

$$\approx 520 \text{ flies.}$$

✓ Checkpoint Audio-video solution in English & Spanish at LarsonPrecalculus.com.

The number of bacteria in a culture is increasing according to the law of exponential growth. After 1 hour there are 100 bacteria, and after 2 hours there are 200 bacteria. How many bacteria will there be after 3 hours?

In living organic material, the ratio of the number of radioactive carbon isotopes (carbon 14) to the number of nonradioactive carbon isotopes (carbon 12) is about 1 to 10^{12}. When organic material dies, its carbon 12 content remains fixed, whereas its radioactive carbon 14 begins to decay with a half-life of about 5700 years. To estimate the age of dead organic material, scientists use the following formula, which denotes the ratio of carbon 14 to carbon 12 present at any time t (in years).

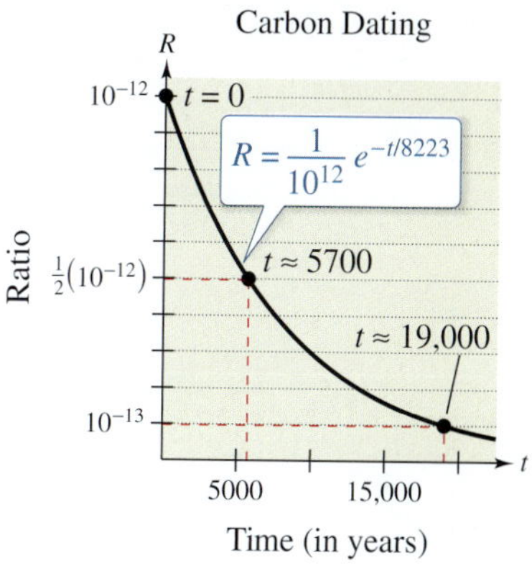

$$R = \frac{1}{10^{12}}e^{-t/8223} \quad \text{Carbon dating model}$$

The graph of R is shown at the right. Note that R decreases as t increases.

EXAMPLE 3 Carbon Dating

Estimate the age of a newly discovered fossil for which the ratio of carbon 14 to carbon 12 is $R = 1/10^{13}$.

Algebraic Solution

In the carbon dating model, substitute the given value of R to obtain the following.

$$\frac{1}{10^{12}}e^{-t/8223} = R \quad \text{Write original model.}$$

$$\frac{e^{-t/8223}}{10^{12}} = \frac{1}{10^{13}} \quad \text{Substitute } \frac{1}{10^{13}} \text{ for } R.$$

$$e^{-t/8223} = \frac{1}{10} \quad \text{Multiply each side by } 10^{12}.$$

$$\ln e^{-t/8223} = \ln\frac{1}{10} \quad \text{Take natural log of each side.}$$

$$-\frac{t}{8223} \approx -2.3026 \quad \text{Inverse Property}$$

$$t \approx 18{,}934 \quad \text{Multiply each side by } -8223.$$

So, to the nearest thousand years, the age of the fossil is about 19,000 years.

Graphical Solution

Use a graphing utility to graph the formula for the ratio of carbon 14 to carbon 12 at any time t as

$$y_1 = \frac{1}{10^{12}}e^{-x/8223}.$$

In the same viewing window, graph $y_2 = 1/10^{13}$.

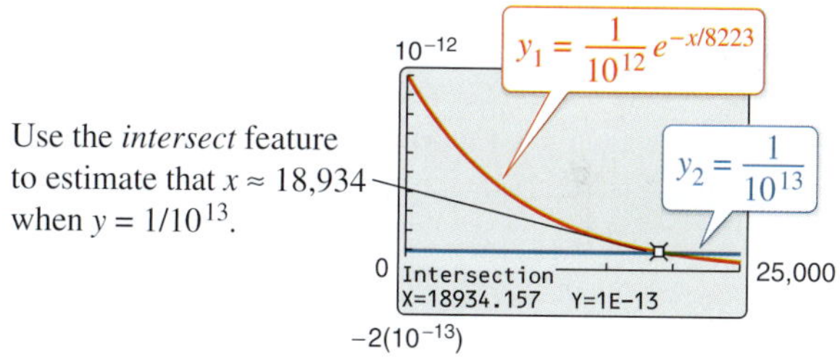

So, to the nearest thousand years, the age of the fossil is about 19,000 years.

Checkpoint *Audio-video solution in English & Spanish at LarsonPrecalculus.com.*

Estimate the age of a newly discovered fossil for which the ratio of carbon 14 to carbon 12 is $R = 1/10^{14}$.

The value of b in the exponential decay model $y = ae^{-bt}$ determines the *decay* of radioactive isotopes. For instance, to find how much of an initial 10 grams of ^{226}Ra isotope with a half-life of 1599 years is left after 500 years, substitute this information into the model $y = ae^{-bt}$.

$$\frac{1}{2}(10) = 10e^{-b(1599)} \Rightarrow \ln\frac{1}{2} = -1599b \Rightarrow b = -\frac{\ln\frac{1}{2}}{1599}$$

Using the value of b found above and $a = 10$, the amount left is

$$y = 10e^{-[-\ln(1/2)/1599](500)} \approx 8.05 \text{ grams.}$$

Gaussian Models

As mentioned at the beginning of this section, Gaussian models are of the form

$$y = ae^{-(x-b)^2/c}.$$

In probability and statistics, this type of model commonly represents populations that are **normally distributed.** The graph of a Gaussian model is called a **bell-shaped curve.** Try graphing the normal distribution with a graphing utility. Can you see why it is called a bell-shaped curve?

For *standard* normal distributions, the model takes the form

$$y = \frac{1}{\sqrt{2\pi}}e^{-x^2/2}.$$

The **average value** of a population can be found from the bell-shaped curve by observing where the maximum y-value of the function occurs. The x-value corresponding to the maximum y-value of the function represents the average value of the independent variable—in this case, x.

EXAMPLE 4 **SAT Scores**

In 2011, the SAT mathematics scores for high school graduates in the United States roughly followed the normal distribution given by

$$y = 0.0034e^{-(x-514)^2/27{,}378}, \quad 200 \le x \le 800$$

where x is the SAT score for mathematics. Sketch the graph of this function. From the graph, estimate the average SAT mathematics score. *(Source: The College Board)*

Solution The graph of the function is shown below. On this bell-shaped curve, the maximum value of the curve represents the average score. From the graph, you can estimate that the average mathematics score for high school graduates in 2011 was 514.

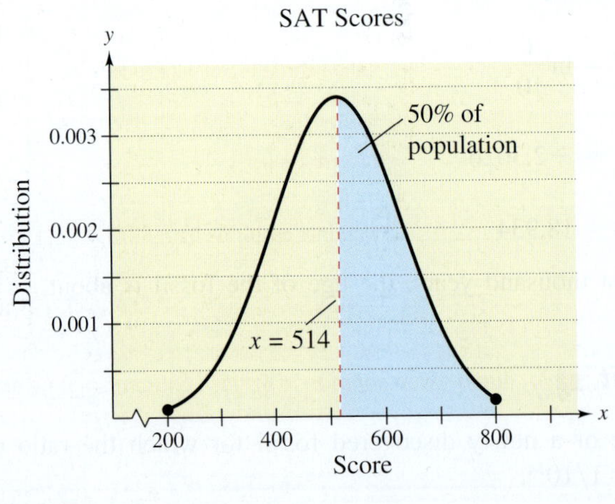

Checkpoint Audio-video solution in English & Spanish at LarsonPrecalculus.com.

In 2011, the SAT critical reading scores for high school graduates in the United States roughly followed the normal distribution given by

$$y = 0.0035e^{-(x-497)^2/25{,}992}, \quad 200 \le x \le 800$$

where x is the SAT score for critical reading. Sketch the graph of this function. From the graph, estimate the average SAT critical reading score. *(Source: The College Board)*

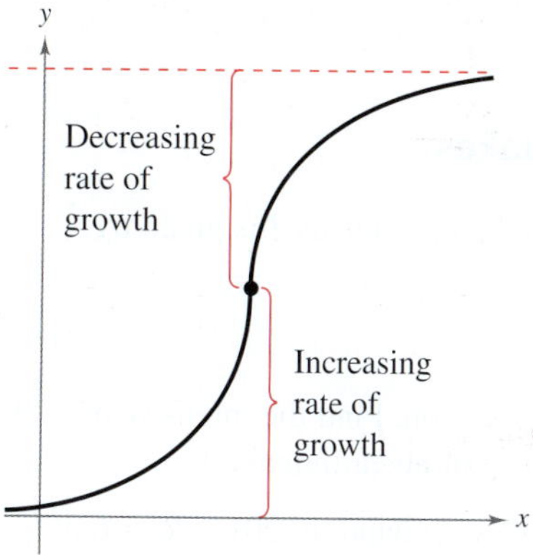

Figure 3.21

Logistic Growth Models

Some populations initially have rapid growth, followed by a declining rate of growth, as indicated by the graph in Figure 3.21. One model for describing this type of growth pattern is the **logistic curve** given by the function

$$y = \frac{a}{1 + be^{-rx}}$$

where y is the population size and x is the time. An example is a bacteria culture that is initially allowed to grow under ideal conditions and then under less favorable conditions that inhibit growth. A logistic growth curve is also called a **sigmoidal curve.**

EXAMPLE 5 Spread of a Virus

On a college campus of 5000 students, one student returns from vacation with a contagious and long-lasting flu virus. The spread of the virus is modeled by

$$y = \frac{5000}{1 + 4999e^{-0.8t}}, \quad t \geq 0$$

where y is the total number of students infected after t days. The college will cancel classes when 40% or more of the students are infected.

a. How many students are infected after 5 days?

b. After how many days will the college cancel classes?

Algebraic Solution

a. After 5 days, the number of students infected is

$$y = \frac{5000}{1 + 4999e^{-0.8(5)}} = \frac{5000}{1 + 4999e^{-4}} \approx 54.$$

b. The college will cancel classes when the number of infected students is $(0.40)(5000) = 2000$.

$$2000 = \frac{5000}{1 + 4999e^{-0.8t}}$$

$$1 + 4999e^{-0.8t} = 2.5$$

$$e^{-0.8t} = \frac{1.5}{4999}$$

$$-0.8t = \ln \frac{1.5}{4999}$$

$$t = -\frac{1}{0.8} \ln \frac{1.5}{4999}$$

$$t \approx 10.14$$

So, after about 10 days, at least 40% of the students will be infected, and the college will cancel classes.

Graphical Solution

a. Use the *value* feature to estimate that $y \approx 54$ when $x = 5$. So, after 5 days, about 54 students are infected.

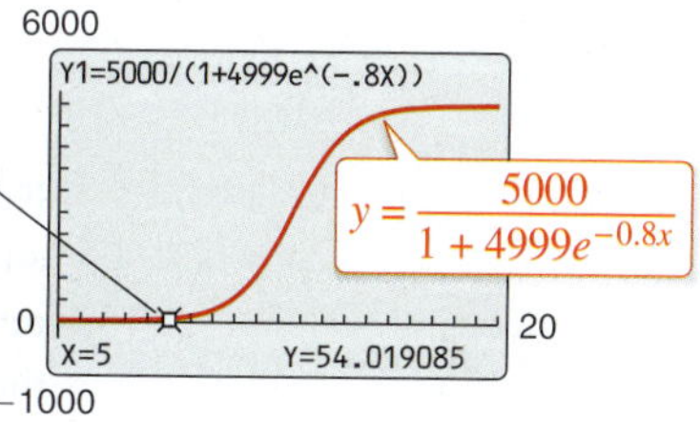

b. The college will cancel classes when the number of infected students is $(0.40)(5000) = 2000$. Use a graphing utility to graph

$$y_1 = \frac{5000}{1 + 4999e^{-0.8x}} \quad \text{and} \quad y_2 = 2000$$

in the same viewing window. Use the *intersect* feature of the graphing utility to find the point of intersection of the graphs.

The point of intersection occurs near $x \approx 10.14$. So, after about 10 days, at least 40% of the students will be infected, and the college will cancel classes.

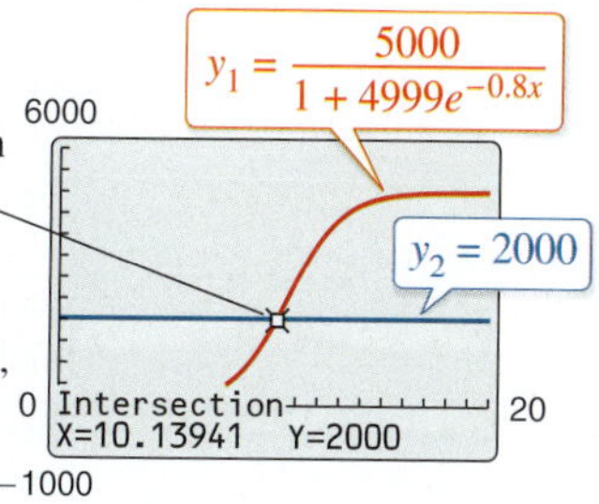

Checkpoint Audio-video solution in English & Spanish at LarsonPrecalculus.com.

In Example 5, after how many days are 250 students infected?

Logarithmic Models

EXAMPLE 6 **Magnitudes of Earthquakes**

On the Richter scale, the magnitude R of an earthquake of intensity I is given by

$$R = \log \frac{I}{I_0}$$

where $I_0 = 1$ is the minimum intensity used for comparison. Find the intensity of each earthquake. (Intensity is a measure of the wave energy of an earthquake.)

a. Alaska in 2012: $R = 4.0$ **b.** Christchurch, New Zealand, in 2011: $R = 6.3$

On February 22, 2011, an earthquake of magnitude 6.3 struck Christchurch, New Zealand. The total economic loss was estimated at 15.5 billion U.S. dollars.

Solution

a. Because $I_0 = 1$ and $R = 4.0$, you have

$4.0 = \log \frac{I}{1}$ Substitute 1 for I_0 and 4.0 for R.

$10^{4.0} = 10^{\log I}$ Exponentiate each side.

$10^{4.0} = I$ Inverse Property

$10{,}000 = I.$ Simplify.

b. For $R = 6.3$, you have

$6.3 = \log \frac{I}{1}$ Substitute 1 for I_0 and 6.3 for R.

$10^{6.3} = 10^{\log I}$ Exponentiate each side.

$10^{6.3} = I$ Inverse Property

$2{,}000{,}000 \approx I.$ Use a calculator.

Note that an increase of 2.3 units on the Richter scale (from 4.0 to 6.3) represents an increase in intensity by a factor of $2{,}000{,}000/10{,}000 = 200$. In other words, the intensity of the earthquake in Christchurch was about 200 times as great as that of the earthquake in Alaska.

✓ Checkpoint *Audio-video solution in English & Spanish at LarsonPrecalculus.com.*

Find the intensities of earthquakes whose magnitudes are (a) $R = 6.0$ and (b) $R = 7.9$.

Summarize (Section 3.5)

1. State the five most common types of models involving exponential and logarithmic functions *(page 238)*.
2. Describe examples of how to use exponential growth and decay functions to model and solve real-life problems *(pages 239–241, Examples 1–3)*.
3. Describe an example of how to use a Gaussian function to model and solve a real-life problem *(page 242, Example 4)*.
4. Describe an example of how to use a logistic growth function to model and solve a real-life problem *(page 243, Example 5)*.
5. Describe an example of how to use a logarithmic function to model and solve a real-life problem *(page 244, Example 6)*.

Darrenp/Shutterstock.com

3.5 Exercises

See CalcChat.com for tutorial help and worked-out solutions to odd-numbered exercises.

Vocabulary: Fill in the blanks.

1. An exponential growth model has the form ________, and an exponential decay model has the form ________.
2. A logarithmic model has the form ________ or ________.
3. In probability and statistics, Gaussian models commonly represent populations that are ________ ________.
4. A logistic growth model has the form ________.

Skills and Applications

Solving for a Variable **In Exercises 5 and 6, (a) solve for P and (b) solve for t.**

5. $A = Pe^{rt}$

6. $A = P\left(1 + \dfrac{r}{n}\right)^{nt}$

Compound Interest **In Exercises 7–12, complete the table assuming continuously compounded interest.**

	Initial Investment	Annual % Rate	Time to Double	Amount After 10 Years
7.	\$1000	3.5%		
8.	\$750	$10\frac{1}{2}$%		
9.	\$750		$7\frac{3}{4}$ yr	
10.	\$500			\$1505.00
11.		4.5%		\$10,000.00
12.			12 yr	\$2000.00

Compound Interest **In Exercises 13 and 14, determine the principal P that must be invested at rate r, compounded monthly, so that \$500,000 will be available for retirement in t years.**

13. $r = 5\%,\ t = 10$

14. $r = 3\frac{1}{2}\%,\ t = 15$

Compound Interest **In Exercises 15 and 16, determine the time necessary for P dollars to double when it is invested at interest rate r compounded (a) annually, (b) monthly, (c) daily, and (d) continuously.**

15. $r = 10\%$

16. $r = 6.5\%$

17. **Compound Interest** Complete the table for the time t (in years) necessary for P dollars to triple when interest is compounded (a) continuously and (b) annually at rate r.

r	2%	4%	6%	8%	10%	12%
t						

18. **Modeling Data** Draw scatter plots of the data in Exercise 17. Use the *regression* feature of a graphing utility to find models for the data.

19. **Comparing Models** If \$1 is invested over a 10-year period, then the balance A, where t represents the time in years, is given by $A = 1 + 0.075[\![t]\!]$ or $A = e^{0.07t}$ depending on whether the interest is simple interest at $7\frac{1}{2}$% or continuous compound interest at 7%. Graph each function on the same set of axes. Which grows at a greater rate? (Remember that $[\![t]\!]$ is the greatest integer function discussed in Section 1.6.)

20. **Comparing Models** If \$1 is invested over a 10-year period, then the balance A, where t represents the time in years, is given by $A = 1 + 0.06[\![t]\!]$ or $A = [1 + (0.055/365)]^{[\![365t]\!]}$ depending on whether the interest is simple interest at 6% or compound interest at $5\frac{1}{2}$% compounded daily. Use a graphing utility to graph each function in the same viewing window. Which grows at a greater rate?

Radioactive Decay **In Exercises 21–24, complete the table for the radioactive isotope.**

	Isotope	Half-life (years)	Initial Quantity	Amount After 1000 Years
21.	^{226}Ra	1599	10 g	
22.	^{14}C	5715	6.5 g	
23.	^{14}C	5715		2 g
24.	^{239}Pu	24,100		0.4 g

Finding an Exponential Model **In Exercises 25–28, find the exponential model that fits the points shown in the graph or table.**

25.

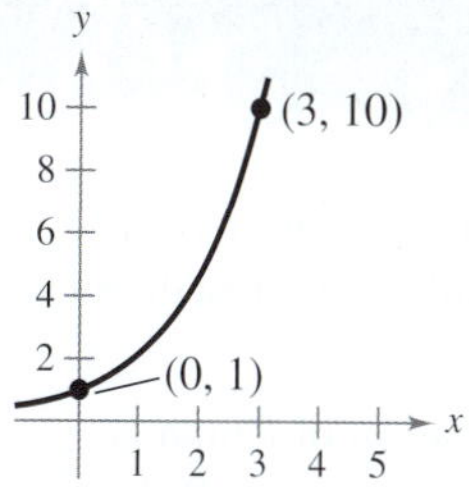

26.

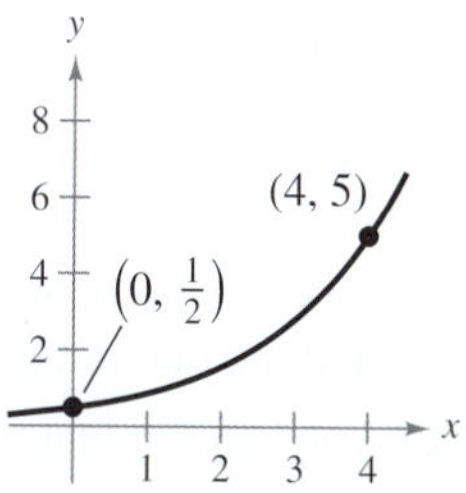

27.

x	0	4
y	5	1

28.

x	0	3
y	1	$\frac{1}{4}$

29. Population The populations P (in thousands) of Horry County, South Carolina, from 1980 through 2010 can be modeled by

$$P = 20.6 + 85.5e^{0.0360t}$$

where t represents the year, with $t = 0$ corresponding to 1980. *(Source: U.S. Census Bureau)*

(a) Use the model to complete the table.

Year	1980	1990	2000	2010
Population				

(b) According to the model, when will the population of Horry County reach 350,000?

(c) Do you think the model is valid for long-term predictions of the population? Explain.

30. Population

The table shows the mid-year populations (in millions) of five countries in 2010 and the projected populations (in millions) for the year 2020. *(Source: U.S. Census Bureau)*

Country	2010	2020
Bulgaria	7.1	6.6
Canada	33.8	36.4
China	1330.1	1384.5
United Kingdom	62.3	65.8
United States	310.2	341.4

(a) Find the exponential growth or decay model $y = ae^{bt}$ or $y = ae^{-bt}$ for the population of each country by letting $t = 10$ correspond to 2010. Use the model to predict the population of each country in 2030.

(b) You can see that the populations of the United States and the United Kingdom are growing at different rates. What constant in the equation $y = ae^{bt}$ gives the growth rate? Discuss the relationship between the different growth rates and the magnitude of the constant.

(c) You can see that the population of China is increasing, whereas the population of Bulgaria is decreasing. What constant in the equation $y = ae^{bt}$ reflects this difference? Explain.

31. Website Growth The number y of hits a new website receives each month can be modeled by $y = 4080e^{kt}$, where t represents the number of months the website has been operating. In the website's third month, there were 10,000 hits. Find the value of k, and use this value to predict the number of hits the website will receive after 24 months.

32. Population The populations P (in thousands) of Tallahassee, Florida, from 2005 through 2010 can be modeled by $P = 319.2e^{kt}$, where t represents the year, with $t = 5$ corresponding to 2005. In 2006, the population of Tallahassee was about 347,000. *(Source: U.S. Census Bureau)*

(a) Find the value of k. Is the population increasing or decreasing? Explain.

(b) Use the model to predict the populations of Tallahassee in 2015 and 2020. Are the results reasonable? Explain.

(c) According to the model, during what year will the population reach 410,000?

33. Bacteria Growth The number of bacteria in a culture is increasing according to the law of exponential growth. After 3 hours there are 100 bacteria, and after 5 hours there are 400 bacteria. How many bacteria will there be after 6 hours?

34. Bacteria Growth The number of bacteria in a culture is increasing according to the law of exponential growth. The initial population is 250 bacteria, and the population after 10 hours is double the population after 1 hour. How many bacteria will there be after 6 hours?

35. Depreciation A laptop computer that costs \$1150 new has a book value of \$550 after 2 years.

(a) Find the linear model $V = mt + b$.

(b) Find the exponential model $V = ae^{kt}$.

(c) Use a graphing utility to graph the two models in the same viewing window. Which model depreciates faster in the first 2 years?

(d) Find the book values of the computer after 1 year and after 3 years using each model.

(e) Explain the advantages and disadvantages of using each model to a buyer and a seller.

36. Learning Curve The management at a plastics factory has found that the maximum number of units a worker can produce in a day is 30. The learning curve for the number N of units produced per day after a new employee has worked t days is modeled by $N = 30(1 - e^{kt})$. After 20 days on the job, a new employee produces 19 units.

(a) Find the learning curve for this employee (first, find the value of k).

(b) How many days should pass before this employee is producing 25 units per day?

Alan Becker/Stone/Getty Images

37. Carbon Dating

(a) The ratio of carbon 14 to carbon 12 in a piece of wood discovered in a cave is $R = 1/8^{14}$. Estimate the age of the piece of wood.

(b) The ratio of carbon 14 to carbon 12 in a piece of paper buried in a tomb is $R = 1/13^{11}$. Estimate the age of the piece of paper.

38. Radioactive Decay Carbon 14 dating assumes that the carbon dioxide on Earth today has the same radioactive content as it did centuries ago. If this is true, then the amount of ^{14}C absorbed by a tree that grew several centuries ago should be the same as the amount of ^{14}C absorbed by a tree growing today. A piece of ancient charcoal contains only 15% as much radioactive carbon as a piece of modern charcoal. How long ago was the tree burned to make the ancient charcoal, assuming that the half-life of ^{14}C is 5715 years?

39. IQ Scores The IQ scores for a sample of a class of returning adult students at a small northeastern college roughly follow the normal distribution

$$y = 0.0266e^{-(x-100)^2/450}, \quad 70 \le x \le 115$$

where x is the IQ score.

(a) Use a graphing utility to graph the function.

(b) From the graph in part (a), estimate the average IQ score of an adult student.

40. Education The amount of time (in hours per week) a student utilizes a math-tutoring center roughly follows the normal distribution

$$y = 0.7979e^{-(x-5.4)^2/0.5}, \quad 4 \le x \le 7$$

where x is the number of hours.

(a) Use a graphing utility to graph the function.

(b) From the graph in part (a), estimate the average number of hours per week a student uses the tutoring center.

41. Cell Sites A cell site is a site where electronic communications equipment is placed in a cellular network for the use of mobile phones. The numbers y of cell sites from 1985 through 2011 can be modeled by

$$y = \frac{269{,}573}{1 + 985e^{-0.308t}}$$

where t represents the year, with $t = 5$ corresponding to 1985. *(Source: CTIA-The Wireless Association)*

(a) Use the model to find the numbers of cell sites in the years 1998, 2003, and 2006.

(b) Use a graphing utility to graph the function.

(c) Use the graph to determine the year in which the number of cell sites reached 250,000.

(d) Confirm your answer to part (c) algebraically.

42. Population The populations P (in thousands) of a city from 2000 through 2010 can be modeled by

$$P = \frac{2632}{1 + 0.083e^{0.050t}}$$

where t represents the year, with $t = 0$ corresponding to 2000.

(a) Use the model to find the populations of the city in the years 2000, 2005, and 2010.

(b) Use a graphing utility to graph the function.

(c) Use the graph to determine the year in which the population will reach 2.2 million.

(d) Confirm your answer to part (c) algebraically.

43. Population Growth A conservation organization released 100 animals of an endangered species into a game preserve. The preserve has a carrying capacity of 1000 animals. The growth of the pack is modeled by the logistic curve

$$p(t) = \frac{1000}{1 + 9e^{-0.1656t}}$$

where t is measured in months (see figure).

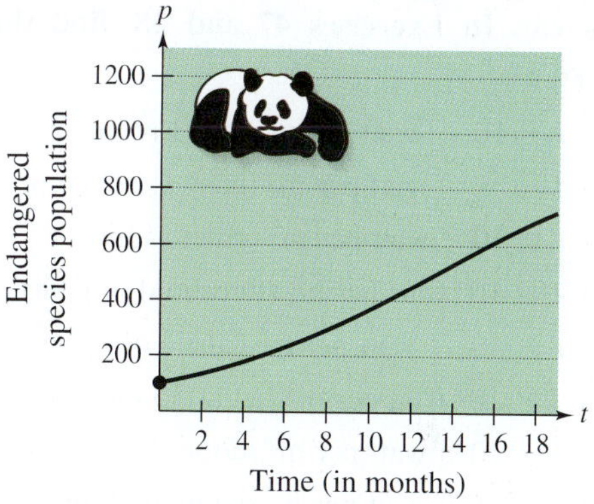

(a) Estimate the population after 5 months.

(b) After how many months is the population 500?

(c) Use a graphing utility to graph the function. Use the graph to determine the horizontal asymptotes, and interpret the meaning of the asymptotes in the context of the problem.

44. Sales After discontinuing all advertising for a tool kit in 2007, the manufacturer noted that sales began to drop according to the model

$$S = \frac{500{,}000}{1 + 0.4e^{kt}}$$

where S represents the number of units sold and $t = 7$ represents 2007. In 2011, 300,000 units were sold.

(a) Complete the model by solving for k.

(b) Estimate sales in 2015.

Geology **In Exercises 45 and 46, use the Richter scale**

$$R = \log \frac{I}{I_0}$$

for measuring the magnitude R of an earthquake.

45. Find the intensity I of an earthquake measuring R on the Richter scale (let $I_0 = 1$).

(a) South Shetland Islands in 2012: $R = 6.6$

(b) Oklahoma in 2011: $R = 5.6$

(c) Papua New Guinea in 2011: $R = 7.1$

46. Find the magnitude R of each earthquake of intensity I (let $I_0 = 1$).

(a) $I = 199{,}500{,}000$

(b) $I = 48{,}275{,}000$

(c) $I = 17{,}000$

Intensity of Sound **In Exercises 47–50, use the following information for determining sound intensity. The level of sound β, in decibels, with an intensity of I, is given by $\beta = 10 \log(I/I_0)$, where I_0 is an intensity of 10^{-12} watt per square meter, corresponding roughly to the faintest sound that can be heard by the human ear. In Exercises 47 and 48, find the level of sound β.**

47. (a) $I = 10^{-10}$ watt per m^2 (quiet room)

(b) $I = 10^{-5}$ watt per m^2 (busy street corner)

(c) $I = 10^{-8}$ watt per m^2 (quiet radio)

(d) $I = 10^{0}$ watt per m^2 (threshold of pain)

48. (a) $I = 10^{-11}$ watt per m^2 (rustle of leaves)

(b) $I = 10^{2}$ watt per m^2 (jet at 30 meters)

(c) $I = 10^{-4}$ watt per m^2 (door slamming)

(d) $I = 10^{-2}$ watt per m^2 (siren at 30 meters)

49. Due to the installation of noise suppression materials, the noise level in an auditorium decreased from 93 to 80 decibels. Find the percent decrease in the intensity level of the noise as a result of the installation of these materials.

50. Due to the installation of a muffler, the noise level of an engine decreased from 88 to 72 decibels. Find the percent decrease in the intensity level of the noise as a result of the installation of the muffler.

pH Levels **In Exercises 51–56, use the acidity model given by $\text{pH} = -\log[H^+]$, where acidity (pH) is a measure of the hydrogen ion concentration $[H^+]$ (measured in moles of hydrogen per liter) of a solution.**

51. Find the pH when $[H^+] = 2.3 \times 10^{-5}$.

52. Find the pH when $[H^+] = 1.13 \times 10^{-5}$.

53. Compute $[H^+]$ for a solution in which pH $= 5.8$.

54. Compute $[H^+]$ for a solution in which pH $= 3.2$.

55. Apple juice has a pH of 2.9 and drinking water has a pH of 8.0. The hydrogen ion concentration of the apple juice is how many times the concentration of drinking water?

56. The pH of a solution decreases by one unit. By what factor does the hydrogen ion concentration increase?

57. **Forensics** At 8:30 A.M., a coroner went to the home of a person who had died during the night. In order to estimate the time of death, the coroner took the person's temperature twice. At 9:00 A.M. the temperature was 85.7°F, and at 11:00 A.M. the temperature was 82.8°F. From these two temperatures, the coroner was able to determine that the time elapsed since death and the body temperature were related by the formula

$$t = -10 \ln \frac{T - 70}{98.6 - 70}$$

where t is the time in hours elapsed since the person died and T is the temperature (in degrees Fahrenheit) of the person's body. (This formula comes from a general cooling principle called *Newton's Law of Cooling*. It uses the assumptions that the person had a normal body temperature of 98.6°F at death and that the room temperature was a constant 70°F.) Use the formula to estimate the time of death of the person.

58. **Home Mortgage** A \$120,000 home mortgage for 30 years at $7\frac{1}{2}\%$ has a monthly payment of \$839.06. Part of the monthly payment covers the interest charge on the unpaid balance, and the remainder of the payment reduces the principal. The amount paid toward the interest is

$$u = M - \left(M - \frac{Pr}{12}\right)\left(1 + \frac{r}{12}\right)^{12t}$$

and the amount paid toward the reduction of the principal is

$$v = \left(M - \frac{Pr}{12}\right)\left(1 + \frac{r}{12}\right)^{12t}.$$

In these formulas, P is the size of the mortgage, r is the interest rate, M is the monthly payment, and t is the time (in years).

(a) Use a graphing utility to graph each function in the same viewing window. (The viewing window should show all 30 years of mortgage payments.)

(b) In the early years of the mortgage, is the greater part of the monthly payment paid toward the interest or the principal? Approximate the time when the monthly payment is evenly divided between interest and principal reduction.

(c) Repeat parts (a) and (b) for a repayment period of 20 years ($M = \$966.71$). What can you conclude?

59. Home Mortgage The total interest u paid on a home mortgage of P dollars at interest rate r for t years is

$$u = P\left[\frac{rt}{1 - \left(\dfrac{1}{1 + r/12}\right)^{12t}} - 1\right].$$

Consider a \$120,000 home mortgage at $7\frac{1}{2}\%$.

(a) Use a graphing utility to graph the total interest function.

(b) Approximate the length of the mortgage for which the total interest paid is the same as the size of the mortgage. Is it possible that some people are paying twice as much in interest charges as the size of the mortgage?

60. Data Analysis The table shows the time t (in seconds) required for a car to attain a speed of s miles per hour from a standing start.

DATA

Speed, s	Time, t
30	3.4
40	5.0
50	7.0
60	9.3
70	12.0
80	15.8
90	20.0

Spreadsheet at LarsonPrecalculus.com

Two models for these data are as follows.

$t_1 = 40.757 + 0.556s - 15.817 \ln s$

$t_2 = 1.2259 + 0.0023s^2$

(a) Use the *regression* feature of a graphing utility to find a linear model t_3 and an exponential model t_4 for the data.

(b) Use the graphing utility to graph the data and each model in the same viewing window.

(c) Create a table comparing the data with estimates obtained from each model.

(d) Use the results of part (c) to find the sum of the absolute values of the differences between the data and the estimated values given by each model. Based on the four sums, which model do you think best fits the data? Explain.

Exploration

True or False? **In Exercises 61–64, determine whether the statement is true or false. Justify your answer.**

61. The domain of a logistic growth function cannot be the set of real numbers.

62. A logistic growth function will always have an x-intercept.

63. The graph of $f(x) = \dfrac{4}{1 + 6e^{-2x}} + 5$ is the graph of $g(x) = \dfrac{4}{1 + 6e^{-2x}}$ shifted to the right five units.

64. The graph of a Gaussian model will never have an x-intercept.

65. Writing Use your school's library, the Internet, or some other reference source to write a paper describing John Napier's work with logarithms.

66. HOW DO YOU SEE IT? Identify each model as exponential growth, exponential decay, Gaussian, linear, logarithmic, logistic growth, quadratic, or none of the above. Explain your reasoning.

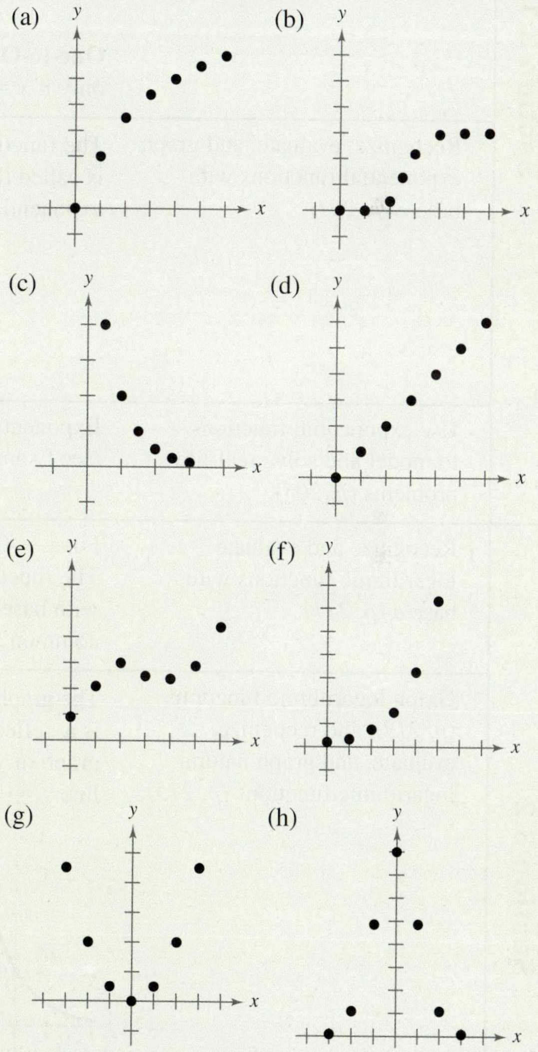

Project: Sales per Share To work an extended application analyzing the sales per share for Kohl's Corporation from 1995 through 2010, visit this text's website at *LarsonPrecalculus.com*. *(Source: Kohl's Corporation)*

Chapter Summary

<table>
<tr><th></th><th>What Did You Learn?</th><th>Explanation/Examples</th><th>Review Exercises</th></tr>
<tr><td rowspan="4">Section 3.1</td><td>Recognize and evaluate exponential functions with base a (p. 200).</td><td>The exponential function f with base a is denoted by $f(x) = a^x$, where $a > 0$, $a \neq 1$, and x is any real number.</td><td>1–6</td></tr>
<tr><td>Graph exponential functions and use the One-to-One Property (p. 201).</td><td>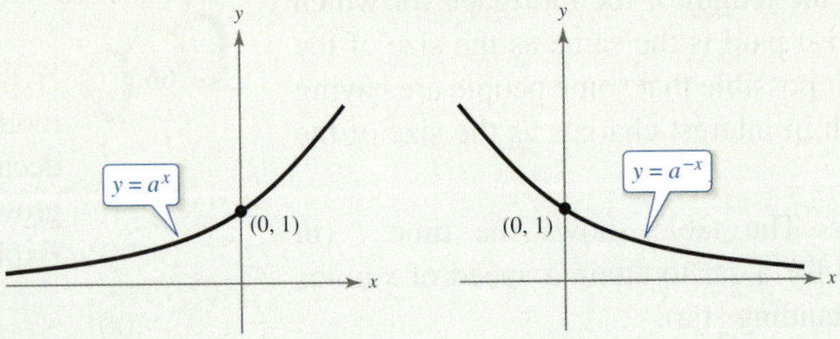

One-to-One Property: For $a > 0$ and $a \neq 1$, $a^x = a^y$ if and only if $x = y$.</td><td>7–20</td></tr>
<tr><td>Recognize, evaluate, and graph exponential functions with base e (p. 204).</td><td>The function $f(x) = e^x$ is called the natural exponential function.
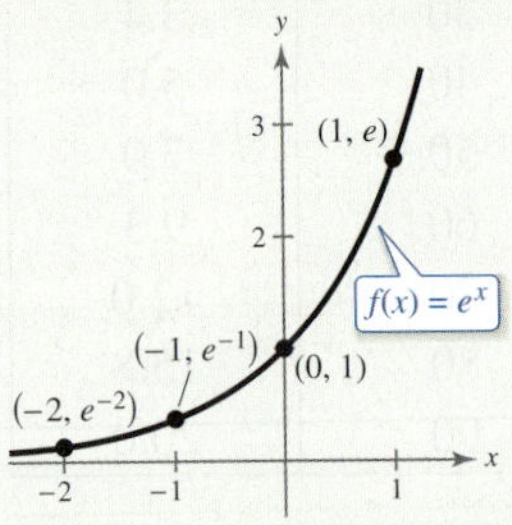
</td><td>21–28</td></tr>
<tr><td>Use exponential functions to model and solve real-life problems (p. 205).</td><td>Exponential functions are used in compound interest formulas (see Example 8) and in radioactive decay models (see Example 9).</td><td>29–32</td></tr>
<tr><td rowspan="3">Section 3.2</td><td>Recognize and evaluate logarithmic functions with base a (p. 211).</td><td>For $x > 0$, $a > 0$, and $a \neq 1$, $y = \log_a x$ if and only if $x = a^y$. The function $f(x) = \log_a x$ is called the logarithmic function with base a. The logarithmic function with base 10 is the common logarithmic function. It is denoted by $\log_{10}$ or log.</td><td>33–44</td></tr>
<tr><td>Graph logarithmic functions (p. 213), and recognize, evaluate, and graph natural logarithmic functions (p. 215).</td><td>The graph of $y = \log_a x$ is a reflection of the graph of $y = a^x$ in the line $y = x$.
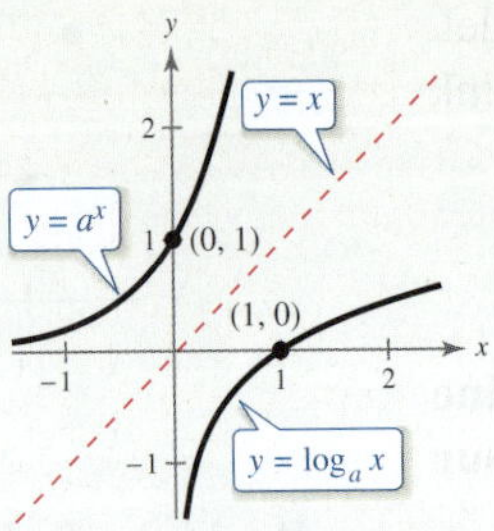

The function $f(x) = \ln x$, $x > 0$, is the natural logarithmic function. Its graph is a reflection of the graph of $f(x) = e^x$ in the line $y = x$.
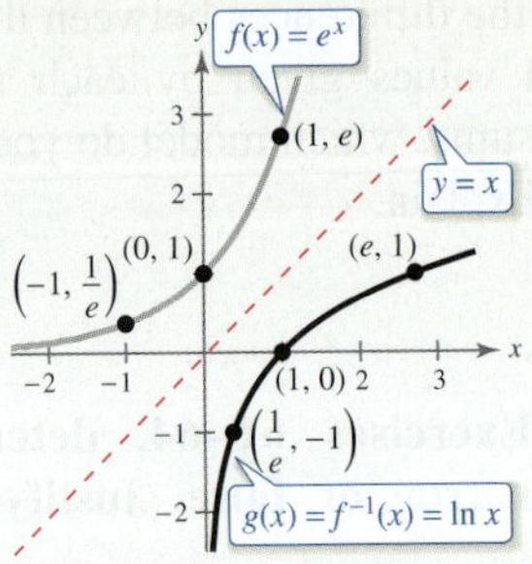
</td><td>45–56</td></tr>
<tr><td>Use logarithmic functions to model and solve real-life problems (p. 217).</td><td>A logarithmic function can model human memory. (See Example 11.)</td><td>57, 58</td></tr>
</table>

Section	What Did You Learn?	Explanation/Examples	Review Exercises
Section 3.3	Use the change-of-base formula to rewrite and evaluate logarithmic expressions *(p. 221)*.	Let a, b, and x be positive real numbers such that $a \neq 1$ and $b \neq 1$. Then $\log_a x$ can be converted to a different base as follows. **Base b**: $\log_a x = \dfrac{\log_b x}{\log_b a}$ **Base 10**: $\log_a x = \dfrac{\log x}{\log a}$ **Base e**: $\log_a x = \dfrac{\ln x}{\ln a}$	59–62
	Use properties of logarithms to evaluate, rewrite, expand, or condense logarithmic expressions *(p. 222)*.	Let a be a positive number such that $a \neq 1$, let n be a real number, and let u and v be positive real numbers. **1. Product Property:** $\log_a(uv) = \log_a u + \log_a v$ $\ln(uv) = \ln u + \ln v$ **2. Quotient Property:** $\log_a(u/v) = \log_a u - \log_a v$ $\ln(u/v) = \ln u - \ln v$ **3. Power Property:** $\log_a u^n = n \log_a u$, $\ln u^n = n \ln u$	63–78
	Use logarithmic functions to model and solve real-life problems *(p. 224)*.	Logarithmic functions can help you find an equation that relates the periods of several planets and their distances from the sun. (See Example 7.)	79, 80
Section 3.4	Solve simple exponential and logarithmic equations *(p. 228)*.	One-to-One Properties and Inverse Properties of exponential or logarithmic functions can help you solve exponential or logarithmic equations.	81–86
	Solve more complicated exponential equations *(p. 229)* and logarithmic equations *(p. 231)*.	To solve more complicated equations, rewrite the equations to allow the use of the One-to-One Properties or Inverse Properties of exponential or logarithmic functions. (See Examples 2–9.)	87–104
	Use exponential and logarithmic equations to model and solve real-life problems *(p. 233)*.	Exponential and logarithmic equations can help you find how long it will take to double an investment (see Example 10) and find the year in which companies reached a given amount of sales (see Example 11).	105, 106
Section 3.5	Recognize the five most common types of models involving exponential and logarithmic functions *(p. 238)*.	**1. Exponential growth model:** $y = ae^{bx}, \quad b > 0$ **2. Exponential decay model:** $y = ae^{-bx}, \quad b > 0$ **3. Gaussian model:** $y = ae^{-(x-b)^2/c}$ **4. Logistic growth model:** $y = \dfrac{a}{1 + be^{-rx}}$ **5. Logarithmic models:** $y = a + b \ln x$, $y = a + b \log x$	107–112
	Use exponential growth and decay functions to model and solve real-life problems *(p. 239)*.	An exponential growth function can help you model a population of fruit flies (see Example 2), and an exponential decay function can help you estimate the age of a fossil (see Example 3).	113, 114
	Use Gaussian functions *(p. 242)*, logistic growth functions *(p. 243)*, and logarithmic functions *(p. 244)* to model and solve real-life problems.	A Gaussian function can help you model SAT mathematics scores for high school graduates. (See Example 4.) A logistic growth function can help you model the spread of a flu virus. (See Example 5.) A logarithmic function can help you find the intensity of an earthquake given its magnitude. (See Example 6.)	115–117

Review Exercises

See CalcChat.com for tutorial help and worked-out solutions to odd-numbered exercises.

3.1 Evaluating an Exponential Function **In Exercises 1–6, evaluate the function at the indicated value of x. Round your result to three decimal places.**

1. $f(x) = 0.3^x, \quad x = 1.5$
2. $f(x) = 30^x, \quad x = \sqrt{3}$
3. $f(x) = 2^{-0.5x}, \quad x = \pi$
4. $f(x) = 1278^{x/5}, \quad x = 1$
5. $f(x) = 7(0.2^x), \quad x = -\sqrt{11}$
6. $f(x) = -14(5^x), \quad x = -0.8$

Transforming the Graph of an Exponential Function **In Exercises 7–10, use the graph of f to describe the transformation that yields the graph of g.**

7. $f(x) = 5^x, \quad g(x) = 5^x + 1$
8. $f(x) = 6^x, \quad g(x) = 6^{x+1}$
9. $f(x) = 3^x, \quad g(x) = 1 - 3^x$
10. $f(x) = \left(\frac{1}{2}\right)^x, \quad g(x) = -\left(\frac{1}{2}\right)^{x+2}$

Graphing an Exponential Function **In Exercises 11–16, use a graphing utility to construct a table of values for the function. Then sketch the graph of the function.**

11. $f(x) = 4^{-x} + 4$
12. $f(x) = 2.65^{x-1}$
13. $f(x) = 5^{x-2} + 4$
14. $f(x) = 2^{x-6} - 5$
15. $f(x) = \left(\frac{1}{2}\right)^{-x} + 3$
16. $f(x) = \left(\frac{1}{8}\right)^{x+2} - 5$

Using the One-to-One Property **In Exercises 17–20, use the One-to-One Property to solve the equation for x.**

17. $\left(\frac{1}{3}\right)^{x-3} = 9$
18. $3^{x+3} = \frac{1}{81}$
19. $e^{3x-5} = e^7$
20. $e^{8-2x} = e^{-3}$

Evaluating the Natural Exponential Function **In Exercises 21–24, evaluate $f(x) = e^x$ at the indicated value of x. Round your result to three decimal places.**

21. $x = 8$
22. $x = \frac{5}{8}$
23. $x = -1.7$
24. $x = 0.278$

Graphing a Natural Exponential Function **In Exercises 25–28, use a graphing utility to construct a table of values for the function. Then sketch the graph of the function.**

25. $h(x) = e^{-x/2}$
26. $h(x) = 2 - e^{-x/2}$
27. $f(x) = e^{x+2}$
28. $s(t) = 4e^{-2/t}, \quad t > 0$

29. **Waiting Times** The average time between incoming calls at a switchboard is 3 minutes. The probability F of waiting less than t minutes until the next incoming call is approximated by the model $F(t) = 1 - e^{-t/3}$. The switchboard has just received a call. Find the probability that the next call will be within
    (a) $\frac{1}{2}$ minute. (b) 2 minutes. (c) 5 minutes.

30. **Depreciation** After t years, the value V of a car that originally cost \$23,970 is given by $V(t) = 23{,}970\left(\frac{3}{4}\right)^t$.
    (a) Use a graphing utility to graph the function.
    (b) Find the value of the car 2 years after it was purchased.
    (c) According to the model, when does the car depreciate most rapidly? Is this realistic? Explain.
    (d) According to the model, when will the car have no value?

Compound Interest **In Exercises 31 and 32, complete the table to determine the balance A for P dollars invested at rate r for t years and compounded n times per year.**

n	1	2	4	12	365	Continuous
A						

31. $P = \$5000, \quad r = 3\%, \quad t = 10$ years
32. $P = \$4500, \quad r = 2.5\%, \quad t = 30$ years

3.2 Writing a Logarithmic Equation **In Exercises 33–36, write the exponential equation in logarithmic form. For example, the logarithmic form of $2^3 = 8$ is $\log_2 8 = 3$.**

33. $3^3 = 27$
34. $25^{3/2} = 125$
35. $e^{0.8} = 2.2255\ldots$
36. $e^0 = 1$

Evaluating a Logarithmic Function **In Exercises 37–40, evaluate the function at the indicated value of x without using a calculator.**

37. $f(x) = \log x, x = 1000$
38. $g(x) = \log_9 x, x = 3$
39. $g(x) = \log_2 x, x = \frac{1}{4}$
40. $f(x) = \log_3 x, x = \frac{1}{81}$

Using the One-to-One Property **In Exercises 41–44, use the One-to-One Property to solve the equation for x.**

41. $\log_4(x + 7) = \log_4 14$
42. $\log_8(3x - 10) = \log_8 5$
43. $\ln(x + 9) = \ln 4$
44. $\ln(2x - 1) = \ln 11$

Sketching the Graph of a Logarithmic Function **In Exercises 45–48, find the domain, x-intercept, and vertical asymptote of the logarithmic function and sketch its graph.**

45. $g(x) = \log_7 x$
46. $f(x) = \log\left(\frac{x}{3}\right)$
47. $f(x) = 4 - \log(x + 5)$
48. $f(x) = \log(x - 3) + 1$

Evaluating a Logarithmic Function on a Calculator **In Exercises 49–52, use a calculator to evaluate the function at the indicated value of x. Round your result to three decimal places, if necessary.**

49. $f(x) = \ln x, \quad x = 22.6$

50. $f(x) = \ln x, \quad x = e^{-12}$

51. $f(x) = \frac{1}{2}\ln x, \quad x = \sqrt{e}$

52. $f(x) = 5\ln x, \quad x = 0.98$

Graphing a Natural Logarithmic Function **In Exercises 53–56, find the domain, x-intercept, and vertical asymptote of the logarithmic function and sketch its graph.**

53. $f(x) = \ln x + 3$

54. $f(x) = \ln(x - 3)$

55. $h(x) = \ln(x^2)$

56. $f(x) = \frac{1}{4}\ln x$

57. Antler Spread The antler spread a (in inches) and shoulder height h (in inches) of an adult male American elk are related by the model

$$h = 116\log(a + 40) - 176.$$

Approximate the shoulder height of a male American elk with an antler spread of 55 inches.

58. Snow Removal The number of miles s of roads cleared of snow is approximated by the model

$$s = 25 - \frac{13\ln(h/12)}{\ln 3}, \quad 2 \le h \le 15$$

where h is the depth of the snow in inches. Use this model to find s when $h = 10$ inches.

3.3 Using the Change-of-Base Formula **In Exercises 59–62, evaluate the logarithm using the change-of-base formula (a) with common logarithms and (b) with natural logarithms. Round your results to three decimal places.**

59. $\log_2 6$

60. $\log_{12} 200$

61. $\log_{1/2} 5$

62. $\log_3 0.28$

Using Properties of Logarithms **In Exercises 63–66, use the properties of logarithms to rewrite and simplify the logarithmic expression.**

63. $\log 18$

64. $\log_2\left(\frac{1}{12}\right)$

65. $\ln 20$

66. $\ln(3e^{-4})$

Expanding a Logarithmic Expression **In Exercises 67–72, use the properties of logarithms to expand the expression as a sum, difference, and/or constant multiple of logarithms. (Assume all variables are positive.)**

67. $\log_5 5x^2$

68. $\log 7x^4$

69. $\log_3 \frac{9}{\sqrt{x}}$

70. $\log_7 \frac{\sqrt[3]{x}}{14}$

71. $\ln x^2y^2z$

72. $\ln\left(\frac{y-1}{4}\right)^2, \quad y > 1$

Condensing a Logarithmic Expression **In Exercises 73–78, condense the expression to the logarithm of a single quantity.**

73. $\log_2 5 + \log_2 x$

74. $\log_6 y - 2\log_6 z$

75. $\ln x - \frac{1}{4}\ln y$

76. $3\ln x + 2\ln(x + 1)$

77. $\frac{1}{2}\log_3 x - 2\log_3(y + 8)$

78. $5\ln(x - 2) - \ln(x + 2) - 3\ln x$

79. Climb Rate The time t (in minutes) for a small plane to climb to an altitude of h feet is modeled by $t = 50\log[18{,}000/(18{,}000 - h)]$, where 18,000 feet is the plane's absolute ceiling.

(a) Determine the domain of the function in the context of the problem.

(b) Use a graphing utility to graph the function and identify any asymptotes.

(c) As the plane approaches its absolute ceiling, what can be said about the time required to increase its altitude?

(d) Find the time for the plane to climb to an altitude of 4000 feet.

80. Human Memory Model Students in a learning theory study took an exam and then retested monthly for 6 months with an equivalent exam. The data obtained in the study are given by the ordered pairs (t, s), where t is the time in months after the initial exam and s is the average score for the class. Use the data to find a logarithmic equation that relates t and s.

$(1, 84.2), (2, 78.4), (3, 72.1),$
$(4, 68.5), (5, 67.1), (6, 65.3)$

3.4 Solving a Simple Equation **In Exercises 81–86, solve for x.**

81. $5^x = 125$

82. $6^x = \frac{1}{216}$

83. $e^x = 3$

84. $\log_6 x = -1$

85. $\ln x = 4$

86. $\ln x = -1.6$

Solving an Exponential Equation **In Exercises 87–90, solve the exponential equation algebraically. Approximate the result to three decimal places.**

87. $e^{4x} = e^{x^2+3}$

88. $e^{3x} = 25$

89. $2^x - 3 = 29$

90. $e^{2x} - 6e^x + 8 = 0$

Graphing and Solving an Exponential Equation **In Exercises 91 and 92, use a graphing utility to graph and solve the equation. Approximate the result to three decimal places. Verify your result algebraically.**

91. $25e^{-0.3x} = 12$

92. $2^x = 3 + x - e^x$

Solving a Logarithmic Equation **In Exercises 93–100, solve the logarithmic equation algebraically. Approximate the result to three decimal places.**

93. $\ln 3x = 8.2$

94. $4 \ln 3x = 15$

95. $\ln x - \ln 3 = 2$

96. $\ln \sqrt{x+8} = 3$

97. $\log_8(x-1) = \log_8(x-2) - \log_8(x+2)$

98. $\log_6(x+2) - \log_6 x = \log_6(x+5)$

99. $\log(1-x) = -1$

100. $\log(-x-4) = 2$

Graphing and Solving a Logarithmic Equation **In Exercises 101–104, use a graphing utility to graph and solve the equation. Approximate the result to three decimal places. Verify your result algebraically.**

101. $2 \ln(x+3) - 3 = 0$

102. $x - 2 \log(x+4) = 0$

103. $6 \log(x^2+1) - x = 0$

104. $3 \ln x + 2 \log x = e^x - 25$

105. Compound Interest You deposit \$8500 in an account that pays 1.5% interest, compounded continuously. How long will it take for the money to triple?

106. Meteorology The speed of the wind S (in miles per hour) near the center of a tornado and the distance d (in miles) the tornado travels are related by the model $S = 93 \log d + 65$. On March 18, 1925, a large tornado struck portions of Missouri, Illinois, and Indiana with a wind speed at the center of about 283 miles per hour. Approximate the distance traveled by this tornado.

3.5 Matching a Function with Its Graph **In Exercises 107–112, match the function with its graph. [The graphs are labeled (a), (b), (c), (d), (e), and (f).]**

(a)

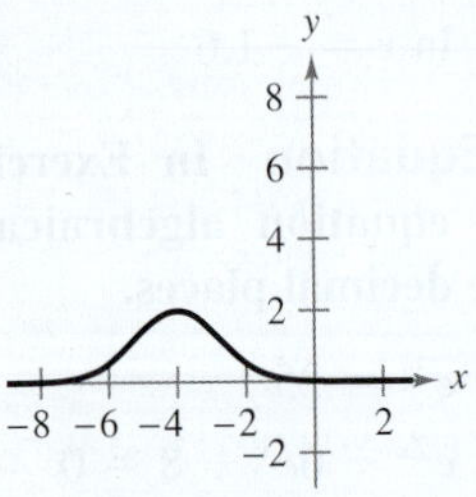

(b)

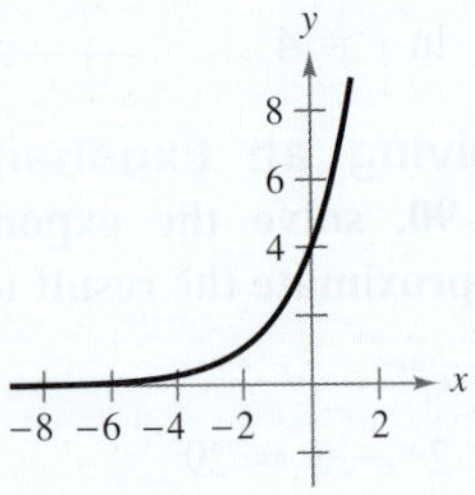

(c)

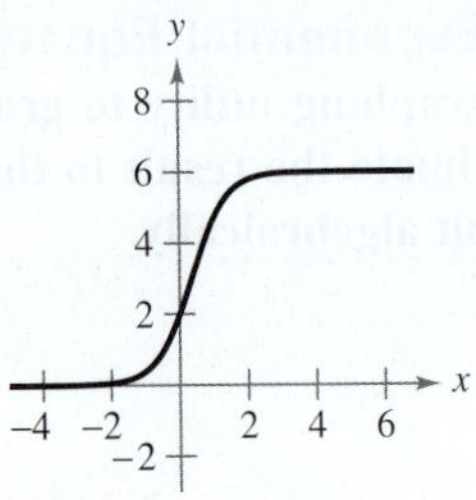

(d)

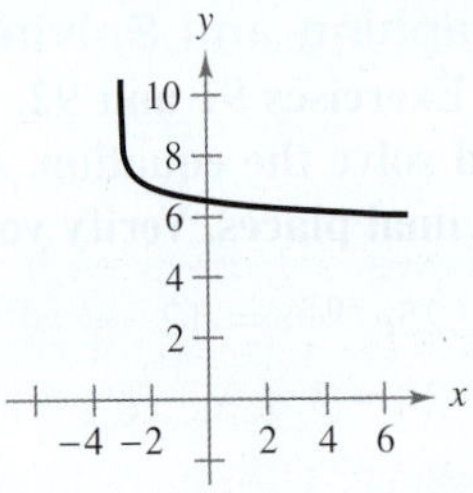

(e)

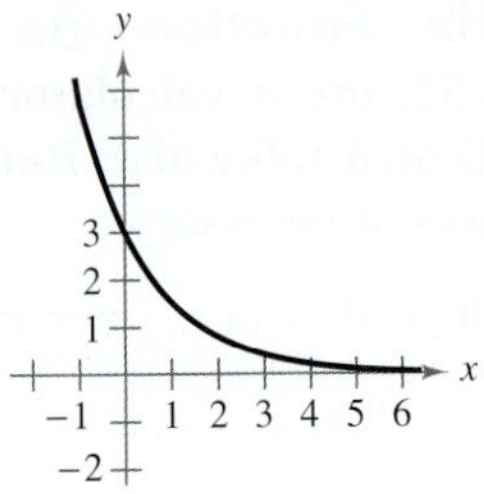

(f)

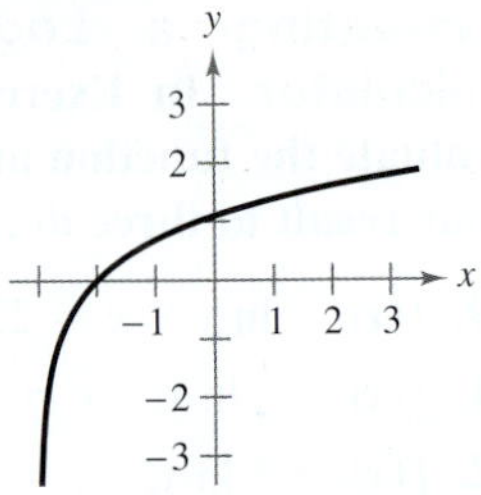

107. $y = 3e^{-2x/3}$ **108.** $y = 4e^{2x/3}$

109. $y = \ln(x+3)$ **110.** $y = 7 - \log(x+3)$

111. $y = 2e^{-(x+4)^2/3}$ **112.** $y = \dfrac{6}{1+2e^{-2x}}$

113. Finding an Exponential Model Find the exponential model $y = ae^{bx}$ that fits the points $(0, 2)$ and $(4, 3)$.

114. Wildlife Population A species of bat is in danger of becoming extinct. Five years ago, the total population of the species was 2000. Two years ago, the total population of the species was 1400. What was the total population of the species one year ago?

115. Test Scores The test scores for a biology test follow a normal distribution modeled by

$$y = 0.0499e^{-(x-71)^2/128}, \quad 40 \le x \le 100$$

where x is the test score. Use a graphing utility to graph the equation and estimate the average test score.

116. Typing Speed In a typing class, the average number N of words per minute typed after t weeks of lessons is

$N = 157/(1 + 5.4e^{-0.12t})$.

Find the time necessary to type (a) 50 words per minute and (b) 75 words per minute.

117. Sound Intensity The relationship between the number of decibels β and the intensity of a sound I in watts per square meter is

$\beta = 10 \log(I/10^{-12})$.

Find I for each decibel level β.

(a) $\beta = 60$ (b) $\beta = 135$ (c) $\beta = 1$

Exploration

118. Graph of an Exponential Function Consider the graph of $y = e^{kt}$. Describe the characteristics of the graph when k is positive and when k is negative.

True or False? **In Exercises 119 and 120, determine whether the equation is true or false. Justify your answer.**

119. $\log_b b^{2x} = 2x$

120. $\ln(x+y) = \ln x + \ln y$

Chapter Test

See CalcChat.com for tutorial help and worked-out solutions to odd-numbered exercises.

Take this test as you would take a test in class. When you are finished, check your work against the answers given in the back of the book.

In Exercises 1–4, evaluate the expression. Round your result to three decimal places.

1. $4.2^{0.6}$
2. $4^{3\pi/2}$
3. $e^{-7/10}$
4. $e^{3.1}$

In Exercises 5–7, construct a table of values for the function. Then sketch the graph of the function.

5. $f(x) = 10^{-x}$
6. $f(x) = -6^{x-2}$
7. $f(x) = 1 - e^{2x}$

8. Evaluate (a) $\log_7 7^{-0.89}$ and (b) $4.6 \ln e^2$.

In Exercises 9–11, find the domain, x-intercept, and vertical asymptote of the logarithmic function and sketch its graph.

9. $f(x) = -\log x - 6$
10. $f(x) = \ln(x - 4)$
11. $f(x) = 1 + \ln(x + 6)$

In Exercises 12–14, evaluate the logarithm using the change-of-base formula. Round your result to three decimal places.

12. $\log_7 44$
13. $\log_{16} 0.63$
14. $\log_{3/4} 24$

In Exercises 15–17, use the properties of logarithms to expand the expression as a sum, difference, and/or constant multiple of logarithms. (Assume all variables are positive.)

15. $\log_2 3a^4$
16. $\ln \dfrac{5\sqrt{x}}{6}$
17. $\log \dfrac{(x-1)^3}{y^2 z}$

In Exercises 18–20, condense the expression to the logarithm of a single quantity.

18. $\log_3 13 + \log_3 y$
19. $4 \ln x - 4 \ln y$
20. $3 \ln x - \ln(x + 3) + 2 \ln y$

In Exercises 21–26, solve the equation algebraically. Approximate the result to three decimal places, if necessary.

21. $5^x = \dfrac{1}{25}$
22. $3e^{-5x} = 132$
23. $\dfrac{1025}{8 + e^{4x}} = 5$
24. $\ln x = \dfrac{1}{2}$
25. $18 + 4 \ln x = 7$
26. $\log x + \log(x - 15) = 2$

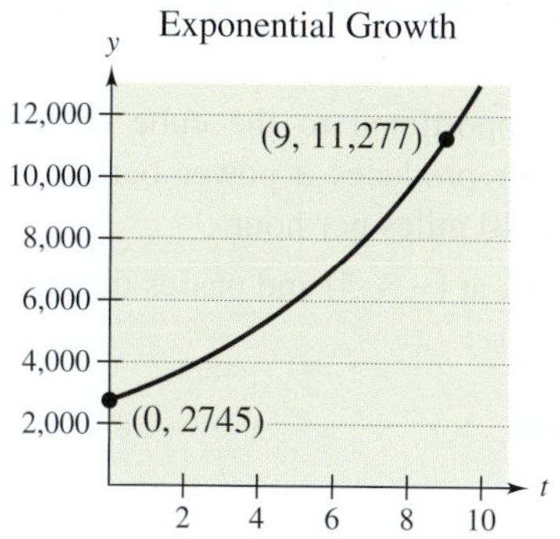

Figure for 27

27. Find the exponential growth model that fits the points shown in the graph.
28. The half-life of radioactive actinium (^{227}Ac) is 21.77 years. What percent of a present amount of radioactive actinium will remain after 19 years?
29. A model that can predict the height H (in centimeters) of a child based on his or her age is $H = 70.228 + 5.104x + 9.222 \ln x$, $\frac{1}{4} \le x \le 6$, where x is the age of the child in years. *(Source: Snapshots of Applications in Mathematics)*
   (a) Construct a table of values for the model. Then sketch the graph of the model.
   (b) Use the graph from part (a) to predict the height of a child when he or she is four years old. Then confirm your prediction algebraically.

Cumulative Test for Chapters 1–3

See CalcChat.com for tutorial help and worked-out solutions to odd-numbered exercises.

Take this test as you would take a test in class. When you are finished, check your work against the answers given in the back of the book.

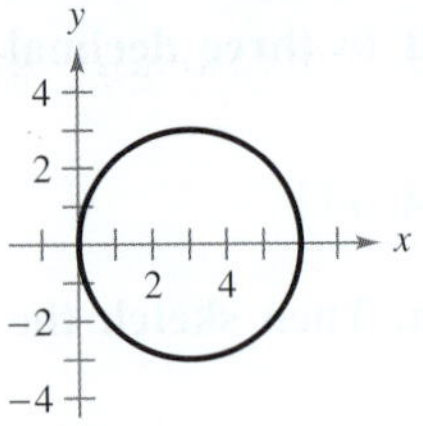

Figure for 6

1. Plot the points $(-2, 5)$ and $(3, -1)$. Find the coordinates of the midpoint of the line segment joining the points and the distance between the points.

In Exercises 2–4, sketch the graph of the equation.

2. $x - 3y + 12 = 0$ **3.** $y = x^2 - 9$ **4.** $y = \sqrt{4 - x}$

5. Find an equation of the line passing through $\left(-\frac{1}{2}, 1\right)$ and $(3, 8)$.

6. Explain why the graph at the left does not represent y as a function of x.

7. Evaluate (if possible) the function $f(x) = \dfrac{x}{x - 2}$ at each specified value of the independent variable.

(a) $f(6)$ (b) $f(2)$ (c) $f(s + 2)$

8. Compare the graph of each function with the graph of $y = \sqrt[3]{x}$. (*Note:* It is not necessary to sketch the graphs.)

(a) $r(x) = \frac{1}{2}\sqrt[3]{x}$ (b) $h(x) = \sqrt[3]{x} + 2$ (c) $g(x) = \sqrt[3]{x + 2}$

In Exercises 9 and 10, find (a) $(f + g)(x)$, (b) $(f - g)(x)$, (c) $(fg)(x)$, and (d) $(f/g)(x)$. What is the domain of f/g?

9. $f(x) = x - 3,\quad g(x) = 4x + 1$

10. $f(x) = \sqrt{x - 1},\quad g(x) = x^2 + 1$

In Exercises 11 and 12, find (a) $f \circ g$ and (b) $g \circ f$. Find the domain of each composite function.

11. $f(x) = 2x^2,\quad g(x) = \sqrt{x + 6}$

12. $f(x) = x - 2,\quad g(x) = |x|$

13. Determine whether $h(x) = -5x + 3$ has an inverse function. If so, find the inverse function.

14. The power P produced by a wind turbine is proportional to the cube of the wind speed S. A wind speed of 27 miles per hour produces a power output of 750 kilowatts. Find the output for a wind speed of 40 miles per hour.

15. Find the quadratic function whose graph has a vertex at $(-8, 5)$ and passes through the point $(-4, -7)$.

In Exercises 16–18, sketch the graph of the function.

16. $h(x) = -(x^2 + 4x)$

17. $f(t) = \frac{1}{4}t(t - 2)^2$

18. $g(s) = s^2 + 4s + 10$

In Exercises 19–21, find all the zeros of the function and write the function as a product of linear factors.

19. $f(x) = x^3 + 2x^2 + 4x + 8$

20. $f(x) = x^4 + 4x^3 - 21x^2$

21. $f(x) = 2x^4 - 11x^3 + 30x^2 - 62x - 40$

22. Use long division to divide $6x^3 - 4x^2$ by $2x^2 + 1$.

23. Use synthetic division to divide $3x^4 + 2x^2 - 5x + 3$ by $x - 2$.

24. Use the Intermediate Value Theorem and the *table* feature of a graphing utility to find intervals one unit in length in which the function $g(x) = x^3 + 3x^2 - 6$ is guaranteed to have a zero. Adjust the table to approximate the real zeros of the function.

In Exercises 25–27, sketch the graph of the rational function. Identify all intercepts and asymptotes.

25. $f(x) = \dfrac{2x}{x^2 + 2x - 3}$

26. $f(x) = \dfrac{x^2 - 4}{x^2 + x - 2}$

27. $f(x) = \dfrac{x^3 - 2x^2 - 9x + 18}{x^2 + 4x + 3}$

In Exercises 28 and 29, solve the inequality. Then graph the solution set.

28. $2x^3 - 18x \le 0$

29. $\dfrac{1}{x + 1} \ge \dfrac{1}{x + 5}$

In Exercises 30 and 31, use the graph of *f* to describe the transformation that yields the graph of *g*.

30. $f(x) = \left(\frac{2}{5}\right)^x, \quad g(x) = -\left(\frac{2}{5}\right)^{-x+3}$

31. $f(x) = 2.2^x, \quad g(x) = -2.2^x + 4$

In Exercises 32–35, use a calculator to evaluate the expression. Round your result to three decimal places.

32. $\log 98$

33. $\log \frac{6}{7}$

34. $\ln \sqrt{31}$

35. $\ln\left(\sqrt{40} - 5\right)$

36. Use the properties of logarithms to expand $\ln\left(\dfrac{x^2 - 16}{x^4}\right)$, where $x > 4$.

37. Condense $2 \ln x - \frac{1}{2}\ln(x + 5)$ to the logarithm of a single quantity.

In Exercises 38–40, solve the equation algebraically. Approximate the result to three decimal places.

38. $6e^{2x} = 72$

39. $e^{2x} - 13e^x + 42 = 0$

40. $\ln\sqrt{x + 2} = 3$

41. The sales S (in billions of dollars) of lottery tickets in the United States from 2000 through 2010 are shown in the table. *(Source: TLF Publications, Inc.)*

(a) Use a graphing utility to create a scatter plot of the data. Let t represent the year, with $t = 0$ corresponding to 2000.

(b) Use the *regression* feature of the graphing utility to find a cubic model for the data.

(c) Use the graphing utility to graph the model in the same viewing window used for the scatter plot. How well does the model fit the data?

(d) Use the model to predict the sales of lottery tickets in 2018. Does your answer seem reasonable? Explain.

DATA

Year	Sales, S
2000	37.2
2001	38.4
2002	42.0
2003	43.5
2004	47.7
2005	47.4
2006	51.6
2007	52.4
2008	53.4
2009	53.1
2010	54.2

Spreadsheet at LarsonPrecalculus.com

Table for 41

42. The number N of bacteria in a culture is given by the model $N = 175e^{kt}$, where t is the time in hours. If $N = 420$ when $t = 8$, then estimate the time required for the population to double in size.

Proofs in Mathematics

Each of the following three properties of logarithms can be proved by using properties of exponential functions.

> **SLIDE RULES**
>
> William Oughtred (1574–1660) invented the slide rule in 1625. The slide rule is a computational device with a sliding portion and a fixed portion. A slide rule enables you to perform multiplication by using the Product Property of Logarithms. There are other slide rules that allow for the calculation of roots and trigonometric functions. Mathematicians and engineers used slide rules until the invention of the hand-held calculator in 1972.

Properties of Logarithms *(p. 222)*

Let a be a positive number such that $a \neq 1$, and let n be a real number. If u and v are positive real numbers, then the following properties are true.

	Logarithm with Base a	Natural Logarithm
1. Product Property:	$\log_a(uv) = \log_a u + \log_a v$	$\ln(uv) = \ln u + \ln v$
2. Quotient Property:	$\log_a \frac{u}{v} = \log_a u - \log_a v$	$\ln \frac{u}{v} = \ln u - \ln v$
3. Power Property:	$\log_a u^n = n \log_a u$	$\ln u^n = n \ln u$

Proof

Let

$$x = \log_a u \quad \text{and} \quad y = \log_a v.$$

The corresponding exponential forms of these two equations are

$$a^x = u \quad \text{and} \quad a^y = v.$$

To prove the Product Property, multiply u and v to obtain

$$\begin{aligned} uv &= a^x a^y \\ &= a^{x+y}. \end{aligned}$$

The corresponding logarithmic form of $uv = a^{x+y}$ is $\log_a(uv) = x + y$. So,

$$\log_a(uv) = \log_a u + \log_a v.$$

To prove the Quotient Property, divide u by v to obtain

$$\begin{aligned} \frac{u}{v} &= \frac{a^x}{a^y} \\ &= a^{x-y}. \end{aligned}$$

The corresponding logarithmic form of $\frac{u}{v} = a^{x-y}$ is $\log_a \frac{u}{v} = x - y$. So,

$$\log_a \frac{u}{v} = \log_a u - \log_a v.$$

To prove the Power Property, substitute a^x for u in the expression $\log_a u^n$, as follows.

$$\begin{aligned} \log_a u^n &= \log_a(a^x)^n && \text{Substitute } a^x \text{ for } u. \\ &= \log_a a^{nx} && \text{Property of Exponents} \\ &= nx && \text{Inverse Property of Logarithms} \\ &= n \log_a u && \text{Substitute } \log_a u \text{ for } x. \end{aligned}$$

So, $\log_a u^n = n \log_a u$. ■

P.S. Problem Solving

1. **Graphical Analysis** Graph the exponential function $y = a^x$ for $a = 0.5$, 1.2, and 2.0. Which of these curves intersects the line $y = x$? Determine all positive numbers a for which the curve $y = a^x$ intersects the line $y = x$.

2. **Graphical Analysis** Use a graphing utility to graph $y_1 = e^x$ and each of the functions $y_2 = x^2$, $y_3 = x^3$, $y_4 = \sqrt{x}$, and $y_5 = |x|$. Which function increases at the greatest rate as x approaches $+\infty$?

3. **Conjecture** Use the result of Exercise 2 to make a conjecture about the rate of growth of $y_1 = e^x$ and $y = x^n$, where n is a natural number and x approaches $+\infty$.

4. **Implication of "Growing Exponentially"** Use the results of Exercises 2 and 3 to describe what is implied when it is stated that a quantity is growing exponentially.

5. **Exponential Function** Given the exponential function

$$f(x) = a^x$$

show that

(a) $f(u + v) = f(u) \cdot f(v)$. (b) $f(2x) = [f(x)]^2$.

6. **Hyperbolic Functions** Given that

$$f(x) = \frac{e^x + e^{-x}}{2} \quad \text{and} \quad g(x) = \frac{e^x - e^{-x}}{2}$$

show that

$$[f(x)]^2 - [g(x)]^2 = 1.$$

7. **Graphical Analysis** Use a graphing utility to compare the graph of the function $y = e^x$ with the graph of each given function. [$n!$ (read "n factorial") is defined as $n! = 1 \cdot 2 \cdot 3 \cdot \cdot \cdot (n - 1) \cdot n$.]

(a) $y_1 = 1 + \dfrac{x}{1!}$

(b) $y_2 = 1 + \dfrac{x}{1!} + \dfrac{x^2}{2!}$

(c) $y_3 = 1 + \dfrac{x}{1!} + \dfrac{x^2}{2!} + \dfrac{x^3}{3!}$

8. **Identifying a Pattern** Identify the pattern of successive polynomials given in Exercise 7. Extend the pattern one more term and compare the graph of the resulting polynomial function with the graph of $y = e^x$. What do you think this pattern implies?

9. **Finding an Inverse Function** Graph the function

$$f(x) = e^x - e^{-x}.$$

From the graph, the function appears to be one-to-one. Assuming that the function has an inverse function, find $f^{-1}(x)$.

10. **Finding a Pattern for an Inverse Function** Find a pattern for $f^{-1}(x)$ when

$$f(x) = \frac{a^x + 1}{a^x - 1}$$

where $a > 0$, $a \neq 1$.

11. **Determining the Equation of a Graph** By observation, determine whether equation (a), (b), or (c) corresponds to the graph. Explain your reasoning.

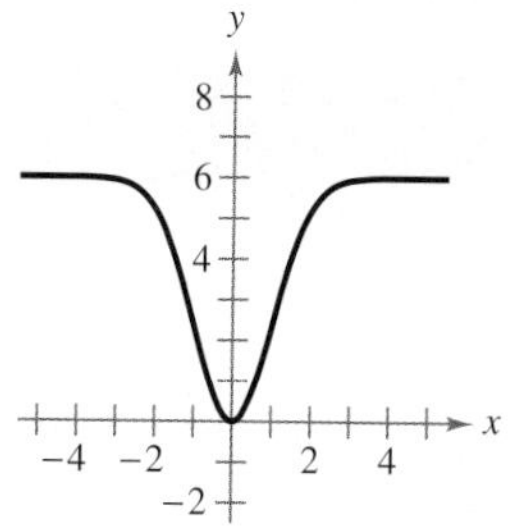

(a) $y = 6e^{-x^2/2}$

(b) $y = \dfrac{6}{1 + e^{-x/2}}$

(c) $y = 6(1 - e^{-x^2/2})$

12. **Simple and Compound Interest** You have two options for investing \$500. The first earns 7% compounded annually, and the second earns 7% simple interest. The figure shows the growth of each investment over a 30-year period.

(a) Identify which graph represents each type of investment. Explain your reasoning.

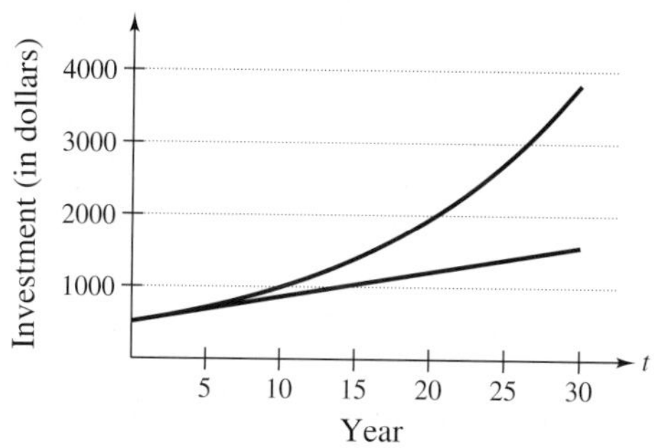

(b) Verify your answer in part (a) by finding the equations that model the investment growth and by graphing the models.

(c) Which option would you choose? Explain your reasoning.

13. **Radioactive Decay** Two different samples of radioactive isotopes are decaying. The isotopes have initial amounts of c_1 and c_2, as well as half-lives of k_1 and k_2, respectively. Find the time t required for the samples to decay to equal amounts.

14. **Bacteria Decay** A lab culture initially contains 500 bacteria. Two hours later, the number of bacteria decreases to 200. Find the exponential decay model of the form

$$B = B_0 a^{kt}$$

that approximates the number of bacteria after t hours.

15. **Colonial Population** The table shows the colonial population estimates of the American colonies from 1700 through 1780. *(Source: U.S. Census Bureau)*

Year	Population
1700	250,900
1710	331,700
1720	466,200
1730	629,400
1740	905,600
1750	1,170,800
1760	1,593,600
1770	2,148,400
1780	2,780,400

DATA Spreadsheet at LarsonPrecalculus.com

In each of the following, let y represent the population in the year t, with $t = 0$ corresponding to 1700.

(a) Use the *regression* feature of a graphing utility to find an exponential model for the data.

(b) Use the *regression* feature of the graphing utility to find a quadratic model for the data.

(c) Use the graphing utility to plot the data and the models from parts (a) and (b) in the same viewing window.

(d) Which model is a better fit for the data? Would you use this model to predict the population of the United States in 2018? Explain your reasoning.

16. **Ratio of Logarithms** Show that

$$\frac{\log_a x}{\log_{a/b} x} = 1 + \log_a \frac{1}{b}.$$

17. **Solving a Logarithmic Equation** Solve

$$(\ln x)^2 = \ln x^2.$$

18. **Graphical Analysis** Use a graphing utility to compare the graph of the function $y = \ln x$ with the graph of each given function.

(a) $y_1 = x - 1$

(b) $y_2 = (x - 1) - \frac{1}{2}(x - 1)^2$

(c) $y_3 = (x - 1) - \frac{1}{2}(x - 1)^2 + \frac{1}{3}(x - 1)^3$

19. **Identifying a Pattern** Identify the pattern of successive polynomials given in Exercise 18. Extend the pattern one more term and compare the graph of the resulting polynomial function with the graph of $y = \ln x$. What do you think the pattern implies?

20. **Finding Slope and y-Intercept** Using

$$y = ab^x \quad \text{and} \quad y = ax^b$$

take the natural logarithm of each side of each equation. What are the slope and y-intercept of the line relating x and $\ln y$ for $y = ab^x$? What are the slope and y-intercept of the line relating $\ln x$ and $\ln y$ for $y = ax^b$?

Ventilation Rate **In Exercises 21 and 22, use the model**

$$y = 80.4 - 11 \ln x, \quad 100 \le x \le 1500$$

which approximates the minimum required ventilation rate in terms of the air space per child in a public school classroom. In the model, x is the air space per child in cubic feet and y is the ventilation rate per child in cubic feet per minute.

21. Use a graphing utility to graph the model and approximate the required ventilation rate when there are 300 cubic feet of air space per child.

22. In a classroom designed for 30 students, the air conditioning system can move 450 cubic feet of air per minute.

(a) Determine the ventilation rate per child in a full classroom.

(b) Estimate the air space required per child.

(c) Determine the minimum number of square feet of floor space required for the room when the ceiling height is 30 feet.

Data Analysis **In Exercises 23–26, (a) use a graphing utility to create a scatter plot of the data, (b) decide whether the data could best be modeled by a linear model, an exponential model, or a logarithmic model, (c) explain why you chose the model you did in part (b), (d) use the *regression* feature of the graphing utility to find the model you chose in part (b) for the data and graph the model with the scatter plot, and (e) determine how well the model you chose fits the data.**

23. (1, 2.0), (1.5, 3.5), (2, 4.0), (4, 5.8), (6, 7.0), (8, 7.8)

24. (1, 4.4), (1.5, 4.7), (2, 5.5), (4, 9.9), (6, 18.1), (8, 33.0)

25. (1, 7.5), (1.5, 7.0), (2, 6.8), (4, 5.0), (6, 3.5), (8, 2.0)

26. (1, 5.0), (1.5, 6.0), (2, 6.4), (4, 7.8), (6, 8.6), (8, 9.0)

4 Trigonometry

Television Coverage *(Exercise 84, page 319)*

Waterslide Design *(Exercise 32, page 337)*

Respiratory Cycle *(Exercise 88, page 308)*

Meteorology *(Exercise 99, page 298)*

Skateboarding *(Example 10, page 285)*

Clockwise from top left, ariadna de raadt/Shutterstock.com; Alexey Bykov/Shutterstock.com; tusharkoley/Shutterstock.com; Vladimir Ivanovich Danilov/Shutterstock; AISPIX by Image Source/Shutterstock.com

4.1 Radian and Degree Measure

Angles can help you model and solve real-life problems. For instance, in Exercise 68 on page 271, you will use angles to find the speed of a bicycle.

- **Describe angles.**
- **Use radian measure.**
- **Use degree measure.**
- **Use angles to model and solve real-life problems.**

Angles

As derived from the Greek language, the word **trigonometry** means "measurement of triangles." Initially, trigonometry dealt with relationships among the sides and angles of triangles and was used in the development of astronomy, navigation, and surveying. With the development of calculus and the physical sciences in the 17th century, a different perspective arose—one that viewed the classic trigonometric relationships as *functions* with the set of real numbers as their domains. Consequently, the applications of trigonometry expanded to include a vast number of physical phenomena, such as sound waves, planetary orbits, vibrating strings, pendulums, and orbits of atomic particles.

This text incorporates *both* perspectives, starting with angles and their measures.

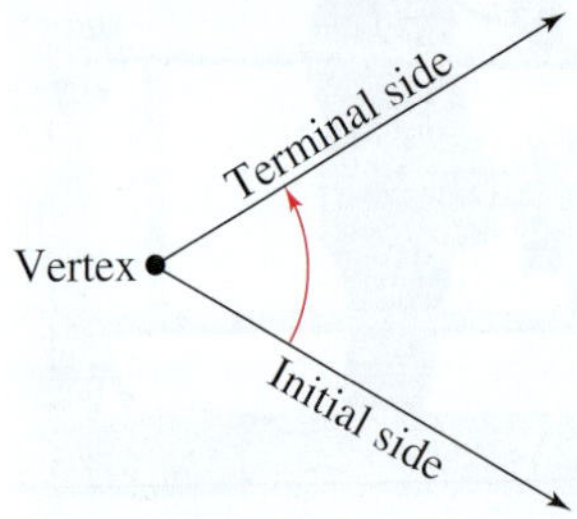

Angle
Figure 4.1

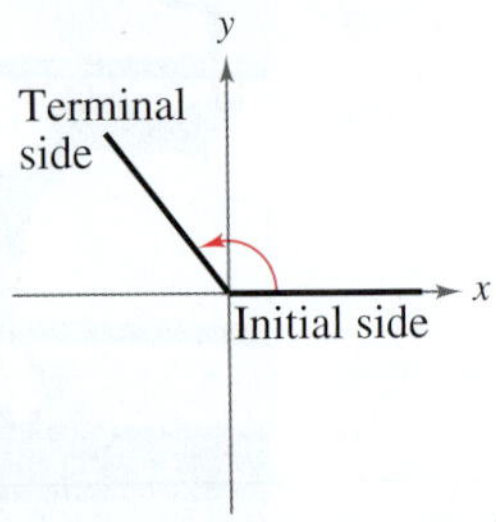

Angle in standard position
Figure 4.2

An **angle** is determined by rotating a ray (half-line) about its endpoint. The starting position of the ray is the **initial side** of the angle, and the position after rotation is the **terminal side,** as shown in Figure 4.1. The endpoint of the ray is the **vertex** of the angle. This perception of an angle fits a coordinate system in which the origin is the vertex and the initial side coincides with the positive x-axis. Such an angle is in **standard position,** as shown in Figure 4.2. Counterclockwise rotation generates **positive angles** and clockwise rotation generates **negative angles,** as shown in Figure 4.3. Angles are labeled with Greek letters such as

α (alpha), β (beta), and θ (theta)

as well as uppercase letters such as

A, B, and C.

In Figure 4.4, note that angles α and β have the same initial and terminal sides. Such angles are **coterminal.**

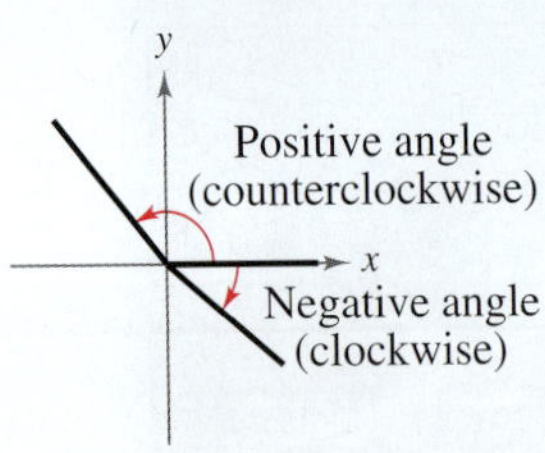

Figure 4.3

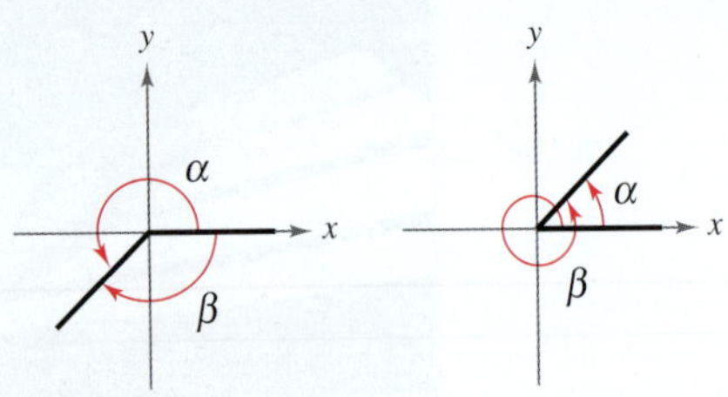

Coterminal angles
Figure 4.4

Paman Aheri - Malaysia Event/Shutterstock.com

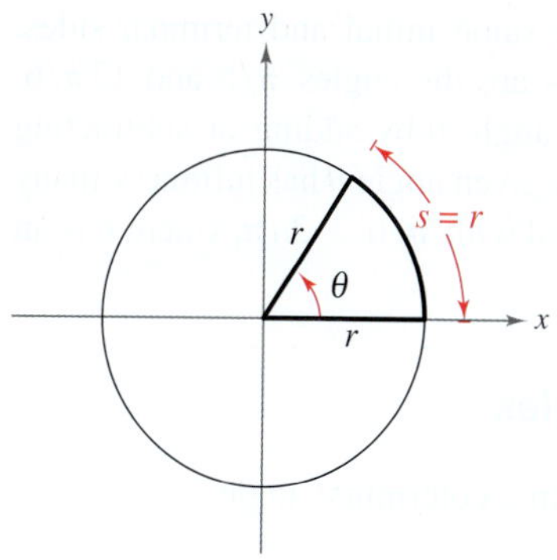

Arc length = radius when $\theta = 1$ radian.
Figure 4.5

Radian Measure

You determine the **measure of an angle** by the amount of rotation from the initial side to the terminal side. One way to measure angles is in *radians*. This type of measure is especially useful in calculus. To define a radian, you can use a **central angle** of a circle, one whose vertex is the center of the circle, as shown in Figure 4.5.

> **Definition of Radian**
>
> One **radian** is the measure of a central angle θ that intercepts an arc s equal in length to the radius r of the circle. See Figure 4.5. Algebraically, this means that
>
> $$\theta = \frac{s}{r}$$
>
> where θ is measured in radians. (Note that $\theta = 1$ when $s = r$.)

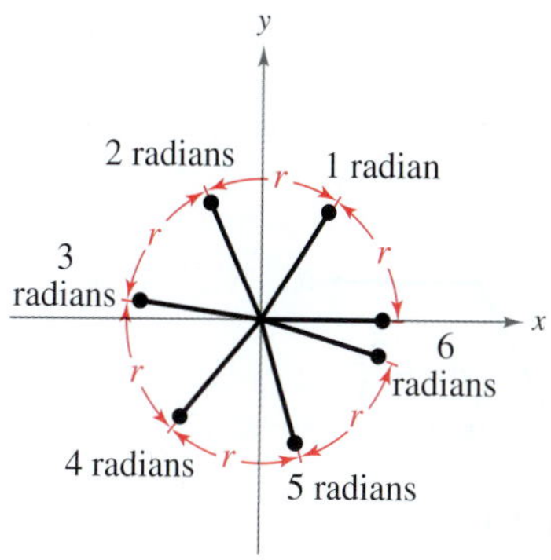

Figure 4.6

Because the circumference of a circle is $2\pi r$ units, it follows that a central angle of one full revolution (counterclockwise) corresponds to an arc length of $s = 2\pi r$. Moreover, because $2\pi \approx 6.28$, there are just over six radius lengths in a full circle, as shown in Figure 4.6. Because the units of measure for s and r are the same, the ratio s/r has no units—it is a real number.

Because the measure of an angle of one full revolution is $s/r = 2\pi r/r = 2\pi$ radians, you can obtain the following.

$$\frac{1}{2}\text{ revolution} = \frac{2\pi}{2} = \pi \text{ radians} \qquad \frac{1}{4}\text{ revolution} = \frac{2\pi}{4} = \frac{\pi}{2}\text{ radians}$$

$$\frac{1}{6}\text{ revolution} = \frac{2\pi}{6} = \frac{\pi}{3}\text{ radians}$$

These and other common angles are shown below.

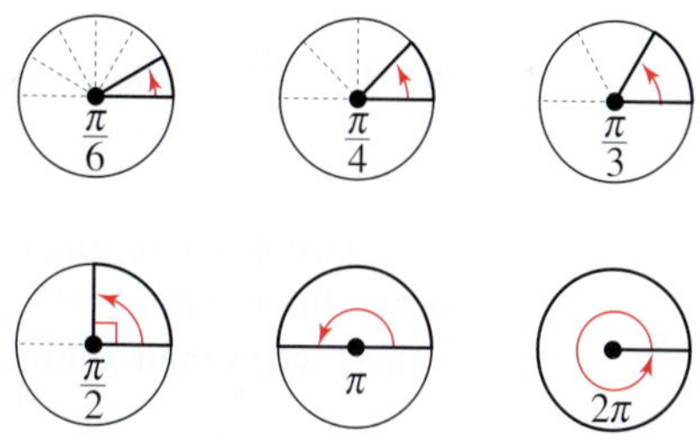

• • **REMARK** The phrase "the terminal side of θ lies in a quadrant" is often abbreviated by the phrase "θ lies in a quadrant." The terminal sides of the "quadrant angles" 0, $\pi/2$, π, and $3\pi/2$ do not lie within quadrants.

Recall that the four quadrants in a coordinate system are numbered I, II, III, and IV. The figure below shows which angles between 0 and 2π lie in each of the four quadrants. Note that angles between 0 and $\pi/2$ are **acute** angles and angles between $\pi/2$ and π are **obtuse** angles.

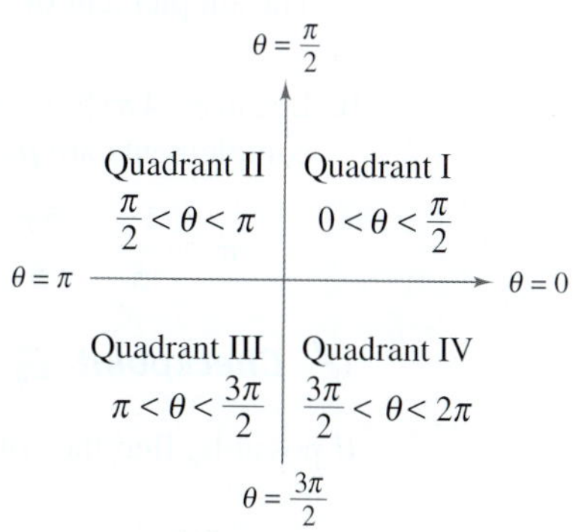

Two angles are coterminal when they have the same initial and terminal sides. For instance, the angles 0 and 2π are coterminal, as are the angles $\pi/6$ and $13\pi/6$. You can find an angle that is coterminal to a given angle θ by adding or subtracting 2π (one revolution), as demonstrated in Example 1. A given angle θ has infinitely many coterminal angles. For instance, $\theta = \pi/6$ is coterminal with $\pi/6 + 2n\pi$, where n is an integer.

EXAMPLE 1 Finding Coterminal Angles

> **ALGEBRA HELP** You can review operations involving fractions in Appendix A.1.

a. For the positive angle $13\pi/6$, subtract 2π to obtain a coterminal angle

$$\frac{13\pi}{6} - 2\pi = \frac{\pi}{6}.$$ See Figure 4.7.

b. For the negative angle $-2\pi/3$, add 2π to obtain a coterminal angle

$$-\frac{2\pi}{3} + 2\pi = \frac{4\pi}{3}.$$ See Figure 4.8.

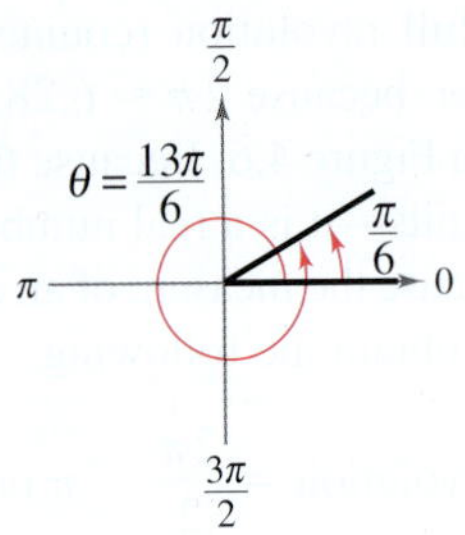

Figure 4.7

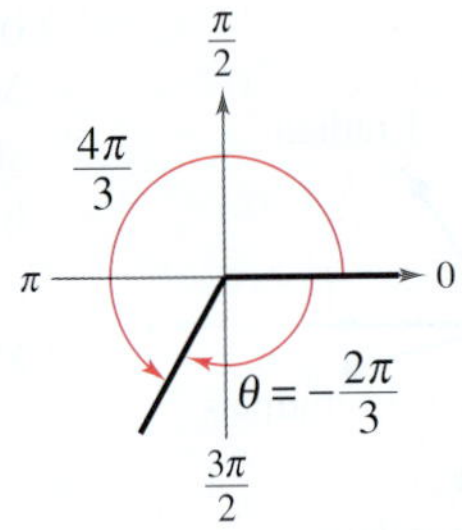

Figure 4.8

✓ Checkpoint Audio-video solution in English & Spanish at LarsonPrecalculus.com.

Determine two coterminal angles (one positive and one negative) for each angle.

a. $\theta = \dfrac{9\pi}{4}$ **b.** $\theta = -\dfrac{\pi}{3}$

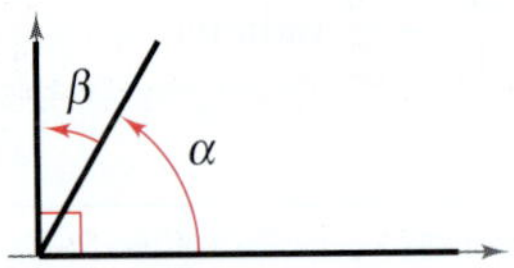

Complementary angles

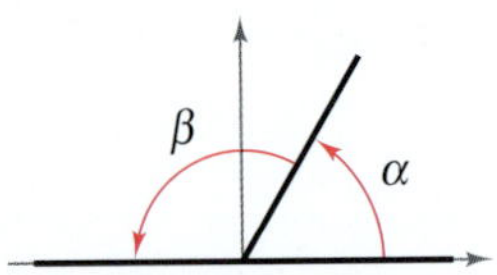

Supplementary angles
Figure 4.9

Two positive angles α and β are **complementary** (complements of each other) when their sum is $\pi/2$. Two positive angles are **supplementary** (supplements of each other) when their sum is π. See Figure 4.9.

EXAMPLE 2 Complementary and Supplementary Angles

a. The complement of $\dfrac{2\pi}{5}$ is $\dfrac{\pi}{2} - \dfrac{2\pi}{5} = \dfrac{5\pi}{10} - \dfrac{4\pi}{10} = \dfrac{\pi}{10}$.

The supplement of $\dfrac{2\pi}{5}$ is $\pi - \dfrac{2\pi}{5} = \dfrac{5\pi}{5} - \dfrac{2\pi}{5} = \dfrac{3\pi}{5}$.

b. Because $4\pi/5$ is greater than $\pi/2$, it has no complement. (Remember that complements are *positive* angles.) The supplement of $4\pi/5$ is

$$\pi - \frac{4\pi}{5} = \frac{5\pi}{5} - \frac{4\pi}{5} = \frac{\pi}{5}.$$

✓ Checkpoint Audio-video solution in English & Spanish at LarsonPrecalculus.com.

If possible, find the complement and the supplement of (a) $\pi/6$ and (b) $5\pi/6$.

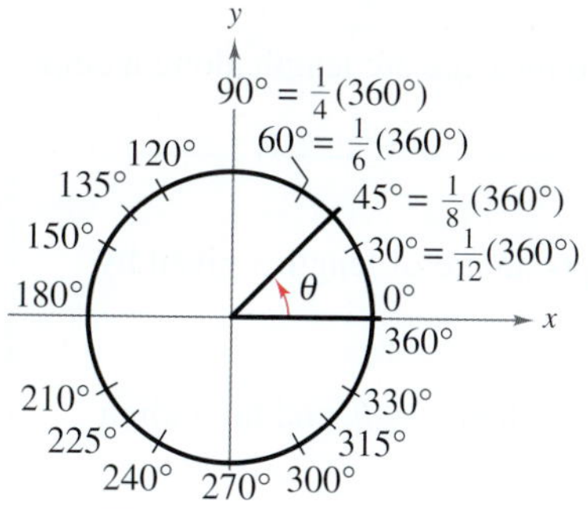

Figure 4.10

Degree Measure

A second way to measure angles is in **degrees,** denoted by the symbol °. A measure of one degree (1°) is equivalent to a rotation of $\frac{1}{360}$ of a complete revolution about the vertex. To measure angles, it is convenient to mark degrees on the circumference of a circle, as shown in Figure 4.10. So, a full revolution (counterclockwise) corresponds to 360°, a half revolution to 180°, a quarter revolution to 90°, and so on.

Because 2π radians corresponds to one complete revolution, degrees and radians are related by the equations

$$360° = 2\pi \text{ rad} \quad \text{and} \quad 180° = \pi \text{ rad}.$$

From the latter equation, you obtain

$$1° = \frac{\pi}{180} \text{ rad} \quad \text{and} \quad 1 \text{ rad} = \left(\frac{180}{\pi}\right)°$$

which lead to the following conversion rules.

Conversions Between Degrees and Radians

1. To convert degrees to radians, multiply degrees by $\dfrac{\pi \text{ rad}}{180°}$.

2. To convert radians to degrees, multiply radians by $\dfrac{180°}{\pi \text{ rad}}$.

To apply these two conversion rules, use the basic relationship π rad = 180°. (See Figure 4.11.)

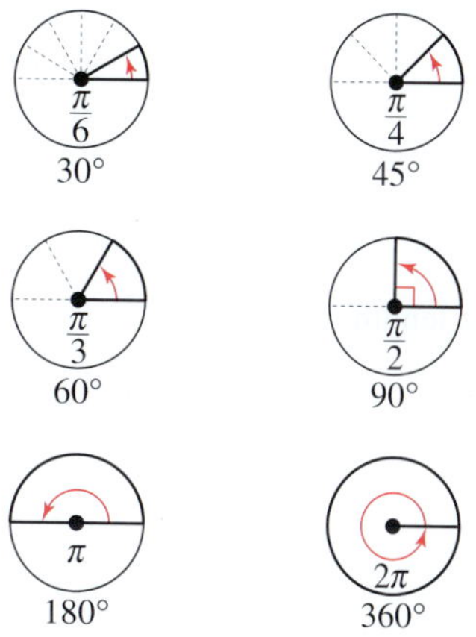

Figure 4.11

TECHNOLOGY

With calculators, it is convenient to use *decimal* degrees to denote fractional parts of degrees. Historically, however, fractional parts of degrees were expressed in *minutes* and *seconds*, using the prime (′) and double prime (″) notations, respectively. That is,

$$1' = \text{one minute} = \tfrac{1}{60}(1°)$$

$$1'' = \text{one second} = \tfrac{1}{3600}(1°).$$

Consequently, an angle of 64 degrees, 32 minutes, and 47 seconds is represented by $\theta = 64° 32' 47''$. Many calculators have special keys for converting an angle in degrees, minutes, and seconds (D° M′ S″) to decimal degree form, and vice versa.

When no units of angle measure are specified, *radian measure is implied.* For instance, $\theta = 2$ implies that $\theta = 2$ radians.

EXAMPLE 3 Converting from Degrees to Radians

a. $135° = (135 \text{ deg})\left(\dfrac{\pi \text{ rad}}{180 \text{ deg}}\right) = \dfrac{3\pi}{4}$ radians Multiply by $\dfrac{\pi \text{ rad}}{180°}$.

b. $540° = (540 \text{ deg})\left(\dfrac{\pi \text{ rad}}{180 \text{ deg}}\right) = 3\pi$ radians Multiply by $\dfrac{\pi \text{ rad}}{180°}$.

Checkpoint Audio-video solution in English & Spanish at LarsonPrecalculus.com.

Rewrite each angle in radian measure as a multiple of π. (Do not use a calculator.)

a. $\theta = 60°$ **b.** $\theta = 320°$

EXAMPLE 4 Converting from Radians to Degrees

a. $-\dfrac{\pi}{2} \text{ rad} = \left(-\dfrac{\pi}{2} \text{ rad}\right)\left(\dfrac{180 \text{ deg}}{\pi \text{ rad}}\right) = -90°$ Multiply by $\dfrac{180°}{\pi \text{ rad}}$.

b. $2 \text{ rad} = (2 \text{ rad})\left(\dfrac{180 \text{ deg}}{\pi \text{ rad}}\right) = \dfrac{360°}{\pi} \approx 114.59°$ Multiply by $\dfrac{180°}{\pi \text{ rad}}$.

Checkpoint Audio-video solution in English & Spanish at LarsonPrecalculus.com.

Rewrite each angle in degree measure. (Do not use a calculator.)

a. $\pi/6$ **b.** $5\pi/3$

Applications

The *radian measure* formula, $\theta = s/r$, can be used to measure arc length along a circle.

Arc Length

For a circle of radius r, a central angle θ intercepts an arc of length s given by

$$s = r\theta \quad \text{Length of circular arc}$$

where θ is measured in radians. Note that if $r = 1$, then $s = \theta$, and the radian measure of θ equals the arc length.

EXAMPLE 5 Finding Arc Length

A circle has a radius of 4 inches. Find the length of the arc intercepted by a central angle of 240°, as shown in Figure 4.12.

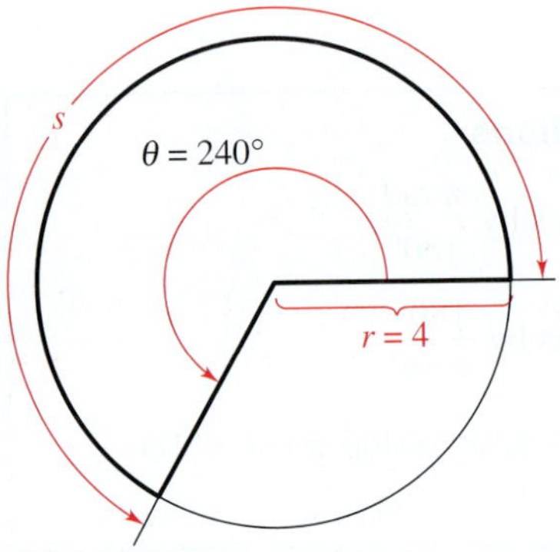

Figure 4.12

Solution To use the formula $s = r\theta$, first convert 240° to radian measure.

$$240° = (240 \text{ deg})\left(\frac{\pi \text{ rad}}{180 \text{ deg}}\right)$$

$$= \frac{4\pi}{3} \text{ radians}$$

Then, using a radius of $r = 4$ inches, you can find the arc length to be

$$s = r\theta$$

$$= 4\left(\frac{4\pi}{3}\right)$$

$$\approx 16.76 \text{ inches.}$$

Note that the units for r determine the units for $r\theta$ because θ is given in radian measure, which has no units.

Checkpoint Audio-video solution in English & Spanish at LarsonPrecalculus.com.

A circle has a radius of 27 inches. Find the length of the arc intercepted by a central angle of 160°.

The formula for the length of a circular arc can help you analyze the motion of a particle moving at a *constant speed* along a circular path.

REMARK Linear speed measures how fast the particle moves, and angular speed measures how fast the angle changes. By dividing each side of the formula for arc length by t, you can establish a relationship between linear speed v and angular speed ω, as shown.

$$s = r\theta$$

$$\frac{s}{t} = \frac{r\theta}{t}$$

$$v = r\omega$$

Linear and Angular Speeds

Consider a particle moving at a constant speed along a circular arc of radius r. If s is the length of the arc traveled in time t, then the **linear speed** v of the particle is

$$\text{Linear speed } v = \frac{\text{arc length}}{\text{time}} = \frac{s}{t}.$$

Moreover, if θ is the angle (in radian measure) corresponding to the arc length s, then the **angular speed** ω (the lowercase Greek letter omega) of the particle is

$$\text{Angular speed } \omega = \frac{\text{central angle}}{\text{time}} = \frac{\theta}{t}.$$

Figure 4.13

EXAMPLE 6 Finding Linear Speed

The second hand of a clock is 10.2 centimeters long, as shown in Figure 4.13. Find the linear speed of the tip of this second hand as it passes around the clock face.

Solution In one revolution, the arc length traveled is

$$\begin{aligned} s &= 2\pi r \\ &= 2\pi(10.2) && \text{Substitute for } r. \\ &= 20.4\pi \text{ centimeters.} \end{aligned}$$

The time required for the second hand to travel this distance is

$$\begin{aligned} t &= 1 \text{ minute} \\ &= 60 \text{ seconds.} \end{aligned}$$

So, the linear speed of the tip of the second hand is

$$\begin{aligned} \text{Linear speed} &= \frac{s}{t} \\ &= \frac{20.4\pi \text{ centimeters}}{60 \text{ seconds}} \\ &\approx 1.068 \text{ centimeters per second.} \end{aligned}$$

✓ **Checkpoint** Audio-video solution in English & Spanish at LarsonPrecalculus.com.

The second hand of a clock is 8 centimeters long. Find the linear speed of the tip of this second hand as it passes around the clock face.

Figure 4.14

EXAMPLE 7 Finding Angular and Linear Speeds

The blades of a wind turbine are 116 feet long (see Figure 4.14). The propeller rotates at 15 revolutions per minute.

a. Find the angular speed of the propeller in radians per minute.

b. Find the linear speed of the tips of the blades.

Solution

a. Because each revolution generates 2π radians, it follows that the propeller turns

$$(15)(2\pi) = 30\pi \text{ radians per minute.}$$

In other words, the angular speed is

$$\text{Angular speed} = \frac{\theta}{t} = \frac{30\pi \text{ radians}}{1 \text{ minute}} = 30\pi \text{ radians per minute.}$$

b. The linear speed is

$$\text{Linear speed} = \frac{s}{t} = \frac{r\theta}{t} = \frac{(116)(30\pi) \text{ feet}}{1 \text{ minute}} \approx 10{,}933 \text{ feet per minute.}$$

✓ **Checkpoint** Audio-video solution in English & Spanish at LarsonPrecalculus.com.

The circular blade on a saw rotates at 2400 revolutions per minute.

a. Find the angular speed of the blade in radians per minute.

b. The blade has a radius of 4 inches. Find the linear speed of a blade tip.

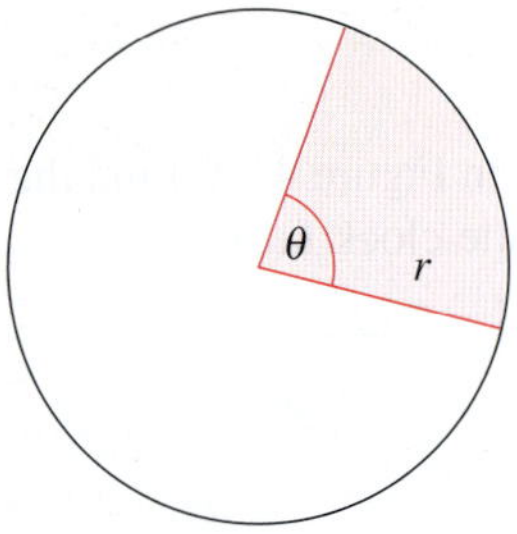

Figure 4.15

A **sector** of a circle is the region bounded by two radii of the circle and their intercepted arc (see Figure 4.15).

> **Area of a Sector of a Circle**
>
> For a circle of radius r, the area A of a sector of the circle with central angle θ is
>
> $$A = \frac{1}{2}r^2\theta$$
>
> where θ is measured in radians.

EXAMPLE 8 Area of a Sector of a Circle

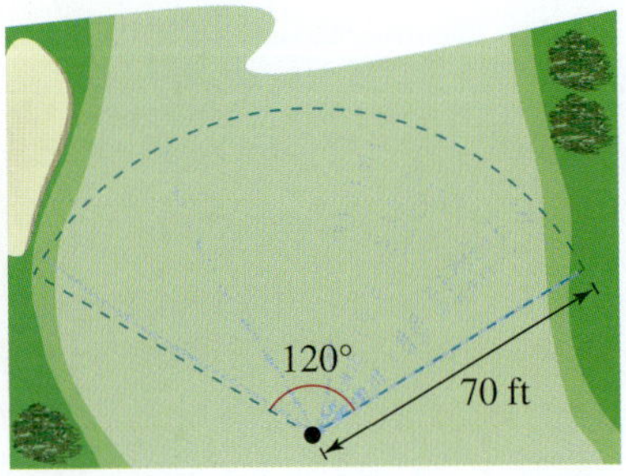

Figure 4.16

A sprinkler on a golf course fairway sprays water over a distance of 70 feet and rotates through an angle of 120° (see Figure 4.16). Find the area of the fairway watered by the sprinkler.

Solution

First convert 120° to radian measure as follows.

$$\theta = 120°$$

$$= (120 \text{ deg})\left(\frac{\pi \text{ rad}}{180 \text{ deg}}\right) \qquad \text{Multiply by } \frac{\pi \text{ rad}}{180°}.$$

$$= \frac{2\pi}{3} \text{ radians}$$

Then, using $\theta = 2\pi/3$ and $r = 70$, the area is

$$A = \frac{1}{2}r^2\theta \qquad \text{Formula for the area of a sector of a circle}$$

$$= \frac{1}{2}(70)^2\left(\frac{2\pi}{3}\right) \qquad \text{Substitute for } r \text{ and } \theta.$$

$$= \frac{4900\pi}{3} \qquad \text{Multiply.}$$

$$\approx 5131 \text{ square feet.} \qquad \text{Simplify.}$$

✓ **Checkpoint** *Audio-video solution in English & Spanish at LarsonPrecalculus.com.*

A sprinkler sprays water over a distance of 40 feet and rotates through an angle of 80°. Find the area watered by the sprinkler.

> **Summarize (Section 4.1)**
>
> **1.** Describe an angle *(page 262)*.
>
> **2.** Describe how to determine the measure of an angle using radians *(page 263)*. For examples involving radian measure, see Examples 1 and 2.
>
> **3.** Describe how to determine the measure of an angle using degrees *(page 265)*. For examples involving degree measure, see Examples 3 and 4.
>
> **4.** Describe examples of how to use angles to model and solve real-life problems *(pages 266–268, Examples 5–8)*.

4.1 Exercises

See CalcChat.com for tutorial help and worked-out solutions to odd-numbered exercises.

Vocabulary: **Fill in the blanks.**

1. Two angles that have the same initial and terminal sides are ________.
2. One ________ is the measure of a central angle that intercepts an arc equal to the radius of the circle.
3. Two positive angles that have a sum of $\pi/2$ are ________ angles, whereas two positive angles that have a sum of π are ________ angles.
4. The angle measure that is equivalent to a rotation of $\frac{1}{360}$ of a complete revolution about an angle's vertex is one ________.
5. The ________ speed of a particle is the ratio of the arc length to the time traveled, and the ________ speed of a particle is the ratio of the central angle to the time traveled.
6. The area A of a sector of a circle with radius r and central angle θ, where θ is measured in radians, is given by the formula ________.

Skills and Applications

Estimating an Angle **In Exercises 7–10, estimate the angle to the nearest one-half radian.**

7. 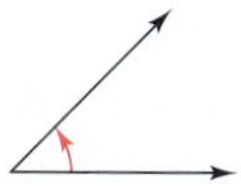

8. 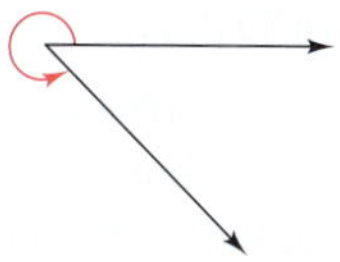

9. 

10. 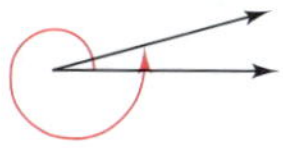

Determining Quadrants **In Exercises 11 and 12, determine the quadrant in which each angle lies.**

11. (a) $\frac{\pi}{4}$ (b) $\frac{5\pi}{4}$

12. (a) $-\frac{\pi}{6}$ (b) $-\frac{11\pi}{9}$

Sketching Angles **In Exercises 13 and 14, sketch each angle in standard position.**

13. (a) $\frac{\pi}{3}$ (b) $-\frac{2\pi}{3}$

14. (a) $\frac{5\pi}{2}$ (b) 4

Finding Coterminal Angles **In Exercises 15 and 16, determine two coterminal angles (one positive and one negative) for each angle. Give your answers in radians.**

15. (a) $\frac{\pi}{6}$ (b) $\frac{7\pi}{6}$

16. (a) $\frac{2\pi}{3}$ (b) $-\frac{9\pi}{4}$

Complementary and Supplementary Angles **In Exercises 17–20, find (if possible) the complement and the supplement of each angle.**

17. (a) $\frac{\pi}{3}$ (b) $\frac{\pi}{4}$

18. (a) $\frac{\pi}{12}$ (b) $\frac{11\pi}{12}$

19. (a) 1 (b) 2

20. (a) 3 (b) 1.5

Estimating an Angle **In Exercises 21–24, estimate the number of degrees in the angle.**

21. 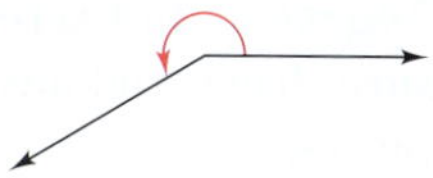

22. 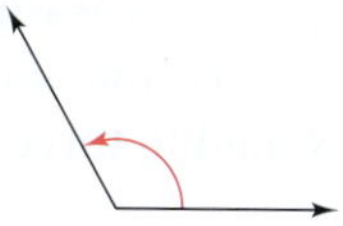

23. 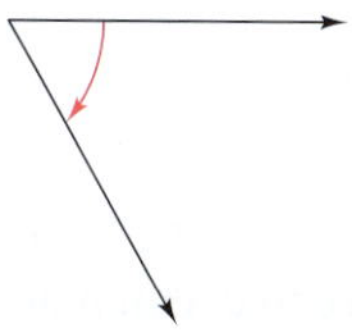

24. 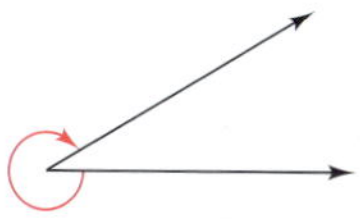

Determining Quadrants **In Exercises 25 and 26, determine the quadrant in which each angle lies.**

25. (a) $130°$ (b) $8.3°$

26. (a) $-132° 50'$ (b) $-3.4°$

Sketching Angles **In Exercises 27 and 28, sketch each angle in standard position.**

27. (a) $270°$ (b) $120°$

28. (a) $-135°$ (b) $-750°$

Finding Coterminal Angles **In Exercises 29 and 30, determine two coterminal angles (one positive and one negative) for each angle. Give your answers in degrees.**

29. (a) $45°$ (b) $-36°$

30. (a) $120°$ (b) $-420°$

Complementary and Supplementary Angles **In Exercises 31–34, find (if possible) the complement and the supplement of each angle.**

31. (a) $18°$ (b) $85°$

32. (a) $46°$ (b) $93°$

33. (a) $150°$ (b) $79°$

34. (a) $130°$ (b) $170°$

Converting from Degrees to Radians In Exercises 35 and 36, rewrite each angle in radian measure as a multiple of π. (Do not use a calculator.)

35. (a) $120°$ (b) $-20°$

36. (a) $-60°$ (b) $144°$

Converting from Radians to Degrees In Exercises 37 and 38, rewrite each angle in degree measure. (Do not use a calculator.)

37. (a) $\frac{3\pi}{2}$ (b) $\frac{7\pi}{6}$

38. (a) $-\frac{7\pi}{12}$ (b) $\frac{5\pi}{4}$

Converting from Degrees to Radians In Exercises 39–42, convert the angle measure from degrees to radians. Round to three decimal places.

39. $45°$ **40.** $-48.27°$

41. $0.54°$ **42.** $345°$

Converting from Radians to Degrees In Exercises 43–46, convert the angle measure from radians to degrees. Round to three decimal places.

43. $\frac{5\pi}{11}$ **44.** $\frac{15\pi}{8}$

45. -4.2π **46.** -0.57

Converting to Decimal Degree Form In Exercises 47 and 48, convert each angle measure to decimal degree form without using a calculator. Then check your answers using a calculator.

47. (a) $54° 45'$ (b) $-128° 30'$

48. (a) $-135° 36''$ (b) $-408° 16' 20''$

Converting to D°M′S″ Form In Exercises 49 and 50, convert each angle measure to degrees, minutes, and seconds without using a calculator. Then check your answers using a calculator.

49. (a) $240.6°$ (b) $-145.8°$

50. (a) $-345.12°$ (b) $-3.58°$

Finding Arc Length In Exercises 51 and 52, find the length of the arc on a circle of radius r intercepted by a central angle θ.

51. $r = 15$ inches, $\theta = 120°$

52. $r = 3$ meters, $\theta = 150°$

Finding the Central Angle In Exercises 53 and 54, find the radian measure of the central angle of a circle of radius r that intercepts an arc of length s.

53. $r = 80$ kilometers, $s = 150$ kilometers

54. $r = 14$ feet, $s = 8$ feet

Finding an Angle In Exercises 55 and 56, use the given arc length and radius to find the angle θ (in radians).

55.

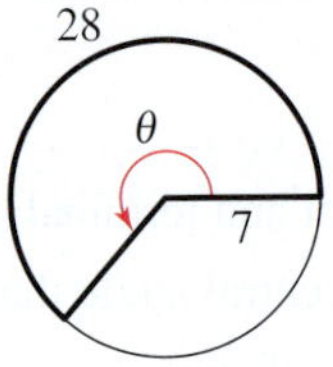

56.

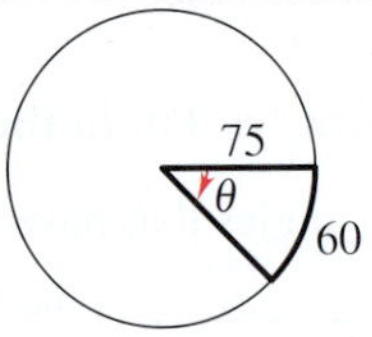

Area of a Sector of a Circle In Exercises 57 and 58, find the area of the sector of a circle of radius r and central angle θ.

57. $r = 12$ millimeters, $\theta = \frac{\pi}{4}$

58. $r = 2.5$ feet, $\theta = 225°$

59. Distance Between Cities Find the distance between Dallas, Texas, whose latitude is $32° 47' 39''$ N, and Omaha, Nebraska, whose latitude is $41° 15' 50''$ N. Assume that Earth is a sphere of radius 4000 miles and that the cities are on the same longitude (Omaha is due north of Dallas).

60. Difference in Latitudes Assuming that Earth is a sphere of radius 6378 kilometers, what is the difference in the latitudes of Lynchburg, Virginia, and Myrtle Beach, South Carolina, where Lynchburg is about 400 kilometers due north of Myrtle Beach?

61. Instrumentation The pointer on a voltmeter is 6 centimeters in length (see figure). Find the number of degrees through which the pointer rotates when it moves 2.5 centimeters on the scale.

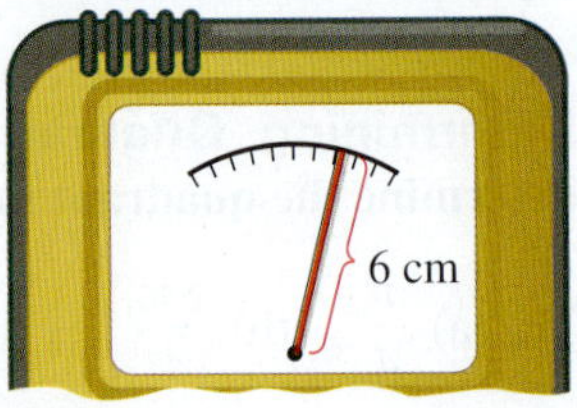

62. Linear Speed A satellite in a circular orbit 1250 kilometers above Earth makes one complete revolution every 110 minutes. Assuming that Earth is a sphere of radius 6378 kilometers, what is the linear speed (in kilometers per minute) of the satellite?

63. Angular and Linear Speeds The circular blade on a saw rotates at 5000 revolutions per minute.

(a) Find the angular speed of the blade in radians per minute.

(b) The blade has a diameter of $7\frac{1}{4}$ inches. Find the linear speed of a blade tip.

64. Angular and Linear Speeds A carousel with a 50-foot diameter makes 4 revolutions per minute.

(a) Find the angular speed of the carousel in radians per minute.

(b) Find the linear speed (in feet per minute) of the platform rim of the carousel.

65. Angular and Linear Speeds A DVD is approximately 12 centimeters in diameter. The drive motor of the DVD player rotates between 200 and 500 revolutions per minute, depending on what track is being read.

(a) Find an interval for the angular speed of the DVD as it rotates.

(b) Find an interval for the linear speed of a point on the outermost track as the DVD rotates.

66. Angular Speed A car is moving at a rate of 65 miles per hour, and the diameter of its wheels is 2 feet.

(a) Find the number of revolutions per minute the wheels are rotating.

(b) Find the angular speed of the wheels in radians per minute.

67. Linear and Angular Speeds A computerized spin balance machine rotates a 25-inch-diameter tire at 480 revolutions per minute.

(a) Find the road speed (in miles per hour) at which the tire is being balanced.

(b) At what rate should the spin balance machine be set so that the tire is being tested for 55 miles per hour?

68. Speed of a Bicycle

The radii of the pedal sprocket, the wheel sprocket, and the wheel of the bicycle in the figure are 4 inches, 2 inches, and 14 inches, respectively. A cyclist is pedaling at a rate of 1 revolution per second.

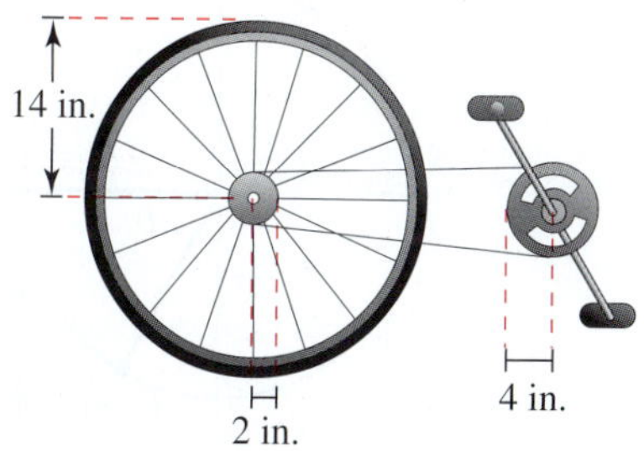

(a) Find the speed of the bicycle in feet per second and miles per hour.

(b) Use your result from part (a) to write a function for the distance d (in miles) a cyclist travels in terms of the number n of revolutions of the pedal sprocket.

(c) Write a function for the distance d (in miles) a cyclist travels in terms of the time t (in seconds). Compare this function with the function from part (b).

69. Area A sprinkler on a golf green sprays water over a distance of 15 meters and rotates through an angle of 140°. Draw a diagram that shows the region that the sprinkler can irrigate. Find the area of the region.

70. Area A car's rear windshield wiper rotates 125°. The total length of the wiper mechanism is 25 inches and wipes the windshield over a distance of 14 inches. Find the area covered by the wiper.

Exploration

True or False? **In Exercises 71–73, determine whether the statement is true or false. Justify your answer.**

71. A measurement of 4 radians corresponds to two complete revolutions from the initial side to the terminal side of an angle.

72. The difference between the measures of two coterminal angles is always a multiple of 360° when expressed in degrees and is always a multiple of 2π radians when expressed in radians.

73. An angle that measures $-1260°$ lies in Quadrant III.

74. HOW DO YOU SEE IT? Determine which angles in the figure are coterminal angles with angle A. Explain your reasoning.

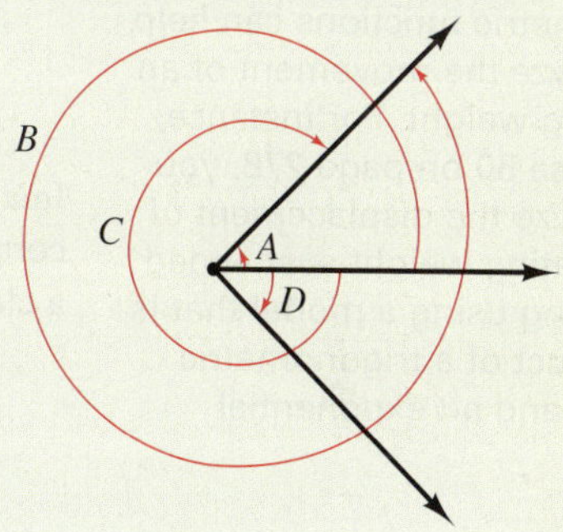

75. Think About It A fan motor turns at a given angular speed. How does the speed of the tips of the blades change when a fan of greater diameter is on the motor? Explain.

76. Think About It Is a degree or a radian the greater unit of measure? Explain.

77. Writing When the radius of a circle increases and the magnitude of a central angle is constant, how does the length of the intercepted arc change? Explain your reasoning.

78. Proof Prove that the area of a circular sector of radius r with central angle θ is

$$A = \frac{1}{2}\theta r^2$$

where θ is measured in radians.

Paman Aheri - Malaysia Event/Shutterstock.com

4.2 Trigonometric Functions: The Unit Circle

- Identify a unit circle and describe its relationship to real numbers.
- Evaluate trigonometric functions using the unit circle.
- Use domain and period to evaluate sine and cosine functions, and use a calculator to evaluate trigonometric functions.

Trigonometric functions can help you analyze the movement of an oscillating weight. For instance, in Exercise 50 on page 278, you will analyze the displacement of an oscillating weight suspended by a spring using a model that is the product of a trigonometric function and an exponential function.

The Unit Circle

The two historical perspectives of trigonometry incorporate different methods for introducing the trigonometric functions. One such perspective follows and is based on the unit circle.

Consider the **unit circle** given by

$x^2 + y^2 = 1$ Unit circle

as shown below.

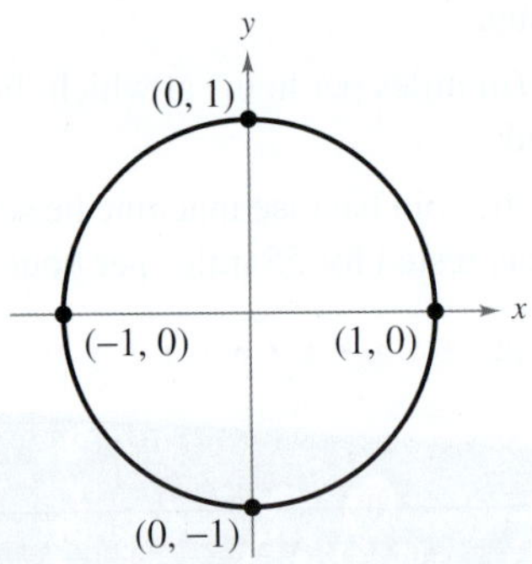

Imagine wrapping the real number line around this circle, with positive numbers corresponding to a counterclockwise wrapping and negative numbers corresponding to a clockwise wrapping, as shown below.

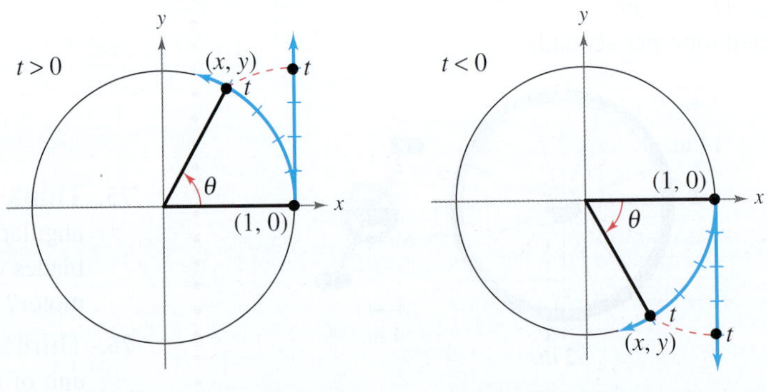

As the real number line wraps around the unit circle, each real number t corresponds to a point (x, y) on the circle. For example, the real number 0 corresponds to the point $(1, 0)$. Moreover, because the unit circle has a circumference of 2π, the real number 2π also corresponds to the point $(1, 0)$.

In general, each real number t also corresponds to a central angle θ (in standard position) whose radian measure is t. With this interpretation of t, the arc length formula

$s = r\theta$ (with $r = 1$)

indicates that the real number t is the (directional) length of the arc intercepted by the angle θ, given in radians.

Richard Megna/Fundamental Photographs

The Trigonometric Functions

From the preceding discussion, it follows that the coordinates x and y are two functions of the real variable t. You can use these coordinates to define the six trigonometric functions of t.

sine **cosecant** **cosine** **secant** **tangent** **cotangent**

These six functions are normally abbreviated sin, csc, cos, sec, tan, and cot, respectively.

> **REMARK** Note that the functions in the second row are the *reciprocals* of the corresponding functions in the first row.

Definitions of Trigonometric Functions

Let t be a real number and let (x, y) be the point on the unit circle corresponding to t.

$$\sin t = y \qquad \cos t = x \qquad \tan t = \frac{y}{x},\ x \neq 0$$

$$\csc t = \frac{1}{y},\ y \neq 0 \qquad \sec t = \frac{1}{x},\ x \neq 0 \qquad \cot t = \frac{x}{y},\ y \neq 0$$

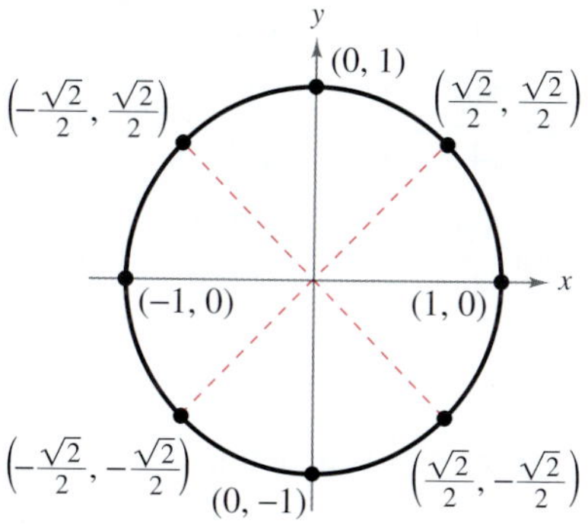

Figure 4.17

In the definitions of the trigonometric functions, note that the tangent and secant are not defined when $x = 0$. For instance, because $t = \pi/2$ corresponds to $(x, y) = (0, 1)$, it follows that $\tan(\pi/2)$ and $\sec(\pi/2)$ are *undefined*. Similarly, the cotangent and cosecant are not defined when $y = 0$. For instance, because $t = 0$ corresponds to $(x, y) = (1, 0)$, cot 0 and csc 0 are *undefined*.

In Figure 4.17, the unit circle is divided into eight equal arcs, corresponding to t-values of

$$0, \frac{\pi}{4}, \frac{\pi}{2}, \frac{3\pi}{4}, \pi, \frac{5\pi}{4}, \frac{3\pi}{2}, \frac{7\pi}{4}, \text{ and } 2\pi.$$

Similarly, in Figure 4.18, the unit circle is divided into 12 equal arcs, corresponding to t-values of

$$0, \frac{\pi}{6}, \frac{\pi}{3}, \frac{\pi}{2}, \frac{2\pi}{3}, \frac{5\pi}{6}, \pi, \frac{7\pi}{6}, \frac{4\pi}{3}, \frac{3\pi}{2}, \frac{5\pi}{3}, \frac{11\pi}{6}, \text{ and } 2\pi.$$

To verify the points on the unit circle in Figure 4.17, note that

$$\left(\frac{\sqrt{2}}{2}, \frac{\sqrt{2}}{2}\right)$$

also lies on the line $y = x$. So, substituting x for y in the equation of the unit circle produces the following.

$$x^2 + x^2 = 1 \implies 2x^2 = 1 \implies x^2 = \frac{1}{2} \implies x = \pm\frac{\sqrt{2}}{2}$$

Because the point is in the first quadrant and $y = x$, you have

$$x = \frac{\sqrt{2}}{2} \quad \text{and} \quad y = \frac{\sqrt{2}}{2}.$$

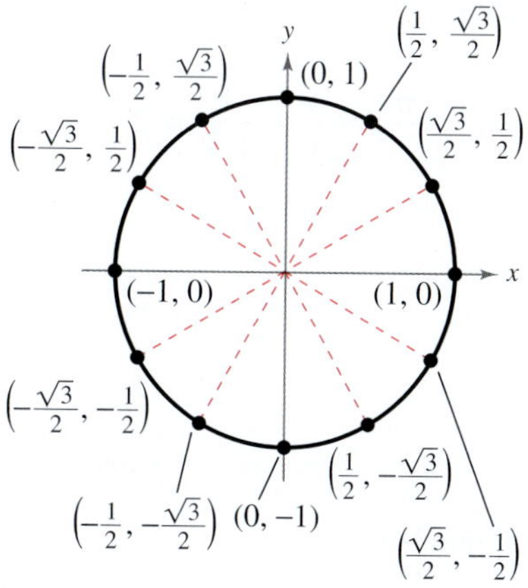

Figure 4.18

You can use similar reasoning to verify the rest of the points in Figure 4.17 and the points in Figure 4.18.

Using the (x, y) coordinates in Figures 4.17 and 4.18, you can evaluate the trigonometric functions for common t-values. Examples 1 and 2 demonstrate this procedure. You should study and learn these exact function values for common t-values because they will help you in later sections to perform calculations.

EXAMPLE 1 Evaluating Trigonometric Functions

ALGEBRA HELP You can review dividing fractions and rationalizing denominators in Appendix A.1 and Appendix A.2, respectively.

Evaluate the six trigonometric functions at each real number.

a. $t = \dfrac{\pi}{6}$ **b.** $t = \dfrac{5\pi}{4}$ **c.** $t = \pi$ **d.** $t = -\dfrac{\pi}{3}$

Solution For each t-value, begin by finding the corresponding point (x, y) on the unit circle. Then use the definitions of trigonometric functions listed on page 273.

a. $t = \pi/6$ corresponds to the point $(x, y) = \left(\sqrt{3}/2, 1/2\right)$.

$$\sin\frac{\pi}{6} = y = \frac{1}{2} \qquad \csc\frac{\pi}{6} = \frac{1}{y} = \frac{1}{1/2} = 2$$

$$\cos\frac{\pi}{6} = x = \frac{\sqrt{3}}{2} \qquad \sec\frac{\pi}{6} = \frac{1}{x} = \frac{2}{\sqrt{3}} = \frac{2\sqrt{3}}{3}$$

$$\tan\frac{\pi}{6} = \frac{y}{x} = \frac{1/2}{\sqrt{3}/2} = \frac{1}{\sqrt{3}} = \frac{\sqrt{3}}{3} \qquad \cot\frac{\pi}{6} = \frac{x}{y} = \frac{\sqrt{3}/2}{1/2} = \sqrt{3}$$

b. $t = 5\pi/4$ corresponds to the point $(x, y) = \left(-\sqrt{2}/2, -\sqrt{2}/2\right)$.

$$\sin\frac{5\pi}{4} = y = -\frac{\sqrt{2}}{2} \qquad \csc\frac{5\pi}{4} = \frac{1}{y} = -\frac{2}{\sqrt{2}} = -\sqrt{2}$$

$$\cos\frac{5\pi}{4} = x = -\frac{\sqrt{2}}{2} \qquad \sec\frac{5\pi}{4} = \frac{1}{x} = -\frac{2}{\sqrt{2}} = -\sqrt{2}$$

$$\tan\frac{5\pi}{4} = \frac{y}{x} = \frac{-\sqrt{2}/2}{-\sqrt{2}/2} = 1 \qquad \cot\frac{5\pi}{4} = \frac{x}{y} = \frac{-\sqrt{2}/2}{-\sqrt{2}/2} = 1$$

c. $t = \pi$ corresponds to the point $(x, y) = (-1, 0)$.

$$\sin\pi = y = 0 \qquad \csc\pi = \frac{1}{y} \text{ is undefined.}$$

$$\cos\pi = x = -1 \qquad \sec\pi = \frac{1}{x} = \frac{1}{-1} = -1$$

$$\tan\pi = \frac{y}{x} = \frac{0}{-1} = 0 \qquad \cot\pi = \frac{x}{y} \text{ is undefined.}$$

d. Moving *clockwise* around the unit circle, it follows that $t = -\pi/3$ corresponds to the point $(x, y) = \left(1/2, -\sqrt{3}/2\right)$.

$$\sin\left(-\frac{\pi}{3}\right) = y = -\frac{\sqrt{3}}{2} \qquad \csc\left(-\frac{\pi}{3}\right) = \frac{1}{y} = -\frac{2}{\sqrt{3}} = -\frac{2\sqrt{3}}{3}$$

$$\cos\left(-\frac{\pi}{3}\right) = x = \frac{1}{2} \qquad \sec\left(-\frac{\pi}{3}\right) = \frac{1}{x} = \frac{1}{1/2} = 2$$

$$\tan\left(-\frac{\pi}{3}\right) = \frac{y}{x} = \frac{-\sqrt{3}/2}{1/2} = -\sqrt{3}$$

$$\cot\left(-\frac{\pi}{3}\right) = \frac{x}{y} = \frac{1/2}{-\sqrt{3}/2} = -\frac{1}{\sqrt{3}} = -\frac{\sqrt{3}}{3}$$

✓ **Checkpoint** Audio-video solution in English & Spanish at LarsonPrecalculus.com.

Evaluate the six trigonometric functions at each real number.

a. $t = \pi/2$ **b.** $t = 0$ **c.** $t = -5\pi/6$ **d.** $t = -3\pi/4$

Domain and Period of Sine and Cosine

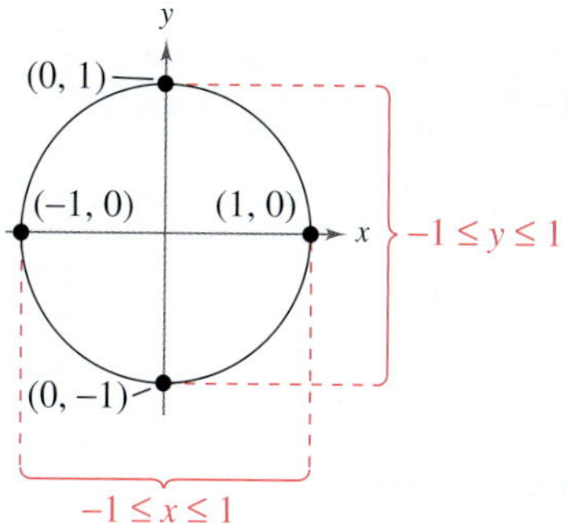

Figure 4.19

The *domain* of the sine and cosine functions is the set of all real numbers. To determine the *range* of these two functions, consider the unit circle shown in Figure 4.19. By definition, $\sin t = y$ and $\cos t = x$. Because (x, y) is on the unit circle, you know that $-1 \le y \le 1$ and $-1 \le x \le 1$. So, the values of sine and cosine also range between -1 and 1.

$$\begin{aligned} -1 \le &\ y \le 1 \\ -1 \le &\ \sin t \le 1 \end{aligned} \quad \text{and} \quad \begin{aligned} -1 \le &\ x \le 1 \\ -1 \le &\ \cos t \le 1 \end{aligned}$$

Adding 2π to each value of t in the interval $[0, 2\pi]$ results in a revolution around the unit circle, as shown below.

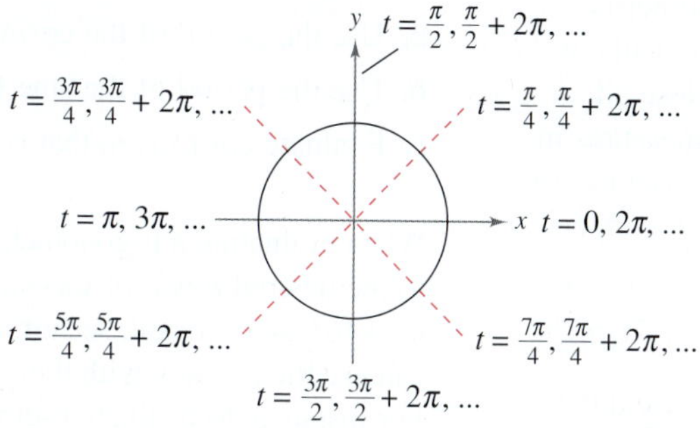

The values of $\sin(t + 2\pi)$ and $\cos(t + 2\pi)$ correspond to those of $\sin t$ and $\cos t$. Similar results can be obtained for repeated revolutions (positive or negative) on the unit circle. This leads to the general result

$$\sin(t + 2\pi n) = \sin t \quad \text{and} \quad \cos(t + 2\pi n) = \cos t$$

for any integer n and real number t. Functions that behave in such a repetitive (or cyclic) manner are called **periodic.**

REMARK From this definition, it follows that the sine and cosine functions are periodic and have a period of 2π. The other four trigonometric functions are also periodic and will be discussed further in Section 4.6.

Definition of Periodic Function

A function f is **periodic** when there exists a positive real number c such that

$$f(t + c) = f(t)$$

for all t in the domain of f. The smallest number c for which f is periodic is called the **period** of f.

Recall from Section 1.5 that a function f is *even* when $f(-t) = f(t)$ and is *odd* when $f(-t) = -f(t)$.

Even and Odd Trigonometric Functions

The cosine and secant functions are *even.*

$$\cos(-t) = \cos t \qquad \sec(-t) = \sec t$$

The sine, cosecant, tangent, and cotangent functions are *odd.*

$$\sin(-t) = -\sin t \qquad \csc(-t) = -\csc t$$

$$\tan(-t) = -\tan t \qquad \cot(-t) = -\cot t$$

EXAMPLE 2 Evaluating Sine and Cosine

a. Because $\frac{13\pi}{6} = 2\pi + \frac{\pi}{6}$, you have $\sin\frac{13\pi}{6} = \sin\left(2\pi + \frac{\pi}{6}\right) = \sin\frac{\pi}{6} = \frac{1}{2}$.

b. Because $-\frac{7\pi}{2} = -4\pi + \frac{\pi}{2}$, you have

$$\cos\left(-\frac{7\pi}{2}\right) = \cos\left(-4\pi + \frac{\pi}{2}\right) = \cos\frac{\pi}{2} = 0.$$

c. For $\sin t = \frac{4}{5}$, $\sin(-t) = -\frac{4}{5}$ because the sine function is odd.

✓ ***Checkpoint*** *Audio-video solution in English & Spanish at LarsonPrecalculus.com.*

a. Use the period of the cosine function to evaluate $\cos(9\pi/2)$.

b. Use the period of the sine function to evaluate $\sin(-7\pi/3)$.

c. Evaluate $\cos t$ given that $\cos(-t) = 0.3$.

> **TECHNOLOGY** When evaluating trigonometric functions with a calculator, remember to enclose all fractional angle measures in parentheses. For instance, to evaluate $\sin t$ for $t = \pi/6$, you should enter
>
> [SIN] [(] [π] [÷] 6 [)] [ENTER].
>
> These keystrokes yield the correct value of 0.5. Note that some calculators automatically place a left parenthesis after trigonometric functions.

When evaluating a trigonometric function with a calculator, you need to set the calculator to the desired *mode* of measurement (*degree* or *radian*). Most calculators do not have keys for the cosecant, secant, and cotangent functions. To evaluate these functions, you can use the [x^{-1}] key with their respective reciprocal functions: sine, cosine, and tangent. For instance, to evaluate $\csc(\pi/8)$, use the fact that

$$\csc\frac{\pi}{8} = \frac{1}{\sin(\pi/8)}$$

and enter the following keystroke sequence in *radian* mode.

[(] [SIN] [(] [π] [÷] 8 [)] [)] [x^{-1}] [ENTER] Display 2.6131259

EXAMPLE 3 Using a Calculator

Function	Mode	Calculator Keystrokes	Display
a. $\sin\frac{2\pi}{3}$	Radian	[SIN] [(] 2 [π] [÷] 3 [)] [ENTER]	0.8660254
b. $\cot 1.5$	Radian	[(] [TAN] [(] 1.5 [)] [)] [x^{-1}] [ENTER]	0.0709148

✓ ***Checkpoint*** *Audio-video solution in English & Spanish at LarsonPrecalculus.com.*

Use a calculator to evaluate (a) $\sin(5\pi/7)$ and (b) $\csc 2.0$.

Summarize (Section 4.2)

1. Explain how to identify a unit circle and describe its relationship to real numbers *(page 272)*.
2. State the unit circle definitions of the trigonometric functions *(page 273)*. For an example of evaluating trigonometric functions using the unit circle, see Example 1.
3. Explain how to use domain and period to evaluate sine and cosine functions *(page 275)* and describe how to use a calculator to evaluate trigonometric functions *(page 276)*. For an example of using period and an odd trigonometric function to evaluate sine and cosine functions, see Example 2. For an example of using a calculator to evaluate trigonometric functions, see Example 3.

4.2 Exercises

See CalcChat.com for tutorial help and worked-out solutions to odd-numbered exercises.

Vocabulary: Fill in the blanks.

1. Each real number t corresponds to a point (x, y) on the ________ ________.
2. A function f is ________ when there exists a positive real number c such that $f(t + c) = f(t)$ for all t in the domain of f.
3. The smallest number c for which a function f is periodic is called the ________ of f.
4. A function f is ________ when $f(-t) = -f(t)$ and ________ when $f(-t) = f(t)$.

Skills and Applications

Determining Values of Trigonometric Functions In Exercises 5–8, determine the exact values of the six trigonometric functions of the real number t.

5.

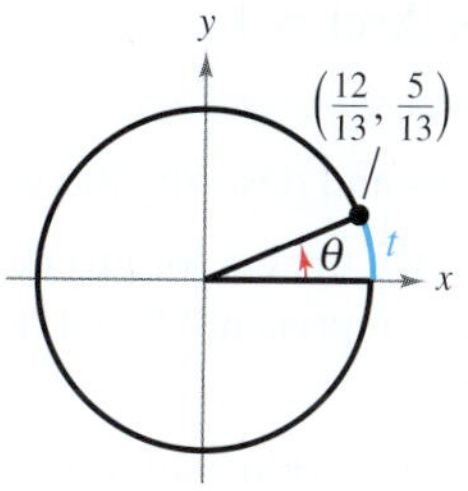

6.

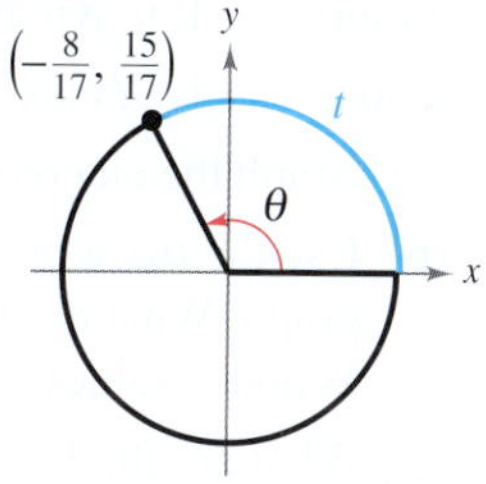

7.

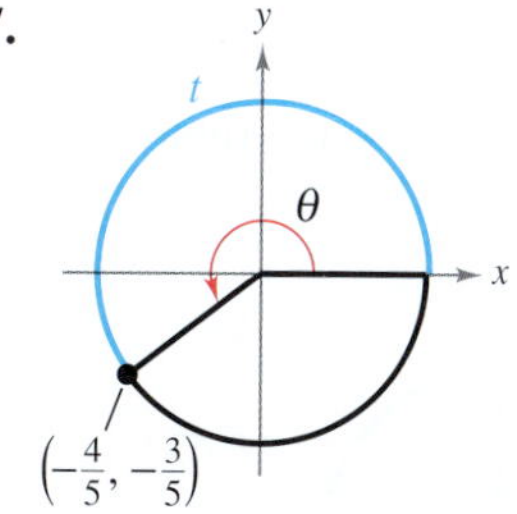

8.

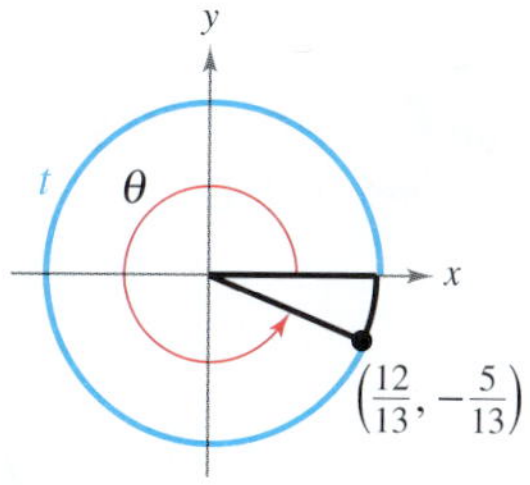

Finding a Point on the Unit Circle In Exercises 9–12, find the point (x, y) on the unit circle that corresponds to the real number t.

9. $t = \pi/2$
10. $t = \pi/4$
11. $t = 5\pi/6$
12. $t = 4\pi/3$

Evaluating Sine, Cosine, and Tangent In Exercises 13–22, evaluate (if possible) the sine, cosine, and tangent at the real number.

13. $t = \dfrac{\pi}{4}$
14. $t = \dfrac{\pi}{3}$
15. $t = -\dfrac{\pi}{6}$
16. $t = -\dfrac{\pi}{4}$
17. $t = -\dfrac{7\pi}{4}$
18. $t = -\dfrac{4\pi}{3}$
19. $t = \dfrac{11\pi}{6}$
20. $t = \dfrac{5\pi}{3}$
21. $t = -\dfrac{3\pi}{2}$
22. $t = -2\pi$

Evaluating Trigonometric Functions In Exercises 23–30, evaluate (if possible) the six trigonometric functions at the real number.

23. $t = 2\pi/3$
24. $t = 5\pi/6$
25. $t = 4\pi/3$
26. $t = 7\pi/4$
27. $t = -5\pi/3$
28. $t = 3\pi/2$
29. $t = -\pi/2$
30. $t = -\pi$

Using Period to Evaluate Sine and Cosine In Exercises 31–36, evaluate the trigonometric function using its period as an aid.

31. $\sin 4\pi$
32. $\cos 3\pi$
33. $\cos \dfrac{7\pi}{3}$
34. $\sin \dfrac{9\pi}{4}$
35. $\sin \dfrac{19\pi}{6}$
36. $\sin\left(-\dfrac{8\pi}{3}\right)$

Using the Value of a Trigonometric Function In Exercises 37–42, use the value of the trigonometric function to evaluate the indicated functions.

37. $\sin t = \frac{1}{2}$
    (a) $\sin(-t)$
    (b) $\csc(-t)$
38. $\sin(-t) = \frac{3}{8}$
    (a) $\sin t$
    (b) $\csc t$
39. $\cos(-t) = -\frac{1}{5}$
    (a) $\cos t$
    (b) $\sec(-t)$
40. $\cos t = -\frac{3}{4}$
    (a) $\cos(-t)$
    (b) $\sec(-t)$
41. $\sin t = \frac{4}{5}$
    (a) $\sin(\pi - t)$
    (b) $\sin(t + \pi)$
42. $\cos t = \frac{4}{5}$
    (a) $\cos(\pi - t)$
    (b) $\cos(t + \pi)$

Using a Calculator In Exercises 43–48, use a calculator to evaluate the trigonometric function. Round your answer to four decimal places. (Be sure the calculator is in the correct mode.)

43. $\tan \pi/3$
44. $\csc 2\pi/3$
45. $\csc 0.8$
46. $\cos(-1.7)$
47. $\sec 1.8$
48. $\cot(-0.9)$

49. Harmonic Motion The displacement from equilibrium of an oscillating weight suspended by a spring is given by

$$y(t) = \frac{1}{4}\cos 6t$$

where y is the displacement (in feet) and t is the time (in seconds). Find the displacement when (a) $t = 0$, (b) $t = \frac{1}{4}$, and (c) $t = \frac{1}{2}$.

50. Harmonic Motion

The displacement from equilibrium of an oscillating weight suspended by a spring and subject to the damping effect of friction is given by $y(t) = \frac{1}{4}e^{-t}\cos 6t$, where y is the displacement (in feet) and t is the time (in seconds).

(a) Complete the table.

t	0	$\frac{1}{4}$	$\frac{1}{2}$	$\frac{3}{4}$	1
y					

(b) Use the *table* feature of a graphing utility to approximate the time when the weight reaches equilibrium.

(c) What appears to happen to the displacement as t increases?

Exploration

True or False? **In Exercises 51–54, determine whether the statement is true or false. Justify your answer.**

51. Because $\sin(-t) = -\sin t$, the sine of a negative angle is a negative number.

52. The real number 0 corresponds to the point $(0, 1)$ on the unit circle.

53. $\tan a = \tan(a - 6\pi)$

54. $\cos\left(-\frac{7\pi}{2}\right) = \cos\left(\pi + \frac{\pi}{2}\right)$

55. Conjecture Let (x_1, y_1) and (x_2, y_2) be points on the unit circle corresponding to $t = t_1$ and $t = \pi - t_1$, respectively.

(a) Identify the symmetry of the points (x_1, y_1) and (x_2, y_2).

(b) Make a conjecture about any relationship between $\sin t_1$ and $\sin(\pi - t_1)$.

(c) Make a conjecture about any relationship between $\cos t_1$ and $\cos(\pi - t_1)$.

56. Using the Unit Circle Use the unit circle to verify that the cosine and secant functions are even and that the sine, cosecant, tangent, and cotangent functions are odd.

57. Verifying Expressions Are Not Equal Verify that $\cos 2t \neq 2\cos t$ by approximating $\cos 1.5$ and $2\cos 0.75$.

58. Verifying Expressions Are Not Equal Verify that $\sin(t_1 + t_2) \neq \sin t_1 + \sin t_2$ by approximating $\sin 0.25$, $\sin 0.75$, and $\sin 1$.

59. Graphical Analysis With a graphing utility in *radian* and *parametric* modes, enter the equations

$$X_{1T} = \cos T \quad \text{and} \quad Y_{1T} = \sin T$$

and use the following settings.

Tmin = 0, Tmax = 6.3, Tstep = 0.1

Xmin = −1.5, Xmax = 1.5, Xscl = 1

Ymin = −1, Ymax = 1, Yscl = 1

(a) Graph the entered equations and describe the graph.

(b) Use the *trace* feature to move the cursor around the graph. What do the t-values represent? What do the x- and y-values represent?

(c) What are the least and greatest values of x and y?

60. HOW DO YOU SEE IT? Use the figure below.

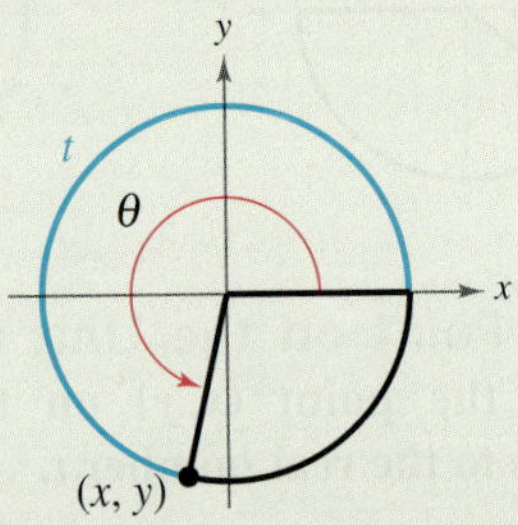

(a) Do all of the trigonometric functions of t exist? Explain your reasoning.

(b) For those trigonometric functions that exist, determine whether the sign of the trigonometric function is positive or negative. Explain your reasoning.

61. Think About It Because $f(t) = \sin t$ is an odd function and $g(t) = \cos t$ is an even function, what can be said about the function $h(t) = f(t)g(t)$?

62. Think About It Because $f(t) = \sin t$ and $g(t) = \tan t$ are odd functions, what can be said about the function $h(t) = f(t)g(t)$?

Richard Megna/Fundamental Photographs

4.3 Right Triangle Trigonometry

Trigonometric functions can help you analyze real-life situations. For instance, in Exercise 76 on page 289, you will use trigonometric functions to find the height of a helium-filled balloon.

- **Evaluate trigonometric functions of acute angles, and use a calculator to evaluate trigonometric functions.**
- **Use the fundamental trigonometric identities.**
- **Use trigonometric functions to model and solve real-life problems.**

The Six Trigonometric Functions

This section introduces the trigonometric functions from a *right triangle* perspective. Consider a right triangle with one acute angle labeled θ, as shown below. Relative to the angle θ, the three sides of the triangle are the **hypotenuse,** the **opposite side** (the side opposite the angle θ), and the **adjacent side** (the side adjacent to the angle θ).

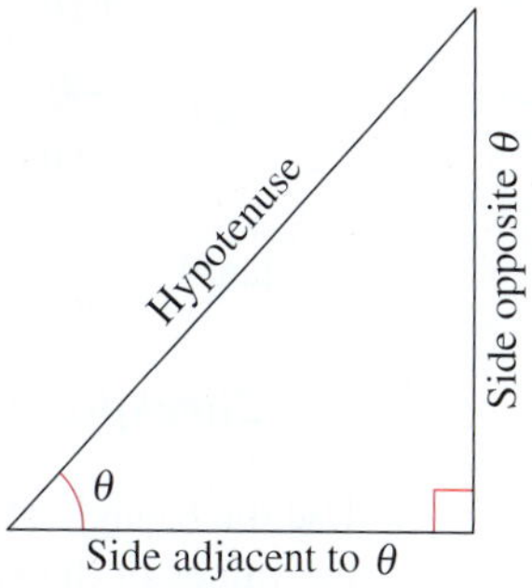

Using the lengths of these three sides, you can form six ratios that define the six trigonometric functions of the acute angle θ.

sine **cosecant** **cosine** **secant** **tangent** **cotangent**

In the following definitions, it is important to see that

$$0° < \theta < 90°$$

(θ lies in the first quadrant) and that for such angles the value of each trigonometric function is *positive.*

Right Triangle Definitions of Trigonometric Functions

Let θ be an *acute* angle of a right triangle. The six trigonometric functions of the angle θ are defined as follows. (Note that the functions in the second row are the *reciprocals* of the corresponding functions in the first row.)

$$\sin\theta = \frac{\text{opp}}{\text{hyp}} \qquad \cos\theta = \frac{\text{adj}}{\text{hyp}} \qquad \tan\theta = \frac{\text{opp}}{\text{adj}}$$

$$\csc\theta = \frac{\text{hyp}}{\text{opp}} \qquad \sec\theta = \frac{\text{hyp}}{\text{adj}} \qquad \cot\theta = \frac{\text{adj}}{\text{opp}}$$

The abbreviations

opp, adj, and hyp

represent the lengths of the three sides of a right triangle.

opp = the length of the side *opposite* θ

adj = the length of the side *adjacent to* θ

hyp = the length of the *hypotenuse*

Scott Cornell/Shutterstock.com

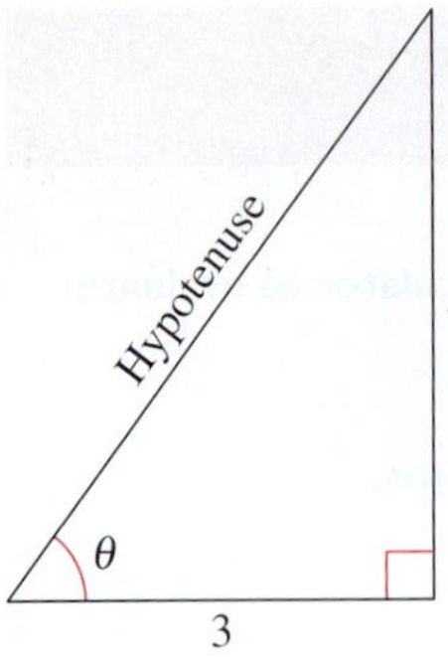

Figure 4.20

EXAMPLE 1 Evaluating Trigonometric Functions

Use the triangle in Figure 4.20 to find the values of the six trigonometric functions of θ.

Solution By the Pythagorean Theorem,

$$(\text{hyp})^2 = (\text{opp})^2 + (\text{adj})^2$$

it follows that

$$\begin{aligned}\text{hyp} &= \sqrt{4^2 + 3^2}\\ &= \sqrt{25}\\ &= 5.\end{aligned}$$

So, the six trigonometric functions of θ are

$$\sin\theta = \frac{\text{opp}}{\text{hyp}} = \frac{4}{5} \qquad \csc\theta = \frac{\text{hyp}}{\text{opp}} = \frac{5}{4}$$

$$\cos\theta = \frac{\text{adj}}{\text{hyp}} = \frac{3}{5} \qquad \sec\theta = \frac{\text{hyp}}{\text{adj}} = \frac{5}{3}$$

$$\tan\theta = \frac{\text{opp}}{\text{adj}} = \frac{4}{3} \qquad \cot\theta = \frac{\text{adj}}{\text{opp}} = \frac{3}{4}.$$

ALGEBRA HELP You can review the Pythagorean Theorem in Section 1.1.

✓ Checkpoint Audio-video solution in English & Spanish at LarsonPrecalculus.com.

Use the triangle at the right to find the values of the six trigonometric functions of θ.

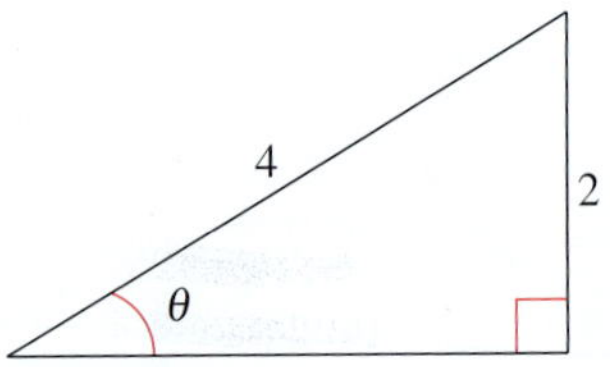

HISTORICAL NOTE

Georg Joachim Rhaeticus (1514–1574) was the leading Teutonic mathematical astronomer of the 16th century. He was the first to define the trigonometric functions as ratios of the sides of a right triangle.

In Example 1, you were given the lengths of two sides of the right triangle, but not the angle θ. Often, you will be asked to find the trigonometric functions of a *given* acute angle θ. To do this, construct a right triangle having θ as one of its angles.

EXAMPLE 2 Evaluating Trigonometric Functions of 45°

Find the values of sin 45°, cos 45°, and tan 45°.

Solution Construct a right triangle having 45° as one of its acute angles, as shown in Figure 4.21. Choose 1 as the length of the adjacent side. From geometry, you know that the other acute angle is also 45°. So, the triangle is isosceles and the length of the opposite side is also 1. Using the Pythagorean Theorem, you find the length of the hypotenuse to be $\sqrt{2}$.

$$\sin 45^\circ = \frac{\text{opp}}{\text{hyp}} = \frac{1}{\sqrt{2}} = \frac{\sqrt{2}}{2}$$

$$\cos 45^\circ = \frac{\text{adj}}{\text{hyp}} = \frac{1}{\sqrt{2}} = \frac{\sqrt{2}}{2}$$

$$\tan 45^\circ = \frac{\text{opp}}{\text{adj}} = \frac{1}{1} = 1$$

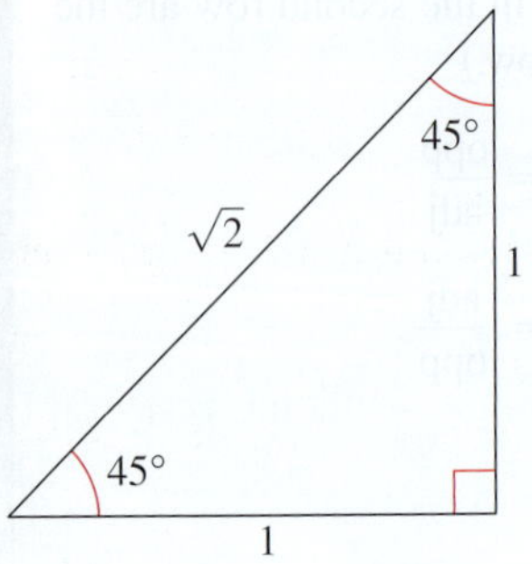

Figure 4.21

✓ Checkpoint Audio-video solution in English & Spanish at LarsonPrecalculus.com.

Find the value of sec 45°.

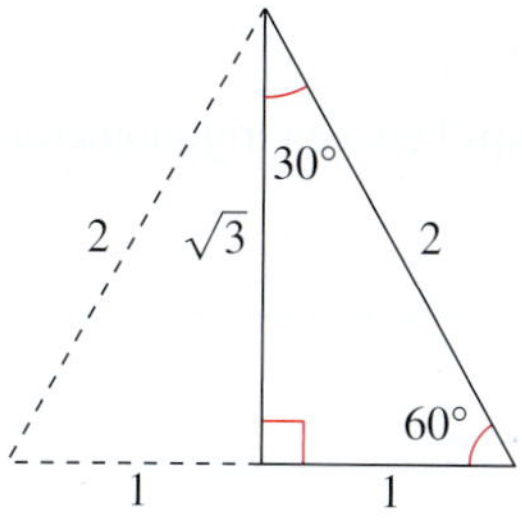

Figure 4.22

REMARK Because the angles 30°, 45°, and 60° ($\pi/6$, $\pi/4$, and $\pi/3$, respectively) occur frequently in trigonometry, you should learn to construct the triangles shown in Figures 4.21 and 4.22.

EXAMPLE 3 **Trigonometric Functions of 30° and 60°**

Use the equilateral triangle shown in Figure 4.22 to find the values of sin 60°, cos 60°, sin 30°, and cos 30°.

Solution For $\theta = 60°$, you have adj = 1, opp = $\sqrt{3}$, and hyp = 2. So,

$$\sin 60° = \frac{\text{opp}}{\text{hyp}} = \frac{\sqrt{3}}{2} \quad \text{and} \quad \cos 60° = \frac{\text{adj}}{\text{hyp}} = \frac{1}{2}.$$

For $\theta = 30°$, adj = $\sqrt{3}$, opp = 1, and hyp = 2. So,

$$\sin 30° = \frac{\text{opp}}{\text{hyp}} = \frac{1}{2} \quad \text{and} \quad \cos 30° = \frac{\text{adj}}{\text{hyp}} = \frac{\sqrt{3}}{2}.$$

✓ **Checkpoint** *Audio-video solution in English & Spanish at LarsonPrecalculus.com.*

Use the equilateral triangle shown in Figure 4.22 to find the values of tan 60° and tan 30°.

Sines, Cosines, and Tangents of Special Angles

$$\sin 30° = \sin\frac{\pi}{6} = \frac{1}{2} \qquad \cos 30° = \cos\frac{\pi}{6} = \frac{\sqrt{3}}{2} \qquad \tan 30° = \tan\frac{\pi}{6} = \frac{\sqrt{3}}{3}$$

$$\sin 45° = \sin\frac{\pi}{4} = \frac{\sqrt{2}}{2} \qquad \cos 45° = \cos\frac{\pi}{4} = \frac{\sqrt{2}}{2} \qquad \tan 45° = \tan\frac{\pi}{4} = 1$$

$$\sin 60° = \sin\frac{\pi}{3} = \frac{\sqrt{3}}{2} \qquad \cos 60° = \cos\frac{\pi}{3} = \frac{1}{2} \qquad \tan 60° = \tan\frac{\pi}{3} = \sqrt{3}$$

REMARK Throughout this text, angles are assumed to be measured in radians unless noted otherwise. For example, sin 1 means the sine of 1 radian and sin 1° means the sine of 1 degree.

In the box, note that $\sin 30° = \frac{1}{2} = \cos 60°$. This occurs because 30° and 60° are complementary angles. In general, it can be shown from the right triangle definitions that *cofunctions of complementary angles are equal*. That is, if θ is an acute angle, then the following relationships are true.

$$\sin(90° - \theta) = \cos\theta \qquad \cos(90° - \theta) = \sin\theta \qquad \tan(90° - \theta) = \cot\theta$$

$$\cot(90° - \theta) = \tan\theta \qquad \sec(90° - \theta) = \csc\theta \qquad \csc(90° - \theta) = \sec\theta$$

To use a calculator to evaluate trigonometric functions of angles measured in degrees, first set the calculator to *degree* mode and then proceed as demonstrated in Section 4.2.

EXAMPLE 4 **Using a Calculator**

Use a calculator to evaluate sec 5° 40′ 12″.

Solution Begin by converting to decimal degree form. $\left[\text{Recall that } 1' = \frac{1}{60}(1°) \text{ and } 1'' = \frac{1}{3600}(1°).\right]$

$$5°\,40'\,12'' = 5° + \left(\frac{40}{60}\right)^{\circ} + \left(\frac{12}{3600}\right)^{\circ} = 5.67°$$

Then, use a calculator to evaluate sec 5.67°.

Function	Calculator Keystrokes	Display
sec 5° 40′ 12″ = sec 5.67°	(COS (5.67)) x⁻¹ ENTER	1.0049166

✓ **Checkpoint** *Audio-video solution in English & Spanish at LarsonPrecalculus.com.*

Use a calculator to evaluate csc 34° 30′ 36″.

Trigonometric Identities

In trigonometry, a great deal of time is spent studying relationships between trigonometric functions (identities).

Fundamental Trigonometric Identities

Reciprocal Identities

$$\sin\theta = \frac{1}{\csc\theta} \qquad \cos\theta = \frac{1}{\sec\theta} \qquad \tan\theta = \frac{1}{\cot\theta}$$

$$\csc\theta = \frac{1}{\sin\theta} \qquad \sec\theta = \frac{1}{\cos\theta} \qquad \cot\theta = \frac{1}{\tan\theta}$$

Quotient Identities

$$\tan\theta = \frac{\sin\theta}{\cos\theta} \qquad \cot\theta = \frac{\cos\theta}{\sin\theta}$$

Pythagorean Identities

$$\sin^2\theta + \cos^2\theta = 1$$

$$1 + \tan^2\theta = \sec^2\theta$$

$$1 + \cot^2\theta = \csc^2\theta$$

Note that $\sin^2\theta$ represents $(\sin\theta)^2$, $\cos^2\theta$ represents $(\cos\theta)^2$, and so on.

EXAMPLE 5 Applying Trigonometric Identities

Let θ be an acute angle such that $\sin\theta = 0.6$. Find the values of (a) $\cos\theta$ and (b) $\tan\theta$ using trigonometric identities.

Solution

a. To find the value of $\cos\theta$, use the Pythagorean identity

$$\sin^2\theta + \cos^2\theta = 1.$$

So, you have

$$(0.6)^2 + \cos^2\theta = 1 \qquad \text{Substitute 0.6 for } \sin\theta.$$

$$\cos^2\theta = 1 - (0.6)^2 \qquad \text{Subtract } (0.6)^2 \text{ from each side.}$$

$$\cos^2\theta = 0.64 \qquad \text{Simplify.}$$

$$\cos\theta = \sqrt{0.64} \qquad \text{Extract positive square root.}$$

$$\cos\theta = 0.8. \qquad \text{Simplify.}$$

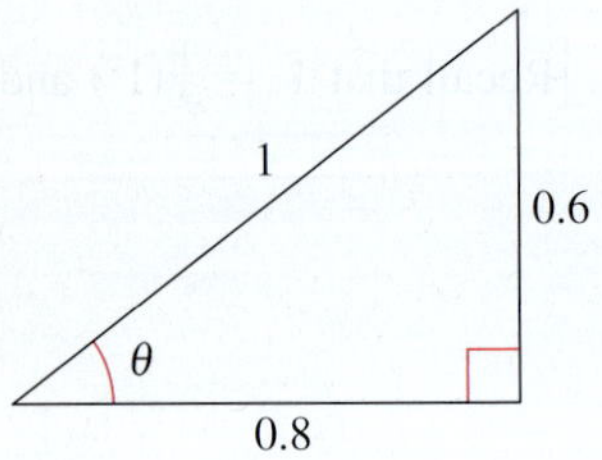

Figure 4.23

b. Now, knowing the sine and cosine of θ, you can find the tangent of θ to be

$$\tan\theta = \frac{\sin\theta}{\cos\theta} = \frac{0.6}{0.8} = 0.75.$$

Use the definitions of $\cos\theta$ and $\tan\theta$ and the triangle shown in Figure 4.23 to check these results.

✓ **Checkpoint** *Audio-video solution in English & Spanish at LarsonPrecalculus.com.*

Let θ be an acute angle such that $\cos\theta = 0.25$. Find the values of (a) $\sin\theta$ and (b) $\tan\theta$ using trigonometric identities.

EXAMPLE 6 **Applying Trigonometric Identities**

Let θ be an acute angle such that $\tan \theta = \frac{1}{3}$. Find the values of (a) $\cot \theta$ and (b) $\sec \theta$ using trigonometric identities.

Solution

a. $\cot \theta = \dfrac{1}{\tan \theta}$ — Reciprocal identity

$= \dfrac{1}{1/3}$

$= 3$

b. $\sec^2 \theta = 1 + \tan^2 \theta$ — Pythagorean identity

$= 1 + (1/3)^2$

$= 10/9$

$\sec \theta = \sqrt{10}/3$

Use the definitions of $\cot \theta$ and $\sec \theta$ and the triangle shown below to check these results.

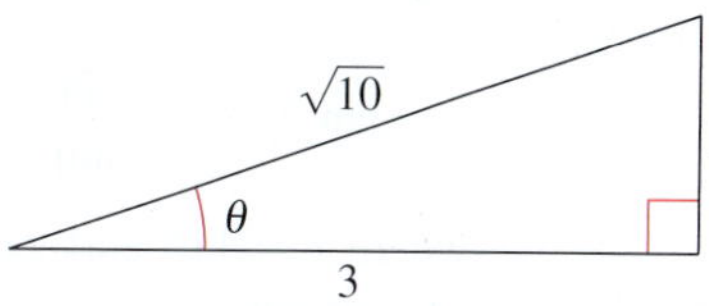

✓ Checkpoint *Audio-video solution in English & Spanish at LarsonPrecalculus.com.*

Let β be an acute angle such that $\tan \beta = 2$. Find the values of (a) $\cot \beta$ and (b) $\sec \beta$ using trigonometric identities.

EXAMPLE 7 **Using Trigonometric Identities**

Use trigonometric identities to transform the left side of the equation into the right side $(0 < \theta < \pi/2)$.

a. $\sin \theta \csc \theta = 1$ **b.** $(\csc \theta + \cot \theta)(\csc \theta - \cot \theta) = 1$

Solution

a. $\sin \theta \csc \theta = \left(\dfrac{1}{\csc \theta}\right)\csc \theta = 1$ — Use a reciprocal identity and simplify.

b. $(\csc \theta + \cot \theta)(\csc \theta - \cot \theta)$

$= \csc^2 \theta - \csc \theta \cot \theta + \csc \theta \cot \theta - \cot^2 \theta$ — FOIL Method

$= \csc^2 \theta - \cot^2 \theta$ — Simplify.

$= 1$ — Pythagorean identity

✓ Checkpoint *Audio-video solution in English & Spanish at LarsonPrecalculus.com.*

Use trigonometric identities to transform the left side of the equation into the right side $(0 < \theta < \pi/2)$.

a. $\tan \theta \csc \theta = \sec \theta$

b. $(\csc \theta + 1)(\csc \theta - 1) = \cot^2 \theta$

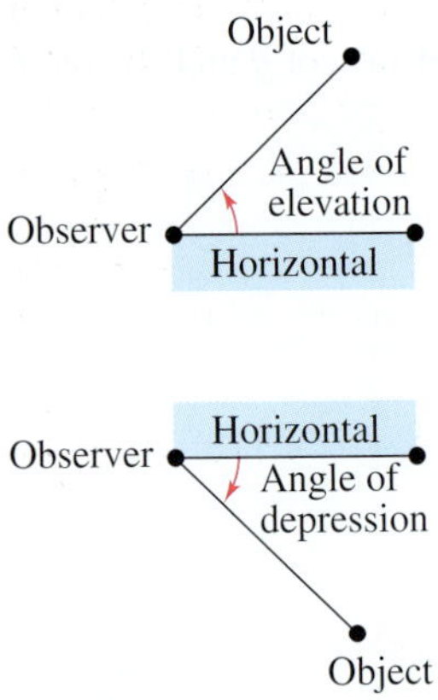

Figure 4.24

Applications Involving Right Triangles

Many applications of trigonometry involve a process called **solving right triangles.** In this type of application, you are usually given one side of a right triangle and one of the acute angles and are asked to find one of the other sides, *or* you are given two sides and are asked to find one of the acute angles.

In Example 8, the angle you are given is the **angle of elevation,** which represents the angle from the horizontal upward to an object. In other applications you may be given the **angle of depression,** which represents the angle from the horizontal downward to an object. (See Figure 4.24.)

EXAMPLE 8 Using Trigonometry to Solve a Right Triangle

A surveyor is standing 115 feet from the base of the Washington Monument, as shown in the figure at the right. The surveyor measures the angle of elevation to the top of the monument as 78.3°. How tall is the Washington Monument?

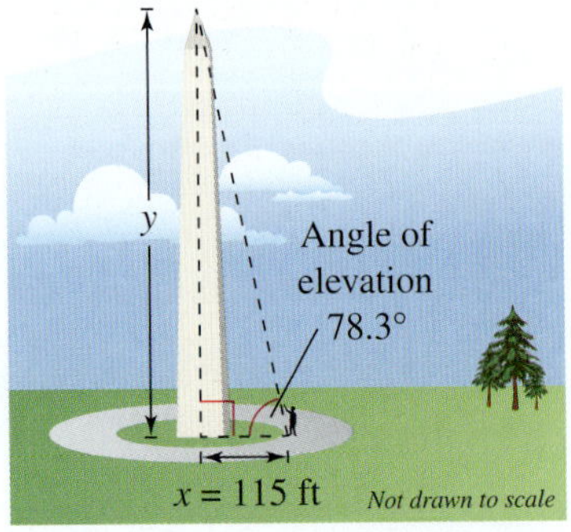

Solution From the figure, you can see that

$$\tan 78.3^\circ = \frac{\text{opp}}{\text{adj}} = \frac{y}{x}$$

where $x = 115$ and y is the height of the monument. So, the height of the Washington Monument is

$$\begin{aligned} y &= x \tan 78.3^\circ \\ &\approx 115(4.82882) \\ &\approx 555 \text{ feet.} \end{aligned}$$

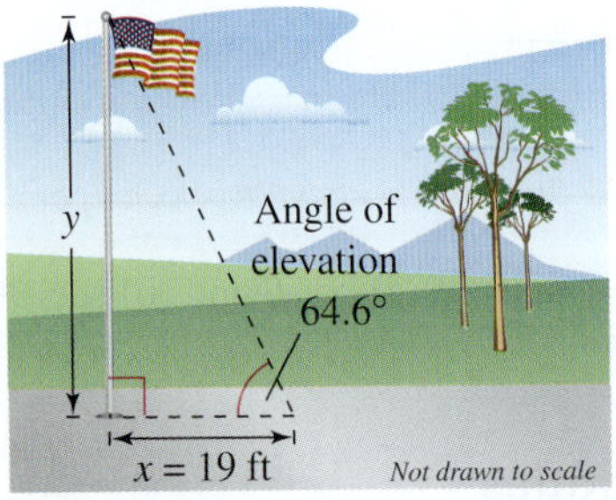

Figure 4.25

✓ ***Checkpoint*** *Audio-video solution in English & Spanish at LarsonPrecalculus.com.*

How tall is the flagpole in Figure 4.25?

EXAMPLE 9 Using Trigonometry to Solve a Right Triangle

A historic lighthouse is 200 yards from a bike path along the edge of a lake. A walkway to the lighthouse is 400 yards long. Find the acute angle θ between the bike path and the walkway, as illustrated in the figure at the right.

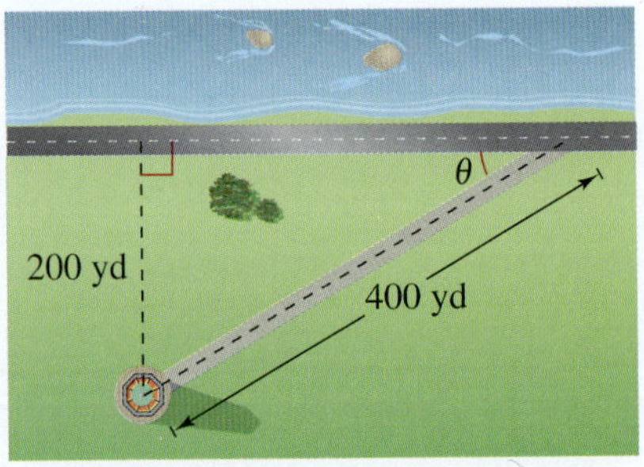

Solution From the figure, you can see that the sine of the angle θ is

$$\sin \theta = \frac{\text{opp}}{\text{hyp}} = \frac{200}{400} = \frac{1}{2}.$$

Now you should recognize that $\theta = 30^\circ$.

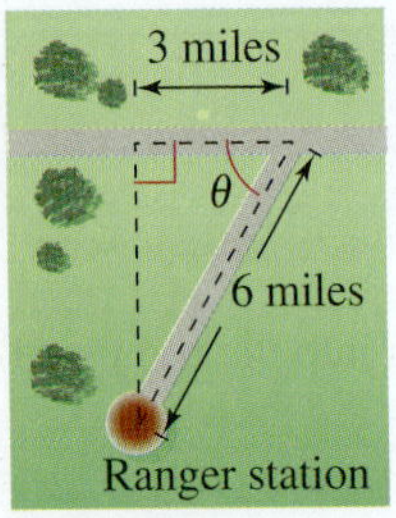

Figure 4.26

✓ ***Checkpoint*** *Audio-video solution in English & Spanish at LarsonPrecalculus.com.*

Find the acute angle θ between the two paths, as illustrated in Figure 4.26.

In Example 9, you were able to recognize that $\theta = 30°$ is the acute angle that satisfies the equation $\sin \theta = \frac{1}{2}$. Suppose, however, that you were given the equation $\sin \theta = 0.6$ and were asked to find the acute angle θ. Because $\sin 30° = \frac{1}{2} = 0.5000$ and $\sin 45° = 1/\sqrt{2} \approx 0.7071$, you might guess that θ lies somewhere between $30°$ and $45°$. In a later section, you will study a method by which a more precise value of θ can be determined.

EXAMPLE 10 Solving a Right Triangle

Skateboarders can go to a skatepark, which is a recreational environment built with many different types of ramps and rails.

Find the length c of the skateboard ramp shown in the figure below. Find the horizontal length a of the ramp.

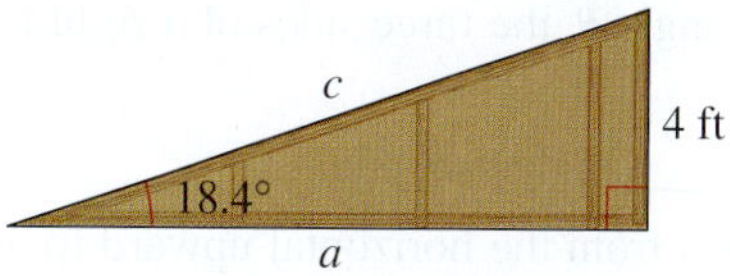

Solution From the figure, you can see that

$$\sin 18.4° = \frac{\text{opp}}{\text{hyp}} = \frac{4}{c}.$$

So, the length of the skateboard ramp is

$$c = \frac{4}{\sin 18.4°} \approx \frac{4}{0.3156} \approx 12.7 \text{ feet}.$$

Also from the figure, you can see that

$$\tan 18.4° = \frac{\text{opp}}{\text{adj}} = \frac{4}{a}.$$

So, the horizontal length is

$$a = \frac{4}{\tan 18.4°} \approx 12.0 \text{ feet}.$$

Checkpoint *Audio-video solution in English & Spanish at LarsonPrecalculus.com.*

Find the length c of the loading ramp shown in the figure below. Find the horizontal length a of the ramp.

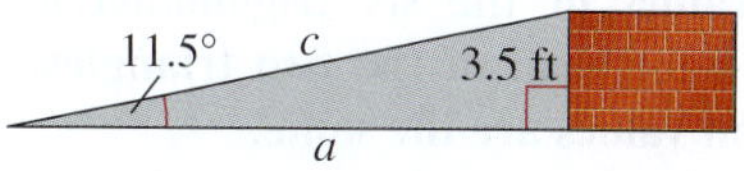

Summarize (Section 4.3)

1. State the right triangle definitions of the trigonometric functions *(page 279)* and describe how to use a calculator to evaluate trigonometric functions *(page 281)*. For examples of evaluating trigonometric functions of acute angles, see Examples 1–3. For an example of using a calculator to evaluate a trigonometric function, see Example 4.
2. List the fundamental trigonometric identities *(page 282)*. For examples of using the fundamental trigonometric identities, see Examples 5–7.
3. Describe examples of how to use trigonometric functions to model and solve real-life problems *(pages 284 and 285, Examples 8–10)*.

Vladimir Ivanovich Danilov/Shutterstock.com

4.3 Exercises

See CalcChat.com for tutorial help and worked-out solutions to odd-numbered exercises.

Vocabulary

1. Match each trigonometric function with its right triangle definition.

(a) sine (b) cosine (c) tangent (d) cosecant (e) secant (f) cotangent

(i) $\dfrac{\text{hypotenuse}}{\text{adjacent}}$ (ii) $\dfrac{\text{adjacent}}{\text{opposite}}$ (iii) $\dfrac{\text{hypotenuse}}{\text{opposite}}$ (iv) $\dfrac{\text{adjacent}}{\text{hypotenuse}}$ (v) $\dfrac{\text{opposite}}{\text{hypotenuse}}$ (vi) $\dfrac{\text{opposite}}{\text{adjacent}}$

In Exercises 2–4, fill in the blanks.

2. Relative to the acute angle θ, the three sides of a right triangle are the ________ side, the ________ side, and the ________.

3. Cofunctions of ________ angles are equal.

4. An angle that measures from the horizontal upward to an object is called the angle of ________, whereas an angle that measures from the horizontal downward to an object is called the angle of ________.

Skills and Applications

Evaluating Trigonometric Functions **In Exercises 5–8, find the exact values of the six trigonometric functions of the angle θ shown in the figure. (Use the Pythagorean Theorem to find the third side of the triangle.)**

5.
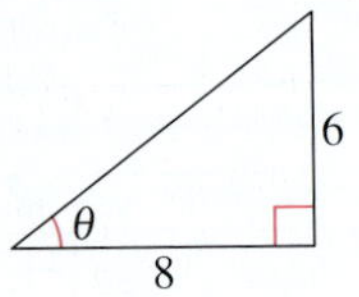

6.
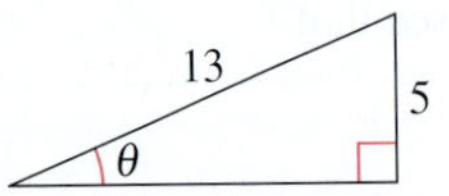

7.
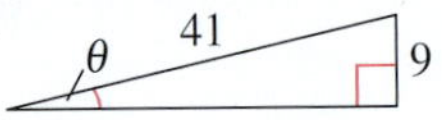

8.
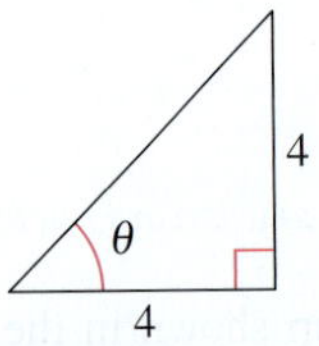

Evaluating Trigonometric Functions **In Exercises 9–12, find the exact values of the six trigonometric functions of the angle θ for each of the two triangles. Explain why the function values are the same.**

9.
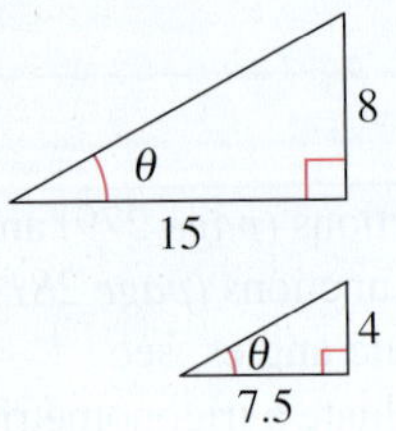

10.
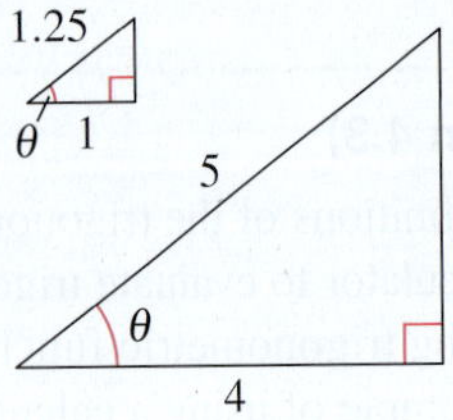

11.
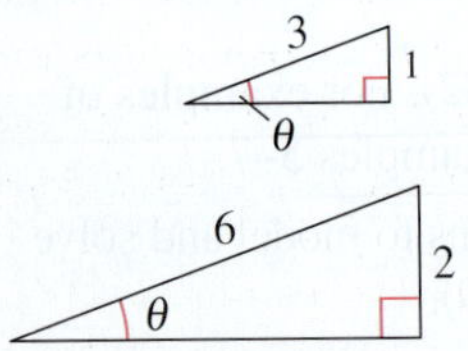

12.
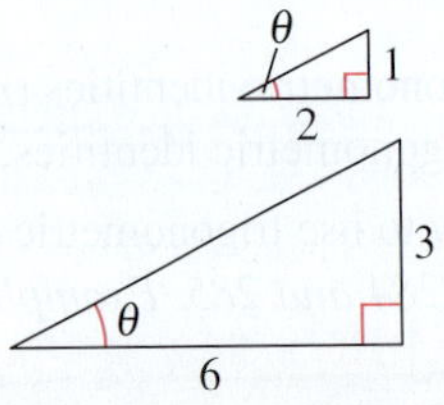

Evaluating Trigonometric Functions **In Exercises 13–20, sketch a right triangle corresponding to the trigonometric function of the acute angle θ. Use the Pythagorean Theorem to determine the third side and then find the other five trigonometric functions of θ.**

13. $\tan\theta = \frac{3}{4}$ **14.** $\cos\theta = \frac{5}{6}$

15. $\sec\theta = \frac{3}{2}$ **16.** $\tan\theta = \frac{4}{5}$

17. $\sin\theta = \frac{1}{5}$ **18.** $\sec\theta = \frac{17}{7}$

19. $\cot\theta = 3$ **20.** $\csc\theta = 9$

Evaluating Trigonometric Functions of 30°, 45°, and 60° **In Exercises 21–30, construct an appropriate triangle to find the missing values. $(0° \le \theta \le 90°, 0 \le \theta \le \pi/2)$**

	Function	θ (deg)	θ (rad)	Function Value
21.	sin	30°		
22.	cos	45°		
23.	sec		$\dfrac{\pi}{4}$	
24.	tan		$\dfrac{\pi}{3}$	
25.	cot			$\dfrac{\sqrt{3}}{3}$
26.	csc			$\sqrt{2}$
27.	csc		$\dfrac{\pi}{6}$	
28.	sin		$\dfrac{\pi}{4}$	
29.	cot			1
30.	tan			$\dfrac{\sqrt{3}}{3}$

Using a Calculator **In Exercises 31–40, use a calculator to evaluate each function. Round your answers to four decimal places. (Be sure the calculator is in the correct mode.)**

31. (a) $\sin 10°$ (b) $\cos 80°$

32. (a) $\tan 23.5°$ (b) $\cot 66.5°$

33. (a) $\sin 16.35°$ (b) $\csc 16.35°$

34. (a) $\cot 79.56°$ (b) $\sec 79.56°$

35. (a) $\cos 4° 50' 15''$ (b) $\sec 4° 50' 15''$

36. (a) $\sec 42° 12'$ (b) $\csc 48° 7'$

37. (a) $\cot 11° 15'$ (b) $\tan 11° 15'$

38. (a) $\sec 56° 8' 10''$ (b) $\cos 56° 8' 10''$

39. (a) $\csc 32° 40' 3''$ (b) $\tan 44° 28' 16''$

40. (a) $\sec\left(\frac{9}{5} \cdot 20 + 32\right)°$ (b) $\cot\left(\frac{9}{5} \cdot 30 + 32\right)°$

Applying Trigonometric Identities **In Exercises 41–46, use the given function value(s) and the trigonometric identities to find the indicated trigonometric functions.**

41. $\sin 60° = \dfrac{\sqrt{3}}{2}$, $\cos 60° = \dfrac{1}{2}$

(a) $\sin 30°$ (b) $\cos 30°$

(c) $\tan 60°$ (d) $\cot 60°$

42. $\sin 30° = \dfrac{1}{2}$, $\tan 30° = \dfrac{\sqrt{3}}{3}$

(a) $\csc 30°$ (b) $\cot 60°$

(c) $\cos 30°$ (d) $\cot 30°$

43. $\cos \theta = \frac{1}{3}$

(a) $\sin \theta$ (b) $\tan \theta$

(c) $\sec \theta$ (d) $\csc(90° - \theta)$

44. $\sec \theta = 5$

(a) $\cos \theta$ (b) $\cot \theta$

(c) $\cot(90° - \theta)$ (d) $\sin \theta$

45. $\cot \alpha = 5$

(a) $\tan \alpha$ (b) $\csc \alpha$

(c) $\cot(90° - \alpha)$ (d) $\cos \alpha$

46. $\cos \beta = \dfrac{\sqrt{7}}{4}$

(a) $\sec \beta$ (b) $\sin \beta$

(c) $\cot \beta$ (d) $\sin(90° - \beta)$

Using Trigonometric Identities **In Exercises 47–56, use trigonometric identities to transform the left side of the equation into the right side $(0 < \theta < \pi/2)$.**

47. $\tan \theta \cot \theta = 1$

48. $\cos \theta \sec \theta = 1$

49. $\tan \alpha \cos \alpha = \sin \alpha$

50. $\cot \alpha \sin \alpha = \cos \alpha$

51. $(1 + \sin \theta)(1 - \sin \theta) = \cos^2 \theta$

52. $(1 + \cos \theta)(1 - \cos \theta) = \sin^2 \theta$

53. $(\sec \theta + \tan \theta)(\sec \theta - \tan \theta) = 1$

54. $\sin^2 \theta - \cos^2 \theta = 2 \sin^2 \theta - 1$

55. $\dfrac{\sin \theta}{\cos \theta} + \dfrac{\cos \theta}{\sin \theta} = \csc \theta \sec \theta$

56. $\dfrac{\tan \beta + \cot \beta}{\tan \beta} = \csc^2 \beta$

Evaluating Trigonometric Functions **In Exercises 57–62, find each value of θ in degrees $(0° < \theta < 90°)$ and radians $(0 < \theta < \pi/2)$ without using a calculator.**

57. (a) $\sin \theta = \frac{1}{2}$ (b) $\csc \theta = 2$

58. (a) $\cos \theta = \dfrac{\sqrt{2}}{2}$ (b) $\tan \theta = 1$

59. (a) $\sec \theta = 2$ (b) $\cot \theta = 1$

60. (a) $\tan \theta = \sqrt{3}$ (b) $\cos \theta = \frac{1}{2}$

61. (a) $\csc \theta = \dfrac{2\sqrt{3}}{3}$ (b) $\sin \theta = \dfrac{\sqrt{2}}{2}$

62. (a) $\cot \theta = \dfrac{\sqrt{3}}{3}$ (b) $\sec \theta = \sqrt{2}$

Finding Side Lengths of a Triangle **In Exercises 63–66, find the exact values of the indicated variables.**

63. Find x and y.

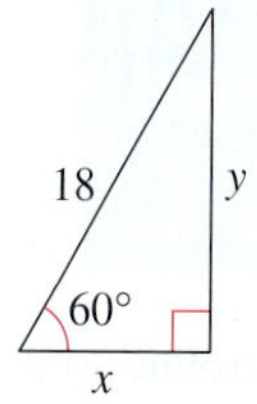

64. Find x and r.

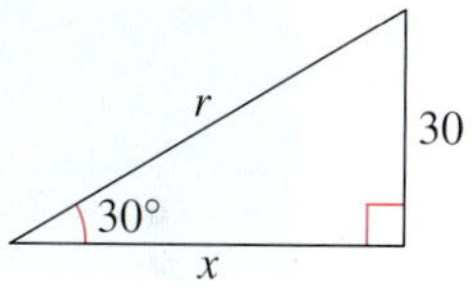

65. Find x and r.

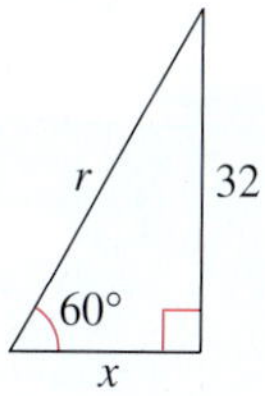

66. Find x and r.

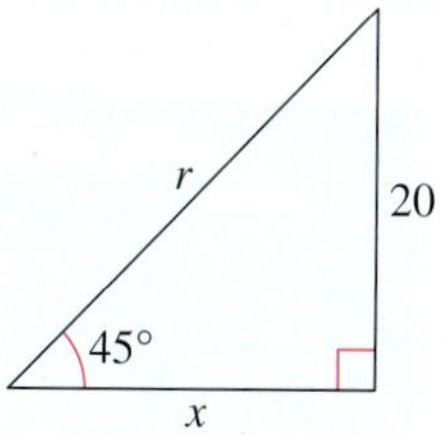

67. Empire State Building You are standing 45 meters from the base of the Empire State Building. You estimate that the angle of elevation to the top of the 86th floor (the observatory) is 82°. The total height of the building is another 123 meters above the 86th floor. What is the approximate height of the building? One of your friends is on the 86th floor. What is the distance between you and your friend?

68. Height A six-foot-tall person walks from the base of a broadcasting tower directly toward the tip of the shadow cast by the tower. When the person is 132 feet from the tower and 3 feet from the tip of the shadow, the person's shadow starts to appear beyond the tower's shadow.

(a) Draw a right triangle that gives a visual representation of the problem. Show the known quantities of the triangle and use a variable to indicate the height of the tower.

(b) Use a trigonometric function to write an equation involving the unknown quantity.

(c) What is the height of the tower?

69. Angle of Elevation You are skiing down a mountain with a vertical height of 1500 feet. The distance from the top of the mountain to the base is 3000 feet. What is the angle of elevation from the base to the top of the mountain?

70. Width of a River A biologist wants to know the width w of a river in order to properly set instruments for studying the pollutants in the water. From point A, the biologist walks downstream 100 feet and sights to point C (see figure). From this sighting, the biologist determines that $\theta = 54°$. How wide is the river?

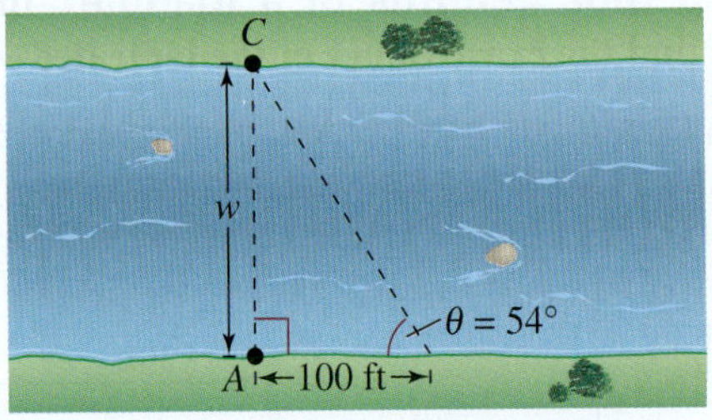

71. Length A guy wire runs from the ground to a cell tower. The wire is attached to the cell tower 150 feet above the ground. The angle formed between the wire and the ground is 43° (see figure).

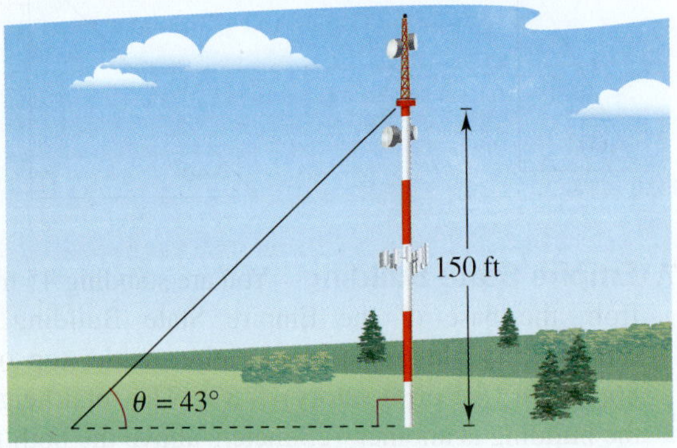

(a) How long is the guy wire?

(b) How far from the base of the tower is the guy wire anchored to the ground?

72. Height of a Mountain In traveling across flat land, you notice a mountain directly in front of you. Its angle of elevation (to the peak) is 3.5°. After you drive 13 miles closer to the mountain, the angle of elevation is 9° (see figure). Approximate the height of the mountain.

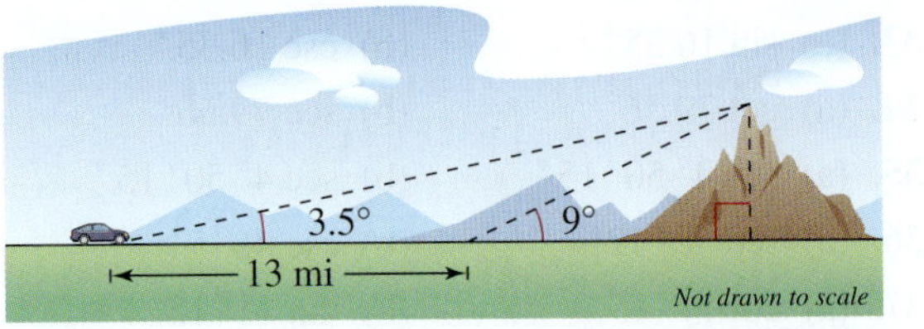

73. Machine Shop Calculations A steel plate has the form of one-fourth of a circle with a radius of 60 centimeters. Two two-centimeter holes are to be drilled in the plate, positioned as shown in the figure. Find the coordinates of the center of each hole.

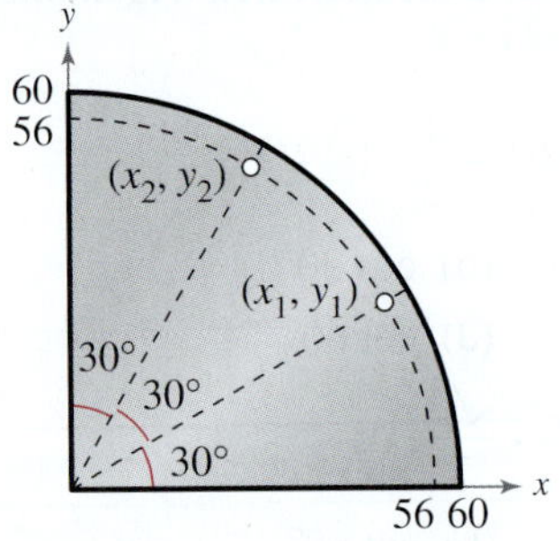

Figure for 73

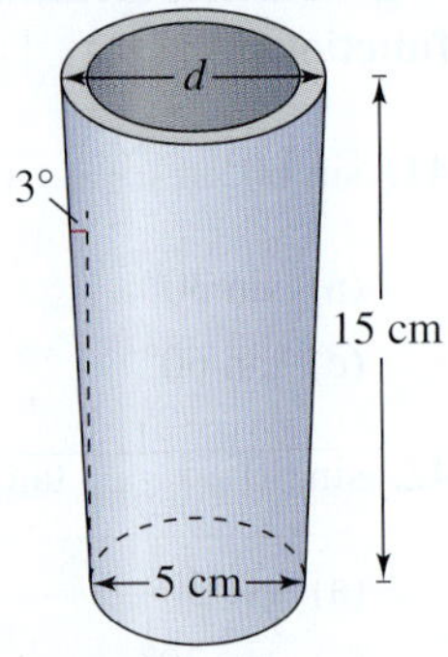

Figure for 74

74. Machine Shop Calculations A tapered shaft has a diameter of 5 centimeters at the small end and is 15 centimeters long (see figure). The taper is 3°. Find the diameter d of the large end of the shaft.

75. Geometry Use a compass to sketch a quarter of a circle of radius 10 centimeters. Using a protractor, construct an angle of 20° in standard position (see figure). Drop a perpendicular line from the point of intersection of the terminal side of the angle and the arc of the circle. By actual measurement, calculate the coordinates (x, y) of the point of intersection and use these measurements to approximate the six trigonometric functions of a 20° angle.

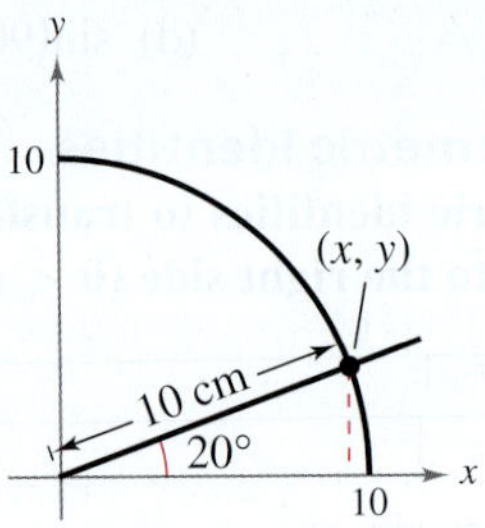

76. Height A 20-meter line is a tether for a helium-filled balloon. Because of a breeze, the line makes an angle of approximately 85° with the ground.

(a) Draw a right triangle that gives a visual representation of the problem. Show the known quantities of the triangle and use a variable to indicate the height of the balloon.

(b) Use a trigonometric function to write and solve an equation for the height of the balloon.

(c) The breeze becomes stronger and the angle the line makes with the ground decreases. How does this affect the triangle you drew in part (a)?

(d) Complete the table, which shows the heights (in meters) of the balloon for decreasing angle measures θ.

Angle, θ	80°	70°	60°	50°
Height				

Angle, θ	40°	30°	20°	10°
Height				

(e) As θ approaches 0°, how does this affect the height of the balloon? Draw a right triangle to explain your reasoning.

77. Johnstown Inclined Plane The Johnstown Inclined Plane in Pennsylvania is one of the longest and steepest hoists in the world. The railway cars travel a distance of 896.5 feet at an angle of approximately 35.4°, rising to a height of 1693.5 feet above sea level.

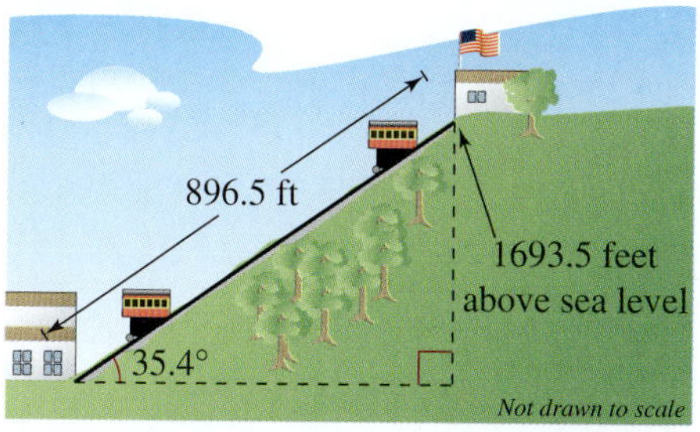

(a) Find the vertical rise of the inclined plane.

(b) Find the elevation of the lower end of the inclined plane.

(c) The cars move up the mountain at a rate of 300 feet per minute. Find the rate at which they rise vertically.

Exploration

78. Writing In right triangle trigonometry, explain why $\sin 30° = \frac{1}{2}$ regardless of the size of the triangle.

True or False? **In Exercises 79–84, determine whether the statement is true or false. Justify your answer.**

79. $\sin 60° \csc 60° = 1$

80. $\sec 30° = \csc 60°$

81. $\sin 45° + \cos 45° = 1$

82. $\cot^2 10° - \csc^2 10° = -1$

83. $\dfrac{\sin 60°}{\sin 30°} = \sin 2°$

84. $\tan[(5°)^2] = \tan^2 5°$

85. Think About It You are given the value of $\tan \theta$. Is it possible to find the value of $\sec \theta$ without finding the measure of θ? Explain.

86. Think About It

(a) Complete the table.

θ	0.1	0.2	0.3	0.4	0.5
$\sin \theta$					

(b) Is θ or $\sin \theta$ greater for θ in the interval $(0, 0.5]$?

(c) As θ approaches 0, how do θ and $\sin \theta$ compare? Explain.

87. Think About It

(a) Complete the table.

θ	0°	18°	36°	54°	72°	90°
$\sin \theta$						
$\cos \theta$						

(b) Discuss the behavior of the sine function for θ in the range from 0° to 90°.

(c) Discuss the behavior of the cosine function for θ in the range from 0° to 90°.

(d) Use the definitions of the sine and cosine functions to explain the results of parts (b) and (c).

88. HOW DO YOU SEE IT? Use the figure below.

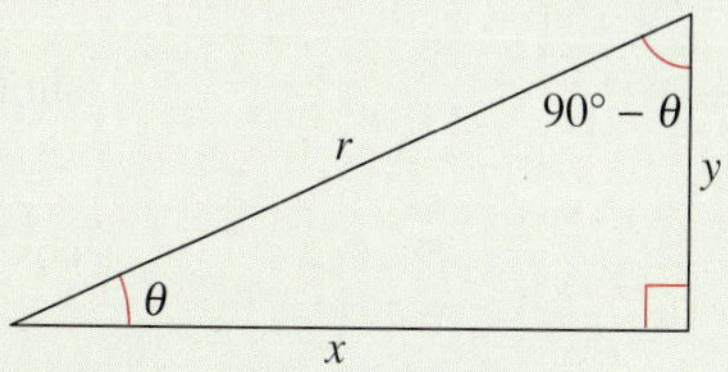

(a) Which side is opposite θ?

(b) Which side is adjacent to $90° - \theta$?

(c) Explain why $\sin \theta = \cos(90° - \theta)$.

Scott Cornell/Shutterstock.com

4.4 Trigonometric Functions of Any Angle

Trigonometric functions can help you model and solve real-life problems. For instance, in Exercise 99 on page 298, you will use trigonometric functions to model the monthly normal temperatures in New York City and Fairbanks, Alaska.

- **Evaluate trigonometric functions of any angle.**
- **Find reference angles.**
- **Evaluate trigonometric functions of real numbers.**

Introduction

In Section 4.3, the definitions of trigonometric functions were restricted to acute angles. In this section, the definitions are extended to cover *any* angle. When θ is an *acute* angle, the definitions here coincide with those given in the preceding section.

Definitions of Trigonometric Functions of Any Angle

Let θ be an angle in standard position with (x, y) a point on the terminal side of θ and $r = \sqrt{x^2 + y^2} \neq 0$.

$$\sin\theta = \frac{y}{r} \qquad \cos\theta = \frac{x}{r}$$

$$\tan\theta = \frac{y}{x},\quad x \neq 0 \qquad \cot\theta = \frac{x}{y},\quad y \neq 0$$

$$\sec\theta = \frac{r}{x},\quad x \neq 0 \qquad \csc\theta = \frac{r}{y},\quad y \neq 0$$

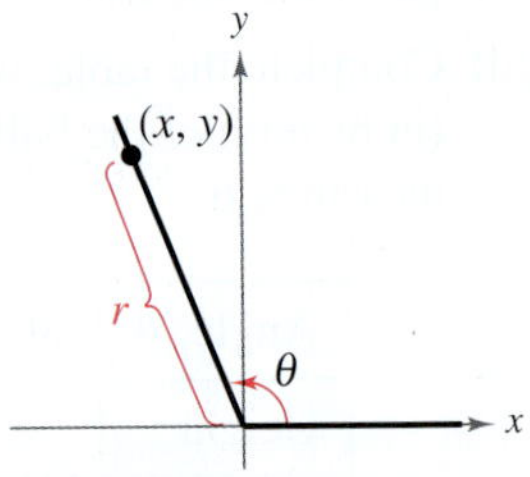

Because $r = \sqrt{x^2 + y^2}$ *cannot* be zero, it follows that the sine and cosine functions are defined for any real value of θ. However, when $x = 0$, the tangent and secant of θ are undefined. For example, the tangent of 90° is undefined. Similarly, when $y = 0$, the cotangent and cosecant of θ are undefined.

EXAMPLE 1 Evaluating Trigonometric Functions

Let $(-3, 4)$ be a point on the terminal side of θ. Find the sine, cosine, and tangent of θ.

Solution Referring to Figure 4.27, you can see that $x = -3$, $y = 4$, and

$$\begin{aligned} r &= \sqrt{x^2 + y^2} \\ &= \sqrt{(-3)^2 + 4^2} \\ &= \sqrt{25} \\ &= 5. \end{aligned}$$

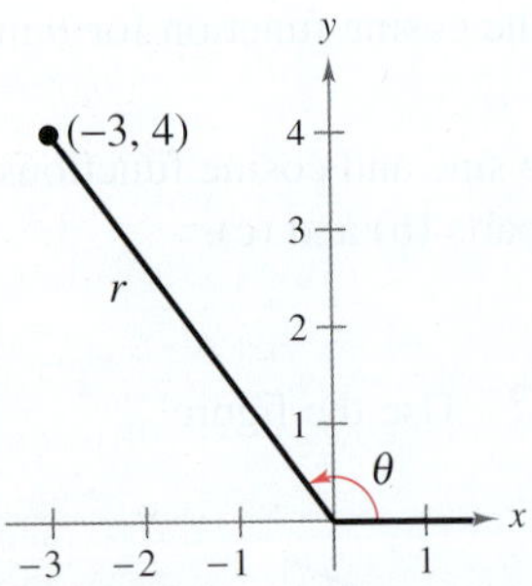

Figure 4.27

So, you have the following.

$$\sin\theta = \frac{y}{r} = \frac{4}{5}$$

$$\cos\theta = \frac{x}{r} = -\frac{3}{5}$$

$$\tan\theta = \frac{y}{x} = -\frac{4}{3}$$

ALGEBRA HELP The formula $r = \sqrt{x^2 + y^2}$ is a result of the Distance Formula. You can review the Distance Formula in Section 1.1.

Checkpoint *Audio-video solution in English & Spanish at LarsonPrecalculus.com.*

Let $(-2, 3)$ be a point on the terminal side of θ. Find the sine, cosine, and tangent of θ.

tusharkoley/Shutterstock.com

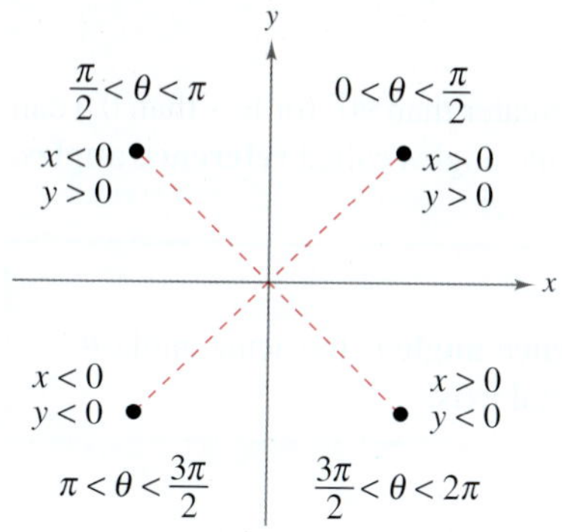

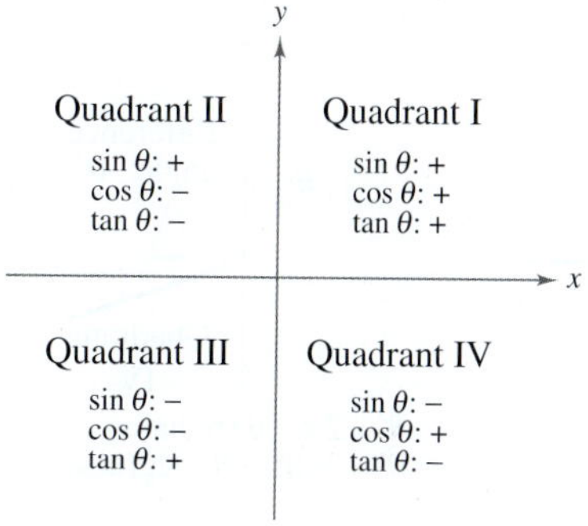

Figure 4.28

The *signs* of the trigonometric functions in the four quadrants can be determined from the definitions of the functions. For instance, because $\cos\theta = x/r$, it follows that $\cos\theta$ is positive wherever $x > 0$, which is in Quadrants I and IV. (Remember, r is always positive.) In a similar manner, you can verify the results shown in Figure 4.28.

EXAMPLE 2 Evaluating Trigonometric Functions

Given $\tan\theta = -\frac{5}{4}$ and $\cos\theta > 0$, find $\sin\theta$ and $\sec\theta$.

Solution Note that θ lies in Quadrant IV because that is the only quadrant in which the tangent is negative and the cosine is positive. Moreover, using

$$\tan\theta = \frac{y}{x} = -\frac{5}{4}$$

and the fact that y is negative in Quadrant IV, you can let $y = -5$ and $x = 4$. So, $r = \sqrt{16 + 25} = \sqrt{41}$ and you have the following.

$$\sin\theta = \frac{y}{r}$$

$$= \frac{-5}{\sqrt{41}} \qquad \text{Exact value}$$

$$\approx -0.7809 \qquad \text{Approximate value}$$

$$\sec\theta = \frac{r}{x}$$

$$= \frac{\sqrt{41}}{4} \qquad \text{Exact value}$$

$$\approx 1.6008 \qquad \text{Approximate value}$$

Checkpoint *Audio-video solution in English & Spanish at LarsonPrecalculus.com.*

Given $\sin\theta = \frac{4}{5}$ and $\tan\theta < 0$, find $\cos\theta$.

EXAMPLE 3 Trigonometric Functions of Quadrant Angles

Evaluate the cosine and tangent functions at the four quadrant angles 0, $\frac{\pi}{2}$, π, and $\frac{3\pi}{2}$.

Solution To begin, choose a point on the terminal side of each angle, as shown in Figure 4.29. For each of the four points, $r = 1$ and you have the following.

$$\cos 0 = \frac{x}{r} = \frac{1}{1} = 1 \qquad \tan 0 = \frac{y}{x} = \frac{0}{1} = 0 \qquad (x, y) = (1, 0)$$

$$\cos\frac{\pi}{2} = \frac{x}{r} = \frac{0}{1} = 0 \qquad \tan\frac{\pi}{2} = \frac{y}{x} = \frac{1}{0} \Rightarrow \text{undefined} \qquad (x, y) = (0, 1)$$

$$\cos\pi = \frac{x}{r} = \frac{-1}{1} = -1 \qquad \tan\pi = \frac{y}{x} = \frac{0}{-1} = 0 \qquad (x, y) = (-1, 0)$$

$$\cos\frac{3\pi}{2} = \frac{x}{r} = \frac{0}{1} = 0 \qquad \tan\frac{3\pi}{2} = \frac{y}{x} = \frac{-1}{0} \Rightarrow \text{undefined} \qquad (x, y) = (0, -1)$$

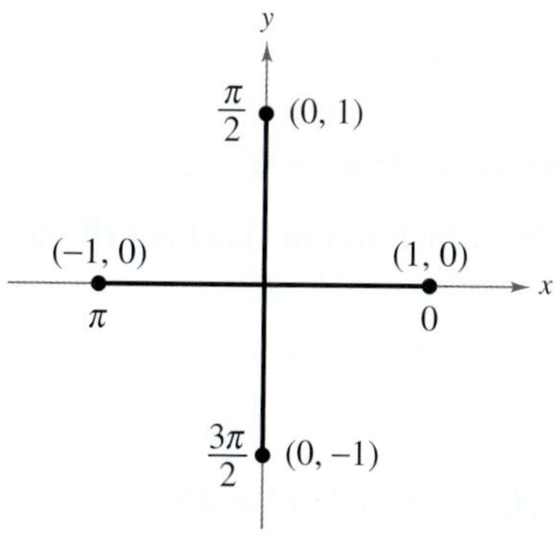

Figure 4.29

Checkpoint *Audio-video solution in English & Spanish at LarsonPrecalculus.com.*

Evaluate the sine and cotangent functions at the quadrant angle $\frac{3\pi}{2}$.

Reference Angles

The values of the trigonometric functions of angles greater than 90° (or less than 0°) can be determined from their values at corresponding acute angles called **reference angles.**

> **Definition of Reference Angle**
>
> Let θ be an angle in standard position. Its **reference angle** is the acute angle θ' formed by the terminal side of θ and the horizontal axis.

The reference angles for θ in Quadrants II, III, and IV are shown below.

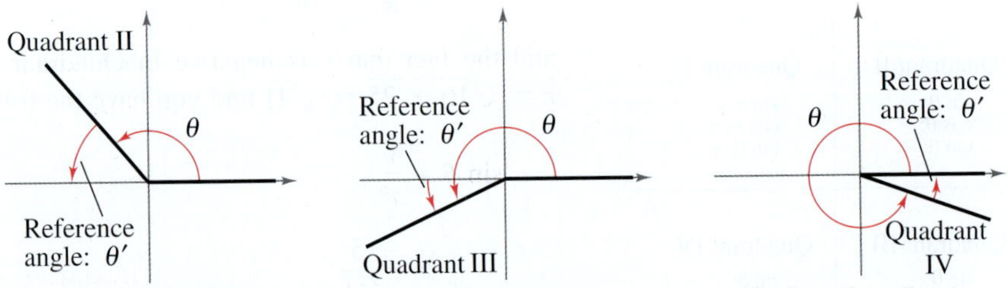

$\theta' = \pi - \theta$ (radians) $\quad$ $\theta' = \theta - \pi$ (radians) $\quad$ $\theta' = 2\pi - \theta$ (radians)

$\theta' = 180° - \theta$ (degrees) $\quad$ $\theta' = \theta - 180°$ (degrees) $\quad$ $\theta' = 360° - \theta$ (degrees)

EXAMPLE 4 Finding Reference Angles

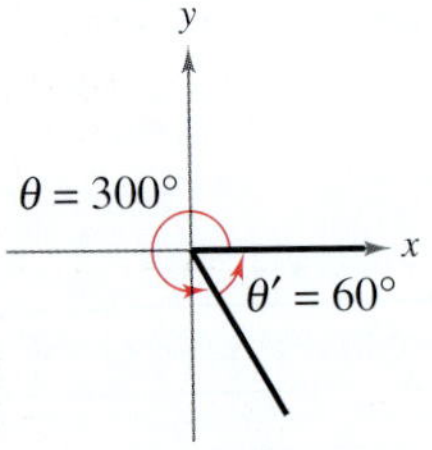

Figure 4.30

Find the reference angle θ'.

a. $\theta = 300°$ **b.** $\theta = 2.3$ **c.** $\theta = -135°$

Solution

a. Because 300° lies in Quadrant IV, the angle it makes with the x-axis is

$$\theta' = 360° - 300°$$
$$= 60°. \quad \text{Degrees}$$

Figure 4.30 shows the angle $\theta = 300°$ and its reference angle $\theta' = 60°$.

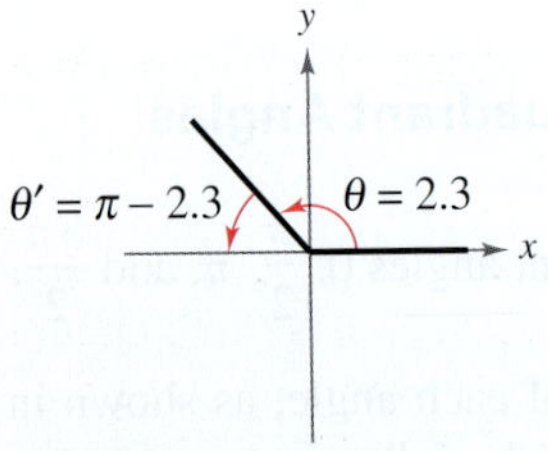

Figure 4.31

b. Because 2.3 lies between $\pi/2 \approx 1.5708$ and $\pi \approx 3.1416$, it follows that it is in Quadrant II and its reference angle is

$$\theta' = \pi - 2.3$$
$$\approx 0.8416. \quad \text{Radians}$$

Figure 4.31 shows the angle $\theta = 2.3$ and its reference angle $\theta' = \pi - 2.3$.

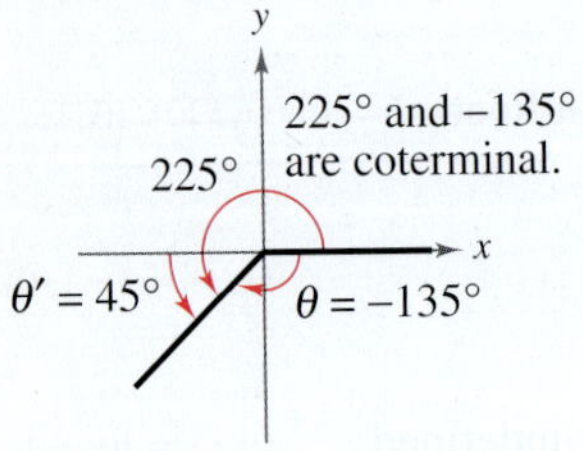

Figure 4.32

c. First, determine that $-135°$ is coterminal with 225°, which lies in Quadrant III. So, the reference angle is

$$\theta' = 225° - 180°$$
$$= 45°. \quad \text{Degrees}$$

Figure 4.32 shows the angle $\theta = -135°$ and its reference angle $\theta' = 45°$.

✓ **Checkpoint** Audio-video solution in English & Spanish at LarsonPrecalculus.com.

Find the reference angle θ'.

a. 213° **b.** $\dfrac{14\pi}{9}$ **c.** $\dfrac{4\pi}{5}$

Trigonometric Functions of Real Numbers

To see how a reference angle is used to evaluate a trigonometric function, consider the point (x, y) on the terminal side of θ, as shown at the right. By definition, you know that

$$\sin \theta = \frac{y}{r}$$

and

$$\tan \theta = \frac{y}{x}.$$

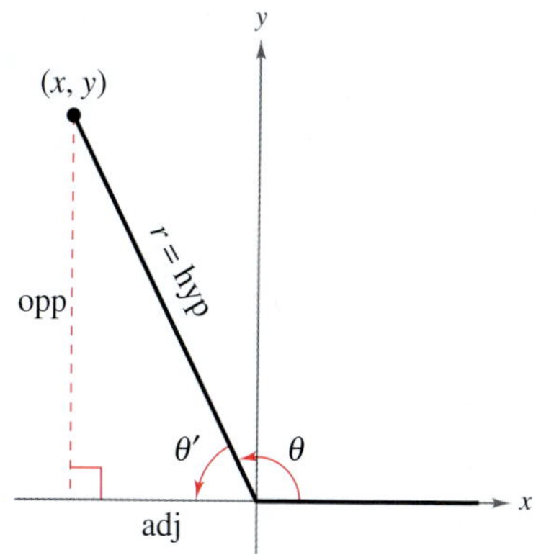

$\text{opp} = |y|, \text{adj} = |x|$

For the right triangle with acute angle θ' and sides of lengths $|x|$ and $|y|$, you have

$$\sin \theta' = \frac{\text{opp}}{\text{hyp}} = \frac{|y|}{r}$$

and

$$\tan \theta' = \frac{\text{opp}}{\text{adj}} = \frac{|y|}{|x|}.$$

So, it follows that $\sin \theta$ and $\sin \theta'$ are equal, *except possibly in sign*. The same is true for $\tan \theta$ and $\tan \theta'$ *and* for the other four trigonometric functions. In all cases, the quadrant in which θ lies determines the sign of the function value.

Evaluating Trigonometric Functions of Any Angle

To find the value of a trigonometric function of any angle θ:

1. Determine the function value of the associated reference angle θ'.
2. Depending on the quadrant in which θ lies, affix the appropriate sign to the function value.

REMARK Learning the table of values at the right is worth the effort because doing so will increase both your efficiency and your confidence. Here is a pattern for the sine function that may help you remember the values.

θ	$0°$	$30°$	$45°$	$60°$	$90°$
$\sin \theta$	$\frac{\sqrt{0}}{2}$	$\frac{\sqrt{1}}{2}$	$\frac{\sqrt{2}}{2}$	$\frac{\sqrt{3}}{2}$	$\frac{\sqrt{4}}{2}$

Reverse the order to get cosine values of the same angles.

By using reference angles and the special angles discussed in the preceding section, you can greatly extend the scope of *exact* trigonometric values. For instance, knowing the function values of 30° means that you know the function values of all angles for which 30° is a reference angle. For convenience, the table below shows the exact values of the sine, cosine, and tangent functions of special angles and quadrant angles.

Trigonometric Values of Common Angles

θ (degrees)	$0°$	$30°$	$45°$	$60°$	$90°$	$180°$	$270°$
θ (radians)	0	$\frac{\pi}{6}$	$\frac{\pi}{4}$	$\frac{\pi}{3}$	$\frac{\pi}{2}$	π	$\frac{3\pi}{2}$
$\sin \theta$	0	$\frac{1}{2}$	$\frac{\sqrt{2}}{2}$	$\frac{\sqrt{3}}{2}$	1	0	-1
$\cos \theta$	1	$\frac{\sqrt{3}}{2}$	$\frac{\sqrt{2}}{2}$	$\frac{1}{2}$	0	-1	0
$\tan \theta$	0	$\frac{\sqrt{3}}{3}$	1	$\sqrt{3}$	Undef.	0	Undef.

EXAMPLE 5 Using Reference Angles

Evaluate each trigonometric function.

a. $\cos \dfrac{4\pi}{3}$ **b.** $\tan(-210°)$ **c.** $\csc \dfrac{11\pi}{4}$

Solution

a. Because $\theta = 4\pi/3$ lies in Quadrant III, the reference angle is

$$\theta' = \frac{4\pi}{3} - \pi = \frac{\pi}{3}$$

as shown in Figure 4.33. Moreover, the cosine is negative in Quadrant III, so

$$\cos \frac{4\pi}{3} = (-) \cos \frac{\pi}{3}$$

$$= -\frac{1}{2}.$$

b. Because $-210° + 360° = 150°$, it follows that $-210°$ is coterminal with the second-quadrant angle $150°$. So, the reference angle is $\theta' = 180° - 150° = 30°$, as shown in Figure 4.34. Finally, because the tangent is negative in Quadrant II, you have

$$\tan(-210°) = (-) \tan 30°$$

$$= -\frac{\sqrt{3}}{3}.$$

c. Because $(11\pi/4) - 2\pi = 3\pi/4$, it follows that $11\pi/4$ is coterminal with the second-quadrant angle $3\pi/4$. So, the reference angle is $\theta' = \pi - (3\pi/4) = \pi/4$, as shown in Figure 4.35. Because the cosecant is positive in Quadrant II, you have

$$\csc \frac{11\pi}{4} = (+) \csc \frac{\pi}{4}$$

$$= \frac{1}{\sin(\pi/4)}$$

$$= \sqrt{2}.$$

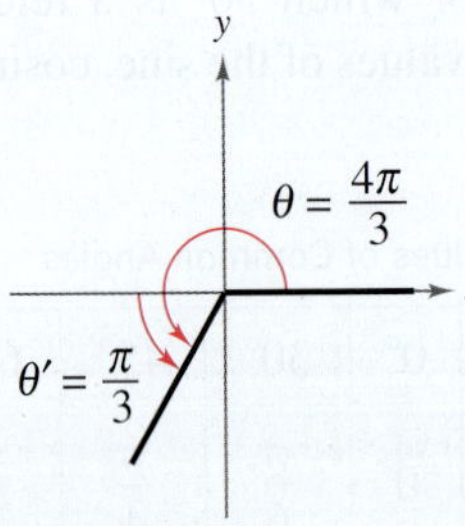

Figure 4.33

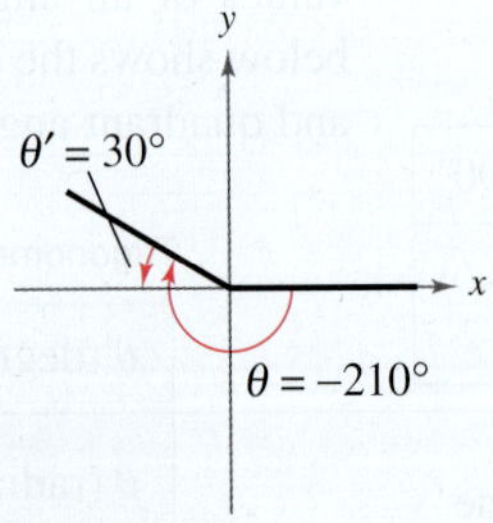

Figure 4.34

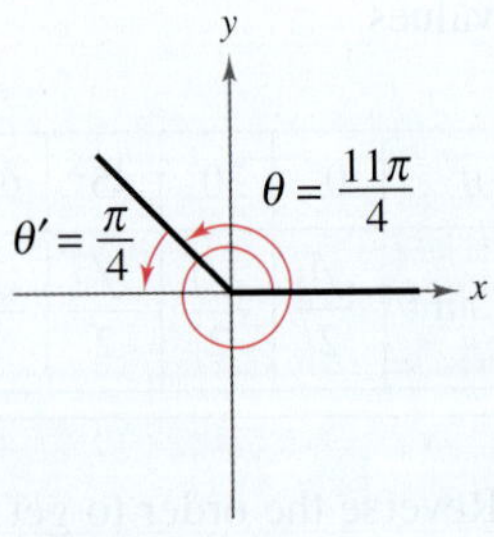

Figure 4.35

✓ Checkpoint Audio-video solution in English & Spanish at LarsonPrecalculus.com.

Evaluate each trigonometric function.

a. $\sin \dfrac{7\pi}{4}$ **b.** $\cos(-120°)$ **c.** $\tan \dfrac{11\pi}{6}$

EXAMPLE 6 Using Trigonometric Identities

Let θ be an angle in Quadrant II such that $\sin\theta = \frac{1}{3}$. Find (a) $\cos\theta$ and (b) $\tan\theta$ by using trigonometric identities.

Solution

a. Using the Pythagorean identity $\sin^2\theta + \cos^2\theta = 1$, you obtain

$$\left(\frac{1}{3}\right)^2 + \cos^2\theta = 1 \quad \Longrightarrow \quad \cos^2\theta = 1 - \frac{1}{9} = \frac{8}{9}.$$

Because $\cos\theta < 0$ in Quadrant II, use the negative root to obtain

$$\cos\theta = -\frac{\sqrt{8}}{\sqrt{9}} = -\frac{2\sqrt{2}}{3}.$$

b. Using the trigonometric identity $\tan\theta = \dfrac{\sin\theta}{\cos\theta}$, you obtain

$$\tan\theta = \frac{1/3}{-2\sqrt{2}/3} = -\frac{1}{2\sqrt{2}} = -\frac{\sqrt{2}}{4}.$$

✓ ***Checkpoint*** *Audio-video solution in English & Spanish at LarsonPrecalculus.com.*

Let θ be an angle in Quadrant III such that $\sin\theta = -\frac{4}{5}$. Find (a) $\cos\theta$ and (b) $\tan\theta$ by using trigonometric identities.

EXAMPLE 7 Using a Calculator

Use a calculator to evaluate each trigonometric function.

a. $\cot 410°$ **b.** $\sin(-7)$ **c.** $\sec\dfrac{\pi}{9}$

Solution

Function	Mode	Calculator Keystrokes	Display
a. $\cot 410°$	Degree	(TAN (410)) x⁻¹ ENTER	0.8390996
b. $\sin(-7)$	Radian	SIN ((−) 7) ENTER	−0.6569866
c. $\sec(\pi/9)$	Radian	(COS (π ÷ 9)) x⁻¹ ENTER	1.0641778

✓ ***Checkpoint*** *Audio-video solution in English & Spanish at LarsonPrecalculus.com.*

Use a calculator to evaluate each trigonometric function.

a. $\tan 119°$ **b.** $\csc 5$ **c.** $\cos\dfrac{\pi}{5}$

Summarize (Section 4.4)

1. State the definitions of the trigonometric functions of any angle *(page 290)*. For examples of evaluating trigonometric functions, see Examples 1–3.
2. Explain how to find a reference angle *(page 292)*. For an example of finding reference angles, see Example 4.
3. Explain how to evaluate a trigonometric function of a real number *(page 293)*. For examples of evaluating trigonometric functions of real numbers, see Examples 5–7.

4.4 Exercises

See **CalcChat.com** for tutorial help and worked-out solutions to odd-numbered exercises.

Vocabulary: Fill in the blanks.

In Exercises 1–6, let θ be an angle in standard position with (x, y) a point on the terminal side of θ and $r = \sqrt{x^2 + y^2} \neq 0$.

1. $\sin\theta =$ ________ **2.** $\frac{r}{y} =$ ________ **3.** $\tan\theta =$ ________

4. $\sec\theta =$ ________ **5.** $\frac{x}{r} =$ ________ **6.** $\frac{x}{y} =$ ________

7. Because $r = \sqrt{x^2 + y^2}$ cannot be ________, the sine and cosine functions are ________ for any real value of θ.

8. The acute positive angle formed by the terminal side of an angle θ and the horizontal axis is called the ________ angle of θ and is denoted by θ'.

Skills and Applications

Evaluating Trigonometric Functions **In Exercises 9–12, determine the exact values of the six trigonometric functions of each angle θ.**

9. (a)

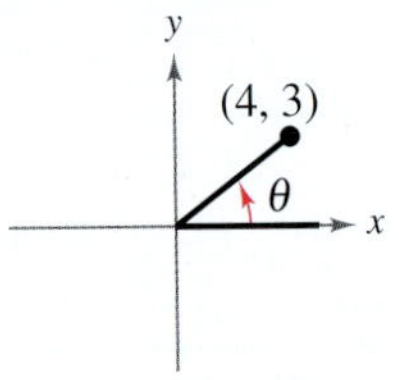

(b)

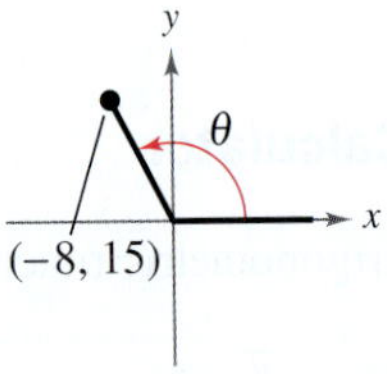

10. (a)

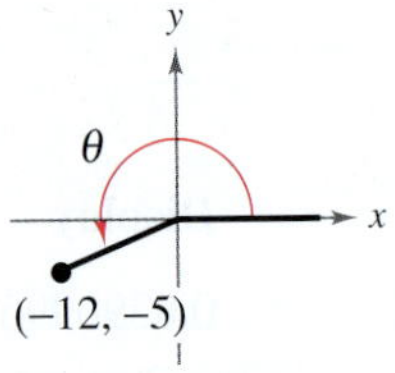

(b)

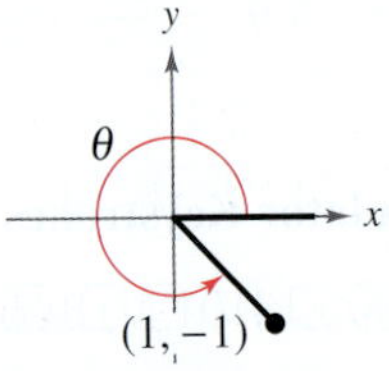

11. (a)

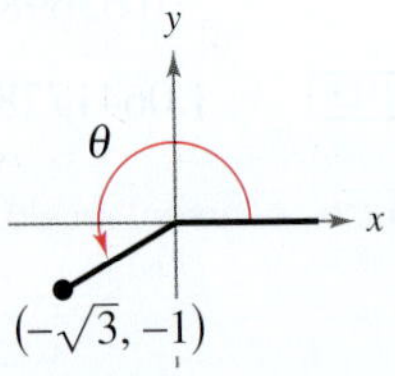

(b)

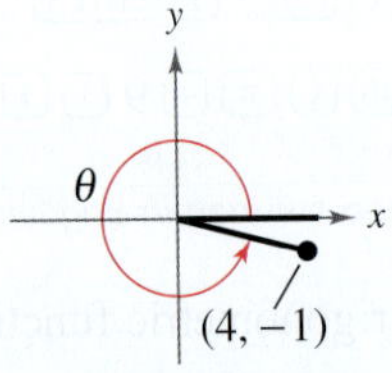

12. (a)

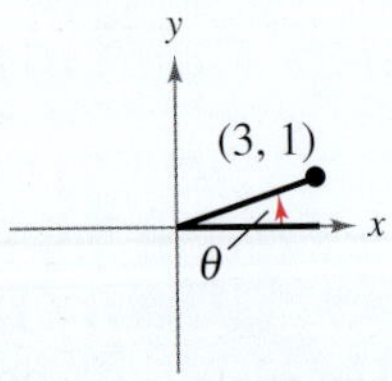

(b)

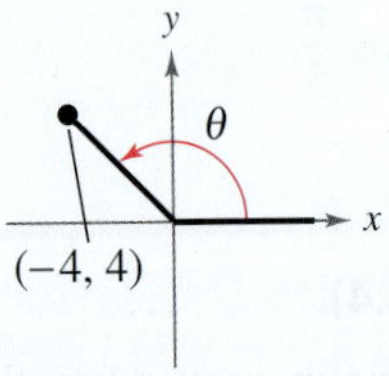

Evaluating Trigonometric Functions **In Exercises 13–18, the point is on the terminal side of an angle in standard position. Determine the exact values of the six trigonometric functions of the angle.**

13. $(5, 12)$ **14.** $(8, 15)$

15. $(-5, -2)$ **16.** $(-4, 10)$

17. $(-5.4, 7.2)$ **18.** $\left(3\frac{1}{2}, -7\frac{3}{4}\right)$

Determining a Quadrant **In Exercises 19–22, state the quadrant in which θ lies.**

19. $\sin\theta > 0$ and $\cos\theta > 0$

20. $\sin\theta < 0$ and $\cos\theta < 0$

21. $\sin\theta > 0$ and $\cos\theta < 0$

22. $\sec\theta > 0$ and $\cot\theta < 0$

Evaluating Trigonometric Functions **In Exercises 23–32, find the values of the six trigonometric functions of θ with the given constraint.**

Function Value	Constraint
23. $\tan\theta = -\frac{15}{8}$	$\sin\theta > 0$
24. $\cos\theta = \frac{8}{17}$	$\tan\theta < 0$
25. $\sin\theta = \frac{3}{5}$	θ lies in Quadrant II.
26. $\cos\theta = -\frac{4}{5}$	θ lies in Quadrant III.
27. $\cot\theta = -3$	$\cos\theta > 0$
28. $\csc\theta = 4$	$\cot\theta < 0$
29. $\sec\theta = -2$	$\sin\theta < 0$
30. $\sin\theta = 0$	$\sec\theta = -1$
31. $\cot\theta$ is undefined.	$\pi/2 \le \theta \le 3\pi/2$
32. $\tan\theta$ is undefined.	$\pi \le \theta \le 2\pi$

An Angle Formed by a Line Through the Origin **In Exercises 33–36, the terminal side of θ lies on the given line in the specified quadrant. Find the values of the six trigonometric functions of θ by finding a point on the line.**

Line	Quadrant
33. $y = -x$	II
34. $y = \frac{1}{3}x$	III
35. $2x - y = 0$	III
36. $4x + 3y = 0$	IV

Trigonometric Function of a Quadrant Angle In Exercises 37–44, evaluate the trigonometric function of the quadrant angle, if possible.

37. $\sin \pi$ **38.** $\csc \frac{3\pi}{2}$

39. $\sec \frac{3\pi}{2}$ **40.** $\sec \pi$

41. $\sin \frac{\pi}{2}$ **42.** $\cot \pi$

43. $\csc \pi$ **44.** $\cot \frac{\pi}{2}$

Finding a Reference Angle In Exercises 45–52, find the reference angle θ' and sketch θ and θ' in standard position.

45. $\theta = 160°$ **46.** $\theta = 309°$

47. $\theta = -125°$ **48.** $\theta = -215°$

49. $\theta = \frac{2\pi}{3}$ **50.** $\theta = \frac{7\pi}{6}$

51. $\theta = 4.8$ **52.** $\theta = 11.6$

Using a Reference Angle In Exercises 53–68, evaluate the sine, cosine, and tangent of the angle without using a calculator.

53. $225°$ **54.** $300°$

55. $750°$ **56.** $-405°$

57. $-840°$ **58.** $510°$

59. $\frac{2\pi}{3}$ **60.** $\frac{3\pi}{4}$

61. $\frac{5\pi}{4}$ **62.** $\frac{7\pi}{6}$

63. $-\frac{\pi}{6}$ **64.** $-\frac{\pi}{2}$

65. $\frac{9\pi}{4}$ **66.** $\frac{10\pi}{3}$

67. $-\frac{3\pi}{2}$ **68.** $-\frac{23\pi}{4}$

Using Trigonometric Identities In Exercises 69–74, use a trigonometric identity to find the indicated value in the specified quadrant.

Function Value	Quadrant	Value
69. $\sin \theta = -\frac{3}{5}$	IV	$\cos \theta$
70. $\cot \theta = -3$	II	$\sin \theta$
71. $\tan \theta = \frac{3}{2}$	III	$\sec \theta$
72. $\csc \theta = -2$	IV	$\cot \theta$
73. $\cos \theta = \frac{5}{8}$	I	$\sec \theta$
74. $\sec \theta = -\frac{9}{4}$	III	$\tan \theta$

Using a Calculator In Exercises 75–90, use a calculator to evaluate the trigonometric function. Round your answer to four decimal places. (Be sure the calculator is in the correct mode.)

75. $\sin 10°$ **76.** $\sec 225°$

77. $\cos(-110°)$ **78.** $\csc(-330°)$

79. $\tan 304°$ **80.** $\cot 178°$

81. $\sec 72°$ **82.** $\tan(-188°)$

83. $\tan 4.5$ **84.** $\cot 1.35$

85. $\tan \frac{\pi}{9}$ **86.** $\tan\left(-\frac{\pi}{9}\right)$

87. $\sin(-0.65)$ **88.** $\sec 0.29$

89. $\cot\left(-\frac{11\pi}{8}\right)$ **90.** $\csc\left(-\frac{15\pi}{14}\right)$

Solving for θ In Exercises 91–96, find two solutions of each equation. Give your answers in degrees ($0° \le \theta < 360°$) and in radians ($0 \le \theta < 2\pi$). Do not use a calculator.

91. (a) $\sin \theta = \frac{1}{2}$ (b) $\sin \theta = -\frac{1}{2}$

92. (a) $\cos \theta = \frac{\sqrt{2}}{2}$ (b) $\cos \theta = -\frac{\sqrt{2}}{2}$

93. (a) $\csc \theta = \frac{2\sqrt{3}}{3}$ (b) $\cot \theta = -1$

94. (a) $\sec \theta = 2$ (b) $\sec \theta = -2$

95. (a) $\tan \theta = 1$ (b) $\cot \theta = -\sqrt{3}$

96. (a) $\sin \theta = \frac{\sqrt{3}}{2}$ (b) $\sin \theta = -\frac{\sqrt{3}}{2}$

97. Distance An airplane, flying at an altitude of 6 miles, is on a flight path that passes directly over an observer (see figure). Let θ be the angle of elevation from the observer to the plane. Find the distance d from the observer to the plane when (a) $\theta = 30°$, (b) $\theta = 90°$, and (c) $\theta = 120°$.

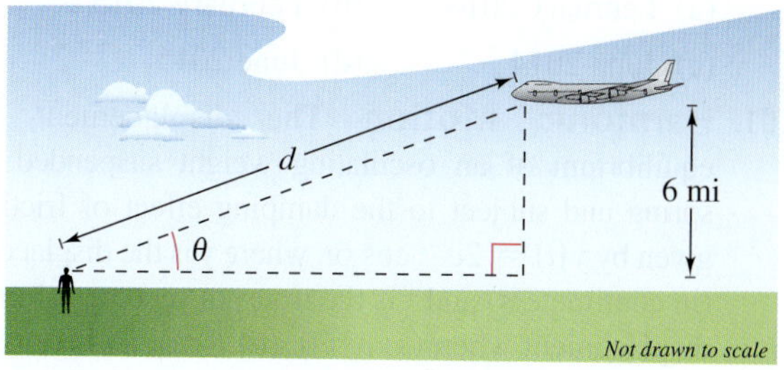

98. Harmonic Motion The displacement from equilibrium of an oscillating weight suspended by a spring is given by $y(t) = 2 \cos 6t$, where y is the displacement (in centimeters) and t is the time (in seconds). Find the displacement when (a) $t = 0$, (b) $t = \frac{1}{4}$, and (c) $t = \frac{1}{2}$.

99. Data Analysis: Meteorology The table shows the monthly normal temperatures (in degrees Fahrenheit) for selected months in New York City (N) and Fairbanks, Alaska (F). *(Source: National Climatic Data Center)*

Month	New York City, N	Fairbanks, F
January	33	−10
April	52	32
July	77	62
October	58	24
December	38	−6

Spreadsheet at LarsonPrecalculus.com

(a) Use the *regression* feature of a graphing utility to find a model of the form $y = a \sin(bt + c) + d$ for each city. Let t represent the month, with $t = 1$ corresponding to January.

(b) Use the models from part (a) to find the monthly normal temperatures for the two cities in February, March, May, June, August, September, and November.

(c) Compare the models for the two cities.

100. Sales A company that produces snowboards forecasts monthly sales over the next 2 years to be

$$S = 23.1 + 0.442t + 4.3 \cos \frac{\pi t}{6}$$

where S is measured in thousands of units and t is the time in months, with $t = 1$ representing January 2014. Predict sales for each of the following months.

(a) February 2014 (b) February 2015

(c) June 2014 (d) June 2015

101. Harmonic Motion The displacement from equilibrium of an oscillating weight suspended by a spring and subject to the damping effect of friction is given by $y(t) = 2e^{-t} \cos 6t$, where y is the displacement (in centimeters) and t is the time (in seconds). Find the displacement when (a) $t = 0$, (b) $t = \frac{1}{4}$, and (c) $t = \frac{1}{2}$.

102. Electric Circuits The current I (in amperes) when 100 volts is applied to a circuit is given by

$$I = 5e^{-2t} \sin t$$

where t is the time (in seconds) after the voltage is applied. Approximate the current at $t = 0.7$ second after the voltage is applied.

Exploration

True or False? **In Exercises 103 and 104, determine whether the statement is true or false. Justify your answer.**

103. In each of the four quadrants, the signs of the secant function and sine function are the same.

104. To find the reference angle for an angle θ (given in degrees), find the integer n such that $0 \le 360°n - \theta \le 360°$. The difference $360°n - \theta$ is the reference angle.

105. Writing Write a short paper explaining to a classmate how to evaluate the six trigonometric functions of any angle θ in standard position. Include an explanation of reference angles and how to use them, the signs of the functions in each of the four quadrants, and the trigonometric values of common angles. Be sure to include figures or diagrams in your paper.

106. HOW DO YOU SEE IT? Consider an angle in standard position with $r = 12$ centimeters, as shown in the figure. Describe the changes in the values of x, y, $\sin \theta$, $\cos \theta$, and $\tan \theta$ as θ increases continuously from 0° to 90°.

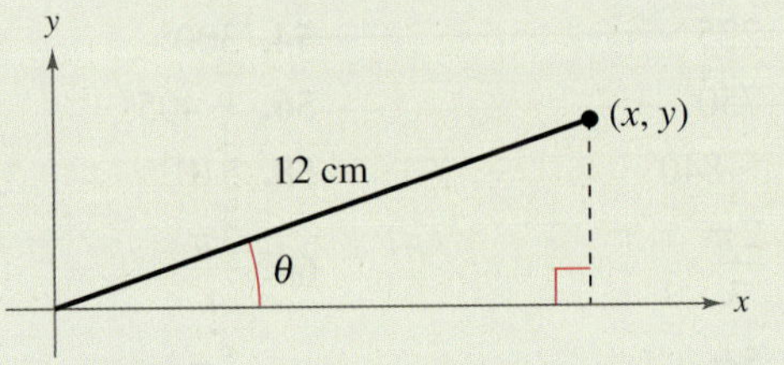

107. Think About It The figure shows point $P(x, y)$ on a unit circle and right triangle OAP.

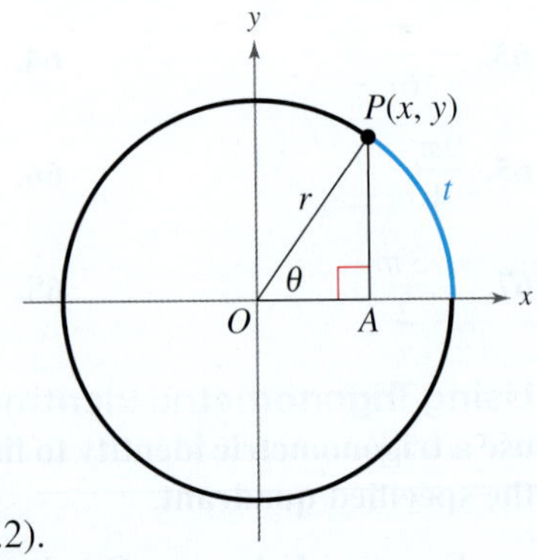

(a) Find $\sin t$ and $\cos t$ using the unit circle definitions of sine and cosine (from Section 4.2).

(b) What is the value of r? Explain.

(c) Use the definitions of sine and cosine given in this section to find $\sin \theta$ and $\cos \theta$. Write your answers in terms of x and y.

(d) Based on your answers to parts (a) and (c), what can you conclude?

tusharkoley/Shutterstock.com

4.5 Graphs of Sine and Cosine Functions

You can use sine and cosine functions in scientific calculations. For instance, in Exercise 88 on page 308, you will use a trigonometric function to model the airflow of your respiratory cycle.

- **Sketch the graphs of basic sine and cosine functions.**
- **Use amplitude and period to help sketch the graphs of sine and cosine functions.**
- **Sketch translations of the graphs of sine and cosine functions.**
- **Use sine and cosine functions to model real-life data.**

Basic Sine and Cosine Curves

In this section, you will study techniques for sketching the graphs of the sine and cosine functions. The graph of the sine function is a **sine curve.** In Figure 4.36, the black portion of the graph represents one period of the function and is called **one cycle** of the sine curve. The gray portion of the graph indicates that the basic sine curve repeats indefinitely to the left and right. The graph of the cosine function is shown in Figure 4.37.

Recall from Section 4.2 that the domain of the sine and cosine functions is the set of all real numbers. Moreover, the range of each function is the interval $[-1, 1]$, and each function has a period of 2π. Do you see how this information is consistent with the basic graphs shown in Figures 4.36 and 4.37?

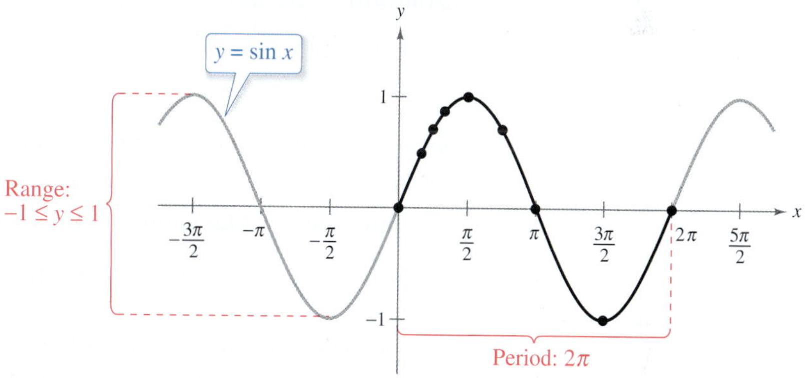

Figure 4.36

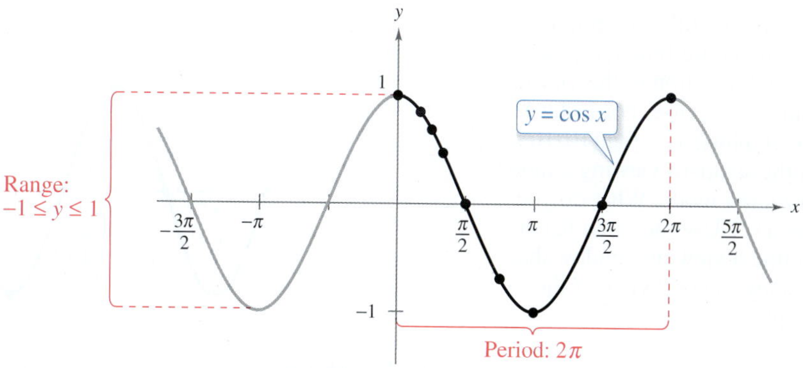

Figure 4.37

Note in Figures 4.36 and 4.37 that the sine curve is symmetric with respect to the *origin*, whereas the cosine curve is symmetric with respect to the *y-axis*. These properties of symmetry follow from the fact that the sine function is odd and the cosine function is even.

AISPIX by Image Source/Shutterstock.com

To sketch the graphs of the basic sine and cosine functions by hand, it helps to note five **key points** in one period of each graph: the *intercepts, maximum points,* and *minimum points* (see below).

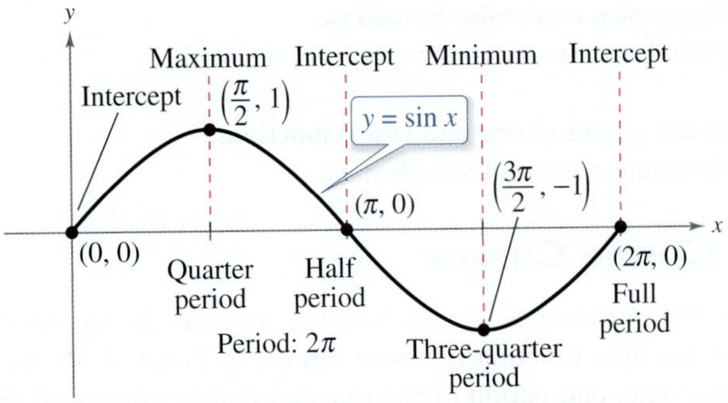

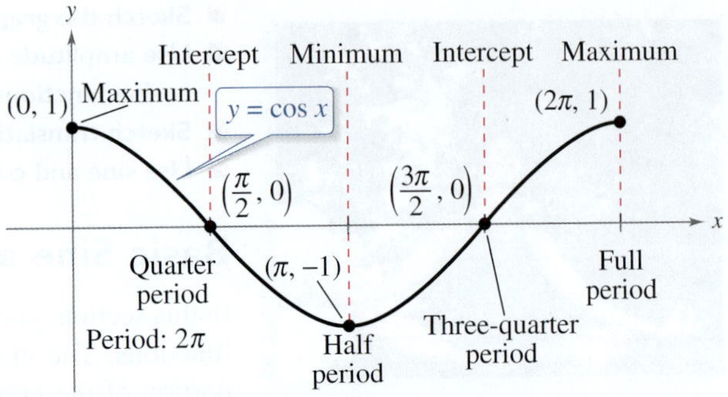

EXAMPLE 1 Using Key Points to Sketch a Sine Curve

Sketch the graph of

$$y = 2 \sin x$$

on the interval $[-\pi, 4\pi]$.

Solution Note that

$$y = 2 \sin x$$
$$= 2(\sin x)$$

indicates that the y-values for the key points will have twice the magnitude of those on the graph of $y = \sin x$. Divide the period 2π into four equal parts to get the key points

Intercept	Maximum	Intercept	Minimum		Intercept
$(0, 0)$,	$\left(\frac{\pi}{2}, 2\right)$,	$(\pi, 0)$,	$\left(\frac{3\pi}{2}, -2\right)$,	and	$(2\pi, 0)$.

By connecting these key points with a smooth curve and extending the curve in both directions over the interval $[-\pi, 4\pi]$, you obtain the graph shown below.

▷ **TECHNOLOGY** When using a graphing utility to graph trigonometric functions, pay special attention to the viewing window you use. For instance, try graphing $y = [\sin(10x)]/10$ in the standard viewing window in *radian* mode. What do you observe? Use the *zoom* feature to find a viewing window that displays a good view of the graph.

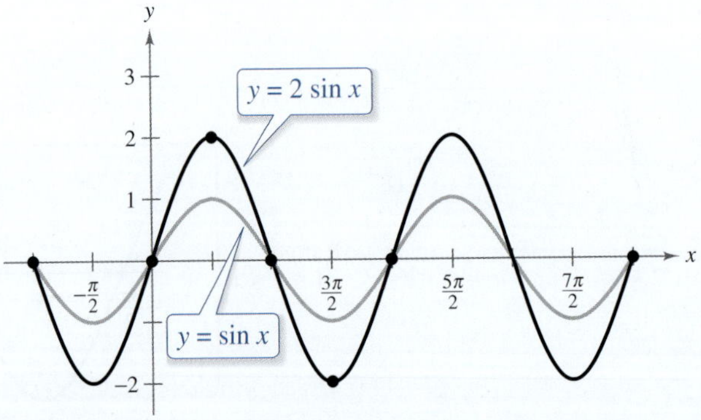

Checkpoint Audio-video solution in English & Spanish at LarsonPrecalculus.com.

Sketch the graph of

$$y = 2 \cos x$$

on the interval $\left[-\frac{\pi}{2}, \frac{9\pi}{2}\right]$. ■

Amplitude and Period

In the rest of this section, you will study the graphic effect of each of the constants a, b, c, and d in equations of the forms

$$y = d + a\sin(bx - c)$$

and

$$y = d + a\cos(bx - c).$$

A quick review of the transformations you studied in Section 1.7 should help in this investigation.

The constant factor a in $y = a\sin x$ acts as a *scaling factor*—a *vertical stretch* or *vertical shrink* of the basic sine curve. When $|a| > 1$, the basic sine curve is stretched, and when $|a| < 1$, the basic sine curve is shrunk. The result is that the graph of $y = a\sin x$ ranges between $-a$ and a instead of between -1 and 1. The absolute value of a is the **amplitude** of the function $y = a\sin x$. The range of the function $y = a\sin x$ for $a > 0$ is $-a \le y \le a$.

> **Definition of Amplitude of Sine and Cosine Curves**
>
> The **amplitude** of $y = a\sin x$ and $y = a\cos x$ represents half the distance between the maximum and minimum values of the function and is given by
>
> $$\text{Amplitude} = |a|.$$

EXAMPLE 2 Scaling: Vertical Shrinking and Stretching

In the same coordinate plane, sketch the graph of each function.

a. $y = \frac{1}{2}\cos x$

b. $y = 3\cos x$

Solution

a. Because the amplitude of $y = \frac{1}{2}\cos x$ is $\frac{1}{2}$, the maximum value is $\frac{1}{2}$ and the minimum value is $-\frac{1}{2}$. Divide one cycle, $0 \le x \le 2\pi$, into four equal parts to get the key points

Maximum	Intercept	Minimum	Intercept		Maximum
$\left(0, \frac{1}{2}\right)$,	$\left(\frac{\pi}{2}, 0\right)$,	$\left(\pi, -\frac{1}{2}\right)$,	$\left(\frac{3\pi}{2}, 0\right)$,	and	$\left(2\pi, \frac{1}{2}\right)$.

b. A similar analysis shows that the amplitude of $y = 3\cos x$ is 3, and the key points are

Maximum	Intercept	Minimum	Intercept		Maximum
$(0, 3)$,	$\left(\frac{\pi}{2}, 0\right)$,	$(\pi, -3)$,	$\left(\frac{3\pi}{2}, 0\right)$,	and	$(2\pi, 3)$.

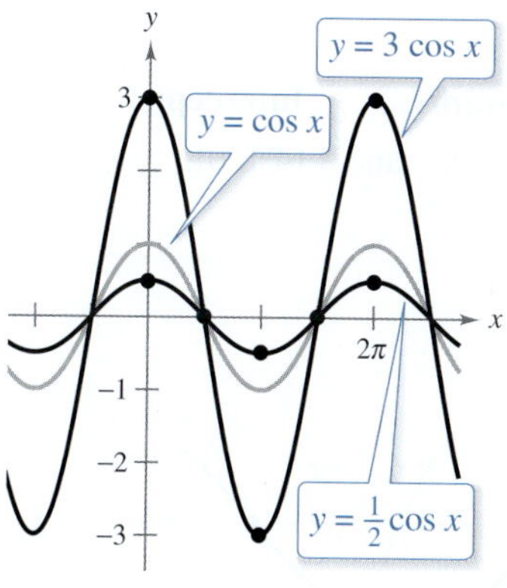

Figure 4.38

The graphs of these two functions are shown in Figure 4.38. Notice that the graph of $y = \frac{1}{2}\cos x$ is a vertical *shrink* of the graph of $y = \cos x$ and the graph of $y = 3\cos x$ is a vertical *stretch* of the graph of $y = \cos x$.

✓ ***Checkpoint*** *Audio-video solution in English & Spanish at LarsonPrecalculus.com.*

In the same coordinate plane, sketch the graph of each function.

a. $y = \frac{1}{3}\sin x$

b. $y = 3\sin x$

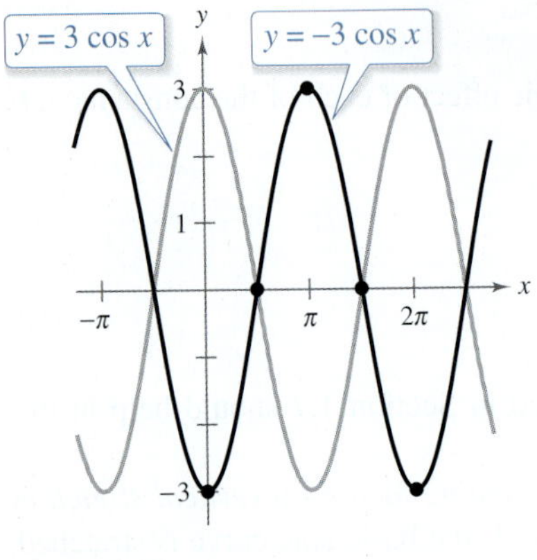

Figure 4.39

You know from Section 1.7 that the graph of $y = -f(x)$ is a **reflection** in the x-axis of the graph of $y = f(x)$. For instance, the graph of $y = -3 \cos x$ is a reflection of the graph of $y = 3 \cos x$, as shown in Figure 4.39.

Because $y = a \sin x$ completes one cycle from $x = 0$ to $x = 2\pi$, it follows that $y = a \sin bx$ completes one cycle from $x = 0$ to $x = 2\pi/b$, where b is a positive real number.

Period of Sine and Cosine Functions

Let b be a positive real number. The **period** of $y = a \sin bx$ and $y = a \cos bx$ is given by

$$\text{Period} = \frac{2\pi}{b}.$$

Note that when $0 < b < 1$, the period of $y = a \sin bx$ is greater than 2π and represents a *horizontal stretching* of the graph of $y = a \sin x$. Similarly, when $b > 1$, the period of $y = a \sin bx$ is less than 2π and represents a *horizontal shrinking* of the graph of $y = a \sin x$. When b is negative, the identities $\sin(-x) = -\sin x$ and $\cos(-x) = \cos x$ are used to rewrite the function.

EXAMPLE 3 Scaling: Horizontal Stretching

Sketch the graph of

$$y = \sin \frac{x}{2}.$$

Solution The amplitude is 1. Moreover, because $b = \frac{1}{2}$, the period is

$$\frac{2\pi}{b} = \frac{2\pi}{\frac{1}{2}} = 4\pi. \qquad \text{Substitute for } b.$$

REMARK In general, to divide a period-interval into four equal parts, successively add "period/4," starting with the left endpoint of the interval. For instance, for the period-interval $[-\pi/6, \pi/2]$ of length $2\pi/3$, you would successively add

$$\frac{2\pi/3}{4} = \frac{\pi}{6}$$

to get $-\pi/6, 0, \pi/6, \pi/3$, and $\pi/2$ as the x-values for the key points on the graph.

Now, divide the period-interval $[0, 4\pi]$ into four equal parts using the values π, 2π, and 3π to obtain the key points

Intercept	Maximum	Intercept	Minimum		Intercept
$(0, 0)$,	$(\pi, 1)$,	$(2\pi, 0)$,	$(3\pi, -1)$,	and	$(4\pi, 0)$.

The graph is shown below.

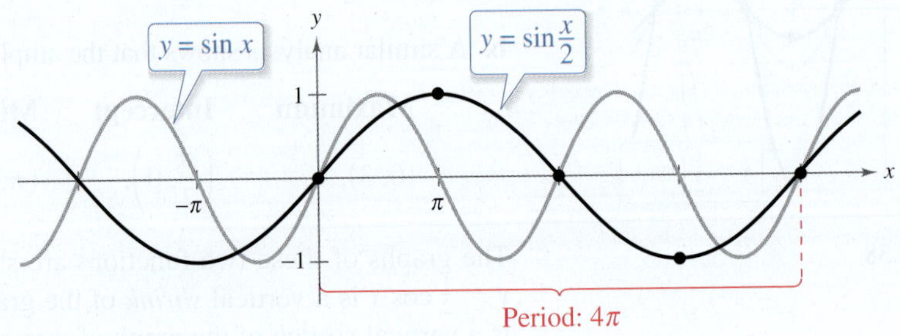

✓ **Checkpoint** Audio-video solution in English & Spanish at LarsonPrecalculus.com.

Sketch the graph of

$$y = \cos \frac{x}{3}.$$

Translations of Sine and Cosine Curves

▷ **ALGEBRA HELP** You can review the techniques for shifting, reflecting, and stretching graphs in Section 1.7.

The constant c in the general equations

$$y = a\sin(bx - c) \quad \text{and} \quad y = a\cos(bx - c)$$

creates *horizontal translations* (shifts) of the basic sine and cosine curves. Comparing $y = a \sin bx$ with $y = a\sin(bx - c)$, you find that the graph of $y = a\sin(bx - c)$ completes one cycle from $bx - c = 0$ to $bx - c = 2\pi$. By solving for x, you can find the interval for one cycle to be

$$\underbrace{\frac{c}{b}}_{\text{Left endpoint}} \le x \le \underbrace{\frac{c}{b} + \frac{2\pi}{b}}_{\text{Right endpoint}}.$$

(Period: $\frac{2\pi}{b}$)

This implies that the period of $y = a\sin(bx - c)$ is $2\pi/b$, and the graph of $y = a \sin bx$ is shifted by an amount c/b. The number c/b is the **phase shift.**

> **Graphs of Sine and Cosine Functions**
>
> The graphs of $y = a\sin(bx - c)$ and $y = a\cos(bx - c)$ have the following characteristics. (Assume $b > 0$.)
>
> $$\text{Amplitude} = |a| \qquad \text{Period} = \frac{2\pi}{b}$$
>
> The left and right endpoints of a one-cycle interval can be determined by solving the equations $bx - c = 0$ and $bx - c = 2\pi$.

EXAMPLE 4 **Horizontal Translation**

Analyze the graph of $y = \frac{1}{2}\sin\left(x - \frac{\pi}{3}\right)$.

Algebraic Solution

The amplitude is $\frac{1}{2}$ and the period is 2π. By solving the equations

$$x - \frac{\pi}{3} = 0 \quad \Longrightarrow \quad x = \frac{\pi}{3}$$

and

$$x - \frac{\pi}{3} = 2\pi \quad \Longrightarrow \quad x = \frac{7\pi}{3}$$

you see that the interval $[\pi/3, 7\pi/3]$ corresponds to one cycle of the graph. Dividing this interval into four equal parts produces the key points

Intercept	Maximum	Intercept	Minimum		Intercept
$\left(\frac{\pi}{3}, 0\right)$,	$\left(\frac{5\pi}{6}, \frac{1}{2}\right)$,	$\left(\frac{4\pi}{3}, 0\right)$,	$\left(\frac{11\pi}{6}, -\frac{1}{2}\right)$,	and	$\left(\frac{7\pi}{3}, 0\right)$.

Graphical Solution

Use a graphing utility set in *radian* mode to graph $y = (1/2)\sin(x - \pi/3)$, as shown below. Use the *minimum, maximum,* and *zero* or *root* features of the graphing utility to approximate the key points $(1.05, 0)$, $(2.62, 0.5)$, $(4.19, 0)$, $(5.76, -0.5)$, and $(7.33, 0)$.

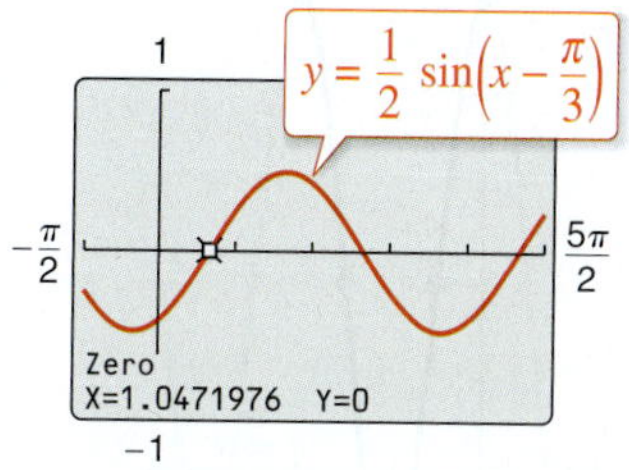

✓ **Checkpoint** Audio-video solution in English & Spanish at LarsonPrecalculus.com.

Analyze the graph of $y = 2\cos\left(x - \frac{\pi}{2}\right)$.

EXAMPLE 5 **Horizontal Translation**

Sketch the graph of

$$y = -3\cos(2\pi x + 4\pi).$$

Solution The amplitude is 3 and the period is $2\pi/2\pi = 1$. By solving the equations

$$2\pi x + 4\pi = 0$$
$$2\pi x = -4\pi$$
$$x = -2$$

and

$$2\pi x + 4\pi = 2\pi$$
$$2\pi x = -2\pi$$
$$x = -1$$

you see that the interval $[-2, -1]$ corresponds to one cycle of the graph. Dividing this interval into four equal parts produces the key points

Minimum	Intercept	Maximum	Intercept		Minimum
$(-2, -3)$,	$\left(-\frac{7}{4}, 0\right)$,	$\left(-\frac{3}{2}, 3\right)$,	$\left(-\frac{5}{4}, 0\right)$,	and	$(-1, -3)$.

The graph is shown in Figure 4.40.

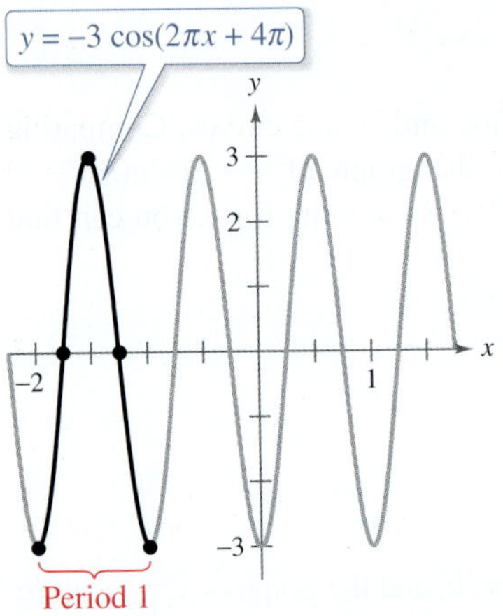

Figure 4.40

Checkpoint Audio-video solution in English & Spanish at LarsonPrecalculus.com.

Sketch the graph of

$$y = -\frac{1}{2}\sin(\pi x + \pi).$$

The final type of transformation is the *vertical translation* caused by the constant d in the equations

$$y = d + a\sin(bx - c) \quad \text{and} \quad y = d + a\cos(bx - c).$$

The shift is d units up for $d > 0$ and d units down for $d < 0$. In other words, the graph oscillates about the horizontal line $y = d$ instead of about the x-axis.

EXAMPLE 6 **Vertical Translation**

Sketch the graph of

$$y = 2 + 3\cos 2x.$$

Solution The amplitude is 3 and the period is π. The key points over the interval $[0, \pi]$ are

$$(0, 5), \quad \left(\frac{\pi}{4}, 2\right), \quad \left(\frac{\pi}{2}, -1\right), \quad \left(\frac{3\pi}{4}, 2\right), \quad \text{and} \quad (\pi, 5).$$

The graph is shown in Figure 4.41. Compared with the graph of $f(x) = 3\cos 2x$, the graph of $y = 2 + 3\cos 2x$ is shifted up two units.

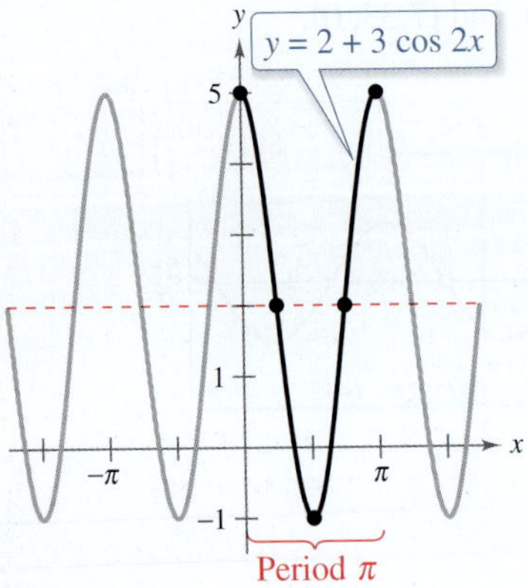

Figure 4.41

Checkpoint Audio-video solution in English & Spanish at LarsonPrecalculus.com.

Sketch the graph of

$$y = 2\cos x - 5.$$

Mathematical Modeling

EXAMPLE 7 Finding a Trigonometric Model

DATA

Time, t	Depth, y
0	3.4
2	8.7
4	11.3
6	9.1
8	3.8
10	0.1
12	1.2

Spreadsheet at LarsonPrecalculus.com

The table shows the depths (in feet) of the water at the end of a dock at various times during the morning, where $t = 0$ corresponds to midnight.

a. Use a trigonometric function to model the data.

b. Find the depths at 9 A.M. and 3 P.M.

c. A boat needs at least 10 feet of water to moor at the dock. During what times in the afternoon can it safely dock?

Solution

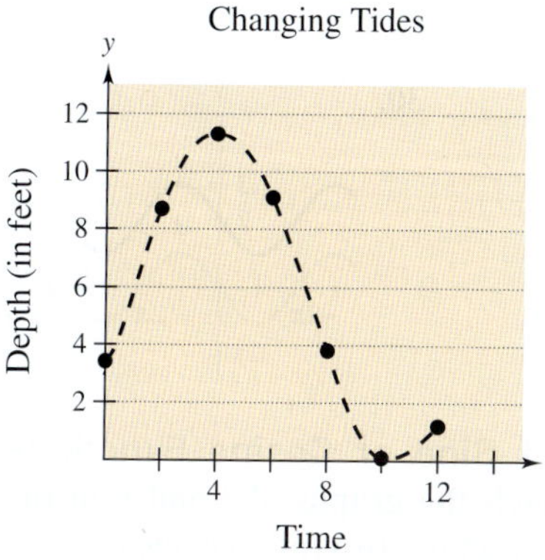

Figure 4.42

a. Begin by graphing the data, as shown in Figure 4.42. You can use either a sine or cosine model. Suppose you use a cosine model of the form $y = a\cos(bt - c) + d$. The difference between the maximum value and minimum value is twice the amplitude of the function. So, the amplitude is

$$a = \tfrac{1}{2}[(\text{maximum depth}) - (\text{minimum depth})] = \tfrac{1}{2}(11.3 - 0.1) = 5.6.$$

The cosine function completes one half of a cycle between the times at which the maximum and minimum depths occur. So, the period p is

$$p = 2[(\text{time of min. depth}) - (\text{time of max. depth})] = 2(10 - 4) = 12$$

which implies that $b = 2\pi/p \approx 0.524$. Because high tide occurs 4 hours after midnight, consider the left endpoint to be $c/b = 4$, so $c \approx 2.094$. Moreover, because the average depth is $\tfrac{1}{2}(11.3 + 0.1) = 5.7$, it follows that $d = 5.7$. So, you can model the depth with the function $y = 5.6\cos(0.524t - 2.094) + 5.7$.

b. The depths at 9 A.M. and 3 P.M. are as follows.

$$y = 5.6\cos(0.524 \cdot 9 - 2.094) + 5.7 \approx 0.84 \text{ foot} \qquad \text{9 A.M.}$$

$$y = 5.6\cos(0.524 \cdot 15 - 2.094) + 5.7 \approx 10.57 \text{ feet} \qquad \text{3 P.M.}$$

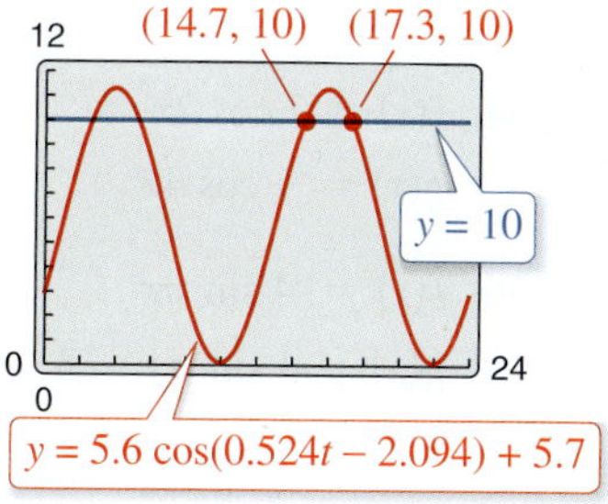

Figure 4.43

c. Using a graphing utility, graph the model with the line $y = 10$. Using the *intersect* feature, you can determine that the depth is at least 10 feet between 2:42 P.M. $(t \approx 14.7)$ and 5:18 P.M. $(t \approx 17.3)$, as shown in Figure 4.43.

✓ ***Checkpoint*** *Audio-video solution in English & Spanish at LarsonPrecalculus.com.*

Find a sine model for the data in Example 7.

Summarize (Section 4.5)

1. Describe how to sketch the graphs of basic sine and cosine functions *(pages 299 and 300)*. For an example of sketching the graph of a sine function, see Example 1.
2. Describe how you can use amplitude and period to help sketch the graphs of sine and cosine functions *(pages 301 and 302)*. For examples of using amplitude and period to sketch graphs of sine and cosine functions, see Examples 2 and 3.
3. Describe how to sketch translations of the graphs of sine and cosine functions *(pages 303 and 304)*. For examples of translating the graphs of sine and cosine functions, see Examples 4–6.
4. Give an example of how to use sine and cosine functions to model real-life data *(page 305, Example 7)*.

4.5 Exercises

See CalcChat.com for tutorial help and worked-out solutions to odd-numbered exercises.

Vocabulary: Fill in the blanks.

1. One period of a sine or cosine function is called one ________ of the sine or cosine curve.
2. The ________ of a sine or cosine curve represents half the distance between the maximum and minimum values of the function.
3. For the function $y = a \sin(bx - c)$, $\frac{c}{b}$ represents the ________ ________ of the graph of the function.
4. For the function $y = d + a \cos(bx - c)$, d represents a ________ ________ of the graph of the function.

Skills and Applications

Finding the Period and Amplitude **In Exercises 5–18, find the period and amplitude.**

5. $y = 2 \sin 5x$
6. $y = 3 \cos 2x$
7. $y = \frac{3}{4} \cos \frac{x}{2}$
8. $y = -3 \sin \frac{x}{3}$
9. $y = \frac{1}{2} \sin \frac{\pi x}{3}$
10. $y = \frac{3}{2} \cos \frac{\pi x}{2}$
11. $y = -4 \sin x$
12. $y = -\cos \frac{2x}{3}$
13. $y = 3 \sin 10x$
14. $y = \frac{1}{5} \sin 6x$
15. $y = \frac{5}{3} \cos \frac{4x}{5}$
16. $y = \frac{5}{2} \cos \frac{x}{4}$
17. $y = \frac{1}{4} \sin 2\pi x$
18. $y = \frac{2}{3} \cos \frac{\pi x}{10}$

Describing the Relationship Between Graphs **In Exercises 19–30, describe the relationship between the graphs of f and g. Consider amplitude, period, and shifts.**

19. $f(x) = \sin x$
    $g(x) = \sin(x - \pi)$
20. $f(x) = \cos x$
    $g(x) = \cos(x + \pi)$
21. $f(x) = \cos 2x$
    $g(x) = -\cos 2x$
22. $f(x) = \sin 3x$
    $g(x) = \sin(-3x)$
23. $f(x) = \cos x$
    $g(x) = \cos 2x$
24. $f(x) = \sin x$
    $g(x) = \sin 3x$
25. $f(x) = \sin 2x$
    $g(x) = 3 + \sin 2x$
26. $f(x) = \cos 4x$
    $g(x) = -2 + \cos 4x$

27.

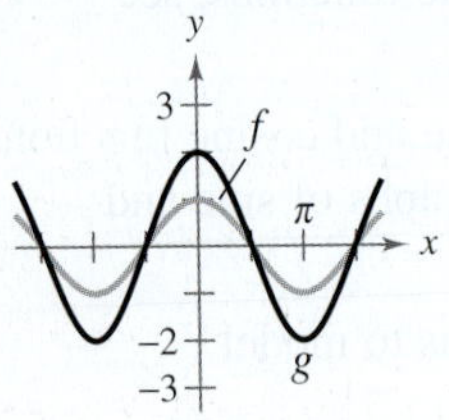

28.

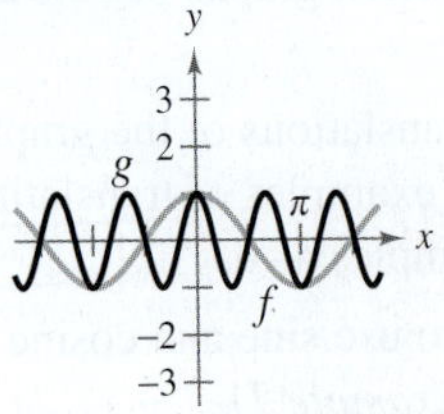

29.

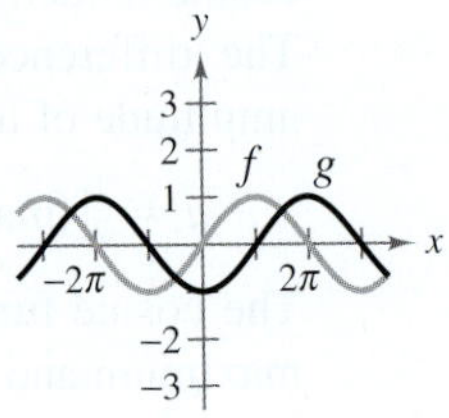

30.

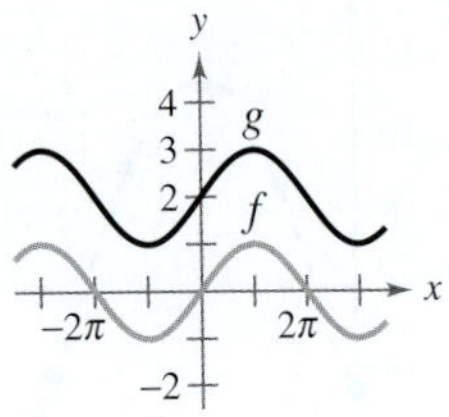

Sketching Graphs of Sine or Cosine Functions **In Exercises 31–38, sketch the graphs of f and g in the same coordinate plane. (Include two full periods.)**

31. $f(x) = -2 \sin x$
    $g(x) = 4 \sin x$
32. $f(x) = \sin x$
    $g(x) = \sin \frac{x}{3}$
33. $f(x) = \cos x$
    $g(x) = 2 + \cos x$
34. $f(x) = 2 \cos 2x$
    $g(x) = -\cos 4x$
35. $f(x) = -\frac{1}{2} \sin \frac{x}{2}$
    $g(x) = 3 - \frac{1}{2} \sin \frac{x}{2}$
36. $f(x) = 4 \sin \pi x$
    $g(x) = 4 \sin \pi x - 3$
37. $f(x) = 2 \cos x$
    $g(x) = 2 \cos(x + \pi)$
38. $f(x) = -\cos x$
    $g(x) = -\cos(x - \pi)$

Sketching the Graph of a Sine or Cosine Function **In Exercises 39–60, sketch the graph of the function. (Include two full periods.)**

39. $y = 5 \sin x$
40. $y = \frac{1}{4} \sin x$
41. $y = \frac{1}{3} \cos x$
42. $y = 4 \cos x$
43. $y = \cos \frac{x}{2}$
44. $y = \sin 4x$
45. $y = \cos 2\pi x$
46. $y = \sin \frac{\pi x}{4}$
47. $y = -\sin \frac{2\pi x}{3}$
48. $y = -10 \cos \frac{\pi x}{6}$
49. $y = 3 \cos(x + \pi)$
50. $y = \sin(x - 2\pi)$

51. $y = \sin\left(x - \dfrac{\pi}{2}\right)$ **52.** $y = 4\cos\left(x + \dfrac{\pi}{4}\right)$

53. $y = 2 - \sin\dfrac{2\pi x}{3}$ **54.** $y = -3 + 5\cos\dfrac{\pi t}{12}$

55. $y = 2 + \frac{1}{10}\cos 60\pi x$ **56.** $y = 2\cos x - 3$

57. $y = 3\cos(x + \pi) - 3$ **58.** $y = -3\cos(6x + \pi)$

59. $y = \dfrac{2}{3}\cos\left(\dfrac{x}{2} - \dfrac{\pi}{4}\right)$ **60.** $y = 4\cos\left(x + \dfrac{\pi}{4}\right) + 4$

Describing a Transformation **In Exercises 61–66, *g* is related to a parent function $f(x) = \sin(x)$ or $f(x) = \cos(x)$. (a) Describe the sequence of transformations from *f* to *g*. (b) Sketch the graph of *g*. (c) Use function notation to write *g* in terms of *f*.**

61. $g(x) = \sin(4x - \pi)$ **62.** $g(x) = \sin(2x + \pi)$

63. $g(x) = \cos(x - \pi) + 2$ **64.** $g(x) = 1 + \cos(x + \pi)$

65. $g(x) = 2\sin(4x - \pi) - 3$ **66.** $g(x) = 4 - \sin(2x + \pi)$

Graphing a Sine or Cosine Function **In Exercises 67–72, use a graphing utility to graph the function. (Include two full periods.) Be sure to choose an appropriate viewing window.**

67. $y = -2\sin(4x + \pi)$ **68.** $y = -4\sin\left(\dfrac{2}{3}x - \dfrac{\pi}{3}\right)$

69. $y = \cos\left(2\pi x - \dfrac{\pi}{2}\right) + 1$

70. $y = 3\cos\left(\dfrac{\pi x}{2} + \dfrac{\pi}{2}\right) - 2$

71. $y = -0.1\sin\left(\dfrac{\pi x}{10} + \pi\right)$ **72.** $y = \dfrac{1}{100}\sin 120\pi t$

Graphical Reasoning **In Exercises 73–76, find *a* and *d* for the function $f(x) = a\cos x + d$ such that the graph of *f* matches the figure.**

73.

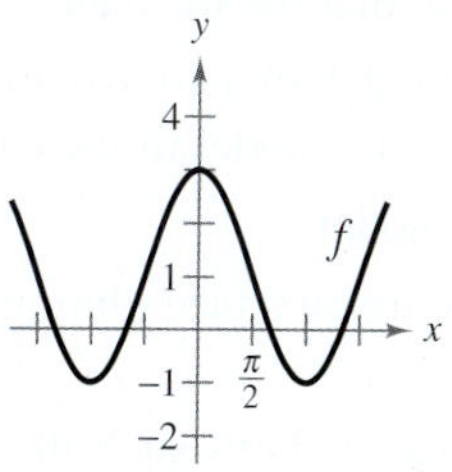

74.

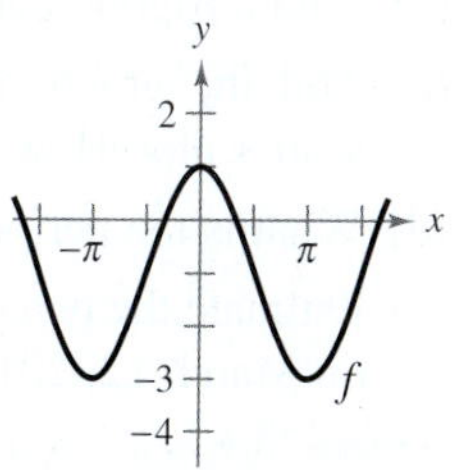

75.

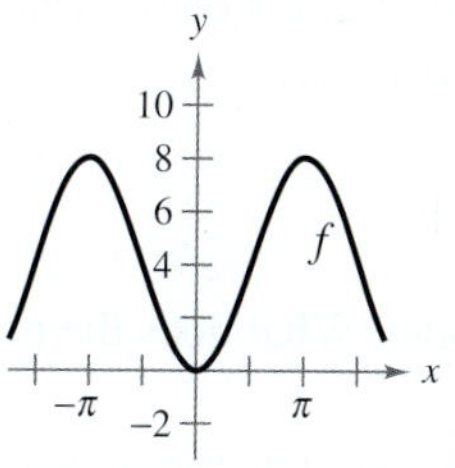

76.

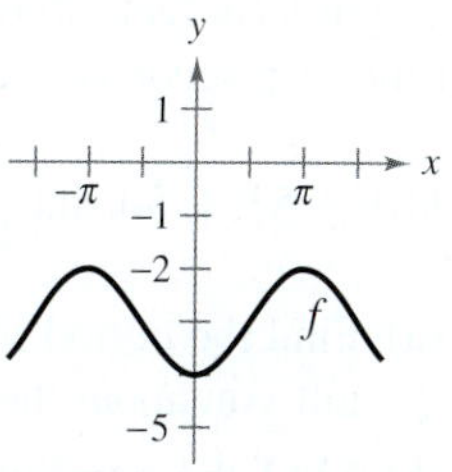

Graphical Reasoning **In Exercises 77–80, find *a*, *b*, and *c* for the function $f(x) = a\sin(bx - c)$ such that the graph of *f* matches the figure.**

77.

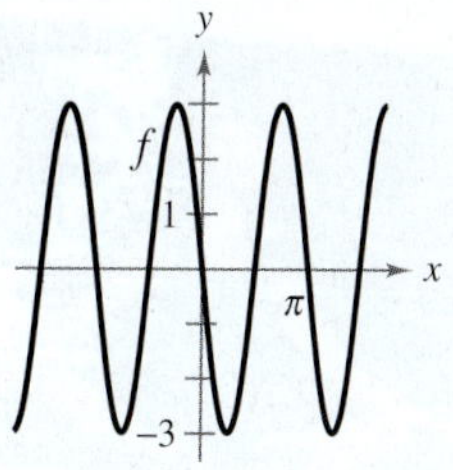

78.

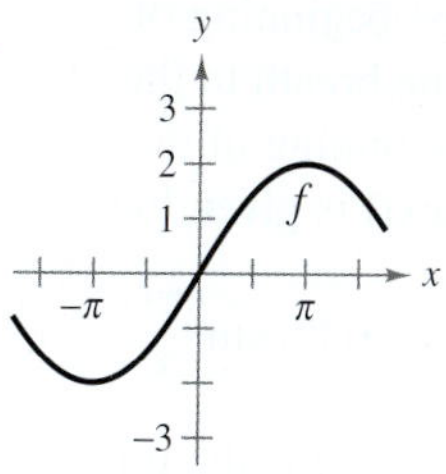

79.

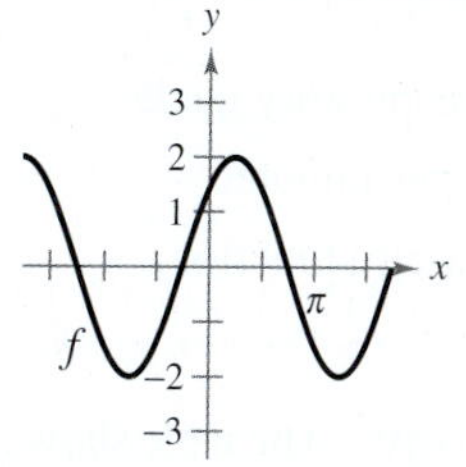

80.

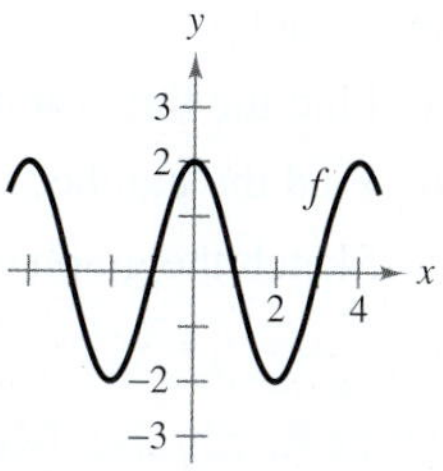

Graphical Analysis **In Exercises 81 and 82, use a graphing utility to graph y_1 and y_2 in the interval $[-2\pi, 2\pi]$. Use the graphs to find real numbers *x* such that $y_1 = y_2$.**

81. $y_1 = \sin x$
$y_2 = -\frac{1}{2}$

82. $y_1 = \cos x$
$y_2 = -1$

Writing an Equation **In Exercises 83–86, write an equation for the function that is described by the given characteristics.**

83. A sine curve with a period of π, an amplitude of 2, a right phase shift of $\pi/2$, and a vertical translation up 1 unit

84. A sine curve with a period of 4π, an amplitude of 3, a left phase shift of $\pi/4$, and a vertical translation down 1 unit

85. A cosine curve with a period of π, an amplitude of 1, a left phase shift of π, and a vertical translation down $\frac{3}{2}$ units

86. A cosine curve with a period of 4π, an amplitude of 3, a right phase shift of $\pi/2$, and a vertical translation up 2 units

87. Respiratory Cycle After exercising for a few minutes, a person has a respiratory cycle for which the velocity of airflow is approximated by

$$v = 1.75\sin\frac{\pi t}{2}$$

where t is the time (in seconds). (Inhalation occurs when $v > 0$, and exhalation occurs when $v < 0$.)

(a) Find the time for one full respiratory cycle.

(b) Find the number of cycles per minute.

(c) Sketch the graph of the velocity function.

88. Respiratory Cycle For a person at rest, the velocity v (in liters per second) of airflow during a respiratory cycle (the time from the beginning of one breath to the beginning of the next) is given by

$$v = 0.85 \sin \frac{\pi t}{3}$$

where t is the time (in seconds).

(a) Find the time for one full respiratory cycle.

(b) Find the number of cycles per minute.

(c) Sketch the graph of the velocity function.

89. Data Analysis: Meteorology The table shows the maximum daily high temperatures in Las Vegas L and International Falls I (in degrees Fahrenheit) for month t, with $t = 1$ corresponding to January. *(Source: National Climatic Data Center)*

DATA

Month, t	Las Vegas, L	International Falls, I
1	57.1	13.8
2	63.0	22.4
3	69.5	34.9
4	78.1	51.5
5	87.8	66.6
6	98.9	74.2
7	104.1	78.6
8	101.8	76.3
9	93.8	64.7
10	80.8	51.7
11	66.0	32.5
12	57.3	18.1

Spreadsheet at LarsonPrecalculus.com

(a) A model for the temperatures in Las Vegas is

$$L(t) = 80.60 + 23.50 \cos\left(\frac{\pi t}{6} - 3.67\right).$$

Find a trigonometric model for International Falls.

(b) Use a graphing utility to graph the data points and the model for the temperatures in Las Vegas. How well does the model fit the data?

(c) Use the graphing utility to graph the data points and the model for the temperatures in International Falls. How well does the model fit the data?

(d) Use the models to estimate the average maximum temperature in each city. Which term of the models did you use? Explain.

(e) What is the period of each model? Are the periods what you expected? Explain.

(f) Which city has the greater variability in temperature throughout the year? Which factor of the models determines this variability? Explain.

90. Health The function

$$P = 100 - 20 \cos \frac{5\pi t}{3}$$

approximates the blood pressure P (in millimeters of mercury) at time t (in seconds) for a person at rest.

(a) Find the period of the function.

(b) Find the number of heartbeats per minute.

91. Piano Tuning When tuning a piano, a technician strikes a tuning fork for the A above middle C and sets up a wave motion that can be approximated by $y = 0.001 \sin 880\pi t$, where t is the time (in seconds).

(a) What is the period of the function?

(b) The frequency f is given by $f = 1/p$. What is the frequency of the note?

92. Data Analysis: Astronomy The percent y (in decimal form) of the moon's face illuminated on day x in the year 2014, where $x = 1$ represents January 1, is shown in the table. *(Source: U.S. Naval Observatory)*

DATA

x	y
1	0.0
8	0.5
16	1.0
24	0.5
30	0.0
37	0.5

Spreadsheet at LarsonPrecalculus.com

(a) Create a scatter plot of the data.

(b) Find a trigonometric model that fits the data.

(c) Add the graph of your model in part (b) to the scatter plot. How well does the model fit the data?

(d) What is the period of the model?

(e) Estimate the percent of the moon's face illuminated on March 12, 2014.

93. Ferris Wheel A Ferris wheel is built such that the height h (in feet) above ground of a seat on the wheel at time t (in seconds) can be modeled by

$$h(t) = 53 + 50 \sin\left(\frac{\pi}{10}t - \frac{\pi}{2}\right).$$

(a) Find the period of the model. What does the period tell you about the ride?

(b) Find the amplitude of the model. What does the amplitude tell you about the ride?

(c) Use a graphing utility to graph one cycle of the model.

AISPIX by Image Source/Shutterstock.com

94. Fuel Consumption The daily consumption C (in gallons) of diesel fuel on a farm is modeled by

$$C = 30.3 + 21.6 \sin\left(\frac{2\pi t}{365} + 10.9\right)$$

where t is the time (in days), with $t = 1$ corresponding to January 1.

(a) What is the period of the model? Is it what you expected? Explain.

(b) What is the average daily fuel consumption? Which term of the model did you use? Explain.

(c) Use a graphing utility to graph the model. Use the graph to approximate the time of the year when consumption exceeds 40 gallons per day.

Exploration

True or False? **In Exercises 95 and 96, determine whether the statement is true or false. Justify your answer.**

95. The graph of the function $f(x) = \sin(x + 2\pi)$ translates the graph of $f(x) = \sin x$ exactly one period to the right so that the two graphs look identical.

96. The function $y = \frac{1}{2}\cos 2x$ has an amplitude that is twice that of the function $y = \cos x$.

Conjecture **In Exercises 97 and 98, graph f and g in the same coordinate plane. Include two full periods. Make a conjecture about the functions.**

97. $f(x) = \sin x, \quad g(x) = \cos\left(x - \frac{\pi}{2}\right)$

98. $f(x) = \sin x, \quad g(x) = -\cos\left(x + \frac{\pi}{2}\right)$

99. Writing Sketch the graph of $y = \cos bx$ for $b = \frac{1}{2}$, 2, and 3. How does the value of b affect the graph? How many complete cycles of the graph of y occur between 0 and 2π for each value of b?

100. Polynomial Approximations Using calculus, it can be shown that the sine and cosine functions can be approximated by the polynomials

$$\sin x \approx x - \frac{x^3}{3!} + \frac{x^5}{5!}$$

and

$$\cos x \approx 1 - \frac{x^2}{2!} + \frac{x^4}{4!}$$

where x is in radians.

(a) Use a graphing utility to graph the sine function and its polynomial approximation in the same viewing window. How do the graphs compare?

(b) Use the graphing utility to graph the cosine function and its polynomial approximation in the same viewing window. How do the graphs compare?

(c) Study the patterns in the polynomial approximations of the sine and cosine functions and predict the next term in each. Then repeat parts (a) and (b). How did the accuracy of the approximations change when an additional term was added?

101. Polynomial Approximations Use the polynomial approximations of the sine and cosine functions in Exercise 100 to approximate the following function values. Compare the results with those given by a calculator. Is the error in the approximation the same in each case? Explain.

(a) $\sin \frac{1}{2}$ (b) $\sin 1$ (c) $\sin \frac{\pi}{6}$

(d) $\cos(-0.5)$ (e) $\cos 1$ (f) $\cos \frac{\pi}{4}$

102. HOW DO YOU SEE IT? The figure below shows the graph of $y = \sin(x - c)$ for $c = -\frac{\pi}{4}$, 0, and $\frac{\pi}{4}$.

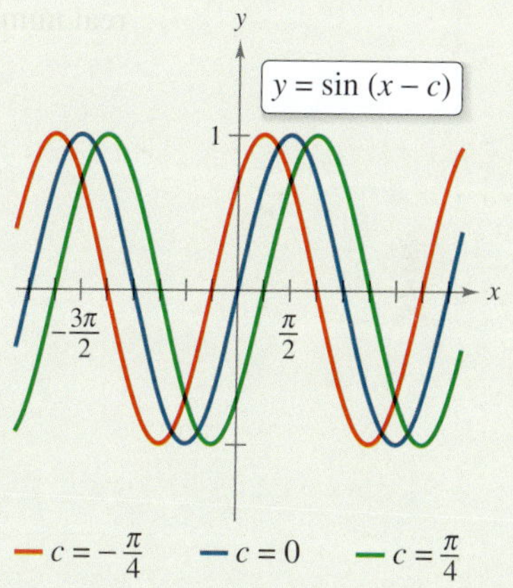

(a) How does the value of c affect the graph?

(b) Which graph is equivalent to that of

$$y = -\cos\left(x + \frac{\pi}{4}\right)?$$

Project: Meteorology To work an extended application analyzing the mean monthly temperature and mean monthly precipitation for Honolulu, Hawaii, visit this text's website at *LarsonPrecalculus.com. (Source: National Climatic Data Center)*

4.6 Graphs of Other Trigonometric Functions

You can use graphs of trigonometric functions to model real-life situations such as the distance from a television camera to a unit in a parade, as in Exercise 84 on page 319.

- **Sketch the graphs of tangent functions.**
- **Sketch the graphs of cotangent functions.**
- **Sketch the graphs of secant and cosecant functions.**
- **Sketch the graphs of damped trigonometric functions.**

Graph of the Tangent Function

Recall that the tangent function is odd. That is, $\tan(-x) = -\tan x$. Consequently, the graph of $y = \tan x$ is symmetric with respect to the origin. You also know from the identity $\tan x = \sin x/\cos x$ that the tangent is undefined for values at which $\cos x = 0$. Two such values are $x = \pm\pi/2 \approx \pm 1.5708$.

x	$-\frac{\pi}{2}$	-1.57	-1.5	$-\frac{\pi}{4}$	0	$\frac{\pi}{4}$	1.5	1.57	$\frac{\pi}{2}$
$\tan x$	Undef.	-1255.8	-14.1	-1	0	1	14.1	1255.8	Undef.

As indicated in the table, $\tan x$ increases without bound as x approaches $\pi/2$ from the left and decreases without bound as x approaches $-\pi/2$ from the right. So, the graph of $y = \tan x$ has *vertical asymptotes* at $x = \pi/2$ and $x = -\pi/2$, as shown below. Moreover, because the period of the tangent function is π, vertical asymptotes also occur at $x = \pi/2 + n\pi$, where n is an integer. The domain of the tangent function is the set of all real numbers other than $x = \pi/2 + n\pi$, and the range is the set of all real numbers.

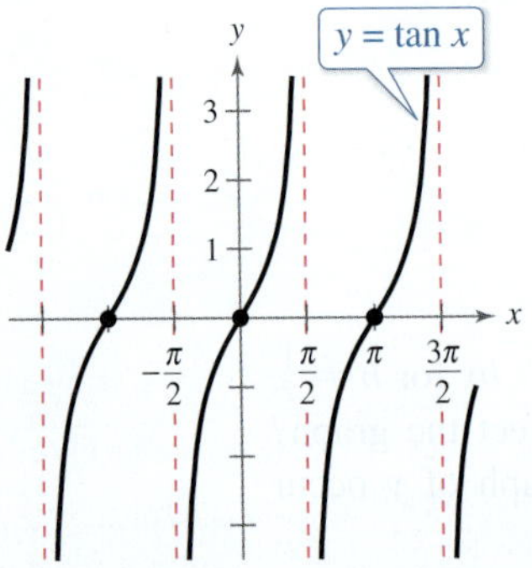

Period: π

Domain: all $x \neq \frac{\pi}{2} + n\pi$

Range: $(-\infty, \infty)$

Vertical asymptotes: $x = \frac{\pi}{2} + n\pi$

Symmetry: origin

ALGEBRA HELP

- You can review odd and even functions in Section 1.5.
- You can review symmetry of a graph in Section 1.2.
- You can review trigonometric identities in Section 4.3.
- You can review asymptotes in Section 2.6.
- You can review domain and range of a function in Section 1.4.
- You can review intercepts of a graph in Section 1.2.

Sketching the graph of $y = a\tan(bx - c)$ is similar to sketching the graph of $y = a\sin(bx - c)$ in that you locate key points that identify the intercepts and asymptotes. Two consecutive vertical asymptotes can be found by solving the equations

$$bx - c = -\frac{\pi}{2} \quad \text{and} \quad bx - c = \frac{\pi}{2}.$$

The midpoint between two consecutive vertical asymptotes is an x-intercept of the graph. The period of the function $y = a\tan(bx - c)$ is the distance between two consecutive vertical asymptotes. The amplitude of a tangent function is not defined. After plotting the asymptotes and the x-intercept, plot a few additional points between the two asymptotes and sketch one cycle. Finally, sketch one or two additional cycles to the left and right.

ariadna de raadt/Shutterstock.com

EXAMPLE 1 Sketching the Graph of a Tangent Function

Sketch the graph of $y = \tan \frac{x}{2}$.

Solution

By solving the equations

$$\frac{x}{2} = -\frac{\pi}{2} \quad \text{and} \quad \frac{x}{2} = \frac{\pi}{2}$$

$$x = -\pi \qquad\qquad x = \pi$$

you can see that two consecutive vertical asymptotes occur at $x = -\pi$ and $x = \pi$. Between these two asymptotes, plot a few points, including the x-intercept, as shown in the table. Three cycles of the graph are shown in Figure 4.44.

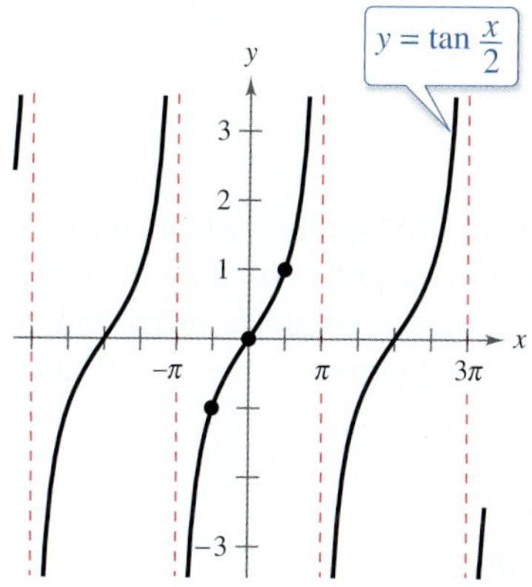

Figure 4.44

x	$-\pi$	$-\frac{\pi}{2}$	0	$\frac{\pi}{2}$	π
$\tan \frac{x}{2}$	Undef.	−1	0	1	Undef.

Checkpoint *Audio-video solution in English & Spanish at LarsonPrecalculus.com.*

Sketch the graph of $y = \tan \frac{x}{4}$.

EXAMPLE 2 Sketching the Graph of a Tangent Function

Sketch the graph of $y = -3 \tan 2x$.

Solution

By solving the equations

$$2x = -\frac{\pi}{2} \quad \text{and} \quad 2x = \frac{\pi}{2}$$

$$x = -\frac{\pi}{4} \qquad\qquad x = \frac{\pi}{4}$$

you can see that two consecutive vertical asymptotes occur at $x = -\pi/4$ and $x = \pi/4$. Between these two asymptotes, plot a few points, including the x-intercept, as shown in the table. Three cycles of the graph are shown in Figure 4.45.

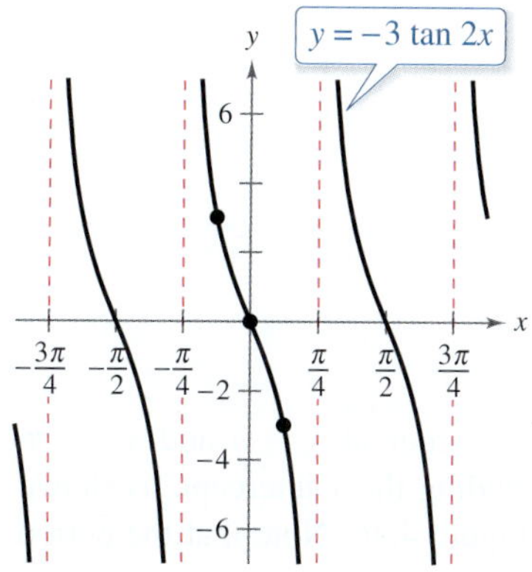

Figure 4.45

x	$-\frac{\pi}{4}$	$-\frac{\pi}{8}$	0	$\frac{\pi}{8}$	$\frac{\pi}{4}$
$-3 \tan 2x$	Undef.	3	0	−3	Undef.

By comparing the graphs in Examples 1 and 2, you can see that the graph of $y = a \tan(bx - c)$ increases between consecutive vertical asymptotes when $a > 0$ and decreases between consecutive vertical asymptotes when $a < 0$. In other words, the graph for $a < 0$ is a reflection in the x-axis of the graph for $a > 0$.

Checkpoint *Audio-video solution in English & Spanish at LarsonPrecalculus.com.*

Sketch the graph of $y = \tan 2x$.

Graph of the Cotangent Function

The graph of the cotangent function is similar to the graph of the tangent function. It also has a period of π. However, from the identity

$$y = \cot x = \frac{\cos x}{\sin x}$$

you can see that the cotangent function has vertical asymptotes when $\sin x$ is zero, which occurs at $x = n\pi$, where n is an integer. The graph of the cotangent function is shown below. Note that two consecutive vertical asymptotes of the graph of $y = a\cot(bx - c)$ can be found by solving the equations $bx - c = 0$ and $bx - c = \pi$.

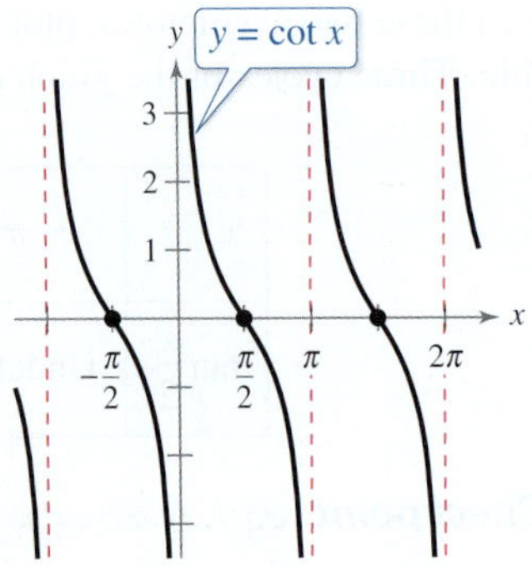

Period: π
Domain: all $x \neq n\pi$
Range: $(-\infty, \infty)$
Vertical asymptotes: $x = n\pi$
Symmetry: origin

▷ **TECHNOLOGY** Some graphing utilities have difficulty graphing trigonometric functions that have vertical asymptotes. Your graphing utility may connect parts of the graphs of tangent, cotangent, secant, and cosecant functions that are not supposed to be connected. To eliminate this problem, change the mode of the graphing utility to *dot* mode.

EXAMPLE 3 **Sketching the Graph of a Cotangent Function**

Sketch the graph of

$$y = 2\cot\frac{x}{3}.$$

Solution

By solving the equations

$$\frac{x}{3} = 0 \quad \text{and} \quad \frac{x}{3} = \pi$$

$$x = 0 \qquad\qquad x = 3\pi$$

you can see that two consecutive vertical asymptotes occur at $x = 0$ and $x = 3\pi$. Between these two asymptotes, plot a few points, including the x-intercept, as shown in the table. Three cycles of the graph are shown in Figure 4.46. Note that the period is 3π, the distance between consecutive asymptotes.

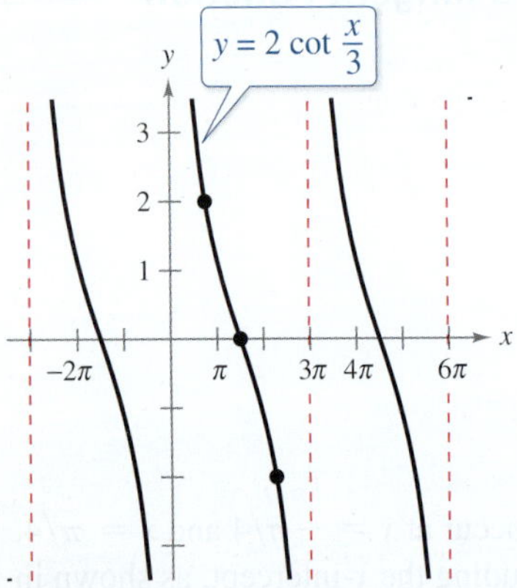

Figure 4.46

x	0	$\frac{3\pi}{4}$	$\frac{3\pi}{2}$	$\frac{9\pi}{4}$	3π
$2\cot\frac{x}{3}$	Undef.	2	0	-2	Undef.

✔ **Checkpoint** Audio-video solution in English & Spanish at LarsonPrecalculus.com.

Sketch the graph of

$$y = \cot\frac{x}{4}.$$

Graphs of the Reciprocal Functions

You can obtain the graphs of the two remaining trigonometric functions from the graphs of the sine and cosine functions using the reciprocal identities

$$\csc x = \frac{1}{\sin x} \quad \text{and} \quad \sec x = \frac{1}{\cos x}.$$

For instance, at a given value of x, the y-coordinate of $\sec x$ is the reciprocal of the y-coordinate of $\cos x$. Of course, when $\cos x = 0$, the reciprocal does not exist. Near such values of x, the behavior of the secant function is similar to that of the tangent function. In other words, the graphs of

$$\tan x = \frac{\sin x}{\cos x} \quad \text{and} \quad \sec x = \frac{1}{\cos x}$$

have vertical asymptotes where $\cos x = 0$—that is, at $x = \pi/2 + n\pi$, where n is an integer. Similarly,

$$\cot x = \frac{\cos x}{\sin x} \quad \text{and} \quad \csc x = \frac{1}{\sin x}$$

have vertical asymptotes where $\sin x = 0$—that is, at $x = n\pi$, where n is an integer.

To sketch the graph of a secant or cosecant function, you should first make a sketch of its reciprocal function. For instance, to sketch the graph of $y = \csc x$, first sketch the graph of $y = \sin x$. Then take reciprocals of the y-coordinates to obtain points on the graph of $y = \csc x$. You can use this procedure to obtain the graphs shown below.

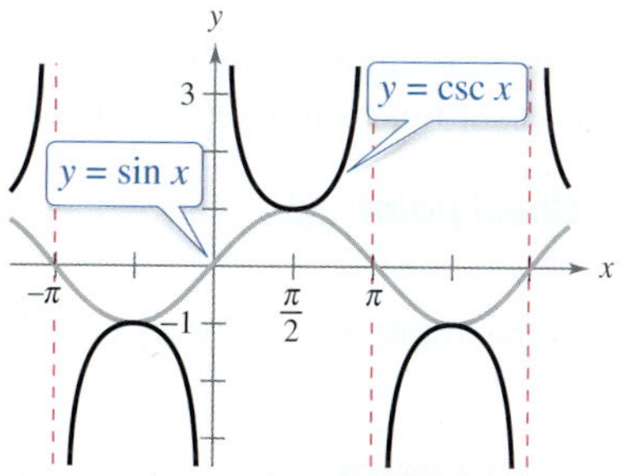

Period: 2π
Domain: all $x \neq n\pi$
Range: $(-\infty, -1] \cup [1, \infty)$
Vertical asymptotes: $x = n\pi$
Symmetry: origin

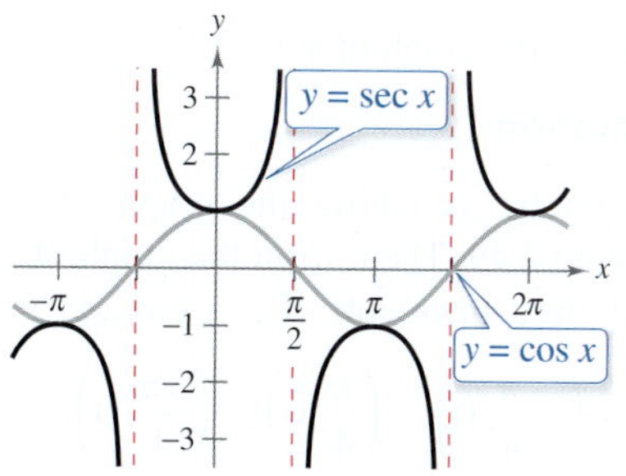

Period: 2π

Domain: all $x \neq \dfrac{\pi}{2} + n\pi$

Range: $(-\infty, -1] \cup [1, \infty)$

Vertical asymptotes: $x = \dfrac{\pi}{2} + n\pi$

Symmetry: y-axis

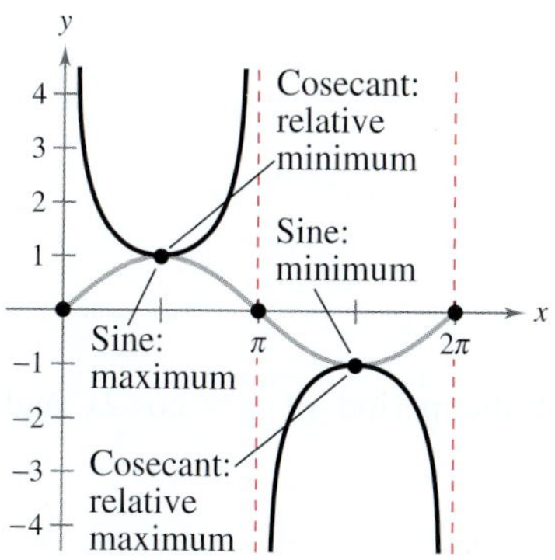

Figure 4.47

In comparing the graphs of the cosecant and secant functions with those of the sine and cosine functions, respectively, note that the "hills" and "valleys" are interchanged. For instance, a hill (or maximum point) on the sine curve corresponds to a valley (a relative minimum) on the cosecant curve, and a valley (or minimum point) on the sine curve corresponds to a hill (a relative maximum) on the cosecant curve, as shown in Figure 4.47. Additionally, x-intercepts of the sine and cosine functions become vertical asymptotes of the cosecant and secant functions, respectively (see Figure 4.47).

EXAMPLE 4 Sketching the Graph of a Cosecant Function

Sketch the graph of $y = 2\csc\left(x + \dfrac{\pi}{4}\right)$.

Solution

Begin by sketching the graph of

$$y = 2\sin\left(x + \frac{\pi}{4}\right).$$

For this function, the amplitude is 2 and the period is 2π. By solving the equations

$$x + \frac{\pi}{4} = 0 \quad \text{and} \quad x + \frac{\pi}{4} = 2\pi$$

$$x = -\frac{\pi}{4} \qquad\qquad x = \frac{7\pi}{4}$$

you can see that one cycle of the sine function corresponds to the interval from $x = -\pi/4$ to $x = 7\pi/4$. The graph of this sine function is represented by the gray curve in Figure 4.48. Because the sine function is zero at the midpoint and endpoints of this interval, the corresponding cosecant function

$$y = 2\csc\left(x + \frac{\pi}{4}\right)$$

$$= 2\left(\frac{1}{\sin[x + (\pi/4)]}\right)$$

has vertical asymptotes at $x = -\pi/4$, $x = 3\pi/4$, $x = 7\pi/4$, and so on. The graph of the cosecant function is represented by the black curve in Figure 4.48.

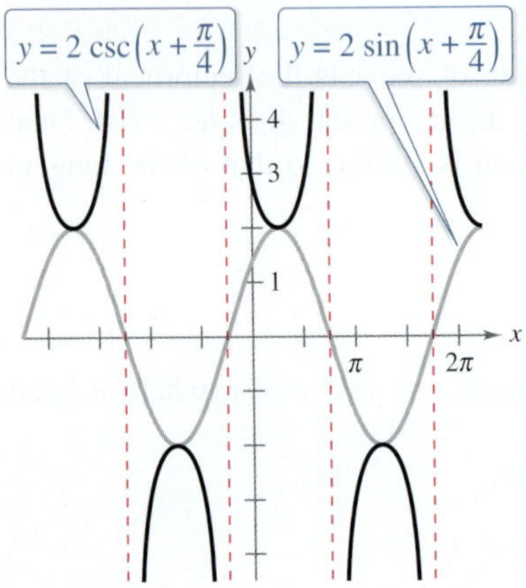

Figure 4.48

Checkpoint Audio-video solution in English & Spanish at LarsonPrecalculus.com.

Sketch the graph of $y = 2\csc\left(x + \dfrac{\pi}{2}\right)$.

EXAMPLE 5 Sketching the Graph of a Secant Function

Sketch the graph of $y = \sec 2x$.

Solution

Begin by sketching the graph of $y = \cos 2x$, as indicated by the gray curve in Figure 4.49. Then, form the graph of $y = \sec 2x$ as the black curve in the figure. Note that the x-intercepts of $y = \cos 2x$

$$\left(-\frac{\pi}{4}, 0\right), \left(\frac{\pi}{4}, 0\right), \left(\frac{3\pi}{4}, 0\right), \ldots$$

correspond to the vertical asymptotes

$$x = -\frac{\pi}{4}, \quad x = \frac{\pi}{4}, \quad x = \frac{3\pi}{4}, \ldots$$

of the graph of $y = \sec 2x$. Moreover, notice that the period of $y = \cos 2x$ and $y = \sec 2x$ is π.

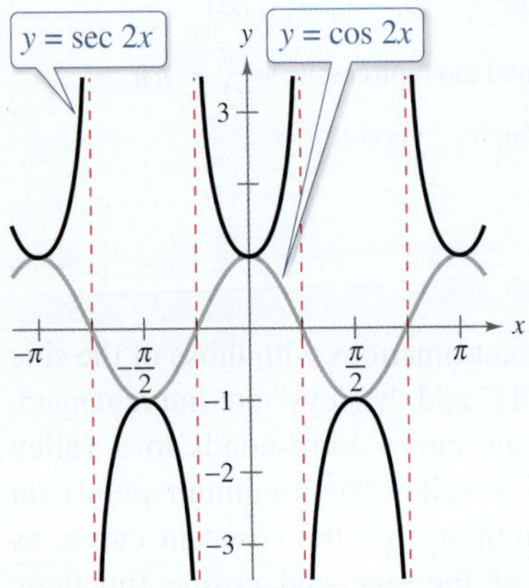

Figure 4.49

Checkpoint Audio-video solution in English & Spanish at LarsonPrecalculus.com.

Sketch the graph of $y = \sec \dfrac{x}{2}$.

Damped Trigonometric Graphs

You can graph a *product* of two functions using properties of the individual functions. For instance, consider the function

$$f(x) = x \sin x$$

as the product of the functions $y = x$ and $y = \sin x$. Using properties of absolute value and the fact that $|\sin x| \le 1$, you have

$$0 \le |x||\sin x| \le |x|.$$

Consequently,

$$-|x| \le x \sin x \le |x|$$

which means that the graph of $f(x) = x \sin x$ lies between the lines $y = -x$ and $y = x$. Furthermore, because

$$f(x) = x \sin x = \pm x \quad \text{at} \quad x = \frac{\pi}{2} + n\pi$$

and

$$f(x) = x \sin x = 0 \quad \text{at} \quad x = n\pi$$

where n is an integer, the graph of f touches the line $y = -x$ or the line $y = x$ at $x = \pi/2 + n\pi$ and has x-intercepts at $x = n\pi$. A sketch of f is shown at the right. In the function $f(x) = x \sin x$, the factor x is called the **damping factor.**

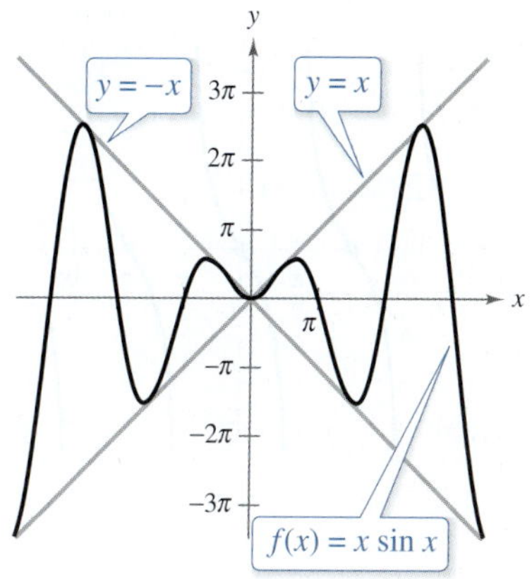

REMARK Do you see why the graph of $f(x) = x \sin x$ touches the lines $y = \pm x$ at $x = \pi/2 + n\pi$ and why the graph has x-intercepts at $x = n\pi$? Recall that the sine function is equal to 1 at $\ldots, -3\pi/2, \pi/2, 5\pi/2, \ldots$ $(x = \pi/2 + 2n\pi)$ and -1 at $\ldots, -\pi/2, 3\pi/2, 7\pi/2, \ldots$ $(x = -\pi/2 + 2n\pi)$ and is equal to 0 at $\ldots, -\pi, 0, \pi, 2\pi, 3\pi, \ldots$ $(x = n\pi)$.

EXAMPLE 6 **Damped Sine Wave**

Sketch the graph of $f(x) = e^{-x} \sin 3x$.

Solution

Consider $f(x)$ as the product of the two functions

$$y = e^{-x} \quad \text{and} \quad y = \sin 3x$$

each of which has the set of real numbers as its domain. For any real number x, you know that $e^{-x} > 0$ and $|\sin 3x| \le 1$. So,

$$e^{-x}|\sin 3x| \le e^{-x}$$

which means that

$$-e^{-x} \le e^{-x} \sin 3x \le e^{-x}.$$

Furthermore, because

$$f(x) = e^{-x} \sin 3x = \pm e^{-x} \quad \text{at} \quad x = \frac{\pi}{6} + \frac{n\pi}{3}$$

and

$$f(x) = e^{-x} \sin 3x = 0 \quad \text{at} \quad x = \frac{n\pi}{3}$$

the graph of f touches the curve $y = -e^{-x}$ or the curve $y = e^{-x}$ at $x = \pi/6 + n\pi/3$ and has intercepts at $x = n\pi/3$. A sketch of f is shown in Figure 4.50.

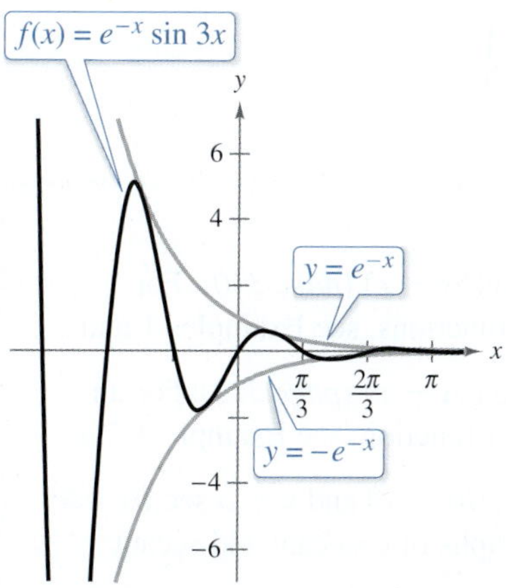

Figure 4.50

✓ **Checkpoint** Audio-video solution in English & Spanish at LarsonPrecalculus.com.

Sketch the graph of $f(x) = e^{x} \sin 4x$.

Below is a summary of the characteristics of the six basic trigonometric functions.

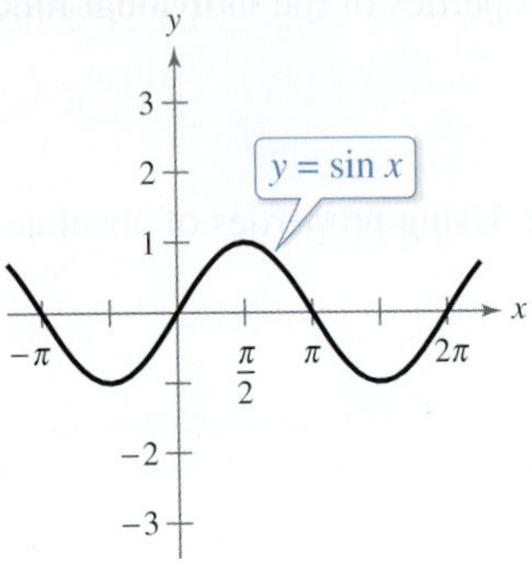

Domain: $(-\infty, \infty)$
Range: $[-1, 1]$
Period: 2π

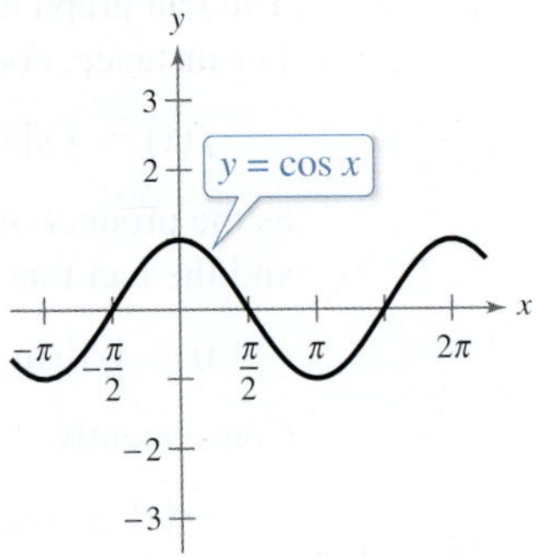

Domain: $(-\infty, \infty)$
Range: $[-1, 1]$
Period: 2π

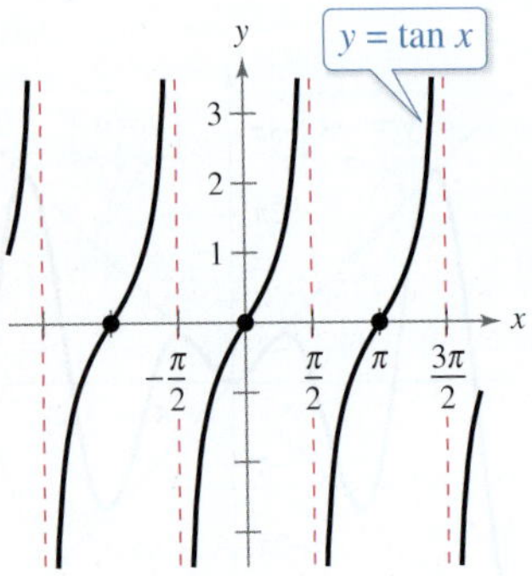

Domain: all $x \neq \frac{\pi}{2} + n\pi$
Range: $(-\infty, \infty)$
Period: π

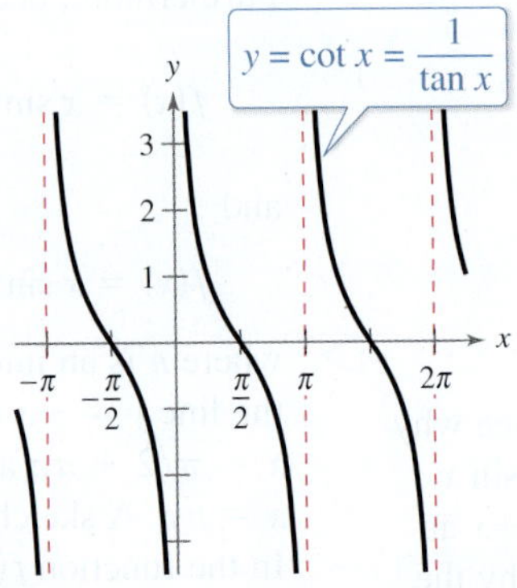

Domain: all $x \neq n\pi$
Range: $(-\infty, \infty)$
Period: π

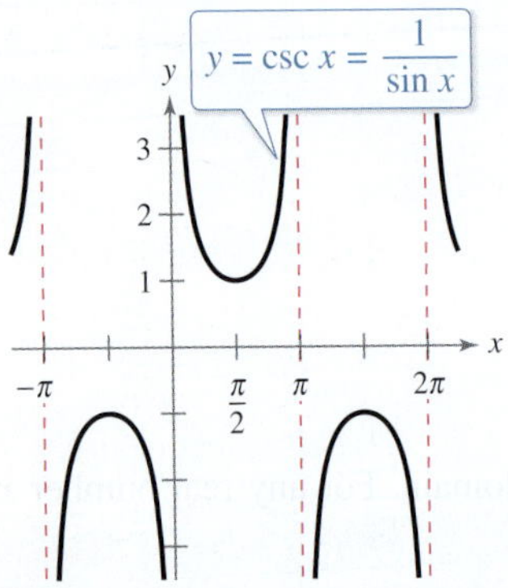

Domain: all $x \neq n\pi$
Range: $(-\infty, -1] \cup [1, \infty)$
Period: 2π

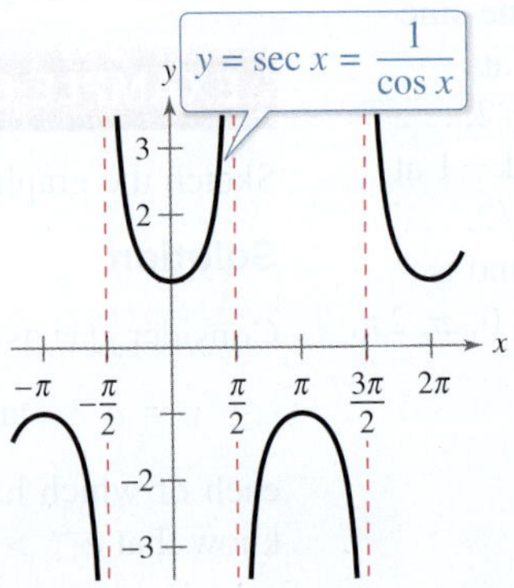

Domain: all $x \neq \frac{\pi}{2} + n\pi$
Range: $(-\infty, -1] \cup [1, \infty)$
Period: 2π

Summarize (Section 4.6)

1. Describe how to sketch the graph of $y = a \tan(bx - c)$ *(page 310)*. For examples of sketching the graphs of tangent functions, see Examples 1 and 2.
2. Describe how to sketch the graph of $y = a \cot(bx - c)$ *(page 312)*. For an example of sketching the graph of a cotangent function, see Example 3.
3. Describe how to sketch the graphs of $y = a \csc(bx - c)$ and $y = a \sec(bx - c)$ *(page 313)*. For examples of sketching the graphs of cosecant and secant functions, see Examples 4 and 5.
4. Describe how to sketch the graph of a damped trigonometric function *(page 315)*. For an example of sketching the graph of a damped trigonometric function, see Example 6.

4.6 Exercises

See CalcChat.com for tutorial help and worked-out solutions to odd-numbered exercises.

Vocabulary: Fill in the blanks.

1. The tangent, cotangent, and cosecant functions are ________ , so the graphs of these functions have symmetry with respect to the ________.
2. The graphs of the tangent, cotangent, secant, and cosecant functions have ________ asymptotes.
3. To sketch the graph of a secant or cosecant function, first make a sketch of its ________ function.
4. For the function $f(x) = g(x) \cdot \sin x$, $g(x)$ is called the ________ factor of the function $f(x)$.
5. The period of $y = \tan x$ is ________.
6. The domain of $y = \cot x$ is all real numbers such that ________.
7. The range of $y = \sec x$ is ________.
8. The period of $y = \csc x$ is ________.

Skills and Applications

Matching **In Exercises 9–14, match the function with its graph. State the period of the function. [The graphs are labeled (a), (b), (c), (d), (e), and (f).]**

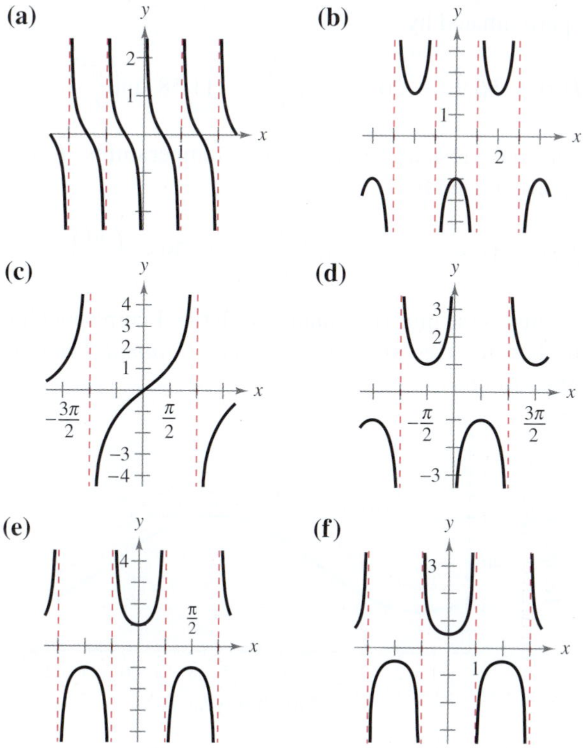

9. $y = \sec 2x$
10. $y = \tan \dfrac{x}{2}$
11. $y = \dfrac{1}{2} \cot \pi x$
12. $y = -\csc x$
13. $y = \dfrac{1}{2} \sec \dfrac{\pi x}{2}$
14. $y = -2 \sec \dfrac{\pi x}{2}$

Sketching the Graph of a Trigonometric Function **In Exercises 15–38, sketch the graph of the function. (Include two full periods.)**

15. $y = \frac{1}{3} \tan x$
16. $y = \tan 4x$
17. $y = -2 \tan 3x$
18. $y = -3 \tan \pi x$
19. $y = -\frac{1}{2} \sec x$
20. $y = \frac{1}{4} \sec x$
21. $y = \csc \pi x$
22. $y = 3 \csc 4x$
23. $y = \frac{1}{2} \sec \pi x$
24. $y = -2 \sec 4x + 2$
25. $y = \csc \dfrac{x}{2}$
26. $y = \csc \dfrac{x}{3}$
27. $y = 3 \cot 2x$
28. $y = 3 \cot \dfrac{\pi x}{2}$
29. $y = 2 \sec 3x$
30. $y = -\frac{1}{2} \tan x$
31. $y = \tan \dfrac{\pi x}{4}$
32. $y = \tan(x + \pi)$
33. $y = 2 \csc(x - \pi)$
34. $y = \csc(2x - \pi)$
35. $y = 2 \sec(x + \pi)$
36. $y = -\sec \pi x + 1$
37. $y = \dfrac{1}{4} \csc\left(x + \dfrac{\pi}{4}\right)$
38. $y = 2 \cot\left(x + \dfrac{\pi}{2}\right)$

Graphing a Trigonometric Function **In Exercises 39–48, use a graphing utility to graph the function. (Include two full periods.)**

39. $y = \tan \dfrac{x}{3}$
40. $y = -\tan 2x$
41. $y = -2 \sec 4x$
42. $y = \sec \pi x$
43. $y = \tan\left(x - \dfrac{\pi}{4}\right)$
44. $y = \dfrac{1}{4} \cot\left(x - \dfrac{\pi}{2}\right)$
45. $y = -\csc(4x - \pi)$
46. $y = 2 \sec(2x - \pi)$
47. $y = 0.1 \tan\left(\dfrac{\pi x}{4} + \dfrac{\pi}{4}\right)$
48. $y = \dfrac{1}{3} \sec\left(\dfrac{\pi x}{2} + \dfrac{\pi}{2}\right)$

Solving a Trigonometric Equation Graphically **In Exercises 49–56, use a graph to solve the equation on the interval $[-2\pi, 2\pi]$.**

49. $\tan x = 1$ **50.** $\tan x = \sqrt{3}$

51. $\cot x = -\dfrac{\sqrt{3}}{3}$ **52.** $\cot x = 1$

53. $\sec x = -2$ **54.** $\sec x = 2$

55. $\csc x = \sqrt{2}$ **56.** $\csc x = -\dfrac{2\sqrt{3}}{3}$

Even and Odd Trigonometric Functions **In Exercises 57–64, use the graph of the function to determine whether the function is even, odd, or neither. Verify your answer algebraically.**

57. $f(x) = \sec x$ **58.** $f(x) = \tan x$

59. $g(x) = \cot x$ **60.** $g(x) = \csc x$

61. $f(x) = x + \tan x$ **62.** $f(x) = x^2 - \sec x$

63. $g(x) = x \csc x$ **64.** $g(x) = x^2 \cot x$

Identifying Damped Trigonometric Functions **In Exercises 65–68, match the function with its graph. Describe the behavior of the function as x approaches zero. [The graphs are labeled (a), (b), (c), and (d).]**

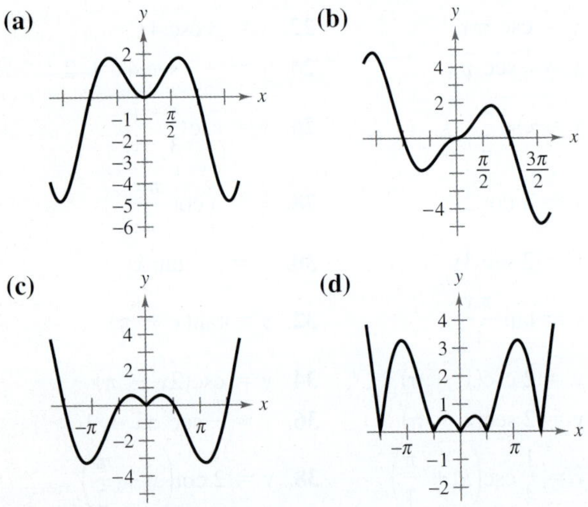

65. $f(x) = |x \cos x|$ **66.** $f(x) = x \sin x$

67. $g(x) = |x| \sin x$ **68.** $g(x) = |x| \cos x$

Conjecture **In Exercises 69–72, graph the functions f and g. Use the graphs to make a conjecture about the relationship between the functions.**

69. $f(x) = \sin x + \cos\left(x + \dfrac{\pi}{2}\right), \quad g(x) = 0$

70. $f(x) = \sin x - \cos\left(x + \dfrac{\pi}{2}\right), \quad g(x) = 2 \sin x$

71. $f(x) = \sin^2 x, \quad g(x) = \frac{1}{2}(1 - \cos 2x)$

72. $f(x) = \cos^2 \dfrac{\pi x}{2}, \quad g(x) = \dfrac{1}{2}(1 + \cos \pi x)$

Analyzing a Damped Trigonometric Graph **In Exercises 73–76, use a graphing utility to graph the function and the damping factor of the function in the same viewing window. Describe the behavior of the function as x increases without bound.**

73. $g(x) = e^{-x^2/2} \sin x$ **74.** $f(x) = e^{-x} \cos x$

75. $f(x) = 2^{-x/4} \cos \pi x$ **76.** $h(x) = 2^{-x^2/4} \sin x$

Analyzing a Trigonometric Graph **In Exercises 77–82, use a graphing utility to graph the function. Describe the behavior of the function as x approaches zero.**

77. $y = \dfrac{6}{x} + \cos x, \quad x > 0$ **78.** $y = \dfrac{4}{x} + \sin 2x, \quad x > 0$

79. $g(x) = \dfrac{\sin x}{x}$ **80.** $f(x) = \dfrac{1 - \cos x}{x}$

81. $f(x) = \sin \dfrac{1}{x}$ **82.** $h(x) = x \sin \dfrac{1}{x}$

83. Meteorology The normal monthly high temperatures H (in degrees Fahrenheit) in Erie, Pennsylvania, are approximated by

$$H(t) = 56.94 - 20.86 \cos\left(\frac{\pi t}{6}\right) - 11.58 \sin\left(\frac{\pi t}{6}\right)$$

and the normal monthly low temperatures L are approximated by

$$L(t) = 41.80 - 17.13 \cos\left(\frac{\pi t}{6}\right) - 13.39 \sin\left(\frac{\pi t}{6}\right)$$

where t is the time (in months), with $t = 1$ corresponding to January (see figure). *(Source: National Climatic Data Center)*

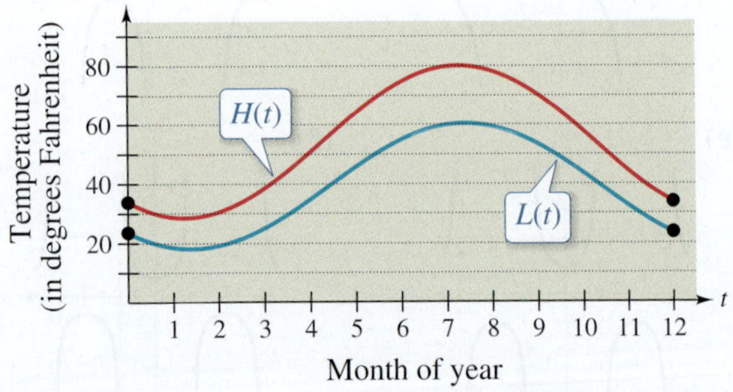

(a) What is the period of each function?

(b) During what part of the year is the difference between the normal high and normal low temperatures greatest? When is it smallest?

(c) The sun is northernmost in the sky around June 21, but the graph shows the warmest temperatures at a later date. Approximate the lag time of the temperatures relative to the position of the sun.

84. Television Coverage A television camera is on a reviewing platform 27 meters from the street on which a parade will be passing from left to right (see figure). Write the distance d from the camera to a particular unit in the parade as a function of the angle x, and graph the function over the interval $-\pi/2 < x < \pi/2$. (Consider x as negative when a unit in the parade approaches from the left.)

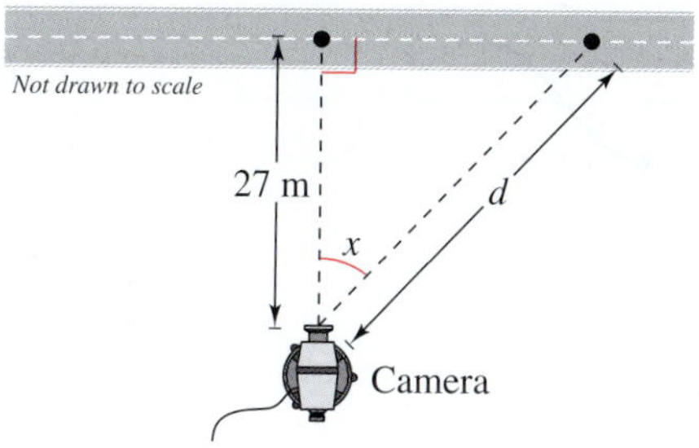

85. Distance A plane flying at an altitude of 7 miles above a radar antenna will pass directly over the radar antenna (see figure). Let d be the ground distance from the antenna to the point directly under the plane and let x be the angle of elevation to the plane from the antenna. (d is positive as the plane approaches the antenna.) Write d as a function of x and graph the function over the interval $0 < x < \pi$.

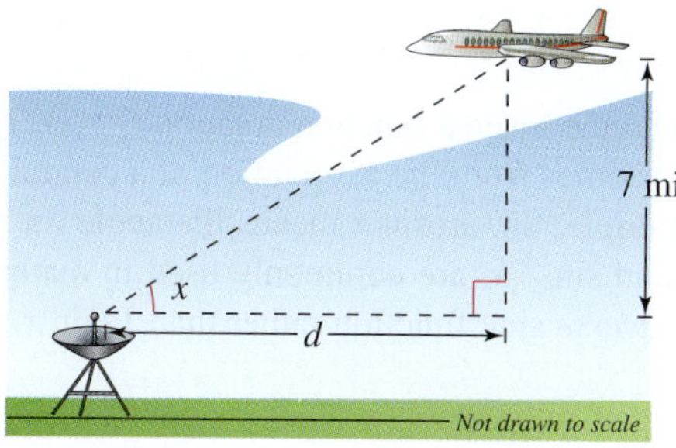

Exploration

True or False? **In Exercises 86 and 87, determine whether the statement is true or false. Justify your answer.**

86. You can obtain the graph of $y = \csc x$ on a calculator by graphing the reciprocal of $y = \sin x$.

87. You can obtain the graph of $y = \sec x$ on a calculator by graphing a translation of the reciprocal of $y = \sin x$.

Graphical Analysis **In Exercises 88 and 89, use a graphing utility to graph the function. Use the graph to determine the behavior of the function as $x \to c$.**

(a) As $x \to 0^+$, the value of $f(x) \to$ ▭.

(b) As $x \to 0^-$, the value of $f(x) \to$ ▭.

(c) As $x \to \pi^+$, the value of $f(x) \to$ ▭.

(d) As $x \to \pi^-$, the value of $f(x) \to$ ▭.

88. $f(x) = \cot x$ **89.** $f(x) = \csc x$

Graphical Analysis **In Exercises 90 and 91, use a graphing utility to graph the function. Use the graph to determine the behavior of the function as $x \to c$.**

(a) $x \to \left(\frac{\pi}{2}\right)^+$ **(b)** $x \to \left(\frac{\pi}{2}\right)^-$

(c) $x \to \left(-\frac{\pi}{2}\right)^+$ **(d)** $x \to \left(-\frac{\pi}{2}\right)^-$

90. $f(x) = \tan x$ **91.** $f(x) = \sec x$

92. HOW DO YOU SEE IT? Determine which function is represented by the graph. Do not use a calculator. Explain your reasoning.

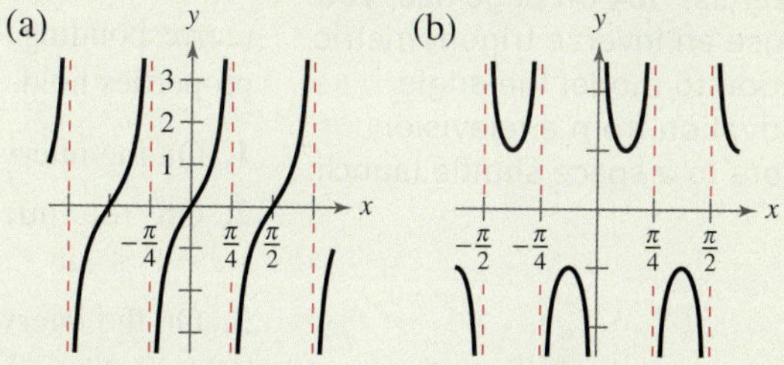

(a)	(b)
(i) $f(x) = \tan 2x$	(i) $f(x) = \sec 4x$
(ii) $f(x) = \tan(x/2)$	(ii) $f(x) = \csc 4x$
(iii) $f(x) = 2 \tan x$	(iii) $f(x) = \csc(x/4)$
(iv) $f(x) = -\tan 2x$	(iv) $f(x) = \sec(x/4)$
(v) $f(x) = -\tan(x/2)$	(v) $f(x) = \csc(4x - \pi)$

93. Think About It Consider the function $f(x) = x - \cos x$.

(a) Use a graphing utility to graph the function and verify that there exists a zero between 0 and 1. Use the graph to approximate the zero.

(b) Starting with $x_0 = 1$, generate a sequence $x_1, x_2, x_3, \ldots,$ where $x_n = \cos(x_{n-1})$. For example,

$$x_0 = 1$$
$$x_1 = \cos(x_0)$$
$$x_2 = \cos(x_1)$$
$$x_3 = \cos(x_2)$$
$$\vdots$$

What value does the sequence approach?

ariadna de raadt/Shutterstock.com

4.7 Inverse Trigonometric Functions

You can use inverse trigonometric functions to model and solve real-life problems. For instance, in Exercise 104 on page 328, you will use an inverse trigonometric function to model the angle of elevation from a television camera to a space shuttle launch.

- **Evaluate and graph the inverse sine function.**
- **Evaluate and graph the other inverse trigonometric functions.**
- **Evaluate the compositions of trigonometric functions.**

Inverse Sine Function

Recall from Section 1.9 that for a function to have an inverse function, it must be one-to-one—that is, it must pass the Horizontal Line Test. From Figure 4.51, you can see that $y = \sin x$ does not pass the test because different values of x yield the same y-value.

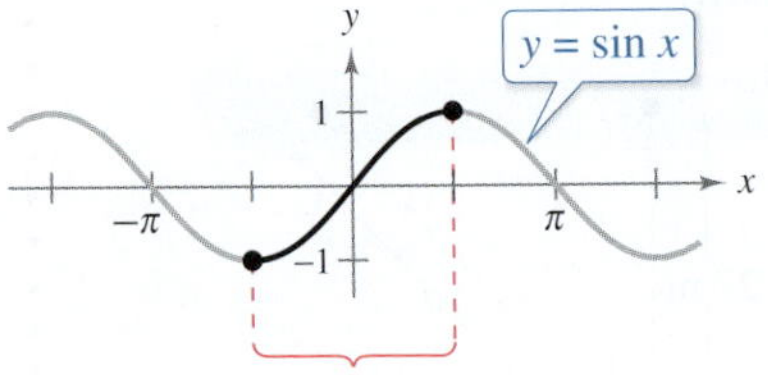

Figure 4.51

However, when you restrict the domain to the interval $-\pi/2 \le x \le \pi/2$ (corresponding to the black portion of the graph in Figure 4.51), the following properties hold.

1. On the interval $[-\pi/2, \pi/2]$, the function $y = \sin x$ is increasing.
2. On the interval $[-\pi/2, \pi/2]$, $y = \sin x$ takes on its full range of values, $-1 \le \sin x \le 1$.
3. On the interval $[-\pi/2, \pi/2]$, $y = \sin x$ is one-to-one.

So, on the restricted domain $-\pi/2 \le x \le \pi/2$, $y = \sin x$ has a unique inverse function called the **inverse sine function.** It is denoted by

$$y = \arcsin x \quad \text{or} \quad y = \sin^{-1} x.$$

The notation $\sin^{-1} x$ is consistent with the inverse function notation $f^{-1}(x)$. The $\arcsin x$ notation (read as "the arcsine of x") comes from the association of a central angle with its intercepted *arc length* on a unit circle. So, $\arcsin x$ means the angle (or arc) whose sine is x. Both notations, $\arcsin x$ and $\sin^{-1} x$, are commonly used in mathematics, so remember that $\sin^{-1} x$ denotes the *inverse* sine function rather than $1/\sin x$. The values of $\arcsin x$ lie in the interval

$$-\frac{\pi}{2} \le \arcsin x \le \frac{\pi}{2}.$$

The graph of $y = \arcsin x$ is shown in Example 2.

REMARK When evaluating the inverse sine function, it helps to remember the phrase "the arcsine of x is the angle (or number) whose sine is x."

Definition of Inverse Sine Function

The **inverse sine function** is defined by

$$y = \arcsin x \quad \text{if and only if} \quad \sin y = x$$

where $-1 \le x \le 1$ and $-\pi/2 \le y \le \pi/2$. The domain of $y = \arcsin x$ is $[-1, 1]$, and the range is $[-\pi/2, \pi/2]$.

NASA

REMARK As with the trigonometric functions, much of the work with the inverse trigonometric functions can be done by *exact* calculations rather than by calculator approximations. Exact calculations help to increase your understanding of the inverse functions by relating them to the right triangle definitions of the trigonometric functions.

EXAMPLE 1 Evaluating the Inverse Sine Function

If possible, find the exact value.

a. $\arcsin\left(-\frac{1}{2}\right)$ **b.** $\sin^{-1}\frac{\sqrt{3}}{2}$ **c.** $\sin^{-1}2$

Solution

a. Because $\sin\left(-\frac{\pi}{6}\right) = -\frac{1}{2}$ and $-\frac{\pi}{6}$ lies in $\left[-\frac{\pi}{2}, \frac{\pi}{2}\right]$, it follows that

$$\arcsin\left(-\frac{1}{2}\right) = -\frac{\pi}{6}.$$ Angle whose sine is $-\frac{1}{2}$

b. Because $\sin\frac{\pi}{3} = \frac{\sqrt{3}}{2}$ and $\frac{\pi}{3}$ lies in $\left[-\frac{\pi}{2}, \frac{\pi}{2}\right]$, it follows that

$$\sin^{-1}\frac{\sqrt{3}}{2} = \frac{\pi}{3}.$$ Angle whose sine is $\sqrt{3}/2$

c. It is not possible to evaluate $y = \sin^{-1}x$ when $x = 2$ because there is no angle whose sine is 2. Remember that the domain of the inverse sine function is $[-1, 1]$.

Checkpoint Audio-video solution in English & Spanish at LarsonPrecalculus.com.

If possible, find the exact value.

a. $\arcsin 1$ **b.** $\sin^{-1}(-2)$

EXAMPLE 2 Graphing the Arcsine Function

Sketch a graph of

$$y = \arcsin x.$$

Solution

By definition, the equations $y = \arcsin x$ and $\sin y = x$ are equivalent for $-\pi/2 \le y \le \pi/2$. So, their graphs are the same. From the interval $[-\pi/2, \pi/2]$, you can assign values to y in the equation $\sin y = x$ to make a table of values. Then plot the points and connect them with a smooth curve.

y	$-\frac{\pi}{2}$	$-\frac{\pi}{4}$	$-\frac{\pi}{6}$	0	$\frac{\pi}{6}$	$\frac{\pi}{4}$	$\frac{\pi}{2}$
$x = \sin y$	-1	$-\frac{\sqrt{2}}{2}$	$-\frac{1}{2}$	0	$\frac{1}{2}$	$\frac{\sqrt{2}}{2}$	1

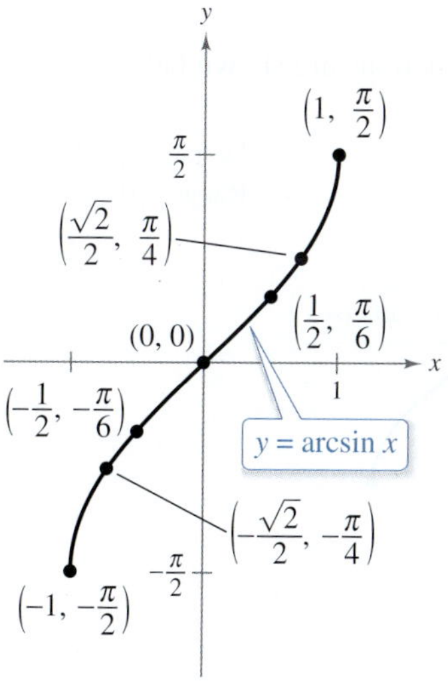

Figure 4.52

The resulting graph of $y = \arcsin x$ is shown in Figure 4.52. Note that it is the reflection (in the line $y = x$) of the black portion of the graph in Figure 4.51. Be sure you see that Figure 4.52 shows the *entire* graph of the inverse sine function. Remember that the domain of $y = \arcsin x$ is the closed interval $[-1, 1]$ and the range is the closed interval $[-\pi/2, \pi/2]$.

Checkpoint Audio-video solution in English & Spanish at LarsonPrecalculus.com.

Use a graphing utility to graph $f(x) = \sin x$, $g(x) = \arcsin x$, and $y = x$ in the same viewing window to verify geometrically that g is the inverse function of f. (Be sure to restrict the domain of f properly.)

Other Inverse Trigonometric Functions

The cosine function is decreasing and one-to-one on the interval $0 \le x \le \pi$, as shown below.

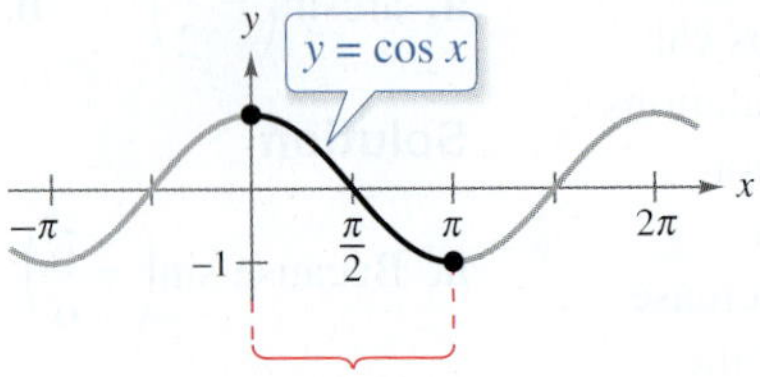

Consequently, on this interval the cosine function has an inverse function—the **inverse cosine function**—denoted by

$$y = \arccos x \quad \text{or} \quad y = \cos^{-1} x.$$

Similarly, you can define an **inverse tangent function** by restricting the domain of $y = \tan x$ to the interval $(-\pi/2, \pi/2)$. The following list summarizes the definitions of the three most common inverse trigonometric functions. The remaining three are defined in Exercises 115–117.

Definitions of the Inverse Trigonometric Functions

Function	Domain	Range
$y = \arcsin x$ if and only if $\sin y = x$	$-1 \le x \le 1$	$-\frac{\pi}{2} \le y \le \frac{\pi}{2}$
$y = \arccos x$ if and only if $\cos y = x$	$-1 \le x \le 1$	$0 \le y \le \pi$
$y = \arctan x$ if and only if $\tan y = x$	$-\infty < x < \infty$	$-\frac{\pi}{2} < y < \frac{\pi}{2}$

The graphs of these three inverse trigonometric functions are shown below.

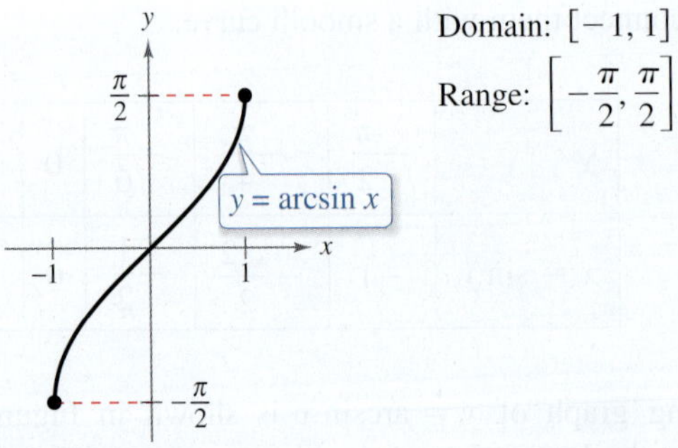

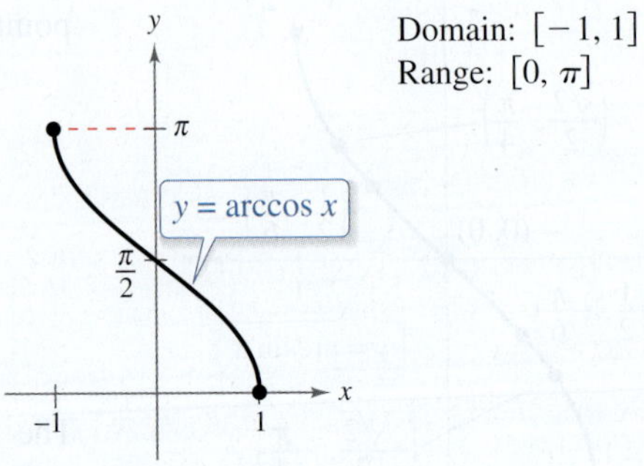

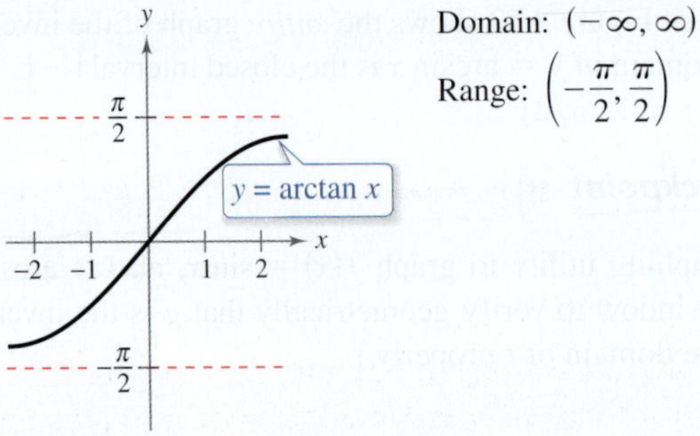

EXAMPLE 3 Evaluating Inverse Trigonometric Functions

Find the exact value.

a. $\arccos \dfrac{\sqrt{2}}{2}$

b. $\arctan 0$

c. $\tan^{-1}(-1)$

Solution

a. Because $\cos(\pi/4) = \sqrt{2}/2$ and $\pi/4$ lies in $[0, \pi]$, it follows that

$$\arccos \frac{\sqrt{2}}{2} = \frac{\pi}{4}. \qquad \text{Angle whose cosine is } \sqrt{2}/2$$

b. Because $\tan 0 = 0$ and 0 lies in $(-\pi/2, \pi/2)$, it follows that

$$\arctan 0 = 0. \qquad \text{Angle whose tangent is } 0$$

c. Because $\tan(-\pi/4) = -1$ and $-\pi/4$ lies in $(-\pi/2, \pi/2)$, it follows that

$$\tan^{-1}(-1) = -\frac{\pi}{4}. \qquad \text{Angle whose tangent is } -1$$

✓ ***Checkpoint*** *Audio-video solution in English & Spanish at LarsonPrecalculus.com.*

Find the exact value of $\cos^{-1}(-1)$.

EXAMPLE 4 Calculators and Inverse Trigonometric Functions

Use a calculator to approximate the value, if possible.

a. $\arctan(-8.45)$

b. $\sin^{-1} 0.2447$

c. $\arccos 2$

Solution

	Function	Mode	Calculator Keystrokes
a.	$\arctan(-8.45)$	Radian	TAN⁻¹ ((–) 8.45) ENTER

From the display, it follows that $\arctan(-8.45) \approx -1.453001$.

	Function	Mode	Calculator Keystrokes
b.	$\sin^{-1} 0.2447$	Radian	SIN⁻¹ (0.2447) ENTER

From the display, it follows that $\sin^{-1} 0.2447 \approx 0.2472103$.

	Function	Mode	Calculator Keystrokes
c.	$\arccos 2$	Radian	COS⁻¹ (2) ENTER

In *radian* mode, the calculator should display an *error message* because the domain of the inverse cosine function is $[-1, 1]$.

REMARK Remember that the domain of the inverse sine function and the inverse cosine function is $[-1, 1]$, as indicated in Example 4(c).

✓ ***Checkpoint*** *Audio-video solution in English & Spanish at LarsonPrecalculus.com.*

Use a calculator to approximate the value, if possible.

a. $\arctan 4.84$

b. $\arcsin(-1.1)$

c. $\arccos(-0.349)$

In Example 4, had you set the calculator to *degree* mode, the displays would have been in degrees rather than in radians. This convention is peculiar to calculators. By definition, the values of inverse trigonometric functions are *always in radians*.

Compositions of Functions

ALGEBRA HELP You can review the composition of functions in Section 1.8.

Recall from Section 1.9 that for all x in the domains of f and f^{-1}, inverse functions have the properties

$$f(f^{-1}(x)) = x \quad \text{and} \quad f^{-1}(f(x)) = x.$$

Inverse Properties of Trigonometric Functions

If $-1 \le x \le 1$ and $-\pi/2 \le y \le \pi/2$, then

$$\sin(\arcsin x) = x \quad \text{and} \quad \arcsin(\sin y) = y.$$

If $-1 \le x \le 1$ and $0 \le y \le \pi$, then

$$\cos(\arccos x) = x \quad \text{and} \quad \arccos(\cos y) = y.$$

If x is a real number and $-\pi/2 < y < \pi/2$, then

$$\tan(\arctan x) = x \quad \text{and} \quad \arctan(\tan y) = y.$$

Keep in mind that these inverse properties do not apply for arbitrary values of x and y. For instance,

$$\arcsin\left(\sin \frac{3\pi}{2}\right) = \arcsin(-1) = -\frac{\pi}{2} \ne \frac{3\pi}{2}.$$

In other words, the property $\arcsin(\sin y) = y$ is not valid for values of y outside the interval $[-\pi/2, \pi/2]$.

EXAMPLE 5 Using Inverse Properties

If possible, find the exact value.

a. $\tan[\arctan(-5)]$ **b.** $\arcsin\left(\sin \frac{5\pi}{3}\right)$ **c.** $\cos(\cos^{-1} \pi)$

Solution

a. Because -5 lies in the domain of the arctangent function, the inverse property applies, and you have

$$\tan[\arctan(-5)] = -5.$$

b. In this case, $5\pi/3$ does not lie in the range of the arcsine function, $-\pi/2 \le y \le \pi/2$. However, $5\pi/3$ is coterminal with

$$\frac{5\pi}{3} - 2\pi = -\frac{\pi}{3}$$

which does lie in the range of the arcsine function, and you have

$$\arcsin\left(\sin \frac{5\pi}{3}\right) = \arcsin\left[\sin\left(-\frac{\pi}{3}\right)\right] = -\frac{\pi}{3}.$$

c. The expression $\cos(\cos^{-1} \pi)$ is not defined because $\cos^{-1} \pi$ is not defined. Remember that the domain of the inverse cosine function is $[-1, 1]$.

✓ ***Checkpoint*** *Audio-video solution in English & Spanish at LarsonPrecalculus.com.*

If possible, find the exact value.

a. $\tan[\tan^{-1}(-14)]$ **b.** $\sin^{-1}\left(\sin \frac{7\pi}{4}\right)$ **c.** $\cos(\arccos 0.54)$

EXAMPLE 6 Evaluating Compositions of Functions

Find the exact value.

a. $\tan\left(\arccos \frac{2}{3}\right)$ **b.** $\cos\left[\arcsin\left(-\frac{3}{5}\right)\right]$

Solution

a. If you let $u = \arccos \frac{2}{3}$, then $\cos u = \frac{2}{3}$. Because the range of the inverse cosine function is the first and second quadrants and $\cos u$ is positive, u is a *first*-quadrant angle. You can sketch and label angle u, as shown in Figure 4.53. Consequently,

$$\tan\left(\arccos \frac{2}{3}\right) = \tan u = \frac{\text{opp}}{\text{adj}} = \frac{\sqrt{5}}{2}.$$

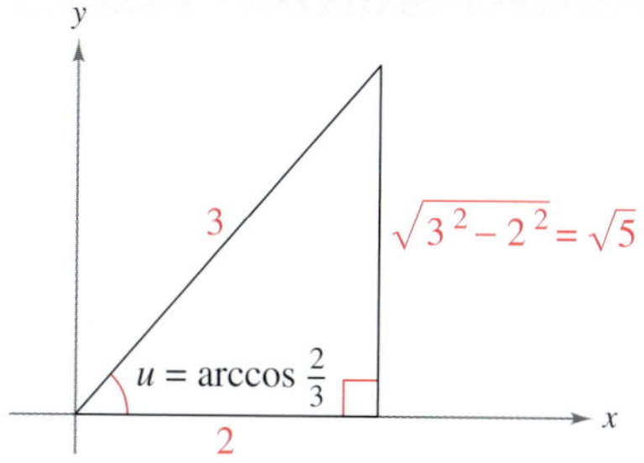

Angle whose cosine is $\frac{2}{3}$

Figure 4.53

b. If you let $u = \arcsin\left(-\frac{3}{5}\right)$, then $\sin u = -\frac{3}{5}$. Because the range of the inverse sine function is the first and fourth quadrants and $\sin u$ is negative, u is a *fourth*-quadrant angle. You can sketch and label angle u, as shown in Figure 4.54. Consequently,

$$\cos\left[\arcsin\left(-\frac{3}{5}\right)\right] = \cos u = \frac{\text{adj}}{\text{hyp}} = \frac{4}{5}.$$

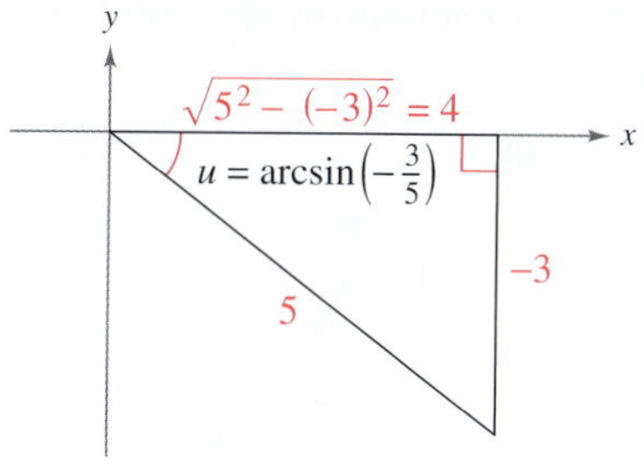

Angle whose sine is $-\frac{3}{5}$

Figure 4.54

✓ **Checkpoint** Audio-video solution in English & Spanish at LarsonPrecalculus.com.

Find the exact value of $\cos\left[\arctan\left(-\frac{3}{4}\right)\right]$.

EXAMPLE 7 Some Problems from Calculus

Write each of the following as an algebraic expression in x.

a. $\sin(\arccos 3x)$, $0 \le x \le \frac{1}{3}$ **b.** $\cot(\arccos 3x)$, $0 \le x < \frac{1}{3}$

Solution

If you let $u = \arccos 3x$, then $\cos u = 3x$, where $-1 \le 3x \le 1$. Because

$$\cos u = \frac{\text{adj}}{\text{hyp}} = \frac{3x}{1}$$

you can sketch a right triangle with acute angle u, as shown in Figure 4.55. From this triangle, you can easily convert each expression to algebraic form.

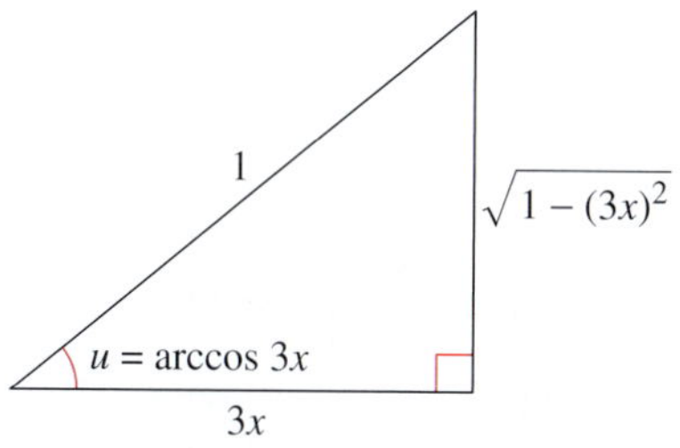

Angle whose cosine is $3x$

Figure 4.55

a. $\sin(\arccos 3x) = \sin u = \dfrac{\text{opp}}{\text{hyp}} = \sqrt{1 - 9x^2}, \quad 0 \le x \le \dfrac{1}{3}$

b. $\cot(\arccos 3x) = \cot u = \dfrac{\text{adj}}{\text{opp}} = \dfrac{3x}{\sqrt{1 - 9x^2}}, \quad 0 \le x < \dfrac{1}{3}$

✓ **Checkpoint** Audio-video solution in English & Spanish at LarsonPrecalculus.com.

Write $\sec(\arctan x)$ as an algebraic expression in x.

Summarize (Section 4.7)

1. State the definition of the inverse sine function *(page 320)*. For examples of evaluating and graphing the inverse sine function, see Examples 1 and 2.
2. State the definitions of the inverse cosine and inverse tangent functions *(page 322)*. For examples of evaluating and graphing inverse trigonometric functions, see Examples 3 and 4.
3. State the inverse properties of trigonometric functions *(page 324)*. For examples involving the compositions of trigonometric functions, see Examples 5–7.

4.7 Exercises

See **CalcChat.com** for tutorial help and worked-out solutions to odd-numbered exercises.

Vocabulary: Fill in the blanks.

	Function	Alternative Notation	Domain	Range
1.	$y = \arcsin x$	________	________	$-\frac{\pi}{2} \le y \le \frac{\pi}{2}$
2.	________	$y = \cos^{-1} x$	$-1 \le x \le 1$	________
3.	$y = \arctan x$	________	________	________

4. Without restrictions, no trigonometric function has an ________ function.

Skills and Applications

Evaluating an Inverse Trigonometric Function

In Exercises 5–18, evaluate the expression without using a calculator.

5. $\arcsin \frac{1}{2}$ **6.** $\arcsin 0$

7. $\arccos \frac{1}{2}$ **8.** $\arccos 0$

9. $\arctan \dfrac{\sqrt{3}}{3}$ **10.** $\arctan 1$

11. $\cos^{-1}\left(-\dfrac{\sqrt{3}}{2}\right)$ **12.** $\sin^{-1}\left(-\dfrac{\sqrt{2}}{2}\right)$

13. $\arctan(-\sqrt{3})$ **14.** $\arctan \sqrt{3}$

15. $\arccos\left(-\dfrac{1}{2}\right)$ **16.** $\arcsin \dfrac{\sqrt{2}}{2}$

17. $\sin^{-1}\left(-\dfrac{\sqrt{3}}{2}\right)$ **18.** $\tan^{-1}\left(-\dfrac{\sqrt{3}}{3}\right)$

Graphing an Inverse Trigonometric Function

In Exercises 19 and 20, use a graphing utility to graph *f*, *g*, and $y = x$ in the same viewing window to verify geometrically that *g* is the inverse function of *f*. (Be sure to restrict the domain of *f* properly.)

19. $f(x) = \cos x,\quad g(x) = \arccos x$

20. $f(x) = \tan x,\quad g(x) = \arctan x$

Calculators and Inverse Trigonometric Functions

In Exercises 21–38, use a calculator to evaluate the expression. Round your result to two decimal places.

21. $\arccos 0.37$ **22.** $\arcsin 0.65$

23. $\arcsin(-0.75)$ **24.** $\arccos(-0.7)$

25. $\arctan(-3)$ **26.** $\arctan 25$

27. $\sin^{-1} 0.31$ **28.** $\cos^{-1} 0.26$

29. $\arccos(-0.41)$ **30.** $\arcsin(-0.125)$

31. $\arctan 0.92$ **32.** $\arctan 2.8$

33. $\arcsin \frac{7}{8}$ **34.** $\arccos\left(-\frac{1}{3}\right)$

35. $\tan^{-1} \frac{19}{4}$ **36.** $\tan^{-1}\left(-\frac{95}{7}\right)$

37. $\tan^{-1}(-\sqrt{372})$ **38.** $\tan^{-1}(-\sqrt{2165})$

Finding Missing Coordinates

In Exercises 39 and 40, determine the missing coordinates of the points on the graph of the function.

39.

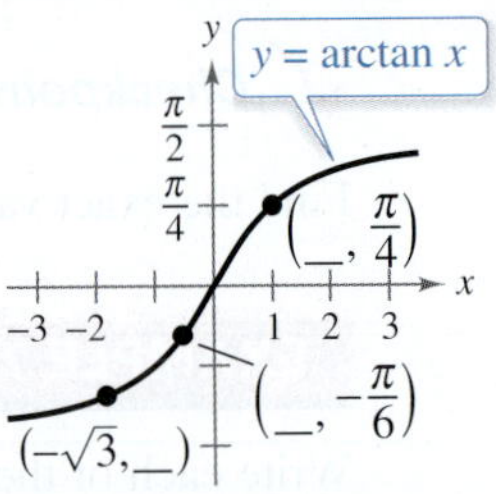

40.

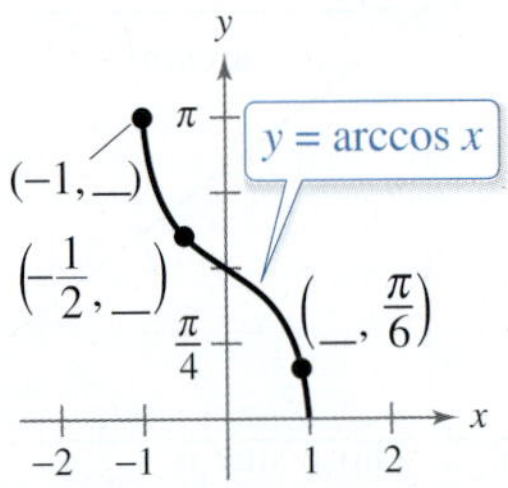

Using an Inverse Trigonometric Function

In Exercises 41–46, use an inverse trigonometric function to write θ as a function of x.

41.

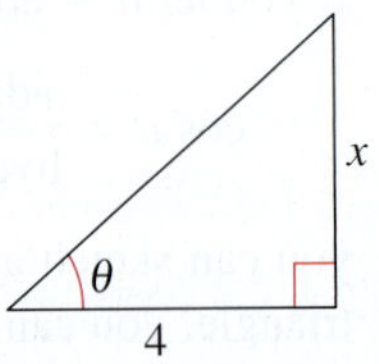

42.

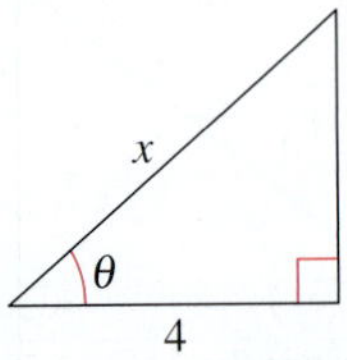

43.

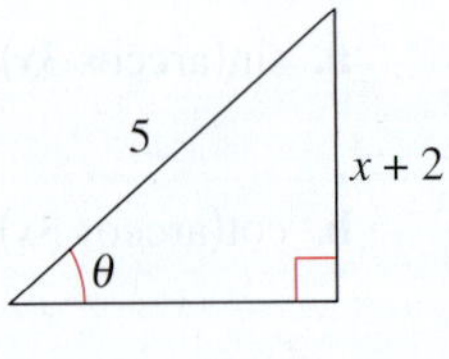

44.

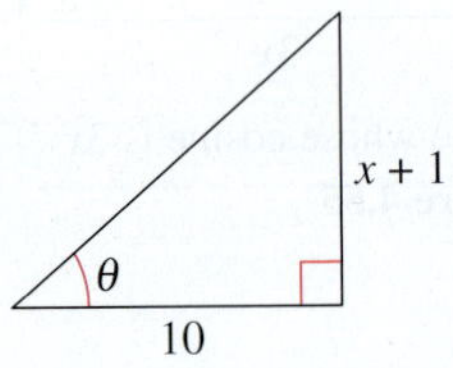

45.

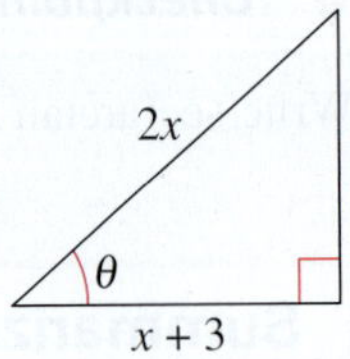

46.

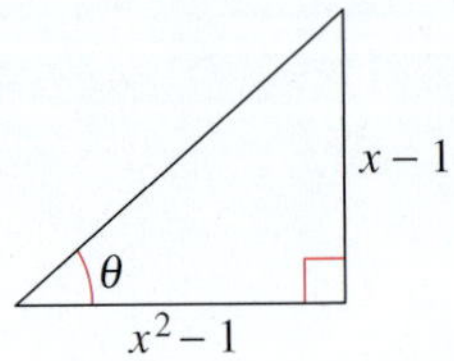

Using Inverse Properties

In Exercises 47–52, use the properties of inverse trigonometric functions to evaluate the expression.

47. $\sin(\arcsin 0.3)$ **48.** $\tan(\arctan 45)$

49. $\cos[\arccos(-0.1)]$ **50.** $\sin[\arcsin(-0.2)]$

51. $\arcsin(\sin 3\pi)$ **52.** $\arccos\left(\cos \dfrac{7\pi}{2}\right)$

Evaluating a Composition of Functions In Exercises 53–64, find the exact value of the expression. (*Hint:* Sketch a right triangle.)

53. $\sin\left(\arctan \frac{3}{4}\right)$
54. $\sec\left(\arcsin \frac{4}{5}\right)$
55. $\cos(\tan^{-1} 2)$
56. $\sin\left(\cos^{-1} \frac{\sqrt{5}}{5}\right)$
57. $\cos\left(\arcsin \frac{5}{13}\right)$
58. $\csc\left[\arctan\left(-\frac{5}{12}\right)\right]$
59. $\sec\left[\arctan\left(-\frac{3}{5}\right)\right]$
60. $\tan\left[\arcsin\left(-\frac{3}{4}\right)\right]$
61. $\sin\left[\arccos\left(-\frac{2}{3}\right)\right]$
62. $\cot\left(\arctan \frac{5}{8}\right)$
63. $\csc\left(\cos^{-1} \frac{\sqrt{3}}{2}\right)$
64. $\sec\left[\sin^{-1}\left(-\frac{\sqrt{2}}{2}\right)\right]$

Writing an Expression In Exercises 65–74, write an algebraic expression that is equivalent to the given expression. (*Hint:* Sketch a right triangle, as demonstrated in Example 7.)

65. $\cot(\arctan x)$
66. $\sin(\arctan x)$
67. $\cos(\arcsin 2x)$
68. $\sec(\arctan 3x)$
69. $\sin(\arccos x)$
70. $\sec[\arcsin(x - 1)]$
71. $\tan\left(\arccos \frac{x}{3}\right)$
72. $\cot\left(\arctan \frac{1}{x}\right)$
73. $\csc\left(\arctan \frac{x}{\sqrt{2}}\right)$
74. $\cos\left(\arcsin \frac{x - h}{r}\right)$

Graphical Analysis In Exercises 75 and 76, use a graphing utility to graph f and g in the same viewing window to verify that the two functions are equal. Explain why they are equal. Identify any asymptotes of the graphs.

75. $f(x) = \sin(\arctan 2x), \quad g(x) = \dfrac{2x}{\sqrt{1 + 4x^2}}$
76. $f(x) = \tan\left(\arccos \dfrac{x}{2}\right), \quad g(x) = \dfrac{\sqrt{4 - x^2}}{x}$

Completing an Equation In Exercises 77–80, complete the equation.

77. $\arctan \dfrac{9}{x} = \arcsin(\ \_\_\_\_\ ), \quad x > 0$
78. $\arcsin \dfrac{\sqrt{36 - x^2}}{6} = \arccos(\ \_\_\_\_\ ), \quad 0 \le x \le 6$
79. $\arccos \dfrac{3}{\sqrt{x^2 - 2x + 10}} = \arcsin(\ \_\_\_\_\ )$
80. $\arccos \dfrac{x - 2}{2} = \arctan(\ \_\_\_\_\ ), \quad 2 < x < 4$

Comparing Graphs In Exercises 81 and 82, sketch a graph of the function and compare the graph of g with the graph of $f(x) = \arcsin x$.

81. $g(x) = \arcsin(x - 1)$
82. $g(x) = \arcsin \dfrac{x}{2}$

Sketching the Graph of a Function In Exercises 83–88, sketch a graph of the function.

83. $y = 2 \arccos x$
84. $g(t) = \arccos(t + 2)$
85. $f(x) = \arctan 2x$
86. $f(x) = \dfrac{\pi}{2} + \arctan x$
87. $h(v) = \arccos \dfrac{v}{2}$
88. $f(x) = \arccos \dfrac{x}{4}$

Graphing an Inverse Trigonometric Function In Exercises 89–94, use a graphing utility to graph the function.

89. $f(x) = 2 \arccos(2x)$
90. $f(x) = \pi \arcsin(4x)$
91. $f(x) = \arctan(2x - 3)$
92. $f(x) = -3 + \arctan(\pi x)$
93. $f(x) = \pi - \sin^{-1}\left(\dfrac{2}{3}\right)$
94. $f(x) = \dfrac{\pi}{2} + \cos^{-1}\left(\dfrac{1}{\pi}\right)$

Using a Trigonometric Identity In Exercises 95 and 96, write the function in terms of the sine function by using the identity

$$A \cos \omega t + B \sin \omega t = \sqrt{A^2 + B^2} \sin\left(\omega t + \arctan \frac{A}{B}\right).$$

Use a graphing utility to graph both forms of the function. What does the graph imply?

95. $f(t) = 3 \cos 2t + 3 \sin 2t$
96. $f(t) = 4 \cos \pi t + 3 \sin \pi t$

Behavior of an Inverse Trigonometric Function In Exercises 97–102, fill in the blank. If not possible, state the reason.

97. As $x \to 1^-$, the value of $\arcsin x \to$ ____.
98. As $x \to 1^-$, the value of $\arccos x \to$ ____.
99. As $x \to \infty$, the value of $\arctan x \to$ ____.
100. As $x \to -1^+$, the value of $\arcsin x \to$ ____.
101. As $x \to -1^+$, the value of $\arccos x \to$ ____.
102. As $x \to -\infty$, the value of $\arctan x \to$ ____.

103. **Docking a Boat** A boat is pulled in by means of a winch located on a dock 5 feet above the deck of the boat (see figure). Let θ be the angle of elevation from the boat to the winch and let s be the length of the rope from the winch to the boat.

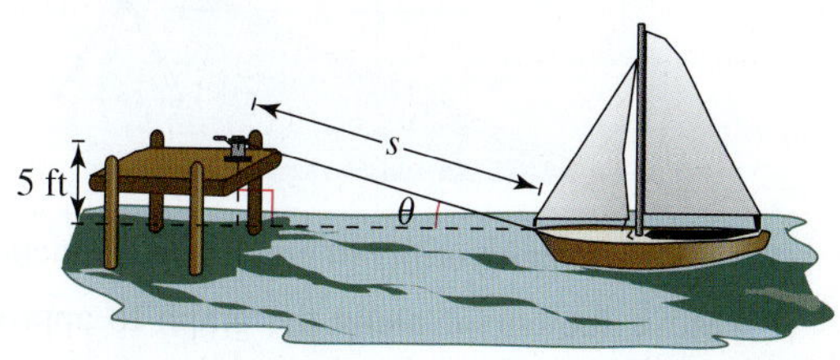

(a) Write θ as a function of s.

(b) Find θ when $s = 40$ feet and $s = 20$ feet.

104. Photography A television camera at ground level is filming the lift-off of a space shuttle at a point 750 meters from the launch pad (see figure). Let θ be the angle of elevation to the shuttle and let s be the height of the shuttle.

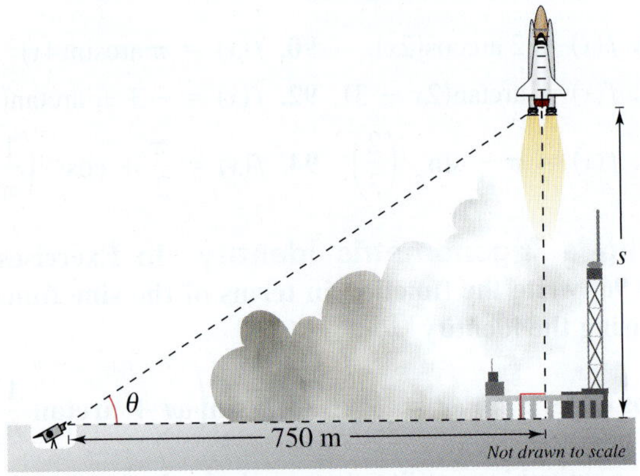

(a) Write θ as a function of s.

(b) Find θ when $s = 300$ meters and $s = 1200$ meters.

105. Photography A photographer is taking a picture of a three-foot-tall painting hung in an art gallery. The camera lens is 1 foot below the lower edge of the painting (see figure). The angle β subtended by the camera lens x feet from the painting is given by

$$\beta = \arctan \frac{3x}{x^2 + 4}, \quad x > 0.$$

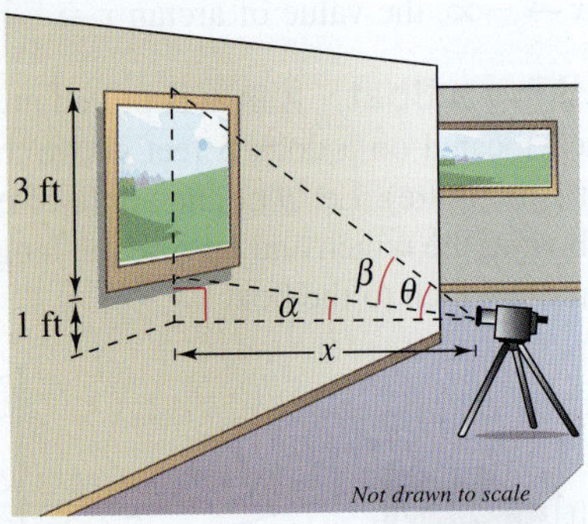

(a) Use a graphing utility to graph β as a function of x.

(b) Move the cursor along the graph to approximate the distance from the picture when β is maximum.

(c) Identify the asymptote of the graph and discuss its meaning in the context of the problem.

106. Granular Angle of Repose Different types of granular substances naturally settle at different angles when stored in cone-shaped piles. This angle θ is called the *angle of repose* (see figure). When rock salt is stored in a cone-shaped pile 11 feet high, the diameter of the pile's base is about 34 feet. *(Source: Bulk-Store Structures, Inc.)*

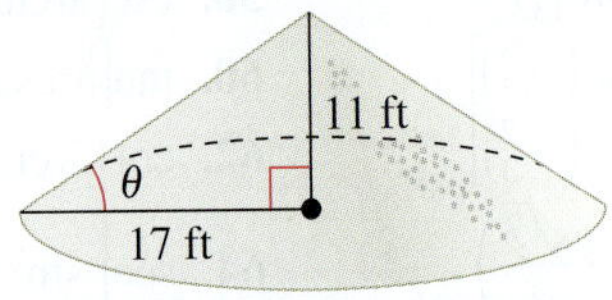

(a) Find the angle of repose for rock salt.

(b) How tall is a pile of rock salt that has a base diameter of 40 feet?

107. Granular Angle of Repose When whole corn is stored in a cone-shaped pile 20 feet high, the diameter of the pile's base is about 82 feet.

(a) Find the angle of repose for whole corn.

(b) How tall is a pile of corn that has a base diameter of 100 feet?

108. Angle of Elevation An airplane flies at an altitude of 6 miles toward a point directly over an observer. Consider θ and x as shown in the figure.

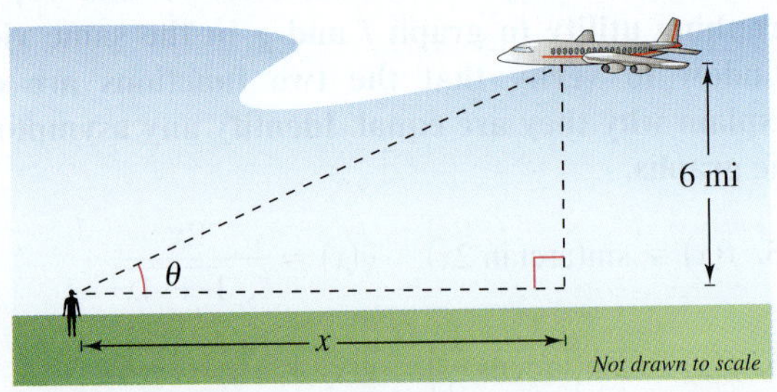

(a) Write θ as a function of x.

(b) Find θ when $x = 7$ miles and $x = 1$ mile.

109. Security Patrol A security car with its spotlight on is parked 20 meters from a warehouse. Consider θ and x as shown in the figure.

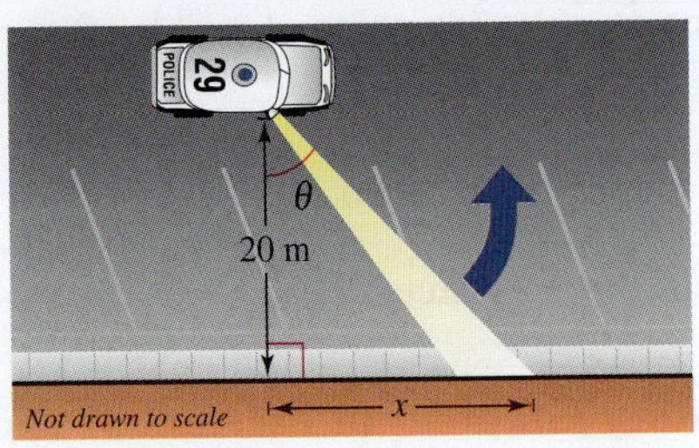

(a) Write θ as a function of x.

(b) Find θ when $x = 5$ meters and $x = 12$ meters.

NASA

Exploration

True or False? **In Exercises 110–113, determine whether the statement is true or false. Justify your answer.**

110. $\sin \dfrac{5\pi}{6} = \dfrac{1}{2} \Rightarrow \arcsin \dfrac{1}{2} = \dfrac{5\pi}{6}$

111. $\tan \dfrac{5\pi}{4} = 1 \Rightarrow \arctan 1 = \dfrac{5\pi}{4}$

112. $\arctan x = \dfrac{\arcsin x}{\arccos x}$ **113.** $\sin^{-1} x = \dfrac{1}{\sin x}$

114. **HOW DO YOU SEE IT?** Use the figure below to determine the value(s) of x for which each statement is true.

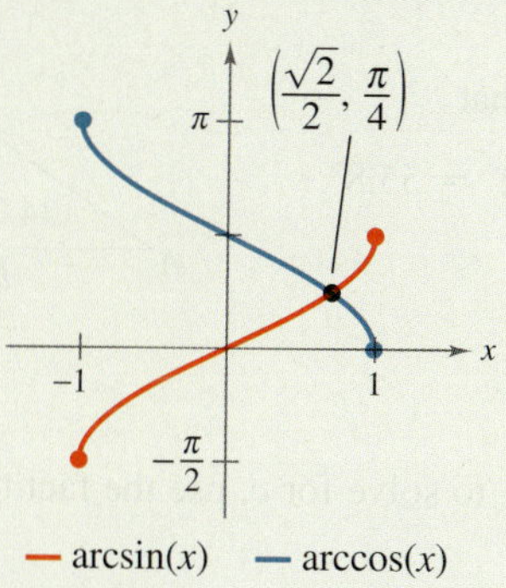

(a) $\arcsin x < \arccos x$

(b) $\arcsin x = \arccos x$

(c) $\arcsin x > \arccos x$

115. Inverse Cotangent Function Define the inverse cotangent function by restricting the domain of the cotangent function to the interval $(0, \pi)$, and sketch the graph of the inverse trigonometric function.

116. Inverse Secant Function Define the inverse secant function by restricting the domain of the secant function to the intervals $[0, \pi/2)$ and $(\pi/2, \pi]$, and sketch the graph of the inverse trigonometric function.

117. Inverse Cosecant Function Define the inverse cosecant function by restricting the domain of the cosecant function to the intervals $[-\pi/2, 0)$ and $(0, \pi/2]$, and sketch the graph of the inverse trigonometric function.

118. Writing Use the results of Exercises 115–117 to explain how to graph (a) the inverse cotangent function, (b) the inverse secant function, and (c) the inverse cosecant function on a graphing utility.

Evaluating an Inverse Trigonometric Function **In Exercises 119–126, use the results of Exercises 115–117 to evaluate the expression without using a calculator.**

119. $\operarcsec \sqrt{2}$ **120.** $\operatorname{arcsec} 1$

121. $\operatorname{arccot}(-1)$ **122.** $\operatorname{arccot}(-\sqrt{3})$

123. $\operatorname{arccsc} 2$ **124.** $\operatorname{arccsc}(-1)$

125. $\operatorname{arccsc}\left(\dfrac{2\sqrt{3}}{3}\right)$ **126.** $\operatorname{arcsec}\left(-\dfrac{2\sqrt{3}}{3}\right)$

Calculators and Inverse Trigonometric Functions **In Exercises 127–134, use the results of Exercises 115–117 and a calculator to approximate the value of the expression. Round your result to two decimal places.**

127. $\operatorname{arcsec} 2.54$ **128.** $\operatorname{arcsec}(-1.52)$

129. $\operatorname{arccot} 5.25$ **130.** $\operatorname{arccot}(-10)$

131. $\operatorname{arccot} \frac{5}{3}$ **132.** $\operatorname{arccot}(-\frac{16}{7})$

133. $\operatorname{arccsc}(-\frac{25}{3})$ **134.** $\operatorname{arccsc}(-12)$

135. Area In calculus, it is shown that the area of the region bounded by the graphs of $y = 0$, $y = 1/(x^2 + 1)$, $x = a$, and $x = b$ is given by

Area $= \arctan b - \arctan a$

(see figure). Find the area for the following values of a and b.

(a) $a = 0, b = 1$ (b) $a = -1, b = 1$

(c) $a = 0, b = 3$ (d) $a = -1, b = 3$

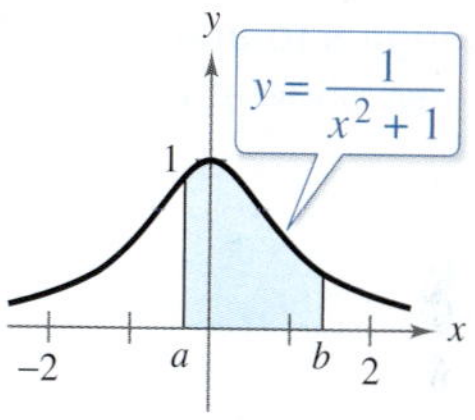

136. Think About It Use a graphing utility to graph the functions $f(x) = \sqrt{x}$ and $g(x) = 6 \arctan x$. For $x > 0$, it appears that $g > f$. Explain why you know that there exists a positive real number a such that $g < f$ for $x > a$. Approximate the number a.

137. Think About It Consider the functions $f(x) = \sin x$ and $f^{-1}(x) = \arcsin x$.

(a) Use a graphing utility to graph the composite functions $f \circ f^{-1}$ and $f^{-1} \circ f$.

(b) Explain why the graphs in part (a) are not the graph of the line $y = x$. Why do the graphs of $f \circ f^{-1}$ and $f^{-1} \circ f$ differ?

138. Proof Prove each identity.

(a) $\arcsin(-x) = -\arcsin x$

(b) $\arctan(-x) = -\arctan x$

(c) $\arctan x + \arctan \dfrac{1}{x} = \dfrac{\pi}{2}, \quad x > 0$

(d) $\arcsin x + \arccos x = \dfrac{\pi}{2}$

(e) $\arcsin x = \arctan \dfrac{x}{\sqrt{1 - x^2}}$

4.8 Applications and Models

- Solve real-life problems involving right triangles.
- Solve real-life problems involving directional bearings.
- Solve real-life problems involving harmonic motion.

Right triangles often occur in real-life situations. For instance, in Exercise 32 on page 337, you will use right triangles to analyze the design of a new slide at a water park.

Applications Involving Right Triangles

In this section, the three angles of a right triangle are denoted by the letters A, B, and C (where C is the right angle), and the lengths of the sides opposite these angles by the letters a, b, and c, respectively (where c is the hypotenuse).

EXAMPLE 1 Solving a Right Triangle

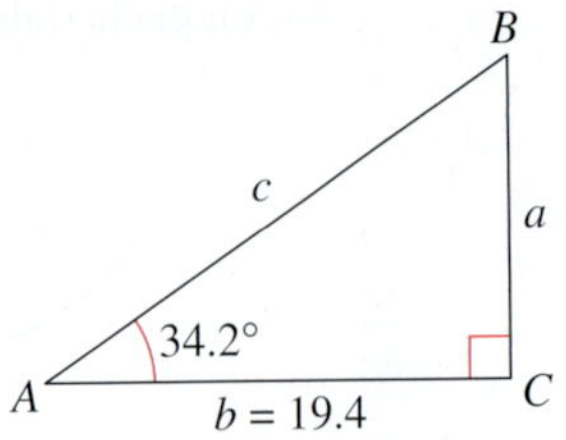

Solve the right triangle shown at the right for all unknown sides and angles.

Solution Because $C = 90°$, it follows that

$$A + B = 90° \quad \text{and} \quad B = 90° - 34.2° = 55.8°.$$

To solve for a, use the fact that

$$\tan A = \frac{\text{opp}}{\text{adj}} = \frac{a}{b} \Rightarrow a = b\tan A.$$

So, $a = 19.4 \tan 34.2° \approx 13.18$. Similarly, to solve for c, use the fact that

$$\cos A = \frac{\text{adj}}{\text{hyp}} = \frac{b}{c} \Rightarrow c = \frac{b}{\cos A}.$$

So, $c = \dfrac{19.4}{\cos 34.2°} \approx 23.46$.

✓ **Checkpoint** Audio-video solution in English & Spanish at LarsonPrecalculus.com.

Solve the right triangle shown in Figure 4.56 for all unknown sides and angles.

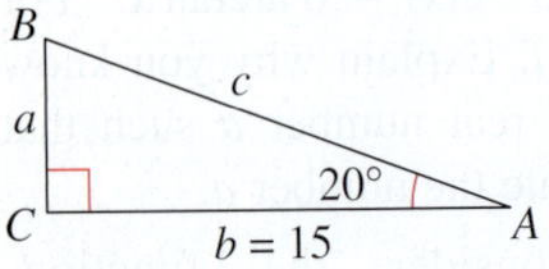

Figure 4.56

EXAMPLE 2 Finding a Side of a Right Triangle

A safety regulation states that the maximum angle of elevation for a rescue ladder is 72°. A fire department's longest ladder is 110 feet. What is the maximum safe rescue height?

Solution A sketch is shown in Figure 4.57. From the equation $\sin A = a/c$, it follows that

$$\begin{aligned} a &= c\sin A \\ &= 110 \sin 72° \\ &\approx 104.6. \end{aligned}$$

So, the maximum safe rescue height is about 104.6 feet above the height of the fire truck.

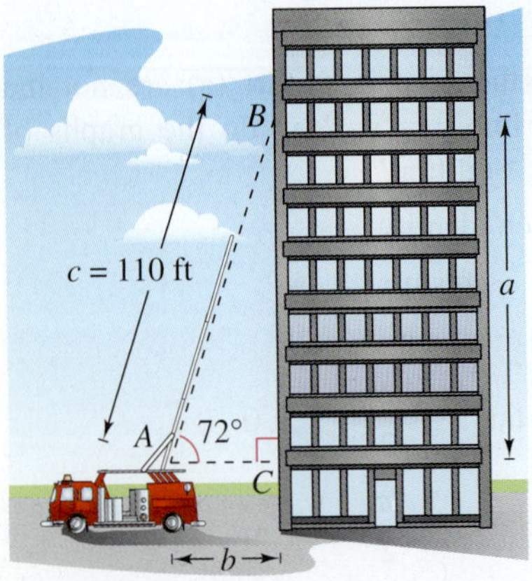

Figure 4.57

✓ **Checkpoint** Audio-video solution in English & Spanish at LarsonPrecalculus.com.

A ladder that is 16 feet long leans against the side of a house. The angle of elevation of the ladder is 80°. Find the height from the top of the ladder to the ground. ■

Alexey Bykov/Shutterstock.com

EXAMPLE 3 Finding a Side of a Right Triangle

At a point 200 feet from the base of a building, the angle of elevation to the *bottom* of a smokestack is 35°, whereas the angle of elevation to the *top* is 53°, as shown in Figure 4.58. Find the height s of the smokestack alone.

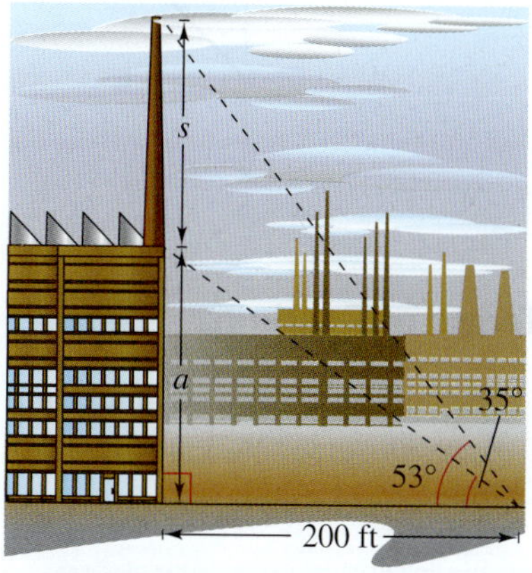

Figure 4.58

Solution

Note from Figure 4.58 that this problem involves two right triangles. For the smaller right triangle, use the fact that

$$\tan 35° = \frac{a}{200}$$

to conclude that the height of the building is

$$a = 200 \tan 35°.$$

For the larger right triangle, use the equation

$$\tan 53° = \frac{a + s}{200}$$

to conclude that $a + s = 200 \tan 53°$. So, the height of the smokestack is

$$s = 200 \tan 53° - a$$

$$= 200 \tan 53° - 200 \tan 35°$$

$$\approx 125.4 \text{ feet.}$$

✓ **Checkpoint** Audio-video solution in English & Spanish at LarsonPrecalculus.com.

At a point 65 feet from the base of a church, the angles of elevation to the bottom of the steeple and the top of the steeple are 35° and 43°, respectively. Find the height of the steeple.

EXAMPLE 4 Finding an Acute Angle of a Right Triangle

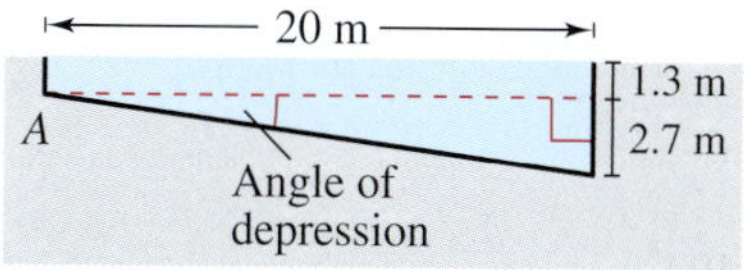

Figure 4.59

A swimming pool is 20 meters long and 12 meters wide. The bottom of the pool is slanted so that the water depth is 1.3 meters at the shallow end and 4 meters at the deep end, as shown in Figure 4.59. Find the angle of depression (in degrees) of the bottom of the pool.

Solution Using the tangent function, you can see that

$$\tan A = \frac{\text{opp}}{\text{adj}}$$

$$= \frac{2.7}{20}$$

$$= 0.135.$$

So, the angle of depression is

$$A = \arctan 0.135$$

$$\approx 0.13419 \text{ radian}$$

$$\approx 7.69°.$$

✓ **Checkpoint** Audio-video solution in English & Spanish at LarsonPrecalculus.com.

From the time a small airplane is 100 feet high and 1600 ground feet from its landing runway, the plane descends in a straight line to the runway. Determine the angle of descent (in degrees) of the plane.

Trigonometry and Bearings

In surveying and navigation, directions can be given in terms of **bearings.** A bearing measures the acute angle that a path or line of sight makes with a fixed north-south line. For instance, the bearing S 35° E, shown below, means 35 degrees east of south.

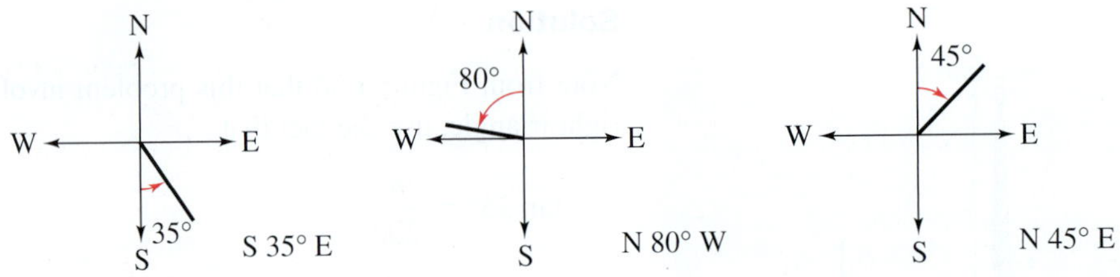

REMARK In *air navigation*, bearings are measured in degrees *clockwise* from north. Examples of air navigation bearings are shown below.

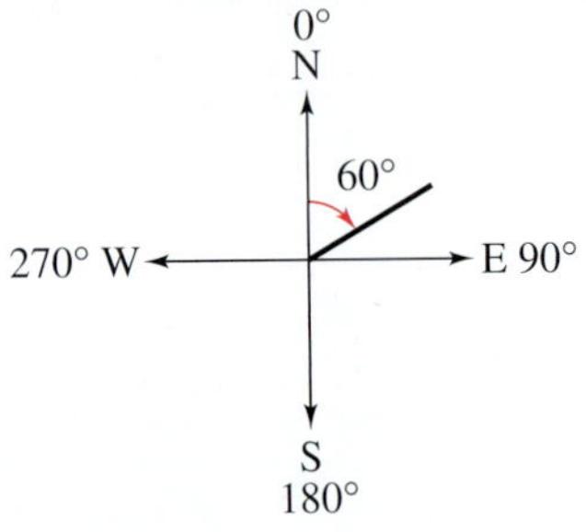

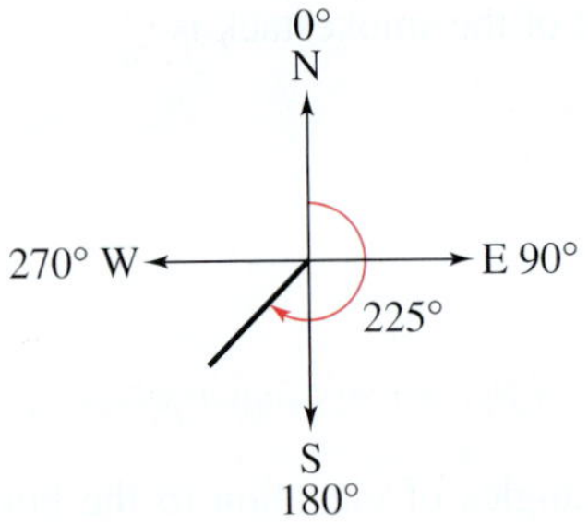

EXAMPLE 5 Finding Directions in Terms of Bearings

A ship leaves port at noon and heads due west at 20 knots, or 20 nautical miles (nm) per hour. At 2 P.M. the ship changes course to N 54° W, as shown below. Find the ship's bearing and distance from the port of departure at 3 P.M.

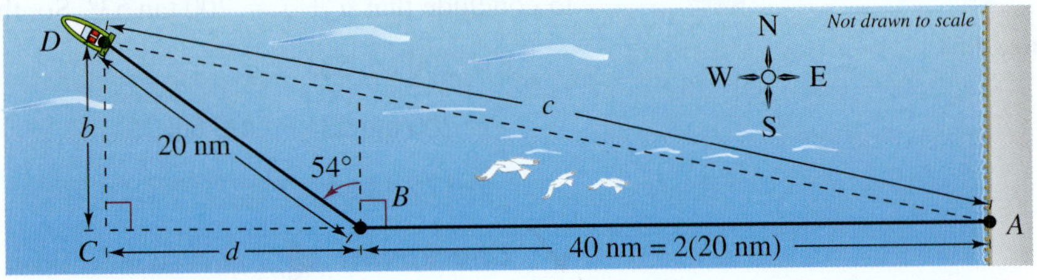

Solution

For triangle *BCD*, you have

$$B = 90° - 54° = 36°.$$

The two sides of this triangle can be determined to be

$$b = 20 \sin 36° \quad \text{and} \quad d = 20 \cos 36°.$$

For triangle *ACD*, you can find angle *A* as follows.

$$\tan A = \frac{b}{d + 40} = \frac{20 \sin 36°}{20 \cos 36° + 40} \approx 0.2092494$$

$$A \approx \arctan 0.2092494 \approx 0.2062732 \text{ radian} \approx 11.82°$$

The angle with the north-south line is $90° - 11.82° = 78.18°$. So, the bearing of the ship is N 78.18° W. Finally, from triangle *ACD*, you have $\sin A = b/c$, which yields

$$c = \frac{b}{\sin A}$$

$$= \frac{20 \sin 36°}{\sin 11.82°}$$

$$\approx 57.4 \text{ nautical miles.}$$ Distance from port

✓ **Checkpoint** Audio-video solution in English & Spanish at LarsonPrecalculus.com.

A sailboat leaves a pier heading due west at 8 knots. After 15 minutes, the sailboat tacks, changing course to N 16° W at 10 knots. Find the sailboat's distance and bearing from the pier after 12 minutes on this course.

Harmonic Motion

The periodic nature of the trigonometric functions is useful for describing the motion of a point on an object that vibrates, oscillates, rotates, or is moved by wave motion.

For example, consider a ball that is bobbing up and down on the end of a spring, as shown in Figure 4.60. Suppose that 10 centimeters is the maximum distance the ball moves vertically upward or downward from its equilibrium (at rest) position. Suppose further that the time it takes for the ball to move from its maximum displacement above zero to its maximum displacement below zero and back again is $t = 4$ seconds. Assuming the ideal conditions of perfect elasticity and no friction or air resistance, the ball would continue to move up and down in a uniform and regular manner.

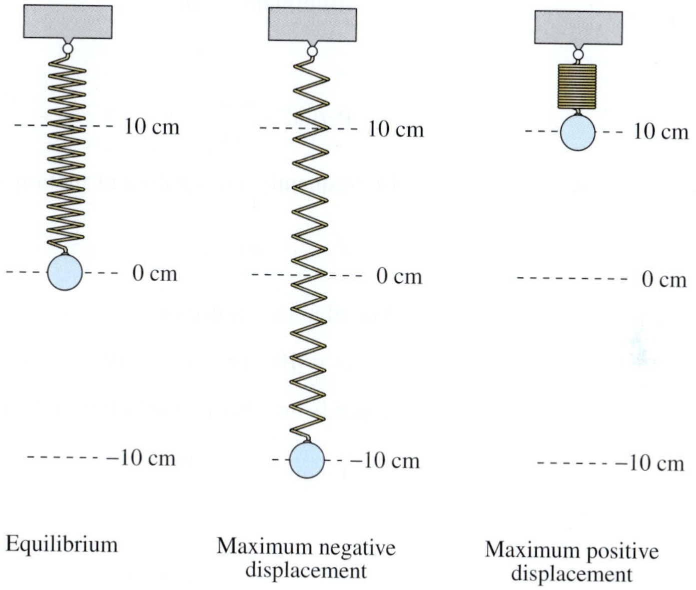

Figure 4.60

From this spring you can conclude that the period (time for one complete cycle) of the motion is

$$\text{Period} = 4 \text{ seconds}$$

its amplitude (maximum displacement from equilibrium) is

$$\text{Amplitude} = 10 \text{ centimeters}$$

and its **frequency** (number of cycles per second) is

$$\text{Frequency} = \frac{1}{4} \text{ cycle per second.}$$

Motion of this nature can be described by a sine or cosine function and is called **simple harmonic motion.**

Definition of Simple Harmonic Motion

A point that moves on a coordinate line is in **simple harmonic motion** when its distance d from the origin at time t is given by either

$$d = a \sin \omega t \quad \text{or} \quad d = a \cos \omega t$$

where a and ω are real numbers such that $\omega > 0$. The motion has amplitude $|a|$, period $\frac{2\pi}{\omega}$, and frequency $\frac{\omega}{2\pi}$.

EXAMPLE 6 Simple Harmonic Motion

Write an equation for the simple harmonic motion of the ball described in Figure 4.60, where the period is 4 seconds. What is the frequency of this harmonic motion?

Solution

Because the spring is at equilibrium ($d = 0$) when $t = 0$, use the equation

$$d = a \sin \omega t.$$

Moreover, because the maximum displacement from zero is 10 and the period is 4, you have the following.

$$\text{Amplitude} = |a| = 10$$

$$\text{Period} = \frac{2\pi}{\omega} = 4 \Longrightarrow \omega = \frac{\pi}{2}$$

Consequently, an equation of motion is

$$d = 10 \sin \frac{\pi}{2} t.$$

Note that the choice of

$$a = 10 \quad \text{or} \quad a = -10$$

depends on whether the ball initially moves up or down. The frequency is

$$\text{Frequency} = \frac{\omega}{2\pi} = \frac{\pi/2}{2\pi} = \frac{1}{4} \text{ cycle per second.}$$

✓ Checkpoint Audio-video solution in English & Spanish at LarsonPrecalculus.com.

Find a model for simple harmonic motion that satisfies the following conditions: $d = 0$ when $t = 0$, the amplitude is 6 centimeters, and the period is 3 seconds. Then find the frequency.

One illustration of the relationship between sine waves and harmonic motion is in the wave motion that results when a stone is dropped into a calm pool of water. The waves move outward in roughly the shape of sine (or cosine) waves, as shown in Figure 4.61. As an example, suppose you are fishing and your fishing bobber is attached so that it does not move horizontally. As the waves move outward from the dropped stone, your fishing bobber will move up and down in simple harmonic motion, as shown in Figure 4.62.

Figure 4.61

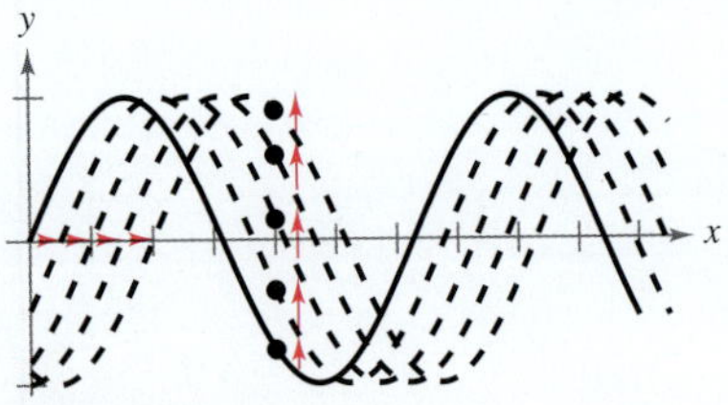

Figure 4.62

EXAMPLE 7 **Simple Harmonic Motion**

Given the equation for simple harmonic motion $d = 6 \cos \frac{3\pi}{4} t$, find (a) the maximum displacement, (b) the frequency, (c) the value of d when $t = 4$, and (d) the least positive value of t for which $d = 0$.

Algebraic Solution

The given equation has the form $d = a \cos \omega t$, with $a = 6$ and $\omega = 3\pi/4$.

a. The maximum displacement (from the point of equilibrium) is given by the amplitude. So, the maximum displacement is 6.

b.
$$\text{Frequency} = \frac{\omega}{2\pi} = \frac{3\pi/4}{2\pi} = \frac{3}{8} \text{ cycle per unit of time}$$

c.
$$d = 6 \cos\left[\frac{3\pi}{4}(4)\right] = 6 \cos 3\pi = 6(-1) = -6$$

d. To find the least positive value of t for which $d = 0$, solve the equation

$$6 \cos \frac{3\pi}{4} t = 0.$$

First divide each side by 6 to obtain

$$\cos \frac{3\pi}{4} t = 0.$$

This equation is satisfied when

$$\frac{3\pi}{4} t = \frac{\pi}{2}, \frac{3\pi}{2}, \frac{5\pi}{2}, \ldots.$$

Multiply these values by $4/(3\pi)$ to obtain

$$t = \frac{2}{3}, 2, \frac{10}{3}, \ldots.$$

So, the least positive value of t is $t = \frac{2}{3}$.

Graphical Solution

Use a graphing utility set in *radian* mode.

a.

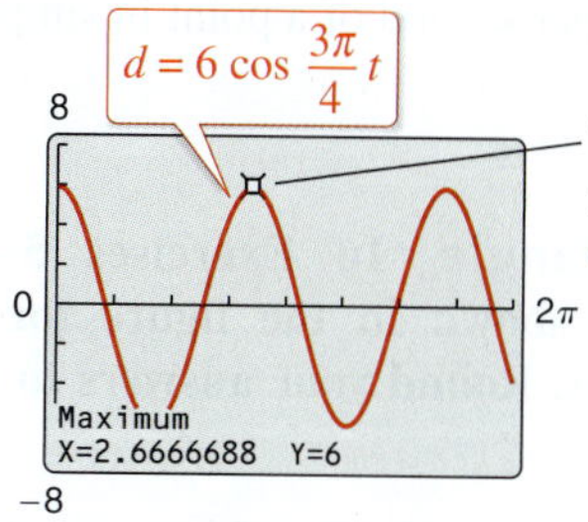

b.

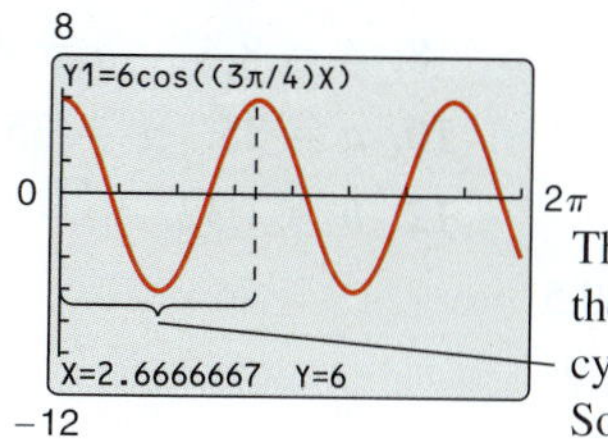

c.

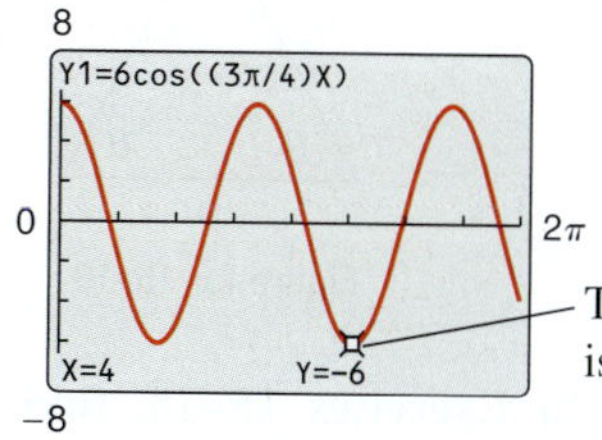

d.

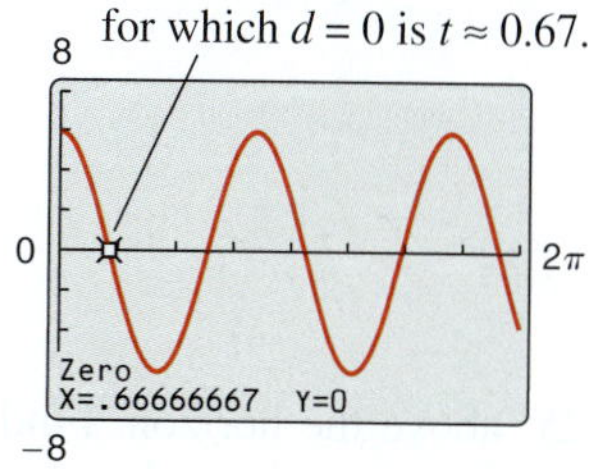

✓ Checkpoint *Audio-video solution in English & Spanish at LarsonPrecalculus.com.*

Rework Example 7 for the equation $d = 4 \cos 6\pi t$.

Summarize **(Section 4.8)**

1. Describe real-life problems that can be solved using right triangles *(pages 330 and 331, Examples 1–4)*.
2. State the definition of a bearing *(page 332, Example 5)*.
3. State the definition of simple harmonic motion *(page 333, Examples 6 and 7)*.

4.8 Exercises

See CalcChat.com for tutorial help and worked-out solutions to odd-numbered exercises.

Vocabulary: Fill in the blanks.

1. A ________ measures the acute angle that a path or line of sight makes with a fixed north-south line.
2. A point that moves on a coordinate line is said to be in simple ________ ________ when its distance d from the origin at time t is given by either $d = a \sin \omega t$ or $d = a \cos \omega t$.
3. The time for one complete cycle of a point in simple harmonic motion is its ________.
4. The number of cycles per second of a point in simple harmonic motion is its ________.

Skills and Applications

Solving a Right Triangle In Exercises 5–14, solve the right triangle shown in the figure for all unknown sides and angles. Round your answers to two decimal places.

5. $A = 30°, \quad b = 3$
6. $B = 54°, \quad c = 15$
7. $B = 71°, \quad b = 24$
8. $A = 8.4°, \quad a = 40.5$
9. $a = 3, \quad b = 4$
10. $a = 25, \quad c = 35$
11. $b = 16, \quad c = 52$
12. $b = 1.32, \quad c = 9.45$
13. $A = 12°15', \quad c = 430.5$
14. $B = 65°12', \quad a = 14.2$

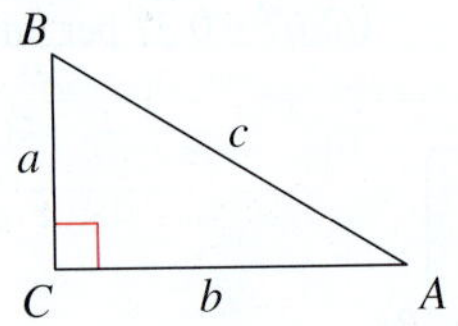

Figure for 5–14

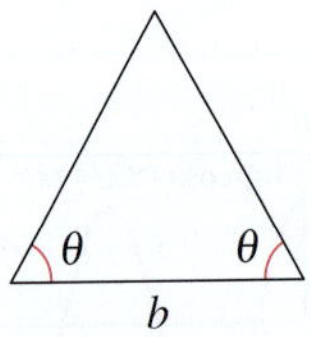

Figure for 15–18

Finding an Altitude In Exercises 15–18, find the altitude of the isosceles triangle shown in the figure. Round your answers to two decimal places.

15. $\theta = 45°, \quad b = 6$
16. $\theta = 18°, \quad b = 10$
17. $\theta = 32°, \quad b = 8$
18. $\theta = 27°, \quad b = 11$

19. **Length** The sun is 25° above the horizon. Find the length of a shadow cast by a building that is 100 feet tall (see figure).

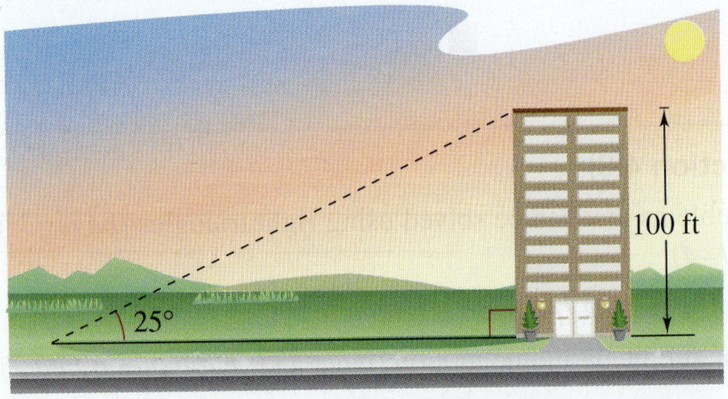

20. **Length** The sun is 20° above the horizon. Find the length of a shadow cast by a park statue that is 12 feet tall.
21. **Height** A ladder that is 20 feet long leans against the side of a house. The angle of elevation of the ladder is 80°. Find the height from the top of the ladder to the ground.
22. **Height** The length of a shadow of a tree is 125 feet when the angle of elevation of the sun is 33°. Approximate the height of the tree.
23. **Height** At a point 50 feet from the base of a church, the angles of elevation to the bottom of the steeple and the top of the steeple are 35° and 47° 40′, respectively. Find the height of the steeple.
24. **Distance** An observer in a lighthouse 350 feet above sea level observes two ships directly offshore. The angles of depression to the ships are 4° and 6.5° (see figure). How far apart are the ships?

25. **Distance** A passenger in an airplane at an altitude of 10 kilometers sees two towns directly to the east of the plane. The angles of depression to the towns are 28° and 55° (see figure). How far apart are the towns?

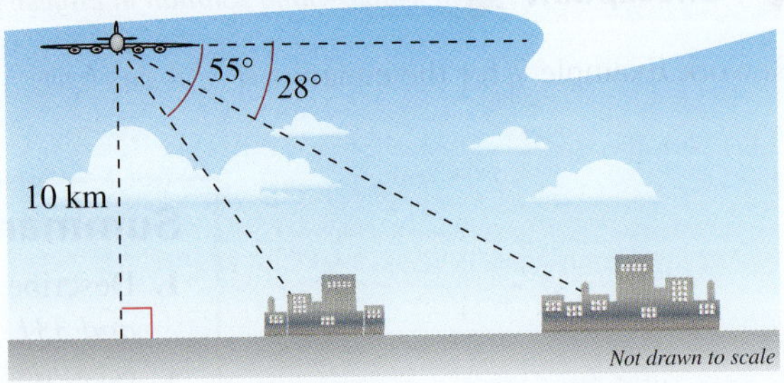

26. Altitude You observe a plane approaching overhead and assume that its speed is 550 miles per hour. The angle of elevation of the plane is 16° at one time and 57° one minute later. Approximate the altitude of the plane.

27. Angle of Elevation An engineer erects a 75-foot cellular telephone tower. Find the angle of elevation to the top of the tower at a point on level ground 50 feet from its base.

28. Angle of Elevation The height of an outdoor basketball backboard is $12\frac{1}{2}$ feet, and the backboard casts a shadow $17\frac{1}{3}$ feet long.

(a) Draw a right triangle that gives a visual representation of the problem. Label the known and unknown quantities.

(b) Use a trigonometric function to write an equation involving the unknown angle of elevation.

(c) Find the angle of elevation of the sun.

29. Angle of Depression A cellular telephone tower that is 150 feet tall is placed on top of a mountain that is 1200 feet above sea level. What is the angle of depression from the top of the tower to a cell phone user who is 5 horizontal miles away and 400 feet above sea level?

30. Angle of Depression A Global Positioning System satellite orbits 12,500 miles above Earth's surface (see figure). Find the angle of depression from the satellite to the horizon. Assume the radius of Earth is 4000 miles.

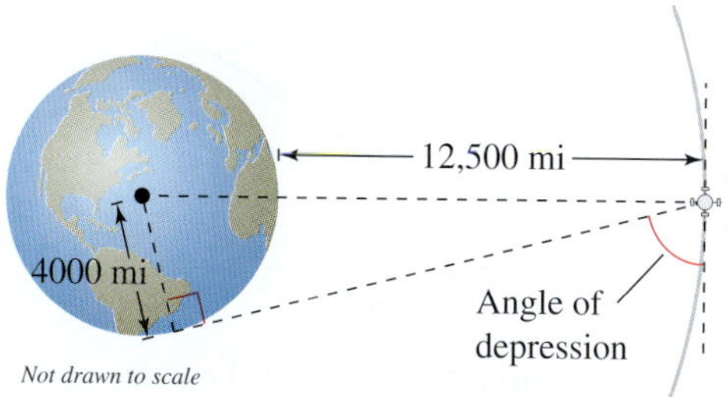

31. Height You are holding one of the tethers attached to the top of a giant character balloon in a parade. Before the start of the parade the balloon is upright and the bottom is floating approximately 20 feet above ground level. You are standing approximately 100 feet ahead of the balloon (see figure).

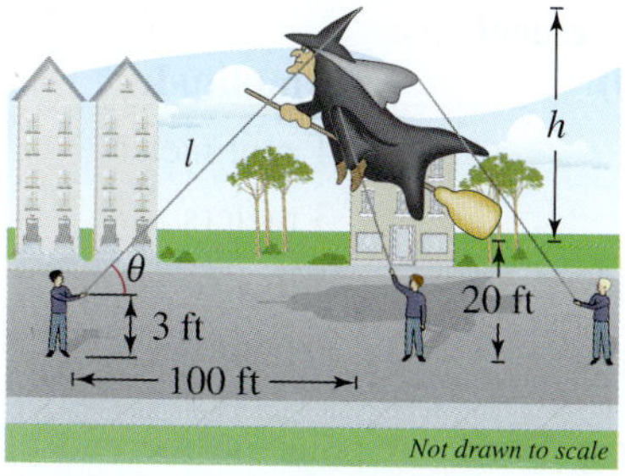

(a) Find an equation for the length l of the tether you are holding in terms of h, the height of the balloon from top to bottom.

(b) Find an equation for the angle of elevation θ from you to the top of the balloon.

(c) The angle of elevation to the top of the balloon is 35°. Find the height h of the balloon.

32. Waterslide Design

The designers of a water park are creating a new slide and have sketched some preliminary drawings. The length of the ladder is 30 feet, and its angle of elevation is 60° (see figure).

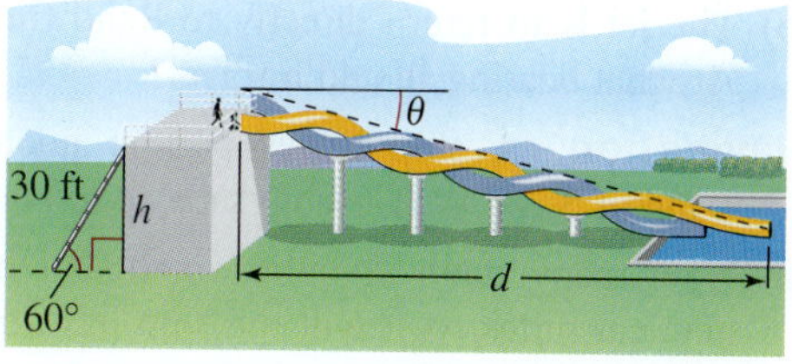

(a) Find the height h of the slide.

(b) Find the angle of depression θ from the top of the slide to the end of the slide at the ground in terms of the horizontal distance d a rider travels.

(c) Safety restrictions require the angle of depression to be no less than 25° and no more than 30°. Find an interval for how far a rider travels horizontally.

33. Speed Enforcement A police department has set up a speed enforcement zone on a straight length of highway. A patrol car is parked parallel to the zone, 200 feet from one end and 150 feet from the other end (see figure).

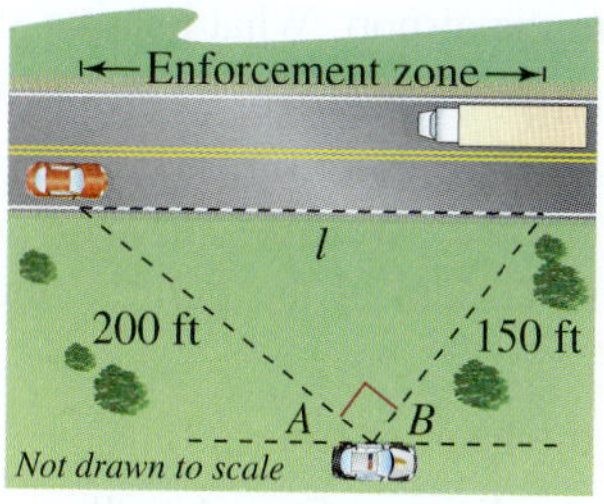

(a) Find the length l of the zone and the measures of the angles A and B (in degrees).

(b) Find the minimum amount of time (in seconds) it takes for a vehicle to pass through the zone without exceeding the posted speed limit of 35 miles per hour.

34. Airplane Ascent During takeoff, an airplane's angle of ascent is 18° and its speed is 275 feet per second.

(a) Find the plane's altitude after 1 minute.

(b) How long will it take for the plane to climb to an altitude of 10,000 feet?

karnizz/Shutterstock.com

35. Navigation An airplane flying at 600 miles per hour has a bearing of 52°. After flying for 1.5 hours, how far north and how far east will the plane have traveled from its point of departure?

36. Navigation A jet leaves Reno, Nevada, and is headed toward Miami, Florida, at a bearing of 100°. The distance between the two cities is approximately 2472 miles.

(a) How far north and how far west is Reno relative to Miami?

(b) The jet is to return directly to Reno from Miami. At what bearing should it travel?

37. Navigation A ship leaves port at noon and has a bearing of S 29° W. The ship sails at 20 knots.

(a) How many nautical miles south and how many nautical miles west will the ship have traveled by 6:00 P.M.?

(b) At 6:00 P.M., the ship changes course to due west. Find the ship's bearing and distance from the port of departure at 7:00 P.M.

38. Navigation A privately owned yacht leaves a dock in Myrtle Beach, South Carolina, and heads toward Freeport in the Bahamas at a bearing of S 1.4° E. The yacht averages a speed of 20 knots over the 428-nautical-mile trip.

(a) How long will it take the yacht to make the trip?

(b) How far east and south is the yacht after 12 hours?

(c) A plane leaves Myrtle Beach to fly to Freeport. What bearing should be taken?

39. Navigation A ship is 45 miles east and 30 miles south of port. The captain wants to sail directly to port. What bearing should be taken?

40. Navigation An airplane is 160 miles north and 85 miles east of an airport. The pilot wants to fly directly to the airport. What bearing should be taken?

41. Surveying A surveyor wants to find the distance across a pond (see figure). The bearing from A to B is N 32° W. The surveyor walks 50 meters from A to C, and at the point C the bearing to B is N 68° W.

(a) Find the bearing from A to C.

(b) Find the distance from A to B.

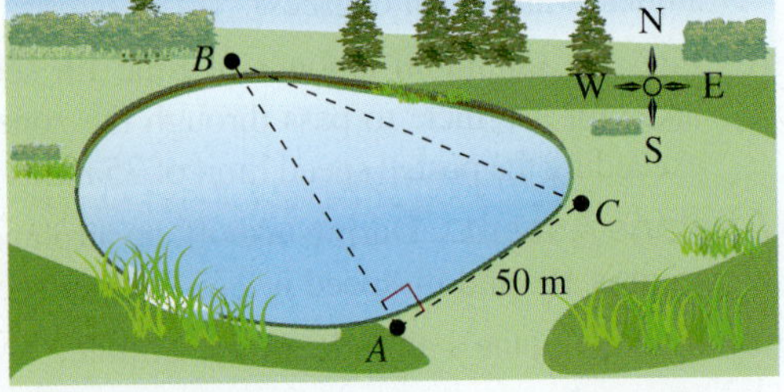

42. Location of a Fire Fire tower A is 30 kilometers due west of fire tower B. A fire is spotted from the towers, and the bearings from A and B are N 76° E and N 56° W, respectively (see figure). Find the distance d of the fire from the line segment AB.

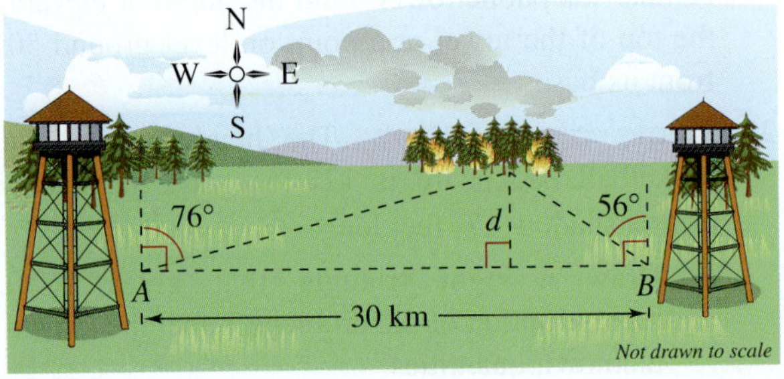

43. Geometry Determine the angle between the diagonal of a cube and the diagonal of its base, as shown in the figure.

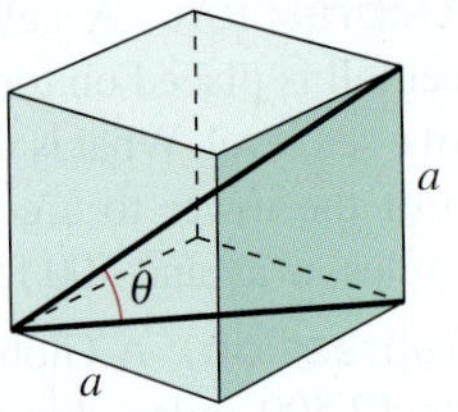

44. Geometry Determine the angle between the diagonal of a cube and its edge, as shown in the figure.

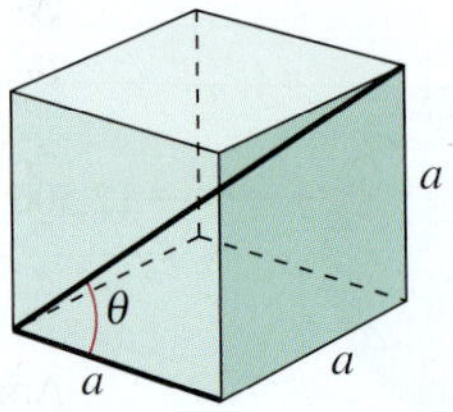

45. Geometry Find the length of the sides of a regular pentagon inscribed in a circle of radius 25 inches.

46. Geometry Find the length of the sides of a regular hexagon inscribed in a circle of radius 25 inches.

Harmonic Motion **In Exercises 47–50, find a model for simple harmonic motion satisfying the specified conditions.**

	Displacement ($t = 0$)	Amplitude	Period
47.	0	4 centimeters	2 seconds
48.	0	3 meters	6 seconds
49.	3 inches	3 inches	1.5 seconds
50.	2 feet	2 feet	10 seconds

51. Tuning Fork A point on the end of a tuning fork moves in simple harmonic motion described by $d = a \sin \omega t$. Find ω given that the tuning fork for middle C has a frequency of 264 vibrations per second.

52. **Wave Motion** A buoy oscillates in simple harmonic motion as waves go past. The buoy moves a total of 3.5 feet from its low point to its high point (see figure), and it returns to its high point every 10 seconds. Write an equation that describes the motion of the buoy where the high point corresponds to the time $t = 0$.

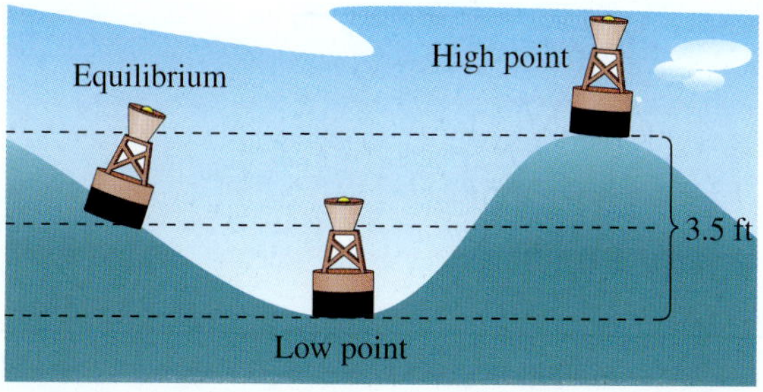

Harmonic Motion **In Exercises 53–56, for the simple harmonic motion described by the trigonometric function, find (a) the maximum displacement, (b) the frequency, (c) the value of d when $t = 5$, and (d) the least positive value of t for which $d = 0$. Use a graphing utility to verify your results.**

53. $d = 9 \cos \frac{6\pi}{5} t$

54. $d = \frac{1}{2} \cos 20\pi t$

55. $d = \frac{1}{4} \sin 6\pi t$

56. $d = \frac{1}{64} \sin 792\pi t$

57. **Oscillation of a Spring** A ball that is bobbing up and down on the end of a spring has a maximum displacement of 3 inches. Its motion (in ideal conditions) is modeled by $y = \frac{1}{4} \cos 16t,\ t > 0$, where y is measured in feet and t is the time in seconds.

(a) Graph the function.

(b) What is the period of the oscillations?

(c) Determine the first time the weight passes the point of equilibrium $(y = 0)$.

58. **Data Analysis** The table shows the average sales S (in millions of dollars) of an outerwear manufacturer for each month t, where $t = 1$ represents January.

Time, t	1	2	3	4	5	6
Sales, S	13.46	11.15	8.00	4.85	2.54	1.70

Time, t	7	8	9	10	11	12
Sales, S	2.54	4.85	8.00	11.15	13.46	14.30

(a) Create a scatter plot of the data.

(b) Find a trigonometric model that fits the data. Graph the model with your scatter plot. How well does the model fit the data?

(c) What is the period of the model? Do you think it is reasonable given the context? Explain your reasoning.

(d) Interpret the meaning of the model's amplitude in the context of the problem.

59. **Data Analysis** The numbers of hours H of daylight in Denver, Colorado, on the 15th of each month are: 1(9.67), 2(10.72), 3(11.92), 4(13.25), 5(14.37), 6(14.97), 7(14.72), 8(13.77), 9(12.48), 10(11.18), 11(10.00), 12(9.38). The month is represented by t, with $t = 1$ corresponding to January. A model for the data is

$$H(t) = 12.13 + 2.77 \sin\left(\frac{\pi t}{6} - 1.60\right).$$

(a) Use a graphing utility to graph the data points and the model in the same viewing window.

(b) What is the period of the model? Is it what you expected? Explain.

(c) What is the amplitude of the model? What does it represent in the context of the problem? Explain.

Exploration

60. **HOW DO YOU SEE IT?** The graph below shows the displacement of an object in simple harmonic motion.

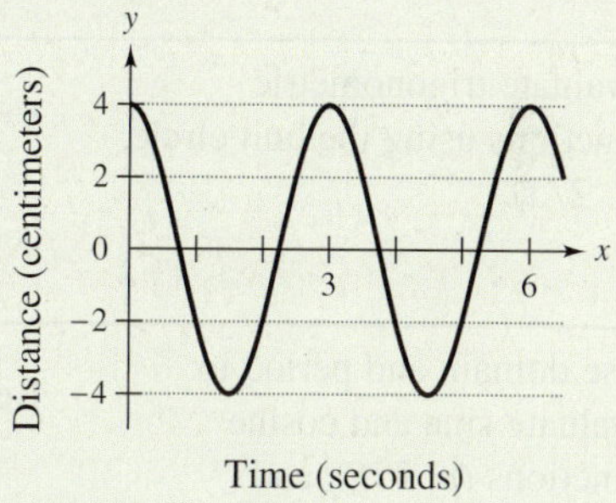

(a) What is the amplitude?

(b) What is the period?

(c) Is the equation of the simple harmonic motion of the form $d = a \sin \omega t$ or $d = a \cos \omega t$?

True or False? **In Exercises 61 and 62, determine whether the statement is true or false. Justify your answer.**

61. The Leaning Tower of Pisa is not vertical, but when you know the angle of elevation θ to the top of the tower as you stand d feet away from it, you can find its height h using the formula $h = d \tan \theta$.

62. The bearing N 24° E means 24 degrees north of east.

Chapter Summary

	What Did You Learn?	Explanation/Examples	Review Exercises
Section 4.1	Describe angles *(p. 262).*	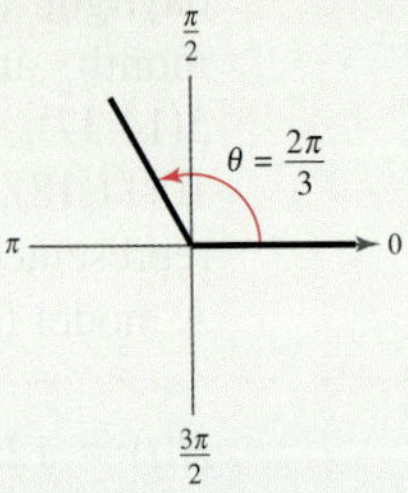 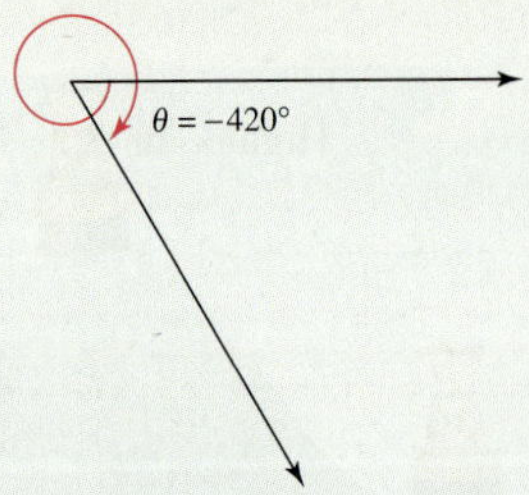	1–4
	Convert between degrees and radians *(p. 265).*	To convert degrees to radians, multiply degrees by $(\pi \text{ rad})/180°$. To convert radians to degrees, multiply radians by $180°/(\pi \text{ rad})$.	5–14
	Use angles to model and solve real-life problems *(p. 266).*	Angles can help you find the length of a circular arc and the area of a sector of a circle. (See Examples 5 and 8.)	15–18
Section 4.2	Identify a unit circle and describe its relationship to real numbers *(p. 272).*	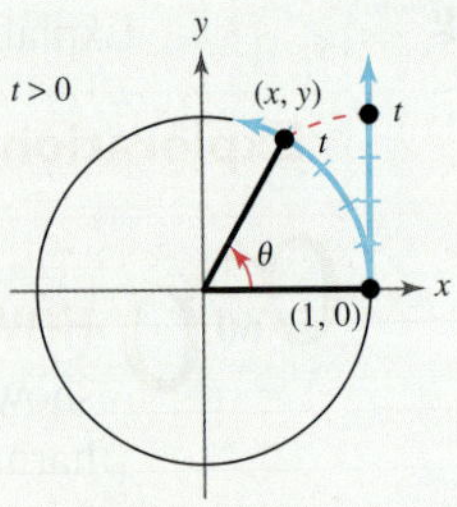 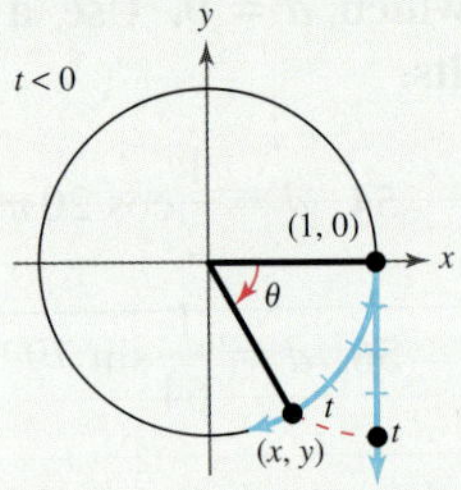	19–22
	Evaluate trigonometric functions using the unit circle *(p. 273).*	$t = \frac{2\pi}{3}$ corresponds to $(x, y) = \left(-\frac{1}{2}, \frac{\sqrt{3}}{2}\right)$. So $\cos\frac{2\pi}{3} = -\frac{1}{2}$, $\sin\frac{2\pi}{3} = \frac{\sqrt{3}}{2}$, and $\tan\frac{2\pi}{3} = -\sqrt{3}$.	23, 24
	Use domain and period to evaluate sine and cosine functions *(p. 275),* and use a calculator to evaluate trigonometric functions *(p. 276).*	Because $\frac{13\pi}{6} = 2\pi + \frac{\pi}{6}$, $\sin\frac{13\pi}{6} = \sin\frac{\pi}{6} = \frac{1}{2}$. $\sin\frac{3\pi}{8} \approx 0.9239$, $\cot(-1.2) \approx -0.3888$	25–32
Section 4.3	Evaluate trigonometric functions of acute angles *(p. 279),* and use a calculator to evaluate trigonometric functions *(p. 281).*	$\sin\theta = \frac{\text{opp}}{\text{hyp}}$, $\cos\theta = \frac{\text{adj}}{\text{hyp}}$, $\tan\theta = \frac{\text{opp}}{\text{adj}}$ $\csc\theta = \frac{\text{hyp}}{\text{opp}}$, $\sec\theta = \frac{\text{hyp}}{\text{adj}}$, $\cot\theta = \frac{\text{adj}}{\text{opp}}$ $\tan 34.7° \approx 0.6924$, $\csc 29°\,15' \approx 2.0466$	33–38
	Use the fundamental trigonometric identities *(p. 282).*	$\sin\theta = \frac{1}{\csc\theta}$, $\tan\theta = \frac{\sin\theta}{\cos\theta}$, $\sin^2\theta + \cos^2\theta = 1$	39, 40
	Use trigonometric functions to model and solve real-life problems *(p. 284).*	Trigonometric functions can help you find the height of a monument, the angle between two paths, and the length of a ramp. (See Examples 8–10.)	41, 42

Section	What Did You Learn?	Explanation/Examples	Review Exercises
Section 4.4	Evaluate trigonometric functions of any angle *(p. 290).*	Let (3, 4) be a point on the terminal side of θ. Then $\sin\theta = \frac{4}{5}$, $\cos\theta = \frac{3}{5}$, and $\tan\theta = \frac{4}{3}$.	43–50
	Find reference angles *(p. 292).*	Let θ be an angle in standard position. Its reference angle is the acute angle θ' formed by the terminal side of θ and the horizontal axis.	51–54
	Evaluate trigonometric functions of real numbers *(p. 293).*	$\cos\frac{7\pi}{3} = \frac{1}{2}$ because $\theta' = \frac{7\pi}{3} - 2\pi = \frac{\pi}{3}$ and $\cos\frac{\pi}{3} = \frac{1}{2}$.	55–62
Section 4.5	Sketch the graphs of sine and cosine functions using amplitude and period *(p. 299).*	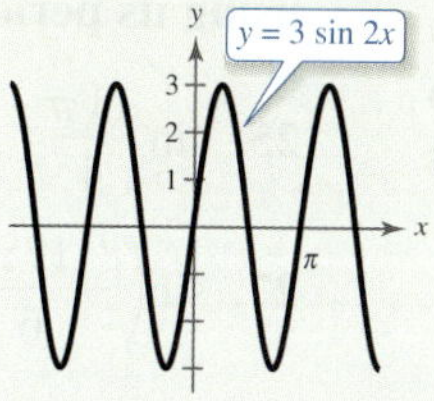 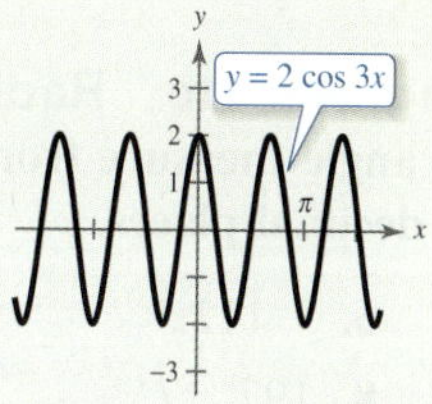	63, 64
	Sketch translations of the graphs of sine and cosine functions *(p. 303).*	For $y = d + a\sin(bx - c)$ and $y = d + a\cos(bx - c)$, the constant c creates a horizontal translation. The constant d creates a vertical translation. (See Examples 4–6.)	65–68
	Use sine and cosine functions to model real-life data *(p. 305).*	A cosine function can help you model the depth of the water at the end of a dock at various times. (See Example 7.)	69, 70
Section 4.6	Sketch the graphs of tangent *(p. 310),* cotangent *(p. 312),* secant *(p. 313),* and cosecant *(p. 313)* functions.	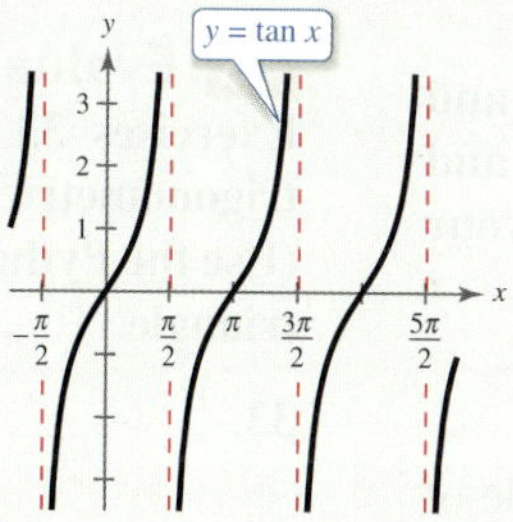 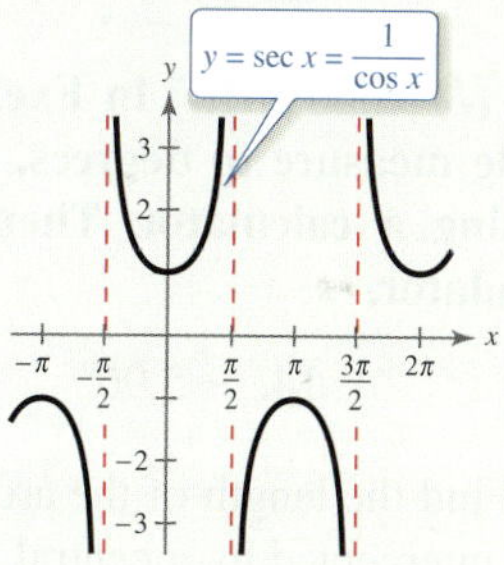	71–74
	Sketch the graphs of damped trigonometric functions *(p. 315).*	In $f(x) = x\cos 2x$, the factor x is called the damping factor.	75, 76
Section 4.7	Evaluate and graph inverse trigonometric functions *(p. 320).*	$\sin^{-1}\frac{\sqrt{3}}{2} = \frac{\pi}{3}$, $\cos^{-1}\left(-\frac{\sqrt{2}}{2}\right) = \frac{3\pi}{4}$, $\tan^{-1}(-1) = -\frac{\pi}{4}$	77–86
	Evaluate the compositions of trigonometric functions *(p. 324).*	$\cos\left[\arctan\left(\frac{5}{12}\right)\right] = \frac{12}{13}$, $\sin(\sin^{-1} 0.4) = 0.4$	87–92
Section 4.8	Solve real-life problems involving right triangles *(p. 330).*	A trigonometric function can help you find the height of a smokestack on top of a building. (See Example 3.)	93, 94
	Solve real-life problems involving directional bearings *(p. 332).*	Trigonometric functions can help you find a ship's bearing and distance from a port at a given time. (See Example 5.)	95
	Solve real-life problems involving harmonic motion *(p. 333).*	Sine or cosine functions can help you describe the motion of an object that vibrates, oscillates, rotates, or is moved by wave motion. (See Examples 6 and 7.)	96

Review Exercises

See CalcChat.com for tutorial help and worked-out solutions to odd-numbered exercises.

4.1 Using Radian or Degree Measure In Exercises 1–4, (a) sketch the angle in standard position, (b) determine the quadrant in which the angle lies, and (c) determine one positive and one negative coterminal angle.

1. $\frac{15\pi}{4}$
2. $-\frac{4\pi}{3}$
3. $-110°$
4. $280°$

Converting from Degrees to Radians In Exercises 5–8, convert the angle measure from degrees to radians. Round to three decimal places.

5. $450°$
6. $-112.5°$
7. $-33° 45'$
8. $197° 17'$

Converting from Radians to Degrees In Exercises 9–12, convert the angle measure from radians to degrees. Round to three decimal places.

9. $\frac{3\pi}{10}$
10. $-\frac{11\pi}{6}$
11. -3.5
12. 5.7

Converting to D° M′ S″ Form In Exercises 13 and 14, convert the angle measure to degrees, minutes, and seconds without using a calculator. Then check your answer using a calculator.

13. $198.4°$
14. $-5.96°$

15. **Arc Length** Find the length of the arc on a circle of radius 20 inches intercepted by a central angle of $138°$.
16. **Phonograph** Phonograph records are vinyl discs that rotate on a turntable. A typical record album is 12 inches in diameter and plays at $33\frac{1}{3}$ revolutions per minute.
    (a) What is the angular speed of a record album?
    (b) What is the linear speed of the outer edge of a record album?
17. **Circular Sector** Find the area of the sector of a circle of radius 18 inches and central angle $\theta = 120°$.
18. **Circular Sector** Find the area of the sector of a circle of radius 6.5 millimeters and central angle $\theta = 5\pi/6$.

4.2 Finding a Point on the Unit Circle In Exercises 19–22, find the point (x, y) on the unit circle that corresponds to the real number t.

19. $t = \frac{2\pi}{3}$
20. $t = \frac{7\pi}{4}$
21. $t = \frac{7\pi}{6}$
22. $t = -\frac{4\pi}{3}$

Evaluating Trigonometric Functions In Exercises 23 and 24, evaluate (if possible) the six trigonometric functions at the real number.

23. $t = \frac{3\pi}{4}$
24. $t = -\frac{2\pi}{3}$

Using Period to Evaluate Sine and Cosine In Exercises 25–28, evaluate the trigonometric function using its period as an aid.

25. $\sin \frac{11\pi}{4}$
26. $\cos 4\pi$
27. $\sin\left(-\frac{17\pi}{6}\right)$
28. $\cos\left(-\frac{13\pi}{3}\right)$

Using a Calculator In Exercises 29–32, use a calculator to evaluate the trigonometric function. Round your answer to four decimal places. (Be sure the calculator is in the correct mode.)

29. $\tan 33$
30. $\csc 10.5$
31. $\sec \frac{12\pi}{5}$
32. $\sin\left(-\frac{\pi}{9}\right)$

4.3 Evaluating Trigonometric Functions In Exercises 33 and 34, find the exact values of the six trigonometric functions of the angle θ shown in the figure. (Use the Pythagorean Theorem to find the third side of the triangle.)

33.
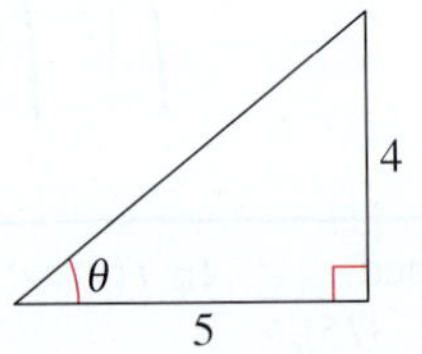

34.
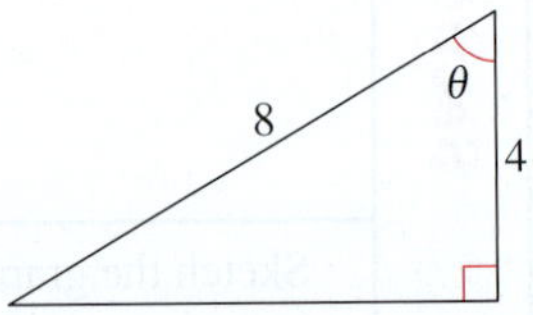

Using a Calculator In Exercises 35–38, use a calculator to evaluate the trigonometric function. Round your answer to four decimal places. (Be sure the calculator is in the correct mode.)

35. $\tan 33°$
36. $\sec 79.3°$
37. $\cot 15° 14'$
38. $\cos 78° 11' 58''$

Applying Trigonometric Identities In Exercises 39 and 40, use the given function value and the trigonometric identities to find the indicated trigonometric functions.

39. $\sin \theta = \frac{1}{3}$ (a) $\csc \theta$ (b) $\cos \theta$ (c) $\sec \theta$ (d) $\tan \theta$
40. $\csc \theta = 5$ (a) $\sin \theta$ (b) $\cot \theta$ (c) $\tan \theta$ (d) $\sec(90° - \theta)$

41. Railroad Grade A train travels 3.5 kilometers on a straight track with a grade of $1^\circ\, 10'$ (see figure). What is the vertical rise of the train in that distance?

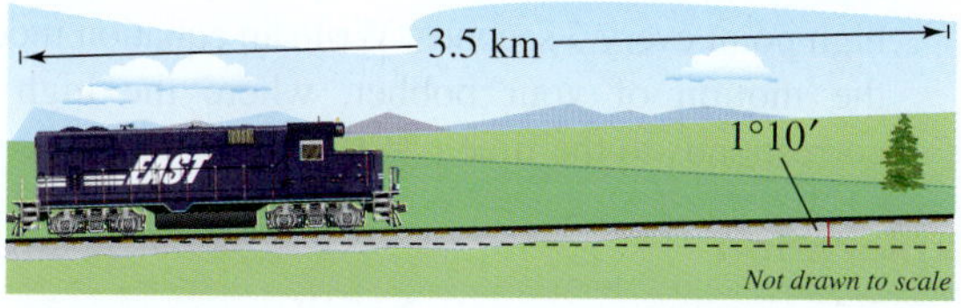

42. Guy Wire A guy wire runs from the ground to the top of a 25-foot telephone pole. The angle formed between the wire and the ground is 52°. How far from the base of the pole is the wire attached to the ground? Assume the pole is perpendicular to the ground.

4.4 Evaluating Trigonometric Functions **In Exercises 43–46, the point is on the terminal side of an angle in standard position. Determine the exact values of the six trigonometric functions of the angle.**

43. $(12, 16)$ **44.** $(3, -4)$

45. $(0.3, 0.4)$ **46.** $\left(-\frac{10}{3}, -\frac{2}{3}\right)$

Evaluating Trigonometric Functions **In Exercises 47–50, find the values of the remaining five trigonometric functions of θ with the given constraint.**

	Function Value	Constraint
47.	$\sec\theta = \frac{6}{5}$	$\tan\theta < 0$
48.	$\csc\theta = \frac{3}{2}$	$\cos\theta < 0$
49.	$\cos\theta = -\frac{2}{5}$	$\sin\theta > 0$
50.	$\sin\theta = -\frac{1}{2}$	$\cos\theta > 0$

Finding a Reference Angle **In Exercises 51–54, find the reference angle θ' and sketch θ and θ' in standard position.**

51. $\theta = 264^\circ$ **52.** $\theta = 635^\circ$

53. $\theta = -6\pi/5$ **54.** $\theta = 17\pi/3$

Using a Reference Angle **In Exercises 55–58, evaluate the sine, cosine, and tangent of the angle without using a calculator.**

55. $\pi/3$ **56.** $-5\pi/4$

57. -150° **58.** 495°

Using a Calculator **In Exercises 59–62, use a calculator to evaluate the trigonometric function. Round your answer to four decimal places. (Be sure the calculator is in the correct mode.)**

59. $\sin 4$

60. $\cot(-4.8)$

61. $\sin(12\pi/5)$

62. $\tan(-25\pi/7)$

4.5 Sketching the Graph of a Sine or Cosine Function **In Exercises 63–68, sketch the graph of the function. (Include two full periods.)**

63. $y = \sin 6x$

64. $f(x) = 5\sin(2x/5)$

65. $y = 5 + \sin x$

66. $y = -4 - \cos \pi x$

67. $g(t) = \frac{5}{2}\sin(t - \pi)$

68. $g(t) = 3\cos(t + \pi)$

69. Sound Waves Sine functions of the form $y = a \sin bx$, where x is measured in seconds, can model sine waves.

(a) Write an equation of a sound wave whose amplitude is 2 and whose period is $\frac{1}{264}$ second.

(b) What is the frequency of the sound wave described in part (a)?

70. Data Analysis: Meteorology The times S of sunset (Greenwich Mean Time) at 40° north latitude on the 15th of each month are: 1(16:59), 2(17:35), 3(18:06), 4(18:38), 5(19:08), 6(19:30), 7(19:28), 8(18:57), 9(18:09), 10(17:21), 11(16:44), 12(16:36). The month is represented by t, with $t = 1$ corresponding to January. A model (in which minutes have been converted to the decimal parts of an hour) for the data is $S(t) = 18.09 + 1.41 \sin[(\pi t/6) + 4.60]$.

(a) Use a graphing utility to graph the data points and the model in the same viewing window.

(b) What is the period of the model? Is it what you expected? Explain.

(c) What is the amplitude of the model? What does it represent in the model? Explain.

4.6 Sketching the Graph of a Trigonometric Function **In Exercises 71–74, sketch the graph of the function. (Include two full periods.)**

71. $f(t) = \tan\left(t + \dfrac{\pi}{2}\right)$

72. $f(x) = \dfrac{1}{2}\cot x$

73. $f(x) = \dfrac{1}{2}\csc\dfrac{x}{2}$

74. $h(t) = \sec\left(t - \dfrac{\pi}{4}\right)$

Analyzing a Damped Trigonometric Graph **In Exercises 75 and 76, use a graphing utility to graph the function and the damping factor of the function in the same viewing window. Describe the behavior of the function as x increases without bound.**

75. $f(x) = x\cos x$ **76.** $g(x) = x^4 \cos x$

4.7 Evaluating an Inverse Trigonometric Function **In Exercises 77–80, evaluate the expression without using a calculator.**

77. $\arcsin(-1)$

78. $\cos^{-1} 1$

79. $\operatorname{arccot} \sqrt{3}$

80. $\operatorname{arcsec}(-\sqrt{2})$

Calculators and Inverse Trigonometric Functions **In Exercises 81–84, use a calculator to evaluate the expression. Round your result to two decimal places.**

81. $\tan^{-1}(-1.5)$

82. $\arccos 0.324$

83. $\operatorname{arccot} 10.5$

84. $\operatorname{arccsc}(-2.01)$

Graphing an Inverse Trigonometric Function **In Exercises 85 and 86, use a graphing utility to graph the function.**

85. $f(x) = \arctan(x/2)$

86. $f(x) = -\arcsin 2x$

Evaluating a Composition of Functions **In Exercises 87–90, find the exact value of the expression. (*Hint:* Sketch a right triangle.)**

87. $\cos(\arctan \frac{3}{4})$

88. $\sec(\tan^{-1} \frac{12}{5})$

89. $\sec[\sin^{-1}(-\frac{1}{4})]$

90. $\cot[\arcsin(-\frac{12}{13})]$

Writing an Expression **In Exercises 91 and 92, write an algebraic expression that is equivalent to the given expression. (*Hint:* Sketch a right triangle.)**

91. $\tan[\arccos(x/2)]$

92. $\sec(\arcsin x)$

4.8

93. **Angle of Elevation** The height of a radio transmission tower is 70 meters, and it casts a shadow of length 30 meters. Draw a right triangle that gives a visual representation of the problem. Label the known and unknown quantities. Then find the angle of elevation of the sun.

94. **Height** Your football has landed at the edge of the roof of your school building. When you are 25 feet from the base of the building, the angle of elevation to your football is 21°. How high off the ground is your football?

95. **Distance** From city A to city B, a plane flies 650 miles at a bearing of 48°. From city B to city C, the plane flies 810 miles at a bearing of 115°. Find the distance from city A to city C and the bearing from city A to city C.

96. **Wave Motion** Your fishing bobber oscillates in simple harmonic motion from the waves in the lake where you fish. Your bobber moves a total of 1.5 inches from its high point to its low point and returns to its high point every 3 seconds. Write an equation modeling the motion of your bobber, where the high point corresponds to the time $t = 0$.

Exploration

True or False? **In Exercises 97 and 98, determine whether the statement is true or false. Justify your answer.**

97. $y = \sin \theta$ is not a function because $\sin 30° = \sin 150°$.

98. Because $\tan 3\pi/4 = -1$, $\arctan(-1) = 3\pi/4$.

99. **Writing** Describe the behavior of $f(\theta) = \sec \theta$ at the zeros of $g(\theta) = \cos \theta$. Explain your reasoning.

100. **Conjecture**

(a) Use a graphing utility to complete the table.

θ	0.1	0.4	0.7	1.0	1.3
$\tan\left(\theta - \frac{\pi}{2}\right)$					
$-\cot \theta$					

(b) Make a conjecture about the relationship between $\tan[\theta - (\pi/2)]$ and $-\cot \theta$.

101. **Writing** When graphing the sine and cosine functions, determining the amplitude is part of the analysis. Explain why this is not true for the other four trigonometric functions.

102. **Oscillation of a Spring** A weight is suspended from a ceiling by a steel spring. The weight is lifted (positive direction) from the equilibrium position and released. The resulting motion of the weight is modeled by $y = Ae^{-kt} \cos bt = \frac{1}{5}e^{-t/10} \cos 6t$, where y is the distance (in feet) from equilibrium and t is the time (in seconds). The figure shows the graph of the function. For each of the following, describe the change in the system without graphing the resulting function.

(a) A is changed from $\frac{1}{5}$ to $\frac{1}{3}$.

(b) k is changed from $\frac{1}{10}$ to $\frac{1}{3}$.

(c) b is changed from 6 to 9.

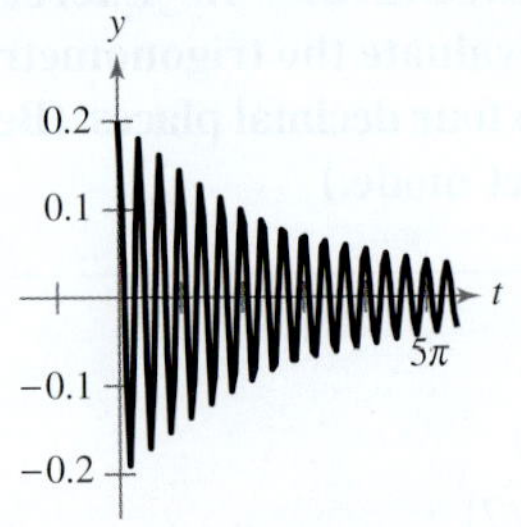

Chapter Test

See CalcChat.com for tutorial help and worked-out solutions to odd-numbered exercises.

Take this test as you would take a test in class. When you are finished, check your work against the answers given in the back of the book.

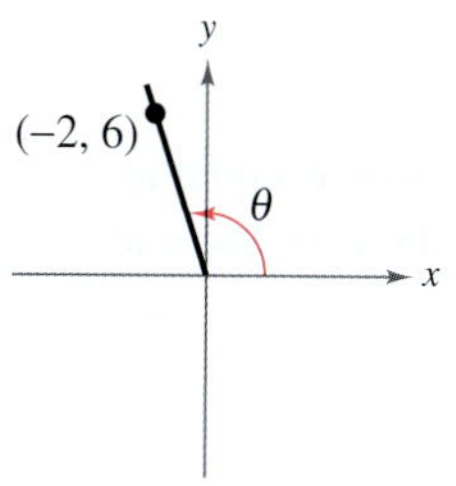

Figure for 4

1. Consider an angle that measures $\frac{5\pi}{4}$ radians.

(a) Sketch the angle in standard position.

(b) Determine two coterminal angles (one positive and one negative).

(c) Convert the angle to degree measure.

2. A truck is moving at a rate of 105 kilometers per hour, and the diameter of each of its wheels is 1 meter. Find the angular speed of the wheels in radians per minute.

3. A water sprinkler sprays water on a lawn over a distance of 25 feet and rotates through an angle of 130°. Find the area of the lawn watered by the sprinkler.

4. Find the exact values of the six trigonometric functions of the angle θ shown in the figure.

5. Given that $\tan \theta = \frac{3}{2}$, find the other five trigonometric functions of θ.

6. Determine the reference angle θ' of the angle $\theta = 205°$ and sketch θ and θ' in standard position.

7. Determine the quadrant in which θ lies when $\sec \theta < 0$ and $\tan \theta > 0$.

8. Find two exact values of θ in degrees $(0 \le \theta < 360°)$ for which $\cos \theta = -\sqrt{3}/2$. (Do not use a calculator.)

9. Use a calculator to approximate two values of θ in radians $(0 \le \theta < 2\pi)$ for which $\csc \theta = 1.030$. Round the results to two decimal places.

In Exercises 10 and 11, find the values of the remaining five trigonometric functions of θ with the given constraint.

10. $\cos \theta = \frac{3}{5}, \quad \tan \theta < 0$

11. $\sec \theta = -\frac{29}{20}, \quad \sin \theta > 0$

In Exercises 12 and 13, sketch the graph of the function. (Include two full periods.)

12. $g(x) = -2 \sin\left(x - \frac{\pi}{4}\right)$

13. $f(\alpha) = \frac{1}{2} \tan 2\alpha$

In Exercises 14 and 15, use a graphing utility to graph the function. If the function is periodic, then find its period.

14. $y = \sin 2\pi x + 2 \cos \pi x$

15. $y = 6e^{-0.12t} \cos(0.25t), \quad 0 \le t \le 32$

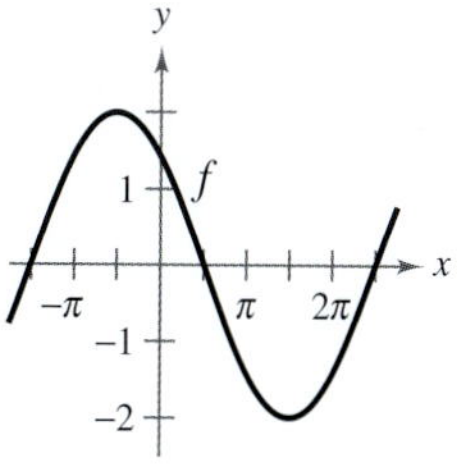

Figure for 16

16. Find a, b, and c for the function $f(x) = a \sin(bx + c)$ such that the graph of f matches the figure.

17. Find the exact value of $\cot\left(\arcsin \frac{3}{8}\right)$ without using a calculator.

18. Graph the function $f(x) = 2 \arcsin\left(\frac{1}{2}x\right)$.

19. A plane is 90 miles south and 110 miles east of London Heathrow Airport. What bearing should be taken to fly directly to the airport?

20. Write the equation for the simple harmonic motion of a ball on a spring that starts at its lowest point of 6 inches below equilibrium, bounces to its maximum height of 6 inches above equilibrium, and returns to its lowest point in a total of 2 seconds.

Proofs in Mathematics

The Pythagorean Theorem

The Pythagorean Theorem is one of the most famous theorems in mathematics. More than 100 different proofs now exist. James A. Garfield, the twentieth president of the United States, developed a proof of the Pythagorean Theorem in 1876. His proof, shown below, involves the fact that a trapezoid can be formed from two congruent right triangles and an isosceles right triangle.

> **The Pythagorean Theorem**
>
> In a right triangle, the sum of the squares of the lengths of the legs is equal to the square of the length of the hypotenuse, where a and b are the legs and c is the hypotenuse.
>
> $$a^2 + b^2 = c^2$$

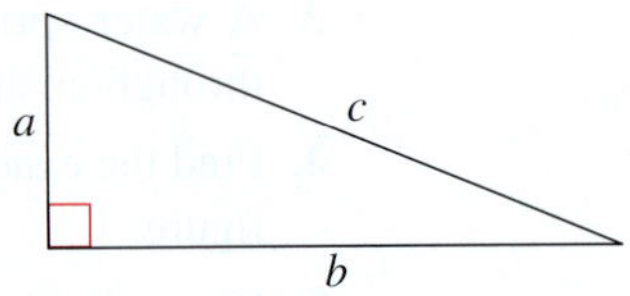

Proof

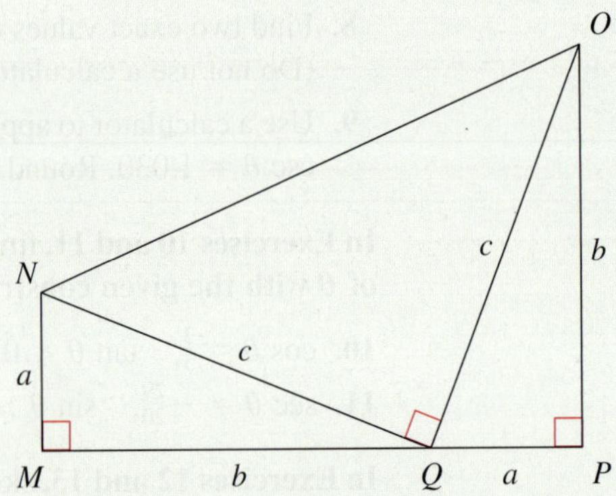

$$\underset{\text{trapezoid } MNOP}{\text{Area of}} = \underset{\triangle MNQ}{\text{Area of}} + \underset{\triangle PQO}{\text{Area of}} + \underset{\triangle NOQ}{\text{Area of}}$$

$$\frac{1}{2}(a + b)(a + b) = \frac{1}{2}ab + \frac{1}{2}ab + \frac{1}{2}c^2$$

$$\frac{1}{2}(a + b)(a + b) = ab + \frac{1}{2}c^2$$

$$(a + b)(a + b) = 2ab + c^2$$

$$a^2 + 2ab + b^2 = 2ab + c^2$$

$$a^2 + b^2 = c^2$$

P.S. Problem Solving

1. **Angle of Rotation** The restaurant at the top of the Space Needle in Seattle, Washington, is circular and has a radius of 47.25 feet. The dining part of the restaurant revolves, making about one complete revolution every 48 minutes. A dinner party, seated at the edge of the revolving restaurant at 6:45 P.M., finishes at 8:57 P.M.

   (a) Find the angle through which the dinner party rotated.

   (b) Find the distance the party traveled during dinner.

2. **Bicycle Gears** A bicycle's gear ratio is the number of times the freewheel turns for every one turn of the chainwheel (see figure). The table shows the numbers of teeth in the freewheel and chainwheel for the first five gears of an 18-speed touring bicycle. The chainwheel completes one rotation for each gear. Find the angle through which the freewheel turns for each gear. Give your answers in both degrees and radians.

DATA

Spreadsheet at LarsonPrecalculus.com

Gear Number	Number of Teeth in Freewheel	Number of Teeth in Chainwheel
1	32	24
2	26	24
3	22	24
4	32	40
5	19	24

3. **Surveying** A surveyor in a helicopter is trying to determine the width of an island, as shown in the figure.

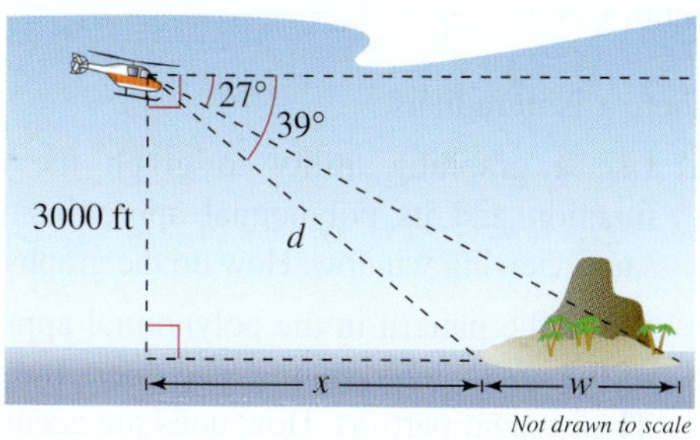

   (a) What is the shortest distance d the helicopter would have to travel to land on the island?

   (b) What is the horizontal distance x the helicopter would have to travel before it would be directly over the nearer end of the island?

   (c) Find the width w of the island. Explain how you found your answer.

4. **Similar Triangles and Trigonometric Functions** Use the figure below.

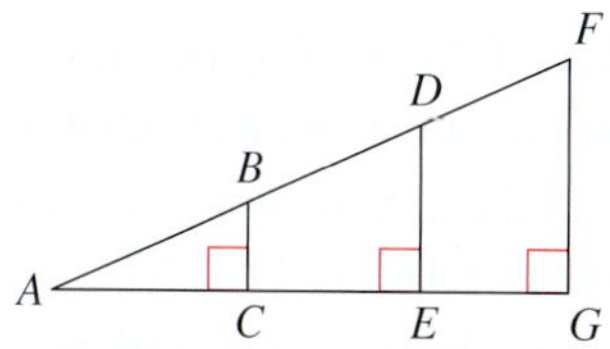

   (a) Explain why $\triangle ABC$, $\triangle ADE$, and $\triangle AFG$ are similar triangles.

   (b) What does similarity imply about the ratios

   $$\frac{BC}{AB}, \quad \frac{DE}{AD}, \quad \text{and} \quad \frac{FG}{AF}?$$

   (c) Does the value of $\sin A$ depend on which triangle from part (a) you use to calculate it? Would using a different right triangle similar to the three given triangles change the value of $\sin A$?

   (d) Do your conclusions from part (c) apply to the other five trigonometric functions? Explain.

5. **Graphical Analysis** Use a graphing utility to graph h, and use the graph to decide whether h is even, odd, or neither.

   (a) $h(x) = \cos^2 x$

   (b) $h(x) = \sin^2 x$

6. **Squares of Even and Odd Functions** Given that f is an even function and g is an odd function, use the results of Exercise 5 to make a conjecture about h, where

   (a) $h(x) = [f(x)]^2$

   (b) $h(x) = [g(x)]^2$.

7. **Height of a Ferris Wheel Car** The model for the height h (in feet) of a Ferris wheel car is

   $$h = 50 + 50 \sin 8\pi t$$

   where t is the time (in minutes). (The Ferris wheel has a radius of 50 feet.) This model yields a height of 50 feet when $t = 0$. Alter the model so that the height of the car is 1 foot when $t = 0$.

8. **Periodic Function** The function f is periodic, with period c. So, $f(t + c) = f(t)$. Are the following statements true? Explain.

   (a) $f(t - 2c) = f(t)$

   (b) $f\left(t + \frac{1}{2}c\right) = f\left(\frac{1}{2}t\right)$

   (c) $f\left(\frac{1}{2}(t + c)\right) = f\left(\frac{1}{2}t\right)$

9. **Blood Pressure** The pressure P (in millimeters of mercury) against the walls of the blood vessels of a patient is modeled by

$$P = 100 - 20\cos\left(\frac{8\pi}{3}t\right)$$

where t is time (in seconds).

(a) Use a graphing utility to graph the model.

(b) What is the period of the model? What does the period tell you about this situation?

(c) What is the amplitude of the model? What does it tell you about this situation?

(d) If one cycle of this model is equivalent to one heartbeat, what is the pulse of this patient?

(e) A physician wants this patient's pulse rate to be 64 beats per minute or less. What should the period be? What should the coefficient of t be?

10. **Biorhythms** A popular theory that attempts to explain the ups and downs of everyday life states that each of us has three cycles, called biorhythms, which begin at birth. These three cycles can be modeled by sine waves.

Physical (23 days): $P = \sin\frac{2\pi t}{23}, \quad t \geq 0$

Emotional (28 days): $E = \sin\frac{2\pi t}{28}, \quad t \geq 0$

Intellectual (33 days): $I = \sin\frac{2\pi t}{33}, \quad t \geq 0$

where t is the number of days since birth. Consider a person who was born on July 20, 1990.

(a) Use a graphing utility to graph the three models in the same viewing window for $7300 \leq t \leq 7380$.

(b) Describe the person's biorhythms during the month of September 2010.

(c) Calculate the person's three energy levels on September 22, 2010.

11. (a) **Graphical Reasoning** Use a graphing utility to graph the functions

$f(x) = 2\cos 2x + 3\sin 3x$ and

$g(x) = 2\cos 2x + 3\sin 4x.$

(b) Use the graphs from part (a) to find the period of each function.

(c) Is the function $h(x) = A\cos\alpha x + B\sin\beta x$, where α and β are positive integers, periodic? Explain your reasoning.

12. **Analyzing Trigonometric Functions** Two trigonometric functions f and g have periods of 2, and their graphs intersect at $x = 5.35$.

(a) Give one positive value of x less than 5.35 and one value of x greater than 5.35 at which the functions have the same value.

(b) Determine one negative value of x at which the graphs intersect.

(c) Is it true that $f(13.35) = g(-4.65)$? Explain your reasoning.

13. **Refraction** When you stand in shallow water and look at an object below the surface of the water, the object will look farther away from you than it really is. This is because when light rays pass between air and water, the water refracts, or bends, the light rays. The index of refraction for water is 1.333. This is the ratio of the sine of θ_1 and the sine of θ_2 (see figure).

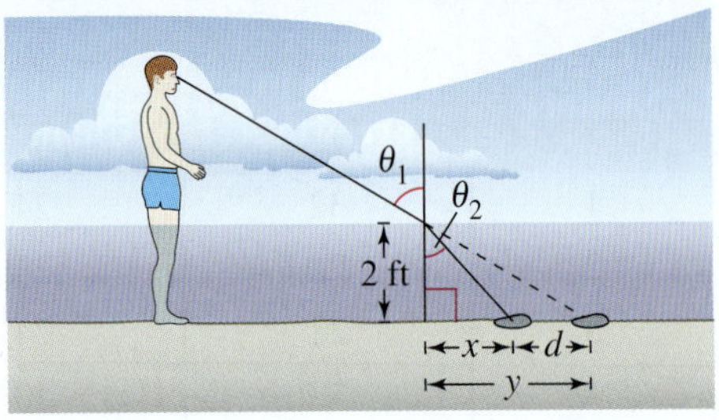

(a) While standing in water that is 2 feet deep, you look at a rock at angle $\theta_1 = 60°$ (measured from a line perpendicular to the surface of the water). Find θ_2.

(b) Find the distances x and y.

(c) Find the distance d between where the rock is and where it appears to be.

(d) What happens to d as you move closer to the rock? Explain your reasoning.

14. **Polynomial Approximation** In calculus, it can be shown that the arctangent function can be approximated by the polynomial

$$\arctan x \approx x - \frac{x^3}{3} + \frac{x^5}{5} - \frac{x^7}{7}$$

where x is in radians.

(a) Use a graphing utility to graph the arctangent function and its polynomial approximation in the same viewing window. How do the graphs compare?

(b) Study the pattern in the polynomial approximation of the arctangent function and guess the next term. Then repeat part (a). How does the accuracy of the approximation change when you add additional terms?

5 Analytic Trigonometry

Projectile Motion *(Example 10, page 388)*

Standing Waves *(page 378)*

Honeycomb Cell *(Example 10, page 370)*

Shadow Length *(Exercise 66, page 363)*

Friction *(Exercise 65, page 356)*

Clockwise from top left, Brian A Jackson/Shutterstock.com; Aspen Photo/Shutterstock.com; maigi/Shutterstock.com; Stocksnapper/Shutterstock.com; LilKar/Shutterstock.com

5.1 Using Fundamental Identities

Fundamental trigonometric identities can help you simplify trigonometric expressions. For instance, in Exercise 65 on page 356, you will use trigonometric identities to simplify an expression for the coefficient of friction.

- **Recognize and write the fundamental trigonometric identities.**
- **Use the fundamental trigonometric identities to evaluate trigonometric functions, simplify trigonometric expressions, and rewrite trigonometric expressions.**

Introduction

In Chapter 4, you studied the basic definitions, properties, graphs, and applications of the individual trigonometric functions. In this chapter, you will learn how to use the fundamental identities to do the following.

1. Evaluate trigonometric functions.
2. Simplify trigonometric expressions.
3. Develop additional trigonometric identities.
4. Solve trigonometric equations.

Fundamental Trigonometric Identities

Reciprocal Identities

$$\sin u = \frac{1}{\csc u} \qquad \cos u = \frac{1}{\sec u} \qquad \tan u = \frac{1}{\cot u}$$

$$\csc u = \frac{1}{\sin u} \qquad \sec u = \frac{1}{\cos u} \qquad \cot u = \frac{1}{\tan u}$$

Quotient Identities

$$\tan u = \frac{\sin u}{\cos u} \qquad \cot u = \frac{\cos u}{\sin u}$$

Pythagorean Identities

$$\sin^2 u + \cos^2 u = 1 \qquad 1 + \tan^2 u = \sec^2 u \qquad 1 + \cot^2 u = \csc^2 u$$

Cofunction Identities

$$\sin\left(\frac{\pi}{2} - u\right) = \cos u \qquad \cos\left(\frac{\pi}{2} - u\right) = \sin u$$

$$\tan\left(\frac{\pi}{2} - u\right) = \cot u \qquad \cot\left(\frac{\pi}{2} - u\right) = \tan u$$

$$\sec\left(\frac{\pi}{2} - u\right) = \csc u \qquad \csc\left(\frac{\pi}{2} - u\right) = \sec u$$

Even/Odd Identities

$$\sin(-u) = -\sin u \qquad \cos(-u) = \cos u \qquad \tan(-u) = -\tan u$$

$$\csc(-u) = -\csc u \qquad \sec(-u) = \sec u \qquad \cot(-u) = -\cot u$$

REMARK You should learn the fundamental trigonometric identities well, because you will use them frequently in trigonometry and they will also appear in calculus. Note that u can be an angle, a real number, or a variable.

Pythagorean identities are sometimes used in radical form such as

$$\sin u = \pm\sqrt{1 - \cos^2 u}$$

or

$$\tan u = \pm\sqrt{\sec^2 u - 1}$$

where the sign depends on the choice of u.

Stocksnapper/Shutterstock.com

Using the Fundamental Identities

One common application of trigonometric identities is to use given values of trigonometric functions to evaluate other trigonometric functions.

EXAMPLE 1 Using Identities to Evaluate a Function

Use the values $\sec u = -\frac{3}{2}$ and $\tan u > 0$ to find the values of all six trigonometric functions.

Solution Using a reciprocal identity, you have

$$\cos u = \frac{1}{\sec u} = \frac{1}{-3/2} = -\frac{2}{3}.$$

Using a Pythagorean identity, you have

$$\sin^2 u = 1 - \cos^2 u \qquad \text{Pythagorean identity}$$
$$= 1 - \left(-\tfrac{2}{3}\right)^2 \qquad \text{Substitute } -\tfrac{2}{3} \text{ for } \cos u.$$
$$= \tfrac{5}{9}. \qquad \text{Simplify.}$$

Because $\sec u < 0$ and $\tan u > 0$, it follows that u lies in Quadrant III. Moreover, because $\sin u$ is negative when u is in Quadrant III, choose the negative root and obtain $\sin u = -\sqrt{5}/3$. Knowing the values of the sine and cosine enables you to find the values of all six trigonometric functions.

$$\sin u = -\frac{\sqrt{5}}{3} \qquad \csc u = \frac{1}{\sin u} = -\frac{3}{\sqrt{5}} = -\frac{3\sqrt{5}}{5}$$

$$\cos u = -\frac{2}{3} \qquad \sec u = \frac{1}{\cos u} = -\frac{3}{2}$$

$$\tan u = \frac{\sin u}{\cos u} = \frac{-\sqrt{5}/3}{-2/3} = \frac{\sqrt{5}}{2} \qquad \cot u = \frac{1}{\tan u} = \frac{2}{\sqrt{5}} = \frac{2\sqrt{5}}{5}$$

✓ Checkpoint Audio-video solution in English & Spanish at LarsonPrecalculus.com.

Use the values $\tan x = \frac{1}{3}$ and $\cos x < 0$ to find the values of all six trigonometric functions.

TECHNOLOGY To use a graphing utility to check the result of Example 2, graph

$$y_1 = \sin x \cos^2 x - \sin x$$

and

$$y_2 = -\sin^3 x$$

in the same viewing window, as shown below. Because Example 2 shows the equivalence algebraically and the two graphs appear to coincide, you can conclude that the expressions are equivalent.

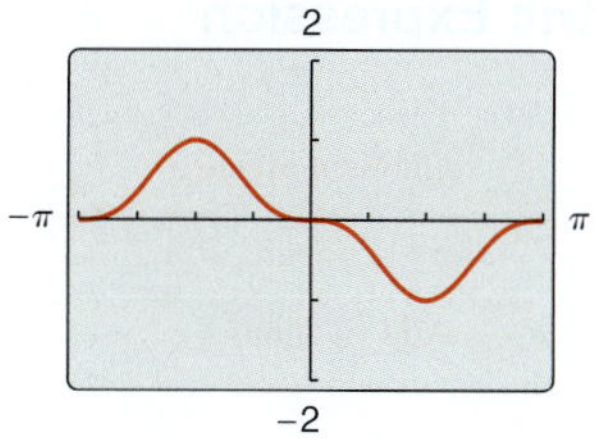

EXAMPLE 2 Simplifying a Trigonometric Expression

Simplify

$$\sin x \cos^2 x - \sin x.$$

Solution First factor out a common monomial factor and then use a fundamental identity.

$$\sin x \cos^2 x - \sin x = \sin x(\cos^2 x - 1) \qquad \text{Factor out common monomial factor.}$$
$$= -\sin x(1 - \cos^2 x) \qquad \text{Factor out } -1.$$
$$= -\sin x(\sin^2 x) \qquad \text{Pythagorean identity}$$
$$= -\sin^3 x \qquad \text{Multiply.}$$

✓ Checkpoint Audio-video solution in English & Spanish at LarsonPrecalculus.com.

Simplify

$$\cos^2 x \csc x - \csc x.$$

When factoring trigonometric expressions, it is helpful to find a special polynomial factoring form that fits the expression, as shown in Example 3.

EXAMPLE 3 **Factoring Trigonometric Expressions**

Factor each expression.

a. $\sec^2 \theta - 1$ **b.** $4 \tan^2 \theta + \tan \theta - 3$

Solution

a. This expression has the form $u^2 - v^2$, which is the difference of two squares. It factors as

$$\sec^2 \theta - 1 = (\sec \theta + 1)(\sec \theta - 1).$$

b. This expression has the polynomial form $ax^2 + bx + c$, and it factors as

$$4 \tan^2 \theta + \tan \theta - 3 = (4 \tan \theta - 3)(\tan \theta + 1).$$

ALGEBRA HELP In Example 3, you need to be able to factor the difference of two squares and factor a trinomial. You can review the techniques for factoring in Appendix A.3.

Checkpoint *Audio-video solution in English & Spanish at LarsonPrecalculus.com.*

Factor each expression.

a. $1 - \cos^2 \theta$ **b.** $2 \csc^2 \theta - 7 \csc \theta + 6$

On occasion, factoring or simplifying can best be done by first rewriting the expression in terms of just *one* trigonometric function or in terms of *sine and cosine only*. Examples 4 and 5, respectively, show these strategies.

EXAMPLE 4 **Factoring a Trigonometric Expression**

Factor $\csc^2 x - \cot x - 3$.

Solution Use the identity $\csc^2 x = 1 + \cot^2 x$ to rewrite the expression.

$$\begin{aligned} \csc^2 x - \cot x - 3 &= (1 + \cot^2 x) - \cot x - 3 && \text{Pythagorean identity} \\ &= \cot^2 x - \cot x - 2 && \text{Combine like terms.} \\ &= (\cot x - 2)(\cot x + 1) && \text{Factor.} \end{aligned}$$

Checkpoint *Audio-video solution in English & Spanish at LarsonPrecalculus.com.*

Factor $\sec^2 x + 3 \tan x + 1$.

REMARK Remember that when adding rational expressions, you must first find the least common denominator (LCD). In Example 5, the LCD is $\sin t$.

EXAMPLE 5 **Simplifying a Trigonometric Expression**

$$\begin{aligned} \sin t + \cot t \cos t &= \sin t + \left(\frac{\cos t}{\sin t}\right) \cos t && \text{Quotient identity} \\ &= \frac{\sin^2 t + \cos^2 t}{\sin t} && \text{Add fractions.} \\ &= \frac{1}{\sin t} && \text{Pythagorean identity} \\ &= \csc t && \text{Reciprocal identity} \end{aligned}$$

Checkpoint *Audio-video solution in English & Spanish at LarsonPrecalculus.com.*

Simplify $\csc x - \cos x \cot x$.

EXAMPLE 6 Adding Trigonometric Expressions

Perform the addition $\dfrac{\sin\theta}{1+\cos\theta}+\dfrac{\cos\theta}{\sin\theta}$ and simplify.

Solution

$$\frac{\sin\theta}{1+\cos\theta}+\frac{\cos\theta}{\sin\theta}=\frac{(\sin\theta)(\sin\theta)+(\cos\theta)(1+\cos\theta)}{(1+\cos\theta)(\sin\theta)}$$

$$=\frac{\sin^2\theta+\cos^2\theta+\cos\theta}{(1+\cos\theta)(\sin\theta)}$$ Multiply.

$$=\frac{\cancel{1+\cos\theta}}{(\cancel{1+\cos\theta})(\sin\theta)}$$ Pythagorean identity: $\sin^2\theta+\cos^2\theta=1$

$$=\frac{1}{\sin\theta}$$ Divide out common factor.

$$=\csc\theta$$ Reciprocal identity

Checkpoint Audio-video solution in English & Spanish at LarsonPrecalculus.com.

Perform the addition $\dfrac{1}{1+\sin\theta}+\dfrac{1}{1-\sin\theta}$ and simplify.

The next two examples involve techniques for rewriting expressions in forms that are used in calculus.

EXAMPLE 7 Rewriting a Trigonometric Expression

Rewrite $\dfrac{1}{1+\sin x}$ so that it is *not* in fractional form.

Solution From the Pythagorean identity

$$\cos^2 x=1-\sin^2 x=(1-\sin x)(1+\sin x)$$

multiplying both the numerator and the denominator by $(1-\sin x)$ will produce a monomial denominator.

$$\frac{1}{1+\sin x}=\frac{1}{1+\sin x}\cdot\frac{1-\sin x}{1-\sin x}$$ Multiply numerator and denominator by $(1-\sin x)$.

$$=\frac{1-\sin x}{1-\sin^2 x}$$ Multiply.

$$=\frac{1-\sin x}{\cos^2 x}$$ Pythagorean identity

$$=\frac{1}{\cos^2 x}-\frac{\sin x}{\cos^2 x}$$ Write as separate fractions.

$$=\frac{1}{\cos^2 x}-\frac{\sin x}{\cos x}\cdot\frac{1}{\cos x}$$ Product of fractions

$$=\sec^2 x-\tan x\sec x$$ Reciprocal and quotient identities

Checkpoint Audio-video solution in English & Spanish at LarsonPrecalculus.com.

Rewrite $\dfrac{\cos^2\theta}{1-\sin\theta}$ so that it is *not* in fractional form.

EXAMPLE 8 Trigonometric Substitution

Use the substitution $x = 2\tan\theta$, $0 < \theta < \pi/2$, to write $\sqrt{4 + x^2}$ as a trigonometric function of θ.

Solution Begin by letting $x = 2\tan\theta$. Then, you obtain

$$\sqrt{4 + x^2} = \sqrt{4 + (2\tan\theta)^2} \quad \text{Substitute } 2\tan\theta \text{ for } x.$$
$$= \sqrt{4 + 4\tan^2\theta} \quad \text{Rule of exponents}$$
$$= \sqrt{4(1 + \tan^2\theta)} \quad \text{Factor.}$$
$$= \sqrt{4\sec^2\theta} \quad \text{Pythagorean identity}$$
$$= 2\sec\theta. \quad \sec\theta > 0 \text{ for } 0 < \theta < \pi/2$$

✓ **Checkpoint** Audio-video solution in English & Spanish at LarsonPrecalculus.com.

Use the substitution $x = 3\sin\theta$, $0 < \theta < \pi/2$, to write $\sqrt{9 - x^2}$ as a trigonometric function of θ.

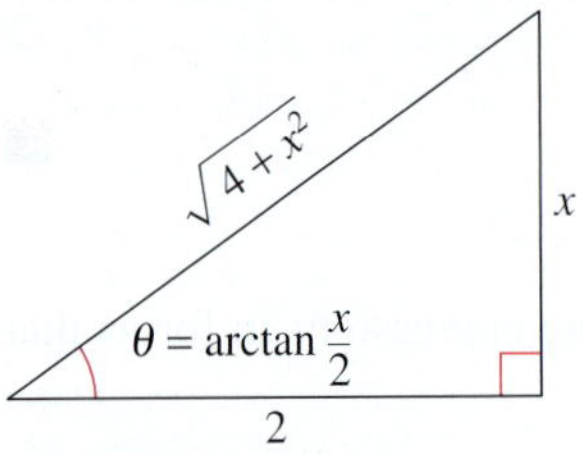

Angle whose tangent is $x/2$
Figure 5.1

Figure 5.1 shows the right triangle illustration of the trigonometric substitution $x = 2\tan\theta$ in Example 8. Use this triangle to check the solution of Example 8, as follows. For $0 < \theta < \pi/2$, you have

$$\text{opp} = x, \quad \text{adj} = 2, \quad \text{and} \quad \text{hyp} = \sqrt{4 + x^2}.$$

With these expressions, you can write

$$\sec\theta = \frac{\text{hyp}}{\text{adj}} = \frac{\sqrt{4 + x^2}}{2}.$$

So, $2\sec\theta = \sqrt{4 + x^2}$, and the solution checks.

EXAMPLE 9 Rewriting a Logarithmic Expression

Rewrite $\ln|\csc\theta| + \ln|\tan\theta|$ as a single logarithm and simplify the result.

Solution

$$\ln|\csc\theta| + \ln|\tan\theta| = \ln|\csc\theta\tan\theta| \quad \text{Product Property of Logarithms}$$
$$= \ln\left|\frac{1}{\sin\theta} \cdot \frac{\sin\theta}{\cos\theta}\right| \quad \text{Reciprocal and quotient identities}$$
$$= \ln\left|\frac{1}{\cos\theta}\right| \quad \text{Simplify.}$$
$$= \ln|\sec\theta| \quad \text{Reciprocal identity}$$

ALGEBRA HELP Recall that for positive real numbers u and v,

$$\ln u + \ln v = \ln(uv).$$

You can review the properties of logarithms in Section 3.3.

✓ **Checkpoint** Audio-video solution in English & Spanish at LarsonPrecalculus.com.

Rewrite $\ln|\sec x| + \ln|\sin x|$ as a single logarithm and simplify the result.

Summarize (Section 5.1)

1. State the fundamental trigonometric identities *(page 350)*.
2. Explain how to use the fundamental trigonometric identities to evaluate trigonometric functions, simplify trigonometric expressions, and rewrite trigonometric expressions *(pages 351–354)*. For examples of these concepts, see Examples 1–9.

5.1 Exercises

See CalcChat.com for tutorial help and worked-out solutions to odd-numbered exercises.

Vocabulary: **Fill in the blank to complete the trigonometric identity.**

1. $\dfrac{\sin u}{\cos u} =$ ______

2. $\dfrac{1}{\csc u} =$ ______

3. $\dfrac{1}{\tan u} =$ ______

4. $\sec\left(\dfrac{\pi}{2} - u\right) =$ ______

5. $1 +$ ______ $= \csc^2 u$

6. $\cot(-u) =$ ______

Skills and Applications

Using Identities to Evaluate a Function **In Exercises 7–14, use the given values to find the values (if possible) of all six trigonometric functions.**

7. $\sin x = \dfrac{1}{2}, \quad \cos x = \dfrac{\sqrt{3}}{2}$

8. $\csc \theta = \dfrac{25}{7}, \quad \tan \theta = \dfrac{7}{24}$

9. $\cos\left(\dfrac{\pi}{2} - x\right) = \dfrac{3}{5}, \quad \cos x = \dfrac{4}{5}$

10. $\sin(-x) = -\dfrac{1}{3}, \quad \tan x = -\dfrac{\sqrt{2}}{4}$

11. $\sec x = 4, \quad \sin x > 0$

12. $\csc \theta = -5, \quad \cos \theta < 0$

13. $\sin \theta = -1, \quad \cot \theta = 0$

14. $\tan \theta$ is undefined, $\quad \sin \theta > 0$

Matching Trigonometric Expressions **In Exercises 15–20, match the trigonometric expression with one of the following.**

(a) $\csc x$ (b) -1 (c) 1

(d) $\sin x \tan x$ (e) $\sec^2 x$ (f) $\sec^2 x + \tan^2 x$

15. $\sec x \cos x$

16. $\cot^2 x - \csc^2 x$

17. $\sec^4 x - \tan^4 x$

18. $\cot x \sec x$

19. $\dfrac{\sec^2 x - 1}{\sin^2 x}$

20. $\dfrac{\cos^2[(\pi/2) - x]}{\cos x}$

Factoring a Trigonometric Expression **In Exercises 21–28, factor the expression and use the fundamental identities to simplify. There is more than one correct form of each answer.**

21. $\tan^2 x - \tan^2 x \sin^2 x$

22. $\sin^2 x \sec^2 x - \sin^2 x$

23. $\dfrac{\sec^2 x - 1}{\sec x - 1}$

24. $\dfrac{\cos x - 2}{\cos^2 x - 4}$

25. $1 - 2\cos^2 x + \cos^4 x$

26. $\sec^4 x - \tan^4 x$

27. $\cot^3 x + \cot^2 x + \cot x + 1$

28. $\sec^3 x - \sec^2 x - \sec x + 1$

Factoring a Trigonometric Expression **In Exercises 29–32, factor the trigonometric expression. There is more than one correct form of each answer.**

29. $3\sin^2 x - 5\sin x - 2$

30. $6\cos^2 x + 5\cos x - 6$

31. $\cot^2 x + \csc x - 1$

32. $\sin^2 x + 3\cos x + 3$

Multiplying Trigonometric Expressions **In Exercises 33 and 34, perform the multiplication and use the fundamental identities to simplify. There is more than one correct form of each answer.**

33. $(\sin x + \cos x)^2$

34. $(2\csc x + 2)(2\csc x - 2)$

Simplifying a Trigonometric Expression **In Exercises 35–44, use the fundamental identities to simplify the expression. There is more than one correct form of each answer.**

35. $\cot \theta \sec \theta$

36. $\tan(-x)\cos x$

37. $\sin \phi(\csc \phi - \sin \phi)$

38. $\cos t(1 + \tan^2 t)$

39. $\dfrac{1 - \sin^2 x}{\csc^2 x - 1}$

40. $\dfrac{\tan \theta \cot \theta}{\sec \theta}$

41. $\cos\left(\dfrac{\pi}{2} - x\right)\sec x$

42. $\dfrac{\cos^2 y}{1 - \sin y}$

43. $\sin \beta \tan \beta + \cos \beta$

44. $\cot u \sin u + \tan u \cos u$

Adding or Subtracting Trigonometric Expressions **In Exercises 45–48, perform the addition or subtraction and use the fundamental identities to simplify. There is more than one correct form of each answer.**

45. $\dfrac{1}{1 + \cos x} + \dfrac{1}{1 - \cos x}$

46. $\dfrac{1}{\sec x + 1} - \dfrac{1}{\sec x - 1}$

47. $\tan x - \dfrac{\sec^2 x}{\tan x}$

48. $\dfrac{\cos x}{1 + \sin x} + \dfrac{1 + \sin x}{\cos x}$

Rewriting a Trigonometric Expression **In Exercises 49 and 50, rewrite the expression so that it is not in fractional form. There is more than one correct form of each answer.**

49. $\dfrac{\sin^2 y}{1 - \cos y}$

50. $\dfrac{5}{\tan x + \sec x}$

EXAMPLE 1 Verifying a Trigonometric Identity

Verify the identity $\dfrac{\sec^2\theta - 1}{\sec^2\theta} = \sin^2\theta$.

REMARK Remember that an identity is only true for all real values in the domain of the variable. For instance, in Example 1 the identity is not true when $\theta = \pi/2$ because $\sec^2\theta$ is not defined when $\theta = \pi/2$.

Solution Start with the left side because it is more complicated.

$$\frac{\sec^2\theta - 1}{\sec^2\theta} = \frac{(\tan^2\theta + 1) - 1}{\sec^2\theta} \quad \text{Pythagorean identity}$$

$$= \frac{\tan^2\theta}{\sec^2\theta} \quad \text{Simplify.}$$

$$= \tan^2\theta(\cos^2\theta) \quad \text{Reciprocal identity}$$

$$= \frac{\sin^2\theta}{(\cos^2\theta)}(\cos^2\theta) \quad \text{Quotient identity}$$

$$= \sin^2\theta \quad \text{Simplify.}$$

Notice that you verify the identity by starting with the left side of the equation (the more complicated side) and using the fundamental trigonometric identities to simplify it until you obtain the right side.

✓ **Checkpoint** Audio-video solution in English & Spanish at LarsonPrecalculus.com.

Verify the identity $\dfrac{\sin^2\theta + \cos^2\theta}{\cos^2\theta\sec^2\theta} = 1$.

There can be more than one way to verify an identity. Here is another way to verify the identity in Example 1.

$$\frac{\sec^2\theta - 1}{\sec^2\theta} = \frac{\sec^2\theta}{\sec^2\theta} - \frac{1}{\sec^2\theta} \quad \text{Write as separate fractions.}$$

$$= 1 - \cos^2\theta \quad \text{Reciprocal identity}$$

$$= \sin^2\theta \quad \text{Pythagorean identity}$$

EXAMPLE 2 Verifying a Trigonometric Identity

Verify the identity $2\sec^2\alpha = \dfrac{1}{1 - \sin\alpha} + \dfrac{1}{1 + \sin\alpha}$.

Algebraic Solution

Start with the right side because it is more complicated.

$$\frac{1}{1 - \sin\alpha} + \frac{1}{1 + \sin\alpha} = \frac{1 + \sin\alpha + 1 - \sin\alpha}{(1 - \sin\alpha)(1 + \sin\alpha)} \quad \text{Add fractions.}$$

$$= \frac{2}{1 - \sin^2\alpha} \quad \text{Simplify.}$$

$$= \frac{2}{\cos^2\alpha} \quad \text{Pythagorean identity}$$

$$= 2\sec^2\alpha \quad \text{Reciprocal identity}$$

Numerical Solution

Use a graphing utility to create a table that shows the values of $y_1 = 2/\cos^2 x$ and $y_2 = 1/(1 - \sin x) + 1/(1 + \sin x)$ for different values of x.

X	Y1	Y2
-.5	2.5969	2.5969
-.25	2.1304	2.1304
0	2	2
.25	2.1304	2.1304
.5	2.5969	2.5969
.75	3.7357	3.7357
1	6.851	6.851

X=-.5

The values for y_1 and y_2 appear to be identical, so the equation appears to be an identity.

✓ **Checkpoint** Audio-video solution in English & Spanish at LarsonPrecalculus.com.

Verify the identity $2\csc^2\beta = \dfrac{1}{1 - \cos\beta} + \dfrac{1}{1 + \cos\beta}$.

In Example 2, you needed to write the Pythagorean identity $\sin^2 u + \cos^2 u = 1$ in the equivalent form $\cos^2 u = 1 - \sin^2 u$. When verifying identities, you may find it useful to write the Pythagorean identities in one of these equivalent forms.

Pythagorean Identities	**Equivalent Forms**
$\sin^2 u + \cos^2 u = 1$	$\sin^2 u = 1 - \cos^2 u$
	$\cos^2 u = 1 - \sin^2 u$
$1 + \tan^2 u = \sec^2 u$	$1 = \sec^2 u - \tan^2 u$
	$\tan^2 u = \sec^2 u - 1$
$1 + \cot^2 u = \csc^2 u$	$1 = \csc^2 u - \cot^2 u$
	$\cot^2 u = \csc^2 u - 1$

EXAMPLE 3 Verifying a Trigonometric Identity

Verify the identity $(\tan^2 x + 1)(\cos^2 x - 1) = -\tan^2 x$.

Algebraic Solution

By applying identities before multiplying, you obtain the following.

$$(\tan^2 x + 1)(\cos^2 x - 1) = (\sec^2 x)(-\sin^2 x) \quad \text{Pythagorean identities}$$

$$= -\frac{\sin^2 x}{\cos^2 x} \quad \text{Reciprocal identity}$$

$$= -\left(\frac{\sin x}{\cos x}\right)^2 \quad \text{Property of exponents}$$

$$= -\tan^2 x \quad \text{Quotient identity}$$

Graphical Solution

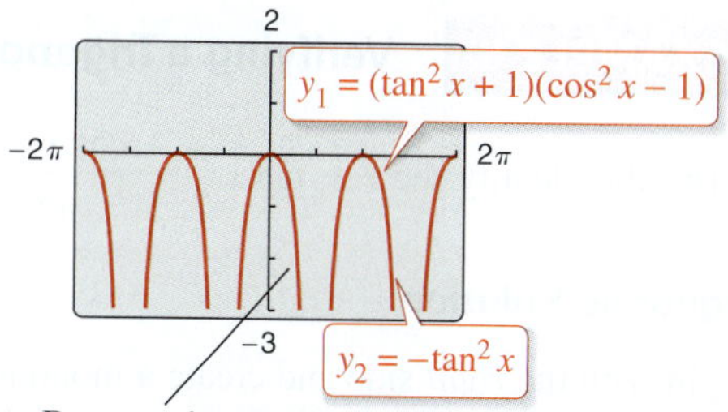

Because the graphs appear to coincide, the given equation appears to be an identity.

✓ **Checkpoint** *Audio-video solution in English & Spanish at LarsonPrecalculus.com.*

Verify the identity $(\sec^2 x - 1)(\sin^2 x - 1) = -\sin^2 x$.

REMARK Although a graphing utility can be useful in helping to verify an identity, you must use algebraic techniques to produce a *valid* proof.

EXAMPLE 4 Converting to Sines and Cosines

Verify the identity $\tan x + \cot x = \sec x \csc x$.

Solution Convert the left side into sines and cosines.

$$\tan x + \cot x = \frac{\sin x}{\cos x} + \frac{\cos x}{\sin x} \quad \text{Quotient identities}$$

$$= \frac{\sin^2 x + \cos^2 x}{\cos x \sin x} \quad \text{Add fractions.}$$

$$= \frac{1}{\cos x \sin x} \quad \text{Pythagorean identity}$$

$$= \frac{1}{\cos x} \cdot \frac{1}{\sin x} \quad \text{Product of fractions}$$

$$= \sec x \csc x \quad \text{Reciprocal identities}$$

✓ **Checkpoint** *Audio-video solution in English & Spanish at LarsonPrecalculus.com.*

Verify the identity $\csc x - \sin x = \cos x \cot x$.

Recall from algebra that *rationalizing the denominator* using conjugates is, on occasion, a powerful simplification technique. A related form of this technique works for simplifying trigonometric expressions as well. For instance, to simplify

$$\frac{1}{1 - \cos x}$$

multiply the numerator and the denominator by $1 + \cos x$.

$$\begin{aligned}\frac{1}{1 - \cos x} &= \frac{1}{1 - \cos x}\left(\frac{1 + \cos x}{1 + \cos x}\right)\\ &= \frac{1 + \cos x}{1 - \cos^2 x}\\ &= \frac{1 + \cos x}{\sin^2 x}\\ &= \csc^2 x(1 + \cos x)\end{aligned}$$

The expression $\csc^2 x(1 + \cos x)$ is considered a simplified form of

$$\frac{1}{1 - \cos x}$$

because $\csc^2 x(1 + \cos x)$ does not contain fractions.

EXAMPLE 5 Verifying a Trigonometric Identity

Verify the identity $\sec x + \tan x = \dfrac{\cos x}{1 - \sin x}$.

Algebraic Solution

Begin with the *right* side and create a monomial denominator by multiplying the numerator and the denominator by $1 + \sin x$.

$$\begin{aligned}\frac{\cos x}{1 - \sin x} &= \frac{\cos x}{1 - \sin x}\left(\frac{1 + \sin x}{1 + \sin x}\right) && \text{Multiply numerator and denominator by } 1 + \sin x.\\ &= \frac{\cos x + \cos x \sin x}{1 - \sin^2 x} && \text{Multiply.}\\ &= \frac{\cos x + \cos x \sin x}{\cos^2 x} && \text{Pythagorean identity}\\ &= \frac{\cos x}{\cos^2 x} + \frac{\cos x \sin x}{\cos^2 x} && \text{Write as separate fractions.}\\ &= \frac{1}{\cos x} + \frac{\sin x}{\cos x} && \text{Simplify.}\\ &= \sec x + \tan x && \text{Identities}\end{aligned}$$

Graphical Solution

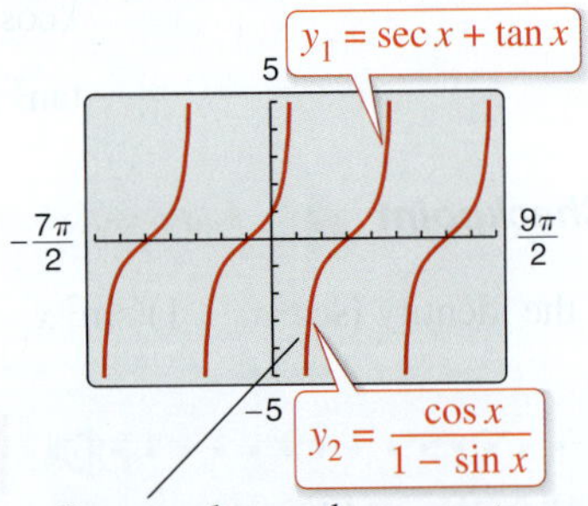

Because the graphs appear to coincide, the given equation appears to be an identity.

Checkpoint *Audio-video solution in English & Spanish at LarsonPrecalculus.com.*

Verify the identity $\csc x + \cot x = \dfrac{\sin x}{1 - \cos x}$.

In Examples 1 through 5, you have been verifying trigonometric identities by working with one side of the equation and converting to the form given on the other side. On occasion, it is practical to work with each side *separately*, to obtain one common form that is equivalent to both sides. This is illustrated in Example 6.

EXAMPLE 6 Working with Each Side Separately

Verify the identity $\dfrac{\cot^2\theta}{1+\csc\theta} = \dfrac{1-\sin\theta}{\sin\theta}$.

Algebraic Solution

Working with the left side, you have

$$\frac{\cot^2\theta}{1+\csc\theta} = \frac{\csc^2\theta - 1}{1+\csc\theta} \qquad \text{Pythagorean identity}$$

$$= \frac{(\csc\theta - 1)\cancel{(\csc\theta + 1)}}{\cancel{1+\csc\theta}} \qquad \text{Factor.}$$

$$= \csc\theta - 1. \qquad \text{Simplify.}$$

Now, simplifying the right side, you have

$$\frac{1-\sin\theta}{\sin\theta} = \frac{1}{\sin\theta} - \frac{\sin\theta}{\sin\theta} = \csc\theta - 1.$$

This verifies the identity because both sides are equal to $\csc\theta - 1$.

Numerical Solution

Use a graphing utility to create a table that shows the values of

$$y_1 = \frac{\cot^2 x}{1+\csc x} \quad \text{and} \quad y_2 = \frac{1-\sin x}{\sin x}$$

for different values of x.

X	Y1	Y2
-.5	-3.086	-3.086
-.25	-5.042	-5.042
0	ERROR	ERROR
.25	3.042	3.042
.5	1.0858	1.0858
.75	.46705	.46705
1	.1884	.1884

X=1

The values for y_1 and y_2 appear to be identical, so the equation appears to be an identity.

✓ **Checkpoint** *Audio-video solution in English & Spanish at LarsonPrecalculus.com.*

Verify the identity $\dfrac{\tan^2\theta}{1+\sec\theta} = \dfrac{1-\cos\theta}{\cos\theta}$.

Example 7 shows powers of trigonometric functions rewritten as more complicated sums of products of trigonometric functions. This is a common procedure used in calculus.

EXAMPLE 7 Two Examples from Calculus

Verify each identity.

a. $\tan^4 x = \tan^2 x \sec^2 x - \tan^2 x$ **b.** $\csc^4 x \cot x = \csc^2 x(\cot x + \cot^3 x)$

Solution

a. $\tan^4 x = (\tan^2 x)(\tan^2 x)$ Write as separate factors.

$= \tan^2 x(\sec^2 x - 1)$ Pythagorean identity

$= \tan^2 x \sec^2 x - \tan^2 x$ Multiply.

b. $\csc^4 x \cot x = \csc^2 x \csc^2 x \cot x$ Write as separate factors.

$= \csc^2 x(1 + \cot^2 x)\cot x$ Pythagorean identity

$= \csc^2 x(\cot x + \cot^3 x)$ Multiply.

✓ **Checkpoint** *Audio-video solution in English & Spanish at LarsonPrecalculus.com.*

Verify each identity.

a. $\tan^3 x = \tan x \sec^2 x - \tan x$ **b.** $\sin^3 x \cos^4 x = (\cos^4 x - \cos^6 x)\sin x$

Summarize (Section 5.2)

1. State the guidelines for verifying trigonometric identities *(page 357)*. For examples of verifying trigonometric identities, see Examples 1–7.

5.2 Exercises

See CalcChat.com for tutorial help and worked-out solutions to odd-numbered exercises.

Vocabulary

In Exercises 1 and 2, fill in the blanks.

1. An equation that is true for all real values in its domain is called an ________.
2. An equation that is true for only some values in its domain is called a ________ ________.

In Exercises 3–8, fill in the blank to complete the fundamental trigonometric identity.

3. $\dfrac{1}{\cot u} = $ ________
4. $\dfrac{\cos u}{\sin u} = $ ________
5. $\sin^2 u + $ ________ $= 1$
6. $\cos\left(\dfrac{\pi}{2} - u\right) = $ ________
7. $\csc(-u) = $ ________
8. $\sec(-u) = $ ________

Skills and Applications

Verifying a Trigonometric Identity **In Exercises 9–50, verify the identity.**

9. $\tan t \cot t = 1$
10. $\sec y \cos y = 1$
11. $\cot^2 y(\sec^2 y - 1) = 1$
12. $\cos x + \sin x \tan x = \sec x$
13. $(1 + \sin \alpha)(1 - \sin \alpha) = \cos^2 \alpha$
14. $\cos^2 \beta - \sin^2 \beta = 2\cos^2 \beta - 1$
15. $\cos^2 \beta - \sin^2 \beta = 1 - 2\sin^2 \beta$
16. $\sin^2 \alpha - \sin^4 \alpha = \cos^2 \alpha - \cos^4 \alpha$
17. $\dfrac{\tan^2 \theta}{\sec \theta} = \sin \theta \tan \theta$
18. $\dfrac{\cot^3 t}{\csc t} = \cos t(\csc^2 t - 1)$
19. $\dfrac{\cot^2 t}{\csc t} = \dfrac{1 - \sin^2 t}{\sin t}$
20. $\dfrac{1}{\tan \beta} + \tan \beta = \dfrac{\sec^2 \beta}{\tan \beta}$
21. $\sin^{1/2} x \cos x - \sin^{5/2} x \cos x = \cos^3 x\sqrt{\sin x}$
22. $\sec^6 x(\sec x \tan x) - \sec^4 x(\sec x \tan x) = \sec^5 x \tan^3 x$
23. $\dfrac{\cot x}{\sec x} = \csc x - \sin x$
24. $\dfrac{\sec \theta - 1}{1 - \cos \theta} = \sec \theta$
25. $\sec x - \cos x = \sin x \tan x$
26. $\sec x(\csc x - 2\sin x) = \cot x - \tan x$
27. $\dfrac{1}{\tan x} + \dfrac{1}{\cot x} = \tan x + \cot x$
28. $\dfrac{1}{\sin x} - \dfrac{1}{\csc x} = \csc x - \sin x$
29. $\dfrac{1 + \sin \theta}{\cos \theta} + \dfrac{\cos \theta}{1 + \sin \theta} = 2\sec \theta$
30. $\dfrac{\cos \theta \cot \theta}{1 - \sin \theta} - 1 = \csc \theta$
31. $\dfrac{1}{\cos x + 1} + \dfrac{1}{\cos x - 1} = -2\csc x \cot x$
32. $\cos x - \dfrac{\cos x}{1 - \tan x} = \dfrac{\sin x \cos x}{\sin x - \cos x}$
33. $\tan\left(\dfrac{\pi}{2} - \theta\right)\tan \theta = 1$
34. $\dfrac{\cos[(\pi/2) - x]}{\sin[(\pi/2) - x]} = \tan x$
35. $\dfrac{\tan x \cot x}{\cos x} = \sec x$
36. $\dfrac{\csc(-x)}{\sec(-x)} = -\cot x$
37. $(1 + \sin y)[1 + \sin(-y)] = \cos^2 y$
38. $\dfrac{\tan x + \tan y}{1 - \tan x \tan y} = \dfrac{\cot x + \cot y}{\cot x \cot y - 1}$
39. $\dfrac{\tan x + \cot y}{\tan x \cot y} = \tan y + \cot x$
40. $\dfrac{\cos x - \cos y}{\sin x + \sin y} + \dfrac{\sin x - \sin y}{\cos x + \cos y} = 0$
41. $\sqrt{\dfrac{1 + \sin \theta}{1 - \sin \theta}} = \dfrac{1 + \sin \theta}{|\cos \theta|}$
42. $\sqrt{\dfrac{1 - \cos \theta}{1 + \cos \theta}} = \dfrac{1 - \cos \theta}{|\sin \theta|}$
43. $\cos^2 \beta + \cos^2\left(\dfrac{\pi}{2} - \beta\right) = 1$
44. $\sec^2 y - \cot^2\left(\dfrac{\pi}{2} - y\right) = 1$
45. $\sin t \csc\left(\dfrac{\pi}{2} - t\right) = \tan t$
46. $\sec^2\left(\dfrac{\pi}{2} - x\right) - 1 = \cot^2 x$
47. $\tan(\sin^{-1} x) = \dfrac{x}{\sqrt{1 - x^2}}$
48. $\cos(\sin^{-1} x) = \sqrt{1 - x^2}$
49. $\tan\left(\sin^{-1}\dfrac{x - 1}{4}\right) = \dfrac{x - 1}{\sqrt{16 - (x - 1)^2}}$
50. $\tan\left(\cos^{-1}\dfrac{x + 1}{2}\right) = \dfrac{\sqrt{4 - (x + 1)^2}}{x + 1}$

Error Analysis **In Exercises 51 and 52, describe the error(s).**

51. $(1 + \tan x)[1 + \cot(-x)]$
$= (1 + \tan x)(1 + \cot x)$
$= 1 + \cot x + \tan x + \tan x \cot x$
$= 1 + \cot x + \tan x + 1$
$= 2 + \cot x + \tan x$

52. $\dfrac{1 + \sec(-\theta)}{\sin(-\theta) + \tan(-\theta)} = \dfrac{1 - \sec\theta}{\sin\theta - \tan\theta}$
$= \dfrac{1 - \sec\theta}{(\sin\theta)[1 - (1/\cos\theta)]}$
$= \dfrac{1 - \sec\theta}{\sin\theta(1 - \sec\theta)}$
$= \dfrac{1}{\sin\theta} = \csc\theta$

Determining Trigonometric Identities **In Exercises 53–58, (a) use a graphing utility to graph each side of the equation to determine whether the equation is an identity, (b) use the *table* feature of the graphing utility to determine whether the equation is an identity, and (c) confirm the results of parts (a) and (b) algebraically.**

53. $(1 + \cot^2 x)(\cos^2 x) = \cot^2 x$

54. $\csc x(\csc x - \sin x) + \dfrac{\sin x - \cos x}{\sin x} + \cot x = \csc^2 x$

55. $2 + \cos^2 x - 3\cos^4 x = \sin^2 x(3 + 2\cos^2 x)$

56. $\tan^4 x + \tan^2 x - 3 = \sec^2 x(4\tan^2 x - 3)$

57. $\dfrac{1 + \cos x}{\sin x} = \dfrac{\sin x}{1 - \cos x}$

58. $\dfrac{\cot\alpha}{\csc\alpha + 1} = \dfrac{\csc\alpha + 1}{\cot\alpha}$

Verifying a Trigonometric Identity **In Exercises 59–62, verify the identity.**

59. $\tan^5 x = \tan^3 x \sec^2 x - \tan^3 x$

60. $\sec^4 x \tan^2 x = (\tan^2 x + \tan^4 x)\sec^2 x$

61. $\cos^3 x \sin^2 x = (\sin^2 x - \sin^4 x)\cos x$

62. $\sin^4 x + \cos^4 x = 1 - 2\cos^2 x + 2\cos^4 x$

Using Cofunction Identities **In Exercises 63 and 64, use the cofunction identities to evaluate the expression without using a calculator.**

63. $\sin^2 25° + \sin^2 65°$

64. $\tan^2 63° + \cot^2 16° - \sec^2 74° - \csc^2 27°$

65. Rate of Change The rate of change of the function $f(x) = \sin x + \csc x$ with respect to change in the variable x is given by the expression $\cos x - \csc x \cot x$. Show that the expression for the rate of change can also be written as $-\cos x \cot^2 x$.

66. Shadow Length The length s of a shadow cast by a vertical gnomon (a device used to tell time) of height h when the angle of the sun above the horizon is θ can be modeled by the equation

$$s = \frac{h\sin(90° - \theta)}{\sin\theta}.$$

(a) Verify that the expression for s is equal to $h\cot\theta$.

(b) Use a graphing utility to complete the table. Let $h = 5$ feet.

θ	15°	30°	45°	60°	75°	90°
s						

(c) Use your table from part (b) to determine the angles of the sun that result in the maximum and minimum lengths of the shadow.

(d) Based on your results from part (c), what time of day do you think it is when the angle of the sun above the horizon is 90°?

Exploration

True or False? **In Exercises 67–69, determine whether the statement is true or false. Justify your answer.**

67. There can be more than one way to verify a trigonometric identity.

68. The equation $\sin^2\theta + \cos^2\theta = 1 + \tan^2\theta$ is an identity because $\sin^2(0) + \cos^2(0) = 1$ and $1 + \tan^2(0) = 1$.

69. $\sin x^2 = \sin^2 x$

70. HOW DO YOU SEE IT? Explain how to use the figure to derive the identity

$$\frac{\sec^2\theta - 1}{\sec^2\theta} = \sin^2\theta$$

given in Example 1.

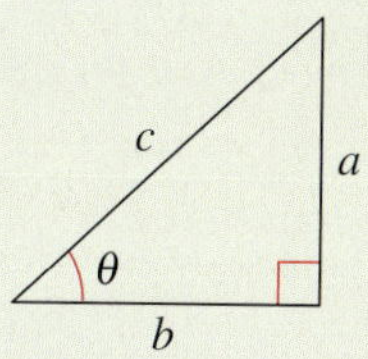

Think About It **In Exercises 71–74, explain why the equation is not an identity and find one value of the variable for which the equation is not true.**

71. $\sin\theta = \sqrt{1 - \cos^2\theta}$

72. $\tan\theta = \sqrt{\sec^2\theta - 1}$

73. $1 - \cos\theta = \sin\theta$

74. $1 + \tan\theta = \sec\theta$

Robert W. Ginn/PhotoEdit

5.3 Solving Trigonometric Equations

Trigonometric equations can help you solve a variety of real-life problems. For instance, in Exercise 94 on page 374, you will solve a trigonometric equation to determine the height above ground of a seat on a Ferris wheel.

- **Use standard algebraic techniques to solve trigonometric equations.**
- **Solve trigonometric equations of quadratic type.**
- **Solve trigonometric equations involving multiple angles.**
- **Use inverse trigonometric functions to solve trigonometric equations.**

Introduction

To solve a trigonometric equation, use standard algebraic techniques (when possible) such as collecting like terms and factoring. Your preliminary goal in solving a trigonometric equation is to *isolate* the trigonometric function on one side of the equation. For example, to solve the equation $2 \sin x = 1$, divide each side by 2 to obtain

$$\sin x = \frac{1}{2}.$$

To solve for x, note in the figure below that the equation $\sin x = \frac{1}{2}$ has solutions $x = \pi/6$ and $x = 5\pi/6$ in the interval $[0, 2\pi)$. Moreover, because $\sin x$ has a period of 2π, there are infinitely many other solutions, which can be written as

$$x = \frac{\pi}{6} + 2n\pi \quad \text{and} \quad x = \frac{5\pi}{6} + 2n\pi \qquad \text{General solution}$$

where n is an integer, as shown below.

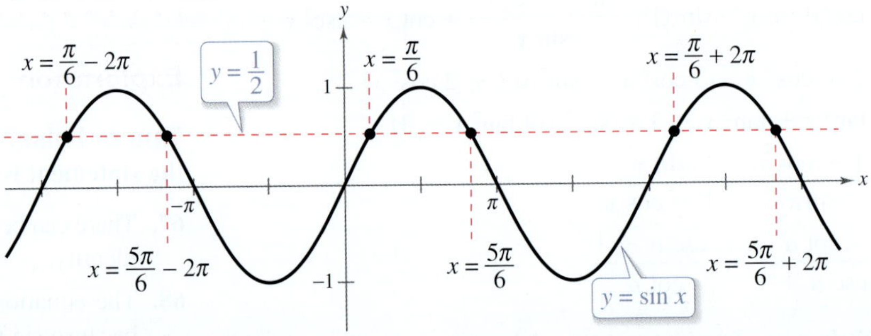

The figure below illustrates another way to show that the equation $\sin x = \frac{1}{2}$ has infinitely many solutions. Any angles that are coterminal with $\pi/6$ or $5\pi/6$ will also be solutions of the equation.

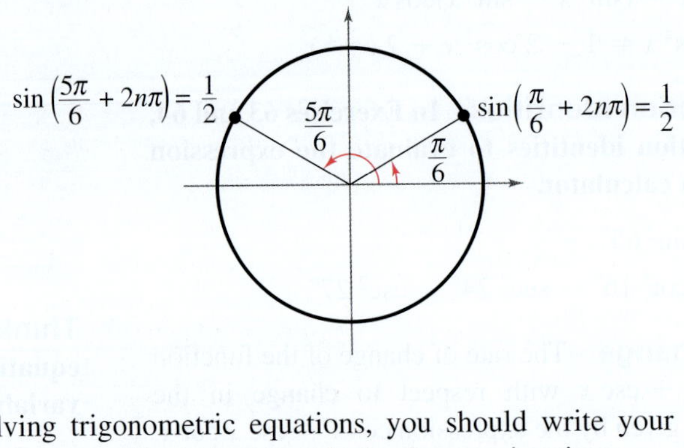

When solving trigonometric equations, you should write your answer(s) using exact values, when possible, rather than decimal approximations.

white coast art/Shutterstock.com

EXAMPLE 1 Collecting Like Terms

Solve

$$\sin x + \sqrt{2} = -\sin x.$$

Solution Begin by isolating $\sin x$ on one side of the equation.

$$\sin x + \sqrt{2} = -\sin x \qquad \text{Write original equation.}$$

$$\sin x + \sin x + \sqrt{2} = 0 \qquad \text{Add } \sin x \text{ to each side.}$$

$$\sin x + \sin x = -\sqrt{2} \qquad \text{Subtract } \sqrt{2} \text{ from each side.}$$

$$2 \sin x = -\sqrt{2} \qquad \text{Combine like terms.}$$

$$\sin x = -\frac{\sqrt{2}}{2} \qquad \text{Divide each side by 2.}$$

Because $\sin x$ has a period of 2π, first find all solutions in the interval $[0, 2\pi)$. These solutions are $x = 5\pi/4$ and $x = 7\pi/4$. Finally, add multiples of 2π to each of these solutions to obtain the general form

$$x = \frac{5\pi}{4} + 2n\pi \quad \text{and} \quad x = \frac{7\pi}{4} + 2n\pi \qquad \text{General solution}$$

where n is an integer.

Checkpoint *Audio-video solution in English & Spanish at LarsonPrecalculus.com.*

Solve $\sin x - \sqrt{2} = -\sin x$.

EXAMPLE 2 Extracting Square Roots

Solve

$$3 \tan^2 x - 1 = 0.$$

Solution Begin by isolating $\tan x$ on one side of the equation.

$$3 \tan^2 x - 1 = 0 \qquad \text{Write original equation.}$$

$$3 \tan^2 x = 1 \qquad \text{Add 1 to each side.}$$

$$\tan^2 x = \frac{1}{3} \qquad \text{Divide each side by 3.}$$

$$\tan x = \pm\frac{1}{\sqrt{3}} \qquad \text{Extract square roots.}$$

$$\tan x = \pm\frac{\sqrt{3}}{3} \qquad \text{Rationalize the denominator.}$$

REMARK When you extract square roots, make sure you account for both the positive and negative solutions.

Because $\tan x$ has a period of π, first find all solutions in the interval $[0, \pi)$. These solutions are $x = \pi/6$ and $x = 5\pi/6$. Finally, add multiples of π to each of these solutions to obtain the general form

$$x = \frac{\pi}{6} + n\pi \quad \text{and} \quad x = \frac{5\pi}{6} + n\pi \qquad \text{General solution}$$

where n is an integer.

Checkpoint *Audio-video solution in English & Spanish at LarsonPrecalculus.com.*

Solve $4 \sin^2 x - 3 = 0$.

The equations in Examples 1 and 2 involved only one trigonometric function. When two or more functions occur in the same equation, collect all terms on one side and try to separate the functions by factoring or by using appropriate identities. This may produce factors that yield no solutions, as illustrated in Example 3.

EXAMPLE 3 Factoring

Solve $\cot x \cos^2 x = 2 \cot x$.

Solution Begin by collecting all terms on one side of the equation and factoring.

$$\cot x \cos^2 x = 2 \cot x$$ Write original equation.

$$\cot x \cos^2 x - 2 \cot x = 0$$ Subtract $2 \cot x$ from each side.

$$\cot x(\cos^2 x - 2) = 0$$ Factor.

By setting each of these factors equal to zero, you obtain

$$\cot x = 0 \quad \text{and} \quad \cos^2 x - 2 = 0$$

$$\cos^2 x = 2$$

$$\cos x = \pm\sqrt{2}.$$

In the interval $(0, \pi)$, the equation $\cot x = 0$ has the solution

$$x = \frac{\pi}{2}.$$

No solution exists for $\cos x = \pm\sqrt{2}$ because $\pm\sqrt{2}$ are outside the range of the cosine function. Because $\cot x$ has a period of π, you obtain the general form of the solution by adding multiples of π to $x = \pi/2$ to get

$$x = \frac{\pi}{2} + n\pi$$ General solution

where n is an integer. Confirm this graphically by sketching the graph of $y = \cot x \cos^2 x - 2 \cot x$, as shown below.

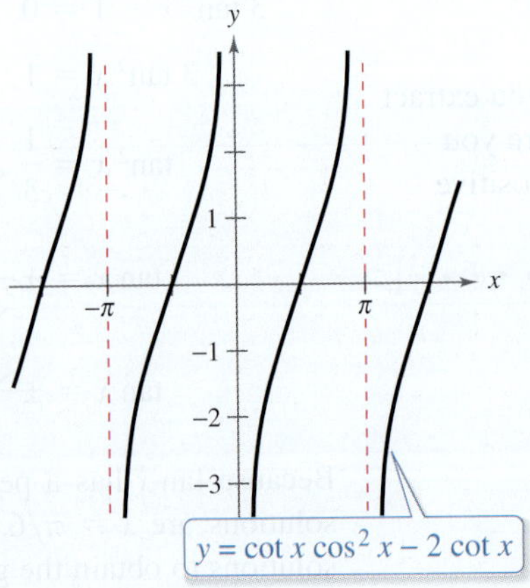

Notice that the x-intercepts occur at

$$-\frac{3\pi}{2}, \quad -\frac{\pi}{2}, \quad \frac{\pi}{2}, \quad \frac{3\pi}{2}$$

and so on. These x-intercepts correspond to the solutions of $\cot x \cos^2 x - 2 \cot x = 0$.

✓ **Checkpoint** Audio-video solution in English & Spanish at LarsonPrecalculus.com.

Solve $\sin^2 x = 2 \sin x$.

ALGEBRA HELP You can review the techniques for solving quadratic equations in Appendix A.5.

Equations of Quadratic Type

Many trigonometric equations are of quadratic type $ax^2 + bx + c = 0$, as shown below. To solve equations of this type, factor the quadratic or, when this is not possible, use the Quadratic Formula.

Quadratic in sin *x*	**Quadratic in sec *x***
$2\sin^2 x - \sin x - 1 = 0$	$\sec^2 x - 3\sec x - 2 = 0$
$2(\sin x)^2 - \sin x - 1 = 0$	$(\sec x)^2 - 3(\sec x) - 2 = 0$

EXAMPLE 4 **Factoring an Equation of Quadratic Type**

Find all solutions of $2\sin^2 x - \sin x - 1 = 0$ in the interval $[0, 2\pi)$.

Algebraic Solution

Treat the equation as a quadratic in $\sin x$ and factor.

$2\sin^2 x - \sin x - 1 = 0$ Write original equation.

$(2\sin x + 1)(\sin x - 1) = 0$ Factor.

Setting each factor equal to zero, you obtain the following solutions in the interval $[0, 2\pi)$.

$2\sin x + 1 = 0$ and $\sin x - 1 = 0$

$\sin x = -\dfrac{1}{2}$ $\sin x = 1$

$x = \dfrac{7\pi}{6}, \dfrac{11\pi}{6}$ $x = \dfrac{\pi}{2}$

Graphical Solution

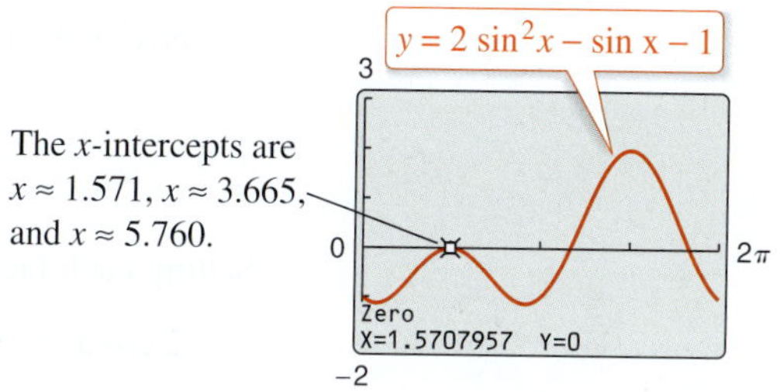

From the above figure, you can conclude that the approximate solutions of $2\sin^2 x - \sin x - 1 = 0$ in the interval $[0, 2\pi)$ are

$$x \approx 1.571 \approx \frac{\pi}{2},\quad x \approx 3.665 \approx \frac{7\pi}{6},\quad \text{and}\quad x \approx 5.760 \approx \frac{11\pi}{6}.$$

✓ **Checkpoint** Audio-video solution in English & Spanish at LarsonPrecalculus.com.

Find all solutions of $2\sin^2 x - 3\sin x + 1 = 0$ in the interval $[0, 2\pi)$.

EXAMPLE 5 **Rewriting with a Single Trigonometric Function**

Solve $2\sin^2 x + 3\cos x - 3 = 0$.

Solution This equation contains both sine and cosine functions. You can rewrite the equation so that it has only cosine functions by using the identity $\sin^2 x = 1 - \cos^2 x$.

$2\sin^2 x + 3\cos x - 3 = 0$ Write original equation.

$2(1 - \cos^2 x) + 3\cos x - 3 = 0$ Pythagorean identity

$2\cos^2 x - 3\cos x + 1 = 0$ Multiply each side by -1.

$(2\cos x - 1)(\cos x - 1) = 0$ Factor.

By setting each factor equal to zero, you can find the solutions in the interval $[0, 2\pi)$ to be $x = 0$, $x = \pi/3$, and $x = 5\pi/3$. Because $\cos x$ has a period of 2π, the general solution is

$$x = 2n\pi,\quad x = \frac{\pi}{3} + 2n\pi,\quad x = \frac{5\pi}{3} + 2n\pi$$ General solution

where n is an integer.

✓ **Checkpoint** Audio-video solution in English & Spanish at LarsonPrecalculus.com.

Solve $3\sec^2 x - 2\tan^2 x - 4 = 0$.

Sometimes you must square each side of an equation to obtain a quadratic, as demonstrated in the next example. Because this procedure can introduce extraneous solutions, you should check any solutions in the original equation to see whether they are valid or extraneous.

REMARK You square each side of the equation in Example 6 because the squares of the sine and cosine functions are related by a Pythagorean identity. The same is true for the squares of the secant and tangent functions and for the squares of the cosecant and cotangent functions.

EXAMPLE 6 Squaring and Converting to Quadratic Type

Find all solutions of $\cos x + 1 = \sin x$ in the interval $[0, 2\pi)$.

Solution It is not clear how to rewrite this equation in terms of a single trigonometric function. Notice what happens when you square each side of the equation.

$$\cos x + 1 = \sin x \quad \text{Write original equation.}$$

$$\cos^2 x + 2\cos x + 1 = \sin^2 x \quad \text{Square each side.}$$

$$\cos^2 x + 2\cos x + 1 = 1 - \cos^2 x \quad \text{Pythagorean identity}$$

$$\cos^2 x + \cos^2 x + 2\cos x + 1 - 1 = 0 \quad \text{Rewrite equation.}$$

$$2\cos^2 x + 2\cos x = 0 \quad \text{Combine like terms.}$$

$$2\cos x(\cos x + 1) = 0 \quad \text{Factor.}$$

Setting each factor equal to zero produces

$$2\cos x = 0 \qquad \text{and} \qquad \cos x + 1 = 0$$

$$\cos x = 0 \qquad\qquad \cos x = -1$$

$$x = \frac{\pi}{2}, \frac{3\pi}{2} \qquad\qquad x = \pi.$$

Because you squared the original equation, check for extraneous solutions.

Check $x = \frac{\pi}{2}$

$$\cos\frac{\pi}{2} + 1 \stackrel{?}{=} \sin\frac{\pi}{2} \quad \text{Substitute } \frac{\pi}{2} \text{ for } x.$$

$$0 + 1 = 1 \quad \text{Solution checks. ✓}$$

Check $x = \frac{3\pi}{2}$

$$\cos\frac{3\pi}{2} + 1 \stackrel{?}{=} \sin\frac{3\pi}{2} \quad \text{Substitute } \frac{3\pi}{2} \text{ for } x.$$

$$0 + 1 \neq -1 \quad \text{Solution does not check.}$$

Check $x = \pi$

$$\cos\pi + 1 \stackrel{?}{=} \sin\pi \quad \text{Substitute } \pi \text{ for } x.$$

$$-1 + 1 = 0 \quad \text{Solution checks. ✓}$$

Of the three possible solutions, $x = 3\pi/2$ is extraneous. So, in the interval $[0, 2\pi)$, the only two solutions are

$$x = \frac{\pi}{2} \quad \text{and} \quad x = \pi.$$

✓ **Checkpoint** *Audio-video solution in English & Spanish at LarsonPrecalculus.com.*

Find all solutions of $\sin x + 1 = \cos x$ in the interval $[0, 2\pi)$.

Functions Involving Multiple Angles

The next two examples involve trigonometric functions of multiple angles of the forms $\cos ku$ and $\tan ku$. To solve equations of these forms, first solve the equation for ku, and then divide your result by k.

EXAMPLE 7 Solving a Multiple-Angle Equation

Solve $2 \cos 3t - 1 = 0$.

Solution

$$2 \cos 3t - 1 = 0 \qquad \text{Write original equation.}$$

$$2 \cos 3t = 1 \qquad \text{Add 1 to each side.}$$

$$\cos 3t = \frac{1}{2} \qquad \text{Divide each side by 2.}$$

In the interval $[0, 2\pi)$, you know that $3t = \pi/3$ and $3t = 5\pi/3$ are the only solutions, so, in general, you have

$$3t = \frac{\pi}{3} + 2n\pi \quad \text{and} \quad 3t = \frac{5\pi}{3} + 2n\pi.$$

Dividing these results by 3, you obtain the general solution

$$t = \frac{\pi}{9} + \frac{2n\pi}{3} \quad \text{and} \quad t = \frac{5\pi}{9} + \frac{2n\pi}{3} \qquad \text{General solution}$$

where n is an integer.

✓ **Checkpoint** *Audio-video solution in English & Spanish at LarsonPrecalculus.com.*

Solve $2 \sin 2t - \sqrt{3} = 0$.

EXAMPLE 8 Solving a Multiple-Angle Equation

$$3 \tan \frac{x}{2} + 3 = 0 \qquad \text{Original equation}$$

$$3 \tan \frac{x}{2} = -3 \qquad \text{Subtract 3 from each side.}$$

$$\tan \frac{x}{2} = -1 \qquad \text{Divide each side by 3.}$$

In the interval $[0, \pi)$, you know that $x/2 = 3\pi/4$ is the only solution, so, in general, you have

$$\frac{x}{2} = \frac{3\pi}{4} + n\pi.$$

Multiplying this result by 2, you obtain the general solution

$$x = \frac{3\pi}{2} + 2n\pi \qquad \text{General solution}$$

where n is an integer.

✓ **Checkpoint** *Audio-video solution in English & Spanish at LarsonPrecalculus.com.*

Solve $2 \tan \frac{x}{2} - 2 = 0$.

Using Inverse Functions

It is possible to find the minimum surface area of a honeycomb cell using a graphing utility or using calculus and the arccosine function.

EXAMPLE 9 Using Inverse Functions

$$\sec^2 x - 2\tan x = 4 \quad \text{Original equation}$$

$$1 + \tan^2 x - 2\tan x - 4 = 0 \quad \text{Pythagorean identity}$$

$$\tan^2 x - 2\tan x - 3 = 0 \quad \text{Combine like terms.}$$

$$(\tan x - 3)(\tan x + 1) = 0 \quad \text{Factor.}$$

Setting each factor equal to zero, you obtain two solutions in the interval $(-\pi/2, \pi/2)$. [Recall that the range of the inverse tangent function is $(-\pi/2, \pi/2)$.]

$$x = \arctan 3 \quad \text{and} \quad x = \arctan(-1) = -\pi/4$$

Finally, because $\tan x$ has a period of π, you add multiples of π to obtain

$$x = \arctan 3 + n\pi \quad \text{and} \quad x = (-\pi/4) + n\pi \quad \text{General solution}$$

where n is an integer. You can use a calculator to approximate the value of arctan 3.

✓ ***Checkpoint*** *Audio-video solution in English & Spanish at LarsonPrecalculus.com.*

Solve $4\tan^2 x + 5\tan x - 6 = 0$.

EXAMPLE 10 Surface Area of a Honeycomb Cell

The surface area S (in square inches) of a honeycomb cell is given by

$$S = 6hs + 1.5s^2\left[\left(\sqrt{3} - \cos\theta\right)/\sin\theta\right], \quad 0 < \theta \le 90^\circ$$

where $h = 2.4$ inches, $s = 0.75$ inch, and θ is the angle shown in Figure 5.2. What value of θ gives the minimum surface area?

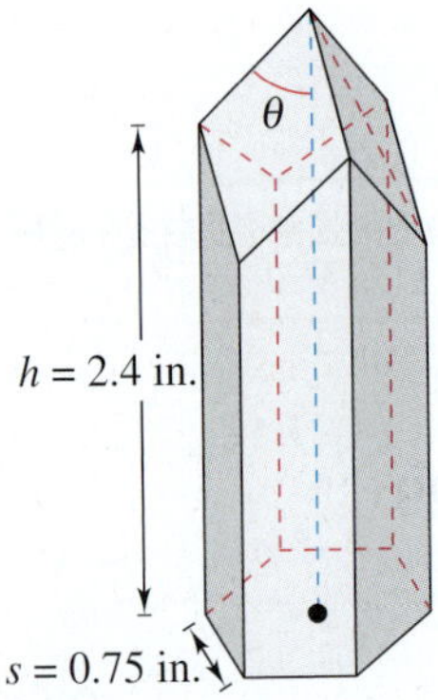

Figure 5.2

Solution Letting $h = 2.4$ and $s = 0.75$, you obtain

$$S = 10.8 + 0.84375\left[\left(\sqrt{3} - \cos\theta\right)/\sin\theta\right].$$

Graph this function using a graphing utility. The minimum point on the graph, which occurs at $\theta \approx 54.7^\circ$, is shown in Figure 5.3. By using calculus, the exact minimum point on the graph can be shown to occur at $\theta = \arccos\left(1/\sqrt{3}\right) \approx 0.9553 \approx 54.7356^\circ$.

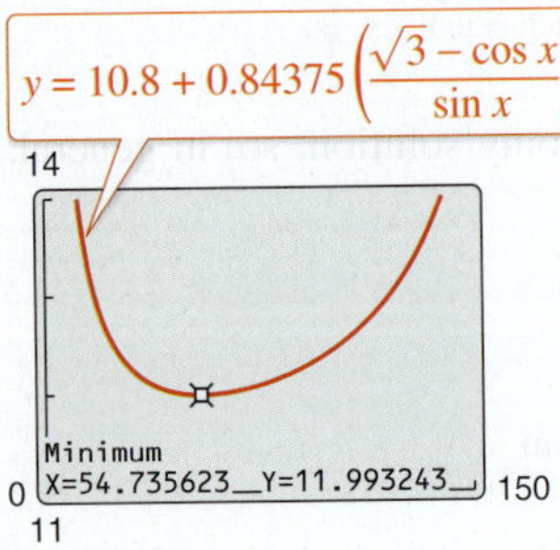

Figure 5.3

✓ ***Checkpoint*** *Audio-video solution in English & Spanish at LarsonPrecalculus.com.*

In Example 10, for what value(s) of θ is the surface area 12 square inches?

Summarize (Section 5.3)

1. Describe how to use standard algebraic techniques to solve trigonometric equations *(page 364)*. For examples of using standard algebraic techniques to solve trigonometric equations, see Examples 1–3.
2. Explain how to solve a trigonometric equation of quadratic type *(page 367)*. For examples of solving trigonometric equations of quadratic type, see Examples 4–6.
3. Explain how to solve a trigonometric equation involving multiple angles *(page 369)*. For examples of solving trigonometric equations involving multiple angles, see Examples 7 and 8.
4. Explain how to use inverse trigonometric functions to solve trigonometric equations *(page 370)*. For examples of using inverse trigonometric functions to solve trigonometric equations, see Examples 9 and 10.

LilKar/Shutterstock.com

5.3 Exercises

See CalcChat.com for tutorial help and worked-out solutions to odd-numbered exercises.

Vocabulary: Fill in the blanks.

1. When solving a trigonometric equation, the preliminary goal is to ________ the trigonometric function involved in the equation.

2. The equation $2 \sin \theta + 1 = 0$ has the solutions $\theta = \frac{7\pi}{6} + 2n\pi$ and $\theta = \frac{11\pi}{6} + 2n\pi$, which are called ________ solutions.

3. The equation $2 \tan^2 x - 3 \tan x + 1 = 0$ is a trigonometric equation that is of ________ type.

4. A solution of an equation that does not satisfy the original equation is called an ________ solution.

Skills and Applications

Verifying Solutions In Exercises 5–10, verify that the x-values are solutions of the equation.

5. $\tan x - \sqrt{3} = 0$
 (a) $x = \frac{\pi}{3}$
 (b) $x = \frac{4\pi}{3}$

6. $\sec x - 2 = 0$
 (a) $x = \frac{\pi}{3}$
 (b) $x = \frac{5\pi}{3}$

7. $3 \tan^2 2x - 1 = 0$
 (a) $x = \frac{\pi}{12}$
 (b) $x = \frac{5\pi}{12}$

8. $2 \cos^2 4x - 1 = 0$
 (a) $x = \frac{\pi}{16}$
 (b) $x = \frac{3\pi}{16}$

9. $2 \sin^2 x - \sin x - 1 = 0$
 (a) $x = \frac{\pi}{2}$
 (b) $x = \frac{7\pi}{6}$

10. $\csc^4 x - 4 \csc^2 x = 0$
 (a) $x = \frac{\pi}{6}$
 (b) $x = \frac{5\pi}{6}$

Solving a Trigonometric Equation In Exercises 11–24, solve the equation.

11. $\sqrt{3} \csc x - 2 = 0$
12. $\tan x + \sqrt{3} = 0$
13. $\cos x + 1 = -\cos x$
14. $3 \sin x + 1 = \sin x$
15. $3 \sec^2 x - 4 = 0$
16. $3 \cot^2 x - 1 = 0$
17. $4 \cos^2 x - 1 = 0$
18. $\sin^2 x = 3 \cos^2 x$
19. $2 \sin^2 2x = 1$
20. $\tan^2 3x = 3$
21. $\tan 3x(\tan x - 1) = 0$
22. $\cos 2x(2 \cos x + 1) = 0$
23. $\sin x(\sin x + 1) = 0$
24. $(2 \sin^2 x - 1)(\tan^2 x - 3) = 0$

Solving a Trigonometric Equation In Exercises 25–38, find all solutions of the equation in the interval $[0, 2\pi)$.

25. $\cos^3 x = \cos x$
26. $\sec^2 x - 1 = 0$
27. $3 \tan^3 x = \tan x$
28. $2 \sin^2 x = 2 + \cos x$
29. $\sec^2 x - \sec x = 2$
30. $\sec x \csc x = 2 \csc x$
31. $2 \sin x + \csc x = 0$
32. $\sin x - 2 = \cos x - 2$
33. $2 \cos^2 x + \cos x - 1 = 0$
34. $2 \sin^2 x + 3 \sin x + 1 = 0$
35. $2 \sec^2 x + \tan^2 x - 3 = 0$
36. $\cos x + \sin x \tan x = 2$
37. $\csc x + \cot x = 1$
38. $\sec x + \tan x = 1$

Solving a Multiple-Angle Equation In Exercises 39–44, solve the multiple-angle equation.

39. $2 \cos 2x - 1 = 0$
40. $2 \sin 2x + \sqrt{3} = 0$
41. $\tan 3x - 1 = 0$
42. $\sec 4x - 2 = 0$
43. $2 \cos \frac{x}{2} - \sqrt{2} = 0$
44. $2 \sin \frac{x}{2} + \sqrt{3} = 0$

Finding x-Intercepts In Exercises 45–48, find the x-intercepts of the graph.

45. $y = \sin \frac{\pi x}{2} + 1$

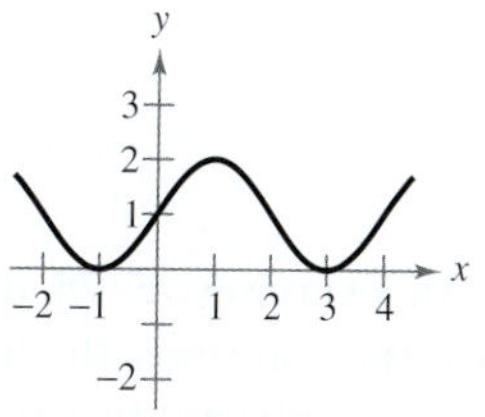

46. $y = \sin \pi x + \cos \pi x$

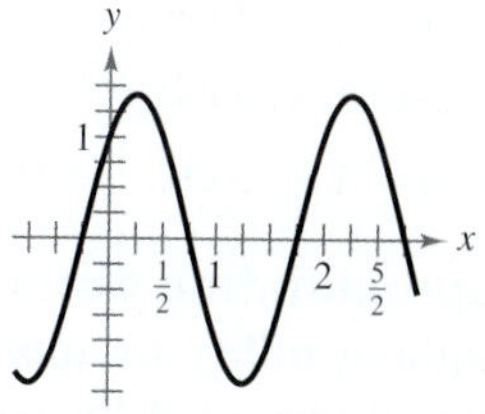

47. $y = \tan^2\left(\frac{\pi x}{6}\right) - 3$

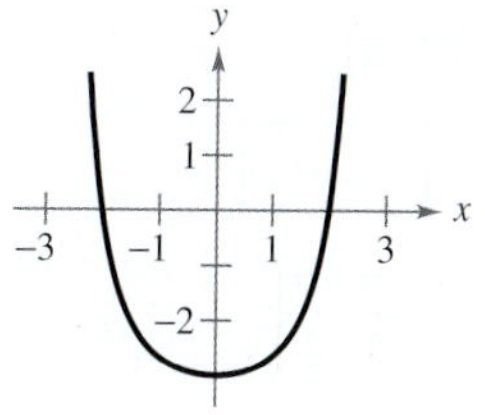

48. $y = \sec^4\left(\frac{\pi x}{8}\right) - 4$

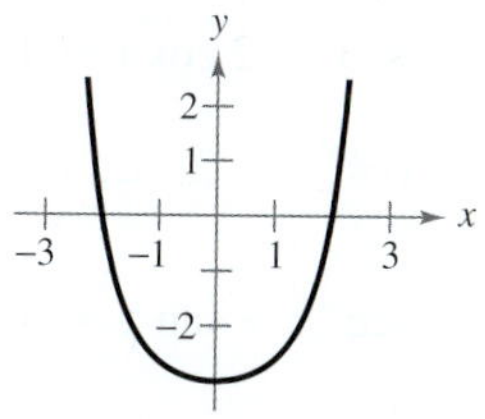

Approximating Solutions **In Exercises 49–58, use a graphing utility to approximate the solutions (to three decimal places) of the equation in the interval $[0, 2\pi)$.**

49. $2 \sin x + \cos x = 0$

50. $4 \sin^3 x + 2 \sin^2 x - 2 \sin x - 1 = 0$

51. $\dfrac{1 + \sin x}{\cos x} + \dfrac{\cos x}{1 + \sin x} = 4$

52. $\dfrac{\cos x \cot x}{1 - \sin x} = 3$

53. $x \tan x - 1 = 0$ **54.** $x \cos x - 1 = 0$

55. $\sec^2 x + 0.5 \tan x - 1 = 0$

56. $\csc^2 x + 0.5 \cot x - 5 = 0$

57. $2 \tan^2 x + 7 \tan x - 15 = 0$

58. $6 \sin^2 x - 7 \sin x + 2 = 0$

Using the Quadratic Formula **In Exercises 59–62, use the Quadratic Formula to solve the equation in the interval $[0, 2\pi)$. Then use a graphing utility to approximate the angle x.**

59. $12 \sin^2 x - 13 \sin x + 3 = 0$

60. $3 \tan^2 x + 4 \tan x - 4 = 0$

61. $\tan^2 x + 3 \tan x + 1 = 0$

62. $4 \cos^2 x - 4 \cos x - 1 = 0$

Using Inverse Functions **In Exercises 63–74, use inverse functions where needed to find all solutions of the equation in the interval $[0, 2\pi)$.**

63. $\tan^2 x + \tan x - 12 = 0$

64. $\tan^2 x - \tan x - 2 = 0$

65. $\sec^2 x - 6 \tan x = -4$

66. $\sec^2 x + \tan x - 3 = 0$

67. $2 \sin^2 x + 5 \cos x = 4$ **68.** $2 \cos^2 x + 7 \sin x = 5$

69. $\cot^2 x - 9 = 0$ **70.** $\cot^2 x - 6 \cot x + 5 = 0$

71. $\sec^2 x - 4 \sec x = 0$

72. $\sec^2 x + 2 \sec x - 8 = 0$

73. $\csc^2 x + 3 \csc x - 4 = 0$

74. $\csc^2 x - 5 \csc x = 0$

Approximating Solutions **In Exercises 75–78, use a graphing utility to approximate the solutions (to three decimal places) of the equation in the given interval.**

75. $3 \tan^2 x + 5 \tan x - 4 = 0, \quad \left[-\dfrac{\pi}{2}, \dfrac{\pi}{2}\right]$

76. $\cos^2 x - 2 \cos x - 1 = 0, \quad [0, \pi]$

77. $4 \cos^2 x - 2 \sin x + 1 = 0, \quad \left[-\dfrac{\pi}{2}, \dfrac{\pi}{2}\right]$

78. $2 \sec^2 x + \tan x - 6 = 0, \quad \left[-\dfrac{\pi}{2}, \dfrac{\pi}{2}\right]$

Approximating Maximum and Minimum Points **In Exercises 79–84, (a) use a graphing utility to graph the function and approximate the maximum and minimum points on the graph in the interval $[0, 2\pi)$, and (b) solve the trigonometric equation and demonstrate that its solutions are the x-coordinates of the maximum and minimum points of f. (Calculus is required to find the trigonometric equation.)**

	Function	Trigonometric Equation
79.	$f(x) = \sin^2 x + \cos x$	$2 \sin x \cos x - \sin x = 0$
80.	$f(x) = \cos^2 x - \sin x$	$-2 \sin x \cos x - \cos x = 0$
81.	$f(x) = \sin x + \cos x$	$\cos x - \sin x = 0$
82.	$f(x) = 2 \sin x + \cos 2x$	$2 \cos x - 4 \sin x \cos x = 0$
83.	$f(x) = \sin x \cos x$	$-\sin^2 x + \cos^2 x = 0$
84.	$f(x) = \sec x + \tan x - x$	$\sec x \tan x + \sec^2 x = 1$

Number of Points of Intersection **In Exercises 85 and 86, use the graph to approximate the number of points of intersection of the graphs of y_1 and y_2.**

85. $y_1 = 2 \sin x$
$y_2 = 3x + 1$

86. $y_1 = 2 \sin x$
$y_2 = \frac{1}{2}x + 1$

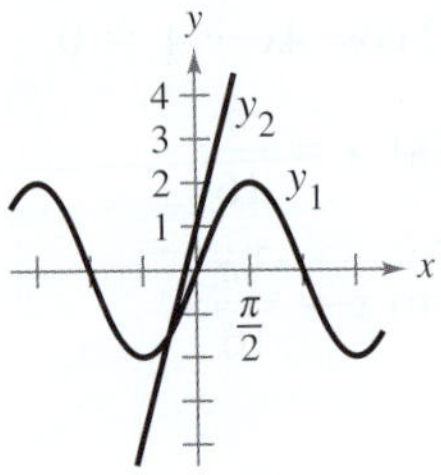

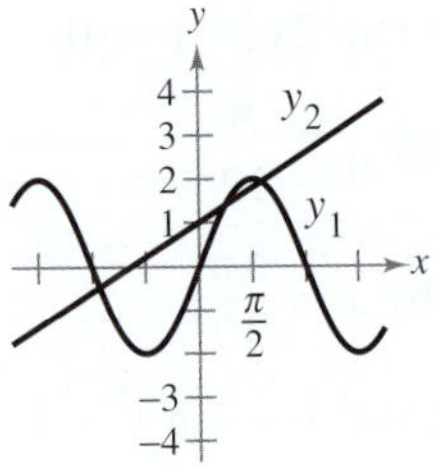

87. Graphical Reasoning Consider the function $f(x) = (\sin x)/x$ and its graph shown in the figure.

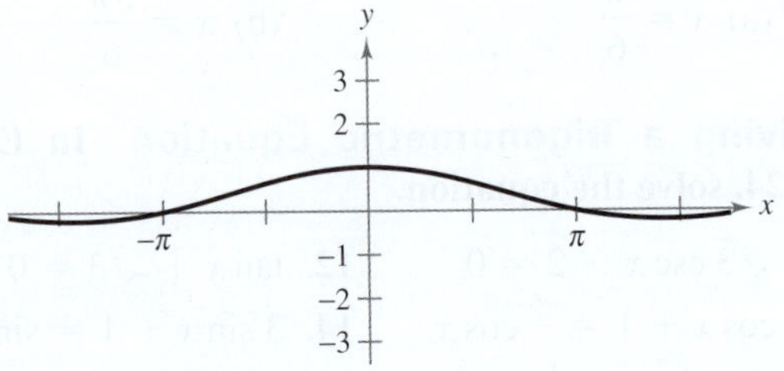

(a) What is the domain of the function?

(b) Identify any symmetry and any asymptotes of the graph.

(c) Describe the behavior of the function as $x \to 0$.

(d) How many solutions does the equation

$$\frac{\sin x}{x} = 0$$

have in the interval $[-8, 8]$? Find the solutions.

88. Graphical Reasoning Consider the function

$$f(x) = \cos \frac{1}{x}$$

and its graph shown in the figure.

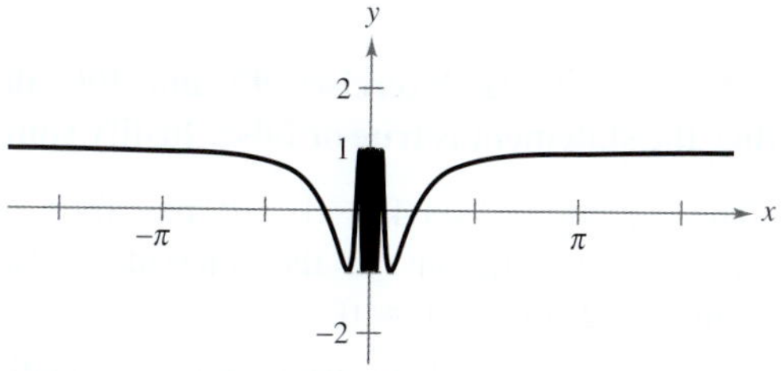

(a) What is the domain of the function?

(b) Identify any symmetry and any asymptotes of the graph.

(c) Describe the behavior of the function as $x \to 0$.

(d) How many solutions does the equation

$$\cos \frac{1}{x} = 0$$

have in the interval $[-1, 1]$? Find the solutions.

(e) Does the equation $\cos(1/x) = 0$ have a greatest solution? If so, then approximate the solution. If not, then explain why.

89. Harmonic Motion A weight is oscillating on the end of a spring (see figure). The position of the weight relative to the point of equilibrium is given by

$$y = \tfrac{1}{12}(\cos 8t - 3 \sin 8t)$$

where y is the displacement (in meters) and t is the time (in seconds). Find the times when the weight is at the point of equilibrium $(y = 0)$ for $0 \le t \le 1$.

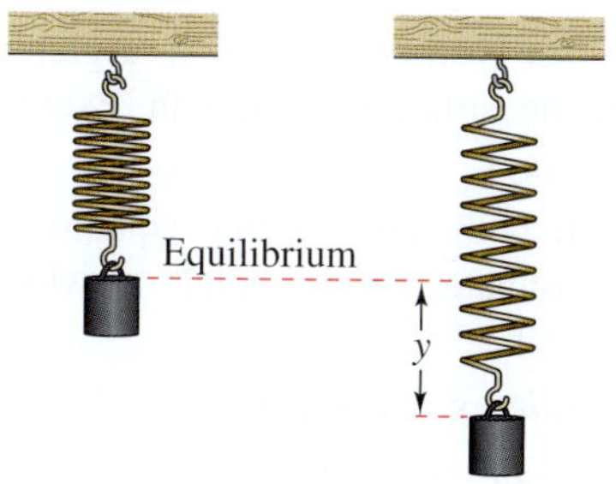

90. Damped Harmonic Motion The displacement from equilibrium of a weight oscillating on the end of a spring is given by

$$y = 1.56e^{-0.22t} \cos 4.9t$$

where y is the displacement (in feet) and t is the time (in seconds). Use a graphing utility to graph the displacement function for $0 \le t \le 10$. Find the time beyond which the displacement does not exceed 1 foot from equilibrium.

91. Sales The monthly sales S (in hundreds of units) of skiing equipment at a sports store are approximated by

$$S = 58.3 + 32.5 \cos \frac{\pi t}{6}$$

where t is the time (in months), with $t = 1$ corresponding to January. Determine the months in which sales exceed 7500 units.

92. Projectile Motion A baseball is hit at an angle of θ with the horizontal and with an initial velocity of $v_0 = 100$ feet per second. An outfielder catches the ball 300 feet from home plate (see figure). Find θ when the range r of a projectile is given by

$$r = \frac{1}{32} v_0^2 \sin 2\theta.$$

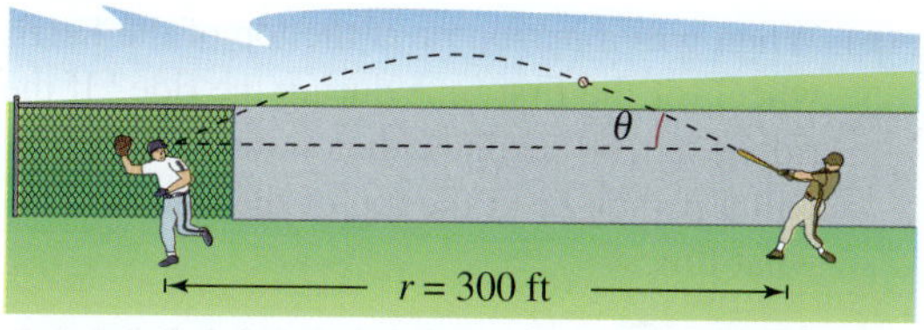

Not drawn to scale

93. Data Analysis: Meteorology The table shows the normal daily high temperatures in Houston H (in degrees Fahrenheit) for month t, with $t = 1$ corresponding to January. *(Source: NOAA)*

DATA — Spreadsheet at LarsonPrecalculus.com

Month, t	Houston, H
1	62.3
2	66.5
3	73.3
4	79.1
5	85.5
6	90.7
7	93.6
8	93.5
9	89.3
10	82.0
11	72.0
12	64.6

(a) Create a scatter plot of the data.

(b) Find a cosine model for the temperatures.

(c) Use a graphing utility to graph the data points and the model for the temperatures. How well does the model fit the data?

(d) What is the overall normal daily high temperature?

(e) Use the graphing utility to describe the months during which the normal daily high temperature is above 86°F and below 86°F.

94. Ferris Wheel The height h (in feet) above ground of a seat on a Ferris wheel at time t (in minutes) can be modeled by

$$h(t) = 53 + 50\sin\left(\frac{\pi}{16}t - \frac{\pi}{2}\right).$$

The wheel makes one revolution every 32 seconds. The ride begins when $t = 0$.

(a) During the first 32 seconds of the ride, when will a person on the Ferris wheel be 53 feet above ground?

(b) When will a person be at the top of the Ferris wheel for the first time during the ride? If the ride lasts 160 seconds, then how many times will a person be at the top of the ride, and at what times?

95. Geometry The area of a rectangle (see figure) inscribed in one arc of the graph of $y = \cos x$ is given by

$$A = 2x\cos x,\ 0 < x < \pi/2.$$

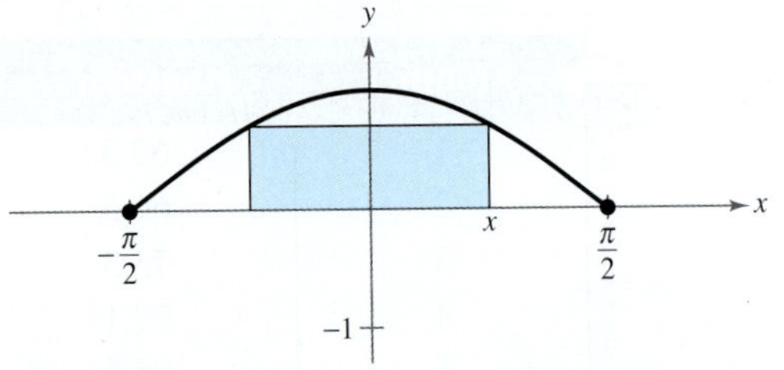

(a) Use a graphing utility to graph the area function, and approximate the area of the largest inscribed rectangle.

(b) Determine the values of x for which $A \geq 1$.

96. Quadratic Approximation Consider the function

$$f(x) = 3\sin(0.6x - 2).$$

(a) Approximate the zero of the function in the interval $[0, 6]$.

(b) A quadratic approximation agreeing with f at $x = 5$ is

$$g(x) = -0.45x^2 + 5.52x - 13.70.$$

Use a graphing utility to graph f and g in the same viewing window. Describe the result.

(c) Use the Quadratic Formula to find the zeros of g. Compare the zero in the interval $[0, 6]$ with the result of part (a).

Fixed Point **In Exercises 97 and 98, find the smallest positive fixed point of the function f. [A *fixed point* of a function f is a real number c such that $f(c) = c$.]**

97. $f(x) = \tan(\pi x/4)$ **98.** $f(x) = \cos x$

Exploration

True or False? **In Exercises 99 and 100, determine whether the statement is true or false. Justify your answer.**

99. The equation $2\sin 4t - 1 = 0$ has four times the number of solutions in the interval $[0, 2\pi)$ as the equation $2\sin t - 1 = 0$.

100. If you correctly solve a trigonometric equation to the statement $\sin x = 3.4$, then you can finish solving the equation by using an inverse function.

101. Think About It Explain what happens when you divide each side of the equation $\cot x\cos^2 x = 2\cot x$ by $\cot x$. Is this a correct method to use when solving equations?

102. HOW DO YOU SEE IT? Explain how to use the figure to solve the equation $2\cos x - 1 = 0$.

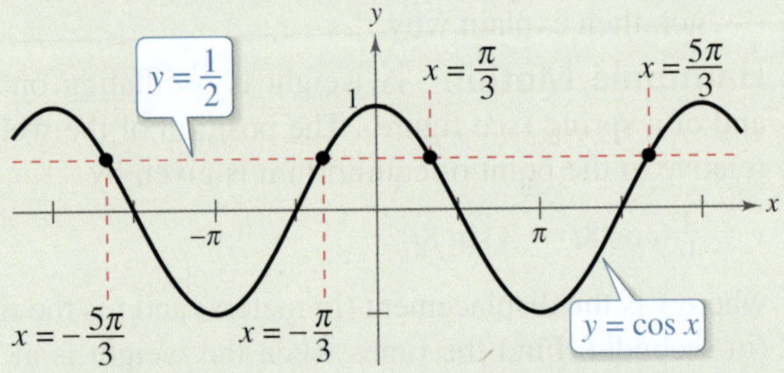

103. Graphical Reasoning Use a graphing utility to confirm the solutions found in Example 6 in two different ways.

(a) Graph both sides of the equation and find the x-coordinates of the points at which the graphs intersect.

Left side: $y = \cos x + 1$

Right side: $y = \sin x$

(b) Graph the equation $y = \cos x + 1 - \sin x$ and find the x-intercepts of the graph. Do both methods produce the same x-values? Which method do you prefer? Explain.

104. Discussion Explain in your own words how knowledge of algebra is important when solving trigonometric equations.

Project: Meteorology To work an extended application analyzing the normal daily high temperatures in Phoenix and in Seattle, visit this text's website at *LarsonPrecalculus.com.* *(Source: NOAA)*

white coast art/Shutterstock.com

5.4 Sum and Difference Formulas

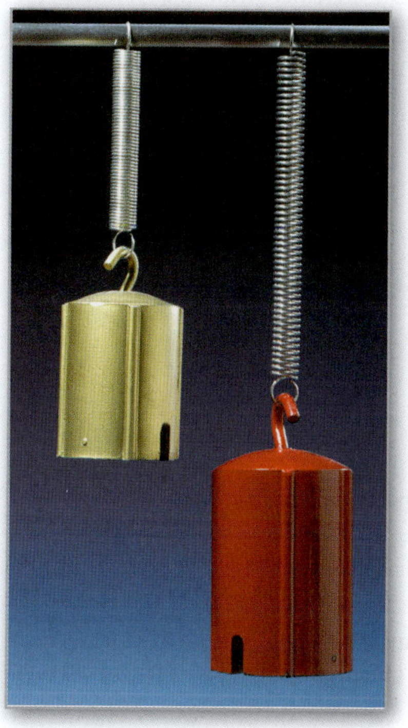

Trigonometric identities enable you to rewrite trigonometric expressions. For instance, in Exercise 79 on page 380, you will use an identity to rewrite a trigonometric expression in a form that helps you analyze a harmonic motion equation.

- **Use sum and difference formulas to evaluate trigonometric functions, verify identities, and solve trigonometric equations.**

Using Sum and Difference Formulas

In this and the following section, you will study the uses of several trigonometric identities and formulas.

Sum and Difference Formulas

$$\sin(u + v) = \sin u \cos v + \cos u \sin v$$

$$\sin(u - v) = \sin u \cos v - \cos u \sin v$$

$$\cos(u + v) = \cos u \cos v - \sin u \sin v$$

$$\cos(u - v) = \cos u \cos v + \sin u \sin v$$

$$\tan(u + v) = \frac{\tan u + \tan v}{1 - \tan u \tan v} \qquad \tan(u - v) = \frac{\tan u - \tan v}{1 + \tan u \tan v}$$

For a proof of the sum and difference formulas for $\cos(u \pm v)$ and $\tan(u \pm v)$, see Proofs in Mathematics on page 396.

Examples 1 and 2 show how **sum and difference formulas** can enable you to find exact values of trigonometric functions involving sums or differences of special angles.

EXAMPLE 1 Evaluating a Trigonometric Function

Find the exact value of $\sin \frac{\pi}{12}$.

Solution To find the *exact* value of $\sin \pi/12$, use the fact that

$$\frac{\pi}{12} = \frac{\pi}{3} - \frac{\pi}{4}.$$

Consequently, the formula for $\sin(u - v)$ yields

$$\sin \frac{\pi}{12} = \sin\left(\frac{\pi}{3} - \frac{\pi}{4}\right)$$

$$= \sin \frac{\pi}{3} \cos \frac{\pi}{4} - \cos \frac{\pi}{3} \sin \frac{\pi}{4}$$

$$= \frac{\sqrt{3}}{2}\left(\frac{\sqrt{2}}{2}\right) - \frac{1}{2}\left(\frac{\sqrt{2}}{2}\right)$$

$$= \frac{\sqrt{6} - \sqrt{2}}{4}.$$

Try checking this result on your calculator. You will find that $\sin \pi/12 \approx 0.259$.

Checkpoint Audio-video solution in English & Spanish at LarsonPrecalculus.com.

Find the exact value of $\cos \frac{\pi}{12}$.

Richard Megna/Fundamental Photographs

REMARK Another way to solve Example 2 is to use the fact that $75° = 120° - 45°$ together with the formula for $\cos(u - v)$.

EXAMPLE 2 Evaluating a Trigonometric Function

Find the exact value of $\cos 75°$.

Solution Using the fact that $75° = 30° + 45°$, together with the formula for $\cos(u + v)$, you obtain

$$\begin{aligned}\cos 75° &= \cos(30° + 45°)\\ &= \cos 30° \cos 45° - \sin 30° \sin 45°\\ &= \frac{\sqrt{3}}{2}\left(\frac{\sqrt{2}}{2}\right) - \frac{1}{2}\left(\frac{\sqrt{2}}{2}\right)\\ &= \frac{\sqrt{6} - \sqrt{2}}{4}.\end{aligned}$$

✓ Checkpoint Audio-video solution in English & Spanish at LarsonPrecalculus.com.

Find the exact value of $\sin 75°$.

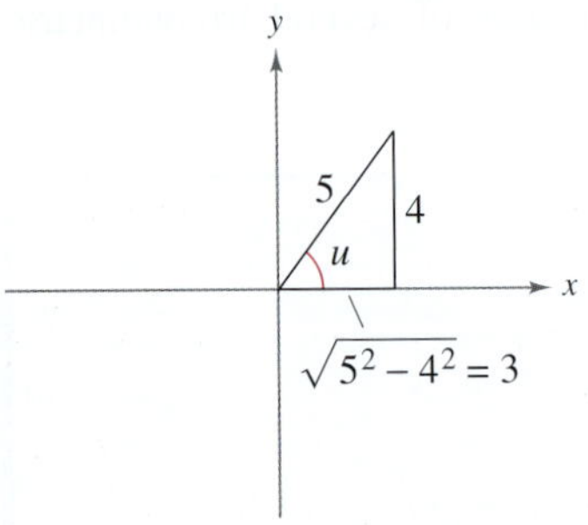

Figure 5.4

EXAMPLE 3 Evaluating a Trigonometric Expression

Find the exact value of $\sin(u + v)$ given $\sin u = 4/5$, where $0 < u < \pi/2$, and $\cos v = -12/13$, where $\pi/2 < v < \pi$.

Solution Because $\sin u = 4/5$ and u is in Quadrant I, $\cos u = 3/5$, as shown in Figure 5.4. Because $\cos v = -12/13$ and v is in Quadrant II, $\sin v = 5/13$, as shown in Figure 5.5. You can find $\sin(u + v)$ as follows.

$$\begin{aligned}\sin(u + v) &= \sin u \cos v + \cos u \sin v\\ &= \frac{4}{5}\left(-\frac{12}{13}\right) + \frac{3}{5}\left(\frac{5}{13}\right)\\ &= -\frac{33}{65}\end{aligned}$$

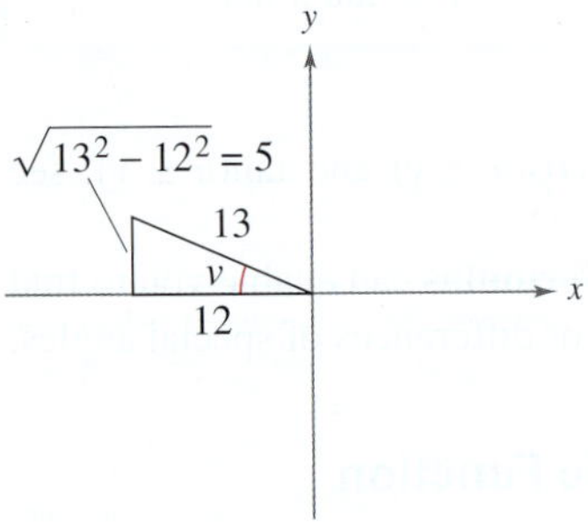

Figure 5.5

✓ Checkpoint Audio-video solution in English & Spanish at LarsonPrecalculus.com.

Find the exact value of $\cos(u + v)$ given $\sin u = 12/13$, where $0 < u < \pi/2$, and $\cos v = -3/5$, where $\pi/2 < v < \pi$.

EXAMPLE 4 An Application of a Sum Formula

Write $\cos(\arctan 1 + \arccos x)$ as an algebraic expression.

Solution This expression fits the formula for $\cos(u + v)$. Figure 5.6 shows angles $u = \arctan 1$ and $v = \arccos x$. So,

$$\begin{aligned}\cos(u + v) &= \cos(\arctan 1)\cos(\arccos x) - \sin(\arctan 1)\sin(\arccos x)\\ &= \frac{1}{\sqrt{2}} \cdot x - \frac{1}{\sqrt{2}} \cdot \sqrt{1 - x^2}\\ &= \frac{x - \sqrt{1 - x^2}}{\sqrt{2}}.\end{aligned}$$

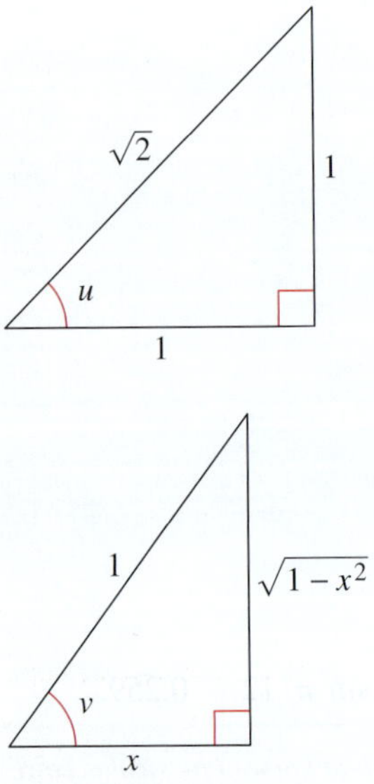

Figure 5.6

✓ Checkpoint Audio-video solution in English & Spanish at LarsonPrecalculus.com.

Write $\sin(\arctan 1 + \arccos x)$ as an algebraic expression.

Hipparchus, considered the most eminent of Greek astronomers, was born about 190 B.C. in Nicaea. He is credited with the invention of trigonometry. He also derived the sum and difference formulas for $\sin(A \pm B)$ and $\cos(A \pm B)$.

EXAMPLE 5 **Proving a Cofunction Identity**

Use a difference formula to prove the cofunction identity $\cos\left(\frac{\pi}{2} - x\right) = \sin x$.

Solution Using the formula for $\cos(u - v)$, you have

$$\begin{aligned}\cos\left(\frac{\pi}{2} - x\right) &= \cos\frac{\pi}{2}\cos x + \sin\frac{\pi}{2}\sin x \\ &= (0)(\cos x) + (1)(\sin x) \\ &= \sin x.\end{aligned}$$

✓ ***Checkpoint*** *Audio-video solution in English & Spanish at LarsonPrecalculus.com.*

Use a difference formula to prove the cofunction identity $\sin\left(x - \frac{\pi}{2}\right) = -\cos x$. ■

Sum and difference formulas can be used to rewrite expressions such as

$$\sin\left(\theta + \frac{n\pi}{2}\right) \quad \text{and} \quad \cos\left(\theta + \frac{n\pi}{2}\right), \quad \text{where } n \text{ is an integer}$$

as expressions involving only $\sin\theta$ or $\cos\theta$. The resulting formulas are called **reduction formulas.**

EXAMPLE 6 **Deriving Reduction Formulas**

Simplify each expression.

a. $\cos\left(\theta - \frac{3\pi}{2}\right)$

b. $\tan(\theta + 3\pi)$

Solution

a. Using the formula for $\cos(u - v)$, you have

$$\begin{aligned}\cos\left(\theta - \frac{3\pi}{2}\right) &= \cos\theta\cos\frac{3\pi}{2} + \sin\theta\sin\frac{3\pi}{2} \\ &= (\cos\theta)(0) + (\sin\theta)(-1) \\ &= -\sin\theta.\end{aligned}$$

b. Using the formula for $\tan(u + v)$, you have

$$\begin{aligned}\tan(\theta + 3\pi) &= \frac{\tan\theta + \tan 3\pi}{1 - \tan\theta\tan 3\pi} \\ &= \frac{\tan\theta + 0}{1 - (\tan\theta)(0)} \\ &= \tan\theta.\end{aligned}$$

✓ ***Checkpoint*** *Audio-video solution in English & Spanish at LarsonPrecalculus.com.*

Simplify each expression.

a. $\sin\left(\frac{3\pi}{2} - \theta\right)$ **b.** $\tan\left(\theta - \frac{\pi}{4}\right)$ ■

Mary Evans Picture Library

EXAMPLE 7 Solving a Trigonometric Equation

Find all solutions of $\sin[x + (\pi/4)] + \sin[x - (\pi/4)] = -1$ in the interval $[0, 2\pi)$.

Algebraic Solution

Using sum and difference formulas, rewrite the equation as

$$\sin x \cos \frac{\pi}{4} + \cos x \sin \frac{\pi}{4} + \sin x \cos \frac{\pi}{4} - \cos x \sin \frac{\pi}{4} = -1$$

$$2 \sin x \cos \frac{\pi}{4} = -1$$

$$2(\sin x)\left(\frac{\sqrt{2}}{2}\right) = -1$$

$$\sin x = -\frac{1}{\sqrt{2}}$$

$$\sin x = -\frac{\sqrt{2}}{2}.$$

So, the only solutions in the interval $[0, 2\pi)$ are $x = 5\pi/4$ and $x = 7\pi/4$.

Graphical Solution

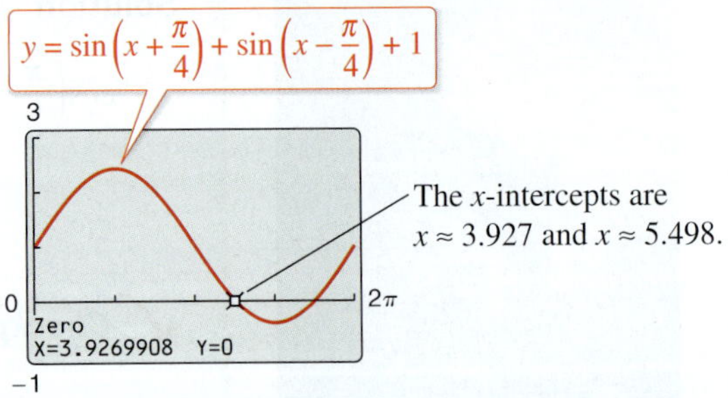

From the above figure, you can conclude that the approximate solutions in the interval $[0, 2\pi)$ are

$$x \approx 3.927 \approx \frac{5\pi}{4} \quad \text{and} \quad x \approx 5.498 \approx \frac{7\pi}{4}.$$

✓ Checkpoint *Audio-video solution in English & Spanish at LarsonPrecalculus.com.*

Find all solutions of $\sin[x + (\pi/2)] + \sin[x - (3\pi/2)] = 1$ in the interval $[0, 2\pi)$.

The next example is an application from calculus.

EXAMPLE 8 An Application from Calculus

Verify that $\dfrac{\sin(x + h) - \sin x}{h} = (\cos x)\left(\dfrac{\sin h}{h}\right) - (\sin x)\left(\dfrac{1 - \cos h}{h}\right)$, where $h \neq 0$.

Solution Using the formula for $\sin(u + v)$, you have

$$\frac{\sin(x + h) - \sin x}{h} = \frac{\sin x \cos h + \cos x \sin h - \sin x}{h}$$

$$= \frac{\cos x \sin h - \sin x(1 - \cos h)}{h}$$

$$= (\cos x)\left(\frac{\sin h}{h}\right) - (\sin x)\left(\frac{1 - \cos h}{h}\right).$$

One application of the sum and difference formulas is in the analysis of standing waves, such as those that can be produced when plucking a guitar string. You will investigate standing waves in Exercise 80.

✓ Checkpoint *Audio-video solution in English & Spanish at LarsonPrecalculus.com.*

Verify that $\dfrac{\cos(x + h) - \cos x}{h} = (\cos x)\left(\dfrac{\cos h - 1}{h}\right) - (\sin x)\left(\dfrac{\sin h}{h}\right)$, where $h \neq 0$.

Summarize (Section 5.4)

1. State the sum and difference formulas for sine, cosine, and tangent *(page 375)*. For examples of using the sum and difference formulas to evaluate trigonometric functions, verify identities, and solve trigonometric equations, see Examples 1–8.

Brian A Jackson/Shutterstock.com

5.4 Exercises

See CalcChat.com for tutorial help and worked-out solutions to odd-numbered exercises.

Vocabulary: Fill in the blank.

1. $\sin(u - v) =$ ________
2. $\cos(u + v) =$ ________
3. $\tan(u + v) =$ ________
4. $\sin(u + v) =$ ________
5. $\cos(u - v) =$ ________
6. $\tan(u - v) =$ ________

Skills and Applications

Evaluating Trigonometric Expressions In Exercises 7–10, find the exact value of each expression.

7. (a) $\cos\left(\frac{\pi}{4} + \frac{\pi}{3}\right)$ (b) $\cos\frac{\pi}{4} + \cos\frac{\pi}{3}$
8. (a) $\sin\left(\frac{7\pi}{6} - \frac{\pi}{3}\right)$ (b) $\sin\frac{7\pi}{6} - \sin\frac{\pi}{3}$
9. (a) $\sin(135° - 30°)$ (b) $\sin 135° - \cos 30°$
10. (a) $\cos(120° + 45°)$ (b) $\cos 120° + \cos 45°$

Evaluating Trigonometric Functions In Exercises 11–26, find the exact values of the sine, cosine, and tangent of the angle.

11. $\frac{11\pi}{12} = \frac{3\pi}{4} + \frac{\pi}{6}$
12. $\frac{7\pi}{12} = \frac{\pi}{3} + \frac{\pi}{4}$
13. $\frac{17\pi}{12} = \frac{9\pi}{4} - \frac{5\pi}{6}$
14. $-\frac{\pi}{12} = \frac{\pi}{6} - \frac{\pi}{4}$
15. $105° = 60° + 45°$
16. $165° = 135° + 30°$
17. $195° = 225° - 30°$
18. $255° = 300° - 45°$
19. $\frac{13\pi}{12}$
20. $-\frac{7\pi}{12}$
21. $-\frac{13\pi}{12}$
22. $\frac{5\pi}{12}$
23. $285°$
24. $-105°$
25. $-165°$
26. $15°$

Rewriting a Trigonometric Expression In Exercises 27–34, write the expression as the sine, cosine, or tangent of an angle.

27. $\sin 3 \cos 1.2 - \cos 3 \sin 1.2$
28. $\cos\frac{\pi}{7}\cos\frac{\pi}{5} - \sin\frac{\pi}{7}\sin\frac{\pi}{5}$
29. $\sin 60° \cos 15° + \cos 60° \sin 15°$
30. $\cos 130° \cos 40° - \sin 130° \sin 40°$
31. $\frac{\tan 45° - \tan 30°}{1 + \tan 45° \tan 30°}$
32. $\frac{\tan 140° - \tan 60°}{1 + \tan 140° \tan 60°}$
33. $\cos 3x \cos 2y + \sin 3x \sin 2y$
34. $\frac{\tan 2x + \tan x}{1 - \tan 2x \tan x}$

Evaluating a Trigonometric Expression In Exercises 35–40, find the exact value of the expression.

35. $\sin\frac{\pi}{12}\cos\frac{\pi}{4} + \cos\frac{\pi}{12}\sin\frac{\pi}{4}$
36. $\cos\frac{\pi}{16}\cos\frac{3\pi}{16} - \sin\frac{\pi}{16}\sin\frac{3\pi}{16}$
37. $\sin 120° \cos 60° - \cos 120° \sin 60°$
38. $\cos 120° \cos 30° + \sin 120° \sin 30°$
39. $\frac{\tan(5\pi/6) - \tan(\pi/6)}{1 + \tan(5\pi/6)\tan(\pi/6)}$
40. $\frac{\tan 25° + \tan 110°}{1 - \tan 25° \tan 110°}$

Evaluating a Trigonometric Expression In Exercises 41–46, find the exact value of the trigonometric expression given that $\sin u = \frac{5}{13}$ and $\cos v = -\frac{3}{5}$. (Both u and v are in Quadrant II.)

41. $\sin(u + v)$
42. $\cos(u - v)$
43. $\tan(u + v)$
44. $\csc(u - v)$
45. $\sec(v - u)$
46. $\cot(u + v)$

Evaluating a Trigonometric Expression In Exercises 47–52, find the exact value of the trigonometric expression given that $\sin u = -\frac{7}{25}$ and $\cos v = -\frac{4}{5}$. (Both u and v are in Quadrant III.)

47. $\cos(u + v)$
48. $\sin(u + v)$
49. $\tan(u - v)$
50. $\cot(v - u)$
51. $\csc(u - v)$
52. $\sec(v - u)$

An Application of a Sum or Difference Formula In Exercises 53–56, write the trigonometric expression as an algebraic expression.

53. $\sin(\arcsin x + \arccos x)$
54. $\sin(\arctan 2x - \arccos x)$
55. $\cos(\arccos x + \arcsin x)$
56. $\cos(\arccos x - \arctan x)$

Proving a Trigonometric Identity In Exercises 57–64, prove the identity.

57. $\sin\left(\frac{\pi}{2} - x\right) = \cos x$

58. $\sin\left(\frac{\pi}{2} + x\right) = \cos x$

59. $\sin\left(\frac{\pi}{6} + x\right) = \frac{1}{2}(\cos x + \sqrt{3}\sin x)$

60. $\cos\left(\frac{5\pi}{4} - x\right) = -\frac{\sqrt{2}}{2}(\cos x + \sin x)$

61. $\cos(\pi - \theta) + \sin\left(\frac{\pi}{2} + \theta\right) = 0$

62. $\tan\left(\frac{\pi}{4} - \theta\right) = \frac{1 - \tan\theta}{1 + \tan\theta}$

63. $\cos(x + y)\cos(x - y) = \cos^2 x - \sin^2 y$

64. $\sin(x + y) + \sin(x - y) = 2\sin x \cos y$

Deriving a Reduction Formula In Exercises 65–68, simplify the expression algebraically and use a graphing utility to confirm your answer graphically.

65. $\cos\left(\frac{3\pi}{2} - x\right)$

66. $\cos(\pi + x)$

67. $\sin\left(\frac{3\pi}{2} + \theta\right)$

68. $\tan(\pi + \theta)$

Solving a Trigonometric Equation In Exercises 69–74, find all solutions of the equation in the interval $[0, 2\pi)$.

69. $\sin(x + \pi) - \sin x + 1 = 0$

70. $\cos(x + \pi) - \cos x - 1 = 0$

71. $\cos\left(x + \frac{\pi}{4}\right) - \cos\left(x - \frac{\pi}{4}\right) = 1$

72. $\sin\left(x + \frac{\pi}{6}\right) - \sin\left(x - \frac{7\pi}{6}\right) = \frac{\sqrt{3}}{2}$

73. $\tan(x + \pi) + 2\sin(x + \pi) = 0$

74. $\sin\left(x + \frac{\pi}{2}\right) - \cos^2 x = 0$

Approximating Solutions In Exercises 75–78, use a graphing utility to approximate the solutions of the equation in the interval $[0, 2\pi)$.

75. $\cos\left(x + \frac{\pi}{4}\right) + \cos\left(x - \frac{\pi}{4}\right) = 1$

76. $\tan(x + \pi) - \cos\left(x + \frac{\pi}{2}\right) = 0$

77. $\sin\left(x + \frac{\pi}{2}\right) + \cos^2 x = 0$

78. $\cos\left(x - \frac{\pi}{2}\right) - \sin^2 x = 0$

79. **Harmonic Motion**

A weight is attached to a spring suspended vertically from a ceiling. When a driving force is applied to the system, the weight moves vertically from its equilibrium position, and this motion is modeled by

$$y = \frac{1}{3}\sin 2t + \frac{1}{4}\cos 2t$$

where y is the distance from equilibrium (in feet) and t is the time (in seconds).

(a) Use the identity

$$a\sin B\theta + b\cos B\theta = \sqrt{a^2 + b^2}\sin(B\theta + C)$$

where $C = \arctan(b/a)$, $a > 0$, to write the model in the form

$$y = \sqrt{a^2 + b^2}\sin(Bt + C).$$

(b) Find the amplitude of the oscillations of the weight.

(c) Find the frequency of the oscillations of the weight.

80. **Standing Waves** The equation of a standing wave is obtained by adding the displacements of two waves traveling in opposite directions (see figure). Assume that each of the waves has amplitude A, period T, and wavelength λ. If the models for these waves are

$$y_1 = A\cos 2\pi\left(\frac{t}{T} - \frac{x}{\lambda}\right) \quad \text{and} \quad y_2 = A\cos 2\pi\left(\frac{t}{T} + \frac{x}{\lambda}\right)$$

then show that

$$y_1 + y_2 = 2A\cos\frac{2\pi t}{T}\cos\frac{2\pi x}{\lambda}.$$

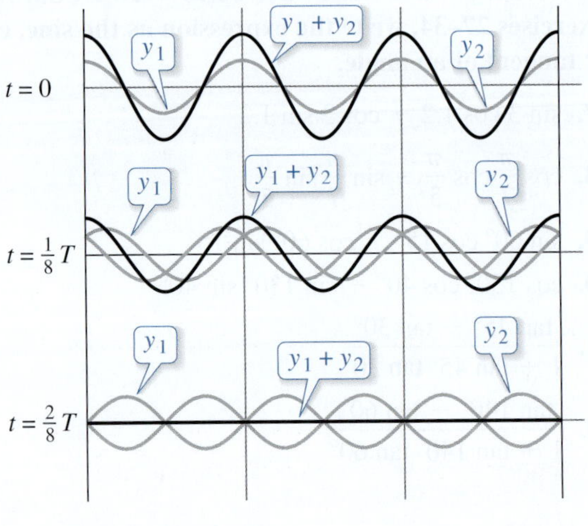

Richard Megna/Fundamental Photographs

Exploration

True or False? **In Exercises 81–84, determine whether the statement is true or false. Justify your answer.**

81. $\sin(u \pm v) = \sin u \cos v \pm \cos u \sin v$

82. $\cos(u \pm v) = \cos u \cos v \pm \sin u \sin v$

83. $\tan\left(x - \frac{\pi}{4}\right) = \frac{\tan x + 1}{1 - \tan x}$

84. $\sin\left(x - \frac{\pi}{2}\right) = -\cos x$

85. **An Application from Calculus** Let $x = \pi/3$ in the identity in Example 8 and define the functions f and g as follows.

$$f(h) = \frac{\sin[(\pi/3) + h] - \sin(\pi/3)}{h}$$

$$g(h) = \cos\frac{\pi}{3}\left(\frac{\sin h}{h}\right) - \sin\frac{\pi}{3}\left(\frac{1 - \cos h}{h}\right)$$

(a) What are the domains of the functions f and g?

(b) Use a graphing utility to complete the table.

h	0.5	0.2	0.1	0.05	0.02	0.01
$f(h)$						
$g(h)$						

(c) Use the graphing utility to graph the functions f and g.

(d) Use the table and the graphs to make a conjecture about the values of the functions f and g as $h \to 0^+$.

86. **HOW DO YOU SEE IT?** Explain how to use the figure to justify each statement.

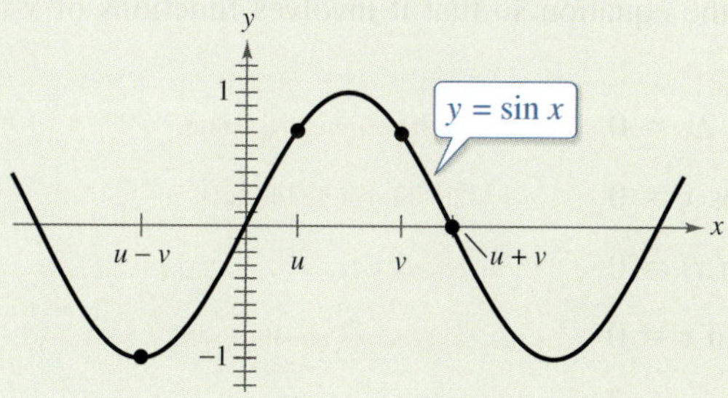

(a) $\sin(u + v) \neq \sin u + \sin v$

(b) $\sin(u - v) \neq \sin u - \sin v$

Verifying an Identity **In Exercises 87–90, verify the identity.**

87. $\cos(n\pi + \theta) = (-1)^n \cos\theta$, n is an integer

88. $\sin(n\pi + \theta) = (-1)^n \sin\theta$, n is an integer

89. $a \sin B\theta + b \cos B\theta = \sqrt{a^2 + b^2}\sin(B\theta + C)$, where $C = \arctan(b/a)$ and $a > 0$

90. $a \sin B\theta + b \cos B\theta = \sqrt{a^2 + b^2}\cos(B\theta - C)$, where $C = \arctan(a/b)$ and $b > 0$

Rewriting a Trigonometric Expression **In Exercises 91–94, use the formulas given in Exercises 89 and 90 to write the trigonometric expression in the following forms.**

(a) $\sqrt{a^2 + b^2}\sin(B\theta + C)$ **(b)** $\sqrt{a^2 + b^2}\cos(B\theta - C)$

91. $\sin\theta + \cos\theta$

92. $3 \sin 2\theta + 4 \cos 2\theta$

93. $12 \sin 3\theta + 5 \cos 3\theta$

94. $\sin 2\theta + \cos 2\theta$

Rewriting a Trigonometric Expression **In Exercises 95 and 96, use the formulas given in Exercises 89 and 90 to write the trigonometric expression in the form $a \sin B\theta + b \cos B\theta$.**

95. $2 \sin\left(\theta + \frac{\pi}{4}\right)$

96. $5 \cos\left(\theta - \frac{\pi}{4}\right)$

Angle Between Two Lines **In Exercises 97 and 98, use the figure, which shows two lines whose equations are $y_1 = m_1x + b_1$ and $y_2 = m_2x + b_2$. Assume that both lines have positive slopes. Derive a formula for the angle between the two lines. Then use your formula to find the angle between the given pair of lines.**

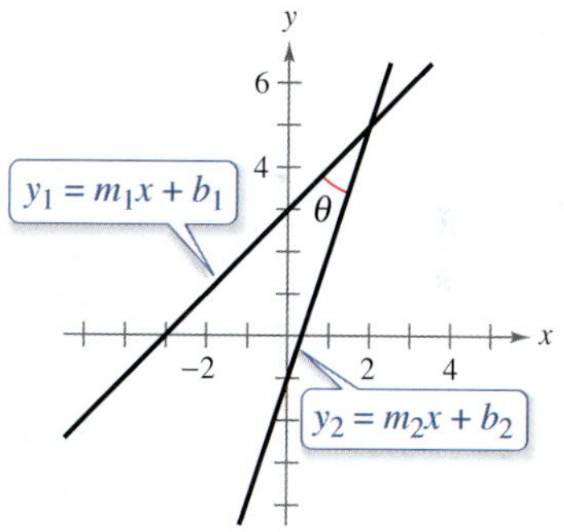

97. $y = x$ and $y = \sqrt{3}x$

98. $y = x$ and $y = \frac{1}{\sqrt{3}}x$

Graphical Reasoning **In Exercises 99 and 100, use a graphing utility to graph y_1 and y_2 in the same viewing window. Use the graphs to determine whether $y_1 = y_2$. Explain your reasoning.**

99. $y_1 = \cos(x + 2)$, $y_2 = \cos x + \cos 2$

100. $y_1 = \sin(x + 4)$, $y_2 = \sin x + \sin 4$

101. **Proof**

(a) Write a proof of the formula for $\sin(u + v)$.

(b) Write a proof of the formula for $\sin(u - v)$.

5.5 Multiple-Angle and Product-to-Sum Formulas

A variety of trigonometric formulas enable you to rewrite trigonometric functions in more convenient forms. For instance, in Exercise 73 on page 390, you will use a half-angle formula to relate the Mach number of a supersonic airplane to the apex angle of the cone formed by the sound waves behind the airplane.

- **Use multiple-angle formulas to rewrite and evaluate trigonometric functions.**
- **Use power-reducing formulas to rewrite and evaluate trigonometric functions.**
- **Use half-angle formulas to rewrite and evaluate trigonometric functions.**
- **Use product-to-sum and sum-to-product formulas to rewrite and evaluate trigonometric functions.**
- **Use trigonometric formulas to rewrite real-life models.**

Multiple-Angle Formulas

In this section, you will study four other categories of trigonometric identities.

1. The first category involves *functions of multiple angles* such as $\sin ku$ and $\cos ku$.
2. The second category involves *squares of trigonometric functions* such as $\sin^2 u$.
3. The third category involves *functions of half-angles* such as $\sin(u/2)$.
4. The fourth category involves *products of trigonometric functions* such as $\sin u \cos v$.

You should learn the **double-angle formulas** because they are used often in trigonometry and calculus. For proofs of these formulas, see Proofs in Mathematics on page 397.

Double-Angle Formulas

$$\sin 2u = 2 \sin u \cos u \qquad \cos 2u = \cos^2 u - \sin^2 u$$

$$\tan 2u = \frac{2 \tan u}{1 - \tan^2 u} \qquad \qquad = 2\cos^2 u - 1$$

$$= 1 - 2 \sin^2 u$$

EXAMPLE 1 Solving a Multiple-Angle Equation

Solve $2 \cos x + \sin 2x = 0$.

Solution Begin by rewriting the equation so that it involves functions of x (rather than $2x$). Then factor and solve.

$$2 \cos x + \sin 2x = 0 \qquad \text{Write original equation.}$$

$$2 \cos x + 2 \sin x \cos x = 0 \qquad \text{Double-angle formula}$$

$$2 \cos x(1 + \sin x) = 0 \qquad \text{Factor.}$$

$$2 \cos x = 0 \quad \text{and} \quad 1 + \sin x = 0 \qquad \text{Set factors equal to zero.}$$

$$x = \frac{\pi}{2}, \frac{3\pi}{2} \qquad x = \frac{3\pi}{2} \qquad \text{Solutions in } [0, 2\pi)$$

So, the general solution is

$$x = \frac{\pi}{2} + 2n\pi \quad \text{and} \quad x = \frac{3\pi}{2} + 2n\pi$$

where n is an integer. Try verifying these solutions graphically.

✓ ***Checkpoint*** *Audio-video solution in English & Spanish at LarsonPrecalculus.com.*

Solve $\cos 2x + \cos x = 0$.

Lukich/Shutterstock.com

EXAMPLE 2 Evaluating Functions Involving Double Angles

Use the following to find $\sin 2\theta$, $\cos 2\theta$, and $\tan 2\theta$.

$$\cos\theta = \frac{5}{13}, \quad \frac{3\pi}{2} < \theta < 2\pi$$

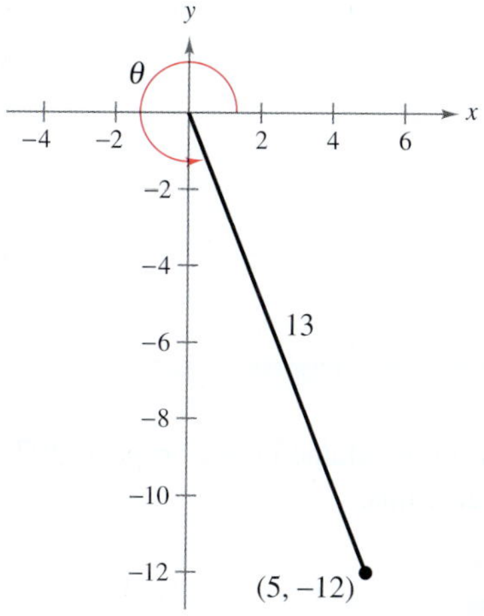

Figure 5.7

Solution From Figure 5.7,

$$\sin\theta = \frac{y}{r} = -\frac{12}{13} \quad \text{and} \quad \tan\theta = \frac{y}{x} = -\frac{12}{5}.$$

Consequently, using each of the double-angle formulas, you can write

$$\sin 2\theta = 2\sin\theta\cos\theta = 2\left(-\frac{12}{13}\right)\left(\frac{5}{13}\right) = -\frac{120}{169}$$

$$\cos 2\theta = 2\cos^2\theta - 1 = 2\left(\frac{25}{169}\right) - 1 = -\frac{119}{169}$$

$$\tan 2\theta = \frac{2\tan\theta}{1-\tan^2\theta} = \frac{2\left(-\frac{12}{5}\right)}{1-\left(-\frac{12}{5}\right)^2} = \frac{120}{119}.$$

✓ Checkpoint *Audio-video solution in English & Spanish at LarsonPrecalculus.com.*

Use the following to find $\sin 2\theta$, $\cos 2\theta$, and $\tan 2\theta$.

$$\sin\theta = \frac{3}{5}, \quad 0 < \theta < \frac{\pi}{2}$$

The double-angle formulas are not restricted to angles 2θ and θ. Other *double* combinations, such as 4θ and 2θ or 6θ and 3θ, are also valid. Here are two examples.

$$\sin 4\theta = 2\sin 2\theta\cos 2\theta \quad \text{and} \quad \cos 6\theta = \cos^2 3\theta - \sin^2 3\theta$$

By using double-angle formulas together with the sum formulas given in the preceding section, you can form other multiple-angle formulas.

EXAMPLE 3 Deriving a Triple-Angle Formula

Rewrite $\sin 3x$ in terms of $\sin x$.

Solution

$\sin 3x = \sin(2x + x)$	Rewrite as a sum.
$= \sin 2x\cos x + \cos 2x\sin x$	Sum formula
$= 2\sin x\cos x\cos x + (1 - 2\sin^2 x)\sin x$	Double-angle formulas
$= 2\sin x\cos^2 x + \sin x - 2\sin^3 x$	Distributive Property
$= 2\sin x(1 - \sin^2 x) + \sin x - 2\sin^3 x$	Pythagorean identity
$= 2\sin x - 2\sin^3 x + \sin x - 2\sin^3 x$	Distributive Property
$= 3\sin x - 4\sin^3 x$	Simplify.

✓ Checkpoint *Audio-video solution in English & Spanish at LarsonPrecalculus.com.*

Rewrite $\cos 3x$ in terms of $\cos x$.

Power-Reducing Formulas

The double-angle formulas can be used to obtain the following **power-reducing formulas.**

Power-Reducing Formulas

$$\sin^2 u = \frac{1 - \cos 2u}{2}$$

$$\cos^2 u = \frac{1 + \cos 2u}{2}$$

$$\tan^2 u = \frac{1 - \cos 2u}{1 + \cos 2u}$$

For a proof of the power-reducing formulas, see Proofs in Mathematics on page 397. Example 4 shows a typical power reduction used in calculus.

EXAMPLE 4 Reducing a Power

Rewrite $\sin^4 x$ in terms of first powers of the cosines of multiple angles.

Solution Note the repeated use of power-reducing formulas.

$$\sin^4 x = (\sin^2 x)^2 \qquad \text{Property of exponents}$$

$$= \left(\frac{1 - \cos 2x}{2}\right)^2 \qquad \text{Power-reducing formula}$$

$$= \frac{1}{4}(1 - 2\cos 2x + \cos^2 2x) \qquad \text{Expand.}$$

$$= \frac{1}{4}\left(1 - 2\cos 2x + \frac{1 + \cos 4x}{2}\right) \qquad \text{Power-reducing formula}$$

$$= \frac{1}{4} - \frac{1}{2}\cos 2x + \frac{1}{8} + \frac{1}{8}\cos 4x \qquad \text{Distributive Property}$$

$$= \frac{3}{8} - \frac{1}{2}\cos 2x + \frac{1}{8}\cos 4x \qquad \text{Simplify.}$$

$$= \frac{1}{8}(3 - 4\cos 2x + \cos 4x) \qquad \text{Factor out common factor.}$$

You can use a graphing utility to check this result, as shown below. Notice that the graphs coincide.

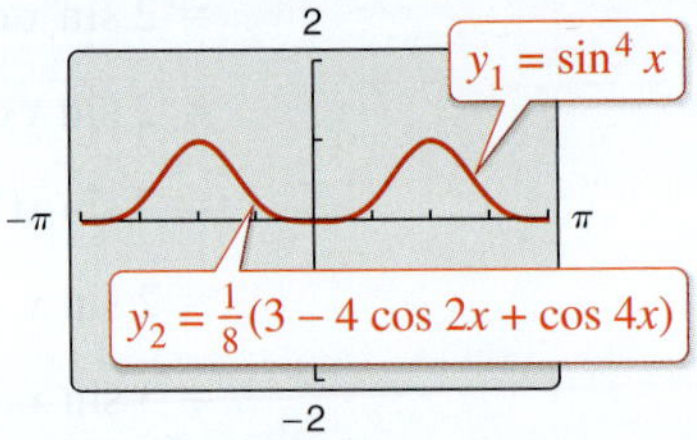

Checkpoint Audio-video solution in English & Spanish at LarsonPrecalculus.com.

Rewrite $\tan^4 x$ in terms of first powers of the cosines of multiple angles.

Half-Angle Formulas

You can derive some useful alternative forms of the power-reducing formulas by replacing u with $u/2$. The results are called **half-angle formulas.**

REMARK To find the exact value of a trigonometric function with an angle measure in D°M′S″ form using a half-angle formula, first convert the angle measure to decimal degree form. Then multiply the resulting angle measure by 2.

Half-Angle Formulas

$$\sin\frac{u}{2} = \pm\sqrt{\frac{1-\cos u}{2}} \qquad \cos\frac{u}{2} = \pm\sqrt{\frac{1+\cos u}{2}}$$

$$\tan\frac{u}{2} = \frac{1-\cos u}{\sin u} = \frac{\sin u}{1+\cos u}$$

The signs of $\sin\frac{u}{2}$ and $\cos\frac{u}{2}$ depend on the quadrant in which $\frac{u}{2}$ lies.

REMARK Use your calculator to verify the result obtained in Example 5. That is, evaluate $\sin 105°$ and $\left(\sqrt{2+\sqrt{3}}\right)/2$. Note that both values are approximately 0.9659258.

EXAMPLE 5 Using a Half-Angle Formula

Find the exact value of $\sin 105°$.

Solution Begin by noting that 105° is half of 210°. Then, using the half-angle formula for $\sin(u/2)$ and the fact that 105° lies in Quadrant II, you have

$$\sin 105° = \sqrt{\frac{1-\cos 210°}{2}} = \sqrt{\frac{1+\left(\sqrt{3}/2\right)}{2}} = \frac{\sqrt{2+\sqrt{3}}}{2}.$$

The positive square root is chosen because $\sin\theta$ is positive in Quadrant II.

Checkpoint Audio-video solution in English & Spanish at LarsonPrecalculus.com.

Find the exact value of $\cos 105°$.

EXAMPLE 6 Solving a Trigonometric Equation

Find all solutions of $1 + \cos^2 x = 2\cos^2\frac{x}{2}$ in the interval $[0, 2\pi)$.

Algebraic Solution

$$1+\cos^2 x = 2\cos^2\frac{x}{2} \qquad \text{Write original equation.}$$

$$1+\cos^2 x = 2\left(\pm\sqrt{\frac{1+\cos x}{2}}\right)^2 \qquad \text{Half-angle formula}$$

$$1+\cos^2 x = 1+\cos x \qquad \text{Simplify.}$$

$$\cos^2 x - \cos x = 0 \qquad \text{Simplify.}$$

$$\cos x(\cos x - 1) = 0 \qquad \text{Factor.}$$

By setting the factors $\cos x$ and $\cos x - 1$ equal to zero, you find that the solutions in the interval $[0, 2\pi)$ are

$$x = \frac{\pi}{2}, \quad x = \frac{3\pi}{2}, \quad \text{and} \quad x = 0.$$

Graphical Solution

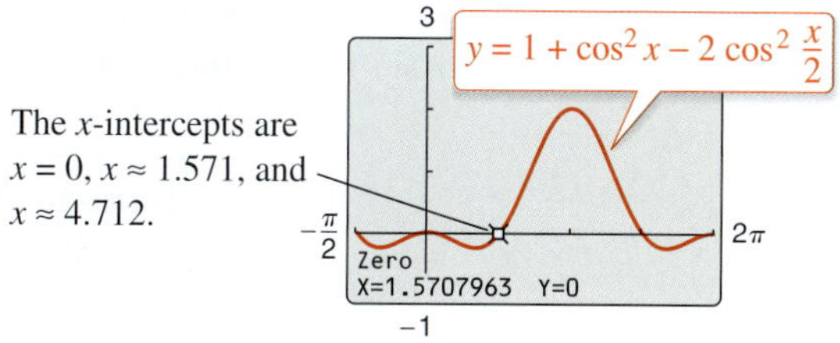

From the above figure, you can conclude that the approximate solutions of $1 + \cos^2 x = 2\cos^2 x/2$ in the interval $[0, 2\pi)$ are

$$x = 0, \quad x \approx 1.571 \approx \frac{\pi}{2}, \quad \text{and} \quad x \approx 4.712 \approx \frac{3\pi}{2}.$$

Checkpoint Audio-video solution in English & Spanish at LarsonPrecalculus.com.

Find all solutions of $\cos^2 x = \sin^2\frac{x}{2}$ in the interval $[0, 2\pi)$.

Product-to-Sum Formulas

Each of the following **product-to-sum formulas** can be verified using the sum and difference formulas discussed in the preceding section.

Product-to-Sum Formulas

$$\sin u \sin v = \frac{1}{2}[\cos(u - v) - \cos(u + v)]$$

$$\cos u \cos v = \frac{1}{2}[\cos(u - v) + \cos(u + v)]$$

$$\sin u \cos v = \frac{1}{2}[\sin(u + v) + \sin(u - v)]$$

$$\cos u \sin v = \frac{1}{2}[\sin(u + v) - \sin(u - v)]$$

Product-to-sum formulas are used in calculus to solve problems involving the products of sines and cosines of two different angles.

EXAMPLE 7 Writing Products as Sums

Rewrite the product $\cos 5x \sin 4x$ as a sum or difference.

Solution Using the appropriate product-to-sum formula, you obtain

$$\cos 5x \sin 4x = \frac{1}{2}[\sin(5x + 4x) - \sin(5x - 4x)]$$

$$= \frac{1}{2}\sin 9x - \frac{1}{2}\sin x.$$

✓ ***Checkpoint*** *Audio-video solution in English & Spanish at LarsonPrecalculus.com.*

Rewrite the product $\sin 5x \cos 3x$ as a sum or difference.

Occasionally, it is useful to reverse the procedure and write a sum of trigonometric functions as a product. This can be accomplished with the following **sum-to-product formulas.**

Sum-to-Product Formulas

$$\sin u + \sin v = 2\sin\left(\frac{u + v}{2}\right)\cos\left(\frac{u - v}{2}\right)$$

$$\sin u - \sin v = 2\cos\left(\frac{u + v}{2}\right)\sin\left(\frac{u - v}{2}\right)$$

$$\cos u + \cos v = 2\cos\left(\frac{u + v}{2}\right)\cos\left(\frac{u - v}{2}\right)$$

$$\cos u - \cos v = -2\sin\left(\frac{u + v}{2}\right)\sin\left(\frac{u - v}{2}\right)$$

For a proof of the sum-to-product formulas, see Proofs in Mathematics on page 398.

EXAMPLE 8 Using a Sum-to-Product Formula

Find the exact value of $\cos 195° + \cos 105°$.

Solution Using the appropriate sum-to-product formula, you obtain

$$\begin{aligned}\cos 195° + \cos 105° &= 2\cos\left(\frac{195° + 105°}{2}\right)\cos\left(\frac{195° - 105°}{2}\right)\\ &= 2\cos 150° \cos 45°\\ &= 2\left(-\frac{\sqrt{3}}{2}\right)\left(\frac{\sqrt{2}}{2}\right)\\ &= -\frac{\sqrt{6}}{2}.\end{aligned}$$

Checkpoint *Audio-video solution in English & Spanish at LarsonPrecalculus.com.*

Find the exact value of $\sin 195° + \sin 105°$.

EXAMPLE 9 Solving a Trigonometric Equation

Solve $\sin 5x + \sin 3x = 0$.

Solution

$$\sin 5x + \sin 3x = 0 \qquad \text{Write original equation.}$$

$$2\sin\left(\frac{5x + 3x}{2}\right)\cos\left(\frac{5x - 3x}{2}\right) = 0 \qquad \text{Sum-to-product formula}$$

$$2\sin 4x \cos x = 0 \qquad \text{Simplify.}$$

By setting the factor $2\sin 4x$ equal to zero, you can find that the solutions in the interval $[0, 2\pi)$ are

$$x = 0, \frac{\pi}{4}, \frac{\pi}{2}, \frac{3\pi}{4}, \pi, \frac{5\pi}{4}, \frac{3\pi}{2}, \frac{7\pi}{4}.$$

The equation $\cos x = 0$ yields no additional solutions, so you can conclude that the solutions are of the form $x = n\pi/4$ where n is an integer. To confirm this graphically, sketch the graph of $y = \sin 5x + \sin 3x$, as shown below.

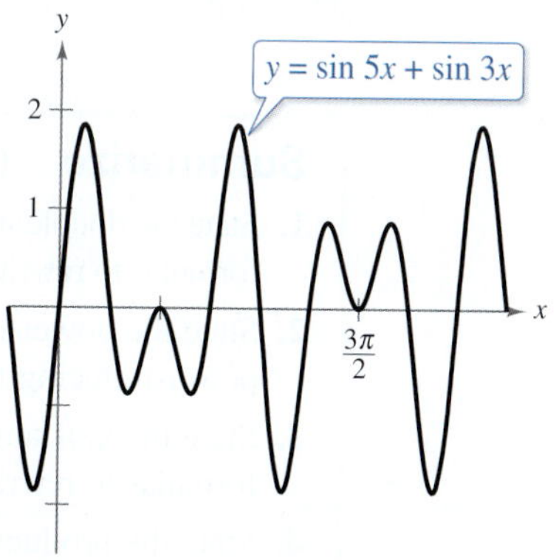

Notice from the graph that the x-intercepts occur at multiples of $\pi/4$.

Checkpoint *Audio-video solution in English & Spanish at LarsonPrecalculus.com.*

Solve $\sin 4x - \sin 2x = 0$.

Application

EXAMPLE 10 Projectile Motion

Kicking a football with an initial velocity of 80 feet per second at an angle of 45° with the horizontal results in a distance traveled of 200 feet.

Ignoring air resistance, the range of a projectile fired at an angle θ with the horizontal and with an initial velocity of v_0 feet per second is given by

$$r = \frac{1}{16}v_0^2 \sin \theta \cos \theta$$

where r is the horizontal distance (in feet) that the projectile travels. A football player can kick a football from ground level with an initial velocity of 80 feet per second.

a. Write the projectile motion model in a simpler form.

b. At what angle must the player kick the football so that the football travels 200 feet?

Solution

a. You can use a double-angle formula to rewrite the projectile motion model as

$$r = \frac{1}{32}v_0^2(2 \sin \theta \cos \theta) \quad \text{Rewrite original projectile motion model.}$$

$$= \frac{1}{32}v_0^2 \sin 2\theta. \quad \text{Rewrite model using a double-angle formula.}$$

b.

$$r = \frac{1}{32}v_0^2 \sin 2\theta \quad \text{Write projectile motion model.}$$

$$200 = \frac{1}{32}(80)^2 \sin 2\theta \quad \text{Substitute 200 for } r \text{ and 80 for } v_0.$$

$$200 = 200 \sin 2\theta \quad \text{Simplify.}$$

$$1 = \sin 2\theta \quad \text{Divide each side by 200.}$$

You know that $2\theta = \pi/2$, so dividing this result by 2 produces $\theta = \pi/4$. Because $\pi/4 = 45°$, the player must kick the football at an angle of 45° so that the football travels 200 feet.

✓ **Checkpoint** *Audio-video solution in English & Spanish at LarsonPrecalculus.com.*

In Example 10, for what angle is the horizontal distance the football travels a maximum?

Summarize (Section 5.5)

1. State the double-angle formulas *(page 382)*. For examples of using multiple-angle formulas to rewrite and evaluate trigonometric functions, see Examples 1–3.
2. State the power-reducing formulas *(page 384)*. For an example of using power-reducing formulas to rewrite a trigonometric function, see Example 4.
3. State the half-angle formulas *(page 385)*. For examples of using half-angle formulas to rewrite and evaluate trigonometric functions, see Examples 5 and 6.
4. State the product-to-sum and sum-to-product formulas *(page 386)*. For an example of using a product-to-sum formula to rewrite a trigonometric function, see Example 7. For examples of using sum-to-product formulas to rewrite and evaluate trigonometric functions, see Examples 8 and 9.
5. Describe an example of how to use a trigonometric formula to rewrite a real-life model *(page 388, Example 10)*.

Aspen Photo/Shutterstock.com

5.5 Exercises

See CalcChat.com for tutorial help and worked-out solutions to odd-numbered exercises.

Vocabulary: Fill in the blank to complete the trigonometric formula.

1. $\sin 2u =$ ________
2. $\cos 2u =$ ________
3. $\dfrac{1 - \cos 2u}{1 + \cos 2u} =$ ________
4. $\sin \dfrac{u}{2} =$ ________
5. $\sin u \cos v =$ ________
6. $\cos u \ - \cos v =$ ________

Skills and Applications

Solving a Multiple-Angle Equation In Exercises 7–14, find the exact solutions of the equation in the interval $[0, 2\pi)$.

7. $\sin 2x - \sin x = 0$
8. $\sin 2x \sin x = \cos x$
9. $\cos 2x - \cos x = 0$
10. $\cos 2x + \sin x = 0$
11. $\sin 4x = -2 \sin 2x$
12. $(\sin 2x + \cos 2x)^2 = 1$
13. $\tan 2x - \cot x = 0$
14. $\tan 2x - 2 \cos x = 0$

Using a Double-Angle Formula In Exercises 15–20, use a double-angle formula to rewrite the expression.

15. $6 \sin x \cos x$
16. $\sin x \cos x$
17. $6 \cos^2 x - 3$
18. $\cos^2 x - \frac{1}{2}$
19. $4 - 8 \sin^2 x$
20. $10 \sin^2 x - 5$

Evaluating Functions Involving Double Angles In Exercises 21–24, find the exact values of $\sin 2u$, $\cos 2u$, and $\tan 2u$ using the double-angle formulas.

21. $\sin u = -3/5, \quad 3\pi/2 < u < 2\pi$
22. $\cos u = -4/5, \quad \pi/2 < u < \pi$
23. $\tan u = 3/5, \quad 0 < u < \pi/2$
24. $\sec u = -2, \quad \pi < u < 3\pi/2$

25. **Deriving a Multiple-Angle Formula** Rewrite $\cos 4x$ in terms of $\cos x$.

26. **Deriving a Multiple-Angle Formula** Rewrite $\tan 3x$ in terms of $\tan x$.

Reducing Powers In Exercises 27–32, use the power-reducing formulas to rewrite the expression in terms of the first power of the cosine.

27. $\cos^4 x$
28. $\sin^4 2x$
29. $\tan^4 2x$
30. $\tan^2 2x \cos^4 2x$
31. $\sin^2 2x \cos^2 2x$
32. $\sin^4 x \cos^2 x$

Using Half-Angle Formulas In Exercises 33–36, use the half-angle formulas to determine the exact values of the sine, cosine, and tangent of the angle.

33. $75°$
34. $67° \, 30'$
35. $\pi/8$
36. $7\pi/12$

Using Half-Angle Formulas In Exercises 37–40, (a) determine the quadrant in which $u/2$ lies, and (b) find the exact values of $\sin(u/2)$, $\cos(u/2)$, and $\tan(u/2)$ using the half-angle formulas.

37. $\cos u = 7/25, \quad 0 < u < \pi/2$
38. $\sin u = 5/13, \quad \pi/2 < u < \pi$
39. $\tan u = -5/12, \quad 3\pi/2 < u < 2\pi$
40. $\cot u = 3, \quad \pi < u < 3\pi/2$

Using Half-Angle Formulas In Exercises 41–44, use the half-angle formulas to simplify the expression.

41. $\sqrt{\dfrac{1 - \cos 6x}{2}}$
42. $\sqrt{\dfrac{1 + \cos 4x}{2}}$
43. $-\sqrt{\dfrac{1 - \cos 8x}{1 + \cos 8x}}$
44. $-\sqrt{\dfrac{1 - \cos(x - 1)}{2}}$

Solving a Trigonometric Equation In Exercises 45–48, find all solutions of the equation in the interval $[0, 2\pi)$. Use a graphing utility to graph the equation and verify the solutions.

45. $\sin \dfrac{x}{2} + \cos x = 0$
46. $\sin \dfrac{x}{2} + \cos x - 1 = 0$
47. $\cos \dfrac{x}{2} - \sin x = 0$
48. $\tan \dfrac{x}{2} - \sin x = 0$

Using Product-to-Sum Formulas In Exercises 49–52, use the product-to-sum formulas to rewrite the product as a sum or difference.

49. $\sin 5\theta \sin 3\theta$
50. $7 \cos(-5\beta) \sin 3\beta$
51. $\cos 2\theta \cos 4\theta$
52. $\sin(x + y) \cos(x - y)$

Using Sum-to-Product Formulas In Exercises 53–56, use the sum-to-product formulas to rewrite the sum or difference as a product.

53. $\sin 5\theta - \sin 3\theta$
54. $\sin 3\theta + \sin \theta$
55. $\cos 6x + \cos 2x$
56. $\cos\left(\theta + \dfrac{\pi}{2}\right) - \cos\left(\theta - \dfrac{\pi}{2}\right)$

Using Sum-to-Product Formulas **In Exercises 57–60, use the sum-to-product formulas to find the exact value of the expression.**

57. $\sin 75^\circ + \sin 15^\circ$

58. $\cos 120^\circ + \cos 60^\circ$

59. $\cos \dfrac{3\pi}{4} - \cos \dfrac{\pi}{4}$

60. $\sin \dfrac{5\pi}{4} - \sin \dfrac{3\pi}{4}$

Solving a Trigonometric Equation **In Exercises 61–64, find all solutions of the equation in the interval $[0, 2\pi)$. Use a graphing utility to graph the equation and verify the solutions.**

61. $\sin 6x + \sin 2x = 0$

62. $\cos 2x - \cos 6x = 0$

63. $\dfrac{\cos 2x}{\sin 3x - \sin x} - 1 = 0$

64. $\sin^2 3x - \sin^2 x = 0$

Verifying a Trigonometric Identity **In Exercises 65–72, verify the identity.**

65. $\csc 2\theta = \dfrac{\csc \theta}{2 \cos \theta}$

66. $\sin \dfrac{\alpha}{3} \cos \dfrac{\alpha}{3} = \dfrac{1}{2} \sin \dfrac{2\alpha}{3}$

67. $1 + \cos 10y = 2 \cos^2 5y$

68. $\cos^4 x - \sin^4 x = \cos 2x$

69. $(\sin x + \cos x)^2 = 1 + \sin 2x$

70. $\tan \dfrac{u}{2} = \csc u - \cot u$

71. $\dfrac{\sin x \pm \sin y}{\cos x + \cos y} = \tan \dfrac{x \pm y}{2}$

72. $\cos\left(\dfrac{\pi}{3} + x\right) + \cos\left(\dfrac{\pi}{3} - x\right) = \cos x$

73. Mach Number The Mach number M of a supersonic airplane is the ratio of its speed to the speed of sound. When an airplane travels faster than the speed of sound, the sound waves form a cone behind the airplane. The Mach number is related to the apex angle θ of the cone by $\sin(\theta/2) = 1/M$.

(a) Use a half-angle formula to rewrite the equation in terms of $\cos \theta$.

(b) Find the angle θ that corresponds to a Mach number of 1.

(c) Find the angle θ that corresponds to a Mach number of 4.5.

(d) The speed of sound is about 760 miles per hour. Determine the speed of an object with the Mach numbers from parts (b) and (c).

74. Projectile Motion The range of a projectile fired at an angle θ with the horizontal and with an initial velocity of v_0 feet per second is

$$r = \frac{1}{32} v_0^2 \sin 2\theta$$

where r is measured in feet. An athlete throws a javelin at 75 feet per second. At what angle must the athlete throw the javelin so that the javelin travels 130 feet?

75. Railroad Track When two railroad tracks merge, the overlapping portions of the tracks are in the shapes of circular arcs (see figure). The radius of each arc r (in feet) and the angle θ are related by

$$\frac{x}{2} = 2r \sin^2 \frac{\theta}{2}.$$

Write a formula for x in terms of $\cos \theta$.

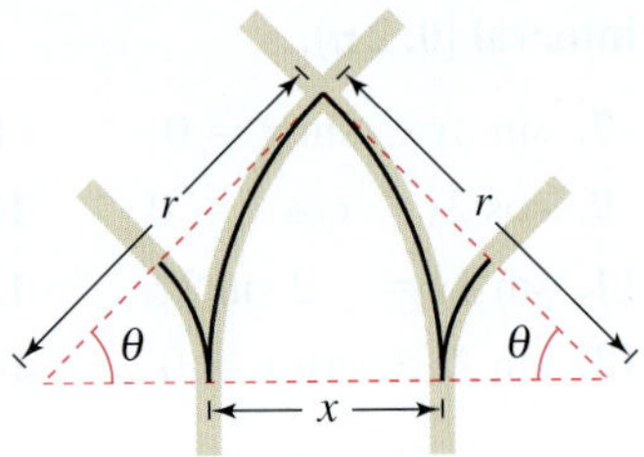

Exploration

76. HOW DO YOU SEE IT? Explain how to use the figure to verify the double-angle formulas (a) $\sin 2u = 2 \sin u \cos u$ and (b) $\cos 2u = \cos^2 u - \sin^2 u$.

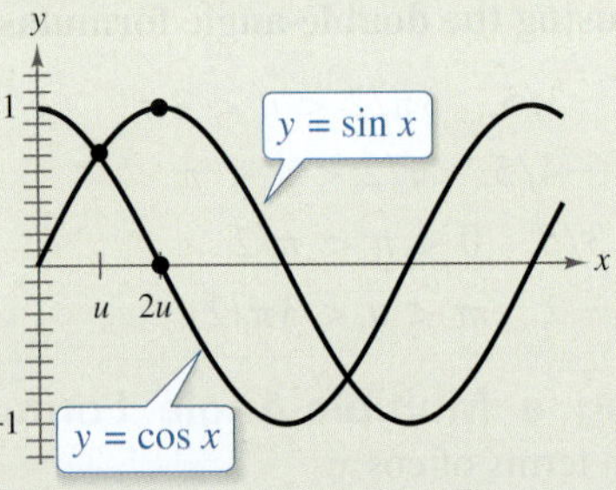

True or False? **In Exercises 77 and 78, determine whether the statement is true or false. Justify your answer.**

77. Because the sine function is an odd function, for a negative number u, $\sin 2u = -2 \sin u \cos u$.

78. $\sin \dfrac{u}{2} = -\sqrt{\dfrac{1 - \cos u}{2}}$ when u is in the second quadrant.

79. Complementary Angles If ϕ and θ are complementary angles, then show that (a) $\sin(\phi - \theta) = \cos 2\theta$ and (b) $\cos(\phi - \theta) = \sin 2\theta$.

Lukich/Shutterstock.com

Chapter Summary

	What Did You Learn?	Explanation/Examples	Review Exercises
Section 5.1	Recognize and write the fundamental trigonometric identities *(p. 350).*	**Reciprocal Identities** $\sin u = 1/\csc u$ $\cos u = 1/\sec u$ $\tan u = 1/\cot u$ $\csc u = 1/\sin u$ $\sec u = 1/\cos u$ $\cot u = 1/\tan u$ **Quotient Identities:** $\tan u = \dfrac{\sin u}{\cos u}$, $\cot u = \dfrac{\cos u}{\sin u}$ **Pythagorean Identities:** $\sin^2 u + \cos^2 u = 1$, $1 + \tan^2 u = \sec^2 u$, $1 + \cot^2 u = \csc^2 u$ **Cofunction Identities** $\sin[(\pi/2) - u] = \cos u$ $\cos[(\pi/2) - u] = \sin u$ $\tan[(\pi/2) - u] = \cot u$ $\cot[(\pi/2) - u] = \tan u$ $\sec[(\pi/2) - u] = \csc u$ $\csc[(\pi/2) - u] = \sec u$ **Even/Odd Identities** $\sin(-u) = -\sin u$ $\cos(-u) = \cos u$ $\tan(-u) = -\tan u$ $\csc(-u) = -\csc u$ $\sec(-u) = \sec u$ $\cot(-u) = -\cot u$	1–4
	Use the fundamental trigonometric identities to evaluate trigonometric functions, simplify trigonometric expressions, and rewrite trigonometric expressions *(p. 351).*	In some cases, when factoring or simplifying trigonometric expressions, it is helpful to rewrite the expression in terms of just *one* trigonometric function or in terms of *sine and cosine only*.	5–18
Section 5.2	Verify trigonometric identities *(p. 357).*	**Guidelines for Verifying Trigonometric Identities** **1.** Work with one side of the equation at a time. **2.** Look to factor an expression, add fractions, square a binomial, or create a monomial denominator. **3.** Look to use the fundamental identities. Note which functions are in the final expression you want. Sines and cosines pair up well, as do secants and tangents, and cosecants and cotangents. **4.** If the preceding guidelines do not help, then try converting all terms to sines and cosines. **5.** Always try *something*.	19–26
Section 5.3	Use standard algebraic techniques to solve trigonometric equations *(p. 364).*	Use standard algebraic techniques (when possible) such as collecting like terms, extracting square roots, and factoring to solve trigonometric equations.	27–32
	Solve trigonometric equations of quadratic type *(p. 367).*	To solve trigonometric equations of quadratic type $ax^2 + bx + c = 0$, factor the quadratic or, when this is not possible, use the Quadratic Formula.	33–36
	Solve trigonometric equations involving multiple angles *(p. 369).*	To solve equations that contain forms such as $\sin ku$ or $\cos ku$, first solve the equation for ku, and then divide your result by k.	37–42
	Use inverse trigonometric functions to solve trigonometric equations *(p. 370).*	After factoring an equation, you may get an equation such as $(\tan x - 3)(\tan x + 1) = 0$. In such cases, use inverse trigonometric functions to solve. (See Example 9.)	43–46

	What Did You Learn?	Explanation/Examples	Review Exercises
Section 5.4	Use sum and difference formulas to evaluate trigonometric functions, verify identities, and solve trigonometric equations *(p. 375)*.	**Sum and Difference Formulas** $\sin(u + v) = \sin u \cos v + \cos u \sin v$ $\sin(u - v) = \sin u \cos v - \cos u \sin v$ $\cos(u + v) = \cos u \cos v - \sin u \sin v$ $\cos(u - v) = \cos u \cos v + \sin u \sin v$ $\tan(u + v) = \dfrac{\tan u + \tan v}{1 - \tan u \tan v}$ $\tan(u - v) = \dfrac{\tan u - \tan v}{1 + \tan u \tan v}$	47–62
Section 5.5	Use multiple-angle formulas to rewrite and evaluate trigonometric functions *(p. 382)*.	**Double-Angle Formulas** $\sin 2u = 2 \sin u \cos u$ $\quad \cos 2u = \cos^2 u - \sin^2 u$ $= 2\cos^2 u - 1$ $\tan 2u = \dfrac{2 \tan u}{1 - \tan^2 u}$ $\quad = 1 - 2\sin^2 u$	63–66
	Use power-reducing formulas to rewrite and evaluate trigonometric functions *(p. 384)*.	**Power-Reducing Formulas** $\sin^2 u = \dfrac{1 - \cos 2u}{2}, \quad \cos^2 u = \dfrac{1 + \cos 2u}{2}$ $\tan^2 u = \dfrac{1 - \cos 2u}{1 + \cos 2u}$	67, 68
	Use half-angle formulas to rewrite and evaluate trigonometric functions *(p. 385)*.	**Half-Angle Formulas** $\sin \dfrac{u}{2} = \pm\sqrt{\dfrac{1 - \cos u}{2}}, \quad \cos \dfrac{u}{2} = \pm\sqrt{\dfrac{1 + \cos u}{2}}$ $\tan \dfrac{u}{2} = \dfrac{1 - \cos u}{\sin u} = \dfrac{\sin u}{1 + \cos u}$ The signs of $\sin(u/2)$ and $\cos(u/2)$ depend on the quadrant in which $u/2$ lies.	69–74
	Use product-to-sum and sum-to-product formulas to rewrite and evaluate trigonometric functions *(p. 386)*.	**Product-to-Sum Formulas** $\sin u \sin v = (1/2)[\cos(u - v) - \cos(u + v)]$ $\cos u \cos v = (1/2)[\cos(u - v) + \cos(u + v)]$ $\sin u \cos v = (1/2)[\sin(u + v) + \sin(u - v)]$ $\cos u \sin v = (1/2)[\sin(u + v) - \sin(u - v)]$ **Sum-to-Product Formulas** $\sin u + \sin v = 2 \sin\left(\dfrac{u + v}{2}\right)\cos\left(\dfrac{u - v}{2}\right)$ $\sin u - \sin v = 2 \cos\left(\dfrac{u + v}{2}\right)\sin\left(\dfrac{u - v}{2}\right)$ $\cos u + \cos v = 2 \cos\left(\dfrac{u + v}{2}\right)\cos\left(\dfrac{u - v}{2}\right)$ $\cos u - \cos v = -2 \sin\left(\dfrac{u + v}{2}\right)\sin\left(\dfrac{u - v}{2}\right)$	75–78
	Use trigonometric formulas to rewrite real-life models *(p. 388)*.	A trigonometric formula can be used to rewrite the projectile motion model $r = (1/16)v_0^2 \sin \theta \cos \theta$. (See Example 10.)	79, 80

Review Exercises

See CalcChat.com for tutorial help and worked-out solutions to odd-numbered exercises.

5.1 Recognizing a Fundamental Identity In Exercises 1–4, name the trigonometric function that is equivalent to the expression.

1. $\dfrac{\sin x}{\cos x}$
2. $\dfrac{1}{\sin x}$
3. $\dfrac{1}{\tan x}$
4. $\sqrt{\cot^2 x + 1}$

Using Identities to Evaluate a Function In Exercises 5 and 6, use the given values and fundamental trigonometric identities to find the values (if possible) of all six trigonometric functions.

5. $\tan\theta = \dfrac{2}{3}$, $\sec\theta = \dfrac{\sqrt{13}}{3}$
6. $\sin\left(\dfrac{\pi}{2} - x\right) = \dfrac{\sqrt{2}}{2}$, $\sin x = -\dfrac{\sqrt{2}}{2}$

Simplifying a Trigonometric Expression In Exercises 7–16, use the fundamental trigonometric identities to simplify the expression. There is more than one correct form of each answer.

7. $\dfrac{1}{\cot^2 x + 1}$
8. $\dfrac{\tan\theta}{1 - \cos^2\theta}$
9. $\tan^2 x(\csc^2 x - 1)$
10. $\cot^2 x(\sin^2 x)$
11. $\dfrac{\cot\left(\dfrac{\pi}{2} - u\right)}{\cos u}$
12. $\dfrac{\sec^2(-\theta)}{\csc^2\theta}$
13. $\cos^2 x + \cos^2 x\cot^2 x$
14. $(\tan x + 1)^2 \cos x$
15. $\dfrac{1}{\csc\theta + 1} - \dfrac{1}{\csc\theta - 1}$
16. $\dfrac{\tan^2 x}{1 + \sec x}$

Trigonometric Substitution In Exercises 17 and 18, use the trigonometric substitution to write the algebraic expression as a trigonometric function of θ, where $0 < \theta < \pi/2$.

17. $\sqrt{25 - x^2}$, $x = 5\sin\theta$
18. $\sqrt{x^2 - 16}$, $x = 4\sec\theta$

5.2 Verifying a Trigonometric Identity In Exercises 19–26, verify the identity.

19. $\cos x(\tan^2 x + 1) = \sec x$
20. $\sec^2 x\cot x - \cot x = \tan x$
21. $\sec\left(\dfrac{\pi}{2} - \theta\right) = \csc\theta$
22. $\cot\left(\dfrac{\pi}{2} - x\right) = \tan x$
23. $\dfrac{1}{\tan\theta\csc\theta} = \cos\theta$
24. $\dfrac{1}{\tan x\csc x\sin x} = \cot x$
25. $\sin^5 x\cos^2 x = (\cos^2 x - 2\cos^4 x + \cos^6 x)\sin x$
26. $\cos^3 x\sin^2 x = (\sin^2 x - \sin^4 x)\cos x$

5.3 Solving a Trigonometric Equation In Exercises 27–32, solve the equation.

27. $\sin x = \sqrt{3} - \sin x$
28. $4\cos\theta = 1 + 2\cos\theta$
29. $3\sqrt{3}\tan u = 3$
30. $\frac{1}{2}\sec x - 1 = 0$
31. $3\csc^2 x = 4$
32. $4\tan^2 u - 1 = \tan^2 u$

Solving a Trigonometric Equation In Exercises 33–42, find all solutions of the equation in the interval $[0, 2\pi)$.

33. $2\cos^2 x - \cos x = 1$
34. $2\cos^2 x + 3\cos x = 0$
35. $\cos^2 x + \sin x = 1$
36. $\sin^2 x + 2\cos x = 2$
37. $2\sin 2x - \sqrt{2} = 0$
38. $2\cos\dfrac{x}{2} + 1 = 0$
39. $3\tan^2\left(\dfrac{x}{3}\right) - 1 = 0$
40. $\sqrt{3}\tan 3x = 0$
41. $\cos 4x(\cos x - 1) = 0$
42. $3\csc^2 5x = -4$

Using Inverse Functions In Exercises 43–46, use inverse functions where needed to find all solutions of the equation in the interval $[0, 2\pi)$.

43. $\tan^2 x - 2\tan x = 0$
44. $2\tan^2 x - 3\tan x = -1$
45. $\tan^2\theta + \tan\theta - 6 = 0$
46. $\sec^2 x + 6\tan x + 4 = 0$

5.4 Evaluating Trigonometric Functions In Exercises 47–50, find the exact values of the sine, cosine, and tangent of the angle.

47. $285° = 315° - 30°$
48. $345° = 300° + 45°$
49. $\dfrac{25\pi}{12} = \dfrac{11\pi}{6} + \dfrac{\pi}{4}$
50. $\dfrac{19\pi}{12} = \dfrac{11\pi}{6} - \dfrac{\pi}{4}$

Rewriting a Trigonometric Expression In Exercises 51 and 52, write the expression as the sine, cosine, or tangent of an angle.

51. $\sin 60° \cos 45° - \cos 60° \sin 45°$
52. $\dfrac{\tan 68° - \tan 115°}{1 + \tan 68° \tan 115°}$

Evaluating a Trigonometric Expression In Exercises 53–56, find the exact value of the trigonometric expression given that $\tan u = \frac{3}{4}$ and $\cos v = -\frac{4}{5}$. (u is in Quadrant I and v is in Quadrant III.)

53. $\sin(u + v)$
54. $\tan(u + v)$
55. $\cos(u - v)$
56. $\sin(u - v)$

Proving a Trigonometric Identity In Exercises 57–60, prove the identity.

57. $\cos\left(x + \frac{\pi}{2}\right) = -\sin x$

58. $\tan\left(x - \frac{\pi}{2}\right) = -\cot x$

59. $\tan(\pi - x) = -\tan x$

60. $\cos 3x = 4\cos^3 x - 3\cos x$

Solving a Trigonometric Equation In Exercises 61 and 62, find all solutions of the equation in the interval $[0, 2\pi)$.

61. $\sin\left(x + \frac{\pi}{4}\right) - \sin\left(x - \frac{\pi}{4}\right) = 1$

62. $\cos\left(x + \frac{\pi}{6}\right) - \cos\left(x - \frac{\pi}{6}\right) = 1$

5.5 Evaluating Functions Involving Double Angles

In Exercises 63 and 64, find the exact values of $\sin 2u$, $\cos 2u$, and $\tan 2u$ using the double-angle formulas.

63. $\sin u = -\frac{4}{5}, \quad \pi < u < 3\pi/2$

64. $\cos u = -2/\sqrt{5}, \quad \pi/2 < u < \pi$

Verifying a Trigonometric Identity In Exercises 65 and 66, use the double-angle formulas to verify the identity algebraically and use a graphing utility to confirm your result graphically.

65. $\sin 4x = 8\cos^3 x \sin x - 4\cos x \sin x$

66. $\tan^2 x = \dfrac{1 - \cos 2x}{1 + \cos 2x}$

Reducing Powers In Exercises 67 and 68, use the power-reducing formulas to rewrite the expression in terms of the first power of the cosine.

67. $\tan^2 2x$

68. $\sin^2 x \tan^2 x$

Using Half-Angle Formulas In Exercises 69 and 70, use the half-angle formulas to determine the exact values of the sine, cosine, and tangent of the angle.

69. $-75°$

70. $\dfrac{19\pi}{12}$

Using Half-Angle Formulas In Exercises 71 and 72, (a) determine the quadrant in which $u/2$ lies, and (b) find the exact values of $\sin(u/2)$, $\cos(u/2)$, and $\tan(u/2)$ using the half-angle formulas.

71. $\tan u = \frac{4}{3}, \quad \pi < u < 3\pi/2$

72. $\cos u = -\frac{2}{7}, \quad \pi/2 < u < \pi$

Using Half-Angle Formulas In Exercises 73 and 74, use the half-angle formulas to simplify the expression.

73. $-\sqrt{\dfrac{1 + \cos 10x}{2}}$

74. $\dfrac{\sin 6x}{1 + \cos 6x}$

Using Product-to-Sum Formulas In Exercises 75 and 76, use the product-to-sum formulas to rewrite the product as a sum or difference.

75. $\cos 4\theta \sin 6\theta$

76. $2\sin 7\theta \cos 3\theta$

Using Sum-to-Product Formulas In Exercises 77 and 78, use the sum-to-product formulas to rewrite the sum or difference as a product.

77. $\cos 6\theta + \cos 5\theta$

78. $\sin\left(x + \frac{\pi}{4}\right) - \sin\left(x - \frac{\pi}{4}\right)$

79. **Projectile Motion** A baseball leaves the hand of a player at first base at an angle of θ with the horizontal and at an initial velocity of $v_0 = 80$ feet per second. A player at second base 100 feet away catches the ball. Find θ when the range r of a projectile is

$$r = \frac{1}{32}v_0^2 \sin 2\theta.$$

80. **Geometry** A trough for feeding cattle is 4 meters long and its cross sections are isosceles triangles with the two equal sides being $\frac{1}{2}$ meter (see figure). The angle between the two sides is θ.

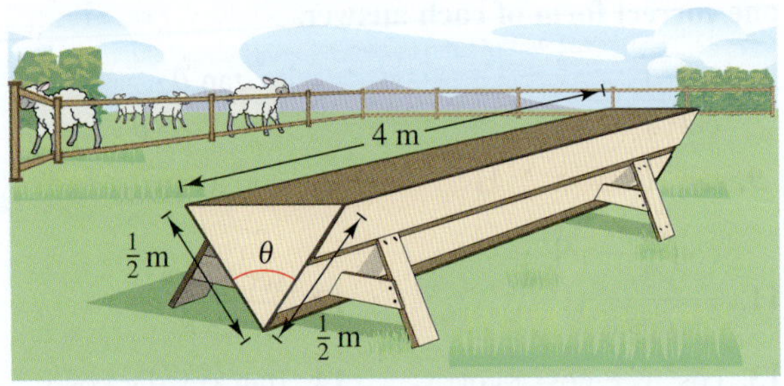

(a) Write the trough's volume as a function of $\theta/2$.

(b) Write the volume of the trough as a function of θ and determine the value of θ such that the volume is maximum.

Exploration

True or False? In Exercises 81–84, determine whether the statement is true or false. Justify your answer.

81. If $\frac{\pi}{2} < \theta < \pi$, then $\cos\frac{\theta}{2} < 0$.

82. $\sin(x + y) = \sin x + \sin y$

83. $4\sin(-x)\cos(-x) = -2\sin 2x$

84. $4\sin 45° \cos 15° = 1 + \sqrt{3}$

85. **Think About It** When a trigonometric equation has an infinite number of solutions, is it true that the equation is an identity? Explain.

Chapter Test

See CalcChat.com for tutorial help and worked-out solutions to odd-numbered exercises.

Take this test as you would take a test in class. When you are finished, check your work against the answers given in the back of the book.

1. When $\tan\theta = \frac{6}{5}$ and $\cos\theta < 0$, evaluate (if possible) all six trigonometric functions of θ.

2. Use the fundamental identities to simplify $\csc^2\beta(1 - \cos^2\beta)$.

3. Factor and simplify $\dfrac{\sec^4 x - \tan^4 x}{\sec^2 x + \tan^2 x}$.

4. Add and simplify $\dfrac{\cos\theta}{\sin\theta} + \dfrac{\sin\theta}{\cos\theta}$.

5. Determine the values of θ, $0 \le \theta < 2\pi$, for which $\tan\theta = -\sqrt{\sec^2\theta - 1}$.

6. Use a graphing utility to graph the functions $y_1 = \cos x + \sin x \tan x$ and $y_2 = \sec x$ in the same viewing window. Make a conjecture about y_1 and y_2. Verify the result algebraically.

In Exercises 7–12, verify the identity.

7. $\sin\theta\sec\theta = \tan\theta$

8. $\sec^2 x\tan^2 x + \sec^2 x = \sec^4 x$

9. $\dfrac{\csc\alpha + \sec\alpha}{\sin\alpha + \cos\alpha} = \cot\alpha + \tan\alpha$

10. $\tan\left(x + \dfrac{\pi}{2}\right) = -\cot x$

11. $\sin(n\pi + \theta) = (-1)^n \sin\theta$, n is an integer.

12. $(\sin x + \cos x)^2 = 1 + \sin 2x$

13. Rewrite $\sin^4 \dfrac{x}{2}$ in terms of the first power of the cosine.

14. Use a half-angle formula to simplify the expression $(\sin 4\theta)/(1 + \cos 4\theta)$.

15. Rewrite $4 \sin 3\theta \cos 2\theta$ as a sum or difference.

16. Rewrite $\cos 3\theta - \cos\theta$ as a product.

In Exercises 17–20, find all solutions of the equation in the interval $[0, 2\pi)$.

17. $\tan^2 x + \tan x = 0$

18. $\sin 2\alpha - \cos\alpha = 0$

19. $4\cos^2 x - 3 = 0$

20. $\csc^2 x - \csc x - 2 = 0$

21. Use a graphing utility to approximate the solutions (to three decimal places) of $5 \sin x - x = 0$ in the interval $[0, 2\pi)$.

22. Find the exact value of $\cos 105°$ using the fact that $105° = 135° - 30°$.

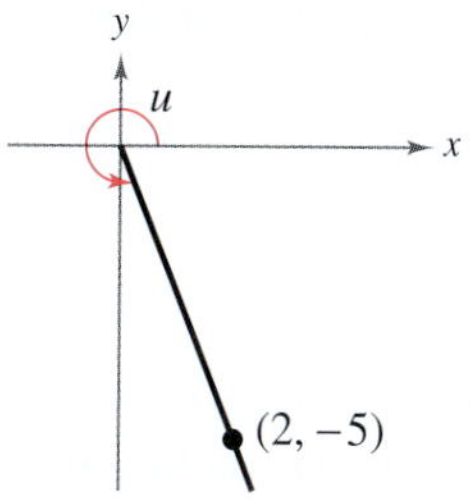

Figure for 23

23. Use the figure to find the exact values of $\sin 2u$, $\cos 2u$, and $\tan 2u$.

24. Cheyenne, Wyoming, has a latitude of 41°N. At this latitude, the position of the sun at sunrise can be modeled by

$$D = 31 \sin\left(\frac{2\pi}{365}t - 1.4\right)$$

where t is the time (in days), with $t = 1$ representing January 1. In this model, D represents the number of degrees north or south of due east that the sun rises. Use a graphing utility to determine the days on which the sun is more than 20° north of due east at sunrise.

25. The heights above ground h_1 and h_2 (in feet) of two people in different seats on a Ferris wheel can be modeled by

$$h_1 = 28 \cos 10t + 38 \quad \text{and} \quad h_2 = 28 \cos\left[10\left(t - \frac{\pi}{6}\right)\right] + 38,\ 0 \le t \le 2$$

where t is the time (in minutes). When are the two people at the same height?

Proofs in Mathematics

Sum and Difference Formulas *(p. 375)*

$\sin(u + v) = \sin u \cos v + \cos u \sin v$

$\sin(u - v) = \sin u \cos v - \cos u \sin v$

$\cos(u + v) = \cos u \cos v - \sin u \sin v$

$\cos(u - v) = \cos u \cos v + \sin u \sin v$

$\tan(u + v) = \dfrac{\tan u + \tan v}{1 - \tan u \tan v}$

$\tan(u - v) = \dfrac{\tan u - \tan v}{1 + \tan u \tan v}$

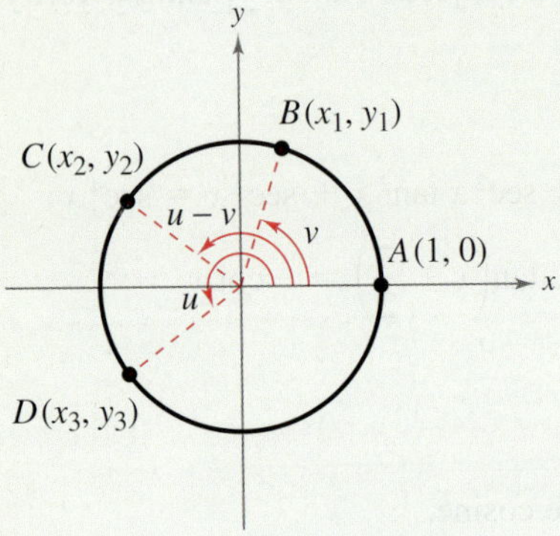

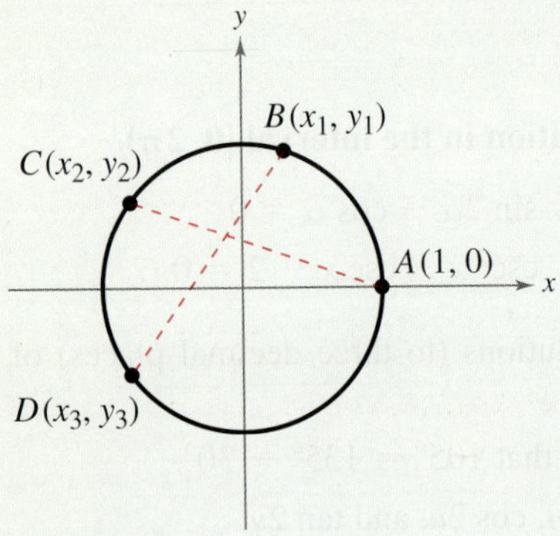

Proof

Use the figures at the left for the proofs of the formulas for $\cos(u \pm v)$. In the top figure, let A be the point $(1, 0)$ and then use u and v to locate the points $B(x_1, y_1)$, $C(x_2, y_2)$, and $D(x_3, y_3)$ on the unit circle. So, $x_i^2 + y_i^2 = 1$ for $i = 1, 2$, and 3. For convenience, assume that $0 < v < u < 2\pi$. In the bottom figure, note that arcs AC and BD have the same length. So, line segments AC and BD are also equal in length, which implies that

$$\sqrt{(x_2 - 1)^2 + (y_2 - 0)^2} = \sqrt{(x_3 - x_1)^2 + (y_3 - y_1)^2}$$

$$x_2^2 - 2x_2 + 1 + y_2^2 = x_3^2 - 2x_1x_3 + x_1^2 + y_3^2 - 2y_1y_3 + y_1^2$$

$$(x_2^2 + y_2^2) + 1 - 2x_2 = (x_3^2 + y_3^2) + (x_1^2 + y_1^2) - 2x_1x_3 - 2y_1y_3$$

$$1 + 1 - 2x_2 = 1 + 1 - 2x_1x_3 - 2y_1y_3$$

$$x_2 = x_3x_1 + y_3y_1.$$

Finally, by substituting the values $x_2 = \cos(u - v)$, $x_3 = \cos u$, $x_1 = \cos v$, $y_3 = \sin u$, and $y_1 = \sin v$, you obtain $\cos(u - v) = \cos u \cos v + \sin u \sin v$. To establish the formula for $\cos(u + v)$, consider $u + v = u - (-v)$ and use the formula just derived to obtain

$$\begin{aligned}\cos(u + v) &= \cos[u - (-v)] \\ &= \cos u \cos(-v) + \sin u \sin(-v) \\ &= \cos u \cos v - \sin u \sin v.\end{aligned}$$

You can use the sum and difference formulas for sine and cosine to prove the formulas for $\tan(u \pm v)$.

$$\tan(u \pm v) = \frac{\sin(u \pm v)}{\cos(u \pm v)} \qquad \text{Quotient identity}$$

$$= \frac{\sin u \cos v \pm \cos u \sin v}{\cos u \cos v \mp \sin u \sin v} \qquad \text{Sum and difference formulas}$$

$$= \frac{\dfrac{\sin u \cos v \pm \cos u \sin v}{\cos u \cos v}}{\dfrac{\cos u \cos v \mp \sin u \sin v}{\cos u \cos v}} \qquad \text{Divide numerator and denominator by } \cos u \cos v.$$

$$= \frac{\dfrac{\sin u \cos v}{\cos u \cos v} \pm \dfrac{\cos u \sin v}{\cos u \cos v}}{\dfrac{\cos u \cos v}{\cos u \cos v} \mp \dfrac{\sin u \sin v}{\cos u \cos v}}$$ Write as separate fractions.

$$= \frac{\dfrac{\sin u}{\cos u} \pm \dfrac{\sin v}{\cos v}}{1 \mp \dfrac{\sin u}{\cos u} \cdot \dfrac{\sin v}{\cos v}}$$ Simplify.

$$= \frac{\tan u \pm \tan v}{1 \mp \tan u \tan v}$$ Quotient identity

TRIGONOMETRY AND ASTRONOMY

Early astronomers used trigonometry to calculate measurements in the universe. For instance, they used trigonometry to calculate the circumference of Earth and the distance from Earth to the moon. Another major accomplishment in astronomy using trigonometry was computing distances to stars.

Double-Angle Formulas *(p. 382)*

$$\sin 2u = 2 \sin u \cos u \qquad \cos 2u = \cos^2 u - \sin^2 u$$

$$\tan 2u = \frac{2 \tan u}{1 - \tan^2 u} \qquad = 2\cos^2 u - 1$$

$$= 1 - 2 \sin^2 u$$

Proof

To prove all three formulas, let $v = u$ in the corresponding sum formulas.

$$\sin 2u = \sin(u + u) = \sin u \cos u + \cos u \sin u = 2 \sin u \cos u$$

$$\cos 2u = \cos(u + u) = \cos u \cos u - \sin u \sin u = \cos^2 u - \sin^2 u$$

$$\tan 2u = \tan(u + u) = \frac{\tan u + \tan u}{1 - \tan u \tan u} = \frac{2 \tan u}{1 - \tan^2 u}$$

Power-Reducing Formulas *(p. 384)*

$$\sin^2 u = \frac{1 - \cos 2u}{2} \qquad \cos^2 u = \frac{1 + \cos 2u}{2} \qquad \tan^2 u = \frac{1 - \cos 2u}{1 + \cos 2u}$$

Proof

To prove the first formula, solve for $\sin^2 u$ in the double-angle formula $\cos 2u = 1 - 2 \sin^2 u$, as follows.

$$\cos 2u = 1 - 2 \sin^2 u$$ Write double-angle formula.

$$2 \sin^2 u = 1 - \cos 2u$$ Subtract $\cos 2u$ from, and add $2 \sin^2 u$ to, each side.

$$\sin^2 u = \frac{1 - \cos 2u}{2}$$ Divide each side by 2.

In a similar way, you can prove the second formula by solving for $\cos^2 u$ in the double-angle formula

$$\cos 2u = 2\cos^2 u - 1.$$

To prove the third formula, use a quotient identity, as follows.

$$\begin{aligned}\tan^2 u &= \frac{\sin^2 u}{\cos^2 u}\\ &= \frac{\dfrac{1-\cos 2u}{2}}{\dfrac{1+\cos 2u}{2}}\\ &= \frac{1-\cos 2u}{1+\cos 2u}\end{aligned}$$

Sum-to-Product Formulas ***(p. 386)***

$$\sin u + \sin v = 2\sin\left(\frac{u+v}{2}\right)\cos\left(\frac{u-v}{2}\right)$$

$$\sin u - \sin v = 2\cos\left(\frac{u+v}{2}\right)\sin\left(\frac{u-v}{2}\right)$$

$$\cos u + \cos v = 2\cos\left(\frac{u+v}{2}\right)\cos\left(\frac{u-v}{2}\right)$$

$$\cos u - \cos v = -2\sin\left(\frac{u+v}{2}\right)\sin\left(\frac{u-v}{2}\right)$$

Proof

To prove the first formula, let $x = u + v$ and $y = u - v$. Then substitute $u = (x + y)/2$ and $v = (x - y)/2$ in the product-to-sum formula.

$$\sin u \cos v = \frac{1}{2}[\sin(u+v) + \sin(u-v)]$$

$$\sin\left(\frac{x+y}{2}\right)\cos\left(\frac{x-y}{2}\right) = \frac{1}{2}(\sin x + \sin y)$$

$$2\sin\left(\frac{x+y}{2}\right)\cos\left(\frac{x-y}{2}\right) = \sin x + \sin y$$

The other sum-to-product formulas can be proved in a similar manner.

P.S. Problem Solving

1. **Writing Trigonometric Functions in Terms of Cosine** Write each of the other trigonometric functions of θ in terms of $\cos\theta$.

2. **Verifying a Trigonometric Identity** Verify that for all integers n,

$$\cos\left[\frac{(2n+1)\pi}{2}\right] = 0.$$

3. **Verifying a Trigonometric Identity** Verify that for all integers n,

$$\sin\left[\frac{(12n+1)\pi}{6}\right] = \frac{1}{2}.$$

4. **Sound Wave** A sound wave is modeled by

$$p(t) = \frac{1}{4\pi}[p_1(t) + 30p_2(t) + p_3(t) + p_5(t) + 30p_6(t)]$$

where $p_n(t) = \frac{1}{n}\sin(524n\pi t)$, and t is the time (in seconds).

(a) Find the sine components $p_n(t)$ and use a graphing utility to graph the components. Then verify the graph of p shown below.

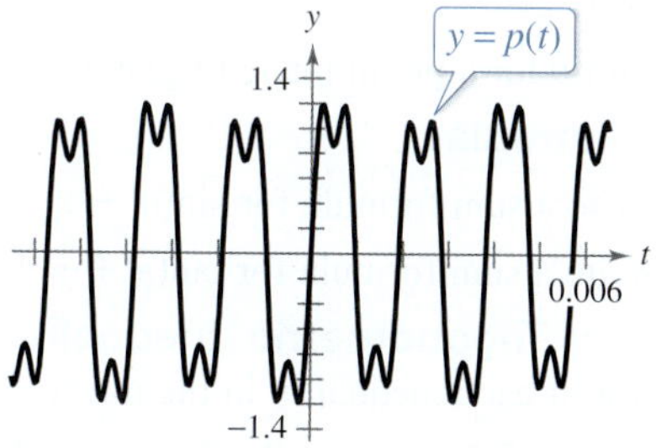

(b) Find the period of each sine component of p. Is p periodic? If so, then what is its period?

(c) Use the graphing utility to find the t-intercepts of the graph of p over one cycle.

(d) Use the graphing utility to approximate the absolute maximum and absolute minimum values of p over one cycle.

5. **Geometry** Three squares of side s are placed side by side (see figure). Make a conjecture about the relationship between the sum $u + v$ and w. Prove your conjecture by using the identity for the tangent of the sum of two angles.

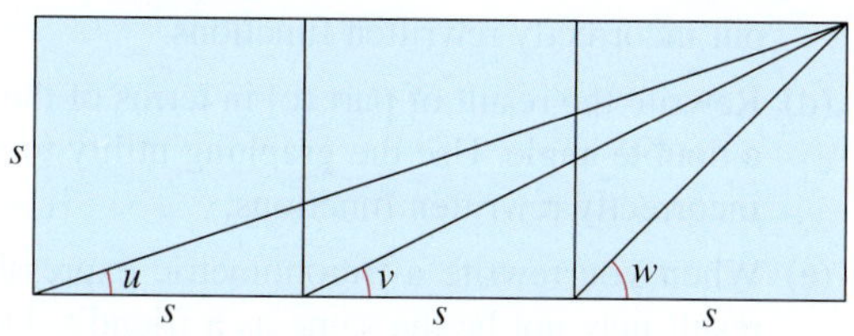

6. **Projectile Motion** The path traveled by an object (neglecting air resistance) that is projected at an initial height of h_0 feet, an initial velocity of v_0 feet per second, and an initial angle θ is given by

$$y = -\frac{16}{v_0^2\cos^2\theta}x^2 + (\tan\theta)x + h_0$$

where x and y are measured in feet. Find a formula for the maximum height of an object projected from ground level at velocity v_0 and angle θ. To do this, find half of the horizontal distance

$$\frac{1}{32}v_0^2\sin 2\theta$$

and then substitute it for x in the general model for the path of a projectile (where $h_0 = 0$).

7. **Geometry** The length of each of the two equal sides of an isosceles triangle is 10 meters (see figure). The angle between the two sides is θ.

(a) Write the area of the triangle as a function of $\theta/2$.

(b) Write the area of the triangle as a function of θ. Determine the value of θ such that the area is a maximum.

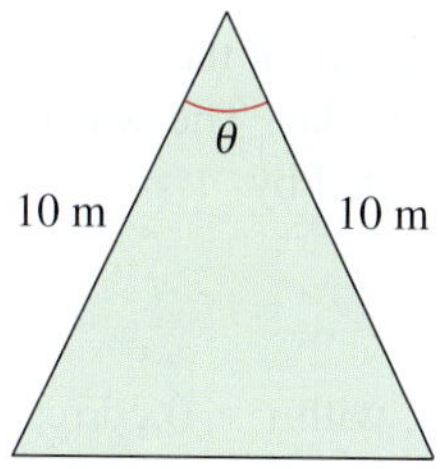

Figure for 7

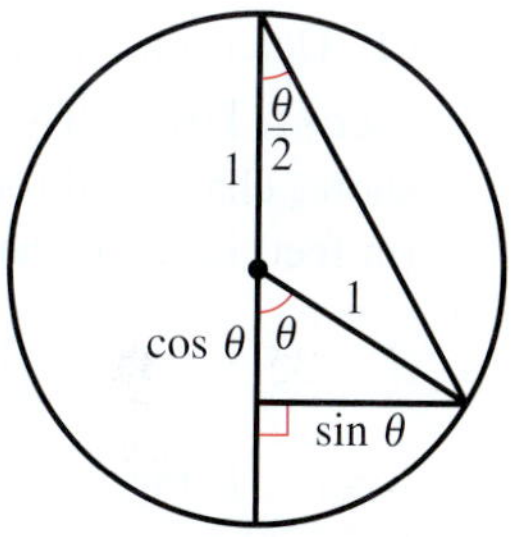

Figure for 8

8. **Geometry** Use the figure to derive the formulas for

$$\sin\frac{\theta}{2},\ \cos\frac{\theta}{2},\ \text{and}\ \tan\frac{\theta}{2}$$

where θ is an acute angle.

9. **Force** The force F (in pounds) on a person's back when he or she bends over at an angle θ is modeled by

$$F = \frac{0.6W\sin(\theta + 90^\circ)}{\sin 12^\circ}$$

where W is the person's weight (in pounds).

(a) Simplify the model.

(b) Use a graphing utility to graph the model, where $W = 185$ and $0^\circ < \theta < 90^\circ$.

(c) At what angle is the force a maximum? At what angle is the force a minimum?

10. Hours of Daylight The number of hours of daylight that occur at any location on Earth depends on the time of year and the latitude of the location. The following equations model the numbers of hours of daylight in Seward, Alaska (60° latitude), and New Orleans, Louisiana (30° latitude).

$$D = 12.2 - 6.4 \cos\left[\frac{\pi(t + 0.2)}{182.6}\right] \quad \text{Seward}$$

$$D = 12.2 - 1.9 \cos\left[\frac{\pi(t + 0.2)}{182.6}\right] \quad \text{New Orleans}$$

In these models, D represents the number of hours of daylight and t represents the day, with $t = 0$ corresponding to January 1.

(a) Use a graphing utility to graph both models in the same viewing window. Use a viewing window of $0 \le t \le 365$.

(b) Find the days of the year on which both cities receive the same amount of daylight.

(c) Which city has the greater variation in the number of daylight hours? Which constant in each model would you use to determine the difference between the greatest and least numbers of hours of daylight?

(d) Determine the period of each model.

11. Ocean Tide The tide, or depth of the ocean near the shore, changes throughout the day. The water depth d (in feet) of a bay can be modeled by

$$d = 35 - 28 \cos \frac{\pi}{6.2} t$$

where t is the time in hours, with $t = 0$ corresponding to 12:00 A.M.

(a) Algebraically find the times at which the high and low tides occur.

(b) If possible, algebraically find the time(s) at which the water depth is 3.5 feet.

(c) Use a graphing utility to verify your results from parts (a) and (b).

12. Piston Heights The heights h (in inches) of pistons 1 and 2 in an automobile engine can be modeled by

$$h_1 = 3.75 \sin 733t + 7.5$$

and

$$h_2 = 3.75 \sin 733(t + 4\pi/3) + 7.5$$

respectively, where t is measured in seconds.

(a) Use a graphing utility to graph the heights of these pistons in the same viewing window for $0 \le t \le 1$.

(b) How often are the pistons at the same height?

13. Index of Refraction The index of refraction n of a transparent material is the ratio of the speed of light in a vacuum to the speed of light in the material. Some common materials and their indices of refraction are air (1.00), water (1.33), and glass (1.50). Triangular prisms are often used to measure the index of refraction based on the formula

$$n = \frac{\sin\left(\frac{\theta}{2} + \frac{\alpha}{2}\right)}{\sin \frac{\theta}{2}}.$$

For the prism shown in the figure, $\alpha = 60°$.

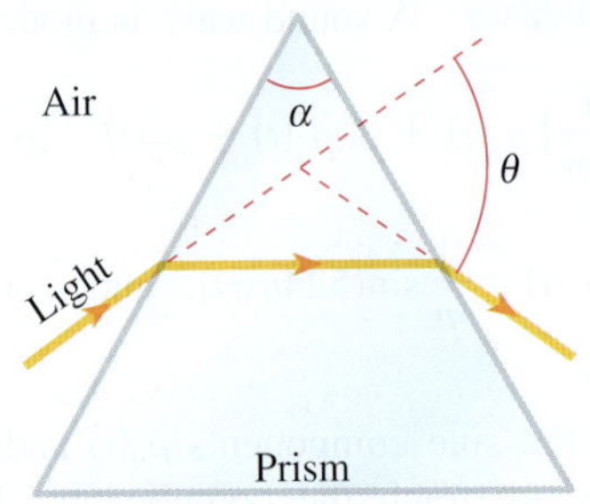

(a) Write the index of refraction as a function of $\cot(\theta/2)$.

(b) Find θ for a prism made of glass.

14. Sum Formulas

(a) Write a sum formula for $\sin(u + v + w)$.

(b) Write a sum formula for $\tan(u + v + w)$.

15. Solving Trigonometric Inequalities Find the solution of each inequality in the interval $[0, 2\pi)$.

(a) $\sin x \ge 0.5$ (b) $\cos x \le -0.5$

(c) $\tan x < \sin x$ (d) $\cos x \ge \sin x$

16. Sum of Fourth Powers Consider the function $f(x) = \sin^4 x + \cos^4 x$.

(a) Use the power-reducing formulas to write the function in terms of cosine to the first power.

(b) Determine another way of rewriting the function. Use a graphing utility to rule out incorrectly rewritten functions.

(c) Add a trigonometric term to the function so that it becomes a perfect square trinomial. Rewrite the function as a perfect square trinomial minus the term that you added. Use the graphing utility to rule out incorrectly rewritten functions.

(d) Rewrite the result of part (c) in terms of the sine of a double angle. Use the graphing utility to rule out incorrectly rewritten functions.

(e) When you rewrite a trigonometric expression, the result may not be the same as a friend's. Does this mean that one of you is wrong? Explain.

6 Additional Topics in Trigonometry

Work *(page 436)*

Electrical Engineering *(Exercise 105, page 449)*

Navigation *(Example 10, page 426)*

Engine Design *(Exercise 56, page 417)*

Surveying *(page 403)*

Clockwise from top left, Vince Clements/Shutterstock.com; auremar/Shutterstock.com; Smart-foto/Shutterstock.com; Daniel Prudek/Shutterstock.com; MC_PP/Shutterstock.com

6.1 Law of Sines

You can use the Law of Sines to solve real-life problems involving oblique triangles. For instance, in Exercise 53 on page 409, you will use the Law of Sines to determine the distance from a boat to the shoreline.

- **Use the Law of Sines to solve oblique triangles (AAS or ASA).**
- **Use the Law of Sines to solve oblique triangles (SSA).**
- **Find the areas of oblique triangles.**
- **Use the Law of Sines to model and solve real-life problems.**

Introduction

In Chapter 4, you studied techniques for solving right triangles. In this section and the next, you will solve **oblique triangles**—triangles that have no right angles. As standard notation, the angles of a triangle are labeled A, B, and C, and their opposite sides are labeled a, b, and c, as shown below.

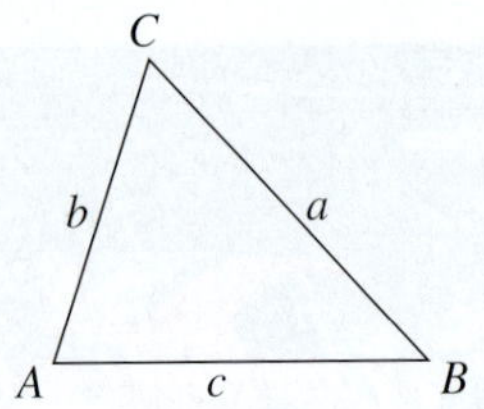

To solve an oblique triangle, you need to know the measure of at least one side and any two other measures of the triangle—either two sides, two angles, or one angle and one side. This breaks down into the following four cases.

1. Two angles and any side (AAS or ASA)
2. Two sides and an angle opposite one of them (SSA)
3. Three sides (SSS)
4. Two sides and their included angle (SAS)

The first two cases can be solved using the **Law of Sines,** whereas the last two cases require the Law of Cosines (see Section 6.2).

Law of Sines

If ABC is a triangle with sides a, b, and c, then

$$\frac{a}{\sin A} = \frac{b}{\sin B} = \frac{c}{\sin C}.$$

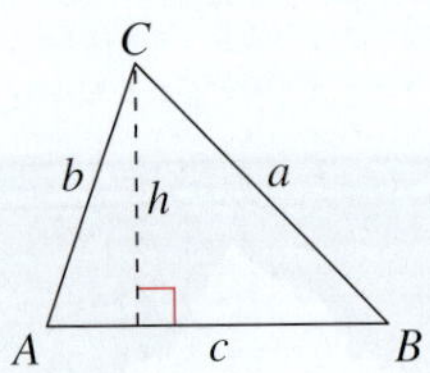

A is acute.

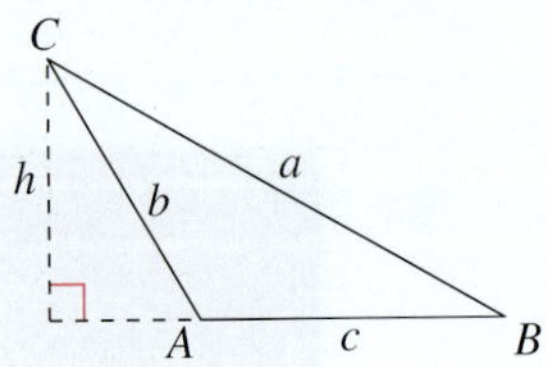

A is obtuse.

The Law of Sines can also be written in the reciprocal form

$$\frac{\sin A}{a} = \frac{\sin B}{b} = \frac{\sin C}{c}.$$

For a proof of the Law of Sines, see Proofs in Mathematics on page 459.

© Owen Franken/CORBIS

EXAMPLE 1 Given Two Angles and One Side—AAS

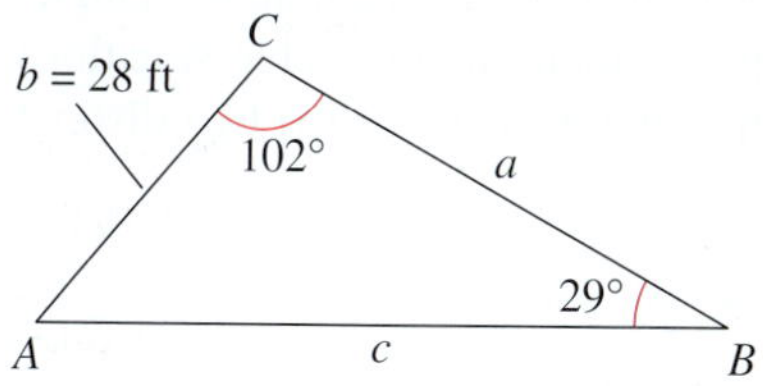

Figure 6.1

For the triangle in Figure 6.1, $C = 102°$, $B = 29°$, and $b = 28$ feet. Find the remaining angle and sides.

Solution The third angle of the triangle is

$$\begin{aligned} A &= 180° - B - C \\ &= 180° - 29° - 102° \\ &= 49°. \end{aligned}$$

By the Law of Sines, you have

$$\frac{a}{\sin A} = \frac{b}{\sin B} = \frac{c}{\sin C}.$$

Using $b = 28$ produces

$$a = \frac{b}{\sin B}(\sin A) = \frac{28}{\sin 29°}(\sin 49°) \approx 43.59 \text{ feet}$$

and

$$c = \frac{b}{\sin B}(\sin C) = \frac{28}{\sin 29°}(\sin 102°) \approx 56.49 \text{ feet.}$$

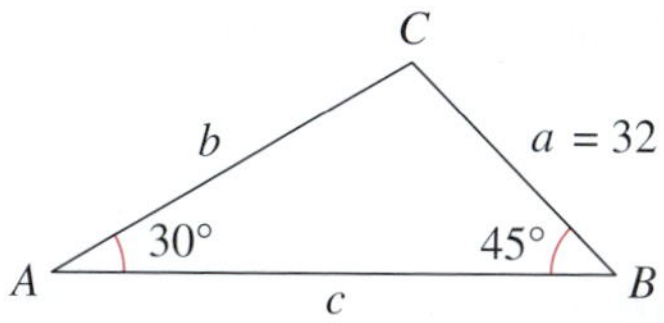

Figure 6.2

✓ ***Checkpoint*** *Audio-video solution in English & Spanish at LarsonPrecalculus.com.*

For the triangle in Figure 6.2, $A = 30°$, $B = 45°$, and $a = 32$. Find the remaining angle and sides.

In the 1850s, surveyors used the Law of Sines to calculate the height of Mount Everest. Their calculation was within 30 feet of the currently accepted value.

EXAMPLE 2 Given Two Angles and One Side—ASA

A pole tilts *toward* the sun at an 8° angle from the vertical, and it casts a 22-foot shadow. The angle of elevation from the tip of the shadow to the top of the pole is 43°. How tall is the pole?

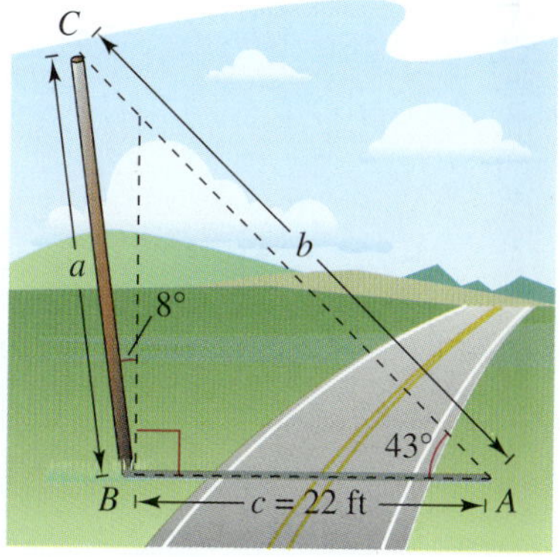

Solution From the figure at the right, note that $A = 43°$ and

$$B = 90° + 8° = 98°.$$

So, the third angle is

$$\begin{aligned} C &= 180° - A - B \\ &= 180° - 43° - 98° \\ &= 39°. \end{aligned}$$

By the Law of Sines, you have

$$\frac{a}{\sin A} = \frac{c}{\sin C}.$$

Because $c = 22$ feet, the height of the pole is

$$a = \frac{c}{\sin C}(\sin A) = \frac{22}{\sin 39°}(\sin 43°) \approx 23.84 \text{ feet.}$$

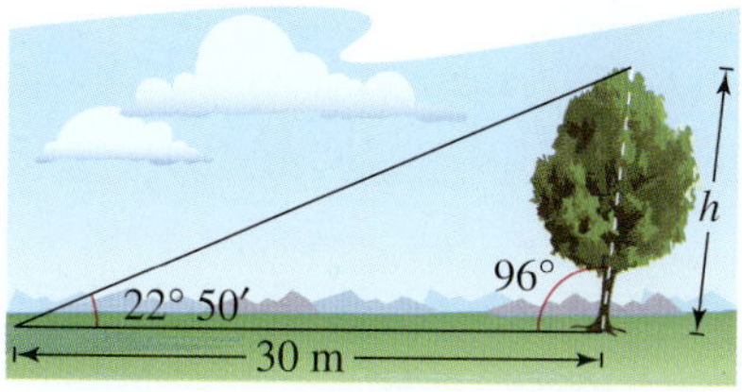

Figure 6.3

✓ ***Checkpoint*** *Audio-video solution in English & Spanish at LarsonPrecalculus.com.*

Find the height of the tree shown in Figure 6.3.

Daniel Prudek/Shutterstock.com

The Ambiguous Case (SSA)

In Examples 1 and 2, you saw that two angles and one side determine a unique triangle. However, if two sides and one opposite angle are given, then three possible situations can occur: (1) no such triangle exists, (2) one such triangle exists, or (3) two distinct triangles may satisfy the conditions.

The Ambiguous Case (SSA)

Consider a triangle in which you are given a, b, and A. $(h = b \sin A)$

	A is acute.	A is acute.	A is acute.	A is acute.	A is obtuse.	A is obtuse.
Sketch						
Necessary condition	$a < h$	$a = h$	$a \geq b$	$h < a < b$	$a \leq b$	$a > b$
Triangles possible	None	One	One	Two	None	One

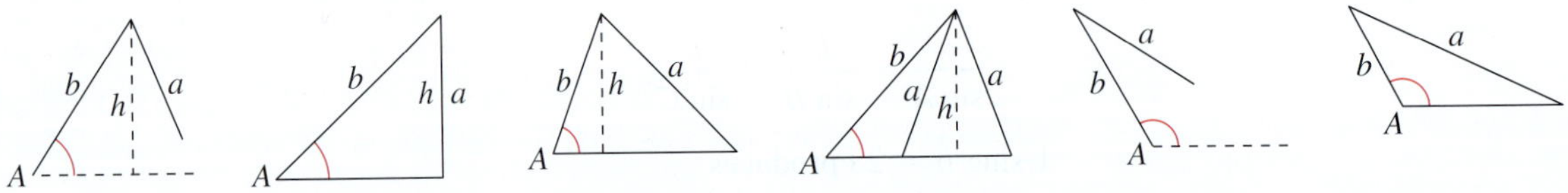

EXAMPLE 3 Single-Solution Case—SSA

For the triangle in Figure 6.4, $a = 22$ inches, $b = 12$ inches, and $A = 42°$. Find the remaining side and angles.

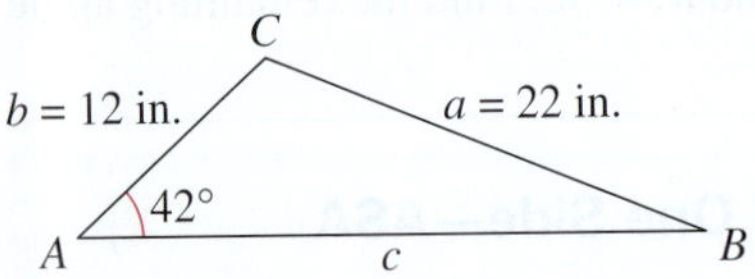

One solution: $a \geq b$

Figure 6.4

Solution By the Law of Sines, you have

$$\frac{\sin B}{b} = \frac{\sin A}{a} \qquad \text{Reciprocal form}$$

$$\sin B = b\left(\frac{\sin A}{a}\right) \qquad \text{Multiply each side by } b.$$

$$\sin B = 12\left(\frac{\sin 42°}{22}\right) \qquad \text{Substitute for } A, a, \text{ and } b.$$

$$B \approx 21.41°.$$

Now, you can determine that

$$C \approx 180° - 42° - 21.41° = 116.59°.$$

Then, the remaining side is

$$\frac{c}{\sin C} = \frac{a}{\sin A}$$

$$c = \frac{a}{\sin A}(\sin C)$$

$$= \frac{22}{\sin 42°}(\sin 116.59°)$$

$$\approx 29.40 \text{ inches.}$$

✓ **Checkpoint** *Audio-video solution in English & Spanish at LarsonPrecalculus.com.*

Given $A = 31°$, $a = 12$, and $b = 5$, find the remaining side and angles of the triangle.

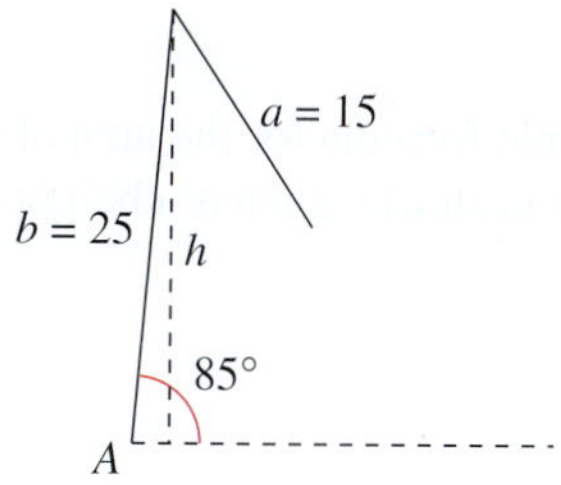

No solution: $a < h$

Figure 6.5

EXAMPLE 4 No-Solution Case—SSA

Show that there is no triangle for which $a = 15$, $b = 25$, and $A = 85°$.

Solution Begin by making the sketch shown in Figure 6.5. From this figure, it appears that no triangle is formed. You can verify this using the Law of Sines.

$$\frac{\sin B}{b} = \frac{\sin A}{a} \quad \text{Reciprocal form}$$

$$\sin B = b\left(\frac{\sin A}{a}\right) \quad \text{Multiply each side by } b.$$

$$\sin B = 25\left(\frac{\sin 85°}{15}\right) \approx 1.6603 > 1$$

This contradicts the fact that

$$|\sin B| \leq 1.$$

So, no triangle can be formed having sides $a = 15$ and $b = 25$ and angle $A = 85°$.

Checkpoint *Audio-video solution in English & Spanish at LarsonPrecalculus.com.*

Show that there is no triangle for which $a = 4$, $b = 14$, and $A = 60°$.

EXAMPLE 5 Two-Solution Case—SSA

Find two triangles for which $a = 12$ meters, $b = 31$ meters, and $A = 20.5°$.

Solution By the Law of Sines, you have

$$\frac{\sin B}{b} = \frac{\sin A}{a} \quad \text{Reciprocal form}$$

$$\sin B = b\left(\frac{\sin A}{a}\right) = 31\left(\frac{\sin 20.5°}{12}\right) \approx 0.9047.$$

There are two angles, $B_1 \approx 64.8°$ and $B_2 \approx 180° - 64.8° = 115.2°$, between $0°$ and $180°$ whose sine is 0.9047. For $B_1 \approx 64.8°$, you obtain

$$C \approx 180° - 20.5° - 64.8° = 94.7°$$

$$c = \frac{a}{\sin A}(\sin C) = \frac{12}{\sin 20.5°}(\sin 94.7°) \approx 34.15 \text{ meters.}$$

For $B_2 \approx 115.2°$, you obtain

$$C \approx 180° - 20.5° - 115.2° = 44.3°$$

$$c = \frac{a}{\sin A}(\sin C) = \frac{12}{\sin 20.5°}(\sin 44.3°) \approx 23.93 \text{ meters.}$$

The resulting triangles are shown below.

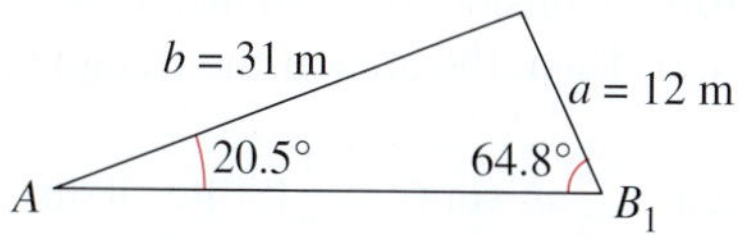

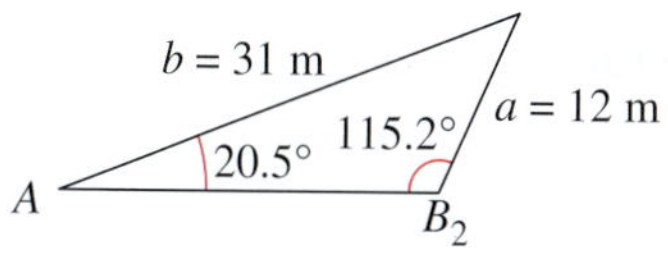

Two solutions: $h < a < b$

Checkpoint *Audio-video solution in English & Spanish at LarsonPrecalculus.com.*

Find two triangles for which $a = 4.5$ feet, $b = 5$ feet, and $A = 58°$.

Area of an Oblique Triangle

The procedure used to prove the Law of Sines leads to a simple formula for the area of an oblique triangle. Referring to the triangles below, note that each triangle has a height of $h = b \sin A$. Consequently, the area of each triangle is

$$\text{Area} = \frac{1}{2}(\text{base})(\text{height})$$
$$= \frac{1}{2}(c)(b \sin A)$$
$$= \frac{1}{2}bc \sin A.$$

By similar arguments, you can develop the formulas

$$\text{Area} = \frac{1}{2}ab \sin C = \frac{1}{2}ac \sin B.$$

REMARK To see how to obtain the height of the obtuse triangle, notice the use of the reference angle $180° - A$ and the difference formula for sine, as follows.

$$\begin{aligned} h &= b \sin(180° - A) \\ &= b(\sin 180° \cos A \\ &\qquad - \cos 180° \sin A) \\ &= b[0 \cdot \cos A - (-1) \cdot \sin A] \\ &= b \sin A \end{aligned}$$

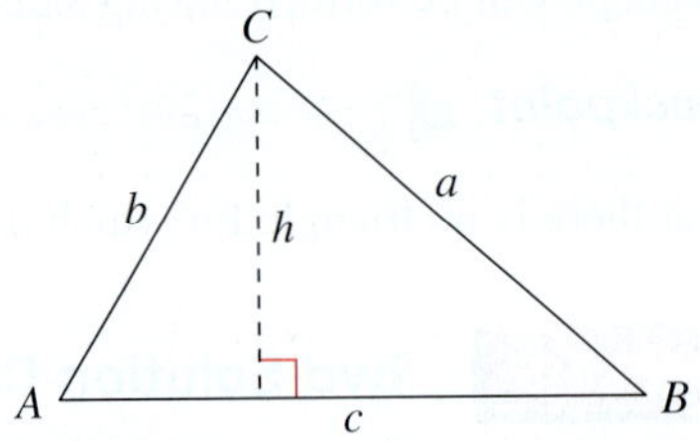

A is acute.

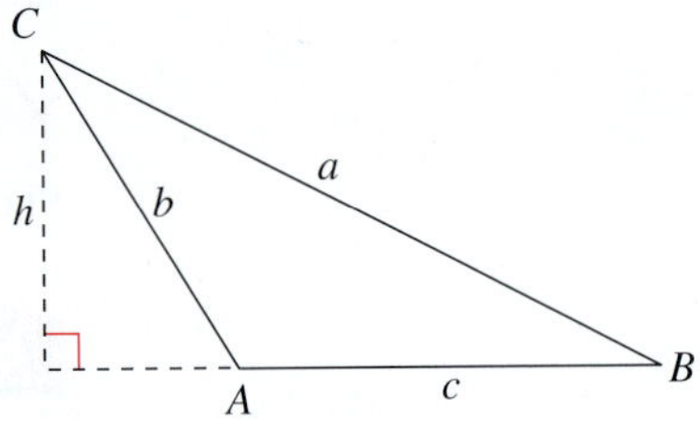

A is obtuse.

> **Area of an Oblique Triangle**
>
> The area of any triangle is one-half the product of the lengths of two sides times the sine of their included angle. That is,
>
> $$\text{Area} = \frac{1}{2}bc \sin A = \frac{1}{2}ab \sin C = \frac{1}{2}ac \sin B.$$

Note that when angle A is 90°, the formula gives the area of a right triangle:

$$\text{Area} = \frac{1}{2}bc(\sin 90°) = \frac{1}{2}bc = \frac{1}{2}(\text{base})(\text{height}). \qquad \sin 90° = 1$$

Similar results are obtained for angles C and B equal to 90°.

EXAMPLE 6 **Finding the Area of a Triangular Lot**

Find the area of a triangular lot having two sides of lengths 90 meters and 52 meters and an included angle of 102°.

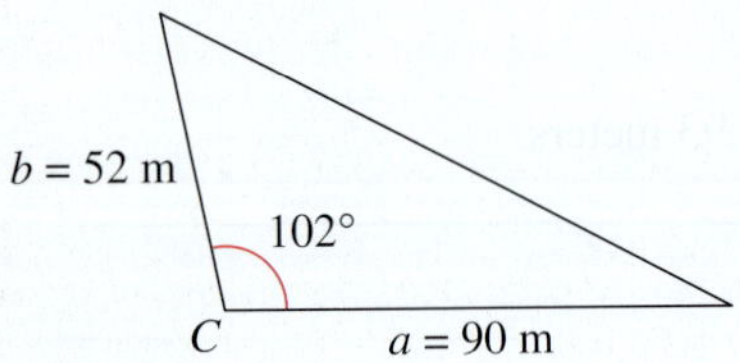

Figure 6.6

Solution Consider $a = 90$ meters, $b = 52$ meters, and angle $C = 102°$, as shown in Figure 6.6. Then, the area of the triangle is

$$\text{Area} = \frac{1}{2}ab \sin C = \frac{1}{2}(90)(52)(\sin 102°) \approx 2289 \text{ square meters.}$$

✓ Checkpoint *Audio-video solution in English & Spanish at LarsonPrecalculus.com.*

Find the area of a triangular lot having two sides of lengths 24 inches and 18 inches and an included angle of 80°.

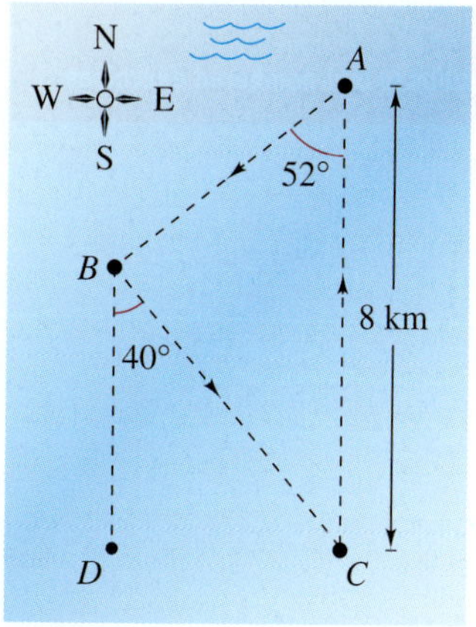

Figure 6.7

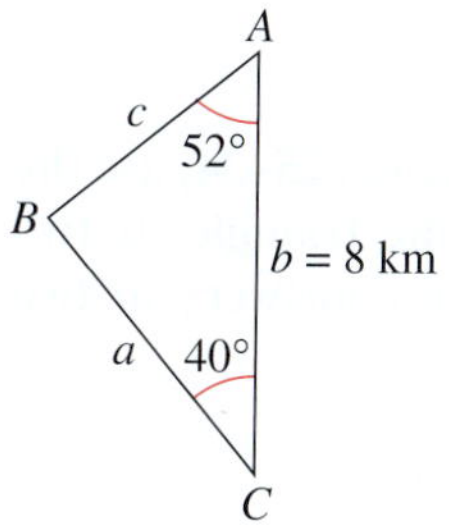

Figure 6.8

Application

EXAMPLE 7 An Application of the Law of Sines

The course for a boat race starts at point A and proceeds in the direction S 52° W to point B, then in the direction S 40° E to point C, and finally back to point A, as shown in Figure 6.7. Point C lies 8 kilometers directly south of point A. Approximate the total distance of the race course.

Solution Because lines BD and AC are parallel, it follows that $\angle BCA \cong \angle CBD$. Consequently, triangle ABC has the measures shown in Figure 6.8. The measure of angle B is $180° - 52° - 40° = 88°$. Using the Law of Sines,

$$\frac{a}{\sin 52°} = \frac{b}{\sin 88°} = \frac{c}{\sin 40°}.$$

Because $b = 8$,

$$a = \frac{8}{\sin 88°}(\sin 52°) \approx 6.31 \quad \text{and} \quad c = \frac{8}{\sin 88°}(\sin 40°) \approx 5.15.$$

The total distance of the course is approximately

$$\text{Length} \approx 8 + 6.31 + 5.15 = 19.46 \text{ kilometers.}$$

✓ **Checkpoint** Audio-video solution in English & Spanish at LarsonPrecalculus.com.

On a small lake, you swim from point A to point B at a bearing of N 28° E, then to point C at a bearing of N 58° W, and finally back to point A, as shown in the figure below. Point C lies 800 meters directly north of point A. Approximate the total distance that you swim.

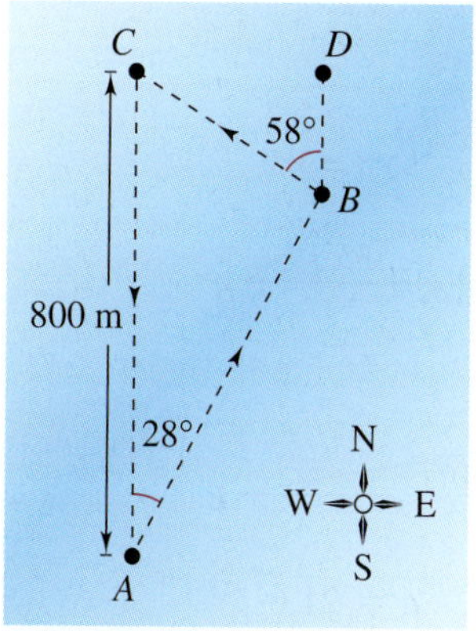

Summarize (Section 6.1)

1. State the Law of Sines *(page 402)*. For examples of using the Law of Sines to solve oblique triangles (AAS or ASA), see Examples 1 and 2.
2. List the necessary conditions and the numbers of possible triangles for the ambiguous case (SSA) *(page 404)*. For examples of using the Law of Sines to solve oblique triangles (SSA), see Examples 3–5.
3. State the formula for the area of an oblique triangle *(page 406)*. For an example of finding the area of an oblique triangle, see Example 6.
4. Describe how you can use the Law of Sines to model and solve a real-life problem *(page 407, Example 7)*.

6.1 Exercises

See **CalcChat.com** for tutorial help and worked-out solutions to odd-numbered exercises.

Vocabulary: Fill in the blanks.

1. An ________ triangle is a triangle that has no right angle.
2. For triangle ABC, the Law of Sines is $\frac{a}{\sin A} = $ ________ $= \frac{c}{\sin C}$.
3. Two ________ and one ________ determine a unique triangle.
4. The area of an oblique triangle is $\frac{1}{2}bc \sin A = \frac{1}{2}ab \sin C = $ ________ .

Skills and Applications

Using the Law of Sines

In Exercises 5–24, use the Law of Sines to solve the triangle. Round your answers to two decimal places.

5.

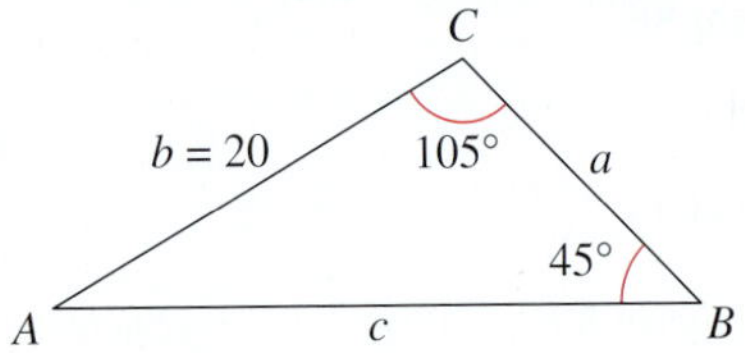

6.

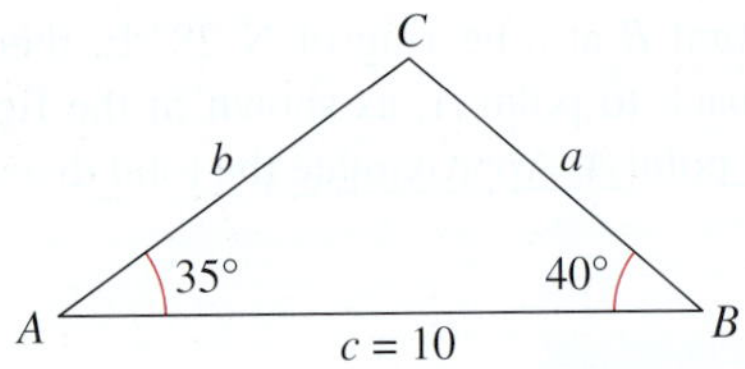

7.

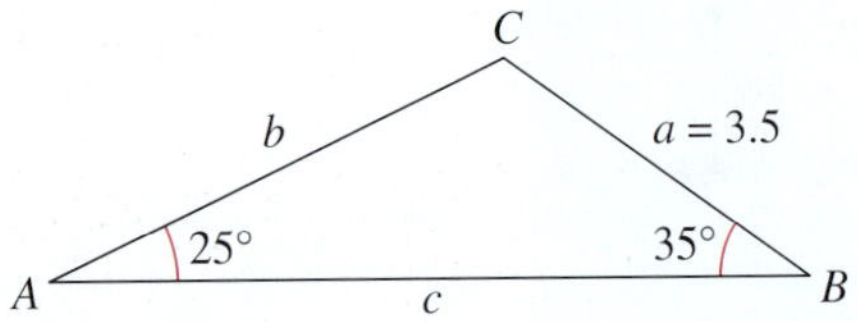

8.

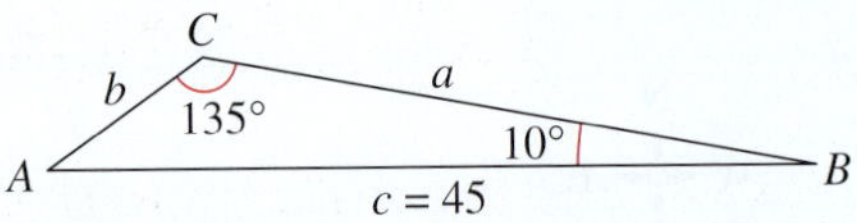

9. $A = 102.4°, \quad C = 16.7°, \quad a = 21.6$
10. $A = 24.3°, \quad C = 54.6°, \quad c = 2.68$
11. $A = 83° 20', \quad C = 54.6°, \quad c = 18.1$
12. $A = 5° 40', \quad B = 8° 15', \quad b = 4.8$
13. $A = 35°, \quad B = 65°, \quad c = 10$
14. $A = 120°, \quad B = 45°, \quad c = 16$
15. $A = 55°, \quad B = 42°, \quad c = \frac{3}{4}$
16. $B = 28°, \quad C = 104°, \quad a = 3\frac{5}{8}$
17. $A = 36°, \quad a = 8, \quad b = 5$
18. $A = 60°, \quad a = 9, \quad c = 10$
19. $B = 15° 30', \quad a = 4.5, \quad b = 6.8$
20. $B = 2° 45', \quad b = 6.2, \quad c = 5.8$
21. $A = 145°, \quad a = 14, \quad b = 4$
22. $A = 100°, \quad a = 125, \quad c = 10$
23. $A = 110° 15', \quad a = 48, \quad b = 16$
24. $C = 95.20°, \quad a = 35, \quad c = 50$

Using the Law of Sines

In Exercises 25–34, use the Law of Sines to solve (if possible) the triangle. If two solutions exist, find both. Round your answers to two decimal places.

25. $A = 110°, \quad a = 125, \quad b = 100$
26. $A = 110°, \quad a = 125, \quad b = 200$
27. $A = 76°, \quad a = 18, \quad b = 20$
28. $A = 76°, \quad a = 34, \quad b = 21$
29. $A = 58°, \quad a = 11.4, \quad b = 12.8$
30. $A = 58°, \quad a = 4.5, \quad b = 12.8$
31. $A = 120°, \quad a = b = 25$
32. $A = 120°, \quad a = 25, \quad b = 24$
33. $A = 45°, \quad a = b = 1$
34. $A = 25° 4', \quad a = 9.5, \quad b = 22$

Using the Law of Sines

In Exercises 35–38, find values for b such that the triangle has (a) one solution, (b) two solutions, and (c) no solution.

35. $A = 36°, \quad a = 5$
36. $A = 60°, \quad a = 10$
37. $A = 10°, \quad a = 10.8$
38. $A = 88°, \quad a = 315.6$

Finding the Area of a Triangle

In Exercises 39–46, find the area of the triangle having the indicated angle and sides.

39. $C = 120°, \quad a = 4, \quad b = 6$
40. $B = 130°, \quad a = 62, \quad c = 20$
41. $A = 150°, \quad b = 8, \quad c = 10$
42. $C = 170°, \quad a = 14, \quad b = 24$
43. $A = 43° 45', \quad b = 57, \quad c = 85$
44. $A = 5° 15', \quad b = 4.5, \quad c = 22$
45. $B = 72° 30', \quad a = 105, \quad c = 64$
46. $C = 84° 30', \quad a = 16, \quad b = 20$

47. Height Because of prevailing winds, a tree grew so that it was leaning 4° from the vertical. At a point 40 meters from the tree, the angle of elevation to the top of the tree is 30° (see figure). Find the height h of the tree.

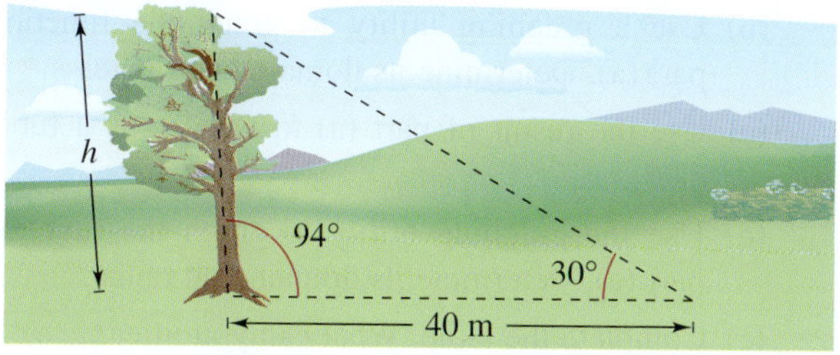

48. Height A flagpole at a right angle to the horizontal is located on a slope that makes an angle of 12° with the horizontal. The flagpole's shadow is 16 meters long and points directly up the slope. The angle of elevation from the tip of the shadow to the sun is 20°.

(a) Draw a triangle to represent the situation. Show the known quantities on the triangle and use a variable to indicate the height of the flagpole.

(b) Write an equation that can be used to find the height of the flagpole.

(c) Find the height of the flagpole.

49. Angle of Elevation A 10-meter utility pole casts a 17-meter shadow directly down a slope when the angle of elevation of the sun is 42° (see figure). Find θ, the angle of elevation of the ground.

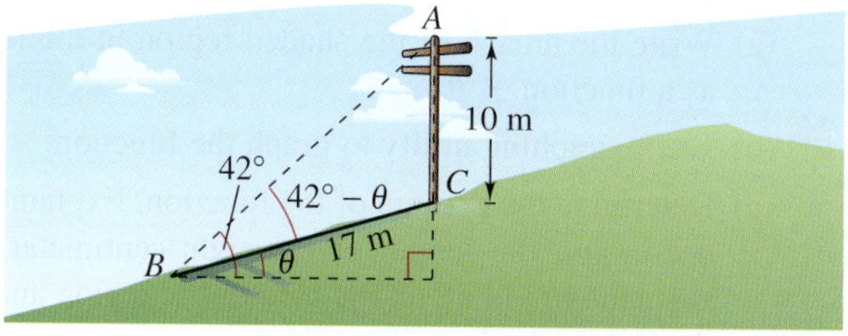

50. Bridge Design A bridge is to be built across a small lake from a gazebo to a dock (see figure). The bearing from the gazebo to the dock is S 41° W. From a tree 100 meters from the gazebo, the bearings to the gazebo and the dock are S 74° E and S 28° E, respectively. Find the distance from the gazebo to the dock.

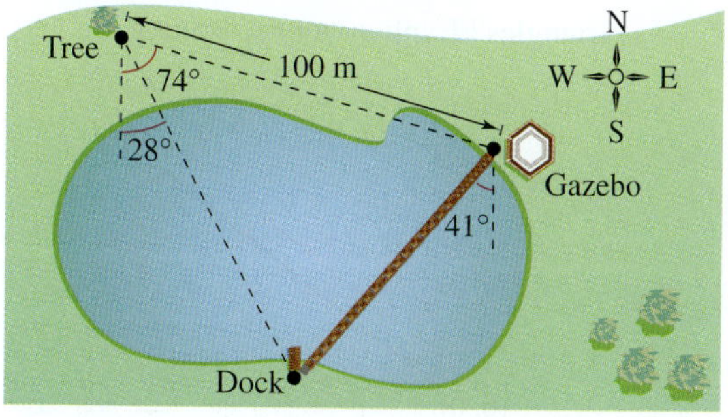

51. Flight Path A plane flies 500 kilometers with a bearing of 316° from Naples to Elgin (see figure). The plane then flies 720 kilometers from Elgin to Canton (Canton is due west of Naples). Find the bearing of the flight from Elgin to Canton.

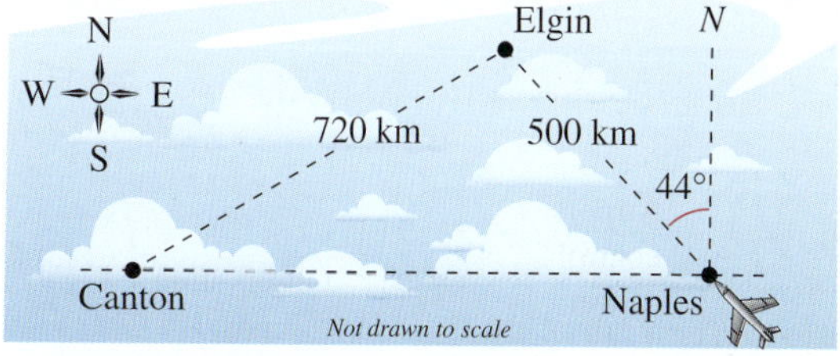

52. Locating a Fire The bearing from the Pine Knob fire tower to the Colt Station fire tower is N 65° E, and the two towers are 30 kilometers apart. A fire spotted by rangers in each tower has a bearing of N 80° E from Pine Knob and S 70° E from Colt Station (see figure). Find the distance of the fire from each tower.

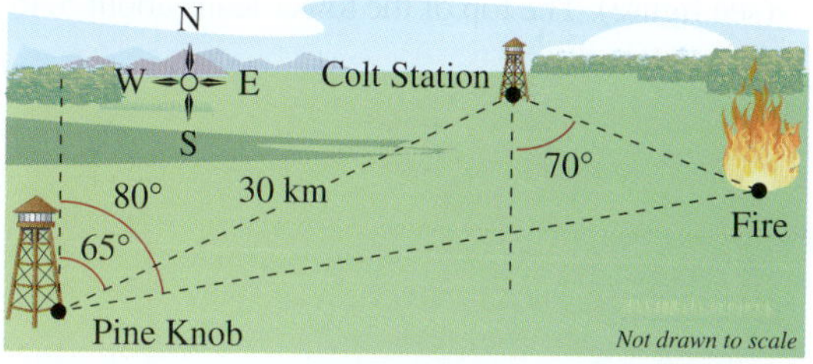

53. Distance A boat is sailing due east parallel to the shoreline at a speed of 10 miles per hour. At a given time, the bearing to the lighthouse is S 70° E, and 15 minutes later the bearing is S 63° E (see figure). The lighthouse is located at the shoreline. What is the distance from the boat to the shoreline?

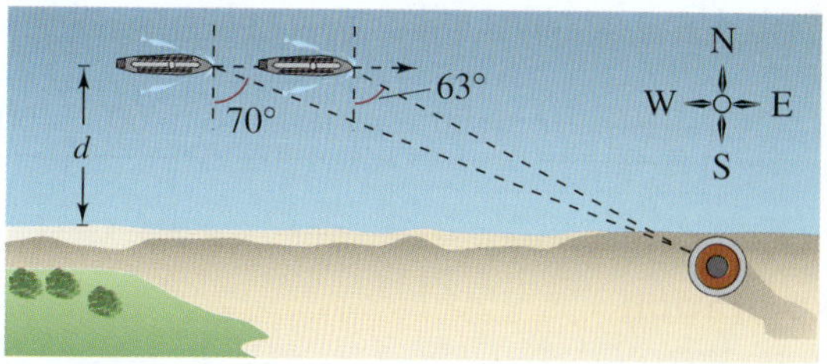

54. Altitude The angles of elevation to an airplane from two points A and B on level ground are 55° and 72°, respectively. The points A and B are 2.2 miles apart, and the airplane is east of both points in the same vertical plane. Find the altitude of the plane.

© Owen Franken/CORBIS

55. Distance The angles of elevation θ and ϕ to an airplane from the airport control tower and from an observation post 2 miles away are being continuously monitored (see figure). Write an equation giving the distance d between the plane and observation post in terms of θ and ϕ.

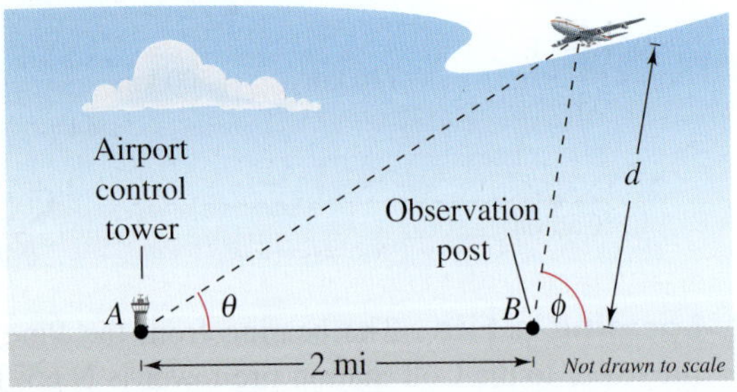

56. The Leaning Tower of Pisa The Leaning Tower of Pisa in Italy leans because it was built on unstable soil—a mixture of clay, sand, and water. The tower is approximately 58.36 meters tall from its foundation (see figure). The top of the tower leans about 5.45 meters off center.

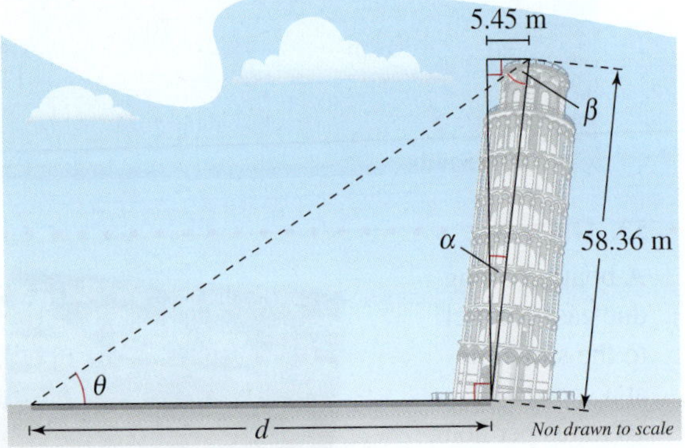

(a) Find the angle of lean α of the tower.

(b) Write β as a function of d and θ, where θ is the angle of elevation to the sun.

(c) Use the Law of Sines to write an equation for the length d of the shadow cast by the tower in terms of θ.

(d) Use a graphing utility to complete the table.

θ	10°	20°	30°	40°	50°	60°
d						

Exploration

True or False? **In Exercises 57–59, determine whether the statement is true or false. Justify your answer.**

57. If a triangle contains an obtuse angle, then it must be oblique.

58. Two angles and one side of a triangle do not necessarily determine a unique triangle.

59. If three sides or three angles of an oblique triangle are known, then the triangle can be solved.

60. Graphical and Numerical Analysis In the figure, α and β are positive angles.

(a) Write α as a function of β.

(b) Use a graphing utility to graph the function in part (a). Determine its domain and range.

(c) Use the result of part (a) to write c as a function of β.

(d) Use the graphing utility to graph the function in part (c). Determine its domain and range.

(e) Complete the table. What can you infer?

β	0.4	0.8	1.2	1.6	2.0	2.4	2.8
α							
c							

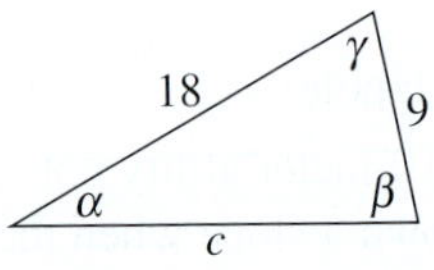

Figure for 60

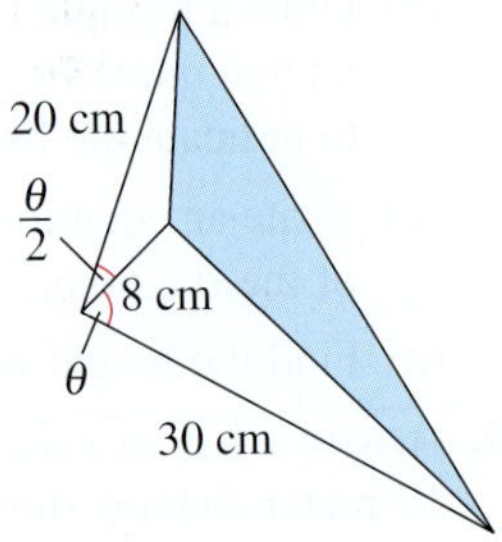

Figure for 61

61. Graphical Analysis

(a) Write the area A of the shaded region in the figure as a function of θ.

(b) Use a graphing utility to graph the function.

(c) Determine the domain of the function. Explain how decreasing the length of the eight-centimeter line segment would affect the area of the region and the domain of the function.

62. HOW DO YOU SEE IT? In the figure, a triangle is to be formed by drawing a line segment of length a from $(4, 3)$ to the positive x-axis. For what value(s) of a can you form (a) one triangle, (b) two triangles, and (c) no triangles? Explain your reasoning.

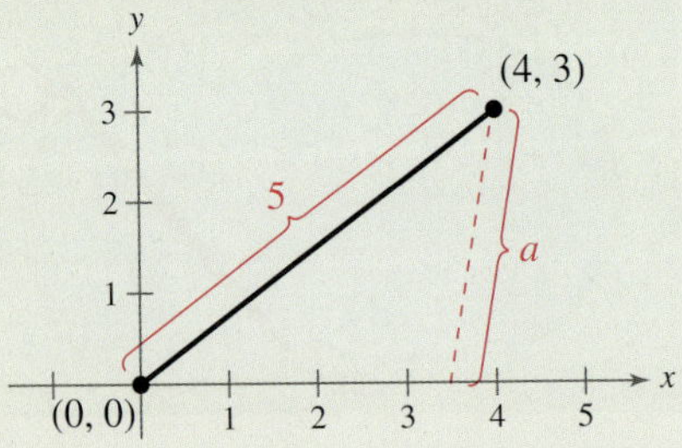

6.2 Law of Cosines

You can use the Law of Cosines to solve real-life problems involving oblique triangles. For instance, in Exercise 56 on page 417, you will use the Law of Cosines to determine the total distance a piston moves in an engine.

- Use the Law of Cosines to solve oblique triangles (SSS or SAS).
- Use the Law of Cosines to model and solve real-life problems.
- Use Heron's Area Formula to find the area of a triangle.

Introduction

Two cases remain in the list of conditions needed to solve an oblique triangle—SSS and SAS. When you are given three sides (SSS), or two sides and their included angle (SAS), none of the ratios in the Law of Sines would be complete. In such cases, you can use the **Law of Cosines.**

Law of Cosines

Standard Form	Alternative Form
$a^2 = b^2 + c^2 - 2bc \cos A$	$\cos A = \dfrac{b^2 + c^2 - a^2}{2bc}$
$b^2 = a^2 + c^2 - 2ac \cos B$	$\cos B = \dfrac{a^2 + c^2 - b^2}{2ac}$
$c^2 = a^2 + b^2 - 2ab \cos C$	$\cos C = \dfrac{a^2 + b^2 - c^2}{2ab}$

For a proof of the Law of Cosines, see Proofs in Mathematics on page 460.

EXAMPLE 1 Three Sides of a Triangle—SSS

Find the three angles of the triangle shown below.

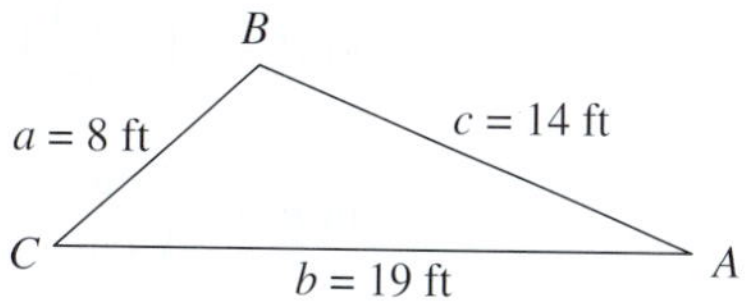

Solution It is a good idea first to find the angle opposite the longest side—side b in this case. Using the alternative form of the Law of Cosines, you find that

$$\cos B = \frac{a^2 + c^2 - b^2}{2ac} = \frac{8^2 + 14^2 - 19^2}{2(8)(14)} \approx -0.45089.$$

Because $\cos B$ is negative, B is an *obtuse* angle given by $B \approx 116.80°$. At this point, it is simpler to use the Law of Sines to determine A.

$$\sin A = a\left(\frac{\sin B}{b}\right) \approx 8\left(\frac{\sin 116.80°}{19}\right) \approx 0.37583$$

Because B is obtuse and a triangle can have at most one obtuse angle, you know that A must be acute. So, $A \approx 22.08°$ and $C \approx 180° - 22.08° - 116.80° = 41.12°$.

✓ **Checkpoint** Audio-video solution in English & Spanish at LarsonPrecalculus.com.

Find the three angles of the triangle whose sides have lengths $a = 6$, $b = 8$, and $c = 12$.

Smart-foto/Shutterstock.com

Do you see why it was wise to find the largest angle *first* in Example 1? Knowing the cosine of an angle, you can determine whether the angle is acute or obtuse. That is,

$\cos \theta > 0$ for $0° < \theta < 90°$ Acute

$\cos \theta < 0$ for $90° < \theta < 180°$. Obtuse

So, in Example 1, once you found that angle B was obtuse, you knew that angles A and C were both acute. Furthermore, if the largest angle is acute, then the remaining two angles are also acute.

REMARK When solving an oblique triangle given three sides, you use the alternative form of the Law of Cosines to solve for an angle. When solving an oblique triangle given two sides and their included angle, you use the standard form of the Law of Cosines to solve for an unknown.

EXAMPLE 2 **Two Sides and the Included Angle—SAS**

Find the remaining angles and side of the triangle shown below.

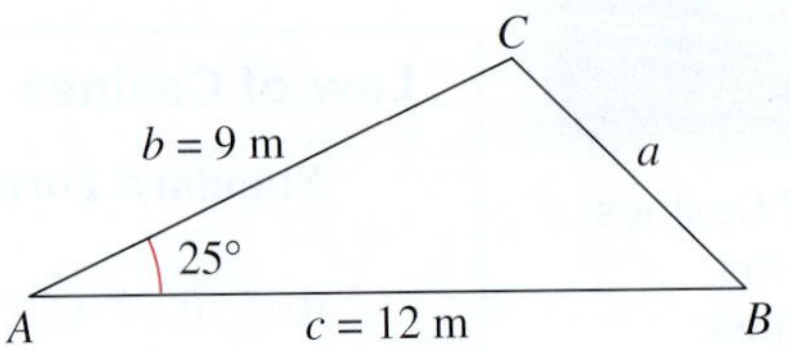

Solution Use the Law of Cosines to find the unknown side a in the figure.

$$a^2 = b^2 + c^2 - 2bc \cos A$$

$$a^2 = 9^2 + 12^2 - 2(9)(12) \cos 25°$$

$$a^2 \approx 29.2375$$

$$a \approx 5.4072$$

Because $a \approx 5.4072$ meters, you now know the ratio $(\sin A)/a$, and you can use the reciprocal form of the Law of Sines to solve for B.

$$\frac{\sin B}{b} = \frac{\sin A}{a}$$ Reciprocal form

$$\sin B = b\left(\frac{\sin A}{a}\right)$$ Multiply each side by b.

$$\sin B \approx 9\left(\frac{\sin 25°}{5.4072}\right)$$ Substitute for A, a, and b.

$$\sin B \approx 0.7034$$ Use a calculator.

There are two angles between 0° and 180° whose sine is 0.7034, $B_1 \approx 44.7°$ and $B_2 \approx 180° - 44.7° = 135.3°$.

For $B_1 \approx 44.7°$,

$$C_1 \approx 180° - 25° - 44.7° = 110.3°.$$

For $B_2 \approx 135.3°$,

$$C_2 \approx 180° - 25° - 135.3° = 19.7°.$$

Because side c is the longest side of the triangle, C must be the largest angle of the triangle. So, $B \approx 44.7°$ and $C \approx 110.3°$.

✓ **Checkpoint** Audio-video solution in English & Spanish at LarsonPrecalculus.com.

Given $A = 80°$, $b = 16$, and $c = 12$, find the remaining angles and side of the triangle.

Applications

EXAMPLE 3 An Application of the Law of Cosines

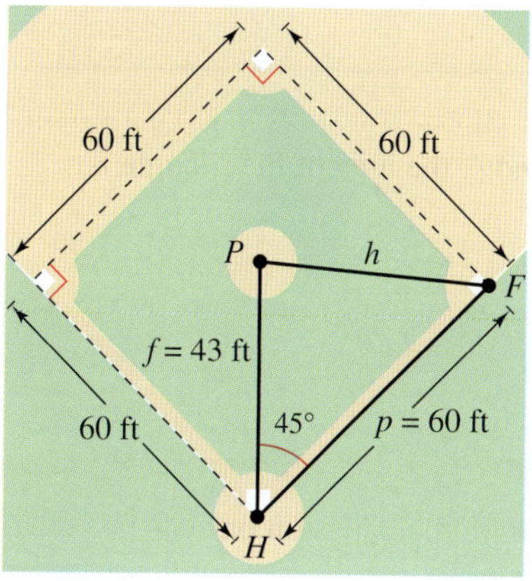

Figure 6.9

The pitcher's mound on a women's softball field is 43 feet from home plate and the distance between the bases is 60 feet, as shown in Figure 6.9. (The pitcher's mound is *not* halfway between home plate and second base.) How far is the pitcher's mound from first base?

Solution In triangle HPF, $H = 45°$ (line HP bisects the right angle at H), $f = 43$, and $p = 60$. Using the Law of Cosines for this SAS case, you have

$$h^2 = f^2 + p^2 - 2fp\cos H$$

$$= 43^2 + 60^2 - 2(43)(60)\cos 45°$$

$$\approx 1800.3.$$

So, the approximate distance from the pitcher's mound to first base is

$$h \approx \sqrt{1800.3} \approx 42.43 \text{ feet.}$$

Checkpoint Audio-video solution in English & Spanish at LarsonPrecalculus.com.

In a softball game, a batter hits a ball to dead center field, a distance of 240 feet from home plate. The center fielder then throws the ball to third base and gets a runner out. The distance between the bases is 60 feet. How far is the center fielder from third base?

EXAMPLE 4 An Application of the Law of Cosines

A ship travels 60 miles due east and then adjusts its course, as shown below. After traveling 80 miles in this new direction, the ship is 139 miles from its point of departure. Describe the bearing from point B to point C.

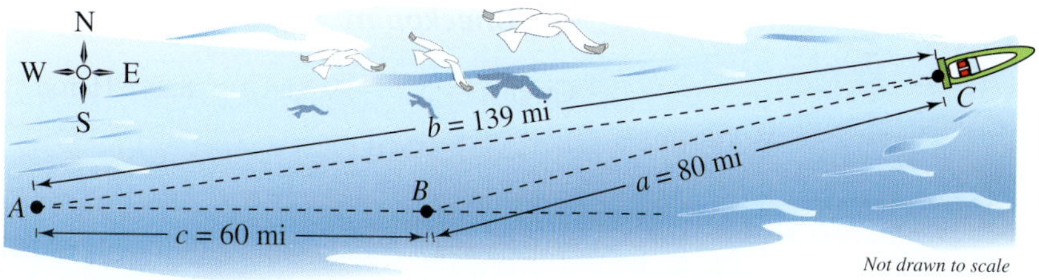

Solution You have $a = 80$, $b = 139$, and $c = 60$. So, using the alternative form of the Law of Cosines, you have

$$\cos B = \frac{a^2 + c^2 - b^2}{2ac}$$

$$= \frac{80^2 + 60^2 - 139^2}{2(80)(60)}$$

$$\approx -0.97094.$$

So, $B \approx 166.15°$, and thus the bearing measured from due north from point B to point C is $166.15° - 90° = 76.15°$, or N 76.15° E.

Checkpoint Audio-video solution in English & Spanish at LarsonPrecalculus.com.

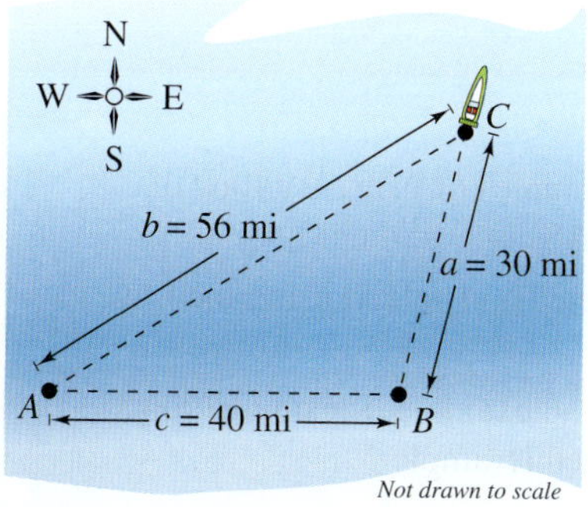

Figure 6.10

A ship travels 40 miles due east and then changes direction, as shown in Figure 6.10. After traveling 30 miles in this new direction, the ship is 56 miles from its point of departure. Describe the bearing from point B to point C.

> **HISTORICAL NOTE**
>
> Heron of Alexandria (ca. 100 B.C.) was a Greek geometer and inventor. His works describe how to find the areas of triangles, quadrilaterals, regular polygons having 3 to 12 sides, and circles, as well as the surface areas and volumes of three-dimensional objects.

Heron's Area Formula

The Law of Cosines can be used to establish the following formula for the area of a triangle. This formula is called **Heron's Area Formula** after the Greek mathematician Heron (ca. 100 B.C.).

> **Heron's Area Formula**
>
> Given any triangle with sides of lengths a, b, and c, the area of the triangle is
>
> $$\text{Area} = \sqrt{s(s-a)(s-b)(s-c)}$$
>
> where
>
> $$s = \frac{a+b+c}{2}.$$

For a proof of Heron's Area Formula, see Proofs in Mathematics on page 461.

EXAMPLE 5 Using Heron's Area Formula

Find the area of a triangle having sides of lengths $a = 43$ meters, $b = 53$ meters, and $c = 72$ meters.

Solution Because $s = (a + b + c)/2 = 168/2 = 84$, Heron's Area Formula yields

$$\begin{aligned}\text{Area} &= \sqrt{s(s-a)(s-b)(s-c)}\\ &= \sqrt{84(84-43)(84-53)(84-72)}\\ &= \sqrt{84(41)(31)(12)}\\ &\approx 1131.89 \text{ square meters.}\end{aligned}$$

✓ ***Checkpoint*** *Audio-video solution in English & Spanish at LarsonPrecalculus.com.*

Given $a = 5$, $b = 9$, and $c = 8$, use Heron's Area Formula to find the area of the triangle.

You have now studied three different formulas for the area of a triangle.

Standard Formula: $\text{Area} = \frac{1}{2}bh$

Oblique Triangle: $\text{Area} = \frac{1}{2}bc \sin A = \frac{1}{2}ab \sin C = \frac{1}{2}ac \sin B$

Heron's Area Formula: $\text{Area} = \sqrt{s(s-a)(s-b)(s-c)}$

> **Summarize (Section 6.2)**
>
> 1. State the Law of Cosines *(page 411)*. For examples of using the Law of Cosines to solve oblique triangles (SSS or SAS), see Examples 1 and 2.
> 2. Describe real-life problems that can be modeled and solved using the Law of Cosines *(page 413, Examples 3 and 4)*.
> 3. State Heron's Area Formula *(page 414)*. For an example of using Heron's Area Formula to find the area of a triangle, see Example 5.

6.2 Exercises

See CalcChat.com for tutorial help and worked-out solutions to odd-numbered exercises.

Vocabulary: Fill in the blanks.

1. When you are given three sides of a triangle, you use the Law of ________ to find the three angles of the triangle.
2. When you are given two angles and any side of a triangle, you use the Law of ________ to solve the triangle.
3. The standard form of the Law of Cosines for $\cos B = \dfrac{a^2 + c^2 - b^2}{2ac}$ is ________ .
4. The Law of Cosines can be used to establish a formula for finding the area of a triangle called ________ ________ Formula.

Skills and Applications

Using the Law of Cosines In Exercises 5–24, use the Law of Cosines to solve the triangle. Round your answers to two decimal places.

5.
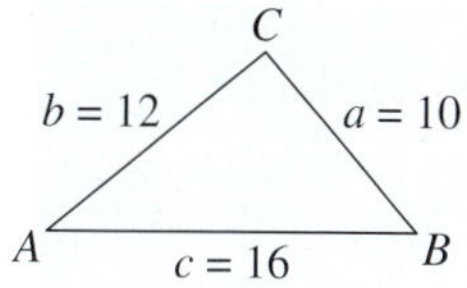

6.
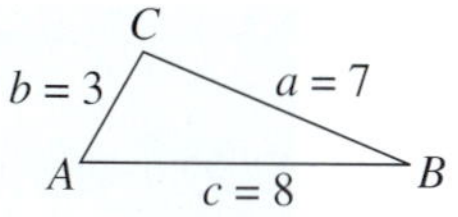

7.
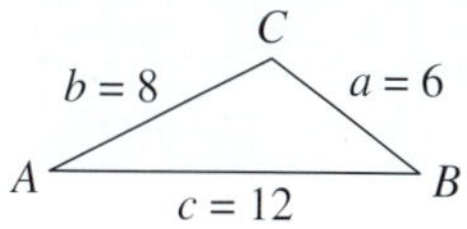

8.
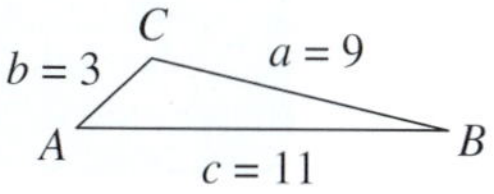

9.
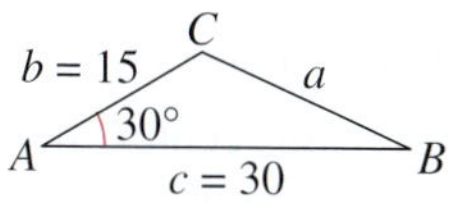

10.
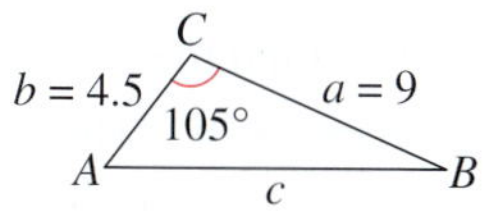

11.
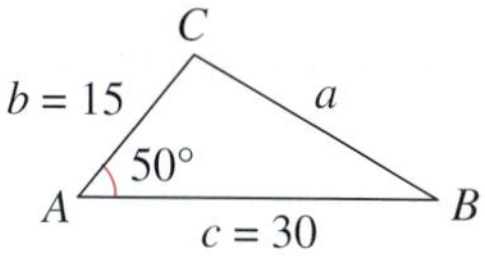

12.
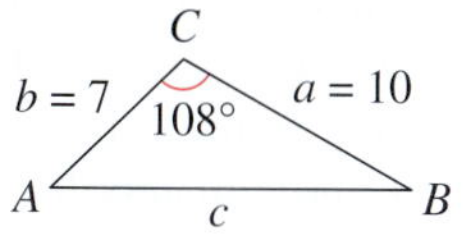

13. $a = 11, \quad b = 15, \quad c = 21$
14. $a = 55, \quad b = 25, \quad c = 72$
15. $a = 75.4, \quad b = 52, \quad c = 52$
16. $a = 1.42, \quad b = 0.75, \quad c = 1.25$
17. $A = 120°, \quad b = 6, \quad c = 7$
18. $A = 48°, \quad b = 3, \quad c = 14$
19. $B = 10° 35', \quad a = 40, \quad c = 30$
20. $B = 75° 20', \quad a = 6.2, \quad c = 9.5$
21. $B = 125° 40', \quad a = 37, \quad c = 37$
22. $C = 15° 15', \quad a = 7.45, \quad b = 2.15$
23. $C = 43°, \quad a = \frac{4}{9}, \quad b = \frac{7}{9}$
24. $C = 101°, \quad a = \frac{3}{8}, \quad b = \frac{3}{4}$

Finding Measures in a Parallelogram In Exercises 25–30, complete the table by solving the parallelogram shown in the figure. (The lengths of the diagonals are given by *c* and *d*.)

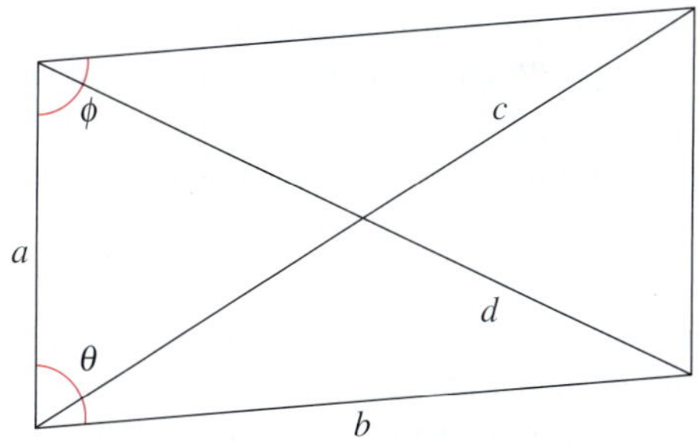

	a	*b*	*c*	*d*	*θ*	*ϕ*
25.	5	8			45°	
26.	25	35				120°
27.	10	14	20			
28.	40	60		80		
29.	15		25	20		
30.		25	50	35		

Solving a Triangle In Exercises 31–36, determine whether the Law of Sines or the Law of Cosines is needed to solve the triangle. Then solve (if possible) the triangle. If two solutions exist, find both. Round your answers to two decimal places.

31. $a = 8, \quad c = 5, \quad B = 40°$
32. $a = 10, \quad b = 12, \quad C = 70°$
33. $A = 24°, \quad a = 4, \quad b = 18$
34. $a = 11, \quad b = 13, \quad c = 7$
35. $A = 42°, \quad B = 35°, \quad c = 1.2$
36. $a = 160, \quad B = 12°, \quad C = 7°$

Using Heron's Area Formula **In Exercises 37–44, use Heron's Area Formula to find the area of the triangle.**

37. $a = 8, \quad b = 12, \quad c = 17$
38. $a = 33, \quad b = 36, \quad c = 25$
39. $a = 2.5, \quad b = 10.2, \quad c = 9$
40. $a = 75.4, \quad b = 52, \quad c = 52$
41. $a = 12.32, \quad b = 8.46, \quad c = 15.05$
42. $a = 3.05, \quad b = 0.75, \quad c = 2.45$
43. $a = 1, \quad b = \frac{1}{2}, \quad c = \frac{3}{4}$
44. $a = \frac{3}{5}, \quad b = \frac{5}{8}, \quad c = \frac{3}{8}$

45. **Navigation** A boat race runs along a triangular course marked by buoys A, B, and C. The race starts with the boats headed west for 3700 meters. The other two sides of the course lie to the north of the first side, and their lengths are 1700 meters and 3000 meters. Draw a figure that gives a visual representation of the situation. Then find the bearings for the last two legs of the race.

46. **Navigation** A plane flies 810 miles from Franklin to Centerville with a bearing of 75°. Then it flies 648 miles from Centerville to Rosemount with a bearing of 32°. Draw a figure that visually represents the situation. Then find the straight-line distance and bearing from Franklin to Rosemount.

47. **Surveying** To approximate the length of a marsh, a surveyor walks 250 meters from point A to point B, then turns 75° and walks 220 meters to point C (see figure). Approximate the length AC of the marsh.

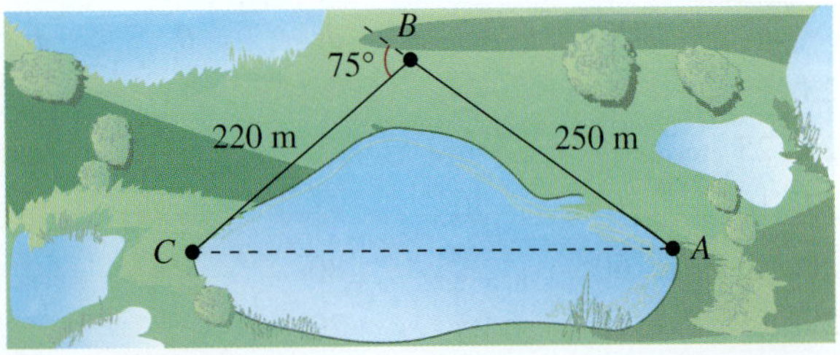

48. **Streetlight Design** Determine the angle θ in the design of the streetlight shown in the figure.

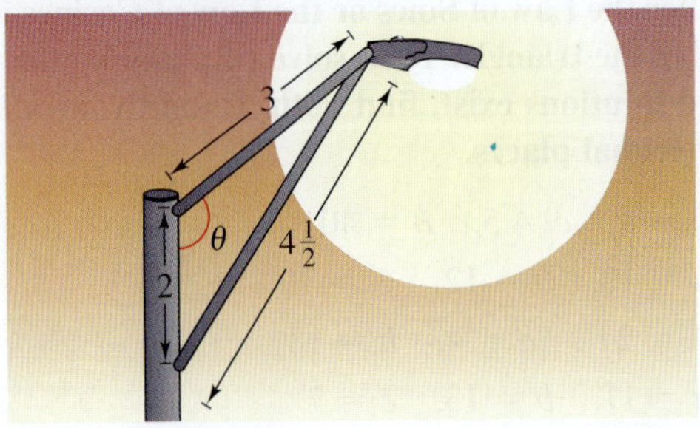

49. **Distance** Two ships leave a port at 9 A.M. One travels at a bearing of N 53° W at 12 miles per hour, and the other travels at a bearing of S 67° W at 16 miles per hour. Approximate how far apart they are at noon that day.

50. **Length** A 100-foot vertical tower is to be erected on the side of a hill that makes a 6° angle with the horizontal (see figure). Find the length of each of the two guy wires that will be anchored 75 feet uphill and downhill from the base of the tower.

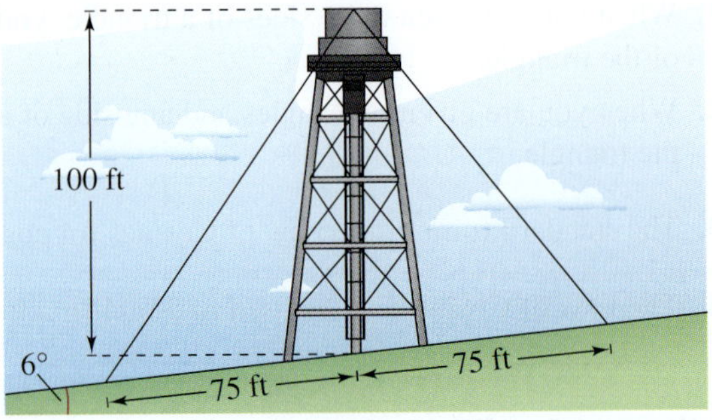

51. **Navigation** On a map, Minneapolis is 165 millimeters due west of Albany, Phoenix is 216 millimeters from Minneapolis, and Phoenix is 368 millimeters from Albany (see figure).

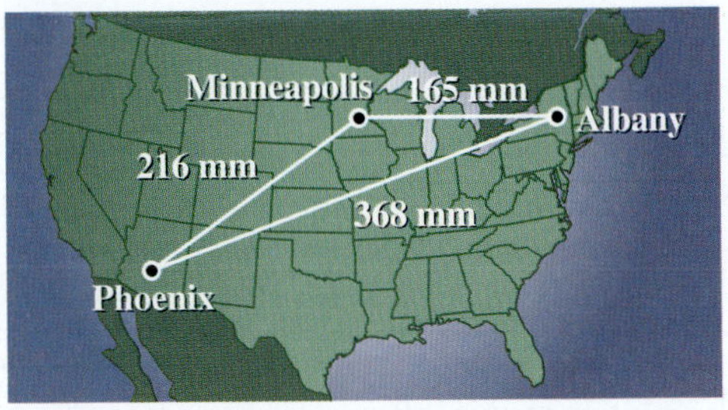

(a) Find the bearing of Minneapolis from Phoenix.

(b) Find the bearing of Albany from Phoenix.

52. **Baseball** The baseball player in center field is playing approximately 330 feet from the television camera that is behind home plate. A batter hits a fly ball that goes to the wall 420 feet from the camera (see figure). The camera turns 8° to follow the play. Approximately how far does the center fielder have to run to make the catch?

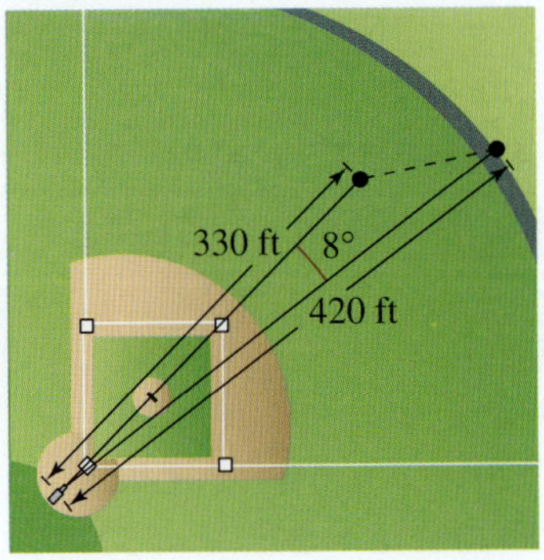

53. **Baseball** On a baseball diamond with 90-foot sides, the pitcher's mound is 60.5 feet from home plate. How far is it from the pitcher's mound to third base?

54. Surveying A triangular parcel of land has 115 meters of frontage, and the other boundaries have lengths of 76 meters and 92 meters. What angles does the frontage make with the two other boundaries?

55. Surveying A triangular parcel of ground has sides of lengths 725 feet, 650 feet, and 575 feet. Find the measure of the largest angle.

56. Engine Design

An engine has a seven-inch connecting rod fastened to a crank (see figure).

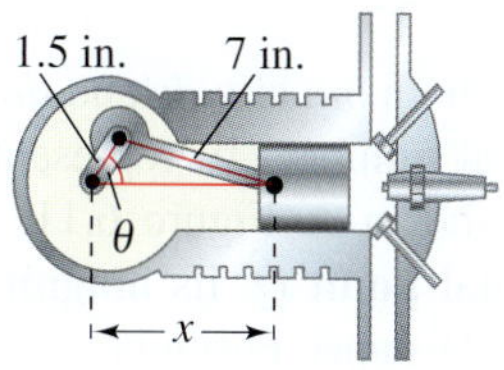

(a) Use the Law of Cosines to write an equation giving the relationship between x and θ.

(b) Write x as a function of θ. (Select the sign that yields positive values of x.)

(c) Use a graphing utility to graph the function in part (b).

(d) Use the graph in part (c) to determine the total distance the piston moves in one cycle.

57. Geometry The lengths of the sides of a triangular parcel of land are approximately 200 feet, 500 feet, and 600 feet. Approximate the area of the parcel.

58. Geometry A parking lot has the shape of a parallelogram (see figure). The lengths of two adjacent sides are 70 meters and 100 meters. The angle between the two sides is 70°. What is the area of the parking lot?

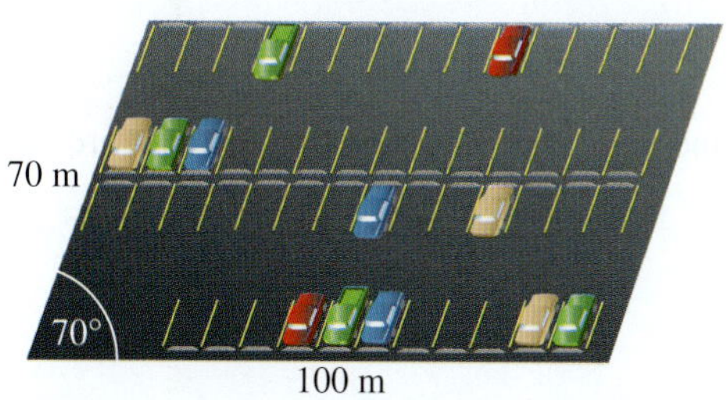

59. Geometry You want to buy a triangular lot measuring 510 yards by 840 yards by 1120 yards. The price of the land is \$2000 per acre. How much does the land cost? (*Hint:* 1 acre = 4840 square yards)

60. Geometry You want to buy a triangular lot measuring 1350 feet by 1860 feet by 2490 feet. The price of the land is \$2200 per acre. How much does the land cost? (*Hint:* 1 acre = 43,560 square feet)

Exploration

True or False? **In Exercises 61 and 62, determine whether the statement is true or false. Justify your answer.**

61. In Heron's Area Formula, s is the average of the lengths of the three sides of the triangle.

62. In addition to SSS and SAS, the Law of Cosines can be used to solve triangles with AAS conditions.

63. Think About It What familiar formula do you obtain when you use the standard form of the Law of Cosines $c^2 = a^2 + b^2 - 2ab \cos C$, and you let $C = 90°$? What is the relationship between the Law of Cosines and this formula?

64. Writing Describe how the Law of Cosines can be used to solve the ambiguous case of the oblique triangle ABC, where $a = 12$ feet, $b = 30$ feet, and $A = 20°$. Is the result the same as when the Law of Sines is used to solve the triangle? Describe the advantages and the disadvantages of each method.

65. Writing In Exercise 64, the Law of Cosines was used to solve a triangle in the two-solution case of SSA. Can the Law of Cosines be used to solve the no-solution and single-solution cases of SSA? Explain.

66. HOW DO YOU SEE IT? Determine whether the Law of Sines or the Law of Cosines is needed to solve the triangle.

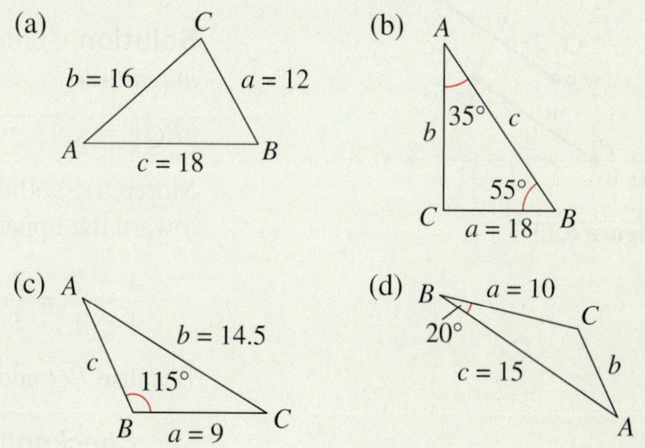

67. Proof Use the Law of Cosines to prove that

$$\frac{1}{2}bc(1 + \cos A) = \frac{a + b + c}{2} \cdot \frac{-a + b + c}{2}.$$

68. Proof Use the Law of Cosines to prove that

$$\frac{1}{2}bc(1 - \cos A) = \frac{a - b + c}{2} \cdot \frac{a + b - c}{2}.$$

Smart-foto/Shutterstock.com

6.3 Vectors in the Plane

You can use vectors to model and solve real-life problems involving magnitude and direction. For instance, in Exercise 102 on page 430, you will use vectors to determine the true direction of a commercial jet.

- **Represent vectors as directed line segments.**
- **Write the component forms of vectors.**
- **Perform basic vector operations and represent them graphically.**
- **Write vectors as linear combinations of unit vectors.**
- **Find the direction angles of vectors.**
- **Use vectors to model and solve real-life problems.**

Introduction

Quantities such as force and velocity involve both *magnitude* and *direction* and cannot be completely characterized by a single real number. To represent such a quantity, you can use a **directed line segment,** as shown in Figure 6.11. The directed line segment $\overrightarrow{PQ}$ has **initial point** P and **terminal point** Q. Its **magnitude** (or length) is denoted by $\|\overrightarrow{PQ}\|$ and can be found using the Distance Formula.

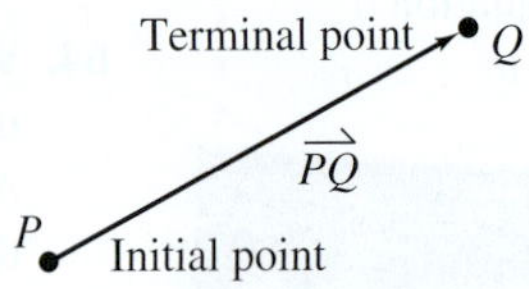

Figure 6.11

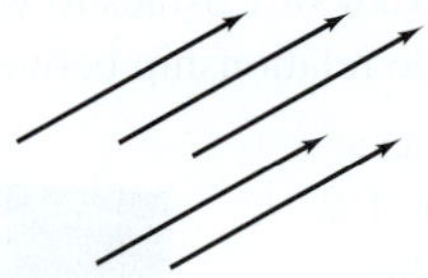

Figure 6.12

Two directed line segments that have the same magnitude and direction are equivalent. For example, the directed line segments in Figure 6.12 are all equivalent. The set of all directed line segments that are equivalent to the directed line segment $\overrightarrow{PQ}$ is a **vector v in the plane,** written $\mathbf{v} = \overrightarrow{PQ}$. Vectors are denoted by lowercase, boldface letters such as **u**, **v**, and **w**.

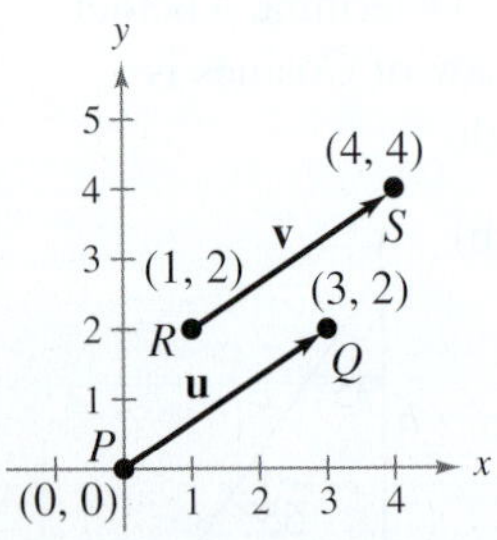

Figure 6.13

EXAMPLE 1 Showing That Two Vectors Are Equivalent

Show that **u** and **v** in Figure 6.13 are equivalent.

Solution From the Distance Formula, it follows that $\overrightarrow{PQ}$ and $\overrightarrow{RS}$ have the *same magnitude.*

$$\|\overrightarrow{PQ}\| = \sqrt{(3-0)^2 + (2-0)^2} = \sqrt{13} \qquad \|\overrightarrow{RS}\| = \sqrt{(4-1)^2 + (4-2)^2} = \sqrt{13}$$

Moreover, both line segments have the *same direction* because they are both directed toward the upper right on lines having a slope of

$$\frac{4-2}{4-1} = \frac{2-0}{3-0} = \frac{2}{3}.$$

Because $\overrightarrow{PQ}$ and $\overrightarrow{RS}$ have the same magnitude and direction, **u** and **v** are equivalent.

✓ ***Checkpoint*** *Audio-video solution in English & Spanish at LarsonPrecalculus.com.*

Show that **u** and **v** in the figure at the right are equivalent.

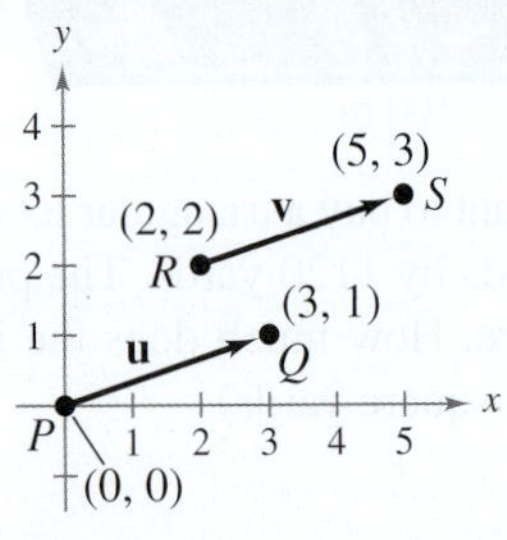

Bill Bachman/Photo Researchers, Inc.

Component Form of a Vector

The directed line segment whose initial point is the origin is often the most convenient representative of a set of equivalent directed line segments. This representative of the vector **v** is in **standard position.**

A vector whose initial point is the origin $(0, 0)$ can be uniquely represented by the coordinates of its terminal point (v_1, v_2). This is the **component form of a vector v,** written as $\mathbf{v} = \langle v_1, v_2 \rangle$. The coordinates v_1 and v_2 are the *components* of **v**. If both the initial point and the terminal point lie at the origin, then **v** is the **zero vector** and is denoted by $\mathbf{0} = \langle 0, 0 \rangle$.

▷ TECHNOLOGY You can graph vectors with a graphing utility by graphing directed line segments. Consult the user's guide for your graphing utility for specific instructions.

Component Form of a Vector

The component form of the vector with initial point $P(p_1, p_2)$ and terminal point $Q(q_1, q_2)$ is given by

$$\overrightarrow{PQ} = \langle q_1 - p_1, q_2 - p_2 \rangle = \langle v_1, v_2 \rangle = \mathbf{v}.$$

The **magnitude** (or length) of **v** is given by

$$\|\mathbf{v}\| = \sqrt{(q_1 - p_1)^2 + (q_2 - p_2)^2} = \sqrt{v_1^2 + v_2^2}.$$

If $\|\mathbf{v}\| = 1$, then **v** is a **unit vector.** Moreover, $\|\mathbf{v}\| = 0$ if and only if **v** is the zero vector **0**.

Two vectors $\mathbf{u} = \langle u_1, u_2 \rangle$ and $\mathbf{v} = \langle v_1, v_2 \rangle$ are *equal* if and only if $u_1 = v_1$ and $u_2 = v_2$. For instance, in Example 1, the vector **u** from $P(0, 0)$ to $Q(3, 2)$ is $\mathbf{u} = \overrightarrow{PQ} = \langle 3 - 0, 2 - 0 \rangle = \langle 3, 2 \rangle$, and the vector **v** from $R(1, 2)$ to $S(4, 4)$ is $\mathbf{v} = \overrightarrow{RS} = \langle 4 - 1, 4 - 2 \rangle = \langle 3, 2 \rangle$.

EXAMPLE 2 Finding the Component Form of a Vector

Find the component form and magnitude of the vector **v** that has initial point $(4, -7)$ and terminal point $(-1, 5)$.

Algebraic Solution

Let

$$P(4, -7) = (p_1, p_2)$$

and

$$Q(-1, 5) = (q_1, q_2).$$

Then, the components of $\mathbf{v} = \langle v_1, v_2 \rangle$ are

$$v_1 = q_1 - p_1 = -1 - 4 = -5$$

$$v_2 = q_2 - p_2 = 5 - (-7) = 12.$$

So, $\mathbf{v} = \langle -5, 12 \rangle$ and the magnitude of **v** is

$$\|\mathbf{v}\| = \sqrt{(-5)^2 + 12^2} = \sqrt{169} = 13.$$

Graphical Solution

Use centimeter graph paper to plot the points $P(4, -7)$ and $Q(-1, 5)$. Carefully sketch the vector **v**. Use the sketch to find the components of $\mathbf{v} = \langle v_1, v_2 \rangle$. Then use a centimeter ruler to find the magnitude of **v**. The figure at the right shows that the components of **v** are $v_1 = -5$ and $v_2 = 12$, so $\mathbf{v} = \langle -5, 12 \rangle$. The figure also shows that the magnitude of **v** is $\|\mathbf{v}\| = 13$.

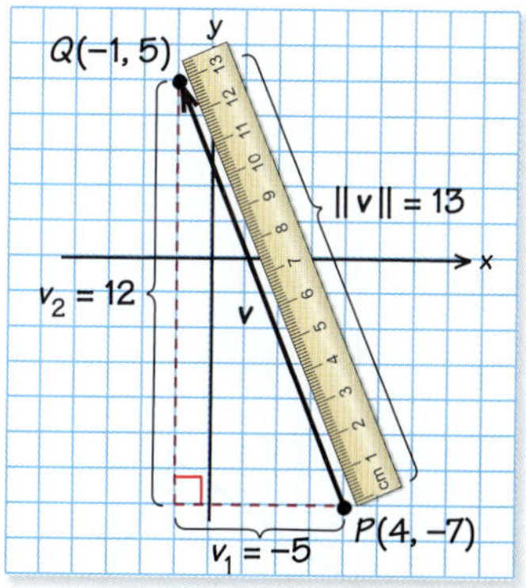

✓ **Checkpoint** Audio-video solution in English & Spanish at LarsonPrecalculus.com.

Find the component form and magnitude of the vector **v** that has initial point $(-2, 3)$ and terminal point $(-7, 9)$.

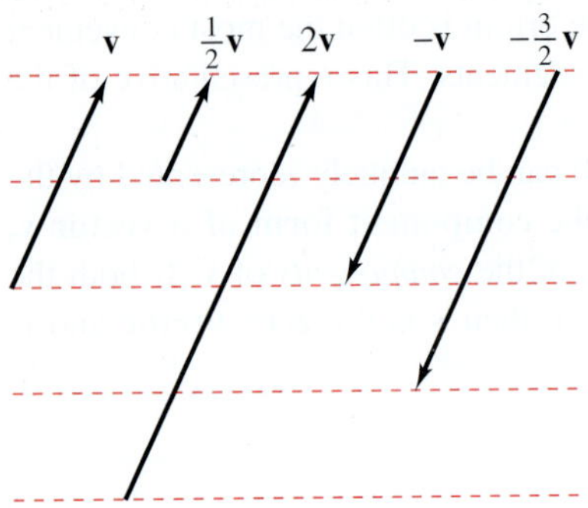

Figure 6.14

Vector Operations

The two basic vector operations are **scalar multiplication** and **vector addition.** In operations with vectors, numbers are usually referred to as **scalars.** In this text, scalars will always be real numbers. Geometrically, the product of a vector $\mathbf{v}$ and a scalar k is the vector that is $|k|$ times as long as $\mathbf{v}$. When k is positive, $k\mathbf{v}$ has the same direction as $\mathbf{v}$, and when k is negative, $k\mathbf{v}$ has the direction opposite that of $\mathbf{v}$, as shown in Figure 6.14.

To add two vectors $\mathbf{u}$ and $\mathbf{v}$ geometrically, first position them (without changing their lengths or directions) so that the initial point of the second vector $\mathbf{v}$ coincides with the terminal point of the first vector $\mathbf{u}$. The sum

$$\mathbf{u} + \mathbf{v}$$

is the vector formed by joining the initial point of the first vector $\mathbf{u}$ with the terminal point of the second vector $\mathbf{v}$, as shown below. This technique is called the **parallelogram law** for vector addition because the vector $\mathbf{u} + \mathbf{v}$, often called the **resultant** of vector addition, is the diagonal of a parallelogram having adjacent sides $\mathbf{u}$ and $\mathbf{v}$.

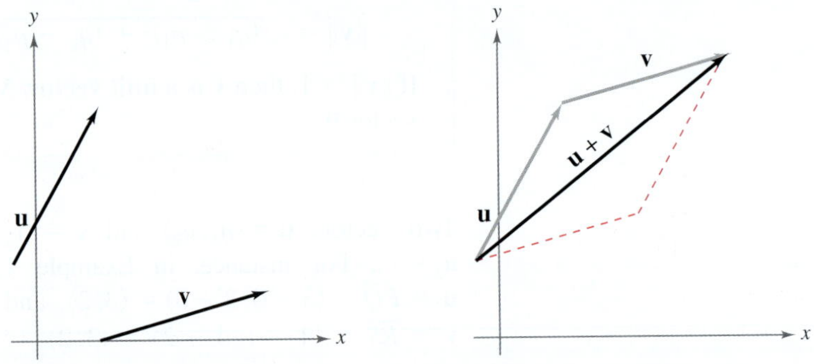

> **Definitions of Vector Addition and Scalar Multiplication**
>
> Let $\mathbf{u} = \langle u_1, u_2\rangle$ and $\mathbf{v} = \langle v_1, v_2\rangle$ be vectors and let k be a scalar (a real number). Then the **sum** of $\mathbf{u}$ and $\mathbf{v}$ is the vector
>
> $$\mathbf{u} + \mathbf{v} = \langle u_1 + v_1, u_2 + v_2\rangle \qquad \text{Sum}$$
>
> and the **scalar multiple** of k times $\mathbf{u}$ is the vector
>
> $$k\mathbf{u} = k\langle u_1, u_2\rangle = \langle ku_1, ku_2\rangle. \qquad \text{Scalar multiple}$$

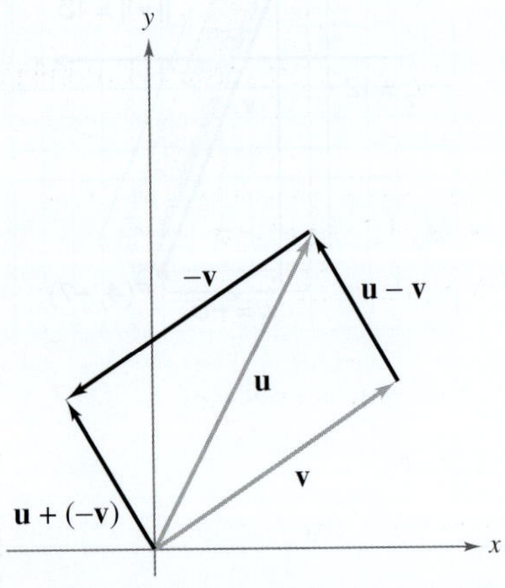

$\mathbf{u} - \mathbf{v} = \mathbf{u} + (-\mathbf{v})$
Figure 6.15

The **negative** of $\mathbf{v} = \langle v_1, v_2\rangle$ is

$$-\mathbf{v} = (-1)\mathbf{v}$$

$$= \langle -v_1, -v_2\rangle \qquad \text{Negative}$$

and the **difference** of $\mathbf{u}$ and $\mathbf{v}$ is

$$\mathbf{u} - \mathbf{v} = \mathbf{u} + (-\mathbf{v}) \qquad \text{Add } (-\mathbf{v})\text{. See Figure 6.15.}$$

$$= \langle u_1 - v_1, u_2 - v_2\rangle. \qquad \text{Difference}$$

To represent $\mathbf{u} - \mathbf{v}$ geometrically, you can use directed line segments with the *same* initial point. The difference $\mathbf{u} - \mathbf{v}$ is the vector from the terminal point of $\mathbf{v}$ to the terminal point of $\mathbf{u}$, which is equal to

$$\mathbf{u} + (-\mathbf{v})$$

as shown in Figure 6.15.

The component definitions of vector addition and scalar multiplication are illustrated in Example 3. In this example, notice that each of the vector operations can be interpreted geometrically.

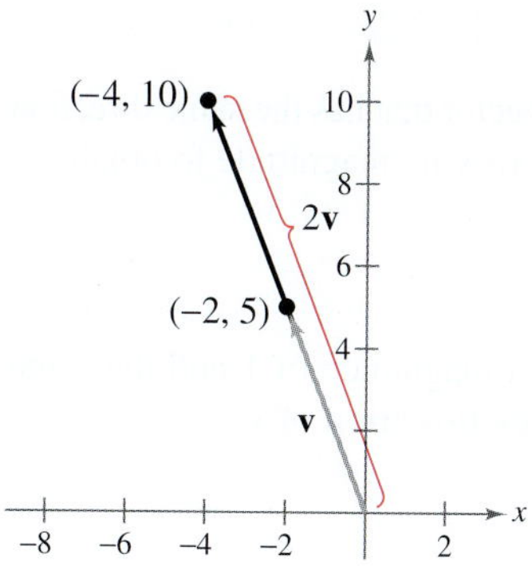

Figure 6.16

EXAMPLE 3 Vector Operations

Let $\mathbf{v} = \langle -2, 5\rangle$ and $\mathbf{w} = \langle 3, 4\rangle$. Find each of the following vectors.

a. $2\mathbf{v}$ **b.** $\mathbf{w} - \mathbf{v}$ **c.** $\mathbf{v} + 2\mathbf{w}$

Solution

a. Because $\mathbf{v} = \langle -2, 5\rangle$, you have

$$2\mathbf{v} = 2\langle -2, 5\rangle = \langle 2(-2), 2(5)\rangle = \langle -4, 10\rangle.$$

A sketch of $2\mathbf{v}$ is shown in Figure 6.16.

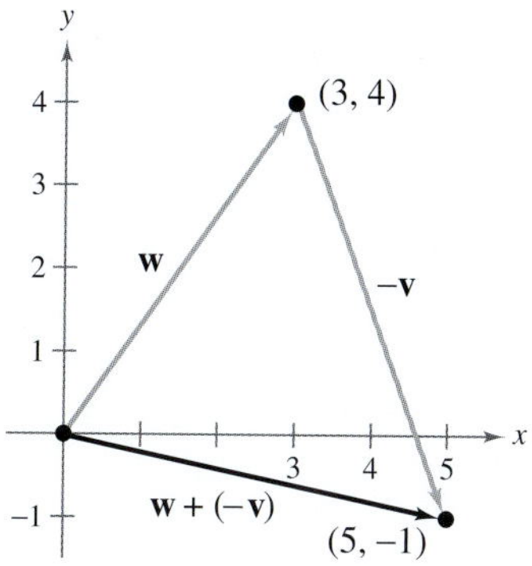

Figure 6.17

b. The difference of $\mathbf{w}$ and $\mathbf{v}$ is

$$\begin{aligned}\mathbf{w} - \mathbf{v} &= \langle 3, 4\rangle - \langle -2, 5\rangle \\ &= \langle 3 - (-2), 4 - 5\rangle \\ &= \langle 5, -1\rangle.\end{aligned}$$

A sketch of $\mathbf{w} - \mathbf{v}$ is shown in Figure 6.17. Note that the figure shows the vector difference $\mathbf{w} - \mathbf{v}$ as the sum $\mathbf{w} + (-\mathbf{v})$.

c. The sum of $\mathbf{v}$ and $2\mathbf{w}$ is

$$\begin{aligned}\mathbf{v} + 2\mathbf{w} &= \langle -2, 5\rangle + 2\langle 3, 4\rangle \\ &= \langle -2, 5\rangle + \langle 2(3), 2(4)\rangle \\ &= \langle -2, 5\rangle + \langle 6, 8\rangle \\ &= \langle -2 + 6, 5 + 8\rangle \\ &= \langle 4, 13\rangle.\end{aligned}$$

A sketch of $\mathbf{v} + 2\mathbf{w}$ is shown in Figure 6.18.

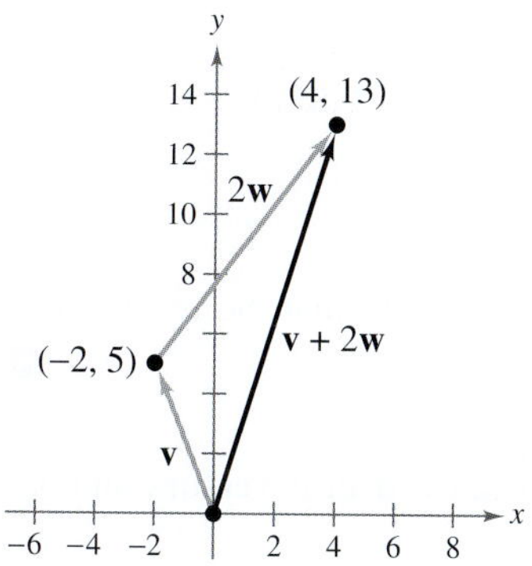

Figure 6.18

✓ ***Checkpoint*** *Audio-video solution in English & Spanish at LarsonPrecalculus.com.*

Let $\mathbf{u} = \langle 1, 4\rangle$ and $\mathbf{v} = \langle 3, 2\rangle$. Find each of the following vectors.

a. $\mathbf{u} + \mathbf{v}$ **b.** $\mathbf{u} - \mathbf{v}$ **c.** $2\mathbf{u} - 3\mathbf{v}$

Vector addition and scalar multiplication share many of the properties of ordinary arithmetic.

REMARK Property 9 can be stated as follows: The magnitude of the vector $c\mathbf{v}$ is the absolute value of c times the magnitude of $\mathbf{v}$.

Properties of Vector Addition and Scalar Multiplication

Let $\mathbf{u}$, $\mathbf{v}$, and $\mathbf{w}$ be vectors and let c and d be scalars. Then the following properties are true.

1. $\mathbf{u} + \mathbf{v} = \mathbf{v} + \mathbf{u}$

2. $(\mathbf{u} + \mathbf{v}) + \mathbf{w} = \mathbf{u} + (\mathbf{v} + \mathbf{w})$

3. $\mathbf{u} + \mathbf{0} = \mathbf{u}$

4. $\mathbf{u} + (-\mathbf{u}) = \mathbf{0}$

5. $c(d\mathbf{u}) = (cd)\mathbf{u}$

6. $(c + d)\mathbf{u} = c\mathbf{u} + d\mathbf{u}$

7. $c(\mathbf{u} + \mathbf{v}) = c\mathbf{u} + c\mathbf{v}$

8. $1(\mathbf{u}) = \mathbf{u}$, $0(\mathbf{u}) = \mathbf{0}$

9. $\|c\mathbf{v}\| = |c|\,\|\mathbf{v}\|$

William Rowan Hamilton (1805–1865), an Irish mathematician, did some of the earliest work with vectors. Hamilton spent many years developing a system of vector-like quantities called quaternions. Although Hamilton was convinced of the benefits of quaternions, the operations he defined did not produce good models for physical phenomena. It was not until the latter half of the nineteenth century that the Scottish physicist James Maxwell (1831–1879) restructured Hamilton's quaternions in a form useful for representing physical quantities such as force, velocity, and acceleration.

Unit Vectors

In many applications of vectors, it is useful to find a unit vector that has the same direction as a given nonzero vector $\mathbf{v}$. To do this, you can divide $\mathbf{v}$ by its magnitude to obtain

$$\mathbf{u} = \text{unit vector} = \frac{\mathbf{v}}{\|\mathbf{v}\|} = \left(\frac{1}{\|\mathbf{v}\|}\right)\mathbf{v}.$$ Unit vector in direction of $\mathbf{v}$

Note that $\mathbf{u}$ is a scalar multiple of $\mathbf{v}$. The vector $\mathbf{u}$ has a magnitude of 1 and the same direction as $\mathbf{v}$. The vector $\mathbf{u}$ is called a **unit vector in the direction of v.**

EXAMPLE 4 Finding a Unit Vector

Find a unit vector in the direction of $\mathbf{v} = \langle -2, 5\rangle$ and verify that the result has a magnitude of 1.

Solution The unit vector in the direction of $\mathbf{v}$ is

$$\begin{aligned}\frac{\mathbf{v}}{\|\mathbf{v}\|} &= \frac{\langle -2, 5\rangle}{\sqrt{(-2)^2 + 5^2}}\\ &= \frac{1}{\sqrt{29}}\langle -2, 5\rangle\\ &= \left\langle \frac{-2}{\sqrt{29}}, \frac{5}{\sqrt{29}}\right\rangle.\end{aligned}$$

This vector has a magnitude of 1 because

$$\begin{aligned}\sqrt{\left(\frac{-2}{\sqrt{29}}\right)^2 + \left(\frac{5}{\sqrt{29}}\right)^2} &= \sqrt{\frac{4}{29} + \frac{25}{29}}\\ &= \sqrt{\frac{29}{29}}\\ &= 1.\end{aligned}$$

Checkpoint Audio-video solution in English & Spanish at LarsonPrecalculus.com.

Find a unit vector $\mathbf{u}$ in the direction of $\mathbf{v} = \langle 6, -1\rangle$ and verify that the result has a magnitude of 1.

The unit vectors $\langle 1, 0\rangle$ and $\langle 0, 1\rangle$ are called the **standard unit vectors** and are denoted by

$$\mathbf{i} = \langle 1, 0\rangle \quad \text{and} \quad \mathbf{j} = \langle 0, 1\rangle$$

as shown in Figure 6.19. (Note that the lowercase letter $\mathbf{i}$ is written in boldface to distinguish it from the imaginary unit $i = \sqrt{-1}$.) These vectors can be used to represent any vector $\mathbf{v} = \langle v_1, v_2\rangle$, as follows.

$$\begin{aligned}\mathbf{v} &= \langle v_1, v_2\rangle\\ &= v_1\langle 1, 0\rangle + v_2\langle 0, 1\rangle\\ &= v_1\mathbf{i} + v_2\mathbf{j}\end{aligned}$$

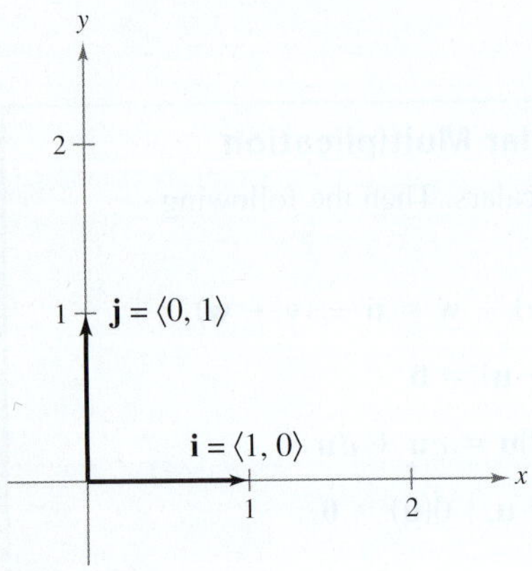

Figure 6.19

The scalars v_1 and v_2 are called the **horizontal** and **vertical components of v,** respectively. The vector sum

$$v_1\mathbf{i} + v_2\mathbf{j}$$

is called a **linear combination** of the vectors $\mathbf{i}$ and $\mathbf{j}$. Any vector in the plane can be written as a linear combination of the standard unit vectors $\mathbf{i}$ and $\mathbf{j}$.

Corbis Images

EXAMPLE 5 Writing a Linear Combination of Unit Vectors

Let **u** be the vector with initial point $(2, -5)$ and terminal point $(-1, 3)$. Write **u** as a linear combination of the standard unit vectors **i** and **j**.

Solution Begin by writing the component form of the vector **u**.

$$\mathbf{u} = \langle -1 - 2, 3 - (-5) \rangle$$
$$= \langle -3, 8 \rangle$$
$$= -3\mathbf{i} + 8\mathbf{j}$$

This result is shown graphically below.

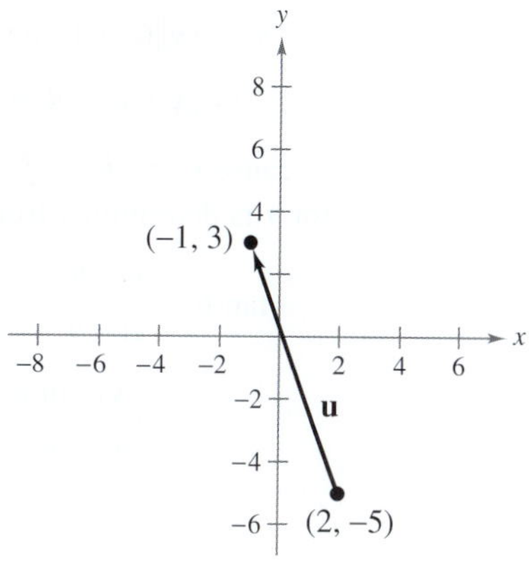

✓ **Checkpoint** Audio-video solution in English & Spanish at LarsonPrecalculus.com.

Let **u** be the vector with initial point $(-2, 6)$ and terminal point $(-8, 3)$. Write **u** as a linear combination of the standard unit vectors **i** and **j**.

EXAMPLE 6 Vector Operations

Let

$$\mathbf{u} = -3\mathbf{i} + 8\mathbf{j} \quad \text{and} \quad \mathbf{v} = 2\mathbf{i} - \mathbf{j}.$$

Find

$$2\mathbf{u} - 3\mathbf{v}.$$

Solution You could solve this problem by converting **u** and **v** to component form. This, however, is not necessary. It is just as easy to perform the operations in unit vector form.

$$2\mathbf{u} - 3\mathbf{v} = 2(-3\mathbf{i} + 8\mathbf{j}) - 3(2\mathbf{i} - \mathbf{j})$$
$$= -6\mathbf{i} + 16\mathbf{j} - 6\mathbf{i} + 3\mathbf{j}$$
$$= -12\mathbf{i} + 19\mathbf{j}$$

✓ **Checkpoint** Audio-video solution in English & Spanish at LarsonPrecalculus.com.

Let

$$\mathbf{u} = \mathbf{i} - 2\mathbf{j} \quad \text{and} \quad \mathbf{v} = -3\mathbf{i} + 2\mathbf{j}.$$

Find

$$5\mathbf{u} - 2\mathbf{v}.$$

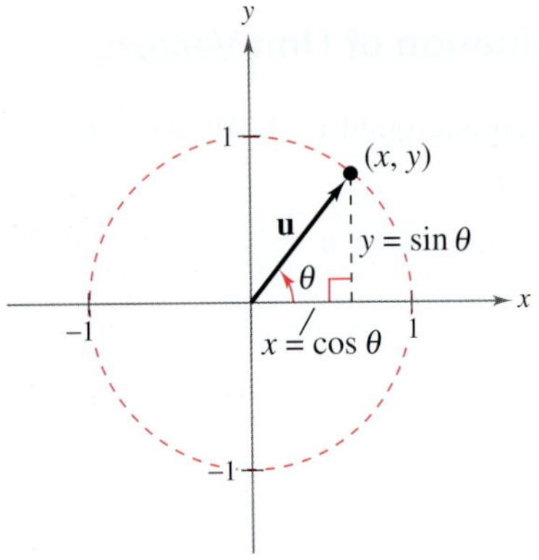

$\|\mathbf{u}\| = 1$
Figure 6.20

Direction Angles

If $\mathbf{u}$ is a *unit vector* such that θ is the angle (measured counterclockwise) from the positive x-axis to $\mathbf{u}$, then the terminal point of $\mathbf{u}$ lies on the unit circle and you have

$$\begin{aligned}\mathbf{u} &= \langle x, y\rangle \\ &= \langle \cos\theta, \sin\theta\rangle \\ &= (\cos\theta)\mathbf{i} + (\sin\theta)\mathbf{j}\end{aligned}$$

as shown in Figure 6.20. The angle θ is the **direction angle** of the vector $\mathbf{u}$.

Suppose that $\mathbf{u}$ is a unit vector with direction angle θ. If $\mathbf{v} = a\mathbf{i} + b\mathbf{j}$ is any vector that makes an angle θ with the positive x-axis, then it has the same direction as $\mathbf{u}$ and you can write

$$\begin{aligned}\mathbf{v} &= \|\mathbf{v}\|\langle \cos\theta, \sin\theta\rangle \\ &= \|\mathbf{v}\|(\cos\theta)\mathbf{i} + \|\mathbf{v}\|(\sin\theta)\mathbf{j}.\end{aligned}$$

Because $\mathbf{v} = a\mathbf{i} + b\mathbf{j} = \|\mathbf{v}\|(\cos\theta)\mathbf{i} + \|\mathbf{v}\|(\sin\theta)\mathbf{j}$, it follows that the direction angle θ for $\mathbf{v}$ is determined from

$$\tan\theta = \frac{\sin\theta}{\cos\theta} \qquad \text{Quotient identity}$$

$$= \frac{\|\mathbf{v}\|\sin\theta}{\|\mathbf{v}\|\cos\theta} \qquad \text{Multiply numerator and denominator by } \|\mathbf{v}\|.$$

$$= \frac{b}{a}. \qquad \text{Simplify.}$$

EXAMPLE 7 Finding Direction Angles of Vectors

Find the direction angle of each vector.

a. $\mathbf{u} = 3\mathbf{i} + 3\mathbf{j}$ **b.** $\mathbf{v} = 3\mathbf{i} - 4\mathbf{j}$

Solution

a. The direction angle is determined from

$$\tan\theta = \frac{b}{a} = \frac{3}{3} = 1.$$

So, $\theta = 45°$, as shown in Figure 6.21.

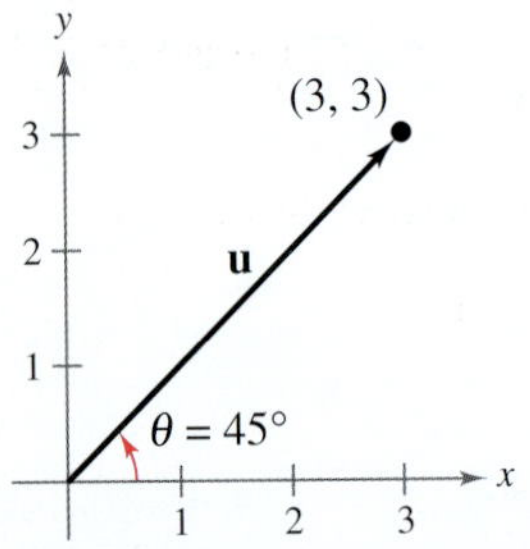

Figure 6.21

b. The direction angle is determined from

$$\tan\theta = \frac{b}{a} = \frac{-4}{3}.$$

Moreover, because $\mathbf{v} = 3\mathbf{i} - 4\mathbf{j}$ lies in Quadrant IV, θ lies in Quadrant IV, and its reference angle is

$$\theta' = \left|\arctan\left(-\frac{4}{3}\right)\right| \approx \left|-0.9273 \text{ radian}\right| \approx |-53.13°| = 53.13°.$$

So, it follows that $\theta \approx 360° - 53.13° = 306.87°$, as shown in Figure 6.22.

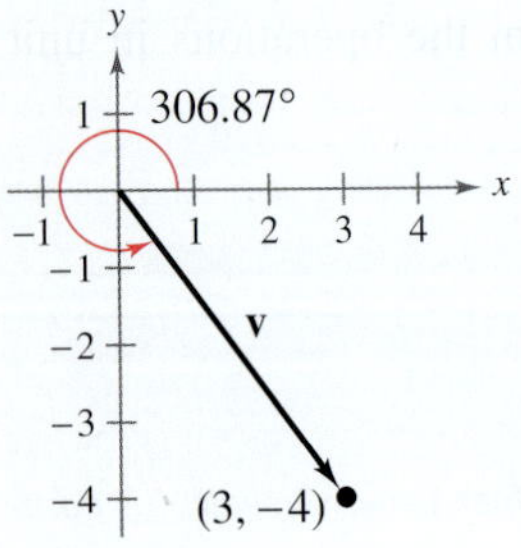

Figure 6.22

✓ **Checkpoint** Audio-video solution in English & Spanish at LarsonPrecalculus.com.

Find the direction angle of each vector.

a. $\mathbf{v} = -6\mathbf{i} + 6\mathbf{j}$ **b.** $\mathbf{v} = -7\mathbf{i} - 4\mathbf{j}$

Applications

EXAMPLE 8 Finding the Component Form of a Vector

Find the component form of the vector that represents the velocity of an airplane descending at a speed of 150 miles per hour at an angle 20° below the horizontal, as shown in Figure 6.23.

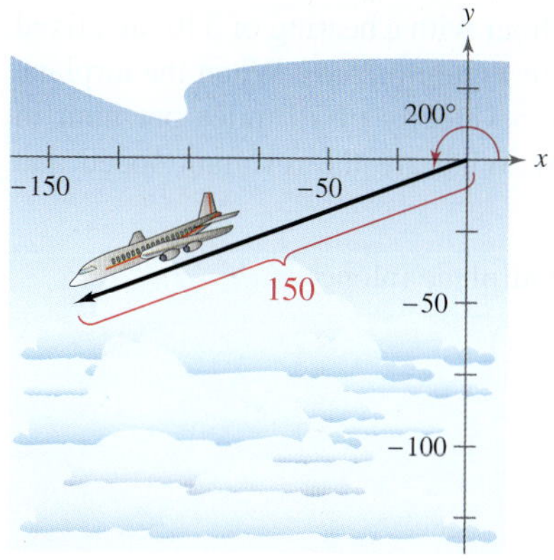

Figure 6.23

Solution The velocity vector **v** has a magnitude of 150 and a direction angle of $\theta = 200°$.

$$\begin{aligned}\mathbf{v} &= \|\mathbf{v}\|(\cos\theta)\mathbf{i} + \|\mathbf{v}\|(\sin\theta)\mathbf{j}\\ &= 150(\cos 200°)\mathbf{i} + 150(\sin 200°)\mathbf{j}\\ &\approx 150(-0.9397)\mathbf{i} + 150(-0.3420)\mathbf{j}\\ &\approx -140.96\mathbf{i} - 51.30\mathbf{j}\\ &= \langle -140.96, -51.30\rangle\end{aligned}$$

You can check that **v** has a magnitude of 150, as follows.

$$\|\mathbf{v}\| \approx \sqrt{(-140.96)^2 + (-51.30)^2} \approx \sqrt{22{,}501.41} \approx 150$$

✓ **Checkpoint** Audio-video solution in English & Spanish at LarsonPrecalculus.com.

Find the component form of the vector that represents the velocity of an airplane descending at a speed of 100 miles per hour at an angle 45° below the horizontal ($\theta = 225°$).

EXAMPLE 9 Using Vectors to Determine Weight

A force of 600 pounds is required to pull a boat and trailer up a ramp inclined at 15° from the horizontal. Find the combined weight of the boat and trailer.

Solution Based on Figure 6.24, you can make the following observations.

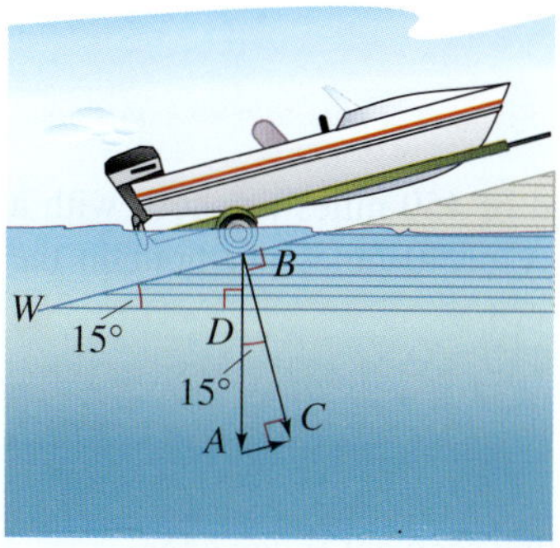

Figure 6.24

$\|\overrightarrow{BA}\|$ = force of gravity = combined weight of boat and trailer

$\|\overrightarrow{BC}\|$ = force against ramp

$\|\overrightarrow{AC}\|$ = force required to move boat up ramp = 600 pounds

By construction, triangles *BWD* and *ABC* are similar. So, angle *ABC* is 15°. In triangle *ABC*, you have

$$\begin{aligned}\sin 15° &= \frac{\|\overrightarrow{AC}\|}{\|\overrightarrow{BA}\|}\\ \sin 15° &= \frac{600}{\|\overrightarrow{BA}\|}\\ \|\overrightarrow{BA}\| &= \frac{600}{\sin 15°}\\ \|\overrightarrow{BA}\| &\approx 2318.\end{aligned}$$

So, the combined weight is approximately 2318 pounds. (In Figure 6.24, note that $\overrightarrow{AC}$ is parallel to the ramp.)

✓ **Checkpoint** Audio-video solution in English & Spanish at LarsonPrecalculus.com.

A force of 500 pounds is required to pull a boat and trailer up a ramp inclined at 12° from the horizontal. Find the combined weight of the boat and trailer.

REMARK Recall from Section 4.8 that in air navigation, bearings can be measured in degrees clockwise from north.

EXAMPLE 10 Using Vectors to Find Speed and Direction

An airplane is traveling at a speed of 500 miles per hour with a bearing of 330° at a fixed altitude with a negligible wind velocity, as shown in Figure 6.25(a). When the airplane reaches a certain point, it encounters a wind with a velocity of 70 miles per hour in the direction N 45° E, as shown in Figure 6.25(b). What are the resultant speed and direction of the airplane?

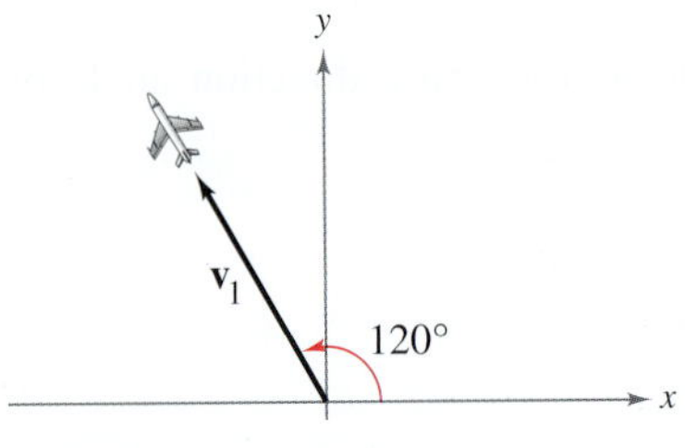

(a)

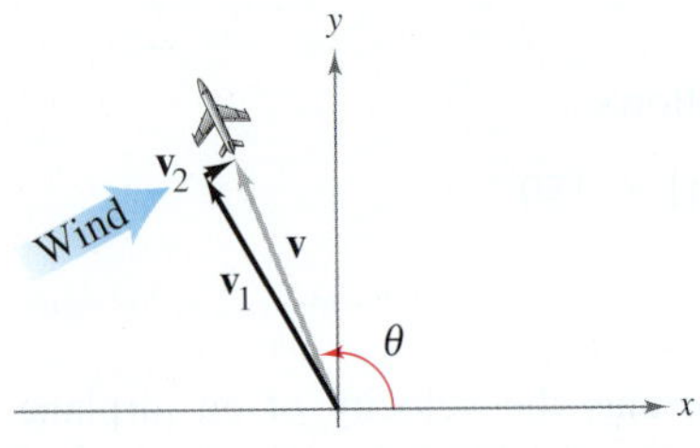

(b)

Figure 6.25

Solution Using Figure 6.25, the velocity of the airplane (alone) is

$$\mathbf{v}_1 = 500\langle \cos 120°, \sin 120° \rangle = \langle -250, 250\sqrt{3} \rangle$$

and the velocity of the wind is

$$\mathbf{v}_2 = 70\langle \cos 45°, \sin 45° \rangle = \langle 35\sqrt{2}, 35\sqrt{2} \rangle.$$

So, the velocity of the airplane (in the wind) is

$$\begin{aligned}\mathbf{v} &= \mathbf{v}_1 + \mathbf{v}_2 \\ &= \langle -250 + 35\sqrt{2}, 250\sqrt{3} + 35\sqrt{2} \rangle \\ &\approx \langle -200.5, 482.5 \rangle\end{aligned}$$

and the resultant speed of the airplane is

$$\|\mathbf{v}\| \approx \sqrt{(-200.5)^2 + (482.5)^2} \approx 522.5 \text{ miles per hour.}$$

Finally, given that θ is the direction angle of the flight path, you have

$$\tan \theta \approx \frac{482.5}{-200.5} \approx -2.4065$$

which implies that

$$\theta \approx 180° - 67.4° = 112.6°.$$

So, the true direction of the airplane is approximately

$$270° + (180° - 112.6°) = 337.4°.$$

Airplanes can take advantage of fast-moving air currents called jet streams to decrease travel time.

✓ Checkpoint *Audio-video solution in English & Spanish at LarsonPrecalculus.com.*

Repeat Example 10 for an airplane traveling at a speed of 450 miles per hour with a bearing of 300° that encounters a wind with a velocity of 40 miles per hour in the direction N 30° E.

Summarize (Section 6.3)

1. Describe how to represent a vector as a directed line segment *(page 418)*. For an example involving vectors represented as directed line segments, see Example 1.
2. Describe how to write a vector in component form *(page 419)*. For an example of finding the component form of a vector, see Example 2.
3. State the definitions of vector addition and scalar multiplication *(page 420)*. For an example of performing vector operations, see Example 3.
4. Describe how to write a vector as a linear combination of unit vectors *(page 422)*. For examples involving unit vectors, see Examples 4–6.
5. Describe how to find the direction angle of a vector *(page 424)*. For an example of finding the direction angles of vectors, see Example 7.
6. Describe real-life situations that can be modeled and solved using vectors *(pages 425 and 426, Examples 8–10)*.

MC_PP/Shutterstock.com

6.3 Exercises

See CalcChat.com for tutorial help and worked-out solutions to odd-numbered exercises.

Vocabulary: Fill in the blanks.

1. A ________ ________ ________ can be used to represent a quantity that involves both magnitude and direction.
2. The directed line segment $\overrightarrow{PQ}$ has ________ point P and ________ point Q.
3. The ________ of the directed line segment $\overrightarrow{PQ}$ is denoted by $\|\overrightarrow{PQ}\|$.
4. The set of all directed line segments that are equivalent to a given directed line segment $\overrightarrow{PQ}$ is a ________ **v** in the plane.
5. In order to show that two vectors are equivalent, you must show that they have the same ________ and the same ________ .
6. The directed line segment whose initial point is the origin is said to be in ________ ________ .
7. A vector that has a magnitude of 1 is called a ________ ________ .
8. The two basic vector operations are scalar ________ and vector ________ .
9. The vector $\mathbf{u} + \mathbf{v}$ is called the ________ of vector addition.
10. The vector sum $v_1\mathbf{i} + v_2\mathbf{j}$ is called a ________ ________ of the vectors **i** and **j**, and the scalars v_1 and v_2 are called the ________ and ________ components of **v**, respectively.

Skills and Applications

Showing That Two Vectors Are Equivalent In Exercises 11 and 12, show that u and v are equivalent.

11.
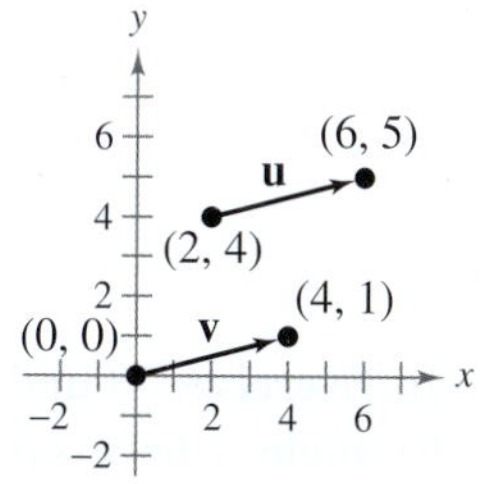

12.
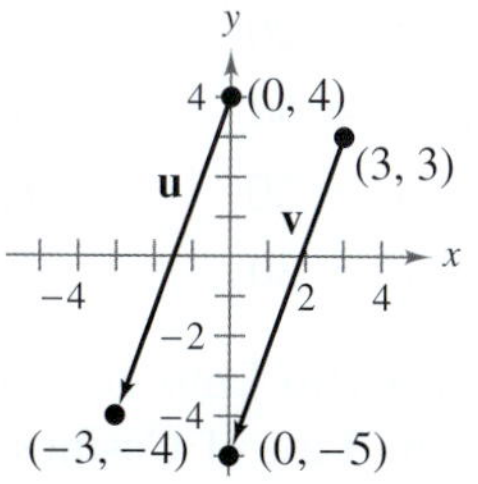

Finding the Component Form of a Vector In Exercises 13–24, find the component form and magnitude of the vector v.

13.
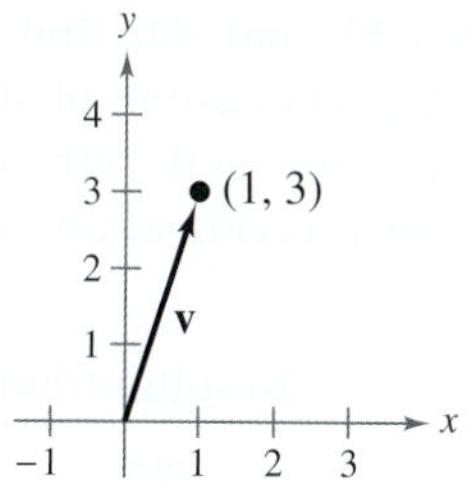

14.
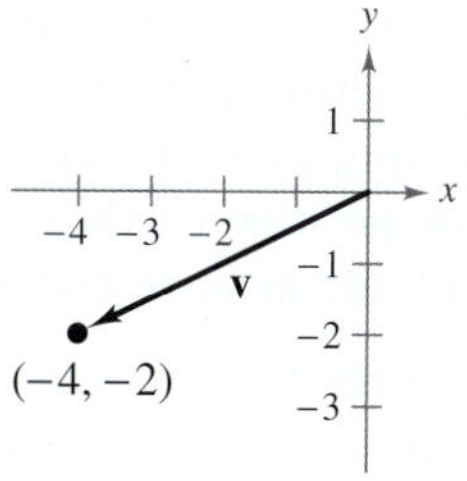

15.
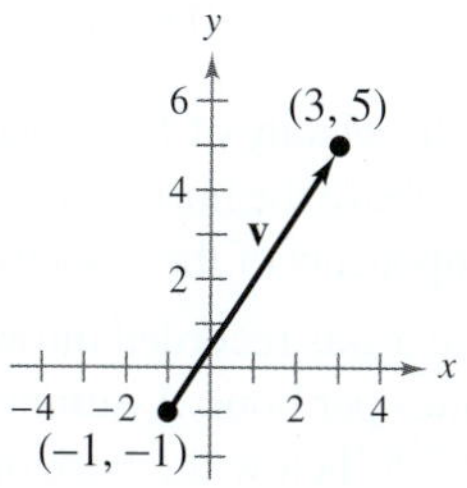

16.
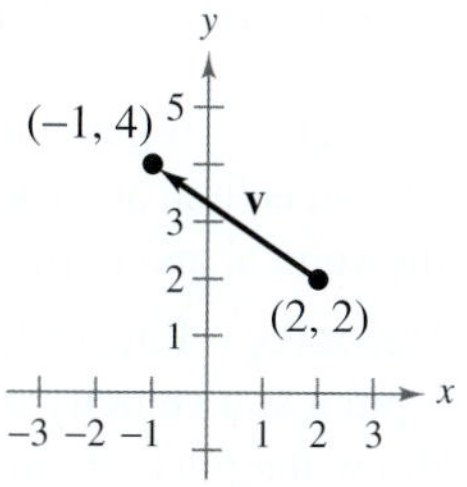

17.
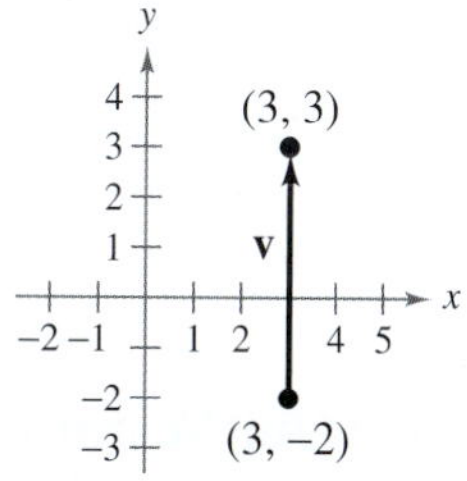

18.
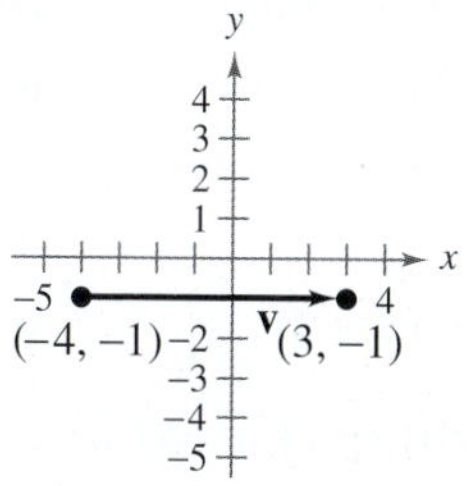

	Initial Point	Terminal Point
19.	$(-3, -5)$	$(5, 1)$
20.	$(-2, 7)$	$(5, -17)$
21.	$(1, 3)$	$(-8, -9)$
22.	$(1, 11)$	$(9, 3)$
23.	$(-1, 5)$	$(15, 12)$
24.	$(-3, 11)$	$(9, 40)$

Sketching the Graph of a Vector In Exercises 25–30, use the figure to sketch a graph of the specified vector. To print an enlarged copy of the graph, go to *MathGraphs.com*.

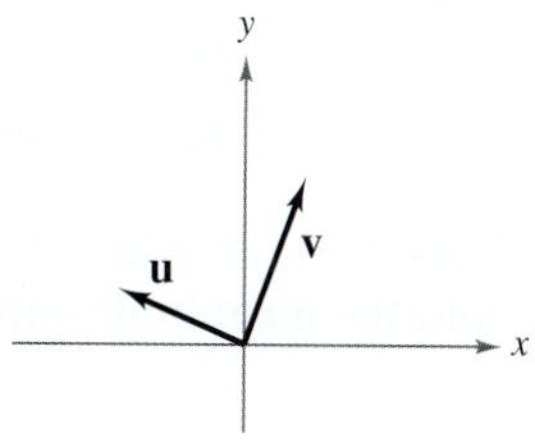

25. $-\mathbf{v}$
26. $5\mathbf{v}$
27. $\mathbf{u} + \mathbf{v}$
28. $\mathbf{u} + 2\mathbf{v}$
29. $\mathbf{u} - \mathbf{v}$
30. $\mathbf{v} - \frac{1}{2}\mathbf{u}$

Vector Operations In Exercises 31–38, find (a) $\mathbf{u} + \mathbf{v}$, (b) $\mathbf{u} - \mathbf{v}$, and (c) $2\mathbf{u} - 3\mathbf{v}$. Then sketch each resultant vector.

31. $\mathbf{u} = \langle 2, 1 \rangle$, $\mathbf{v} = \langle 1, 3 \rangle$ **32.** $\mathbf{u} = \langle 2, 3 \rangle$, $\mathbf{v} = \langle 4, 0 \rangle$

33. $\mathbf{u} = \langle -5, 3 \rangle$, $\mathbf{v} = \langle 0, 0 \rangle$ **34.** $\mathbf{u} = \langle 0, 0 \rangle$, $\mathbf{v} = \langle 2, 1 \rangle$

35. $\mathbf{u} = \mathbf{i} + \mathbf{j}$, $\mathbf{v} = 2\mathbf{i} - 3\mathbf{j}$

36. $\mathbf{u} = -2\mathbf{i} + \mathbf{j}$, $\mathbf{v} = 3\mathbf{j}$

37. $\mathbf{u} = 2\mathbf{i}$, $\mathbf{v} = \mathbf{j}$ **38.** $\mathbf{u} = 2\mathbf{j}$, $\mathbf{v} = 3\mathbf{i}$

Finding a Unit Vector In Exercises 39–48, find a unit vector in the direction of the given vector. Verify that the result has a magnitude of 1.

39. $\mathbf{u} = \langle 3, 0 \rangle$ **40.** $\mathbf{u} = \langle 0, -2 \rangle$

41. $\mathbf{v} = \langle -2, 2 \rangle$ **42.** $\mathbf{v} = \langle 5, -12 \rangle$

43. $\mathbf{v} = \mathbf{i} + \mathbf{j}$ **44.** $\mathbf{v} = 6\mathbf{i} - 2\mathbf{j}$

45. $\mathbf{w} = 4\mathbf{j}$ **46.** $\mathbf{w} = -6\mathbf{i}$

47. $\mathbf{w} = \mathbf{i} - 2\mathbf{j}$ **48.** $\mathbf{w} = 7\mathbf{j} - 3\mathbf{i}$

Finding a Vector In Exercises 49–52, find the vector v with the given magnitude and the same direction as u.

	Magnitude	Direction
49.	$\|\mathbf{v}\| = 10$	$\mathbf{u} = \langle -3, 4 \rangle$
50.	$\|\mathbf{v}\| = 3$	$\mathbf{u} = \langle -12, -5 \rangle$
51.	$\|\mathbf{v}\| = 9$	$\mathbf{u} = \langle 2, 5 \rangle$
52.	$\|\mathbf{v}\| = 8$	$\mathbf{u} = \langle 3, 3 \rangle$

Writing a Linear Combination of Unit Vectors In Exercises 53–56, the initial and terminal points of a vector are given. Write the vector as a linear combination of the standard unit vectors i and j.

	Initial Point	Terminal Point
53.	$(-2, 1)$	$(3, -2)$
54.	$(0, -2)$	$(3, 6)$
55.	$(-6, 4)$	$(0, 1)$
56.	$(-1, -5)$	$(2, 3)$

Vector Operations In Exercises 57–62, find the component form of v and sketch the specified vector operations geometrically, where $\mathbf{u} = 2\mathbf{i} - \mathbf{j}$ and $\mathbf{w} = \mathbf{i} + 2\mathbf{j}$.

57. $\mathbf{v} = \frac{3}{2}\mathbf{u}$ **58.** $\mathbf{v} = \frac{3}{4}\mathbf{w}$

59. $\mathbf{v} = \mathbf{u} + 2\mathbf{w}$ **60.** $\mathbf{v} = -\mathbf{u} + \mathbf{w}$

61. $\mathbf{v} = \frac{1}{2}(3\mathbf{u} + \mathbf{w})$ **62.** $\mathbf{v} = \mathbf{u} - 2\mathbf{w}$

Finding the Direction Angle of a Vector In Exercises 63–66, find the magnitude and direction angle of the vector v.

63. $\mathbf{v} = 6\mathbf{i} - 6\mathbf{j}$ **64.** $\mathbf{v} = -5\mathbf{i} + 4\mathbf{j}$

65. $\mathbf{v} = 3(\cos 60^\circ\mathbf{i} + \sin 60^\circ\mathbf{j})$

66. $\mathbf{v} = 8(\cos 135^\circ\mathbf{i} + \sin 135^\circ\mathbf{j})$

Finding the Component Form of a Vector In Exercises 67–74, find the component form of v given its magnitude and the angle it makes with the positive x-axis. Sketch v.

	Magnitude	Angle
67.	$\|\mathbf{v}\| = 3$	$\theta = 0^\circ$
68.	$\|\mathbf{v}\| = 1$	$\theta = 45^\circ$
69.	$\|\mathbf{v}\| = \frac{7}{2}$	$\theta = 150^\circ$
70.	$\|\mathbf{v}\| = \frac{3}{4}$	$\theta = 150^\circ$
71.	$\|\mathbf{v}\| = 2\sqrt{3}$	$\theta = 45^\circ$
72.	$\|\mathbf{v}\| = 4\sqrt{3}$	$\theta = 90^\circ$
73.	$\|\mathbf{v}\| = 3$	$\mathbf{v}$ in the direction $3\mathbf{i} + 4\mathbf{j}$
74.	$\|\mathbf{v}\| = 2$	$\mathbf{v}$ in the direction $\mathbf{i} + 3\mathbf{j}$

Finding the Component Form of a Vector In Exercises 75–78, find the component form of the sum of u and v with direction angles $\theta_{\mathbf{u}}$ and $\theta_{\mathbf{v}}$.

	Magnitude	Angle
75.	$\|\mathbf{u}\| = 5$	$\theta_{\mathbf{u}} = 0^\circ$
	$\|\mathbf{v}\| = 5$	$\theta_{\mathbf{v}} = 90^\circ$
76.	$\|\mathbf{u}\| = 4$	$\theta_{\mathbf{u}} = 60^\circ$
	$\|\mathbf{v}\| = 4$	$\theta_{\mathbf{v}} = 90^\circ$
77.	$\|\mathbf{u}\| = 20$	$\theta_{\mathbf{u}} = 45^\circ$
	$\|\mathbf{v}\| = 50$	$\theta_{\mathbf{v}} = 180^\circ$
78.	$\|\mathbf{u}\| = 50$	$\theta_{\mathbf{u}} = 30^\circ$
	$\|\mathbf{v}\| = 30$	$\theta_{\mathbf{v}} = 110^\circ$

Using the Law of Cosines In Exercises 79 and 80, use the Law of Cosines to find the angle α between the vectors. (Assume $0^\circ \le \alpha \le 180^\circ$.)

79. $\mathbf{v} = \mathbf{i} + \mathbf{j}$, $\mathbf{w} = 2\mathbf{i} - 2\mathbf{j}$

80. $\mathbf{v} = \mathbf{i} + 2\mathbf{j}$, $\mathbf{w} = 2\mathbf{i} - \mathbf{j}$

Resultant Force In Exercises 81 and 82, find the angle between the forces given the magnitude of their resultant. (*Hint:* Write force 1 as a vector in the direction of the positive x-axis and force 2 as a vector at an angle θ with the positive x-axis.)

	Force 1	Force 2	Resultant Force
81.	45 pounds	60 pounds	90 pounds
82.	3000 pounds	1000 pounds	3750 pounds

83. Velocity A gun with a muzzle velocity of 1200 feet per second is fired at an angle of 6° above the horizontal. Find the vertical and horizontal components of the velocity.

84. Velocity Pitcher Joel Zumaya was recorded throwing a pitch at a velocity of 104 miles per hour. Assuming he threw the pitch at an angle of 3.5° below the horizontal, find the vertical and horizontal components of the velocity. *(Source: Damon Lichtenwalner, Baseball Info Solutions)*

85. Resultant Force Forces with magnitudes of 125 newtons and 300 newtons act on a hook (see figure). The angle between the two forces is 45°. Find the direction and magnitude of the resultant of these forces.

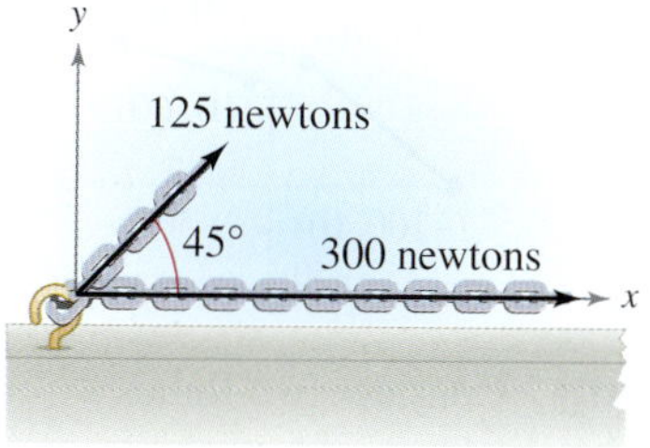

Figure for 85

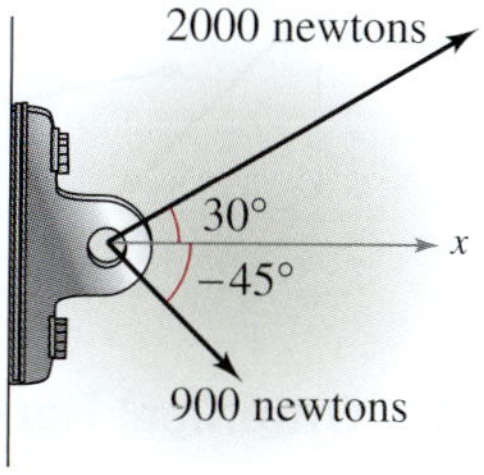

Figure for 86

86. Resultant Force Forces with magnitudes of 2000 newtons and 900 newtons act on a machine part at angles of 30° and −45°, respectively, with the x-axis (see figure). Find the direction and magnitude of the resultant of these forces.

87. Resultant Force Three forces with magnitudes of 75 pounds, 100 pounds, and 125 pounds act on an object at angles of 30°, 45°, and 120°, respectively, with the positive x-axis. Find the direction and magnitude of the resultant of these forces.

88. Resultant Force Three forces with magnitudes of 70 pounds, 40 pounds, and 60 pounds act on an object at angles of −30°, 45°, and 135°, respectively, with the positive x-axis. Find the direction and magnitude of the resultant of these forces.

89. Cable Tension The cranes shown in the figure are lifting an object that weighs 20,240 pounds. Find the tension in the cable of each crane.

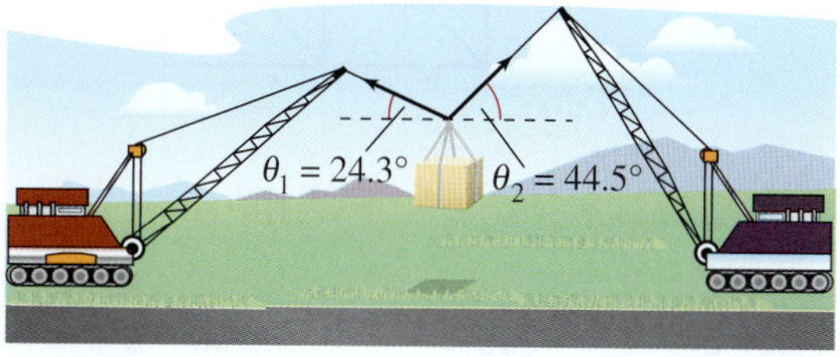

90. Cable Tension Repeat Exercise 89 for $\theta_1 = 35.6°$ and $\theta_2 = 40.4°$.

Cable Tension In Exercises 91 and 92, use the figure to determine the tension in each cable supporting the load.

91.

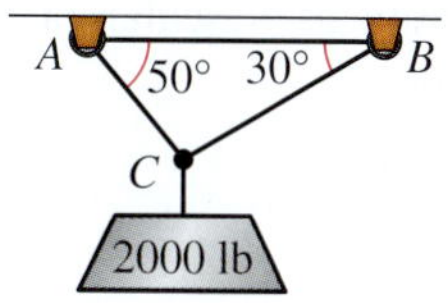

92.

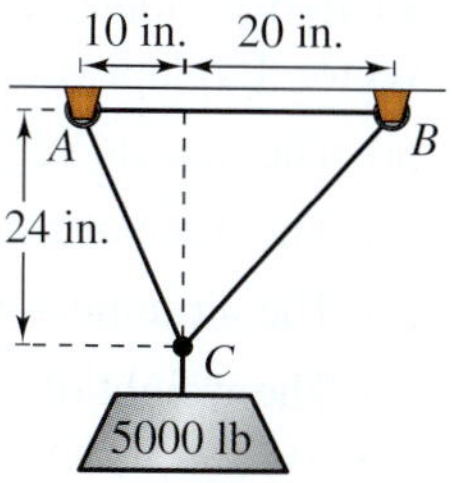

93. Tow Line Tension A loaded barge is being towed by two tugboats, and the magnitude of the resultant is 6000 pounds directed along the axis of the barge (see figure). Find the tension in the tow lines when they each make an 18° angle with the axis of the barge.

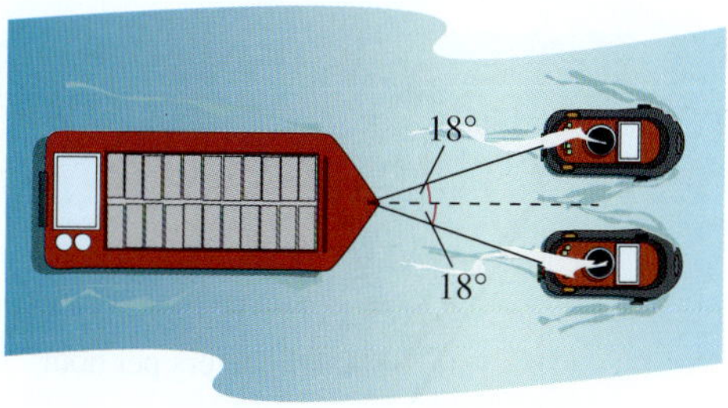

94. Rope Tension To carry a 100-pound cylindrical weight, two people lift on the ends of short ropes that are tied to an eyelet on the top center of the cylinder. Each rope makes a 20° angle with the vertical. Draw a figure that gives a visual representation of the situation. Then find the tension in the ropes.

Inclined Ramp In Exercises 95–98, a force of F pounds is required to pull an object weighing W pounds up a ramp inclined at θ degrees from the horizontal.

95. Find F when $W = 100$ pounds and $\theta = 12°$.

96. Find W when $F = 600$ pounds and $\theta = 14°$.

97. Find θ when $F = 5000$ pounds and $W = 15{,}000$ pounds.

98. Find F when $W = 5000$ pounds and $\theta = 26°$.

99. Work A heavy object is pulled 30 feet across a floor, using a force of 100 pounds. The force is exerted at an angle of 50° above the horizontal (see figure). Find the work done. (Use the formula for work, $W = FD$, where F is the component of the force in the direction of motion and D is the distance.)

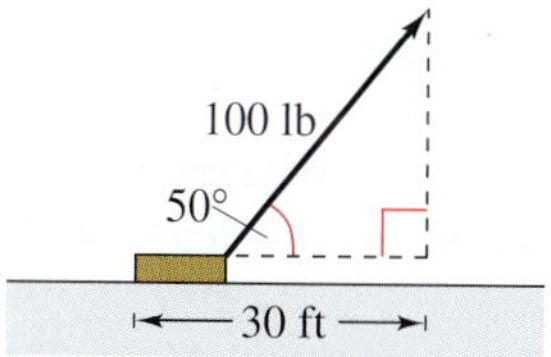

100. Rope Tension A tetherball weighing 1 pound is pulled outward from the pole by a horizontal force **u** until the rope makes a 45° angle with the pole (see figure). Determine the resulting tension in the rope and the magnitude of **u**.

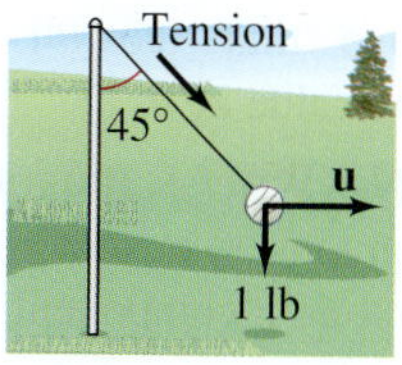

101. Navigation An airplane is flying in the direction of 148° with an airspeed of 875 kilometers per hour. Because of the wind, its groundspeed and direction are 800 kilometers per hour and 140°, respectively (see figure). Find the direction and speed of the wind.

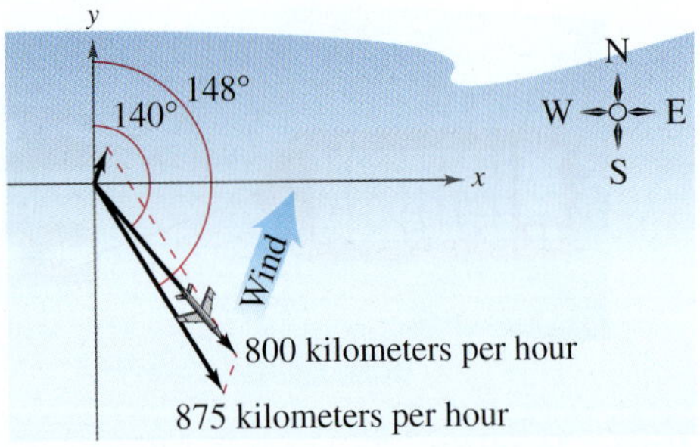

102. Navigation

A commercial jet is flying from Miami to Seattle. The jet's velocity with respect to the air is 580 miles per hour, and its bearing is 332°. The wind, at the altitude of the plane, is blowing from the southwest with a velocity of 60 miles per hour.

(a) Draw a figure that gives a visual representation of the situation.

(b) Write the velocity of the wind as a vector in component form.

(c) Write the velocity of the jet relative to the air in component form.

(d) What is the speed of the jet with respect to the ground?

(e) What is the true direction of the jet?

Exploration

True or False? **In Exercises 103–106, determine whether the statement is true or false. Justify your answer.**

103. If $\mathbf{u}$ and $\mathbf{v}$ have the same magnitude and direction, then $\mathbf{u}$ and $\mathbf{v}$ are equivalent.

104. If $\mathbf{u}$ is a unit vector in the direction of $\mathbf{v}$, then $\mathbf{v} = \|\mathbf{v}\|\mathbf{u}$.

105. If $\mathbf{v} = a\mathbf{i} + b\mathbf{j} = \mathbf{0}$, then $a = -b$.

106. If $\mathbf{u} = a\mathbf{i} + b\mathbf{j}$ is a unit vector, then $a^2 + b^2 = 1$.

107. Proof Prove that $(\cos\theta)\mathbf{i} + (\sin\theta)\mathbf{j}$ is a unit vector for any value of θ.

108. Technology Write a program for your graphing utility that graphs two vectors and their difference given the vectors in component form.

Finding the Difference of Two Vectors **In Exercises 109 and 110, use the program in Exercise 108 to find the difference of the vectors shown in the figure.**

109.

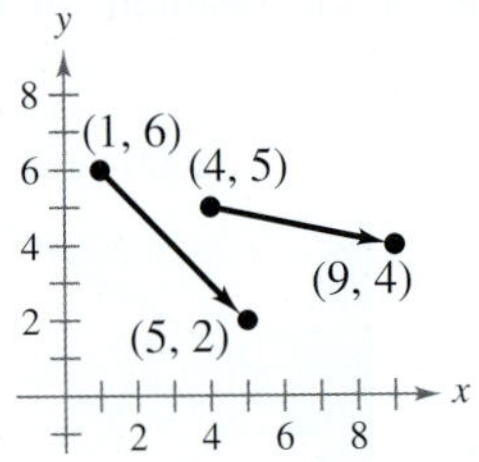

110.

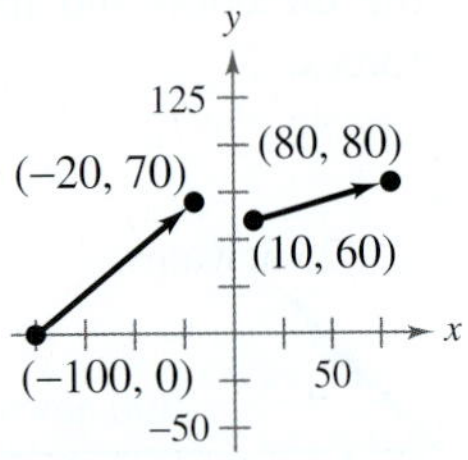

111. Graphical Reasoning Consider two forces

$\mathbf{F}_1 = \langle 10, 0\rangle$ and $\mathbf{F}_2 = 5\langle\cos\theta, \sin\theta\rangle$.

(a) Find $\|\mathbf{F}_1 + \mathbf{F}_2\|$ as a function of θ.

(b) Use a graphing utility to graph the function in part (a) for $0 \le \theta < 2\pi$.

(c) Use the graph in part (b) to determine the range of the function. What is its maximum, and for what value of θ does it occur? What is its minimum, and for what value of θ does it occur?

(d) Explain why the magnitude of the resultant is never 0.

112. HOW DO YOU SEE IT? Use the figure to determine whether each statement is true or false. Justify your answer.

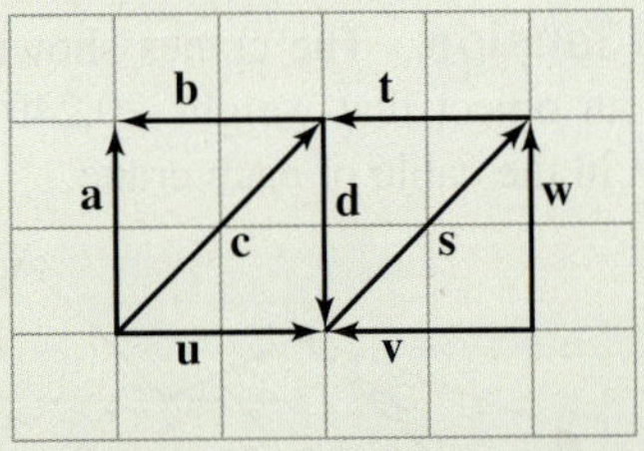

(a) $\mathbf{a} = -\mathbf{d}$

(b) $\mathbf{c} = \mathbf{s}$

(c) $\mathbf{a} + \mathbf{u} = \mathbf{c}$

(d) $\mathbf{v} + \mathbf{w} = -\mathbf{s}$

(e) $\mathbf{a} + \mathbf{w} = -2\mathbf{d}$

(f) $\mathbf{a} + \mathbf{d} = \mathbf{0}$

(g) $\mathbf{u} - \mathbf{v} = -2(\mathbf{b} + \mathbf{t})$

(h) $\mathbf{t} - \mathbf{w} = \mathbf{b} - \mathbf{a}$

113. Writing Give geometric descriptions of the operations of addition of vectors and multiplication of a vector by a scalar.

114. Writing Identify the quantity as a scalar or as a vector. Explain your reasoning.

(a) The muzzle velocity of a bullet

(b) The price of a company's stock

(c) The air temperature in a room

(d) The weight of an automobile

Bill Bachman/Photo Researchers, Inc.

6.4 Vectors and Dot Products

You can use the dot product of two vectors to solve real-life problems involving two vector quantities. For instance, in Exercise 76 on page 438, you will use the dot product to find the force necessary to keep a sport utility vehicle from rolling down a hill.

- **Find the dot product of two vectors and use the properties of the dot product.**
- **Find the angle between two vectors and determine whether two vectors are orthogonal.**
- **Write a vector as the sum of two vector components.**
- **Use vectors to find the work done by a force.**

The Dot Product of Two Vectors

So far you have studied two vector operations—vector addition and multiplication by a scalar—each of which yields another vector. In this section, you will study a third vector operation, the **dot product.** This operation yields a scalar, rather than a vector.

Definition of the Dot Product

The **dot product** of $\mathbf{u} = \langle u_1, u_2 \rangle$ and $\mathbf{v} = \langle v_1, v_2 \rangle$ is

$$\mathbf{u} \cdot \mathbf{v} = u_1v_1 + u_2v_2.$$

Properties of the Dot Product

Let $\mathbf{u}$, $\mathbf{v}$, and $\mathbf{w}$ be vectors in the plane or in space and let c be a scalar.

1. $\mathbf{u} \cdot \mathbf{v} = \mathbf{v} \cdot \mathbf{u}$
2. $\mathbf{0} \cdot \mathbf{v} = 0$
3. $\mathbf{u} \cdot (\mathbf{v} + \mathbf{w}) = \mathbf{u} \cdot \mathbf{v} + \mathbf{u} \cdot \mathbf{w}$
4. $\mathbf{v} \cdot \mathbf{v} = \|\mathbf{v}\|^2$
5. $c(\mathbf{u} \cdot \mathbf{v}) = c\mathbf{u} \cdot \mathbf{v} = \mathbf{u} \cdot c\mathbf{v}$

For proofs of the properties of the dot product, see Proofs in Mathematics on page 462.

REMARK In Example 1, be sure you see that the dot product of two vectors is a scalar (a real number), not a vector. Moreover, notice that the dot product can be positive, zero, or negative.

EXAMPLE 1 Finding Dot Products

a. $\langle 4, 5 \rangle \cdot \langle 2, 3 \rangle = 4(2) + 5(3)$

$= 8 + 15$

$= 23$

b. $\langle 2, -1 \rangle \cdot \langle 1, 2 \rangle = 2(1) + (-1)(2)$

$= 2 - 2$

$= 0$

c. $\langle 0, 3 \rangle \cdot \langle 4, -2 \rangle = 0(4) + 3(-2)$

$= 0 - 6$

$= -6$

✓ ***Checkpoint*** *Audio-video solution in English & Spanish at LarsonPrecalculus.com.*

Find the dot product of $\mathbf{u} = \langle 3, 4 \rangle$ and $\mathbf{v} = \langle 2, -3 \rangle$.

Anthony Berenyi/Shutterstock.com

EXAMPLE 2 **Using Properties of Dot Products**

Let $\mathbf{u} = \langle -1, 3 \rangle$, $\mathbf{v} = \langle 2, -4 \rangle$, and $\mathbf{w} = \langle 1, -2 \rangle$. Use the vectors and the properties of the dot product to find each quantity.

a. $(\mathbf{u} \cdot \mathbf{v})\mathbf{w}$ **b.** $\mathbf{u} \cdot 2\mathbf{v}$ **c.** $\|\mathbf{u}\|$

Solution Begin by finding the dot product of **u** and **v** and the dot product of **u** and **u**.

$$\mathbf{u} \cdot \mathbf{v} = \langle -1, 3 \rangle \cdot \langle 2, -4 \rangle = -1(2) + 3(-4) = -14$$

$$\mathbf{u} \cdot \mathbf{u} = \langle -1, 3 \rangle \cdot \langle -1, 3 \rangle = -1(-1) + 3(3) = 10$$

a. $(\mathbf{u} \cdot \mathbf{v})\mathbf{w} = -14\langle 1, -2 \rangle = \langle -14, 28 \rangle$

b. $\mathbf{u} \cdot 2\mathbf{v} = 2(\mathbf{u} \cdot \mathbf{v}) = 2(-14) = -28$

c. Because $\|\mathbf{u}\|^2 = \mathbf{u} \cdot \mathbf{u} = 10$, it follows that $\|\mathbf{u}\| = \sqrt{\mathbf{u} \cdot \mathbf{u}} = \sqrt{10}$.

In Example 2, notice that the product in part (a) is a vector, whereas the product in part (b) is a scalar. Can you see why?

✓ ***Checkpoint*** *Audio-video solution in English & Spanish at LarsonPrecalculus.com.*

Let $\mathbf{u} = \langle 3, 4 \rangle$ and $\mathbf{v} = \langle -2, 6 \rangle$. Use the vectors and the properties of the dot product to find each quantity.

a. $(\mathbf{u} \cdot \mathbf{v})\mathbf{v}$ **b.** $\mathbf{u} \cdot (\mathbf{u} + \mathbf{v})$ **c.** $\|\mathbf{v}\|$

The Angle Between Two Vectors

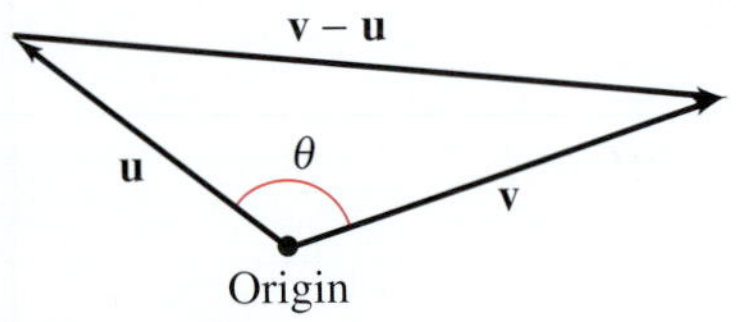

Figure 6.26

The **angle between two nonzero vectors** is the angle θ, $0 \le \theta \le \pi$, between their respective standard position vectors, as shown in Figure 6.26. This angle can be found using the dot product.

Angle Between Two Vectors

If θ is the angle between two nonzero vectors **u** and **v**, then

$$\cos \theta = \frac{\mathbf{u} \cdot \mathbf{v}}{\|\mathbf{u}\| \, \|\mathbf{v}\|}.$$

For a proof of the angle between two vectors, see Proofs in Mathematics on page 462.

EXAMPLE 3 **Finding the Angle Between Two Vectors**

Find the angle θ between $\mathbf{u} = \langle 4, 3 \rangle$ and $\mathbf{v} = \langle 3, 5 \rangle$ (see Figure 6.27).

Solution

$$\cos \theta = \frac{\mathbf{u} \cdot \mathbf{v}}{\|\mathbf{u}\| \, \|\mathbf{v}\|} = \frac{\langle 4, 3 \rangle \cdot \langle 3, 5 \rangle}{\|\langle 4, 3 \rangle\| \, \|\langle 3, 5 \rangle\|} = \frac{4(3) + 3(5)}{\sqrt{4^2 + 3^2}\sqrt{3^2 + 5^2}} = \frac{27}{5\sqrt{34}}$$

This implies that the angle between the two vectors is

$$\theta = \cos^{-1} \frac{27}{5\sqrt{34}} \approx 0.3869 \text{ radian.}$$ Use a calculator.

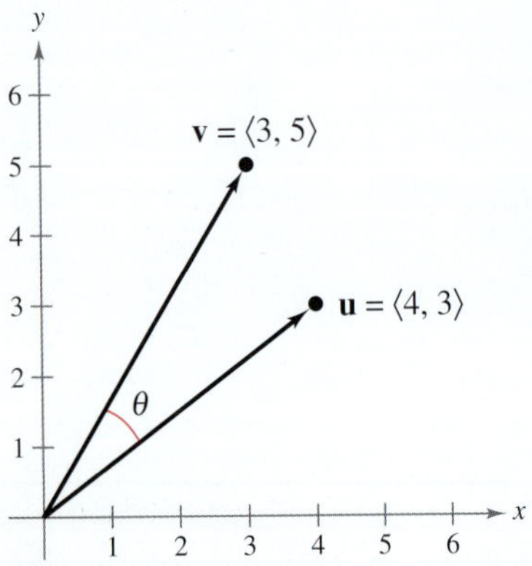

Figure 6.27

✓ ***Checkpoint*** *Audio-video solution in English & Spanish at LarsonPrecalculus.com.*

Find the angle θ between $\mathbf{u} = \langle 2, 1 \rangle$ and $\mathbf{v} = \langle 1, 3 \rangle$.

Rewriting the expression for the angle between two vectors in the form

$$\mathbf{u} \cdot \mathbf{v} = \|\mathbf{u}\| \, \|\mathbf{v}\| \cos \theta \qquad \text{Alternative form of dot product}$$

produces an alternative way to calculate the dot product. From this form, you can see that because $\|\mathbf{u}\|$ and $\|\mathbf{v}\|$ are always positive, $\mathbf{u} \cdot \mathbf{v}$ and $\cos \theta$ will always have the same sign. The five possible orientations of two vectors are shown below.

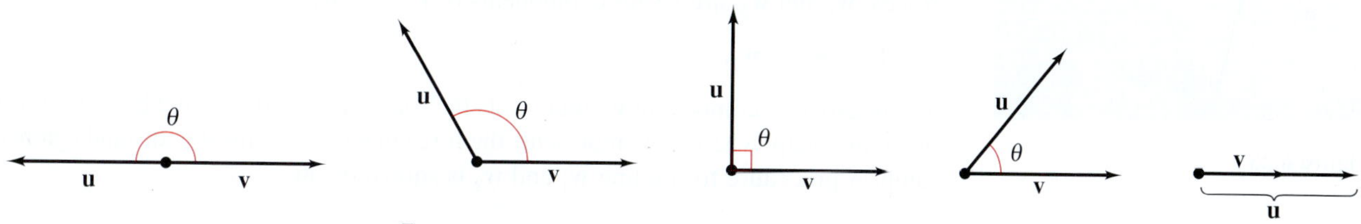

$\theta = \pi$	$\frac{\pi}{2} < \theta < \pi$	$\theta = \frac{\pi}{2}$	$0 < \theta < \frac{\pi}{2}$	$\theta = 0$
$\cos \theta = -1$	$-1 < \cos \theta < 0$	$\cos \theta = 0$	$0 < \cos \theta < 1$	$\cos \theta = 1$
Opposite direction	Obtuse angle	90° angle	Acute angle	Same direction

Definition of Orthogonal Vectors

The vectors **u** and **v** are **orthogonal** if and only if $\mathbf{u} \cdot \mathbf{v} = 0$.

The terms *orthogonal* and *perpendicular* mean essentially the same thing—meeting at right angles. Note that the zero vector is orthogonal to every vector **u**, because $\mathbf{0} \cdot \mathbf{u} = 0$.

▷ **TECHNOLOGY** The graphing utility program, Finding the Angle Between Two Vectors, found on the website for this text at *LarsonPrecalculus.com*, graphs two vectors $\mathbf{u} = \langle a, b \rangle$ and $\mathbf{v} = \langle c, d \rangle$ in standard position and finds the measure of the angle between them. Use the program to verify the solutions for Examples 3 and 4.

EXAMPLE 4 Determining Orthogonal Vectors

Are the vectors $\mathbf{u} = \langle 2, -3 \rangle$ and $\mathbf{v} = \langle 6, 4 \rangle$ orthogonal?

Solution Find the dot product of the two vectors.

$$\begin{aligned} \mathbf{u} \cdot \mathbf{v} &= \langle 2, -3 \rangle \cdot \langle 6, 4 \rangle \\ &= 2(6) + (-3)(4) \\ &= 0 \end{aligned}$$

Because the dot product is 0, the two vectors are orthogonal (see below).

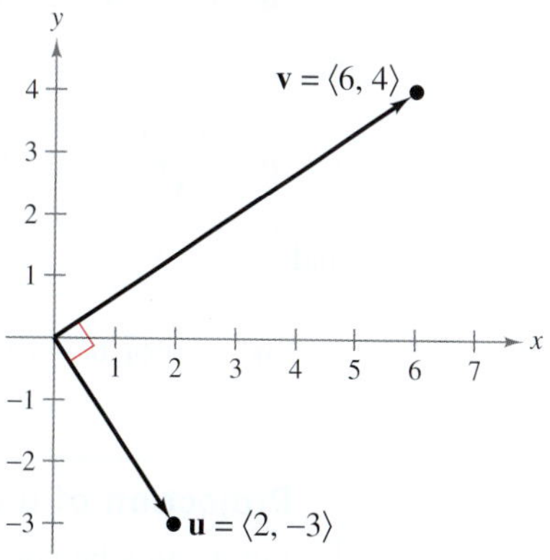

✓ **Checkpoint** *Audio-video solution in English & Spanish at LarsonPrecalculus.com.*

Are the vectors $\mathbf{u} = \langle 6, 10 \rangle$ and $\mathbf{v} = \left\langle -\frac{1}{3}, \frac{1}{5} \right\rangle$ orthogonal?

Finding Vector Components

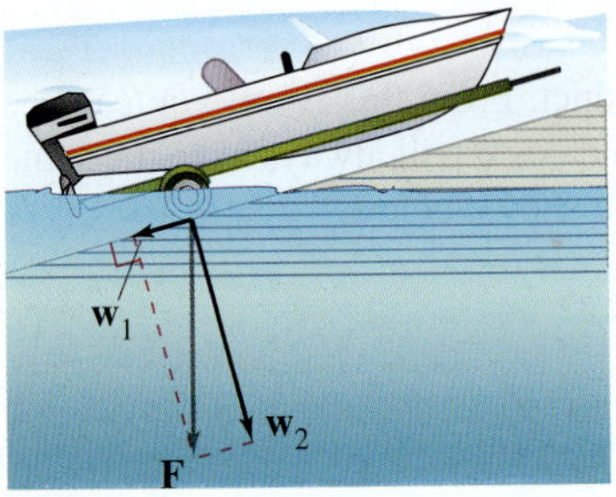

Figure 6.28

You have already seen applications in which two vectors are added to produce a resultant vector. Many applications in physics and engineering pose the reverse problem—decomposing a given vector into the sum of two **vector components.**

Consider a boat on an inclined ramp, as shown in Figure 6.28. The force $\mathbf{F}$ due to gravity pulls the boat *down* the ramp and *against* the ramp. These two orthogonal forces, $\mathbf{w}_1$ and $\mathbf{w}_2$, are vector components of $\mathbf{F}$. That is,

$$\mathbf{F} = \mathbf{w}_1 + \mathbf{w}_2. \qquad \text{Vector components of } \mathbf{F}$$

The negative of component $\mathbf{w}_1$ represents the force needed to keep the boat from rolling down the ramp, whereas $\mathbf{w}_2$ represents the force that the tires must withstand against the ramp. A procedure for finding $\mathbf{w}_1$ and $\mathbf{w}_2$ is shown below.

Definition of Vector Components

Let $\mathbf{u}$ and $\mathbf{v}$ be nonzero vectors such that

$$\mathbf{u} = \mathbf{w}_1 + \mathbf{w}_2$$

where $\mathbf{w}_1$ and $\mathbf{w}_2$ are orthogonal and $\mathbf{w}_1$ is parallel to (or a scalar multiple of) $\mathbf{v}$, as shown in Figure 6.29. The vectors $\mathbf{w}_1$ and $\mathbf{w}_2$ are called **vector components** of $\mathbf{u}$. The vector $\mathbf{w}_1$ is the **projection** of $\mathbf{u}$ onto $\mathbf{v}$ and is denoted by

$$\mathbf{w}_1 = \text{proj}_{\mathbf{v}}\mathbf{u}.$$

The vector $\mathbf{w}_2$ is given by

$$\mathbf{w}_2 = \mathbf{u} - \mathbf{w}_1.$$

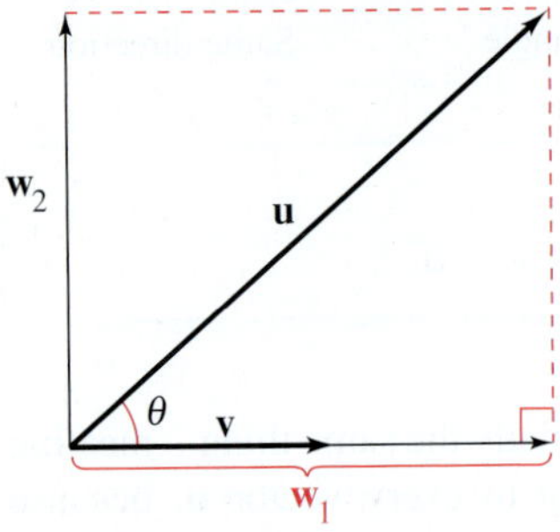

θ is acute.

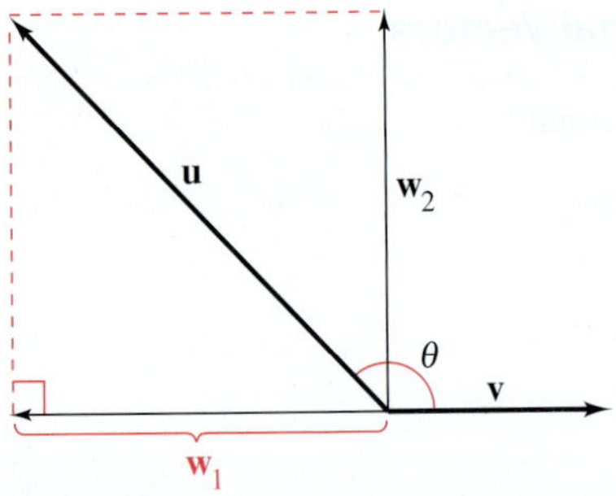

θ is obtuse.

Figure 6.29

From the definition of vector components, you can see that it is easy to find the component $\mathbf{w}_2$ once you have found the projection of $\mathbf{u}$ onto $\mathbf{v}$. To find the projection, you can use the dot product, as follows.

$$\mathbf{u} = \mathbf{w}_1 + \mathbf{w}_2$$

$$\mathbf{u} = c\mathbf{v} + \mathbf{w}_2 \qquad \mathbf{w}_1 \text{ is a scalar multiple of } \mathbf{v}.$$

$$\mathbf{u} \cdot \mathbf{v} = (c\mathbf{v} + \mathbf{w}_2) \cdot \mathbf{v} \qquad \text{Take dot product of each side with } \mathbf{v}.$$

$$\mathbf{u} \cdot \mathbf{v} = c\mathbf{v} \cdot \mathbf{v} + \mathbf{w}_2 \cdot \mathbf{v}$$

$$\mathbf{u} \cdot \mathbf{v} = c\|\mathbf{v}\|^2 + 0 \qquad \mathbf{w}_2 \text{ and } \mathbf{v} \text{ are orthogonal.}$$

So,

$$c = \frac{\mathbf{u} \cdot \mathbf{v}}{\|\mathbf{v}\|^2}$$

and

$$\mathbf{w}_1 = \text{proj}_{\mathbf{v}}\mathbf{u} = c\mathbf{v} = \left(\frac{\mathbf{u} \cdot \mathbf{v}}{\|\mathbf{v}\|^2}\right)\mathbf{v}.$$

Projection of u onto v

Let $\mathbf{u}$ and $\mathbf{v}$ be nonzero vectors. The projection of $\mathbf{u}$ onto $\mathbf{v}$ is

$$\text{proj}_{\mathbf{v}}\mathbf{u} = \left(\frac{\mathbf{u} \cdot \mathbf{v}}{\|\mathbf{v}\|^2}\right)\mathbf{v}.$$

EXAMPLE 5 Decomposing a Vector into Components

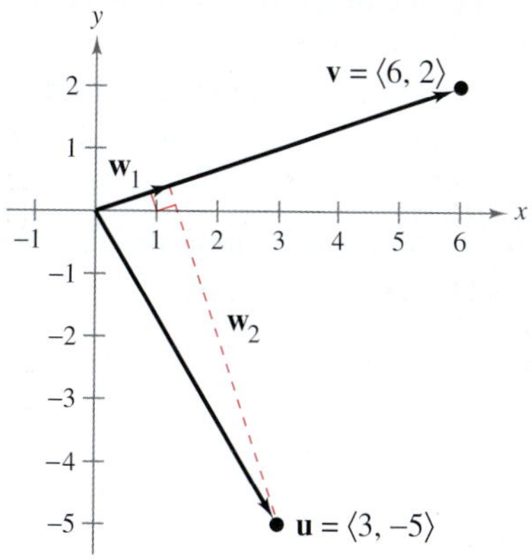

Figure 6.30

Find the projection of $\mathbf{u} = \langle 3, -5 \rangle$ onto $\mathbf{v} = \langle 6, 2 \rangle$. Then write $\mathbf{u}$ as the sum of two orthogonal vectors, one of which is $\text{proj}_{\mathbf{v}}\mathbf{u}$.

Solution The projection of $\mathbf{u}$ onto $\mathbf{v}$ is

$$\mathbf{w}_1 = \text{proj}_{\mathbf{v}}\mathbf{u} = \left(\frac{\mathbf{u} \cdot \mathbf{v}}{\|\mathbf{v}\|^2}\right)\mathbf{v} = \left(\frac{8}{40}\right)\langle 6, 2 \rangle = \left\langle \frac{6}{5}, \frac{2}{5} \right\rangle$$

as shown in Figure 6.30. The other component, $\mathbf{w}_2$, is

$$\mathbf{w}_2 = \mathbf{u} - \mathbf{w}_1 = \langle 3, -5 \rangle - \left\langle \frac{6}{5}, \frac{2}{5} \right\rangle = \left\langle \frac{9}{5}, -\frac{27}{5} \right\rangle.$$

So,

$$\mathbf{u} = \mathbf{w}_1 + \mathbf{w}_2 = \left\langle \frac{6}{5}, \frac{2}{5} \right\rangle + \left\langle \frac{9}{5}, -\frac{27}{5} \right\rangle = \langle 3, -5 \rangle.$$

✓ ***Checkpoint*** *Audio-video solution in English & Spanish at LarsonPrecalculus.com.*

Find the projection of $\mathbf{u} = \langle 3, 4 \rangle$ onto $\mathbf{v} = \langle 8, 2 \rangle$. Then write $\mathbf{u}$ as the sum of two orthogonal vectors, one of which is $\text{proj}_{\mathbf{v}}\mathbf{u}$.

EXAMPLE 6 Finding a Force

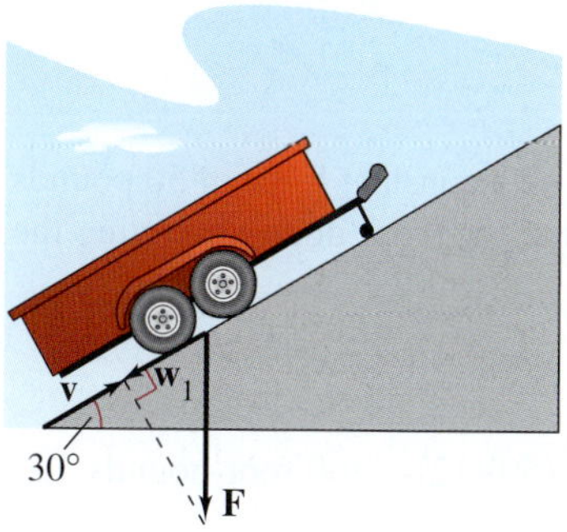

Figure 6.31

A 200-pound cart sits on a ramp inclined at 30°, as shown in Figure 6.31. What force is required to keep the cart from rolling down the ramp?

Solution Because the force due to gravity is vertical and downward, you can represent the gravitational force by the vector

$$\mathbf{F} = -200\mathbf{j}. \qquad \text{Force due to gravity}$$

To find the force required to keep the cart from rolling down the ramp, project $\mathbf{F}$ onto a unit vector $\mathbf{v}$ in the direction of the ramp, as follows.

$$\mathbf{v} = (\cos 30°)\mathbf{i} + (\sin 30°)\mathbf{j}$$
$$= \frac{\sqrt{3}}{2}\mathbf{i} + \frac{1}{2}\mathbf{j} \qquad \text{Unit vector along ramp}$$

So, the projection of $\mathbf{F}$ onto $\mathbf{v}$ is

$$\mathbf{w}_1 = \text{proj}_{\mathbf{v}}\mathbf{F}$$
$$= \left(\frac{\mathbf{F} \cdot \mathbf{v}}{\|\mathbf{v}\|^2}\right)\mathbf{v}$$
$$= (\mathbf{F} \cdot \mathbf{v})\mathbf{v}$$
$$= (-200)\left(\frac{1}{2}\right)\mathbf{v}$$
$$= -100\left(\frac{\sqrt{3}}{2}\mathbf{i} + \frac{1}{2}\mathbf{j}\right).$$

The magnitude of this force is 100. So, a force of 100 pounds is required to keep the cart from rolling down the ramp.

✓ ***Checkpoint*** *Audio-video solution in English & Spanish at LarsonPrecalculus.com.*

Rework Example 6 for a 150-pound cart sitting on a ramp inclined at 15°. ■

Work

The work W done by a *constant* force $\mathbf{F}$ acting along the line of motion of an object is given by

$$W = (\text{magnitude of force})(\text{distance})$$
$$= \|\mathbf{F}\| \, \|\overrightarrow{PQ}\|$$

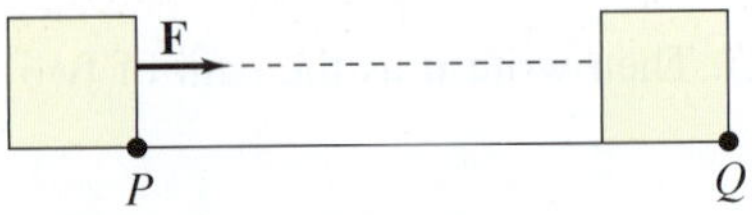

Force acts along the line of motion.

Figure 6.32

as shown in Figure 6.32. When the constant force $\mathbf{F}$ is not directed along the line of motion, as shown in Figure 6.33, the work W done by the force is given by

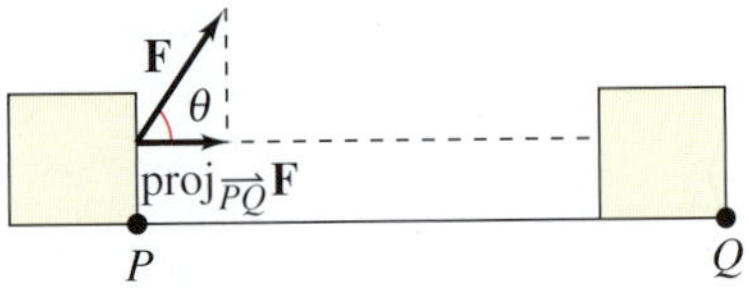

Force acts at angle θ with the line of motion.

Figure 6.33

$$W = \|\text{proj}_{\overrightarrow{PQ}} \mathbf{F}\| \, \|\overrightarrow{PQ}\| \qquad \text{Projection form for work}$$
$$= (\cos \theta)\|\mathbf{F}\| \, \|\overrightarrow{PQ}\| \qquad \|\text{proj}_{\overrightarrow{PQ}} \mathbf{F}\| = (\cos \theta)\|\mathbf{F}\|$$
$$= \mathbf{F} \cdot \overrightarrow{PQ}. \qquad \text{Dot product form for work}$$

This notion of work is summarized in the following definition.

Definition of Work

The **work** W done by a constant force $\mathbf{F}$ as its point of application moves along the vector $\overrightarrow{PQ}$ is given by either of the following.

1. $W = \|\text{proj}_{\overrightarrow{PQ}}\mathbf{F}\| \, \|\overrightarrow{PQ}\|$ — Projection form
2. $W = \mathbf{F} \cdot \overrightarrow{PQ}$ — Dot product form

EXAMPLE 7 Finding Work

To close a sliding barn door, a person pulls on a rope with a constant force of 50 pounds at a constant angle of 60°, as shown in Figure 6.34. Find the work done in moving the barn door 12 feet to its closed position.

Figure 6.34

Solution Using a projection, you can calculate the work as follows.

$$W = \|\text{proj}_{\overrightarrow{PQ}} \mathbf{F}\| \, \|\overrightarrow{PQ}\| = (\cos 60°)\|\mathbf{F}\| \, \|\overrightarrow{PQ}\| = \frac{1}{2}(50)(12) = 300 \text{ foot-pounds}$$

So, the work done is 300 foot-pounds. You can verify this result by finding the vectors $\mathbf{F}$ and $\overrightarrow{PQ}$ and calculating their dot product.

✓ ***Checkpoint*** *Audio-video solution in English & Spanish at LarsonPrecalculus.com.*

A wagon is pulled by exerting a force of 35 pounds on a handle that makes a 30° angle with the horizontal. Find the work done in pulling the wagon 40 feet.

Work is done only when an object is moved. It does not matter how much force is applied—if an object does not move, then no work has been done.

Summarize (Section 6.4)

1. State the definition of the dot product *(page 431)*. For examples of finding dot products and using the properties of the dot product, see Examples 1 and 2.
2. Describe how to find the angle between two vectors *(page 432)*. For examples involving the angle between two vectors, see Examples 3 and 4.
3. Describe how to decompose a vector into components *(page 434)*. For examples involving vector components, see Examples 5 and 6.
4. State the definition of work *(page 436)*. For an example of finding the work done by a constant force, see Example 7.

Vince Clements/Shutterstock.com

6.4 Exercises

See CalcChat.com for tutorial help and worked-out solutions to odd-numbered exercises.

Vocabulary: Fill in the blanks.

1. The ________ ________ of two vectors yields a scalar, rather than a vector.
2. The dot product of $\mathbf{u} = \langle u_1, u_2 \rangle$ and $\mathbf{v} = \langle v_1, v_2 \rangle$ is $\mathbf{u} \cdot \mathbf{v} =$ ________ .
3. If θ is the angle between two nonzero vectors $\mathbf{u}$ and $\mathbf{v}$, then $\cos \theta =$ ________ .
4. The vectors $\mathbf{u}$ and $\mathbf{v}$ are ________ when $\mathbf{u} \cdot \mathbf{v} = 0$.
5. The projection of $\mathbf{u}$ onto $\mathbf{v}$ is given by $\text{proj}_{\mathbf{v}}\mathbf{u} =$ ________ .
6. The work W done by a constant force $\mathbf{F}$ as its point of application moves along the vector $\overrightarrow{PQ}$ is given by $W =$ ________ or $W =$ ________ .

Skills and Applications

Finding a Dot Product In Exercises 7–14, find $\mathbf{u} \cdot \mathbf{v}$.

7. $\mathbf{u} = \langle 7, 1 \rangle$, $\mathbf{v} = \langle -3, 2 \rangle$
8. $\mathbf{u} = \langle 6, 10 \rangle$, $\mathbf{v} = \langle -2, 3 \rangle$
9. $\mathbf{u} = \langle -4, 1 \rangle$, $\mathbf{v} = \langle 2, -3 \rangle$
10. $\mathbf{u} = \langle -2, 5 \rangle$, $\mathbf{v} = \langle -1, -8 \rangle$
11. $\mathbf{u} = 4\mathbf{i} - 2\mathbf{j}$, $\mathbf{v} = \mathbf{i} - \mathbf{j}$
12. $\mathbf{u} = 3\mathbf{i} + 4\mathbf{j}$, $\mathbf{v} = 7\mathbf{i} - 2\mathbf{j}$
13. $\mathbf{u} = 3\mathbf{i} + 2\mathbf{j}$, $\mathbf{v} = -2\mathbf{i} - 3\mathbf{j}$
14. $\mathbf{u} = \mathbf{i} - 2\mathbf{j}$, $\mathbf{v} = -2\mathbf{i} + \mathbf{j}$

Using Properties of Dot Products In Exercises 15–24, use the vectors $\mathbf{u} = \langle 3, 3 \rangle$, $\mathbf{v} = \langle -4, 2 \rangle$, and $\mathbf{w} = \langle 3, -1 \rangle$ to find the indicated quantity. State whether the result is a vector or a scalar.

15. $\mathbf{u} \cdot \mathbf{u}$
16. $3\mathbf{u} \cdot \mathbf{v}$
17. $(\mathbf{u} \cdot \mathbf{v})\mathbf{v}$
18. $(\mathbf{v} \cdot \mathbf{u})\mathbf{w}$
19. $(3\mathbf{w} \cdot \mathbf{v})\mathbf{u}$
20. $(\mathbf{u} \cdot 2\mathbf{v})\mathbf{w}$
21. $\|\mathbf{w}\| - 1$
22. $2 - \|\mathbf{u}\|$
23. $(\mathbf{u} \cdot \mathbf{v}) - (\mathbf{u} \cdot \mathbf{w})$
24. $(\mathbf{v} \cdot \mathbf{u}) - (\mathbf{w} \cdot \mathbf{v})$

Finding the Magnitude of a Vector In Exercises 25–30, use the dot product to find the magnitude of $\mathbf{u}$.

25. $\mathbf{u} = \langle -8, 15 \rangle$
26. $\mathbf{u} = \langle 4, -6 \rangle$
27. $\mathbf{u} = 20\mathbf{i} + 25\mathbf{j}$
28. $\mathbf{u} = 12\mathbf{i} - 16\mathbf{j}$
29. $\mathbf{u} = 6\mathbf{j}$
30. $\mathbf{u} = -21\mathbf{i}$

Finding the Angle Between Two Vectors In Exercises 31–40, find the angle θ between the vectors.

31. $\mathbf{u} = \langle 1, 0 \rangle$, $\mathbf{v} = \langle 0, -2 \rangle$
32. $\mathbf{u} = \langle 3, 2 \rangle$, $\mathbf{v} = \langle 4, 0 \rangle$
33. $\mathbf{u} = 3\mathbf{i} + 4\mathbf{j}$, $\mathbf{v} = -2\mathbf{j}$
34. $\mathbf{u} = 2\mathbf{i} - 3\mathbf{j}$, $\mathbf{v} = \mathbf{i} - 2\mathbf{j}$
35. $\mathbf{u} = 2\mathbf{i} - \mathbf{j}$, $\mathbf{v} = 6\mathbf{i} + 4\mathbf{j}$
36. $\mathbf{u} = -6\mathbf{i} - 3\mathbf{j}$, $\mathbf{v} = -8\mathbf{i} + 4\mathbf{j}$
37. $\mathbf{u} = 5\mathbf{i} + 5\mathbf{j}$, $\mathbf{v} = -6\mathbf{i} + 6\mathbf{j}$
38. $\mathbf{u} = 2\mathbf{i} - 3\mathbf{j}$, $\mathbf{v} = 4\mathbf{i} + 3\mathbf{j}$
39. $\mathbf{u} = \cos\left(\frac{\pi}{3}\right)\mathbf{i} + \sin\left(\frac{\pi}{3}\right)\mathbf{j}$, $\mathbf{v} = \cos\left(\frac{3\pi}{4}\right)\mathbf{i} + \sin\left(\frac{3\pi}{4}\right)\mathbf{j}$
40. $\mathbf{u} = \cos\left(\frac{\pi}{4}\right)\mathbf{i} + \sin\left(\frac{\pi}{4}\right)\mathbf{j}$, $\mathbf{v} = \cos\left(\frac{\pi}{2}\right)\mathbf{i} + \sin\left(\frac{\pi}{2}\right)\mathbf{j}$

Finding the Angle Between Two Vectors In Exercises 41–44, graph the vectors and find the degree measure of the angle θ between the vectors.

41. $\mathbf{u} = 3\mathbf{i} + 4\mathbf{j}$, $\mathbf{v} = -7\mathbf{i} + 5\mathbf{j}$
42. $\mathbf{u} = 6\mathbf{i} + 3\mathbf{j}$, $\mathbf{v} = -4\mathbf{i} + 4\mathbf{j}$
43. $\mathbf{u} = 5\mathbf{i} + 5\mathbf{j}$, $\mathbf{v} = -8\mathbf{i} + 8\mathbf{j}$
44. $\mathbf{u} = 2\mathbf{i} - 3\mathbf{j}$, $\mathbf{v} = 8\mathbf{i} + 3\mathbf{j}$

Finding the Angles in a Triangle In Exercises 45–48, use vectors to find the interior angles of the triangle with the given vertices.

45. $(1, 2), (3, 4), (2, 5)$
46. $(-3, -4), (1, 7), (8, 2)$
47. $(-3, 0), (2, 2), (0, 6)$
48. $(-3, 5), (-1, 9), (7, 9)$

Using the Angle Between Two Vectors In Exercises 49–52, find $\mathbf{u} \cdot \mathbf{v}$, where θ is the angle between $\mathbf{u}$ and $\mathbf{v}$.

49. $\|\mathbf{u}\| = 4, \|\mathbf{v}\| = 10, \theta = \frac{2\pi}{3}$
50. $\|\mathbf{u}\| = 100, \|\mathbf{v}\| = 250, \theta = \frac{\pi}{6}$
51. $\|\mathbf{u}\| = 9, \|\mathbf{v}\| = 36, \theta = \frac{3\pi}{4}$
52. $\|\mathbf{u}\| = 4, \|\mathbf{v}\| = 12, \theta = \frac{\pi}{3}$

Determining Orthogonal Vectors **In Exercises 53–58, determine whether u and v are orthogonal.**

53. $\mathbf{u} = \langle -12, 30 \rangle$
$\mathbf{v} = \langle \frac{1}{2}, -\frac{5}{4} \rangle$

54. $\mathbf{u} = \langle 3, 15 \rangle$
$\mathbf{v} = \langle -1, 5 \rangle$

55. $\mathbf{u} = \frac{1}{4}(3\mathbf{i} - \mathbf{j})$
$\mathbf{v} = 5\mathbf{i} + 6\mathbf{j}$

56. $\mathbf{u} = \mathbf{i}$
$\mathbf{v} = -2\mathbf{i} + 2\mathbf{j}$

57. $\mathbf{u} = 2\mathbf{i} - 2\mathbf{j}$
$\mathbf{v} = -\mathbf{i} - \mathbf{j}$

58. $\mathbf{u} = \langle \cos\theta, \sin\theta \rangle$
$\mathbf{v} = \langle \sin\theta, -\cos\theta \rangle$

Decomposing a Vector into Components **In Exercises 59–62, find the projection of u onto v. Then write u as the sum of two orthogonal vectors, one of which is $\text{proj}_{\mathbf{v}}\mathbf{u}$.**

59. $\mathbf{u} = \langle 2, 2 \rangle$
$\mathbf{v} = \langle 6, 1 \rangle$

60. $\mathbf{u} = \langle 4, 2 \rangle$
$\mathbf{v} = \langle 1, -2 \rangle$

61. $\mathbf{u} = \langle 0, 3 \rangle$
$\mathbf{v} = \langle 2, 15 \rangle$

62. $\mathbf{u} = \langle -3, -2 \rangle$
$\mathbf{v} = \langle -4, -1 \rangle$

Finding the Projection of u onto v **In Exercises 63–66, use the graph to find the projection of u onto v. (The coordinates of the terminal points of the vectors in standard position are given.) Use the formula for the projection of u onto v to verify your result.**

63.
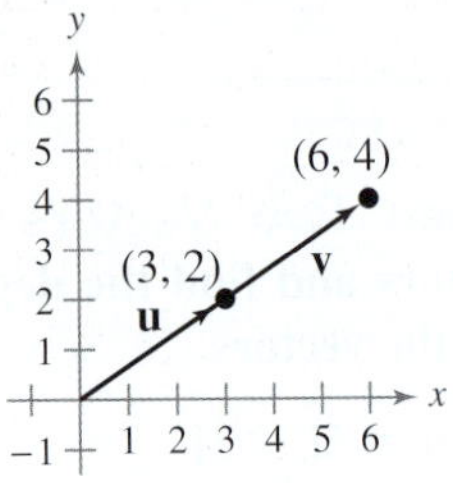

64.
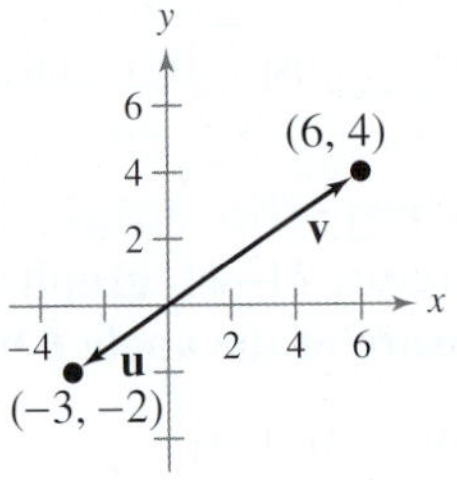

65.
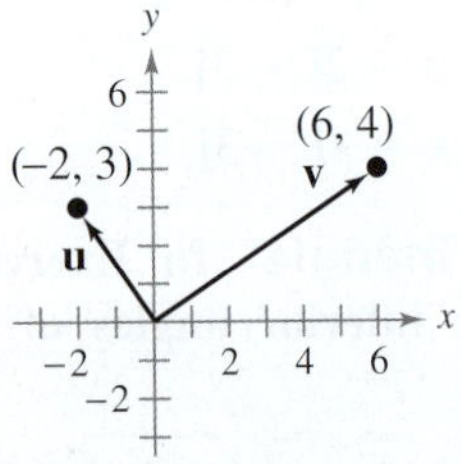

66.
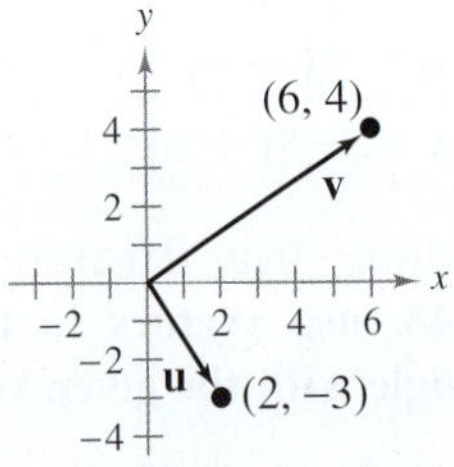

Finding Orthogonal Vectors **In Exercises 67–70, find two vectors in opposite directions that are orthogonal to the vector u. (There are many correct answers.)**

67. $\mathbf{u} = \langle 3, 5 \rangle$

68. $\mathbf{u} = \langle -8, 3 \rangle$

69. $\mathbf{u} = \frac{1}{2}\mathbf{i} - \frac{2}{3}\mathbf{j}$

70. $\mathbf{u} = -\frac{5}{2}\mathbf{i} - 3\mathbf{j}$

Work **In Exercises 71 and 72, find the work done in moving a particle from P to Q when the magnitude and direction of the force are given by v.**

71. $P(0, 0)$, $Q(4, 7)$, $\mathbf{v} = \langle 1, 4 \rangle$

72. $P(1, 3)$, $Q(-3, 5)$, $\mathbf{v} = -2\mathbf{i} + 3\mathbf{j}$

Anthony Berenyi/Shutterstock.com

73. **Business** The vector $\mathbf{u} = \langle 1225, 2445 \rangle$ gives the numbers of hours worked by employees of a temp agency at two pay levels. The vector $\mathbf{v} = \langle 12.20, 8.50 \rangle$ gives the hourly wage (in dollars) paid at each level, respectively.

(a) Find the dot product $\mathbf{u} \cdot \mathbf{v}$ and explain its meaning in the context of the problem.

(b) Identify the vector operation used to increase wages by 2%.

74. **Revenue** The vector $\mathbf{u} = \langle 3140, 2750 \rangle$ gives the numbers of hamburgers and hot dogs, respectively, sold at a fast-food stand in one month. The vector $\mathbf{v} = \langle 2.25, 1.75 \rangle$ gives the prices (in dollars) of the food items.

(a) Find the dot product $\mathbf{u} \cdot \mathbf{v}$ and interpret the result in the context of the problem.

(b) Identify the vector operation used to increase the prices by 2.5%.

75. **Braking Load** A truck with a gross weight of 30,000 pounds is parked on a slope of $d°$ (see figure). Assume that the only force to overcome is the force of gravity.

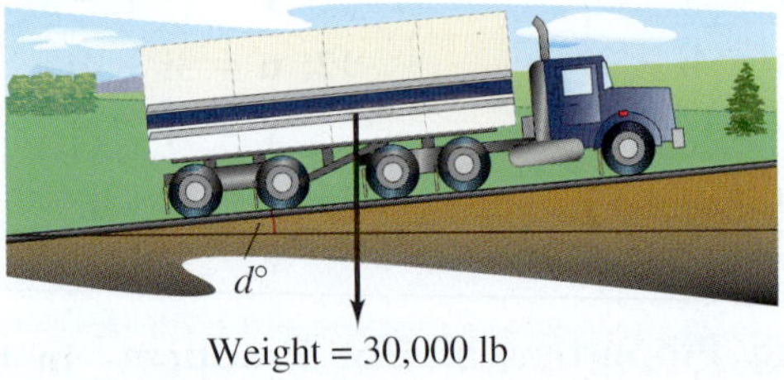

(a) Find the force required to keep the truck from rolling down the hill in terms of the slope d.

(b) Use a graphing utility to complete the table.

d	0°	1°	2°	3°	4°	5°
Force						

d	6°	7°	8°	9°	10°
Force					

(c) Find the force perpendicular to the hill when $d = 5°$.

76. **Braking Load**

A sport utility vehicle with a gross weight of 5400 pounds is parked on a slope of 10°. Assume that the only force to overcome is the force of gravity. Find the force required to keep the vehicle from rolling down the hill. Find the force perpendicular to the hill.

77. Work Determine the work done by a person lifting a 245-newton bag of sugar 3 meters.

78. Work Determine the work done by a crane lifting a 2400-pound car 5 feet.

79. Work A force of 45 pounds, exerted at an angle of 30° with the horizontal, is required to slide a table across a floor. Determine the work done in sliding the table 20 feet.

80. Work A force of 50 pounds, exerted at an angle of 25° with the horizontal, is required to slide a desk across a floor. Determine the work done in sliding the desk 15 feet.

81. Work A tractor pulls a log 800 meters, and the tension in the cable connecting the tractor and log is approximately 15,691 newtons. The direction of the force is 35° above the horizontal. Approximate the work done in pulling the log.

82. Work One of the events in a strength competition is to pull a cement block 100 feet. One competitor pulls the block by exerting a force of 250 pounds on a rope attached to the block at an angle of 30° with the horizontal (see figure). Find the work done in pulling the block.

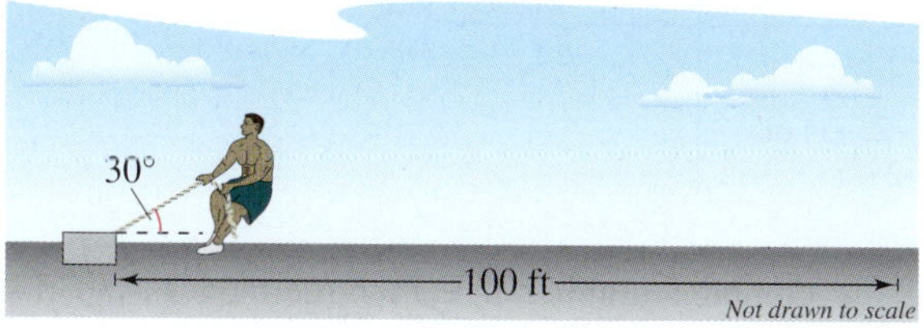

83. Work A toy wagon is pulled by exerting a force of 25 pounds on a handle that makes a 20° angle with the horizontal (see figure). Find the work done in pulling the wagon 50 feet.

84. Work A ski patroller pulls a rescue toboggan across a flat snow surface by exerting a force of 35 pounds on a handle that makes an angle of 22° with the horizontal. Find the work done in pulling the toboggan 200 feet.

85. Programming Given vectors **u** and **v** in component form, write a program for your graphing utility in which the output is (a) $\|\mathbf{u}\|$, (b) $\|\mathbf{v}\|$, and (c) the angle between **u** and **v**.

86. Programming Use the program you wrote in Exercise 85 to find the angle between the given vectors.

(a) $\mathbf{u} = \langle 8, -4 \rangle$ and $\mathbf{v} = \langle 2, 5 \rangle$

(b) $\mathbf{u} = \langle 2, -6 \rangle$ and $\mathbf{v} = \langle 4, 1 \rangle$

87. Programming Given vectors **u** and **v** in component form, write a program for your graphing utility in which the output is the component form of the projection of **u** onto **v**.

88. Programming Use the program you wrote in Exercise 87 to find the projection of **u** onto **v** for the given vectors.

(a) $\mathbf{u} = \langle 5, 6 \rangle$ and $\mathbf{v} = \langle -1, 3 \rangle$

(b) $\mathbf{u} = \langle 3, -2 \rangle$ and $\mathbf{v} = \langle -2, 1 \rangle$

Exploration

True or False? **In Exercises 89 and 90, determine whether the statement is true or false. Justify your answer.**

89. The work W done by a constant force **F** acting along the line of motion of an object is represented by a vector.

90. A sliding door moves along the line of vector $\overrightarrow{PQ}$. If a force is applied to the door along a vector that is orthogonal to $\overrightarrow{PQ}$, then no work is done.

91. Proof Use vectors to prove that the diagonals of a rhombus are perpendicular.

92. HOW DO YOU SEE IT? What is known about θ, the angle between two nonzero vectors **u** and **v**, under each condition (see figure)?

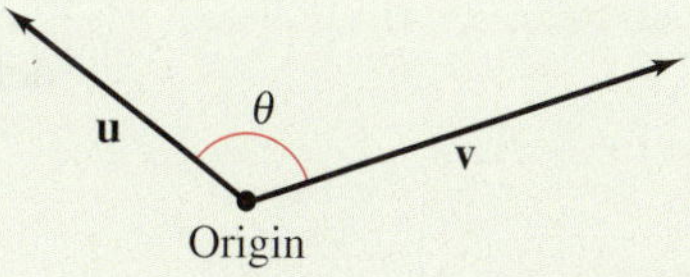

(a) $\mathbf{u} \cdot \mathbf{v} = 0$ (b) $\mathbf{u} \cdot \mathbf{v} > 0$ (c) $\mathbf{u} \cdot \mathbf{v} < 0$

93. Think About It What can be said about the vectors **u** and **v** under each condition?

(a) The projection of **u** onto **v** equals **u**.

(b) The projection of **u** onto **v** equals **0**.

94. Proof Prove the following.

$$\|\mathbf{u} - \mathbf{v}\|^2 = \|\mathbf{u}\|^2 + \|\mathbf{v}\|^2 - 2\mathbf{u} \cdot \mathbf{v}$$

6.5 Trigonometric Form of a Complex Number

You can use the trigonometric form of a complex number to perform operations with complex numbers. For instance, in Exercise 105 on page 449, you will use the trigonometric forms of complex numbers to find the voltage of an alternating current circuit.

- **Plot complex numbers in the complex plane and find absolute values of complex numbers.**
- **Write the trigonometric forms of complex numbers.**
- **Multiply and divide complex numbers written in trigonometric form.**
- **Use DeMoivre's Theorem to find powers of complex numbers.**
- **Find *n*th roots of complex numbers.**

The Complex Plane

Just as real numbers can be represented by points on the real number line, you can represent a complex number $z = a + bi$ as the point (a, b) in a coordinate plane (the **complex plane**). The horizontal axis is called the **real axis** and the vertical axis is called the **imaginary axis,** as shown below.

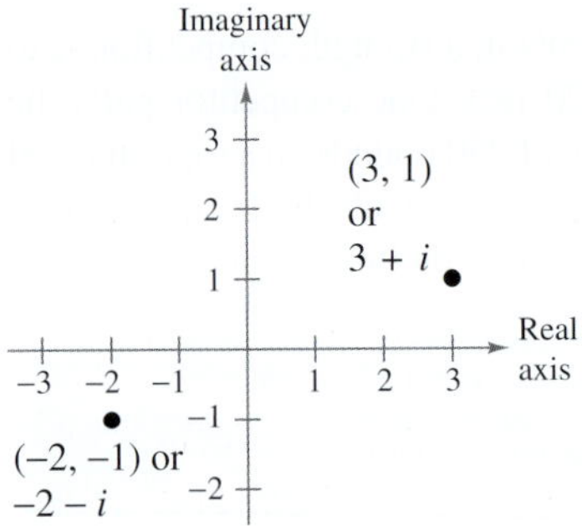

The **absolute value** of the complex number $a + bi$ is defined as the distance between the origin $(0, 0)$ and the point (a, b).

Definition of the Absolute Value of a Complex Number

The **absolute value** of the complex number $z = a + bi$ is

$$|a + bi| = \sqrt{a^2 + b^2}.$$

When the complex number $a + bi$ is a real number (that is, when $b = 0$), this definition agrees with that given for the absolute value of a real number

$$|a + 0i| = \sqrt{a^2 + 0^2}$$
$$= |a|.$$

EXAMPLE 1 **Finding the Absolute Value of a Complex Number**

Plot $z = -2 + 5i$ and find its absolute value.

Solution The number is plotted in Figure 6.35. It has an absolute value of

$$|z| = \sqrt{(-2)^2 + 5^2}$$
$$= \sqrt{29}.$$

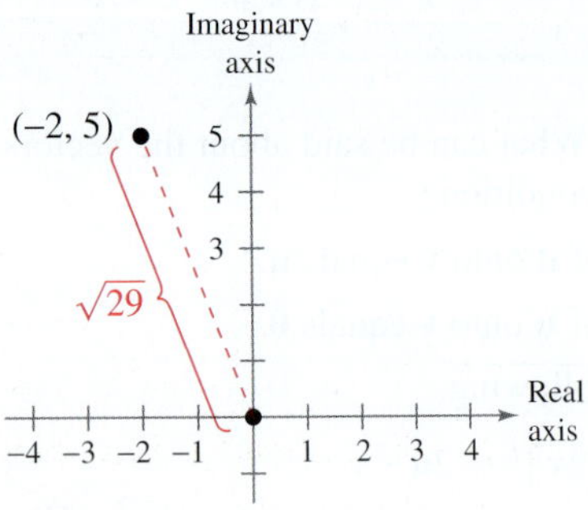

Figure 6.35

✓ ***Checkpoint*** *Audio-video solution in English & Spanish at LarsonPrecalculus.com.*

Plot $z = 3 - 4i$ and find its absolute value.

auremar/Shutterstock.com

Trigonometric Form of a Complex Number

In Section 2.4, you learned how to add, subtract, multiply, and divide complex numbers. To work effectively with *powers* and *roots* of complex numbers, it is helpful to write complex numbers in trigonometric form. In Figure 6.36, consider the nonzero complex number

$$a + bi.$$

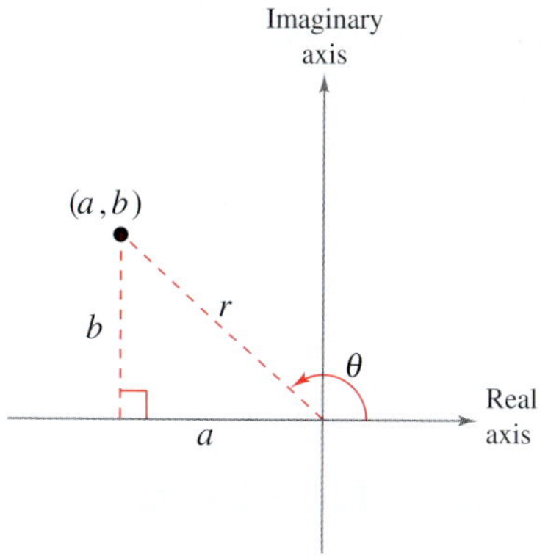

Figure 6.36

By letting θ be the angle from the positive real axis (measured counterclockwise) to the line segment connecting the origin and the point (a, b), you can write

$$a = r\cos\theta \quad \text{and} \quad b = r\sin\theta$$

where

$$r = \sqrt{a^2 + b^2}.$$

Consequently, you have

$$a + bi = (r\cos\theta) + (r\sin\theta)i$$

from which you can obtain the **trigonometric form of a complex number.**

> **Trigonometric Form of a Complex Number**
>
> The **trigonometric form** of the complex number $z = a + bi$ is
>
> $$z = r(\cos\theta + i\sin\theta)$$
>
> where $a = r\cos\theta$, $b = r\sin\theta$, $r = \sqrt{a^2 + b^2}$, and $\tan\theta = b/a$. The number r is the **modulus** of z, and θ is called an **argument** of z.

REMARK When θ is restricted to the interval $0 \le \theta < 2\pi$, use the following guidelines. When a complex number lies in Quadrant I, $\theta = \arctan(b/a)$. When a complex number lies in Quadrant II or Quadrant III, $\theta = \pi + \arctan(b/a)$. When a complex number lies in Quadrant IV, $\theta = 2\pi + \arctan(b/a)$.

The trigonometric form of a complex number is also called the *polar form*. Because there are infinitely many choices for θ, the trigonometric form of a complex number is not unique. Normally, θ is restricted to the interval $0 \le \theta < 2\pi$, although on occasion it is convenient to use $\theta < 0$.

EXAMPLE 2 Trigonometric Form of a Complex Number

Write the complex number $z = -2 - 2\sqrt{3}i$ in trigonometric form.

Solution The absolute value of z is

$$r = \left|-2 - 2\sqrt{3}i\right| = \sqrt{(-2)^2 + \left(-2\sqrt{3}\right)^2} = \sqrt{16} = 4$$

and the argument θ is determined from

$$\tan\theta = \frac{b}{a} = \frac{-2\sqrt{3}}{-2} = \sqrt{3}.$$

Because $z = -2 - 2\sqrt{3}i$ lies in Quadrant III, as shown in Figure 6.37,

$$\theta = \pi + \arctan\sqrt{3} = \pi + \frac{\pi}{3} = \frac{4\pi}{3}.$$

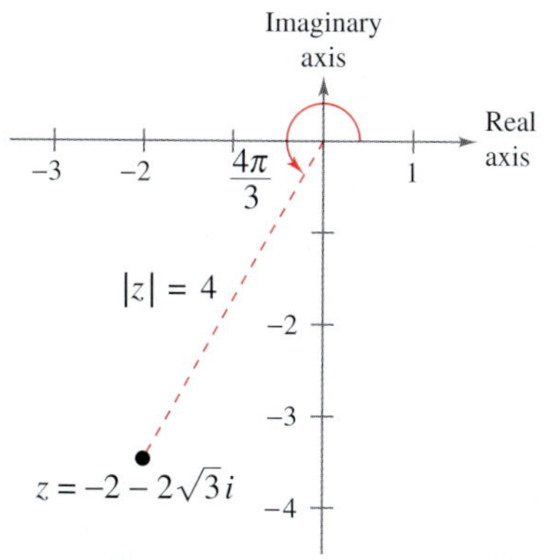

Figure 6.37

So, the trigonometric form is

$$z = r(\cos\theta + i\sin\theta) = 4\left(\cos\frac{4\pi}{3} + i\sin\frac{4\pi}{3}\right).$$

✔ **Checkpoint** *Audio-video solution in English & Spanish at LarsonPrecalculus.com.*

Write the complex number $z = 6 - 6i$ in trigonometric form.

EXAMPLE 3 **Writing a Complex Number in Standard Form**

Write $z = \sqrt{8}[\cos(-\pi/3) + i\sin(-\pi/3)]$ in standard form $a + bi$.

Solution Because $\cos(-\pi/3) = 1/2$ and $\sin(-\pi/3) = -\sqrt{3}/2$, you can write

$$z = \sqrt{8}\left[\cos\left(-\frac{\pi}{3}\right) + i\sin\left(-\frac{\pi}{3}\right)\right] = 2\sqrt{2}\left(\frac{1}{2} - \frac{\sqrt{3}}{2}i\right) = \sqrt{2} - \sqrt{6}i.$$

✓ Checkpoint *Audio-video solution in English & Spanish at LarsonPrecalculus.com.*

Write $z = 8[\cos(2\pi/3) + i\sin(2\pi/3)]$ in standard form $a + bi$.

TECHNOLOGY A graphing utility can be used to convert a complex number in trigonometric (or polar) form to standard form. For specific keystrokes, see the user's manual for your graphing utility.

Multiplication and Division of Complex Numbers

The trigonometric form adapts nicely to multiplication and division of complex numbers. Suppose you are given two complex numbers $z_1 = r_1(\cos\theta_1 + i\sin\theta_1)$ and $z_2 = r_2(\cos\theta_2 + i\sin\theta_2)$. The product of z_1 and z_2 is

$$\begin{aligned} z_1z_2 &= r_1r_2(\cos\theta_1 + i\sin\theta_1)(\cos\theta_2 + i\sin\theta_2) \\ &= r_1r_2[(\cos\theta_1\cos\theta_2 - \sin\theta_1\sin\theta_2) + i(\sin\theta_1\cos\theta_2 + \cos\theta_1\sin\theta_2)]. \end{aligned}$$

Using the sum and difference formulas for cosine and sine, you can rewrite this equation as

$$z_1z_2 = r_1r_2[\cos(\theta_1 + \theta_2) + i\sin(\theta_1 + \theta_2)].$$

This establishes the first part of the following rule. The second part is left for you to verify (see Exercise 108).

REMARK Note that this rule says that to *multiply* two complex numbers you multiply moduli and add arguments, whereas to *divide* two complex numbers you divide moduli and subtract arguments.

Product and Quotient of Two Complex Numbers

Let $z_1 = r_1(\cos\theta_1 + i\sin\theta_1)$ and $z_2 = r_2(\cos\theta_2 + i\sin\theta_2)$ be complex numbers.

$$z_1z_2 = r_1r_2[\cos(\theta_1 + \theta_2) + i\sin(\theta_1 + \theta_2)] \quad \text{Product}$$

$$\frac{z_1}{z_2} = \frac{r_1}{r_2}[\cos(\theta_1 - \theta_2) + i\sin(\theta_1 - \theta_2)], \quad z_2 \neq 0 \quad \text{Quotient}$$

EXAMPLE 4 **Multiplying Complex Numbers**

Find the product z_1z_2 of $z_1 = 3\left(\cos\frac{\pi}{4} + i\sin\frac{\pi}{4}\right)$ and $z_2 = 2\left(\cos\frac{3\pi}{4} + i\sin\frac{3\pi}{4}\right)$.

Solution

$$\begin{aligned} z_1z_2 &= 3\left(\cos\frac{\pi}{4} + i\sin\frac{\pi}{4}\right) \cdot 2\left(\cos\frac{3\pi}{4} + i\sin\frac{3\pi}{4}\right) \\ &= 6\left[\cos\left(\frac{\pi}{4} + \frac{3\pi}{4}\right) + i\sin\left(\frac{\pi}{4} + \frac{3\pi}{4}\right)\right] \\ &= 6(\cos\pi + i\sin\pi) \\ &= 6[-1 + i(0)] \\ &= -6 \end{aligned}$$

✓ Checkpoint *Audio-video solution in English & Spanish at LarsonPrecalculus.com.*

Find the product z_1z_2 of $z_1 = 2\left(\cos\frac{5\pi}{6} + i\sin\frac{5\pi}{6}\right)$ and $z_2 = 5\left(\cos\frac{7\pi}{6} + i\sin\frac{7\pi}{6}\right)$.

EXAMPLE 5 Multiplying Complex Numbers

Find the product z_1z_2 of the complex numbers.

$$z_1 = 2\left(\cos\frac{2\pi}{3} + i\sin\frac{2\pi}{3}\right) \qquad z_2 = 8\left(\cos\frac{11\pi}{6} + i\sin\frac{11\pi}{6}\right)$$

Solution

$$\begin{aligned}
z_1z_2 &= 2\left(\cos\frac{2\pi}{3} + i\sin\frac{2\pi}{3}\right) \cdot 8\left(\cos\frac{11\pi}{6} + i\sin\frac{11\pi}{6}\right) \\
&= 16\left[\cos\left(\frac{2\pi}{3} + \frac{11\pi}{6}\right) + i\sin\left(\frac{2\pi}{3} + \frac{11\pi}{6}\right)\right] && \text{Multiply moduli and add arguments.} \\
&= 16\left(\cos\frac{5\pi}{2} + i\sin\frac{5\pi}{2}\right) \\
&= 16\left(\cos\frac{\pi}{2} + i\sin\frac{\pi}{2}\right) && \frac{5\pi}{2} \text{ and } \frac{\pi}{2} \text{ are coterminal.} \\
&= 16[0 + i(1)] \\
&= 16i
\end{aligned}$$

You can check this result by first converting the complex numbers to the standard forms $z_1 = -1 + \sqrt{3}i$ and $z_2 = 4\sqrt{3} - 4i$ and then multiplying algebraically, as in Section 2.4.

$$\begin{aligned}
z_1z_2 &= \left(-1 + \sqrt{3}i\right)\left(4\sqrt{3} - 4i\right) \\
&= -4\sqrt{3} + 4i + 12i + 4\sqrt{3} \\
&= 16i
\end{aligned}$$

✓ ***Checkpoint*** *Audio-video solution in English & Spanish at LarsonPrecalculus.com.*

Find the product z_1z_2 of the complex numbers.

$$z_1 = 3\left(\cos\frac{\pi}{3} + i\sin\frac{\pi}{3}\right) \qquad z_2 = 4\left(\cos\frac{\pi}{6} + i\sin\frac{\pi}{6}\right)$$

▷ **TECHNOLOGY**
Some graphing utilities can multiply and divide complex numbers in trigonometric form. If you have access to such a graphing utility, use it to find z_1z_2 and z_1/z_2 in Examples 5 and 6.

EXAMPLE 6 Dividing Complex Numbers

Find the quotient z_1/z_2 of the complex numbers.

$$z_1 = 24(\cos 300^\circ + i\sin 300^\circ) \qquad z_2 = 8(\cos 75^\circ + i\sin 75^\circ)$$

Solution

$$\begin{aligned}
\frac{z_1}{z_2} &= \frac{24(\cos 300^\circ + i\sin 300^\circ)}{8(\cos 75^\circ + i\sin 75^\circ)} \\
&= 3[\cos(300^\circ - 75^\circ) + i\sin(300^\circ - 75^\circ)] && \text{Divide moduli and subtract arguments.} \\
&= 3(\cos 225^\circ + i\sin 225^\circ) \\
&= 3\left[-\frac{\sqrt{2}}{2} + i\left(-\frac{\sqrt{2}}{2}\right)\right] \\
&= -\frac{3\sqrt{2}}{2} - \frac{3\sqrt{2}}{2}i
\end{aligned}$$

✓ ***Checkpoint*** *Audio-video solution in English & Spanish at LarsonPrecalculus.com.*

Find the quotient z_1/z_2 of the complex numbers.

$$z_1 = \cos 40^\circ + i\sin 40^\circ \qquad z_2 = \cos 10^\circ + i\sin 10^\circ$$

Powers of Complex Numbers

The trigonometric form of a complex number is used to raise a complex number to a power. To accomplish this, consider repeated use of the multiplication rule.

$$z = r(\cos\theta + i\sin\theta)$$
$$z^2 = r(\cos\theta + i\sin\theta)r(\cos\theta + i\sin\theta) = r^2(\cos 2\theta + i\sin 2\theta)$$
$$z^3 = r^2(\cos 2\theta + i\sin 2\theta)r(\cos\theta + i\sin\theta) = r^3(\cos 3\theta + i\sin 3\theta)$$
$$z^4 = r^4(\cos 4\theta + i\sin 4\theta)$$
$$z^5 = r^5(\cos 5\theta + i\sin 5\theta)$$
$$\vdots$$

This pattern leads to **DeMoivre's Theorem,** which is named after the French mathematician Abraham DeMoivre (1667–1754).

Abraham DeMoivre (1667–1754) is remembered for his work in probability theory and DeMoivre's Theorem. His book *The Doctrine of Chances* (published in 1718) includes the theory of recurring series and the theory of partial fractions.

> **DeMoivre's Theorem**
>
> If $z = r(\cos\theta + i\sin\theta)$ is a complex number and n is a positive integer, then
>
> $$z^n = [r(\cos\theta + i\sin\theta)]^n = r^n(\cos n\theta + i\sin n\theta).$$

EXAMPLE 7 **Finding a Power of a Complex Number**

Use DeMoivre's Theorem to find $\left(-1 + \sqrt{3}i\right)^{12}$.

Solution The absolute value of $z = -1 + \sqrt{3}i$ is

$$r = \left|-1 + \sqrt{3}i\right| = \sqrt{(-1)^2 + \left(\sqrt{3}\right)^2} = 2$$

and the argument θ is given by $\tan\theta = \sqrt{3}/(-1)$. Because $z = -1 + \sqrt{3}i$ lies in Quadrant II,

$$\theta = \pi + \arctan\frac{\sqrt{3}}{-1} = \pi + \left(-\frac{\pi}{3}\right) = \frac{2\pi}{3}.$$

So, the trigonometric form is

$$z = -1 + \sqrt{3}i = 2\left(\cos\frac{2\pi}{3} + i\sin\frac{2\pi}{3}\right).$$

Then, by DeMoivre's Theorem, you have

$$\begin{aligned}\left(-1 + \sqrt{3}i\right)^{12} &= \left[2\left(\cos\frac{2\pi}{3} + i\sin\frac{2\pi}{3}\right)\right]^{12}\\ &= 2^{12}\left[\cos\frac{12(2\pi)}{3} + i\sin\frac{12(2\pi)}{3}\right]\\ &= 4096(\cos 8\pi + i\sin 8\pi)\\ &= 4096(1 + 0)\\ &= 4096.\end{aligned}$$

✓ **Checkpoint** *Audio-video solution in English & Spanish at LarsonPrecalculus.com.*

Use DeMoivre's Theorem to find $(-1 - i)^4$.

Alamy

Roots of Complex Numbers

Recall that a consequence of the Fundamental Theorem of Algebra is that a polynomial equation of degree n has n solutions in the complex number system. So, the equation $x^6 = 1$ has six solutions, and in this particular case you can find the six solutions by factoring and using the Quadratic Formula.

$$x^6 - 1 = 0$$
$$(x^3 - 1)(x^3 + 1) = 0$$
$$(x - 1)(x^2 + x + 1)(x + 1)(x^2 - x + 1) = 0$$

Consequently, the solutions are

$$x = \pm 1, \quad x = \frac{-1 \pm \sqrt{3}i}{2}, \quad \text{and} \quad x = \frac{1 \pm \sqrt{3}i}{2}.$$

Each of these numbers is a sixth root of 1. In general, an ***n*th root of a complex number** is defined as follows.

> **Definition of an *n*th Root of a Complex Number**
>
> The complex number $u = a + bi$ is an ***n*th root** of the complex number z when
>
> $$z = u^n$$
> $$= (a + bi)^n.$$

To find a formula for an nth root of a complex number, let u be an nth root of z, where

$$u = s(\cos \beta + i \sin \beta)$$

and

$$z = r(\cos \theta + i \sin \theta).$$

By DeMoivre's Theorem and the fact that $u^n = z$, you have

$$s^n(\cos n\beta + i \sin n\beta) = r(\cos \theta + i \sin \theta).$$

Taking the absolute value of each side of this equation, it follows that $s^n = r$. Substituting back into the previous equation and dividing by r, you get

$$\cos n\beta + i \sin n\beta = \cos \theta + i \sin \theta.$$

So, it follows that

$$\cos n\beta = \cos \theta$$

and

$$\sin n\beta = \sin \theta.$$

Because both sine and cosine have a period of 2π, these last two equations have solutions if and only if the angles differ by a multiple of 2π. Consequently, there must exist an integer k such that

$$n\beta = \theta + 2\pi k$$
$$\beta = \frac{\theta + 2\pi k}{n}.$$

By substituting this value of β into the trigonometric form of u, you get the result stated on the following page.

Finding nth Roots of a Complex Number

For a positive integer n, the complex number $z = r(\cos\theta + i\sin\theta)$ has exactly n distinct nth roots given by

$$z_k = \sqrt[n]{r}\left(\cos\frac{\theta + 2\pi k}{n} + i\sin\frac{\theta + 2\pi k}{n}\right)$$

where $k = 0, 1, 2, \ldots, n - 1$.

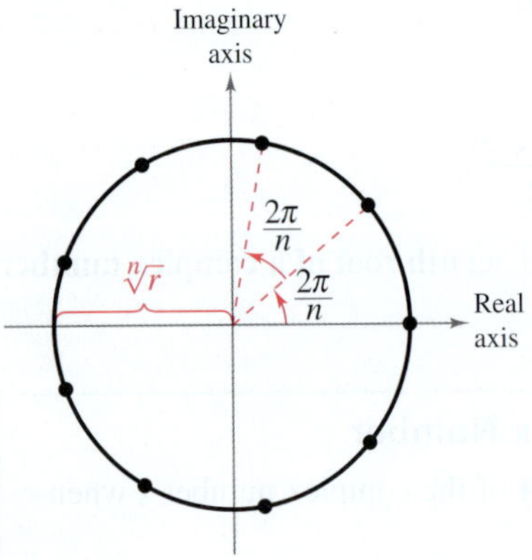

Figure 6.38

When $k > n - 1$, the roots begin to repeat. For instance, when $k = n$, the angle

$$\frac{\theta + 2\pi n}{n} = \frac{\theta}{n} + 2\pi$$

is coterminal with θ/n, which is also obtained when $k = 0$.

The formula for the nth roots of a complex number z has a nice geometrical interpretation, as shown in Figure 6.38. Note that because the nth roots of z all have the same magnitude $\sqrt[n]{r}$, they all lie on a circle of radius $\sqrt[n]{r}$ with center at the origin. Furthermore, because successive nth roots have arguments that differ by $2\pi/n$, the n roots are equally spaced around the circle.

You have already found the sixth roots of 1 by factoring and using the Quadratic Formula. Example 8 shows how you can solve the same problem with the formula for nth roots.

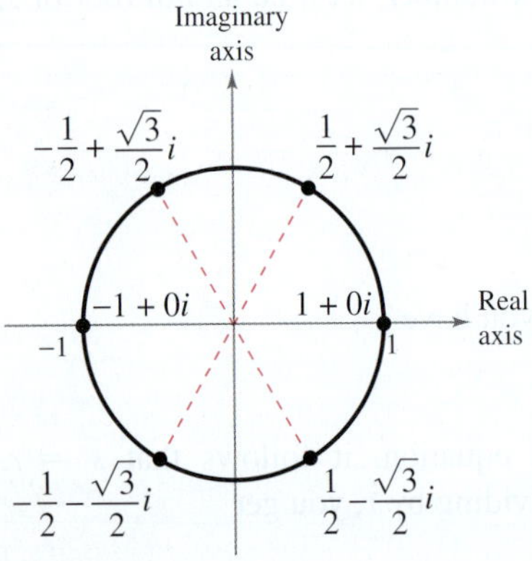

Figure 6.39

EXAMPLE 8 **Finding the nth Roots of a Real Number**

Find all sixth roots of 1.

Solution First, write 1 in the trigonometric form $z = 1(\cos 0 + i\sin 0)$. Then, by the nth root formula, with $n = 6$ and $r = 1$, the roots have the form

$$z_k = \sqrt[6]{1}\left(\cos\frac{0 + 2\pi k}{6} + i\sin\frac{0 + 2\pi k}{6}\right) = \cos\frac{\pi k}{3} + i\sin\frac{\pi k}{3}.$$

So, for $k = 0, 1, 2, 3, 4$, and 5, the sixth roots are as follows. (See Figure 6.39.)

$$z_0 = \cos 0 + i\sin 0 = 1$$

$$z_1 = \cos\frac{\pi}{3} + i\sin\frac{\pi}{3} = \frac{1}{2} + \frac{\sqrt{3}}{2}i$$ Increment by $\frac{2\pi}{n} = \frac{2\pi}{6} = \frac{\pi}{3}$

$$z_2 = \cos\frac{2\pi}{3} + i\sin\frac{2\pi}{3} = -\frac{1}{2} + \frac{\sqrt{3}}{2}i$$

$$z_3 = \cos\pi + i\sin\pi = -1$$

$$z_4 = \cos\frac{4\pi}{3} + i\sin\frac{4\pi}{3} = -\frac{1}{2} - \frac{\sqrt{3}}{2}i$$

$$z_5 = \cos\frac{5\pi}{3} + i\sin\frac{5\pi}{3} = \frac{1}{2} - \frac{\sqrt{3}}{2}i$$

✓ ***Checkpoint*** *Audio-video solution in English & Spanish at LarsonPrecalculus.com.*

Find all fourth roots of 1.

In Figure 6.39, notice that the roots obtained in Example 8 all have a magnitude of 1 and are equally spaced around the unit circle. Also notice that the complex roots occur in conjugate pairs, as discussed in Section 2.5. The n distinct nth roots of 1 are called the ***n*th roots of unity.**

EXAMPLE 9 Finding the nth Roots of a Complex Number

Find the three cube roots of $z = -2 + 2i$.

Solution The absolute value of z is

$$r = |-2 + 2i| = \sqrt{(-2)^2 + 2^2} = \sqrt{8}$$

and the argument θ is given by

$$\tan \theta = \frac{b}{a} = \frac{2}{-2} = -1.$$

Because z lies in Quadrant II, the trigonometric form of z is

$$z = -2 + 2i$$
$$= \sqrt{8}\,(\cos 135^\circ + i \sin 135^\circ). \qquad \theta = \pi + \arctan(-1) = 3\pi/4 = 135^\circ$$

By the formula for nth roots, the cube roots have the form

$$z_k = \sqrt[6]{8}\left(\cos \frac{135^\circ + 360^\circ k}{3} + i \sin \frac{135^\circ + 360^\circ k}{3}\right).$$

REMARK In Example 9, because $r = \sqrt{8}$, it follows that

$$\sqrt[n]{r} = \sqrt[3]{\sqrt{8}} = (8^{1/2})^{1/3} = 8^{1/6} = \sqrt[6]{8}.$$

Finally, for $k = 0$, 1, and 2, you obtain the roots

$$z_0 = \sqrt[6]{8}\left(\cos \frac{135^\circ + 360^\circ(0)}{3} + i \sin \frac{135^\circ + 360^\circ(0)}{3}\right) = \sqrt{2}(\cos 45^\circ + i \sin 45^\circ)$$
$$= 1 + i$$

$$z_1 = \sqrt[6]{8}\left(\cos \frac{135^\circ + 360^\circ(1)}{3} + i \sin \frac{135^\circ + 360^\circ(1)}{3}\right) = \sqrt{2}(\cos 165^\circ + i \sin 165^\circ)$$
$$\approx -1.3660 + 0.3660i$$

$$z_2 = \sqrt[6]{8}\left(\cos \frac{135^\circ + 360^\circ(2)}{3} + i \sin \frac{135^\circ + 360^\circ(2)}{3}\right) = \sqrt{2}(\cos 285^\circ + i \sin 285^\circ)$$
$$\approx 0.3660 - 1.3660i.$$

See Figure 6.40.

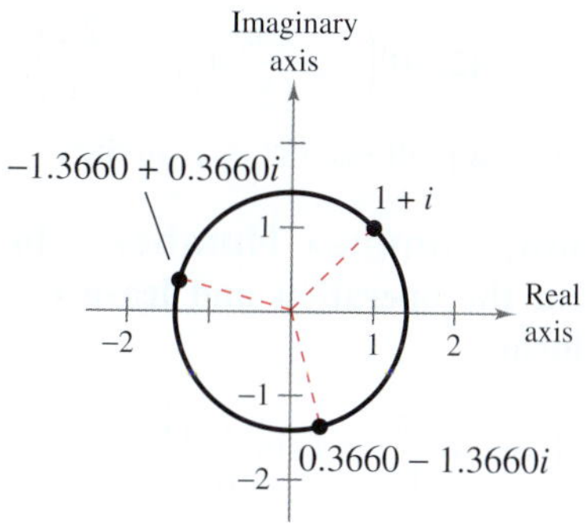

Figure 6.40

Checkpoint Audio-video solution in English & Spanish at LarsonPrecalculus.com.

Find the three cube roots of $z = -6 + 6i$.

Summarize (Section 6.5)

1. State the definition of the absolute value of a complex number *(page 440)*. For an example of finding the absolute value of a complex number, see Example 1.
2. State the definition of the trigonometric form of a complex number *(page 441)*. For examples of writing complex numbers in trigonometric form and standard form, see Examples 2 and 3.
3. Describe how to multiply and divide complex numbers written in trigonometric form *(page 442)*. For examples of multiplying and dividing complex numbers written in trigonometric form, see Examples 4–6.
4. State DeMoivre's Theorem *(page 444)*. For an example of using DeMoivre's Theorem to find a power of a complex number, see Example 7.
5. Describe how to find the nth roots of a complex number *(pages 445 and 446)*. For examples of finding nth roots, see Examples 8 and 9.

6.5 Exercises

See CalcChat.com for tutorial help and worked-out solutions to odd-numbered exercises.

Vocabulary: Fill in the blanks.

1. The ________ ________ of a complex number $a + bi$ is the distance between the origin $(0, 0)$ and the point (a, b).
2. The ________ ________ of a complex number $z = a + bi$ is given by $z = r(\cos\theta + i\sin\theta)$, where r is the ________ of z and θ is the ________ of z.
3. ________ Theorem states that if $z = r(\cos\theta + i\sin\theta)$ is a complex number and n is a positive integer, then $z^n = r^n(\cos n\theta + i\sin n\theta)$.
4. The complex number $u = a + bi$ is an ________ ________ of the complex number z when $z = u^n = (a + bi)^n$.

Skills and Applications

Finding the Absolute Value of a Complex Number In Exercises 5–10, plot the complex number and find its absolute value.

5. $-6 + 8i$
6. $5 - 12i$
7. $-7i$
8. -7
9. $4 - 6i$
10. $-8 + 3i$

Trigonometric Form of a Complex Number In Exercises 11–30, represent the complex number graphically. Then write the trigonometric form of the number.

11. $1 + i$
12. $5 - 5i$
13. $1 - \sqrt{3}i$
14. $4 - 4\sqrt{3}i$
15. $-2(1 + \sqrt{3}i)$
16. $\frac{5}{2}(\sqrt{3} - i)$
17. $-5i$
18. $12i$
19. $-7 + 4i$
20. $3 - i$
21. 2
22. 4
23. $2\sqrt{2} - i$
24. $-3 - i$
25. $5 + 2i$
26. $8 + 3i$
27. $3 + \sqrt{3}i$
28. $3\sqrt{2} - 7i$
29. $-8 - 5\sqrt{3}i$
30. $-9 - 2\sqrt{10}i$

Writing a Complex Number in Standard Form In Exercises 31–40, write the standard form of the complex number. Then represent the complex number graphically.

31. $2(\cos 60° + i\sin 60°)$
32. $5(\cos 135° + i\sin 135°)$
33. $\sqrt{48}[\cos(-30°) + i\sin(-30°)]$
34. $\sqrt{8}(\cos 225° + i\sin 225°)$
35. $\frac{9}{4}\left(\cos\frac{3\pi}{4} + i\sin\frac{3\pi}{4}\right)$
36. $6\left(\cos\frac{5\pi}{12} + i\sin\frac{5\pi}{12}\right)$
37. $7(\cos 0 + i\sin 0)$
38. $8\left(\cos\frac{\pi}{2} + i\sin\frac{\pi}{2}\right)$
39. $5[\cos(198° 45') + i\sin(198° 45')]$
40. $9.75[\cos(280° 30') + i\sin(280° 30')]$

Writing a Complex Number in Standard Form In Exercises 41–44, use a graphing utility to write the complex number in standard form.

41. $5\left(\cos\frac{\pi}{9} + i\sin\frac{\pi}{9}\right)$
42. $10\left(\cos\frac{2\pi}{5} + i\sin\frac{2\pi}{5}\right)$
43. $2(\cos 155° + i\sin 155°)$
44. $9(\cos 58° + i\sin 58°)$

Multiplying or Dividing Complex Numbers In Exercises 45–56, perform the operation and leave the result in trigonometric form.

45. $\left[2\left(\cos\frac{\pi}{4} + i\sin\frac{\pi}{4}\right)\right]\left[6\left(\cos\frac{\pi}{12} + i\sin\frac{\pi}{12}\right)\right]$
46. $\left[\frac{3}{4}\left(\cos\frac{\pi}{3} + i\sin\frac{\pi}{3}\right)\right]\left[4\left(\cos\frac{3\pi}{4} + i\sin\frac{3\pi}{4}\right)\right]$
47. $\left[\frac{5}{3}(\cos 120° + i\sin 120°)\right]\left[\frac{2}{3}(\cos 30° + i\sin 30°)\right]$
48. $\left[\frac{1}{2}(\cos 100° + i\sin 100°)\right]\left[\frac{4}{5}(\cos 300° + i\sin 300°)\right]$
49. $(\cos 80° + i\sin 80°)(\cos 330° + i\sin 330°)$
50. $(\cos 5° + i\sin 5°)(\cos 20° + i\sin 20°)$
51. $\dfrac{3(\cos 50° + i\sin 50°)}{9(\cos 20° + i\sin 20°)}$
52. $\dfrac{\cos 120° + i\sin 120°}{2(\cos 40° + i\sin 40°)}$
53. $\dfrac{\cos\pi + i\sin\pi}{\cos(\pi/3) + i\sin(\pi/3)}$
54. $\dfrac{5(\cos 4.3 + i\sin 4.3)}{4(\cos 2.1 + i\sin 2.1)}$
55. $\dfrac{12(\cos 92° + i\sin 92°)}{2(\cos 122° + i\sin 122°)}$
56. $\dfrac{6(\cos 40° + i\sin 40°)}{7(\cos 100° + i\sin 100°)}$

Multiplying or Dividing Complex Numbers In Exercises 57–62, (a) write the trigonometric forms of the complex numbers, (b) perform the indicated operation using the trigonometric forms, and (c) perform the indicated operation using the standard forms, and check your result with that of part (b).

57. $(2 + 2i)(1 - i)$
58. $(\sqrt{3} + i)(1 + i)$
59. $-2i(1 + i)$
60. $3i(1 - \sqrt{2}i)$
61. $\dfrac{3 + 4i}{1 - \sqrt{3}i}$
62. $\dfrac{1 + \sqrt{3}i}{6 - 3i}$

Graphing Powers of a Complex Number In Exercises 63 and 64, represent the powers z, z^2, z^3, and z^4 graphically. Describe the pattern.

63. $z = \frac{\sqrt{2}}{2}(1 + i)$ **64.** $z = \frac{1}{2}(1 + \sqrt{3}i)$

Finding a Power of a Complex Number In Exercises 65–80, use DeMoivre's Theorem to find the indicated power of the complex number. Write the result in standard form.

65. $(1 + i)^5$ **66.** $(2 + 2i)^6$

67. $(-1 + i)^6$ **68.** $(3 - 2i)^8$

69. $2(\sqrt{3} + i)^{10}$ **70.** $4(1 - \sqrt{3}i)^3$

71. $[5(\cos 20° + i \sin 20°)]^3$

72. $[3(\cos 60° + i \sin 60°)]^4$

73. $\left(\cos \frac{\pi}{4} + i \sin \frac{\pi}{4}\right)^{12}$ **74.** $\left[2\left(\cos \frac{\pi}{2} + i \sin \frac{\pi}{2}\right)\right]^8$

75. $[5(\cos 3.2 + i \sin 3.2)]^4$ **76.** $(\cos 0 + i \sin 0)^{20}$

77. $(3 - 2i)^5$ **78.** $(\sqrt{5} - 4i)^3$

79. $[3(\cos 15° + i \sin 15°)]^4$ **80.** $\left[2\left(\cos \frac{\pi}{8} + i \sin \frac{\pi}{8}\right)\right]^6$

Finding the *n*th Roots of a Complex Number In Exercises 81–96, (a) use the formula on page 446 to find the indicated roots of the complex number, (b) represent each of the roots graphically, and (c) write each of the roots in standard form.

81. Square roots of $5(\cos 120° + i \sin 120°)$

82. Square roots of $16(\cos 60° + i \sin 60°)$

83. Cube roots of $8\left(\cos \frac{2\pi}{3} + i \sin \frac{2\pi}{3}\right)$

84. Fifth roots of $32\left(\cos \frac{5\pi}{6} + i \sin \frac{5\pi}{6}\right)$

85. Cube roots of $-\frac{125}{2}(1 + \sqrt{3}i)$

86. Cube roots of $-4\sqrt{2}(-1 + i)$

87. Square roots of $-25i$ **88.** Fourth roots of $625i$

89. Fourth roots of 16 **90.** Fourth roots of i

91. Fifth roots of 1 **92.** Cube roots of 1000

93. Cube roots of -125 **94.** Fourth roots of -4

95. Fifth roots of $4(1 - i)$ **96.** Sixth roots of $64i$

Solving an Equation In Exercises 97–104, use the formula on page 446 to find all solutions of the equation and represent the solutions graphically.

97. $x^4 + i = 0$ **98.** $x^3 + 1 = 0$

99. $x^5 + 243 = 0$ **100.** $x^3 - 27 = 0$

101. $x^4 + 16i = 0$ **102.** $x^6 + 64i = 0$

103. $x^3 - (1 - i) = 0$ **104.** $x^4 + (1 + i) = 0$

auremar/Shutterstock.com

105. Electrical Engineering

Ohm's law for alternating current circuits is $E = I \cdot Z$, where E is the voltage in volts, I is the current in amperes, and Z is the impedance in ohms. Each variable is a complex number.

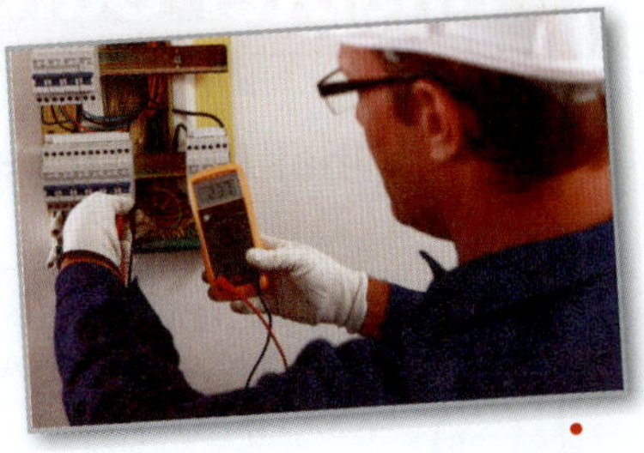

(a) Write E in trigonometric form when $I = 6(\cos 41° + i \sin 41°)$ amperes and $Z = 4[\cos(-11°) + i \sin(-11°)]$ ohms.

(b) Write the voltage from part (a) in standard form.

(c) A voltmeter measures the magnitude of the voltage in a circuit. What would be the reading on a voltmeter for the circuit described in part (a)?

Exploration

True or False? In Exercises 106 and 107, determine whether the statement is true or false. Justify your answer.

106. Geometrically, the nth roots of any complex number z are all equally spaced around the unit circle.

107. The product of two complex numbers is zero only when the modulus of one (or both) of the complex numbers is zero.

108. Quotient of Two Complex Numbers Given two complex numbers $z_1 = r_1(\cos \theta_1 + i \sin \theta_1)$ and $z_2 = r_2(\cos \theta_2 + i \sin \theta_2)$, $z_2 \neq 0$, show that

$$\frac{z_1}{z_2} = \frac{r_1}{r_2}[\cos(\theta_1 - \theta_2) + i \sin(\theta_1 - \theta_2)].$$

109. Reasoning Show that $\bar{z} = r[\cos(-\theta) + i \sin(-\theta)]$ is the complex conjugate of $z = r(\cos \theta + i \sin \theta)$.

110. Reasoning Use the trigonometric forms of z and $\bar{z}$ in Exercise 109 to find (a) $z\bar{z}$ and (b) $z/\bar{z}$, $\bar{z} \neq 0$.

111. Think About It Show that the negative of $z = r(\cos \theta + i \sin \theta)$ is

$$-z = r[\cos(\theta + \pi) + i \sin(\theta + \pi)].$$

112. HOW DO YOU SEE IT? One of the fourth roots of a complex number z is shown in the figure.

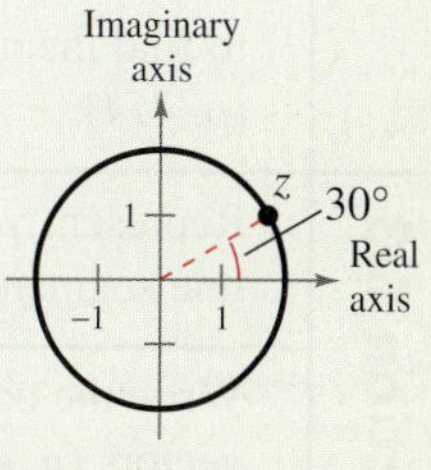

(a) How many roots are not shown?

(b) Describe the other roots.

Chapter Summary

	What Did You Learn?	Explanation/Examples	Review Exercises
Section 6.1	Use the Law of Sines to solve oblique triangles (AAS or ASA) *(p. 402).*	**Law of Sines** If ABC is a triangle with sides a, b, and c, then $\dfrac{a}{\sin A} = \dfrac{b}{\sin B} = \dfrac{c}{\sin C}.$ 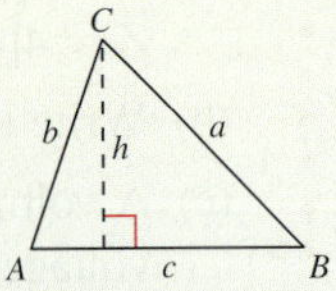 A is acute. 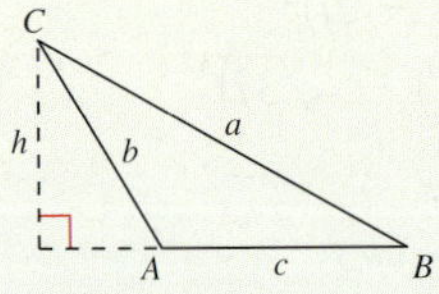 A is obtuse.	1–12
	Use the Law of Sines to solve oblique triangles (SSA) *(p. 404).*	If two sides and one opposite angle are given, then three possible situations can occur: (1) no such triangle exists (see Example 4), (2) one such triangle exists (see Example 3), or (3) two distinct triangles may satisfy the conditions (see Example 5).	1–12, 31–34
	Find the areas of oblique triangles *(p. 406).*	The area of any triangle is one-half the product of the lengths of two sides times the sine of their included angle. That is, Area $= \frac{1}{2}bc \sin A = \frac{1}{2}ab \sin C = \frac{1}{2}ac \sin B$.	13–16
	Use the Law of Sines to model and solve real-life problems *(p. 407).*	You can use the Law of Sines to approximate the total distance of a boat race course. (See Example 7.)	17–20
Section 6.2	Use the Law of Cosines to solve oblique triangles (SSS or SAS) *(p. 411).*	**Law of Cosines** **Standard Form** $a^2 = b^2 + c^2 - 2bc \cos A$ $b^2 = a^2 + c^2 - 2ac \cos B$ $c^2 = a^2 + b^2 - 2ab \cos C$ **Alternative Form** $\cos A = \dfrac{b^2 + c^2 - a^2}{2bc}$ $\cos B = \dfrac{a^2 + c^2 - b^2}{2ac}$ $\cos C = \dfrac{a^2 + b^2 - c^2}{2ab}$	21–34
	Use the Law of Cosines to model and solve real-life problems *(p. 413).*	You can use the Law of Cosines to find the distance between the pitcher's mound and first base on a women's softball field. (See Example 3.)	35–38
	Use Heron's Area Formula to find the area of a triangle *(p. 414).*	**Heron's Area Formula:** Given any triangle with sides of lengths a, b, and c, the area of the triangle is Area $= \sqrt{s(s-a)(s-b)(s-c)}$, where $s = (a + b + c)/2$.	39–42
Section 6.3	Represent vectors as directed line segments *(p. 418).*	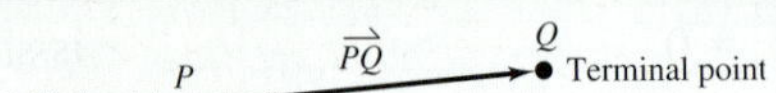	43, 44
	Write the component forms of vectors *(p. 419).*	The component form of the vector with initial point $P(p_1, p_2)$ and terminal point $Q(q_1, q_2)$ is given by $\overrightarrow{PQ} = \langle q_1 - p_1, q_2 - p_2 \rangle = \langle v_1, v_2 \rangle = \mathbf{v}$.	45–50

	What Did You Learn?	Explanation/Examples	Review Exercises
Section 6.3	Perform basic vector operations and represent them graphically *(p. 420)*.	Let $\mathbf{u} = \langle u_1, u_2 \rangle$ and $\mathbf{v} = \langle v_1, v_2 \rangle$ be vectors and let k be a scalar (a real number). $\mathbf{u} + \mathbf{v} = \langle u_1 + v_1, u_2 + v_2 \rangle$ $k\mathbf{u} = \langle ku_1, ku_2 \rangle$ $-\mathbf{v} = \langle -v_1, -v_2 \rangle$ $\mathbf{u} - \mathbf{v} = \langle u_1 - v_1, u_2 - v_2 \rangle$	51–64
	Write vectors as linear combinations of unit vectors *(p. 422)*.	The vector sum $\mathbf{v} = \langle v_1, v_2 \rangle = v_1\langle 1, 0 \rangle + v_2\langle 0, 1 \rangle = v_1\mathbf{i} + v_2\mathbf{j}$ is a linear combination of the vectors **i** and **j**.	65–68
	Find the direction angles of vectors *(p. 424)*.	If $\mathbf{u} = 2\mathbf{i} + 2\mathbf{j}$, then the direction angle is determined from $\tan\theta = 2/2 = 1$. So, $\theta = 45°$.	69–74
	Use vectors to model and solve real-life problems *(p. 425)*.	You can use vectors to find the resultant speed and direction of an airplane. (See Example 10.)	75–78
Section 6.4	Find the dot product of two vectors and use the properties of the dot product *(p. 431)*.	The dot product of $\mathbf{u} = \langle u_1, u_2 \rangle$ and $\mathbf{v} = \langle v_1, v_2 \rangle$ is $\mathbf{u} \cdot \mathbf{v} = u_1v_1 + u_2v_2$.	79–90
	Find the angle between two vectors and determine whether two vectors are orthogonal *(p. 432)*.	If θ is the angle between two nonzero vectors **u** and **v**, then $\cos\theta = \dfrac{\mathbf{u} \cdot \mathbf{v}}{\lVert\mathbf{u}\rVert\,\lVert\mathbf{v}\rVert}$. Vectors **u** and **v** are orthogonal if and only if $\mathbf{u} \cdot \mathbf{v} = 0$.	91–98
	Write a vector as the sum of two vector components *(p. 434)*.	Many applications in physics and engineering require the decomposition of a given vector into the sum of two vector components. (See Example 6.)	99–102
	Use vectors to find the work done by a force *(p. 436)*.	The work W done by a constant force **F** as its point of application moves along the vector $\overrightarrow{PQ}$ is given by either of the following. **1.** $W = \lVert\text{proj}_{\overrightarrow{PQ}}\mathbf{F}\rVert\,\lVert\overrightarrow{PQ}\rVert$ **2.** $W = \mathbf{F} \cdot \overrightarrow{PQ}$	103–106
Section 6.5	Plot complex numbers in the complex plane and find absolute values of complex numbers *(p. 440)*.	A complex number $z = a + bi$ can be represented as the point (a, b) in the complex plane. The horizontal axis is the real axis and the vertical axis is the imaginary axis. The absolute value of $z = a + bi$ is $\lvert a + bi \rvert = \sqrt{a^2 + b^2}$.	107–112
	Write the trigonometric forms of complex numbers *(p. 441)*.	The trigonometric form of the complex number $z = a + bi$ is $z = r(\cos\theta + i\sin\theta)$ where $a = r\cos\theta$, $b = r\sin\theta$, $r = \sqrt{a^2 + b^2}$, and $\tan\theta = b/a$.	113–118
	Multiply and divide complex numbers written in trigonometric form *(p. 442)*.	Let $z_1 = r_1(\cos\theta_1 + i\sin\theta_1)$ and $z_2 = r_2(\cos\theta_2 + i\sin\theta_2)$ be complex numbers. $z_1z_2 = r_1r_2[\cos(\theta_1 + \theta_2) + i\sin(\theta_1 + \theta_2)]$ $z_1/z_2 = (r_1/r_2)[\cos(\theta_1 - \theta_2) + i\sin(\theta_1 - \theta_2)],\quad z_2 \neq 0$	119, 120
	Use DeMoivre's Theorem to find powers of complex numbers *(p. 444)*.	**DeMoivre's Theorem:** If $z = r(\cos\theta + i\sin\theta)$ is a complex number and n is a positive integer, then $z^n = [r(\cos\theta + i\sin\theta)]^n = r^n(\cos n\theta + i\sin n\theta)$.	121–124
	Find nth roots of complex numbers *(p. 445)*.	The complex number $u = a + bi$ is an nth root of the complex number z when $z = u^n = (a + bi)^n$.	125–134

Review Exercises

See CalcChat.com for tutorial help and worked-out solutions to odd-numbered exercises.

6.1 Using the Law of Sines

In Exercises 1–12, use the Law of Sines to solve (if possible) the triangle. If two solutions exist, find both. Round your answers to two decimal places.

1.

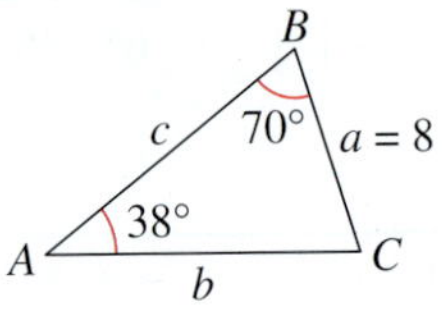

2.

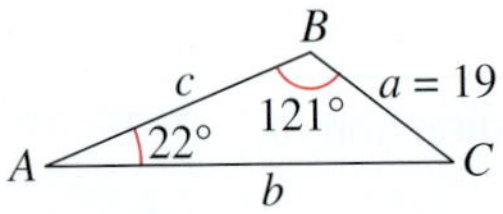

3. $B = 72°,\ C = 82°,\ b = 54$
4. $B = 10°,\ C = 20°,\ c = 33$
5. $A = 16°,\ B = 98°,\ c = 8.4$
6. $A = 95°,\ B = 45°,\ c = 104.8$
7. $A = 24°,\ C = 48°,\ b = 27.5$
8. $B = 64°,\ C = 36°,\ a = 367$
9. $B = 150°,\ b = 30,\ c = 10$
10. $B = 150°,\ a = 10,\ b = 3$
11. $A = 75°,\ a = 51.2,\ b = 33.7$
12. $B = 25°,\ a = 6.2,\ b = 4$

Finding the Area of a Triangle

In Exercises 13–16, find the area of the triangle having the indicated angle and sides.

13. $A = 33°,\ b = 7,\ c = 10$
14. $B = 80°,\ a = 4,\ c = 8$
15. $C = 119°,\ a = 18,\ b = 6$
16. $A = 11°,\ b = 22,\ c = 21$

17. **Height** From a certain distance, the angle of elevation to the top of a building is 17°. At a point 50 meters closer to the building, the angle of elevation is 31°. Approximate the height of the building.

18. **Geometry** Find the length of the side w of the parallelogram.

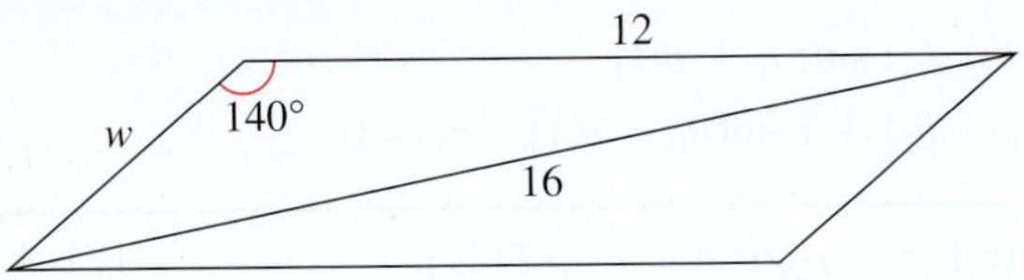

19. **Height** A tree stands on a hillside of slope 28° from the horizontal. From a point 75 feet down the hill, the angle of elevation to the top of the tree is 45° (see figure). Find the height of the tree.

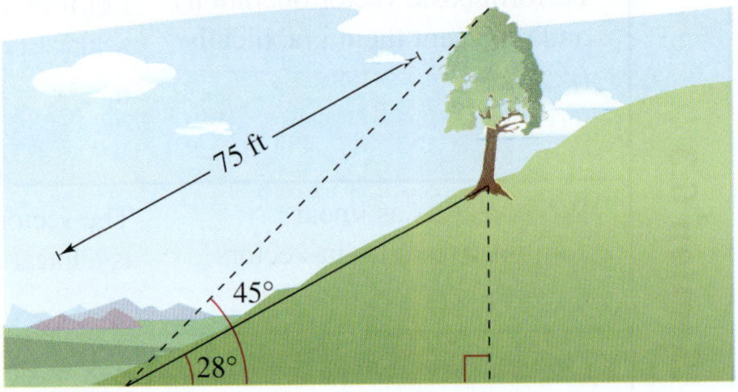

Figure for 19

20. **River Width** A surveyor finds that a tree on the opposite bank of a river flowing due east has a bearing of N 22° 30′ E from a certain point and a bearing of N 15° W from a point 400 feet downstream. Find the width of the river.

6.2 Using the Law of Cosines

In Exercises 21–30, use the Law of Cosines to solve the triangle. Round your answers to two decimal places.

21.

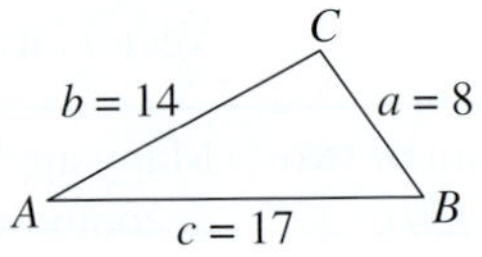

22.

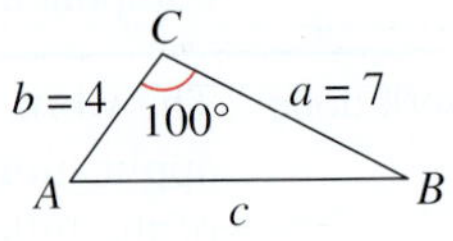

23. $a = 6,\ b = 9,\ c = 14$
24. $a = 75,\ b = 50,\ c = 110$
25. $a = 2.5,\ b = 5.0,\ c = 4.5$
26. $a = 16.4,\ b = 8.8,\ c = 12.2$
27. $B = 108°,\ a = 11,\ c = 11$
28. $B = 150°,\ a = 10,\ c = 20$
29. $C = 43°,\ a = 22.5,\ b = 31.4$
30. $A = 62°,\ b = 11.34,\ c = 19.52$

Solving a Triangle

In Exercises 31–34, determine whether the Law of Sines or the Law of Cosines is needed to solve the triangle. Then solve (if possible) the triangle. If two solutions exist, find both. Round your answers to two decimal places.

31. $b = 9,\ c = 13,\ C = 64°$
32. $a = 4,\ c = 5,\ B = 52°$
33. $a = 13,\ b = 15,\ c = 24$
34. $A = 44°,\ B = 31°,\ c = 2.8$

35. **Geometry** The lengths of the diagonals of a parallelogram are 10 feet and 16 feet. Find the lengths of the sides of the parallelogram when the diagonals intersect at an angle of 28°.

36. **Geometry** The lengths of the diagonals of a parallelogram are 30 meters and 40 meters. Find the lengths of the sides of the parallelogram when the diagonals intersect at an angle of 34°.

37. **Surveying** To approximate the length of a marsh, a surveyor walks 425 meters from point A to point B. Then the surveyor turns 65° and walks 300 meters to point C (see figure). Find the length AC of the marsh.

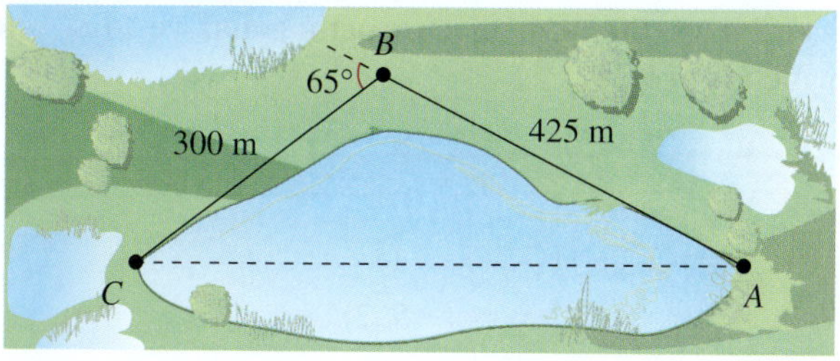

38. **Navigation** Two planes leave an airport at approximately the same time. One is flying 425 miles per hour at a bearing of 355°, and the other is flying 530 miles per hour at a bearing of 67°. Draw a figure that gives a visual representation of the situation and determine the distance between the planes after they have flown for 2 hours.

Using Heron's Area Formula **In Exercises 39–42, use Heron's Area Formula to find the area of the triangle.**

39. $a = 3,\ b = 6,\ c = 8$

40. $a = 15,\ b = 8,\ c = 10$

41. $a = 12.3,\ b = 15.8,\ c = 3.7$

42. $a = \frac{4}{5},\ b = \frac{3}{4},\ c = \frac{5}{8}$

6.3 Showing That Two Vectors Are Equivalent

In Exercises 43 and 44, show that u and v are equivalent.

43.

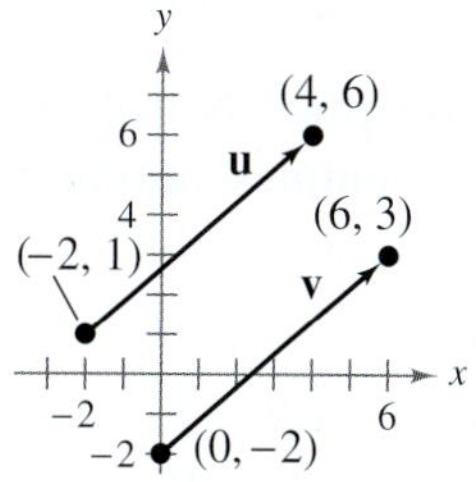

44.

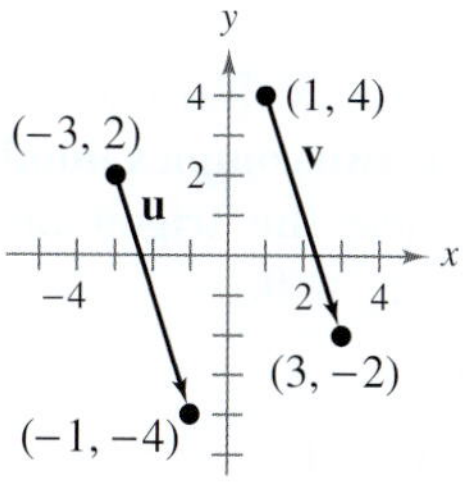

Finding the Component Form of a Vector **In Exercises 45–50, find the component form of the vector v satisfying the given conditions.**

45.

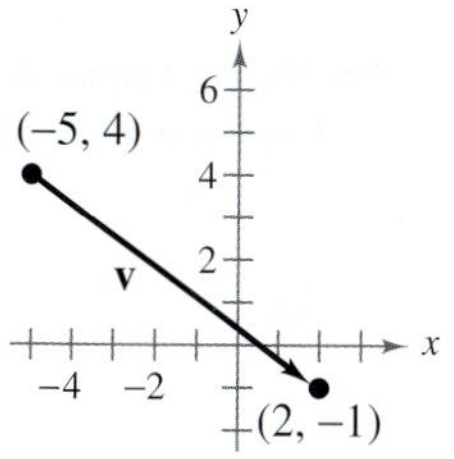

46.

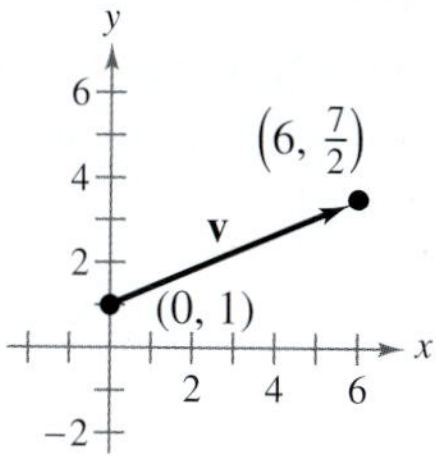

47. Initial point: (0, 10); terminal point: (7, 3)

48. Initial point: (1, 5); terminal point: (15, 9)

49. $\|\mathbf{v}\| = 8,\quad \theta = 120°$

50. $\|\mathbf{v}\| = \frac{1}{2},\quad \theta = 225°$

Vector Operations **In Exercises 51–58, find (a) u + v, (b) u − v, (c) 4u, and (d) 3v + 5u. Then sketch each resultant vector.**

51. $\mathbf{u} = \langle -1, -3\rangle,\ \mathbf{v} = \langle -3, 6\rangle$

52. $\mathbf{u} = \langle 4, 5\rangle,\ \mathbf{v} = \langle 0, -1\rangle$

53. $\mathbf{u} = \langle -5, 2\rangle,\ \mathbf{v} = \langle 4, 4\rangle$

54. $\mathbf{u} = \langle 1, -8\rangle,\ \mathbf{v} = \langle 3, -2\rangle$

55. $\mathbf{u} = 2\mathbf{i} - \mathbf{j},\ \mathbf{v} = 5\mathbf{i} + 3\mathbf{j}$

56. $\mathbf{u} = -7\mathbf{i} - 3\mathbf{j},\ \mathbf{v} = 4\mathbf{i} - \mathbf{j}$

57. $\mathbf{u} = 4\mathbf{i},\ \mathbf{v} = -\mathbf{i} + 6\mathbf{j}$

58. $\mathbf{u} = -6\mathbf{j},\ \mathbf{v} = \mathbf{i} + \mathbf{j}$

Vector Operations **In Exercises 59–64, find the component form of w and sketch the specified vector operations geometrically, where $\mathbf{u} = 6\mathbf{i} - 5\mathbf{j}$ and $\mathbf{v} = 10\mathbf{i} + 3\mathbf{j}$.**

59. $\mathbf{w} = 3\mathbf{v}$

60. $\mathbf{w} = \frac{1}{2}\mathbf{v}$

61. $\mathbf{w} = 2\mathbf{u} + \mathbf{v}$

62. $\mathbf{w} = 4\mathbf{u} - 5\mathbf{v}$

63. $\mathbf{w} = 5\mathbf{u} - 4\mathbf{v}$

64. $\mathbf{w} = -3\mathbf{u} + 2\mathbf{v}$

Writing a Linear Combination of Unit Vectors **In Exercises 65–68, the initial and terminal points of a vector are given. Write the vector as a linear combination of the standard unit vectors i and j.**

	Initial Point	Terminal Point
65.	(2, 3)	(1, 8)
66.	(4, −2)	(−2, −10)
67.	(3, 4)	(9, 8)
68.	(−2, 7)	(5, −9)

Finding the Direction Angle of a Vector **In Exercises 69–74, find the magnitude and direction angle of the vector v.**

69. $\mathbf{v} = 7(\cos 60°\mathbf{i} + \sin 60°\mathbf{j})$

70. $\mathbf{v} = 3(\cos 150°\mathbf{i} + \sin 150°\mathbf{j})$

71. $\mathbf{v} = 5\mathbf{i} + 4\mathbf{j}$

72. $\mathbf{v} = -4\mathbf{i} + 7\mathbf{j}$

73. $\mathbf{v} = -3\mathbf{i} - 3\mathbf{j}$

74. $\mathbf{v} = 8\mathbf{i} - \mathbf{j}$

75. Resultant Force Forces with magnitudes of 85 pounds and 50 pounds act on a single point. The angle between the forces is 15°. Describe the resultant force.

76. Rope Tension A 180-pound weight is supported by two ropes, as shown in the figure. Find the tension in each rope.

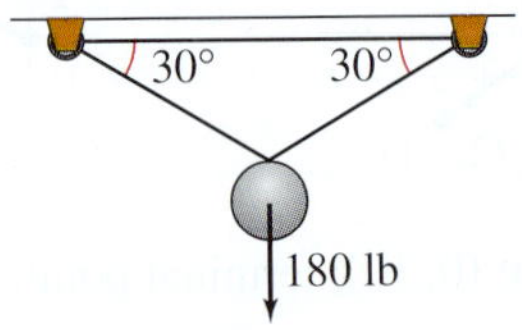

77. Navigation An airplane has an airspeed of 430 miles per hour at a bearing of 135°. The wind velocity is 35 miles per hour in the direction of N 30° E. Find the resultant speed and direction of the airplane.

78. Navigation An airplane has an airspeed of 724 kilometers per hour at a bearing of 30°. The wind velocity is 32 kilometers per hour from the west. Find the resultant speed and direction of the airplane.

6.4 Finding a Dot Product **In Exercises 79–82, find the dot product of u and v.**

79. $\mathbf{u} = \langle 6, 7 \rangle$
$\mathbf{v} = \langle -3, 9 \rangle$

80. $\mathbf{u} = \langle -7, 12 \rangle$
$\mathbf{v} = \langle -4, -14 \rangle$

81. $\mathbf{u} = 3\mathbf{i} + 7\mathbf{j}$
$\mathbf{v} = 11\mathbf{i} - 5\mathbf{j}$

82. $\mathbf{u} = -7\mathbf{i} + 2\mathbf{j}$
$\mathbf{v} = 16\mathbf{i} - 12\mathbf{j}$

Using Properties of Dot Products **In Exercises 83–90, use the vectors $\mathbf{u} = \langle -4, 2 \rangle$ and $\mathbf{v} = \langle 5, 1 \rangle$ to find the indicated quantity. State whether the result is a vector or a scalar.**

83. $2\mathbf{u} \cdot \mathbf{u}$ **84.** $3\mathbf{u} \cdot \mathbf{v}$

85. $4 - \|\mathbf{u}\|$ **86.** $\|\mathbf{v}\|^2$

87. $\mathbf{u}(\mathbf{u} \cdot \mathbf{v})$ **88.** $(\mathbf{u} \cdot \mathbf{v})\mathbf{v}$

89. $(\mathbf{u} \cdot \mathbf{u}) - (\mathbf{u} \cdot \mathbf{v})$

90. $(\mathbf{v} \cdot \mathbf{v}) - (\mathbf{v} \cdot \mathbf{u})$

Finding the Angle Between Two Vectors **In Exercises 91–94, find the angle θ between the vectors.**

91. $\mathbf{u} = \cos \frac{7\pi}{4}\mathbf{i} + \sin \frac{7\pi}{4}\mathbf{j}$

$\mathbf{v} = \cos \frac{5\pi}{6}\mathbf{i} + \sin \frac{5\pi}{6}\mathbf{j}$

92. $\mathbf{u} = \cos 45°\mathbf{i} + \sin 45°\mathbf{j}$
$\mathbf{v} = \cos 300°\mathbf{i} + \sin 300°\mathbf{j}$

93. $\mathbf{u} = \langle 2\sqrt{2}, -4 \rangle$, $\mathbf{v} = \langle -\sqrt{2}, 1 \rangle$

94. $\mathbf{u} = \langle 3, \sqrt{3} \rangle$, $\mathbf{v} = \langle 4, 3\sqrt{3} \rangle$

Determining Orthogonal Vectors **In Exercises 95–98, determine whether u and v are orthogonal.**

95. $\mathbf{u} = \langle -3, 8 \rangle$
$\mathbf{v} = \langle 8, 3 \rangle$

96. $\mathbf{u} = \langle \frac{1}{4}, -\frac{1}{2} \rangle$
$\mathbf{v} = \langle -2, 4 \rangle$

97. $\mathbf{u} = -\mathbf{i}$
$\mathbf{v} = \mathbf{i} + 2\mathbf{j}$

98. $\mathbf{u} = -2\mathbf{i} + \mathbf{j}$
$\mathbf{v} = 3\mathbf{i} + 6\mathbf{j}$

Decomposing a Vector into Components **In Exercises 99–102, find the projection of u onto v. Then write u as the sum of two orthogonal vectors, one of which is $\text{proj}_{\mathbf{v}}\mathbf{u}$.**

99. $\mathbf{u} = \langle -4, 3 \rangle$, $\mathbf{v} = \langle -8, -2 \rangle$

100. $\mathbf{u} = \langle 5, 6 \rangle$, $\mathbf{v} = \langle 10, 0 \rangle$

101. $\mathbf{u} = \langle 2, 7 \rangle$, $\mathbf{v} = \langle 1, -1 \rangle$

102. $\mathbf{u} = \langle -3, 5 \rangle$, $\mathbf{v} = \langle -5, 2 \rangle$

Work **In Exercises 103 and 104, find the work done in moving a particle from P to Q when the magnitude and direction of the force are given by v.**

103. $P(5, 3)$, $Q(8, 9)$, $\mathbf{v} = \langle 2, 7 \rangle$

104. $P(-2, -9)$, $Q(-12, 8)$, $\mathbf{v} = 3\mathbf{i} - 6\mathbf{j}$

105. Work Determine the work done (in foot-pounds) by a crane lifting an 18,000-pound truck 48 inches.

106. Work A mover exerts a horizontal force of 25 pounds on a crate as it is pushed up a ramp that is 12 feet long and inclined at an angle of 20° above the horizontal. Find the work done in pushing the crate.

6.5 Finding the Absolute Value of a Complex Number **In Exercises 107–112, plot the complex number and find its absolute value.**

107. $7i$ **108.** $-6i$

109. $5 + 3i$ **110.** $-10 - 4i$

111. $\sqrt{2} - \sqrt{2}i$ **112.** $-\sqrt{2} + \sqrt{2}i$

Trigonometric Form of a Complex Number **In Exercises 113–118, write the complex number in trigonometric form.**

113. $4i$ **114.** -7

115. $7 - 7i$ **116.** $5 + 12i$

117. $-5 - 12i$ **118.** $-3\sqrt{3} + 3i$

Multiplying or Dividing Complex Numbers **In Exercises 119 and 120, (a) write the two complex numbers in trigonometric form and (b) use the trigonometric forms to find z_1z_2 and z_1/z_2, where $z_2 \neq 0$.**

119. $z_1 = 2\sqrt{3} - 2i$, $z_2 = -10i$

120. $z_1 = -3(1 + i)$, $z_2 = 2(\sqrt{3} + i)$

Finding a Power of a Complex Number **In Exercises 121–124, use DeMoivre's Theorem to find the indicated power of the complex number. Write the result in standard form.**

121. $\left[5\left(\cos\frac{\pi}{12} + i\sin\frac{\pi}{12}\right)\right]^4$

122. $\left[2\left(\cos\frac{4\pi}{15} + i\sin\frac{4\pi}{15}\right)\right]^5$

123. $(2 + 3i)^6$

124. $(1 - i)^8$

Finding the *n*th Roots of a Complex Number **In Exercises 125–128, (a) use the formula on page 446 to find the indicated roots of the complex number, (b) represent each of the roots graphically, and (c) write each of the roots in standard form.**

125. Sixth roots of $-729i$

126. Fourth roots of $256i$

127. Cube roots of 8

128. Fifth roots of -1024

Solving an Equation **In Exercises 129–134, use the formula on page 446 to find all solutions of the equation and represent the solutions graphically.**

129. $x^4 + 81 = 0$

130. $x^5 - 32 = 0$

131. $x^3 + 8i = 0$

132. $x^4 - 64i = 0$

133. $x^5 + x^3 - x^2 - 1 = 0$

134. $x^5 + 4x^3 - 8x^2 - 32 = 0$

Exploration

True or False? **In Exercises 135–137, determine whether the statement is true or false. Justify your answer.**

135. The Law of Sines is true when one of the angles in the triangle is a right angle.

136. When the Law of Sines is used, the solution is always unique.

137. $x = \sqrt{3} + i$ is a solution of the equation $x^2 - 8i = 0$.

138. Law of Sines State the Law of Sines from memory.

139. Law of Cosines State the Law of Cosines from memory.

140. Reasoning What characterizes a vector in the plane?

141. Think About It Which vectors in the figure appear to be equivalent?

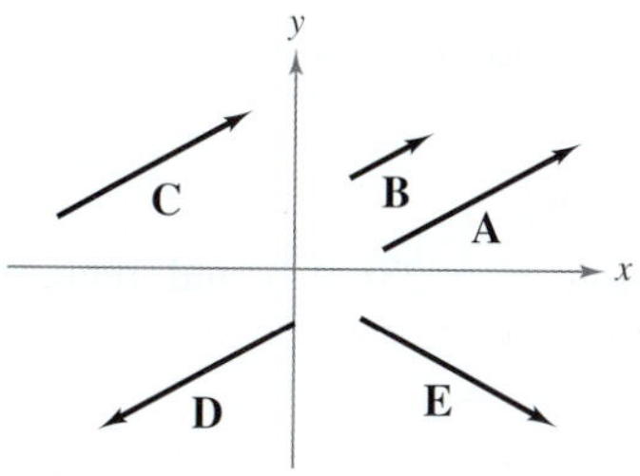

142. Think About It The vectors **u** and **v** have the same magnitudes in the two figures. In which figure will the magnitude of the sum be greater? Give a reason for your answer.

(a)

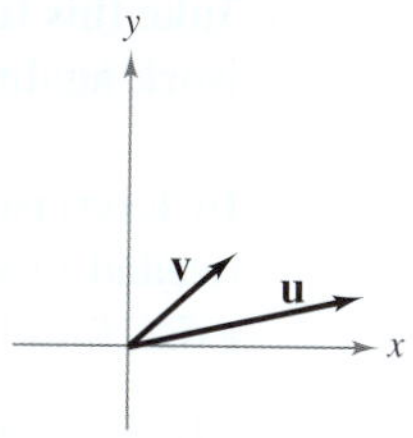

(b)

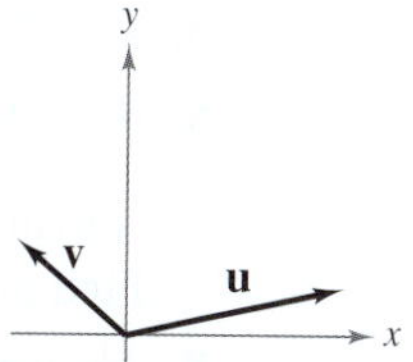

143. Geometry Describe geometrically the scalar multiple $k\mathbf{u}$ of the vector **u**, for $k > 0$ and $k < 0$.

144. Geometry Describe geometrically the sum of the vectors **u** and **v**.

Graphical Reasoning **In Exercises 145–148, use the graph of the roots of a complex number.**

(a) Write each of the roots in trigonometric form.

(b) Identify the complex number whose roots are given. Use a graphing utility to verify your results.

145.

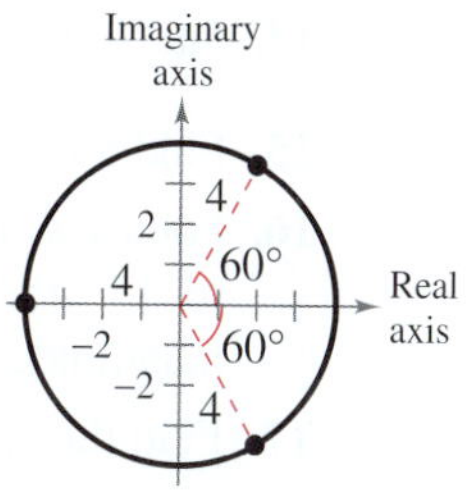

146.

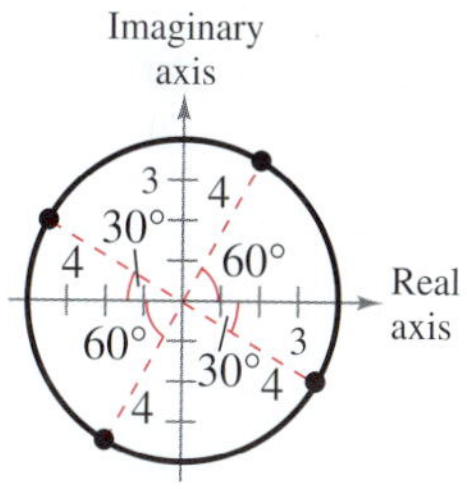

147.

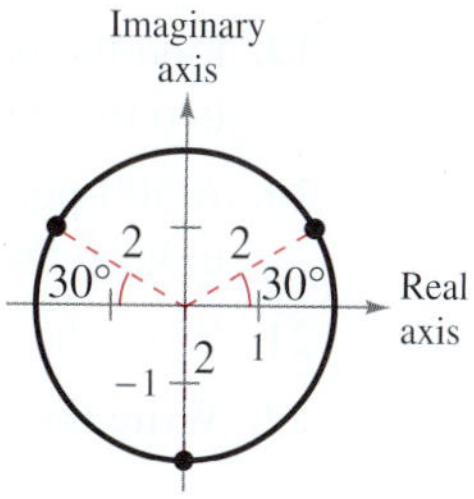

148.

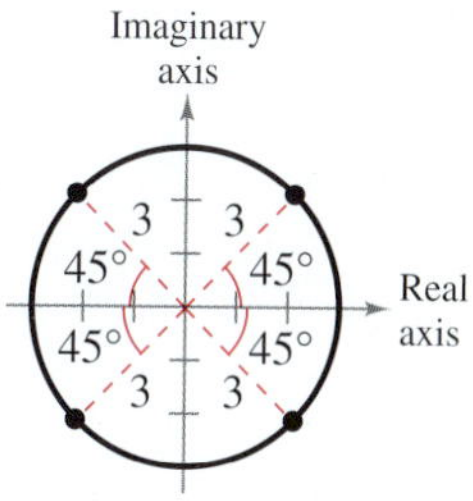

149. Graphical Reasoning The figure shows z_1 and z_2. Describe z_1z_2 and z_1/z_2.

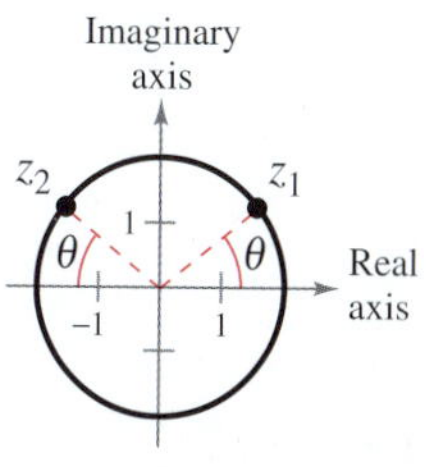

Chapter Test

See CalcChat.com for tutorial help and worked-out solutions to odd-numbered exercises.

Take this test as you would take a test in class. When you are finished, check your work against the answers given in the back of the book.

In Exercises 1–6, determine whether the Law of Sines or the Law of Cosines is needed to solve the triangle. Then solve (if possible) the triangle. If two solutions exist, find both. Round your answers to two decimal places.

1. $A = 24°,\ B = 68°,\ a = 12.2$

2. $B = 110°,\ C = 28°,\ a = 15.6$

3. $A = 24°,\ a = 11.2,\ b = 13.4$

4. $a = 4.0,\ b = 7.3,\ c = 12.4$

5. $B = 100°,\ a = 15,\ b = 23$

6. $C = 121°,\ a = 34,\ b = 55$

7. A triangular parcel of land has borders of lengths 60 meters, 70 meters, and 82 meters. Find the area of the parcel of land.

8. An airplane flies 370 miles from point A to point B with a bearing of 24°. It then flies 240 miles from point B to point C with a bearing of 37° (see figure). Find the distance and bearing from point A to point C.

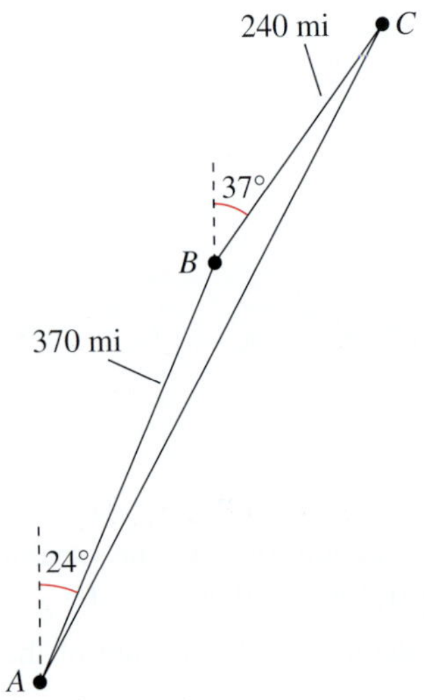

Figure for 8

In Exercises 9 and 10, find the component form of the vector v satisfying the given conditions.

9. Initial point of **v**: $(-3, 7)$; terminal point of **v**: $(11, -16)$

10. Magnitude of **v**: $\|\mathbf{v}\| = 12$; direction of **v**: $\mathbf{u} = \langle 3, -5 \rangle$

In Exercises 11–14, $\mathbf{u} = \langle 2, 7 \rangle$ and $\mathbf{v} = \langle -6, 5 \rangle$. Find the resultant vector and sketch its graph.

11. $\mathbf{u} + \mathbf{v}$

12. $\mathbf{u} - \mathbf{v}$

13. $5\mathbf{u} - 3\mathbf{v}$

14. $4\mathbf{u} + 2\mathbf{v}$

15. Find a unit vector in the direction of $\mathbf{u} = \langle 24, -7 \rangle$.

16. Forces with magnitudes of 250 pounds and 130 pounds act on an object at angles of 45° and $-60°$, respectively, with the positive x-axis. Find the direction and magnitude of the resultant of these forces.

17. Find the angle between the vectors $\mathbf{u} = \langle -1, 5 \rangle$ and $\mathbf{v} = \langle 3, -2 \rangle$.

18. Are the vectors $\mathbf{u} = \langle 6, -10 \rangle$ and $\mathbf{v} = \langle 5, 3 \rangle$ orthogonal?

19. Find the projection of $\mathbf{u} = \langle 6, 7 \rangle$ onto $\mathbf{v} = \langle -5, -1 \rangle$. Then write **u** as the sum of two orthogonal vectors, one of which is $\text{proj}_{\mathbf{v}}\mathbf{u}$.

20. A 500-pound motorcycle is headed up a hill inclined at 12°. What force is required to keep the motorcycle from rolling down the hill when stopped at a red light?

21. Write the complex number $z = 4 - 4i$ in trigonometric form.

22. Write the complex number $z = 6(\cos 120° + i \sin 120°)$ in standard form.

In Exercises 23 and 24, use DeMoivre's Theorem to find the indicated power of the complex number. Write the result in standard form.

23. $\left[3\left(\cos \frac{7\pi}{6} + i \sin \frac{7\pi}{6}\right)\right]^8$

24. $(3 - 3i)^6$

25. Find the fourth roots of $256(1 + \sqrt{3}i)$.

26. Find all solutions of the equation $x^3 - 27i = 0$ and represent the solutions graphically.

Cumulative Test for Chapters 4–6

See CalcChat.com for tutorial help and worked-out solutions to odd-numbered exercises.

Take this test as you would take a test in class. When you are finished, check your work against the answers given in the back of the book.

1. Consider the angle $\theta = -120°$.

(a) Sketch the angle in standard position.

(b) Determine a coterminal angle in the interval $[0°, 360°)$.

(c) Rewrite the angle in radian measure as a multiple of π.

(d) Find the reference angle θ'.

(e) Find the exact values of the six trigonometric functions of θ.

2. Convert the angle $\theta = -1.45$ radians to degrees. Round the answer to one decimal place.

3. Find $\cos \theta$ when $\tan \theta = -\frac{21}{20}$ and $\sin \theta < 0$.

In Exercises 4–6, sketch the graph of the function. (Include two full periods.)

4. $f(x) = 3 - 2 \sin \pi x$ **5.** $g(x) = \dfrac{1}{2} \tan\left(x - \dfrac{\pi}{2}\right)$ **6.** $h(x) = -\sec(x + \pi)$

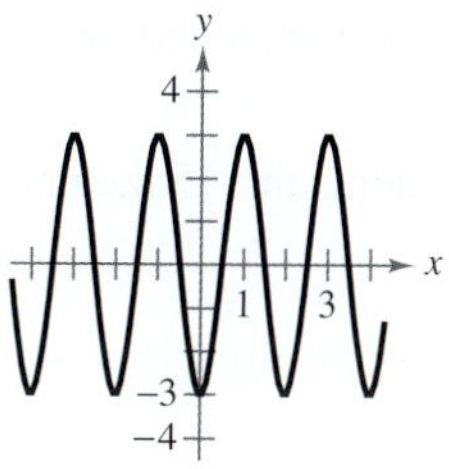

Figure for 7

7. Find a, b, and c such that the graph of the function $h(x) = a \cos(bx + c)$ matches the graph shown in the figure.

8. Sketch the graph of the function $f(x) = \frac{1}{2}x \sin x$ on the interval $-3\pi \le x \le 3\pi$.

In Exercises 9 and 10, find the exact value of the expression without using a calculator.

9. $\tan(\arctan 4.9)$ **10.** $\tan\left(\arcsin \frac{3}{5}\right)$

11. Write an algebraic expression equivalent to $\sin(\arccos 2x)$.

12. Use the fundamental identities to simplify: $\cos\left(\dfrac{\pi}{2} - x\right) \csc x$.

13. Subtract and simplify: $\dfrac{\sin \theta - 1}{\cos \theta} - \dfrac{\cos \theta}{\sin \theta - 1}$.

In Exercises 14–16, verify the identity.

14. $\cot^2 \alpha(\sec^2 \alpha - 1) = 1$

15. $\sin(x + y) \sin(x - y) = \sin^2 x - \sin^2 y$

16. $\sin^2 x \cos^2 x = \frac{1}{8}(1 - \cos 4x)$

In Exercises 17 and 18, find all solutions of the equation in the interval $[0, 2\pi)$.

17. $2 \cos^2 \beta - \cos \beta = 0$

18. $3 \tan \theta - \cot \theta = 0$

19. Use the Quadratic Formula to solve the equation in the interval $[0, 2\pi)$: $\sin^2 x + 2 \sin x + 1 = 0$.

20. Given that $\sin u = \frac{12}{13}$, $\cos v = \frac{3}{5}$, and angles u and v are both in Quadrant I, find $\tan(u - v)$.

21. Given that $\tan \theta = \frac{1}{2}$, find the exact value of $\tan(2\theta)$.

22. Given that $\tan \theta = \dfrac{4}{3}$, find the exact value of $\sin \dfrac{\theta}{2}$.

23. Write the product $5 \sin \frac{3\pi}{4} \cdot \cos \frac{7\pi}{4}$ as a sum or difference.

24. Write $\cos 9x - \cos 7x$ as a product.

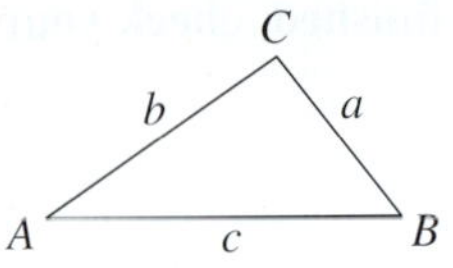

Figure for 25–30

In Exercises 25–30, determine whether the Law of Sines or the Law of Cosines is needed to solve the triangle at the left, then solve the triangle. Round your answers to two decimal places.

25. $A = 30°,\ a = 9,\ b = 8$

26. $A = 30°,\ b = 8,\ c = 10$

27. $A = 30°,\ C = 90°,\ b = 10$

28. $a = 4.7,\ b = 8.1,\ c = 10.3$

29. $A = 45°,\ B = 26°,\ c = 20$

30. $a = 1.2,\ b = 10,\ C = 80°$

31. Two sides of a triangle have lengths 7 inches and 12 inches. Their included angle measures 99°. Find the area of the triangle.

32. Find the area of a triangle with sides of lengths 30 meters, 41 meters, and 45 meters.

33. Write the vector

$$\mathbf{u} = \langle 7, 8 \rangle$$

as a linear combination of the standard unit vectors **i** and **j**.

34. Find a unit vector in the direction of $\mathbf{v} = \mathbf{i} + \mathbf{j}$.

35. Find $\mathbf{u} \cdot \mathbf{v}$ for $\mathbf{u} = 3\mathbf{i} + 4\mathbf{j}$ and $\mathbf{v} = \mathbf{i} - 2\mathbf{j}$.

36. Find the projection of $\mathbf{u} = \langle 8, -2 \rangle$ onto $\mathbf{v} = \langle 1, 5 \rangle$. Then write **u** as the sum of two orthogonal vectors, one of which is $\text{proj}_{\mathbf{v}}\mathbf{u}$.

37. Write the complex number $-2 + 2i$ in trigonometric form.

38. Find the product of $[4(\cos 30° + i \sin 30°)]$ and $[6(\cos 120° + i \sin 120°)]$. Write the answer in standard form.

39. Find the three cube roots of 1.

40. Find all solutions of the equation $x^5 + 243 = 0$.

41. A ceiling fan with 21-inch blades makes 63 revolutions per minute. Find the angular speed of the fan in radians per minute. Find the linear speed of the tips of the blades in inches per minute.

42. Find the area of the sector of a circle with a radius of 12 yards and a central angle of 105°.

43. From a point 200 feet from a flagpole, the angles of elevation to the bottom and top of the flag are 16° 45′ and 18°, respectively. Approximate the height of the flag to the nearest foot.

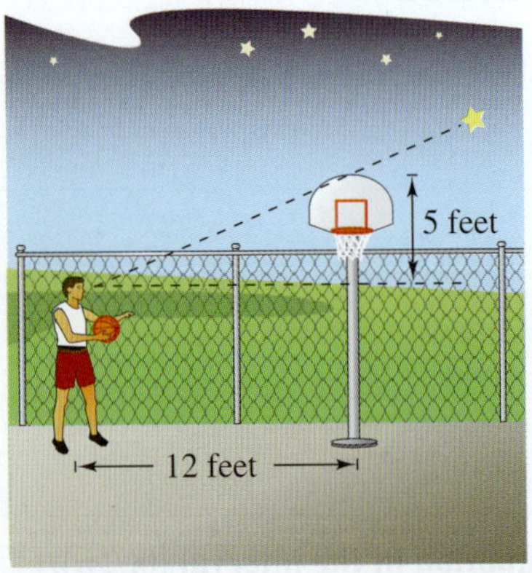

Figure for 44

44. To determine the angle of elevation of a star in the sky, you get the star and the top of the backboard of a basketball hoop that is 5 feet higher than your eyes in your line of vision (see figure). Your horizontal distance from the backboard is 12 feet. What is the angle of elevation of the star?

45. Find a model for a particle in simple harmonic motion with a displacement (at $t = 0$) of 4 inches, an amplitude of 4 inches, and a period of 8 seconds.

46. An airplane has an airspeed of 500 kilometers per hour at a bearing of 30°. The wind velocity is 50 kilometers per hour in the direction of N 60° E. Find the resultant speed and direction of the airplane.

47. A force of 85 pounds, exerted at an angle of 60° with the horizontal, is required to slide an object across a floor. Determine the work done in sliding the object 10 feet.

Proofs in Mathematics

LAW OF TANGENTS

Besides the Law of Sines and the Law of Cosines, there is also a Law of Tangents, which was developed by Francois Viète (1540–1603). The Law of Tangents follows from the Law of Sines and the sum-to-product formulas for sine and is defined as follows.

$$\frac{a+b}{a-b} = \frac{\tan[(A+B)/2]}{\tan[(A-B)/2]}$$

The Law of Tangents can be used to solve a triangle when two sides and the included angle are given (SAS). Before calculators were invented, the Law of Tangents was used to solve the SAS case instead of the Law of Cosines because computation with a table of tangent values was easier.

Law of Sines *(p. 402)*

If ABC is a triangle with sides a, b, and c, then

$$\frac{a}{\sin A} = \frac{b}{\sin B} = \frac{c}{\sin C}.$$

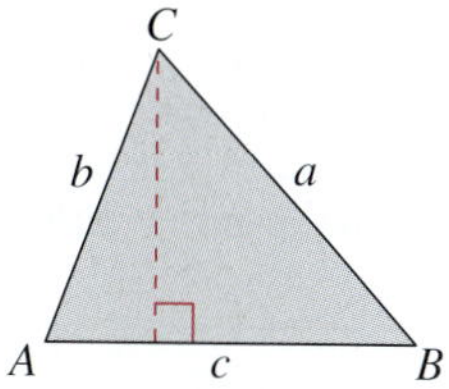

A is acute.

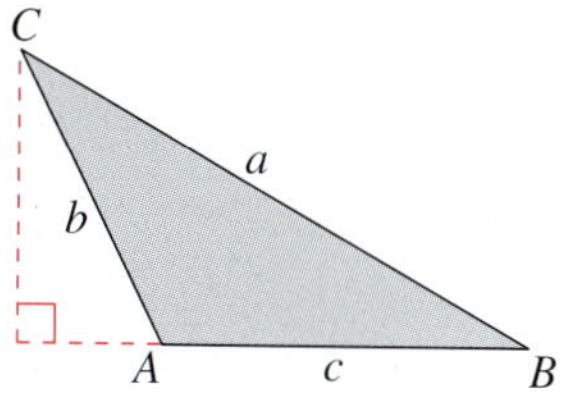

A is obtuse.

Proof

Let h be the altitude of either triangle in the figure above. Then you have

$$\sin A = \frac{h}{b} \quad \text{or} \quad h = b \sin A$$

$$\sin B = \frac{h}{a} \quad \text{or} \quad h = a \sin B.$$

Equating these two values of h, you have

$$a \sin B = b \sin A \quad \text{or} \quad \frac{a}{\sin A} = \frac{b}{\sin B}.$$

Note that $\sin A \neq 0$ and $\sin B \neq 0$ because no angle of a triangle can have a measure of $0°$ or $180°$. In a similar manner, construct an altitude h from vertex B to side AC (extended in the obtuse triangle), as shown at the left. Then you have

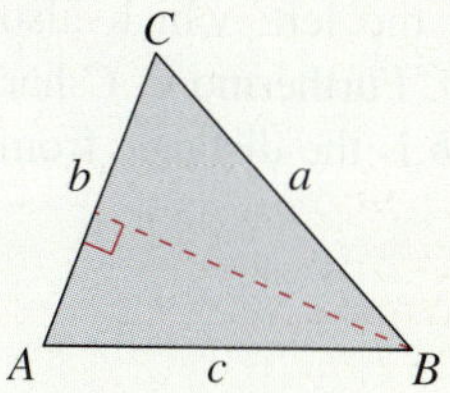

A is acute.

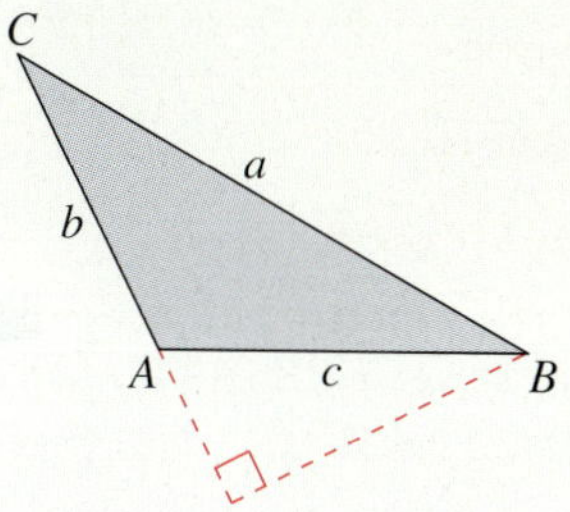

A is obtuse.

$$\sin A = \frac{h}{c} \quad \text{or} \quad h = c \sin A$$

$$\sin C = \frac{h}{a} \quad \text{or} \quad h = a \sin C.$$

Equating these two values of h, you have

$$a \sin C = c \sin A \quad \text{or} \quad \frac{a}{\sin A} = \frac{c}{\sin C}.$$

By the Transitive Property of Equality, you know that

$$\frac{a}{\sin A} = \frac{b}{\sin B} = \frac{c}{\sin C}.$$

So, the Law of Sines is established. ∎

Law of Cosines *(p. 411)*

Standard Form	**Alternative Form**
$a^2 = b^2 + c^2 - 2bc\cos A$	$\cos A = \dfrac{b^2 + c^2 - a^2}{2bc}$
$b^2 = a^2 + c^2 - 2ac\cos B$	$\cos B = \dfrac{a^2 + c^2 - b^2}{2ac}$
$c^2 = a^2 + b^2 - 2ab\cos C$	$\cos C = \dfrac{a^2 + b^2 - c^2}{2ab}$

Proof

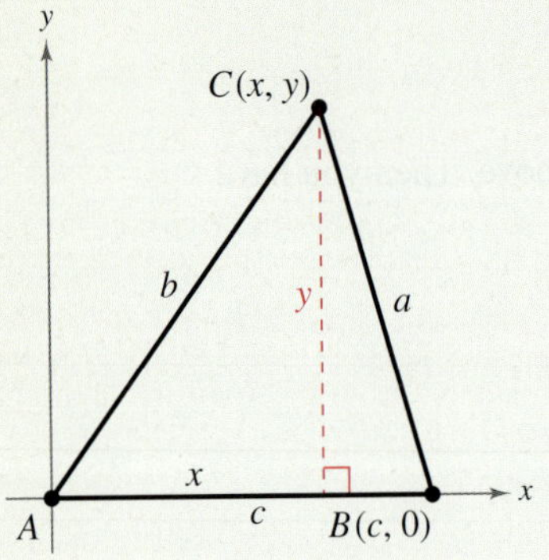

To prove the first formula, consider the top triangle at the left, which has three acute angles. Note that vertex B has coordinates $(c, 0)$. Furthermore, C has coordinates (x, y), where $x = b\cos A$ and $y = b\sin A$. Because a is the distance from vertex C to vertex B, it follows that

$$a = \sqrt{(x-c)^2 + (y-0)^2} \qquad \text{Distance Formula}$$

$$a^2 = (x-c)^2 + (y-0)^2 \qquad \text{Square each side.}$$

$$a^2 = (b\cos A - c)^2 + (b\sin A)^2 \qquad \text{Substitute for } x \text{ and } y.$$

$$a^2 = b^2\cos^2 A - 2bc\cos A + c^2 + b^2\sin^2 A \qquad \text{Expand.}$$

$$a^2 = b^2(\sin^2 A + \cos^2 A) + c^2 - 2bc\cos A \qquad \text{Factor out } b^2.$$

$$a^2 = b^2 + c^2 - 2bc\cos A. \qquad \sin^2 A + \cos^2 A = 1$$

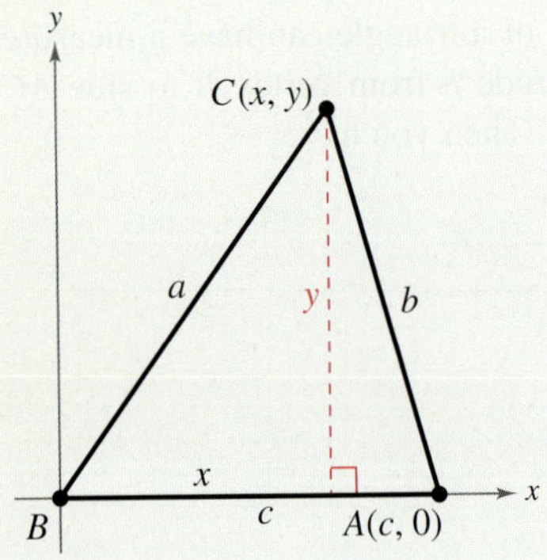

To prove the second formula, consider the bottom triangle at the left, which also has three acute angles. Note that vertex A has coordinates $(c, 0)$. Furthermore, C has coordinates (x, y), where $x = a\cos B$ and $y = a\sin B$. Because b is the distance from vertex C to vertex A, it follows that

$$b = \sqrt{(x-c)^2 + (y-0)^2} \qquad \text{Distance Formula}$$

$$b^2 = (x-c)^2 + (y-0)^2 \qquad \text{Square each side.}$$

$$b^2 = (a\cos B - c)^2 + (a\sin B)^2 \qquad \text{Substitute for } x \text{ and } y.$$

$$b^2 = a^2\cos^2 B - 2ac\cos B + c^2 + a^2\sin^2 B \qquad \text{Expand.}$$

$$b^2 = a^2(\sin^2 B + \cos^2 B) + c^2 - 2ac\cos B \qquad \text{Factor out } a^2.$$

$$b^2 = a^2 + c^2 - 2ac\cos B. \qquad \sin^2 B + \cos^2 B = 1$$

A similar argument is used to establish the third formula. ■

Heron's Area Formula *(p. 414)*

Given any triangle with sides of lengths a, b, and c, the area of the triangle is

$$\text{Area} = \sqrt{s(s-a)(s-b)(s-c)}$$

where $s = \dfrac{a+b+c}{2}$.

Proof

From Section 6.1, you know that

$$\text{Area} = \frac{1}{2}bc \sin A \qquad \text{Formula for the area of an oblique triangle}$$

$$(\text{Area})^2 = \frac{1}{4}b^2c^2 \sin^2 A \qquad \text{Square each side.}$$

$$\text{Area} = \sqrt{\frac{1}{4}b^2c^2 \sin^2 A} \qquad \text{Take the square root of each side.}$$

$$= \sqrt{\frac{1}{4}b^2c^2(1-\cos^2 A)} \qquad \text{Pythagorean identity}$$

$$= \sqrt{\left[\frac{1}{2}bc(1+\cos A)\right]\left[\frac{1}{2}bc(1-\cos A)\right]}. \qquad \text{Factor.}$$

Using the Law of Cosines, you can show that

$$\frac{1}{2}bc(1+\cos A) = \frac{a+b+c}{2} \cdot \frac{-a+b+c}{2}$$

and

$$\frac{1}{2}bc(1-\cos A) = \frac{a-b+c}{2} \cdot \frac{a+b-c}{2}.$$

Letting $s = (a+b+c)/2$, these two equations can be rewritten as

$$\frac{1}{2}bc(1+\cos A) = s(s-a)$$

and

$$\frac{1}{2}bc(1-\cos A) = (s-b)(s-c).$$

By substituting into the last formula for area, you can conclude that

$$\text{Area} = \sqrt{s(s-a)(s-b)(s-c)}.$$

Properties of the Dot Product *(p. 431)*

Let **u**, **v**, and **w** be vectors in the plane or in space and let c be a scalar.

1. $\mathbf{u} \cdot \mathbf{v} = \mathbf{v} \cdot \mathbf{u}$
2. $\mathbf{0} \cdot \mathbf{v} = 0$
3. $\mathbf{u} \cdot (\mathbf{v} + \mathbf{w}) = \mathbf{u} \cdot \mathbf{v} + \mathbf{u} \cdot \mathbf{w}$
4. $\mathbf{v} \cdot \mathbf{v} = \|\mathbf{v}\|^2$
5. $c(\mathbf{u} \cdot \mathbf{v}) = c\mathbf{u} \cdot \mathbf{v} = \mathbf{u} \cdot c\mathbf{v}$

Proof

Let $\mathbf{u} = \langle u_1, u_2 \rangle$, $\mathbf{v} = \langle v_1, v_2 \rangle$, $\mathbf{w} = \langle w_1, w_2 \rangle$, $\mathbf{0} = \langle 0, 0 \rangle$, and let c be a scalar.

1. $\mathbf{u} \cdot \mathbf{v} = u_1v_1 + u_2v_2 = v_1u_1 + v_2u_2 = \mathbf{v} \cdot \mathbf{u}$

2. $\mathbf{0} \cdot \mathbf{v} = 0 \cdot v_1 + 0 \cdot v_2 = 0$

3. $$\begin{aligned}\mathbf{u} \cdot (\mathbf{v} + \mathbf{w}) &= \mathbf{u} \cdot \langle v_1 + w_1, v_2 + w_2 \rangle \\ &= u_1(v_1 + w_1) + u_2(v_2 + w_2) \\ &= u_1v_1 + u_1w_1 + u_2v_2 + u_2w_2 \\ &= (u_1v_1 + u_2v_2) + (u_1w_1 + u_2w_2) = \mathbf{u} \cdot \mathbf{v} + \mathbf{u} \cdot \mathbf{w}\end{aligned}$$

4. $\mathbf{v} \cdot \mathbf{v} = v_1^2 + v_2^2 = \left(\sqrt{v_1^2 + v_2^2}\right)^2 = \|\mathbf{v}\|^2$

5. $$\begin{aligned}c(\mathbf{u} \cdot \mathbf{v}) &= c(\langle u_1, u_2 \rangle \cdot \langle v_1, v_2 \rangle) \\ &= c(u_1v_1 + u_2v_2) \\ &= (cu_1)v_1 + (cu_2)v_2 \\ &= \langle cu_1, cu_2 \rangle \cdot \langle v_1, v_2 \rangle \\ &= c\mathbf{u} \cdot \mathbf{v}\end{aligned}$$

Angle Between Two Vectors *(p. 432)*

If θ is the angle between two nonzero vectors **u** and **v**, then $\cos\theta = \dfrac{\mathbf{u} \cdot \mathbf{v}}{\|\mathbf{u}\|\,\|\mathbf{v}\|}$.

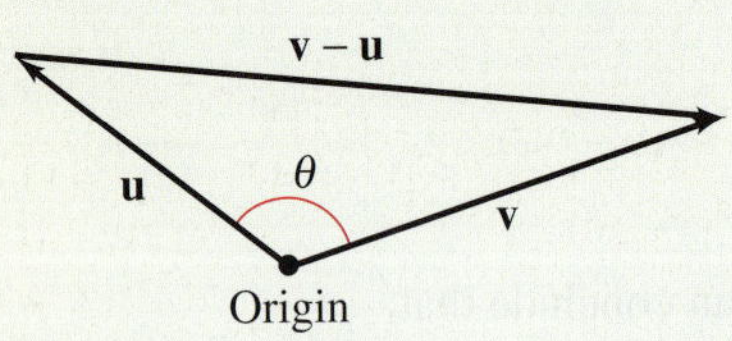

Proof

Consider the triangle determined by vectors **u**, **v**, and **v** − **u**, as shown at the left. By the Law of Cosines, you can write

$$\begin{aligned}\|\mathbf{v} - \mathbf{u}\|^2 &= \|\mathbf{u}\|^2 + \|\mathbf{v}\|^2 - 2\|\mathbf{u}\|\,\|\mathbf{v}\| \cos\theta \\ (\mathbf{v} - \mathbf{u}) \cdot (\mathbf{v} - \mathbf{u}) &= \|\mathbf{u}\|^2 + \|\mathbf{v}\|^2 - 2\|\mathbf{u}\|\,\|\mathbf{v}\| \cos\theta \\ (\mathbf{v} - \mathbf{u}) \cdot \mathbf{v} - (\mathbf{v} - \mathbf{u}) \cdot \mathbf{u} &= \|\mathbf{u}\|^2 + \|\mathbf{v}\|^2 - 2\|\mathbf{u}\|\,\|\mathbf{v}\| \cos\theta \\ \mathbf{v} \cdot \mathbf{v} - \mathbf{u} \cdot \mathbf{v} - \mathbf{v} \cdot \mathbf{u} + \mathbf{u} \cdot \mathbf{u} &= \|\mathbf{u}\|^2 + \|\mathbf{v}\|^2 - 2\|\mathbf{u}\|\,\|\mathbf{v}\| \cos\theta \\ \|\mathbf{v}\|^2 - 2\mathbf{u} \cdot \mathbf{v} + \|\mathbf{u}\|^2 &= \|\mathbf{u}\|^2 + \|\mathbf{v}\|^2 - 2\|\mathbf{u}\|\,\|\mathbf{v}\| \cos\theta \\ \cos\theta &= \frac{\mathbf{u} \cdot \mathbf{v}}{\|\mathbf{u}\|\,\|\mathbf{v}\|}.\end{aligned}$$

P.S. Problem Solving

1. **Distance** In the figure, a beam of light is directed at the blue mirror, reflected to the red mirror, and then reflected back to the blue mirror. Find PT, the distance that the light travels from the red mirror back to the blue mirror.

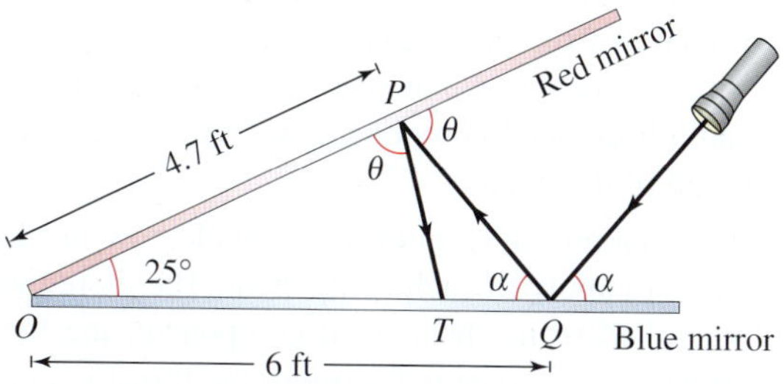

2. **Correcting a Course** A triathlete sets a course to swim S 25° E from a point on shore to a buoy $\frac{3}{4}$ mile away. After swimming 300 yards through a strong current, the triathlete is off course at a bearing of S 35° E. Find the bearing and distance the triathlete needs to swim to correct her course.

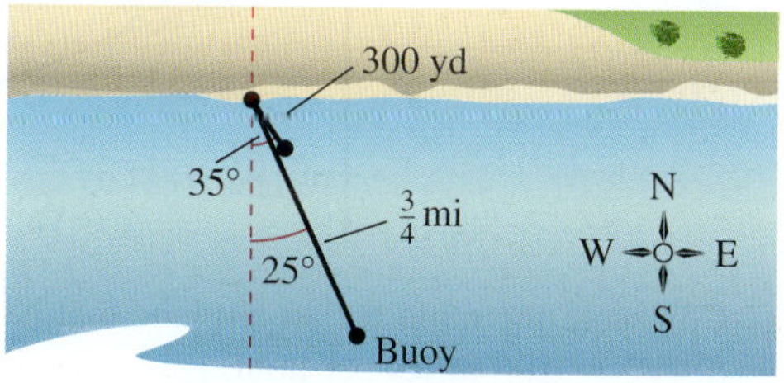

3. **Locating Lost Hikers** A group of hikers is lost in a national park. Two ranger stations have received an emergency SOS signal from the hikers. Station B is 75 miles due east of station A. The bearing from station A to the signal is S 60° E and the bearing from station B to the signal is S 75° W.
   (a) Draw a diagram that gives a visual representation of the problem.
   (b) Find the distance from each station to the SOS signal.
   (c) A rescue party is in the park 20 miles from station A at a bearing of S 80° E. Find the distance and the bearing the rescue party must travel to reach the lost hikers.

4. **Seeding a Courtyard** You are seeding a triangular courtyard. One side of the courtyard is 52 feet long and another side is 46 feet long. The angle opposite the 52-foot side is 65°.
   (a) Draw a diagram that gives a visual representation of the situation.
   (b) How long is the third side of the courtyard?
   (c) One bag of grass seed covers an area of 50 square feet. How many bags of grass seed will you need to cover the courtyard?

5. **Finding Magnitudes** For each pair of vectors, find the following.
   (i) $\|\mathbf{u}\|$
   (ii) $\|\mathbf{v}\|$
   (iii) $\|\mathbf{u} + \mathbf{v}\|$
   (iv) $\left\|\frac{\mathbf{u}}{\|\mathbf{u}\|}\right\|$
   (v) $\left\|\frac{\mathbf{v}}{\|\mathbf{v}\|}\right\|$
   (vi) $\left\|\frac{\mathbf{u} + \mathbf{v}}{\|\mathbf{u} + \mathbf{v}\|}\right\|$
   (a) $\mathbf{u} = \langle 1, -1 \rangle$
   $\mathbf{v} = \langle -1, 2 \rangle$
   (b) $\mathbf{u} = \langle 0, 1 \rangle$
   $\mathbf{v} = \langle 3, -3 \rangle$
   (c) $\mathbf{u} = \left\langle 1, \frac{1}{2} \right\rangle$
   $\mathbf{v} = \langle 2, 3 \rangle$
   (d) $\mathbf{u} = \langle 2, -4 \rangle$
   $\mathbf{v} = \langle 5, 5 \rangle$

6. **Writing a Vector in Terms of Other Vectors** Write the vector $\mathbf{w}$ in terms of $\mathbf{u}$ and $\mathbf{v}$, given that the terminal point of $\mathbf{w}$ bisects the line segment (see figure).

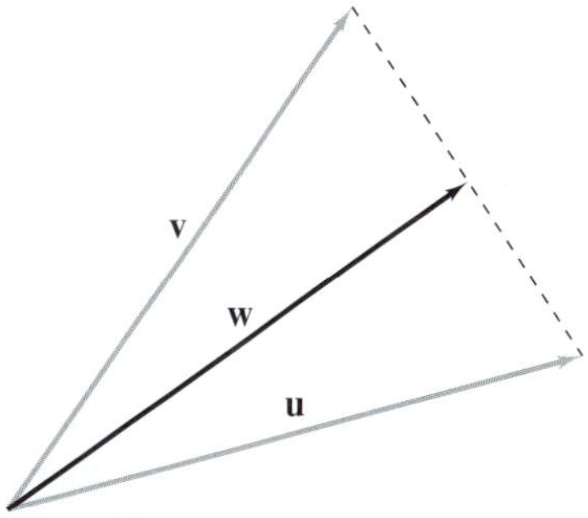

7. **Proof** Prove that if $\mathbf{u}$ is orthogonal to $\mathbf{v}$ and $\mathbf{w}$, then $\mathbf{u}$ is orthogonal to

   $c\mathbf{v} + d\mathbf{w}$

   for any scalars c and d.

8. **Comparing Work** Two forces of the same magnitude $\mathbf{F}_1$ and $\mathbf{F}_2$ act at angles θ_1 and θ_2, respectively. Use a diagram to compare the work done by $\mathbf{F}_1$ with the work done by $\mathbf{F}_2$ in moving along the vector PQ when
   (a) $\theta_1 = -\theta_2$
   (b) $\theta_1 = 60°$ and $\theta_2 = 30°$.

9. **Skydiving** A skydiver is falling at a constant downward velocity of 120 miles per hour. In the figure, vector **u** represents the skydiver's velocity. A steady breeze pushes the skydiver to the east at 40 miles per hour. Vector **v** represents the wind velocity.

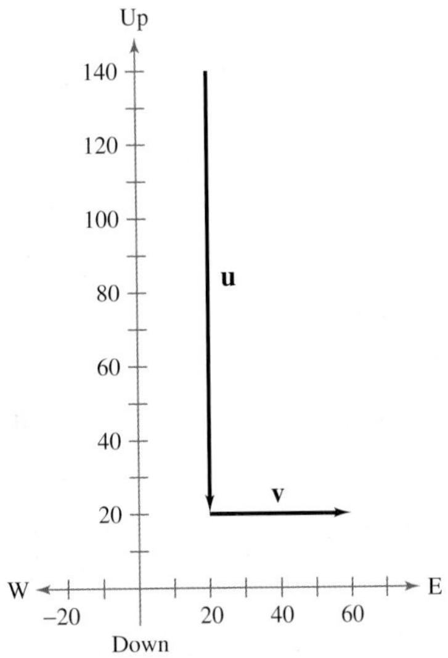

(a) Write the vectors **u** and **v** in component form.

(b) Let

$\mathbf{s} = \mathbf{u} + \mathbf{v}$.

Use the figure to sketch **s**. To print an enlarged copy of the graph, go to *MathGraphs.com*.

(c) Find the magnitude of **s**. What information does the magnitude give you about the skydiver's fall?

(d) If there were no wind, then the skydiver would fall in a path perpendicular to the ground. At what angle to the ground is the path of the skydiver when affected by the 40-mile-per-hour wind from due west?

(e) The skydiver is blown to the west at 30 miles per hour. Draw a new figure that gives a visual representation of the problem and find the skydiver's new velocity.

10. **Speed and Velocity of an Airplane** Four basic forces are in action during flight: weight, lift, thrust, and drag. To fly through the air, an object must overcome its own *weight*. To do this, it must create an upward force called *lift*. To generate lift, a forward motion called *thrust* is needed. The thrust must be great enough to overcome air resistance, which is called *drag*.

For a commercial jet aircraft, a quick climb is important to maximize efficiency because the performance of an aircraft at high altitudes is enhanced. In addition, it is necessary to clear obstacles such as buildings and mountains and to reduce noise in residential areas. In the diagram, the angle θ is called the climb angle. The velocity of the plane can be represented by a vector **v** with a vertical component $\|\mathbf{v}\| \sin \theta$ (called climb speed) and a horizontal component $\|\mathbf{v}\| \cos \theta$, where $\|\mathbf{v}\|$ is the speed of the plane.

When taking off, a pilot must decide how much of the thrust to apply to each component. The more the thrust is applied to the horizontal component, the faster the airplane gains speed. The more the thrust is applied to the vertical component, the quicker the airplane climbs.

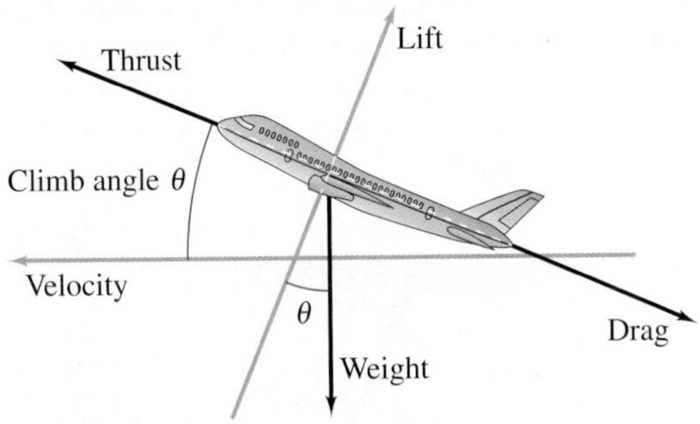

(a) Complete the table for an airplane that has a speed of $\|\mathbf{v}\| = 100$ miles per hour.

θ	0.5°	1.0°	1.5°	2.0°	2.5°	3.0°
$\|\mathbf{v}\| \sin \theta$						
$\|\mathbf{v}\| \cos \theta$						

(b) Does an airplane's speed equal the sum of the vertical and horizontal components of its velocity? If not, how could you find the speed of an airplane whose velocity components were known?

(c) Use the result of part (b) to find the speed of an airplane with the given velocity components.

(i) $\|\mathbf{v}\| \sin \theta = 5.235$ miles per hour

$\|\mathbf{v}\| \cos \theta = 149.909$ miles per hour

(ii) $\|\mathbf{v}\| \sin \theta = 10.463$ miles per hour

$\|\mathbf{v}\| \cos \theta = 149.634$ miles per hour

7 Systems of Equations and Inequalities

Nutrition (Example 9, page 514)

Waste Water (Exercise 59, page 507)

Global Positioning System (page 493)

Fuel Mixture (Exercise 50, page 485)

Traffic Flow (page 467)

Clockwise from top left, tlorna/Shutterstock.com; Kinetic Imagery/Shutterstock.com; Vlad61/Shutterstock.com; Taras Vyshnya/Shutterstock.com; edobric/Shutterstock.com

7.1 Linear and Nonlinear Systems of Equations

Graphs of systems of equations can help you solve real-life problems. For instance, in Exercise 64 on page 475, you will use the graph of a system of equations to approximate when the consumption of wind energy surpassed the consumption of geothermal energy.

- **Use the method of substitution to solve systems of linear equations in two variables.**
- **Use the method of substitution to solve systems of nonlinear equations in two variables.**
- **Use a graphical approach to solve systems of equations in two variables.**
- **Use systems of equations to model and solve real-life problems.**

The Method of Substitution

Up to this point in the text, most problems have involved either a function of one variable or a single equation in two variables. However, many problems in science, business, and engineering involve two or more equations in two or more variables. To solve such a problem, you need to find solutions of a **system of equations.** Here is an example of a system of two equations in two unknowns.

$$\begin{cases} 2x + y = 5 & \text{Equation 1} \\ 3x - 2y = 4 & \text{Equation 2} \end{cases}$$

A **solution** of this system is an ordered pair that satisfies each equation in the system. Finding the set of all solutions is called **solving the system of equations.** For instance, the ordered pair $(2, 1)$ is a solution of this system. To check this, substitute 2 for x and 1 for y in *each* equation.

Check (2, 1) in Equation 1 and Equation 2:

$2x + y = 5$ Write Equation 1.

$2(2) + 1 \stackrel{?}{=} 5$ Substitute 2 for x and 1 for y.

$4 + 1 = 5$ Solution checks in Equation 1. ✓

$3x - 2y = 4$ Write Equation 2.

$3(2) - 2(1) \stackrel{?}{=} 4$ Substitute 2 for x and 1 for y.

$6 - 2 = 4$ Solution checks in Equation 2. ✓

In this chapter, you will study four ways to solve systems of equations, beginning with the **method of substitution.**

Method	Section	Type of System
1. Substitution	7.1	Linear or nonlinear, two variables
2. Graphical method	7.1	Linear or nonlinear, two variables
3. Elimination	7.2	Linear, two variables
4. Gaussian elimination	7.3	Linear, three or more variables

Method of Substitution

1. *Solve* one of the equations for one variable in terms of the other.
2. *Substitute* the expression found in Step 1 into the other equation to obtain an equation in one variable.
3. *Solve* the equation obtained in Step 2.
4. *Back-substitute* the value obtained in Step 3 into the expression obtained in Step 1 to find the value of the other variable.
5. *Check* that the solution satisfies *each* of the original equations.

ssuaphotos/Shutterstock.com

Systems of equations have a wide variety of real-life applications. For instance, researchers in Italy studying the acoustical noise levels from vehicular traffic at a busy three-way intersection used a system of equations to model the traffic flow at the intersection. To help formulate the system of equations, "operators" stationed themselves at various locations along the intersection and counted the numbers of vehicles that passed them. *(Source: Proceedings of the 11th WSEAS International Conference on Acoustics & Music: Theory & Applications)*

EXAMPLE 1 Solving a System of Equations by Substitution

Solve the system of equations.

$$\begin{cases} x + y = 4 & \text{Equation 1} \\ x - y = 2 & \text{Equation 2} \end{cases}$$

Solution Begin by solving for y in Equation 1.

$$y = 4 - x \qquad \text{Solve for } y \text{ in Equation 1.}$$

Next, substitute this expression for y into Equation 2 and solve the resulting single-variable equation for x.

$$x - y = 2 \qquad \text{Write Equation 2.}$$

$$x - (4 - x) = 2 \qquad \text{Substitute } 4 - x \text{ for } y.$$

$$x - 4 + x = 2 \qquad \text{Distributive Property}$$

$$2x = 6 \qquad \text{Combine like terms.}$$

$$x = 3 \qquad \text{Divide each side by 2.}$$

Finally, solve for y by *back-substituting* $x = 3$ into the equation $y = 4 - x$, to obtain

$$y = 4 - x \qquad \text{Write revised Equation 1.}$$

$$y = 4 - 3 \qquad \text{Substitute 3 for } x.$$

$$y = 1. \qquad \text{Solve for } y.$$

The solution is the ordered pair

$(3, 1)$.

Check this solution as follows.

REMARK Because many steps are required to solve a system of equations, it is very easy to make errors in arithmetic. So, you should always check your solution by substituting it into *each* equation in the original system.

Check

Substitute $(3, 1)$ into Equation 1:

$$x + y = 4 \qquad \text{Write Equation 1.}$$

$$3 + 1 \stackrel{?}{=} 4 \qquad \text{Substitute for } x \text{ and } y.$$

$$4 = 4 \qquad \text{Solution checks in Equation 1.} \checkmark$$

Substitute $(3, 1)$ into Equation 2:

$$x - y = 2 \qquad \text{Write Equation 2.}$$

$$3 - 1 \stackrel{?}{=} 2 \qquad \text{Substitute for } x \text{ and } y.$$

$$2 = 2 \qquad \text{Solution checks in Equation 2.} \checkmark$$

Because $(3, 1)$ satisfies both equations in the system, it is a solution of the system of equations.

✓ **Checkpoint** *Audio-video solution in English & Spanish at LarsonPrecalculus.com.*

Solve the system of equations.

$$\begin{cases} x - y = 0 \\ 5x - 3y = 6 \end{cases}$$

The term *back-substitution* implies that you work *backwards*. First you solve for one of the variables, and then you substitute that value *back* into one of the equations in the system to find the value of the other variable.

Taras Vyshnya/Shutterstock.com

EXAMPLE 2 Solving a System by Substitution

A total of \$12,000 is invested in two funds paying 5% and 3% simple interest. (Recall that the formula for simple interest is $I = Prt$, where P is the principal, r is the annual interest rate, and t is the time.) The yearly interest is \$500. How much is invested at each rate?

Solution

Verbal Model: [Amount in 5% fund] + [Amount in 3% fund] = [Total investment]

[Interest for 5% fund] + [Interest for 3% fund] = [Total interest]

Labels: Amount in 5% fund $= x$ (dollars)
Interest for 5% fund $= 0.05x$ (dollars)
Amount in 3% fund $= y$ (dollars)
Interest for 3% fund $= 0.03y$ (dollars)
Total investment $= 12{,}000$ (dollars)
Total interest $= 500$ (dollars)

System: $\begin{cases} x + y = 12{,}000 \\ 0.05x + 0.03y = 500 \end{cases}$ Equation 1, Equation 2

To begin, it is convenient to multiply each side of Equation 2 by 100. This eliminates the need to work with decimals.

$100(0.05x + 0.03y) = 100(500)$ Multiply each side of Equation 2 by 100.

$5x + 3y = 50{,}000$ Revised Equation 2

••REMARK When using the method of substitution, it does not matter which variable you choose to solve for first. Whether you solve for y first or x first, you will obtain the same solution. You should choose the variable and equation that make your work easier. For instance, in Example 2, solving for x in Equation 1 is easier than solving for x in Equation 2.

To solve this system, you can solve for x in Equation 1.

$x = 12{,}000 - y$ Revised Equation 1

Then, substitute this expression for x into revised Equation 2 and solve the resulting equation for y.

$5x + 3y = 50{,}000$ Write revised Equation 2.

$5(12{,}000 - y) + 3y = 50{,}000$ Substitute $12{,}000 - y$ for x.

$60{,}000 - 5y + 3y = 50{,}000$ Distributive Property

$-2y = -10{,}000$ Combine like terms.

$y = 5000$ Divide each side by -2.

Next, back-substitute $y = 5000$ to solve for x.

$x = 12{,}000 - y$ Write revised Equation 1.

$x = 12{,}000 - 5000$ Substitute 5000 for y.

$x = 7000$ Subtract.

The solution is (7000, 5000). So, \$7000 is invested at 5% and \$5000 is invested at 3%. Check this in the original system.

TECHNOLOGY One way to check the answers you obtain in this section is to use a graphing utility. For instance, graph the two equations in Example 2

$y_1 = 12{,}000 - x$

$y_2 = \dfrac{500 - 0.05x}{0.03}$

and find the point of intersection. Does this point agree with the solution obtained at the right?

✓ **Checkpoint** *Audio-video solution in English & Spanish at LarsonPrecalculus.com.*

A total of \$25,000 is invested in two funds paying 6.5% and 8.5% simple interest. The yearly interest is \$2000. How much is invested at each rate?

Nonlinear Systems of Equations

The equations in Examples 1 and 2 are linear. The method of substitution can also be used to solve systems in which one or both of the equations are nonlinear.

EXAMPLE 3 Substitution: Two-Solution Case

Solve the system of equations.

$$\begin{cases} 3x^2 + 4x - y = 7 & \text{Equation 1} \\ 2x - y = -1 & \text{Equation 2} \end{cases}$$

Solution Begin by solving for y in Equation 2 to obtain $y = 2x + 1$. Next, substitute this expression for y into Equation 1 and solve for x.

$3x^2 + 4x - (2x + 1) = 7$ — Substitute $2x + 1$ for y in Equation 1.

$3x^2 + 2x - 1 = 7$ — Simplify.

$3x^2 + 2x - 8 = 0$ — Write in general form.

$(3x - 4)(x + 2) = 0$ — Factor.

$x = \frac{4}{3}, -2$ — Solve for x.

Back-substituting these values of x to solve for the corresponding values of y produces the solutions $\left(\frac{4}{3}, \frac{11}{3}\right)$ and $(-2, -3)$. Check these in the original system.

▷ **ALGEBRA HELP** You can review the techniques for factoring in Appendix A.3.

✓ **Checkpoint** Audio-video solution in English & Spanish at LarsonPrecalculus.com.

Solve the system of equations.

$$\begin{cases} -2x + y = 5 \\ x^2 - y + 3x = 1 \end{cases}$$

EXAMPLE 4 Substitution: No-Real-Solution Case

Solve the system of equations.

$$\begin{cases} -x + y = 4 & \text{Equation 1} \\ x^2 + y = 3 & \text{Equation 2} \end{cases}$$

Solution Begin by solving for y in Equation 1 to obtain $y = x + 4$. Next, substitute this expression for y into Equation 2 and solve for x.

$x^2 + (x + 4) = 3$ — Substitute $x + 4$ for y in Equation 2.

$x^2 + x + 1 = 0$ — Write in general form.

$x = \frac{-1 \pm \sqrt{-3}}{2}$ — Use the Quadratic Formula.

Because the discriminant is negative, the equation

$$x^2 + x + 1 = 0$$

has no (real) solution. So, the original system has no (real) solution.

✓ **Checkpoint** Audio-video solution in English & Spanish at LarsonPrecalculus.com.

Solve the system of equations.

$$\begin{cases} 2x - y = -3 \\ 2x^2 + 4x - y^2 = 0 \end{cases}$$

▷ **TECHNOLOGY** Most graphing utilities have built-in features that approximate the point(s) of intersection of two graphs. Use a graphing utility to find the points of intersection of the graphs in Figures 7.1 through 7.3. Be sure to adjust the viewing window so that you see all the points of intersection.

▷ **ALGEBRA HELP** You can review the techniques for graphing equations in Section 1.2.

Graphical Approach to Finding Solutions

Notice from Examples 2, 3, and 4 that a system of two equations in two unknowns can have exactly one solution, more than one solution, or no solution. By using a **graphical method,** you can gain insight about the number of solutions and the location(s) of the solution(s) of a system of equations by graphing each of the equations in the same coordinate plane. The solutions of the system correspond to the **points of intersection** of the graphs. For instance, the two equations in Figure 7.1 graph as two lines with a *single point* of intersection; the two equations in Figure 7.2 graph as a parabola and a line with *two points* of intersection; and the two equations in Figure 7.3 graph as a parabola and a line with *no points* of intersection.

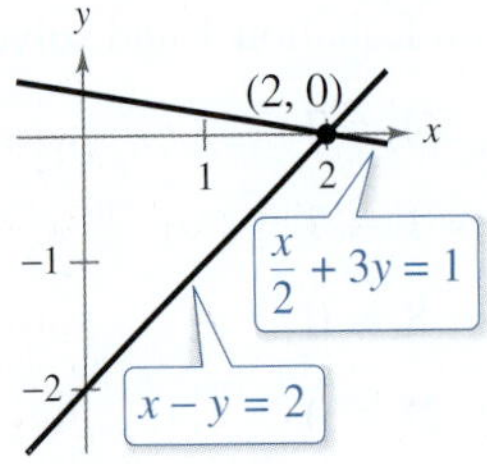

One intersection point
Figure 7.1

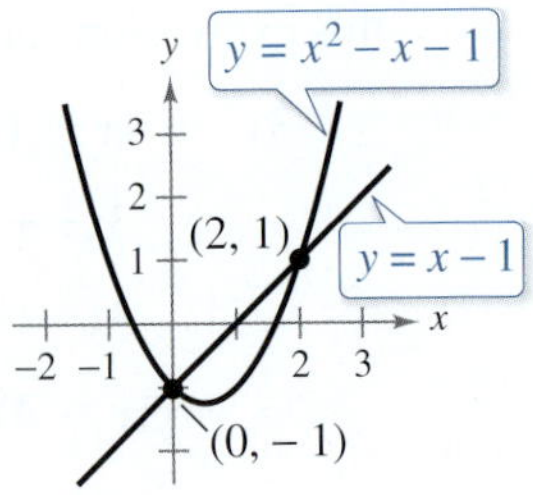

Two intersection points
Figure 7.2

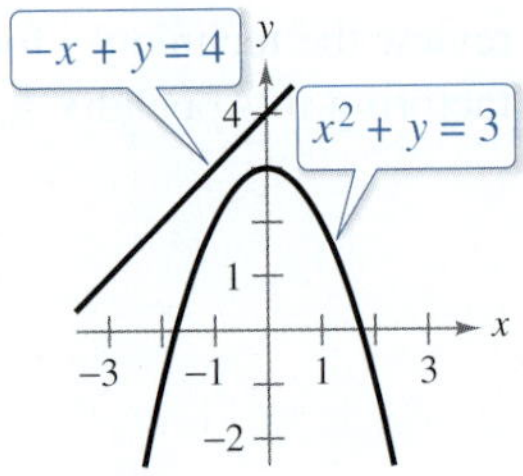

No intersection points
Figure 7.3

EXAMPLE 5 Solving a System of Equations Graphically

Solve the system of equations.

$$\begin{cases} y = \ln x & \text{Equation 1} \\ x + y = 1 & \text{Equation 2} \end{cases}$$

Solution There is only one point of intersection of the graphs of the two equations, and (1, 0) is the solution point (see Figure 7.4). Check this solution as follows.

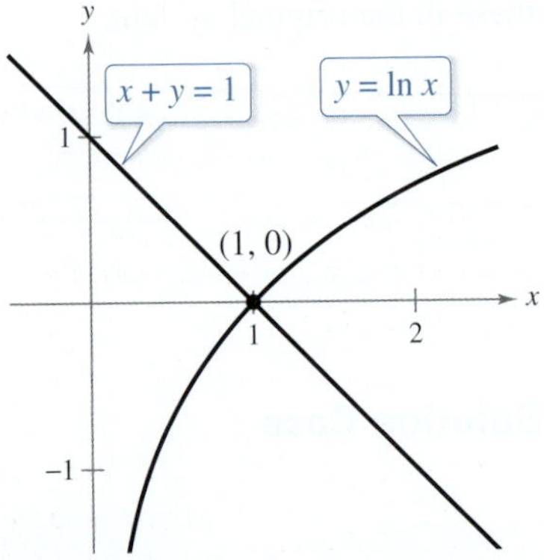

Figure 7.4

Check (1, 0) in Equation 1:

$y = \ln x$ Write Equation 1.

$0 = \ln 1$ Substitute for x and y.

$0 = 0$ Solution checks in Equation 1. ✓

Check (1, 0) in Equation 2:

$x + y = 1$ Write Equation 2.

$1 + 0 = 1$ Substitute for x and y.

$1 = 1$ Solution checks in Equation 2. ✓

✓ ***Checkpoint*** *Audio-video solution in English & Spanish at LarsonPrecalculus.com.*

Solve the system of equations.

$$\begin{cases} y = 3 - \log x \\ -2x + y = 1 \end{cases}$$

Example 5 shows the benefit of a graphical approach to solving systems of equations in two variables. Notice that by trying only the substitution method in Example 5, you would obtain the equation $x + \ln x = 1$. It would be difficult to solve this equation for x using standard algebraic techniques.

Applications

The total cost C of producing x units of a product typically has two components—the initial cost and the cost per unit. When enough units have been sold so that the total revenue R equals the total cost C, the sales are said to have reached the **break-even point.** You will find that the break-even point corresponds to the point of intersection of the cost and revenue curves.

EXAMPLE 6 Break-Even Analysis

A shoe company invests \$300,000 in equipment to produce a new line of athletic footwear. Each pair of shoes costs \$5 to produce and sells for \$60. How many pairs of shoes must the company sell to break even?

Algebraic Solution

The total cost of producing x units is

Total cost = Cost per unit · Number of units + Initial cost

$$C = 5x + 300{,}000. \qquad \text{Equation 1}$$

The revenue obtained by selling x units is

Total revenue = Price per unit · Number of units

$$R = 60x. \qquad \text{Equation 2}$$

Because the break-even point occurs when $R = C$, you have $C = 60x$, and the system of equations to solve is

$$\begin{cases} C = 5x + 300{,}000 \\ C = 60x \end{cases}.$$

Solve by substitution.

$60x = 5x + 300{,}000$ Substitute $60x$ for C in Equation 1.

$55x = 300{,}000$ Subtract $5x$ from each side.

$x \approx 5455$ Divide each side by 55.

So, the company must sell about 5455 pairs of shoes to break even.

Graphical Solution

The total cost of producing x units is

Total cost = Cost per unit · Number of units + Initial cost

$$C = 5x + 300{,}000. \qquad \text{Equation 1}$$

The revenue obtained by selling x units is

Total revenue = Price per unit · Number of units

$$R = 60x. \qquad \text{Equation 2}$$

Because the break-even point occurs when $R = C$, you have $C = 60x$, and the system of equations to solve is

$$\begin{cases} C = 5x + 300{,}000 \\ C = 60x \end{cases}.$$

Use a graphing utility to graph $y_1 = 5x + 300{,}000$ and $y_2 = 60x$ in the same viewing window.

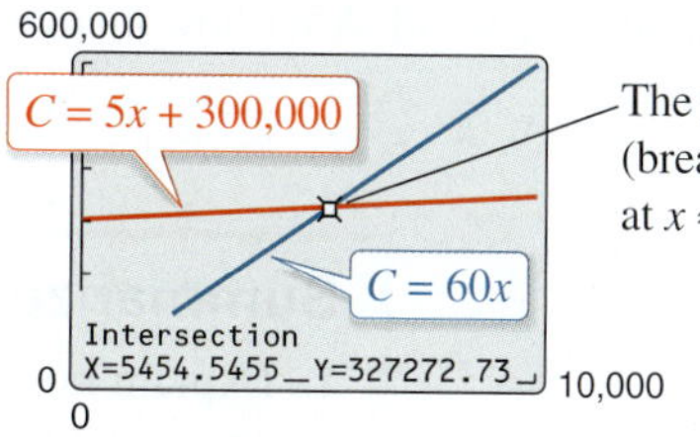

The point of intersection (break-even point) occurs at $x \approx 5455$.

So, the company must sell about 5455 pairs of shoes to break even.

✓ Checkpoint *Audio-video solution in English & Spanish at LarsonPrecalculus.com.*

In Example 6, each pair of shoes costs \$12 to produce. How many pairs of shoes must the company sell to break even?

Another way to view the solution in Example 6 is to consider the profit function $P = R - C$. The break-even point occurs when the profit is 0, which is the same as saying that $R = C$.

EXAMPLE 7 Movie Ticket Sales

The weekly ticket sales for a new comedy movie decreased each week. At the same time, the weekly ticket sales for a new drama movie increased each week. Models that approximate the weekly ticket sales S (in millions of dollars) for the movies are

$$\begin{cases} S = 60 - 8x & \text{Comedy} \\ S = 10 + 4.5x & \text{Drama} \end{cases}$$

where x represents the number of weeks each movie was in theaters, with $x = 0$ corresponding to the opening weekend. After how many weeks will the ticket sales for the two movies be equal?

Algebraic Solution

Because both equations are already solved for S in terms of x, substitute either expression for S into the other equation and solve for x.

$10 + 4.5x = 60 - 8x$ Substitute for S in Equation 1.

$4.5x + 8x = 60 - 10$ Add $8x$ and -10 to each side.

$12.5x = 50$ Combine like terms.

$x = 4$ Divide each side by 12.5.

So, the weekly ticket sales for the two movies will be equal after 4 weeks.

Numerical Solution

You can create a table of values for each model to determine when the ticket sales for the two movies will be equal.

Number of Weeks, x	0	1	2	3	4	5	6
Sales, S (comedy)	60	52	44	36	28	20	12
Sales, S (drama)	10	14.5	19	23.5	28	32.5	37

So, from the above table, the weekly ticket sales for the two movies will be equal after 4 weeks.

✓ Checkpoint Audio-video solution in English & Spanish at LarsonPrecalculus.com.

The weekly ticket sales for a new animated movie decreased each week. At the same time, the weekly ticket sales for a new horror movie increased each week. Models that approximate the weekly ticket sales S (in millions of dollars) for the movies are

$$\begin{cases} S = 108 - 9.4x & \text{Animated} \\ S = 16 + 9x & \text{Horror} \end{cases}$$

where x represents the number of weeks each movie was in theaters, with $x = 0$ corresponding to the opening weekend. After how many weeks will the ticket sales for the two movies be equal?

Summarize (Section 7.1)

1. Explain how to use the method of substitution to solve a system of linear equations in two variables *(page 466)*. For examples of using the method of substitution to solve systems of linear equations in two variables, see Examples 1 and 2.
2. Explain how to use the method of substitution to solve a system of nonlinear equations in two variables *(page 469)*. For examples of using the method of substitution to solve systems of nonlinear equations in two variables, see Examples 3 and 4.
3. Explain how to use a graphical approach to solve a system of equations in two variables *(page 470)*. For an example of using a graphical approach to solve a system of equations in two variables, see Example 5.
4. Describe examples of how to use systems of equations to model and solve real-life problems *(pages 471 and 472, Examples 6 and 7)*.

7.1 Exercises

See CalcChat.com for tutorial help and worked-out solutions to odd-numbered exercises.

Vocabulary: Fill in the blanks.

1. A ________ of a system of equations is an ordered pair that satisfies each equation in the system.
2. The first step in solving a system of equations by the method of ________ is to solve one of the equations for one variable in terms of the other variable.
3. Graphically, the solution of a system of two equations is the ________ of ________ of the graphs of the two equations.
4. In business applications, the point at which the revenue equals costs is called the ________ point.

Skills and Applications

Checking Solutions **In Exercises 5 and 6, determine whether each ordered pair is a solution of the system.**

5. $\begin{cases} 2x - y = 4 \\ 8x + y = -9 \end{cases}$ (a) $(0, -4)$ (b) $(-2, 7)$ (c) $\left(\frac{3}{2}, -1\right)$ (d) $\left(-\frac{1}{2}, -5\right)$

6. $\begin{cases} 4x^2 + y = 3 \\ -x - y = 11 \end{cases}$ (a) $(2, -13)$ (b) $(2, -9)$ (c) $\left(-\frac{3}{2}, -\frac{31}{3}\right)$ (d) $\left(-\frac{7}{4}, -\frac{37}{4}\right)$

Solving a System by Substitution **In Exercises 7–14, solve the system by the method of substitution. Check your solution(s) graphically.**

7. $\begin{cases} 2x + y = 6 \\ -x + y = 0 \end{cases}$

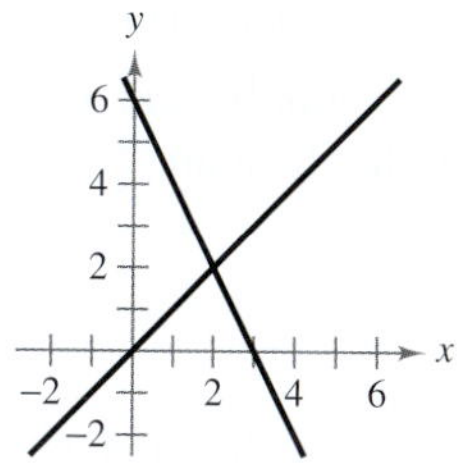

8. $\begin{cases} x - 4y = -11 \\ x + 3y = 3 \end{cases}$

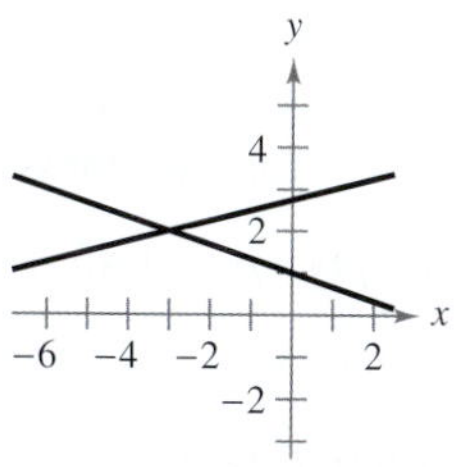

9. $\begin{cases} x - y = -4 \\ x^2 - y = -2 \end{cases}$

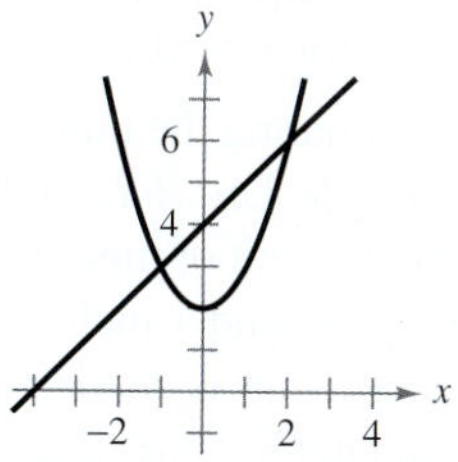

10. $\begin{cases} 3x + y = 2 \\ x^3 - 2 + y = 0 \end{cases}$

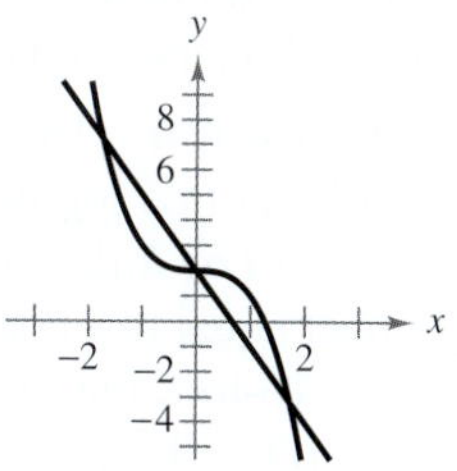

11. $\begin{cases} -\frac{1}{2}x + y = -\frac{5}{2} \\ x^2 + y^2 = 25 \end{cases}$

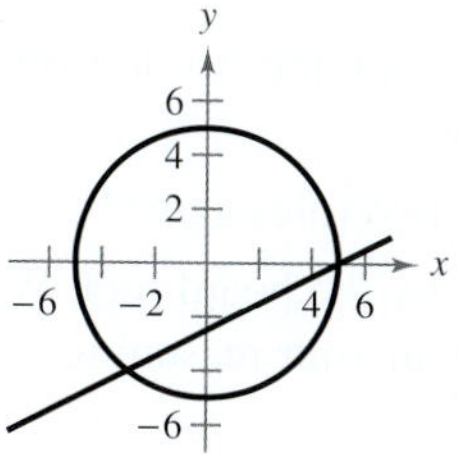

12. $\begin{cases} x + y = 0 \\ x^3 - 5x - y = 0 \end{cases}$

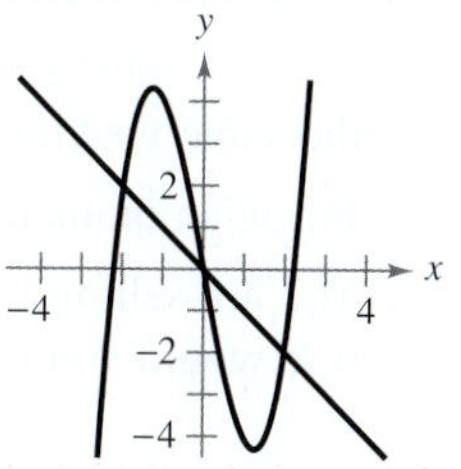

13. $\begin{cases} x^2 + y = 0 \\ x^2 - 4x - y = 0 \end{cases}$

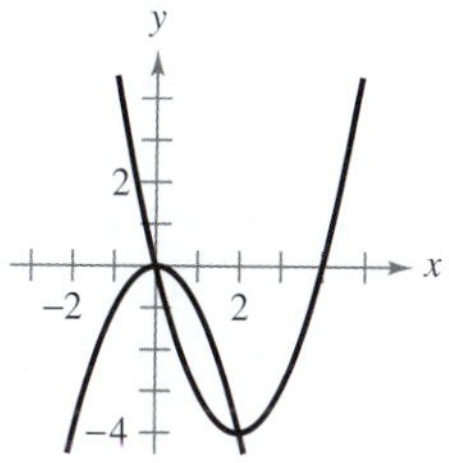

14. $\begin{cases} y = x^3 - 3x^2 + 1 \\ y = x^2 - 3x + 1 \end{cases}$

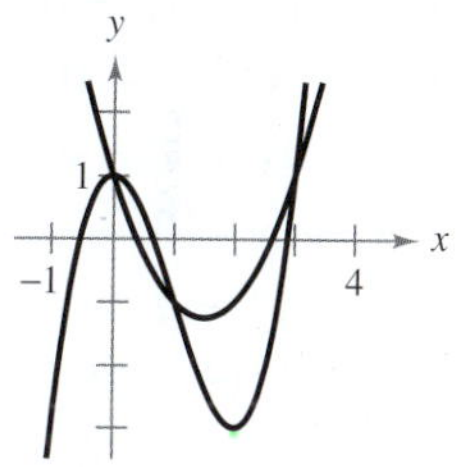

Solving a System by Substitution **In Exercises 15–24, solve the system by the method of substitution.**

15. $\begin{cases} x - y = 2 \\ 6x - 5y = 16 \end{cases}$

16. $\begin{cases} x + 4y = 3 \\ 2x - 7y = -24 \end{cases}$

17. $\begin{cases} 2x - y + 2 = 0 \\ 4x + y - 5 = 0 \end{cases}$

18. $\begin{cases} 6x - 3y - 4 = 0 \\ x + 2y - 4 = 0 \end{cases}$

19. $\begin{cases} 1.5x + 0.8y = 2.3 \\ 0.3x - 0.2y = 0.1 \end{cases}$

20. $\begin{cases} 0.5x + 3.2y = 9.0 \\ 0.2x - 1.6y = -3.6 \end{cases}$

21. $\begin{cases} \frac{1}{5}x + \frac{1}{2}y = 8 \\ x + y = 20 \end{cases}$

22. $\begin{cases} \frac{1}{2}x + \frac{3}{4}y = 10 \\ \frac{3}{4}x - y = 4 \end{cases}$

23. $\begin{cases} 6x + 5y = -3 \\ -x - \frac{5}{6}y = -7 \end{cases}$

24. $\begin{cases} -\frac{2}{3}x + y = 2 \\ 2x - 3y = 6 \end{cases}$

Solving a System by Substitution **In Exercises 25–28, you are given the total annual interest earned from a total of \$12,000 invested in two funds paying the given rates of simple interest. Write and solve a system of equations to find the amount invested at each rate.**

	Annual Interest	Rate 1	Rate 2
25.	\$500	3%	5%
26.	\$630	4%	6%
27.	\$396	2.8%	3.8%
28.	\$254	1.75%	2.25%

Solving a System with a Nonlinear Equation **In Exercises 29–32, solve the system by the method of substitution.**

29. $\begin{cases} x^2 - y = 0 \\ 2x + y = 0 \end{cases}$
30. $\begin{cases} x - 2y = 0 \\ 3x - y^2 = 0 \end{cases}$
31. $\begin{cases} x - y = -1 \\ x^2 - y = -4 \end{cases}$
32. $\begin{cases} y = -x \\ y = x^3 + 3x^2 + 2x \end{cases}$

Solving a System of Equations Graphically **In Exercises 33–44, solve the system graphically.**

33. $\begin{cases} -x + 2y = -2 \\ 3x + y = 20 \end{cases}$
34. $\begin{cases} x + y = 0 \\ 3x - 2y = 5 \end{cases}$
35. $\begin{cases} x - 3y = -3 \\ 5x + 3y = -6 \end{cases}$
36. $\begin{cases} -x + 2y = -7 \\ x - y = 2 \end{cases}$
37. $\begin{cases} x + y = 4 \\ x^2 + y^2 - 4x = 0 \end{cases}$
38. $\begin{cases} -x + y = 3 \\ x^2 - 6x - 27 + y^2 = 0 \end{cases}$
39. $\begin{cases} x - y + 3 = 0 \\ x^2 - 4x + 7 = y \end{cases}$
40. $\begin{cases} y^2 - 4x + 11 = 0 \\ -\frac{1}{2}x + y = -\frac{1}{2} \end{cases}$
41. $\begin{cases} 3x - 2y = 0 \\ x^2 - y^2 = 4 \end{cases}$
42. $\begin{cases} 2x - y + 3 = 0 \\ x^2 + y^2 - 4x = 0 \end{cases}$
43. $\begin{cases} x^2 + y^2 = 25 \\ 3x^2 - 16y = 0 \end{cases}$
44. $\begin{cases} x^2 + y^2 = 25 \\ (x - 8)^2 + y^2 = 41 \end{cases}$

Solving a System of Equations Graphically **In Exercises 45–48, use a graphing utility to solve the system of equations. Find the solution(s) accurate to two decimal places.**

45. $\begin{cases} y = e^x \\ x - y + 1 = 0 \end{cases}$
46. $\begin{cases} y = -4e^{-x} \\ y + 3x + 8 = 0 \end{cases}$
47. $\begin{cases} y + 2 = \ln(x - 1) \\ 3y + 2x = 9 \end{cases}$
48. $\begin{cases} x^2 + y^2 = 4 \\ 2x^2 - y = 2 \end{cases}$

Choosing a Solution Method **In Exercises 49–56, solve the system graphically or algebraically. Explain your choice of method.**

49. $\begin{cases} y = 2x \\ y = x^2 + 1 \end{cases}$
50. $\begin{cases} x^2 + y^2 = 25 \\ 2x + y = 10 \end{cases}$
51. $\begin{cases} x - 2y = 4 \\ x^2 - y = 0 \end{cases}$
52. $\begin{cases} y = x^3 - 2x^2 + x - 1 \\ y = -x^2 + 3x - 1 \end{cases}$
53. $\begin{cases} y - e^{-x} = 1 \\ y - \ln x = 3 \end{cases}$
54. $\begin{cases} x^2 + y = 4 \\ e^x - y = 0 \end{cases}$
55. $\begin{cases} xy - 1 = 0 \\ 2x - 4y + 7 = 0 \end{cases}$
56. $\begin{cases} x - 2y = 1 \\ y = \sqrt{x - 1} \end{cases}$

Break-Even Analysis **In Exercises 57 and 58, find the sales necessary to break even ($R = C$) for the total cost C of producing x units and the revenue R obtained by selling x units. (Round to the nearest whole unit.)**

57. $C = 8650x + 250{,}000, \quad R = 9950x$
58. $C = 5.5\sqrt{x} + 10{,}000, \quad R = 3.29x$

59. **Break-Even Analysis** A small software company invests \$16,000 to produce a software package that will sell for \$55.95. Each unit costs \$9.45 to produce.
    (a) How many units must the company sell to break even?
    (b) How many units must the company sell to make a profit of \$100,000?

60. **Choice of Two Jobs** You receive two sales job offers. One company offers a straight commission of 6% of sales. The other company offers a salary of \$500 per week plus 3% of sales. How much would you have to sell in a week in order to make the straight commission job offer better?

61. **DVD Rentals** The number of rentals of a newly released DVD of a horror film at a movie rental store decreased each week. At the same time, the number of rentals of a newly released DVD of a comedy film increased each week. Models that approximate the numbers N of DVDs rented are

$$\begin{cases} N = 360 - 24x & \text{Horror film} \\ N = 24 + 18x & \text{Comedy film} \end{cases}$$

where x represents the week, with $x = 1$ corresponding to the first week of release.
    (a) After how many weeks will the numbers of DVDs rented for the two films be equal?
    (b) Use a table to solve the system of equations numerically. Compare your result with that of part (a).

62. **Supply and Demand** The supply and demand curves for a business dealing with wheat are

Supply: $p = 1.45 + 0.00014x^2$

Demand: $p = (2.388 - 0.007x)^2$

where p is the price in dollars per bushel and x is the quantity in bushels per day. Use a graphing utility to graph the supply and demand equations and find the market equilibrium. (The market equilibrium is the point of intersection of the graphs for $x > 0$.)

63. **Log Volume** Two rules for estimating the number of board feet in a log include the *Doyle Log Rule* and the *Scribner Log Rule*. (A board foot is a unit of measure for lumber equal to a board 1 foot square and 1 inch thick.) For a 16-foot log, the Doyle Log Rule is modeled by $V_1 = (D - 4)^2, 5 \le D \le 40$, and the Scribner Log Rule is modeled by $V_2 = 0.79D^2 - 2D - 4, 5 \le D \le 40$, where D is the diameter (in inches) of the log and V is its volume (in board feet).
    (a) Use a graphing utility to graph the two log rules in the same viewing window.
    (b) For what diameter do the two rules agree?
    (c) You are selling large logs by the board foot. Which rule would you use? Explain your reasoning.

64. Data Analysis: Renewable Energy The table shows the consumption C (in trillions of Btus) of geothermal energy and wind energy in the United States from 2001 through 2009. *(Source: U.S. Energy Information Administration)*

DATA

Year	Geothermal, C	Wind, C
2001	164	70
2002	171	105
2003	175	115
2004	178	142
2005	181	178
2006	181	264
2007	186	341
2008	192	546
2009	200	721

Spreadsheet at LarsonPrecalculus.com

(a) Use a graphing utility to find a cubic model for the geothermal energy consumption data and a quadratic model for the wind energy consumption data. Let t represent the year, with $t = 1$ corresponding to 2001.

(b) Use the graphing utility to graph the data and the two models in the same viewing window.

(c) Use the graph from part (b) to approximate the point of intersection of the graphs of the models. Interpret your answer in the context of the problem.

(d) Describe the behavior of each model. Do you think the models can accurately predict consumption of geothermal energy and wind energy in the United States for future years? Explain.

(e) Use your school's library, the Internet, or some other reference source to research the advantages and disadvantages of using renewable energy.

Geometry **In Exercises 65 and 66, find the dimensions of the rectangle meeting the specified conditions.**

65. The perimeter is 56 meters and the length is 4 meters greater than the width.

66. The perimeter is 42 inches and the width is three-fourths the length.

67. Geometry What are the dimensions of a rectangular tract of land when its perimeter is 44 kilometers and its area is 120 square kilometers?

68. Geometry What are the dimensions of a right triangle with a two-inch hypotenuse and an area of 1 square inch?

Exploration

True or False? **In Exercises 69 and 70, determine whether the statement is true or false. Justify your answer.**

69. In order to solve a system of equations by substitution, you must always solve for y in one of the two equations and then back-substitute.

70. If a system consists of a parabola and a circle, then the system can have at most two solutions.

71. Think About It When solving a system of equations by substitution, how do you recognize that the system has no solution?

72. HOW DO YOU SEE IT? The cost C of producing x units and the revenue R obtained by selling x units are shown in the figure.

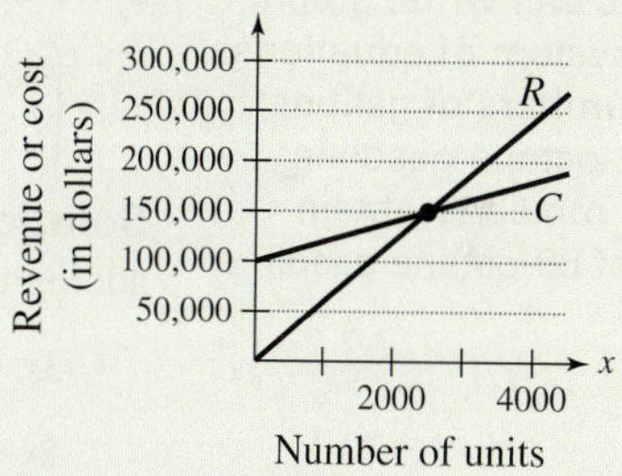

(a) Estimate the point of intersection. What does this point represent?

(b) Use the figure to identify the x-values that correspond to (i) an overall loss and (ii) a profit. Explain your reasoning.

73. Think About It Consider the system of equations

$$\begin{cases} ax + by = c \\ dx + ey = f \end{cases}.$$

(a) Find values for a, b, c, d, e, and f so that the system has one distinct solution. (There is more than one correct answer.)

(b) Explain how to solve the system in part (a) by the method of substitution and graphically.

(c) Write a brief paragraph describing any advantages of the method of substitution over the graphical method of solving a system of equations.

ssuaphotos/Shutterstock.com

7.2 Two-Variable Linear Systems

Systems of equations in two variables can help you model and solve real-life problems. For instance, in Exercise 50 on page 485, you will write, graph, and solve a system of equations to find the numbers of gallons of 87- and 92-octane gasoline that must be mixed to obtain 500 gallons of 89-octane gasoline.

- **Use the method of elimination to solve systems of linear equations in two variables.**
- **Interpret graphically the numbers of solutions of systems of linear equations in two variables.**
- **Use systems of linear equations in two variables to model and solve real-life problems.**

The Method of Elimination

In Section 7.1, you studied two methods for solving a system of equations: substitution and graphing. Now, you will study the **method of elimination.** The key step in this method is to obtain, for one of the variables, coefficients that differ only in sign so that *adding* the equations eliminates the variable.

$$\begin{aligned} 3x + 5y &= 7 && \text{Equation 1} \\ -3x - 2y &= -1 && \text{Equation 2} \\ 3y &= 6 && \text{Add equations.} \end{aligned}$$

Note that by adding the two equations, you eliminate the x-terms and obtain a single equation in y. Solving this equation for y produces $y = 2$, which you can then back-substitute into one of the original equations to solve for x.

EXAMPLE 1 Solving a System of Equations by Elimination

Solve the system of linear equations.

$$\begin{cases} 3x + 2y = 4 & \text{Equation 1} \\ 5x - 2y = 12 & \text{Equation 2} \end{cases}$$

Solution Because the coefficients of y differ only in sign, eliminate the y-terms by adding the two equations.

$$\begin{aligned} 3x + 2y &= 4 && \text{Write Equation 1.} \\ 5x - 2y &= 12 && \text{Write Equation 2.} \\ 8x &= 16 && \text{Add equations.} \\ x &= 2 && \text{Solve for } x. \end{aligned}$$

Solve for y by back-substituting $x = 2$ into Equation 1.

$$\begin{aligned} 3x + 2y &= 4 && \text{Write Equation 1.} \\ 3(2) + 2y &= 4 && \text{Substitute 2 for } x. \\ y &= -1 && \text{Solve for } y. \end{aligned}$$

The solution is $(2, -1)$. Check this in the original system, as follows.

Check

$$\begin{aligned} 3(2) + 2(-1) &= 4 && \text{Solution checks in Equation 1. ✓} \\ 5(2) - 2(-1) &= 12 && \text{Solution checks in Equation 2. ✓} \end{aligned}$$

✓ ***Checkpoint*** *Audio-video solution in English & Spanish at LarsonPrecalculus.com.*

Solve the system of linear equations.

$$\begin{cases} 2x + y = 4 \\ 2x - y = -1 \end{cases}$$

Vlad61/Shutterstock.com

Method of Elimination

To use the **method of elimination** to solve a system of two linear equations in x and y, perform the following steps.

1. *Obtain coefficients* for x (or y) that differ only in sign by multiplying all terms of one or both equations by suitably chosen constants.
2. *Add* the equations to eliminate one variable.
3. *Solve* the equation obtained in Step 2.
4. *Back-substitute* the value obtained in Step 3 into either of the original equations and solve for the other variable.
5. *Check* that the solution satisfies *each* of the original equations.

EXAMPLE 2 Solving a System of Equations by Elimination

Solve the system of linear equations.

$$\begin{cases} 2x - 4y = -7 & \text{Equation 1} \\ 5x + y = -1 & \text{Equation 2} \end{cases}$$

Solution To obtain coefficients that differ only in sign, multiply Equation 2 by 4.

$$2x - 4y = -7 \quad \Rightarrow \quad 2x - 4y = -7 \qquad \text{Write Equation 1.}$$
$$5x + y = -1 \quad \Rightarrow \quad 20x + 4y = -4 \qquad \text{Multiply Equation 2 by 4.}$$
$$22x = -11 \qquad \text{Add equations.}$$
$$x = -\tfrac{1}{2} \qquad \text{Solve for } x.$$

Solve for y by back-substituting $x = -\frac{1}{2}$ into Equation 1.

$$2x - 4y = -7 \qquad \text{Write Equation 1.}$$
$$2\left(-\tfrac{1}{2}\right) - 4y = -7 \qquad \text{Substitute } -\tfrac{1}{2} \text{ for } x.$$
$$-4y = -6 \qquad \text{Simplify.}$$
$$y = \tfrac{3}{2} \qquad \text{Solve for } y.$$

The solution is $\left(-\frac{1}{2}, \frac{3}{2}\right)$. Check this in the original system, as follows.

Check

$$2x - 4y = -7 \qquad \text{Write Equation 1.}$$
$$2\left(-\tfrac{1}{2}\right) - 4\left(\tfrac{3}{2}\right) \stackrel{?}{=} -7 \qquad \text{Substitute for } x \text{ and } y.$$
$$-1 - 6 = -7 \qquad \text{Solution checks in Equation 1. ✓}$$
$$5x + y = -1 \qquad \text{Write Equation 2.}$$
$$5\left(-\tfrac{1}{2}\right) + \tfrac{3}{2} \stackrel{?}{=} -1 \qquad \text{Substitute for } x \text{ and } y.$$
$$-\tfrac{5}{2} + \tfrac{3}{2} = -1 \qquad \text{Solution checks in Equation 2. ✓}$$

✓ ***Checkpoint*** *Audio-video solution in English & Spanish at LarsonPrecalculus.com.*

Solve the system of linear equations.

$$\begin{cases} 2x + 3y = 17 \\ 5x - y = 17 \end{cases}$$

In Example 2, the two systems of linear equations (the original system and the system obtained by multiplying by a constant)

$$\begin{cases} 2x - 4y = -7 \\ 5x + \ \ y = -1 \end{cases} \quad \text{and} \quad \begin{cases} 2x - 4y = -7 \\ 20x + 4y = -4 \end{cases}$$

are called **equivalent systems** because they have precisely the same solution set. The operations that can be performed on a system of linear equations to produce an equivalent system are (1) interchanging any two equations, (2) multiplying an equation by a nonzero constant, and (3) adding a multiple of one equation to any other equation in the system.

EXAMPLE 3 Solving the System of Equations by Elimination

Solve the system of linear equations.

$$\begin{cases} 5x + 3y = 9 & \text{Equation 1} \\ 2x - 4y = 14 & \text{Equation 2} \end{cases}$$

Algebraic Solution

You can obtain coefficients that differ only in sign by multiplying Equation 1 by 4 and multiplying Equation 2 by 3.

$5x + 3y = 9 \Rightarrow 20x + 12y = 36$ Multiply Equation 1 by 4.

$2x - 4y = 14 \Rightarrow 6x - 12y = 42$ Multiply Equation 2 by 3.

$26x = 78$ Add equations.

$x = 3$ Solve for x.

Solve for y by back-substituting $x = 3$ into Equation 2.

$2x - 4y = 14$ Write Equation 2.

$2(3) - 4y = 14$ Substitute 3 for x.

$-4y = 8$ Simplify.

$y = -2$ Solve for y.

The solution is

$(3, -2)$.

Check this in the original system.

Graphical Solution

Solve each equation for y and use a graphing utility to graph the equations in the same viewing window.

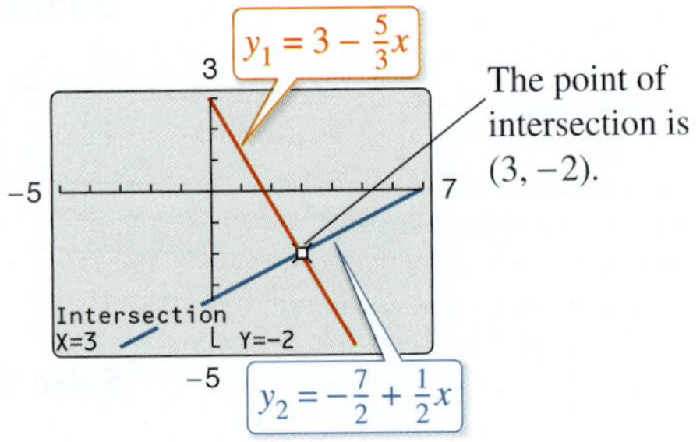

From the figure, the solution is

$(3, -2)$.

Check this in the original system.

✓ Checkpoint Audio-video solution in English & Spanish at LarsonPrecalculus.com.

Solve the system of linear equations.

$$\begin{cases} 3x + 2y = 7 \\ 2x + 5y = 1 \end{cases}$$

Check the solution from Example 3, as follows.

$5(3) + 3(-2) \stackrel{?}{=} 9$ Substitute 3 for x and -2 for y in Equation 1.

$15 - 6 = 9$ Solution checks in Equation 1. ✓

$2(3) - 4(-2) \stackrel{?}{=} 14$ Substitute 3 for x and -2 for y in Equation 2.

$6 + 8 = 14$ Solution checks in Equation 2. ✓

Keep in mind that the terminology and methods discussed in this section apply only to systems of *linear* equations.

▷ **TECHNOLOGY** The general solution of the linear system

$$\begin{cases} ax + by = c \\ dx + ey = f \end{cases}$$

is

$$x = \frac{ce - bf}{ae - bd}$$

and

$$y = \frac{af - cd}{ae - bd}.$$

If $ae - bd = 0$, then the system does not have a unique solution. A graphing utility program (called Systems of Linear Equations) for solving such a system can be found at the website for this text at *LarsonPrecalculus.com*. Try using the program to solve the system in Example 4.

Example 4 illustrates a strategy for solving a system of linear equations that has decimal coefficients.

EXAMPLE 4 A Linear System Having Decimal Coefficients

Solve the system of linear equations.

$$\begin{cases} 0.02x - 0.05y = -0.38 \\ 0.03x + 0.04y = 1.04 \end{cases}$$ Equation 1, Equation 2

Solution Because the coefficients in this system have two decimal places, you can begin by multiplying each equation by 100. This produces a system in which the coefficients are all integers.

$$\begin{cases} 2x - 5y = -38 \\ 3x + 4y = 104 \end{cases}$$ Revised Equation 1, Revised Equation 2

Now, to obtain coefficients that differ only in sign, multiply Equation 1 by 3 and multiply Equation 2 by -2.

$2x - 5y = -38 \Rightarrow 6x - 15y = -114$ Multiply Equation 1 by 3.

$3x + 4y = 104 \Rightarrow -6x - 8y = -208$ Multiply Equation 2 by -2.

$-23y = -322$ Add equations.

So,

$$y = \frac{-322}{-23} = 14.$$

Back-substituting $y = 14$ into revised Equation 2 produces the following.

$3x + 4y = 104$ Write revised Equation 2.

$3x + 4(14) = 104$ Substitute 14 for y.

$3x = 48$ Simplify.

$x = 16$ Solve for x.

The solution is

$(16, 14).$

Check this in the original system, as follows.

Check

$0.02x - 0.05y = -0.38$ Write Equation 1.

$0.02(16) - 0.05(14) \stackrel{?}{=} -0.38$ Substitute for x and y.

$0.32 - 0.70 = -0.38$ Solution checks in Equation 1. ✓

$0.03x + 0.04y = 1.04$ Write Equation 2.

$0.03(16) + 0.04(14) \stackrel{?}{=} 1.04$ Substitute for x and y.

$0.48 + 0.56 = 1.04$ Solution checks in Equation 2. ✓

✓ **Checkpoint** Audio-video solution in English & Spanish at LarsonPrecalculus.com.

Solve the system of linear equations.

$$\begin{cases} 0.03x + 0.04y = 0.75 \\ 0.02x + 0.06y = 0.90 \end{cases}$$

Graphical Interpretation of Solutions

It is possible for a *general* system of equations to have exactly one solution, two or more solutions, or no solution. If a system of *linear* equations has two different solutions, then it must have an *infinite* number of solutions.

Graphical Interpretations of Solutions

For a system of two linear equations in two variables, the number of solutions is one of the following.

Number of Solutions	Graphical Interpretation	Slopes of Lines
1. Exactly one solution	The two lines intersect at one point.	The slopes of the two lines are not equal.
2. Infinitely many solutions	The two lines coincide (are identical).	The slopes of the two lines are equal.
3. No solution	The two lines are parallel.	The slopes of the two lines are equal.

A system of linear equations is **consistent** when it has at least one solution. A consistent system with exactly one solution is *independent*, whereas a consistent system with infinitely many solutions is *dependent*. A system is **inconsistent** when it has no solution.

REMARK A comparison of the slopes of two lines gives useful information about the number of solutions of the corresponding system of equations. To solve a system of equations graphically, it helps to begin by writing the equations in slope-intercept form. Try doing this for the systems in Example 5.

EXAMPLE 5 **Recognizing Graphs of Linear Systems**

Match each system of linear equations with its graph. Describe the number of solutions and state whether the system is consistent or inconsistent.

a. $\begin{cases} 2x - 3y = 3 \\ -4x + 6y = 6 \end{cases}$ **b.** $\begin{cases} 2x - 3y = 3 \\ x + 2y = 5 \end{cases}$ **c.** $\begin{cases} 2x - 3y = 3 \\ -4x + 6y = -6 \end{cases}$

i.

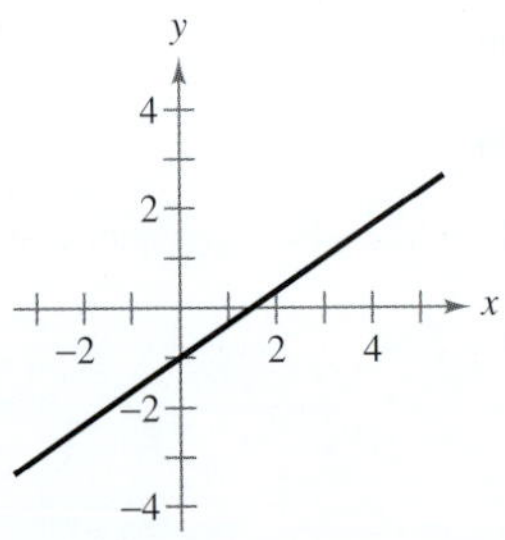

ii.

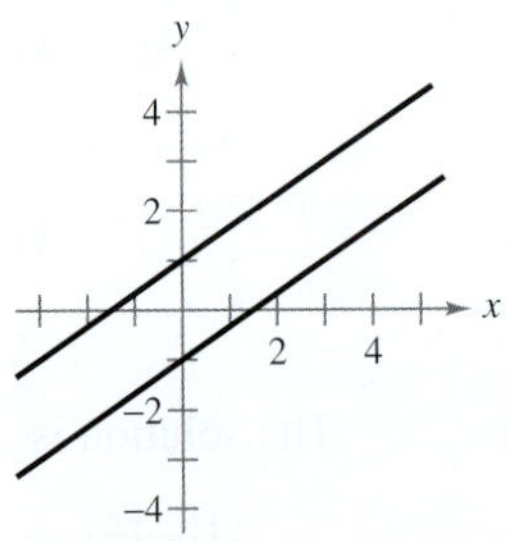

iii.

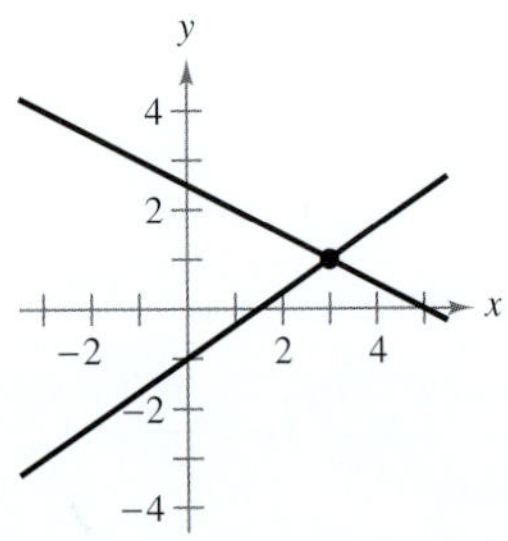

Solution

a. The graph of system (a) is a pair of parallel lines (ii). The lines have no point of intersection, so the system has no solution. The system is inconsistent.

b. The graph of system (b) is a pair of intersecting lines (iii). The lines have one point of intersection, so the system has exactly one solution. The system is consistent.

c. The graph of system (c) is a pair of lines that coincide (i). The lines have infinitely many points of intersection, so the system has infinitely many solutions. The system is consistent.

✓ ***Checkpoint*** *Audio-video solution in English & Spanish at LarsonPrecalculus.com.*

Sketch the graph of the system of linear equations. Then describe the number of solutions and state whether the system is consistent or inconsistent.

$$\begin{cases} -2x + 3y = 6 \\ 4x - 6y = -9 \end{cases}$$

In Examples 6 and 7, note how you can use the method of elimination to determine that a system of linear equations has no solution or infinitely many solutions.

EXAMPLE 6 No-Solution Case: Method of Elimination

Solve the system of linear equations.

$$\begin{cases} x - 2y = 3 & \text{Equation 1} \\ -2x + 4y = 1 & \text{Equation 2} \end{cases}$$

Solution To obtain coefficients that differ only in sign, multiply Equation 1 by 2.

$$\begin{array}{rcrl} x - 2y = 3 & \Rightarrow & 2x - 4y = 6 & \text{Multiply Equation 1 by 2.} \\ \underline{-2x + 4y = 1} & \Rightarrow & \underline{-2x + 4y = 1} & \text{Write Equation 2.} \\ & & 0 = 7 & \text{Add equations.} \end{array}$$

Because there are no values of x and y for which $0 = 7$, you can conclude that the system is inconsistent and has no solution. Figure 7.5 shows the lines corresponding to the two equations in this system. Note that the two lines are parallel, so they have no point of intersection.

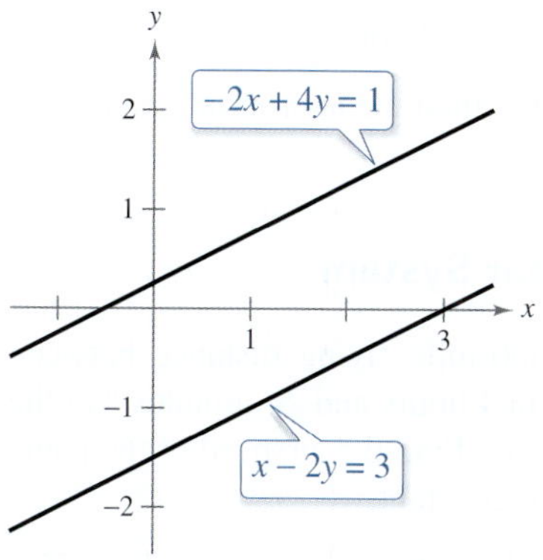

Figure 7.5

✓ **Checkpoint** *Audio-video solution in English & Spanish at LarsonPrecalculus.com.*

Solve the system of linear equations.

$$\begin{cases} 6x - 5y = 3 \\ -12x + 10y = 5 \end{cases}$$

In Example 6, note that the occurrence of a false statement, such as $0 = 7$, indicates that the system has no solution. In the next example, note that the occurrence of a statement that is true for all values of the variables, such as $0 = 0$, indicates that the system has infinitely many solutions.

EXAMPLE 7 Many-Solution Case: Method of Elimination

Solve the system of linear equations.

$$\begin{cases} 2x - y = 1 & \text{Equation 1} \\ 4x - 2y = 2 & \text{Equation 2} \end{cases}$$

Solution To obtain coefficients that differ only in sign, multiply Equation 1 by -2.

$$\begin{array}{rcrl} 2x - y = 1 & \Rightarrow & -4x + 2y = -2 & \text{Multiply Equation 1 by } -2. \\ \underline{4x - 2y = 2} & \Rightarrow & \underline{4x - 2y = 2} & \text{Write Equation 2.} \\ & & 0 = 0 & \text{Add equations.} \end{array}$$

Because the two equations are equivalent (have the same solution set), the system has infinitely many solutions. The solution set consists of all points (x, y) lying on the line $2x - y = 1$, as shown in Figure 7.6. Letting $x = a$, where a is any real number, the solutions of the system are $(a, 2a - 1)$.

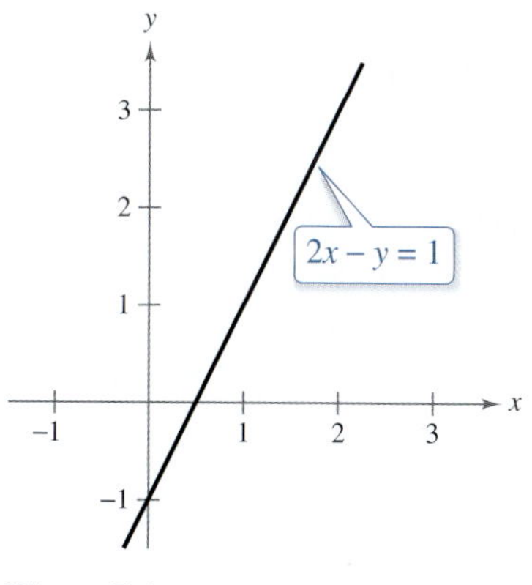

Figure 7.6

✓ **Checkpoint** *Audio-video solution in English & Spanish at LarsonPrecalculus.com.*

Solve the system of linear equations.

$$\begin{cases} \frac{1}{2}x - \frac{1}{8}y = -\frac{3}{8} \\ -4x + y = 3 \end{cases}$$

Applications

At this point, you may be asking the question "How can I tell whether I can solve an application problem using a system of linear equations?" The answer comes from the following considerations.

1. Does the problem involve more than one unknown quantity?
2. Are there two (or more) equations or conditions to be satisfied?

If the answer to one or both of these questions is "yes," then the appropriate model for the problem may be a system of linear equations.

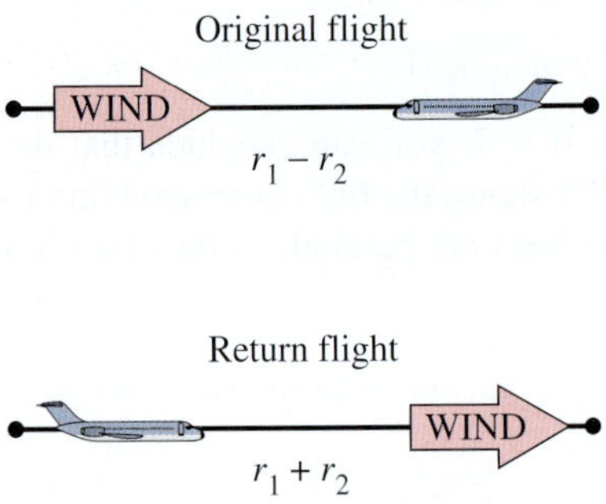

Figure 7.7

EXAMPLE 8 An Application of a Linear System

An airplane flying into a headwind travels the 2000-mile flying distance between Chicopee, Massachusetts, and Salt Lake City, Utah, in 4 hours and 24 minutes. On the return flight, the airplane travels this distance in 4 hours. Find the airspeed of the plane and the speed of the wind, assuming that both remain constant.

Solution The two unknown quantities are the speeds of the wind and the plane. If r_1 is the speed of the plane and r_2 is the speed of the wind, then

$$r_1 - r_2 = \text{speed of the plane against the wind}$$

$$r_1 + r_2 = \text{speed of the plane with the wind}$$

as shown in Figure 7.7. Using the formula

$$\text{distance} = (\text{rate})(\text{time})$$

for these two speeds, you obtain the following equations.

$$2000 = (r_1 - r_2)\left(4 + \frac{24}{60}\right)$$

$$2000 = (r_1 + r_2)(4)$$

These two equations simplify as follows.

$$\begin{cases} 5000 = 11r_1 - 11r_2 & \text{Equation 1} \\ 500 = r_1 + r_2 & \text{Equation 2} \end{cases}$$

To solve this system by elimination, multiply Equation 2 by 11.

$$5000 = 11r_1 - 11r_2 \quad \Longrightarrow \quad 5000 = 11r_1 - 11r_2 \qquad \text{Write Equation 1.}$$

$$500 = r_1 + r_2 \quad \Longrightarrow \quad 5500 = 11r_1 + 11r_2 \qquad \text{Multiply Equation 2 by 11.}$$

$$10{,}500 = 22r_1 \qquad \text{Add equations.}$$

So,

$$r_1 = \frac{10{,}500}{22} = \frac{5250}{11} \approx 477.27 \text{ miles per hour} \qquad \text{Speed of plane}$$

and

$$r_2 = 500 - \frac{5250}{11} = \frac{250}{11} \approx 22.73 \text{ miles per hour.} \qquad \text{Speed of wind}$$

Check this solution in the original statement of the problem.

✓ **Checkpoint** Audio-video solution in English & Spanish at LarsonPrecalculus.com.

In Example 8, the return flight takes 4 hours and 6 minutes. Find the airspeed of the plane and the speed of the wind, assuming that both remain constant.

In a free market, the demands for many products are related to the prices of the products. As the prices decrease, the demands by consumers increase and the amounts that producers are able or willing to supply decrease.

EXAMPLE 9 Finding the Equilibrium Point

The demand and supply equations for a video game console are

$$\begin{cases} p = 180 - 0.00001x & \text{Demand equation} \\ p = 90 + 0.00002x & \text{Supply equation} \end{cases}$$

where p is the price per unit (in dollars) and x is the number of units. Find the equilibrium point for this market. **The equilibrium point** is the price p and number of units x that satisfy both the demand and supply equations.

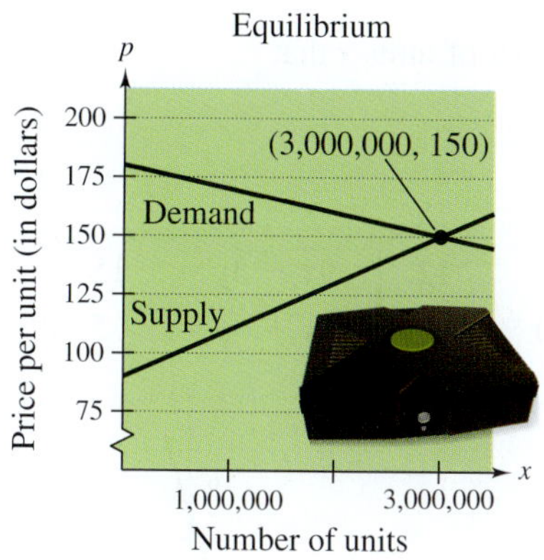

Figure 7.8

Solution Because p is written in terms of x, begin by substituting the value of p given in the supply equation into the demand equation.

$$p = 180 - 0.00001x \qquad \text{Write demand equation.}$$

$$90 + 0.00002x = 180 - 0.00001x \qquad \text{Substitute } 90 + 0.00002x \text{ for } p.$$

$$0.00003x = 90 \qquad \text{Combine like terms.}$$

$$x = 3{,}000{,}000 \qquad \text{Solve for } x.$$

So, the equilibrium point occurs when the demand and supply are each 3 million units. (See Figure 7.8.) Obtain the price that corresponds to this x-value by back-substituting $x = 3{,}000{,}000$ into either of the original equations. For instance, back-substituting into the demand equation produces

$$\begin{aligned} p &= 180 - 0.00001(3{,}000{,}000) \\ &= 180 - 30 \\ &= \$150. \end{aligned}$$

The solution is (3,000,000, 150). Check this by substituting (3,000,000, 150) into the demand and supply equations.

Checkpoint *Audio-video solution in English & Spanish at LarsonPrecalculus.com.*

The demand and supply equations for a flat-screen television are

$$\begin{cases} p = 567 - 0.00002x & \text{Demand equation} \\ p = 492 + 0.00003x & \text{Supply equation} \end{cases}$$

where p is the price per unit (in dollars) and x is the number of units. Find the equilibrium point for this market.

Summarize (Section 7.2)

1. Explain how to use the method of elimination to solve a system of linear equations in two variables *(page 476)*. For examples of using the method of elimination to solve systems of linear equations in two variables, see Examples 1–4.
2. Explain how to interpret graphically the numbers of solutions of systems of linear equations in two variables *(page 480)*. For examples of interpreting graphically the numbers of solutions of systems of linear equations in two variables, see Examples 5–7.
3. Describe examples of how to use systems of linear equations in two variables to model and solve real-life problems *(pages 482 and 483, Examples 8 and 9)*.

7.2 Exercises

See **CalcChat.com** for tutorial help and worked-out solutions to odd-numbered exercises.

Vocabulary: Fill in the blanks.

1. The first step in solving a system of equations by the method of ________ is to obtain coefficients for x (or y) that differ only in sign.
2. Two systems of equations that have the same solution set are called ________ systems.
3. A system of linear equations that has at least one solution is called ________, whereas a system of linear equations that has no solution is called ________.
4. In business applications, the ________ ________ is defined as the price p and the number of units x that satisfy both the demand and supply equations.

Skills and Applications

Solving a System by Elimination **In Exercises 5–12, solve the system by the method of elimination. Label each line with its equation. To print an enlarged copy of the graph, go to *MathGraphs.com*.**

5. $\begin{cases} 2x + y = 5 \\ x - y = 1 \end{cases}$

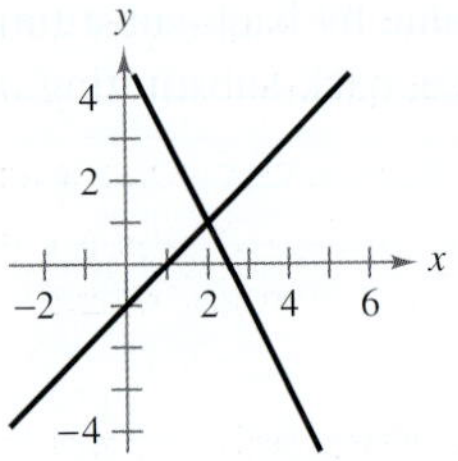

6. $\begin{cases} x + 3y = 1 \\ -x + 2y = 4 \end{cases}$

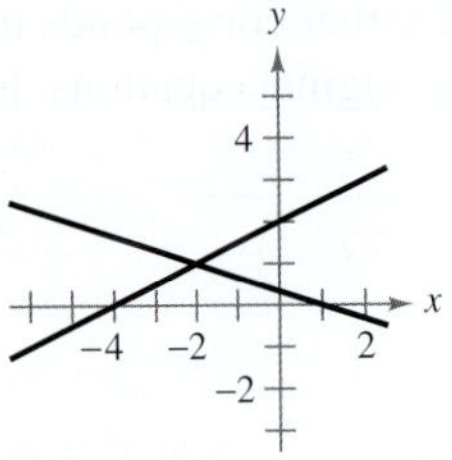

7. $\begin{cases} x + y = 0 \\ 3x + 2y = 1 \end{cases}$

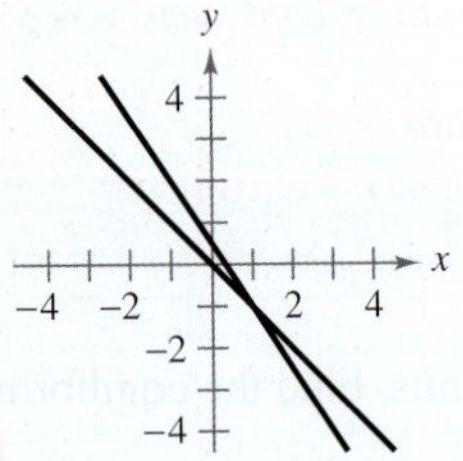

8. $\begin{cases} 2x - y = -3 \\ 4x + 3y = -21 \end{cases}$

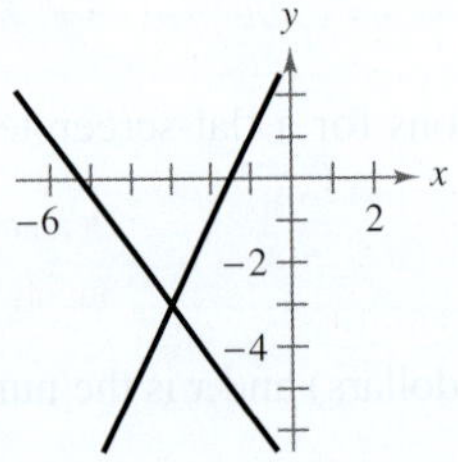

9. $\begin{cases} x - y = 2 \\ -2x + 2y = 5 \end{cases}$

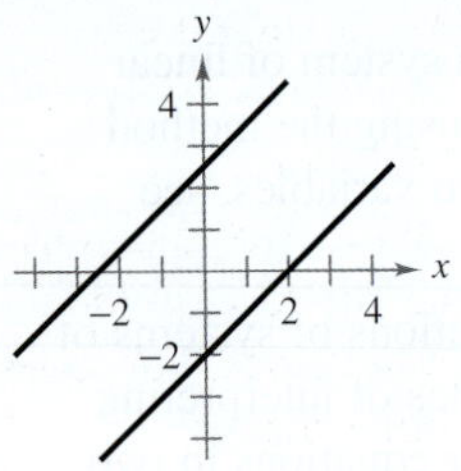

10. $\begin{cases} 3x + 2y = 3 \\ 6x + 4y = 14 \end{cases}$

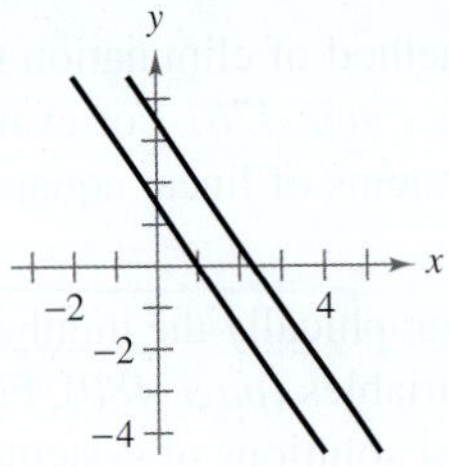

11. $\begin{cases} 3x - 2y = 5 \\ -6x + 4y = -10 \end{cases}$

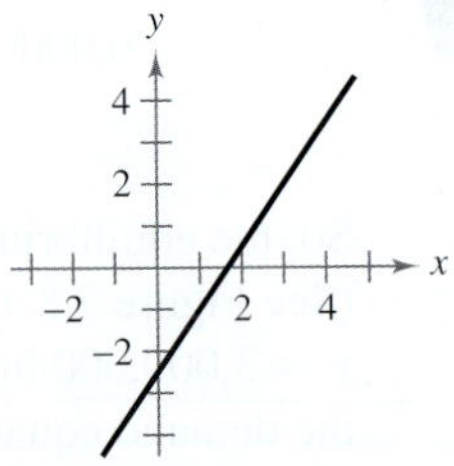

12. $\begin{cases} 9x - 3y = -15 \\ -3x + y = 5 \end{cases}$

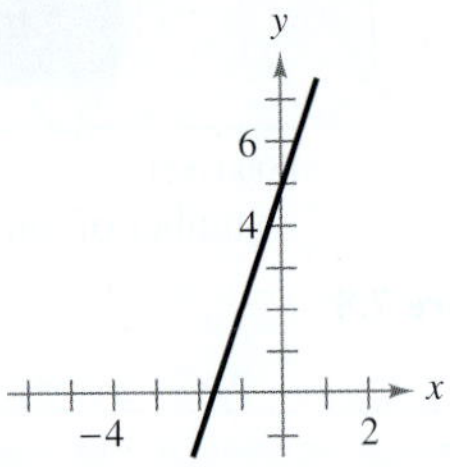

Solving a System by Elimination **In Exercises 13–30, solve the system by the method of elimination and check any solutions algebraically.**

13. $\begin{cases} x + 2y = 6 \\ x - 2y = 2 \end{cases}$

14. $\begin{cases} 3x - 5y = 8 \\ 2x + 5y = 22 \end{cases}$

15. $\begin{cases} 5x + 3y = 6 \\ 3x - y = 5 \end{cases}$

16. $\begin{cases} x + 5y = 10 \\ 3x - 10y = -5 \end{cases}$

17. $\begin{cases} 3x + 2y = 10 \\ 2x + 5y = 3 \end{cases}$

18. $\begin{cases} 2r + 4s = 5 \\ 16r + 50s = 55 \end{cases}$

19. $\begin{cases} 5u + 6v = 24 \\ 3u + 5v = 18 \end{cases}$

20. $\begin{cases} 3x + 11y = 4 \\ -2x - 5y = 9 \end{cases}$

21. $\begin{cases} \frac{9}{5}x + \frac{6}{5}y = 4 \\ 9x + 6y = 3 \end{cases}$

22. $\begin{cases} \frac{3}{4}x + y = \frac{1}{8} \\ \frac{9}{4}x + 3y = \frac{3}{8} \end{cases}$

23. $\begin{cases} -5x + 6y = -3 \\ 20x - 24y = 12 \end{cases}$

24. $\begin{cases} 7x + 8y = 6 \\ -14x - 16y = -12 \end{cases}$

25. $\begin{cases} 0.2x - 0.5y = -27.8 \\ 0.3x + 0.4y = 68.7 \end{cases}$

26. $\begin{cases} 0.05x - 0.03y = 0.21 \\ 0.07x + 0.02y = 0.16 \end{cases}$

27. $\begin{cases} 4b + 3m = 3 \\ 3b + 11m = 13 \end{cases}$

28. $\begin{cases} 2x + 5y = 8 \\ 5x + 8y = 10 \end{cases}$

29. $\begin{cases} \dfrac{x+3}{4} + \dfrac{y-1}{3} = 1 \\ 2x - y = 12 \end{cases}$

30. $\begin{cases} \dfrac{x-1}{2} + \dfrac{y+2}{3} = 4 \\ x - 2y = 5 \end{cases}$

Matching a System with Its Graph **In Exercises 31–34, match the system of linear equations with its graph. Describe the number of solutions and state whether the system is consistent or inconsistent. [The graphs are labeled (a), (b), (c) and (d).]**

(a)

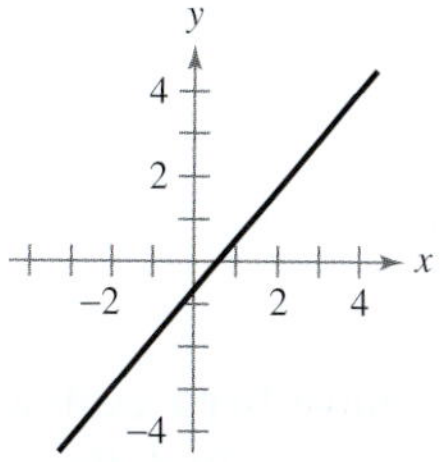

(b)

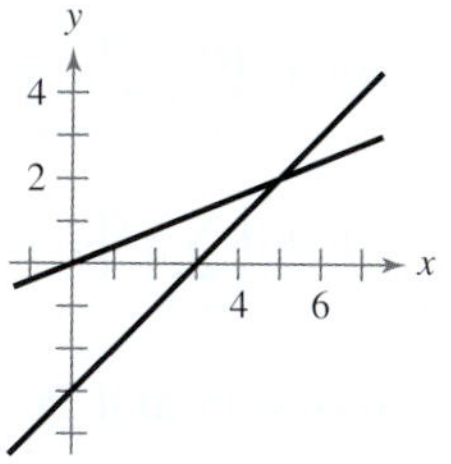

(c)

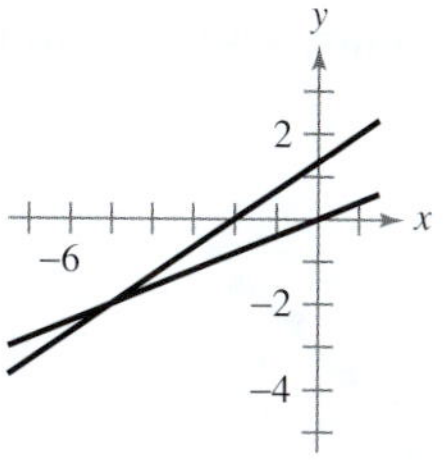

(d)

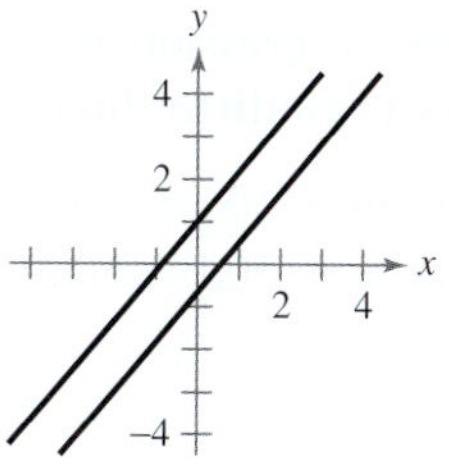

31. $\begin{cases} 2x - 5y = 0 \\ x - y = 3 \end{cases}$

32. $\begin{cases} 2x - 5y = 0 \\ 2x - 3y = -4 \end{cases}$

33. $\begin{cases} -7x + 6y = -4 \\ 14x - 12y = 8 \end{cases}$

34. $\begin{cases} 7x - 6y = -6 \\ 7x + 6y = -4 \end{cases}$

Solving a System **In Exercises 35–40, use any method to solve the system.**

35. $\begin{cases} 3x - 5y = 7 \\ 2x + y = 9 \end{cases}$

36. $\begin{cases} -x + 3y = 17 \\ 4x + 3y = 7 \end{cases}$

37. $\begin{cases} y = 2x - 5 \\ y = 5x - 11 \end{cases}$

38. $\begin{cases} 7x + 3y = 16 \\ y = x + 2 \end{cases}$

39. $\begin{cases} x - 5y = 21 \\ 6x + 5y = 21 \end{cases}$

40. $\begin{cases} 4x - 3y = 6 \\ -5x + 7y = -1 \end{cases}$

41. Airplane Speed An airplane flying into a headwind travels the 1800-mile flying distance between Pittsburgh, Pennsylvania, and Phoenix, Arizona, in 3 hours and 36 minutes. On the return flight, the airplane travels this distance in 3 hours. Find the airspeed of the plane and the speed of the wind, assuming that both remain constant.

42. Airplane Speed Two planes start from Los Angeles International Airport and fly in opposite directions. The second plane starts $\frac{1}{2}$ hour after the first plane, but its speed is 80 kilometers per hour faster. Find the airspeed of each plane when 2 hours after the first plane departs the planes are 3200 kilometers apart.

43. Nutrition Two cheeseburgers and one small order of French fries from a fast-food restaurant contain a total of 830 calories. Three cheeseburgers and two small orders of French fries contain a total of 1360 calories. Find the caloric content of each item.

44. Nutrition One eight-ounce glass of apple juice and one eight-ounce glass of orange juice contain a total of 179.2 milligrams of vitamin C. Two eight-ounce glasses of apple juice and three eight-ounce glasses of orange juice contain a total of 442.1 milligrams of vitamin C. How much vitamin C is in an eight-ounce glass of each type of juice?

Finding the Equilibrium Point **In Exercises 45–48, find the equilibrium point of the demand and supply equations.**

	Demand	Supply
45.	$p = 500 - 0.4x$	$p = 380 + 0.1x$
46.	$p = 100 - 0.05x$	$p = 25 + 0.1x$
47.	$p = 140 - 0.00002x$	$p = 80 + 0.00001x$
48.	$p = 400 - 0.0002x$	$p = 225 + 0.0005x$

49. Acid Mixture Thirty liters of a 40% acid solution is obtained by mixing a 25% solution with a 50% solution.

(a) Write a system of equations in which one equation represents the amount of final mixture required and the other represents the percent of acid in the final mixture. Let x and y represent the amounts of the 25% and 50% solutions, respectively.

(b) Use a graphing utility to graph the two equations in part (a) in the same viewing window. As the amount of the 25% solution increases, how does the amount of the 50% solution change?

(c) How much of each solution is required to obtain the specified concentration of the final mixture?

50. Fuel Mixture

Five hundred gallons of 89-octane gasoline is obtained by mixing 87-octane gasoline with 92-octane gasoline.

(a) Write a system of equations in which one equation represents the amount of final mixture required and the other represents the amounts of 87- and 92-octane gasolines in the final mixture. Let x and y represent the numbers of gallons of 87-octane and 92-octane gasolines, respectively.

(b) Use a graphing utility to graph the two equations in part (a) in the same viewing window. As the amount of 87-octane gasoline increases, how does the amount of 92-octane gasoline change?

(c) How much of each type of gasoline is required to obtain the 500 gallons of 89-octane gasoline?

Vlad61/Shutterstock.com

51. Investment Portfolio A total of \$24,000 is invested in two corporate bonds that pay 3.5% and 5% simple interest. The investor wants an annual interest income of \$930 from the investments. What amount should be invested in the 3.5% bond?

52. Investment Portfolio A total of \$32,000 is invested in two municipal bonds that pay 5.75% and 6.25% simple interest. The investor wants an annual interest income of \$1900 from the investments. What amount should be invested in the 5.75% bond?

53. Prescriptions The numbers of prescriptions P (in thousands) filled at two pharmacies from 2009 through 2013 are shown in the table.

Year	Pharmacy A	Pharmacy B
2009	19.2	20.4
2010	19.6	20.8
2011	20.0	21.1
2012	20.6	21.5
2013	21.3	22.0

(a) Use a graphing utility to create a scatter plot of the data for pharmacy A and find a linear model. Let t represent the year, with $t = 9$ corresponding to 2009. Repeat the procedure for pharmacy B.

(b) Assuming that the numbers for the given five years are representative of future years, will the number of prescriptions filled at pharmacy A ever exceed the number of prescriptions filled at pharmacy B? If so, then when?

54. Data Analysis A store manager wants to know the demand for a product as a function of the price. The table shows the daily sales y for different prices x of the product.

Price, x	Demand, y
\$1.00	45
\$1.20	37
\$1.50	23

(a) Find the least squares regression line $y = ax + b$ for the data by solving the system for a and b.

$$\begin{cases} 3.00b + 3.70a = 105.00 \\ 3.70b + 4.69a = 123.90 \end{cases}$$

(b) Use a graphing utility to confirm the result of part (a).

(c) Use the linear model from part (a) to predict the demand when the price is \$1.75.

Fitting a Line to Data **To find the least squares regression line $y = ax + b$ for a set of points**

$$(x_1, y_1), (x_2, y_2), \ldots, (x_n, y_n)$$

you can solve the following system for a and b.

$$\begin{cases} nb + \left(\sum_{i=1}^{n} x_i\right)a = \left(\sum_{i=1}^{n} y_i\right) \\ \left(\sum_{i=1}^{n} x_i\right)b + \left(\sum_{i=1}^{n} x_i^2\right)a = \left(\sum_{i=1}^{n} x_i y_i\right) \end{cases}$$

In Exercises 55 and 56, the sums have been evaluated. Solve the given system for a and b to find the least squares regression line for the points. Use a graphing utility to confirm the result.

55. $\begin{cases} 5b + 10a = 20.2 \\ 10b + 30a = 50.1 \end{cases}$

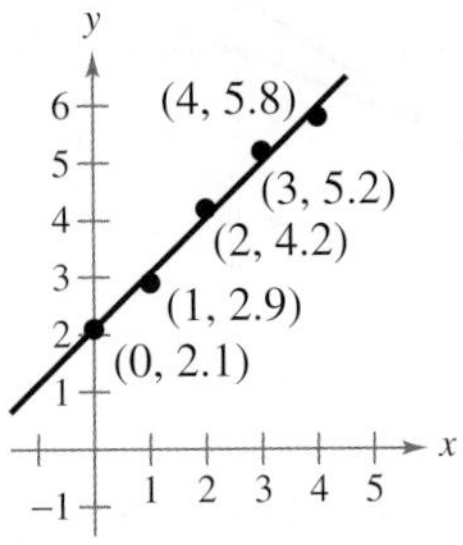

56. $\begin{cases} 6b + 15a = 23.6 \\ 15b + 55a = 48.8 \end{cases}$

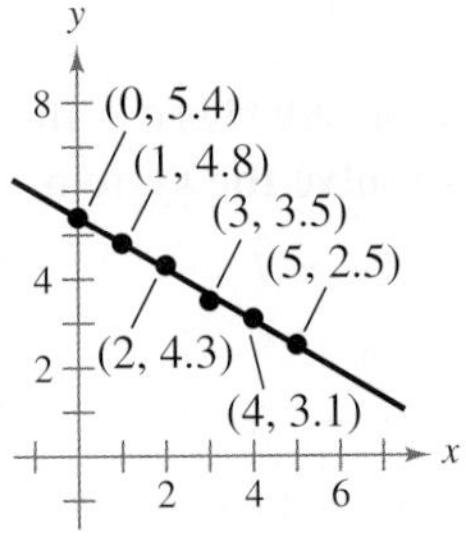

57. Data Analysis An agricultural scientist used four test plots to determine the relationship between wheat yield y (in bushels per acre) and the amount of fertilizer x (in hundreds of pounds per acre). The table shows the results.

Fertilizer, x	Yield, y
1.0	32
1.5	41
2.0	48
2.5	53

(a) Find the least squares regression line $y = ax + b$ for the data by solving the system for a and b.

$$\begin{cases} 4b + 7.0a = 174 \\ 7b + 13.5a = 322 \end{cases}$$

(b) Use the linear model from part (a) to estimate the yield for a fertilizer application of 160 pounds per acre.

58. Defense Department Outlays The table shows the total national outlays y for defense functions (in billions of dollars) for the years 2004 through 2011. *(Source: U.S. Office of Management and Budget)*

DATA

Year	Outlays, y
2004	455.8
2005	495.3
2006	521.8
2007	551.3
2008	616.1
2009	661.0
2010	693.6
2011	705.6

Spreadsheet at LarsonPrecalculus.com

(a) Find the least squares regression line $y = at + b$ for the data, where t represents the year with $t = 4$ corresponding to 2004, by solving the system for a and b.

$$\begin{cases} 8b + 60a = 4700.5 \\ 60b + 492a = 36{,}865.0 \end{cases}$$

(b) Use the *regression* feature of a graphing utility to confirm the result of part (a).

(c) Use the linear model to create a table of estimated values of y. Compare the estimated values with the actual data.

(d) Use the linear model to estimate the total national outlay for 2012.

(e) Use the Internet, your school's library, or some other reference source to find the total national outlay for 2012. How does this value compare with your answer in part (d)?

(f) Is the linear model valid for long-term predictions of total national outlays? Explain.

Exploration

True or False? **In Exercises 59 and 60, determine whether the statement is true or false. Justify your answer.**

59. If two lines do not have exactly one point of intersection, then they must be parallel.

60. Solving a system of equations graphically will always give an exact solution.

Finding the Value of a Constant **In Exercises 61 and 62, find the value of k such that the system of linear equations is inconsistent.**

61. $\begin{cases} 4x - 8y = -3 \\ 2x + ky = 16 \end{cases}$ **62.** $\begin{cases} 15x + 3y = 6 \\ -10x + ky = 9 \end{cases}$

63. Writing Briefly explain whether it is possible for a consistent system of linear equations to have exactly two solutions.

64. Think About It Give examples of a system of linear equations that has (a) no solution and (b) an infinite number of solutions.

65. Comparing Methods Use the method of substitution to solve the system in Example 1. Is the method of substitution or the method of elimination easier? Explain.

66. HOW DO YOU SEE IT? Use the graphs of the two equations shown below.

(a) Describe the graphs of the two equations.

(b) Can you conclude that the system of equations shown in the graph is inconsistent? Explain.

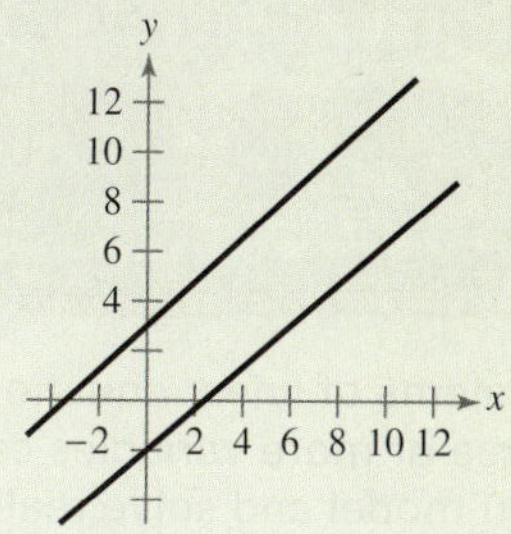

Think About It **In Exercises 67 and 68, the graphs of the two equations appear to be parallel. Yet, when you solve the system algebraically, you find that the system does have a solution. Find the solution and explain why it does not appear on the portion of the graph shown.**

67. $\begin{cases} 100y - x = 200 \\ 99y - x = -198 \end{cases}$ **68.** $\begin{cases} 21x - 20y = 0 \\ 13x - 12y = 120 \end{cases}$

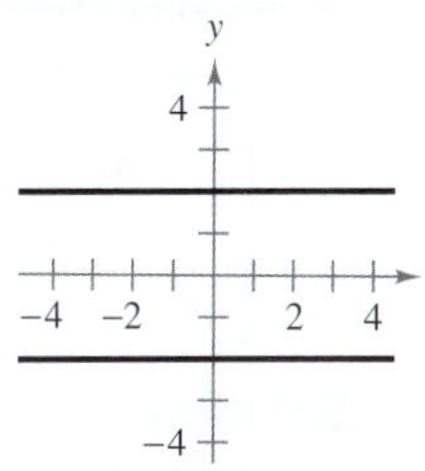

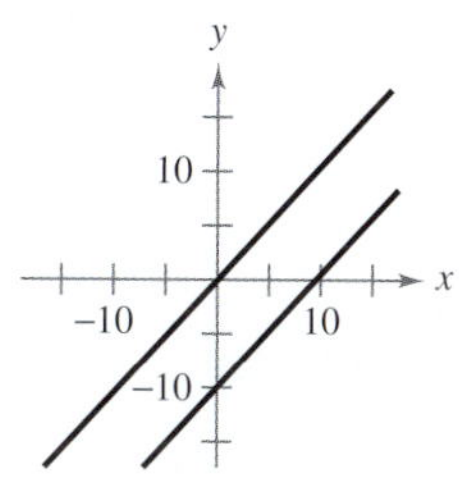

Advanced Applications **In Exercises 69 and 70, solve the system of equations for u and v. While solving for these variables, consider the transcendental functions as constants. (Systems of this type appear in a course in differential equations.)**

69. $\begin{cases} u \sin x + v \cos x = 0 \\ u \cos x - v \sin x = \sec x \end{cases}$

70. $\begin{cases} u \cos 2x + v \sin 2x = 0 \\ u(-2 \sin 2x) + v(2 \cos 2x) = \csc 2x \end{cases}$

Project: College Expenses To work an extended application analyzing the average undergraduate tuition, room, and board charges at private degree-granting institutions in the United States from 1990 through 2010, visit this text's website at *LarsonPrecalculus.com.* *(Source: U.S. Department of Education)*

7.3 Multivariable Linear Systems

Systems of linear equations in three or more variables can help you model and solve real-life problems. For instance, in Exercise 72 on page 499, you will use a system of linear equations in three variables to analyze the reproductive rates of deer in a wildlife preserve.

- Use back-substitution to solve linear systems in row-echelon form.
- Use Gaussian elimination to solve systems of linear equations.
- Solve nonsquare systems of linear equations.
- Use systems of linear equations in three or more variables to model and solve real-life problems.

Row-Echelon Form and Back-Substitution

The method of elimination can be applied to a system of linear equations in more than two variables. In fact, this method adapts to computer use for solving linear systems with dozens of variables.

When using the method of elimination to solve a system of linear equations, the goal is to rewrite the system in a form to which back-substitution can be applied. To see how this works, consider the following two systems of linear equations.

$$\begin{cases} x - 2y + 3z = 9 \\ -x + 3y = -4 \\ 2x - 5y + 5z = 17 \end{cases}$$ System of three linear equations in three variables (See Example 3.)

$$\begin{cases} x - 2y + 3z = 9 \\ y + 3z = 5 \\ z = 2 \end{cases}$$ Equivalent system in row-echelon form (See Example 1.)

The second system is in **row-echelon form,** which means that it has a "stair-step" pattern with leading coefficients of 1. After comparing the two systems, it should be clear that it is easier to solve the system in row-echelon form, using back-substitution.

EXAMPLE 1 Using Back-Substitution in Row-Echelon Form

Solve the system of linear equations.

$$\begin{cases} x - 2y + 3z = 9 & \text{Equation 1} \\ y + 3z = 5 & \text{Equation 2} \\ z = 2 & \text{Equation 3} \end{cases}$$

Solution From Equation 3, you know the value of z. To solve for y, back-substitute $z = 2$ into Equation 2 to obtain

$y + 3(2) = 5$ Substitute 2 for z.

$y = -1.$ Solve for y.

Then back-substitute $y = -1$ and $z = 2$ into Equation 1 to obtain

$x - 2(-1) + 3(2) = 9$ Substitute -1 for y and 2 for z.

$x = 1.$ Solve for x.

The solution is $x = 1$, $y = -1$, and $z = 2$, which can be written as the **ordered triple** $(1, -1, 2)$. Check this in the original system of equations.

✓ **Checkpoint** Audio-video solution in English & Spanish at LarsonPrecalculus.com.

Solve the system of linear equations.

$$\begin{cases} 2x - y + 5z = 22 \\ y + 3z = 6 \\ z = 3 \end{cases}$$

Bruce MacQueen/Shutterstock.com

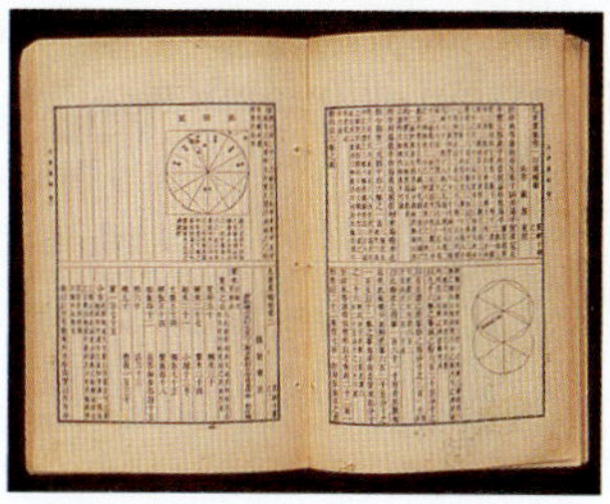

One of the most influential Chinese mathematics books was the *Chui-chang suan-shu* or *Nine Chapters on the Mathematical Art* (written in approximately 250 B.C.). Chapter Eight of the *Nine Chapters* contained solutions of systems of linear equations such as

$$\begin{cases} 3x + 2y + z = 39 \\ 2x + 3y + z = 34 \\ x + 2y + 3z = 26 \end{cases}$$

using positive and negative numbers. This system was solved using column operations on a matrix. Matrices (plural for matrix) are discussed in the next chapter.

Gaussian Elimination

Two systems of equations are *equivalent* when they have the same solution set. To solve a system that is not in row-echelon form, first convert it to an *equivalent* system that is in row-echelon form by using the following operations.

Operations That Produce Equivalent Systems

Each of the following **row operations** on a system of linear equations produces an *equivalent* system of linear equations.

1. Interchange two equations.
2. Multiply one of the equations by a nonzero constant.
3. Add a multiple of one of the equations to another equation to replace the latter equation.

To see how to use row operations, take another look at the method of elimination, as applied to a system of two linear equations.

EXAMPLE 2 Using Row Operations to Solve a System

Solve the system of linear equations.

$$\begin{cases} 3x - 2y = -1 \\ x - y = 0 \end{cases}$$ Equation 1, Equation 2

Solution There are two strategies that seem reasonable: eliminate the variable x or eliminate the variable y. The following steps show how to eliminate x.

$$\begin{cases} x - y = 0 \\ 3x - 2y = -1 \end{cases}$$ Interchange the two equations in the system.

$$\begin{cases} -3x + 3y = 0 \\ 3x - 2y = -1 \end{cases}$$ Multiply the first equation by -3.

$$\begin{array}{r} -3x + 3y = 0 \\ \underline{3x - 2y = -1} \\ y = -1 \end{array}$$ Add the multiple of the first equation to the second equation to obtain a new second equation.

$$\begin{cases} x - y = 0 \\ y = -1 \end{cases}$$ New system in row-echelon form

Notice in the first step that interchanging rows is a way of obtaining a leading coefficient of 1. Now back-substitute $y = -1$ into Equation 2 and solve for x.

$x - (-1) = 0$ Substitute -1 for y.

$x = -1$ Solve for x.

The solution is $x = -1$ and $y = -1$, which can be written as the ordered pair $(-1, -1)$.

Checkpoint Audio-video solution in English & Spanish at LarsonPrecalculus.com.

Solve the system of linear equations.

$$\begin{cases} 2x + y = 3 \\ x + 2y = 3 \end{cases}$$

Christopher Lui/China Stock

Rewriting a system of linear equations in row-echelon form usually involves a chain of equivalent systems, which you obtain by using one of the three basic row operations listed on the previous page. This process is called **Gaussian elimination,** after the German mathematician Carl Friedrich Gauss (1777–1855).

EXAMPLE 3 Using Gaussian Elimination to Solve a System

Solve the system of linear equations.

$$\begin{cases} x - 2y + 3z = 9 & \text{Equation 1} \\ -x + 3y = -4 & \text{Equation 2} \\ 2x - 5y + 5z = 17 & \text{Equation 3} \end{cases}$$

REMARK Arithmetic errors are common when performing row operations. You should note the operation performed in each step to make checking your work easier.

Solution Because the leading coefficient of the first equation is 1, begin by keeping the x in the upper left position and eliminating the other x-terms from the first column.

$$\begin{array}{rl} x - 2y + 3z = 9 & \text{Write Equation 1.} \\ -x + 3y \qquad = -4 & \text{Write Equation 2.} \\ \hline y + 3z = 5 & \text{Add Equation 1 to Equation 2.} \end{array}$$

$$\begin{cases} x - 2y + 3z = 9 \\ y + 3z = 5 \\ 2x - 5y + 5z = 17 \end{cases}$$

Adding the first equation to the second equation produces a new second equation.

$$\begin{array}{rl} -2x + 4y - 6z = -18 & \text{Multiply Equation 1 by } -2. \\ 2x - 5y + 5z = 17 & \text{Write Equation 3.} \\ \hline -y - z = -1 & \text{Add revised Equation 1 to Equation 3.} \end{array}$$

$$\begin{cases} x - 2y + 3z = 9 \\ y + 3z = 5 \\ -y - z = -1 \end{cases}$$

Adding -2 times the first equation to the third equation produces a new third equation.

Now that you have eliminated all but the x in the upper position of the first column, work on the second column. (You need to eliminate y from the third equation.)

$$\begin{cases} x - 2y + 3z = 9 \\ y + 3z = 5 \\ 2z = 4 \end{cases}$$

Adding the second equation to the third equation produces a new third equation.

Finally, you need a coefficient of 1 for z in the third equation.

$$\begin{cases} x - 2y + 3z = 9 \\ y + 3z = 5 \\ z = 2 \end{cases}$$

Multiplying the third equation by $\frac{1}{2}$ produces a new third equation.

This is the same system in Example 1, and, as in that example, the solution is

$$x = 1, \quad y = -1, \quad \text{and} \quad z = 2.$$

✓ Checkpoint *Audio-video solution in English & Spanish at LarsonPrecalculus.com.*

Solve the system of linear equations.

$$\begin{cases} x + y + z = 6 \\ 2x - y + z = 3 \\ 3x + y - z = 2 \end{cases}$$

The next example involves an inconsistent system—one that has no solution. The key to recognizing an inconsistent system is that at some stage in the elimination process, you obtain a false statement such as $0 = -2$.

EXAMPLE 4 **An Inconsistent System**

Solve the system of linear equations.

$$\begin{cases} x - 3y + z = 1 & \text{Equation 1} \\ 2x - y - 2z = 2 & \text{Equation 2} \\ x + 2y - 3z = -1 & \text{Equation 3} \end{cases}$$

Solution

$$\begin{cases} x - 3y + z = 1 \\ 5y - 4z = 0 \\ x + 2y - 3z = -1 \end{cases}$$

Adding -2 times the first equation to the second equation produces a new second equation.

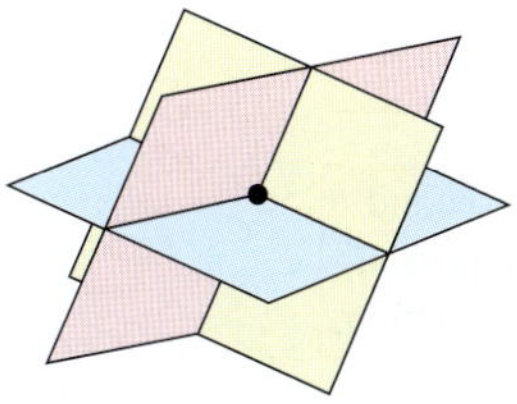

Solution: one point
Figure 7.9

$$\begin{cases} x - 3y + z = 1 \\ 5y - 4z = 0 \\ 5y - 4z = -2 \end{cases}$$

Adding -1 times the first equation to the third equation produces a new third equation.

$$\begin{cases} x - 3y + z = 1 \\ 5y - 4z = 0 \\ 0 = -2 \end{cases}$$

Adding -1 times the second equation to the third equation produces a new third equation.

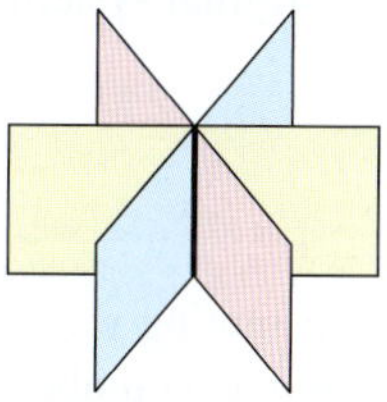

Solution: one line
Figure 7.10

Because $0 = -2$ is a false statement, this system is inconsistent and has no solution. Moreover, because this system is equivalent to the original system, the original system also has no solution.

✓ ***Checkpoint*** *Audio-video solution in English & Spanish at LarsonPrecalculus.com.*

Solve the system of linear equations.

$$\begin{cases} x + y - 2z = 3 \\ 3x - 2y + 4z = 1 \\ 2x - 3y + 6z = 8 \end{cases}$$

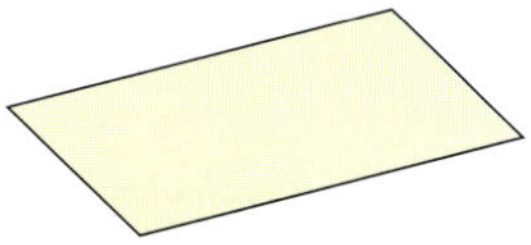

Solution: one plane
Figure 7.11

As with a system of linear equations in two variables, the solution(s) of a system of linear equations in more than two variables must fall into one of three categories.

The Number of Solutions of a Linear System

For a system of linear equations, exactly one of the following is true.

1. There is exactly one solution.
2. There are infinitely many solutions.
3. There is no solution.

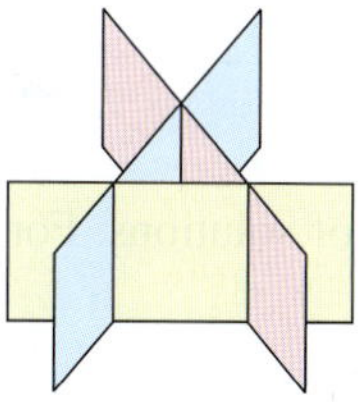

Solution: none
Figure 7.12

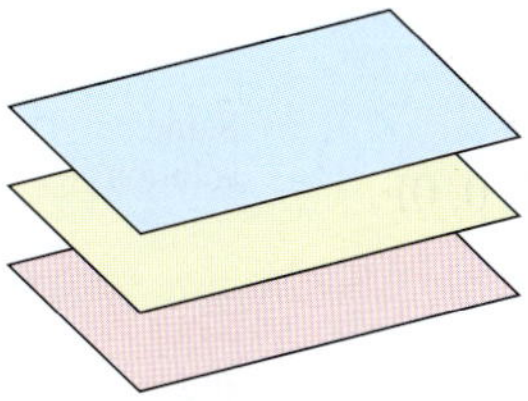

Solution: none
Figure 7.13

In Section 7.2, you learned that a system of two linear equations in two variables can be represented graphically as a pair of lines that are intersecting, coincident, or parallel. A system of three linear equations in three variables has a similar graphical representation—it can be represented as three planes in space that intersect in one point (exactly one solution, see Figure 7.9), intersect in a line or a plane (infinitely many solutions, see Figures 7.10 and 7.11), or have no points common to all three planes (no solution, see Figures 7.12 and 7.13).

EXAMPLE 5 A System with Infinitely Many Solutions

Solve the system of linear equations.

$$\begin{cases} x + y - 3z = -1 & \text{Equation 1} \\ y - z = 0 & \text{Equation 2} \\ -x + 2y = 1 & \text{Equation 3} \end{cases}$$

Solution

$$\begin{cases} x + y - 3z = -1 \\ y - z = 0 \\ 3y - 3z = 0 \end{cases}$$

Adding the first equation to the third equation produces a new third equation.

$$\begin{cases} x + y - 3z = -1 \\ y - z = 0 \\ 0 = 0 \end{cases}$$

Adding -3 times the second equation to the third equation produces a new third equation.

This result means that Equation 3 depends on Equations 1 and 2 in the sense that it gives no additional information about the variables. Because $0 = 0$ is a true statement, this system has infinitely many solutions. However, it is incorrect to say that the solution is "infinite." You must also specify the correct form of the solution. So, the original system is equivalent to the system

$$\begin{cases} x + y - 3z = -1 \\ y - z = 0 \end{cases}.$$

In the second equation, solve for y in terms of z to obtain $y = z$. Back-substituting $y = z$ in the first equation produces $x = 2z - 1$. Finally, letting $z = a$, where a is a real number, the solutions of the given system are all of the form $x = 2a - 1$, $y = a$, and $z = a$. So, every ordered triple of the form

$$(2a - 1, a, a) \qquad a \text{ is a real number.}$$

is a solution of the system.

✓ **Checkpoint** Audio-video solution in English & Spanish at LarsonPrecalculus.com.

Solve the system of linear equations.

$$\begin{cases} x + 2y - 7z = -4 \\ 2x + 3y + z = 5 \\ 3x + 7y - 36z = -25 \end{cases}$$

In Example 5, there are other ways to write the same infinite set of solutions. For instance, letting $x = b$, the solutions could have been written as

$$\left(b, \tfrac{1}{2}(b + 1), \tfrac{1}{2}(b + 1)\right). \qquad b \text{ is a real number.}$$

To convince yourself that this description produces the same set of solutions, consider the following.

Substitution	**Solution**	
$a = 0$	$(2(0) - 1, 0, 0) = (-1, 0, 0)$	Same solution
$b = -1$	$\left(-1, \tfrac{1}{2}(-1 + 1), \tfrac{1}{2}(-1 + 1)\right) = (-1, 0, 0)$	
$a = 1$	$(2(1) - 1, 1, 1) = (1, 1, 1)$	Same solution
$b = 1$	$\left(1, \tfrac{1}{2}(1 + 1), \tfrac{1}{2}(1 + 1)\right) = (1, 1, 1)$	
$a = 2$	$(2(2) - 1, 2, 2) = (3, 2, 2)$	Same solution
$b = 3$	$\left(3, \tfrac{1}{2}(3 + 1), \tfrac{1}{2}(3 + 1)\right) = (3, 2, 2)$	

The Global Positioning System (GPS) is a network of 24 satellites originally developed by the U.S. military as a navigational tool. Civilian applications of GPS receivers include determining directions, locating vessels lost at sea, and monitoring earthquakes. A GPS receiver works by using satellite readings to calculate its location. In a simplified mathematical model, nonsquare systems of three linear equations in four variables (three dimensions and time) determine the coordinates of the receiver as a function of time.

Nonsquare Systems

So far, each system of linear equations you have looked at has been *square*, which means that the number of equations is equal to the number of variables. In a **nonsquare** system, the number of equations differs from the number of variables. A system of linear equations cannot have a unique solution unless there are at least as many equations as there are variables in the system.

EXAMPLE 6 A System with Fewer Equations than Variables

Solve the system of linear equations.

$$\begin{cases} x - 2y + z = 2 & \text{Equation 1} \\ 2x - y - z = 1 & \text{Equation 2} \end{cases}$$

Solution Begin by rewriting the system in row-echelon form.

$$\begin{cases} x - 2y + z = 2 \\ 3y - 3z = -3 \end{cases}$$

Adding -2 times the first equation to the second equation produces a new second equation.

$$\begin{cases} x - 2y + z = 2 \\ y - z = -1 \end{cases}$$

Multiplying the second equation by $\frac{1}{3}$ produces a new second equation.

Solve for y in terms of z to obtain

$$y = z - 1.$$

Solve for x by back-substituting $y = z - 1$ into Equation 1.

$$x - 2y + z = 2 \quad \text{Write Equation 1.}$$

$$x - 2(z - 1) + z = 2 \quad \text{Substitute } z - 1 \text{ for } y.$$

$$x - 2z + 2 + z = 2 \quad \text{Distributive Property}$$

$$x = z \quad \text{Solve for } x.$$

Finally, by letting $z = a$, where a is a real number, you have the solution

$$x = a, \quad y = a - 1, \quad \text{and} \quad z = a.$$

So, every ordered triple of the form

$$(a, a - 1, a) \quad a \text{ is a real number.}$$

is a solution of the system. Because there were originally three variables and only two equations, the system cannot have a unique solution.

✓ **Checkpoint** Audio-video solution in English & Spanish at LarsonPrecalculus.com.

Solve the system of linear equations.

$$\begin{cases} x - y + 4z = 3 \\ 4x - z = 0 \end{cases}$$

In Example 6, try choosing some values of a to obtain different solutions of the system, such as

$$(1, 0, 1), \quad (2, 1, 2), \quad \text{and} \quad (3, 2, 3).$$

Then check each ordered triple in the original system to verify that it is a solution of the system.

edobric/Shutterstock.com

Applications

EXAMPLE 7 Vertical Motion

The height at time t of an object that is moving in a (vertical) line with constant acceleration a is given by the **position equation**

$$s = \frac{1}{2}at^2 + v_0t + s_0$$

where s is the height in feet, a is the acceleration in feet per second squared, t is the time in seconds, v_0 is the initial velocity (at $t = 0$), and s_0 is the initial height. Find the values of a, v_0, and s_0 when $s = 52$ at $t = 1$, $s = 52$ at $t = 2$, and $s = 20$ at $t = 3$, and interpret the result. (See Figure 7.14.)

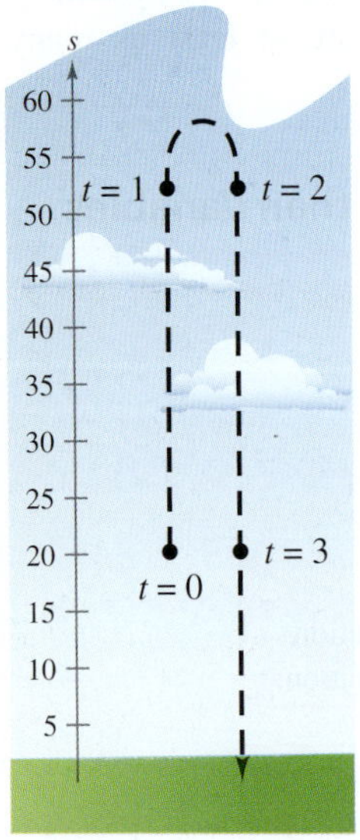

Figure 7.14

Solution By substituting the three values of t and s into the position equation, you can obtain three linear equations in a, v_0, and s_0.

When $t = 1$: $\frac{1}{2}a(1)^2 + v_0(1) + s_0 = 52$ ⇨ $a + 2v_0 + 2s_0 = 104$

When $t = 2$: $\frac{1}{2}a(2)^2 + v_0(2) + s_0 = 52$ ⇨ $2a + 2v_0 + s_0 = 52$

When $t = 3$: $\frac{1}{2}a(3)^2 + v_0(3) + s_0 = 20$ ⇨ $9a + 6v_0 + 2s_0 = 40$

This produces the following system of linear equations.

$$\begin{cases} a + 2v_0 + 2s_0 = 104 \\ 2a + 2v_0 + s_0 = 52 \\ 9a + 6v_0 + 2s_0 = 40 \end{cases}$$

Now solve the system using Gaussian elimination.

$$\begin{cases} a + 2v_0 + 2s_0 = 104 \\ -2v_0 - 3s_0 = -156 \\ 9a + 6v_0 + 2s_0 = 40 \end{cases}$$

Adding -2 times the first equation to the second equation produces a new second equation.

$$\begin{cases} a + 2v_0 + 2s_0 = 104 \\ -2v_0 - 3s_0 = -156 \\ -12v_0 - 16s_0 = -896 \end{cases}$$

Adding -9 times the first equation to the third equation produces a new third equation.

$$\begin{cases} a + 2v_0 + 2s_0 = 104 \\ -2v_0 - 3s_0 = -156 \\ 2s_0 = 40 \end{cases}$$

Adding -6 times the second equation to the third equation produces a new third equation.

$$\begin{cases} a + 2v_0 + 2s_0 = 104 \\ v_0 + \frac{3}{2}s_0 = 78 \\ s_0 = 20 \end{cases}$$

Multiplying the second equation by $-\frac{1}{2}$ produces a new second equation and multiplying the third equation by $\frac{1}{2}$ produces a new third equation.

So, the solution of this system is $a = -32$, $v_0 = 48$, and $s_0 = 20$, which can be written as $(-32, 48, 20)$. This solution results in a position equation of $s = -16t^2 + 48t + 20$ and implies that the object was thrown upward at a velocity of 48 feet per second from a height of 20 feet.

✔ **Checkpoint** *Audio-video solution in English & Spanish at LarsonPrecalculus.com.*

For the position equation

$$s = \frac{1}{2}at^2 + v_0t + s_0$$

given in Example 7, find the values of a, v_0, and s_0 when $s = 104$ at $t = 1$, $s = 76$ at $t = 2$, and $s = 16$ at $t = 3$, and interpret the result.

EXAMPLE 8 Data Analysis: Curve-Fitting

Find a quadratic equation $y = ax^2 + bx + c$ whose graph passes through the points $(-1, 3)$, $(1, 1)$, and $(2, 6)$.

Solution Because the graph of $y = ax^2 + bx + c$ passes through the points $(-1, 3)$, $(1, 1)$, and $(2, 6)$, you can write the following.

When $x = -1, y = 3$: $a(-1)^2 + b(-1) + c = 3$

When $x = 1, y = 1$: $a(1)^2 + b(1) + c = 1$

When $x = 2, y = 6$: $a(2)^2 + b(2) + c = 6$

This produces the following system of linear equations.

$$\begin{cases} a - b + c = 3 & \text{Equation 1} \\ a + b + c = 1 & \text{Equation 2} \\ 4a + 2b + c = 6 & \text{Equation 3} \end{cases}$$

The solution of this system is $a = 2$, $b = -1$, and $c = 0$. So, the equation of the parabola is

$$y = 2x^2 - x$$

as shown below.

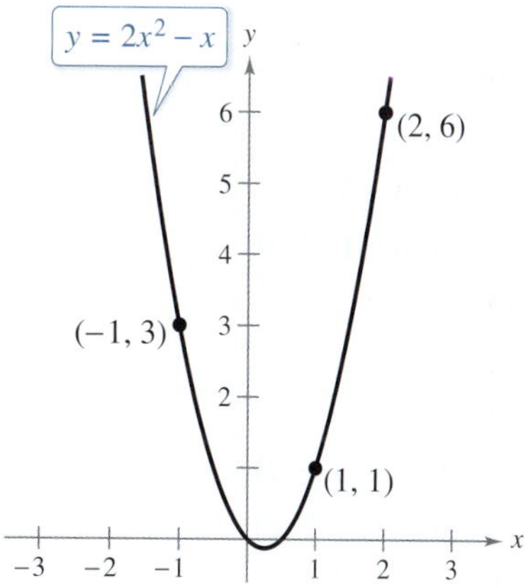

✓ **Checkpoint** *Audio-video solution in English & Spanish at LarsonPrecalculus.com.*

Find a quadratic equation $y = ax^2 + bx + c$ whose graph passes through the points $(0, 0)$, $(3, -3)$, and $(6, 0)$.

Summarize (Section 7.3)

1. State the definition of row-echelon form *(page 488)*. For an example of solving a linear system in row-echelon form, see Example 1.
2. Describe the process of Gaussian elimination *(pages 489 and 490)*. For examples of using Gaussian elimination to solve systems of linear equations, see Examples 2–5.
3. Explain the difference between a square system of linear equations and a nonsquare system of linear equations *(page 493)*. For an example of solving a nonsquare system of linear equations, see Example 6.
4. Describe examples of how to use systems of linear equations in three or more variables to model and solve real-life problems *(pages 494 and 495, Examples 7 and 8)*.

7.3 Exercises

See CalcChat.com for tutorial help and worked-out solutions to odd-numbered exercises.

Vocabulary: Fill in the blanks.

1. A system of equations in ________ form has a "stair-step" pattern with leading coefficients of 1.
2. A solution of a system of three linear equations in three unknowns can be written as an ________ ________, which has the form (x, y, z).
3. The process used to write a system of linear equations in row-echelon form is called ________ elimination.
4. Interchanging two equations of a system of linear equations is a ________ ________ that produces an equivalent system.
5. A system of equations is called ________ when the number of equations differs from the number of variables in the system.
6. The equation $s = \frac{1}{2}at^2 + v_0t + s_0$ is called the ________ equation, and it models the height s of an object at time t that is moving in a vertical line with a constant acceleration a.

Skills and Applications

Checking Solutions **In Exercises 7–10, determine whether each ordered triple is a solution of the system of equations.**

7. $\begin{cases} 6x - y + z = -1 \\ 4x - 3z = -19 \\ 2y + 5z = 25 \end{cases}$

(a) $(2, 0, -2)$ (b) $(-3, 0, 5)$
(c) $(0, -1, 4)$ (d) $(-1, 0, 5)$

8. $\begin{cases} 3x + 4y - z = 17 \\ 5x - y + 2z = -2 \\ 2x - 3y + 7z = -21 \end{cases}$

(a) $(3, -1, 2)$ (b) $(1, 3, -2)$
(c) $(4, 1, -3)$ (d) $(1, -2, 2)$

9. $\begin{cases} 4x + y - z = 0 \\ -8x - 6y + z = -\frac{7}{4} \\ 3x - y = -\frac{9}{4} \end{cases}$

(a) $\left(\frac{1}{2}, -\frac{3}{4}, -\frac{7}{4}\right)$ (b) $\left(-\frac{3}{2}, \frac{5}{4}, -\frac{5}{4}\right)$
(c) $\left(-\frac{1}{2}, \frac{3}{4}, -\frac{5}{4}\right)$ (d) $\left(-\frac{1}{2}, \frac{1}{6}, -\frac{3}{4}\right)$

10. $\begin{cases} -4x - y - 8z = -6 \\ y + z = 0 \\ 4x - 7y = 6 \end{cases}$

(a) $(-2, -2, 2)$ (b) $\left(-\frac{33}{2}, -10, 10\right)$
(c) $\left(\frac{1}{8}, -\frac{1}{2}, \frac{1}{2}\right)$ (d) $\left(-\frac{11}{2}, -4, 4\right)$

Using Back-Substitution **In Exercises 11–16, use back-substitution to solve the system of linear equations.**

11. $\begin{cases} 2x - y + 5z = 24 \\ y + 2z = 6 \\ z = 8 \end{cases}$

12. $\begin{cases} 4x - 3y - 2z = 21 \\ 6y - 5z = -8 \\ z = -2 \end{cases}$

13. $\begin{cases} 2x + y - 3z = 10 \\ y + z = 12 \\ z = 2 \end{cases}$

14. $\begin{cases} x - y + 2z = 22 \\ 3y - 8z = -9 \\ z = -3 \end{cases}$

15. $\begin{cases} 4x - 2y + z = 8 \\ -y + z = 4 \\ z = 11 \end{cases}$

16. $\begin{cases} 5x - 8z = 22 \\ 3y - 5z = 10 \\ z = -4 \end{cases}$

Performing Row Operations **In Exercises 17 and 18, perform the row operation and write the equivalent system.**

17. Add Equation 1 to Equation 2.

$\begin{cases} x - 2y + 3z = 5 & \text{Equation 1} \\ -x + 3y - 5z = 4 & \text{Equation 2} \\ 2x - 3z = 0 & \text{Equation 3} \end{cases}$

What did this operation accomplish?

18. Add -2 times Equation 1 to Equation 3.

$\begin{cases} x - 2y + 3z = 5 & \text{Equation 1} \\ -x + 3y - 5z = 4 & \text{Equation 2} \\ 2x - 3z = 0 & \text{Equation 3} \end{cases}$

What did this operation accomplish?

Solving a System of Linear Equations **In Exercises 19–24, solve the system of linear equations and check any solutions algebraically.**

19. $\begin{cases} -2x + 3y = 10 \\ x + y = 0 \end{cases}$

20. $\begin{cases} 2x - y = 0 \\ x - y = 7 \end{cases}$

21. $\begin{cases} 3x - y = 9 \\ x - 2y = -2 \end{cases}$

22. $\begin{cases} x + 2y = 1 \\ 5x - 4y = -23 \end{cases}$

23. $\begin{cases} 1.5x + 0.8y = -0.1 \\ -0.3x + 0.2y = -0.7 \end{cases}$

24. $\begin{cases} \frac{1}{5}x + \frac{1}{2}y = -13 \\ x + y = -35 \end{cases}$

Solving a System of Linear Equations In Exercises 25–46, solve the system of linear equations and check any solutions algebraically.

25. $\begin{cases} x + y + z = 7 \\ 2x - y + z = 9 \\ 3x - z = 10 \end{cases}$

26. $\begin{cases} x + y + z = 5 \\ x - 2y + 4z = 13 \\ 3y + 4z = 13 \end{cases}$

27. $\begin{cases} 2x + 2z = 2 \\ 5x + 3y = 4 \\ 3y - 4z = 4 \end{cases}$

28. $\begin{cases} 2x + 4y + z = 1 \\ x - 2y - 3z = 2 \\ x + y - z = -1 \end{cases}$

29. $\begin{cases} 2x + y - z = 7 \\ x - 2y + 2z = -9 \\ 3x - y + z = 5 \end{cases}$

30. $\begin{cases} 5x - 3y + 2z = 3 \\ 2x + 4y - z = 7 \\ x - 11y + 4z = 3 \end{cases}$

31. $\begin{cases} 3x - 5y + 5z = 1 \\ 5x - 2y + 3z = 0 \\ 7x - y + 3z = 0 \end{cases}$

32. $\begin{cases} 2x + y + 3z = 1 \\ 2x + 6y + 8z = 3 \\ 6x + 8y + 18z = 5 \end{cases}$

33. $\begin{cases} 2x + 3y = 0 \\ 4x + 3y - z = 0 \\ 8x + 3y + 3z = 0 \end{cases}$

34. $\begin{cases} 4x + 3y + 17z = 0 \\ 5x + 4y + 22z = 0 \\ 4x + 2y + 19z = 0 \end{cases}$

35. $\begin{cases} x + 4z = 1 \\ x + y + 10z = 10 \\ 2x - y + 2z = -5 \end{cases}$

36. $\begin{cases} 2x - 2y - 6z = -4 \\ -3x + 2y + 6z = 1 \\ x - y - 5z = -3 \end{cases}$

37. $\begin{cases} 3x - 3y + 6z = 6 \\ x + 2y - z = 5 \\ 5x - 8y + 13z = 7 \end{cases}$

38. $\begin{cases} x + 2z = 5 \\ 3x - y - z = 1 \\ 6x - y + 5z = 16 \end{cases}$

39. $\begin{cases} x - 2y + 5z = 2 \\ 4x - z = 0 \end{cases}$

40. $\begin{cases} x - 3y + 2z = 18 \\ 5x - 13y + 12z = 80 \end{cases}$

41. $\begin{cases} x + 2y - 7z = -4 \\ 2x + y + z = 13 \\ 3x + 9y - 36z = -33 \end{cases}$

42. $\begin{cases} 2x + y - 3z = 4 \\ 4x + 2z = 10 \\ -2x + 3y - 13z = -8 \end{cases}$

43. $\begin{cases} 2x - 3y + z = -2 \\ -4x + 9y = 7 \end{cases}$

44. $\begin{cases} 2x + 3y + 3z = 7 \\ 4x + 18y + 15z = 44 \end{cases}$

45. $\begin{cases} x + 3w = 4 \\ 2y - z - w = 0 \\ 3y - 2w = 1 \\ 2x - y + 4z = 5 \end{cases}$

46. $\begin{cases} x + y + z + w = 6 \\ 2x + 3y - w = 0 \\ -3x + 4y + z + 2w = 4 \\ x + 2y - z + w = 0 \end{cases}$

Vertical Motion In Exercises 47 and 48, an object moving vertically is at the given heights at the specified times. Find the position equation $s = \frac{1}{2}at^2 + v_0t + s_0$ for the object.

47. At $t = 1$ second, $s = 128$ feet
At $t = 2$ seconds, $s = 80$ feet
At $t = 3$ seconds, $s = 0$ feet

48. At $t = 1$ second, $s = 132$ feet
At $t = 2$ seconds, $s = 100$ feet
At $t = 3$ seconds, $s = 36$ feet

Finding the Equation of a Parabola In Exercises 49–54, find the equation of the parabola

$y = ax^2 + bx + c$

that passes through the points. To verify your result, use a graphing utility to plot the points and graph the parabola.

49. $(0, 0), (2, -2), (4, 0)$

50. $(0, 3), (1, 4), (2, 3)$

51. $(2, 0), (3, -1), (4, 0)$

52. $(1, 3), (2, 2), (3, -3)$

53. $\left(\frac{1}{2}, 1\right), (1, 3), (2, 13)$

54. $(-2, -3), (-1, 0), \left(\frac{1}{2}, -3\right)$

Finding the Equation of a Circle In Exercises 55–58, find the equation of the circle

$x^2 + y^2 + Dx + Ey + F = 0$

that passes through the points. To verify your result, use a graphing utility to plot the points and graph the circle.

55. $(0, 0), (5, 5), (10, 0)$

56. $(0, 0), (0, 6), (3, 3)$

57. $(-3, -1), (2, 4), (-6, 8)$

58. $(0, 0), (0, -2), (3, 0)$

59. **Sports** In Super Bowl I, on January 15, 1967, the Green Bay Packers defeated the Kansas City Chiefs by a score of 35 to 10. The total points scored came from a combination of touchdowns, extra-point kicks, and field goals, worth 6, 1, and 3 points, respectively. The numbers of touchdowns and extra-point kicks were equal. There were six times as many touchdowns as field goals. Find the numbers of touchdowns, extra-point kicks, and field goals scored. *(Source: National Football League)*

60. **Agriculture** A mixture of 5 pounds of fertilizer A, 13 pounds of fertilizer B, and 4 pounds of fertilizer C provides the optimal nutrients for a plant. Commercial brand X contains equal parts of fertilizer B and fertilizer C. Commercial brand Y contains one part of fertilizer A and two parts of fertilizer B. Commercial brand Z contains two parts of fertilizer A, five parts of fertilizer B, and two parts of fertilizer C. How much of each fertilizer brand is needed to obtain the desired mixture?

61. Finance A small corporation borrowed \$775,000 to expand its clothing line. Some of the money was borrowed at 8%, some at 9%, and some at 10%. How much was borrowed at each rate when the annual interest owed was \$67,500 and the amount borrowed at 8% was four times the amount borrowed at 10%?

62. Advertising A health insurance company advertises on television, on radio, and in the local newspaper. The marketing department has an advertising budget of \$42,000 per month. A television ad costs \$1000, a radio ad costs \$200, and a newspaper ad costs \$500. The department wants to run 60 ads per month and have as many television ads as radio and newspaper ads combined. How many of each type of ad can the department run each month?

Geometry In Exercises 63 and 64, find the values of *x*, *y*, and *z* in the figure.

63.

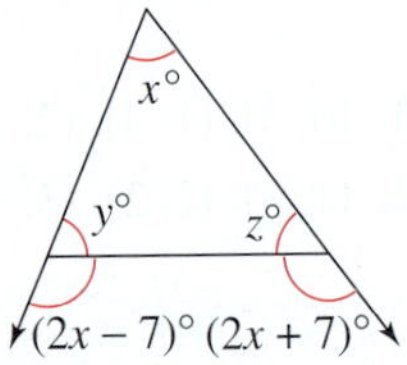

64.

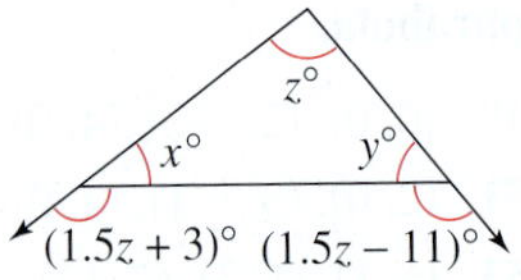

65. Geometry The perimeter of a triangle is 180 feet. The longest side of the triangle is 9 feet shorter than twice the shortest side. The sum of the lengths of the two shorter sides is 30 feet more than the length of the longest side. Find the lengths of the sides of the triangle.

66. Acid Mixture A chemist needs 10 liters of a 25% acid solution. The solution is to be mixed from three solutions whose concentrations are 10%, 20%, and 50%. How many liters of each solution will satisfy each condition?

(a) Use 2 liters of the 50% solution.

(b) Use as little as possible of the 50% solution.

(c) Use as much as possible of the 50% solution.

67. Electrical Network Applying Kirchhoff's Laws to the electrical network in the figure, the currents I_1, I_2, and I_3 are the solution of the system

$$\begin{cases} I_1 - I_2 + I_3 = 0 \\ 3I_1 + 2I_2 = 7. \\ 2I_2 + 4I_3 = 8 \end{cases}$$

Find the currents.

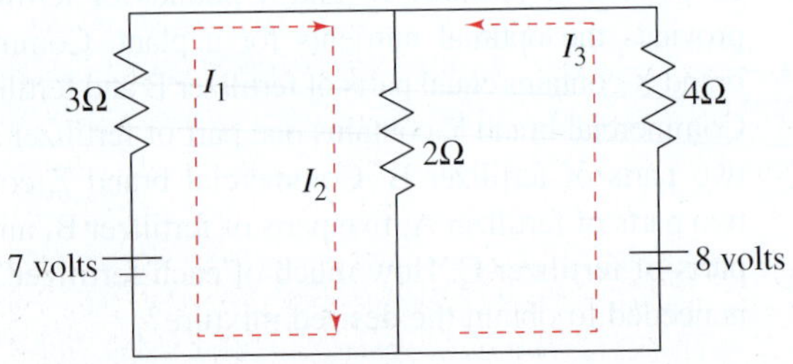

68. Pulley System A system of pulleys is loaded with 128-pound and 32-pound weights (see figure). The tensions t_1 and t_2 in the ropes and the acceleration a of the 32-pound weight are found by solving the system of equations

$$\begin{cases} t_1 - 2t_2 = 0 \\ t_1 - 2a = 128 \\ t_2 + a = 32 \end{cases}$$

where t_1 and t_2 are in pounds and a is in feet per second squared. Solve this system.

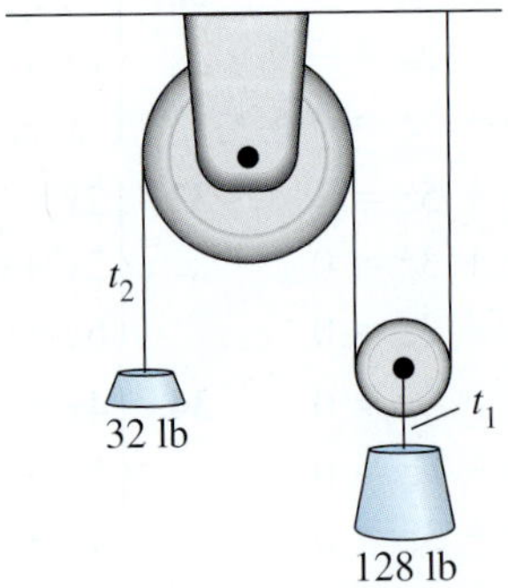

Fitting a Parabola To find the least squares regression parabola $y = ax^2 + bx + c$ for a set of points

$$(x_1, y_1), (x_2, y_2), \ldots, (x_n, y_n)$$

you can solve the following system of linear equations for *a*, *b*, and *c*.

$$\begin{cases} nc + \left(\sum_{i=1}^{n} x_i\right)b + \left(\sum_{i=1}^{n} x_i^2\right)a = \sum_{i=1}^{n} y_i \\ \left(\sum_{i=1}^{n} x_i\right)c + \left(\sum_{i=1}^{n} x_i^2\right)b + \left(\sum_{i=1}^{n} x_i^3\right)a = \sum_{i=1}^{n} x_i y_i \\ \left(\sum_{i=1}^{n} x_i^2\right)c + \left(\sum_{i=1}^{n} x_i^3\right)b + \left(\sum_{i=1}^{n} x_i^4\right)a = \sum_{i=1}^{n} x_i^2 y_i \end{cases}$$

In Exercises 69 and 70, the sums have been evaluated. Solve the given system for *a*, *b*, and *c* to find the least squares regression parabola for the points. Use a graphing utility to confirm the result.

69. $\begin{cases} 4c + 9b + 29a = 20 \\ 9c + 29b + 99a = 70 \\ 29c + 99b + 353a = 254 \end{cases}$

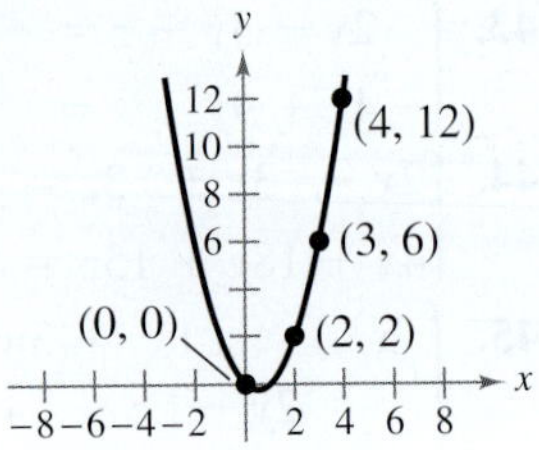

70. $\begin{cases} 4c + 40a = 19 \\ 40b = -12 \\ 40c + 544a = 160 \end{cases}$

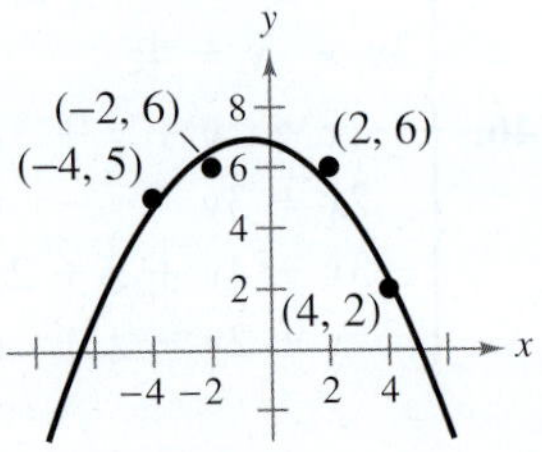

71. Data Analysis: Stopping Distance In testing a new automobile braking system, engineers recorded the speed x (in miles per hour) and the stopping distance y (in feet) in the following table.

Speed, x	30	40	50
Stopping Distance, y	55	105	188

(a) Find the least squares regression parabola $y = ax^2 + bx + c$ for the data by solving the system.

$$\begin{cases} 3c + 120b + 5000a = 348 \\ 120c + 5000b + 216{,}000a = 15{,}250 \\ 5000c + 216{,}000b + 9{,}620{,}000a = 687{,}500 \end{cases}$$

(b) Graph the parabola and the data on the same set of axes.

(c) Use the model to estimate the stopping distance when the speed is 70 miles per hour.

72. Data Analysis: Wildlife

A wildlife management team studied the reproductive rates of deer in three tracts of a wildlife preserve. Each tract contained 5 acres. In each tract, the number of females x, and the percent of females y, that had offspring the following year were recorded. The table shows the results.

Number, x	100	120	140
Percent, y	75	68	55

(a) Use the data to create a system of linear equations. Then find the least squares regression parabola for the data by solving the system.

(b) Use a graphing utility to graph the parabola and the data in the same viewing window.

(c) Use the model to estimate the percent of females that had offspring when there were 170 females.

(d) Use the model to estimate the number of females when 40% of the females had offspring.

Advanced Applications **In Exercises 73 and 74, find values of x, y, and λ that satisfy the system. These systems arise in certain optimization problems in calculus, and λ is called a Lagrange multiplier.**

73. $\begin{cases} 2x - 2x\lambda = 0 \\ -2y + \lambda = 0 \\ y - x^2 = 0 \end{cases}$

74. $\begin{cases} 2 + 2y + 2\lambda = 0 \\ 2x + 1 + \lambda = 0 \\ 2x + y - 100 = 0 \end{cases}$

Exploration

True or False? **In Exercises 75 and 76, determine whether the statement is true or false. Justify your answer.**

75. The system

$$\begin{cases} x + 3y - 6z = -16 \\ 2y - z = -1 \\ z = 3 \end{cases}$$

is in row-echelon form.

76. If a system of three linear equations is inconsistent, then its graph has no points common to all three equations.

77. Think About It Are the following two systems of equations equivalent? Give reasons for your answer.

$$\begin{cases} x + 3y - z = 6 \\ 2x - y + 2z = 1 \\ 3x + 2y - z = 2 \end{cases}$$

$$\begin{cases} x + 3y - z = 6 \\ -7y + 4z = 1 \\ -7y - 4z = -16 \end{cases}$$

78. HOW DO YOU SEE IT? The number of sides x and the combined number of sides and diagonals y for each of three regular polygons are shown below. Write a system of linear equations to find an equation of the form $y = ax^2 + bx + c$ that represents the relationship between x and y for the three polygons.

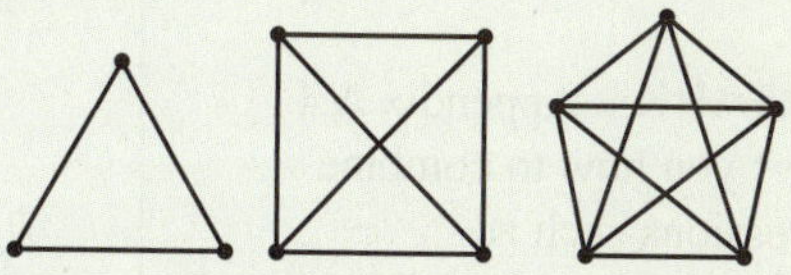

$x = 3, y = 3$ $x = 4, y = 6$ $x = 5, y = 10$

Finding Systems of Linear Equations **In Exercises 79–82, find two systems of linear equations that have the ordered triple as a solution. (There are many correct answers.)**

79. $(3, -4, 2)$

80. $(-5, -2, 1)$

81. $\left(-6, -\frac{1}{2}, -\frac{7}{4}\right)$

82. $\left(-\frac{3}{2}, 4, -7\right)$

Project: Earnings per Share To work an extended application analyzing the earnings per share for Wal-Mart Stores, Inc., from 1995 through 2010, visit this text's website at *LarsonPrecalculus.com.* *(Source: Wal-Mart Stores, Inc.)*

Bruce MacQueen/Shutterstock.com

7.4 Partial Fractions

Partial fractions can help you analyze the behavior of a rational function. For instance, in Exercise 59 on page 507, you will use partial fractions to analyze the predicted cost for a company to remove a percentage of a chemical from its waste water.

- **Recognize partial fraction decompositions of rational expressions.**
- **Find partial fraction decompositions of rational expressions.**

Introduction

In this section, you will learn to write a rational expression as the sum of two or more simpler rational expressions. For example, the rational expression

$$\frac{x+7}{x^2-x-6}$$

can be written as the sum of two fractions with first-degree denominators. That is,

$$\frac{x+7}{x^2-x-6} = \underbrace{\overbrace{\underbrace{\frac{2}{x-3}}_{\text{Partial fraction}} + \underbrace{\frac{-1}{x+2}}_{\text{Partial fraction}}}^{\text{Partial fraction decomposition of } \frac{x+7}{x^2-x-6}}}.$$

Each fraction on the right side of the equation is a **partial fraction,** and together they make up the **partial fraction decomposition** of the left side.

ALGEBRA HELP You can review how to find the degree of a polynomial (such as $x - 3$ and $x + 2$) in Appendix A.3.

REMARK Appendix A.4 shows you how to combine expressions such as

$$\frac{1}{x-2} + \frac{-1}{x+3} = \frac{5}{(x-2)(x+3)}.$$

The method of partial fraction decomposition shows you how to reverse this process and write

$$\frac{5}{(x-2)(x+3)} = \frac{1}{x-2} + \frac{-1}{x+3}.$$

Decomposition of $N(x)/D(x)$ into Partial Fractions

1. *Divide when improper:* When $N(x)/D(x)$ is an improper fraction [degree of $N(x) \geq$ degree of $D(x)$], divide the denominator into the numerator to obtain

$$\frac{N(x)}{D(x)} = (\text{polynomial}) + \frac{N_1(x)}{D(x)}$$

and apply Steps 2, 3, and 4 below to the proper rational expression

$$\frac{N_1(x)}{D(x)}.$$

Note that $N_1(x)$ is the remainder from the division of $N(x)$ by $D(x)$.

2. *Factor the denominator:* Completely factor the denominator into factors of the form

$$(px+q)^m \quad \text{and} \quad (ax^2+bx+c)^n$$

where $(ax^2 + bx + c)$ is irreducible.

3. *Linear factors:* For *each* factor of the form $(px + q)^m$, the partial fraction decomposition must include the following sum of m fractions.

$$\frac{A_1}{(px+q)} + \frac{A_2}{(px+q)^2} + \cdots + \frac{A_m}{(px+q)^m}$$

4. *Quadratic factors:* For *each* factor of the form $(ax^2 + bx + c)^n$, the partial fraction decomposition must include the following sum of n fractions.

$$\frac{B_1x+C_1}{ax^2+bx+c} + \frac{B_2x+C_2}{(ax^2+bx+c)^2} + \cdots + \frac{B_nx+C_n}{(ax^2+bx+c)^n}$$

tlorna/Shutterstock.com

Partial Fraction Decomposition

The following examples demonstrate algebraic techniques for determining the constants in the numerators of partial fractions. Note that the techniques vary slightly, depending on the type of factors of the denominator: linear or quadratic, distinct or repeated.

EXAMPLE 1 Distinct Linear Factors

Write the partial fraction decomposition of

$$\frac{x+7}{x^2-x-6}.$$

Solution The expression is proper, so you should begin by factoring the denominator. Because

$$x^2-x-6=(x-3)(x+2)$$

you should include one partial fraction with a constant numerator for each linear factor of the denominator. Write the form of the decomposition as follows.

$$\frac{x+7}{x^2-x-6}=\frac{A}{x-3}+\frac{B}{x+2} \qquad \text{Write form of decomposition.}$$

Multiplying each side of this equation by the least common denominator, $(x-3)(x+2)$, leads to the **basic equation**

$$x+7=A(x+2)+B(x-3). \qquad \text{Basic equation}$$

Because this equation is true for all x, substitute any *convenient* values of x that will help determine the constants A and B. Values of x that are especially convenient are those that make the factors $(x+2)$ and $(x-3)$ equal to zero. For instance, to solve for B, let $x=-2$. Then

$$-2+7=A(-2+2)+B(-2-3) \qquad \text{Substitute } -2 \text{ for } x.$$

$$5=A(0)+B(-5)$$

$$5=-5B$$

$$-1=B.$$

To solve for A, let $x=3$ and obtain

$$3+7=A(3+2)+B(3-3) \qquad \text{Substitute 3 for } x.$$

$$10=A(5)+B(0)$$

$$10=5A$$

$$2=A.$$

So, the partial fraction decomposition is

$$\frac{x+7}{x^2-x-6}=\frac{2}{x-3}+\frac{-1}{x+2}.$$

Check this result by combining the two partial fractions on the right side of the equation, or by using your graphing utility.

Checkpoint Audio-video solution in English & Spanish at LarsonPrecalculus.com.

Write the partial fraction decomposition of

$$\frac{x+5}{2x^2-x-1}.$$

TECHNOLOGY To use a graphing utility to check the decomposition found in Example 1, graph

$$y_1=\frac{x+7}{x^2-x-6}$$

and

$$y_2=\frac{2}{x-3}+\frac{-1}{x+2}$$

in the same viewing window. The graphs should be identical, as shown below.

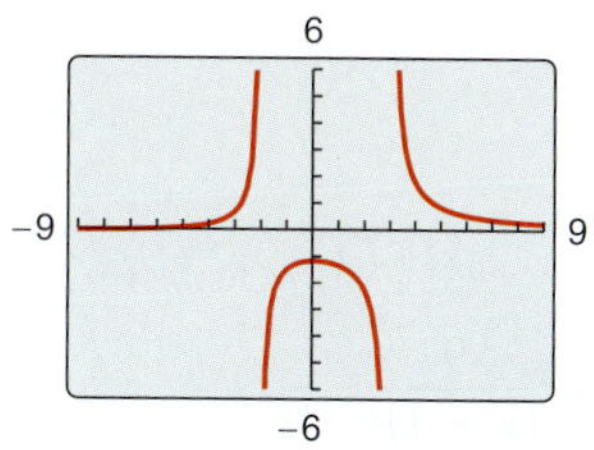

The next example shows how to find the partial fraction decomposition of a rational expression whose denominator has a *repeated* linear factor.

EXAMPLE 2 Repeated Linear Factors

Write the partial fraction decomposition of $\dfrac{x^4 + 2x^3 + 6x^2 + 20x + 6}{x^3 + 2x^2 + x}$.

ALGEBRA HELP You can review long division of polynomials in Section 2.3. You can review factoring of polynomials in Appendix A.3.

Solution This rational expression is improper, so you should begin by dividing the numerator by the denominator to obtain

$$x + \frac{5x^2 + 20x + 6}{x^3 + 2x^2 + x}.$$

Because the denominator of the remainder factors as

$$x^3 + 2x^2 + x = x(x^2 + 2x + 1) = x(x + 1)^2$$

you should include one partial fraction with a constant numerator for each power of x and $(x + 1)$, and write the form of the decomposition as follows.

$$\frac{5x^2 + 20x + 6}{x(x + 1)^2} = \frac{A}{x} + \frac{B}{x + 1} + \frac{C}{(x + 1)^2} \qquad \text{Write form of decomposition.}$$

Multiplying each side by the LCD, $x(x + 1)^2$, leads to the basic equation

$$5x^2 + 20x + 6 = A(x + 1)^2 + Bx(x + 1) + Cx. \qquad \text{Basic equation}$$

REMARK To obtain the basic equation, be sure to multiply *each* fraction in the form of the decomposition by the LCD.

Letting $x = -1$ eliminates the A- and B-terms and yields

$$5(-1)^2 + 20(-1) + 6 = A(-1 + 1)^2 + B(-1)(-1 + 1) + C(-1)$$
$$5 - 20 + 6 = 0 + 0 - C$$
$$C = 9.$$

Letting $x = 0$ eliminates the B- and C-terms and yields

$$5(0)^2 + 20(0) + 6 = A(0 + 1)^2 + B(0)(0 + 1) + C(0)$$
$$6 = A(1) + 0 + 0$$
$$6 = A.$$

At this point, you have exhausted the most convenient values of x, so to find the value of B, use *any other value* of x along with the known values of A and C. So, using $x = 1$, $A = 6$, and $C = 9$,

$$5(1)^2 + 20(1) + 6 = 6(1 + 1)^2 + B(1)(1 + 1) + 9(1)$$
$$31 = 6(4) + 2B + 9$$
$$-2 = 2B$$
$$-1 = B.$$

So, the partial fraction decomposition is

$$\frac{x^4 + 2x^3 + 6x^2 + 20x + 6}{x^3 + 2x^2 + x} = x + \frac{6}{x} + \frac{-1}{x + 1} + \frac{9}{(x + 1)^2}.$$

✓ Checkpoint Audio-video solution in English & Spanish at LarsonPrecalculus.com.

Write the partial fraction decomposition of

$$\frac{x^4 + x^3 + x + 4}{x^3 + x^2}.$$

The procedure used to solve for the constants in Examples 1 and 2 works well when the factors of the denominator are linear. However, when the denominator contains irreducible quadratic factors, you should use a different procedure, which involves writing the right side of the basic equation in polynomial form, *equating the coefficients* of like terms, and using a system of equations to solve for the coefficients.

EXAMPLE 3 Distinct Linear and Quadratic Factors

Write the partial fraction decomposition of

$$\frac{3x^2 + 4x + 4}{x^3 + 4x}.$$

John Bernoulli (1667–1748), a Swiss mathematician, introduced the method of partial fractions and was instrumental in the early development of calculus. Bernoulli was a professor at the University of Basel and taught many outstanding students, the most famous of whom was Leonhard Euler.

Solution This expression is proper, so begin by factoring the denominator. Because the denominator factors as

$$x^3 + 4x = x(x^2 + 4)$$

you should include one partial fraction with a constant numerator and one partial fraction with a linear numerator, and write the form of the decomposition as follows.

$$\frac{3x^2 + 4x + 4}{x^3 + 4x} = \frac{A}{x} + \frac{Bx + C}{x^2 + 4}$$ Write form of decomposition.

Multiplying each side by the LCD, $x(x^2 + 4)$, yields the basic equation

$$3x^2 + 4x + 4 = A(x^2 + 4) + (Bx + C)x.$$ Basic equation

Expanding this basic equation and collecting like terms produces

$$3x^2 + 4x + 4 = Ax^2 + 4A + Bx^2 + Cx$$

$$= (A + B)x^2 + Cx + 4A.$$ Polynomial form

Finally, because two polynomials are equal if and only if the coefficients of like terms are equal, equate the coefficients of like terms on opposite sides of the equation.

$$3x^2 + 4x + 4 = (A + B)x^2 + Cx + 4A$$ Equate coefficients of like terms.

Now write the following system of linear equations.

$$\begin{cases} A + B \quad\quad = 3 \\ \quad\quad\quad C = 4 \\ 4A \quad\quad\quad = 4 \end{cases}$$

Equation 1

Equation 2

Equation 3

From this system, you can see that

$$A = 1 \quad \text{and} \quad C = 4.$$

Moreover, back-substituting $A = 1$ into Equation 1 yields

$$1 + B = 3 \Longrightarrow B = 2.$$

So, the partial fraction decomposition is

$$\frac{3x^2 + 4x + 4}{x^3 + 4x} = \frac{1}{x} + \frac{2x + 4}{x^2 + 4}.$$

✓ ***Checkpoint*** *Audio-video solution in English & Spanish at LarsonPrecalculus.com.*

Write the partial fraction decomposition of

$$\frac{2x^2 - 5}{x^3 + x}.$$

Mary Evans Picture Library

The next example shows how to find the partial fraction decomposition of a rational expression whose denominator has a *repeated* quadratic factor.

EXAMPLE 4 Repeated Quadratic Factors

Write the partial fraction decomposition of

$$\frac{8x^3 + 13x}{(x^2 + 2)^2}.$$

Solution Include one partial fraction with a linear numerator for each power of $(x^2 + 2)$.

$$\frac{8x^3 + 13x}{(x^2 + 2)^2} = \frac{Ax + B}{x^2 + 2} + \frac{Cx + D}{(x^2 + 2)^2} \qquad \text{Write form of decomposition.}$$

Multiplying each side by the LCD, $(x^2 + 2)^2$, yields the basic equation

$$\begin{aligned} 8x^3 + 13x &= (Ax + B)(x^2 + 2) + Cx + D && \text{Basic equation} \\ &= Ax^3 + 2Ax + Bx^2 + 2B + Cx + D \\ &= Ax^3 + Bx^2 + (2A + C)x + (2B + D). && \text{Polynomial form} \end{aligned}$$

Equating coefficients of like terms on opposite sides of the equation

$$8x^3 + 0x^2 + 13x + 0 = Ax^3 + Bx^2 + (2A + C)x + (2B + D)$$

produces the following system of linear equations.

$$\begin{cases} A \qquad\qquad\quad = 8 & \text{Equation 1} \\ \quad B \qquad\qquad = 0 & \text{Equation 2} \\ 2A + \qquad C \quad = 13 & \text{Equation 3} \\ \quad 2B + \qquad D = 0 & \text{Equation 4} \end{cases}$$

Use the values $A = 8$ and $B = 0$ to obtain the following.

$$2(8) + C = 13 \qquad \text{Substitute 8 for } A \text{ in Equation 3.}$$

$$C = -3$$

$$2(0) + D = 0 \qquad \text{Substitute 0 for } B \text{ in Equation 4.}$$

$$D = 0$$

So, using

$$A = 8, \quad B = 0, \quad C = -3, \quad \text{and} \quad D = 0$$

the partial fraction decomposition is

$$\frac{8x^3 + 13x}{(x^2 + 2)^2} = \frac{8x}{x^2 + 2} + \frac{-3x}{(x^2 + 2)^2}.$$

Check this result by combining the two partial fractions on the right side of the equation, or by using your graphing utility.

✓ **Checkpoint** *Audio-video solution in English & Spanish at LarsonPrecalculus.com.*

Write the partial fraction decomposition of

$$\frac{x^3 + 3x^2 - 2x + 7}{(x^2 + 4)^2}.$$

EXAMPLE 5 **Repeated Linear and Quadratic Factors**

Write the partial fraction decomposition of $\dfrac{x + 5}{x^2(x^2 + 1)^2}$.

Solution Include one partial fraction with a constant numerator for each power of x and one partial fraction with a linear numerator for each power of $(x^2 + 1)$.

$$\frac{x + 5}{x^2(x^2 + 1)^2} = \frac{A}{x} + \frac{B}{x^2} + \frac{Cx + D}{x^2 + 1} + \frac{Ex + F}{(x^2 + 1)^2}$$ Write form of decomposition.

Multiplying each side by the LCD, $x^2(x^2 + 1)^2$, yields the basic equation

$$x + 5 = Ax(x^2 + 1)^2 + B(x^2 + 1)^2 + (Cx + D)x^2(x^2 + 1) + (Ex + F)x^2$$ Basic equation

$$= (A + C)x^5 + (B + D)x^4 + (2A + C + E)x^3 + (2B + D + F)x^2 + Ax + B.$$

Verify, by equating coefficients of like terms on opposite sides of the equation and solving the resulting system of linear equations, that $A = 1$, $B = 5$, $C = -1$, $D = -5$, $E = -1$, and $F = -5$, and that the partial fraction decomposition is

$$\frac{x + 5}{x^2(x^2 + 1)^2} = \frac{1}{x} + \frac{5}{x^2} - \frac{x + 5}{x^2 + 1} - \frac{x + 5}{(x^2 + 1)^2}.$$

✓ **Checkpoint** *Audio-video solution in English & Spanish at LarsonPrecalculus.com.*

Write the partial fraction decomposition of $\dfrac{4x - 8}{x^2(x^2 + 2)^2}$.

Guidelines for Solving the Basic Equation

Linear Factors

1. Substitute the *zeros* of the distinct linear factors into the basic equation.
2. For repeated linear factors, use the coefficients determined in Step 1 to rewrite the basic equation. Then substitute *other* convenient values of x and solve for the remaining coefficients.

Quadratic Factors

1. Expand the basic equation.
2. Collect terms according to powers of x.
3. Equate the coefficients of like terms to obtain equations involving A, B, C, and so on.
4. Use a system of linear equations to solve for A, B, C, and so on.

Keep in mind that for *improper* rational expressions, you must first divide before applying partial fraction decomposition.

Summarize (Section 7.4)

1. Explain what is meant by the partial fraction decomposition of a rational expression *(page 500)*.
2. Explain how to find the partial fraction decomposition of a rational expression *(pages 500–505)*. For examples of finding partial fraction decompositions of rational expressions, see Examples 1–5.

7.4 Exercises

See CalcChat.com for tutorial help and worked-out solutions to odd-numbered exercises.

Vocabulary: Fill in the blanks.

1. The process of writing a rational expression as the sum of two or more simpler rational expressions is called ________ ________ ________.
2. If the degree of the numerator of a rational expression is greater than or equal to the degree of the denominator, then the fraction is called ________.
3. Each fraction on the right side of the equation $\dfrac{x-1}{x^2-8x+15} = \dfrac{-1}{x-3} + \dfrac{2}{x-5}$ is a ________ ________.
4. You obtain the ________ ________ after multiplying each side of the partial fraction decomposition form by the least common denominator.

Skills and Applications

Matching **In Exercises 5–8, match the rational expression with the form of its decomposition. [The decompositions are labeled (a), (b), (c), and (d).]**

(a) $\dfrac{A}{x} + \dfrac{B}{x+2} + \dfrac{C}{x-2}$ (b) $\dfrac{A}{x} + \dfrac{B}{x-4}$

(c) $\dfrac{A}{x} + \dfrac{B}{x^2} + \dfrac{C}{x-4}$ (d) $\dfrac{A}{x} + \dfrac{Bx+C}{x^2+4}$

5. $\dfrac{3x-1}{x(x-4)}$
6. $\dfrac{3x-1}{x^2(x-4)}$
7. $\dfrac{3x-1}{x(x^2+4)}$
8. $\dfrac{3x-1}{x(x^2-4)}$

Writing the Form of the Decomposition **In Exercises 9–16, write the form of the partial fraction decomposition of the rational expression. Do not solve for the constants.**

9. $\dfrac{3}{x^2-2x}$
10. $\dfrac{x-2}{x^2+4x+3}$
11. $\dfrac{9}{x^3-7x^2}$
12. $\dfrac{2x-3}{x^3+10x}$
13. $\dfrac{4x^2+3}{(x-5)^3}$
14. $\dfrac{6x+5}{(x+2)^4}$
15. $\dfrac{x-1}{x(x^2+1)^2}$
16. $\dfrac{x+4}{x^2(3x-1)^2}$

Writing the Partial Fraction Decomposition **In Exercises 17–42, write the partial fraction decomposition of the rational expression. Check your result algebraically.**

17. $\dfrac{1}{x^2+x}$
18. $\dfrac{3}{x^2-3x}$
19. $\dfrac{1}{2x^2+x}$
20. $\dfrac{5}{x^2+x-6}$
21. $\dfrac{3}{x^2+x-2}$
22. $\dfrac{x+1}{x^2-x-6}$
23. $\dfrac{1}{x^2-1}$
24. $\dfrac{1}{4x^2-9}$
25. $\dfrac{x^2+12x+12}{x^3-4x}$
26. $\dfrac{x+2}{x(x^2-9)}$
27. $\dfrac{3x}{(x-3)^2}$
28. $\dfrac{2x-3}{(x-1)^2}$
29. $\dfrac{4x^2+2x-1}{x^2(x+1)}$
30. $\dfrac{6x^2+1}{x^2(x-1)^2}$
31. $\dfrac{x^2+2x+3}{x^3+x}$
32. $\dfrac{2x}{x^3-1}$
33. $\dfrac{x}{x^3-x^2-2x+2}$
34. $\dfrac{x+6}{x^3-3x^2-4x+12}$
35. $\dfrac{2x^2+x+8}{(x^2+4)^2}$
36. $\dfrac{x^2}{x^4-2x^2-8}$
37. $\dfrac{x}{16x^4-1}$
38. $\dfrac{3}{x^4+x}$
39. $\dfrac{x^2+5}{(x+1)(x^2-2x+3)}$
40. $\dfrac{x^2-4x+7}{(x+1)(x^2-2x+3)}$
41. $\dfrac{8x-12}{x^2(x^2+2)^2}$
42. $\dfrac{x+1}{x^3(x^2+1)^2}$

Improper Rational Expression Decomposition **In Exercises 43–50, write the partial fraction decomposition of the improper rational expression.**

43. $\dfrac{x^2-x}{x^2+x+1}$
44. $\dfrac{x^2-4x}{x^2+x+6}$
45. $\dfrac{2x^3-x^2+x+5}{x^2+3x+2}$
46. $\dfrac{x^3+2x^2-x+1}{x^2+3x-4}$
47. $\dfrac{x^4}{(x-1)^3}$
48. $\dfrac{16x^4}{(2x-1)^3}$
49. $\dfrac{x^4+2x^3+4x^2+8x+2}{x^3+2x^2+x}$
50. $\dfrac{2x^4+8x^3+7x^2-7x-12}{x^3+4x^2+4x}$

Writing the Partial Fraction Decomposition **In Exercises 51–58, write the partial fraction decomposition of the rational expression. Use a graphing utility to check your result.**

51. $\dfrac{5-x}{2x^2+x-1}$

52. $\dfrac{3x^2-7x-2}{x^3-x}$

53. $\dfrac{4x^2-1}{2x(x+1)^2}$

54. $\dfrac{3x+1}{2x^3+3x^2}$

55. $\dfrac{x^2+x+2}{(x^2+2)^2}$

56. $\dfrac{x^3}{(x+2)^2(x-2)^2}$

57. $\dfrac{2x^3-4x^2-15x+5}{x^2-2x-8}$

58. $\dfrac{x^3-x+3}{x^2+x-2}$

59. Waste Water The predicted cost C (in thousands of dollars) for a company to remove $p\%$ of a chemical from its waste water is given by the model

$$C = \frac{120p}{10{,}000 - p^2}, \quad 0 \le p < 100.$$

Write the partial fraction decomposition for the rational function. Verify your result by using a graphing utility to create a table comparing the original function with the partial fractions.

60. Thermodynamics The magnitude of the range R of exhaust temperatures (in degrees Fahrenheit) in an experimental diesel engine is approximated by the model

$$R = \frac{5000(4-3x)}{(11-7x)(7-4x)}, \quad 0 < x \le 1$$

where x is the relative load (in foot-pounds).

(a) Write the partial fraction decomposition of the equation.

(b) The decomposition in part (a) is the difference of two fractions. The absolute values of the terms give the expected maximum and minimum temperatures of the exhaust gases for different loads.

$\text{Ymax} = |\text{1st term}| \qquad \text{Ymin} = |\text{2nd term}|$

Write the equations for Ymax and Ymin.

(c) Use a graphing utility to graph each equation from part (b) in the same viewing window.

(d) Determine the expected maximum and minimum temperatures for a relative load of 0.5.

Exploration

True or False? **In Exercises 61–63, determine whether the statement is true or false. Justify your answer.**

61. For the rational expression $\dfrac{x}{(x+10)(x-10)^2}$, the partial fraction decomposition is of the form $\dfrac{A}{x+10} + \dfrac{B}{(x-10)^2}$.

62. For the rational expression $\dfrac{2x+3}{x^2(x+2)^2}$, the partial fraction decomposition is of the form $\dfrac{Ax+B}{x^2} + \dfrac{Cx+D}{(x+2)^2}$.

63. When writing the partial fraction decomposition of the expression $\dfrac{x^3+x-2}{x^2-5x-14}$, the first step is to divide the numerator by the denominator.

64. HOW DO YOU SEE IT? Identify the graph of the rational function and the graph representing each partial fraction of its partial fraction decomposition. Then state any relationship between the vertical asymptotes of the graph of the rational function and the vertical asymptotes of the graphs representing the partial fractions of the decomposition. To print an enlarged copy of the graph, go to *MathGraphs.com.*

(a) $y = \dfrac{x-12}{x(x-4)} = \dfrac{3}{x} - \dfrac{2}{x-4}$

(b) $y = \dfrac{2(4x-3)}{x^2-9} = \dfrac{3}{x-3} + \dfrac{5}{x+3}$

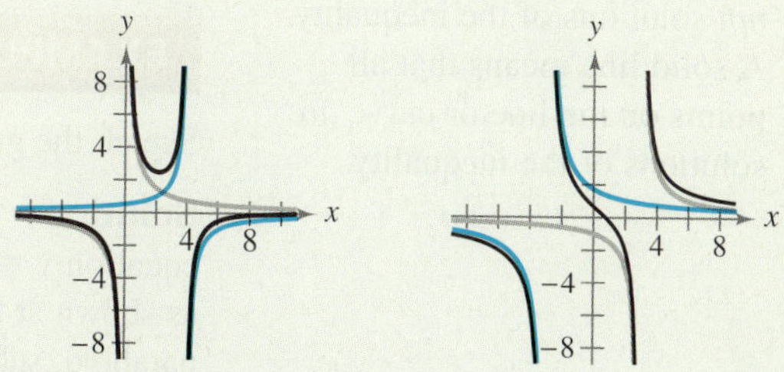

Writing the Partial Fraction Decomposition **In Exercises 65–68, write the partial fraction decomposition of the rational expression. Then assign a value to the constant a to check the result algebraically and graphically.**

65. $\dfrac{1}{a^2-x^2}$

66. $\dfrac{1}{x(x+a)}$

67. $\dfrac{1}{y(a-y)}$

68. $\dfrac{1}{(x+1)(a-x)}$

69. Writing Describe two ways of solving for the constants in a partial fraction decomposition.

tlorna/Shutterstock.com

7.5 Systems of Inequalities

Systems of inequalities in two variables can help you model and solve real-life problems. For instance, in Exercise 70 on page 516, you will use a system of inequalities to analyze the nutritional content of a dietary supplement.

- Sketch the graphs of inequalities in two variables.
- Solve systems of inequalities.
- Use systems of inequalities in two variables to model and solve real-life problems.

The Graph of an Inequality

The statements

$$3x - 2y < 6 \quad \text{and} \quad 2x^2 + 3y^2 \geq 6$$

are inequalities in two variables. An ordered pair (a, b) is a **solution of an inequality** in x and y when the inequality is true after a and b are substituted for x and y, respectively. The **graph of an inequality** is the collection of all solutions of the inequality. To sketch the graph of an inequality, begin by sketching the graph of the *corresponding equation*. The graph of the equation will usually separate the plane into two or more regions. In each such region, one of the following must be true.

1. *All* points in the region are solutions of the inequality.
2. *No* point in the region is a solution of the inequality.

So, you can determine whether the points in an entire region satisfy the inequality by testing *one* point in the region.

Sketching the Graph of an Inequality in Two Variables

1. Replace the inequality sign by an equal sign and sketch the graph of the resulting equation. (Use a dashed line for $<$ or $>$ and a solid line for $\leq$ or $\geq$.)
2. Test one point in each of the regions formed by the graph in Step 1. If the point satisfies the inequality, then shade the entire region to denote that every point in the region satisfies the inequality.

REMARK Be careful when you are sketching the graph of an inequality in two variables. A dashed line means that all points on the line or curve *are not* solutions of the inequality. A solid line means that all points on the line or curve *are* solutions of the inequality.

EXAMPLE 1 Sketching the Graph of an Inequality

Sketch the graph of $y \geq x^2 - 1$.

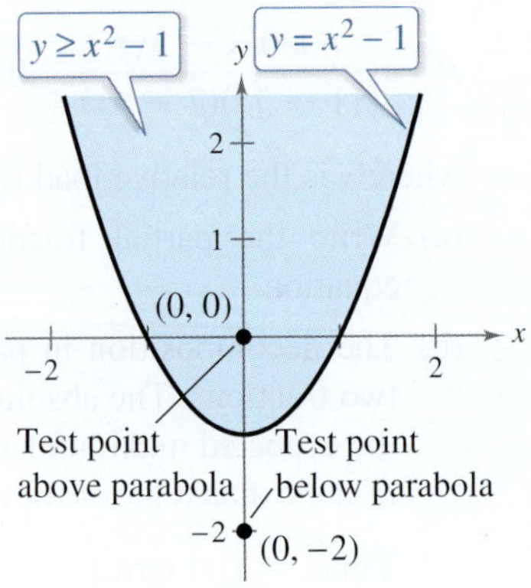

Solution Begin by graphing the corresponding equation $y = x^2 - 1$, which is a parabola, as shown at the right. Test a point *above* the parabola, such as $(0, 0)$, and a point *below* the parabola, such as $(0, -2)$.

The points that satisfy the inequality are those lying above (or on) the parabola.

$(0, 0)$: $0 \stackrel{?}{\geq} 0^2 - 1$

$0 \geq -1$ $\quad$ $(0, 0)$ is a solution.

$(0, -2)$: $-2 \stackrel{?}{\geq} 0^2 - 1$

$-2 \not\geq -1$ $\quad$ $(0, -2)$ is not a solution.

✓ **Checkpoint** *Audio-video solution in English & Spanish at LarsonPrecalculus.com.*

Sketch the graph of $(x + 2)^2 + (y - 2)^2 < 16$.

alexskopje/Shutterstock.com

The inequality in Example 1 is a nonlinear inequality in two variables. Most of the following examples involve **linear inequalities** such as $ax + by < c$ (where a and b are not both zero). The graph of a linear inequality is a half-plane lying on one side of the line $ax + by = c$.

EXAMPLE 2 **Sketching the Graph of a Linear Inequality**

Sketch the graph of each linear inequality.

a. $x > -2$ **b.** $y \leq 3$

Solution

a. The graph of the corresponding equation $x = -2$ is a vertical line. The points that satisfy the inequality $x > -2$ are those lying to the right of this line, as shown in Figure 7.15.

b. The graph of the corresponding equation $y = 3$ is a horizontal line. The points that satisfy the inequality $y \leq 3$ are those lying below (or on) this line, as shown in Figure 7.16.

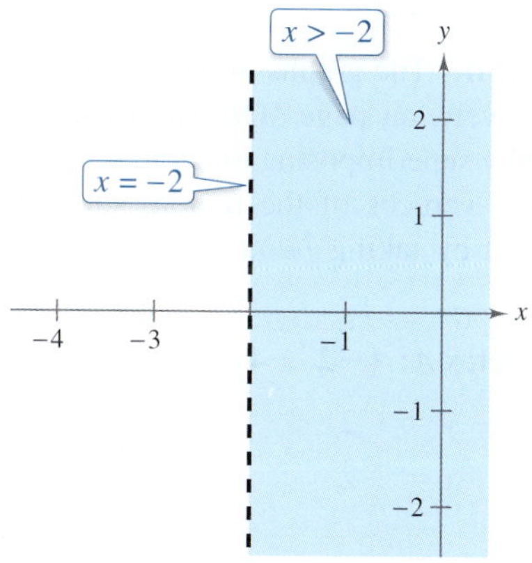

Figure 7.15

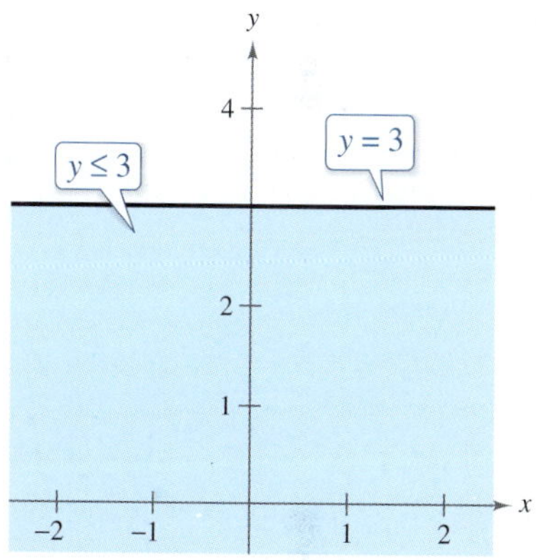

Figure 7.16

Checkpoint Audio-video solution in English & Spanish at LarsonPrecalculus.com.

Sketch the graph of $x \geq 3$.

EXAMPLE 3 **Sketching the Graph of a Linear Inequality**

Sketch the graph of $x - y < 2$.

Solution The graph of the corresponding equation $x - y = 2$ is a line, as shown in Figure 7.17. Because the origin $(0, 0)$ satisfies the inequality, the graph consists of the half-plane lying above the line. (Try checking a point below the line. Regardless of which point you choose, you will see that it does not satisfy the inequality.)

Checkpoint Audio-video solution in English & Spanish at LarsonPrecalculus.com.

Sketch the graph of $x + y > -2$.

To graph a linear inequality, it can help to write the inequality in slope-intercept form. For instance, by writing $x - y < 2$ in the form

$$y > x - 2$$

you can see that the solution points lie *above* the line $x - y = 2$ (or $y = x - 2$), as shown in Figure 7.17.

TECHNOLOGY A graphing utility can be used to graph an inequality or a system of inequalities. For instance, to graph $y \geq x - 2$, enter $y = x - 2$ and use the *shade* feature of the graphing utility to shade the correct part of the graph. You should obtain the graph shown below. Consult the user's guide for your graphing utility for specific keystrokes.

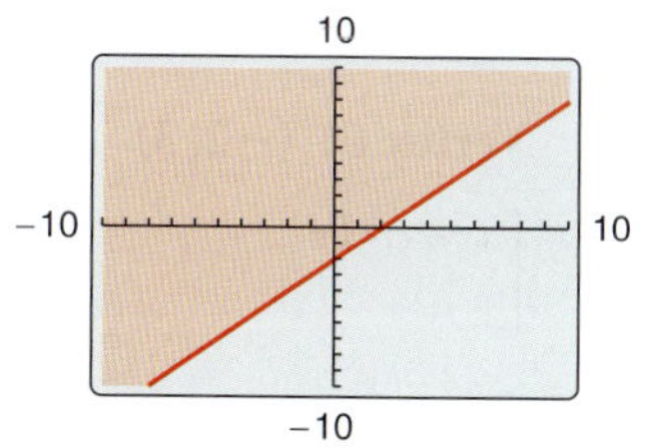

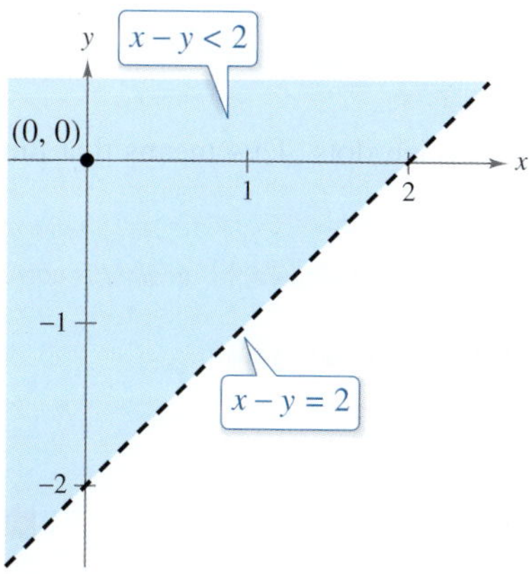

Figure 7.17

Systems of Inequalities

Many practical problems in business, science, and engineering involve systems of linear inequalities. A **solution** of a system of inequalities in x and y is a point

(x, y)

that satisfies each inequality in the system.

To sketch the graph of a system of inequalities in two variables, first sketch the graph of each individual inequality (on the same coordinate system) and then find the region that is *common* to every graph in the system. This region represents the **solution set** of the system. For systems of *linear inequalities,* it is helpful to find the vertices of the solution region.

EXAMPLE 4 Solving a System of Inequalities

Sketch the graph (and label the vertices) of the solution set of the system.

$$\begin{cases} x - y < 2 & \text{Inequality 1} \\ x > -2 & \text{Inequality 2} \\ y \leq 3 & \text{Inequality 3} \end{cases}$$

Solution The graphs of these inequalities are shown in Figures 7.17, 7.15, and 7.16, respectively, on page 509. The triangular region common to all three graphs can be found by superimposing the graphs on the same coordinate system, as shown below. To find the vertices of the region, solve the three systems of corresponding equations obtained by taking *pairs* of equations representing the boundaries of the individual regions.

Vertex A: $(-2, -4)$

$$\begin{cases} x - y = 2 \\ x = -2 \end{cases}$$

Vertex B: $(5, 3)$

$$\begin{cases} x - y = 2 \\ y = 3 \end{cases}$$

Vertex C: $(-2, 3)$

$$\begin{cases} x = -2 \\ y = 3 \end{cases}$$

• • **REMARK** Using a different colored pencil to shade the solution of each inequality in a system will make identifying the solution of the system of inequalities easier.

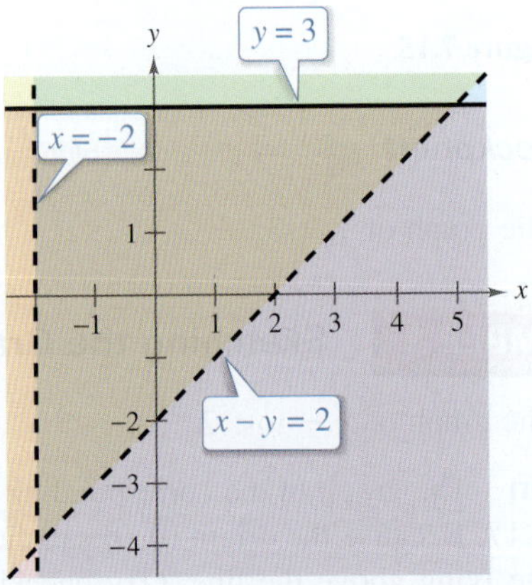

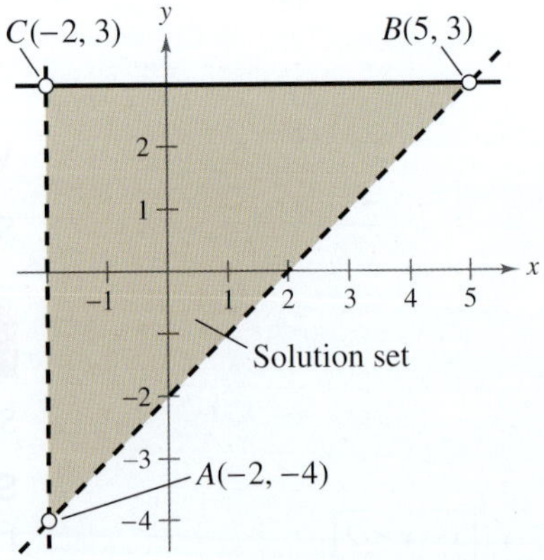

Note that the vertices of the region are represented by open dots. This means that the vertices *are not* solutions of the system of inequalities.

Checkpoint Audio-video solution in English & Spanish at LarsonPrecalculus.com.

Sketch the graph (and label the vertices) of the solution set of the system.

$$\begin{cases} x + y \geq 1 \\ -x + y \geq 1 \\ y \leq 2 \end{cases}$$

For the triangular region shown on page 510, each point of intersection of a pair of boundary lines corresponds to a vertex. With more complicated regions, two border lines can sometimes intersect at a point that is not a vertex of the region, as shown below. To keep track of which points of intersection are actually vertices of the region, you should sketch the region and refer to your sketch as you find each point of intersection.

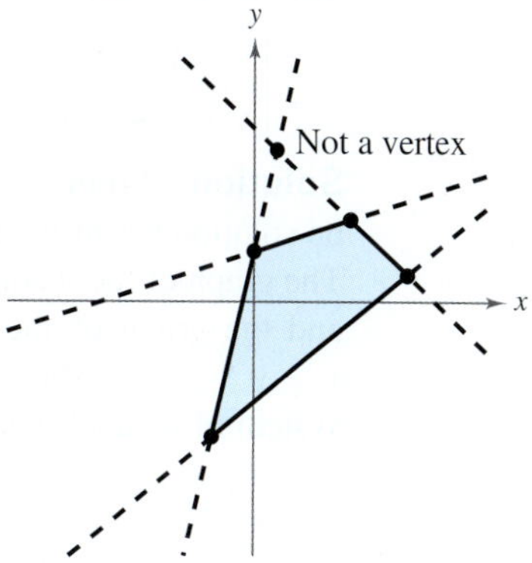

EXAMPLE 5 **Solving a System of Inequalities**

Sketch the region containing all points that satisfy the system of inequalities.

$$\begin{cases} x^2 - y \leq 1 \\ -x + y \leq 1 \end{cases}$$ Inequality 1, Inequality 2

Solution As shown in the figure below, the points that satisfy the inequality

$x^2 - y \leq 1$ Inequality 1

are the points lying above (or on) the parabola

$y = x^2 - 1.$ Parabola

The points satisfying the inequality

$-x + y \leq 1$ Inequality 2

are the points lying below (or on) the line

$y = x + 1.$ Line

To find the points of intersection of the parabola and the line, solve the system of corresponding equations.

$$\begin{cases} x^2 - y = 1 \\ -x + y = 1 \end{cases}$$

Using the method of substitution, you can find the solutions to be $(-1, 0)$ and $(2, 3)$. So, the region containing all points that satisfy the system is indicated by the shaded region at the right.

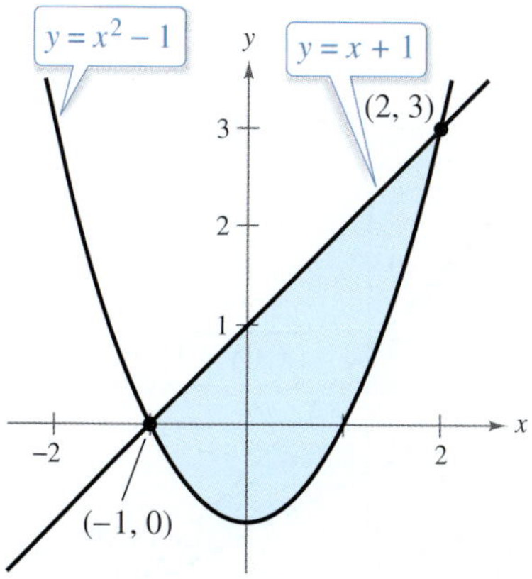

✓ **Checkpoint** Audio-video solution in English & Spanish at LarsonPrecalculus.com.

Sketch the region containing all points that satisfy the system of inequalities.

$$\begin{cases} x - y^2 > 0 \\ x + y < 2 \end{cases}$$

When solving a system of inequalities, you should be aware that the system might have no solution *or* its graph might be an unbounded region in the plane. Examples 6 and 7 show these two possibilities.

EXAMPLE 6 A System with No Solution

Sketch the solution set of the system of inequalities.

$$\begin{cases} x + y > 3 & \text{Inequality 1} \\ x + y < -1 & \text{Inequality 2} \end{cases}$$

Solution From the way the system is written, it should be clear that the system has no solution because the quantity $(x + y)$ cannot be both less than -1 and greater than 3. The graph of the inequality $x + y > 3$ is the half-plane lying above the line $x + y = 3$, and the graph of the inequality $x + y < -1$ is the half-plane lying below the line $x + y = -1$, as shown below. These two half-planes have no points in common. So, the system of inequalities has no solution.

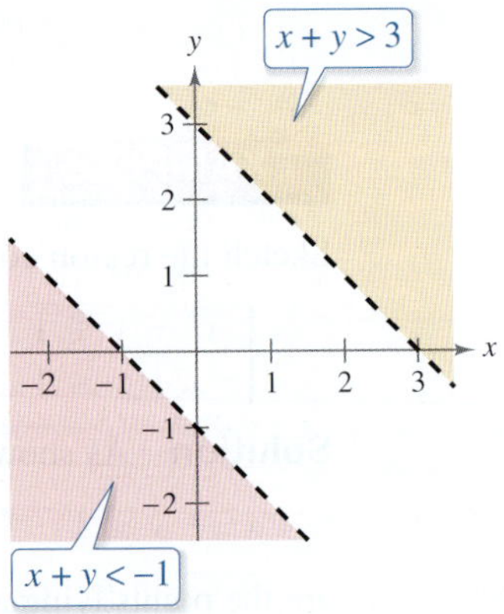

Checkpoint Audio-video solution in English & Spanish at LarsonPrecalculus.com.

Sketch the solution set of the system of inequalities.

$$\begin{cases} 2x - y < -3 \\ 2x - y > \quad 1 \end{cases}$$

EXAMPLE 7 An Unbounded Solution Set

Sketch the solution set of the system of inequalities.

$$\begin{cases} x + y < 3 & \text{Inequality 1} \\ x + 2y > 3 & \text{Inequality 2} \end{cases}$$

Solution The graph of the inequality $x + y < 3$ is the half-plane that lies below the line $x + y = 3$. The graph of the inequality $x + 2y > 3$ is the half-plane that lies above the line $x + 2y = 3$. The intersection of these two half-planes is an *infinite wedge* that has a vertex at $(3, 0)$, as shown in Figure 7.18. So, the solution set of the system of inequalities is unbounded.

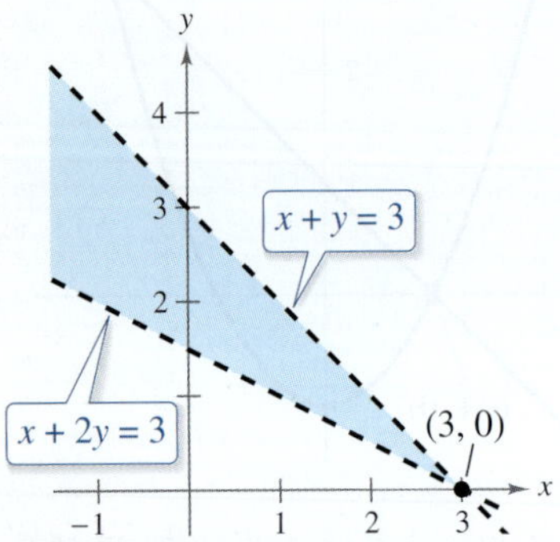

Figure 7.18

Checkpoint Audio-video solution in English & Spanish at LarsonPrecalculus.com.

Sketch the solution set of the system of inequalities.

$$\begin{cases} x^2 - y < \quad 0 \\ x - y < -2 \end{cases}$$

Applications

Example 9 in Section 7.2 discussed the *equilibrium point* for a system of demand and supply equations. The next example discusses two related concepts that economists call **consumer surplus** and **producer surplus.** As shown in Figure 7.19, the consumer surplus is defined as the area of the region formed by the *demand* curve, the horizontal line passing through the equilibrium point, and the p-axis. Similarly, the producer surplus is defined as the area of the region formed by the *supply* curve, the horizontal line passing through the equilibrium point, and the p-axis. The consumer surplus is a measure of the amount that consumers would have been willing to pay *above what they actually paid,* whereas the producer surplus is a measure of the amount that producers would have been willing to receive *below what they actually received.*

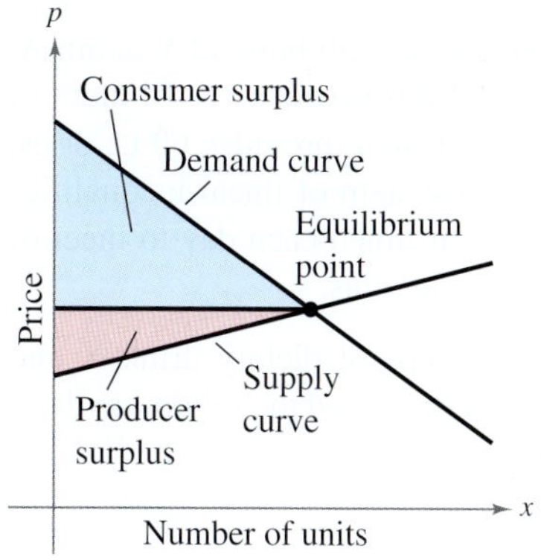

Figure 7.19

EXAMPLE 8 Consumer Surplus and Producer Surplus

The demand and supply equations for a new type of video game console are

$$\begin{cases} p = 180 - 0.00001x & \text{Demand equation} \\ p = 90 + 0.00002x & \text{Supply equation} \end{cases}$$

where p is the price per unit (in dollars) and x is the number of units. Find the consumer surplus and producer surplus for these two equations.

Solution Begin by finding the equilibrium point (when supply and demand are equal) by solving the equation

$$90 + 0.00002x = 180 - 0.00001x.$$

In Example 9 in Section 7.2, you saw that the solution is $x = 3{,}000{,}000$ units, which corresponds to an equilibrium price of $p = \$150$. So, the consumer surplus and producer surplus are the areas of the following triangular regions.

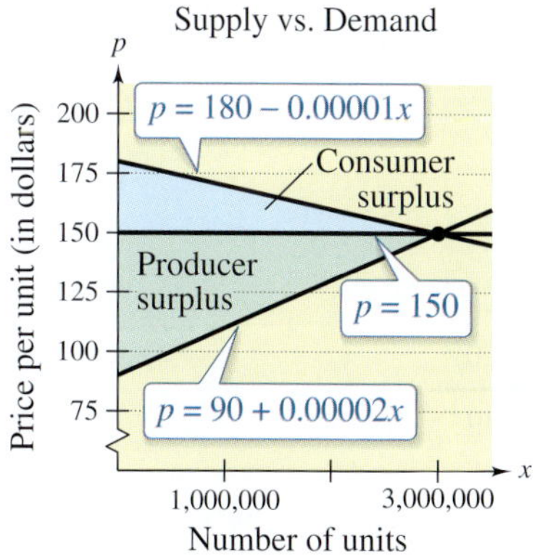

Figure 7.20

Consumer Surplus

$$\begin{cases} p \le 180 - 0.00001x \\ p \ge 150 \\ x \ge 0 \end{cases}$$

Producer Surplus

$$\begin{cases} p \ge 90 + 0.00002x \\ p \le 150 \\ x \ge 0 \end{cases}$$

The consumer and producer surpluses are the areas of the shaded triangles shown in Figure 7.20.

$$\text{Consumer surplus} = \tfrac{1}{2}(\text{base})(\text{height})$$
$$= \tfrac{1}{2}(3{,}000{,}000)(30)$$
$$= \$45{,}000{,}000$$

$$\text{Producer surplus} = \tfrac{1}{2}(\text{base})(\text{height})$$
$$= \tfrac{1}{2}(3{,}000{,}000)(60)$$
$$= \$90{,}000{,}000$$

✓ Checkpoint Audio-video solution in English & Spanish at LarsonPrecalculus.com.

The demand and supply equations for a flat-screen television are

$$\begin{cases} p = 567 - 0.00002x & \text{Demand equation} \\ p = 492 + 0.00003x & \text{Supply equation} \end{cases}$$

where p is the price per unit (in dollars) and x is the number of units. Find the consumer surplus and producer surplus for these two equations.

A system of linear inequalities can help you determine amounts of dietary drinks to consume to meet or exceed daily requirements for calories and vitamins.

EXAMPLE 9 **Nutrition**

The liquid portion of a diet is to provide at least 300 calories, 36 units of vitamin A, and 90 units of vitamin C. A cup of dietary drink X provides 60 calories, 12 units of vitamin A, and 10 units of vitamin C. A cup of dietary drink Y provides 60 calories, 6 units of vitamin A, and 30 units of vitamin C. Set up a system of linear inequalities that describes how many cups of each drink should be consumed each day to meet or exceed the minimum daily requirements for calories and vitamins.

Solution Begin by letting x represent the number of cups of dietary drink X and y represent the number of cups of dietary drink Y. To meet or exceed the minimum daily requirements, the following inequalities must be satisfied.

$$\begin{cases} 60x + 60y \geq 300 & \text{Calories} \\ 12x + 6y \geq 36 & \text{Vitamin A} \\ 10x + 30y \geq 90 & \text{Vitamin C} \\ x \geq 0 & \\ y \geq 0 & \end{cases}$$

The last two inequalities are included because x and y cannot be negative. The graph of this system of inequalities is shown below. (More is said about this application in Example 6 in Section 7.6.)

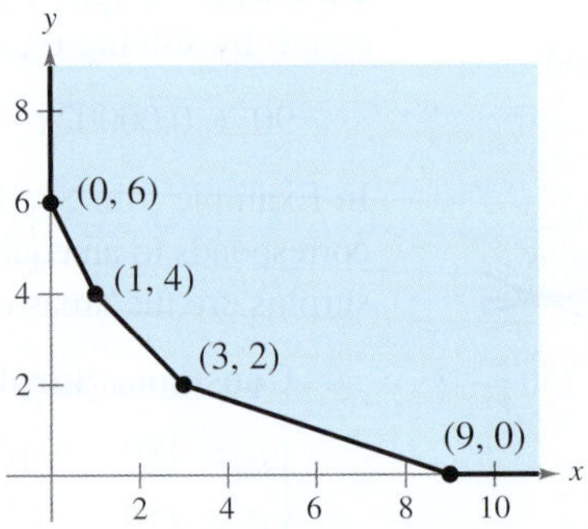

✓ ***Checkpoint*** *Audio-video solution in English & Spanish at LarsonPrecalculus.com.*

A pet supply company mixes two brands of dry dog food. A bag of brand X contains 8 units of nutrient A, 1 unit of nutrient B, and 2 units of nutrient C. A bag of brand Y contains 2 units of nutrient A, 1 unit of nutrient B, and 7 units of nutrient C. The minimum required amounts of nutrients A, B, and C are 16 units, 5 units, and 20 units, respectively. Set up a system of linear inequalities that describes how many bags of each brand should be mixed to meet or exceed the minimum required amounts of nutrients.

Summarize (Section 7.5)

1. Explain how to sketch the graph of an inequality in two variables *(page 508)*. For examples of sketching the graphs of inequalities in two variables, see Examples 1–3.
2. Explain how to solve a system of inequalities *(page 510)*. For examples of solving systems of inequalities, see Examples 4–7.
3. Describe examples of how to use systems of inequalities in two variables to model and solve real-life problems *(pages 513 and 514, Examples 8 and 9)*.

Kinetic Imagery/Shutterstock.com

7.5 Exercises

See CalcChat.com for tutorial help and worked-out solutions to odd-numbered exercises.

Vocabulary: **Fill in the blanks.**

1. An ordered pair (a, b) is a ________ of an inequality in x and y when the inequality is true after a and b are substituted for x and y, respectively.
2. The ________ of an inequality is the collection of all solutions of the inequality.
3. A ________ of a system of inequalities in x and y is a point (x, y) that satisfies each inequality in the system.
4. A ________ ________ of a system of inequalities in two variables is the region common to the graphs of every inequality in the system.

Skills and Applications

Graphing an Inequality **In Exercises 5–18, sketch the graph of the inequality.**

5. $y < 5 - x^2$
6. $y^2 - x < 0$
7. $x \geq 6$
8. $x < -4$
9. $y > -7$
10. $10 \geq y$
11. $y < 2 - x$
12. $y > 4x - 3$
13. $2y - x \geq 4$
14. $5x + 3y \geq -15$
15. $(x + 1)^2 + (y - 2)^2 < 9$
16. $(x - 1)^2 + (y - 4)^2 > 9$
17. $y \leq \dfrac{1}{1 + x^2}$
18. $y > \dfrac{-15}{x^2 + x + 4}$

Graphing an Inequality **In Exercises 19–28, use a graphing utility to graph the inequality.**

19. $y < \ln x$
20. $y \geq -2 - \ln(x + 3)$
21. $y < 4^{-x-5}$
22. $y \leq 2^{2x-0.5} - 7$
23. $y \leq 6 - \frac{3}{2}x$
24. $y < -3.8x + 1.1$
25. $x^2 + 5y - 10 \leq 0$
26. $2x^2 - y - 3 > 0$
27. $\frac{5}{2}y - 3x^2 - 6 \geq 0$
28. $-\frac{1}{10}x^2 - \frac{3}{8}y < -\frac{1}{4}$

Writing an Inequality **In Exercises 29–32, write an inequality for the shaded region shown in the figure.**

29.

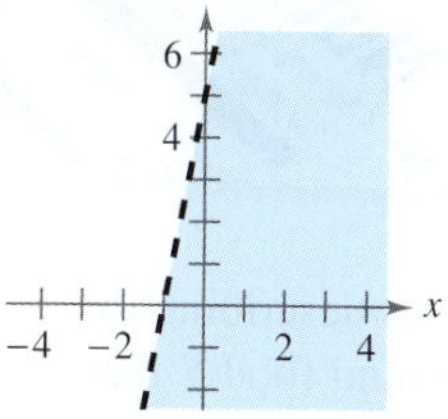

30.

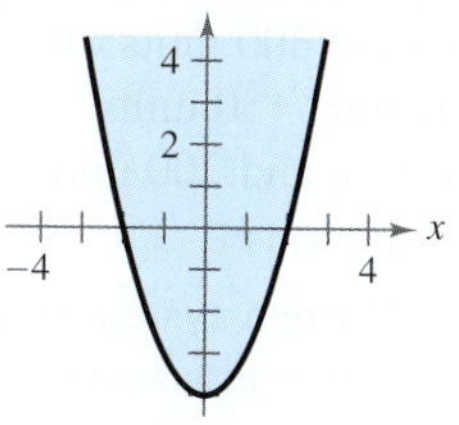

31.

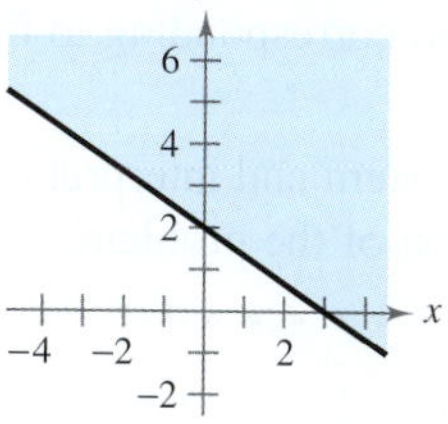

32.

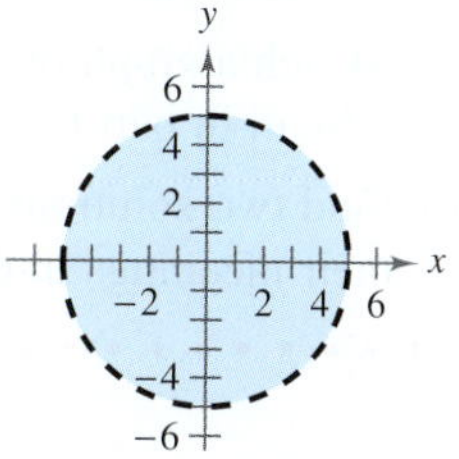

Solving a System of Inequalities **In Exercises 33–46, sketch the graph (and label the vertices) of the solution set of the system of inequalities.**

33. $\begin{cases} x + y \leq 1 \\ -x + y \leq 1 \\ y \geq 0 \end{cases}$
34. $\begin{cases} 3x + 4y < 12 \\ x > 0 \\ y > 0 \end{cases}$
35. $\begin{cases} x^2 + y \leq 7 \\ x \geq -2 \\ y \geq 0 \end{cases}$
36. $\begin{cases} 4x^2 + y \geq 2 \\ x \leq 1 \\ y \leq 1 \end{cases}$
37. $\begin{cases} 2x + y > 2 \\ 6x + 3y < 2 \end{cases}$
38. $\begin{cases} x - 7y > -36 \\ 5x + 2y > 5 \\ 6x - 5y > 6 \end{cases}$
39. $\begin{cases} -3x + 2y < 6 \\ x - 4y > -2 \\ 2x + y < 3 \end{cases}$
40. $\begin{cases} x - 2y < -6 \\ 5x - 3y > -9 \end{cases}$
41. $\begin{cases} x > y^2 \\ x < y + 2 \end{cases}$
42. $\begin{cases} x - y^2 > 0 \\ x - y > 2 \end{cases}$
43. $\begin{cases} x^2 + y^2 \leq 36 \\ x^2 + y^2 \geq 9 \end{cases}$
44. $\begin{cases} x^2 + y^2 \leq 25 \\ 4x - 3y \leq 0 \end{cases}$
45. $\begin{cases} 3x + 4 \geq y^2 \\ x - y < 0 \end{cases}$
46. $\begin{cases} x < 2y - y^2 \\ 0 < x + y \end{cases}$

Solving a System of Inequalities **In Exercises 47–52, use a graphing utility to graph the solution set of the system of inequalities.**

47. $\begin{cases} y \leq \sqrt{3x} + 1 \\ y \geq x^2 + 1 \end{cases}$
48. $\begin{cases} y < -x^2 + 2x + 3 \\ y > x^2 - 4x + 3 \end{cases}$
49. $\begin{cases} y < x^3 - 2x + 1 \\ y > -2x \\ x \leq 1 \end{cases}$
50. $\begin{cases} y \geq x^4 - 2x^2 + 1 \\ y \leq 1 - x^2 \end{cases}$
51. $\begin{cases} x^2y \geq 1 \\ 0 < x \leq 4 \\ y \leq 4 \end{cases}$
52. $\begin{cases} y \leq e^{-x^2/2} \\ y \geq 0 \\ -2 \leq x \leq 2 \end{cases}$

Writing a System of Inequalities In Exercises 53–60, write a system of inequalities to describe the region.

53.

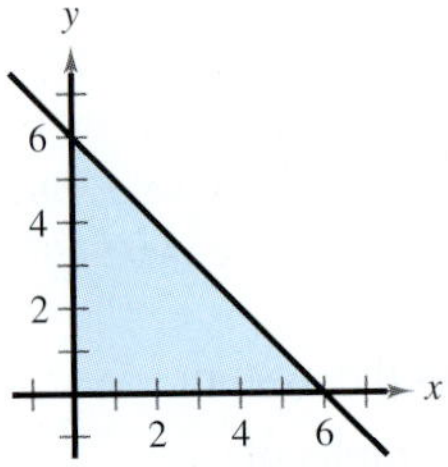

54.

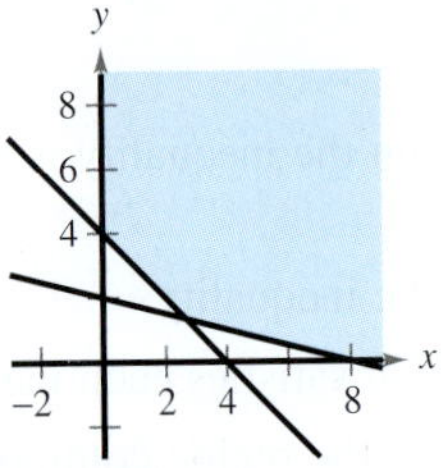

55.

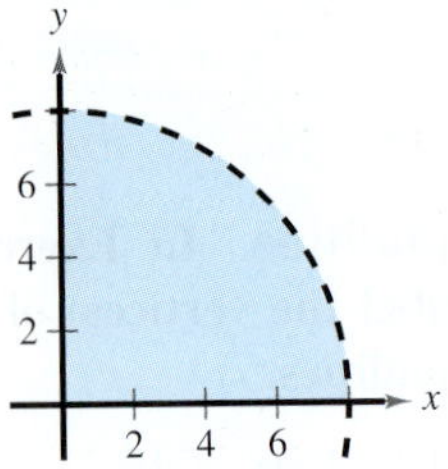

56.

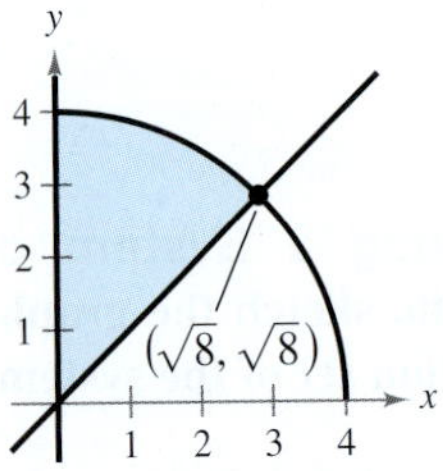

57. Rectangle: vertices at $(4, 3)$, $(9, 3)$, $(9, 9)$, $(4, 9)$

58. Parallelogram: vertices at $(0, 0)$, $(4, 0)$, $(1, 4)$, $(5, 4)$

59. Triangle: vertices at $(0, 0)$, $(6, 0)$, $(1, 5)$

60. Triangle: vertices at $(-1, 0)$, $(1, 0)$, $(0, 1)$

Consumer Surplus and Producer Surplus In Exercises 61–64, (a) graph the systems representing the consumer surplus and producer surplus for the supply and demand equations and (b) find the consumer surplus and producer surplus.

	Demand	**Supply**
61.	$p = 50 - 0.5x$	$p = 0.125x$
62.	$p = 100 - 0.05x$	$p = 25 + 0.1x$
63.	$p = 140 - 0.00002x$	$p = 80 + 0.00001x$
64.	$p = 400 - 0.0002x$	$p = 225 + 0.0005x$

65. Production A furniture company produces tables and chairs. Each table requires 1 hour in the assembly center and $1\frac{1}{3}$ hours in the finishing center. Each chair requires $1\frac{1}{2}$ hours in the assembly center and $1\frac{1}{2}$ hours in the finishing center. The assembly center is available 12 hours per day, and the finishing center is available 15 hours per day. Find and graph a system of inequalities describing all possible production levels.

66. Inventory A store sells two models of laptop computers. Because of the demand, the store stocks at least twice as many units of model A as of model B. The costs to the store for the two models are \$800 and \$1200, respectively. The management does not want more than \$20,000 in computer inventory at any one time, and it wants at least four model A laptop computers and two model B laptop computers in inventory at all times. Find and graph a system of inequalities describing all possible inventory levels.

67. Investment Analysis A person plans to invest up to \$20,000 in two different interest-bearing accounts. Each account is to contain at least \$5000. Moreover, the amount in one account should be at least twice the amount in the other account. Find and graph a system of inequalities to describe the various amounts that can be deposited in each account.

68. Ticket Sales For a concert event, there are \$30 reserved seat tickets and \$20 general admission tickets. There are 2000 reserved seats available, and fire regulations limit the number of paid ticket holders to 3000. The promoter must take in at least \$75,000 in ticket sales. Find and graph a system of inequalities describing the different numbers of tickets that can be sold.

69. Truck Scheduling A small company that manufactures two models of exercise machines has an order for 15 units of the standard model and 16 units of the deluxe model. The company has trucks of two different sizes that can haul the products, as shown in the table.

Truck	Standard	Deluxe
Large	6	3
Medium	4	6

Find and graph a system of inequalities describing the numbers of trucks of each size that are needed to deliver the order.

70. Nutrition

A dietitian designs a special dietary supplement using two different foods. Each ounce of food X contains 20 units of calcium, 15 units of iron, and 10 units of vitamin B. Each ounce of food Y contains 10 units of calcium, 10 units of iron, and 20 units of vitamin B. The minimum daily requirements of the diet are 300 units of calcium, 150 units of iron, and 200 units of vitamin B.

(a) Write a system of inequalities describing the different amounts of food X and food Y that can be used.

(b) Sketch a graph of the region corresponding to the system in part (a).

(c) Find two solutions of the system and interpret their meanings in the context of the problem.

alexskopje/Shutterstock.com

71. Health A person's maximum heart rate is $220 - x$, where x is the person's age in years for $20 \le x \le 70$. The American Heart Association recommends that when a person exercises, the person should strive for a heart rate that is at least 50% of the maximum and at most 85% of the maximum. *(Source: American Heart Association)*

(a) Write a system of inequalities that describes the exercise target heart rate region.

(b) Sketch a graph of the region in part (a).

(c) Find two solutions to the system and interpret their meanings in the context of the problem.

72. Physical Fitness Facility An indoor running track is to be constructed with a space for exercise equipment inside the track (see figure). The track must be at least 125 meters long, and the exercise space must have an area of at least 500 square meters.

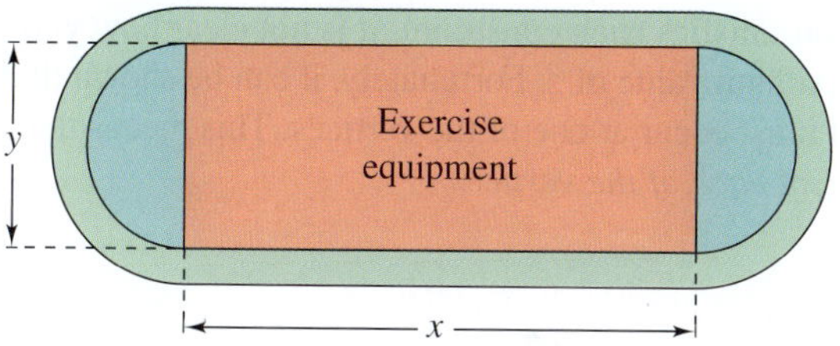

(a) Find a system of inequalities describing the requirements of the facility.

(b) Graph the system from part (a).

Exploration

True or False? **In Exercises 73 and 74, determine whether the statement is true or false. Justify your answer.**

73. The area of the figure defined by the system

$$\begin{cases} x \ge -3 \\ x \le 6 \\ y \le 5 \\ y \ge -6 \end{cases}$$

is 99 square units.

74. The following graph shows the solution of the system

$$\begin{cases} y \le 6 \\ -4x - 9y > 6. \\ 3x + y^2 \ge 2 \end{cases}$$

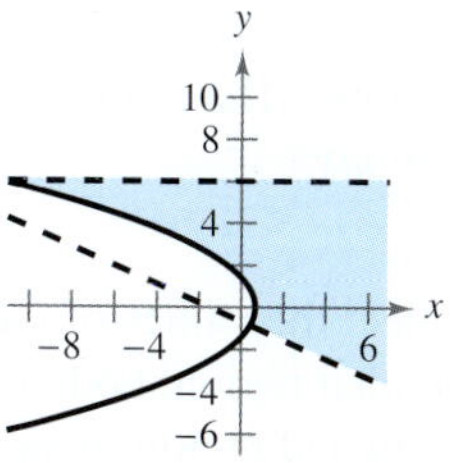

75. Think About It After graphing the boundary of the inequality $x + y < 3$, explain how you decide on which side of the boundary the solution set of the inequality lies.

76. HOW DO YOU SEE IT? Match the system of inequalities with the graph of its solution. [The graphs are labeled (i), (ii), (iii), and (iv).]

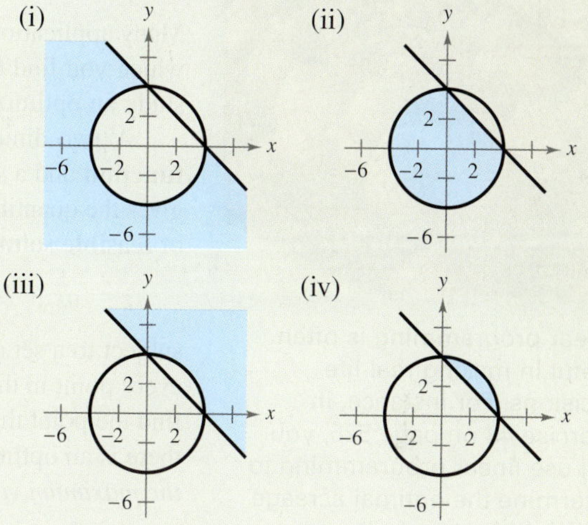

(a) $\begin{cases} x^2 + y^2 \le 16 \\ x + y \ge 4 \end{cases}$ (b) $\begin{cases} x^2 + y^2 \le 16 \\ x + y \le 4 \end{cases}$

(c) $\begin{cases} x^2 + y^2 \ge 16 \\ x + y \ge 4 \end{cases}$ (d) $\begin{cases} x^2 + y^2 \ge 16 \\ x + y \le 4 \end{cases}$

77. Graphical Reasoning Two concentric circles have radii x and y, where $y > x$. The area between the circles is at least 10 square units.

(a) Find a system of inequalities describing the constraints on the circles.

(b) Use a graphing utility to graph the system of inequalities in part (a). Graph the line $y = x$ in the same viewing window.

(c) Identify the graph of the line in relation to the boundary of the inequality. Explain its meaning in the context of the problem.

78. Changing the Inequality Symbol The graph of the solution of the inequality $x + 2y < 6$ is shown in the figure. Describe how the solution set would change for each of the following.

(a) $x + 2y \le 6$ (b) $x + 2y > 6$

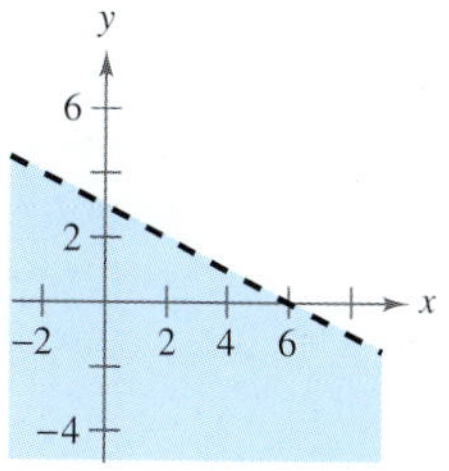

7.6 Linear Programming

Linear programming is often useful in making real-life decisions. For instance, in Exercise 43 on page 526, you will use linear programming to determine the optimal acreage and yield for two fruit crops.

- Solve linear programming problems.
- Use linear programming to model and solve real-life problems.

Linear Programming: A Graphical Approach

Many applications in business and economics involve a process called **optimization,** in which you find the minimum or maximum value of a quantity. In this section, you will study an optimization strategy called **linear programming.**

A two-dimensional linear programming problem consists of a linear **objective function** and a system of linear inequalities called **constraints.** The objective function gives the quantity to be maximized (or minimized), and the constraints determine the set of **feasible solutions.** For instance, suppose you want to maximize the value of

$$z = ax + by \qquad \text{Objective function}$$

subject to a set of constraints that determines the shaded region shown below. Because every point in the shaded region satisfies each constraint, it is not clear how you should find the point that yields a maximum value of z. Fortunately, it can be shown that when there is an optimal solution, it must occur at one of the vertices. This means that *to find the maximum value of z, test z at each of the vertices.*

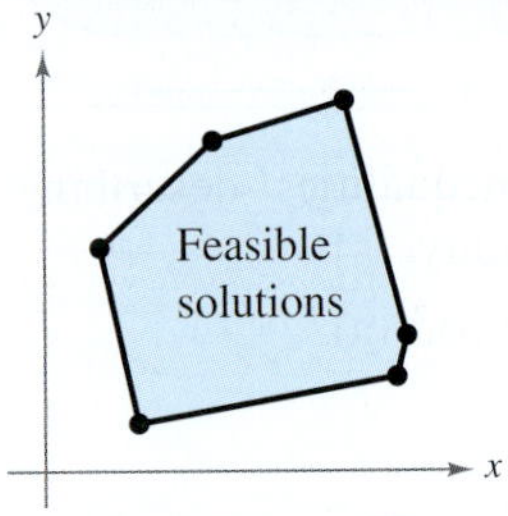

Optimal Solution of a Linear Programming Problem

If a linear programming problem has a solution, then it must occur at a vertex of the set of feasible solutions. If there is more than one solution, then at least one solution must occur at such a vertex. In either case, the value of the objective function is unique.

A linear programming problem can include hundreds, and sometimes even thousands, of variables. However, in this section, you will solve linear programming problems that involve only two variables. The guidelines for solving a linear programming problem in two variables are listed below.

Solving a Linear Programming Problem

1. Sketch the region corresponding to the system of constraints. (The points inside or on the boundary of the region are *feasible solutions.*)
2. Find the vertices of the region.
3. Test the objective function at each of the vertices and select the values of the variables that optimize the objective function. For a bounded region, both a minimum and a maximum value will exist. (For an unbounded region, *if* an optimal solution exists, then it will occur at a vertex.)

Alison McC/Shutterstock.com

EXAMPLE 1 Solving a Linear Programming Problem

Find the maximum value of

$$z = 3x + 2y \quad \text{Objective function}$$

subject to the following constraints.

$$\left.\begin{aligned} x &\geq 0 \\ y &\geq 0 \\ x + 2y &\leq 4 \\ x - y &\leq 1 \end{aligned}\right\} \quad \text{Constraints}$$

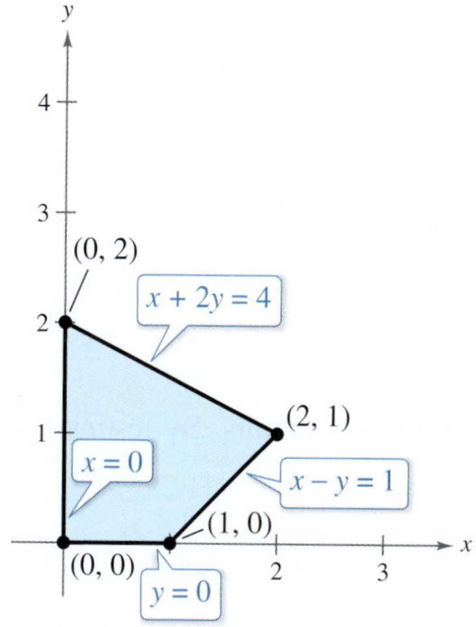

Figure 7.21

Solution The constraints form the region shown in Figure 7.21. At the four vertices of this region, the objective function has the following values.

At (0, 0): $z = 3(0) + 2(0) = 0$

At (0, 2): $z = 3(0) + 2(2) = 4$

At (2, 1): $z = 3(2) + 2(1) = 8$ Maximum value of z

At (1, 0): $z = 3(1) + 2(0) = 3$

So, the maximum value of z is 8, and this occurs when $x = 2$ and $y = 1$.

Checkpoint *Audio-video solution in English & Spanish at LarsonPrecalculus.com.*

Find the maximum value of

$$z = 4x + 5y$$

subject to the following constraints.

$$\begin{aligned} x &\geq 0 \\ y &\geq 0 \\ x + y &\leq 6 \end{aligned}$$

In Example 1, try testing some of the *interior* points in the region. You will see that the corresponding values of z are less than 8. Here are some examples.

At (1, 1): $z = 3(1) + 2(1) = 5$

At $\left(1, \frac{1}{2}\right)$: $z = 3(1) + 2\left(\frac{1}{2}\right) = 4$

At $\left(\frac{1}{2}, \frac{3}{2}\right)$: $z = 3\left(\frac{1}{2}\right) + 2\left(\frac{3}{2}\right) = \frac{9}{2}$

At $\left(\frac{3}{2}, 1\right)$: $z = 3\left(\frac{3}{2}\right) + 2(1) = \frac{13}{2}$

To see why the maximum value of the objective function in Example 1 must occur at a vertex, consider writing the objective function in slope-intercept form

$$y = -\frac{3}{2}x + \frac{z}{2} \quad \text{Family of lines}$$

where

$$\frac{z}{2}$$

is the y-intercept of the objective function. This equation represents a family of lines, each of slope $-\frac{3}{2}$. Of these infinitely many lines, you want the one that has the largest z-value while still intersecting the region determined by the constraints. In other words, of all the lines whose slope is $-\frac{3}{2}$, you want the one that has the largest y-intercept *and* intersects the given region, as shown in Figure 7.22. Notice from the graph that such a line will pass through one (or more) of the vertices of the region.

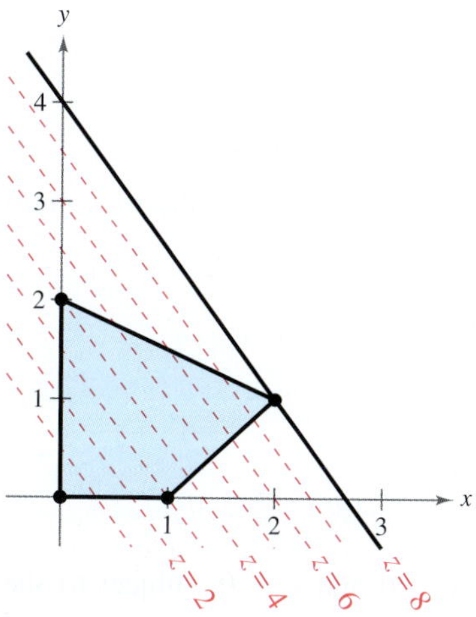

Figure 7.22

The next example shows that the same basic procedure can be used to solve a problem in which the objective function is to be *minimized.*

EXAMPLE 2 Minimizing an Objective Function

Find the minimum value of

$$z = 5x + 7y \qquad \text{Objective function}$$

where $x \geq 0$ and $y \geq 0$, subject to the following constraints.

$$\left.\begin{aligned} 2x + 3y &\geq 6 \\ 3x - y &\leq 15 \\ -x + y &\leq 4 \\ 2x + 5y &\leq 27 \end{aligned}\right\} \quad \text{Constraints}$$

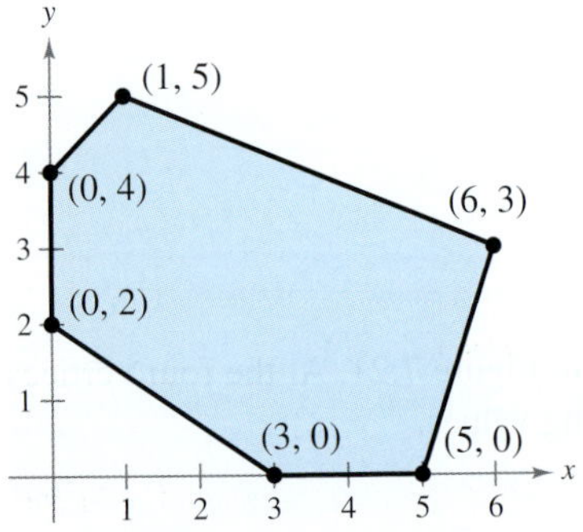

Figure 7.23

Solution Figure 7.23 shows the region bounded by the constraints. By testing the objective function at each vertex, you obtain the following.

At (0, 2): $z = 5(0) + 7(2) = 14$ Minimum value of z

At (0, 4): $z = 5(0) + 7(4) = 28$

At (1, 5): $z = 5(1) + 7(5) = 40$

At (6, 3): $z = 5(6) + 7(3) = 51$

At (5, 0): $z = 5(5) + 7(0) = 25$

At (3, 0): $z = 5(3) + 7(0) = 15$

So, the minimum value of z is 14, and this occurs when $x = 0$ and $y = 2$.

✓ **Checkpoint** Audio-video solution in English & Spanish at LarsonPrecalculus.com.

Find the minimum value of

$$z = 12x + 8y$$

where $x \geq 0$ and $y \geq 0$, subject to the following constraints.

$$\begin{aligned} 5x + 6y &\leq 420 \\ -x + 6y &\leq 240 \\ -2x + y &\geq -100 \end{aligned}$$

EXAMPLE 3 Maximizing an Objective Function

Find the maximum value of

$$z = 5x + 7y \qquad \text{Objective function}$$

where $x \geq 0$ and $y \geq 0$, subject to the constraints given in Example 2.

Solution Using the values of z at the vertices shown in Example 2, the maximum value of z is

$$z = 5(6) + 7(3)$$
$$= 51$$

and occurs when $x = 6$ and $y = 3$.

✓ **Checkpoint** Audio-video solution in English & Spanish at LarsonPrecalculus.com.

Find the maximum value of $z = 12x + 8y$, where $x \geq 0$ and $y \geq 0$, subject to the constraints given in the Checkpoint with Example 2.

George Dantzig (1914–2005) was the first to propose the simplex method, or linear programming, in 1947. This technique defined the steps needed to find the optimal solution to a complex multivariable problem.

Edward W. Souza/News Services/Stanford University

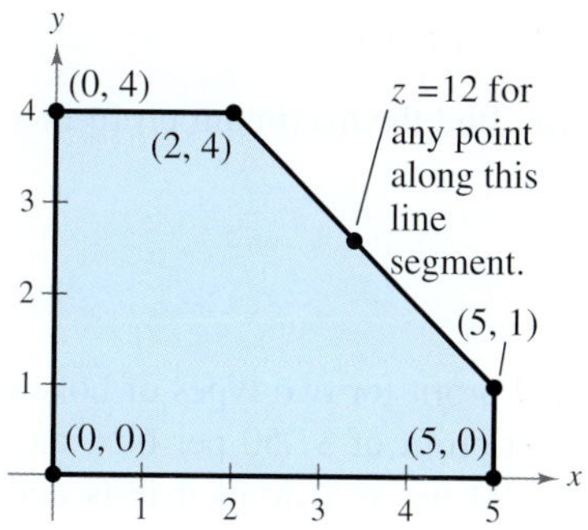

Figure 7.24

It is possible for the maximum (or minimum) value in a linear programming problem to occur at *two* different vertices. For instance, at the vertices of the region shown in Figure 7.24, the objective function

$$z = 2x + 2y \quad \text{Objective function}$$

has the following values.

$$\text{At } (0, 0): \quad z = 2(0) + 2(0) = 0$$

$$\text{At } (0, 4): \quad z = 2(0) + 2(4) = 8$$

$$\text{At } (2, 4): \quad z = 2(2) + 2(4) = 12 \quad \text{Maximum value of } z$$

$$\text{At } (5, 1): \quad z = 2(5) + 2(1) = 12 \quad \text{Maximum value of } z$$

$$\text{At } (5, 0): \quad z = 2(5) + 2(0) = 10$$

In this case, the objective function has a maximum value not only at the vertices (2, 4) and (5, 1); it also has a maximum value (of 12) at *any point on the line segment connecting these two vertices.* Note that the objective function in slope-intercept form $y = -x + \frac{1}{2}z$ has the same slope as the line through the vertices (2, 4) and (5, 1).

Some linear programming problems have no optimal solutions. This can occur when the region determined by the constraints is *unbounded*. Example 4 illustrates such a problem.

ALGEBRA HELP The slope m of the nonvertical line through the points (x_1, y_1) and (x_2, y_2) is

$$m = \frac{y_2 - y_1}{x_2 - x_1}$$

where $x_1 \neq x_2$.

EXAMPLE 4 An Unbounded Region

Find the maximum value of

$$z = 4x + 2y \quad \text{Objective function}$$

where $x \geq 0$ and $y \geq 0$, subject to the following constraints.

$$\left.\begin{aligned} x + 2y &\geq 4 \\ 3x + y &\geq 7 \\ -x + 2y &\leq 7 \end{aligned}\right\} \quad \text{Constraints}$$

Solution Figure 7.25 shows the region determined by the constraints. For this unbounded region, there is no maximum value of z. To see this, note that the point $(x, 0)$ lies in the region for all values of $x \geq 4$. Substituting this point into the objective function, you get

$$z = 4(x) + 2(0) = 4x.$$

By choosing x to be large, you can obtain values of z that are as large as you want. So, there is no maximum value of z. However, there *is* a minimum value of z.

$$\text{At } (1, 4): \quad z = 4(1) + 2(4) = 12$$

$$\text{At } (2, 1): \quad z = 4(2) + 2(1) = 10 \quad \text{Minimum value of } z$$

$$\text{At } (4, 0): \quad z = 4(4) + 2(0) = 16$$

So, the minimum value of z is 10, and this occurs when $x = 2$ and $y = 1$.

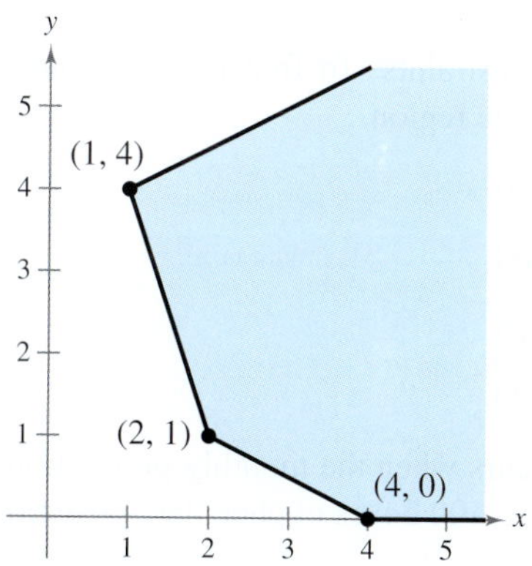

Figure 7.25

Checkpoint Audio-video solution in English & Spanish at LarsonPrecalculus.com.

Find the minimum value of

$$z = 3x + 7y$$

where $x \geq 0$ and $y \geq 0$, subject to the following constraints.

$$\begin{aligned} x + y &\geq 8 \\ 3x + 5y &\geq 30 \end{aligned}$$

Applications

Example 5 shows how linear programming can help you find the maximum profit in a business application.

EXAMPLE 5 Optimal Profit

A candy manufacturer wants to maximize the combined profit for two types of boxed chocolates. A box of chocolate-covered creams yields a profit of $1.50 per box, and a box of chocolate-covered nuts yields a profit of $2.00 per box. Market tests and available resources have indicated the following constraints.

1. The combined production level should not exceed 1200 boxes per month.
2. The demand for a box of chocolate-covered nuts is no more than half the demand for a box of chocolate-covered creams.
3. The production level for chocolate-covered creams should be less than or equal to 600 boxes plus three times the production level for chocolate-covered nuts.

What is the maximum monthly profit? How many boxes of each type should be produced per month to yield the maximum profit?

Solution Let x be the number of boxes of chocolate-covered creams and let y be the number of boxes of chocolate-covered nuts. So, the objective function (for the combined profit) is

$$P = 1.5x + 2y. \qquad \text{Objective function}$$

The three constraints translate into the following linear inequalities.

1. $x + y \le 1200 \quad \Rightarrow \quad x + y \le 1200$
2. $y \le \frac{1}{2}x \quad \Rightarrow \quad -x + 2y \le 0$
3. $x \le 600 + 3y \quad \Rightarrow \quad x - 3y \le 600$

Because neither x nor y can be negative, you also have the two additional constraints of

$$x \ge 0 \quad \text{and} \quad y \ge 0.$$

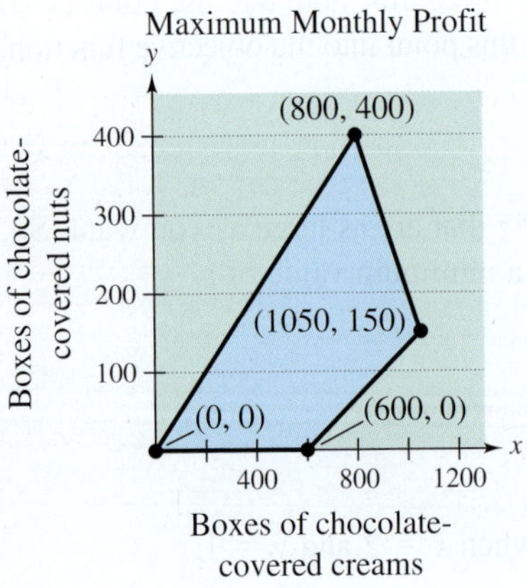

Figure 7.26

Figure 7.26 shows the region determined by the constraints. To find the maximum monthly profit, test the values of P at the vertices of the region.

At (0, 0): $P = 1.5(0) + 2(0) = 0$

At (800, 400): $P = 1.5(800) + 2(400) = 2000$ Maximum profit

At (1050, 150): $P = 1.5(1050) + 2(150) = 1875$

At (600, 0): $P = 1.5(600) + 2(0) = 900$

So, the maximum monthly profit is $2000, and it occurs when the monthly production consists of 800 boxes of chocolate-covered creams and 400 boxes of chocolate-covered nuts.

✓ **Checkpoint** *Audio-video solution in English & Spanish at LarsonPrecalculus.com.*

In Example 5, the candy manufacturer improves the production of chocolate-covered creams so that the profit is $2.50 per box. What is the maximum monthly profit? How many boxes of each type should be produced per month to yield the maximum profit?

Example 6 shows how linear programming can help you find the optimal cost in a real-life application.

EXAMPLE 6 **Optimal Cost**

The liquid portion of a diet is to provide at least 300 calories, 36 units of vitamin A, and 90 units of vitamin C. A cup of dietary drink X costs \$0.12 and provides 60 calories, 12 units of vitamin A, and 10 units of vitamin C. A cup of dietary drink Y costs \$0.15 and provides 60 calories, 6 units of vitamin A, and 30 units of vitamin C. How many cups of each drink should be consumed each day to obtain an optimal cost and still meet the daily requirements?

Solution As in Example 9 in Section 7.5, let x be the number of cups of dietary drink X and let y be the number of cups of dietary drink Y.

$$\left.\begin{array}{ll} \text{For calories:} & 60x + 60y \geq 300 \\ \text{For vitamin A:} & 12x + 6y \geq 36 \\ \text{For vitamin C:} & 10x + 30y \geq 90 \\ & x \geq 0 \\ & y \geq 0 \end{array}\right\} \quad \text{Constraints}$$

The cost C is given by $C = 0.12x + 0.15y$. Objective function

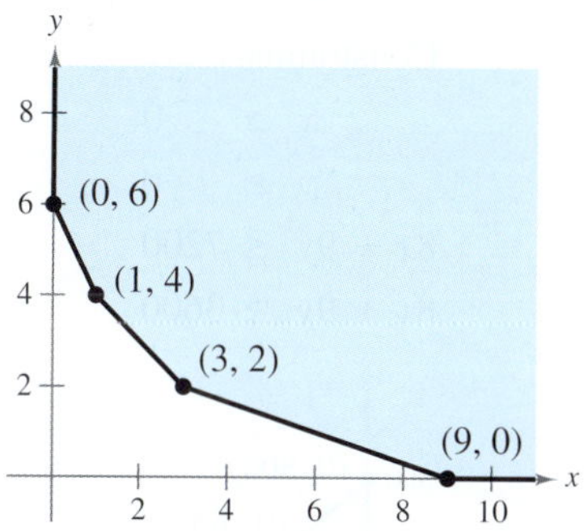

Figure 7.27

Figure 7.27 shows the graph of the region corresponding to the constraints. Because you want to incur as little cost as possible, you want to determine the *minimum* cost. To determine the minimum cost, test C at each vertex of the region.

At $(0, 6)$: $C = 0.12(0) + 0.15(6) = 0.90$

At $(1, 4)$: $C = 0.12(1) + 0.15(4) = 0.72$

At $(3, 2)$: $C = 0.12(3) + 0.15(2) = 0.66$ Minimum value of C

At $(9, 0)$: $C = 0.12(9) + 0.15(0) = 1.08$

So, the minimum cost is \$0.66 per day, and this occurs when 3 cups of drink X and 2 cups of drink Y are consumed each day.

✓ ***Checkpoint*** *Audio-video solution in English & Spanish at LarsonPrecalculus.com.*

A pet supply company mixes two brands of dry dog food. Brand X costs \$15 per bag and contains eight units of nutrient A, one unit of nutrient B, and two units of nutrient C. Brand Y costs \$30 per bag and contains two units of nutrient A, one unit of nutrient B, and seven units of nutrient C. The minimum required amounts of nutrients A, B, and C are 16 units, 5 units, and 20 units, respectively. How many bags of each brand should be mixed to obtain an optimal cost and still meet the minimum required amounts of nutrients?

Summarize (Section 7.6)

1. State the guidelines for solving a linear programming problem *(page 518)*. For an example of solving a linear programming problem, see Example 1.
2. Explain how to find the minimum and maximum values of an objective function of a linear programming problem *(pages 519 and 520)*. For an example of minimizing an objective function, see Example 2. For examples of maximizing objective functions, see Examples 1 and 3.
3. Describe a situation in which a linear programming problem has no optimal solution *(page 521)*. For an example of solving a linear programming problem that has no optimal solution, see Example 4.
4. Describe real-life examples of how to use linear programming to find the maximum profit in a business application *(page 522, Example 5)* and the optimal cost of diet drinks *(page 523, Example 6)*.

7.6 Exercises

See CalcChat.com for tutorial help and worked-out solutions to odd-numbered exercises.

Vocabulary: Fill in the blanks.

1. In the process called ________, you find the maximum or minimum value of a quantity.
2. One type of optimization strategy is called ________ ________.
3. The ________ function of a linear programming problem gives the quantity to be maximized or minimized.
4. The ________ of a linear programming problem determine the set of ________ ________.
5. The feasible solutions are ________ or ________ the boundary of the region corresponding to a system of constraints.
6. If a linear programming problem has a solution, then it must occur at a ________ of the set of feasible solutions.

Skills and Applications

Solving a Linear Programming Problem **In Exercises 7–12, find the minimum and maximum values of the objective function and where they occur, subject to the indicated constraints. (For each exercise, the graph of the region determined by the constraints is provided.)**

7. Objective function:

$z = 4x + 3y$

Constraints:

$x \geq 0$

$y \geq 0$

$x + y \leq 5$

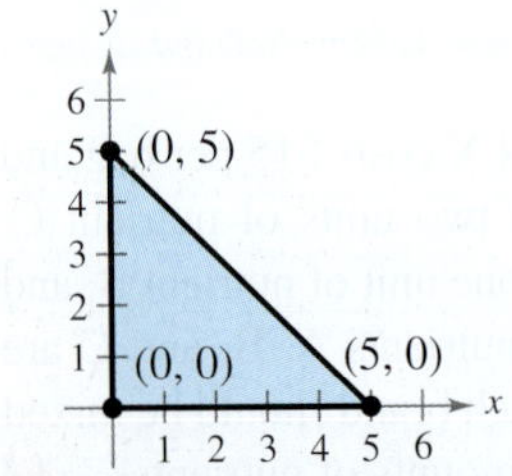

8. Objective function:

$z = 2x + 8y$

Constraints:

$x \geq 0$

$y \geq 0$

$2x + y \leq 4$

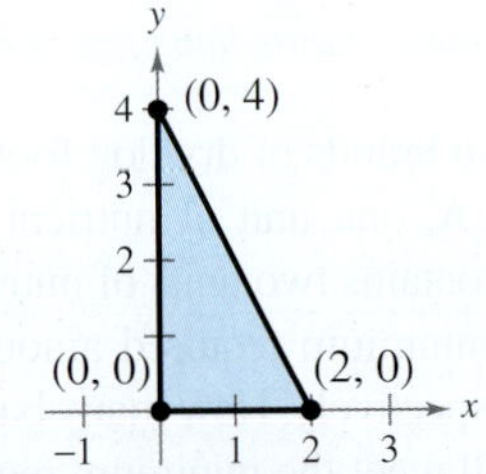

9. Objective function:

$z = 2x + 5y$

Constraints:

$x \geq 0$

$y \geq 0$

$x + 3y \leq 15$

$4x + y \leq 16$

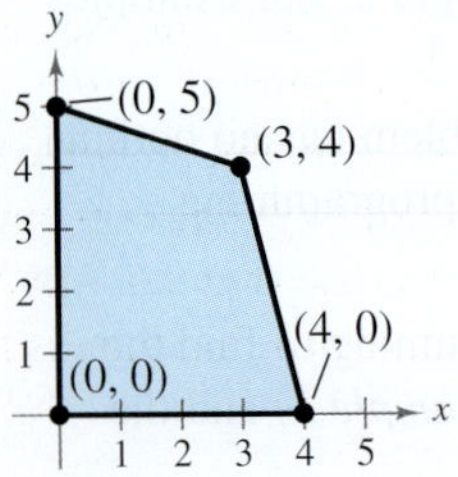

10. Objective function:

$z = 4x + 5y$

Constraints:

$x \geq 0$

$2x + 3y \geq 6$

$3x - y \leq 9$

$x + 4y \leq 16$

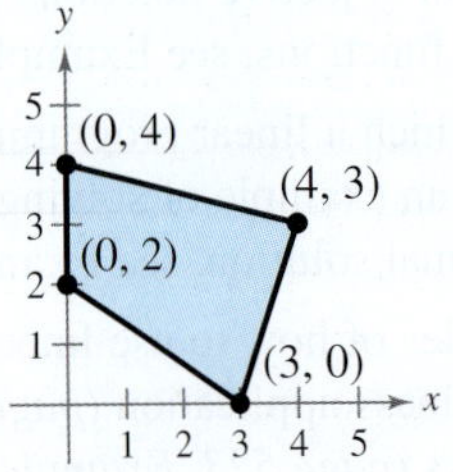

11. Objective function:

$z = 10x + 7y$

Constraints:

$0 \leq x \leq 60$

$0 \leq y \leq 45$

$5x + 6y \leq 420$

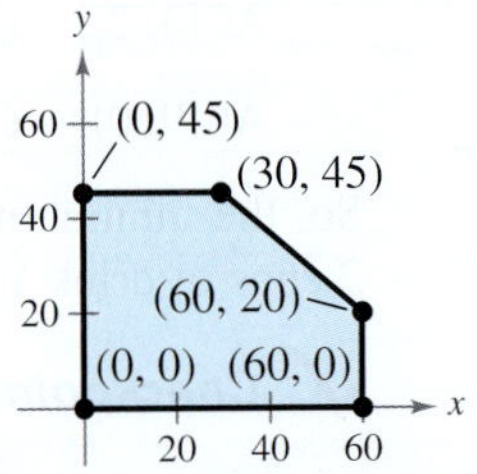

12. Objective function:

$z = 40x + 45y$

Constraints:

$x \geq 0$

$y \geq 0$

$8x + 9y \leq 7200$

$8x + 9y \geq 3600$

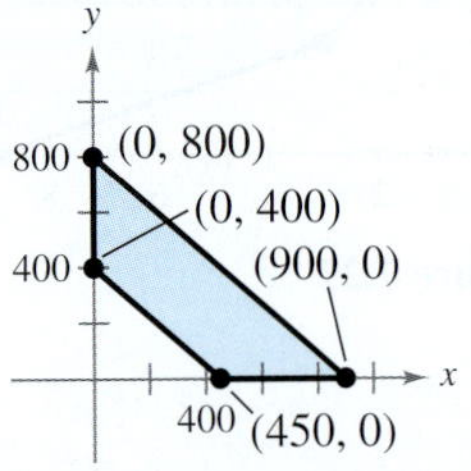

Solving a Linear Programming Problem **In Exercises 13–16, sketch the region determined by the constraints. Then find the minimum and maximum values of the objective function (if possible) and where they occur, subject to the indicated constraints.**

13. Objective function:

$z = 3x + 2y$

Constraints:

$x \geq 0$

$y \geq 0$

$5x + 2y \leq 20$

$5x + y \geq 10$

14. Objective function:

$z = 5x + \frac{1}{2}y$

Constraints:

$x \geq 0$

$y \geq 0$

$\frac{1}{2}x + y \leq 8$

$x + \frac{1}{2}y \geq 4$

15. Objective function:

$z = 4x + 5y$

Constraints:

$x \geq 0$

$y \geq 0$

$x + y \geq 8$

$3x + 5y \geq 30$

16. Objective function:

$z = 5x + 4y$

Constraints:

$x \geq 0$

$y \geq 0$

$2x + 2y \geq 10$

$x + 2y \geq 6$

Solving a Linear Programming Problem In Exercises 17–20, use a graphing utility to graph the region determined by the constraints. Then find the minimum and maximum values of the objective function and where they occur, subject to the constraints.

17. Objective function:

$z = 3x + y$

Constraints:

$x \geq 0$

$y \geq 0$

$x + 4y \leq 60$

$3x + 2y \geq 48$

18. Objective function:

$z = x$

Constraints:

$x \geq 0$

$y \geq 0$

$2x + 3y \leq 60$

$2x + y \leq 28$

$4x + y \leq 48$

19. Objective function:

$z = x + 4y$

Constraints:

(See Exercise 17.)

20. Objective function:

$z = y$

Constraints:

(See Exercise 18.)

Finding Minimum and Maximum Values In Exercises 21–24, find the minimum and maximum values of the objective function and where they occur, subject to the constraints $x \geq 0$, $y \geq 0$, $3x + y \leq 15$, and $4x + 3y \leq 30$.

21. $z = 2x + y$

22. $z = 5x + y$

23. $z = x + y$

24. $z = 3x + y$

Finding Minimum and Maximum Values In Exercises 25–28, find the minimum and maximum values of the objective function and where they occur, subject to the constraints $x \geq 0$, $y \geq 0$, $x + 4y \leq 20$, $x + y \leq 18$, and $2x + 2y \leq 21$.

25. $z = x + 5y$

26. $z = 2x + 4y$

27. $z = 4x + 5y$

28. $z = 4x + y$

Describing an Unusual Characteristic In Exercises 29–36, the linear programming problem has an unusual characteristic. Sketch a graph of the solution region for the problem and describe the unusual characteristic. Find the minimum and maximum values of the objective function (if possible) and where they occur.

29. Objective function:

$z = 2.5x + y$

Constraints:

$x \geq 0$

$y \geq 0$

$3x + 5y \leq 15$

$5x + 2y \leq 10$

30. Objective function:

$z = x + y$

Constraints:

$x \geq 0$

$y \geq 0$

$-x + y \leq 1$

$-x + 2y \leq 4$

31. Objective function:

$z = -x + 2y$

Constraints:

$x \geq 0$

$y \geq 0$

$x \leq 10$

$x + y \leq 7$

32. Objective function:

$z = x + y$

Constraints:

$x \geq 0$

$y \geq 0$

$-x + y \leq 0$

$-3x + y \geq 3$

33. Objective function:

$z = 3x + 4y$

Constraints:

$x \geq 0$

$y \geq 0$

$x + y \leq 1$

$2x + y \leq 4$

34. Objective function:

$z = x + 2y$

Constraints:

$x \geq 0$

$y \geq 0$

$x + 2y \leq 4$

$2x + y \leq 4$

35. Objective function:

$z = x + y$

Constraints:

(See Exercise 33.)

36. Objective function:

$z = 2x + y$

Constraints:

(See Exercise 34.)

37. Optimal Profit A merchant plans to sell two models of MP3 players at prices of \$225 and \$250. The \$225 model yields a profit of \$30 per unit and the \$250 model yields a profit of \$31 per unit. The merchant estimates that the total monthly demand will not exceed 275 units. The merchant does not want to invest more than \$63,000 in inventory for these products. What is the optimal inventory level for each model? What is the optimal profit?

38. Optimal Profit A manufacturer produces two models of elliptical cross-training exercise machines. The times for assembling, finishing, and packaging model X are 3 hours, 3 hours, and 0.8 hour, respectively. The times for model Y are 4 hours, 2.5 hours, and 0.4 hour. The total times available for assembling, finishing, and packaging are 6000 hours, 4200 hours, and 950 hours, respectively. The profits per unit are \$300 for model X and \$375 for model Y. What is the optimal production level for each model? What is the optimal profit?

39. Optimal Cost An animal shelter mixes two brands of dry dog food. Brand X costs \$25 per bag and contains two units of nutrient A, two units of nutrient B, and two units of nutrient C. Brand Y costs \$20 per bag and contains one unit of nutrient A, nine units of nutrient B, and three units of nutrient C. The minimum required amounts of nutrients A, B, and C are 12 units, 36 units, and 24 units, respectively. What is the optimal number of bags of each brand that should be mixed? What is the optimal cost?

40. Optimal Cost A humanitarian agency can use two models of vehicles for a refugee rescue mission. Each model A vehicle costs \$1000 and each model B vehicle costs \$1500. Mission strategies and objectives indicate the following constraints.

- The agency must use a total of at least 20 vehicles.
- A model A vehicle can hold 45 boxes of supplies. A model B vehicle can hold 30 boxes of supplies. The agency must deliver at least 690 boxes of supplies to the refugee camp.
- A model A vehicle can hold 20 refugees. A model B vehicle can hold 32 refugees. The agency must rescue at least 520 refugees.

What is the optimal number of vehicles of each model that should be used? What is the optimal cost?

41. Optimal Revenue An accounting firm has 780 hours of staff time and 272 hours of reviewing time available each week. The firm charges \$1600 for an audit and \$250 for a tax return. Each audit requires 60 hours of staff time and 16 hours of review time. Each tax return requires 10 hours of staff time and 4 hours of review time. What numbers of audits and tax returns will yield an optimal revenue? What is the optimal revenue?

42. Optimal Revenue The accounting firm in Exercise 41 lowers its charge for an audit to \$1400. What numbers of audits and tax returns will yield an optimal revenue? What is the optimal revenue?

43. Optimal Yield

A fruit grower raises crops A and B. The yield is 300 bushels per acre for crop A and 500 bushels per acre for crop B. Research and available resources indicate the following constraints.

- The fruit grower has 150 acres of land for raising the crops.
- It takes 1 day to trim an acre of crop A and 2 days to trim an acre of crop B, and there are 240 days per year available for trimming.
- It takes 0.3 day to pick an acre of crop A and 0.1 day to pick an acre of crop B, and there are 30 days per year available for picking.

What is the optimal acreage for each fruit? What is the optimal yield?

44. Optimal Profit In Exercise 43, the profit is \$185 per acre for crop A and \$245 per acre for crop B. What is the optimal profit?

45. Media Selection A company has budgeted a maximum of \$1,000,000 for national advertising of an allergy medication. Each minute of television time costs \$100,000 and each one-page newspaper ad costs \$20,000. Each television ad is expected to be viewed by 20 million viewers, and each newspaper ad is expected to be seen by 5 million readers. The company's market research department recommends that at most 80% of the advertising budget be spent on television ads. What is the optimal amount that should be spent on each type of ad? What is the optimal total audience?

46. Investment Portfolio An investor has up to \$450,000 to invest in two types of investments. Type A pays 6% annually and type B pays 10% annually. To have a well-balanced portfolio, the investor imposes the following conditions. At least one-half of the total portfolio is to be allocated to type A investments and at least one-fourth of the portfolio is to be allocated to type B investments. What is the optimal amount that should be invested in each type of investment? What is the optimal return?

Exploration

True or False? **In Exercises 47–49, determine whether the statement is true or false. Justify your answer.**

47. If an objective function has a maximum value at the vertices $(4, 7)$ and $(8, 3)$, then it also has a maximum value at the points $(4.5, 6.5)$ and $(7.8, 3.2)$.

48. If an objective function has a minimum value at the vertex $(20, 0)$, then it also has a minimum value at the point $(0, 0)$.

49. When solving a linear programming problem, if the objective function has a maximum value at more than one vertex, then there are an infinite number of points that will produce the maximum value.

50. HOW DO YOU SEE IT? Using the constraint region shown below, determine which of the following objective functions has (a) a maximum at vertex A, (b) a maximum at vertex B, (c) a maximum at vertex C, and (d) a minimum at vertex C.

(i) $z = 2x + y$

(ii) $z = 2x - y$

(iii) $z = -x + 2y$

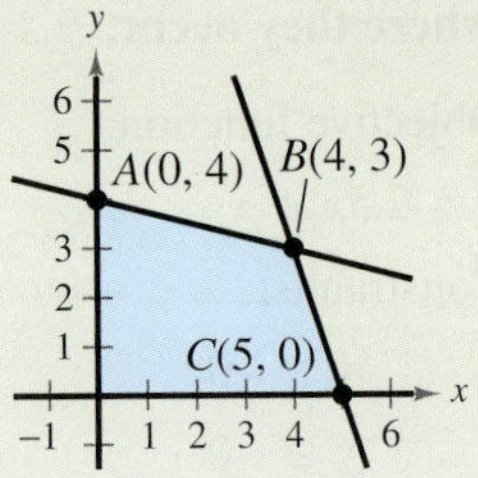

Alison McC/Shutterstock.com

Chapter Summary

Section	What Did You Learn?	Explanation/Examples	Review Exercises
Section 7.1	Use the method of substitution to solve systems of linear equations in two variables *(p. 466).*	**Method of Substitution** **1.** *Solve* one of the equations for one variable in terms of the other. **2.** *Substitute* the expression found in Step 1 into the other equation to obtain an equation in one variable. **3.** *Solve* the equation obtained in Step 2. **4.** *Back-substitute* the value obtained in Step 3 into the expression obtained in Step 1 to find the value of the other variable. **5.** *Check* that the solution satisfies *each* of the original equations.	1–6
	Use the method of substitution to solve systems of nonlinear equations in two variables *(p. 469).*	The method of substitution (see steps above) can be used to solve systems in which one or both of the equations are nonlinear. (See Examples 3 and 4.)	7–10
	Use a graphical approach to solve systems of equations in two variables *(p. 470).*	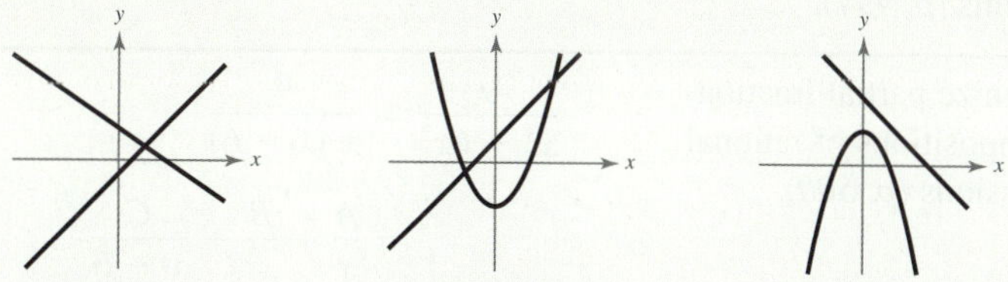 One intersection point — Two intersection points — No intersection points	11–18
	Use systems of equations to model and solve real-life problems *(p. 471).*	A system of equations can help you find the break-even point for a company. (See Example 6.)	19–22
Section 7.2	Use the method of elimination to solve systems of linear equations in two variables *(p. 476).*	**Method of Elimination** **1.** *Obtain coefficients* for x (or y) that differ only in sign. **2.** *Add* the equations to eliminate one variable. **3.** *Solve* the equation obtained in Step 2. **4.** *Back-substitute* the value obtained in Step 3 into either of the original equations and solve for the other variable. **5.** *Check* that the solution satisfies *each* of the original equations.	23–28
	Interpret graphically the numbers of solutions of systems of linear equations in two variables *(p. 480).*	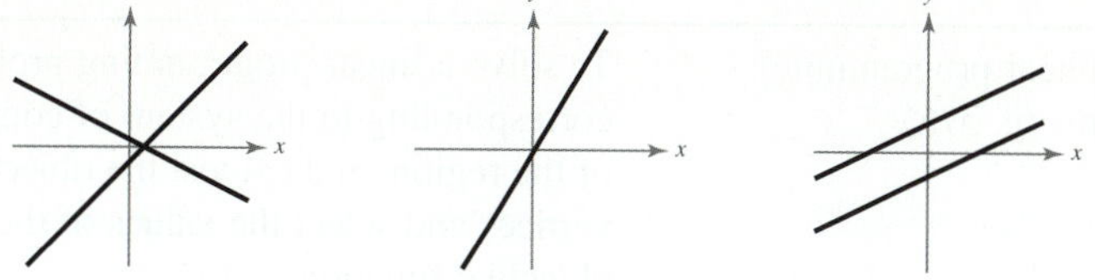 Exactly one solution — Infinitely many solutions — No solution	29–32
	Use systems of linear equations in two variables to model and solve real-life problems *(p. 482).*	A system of linear equations in two variables can help you find the equilibrium point for a market. (See Example 9.)	33, 34

	What Did You Learn?	Explanation/Examples	Review Exercises
Section 7.3	Use back-substitution to solve linear systems in row-echelon form *(p. 488).*	$\begin{cases} x - 2y + 3z = 9 \\ -x + 3y = -4 \\ 2x - 5y + 5z = 17 \end{cases} \Rightarrow$ **Row-Echelon Form** $\begin{cases} x - 2y + 3z = 9 \\ y + 3z = 5 \\ z = 2 \end{cases}$	35–38
	Use Gaussian elimination to solve systems of linear equations *(p. 489).*	To produce an equivalent system of linear equations, use row operations by (1) interchanging two equations, (2) multiplying one equation by a nonzero constant, or (3) adding a multiple of one equation to another equation to replace the latter equation.	39–42
	Solve nonsquare systems of linear equations *(p. 493).*	In a nonsquare system, the number of equations differs from the number of variables. A system of linear equations cannot have a unique solution unless there are at least as many equations as there are variables.	43, 44
	Use systems of linear equations in three or more variables to model and solve real-life problems *(p. 494).*	A system of linear equations in three variables can help you find the position equation of an object that is moving in a (vertical) line with constant acceleration. (See Example 7.)	45–54
Section 7.4	Recognize partial fraction decompositions of rational expressions *(p. 500).*	$\frac{9}{x^3 - 6x^2} = \frac{9}{x^2(x - 6)} = \frac{A}{x} + \frac{B}{x^2} + \frac{C}{x - 6}$	55–58
	Find partial fraction decompositions of rational expressions *(p. 501).*	The techniques used for determining constants in the numerators of partial fractions vary slightly, depending on the type of factors of the denominator: linear or quadratic, distinct or repeated.	59–66
Section 7.5	Sketch the graphs of inequalities in two variables *(p. 508)*, and solve systems of inequalities *(p. 510).*	$\begin{cases} x^2 + y \le 5 \\ x \ge -1 \\ y \ge 0 \end{cases}$ 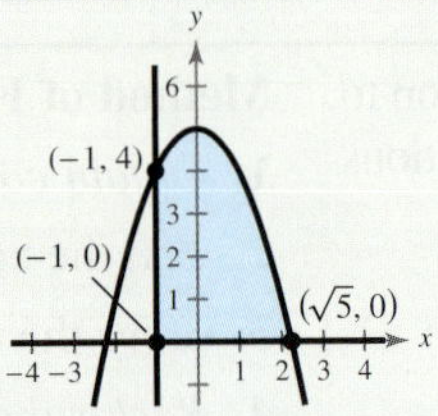	67–78
	Use systems of inequalities in two variables to model and solve real-life problems *(p. 513).*	A system of inequalities in two variables can help you find the consumer surplus and producer surplus for given demand and supply equations. (See Example 8.)	79–84
Section 7.6	Solve linear programming problems *(p. 518).*	To solve a linear programming problem, (1) sketch the region corresponding to the system of constraints, (2) find the vertices of the region, and (3) test the objective function at each of the vertices and select the values of the variables that optimize the objective function.	85–88
	Use linear programming to model and solve real-life problems *(p. 522).*	Linear programming can help you find the maximum profit in business applications. (See Example 5.)	89, 90

Review Exercises

See CalcChat.com for tutorial help and worked-out solutions to odd-numbered exercises.

7.1 Solving a System by Substitution **In Exercises 1–10, solve the system by the method of substitution.**

1. $\begin{cases} x + y = 2 \\ x - y = 0 \end{cases}$
2. $\begin{cases} 2x - 3y = 3 \\ x - y = 0 \end{cases}$
3. $\begin{cases} 4x - y - 1 = 0 \\ 8x + y - 17 = 0 \end{cases}$
4. $\begin{cases} 10x + 6y + 14 = 0 \\ x + 9y + 7 = 0 \end{cases}$
5. $\begin{cases} 0.5x + y = 0.75 \\ 1.25x - 4.5y = -2.5 \end{cases}$
6. $\begin{cases} -x + \frac{2}{5}y = \frac{3}{5} \\ -x + \frac{1}{5}y = -\frac{4}{5} \end{cases}$
7. $\begin{cases} x^2 - y^2 = 9 \\ x - y = 1 \end{cases}$
8. $\begin{cases} x^2 + y^2 = 169 \\ 3x + 2y = 39 \end{cases}$
9. $\begin{cases} y = 2x^2 \\ y = x^4 - 2x^2 \end{cases}$
10. $\begin{cases} x = y + 3 \\ x = y^2 + 1 \end{cases}$

Solving a System of Equations Graphically **In Exercises 11–14, solve the system graphically.**

11. $\begin{cases} 2x - y = 10 \\ x + 5y = -6 \end{cases}$
12. $\begin{cases} 8x - 3y = -3 \\ 2x + 5y = 28 \end{cases}$
13. $\begin{cases} y = 2x^2 - 4x + 1 \\ y = x^2 - 4x + 3 \end{cases}$
14. $\begin{cases} y^2 - 2y + x = 0 \\ x + y = 0 \end{cases}$

Solving a System of Equations Graphically **In Exercises 15–18, use a graphing utility to solve the systems of equations. Find the solution(s) accurate to two decimal places.**

15. $\begin{cases} y = -2e^{-x} \\ 2e^{x} + y = 0 \end{cases}$
16. $\begin{cases} x^2 + y^2 = 100 \\ 2x - 3y = -12 \end{cases}$
17. $\begin{cases} y = 2 + \log x \\ y = \frac{3}{4}x + 5 \end{cases}$
18. $\begin{cases} y = \ln(x - 1) - 3 \\ y = 4 - \frac{1}{2}x \end{cases}$

19. **Body Mass Index** Body Mass Index (BMI) is a measure of body fat based on height and weight. The 85th percentile BMI for females, ages 9 to 20, is growing more slowly than that for males of the same age range. Models that represent the 85th percentile BMI for males and females, ages 9 to 20, are

$$\begin{cases} B = 0.77a + 11.7 & \text{Males} \\ B = 0.68a + 13.5 & \text{Females} \end{cases}$$

where B is the BMI (kg/m^2) and a represents the age, with $a = 9$ corresponding to 9 years old. Use a graphing utility to determine whether the BMI for males exceeds the BMI for females. *(Source: Centers for Disease Control and Prevention)*

20. **Choice of Two Jobs** You receive two sales job offers. One company offers an annual salary of \$55,000 plus a year-end bonus of 1.5% of your total sales. The other company offers an annual salary of \$52,000 plus a year-end bonus of 2% of your total sales. What amount of sales will make the second job offer better? Explain.

21. **Geometry** The perimeter of a rectangle is 68 feet and its width is $\frac{8}{9}$ times its length. Find the dimensions of the rectangle.

22. **Geometry** The perimeter of a rectangle is 40 inches. The area of the rectangle is 96 square inches. Find the dimensions of the rectangle.

7.2 Solving a System by Elimination **In Exercises 23–28, solve the system by the method of elimination and check any solutions algebraically.**

23. $\begin{cases} 2x - y = 2 \\ 6x + 8y = 39 \end{cases}$
24. $\begin{cases} 12x + 42y = -17 \\ 30x - 18y = 19 \end{cases}$
25. $\begin{cases} 3x - 2y = 0 \\ 3x + 2(y + 5) = 10 \end{cases}$
26. $\begin{cases} 7x + 12y = 63 \\ 2x + 3(y + 2) = 21 \end{cases}$
27. $\begin{cases} 1.25x - 2y = 3.5 \\ 5x - 8y = 14 \end{cases}$
28. $\begin{cases} 1.5x + 2.5y = 8.5 \\ 6x + 10y = 24 \end{cases}$

Matching a System with Its Graph **In Exercises 29–32, match the system of linear equations with its graph. Describe the number of solutions and state whether the system is consistent or inconsistent. [The graphs are labeled (a), (b), (c), and (d).]**

(a)

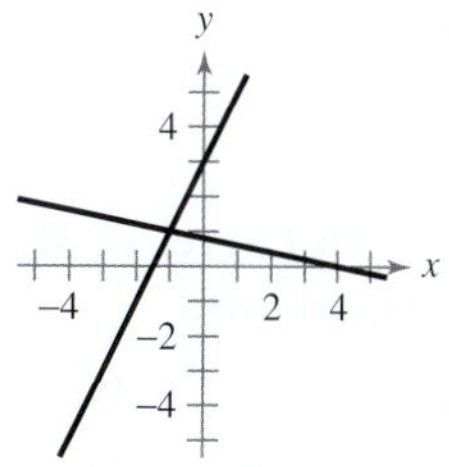

(b)

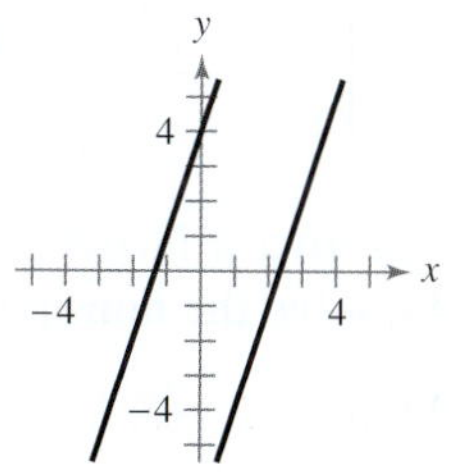

(c)

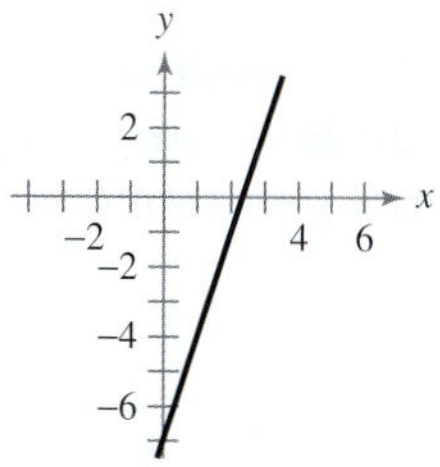

(d)

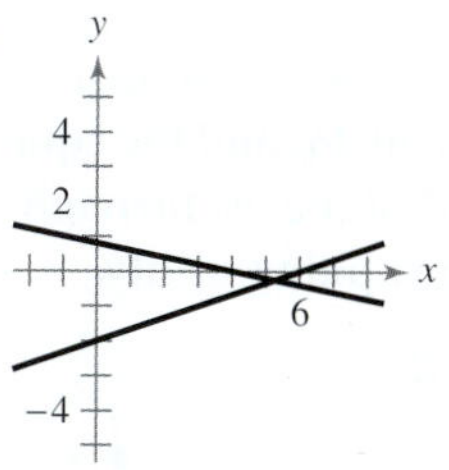

29. $\begin{cases} x + 5y = 4 \\ x - 3y = 6 \end{cases}$
30. $\begin{cases} -3x + y = -7 \\ 9x - 3y = 21 \end{cases}$
31. $\begin{cases} 3x - y = 7 \\ -6x + 2y = 8 \end{cases}$
32. $\begin{cases} 2x - y = -3 \\ x + 5y = 4 \end{cases}$

Finding the Equilibrium Point **In Exercises 33 and 34, find the equilibrium point of the demand and supply equations.**

	Demand	Supply
33.	$p = 43 - 0.0002x$	$p = 22 + 0.00001x$
34.	$p = 120 - 0.0001x$	$p = 45 + 0.0002x$

7.3 Using Back-Substitution **In Exercises 35–38, use back-substitution to solve the system of linear equations.**

35. $\begin{cases} x - 4y + 3z = 3 \\ -y + z = -1 \\ z = -5 \end{cases}$

36. $\begin{cases} x - 7y + 8z = 85 \\ y - 9z = -35 \\ z = 3 \end{cases}$

37. $\begin{cases} 4x - 3y - 2z = -65 \\ 8y - 7z = -14 \\ z = 10 \end{cases}$

38. $\begin{cases} 5x - 7z = 9 \\ 3y - 8z = -4 \\ z = -7 \end{cases}$

Using Gaussian Elimination **In Exercises 39–42, use Gaussian elimination to solve the system of linear equations and check any solutions algebraically.**

39. $\begin{cases} x + 2y + 6z = 4 \\ -3x + 2y - z = -4 \\ 4x + 2z = 16 \end{cases}$

40. $\begin{cases} x - 2y + z = -6 \\ 2x - 3y = -7 \\ -x + 3y - 3z = 11 \end{cases}$

41. $\begin{cases} 2x + 6z = -9 \\ 3x - 2y + 11z = -16 \\ 3x - y + 7z = -11 \end{cases}$

42. $\begin{cases} x + 4w = 1 \\ 3y + z - w = 4 \\ 2y - 3w = 2 \\ 4x - y + 2z = 5 \end{cases}$

Solving a Nonsquare System **In Exercises 43 and 44, solve the nonsquare system of equations.**

43. $\begin{cases} 5x - 12y + 7z = 16 \\ 3x - 7y + 4z = 9 \end{cases}$

44. $\begin{cases} 2x + 5y - 19z = 34 \\ 3x + 8y - 31z = 54 \end{cases}$

Finding the Equation of a Parabola **In Exercises 45 and 46, find the equation of the parabola $y = ax^2 + bx + c$ that passes through the points. Verify your result using a graphing utility.**

45.

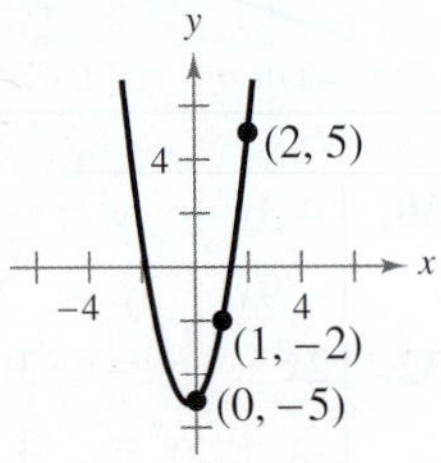

46.

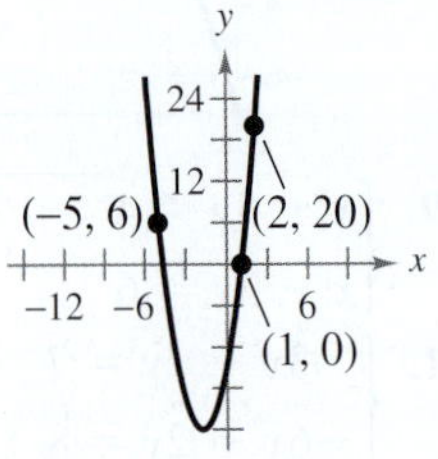

Finding the Equation of a Circle **In Exercises 47 and 48, find the equation of the circle**

$$x^2 + y^2 + Dx + Ey + F = 0$$

that passes through the points. Verify your result using a graphing utility.

47.

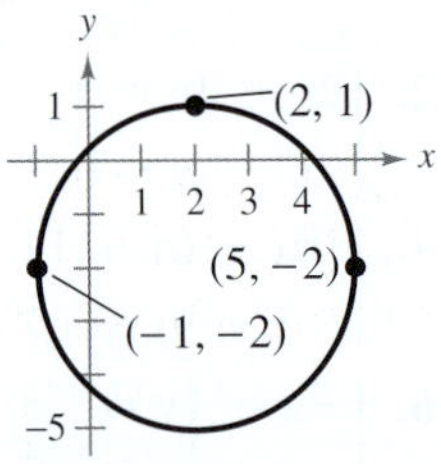

48.

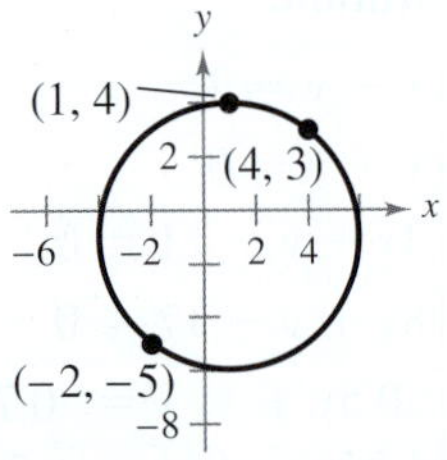

49. Data Analysis: E-Commerce The table shows the retail e-commerce sales y (in billions of dollars) in the United States from 2007 through 2009. *(Source: U.S. Census Bureau)*

Year	2007	2008	2009
Retail E-Commerce Sales, y	126.7	141.9	145.2

(a) Find the least squares regression parabola $y = at^2 + bt + c$ for the data, where t represents the year with $t = 7$ corresponding to 2007, by solving the system for a, b, and c.

$$\begin{cases} 3c + 24b + 194a = 413.8 \\ 24c + 194b + 1584a = 3328.9 \\ 194c + 1584b + 13{,}058a = 27{,}051.1 \end{cases}$$

(b) Use a graphing utility to graph the parabola and the data in the same viewing window. How well does the model fit the data?

(c) Use the model to estimate the retail e-commerce sales in the United States in 2015. Does your answer seem reasonable?

50. Agriculture A mixture of 6 gallons of chemical A, 8 gallons of chemical B, and 13 gallons of chemical C is required to kill a destructive crop insect. Commercial spray X contains 1, 2, and 2 parts, respectively, of these chemicals. Commercial spray Y contains only chemical C. Commercial spray Z contains chemicals A, B, and C in equal amounts. How much of each type of commercial spray gives the desired mixture?

51. Investment Analysis An inheritance of \$40,000 was divided among three investments yielding \$3500 in interest per year. The interest rates for the three investments were 7%, 9%, and 11%. Find the amount placed in each investment when the second and third were \$3000 and \$5000 less than the first, respectively.

52. Sports The Old Course at St Andrews Links in St Andrews, Scotland, is one of the oldest golf courses in the world. It is an 18-hole course that consists of par-3 holes, par-4 holes, and par-5 holes. There are seven times as many par-4 holes as par-5 holes, and the sum of the numbers of par-3 and par-5 holes is four. Find the numbers of par-3, par-4, and par-5 holes on the course. *(Source: St Andrews Links Trust)*

Vertical Motion **In Exercises 53 and 54, an object moving vertically is at the given heights at the specified times. Find the position equation**

$$s = \frac{1}{2}at^2 + v_0t + s_0$$

for the object.

53. At $t = 1$ second, $s = 134$ feet
At $t = 2$ seconds, $s = 86$ feet
At $t = 3$ seconds, $s = 6$ feet

54. At $t = 1$ second, $s = 184$ feet
At $t = 2$ seconds, $s = 116$ feet
At $t = 3$ seconds, $s = 16$ feet

7.4 Writing the Form of the Decomposition **In Exercises 55–58, write the form of the partial fraction decomposition of the rational expression. Do not solve for the constants.**

55. $\dfrac{3}{x^2 + 20x}$ **56.** $\dfrac{x - 8}{x^2 - 3x - 28}$

57. $\dfrac{3x - 4}{x^3 - 5x^2}$ **58.** $\dfrac{x - 2}{x(x^2 + 2)^2}$

Writing the Partial Fraction Decomposition **In Exercises 59–66, write the partial fraction decomposition of the rational expression. Check your result algebraically.**

59. $\dfrac{4 - x}{x^2 + 6x + 8}$ **60.** $\dfrac{-x}{x^2 + 3x + 2}$

61. $\dfrac{x^2}{x^2 + 2x - 15}$ **62.** $\dfrac{9}{x^2 - 9}$

63. $\dfrac{x^2 + 2x}{x^3 - x^2 + x - 1}$ **64.** $\dfrac{4x}{3(x - 1)^2}$

65. $\dfrac{3x^2 + 4x}{(x^2 + 1)^2}$ **66.** $\dfrac{4x^2}{(x - 1)(x^2 + 1)}$

7.5 Graphing an Inequality **In Exercises 67–72, sketch the graph of the inequality.**

67. $y \le 5 - \frac{1}{2}x$ **68.** $3y - x \ge 7$

69. $y - 4x^2 > -1$ **70.** $y \le \dfrac{3}{x^2 + 2}$

71. $(x - 1)^2 + (y - 3)^2 < 16$

72. $x^2 + (y + 5)^2 > 1$

Solving a System of Inequalities **In Exercises 73–78, sketch the graph (and label the vertices) of the solution set of the system of inequalities.**

73. $\begin{cases} x + 2y \le 160 \\ 3x + y \le 180 \\ x \ge 0 \\ y \ge 0 \end{cases}$ **74.** $\begin{cases} 3x + 2y \ge 24 \\ x + 2y \ge 12 \\ 2 \le x \le 15 \\ y \le 15 \end{cases}$

75. $\begin{cases} y < x + 1 \\ y > x^2 - 1 \end{cases}$ **76.** $\begin{cases} y \le 6 - 2x - x^2 \\ y \ge x + 6 \end{cases}$

77. $\begin{cases} 2x - 3y \ge 0 \\ 2x - y \le 8 \\ y \ge 0 \end{cases}$ **78.** $\begin{cases} x^2 + y^2 \le 9 \\ (x - 3)^2 + y^2 \le 9 \end{cases}$

Consumer Surplus and Producer Surplus **In Exercises 79 and 80, (a) graph the systems representing the consumer surplus and producer surplus for the supply and demand equations and (b) find the consumer surplus and producer surplus.**

	Demand	Supply
79.	$p = 160 - 0.0001x$	$p = 70 + 0.0002x$
80.	$p = 130 - 0.0002x$	$p = 30 + 0.0003x$

81. Geometry Write a system of inequalities to describe the region of a rectangle with vertices at (3, 1), (7, 1), (7, 10), and (3, 10).

82. Data Analysis: Sales The table shows the sales y (in millions of dollars) for Aéropostale, Inc., from 2003 through 2010. *(Source: Aéropostale, Inc.)*

DATA

Spreadsheet at LarsonPrecalculus.com

Year	Sales, y
2003	734.9
2004	964.2
2005	1204.3
2006	1413.2
2007	1590.9
2008	1885.5
2009	2230.1
2010	2400.4

(a) Use the *regression* feature of a graphing utility to find a linear model for the data. Let t represent the year, with $t = 3$ corresponding to 2003.

(b) The total sales for Aéropostale during this eight-year period can be approximated by finding the area of the trapezoid bounded by the linear model you found in part (a) and the lines $y = 0$, $t = 2.5$, and $t = 10.5$. Use the graphing utility to graph this region.

(c) Use the formula for the area of a trapezoid to approximate the total sales for Aéropostale.

83. Inventory Costs A warehouse operator has 24,000 square feet of floor space in which to store two products. Each unit of product I requires 20 square feet of floor space and costs $12 per day to store. Each unit of product II requires 30 square feet of floor space and costs $8 per day to store. The total storage cost per day cannot exceed $12,400. Find and graph a system of inequalities describing all possible inventory levels.

84. Nutrition A dietitian designs a special dietary supplement using two different foods. Each ounce of food X contains 12 units of calcium, 10 units of iron, and 20 units of vitamin B. Each ounce of food Y contains 15 units of calcium, 20 units of iron, and 12 units of vitamin B. The minimum daily requirements of the diet are 300 units of calcium, 280 units of iron, and 300 units of vitamin B.

(a) Write a system of inequalities describing the different amounts of food X and food Y that can be used.

(b) Sketch a graph of the region in part (a).

(c) Find two solutions to the system and interpret their meanings in the context of the problem.

7.6 Solving a Linear Programming Problem In Exercises 85–88, sketch the region determined by the constraints. Then find the minimum and maximum values of the objective function (if possible) and where they occur, subject to the indicated constraints.

85. Objective function:

$z = 3x + 4y$

Constraints:

$$\begin{aligned} x &\geq 0 \\ y &\geq 0 \\ 2x + 5y &\leq 50 \\ 4x + y &\leq 28 \end{aligned}$$

86. Objective function:

$z = 10x + 7y$

Constraints:

$$\begin{aligned} x &\geq 0 \\ y &\geq 0 \\ 2x + y &\geq 100 \\ x + y &\geq 75 \end{aligned}$$

87. Objective function:

$z = 1.75x + 2.25y$

Constraints:

$$\begin{aligned} x &\geq 0 \\ y &\geq 0 \\ 2x + y &\geq 25 \\ 3x + 2y &\geq 45 \end{aligned}$$

88. Objective function:

$z = 50x + 70y$

Constraints:

$$\begin{aligned} x &\geq 0 \\ y &\geq 0 \\ x + 2y &\leq 1500 \\ 5x + 2y &\leq 3500 \end{aligned}$$

89. Optimal Revenue A student is working part time as a hairdresser to pay college expenses. The student may work no more than 24 hours per week. Haircuts cost $25 and require an average of 20 minutes, and permanents cost $70 and require an average of 1 hour and 10 minutes. How many haircuts and/or permanents will yield an optimal revenue? What is the optimal revenue?

90. Optimal Profit A manufacturer produces two models of bicycles. The table shows the times (in hours) required for assembling, painting, and packaging each model.

Process	Hours, Model A	Hours, Model B
Assembling	2	2.5
Painting	4	1
Packaging	1	0.75

The total times available for assembling, painting, and packaging are 4000 hours, 4800 hours, and 1500 hours, respectively. The profits per unit are $45 for model A and $50 for model B. What is the optimal production level for each model? What is the optimal profit?

Exploration

True or False? In Exercises 91 and 92, determine whether the statement is true or false. Justify your answer.

91. The system

$$\begin{cases} y \leq 5 \\ y \geq -2 \\ y \leq \frac{7}{2}x - 9 \\ y \leq -\frac{7}{2}x + 26 \end{cases}$$

represents the region covered by an isosceles trapezoid.

92. It is possible for an objective function of a linear programming problem to have exactly 10 maximum value points.

Finding a System of Linear Equations In Exercises 93–96, find a system of linear equations having the ordered pair as a solution. (There are many correct answers.)

93. $(-8, 10)$ **94.** $(5, -4)$

95. $\left(\frac{4}{3}, 3\right)$ **96.** $\left(-2, \frac{11}{5}\right)$

Finding a System of Linear Equations In Exercises 97–100, find a system of linear equations having the ordered triple as a solution. (There are many correct answers.)

97. $(4, -1, 3)$ **98.** $(-3, 5, 6)$

99. $\left(5, \frac{3}{2}, 2\right)$ **100.** $\left(-\frac{1}{2}, -2, -\frac{3}{4}\right)$

101. Writing Explain what is meant by an inconsistent system of linear equations.

102. Graphical Reasoning How can you tell graphically that a system of linear equations in two variables has no solution? Give an example.

Chapter Test

See CalcChat.com for tutorial help and worked-out solutions to odd-numbered exercises.

Take this test as you would take a test in class. When you are finished, check your work against the answers given in the back of the book.

In Exercises 1–3, solve the system by the method of substitution.

1. $\begin{cases} x + y = -9 \\ 5x - 8y = 20 \end{cases}$ **2.** $\begin{cases} y = x - 1 \\ y = (x - 1)^3 \end{cases}$ **3.** $\begin{cases} 2x - y^2 = 0 \\ x - y = 4 \end{cases}$

In Exercises 4–6, solve the system graphically.

4. $\begin{cases} 3x - 6y = 0 \\ 3x + 6y = 18 \end{cases}$ **5.** $\begin{cases} y = 9 - x^2 \\ y = x + 3 \end{cases}$ **6.** $\begin{cases} y - \ln x = 12 \\ 7x - 2y + 11 = -6 \end{cases}$

In Exercises 7 and 8, solve the system by the method of elimination.

7. $\begin{cases} 3x + 4y = -26 \\ 7x - 5y = 11 \end{cases}$ **8.** $\begin{cases} 1.4x - y = 17 \\ 0.8x + 6y = -10 \end{cases}$

In Exercises 9 and 10, solve the system of linear equations and check any solutions algebraically.

9. $\begin{cases} x - 2y + 3z = 11 \\ 2x - z = 3 \\ 3y + z = -8 \end{cases}$ **10.** $\begin{cases} 3x + 2y + z = 17 \\ -x + y + z = 4 \\ x - y - z = 3 \end{cases}$

In Exercises 11–14, write the partial fraction decomposition of the rational expression. Check your result algebraically.

11. $\dfrac{2x + 5}{x^2 - x - 2}$ **12.** $\dfrac{3x^2 - 2x + 4}{x^2(2 - x)}$ **13.** $\dfrac{x^2 + 5}{x^3 - x}$ **14.** $\dfrac{x^2 - 4}{x^3 + 2x}$

In Exercises 15–17, sketch the graph (and label the vertices) of the solution set of the system of inequalities.

15. $\begin{cases} 2x + y \le 4 \\ 2x - y \ge 0 \\ x \ge 0 \end{cases}$ **16.** $\begin{cases} y < -x^2 + x + 4 \\ y > 4x \end{cases}$ **17.** $\begin{cases} x^2 + y^2 \le 36 \\ x \ge 2 \\ y \ge -4 \end{cases}$

18. Find the minimum and maximum values of the objective function $z = 20x + 12y$ and where they occur, subject to the following constraints.

$$\left.\begin{aligned} x &\ge 0 \\ y &\ge 0 \\ x + 4y &\le 32 \\ 3x + 2y &\le 36 \end{aligned}\right\} \quad \text{Constraints}$$

19. A total of \$50,000 is invested in two funds that pay 4% and 5.5% simple interest. The yearly interest is \$2390. How much is invested at each rate?

20. Find the equation of the parabola $y = ax^2 + bx + c$ that passes through the points $(0, 6)$, $(-2, 2)$, and $(3, \frac{9}{2})$.

21. A manufacturer produces two models of television stands. The table at the left shows the times (in hours) required for assembling, staining, and packaging the two models. The total times available for assembling, staining, and packaging are 4000 hours, 8950 hours, and 2650 hours, respectively. The profits per unit are \$30 for model I and \$40 for model II. What is the optimal inventory level for each model? What is the optimal profit?

	Model I	Model II
Assembling	0.5	0.75
Staining	2.0	1.5
Packaging	0.5	0.5

Table for 21

Proofs in Mathematics

An **indirect proof** can be useful in proving statements of the form "p implies q." Recall that the conditional statement $p \to q$ is false only when p is true and q is false. To prove a conditional statement indirectly, assume that p is true and q is false. If this assumption leads to an impossibility, then you have proved that the conditional statement is true. An indirect proof is also called a **proof by contradiction.**

To use an indirect proof to prove the following conditional statement,

"If a is a positive integer and a^2 is divisible by 2, then a is divisible by 2,"

proceed as follows. First, assume that p, "a is a positive integer and a^2 is divisible by 2," is true and q, "a is divisible by 2," is false. This means that a is not divisible by 2. If so, then a is odd and can be written as $a = 2n + 1$, where n is an integer.

$$a = 2n + 1 \qquad \text{Definition of an odd integer}$$

$$a^2 = 4n^2 + 4n + 1 \qquad \text{Square each side.}$$

$$a^2 = 2(2n^2 + 2n) + 1 \qquad \text{Distributive Property}$$

So, by the definition of an odd integer, a^2 is odd. This contradicts the assumption, and you can conclude that a is divisible by 2.

EXAMPLE **Using an Indirect Proof**

Use an indirect proof to prove that $\sqrt{2}$ is an irrational number.

Solution Begin by assuming that $\sqrt{2}$ is *not* an irrational number. Then $\sqrt{2}$ can be written as the quotient of two integers a and $b (b \neq 0)$ that have no common factors.

$$\sqrt{2} = \frac{a}{b} \qquad \text{Assume that } \sqrt{2} \text{ is a rational number.}$$

$$2 = \frac{a^2}{b^2} \qquad \text{Square each side.}$$

$$2b^2 = a^2 \qquad \text{Multiply each side by } b^2.$$

This implies that 2 is a factor of a^2. So, 2 is also a factor of a, and a can be written as $2c$, where c is an integer.

$$2b^2 = (2c)^2 \qquad \text{Substitute } 2c \text{ for } a.$$

$$2b^2 = 4c^2 \qquad \text{Simplify.}$$

$$b^2 = 2c^2 \qquad \text{Divide each side by 2.}$$

This implies that 2 is a factor of b^2 and also a factor of b. So, 2 is a factor of both a and b. This contradicts the assumption that a and b have no common factors. So, you can conclude that $\sqrt{2}$ is an irrational number. ■

P.S. Problem Solving

1. **Geometry** A theorem from geometry states that if a triangle is inscribed in a circle such that one side of the triangle is a diameter of the circle, then the triangle is a right triangle. Show that this theorem is true for the circle

$x^2 + y^2 = 100$

and the triangle formed by the lines

$y = 0$, $y = \frac{1}{2}x + 5$, and $y = -2x + 20$.

2. **Finding Values of Constants** Find values of k_1 and k_2 such that the system of equations has an infinite number of solutions.

$$\begin{cases} 3x - 5y = 8 \\ 2x + k_1 y = k_2 \end{cases}$$

3. **Finding Conditions on Constants** Consider the following system of linear equations in x and y.

$$\begin{cases} ax + by = e \\ cx + dy = f \end{cases}$$

Under what condition(s) will the system have exactly one solution?

4. **Finding Values of Constants** Find values of a, b, and c (if possible) such that the system of linear equations has (a) a unique solution, (b) no solution, and (c) an infinite number of solutions.

$$\begin{cases} x + y = 2 \\ y + z = 2 \\ x + z = 2 \\ ax + by + cz = 0 \end{cases}$$

5. **Graphical Analysis** Graph the lines determined by each system of linear equations. Then use Gaussian elimination to solve each system. At each step of the elimination process, graph the corresponding lines. What do you observe?

(a) $\begin{cases} x - 4y = -3 \\ 5x - 6y = 13 \end{cases}$

(b) $\begin{cases} 2x - 3y = 7 \\ -4x + 6y = -14 \end{cases}$

6. **Maximum Numbers of Solutions** A system of two equations in two unknowns has a finite number of solutions. Determine the maximum number of solutions of the system satisfying each condition.

(a) Both equations are linear.

(b) One equation is linear and the other is quadratic.

(c) Both equations are quadratic.

7. **Vietnam Veterans Memorial** The Vietnam Veterans Memorial (or "The Wall") in Washington, D.C., was designed by Maya Ying Lin when she was a student at Yale University. This monument has two vertical, triangular sections of black granite with a common side (see figure). The bottom of each section is level with the ground. The tops of the two sections can be approximately modeled by the equations

$-2x + 50y = 505$

and

$2x + 50y = 505$

when the x-axis is superimposed at the base of the wall. Each unit in the coordinate system represents 1 foot. How high is the memorial at the point where the two sections meet? How long is each section?

8. **Finding Atomic Weights** Weights of atoms and molecules are measured in atomic mass units (u). A molecule of C_2H_6 (ethane) is made up of two carbon atoms and six hydrogen atoms and weighs 30.070 u. A molecule of C_3H_8 (propane) is made up of three carbon atoms and eight hydrogen atoms and weighs 44.097 u. Find the weights of a carbon atom and a hydrogen atom.

9. **DVD Connector Cables** Connecting a DVD player to a television set requires a cable with special connectors at both ends. You buy a six-foot cable for \$15.50 and a three-foot cable for \$10.25. Assuming that the cost of a cable is the sum of the cost of the two connectors and the cost of the cable itself, what is the cost of a four-foot cable? Explain your reasoning.

10. **Distance** A hotel 35 miles from an airport runs a shuttle service to and from the airport. The 9:00 A.M. bus leaves for the airport traveling at 30 miles per hour. The 9:15 A.M. bus leaves for the airport traveling at 40 miles per hour. Write a system of linear equations that represents distance as a function of time for the buses. Graph and solve the system. How far from the airport will the 9:15 A.M. bus catch up to the 9:00 A.M. bus?

11. Systems with Rational Expressions Solve each system of equations by letting $X = 1/x$, $Y = 1/y$, and $Z = 1/z$.

(a) $\begin{cases} \dfrac{12}{x} - \dfrac{12}{y} = 7 \\ \dfrac{3}{x} + \dfrac{4}{y} = 0 \end{cases}$

(b) $\begin{cases} \dfrac{2}{x} + \dfrac{1}{y} - \dfrac{3}{z} = 4 \\ \dfrac{4}{x} \qquad + \dfrac{2}{z} = 10 \\ -\dfrac{2}{x} + \dfrac{3}{y} - \dfrac{13}{z} = -8 \end{cases}$

12. Finding Values of Constants What values should be given to a, b, and c so that the following linear system has $(-1, 2, -3)$ as its only solution?

$$\begin{cases} x + 2y - 3z = a & \text{Equation 1} \\ -x - y + z = b & \text{Equation 2} \\ 2x + 3y - 2z = c & \text{Equation 3} \end{cases}$$

13. System of Linear Equations The following system has one solution: $x = 1$, $y = -1$, and $z = 2$.

$$\begin{cases} 4x - 2y + 5z = 16 & \text{Equation 1} \\ x + y \qquad = 0 & \text{Equation 2} \\ -x - 3y + 2z = 6 & \text{Equation 3} \end{cases}$$

Solve the system given by (a) Equation 1 and Equation 2, (b) Equation 1 and Equation 3, and (c) Equation 2 and Equation 3. (d) How many solutions does each of these systems have?

14. System of Linear Equations Solve the system of linear equations algebraically.

$$\begin{cases} x_1 - x_2 + 2x_3 + 2x_4 + 6x_5 = 6 \\ 3x_1 - 2x_2 + 4x_3 + 4x_4 + 12x_5 = 14 \\ \quad - x_2 - x_3 - x_4 - 3x_5 = -3 \\ 2x_1 - 2x_2 + 4x_3 + 5x_4 + 15x_5 = 10 \\ 2x_1 - 2x_2 + 4x_3 + 4x_4 + 13x_5 = 13 \end{cases}$$

15. Biology Each day, an average adult moose can process about 32 kilograms of terrestrial vegetation (twigs and leaves) and aquatic vegetation. From this food, it needs to obtain about 1.9 grams of sodium and 11,000 calories of energy. Aquatic vegetation has about 0.15 gram of sodium per kilogram and about 193 calories of energy per kilogram, whereas terrestrial vegetation has minimal sodium and about four times as much energy as aquatic vegetation. Write and graph a system of inequalities that describes the amounts t and a of terrestrial and aquatic vegetation, respectively, for the daily diet of an average adult moose. *(Source: Biology by Numbers)*

16. Height and Weight For a healthy person who is 4 feet 10 inches tall, the recommended minimum weight is about 91 pounds and increases by about 3.6 pounds for each additional inch of height. The recommended maximum weight is about 115 pounds and increases by about 4.5 pounds for each additional inch of height. *(Source: National Institutes of Health)*

(a) Let x be the number of inches by which a person's height exceeds 4 feet 10 inches and let y be the person's weight in pounds. Write a system of inequalities that describes the possible values of x and y for a healthy person.

(b) Use a graphing utility to graph the system of inequalities from part (a).

(c) What is the recommended weight range for a healthy person who is 6 feet tall?

17. Cholesterol Cholesterol in human blood is necessary, but too much can lead to health problems. There are three main types of cholesterol: HDL (high-density lipoproteins), LDL (low-density lipoproteins), and VLDL (very low-density lipoproteins). HDL is considered "good" cholesterol; LDL and VLDL are considered "bad" cholesterol.

A standard fasting cholesterol blood test measures total cholesterol, HDL cholesterol, and triglycerides. These numbers are used to estimate LDL and VLDL, which are difficult to measure directly. It is recommended that your combined LDL/VLDL cholesterol level be less than 130 milligrams per deciliter, your HDL cholesterol level be at least 40 milligrams per deciliter, and your total cholesterol level be no more than 200 milligrams per deciliter. *(Source: American Heart Association)*

(a) Write a system of linear inequalities for the recommended cholesterol levels. Let x represent the HDL cholesterol level, and let y represent the combined LDL/VLDL cholesterol level.

(b) Graph the system of inequalities from part (a). Label any vertices of the solution region.

(c) Are the following cholesterol levels within recommendations? Explain your reasoning.

LDL/VLDL: 120 milligrams per deciliter

HDL: 90 milligrams per deciliter

Total: 210 milligrams per deciliter

(d) Give an example of cholesterol levels in which the LDL/VLDL cholesterol level is too high but the HDL cholesterol level is acceptable.

(e) Another recommendation is that the ratio of total cholesterol to HDL cholesterol be less than 4 (that is, less than 4 to 1). Find a point in the solution region from part (b) that meets this recommendation, and explain why it meets the recommendation.

8 Matrices and Determinants

Sudoku *(page 578)*

Data Encryption *(page 588)*

Beam Deflection *(page 566)*

Flight Crew Scheduling *(page 559)*

Health and Wellness *(Exercise 101, page 550)*

Clockwise from top left, James E. Knopf/Shutterstock.com; Andrea Danti/Shutterstock.com; © Sean Locke/iStockphoto.com; Tyler Olson/Shutterstock.com; nostal6ie/Shutterstock.com

8.1 Matrices and Systems of Equations

You can use matrices to solve systems of linear equations in two or more variables. For instance, in Exercise 101 on page 550, you will use a matrix to find a model for the numbers of new cases of a waterborne disease in a small city.

- **Write matrices and identify their orders.**
- **Perform elementary row operations on matrices.**
- **Use matrices and Gaussian elimination to solve systems of linear equations.**
- **Use matrices and Gauss-Jordan elimination to solve systems of linear equations.**

Matrices

In this section, you will study a streamlined technique for solving systems of linear equations. This technique involves the use of a rectangular array of real numbers called a **matrix.** The plural of matrix is *matrices.*

Definition of Matrix

If m and n are positive integers, then an $m \times n$ (read "m by n") matrix is a rectangular array

$$\begin{matrix} & \text{Column 1} & \text{Column 2} & \text{Column 3} & \ldots & \text{Column } n \\ \text{Row 1} \\ \text{Row 2} \\ \text{Row 3} \\ \vdots \\ \text{Row } m \end{matrix} \begin{bmatrix} a_{11} & a_{12} & a_{13} & \cdots & a_{1n} \\ a_{21} & a_{22} & a_{23} & \cdots & a_{2n} \\ a_{31} & a_{32} & a_{33} & \cdots & a_{3n} \\ \vdots & \vdots & \vdots & & \vdots \\ a_{m1} & a_{m2} & a_{m3} & \cdots & a_{mn} \end{bmatrix}$$

in which each **entry** a_{ij} of the matrix is a number. An $m \times n$ matrix has m rows and n columns. Matrices are usually denoted by capital letters.

The entry in the ith row and jth column is denoted by the *double subscript* notation a_{ij}. For instance, a_{23} refers to the entry in the second row, third column. A matrix that has only one row is called a **row matrix,** and a matrix that has only one column is called a **column matrix.** A matrix having m rows and n columns is said to be of **order** $m \times n$. If $m = n$, then the matrix is **square** of order $m \times m$ (or $n \times n$). For a square matrix, the entries $a_{11}, a_{22}, a_{33}, \ldots$ are the **main diagonal** entries.

EXAMPLE 1 Orders of Matrices

Determine the order of each matrix.

a. $[2]$ **b.** $\begin{bmatrix} 1 & -3 & 0 & \frac{1}{2} \end{bmatrix}$ **c.** $\begin{bmatrix} 0 & 0 \\ 0 & 0 \end{bmatrix}$ **d.** $\begin{bmatrix} 5 & 0 \\ 2 & -2 \\ -7 & 4 \end{bmatrix}$

Solution

a. This matrix has *one* row and *one* column. The order of the matrix is 1×1.

b. This matrix has *one* row and *four* columns. The order of the matrix is 1×4.

c. This matrix has *two* rows and *two* columns. The order of the matrix is 2×2.

d. This matrix has *three* rows and *two* columns. The order of the matrix is 3×2.

✓ ***Checkpoint*** *Audio-video solution in English & Spanish at LarsonPrecalculus.com.*

Determine the order of the matrix $\begin{bmatrix} 14 & 7 & 10 \\ -2 & -3 & -8 \end{bmatrix}$.

Tyler Olson/Shutterstock.com

A matrix derived from a system of linear equations (each written in standard form with the constant term on the right) is the **augmented matrix** of the system. Moreover, the matrix derived from the coefficients of the system (but not including the constant terms) is the **coefficient matrix** of the system.

System: $\begin{cases} x - 4y + 3z = 5 \\ -x + 3y - z = -3 \\ 2x - 4z = 6 \end{cases}$

Augmented Matrix: $\begin{bmatrix} 1 & -4 & 3 & \vdots & 5 \\ -1 & 3 & -1 & \vdots & -3 \\ 2 & 0 & -4 & \vdots & 6 \end{bmatrix}$ *Coefficient Matrix:* $\begin{bmatrix} 1 & -4 & 3 \\ -1 & 3 & -1 \\ 2 & 0 & -4 \end{bmatrix}$

REMARK The vertical dots in an augmented matrix separate the coefficients of the linear system from the constant terms.

Note the use of 0 for the coefficient of the missing y-variable in the third equation, and also note the fourth column of constant terms in the augmented matrix.

When forming either the coefficient matrix or the augmented matrix of a system, you should begin by vertically aligning the variables in the equations and using zeros for the coefficients of the missing variables.

EXAMPLE 2 Writing an Augmented Matrix

Write the augmented matrix for the system of linear equations.

$$\begin{cases} x + 3y - w = 9 \\ -y + 4z + 2w = -2 \\ x - 5z - 6w = 0 \\ 2x + 4y - 3z = 4 \end{cases}$$

What is the order of the augmented matrix?

Solution

Begin by rewriting the linear system and aligning the variables.

$$\begin{cases} x + 3y \qquad\;\; - w = 9 \\ \quad -y + 4z + 2w = -2 \\ x \qquad - 5z - 6w = 0 \\ 2x + 4y - 3z \qquad = 4 \end{cases}$$

Next, use the coefficients and constant terms as the matrix entries. Include zeros for the coefficients of the missing variables.

$$\begin{matrix} R_1 \\ R_2 \\ R_3 \\ R_4 \end{matrix} \begin{bmatrix} 1 & 3 & 0 & -1 & \vdots & 9 \\ 0 & -1 & 4 & 2 & \vdots & -2 \\ 1 & 0 & -5 & -6 & \vdots & 0 \\ 2 & 4 & -3 & 0 & \vdots & 4 \end{bmatrix}$$

The augmented matrix has four rows and five columns, so it is a 4×5 matrix. The notation R_n is used to designate each row in the matrix. For example, Row 1 is represented by R_1.

Checkpoint Audio-video solution in English & Spanish at LarsonPrecalculus.com.

Write the augmented matrix for the system of linear equations.

$$\begin{cases} x + y + z = 2 \\ 2x - y + 3z = -1 \\ -x + 2y - z = 4 \end{cases}$$

Elementary Row Operations

In Section 7.3, you studied three operations that can be used on a system of linear equations to produce an equivalent system.

1. Interchange two equations.
2. Multiply an equation by a nonzero constant.
3. Add a multiple of an equation to another equation.

In matrix terminology, these three operations correspond to **elementary row operations.** An elementary row operation on an augmented matrix of a given system of linear equations produces a new augmented matrix corresponding to a new (but equivalent) system of linear equations. Two matrices are **row-equivalent** when one can be obtained from the other by a sequence of elementary row operations.

REMARK Although elementary row operations are simple to perform, they involve a lot of arithmetic. Because it is easy to make a mistake, you should get in the habit of noting the elementary row operations performed in each step so that you can go back and check your work.

Elementary Row Operations

Operation	Notation
1. Interchange two rows.	$R_a \leftrightarrow R_b$
2. Multiply a row by a nonzero constant.	$cR_a \quad (c \neq 0)$
3. Add a multiple of a row to another row.	$cR_a + R_b$

EXAMPLE 3 Elementary Row Operations

a. Interchange the first and second rows of the original matrix.

Original Matrix

$$\begin{bmatrix} 0 & 1 & 3 & 4 \\ -1 & 2 & 0 & 3 \\ 2 & -3 & 4 & 1 \end{bmatrix}$$

New Row-Equivalent Matrix

$$\begin{matrix} R_2 \\ R_1 \\ \end{matrix} \begin{bmatrix} -1 & 2 & 0 & 3 \\ 0 & 1 & 3 & 4 \\ 2 & -3 & 4 & 1 \end{bmatrix}$$

b. Multiply the first row of the original matrix by $\frac{1}{2}$.

Original Matrix

$$\begin{bmatrix} 2 & -4 & 6 & -2 \\ 1 & 3 & -3 & 0 \\ 5 & -2 & 1 & 2 \end{bmatrix}$$

New Row-Equivalent Matrix

$$\begin{matrix} \frac{1}{2}R_1 \rightarrow \\ \\ \end{matrix} \begin{bmatrix} 1 & -2 & 3 & -1 \\ 1 & 3 & -3 & 0 \\ 5 & -2 & 1 & 2 \end{bmatrix}$$

c. Add -2 times the first row of the original matrix to the third row.

Original Matrix

$$\begin{bmatrix} 1 & 2 & -4 & 3 \\ 0 & 3 & -2 & -1 \\ 2 & 1 & 5 & -2 \end{bmatrix}$$

New Row-Equivalent Matrix

$$\begin{matrix} \\ \\ -2R_1 + R_3 \rightarrow \end{matrix} \begin{bmatrix} 1 & 2 & -4 & 3 \\ 0 & 3 & -2 & -1 \\ 0 & -3 & 13 & -8 \end{bmatrix}$$

Note that the elementary row operation is written beside the row that is *changed*.

TECHNOLOGY Most graphing utilities can perform elementary row operations on matrices. Consult the user's guide for your graphing utility for specific keystrokes.

After performing a row operation, the new row-equivalent matrix that is displayed on your graphing utility is stored in the *answer* variable. You should use the *answer* variable and not the original matrix for subsequent row operations.

✓ **Checkpoint** Audio-video solution in English & Spanish at LarsonPrecalculus.com.

Identify the elementary row operation being performed to obtain the new row-equivalent matrix.

Original Matrix

$$\begin{bmatrix} 1 & 0 & 2 \\ 3 & 1 & 7 \\ 2 & -6 & 14 \end{bmatrix}$$

New Row-Equivalent Matrix

$$\begin{bmatrix} 1 & 0 & 2 \\ 0 & 1 & 1 \\ 2 & -6 & 14 \end{bmatrix}$$

Gaussian Elimination with Back-Substitution

In Example 3 in Section 7.3, you used Gaussian elimination with back-substitution to solve a system of linear equations. The next example demonstrates the matrix version of Gaussian elimination. The two methods are essentially the same. The basic difference is that with matrices you do not need to keep writing the variables.

EXAMPLE 4 Comparing Linear Systems and Matrix Operations

Linear System

$$\begin{cases} x - 2y + 3z = 9 \\ -x + 3y = -4 \\ 2x - 5y + 5z = 17 \end{cases}$$

Associated Augmented Matrix

$$\begin{bmatrix} 1 & -2 & 3 & \vdots & 9 \\ -1 & 3 & 0 & \vdots & -4 \\ 2 & -5 & 5 & \vdots & 17 \end{bmatrix}$$

Add the first equation to the second equation.

$$\begin{cases} x - 2y + 3z = 9 \\ y + 3z = 5 \\ 2x - 5y + 5z = 17 \end{cases}$$

Add the first row to the second row ($R_1 + R_2$).

$$\begin{matrix} \\ R_1 + R_2 \rightarrow \\ \\ \end{matrix}\begin{bmatrix} 1 & -2 & 3 & \vdots & 9 \\ 0 & 1 & 3 & \vdots & 5 \\ 2 & -5 & 5 & \vdots & 17 \end{bmatrix}$$

Add -2 times the first equation to the third equation.

$$\begin{cases} x - 2y + 3z = 9 \\ y + 3z = 5 \\ -y - z = -1 \end{cases}$$

Add -2 times the first row to the third row ($-2R_1 + R_3$).

$$\begin{matrix} \\ \\ -2R_1 + R_3 \rightarrow \end{matrix}\begin{bmatrix} 1 & -2 & 3 & \vdots & 9 \\ 0 & 1 & 3 & \vdots & 5 \\ 0 & -1 & -1 & \vdots & -1 \end{bmatrix}$$

Add the second equation to the third equation.

$$\begin{cases} x - 2y + 3z = 9 \\ y + 3z = 5 \\ 2z = 4 \end{cases}$$

Add the second row to the third row ($R_2 + R_3$).

$$\begin{matrix} \\ \\ R_2 + R_3 \rightarrow \end{matrix}\begin{bmatrix} 1 & -2 & 3 & \vdots & 9 \\ 0 & 1 & 3 & \vdots & 5 \\ 0 & 0 & 2 & \vdots & 4 \end{bmatrix}$$

Multiply the third equation by $\frac{1}{2}$.

$$\begin{cases} x - 2y + 3z = 9 \\ y + 3z = 5 \\ z = 2 \end{cases}$$

Multiply the third row by $\frac{1}{2}$ $\left(\frac{1}{2}R_3\right)$.

$$\begin{matrix} \\ \\ \frac{1}{2}R_3 \rightarrow \end{matrix}\begin{bmatrix} 1 & -2 & 3 & \vdots & 9 \\ 0 & 1 & 3 & \vdots & 5 \\ 0 & 0 & 1 & \vdots & 2 \end{bmatrix}$$

At this point, you can use back-substitution to find x and y.

$y + 3(2) = 5$ Substitute 2 for z.

$y = -1$ Solve for y.

$x - 2(-1) + 3(2) = 9$ Substitute -1 for y and 2 for z.

$x = 1$ Solve for x.

The solution is $x = 1$, $y = -1$, and $z = 2$.

> **REMARK** Remember that you should check a solution by substituting the values of x, y, and z into each equation of the original system. For example, you can check the solution to Example 4 as follows.
>
> Equation 1:
> $1 - 2(-1) + 3(2) = 9$ ✓
>
> Equation 2:
> $-1 + 3(-1) = -4$ ✓
>
> Equation 3:
> $2(1) - 5(-1) + 5(2) = 17$ ✓

✓ **Checkpoint** Audio-video solution in English & Spanish at LarsonPrecalculus.com.

Compare solving the linear system below to solving it using its associated augmented matrix.

$$\begin{cases} 2x + y - z = -3 \\ 4x - 2y + 2z = -2 \\ -6x + 5y + 4z = 10 \end{cases}$$

The last matrix in Example 4 is said to be in **row-echelon form.** The term *echelon* refers to the stair-step pattern formed by the nonzero entries of the matrix. To be in this form, a matrix must have the following properties.

Row-Echelon Form and Reduced Row-Echelon Form

A matrix in **row-echelon form** has the following properties.

1. Any rows consisting entirely of zeros occur at the bottom of the matrix.
2. For each row that does not consist entirely of zeros, the first nonzero entry is 1 (called a **leading 1**).
3. For two successive (nonzero) rows, the leading 1 in the higher row is farther to the left than the leading 1 in the lower row.

A matrix in *row-echelon form* is in **reduced row-echelon form** when every column that has a leading 1 has zeros in every position above and below its leading 1.

It is worth noting that the row-echelon form of a matrix is not unique. That is, two different sequences of elementary row operations may yield different row-echelon forms. However, the *reduced* row-echelon form of a given matrix is unique.

EXAMPLE 5 Row-Echelon Form

Determine whether each matrix is in row-echelon form. If it is, determine whether the matrix is in reduced row-echelon form.

a. $\begin{bmatrix} 1 & 2 & -1 & 4 \\ 0 & 1 & 0 & 3 \\ 0 & 0 & 1 & -2 \end{bmatrix}$ **b.** $\begin{bmatrix} 1 & 2 & -1 & 2 \\ 0 & 0 & 0 & 0 \\ 0 & 1 & 2 & -4 \end{bmatrix}$

c. $\begin{bmatrix} 1 & -5 & 2 & -1 & 3 \\ 0 & 0 & 1 & 3 & -2 \\ 0 & 0 & 0 & 1 & 4 \\ 0 & 0 & 0 & 0 & 1 \end{bmatrix}$ **d.** $\begin{bmatrix} 1 & 0 & 0 & -1 \\ 0 & 1 & 0 & 2 \\ 0 & 0 & 1 & 3 \\ 0 & 0 & 0 & 0 \end{bmatrix}$

e. $\begin{bmatrix} 1 & 2 & -3 & 4 \\ 0 & 2 & 1 & -1 \\ 0 & 0 & 1 & -3 \end{bmatrix}$ **f.** $\begin{bmatrix} 0 & 1 & 0 & 5 \\ 0 & 0 & 1 & 3 \\ 0 & 0 & 0 & 0 \end{bmatrix}$

Solution The matrices in (a), (c), (d), and (f) are in row-echelon form. The matrices in (d) and (f) are in *reduced* row-echelon form because every column that has a leading 1 has zeros in every position above and below its leading 1. The matrix in (b) is not in row-echelon form because a row of all zeros does not occur at the bottom of the matrix. The matrix in (e) is not in row-echelon form because the first nonzero entry in Row 2 is not a leading 1.

✓ ***Checkpoint*** *Audio-video solution in English & Spanish at LarsonPrecalculus.com.*

Determine whether the matrix is in row-echelon form. If it is, determine whether it is in reduced row-echelon form.

$$\begin{bmatrix} 1 & 0 & -2 & 4 \\ 0 & 1 & 11 & 3 \\ 0 & 0 & 0 & 0 \end{bmatrix}$$

Every matrix is row-equivalent to a matrix in row-echelon form. For instance, in Example 5, you can change the matrix in part (e) to row-echelon form by multiplying its second row by $\frac{1}{2}$.

Gaussian elimination with back-substitution works well for solving systems of linear equations by hand or with a computer. For this algorithm, the order in which the elementary row operations are performed is important. You should operate from left to right by columns, using elementary row operations to obtain zeros in all entries directly below the leading 1's.

EXAMPLE 6 **Gaussian Elimination with Back-Substitution**

Solve the system $\begin{cases} y + z - 2w = -3 \\ x + 2y - z = 2 \\ 2x + 4y + z - 3w = -2 \\ x - 4y - 7z - w = -19 \end{cases}$.

Solution

$$\begin{bmatrix} 0 & 1 & 1 & -2 & \vdots & -3 \\ 1 & 2 & -1 & 0 & \vdots & 2 \\ 2 & 4 & 1 & -3 & \vdots & -2 \\ 1 & -4 & -7 & -1 & \vdots & -19 \end{bmatrix}$$ Write augmented matrix.

$$\begin{matrix} R_2 \\ R_1 \\ \\ \\ \end{matrix} \begin{bmatrix} 1 & 2 & -1 & 0 & \vdots & 2 \\ 0 & 1 & 1 & -2 & \vdots & -3 \\ 2 & 4 & 1 & -3 & \vdots & -2 \\ 1 & -4 & -7 & -1 & \vdots & -19 \end{bmatrix}$$ Interchange R_1 and R_2 so first column has leading 1 in upper left corner.

$$\begin{matrix} \\ \\ -2R_1 + R_3 \rightarrow \\ -R_1 + R_4 \rightarrow \end{matrix} \begin{bmatrix} 1 & 2 & -1 & 0 & \vdots & 2 \\ 0 & 1 & 1 & -2 & \vdots & -3 \\ 0 & 0 & 3 & -3 & \vdots & -6 \\ 0 & -6 & -6 & -1 & \vdots & -21 \end{bmatrix}$$ Perform operations on R_3 and R_4 so first column has zeros below its leading 1.

$$\begin{matrix} \\ \\ \\ 6R_2 + R_4 \rightarrow \end{matrix} \begin{bmatrix} 1 & 2 & -1 & 0 & \vdots & 2 \\ 0 & 1 & 1 & -2 & \vdots & -3 \\ 0 & 0 & 3 & -3 & \vdots & -6 \\ 0 & 0 & 0 & -13 & \vdots & -39 \end{bmatrix}$$ Perform operations on R_4 so second column has zeros below its leading 1.

$$\begin{matrix} \\ \\ \frac{1}{3}R_3 \rightarrow \\ -\frac{1}{13}R_4 \rightarrow \end{matrix} \begin{bmatrix} 1 & 2 & -1 & 0 & \vdots & 2 \\ 0 & 1 & 1 & -2 & \vdots & -3 \\ 0 & 0 & 1 & -1 & \vdots & -2 \\ 0 & 0 & 0 & 1 & \vdots & 3 \end{bmatrix}$$ Perform operations on R_3 and R_4 so third and fourth columns have leading 1's.

The matrix is now in row-echelon form, and the corresponding system is

$$\begin{cases} x + 2y - z = 2 \\ y + z - 2w = -3 \\ z - w = -2 \\ w = 3 \end{cases}.$$

Using back-substitution, you can determine that the solution is

$x = -1$, $y = 2$, $z = 1$, and $w = 3$.

Checkpoint Audio-video solution in English & Spanish at LarsonPrecalculus.com.

Solve the system $\begin{cases} -3x + 5y + 3z = -19 \\ 3x + 4y + 4z = 8 \\ 4x - 8y - 6z = 26 \end{cases}$.

The procedure for using Gaussian elimination with back-substitution is summarized below.

Gaussian Elimination with Back-Substitution

1. Write the augmented matrix of the system of linear equations.
2. Use elementary row operations to rewrite the augmented matrix in row-echelon form.
3. Write the system of linear equations corresponding to the matrix in row-echelon form and use back-substitution to find the solution.

When solving a system of linear equations, remember that it is possible for the system to have no solution. If, in the elimination process, you obtain a row of all zeros except for the last entry, then the system has no solution, or is *inconsistent*.

EXAMPLE 7 **A System with No Solution**

Solve the system $\begin{cases} x - y + 2z = 4 \\ x + z = 6 \\ 2x - 3y + 5z = 4 \\ 3x + 2y - z = 1 \end{cases}$.

Solution

$$\begin{bmatrix} 1 & -1 & 2 & \vdots & 4 \\ 1 & 0 & 1 & \vdots & 6 \\ 2 & -3 & 5 & \vdots & 4 \\ 3 & 2 & -1 & \vdots & 1 \end{bmatrix}$$ Write augmented matrix.

$$\begin{array}{r} \\ -R_1 + R_2 \rightarrow \\ -2R_1 + R_3 \rightarrow \\ -3R_1 + R_4 \rightarrow \end{array} \begin{bmatrix} 1 & -1 & 2 & \vdots & 4 \\ 0 & 1 & -1 & \vdots & 2 \\ 0 & -1 & 1 & \vdots & -4 \\ 0 & 5 & -7 & \vdots & -11 \end{bmatrix}$$ Perform row operations.

$$\begin{array}{r} \\ \\ R_2 + R_3 \rightarrow \\ \\ \end{array} \begin{bmatrix} 1 & -1 & 2 & \vdots & 4 \\ 0 & 1 & -1 & \vdots & 2 \\ 0 & 0 & 0 & \vdots & -2 \\ 0 & 5 & -7 & \vdots & -11 \end{bmatrix}$$ Perform row operations.

Note that the third row of this matrix consists entirely of zeros except for the last entry. This means that the original system of linear equations is inconsistent. You can see why this is true by converting back to a system of linear equations.

$$\begin{cases} x - y + 2z = 4 \\ y - z = 2 \\ 0 = -2 \\ 5y - 7z = -11 \end{cases}$$

Because the third equation is not possible, the system has no solution.

✓ **Checkpoint** *Audio-video solution in English & Spanish at LarsonPrecalculus.com.*

Solve the system $\begin{cases} x + y + z = 1 \\ x + 2y + 2z = 2. \\ x - y - z = 1 \end{cases}$

Gauss-Jordan Elimination

With Gaussian elimination, elementary row operations are applied to a matrix to obtain a (row-equivalent) row-echelon form of the matrix. A second method of elimination, called **Gauss-Jordan elimination,** after Carl Friedrich Gauss and Wilhelm Jordan (1842–1899), continues the reduction process until a *reduced* row-echelon form is obtained. This procedure is demonstrated in Example 8.

> **TECHNOLOGY** For a demonstration of a graphical approach to Gauss-Jordan elimination on a 2×3 matrix, see the Visualizing Row Operations Program available for several models of graphing calculators at the website for this text at *LarsonPrecalculus.com.*

EXAMPLE 8 Gauss-Jordan Elimination

Use Gauss-Jordan elimination to solve the system $\begin{cases} x - 2y + 3z = 9 \\ -x + 3y = -4. \\ 2x - 5y + 5z = 17 \end{cases}$

Solution In Example 4, Gaussian elimination was used to obtain the row-echelon form of the linear system above.

$$\begin{bmatrix} 1 & -2 & 3 & \vdots & 9 \\ 0 & 1 & 3 & \vdots & 5 \\ 0 & 0 & 1 & \vdots & 2 \end{bmatrix}$$

Now, apply elementary row operations until you obtain zeros above each of the leading 1's, as follows.

$$\begin{matrix} 2R_2 + R_1 \rightarrow \\ \\ \\ \end{matrix} \begin{bmatrix} 1 & 0 & 9 & \vdots & 19 \\ 0 & 1 & 3 & \vdots & 5 \\ 0 & 0 & 1 & \vdots & 2 \end{bmatrix}$$

Perform operations on R_1 so second column has a zero above its leading 1.

$$\begin{matrix} -9R_3 + R_1 \rightarrow \\ -3R_3 + R_2 \rightarrow \\ \\ \end{matrix} \begin{bmatrix} 1 & 0 & 0 & \vdots & 1 \\ 0 & 1 & 0 & \vdots & -1 \\ 0 & 0 & 1 & \vdots & 2 \end{bmatrix}$$

Perform operations on R_1 and R_2 so third column has zeros above its leading 1.

The matrix is now in reduced row-echelon form. Converting back to a system of linear equations, you have

$$\begin{cases} x = 1 \\ y = -1. \\ z = 2 \end{cases}$$

> **REMARK** The advantage of using Gauss-Jordan elimination to solve a system of linear equations is that the solution of the system is easily found without using back-substitution, as illustrated in Example 8.

Now you can simply read the solution, $x = 1$, $y = -1$, and $z = 2$, which can be written as the ordered triple $(1, -1, 2)$.

Checkpoint Audio-video solution in English & Spanish at LarsonPrecalculus.com.

Use Gauss-Jordan elimination to solve the system $\begin{cases} -3x + 7y + 2z = 1 \\ -5x + 3y - 5z = -8. \\ 2x - 2y - 3z = 15 \end{cases}$

The elimination procedures described in this section sometimes result in fractional coefficients. For instance, in the elimination procedure for the system

$$\begin{cases} 2x - 5y + 5z = 17 \\ 3x - 2y + 3z = 11 \\ -3x + 3y = -6 \end{cases}$$

you may be inclined to multiply the first row by $\frac{1}{2}$ to produce a leading 1, which will result in working with fractional coefficients. You can sometimes avoid fractions by judiciously choosing the order in which you apply elementary row operations.

EXAMPLE 9 A System with an Infinite Number of Solutions

Solve the system $\begin{cases} 2x + 4y - 2z = 0 \\ 3x + 5y \quad\quad = 1 \end{cases}$.

Solution

$$\begin{bmatrix} 2 & 4 & -2 & \vdots & 0 \\ 3 & 5 & 0 & \vdots & 1 \end{bmatrix}$$

$$\tfrac{1}{2}R_1 \rightarrow \begin{bmatrix} 1 & 2 & -1 & \vdots & 0 \\ 3 & 5 & 0 & \vdots & 1 \end{bmatrix}$$

$$-3R_1 + R_2 \rightarrow \begin{bmatrix} 1 & 2 & -1 & \vdots & 0 \\ 0 & -1 & 3 & \vdots & 1 \end{bmatrix}$$

$$-R_2 \rightarrow \begin{bmatrix} 1 & 2 & -1 & \vdots & 0 \\ 0 & 1 & -3 & \vdots & -1 \end{bmatrix}$$

$$-2R_2 + R_1 \rightarrow \begin{bmatrix} 1 & 0 & 5 & \vdots & 2 \\ 0 & 1 & -3 & \vdots & -1 \end{bmatrix}$$

The corresponding system of equations is

$$\begin{cases} x + 5z = 2 \\ y - 3z = -1 \end{cases}.$$

Solving for x and y in terms of z, you have

$$x = -5z + 2 \quad \text{and} \quad y = 3z - 1.$$

To write a solution of the system that does not use any of the three variables of the system, let a represent any real number and let $z = a$. Substituting a for z in the equations for x and y, you have

$$x = -5z + 2 = -5a + 2 \quad \text{and} \quad y = 3z - 1 = 3a - 1.$$

So, the solution set can be written as an ordered triple of the form

$$(-5a + 2, 3a - 1, a)$$

where a is any real number. Remember that a solution set of this form represents an infinite number of solutions. Try substituting values for a to obtain a few solutions. Then check each solution in the original system of equations.

✓ Checkpoint *Audio-video solution in English & Spanish at LarsonPrecalculus.com.*

Solve the system $\begin{cases} 2x - 6y + 6z = 46 \\ 2x - 3y \quad\quad = 31 \end{cases}$.

Summarize (Section 8.1)

1. State the definition of a matrix *(page 538)*. For examples of writing matrices and determining their orders, see Examples 1 and 2.
2. List the elementary row operations *(page 540)*. For examples of performing elementary row operations, see Examples 3 and 4.
3. State the definitions of row-echelon form and reduced row-echelon form *(page 542)*. For an example of matrices in these forms, see Example 5.
4. Describe Gaussian elimination with back-substitution *(pages 543 and 544)*. For examples of using this procedure, see Examples 6 and 7.
5. Describe Gauss-Jordan elimination *(page 545)*. For examples of using this procedure, see Examples 8 and 9.

8.1 Exercises

See CalcChat.com for tutorial help and worked-out solutions to odd-numbered exercises.

Vocabulary: Fill in the blanks.

1. A rectangular array of real numbers that can be used to solve a system of linear equations is called a ________.
2. A matrix is ________ when the number of rows equals the number of columns.
3. For a square matrix, the entries $a_{11}, a_{22}, a_{33}, \ldots$ are the ________ ________ entries.
4. A matrix with only one row is called a ________ matrix, and a matrix with only one column is called a ________ matrix.
5. The matrix derived from a system of linear equations is called the ________ matrix of the system.
6. The matrix derived from the coefficients of a system of linear equations is called the ________ matrix of the system.
7. Two matrices are called ________ when one of the matrices can be obtained from the other by a sequence of elementary row operations.
8. A matrix in row-echelon form is in ________ ________ ________ when every column that has a leading 1 has zeros in every position above and below its leading 1.

Skills and Applications

Order of a Matrix In Exercises 9–16, determine the order of the matrix.

9. $\begin{bmatrix} 7 & 0 \end{bmatrix}$

10. $\begin{bmatrix} 5 & -3 & 8 & 7 \end{bmatrix}$

11. $\begin{bmatrix} 2 \\ 36 \\ 3 \end{bmatrix}$

12. $\begin{bmatrix} -3 & 7 & 15 & 0 \\ 0 & 0 & 3 & 3 \\ 1 & 1 & 6 & 7 \end{bmatrix}$

13. $\begin{bmatrix} 33 & 45 \\ -9 & 20 \end{bmatrix}$

14. $\begin{bmatrix} -7 & 6 & 4 \\ 0 & -5 & 1 \end{bmatrix}$

15. $\begin{bmatrix} 1 & 6 & -1 \\ 8 & 0 & 3 \\ 3 & -9 & 9 \end{bmatrix}$

16. $\begin{bmatrix} 3 & -1 \\ 4 & 1 \\ -5 & 9 \end{bmatrix}$

Writing an Augmented Matrix In Exercises 17–22, write the augmented matrix for the system of linear equations.

17. $\begin{cases} 4x - 3y = -5 \\ -x + 3y = 12 \end{cases}$

18. $\begin{cases} 7x + 4y = 22 \\ 5x - 9y = 15 \end{cases}$

19. $\begin{cases} x + 10y - 2z = 2 \\ 5x - 3y + 4z = 0 \\ 2x + y = 6 \end{cases}$

20. $\begin{cases} -x - 8y + 5z = 8 \\ -7x - 15z = -38 \\ 3x - y + 8z = 20 \end{cases}$

21. $\begin{cases} 7x - 5y + z = 13 \\ 19x - 8z = 10 \end{cases}$

22. $\begin{cases} 9x + 2y - 3z = 20 \\ -25y + 11z = -5 \end{cases}$

Writing a System of Equations In Exercises 23–28, write the system of linear equations represented by the augmented matrix. (Use variables x, y, z, and w, if applicable.)

23. $\left[\begin{array}{cc:c} 1 & 2 & 7 \\ 2 & -3 & 4 \end{array}\right]$

24. $\left[\begin{array}{cc:c} 7 & -5 & 0 \\ 8 & 3 & -2 \end{array}\right]$

25. $\left[\begin{array}{ccc:c} 2 & 0 & 5 & -12 \\ 0 & 1 & -2 & 7 \\ 6 & 3 & 0 & 2 \end{array}\right]$

26. $\left[\begin{array}{ccc:c} 4 & -5 & -1 & 18 \\ -11 & 0 & 6 & 25 \\ 3 & 8 & 0 & -29 \end{array}\right]$

27. $\left[\begin{array}{cccc:c} 9 & 12 & 3 & 0 & 0 \\ -2 & 18 & 5 & 2 & 10 \\ 1 & 7 & -8 & 0 & -4 \\ 3 & 0 & 2 & 0 & -10 \end{array}\right]$

28. $\left[\begin{array}{cccc:c} 6 & 2 & -1 & -5 & -25 \\ -1 & 0 & 7 & 3 & 7 \\ 4 & -1 & -10 & 6 & 23 \\ 0 & 8 & 1 & -11 & -21 \end{array}\right]$

Identifying an Elementary Row Operation In Exercises 29–32, identify the elementary row operation(s) being performed to obtain the new row-equivalent matrix.

	Original Matrix	New Row-Equivalent Matrix
29.	$\begin{bmatrix} -2 & 5 & 1 \\ 3 & -1 & -8 \end{bmatrix}$	$\begin{bmatrix} 13 & 0 & -39 \\ 3 & -1 & -8 \end{bmatrix}$
30.	$\begin{bmatrix} 3 & -1 & -4 \\ -4 & 3 & 7 \end{bmatrix}$	$\begin{bmatrix} 3 & -1 & -4 \\ 5 & 0 & -5 \end{bmatrix}$
31.	$\begin{bmatrix} 0 & -1 & -5 & 5 \\ -1 & 3 & -7 & 6 \\ 4 & -5 & 1 & 3 \end{bmatrix}$	$\begin{bmatrix} -1 & 3 & -7 & 6 \\ 0 & -1 & -5 & 5 \\ 0 & 7 & -27 & 27 \end{bmatrix}$
32.	$\begin{bmatrix} -1 & -2 & 3 & -2 \\ 2 & -5 & 1 & -7 \\ 5 & 4 & -7 & 6 \end{bmatrix}$	$\begin{bmatrix} -1 & -2 & 3 & -2 \\ 0 & -9 & 7 & -11 \\ 0 & -6 & 8 & -4 \end{bmatrix}$

Elementary Row Operations **In Exercises 33–40, fill in the blank(s) using elementary row operations to form a row-equivalent matrix.**

33. $\begin{bmatrix} 1 & 4 & 3 \\ 2 & 10 & 5 \end{bmatrix}$

$\begin{bmatrix} 1 & 4 & 3 \\ 0 & \square & -1 \end{bmatrix}$

34. $\begin{bmatrix} 3 & 6 & 8 \\ 4 & -3 & 6 \end{bmatrix}$

$\begin{bmatrix} 1 & \square & \frac{8}{3} \\ 4 & -3 & 6 \end{bmatrix}$

35. $\begin{bmatrix} 1 & 1 & 1 \\ 5 & -2 & 4 \end{bmatrix}$

$\begin{bmatrix} 1 & 1 & 1 \\ 0 & \square & -1 \end{bmatrix}$

36. $\begin{bmatrix} -3 & 3 & 12 \\ 18 & -8 & 4 \end{bmatrix}$

$\begin{bmatrix} 1 & -1 & \square \\ 18 & -8 & 4 \end{bmatrix}$

37. $\begin{bmatrix} 1 & 5 & 4 & -1 \\ 0 & 1 & -2 & 2 \\ 0 & 0 & 1 & -7 \end{bmatrix}$

$\begin{bmatrix} 1 & 0 & \square & \square \\ 0 & 1 & -2 & 2 \\ 0 & 0 & 1 & -7 \end{bmatrix}$

38. $\begin{bmatrix} 1 & 0 & 6 & 1 \\ 0 & -1 & 0 & 7 \\ 0 & 0 & -1 & 3 \end{bmatrix}$

$\begin{bmatrix} 1 & 0 & 6 & 1 \\ 0 & 1 & 0 & \square \\ 0 & 0 & 1 & \square \end{bmatrix}$

39. $\begin{bmatrix} 1 & 1 & 4 & -1 \\ 3 & 8 & 10 & 3 \\ -2 & 1 & 12 & 6 \end{bmatrix}$

$\begin{bmatrix} 1 & 1 & 4 & -1 \\ 0 & 5 & \square & \square \\ 0 & 3 & \square & \square \end{bmatrix}$

$\begin{bmatrix} 1 & 1 & 4 & -1 \\ 0 & 1 & -\frac{2}{5} & \frac{6}{5} \\ 0 & 3 & \square & \square \end{bmatrix}$

40. $\begin{bmatrix} 2 & 4 & 8 & 3 \\ 1 & -1 & -3 & 2 \\ 2 & 6 & 4 & 9 \end{bmatrix}$

$\begin{bmatrix} 1 & \square & \square & \square \\ 1 & -1 & -3 & 2 \\ 2 & 6 & 4 & 9 \end{bmatrix}$

$\begin{bmatrix} 1 & 2 & 4 & \frac{3}{2} \\ 0 & \square & -7 & \frac{1}{2} \\ 0 & 2 & \square & \square \end{bmatrix}$

Comparing Linear Systems and Matrix Operations **In Exercises 41 and 42, (a) perform the row operations to solve the augmented matrix, (b) write and solve the system of linear equations represented by the augmented matrix, and (c) compare the two solution methods. Which do you prefer?**

41. $\begin{bmatrix} -3 & 4 & \vdots & 22 \\ 6 & -4 & \vdots & -28 \end{bmatrix}$

(i) Add R_2 to R_1.

(ii) Add -2 times R_1 to R_2.

(iii) Multiply R_2 by $-\frac{1}{4}$.

(iv) Multiply R_1 by $\frac{1}{3}$.

42. $\begin{bmatrix} 7 & 13 & 1 & \vdots & -4 \\ -3 & -5 & -1 & \vdots & -4 \\ 3 & 6 & 1 & \vdots & -2 \end{bmatrix}$

(i) Add R_2 to R_1.

(ii) Multiply R_1 by $\frac{1}{4}$.

(iii) Add R_3 to R_2.

(iv) Add -3 times R_1 to R_3.

(v) Add -2 times R_2 to R_1.

Row-Echelon Form **In Exercises 43–46, determine whether the matrix is in row-echelon form. If it is, determine whether it is in reduced row-echelon form.**

43. $\begin{bmatrix} 1 & 0 & 0 & 0 \\ 0 & 1 & 1 & 5 \\ 0 & 0 & 0 & 0 \end{bmatrix}$

44. $\begin{bmatrix} 1 & 3 & 0 & 0 \\ 0 & 0 & 1 & 8 \\ 0 & 0 & 0 & 0 \end{bmatrix}$

45. $\begin{bmatrix} 1 & 0 & 0 & 1 \\ 0 & 1 & 0 & -1 \\ 0 & 0 & 0 & 2 \end{bmatrix}$

46. $\begin{bmatrix} 1 & 0 & 1 & 0 \\ 0 & 1 & 0 & 2 \\ 0 & 0 & 1 & 0 \end{bmatrix}$

Writing a Matrix in Row-Echelon Form **In Exercises 47–50, write the matrix in row-echelon form. (Remember that the row-echelon form of a matrix is not unique.)**

47. $\begin{bmatrix} 1 & 1 & 0 & 5 \\ -2 & -1 & 2 & -10 \\ 3 & 6 & 7 & 14 \end{bmatrix}$

48. $\begin{bmatrix} 1 & 2 & -1 & 3 \\ 3 & 7 & -5 & 14 \\ -2 & -1 & -3 & 8 \end{bmatrix}$

49. $\begin{bmatrix} 1 & -1 & -1 & 1 \\ 5 & -4 & 1 & 8 \\ -6 & 8 & 18 & 0 \end{bmatrix}$

50. $\begin{bmatrix} 1 & -3 & 0 & -7 \\ -3 & 10 & 1 & 23 \\ 4 & -10 & 2 & -24 \end{bmatrix}$

Using a Graphing Utility **In Exercises 51–56, use the matrix capabilities of a graphing utility to write the matrix in *reduced* row-echelon form.**

51. $\begin{bmatrix} 3 & 3 & 3 \\ -1 & 0 & -4 \\ 2 & 4 & -2 \end{bmatrix}$

52. $\begin{bmatrix} 1 & 3 & 2 \\ 5 & 15 & 9 \\ 2 & 6 & 10 \end{bmatrix}$

53. $\begin{bmatrix} 1 & 2 & 3 & -5 \\ 1 & 2 & 4 & -9 \\ -2 & -4 & -4 & 3 \\ 4 & 8 & 11 & -14 \end{bmatrix}$

54. $\begin{bmatrix} -2 & 3 & -1 & -2 \\ 4 & -2 & 5 & 8 \\ 1 & 5 & -2 & 0 \\ 3 & 8 & -10 & -30 \end{bmatrix}$

55. $\begin{bmatrix} -3 & 5 & 1 & 12 \\ 1 & -1 & 1 & 4 \end{bmatrix}$

56. $\begin{bmatrix} 5 & 1 & 2 & 4 \\ -1 & 5 & 10 & -32 \end{bmatrix}$

Using Back-Substitution **In Exercises 57–60, write the system of linear equations represented by the augmented matrix. Then use back-substitution to solve. (Use variables x, y, and z, if applicable.)**

57. $\begin{bmatrix} 1 & -2 & \vdots & 4 \\ 0 & 1 & \vdots & -3 \end{bmatrix}$

58. $\begin{bmatrix} 1 & 5 & \vdots & 0 \\ 0 & 1 & \vdots & -1 \end{bmatrix}$

59. $\begin{bmatrix} 1 & -1 & 2 & \vdots & 4 \\ 0 & 1 & -1 & \vdots & 2 \\ 0 & 0 & 1 & \vdots & -2 \end{bmatrix}$

60. $\begin{bmatrix} 1 & 2 & -2 & \vdots & -1 \\ 0 & 1 & 1 & \vdots & 9 \\ 0 & 0 & 1 & \vdots & -3 \end{bmatrix}$

Interpreting Reduced Row-Echelon Form In Exercises 61–64, an augmented matrix that represents a system of linear equations (in variables x, y, and z, if applicable) has been reduced using Gauss-Jordan elimination. Write the solution represented by the augmented matrix.

61. $\begin{bmatrix} 1 & 0 & \vdots & 3 \\ 0 & 1 & \vdots & -4 \end{bmatrix}$

62. $\begin{bmatrix} 1 & 0 & \vdots & -6 \\ 0 & 1 & \vdots & 10 \end{bmatrix}$

63. $\begin{bmatrix} 1 & 0 & 0 & \vdots & -4 \\ 0 & 1 & 0 & \vdots & -10 \\ 0 & 0 & 1 & \vdots & 4 \end{bmatrix}$

64. $\begin{bmatrix} 1 & 0 & 0 & \vdots & 5 \\ 0 & 1 & 0 & \vdots & -3 \\ 0 & 0 & 1 & \vdots & 0 \end{bmatrix}$

Gaussian Elimination with Back-Substitution In Exercises 65–74, use matrices to solve the system of equations (if possible). Use Gaussian elimination with back-substitution.

65. $\begin{cases} x + 2y = 7 \\ 2x + y = 8 \end{cases}$

66. $\begin{cases} 2x + 6y = 16 \\ 2x + 3y = 7 \end{cases}$

67. $\begin{cases} 3x - 2y = -27 \\ x + 3y = 13 \end{cases}$

68. $\begin{cases} -x + y = 4 \\ 2x - 4y = -34 \end{cases}$

69. $\begin{cases} x + 2y - 3z = -28 \\ 4y + 2z = 0 \\ -x + y - z = -5 \end{cases}$

70. $\begin{cases} 3x - 2y + z = 15 \\ -x + y + 2z = -10 \\ x - y - 4z = 14 \end{cases}$

71. $\begin{cases} -x + y = -22 \\ 3x + 4y = 4 \\ 4x - 8y = 32 \end{cases}$

72. $\begin{cases} x + 2y = 0 \\ x + y = 6 \\ 3x - 2y = 8 \end{cases}$

73. $\begin{cases} 3x + 2y - z + w = 0 \\ x - y + 4z + 2w = 25 \\ -2x + y + 2z - w = 2 \\ x + y + z + w = 6 \end{cases}$

74. $\begin{cases} x - 4y + 3z - 2w = 9 \\ 3x - 2y + z - 4w = -13 \\ -4x + 3y - 2z + w = -4 \\ -2x + y - 4z + 3w = -10 \end{cases}$

Gauss-Jordan Elimination In Exercises 75–84, use matrices to solve the system of equations (if possible). Use Gauss-Jordan elimination.

75. $\begin{cases} -2x + 6y = -22 \\ x + 2y = -9 \end{cases}$

76. $\begin{cases} 5x - 5y = -5 \\ -2x - 3y = 7 \end{cases}$

77. $\begin{cases} 8x - 4y = 7 \\ 5x + 2y = 1 \end{cases}$

78. $\begin{cases} x - 3y = 5 \\ -2x + 6y = -10 \end{cases}$

79. $\begin{cases} x + 2y + z = 8 \\ 3x + 7y + 6z = 26 \end{cases}$

80. $\begin{cases} x + y + 4z = 5 \\ 2x + y - z = 9 \end{cases}$

81. $\begin{cases} x - 3z = -2 \\ 3x + y - 2z = 5 \\ 2x + 2y + z = 4 \end{cases}$

82. $\begin{cases} 2x - y + 3z = 24 \\ 2y - z = 14 \\ 7x - 5y = 6 \end{cases}$

83. $\begin{cases} -x + y - z = -14 \\ 2x - y + z = 21 \\ 3x + 2y + z = 19 \end{cases}$

84. $\begin{cases} 2x + 2y - z = 2 \\ x - 3y + z = -28 \\ -x + y = 14 \end{cases}$

Using a Graphing Utility In Exercises 85–90, use the matrix capabilities of a graphing utility to write the augmented matrix corresponding to the system of equations in reduced row-echelon form. Then solve the system.

85. $\begin{cases} 3x + 3y + 12z = 6 \\ x + y + 4z = 2 \\ 2x + 5y + 20z = 10 \\ -x + 2y + 8z = 4 \end{cases}$

86. $\begin{cases} 2x + 10y + 2z = 6 \\ x + 5y + 2z = 6 \\ x + 5y + z = 3 \\ -3x - 15y - 3z = -9 \end{cases}$

87. $\begin{cases} 2x + y - z + 2w = -6 \\ 3x + 4y + w = 1 \\ x + 5y + 2z + 6w = -3 \\ 5x + 2y - z - w = 3 \end{cases}$

88. $\begin{cases} x + 2y + 2z + 4w = 11 \\ 3x + 6y + 5z + 12w = 30 \\ x + 3y - 3z + 2w = -5 \\ 6x - y - z + w = -9 \end{cases}$

89. $\begin{cases} x + y + z + w = 0 \\ 2x + 3y + z - 2w = 0 \\ 3x + 5y + z = 0 \end{cases}$

90. $\begin{cases} x + 2y + z + 3w = 0 \\ x - y + w = 0 \\ y - z + 2w = 0 \end{cases}$

Comparing Solutions of Two Systems In Exercises 91–94, determine whether the two systems of linear equations yield the same solution. If so, find the solution using matrices.

91. (a) $\begin{cases} x - 2y + z = -6 \\ y - 5z = 16 \\ z = -3 \end{cases}$ (b) $\begin{cases} x + y - 2z = 6 \\ y + 3z = -8 \\ z = -3 \end{cases}$

92. (a) $\begin{cases} x - 3y + 4z = -11 \\ y - z = -4 \\ z = 2 \end{cases}$ (b) $\begin{cases} x + 4y = -11 \\ y + 3z = 4 \\ z = 2 \end{cases}$

93. (a) $\begin{cases} x - 4y + 5z = 27 \\ y - 7z = -54 \\ z = 8 \end{cases}$ (b) $\begin{cases} x - 6y + z = 15 \\ y + 5z = 42 \\ z = 8 \end{cases}$

94. (a) $\begin{cases} x + 3y - z = 19 \\ y + 6z = -18 \\ z = -4 \end{cases}$ (b) $\begin{cases} x - y + 3z = -15 \\ y - 2z = 14 \\ z = -4 \end{cases}$

Curve Fitting In Exercises 95–100, use a system of equations to find the quadratic function $f(x) = ax^2 + bx + c$ that satisfies the given conditions. Solve the system using matrices.

95. $f(1) = 1, f(2) = -1, f(3) = -5$

96. $f(1) = 2, f(2) = 9, f(3) = 20$

97. $f(-2) = -15, f(-1) = 7, f(1) = -3$

98. $f(-2) = -3, f(1) = -3, f(2) = -11$

99. $f(1) = 8, f(2) = 13, f(3) = 20$

100. $f(1) = 9, f(2) = 8, f(3) = 5$

101. Health and Wellness

From 2000 through 2011, the numbers of new cases of a waterborne disease in a small city increased in a pattern that was approximately linear (see figure). Find the least squares regression line

$$y = at + b$$

for the data shown in the figure by solving the following system using matrices. Let t represent the year, with $t = 0$ corresponding to 2000.

$$\begin{cases} 12b + 66a = 831 \\ 66b + 506a = 5643 \end{cases}$$

Use the result to predict the number of new cases of the waterborne disease in 2014. Is the estimate reasonable? Explain.

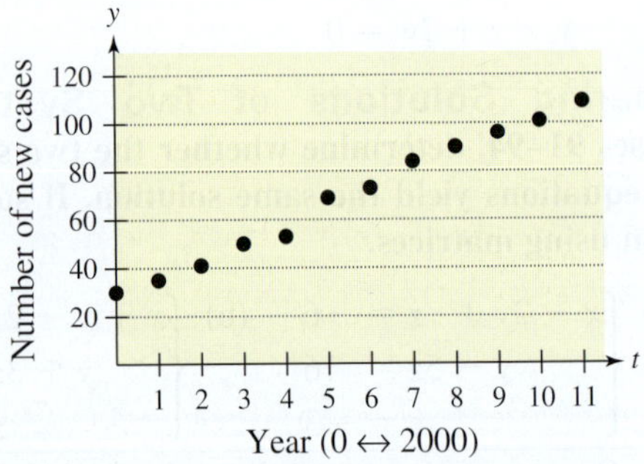

102. Breeding Facility A city zoo borrowed \$2,000,000 at simple annual interest to construct a breeding facility. Some of the money was borrowed at 8%, some at 9%, and some at 12%. Use a system of linear equations to determine how much was borrowed at each rate given that the total annual interest was \$186,000 and the amount borrowed at 8% was twice the amount borrowed at 12%. Solve the system of linear equations using matrices.

103. Museum A natural history museum borrowed \$2,000,000 at simple annual interest to purchase new exhibits. Some of the money was borrowed at 7%, some at 8.5%, and some at 9.5%. Use a system of linear equations to determine how much was borrowed at each rate given that the total annual interest was \$169,750 and the amount borrowed at 8.5% was four times the amount borrowed at 9.5%. Solve the system of linear equations using matrices.

104. Mathematical Modeling A video of the path of a ball thrown by a baseball player was analyzed with a grid covering the TV screen. The tape was paused three times, and the position of the ball was measured each time. The coordinates obtained are shown in the table. (x and y are measured in feet.)

Horizontal Distance, x	0	15	30
Height, y	5.0	9.6	12.4

(a) Use a system of equations to find the equation of the parabola $y = ax^2 + bx + c$ that passes through the three points. Solve the system using matrices.

(b) Use a graphing utility to graph the parabola.

(c) Graphically approximate the maximum height of the ball and the point at which the ball struck the ground.

(d) Analytically find the maximum height of the ball and the point at which the ball struck the ground.

(e) Compare your results from parts (c) and (d).

Exploration

True or False? In Exercises 105 and 106, determine whether the statement is true or false. Justify your answer.

105. $\begin{bmatrix} 5 & 0 & -2 & 7 \\ -1 & 3 & -6 & 0 \end{bmatrix}$ is a 4×2 matrix.

106. The method of Gaussian elimination reduces a matrix until a reduced row-echelon form is obtained.

107. Think About It What is the relationship between the three elementary row operations performed on an augmented matrix and the operations that lead to equivalent systems of equations?

108. HOW DO YOU SEE IT? Determine whether the matrix below is in row-echelon form, reduced row-echelon form, or neither when it satisfies the given conditions.

$$\begin{bmatrix} 1 & b \\ c & 1 \end{bmatrix}$$

(a) $b = 0, c = 0$ (b) $b \neq 0, c = 0$

(c) $b = 0, c \neq 0$ (d) $b \neq 0, c \neq 0$

Tyler Olson/Shutterstock.com

8.2 Operations with Matrices

You can use matrix operations to model and solve real-life problems. For instance, in Exercise 74 on page 564, you will use matrix multiplication to determine the numbers of calories burned by individuals of different body weights while performing different types of exercises.

- **Decide whether two matrices are equal.**
- **Add and subtract matrices, and multiply matrices by scalars.**
- **Multiply two matrices.**
- **Use matrix operations to model and solve real-life problems.**

Equality of Matrices

In Section 8.1, you used matrices to solve systems of linear equations. There is a rich mathematical theory of matrices, and its applications are numerous. This section and the next two introduce some fundamentals of matrix theory. It is standard mathematical convention to represent matrices in any of the following three ways.

Representation of Matrices

1. A matrix can be denoted by an uppercase letter such as A, B, or C.
2. A matrix can be denoted by a representative element enclosed in brackets, such as $[a_{ij}]$, $[b_{ij}]$, or $[c_{ij}]$.
3. A matrix can be denoted by a rectangular array of numbers such as

$$A = [a_{ij}] = \begin{bmatrix} a_{11} & a_{12} & a_{13} & \cdots & a_{1n} \\ a_{21} & a_{22} & a_{23} & \cdots & a_{2n} \\ a_{31} & a_{32} & a_{33} & \cdots & a_{3n} \\ \vdots & \vdots & \vdots & & \vdots \\ a_{m1} & a_{m2} & a_{m3} & \cdots & a_{mn} \end{bmatrix}.$$

Two matrices $A = [a_{ij}]$ and $B = [b_{ij}]$ are **equal** when they have the same order $(m \times n)$ and $a_{ij} = b_{ij}$ for $1 \le i \le m$ and $1 \le j \le n$. In other words, two matrices are equal when their corresponding entries are equal.

EXAMPLE 1 Equality of Matrices

Solve for a_{11}, a_{12}, a_{21}, and a_{22} in the following matrix equation.

$$\begin{bmatrix} a_{11} & a_{12} \\ a_{21} & a_{22} \end{bmatrix} = \begin{bmatrix} 2 & -1 \\ -3 & 0 \end{bmatrix}$$

Solution Because two matrices are equal when their corresponding entries are equal, you can conclude that $a_{11} = 2$, $a_{12} = -1$, $a_{21} = -3$, and $a_{22} = 0$.

✓ ***Checkpoint*** *Audio-video solution in English & Spanish at LarsonPrecalculus.com.*

Solve for a_{11}, a_{12}, a_{21}, and a_{22} in the following matrix equation.

$$\begin{bmatrix} a_{11} & a_{12} \\ a_{21} & a_{22} \end{bmatrix} = \begin{bmatrix} 6 & 3 \\ -2 & 4 \end{bmatrix}$$

Be sure you see that for two matrices to be equal, they must have the same order *and* their corresponding entries must be equal. For instance,

$$\begin{bmatrix} 2 & -1 \\ \sqrt{4} & \frac{1}{2} \end{bmatrix} = \begin{bmatrix} 2 & -1 \\ 2 & 0.5 \end{bmatrix} \quad \text{but} \quad \begin{bmatrix} 2 & -1 & 0 \\ 3 & 4 & 0 \end{bmatrix} \neq \begin{bmatrix} 2 & -1 \\ 3 & 4 \end{bmatrix}.$$

Simone van den Berg/Shutterstock.com

Matrix Addition and Scalar Multiplication

In this section, three basic matrix operations will be covered. The first two are matrix addition and scalar multiplication. With matrix addition, you can add two matrices (of the same order) by adding their corresponding entries.

HISTORICAL NOTE

Arthur Cayley (1821–1895), a British mathematician, invented matrices around 1858. Cayley was a Cambridge University graduate and a lawyer by profession. His groundbreaking work on matrices was begun as he studied the theory of transformations. Cayley also was instrumental in the development of determinants. Cayley and two American mathematicians, Benjamin Peirce (1809–1880) and his son Charles S. Peirce (1839–1914), are credited with developing "matrix algebra."

Definition of Matrix Addition

If $A = [a_{ij}]$ and $B = [b_{ij}]$ are matrices of order $m \times n$, then their sum is the $m \times n$ matrix given by

$$A + B = [a_{ij} + b_{ij}].$$

The sum of two matrices of different orders is undefined.

EXAMPLE 2 Addition of Matrices

a. $\begin{bmatrix} -1 & 2 \\ 0 & 1 \end{bmatrix} + \begin{bmatrix} 1 & 3 \\ -1 & 2 \end{bmatrix} = \begin{bmatrix} -1+1 & 2+3 \\ 0+(-1) & 1+2 \end{bmatrix} = \begin{bmatrix} 0 & 5 \\ -1 & 3 \end{bmatrix}$

b. $\begin{bmatrix} 0 & 1 & -2 \\ 1 & 2 & 3 \end{bmatrix} + \begin{bmatrix} 0 & 0 & 0 \\ 0 & 0 & 0 \end{bmatrix} = \begin{bmatrix} 0 & 1 & -2 \\ 1 & 2 & 3 \end{bmatrix}$

c. $\begin{bmatrix} 1 \\ -3 \\ -2 \end{bmatrix} + \begin{bmatrix} -1 \\ 3 \\ 2 \end{bmatrix} = \begin{bmatrix} 0 \\ 0 \\ 0 \end{bmatrix}$

d. The sum of

$$A = \begin{bmatrix} 2 & 1 & 0 \\ 4 & 0 & -1 \\ 3 & -2 & 2 \end{bmatrix}$$

and

$$B = \begin{bmatrix} 0 & 1 \\ -1 & 3 \\ 2 & 4 \end{bmatrix}$$

is undefined because A is of order 3×3 and B is of order 3×2.

✓ **Checkpoint** *Audio-video solution in English & Spanish at LarsonPrecalculus.com.*

Evaluate the expression.

$$\begin{bmatrix} 4 & -1 \\ 2 & -3 \end{bmatrix} + \begin{bmatrix} 2 & -1 \\ 0 & 6 \end{bmatrix}$$

In operations with matrices, numbers are usually referred to as **scalars.** In this text, scalars will always be real numbers. You can multiply a matrix A by a scalar c by multiplying each entry in A by c.

Definition of Scalar Multiplication

If $A = [a_{ij}]$ is an $m \times n$ matrix and c is a scalar, then the **scalar multiple** of A by c is the $m \times n$ matrix given by

$$cA = [ca_{ij}].$$

Bettmann/CORBIS

The symbol $-A$ represents the negation of A, which is the scalar product $(-1)A$. Moreover, if A and B are of the same order, then $A - B$ represents the sum of A and $(-1)B$. That is,

$$A - B = A + (-1)B. \qquad \text{Subtraction of matrices}$$

EXAMPLE 3 Scalar Multiplication and Matrix Subtraction

For the following matrices, find (a) $3A$, (b) $-B$, and (c) $3A - B$.

$$A = \begin{bmatrix} 2 & 2 & 4 \\ -3 & 0 & -1 \\ 2 & 1 & 2 \end{bmatrix} \quad \text{and} \quad B = \begin{bmatrix} 2 & 0 & 0 \\ 1 & -4 & 3 \\ -1 & 3 & 2 \end{bmatrix}$$

Solution

a. $3A = 3\begin{bmatrix} 2 & 2 & 4 \\ -3 & 0 & -1 \\ 2 & 1 & 2 \end{bmatrix}$ Scalar multiplication

$= \begin{bmatrix} 3(2) & 3(2) & 3(4) \\ 3(-3) & 3(0) & 3(-1) \\ 3(2) & 3(1) & 3(2) \end{bmatrix}$ Multiply each entry by 3.

$= \begin{bmatrix} 6 & 6 & 12 \\ -9 & 0 & -3 \\ 6 & 3 & 6 \end{bmatrix}$ Simplify.

b. $-B = (-1)\begin{bmatrix} 2 & 0 & 0 \\ 1 & -4 & 3 \\ -1 & 3 & 2 \end{bmatrix}$ Definition of negation

$= \begin{bmatrix} -2 & 0 & 0 \\ -1 & 4 & -3 \\ 1 & -3 & -2 \end{bmatrix}$ Multiply each entry by -1.

c. $3A - B = \begin{bmatrix} 6 & 6 & 12 \\ -9 & 0 & -3 \\ 6 & 3 & 6 \end{bmatrix} - \begin{bmatrix} 2 & 0 & 0 \\ 1 & -4 & 3 \\ -1 & 3 & 2 \end{bmatrix}$ Perform scalar multiplication first.

$= \begin{bmatrix} 4 & 6 & 12 \\ -10 & 4 & -6 \\ 7 & 0 & 4 \end{bmatrix}$ Subtract corresponding entries.

REMARK The order of operations for matrix expressions is similar to that for real numbers. In particular, you perform scalar multiplication before matrix addition and subtraction, as shown in Example 3(c).

Checkpoint *Audio-video solution in English & Spanish at LarsonPrecalculus.com.*

For the following matrices, find (a) $A + B$, (b) $A - B$, (c) $3A$, and (d) $3A - 2B$.

$$A = \begin{bmatrix} 4 & -1 \\ 0 & 4 \\ -3 & 8 \end{bmatrix}, \quad B = \begin{bmatrix} 0 & 4 \\ -1 & 3 \\ 1 & 7 \end{bmatrix}$$

It is often convenient to rewrite the scalar multiple cA by factoring c out of every entry in the matrix. For instance, in the following example, the scalar $\frac{1}{2}$ has been factored out of the matrix.

$$\begin{bmatrix} \frac{1}{2} & -\frac{3}{2} \\ \frac{5}{2} & \frac{1}{2} \end{bmatrix} = \begin{bmatrix} \frac{1}{2}(1) & \frac{1}{2}(-3) \\ \frac{1}{2}(5) & \frac{1}{2}(1) \end{bmatrix} = \frac{1}{2}\begin{bmatrix} 1 & -3 \\ 5 & 1 \end{bmatrix}$$

▷ **ALGEBRA HELP** You can review the properties of addition and multiplication of real numbers (and other properties of real numbers) in Appendix A.1.

The properties of matrix addition and scalar multiplication are similar to those of addition and multiplication of real numbers.

Properties of Matrix Addition and Scalar Multiplication

Let A, B, and C be $m \times n$ matrices and let c and d be scalars.

1. $A + B = B + A$ — Commutative Property of Matrix Addition
2. $A + (B + C) = (A + B) + C$ — Associative Property of Matrix Addition
3. $(cd)A = c(dA)$ — Associative Property of Scalar Multiplication
4. $1A = A$ — Scalar Identity Property
5. $c(A + B) = cA + cB$ — Distributive Property
6. $(c + d)A = cA + dA$ — Distributive Property

Note that the Associative Property of Matrix Addition allows you to write expressions such as $A + B + C$ without ambiguity because the same sum occurs no matter how the matrices are grouped. This same reasoning applies to sums of four or more matrices.

▷ **TECHNOLOGY** Most graphing utilities have the capability of performing matrix operations. Consult the user's guide for your graphing utility for specific keystrokes. Try using a graphing utility to find the sum of the matrices

$$A = \begin{bmatrix} 2 & -3 \\ -1 & 0 \end{bmatrix}$$

and

$$B = \begin{bmatrix} -1 & 4 \\ 2 & -5 \end{bmatrix}.$$

EXAMPLE 4 Addition of More than Two Matrices

By adding corresponding entries, you obtain the following sum of four matrices.

$$\begin{bmatrix} 1 \\ 2 \\ -3 \end{bmatrix} + \begin{bmatrix} -1 \\ -1 \\ 2 \end{bmatrix} + \begin{bmatrix} 0 \\ 1 \\ 4 \end{bmatrix} + \begin{bmatrix} 2 \\ -3 \\ -2 \end{bmatrix} = \begin{bmatrix} 2 \\ -1 \\ 1 \end{bmatrix}$$

✓ **Checkpoint** Audio-video solution in English & Spanish at LarsonPrecalculus.com.

Evaluate the expression.

$$\begin{bmatrix} 3 & -8 \\ 0 & 2 \end{bmatrix} + \begin{bmatrix} -2 & 3 \\ 6 & -5 \end{bmatrix} + \begin{bmatrix} 0 & 7 \\ 4 & -1 \end{bmatrix}$$

EXAMPLE 5 Using the Distributive Property

$$3\left(\begin{bmatrix} -2 & 0 \\ 4 & 1 \end{bmatrix} + \begin{bmatrix} 4 & -2 \\ 3 & 7 \end{bmatrix}\right) = 3\begin{bmatrix} -2 & 0 \\ 4 & 1 \end{bmatrix} + 3\begin{bmatrix} 4 & -2 \\ 3 & 7 \end{bmatrix}$$

$$= \begin{bmatrix} -6 & 0 \\ 12 & 3 \end{bmatrix} + \begin{bmatrix} 12 & -6 \\ 9 & 21 \end{bmatrix}$$

$$= \begin{bmatrix} 6 & -6 \\ 21 & 24 \end{bmatrix}$$

✓ **Checkpoint** Audio-video solution in English & Spanish at LarsonPrecalculus.com.

Evaluate the expression using the Distributive Property.

$$2\left(\begin{bmatrix} 1 & 3 \\ -2 & 2 \end{bmatrix} + \begin{bmatrix} -4 & 0 \\ -3 & 1 \end{bmatrix}\right)$$

In Example 5, you could add the two matrices first and then multiply the resulting matrix by 3. The result would be the same.

One important property of addition of real numbers is that the number 0 is the additive identity. That is, $c + 0 = c$ for any real number c. For matrices, a similar property holds. That is, if A is an $m \times n$ matrix and O is the $m \times n$ **zero matrix** consisting entirely of zeros, then $A + O = A$.

In other words, O is the **additive identity** for the set of all $m \times n$ matrices. For example, the following matrices are the additive identities for the sets of all 2×3 and 2×2 matrices.

$$O = \underbrace{\begin{bmatrix} 0 & 0 & 0 \\ 0 & 0 & 0 \end{bmatrix}}_{2 \times 3 \text{ zero matrix}} \quad \text{and} \quad O = \underbrace{\begin{bmatrix} 0 & 0 \\ 0 & 0 \end{bmatrix}}_{2 \times 2 \text{ zero matrix}}$$

The algebra of real numbers and the algebra of matrices have many similarities. For example, compare the following solutions.

> **REMARK** Remember that matrices are denoted by capital letters. So, when you solve for X, you are solving for a *matrix* that makes the matrix equation true.

Real Numbers (Solve for x.)	**$m \times n$ Matrices (Solve for X.)**
$x + a = b$	$X + A = B$
$x + a + (-a) = b + (-a)$	$X + A + (-A) = B + (-A)$
$x + 0 = b - a$	$X + O = B - A$
$x = b - a$	$X = B - A$

The algebra of real numbers and the algebra of matrices also have important differences (see Example 10 and Exercises 77–82).

EXAMPLE 6 Solving a Matrix Equation

Solve for X in the equation $3X + A = B$, where

$$A = \begin{bmatrix} 1 & -2 \\ 0 & 3 \end{bmatrix} \quad \text{and} \quad B = \begin{bmatrix} -3 & 4 \\ 2 & 1 \end{bmatrix}.$$

Solution Begin by solving the matrix equation for X to obtain

$$3X + A = B$$

$$3X = B - A$$

$$X = \frac{1}{3}(B - A).$$

Now, using the matrices A and B, you have

$$X = \frac{1}{3}\left(\begin{bmatrix} -3 & 4 \\ 2 & 1 \end{bmatrix} - \begin{bmatrix} 1 & -2 \\ 0 & 3 \end{bmatrix}\right) \qquad \text{Substitute the matrices.}$$

$$= \frac{1}{3}\begin{bmatrix} -4 & 6 \\ 2 & -2 \end{bmatrix} \qquad \text{Subtract matrix } A \text{ from matrix } B.$$

$$= \begin{bmatrix} -\frac{4}{3} & 2 \\ \frac{2}{3} & -\frac{2}{3} \end{bmatrix}. \qquad \text{Multiply the resulting matrix by } \frac{1}{3}.$$

✓ **Checkpoint** *Audio-video solution in English & Spanish at LarsonPrecalculus.com.*

Solve for X in the equation $2X - A = B$, where

$$A = \begin{bmatrix} 6 & 1 \\ 0 & 3 \end{bmatrix} \quad \text{and} \quad B = \begin{bmatrix} 4 & -1 \\ -2 & 5 \end{bmatrix}.$$

Matrix Multiplication

Another basic matrix operation is **matrix multiplication.** At first glance, the definition may seem unusual. You will see later, however, that this definition of the product of two matrices has many practical applications.

> **Definition of Matrix Multiplication**
>
> If $A = [a_{ij}]$ is an $m \times n$ matrix and $B = [b_{ij}]$ is an $n \times p$ matrix, then the product AB is an $m \times p$ matrix
>
> $$AB = [c_{ij}]$$
>
> where $c_{ij} = a_{i1}b_{1j} + a_{i2}b_{2j} + a_{i3}b_{3j} + \cdot \cdot \cdot + a_{in}b_{nj}$.

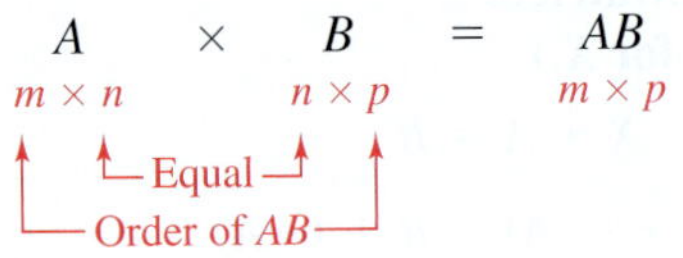

The definition of matrix multiplication indicates a *row-by-column* multiplication, where the entry in the ith row and jth column of the product AB is obtained by multiplying the entries in the ith row of A by the corresponding entries in the jth column of B and then adding the results. So, for the product of two matrices to be defined, the number of columns of the first matrix must equal the number of rows of the second matrix. That is, the middle two indices must be the same. The outside two indices give the order of the product, as shown at the left. The general pattern for matrix multiplication is as follows.

$$\begin{bmatrix} a_{11} & a_{12} & a_{13} & \cdots & a_{1n} \\ a_{21} & a_{22} & a_{23} & \cdots & a_{2n} \\ a_{31} & a_{32} & a_{33} & \cdots & a_{3n} \\ \vdots & \vdots & \vdots & & \vdots \\ a_{i1} & a_{i2} & a_{i3} & \cdots & a_{in} \\ \vdots & \vdots & \vdots & & \vdots \\ a_{m1} & a_{m2} & a_{m3} & \cdots & a_{mn} \end{bmatrix} \begin{bmatrix} b_{11} & b_{12} & \cdots & b_{1j} & \cdots & b_{1p} \\ b_{21} & b_{22} & \cdots & b_{2j} & \cdots & b_{2p} \\ b_{31} & b_{32} & \cdots & b_{3j} & \cdots & b_{3p} \\ \vdots & \vdots & & \vdots & & \vdots \\ b_{n1} & b_{n2} & \cdots & b_{nj} & \cdots & b_{np} \end{bmatrix} = \begin{bmatrix} c_{11} & c_{12} & \cdots & c_{1j} & \cdots & c_{1p} \\ c_{21} & c_{22} & \cdots & c_{2j} & \cdots & c_{2p} \\ \vdots & \vdots & & \vdots & & \vdots \\ c_{i1} & c_{i2} & \cdots & c_{ij} & \cdots & c_{ip} \\ \vdots & \vdots & & \vdots & & \vdots \\ c_{m1} & c_{m2} & \cdots & c_{mj} & \cdots & c_{mp} \end{bmatrix}$$

$$a_{i1}b_{1j} + a_{i2}b_{2j} + a_{i3}b_{3j} + \cdot \cdot \cdot + a_{in}b_{nj} = c_{ij}$$

EXAMPLE 7 **Finding the Product of Two Matrices**

Find the product AB using $A = \begin{bmatrix} -1 & 3 \\ 4 & -2 \\ 5 & 0 \end{bmatrix}$ and $B = \begin{bmatrix} -3 & 2 \\ -4 & 1 \end{bmatrix}$.

Solution To find the entries of the product, multiply each row of A by each column of B.

$$AB = \begin{bmatrix} -1 & 3 \\ 4 & -2 \\ 5 & 0 \end{bmatrix} \begin{bmatrix} -3 & 2 \\ -4 & 1 \end{bmatrix}$$

$$= \begin{bmatrix} (-1)(-3) + (3)(-4) & (-1)(2) + (3)(1) \\ (4)(-3) + (-2)(-4) & (4)(2) + (-2)(1) \\ (5)(-3) + (0)(-4) & (5)(2) + (0)(1) \end{bmatrix}$$

$$= \begin{bmatrix} -9 & 1 \\ -4 & 6 \\ -15 & 10 \end{bmatrix}$$

REMARK In Example 7, the product AB is defined because the number of columns of A is equal to the number of rows of B. Also, note that the product AB has order 3×2.

✓ ***Checkpoint*** *Audio-video solution in English & Spanish at LarsonPrecalculus.com.*

Find the product AB using $A = \begin{bmatrix} -1 & 4 \\ 2 & 0 \\ 1 & 2 \end{bmatrix}$ and $B = \begin{bmatrix} 1 & -2 \\ 0 & 7 \end{bmatrix}$.

EXAMPLE 8 Finding the Product of Two Matrices

Find the product AB using $A = \begin{bmatrix} 1 & 0 & 3 \\ 2 & -1 & -2 \end{bmatrix}$ and $B = \begin{bmatrix} -2 & 4 \\ 1 & 0 \\ -1 & 1 \end{bmatrix}$.

Solution Note that the order of A is 2×3 and the order of B is 3×2. So, the product AB has order 2×2.

$$AB = \begin{bmatrix} 1 & 0 & 3 \\ 2 & -1 & -2 \end{bmatrix}\begin{bmatrix} -2 & 4 \\ 1 & 0 \\ -1 & 1 \end{bmatrix}$$

$$= \begin{bmatrix} 1(-2) + 0(1) + 3(-1) & 1(4) + 0(0) + 3(1) \\ 2(-2) + (-1)(1) + (-2)(-1) & 2(4) + (-1)(0) + (-2)(1) \end{bmatrix}$$

$$= \begin{bmatrix} -5 & 7 \\ -3 & 6 \end{bmatrix}$$

✓ Checkpoint *Audio-video solution in English & Spanish at LarsonPrecalculus.com.*

Find the product AB using $A = \begin{bmatrix} 0 & 4 & -3 \\ 2 & 1 & 7 \\ 3 & -2 & 1 \end{bmatrix}$ and $B = \begin{bmatrix} -2 & 0 \\ 0 & -4 \\ 1 & 2 \end{bmatrix}$.

EXAMPLE 9 Patterns in Matrix Multiplication

a. $\underset{2 \times 2}{\begin{bmatrix} 3 & 4 \\ -2 & 5 \end{bmatrix}}\underset{2 \times 2}{\begin{bmatrix} 1 & 0 \\ 0 & 1 \end{bmatrix}} = \underset{2 \times 2}{\begin{bmatrix} 3 & 4 \\ -2 & 5 \end{bmatrix}}$

b. $\underset{3 \times 3}{\begin{bmatrix} 6 & 2 & 0 \\ 3 & -1 & 2 \\ 1 & 4 & 6 \end{bmatrix}}\underset{3 \times 1}{\begin{bmatrix} 1 \\ 2 \\ -3 \end{bmatrix}} = \underset{3 \times 1}{\begin{bmatrix} 10 \\ -5 \\ -9 \end{bmatrix}}$

c. The product AB for the following matrices is not defined.

$$A = \underset{3 \times 2}{\begin{bmatrix} -2 & 1 \\ 1 & -3 \\ 1 & 4 \end{bmatrix}} \quad \text{and} \quad B = \underset{3 \times 4}{\begin{bmatrix} -2 & 3 & 1 & 4 \\ 0 & 1 & -1 & 2 \\ 2 & -1 & 0 & 1 \end{bmatrix}}$$

✓ Checkpoint *Audio-video solution in English & Spanish at LarsonPrecalculus.com.*

Find, if possible, the product AB using $A = \begin{bmatrix} 3 & 1 & 2 \\ 7 & 0 & -2 \end{bmatrix}$ and $B = \begin{bmatrix} 6 & 4 \\ 2 & -1 \end{bmatrix}$.

REMARK In Example 10, note that the two products are different. Even when both AB and BA are defined, matrix multiplication is not, in general, commutative. That is, for most matrices, $AB \neq BA$. This is one way in which the algebra of real numbers and the algebra of matrices differ.

EXAMPLE 10 Patterns in Matrix Multiplication

a. $\underset{1 \times 3}{\begin{bmatrix} 1 & -2 & -3 \end{bmatrix}}\underset{3 \times 1}{\begin{bmatrix} 2 \\ -1 \\ 1 \end{bmatrix}} = \underset{1 \times 1}{\begin{bmatrix} 1 \end{bmatrix}}$

b. $\underset{3 \times 1}{\begin{bmatrix} 2 \\ -1 \\ 1 \end{bmatrix}}\underset{1 \times 3}{\begin{bmatrix} 1 & -2 & -3 \end{bmatrix}} = \underset{3 \times 3}{\begin{bmatrix} 2 & -4 & -6 \\ -1 & 2 & 3 \\ 1 & -2 & -3 \end{bmatrix}}$

✓ Checkpoint *Audio-video solution in English & Spanish at LarsonPrecalculus.com.*

Find AB and BA using $A = \begin{bmatrix} 3 & -1 \end{bmatrix}$ and $B = \begin{bmatrix} 1 \\ -3 \end{bmatrix}$.

EXAMPLE 11 Squaring a Matrix

Find A^2, where $A = \begin{bmatrix} 3 & 1 \\ -1 & 2 \end{bmatrix}$. (*Note:* $A^2 = AA$.)

Solution

$$A^2 = AA = \begin{bmatrix} 3 & 1 \\ -1 & 2 \end{bmatrix}\begin{bmatrix} 3 & 1 \\ -1 & 2 \end{bmatrix} = \begin{bmatrix} 3(3) + 1(-1) & 3(1) + 1(2) \\ -1(3) + 2(-1) & -1(1) + 2(2) \end{bmatrix}$$

$$= \begin{bmatrix} 8 & 5 \\ -5 & 3 \end{bmatrix}$$

✓ ***Checkpoint*** *Audio-video solution in English & Spanish at LarsonPrecalculus.com.*

Find A^2, where $A = \begin{bmatrix} 2 & 1 \\ 3 & -2 \end{bmatrix}$.

Properties of Matrix Multiplication

Let A, B, and C be matrices and let c be a scalar.

1. $A(BC) = (AB)C$ — Associative Property of Matrix Multiplication
2. $A(B + C) = AB + AC$ — Distributive Property
3. $(A + B)C = AC + BC$ — Distributive Property
4. $c(AB) = (cA)B = A(cB)$ — Associative Property of Scalar Multiplication

Definition of Identity Matrix

The $n \times n$ matrix that consists of 1's on its main diagonal and 0's elsewhere is called the **identity matrix of order $n \times n$** and is denoted by

$$I_n = \begin{bmatrix} 1 & 0 & 0 & \dots & 0 \\ 0 & 1 & 0 & \dots & 0 \\ 0 & 0 & 1 & \dots & 0 \\ \vdots & \vdots & \vdots & & \vdots \\ 0 & 0 & 0 & \dots & 1 \end{bmatrix}.$$ Identity matrix

Note that an identity matrix must be *square*. When the order is understood to be $n \times n$, you can denote I_n simply by I.

If A is an $n \times n$ matrix, then the identity matrix has the property that $AI_n = A$ and $I_nA = A$. For example,

$$\begin{bmatrix} 3 & -2 & 5 \\ 1 & 0 & 4 \\ -1 & 2 & -3 \end{bmatrix}\begin{bmatrix} 1 & 0 & 0 \\ 0 & 1 & 0 \\ 0 & 0 & 1 \end{bmatrix} = \begin{bmatrix} 3 & -2 & 5 \\ 1 & 0 & 4 \\ -1 & 2 & -3 \end{bmatrix}$$ $AI = A$

and

$$\begin{bmatrix} 1 & 0 & 0 \\ 0 & 1 & 0 \\ 0 & 0 & 1 \end{bmatrix}\begin{bmatrix} 3 & -2 & 5 \\ 1 & 0 & 4 \\ -1 & 2 & -3 \end{bmatrix} = \begin{bmatrix} 3 & -2 & 5 \\ 1 & 0 & 4 \\ -1 & 2 & -3 \end{bmatrix}.$$ $IA = A$

Many real-life applications of linear systems involve enormous numbers of equations and variables. For example, a flight crew scheduling problem for American Airlines required the manipulation of matrices with 837 rows and 12,753,313 columns. *(Source: Very Large-Scale Linear Programming. A Case Study in Combining Interior Point and Simplex Methods, Bixby, Robert E., et al., Operations Research, 40, no. 5)*

Applications

Matrix multiplication can be used to represent a system of linear equations. Note how the system below can be written as the matrix equation $AX = B$, where A is the *coefficient matrix* of the system and X and B are column matrices. The column matrix B is also called a *constant matrix*. Its entries are the constant terms in the system of equations.

System

$$\begin{cases} a_{11}x_1 + a_{12}x_2 + a_{13}x_3 = b_1 \\ a_{21}x_1 + a_{22}x_2 + a_{23}x_3 = b_2 \\ a_{31}x_1 + a_{32}x_2 + a_{33}x_3 = b_3 \end{cases}$$

Matrix Equation $AX = B$

$$\underbrace{\begin{bmatrix} a_{11} & a_{12} & a_{13} \\ a_{21} & a_{22} & a_{23} \\ a_{31} & a_{32} & a_{33} \end{bmatrix}}_{A} \times \underbrace{\begin{bmatrix} x_1 \\ x_2 \\ x_3 \end{bmatrix}}_{X} = \underbrace{\begin{bmatrix} b_1 \\ b_2 \\ b_3 \end{bmatrix}}_{B}$$

In Example 12, $[A \vdots B]$ represents the augmented matrix formed when matrix B is *adjoined* to matrix A. Also, $[I \vdots X]$ represents the reduced row-echelon form of the augmented matrix that yields the solution of the system.

EXAMPLE 12 Solving a System of Linear Equations

For the system of linear equations, (a) write the system as a matrix equation, $AX = B$, and (b) use Gauss-Jordan elimination on $[A \vdots B]$ to solve for the matrix X.

$$\begin{cases} x_1 - 2x_2 + x_3 = -4 \\ x_2 + 2x_3 = 4 \\ 2x_1 + 3x_2 - 2x_3 = 2 \end{cases}$$

Solution

a. In matrix form, $AX = B$, the system can be written as follows.

$$\begin{bmatrix} 1 & -2 & 1 \\ 0 & 1 & 2 \\ 2 & 3 & -2 \end{bmatrix}\begin{bmatrix} x_1 \\ x_2 \\ x_3 \end{bmatrix} = \begin{bmatrix} -4 \\ 4 \\ 2 \end{bmatrix}$$

b. The augmented matrix is formed by adjoining matrix B to matrix A.

$$[A \vdots B] = \left[\begin{array}{ccc|c} 1 & -2 & 1 & -4 \\ 0 & 1 & 2 & 4 \\ 2 & 3 & -2 & 2 \end{array}\right]$$

Using Gauss-Jordan elimination, you can rewrite this matrix as

$$[I \vdots X] = \left[\begin{array}{ccc|c} 1 & 0 & 0 & -1 \\ 0 & 1 & 0 & 2 \\ 0 & 0 & 1 & 1 \end{array}\right].$$

So, the solution of the matrix equation is

$$X = \begin{bmatrix} x_1 \\ x_2 \\ x_3 \end{bmatrix} = \begin{bmatrix} -1 \\ 2 \\ 1 \end{bmatrix}.$$

Checkpoint *Audio-video solution in English & Spanish at LarsonPrecalculus.com.*

For the system of linear equations, (a) write the system as a matrix equation, $AX = B$, and (b) use Gauss-Jordan elimination on $[A \vdots B]$ to solve for the matrix X.

$$\begin{cases} -2x_1 - 3x_2 = -4 \\ 6x_1 + x_2 = -36 \end{cases}$$

© Sean Locke/iStockphoto.com

EXAMPLE 13 Softball Team Expenses

Two softball teams submit equipment lists to their sponsors.

Equipment	Women's Team	Men's Team
Bats	12	15
Balls	45	38
Gloves	15	17

Each bat costs \$80, each ball costs \$6, and each glove costs \$60. Use matrices to find the total cost of equipment for each team.

Solution The equipment lists E and the costs per item C can be written in matrix form as

$$E = \begin{bmatrix} 12 & 15 \\ 45 & 38 \\ 15 & 17 \end{bmatrix}$$

and

$$C = [80 \quad 6 \quad 60].$$

The total cost of equipment for each team is given by the product

$$CE = [80 \quad 6 \quad 60]\begin{bmatrix} 12 & 15 \\ 45 & 38 \\ 15 & 17 \end{bmatrix}$$

$$= [80(12) + 6(45) + 60(15) \quad 80(15) + 6(38) + 60(17)]$$

$$= [2130 \quad 2448].$$

So, the total cost of equipment for the women's team is \$2130, and the total cost of equipment for the men's team is \$2448.

REMARK Notice in Example 13 that you cannot find the total cost using the product EC because EC is not defined. That is, the number of columns of E (2 columns) does not equal the number of rows of C (1 row).

Checkpoint *Audio-video solution in English & Spanish at LarsonPrecalculus.com.*

Repeat Example 13 when each bat costs \$100, each ball costs \$7, and each glove costs \$65.

Summarize (Section 8.2)

1. State the conditions under which two matrices are equal *(page 551)*. For an example of the equality of matrices, see Example 1.
2. State the definition of matrix addition *(page 552)*. For an example of matrix addition, see Example 2.
3. State the definition of scalar multiplication *(page 552)*. For an example of scalar multiplication, see Example 3.
4. List the properties of matrix addition and scalar multiplication *(page 554)*. For examples of using these properties, see Examples 4, 5, and 6.
5. State the definition of matrix multiplication *(page 556)*. For examples of matrix multiplication, see Examples 7–11.
6. Describe applications of matrix multiplication *(pages 559 and 560, Examples 12 and 13)*.

8.2 Exercises

See CalcChat.com for tutorial help and worked-out solutions to odd-numbered exercises.

Vocabulary: In Exercises 1–4, fill in the blanks.

1. Two matrices are ________ when their corresponding entries are equal.
2. When performing matrix operations, real numbers are often referred to as ________.
3. A matrix consisting entirely of zeros is called a ________ matrix and is denoted by ________.
4. The $n \times n$ matrix consisting of 1's on its main diagonal and 0's elsewhere is called the ________ matrix of order $n \times n$.

Skills and Applications

Equality of Matrices In Exercises 5–8, find x and y.

5. $\begin{bmatrix} x & -2 \\ 7 & y \end{bmatrix} = \begin{bmatrix} -4 & -2 \\ 7 & 22 \end{bmatrix}$

6. $\begin{bmatrix} -5 & x \\ y & 8 \end{bmatrix} = \begin{bmatrix} -5 & 13 \\ 12 & 8 \end{bmatrix}$

7. $\begin{bmatrix} 16 & 4 & 5 & 4 \\ -3 & 13 & 15 & 6 \\ 0 & 2 & 4 & 0 \end{bmatrix} = \begin{bmatrix} 16 & 4 & 2x+1 & 4 \\ -3 & 13 & 15 & 3x \\ 0 & 2 & 3y-5 & 0 \end{bmatrix}$

8. $\begin{bmatrix} x+2 & 8 & -3 \\ 1 & 2y & 2x \\ 7 & -2 & y+2 \end{bmatrix} = \begin{bmatrix} 2x+6 & 8 & -3 \\ 1 & 18 & -8 \\ 7 & -2 & 11 \end{bmatrix}$

Operations with Matrices In Exercises 9–16, if possible, find (a) $A + B$, (b) $A - B$, (c) $3A$, and (d) $3A - 2B$.

9. $A = \begin{bmatrix} 1 & -1 \\ 2 & -1 \end{bmatrix}$, $B = \begin{bmatrix} 2 & -1 \\ -1 & 8 \end{bmatrix}$

10. $A = \begin{bmatrix} 1 & 2 \\ 2 & 1 \end{bmatrix}$, $B = \begin{bmatrix} -3 & -2 \\ 4 & 2 \end{bmatrix}$

11. $A = \begin{bmatrix} 8 & -1 \\ 2 & 3 \\ -4 & 5 \end{bmatrix}$, $B = \begin{bmatrix} 1 & 6 \\ -1 & -5 \\ 1 & 10 \end{bmatrix}$

12. $A = \begin{bmatrix} 1 & -1 & 3 \\ 0 & 6 & 9 \end{bmatrix}$, $B = \begin{bmatrix} -2 & 0 & -5 \\ -3 & 4 & -7 \end{bmatrix}$

13. $A = \begin{bmatrix} 4 & 5 & -1 & 3 & 4 \\ 1 & 2 & -2 & -1 & 0 \end{bmatrix}$,
$B = \begin{bmatrix} 1 & 0 & -1 & 1 & 0 \\ -6 & 8 & 2 & -3 & -7 \end{bmatrix}$

14. $A = \begin{bmatrix} -1 & 4 & 0 \\ 3 & -2 & 2 \\ 5 & 4 & -1 \\ 0 & 8 & -6 \\ -4 & -1 & 0 \end{bmatrix}$, $B = \begin{bmatrix} -3 & 5 & 1 \\ 2 & -4 & -7 \\ 10 & -9 & -1 \\ 3 & 2 & -4 \\ 0 & 1 & -2 \end{bmatrix}$

15. $A = \begin{bmatrix} 6 & 0 & 3 \\ -1 & -4 & 0 \end{bmatrix}$, $B = \begin{bmatrix} 8 & -1 \\ 4 & -3 \end{bmatrix}$

16. $A = \begin{bmatrix} 3 \\ 2 \\ -1 \end{bmatrix}$, $B = \begin{bmatrix} -4 & 6 & 2 \end{bmatrix}$

Evaluating an Expression In Exercises 17–22, evaluate the expression.

17. $\begin{bmatrix} -5 & 0 \\ 3 & -6 \end{bmatrix} + \begin{bmatrix} 7 & 1 \\ -2 & -1 \end{bmatrix} + \begin{bmatrix} -10 & -8 \\ 14 & 6 \end{bmatrix}$

18. $\begin{bmatrix} 6 & 8 \\ -1 & 0 \end{bmatrix} + \begin{bmatrix} 0 & 5 \\ -3 & -1 \end{bmatrix} + \begin{bmatrix} -11 & -7 \\ 2 & -1 \end{bmatrix}$

19. $4\left(\begin{bmatrix} -4 & 0 & 1 \\ 0 & 2 & 3 \end{bmatrix} - \begin{bmatrix} 2 & 1 & -2 \\ 3 & -6 & 0 \end{bmatrix}\right)$

20. $\frac{1}{2}(\begin{bmatrix} 5 & -2 & 4 & 0 \end{bmatrix} + \begin{bmatrix} 14 & 6 & -18 & 9 \end{bmatrix})$

21. $-3\left(\begin{bmatrix} 0 & -3 \\ 7 & 2 \end{bmatrix} + \begin{bmatrix} -6 & 3 \\ 8 & 1 \end{bmatrix}\right) - 2\begin{bmatrix} 4 & -4 \\ 7 & -9 \end{bmatrix}$

22. $-1\begin{bmatrix} 4 & 11 \\ -2 & -1 \\ 9 & 3 \end{bmatrix} + \frac{1}{6}\left(\begin{bmatrix} -5 & -1 \\ 3 & 4 \\ 0 & 13 \end{bmatrix} + \begin{bmatrix} 7 & 5 \\ -9 & -1 \\ 6 & -1 \end{bmatrix}\right)$

Operations with Matrices In Exercises 23–26, use the matrix capabilities of a graphing utility to evaluate the expression.

23. $\frac{11}{25}\begin{bmatrix} 2 & 5 \\ -1 & -4 \end{bmatrix} + 6\begin{bmatrix} -3 & 0 \\ 2 & 2 \end{bmatrix}$

24. $55\left(\begin{bmatrix} 14 & -11 \\ -22 & 19 \end{bmatrix} + \begin{bmatrix} -22 & 20 \\ 13 & 6 \end{bmatrix}\right)$

25. $-1\begin{bmatrix} 3.211 & 6.829 \\ -1.004 & 4.914 \\ 0.055 & -3.889 \end{bmatrix} - \begin{bmatrix} -1.630 & -3.090 \\ 5.256 & 8.335 \\ -9.768 & 4.251 \end{bmatrix}$

26. $-1\begin{bmatrix} 10 & 15 \\ -20 & 10 \\ 12 & 4 \end{bmatrix} + \frac{1}{8}\left(\begin{bmatrix} -13 & 11 \\ 7 & 0 \\ 6 & 9 \end{bmatrix} + \begin{bmatrix} -3 & 13 \\ -3 & 8 \\ -14 & 15 \end{bmatrix}\right)$

Solving a Matrix Equation In Exercises 27–34, solve for X in the equation, where

$$A = \begin{bmatrix} -2 & -1 \\ 1 & 0 \\ 3 & -4 \end{bmatrix} \quad \text{and} \quad B = \begin{bmatrix} 0 & 3 \\ 2 & 0 \\ -4 & -1 \end{bmatrix}.$$

27. $X = 2A + 2B$
28. $X = 3A - 2B$
29. $2X = A + B$
30. $2X = 2A - B$
31. $2X + 3A = B$
32. $3X - 4A = 2B$
33. $2A + 4B = -2X$
34. $5A - 6B = -3X$

Finding the Product of Two Matrices **In Exercises 35–42, if possible, find AB and state the order of the result.**

35. $A = \begin{bmatrix} 2 & 1 \\ -3 & 4 \\ 1 & 6 \end{bmatrix}, \quad B = \begin{bmatrix} 0 & -1 & 0 \\ 4 & 0 & 2 \\ 8 & -1 & 7 \end{bmatrix}$

36. $A = \begin{bmatrix} 0 & -1 & 2 \\ 6 & 0 & 3 \\ 7 & -1 & 8 \end{bmatrix}, \quad B = \begin{bmatrix} 2 & -1 \\ 4 & -5 \\ 1 & 6 \end{bmatrix}$

37. $A = \begin{bmatrix} -1 & 6 \\ -4 & 5 \\ 0 & 3 \end{bmatrix}, \quad B = \begin{bmatrix} 2 & 3 \\ 0 & 9 \end{bmatrix}$

38. $A = \begin{bmatrix} 1 & 0 & 0 \\ 0 & 4 & 0 \\ 0 & 0 & -2 \end{bmatrix}, \quad B = \begin{bmatrix} 3 & 0 & 0 \\ 0 & -1 & 0 \\ 0 & 0 & 5 \end{bmatrix}$

39. $A = \begin{bmatrix} 5 & 0 & 0 \\ 0 & -8 & 0 \\ 0 & 0 & 7 \end{bmatrix}, \quad B = \begin{bmatrix} \frac{1}{5} & 0 & 0 \\ 0 & -\frac{1}{8} & 0 \\ 0 & 0 & \frac{1}{2} \end{bmatrix}$

40. $A = \begin{bmatrix} 0 & 0 & 5 \\ 0 & 0 & -3 \\ 0 & 0 & 4 \end{bmatrix}, \quad B = \begin{bmatrix} 6 & -11 & 4 \\ 8 & 16 & 4 \\ 0 & 0 & 0 \end{bmatrix}$

41. $A = \begin{bmatrix} 10 \\ 12 \end{bmatrix}, \quad B = \begin{bmatrix} 6 & -2 & 1 & 6 \end{bmatrix}$

42. $A = \begin{bmatrix} 1 & 0 & 3 & -2 \\ 6 & 13 & 8 & -17 \end{bmatrix}, \quad B = \begin{bmatrix} 1 & 6 \\ 4 & 2 \end{bmatrix}$

Matrix Multiplication **In Exercises 43–46, use the matrix capabilities of a graphing utility to find AB, if possible.**

43. $A = \begin{bmatrix} 7 & 5 & -4 \\ -2 & 5 & 1 \\ 10 & -4 & -7 \end{bmatrix}, \quad B = \begin{bmatrix} 2 & -2 & 3 \\ 8 & 1 & 4 \\ -4 & 2 & -8 \end{bmatrix}$

44. $A = \begin{bmatrix} 11 & -12 & 4 \\ 14 & 10 & 12 \\ 6 & -2 & 9 \end{bmatrix}, \quad B = \begin{bmatrix} 12 & 10 \\ -5 & 12 \\ 15 & 16 \end{bmatrix}$

45. $A = \begin{bmatrix} -3 & 8 & -6 & 8 \\ -12 & 15 & 9 & 6 \\ 5 & -1 & 1 & 5 \end{bmatrix}, \quad B = \begin{bmatrix} 3 & 1 & 6 \\ 24 & 15 & 14 \\ 16 & 10 & 21 \\ 8 & -4 & 10 \end{bmatrix}$

46. $A = \begin{bmatrix} -2 & 4 & 8 \\ 21 & 5 & 6 \\ 13 & 2 & 6 \end{bmatrix}, \quad B = \begin{bmatrix} 2 & 0 \\ -7 & 15 \\ 32 & 14 \\ 0.5 & 1.6 \end{bmatrix}$

Operations with Matrices **In Exercises 47–52, if possible, find (a) AB, (b) BA, and (c) A^2.**

47. $A = \begin{bmatrix} 1 & 2 \\ 4 & 2 \end{bmatrix}, \quad B = \begin{bmatrix} 2 & -1 \\ -1 & 8 \end{bmatrix}$

48. $A = \begin{bmatrix} 6 & 3 \\ -2 & -4 \end{bmatrix}, \quad B = \begin{bmatrix} -2 & 0 \\ 2 & 4 \end{bmatrix}$

49. $A = \begin{bmatrix} 3 & -1 \\ 1 & 3 \end{bmatrix}, \quad B = \begin{bmatrix} 1 & -3 \\ 3 & 1 \end{bmatrix}$

50. $A = \begin{bmatrix} 1 & -1 \\ 1 & 1 \end{bmatrix}, \quad B = \begin{bmatrix} 1 & 3 \\ -3 & 1 \end{bmatrix}$

51. $A = \begin{bmatrix} 7 \\ 8 \\ -1 \end{bmatrix}, \quad B = \begin{bmatrix} 1 & 1 & 2 \end{bmatrix}$

52. $A = \begin{bmatrix} 3 & 2 & 1 \end{bmatrix}, \quad B = \begin{bmatrix} 2 \\ 3 \\ 0 \end{bmatrix}$

Operations with Matrices **In Exercises 53–56, evaluate the expression. Use the matrix capabilities of a graphing utility to verify your answer.**

53. $\begin{bmatrix} 3 & 1 \\ 0 & -2 \end{bmatrix}\begin{bmatrix} 1 & 0 \\ -2 & 2 \end{bmatrix}\begin{bmatrix} 1 & 0 \\ 2 & 4 \end{bmatrix}$

54. $-3\left(\begin{bmatrix} 6 & 5 & -1 \\ 1 & -2 & 0 \end{bmatrix}\begin{bmatrix} 0 & 3 \\ -1 & -3 \\ 4 & 1 \end{bmatrix}\right)$

55. $\begin{bmatrix} 0 & 2 & -2 \\ 4 & 1 & 2 \end{bmatrix}\left(\begin{bmatrix} 4 & 0 \\ 0 & -1 \\ -1 & 2 \end{bmatrix} + \begin{bmatrix} -2 & 3 \\ -3 & 5 \\ 0 & -3 \end{bmatrix}\right)$

56. $\begin{bmatrix} 3 \\ -1 \\ 5 \\ 7 \end{bmatrix}\left(\begin{bmatrix} 5 & -6 \end{bmatrix} + \begin{bmatrix} 7 & -1 \end{bmatrix} + \begin{bmatrix} -8 & 9 \end{bmatrix}\right)$

Solving a System of Linear Equations **In Exercises 57–64, (a) write the system of linear equations as a matrix equation, $AX = B$, and (b) use Gauss-Jordan elimination on $[A \vdots B]$ to solve for the matrix X.**

57. $\begin{cases} -x_1 + x_2 = 4 \\ -2x_1 + x_2 = 0 \end{cases}$

58. $\begin{cases} 2x_1 + 3x_2 = 5 \\ x_1 + 4x_2 = 10 \end{cases}$

59. $\begin{cases} -2x_1 - 3x_2 = -4 \\ 6x_1 + x_2 = -36 \end{cases}$

60. $\begin{cases} -4x_1 + 9x_2 = -13 \\ x_1 - 3x_2 = 12 \end{cases}$

61. $\begin{cases} x_1 - 2x_2 + 3x_3 = 9 \\ -x_1 + 3x_2 - x_3 = -6 \\ 2x_1 - 5x_2 + 5x_3 = 17 \end{cases}$

62. $\begin{cases} x_1 + x_2 - 3x_3 = -1 \\ -x_1 + 2x_2 = 1 \\ x_1 - x_2 + x_3 = 2 \end{cases}$

63. $\begin{cases} x_1 - 5x_2 + 2x_3 = -20 \\ -3x_1 + x_2 - x_3 = 8 \\ -2x_2 + 5x_3 = -16 \end{cases}$

64. $\begin{cases} x_1 - x_2 + 4x_3 = 17 \\ x_1 + 3x_2 = -11 \\ -6x_2 + 5x_3 = 40 \end{cases}$

8.3 The Inverse of a Square Matrix

You can use inverse matrices to model and solve real-life problems. For instance, in Exercises 65–68 on pages 572 and 573, you will use an inverse matrix to find the currents in a circuit.

- **Verify that two matrices are inverses of each other.**
- **Use Gauss-Jordan elimination to find the inverses of matrices.**
- **Use a formula to find the inverses of 2×2 matrices.**
- **Use inverse matrices to solve systems of linear equations.**

The Inverse of a Matrix

This section further develops the algebra of matrices. To begin, consider the real number equation $ax = b$. To solve this equation for x, multiply each side of the equation by a^{-1} (provided that $a \neq 0$).

$$ax = b$$

$$(a^{-1}a)x = a^{-1}b$$

$$(1)x = a^{-1}b$$

$$x = a^{-1}b$$

The number a^{-1} is called the *multiplicative inverse* of a because $a^{-1}a = 1$. The definition of the multiplicative **inverse of a matrix** is similar.

Definition of the Inverse of a Square Matrix

Let A be an $n \times n$ matrix and let I_n be the $n \times n$ identity matrix. If there exists a matrix A^{-1} such that

$$AA^{-1} = I_n = A^{-1}A$$

then A^{-1} is called the **inverse** of A. The symbol A^{-1} is read "A inverse."

EXAMPLE 1 The Inverse of a Matrix

Show that B is the inverse of A, where

$$A = \begin{bmatrix} -1 & 2 \\ -1 & 1 \end{bmatrix} \quad \text{and} \quad B = \begin{bmatrix} 1 & -2 \\ 1 & -1 \end{bmatrix}.$$

Solution To show that B is the inverse of A, show that $AB = I = BA$, as follows.

$$AB = \begin{bmatrix} -1 & 2 \\ -1 & 1 \end{bmatrix}\begin{bmatrix} 1 & -2 \\ 1 & -1 \end{bmatrix} = \begin{bmatrix} -1+2 & 2-2 \\ -1+1 & 2-1 \end{bmatrix} = \begin{bmatrix} 1 & 0 \\ 0 & 1 \end{bmatrix}$$

$$BA = \begin{bmatrix} 1 & -2 \\ 1 & -1 \end{bmatrix}\begin{bmatrix} -1 & 2 \\ -1 & 1 \end{bmatrix} = \begin{bmatrix} -1+2 & 2-2 \\ -1+1 & 2-1 \end{bmatrix} = \begin{bmatrix} 1 & 0 \\ 0 & 1 \end{bmatrix}$$

As you can see, $AB = I = BA$. This is an example of a square matrix that has an inverse. Note that not all square matrices have inverses.

✓ **Checkpoint** *Audio-video solution in English & Spanish at LarsonPrecalculus.com.*

Show that B is the inverse of A, where

$$A = \begin{bmatrix} 2 & -1 \\ -3 & 1 \end{bmatrix} \quad \text{and} \quad B = \begin{bmatrix} -1 & -1 \\ -3 & -2 \end{bmatrix}.$$

Recall that it is not always true that $AB = BA$, even when both products are defined. However, if A and B are both square matrices and $AB = I_n$, then it can be shown that $BA = I_n$. So, in Example 1, you need only to check that $AB = I_2$.

Ingrid Prats/Shutterstock.com

One real-life application of inverse matrices is in the study of beam deflection. In a simply supported elastic beam subjected to multiple forces, deflection **d** is related to force **w** by the matrix equation

$$\mathbf{d} = F\mathbf{w}$$

where F is a *flexibility matrix* whose entries depend on the material of the beam. The inverse of the flexibility matrix, F^{-1}, is called the *stiffness matrix*.

Finding Inverse Matrices

If a matrix A has an inverse, then A is called **invertible** (or **nonsingular**); otherwise, A is called **singular.** A nonsquare matrix cannot have an inverse. To see this, note that when A is of order $m \times n$ and B is of order $n \times m$ (where $m \neq n$), the products AB and BA are of different orders and so cannot be equal to each other. Not all square matrices have inverses (see the matrix at the bottom of page 568). When, however, a matrix does have an inverse, that inverse is unique. Example 2 shows how to use a system of equations to find the inverse of a matrix.

EXAMPLE 2 Finding the Inverse of a Matrix

Find the inverse of

$$A = \begin{bmatrix} 1 & 4 \\ -1 & -3 \end{bmatrix}.$$

Solution To find the inverse of A, solve the matrix equation $AX = I$ for X.

$$\overset{A}{\begin{bmatrix} 1 & 4 \\ -1 & -3 \end{bmatrix}} \overset{X}{\begin{bmatrix} x_{11} & x_{12} \\ x_{21} & x_{22} \end{bmatrix}} \overset{=}{=} \overset{I}{\begin{bmatrix} 1 & 0 \\ 0 & 1 \end{bmatrix}}$$ Write matrix equation.

$$\begin{bmatrix} x_{11} + 4x_{21} & x_{12} + 4x_{22} \\ -x_{11} - 3x_{21} & -x_{12} - 3x_{22} \end{bmatrix} = \begin{bmatrix} 1 & 0 \\ 0 & 1 \end{bmatrix}$$ Multiply.

Equating corresponding entries, you obtain two systems of linear equations.

$$\begin{cases} x_{11} + 4x_{21} = 1 \\ -x_{11} - 3x_{21} = 0 \end{cases} \qquad \begin{cases} x_{12} + 4x_{22} = 0 \\ -x_{12} - 3x_{22} = 1 \end{cases}$$

Solve the first system using elementary row operations to determine that

$$x_{11} = -3 \quad \text{and} \quad x_{21} = 1.$$

From the second system, you can determine that

$$x_{12} = -4 \quad \text{and} \quad x_{22} = 1.$$

So, the inverse of A is

$$X = A^{-1} = \begin{bmatrix} -3 & -4 \\ 1 & 1 \end{bmatrix}.$$

You can use matrix multiplication to check this result.

Check

$$AA^{-1} = \begin{bmatrix} 1 & 4 \\ -1 & -3 \end{bmatrix}\begin{bmatrix} -3 & -4 \\ 1 & 1 \end{bmatrix} = \begin{bmatrix} 1 & 0 \\ 0 & 1 \end{bmatrix} \checkmark$$

$$A^{-1}A = \begin{bmatrix} -3 & -4 \\ 1 & 1 \end{bmatrix}\begin{bmatrix} 1 & 4 \\ -1 & -3 \end{bmatrix} = \begin{bmatrix} 1 & 0 \\ 0 & 1 \end{bmatrix} \checkmark$$

✓ ***Checkpoint*** *Audio-video solution in English & Spanish at LarsonPrecalculus.com.*

Find the inverse of

$$A = \begin{bmatrix} 1 & -2 \\ -1 & 3 \end{bmatrix}.$$

nostal6ie/Shutterstock.com

In Example 2, note that the two systems of linear equations have the *same coefficient matrix A*. Rather than solve the two systems represented by

$$\begin{bmatrix} 1 & 4 & \vdots & 1 \\ -1 & -3 & \vdots & 0 \end{bmatrix}$$

and

$$\begin{bmatrix} 1 & 4 & \vdots & 0 \\ -1 & -3 & \vdots & 1 \end{bmatrix}$$

separately, you can solve them *simultaneously* by *adjoining* the identity matrix to the coefficient matrix to obtain

$$\overset{A}{\begin{bmatrix} 1 & 4 \\ -1 & -3 \end{bmatrix}} \begin{matrix} \vdots \\ \vdots \end{matrix} \overset{I}{\begin{bmatrix} 1 & 0 \\ 0 & 1 \end{bmatrix}}.$$

This "doubly augmented" matrix can be represented as

$[A \ \vdots \ I]$.

By applying Gauss-Jordan elimination to this matrix, you can solve *both* systems with a single elimination process.

▷ **TECHNOLOGY** Most graphing utilities can find the inverse of a square matrix. To do so, you may have to use the inverse key x^{-1}. Consult the user's guide for your graphing utility for specific keystrokes.

$$\begin{bmatrix} 1 & 4 & \vdots & 1 & 0 \\ -1 & -3 & \vdots & 0 & 1 \end{bmatrix}$$

$$R_1 + R_2 \rightarrow \begin{bmatrix} 1 & 4 & \vdots & 1 & 0 \\ 0 & 1 & \vdots & 1 & 1 \end{bmatrix}$$

$$-4R_2 + R_1 \rightarrow \begin{bmatrix} 1 & 0 & \vdots & -3 & -4 \\ 0 & 1 & \vdots & 1 & 1 \end{bmatrix}$$

So, from the "doubly augmented" matrix $[A \ \vdots \ I]$, you obtain the matrix $[I \ \vdots \ A^{-1}]$.

$$\overset{A}{\begin{bmatrix} 1 & 4 \\ -1 & -3 \end{bmatrix}} \begin{matrix} \vdots \\ \vdots \end{matrix} \overset{I}{\begin{bmatrix} 1 & 0 \\ 0 & 1 \end{bmatrix}} \Rightarrow \overset{I}{\begin{bmatrix} 1 & 0 \\ 0 & 1 \end{bmatrix}} \begin{matrix} \vdots \\ \vdots \end{matrix} \overset{A^{-1}}{\begin{bmatrix} -3 & -4 \\ 1 & 1 \end{bmatrix}}$$

This procedure (or algorithm) works for any square matrix that has an inverse.

Finding an Inverse Matrix

Let A be a square matrix of order $n \times n$.

1. Write the $n \times 2n$ matrix that consists of the given matrix A on the left and the $n \times n$ identity matrix I on the right to obtain

   $[A \ \vdots \ I]$.

2. If possible, row reduce A to I using elementary row operations on the *entire* matrix

   $[A \ \vdots \ I]$.

   The result will be the matrix

   $[I \ \vdots \ A^{-1}]$.

   If this is not possible, then A is not invertible.

3. Check your work by multiplying to see that

   $AA^{-1} = I = A^{-1}A$.

EXAMPLE 3 Finding the Inverse of a Matrix

Find the inverse of

$$A = \begin{bmatrix} 1 & -1 & 0 \\ 1 & 0 & -1 \\ 6 & -2 & -3 \end{bmatrix}.$$

Solution Begin by adjoining the identity matrix to A to form the matrix

$$[A \ \vdots \ I] = \left[\begin{array}{ccc:ccc} 1 & -1 & 0 & 1 & 0 & 0 \\ 1 & 0 & -1 & 0 & 1 & 0 \\ 6 & -2 & -3 & 0 & 0 & 1 \end{array}\right].$$

Use elementary row operations to obtain the form $[I \ \vdots \ A^{-1}]$, as follows.

$$\begin{array}{r} \\ -R_1 + R_2 \to \\ -6R_1 + R_3 \to \end{array} \left[\begin{array}{ccc:ccc} 1 & -1 & 0 & 1 & 0 & 0 \\ 0 & 1 & -1 & -1 & 1 & 0 \\ 0 & 4 & -3 & -6 & 0 & 1 \end{array}\right]$$

$$\begin{array}{r} R_2 + R_1 \to \\ \\ -4R_2 + R_3 \to \end{array} \left[\begin{array}{ccc:ccc} 1 & 0 & -1 & 0 & 1 & 0 \\ 0 & 1 & -1 & -1 & 1 & 0 \\ 0 & 0 & 1 & -2 & -4 & 1 \end{array}\right]$$

$$\begin{array}{r} R_3 + R_1 \to \\ R_3 + R_2 \to \\ \\ \end{array} \left[\begin{array}{ccc:ccc} 1 & 0 & 0 & -2 & -3 & 1 \\ 0 & 1 & 0 & -3 & -3 & 1 \\ 0 & 0 & 1 & -2 & -4 & 1 \end{array}\right] = [I \ \vdots \ A^{-1}]$$

So, the matrix A is invertible and its inverse is $A^{-1} = \begin{bmatrix} -2 & -3 & 1 \\ -3 & -3 & 1 \\ -2 & -4 & 1 \end{bmatrix}$.

Confirm this result by multiplying A and A^{-1} to obtain I, as follows.

Check

$$AA^{-1} = \begin{bmatrix} 1 & -1 & 0 \\ 1 & 0 & -1 \\ 6 & -2 & -3 \end{bmatrix} \begin{bmatrix} -2 & -3 & 1 \\ -3 & -3 & 1 \\ -2 & -4 & 1 \end{bmatrix} = \begin{bmatrix} 1 & 0 & 0 \\ 0 & 1 & 0 \\ 0 & 0 & 1 \end{bmatrix} = I$$

REMARK Be sure to check your solution because it is easy to make algebraic errors when using elementary row operations.

✓ ***Checkpoint*** *Audio-video solution in English & Spanish at LarsonPrecalculus.com.*

Find the inverse of

$$A = \begin{bmatrix} 1 & -2 & -1 \\ 0 & -1 & 2 \\ 1 & -2 & 0 \end{bmatrix}.$$

The process shown in Example 3 applies to any $n \times n$ matrix A. When using this algorithm, if the matrix A does not reduce to the identity matrix, then A does not have an inverse. For instance, the following matrix has no inverse.

$$A = \begin{bmatrix} 1 & 2 & 0 \\ 3 & -1 & 2 \\ -2 & 3 & -2 \end{bmatrix}$$

To confirm that matrix A above has no inverse, adjoin the identity matrix to A to form $[A \ \vdots \ I]$ and perform elementary row operations on the matrix. After doing so, you will see that it is impossible to obtain the identity matrix I on the left. So, A is not invertible.

The Inverse of a 2 × 2 Matrix

Using Gauss-Jordan elimination to find the inverse of a matrix works well (even as a computer technique) for matrices of order 3×3 or greater. For 2×2 matrices, however, many people prefer to use a formula for the inverse rather than Gauss-Jordan elimination. This simple formula, which works *only* for 2×2 matrices, is explained as follows. If A is a 2×2 matrix given by

$$A = \begin{bmatrix} a & b \\ c & d \end{bmatrix}$$

then A is invertible if and only if $ad - bc \neq 0$. Moreover, if $ad - bc \neq 0$, then the inverse is given by

$$A^{-1} = \frac{1}{ad - bc}\begin{bmatrix} d & -b \\ -c & a \end{bmatrix}.$$ Formula for the inverse of a 2 × 2 matrix

The denominator $ad - bc$ is called the **determinant** of the 2×2 matrix A. You will study determinants in the next section.

EXAMPLE 4 Finding the Inverse of a 2 × 2 Matrix

If possible, find the inverse of each matrix.

a. $A = \begin{bmatrix} 3 & -1 \\ -2 & 2 \end{bmatrix}$ **b.** $B = \begin{bmatrix} 3 & -1 \\ -6 & 2 \end{bmatrix}$

Solution

a. For the matrix A, apply the formula for the inverse of a 2×2 matrix to obtain

$$ad - bc = 3(2) - (-1)(-2) = 4.$$

Because this quantity is not zero, the matrix is invertible. The inverse is formed by interchanging the entries on the main diagonal, changing the signs of the other two entries, and multiplying by the scalar $\frac{1}{4}$, as follows.

$$A^{-1} = \frac{1}{ad - bc}\begin{bmatrix} d & -b \\ -c & a \end{bmatrix}$$ Formula for the inverse of a 2 × 2 matrix

$$= \frac{1}{4}\begin{bmatrix} 2 & 1 \\ 2 & 3 \end{bmatrix}$$ Substitute for a, b, c, d, and the determinant.

$$= \begin{bmatrix} \frac{1}{4}(2) & \frac{1}{4}(1) \\ \frac{1}{4}(2) & \frac{1}{4}(3) \end{bmatrix}$$ Multiply by the scalar $\frac{1}{4}$.

$$= \begin{bmatrix} \frac{1}{2} & \frac{1}{4} \\ \frac{1}{2} & \frac{3}{4} \end{bmatrix}$$ Simplify.

b. For the matrix B, you have

$$ad - bc = 3(2) - (-1)(-6) = 0.$$

Because $ad - bc = 0$, B is not invertible.

✓ ***Checkpoint*** *Audio-video solution in English & Spanish at LarsonPrecalculus.com.*

If possible, find the inverse of $A = \begin{bmatrix} 5 & -1 \\ 3 & 4 \end{bmatrix}$.

Systems of Linear Equations

You know that a system of linear equations can have exactly one solution, infinitely many solutions, or no solution. If the coefficient matrix A of a *square* system (a system that has the same number of equations as variables) is invertible, then the system has a unique solution, which is defined as follows.

> **A System of Equations with a Unique Solution**
>
> If A is an invertible matrix, then the system of linear equations represented by $AX = B$ has a unique solution given by $X = A^{-1}B$.

EXAMPLE 5 **Solving a System Using an Inverse Matrix**

Use an inverse matrix to solve the system

$$\begin{cases} x + y + z = 10{,}000 \\ 0.06x + 0.075y + 0.095z = 730. \\ x - 2z = 0 \end{cases}$$

Solution Begin by writing the system in the matrix form $AX = B$.

$$\begin{bmatrix} 1 & 1 & 1 \\ 0.06 & 0.075 & 0.095 \\ 1 & 0 & -2 \end{bmatrix}\begin{bmatrix} x \\ y \\ z \end{bmatrix} = \begin{bmatrix} 10{,}000 \\ 730 \\ 0 \end{bmatrix}$$

Then, use Gauss-Jordan elimination to find A^{-1}.

$$A^{-1} = \begin{bmatrix} 15 & -200 & -2 \\ -21.5 & 300 & 3.5 \\ 7.5 & -100 & -1.5 \end{bmatrix}$$

Finally, multiply B by A^{-1} on the left to obtain the solution.

$$X = A^{-1}B = \begin{bmatrix} 15 & -200 & -2 \\ -21.5 & 300 & 3.5 \\ 7.5 & -100 & -1.5 \end{bmatrix}\begin{bmatrix} 10{,}000 \\ 730 \\ 0 \end{bmatrix} = \begin{bmatrix} 4000 \\ 4000 \\ 2000 \end{bmatrix}$$

The solution of the system is $x = 4000$, $y = 4000$, and $z = 2000$.

TECHNOLOGY To solve a system of equations with a graphing utility, enter the matrices A and B in the matrix editor. Then, using the inverse key, solve for X.

A [x^{-1}] B [ENTER]

The screen will display the solution, matrix X.

✓ ***Checkpoint*** *Audio-video solution in English & Spanish at LarsonPrecalculus.com.*

Use an inverse matrix to solve the system $\begin{cases} 2x + 3y + z = -1 \\ 3x + 3y + z = 1. \\ 2x + 4y + z = -2 \end{cases}$

> **Summarize** **(Section 8.3)**
>
> 1. State the definition of the inverse of a square matrix *(page 565)*. For an example of how to show that a matrix is the inverse of another matrix, see Example 1.
> 2. Explain how to find an inverse matrix *(pages 566 and 567)*. For examples of finding inverse matrices, see Examples 2 and 3.
> 3. Give the formula for finding the inverse of a 2×2 matrix *(page 569)*. For an example of using this formula, see Example 4.
> 4. Describe how to use an inverse matrix to solve a system of linear equations *(page 570)*. For an example of using an inverse matrix to solve a system of linear equations, see Example 5.

8.3 Exercises

See CalcChat.com for tutorial help and worked-out solutions to odd-numbered exercises.

Vocabulary: Fill in the blanks.

1. In a ________ matrix, the number of rows equals the number of columns.
2. If there exists an $n \times n$ matrix A^{-1} such that $AA^{-1} = I_n = A^{-1}A$, then A^{-1} is called the ________ of A.
3. If a matrix A has an inverse, then it is called invertible or ________; if it does not have an inverse, then it is called ________.
4. If A is an invertible matrix, then the system of linear equations represented by $AX = B$ has a unique solution given by $X =$ ________.

Skills and Applications

The Inverse of a Matrix In Exercises 5–12, show that B is the inverse of A.

5. $A = \begin{bmatrix} 2 & 1 \\ 5 & 3 \end{bmatrix}, B = \begin{bmatrix} 3 & -1 \\ -5 & 2 \end{bmatrix}$

6. $A = \begin{bmatrix} 1 & -1 \\ -1 & 2 \end{bmatrix}, B = \begin{bmatrix} 2 & 1 \\ 1 & 1 \end{bmatrix}$

7. $A = \begin{bmatrix} 1 & 2 \\ 3 & 4 \end{bmatrix}, B = \frac{1}{2}\begin{bmatrix} -4 & 2 \\ 3 & -1 \end{bmatrix}$

8. $A = \begin{bmatrix} 1 & -1 \\ 2 & 3 \end{bmatrix}, B = \frac{1}{5}\begin{bmatrix} 3 & 1 \\ -2 & 1 \end{bmatrix}$

9. $A = \begin{bmatrix} 2 & -17 & 11 \\ -1 & 11 & -7 \\ 0 & 3 & -2 \end{bmatrix}, B = \begin{bmatrix} 1 & 1 & 2 \\ 2 & 4 & -3 \\ 3 & 6 & -5 \end{bmatrix}$

10. $A = \begin{bmatrix} -4 & 1 & 5 \\ -1 & 2 & 4 \\ 0 & -1 & -1 \end{bmatrix}, B = \frac{1}{4}\begin{bmatrix} -2 & 4 & 6 \\ 1 & -4 & -11 \\ -1 & 4 & 7 \end{bmatrix}$

11. $A = \begin{bmatrix} 2 & 0 & 2 & 1 \\ 3 & 0 & 0 & 1 \\ -1 & 1 & -2 & 1 \\ 3 & -1 & 1 & 0 \end{bmatrix},$

$B = \frac{1}{3}\begin{bmatrix} -1 & 3 & -2 & -2 \\ -2 & 9 & -7 & -10 \\ 1 & 0 & -1 & -1 \\ 3 & -6 & 6 & 6 \end{bmatrix}$

12. $A = \begin{bmatrix} -1 & 1 & 0 & -1 \\ 1 & -1 & 1 & 0 \\ -1 & 1 & 2 & 0 \\ 0 & -1 & 1 & 1 \end{bmatrix},$

$B = \frac{1}{3}\begin{bmatrix} -3 & 1 & 1 & -3 \\ -3 & -1 & 2 & -3 \\ 0 & 1 & 1 & 0 \\ -3 & -2 & 1 & 0 \end{bmatrix}$

Finding the Inverse of a Matrix In Exercises 13–24, find the inverse of the matrix (if it exists).

13. $\begin{bmatrix} 2 & 0 \\ 0 & 3 \end{bmatrix}$

14. $\begin{bmatrix} 1 & 2 \\ 3 & 7 \end{bmatrix}$

15. $\begin{bmatrix} 1 & -2 \\ 2 & -3 \end{bmatrix}$

16. $\begin{bmatrix} -7 & 33 \\ 4 & -19 \end{bmatrix}$

17. $\begin{bmatrix} 3 & 1 \\ 4 & 2 \end{bmatrix}$

18. $\begin{bmatrix} 4 & -1 \\ -3 & 1 \end{bmatrix}$

19. $\begin{bmatrix} 1 & 1 & 1 \\ 3 & 5 & 4 \\ 3 & 6 & 5 \end{bmatrix}$

20. $\begin{bmatrix} 1 & 2 & 2 \\ 3 & 7 & 9 \\ -1 & -4 & -7 \end{bmatrix}$

21. $\begin{bmatrix} -5 & 0 & 0 \\ 2 & 0 & 0 \\ -1 & 5 & 7 \end{bmatrix}$

22. $\begin{bmatrix} 1 & 0 & 0 \\ 3 & 0 & 0 \\ 2 & 5 & 5 \end{bmatrix}$

23. $\begin{bmatrix} -8 & 0 & 0 & 0 \\ 0 & 1 & 0 & 0 \\ 0 & 0 & 4 & 0 \\ 0 & 0 & 0 & -5 \end{bmatrix}$

24. $\begin{bmatrix} 1 & 3 & -2 & 0 \\ 0 & 2 & 4 & 6 \\ 0 & 0 & -2 & 1 \\ 0 & 0 & 0 & 5 \end{bmatrix}$

Finding the Inverse of a Matrix In Exercises 25–34, use the matrix capabilities of a graphing utility to find the inverse of the matrix (if it exists).

25. $\begin{bmatrix} 1 & 2 & -1 \\ 3 & 7 & -10 \\ -5 & -7 & -15 \end{bmatrix}$

26. $\begin{bmatrix} 10 & 5 & -7 \\ -5 & 1 & 4 \\ 3 & 2 & -2 \end{bmatrix}$

27. $\begin{bmatrix} 1 & 1 & 2 \\ 3 & 1 & 0 \\ -2 & 0 & 3 \end{bmatrix}$

28. $\begin{bmatrix} 3 & 2 & 2 \\ 2 & 2 & 2 \\ -4 & 4 & 3 \end{bmatrix}$

29. $\begin{bmatrix} -\frac{1}{2} & \frac{3}{4} & \frac{1}{4} \\ 1 & 0 & -\frac{3}{2} \\ 0 & -1 & \frac{1}{2} \end{bmatrix}$

30. $\begin{bmatrix} -\frac{5}{6} & \frac{1}{3} & \frac{11}{6} \\ 0 & \frac{2}{3} & 2 \\ 1 & -\frac{1}{2} & -\frac{5}{2} \end{bmatrix}$

31. $\begin{bmatrix} 0.1 & 0.2 & 0.3 \\ -0.3 & 0.2 & 0.2 \\ 0.5 & 0.4 & 0.4 \end{bmatrix}$

32. $\begin{bmatrix} 0.6 & 0 & -0.3 \\ 0.7 & -1 & 0.2 \\ 1 & 0 & -0.9 \end{bmatrix}$

33. $\begin{bmatrix} -1 & 0 & 1 & 0 \\ 0 & 2 & 0 & -1 \\ 2 & 0 & -1 & 0 \\ 0 & -1 & 0 & 1 \end{bmatrix}$

34. $\begin{bmatrix} 1 & -2 & -1 & -2 \\ 3 & -5 & -2 & -3 \\ 2 & -5 & -2 & -5 \\ -1 & 4 & 4 & 11 \end{bmatrix}$

Finding the Inverse of a 2 × 2 Matrix **In Exercises 35–40, use the formula on page 569 to find the inverse of the 2 × 2 matrix (if it exists).**

35. $\begin{bmatrix} 2 & 3 \\ -1 & 5 \end{bmatrix}$

36. $\begin{bmatrix} 1 & -2 \\ -3 & 2 \end{bmatrix}$

37. $\begin{bmatrix} -4 & -6 \\ 2 & 3 \end{bmatrix}$

38. $\begin{bmatrix} -12 & 3 \\ 5 & -2 \end{bmatrix}$

39. $\begin{bmatrix} \frac{7}{2} & -\frac{3}{4} \\ \frac{1}{5} & \frac{4}{5} \end{bmatrix}$

40. $\begin{bmatrix} -\frac{1}{4} & \frac{9}{4} \\ \frac{5}{3} & \frac{8}{9} \end{bmatrix}$

Solving a System Using an Inverse Matrix **In Exercises 41–44, use the inverse matrix found in Exercise 15 to solve the system of linear equations.**

41. $\begin{cases} x - 2y = 5 \\ 2x - 3y = 10 \end{cases}$

42. $\begin{cases} x - 2y = 0 \\ 2x - 3y = 3 \end{cases}$

43. $\begin{cases} x - 2y = 4 \\ 2x - 3y = 2 \end{cases}$

44. $\begin{cases} x - 2y = 1 \\ 2x - 3y = -2 \end{cases}$

Solving a System Using an Inverse Matrix **In Exercises 45 and 46, use the inverse matrix found in Exercise 19 to solve the system of linear equations.**

45. $\begin{cases} x + y + z = 0 \\ 3x + 5y + 4z = 5 \\ 3x + 6y + 5z = 2 \end{cases}$

46. $\begin{cases} x + y + z = -1 \\ 3x + 5y + 4z = 2 \\ 3x + 6y + 5z = 0 \end{cases}$

Solving a System Using an Inverse Matrix **In Exercises 47 and 48, use the inverse matrix found in Exercise 34 to solve the system of linear equations.**

47. $\begin{cases} x_1 - 2x_2 - x_3 - 2x_4 = 0 \\ 3x_1 - 5x_2 - 2x_3 - 3x_4 = 1 \\ 2x_1 - 5x_2 - 2x_3 - 5x_4 = -1 \\ -x_1 + 4x_2 + 4x_3 + 11x_4 - 2 \end{cases}$

48. $\begin{cases} x_1 - 2x_2 - x_3 - 2x_4 = 1 \\ 3x_1 - 5x_2 - 2x_3 - 3x_4 = -2 \\ 2x_1 - 5x_2 - 2x_3 - 5x_4 = 0 \\ -x_1 + 4x_2 + 4x_3 + 11x_4 = -3 \end{cases}$

Solving a System Using an Inverse Matrix **In Exercises 49–56, use an inverse matrix to solve (if possible) the system of linear equations.**

49. $\begin{cases} 3x + 4y = -2 \\ 5x + 3y = 4 \end{cases}$

50. $\begin{cases} 18x + 12y = 13 \\ 30x + 24y = 23 \end{cases}$

51. $\begin{cases} -0.4x + 0.8y = 1.6 \\ 2x - 4y = 5 \end{cases}$

52. $\begin{cases} 0.2x - 0.6y = 2.4 \\ -x + 1.4y = -8.8 \end{cases}$

53. $\begin{cases} -\frac{1}{4}x + \frac{3}{8}y = -2 \\ \frac{3}{2}x + \frac{3}{4}y = -12 \end{cases}$

54. $\begin{cases} \frac{5}{6}x - y = -20 \\ \frac{4}{3}x - \frac{7}{2}y = -51 \end{cases}$

55. $\begin{cases} 4x - y + z = -5 \\ 2x + 2y + 3z = 10 \\ 5x - 2y + 6z = 1 \end{cases}$

56. $\begin{cases} 4x - 2y + 3z = -2 \\ 2x + 2y + 5z = 16 \\ 8x - 5y - 2z = 4 \end{cases}$

Using a Graphing Utility **In Exercises 57–60, use the matrix capabilities of a graphing utility to solve (if possible) the system of linear equations.**

57. $\begin{cases} 5x - 3y + 2z = 2 \\ 2x + 2y - 3z = 3 \\ x - 7y + 8z = -4 \end{cases}$

58. $\begin{cases} 2x + 3y + 5z = 4 \\ 3x + 5y + 9z = 7 \\ 5x + 9y + 17z = 13 \end{cases}$

59. $\begin{cases} 3x - 2y + z = -29 \\ -4x + y - 3z = 37 \\ x - 5y + z = -24 \end{cases}$

60. $\begin{cases} -8x + 7y - 10z = -151 \\ 12x + 3y - 5z = 86 \\ 15x - 9y + 2z = 187 \end{cases}$

Investment Portfolio **In Exercises 61–64, you invest in AAA-rated bonds, A-rated bonds, and B-rated bonds. The average yields are 6.5% on AAA bonds, 7% on A bonds, and 9% on B bonds. You invest twice as much in B bonds as in A bonds. Let x, y, and z represent the amounts invested in AAA, A, and B bonds, respectively.**

$$\begin{cases} x + y + z = \text{(total investment)} \\ 0.065x + 0.07y + 0.09z = \text{(annual return)} \\ 2y - z = 0 \end{cases}$$

Use the inverse of the coefficient matrix of this system to find the amount invested in each type of bond.

	Total Investment	Annual Return
61.	$10,000	$705
62.	$10,000	$760
63.	$12,000	$835
64.	$500,000	$38,000

Circuit Analysis

In Exercises 65–68, consider the circuit shown in the figure. The currents I_1, I_2, and I_3, in amperes, are the solution of the system of linear equations.

$$\begin{cases} 2I_1 + 4I_3 = E_1 \\ I_2 + 4I_3 = E_2 \\ I_1 + I_2 - I_3 = 0 \end{cases}$$

where E_1 and E_2 are voltages. Use the inverse of the coefficient matrix of this system to find the unknown currents for the given voltages.

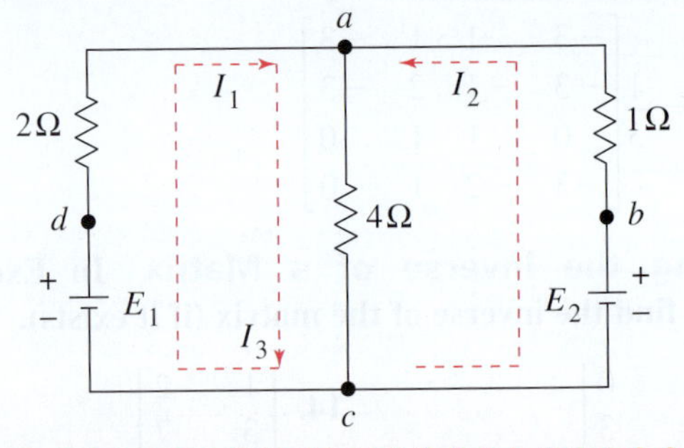

Circuit Analysis —continued—

65. $E_1 = 15$ volts, $E_2 = 17$ volts

66. $E_1 = 10$ volts, $E_2 = 10$ volts

67. $E_1 = 28$ volts, $E_2 = 21$ volts

68. $E_1 = 24$ volts, $E_2 = 23$ volts

Raw Materials **In Exercises 69 and 70, find the numbers of bags of potting soil that a company can produce for seedlings, general potting, and hardwood plants with the given amounts of raw materials. The raw materials used in one bag of each type of potting soil are shown below.**

	Sand	Loam	Peat Moss
Seedlings	2 units	1 units	1 unit
General	1 unit	2 units	1 unit
Hardwoods	2 units	2 units	2 units

69. 500 units of sand
500 units of loam
400 units of peat moss

70. 500 units of sand
750 units of loam
450 units of peat moss

71. Floral Design A florist is creating 10 centerpieces for the tables at a wedding reception. Roses cost \$2.50 each, lilies cost \$4 each, and irises cost \$2 each. The customer has a budget of \$300 allocated for the centerpieces and wants each centerpiece to contain 12 flowers, with twice as many roses as the number of irises and lilies combined.

(a) Write a system of linear equations that represents the situation.

(b) Write a matrix equation that corresponds to your system.

(c) Solve your system of linear equations using an inverse matrix. Find the number of flowers of each type that the florist can use to create the 10 centerpieces.

72. International Travel The table shows the numbers of international travelers y (in thousands) to the United States from South America from 2008 through 2010. *(Source: U.S. Department of Commerce)*

Year	Travelers, y (in thousands)
2008	2556
2009	2742
2010	3250

(a) The data can be modeled by the quadratic function $y = at^2 + bt + c$. Create a system of linear equations for the data. Let t represent the year, with $t = 8$ corresponding to 2008.

(b) Use the matrix capabilities of a graphing utility to find the inverse matrix to solve the system from part (a) and find the least squares regression parabola $y = at^2 + bt + c$.

(c) Use the graphing utility to graph the parabola with the data.

(d) Do you believe the model is a reasonable predictor of future numbers of travelers? Explain.

Exploration

True or False? **In Exercises 73 and 74, determine whether the statement is true or false. Justify your answer.**

73. Multiplication of an invertible matrix and its inverse is commutative.

74. When the product of two square matrices is the identity matrix, the matrices are inverses of one another.

75. Writing Explain how to determine whether the inverse of a 2×2 matrix exists. If so, explain how to find the inverse.

76. Writing Explain in your own words how to write a system of three linear equations in three variables as a matrix equation, $AX = B$, as well as how to solve the system using an inverse matrix.

77. Conjecture Consider matrices of the form

$$A = \begin{bmatrix} a_{11} & 0 & 0 & 0 & \cdots & 0 \\ 0 & a_{22} & 0 & 0 & \cdots & 0 \\ 0 & 0 & a_{33} & 0 & \cdots & 0 \\ \vdots & \vdots & \vdots & \vdots & \cdots & \vdots \\ 0 & 0 & 0 & 0 & \cdots & a_{nn} \end{bmatrix}.$$

(a) Write a 2×2 matrix and a 3×3 matrix in the form of A. Find the inverse of each.

(b) Use the result of part (a) to make a conjecture about the inverses of matrices in the form of A.

78. HOW DO YOU SEE IT? Let A be the 2×2 matrix

$$A = \begin{bmatrix} x & y \\ 0 & z \end{bmatrix}.$$

Use the determinant of A to state the conditions for which (a) A^{-1} exists and (b) $A^{-1} = A$.

Project: Viewing Television To work an extended application analyzing the average amounts of time spent viewing television in the United States, visit this text's website at *LarsonPrecalculus.com*. *(Source: The Nielsen Company)*

Ingrid Prats/Shutterstock.com

8.4 The Determinant of a Square Matrix

Determinants are often used in other branches of mathematics. For instance, Exercises 89–94 on page 581 show some types of determinants that are useful when changes of variables are made in calculus.

- **Find the determinants of 2 × 2 matrices.**
- **Find minors and cofactors of square matrices.**
- **Find the determinants of square matrices.**

The Determinant of a 2 × 2 Matrix

Every *square* matrix can be associated with a real number called its **determinant.** Determinants have many uses, and several will be discussed in this and the next section. Historically, the use of determinants arose from special number patterns that occur when systems of linear equations are solved. For instance, the system

$$\begin{cases} a_1x + b_1y = c_1 \\ a_2x + b_2y = c_2 \end{cases}$$

has a solution

$$x = \frac{c_1b_2 - c_2b_1}{a_1b_2 - a_2b_1} \quad \text{and} \quad y = \frac{a_1c_2 - a_2c_1}{a_1b_2 - a_2b_1}$$

provided that $a_1b_2 - a_2b_1 \neq 0$. Note that the denominators of the two fractions are the same. This denominator is called the *determinant* of the coefficient matrix of the system.

Coefficient Matrix	**Determinant**
$A = \begin{bmatrix} a_1 & b_1 \\ a_2 & b_2 \end{bmatrix}$	$\det(A) = a_1b_2 - a_2b_1$

The determinant of the matrix A can also be denoted by vertical bars on both sides of the matrix, as indicated in the following definition.

Definition of the Determinant of a 2 × 2 Matrix

The **determinant** of the matrix

$$A = \begin{bmatrix} a_1 & b_1 \\ a_2 & b_2 \end{bmatrix}$$

is given by

$$\det(A) = |A| = \begin{vmatrix} a_1 & b_1 \\ a_2 & b_2 \end{vmatrix} = a_1b_2 - a_2b_1.$$

In this text, $\det(A)$ and $|A|$ are used interchangeably to represent the determinant of A. Although vertical bars are also used to denote the absolute value of a real number, the context will show which use is intended.

A convenient method for remembering the formula for the determinant of a 2 × 2 matrix is shown in the following diagram.

$$\det(A) = \begin{vmatrix} a_1 & b_1 \\ a_2 & b_2 \end{vmatrix} = a_1b_2 - a_2b_1$$

Note that the determinant is the difference of the products of the two diagonals of the matrix.

lightpoet/Shutterstock.com

In Example 1, you will see that the determinant of a matrix can be positive, zero, or negative.

EXAMPLE 1 The Determinant of a 2×2 Matrix

Find the determinant of each matrix.

a. $A = \begin{bmatrix} 2 & -3 \\ 1 & 2 \end{bmatrix}$

b. $B = \begin{bmatrix} 2 & 1 \\ 4 & 2 \end{bmatrix}$

c. $C = \begin{bmatrix} 0 & \frac{3}{2} \\ 2 & 4 \end{bmatrix}$

Solution

a. $\det(A) = \begin{vmatrix} 2 & -3 \\ 1 & 2 \end{vmatrix}$

$= 2(2) - 1(-3)$

$= 4 + 3$

$= 7$

b. $\det(B) = \begin{vmatrix} 2 & 1 \\ 4 & 2 \end{vmatrix}$

$= 2(2) - 4(1)$

$= 4 - 4$

$= 0$

c. $\det(C) = \begin{vmatrix} 0 & \frac{3}{2} \\ 2 & 4 \end{vmatrix}$

$= 0(4) - 2\left(\frac{3}{2}\right)$

$= 0 - 3$

$= -3$

✓ Checkpoint *Audio-video solution in English & Spanish at LarsonPrecalculus.com.*

Find the determinant of each matrix.

a. $A = \begin{bmatrix} 1 & 2 \\ 3 & -1 \end{bmatrix}$ **b.** $B = \begin{bmatrix} 5 & 0 \\ -4 & 2 \end{bmatrix}$ **c.** $C = \begin{bmatrix} 3 & 6 \\ 2 & 4 \end{bmatrix}$

The determinant of a matrix of order 1×1 is defined simply as the entry of the matrix. For instance, if $A = [-2]$, then $\det(A) = -2$.

TECHNOLOGY Most graphing utilities can evaluate the determinant of a matrix. For instance, you can evaluate the determinant of

$$A = \begin{bmatrix} 2 & -3 \\ 1 & 2 \end{bmatrix}$$

by entering the matrix as $[A]$ and then choosing the *determinant* feature. The result should be 7, as in Example 1(a). Try evaluating the determinants of other matrices. Consult the user's guide for your graphing utility for specific keystrokes.

Minors and Cofactors

To define the determinant of a square matrix of order 3×3 or higher, it is convenient to introduce the concepts of **minors** and **cofactors.**

Sign Pattern for Cofactors

$$\begin{bmatrix} + & - & + \\ - & + & - \\ + & - & + \end{bmatrix}$$

3×3 matrix

$$\begin{bmatrix} + & - & + & - \\ - & + & - & + \\ + & - & + & - \\ - & + & - & + \end{bmatrix}$$

4×4 matrix

$$\begin{bmatrix} + & - & + & - & + & \cdots \\ - & + & - & + & - & \cdots \\ + & - & + & - & + & \cdots \\ - & + & - & + & - & \cdots \\ + & - & + & - & + & \cdots \\ \vdots & \vdots & \vdots & \vdots & \vdots & \end{bmatrix}$$

$n \times n$ matrix

Minors and Cofactors of a Square Matrix

If A is a square matrix, then the **minor** M_{ij} of the entry a_{ij} is the determinant of the matrix obtained by deleting the ith row and jth column of A. The **cofactor** C_{ij} of the entry a_{ij} is

$$C_{ij} = (-1)^{i+j}M_{ij}.$$

In the sign pattern for cofactors at the left, notice that *odd* positions (where $i + j$ is odd) have negative signs and *even* positions (where $i + j$ is even) have positive signs.

EXAMPLE 2 Finding the Minors and Cofactors of a Matrix

Find all the minors and cofactors of

$$A = \begin{bmatrix} 0 & 2 & 1 \\ 3 & -1 & 2 \\ 4 & 0 & 1 \end{bmatrix}.$$

Solution To find the minor M_{11}, delete the first row and first column of A and evaluate the determinant of the resulting matrix.

$$\begin{bmatrix} 0 & 2 & 1 \\ 3 & -1 & 2 \\ 4 & 0 & 1 \end{bmatrix}, \quad M_{11} = \begin{vmatrix} -1 & 2 \\ 0 & 1 \end{vmatrix} = -1(1) - 0(2) = -1$$

Similarly, to find M_{12}, delete the first row and second column.

$$\begin{bmatrix} 0 & 2 & 1 \\ 3 & -1 & 2 \\ 4 & 0 & 1 \end{bmatrix}, \quad M_{12} = \begin{vmatrix} 3 & 2 \\ 4 & 1 \end{vmatrix} = 3(1) - 4(2) = -5$$

Continuing this pattern, you obtain the minors.

$$\begin{array}{lll} M_{11} = -1 & M_{12} = -5 & M_{13} = 4 \\ M_{21} = 2 & M_{22} = -4 & M_{23} = -8 \\ M_{31} = 5 & M_{32} = -3 & M_{33} = -6 \end{array}$$

Now, to find the cofactors, combine these minors with the checkerboard pattern of signs for a 3×3 matrix shown at the upper left.

$$\begin{array}{lll} C_{11} = -1 & C_{12} = 5 & C_{13} = 4 \\ C_{21} = -2 & C_{22} = -4 & C_{23} = 8 \\ C_{31} = 5 & C_{32} = 3 & C_{33} = -6 \end{array}$$

✓ **Checkpoint** *Audio-video solution in English & Spanish at LarsonPrecalculus.com.*

Find all the minors and cofactors of

$$A = \begin{bmatrix} 1 & 2 & 3 \\ 0 & -1 & 5 \\ 2 & 1 & 4 \end{bmatrix}.$$

The Determinant of a Square Matrix

The definition below is called *inductive* because it uses determinants of matrices of order $n - 1$ to define determinants of matrices of order n.

> **Determinant of a Square Matrix**
>
> If A is a square matrix (of order 2×2 or greater), then the determinant of A is the sum of the entries in any row (or column) of A multiplied by their respective cofactors. For instance, expanding along the first row yields
>
> $$|A| = a_{11}C_{11} + a_{12}C_{12} + \cdot \cdot \cdot + a_{1n}C_{1n}.$$
>
> Applying this definition to find a determinant is called **expanding by cofactors.**

Try checking that for a 2×2 matrix

$$A = \begin{bmatrix} a_1 & b_1 \\ a_2 & b_2 \end{bmatrix}$$

this definition of the determinant yields

$$|A| = a_1b_2 - a_2b_1$$

as previously defined.

EXAMPLE 3 The Determinant of a 3 × 3 Matrix

Find the determinant of $A = \begin{bmatrix} 0 & 2 & 1 \\ 3 & -1 & 2 \\ 4 & 0 & 1 \end{bmatrix}$.

Solution Note that this is the same matrix that was given in Example 2. There you found the cofactors of the entries in the first row to be

$$C_{11} = -1, \quad C_{12} = 5, \quad \text{and} \quad C_{13} = 4.$$

So, by the definition of a determinant, you have

$$\begin{aligned} |A| &= a_{11}C_{11} + a_{12}C_{12} + a_{13}C_{13} && \text{First-row expansion} \\ &= 0(-1) + 2(5) + 1(4) \\ &= 14. \end{aligned}$$

✓ **Checkpoint** *Audio-video solution in English & Spanish at LarsonPrecalculus.com.*

Find the determinant of $A = \begin{bmatrix} 3 & 4 & -2 \\ 3 & 5 & 0 \\ -1 & 4 & 1 \end{bmatrix}$.

In Example 3, the determinant was found by expanding by cofactors in the first row. You could have used any row or column. For instance, you could have expanded along the second row to obtain

$$\begin{aligned} |A| &= a_{21}C_{21} + a_{22}C_{22} + a_{23}C_{23} && \text{Second-row expansion} \\ &= 3(-2) + (-1)(-4) + 2(8) \\ &= 14. \end{aligned}$$

In the number-placement puzzle Sudoku, the object is to fill out a partially completed 9×9 grid of boxes with numbers from 1 to 9 so that each column, row, and 3×3 sub-grid contains each of these numbers without repetition. For a completed Sudoku grid to be valid, no two rows (or columns) will have the numbers in the same order. If this should happen in a row or column, then the determinant of the matrix formed by the numbers in the grid will be zero.

When expanding by cofactors, you do not need to find cofactors of zero entries, because zero times its cofactor is zero. So, the row (or column) containing the most zeros is usually the best choice for expansion by cofactors. This is demonstrated in the next example.

EXAMPLE 4 The Determinant of a 4 × 4 Matrix

Find the determinant of $A = \begin{bmatrix} 1 & -2 & 3 & 0 \\ -1 & 1 & 0 & 2 \\ 0 & 2 & 0 & 3 \\ 3 & 4 & 0 & 2 \end{bmatrix}$.

Solution After inspecting this matrix, you can see that three of the entries in the third column are zeros. So, you can eliminate some of the work in the expansion by using the third column.

$$|A| = 3(C_{13}) + 0(C_{23}) + 0(C_{33}) + 0(C_{43})$$

Because C_{23}, C_{33}, and C_{43} have zero coefficients, you need only find the cofactor C_{13}. To do this, delete the first row and third column of A and evaluate the determinant of the resulting matrix.

$$C_{13} = (-1)^{1+3}\begin{vmatrix} -1 & 1 & 2 \\ 0 & 2 & 3 \\ 3 & 4 & 2 \end{vmatrix}$$ Delete 1st row and 3rd column.

$$= \begin{vmatrix} -1 & 1 & 2 \\ 0 & 2 & 3 \\ 3 & 4 & 2 \end{vmatrix}$$ Simplify.

Expanding by cofactors in the second row yields

$$\begin{aligned} C_{13} &= 0(-1)^3\begin{vmatrix} 1 & 2 \\ 4 & 2 \end{vmatrix} + 2(-1)^4\begin{vmatrix} -1 & 2 \\ 3 & 2 \end{vmatrix} + 3(-1)^5\begin{vmatrix} -1 & 1 \\ 3 & 4 \end{vmatrix} \\ &= 0 + 2(1)(-8) + 3(-1)(-7) \\ &= 5. \end{aligned}$$

So, $|A| = 3C_{13} = 3(5) = 15$.

✓ Checkpoint Audio-video solution in English & Spanish at LarsonPrecalculus.com.

Find the determinant of $A = \begin{bmatrix} 2 & 6 & -4 & 2 \\ 2 & -2 & 3 & 6 \\ 1 & 5 & 0 & 1 \\ 3 & 1 & 0 & -5 \end{bmatrix}$.

Summarize (Section 8.4)

1. State the definition of the determinant of a 2×2 matrix *(page 574)*. For an example of finding the determinants of 2×2 matrices, see Example 1.
2. State the definitions of minors and cofactors of a square matrix *(page 576)*. For an example of finding the minors and cofactors of a square matrix, see Example 2.
3. State the definition of the determinant of a square matrix using expansion by cofactors *(page 577)*. For examples of finding determinants using this definition, see Examples 3 and 4.

James E. Knopf/Shutterstock.com

8.4 Exercises

See CalcChat.com for tutorial help and worked-out solutions to odd-numbered exercises.

Vocabulary: Fill in the blanks.

1. Both $\det(A)$ and $|A|$ represent the ________ of the matrix A.
2. The ________ M_{ij} of the entry a_{ij} is the determinant of the matrix obtained by deleting the ith row and jth column of the square matrix A.
3. The ________ C_{ij} of the entry a_{ij} of the square matrix A is given by $(-1)^{i+j}M_{ij}$.
4. The method of finding the determinant of a matrix of order 2×2 or greater is ________ by ________.

Skills and Applications

Finding the Determinant of a Matrix In Exercises 5–22, find the determinant of the matrix.

5. $[4]$
6. $[-10]$
7. $\begin{bmatrix} 8 & 4 \\ 2 & 3 \end{bmatrix}$
8. $\begin{bmatrix} -9 & 0 \\ 6 & 2 \end{bmatrix}$
9. $\begin{bmatrix} 6 & 2 \\ -5 & 3 \end{bmatrix}$
10. $\begin{bmatrix} 3 & -3 \\ 4 & -8 \end{bmatrix}$
11. $\begin{bmatrix} -7 & 0 \\ 3 & 0 \end{bmatrix}$
12. $\begin{bmatrix} 4 & -3 \\ 0 & 0 \end{bmatrix}$
13. $\begin{bmatrix} 2 & 6 \\ 0 & 3 \end{bmatrix}$
14. $\begin{bmatrix} 2 & -3 \\ -6 & 9 \end{bmatrix}$
15. $\begin{bmatrix} -3 & -2 \\ -6 & -1 \end{bmatrix}$
16. $\begin{bmatrix} 4 & 7 \\ -2 & 5 \end{bmatrix}$
17. $\begin{bmatrix} -2 & -7 \\ -3 & 1 \end{bmatrix}$
18. $\begin{bmatrix} 2 & -5 \\ -4 & -1 \end{bmatrix}$
19. $\begin{bmatrix} -7 & 6 \\ \frac{1}{2} & 3 \end{bmatrix}$
20. $\begin{bmatrix} 0 & 6 \\ -3 & 2 \end{bmatrix}$
21. $\begin{bmatrix} -\frac{1}{2} & \frac{1}{3} \\ -6 & \frac{1}{3} \end{bmatrix}$
22. $\begin{bmatrix} \frac{2}{3} & \frac{4}{3} \\ -1 & -\frac{1}{3} \end{bmatrix}$

Using a Graphing Utility In Exercises 23–30, use the matrix capabilities of a graphing utility to find the determinant of the matrix.

23. $\begin{bmatrix} 3 & 4 \\ -2 & 1 \end{bmatrix}$
24. $\begin{bmatrix} 5 & -9 \\ 7 & 16 \end{bmatrix}$
25. $\begin{bmatrix} 19 & -20 \\ 43 & 56 \end{bmatrix}$
26. $\begin{bmatrix} 101 & 197 \\ -253 & 172 \end{bmatrix}$
27. $\begin{bmatrix} 0.3 & 0.2 \\ 0.2 & 0.2 \end{bmatrix}$
28. $\begin{bmatrix} 0.1 & 0.2 \\ -0.3 & 0.2 \end{bmatrix}$
29. $\begin{bmatrix} 0.9 & 0.7 \\ -0.1 & 0.3 \end{bmatrix}$
30. $\begin{bmatrix} 0.1 & 0.1 \\ 7.5 & 6.2 \end{bmatrix}$

Finding the Minors and Cofactors of a Matrix In Exercises 31–38, find all the (a) minors and (b) cofactors of the matrix.

31. $\begin{bmatrix} 4 & 5 \\ 3 & -6 \end{bmatrix}$
32. $\begin{bmatrix} 0 & 10 \\ 3 & -4 \end{bmatrix}$
33. $\begin{bmatrix} 3 & 1 \\ -2 & -4 \end{bmatrix}$
34. $\begin{bmatrix} -6 & 5 \\ 7 & -2 \end{bmatrix}$
35. $\begin{bmatrix} 4 & 0 & 2 \\ -3 & 2 & 1 \\ 1 & -1 & 1 \end{bmatrix}$
36. $\begin{bmatrix} 1 & -1 & 0 \\ 3 & 2 & 5 \\ 4 & -6 & 4 \end{bmatrix}$
37. $\begin{bmatrix} -4 & 6 & 3 \\ 7 & -2 & 8 \\ 1 & 0 & -5 \end{bmatrix}$
38. $\begin{bmatrix} -2 & 9 & 4 \\ 7 & -6 & 0 \\ 6 & 7 & -6 \end{bmatrix}$

Finding the Determinant of a Matrix In Exercises 39–46, find the determinant of the matrix. Expand by cofactors using the indicated row or column.

39. $\begin{bmatrix} -3 & 2 & 1 \\ 4 & 5 & 6 \\ 2 & -3 & 1 \end{bmatrix}$
(a) Row 1
(b) Column 2

40. $\begin{bmatrix} -3 & 4 & 2 \\ 6 & 3 & 1 \\ 4 & -7 & -8 \end{bmatrix}$
(a) Row 2
(b) Column 3

41. $\begin{bmatrix} 5 & 0 & -3 \\ 0 & 12 & 4 \\ 1 & 6 & 3 \end{bmatrix}$
(a) Row 2
(b) Column 2

42. $\begin{bmatrix} 10 & -5 & 5 \\ 30 & 0 & 10 \\ 0 & 10 & 1 \end{bmatrix}$
(a) Row 3
(b) Column 1

43. $\begin{bmatrix} -2 & 4 & 7 & 1 \\ 3 & 0 & 0 & 0 \\ 8 & 5 & 10 & 5 \\ 6 & 0 & 5 & 0 \end{bmatrix}$
(a) Row 2
(b) Column 4

44. $\begin{bmatrix} 7 & 0 & 0 & -6 \\ 6 & 0 & 1 & -2 \\ 1 & -2 & 3 & 2 \\ -3 & 0 & -1 & 4 \end{bmatrix}$
(a) Row 4
(b) Column 2

45. $\begin{bmatrix} 6 & 0 & -3 & 5 \\ 4 & 13 & 6 & -8 \\ -1 & 0 & 7 & 4 \\ 8 & 6 & 0 & 2 \end{bmatrix}$
(a) Row 2
(b) Column 2

46. $\begin{bmatrix} 10 & 8 & 3 & -7 \\ 4 & 0 & 5 & -6 \\ 0 & 3 & 2 & 7 \\ 1 & 0 & -3 & 2 \end{bmatrix}$
(a) Row 3
(b) Column 1

Finding the Determinant of a Matrix **In Exercises 47–62, find the determinant of the matrix. Expand by cofactors using the row or column that appears to make the computations easiest.**

47. $\begin{bmatrix} 2 & -1 & 0 \\ 4 & 2 & 1 \\ 4 & 2 & 1 \end{bmatrix}$

48. $\begin{bmatrix} -2 & 2 & 3 \\ 1 & -1 & 0 \\ 0 & 1 & 4 \end{bmatrix}$

49. $\begin{bmatrix} 6 & 3 & -7 \\ 0 & 0 & 0 \\ 4 & -6 & 3 \end{bmatrix}$

50. $\begin{bmatrix} 1 & 1 & 2 \\ 3 & 1 & 0 \\ -2 & 0 & 3 \end{bmatrix}$

51. $\begin{bmatrix} -1 & 8 & -3 \\ 0 & 3 & -6 \\ 0 & 0 & 3 \end{bmatrix}$

52. $\begin{bmatrix} 1 & 0 & 0 \\ -1 & -1 & 0 \\ 4 & 11 & 5 \end{bmatrix}$

53. $\begin{bmatrix} 1 & 4 & -2 \\ 3 & 2 & 0 \\ -1 & 4 & 3 \end{bmatrix}$

54. $\begin{bmatrix} 2 & -1 & 3 \\ 1 & 4 & 4 \\ 1 & 0 & 2 \end{bmatrix}$

55. $\begin{bmatrix} 2 & 4 & 6 \\ 0 & 3 & 1 \\ 0 & 0 & -5 \end{bmatrix}$

56. $\begin{bmatrix} -3 & 0 & 0 \\ 7 & 11 & 0 \\ 1 & 2 & 2 \end{bmatrix}$

57. $\begin{bmatrix} 2 & 6 & 6 & 2 \\ 2 & 7 & 3 & 6 \\ 1 & 5 & 0 & 1 \\ 3 & 7 & 0 & 7 \end{bmatrix}$

58. $\begin{bmatrix} 3 & 6 & -5 & 4 \\ -2 & 0 & 6 & 0 \\ 1 & 1 & 2 & 2 \\ 0 & 3 & -1 & -1 \end{bmatrix}$

59. $\begin{bmatrix} 5 & 3 & 0 & 6 \\ 4 & 6 & 4 & 12 \\ 0 & 2 & -3 & 4 \\ 0 & 1 & -2 & 2 \end{bmatrix}$

60. $\begin{bmatrix} 1 & 4 & 3 & 2 \\ -5 & 6 & 2 & 1 \\ 0 & 0 & 0 & 0 \\ 3 & -2 & 1 & 5 \end{bmatrix}$

61. $\begin{bmatrix} 3 & 2 & 4 & -1 & 5 \\ -2 & 0 & 1 & 3 & 2 \\ 1 & 0 & 0 & 4 & 0 \\ 6 & 0 & 2 & -1 & 0 \\ 3 & 0 & 5 & 1 & 0 \end{bmatrix}$

62. $\begin{bmatrix} 5 & 2 & 0 & 0 & -2 \\ 0 & 1 & 4 & 3 & 2 \\ 0 & 0 & 2 & 6 & 3 \\ 0 & 0 & 3 & 4 & 1 \\ 0 & 0 & 0 & 0 & 2 \end{bmatrix}$

Using a Graphing Utility **In Exercises 63–66, use the matrix capabilities of a graphing utility to evaluate the determinant.**

63. $\begin{vmatrix} 3 & 8 & -7 \\ 0 & -5 & 4 \\ 8 & 1 & 6 \end{vmatrix}$

64. $\begin{vmatrix} 5 & -8 & 0 \\ 9 & 7 & 4 \\ -8 & 7 & 1 \end{vmatrix}$

65. $\begin{vmatrix} 1 & -1 & 8 & 4 \\ 2 & 6 & 0 & -4 \\ 2 & 0 & 2 & 6 \\ 0 & 2 & 8 & 0 \end{vmatrix}$

66. $\begin{vmatrix} 0 & -3 & 8 & 2 \\ 8 & 1 & -1 & 6 \\ -4 & 6 & 0 & 9 \\ -7 & 0 & 0 & 14 \end{vmatrix}$

The Determinant of a Matrix Product **In Exercises 67–74, find (a) $|A|$, (b) $|B|$, (c) AB, and (d) $|AB|$.**

67. $A = \begin{bmatrix} -1 & 0 \\ 0 & 3 \end{bmatrix}, \quad B = \begin{bmatrix} 2 & 0 \\ 0 & -1 \end{bmatrix}$

68. $A = \begin{bmatrix} -2 & 1 \\ 4 & -2 \end{bmatrix}, \quad B = \begin{bmatrix} 1 & 2 \\ 0 & -1 \end{bmatrix}$

69. $A = \begin{bmatrix} 4 & 0 \\ 3 & -2 \end{bmatrix}, \quad B = \begin{bmatrix} -1 & 1 \\ -2 & 2 \end{bmatrix}$

70. $A = \begin{bmatrix} 5 & 4 \\ 3 & -1 \end{bmatrix}, \quad B = \begin{bmatrix} 0 & 6 \\ 1 & -2 \end{bmatrix}$

71. $A = \begin{bmatrix} 0 & 1 & 2 \\ -3 & -2 & 1 \\ 0 & 4 & 1 \end{bmatrix}, \quad B = \begin{bmatrix} 3 & -2 & 0 \\ 1 & -1 & 2 \\ 3 & 1 & 1 \end{bmatrix}$

72. $A = \begin{bmatrix} 3 & 2 & 0 \\ -1 & -3 & 4 \\ -2 & 0 & 1 \end{bmatrix}, \quad B = \begin{bmatrix} -3 & 0 & 1 \\ 0 & 2 & -1 \\ -2 & -1 & 1 \end{bmatrix}$

73. $A = \begin{bmatrix} -1 & 2 & 1 \\ 1 & 0 & 1 \\ 0 & 1 & 0 \end{bmatrix}, \quad B = \begin{bmatrix} -1 & 0 & 0 \\ 0 & 2 & 0 \\ 0 & 0 & 3 \end{bmatrix}$

74. $A = \begin{bmatrix} 2 & 0 & 1 \\ 1 & -1 & 2 \\ 3 & 1 & 0 \end{bmatrix}, \quad B = \begin{bmatrix} 2 & -1 & 4 \\ 0 & 1 & 3 \\ 3 & -2 & 1 \end{bmatrix}$

Verifying an Equation **In Exercises 75–80, evaluate the determinant(s) to verify the equation.**

75. $\begin{vmatrix} w & x \\ y & z \end{vmatrix} = -\begin{vmatrix} y & z \\ w & x \end{vmatrix}$

76. $\begin{vmatrix} w & cx \\ y & cz \end{vmatrix} = c\begin{vmatrix} w & x \\ y & z \end{vmatrix}$

77. $\begin{vmatrix} w & x \\ y & z \end{vmatrix} = \begin{vmatrix} w & x + cw \\ y & z + cy \end{vmatrix}$

78. $\begin{vmatrix} w & x \\ cw & cx \end{vmatrix} = 0$

79. $\begin{vmatrix} 1 & x & x^2 \\ 1 & y & y^2 \\ 1 & z & z^2 \end{vmatrix} = (y - x)(z - x)(z - y)$

80. $\begin{vmatrix} a + b & a & a \\ a & a + b & a \\ a & a & a + b \end{vmatrix} = b^2(3a + b)$

Solving an Equation **In Exercises 81–88, solve for x.**

81. $\begin{vmatrix} x & 2 \\ 1 & x \end{vmatrix} = 2$

82. $\begin{vmatrix} x & 4 \\ -1 & x \end{vmatrix} = 20$

83. $\begin{vmatrix} x & 1 \\ 2 & x - 2 \end{vmatrix} = -1$

84. $\begin{vmatrix} x + 1 & 2 \\ -1 & x \end{vmatrix} = 4$

85. $\begin{vmatrix} x - 1 & 2 \\ 3 & x - 2 \end{vmatrix} = 0$

86. $\begin{vmatrix} x - 2 & -1 \\ -3 & x \end{vmatrix} = 0$

87. $\begin{vmatrix} x + 3 & 2 \\ 1 & x + 2 \end{vmatrix} = 0$

88. $\begin{vmatrix} x + 4 & -2 \\ 7 & x - 5 \end{vmatrix} = 0$

Entries Involving Expressions In Exercises 89–94, evaluate the determinant in which the entries are functions. Determinants of this type occur when changes of variables are made in calculus.

89. $\begin{vmatrix} 4u & -1 \\ -1 & 2v \end{vmatrix}$

90. $\begin{vmatrix} 3x^2 & -3y^2 \\ 1 & 1 \end{vmatrix}$

91. $\begin{vmatrix} e^{2x} & e^{3x} \\ 2e^{2x} & 3e^{3x} \end{vmatrix}$

92. $\begin{vmatrix} e^{-x} & xe^{-x} \\ -e^{-x} & (1-x)e^{-x} \end{vmatrix}$

93. $\begin{vmatrix} x & \ln x \\ 1 & 1/x \end{vmatrix}$

94. $\begin{vmatrix} x & x\ln x \\ 1 & 1+\ln x \end{vmatrix}$

Exploration

True or False? In Exercises 95 and 96, determine whether the statement is true or false. Justify your answer.

95. If a square matrix has an entire row of zeros, then the determinant will always be zero.

96. If two columns of a square matrix are the same, then the determinant of the matrix will be zero.

97. **Providing a Counterexample** Find square matrices A and B to demonstrate that $|A+B| \neq |A| + |B|$.

98. **Conjecture** Consider square matrices in which the entries are consecutive integers. An example of such a matrix is

$$\begin{bmatrix} 4 & 5 & 6 \\ 7 & 8 & 9 \\ 10 & 11 & 12 \end{bmatrix}.$$

(a) Use a graphing utility to evaluate the determinants of four matrices of this type. Make a conjecture based on the results.

(b) Verify your conjecture.

99. **Writing** Write a brief paragraph explaining the difference between a square matrix and its determinant.

100. **Think About It** Let A be a 3×3 matrix such that $|A| = 5$. Is it possible to find $|2A|$? Explain.

Properties of Determinants In Exercises 101–103, a property of determinants is given (A and B are square matrices). State how the property has been applied to the given determinants and use a graphing utility to verify the results.

101. If B is obtained from A by interchanging two rows of A or interchanging two columns of A, then $|B| = -|A|$.

(a) $\begin{vmatrix} 1 & 3 & 4 \\ -7 & 2 & -5 \\ 6 & 1 & 2 \end{vmatrix} = -\begin{vmatrix} 1 & 4 & 3 \\ -7 & -5 & 2 \\ 6 & 2 & 1 \end{vmatrix}$

(b) $\begin{vmatrix} 1 & 3 & 4 \\ -2 & 2 & 0 \\ 1 & 6 & 2 \end{vmatrix} = -\begin{vmatrix} 1 & 6 & 2 \\ -2 & 2 & 0 \\ 1 & 3 & 4 \end{vmatrix}$

102. If B is obtained from A by adding a multiple of a row of A to another row of A or by adding a multiple of a column of A to another column of A, then $|B| = |A|$.

(a) $\begin{vmatrix} 1 & -3 \\ 5 & 2 \end{vmatrix} = \begin{vmatrix} 1 & -3 \\ 0 & 17 \end{vmatrix}$

(b) $\begin{vmatrix} 5 & 4 & 2 \\ 2 & -3 & 4 \\ 7 & 6 & 3 \end{vmatrix} = \begin{vmatrix} 1 & 10 & -6 \\ 2 & -3 & 4 \\ 7 & 6 & 3 \end{vmatrix}$

103. If B is obtained from A by multiplying a row by a nonzero constant c or by multiplying a column by a nonzero constant c, then $|B| = c|A|$.

(a) $\begin{vmatrix} 5 & 10 \\ 2 & -3 \end{vmatrix} = 5\begin{vmatrix} 1 & 2 \\ 2 & -3 \end{vmatrix}$

(b) $\begin{vmatrix} 1 & 8 & -3 \\ 3 & -12 & 6 \\ 7 & 4 & 9 \end{vmatrix} = 12\begin{vmatrix} 1 & 2 & -1 \\ 3 & -3 & 2 \\ 7 & 1 & 3 \end{vmatrix}$

104. **HOW DO YOU SEE IT?** Explain why the determinant of the matrix is equal to zero.

(a) $\begin{bmatrix} 3 & 4 & -2 & 7 \\ 1 & 3 & -1 & 2 \\ 0 & 5 & 7 & 1 \\ 1 & 3 & -1 & 2 \end{bmatrix}$ (b) $\begin{bmatrix} 3 & 2 & -1 \\ -6 & -4 & 2 \\ 5 & -7 & 9 \end{bmatrix}$

(c) $\begin{bmatrix} 2 & -4 & 5 \\ 1 & -2 & 3 \\ 0 & 0 & 0 \end{bmatrix}$ (d) $\begin{bmatrix} 4 & -4 & 5 & 7 \\ 2 & -2 & 3 & 1 \\ 4 & -4 & 5 & 7 \\ 6 & 1 & -3 & -3 \end{bmatrix}$

105. **Conjecture** A **diagonal matrix** is a square matrix with all zero entries above and below its main diagonal. Evaluate the determinant of each diagonal matrix. Make a conjecture based on your results.

(a) $\begin{bmatrix} 7 & 0 \\ 0 & 4 \end{bmatrix}$ (b) $\begin{bmatrix} -1 & 0 & 0 \\ 0 & 5 & 0 \\ 0 & 0 & 2 \end{bmatrix}$ (c) $\begin{bmatrix} 2 & 0 & 0 & 0 \\ 0 & -2 & 0 & 0 \\ 0 & 0 & 1 & 0 \\ 0 & 0 & 0 & 3 \end{bmatrix}$

lightpoet/Shutterstock.com

8.5 Applications of Matrices and Determinants

You can use determinants to solve real-life problems. For instance, in Exercise 33 on page 591, you will use a determinant to find the area of a region of forest infested with gypsy moths.

- **Use Cramer's Rule to solve systems of linear equations.**
- **Use determinants to find the areas of triangles.**
- **Use a determinant to test for collinear points and find an equation of a line passing through two points.**
- **Use matrices to encode and decode messages.**

Cramer's Rule

So far, you have studied three methods for solving a system of linear equations: substitution, elimination with equations, and elimination with matrices. In this section, you will study one more method, **Cramer's Rule,** named after Gabriel Cramer (1704–1752). This rule uses determinants to write the solution of a system of linear equations. To see how Cramer's Rule works, take another look at the solution described at the beginning of Section 8.4. There, it was pointed out that the system

$$\begin{cases} a_1x + b_1y = c_1 \\ a_2x + b_2y = c_2 \end{cases}$$

has a solution

$$x = \frac{c_1b_2 - c_2b_1}{a_1b_2 - a_2b_1} \quad \text{and} \quad y = \frac{a_1c_2 - a_2c_1}{a_1b_2 - a_2b_1}$$

provided that

$$a_1b_2 - a_2b_1 \neq 0.$$

Each numerator and denominator in this solution can be expressed as a determinant, as follows.

$$x = \frac{c_1b_2 - c_2b_1}{a_1b_2 - a_2b_1} = \frac{\begin{vmatrix} c_1 & b_1 \\ c_2 & b_2 \end{vmatrix}}{\begin{vmatrix} a_1 & b_1 \\ a_2 & b_2 \end{vmatrix}} \qquad y = \frac{a_1c_2 - a_2c_1}{a_1b_2 - a_2b_1} = \frac{\begin{vmatrix} a_1 & c_1 \\ a_2 & c_2 \end{vmatrix}}{\begin{vmatrix} a_1 & b_1 \\ a_2 & b_2 \end{vmatrix}}$$

Relative to the original system, the denominators for x and y are simply the determinant of the *coefficient* matrix of the system. This determinant is denoted by D. The numerators for x and y are denoted by D_x and D_y, respectively. They are formed by using the column of constants as replacements for the coefficients of x and y, as follows.

Coefficient Matrix	D	D_x	D_y
$\begin{bmatrix} a_1 & b_1 \\ a_2 & b_2 \end{bmatrix}$	$\begin{vmatrix} a_1 & b_1 \\ a_2 & b_2 \end{vmatrix}$	$\begin{vmatrix} c_1 & b_1 \\ c_2 & b_2 \end{vmatrix}$	$\begin{vmatrix} a_1 & c_1 \\ a_2 & c_2 \end{vmatrix}$

For example, given the system

$$\begin{cases} 2x - 5y = 3 \\ -4x + 3y = 8 \end{cases}$$

the coefficient matrix, D, D_x, and D_y are as follows.

Coefficient Matrix	D	D_x	D_y
$\begin{bmatrix} 2 & -5 \\ -4 & 3 \end{bmatrix}$	$\begin{vmatrix} 2 & -5 \\ -4 & 3 \end{vmatrix}$	$\begin{vmatrix} 3 & -5 \\ 8 & 3 \end{vmatrix}$	$\begin{vmatrix} 2 & 3 \\ -4 & 8 \end{vmatrix}$

Martynova Anna/Shutterstock.com

Cramer's Rule generalizes easily to systems of n equations in n variables. The value of each variable is given as the quotient of two determinants. The denominator is the determinant of the coefficient matrix, and the numerator is the determinant of the matrix formed by replacing the column corresponding to the variable (being solved for) with the column representing the constants. For instance, the solution for x_3 in the following system is shown.

$$\begin{cases} a_{11}x_1 + a_{12}x_2 + a_{13}x_3 = b_1 \\ a_{21}x_1 + a_{22}x_2 + a_{23}x_3 = b_2 \\ a_{31}x_1 + a_{32}x_2 + a_{33}x_3 = b_3 \end{cases} \qquad x_3 = \frac{|A_3|}{|A|} = \frac{\begin{vmatrix} a_{11} & a_{12} & b_1 \\ a_{21} & a_{22} & b_2 \\ a_{31} & a_{32} & b_3 \end{vmatrix}}{\begin{vmatrix} a_{11} & a_{12} & a_{13} \\ a_{21} & a_{22} & a_{23} \\ a_{31} & a_{32} & a_{33} \end{vmatrix}}$$

Cramer's Rule

If a system of n linear equations in n variables has a coefficient matrix A with a nonzero determinant $|A|$, then the solution of the system is

$$x_1 = \frac{|A_1|}{|A|}, \quad x_2 = \frac{|A_2|}{|A|}, \quad . . . \quad , \quad x_n = \frac{|A_n|}{|A|}$$

where the ith column of A_i is the column of constants in the system of equations. If the determinant of the coefficient matrix is zero, then the system has either no solution or infinitely many solutions.

EXAMPLE 1 Using Cramer's Rule for a 2 × 2 System

Use Cramer's Rule to solve the system

$$\begin{cases} 4x - 2y = 10 \\ 3x - 5y = 11 \end{cases}.$$

Solution To begin, find the determinant of the coefficient matrix.

$$D = \begin{vmatrix} 4 & -2 \\ 3 & -5 \end{vmatrix} = -20 - (-6) = -14$$

Because this determinant is not zero, you can apply Cramer's Rule.

$$x = \frac{D_x}{D} = \frac{\begin{vmatrix} 10 & -2 \\ 11 & -5 \end{vmatrix}}{-14} = \frac{-50 - (-22)}{-14} = \frac{-28}{-14} = 2$$

$$y = \frac{D_y}{D} = \frac{\begin{vmatrix} 4 & 10 \\ 3 & 11 \end{vmatrix}}{-14} = \frac{44 - 30}{-14} = \frac{14}{-14} = -1$$

So, the solution is $x = 2$ and $y = -1$. Check this in the original system.

Checkpoint Audio-video solution in English & Spanish at LarsonPrecalculus.com.

Use Cramer's Rule to solve the system

$$\begin{cases} 3x + 4y = 1 \\ 5x + 3y = 9 \end{cases}.$$

EXAMPLE 2 Using Cramer's Rule for a 3 × 3 System

Use Cramer's Rule to solve the system $\begin{cases} -x + 2y - 3z = 1 \\ 2x \qquad\quad + z = 0. \\ 3x - 4y + 4z = 2 \end{cases}$

Solution To find the determinant of the coefficient matrix

$$\begin{bmatrix} -1 & 2 & -3 \\ 2 & 0 & 1 \\ 3 & -4 & 4 \end{bmatrix}$$

expand along the second row, as follows.

$$D = 2(-1)^3\begin{vmatrix} 2 & -3 \\ -4 & 4 \end{vmatrix} + 0(-1)^4\begin{vmatrix} -1 & -3 \\ 3 & 4 \end{vmatrix} + 1(-1)^5\begin{vmatrix} -1 & 2 \\ 3 & -4 \end{vmatrix}$$

$$= -2(-4) + 0 - 1(-2)$$

$$= 10$$

Because this determinant is not zero, you can apply Cramer's Rule.

$$x = \frac{D_x}{D} = \frac{\begin{vmatrix} 1 & 2 & -3 \\ 0 & 0 & 1 \\ 2 & -4 & 4 \end{vmatrix}}{10} = \frac{8}{10} = \frac{4}{5}$$

$$y = \frac{D_y}{D} = \frac{\begin{vmatrix} -1 & 1 & -3 \\ 2 & 0 & 1 \\ 3 & 2 & 4 \end{vmatrix}}{10} = \frac{-15}{10} = -\frac{3}{2}$$

$$z = \frac{D_z}{D} = \frac{\begin{vmatrix} -1 & 2 & 1 \\ 2 & 0 & 0 \\ 3 & -4 & 2 \end{vmatrix}}{10} = \frac{-16}{10} = -\frac{8}{5}$$

The solution is $\left(\frac{4}{5}, -\frac{3}{2}, -\frac{8}{5}\right)$. Check this in the original system, as follows.

Check

$-\left(\frac{4}{5}\right) + 2\left(-\frac{3}{2}\right) - 3\left(-\frac{8}{5}\right) \stackrel{?}{=} 1$ Substitute into Equation 1.

$-\frac{4}{5} - 3 + \frac{24}{5} = 1$ Equation 1 checks. ✓

$2\left(\frac{4}{5}\right) + \left(-\frac{8}{5}\right) \stackrel{?}{=} 0$ Substitute into Equation 2.

$\frac{8}{5} - \frac{8}{5} = 0$ Equation 2 checks. ✓

$3\left(\frac{4}{5}\right) - 4\left(-\frac{3}{2}\right) + 4\left(-\frac{8}{5}\right) \stackrel{?}{=} 2$ Substitute into Equation 3.

$\frac{12}{5} + 6 - \frac{32}{5} = 2$ Equation 3 checks. ✓

✓ **Checkpoint** Audio-video solution in English & Spanish at LarsonPrecalculus.com.

Use Cramer's Rule to solve the system $\begin{cases} 4x - y + z = 12 \\ 2x + 2y + 3z = 1. \\ 5x - 2y + 6z = 22 \end{cases}$

Remember that Cramer's Rule does not apply when the determinant of the coefficient matrix is zero. This would create division by zero, which is undefined.

Area of a Triangle

Another application of matrices and determinants is finding the area of a triangle whose vertices are given as points in a coordinate plane.

> **Area of a Triangle**
>
> The area of a triangle with vertices (x_1, y_1), (x_2, y_2), and (x_3, y_3) is
>
> $$\text{Area} = \pm\frac{1}{2}\begin{vmatrix} x_1 & y_1 & 1 \\ x_2 & y_2 & 1 \\ x_3 & y_3 & 1 \end{vmatrix}$$
>
> where the symbol $\pm$ indicates that the appropriate sign should be chosen to yield a positive area.

EXAMPLE 3 Finding the Area of a Triangle

Find the area of a triangle whose vertices are $(1, 0)$, $(2, 2)$, and $(4, 3)$, as shown below.

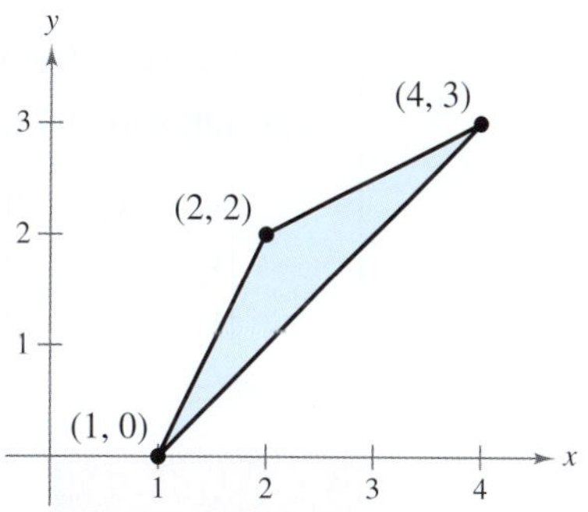

Solution Let $(x_1, y_1) = (1, 0)$, $(x_2, y_2) = (2, 2)$, and $(x_3, y_3) = (4, 3)$. Then, to find the area of the triangle, evaluate the determinant.

$$\begin{vmatrix} x_1 & y_1 & 1 \\ x_2 & y_2 & 1 \\ x_3 & y_3 & 1 \end{vmatrix} = \begin{vmatrix} 1 & 0 & 1 \\ 2 & 2 & 1 \\ 4 & 3 & 1 \end{vmatrix}$$

$$= 1(-1)^2\begin{vmatrix} 2 & 1 \\ 3 & 1 \end{vmatrix} + 0(-1)^3\begin{vmatrix} 2 & 1 \\ 4 & 1 \end{vmatrix} + 1(-1)^4\begin{vmatrix} 2 & 2 \\ 4 & 3 \end{vmatrix}$$

$$= 1(-1) + 0 + 1(-2)$$

$$= -3$$

Using this value, you can conclude that the area of the triangle is

$$\text{Area} = -\frac{1}{2}\begin{vmatrix} 1 & 0 & 1 \\ 2 & 2 & 1 \\ 4 & 3 & 1 \end{vmatrix}$$ Choose $(-)$ so that the area is positive.

$$= -\frac{1}{2}(-3)$$

$$= \frac{3}{2} \text{ square units.}$$

✓ **Checkpoint** *Audio-video solution in English & Spanish at LarsonPrecalculus.com.*

Find the area of a triangle whose vertices are $(0, 0)$, $(4, 1)$, and $(2, 5)$.

Lines in a Plane

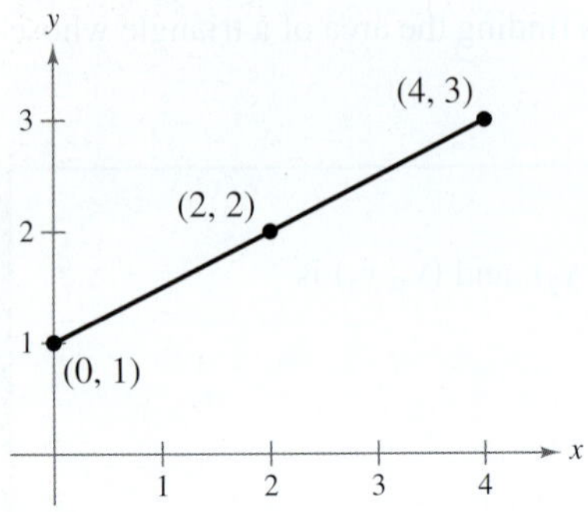

Figure 8.1

Suppose the three points in Example 3 had been on the same line. What would have happened had the area formula been applied to three such points? The answer is that the determinant would have been zero. Consider, for instance, the three collinear points (0, 1), (2, 2), and (4, 3), as shown in Figure 8.1. The area of the "triangle" that has these three points as vertices is

$$\frac{1}{2}\begin{vmatrix} 0 & 1 & 1 \\ 2 & 2 & 1 \\ 4 & 3 & 1 \end{vmatrix} = \frac{1}{2}\left[0(-1)^2\begin{vmatrix} 2 & 1 \\ 3 & 1 \end{vmatrix} + 1(-1)^3\begin{vmatrix} 2 & 1 \\ 4 & 1 \end{vmatrix} + 1(-1)^4\begin{vmatrix} 2 & 2 \\ 4 & 3 \end{vmatrix}\right]$$

$$= \frac{1}{2}[0 - 1(-2) + 1(-2)]$$

$$= 0.$$

This result is generalized as follows.

Test for Collinear Points

Three points

$$(x_1, y_1), \quad (x_2, y_2), \quad \text{and} \quad (x_3, y_3)$$

are **collinear** (lie on the same line) if and only if

$$\begin{vmatrix} x_1 & y_1 & 1 \\ x_2 & y_2 & 1 \\ x_3 & y_3 & 1 \end{vmatrix} = 0.$$

EXAMPLE 4 Testing for Collinear Points

Determine whether the points $(-2, -2)$, $(1, 1)$, and $(7, 5)$ are collinear. (See Figure 8.2.)

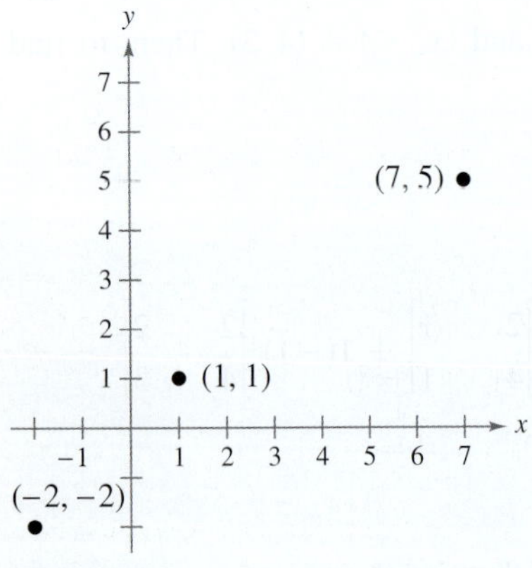

Figure 8.2

Solution Letting $(x_1, y_1) = (-2, -2)$, $(x_2, y_2) = (1, 1)$, and $(x_3, y_3) = (7, 5)$, you have

$$\begin{vmatrix} x_1 & y_1 & 1 \\ x_2 & y_2 & 1 \\ x_3 & y_3 & 1 \end{vmatrix} = \begin{vmatrix} -2 & -2 & 1 \\ 1 & 1 & 1 \\ 7 & 5 & 1 \end{vmatrix}$$

$$= -2(-1)^2\begin{vmatrix} 1 & 1 \\ 5 & 1 \end{vmatrix} + (-2)(-1)^3\begin{vmatrix} 1 & 1 \\ 7 & 1 \end{vmatrix} + 1(-1)^4\begin{vmatrix} 1 & 1 \\ 7 & 5 \end{vmatrix}$$

$$= -2(-4) + 2(-6) + 1(-2)$$

$$= -6.$$

Because the value of this determinant is *not* zero, you can conclude that the three points do not lie on the same line. Moreover, the area of the triangle with vertices at these points is $\left(-\frac{1}{2}\right)(-6) = 3$ square units.

✓ **Checkpoint** Audio-video solution in English & Spanish at LarsonPrecalculus.com.

Determine whether the points

$$(-2, 4), \quad (3, -1), \quad \text{and} \quad (6, -4)$$

are collinear.

The test for collinear points can be adapted for another use. Given two points on a rectangular coordinate system, you can find an equation of the line passing through the two points.

Two-Point Form of the Equation of a Line

An equation of the line passing through the distinct points (x_1, y_1) and (x_2, y_2) is given by

$$\begin{vmatrix} x & y & 1 \\ x_1 & y_1 & 1 \\ x_2 & y_2 & 1 \end{vmatrix} = 0.$$

EXAMPLE 5 **Finding an Equation of a Line**

Find an equation of the line passing through the two points $(2, 4)$ and $(-1, 3)$, as shown below.

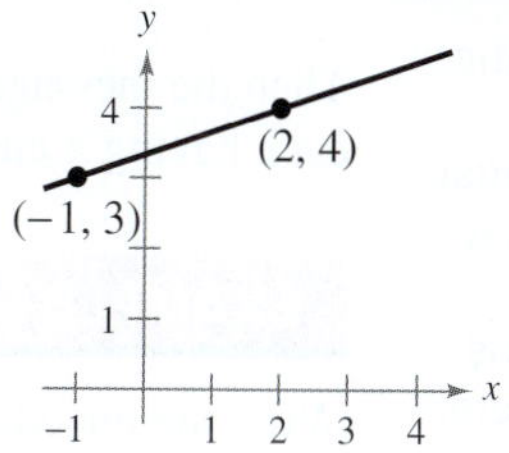

Solution Let $(x_1, y_1) = (2, 4)$ and $(x_2, y_2) = (-1, 3)$. Applying the determinant formula for the equation of a line produces

$$\begin{vmatrix} x & y & 1 \\ 2 & 4 & 1 \\ -1 & 3 & 1 \end{vmatrix} = 0.$$

To evaluate this determinant, you can expand by cofactors along the first row to find an equation of the line.

$$x(-1)^2\begin{vmatrix} 4 & 1 \\ 3 & 1 \end{vmatrix} + y(-1)^3\begin{vmatrix} 2 & 1 \\ -1 & 1 \end{vmatrix} + 1(-1)^4\begin{vmatrix} 2 & 4 \\ -1 & 3 \end{vmatrix} = 0$$

$$x(1) - y(3) + (1)(10) = 0$$

$$x - 3y + 10 = 0$$

✓ **Checkpoint** *Audio-video solution in English & Spanish at LarsonPrecalculus.com.*

Find an equation of the line passing through the two points $(-3, -1)$ and $(3, 5)$. ■

Note that this method of finding the equation of a line works for all lines, including horizontal and vertical lines. For instance, the equation of the vertical line passing through $(2, 0)$ and $(2, 2)$ is

$$\begin{vmatrix} x & y & 1 \\ 2 & 0 & 1 \\ 2 & 2 & 1 \end{vmatrix} = 0$$

$$4 - 2x = 0$$

$$x = 2.$$

Because of the heavy use of the Internet to conduct business, Internet security is of the utmost importance. If a malicious party should receive confidential information such as passwords, personal identification numbers, credit card numbers, social security numbers, bank account details, or corporate secrets, the effects can be damaging. To protect the confidentiality and integrity of such information, the most popular forms of Internet security use data *encryption,* the process of encoding information so that the only way to decode it, apart from a brute force "exhaustion attack," is to use a *key.* Data encryption technology uses algorithms based on the material presented here, but on a much more sophisticated level, to prevent malicious parties from discovering the key.

Cryptography

A **cryptogram** is a message written according to a secret code. (The Greek word *kryptos* means "hidden.") Matrix multiplication can be used to encode and decode messages. To begin, you need to assign a number to each letter in the alphabet (with 0 assigned to a blank space), as follows.

0 = _	9 = I	18 = R
1 = A	10 = J	19 = S
2 = B	11 = K	20 = T
3 = C	12 = L	21 = U
4 = D	13 = M	22 = V
5 = E	14 = N	23 = W
6 = F	15 = O	24 = X
7 = G	16 = P	25 = Y
8 = H	17 = Q	26 = Z

Then the message is converted to numbers and partitioned into **uncoded row matrices,** each having n entries, as demonstrated in Example 6.

EXAMPLE 6 Forming Uncoded Row Matrices

Write the uncoded row matrices of order 1×3 for the message

MEET ME MONDAY.

Solution Partitioning the message (including blank spaces, but ignoring punctuation) into groups of three produces the following uncoded row matrices.

$$[13 \quad 5 \quad 5] \quad [20 \quad 0 \quad 13] \quad [5 \quad 0 \quad 13] \quad [15 \quad 14 \quad 4] \quad [1 \quad 25 \quad 0]$$

M E E T _ M E _ M O N D A Y _

Note that a blank space is used to fill out the last uncoded row matrix.

✓ **Checkpoint** *Audio-video solution in English & Spanish at LarsonPrecalculus.com.*

Write the uncoded row matrices of order 1×3 for the message

OWLS ARE NOCTURNAL.

To encode a message, use the techniques demonstrated in Section 8.3 to choose an $n \times n$ invertible matrix such as

$$A = \begin{bmatrix} 1 & -2 & 2 \\ -1 & 1 & 3 \\ 1 & -1 & -4 \end{bmatrix}$$

and multiply the uncoded row matrices by A (on the right) to obtain **coded row matrices.** Here is an example.

Uncoded Matrix	**Encoding Matrix A**	**Coded Matrix**

$$[13 \quad 5 \quad 5] \begin{bmatrix} 1 & -2 & 2 \\ -1 & 1 & 3 \\ 1 & -1 & -4 \end{bmatrix} = [13 \quad -26 \quad 21]$$

Andrea Danti/Shutterstock.com

EXAMPLE 7 Encoding a Message

Use the following invertible matrix to encode the message MEET ME MONDAY.

$$A = \begin{bmatrix} 1 & -2 & 2 \\ -1 & 1 & 3 \\ 1 & -1 & -4 \end{bmatrix}$$

Solution The coded row matrices are obtained by multiplying each of the uncoded row matrices found in Example 6 by the matrix A, as follows.

Uncoded Matrix Encoding Matrix A Coded Matrix

$$\begin{bmatrix} 13 & 5 & 5 \end{bmatrix} \begin{bmatrix} 1 & -2 & 2 \\ -1 & 1 & 3 \\ 1 & -1 & -4 \end{bmatrix} = \begin{bmatrix} 13 & -26 & 21 \end{bmatrix}$$

$$\begin{bmatrix} 20 & 0 & 13 \end{bmatrix} \begin{bmatrix} 1 & -2 & 2 \\ -1 & 1 & 3 \\ 1 & -1 & -4 \end{bmatrix} = \begin{bmatrix} 33 & -53 & -12 \end{bmatrix}$$

$$\begin{bmatrix} 5 & 0 & 13 \end{bmatrix} \begin{bmatrix} 1 & -2 & 2 \\ -1 & 1 & 3 \\ 1 & -1 & -4 \end{bmatrix} = \begin{bmatrix} 18 & -23 & -42 \end{bmatrix}$$

$$\begin{bmatrix} 15 & 14 & 4 \end{bmatrix} \begin{bmatrix} 1 & -2 & 2 \\ -1 & 1 & 3 \\ 1 & -1 & -4 \end{bmatrix} = \begin{bmatrix} 5 & -20 & 56 \end{bmatrix}$$

$$\begin{bmatrix} 1 & 25 & 0 \end{bmatrix} \begin{bmatrix} 1 & -2 & 2 \\ -1 & 1 & 3 \\ 1 & -1 & -4 \end{bmatrix} = \begin{bmatrix} -24 & 23 & 77 \end{bmatrix}$$

So, the sequence of coded row matrices is

$$[13 \ -26 \ 21][33 \ -53 \ -12][18 \ -23 \ -42][5 \ -20 \ 56][-24 \ 23 \ 77].$$

Finally, removing the matrix notation produces the following cryptogram.

13 −26 21 33 −53 −12 18 −23 −42 5 −20 56 −24 23 77

✓ Checkpoint *Audio-video solution in English & Spanish at LarsonPrecalculus.com.*

Use the following invertible matrix to encode the message OWLS ARE NOCTURNAL.

$$A = \begin{bmatrix} 1 & -1 & 0 \\ 1 & 0 & -1 \\ 6 & -2 & -3 \end{bmatrix}$$

For those who do not know the encoding matrix A, decoding the cryptogram found in Example 7 is difficult. But for an authorized receiver who knows the encoding matrix A, decoding is simple. The receiver just needs to multiply the coded row matrices by A^{-1} (on the right) to retrieve the uncoded row matrices. Here is an example.

$$\underbrace{\begin{bmatrix} 13 & -26 & 21 \end{bmatrix}}_{\text{Coded}} \underbrace{\begin{bmatrix} -1 & -10 & -8 \\ -1 & -6 & -5 \\ 0 & -1 & -1 \end{bmatrix}}_{A^{-1}} = \underbrace{\begin{bmatrix} 13 & 5 & 5 \end{bmatrix}}_{\text{Uncoded}}$$

HISTORICAL NOTE

During World War II, Navajo soldiers created a code using their native language to send messages between battalions. Native words were assigned to represent characters in the English alphabet, and the soldiers created a number of expressions for important military terms, such as *iron-fish* to mean *submarine*. Without the Navajo Code Talkers, the Second World War might have had a very different outcome.

EXAMPLE 8 Decoding a Message

Use the inverse of the matrix A in Example 7 to decode the cryptogram

$$13 \quad -26 \quad 21 \quad 33 \quad -53 \quad -12 \quad 18 \quad -23 \quad -42 \quad 5 \quad -20 \quad 56 \quad -24 \quad 23 \quad 77.$$

Solution First find the decoding matrix A^{-1} by using the techniques demonstrated in Section 8.3. Then partition the message into groups of three to form the coded row matrices. Finally, multiply each coded row matrix by A^{-1} (on the right).

Coded Matrix **Decoding Matrix A^{-1}** **Decoded Matrix**

$$[13 \quad -26 \quad 21]\begin{bmatrix}-1 & -10 & -8\\-1 & -6 & -5\\0 & -1 & -1\end{bmatrix} = [13 \quad 5 \quad 5]$$

$$[33 \quad -53 \quad -12]\begin{bmatrix}-1 & -10 & -8\\-1 & -6 & -5\\0 & -1 & -1\end{bmatrix} = [20 \quad 0 \quad 13]$$

$$[18 \quad -23 \quad -42]\begin{bmatrix}-1 & -10 & -8\\-1 & -6 & -5\\0 & -1 & -1\end{bmatrix} = [5 \quad 0 \quad 13]$$

$$[5 \quad -20 \quad 56]\begin{bmatrix}-1 & -10 & -8\\-1 & -6 & -5\\0 & -1 & -1\end{bmatrix} = [15 \quad 14 \quad 4]$$

$$[-24 \quad 23 \quad 77]\begin{bmatrix}-1 & -10 & -8\\-1 & -6 & -5\\0 & -1 & -1\end{bmatrix} = [1 \quad 25 \quad 0]$$

So, the message is as follows.

$$[13 \quad 5 \quad 5] \; [20 \quad 0 \quad 13] \; [5 \quad 0 \quad 13] \; [15 \quad 14 \quad 4] \; [1 \quad 25 \quad 0]$$

M E E T M E M O N D A Y

✓ **Checkpoint** Audio-video solution in English & Spanish at LarsonPrecalculus.com.

Use the inverse of the matrix A in the Checkpoint with Example 7 to decode the cryptogram

$$110 \quad -39 \quad -59 \quad 25 \quad -21 \quad -3 \quad 23 \quad -18 \quad -5 \quad 47 \quad -20 \quad -24$$
$$149 \quad -56 \quad -75 \quad 87 \quad -38 \quad -37.$$

Summarize (Section 8.5)

1. State Cramer's Rule *(page 583)*. For examples of using Cramer's Rule to solve systems of linear equations, see Examples 1 and 2.
2. State the formula for finding the area of a triangle using a determinant *(page 585)*. For an example of finding the area of a triangle, see Example 3.
3. State the test for collinear points *(page 586)*. For an example of testing for collinear points, see Example 4.
4. Explain how to use a determinant to find an equation of a line *(page 587)*. For an example of finding an equation of a line, see Example 5.
5. Describe how to use matrices to encode and decode a message *(pages 588–590, Examples 6–8)*.

Bettmann/Corbis

8.5 Exercises

See CalcChat.com for tutorial help and worked-out solutions to odd-numbered exercises.

Vocabulary: Fill in the blanks.

1. The method of using determinants to solve a system of linear equations is called ________ ________.
2. Three points are ________ when the points lie on the same line.
3. The area A of a triangle with vertices (x_1, y_1), (x_2, y_2), and (x_3, y_3) is given by ________.
4. A message written according to a secret code is called a ________.
5. To encode a message, choose an invertible matrix A and multiply the ________ row matrices by A (on the right) to obtain ________ row matrices.
6. If a message is encoded using an invertible matrix A, then the message can be decoded by multiplying the coded row matrices by ________ (on the right).

Skills and Applications

Using Cramer's Rule **In Exercises 7–16, use Cramer's Rule to solve (if possible) the system of equations.**

7. $\begin{cases} -7x + 11y = -1 \\ 3x - 9y = 9 \end{cases}$

8. $\begin{cases} 4x - 3y = -10 \\ 6x + 9y = 12 \end{cases}$

9. $\begin{cases} 3x + 2y = -2 \\ 6x + 4y = 4 \end{cases}$

10. $\begin{cases} 6x - 5y = 17 \\ -13x + 3y = -76 \end{cases}$

11. $\begin{cases} -0.4x + 0.8y = 1.6 \\ 0.2x + 0.3y = 2.2 \end{cases}$

12. $\begin{cases} 2.4x - 1.3y = 14.63 \\ 4.6x + 0.5y = -11.51 \end{cases}$

13. $\begin{cases} 4x - y + z = -5 \\ 2x + 2y + 3z = 10 \\ 5x - 2y + 6z = 1 \end{cases}$

14. $\begin{cases} 4x - 2y + 3z = -2 \\ 2x + 2y + 5z = 16 \\ 8x - 5y - 2z = 4 \end{cases}$

15. $\begin{cases} x + 2y + 3z = -3 \\ -2x + y - z = 6 \\ 3x - 3y + 2z = -11 \end{cases}$

16. $\begin{cases} 5x - 4y + z = -14 \\ -x + 2y - 2z = 10 \\ 3x + y + z = 1 \end{cases}$

Finding the Area of a Triangle **In Exercises 17–28, use a determinant to find the area with the given vertices.**

17. 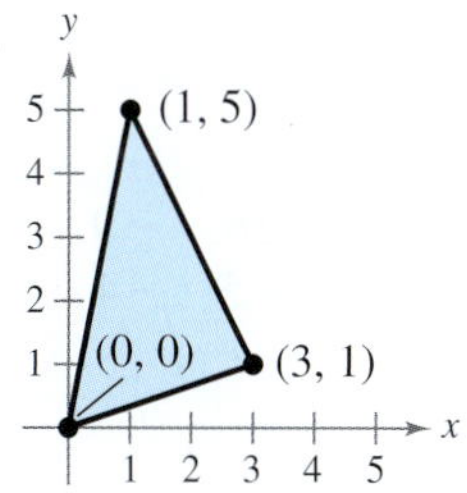

18. 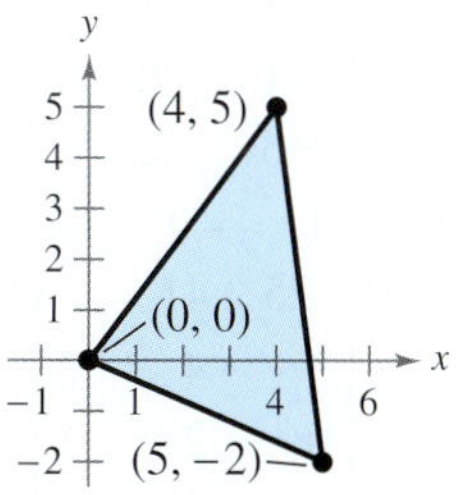

19. 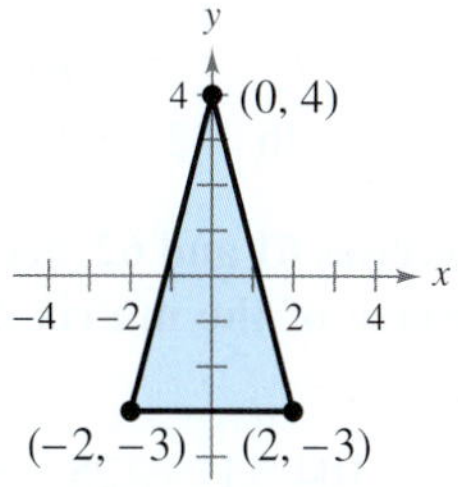

20. 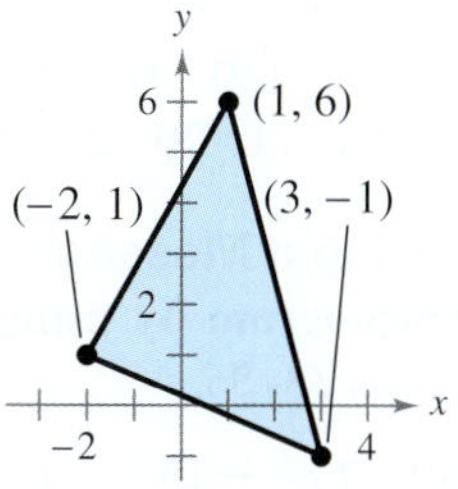

21. $(0, \frac{1}{2}), (4, 3), (\frac{5}{2}, 0)$
22. $(-4, -5), (6, 10), (6, -1)$
23. $(-2, 4), (2, 3), (-1, 5)$
24. $(0, -2), (-1, 4), (3, 5)$
25. $(-3, 5), (2, 6), (3, -5)$
26. $(-2, 4), (1, 5), (3, -2)$
27. $(-4, 2), (0, \frac{7}{2}), (3, -\frac{1}{2})$
28. $(\frac{9}{2}, 0), (2, 6), (0, -\frac{3}{2})$

Finding a Coordinate **In Exercises 29 and 30, find a value of y such that the triangle with the given vertices has an area of 4 square units.**

29. $(-5, 1), (0, 2), (-2, y)$
30. $(-4, 2), (-3, 5), (-1, y)$

Finding a Coordinate **In Exercises 31 and 32, find a value of y such that the triangle with the given vertices has an area of 6 square units.**

31. $(-2, -3), (1, -1), (-8, y)$
32. $(1, 0), (5, -3), (-3, y)$

33. Area of Infestation

A large region of forest has been infested with gypsy moths. The region is roughly triangular, as shown in the figure. From the northernmost vertex A of the region, the distances to the other vertices are 25 miles south and 10 miles east (for vertex B), and 20 miles south and 28 miles east (for vertex C). Use a graphing utility to approximate the number of square miles in this region.

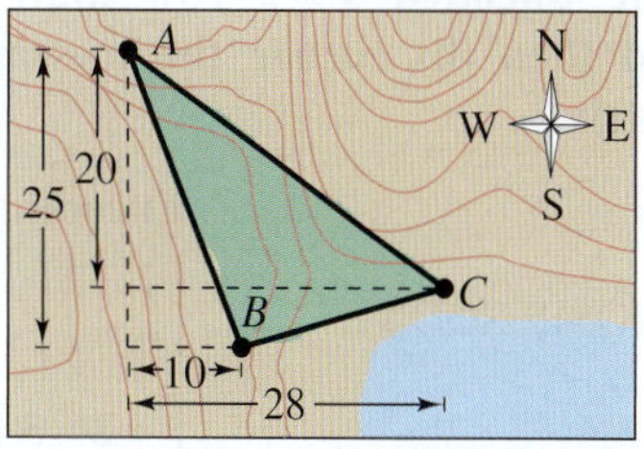

Martynova Anna/Shutterstock.com

34. Botany A botanist is studying the plants growing in a triangular tract of land, as shown in the figure. To estimate the number of square feet in the tract, the botanist starts at one vertex, walks 65 feet east and 50 feet north to the second vertex, and then walks 85 feet west and 30 feet north to the third vertex. Use a graphing utility to determine how many square feet there are in the tract of land.

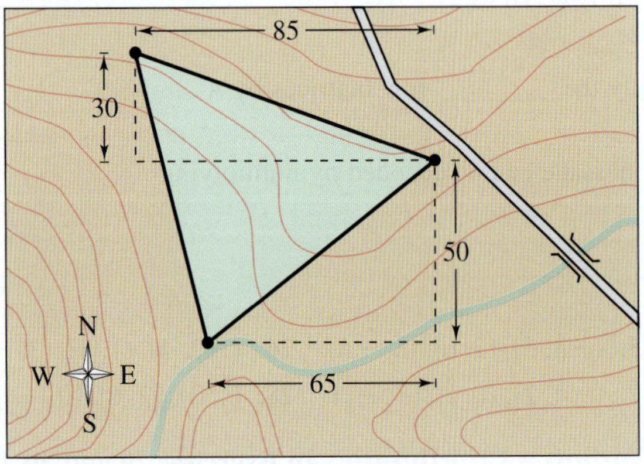

Testing for Collinear Points **In Exercises 35–40, use a determinant to determine whether the points are collinear.**

35. $(3, -1), (0, -3), (12, 5)$

36. $(3, -5), (6, 1), (4, 2)$

37. $\left(2, -\frac{1}{2}\right), (-4, 4), (6, -3)$

38. $\left(0, \frac{1}{2}\right), (2, -1), \left(-4, \frac{7}{2}\right)$

39. $(0, 2), (1, 2.4), (-1, 1.6)$

40. $(2, 3), (3, 3.5), (-1, 2)$

Finding a Coordinate **In Exercises 41 and 42, find y such that the points are collinear.**

41. $(2, -5), (4, y), (5, -2)$ **42.** $(-6, 2), (-5, y), (-3, 5)$

Finding an Equation of a Line **In Exercises 43–48, use a determinant to find an equation of the line passing through the points.**

43. $(0, 0), (5, 3)$ **44.** $(0, 0), (-2, 2)$

45. $(-4, 3), (2, 1)$ **46.** $(10, 7), (-2, -7)$

47. $\left(-\frac{1}{2}, 3\right), \left(\frac{5}{2}, 1\right)$ **48.** $\left(\frac{2}{3}, 4\right), (6, 12)$

Encoding a Message **In Exercises 49 and 50, (a) write the uncoded 1×2 row matrices for the message, and then (b) encode the message using the encoding matrix.**

Message	Encoding Matrix
49. COME HOME SOON	$\begin{bmatrix} 1 & 2 \\ 3 & 5 \end{bmatrix}$
50. HELP IS ON THE WAY	$\begin{bmatrix} -2 & 3 \\ -1 & 1 \end{bmatrix}$

Encoding a Message **In Exercises 51 and 52, (a) write the uncoded 1×3 row matrices for the message, and then (b) encode the message using the encoding matrix.**

Message	Encoding Matrix
51. CALL ME TOMORROW	$\begin{bmatrix} 1 & -1 & 0 \\ 1 & 0 & -1 \\ -6 & 2 & 3 \end{bmatrix}$
52. PLEASE SEND MONEY	$\begin{bmatrix} 4 & 2 & 1 \\ -3 & -3 & -1 \\ 3 & 2 & 1 \end{bmatrix}$

Encoding a Message **In Exercises 53–56, write a cryptogram for the message using the matrix**

$$A = \begin{bmatrix} 1 & 2 & 2 \\ 3 & 7 & 9 \\ -1 & -4 & -7 \end{bmatrix}.$$

53. LANDING SUCCESSFUL

54. ICEBERG DEAD AHEAD

55. HAPPY BIRTHDAY

56. OPERATION OVERLOAD

Decoding a Message **In Exercises 57–60, use A^{-1} to decode the cryptogram.**

57. $A = \begin{bmatrix} 1 & 2 \\ 3 & 5 \end{bmatrix}$

11 21 64 112 25 50 29 53 23 46 40 75 55 92

58. $A = \begin{bmatrix} 2 & 3 \\ 3 & 4 \end{bmatrix}$

85 120 6 8 10 15 84 117 42 56 90 125 60 80 30 45 19 26

59. $A = \begin{bmatrix} 1 & -1 & 0 \\ 1 & 0 & -1 \\ -6 & 2 & 3 \end{bmatrix}$

9 −1 −9 38 −19 −19 28 −9 −19 −80 25 41 −64 21 31 9 −5 −4

60. $A = \begin{bmatrix} 3 & -4 & 2 \\ 0 & 2 & 1 \\ 4 & -5 & 3 \end{bmatrix}$

112 −140 83 19 −25 13 72 −76 61 95 −118 71 20 21 38 35 −23 36 42 −48 32

Decoding a Message **In Exercises 61 and 62, decode the cryptogram by using the inverse of the matrix A in Exercises 53–56.**

61. 20 17 −15 −12 −56 −104 1 −25 −65 62 143 181

62. 13 −9 −59 61 112 106 −17 −73 −131 11 24 29 65 144 172

63. Decoding a Message The following cryptogram was encoded with a 2×2 matrix.

8 21 −15 −10 −13 −13 5 10 5 25 5 19
−1 6 20 40 −18 −18 1 16

The last word of the message is _RON. What is the message?

64. Decoding a Message The following cryptogram was encoded with a 2×2 matrix.

5 2 25 11 −2 −7 −15 −15 32 14 −8
−13 38 19 −19 −19 37 16

The last word of the message is _SUE. What is the message?

65. Circuit Analysis Consider the circuit in the figure. The currents I_1, I_2, and I_3 in amperes are given by the solution of the system of linear equations.

$$\begin{cases} 4I_1 + 8I_3 = 2 \\ 2I_2 + 8I_3 = 6 \\ I_1 + I_2 - I_3 = 0. \end{cases}$$

Use Cramer's Rule to find the three currents.

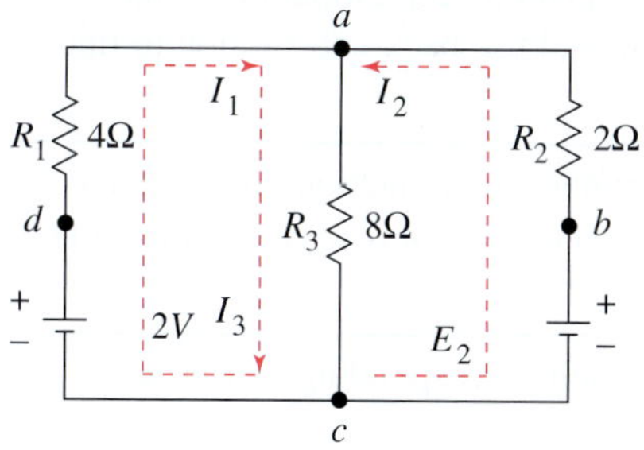

66. Pulley System A system of pulleys that is assumed frictionless and without mass is loaded with 192-pound and 64-pound weights (see figure). The tensions t_1 and t_2 in the ropes and the acceleration a of the 64-pound weight are found by solving the system

$$\begin{cases} t_1 - 2t_2 = 0 \\ t_1 - 3a = 192 \\ t_2 + 2a = 64 \end{cases}$$

where t_1 and t_2 are measured in pounds and a is in feet per second squared. Use Cramer's Rule to find t_1, t_2, and a.

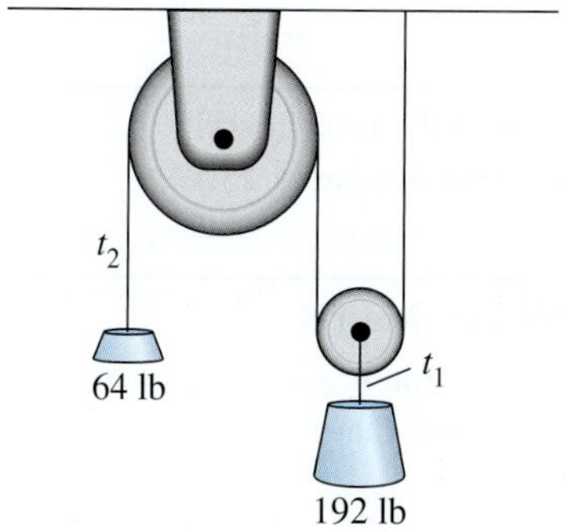

Exploration

True or False? **In Exercises 67–70, determine whether the statement is true or false. Justify your answer.**

67. In Cramer's Rule, the numerator is the determinant of the coefficient matrix.

68. You cannot use Cramer's Rule to solve a system of linear equations when the determinant of the coefficient matrix is zero.

69. In a system of linear equations, when the determinant of the coefficient matrix is zero, the system has no solution.

70. The points $(-5, -13)$, $(0, 2)$, and $(3, 11)$ are collinear.

71. Writing Use your school's library, the Internet, or some other reference source to research a few current real-life uses of cryptography. Write a short summary of these uses. Include a description of how messages are encoded and decoded in each case.

72. Writing

(a) State Cramer's Rule for solving a system of linear equations.

(b) At this point in the text, you have learned several methods for solving systems of linear equations. Briefly describe which method(s) you find easiest to use and which method(s) you find most difficult to use.

73. Reasoning Use determinants to find the area of a triangle with vertices $(3, -1)$, $(7, -1)$, and $(7, 5)$. Confirm your answer by plotting the points in a coordinate plane and using the formula

$$\text{Area} = \tfrac{1}{2}(\text{base})(\text{height}).$$

74. **HOW DO YOU SEE IT?** At this point in the text, you have learned several methods for finding an equation of a line that passes through two given points. Briefly describe the methods that can be used to find the equation of the line that passes through the two points shown. Discuss the advantages and disadvantages of each method.

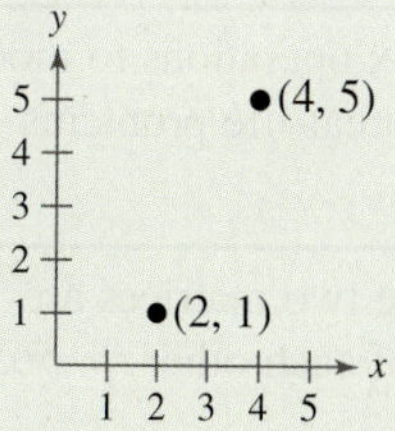

Chapter Summary

Section	What Did You Learn?	Explanation/Examples	Review Exercises
Section 8.1	Write matrices and identify their orders *(p. 538).*	$\begin{bmatrix} -1 & 1 \\ 4 & 7 \end{bmatrix}$ 2×2; $[-2 \quad 3 \quad 0]$ 1×3; $\begin{bmatrix} 4 & -3 \\ 5 & 0 \\ -2 & 1 \end{bmatrix}$ 3×2; $\begin{bmatrix} 8 \\ -8 \end{bmatrix}$ 2×1	1–8
	Perform elementary row operations on matrices *(p. 540).*	**Elementary Row Operations** **1.** Interchange two rows. **2.** Multiply a row by a nonzero constant. **3.** Add a multiple of a row to another row.	9, 10
	Use matrices and Gaussian elimination to solve systems of linear equations *(p. 541).*	**Gaussian Elimination with Back-Substitution** **1.** Write the augmented matrix of the system of linear equations. **2.** Use elementary row operations to rewrite the augmented matrix in row-echelon form. **3.** Write the system of linear equations corresponding to the matrix in row-echelon form and use back-substitution to find the solution.	11–28
	Use matrices and Gauss-Jordan elimination to solve systems of linear equations *(p. 545).*	Gauss-Jordan elimination continues the reduction process on a matrix in row-echelon form until a *reduced* row-echelon form is obtained. (See Example 8.)	29–36
Section 8.2	Decide whether two matrices are equal *(p. 551).*	Two matrices are equal when their corresponding entries are equal.	37–40
	Add and subtract matrices, and multiply matrices by scalars *(p. 552).*	**Definition of Matrix Addition** If $A = [a_{ij}]$ and $B = [b_{ij}]$ are matrices of order $m \times n$, then their sum is the $m \times n$ matrix given by $A + B = [a_{ij} + b_{ij}]$. **Definition of Scalar Multiplication** If $A = [a_{ij}]$ is an $m \times n$ matrix and c is a scalar, then the scalar multiple of A by c is the $m \times n$ matrix given by $cA = [ca_{ij}]$.	41–54
	Multiply two matrices *(p. 556).*	**Definition of Matrix Multiplication** If $A = [a_{ij}]$ is an $m \times n$ matrix and $B = [b_{ij}]$ is an $n \times p$ matrix, then the product AB is an $m \times p$ matrix $AB = [c_{ij}]$ where $c_{ij} = a_{i1}b_{1j} + a_{i2}b_{2j} + a_{i3}b_{3j} + \cdot \cdot \cdot + a_{in}b_{nj}$.	55–66
	Use matrix operations to model and solve real-life problems *(p. 559).*	You can use matrix operations to find the total cost of equipment for two softball teams. (See Example 13.)	67–70
Section 8.3	Verify that two matrices are inverses of each other *(p. 565).*	**Definition of the Inverse of a Square Matrix** Let A be an $n \times n$ matrix and let I_n be the $n \times n$ identity matrix. If there exists a matrix A^{-1} such that $AA^{-1} = I_n = A^{-1}A$ then A^{-1} is the inverse of A.	71–74

	What Did You Learn?	Explanation/Examples	Review Exercises
Section 8.3	Use Gauss-Jordan elimination to find the inverses of matrices *(p. 566)*.	**Finding an Inverse Matrix** Let A be a square matrix of order $n \times n$. **1.** Write the $n \times 2n$ matrix that consists of the given matrix A on the left and the $n \times n$ identity matrix I on the right to obtain $[A \,\vdots\, I]$. **2.** If possible, row reduce A to I using elementary row operations on the *entire* matrix $[A \,\vdots\, I]$. The result will be the matrix $[I \,\vdots\, A^{-1}]$. If this is not possible, then A is not invertible. **3.** Check your work by multiplying to see that $AA^{-1} = I = A^{-1}A$.	75–82
	Use a formula to find the inverses of 2×2 matrices *(p. 569)*.	If $A = \begin{bmatrix} a & b \\ c & d \end{bmatrix}$ and $ad - bc \neq 0$, then $A^{-1} = \dfrac{1}{ad - bc}\begin{bmatrix} d & -b \\ -c & a \end{bmatrix}$.	83–90
	Use inverse matrices to solve systems of linear equations *(p. 570)*.	If A is an invertible matrix, then the system of linear equations represented by $AX = B$ has a unique solution given by $X = A^{-1}B$.	91–108
Section 8.4	Find the determinants of 2×2 matrices *(p. 574)*.	The determinant of the matrix $A = \begin{bmatrix} a_1 & b_1 \\ a_2 & b_2 \end{bmatrix}$ is given by $\det(A) = \lvert A \rvert = \begin{vmatrix} a_1 & b_1 \\ a_2 & b_2 \end{vmatrix} = a_1b_2 - a_2b_1$.	109–112
	Find minors and cofactors of square matrices *(p. 576)*.	If A is a square matrix, then the minor M_{ij} of the entry a_{ij} is the determinant of the matrix obtained by deleting the ith row and jth column of A. The cofactor C_{ij} of the entry a_{ij} is $C_{ij} = (-1)^{i+j}M_{ij}$.	113–116
	Find the determinants of square matrices *(p. 577)*.	If A is a square matrix (of order 2×2 or greater), then the determinant of A is the sum of the entries in any row (or column) of A multiplied by their respective cofactors.	117–126
Section 8.5	Use Cramer's Rule to solve systems of linear equations *(p. 582)*.	Cramer's Rule uses determinants to write the solution of a system of linear equations.	127–130
	Use determinants to find the areas of triangles *(p. 585)*.	The area of a triangle with vertices (x_1, y_1), (x_2, y_2), and (x_3, y_3) is $\text{Area} = \pm\frac{1}{2}\begin{vmatrix} x_1 & y_1 & 1 \\ x_2 & y_2 & 1 \\ x_3 & y_3 & 1 \end{vmatrix}$ where the symbol $\pm$ indicates that the appropriate sign should be chosen to yield a positive area.	131–134
	Use a determinant to test for collinear points and find an equation of a line passing through two points *(p. 586)*.	Three points (x_1, y_1), (x_2, y_2), and (x_3, y_3) are collinear (lie on the same line) if and only if $\begin{vmatrix} x_1 & y_1 & 1 \\ x_2 & y_2 & 1 \\ x_3 & y_3 & 1 \end{vmatrix} = 0$.	135–140
	Use matrices to encode and decode messages *(p. 588)*.	You can use the inverse of a matrix to decode a cryptogram. (See Example 8.)	141–144

Review Exercises

See CalcChat.com for tutorial help and worked-out solutions to odd-numbered exercises.

8.1 Order of a Matrix **In Exercises 1–4, determine the order of the matrix.**

1. $\begin{bmatrix} -4 \\ 0 \\ 5 \end{bmatrix}$

2. $\begin{bmatrix} 3 & -1 & 0 & 6 \\ -2 & 7 & 1 & 4 \end{bmatrix}$

3. $[3]$

4. $[6 \quad 2 \quad -5 \quad 8 \quad 0]$

Writing an Augmented Matrix **In Exercises 5 and 6, write the augmented matrix for the system of linear equations.**

5. $\begin{cases} 3x - 10y = 15 \\ 5x + 4y = 22 \end{cases}$

6. $\begin{cases} 8x - 7y + 4z = 12 \\ 3x - 5y + 2z = 20 \end{cases}$

Writing a System of Equations **In Exercises 7 and 8, write the system of linear equations represented by the augmented matrix. (Use variables *x*, *y*, *z*, and *w*, if applicable.)**

7. $\left[\begin{array}{ccc:c} 5 & 1 & 7 & -9 \\ 4 & 2 & 0 & 10 \\ 9 & 4 & 2 & 3 \end{array}\right]$

8. $\left[\begin{array}{cccc:c} 13 & 16 & 7 & 3 & 2 \\ 4 & 10 & -4 & 3 & -1 \end{array}\right]$

Writing a Matrix in Row-Echelon Form **In Exercises 9 and 10, write the matrix in row-echelon form. (Remember that the row-echelon form of a matrix is not unique.)**

9. $\begin{bmatrix} 0 & 1 & 1 \\ 1 & 2 & 3 \\ 2 & 2 & 2 \end{bmatrix}$

10. $\begin{bmatrix} 4 & 8 & 16 \\ 3 & -1 & 2 \\ -2 & 10 & 12 \end{bmatrix}$

Using Back-Substitution **In Exercises 11–14, write the system of linear equations represented by the augmented matrix. Then use back-substitution to solve the system. (Use variables *x*, *y*, and *z*, if applicable.)**

11. $\left[\begin{array}{ccc:c} 1 & 2 & 3 & 9 \\ 0 & 1 & -2 & 2 \\ 0 & 0 & 1 & 0 \end{array}\right]$

12. $\left[\begin{array}{ccc:c} 1 & 3 & -9 & 4 \\ 0 & 1 & -1 & 10 \\ 0 & 0 & 1 & -2 \end{array}\right]$

13. $\left[\begin{array}{ccc:c} 1 & -5 & 4 & 1 \\ 0 & 1 & 2 & 3 \\ 0 & 0 & 1 & 4 \end{array}\right]$

14. $\left[\begin{array}{ccc:c} 1 & -8 & 0 & -2 \\ 0 & 1 & -1 & -7 \\ 0 & 0 & 1 & 1 \end{array}\right]$

Gaussian Elimination with Back-Substitution **In Exercises 15–28, use matrices to solve the system of equations (if possible). Use Gaussian elimination with back-substitution.**

15. $\begin{cases} 5x + 4y = 2 \\ -x + y = -22 \end{cases}$

16. $\begin{cases} 2x - 5y = 2 \\ 3x - 7y = 1 \end{cases}$

17. $\begin{cases} 0.3x - 0.1y = -0.13 \\ 0.2x - 0.3y = -0.25 \end{cases}$

18. $\begin{cases} 0.2x - 0.1y = 0.07 \\ 0.4x - 0.5y = -0.01 \end{cases}$

19. $\begin{cases} -x + 2y = 3 \\ 2x - 4y = 6 \end{cases}$

20. $\begin{cases} -x + 2y = 3 \\ 2x - 4y = -6 \end{cases}$

21. $\begin{cases} x - 2y + z = 7 \\ 2x + y - 2z = -4 \\ -x + 3y + 2z = -3 \end{cases}$

22. $\begin{cases} x - 2y + z = 4 \\ 2x + y - 2z = -24 \\ -x + 3y + 2z = 20 \end{cases}$

23. $\begin{cases} 2x + y + 2z = 4 \\ 2x + 2y = 5 \\ 2x - y + 6z = 2 \end{cases}$

24. $\begin{cases} x + 2y + 6z = 1 \\ 2x + 5y + 15z = 4 \\ 3x + y + 3z = -6 \end{cases}$

25. $\begin{cases} 2x + 3y + z = 10 \\ 2x - 3y - 3z = 22 \\ 4x - 2y + 3z = -2 \end{cases}$

26. $\begin{cases} 2x + 3y + 3z = 3 \\ 6x + 6y + 12z = 13 \\ 12x + 9y - z = 2 \end{cases}$

27. $\begin{cases} 2x + y + z = 6 \\ -2y + 3z - w = 9 \\ 3x + 3y - 2z - 2w = -11 \\ x + z + 3w = 14 \end{cases}$

28. $\begin{cases} x + 2y + w = 3 \\ -3y + 3z = 0 \\ 4x + 4y + z + 2w = 0 \\ 2x + z = 3 \end{cases}$

Gauss-Jordan Elimination **In Exercises 29–34, use matrices to solve the system of equations (if possible). Use Gauss-Jordan elimination.**

29. $\begin{cases} x + 2y - z = 3 \\ x - y - z = -3 \\ 2x + y + 3z = 10 \end{cases}$

30. $\begin{cases} x - 3y + z = 2 \\ 3x - y - z = -6 \\ -x + y - 3z = -2 \end{cases}$

31. $\begin{cases} -x + y + 2z = 1 \\ 2x + 3y + z = -2 \\ 5x + 4y + 2z = 4 \end{cases}$

32. $\begin{cases} 4x + 4y + 4z = 5 \\ 4x - 2y - 8z = 1 \\ 5x + 3y + 8z = 6 \end{cases}$

33. $\begin{cases} 2x - y + 9z = -8 \\ -x - 3y + 4z = -15 \\ 5x + 2y - z = 17 \end{cases}$

34. $\begin{cases} -3x + y + 7z = -20 \\ 5x - 2y - z = 34 \\ -x + y + 4z = -8 \end{cases}$

Using a Graphing Utility In Exercises 35 and 36, use the matrix capabilities of a graphing utility to write the augmented matrix corresponding to the system of equations in reduced row-echelon form. Then solve the system.

35. $\begin{cases} 3x - y + 5z - 2w = -44 \\ x + 6y + 4z - w = 1 \\ 5x - y + z + 3w = -15 \\ 4y - z - 8w = 58 \end{cases}$

36. $\begin{cases} 4x + 12y + 2z = 20 \\ x + 6y + 4z = 12 \\ x + 6y + z = 8 \\ -2x - 10y - 2z = -10 \end{cases}$

8.2 Equality of Matrices In Exercises 37–40, find x and y.

37. $\begin{bmatrix} -1 & x \\ y & 9 \end{bmatrix} = \begin{bmatrix} -1 & 12 \\ -7 & 9 \end{bmatrix}$

38. $\begin{bmatrix} -1 & 0 \\ x & 5 \\ -4 & y \end{bmatrix} = \begin{bmatrix} -1 & 0 \\ 8 & 5 \\ -4 & 0 \end{bmatrix}$

39. $\begin{bmatrix} x+3 & -4 & 4y \\ 0 & -3 & 2 \\ -2 & y+5 & 6x \end{bmatrix} = \begin{bmatrix} 5x-1 & -4 & 44 \\ 0 & -3 & 2 \\ -2 & 16 & 6 \end{bmatrix}$

40. $\begin{bmatrix} -9 & 4 & 2 & -5 \\ 0 & -3 & 7 & -4 \\ 6 & -1 & 1 & 0 \end{bmatrix} = \begin{bmatrix} -9 & 4 & x-10 & -5 \\ 0 & -3 & 7 & 2y \\ \frac{1}{2}x & -1 & 1 & 0 \end{bmatrix}$

Operations with Matrices In Exercises 41–44, if possible, find (a) $A + B$, (b) $A - B$, (c) $4A$, and (d) $A + 3B$.

41. $A = \begin{bmatrix} 2 & -2 \\ 3 & 5 \end{bmatrix}, \quad B = \begin{bmatrix} -3 & 10 \\ 12 & 8 \end{bmatrix}$

42. $A = \begin{bmatrix} 4 & 3 \\ -6 & 1 \\ 10 & 1 \end{bmatrix}, \quad B = \begin{bmatrix} 3 & 11 \\ 15 & 25 \\ 20 & 29 \end{bmatrix}$

43. $A = \begin{bmatrix} 5 & 4 \\ -7 & 2 \\ 11 & 2 \end{bmatrix}, \quad B = \begin{bmatrix} 0 & 3 \\ 4 & 12 \\ 20 & 40 \end{bmatrix}$

44. $A = \begin{bmatrix} 6 & -5 & 7 \end{bmatrix}, \quad B = \begin{bmatrix} -1 \\ 4 \\ 8 \end{bmatrix}$

Evaluating an Expression In Exercises 45–50, evaluate the expression. If it is not possible, explain why.

45. $\begin{bmatrix} 7 & 3 \\ -1 & 5 \end{bmatrix} + \begin{bmatrix} 10 & -20 \\ 14 & -3 \end{bmatrix}$

46. $\begin{bmatrix} -11 & 16 & 19 \\ -7 & -2 & 1 \end{bmatrix} - \begin{bmatrix} 6 & 0 \\ 8 & -4 \\ -2 & 10 \end{bmatrix}$

47. $-2\begin{bmatrix} 1 & 2 \\ 5 & -4 \\ 6 & 0 \end{bmatrix} + 8\begin{bmatrix} 7 & 1 \\ 1 & 2 \\ 1 & 4 \end{bmatrix}$

48. $-\begin{bmatrix} 8 & -1 & 8 \\ -2 & 4 & 12 \\ 0 & -6 & 0 \end{bmatrix} - 5\begin{bmatrix} -2 & 0 & -4 \\ 3 & -1 & 1 \\ 6 & 12 & -8 \end{bmatrix}$

49. $3\begin{bmatrix} 8 & -2 & 5 \\ 1 & 3 & -1 \end{bmatrix} + 6\begin{bmatrix} 4 & -2 & -3 \\ 2 & 7 & 6 \end{bmatrix}$

50. $-5\begin{bmatrix} 2 & 0 \\ 7 & -2 \\ 8 & 2 \end{bmatrix} + 4\begin{bmatrix} 4 & -2 \\ 6 & 11 \\ -1 & 3 \end{bmatrix}$

Solving a Matrix Equation In Exercises 51–54, solve for X in the equation, where

$$A = \begin{bmatrix} -4 & 0 \\ 1 & -5 \\ -3 & 2 \end{bmatrix} \quad \text{and} \quad B = \begin{bmatrix} 1 & 2 \\ -2 & 1 \\ 4 & 4 \end{bmatrix}.$$

51. $X = 2A - 3B$

52. $6X = 4A + 3B$

53. $3X + 2A = B$

54. $2A - 5B = 3X$

Finding the Product of Two Matrices In Exercises 55–58, find AB, if possible.

55. $A = \begin{bmatrix} 2 & 2 \\ 3 & 5 \end{bmatrix}, \quad B = \begin{bmatrix} -3 & 10 \\ 12 & 8 \end{bmatrix}$

56. $A = \begin{bmatrix} 5 & 4 \\ -7 & 2 \\ 11 & 2 \end{bmatrix}, \quad B = \begin{bmatrix} 4 & 12 \\ 20 & 40 \\ 15 & 30 \end{bmatrix}$

57. $A = \begin{bmatrix} 5 & 4 \\ -7 & 2 \\ 11 & 2 \end{bmatrix}, \quad B = \begin{bmatrix} 4 & 12 \\ 20 & 40 \end{bmatrix}$

58. $A = \begin{bmatrix} 6 & -5 & 7 \end{bmatrix}, \quad B = \begin{bmatrix} -1 \\ 4 \\ 8 \end{bmatrix}$

Evaluating an Expression In Exercises 59–62, evaluate the expression. If it is not possible, explain why.

59. $\begin{bmatrix} 1 & 2 \\ 5 & -4 \\ 6 & 0 \end{bmatrix}\begin{bmatrix} 6 & -2 & 8 \\ 4 & 0 & 0 \end{bmatrix}$

60. $\begin{bmatrix} 1 & 5 & 6 \\ 2 & -4 & 0 \end{bmatrix}\begin{bmatrix} 6 & -2 & 8 \\ 4 & 0 & 0 \end{bmatrix}$

61. $\begin{bmatrix} 1 & 5 & 6 \\ 2 & -4 & 0 \end{bmatrix}\begin{bmatrix} 6 & 4 \\ -2 & 0 \\ 8 & 0 \end{bmatrix}$

62. $\begin{bmatrix} 1 & 3 & 2 \\ 0 & 2 & -4 \\ 0 & 0 & 3 \end{bmatrix}\begin{bmatrix} 4 & -3 & 2 \\ 0 & 3 & -1 \\ 0 & 0 & 2 \end{bmatrix}$

Matrix Multiplication **In Exercises 63–66, use the matrix capabilities of a graphing utility to find the product, if possible.**

63. $\begin{bmatrix} 4 & 1 \\ 11 & -7 \\ 12 & 3 \end{bmatrix}\begin{bmatrix} 3 & -5 & 6 \\ 2 & -2 & -2 \end{bmatrix}$

64. $\begin{bmatrix} -2 & 3 & 10 \\ 4 & -2 & 2 \end{bmatrix}\begin{bmatrix} 1 & 1 \\ -5 & 2 \\ 3 & 2 \end{bmatrix}$

65. $\begin{bmatrix} 1 & 2 & -1 \\ 0 & 4 & -2 \\ 1 & 1 & 3 \end{bmatrix}\begin{bmatrix} 1 & -1 & 2 \end{bmatrix}$

66. $\begin{bmatrix} 4 & -2 & 6 \end{bmatrix}\begin{bmatrix} -2 & 1 \\ 0 & -3 \\ 2 & 0 \end{bmatrix}$

67. **Manufacturing** A tire corporation has three factories, each of which manufactures two models of tires. The production levels are represented by A.

$$A = \overset{\text{Factory}}{\begin{matrix} 1 & 2 & 3 \end{matrix}} \atop \begin{bmatrix} 80 & 120 & 140 \\ 40 & 100 & 80 \end{bmatrix}\begin{matrix} \text{A} \\ \text{B} \end{matrix}\Bigg\}\ \text{Model}$$

Find the production levels when production is decreased by 5%.

68. **Manufacturing** A power tool company has four manufacturing plants, each of which produces three types of cordless power tools. The production levels are represented by A.

$$A = \overset{\text{Plant: } 1 \quad 2 \quad 3 \quad 4}{\begin{bmatrix} 80 & 70 & 90 & 40 \\ 50 & 30 & 80 & 20 \\ 90 & 60 & 100 & 50 \end{bmatrix}}\begin{matrix} \text{A} \\ \text{B} \\ \text{C} \end{matrix}\Bigg\}\ \text{Type}$$

Find the production levels when production is increased by 20%.

69. **Manufacturing** An electronics manufacturing company produces three different models of headphones that are shipped to two warehouses. The shipment levels are represented by A.

$$A = \overset{\text{Warehouse: } 1 \quad 2}{\begin{bmatrix} 8200 & 7400 \\ 6500 & 9800 \\ 5400 & 4800 \end{bmatrix}}\begin{matrix} \text{A} \\ \text{B} \\ \text{C} \end{matrix}\Bigg\}\ \text{Model}$$

The prices per unit are represented by the matrix

$B = [\$79.99 \quad \$109.95 \quad \$189.99]$.

Compute BA and interpret the result.

70. **Cell Phone Charges** The pay-as-you-go charges (in dollars per minute) of two cellular telephone companies for calls inside the coverage area, regional roaming calls, and calls outside the coverage area are represented by C.

$$C = \overset{\text{Company: A} \quad \text{B}}{\begin{bmatrix} 0.07 & 0.095 \\ 0.10 & 0.08 \\ 0.28 & 0.25 \end{bmatrix}}\begin{matrix} \text{Inside} \\ \text{Regional roaming} \\ \text{Outside} \end{matrix}\Bigg\}\ \text{Coverage area}$$

The numbers of minutes you plan to use in the coverage areas per month are represented by the matrix

$T = [120 \quad 80 \quad 20]$.

Compute TC and interpret the result.

8.3 The Inverse of a Matrix **In Exercises 71–74, show that B is the inverse of A.**

71. $A = \begin{bmatrix} -4 & -1 \\ 7 & 2 \end{bmatrix}$, $B = \begin{bmatrix} -2 & -1 \\ 7 & 4 \end{bmatrix}$

72. $A = \begin{bmatrix} 5 & -1 \\ 11 & -2 \end{bmatrix}$, $B = \begin{bmatrix} -2 & 1 \\ -11 & 5 \end{bmatrix}$

73. $A = \begin{bmatrix} 1 & 1 & 0 \\ 1 & 0 & 1 \\ 6 & 2 & 3 \end{bmatrix}$, $B = \begin{bmatrix} -2 & -3 & 1 \\ 3 & 3 & -1 \\ 2 & 4 & -1 \end{bmatrix}$

74. $A = \begin{bmatrix} 1 & -1 & 0 \\ -1 & 0 & -1 \\ 8 & -4 & 2 \end{bmatrix}$, $B = \begin{bmatrix} -2 & 1 & \frac{1}{2} \\ -3 & 1 & \frac{1}{2} \\ 2 & -2 & -\frac{1}{2} \end{bmatrix}$

Finding the Inverse of a Matrix **In Exercises 75–78, find the inverse of the matrix (if it exists).**

75. $\begin{bmatrix} -6 & 5 \\ -5 & 4 \end{bmatrix}$

76. $\begin{bmatrix} -3 & -5 \\ 2 & 3 \end{bmatrix}$

77. $\begin{bmatrix} 2 & 0 & 3 \\ -1 & 1 & 1 \\ 2 & -2 & 1 \end{bmatrix}$

78. $\begin{bmatrix} 0 & -2 & 1 \\ -5 & -2 & -3 \\ 7 & 3 & 4 \end{bmatrix}$

Finding the Inverse of a Matrix **In Exercises 79–82, use the matrix capabilities of a graphing utility to find the inverse of the matrix (if it exists).**

79. $\begin{bmatrix} -1 & -2 & -2 \\ 3 & 7 & 9 \\ 1 & 4 & 7 \end{bmatrix}$

80. $\begin{bmatrix} 1 & 4 & 6 \\ 2 & -3 & 1 \\ -1 & 18 & 16 \end{bmatrix}$

81. $\begin{bmatrix} 1 & 3 & 1 & 6 \\ 4 & 4 & 2 & 6 \\ 3 & 4 & 1 & 2 \\ -1 & 2 & -1 & -2 \end{bmatrix}$

82. $\begin{bmatrix} 8 & 0 & 2 & 8 \\ 4 & -2 & 0 & -2 \\ 1 & 2 & 1 & 4 \\ -1 & 4 & 1 & 1 \end{bmatrix}$

Finding the Inverse of a 2 × 2 Matrix In Exercises 83–90, use the formula on page 569 to find the inverse of the 2 × 2 matrix (if it exists).

83. $\begin{bmatrix} -7 & 2 \\ -8 & 2 \end{bmatrix}$

84. $\begin{bmatrix} 10 & 4 \\ 7 & 3 \end{bmatrix}$

85. $\begin{bmatrix} -12 & 6 \\ 10 & -5 \end{bmatrix}$

86. $\begin{bmatrix} -18 & -15 \\ -6 & -5 \end{bmatrix}$

87. $\begin{bmatrix} -\frac{1}{2} & 20 \\ \frac{3}{10} & -6 \end{bmatrix}$

88. $\begin{bmatrix} -\frac{3}{4} & \frac{5}{2} \\ -\frac{4}{5} & -\frac{8}{3} \end{bmatrix}$

89. $\begin{bmatrix} 0.5 & 0.1 \\ -0.2 & -0.4 \end{bmatrix}$

90. $\begin{bmatrix} 1.6 & -3.2 \\ 1.2 & -2.4 \end{bmatrix}$

Solving a System Using an Inverse Matrix In Exercises 91–102, use an inverse matrix to solve (if possible) the system of linear equations.

91. $\begin{cases} -x + 4y = 8 \\ 2x - 7y = -5 \end{cases}$

92. $\begin{cases} 5x - y = 13 \\ -9x + 2y = -24 \end{cases}$

93. $\begin{cases} -3x + 10y = 8 \\ 5x - 17y = -13 \end{cases}$

94. $\begin{cases} 4x - 2y = -10 \\ -19x + 9y = 47 \end{cases}$

95. $\begin{cases} \frac{1}{2}x + \frac{1}{3}y = 2 \\ -3x + 2y = 0 \end{cases}$

96. $\begin{cases} -\frac{5}{6}x + \frac{3}{8}y = -2 \\ 4x - 3y = 0 \end{cases}$

97. $\begin{cases} 0.3x + 0.7y = 10.2 \\ 0.4x + 0.6y = 7.6 \end{cases}$

98. $\begin{cases} 3.5x - 4.5y = 8 \\ 2.5x - 7.5y = 25 \end{cases}$

99. $\begin{cases} 3x + 2y - z = 6 \\ x - y + 2z = -1 \\ 5x + y + z = 7 \end{cases}$

100. $\begin{cases} 4x + 5y - 6z = -6 \\ 3x + 2y + 2z = 8 \\ 2x + y + z = 3 \end{cases}$

101. $\begin{cases} -2x + y + 2z = -13 \\ -x - 4y + z = -11 \\ -y - z = 0 \end{cases}$

102. $\begin{cases} 3x - y + 5z = -14 \\ -x + y + 6z = 8 \\ -8x + 4y - z = 44 \end{cases}$

Using a Graphing Utility In Exercises 103–108, use the matrix capabilities of a graphing utility to solve (if possible) the system of linear equations.

103. $\begin{cases} x + 2y = -1 \\ 3x + 4y = -5 \end{cases}$

104. $\begin{cases} x + 3y = 23 \\ -6x + 2y = -18 \end{cases}$

105. $\begin{cases} \frac{6}{5}x - \frac{4}{7}y = \frac{6}{5} \\ -\frac{12}{5}x + \frac{12}{7}y = -\frac{17}{5} \end{cases}$

106. $\begin{cases} 5x + 10y = 7 \\ 2x + y = -98 \end{cases}$

107. $\begin{cases} -3x - 3y - 4z = 2 \\ y + z = -1 \\ 4x + 3y + 4z = -1 \end{cases}$

108. $\begin{cases} x - 3y - 2z = 8 \\ -2x + 7y + 3z = -19 \\ x - y - 3z = 3 \end{cases}$

8.4 Finding the Determinant of a Matrix In Exercises 109–112, find the determinant of the matrix.

109. $\begin{bmatrix} 8 & 5 \\ 2 & -4 \end{bmatrix}$

110. $\begin{bmatrix} -9 & 11 \\ 7 & -4 \end{bmatrix}$

111. $\begin{bmatrix} 50 & -30 \\ 10 & 5 \end{bmatrix}$

112. $\begin{bmatrix} 14 & -24 \\ 12 & -15 \end{bmatrix}$

Finding the Minors and Cofactors of a Matrix In Exercises 113–116, find all the (a) minors and (b) cofactors of the matrix.

113. $\begin{bmatrix} 2 & -1 \\ 7 & 4 \end{bmatrix}$

114. $\begin{bmatrix} 3 & 6 \\ 5 & -4 \end{bmatrix}$

115. $\begin{bmatrix} 3 & 2 & -1 \\ -2 & 5 & 0 \\ 1 & 8 & 6 \end{bmatrix}$

116. $\begin{bmatrix} 8 & 3 & 4 \\ 6 & 5 & -9 \\ -4 & 1 & 2 \end{bmatrix}$

Finding the Determinant of a Matrix In Exercises 117–126, find the determinant of the matrix. Expand by cofactors using the row or column that appears to make the computations easiest.

117. $\begin{bmatrix} -2 & 0 & 0 \\ 2 & -1 & 0 \\ -1 & 1 & -3 \end{bmatrix}$

118. $\begin{bmatrix} 0 & 1 & -2 \\ 0 & 1 & 2 \\ -1 & -1 & 3 \end{bmatrix}$

119. $\begin{bmatrix} 4 & 1 & -1 \\ 2 & 3 & 2 \\ 1 & -1 & 0 \end{bmatrix}$

120. $\begin{bmatrix} -1 & -2 & 1 \\ 2 & 3 & 0 \\ -5 & -1 & 3 \end{bmatrix}$

121. $\begin{bmatrix} -2 & 4 & 1 \\ -6 & 0 & 2 \\ 5 & 3 & 4 \end{bmatrix}$

122. $\begin{bmatrix} 1 & 1 & 4 \\ -4 & 1 & 2 \\ 0 & 1 & -1 \end{bmatrix}$

123. $\begin{bmatrix} 1 & 2 & -1 & 0 \\ 1 & 2 & -4 & 1 \\ 2 & -4 & -3 & 1 \\ 2 & 0 & 0 & 0 \end{bmatrix}$

124. $\begin{bmatrix} 1 & -2 & 1 & 2 \\ 4 & 1 & 4 & 1 \\ 2 & 3 & 3 & 0 \\ 0 & -2 & -4 & 2 \end{bmatrix}$

125. $\begin{bmatrix} 3 & 0 & -4 & 0 \\ 0 & 8 & 1 & 2 \\ 6 & 1 & 8 & 2 \\ 0 & 3 & -4 & 1 \end{bmatrix}$

126. $\begin{bmatrix} -5 & 6 & 0 & 0 \\ 0 & 1 & -1 & 2 \\ -3 & 4 & -5 & 1 \\ 1 & 6 & 0 & 3 \end{bmatrix}$

8.5 Using Cramer's Rule In Exercises 127–130, use Cramer's Rule to solve (if possible) the system of equations.

127. $\begin{cases} 5x - 2y = 6 \\ -11x + 3y = -23 \end{cases}$

128. $\begin{cases} 3x + 8y = -7 \\ 9x - 5y = 37 \end{cases}$

129. $\begin{cases} -2x + 3y - 5z = -11 \\ 4x - y + z = -3 \\ -x - 4y + 6z = 15 \end{cases}$

130. $\begin{cases} 5x - 2y + z = 15 \\ 3x - 3y - z = -7 \\ 2x - y - 7z = -3 \end{cases}$

Finding the Area of a Triangle In Exercises 131–134, use a determinant to find the area of the triangle with the given vertices.

131.

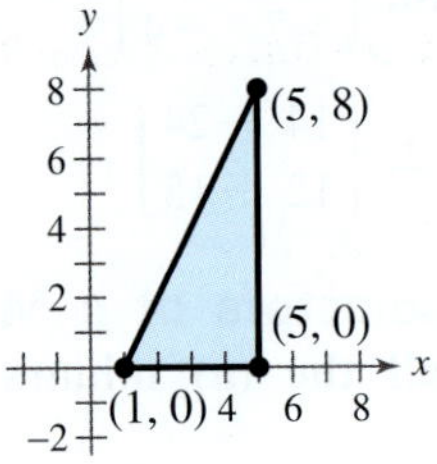

132.

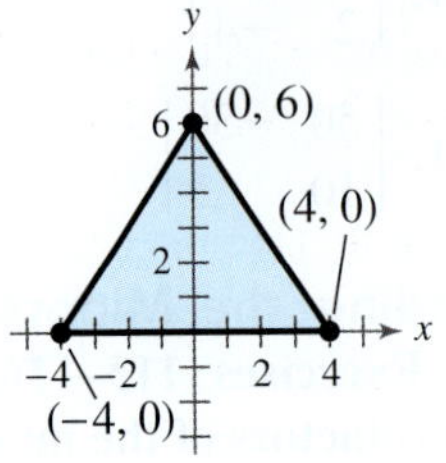

133.

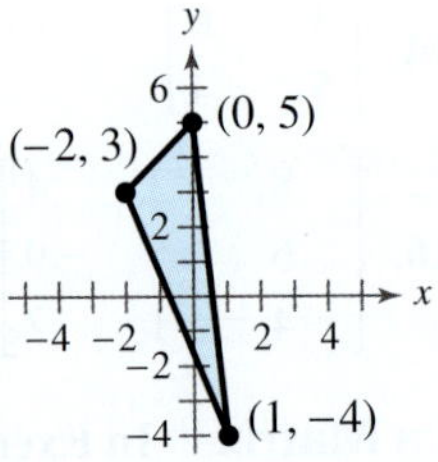

134.

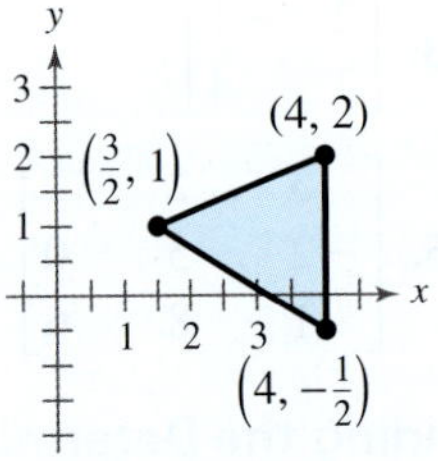

Testing for Collinear Points In Exercises 135 and 136, use a determinant to determine whether the points are collinear.

135. $(-1, 7), (3, -9), (-3, 15)$

136. $(0, -5), (-2, -6), (8, -1)$

Finding an Equation of a Line In Exercises 137–140, use a determinant to find an equation of the line passing through the points.

137. $(-4, 0), (4, 4)$ **138.** $(2, 5), (6, -1)$

139. $\left(-\frac{5}{2}, 3\right), \left(\frac{7}{2}, 1\right)$ **140.** $(-0.8, 0.2), (0.7, 3.2)$

Encoding a Message In Exercises 141 and 142, (a) write the uncoded 1×3 row matrices for the message and then (b) encode the message using the encoding matrix.

	Message	Encoding Matrix
141.	LOOK OUT BELOW	$\begin{bmatrix} 2 & -2 & 0 \\ 3 & 0 & -3 \\ -6 & 2 & 3 \end{bmatrix}$
142.	HEAD DUE WEST	$\begin{bmatrix} 1 & 2 & 2 \\ 3 & 7 & 9 \\ -1 & -4 & -7 \end{bmatrix}$

Decoding a Message In Exercises 143 and 144, decode the cryptogram by using the inverse of the matrix

$$A = \begin{bmatrix} -5 & 4 & -3 \\ 10 & -7 & 6 \\ 8 & -6 & 5 \end{bmatrix}.$$

143. -5 11 -2 370 -265 225 -57 48 -33 32 -15 20 245 -171 147

144. 145 -105 92 264 -188 160 23 -16 15 129 -84 78 -9 8 -5 159 -118 100 219 -152 133 370 -265 225 -105 84 -63

Exploration

True or False? In Exercises 145 and 146, determine whether the statement is true or false. Justify your answer.

145. It is possible to find the determinant of a 4×5 matrix.

146. $\begin{vmatrix} a_{11} & a_{12} & a_{13} \\ a_{21} & a_{22} & a_{23} \\ a_{31} + c_1 & a_{32} + c_2 & a_{33} + c_3 \end{vmatrix} = \begin{vmatrix} a_{11} & a_{12} & a_{13} \\ a_{21} & a_{22} & a_{23} \\ a_{31} & a_{32} & a_{33} \end{vmatrix} + \begin{vmatrix} a_{11} & a_{12} & a_{13} \\ a_{21} & a_{22} & a_{23} \\ c_1 & c_2 & c_3 \end{vmatrix}$

147. Using a Graphing Utility Use the matrix capabilities of a graphing utility to find the inverse of the matrix

$$A = \begin{bmatrix} 1 & -3 \\ -2 & 6 \end{bmatrix}.$$

What message appears on the screen? Why does the graphing utility display this message?

148. Invertible Matrices Under what conditions does a matrix have an inverse?

149. Writing What is meant by the cofactor of an entry of a matrix? How are cofactors used to find the determinant of the matrix?

150. Think About It Three people were asked to solve a system of equations using an augmented matrix. Each person reduced the matrix to row-echelon form. The reduced matrices were

$$\begin{bmatrix} 1 & 2 & \vdots & 3 \\ 0 & 1 & \vdots & 1 \end{bmatrix},$$

$$\begin{bmatrix} 1 & 0 & \vdots & 1 \\ 0 & 1 & \vdots & 1 \end{bmatrix},$$

and

$$\begin{bmatrix} 1 & 2 & \vdots & 3 \\ 0 & 0 & \vdots & 0 \end{bmatrix}.$$

Can all three be right? Explain.

151. Think About It Describe the row-echelon form of an augmented matrix that corresponds to a system of linear equations that has a unique solution.

152. Solving an Equation Solve the equation for λ.

$$\begin{vmatrix} 2 - \lambda & 5 \\ 3 & -8 - \lambda \end{vmatrix} = 0$$

Chapter Test

See CalcChat.com for tutorial help and worked-out solutions to odd-numbered exercises.

Take this test as you would take a test in class. When you are finished, check your work against the answers given in the back of the book.

In Exercises 1 and 2, write the matrix in reduced row-echelon form.

1. $\begin{bmatrix} 1 & -1 & 5 \\ 6 & 2 & 3 \\ 5 & 3 & -3 \end{bmatrix}$ **2.** $\begin{bmatrix} 1 & 0 & -1 & 2 \\ -1 & 1 & 1 & -3 \\ 1 & 1 & -1 & 1 \\ 3 & 2 & -3 & 4 \end{bmatrix}$

3. Write the augmented matrix for the system of equations and solve the system.

$$\begin{cases} 4x + 3y - 2z = 14 \\ -x - y + 2z = -5 \\ 3x + y - 4z = 8 \end{cases}$$

4. Find (a) $A - B$, (b) $3A$, (c) $3A - 2B$, and (d) AB (if possible).

$$A = \begin{bmatrix} 6 & 5 \\ -5 & -5 \end{bmatrix}, \quad B = \begin{bmatrix} 5 & 0 \\ -5 & -1 \end{bmatrix}$$

In Exercises 5 and 6, find the inverse of the matrix (if it exists).

5. $\begin{bmatrix} -4 & 3 \\ 5 & -2 \end{bmatrix}$ **6.** $\begin{bmatrix} -2 & 4 & -6 \\ 2 & 1 & 0 \\ 4 & -2 & 5 \end{bmatrix}$

7. Use the result of Exercise 5 to solve the system.

$$\begin{cases} -4x + 3y = 6 \\ 5x - 2y = 24 \end{cases}$$

In Exercises 8–10, find the determinant of the matrix.

8. $\begin{bmatrix} -6 & 4 \\ 10 & 12 \end{bmatrix}$ **9.** $\begin{bmatrix} \frac{5}{2} & \frac{13}{4} \\ -8 & \frac{6}{5} \end{bmatrix}$ **10.** $\begin{bmatrix} 6 & -7 & 2 \\ 3 & -2 & 0 \\ 1 & 5 & 1 \end{bmatrix}$

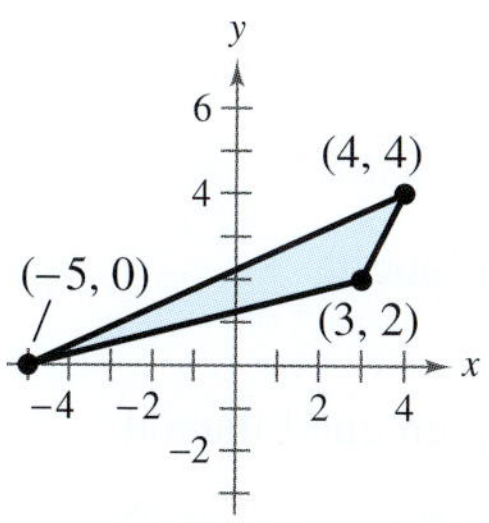

Figure for 13

In Exercises 11 and 12, use Cramer's Rule to solve (if possible) the system of equations.

11. $\begin{cases} 7x + 6y = 9 \\ -2x - 11y = -49 \end{cases}$ **12.** $\begin{cases} 6x - y + 2z = -4 \\ -2x + 3y - z = 10 \\ 4x - 4y + z = -18 \end{cases}$

13. Use a determinant to find the area of the triangle at the left.

14. Write the uncoded 1×3 row matrices for the message KNOCK ON WOOD. Then encode the message using the encoding matrix A below.

$$A = \begin{bmatrix} 1 & -1 & 0 \\ 1 & 0 & -1 \\ 6 & -2 & -3 \end{bmatrix}$$

15. One hundred liters of a 50% solution is obtained by mixing a 60% solution with a 20% solution. How many liters of each solution must be used to obtain the desired mixture?

Proofs in Mathematics

A proof without words is a picture or diagram that gives a visual understanding of why a theorem or statement is true. It can also provide a starting point for writing a formal proof. The following proof shows that a 2×2 determinant is the area of a parallelogram.

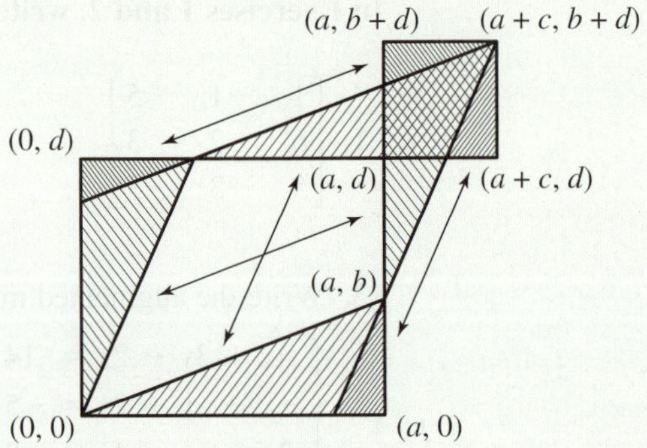

$$\begin{vmatrix} a & b \\ c & d \end{vmatrix} = ad - bc = \|\text{▭}\| - \|\text{□}\| = \|\text{▱}\|$$

The following is a color-coded version of the proof along with a brief explanation of why this proof works.

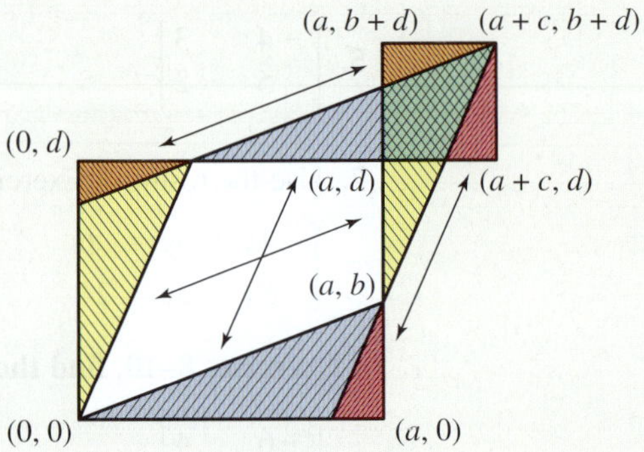

$$\begin{vmatrix} a & b \\ c & d \end{vmatrix} = ad - bc = \|\text{▭}\| - \|\text{□}\| = \|\text{▱}\|$$

Area of ▭ = Area of orange △ + Area of yellow △ + Area of blue △ + Area of pink △ + Area of white quadrilateral

Area of □ = Area of orange △ + Area of pink △ + Area of green quadrilateral

Area of ▱ = Area of white quadrilateral + Area of blue △ + Area of yellow △ − Area of green quadrilateral

= Area of ▭ − Area of □

From "Proof Without Words" by Solomon W. Golomb, *Mathematics Magazine,* March 1985, Vol. 58, No. 2, pg. 107. Reprinted with permission.

P.S. Problem Solving

1. **Multiplying by a Transformation Matrix** The columns of matrix T show the coordinates of the vertices of a triangle. Matrix A is a transformation matrix.

$$A = \begin{bmatrix} 0 & -1 \\ 1 & 0 \end{bmatrix} \qquad T = \begin{bmatrix} 1 & 2 & 3 \\ 1 & 4 & 2 \end{bmatrix}$$

(a) Find AT and AAT. Then sketch the original triangle and the two images of the triangle. What transformation does A represent?

(b) Given the triangle determined by AAT, describe the transformation that produces the triangle determined by AT and then the triangle determined by T.

2. **Population** The matrices show the number of people (in thousands) who lived in each region of the United States in 2000 and the number of people (in thousands) who lived in each region in 2010. The regional populations are separated into three age groups. *(Source: U.S. Census Bureau)*

2000

	0–17	18–64	65+
Northeast	13,048	33,174	7,372
Midwest	16,648	39,486	8,259
South	25,567	62,232	12,438
West	17,031	39,244	6,922

2010

	0–17	18–64	65+
Northeast	12,333	35,179	7,805
Midwest	16,128	41,777	9,022
South	27,789	71,873	14,894
West	17,931	45,467	8,547

(a) The total population in 2000 was approximately 281,422,000 and the total population in 2010 was approximately 308,746,000. Rewrite the matrices to give the information as percents of the total population.

(b) Write a matrix that gives the change in the percent of the population in each region and age group from 2000 to 2010.

(c) Based on the result of part (b), which region(s) and age group(s) had percents that decreased from 2000 to 2010?

3. **Determining Whether Matrices are Idempotent** Determine whether the matrix is idempotent. A square matrix is **idempotent** when $A^2 = A$.

(a) $\begin{bmatrix} 1 & 0 \\ 0 & 0 \end{bmatrix}$ (b) $\begin{bmatrix} 0 & 1 \\ 1 & 0 \end{bmatrix}$

(c) $\begin{bmatrix} 2 & 3 \\ -1 & -2 \end{bmatrix}$ (d) $\begin{bmatrix} 2 & 3 \\ 1 & 2 \end{bmatrix}$

4. **Quadratic Matrix Equation** Let $A = \begin{bmatrix} 1 & 2 \\ -2 & 1 \end{bmatrix}$.

(a) Show that $A^2 - 2A + 5I = O$, where I is the identity matrix of order 2×2.

(b) Show that $A^{-1} = \frac{1}{5}(2I - A)$.

(c) Show in general that for any square matrix satisfying

$$A^2 - 2A + 5I = O$$

the inverse of A is given by

$$A^{-1} = \tfrac{1}{5}(2I - A).$$

5. **Satellite Television** Two competing companies offer satellite television to a city with 100,000 households. Gold Satellite System has 25,000 subscribers and Galaxy Satellite Network has 30,000 subscribers. (The other 45,000 households do not subscribe.) The percent changes in satellite subscriptions each year are shown in the matrix below.

		Percent Changes		
		From Gold	From Galaxy	From Nonsubscriber
Percent Changes	To Gold	0.70	0.15	0.15
	To Galaxy	0.20	0.80	0.15
	To Nonsubscriber	0.10	0.05	0.70

(a) Find the number of subscribers each company will have in 1 year using matrix multiplication. Explain how you obtained your answer.

(b) Find the number of subscribers each company will have in 2 years using matrix multiplication. Explain how you obtained your answer.

(c) Find the number of subscribers each company will have in 3 years using matrix multiplication. Explain how you obtained your answer.

(d) What is happening to the number of subscribers to each company? What is happening to the number of nonsubscribers?

6. **The Transpose of a Matrix** The **transpose** of a matrix, denoted A^T, is formed by writing its columns as rows. Find the transpose of each matrix and verify that $(AB)^T = B^TA^T$.

$$A = \begin{bmatrix} -1 & 1 & -2 \\ 2 & 0 & 1 \end{bmatrix}, \quad B = \begin{bmatrix} -3 & 0 \\ 1 & 2 \\ 1 & -1 \end{bmatrix}$$

7. **Finding a Value** Find x such that the matrix is equal to its own inverse.

$$A = \begin{bmatrix} 3 & x \\ -2 & -3 \end{bmatrix}$$

8. Finding a Value Find x such that the matrix is singular.

$$A = \begin{bmatrix} 4 & x \\ -2 & -3 \end{bmatrix}$$

9. Finding a Matrix Find a singular 2×2 matrix satisfying $A^2 = A$.

10. Verifying an Equation Verify the following equation.

$$\begin{vmatrix} 1 & 1 & 1 \\ a & b & c \\ a^2 & b^2 & c^2 \end{vmatrix} = (a - b)(b - c)(c - a)$$

11. Verifying an Equation Verify the following equation.

$$\begin{vmatrix} 1 & 1 & 1 \\ a & b & c \\ a^3 & b^3 & c^3 \end{vmatrix} = (a - b)(b - c)(c - a)(a + b + c)$$

12. Verifying an Equation Verify the following equation.

$$\begin{vmatrix} x & 0 & c \\ -1 & x & b \\ 0 & -1 & a \end{vmatrix} = ax^2 + bx + c$$

13. Finding a Determinant Use the equation given in Exercise 12 as a model to find a determinant that is equal to $ax^3 + bx^2 + cx + d$.

14. Finding Atomic Masses The atomic masses of three compounds are shown in the table. Use a linear system and Cramer's Rule to find the atomic masses of sulfur (S), nitrogen (N), and fluorine (F).

DATA

Compound	Formula	Atomic Mass
Tetrasulfur tetranitride	S_4N_4	184
Sulfur hexafluoride	SF_6	146
Dinitrogen tetrafluoride	N_2F_4	104

Spreadsheet at LarsonPrecalculus.com

15. Finding the Costs of Items A walkway lighting package includes a transformer, a certain length of wire, and a certain number of lights on the wire. The price of each lighting package depends on the length of wire and the number of lights on the wire. Use the following information to find the cost of a transformer, the cost per foot of wire, and the cost of a light. Assume that the cost of each item is the same in each lighting package.

- A package that contains a transformer, 25 feet of wire, and 5 lights costs \$20.
- A package that contains a transformer, 50 feet of wire, and 15 lights costs \$35.
- A package that contains a transformer, 100 feet of wire, and 20 lights costs \$50.

16. Decoding a Message Use the inverse of matrix A to decode the cryptogram.

$$A = \begin{bmatrix} 1 & -2 & 2 \\ 1 & 1 & -3 \\ 1 & -1 & 4 \end{bmatrix}$$

23 13 −34 31 −34 63 25 −17 61
24 14 −37 41 −17 −8 20 −29 40
38 −56 116 13 −11 1 22 −3 −6
41 −53 85 28 −32 16

17. Decoding a Message A code breaker intercepted the encoded message below.

45 −35 38 −30 18 −18 35 −30 81 −60
42 −28 75 −55 2 −2 22 −21 15 −10

Let $A^{-1} = \begin{bmatrix} w & x \\ y & z \end{bmatrix}$.

(a) You know that $[45 \quad -35]A^{-1} = [10 \quad 15]$ and that $[38 \quad -30]A^{-1} = [8 \quad 14]$, where A^{-1} is the inverse of the encoding matrix A. Write and solve two systems of equations to find w, x, y, and z.

(b) Decode the message.

18. Conjecture Let

$$A = \begin{bmatrix} 6 & 4 & 1 \\ 0 & 2 & 3 \\ 1 & 1 & 2 \end{bmatrix}.$$

Use a graphing utility to find A^{-1}. Compare $|A^{-1}|$ with $|A|$. Make a conjecture about the determinant of the inverse of a matrix.

19. Finding the Determinant of a Matrix Let A be an $n \times n$ matrix each of whose rows adds up to zero. Find $|A|$.

20. Conjecture Consider matrices of the form

$$A = \begin{bmatrix} 0 & a_{12} & a_{13} & a_{14} & \cdots & a_{1n} \\ 0 & 0 & a_{23} & a_{24} & \cdots & a_{2n} \\ 0 & 0 & 0 & a_{34} & \cdots & a_{3n} \\ \vdots & \vdots & \vdots & \vdots & & \vdots \\ 0 & 0 & 0 & 0 & \cdots & a_{(n-1)n} \\ 0 & 0 & 0 & 0 & \cdots & 0 \end{bmatrix}.$$

(a) Write a 2×2 matrix and a 3×3 matrix in the form of A.

(b) Use a graphing utility to raise each of the matrices to higher powers. Describe the result.

(c) Use the result of part (b) to make a conjecture about powers of A when A is a 4×4 matrix. Use the graphing utility to test your conjecture.

(d) Use the results of parts (b) and (c) to make a conjecture about powers of A when A is an $n \times n$ matrix.

9 Sequences, Series, and Probability

Horse Racing *(Example 6, page 655)*

Tossing Dice *(Example 3, page 664)*

Electricity *(Exercise 92, page 651)*

Dominoes *(page 635)*

AIDS Cases *(Exercise 98, page 615)*

Clockwise from top left, Mikhail Pogosov/Shutterstock.com; Sergey Mironov/Shutterstock.com; © mark wragg/istockphoto.com; michaeljung/Shutterstock.com; wang song/Shutterstock.com

9.1 Sequences and Series

Sequences and series can help you model real-life problems. For instance, in Exercise 98 on page 615, a sequence models the numbers of AIDS cases reported from 2003 through 2010.

- **Use sequence notation to write the terms of sequences.**
- **Use factorial notation.**
- **Use summation notation to write sums.**
- **Find the sums of series.**
- **Use sequences and series to model and solve real-life problems.**

Sequences

In mathematics, the word *sequence* is used in much the same way as in ordinary English. Saying that a collection is listed in *sequence* means that it is ordered so that it has a first member, a second member, a third member, and so on. Two examples are $1, 2, 3, 4, \ldots$ and $1, 3, 5, 7, \ldots$.

Mathematically, you can think of a sequence as a *function* whose domain is the set of positive integers. Rather than using function notation, however, sequences are usually written using subscript notation, as indicated in the following definition.

Definition of Sequence

An **infinite sequence** is a function whose domain is the set of positive integers. The function values

$$a_1, a_2, a_3, a_4, \ldots, a_n, \ldots$$

are the **terms** of the sequence. When the domain of the function consists of the first n positive integers only, the sequence is a **finite sequence.**

On occasion, it is convenient to begin subscripting a sequence with 0 instead of 1 so that the terms of the sequence become

$$a_0, a_1, a_2, a_3, \ldots.$$

When this is the case, the domain includes 0.

REMARK The subscripts of a sequence make up the domain of the sequence and serve to identify the locations of terms within the sequence. For example, a_4 is the fourth term of the sequence, and a_n is the nth term of the sequence. Any variable can be a subscript. The most commonly used variable subscripts in sequence and series notation are i, j, k, and n.

EXAMPLE 1 Writing the Terms of a Sequence

a. The first four terms of the sequence given by $a_n = 3n - 2$ are

$a_1 = 3(1) - 2 = 1$ 1st term

$a_2 = 3(2) - 2 = 4$ 2nd term

$a_3 = 3(3) - 2 = 7$ 3rd term

$a_4 = 3(4) - 2 = 10.$ 4th term

b. The first four terms of the sequence given by $a_n = 3 + (-1)^n$ are

$a_1 = 3 + (-1)^1 = 3 - 1 = 2$ 1st term

$a_2 = 3 + (-1)^2 = 3 + 1 = 4$ 2nd term

$a_3 = 3 + (-1)^3 = 3 - 1 = 2$ 3rd term

$a_4 = 3 + (-1)^4 = 3 + 1 = 4.$ 4th term

✓ ***Checkpoint*** *Audio-video solution in English & Spanish at LarsonPrecalculus.com.*

Write the first four terms of the sequence given by $a_n = 2n + 1$.

michaeljung/Shutterstock.com

REMARK Write the first four terms of the sequence whose nth term is

$$a_n = \frac{(-1)^{n+1}}{2n+1}.$$

Are they the same as the first four terms of the sequence in Example 2? If not, then how do they differ?

EXAMPLE 2 A Sequence Whose Terms Alternate in Sign

Write the first four terms of the sequence given by $a_n = \dfrac{(-1)^n}{2n+1}$.

Solution The first four terms of the sequence are as follows.

$$a_1 = \frac{(-1)^1}{2(1)+1} = \frac{-1}{2+1} = -\frac{1}{3} \qquad \text{1st term}$$

$$a_2 = \frac{(-1)^2}{2(2)+1} = \frac{1}{4+1} = \frac{1}{5} \qquad \text{2nd term}$$

$$a_3 = \frac{(-1)^3}{2(3)+1} = \frac{-1}{6+1} = -\frac{1}{7} \qquad \text{3rd term}$$

$$a_4 = \frac{(-1)^4}{2(4)+1} = \frac{1}{8+1} = \frac{1}{9} \qquad \text{4th term}$$

✓ **Checkpoint** *Audio-video solution in English & Spanish at LarsonPrecalculus.com.*

Write the first four terms of the sequence given by $a_n = \dfrac{2 + (-1)^n}{n}$.

Simply listing the first few terms is not sufficient to define a unique sequence—the nth term *must be given*. To see this, consider the following sequences, both of which have the same first three terms.

$$\frac{1}{2}, \frac{1}{4}, \frac{1}{8}, \frac{1}{16}, \ldots, \frac{1}{2^n}, \ldots$$

$$\frac{1}{2}, \frac{1}{4}, \frac{1}{8}, \frac{1}{15}, \ldots, \frac{6}{(n+1)(n^2-n+6)}, \ldots$$

TECHNOLOGY To graph a sequence using a graphing utility, set the mode to *sequence* and *dot* and enter the sequence. The graph of the sequence in Example 3(a) is below. To identify the terms, use the *trace* feature or *value* feature.

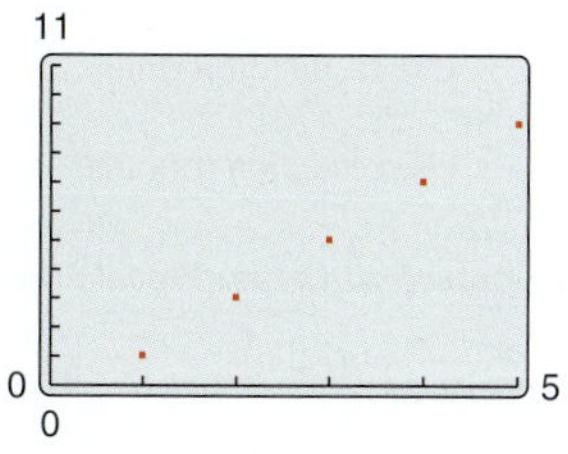

EXAMPLE 3 Finding the nth Term of a Sequence

Write an expression for the apparent nth term (a_n) of each sequence.

a. $1, 3, 5, 7, \ldots$ **b.** $2, -5, 10, -17, \ldots$

Solution

a. n: 1 2 3 4 . . . n

Terms: 1 3 5 7 . . . a_n

Apparent pattern: Each term is 1 less than twice n, which implies that

$$a_n = 2n - 1.$$

b. n: 1 2 3 4 . . . n

Terms: 2 −5 10 −17 . . . a_n

Apparent pattern: The absolute value of each term is 1 more than the square of n, and the terms have alternating signs, with those in the even positions being negative. This implies that

$$a_n = (-1)^{n+1}(n^2 + 1).$$

✓ **Checkpoint** *Audio-video solution in English & Spanish at LarsonPrecalculus.com.*

Write an expression for the apparent nth term (a_n) of each sequence.

a. $1, 5, 9, 13, \ldots$ **b.** $2, -4, 6, -8, \ldots$

Some sequences are defined **recursively.** To define a sequence recursively, you need to be given one or more of the first few terms. All other terms of the sequence are then defined using previous terms.

EXAMPLE 4 **A Recursive Sequence**

Write the first five terms of the sequence defined recursively as

$$a_1 = 3, \quad a_k = 2a_{k-1} + 1, \quad \text{where } k \geq 2.$$

Solution

$a_1 = 3$ 1st term is given.

$a_2 = 2a_{2-1} + 1 = 2a_1 + 1 = 2(3) + 1 = 7$ Use recursion formula.

$a_3 = 2a_{3-1} + 1 = 2a_2 + 1 = 2(7) + 1 = 15$ Use recursion formula.

$a_4 = 2a_{4-1} + 1 = 2a_3 + 1 = 2(15) + 1 = 31$ Use recursion formula.

$a_5 = 2a_{5-1} + 1 = 2a_4 + 1 = 2(31) + 1 = 63$ Use recursion formula.

✓ ***Checkpoint*** *Audio-video solution in English & Spanish at LarsonPrecalculus.com.*

Write the first five terms of the sequence defined recursively as

$$a_1 = 6, \quad a_{k+1} = a_k + 1, \quad \text{where} \quad k \geq 2.$$

In the next example, you will study a well-known recursive sequence, the Fibonacci sequence.

EXAMPLE 5 **The Fibonacci Sequence: A Recursive Sequence**

The Fibonacci sequence is defined recursively, as follows.

$$a_0 = 1, \quad a_1 = 1, \quad a_k = a_{k-2} + a_{k-1}, \quad \text{where} \quad k \geq 2$$

Write the first six terms of this sequence.

Solution

$a_0 = 1$ 0th term is given.

$a_1 = 1$ 1st term is given.

$a_2 = a_{2-2} + a_{2-1} = a_0 + a_1 = 1 + 1 = 2$ Use recursion formula.

$a_3 = a_{3-2} + a_{3-1} = a_1 + a_2 = 1 + 2 = 3$ Use recursion formula.

$a_4 = a_{4-2} + a_{4-1} = a_2 + a_3 = 2 + 3 = 5$ Use recursion formula.

$a_5 = a_{5-2} + a_{5-1} = a_3 + a_4 = 3 + 5 = 8$ Use recursion formula.

✓ ***Checkpoint*** *Audio-video solution in English & Spanish at LarsonPrecalculus.com.*

Write the first five terms of the sequence defined recursively as

$$a_0 = 1, \quad a_1 = 3, \quad a_k = a_{k-2} + a_{k-1}, \quad \text{where} \quad k \geq 2.$$

Factorial Notation

Some very important sequences in mathematics involve terms that are defined with special types of products called **factorials.**

> **Definition of Factorial**
>
> If n is a positive integer, then ***n* factorial** is defined as
>
> $$n! = 1 \cdot 2 \cdot 3 \cdot 4 \cdot \cdot \cdot (n-1) \cdot n.$$
>
> As a special case, zero factorial is defined as $0! = 1$.

REMARK The value of n does not have to be very large before the value of $n!$ becomes extremely large. For instance, $10! = 3{,}628{,}800$.

Notice that $0! = 1$ and $1! = 1$. Here are some other values of $n!$.

$$2! = 1 \cdot 2 = 2 \qquad 3! = 1 \cdot 2 \cdot 3 = 6 \qquad 4! = 1 \cdot 2 \cdot 3 \cdot 4 = 24$$

Factorials follow the same conventions for order of operations as do exponents. For instance,

$$2n! = 2(n!) = 2(1 \cdot 2 \cdot 3 \cdot 4 \cdot \cdot \cdot n), \quad \text{whereas} \quad (2n)! = 1 \cdot 2 \cdot 3 \cdot 4 \cdot \cdot \cdot 2n.$$

EXAMPLE 6 Writing the Terms of a Sequence Involving Factorials

Write the first five terms of the sequence given by $a_n = \dfrac{2^n}{n!}$. Begin with $n = 0$.

Algebraic Solution

$a_0 = \dfrac{2^0}{0!} = \dfrac{1}{1} = 1$ 0th term

$a_1 = \dfrac{2^1}{1!} = \dfrac{2}{1} = 2$ 1st term

$a_2 = \dfrac{2^2}{2!} = \dfrac{4}{2} = 2$ 2nd term

$a_3 = \dfrac{2^3}{3!} = \dfrac{8}{6} = \dfrac{4}{3}$ 3rd term

$a_4 = \dfrac{2^4}{4!} = \dfrac{16}{24} = \dfrac{2}{3}$ 4th term

Graphical Solution

Using a graphing utility set to *dot* and *sequence* modes, enter the sequence. Next, graph the sequence. You can estimate the first five terms of the sequence as follows.

$u_0 = 1$

$u_1 = 2$

$u_2 = 2$

$u_3 \approx 1.333 \approx \frac{4}{3}$

$u_4 \approx 0.667 \approx \frac{2}{3}$

Use the *trace* feature to approximate the first five terms.

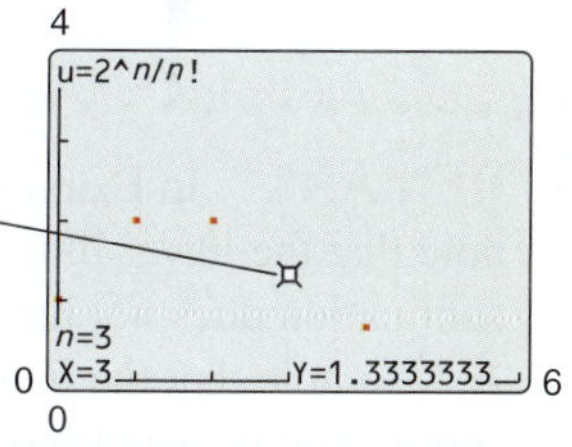

Checkpoint Audio-video solution in English & Spanish at LarsonPrecalculus.com.

Write the first five terms of the sequence given by $a_n = \dfrac{3^n + 1}{n!}$. Begin with $n = 0$.

When working with fractions involving factorials, you will often be able to reduce the fractions to simplify the computations.

EXAMPLE 7 Simplifying Factorial Expressions

a. $\dfrac{8!}{2! \cdot 6!} = \dfrac{1 \cdot 2 \cdot \cancel{3} \cdot \cancel{4} \cdot \cancel{5} \cdot \cancel{6} \cdot 7 \cdot 8}{1 \cdot 2 \cdot \cancel{1} \cdot \cancel{2} \cdot \cancel{3} \cdot \cancel{4} \cdot \cancel{5} \cdot \cancel{6}} = \dfrac{7 \cdot 8}{2} = 28$

b. $\dfrac{n!}{(n-1)!} = \dfrac{\cancel{1 \cdot 2 \cdot 3 \cdots (n-1)} \cdot n}{\cancel{1 \cdot 2 \cdot 3 \cdots (n-1)}} = n$

ALGEBRA HELP You can also simplify the expression in Example 7(a) as follows.

$$\frac{8!}{2! \cdot 6!} = \frac{8 \cdot 7 \cdot \cancel{6!}}{2! \cdot \cancel{6!}} = \frac{8 \cdot 7}{2 \cdot 1} = 28$$

Checkpoint Audio-video solution in English & Spanish at LarsonPrecalculus.com.

Simplify the factorial expression $\dfrac{4!(n+1)!}{3!n!}$.

TECHNOLOGY Most graphing utilities can sum the first n terms of a sequence. Check your user's guide for a *sum sequence* feature or a *series* feature.

Summation Notation

A convenient notation for the sum of the terms of a finite sequence is called **summation notation** or **sigma notation.** It involves the use of the uppercase Greek letter sigma, written as Σ.

Definition of Summation Notation

The sum of the first n terms of a sequence is represented by

$$\sum_{i=1}^{n} a_i = a_1 + a_2 + a_3 + a_4 + \cdot \cdot \cdot + a_n$$

where i is called the **index of summation,** n is the **upper limit of summation,** and 1 is the **lower limit of summation.**

REMARK Summation notation is an instruction to add the terms of a sequence. From the definition at the right, the upper limit of summation tells you the last term of the sum. Summation notation helps you generate the terms of the sequence prior to finding the sum.

EXAMPLE 8 Summation Notation for a Sum

a. $\displaystyle\sum_{i=1}^{5} 3i = 3(1) + 3(2) + 3(3) + 3(4) + 3(5)$

$= 45$

b. $\displaystyle\sum_{k=3}^{6} (1 + k^2) = (1 + 3^2) + (1 + 4^2) + (1 + 5^2) + (1 + 6^2)$

$= 10 + 17 + 26 + 37$

$= 90$

c. $\displaystyle\sum_{i=0}^{8} \frac{1}{i!} = \frac{1}{0!} + \frac{1}{1!} + \frac{1}{2!} + \frac{1}{3!} + \frac{1}{4!} + \frac{1}{5!} + \frac{1}{6!} + \frac{1}{7!} + \frac{1}{8!}$

$\displaystyle = 1 + 1 + \frac{1}{2} + \frac{1}{6} + \frac{1}{24} + \frac{1}{120} + \frac{1}{720} + \frac{1}{5040} + \frac{1}{40{,}320}$

≈ 2.71828

For this summation, note that the sum is very close to the irrational number

$e \approx 2.718281828.$

It can be shown that as more terms of the sequence whose nth term is $1/n!$ are added, the sum becomes closer and closer to e.

REMARK In Example 8, note that the lower limit of a summation does not have to be 1. Also note that the index of summation does not have to be the letter i. For instance, in part (b), the letter k is the index of summation.

✓ **Checkpoint** Audio-video solution in English & Spanish at LarsonPrecalculus.com.

Find the sum $\displaystyle\sum_{i=1}^{4} (4i + 1)$.

Properties of Sums

1. $\displaystyle\sum_{i=1}^{n} c = cn$, c is a constant.

2. $\displaystyle\sum_{i=1}^{n} ca_i = c\sum_{i=1}^{n} a_i$, c is a constant.

3. $\displaystyle\sum_{i=1}^{n} (a_i + b_i) = \sum_{i=1}^{n} a_i + \sum_{i=1}^{n} b_i$

4. $\displaystyle\sum_{i=1}^{n} (a_i - b_i) = \sum_{i=1}^{n} a_i - \sum_{i=1}^{n} b_i$

For proofs of these properties, see Proofs in Mathematics on page 682.

Series

Many applications involve the sum of the terms of a finite or infinite sequence. Such a sum is called a **series.**

Definition of Series

Consider the infinite sequence $a_1, a_2, a_3, \ldots, a_i, \ldots$.

1. The sum of the first n terms of the sequence is called a **finite series** or the ***n*th partial sum** of the sequence and is denoted by

$$a_1 + a_2 + a_3 + \cdots + a_n = \sum_{i=1}^{n} a_i.$$

2. The sum of all the terms of the infinite sequence is called an **infinite series** and is denoted by

$$a_1 + a_2 + a_3 + \cdots + a_i + \cdots = \sum_{i=1}^{\infty} a_i.$$

EXAMPLE 9 **Finding the Sum of a Series**

For the series

$$\sum_{i=1}^{\infty} \frac{3}{10^i}$$

find (a) the third partial sum and (b) the sum.

Solution

a. The third partial sum is

$$\sum_{i=1}^{3} \frac{3}{10^i} = \frac{3}{10^1} + \frac{3}{10^2} + \frac{3}{10^3}$$

$$= 0.3 + 0.03 + 0.003$$

$$= 0.333.$$

b. The sum of the series is

$$\sum_{i=1}^{\infty} \frac{3}{10^i} = \frac{3}{10^1} + \frac{3}{10^2} + \frac{3}{10^3} + \frac{3}{10^4} + \frac{3}{10^5} + \cdots$$

$$= 0.3 + 0.03 + 0.003 + 0.0003 + 0.00003 + \cdots$$

$$= 0.33333\ldots$$

$$= \frac{1}{3}.$$

✓ ***Checkpoint*** *Audio-video solution in English & Spanish at LarsonPrecalculus.com.*

For the series

$$\sum_{i=1}^{\infty} \frac{5}{10^i}$$

find (a) the fourth partial sum and (b) the sum.

Notice in Example 9(b) that the sum of an infinite series can be a finite number.

Application

Sequences have many applications in business and science. Example 10 illustrates such an application.

EXAMPLE 10 Compound Interest

An investor deposits \$5000 in an account that earns 3% interest compounded quarterly. The balance in the account after n quarters is given by

$$A_n = 5000\left(1 + \frac{0.03}{4}\right)^n, \quad n = 0, 1, 2, \ldots .$$

a. Write the first three terms of the sequence.

b. Find the balance in the account after 10 years by computing the 40th term of the sequence.

Solution

a. The first three terms of the sequence are as follows.

$$A_0 = 5000\left(1 + \frac{0.03}{4}\right)^0 = \$5000.00$$ Original deposit

$$A_1 = 5000\left(1 + \frac{0.03}{4}\right)^1 = \$5037.50$$ First-quarter balance

$$A_2 = 5000\left(1 + \frac{0.03}{4}\right)^2 \approx \$5075.28$$ Second-quarter balance

b. The 40th term of the sequence is

$$A_{40} = 5000\left(1 + \frac{0.03}{4}\right)^{40} \approx \$6741.74.$$ Ten-year balance

✓ **Checkpoint** Audio-video solution in English & Spanish at LarsonPrecalculus.com.

An investor deposits \$1000 in an account that earns 3% interest compounded monthly. The balance in the account after n months is given by

$$A_n = 1000\left(1 + \frac{0.03}{12}\right)^n, \quad n = 0, 1, 2, \ldots .$$

a. Write the first three terms of the sequence.

b. Find the balance in the account after four years by computing the 48th term of the sequence.

Summarize (Section 9.1)

1. State the definition of a sequence *(page 606)*. For examples of writing the terms of sequences, see Examples 1–5.
2. State the definition of a factorial *(page 609)*. For examples of using factorial notation, see Examples 6 and 7.
3. State the definition of summation notation *(page 610)*. For an example of using summation notation, see Example 8.
4. State the definition of a series *(page 611)*. For an example of finding the sum of a series, see Example 9.
5. Describe an example of how to use a sequence to model and solve a real-life problem *(page 612, Example 10)*.

9.1 Exercises

See CalcChat.com for tutorial help and worked-out solutions to odd-numbered exercises.

Vocabulary: Fill in the blanks.

1. An ________ ________ is a function whose domain is the set of positive integers.
2. A sequence is a ________ sequence when the domain of the function consists only of the first n positive integers.
3. If you are given one or more of the first few terms of a sequence, and all other terms of the sequence are defined using previous terms, then the sequence is said to be defined ________.
4. If n is a positive integer, then n ________ is defined as $n! = 1 \cdot 2 \cdot 3 \cdot 4 \cdots (n-1) \cdot n$.
5. For the sum $\sum_{i=1}^{n} a_i$, i is called the ________ of summation, n is the ________ limit of summation, and 1 is the ________ limit of summation.
6. The sum of the terms of a finite or infinite sequence is called a ________.

Skills and Applications

Writing the Terms of a Sequence In Exercises 7–22, write the first five terms of the sequence. (Assume that n begins with 1.)

7. $a_n = 4n - 7$
8. $a_n = 2 - \dfrac{1}{3^n}$
9. $a_n = (-2)^n$
10. $a_n = \left(\frac{1}{2}\right)^n$
11. $a_n = \dfrac{n}{n+2}$
12. $a_n = \dfrac{6n}{3n^2 - 1}$
13. $a_n = \dfrac{1 + (-1)^n}{n}$
14. $a_n = \dfrac{(-1)^n}{n^2}$
15. $a_n = \dfrac{2^n}{3^n}$
16. $a_n = \dfrac{1}{n^{3/2}}$
17. $a_n = \frac{2}{3}$
18. $a_n = 1 + (-1)^n$
19. $a_n = n(n-1)(n-2)$
20. $a_n = n(n^2 - 6)$
21. $a_n = (-1)^n\left(\dfrac{n}{n+1}\right)$
22. $a_n = \dfrac{(-1)^{n+1}}{n^2 + 1}$

Finding a Term of a Sequence In Exercises 23–26, find the indicated term of the sequence.

23. $a_n = (-1)^n(3n - 2)$
    $a_{25} = \square$
24. $a_n = (-1)^{n-1}[n(n-1)]$
    $a_{16} = \square$
25. $a_n = \dfrac{4n}{2n^2 - 3}$
    $a_{11} = \square$
26. $a_n = \dfrac{4n^2 - n + 3}{n(n-1)(n+2)}$
    $a_{13} = \square$

Graphing the Terms of a Sequence In Exercises 27–32, use a graphing utility to graph the first 10 terms of the sequence. (Assume that n begins with 1.)

27. $a_n = \dfrac{2}{3}n$
28. $a_n = 2 - \dfrac{4}{n}$
29. $a_n = 16(-0.5)^{n-1}$
30. $a_n = 8(0.75)^{n-1}$
31. $a_n = \dfrac{2n}{n+1}$
32. $a_n = \dfrac{3n^2}{n^2 + 1}$

Matching a Sequence with a Graph In Exercises 33–36, match the sequence with the graph of its first 10 terms. [The graphs are labeled (a), (b), (c), and (d).]

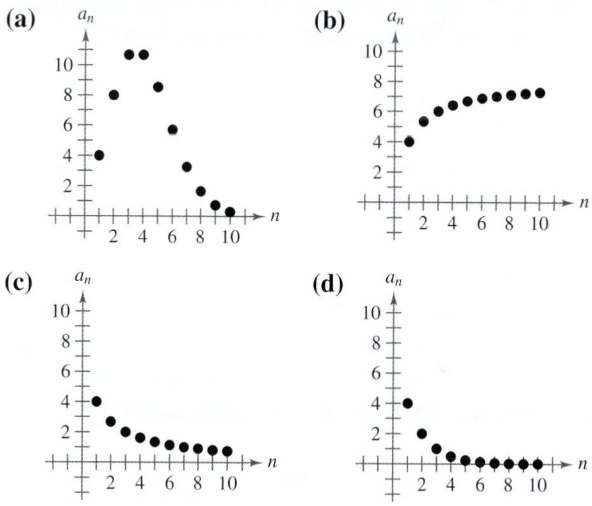

33. $a_n = \dfrac{8}{n+1}$
34. $a_n = \dfrac{8n}{n+1}$
35. $a_n = 4(0.5)^{n-1}$
36. $a_n = \dfrac{4^n}{n!}$

Finding the nth Term of a Sequence In Exercises 37–48, write an expression for the apparent nth term (a_n) of the sequence. (Assume that n begins with 1.)

37. 3, 7, 11, 15, 19, . . .
38. 0, 3, 8, 15, 24, . . .
39. $-\frac{2}{3}, \frac{3}{4}, -\frac{4}{5}, \frac{5}{6}, -\frac{6}{7}, \ldots$
40. $\frac{1}{2}, -\frac{1}{4}, \frac{1}{8}, -\frac{1}{16}, \ldots$
41. $\frac{2}{1}, \frac{3}{3}, \frac{4}{5}, \frac{5}{7}, \frac{6}{9}, \ldots$
42. $\frac{1}{3}, \frac{2}{9}, \frac{4}{27}, \frac{8}{81}, \ldots$
43. $1, \frac{1}{4}, \frac{1}{9}, \frac{1}{16}, \frac{1}{25}, \ldots$
44. $1, \frac{1}{2}, \frac{1}{6}, \frac{1}{24}, \frac{1}{120}, \ldots$
45. $1, -1, 1, -1, 1, \ldots$
46. 1, 3, 1, 3, 1, . . .
47. $1, 3, \dfrac{3^2}{2}, \dfrac{3^3}{6}, \dfrac{3^4}{24}, \dfrac{3^5}{120}, \ldots$
48. $1 + \frac{1}{2}, 1 + \frac{3}{4}, 1 + \frac{7}{8}, 1 + \frac{15}{16}, 1 + \frac{31}{32}, \ldots$

Writing the Terms of a Recursive Sequence In Exercises 49–52, write the first five terms of the sequence defined recursively.

49. $a_1 = 28, \quad a_{k+1} = a_k - 4$
50. $a_1 = 3, \quad a_{k+1} = 2(a_k - 1)$
51. $a_0 = 1, \quad a_1 = 2, \quad a_k = a_{k-2} + \frac{1}{2}a_{k-1}$
52. $a_0 = -1, \quad a_1 = 1, \quad a_k = a_{k-2} + a_{k-1}$

Writing the *n*th Term of a Recursive Sequence In Exercises 53–56, write the first five terms of the sequence defined recursively. Use the pattern to write the *n*th term of the sequence as a function of *n*.

53. $a_1 = 6, \quad a_{k+1} = a_k + 2$
54. $a_1 = 25, \quad a_{k+1} = a_k - 5$
55. $a_1 = 81, \quad a_{k+1} = \frac{1}{3}a_k$
56. $a_1 = 14, \quad a_{k+1} = (-2)a_k$

Fibonacci Sequence In Exercises 57 and 58, use the Fibonacci sequence. (See Example 5.)

57. Write the first 12 terms of the Fibonacci sequence a_n and the first 10 terms of the sequence given by

$$b_n = \frac{a_{n+1}}{a_n}, \quad n \geq 1.$$

58. Using the definition for b_n in Exercise 57, show that b_n can be defined recursively by

$$b_n = 1 + \frac{1}{b_{n-1}}.$$

Writing the Terms of a Sequence Involving Factorials In Exercises 59–62, write the first five terms of the sequence. (Assume that *n* begins with 0.)

59. $a_n = \dfrac{5}{n!}$
60. $a_n = \dfrac{n!}{2n+1}$
61. $a_n = \dfrac{1}{(n+1)!}$
62. $a_n = \dfrac{(-1)^{2n+1}}{(2n+1)!}$

Simplifying a Factorial Expression In Exercises 63–66, simplify the factorial expression.

63. $\dfrac{4!}{6!}$
64. $\dfrac{12!}{4! \cdot 8!}$
65. $\dfrac{(n+1)!}{n!}$
66. $\dfrac{(2n-1)!}{(2n+1)!}$

Finding a Sum In Exercises 67–74, find the sum.

67. $\displaystyle\sum_{i=1}^{5} (2i + 1)$
68. $\displaystyle\sum_{j=3}^{5} \frac{1}{j^2 - 3}$
69. $\displaystyle\sum_{k=1}^{4} 10$
70. $\displaystyle\sum_{i=0}^{4} i^2$
71. $\displaystyle\sum_{k=2}^{5} (k+1)^2(k-3)$
72. $\displaystyle\sum_{i=1}^{4} [(i-1)^2 + (i+1)^3]$
73. $\displaystyle\sum_{i=1}^{4} 2^i$
74. $\displaystyle\sum_{j=0}^{4} (-2)^j$

Finding a Sum In Exercises 75–78, use a graphing utility to find the sum.

75. $\displaystyle\sum_{n=0}^{5} \frac{1}{2n+1}$
76. $\displaystyle\sum_{k=0}^{4} \frac{(-1)^k}{k+1}$
77. $\displaystyle\sum_{k=0}^{4} \frac{(-1)^k}{k!}$
78. $\displaystyle\sum_{n=0}^{25} \frac{1}{4^n}$

Using Sigma Notation to Write a Sum In Exercises 79–88, use sigma notation to write the sum.

79. $\dfrac{1}{3(1)} + \dfrac{1}{3(2)} + \dfrac{1}{3(3)} + \cdots + \dfrac{1}{3(9)}$
80. $\dfrac{5}{1+1} + \dfrac{5}{1+2} + \dfrac{5}{1+3} + \cdots + \dfrac{5}{1+15}$
81. $\left[2\left(\frac{1}{8}\right) + 3\right] + \left[2\left(\frac{2}{8}\right) + 3\right] + \cdots + \left[2\left(\frac{8}{8}\right) + 3\right]$
82. $\left[1 - \left(\frac{1}{6}\right)^2\right] + \left[1 - \left(\frac{2}{6}\right)^2\right] + \cdots + \left[1 - \left(\frac{6}{6}\right)^2\right]$
83. $3 - 9 + 27 - 81 + 243 - 729$
84. $1 - \frac{1}{2} + \frac{1}{4} - \frac{1}{8} + \cdots - \frac{1}{128}$
85. $\dfrac{1}{1^2} - \dfrac{1}{2^2} + \dfrac{1}{3^2} - \dfrac{1}{4^2} + \cdots - \dfrac{1}{20^2}$
86. $\dfrac{1}{1 \cdot 3} + \dfrac{1}{2 \cdot 4} + \dfrac{1}{3 \cdot 5} + \cdots + \dfrac{1}{10 \cdot 12}$
87. $\frac{1}{4} + \frac{3}{8} + \frac{7}{16} + \frac{15}{32} + \frac{31}{64}$
88. $\frac{1}{2} + \frac{2}{4} + \frac{6}{8} + \frac{24}{16} + \frac{120}{32} + \frac{720}{64}$

Finding a Partial Sum of a Series In Exercises 89–92, find the indicated partial sum of the series.

89. $\displaystyle\sum_{i=1}^{\infty} 5\left(\frac{1}{2}\right)^i$ Fourth partial sum
90. $\displaystyle\sum_{i=1}^{\infty} 2\left(\frac{1}{3}\right)^i$ Fifth partial sum
91. $\displaystyle\sum_{n=1}^{\infty} 4\left(-\frac{1}{2}\right)^n$ Third partial sum
92. $\displaystyle\sum_{n=1}^{\infty} 8\left(-\frac{1}{4}\right)^n$ Fourth partial sum

Finding the Sum of an Infinite Series In Exercises 93–96, find the sum of the infinite series.

93. $\displaystyle\sum_{i=1}^{\infty} \frac{6}{10^i}$
94. $\displaystyle\sum_{k=1}^{\infty} \left(\frac{1}{10}\right)^k$
95. $\displaystyle\sum_{k=1}^{\infty} 7\left(\frac{1}{10}\right)^k$
96. $\displaystyle\sum_{i=1}^{\infty} \frac{2}{10^i}$

97. Compound Interest An investor deposits \$10,000 in an account that earns 3.5% interest compounded quarterly. The balance in the account after n quarters is given by

$$A_n = 10{,}000\left(1 + \frac{0.035}{4}\right)^n, \quad n = 1, 2, 3, \ldots.$$

(a) Write the first eight terms of the sequence.

(b) Find the balance in the account after 10 years by computing the 40th term of the sequence.

(c) Is the balance after 20 years twice the balance after 10 years? Explain.

98. AIDS Cases

The numbers a_n (in thousands) of AIDS cases reported from 2003 through 2010 can be approximated by

$a_n = -0.0126n^3 + 0.391n^2 - 4.21n + 48.5,$

$n = 3, 4, \ldots, 10$

where n is the year, with $n = 3$ corresponding to 2003. *(Source: U.S. Centers for Disease Control and Prevention)*

(a) Write the terms of this finite sequence. Use a graphing utility to construct a bar graph that represents the sequence.

(b) What does the graph in part (a) say about reported cases of AIDS?

Exploration

True or False? **In Exercises 99 and 100, determine whether the statement is true or false. Justify your answer.**

99. $\displaystyle\sum_{i=1}^{4}(i^2 + 2i) = \sum_{i=1}^{4} i^2 + 2\sum_{i=1}^{4} i$

100. $\displaystyle\sum_{j=1}^{4} 2^j = \sum_{j=3}^{6} 2^{j-2}$

Arithmetic Mean **In Exercises 101–103, use the following definition of the arithmetic mean $\bar{x}$ of a set of n measurements $x_1, x_2, x_3, \ldots, x_n$.**

$$\bar{x} = \frac{1}{n}\sum_{i=1}^{n} x_i$$

101. Find the arithmetic mean of the six checking account balances \$327.15, \$785.69, \$433.04, \$265.38, \$604.12, and \$590.30. Use the statistical capabilities of a graphing utility to verify your result.

102. Proof Prove that $\displaystyle\sum_{i=1}^{n}(x_i - \bar{x}) = 0.$

103. Proof Prove that $\displaystyle\sum_{i=1}^{n}(x_i - \bar{x})^2 = \sum_{i=1}^{n} x_i^2 - \frac{1}{n}\left(\sum_{i=1}^{n} x_i\right)^2.$

104. HOW DO YOU SEE IT? The graph represents the first 10 terms of a sequence. Complete each expression for the apparent nth term a_n of the sequence. Which expressions are appropriate to represent the cost a_n to buy n MP3 songs at a cost of \$1 per song? Explain.

(a) $a_n = 1 \cdot \square$

(b) $a_n = \dfrac{\square\,!}{(n-1)!}$

(c) $a_n = \displaystyle\sum_{k=1}^{n} \square$

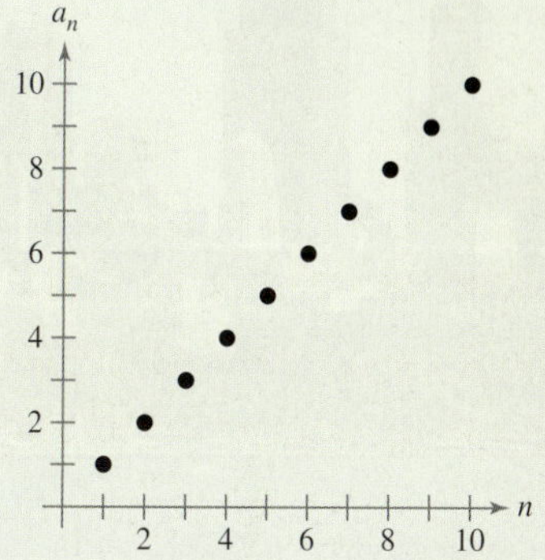

Finding the Terms of a Sequence **In Exercises 105 and 106, find the first five terms of the sequence.**

105. $a_n = \dfrac{x^n}{n!}$

106. $a_n = \dfrac{(-1)^n x^{2n+1}}{2n+1}$

107. Cube A $3 \times 3 \times 3$ cube is made up of 27 unit cubes (a unit cube has a length, width, and height of 1 unit), and only the faces of each cube that are visible are painted blue, as shown in the figure.

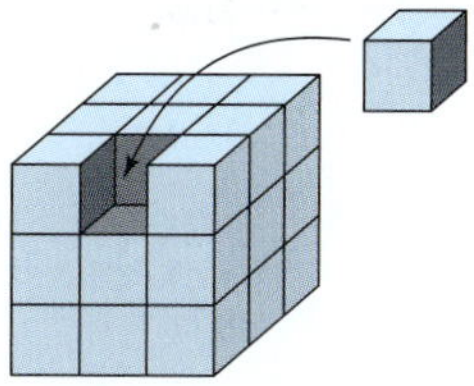

(a) Complete the table to determine how many unit cubes of the $3 \times 3 \times 3$ cube have 0 blue faces, 1 blue face, 2 blue faces, and 3 blue faces.

Number of Blue Cube Faces	0	1	2	3
$3 \times 3 \times 3$				

(b) Repeat part (a) for a $4 \times 4 \times 4$ cube, a $5 \times 5 \times 5$ cube, and a $6 \times 6 \times 6$ cube.

(c) What type of pattern do you observe?

(d) Write formulas you could use to repeat part (a) for an $n \times n \times n$ cube.

michaeljung/Shutterstock.com

9.2 Arithmetic Sequences and Partial Sums

- **Recognize, write, and find the *n*th terms of arithmetic sequences.**
- **Find *n*th partial sums of arithmetic sequences.**
- **Use arithmetic sequences to model and solve real-life problems.**

Arithmetic sequences have many real-life applications. For instance, in Exercise 81 on page 623, you will use an arithmetic sequence to determine how far an object falls in 7 seconds from the top of the Willis Tower in Chicago.

Arithmetic Sequences

A sequence whose consecutive terms have a common difference is called an **arithmetic sequence.**

> **Definition of Arithmetic Sequence**
>
> A sequence is **arithmetic** when the differences between consecutive terms are the same. So, the sequence
>
> $$a_1, a_2, a_3, a_4, \ldots, a_n, \ldots$$
>
> is arithmetic when there is a number d such that
>
> $$a_2 - a_1 = a_3 - a_2 = a_4 - a_3 = \cdots = d.$$
>
> The number d is the **common difference** of the arithmetic sequence.

EXAMPLE 1 **Examples of Arithmetic Sequences**

a. The sequence whose nth term is $4n + 3$ is arithmetic. For this sequence, the common difference between consecutive terms is 4.

$$\underbrace{7, 11}_{11 - 7 = 4}, 15, 19, \ldots, 4n + 3, \ldots \qquad \text{Begin with } n = 1.$$

b. The sequence whose nth term is $7 - 5n$ is arithmetic. For this sequence, the common difference between consecutive terms is -5.

$$\underbrace{2, -3}_{-3 - 2 = -5}, -8, -13, \ldots, 7 - 5n, \ldots \qquad \text{Begin with } n = 1.$$

c. The sequence whose nth term is $\frac{1}{4}(n + 3)$ is arithmetic. For this sequence, the common difference between consecutive terms is $\frac{1}{4}$.

$$\underbrace{1, \frac{5}{4}}_{\frac{5}{4} - 1 = \frac{1}{4}}, \frac{3}{2}, \frac{7}{4}, \ldots, \frac{n + 3}{4}, \ldots \qquad \text{Begin with } n = 1.$$

✓ **Checkpoint** Audio-video solution in English & Spanish at LarsonPrecalculus.com.

Write the first four terms of the arithmetic sequence whose nth term is $3n - 1$. Then find the common difference between consecutive terms.

The sequence 1, 4, 9, 16, . . . , whose nth term is n^2, is *not* arithmetic. The difference between the first two terms is

$$a_2 - a_1 = 4 - 1 = 3$$

but the difference between the second and third terms is

$$a_3 - a_2 = 9 - 4 = 5.$$

Eugene Moerman/Shutterstock.com

The nth term of an arithmetic sequence can be derived from the following pattern.

$a_1 = a_1$ 1st term

$a_2 = a_1 + d$ 2nd term

$a_3 = a_1 + 2d$ 3rd term

$a_4 = a_1 + 3d$ 4th term

$a_5 = a_1 + 4d$ 5th term

1 less

$\vdots$

$a_n = a_1 + (n - 1)d$ nth term

1 less

The following definition summarizes this result.

The nth Term of an Arithmetic Sequence

The nth term of an arithmetic sequence has the form

$$a_n = a_1 + (n - 1)d$$

where d is the common difference between consecutive terms of the sequence and a_1 is the first term.

EXAMPLE 2 Finding the nth Term

Find a formula for the nth term of the arithmetic sequence whose common difference is 3 and whose first term is 2.

Solution You know that the formula for the nth term is of the form $a_n = a_1 + (n - 1)d$. Moreover, because the common difference is $d = 3$ and the first term is $a_1 = 2$, the formula must have the form

$$a_n = 2 + 3(n - 1).$$ Substitute 2 for a_1 and 3 for d.

So, the formula for the nth term is $a_n = 3n - 1$. The sequence therefore has the following form.

$$2, 5, 8, 11, 14, \ldots, 3n - 1, \ldots$$

The figure below shows a graph of the first 15 terms of the sequence. Notice that the points lie on a line. This makes sense because a_n is a linear function of n. In other words, the terms "arithmetic" and "linear" are closely connected.

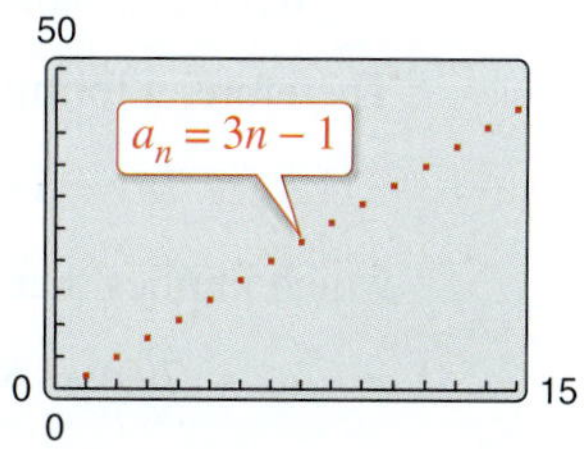

Checkpoint Audio-video solution in English & Spanish at LarsonPrecalculus.com.

Find a formula for the nth term of the arithmetic sequence whose common difference is 5 and whose first term is -1.

REMARK You can find a_1 in Example 3 by using the nth term of an arithmetic sequence, as follows.

$$a_n = a_1 + (n - 1)d$$
$$a_4 = a_1 + (4 - 1)d$$
$$20 = a_1 + (4 - 1)5$$
$$20 = a_1 + 15$$
$$5 = a_1$$

EXAMPLE 3 **Writing the Terms of an Arithmetic Sequence**

The fourth term of an arithmetic sequence is 20, and the 13th term is 65. Write the first 11 terms of this sequence.

Solution You know that $a_4 = 20$ and $a_{13} = 65$. So, you must add the common difference d nine times to the fourth term to obtain the 13th term. Therefore, the fourth and 13th terms of the sequence are related by

$$a_{13} = a_4 + 9d. \qquad a_4 \text{ and } a_{13} \text{ are nine terms apart.}$$

Using $a_4 = 20$ and $a_{13} = 65$, solve for d to find that $d = 5$, which implies that the sequence is as follows.

a_1	a_2	a_3	a_4	a_5	a_6	a_7	a_8	a_9	a_{10}	a_{11} . . .
5	10	15	20	25	30	35	40	45	50	55 . . .

✓ **Checkpoint** Audio-video solution in English & Spanish at LarsonPrecalculus.com.

The eighth term of an arithmetic sequence is 25, and the 12th term is 41. Write the first 11 terms of this sequence.

When you know the nth term of an arithmetic sequence *and* you know the common difference of the sequence, you can find the $(n + 1)$th term by using the *recursion formula*

$$a_{n+1} = a_n + d. \qquad \text{Recursion formula}$$

With this formula, you can find any term of an arithmetic sequence, *provided* that you know the preceding term. For instance, when you know the first term, you can find the second term. Then, knowing the second term, you can find the third term, and so on.

EXAMPLE 4 **Using a Recursion Formula**

Find the ninth term of the arithmetic sequence that begins with 2 and 9.

Solution For this sequence, the common difference is

$$d = 9 - 2 - 7.$$

There are two ways to find the ninth term. One way is to write the first nine terms (by repeatedly adding 7).

2, 9, 16, 23, 30, 37, 44, 51, 58

Another way to find the ninth term is to first find a formula for the nth term. Because the common difference is $d = 7$ and the first term is $a_1 = 2$, the formula must have the form

$$a_n = 2 + 7(n - 1). \qquad \text{Substitute 2 for } a_1 \text{ and 7 for } d.$$

Therefore, a formula for the nth term is

$$a_n = 7n - 5$$

which implies that the ninth term is

$$a_9 = 7(9) - 5$$
$$= 58.$$

✓ **Checkpoint** Audio-video solution in English & Spanish at LarsonPrecalculus.com.

Find the tenth term of the arithmetic sequence that begins with 7 and 15.

The Sum of a Finite Arithmetic Sequence

There is a formula for the *sum* of a finite arithmetic sequence.

REMARK Note that this formula works only for *arithmetic* sequences.

> **The Sum of a Finite Arithmetic Sequence**
>
> The sum of a finite arithmetic sequence with n terms is $S_n = \frac{n}{2}(a_1 + a_n)$.

For a proof of this formula for the sum of a finite arithmetic sequence, see Proofs in Mathematics on page 683.

EXAMPLE 5 Sum of a Finite Arithmetic Sequence

Find the sum: $1 + 3 + 5 + 7 + 9 + 11 + 13 + 15 + 17 + 19$.

Solution To begin, notice that the sequence is arithmetic (with a common difference of 2). Moreover, the sequence has 10 terms. So, the sum of the sequence is

$$S_n = \frac{n}{2}(a_1 + a_n) \qquad \text{Sum of a finite arithmetic sequence}$$

$$= \frac{10}{2}(1 + 19) \qquad \text{Substitute 10 for } n, \text{ 1 for } a_1, \text{ and 19 for } a_n.$$

$$= 100. \qquad \text{Simplify.}$$

✓ **Checkpoint** Audio-video solution in English & Spanish at LarsonPrecalculus.com.

Find the sum: $40 + 37 + 34 + 31 + 28 + 25 + 22$.

A teacher of Carl Friedrich Gauss (1777–1855) asked him to add all the integers from 1 to 100. When Gauss returned with the correct answer after only a few moments, the teacher could only look at him in astounded silence. This is what Gauss did:

$$\begin{aligned} S_n &= 1 + 2 + 3 + \cdots + 100 \\ S_n &= 100 + 99 + 98 + \cdots + 1 \\ \hline 2S_n &= 101 + 101 + 101 + \cdots + 101 \\ S_n &= \frac{100 \times 101}{2} = 5050 \end{aligned}$$

EXAMPLE 6 Sum of a Finite Arithmetic Sequence

Find the sum of the integers (a) from 1 to 100 and (b) from 1 to N.

Solution

a. The integers from 1 to 100 form an arithmetic sequence that has 100 terms. So, you can use the formula for the sum of a finite arithmetic sequence, as follows.

$$S_n = 1 + 2 + 3 + 4 + 5 + 6 + \cdots + 99 + 100$$

$$= \frac{n}{2}(a_1 + a_n) \qquad \text{Sum of a finite arithmetic sequence}$$

$$= \frac{100}{2}(1 + 100) \qquad \text{Substitute 100 for } n, \text{ 1 for } a_1, \text{ and 100 for } a_n.$$

$$= 5050 \qquad \text{Simplify.}$$

b. $S_n = 1 + 2 + 3 + 4 + \cdots + N$

$$= \frac{n}{2}(a_1 + a_n) \qquad \text{Sum of a finite arithmetic sequence}$$

$$= \frac{N}{2}(1 + N) \qquad \text{Substitute } N \text{ for } n, \text{ 1 for } a_1, \text{ and } N \text{ for } a_n.$$

✓ **Checkpoint** Audio-video solution in English & Spanish at LarsonPrecalculus.com.

Find the sum of the integers (a) from 1 to 35 and (b) from 1 to $2N$.

Bettmann/Corbis

The sum of the first n terms of an infinite sequence is the *nth partial sum*. The nth partial sum can be found by using the formula for the sum of a finite arithmetic sequence.

EXAMPLE 7 Partial Sum of an Arithmetic Sequence

Find the 150th partial sum of the arithmetic sequence

5, 16, 27, 38, 49, ...

Solution For this arithmetic sequence, $a_1 = 5$ and $d = 16 - 5 = 11$. So,

$$a_n = 5 + 11(n - 1)$$

and the nth term is

$$a_n = 11n - 6.$$

Therefore, $a_{150} = 11(150) - 6 = 1644$, and the sum of the first 150 terms is

$$S_{150} = \frac{n}{2}(a_1 + a_{150}) \qquad \text{nth partial sum formula}$$

$$= \frac{150}{2}(5 + 1644) \qquad \text{Substitute 150 for } n\text{, 5 for } a_1\text{, and 1644 for } a_{150}.$$

$$= 75(1649) \qquad \text{Simplify.}$$

$$= 123{,}675. \qquad \text{nth partial sum}$$

✓ **Checkpoint** Audio-video solution in English & Spanish at LarsonPrecalculus.com.

Find the 120th partial sum of the arithmetic sequence

6, 12, 18, 24, 30, ...

EXAMPLE 8 Partial Sum of an Arithmetic Sequence

Find the 16th partial sum of the arithmetic sequence

100, 95, 90, 85, 80, ...

Solution For this arithmetic sequence, $a_1 = 100$ and $d = 95 - 100 = -5$. So,

$$a_n = 100 + (-5)(n - 1)$$

and the nth term is

$$a_n = -5n + 105.$$

Therefore, $a_{16} = -5(16) + 105 = 25$, and the sum of the first 16 terms is

$$S_{16} = \frac{n}{2}(a_1 + a_{16}) \qquad \text{nth partial sum formula}$$

$$= \frac{16}{2}(100 + 25) \qquad \text{Substitute 16 for } n\text{, 100 for } a_1\text{, and 25 for } a_{16}.$$

$$= 8(125) \qquad \text{Simplify.}$$

$$= 1000. \qquad \text{nth partial sum}$$

✓ **Checkpoint** Audio-video solution in English & Spanish at LarsonPrecalculus.com.

Find the 30th partial sum of the arithmetic sequence

78, 76, 74, 72, 70, ...

Application

EXAMPLE 9 **Total Sales**

A small business sells \$10,000 worth of skin care products during its first year. The owner of the business has set a goal of increasing annual sales by \$7500 each year for 9 years. Assuming that this goal is met, find the total sales during the first 10 years this business is in operation.

Solution The annual sales form an arithmetic sequence in which

$$a_1 = 10{,}000 \quad \text{and} \quad d = 7500.$$

So,

$$a_n = 10{,}000 + 7500(n - 1)$$

and the nth term of the sequence is

$$a_n = 7500n + 2500.$$

Therefore, the 10th term of the sequence is

$$a_{10} = 7500(10) + 2500$$

$$= 77{,}500. \qquad \text{See figure.}$$

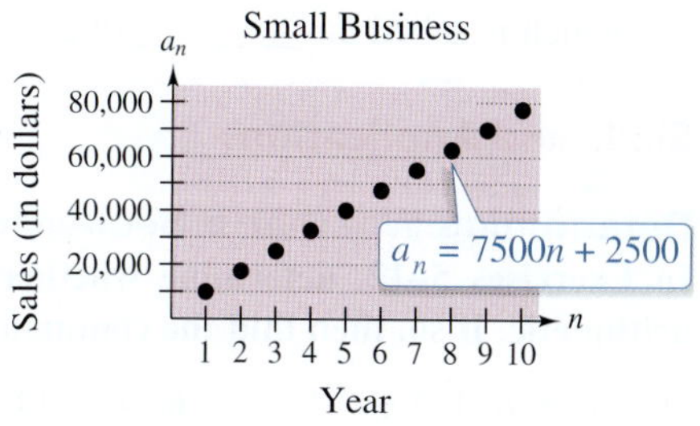

The sum of the first 10 terms of the sequence is

$$S_{10} = \frac{n}{2}(a_1 + a_{10}) \qquad n\text{th partial sum formula}$$

$$= \frac{10}{2}(10{,}000 + 77{,}500) \qquad \text{Substitute 10 for } n\text{, 10,000 for } a_1\text{, and 77,500 for } a_{10}.$$

$$= 5(87{,}500) \qquad \text{Simplify.}$$

$$= 437{,}500. \qquad \text{Multiply.}$$

So, the total sales for the first 10 years will be \$437,500.

✓ ***Checkpoint*** *Audio-video solution in English & Spanish at LarsonPrecalculus.com.*

A company sells \$160,000 worth of printing paper during its first year. The sales manager has set a goal of increasing annual sales of printing paper by \$20,000 each year for 9 years. Assuming that this goal is met, find the total sales of printing paper during the first 10 years this company is in operation.

Summarize (Section 9.2)

1. State the definition of an arithmetic sequence *(page 616)*, and state the formula for the nth term of an arithmetic sequence *(page 617)*. For examples of recognizing, writing, and finding the nth terms of arithmetic sequences, see Examples 1–4.
2. State the formula for the sum of a finite arithmetic sequence and explain how to use it to find an nth partial sum *(pages 619 and 620)*. For examples of finding sums of arithmetic sequences, see Examples 5–8.
3. Describe an example of how to use an arithmetic sequence to model and solve a real-life problem *(page 621, Example 9)*.

9.2 Exercises

See CalcChat.com for tutorial help and worked-out solutions to odd-numbered exercises.

Vocabulary: Fill in the blanks.

1. A sequence is called an ________ sequence when the differences between consecutive terms are the same. This difference is called the ________ difference.
2. The nth term of an arithmetic sequence has the form ________.
3. When you know the nth term of an arithmetic sequence *and* you know the common difference of the sequence, you can find the $(n + 1)$th term by using the ________ formula $a_{n+1} = a_n + d$.
4. You can use the formula $S_n = \frac{n}{2}(a_1 + a_n)$ to find the sum of the first n terms of an arithmetic sequence, which is called the ________ of a ________ ________ ________.

Skills and Applications

Determining Whether a Sequence Is Arithmetic

In Exercises 5–12, determine whether the sequence is arithmetic. If so, then find the common difference.

5. 10, 8, 6, 4, 2, . . .
6. 4, 9, 14, 19, 24, . . .
7. 1, 2, 4, 8, 16, . . .
8. 80, 40, 20, 10, 5, . . .
9. $\frac{9}{4}, 2, \frac{7}{4}, \frac{3}{2}, \frac{5}{4}, \ldots$
10. 5.3, 5.7, 6.1, 6.5, 6.9, . . .
11. ln 1, ln 2, ln 3, ln 4, ln 5, . . .
12. $1^2, 2^2, 3^2, 4^2, 5^2, \ldots$

Writing the Terms of a Sequence

In Exercises 13–20, write the first five terms of the sequence. Determine whether the sequence is arithmetic. If so, then find the common difference. (Assume that n begins with 1.)

13. $a_n = 5 + 3n$
14. $a_n = 100 - 3n$
15. $a_n = 3 - 4(n - 2)$
16. $a_n = 1 + (n - 1)4$
17. $a_n = (-1)^n$
18. $a_n = 2^{n-1}$
19. $a_n = \frac{(-1)^n 3}{n}$
20. $a_n = (2^n)n$

Finding the nth Term

In Exercises 21–30, find a formula for a_n for the arithmetic sequence.

21. $a_1 = 1, d = 3$
22. $a_1 = 15, d = 4$
23. $a_1 = 100, d = -8$
24. $a_1 = 0, d = -\frac{2}{3}$
25. $4, \frac{3}{2}, -1, -\frac{7}{2}, \ldots$
26. $10, 5, 0, -5, -10, \ldots$
27. $a_1 = 5, a_4 = 15$
28. $a_1 = -4, a_5 = 16$
29. $a_3 = 94, a_6 = 85$
30. $a_5 = 190, a_{10} = 115$

Writing the Terms of an Arithmetic Sequence

In Exercises 31–38, write the first five terms of the arithmetic sequence.

31. $a_1 = 5, d = 6$
32. $a_1 = 5, d = -\frac{3}{4}$
33. $a_1 = -\frac{13}{5}, d = -\frac{2}{5}$
34. $a_1 = 16.5, d = 0.25$
35. $a_1 = 2, a_{12} = 46$
36. $a_4 = 16, a_{10} = 46$
37. $a_8 = 26, a_{12} = 42$
38. $a_3 = 19, a_{15} = -1.7$

Writing the Terms of an Arithmetic Sequence

In Exercises 39–42, write the first five terms of the arithmetic sequence defined recursively.

39. $a_1 = 15, a_{n+1} = a_n + 4$
40. $a_1 = 200, a_{n+1} = a_n - 10$
41. $a_1 = \frac{5}{8}, a_{n+1} = a_n - \frac{1}{8}$
42. $a_1 = 0.375, a_{n+1} = a_n + 0.25$

Using a Recursion Formula

In Exercises 43–46, the first two terms of the arithmetic sequence are given. Find the missing term.

43. $a_1 = 5, a_2 = 11, a_{10} =$ ____
44. $a_1 = 3, a_2 = 13, a_9 =$ ____
45. $a_1 = 4.2, a_2 = 6.6, a_7 =$ ____
46. $a_1 = -0.7, a_2 = -13.8, a_8 =$ ____

Sum of a Finite Arithmetic Sequence

In Exercises 47–52, find the sum of the finite arithmetic sequence.

47. $2 + 4 + 6 + 8 + 10 + 12 + 14 + 16 + 18 + 20$
48. $1 + 4 + 7 + 10 + 13 + 16 + 19$
49. $-1 + (-3) + (-5) + (-7) + (-9)$
50. $-5 + (-3) + (-1) + 1 + 3 + 5$
51. Sum of the first 100 positive odd integers
52. Sum of the integers from -100 to 30

Partial Sum of an Arithmetic Sequence

In Exercises 53–58, find the indicated nth partial sum of the arithmetic sequence.

53. $8, 20, 32, 44, \ldots, \quad n = 10$
54. $-6, -2, 2, 6, \ldots, \quad n = 50$
55. $4.2, 3.7, 3.2, 2.7, \ldots, \quad n = 12$
56. $75, 70, 65, 60, \ldots, \quad n = 25$
57. $a_1 = 100, \quad a_{25} = 220, \quad n = 25$
58. $a_1 = 15, \quad a_{100} = 307, \quad n = 100$

Finding a Partial Sum **In Exercises 59–64, find the partial sum.**

59. $\sum_{n=1}^{50} n$ **60.** $\sum_{n=51}^{100} 7n$

61. $\sum_{n=11}^{30} n - \sum_{n=1}^{10} n$ **62.** $\sum_{n=51}^{100} n - \sum_{n=1}^{50} n$

63. $\sum_{n=1}^{500} (n + 8)$ **64.** $\sum_{n=1}^{250} (1000 - n)$

Matching an Arithmetic Sequence with Its Graph **In Exercises 65–68, match the arithmetic sequence with its graph. [The graphs are labeled (a), (b), (c), and (d).]**

(a)
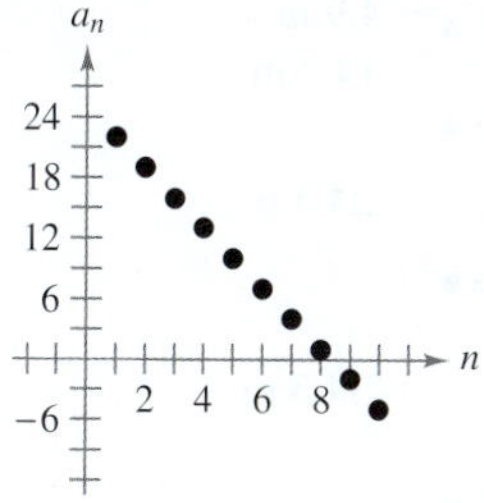

(b)
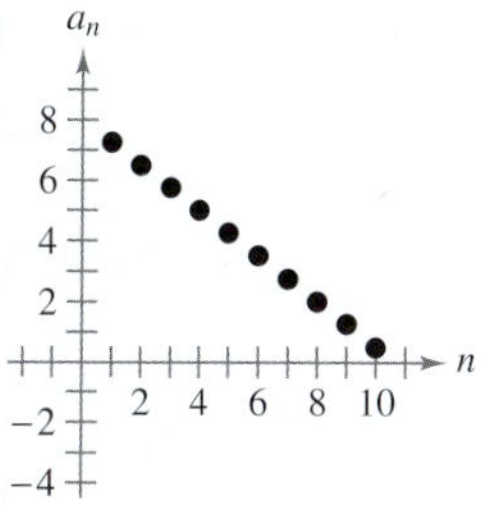

(c)
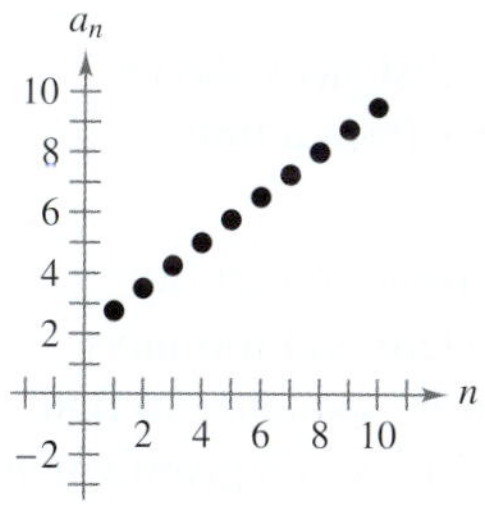

(d)
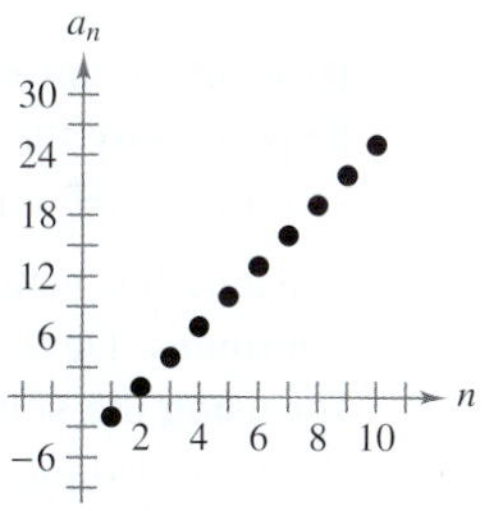

65. $a_n = -\frac{3}{4}n + 8$ **66.** $a_n = 3n - 5$

67. $a_n = 2 + \frac{3}{4}n$ **68.** $a_n = 25 - 3n$

Graphing the Terms of a Sequence **In Exercises 69–72, use a graphing utility to graph the first 10 terms of the sequence. (Assume that *n* begins with 1.)**

69. $a_n = 15 - \frac{3}{2}n$

70. $a_n = -5 + 2n$

71. $a_n = 0.2n + 3$

72. $a_n = -0.3n + 8$

Finding a Partial Sum **In Exercises 73–76, use a graphing utility to find the partial sum.**

73. $\sum_{n=0}^{50} (50 - 2n)$ **74.** $\sum_{n=1}^{100} \frac{n + 1}{2}$

75. $\sum_{i=1}^{60} \left(250 - \frac{2}{5}i\right)$ **76.** $\sum_{j=1}^{200} (10.5 + 0.025j)$

77. Seating Capacity Determine the seating capacity of an auditorium with 36 rows of seats when there are 15 seats in the first row, 18 seats in the second row, 21 seats in the third row, and so on.

Eugene Moerman/Shutterstock.com

78. Brick Pattern A triangular brick wall is made by cutting some bricks in half to use in the first column of every other row (see figure). The wall has 28 rows. The top row is one-half brick wide and the bottom row is 14 bricks wide. How many bricks are in the finished wall?

Job Offer **In Exercises 79 and 80, consider a job offer with the given starting salary and the given annual raise.**

(a) Determine the salary during the sixth year of employment.

(b) Determine the total compensation from the company through six full years of employment.

	Starting Salary	Annual Raise
79.	\$32,500	\$1500
80.	\$36,800	\$1750

81. Falling Object An object with negligible air resistance is dropped from the top of the Willis Tower in Chicago at a height of 1450 feet. During the first second of fall, the object falls 16 feet; during the second second, it falls 48 feet; during the third second, it falls 80 feet; during the fourth second, it falls 112 feet. If this arithmetic pattern continues, then how many feet will the object fall in 7 seconds?

82. Prize Money A county fair is holding a baked goods competition in which the top eight bakers receive cash prizes. First place receives a cash prize of \$200, second place receives \$175, third place receives \$150, and so on.

(a) Write a sequence a_n that represents the cash prize awarded in terms of the place n in which the baked good places.

(b) Find the total amount of prize money awarded at the competition.

83. Total Sales An entrepreneur sells \$15,000 worth of sports memorabilia during one year and sets a goal of increasing annual sales by \$5000 each year for 9 years. Assuming that the entrepreneur meets this goal, find the total sales during the first 10 years of this business. What kinds of economic factors could prevent the business from meeting its goals?

84. Borrowing Money You borrow \$5000 from your parents to purchase a used car. The arrangements of the loan are such that you make payments of \$250 per month plus 1% interest on the unpaid balance.

(a) Find the first year's monthly payments and the unpaid balance after each month.

(b) Find the total amount of interest paid over the term of the loan.

85. Data Analysis: Sales The table shows the sales a_n (in billions of dollars) for Eli Lilly and Co. from 2004 through 2011. *(Source: Eli Lilly and Co.)*

DATA

Year	Sales, a_n
2004	13.9
2005	14.6
2006	15.7
2007	18.6
2008	20.4
2009	21.8
2010	23.1
2011	24.3

Spreadsheet at LarsonPrecalculus.com

(a) Construct a bar graph showing the annual sales from 2004 through 2011.

(b) Find an arithmetic sequence that models the data. Let n represent the year, with $n = 4$ corresponding to 2004.

(c) Use a graphing utility to graph the terms of the finite sequence you found in part (a).

(d) Use summation notation to represent the *total* sales from 2004 through 2011. Find the total sales.

Exploration

86. Writing Explain how to use the first two terms of an arithmetic sequence to find the nth term.

True or False? **In Exercises 87 and 88, determine whether the statement is true or false. Justify your answer.**

87. Given an arithmetic sequence for which only the first two terms are known, it is possible to find the nth term.

88. If the only known information about a finite arithmetic sequence is its first term and its last term, then it is possible to find the sum of the sequence.

Finding the Terms of a Sequence **In Exercises 89 and 90, find the first 10 terms of the sequence.**

89. $a_1 = x, d = 2x$ **90.** $a_1 = -y, d = 5y$

91. Comparing Graphs of a Sequence and a Line

(a) Graph the first 10 terms of the arithmetic sequence $a_n = 2 + 3n$.

(b) Graph the equation of the line $y = 3x + 2$.

(c) Discuss any differences between the graph of $a_n = 2 + 3n$ and the graph of $y = 3x + 2$.

(d) Compare the slope of the line in part (b) with the common difference of the sequence in part (a). What can you conclude about the slope of a line and the common difference of an arithmetic sequence?

92. HOW DO YOU SEE IT? A steel ball with negligible air resistance is dropped from a plane. The figure shows the distance that the ball falls during each of the first four seconds after it is dropped.

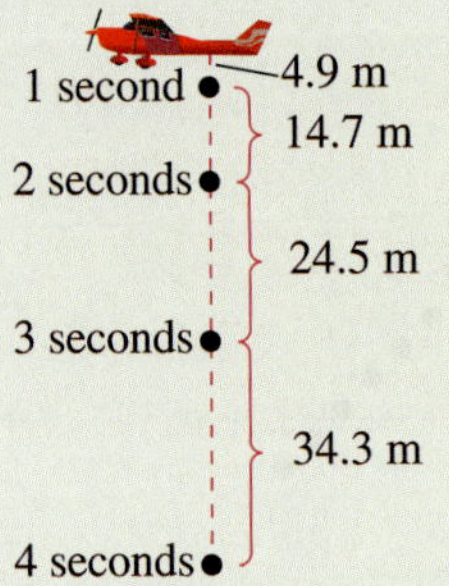

(a) Describe a pattern of the distances shown. Explain why the distances form a finite arithmetic sequence.

(b) Assume the pattern described in part (a) continues. Describe the steps and formulas for using the sum of a finite sequence to find the total distance the ball falls in a given whole number of seconds n.

93. Pattern Recognition

(a) Compute the following sums of consecutive positive odd integers.

$1 + 3 = \square$

$1 + 3 + 5 = \square$

$1 + 3 + 5 + 7 = \square$

$1 + 3 + 5 + 7 + 9 = \square$

$1 + 3 + 5 + 7 + 9 + 11 = \square$

(b) Use the sums in part (a) to make a conjecture about the sums of consecutive positive odd integers. Check your conjecture for the sum

$1 + 3 + 5 + 7 + 9 + 11 + 13 = \square$.

(c) Verify your conjecture algebraically.

Project: Municipal Waste To work an extended application analyzing the amounts of municipal waste recovered in the United States from 1991 through 2010, visit this text's website at *LarsonPrecalculus.com.* *(Source: U.S. Environmental Protection Agency)*

9.3 Geometric Sequences and Series

Geometric sequences can help you model and solve real-life problems. For instance, in Exercise 86 on page 632, you will use a geometric sequence to model the populations of China from 2004 through 2010.

- Recognize, write, and find the *n*th terms of geometric sequences.
- Find the sum of a finite geometric sequence.
- Find the sum of an infinite geometric series.
- Use geometric sequences to model and solve real-life problems.

Geometric Sequences

In Section 9.2, you learned that a sequence whose consecutive terms have a common *difference* is an arithmetic sequence. In this section, you will study another important type of sequence called a **geometric sequence.** Consecutive terms of a geometric sequence have a common *ratio.*

> **Definition of Geometric Sequence**
>
> A sequence is **geometric** when the ratios of consecutive terms are the same. So, the sequence $a_1, a_2, a_3, a_4, \ldots, a_n, \ldots$ is geometric when there is a number r such that
>
> $$\frac{a_2}{a_1} = \frac{a_3}{a_2} = \frac{a_4}{a_3} = \cdots = r, \quad r \neq 0.$$
>
> The number r is the **common ratio** of the geometric sequence.

REMARK Be sure you understand that a sequence such as $1, 4, 9, 16, \ldots$, whose nth term is n^2, is *not* geometric. The ratio of the second term to the first term is

$$\frac{a_2}{a_1} = \frac{4}{1} = 4$$

but the ratio of the third term to the second term is

$$\frac{a_3}{a_2} = \frac{9}{4}.$$

EXAMPLE 1 Examples of Geometric Sequences

a. The sequence whose nth term is 2^n is geometric. For this sequence, the common ratio of consecutive terms is 2.

$$2, 4, 8, 16, \ldots, 2^n, \ldots \qquad \text{Begin with } n = 1.$$

$\frac{4}{2} = 2$

b. The sequence whose nth term is $4(3^n)$ is geometric. For this sequence, the common ratio of consecutive terms is 3.

$$12, 36, 108, 324, \ldots, 4(3^n), \ldots \qquad \text{Begin with } n = 1.$$

$\frac{36}{12} = 3$

c. The sequence whose nth term is $\left(-\frac{1}{3}\right)^n$ is geometric. For this sequence, the common ratio of consecutive terms is $-\frac{1}{3}$.

$$-\frac{1}{3}, \frac{1}{9}, -\frac{1}{27}, \frac{1}{81}, \ldots, \left(-\frac{1}{3}\right)^n, \ldots \qquad \text{Begin with } n = 1.$$

$\frac{1/9}{-1/3} = -\frac{1}{3}$

✓ Checkpoint Audio-video solution in English & Spanish at LarsonPrecalculus.com.

Write the first four terms of the geometric sequence whose nth term is $6(-2)^n$. Then find the common ratio of the consecutive terms.

In Example 1, notice that each of the geometric sequences has an nth term that is of the form ar^n, where the common ratio of the sequence is r. A geometric sequence may be thought of as an exponential function whose domain is the set of natural numbers.

TonyV3112/Shutterstock.com

The *n*th Term of a Geometric Sequence

The *n*th term of a geometric sequence has the form

$$a_n = a_1 r^{n-1}$$

where r is the common ratio of consecutive terms of the sequence. So, every geometric sequence can be written in the following form.

$$a_1,\ a_2,\ a_3,\ a_4,\ a_5,\ \ldots\ldots,\ a_n,\ \ldots$$

$$\downarrow\quad\downarrow\quad\downarrow\quad\downarrow\quad\downarrow\qquad\qquad\downarrow$$

$$a_1,\ a_1r,\ a_1r^2,\ a_1r^3,\ a_1r^4,\ \ldots,\ a_1r^{n-1},\ \ldots$$

When you know the *n*th term of a geometric sequence, you can find the $(n + 1)$th term by multiplying by r. That is, $a_{n+1} = a_n r$.

EXAMPLE 2 Writing the Terms of a Geometric Sequence

Write the first five terms of the geometric sequence whose first term is $a_1 = 3$ and whose common ratio is $r = 2$. Then graph the terms on a set of coordinate axes.

Solution Starting with 3, repeatedly multiply by 2 to obtain the following.

$a_1 = 3$ 1st term

$a_2 = 3(2^1) = 6$ 2nd term

$a_3 = 3(2^2) = 12$ 3rd term

$a_4 = 3(2^3) = 24$ 4th term

$a_5 = 3(2^4) = 48$ 5th term

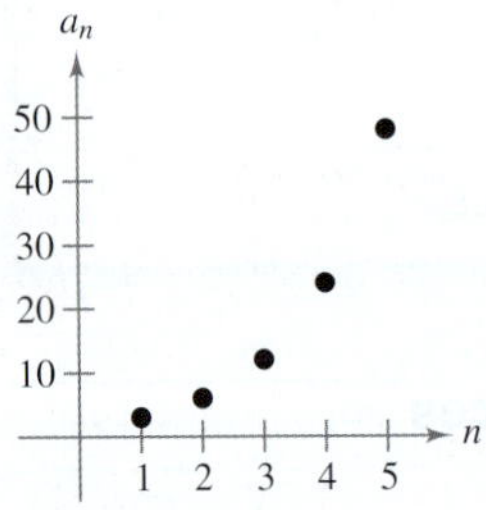

Figure 9.1

Figure 9.1 shows the first five terms of this geometric sequence.

Checkpoint Audio-video solution in English & Spanish at LarsonPrecalculus.com.

Write the first five terms of the geometric sequence whose first term is $a_1 = 2$ and whose common ratio is $r = 4$. Then graph the terms on a set of coordinate axes.

EXAMPLE 3 Finding a Term of a Geometric Sequence

Find the 15th term of the geometric sequence whose first term is 20 and whose common ratio is 1.05.

Algebraic Solution

$a_n = a_1 r^{n-1}$ Formula for *n*th term of a geometric sequence

$a_{15} = 20(1.05)^{15-1}$ Substitute 20 for a_1, 1.05 for r, and 15 for n.

≈ 39.60 Use a calculator.

Numerical Solution

For this sequence, $r = 1.05$ and $a_1 = 20$. So, $a_n = 20(1.05)^{n-1}$. Use a graphing utility to create a table that shows the terms of the sequence.

n	u(n)
9	29.549
10	31.027
11	32.578
12	34.207
13	35.917
14	37.713
15	39.599

u(n)=39.59863199

The number in the 15th row is the 15th term of the sequence.

So, $a_{15} \approx 39.60$.

Checkpoint Audio-video solution in English & Spanish at LarsonPrecalculus.com.

Find the 12th term of the geometric sequence whose first term is 14 and whose common ratio is 1.2.

EXAMPLE 4 Finding a Term of a Geometric Sequence

Find the 12th term of the geometric sequence

$5, 15, 45, \ldots.$

Solution The common ratio of this sequence is $r = 15/5 = 3$. Because the first term is $a_1 = 5$, the 12th term $(n = 12)$ is

$a_n = a_1 r^{n-1}$ Formula for nth term of a geometric sequence

$a_{12} = 5(3)^{12-1}$ Substitute 5 for a_1, 3 for r, and 12 for n.

$= 5(177{,}147)$ Use a calculator.

$= 885{,}735.$ Multiply.

Checkpoint *Audio-video solution in English & Spanish at LarsonPrecalculus.com.*

Find the 12th term of the geometric sequence

$4, 20, 100, \ldots.$

When you know *any* two terms of a geometric sequence, you can use that information to find *any other* term of the sequence.

EXAMPLE 5 Finding a Term of a Geometric Sequence

The fourth term of a geometric sequence is 125, and the 10th term is 125/64. Find the 14th term. (Assume that the terms of the sequence are positive.)

Solution The 10th term is related to the fourth term by the equation

$a_{10} = a_4 r^6$ Multiply fourth term by r^{10-4}.

Because $a_{10} = 125/64$ and $a_4 = 125$, you can solve for r as follows.

$\frac{125}{64} = 125r^6$ Substitute $\frac{125}{64}$ for a_{10} and 125 for a_4.

$\frac{1}{64} = r^6$ Divide each side by 125.

$\frac{1}{2} = r$ Take the sixth root of each side.

You can obtain the 14th term by multiplying the 10th term by r^4.

$a_{14} = a_{10} r^4$ Multiply the 10th term by r^{14-10}.

$= \frac{125}{64}\left(\frac{1}{2}\right)^4$ Substitute $\frac{125}{64}$ for a_{10} and $\frac{1}{2}$ for r.

$= \frac{125}{64}\left(\frac{1}{16}\right)$ Evaluate power.

$= \frac{125}{1024}$ Multiply fractions.

ALGEBRA HELP Remember that r is the common ratio of consecutive terms of a geometric sequence. So, in Example 5

$$\begin{aligned} a_{10} &= a_1 r^9 \\ &= a_1 \cdot r \cdot r \cdot r \cdot r^6 \\ &= a_1 \cdot \frac{a_2}{a_1} \cdot \frac{a_3}{a_2} \cdot \frac{a_4}{a_3} \cdot r^6 \\ &= a_4 r^6. \end{aligned}$$

Checkpoint *Audio-video solution in English & Spanish at LarsonPrecalculus.com.*

The second term of a geometric sequence is 6, and the fifth term is 81/4. Find the eighth term. (Assume that the terms of the sequence are positive.)

The Sum of a Finite Geometric Sequence

The formula for the sum of a *finite* geometric sequence is as follows.

The Sum of a Finite Geometric Sequence

The sum of the finite geometric sequence

$$a_1,\ a_1r,\ a_1r^2,\ a_1r^3,\ a_1r^4,\ .\ .\ .\ ,\ a_1r^{n-1}$$

with common ratio $r \neq 1$ is given by $S_n = \sum_{i=1}^{n} a_1r^{i-1} = a_1\left(\frac{1-r^n}{1-r}\right)$.

For a proof of this formula for the sum of a finite geometric sequence, see Proofs in Mathematics on page 683.

EXAMPLE 6 **Sum of a Finite Geometric Sequence**

Find the sum $\sum_{i=1}^{12} 4(0.3)^{i-1}$.

Solution You have

$$\sum_{i=1}^{12} 4(0.3)^{i-1} = 4(0.3)^0 + 4(0.3)^1 + 4(0.3)^2 + \cdot\cdot\cdot + 4(0.3)^{11}.$$

Now, $a_1 = 4$, $r = 0.3$, and $n = 12$, so applying the formula for the sum of a finite geometric sequence, you obtain

$$S_n = a_1\left(\frac{1-r^n}{1-r}\right) \qquad \text{Sum of a finite geometric sequence}$$

$$\sum_{i=1}^{12} 4(0.3)^{i-1} = 4\left[\frac{1-(0.3)^{12}}{1-0.3}\right] \qquad \text{Substitute 4 for } a_1\text{, 0.3 for } r\text{, and 12 for } n.$$

$$\approx 5.714. \qquad \text{Use a calculator.}$$

✓ ***Checkpoint*** *Audio-video solution in English & Spanish at LarsonPrecalculus.com.*

Find the sum $\sum_{i=1}^{10} 2(0.25)^{i-1}$.

When using the formula for the sum of a finite geometric sequence, be careful to check that the sum is of the form

$$\sum_{i=1}^{n} a_1r^{i-1}. \qquad \text{Exponent for } r \text{ is } i-1.$$

For a sum that is not of this form, you must adjust the formula. For instance, if the sum in Example 6 were $\sum_{i=1}^{12} 4(0.3)^i$, then you would evaluate the sum as follows.

$$\sum_{i=1}^{12} 4(0.3)^i = 4(0.3) + 4(0.3)^2 + 4(0.3)^3 + \cdot\cdot\cdot + 4(0.3)^{12}$$

$$= 4(0.3) + [4(0.3)](0.3) + [4(0.3)](0.3)^2 + \cdot\cdot\cdot + [4(0.3)](0.3)^{11}$$

$$= 4(0.3)\left[\frac{1-(0.3)^{12}}{1-0.3}\right] \qquad a_1 = 4(0.3),\ r = 0.3,\ n = 12$$

$$\approx 1.714$$

Geometric Series

The summation of the terms of an infinite geometric *sequence* is called an **infinite geometric series** or simply a **geometric series.**

The formula for the sum of a *finite* geometric *sequence* can, depending on the value of r, be extended to produce a formula for the sum of an *infinite* geometric *series.* Specifically, if the common ratio r has the property that $|r| < 1$, then it can be shown that r^n approaches zero as n increases without bound. Consequently,

$$a_1\left(\frac{1 - r^n}{1 - r}\right) \longrightarrow a_1\left(\frac{1 - 0}{1 - r}\right) \quad \text{as} \quad n \longrightarrow \infty.$$

The following summarizes this result.

> **The Sum of an Infinite Geometric Series**
>
> If $|r| < 1$, then the infinite geometric series
>
> $$a_1 + a_1 r + a_1 r^2 + a_1 r^3 + \cdots + a_1 r^{n-1} + \cdots$$
>
> has the sum
>
> $$S = \sum_{i=0}^{\infty} a_1 r^i = \frac{a_1}{1 - r}.$$

Note that when $|r| \geq 1$, the series does not have a sum.

EXAMPLE 7 Finding the Sum of an Infinite Geometric Series

Find each sum.

a. $\displaystyle\sum_{n=0}^{\infty} 4(0.6)^n$

b. $3 + 0.3 + 0.03 + 0.003 + \cdots$

Solution

a.
$$\sum_{n=0}^{\infty} 4(0.6)^n = 4 + 4(0.6) + 4(0.6)^2 + 4(0.6)^3 + \cdots + 4(0.6)^n + \cdots$$
$$= \frac{4}{1 - 0.6} \qquad \frac{a_1}{1 - r}$$
$$= 10$$

b.
$$3 + 0.3 + 0.03 + 0.003 + \cdots = 3 + 3(0.1) + 3(0.1)^2 + 3(0.1)^3 + \cdots$$
$$= \frac{3}{1 - 0.1} \qquad \frac{a_1}{1 - r}$$
$$= \frac{10}{3}$$
$$\approx 3.33$$

Checkpoint *Audio-video solution in English & Spanish at LarsonPrecalculus.com.*

Find each sum.

a. $\displaystyle\sum_{n=0}^{\infty} 5(0.5)^n$

b. $5 + 1 + 0.2 + 0.04 + \cdots$

Application

EXAMPLE 8 Increasing Annuity

An investor deposits \$50 on the first day of each month in an account that pays 3% interest, compounded monthly. What is the balance at the end of 2 years? (This type of investment plan is called an **increasing annuity.**)

Solution To find the balance in the account after 24 months, consider each of the 24 deposits separately. The first deposit will gain interest for 24 months, and its balance will be

$$A_{24} = 50\left(1 + \frac{0.03}{12}\right)^{24}$$

$$= 50(1.0025)^{24}.$$

The second deposit will gain interest for 23 months, and its balance will be

$$A_{23} = 50\left(1 + \frac{0.03}{12}\right)^{23}$$

$$= 50(1.0025)^{23}.$$

The last deposit will gain interest for only 1 month, and its balance will be

$$A_1 = 50\left(1 + \frac{0.03}{12}\right)^{1}$$

$$= 50(1.0025).$$

The total balance in the annuity will be the sum of the balances of the 24 deposits. Using the formula for the sum of a finite geometric sequence, with $A_1 = 50(1.0025)$, $r = 1.0025$, and $n = 24$, you have

$$S_n = A_1\left(\frac{1 - r^n}{1 - r}\right) \quad \text{Sum of a finite geometric sequence}$$

$$S_{24} = 50(1.0025)\left[\frac{1 - (1.0025)^{24}}{1 - 1.0025}\right] \quad \text{Substitute } 50(1.0025) \text{ for } A_1,\ 1.0025 \text{ for } r, \text{ and } 24 \text{ for } n.$$

$$\approx \$1238.23. \quad \text{Use a calculator.}$$

REMARK Recall from Section 3.1 that the formula for compound interest (for n compoundings per year) is

$$A = P\left(1 + \frac{r}{n}\right)^{nt}.$$

So, in Example 8, \$50 is the principal P, 0.03 is the annual interest rate r, 12 is the number n of compoundings per year, and 2 is the time t in years. When you substitute these values into the formula, you obtain

$$A = 50\left(1 + \frac{0.03}{12}\right)^{12(2)}$$

$$= 50\left(1 + \frac{0.03}{12}\right)^{24}.$$

✓ **Checkpoint** *Audio-video solution in English & Spanish at LarsonPrecalculus.com.*

An investor deposits \$70 on the first day of each month in an account that pays 2% interest, compounded monthly. What is the balance at the end of 4 years?

Summarize (Section 9.3)

1. State the definition of a geometric sequence *(page 625)* and state the formula for the nth term of a geometric sequence *(page 626)*. For examples of recognizing, writing, and finding the nth terms of geometric sequences, see Examples 1–5.
2. State the formula for the sum of a finite geometric sequence *(page 628)*. For an example of finding the sum of a finite geometric sequence, see Example 6.
3. State the formula for the sum of an infinite geometric series *(page 629)*. For an example of finding the sums of infinite geometric series, see Example 7.
4. Describe an example of how to use a geometric series to model and solve a real-life problem *(page 630, Example 8)*.

9.3 Exercises

See CalcChat.com for tutorial help and worked-out solutions to odd-numbered exercises.

Vocabulary: Fill in the blanks.

1. A sequence is called a ________ sequence when the ratios of consecutive terms are the same. This ratio is called the ________ ratio.
2. The nth term of a geometric sequence has the form ________.
3. The sum of a finite geometric sequence with common ratio $r \neq 1$ is given by ________.
4. The sum of the terms of an infinite geometric sequence is called a ________ ________.

Skills and Applications

Determining Whether a Sequence Is Geometric In Exercises 5–12, determine whether the sequence is geometric. If so, then find the common ratio.

5. $2, 10, 50, 250, \ldots$
6. $3, 12, 21, 30, \ldots$
7. $\frac{1}{8}, \frac{1}{4}, \frac{1}{2}, 1, \ldots$
8. $9, -6, 4, -\frac{8}{3}, \ldots$
9. $1, \frac{1}{2}, \frac{1}{3}, \frac{1}{4}, \ldots$
10. $5, 1, 0.2, 0.04, \ldots$
11. $1, -\sqrt{7}, 7, -7\sqrt{7}, \ldots$
12. $2, \dfrac{4}{\sqrt{3}}, \dfrac{8}{3}, \dfrac{16}{3\sqrt{3}}, \ldots$

Writing the Terms of a Geometric Sequence In Exercises 13–22, write the first five terms of the geometric sequence.

13. $a_1 = 4, r = 3$
14. $a_1 = 8, r = 2$
15. $a_1 = 1, r = \frac{1}{2}$
16. $a_1 = 6, r = -\frac{1}{4}$
17. $a_1 = 1, r = e$
18. $a_1 = 2, r = \pi$
19. $a_1 = 3, r = \sqrt{5}$
20. $a_1 = 4, r = -\dfrac{1}{\sqrt{2}}$
21. $a_1 = 2, r = \dfrac{x}{4}$
22. $a_1 = 5, r = 2x$

Writing the *n*th Term of a Geometric Sequence In Exercises 23–28, write the first five terms of the geometric sequence. Determine the common ratio and write the nth term of the sequence as a function of n.

23. $a_1 = 64, \quad a_{k+1} = \frac{1}{2}a_k$
24. $a_1 = 81, \quad a_{k+1} = \frac{1}{3}a_k$
25. $a_1 = 9, \quad a_{k+1} = 2a_k$
26. $a_1 = 5, \quad a_{k+1} = -2a_k$
27. $a_1 = 6, \quad a_{k+1} = -\frac{3}{2}a_k$
28. $a_1 = 80, \quad a_{k+1} = -\frac{1}{2}a_k$

Finding a Term of a Geometric Sequence In Exercises 29–38, write an expression for the nth term of the geometric sequence. Then find the indicated term.

29. $a_1 = 4, r = \frac{1}{2}, n = 10$
30. $a_1 = 5, r = \frac{7}{2}, n = 8$
31. $a_1 = 6, r = -\frac{1}{3}, n = 12$
32. $a_1 = 64, r = -\frac{1}{4}, n = 10$
33. $a_1 = 100, r = e^x, n = 9$
34. $a_1 = 1, r = e^{-x}, n = 4$
35. $a_1 = 1, r = \sqrt{2}, n = 12$
36. $a_1 = 1, r = \sqrt{3}, n = 8$
37. $a_1 = 500, r = 1.02, n = 40$
38. $a_1 = 1000, r = 1.005, n = 60$

Finding a Term of a Geometric Sequence In Exercises 39–46, find the indicated term of the geometric sequence.

39. 9th term: $11, 33, 99, \ldots$
40. 7th term: $3, 36, 432, \ldots$
41. 8th term: $\frac{1}{2}, -\frac{1}{8}, \frac{1}{32}, -\frac{1}{128}, \ldots$
42. 7th term: $\frac{8}{5}, -\frac{16}{25}, \frac{32}{125}, -\frac{64}{625}, \ldots$
43. 3rd term: $a_1 = 16, a_4 = \frac{27}{4}$
44. 1st term: $a_2 = 3, a_5 = \frac{3}{64}$
45. 6th term: $a_4 = -18, a_7 = \frac{2}{3}$
46. 5th term: $a_2 = 2, a_3 = -\sqrt{2}$

Matching a Geometric Sequence with Its Graph In Exercises 47–50, match the geometric sequence with its graph. [The graphs are labeled (a), (b), (c), and (d).]

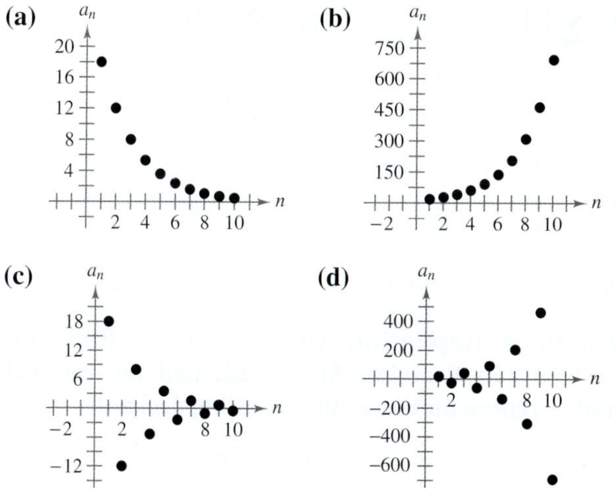

47. $a_n = 18\left(\frac{2}{3}\right)^{n-1}$
48. $a_n = 18\left(-\frac{2}{3}\right)^{n-1}$
49. $a_n = 18\left(\frac{3}{2}\right)^{n-1}$
50. $a_n = 18\left(-\frac{3}{2}\right)^{n-1}$

Graphing the Terms of a Sequence In Exercises 51–54, use a graphing utility to graph the first 10 terms of the sequence.

51. $a_n = 10(1.5)^{n-1}$
52. $a_n = 12(-0.4)^{n-1}$
53. $a_n = 20(-1.25)^{n-1}$
54. $a_n = 2(1.3)^{n-1}$

Sum of a Finite Geometric Sequence **In Exercises 55–68, find the sum of the finite geometric sequence.**

55. $\sum_{n=1}^{7} 4^{n-1}$

56. $\sum_{n=1}^{10} \left(\frac{3}{2}\right)^{n-1}$

57. $\sum_{n=1}^{6} (-7)^{n-1}$

58. $\sum_{n=1}^{8} 5\left(-\frac{5}{2}\right)^{n-1}$

59. $\sum_{i=1}^{7} 64\left(-\frac{1}{2}\right)^{i-1}$

60. $\sum_{i=1}^{12} 16\left(\frac{1}{2}\right)^{i-1}$

61. $\sum_{n=0}^{20} 3\left(\frac{3}{2}\right)^{n}$

62. $\sum_{n=0}^{40} 5\left(\frac{3}{5}\right)^{n}$

63. $\sum_{n=0}^{20} 10\left(\frac{1}{5}\right)^{n}$

64. $\sum_{n=0}^{6} 500(1.04)^{n}$

65. $\sum_{n=0}^{40} 2\left(-\frac{1}{4}\right)^{n}$

66. $\sum_{n=0}^{50} 10\left(\frac{2}{3}\right)^{n-1}$

67. $\sum_{i=1}^{100} 15\left(\frac{2}{3}\right)^{i-1}$

68. $\sum_{i=1}^{10} 8\left(-\frac{1}{4}\right)^{i-1}$

Using Summation Notation **In Exercises 69–72, use summation notation to write the sum.**

69. $10 + 30 + 90 + \cdots + 7290$

70. $15 - 3 + \frac{3}{5} - \cdots - \frac{3}{625}$

71. $0.1 + 0.4 + 1.6 + \cdots + 102.4$

72. $32 + 24 + 18 + \cdots + 10.125$

Sum of an Infinite Geometric Series **In Exercises 73–80, find the sum of the infinite geometric series.**

73. $\sum_{n=0}^{\infty} \left(\frac{1}{2}\right)^{n}$

74. $\sum_{n=0}^{\infty} \left(-\frac{1}{2}\right)^{n}$

75. $\sum_{n=0}^{\infty} 2\left(-\frac{2}{3}\right)^{n}$

76. $\sum_{n=0}^{\infty} 4(0.2)^{n}$

77. $8 + 6 + \frac{9}{2} + \frac{27}{8} + \cdots$

78. $9 + 6 + 4 + \frac{8}{3} + \cdots$

79. $\frac{1}{9} - \frac{1}{3} + 1 - 3 + \cdots$

80. $-\frac{125}{36} + \frac{25}{6} - 5 + 6 - \cdots$

Writing a Repeating Decimal as a Rational Number **In Exercises 81 and 82, find the rational number representation of the repeating decimal.**

81. $0.\overline{36}$

82. $0.3\overline{18}$

Graphical Reasoning **In Exercises 83 and 84, use a graphing utility to graph the function. Identify the horizontal asymptote of the graph and determine its relationship to the sum.**

83. $f(x) = 6\left[\frac{1 - (0.5)^x}{1 - (0.5)}\right], \quad \sum_{n=0}^{\infty} 6\left(\frac{1}{2}\right)^n$

84. $f(x) = 2\left[\frac{1 - (0.8)^x}{1 - (0.8)}\right], \quad \sum_{n=0}^{\infty} 2\left(\frac{4}{5}\right)^n$

85. **Depreciation** A tool and die company buys a machine for \$175,000 and it depreciates at a rate of 30% per year. (In other words, at the end of each year the depreciated value is 70% of what it was at the beginning of the year.) Find the depreciated value of the machine after 5 full years.

86. **Data Analysis: Population** The table shows the mid-year populations a_n of China (in millions) from 2004 through 2010. *(Source: U.S. Census Bureau)*

Spreadsheet at LarsonPrecalculus.com

Year	Population, a_n
2004	1291.0
2005	1297.8
2006	1304.3
2007	1310.6
2008	1317.1
2009	1323.6
2010	1330.1

(a) Use the *exponential regression* feature of a graphing utility to find a geometric sequence that models the data. Let n represent the year, with $n = 4$ corresponding to 2004.

(b) Use the sequence from part (a) to describe the rate at which the population of China is growing.

(c) Use the sequence from part (a) to predict the population of China in 2017. The U.S. Census Bureau predicts the population of China will be 1372.1 million in 2017. How does this value compare with your prediction?

(d) Use the sequence from part (a) to determine when the population of China will reach 1.4 billion.

87. **Annuity** An investor deposits P dollars on the first day of each month in an account with an annual interest rate r, compounded monthly. The balance A after t years is

$$A = P\left(1 + \frac{r}{12}\right) + P\left(1 + \frac{r}{12}\right)^2 + \cdots + P\left(1 + \frac{r}{12}\right)^{12t}.$$

Show that the balance is

$$A = P\left[\left(1 + \frac{r}{12}\right)^{12t} - 1\right]\left(1 + \frac{12}{r}\right).$$

TonyV3112/Shutterstock.com

88. Annuity An investor deposits \$100 on the first day of each month in an account that pays 2% interest, compounded monthly. The balance A in the account at the end of 5 years is

$$A = 100\left(1 + \frac{0.02}{12}\right)^1 + \cdot \cdot \cdot + 100\left(1 + \frac{0.02}{12}\right)^{60}.$$

Use the result of Exercise 87 to find A.

Multiplier Effect In Exercises 89 and 90, use the following information. A state government gives property owners a tax rebate with the anticipation that each property owner will spend approximately $p\%$ of the rebate, and in turn each recipient of this amount will spend $p\%$ of what he or she receives, and so on. Economists refer to this exchange of money and its circulation within the economy as the "multiplier effect." The multiplier effect operates on the idea that the expenditures of one individual become the income of another individual. For the given tax rebate, find the total amount put back into the state's economy, assuming that this effect continues without end.

	Tax rebate	$p\%$
89.	\$400	75%
90.	\$600	72.5%

91. Geometry The sides of a square are 27 inches in length. New squares are formed by dividing the original square into nine squares. The center square is then shaded (see figure). If this process is repeated three more times, then determine the total area of the shaded region.

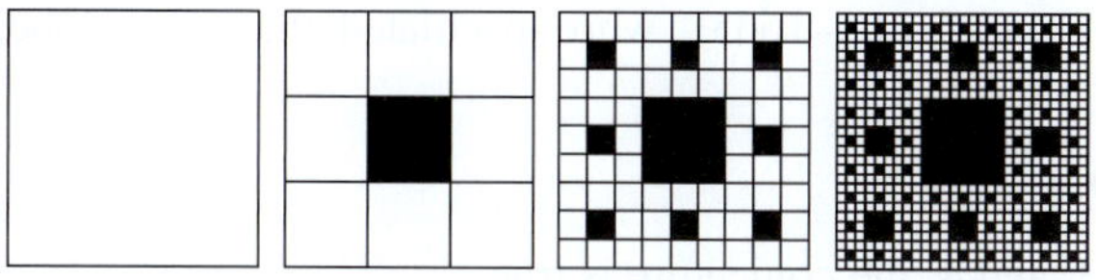

92. Distance A ball is dropped from a height of 6 feet and begins bouncing as shown in the figure. The height of each bounce is three-fourths the height of the previous bounce. Find the total vertical distance the ball travels before coming to rest.

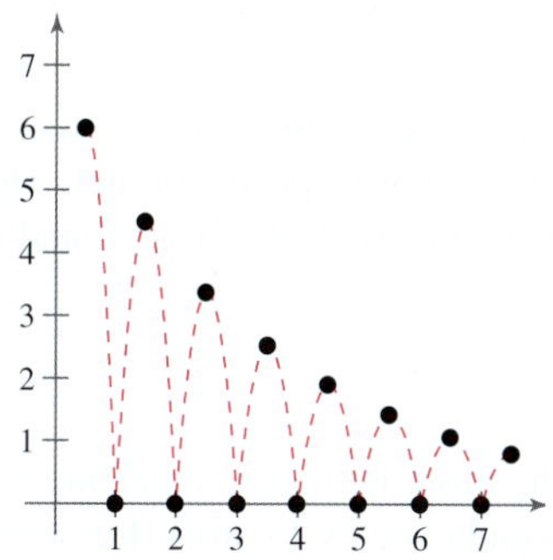

93. Salary An investment firm has a job opening with a salary of \$45,000 for the first year. During the next 39 years, there is a 5% raise each year. Find the total compensation over the 40-year period.

Exploration

94. HOW DO YOU SEE IT? Use the figures shown below.

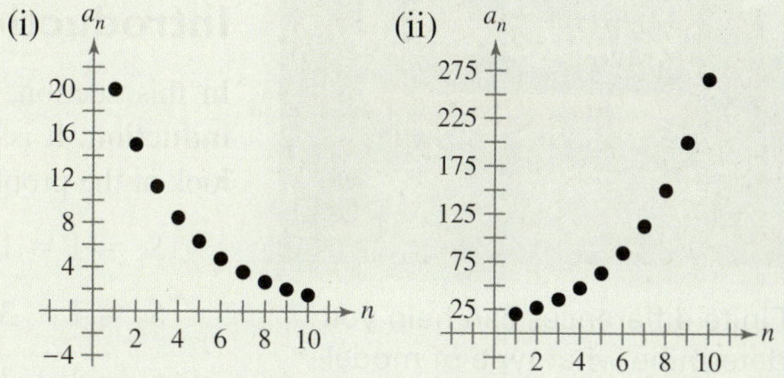

(a) Without performing any calculations, determine which figure shows the terms of the sequence $a_n = 20\left(\frac{4}{3}\right)^{n-1}$ and which shows the terms of $a_n = 20\left(\frac{3}{4}\right)^{n-1}$. Explain your reasoning.

(b) Which sequence has terms that can be summed? Explain your reasoning.

True or False? In Exercises 95 and 96, determine whether the statement is true or false. Justify your answer.

95. A sequence is geometric when the ratios of consecutive differences of consecutive terms are the same.

96. To find the nth term of a geometric sequence, multiply its common ratio by the first term of the sequence raised to the $(n - 1)$th power.

97. Graphical Reasoning Consider the graph of

$$y = \left(\frac{1 - r^x}{1 - r}\right).$$

(a) Use a graphing utility to graph y for $r = \frac{1}{2}, \frac{2}{3}$, and $\frac{4}{5}$. What happens as $x \to \infty$?

(b) Use the graphing utility to graph y for $r = 1.5$, 2, and 3. What happens as $x \to \infty$?

98. Writing Write a brief paragraph explaining why the terms of a geometric sequence decrease in magnitude when $-1 < r < 1$.

Project: Housing Vacancies To work an extended application analyzing the numbers of vacant houses in the United States from 1990 through 2011, visit this text's website at *LarsonPrecalculus.com.* *(Source: U.S. Census Bureau)*

9.4 Mathematical Induction

Finite differences can help you determine what type of model to use to represent a sequence. For instance, in Exercise 75 on page 643, you will use finite differences to find a model that represents the numbers of Alaskan residents from 2005 through 2010.

- **Use mathematical induction to prove statements involving a positive integer *n*.**
- **Use pattern recognition and mathematical induction to write a formula for the *n*th term of a sequence.**
- **Find the sums of powers of integers.**
- **Find finite differences of sequences.**

Introduction

In this section, you will study a form of mathematical proof called **mathematical induction.** It is important that you see clearly the logical need for it, so take a closer look at the problem discussed in Example 5 in Section 9.2.

$$S_1 = 1 = 1^2$$

$$S_2 = 1 + 3 = 2^2$$

$$S_3 = 1 + 3 + 5 = 3^2$$

$$S_4 = 1 + 3 + 5 + 7 = 4^2$$

$$S_5 = 1 + 3 + 5 + 7 + 9 = 5^2$$

$$S_6 = 1 + 3 + 5 + 7 + 9 + 11 = 6^2$$

Judging from the pattern formed by these first six sums, it appears that the sum of the first n odd integers is

$$S_n = 1 + 3 + 5 + 7 + 9 + 11 + \cdot \cdot \cdot + (2n - 1) = n^2.$$

Although this particular formula *is* valid, it is important for you to see that recognizing a pattern and then simply *jumping to the conclusion* that the pattern must be true for all values of n is *not* a logically valid method of proof. There are many examples in which a pattern appears to be developing for small values of n, but then at some point the pattern fails. One of the most famous cases of this was the conjecture by the French mathematician Pierre de Fermat (1601–1665), who speculated that all numbers of the form

$$F_n = 2^{2^n} + 1, \quad n = 0, 1, 2, \ldots$$

are prime. For $n = 0, 1, 2, 3,$ and 4, the conjecture is true.

$$F_0 = 3$$

$$F_1 = 5$$

$$F_2 = 17$$

$$F_3 = 257$$

$$F_4 = 65{,}537$$

The size of the next Fermat number ($F_5 = 4{,}294{,}967{,}297$) is so great that it was difficult for Fermat to determine whether it was prime or not. However, another well-known mathematician, Leonhard Euler (1707–1783), later found the factorization

$$F_5 = 4{,}294{,}967{,}297$$

$$= 641(6{,}700{,}417)$$

which proved that F_5 is not prime and therefore Fermat's conjecture was false.

Just because a rule, pattern, or formula seems to work for several values of n, you cannot simply decide that it is valid for all values of n without going through a *legitimate proof.* Mathematical induction is one method of proof.

Mark Herreid/Shutterstock.com

REMARK It is important to recognize that in order to prove a statement by induction, both parts of the Principle of Mathematical Induction are necessary.

The Principle of Mathematical Induction

Let P_n be a statement involving the positive integer n. If

1. P_1 is true, and
2. for every positive integer k, the truth of P_k implies the truth of P_{k+1}

then the statement P_n must be true for all positive integers n.

To apply the Principle of Mathematical Induction, you need to be able to determine the statement P_{k+1} for a given statement P_k. To determine P_{k+1}, substitute the quantity $k + 1$ for k in the statement P_k.

EXAMPLE 1 A Preliminary Example

Find the statement P_{k+1} for each given statement P_k.

a. $P_k:\ S_k = \dfrac{k^2(k+1)^2}{4}$

b. $P_k:\ S_k = 1 + 5 + 9 + \cdot \cdot \cdot + [4(k-1) - 3] + (4k - 3)$

c. $P_k:\ k + 3 < 5k^2$

d. $P_k:\ 3^k \geq 2k + 1$

Solution

a. $P_{k+1}:\ S_{k+1} = \dfrac{(k+1)^2(k+1+1)^2}{4}$ Replace k with $k + 1$.

$= \dfrac{(k+1)^2(k+2)^2}{4}$ Simplify.

b. $P_{k+1}:\ S_{k+1} = 1 + 5 + 9 + \cdot \cdot \cdot + \{4[(k+1) - 1] - 3\} + [4(k+1) - 3]$

$= 1 + 5 + 9 + \cdot \cdot \cdot + (4k - 3) + (4k + 1)$

c. $P_{k+1}:\ (k+1) + 3 < 5(k+1)^2$

$k + 4 < 5(k^2 + 2k + 1)$

d. $P_{k+1}:\ 3^{k+1} \geq 2(k+1) + 1$

$3^{k+1} \geq 2k + 3$

✓ **Checkpoint** Audio-video solution in English & Spanish at LarsonPrecalculus.com.

Find the statement P_{k+1} for each given statement P_k.

a. $P_k: S_k = \dfrac{6}{k(k+3)}$ **b.** $P_k: k + 2 \leq 3(k-1)^2$ **c.** $P_k: 2^{4k-2} + 1 > 5k$

An unending line of dominoes can illustrate why the Principle of Mathematical Induction works.

A well-known illustration used to explain why the Principle of Mathematical Induction works is that of an unending line of dominoes. If the line actually contains infinitely many dominoes, then it is clear that you could not knock the entire line down by knocking down only *one domino* at a time. However, suppose it were true that each domino would knock down the next one as it fell. Then you could knock them all down by pushing the first one and starting a chain reaction. Mathematical induction works in the same way. If the truth of P_k implies the truth of P_{k+1} and if P_1 is true, then the chain reaction proceeds as follows: P_1 implies P_2, P_2 implies P_3, P_3 implies P_4, and so on.

© mark wragg/istockphoto.com

When using mathematical induction to prove a *summation* formula (such as the one in Example 2), it is helpful to think of S_{k+1} as $S_{k+1} = S_k + a_{k+1}$, where a_{k+1} is the $(k + 1)$th term of the original sum.

EXAMPLE 2 Using Mathematical Induction

Use mathematical induction to prove the formula

$$\begin{aligned} S_n &= 1 + 3 + 5 + 7 + \cdot \cdot \cdot + (2n - 1) \\ &= n^2 \end{aligned}$$

for all integers $n \geq 1$.

Solution Mathematical induction consists of two distinct parts.

1. First, you must show that the formula is true when $n = 1$. When $n = 1$, the formula is valid, because

$$\begin{aligned} S_1 &= 1 \\ &= 1^2. \end{aligned}$$

2. The second part of mathematical induction has two steps. The first step is to *assume* that the formula is valid for some integer k. The second step is to use this assumption to prove that the formula is valid for the *next* integer, $k + 1$. Assuming that the formula

$$\begin{aligned} S_k &= 1 + 3 + 5 + 7 + \cdot \cdot \cdot + (2k - 1) \\ &= k^2 \end{aligned}$$

is true, you must show that the formula $S_{k+1} = (k + 1)^2$ is true.

$$\begin{aligned} S_{k+1} &= 1 + 3 + 5 + 7 + \cdot \cdot \cdot + (2k - 1) \\ &\qquad + [2(k + 1) - 1] && S_{k+1} = S_k + a_{k+1} \\ &= [1 + 3 + 5 + 7 + \cdot \cdot \cdot + (2k - 1)] + (2k + 2 - 1) \\ &= S_k + (2k + 1) && \text{Group terms to form } S_k. \\ &= k^2 + 2k + 1 && \text{Substitute } k^2 \text{ for } S_k. \\ &= (k + 1)^2 \end{aligned}$$

Combining the results of parts (1) and (2), you can conclude by mathematical induction that the formula is valid for all integers $n \geq 1$.

✓ Checkpoint *Audio-video solution in English & Spanish at LarsonPrecalculus.com.*

Use mathematical induction to prove the formula

$$S_n = 5 + 7 + 9 + 11 + \cdot \cdot \cdot + (2n + 3) = n(n + 4)$$

for all integers $n \geq 1$.

It occasionally happens that a statement involving natural numbers is not true for the first $k - 1$ positive integers but is true for all values of $n \geq k$. In these instances, you use a slight variation of the Principle of Mathematical Induction in which you verify P_k rather than P_1. This variation is called the *Extended Principle of Mathematical Induction.* To see the validity of this, note in the unending line of dominoes discussed on page 635 that all but the first $k - 1$ dominoes can be knocked down by knocking over the kth domino. This suggests that you can prove a statement P_n to be true for $n \geq k$ by showing that P_k is true and that P_k implies P_{k+1}. In Exercises 25–30 of this section, you will apply this extension of mathematical induction.

EXAMPLE 3 Using Mathematical Induction

Use mathematical induction to prove the formula

$$S_n = 1^2 + 2^2 + 3^2 + 4^2 + \cdots + n^2$$
$$= \frac{n(n+1)(2n+1)}{6}$$

for all integers $n \geq 1$.

Solution

1. When $n = 1$, the formula is valid, because

$$S_1 = 1^2 = \frac{1(2)(3)}{6}.$$

2. Assuming that

$$S_k = 1^2 + 2^2 + 3^2 + 4^2 + \cdots + k^2 = \frac{k(k+1)(2k+1)}{6} \qquad a_k = k^2$$

you must show that

$$S_{k+1} = \frac{(k+1)(k+1+1)[2(k+1)+1]}{6} = \frac{(k+1)(k+2)(2k+3)}{6}.$$

To do this, write the following.

REMARK Remember that when adding rational expressions, you must first find a common denominator. Example 3 uses the *least* common denominator of 6.

$$S_{k+1} = S_k + a_{k+1}$$

$$= (1^2 + 2^2 + 3^2 + 4^2 + \cdots + k^2) + (k+1)^2 \qquad \text{Substitute for } S_k,\ a_{k+1} = (k+1)^2.$$

$$= \frac{k(k+1)(2k+1)}{6} + (k+1)^2 \qquad \text{By assumption}$$

$$= \frac{k(k+1)(2k+1) + 6(k+1)^2}{6} \qquad \text{Combine fractions.}$$

$$= \frac{(k+1)[k(2k+1) + 6(k+1)]}{6} \qquad \text{Factor.}$$

$$= \frac{(k+1)(2k^2 + 7k + 6)}{6} \qquad \text{Simplify.}$$

$$= \frac{(k+1)(k+2)(2k+3)}{6} \qquad S_k \text{ implies } S_{k+1}.$$

Combining the results of parts (1) and (2), you can conclude by mathematical induction that the formula is valid for all positive integers n.

Checkpoint Audio-video solution in English & Spanish at LarsonPrecalculus.com.

Use mathematical induction to prove the formula

$$S_n = 1(1-1) + 2(2-1) + 3(3-1) + \cdots + n(n-1) = \frac{n(n-1)(n+1)}{3}$$

for all integers $n \geq 1$.

When proving a formula using mathematical induction, the only statement that you *need* to verify is P_1. As a check, however, it is a good idea to try verifying some of the other statements. For instance, in Example 3, try verifying P_2 and P_3.

EXAMPLE 4 Proving an Inequality

Prove that

$$n < 2^n$$

for all positive integers n.

Solution

1. For $n = 1$ and $n = 2$, the statement is true because

$$1 < 2^1 \quad \text{and} \quad 2 < 2^2.$$

2. Assuming that

$$k < 2^k$$

you need to show that $k + 1 < 2^{k+1}$. For $n = k$, you have

$$2^{k+1} = 2(2^k) > 2(k) = 2k. \qquad \text{By assumption}$$

Because $2k = k + k > k + 1$ for all $k > 1$, it follows that

$$2^{k+1} > 2k > k + 1 \quad \text{or} \quad k + 1 < 2^{k+1}.$$

Combining the results of parts (1) and (2), you can conclude by mathematical induction that $n < 2^n$ for all integers $n \geq 1$.

REMARK To check a result that you have proved by mathematical induction, it helps to list the statement for several values of n. For instance, in Example 4, you could list

$$1 < 2^1 = 2, \quad 2 < 2^2 = 4,$$
$$3 < 2^3 = 8, \quad 4 < 2^4 = 16,$$
$$5 < 2^5 = 32, \quad 6 < 2^6 = 64.$$

From this list, your intuition confirms that the statement $n < 2^n$ is reasonable.

✓ **Checkpoint** Audio-video solution in English & Spanish at LarsonPrecalculus.com.

Prove that

$$n! \geq n$$

for all positive integers n.

EXAMPLE 5 Proving a Factor

Prove that 3 is a factor of $4^n - 1$ for all positive integers n.

Solution

1. For $n = 1$, the statement is true because

$$4^1 - 1 = 3.$$

So, 3 is a factor.

2. Assuming that 3 is a factor of $4^k - 1$, you must show that 3 is a factor of $4^{k+1} - 1$. To do this, write the following.

$$4^{k+1} - 1 = 4^{k+1} - 4^k + 4^k - 1 \qquad \text{Subtract and add } 4^k.$$
$$= 4^k(4 - 1) + (4^k - 1) \qquad \text{Regroup terms.}$$
$$= 4^k \cdot 3 + (4^k - 1) \qquad \text{Simplify.}$$

Because 3 is a factor of $4^k \cdot 3$ and 3 is also a factor of $4^k - 1$, it follows that 3 is a factor of $4^{k+1} - 1$. Combining the results of parts (1) and (2), you can conclude by mathematical induction that 3 is a factor of $4^n - 1$ for all positive integers n.

✓ **Checkpoint** Audio-video solution in English & Spanish at LarsonPrecalculus.com.

Prove that 2 is a factor of $3^n + 1$ for all positive integers n.

Pattern Recognition

Although choosing a formula on the basis of a few observations does *not* guarantee the validity of the formula, pattern recognition *is* important. Once you have a pattern or formula that you think works, try using mathematical induction to prove your formula.

> **Finding a Formula for the *n*th Term of a Sequence**
>
> To find a formula for the nth term of a sequence, consider these guidelines.
>
> **1.** Calculate the first several terms of the sequence. It is often a good idea to write the terms in both simplified and factored forms.
>
> **2.** Try to find a recognizable pattern for the terms and write a formula for the nth term of the sequence. This is your *hypothesis* or *conjecture*. You might try computing one or two more terms in the sequence to test your hypothesis.
>
> **3.** Use mathematical induction to prove your hypothesis.

EXAMPLE 6 Finding a Formula for a Finite Sum

Find a formula for the finite sum and prove its validity.

$$\frac{1}{1\cdot 2}+\frac{1}{2\cdot 3}+\frac{1}{3\cdot 4}+\frac{1}{4\cdot 5}+\cdot\cdot\cdot+\frac{1}{n(n+1)}$$

Solution Begin by writing the first few sums.

$$S_1=\frac{1}{1\cdot 2}=\frac{1}{2}=\frac{1}{1+1}$$

$$S_2=\frac{1}{1\cdot 2}+\frac{1}{2\cdot 3}=\frac{4}{6}=\frac{2}{3}=\frac{2}{2+1}$$

$$S_3=\frac{1}{1\cdot 2}+\frac{1}{2\cdot 3}+\frac{1}{3\cdot 4}=\frac{9}{12}=\frac{3}{4}=\frac{3}{3+1}$$

From this sequence, it appears that the formula for the kth sum is

$$S_k=\frac{1}{1\cdot 2}+\frac{1}{2\cdot 3}+\frac{1}{3\cdot 4}+\frac{1}{4\cdot 5}+\cdot\cdot\cdot+\frac{1}{k(k+1)}=\frac{k}{k+1}.$$

To prove the validity of this hypothesis, use mathematical induction. Note that you have already verified the formula for $n = 1$, so begin by assuming that the formula is valid for $n = k$ and trying to show that it is valid for $n = k + 1$.

$$S_{k+1}=\left[\frac{1}{1\cdot 2}+\frac{1}{2\cdot 3}+\frac{1}{3\cdot 4}+\frac{1}{4\cdot 5}+\cdot\cdot\cdot+\frac{1}{k(k+1)}\right]+\frac{1}{(k+1)(k+2)}$$

$$=\frac{k}{k+1}+\frac{1}{(k+1)(k+2)} \qquad \text{By assumption}$$

$$=\frac{k(k+2)+1}{(k+1)(k+2)}=\frac{k^2+2k+1}{(k+1)(k+2)}=\frac{(k+1)^2}{(k+1)(k+2)}=\frac{k+1}{k+2}$$

So, by mathematical induction the hypothesis is valid.

✓ **Checkpoint** *Audio-video solution in English & Spanish at LarsonPrecalculus.com.*

Find a formula for the finite sum and prove its validity.

$$3+7+11+15+\cdot\cdot\cdot+4n-1$$

Sums of Powers of Integers

The formula in Example 3 is one of a collection of useful summation formulas. This and other formulas dealing with the sums of various powers of the first n positive integers are as follows.

Sums of Powers of Integers

1. $1 + 2 + 3 + 4 + \cdots + n = \dfrac{n(n+1)}{2}$

2. $1^2 + 2^2 + 3^2 + 4^2 + \cdots + n^2 = \dfrac{n(n+1)(2n+1)}{6}$

3. $1^3 + 2^3 + 3^3 + 4^3 + \cdots + n^3 = \dfrac{n^2(n+1)^2}{4}$

4. $1^4 + 2^4 + 3^4 + 4^4 + \cdots + n^4 = \dfrac{n(n+1)(2n+1)(3n^2+3n-1)}{30}$

5. $1^5 + 2^5 + 3^5 + 4^5 + \cdots + n^5 = \dfrac{n^2(n+1)^2(2n^2+2n-1)}{12}$

EXAMPLE 7 Finding Sums

Find each sum.

a. $\displaystyle\sum_{i=1}^{7} i^3 = 1^3 + 2^3 + 3^3 + 4^3 + 5^3 + 6^3 + 7^3$ **b.** $\displaystyle\sum_{i=1}^{4} (6i - 4i^2)$

Solution

a. Using the formula for the sum of the cubes of the first n positive integers, you obtain

$$\begin{aligned}\sum_{i=1}^{7} i^3 &= 1^3 + 2^3 + 3^3 + 4^3 + 5^3 + 6^3 + 7^3 \\ &= \frac{7^2(7+1)^2}{4} && \text{Formula 3} \\ &= \frac{49(64)}{4} \\ &= 784.\end{aligned}$$

b.
$$\begin{aligned}\sum_{i=1}^{4} (6i - 4i^2) &= \sum_{i=1}^{4} 6i - \sum_{i=1}^{4} 4i^2 \\ &= 6\sum_{i=1}^{4} i - 4\sum_{i=1}^{4} i^2 \\ &= 6\left[\frac{4(4+1)}{2}\right] - 4\left[\frac{4(4+1)(2 \cdot 4 + 1)}{6}\right] && \text{Formulas 1 and 2} \\ &= 6(10) - 4(30) \\ &= -60\end{aligned}$$

✓ **Checkpoint** *Audio-video solution in English & Spanish at LarsonPrecalculus.com.*

Find each sum.

a. $\displaystyle\sum_{i=1}^{20} i$ **b.** $\displaystyle\sum_{i=1}^{5} (2i^2 + 3i^3)$

Finite Differences

The **first differences** of a sequence are found by subtracting consecutive terms. The **second differences** are found by subtracting consecutive first differences. The first and second differences of the sequence 3, 5, 8, 12, 17, 23, . . . are as follows.

n:	1		2		3		4		5		6
a_n:	3		5		8		12		17		23
First differences:		2		3		4		5		6	
Second differences:			1		1		1		1		

For this sequence, the second differences are all the same. When this happens, the sequence has a perfect *quadratic* model. When the first differences are all the same, the sequence has a perfect *linear* model. That is, it is arithmetic.

EXAMPLE 8 Finding a Quadratic Model

Find the quadratic model for the sequence

3, 5, 8, 12, 17, 23, ...

Solution You know from the second differences shown above that the model is quadratic and has the form $a_n = an^2 + bn + c$. By substituting 1, 2, and 3 for n, you can obtain a system of three linear equations in three variables.

$$a_1 = a(1)^2 + b(1) + c = 3 \qquad \text{Substitute 1 for } n.$$

$$a_2 = a(2)^2 + b(2) + c = 5 \qquad \text{Substitute 2 for } n.$$

$$a_3 = a(3)^2 + b(3) + c = 8 \qquad \text{Substitute 3 for } n.$$

You now have a system of three equations in a, b, and c.

$$\begin{cases} a + b + c = 3 & \text{Equation 1} \\ 4a + 2b + c = 5 & \text{Equation 2} \\ 9a + 3b + c = 8 & \text{Equation 3} \end{cases}$$

Using the techniques discussed in Chapter 7, you can find the solution to be $a = \frac{1}{2}$, $b = \frac{1}{2}$, and $c = 2$. So, the quadratic model is $a_n = \frac{1}{2}n^2 + \frac{1}{2}n + 2$. Try checking the values of a_1, a_2, and a_3.

✓ Checkpoint *Audio-video solution in English & Spanish at LarsonPrecalculus.com.*

Find a quadratic model for the sequence −2, 0, 4, 10, 18, 28, ...

Summarize (Section 9.4)

1. State the Principle of Mathematical Induction *(page 635)*. For examples of using mathematical induction to prove statements involving a positive integer n, see Examples 2–5.
2. Explain how to use pattern recognition and mathematical induction to write a formula for the nth term of a sequence *(page 639)*. For an example of using pattern recognition and mathematical induction to write a formula for the nth term of a sequence, see Example 6.
3. State the formulas for the sums of powers of integers *(page 640)*. For an example of finding sums of powers of integers, see Example 7.
4. Explain how to find finite differences of sequences *(page 641)*. For an example of using finite differences to find a quadratic model, see Example 8.

9.4 Exercises

See CalcChat.com for tutorial help and worked-out solutions to odd-numbered exercises.

Vocabulary: Fill in the blanks.

1. The first step in proving a formula by ________ ________ is to show that the formula is true when $n = 1$.
2. To find the ________ differences of a sequence, subtract consecutive terms.
3. A sequence is an ________ sequence when the first differences are all the same nonzero number.
4. If the ________ differences of a sequence are all the same nonzero number, then the sequence has a perfect quadratic model.

Skills and Applications

Finding P_{k+1} Given P_k In Exercises 5–10, find P_{k+1} for the given P_k.

5. $P_k = \dfrac{5}{k(k+1)}$
6. $P_k = \dfrac{1}{2(k+2)}$
7. $P_k = \dfrac{k^2(k+3)^2}{6}$
8. $P_k = \dfrac{k}{3}(2k+1)$
9. $P_k = \dfrac{3}{(k+2)(k+3)}$
10. $P_k = \dfrac{k^2}{2(k+1)^2}$

Using Mathematical Induction In Exercises 11–24, use mathematical induction to prove the formula for every positive integer n.

11. $2 + 4 + 6 + 8 + \cdots + 2n = n(n+1)$
12. $3 + 6 + 9 + 12 + \cdots + 3n = \dfrac{3n}{2}(n+1)$
13. $2 + 7 + 12 + 17 + \cdots + (5n-3) = \dfrac{n}{2}(5n-1)$
14. $1 + 4 + 7 + 10 + \cdots + (3n-2) = \dfrac{n}{2}(3n-1)$
15. $1 + 2 + 2^2 + 2^3 + \cdots + 2^{n-1} = 2^n - 1$
16. $2(1 + 3 + 3^2 + 3^3 + \cdots + 3^{n-1}) = 3^n - 1$
17. $1 + 2 + 3 + 4 + \cdots + n = \dfrac{n(n+1)}{2}$
18. $1^3 + 2^3 + 3^3 + 4^3 + \cdots + n^3 = \dfrac{n^2(n+1)^2}{4}$
19. $1^2 + 3^2 + 5^2 + \cdots + (2n-1)^2 = \dfrac{n(2n-1)(2n+1)}{3}$
20. $\left(1 + \dfrac{1}{1}\right)\left(1 + \dfrac{1}{2}\right)\left(1 + \dfrac{1}{3}\right)\cdots\left(1 + \dfrac{1}{n}\right) = n + 1$
21. $\displaystyle\sum_{i=1}^{n} i^5 = \dfrac{n^2(n+1)^2(2n^2+2n-1)}{12}$
22. $\displaystyle\sum_{i=1}^{n} i^4 = \dfrac{n(n+1)(2n+1)(3n^2+3n-1)}{30}$
23. $\displaystyle\sum_{i=1}^{n} i(i+1) = \dfrac{n(n+1)(n+2)}{3}$
24. $\displaystyle\sum_{i=1}^{n} \dfrac{1}{(2i-1)(2i+1)} = \dfrac{n}{2n+1}$

Proving an Inequality In Exercises 25–30, use mathematical induction to prove the inequality for the indicated integer values of n.

25. $n! > 2^n, \quad n \ge 4$
26. $\left(\frac{4}{3}\right)^n > n, \quad n \ge 7$
27. $\dfrac{1}{\sqrt{1}} + \dfrac{1}{\sqrt{2}} + \dfrac{1}{\sqrt{3}} + \cdots + \dfrac{1}{\sqrt{n}} > \sqrt{n}, \quad n \ge 2$
28. $\left(\dfrac{x}{y}\right)^{n+1} < \left(\dfrac{x}{y}\right)^n, \quad n \ge 1$ and $0 < x < y$
29. $(1 + a)^n \ge na, \quad n \ge 1$ and $a > 0$
30. $2n^2 > (n+1)^2, \quad n \ge 3$

Proving a Property In Exercises 31–40, use mathematical induction to prove the property for all positive integers n.

31. $(ab)^n = a^n b^n$
32. $\left(\dfrac{a}{b}\right)^n = \dfrac{a^n}{b^n}$
33. If $x_1 \ne 0, x_2 \ne 0, \ldots, x_n \ne 0$, then $(x_1 x_2 x_3 \cdots x_n)^{-1} = x_1^{-1} x_2^{-1} x_3^{-1} \cdots x_n^{-1}$.
34. If $x_1 > 0, x_2 > 0, \ldots, x_n > 0$, then $\ln(x_1 x_2 \cdots x_n) = \ln x_1 + \ln x_2 + \cdots + \ln x_n$.
35. Generalized Distributive Law: $x(y_1 + y_2 + \cdots + y_n) = xy_1 + xy_2 + \cdots + xy_n$
36. $(a + bi)^n$ and $(a - bi)^n$ are complex conjugates for all $n \ge 1$.
37. A factor of $(n^3 + 3n^2 + 2n)$ is 3.
38. A factor of $(n^4 - n + 4)$ is 2.
39. A factor of $(2^{2n+1} + 1)$ is 3.
40. A factor of $(2^{2n-1} + 3^{2n-1})$ is 5.

Finding a Formula for a Sum In Exercises 41–44, use mathematical induction to find a formula for the sum of the first n terms of the sequence.

41. $1, 5, 9, 13, \ldots$
42. $3, -\frac{9}{2}, \frac{27}{4}, -\frac{81}{8}, \ldots$
43. $\dfrac{1}{4}, \dfrac{1}{12}, \dfrac{1}{24}, \dfrac{1}{40}, \ldots, \dfrac{1}{2n(n+1)}, \ldots$
44. $\dfrac{1}{2 \cdot 3}, \dfrac{1}{3 \cdot 4}, \dfrac{1}{4 \cdot 5}, \dfrac{1}{5 \cdot 6}, \ldots, \dfrac{1}{(n+1)(n+2)}, \ldots$

Finding a Sum **In Exercises 45–54, find the sum using the formulas for the sums of powers of integers.**

45. $\sum_{n=1}^{15} n$ **46.** $\sum_{n=1}^{30} n$

47. $\sum_{n=1}^{6} n^2$ **48.** $\sum_{n=1}^{10} n^3$

49. $\sum_{n=1}^{5} n^4$ **50.** $\sum_{n=1}^{8} n^5$

51. $\sum_{n=1}^{6} (n^2 - n)$ **52.** $\sum_{n=1}^{20} (n^3 - n)$

53. $\sum_{i=1}^{6} (6i - 8i^3)$ **54.** $\sum_{j=1}^{10} \left(3 - \frac{1}{2}j + \frac{1}{2}j^2\right)$

Finding a Linear or Quadratic Model **In Exercises 55–60, decide whether the sequence can be represented perfectly by a linear or a quadratic model. If so, then find the model.**

55. 5, 13, 21, 29, 37, 45, . . .

56. 2, 9, 16, 23, 30, 37, . . .

57. 6, 15, 30, 51, 78, 111, . . .

58. 0, 6, 16, 30, 48, 70, . . .

59. −2, 1, 6, 13, 22, 33, . . .

60. −1, 8, 23, 44, 71, 104, . . .

Linear Model, Quadratic Model, or Neither? **In Exercises 61–68, write the first six terms of the sequence beginning with the given term. Then calculate the first and second differences of the sequence. State whether the sequence has a perfect linear model, a perfect quadratic model, or neither.**

61. $a_1 = 0$, $a_n = a_{n-1} + 3$ **62.** $a_1 = 2$, $a_n = a_{n-1} + 2$

63. $a_1 = 3$, $a_n = a_{n-1} - n$ **64.** $a_2 = -3$, $a_n = -2a_{n-1}$

65. $a_0 = 2$, $a_n = (a_{n-1})^2$ **66.** $a_0 = 0$, $a_n = a_{n-1} + n$

67. $a_1 = 2$, $a_n = n - a_{n-1}$ **68.** $a_1 = 0$, $a_n = a_{n-1} + 2n$

Finding a Quadratic Model **In Exercises 69–74, find a quadratic model for the sequence with the indicated terms.**

69. $a_0 = 3,\ a_1 = 3,\ a_4 = 15$

70. $a_0 = 7,\ a_1 = 6,\ a_3 = 10$

71. $a_0 = -3,\ a_2 = 1,\ a_4 = 9$

72. $a_0 = 3,\ a_2 = 0,\ a_6 = 36$

73. $a_1 = 0,\ a_2 = 8,\ a_4 = 30$

74. $a_0 = -3,\ a_2 = -5,\ a_6 = -57$

Mark Herreid/Shutterstock.com

75. Data Analysis: Residents The table shows the numbers a_n (in thousands) of Alaskan residents from 2005 through 2010. *(Source: U.S. Census Bureau)*

DATA — Spreadsheet at LarsonPrecalculus.com

Year	Number of Residents, a_n
2005	664
2006	671
2007	676
2008	682
2009	691
2010	705

(a) Find the first differences of the data shown in the table. Then find a linear model that approximates the data. Let n represent the year, with $n = 5$ corresponding to 2005.

(b) Use a graphing utility to find a linear model for the data. Compare this model with the model from part (a).

(c) Use the models found in parts (a) and (b) to predict the number of residents in 2016. How do these values compare?

Exploration

76. HOW DO YOU SEE IT? Find a formula for the sum of the angles (in degrees) of a regular polygon. Then use mathematical induction to prove this formula for a general n-sided polygon.

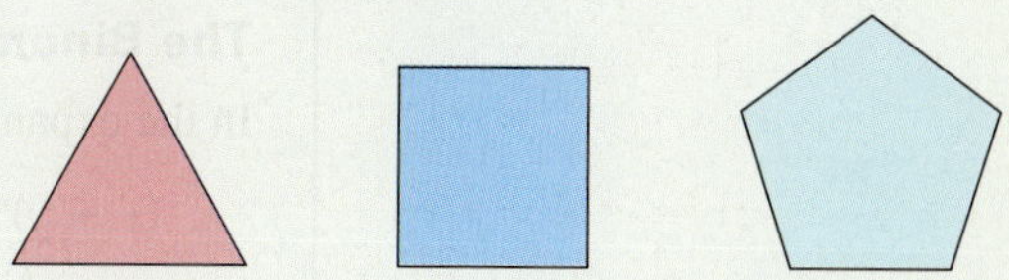

Equilateral triangle (180°) Square (360°) Regular pentagon (540°)

True or False? **In Exercises 77 and 78, determine whether the statement is true or false. Justify your answer.**

77. If the statement P_k is true and P_k implies P_{k+1}, then P_1 is also true.

78. A sequence with n terms has $n - 1$ second differences.

9.5 The Binomial Theorem

Binomial coefficients can help you model and solve real-life problems. For instance, in Exercise 92 on page 651, you will use binomial coefficients to write the expansion of a model that represents the average prices of residential electricity in the United States from 2003 through 2010.

- **Use the Binomial Theorem to calculate binomial coefficients.**
- **Use Pascal's Triangle to calculate binomial coefficients.**
- **Use binomial coefficients to write binomial expansions.**

Binomial Coefficients

Recall that a *binomial* is a polynomial that has two terms. In this section, you will study a formula that provides a quick method of raising a binomial to a power. To begin, look at the expansion of

$$(x + y)^n$$

for several values of n.

$$(x + y)^0 = 1$$
$$(x + y)^1 = x + y$$
$$(x + y)^2 = x^2 + 2xy + y^2$$
$$(x + y)^3 = x^3 + 3x^2y + 3xy^2 + y^3$$
$$(x + y)^4 = x^4 + 4x^3y + 6x^2y^2 + 4xy^3 + y^4$$
$$(x + y)^5 = x^5 + 5x^4y + 10x^3y^2 + 10x^2y^3 + 5xy^4 + y^5$$

There are several observations you can make about these expansions.

1. In each expansion, there are $n + 1$ terms.
2. In each expansion, x and y have symmetrical roles. The powers of x decrease by 1 in successive terms, whereas the powers of y increase by 1.
3. The sum of the powers of each term is n. For instance, in the expansion of $(x + y)^5$, the sum of the powers of each term is 5.

$$(x + y)^5 = x^5 + 5\overbrace{x^4y^1}^{4+1=5} + 10\overbrace{x^3y^2}^{3+2=5} + 10x^2y^3 + 5x^1y^4 + y^5$$

4. The coefficients increase and then decrease in a symmetric pattern.

The coefficients of a binomial expansion are called **binomial coefficients.** To find them, you can use the **Binomial Theorem.**

> **The Binomial Theorem**
>
> In the expansion of $(x + y)^n$
>
> $$(x + y)^n = x^n + nx^{n-1}y + \cdots + {}_nC_r\, x^{n-r}y^r + \cdots + nxy^{n-1} + y^n$$
>
> the coefficient of $x^{n-r}y^r$ is
>
> $${}_nC_r = \frac{n!}{(n - r)!r!}.$$
>
> The symbol $\binom{n}{r}$ is often used in place of ${}_nC_r$ to denote binomial coefficients.

For a proof of the Binomial Theorem, see Proofs in Mathematics on page 684.

wang song/Shutterstock.com

TECHNOLOGY

Most graphing calculators have the capability of evaluating ${}_nC_r$. If yours does, then evaluate ${}_8C_5$. You should get an answer of 56.

EXAMPLE 1 **Finding Binomial Coefficients**

Find each binomial coefficient.

a. ${}_8C_2$ **b.** $\binom{10}{3}$ **c.** ${}_7C_0$ **d.** $\binom{8}{8}$

Solution

$$\textbf{a. } {}_8C_2 = \frac{8!}{6! \cdot 2!} = \frac{(8 \cdot 7) \cdot \cancel{6!}}{\cancel{6!} \cdot 2!} = \frac{8 \cdot 7}{2 \cdot 1} = 28$$

$$\textbf{b. } \binom{10}{3} = \frac{10!}{7! \cdot 3!} = \frac{(10 \cdot 9 \cdot 8) \cdot \cancel{7!}}{\cancel{7!} \cdot 3!} = \frac{10 \cdot 9 \cdot 8}{3 \cdot 2 \cdot 1} = 120$$

$$\textbf{c. } {}_7C_0 = \frac{\cancel{7!}}{\cancel{7!} \cdot 0!} = 1 \qquad \textbf{d. } \binom{8}{8} = \frac{\cancel{8!}}{0! \cdot \cancel{8!}} = 1$$

✓ ***Checkpoint*** *Audio-video solution in English & Spanish at LarsonPrecalculus.com.*

Find each binomial coefficient.

a. $\binom{11}{5}$ **b.** ${}_9C_2$ **c.** $\binom{5}{0}$ **d.** ${}_{15}C_{15}$

When $r \neq 0$ and $r \neq n$, as in parts (a) and (b) above, there is a pattern for evaluating binomial coefficients that works because there will always be factorial terms that divide out from the expression.

$$ {}_8C_2 = \frac{\overbrace{8 \cdot 7}^{\text{2 factors}}}{\underbrace{2 \cdot 1}_{\text{2 factors}}} \quad \text{and} \quad \binom{10}{3} = \frac{\overbrace{10 \cdot 9 \cdot 8}^{\text{3 factors}}}{\underbrace{3 \cdot 2 \cdot 1}_{\text{3 factors}}}$$

EXAMPLE 2 **Finding Binomial Coefficients**

$$\textbf{a. } {}_7C_3 = \frac{7 \cdot \cancel{6} \cdot 5}{\cancel{3} \cdot \cancel{2} \cdot 1} = 35$$

$$\textbf{b. } \binom{7}{4} = \frac{7 \cdot \cancel{6} \cdot 5 \cdot \cancel{4}}{\cancel{4} \cdot \cancel{3} \cdot \cancel{2} \cdot 1} = 35$$

$$\textbf{c. } {}_{12}C_1 = \frac{12}{1} = 12$$

$$\textbf{d. } \binom{12}{11} = \frac{12!}{1! \cdot 11!} = \frac{(12) \cdot \cancel{11!}}{1! \cdot \cancel{11!}} = \frac{12}{1} = 12$$

✓ ***Checkpoint*** *Audio-video solution in English & Spanish at LarsonPrecalculus.com.*

Find each binomial coefficient.

a. ${}_7C_5$ **b.** $\binom{7}{2}$ **c.** ${}_{14}C_{13}$ **d.** $\binom{14}{1}$

It is not a coincidence that the results in parts (a) and (b) of Example 2 are the same and that the results in parts (c) and (d) are the same. In general, it is true that ${}_nC_r = {}_nC_{n-r}$. This shows the symmetric property of binomial coefficients identified earlier.

Pascal's Triangle

There is a convenient way to remember the pattern for binomial coefficients. By arranging the coefficients in a triangular pattern, you obtain the following array, which is called **Pascal's Triangle.** This triangle is named after the famous French mathematician Blaise Pascal (1623–1662).

```
                        1
                     1     1
                  1     2     1
               1     3     3     1
            1     4     6     4     1
         1     5    10    10     5     1           4 + 6 = 10
      1     6    15    20    15     6     1
   1     7    21    35    35    21     7     1     15 + 6 = 21
```

The first and last numbers in each row of Pascal's Triangle are 1. Every other number in each row is formed by adding the two numbers immediately above the number. Pascal noticed that numbers in this triangle are precisely the same numbers that are the coefficients of binomial expansions, as follows.

$$(x + y)^0 = 1 \qquad \text{0th row}$$

$$(x + y)^1 = 1x + 1y \qquad \text{1st row}$$

$$(x + y)^2 = 1x^2 + 2xy + 1y^2 \qquad \text{2nd row}$$

$$(x + y)^3 = 1x^3 + 3x^2y + 3xy^2 + 1y^3 \qquad \text{3rd row}$$

$$(x + y)^4 = 1x^4 + 4x^3y + 6x^2y^2 + 4xy^3 + 1y^4 \qquad \vdots$$

$$(x + y)^5 = 1x^5 + 5x^4y + 10x^3y^2 + 10x^2y^3 + 5xy^4 + 1y^5$$

$$(x + y)^6 = 1x^6 + 6x^5y + 15x^4y^2 + 20x^3y^3 + 15x^2y^4 + 6xy^5 + 1y^6$$

$$(x + y)^7 = 1x^7 + 7x^6y + 21x^5y^2 + 35x^4y^3 + 35x^3y^4 + 21x^2y^5 + 7xy^6 + 1y^7$$

The top row in Pascal's Triangle is called the *zeroth row* because it corresponds to the binomial expansion $(x + y)^0 = 1$. Similarly, the next row is called the *first row* because it corresponds to the binomial expansion

$$(x + y)^1 = 1(x) + 1(y).$$

In general, the *nth row* in Pascal's Triangle gives the coefficients of $(x + y)^n$.

EXAMPLE 3 Using Pascal's Triangle

Use the seventh row of Pascal's Triangle to find the binomial coefficients.

$${}_8C_0, {}_8C_1, {}_8C_2, {}_8C_3, {}_8C_4, {}_8C_5, {}_8C_6, {}_8C_7, {}_8C_8$$

Solution

```
   1     7     21    35    35    21     7     1
1     8     28    56    70    56    28     8     1
|     |     |     |     |     |     |     |     |
8C0   8C1   8C2   8C3   8C4   8C5   8C6   8C7   8C8
```

✓ ***Checkpoint*** *Audio-video solution in English & Spanish at LarsonPrecalculus.com.*

Use the eighth row of Pascal's Triangle to find the binomial coefficients.

$${}_9C_0, {}_9C_1, {}_9C_2, {}_9C_3, {}_9C_4, {}_9C_5, {}_9C_6, {}_9C_7, {}_9C_8, {}_9C_9$$

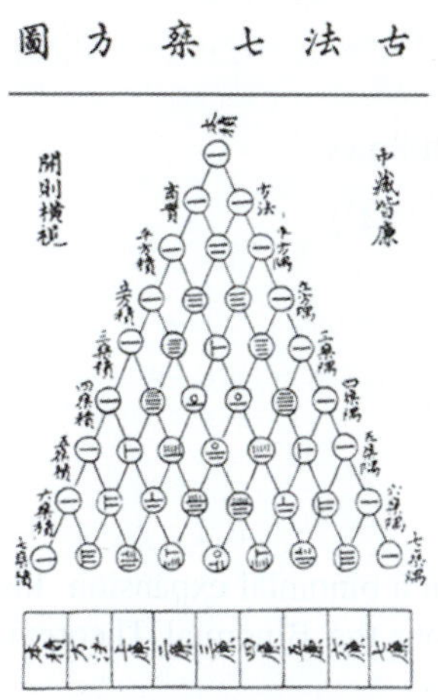

PRECIOUS MIRROR OF THE FOUR ELEMENTS

Eastern cultures were familiar with "Pascal's" Triangle and forms of the Binomial Theorem prior to the Western "discovery" of the theorem. A Chinese text entitled *Precious Mirror of the Four Elements* contains a triangle of binomial expansions through the eighth power.

Binomial Expansions

As mentioned at the beginning of this section, when you write the coefficients for a binomial that is raised to a power, you are **expanding a binomial.** The formula for binomial coefficients and Pascal's Triangle give you a relatively easy way to expand binomials, as demonstrated in the next four examples.

EXAMPLE 4 Expanding a Binomial

Write the expansion of the expression

$(x + 1)^3.$

Solution The binomial coefficients from the third row of Pascal's Triangle are

1, 3, 3, 1.

So, the expansion is as follows.

$$\begin{aligned}(x + 1)^3 &= (1)x^3 + (3)x^2(1) + (3)x(1^2) + (1)(1^3)\\ &= x^3 + 3x^2 + 3x + 1\end{aligned}$$

✓ ***Checkpoint*** *Audio-video solution in English & Spanish at LarsonPrecalculus.com.*

Write the expansion of the expression

$(x + 2)^4.$

To expand binomials representing *differences* rather than sums, you alternate signs. Here are two examples.

$$(x - 1)^3 = x^3 - 3x^2 + 3x - 1$$

$$(x - 1)^4 = x^4 - 4x^3 + 6x^2 - 4x + 1$$

ALGEBRA HELP The solutions to Example 5 use the property of exponents

$$(ab)^m = a^m b^m.$$

For instance, in Example 5(a),

$$(2x)^4 = 2^4x^4 = 16x^4.$$

You can review properties of exponents in Appendix A.2.

EXAMPLE 5 Expanding a Binomial

Write the expansion of each expression.

a. $(2x - 3)^4$

b. $(x - 2y)^4$

Solution The binomial coefficients from the fourth row of Pascal's Triangle are

1, 4, 6, 4, 1.

So, the expansions are as follows.

a. $$\begin{aligned}(2x - 3)^4 &= (1)(2x)^4 - (4)(2x)^3(3) + (6)(2x)^2(3^2) - (4)(2x)(3^3) + (1)(3^4)\\ &= 16x^4 - 96x^3 + 216x^2 - 216x + 81\end{aligned}$$

b. $$\begin{aligned}(x - 2y)^4 &= (1)x^4 - (4)x^3(2y) + (6)x^2(2y)^2 - (4)x(2y)^3 + (1)(2y)^4\\ &= x^4 - 8x^3y + 24x^2y^2 - 32xy^3 + 16y^4\end{aligned}$$

✓ ***Checkpoint*** *Audio-video solution in English & Spanish at LarsonPrecalculus.com.*

Write the expansion of each expression.

a. $(y - 2)^4$

b. $(2x - y)^5$

▷ **TECHNOLOGY** Use a graphing utility to check the expansion in Example 6. Graph the original binomial expression and the expansion in the same viewing window. The graphs should coincide, as shown below.

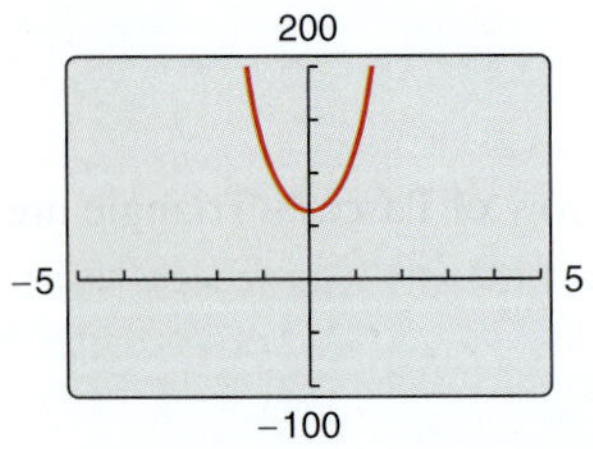

EXAMPLE 6 Expanding a Binomial

Write the expansion of $(x^2 + 4)^3$.

Solution Use the third row of Pascal's Triangle, as follows.

$$(x^2 + 4)^3 = (1)(x^2)^3 + (3)(x^2)^2(4) + (3)x^2(4^2) + (1)(4^3)$$
$$= x^6 + 12x^4 + 48x^2 + 64$$

✓ ***Checkpoint*** *Audio-video solution in English & Spanish at LarsonPrecalculus.com.*

Write the expansion of $(5 + y^2)^3$.

Sometimes you will need to find a specific term in a binomial expansion. Instead of writing the entire expansion, use the fact that, from the Binomial Theorem, the $(r + 1)$th term is ${}_nC_r x^{n-r} y^r$.

EXAMPLE 7 Finding a Term or Coefficient

a. Find the sixth term of $(a + 2b)^8$.

b. Find the coefficient of the term a^6b^5 in the expansion of $(3a - 2b)^{11}$.

Solution

a. Remember that the formula is for the $(r + 1)$th term, so r is one less than the number of the term you need. So, to find the sixth term in this binomial expansion, use $r = 5$, $n = 8$, $x = a$, and $y = 2b$, as shown.

$${}_8C_5 a^{8-5}(2b)^5 = 56 \cdot a^3 \cdot (2b)^5$$
$$= 56(2^5)a^3b^5$$
$$= 1792a^3b^5$$

b. In this case, $n = 11$, $r = 5$, $x = 3a$, and $y = -2b$. Substitute these values to obtain

$${}_nC_r x^{n-r} y^r = {}_{11}C_5(3a)^6(-2b)^5$$
$$= (462)(729a^6)(-32b^5)$$
$$= -10{,}777{,}536a^6b^5.$$

So, the coefficient is $-10{,}777{,}536$.

✓ ***Checkpoint*** *Audio-video solution in English & Spanish at LarsonPrecalculus.com.*

a. Find the fifth term of $(a + 2b)^8$.

b. Find the coefficient of the term a^4b^7 in the expansion of $(3a - 2b)^{11}$.

Summarize (Section 9.5)

1. State the Binomial Theorem *(page 644)*. For examples of using the Binomial Theorem to calculate binomial coefficients, see Examples 1 and 2.
2. Explain how to use Pascal's Triangle to calculate binomial coefficients *(page 646)*. For an example of using Pascal's Triangle to calculate binomial coefficients, see Example 3.
3. Explain how to use binomial coefficients to write a binomial expansion *(page 647)*. For examples of using binomial coefficients to write binomial expansions, see Examples 4–6.

9.5 Exercises

See CalcChat.com for tutorial help and worked-out solutions to odd-numbered exercises.

Vocabulary: **Fill in the blanks.**

1. The coefficients of a binomial expansion are called ________ ________.
2. To find binomial coefficients, you can use the ________ ________ or ________ ________.
3. The symbol used to denote a binomial coefficient is ________ or ________.
4. When you write the coefficients for a binomial that is raised to a power, you are ________ a ________.

Skills and Applications

Finding a Binomial Coefficient **In Exercises 5–14, find the binomial coefficient.**

5. ${}_5C_3$
6. ${}_8C_6$
7. ${}_{12}C_0$
8. ${}_{20}C_{20}$
9. ${}_{20}C_{15}$
10. ${}_{12}C_5$
11. $\binom{10}{4}$
12. $\binom{10}{6}$
13. $\binom{100}{98}$
14. $\binom{100}{2}$

Using Pascal's Triangle **In Exercises 15–18, evaluate using Pascal's Triangle.**

15. $\binom{6}{5}$
16. $\binom{5}{2}$
17. ${}_7C_4$
18. ${}_{10}C_2$

Expanding an Expression **In Exercises 19–40, use the Binomial Theorem to expand and simplify the expression.**

19. $(x + 1)^4$
20. $(x + 1)^6$
21. $(a + 6)^4$
22. $(a + 5)^5$
23. $(y - 4)^3$
24. $(y - 2)^5$
25. $(x + y)^5$
26. $(c + d)^3$
27. $(2x + y)^3$
28. $(7a + b)^3$
29. $(r + 3s)^6$
30. $(x + 2y)^4$
31. $(3a - 4b)^5$
32. $(2x - 5y)^5$
33. $(x^2 + y^2)^4$
34. $(x^2 + y^2)^6$
35. $\left(\frac{1}{x} + y\right)^5$
36. $\left(\frac{1}{x} + 2y\right)^6$
37. $\left(\frac{2}{x} - y\right)^4$
38. $\left(\frac{2}{x} - 3y\right)^5$
39. $2(x - 3)^4 + 5(x - 3)^2$
40. $(4x - 1)^3 - 2(4x - 1)^4$

Expanding a Binomial **In Exercises 41–44, expand the binomial by using Pascal's Triangle to determine the coefficients.**

41. $(2t - s)^5$
42. $(3 - 2z)^4$
43. $(x + 2y)^5$
44. $(3v + 2)^6$

Finding a Term in a Binomial Expansion **In Exercises 45–52, find the specified *n*th term in the expansion of the binomial.**

45. $(x + y)^{10}, \quad n = 4$
46. $(x - y)^6, \quad n = 7$
47. $(x - 6y)^5, \quad n = 3$
48. $(x - 10z)^7, \quad n = 4$
49. $(4x + 3y)^9, \quad n = 8$
50. $(5a + 6b)^5, \quad n = 5$
51. $(10x - 3y)^{12}, \quad n = 10$
52. $(7x + 2y)^{15}, \quad n = 7$

Finding a Coefficient **In Exercises 53–60, find the coefficient *a* of the term in the expansion of the binomial.**

	Binomial	Term
53.	$(x + 3)^{12}$	ax^5
54.	$(x^2 + 3)^{12}$	ax^8
55.	$(4x - y)^{10}$	ax^2y^8
56.	$(x - 2y)^{10}$	ax^8y^2
57.	$(2x - 5y)^9$	ax^4y^5
58.	$(3x - 4y)^8$	ax^6y^2
59.	$(x^2 + y)^{10}$	ax^8y^6
60.	$(z^2 - t)^{10}$	az^4t^8

Expanding an Expression **In Exercises 61–66, use the Binomial Theorem to expand and simplify the expression.**

61. $(\sqrt{x} + 5)^3$
62. $(2\sqrt{t} - 1)^3$
63. $(x^{2/3} - y^{1/3})^3$
64. $(u^{3/5} + 2)^5$
65. $(3\sqrt{t} + \sqrt[4]{t})^4$
66. $(x^{3/4} - 2x^{5/4})^4$

Simplifying a Difference Quotient **In Exercises 67–72, simplify the difference quotient, using the Binomial Theorem if necessary.**

$$\frac{f(x + h) - f(x)}{h} \qquad \text{Difference quotient}$$

67. $f(x) = x^3$
68. $f(x) = x^4$
69. $f(x) = x^6$
70. $f(x) = x^8$
71. $f(x) = \sqrt{x}$
72. $f(x) = \frac{1}{x}$

Expanding a Complex Number **In Exercises 73–78, use the Binomial Theorem to expand the complex number. Simplify your result.**

73. $(1 + i)^4$

74. $(2 - i)^5$

75. $(2 - 3i)^6$

76. $\left(5 + \sqrt{-9}\right)^3$

77. $\left(-\dfrac{1}{2} + \dfrac{\sqrt{3}}{2}i\right)^3$

78. $\left(5 - \sqrt{3}i\right)^4$

Approximation **In Exercises 79–82, use the Binomial Theorem to approximate the quantity accurate to three decimal places. For example, in Exercise 79, use the expansion**

$$\mathbf{(1.02)^8 = (1 + 0.02)^8}$$
$$\mathbf{= 1 + 8(0.02) + 28(0.02)^2 + \cdot \cdot \cdot + (0.02)^8.}$$

79. $(1.02)^8$

80. $(2.005)^{10}$

81. $(2.99)^{12}$

82. $(1.98)^9$

Graphical Reasoning **In Exercises 83 and 84, use a graphing utility to graph f and g in the same viewing window. What is the relationship between the two graphs? Use the Binomial Theorem to write the polynomial function g in standard form.**

83. $f(x) = x^3 - 4x, \quad g(x) = f(x + 4)$

84. $f(x) = -x^4 + 4x^2 - 1, \quad g(x) = f(x - 3)$

Probability **In Exercises 85–88, consider n independent trials of an experiment in which each trial has two possible outcomes: "success" or "failure." The probability of a success on each trial is p, and the probability of a failure is $q = 1 - p$. In this context, the term ${}_nC_k p^k q^{n-k}$ in the expansion of $(p + q)^n$ gives the probability of k successes in the n trials of the experiment.**

85. You toss a fair coin seven times. To find the probability of obtaining four heads, evaluate the term

$${}_7C_4\left(\tfrac{1}{2}\right)^4\left(\tfrac{1}{2}\right)^3$$

in the expansion of $\left(\frac{1}{2} + \frac{1}{2}\right)^7$.

86. The probability of a baseball player getting a hit during any given time at bat is $\frac{1}{4}$. To find the probability that the player gets three hits during the next 10 times at bat, evaluate the term

$${}_{10}C_3\left(\tfrac{1}{4}\right)^3\left(\tfrac{3}{4}\right)^7$$

in the expansion of $\left(\frac{1}{4} + \frac{3}{4}\right)^{10}$.

87. The probability of a sales representative making a sale with any one customer is $\frac{1}{3}$. The sales representative makes eight contacts a day. To find the probability of making four sales, evaluate the term

$${}_8C_4\left(\tfrac{1}{3}\right)^4\left(\tfrac{2}{3}\right)^4$$

in the expansion of $\left(\frac{1}{3} + \frac{2}{3}\right)^8$.

88. To find the probability that the sales representative in Exercise 87 makes four sales when the probability of a sale with any one customer is $\frac{1}{2}$, evaluate the term

$${}_8C_4\left(\tfrac{1}{2}\right)^4\left(\tfrac{1}{2}\right)^4$$

in the expansion of $\left(\frac{1}{2} + \frac{1}{2}\right)^8$.

89. Finding a Pattern Describe the pattern formed by the sums of the numbers along the diagonal line segments shown in Pascal's Triangle (see figure).

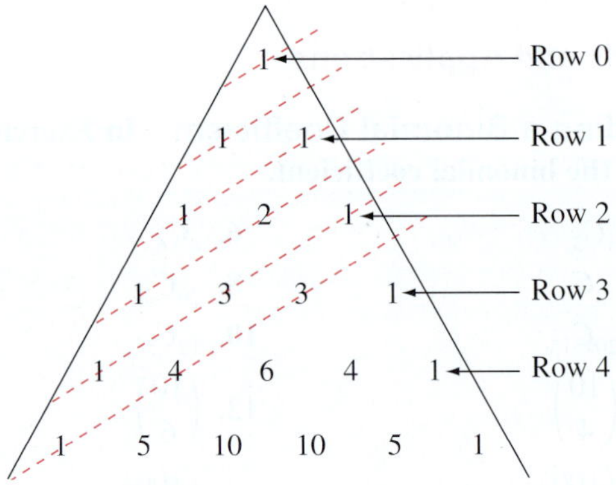

90. Finding a Pattern Use each of the encircled groups of numbers in the figure to form a 2×2 matrix. Find the determinant of each matrix. Describe the pattern.

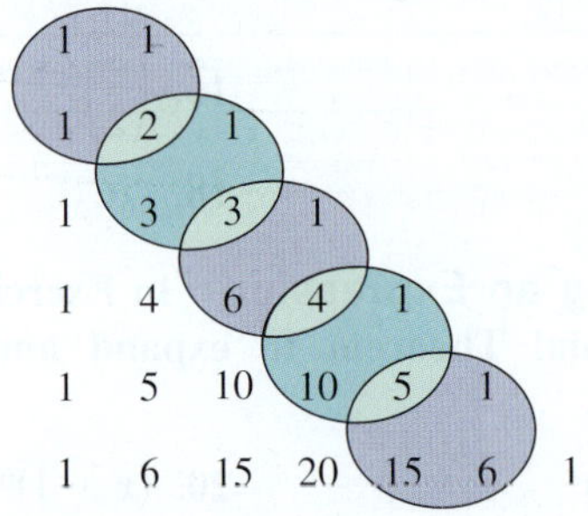

91. Child Support The amounts $f(t)$ (in billions of dollars) of child support collected in the United States from 2002 through 2009 can be approximated by the model

$$f(t) = -0.009t^2 + 1.05t + 18.0, \quad 2 \le t \le 9$$

where t represents the year, with $t = 2$ corresponding to 2002. *(Source: U.S. Department of Health and Human Services)*

(a) You want to adjust the model so that $t = 2$ corresponds to 2007 rather than 2002. To do this, you shift the graph of f five units *to the left* to obtain $g(t) = f(t + 5)$. Use binomial coefficients to write $g(t)$ in standard form.

(b) Use a graphing utility to graph f and g in the same viewing window.

(c) Use the graphs to estimate when the child support collections exceeded \$25 billion.

92. Data Analysis: Electricity The table shows the average prices $f(t)$ (in cents per kilowatt hour) of residential electricity in the United States from 2003 through 2010. *(Source: U.S. Energy Information Administration)*

DATA

Year	Average Price, $f(t)$
2003	8.72
2004	8.95
2005	9.45
2006	10.40
2007	10.65
2008	11.26
2009	11.51
2010	11.58

Spreadsheet at LarsonPrecalculus.com

(a) Use the *regression* feature of a graphing utility to find a cubic model for the data. Let t represent the year, with $t = 3$ corresponding to 2003.

(b) Use the graphing utility to plot the data and the model in the same viewing window.

(c) You want to adjust the model so that $t = 3$ corresponds to 2008 rather than 2003. To do this, you shift the graph of f five units *to the left* to obtain $g(t) = f(t + 5)$. Use binomial coefficients to write $g(t)$ in standard form.

(d) Use the graphing utility to graph g in the same viewing window as f.

(e) Use both models to predict the average price in 2011. Do you obtain the same answer?

(f) Do your answers to part (e) seem reasonable? Explain.

(g) What factors do you think may have contributed to the change in the average price?

Exploration

True or False? **In Exercises 93 and 94, determine whether the statement is true or false. Justify your answer.**

93. The Binomial Theorem could be used to produce each row of Pascal's Triangle.

94. A binomial that represents a difference cannot always be accurately expanded using the Binomial Theorem.

95. Writing In your own words, explain how to form the rows of Pascal's Triangle.

96. Forming Rows of Pascal's Triangle Form rows 8–10 of Pascal's Triangle.

97. Graphical Reasoning Which two functions have identical graphs, and why? Use a graphing utility to graph the functions in the given order and in the same viewing window. Compare the graphs.

(a) $f(x) = (1 - x)^3$

(b) $g(x) = 1 - x^3$

(c) $h(x) = 1 + 3x + 3x^2 + x^3$

(d) $k(x) = 1 - 3x + 3x^2 - x^3$

(e) $p(x) = 1 + 3x - 3x^2 + x^3$

98. HOW DO YOU SEE IT? The expansions of $(x + y)^4$, $(x + y)^5$, and $(x + y)^6$ are as follows.

$$(x + y)^4 = 1x^4 + 4x^3y + 6x^2y^2 + 4xy^3 + 1y^4$$

$$(x + y)^5 = 1x^5 + 5x^4y + 10x^3y^2 + 10x^2y^3 + 5xy^4 + 1y^5$$

$$(x + y)^6 = 1x^6 + 6x^5y + 15x^4y^2 + 20x^3y^3 + 15x^2y^4 + 6xy^5 + 1y^6$$

(a) Explain how the exponent of a binomial is related to the number of terms in its expansion.

(b) How many terms are in the expansion of $(x + y)^n$?

Proof **In Exercises 99–102, prove the property for all integers r and n where $0 \le r \le n$.**

99. ${}_nC_r = {}_nC_{n-r}$

100. ${}_nC_0 - {}_nC_1 + {}_nC_2 - \cdots \pm {}_nC_n = 0$

101. ${}_{n+1}C_r = {}_nC_r + {}_nC_{r-1}$

102. The sum of the numbers in the nth row of Pascal's Triangle is 2^n.

103. Binomial Coefficients and Pascal's Triangle Complete the table and describe the result.

n	r	${}_nC_r$	${}_nC_{n-r}$
9	5		
7	1		
12	4		
6	0		
10	7		

What characteristic of Pascal's Triangle does this table illustrate?

wang song/Shutterstock.com

9.6 Counting Principles

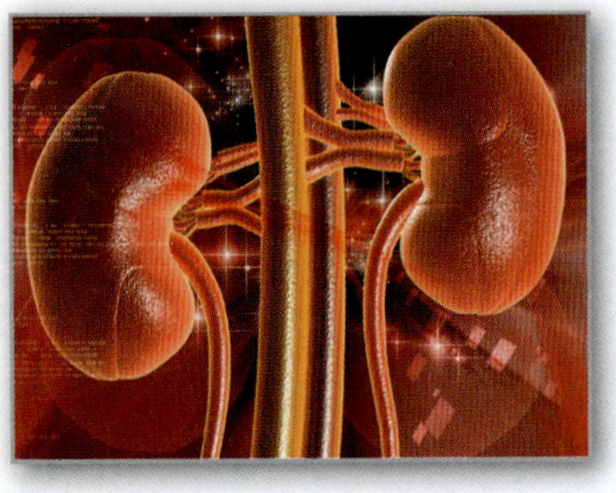

Counting principles can help you solve counting problems that occur in real life. For instance, in Exercise 39 on page 660, you will use counting principles to determine the numbers of possible orders there are for best match, second-best match, and third-best match kidney donors.

- **Solve simple counting problems.**
- **Use the Fundamental Counting Principle to solve counting problems.**
- **Use permutations to solve counting problems.**
- **Use combinations to solve counting problems.**

Simple Counting Problems

This section and Section 9.7 present a brief introduction to some of the basic counting principles and their applications to probability. In Section 9.7, you will see that much of probability has to do with counting the number of ways an event can occur. The following two examples describe simple counting problems.

EXAMPLE 1 Selecting Pairs of Numbers at Random

You place eight pieces of paper, numbered from 1 to 8, in a box. You draw one piece of paper at random from the box, record its number, and *replace* the paper in the box. Then, you draw a second piece of paper at random from the box and record its number. Finally, you add the two numbers. How many different ways can you obtain a sum of 12?

Solution To solve this problem, count the different ways to obtain a sum of 12 using two numbers from 1 to 8.

First number	4	5	6	7	8
Second number	8	7	6	5	4

So, a sum of 12 can occur in five different ways.

✓ **Checkpoint** *Audio-video solution in English & Spanish at LarsonPrecalculus.com.*

In Example 1, how many different ways can you obtain a sum of 14?

EXAMPLE 2 Selecting Pairs of Numbers at Random

You place eight pieces of paper, numbered from 1 to 8, in a box. You draw one piece of paper at random from the box, record its number, and *do not* replace the paper in the box. Then, you draw a second piece of paper at random from the box and record its number. Finally, you add the two numbers. How many different ways can you obtain a sum of 12?

Solution To solve this problem, count the different ways to obtain a sum of 12 *using two different numbers* from 1 to 8.

First number	4	5	7	8
Second number	8	7	5	4

So, a sum of 12 can occur in four different ways.

✓ **Checkpoint** *Audio-video solution in English & Spanish at LarsonPrecalculus.com.*

Repeat Example 2 for drawing *three* pieces of paper.

The difference between the counting problems in Examples 1 and 2 can be described by saying that the random selection in Example 1 occurs **with replacement,** whereas the random selection in Example 2 occurs **without replacement,** which eliminates the possibility of choosing two 6's.

Kannanimages/Shutterstock.com

The Fundamental Counting Principle

Examples 1 and 2 describe simple counting problems in which you can *list* each possible way that an event can occur. When it is possible, this is always the best way to solve a counting problem. However, some events can occur in so many different ways that it is not feasible to write out the entire list. In such cases, you must rely on formulas and counting principles. The most important of these is the **Fundamental Counting Principle.**

> **Fundamental Counting Principle**
>
> Let E_1 and E_2 be two events. The first event E_1 can occur in m_1 different ways. After E_1 has occurred, E_2 can occur in m_2 different ways. The number of ways that the two events can occur is $m_1 \cdot m_2$.

The Fundamental Counting Principle can be extended to three or more events. For instance, the number of ways that three events E_1, E_2, and E_3 can occur is $m_1 \cdot m_2 \cdot m_3$.

EXAMPLE 3 Using the Fundamental Counting Principle

How many different pairs of letters from the English alphabet are possible?

Solution There are two events in this situation. The first event is the choice of the first letter, and the second event is the choice of the second letter. Because the English alphabet contains 26 letters, it follows that the number of two-letter pairs is $26 \cdot 26 = 676$.

✓ ***Checkpoint*** *Audio-video solution in English & Spanish at LarsonPrecalculus.com.*

A combination lock will open when you select the right choice of three numbers (from 1 to 30, inclusive). How many different lock combinations are possible?

EXAMPLE 4 Using the Fundamental Counting Principle

Telephone numbers in the United States have 10 digits. The first three digits are the *area code* and the next seven digits are the *local telephone number*. How many different telephone numbers are possible within each area code? (Note that a local telephone number cannot begin with 0 or 1.)

Solution Because the first digit of a local telephone number cannot be 0 or 1, there are only eight choices for the first digit. For each of the other six digits, there are 10 choices.

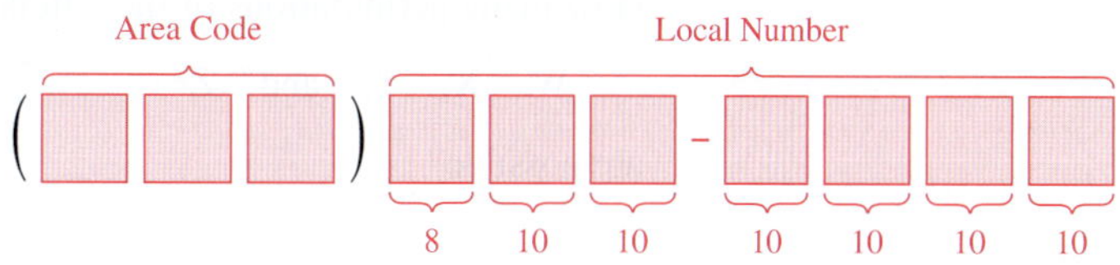

So, the number of telephone numbers that are possible within each area code is

$$8 \cdot 10 \cdot 10 \cdot 10 \cdot 10 \cdot 10 \cdot 10 = 8{,}000{,}000.$$

✓ ***Checkpoint*** *Audio-video solution in English & Spanish at LarsonPrecalculus.com.*

A product's catalog number is made up of one letter from the English alphabet followed by a five-digit number. How many different catalog numbers are possible? ■

Permutations

One important application of the Fundamental Counting Principle is in determining the number of ways that n elements can be arranged (in order). An ordering of n elements is called a **permutation** of the elements.

> **Definition of Permutation**
>
> A **permutation** of n different elements is an ordering of the elements such that one element is first, one is second, one is third, and so on.

EXAMPLE 5 Finding the Number of Permutations

How many permutations of the letters

A, B, C, D, E, and F

are possible?

Solution Consider the following reasoning.

First position: Any of the *six* letters

Second position: Any of the remaining *five* letters

Third position: Any of the remaining *four* letters

Fourth position: Any of the remaining *three* letters

Fifth position: Either of the remaining *two* letters

Sixth position: The *one* remaining letter

So, the numbers of choices for the six positions are as follows.

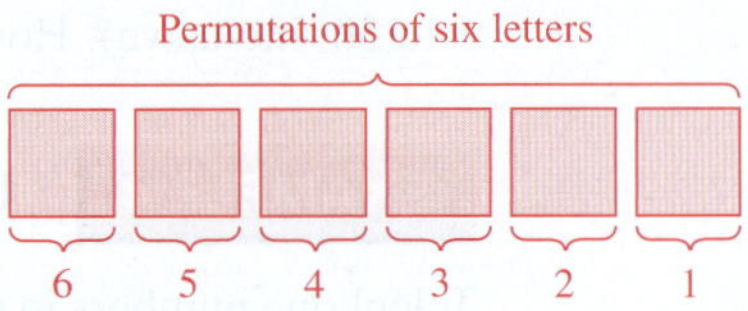

The total number of permutations of the six letters is

$$6! = 6 \cdot 5 \cdot 4 \cdot 3 \cdot 2 \cdot 1$$
$$= 720.$$

✓ **Checkpoint** *Audio-video solution in English & Spanish at LarsonPrecalculus.com.*

How many permutations of the letters

W, X, Y, and Z

are possible?

> **Number of Permutations of n Elements**
>
> The number of permutations of n elements is
>
> $$n \cdot (n-1) \cdot \cdot \cdot 4 \cdot 3 \cdot 2 \cdot 1 = n!.$$
>
> In other words, there are $n!$ different ways that n elements can be ordered.

Eleven thoroughbred racehorses hold the title of Triple Crown winner for winning the Kentucky Derby, the Preakness, and the Belmont Stakes in the same year. Forty-nine horses have won two out of the three races.

EXAMPLE 6 **Counting Horse Race Finishes**

Eight horses are running in a race. In how many different ways can these horses come in first, second, and third? (Assume that there are no ties.)

Solution Here are the different possibilities.

Win (first position): *Eight* choices

Place (second position): *Seven* choices

Show (third position): *Six* choices

Using the Fundamental Counting Principle, multiply these three numbers to obtain the following.

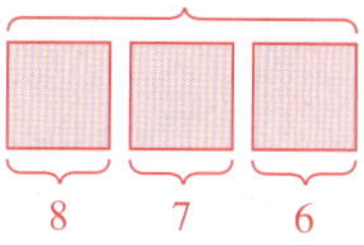

So, there are

$$8 \cdot 7 \cdot 6 = 336$$

different orders.

Checkpoint *Audio-video solution in English & Spanish at LarsonPrecalculus.com.*

A coin club has five members. In how many ways can there be a president and a vice-president?

It is useful, on occasion, to order a *subset* of a collection of elements rather than the entire collection. For example, you might want to order r elements out of a collection of n elements. Such an ordering is called a **permutation of n elements taken r at a time.**

TECHNOLOGY

Most graphing calculators have the capability of evaluating ${}_nP_r$. If yours does, then evaluate ${}_8P_5$. You should get an answer of 6720.

Permutations of n Elements Taken r at a Time

The number of permutations of n elements taken r at a time is

$${}_nP_r = \frac{n!}{(n-r)!} = n(n-1)(n-2) \cdot \cdot \cdot (n-r+1).$$

Using this formula, you can rework Example 6 to find that the number of permutations of eight horses taken three at a time is

$$\begin{aligned}{}_8P_3 &= \frac{8!}{(8-3)!}\\ &= \frac{8!}{5!}\\ &= \frac{8 \cdot 7 \cdot 6 \cdot \cancel{5!}}{\cancel{5!}}\\ &= 336\end{aligned}$$

which is the same answer obtained in the example.

Mikhail Pogosov/Shutterstock.com

Remember that for permutations, order is important. So, to find the possible permutations of the letters A, B, C, and D taken three at a time, you count the permutations (A, B, D) and (B, A, D) as different because the *order* of the elements is different.

Suppose, however, that you want to find the possible permutations of the letters A, A, B, and C. The total number of permutations of the four letters would be ${}_4P_4 = 4!$. However, not all of these arrangements would be *distinguishable* because there are two A's in the list. To find the number of distinguishable permutations, you can use the following formula.

Distinguishable Permutations

Suppose a set of n objects has n_1 of one kind of object, n_2 of a second kind, n_3 of a third kind, and so on, with

$$n = n_1 + n_2 + n_3 + \cdot \cdot \cdot + n_k.$$

Then the number of **distinguishable permutations** of the n objects is

$$\frac{n!}{n_1! \cdot n_2! \cdot n_3! \cdot \cdot \cdot \cdot \cdot n_k!}.$$

EXAMPLE 7 **Distinguishable Permutations**

In how many distinguishable ways can the letters in BANANA be written?

Solution This word has six letters, of which three are A's, two are N's, and one is a B. So, the number of distinguishable ways the letters can be written is

$$\frac{n!}{n_1! \cdot n_2! \cdot n_3!} = \frac{6!}{3! \cdot 2! \cdot 1!}$$

$$= \frac{6 \cdot 5 \cdot 4 \cdot \cancel{3!}}{\cancel{3!} \cdot 2!}$$

$$= 60.$$

The 60 different distinguishable permutations are as follows.

AAABNN	AAANBN	AAANNB	AABANN
AABNAN	AABNNA	AANABN	AANANB
AANBAN	AANBNA	AANNAB	AANNBA
ABAANN	ABANAN	ABANNA	ABNAAN
ABNANA	ABNNAA	ANAABN	ANAANB
ANABAN	ANABNA	ANANAB	ANANBA
ANBAAN	ANBANA	ANBNAA	ANNAAB
ANNABA	ANNBAA	BAAANN	BAANAN
BAANNA	BANAAN	BANANA	BANNAA
BNAAAN	BNAANA	BNANAA	BNNAAA
NAAABN	NAAANB	NAABAN	NAABNA
NAANAB	NAANBA	NABAAN	NABANA
NABNAA	NANAAB	NANABA	NANBAA
NBAAAN	NBAANA	NBANAA	NBNAAA
NNAAAB	NNAABA	NNABAA	NNBAAA

✓ **Checkpoint** Audio-video solution in English & Spanish at LarsonPrecalculus.com.

In how many distinguishable ways can the letters in MITOSIS be written?

Combinations

When you count the number of possible permutations of a set of elements, order is important. As a final topic in this section, you will look at a method of selecting subsets of a larger set in which order is *not* important. Such subsets are called **combinations of *n* elements taken *r* at a time.** For instance, the combinations

$\{A, B, C\}$ and $\{B, A, C\}$

are equivalent because both sets contain the same three elements, and the order in which the elements are listed is not important. So, you would count only one of the two sets. A common example of how a combination occurs is in a card game in which players are free to reorder the cards after they have been dealt.

EXAMPLE 8 Combinations of *n* Elements Taken *r* at a Time

In how many different ways can three letters be chosen from the letters

A, B, C, D, and E?

(The order of the three letters is not important.)

Solution The following subsets represent the different combinations of three letters that can be chosen from the five letters.

$\{A, B, C\}$ $\{A, B, D\}$
$\{A, B, E\}$ $\{A, C, D\}$
$\{A, C, E\}$ $\{A, D, E\}$
$\{B, C, D\}$ $\{B, C, E\}$
$\{B, D, E\}$ $\{C, D, E\}$

From this list, you can conclude that there are 10 different ways that three letters can be chosen from five letters.

Checkpoint *Audio-video solution in English & Spanish at LarsonPrecalculus.com.*

In how many different ways can two letters be chosen from the letters A, B, C, D, E, F, and G? (The order of the two letters is not important.)

Combinations of *n* Elements Taken *r* at a Time

The number of combinations of n elements taken r at a time is

$$_nC_r = \frac{n!}{(n-r)!r!}$$

which is equivalent to $_nC_r = \frac{_nP_r}{r!}$.

Note that the formula for $_nC_r$ is the same one given for binomial coefficients. To see how to use this formula, solve the counting problem in Example 8. In that problem, you want to find the number of combinations of five elements taken three at a time. So, $n = 5$, $r = 3$, and the number of combinations is

$$_5C_3 = \frac{5!}{2!3!} = \frac{5 \cdot \overset{2}{\cancel{4}} \cdot \cancel{3!}}{\cancel{2} \cdot 1 \cdot \cancel{3!}} = 10$$

which is the same answer obtained in Example 8.

A ♥	A ♦	A ♣	A ♠
2 ♥	2 ♦	2 ♣	2 ♠
3 ♥	3 ♦	3 ♣	3 ♠
4 ♥	4 ♦	4 ♣	4 ♠
5 ♥	5 ♦	5 ♣	5 ♠
6 ♥	6 ♦	6 ♣	6 ♠
7 ♥	7 ♦	7 ♣	7 ♠
8 ♥	8 ♦	8 ♣	8 ♠
9 ♥	9 ♦	9 ♣	9 ♠
10 ♥	10 ♦	10 ♣	10 ♠
J ♥	J ♦	J ♣	J ♠
Q ♥	Q ♦	Q ♣	Q ♠
K ♥	K ♦	K ♣	K ♠

Ranks and suits in a standard deck of playing cards
Figure 9.2

EXAMPLE 9 Counting Card Hands

A standard poker hand consists of five cards dealt from a deck of 52 (see Figure 9.2). How many different poker hands are possible? (After the cards are dealt, the player may reorder them, and so order is not important.)

Solution To find the number of different poker hands, use the formula for the number of combinations of 52 elements taken five at a time, as follows.

$$\begin{aligned} {}_{52}C_5 &= \frac{52!}{(52-5)!5!} \\ &= \frac{52!}{47!5!} \\ &= \frac{52 \cdot 51 \cdot 50 \cdot 49 \cdot 48 \cdot \cancel{47!}}{5 \cdot 4 \cdot 3 \cdot 2 \cdot 1 \cdot \cancel{47!}} \\ &= 2{,}598{,}960 \end{aligned}$$

✓ **Checkpoint** *Audio-video solution in English & Spanish at LarsonPrecalculus.com.*

In three-card poker, each player is dealt three cards from a deck of 52. How many different three-card poker hands are possible? (Order is not important.)

EXAMPLE 10 Forming a Team

You are forming a 12-member swim team from 10 girls and 15 boys. The team must consist of five girls and seven boys. How many different 12-member teams are possible?

Solution There are ${}_{10}C_5$ ways of choosing five girls. There are ${}_{15}C_7$ ways of choosing seven boys. By the Fundamental Counting Principle, there are ${}_{10}C_5 \cdot {}_{15}C_7$ ways of choosing five girls and seven boys.

$${}_{10}C_5 \cdot {}_{15}C_7 = \frac{10!}{5! \cdot 5!} \cdot \frac{15!}{8! \cdot 7!} = 252 \cdot 6435 = 1{,}621{,}620$$

So, there are 1,621,620 12-member swim teams possible.

✓ **Checkpoint** *Audio-video solution in English & Spanish at LarsonPrecalculus.com.*

In Example 10, the team must consist of six boys and six girls. How many different 12-member teams are possible?

REMARK When solving problems involving counting principles, you need to be able to distinguish among the various counting principles in order to determine which is necessary to solve the problem correctly. To do this, ask yourself the following questions.

1. Is the order of the elements important? *Permutation*
2. Are the chosen elements a subset of a larger set in which order is not important? *Combination*
3. Does the problem involve two or more separate events? *Fundamental Counting Principle*

Summarize (Section 9.6)

1. Explain how to solve a simple counting problem *(page 652)*. For examples of solving simple counting problems, see Examples 1 and 2.
2. State the Fundamental Counting Principle *(page 653)*. For examples of using the Fundamental Counting Principle to solve counting problems, see Examples 3 and 4.
3. Explain how to find the number of permutations of n elements *(page 654)*, the number of permutations of n elements taken r at a time *(page 655)*, and the number of distinguishable permutations *(page 656)*. For examples of using permutations to solve counting problems, see Examples 5–7.
4. Explain how to find the number of combinations of n elements taken r at a time *(page 657)*. For examples of using combinations to solve counting problems, see Examples 8–10.

9.6 Exercises

See CalcChat.com for tutorial help and worked-out solutions to odd-numbered exercises.

Vocabulary: Fill in the blanks.

1. The ________ ________ ________ states that when there are m_1 different ways for one event to occur and m_2 different ways for a second event to occur, there are $m_1 \cdot m_2$ ways for both events to occur.
2. An ordering of n elements is called a ________ of the elements.
3. The number of permutations of n elements taken r at a time is given by ________.
4. The number of ________ ________ of n objects is given by $\dfrac{n!}{n_1!n_2!n_3! \cdot \cdot \cdot n_k!}$.
5. When selecting subsets of a larger set in which order is not important, you are finding the number of ________ of n elements taken r at a time.
6. The number of combinations of n elements taken r at a time is given by ________.

Skills and Applications

Random Selection **In Exercises 7–14, determine the number of ways a computer can randomly generate one or more such integers from 1 through 12.**

7. An odd integer
8. An even integer
9. A prime integer
10. An integer that is greater than 9
11. An integer that is divisible by 4
12. An integer that is divisible by 3
13. Two *distinct* integers whose sum is 9
14. Two *distinct* integers whose sum is 8

15. **Entertainment Systems** A customer can choose one of three amplifiers, one of two compact disc players, and one of five speaker models for an entertainment system. Determine the number of possible system configurations.
16. **Job Applicants** A small college needs two additional faculty members: a chemist and a statistician. There are five applicants for the chemistry position and three applicants for the statistics position. In how many ways can the college fill these positions?
17. **Course Schedule** A college student is preparing a course schedule for the next semester. The student may select one of two mathematics courses, one of three science courses, and one of five courses from the social sciences and humanities. How many schedules are possible?
18. **Aircraft Boarding** Eight people are boarding an aircraft. Two have tickets for first class and board before those in the economy class. In how many ways can the eight people board the aircraft?
19. **True-False Exam** In how many ways can you answer a six-question true-false exam? (Assume that you do not omit any questions.)
20. **True-False Exam** In how many ways can you answer a 12-question true-false exam? (Assume that you do not omit any questions.)
21. **License Plate Numbers** In the state of Pennsylvania, each standard automobile license plate number consists of three letters followed by a four-digit number. How many distinct license plate numbers can be formed in Pennsylvania?
22. **License Plate Numbers** In a certain state, each automobile license plate number consists of two letters followed by a four-digit number. To avoid confusion between "O" and "zero" and between "I" and "one," the letters "O" and "I" are not used. How many distinct license plate numbers can be formed in this state?
23. **Three-Digit Numbers** How many three-digit numbers can you form under each condition?
    (a) The leading digit cannot be zero.
    (b) The leading digit cannot be zero and no repetition of digits is allowed.
    (c) The leading digit cannot be zero and the number must be a multiple of 5.
    (d) The number is at least 400.
24. **Four-Digit Numbers** How many four-digit numbers can you form under each condition?
    (a) The leading digit cannot be zero.
    (b) The leading digit cannot be zero and no repetition of digits is allowed.
    (c) The leading digit cannot be zero and the number must be less than 5000.
    (d) The leading digit cannot be zero and the number must be even.
25. **Combination Lock** A combination lock will open when you select the right choice of three numbers (from 1 to 40, inclusive). How many different lock combinations are possible?

26. Combination Lock A combination lock will open when you select the right choice of three numbers (from 1 to 50, inclusive). How many different lock combinations are possible?

27. Concert Seats Four couples have reserved seats in one row for a concert. In how many different ways can they sit when

(a) there are no seating restrictions?

(b) the two members of each couple wish to sit together?

28. Single File In how many orders can four girls and four boys walk through a doorway single file when

(a) there are no restrictions?

(b) the girls walk through before the boys?

29. Posing for a Photograph In how many ways can five children posing for a photograph line up in a row?

30. Riding in a Car In how many ways can six people sit in a six-passenger car?

Evaluating $_nP_r$ **In Exercises 31–34, evaluate $_nP_r$.**

31. $_4P_4$ **32.** $_8P_3$

33. $_{20}P_2$ **34.** $_5P_4$

Evaluating $_nP_r$ **In Exercises 35–38, evaluate $_nP_r$ using a graphing utility.**

35. $_{20}P_5$ **36.** $_{100}P_5$

37. $_{100}P_3$ **38.** $_{10}P_8$

39. Kidney Donors

A patient with end-stage kidney disease has nine family members who are potential kidney donors. How many possible orders are there for a best match, a second-best match, and a third-best match?

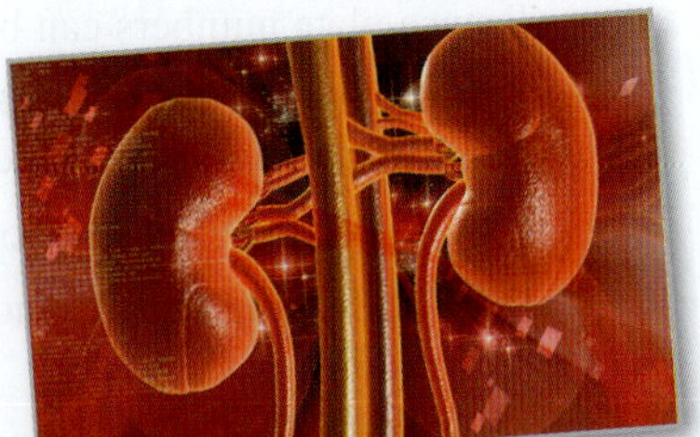

40. Choosing Officers From a pool of 12 candidates, the offices of president, vice-president, secretary, and treasurer will be filled. In how many different ways can the offices be filled?

41. Batting Order A baseball coach is creating a nine-player batting order by selecting from a team of 15 players. How many different batting orders are possible?

42. Athletics Eight sprinters have qualified for the finals in the 100-meter dash at the NCAA national track meet. In how many ways can the sprinters come in first, second, and third? (Assume there are no ties.)

Number of Distinguishable Permutations **In Exercises 43–46, find the number of distinguishable permutations of the group of letters.**

43. A, A, G, E, E, E, M **44.** B, B, B, T, T, T, T, T

45. A, L, G, E, B, R, A **46.** M, I, S, S, I, S, S, I, P, P, I

47. Writing Permutations Write all permutations of the letters A, B, C, and D.

48. Writing Permutations Write all permutations of the letters A, B, C, and D when letters B and C must remain between A and D.

Evaluating $_nC_r$ **In Exercises 49–52, evaluate $_nC_r$ using the formula from this section.**

49. $_5C_2$ **50.** $_6C_3$

51. $_4C_1$ **52.** $_{25}C_0$

Evaluating $_nC_r$ **In Exercises 53–56, evaluate $_nC_r$ using a graphing utility.**

53. $_{20}C_4$ **54.** $_{10}C_7$

55. $_{42}C_5$ **56.** $_{50}C_6$

57. Writing Combinations Write all combinations of two letters that you can form from the letters A, B, C, D, E, and F. (The order of the two letters is not important.)

58. Forming an Experimental Group In order to conduct an experiment, researchers randomly select five students from a class of 20. How many different groups of five students are possible?

59. Jury Selection In how many different ways can a jury of 12 people be randomly selected from a group of 40 people?

60. Committee Members A U.S. Senate Committee has 14 members. Assuming party affiliation is not a factor in selection, how many different committees are possible from the 100 U.S. senators?

61. Lottery Choices In the Massachusetts Mass Cash game, a player randomly chooses five distinct numbers from 1 to 35. In how many ways can a player select the five numbers?

62. Lottery Choices In the Louisiana Lotto game, a player randomly chooses six distinct numbers from 1 to 40. In how many ways can a player select the six numbers?

63. Defective Units A shipment of 25 television sets contains three defective units. In how many ways can a vending company purchase four of these units and receive (a) all good units, (b) two good units, and (c) at least two good units?

64. Interpersonal Relationships The complexity of interpersonal relationships increases dramatically as the size of a group increases. Determine the numbers of different two-person relationships in groups of people of sizes (a) 3, (b) 8, (c) 12, and (d) 20.

Kannanimages/Shutterstock.com

65. Poker Hand You are dealt five cards from a standard deck of 52 playing cards. In how many ways can you get (a) a full house and (b) a five-card combination containing two jacks and three aces? (A full house consists of three of one kind and two of another. For example, A-A-A-5-5 and K-K-K-10-10 are full houses.)

66. Job Applicants An employer interviews 12 people for four openings at a company. Five of the 12 people are women. All 12 applicants are qualified. In how many ways can the employer fill the four positions when (a) the selection is random and (b) exactly two selections are women?

67. Forming a Committee A local college is forming a six-member research committee having one administrator, three faculty members, and two students. There are seven administrators, 12 faculty members, and 20 students in contention for the committee. How many six-member committees are possible?

68. Law Enforcement A police department uses computer imaging to create digital photographs of alleged perpetrators from eyewitness accounts. One software package contains 195 hairlines, 99 sets of eyes and eyebrows, 89 noses, 105 mouths, and 74 chins and cheek structures.

(a) Find the possible number of different faces that the software could create.

(b) An eyewitness can clearly recall the hairline and eyes and eyebrows of a suspect. How many different faces can be produced with this information?

Geometry **In Exercises 69–72, find the number of diagonals of the polygon. (A line segment connecting any two nonadjacent vertices is called a *diagonal* of the polygon.)**

69. Pentagon

70. Hexagon

71. Octagon

72. Decagon (10 sides)

73. Geometry Three points that are not collinear determine three lines. How many lines are determined by nine points, no three of which are collinear?

74. Lottery Powerball is a lottery game that is operated by the Multi-State Lottery Association and is played in 42 states, Washington D.C., and the U.S. Virgin Islands. The game is played by drawing five white balls out of a drum of 59 white balls (numbered 1–59) and one red powerball out of a drum of 35 red balls (numbered 1–35). The jackpot is won by matching all five white balls in any order and the red powerball.

(a) Find the possible number of winning Powerball numbers.

(b) Find the possible number of winning Powerball numbers when you win the jackpot by matching all five white balls in order and the red powerball.

Solving an Equation **In Exercises 75–82, solve for n.**

75. $4 \cdot {}_{n+1}P_2 = {}_{n+2}P_3$

76. $5 \cdot {}_{n-1}P_1 = {}_nP_2$

77. ${}_{n+1}P_3 = 4 \cdot {}_nP_2$

78. ${}_{n+2}P_3 = 6 \cdot {}_{n+2}P_1$

79. $14 \cdot {}_nP_3 = {}_{n+2}P_4$

80. ${}_nP_5 = 18 \cdot {}_{n-2}P_4$

81. ${}_nP_4 = 10 \cdot {}_{n-1}P_3$

82. ${}_nP_6 = 12 \cdot {}_{n-1}P_5$

Exploration

True or False? **In Exercises 83 and 84, determine whether the statement is true or false. Justify your answer.**

83. The number of letter pairs that can be formed in any order from any two of the first 13 letters in the alphabet (A–M) is an example of a permutation.

84. The number of permutations of n elements can be determined by using the Fundamental Counting Principle.

85. Think About It Without calculating the numbers, determine which of the following is greater. Explain.

(a) The number of combinations of 10 elements taken six at a time

(b) The number of permutations of 10 elements taken six at a time

86. HOW DO YOU SEE IT? Without calculating, determine whether the value of ${}_nP_r$ is greater than the value of ${}_nC_r$ for the values of n and r given in the table. Complete the table using yes (Y) or no (N). Is the value of ${}_nP_r$ always greater than the value of ${}_nC_r$? Explain.

n \ r	0	1	2	3	4	5	6	7
1								
2								
3								
4								
5								
6								
7								

Proof **In Exercises 87–90, prove the identity.**

87. ${}_nP_{n-1} = {}_nP_n$

88. ${}_nC_n = {}_nC_0$

89. ${}_nC_{n-1} = {}_nC_1$

90. ${}_nC_r = \dfrac{{}_nP_r}{r!}$

91. Think About It Can your graphing utility evaluate ${}_{100}P_{80}$? If not, then explain why.

9.7 Probability

Probability applies to many real-life applications. For instance, in Exercise 59 on page 672, you will calculate probabilities that relate to a communication network and an independent backup system for a space vehicle.

- **Find the probabilities of events.**
- **Find the probabilities of mutually exclusive events.**
- **Find the probabilities of independent events.**
- **Find the probability of the complement of an event.**

The Probability of an Event

Any happening for which the result is uncertain is called an **experiment.** The possible results of the experiment are **outcomes,** the set of all possible outcomes of the experiment is the **sample space** of the experiment, and any subcollection of a sample space is an **event.**

For instance, when you toss a six-sided die, the numbers 1 through 6 can represent the sample space. For this experiment, each of the outcomes is *equally likely.*

To describe sample spaces in such a way that each outcome is equally likely, you must sometimes distinguish between or among various outcomes in ways that appear artificial. Example 1 illustrates such a situation.

EXAMPLE 1 **Finding a Sample Space**

Find the sample space for each of the following experiments.

a. You toss one coin.

b. You toss two coins.

c. You toss three coins.

Solution

a. Because the coin will land either heads up (denoted by H) or tails up (denoted by T), the sample space is

$$S = \{H, T\}.$$

b. Because either coin can land heads up or tails up, the possible outcomes are as follows.

HH = heads up on both coins

HT = heads up on first coin and tails up on second coin

TH = tails up on first coin and heads up on second coin

TT = tails up on both coins

So, the sample space is

$$S = \{HH, HT, TH, TT\}.$$

Note that this list distinguishes between the two cases HT and TH, even though these two outcomes appear to be similar.

c. Following the notation of part (b), the sample space is

$$S = \{HHH, HHT, HTH, HTT, THH, THT, TTH, TTT\}.$$

Note that this list distinguishes among the cases HHT, HTH, and THH, and among the cases HTT, THT, and TTH.

Checkpoint *Audio-video solution in English & Spanish at LarsonPrecalculus.com.*

Find the sample space for the following experiment.

You toss a coin twice and a six-sided die once.

1971yes/Shutterstock.com

To calculate the probability of an event, count the number of outcomes in the event and in the sample space. The *number of outcomes* in event E is denoted by $n(E)$, and the number of outcomes in the sample space S is denoted by $n(S)$. The probability that event E will occur is given by $n(E)/n(S)$.

The Probability of an Event

If an event E has $n(E)$ equally likely outcomes and its sample space S has $n(S)$ equally likely outcomes, then the **probability** of event E is

$$P(E) = \frac{n(E)}{n(S)}.$$

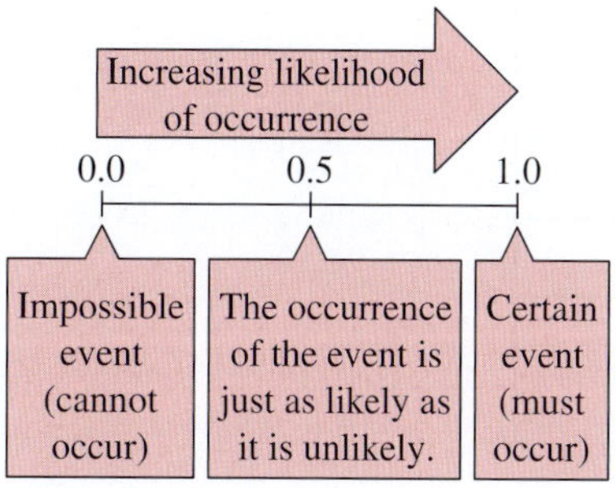

Figure 9.3

Because the number of outcomes in an event must be less than or equal to the number of outcomes in the sample space, the probability of an event must be a number between 0 and 1. That is,

$$0 \le P(E) \le 1$$

as indicated in Figure 9.3. If $P(E) = 0$, then event E *cannot occur,* and E is called an **impossible event.** If $P(E) = 1$, then event E *must occur,* and E is called a **certain event.**

EXAMPLE 2 Finding the Probability of an Event

a. You toss two coins. What is the probability that both land heads up?

b. You draw one card at random from a standard deck of playing cards. What is the probability that it is an ace?

Solution

a. Following the procedure in Example 1(b), let

$$E = \{HH\}$$

and

$$S = \{HH, HT, TH, TT\}.$$

The probability of getting two heads is

$$P(E) = \frac{n(E)}{n(S)} = \frac{1}{4}.$$

REMARK You can write a probability as a fraction, a decimal, or a percent. For instance, in Example 2(a), the probability of getting two heads can be written as $\frac{1}{4}$, 0.25, or 25%.

b. Because there are 52 cards in a standard deck of playing cards and there are four aces (one in each suit), the probability of drawing an ace is

$$P(E) = \frac{n(E)}{n(S)}$$

$$= \frac{4}{52}$$

$$= \frac{1}{13}.$$

Checkpoint Audio-video solution in English & Spanish at LarsonPrecalculus.com.

a. You toss three coins. What is the probability that all three land tails up?

b. You draw one card at random from a standard deck of playing cards. What is the probability that it is a diamond?

As shown in Example 3, when you toss two six-sided dice, the probability that the total of the two dice is 7 is $\frac{1}{6}$.

EXAMPLE 3 **Finding the Probability of an Event**

You toss two six-sided dice. What is the probability that the total of the two dice is 7?

Solution Because there are six possible outcomes on each die, use the Fundamental Counting Principle to conclude that there are $6 \cdot 6$ or 36 different outcomes when you toss two dice. To find the probability of rolling a total of 7, you must first count the number of ways in which this can occur.

First Die	Second Die
1	6
2	5
3	4
4	3
5	2
6	1

So, a total of 7 can be rolled in six ways, which means that the probability of rolling a 7 is

$$P(E) = \frac{n(E)}{n(S)} = \frac{6}{36} = \frac{1}{6}.$$

✓ **Checkpoint** *Audio-video solution in English & Spanish at LarsonPrecalculus.com.*

You toss two six-sided dice. What is the probability that the total of the two dice is 5?

EXAMPLE 4 **Finding the Probability of an Event**

Twelve-sided dice, as shown in Figure 9.4, can be constructed (in the shape of regular dodecahedrons) such that each of the numbers from 1 to 6 appears twice on each die. Prove that these dice can be used in any game requiring ordinary six-sided dice without changing the probabilities of the various events.

Solution For an ordinary six-sided die, each of the numbers

1, 2, 3, 4, 5, and 6

occurs only once, so the probability of rolling any particular number is

$$P(E) = \frac{n(E)}{n(S)} = \frac{1}{6}.$$

For one of the 12-sided dice, each number occurs twice, so the probability of rolling any particular number is

$$P(E) = \frac{n(E)}{n(S)} = \frac{2}{12} = \frac{1}{6}.$$

Figure 9.4

✓ **Checkpoint** *Audio-video solution in English & Spanish at LarsonPrecalculus.com.*

Show that the probability of drawing a club at random from a standard deck of playing cards is the same as the probability of drawing the ace of hearts at random from a set consisting of the aces of hearts, diamonds, clubs, and spades. ■

Sergey Mironov/Shutterstock.com

EXAMPLE 5 **Random Selection**

The following figure shows the numbers of colleges and universities in various regions of the United States in 2010. What is the probability that an institution selected at random is in one of the three southern regions? *(Source: National Center for Education Statistics)*

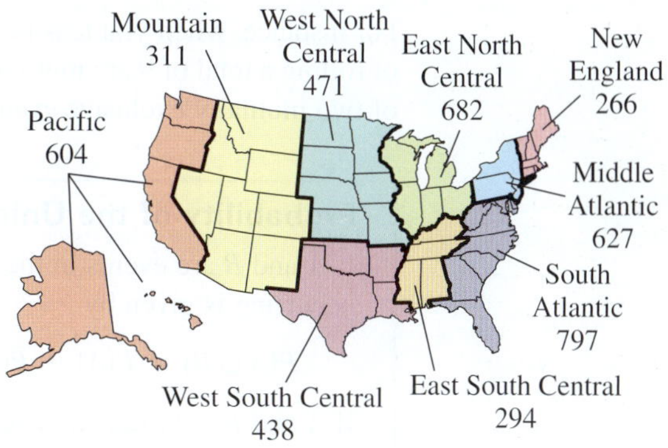

Solution From the figure, the total number of colleges and universities is 4490. Because there are $797 + 294 + 438 = 1529$ colleges and universities in the three southern regions, the probability that the institution is in one of these regions is

$$P(E) = \frac{n(E)}{n(S)} = \frac{1529}{4490} \approx 0.341.$$

Checkpoint Audio-video solution in English & Spanish at LarsonPrecalculus.com.

In Example 5, what is the probability that the randomly selected institution is in the Pacific region?

EXAMPLE 6 **The Probability of Winning a Lottery**

In Arizona's The Pick game, a player chooses six different numbers from 1 to 44. If these six numbers match the six numbers drawn (in any order) by the lottery commission, then the player wins (or shares) the top prize. What is the probability of winning the top prize when the player buys one ticket?

Solution To find the number of elements in the sample space, use the formula for the number of combinations of 44 elements taken six at a time.

$$\begin{aligned} n(S) &= {}_{44}C_6 \\ &= \frac{44 \cdot 43 \cdot 42 \cdot 41 \cdot 40 \cdot 39}{6 \cdot 5 \cdot 4 \cdot 3 \cdot 2 \cdot 1} \\ &= 7{,}059{,}052 \end{aligned}$$

When a player buys one ticket, the probability of winning is

$$P(E) = \frac{1}{7{,}059{,}052}.$$

Checkpoint Audio-video solution in English & Spanish at LarsonPrecalculus.com.

In Pennsylvania's Cash 5 game, a player chooses five different numbers from 1 to 43. If these five numbers match the five numbers drawn (in any order) by the lottery commission, then the player wins (or shares) the top prize. What is the probability of winning the top prize when the player buys one ticket?

Mutually Exclusive Events

Two events A and B (from the same sample space) are **mutually exclusive** when A and B have no outcomes in common. In the terminology of sets, the intersection of A and B is the empty set, which implies that

$$P(A \cap B) = 0.$$

For instance, when you toss two dice, the event A of rolling a total of 6 and the event B of rolling a total of 9 are mutually exclusive. To find the probability that one or the other of two mutually exclusive events will occur, *add* their individual probabilities.

> **Probability of the Union of Two Events**
>
> If A and B are events in the same sample space, then the probability of A or B occurring is given by
>
> $$P(A \cup B) = P(A) + P(B) - P(A \cap B).$$
>
> If A and B are mutually exclusive, then
>
> $$P(A \cup B) = P(A) + P(B).$$

EXAMPLE 7 The Probability of a Union of Events

You draw one card at random from a standard deck of 52 playing cards. What is the probability that the card is either a heart or a face card?

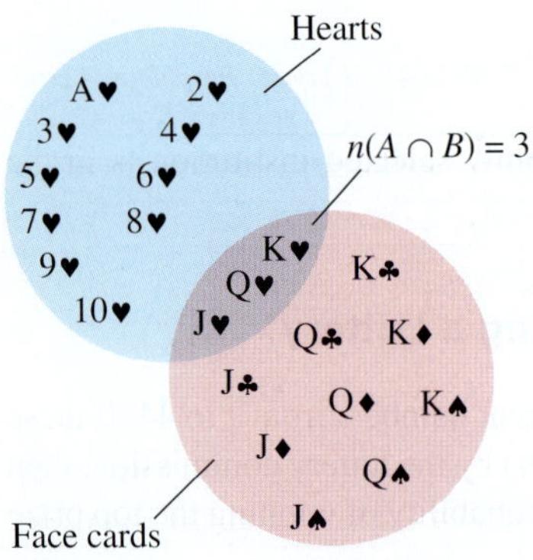

Figure 9.5

Solution Because the deck has 13 hearts, the probability of drawing a heart (event A) is

$$P(A) = \frac{13}{52}.$$

Similarly, because the deck has 12 face cards, the probability of drawing a face card (event B) is

$$P(B) = \frac{12}{52}.$$

Because three of the cards are hearts *and* face cards (see Figure 9.5), it follows that

$$P(A \cap B) = \frac{3}{52}.$$

Finally, applying the formula for the probability of the union of two events, the probability of drawing a heart or a face card is

$$\begin{aligned} P(A \cup B) &= P(A) + P(B) - P(A \cap B) \\ &= \frac{13}{52} + \frac{12}{52} - \frac{3}{52} \\ &= \frac{22}{52} \\ &\approx 0.423. \end{aligned}$$

✓ Checkpoint *Audio-video solution in English & Spanish at LarsonPrecalculus.com.*

You draw one card at random from a standard deck of 52 playing cards. What is the probability that the card is either an ace or a spade?

EXAMPLE 8 Probability of Mutually Exclusive Events

The human resources department of a company has compiled data showing the number of years of service for each employee. The table shows the results.

DATA

Years of Service	Number of Employees
0–4	157
5–9	89
10–14	74
15–19	63
20–24	42
25–29	38
30–34	35
35–39	21
40–44	8
45 or more	2

Spreadsheet at LarsonPrecalculus.com

a. What is the probability that an employee chosen at random has 4 or fewer years of service?

b. What is the probability that an employee chosen at random has 9 or fewer years of service?

Solution

a. To begin, add the number of employees to find that the total is 529. Next, let event A represent choosing an employee with 0 to 4 years of service. Then the probability of choosing an employee who has 4 or fewer years of service is

$$P(A) = \frac{157}{529}$$
$$\approx 0.297.$$

b. Let event B represent choosing an employee with 5 to 9 years of service. Then

$$P(B) = \frac{89}{529}.$$

Because event A from part (a) and event B have no outcomes in common, these two events are mutually exclusive and

$$P(A \cup B) = P(A) + P(B)$$
$$= \frac{157}{529} + \frac{89}{529}$$
$$= \frac{246}{529}$$
$$\approx 0.465.$$

So, the probability of choosing an employee who has 9 or fewer years of service is about 0.465.

✓ **Checkpoint** Audio-video solution in English & Spanish at LarsonPrecalculus.com.

In Example 8, what is the probability that an employee chosen at random has 30 or more years of service?

Independent Events

Two events are **independent** when the occurrence of one has no effect on the occurrence of the other. For instance, rolling a total of 12 with two six-sided dice has no effect on the outcome of future rolls of the dice. To find the probability that two independent events will occur, *multiply* the probabilities of each.

Probability of Independent Events

If A and B are independent events, then the probability that both A and B will occur is

$$P(A \text{ and } B) = P(A) \cdot P(B).$$

EXAMPLE 9 Probability of Independent Events

A random number generator on a computer selects three integers from 1 to 20. What is the probability that all three numbers are less than or equal to 5?

Solution The probability of selecting a number from 1 to 5 is

$$P(A) = \frac{5}{20} = \frac{1}{4}.$$

So, the probability that all three numbers are less than or equal to 5 is

$$P(A) \cdot P(A) \cdot P(A) = \left(\frac{1}{4}\right)\left(\frac{1}{4}\right)\left(\frac{1}{4}\right) = \frac{1}{64}.$$

✓ ***Checkpoint*** *Audio-video solution in English & Spanish at LarsonPrecalculus.com.*

A random number generator on a computer selects two integers from 1 to 30. What is the probability that both numbers are less than 12?

EXAMPLE 10 Probability of Independent Events

In 2012, approximately 27% of the population of the United States got their news from mobile devices. In a survey, researchers selected 10 people at random from the population. What is the probability that all 10 got their news from mobile devices? *(Source: Pew Research Center)*

Solution Let A represent choosing a person who gets his or her news from a mobile device. The probability of choosing a person who got his or her news from a mobile device is 0.27, the probability of choosing a second person who got his or her news from a mobile device is 0.27, and so on. Because these events are independent, the probability that all 10 people got their news from mobile devices is

$$[P(A)]^{10} = (0.27)^{10} \approx 0.000002059.$$

✓ ***Checkpoint*** *Audio-video solution in English & Spanish at LarsonPrecalculus.com.*

In Example 10, researchers selected five people at random from the population. What is the probability that all five got their news from mobile devices?

The Complement of an Event

The **complement of an event** A is the collection of all outcomes in the sample space that are *not* in A. The complement of event A is denoted by A'. Because $P(A \text{ or } A') = 1$ and because A and A' are mutually exclusive, it follows that $P(A) + P(A') = 1$. So, the probability of A' is

$$P(A') = 1 - P(A).$$

For instance, if the probability of *winning* a certain game is $P(A) = \frac{1}{4}$, then the probability of *losing* the game is $P(A') = 1 - \frac{1}{4} = \frac{3}{4}$.

Probability of a Complement

Let A be an event and let A' be its complement. If the probability of A is $P(A)$, then the probability of the complement is

$$P(A') = 1 - P(A).$$

EXAMPLE 11 Finding the Probability of a Complement

A manufacturer has determined that a machine averages one faulty unit for every 1000 it produces. What is the probability that an order of 200 units will have one or more faulty units?

Solution To solve this problem as stated, you would need to find the probabilities of having exactly one faulty unit, exactly two faulty units, exactly three faulty units, and so on. However, using complements, you can find the probability that all units are perfect and then subtract this value from 1. Because the probability that any given unit is perfect is 999/1000, the probability that all 200 units are perfect is

$$P(A) = \left(\frac{999}{1000}\right)^{200} \approx 0.819.$$

So, the probability that at least one unit is faulty is

$$P(A') = 1 - P(A) \approx 1 - 0.819 = 0.181.$$

✓ **Checkpoint** *Audio-video solution in English & Spanish at LarsonPrecalculus.com.*

A manufacturer has determined that a machine averages one faulty unit for every 500 it produces. What is the probability that an order of 300 units will have one or more faulty units?

Summarize (Section 9.7)

1. State the definition of the probability of an event *(page 663)*. For examples of finding the probabilities of events, see Examples 2–6.
2. State the definition of mutually exclusive events and explain how to find the probability of the union of two events *(page 666)*. For examples of finding the probabilities of the unions of two events, see Examples 7 and 8.
3. State the definition of, and explain how to find the probability of, independent events *(page 668)*. For examples of finding the probabilities of independent events, see Examples 9 and 10.
4. State the definition of, and explain how to find the probability of, the complement of an event *(page 669)*. For an example of finding the probability of the complement of an event, see Example 11.

9.7 Exercises

See CalcChat.com for tutorial help and worked-out solutions to odd-numbered exercises.

Vocabulary

In Exercises 1–7, fill in the blanks.

1. An ________ is any happening for which the result is uncertain, and the possible results are called ________.
2. The set of all possible outcomes of an experiment is called the ________ ________.
3. To determine the ________ of an event, you can use the formula $P(E) = \dfrac{n(E)}{n(S)}$, where $n(E)$ is the number of outcomes in the event and $n(S)$ is the number of outcomes in the sample space.
4. If $P(E) = 0$, then E is an ________ event, and if $P(E) = 1$, then E is a ________ event.
5. If two events from the same sample space have no outcomes in common, then the two events are ________ ________.
6. If the occurrence of one event has no effect on the occurrence of a second event, then the events are ________.
7. The ________ of an event A is the collection of all outcomes in the sample space that are not in A.
8. Match the probability formula with the correct probability name.

(a) Probability of the union of two events	(i) $P(A \cup B) = P(A) + P(B)$
(b) Probability of mutually exclusive events	(ii) $P(A') = 1 - P(A)$
(c) Probability of independent events	(iii) $P(A \cup B) = P(A) + P(B) - P(A \cap B)$
(d) Probability of a complement	(iv) $P(A \text{ and } B) = P(A) \cdot P(B)$

Skills and Applications

Finding a Sample Space **In Exercises 9–14, find the sample space for the experiment.**

9. You toss a coin and a six-sided die.
10. You toss a six-sided die twice and record the sum of the results.
11. A taste tester ranks three varieties of yogurt, A, B, and C, according to preference.
12. You select two marbles (without replacement) from a bag containing two red marbles, two blue marbles, and one yellow marble. You record the color of each marble.
13. Two county supervisors are selected from five supervisors, A, B, C, D, and E, to study a recycling plan.
14. A sales representative makes presentations about a product in three homes per day. In each home, there may be a sale (denote by S) or there may be no sale (denote by F).

Tossing a Coin **In Exercises 15–20, find the probability for the experiment of tossing a coin three times. Use the sample space**

$S = \{HHH, HHT, HTH, HTT, THH, THT, TTH, TTT\}$.

15. The probability of getting exactly one tail
16. The probability of getting exactly two tails
17. The probability of getting a head on the first toss
18. The probability of getting a tail on the last toss
19. The probability of getting at least one head
20. The probability of getting at least two heads

Drawing a Card **In Exercises 21–24, find the probability for the experiment of drawing a card at random from a standard deck of 52 playing cards.**

21. The card is a face card.
22. The card is not a face card.
23. The card is a red face card.
24. The card is a 9 or lower. (Aces are low.)

Tossing a Die **In Exercises 25–30, find the probability for the experiment of tossing a six-sided die twice.**

25. The sum is 6.
26. The sum is at least 8.
27. The sum is less than 11.
28. The sum is 2, 3, or 12.
29. The sum is odd and no more than 7.
30. The sum is odd or prime.

Drawing Marbles **In Exercises 31–34, find the probability for the experiment of drawing two marbles at random (without replacement) from a bag containing one green, two yellow, and three red marbles.**

31. Both marbles are red.
32. Both marbles are yellow.
33. Neither marble is yellow.
34. The marbles are different colors.

35. Graphical Reasoning In 2011, there were approximately 13.75 million unemployed workers in the United States. The circle graph shows the age profile of these unemployed workers. *(Source: U.S. Bureau of Labor Statistics)*

Ages of Unemployed Workers

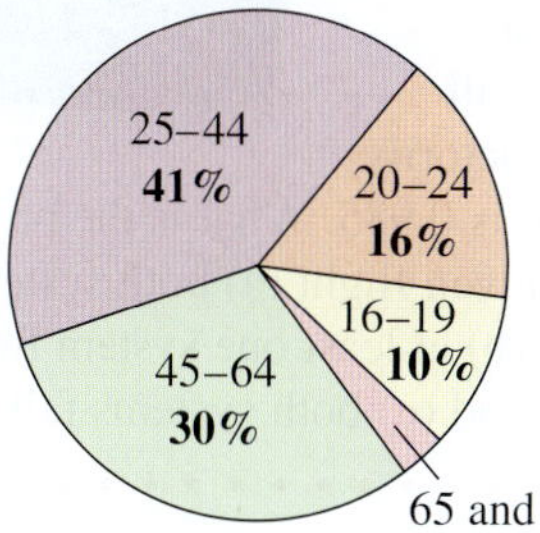

(a) Estimate the number of unemployed workers in the 16–19 age group.

(b) What is the probability that a person selected at random from the population of unemployed workers is in the 25–44 age group?

(c) What is the probability that a person selected at random from the population of unemployed workers is in the 45–64 age group?

(d) What is the probability that a person selected at random from the population of unemployed workers is 45 or older?

36. Data Analysis An independent polling organization interviews one hundred college students to determine their political party affiliations and whether they favor a balanced-budget amendment to the Constitution. The table lists the results of the study. In the table, D represents Democrat and R represents Republican.

	Favor	Not Favor	Unsure	Total
D	23	25	7	55
R	32	9	4	45
Total	55	34	11	100

Find the probability that a person selected at random from the sample is as described.

(a) A person who does not favor the amendment

(b) A Republican

(c) A Democrat who favors the amendment

37. Education In a high school graduating class of 128 students, 52 are on the honor roll. Of these, 48 are going on to college; of the other 76 students, 56 are going on to college. What is the probability that a student selected at random from the class is (a) going to college, (b) not going to college, and (c) not going to college and on the honor roll?

38. Alumni Association A college sends a survey to members of the class of 2012. Of the 1254 people who graduated that year, 672 are women, of whom 124 went on to graduate school. Of the 582 male graduates, 198 went on to graduate school. What is the probability that a class of 2012 alumnus selected at random is (a) female, (b) male, and (c) female and did not attend graduate school?

39. Winning an Election Three people are running for president of a class. The results of a poll indicate that the first candidate has an estimated 37% chance of winning and the second candidate has an estimated 44% chance of winning. What is the probability that the third candidate will win?

40. Payroll Error The employees of a company work in six departments: 31 are in sales, 54 are in research, 42 are in marketing, 20 are in engineering, 47 are in finance, and 58 are in production. The payroll department loses one employee's paycheck. What is the probability that the employee works in the research department?

Using Counting Principles In Exercises 41–48, the sample spaces are large and you should use the counting principles discussed in Section 9.6.

41. A class receives a list of 20 study problems, from which 10 will be part of an upcoming exam. A student knows how to solve 15 of the problems. Find the probability that the student will be able to answer (a) all 10 questions on the exam, (b) exactly eight questions on the exam, and (c) at least nine questions on the exam.

42. A payroll department addresses five paychecks and envelopes to five different people and randomly inserts the paychecks into the envelopes. What is the probability that (a) exactly one paycheck is inserted in the correct envelope and (b) at least one paycheck is in the correct envelope?

43. On a game show, you are given five digits to arrange in the proper order to form the price of a car. If you are correct, then you win the car. What is the probability of winning, given the following conditions?

(a) You guess the position of each digit.

(b) You know the first digit and guess the positions of the other digits.

44. The deck for a card game is made up of 108 cards. Twenty-five each are red, yellow, blue, and green, and eight are wild cards. Each player is randomly dealt a seven-card hand.

(a) What is the probability that a hand will contain exactly two wild cards?

(b) What is the probability that a hand will contain two wild cards, two red cards, and three blue cards?

45. You draw one card at random from a standard deck of 52 playing cards. Find the probability that (a) the card is an even-numbered card, (b) the card is a heart or a diamond, and (c) the card is a nine or a face card.

46. You draw five cards at random from a standard deck of 52 playing cards. What is the probability that the hand drawn is a full house? (A full house is a hand that consists of two of one kind and three of another kind.)

47. A shipment of 12 microwave ovens contains three defective units. A vending company has ordered four units, and because each has identical packaging, the selection will be random. What is the probability that (a) all four units are good, (b) exactly two units are good, and (c) at least two units are good?

48. ATM personal identification number (PIN) codes typically consist of four-digit sequences of numbers. Find the probability that if you forget your PIN, then you can guess the correct sequence (a) at random and (b) when you recall the first two digits.

49. Random Number Generator A random number generator on a computer selects two integers from 1 through 40. What is the probability that (a) both numbers are even, (b) one number is even and one number is odd, (c) both numbers are less than 30, and (d) the same number is selected twice?

50. Flexible Work Hours In a survey, people were asked whether they would prefer to work flexible hours—even when it meant slower career advancement—so they could spend more time with their families. The figure shows the results of the survey. What is the probability that three people chosen at random would prefer flexible work hours?

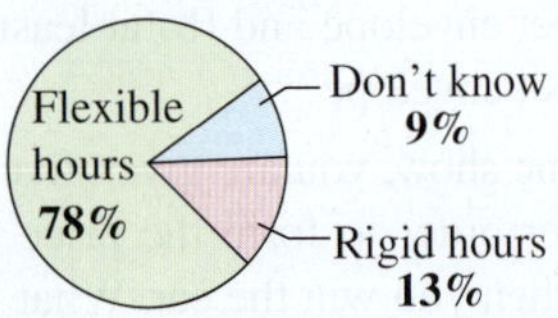

Finding the Probability of a Complement In Exercises 51–54, you are given the probability that an event *will* happen. Find the probability that the event *will not* happen.

51. $P(E) = 0.87$

52. $P(E) = 0.36$

53. $P(E) = \frac{1}{4}$

54. $P(E) = \frac{2}{3}$

Finding the Probability of a Complement In Exercises 55–58, you are given the probability that an event *will not* happen. Find the probability that the event *will* happen.

55. $P(E') = 0.23$

56. $P(E') = 0.92$

57. $P(E') = \frac{17}{35}$

58. $P(E') = \frac{61}{100}$

59. Backup System A space vehicle has an independent backup system for one of its communication networks. The probability that either system will function satisfactorily during a flight is 0.985. What is the probability that during a given flight (a) both systems function satisfactorily, (b) at least one system functions satisfactorily, and (c) both systems fail?

60. Backup Vehicle A fire company keeps two rescue vehicles. Because of the demand on the vehicles and the chance of mechanical failure, the probability that a specific vehicle is available when needed is 90%. The availability of one vehicle is independent of the availability of the other. Find the probability that (a) both vehicles are available at a given time, (b) neither vehicle is available at a given time, and (c) at least one vehicle is available at a given time.

61. Roulette American roulette is a game in which a wheel turns on a spindle and is divided into 38 pockets. Thirty-six of the pockets are numbered 1–36, of which half are red and half are black. Two of the pockets are green and are numbered 0 and 00 (see figure). The dealer spins the wheel and a small ball in opposite directions. As the ball slows to a stop, it has an equal probability of landing in any of the numbered pockets.

(a) Find the probability of landing in the number 00 pocket.

(b) Find the probability of landing in a red pocket.

(c) Find the probability of landing in a green pocket or a black pocket.

(d) Find the probability of landing in the number 14 pocket on two consecutive spins.

(e) Find the probability of landing in a red pocket on three consecutive spins.

1971yes/Shutterstock.com

62. A Boy or a Girl? Assume that the probability of the birth of a child of a particular sex is 50%. In a family with four children, what is the probability that (a) all the children are boys, (b) all the children are the same sex, and (c) there is at least one boy?

63. Geometry You and a friend agree to meet at your favorite fast-food restaurant between 5:00 P.M. and 6:00 P.M. The one who arrives first will wait 15 minutes for the other, and then will leave (see figure). What is the probability that the two of you will actually meet, assuming that your arrival times are random within the hour?

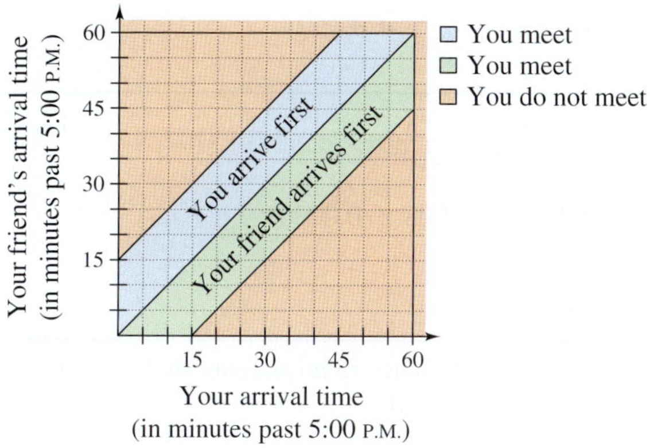

64. Estimating π You drop a coin of diameter d onto a paper that contains a grid of squares d units on a side (see figure).

(a) Find the probability that the coin covers a vertex of one of the squares on the grid.

(b) Perform the experiment 100 times and use the results to approximate π.

Exploration

True or False? In Exercises 65 and 66, determine whether the statement is true or false. Justify your answer.

65. If A and B are independent events with nonzero probabilities, then A can occur when B occurs.

66. Rolling a number less than 3 on a normal six-sided die has a probability of $\frac{1}{3}$. The complement of this event is to roll a number greater than 3, and its probability is $\frac{1}{2}$.

67. Pattern Recognition Consider a group of n people.

(a) Explain why the following pattern gives the probabilities that the n people have distinct birthdays.

$$n = 2: \quad \frac{365}{365} \cdot \frac{364}{365} = \frac{365 \cdot 364}{365^2}$$

$$n = 3: \quad \frac{365}{365} \cdot \frac{364}{365} \cdot \frac{363}{365} = \frac{365 \cdot 364 \cdot 363}{365^3}$$

(b) Use the pattern in part (a) to write an expression for the probability that $n = 4$ people have distinct birthdays.

(c) Let P_n be the probability that the n people have distinct birthdays. Verify that this probability can be obtained recursively by

$$P_1 = 1 \text{ and } P_n = \frac{365 - (n - 1)}{365} P_{n-1}.$$

(d) Explain why $Q_n = 1 - P_n$ gives the probability that at least two people in a group of n people have the same birthday.

(e) Use the results of parts (c) and (d) to complete the table.

n	10	15	20	23	30	40	50
P_n							
Q_n							

(f) How many people must be in a group so that the probability of at least two of them having the same birthday is greater than $\frac{1}{2}$? Explain.

68. HOW DO YOU SEE IT? The circle graphs show the percents of undergraduate students by class level at two colleges. A student is chosen at random from the combined undergraduate population of the two colleges. The probability that the student is a freshman, sophomore, or junior is 81%. Which college has a greater number of undergraduate students? Explain your reasoning.

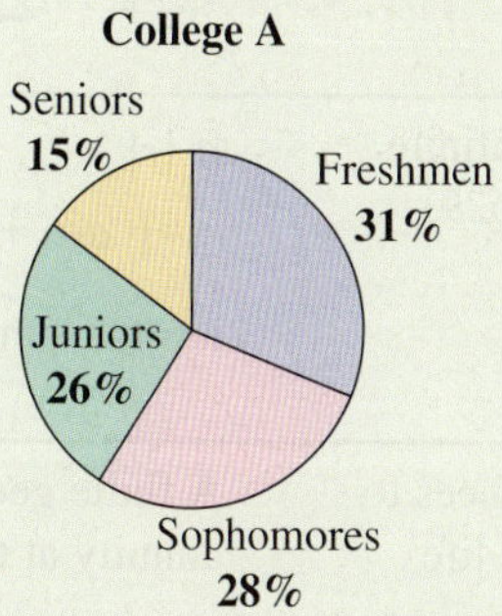

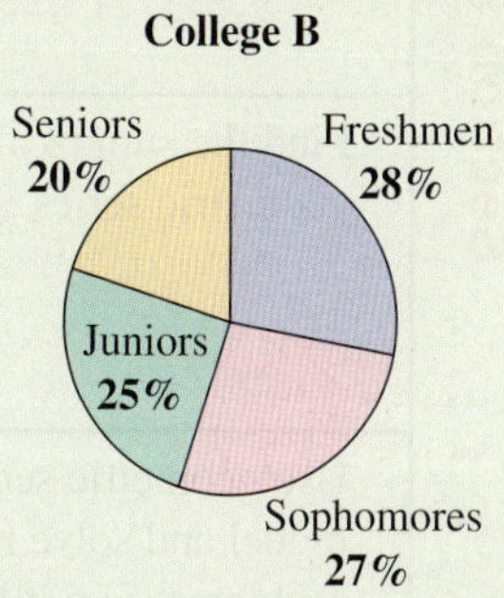

Chapter Summary

	What Did You Learn?	Explanation/Examples	Review Exercises
Section 9.1	Use sequence notation to write the terms of sequences *(p. 606).*	$a_n = 7n - 4;\quad a_1 = 7(1) - 4 = 3,\ a_2 = 7(2) - 4 = 10,$ $a_3 = 7(3) - 4 = 17,\ a_4 = 7(4) - 4 = 24$	1–8
	Use factorial notation *(p. 608).*	If n is a positive integer, then $n! = 1 \cdot 2 \cdot 3 \cdot 4 \cdot \cdot \cdot (n-1) \cdot n.$	9–12
	Use summation notation to write sums *(p. 610).*	The sum of the first n terms of a sequence is represented by $\sum_{i=1}^{n} a_i = a_1 + a_2 + a_3 + a_4 + \cdot \cdot \cdot + a_n.$	13–16
	Find the sums of series *(p. 611).*	$\sum_{i=1}^{\infty} \frac{8}{10^i} = \frac{8}{10^1} + \frac{8}{10^2} + \frac{8}{10^3} + \frac{8}{10^4} + \frac{8}{10^5} + \cdot \cdot \cdot$ $= 0.8 + 0.08 + 0.008 + 0.0008 + 0.00008 + \cdot \cdot \cdot$ $= 0.88888\ldots = \frac{8}{9}$	17, 18
	Use sequences and series to model and solve real-life problems *(p. 612).*	A sequence can help you model the balance in an account that earns compound interest. (See Example 10.)	19, 20
Section 9.2	Recognize, write, and find the *n*th terms of arithmetic sequences *(p. 616).*	$a_n = 9n + 5;\quad a_1 = 9(1) + 5 = 14,\ a_2 = 9(2) + 5 = 23,$ $a_3 = 9(3) + 5 = 32,\ a_4 = 9(4) + 5 = 41$	21–30
	Find *n*th partial sums of arithmetic sequences *(p. 619).*	The sum of a finite arithmetic sequence with n terms is $S_n = (n/2)(a_1 + a_n).$	31–36
	Use arithmetic sequences to model and solve real-life problems *(p. 621).*	An arithmetic sequence can help you find the total sales of a small business. (See Example 9.)	37, 38
Section 9.3	Recognize, write, and find the *n*th terms of geometric sequences *(p. 625).*	$a_n = 3(4^n);\quad a_1 = 3(4^1) = 12,\ a_2 = 3(4^2) = 48,$ $a_3 = 3(4^3) = 192,\ a_4 = 3(4^4) = 768$	39–50
	Find the sum of a finite geometric sequence *(p. 628).*	The sum of the finite geometric sequence $a_1, a_1 r, a_1 r^2, \ldots, a_1 r^{n-1}$ with common ratio $r \neq 1$ is given by $S_n = \sum_{i=1}^{n} a_1 r^{i-1} = a_1\left(\frac{1 - r^n}{1 - r}\right).$	51–58
	Find the sum of an infinite geometric series *(p. 629).*	If $\lvert r \rvert < 1$, then the infinite geometric series $a_1 + a_1 r + a_1 r^2 + \cdot \cdot \cdot + a_1 r^{n-1} + \cdot \cdot \cdot$ has the sum $S = \sum_{i=0}^{\infty} a_1 r^i = \frac{a_1}{1 - r}.$	59–62
	Use geometric sequences to model and solve real-life problems *(p. 630).*	A finite geometric sequence can help you find the balance in an annuity at the end of two years. (See Example 8.)	63, 64

	What Did You Learn?	Explanation/Examples	Review Exercises
Section 9.4	Use mathematical induction to prove statements involving a positive integer n *(p. 634).*	Let P_n be a statement involving the positive integer n. If (1) P_1 is true, and (2) for every positive integer k, the truth of P_k implies the truth of P_{k+1}, then the statement P_n must be true for all positive integers n.	65–68
	Use pattern recognition and mathematical induction to write the nth term of a sequence *(p. 639).*	To find a formula for the nth term of a sequence, (1) calculate the first several terms of the sequence, (2) try to find a pattern for the terms and write a formula (hypothesis) for the nth term of the sequence, and (3) use mathematical induction to prove your hypothesis.	69–72
	Find the sums of powers of integers *(p. 640).*	$\sum_{i=1}^{8} i^2 = \frac{n(n+1)(2n+1)}{6} = \frac{8(8+1)(16+1)}{6} = 204$	73, 74
	Find finite differences of sequences *(p. 641).*	The first differences of a sequence are found by subtracting consecutive terms. The second differences are found by subtracting consecutive first differences.	75, 76
Section 9.5	Use the Binomial Theorem to calculate binomial coefficients *(p. 644).*	**The Binomial Theorem:** In the expansion of $(x + y)^n = x^n + nx^{n-1}y + \cdot \cdot \cdot + {}_nC_r x^{n-r}y^r + \cdot \cdot \cdot + nxy^{n-1} + y^n$, the coefficient of $x^{n-r}y^r$ is ${}_nC_r = \frac{n!}{(n-r)!r!}$.	77, 78
	Use Pascal's Triangle to calculate binomial coefficients *(p. 646).*	First several rows of Pascal's Triangle: 1 1 1 1 2 1 1 3 3 1 1 4 6 4 1	79, 80
	Use binomial coefficients to write binomial expansions *(p. 647).*	$(x + 1)^3 = x^3 + 3x^2 + 3x + 1$ $(x - 1)^4 = x^4 - 4x^3 + 6x^2 - 4x + 1$	81–84
Section 9.6	Solve simple counting problems *(p. 652).*	A computer randomly generates an integer from 1 through 15. The computer can generate an integer that is divisible by 3 in 5 ways (3, 6, 9, 12, and 15).	85, 86
	Use the Fundamental Counting Principle to solve counting problems *(p. 653).*	**Fundamental Counting Principle:** Let E_1 and E_2 be two events. The first event E_1 can occur in m_1 different ways. After E_1 has occurred, E_2 can occur in m_2 different ways. The number of ways that the two events can occur is $m_1 \cdot m_2$.	87, 88
	Use permutations *(p. 654)* and combinations *(p. 657)* to solve counting problems.	The number of permutations of n elements taken r at a time is ${}_nP_r = n!/(n-r)!$. The number of combinations of n elements taken r at a time is ${}_nC_r = n!/[(n-r)!r!]$, or ${}_nC_r = {}_nP_r/r!$.	89–92
Section 9.7	Find the probabilities of events *(p. 662).*	If an event E has $n(E)$ equally likely outcomes and its sample space S has $n(S)$ equally likely outcomes, then the probability of event E is $P(E) = n(E)/n(S)$.	93, 94
	Find the probabilities of mutually exclusive events *(p. 666)* and independent events *(p. 668),* and find the probability of the complement of an event *(p. 669).*	If A and B are events in the same sample space, then the probability of A or B occurring is $P(A \cup B) = P(A) + P(B) - P(A \cap B)$. If A and B are mutually exclusive, then $P(A \cup B) = P(A) + P(B)$. If A and B are independent, then $P(A \text{ and } B) = P(A) \cdot P(B)$. The probability of the complement is $P(A') = 1 - P(A)$.	95–100

Review Exercises

See CalcChat.com for tutorial help and worked-out solutions to odd-numbered exercises.

9.1 Writing the Terms of a Sequence **In Exercises 1–4, write the first five terms of the sequence. (Assume that n begins with 1.)**

1. $a_n = 2 + \frac{6}{n}$
2. $a_n = \frac{(-1)^n 5n}{2n-1}$
3. $a_n = \frac{72}{n!}$
4. $a_n = n(n-1)$

Finding the nth Term of a Sequence **In Exercises 5–8, write an expression for the apparent nth term (a_n) of the sequence. (Assume that n begins with 1.)**

5. $-2, 2, -2, 2, -2, \ldots$
6. $-1, 2, 7, 14, 23, \ldots$
7. $4, 2, \frac{4}{3}, 1, \frac{4}{5}, \ldots$
8. $1, -\frac{1}{2}, \frac{1}{3}, -\frac{1}{4}, \frac{1}{5}, \ldots$

Simplifying a Factorial Expression **In Exercises 9–12, simplify the factorial expression.**

9. $9!$
10. $4! \cdot 0!$
11. $\frac{3! \cdot 5!}{6!}$
12. $\frac{7! \cdot 6!}{6! \cdot 8!}$

Finding a Sum **In Exercises 13 and 14, find the sum.**

13. $\sum_{j=1}^{4} \frac{6}{j^2}$
14. $\sum_{k=1}^{10} 2k^3$

Using Sigma Notation to Write a Sum **In Exercises 15 and 16, use sigma notation to write the sum.**

15. $\frac{1}{2(1)} + \frac{1}{2(2)} + \frac{1}{2(3)} + \cdots + \frac{1}{2(20)}$
16. $\frac{1}{2} + \frac{2}{3} + \frac{3}{4} + \cdots + \frac{9}{10}$

Finding the Sum of an Infinite Series **In Exercises 17 and 18, find the sum of the infinite series.**

17. $\sum_{i=1}^{\infty} \frac{4}{10^i}$
18. $\sum_{k=1}^{\infty} 2\left(\frac{1}{100}\right)^k$

19. **Compound Interest** An investor deposits \$10,000 in an account that earns 2.25% interest compounded monthly. The balance in the account after n months is given by

$$A_n = 10{,}000\left(1 + \frac{0.0225}{12}\right)^n, \quad n = 1, 2, 3, \ldots$$

(a) Write the first 10 terms of the sequence.

(b) Find the balance in the account after 10 years by computing the 120th term of the sequence.

20. **Lottery Ticket Sales** The total sales a_n (in billions of dollars) of lottery tickets in the United States from 2001 through 2010 can be approximated by the model

$$a_n = -0.18n^2 + 3.7n + 35, \quad n = 1, 2, \ldots, 10$$

where n is the year, with $n = 1$ corresponding to 2001. Write the terms of this finite sequence. Use a graphing utility to construct a bar graph that represents the sequence. *(Source: TLF Publications, Inc.)*

9.2 Determining Whether a Sequence Is Arithmetic **In Exercises 21–24, determine whether the sequence is arithmetic. If so, then find the common difference.**

21. $6, -1, -8, -15, -22, \ldots$
22. $0, 1, 3, 6, 10, \ldots$
23. $\frac{1}{2}, 1, \frac{3}{2}, 2, \frac{5}{2}, \ldots$
24. $1, \frac{15}{16}, \frac{7}{8}, \frac{13}{16}, \frac{3}{4}, \ldots$

Finding the nth Term **In Exercises 25–28, find a formula for a_n for the arithmetic sequence.**

25. $a_1 = 7, d = 12$
26. $a_1 = 28, d = -5$
27. $a_2 = 93, a_6 = 65$
28. $a_7 = 8, a_{13} = 6$

Writing the Terms of an Arithmetic Sequence **In Exercises 29 and 30, write the first five terms of the arithmetic sequence.**

29. $a_1 = 3, d = 11$
30. $a_1 = 25, a_{k+1} = a_k + 3$

31. **Sum of a Finite Arithmetic Sequence** Find the sum of the first 100 positive multiples of 7.

32. **Sum of a Finite Arithmetic Sequence** Find the sum of the integers from 40 to 90 (inclusive).

Finding a Partial Sum **In Exercises 33–36, find the partial sum.**

33. $\sum_{j=1}^{10} (2j - 3)$
34. $\sum_{j=1}^{8} (20 - 3j)$
35. $\sum_{k=1}^{11} \left(\frac{2}{3}k + 4\right)$
36. $\sum_{k=1}^{25} \left(\frac{3k+1}{4}\right)$

37. **Job Offer** The starting salary for a job is \$43,800 with a guaranteed increase of \$1950 per year. Determine (a) the salary during the fifth year and (b) the total compensation through five full years of employment.

38. **Baling Hay** In the first two trips baling hay around a large field, a farmer obtains 123 bales and 112 bales, respectively. Because each round gets shorter, the farmer estimates that the same pattern will continue. Estimate the total number of bales made after the farmer takes another six trips around the field.

9.3 Determining Whether a Sequence Is Geometric **In Exercises 39–42, determine whether the sequence is geometric. If so, then find the common ratio.**

39. $6, 12, 24, 48, \ldots$

40. $54, -18, 6, -2, \ldots$

41. $\frac{1}{5}, -\frac{3}{5}, \frac{9}{5}, -\frac{27}{5}, \ldots$

42. $\frac{1}{4}, \frac{2}{5}, \frac{3}{6}, \frac{4}{7}, \ldots$

Writing the Terms of a Geometric Sequence **In Exercises 43–46, write the first five terms of the geometric sequence.**

43. $a_1 = 2,\ r = 15$

44. $a_1 = 4,\ r = -\frac{1}{4}$

45. $a_1 = 9,\ a_3 = 4$

46. $a_1 = 2,\ a_3 = 12$

Finding a Term of a Geometric Sequence **In Exercises 47–50, write an expression for the *n*th term of the geometric sequence. Then find the 10th term of the sequence.**

47. $a_1 = 18, a_2 = -9$

48. $a_3 = 6, a_4 = 1$

49. $a_1 = 100, r = 1.05$

50. $a_1 = 5, r = 0.2$

Sum of a Finite Geometric Sequence **In Exercises 51–56, find the sum of the finite geometric sequence.**

51. $\sum_{i=1}^{7} 2^{i-1}$

52. $\sum_{i=1}^{5} 3^{i-1}$

53. $\sum_{i=1}^{4} \left(\frac{1}{2}\right)^i$

54. $\sum_{i=1}^{6} \left(\frac{1}{3}\right)^{i-1}$

55. $\sum_{i=1}^{5} (2)^{i-1}$

56. $\sum_{i=1}^{4} 6(3)^i$

Sum of a Finite Geometric Sequence **In Exercises 57 and 58, use a graphing utility to find the sum of the finite geometric sequence.**

57. $\sum_{i=1}^{10} 10\left(\frac{3}{5}\right)^{i-1}$

58. $\sum_{i=1}^{15} 20(0.2)^{i-1}$

Sum of an Infinite Geometric Sequence **In Exercises 59–62, find the sum of the infinite geometric series.**

59. $\sum_{i=0}^{\infty} \left(\frac{7}{8}\right)^i$

60. $\sum_{i=0}^{\infty} (0.5)^i$

61. $\sum_{k=1}^{\infty} 4\left(\frac{2}{3}\right)^{k-1}$

62. $\sum_{k=1}^{\infty} 1.3\left(\frac{1}{10}\right)^{k-1}$

63. Depreciation A paper manufacturer buys a machine for \$120,000. During the next 5 years, it will depreciate at a rate of 30% per year. (In other words, at the end of each year the depreciated value will be 70% of what it was at the beginning of the year.)

(a) Find the formula for the *n*th term of a geometric sequence that gives the value of the machine *t* full years after it was purchased.

(b) Find the depreciated value of the machine after 5 full years.

64. Annuity An investor deposits \$800 in an account on the first day of each month for 10 years. The account pays 3%, compounded monthly. What will the balance be at the end of 10 years?

9.4 Using Mathematical Induction **In Exercises 65–68, use mathematical induction to prove the formula for every positive integer *n*.**

65. $3 + 5 + 7 + \cdot \cdot \cdot + (2n + 1) = n(n + 2)$

66. $1 + \frac{3}{2} + 2 + \frac{5}{2} + \cdot \cdot \cdot + \frac{1}{2}(n + 1) = \frac{n}{4}(n + 3)$

67. $\sum_{i=0}^{n-1} ar^i = \frac{a(1 - r^n)}{1 - r}$

68. $\sum_{k=0}^{n-1} (a + kd) = \frac{n}{2}[2a + (n - 1)d]$

Finding a Formula for a Sum **In Exercises 69–72, use mathematical induction to find a formula for the sum of the first *n* terms of the sequence.**

69. $9, 13, 17, 21, \ldots$

70. $68, 60, 52, 44, \ldots$

71. $1, \frac{3}{5}, \frac{9}{25}, \frac{27}{125}, \ldots$

72. $12, -1, \frac{1}{12}, -\frac{1}{144}, \ldots$

Finding a Sum **In Exercises 73 and 74, find the sum using the formulas for the sums of powers of integers.**

73. $\sum_{n=1}^{75} n$

74. $\sum_{n=1}^{6} (n^5 - n^2)$

Linear Model, Quadratic Model, or Neither? **In Exercises 75 and 76, write the first five terms of the sequence beginning with the given term. Then calculate the first and second differences of the sequence. State whether the sequence has a perfect linear model, a perfect quadratic model, or neither.**

75. $a_1 = 5$
$a_n = a_{n-1} + 5$

76. $a_1 = -3$
$a_n = a_{n-1} - 2n$

9.5 Finding a Binomial Coefficient **In Exercises 77 and 78, find the binomial coefficient.**

77. ${}_6C_4$

78. ${}_{12}C_3$

Using Pascal's Triangle **In Exercises 79 and 80, evaluate using Pascal's Triangle.**

79. $\binom{7}{2}$

80. $\binom{10}{4}$

Expanding an Expression **In Exercises 81–84, use the Binomial Theorem to expand and simplify the expression. (Remember that $i = \sqrt{-1}$.)**

81. $(x + 4)^4$

82. $(a - 3b)^5$

83. $(5 + 2i)^4$

84. $(4 - 5i)^3$

9.6

85. **Numbers in a Hat** You place slips of paper numbered 1 through 14 in a hat. In how many ways can you draw two numbers at random with replacement that total 12?

86. **Shopping** A customer in an electronics store can choose one of six speaker systems, one of five DVD players, and one of six flat screen televisions to design a home theater system. How many systems can the customer design?

87. **Telephone Numbers** All of the land line telephone numbers in a small town use the same three-digit prefix. How many different telephone numbers are possible by changing only the last four digits?

88. **Course Schedule** A college student is preparing a course schedule for the next semester. The student may select one of three mathematics courses, one of four science courses, and one of six history courses. How many schedules are possible?

89. **Genetics** A geneticist is using gel electrophoresis to analyze five DNA samples. The geneticist treats each sample with a different restriction enzyme and then injects it into one of five wells formed in a bed of gel. In how many orders can the geneticist inject the five samples into the wells?

90. **Race** There are 10 bicyclists entered in a race. In how many different ways could the top 3 places be decided?

91. **Jury Selection** A group of potential jurors has been narrowed down to 32 people. In how many ways can a jury of 12 people be selected?

92. **Menu Choices** A local sub shop offers five different breads, four different meats, three different cheeses, and six different vegetables. You can choose one bread and any number of the other items. Find the total number of combinations of sandwiches possible.

9.7

93. **Apparel** A man has five pairs of socks, of which no two pairs are the same color. He randomly selects two socks from a drawer. What is the probability that he gets a matched pair?

94. **Bookshelf Order** A child returns a five-volume set of books to a bookshelf. The child is not able to read, and so cannot distinguish one volume from another. What is the probability that the child shelves the books in the correct order?

95. **Students by Class** At a university, 31% of the students are freshmen, 26% are sophomores, 25% are juniors, and 18% are seniors. One student receives a cash scholarship randomly by lottery. What is the probability that the scholarship winner is

(a) a junior or senior?

(b) a freshman, sophomore, or junior?

96. **Data Analysis** Interviewers asked a sample of college students, faculty members, and administrators whether they favored a proposed increase in the annual activity fee to enhance student life on campus. The table lists the results.

	Students	Faculty	Admin.	Total
Favor	237	37	18	292
Oppose	163	38	7	208
Total	400	75	25	500

Find the probability that a person selected at random from the sample is as described.

(a) Not in favor of the proposal

(b) A student

(c) A faculty member in favor of the proposal

97. **Tossing a Die** You toss a six-sided die four times. What is the probability of getting a 5 on each roll?

98. **Tossing a Die** You toss a six-sided die six times. What is the probability of getting each number exactly once?

99. **Drawing a Card** You randomly draw a card from a standard deck of 52 playing cards. What is the probability that the card is not a club?

100. **Tossing a Coin** Find the probability of obtaining at least one tail when you toss a coin five times.

Exploration

True or False? **In Exercises 101–104, determine whether the statement is true or false. Justify your answer.**

101. $\dfrac{(n+2)!}{n!} = \dfrac{n+2}{n}$

102. $\displaystyle\sum_{i=1}^{5}(i^3 + 2i) = \sum_{i=1}^{5} i^3 + \sum_{i=1}^{5} 2i$

103. $\displaystyle\sum_{k=1}^{8} 3k = 3\sum_{k=1}^{8} k$

104. $\displaystyle\sum_{j=1}^{6} 2^j = \sum_{j=3}^{8} 2^{j-2}$

105. **Think About It** An infinite sequence is a function. What is the domain of the function?

106. **Think About It** How do the two sequences differ?

(a) $a_n = \dfrac{(-1)^n}{n}$ (b) $a_n = \dfrac{(-1)^{n+1}}{n}$

107. **Writing** Explain what is meant by a recursion formula.

108. **Writing** Write a brief paragraph explaining how to identify the graph of an arithmetic sequence and the graph of a geometric sequence.

Chapter Test

See CalcChat.com for tutorial help and worked-out solutions to odd-numbered exercises.

Take this test as you would take a test in class. When you are finished, check your work against the answers given in the back of the book.

1. Write the first five terms of the sequence $a_n = \frac{(-1)^n}{3n + 2}$. (Assume that n begins with 1.)

2. Write an expression for the apparent nth term (a_n) of the sequence.

   $\frac{3}{1!}, \frac{4}{2!}, \frac{5}{3!}, \frac{6}{4!}, \frac{7}{5!}, \ldots$

3. Write the next three terms of the series. Then find the sixth partial sum of the series.

   $8 + 21 + 34 + 47 + \cdots$

4. The fifth term of an arithmetic sequence is 5.4, and the 12th term is 11.0. Find the nth term.

5. The second term of a geometric sequence is 28, and the sixth term is 7168. Find the nth term.

6. Write the first five terms of the geometric sequence $a_n = 5(2)^{n-1}$. (Assume that n begins with 1.)

In Exercises 7–9, find the sum.

7. $\sum_{i=1}^{50} (2i^2 + 5)$

8. $\sum_{n=1}^{9} (12n - 7)$

9. $\sum_{i=1}^{\infty} 4\left(\frac{1}{2}\right)^i$

10. Use mathematical induction to prove the formula for every positive integer n.

    $$5 + 10 + 15 + \cdots + 5n = \frac{5n(n + 1)}{2}$$

11. Use the Binomial Theorem to expand and simplify (a) $(x + 6y)^4$ and (b) $3(x - 2)^5 + 4(x - 2)^3$.

12. Find the coefficient of the term a^4b^3 in the expansion of $(3a - 2b)^7$.

In Exercises 13 and 14, evaluate each expression.

13. (a) ${}_9P_2$ (b) ${}_{70}P_3$

14. (a) ${}_{11}C_4$ (b) ${}_{66}C_4$

15. How many distinct license plates can be issued consisting of one letter followed by a three-digit number?

16. Eight people are going for a ride in a boat that seats eight people. One person will drive, and only three of the remaining people are willing to ride in the two bow seats. How many seating arrangements are possible?

17. You attend a karaoke night and hope to hear your favorite song. The karaoke song book has 300 different songs (your favorite song is among them). Assuming that the singers are equally likely to pick any song and no song repeats, what is the probability that your favorite song is one of the 20 that you hear that night?

18. You are with three of your friends at a party. Names of all of the 30 guests are placed in a hat and drawn randomly to award four door prizes. Each guest can win only one prize. What is the probability that you and your friends win all four prizes?

19. The weather report calls for a 90% chance of snow. According to this report, what is the probability that it will *not* snow?

Cumulative Test for Chapters 7–9

See CalcChat.com for tutorial help and worked-out solutions to odd-numbered exercises.

Take this test as you would take a test in class. When you are finished, check your work against the answers given in the back of the book.

In Exercises 1–4, solve the system by the specified method.

1. Substitution
$$\begin{cases} y = 3 - x^2 \\ 2(y - 2) = x - 1 \end{cases}$$

2. Elimination
$$\begin{cases} x + 3y = -6 \\ 2x + 4y = -10 \end{cases}$$

3. Elimination
$$\begin{cases} -2x + 4y - z = -16 \\ x - 2y + 2z = 5 \\ x - 3y - z = 13 \end{cases}$$

4. Gauss-Jordan Elimination
$$\begin{cases} x + 3y - 2z = -7 \\ -2x + y - z = -5 \\ 4x + y + z = 3 \end{cases}$$

5. A custom-blend bird seed is to be mixed from seed mixtures costing \$0.75 per pound and \$1.25 per pound. How many pounds of each seed mixture are used to make 200 pounds of custom-blend bird seed costing \$0.95 per pound?

6. Find the equation of the parabola $y = ax^2 + bx + c$ passing through the points $(0, 6)$, $(2, 3)$, and $(4, 2)$.

In Exercises 7 and 8, sketch the graph (and label the vertices) of the solution set of the system of inequalities.

7. $\begin{cases} 2x + y \geq -3 \\ x - 3y \leq 2 \end{cases}$

8. $\begin{cases} x - y > 6 \\ 5x + 2y < 10 \end{cases}$

9. Sketch the region determined by the constraints. Then find the minimum and maximum values of the objective function $z = 3x + 2y$ and where they occur, subject to the indicated constraints.

$$\begin{aligned} x + 4y &\leq 20 \\ 2x + y &\leq 12 \\ x &\geq 0 \\ y &\geq 0 \end{aligned}$$

$$\begin{cases} -x + 2y - z = 9 \\ 2x - y + 2z = -9 \\ 3x + 3y - 4z = 7 \end{cases}$$

System for 10 and 11

In Exercises 10 and 11, use the system of linear equations shown at the left.

10. Write the augmented matrix for the system.

11. Solve the system using the matrix found in Exercise 10 and Gauss-Jordan elimination.

In Exercises 12–17, perform the operations using the following matrices.

$$A = \begin{bmatrix} 3 & 0 \\ -1 & 4 \end{bmatrix}, \quad B = \begin{bmatrix} -2 & 5 \\ 0 & -1 \end{bmatrix}$$

12. $A + B$

13. $-8B$

14. $2A - 5B$

15. AB

16. A^2

17. $BA - B^2$

$$\begin{bmatrix} 7 & 1 & 0 \\ -2 & 4 & -1 \\ 3 & 8 & 5 \end{bmatrix}$$

Matrix for 19

18. Find the inverse of the matrix (if it exists): $\begin{bmatrix} 1 & 2 & -1 \\ 3 & 7 & -10 \\ -5 & -7 & -15 \end{bmatrix}$.

19. Find the determinant of the matrix shown at the left.

Age group	Gym shoes	Jogging shoes	Walking shoes
14–17	0.079	0.064	0.029
18–24	0.050	0.060	0.022
25–34	0.103	0.259	0.085

Matrix for 20

20. The percents (in decimal form) of the total amounts spent on three types of footwear in a recent year are shown in the matrix at the left. The total amounts (in millions of dollars) spent by the age groups on the three types of footwear were \$479.88 (14–17 age group), \$365.88 (18–24 age group), and \$1248.89 (25–34 age group). How many dollars worth of gym shoes, jogging shoes, and walking shoes were sold that year? *(Source: National Sporting Goods Association)*

In Exercises 21 and 22, use Cramer's Rule to solve the system of equations.

21. $\begin{cases} 8x - 3y = -52 \\ 3x + 5y = 5 \end{cases}$

22. $\begin{cases} 5x + 4y + 3z = 7 \\ -3x - 8y + 7z = -9 \\ 7x - 5y - 6z = -53 \end{cases}$

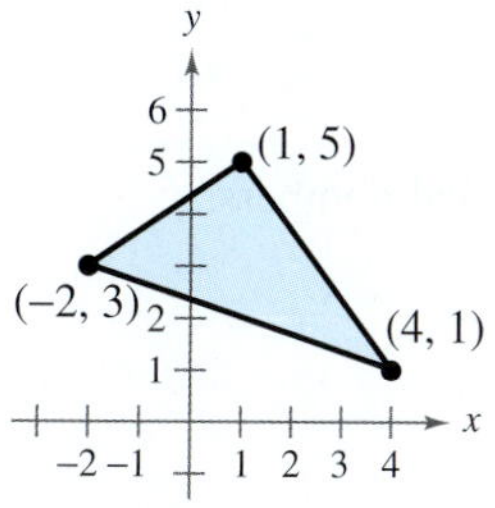

Figure for 23

23. Use a determinant to find the area of the triangle shown at the left.

24. Write the first five terms of the sequence $a_n = \dfrac{(-1)^{n+1}}{2n+3}$. (Assume that n begins with 1.)

25. Write an expression for the apparent nth term (a_n) of the sequence.

$$\frac{2!}{4}, \frac{3!}{5}, \frac{4!}{6}, \frac{5!}{7}, \frac{6!}{8}, \ldots$$

26. Find the 16th partial sum of the arithmetic sequence 6, 18, 30, 42, ...

27. The sixth term of an arithmetic sequence is 20.6, and the ninth term is 30.2.

(a) Find the 20th term.

(b) Find the nth term.

28. Write the first five terms of the sequence $a_n = 3(2)^{n-1}$. (Assume that n begins with 1.)

29. Find the sum: $\displaystyle\sum_{i=0}^{\infty} 1.3\left(\frac{1}{10}\right)^{i-1}$.

30. Use mathematical induction to prove the inequality

$$(n+1)! > 2^n, \quad n \geq 2.$$

31. Use the Binomial Theorem to expand and simplify $(w - 9)^4$.

In Exercises 32–35, evaluate the expression.

32. ${}_{14}P_3$

33. ${}_{25}P_2$

34. $\dbinom{8}{4}$

35. ${}_{11}C_6$

In Exercises 36 and 37, find the number of distinguishable permutations of the group of letters.

36. B, A, S, K, E, T, B, A, L, L

37. A, N, T, A, R, C, T, I, C, A

38. A personnel manager at a department store has 10 applicants to fill three different sales positions. In how many ways can this be done, assuming that all the applicants are qualified for any of the three positions?

39. On a game show, contestants must arrange the digits 3, 4, and 5 in the proper order to form the price of an appliance. When the contestant arranges the digits correctly, he or she wins the appliance. What is the probability of winning when the contestant knows that the price is at least \$400?

Proofs in Mathematics

Properties of Sums *(p. 610)*

1. $\sum_{i=1}^{n} c = cn, \quad c$ is a constant.

2. $\sum_{i=1}^{n} ca_i = c\sum_{i=1}^{n} a_i, \quad c$ is a constant.

3. $\sum_{i=1}^{n} (a_i + b_i) = \sum_{i=1}^{n} a_i + \sum_{i=1}^{n} b_i$

4. $\sum_{i=1}^{n} (a_i - b_i) = \sum_{i=1}^{n} a_i - \sum_{i=1}^{n} b_i$

INFINITE SERIES

People considered the study of infinite series a novelty in the fourteenth century. Logician Richard Suiseth, whose nickname was Calculator, solved this problem.

If throughout the first half of a given time interval a variation continues at a certain intensity; throughout the next quarter of the interval at double the intensity; throughout the following eighth at triple the intensity and so ad infinitum; The average intensity for the whole interval will be the intensity of the variation during the second subinterval (or double the intensity).

This is the same as saying that the sum of the infinite series

$$\frac{1}{2} + \frac{2}{4} + \frac{3}{8} + \cdots + \frac{n}{2^n} + \cdots$$

is 2.

Proof

Each of these properties follows directly from the properties of real numbers.

1. $\sum_{i=1}^{n} c = c + c + c + \cdots + c = cn \qquad n$ terms

The proof of Property 2 uses the Distributive Property.

2. $$\sum_{i=1}^{n} ca_i = ca_1 + ca_2 + ca_3 + \cdots + ca_n$$
$$= c(a_1 + a_2 + a_3 + \cdots + a_n)$$
$$= c\sum_{i=1}^{n} a_i$$

The proof of Property 3 uses the Commutative and Associative Properties of Addition.

3. $$\sum_{i=1}^{n} (a_i + b_i) = (a_1 + b_1) + (a_2 + b_2) + (a_3 + b_3) + \cdots + (a_n + b_n)$$
$$= (a_1 + a_2 + a_3 + \cdots + a_n) + (b_1 + b_2 + b_3 + \cdots + b_n)$$
$$= \sum_{i=1}^{n} a_i + \sum_{i=1}^{n} b_i$$

The proof of Property 4 uses the Commutative and Associative Properties of Addition and the Distributive Property.

4. $$\sum_{i=1}^{n} (a_i - b_i) = (a_1 - b_1) + (a_2 - b_2) + (a_3 - b_3) + \cdots + (a_n - b_n)$$
$$= (a_1 + a_2 + a_3 + \cdots + a_n) + (-b_1 - b_2 - b_3 - \cdots - b_n)$$
$$= (a_1 + a_2 + a_3 + \cdots + a_n) - (b_1 + b_2 + b_3 + \cdots + b_n)$$
$$= \sum_{i=1}^{n} a_i - \sum_{i=1}^{n} b_i$$

The Sum of a Finite Arithmetic Sequence *(p. 619)*

The sum of a finite arithmetic sequence with n terms is

$$S_n = \frac{n}{2}(a_1 + a_n).$$

Proof

Begin by generating the terms of the arithmetic sequence in two ways. In the first way, repeatedly add d to the first term to obtain

$$\begin{aligned} S_n &= a_1 + a_2 + a_3 + \cdots + a_{n-2} + a_{n-1} + a_n \\ &= a_1 + [a_1 + d] + [a_1 + 2d] + \cdots + [a_1 + (n-1)d]. \end{aligned}$$

In the second way, repeatedly subtract d from the nth term to obtain

$$\begin{aligned} S_n &= a_n + a_{n-1} + a_{n-2} + \cdots + a_3 + a_2 + a_1 \\ &= a_n + [a_n - d] + [a_n - 2d] + \cdots + [a_n - (n-1)d]. \end{aligned}$$

If you add these two versions of S_n, then the multiples of d sum to zero and you obtain

$$2S_n = (a_1 + a_n) + (a_1 + a_n) + (a_1 + a_n) + \cdots + (a_1 + a_n) \qquad n \text{ terms}$$

$$2S_n = n(a_1 + a_n)$$

$$S_n = \frac{n}{2}(a_1 + a_n).$$

■

The Sum of a Finite Geometric Sequence *(p. 628)*

The sum of the finite geometric sequence

$$a_1,\ a_1r,\ a_1r^2,\ a_1r^3,\ a_1r^4,\ \ldots,\ a_1r^{n-1}$$

with common ratio $r \neq 1$ is given by $S_n = \sum_{i=1}^{n} a_1r^{i-1} = a_1\left(\frac{1 - r^n}{1 - r}\right)$.

Proof

$$S_n = a_1 + a_1r + a_1r^2 + \cdots + a_1r^{n-2} + a_1r^{n-1}$$

$$rS_n = a_1r + a_1r^2 + a_1r^3 + \cdots + a_1r^{n-1} + a_1r^n \qquad \text{Multiply by } r.$$

Subtracting the second equation from the first yields

$$S_n - rS_n = a_1 - a_1r^n.$$

So, $S_n(1 - r) = a_1(1 - r^n)$, and, because $r \neq 1$, you have $S_n = a_1\left(\frac{1 - r^n}{1 - r}\right)$. ■

The Binomial Theorem *(p. 644)*

In the expansion of $(x + y)^n$

$$(x + y)^n = x^n + nx^{n-1}y + \cdot \cdot \cdot + {}_nC_r\, x^{n-r}y^r + \cdot \cdot \cdot + nxy^{n-1} + y^n$$

the coefficient of $x^{n-r}y^r$ is

$${}_nC_r = \frac{n!}{(n-r)!r!}.$$

Proof

The Binomial Theorem can be proved quite nicely using mathematical induction. The steps are straightforward but look a little messy, so only an outline of the proof follows.

1. For $n = 1$, you have $(x + y)^1 = x^1 + y^1 = {}_1C_0 x + {}_1C_1 y$, and the formula is valid.

2. Assuming that the formula is true for $n = k$, the coefficient of $x^{k-r}y^r$ is

$${}_kC_r = \frac{k!}{(k-r)!r!} = \frac{k(k-1)(k-2)\cdot\cdot\cdot(k-r+1)}{r!}.$$

To show that the formula is true for $n = k + 1$, look at the coefficient of $x^{k+1-r}y^r$ in the expansion of

$$(x + y)^{k+1} = (x + y)^k(x + y).$$

On the right-hand side, the term involving $x^{k+1-r}y^r$ is the sum of two products.

$$\begin{aligned}
&({}_kC_r x^{k-r}y^r)(x) + ({}_kC_{r-1}x^{k+1-r}y^{r-1})(y) \\
&\quad= \left[\frac{k!}{(k-r)!r!} + \frac{k!}{(k+1-r)!(r-1)!}\right]x^{k+1-r}y^r \\
&\quad= \left[\frac{(k+1-r)k!}{(k+1-r)!r!} + \frac{k!r}{(k+1-r)!r!}\right]x^{k+1-r}y^r \\
&\quad= \left[\frac{k!(k+1-r+r)}{(k+1-r)!r!}\right]x^{k+1-r}y^r \\
&\quad= \left[\frac{(k+1)!}{(k+1-r)!r!}\right]x^{k+1-r}y^r \\
&\quad= {}_{k+1}C_r x^{k+1-r}y^r
\end{aligned}$$

So, by mathematical induction, the Binomial Theorem is valid for all positive integers n. ■

P.S. Problem Solving

1. **Decreasing Sequence** Consider the sequence

$$a_n = \frac{n+1}{n^2+1}.$$

(a) Use a graphing utility to graph the first 10 terms of the sequence.

(b) Use the graph from part (a) to estimate the value of a_n as n approaches infinity.

(c) Complete the table.

n	1	10	100	1000	10,000
a_n					

(d) Use the table from part (c) to determine (if possible) the value of a_n as n approaches infinity.

2. **Alternating Sequence** Consider the sequence

$$a_n = 3 + (-1)^n.$$

(a) Use a graphing utility to graph the first 10 terms of the sequence.

(b) Use the graph from part (a) to describe the behavior of the graph of the sequence.

(c) Complete the table.

n	1	10	101	1000	10,001
a_n					

(d) Use the table from part (c) to determine (if possible) the value of a_n as n approaches infinity.

3. **Greek Mythology** Can the Greek hero Achilles, running at 20 feet per second, ever catch a tortoise, starting 20 feet ahead of Achilles and running at 10 feet per second? The Greek mathematician Zeno said no. When Achilles runs 20 feet, the tortoise will be 10 feet ahead. Then, when Achilles runs 10 feet, the tortoise will be 5 feet ahead. Achilles will keep cutting the distance in half but will never catch the tortoise. The table shows Zeno's reasoning. From the table, you can see that both the distances and the times required to achieve them form infinite geometric series. Using the table, show that both series have finite sums. What do these sums represent?

DATA

Distance (in feet)	Time (in seconds)
20	1
10	0.5
5	0.25
2.5	0.125
1.25	0.0625
0.625	0.03125

Spreadsheet at LarsonPrecalculus.com

4. **Conjecture** Let $x_0 = 1$ and consider the sequence x_n given by

$$x_n = \frac{1}{2}x_{n-1} + \frac{1}{x_{n-1}}, \quad n = 1, 2, \ldots$$

Use a graphing utility to compute the first 10 terms of the sequence and make a conjecture about the value of x_n as n approaches infinity.

5. **Operations on an Arithmetic Sequence** The following operations are performed on each term of an arithmetic sequence. Determine whether the resulting sequence is arithmetic. If so, then state the common difference.

(a) A constant C is added to each term.

(b) Each term is multiplied by a nonzero constant C.

(c) Each term is squared.

6. **Sequences of Powers** The following sequence of perfect squares is not arithmetic.

$1, 4, 9, 16, 25, 36, 49, 64, 81, \ldots$

However, you can form a related sequence that is arithmetic by finding the differences of consecutive terms.

(a) Write the first eight terms of the related arithmetic sequence described above. What is the nth term of this sequence?

(b) Describe how you can find an arithmetic sequence that is related to the following sequence of perfect cubes.

$1, 8, 27, 64, 125, 216, 343, 512, 729, \ldots$

(c) Write the first seven terms of the related sequence in part (b) and find the nth term of the sequence.

(d) Describe how you can find the arithmetic sequence that is related to the following sequence of perfect fourth powers.

$1, 16, 81, 256, 625, 1296, 2401, 4096, 6561, \ldots$

(e) Write the first six terms of the related sequence in part (d) and find the nth term of the sequence.

7. **Piecewise-Defined Sequence** You can define a sequence using a piecewise formula. The following is an example of a piecewise-defined sequence.

$$a_1 = 7, \; a_n = \begin{cases} \dfrac{a_{n-1}}{2}, & \text{when } a_{n-1} \text{ is even.} \\ 3a_{n-1} + 1, & \text{when } a_{n-1} \text{ is odd.} \end{cases}$$

(a) Write the first 20 terms of the sequence.

(b) Write the first 10 terms of the sequences for which $a_1 = 4$, $a_1 = 5$, and $a_1 = 12$ (using a_n as defined above). What conclusion can you make about the behavior of each sequence?

8. **Fibonacci Sequence** Let $f_1, f_2, \ldots, f_n, \ldots$ be the Fibonacci sequence.

(a) Use mathematical induction to prove that

$$f_1 + f_2 + \cdots + f_n = f_{n+2} - 1.$$

(b) Find the sum of the first 20 terms of the Fibonacci sequence.

9. **Pentagonal Numbers** The numbers 1, 5, 12, 22, 35, 51, . . . are called pentagonal numbers because they represent the numbers of dots used to make pentagons, as shown below. Use mathematical induction to prove that the nth pentagonal number P_n is given by

$$P_n = \frac{n(3n - 1)}{2}.$$

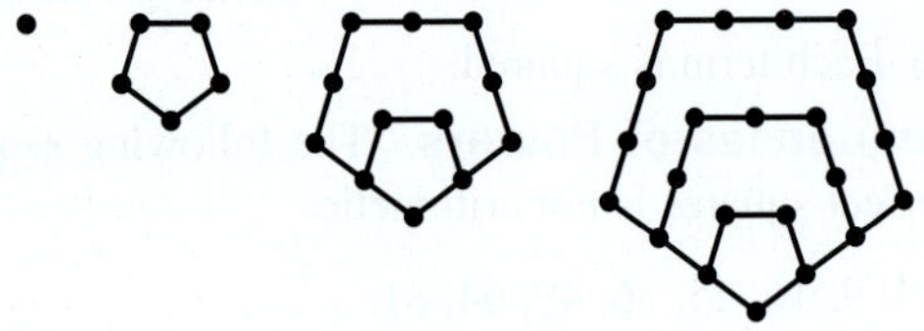

10. **Think About It** What conclusion can be drawn from the following information about the sequence of statements P_n?

(a) P_3 is true and P_k implies P_{k+1}.

(b) $P_1, P_2, P_3, \ldots, P_{50}$ are all true.

(c) P_1, P_2, and P_3 are all true, but the truth of P_k does not imply that P_{k+1} is true.

(d) P_2 is true and P_{2k} implies P_{2k+2}.

11. **Sierpinski Triangle** Recall that a *fractal* is a geometric figure that consists of a pattern that is repeated infinitely on a smaller and smaller scale. A well-known fractal is called the *Sierpinski Triangle*. In the first stage, the midpoints of the three sides are used to create the vertices of a new triangle, which is then removed, leaving three triangles. The figure below shows the first two stages. Note that each remaining triangle is similar to the original triangle. Assume that the length of each side of the original triangle is one unit. Write a formula that describes the side length of the triangles generated in the nth stage. Write a formula for the area of the triangles generated in the nth stage.

12. **Job Offer** You work for a company that pays \$0.01 the first day, \$0.02 the second day, \$0.04 the third day, and so on. If the daily wage keeps doubling, then what would your total income be for working 30 days?

13. **Multiple Choice** A multiple choice question has five possible answers. You know that the answer is not B or D, but you are not sure about answers A, C, and E. What is the probability that you will get the right answer when you take a guess?

14. **Throwing a Dart** You throw a dart at the circular target shown below. The dart is equally likely to hit any point inside the target. What is the probability that it hits the region outside the triangle?

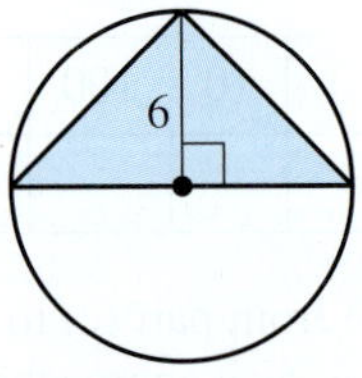

15. **Odds** The odds in favor of an event occurring are the ratio of the probability that the event will occur to the probability that the event will not occur. The reciprocal of this ratio represents the odds against the event occurring.

(a) Six of the marbles in a bag are red. The odds against choosing a red marble are 4 to 1. How many marbles are in the bag?

(b) A bag contains three blue marbles and seven yellow marbles. What are the odds in favor of choosing a blue marble? What are the odds against choosing a blue marble?

(c) Write a formula for converting the odds in favor of an event to the probability of the event.

(d) Write a formula for converting the probability of an event to the odds in favor of the event.

16. **Expected Value** An event A has n possible outcomes, which have the values $x_1, x_2, \ldots, x_n$. The probabilities of the n outcomes occurring are $p_1, p_2, \ldots, p_n$. The **expected value** V of an event A is the sum of the products of the outcomes' probabilities and their values,

$$V = p_1x_1 + p_2x_2 + \cdots + p_nx_n.$$

(a) To win California's Super Lotto Plus game, you must match five different numbers chosen from the numbers 1 to 47, plus one MEGA number chosen from the numbers 1 to 27. You purchase a ticket for \$1. If the jackpot for the next drawing is \$12,000,000, then what is the expected value of the ticket?

(b) You are playing a dice game in which you need to score 60 points to win. On each turn, you toss two six-sided dice. Your score for the turn is 0 when the dice do not show the same number. Your score for the turn is the product of the numbers on the dice when they do show the same number. What is the expected value of each turn? How many turns will it take on average to score 60 points?

10 Topics in Analytic Geometry

Satellite Orbit *(Exercise 64, page 760)*

Microphone Pickup Pattern *(Exercise 69, page 754)*

Nuclear Cooling Towers *(page 714)*

Halley's Comet *(page 708)*

Radio Telescopes *(page 698)*

Clockwise from top left, nikkytok/Shutterstock.com; Cristi Matei/Shutterstock.com, Digital Vision/Getty Images; John A Davis/Shutterstock.com; Malyshev Maksim/Shutterstock.com

10.1 Lines

You can use the inclination of a line to measure heights indirectly. For instance, in Exercise 88 on page 694, you will use the inclination of a line to determine the change in elevation from the base to the top of the Falls Incline Railway in Niagara Falls, Ontario, Canada.

- **Find the inclination of a line.**
- **Find the angle between two lines.**
- **Find the distance between a point and a line.**

Inclination of a Line

In Section 1.3, the slope of a line was described as the ratio of the change in y to the change in x. In this section, you will look at the slope of a line in terms of the angle of inclination of the line.

Every nonhorizontal line must intersect the x-axis. The angle formed by such an intersection determines the **inclination** of the line, as specified in the following definition.

> **Definition of Inclination**
>
> The **inclination** of a nonhorizontal line is the positive angle θ (less than π) measured counterclockwise from the x-axis to the line. (See figures below.)

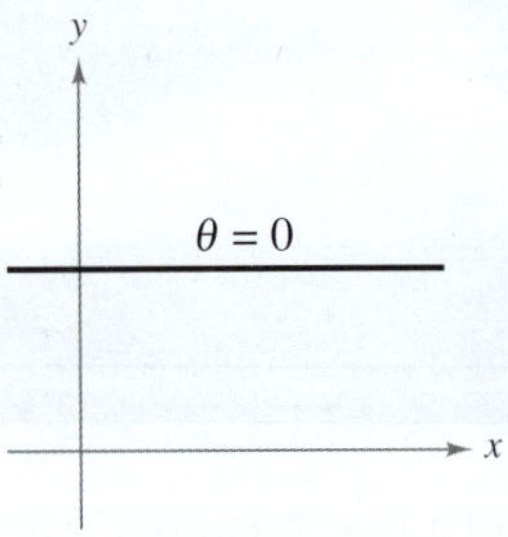

Horizontal Line

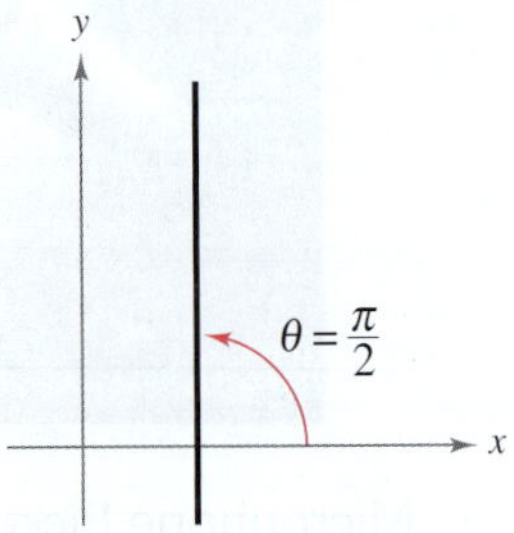

Vertical Line

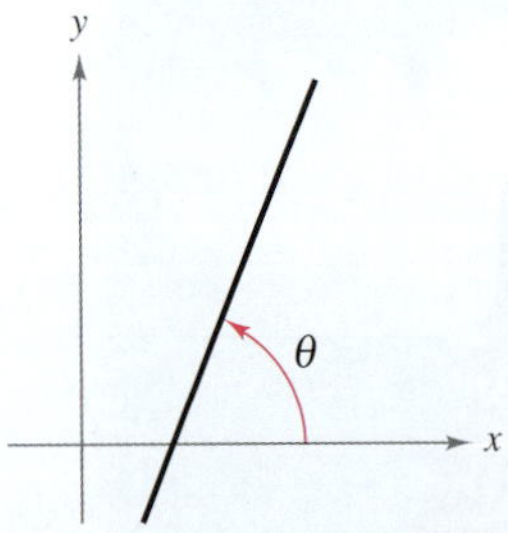

Acute Angle

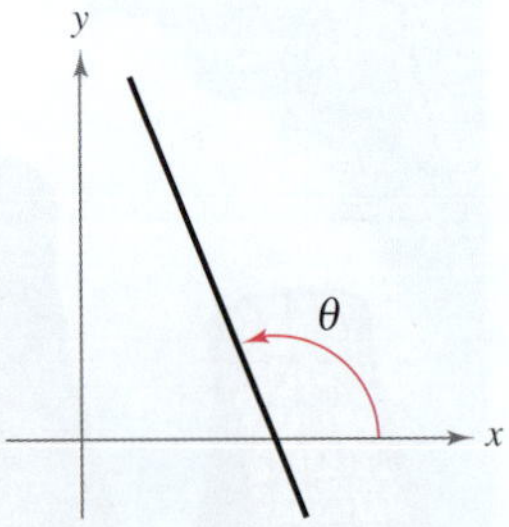

Obtuse Angle

The inclination of a line is related to its slope in the following manner.

> **Inclination and Slope**
>
> If a nonvertical line has inclination θ and slope m, then
>
> $m = \tan\theta.$

For a proof of this relation between inclination and slope, see Proofs in Mathematics on page 768.

Note that if $m \geq 0$, then $\theta = \arctan m$ because $0 \leq \theta < \pi/2$. On the other hand, if $m < 0$, then $\theta = \pi + \arctan m$ because $\pi/2 < \theta < \pi$.

kan_khampanya/Shutterstock.com

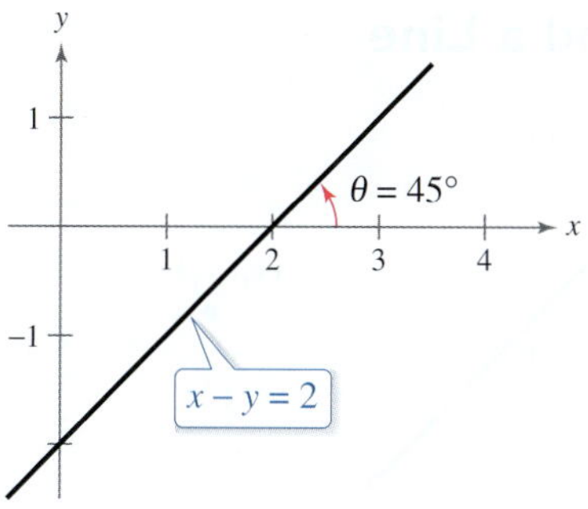

(a)

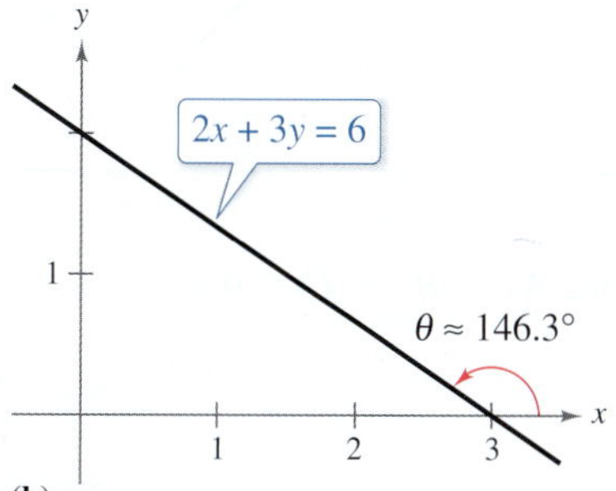

(b)

Figure 10.1

EXAMPLE 1 Finding the Inclination of a Line

Find the inclination of (a) $x - y = 2$ and (b) $2x + 3y = 6$.

Solution

a. The slope of this line is $m = 1$. So, its inclination is determined from $\tan\theta = 1$. Note that $m \geq 0$. This means that

$$\theta = \arctan 1 = \pi/4 \text{ radian} = 45°$$

as shown in Figure 10.1(a).

b. The slope of this line is $m = -\frac{2}{3}$. So, its inclination is determined from $\tan\theta = -\frac{2}{3}$. Note that $m < 0$. This means that

$$\theta = \pi + \arctan\left(-\tfrac{2}{3}\right) \approx \pi + (-0.5880) \approx 2.5536 \text{ radians} \approx 146.3°$$

as shown in Figure 10.1(b).

✓ Checkpoint Audio-video solution in English & Spanish at LarsonPrecalculus.com.

Find the inclination of $4x - 5y = 7$.

The Angle Between Two Lines

When two distinct lines intersect and are nonperpendicular, their intersection forms two pairs of opposite angles. One pair is acute and the other pair is obtuse. The smaller of these angles is the **angle between the two lines.** If two lines have inclinations θ_1 and θ_2, where $\theta_1 < \theta_2$ and $\theta_2 - \theta_1 < \pi/2$, then the angle between the two lines is $\theta = \theta_2 - \theta_1$, as shown in Figure 10.2. You can use the formula for the tangent of the difference of two angles

$$\tan\theta = \tan(\theta_2 - \theta_1) = \frac{\tan\theta_2 - \tan\theta_1}{1 + \tan\theta_1 \tan\theta_2}$$

to obtain the formula for the angle between two lines.

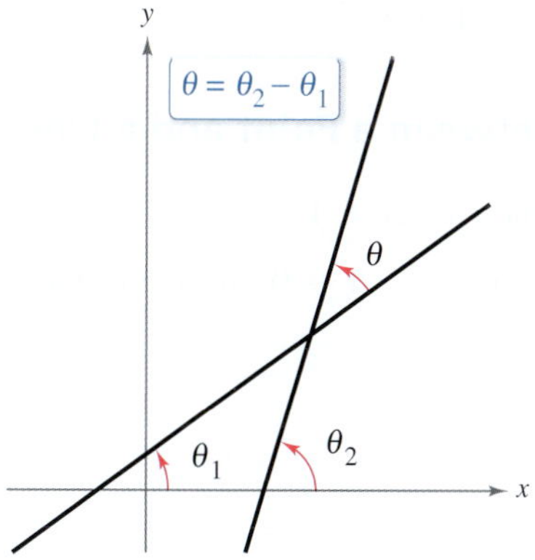

Figure 10.2

> **Angle Between Two Lines**
>
> If two nonperpendicular lines have slopes m_1 and m_2, then the tangent of the angle between the two lines is
>
> $$\tan\theta = \left|\frac{m_2 - m_1}{1 + m_1 m_2}\right|.$$

EXAMPLE 2 Finding the Angle Between Two Lines

Find the angle between $2x - y = 4$ and $3x + 4y = 12$.

Solution The two lines have slopes of $m_1 = 2$ and $m_2 = -\frac{3}{4}$, respectively. So, the tangent of the angle between the two lines is

$$\tan\theta = \left|\frac{m_2 - m_1}{1 + m_1 m_2}\right| = \left|\frac{(-3/4) - 2}{1 + (2)(-3/4)}\right| = \left|\frac{-11/4}{-2/4}\right| = \frac{11}{2}.$$

Finally, you can conclude that the angle is $\theta = \arctan\frac{11}{2} \approx 1.3909$ radians $\approx 79.7°$ as shown in Figure 10.3.

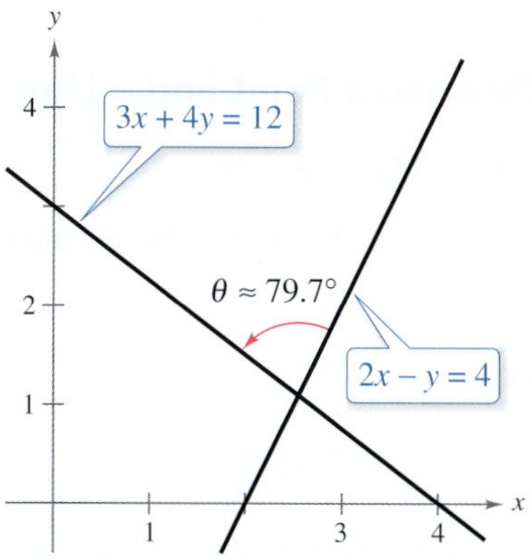

Figure 10.3

✓ Checkpoint Audio-video solution in English & Spanish at LarsonPrecalculus.com.

Find the angle between $4x - 5y + 10 = 0$ and $3x + 2y + 5 = 0$.

The Distance Between a Point and a Line

Finding the distance between a line and a point not on the line is an application of perpendicular lines. This distance is defined as the length of the perpendicular line segment joining the point and the line, as shown at the right.

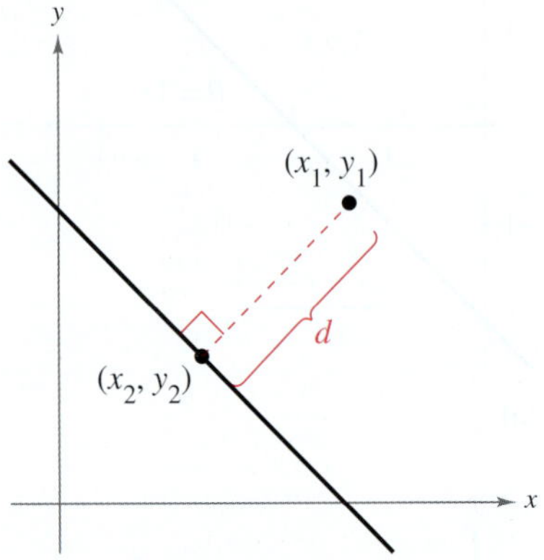

> **Distance Between a Point and a Line**
>
> The distance between the point (x_1, y_1) and the line $Ax + By + C = 0$ is
>
> $$d = \frac{|Ax_1 + By_1 + C|}{\sqrt{A^2 + B^2}}.$$

Remember that the values of A, B, and C in this distance formula correspond to the general equation of a line, $Ax + By + C = 0$. For a proof of this formula for the distance between a point and a line, see Proofs in Mathematics on page 768.

EXAMPLE 3 Finding the Distance Between a Point and a Line

Find the distance between the point $(4, 1)$ and the line $y = 2x + 1$.

Solution The general form of the equation is $-2x + y - 1 = 0$. So, the distance between the point and the line is

$$d = \frac{|-2(4) + 1(1) + (-1)|}{\sqrt{(-2)^2 + 1^2}} = \frac{8}{\sqrt{5}} \approx 3.58 \text{ units.}$$

The line and the point are shown in Figure 10.4.

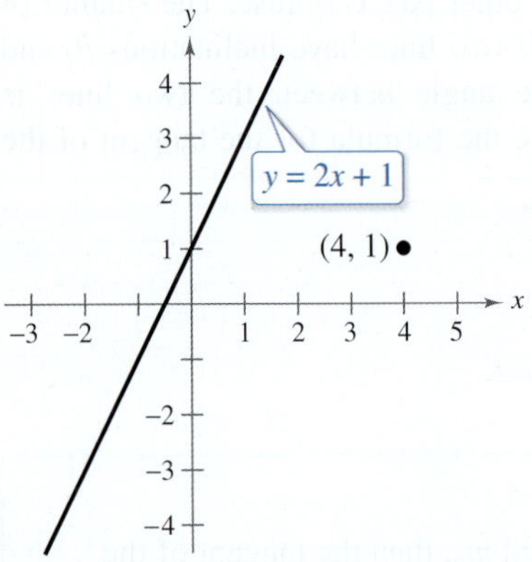

Figure 10.4

✓ **Checkpoint** Audio-video solution in English & Spanish at LarsonPrecalculus.com.

Find the distance between the point $(5, -1)$ and the line $y = -3x + 2$.

EXAMPLE 4 Finding the Distance Between a Point and a Line

Find the distance between the point $(2, -1)$ and the line $-5x + 4y = 8$.

Solution The general form of the equation is $-5x + 4y - 8 = 0$. So, the distance between the point and the line is

$$d = \frac{|-5(2) + 4(-1) + (-8)|}{\sqrt{(-5)^2 + 4^2}} = \frac{22}{\sqrt{41}} \approx 3.44 \text{ units.}$$

The line and the point are shown in Figure 10.5.

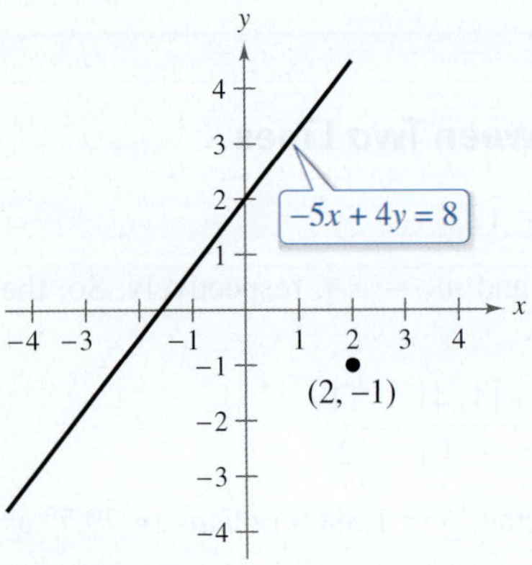

Figure 10.5

✓ **Checkpoint** Audio-video solution in English & Spanish at LarsonPrecalculus.com.

Find the distance between the point $(3, 2)$ and the line $-3x + 5y = -2$.

EXAMPLE 5 An Application of Two Distance Formulas

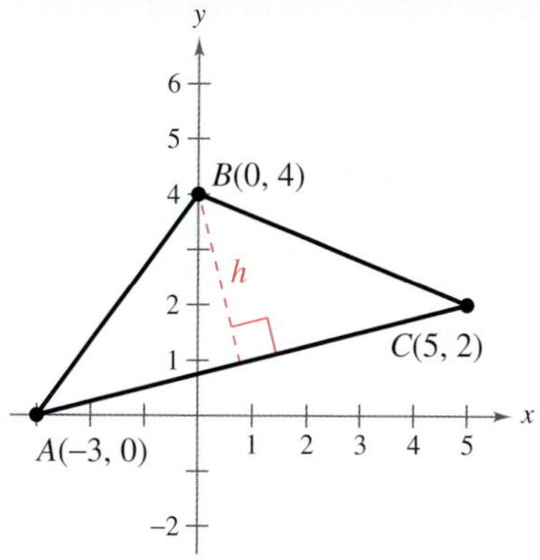

Figure 10.6

Figure 10.6 shows a triangle with vertices $A(-3, 0)$, $B(0, 4)$, and $C(5, 2)$.

a. Find the altitude h from vertex B to side AC.

b. Find the area of the triangle.

Solution

a. To find the altitude, use the formula for the distance between line AC and the point $(0, 4)$. The equation of line AC is obtained as follows.

$$\textit{Slope: } m = \frac{2-0}{5-(-3)} = \frac{2}{8} = \frac{1}{4}$$

$$\textit{Equation: } \quad y - 0 = \frac{1}{4}(x+3) \qquad \text{Point-slope form}$$

$$4y = x + 3 \qquad \text{Multiply each side by 4.}$$

$$x - 4y + 3 = 0 \qquad \text{General form}$$

So, the distance between this line and the point $(0, 4)$ is

$$\text{Altitude} = h = \frac{|1(0) + (-4)(4) + 3|}{\sqrt{1^2 + (-4)^2}} = \frac{13}{\sqrt{17}} \text{ units.}$$

b. Using the formula for the distance between two points, you can find the length of the base AC to be

$$b = \sqrt{[5-(-3)]^2 + (2-0)^2} \qquad \text{Distance Formula}$$

$$= \sqrt{8^2 + 2^2} \qquad \text{Simplify.}$$

$$= 2\sqrt{17} \text{ units.} \qquad \text{Simplify.}$$

Finally, the area of the triangle in Figure 10.6 is

$$A = \frac{1}{2}bh \qquad \text{Formula for the area of a triangle}$$

$$= \frac{1}{2}\left(2\sqrt{17}\right)\left(\frac{13}{\sqrt{17}}\right) \qquad \text{Substitute for } b \text{ and } h.$$

$$= 13 \text{ square units.} \qquad \text{Simplify.}$$

✓ ***Checkpoint*** *Audio-video solution in English & Spanish at LarsonPrecalculus.com.*

A triangle has vertices $A(-2, 0)$, $B(0, 5)$, and $C(4, 3)$.

a. Find the altitude from vertex B to side AC.

b. Find the area of the triangle.

Summarize (Section 10.1)

1. State the definition of the inclination of a line *(page 688)*. For an example of finding the inclinations of lines, see Example 1.
2. Describe how to find the angle between two lines *(page 689)*. For an example of finding the angle between two lines, see Example 2.
3. Describe how to find the distance between a point and a line *(page 690)*. For examples of finding the distances between points and lines, see Examples 3–5.

10.1 Exercises

See CalcChat.com for tutorial help and worked-out solutions to odd-numbered exercises.

Vocabulary: Fill in the blanks.

1. The ________ of a nonhorizontal line is the positive angle θ (less than π) measured counterclockwise from the x-axis to the line.
2. If a nonvertical line has inclination θ and slope m, then $m =$ ________ .
3. If two nonperpendicular lines have slopes m_1 and m_2, then the angle between the two lines is $\tan\theta =$ ________ .
4. The distance between the point (x_1, y_1) and the line $Ax + By + C = 0$ is given by $d =$ ________ .

Skills and Applications

Finding the Slope of a Line In Exercises 5–16, find the slope of the line with inclination θ.

5. $\theta = \dfrac{\pi}{6}$ radian
6. $\theta = \dfrac{\pi}{4}$ radian
7. $\theta = \dfrac{3\pi}{4}$ radians
8. $\theta = \dfrac{2\pi}{3}$ radians
9. $\theta = \dfrac{\pi}{3}$ radians
10. $\theta = \dfrac{5\pi}{6}$ radians
11. $\theta = 0.26$ radian
12. $\theta = 0.74$ radian
13. $\theta = 1.27$ radians
14. $\theta = 1.35$ radians
15. $\theta = 1.81$ radians
16. $\theta = 2.88$ radians

Finding the Inclination of a Line In Exercises 17–24, find the inclination θ (in radians and degrees) of the line with slope m.

17. $m = -1$
18. $m = -2$
19. $m = 1$
20. $m = 2$
21. $m = \frac{3}{4}$
22. $m = \frac{1}{2}$
23. $m = -\frac{5}{2}$
24. $m = -\frac{7}{9}$

Finding the Inclination of a Line In Exercises 25–34, find the inclination θ (in radians and degrees) of the line passing through the points.

25. $(\sqrt{3}, 2), (0, 1)$
26. $(1, 2\sqrt{3}), (0, \sqrt{3})$
27. $(-\sqrt{3}, -1), (0, -2)$
28. $(3, \sqrt{3}), (6, -2\sqrt{3})$
29. $(6, 1), (10, 8)$
30. $(12, 8), (-4, -3)$
31. $(-2, 20), (10, 0)$
32. $(0, 100), (50, 0)$
33. $\left(\frac{1}{4}, \frac{3}{2}\right), \left(\frac{1}{3}, \frac{1}{2}\right)$
34. $\left(\frac{2}{5}, -\frac{3}{4}\right), \left(-\frac{11}{10}, -\frac{1}{4}\right)$

Finding the Inclination of a Line In Exercises 35–44, find the inclination θ (in radians and degrees) of the line.

35. $2x + 2y - 5 = 0$
36. $x - \sqrt{3}y + 1 = 0$
37. $3x - 3y + 1 = 0$
38. $\sqrt{3}x - y + 2 = 0$
39. $x + \sqrt{3}y + 2 = 0$
40. $-2\sqrt{3}x - 2y = 0$
41. $6x - 2y + 8 = 0$
42. $2x - 6y - 12 = 0$
43. $4x + 5y - 9 = 0$
44. $5x + 3y = 0$

Finding the Angle Between Two Lines In Exercises 45–54, find the angle θ (in radians and degrees) between the lines.

45. $3x + y = 3$
    $x - y = 2$

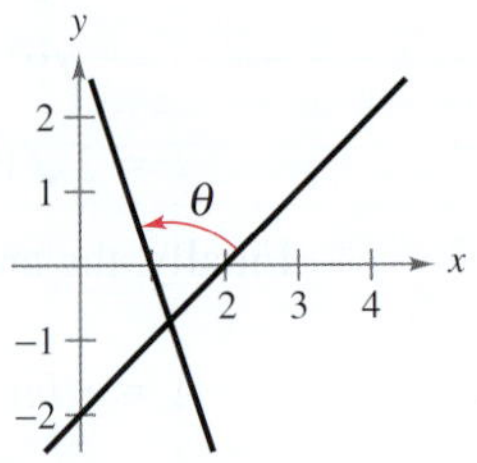

46. $x + 3y = 2$
    $x - 2y = -3$

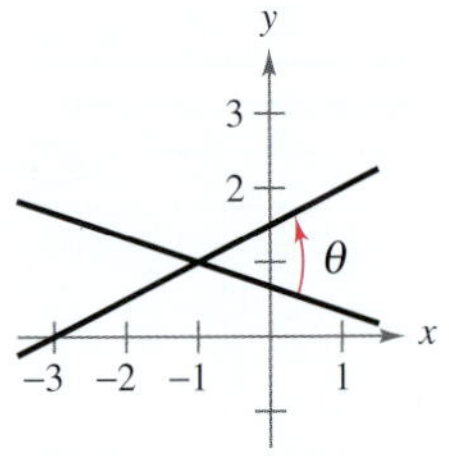

47. $x - y = 0$
    $3x - 2y = -1$

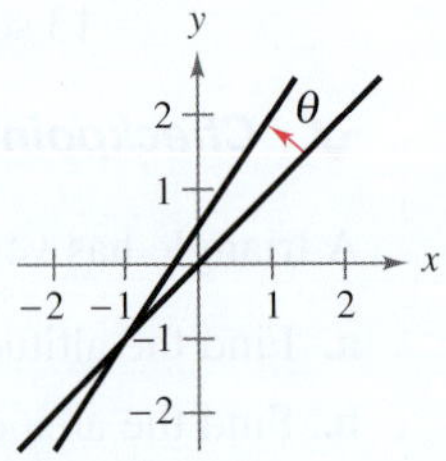

48. $2x - y = 2$
    $4x + 3y = 24$

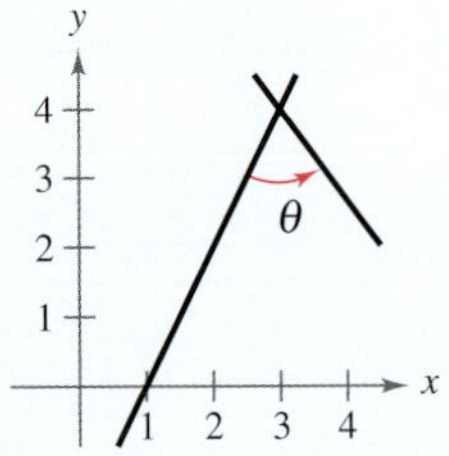

49. $x - 2y = 7$
    $6x + 2y = 5$
50. $5x + 2y = 16$
    $3x - 5y = -1$
51. $x + 2y = 8$
    $x - 2y = 2$
52. $3x - 5y = 3$
    $3x + 5y = 12$
53. $0.05x - 0.03y = 0.21$
    $0.07x + 0.02y = 0.16$
54. $0.02x - 0.05y = -0.19$
    $0.03x + 0.04y = 0.52$

Angle Measurement **In Exercises 55–58, find the slope of each side of the triangle, and use the slopes to find the measures of the interior angles.**

55.

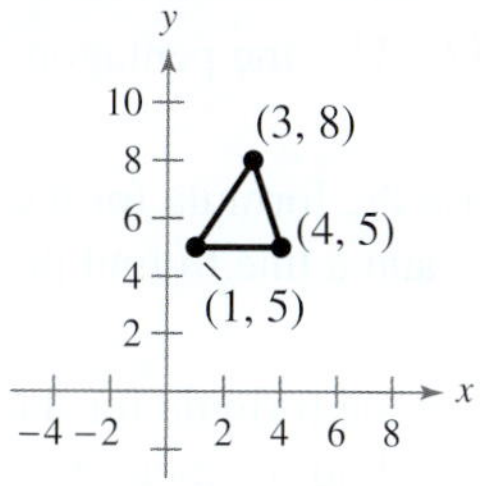

56.

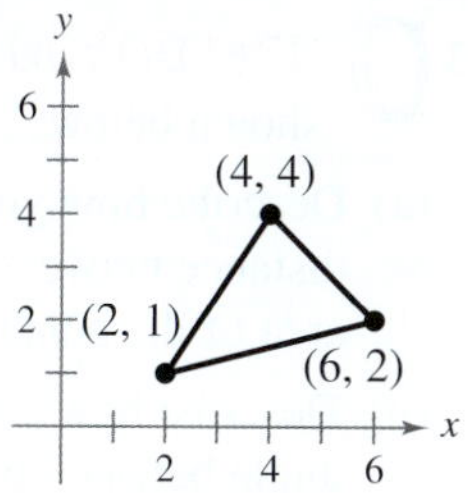

57.

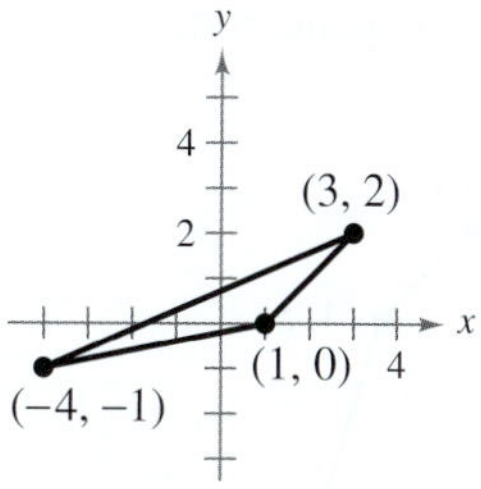

58.

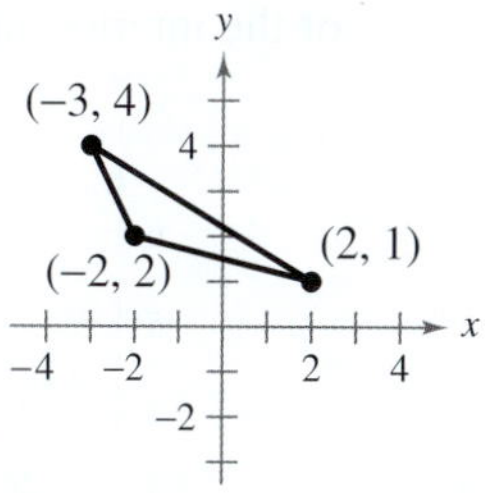

Finding the Distance Between a Point and a Line **In Exercises 59–74, find the distance between the point and the line.**

	Point	Line
59.	$(1, 1)$	$y = x + 1$
60.	$(2, 1)$	$y = x + 2$
61.	$(3, 2)$	$y = 2x - 1$
62.	$(1, 4)$	$y = 4x + 2$
63.	$(-2, 6)$	$y = -x + 5$
64.	$(4, -4)$	$y = -2x - 3$
65.	$(1, -3)$	$y = 2x - 5$
66.	$(-2, 8)$	$y = -3x + 2$
67.	$(2, 3)$	$3x + y = 1$
68.	$(2, 1)$	$-2x + y = 2$
69.	$(6, 2)$	$-3x + 4y = -5$
70.	$(1, -3)$	$4x - 3y = -7$
71.	$(-1, 2)$	$5x + 3y = -4$
72.	$(2, -3)$	$4x - 5y = -2$
73.	$(-1, -5)$	$6x + 3y = 3$
74.	$(-5, -3)$	$-2x - 6y = 7$

An Application of Two Distance Formulas **In Exercises 75–80, the points represent the vertices of a triangle. (a) Draw triangle ABC in the coordinate plane, (b) find the altitude from vertex B of the triangle to side AC, and (c) find the area of the triangle.**

75. $A(-1, 0), B(0, 3), C(3, 1)$

76. $A(-4, 0), B(0, 5), C(3, 3)$

77. $A(-3, 0), B(0, -2), C(2, 3)$

78. $A(-2, 0), B(0, -3), C(5, 1)$

79. $A(1, 1), B(2, 4), C(3, 5)$

80. $A(-3, -2), B(-1, -4), C(3, -1)$

Finding the Distance Between Parallel Lines **In Exercises 81 and 82, find the distance between the parallel lines.**

81. $x + y = 1$
$x + y = 5$

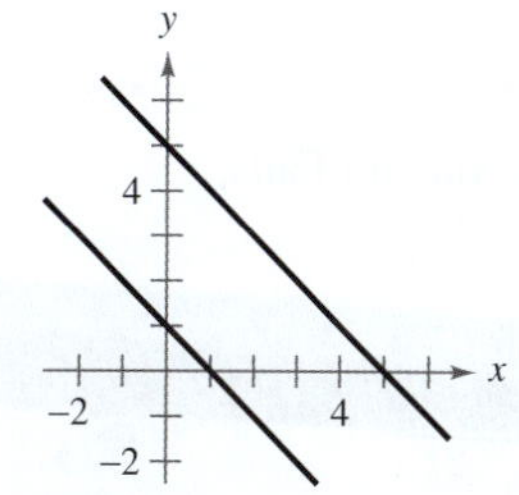

82. $3x - 4y = 1$
$3x - 4y = 10$

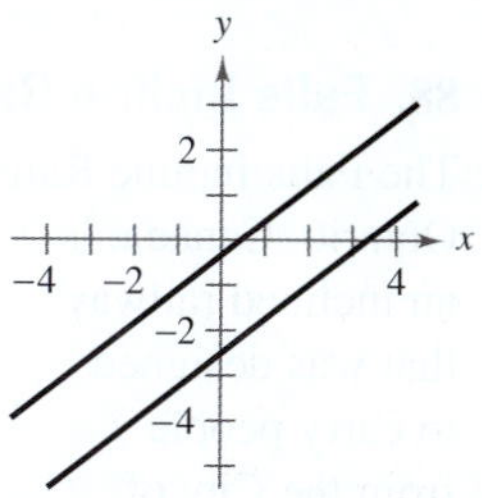

83. **Road Grade** A straight road rises with an inclination of 0.10 radian from the horizontal (see figure). Find the slope of the road and the change in elevation over a two-mile stretch of the road.

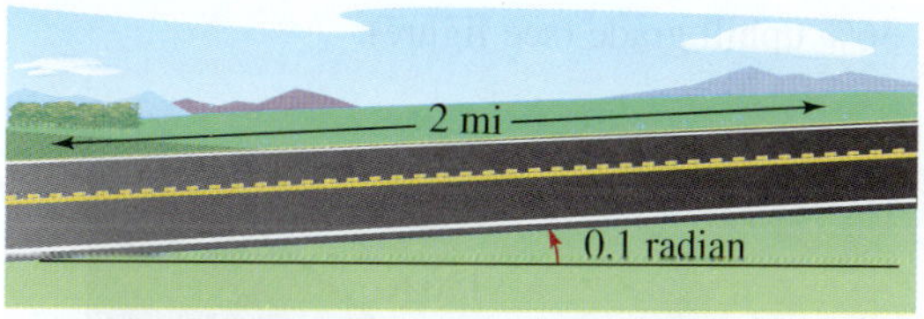

84. **Road Grade** A straight road rises with an inclination of 0.20 radian from the horizontal. Find the slope of the road and the change in elevation over a one-mile stretch of the road.

85. **Pitch of a Roof** A roof has a rise of 3 feet for every horizontal change of 5 feet (see figure). Find the inclination of the roof.

86. **Conveyor Design** A moving conveyor is built so that it rises 1 meter for each 3 meters of horizontal travel.

(a) Draw a diagram that gives a visual representation of the problem.

(b) Find the inclination of the conveyor.

(c) The conveyor runs between two floors in a factory. The distance between the floors is 5 meters. Find the length of the conveyor.

87. Truss Find the angles α and β shown in the drawing of the roof truss.

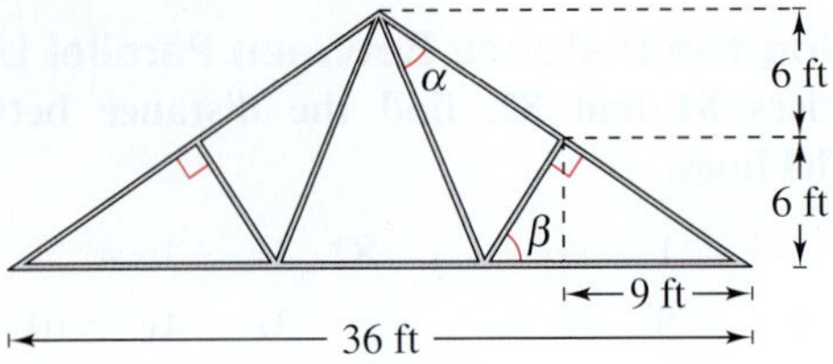

88. Falls Incline Railway

The Falls Incline Railway in Niagara Falls, Ontario, Canada, is an inclined railway that was designed to carry people from the City of Niagara Falls to Queen Victoria Park. The railway is approximately 170 feet long with a 36% uphill grade (see figure).

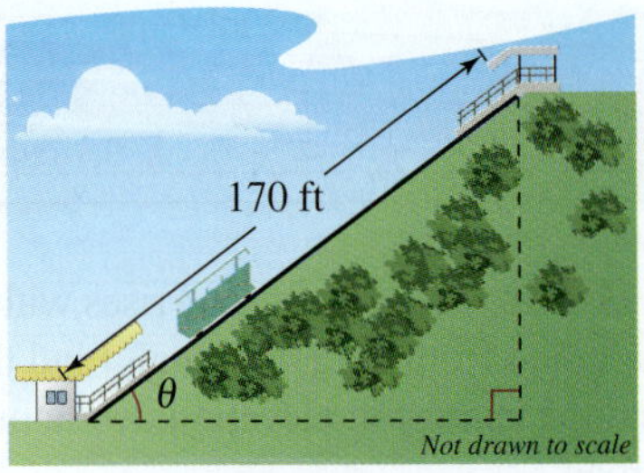

(a) Find the inclination θ of the railway.

(b) Find the change in elevation from the base to the top of the railway.

(c) Using the origin of a rectangular coordinate system as the base of the inclined plane, find the equation of the line that models the railway track.

(d) Sketch a graph of the equation you found in part (c).

Exploration

True or False? **In Exercises 89–91, determine whether the statement is true or false. Justify your answer.**

89. A line that has an inclination greater than $\pi/2$ radians has a negative slope.

90. To find the angle between two lines whose angles of inclination θ_1 and θ_2 are known, substitute θ_1 and θ_2 for m_1 and m_2, respectively, in the formula for the angle between two lines.

91. The inclination of a line is the angle between the line and the x-axis.

92. HOW DO YOU SEE IT? Use the pentagon shown below.

(a) Describe how you can use the formula for the distance between a point and a line to find the area of the pentagon.

(b) Describe how you can use the formula for the angle between two lines to find the measures of the interior angles of the pentagon.

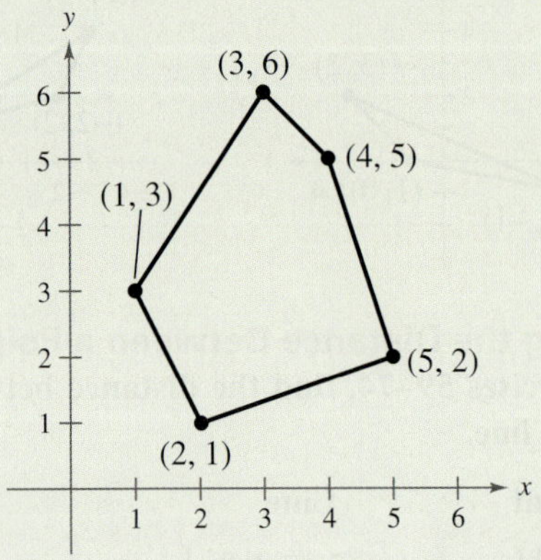

93. Think About It Consider a line with slope m and y-intercept $(0, 4)$.

(a) Write the distance d between the origin and the line as a function of m.

(b) Graph the function in part (a).

(c) Find the slope that yields the maximum distance between the origin and the line.

(d) Find the asymptote of the graph in part (b) and interpret its meaning in the context of the problem.

94. Think About It Consider a line with slope m and y-intercept $(0, 4)$.

(a) Write the distance d between the point $(3, 1)$ and the line as a function of m.

(b) Graph the function in part (a).

(c) Find the slope that yields the maximum distance between the point and the line.

(d) Is it possible for the distance to be 0? If so, what is the slope of the line that yields a distance of 0?

(e) Find the asymptote of the graph in part (b) and interpret its meaning in the context of the problem.

95. Writing Explain why the inclination of a line can be an angle that is greater than $\pi/2$, but the angle between two lines cannot be greater than $\pi/2$.

kan_khampanya/Shutterstock.com

10.2 Introduction to Conics: Parabolas

You can use parabolas to model and solve many types of real-life problems. For instance, in Exercise 72 on page 702, you will use a parabola to model the cables of the Golden Gate Bridge.

- **Recognize a conic as the intersection of a plane and a double-napped cone.**
- **Write equations of parabolas in standard form.**
- **Use the reflective property of parabolas to solve real-life problems.**

Conics

Conic sections were discovered during the classical Greek period, 600 to 300 B.C. This early Greek study was largely concerned with the geometric properties of conics. It was not until the early 17th century that the broad applicability of conics became apparent and played a prominent role in the early development of calculus.

A **conic section** (or simply **conic**) is the intersection of a plane and a double-napped cone. Notice in Figure 10.7 that in the formation of the four basic conics, the intersecting plane does not pass through the vertex of the cone. When the plane does pass through the vertex, the resulting figure is a **degenerate conic,** as shown in Figure 10.8.

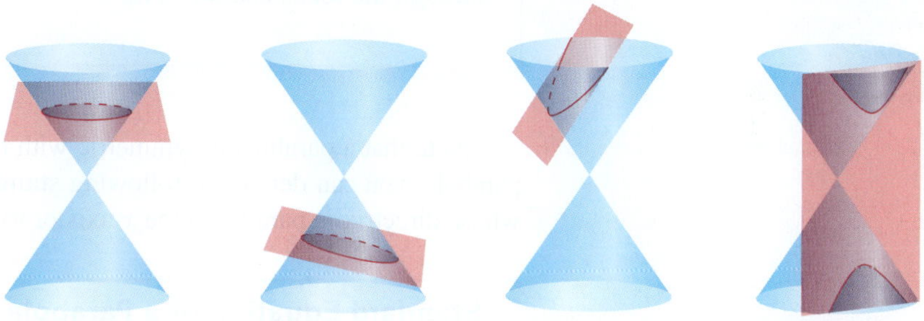

Figure 10.7 Basic Conics

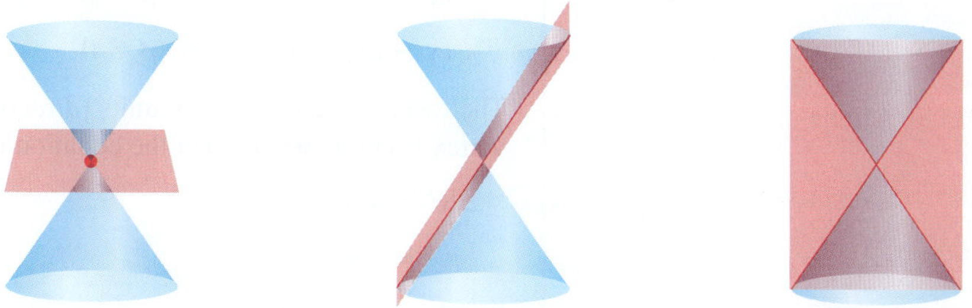

Figure 10.8 Degenerate Conics

There are several ways to approach the study of conics. You could begin by defining conics in terms of the intersections of planes and cones, as the Greeks did, or you could define them algebraically, in terms of the general second-degree equation

$$Ax^2 + Bxy + Cy^2 + Dx + Ey + F = 0.$$

However, you will study a third approach, in which each of the conics is defined as a **locus** (collection) of points satisfying a geometric property. For example, in Section 1.2, you saw how the definition of a circle as the set of all points (x, y) that are equidistant from a fixed point (h, k) led to the standard form of the equation of a circle

$$(x - h)^2 + (y - k)^2 = r^2. \quad \text{Equation of circle}$$

topseller/Shutterstock.com

Parabolas

In Section 2.1, you learned that the graph of the quadratic function

$$f(x) = ax^2 + bx + c$$

is a parabola that opens upward or downward. The following definition of a parabola is more general in the sense that it is independent of the orientation of the parabola.

Definition of Parabola

A **parabola** is the set of all points (x, y) in a plane that are equidistant from a fixed line, called the **directrix,** and a fixed point, called the **focus,** not on the line. (See figure.) The **vertex** is the midpoint between the focus and the directrix. The **axis** of the parabola is the line passing through the focus and the vertex.

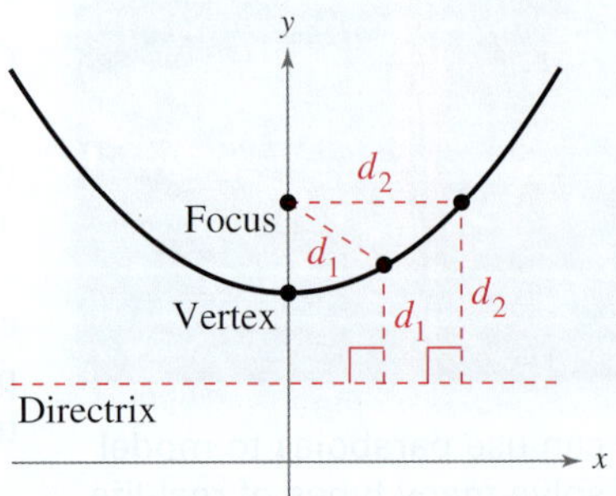

Note that a parabola is symmetric with respect to its axis. Using the definition of a parabola, you can derive the following **standard form of the equation of a parabola** whose directrix is parallel to the x-axis or to the y-axis.

Standard Equation of a Parabola

The **standard form of the equation of a parabola** with vertex at (h, k) is as follows.

$(x - h)^2 = 4p(y - k), \ p \neq 0$ Vertical axis; directrix: $y = k - p$

$(y - k)^2 = 4p(x - h), \ p \neq 0$ Horizontal axis; directrix: $x = h - p$

The focus lies on the axis p units (*directed distance*) from the vertex. If the vertex is at the origin, then the equation takes one of the following forms.

$x^2 = 4py$ Vertical axis

$y^2 = 4px$ Horizontal axis

See the figures below.

For a proof of the standard form of the equation of a parabola, see Proofs in Mathematics on page 769.

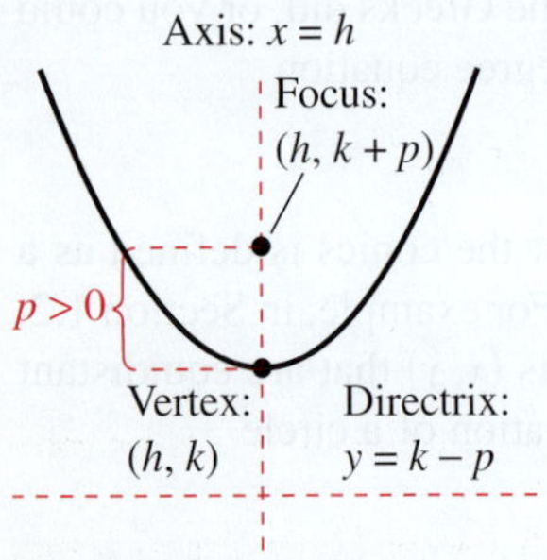

$(x - h)^2 = 4p(y - k)$
Vertical axis: $p > 0$

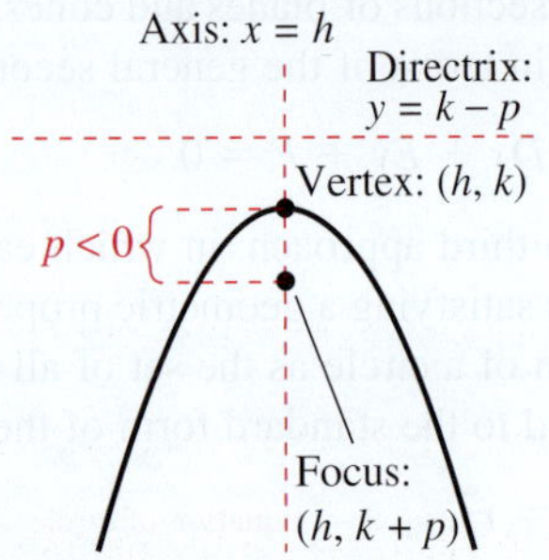

$(x - h)^2 = 4p(y - k)$
Vertical axis: $p < 0$

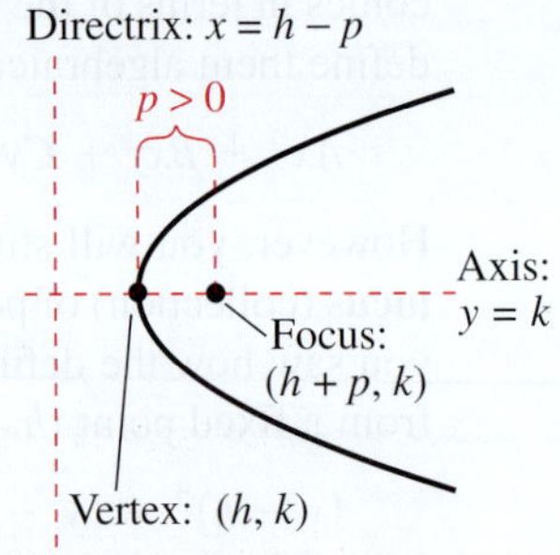

$(y - k)^2 = 4p(x - h)$
Horizontal axis: $p > 0$

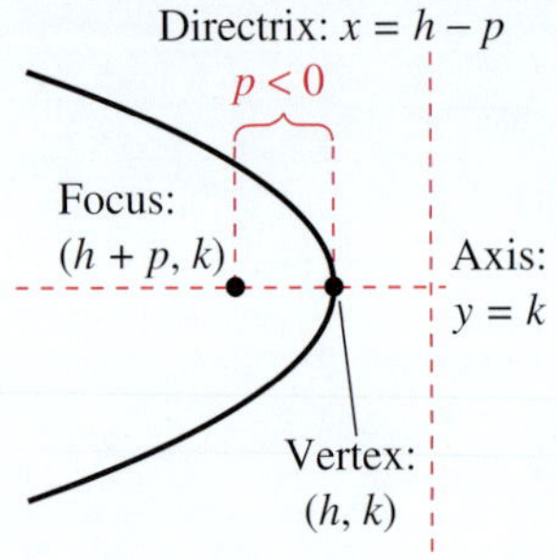

$(y - k)^2 = 4p(x - h)$
Horizontal axis: $p < 0$

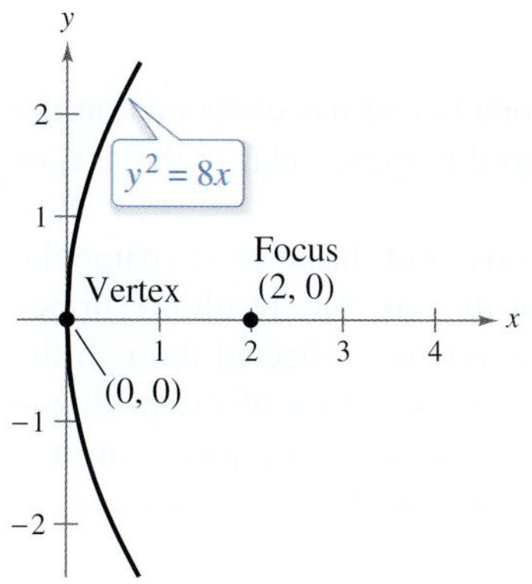

Figure 10.9

EXAMPLE 1 **Finding the Standard Equation of a Parabola**

Find the standard form of the equation of the parabola with vertex at the origin and focus (2, 0).

Solution The axis of the parabola is horizontal, passing through (0, 0) and (2, 0), as shown in Figure 10.9. The standard form is $y^2 = 4px$, where $p = 2$. So, the equation is $y^2 = 8x$. You can use a graphing utility to confirm this equation. To do this, let $y_1 = \sqrt{8x}$ to graph the upper portion and let $y_2 = -\sqrt{8x}$ to graph the lower portion of the parabola.

✓ **Checkpoint** Audio-video solution in English & Spanish at LarsonPrecalculus.com.

Find the standard form of the equation of the parabola with vertex at the origin and focus $\left(0, \frac{3}{8}\right)$.

▷ **ALGEBRA HELP** The technique of completing the square is used to write the equation in Example 2 in standard form. You can review completing the square in Appendix A.5.

EXAMPLE 2 **Finding the Focus of a Parabola**

Find the focus of the parabola $y = -\frac{1}{2}x^2 - x + \frac{1}{2}$.

Solution Convert to standard form by completing the square.

$$y = -\frac{1}{2}x^2 - x + \frac{1}{2} \quad \text{Write original equation.}$$

$$-2y = x^2 + 2x - 1 \quad \text{Multiply each side by } -2.$$

$$1 - 2y = x^2 + 2x \quad \text{Add 1 to each side.}$$

$$1 + 1 - 2y = x^2 + 2x + 1 \quad \text{Complete the square.}$$

$$2 - 2y = x^2 + 2x + 1 \quad \text{Combine like terms.}$$

$$-2(y - 1) = (x + 1)^2 \quad \text{Standard form}$$

Comparing this equation with

$$(x - h)^2 = 4p(y - k)$$

you can conclude that $h = -1$, $k = 1$, and $p = -\frac{1}{2}$. Because p is negative, the parabola opens downward, as shown in Figure 10.10. So, the focus of the parabola is $(h, k + p) = \left(-1, \frac{1}{2}\right)$.

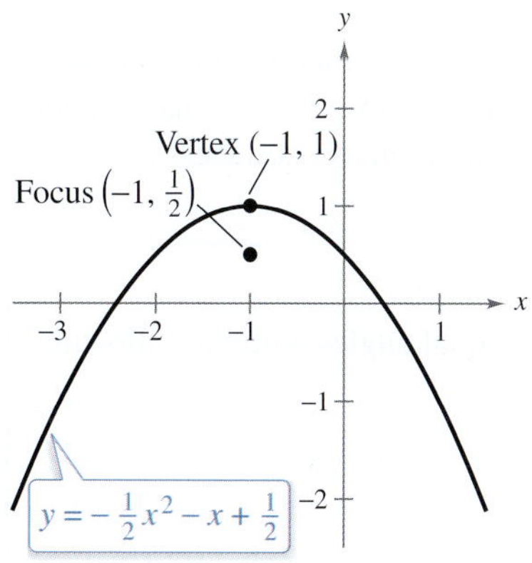

Figure 10.10

✓ **Checkpoint** Audio-video solution in English & Spanish at LarsonPrecalculus.com.

Find the focus of the parabola $x = \frac{1}{4}y^2 + \frac{3}{2}y + \frac{13}{4}$.

EXAMPLE 3 **Finding the Standard Equation of a Parabola**

Find the standard form of the equation of the parabola with vertex (2, 1) and focus (2, 4).

Solution Because the axis of the parabola is vertical, passing through (2, 1) and (2, 4), consider the equation

$$(x - h)^2 = 4p(y - k)$$

where $h = 2$, $k = 1$, and $p = 4 - 1 = 3$. So, the standard form is

$$(x - 2)^2 = 12(y - 1).$$

The graph of this parabola is shown in Figure 10.11.

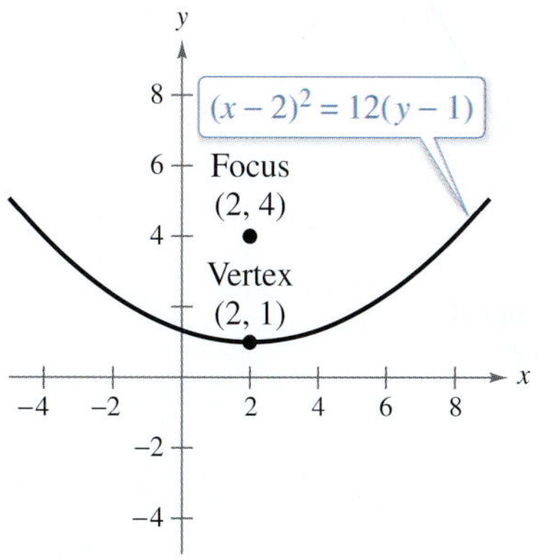

Figure 10.11

✓ **Checkpoint** Audio-video solution in English & Spanish at LarsonPrecalculus.com.

Find the standard form of the equation of the parabola with vertex (2, −3) and focus (4, −3).

One important application of parabolas is in astronomy. Radio telescopes use parabolic dishes to collect radio waves from space.

Application

A line segment that passes through the focus of a parabola and has endpoints on the parabola is called a **focal chord.** The specific focal chord perpendicular to the axis of the parabola is called the **latus rectum.**

Parabolas occur in a wide variety of applications. For instance, a parabolic reflector can be formed by revolving a parabola about its axis. The resulting surface has the property that all incoming rays parallel to the axis are reflected through the focus of the parabola. This is the principle behind the construction of the parabolic mirrors used in reflecting telescopes. Conversely, the light rays emanating from the focus of a parabolic reflector used in a flashlight are all parallel to one another, as shown below.

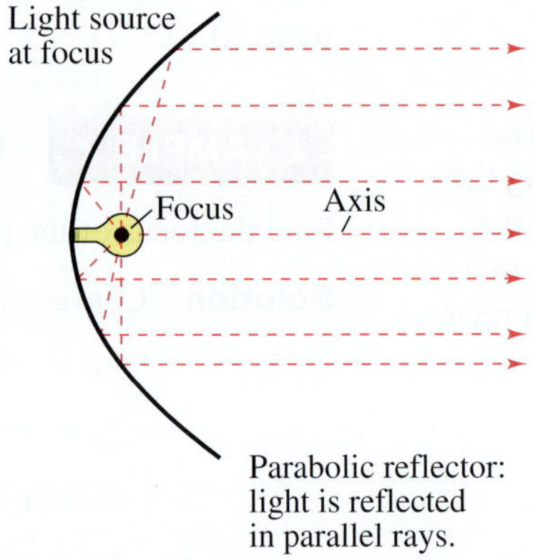

A line is **tangent** to a parabola at a point on the parabola when the line intersects, but does not cross, the parabola at the point. Tangent lines to parabolas have special properties related to the use of parabolas in constructing reflective surfaces.

Reflective Property of a Parabola

The tangent line to a parabola at a point P makes equal angles with the following two lines (see figure below).

1. The line passing through P and the focus
2. The axis of the parabola

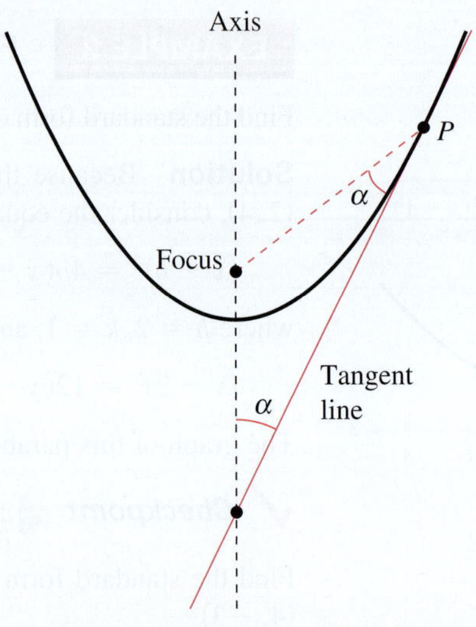

John A Davis/Shutterstock.com

EXAMPLE 4 Finding the Tangent Line at a Point on a Parabola

Find the equation of the tangent line to the parabola $y = x^2$ at the point $(1, 1)$.

Solution For this parabola, $p = \frac{1}{4}$ and the focus is $\left(0, \frac{1}{4}\right)$, as shown in the figure below.

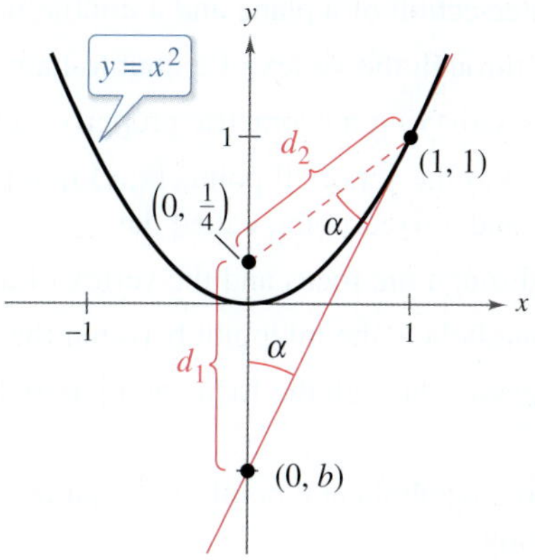

You can find the y-intercept $(0, b)$ of the tangent line by equating the lengths of the two sides of the isosceles triangle shown in the figure:

$$d_1 = \frac{1}{4} - b$$

and

$$d_2 = \sqrt{(1 - 0)^2 + \left(1 - \frac{1}{4}\right)^2} = \frac{5}{4}.$$

Note that $d_1 = \frac{1}{4} - b$ rather than $b - \frac{1}{4}$. The order of subtraction for the distance is important because the distance must be positive. Setting $d_1 = d_2$ produces

$$\frac{1}{4} - b = \frac{5}{4}$$

$$b = -1.$$

So, the slope of the tangent line is

$$m = \frac{1 - (-1)}{1 - 0} = 2$$

and the equation of the tangent line in slope-intercept form is

$$y = 2x - 1.$$

✓ Checkpoint *Audio-video solution in English & Spanish at LarsonPrecalculus.com.*

Find the equation of the tangent line to the parabola $y = 3x^2$ at the point $(1, 3)$.

TECHNOLOGY Use a graphing utility to confirm the result of Example 4. By graphing

$$y_1 = x^2 \text{ and } y_2 = 2x - 1$$

in the same viewing window, you should be able to see that the line touches the parabola at the point $(1, 1)$.

ALGEBRA HELP You can review techniques for writing linear equations in Section 1.3.

Summarize (Section 10.2)

1. List the four basic conic sections *(page 695)*.
2. Define the standard form of the equation of a parabola *(page 696)*. For examples of writing the equations of parabolas in standard form, see Examples 1–3.
3. Describe an application involving the reflective property of parabolas *(page 699, Example 4)*.

10.2 Exercises

See **CalcChat.com** for tutorial help and worked-out solutions to odd-numbered exercises.

Vocabulary: Fill in the blanks.

1. A ________ is the intersection of a plane and a double-napped cone.
2. When a plane passes through the vertex of a double-napped cone, the intersection is a ________ ________.
3. A collection of points satisfying a geometric property can also be referred to as a ________ of points.
4. A ________ is defined as the set of all points (x, y) in a plane that are equidistant from a fixed line, called the ________, and a fixed point, called the ________, not on the line.
5. The line that passes through the focus and the vertex of a parabola is called the ________ of the parabola.
6. The ________ of a parabola is the midpoint between the focus and the directrix.
7. A line segment that passes through the focus of a parabola and has endpoints on the parabola is called a ________ ________.
8. A line is ________ to a parabola at a point on the parabola when the line intersects, but does not cross, the parabola at the point.

Skills and Applications

Matching **In Exercises 9–14, match the equation with its graph. [The graphs are labeled (a), (b), (c), (d), (e), and (f).]**

(a)

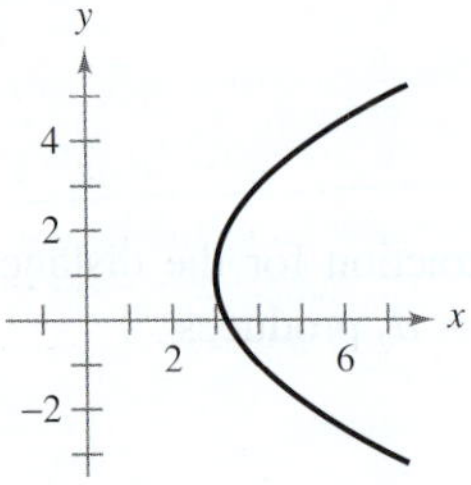

(b)

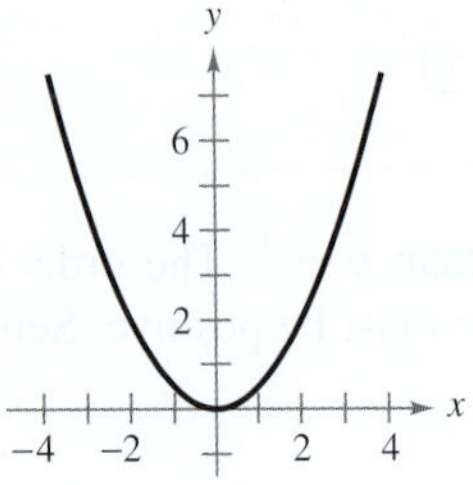

(c)

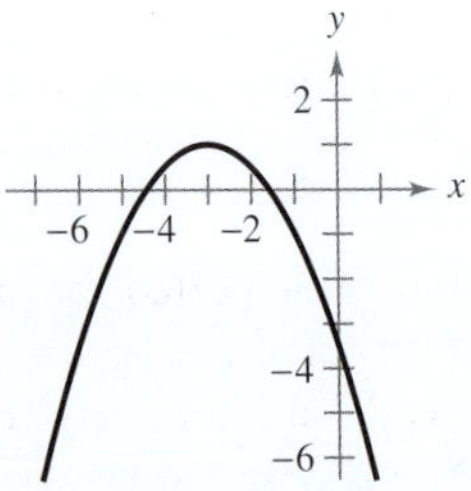

(d)

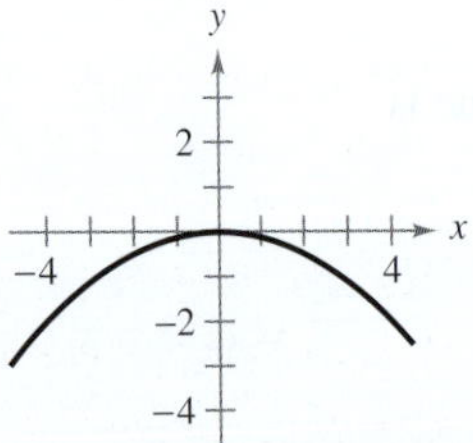

(e)

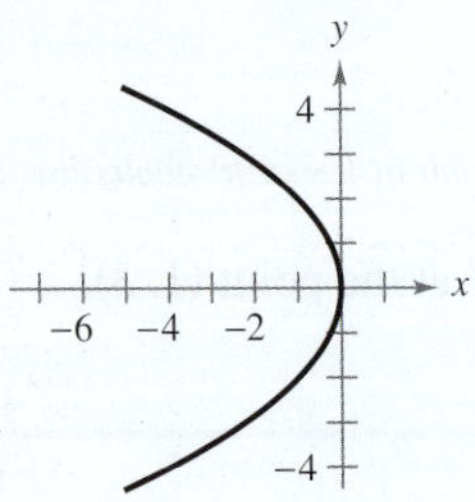

(f)

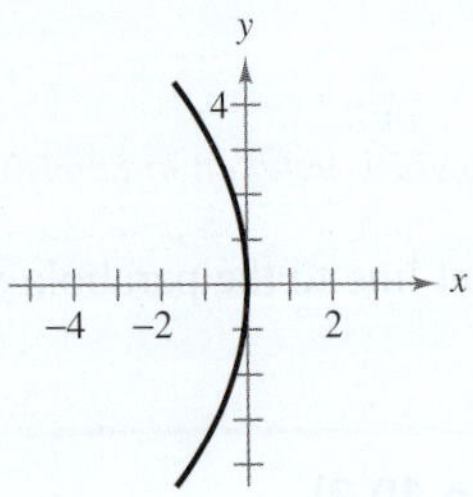

9. $y^2 = -4x$
10. $x^2 = 2y$
11. $x^2 = -8y$
12. $y^2 = -12x$
13. $(y - 1)^2 = 4(x - 3)$
14. $(x + 3)^2 = -2(y - 1)$

Finding the Standard Equation of a Parabola **In Exercises 15–28, find the standard form of the equation of the parabola with the given characteristic(s) and vertex at the origin.**

15.

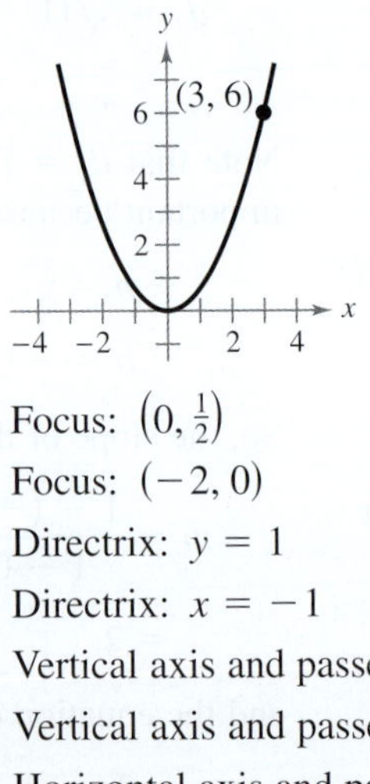

16.

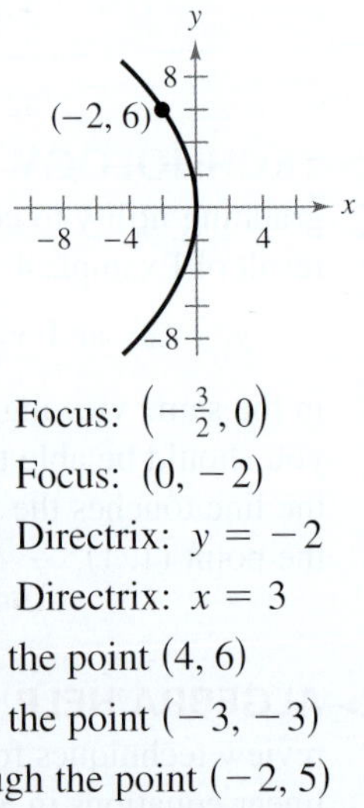

17. Focus: $\left(0, \frac{1}{2}\right)$
18. Focus: $\left(-\frac{3}{2}, 0\right)$
19. Focus: $(-2, 0)$
20. Focus: $(0, -2)$
21. Directrix: $y = 1$
22. Directrix: $y = -2$
23. Directrix: $x = -1$
24. Directrix: $x = 3$
25. Vertical axis and passes through the point $(4, 6)$
26. Vertical axis and passes through the point $(-3, -3)$
27. Horizontal axis and passes through the point $(-2, 5)$
28. Horizontal axis and passes through the point $(3, -2)$

Finding the Vertex, Focus, and Directrix of a Parabola **In Exercises 29–42, find the vertex, focus, and directrix of the parabola. Then sketch the parabola.**

29. $y = \frac{1}{2}x^2$
30. $y = -2x^2$
31. $y^2 = -6x$
32. $y^2 = 3x$
33. $x^2 + 6y = 0$
34. $x + y^2 = 0$
35. $(x - 1)^2 + 8(y + 2) = 0$
36. $(x + 5) + (y - 1)^2 = 0$
37. $(x + 3)^2 = 4\left(y - \frac{3}{2}\right)$
38. $\left(x + \frac{1}{2}\right)^2 = 4(y - 1)$
39. $y = \frac{1}{4}(x^2 - 2x + 5)$
40. $x = \frac{1}{4}(y^2 + 2y + 33)$
41. $y^2 + 6y + 8x + 25 = 0$
42. $y^2 - 4y - 4x = 0$

Finding the Vertex, Focus, and Directrix of a Parabola **In Exercises 43–46, find the vertex, focus, and directrix of the parabola. Use a graphing utility to graph the parabola.**

43. $x^2 + 4x + 6y - 2 = 0$

44. $x^2 - 2x + 8y + 9 = 0$

45. $y^2 + x + y = 0$

46. $y^2 - 4x - 4 = 0$

Finding the Standard Equation of a Parabola **In Exercises 47–56, find the standard form of the equation of the parabola with the given characteristics.**

47.

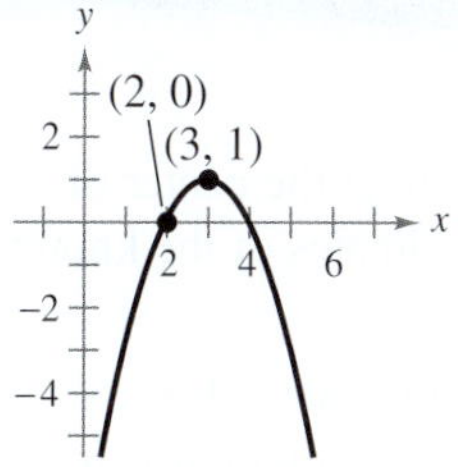

48.

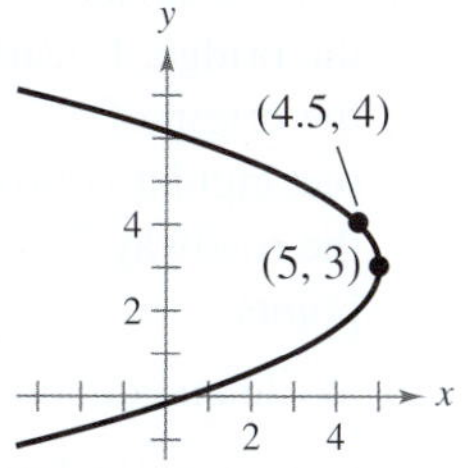

49.

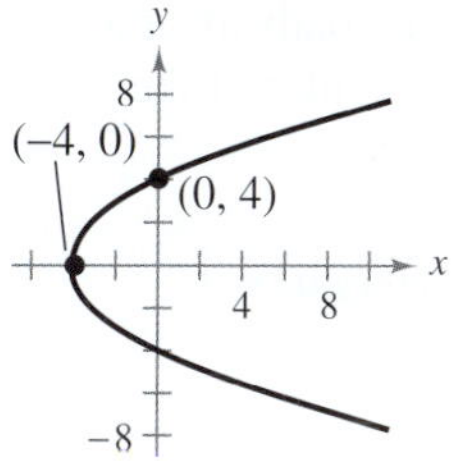

50.

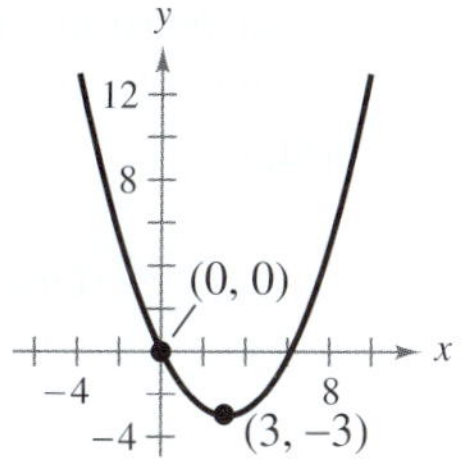

51. Vertex: $(4, 3)$; focus: $(6, 3)$

52. Vertex: $(-1, 2)$; focus: $(-1, 0)$

53. Vertex: $(0, 2)$; directrix: $y = 4$

54. Vertex: $(1, 2)$; directrix: $y = -1$

55. Focus: $(2, 2)$; directrix: $x = -2$

56. Focus: $(0, 0)$; directrix: $y = 8$

Finding the Tangent Line at a Point on a Parabola **In Exercises 57–60, find the equation of the tangent line to the parabola at the given point.**

57. $x^2 = 2y$, $(4, 8)$

58. $x^2 = 2y$, $\left(-3, \frac{9}{2}\right)$

59. $y = -2x^2$, $(-1, -2)$

60. $y = -2x^2$, $(2, -8)$

61. Highway Design Highway engineers design a parabolic curve for an entrance ramp from a straight street to an interstate highway (see figure). Find an equation of the parabola.

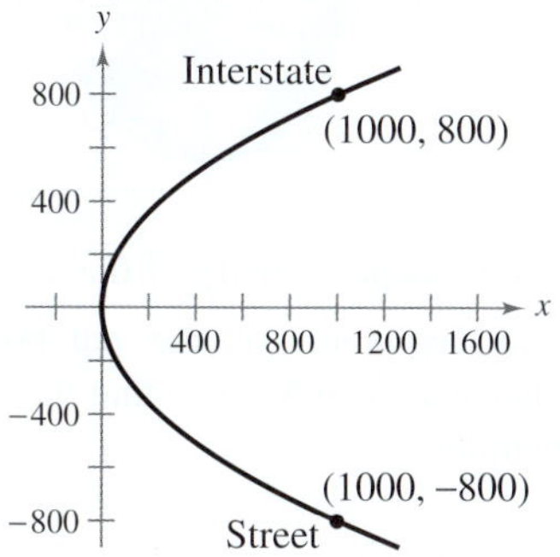

62. Road Design Roads are often designed with parabolic surfaces to allow rain to drain off. A particular road that is 32 feet wide is 0.4 foot higher in the center than it is on the sides (see figure).

(a) Find an equation of the parabola with its vertex at the origin that models the road surface.

(b) How far from the center of the road is the road surface 0.1 foot lower than in the middle?

63. Flashlight The light bulb in a flashlight is at the focus of a parabolic reflector, 1.5 centimeters from the vertex of the reflector (see figure). Write an equation for a cross section of the flashlight's reflector with its focus on the positive x-axis and its vertex at the origin.

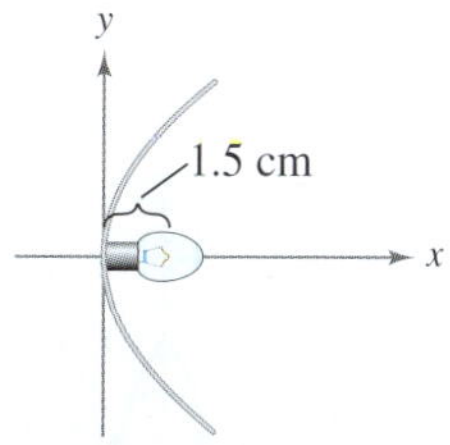

Figure for 63

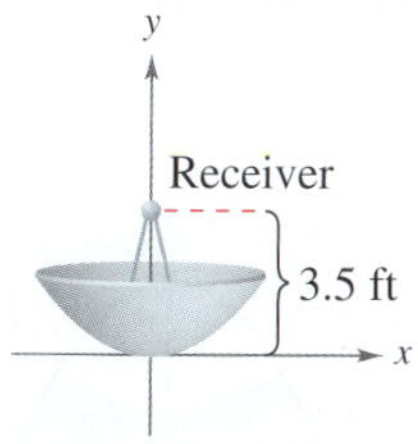

Figure for 64

64. Satellite Antenna Write an equation for a cross section of the parabolic satellite dish antenna shown in the figure.

65. Beam Deflection A simply supported beam is 64 feet long and has a load at the center (see figure). The deflection (bending) of the beam at its center is 1 inch. The shape of the deflected beam is parabolic.

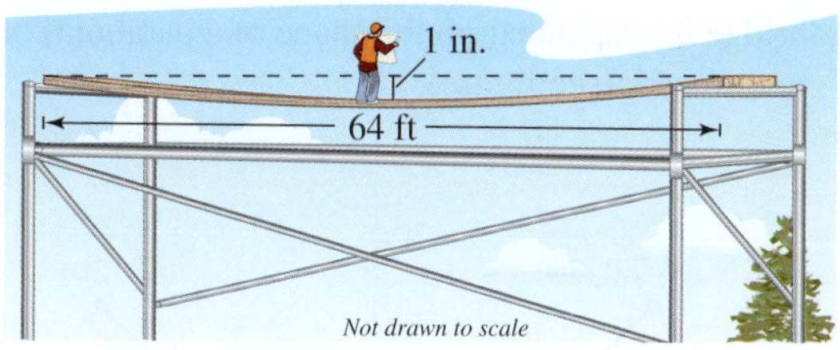

(a) Find an equation of the parabola with its vertex at the origin that models the shape of the beam.

(b) How far from the center of the beam is the deflection equal to $\frac{1}{2}$ inch?

66. Beam Deflection Repeat Exercise 65 when the length of the beam is 36 feet and the deflection of the beam at its center is 2 inches.

67. Fluid Flow Water is flowing from a horizontal pipe 48 feet above the ground. The falling stream of water has the shape of a parabola whose vertex $(0, 48)$ is at the end of the pipe (see figure). The stream of water strikes the ground at the point $(10\sqrt{3}, 0)$. Find the equation of the path taken by the water.

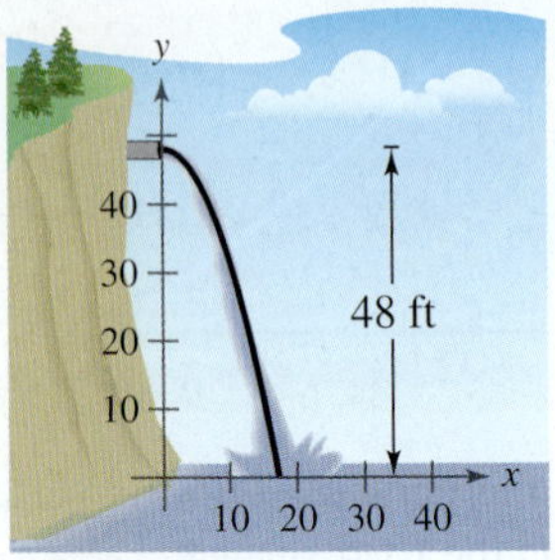

Figure for 67

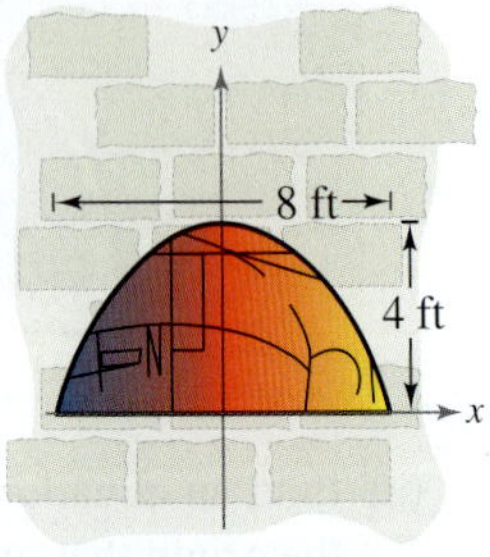

Figure for 68

68. Window Design A church window (see figure) is bounded above by a parabola. Find the equation of the parabola.

69. Archway A parabolic archway (see figure) is 12 meters high at the vertex. At a height of 10 meters, the width of the archway is 8 meters. How wide is the archway at ground level?

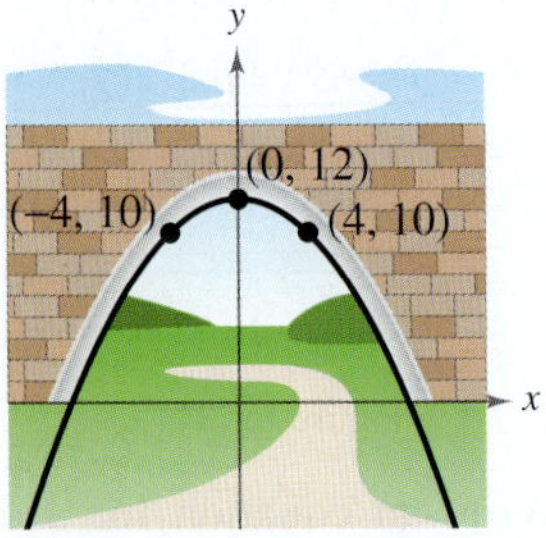

Figure for 69

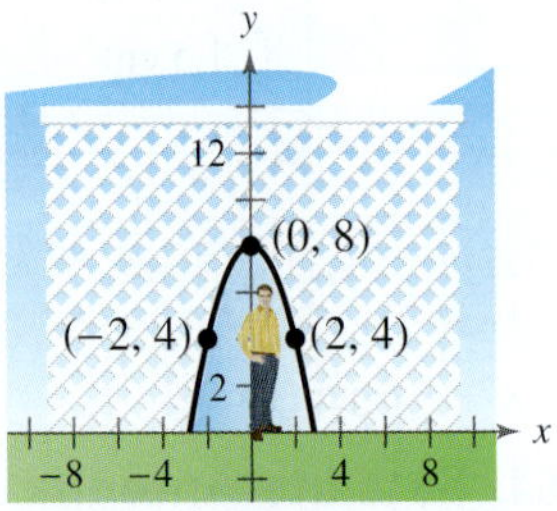

Figure for 70

70. Lattice Arch A parabolic lattice arch is 8 feet high at the vertex. At a height of 4 feet, the width of the lattice arch is 4 feet (see figure). How wide is the lattice arch at ground level?

71. Suspension Bridge Each cable of a suspension bridge is suspended (in the shape of a parabola) between two towers (see figure).

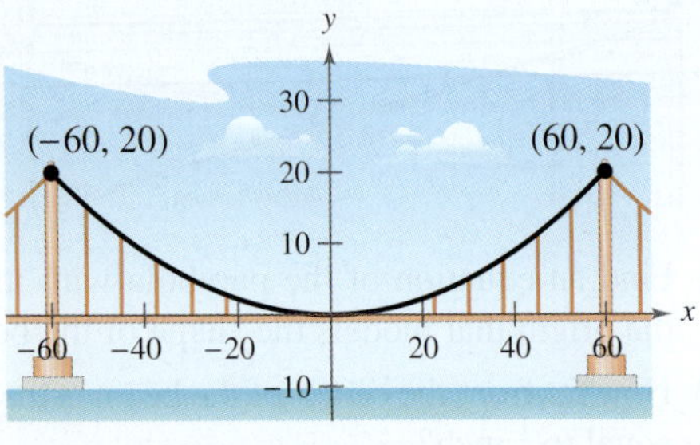

(a) Find the coordinates of the focus.

(b) Write an equation that models the cables.

72. Suspension Bridge Each cable of the Golden Gate Bridge is suspended (in the shape of a parabola) between two towers that are 1280 meters apart. The top of each tower is 152 meters above the roadway. The cables touch the roadway midway between the towers.

(a) Draw a sketch of the bridge. Locate the origin of a rectangular coordinate system at the center of the roadway. Label the coordinates of the known points.

(b) Write an equation that models the cables.

(c) Complete the table by finding the height y of the suspension cables over the roadway at a distance of x meters from the center of the bridge.

Distance, x	Height, y
0	
100	
250	
400	
500	

73. Weather Satellite Orbit A weather satellite in a 100-mile-high circular orbit around Earth has a velocity of approximately 17,500 miles per hour. When this velocity is multiplied by $\sqrt{2}$, the weather satellite has the minimum velocity necessary to escape Earth's gravity and follow a parabolic path with the center of Earth as the focus (see figure).

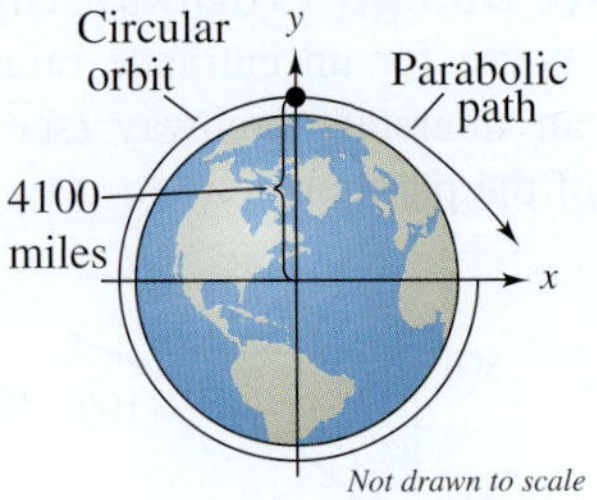

(a) Find the escape velocity of the weather satellite.

(b) Find an equation of the parabolic path of the weather satellite. (Assume that the radius of Earth is 4000 miles.)

topseller/Shutterstock.com

74. Path of a Softball The path of a softball is modeled by

$$-12.5(y - 7.125) = (x - 6.25)^2$$

where the coordinates x and y are measured in feet, with $x = 0$ corresponding to the position from which the ball was thrown.

(a) Use a graphing utility to graph the trajectory of the softball.

(b) Use the *trace* feature of the graphing utility to approximate the highest point and the range of the trajectory.

Projectile Motion In Exercises 75 and 76, consider the path of a projectile projected horizontally with a velocity of v feet per second at a height of s feet, where the model for the path is

$$x^2 = -\frac{v^2}{16}(y - s).$$

In this model (in which air resistance is disregarded), y is the height (in feet) of the projectile and x is the horizontal distance (in feet) the projectile travels.

75. A ball is thrown from the top of a 100-foot tower with a velocity of 28 feet per second.

(a) Find the equation of the parabolic path.

(b) How far does the ball travel horizontally before striking the ground?

76. A cargo plane is flying at an altitude of 500 feet and a speed of 135 miles per hour. A supply crate is dropped from the plane. How many *feet* will the crate travel horizontally before it hits the ground?

Exploration

True or False? In Exercises 77 and 78, determine whether the statement is true or false. Justify your answer.

77. It is possible for a parabola to intersect its directrix.

78. If the vertex and focus of a parabola are on a horizontal line, then the directrix of the parabola is vertical.

79. Slope of a Tangent Line Let (x_1, y_1) be the coordinates of a point on the parabola $x^2 = 4py$. The equation of the line tangent to the parabola at the point is

$$y - y_1 = \frac{x_1}{2p}(x - x_1).$$

What is the slope of the tangent line?

80. Think About It Explain what each of the following equations represents, and how equations (a) and (b) are equivalent.

(a) $y = a(x - h)^2 + k, \quad a \neq 0$

(b) $(x - h)^2 = 4p(y - k), \quad p \neq 0$

(c) $(y - k)^2 = 4p(x - h), \quad p \neq 0$

81. Think About It The equation $x^2 + y^2 = 0$ is a degenerate conic. Sketch the graph of this equation and identify the degenerate conic. Describe the intersection of the plane and the double-napped cone for this particular conic.

82. HOW DO YOU SEE IT?

In parts (a)–(d), describe how a plane could intersect the double-napped cone to form the conic section (see figure).

(a) Circle (b) Ellipse

(c) Parabola (d) Hyperbola

83. Graphical Reasoning Consider the parabola $x^2 = 4py$.

(a) Use a graphing utility to graph the parabola for $p = 1$, $p = 2$, $p = 3$, and $p = 4$. Describe the effect on the graph when p increases.

(b) Locate the focus for each parabola in part (a).

(c) For each parabola in part (a), find the length of the latus rectum (see figure). How can the length of the latus rectum be determined directly from the standard form of the equation of the parabola?

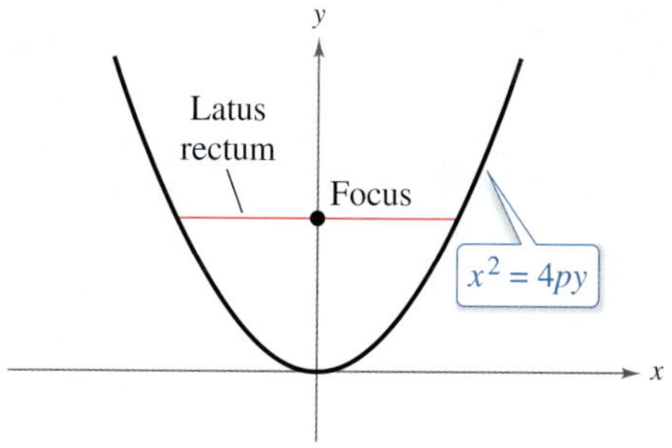

(d) Explain how the result of part (c) can be used as a sketching aid when graphing parabolas.

84. Geometry The area of the shaded region in the figure is $A = \frac{8}{3}p^{1/2}b^{3/2}$.

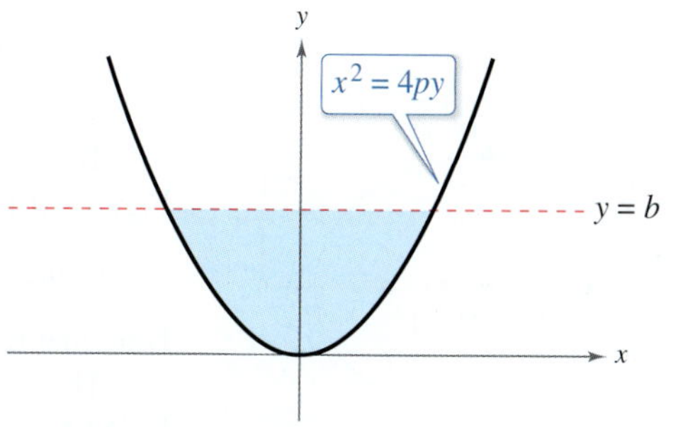

(a) Find the area when $p = 2$ and $b = 4$.

(b) Give a geometric explanation of why the area approaches 0 as p approaches 0.

10.3 Ellipses

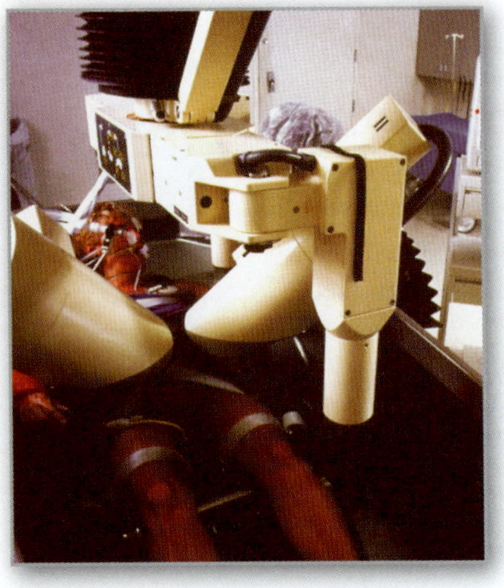

Ellipses have many real-life applications. For instance, Exercise 57 on page 711 shows how the focal properties of an ellipse are used by a lithotripter machine to break up kidney stones.

- **Write equations of ellipses in standard form and graph ellipses.**
- **Use properties of ellipses to model and solve real-life problems.**
- **Find eccentricities of ellipses.**

Introduction

The second type of conic is called an **ellipse.** It is defined as follows.

> **Definition of Ellipse**
>
> An **ellipse** is the set of all points (x, y) in a plane, the sum of whose distances from two distinct fixed points, called **foci,** is constant. See Figure 10.12.

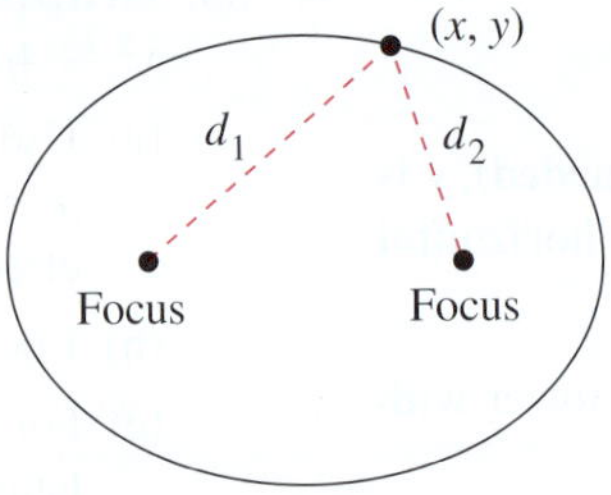

$d_1 + d_2$ is constant.
Figure 10.12

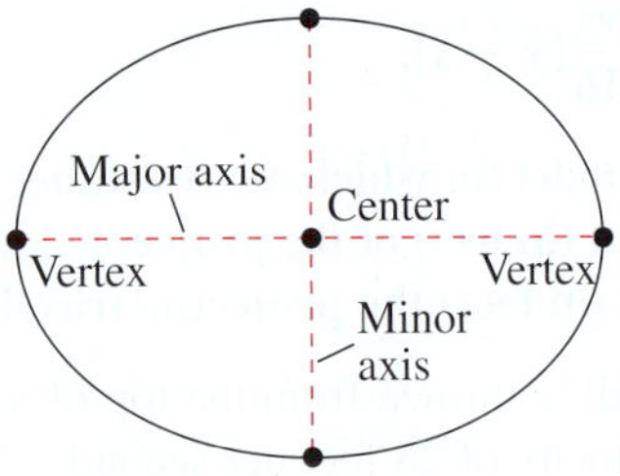

Figure 10.13

The line through the foci intersects the ellipse at two points called **vertices.** The chord joining the vertices is the **major axis,** and its midpoint is the **center** of the ellipse. The chord perpendicular to the major axis at the center is the **minor axis** of the ellipse. See Figure 10.13.

You can visualize the definition of an ellipse by imagining two thumbtacks placed at the foci, as shown below. If the ends of a fixed length of string are fastened to the thumbtacks and the string is drawn taut with a pencil, then the path traced by the pencil will be an ellipse.

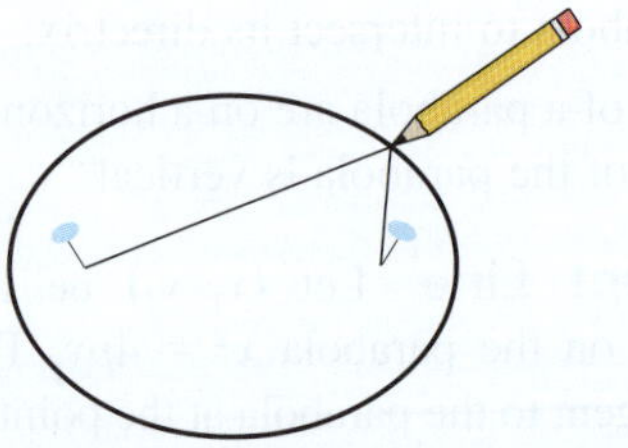

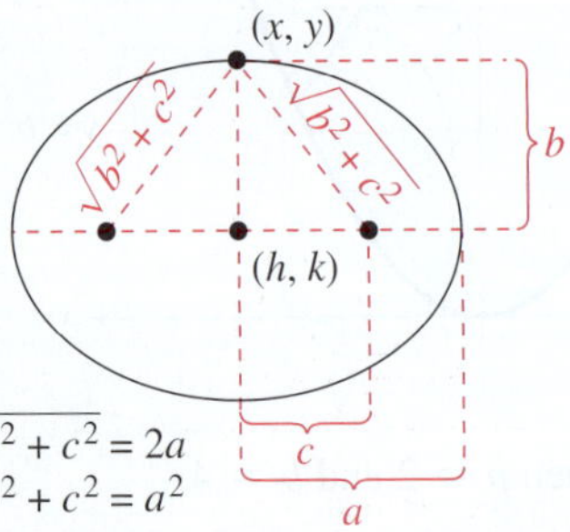

$2\sqrt{b^2 + c^2} = 2a$
$b^2 + c^2 = a^2$

Figure 10.14

To derive the standard form of the equation of an ellipse, consider the ellipse in Figure 10.14 with the following points.

Center: (h, k) Vertices: $(h \pm a, k)$ Foci: $(h \pm c, k)$

Note that the center is the midpoint of the segment joining the foci.

The sum of the distances from any point on the ellipse to the two foci is constant. Using a vertex point, this constant sum is

$$(a + c) + (a - c) = 2a \qquad \text{Length of major axis}$$

or simply the length of the major axis.

Southern Illinois University/Photo Researchers/Getty Images

Now, if you let (x, y) be *any* point on the ellipse, then the sum of the distances between (x, y) and the two foci must also be $2a$. That is,

$$\sqrt{[x-(h-c)]^2+(y-k)^2}+\sqrt{[x-(h+c)]^2+(y-k)^2}=2a$$

which, after expanding and regrouping, reduces to

$$(a^2-c^2)(x-h)^2+a^2(y-k)^2=a^2(a^2-c^2).$$

Finally, in Figure 10.14, you can see that

$$b^2=a^2-c^2$$

which implies that the equation of the ellipse is

$$b^2(x-h)^2+a^2(y-k)^2=a^2b^2$$

$$\frac{(x-h)^2}{a^2}+\frac{(y-k)^2}{b^2}=1.$$

You would obtain a similar equation in the derivation by starting with a vertical major axis. Both results are summarized as follows.

REMARK Consider the equation of the ellipse

$$\frac{(x-h)^2}{a^2}+\frac{(y-k)^2}{b^2}=1.$$

If you let $a = b$, then the equation can be rewritten as

$$(x-h)^2+(y-k)^2=a^2$$

which is the standard form of the equation of a circle with radius $r = a$ (see Section 1.2). Geometrically, when $a = b$ for an ellipse, the major and minor axes are of equal length, and so the graph is a circle.

Standard Equation of an Ellipse

The **standard form of the equation of an ellipse** with center (h, k) and major and minor axes of lengths $2a$ and $2b$, respectively, where $0 < b < a$, is

$$\frac{(x-h)^2}{a^2}+\frac{(y-k)^2}{b^2}=1 \qquad \text{Major axis is horizontal.}$$

$$\frac{(x-h)^2}{b^2}+\frac{(y-k)^2}{a^2}=1. \qquad \text{Major axis is vertical.}$$

The foci lie on the major axis, c units from the center, with

$$c^2=a^2-b^2.$$

If the center is at the origin, then the equation takes one of the following forms.

$$\frac{x^2}{a^2}+\frac{y^2}{b^2}=1 \qquad \text{Major axis is horizontal.}$$

$$\frac{x^2}{b^2}+\frac{y^2}{a^2}=1 \qquad \text{Major axis is vertical.}$$

Both the horizontal and vertical orientations for an ellipse are shown below.

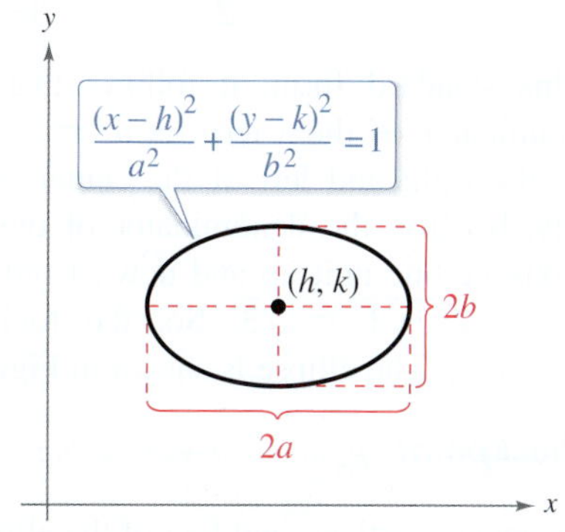

Major axis is horizontal.

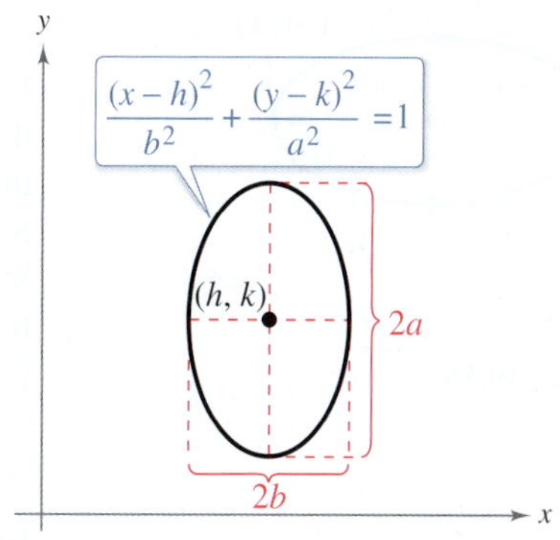

Major axis is vertical.

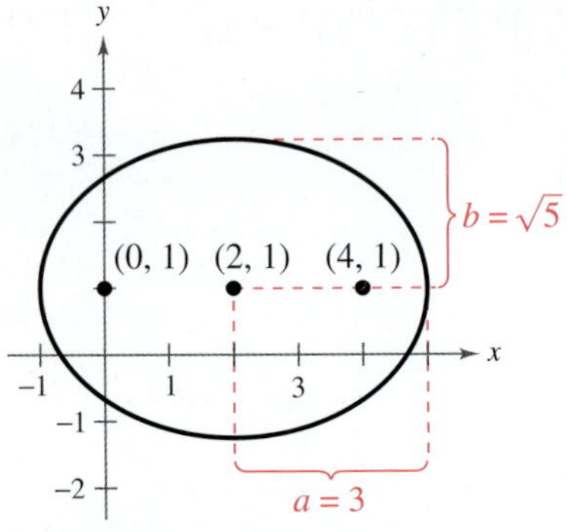

Figure 10.15

EXAMPLE 1 Finding the Standard Equation of an Ellipse

Find the standard form of the equation of the ellipse having foci (0, 1) and (4, 1) and a major axis of length 6, as shown in Figure 10.15.

Solution Because the foci occur at (0, 1) and (4, 1), the center of the ellipse is (2, 1) and the distance from the center to one of the foci is $c = 2$. Because $2a = 6$, you know that $a = 3$. Now, from $c^2 = a^2 - b^2$, you have

$$b = \sqrt{a^2 - c^2} = \sqrt{3^2 - 2^2} = \sqrt{5}.$$

Because the major axis is horizontal, the standard equation is

$$\frac{(x-2)^2}{3^2} + \frac{(y-1)^2}{(\sqrt{5})^2} = 1.$$

This equation simplifies to

$$\frac{(x-2)^2}{9} + \frac{(y-1)^2}{5} = 1.$$

✓ **Checkpoint** *Audio-video solution in English & Spanish at LarsonPrecalculus.com.*

Find the standard form of the equation of the ellipse having foci (2, 0) and (2, 6) and a major axis of length 8.

EXAMPLE 2 Sketching an Ellipse

Find the center, vertices, and foci of the ellipse $x^2 + 4y^2 + 6x - 8y + 9 = 0$. Then sketch the ellipse.

Solution Begin by writing the original equation in standard form. In the fourth step, note that 9 and 4 are added to *both* sides of the equation when completing the squares.

$$x^2 + 4y^2 + 6x - 8y + 9 = 0 \quad \text{Write original equation.}$$

$$(x^2 + 6x + \square) + (4y^2 - 8y + \square) = -9 \quad \text{Group terms.}$$

$$(x^2 + 6x + \square) + 4(y^2 - 2y + \square) = -9 \quad \text{Factor 4 out of } y\text{-terms.}$$

$$(x^2 + 6x + 9) + 4(y^2 - 2y + 1) = -9 + 9 + 4(1)$$

$$(x+3)^2 + 4(y-1)^2 = 4 \quad \text{Write in completed square form.}$$

$$\frac{(x+3)^2}{4} + \frac{(y-1)^2}{1} = 1 \quad \text{Divide each side by 4.}$$

$$\frac{(x+3)^2}{2^2} + \frac{(y-1)^2}{1^2} = 1 \quad \text{Write in standard form.}$$

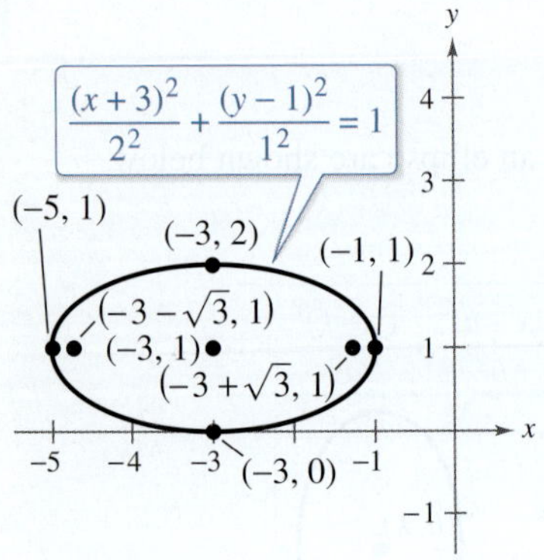

Figure 10.16

From this standard form, it follows that the center is $(h, k) = (-3, 1)$. Because the denominator of the x-term is $a^2 = 2^2$, the endpoints of the major axis lie two units to the right and left of the center. So, the vertices are $(-5, 1)$ and $(-1, 1)$. Similarly, because the denominator of the y-term is $b^2 = 1^2$, the endpoints of the minor axis lie one unit up and down from the center. Now, from $c^2 = a^2 - b^2$, you have $c = \sqrt{2^2 - 1^2} = \sqrt{3}$. So, the foci of the ellipse are $(-3 - \sqrt{3}, 1)$ and $(-3 + \sqrt{3}, 1)$. The ellipse is shown in Figure 10.16.

✓ **Checkpoint** *Audio-video solution in English & Spanish at LarsonPrecalculus.com.*

Find the center, vertices, and foci of the ellipse $9x^2 + 4y^2 + 36x - 8y + 4 = 0$. Then sketch the ellipse.

EXAMPLE 3 Sketching an Ellipse

Find the center, vertices, and foci of the ellipse $4x^2 + y^2 - 8x + 4y - 8 = 0$. Then sketch the ellipse.

Solution By completing the square, you can write the original equation in standard form.

$$4x^2 + y^2 - 8x + 4y - 8 = 0 \quad \text{Write original equation.}$$

$$(4x^2 - 8x + \square) + (y^2 + 4y + \square) = 8 \quad \text{Group terms.}$$

$$4(x^2 - 2x + \square) + (y^2 + 4y + \square) = 8 \quad \text{Factor 4 out of } x\text{-terms.}$$

$$4(x^2 - 2x + 1) + (y^2 + 4y + 4) = 8 + 4(1) + 4$$

$$4(x - 1)^2 + (y + 2)^2 = 16 \quad \text{Write in completed square form.}$$

$$\frac{(x - 1)^2}{4} + \frac{(y + 2)^2}{16} = 1 \quad \text{Divide each side by 16.}$$

$$\frac{(x - 1)^2}{2^2} + \frac{(y + 2)^2}{4^2} = 1 \quad \text{Write in standard form.}$$

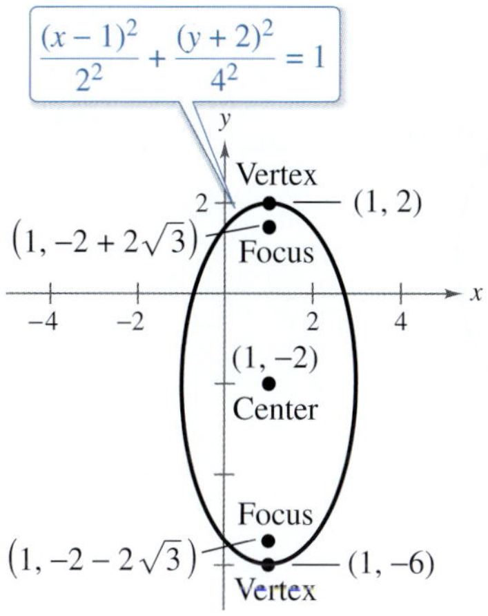

Figure 10.17

The major axis is vertical, where $h = 1$, $k = -2$, $a = 4$, $b = 2$, and

$$c = \sqrt{a^2 - b^2} = \sqrt{16 - 4} = \sqrt{12} = 2\sqrt{3}.$$

So, you have the following.

Center: $(1, -2)$ Vertices: $(1, -6)$, $(1, 2)$ Foci: $(1, -2 - 2\sqrt{3})$, $(1, -2 + 2\sqrt{3})$

The ellipse is shown in Figure 10.17.

Checkpoint *Audio-video solution in English & Spanish at LarsonPrecalculus.com.*

Find the center, vertices, and foci of the ellipse $5x^2 + 9y^2 + 10x - 54y + 41 = 0$. Then sketch the ellipse.

TECHNOLOGY You can use a graphing utility to graph an ellipse by graphing the upper and lower portions in the same viewing window. For instance, to graph the ellipse in Example 3, first solve for y to get

$$y_1 = -2 + 4\sqrt{1 - \frac{(x - 1)^2}{4}}$$

and

$$y_2 = -2 - 4\sqrt{1 - \frac{(x - 1)^2}{4}}.$$

Use a viewing window in which $-6 \le x \le 9$ and $-7 \le y \le 3$. You should obtain the graph shown below.

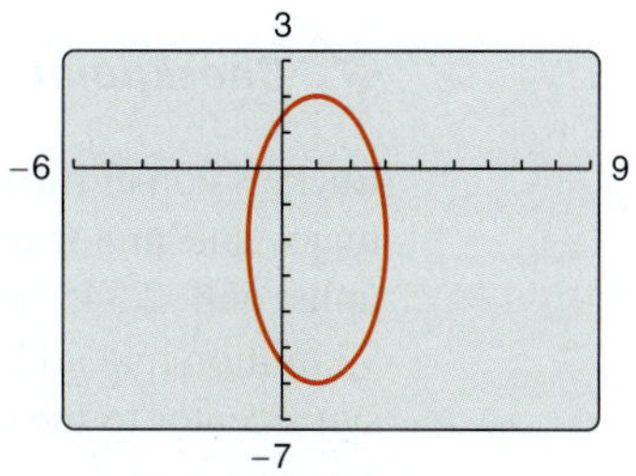

In Exercise 58, you will investigate the elliptical orbit of Halley's comet about the sun. Halley's comet is visible from Earth approximately every 75 years. The comet's latest appearance was in 1986.

REMARK Note in Example 4 and in the figure that Earth is *not* the center of the moon's orbit.

Application

Ellipses have many practical and aesthetic uses. For instance, machine gears, supporting arches, and acoustic designs often involve elliptical shapes. The orbits of satellites and planets are also ellipses. Example 4 investigates the elliptical orbit of the moon about Earth.

EXAMPLE 4 An Application Involving an Elliptical Orbit

The moon travels about Earth in an elliptical orbit with Earth at one focus, as shown below. The major and minor axes of the orbit have lengths of 768,800 kilometers and 767,640 kilometers, respectively. Find the greatest (*apogee*) and least (*perigee*) distances from Earth's center to the moon's center.

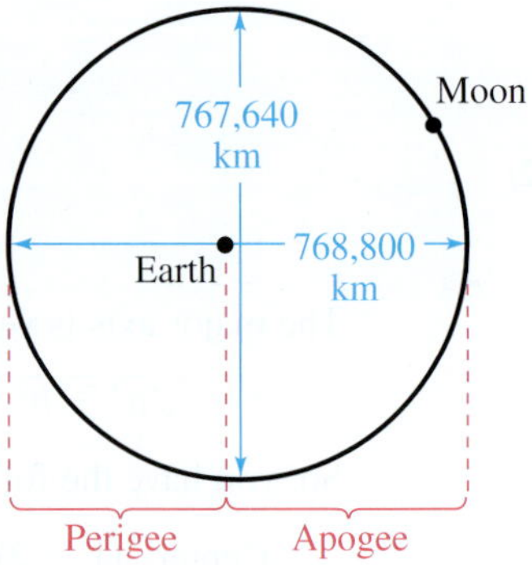

Solution Because

$$2a = 768{,}800 \quad \text{and} \quad 2b = 767{,}640$$

you have

$$a = 384{,}400 \quad \text{and} \quad b = 383{,}820$$

which implies that

$$\begin{aligned} c &= \sqrt{a^2 - b^2} \\ &= \sqrt{384{,}400^2 - 383{,}820^2} \\ &\approx 21{,}108. \end{aligned}$$

So, the greatest distance between the center of Earth and the center of the moon is

$$\begin{aligned} a + c &\approx 384{,}400 + 21{,}108 \\ &= 405{,}508 \text{ kilometers} \end{aligned}$$

and the least distance is

$$\begin{aligned} a - c &\approx 384{,}400 - 21{,}108 \\ &= 363{,}292 \text{ kilometers.} \end{aligned}$$

✓ **Checkpoint** *Audio-video solution in English & Spanish at LarsonPrecalculus.com.*

Encke's comet travels about the sun in an elliptical orbit with the sun at one focus. The major and minor axes of the orbit have lengths of approximately 4.429 astronomical units and 2.345 astronomical units, respectively. (An astronomical unit is about 93 million miles.) Find the greatest (*aphelion*) and least (*perihelion*) distances from the sun's center to the comet's center.

Digital Vision/Getty Images

Eccentricity

One of the reasons it was difficult for early astronomers to detect that the orbits of the planets are ellipses is that the foci of the planetary orbits are relatively close to their centers, and so the orbits are nearly circular. To measure the ovalness of an ellipse, you can use the concept of **eccentricity.**

> **Definition of Eccentricity**
>
> The eccentricity e of an ellipse is given by the ratio $e = \frac{c}{a}$.

Note that $0 < e < 1$ for *every* ellipse.

To see how this ratio is used to describe the shape of an ellipse, note that because the foci of an ellipse are located along the major axis between the vertices and the center, it follows that

$$0 < c < a.$$

For an ellipse that is nearly circular, the foci are close to the center and the ratio c/a is close to 0, as shown in Figure 10.18. On the other hand, for an elongated ellipse, the foci are close to the vertices and the ratio c/a is close to 1, as shown in Figure 10.19.

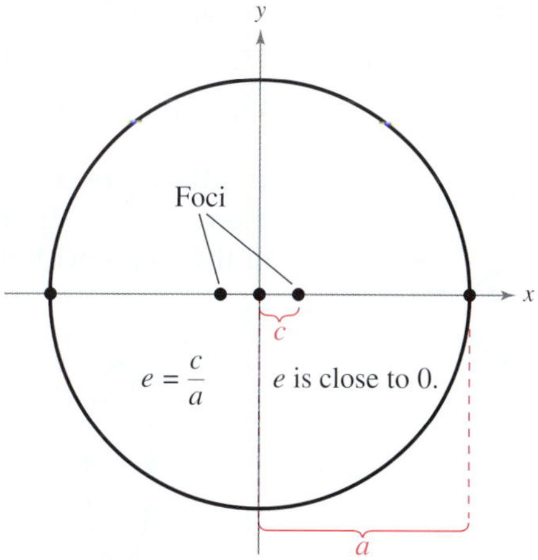

Figure 10.18

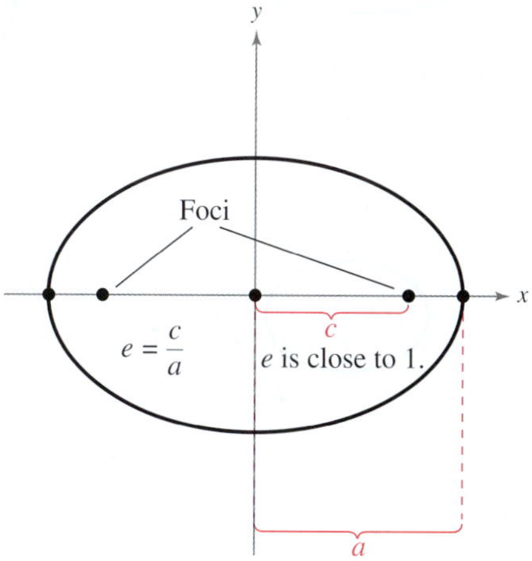

Figure 10.19

The time it takes Saturn to orbit the sun is about 29.4 Earth years.

The orbit of the moon has an eccentricity of $e \approx 0.0549$, and the eccentricities of the eight planetary orbits are as follows.

Mercury:	$e \approx 0.2056$	Jupiter:	$e \approx 0.0484$
Venus:	$e \approx 0.0068$	Saturn:	$e \approx 0.0542$
Earth:	$e \approx 0.0167$	Uranus:	$e \approx 0.0472$
Mars:	$e \approx 0.0934$	Neptune:	$e \approx 0.0086$

Summarize (Section 10.3)

1. State the definition of an ellipse *(page 704)*. For examples involving the equations and graphs of ellipses, see Examples 1–3.
2. Describe a real-life application of an ellipse *(page 708, Example 4)*.
3. State the definition of the eccentricity of an ellipse *(page 709)*.

Ozja/Shutterstock.com

10.3 Exercises

See CalcChat.com for tutorial help and worked-out solutions to odd-numbered exercises.

Vocabulary: Fill in the blanks.

1. An ________ is the set of all points (x, y) in a plane, the sum of whose distances from two distinct fixed points, called ________, is constant.
2. The chord joining the vertices of an ellipse is called the ________ ________, and its midpoint is the ________ of the ellipse.
3. The chord perpendicular to the major axis at the center of the ellipse is called the ________ ________ of the ellipse.
4. The concept of ________ is used to measure the ovalness of an ellipse.

Skills and Applications

Matching In Exercises 5–8, match the equation with its graph. [The graphs are labeled (a), (b), (c), and (d).]

(a)

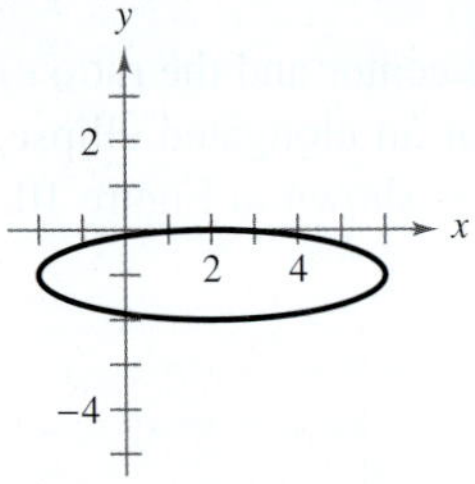

(b)

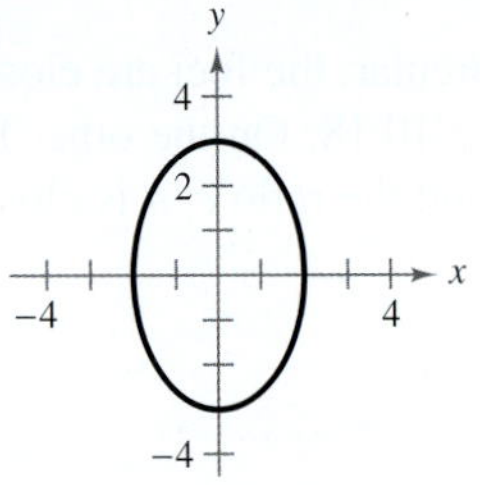

(c)

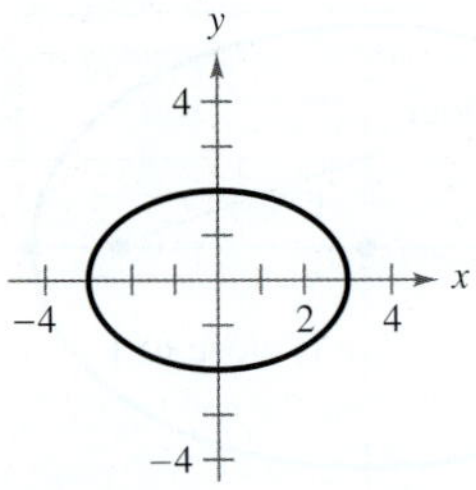

(d)

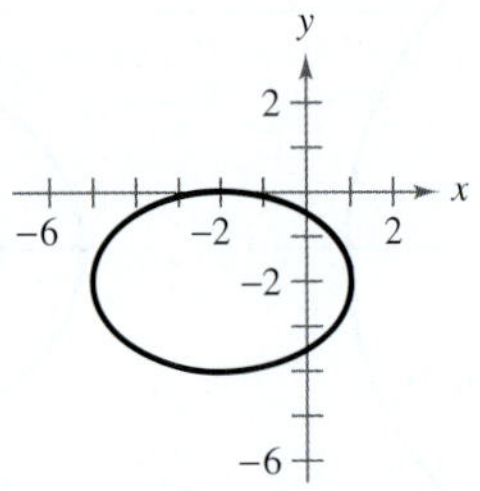

5. $\frac{x^2}{4} + \frac{y^2}{9} = 1$

6. $\frac{x^2}{9} + \frac{y^2}{4} = 1$

7. $\frac{(x-2)^2}{16} + (y+1)^2 = 1$

8. $\frac{(x+2)^2}{9} + \frac{(y+2)^2}{4} = 1$

An Ellipse Centered at the Origin In Exercises 9–18, find the standard form of the equation of the ellipse with the given characteristics and center at the origin.

9.

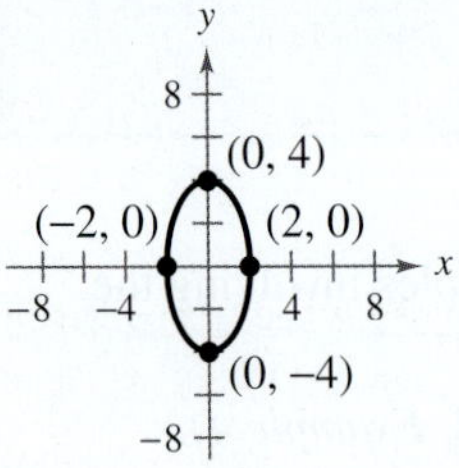

10.

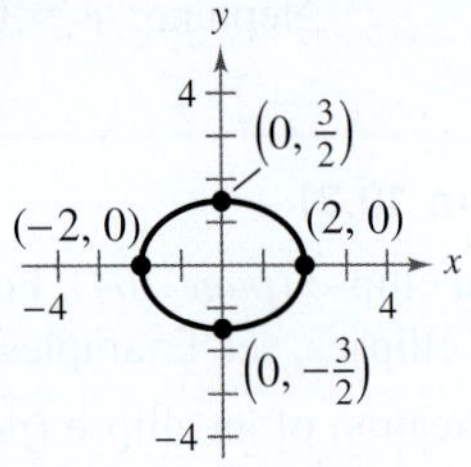

11. Vertices: $(\pm 7, 0)$; foci: $(\pm 2, 0)$
12. Vertices: $(0, \pm 8)$; foci: $(0, \pm 4)$
13. Foci: $(\pm 5, 0)$; major axis of length 14
14. Foci: $(\pm 2, 0)$; major axis of length 10
15. Vertical major axis; passes through the points $(0, 6)$ and $(3, 0)$
16. Horizontal major axis; passes through the points $(5, 0)$ and $(0, 2)$
17. Vertices: $(\pm 6, 0)$; passes through the point $(4, 1)$
18. Vertices: $(0, \pm 5)$; passes through the point $(4, 2)$

Finding the Standard Equation of an Ellipse In Exercises 19–32, find the standard form of the equation of the ellipse with the given characteristics.

19.

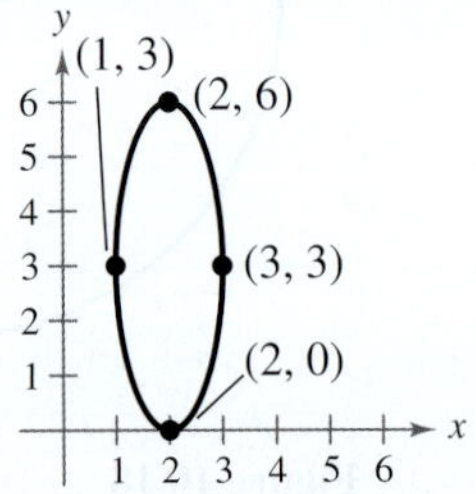

20.

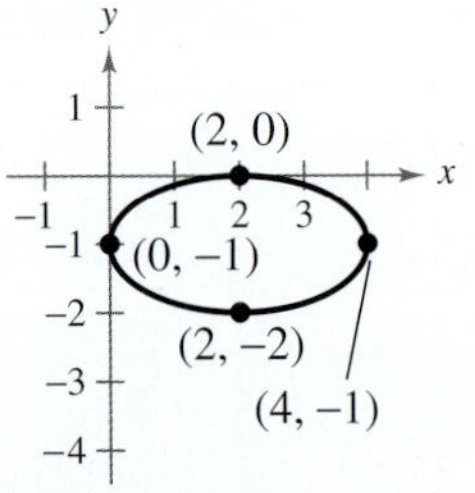

21. Vertices: $(0, 2)$, $(8, 2)$; minor axis of length 2
22. Vertices: $(3, 0)$, $(3, 10)$; minor axis of length 4
23. Foci: $(0, 0)$, $(4, 0)$; major axis of length 6
24. Foci: $(0, 0)$, $(0, 8)$; major axis of length 16
25. Center: $(1, 3)$; vertex: $(-2, 3)$; minor axis of length 4
26. Center: $(2, -1)$; vertex: $(2, \frac{1}{2})$; minor axis of length 2
27. Center: $(0, 2)$; $a = 4c$; vertices: $(0, -6)$, $(0, 10)$
28. Center: $(0, 4)$; $a = 2c$; vertices: $(-4, 4)$, $(4, 4)$
29. Center: $(3, 2)$; $a = 3c$; foci: $(1, 2)$, $(5, 2)$
30. Center: $(1, -5)$; $a = 5c$; foci: $(1, -6)$, $(1, -4)$
31. Vertices: $(0, 2)$, $(4, 2)$; endpoints of the minor axis: $(2, 3)$, $(2, 1)$
32. Vertices: $(5, 0)$, $(5, 12)$; endpoints of the minor axis: $(1, 6)$, $(9, 6)$

Sketching an Ellipse **In Exercises 33–48, find the center, vertices, foci, and eccentricity of the ellipse. Then sketch the ellipse.**

33. $\frac{x^2}{25} + \frac{y^2}{16} = 1$

34. $\frac{x^2}{16} + \frac{y^2}{81} = 1$

35. $\frac{x^2}{5} + \frac{y^2}{9} = 1$

36. $\frac{x^2}{64} + \frac{y^2}{28} = 1$

37. $\frac{(x-4)^2}{16} + \frac{(y+1)^2}{25} = 1$

38. $\frac{(x+3)^2}{12} + \frac{(y-2)^2}{16} = 1$

39. $\frac{(x+5)^2}{9/4} + (y-1)^2 = 1$

40. $(x+2)^2 + \frac{(y+4)^2}{1/4} = 1$

41. $9x^2 + 4y^2 + 36x - 24y + 36 = 0$

42. $9x^2 + 4y^2 - 54x + 40y + 37 = 0$

43. $x^2 + 5y^2 - 8x - 30y - 39 = 0$

44. $3x^2 + y^2 + 18x - 2y - 8 = 0$

45. $6x^2 + 2y^2 + 18x - 10y + 2 = 0$

46. $x^2 + 4y^2 - 6x + 20y - 2 = 0$

47. $16x^2 + 25y^2 - 32x + 50y + 16 = 0$

48. $9x^2 + 25y^2 - 36x - 50y + 60 = 0$

Graphing an Ellipse **In Exercises 49–52, use a graphing utility to graph the ellipse. Find the center, foci, and vertices. (Recall that it may be necessary to solve the equation for *y* and obtain two equations.)**

49. $5x^2 + 3y^2 = 15$

50. $3x^2 + 4y^2 = 12$

51. $12x^2 + 20y^2 - 12x + 40y - 37 = 0$

52. $36x^2 + 9y^2 + 48x - 36y - 72 = 0$

53. Using Eccentricity Find an equation of the ellipse with vertices $(\pm 5, 0)$ and eccentricity $e = \frac{3}{5}$.

54. Using Eccentricity Find an equation of the ellipse with vertices $(0, \pm 8)$ and eccentricity $e = \frac{1}{2}$.

55. Architecture Statuary Hall is an elliptical room in the United States Capitol Building in Washington, D.C. The room is also referred to as the Whispering Gallery because a person standing at one focus of the room can hear even a whisper spoken by a person standing at the other focus. The dimensions of Statuary Hall are 46 feet wide by 97 feet long. (a) Find an equation of the shape of the floor surface of the hall. (b) Determine the distance between the foci.

Southern Illinois University/Photo Researchers/Getty Images

56. Architecture A contractor plans to construct a semielliptical fireplace arch with an opening that is 2 feet high at the center and 6 feet wide along the base (see figure). The contractor draws the outline of the ellipse on the wall using tacks as described on page 704. Determine the required positions of the tacks and the length of the string.

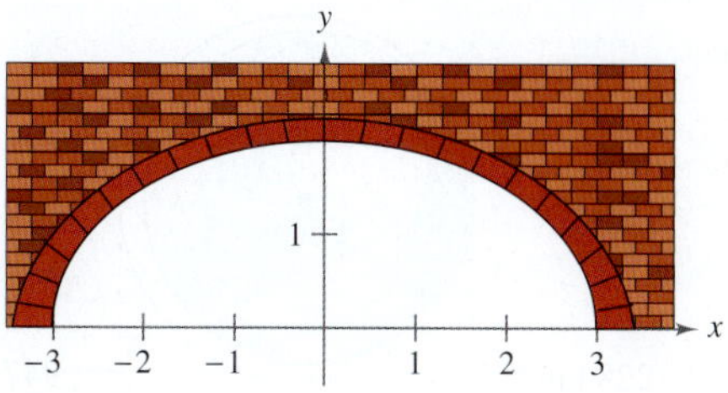

57. Lithotripter

A lithotripter machine uses an elliptical reflector to break up kidney stones nonsurgically. A spark plug in the reflector generates energy waves at one focus of an ellipse. The reflector directs these waves toward the kidney stone, positioned at the other focus of the ellipse, with enough energy to break up the stone (see figure). The lengths of the major and minor axes of the ellipse are 280 millimeters and 160 millimeters, respectively. How far is the spark plug from the kidney stone?

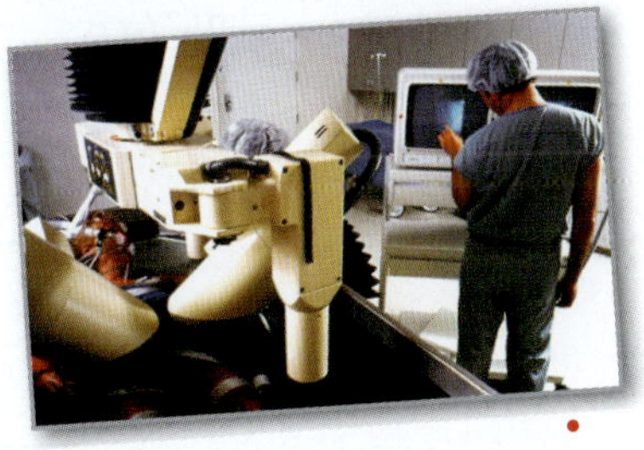

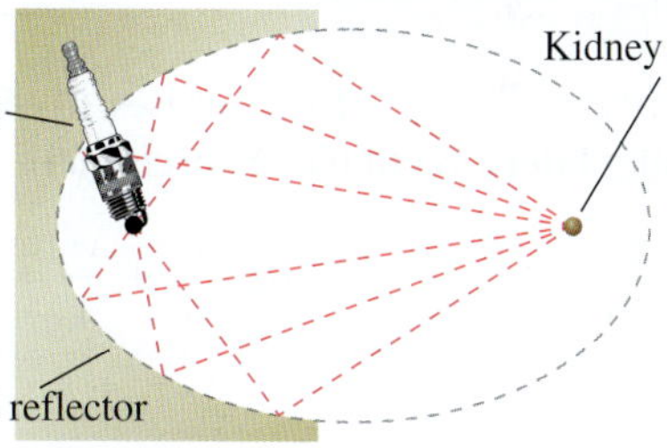

58. Comet Orbit Halley's comet has an elliptical orbit with the sun at one focus. The eccentricity of the orbit is approximately 0.967. The length of the major axis of the orbit is approximately 35.88 astronomical units. (An astronomical unit is about 93 million miles.)

(a) Find an equation of the orbit. Place the center of the orbit at the origin and place the major axis on the *x*-axis.

(b) Use a graphing utility to graph the equation of the orbit.

(c) Find the greatest (aphelion) and least (perihelion) distances from the sun's center to the comet's center.

59. Satellite Orbit The first artificial satellite to orbit Earth was Sputnik I (launched by the former Soviet Union in 1957). Its highest point above Earth's surface was 947 kilometers, and its lowest point was 228 kilometers (see figure). The center of Earth was at one focus of the elliptical orbit, and the radius of Earth is 6378 kilometers. Find the eccentricity of the orbit.

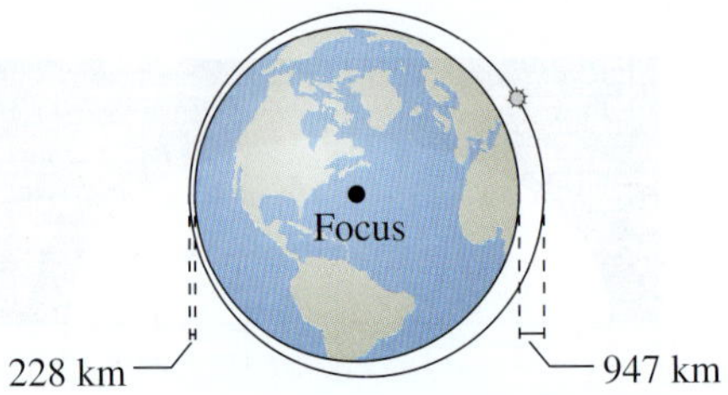

60. Geometry A line segment through a focus of an ellipse with endpoints on the ellipse and perpendicular to the major axis is called a **latus rectum** of the ellipse. Therefore, an ellipse has two latera recta. Knowing the length of the latera recta is helpful in sketching an ellipse because this information yields other points on the ellipse (see figure). Show that the length of each latus rectum is $2b^2/a$.

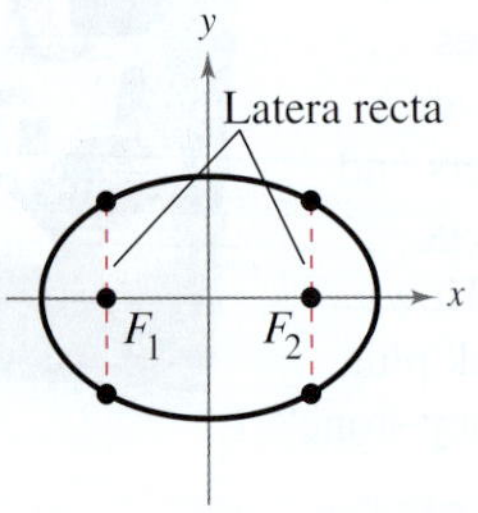

Using Latera Recta **In Exercises 61–64, sketch the ellipse using latera recta (see Exercise 60).**

61. $\dfrac{x^2}{9} + \dfrac{y^2}{16} = 1$

62. $\dfrac{x^2}{4} + \dfrac{y^2}{1} = 1$

63. $5x^2 + 3y^2 = 15$

64. $9x^2 + 4y^2 = 36$

Exploration

True or False? **In Exercises 65 and 66, determine whether the statement is true or false. Justify your answer.**

65. The graph of $x^2 + 4y^4 - 4 = 0$ is an ellipse.

66. It is easier to distinguish the graph of an ellipse from the graph of a circle when the eccentricity of the ellipse is close to 1.

67. Think About It Find the equation of an ellipse such that for any point on the ellipse, the sum of the distances from the point to the points $(2, 2)$ and $(10, 2)$ is 36.

68. Think About It At the beginning of this section, you learned that an ellipse can be drawn using two thumbtacks, a string of fixed length (greater than the distance between the two tacks), and a pencil. If the ends of the string are fastened at the tacks and the string is drawn taut with a pencil, then the path traced by the pencil is an ellipse.

(a) What is the length of the string in terms of a?

(b) Explain why the path is an ellipse.

69. Conjecture Consider the ellipse

$$\frac{x^2}{a^2} + \frac{y^2}{b^2} = 1, \quad a + b = 20.$$

(a) The area of the ellipse is given by $A = \pi ab$. Write the area of the ellipse as a function of a.

(b) Find the equation of an ellipse with an area of 264 square centimeters.

(c) Complete the table using your equation from part (a). Then make a conjecture about the shape of the ellipse with maximum area.

a	8	9	10	11	12	13
A						

(d) Use a graphing utility to graph the area function and use the graph to support your conjecture in part (c).

70. HOW DO YOU SEE IT? Without performing any calculations, order the eccentricities of the ellipses from least to greatest.

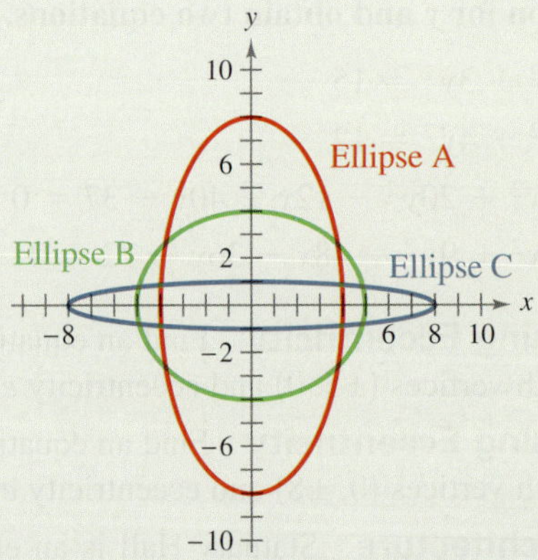

71. Proof Show that $a^2 = b^2 + c^2$ for the ellipse

$$\frac{x^2}{a^2} + \frac{y^2}{b^2} = 1$$

where $a > 0$, $b > 0$, and the distance from the center of the ellipse $(0, 0)$ to a focus is c.

10.4 Hyperbolas

You can use hyperbolas to model and solve many types of real-life problems. For instance, in Exercise 51 on page 721, you will see how you can use hyperbolas in long distance radio navigation for aircraft and ships.

- **Write equations of hyperbolas in standard form.**
- **Find asymptotes of and graph hyperbolas.**
- **Use properties of hyperbolas to solve real-life problems.**
- **Classify conics from their general equations.**

Introduction

The definition of a **hyperbola** is similar to that of an ellipse. For an ellipse, the *sum* of the distances between the foci and a point on the ellipse is fixed. For a hyperbola, the absolute value of the *difference* of the distances between the foci and a point on the hyperbola is fixed.

Definition of Hyperbola

A **hyperbola** is the set of all points (x, y) in a plane for which the absolute value of the difference of the distances from two distinct fixed points, called **foci,** is constant. See Figure 10.20.

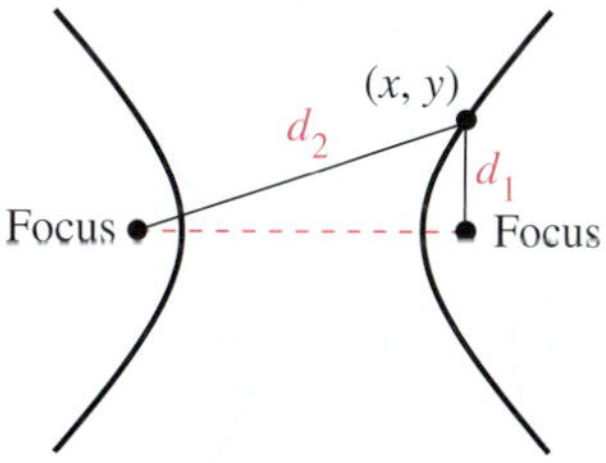

$|d_2 - d_1|$ is constant.

Figure 10.20

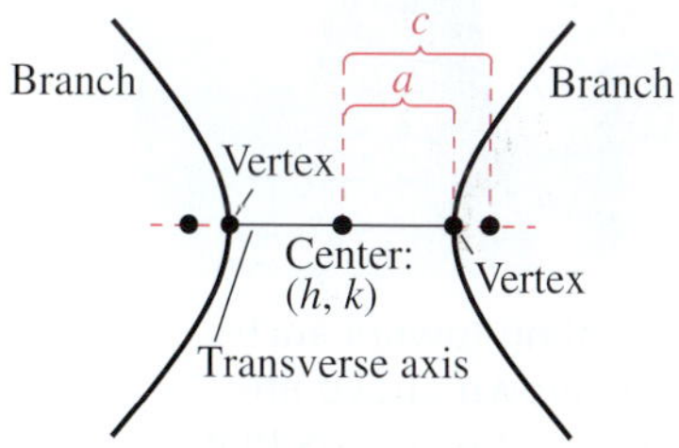

Figure 10.21

The graph of a hyperbola has two disconnected parts called **branches.** The line through the two foci intersects the hyperbola at two points called the **vertices.** The line segment connecting the vertices is the **transverse axis,** and the midpoint of the transverse axis is the **center** of the hyperbola.

Consider the hyperbola in Figure 10.21 with the following points.

Center: (h, k) Vertices: $(h \pm a, k)$ Foci: $(h \pm c, k)$

Note that the center is the midpoint of the segment joining the foci.

The absolute value of the difference of the distances from *any* point on the hyperbola to the two foci is constant. Using a vertex, this constant value is

$$\begin{aligned} |[2a + (c - a)] - (c - a)| &= |2a| \\ &= 2a \end{aligned}$$ Length of transverse axis

or simply the length of the transverse axis. Now, if you let (x, y) be *any* point on the hyperbola, then

$$|d_2 - d_1| = 2a.$$

You would obtain the same result for a hyperbola with a vertical transverse axis.

The development of the standard form of the equation of a hyperbola is similar to that of an ellipse. Note in the definition on the next page that a, b, and c are related differently for hyperbolas than for ellipses.

Melissa Madia/Shutterstock.com

Standard Equation of a Hyperbola

The **standard form of the equation of a hyperbola** with center (h, k) is

$$\frac{(x-h)^2}{a^2} - \frac{(y-k)^2}{b^2} = 1 \qquad \text{Transverse axis is horizontal.}$$

$$\frac{(y-k)^2}{a^2} - \frac{(x-h)^2}{b^2} = 1. \qquad \text{Transverse axis is vertical.}$$

The vertices are a units from the center, and the foci are c units from the center. Moreover, $c^2 = a^2 + b^2$. If the center of the hyperbola is at the origin, then the equation takes one of the following forms.

$$\frac{x^2}{a^2} - \frac{y^2}{b^2} = 1 \qquad \text{Transverse axis is horizontal.}$$

$$\frac{y^2}{a^2} - \frac{x^2}{b^2} = 1 \qquad \text{Transverse axis is vertical.}$$

Nuclear cooling towers such as the one shown above are hyperboloids. This means that they have hyperbolic vertical cross sections.

Both the horizontal and vertical orientations for a hyperbola are shown below.

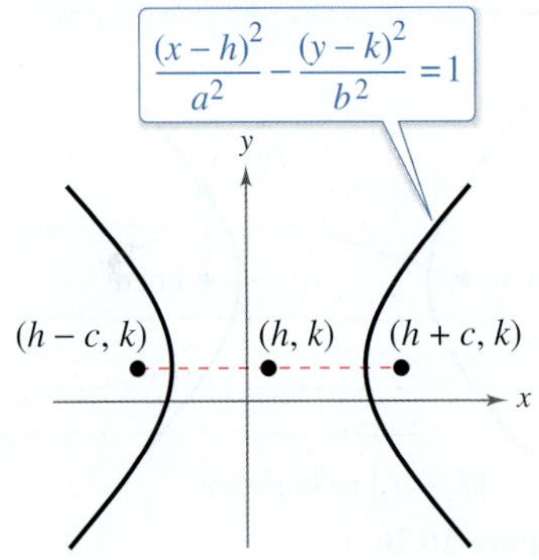

Transverse axis is horizontal.

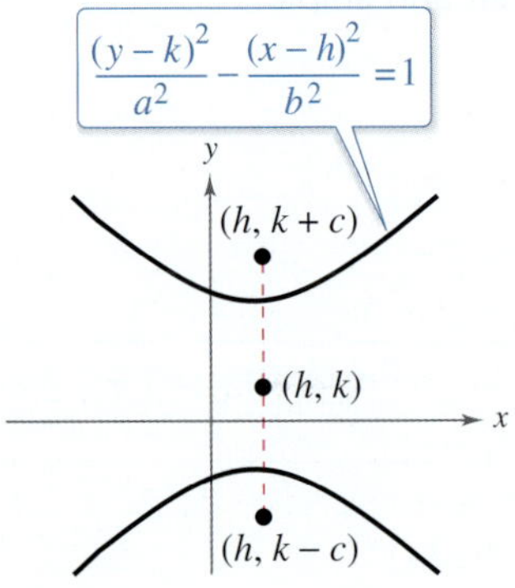

Transverse axis is vertical.

EXAMPLE 1 **Finding the Standard Equation of a Hyperbola**

Find the standard form of the equation of the hyperbola with foci $(-1, 2)$ and $(5, 2)$ and vertices $(0, 2)$ and $(4, 2)$.

Solution By the Midpoint Formula, the center of the hyperbola occurs at the point $(2, 2)$. Furthermore, $c = 5 - 2 = 3$ and $a = 4 - 2 = 2$, and it follows that

$$b = \sqrt{c^2 - a^2} = \sqrt{3^2 - 2^2} = \sqrt{9 - 4} = \sqrt{5}.$$

So, the hyperbola has a horizontal transverse axis and the standard form of the equation is

$$\frac{(x-2)^2}{2^2} - \frac{(y-2)^2}{(\sqrt{5})^2} = 1. \qquad \text{See Figure 10.22.}$$

This equation simplifies to

$$\frac{(x-2)^2}{4} - \frac{(y-2)^2}{5} = 1.$$

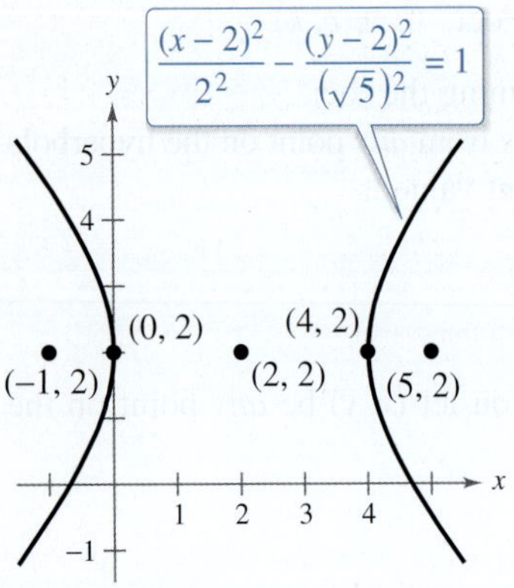

Figure 10.22

✓ ***Checkpoint*** *Audio-video solution in English & Spanish at LarsonPrecalculus.com.*

Find the standard form of the equation of the hyperbola with foci $(2, -5)$ and $(2, 3)$ and vertices $(2, -4)$ and $(2, 2)$.

Malyshev Maksim/Shutterstock.com

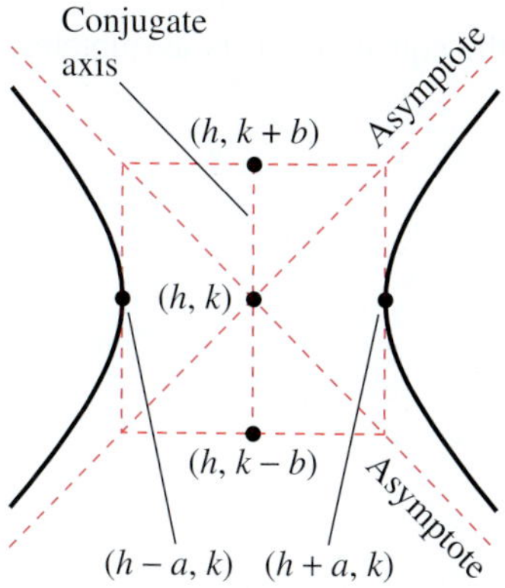

Figure 10.23

Asymptotes of a Hyperbola

Each hyperbola has two **asymptotes** that intersect at the center of the hyperbola, as shown in Figure 10.23. The asymptotes pass through the vertices of a rectangle of dimensions $2a$ by $2b$, with its center at (h, k). The line segment of length $2b$ joining $(h, k + b)$ and $(h, k - b)$ $[$or $(h + b, k)$ and $(h - b, k)]$ is the **conjugate axis** of the hyperbola.

Asymptotes of a Hyperbola

The equations of the asymptotes of a hyperbola are

$$y = k \pm \frac{b}{a}(x - h) \qquad \text{Transverse axis is horizontal.}$$

$$y = k \pm \frac{a}{b}(x - h). \qquad \text{Transverse axis is vertical.}$$

EXAMPLE 2 **Using Asymptotes to Sketch a Hyperbola**

Sketch the hyperbola $4x^2 - y^2 = 16$.

Algebraic Solution

Divide each side of the original equation by 16, and write the equation in standard form.

$$\frac{x^2}{2^2} - \frac{y^2}{4^2} = 1 \qquad \text{Write in standard form.}$$

From this, you can conclude that $a = 2$, $b = 4$, and the transverse axis is horizontal. So, the vertices occur at $(-2, 0)$ and $(2, 0)$, and the endpoints of the conjugate axis occur at $(0, -4)$ and $(0, 4)$. Using these four points, you are able to sketch the rectangle shown in Figure 10.24. Now, from $c^2 = a^2 + b^2$, you have $c = \sqrt{2^2 + 4^2} = \sqrt{20} = 2\sqrt{5}$. So, the foci of the hyperbola are $(-2\sqrt{5}, 0)$ and $(2\sqrt{5}, 0)$. Finally, by drawing the asymptotes through the corners of this rectangle, you can complete the sketch shown in Figure 10.25. Note that the asymptotes are $y = 2x$ and $y = -2x$.

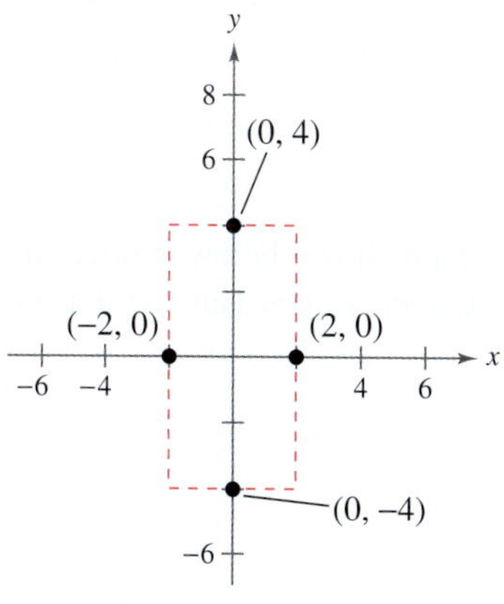

Figure 10.24

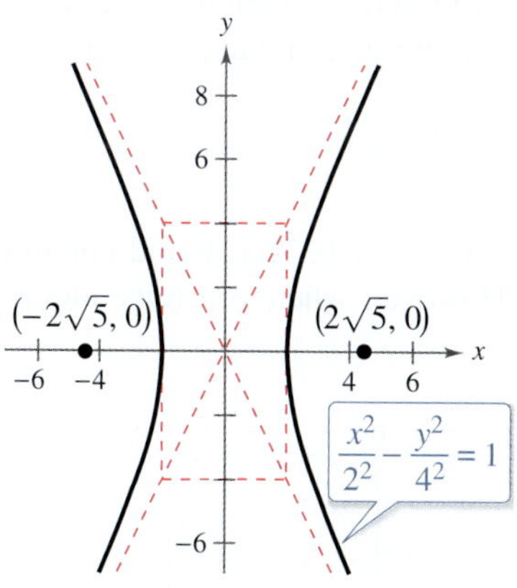

Figure 10.25

Graphical Solution

Solve the equation of the hyperbola for y, as follows.

$$4x^2 - y^2 = 16$$

$$4x^2 - 16 = y^2$$

$$\pm\sqrt{4x^2 - 16} = y$$

Then use a graphing utility to graph

$$y_1 = \sqrt{4x^2 - 16} \quad \text{and} \quad y_2 = -\sqrt{4x^2 - 16}$$

in the same viewing window, as shown below. Be sure to use a square setting.

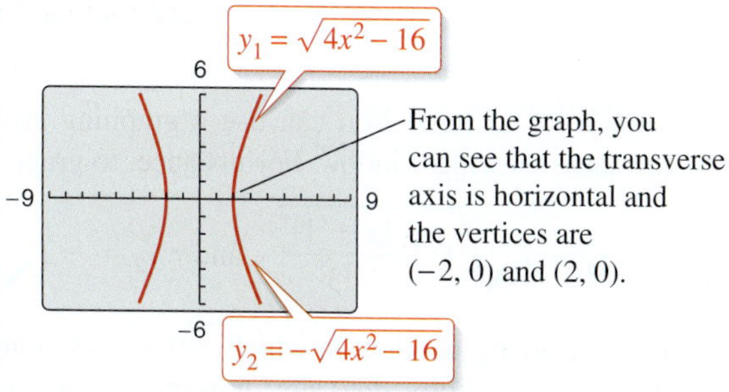

✔ **Checkpoint** *Audio-video solution in English & Spanish at LarsonPrecalculus.com.*

Sketch the hyperbola $4y^2 - 9x^2 = 36$.

EXAMPLE 3 Finding the Asymptotes of a Hyperbola

Sketch the hyperbola $4x^2 - 3y^2 + 8x + 16 = 0$, find the equations of its asymptotes, and find the foci.

Solution

$$4x^2 - 3y^2 + 8x + 16 = 0 \qquad \text{Write original equation.}$$

$$(4x^2 + 8x) - 3y^2 = -16 \qquad \text{Group terms.}$$

$$4(x^2 + 2x) - 3y^2 = -16 \qquad \text{Factor 4 out of } x\text{-terms.}$$

$$4(x^2 + 2x + 1) - 3y^2 = -16 + 4(1) \qquad \text{Complete the square.}$$

$$4(x + 1)^2 - 3y^2 = -12 \qquad \text{Write in completed square form.}$$

$$-\frac{(x + 1)^2}{3} + \frac{y^2}{4} = 1 \qquad \text{Divide each side by } -12.$$

$$\frac{y^2}{2^2} - \frac{(x + 1)^2}{(\sqrt{3})^2} = 1 \qquad \text{Write in standard form.}$$

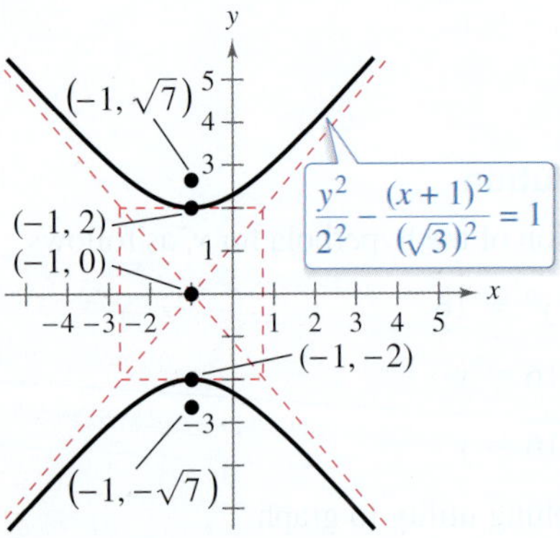

Figure 10.26

From this equation you can conclude that the hyperbola has a vertical transverse axis, is centered at $(-1, 0)$, has vertices $(-1, 2)$ and $(-1, -2)$, and has a conjugate axis with endpoints $(-1 - \sqrt{3}, 0)$ and $(-1 + \sqrt{3}, 0)$. To sketch the hyperbola, draw a rectangle through these four points. The asymptotes are the lines passing through the corners of the rectangle. Using $a = 2$ and $b = \sqrt{3}$, you can conclude that the equations of the asymptotes are

$$y = \frac{2}{\sqrt{3}}(x + 1) \quad \text{and} \quad y = -\frac{2}{\sqrt{3}}(x + 1).$$

Finally, you can determine the foci by using the equation $c^2 = a^2 + b^2$. So, you have $c = \sqrt{2^2 + (\sqrt{3})^2} = \sqrt{7}$, and the foci are $(-1, \sqrt{7})$ and $(-1, -\sqrt{7})$. The hyperbola is shown in Figure 10.26.

Checkpoint *Audio-video solution in English & Spanish at LarsonPrecalculus.com.*

Sketch the hyperbola $9x^2 - 4y^2 + 8y - 40 = 0$, find the equations of its asymptotes, and find the foci.

TECHNOLOGY You can use a graphing utility to graph a hyperbola by graphing the upper and lower portions in the same viewing window. For instance, to graph the hyperbola in Example 3, first solve for y to get

$$y_1 = 2\sqrt{1 + \frac{(x + 1)^2}{3}} \quad \text{and} \quad y_2 = -2\sqrt{1 + \frac{(x + 1)^2}{3}}.$$

Use a viewing window in which $-9 \le x \le 9$ and $-6 \le y \le 6$. You should obtain the graph shown below. Notice that the graphing utility does not draw the asymptotes. However, when you trace along the branches, you can see that the values of the hyperbola approach the asymptotes.

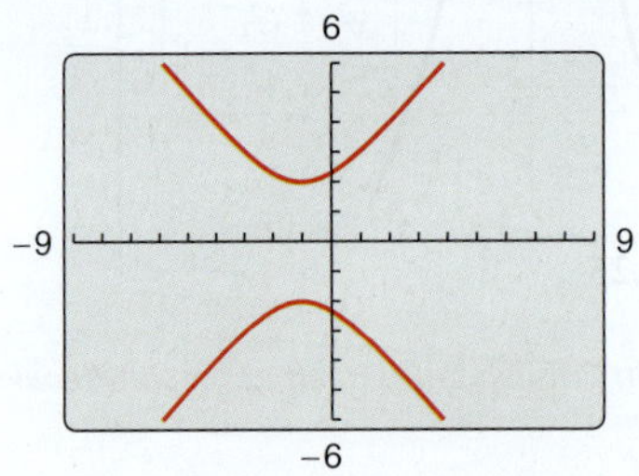

EXAMPLE 4 Using Asymptotes to Find the Standard Equation

Find the standard form of the equation of the hyperbola having vertices $(3, -5)$ and $(3, 1)$ and having asymptotes

$$y = 2x - 8 \quad \text{and} \quad y = -2x + 4$$

as shown in Figure 10.27.

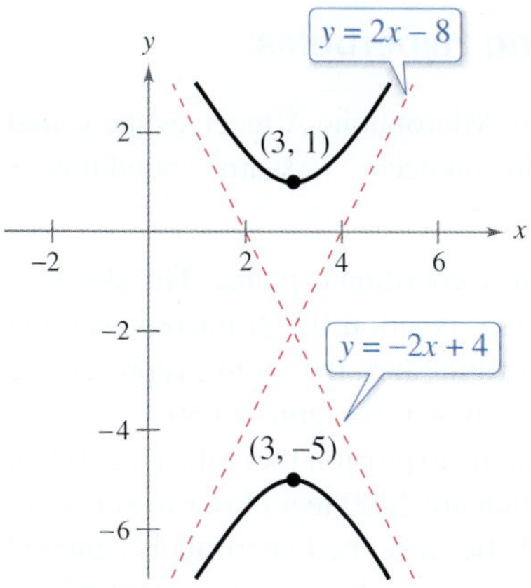

Figure 10.27

Solution By the Midpoint Formula, the center of the hyperbola is $(3, -2)$. Furthermore, the hyperbola has a vertical transverse axis with $a = 3$. From the original equations, you can determine the slopes of the asymptotes to be

$$m_1 = 2 = \frac{a}{b} \quad \text{and} \quad m_2 = -2 = -\frac{a}{b}$$

and, because $a = 3$, you can conclude

$$2 = \frac{a}{b} \implies 2 = \frac{3}{b} \implies b = \frac{3}{2}.$$

So, the standard form of the equation of the hyperbola is

$$\frac{(y+2)^2}{3^2} - \frac{(x-3)^2}{\left(\frac{3}{2}\right)^2} = 1.$$

Checkpoint Audio-video solution in English & Spanish at LarsonPrecalculus.com.

Find the standard form of the equation of the hyperbola having vertices $(3, 2)$ and $(9, 2)$ and having asymptotes

$$y = -2 + \frac{2}{3}x \quad \text{and} \quad y = 6 - \frac{2}{3}x.$$

As with ellipses, the *eccentricity* of a hyperbola is

$$e = \frac{c}{a} \qquad \text{Eccentricity}$$

and because $c > a$, it follows that $e > 1$. When the eccentricity is large, the branches of the hyperbola are nearly flat, as shown in Figure 10.28. When the eccentricity is close to 1, the branches of the hyperbola are more narrow, as shown in Figure 10.29.

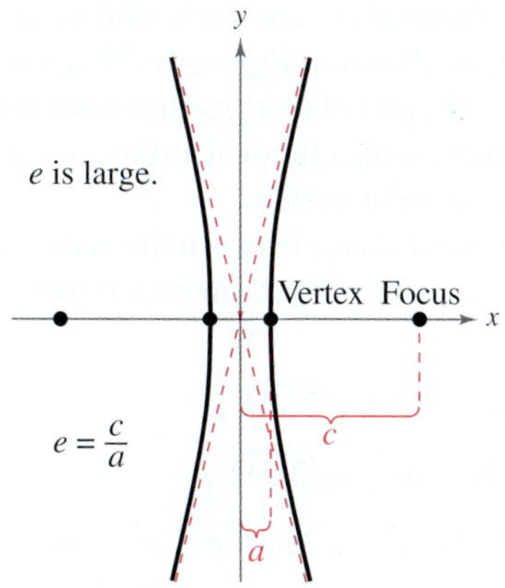

Figure 10.28

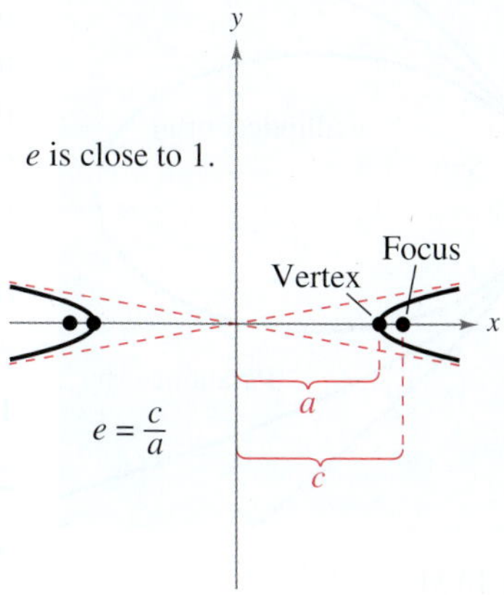

Figure 10.29

Applications

REMARK This application was developed during World War II. It shows how the properties of hyperbolas can be used in radar and other detection systems.

EXAMPLE 5 An Application Involving Hyperbolas

Two microphones, 1 mile apart, record an explosion. Microphone A receives the sound 2 seconds before microphone B. Where did the explosion occur? (Assume sound travels at 1100 feet per second.)

Solution Begin by representing the situation in a coordinate plane. The distance between the microphones is 1 mile, or 5280 feet. So, position the point representing microphone A 2640 units to the right of the origin and the point representing microphone B 2640 units to the left of the origin, as shown in Figure 10.30.

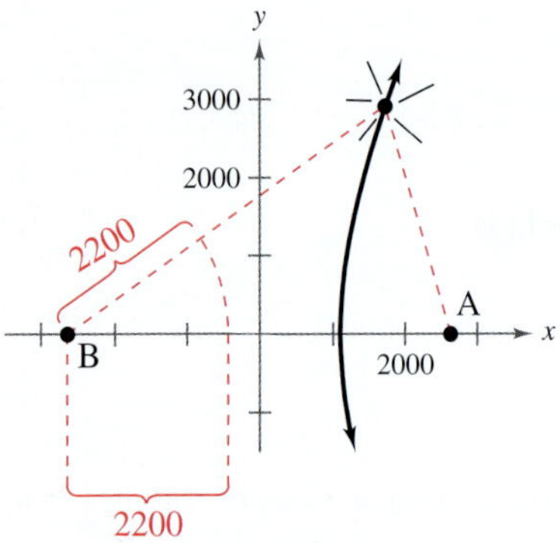

Figure 10.30

Assuming sound travels at 1100 feet per second, the explosion took place 2200 feet farther from B than from A. The locus of all points that are 2200 feet closer to A than to B is one branch of a hyperbola with foci at A and B. Because the hyperbola is centered at the origin and has a horizontal transverse axis, the standard form of its equation is

$$\frac{x^2}{a^2} - \frac{y^2}{b^2} = 1.$$

Because the foci are 2640 units from the center, $c = 2640$. Let d_A and d_B be the distances of any point on the hyperbola from the foci at A and B, respectively. From page 713, you have

$$|d_B - d_A| = 2a$$

$$|2200| = 2a \qquad \text{The points are 2200 feet closer to A than to B.}$$

$$1100 = a. \qquad \text{Divide each side by 2.}$$

So, $b^2 = c^2 - a^2 = 2640^2 - 1100^2 = 5{,}759{,}600$, and you can conclude that the explosion occurred somewhere on the right branch of the hyperbola

$$\frac{x^2}{1{,}210{,}000} - \frac{y^2}{5{,}759{,}600} = 1.$$

Checkpoint *Audio-video solution in English & Spanish at LarsonPrecalculus.com.*

Repeat Example 5 when microphone A receives the sound 4 seconds before microphone B.

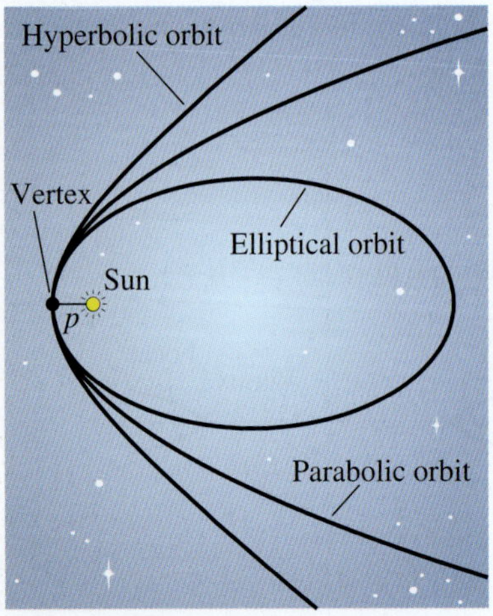

Figure 10.31

Another interesting application of conic sections involves the orbits of comets in our solar system. Of the 610 comets identified prior to 1970, 245 have elliptical orbits, 295 have parabolic orbits, and 70 have hyperbolic orbits. The center of the sun is a focus of each of these orbits, and each orbit has a vertex at the point where the comet is closest to the sun, as shown in Figure 10.31. Undoubtedly, there have been many comets with parabolic or hyperbolic orbits that were not identified. We only get to see such comets *once*. Comets with elliptical orbits, such as Halley's comet, are the only ones that remain in our solar system.

If p is the distance between the vertex and the focus (in meters), and v is the velocity of the comet at the vertex (in meters per second), then the type of orbit is determined as follows.

1. Ellipse: $v < \sqrt{2GM/p}$
2. Parabola: $v = \sqrt{2GM/p}$
3. Hyperbola: $v > \sqrt{2GM/p}$

In each of these relations, $M = 1.989 \times 10^{30}$ kilograms (the mass of the sun) and $G \approx 6.67 \times 10^{-11}$ cubic meter per kilogram-second squared (the universal gravitational constant).

General Equations of Conics

Classifying a Conic from Its General Equation

The graph of $Ax^2 + Cy^2 + Dx + Ey + F = 0$ is one of the following.

1. *Circle:* $A = C$ — $A \neq 0$
2. *Parabola:* $AC = 0$ — $A = 0$ or $C = 0$, but not both.
3. *Ellipse:* $AC > 0$ — A and C have like signs.
4. *Hyperbola:* $AC < 0$ — A and C have unlike signs.

The test above is valid when the graph is a conic. The test does not apply to equations such as $x^2 + y^2 = -1$, whose graph is not a conic.

EXAMPLE 6 Classifying Conics from General Equations

a. For the equation $4x^2 - 9x + y - 5 = 0$, you have

$$AC = 4(0) = 0. \qquad \text{Parabola}$$

So, the graph is a parabola.

b. For the equation $4x^2 - y^2 + 8x - 6y + 4 = 0$, you have

$$AC = 4(-1) < 0. \qquad \text{Hyperbola}$$

So, the graph is a hyperbola.

c. For the equation $2x^2 + 4y^2 - 4x + 12y = 0$, you have

$$AC = 2(4) > 0. \qquad \text{Ellipse}$$

So, the graph is an ellipse.

d. For the equation $2x^2 + 2y^2 - 8x + 12y + 2 = 0$, you have

$$A = C = 2. \qquad \text{Circle}$$

So, the graph is a circle.

Caroline Herschel (1750–1848) was the first woman to be credited with detecting a new comet. During her long life, this English astronomer discovered a total of eight new comets.

✓ Checkpoint Audio-video solution in English & Spanish at LarsonPrecalculus.com.

Classify the graph of each equation.

a. $3x^2 + 3y^2 - 6x + 6y + 5 = 0$

b. $2x^2 - 4y^2 + 4x + 8y - 3 = 0$

c. $3x^2 + y^2 + 6x - 2y + 3 = 0$

d. $2x^2 + 4x + y - 2 = 0$

Summarize (Section 10.4)

1. State the definition of a hyperbola *(page 713)*. For an example of finding the standard form of the equation of a hyperbola, see Example 1.
2. Write the equations of the asymptotes of hyperbolas with horizontal and vertical transverse axes *(page 715)*. For examples involving the asymptotes of hyperbolas, see Examples 2–4.
3. Describe a real-life application of a hyperbola *(page 718, Example 5)*.
4. Describe how to classify a conic from its general equation *(page 719)*. For an example of classifying conics, see Example 6.

© The Art Gallery Collection/Alamy

10.4 Exercises

See CalcChat.com for tutorial help and worked-out solutions to odd-numbered exercises.

Vocabulary: Fill in the blanks.

1. A ________ is the set of all points (x, y) in a plane for which the absolute value of the difference of the distances from two distinct fixed points, called ________, is constant.
2. The graph of a hyperbola has two disconnected parts called ________.
3. The line segment connecting the vertices of a hyperbola is called the ________ ________, and the midpoint of the line segment is the ________ of the hyperbola.
4. Each hyperbola has two ________ that intersect at the center of the hyperbola.

Skills and Applications

Matching In Exercises 5–8, match the equation with its graph. [The graphs are labeled (a), (b), (c), and (d).]

(a)

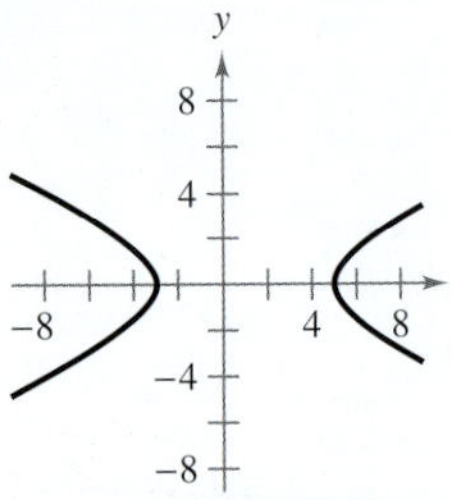

(b)

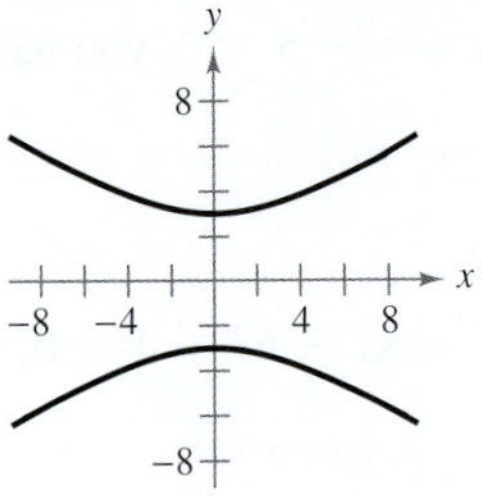

(c)

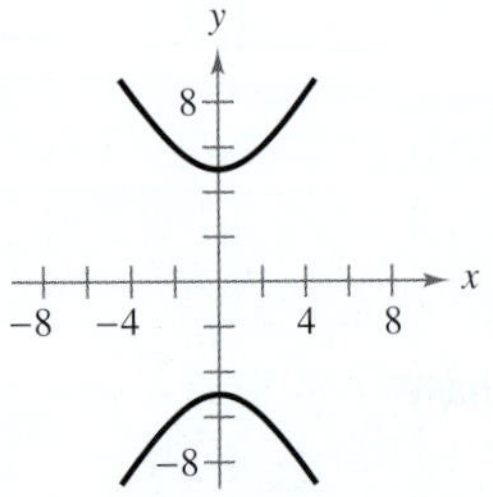

(d)

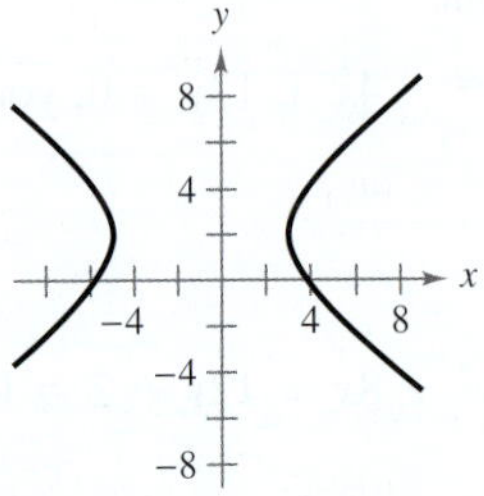

5. $\frac{y^2}{9} - \frac{x^2}{25} = 1$
6. $\frac{y^2}{25} - \frac{x^2}{9} = 1$
7. $\frac{(x-1)^2}{16} - \frac{y^2}{4} = 1$
8. $\frac{(x+1)^2}{16} - \frac{(y-2)^2}{9} = 1$

Finding the Standard Equation of a Hyperbola In Exercises 9–18, find the standard form of the equation of the hyperbola with the given characteristics.

9. Vertices: $(0, \pm 2)$; foci: $(0, \pm 4)$
10. Vertices: $(\pm 4, 0)$; foci: $(\pm 6, 0)$
11. Vertices: $(2, 0)$, $(6, 0)$; foci: $(0, 0)$, $(8, 0)$
12. Vertices: $(2, 3)$, $(2, -3)$; foci: $(2, 6)$, $(2, -6)$
13. Vertices: $(4, 1)$, $(4, 9)$; foci: $(4, 0)$, $(4, 10)$
14. Vertices: $(-2, 1)$, $(2, 1)$; foci: $(-3, 1)$, $(3, 1)$
15. Vertices: $(2, 3)$, $(2, -3)$; passes through the point $(0, 5)$
16. Vertices: $(-2, 1)$, $(2, 1)$; passes through the point $(5, 4)$
17. Vertices: $(0, 4)$, $(0, 0)$; passes through the point $(\sqrt{5}, -1)$
18. Vertices: $(1, 2)$, $(1, -2)$; passes through the point $(0, \sqrt{5})$

Sketching a Hyperbola In Exercises 19–32, find the center, vertices, foci, and the equations of the asymptotes of the hyperbola. Then sketch the hyperbola using the asymptotes as an aid.

19. $x^2 - y^2 = 1$
20. $\frac{x^2}{9} - \frac{y^2}{25} = 1$
21. $\frac{y^2}{25} - \frac{x^2}{81} = 1$
22. $\frac{x^2}{36} - \frac{y^2}{4} = 1$
23. $\frac{y^2}{1} - \frac{x^2}{4} = 1$
24. $\frac{y^2}{9} - \frac{x^2}{1} = 1$
25. $\frac{(x-1)^2}{4} - \frac{(y+2)^2}{1} = 1$
26. $\frac{(x+3)^2}{144} - \frac{(y-2)^2}{25} = 1$
27. $\frac{(y+6)^2}{1/9} - \frac{(x-2)^2}{1/4} = 1$
28. $\frac{(y-1)^2}{1/4} - \frac{(x+3)^2}{1/16} = 1$
29. $9x^2 - y^2 - 36x - 6y + 18 = 0$
30. $x^2 - 9y^2 + 36y - 72 = 0$
31. $x^2 - 9y^2 + 2x - 54y - 80 = 0$
32. $16y^2 - x^2 + 2x + 64y + 63 = 0$

Graphing a Hyperbola In Exercises 33–38, find the center, vertices, foci, and the equations of the asymptotes of the hyperbola. Use a graphing utility to graph the hyperbola and its asymptotes.

33. $2x^2 - 3y^2 = 6$
34. $6y^2 - 3x^2 = 18$
35. $4x^2 - 9y^2 = 36$
36. $25x^2 - 4y^2 = 100$
37. $9y^2 - x^2 + 2x + 54y + 62 = 0$
38. $9x^2 - y^2 + 54x + 10y + 55 = 0$

Finding the Standard Equation **In Exercises 39–46, find the standard form of the equation of the hyperbola with the given characteristics.**

39. Vertices: $(\pm 1, 0)$; asymptotes: $y = \pm 5x$

40. Vertices: $(0, \pm 3)$; asymptotes: $y = \pm 3x$

41. Foci: $(0, \pm 8)$; asymptotes: $y = \pm 4x$

42. Foci: $(\pm 10, 0)$; asymptotes: $y = \pm \frac{3}{4}x$

43. Vertices: $(1, 2)$, $(3, 2)$;
asymptotes: $y = x$, $y = 4 - x$

44. Vertices: $(3, 0)$, $(3, 6)$;
asymptotes: $y = 6 - x$, $y = x$

45. Vertices: $(0, 2)$, $(6, 2)$;
asymptotes: $y = \frac{2}{3}x$, $y = 4 - \frac{2}{3}x$

46. Vertices: $(3, 0)$, $(3, 4)$;
asymptotes: $y = \frac{2}{3}x$, $y = 4 - \frac{2}{3}x$

47. Art A sculpture has a hyperbolic cross section (see figure).

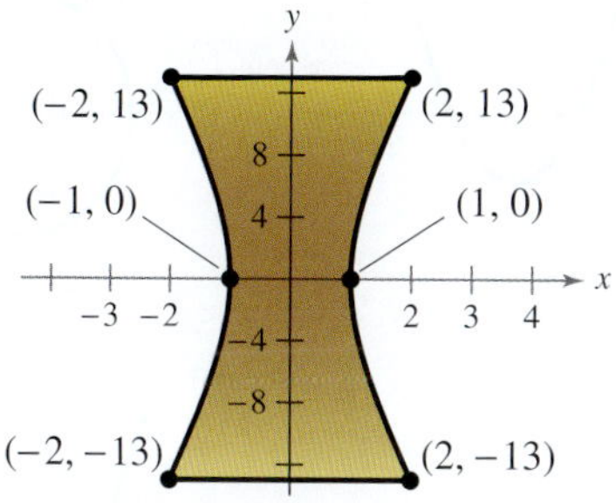

(a) Write an equation that models the curved sides of the sculpture.

(b) Each unit in the coordinate plane represents 1 foot. Find the width of the sculpture at a height of 5 feet.

48. Pendulum The base for a pendulum of a clock has the shape of a hyperbola (see figure).

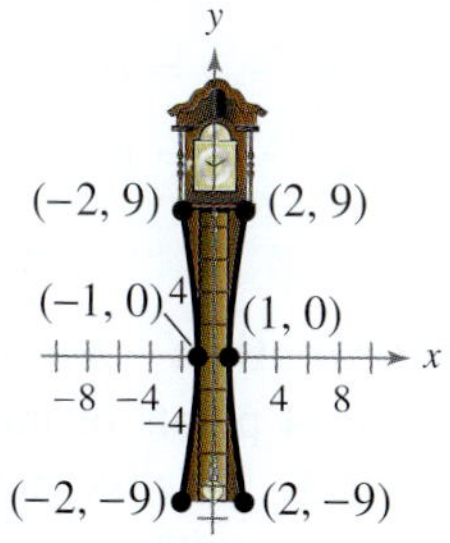

(a) Write an equation of the cross section of the base.

(b) Each unit in the coordinate plane represents $\frac{1}{2}$ foot. Find the width of the base of the pendulum 4 inches from the bottom.

49. Sound Location You and a friend live 4 miles apart. You hear a clap of thunder from lightning. Your friend hears the thunder 18 seconds later. Where did the lightning occur? (Assume sound travels at 1100 feet per second.)

50. Sound Location Three listening stations located at $(3300, 0)$, $(3300, 1100)$, and $(-3300, 0)$ monitor an explosion. The last two stations detect the explosion 1 second and 4 seconds after the first, respectively. Determine the coordinates of the explosion. (Assume that the coordinate system is measured in feet and that sound travels at 1100 feet per second.)

51. Navigation

Long distance radio navigation for aircraft and ships uses synchronized pulses transmitted by widely separated transmitting stations. These pulses travel at the speed of light (186,000 miles per second). The difference in the times of arrival of these pulses at an aircraft or ship is constant on a hyperbola having the transmitting stations as foci.

Assume that two stations, 300 miles apart, are positioned on a rectangular coordinate system at points with coordinates $(-150, 0)$ and $(150, 0)$, and that a ship is traveling on a hyperbolic path with coordinates $(x, 75)$ (see figure).

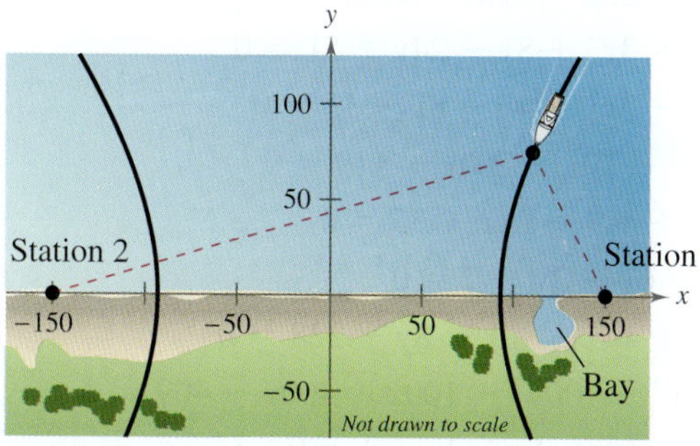

(a) Find the x-coordinate of the position of the ship when the time difference between the pulses from the transmitting stations is 1000 microseconds (0.001 second).

(b) Determine the distance between the ship and station 1 when the ship reaches the shore.

(c) The ship wants to enter a bay located between the two stations. The bay is 30 miles from station 1. What should be the time difference between the pulses?

(d) The ship is 60 miles offshore when the time difference in part (c) is obtained. What is the position of the ship?

Melissa Madia/Shutterstock.com

52. Hyperbolic Mirror A hyperbolic mirror (used in some telescopes) has the property that a light ray directed at a focus will be reflected to the other focus. The focus of a hyperbolic mirror (see figure) has coordinates (24, 0). Find the vertex of the mirror given that the mount at the top edge of the mirror has coordinates (24, 24).

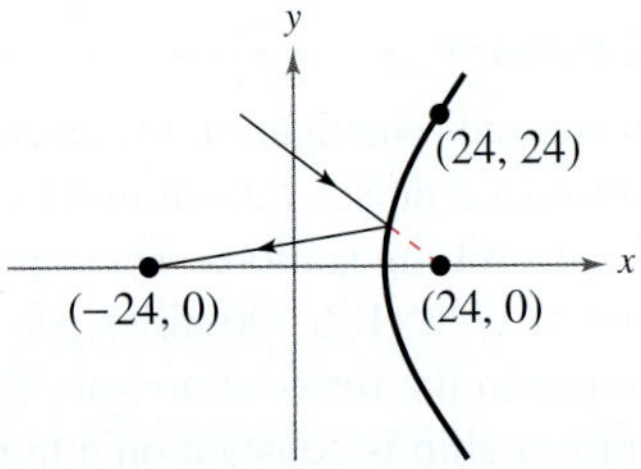

Classifying a Conic from a General Equation **In Exercises 53–68, classify the graph of the equation as a circle, a parabola, an ellipse, or a hyperbola.**

53. $9x^2 + 4y^2 - 18x + 16y - 119 = 0$
54. $x^2 + y^2 - 4x - 6y - 23 = 0$
55. $4x^2 - y^2 - 4x - 3 = 0$
56. $y^2 - 6y - 4x + 21 = 0$
57. $y^2 - 4x^2 + 4x - 2y - 4 = 0$
58. $x^2 + y^2 - 4x + 6y - 3 = 0$
59. $y^2 + 12x + 4y + 28 = 0$
60. $4x^2 + 25y^2 + 16x + 250y + 541 = 0$
61. $4x^2 + 3y^2 + 8x - 24y + 51 = 0$
62. $4y^2 - 2x^2 - 4y - 8x - 15 = 0$
63. $25x^2 - 10x - 200y - 119 = 0$
64. $4y^2 + 4x^2 - 24x + 35 = 0$
65. $x^2 - 6x - 2y + 7 = 0$
66. $9x^2 + 4y^2 - 90x + 8y + 228 = 0$
67. $100x^2 + 100y^2 - 100x + 400y + 409 = 0$
68. $4x^2 - y^2 + 4x + 2y - 1 = 0$

Exploration

True or False? **In Exercises 69–72, determine whether the statement is true or false. Justify your answer.**

69. In the standard form of the equation of a hyperbola, the larger the ratio of b to a, the larger the eccentricity of the hyperbola.

70. In the standard form of the equation of a hyperbola, the trivial solution of two intersecting lines occurs when $b = 0$.

71. If $D \neq 0$ and $E \neq 0$, then the graph of

$$x^2 - y^2 + Dx + Ey = 0$$

is a hyperbola.

72. If the asymptotes of the hyperbola $\frac{x^2}{a^2} - \frac{y^2}{b^2} = 1$, where $a, b > 0$, intersect at right angles, then $a = b$.

73. Think About It Consider a hyperbola centered at the origin with a horizontal transverse axis. Use the definition of a hyperbola to derive its standard form.

74. Writing Explain how the central rectangle of a hyperbola can be used to sketch its asymptotes.

75. Think About It Change the equation of the hyperbola so that its graph is the bottom half of the hyperbola.

$$9x^2 - 54x - 4y^2 + 8y + 41 = 0$$

76. HOW DO YOU SEE IT? Match each equation with its graph.

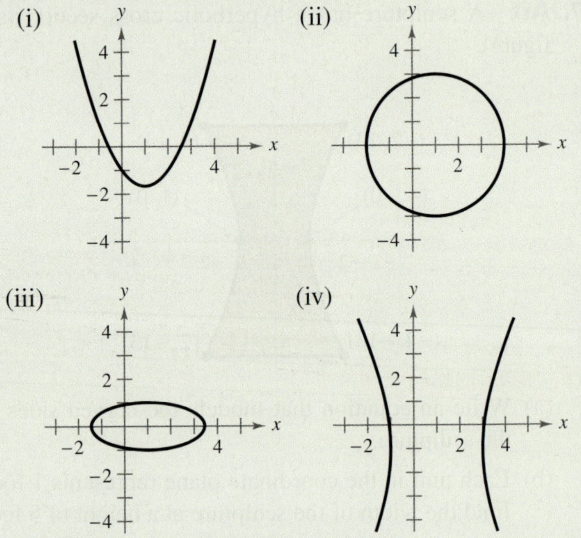

(a) $4x^2 - y^2 - 8x - 2y - 13 = 0$
(b) $x^2 + y^2 - 2x - 8 = 0$
(c) $2x^2 - 4x - 3y - 3 = 0$
(d) $x^2 + 6y^2 - 2x - 5 = 0$

77. Points of Intersection A circle and a parabola can have 0, 1, 2, 3, or 4 points of intersection. Sketch the circle $x^2 + y^2 = 4$. Discuss how this circle could intersect a parabola with an equation of the form $y = x^2 + C$. Then find the values of C for each of the five cases described below. Use a graphing utility to verify your results.

(a) No points of intersection
(b) One point of intersection
(c) Two points of intersection
(d) Three points of intersection
(e) Four points of intersection

10.5 Rotation of Conics

Rotated conics can help you model and solve real-life problems. For instance, in Exercise 63 on page 730, you will use a rotated parabola to model the cross section of a satellite dish.

- **Rotate the coordinate axes to eliminate the *xy*-term in equations of conics.**
- **Use the discriminant to classify conics.**

Rotation

The equation of a conic with axes parallel to one of the coordinate axes has a standard form that can be written in the general form

$$Ax^2 + Cy^2 + Dx + Ey + F = 0. \qquad \text{Horizontal or vertical axis}$$

Conics whose axes are rotated so that they are not parallel to either the x-axis or the y-axis have general equations that contain an xy-term.

$$Ax^2 + Bxy + Cy^2 + Dx + Ey + F = 0 \qquad \text{Equation in } xy\text{-plane}$$

To eliminate this xy-term, you can use a procedure called **rotation of axes.** The objective is to rotate the x- and y-axes until they are parallel to the axes of the conic. The rotated axes are denoted as the x'-axis and the y'-axis, as shown in Figure 10.32. After the rotation, the equation of the conic in the new $x'y'$-plane will have the form

$$A'(x')^2 + C'(y')^2 + D'x' + E'y' + F' = 0. \qquad \text{Equation in } x'y'\text{-plane}$$

Because this equation has no $x'y'$-term, you can obtain a standard form by completing the square. The following theorem identifies how much to rotate the axes to eliminate the xy-term and also the equations for determining the new coefficients A', C', D', E', and F'.

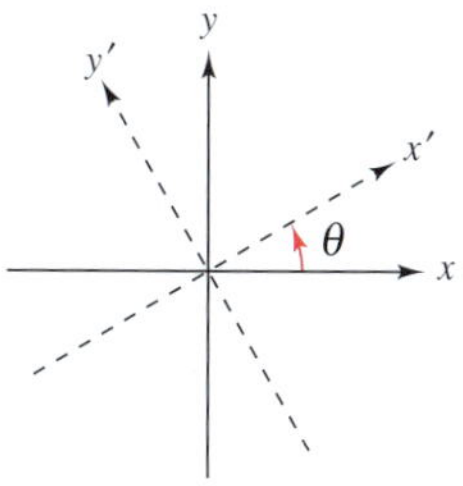

Figure 10.32

Rotation of Axes to Eliminate an *xy*-Term

The general second-degree equation

$$Ax^2 + Bxy + Cy^2 + Dx + Ey + F = 0$$

can be rewritten as

$$A'(x')^2 + C'(y')^2 + D'x' + E'y' + F' = 0$$

by rotating the coordinate axes through an angle θ, where

$$\cot 2\theta = \frac{A - C}{B}.$$

The coefficients of the new equation are obtained by making the substitutions

$$x = x' \cos \theta - y' \sin \theta$$

and

$$y = x' \sin \theta + y' \cos \theta.$$

Remember that the substitutions

$$x = x' \cos \theta - y' \sin \theta$$

and

$$y = x' \sin \theta + y' \cos \theta$$

were developed to eliminate the $x'y'$-term in the rotated system. You can use this as a check of your work. In other words, when your final equation contains an $x'y'$-term, you know that you have made a mistake.

Alexey Gostev/Shutterstock.com

EXAMPLE 1 Rotation of Axes for a Hyperbola

Rotate the axes to eliminate the xy-term in the equation $xy - 1 = 0$. Then write the equation in standard form and sketch its graph.

Solution Because $A = 0$, $B = 1$, and $C = 0$, you have

$$\cot 2\theta = \frac{A - C}{B} = 0 \implies 2\theta = \frac{\pi}{2} \implies \theta = \frac{\pi}{4}$$

which implies that

$$\begin{aligned} x &= x'\cos\frac{\pi}{4} - y'\sin\frac{\pi}{4} \\ &= x'\left(\frac{1}{\sqrt{2}}\right) - y'\left(\frac{1}{\sqrt{2}}\right) \\ &= \frac{x' - y'}{\sqrt{2}} \end{aligned}$$

and

$$\begin{aligned} y &= x'\sin\frac{\pi}{4} + y'\cos\frac{\pi}{4} \\ &= x'\left(\frac{1}{\sqrt{2}}\right) + y'\left(\frac{1}{\sqrt{2}}\right) \\ &= \frac{x' + y'}{\sqrt{2}}. \end{aligned}$$

The equation in the $x'y'$-system is obtained by substituting these expressions in the original equation.

$$xy - 1 = 0$$

$$\left(\frac{x' - y'}{\sqrt{2}}\right)\left(\frac{x' + y'}{\sqrt{2}}\right) - 1 = 0$$

$$\frac{(x')^2 - (y')^2}{2} - 1 = 0$$

$$\frac{(x')^2}{(\sqrt{2})^2} - \frac{(y')^2}{(\sqrt{2})^2} = 1 \qquad \text{Write in standard form.}$$

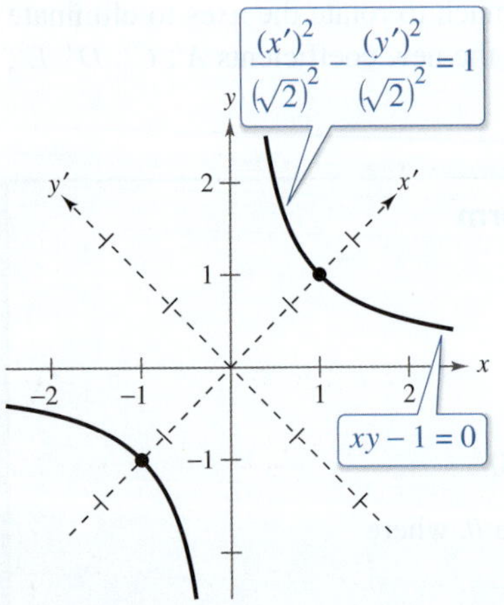

Vertices:
In $x'y'$-system: $(\sqrt{2}, 0), (-\sqrt{2}, 0)$
In xy-system: $(1, 1), (-1, -1)$
Figure 10.33

In the $x'y'$-system, this is a hyperbola centered at the origin with vertices at $(\pm\sqrt{2}, 0)$, as shown in Figure 10.33. To find the coordinates of the vertices in the xy-system, substitute the coordinates $(\pm\sqrt{2}, 0)$ in the equations

$$x = \frac{x' - y'}{\sqrt{2}} \quad \text{and} \quad y = \frac{x' + y'}{\sqrt{2}}.$$

This substitution yields the vertices $(1, 1)$ and $(-1, -1)$ in the xy-system. Note also that the asymptotes of the hyperbola have equations

$$y' = \pm x'$$

which correspond to the original x- and y-axes.

Checkpoint *Audio-video solution in English & Spanish at LarsonPrecalculus.com.*

Rotate the axes to eliminate the xy-term in the equation $xy + 6 = 0$. Then write the equation in standard form and sketch its graph.

EXAMPLE 2 Rotation of Axes for an Ellipse

Rotate the axes to eliminate the xy-term in the equation

$$7x^2 - 6\sqrt{3}xy + 13y^2 - 16 = 0.$$

Then write the equation in standard form and sketch its graph.

Solution Because $A = 7$, $B = -6\sqrt{3}$, and $C = 13$, you have

$$\cot 2\theta = \frac{A - C}{B} = \frac{7 - 13}{-6\sqrt{3}} = \frac{1}{\sqrt{3}}$$

which implies that $\theta = \pi/6$. The equation in the $x'y'$-system is obtained by making the substitutions

$$\begin{aligned} x &= x'\cos\frac{\pi}{6} - y'\sin\frac{\pi}{6} \\ &= x'\left(\frac{\sqrt{3}}{2}\right) - y'\left(\frac{1}{2}\right) \\ &= \frac{\sqrt{3}x' - y'}{2} \end{aligned}$$

and

$$\begin{aligned} y &= x'\sin\frac{\pi}{6} + y'\cos\frac{\pi}{6} \\ &= x'\left(\frac{1}{2}\right) + y'\left(\frac{\sqrt{3}}{2}\right) \\ &= \frac{x' + \sqrt{3}y'}{2} \end{aligned}$$

in the original equation. So, you have

$$7x^2 - 6\sqrt{3}xy + 13y^2 - 16 = 0$$

$$7\left(\frac{\sqrt{3}x' - y'}{2}\right)^2 - 6\sqrt{3}\left(\frac{\sqrt{3}x' - y'}{2}\right)\left(\frac{x' + \sqrt{3}y'}{2}\right) + 13\left(\frac{x' + \sqrt{3}y'}{2}\right)^2 - 16 = 0$$

which simplifies to

$$\begin{aligned} 4(x')^2 + 16(y')^2 - 16 &= 0 \\ 4(x')^2 + 16(y')^2 &= 16 \\ \frac{(x')^2}{4} + \frac{(y')^2}{1} &= 1 \\ \frac{(x')^2}{2^2} + \frac{(y')^2}{1^2} &= 1. \end{aligned}$$

Write in standard form.

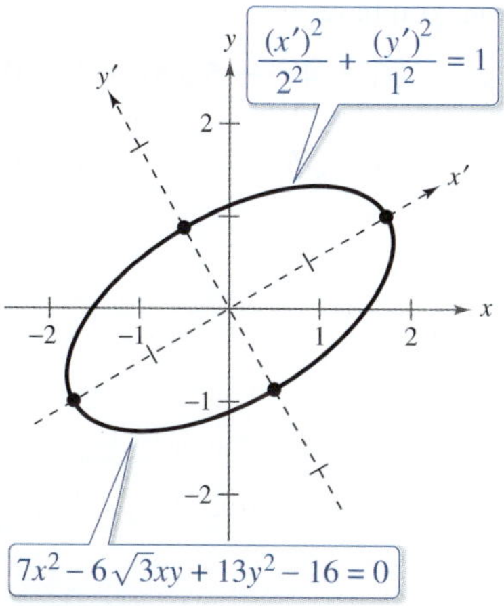

Vertices:
In $x'y'$-system: $(\pm 2, 0)$
In xy-system: $(\sqrt{3}, 1), (-\sqrt{3}, -1)$
Figure 10.34

This is the equation of an ellipse centered at the origin with vertices $(\pm 2, 0)$ in the $x'y'$-system, as shown in Figure 10.34.

Checkpoint *Audio-video solution in English & Spanish at LarsonPrecalculus.com.*

Rotate the axes to eliminate the xy-term in the equation

$$12x^2 + 16\sqrt{3}xy + 28y^2 - 36 = 0.$$

Then write the equation in standard form and sketch its graph.

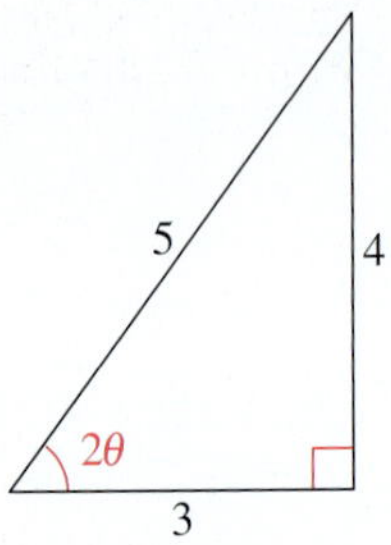

Figure 10.35

EXAMPLE 3 Rotation of Axes for a Parabola

Rotate the axes to eliminate the xy-term in the equation $x^2 - 4xy + 4y^2 + 5\sqrt{5}y + 1 = 0$. Then write the equation in standard form and sketch its graph.

Solution Because $A = 1$, $B = -4$, and $C = 4$, you have

$$\cot 2\theta = \frac{A - C}{B} = \frac{1 - 4}{-4} = \frac{3}{4}.$$

Using this information, draw a right triangle as shown in Figure 10.35. From the figure, you can see that $\cos 2\theta = \frac{3}{5}$. To find the values of $\sin \theta$ and $\cos \theta$, you can use the half-angle formulas in the forms

$$\sin \theta = \sqrt{\frac{1 - \cos 2\theta}{2}} \quad \text{and} \quad \cos \theta = \sqrt{\frac{1 + \cos 2\theta}{2}}.$$

So,

$$\sin \theta = \sqrt{\frac{1 - \cos 2\theta}{2}} = \sqrt{\frac{1 - \frac{3}{5}}{2}} = \sqrt{\frac{1}{5}} = \frac{1}{\sqrt{5}}$$

$$\cos \theta = \sqrt{\frac{1 + \cos 2\theta}{2}} = \sqrt{\frac{1 + \frac{3}{5}}{2}} = \sqrt{\frac{4}{5}} = \frac{2}{\sqrt{5}}.$$

Consequently, you use the substitutions

$$x = x'\cos \theta - y'\sin \theta = x'\left(\frac{2}{\sqrt{5}}\right) - y'\left(\frac{1}{\sqrt{5}}\right) = \frac{2x' - y'}{\sqrt{5}}$$

and

$$y = x'\sin \theta + y'\cos \theta = x'\left(\frac{1}{\sqrt{5}}\right) + y'\left(\frac{2}{\sqrt{5}}\right) = \frac{x' + 2y'}{\sqrt{5}}.$$

Substituting these expressions in the original equation, you have

$$x^2 - 4xy + 4y^2 + 5\sqrt{5}y + 1 = 0$$

$$\left(\frac{2x' - y'}{\sqrt{5}}\right)^2 - 4\left(\frac{2x' - y'}{\sqrt{5}}\right)\left(\frac{x' + 2y'}{\sqrt{5}}\right) + 4\left(\frac{x' + 2y'}{\sqrt{5}}\right)^2 + 5\sqrt{5}\left(\frac{x' + 2y'}{\sqrt{5}}\right) + 1 = 0$$

which simplifies as follows.

$$5(y')^2 + 5x' + 10y' + 1 = 0$$

$$5[(y')^2 + 2y'] = -5x' - 1 \qquad \text{Group terms.}$$

$$5[(y')^2 + 2y' + 1] = -5x' - 1 + 5(1) \qquad \text{Complete the square.}$$

$$5(y' + 1)^2 = -5x' + 4 \qquad \text{Write in completed square form.}$$

$$(y' + 1)^2 = (-1)\left(x' - \frac{4}{5}\right) \qquad \text{Write in standard form.}$$

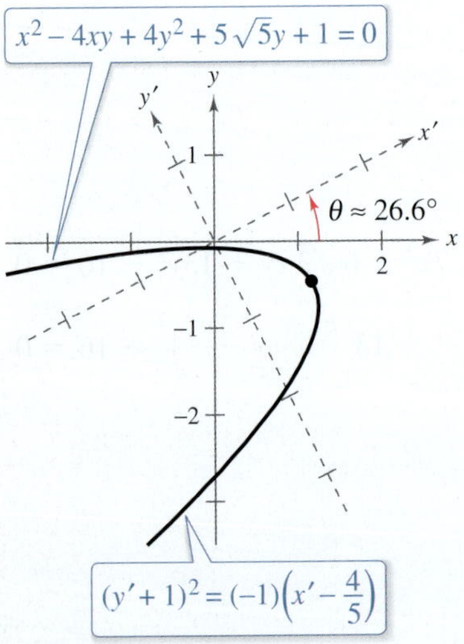

Vertex:

In $x'y'$-system: $\left(\frac{4}{5}, -1\right)$

In xy-system: $\left(\frac{13}{5\sqrt{5}}, -\frac{6}{5\sqrt{5}}\right)$

Figure 10.36

The graph of this equation is a parabola with vertex $\left(\frac{4}{5}, -1\right)$ in the $x'y'$-system. Its axis is parallel to the x'-axis in the $x'y'$-system, and because $\sin \theta = 1/\sqrt{5}$, $\theta \approx 26.6°$, as shown in Figure 10.36.

✓ *Checkpoint* *Audio-video solution in English & Spanish at LarsonPrecalculus.com.*

Rotate the axes to eliminate the xy-term in the equation

$$4x^2 + 4xy + y^2 - 2\sqrt{5}x + 4\sqrt{5}y - 30 = 0.$$

Then write the equation in standard form and sketch its graph.

Invariants Under Rotation

In the rotation of axes theorem listed at the beginning of this section, note that the constant term is the same in both equations, $F' = F$. Such quantities are **invariant under rotation.** The next theorem lists some other rotation invariants.

Rotation Invariants

The rotation of the coordinate axes through an angle θ that transforms the equation $Ax^2 + Bxy + Cy^2 + Dx + Ey + F = 0$ into the form

$$A'(x')^2 + C'(y')^2 + D'x' + E'y' + F' = 0$$

has the following rotation invariants.

1. $F = F'$
2. $A + C = A' + C'$
3. $B^2 - 4AC = (B')^2 - 4A'C'$

REMARK When there is an xy-term in the equation of a conic, you should realize that the conic is rotated. Before rotating the axes, you should use the discriminant to classify the conic.

You can use the results of this theorem to classify the graph of a second-degree equation *with* an xy-term in much the same way you do for a second-degree equation *without* an xy-term. Note that because $B' = 0$, the invariant $B^2 - 4AC$ reduces to

$$B^2 - 4AC = -4A'C'. \qquad \text{Discriminant}$$

This quantity is called the **discriminant** of the equation

$$Ax^2 + Bxy + Cy^2 + Dx + Ey + F = 0.$$

Now, from the classification procedure given in Section 10.4, you know that the value of $A'C'$ determines the type of graph for the equation

$$A'(x')^2 + C'(y')^2 + D'x' + E'y' + F' = 0.$$

Consequently, the value of $B^2 - 4AC$ will determine the type of graph for the original equation, as given in the following classification.

Classification of Conics by the Discriminant

The graph of the equation $Ax^2 + Bxy + Cy^2 + Dx + Ey + F = 0$ is, except in degenerate cases, determined by its discriminant as follows.

1. *Ellipse or circle:* $B^2 - 4AC < 0$
2. *Parabola:* $B^2 - 4AC = 0$
3. *Hyperbola:* $B^2 - 4AC > 0$

For example, in the general equation

$$3x^2 + 7xy + 5y^2 - 6x - 7y + 15 = 0$$

you have $A = 3$, $B = 7$, and $C = 5$. So, the discriminant is

$$\begin{aligned} B^2 - 4AC &= 7^2 - 4(3)(5) \\ &= 49 - 60 \\ &= -11. \end{aligned}$$

Because $-11 < 0$, the graph of the equation is an ellipse or a circle.

EXAMPLE 4 Rotation and Graphing Utilities

For each equation, classify the graph of the equation, use the Quadratic Formula to solve for y, and then use a graphing utility to graph the equation.

a. $2x^2 - 3xy + 2y^2 - 2x = 0$ **b.** $x^2 - 6xy + 9y^2 - 2y + 1 = 0$

c. $3x^2 + 8xy + 4y^2 - 7 = 0$

Solution

a. Because $B^2 - 4AC = 9 - 16 < 0$, the graph is a circle or an ellipse.

$$2x^2 - 3xy + 2y^2 - 2x = 0 \quad \text{Write original equation.}$$

$$2y^2 - 3xy + (2x^2 - 2x) = 0 \quad \text{Quadratic form } ay^2 + by + c = 0$$

$$y = \frac{-(-3x) \pm \sqrt{(-3x)^2 - 4(2)(2x^2 - 2x)}}{2(2)}$$

$$y = \frac{3x \pm \sqrt{x(16 - 7x)}}{4}$$

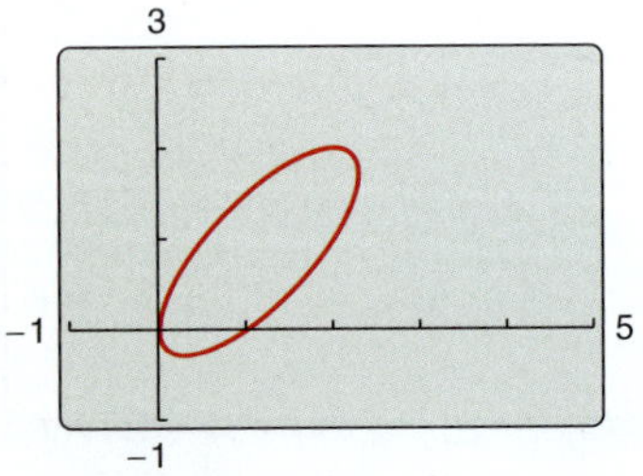

Figure 10.37

Graph both of the equations to obtain the ellipse shown in Figure 10.37.

$$y_1 = \frac{3x + \sqrt{x(16 - 7x)}}{4} \quad \text{Top half of ellipse}$$

$$y_2 = \frac{3x - \sqrt{x(16 - 7x)}}{4} \quad \text{Bottom half of ellipse}$$

b. Because $B^2 - 4AC = 36 - 36 = 0$, the graph is a parabola.

$$x^2 - 6xy + 9y^2 - 2y + 1 = 0 \quad \text{Write original equation.}$$

$$9y^2 - (6x + 2)y + (x^2 + 1) = 0 \quad \text{Quadratic form } ay^2 + by + c = 0$$

$$y = \frac{(6x + 2) \pm \sqrt{(6x + 2)^2 - 4(9)(x^2 + 1)}}{2(9)}$$

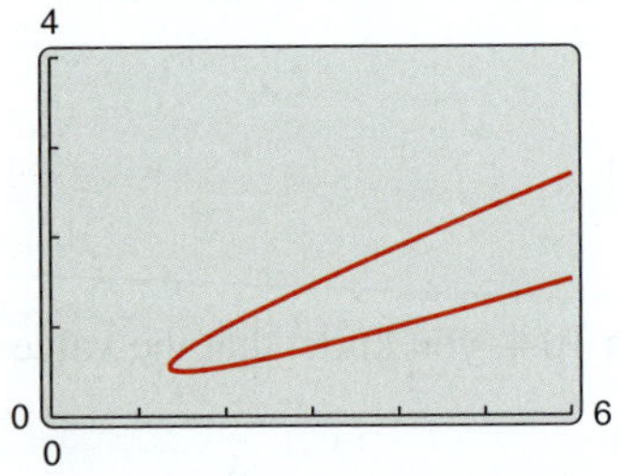

Figure 10.38

Graph both of the equations to obtain the parabola shown in Figure 10.38.

c. Because $B^2 - 4AC = 64 - 48 > 0$, the graph is a hyperbola.

$$3x^2 + 8xy + 4y^2 - 7 = 0 \quad \text{Write original equation.}$$

$$4y^2 + 8xy + (3x^2 - 7) = 0 \quad \text{Quadratic form } ay^2 + by + c = 0$$

$$y = \frac{-8x \pm \sqrt{(8x)^2 - 4(4)(3x^2 - 7)}}{2(4)}$$

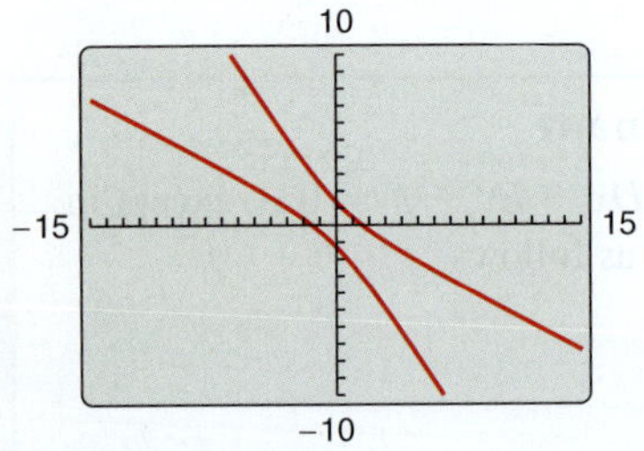

Figure 10.39

Graph both of the equations to obtain the hyperbola shown in Figure 10.39.

Checkpoint *Audio-video solution in English & Spanish at LarsonPrecalculus.com.*

Classify the graph of the equation $2x^2 - 8xy + 8y^2 + 3x + 5 = 0$, use the Quadratic Formula to solve for y, and then use a graphing utility to graph the equation.

Summarize (Section 10.5)

1. Describe how to rotate coordinate axes to eliminate the xy-term in equations of conics *(page 723)*. For examples of rotating coordinate axes to eliminate the xy-term in equations of conics, see Examples 1–3.
2. Describe how to use the discriminant to classify conics *(page 727)*. For an example of using the discriminant to classify conics, see Example 4.

10.5 Exercises

See CalcChat.com for tutorial help and worked-out solutions to odd-numbered exercises.

Vocabulary: Fill in the blanks.

1. The procedure used to eliminate the xy-term in a general second-degree equation is called ________ of ________.
2. After rotating the coordinate axes through an angle θ, the general second-degree equation in the new $x'y'$-plane will have the form ________.
3. Quantities that are equal in both the original equation of a conic and the equation of the rotated conic are ________ ________ ________.
4. The quantity $B^2 - 4AC$ is called the ________ of the equation $Ax^2 + Bxy + Cy^2 + Dx + Ey + F = 0$.

Skills and Applications

Finding a Point in a Rotated Coordinate System **In Exercises 5–12, the $x'y'$-coordinate system has been rotated θ degrees from the xy-coordinate system. The coordinates of a point in the xy-coordinate system are given. Find the coordinates of the point in the rotated coordinate system.**

5. $\theta = 90°, (0, 3)$
6. $\theta = 90°, (2, 2)$
7. $\theta = 30°, (1, 3)$
8. $\theta = 30°, (2, 4)$
9. $\theta = 45°, (2, 1)$
10. $\theta = 45°, (4, 4)$
11. $\theta = 60°, (1, 2)$
12. $\theta = 60°, (3, 1)$

Rotation of Axes **In Exercises 13–24, rotate the axes to eliminate the xy-term in the equation. Then write the equation in standard form. Sketch the graph of the resulting equation, showing both sets of axes.**

13. $xy + 1 = 0$
14. $xy - 4 = 0$
15. $xy + 2x - y + 4 = 0$
16. $xy - 8x - 4y = 0$
17. $5x^2 - 6xy + 5y^2 - 12 = 0$
18. $2x^2 + xy + 2y^2 - 8 = 0$
19. $13x^2 + 6\sqrt{3}xy + 7y^2 - 16 = 0$
20. $7x^2 - 6\sqrt{3}xy + 13y^2 - 64 = 0$
21. $x^2 + 2xy + y^2 + \sqrt{2}x - \sqrt{2}y = 0$
22. $3x^2 - 2\sqrt{3}xy + y^2 + 2x + 2\sqrt{3}y = 0$
23. $9x^2 + 24xy + 16y^2 + 90x - 130y = 0$
24. $9x^2 + 24xy + 16y^2 + 80x - 60y = 0$

Using a Graphing Utility **In Exercises 25–30, use a graphing utility to graph the conic. Determine the angle θ through which the axes are rotated. Explain how you used the graphing utility to obtain the graph.**

25. $x^2 - 4xy + 2y^2 = 6$
26. $17x^2 + 32xy - 7y^2 = 75$
27. $40x^2 + 36xy + 25y^2 = 52$
28. $24x^2 + 18xy + 12y^2 = 34$
29. $2x^2 + 4xy + 2y^2 + \sqrt{26}x + 3y = -15$
30. $4x^2 - 12xy + 9y^2 + \sqrt{6}x - 29y = 91$

Matching **In Exercises 31–36, match the graph with its equation. [The graphs are labeled (a), (b), (c), (d), (e), and (f).]**

(a)
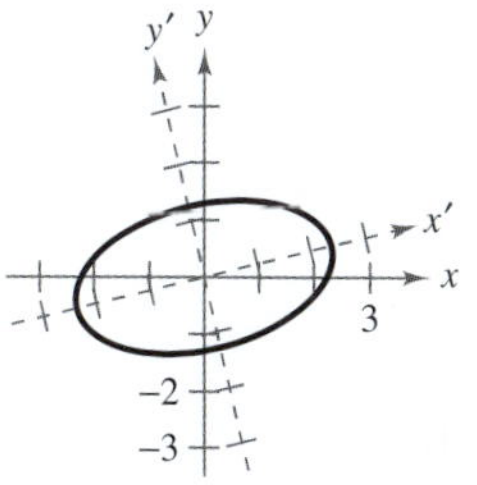

(b)
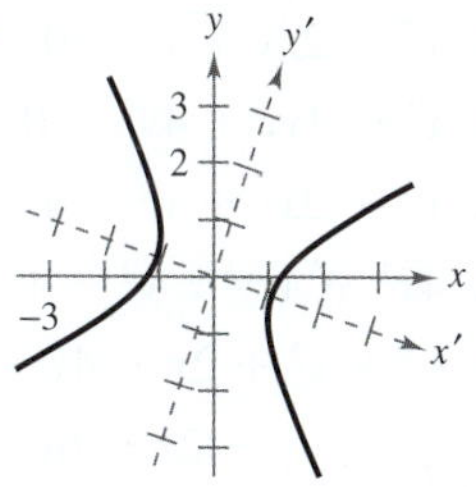

(c)
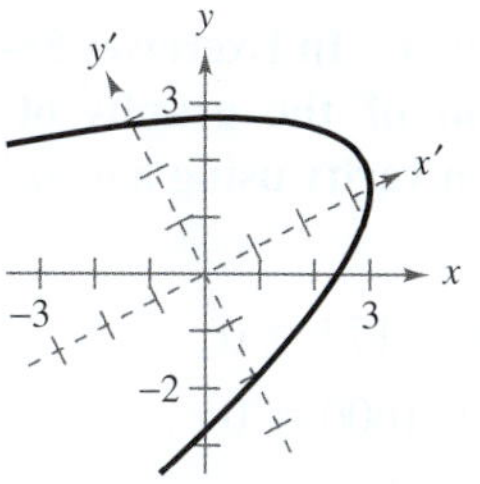

(d)
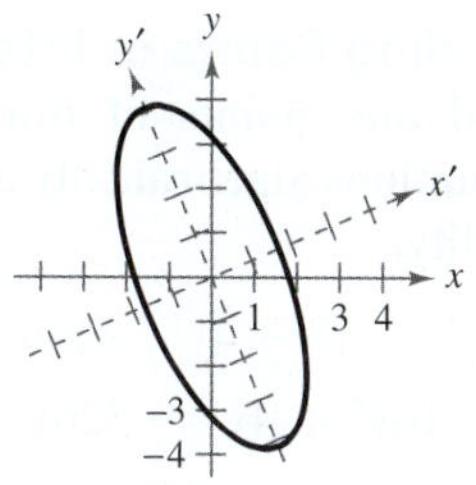

(e)
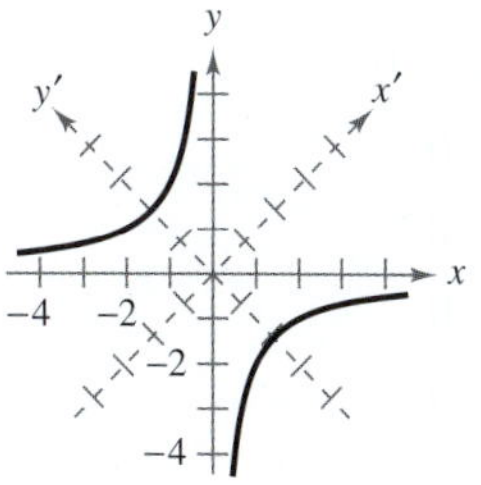

(f)
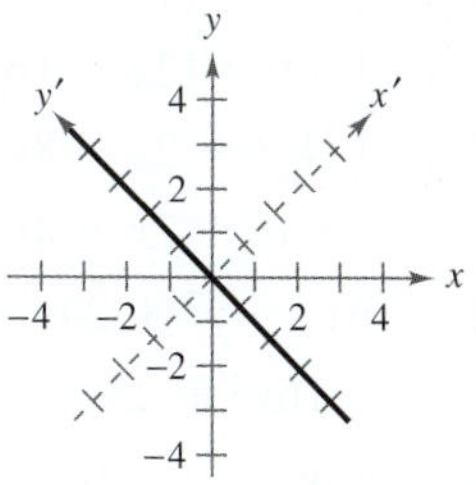

31. $xy + 2 = 0$
32. $x^2 + 2xy + y^2 = 0$
33. $-2x^2 + 3xy + 2y^2 + 3 = 0$
34. $x^2 - xy + 3y^2 - 5 = 0$
35. $3x^2 + 2xy + y^2 - 10 = 0$
36. $x^2 - 4xy + 4y^2 + 10x - 30 = 0$

Rotation and Graphing Utilities **In Exercises 37–44, (a) use the discriminant to classify the graph of the equation, (b) use the Quadratic Formula to solve for y, and (c) use a graphing utility to graph the equation.**

37. $16x^2 - 8xy + y^2 - 10x + 5y = 0$

38. $x^2 - 4xy - 2y^2 - 6 = 0$

39. $12x^2 - 6xy + 7y^2 - 45 = 0$

40. $2x^2 + 4xy + 5y^2 + 3x - 4y - 20 = 0$

41. $x^2 - 6xy - 5y^2 + 4x - 22 = 0$

42. $36x^2 - 60xy + 25y^2 + 9y = 0$

43. $x^2 + 4xy + 4y^2 - 5x - y - 3 = 0$

44. $x^2 + xy + 4y^2 + x + y - 4 = 0$

Sketching the Graph of a Degenerate Conic **In Exercises 45–54, sketch (if possible) the graph of the degenerate conic.**

45. $y^2 - 16x^2 = 0$

46. $y^2 - 25x^2 = 0$

47. $15x^2 - 2xy - y^2 = 0$

48. $32x^2 - 4xy - y^2 = 0$

49. $x^2 - 2xy + y^2 = 0$

50. $x^2 + 4xy + 4y^2 = 0$

51. $x^2 + 2xy + y^2 - 1 = 0$

52. $4x^2 + 4xy + y^2 - 1 = 0$

53. $x^2 + y^2 + 2x - 4y + 5 = 0$

54. $x^2 + y^2 - 2x + 6y + 10 = 0$

Finding Points of Intersection **In Exercises 55–62, find any points of intersection of the graphs of the equations algebraically and then verify using a graphing utility.**

55. $x^2 - 4y^2 - 20x - 64y - 172 = 0$
$16x^2 + 4y^2 - 320x + 64y + 1600 = 0$

56. $x^2 - y^2 - 12x + 16y - 64 = 0$
$x^2 + y^2 - 12x - 16y + 64 = 0$

57. $x^2 + 4y^2 - 2x - 8y + 1 = 0$
$-x^2 + 2x - 4y - 1 = 0$

58. $-16x^2 - y^2 + 24y - 80 = 0$
$16x^2 + 25y^2 - 400 = 0$

59. $x^2 + y^2 - 4 = 0$
$3x - y^2 = 0$

60. $4x^2 + 9y^2 - 36y = 0$
$x^2 + 9y - 27 = 0$

61. $-x^2 + y^2 + 4x - 6y + 4 = 0$
$x^2 + y^2 - 4x - 6y + 12 = 0$

62. $xy + x - 2y + 3 = 0$
$x^2 + 4y^2 - 9 = 0$

Alexey Gostev/Shutterstock.com

63. Satellite Dish

The parabolic cross section of a satellite dish can be modeled by a portion of the graph of the equation

$$x^2 - 2xy - 27\sqrt{2}x + y^2 + 9\sqrt{2}y + 378 = 0$$

where all measurements are in feet.

(a) Rotate the axes to eliminate the xy-term in the equation. Then write the equation in standard form.

(b) A receiver is located at the focus of the cross section. Find the distance from the vertex of the cross section to the receiver.

Exploration

64. Rotating a Circle Show that the equation

$$x^2 + y^2 = r^2$$

is invariant under rotation of axes.

True or False? **In Exercises 65 and 66, determine whether the statement is true or false. Justify your answer.**

65. The graph of the equation

$$x^2 + xy + ky^2 + 6x + 10 = 0$$

where k is any constant less than $\frac{1}{4}$, is a hyperbola.

66. After a rotation of axes is used to eliminate the xy-term from an equation of the form

$$Ax^2 + Bxy + Cy^2 + Dx + Ey + F = 0$$

the coefficients of the x^2- and y^2-terms remain A and C, respectively.

67. Finding Lengths of Axes Find the lengths of the major and minor axes of the ellipse in Exercise 19.

68. HOW DO YOU SEE IT? Match the graph with the discriminant of its corresponding equation.

(a) -7

(b) 0

(c) 1

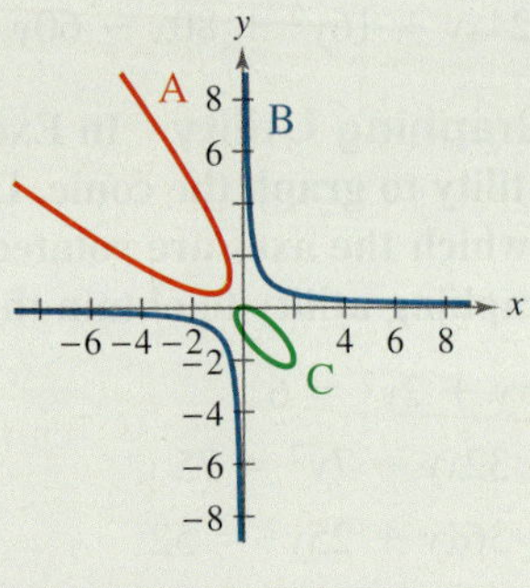

10.6 Parametric Equations

Parametric equations are useful for modeling the path of an object. For instance, in Exercise 95 on page 739, you will use a set of parametric equations to model the path of a baseball.

- **Evaluate sets of parametric equations for given values of the parameter.**
- **Sketch curves that are represented by sets of parametric equations.**
- **Rewrite sets of parametric equations as single rectangular equations by eliminating the parameter.**
- **Find sets of parametric equations for graphs.**

Plane Curves

Up to this point, you have been representing a graph by a single equation involving *two* variables such as x and y. In this section, you will study situations in which it is useful to introduce a *third* variable to represent a curve in the plane.

To see the usefulness of this procedure, consider the path of an object that is propelled into the air at an angle of 45°. When the initial velocity of the object is 48 feet per second, it can be shown that the object follows the parabolic path

$$y = -\frac{x^2}{72} + x. \qquad \text{Rectangular equation}$$

However, this equation does not tell the whole story. Although it does tell you *where* the object has been, it does not tell you *when* the object was at a given point (x, y) on the path. To determine this time, you can introduce a third variable t, called a **parameter.** It is possible to write both x and y as functions of t to obtain the **parametric equations**

$$x = 24\sqrt{2}t \qquad \text{Parametric equation for } x$$

$$y = -16t^2 + 24\sqrt{2}t. \qquad \text{Parametric equation for } y$$

From this set of equations you can determine that at time $t = 0$, the object is at the point $(0, 0)$. Similarly, at time $t = 1$, the object is at the point $(24\sqrt{2}, 24\sqrt{2} - 16)$, and so on, as shown below.

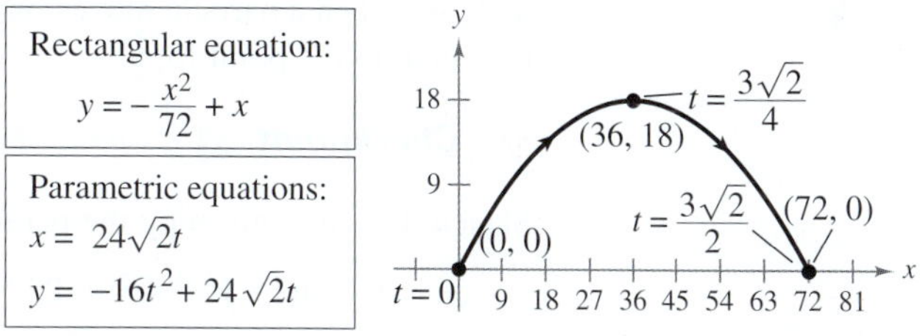

Curvilinear Motion: Two Variables for Position, One Variable for Time

For this particular motion problem, x and y are continuous functions of t, and the resulting path is a **plane curve.** (Recall that a *continuous function* is one whose graph has no breaks, holes, or gaps.)

Definition of Plane Curve

If f and g are continuous functions of t on an interval I, then the set of ordered pairs $(f(t), g(t))$ is a **plane curve** C. The equations

$$x = f(t) \quad \text{and} \quad y = g(t)$$

are **parametric equations** for C, and t is the **parameter.**

Randy Faris/Corbis Super RF/Alamy

Sketching a Plane Curve

When sketching a curve represented by a pair of parametric equations, you still plot points in the xy-plane. Each set of coordinates (x, y) is determined from a value chosen for the parameter t. Plotting the resulting points in the order of *increasing* values of t traces the curve in a specific direction. This is called the **orientation** of the curve.

EXAMPLE 1 **Sketching a Curve**

Sketch the curve given by the parametric equations

$$x = t^2 - 4 \quad \text{and} \quad y = \frac{t}{2}, \quad -2 \le t \le 3.$$

REMARK When using a value of t to find x, be sure to use the same value of t to find the corresponding value of y. Organizing your results in a table, as shown in Example 1, can be helpful.

Solution Using values of t in the specified interval, the parametric equations yield the points (x, y) shown in the table.

t	x	y
-2	0	-1
-1	-3	$-\frac{1}{2}$
0	-4	0
1	-3	$\frac{1}{2}$
2	0	1
3	5	$\frac{3}{2}$

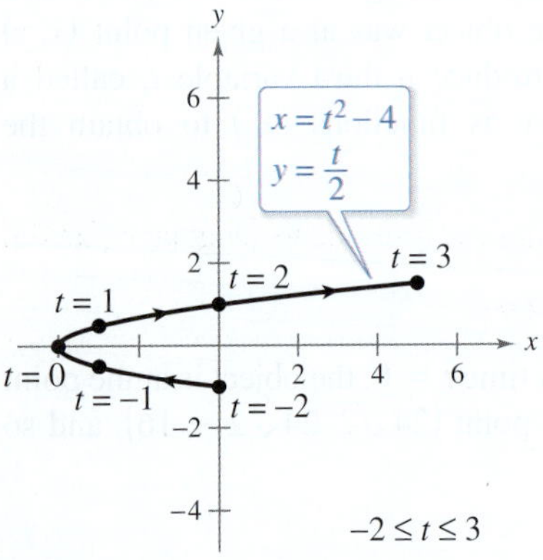

Figure 10.40

By plotting these points in the order of increasing t, you obtain the curve shown in Figure 10.40. The arrows on the curve indicate its orientation as t increases from -2 to 3. So, when a particle moves on this curve, it starts at $(0, -1)$ and then moves along the curve to the point $\left(5, \frac{3}{2}\right)$.

✓ **Checkpoint** *Audio-video solution in English & Spanish at LarsonPrecalculus.com.*

Sketch the curve given by the parametric equations

$$x = 2t \quad \text{and} \quad y = 4t^2 + 2, \quad -2 \le t \le 2.$$

Note that the graph shown in Figure 10.40 does not define y as a function of x. This points out one benefit of parametric equations—they can be used to represent graphs that are more general than graphs of functions.

Two different sets of parametric equations can have the same graph. For example, the set of parametric equations

$$x = 4t^2 - 4 \text{ and } y = t, \quad -1 \le t \le \frac{3}{2}$$

has the same graph as the set of parametric equations given in Example 1. However, by comparing the values of t in Figures 10.40 and 10.41, you can see that the second graph is traced out more *rapidly* (considering t as time) than the first graph. So, in applications, different parametric representations can be used to represent various *speeds* at which objects travel along a given path.

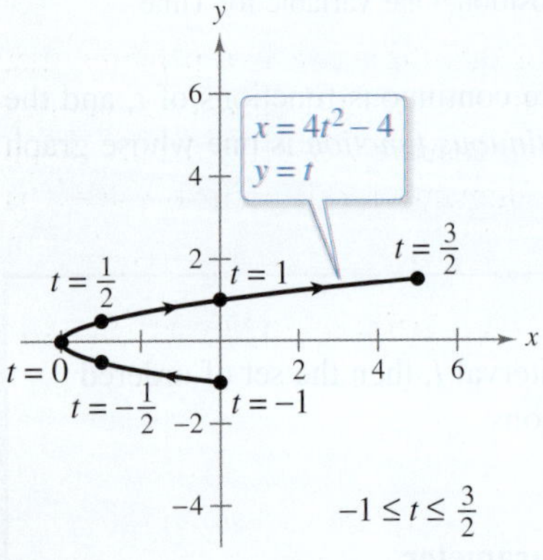

Figure 10.41

Eliminating the Parameter

Example 1 uses simple point plotting to sketch the curve. This tedious process can sometimes be simplified by finding a rectangular equation (in x and y) that has the same graph. This process is called **eliminating the parameter.**

Parametric equations	Solve for t in one equation.	Substitute into other equation.	Rectangular equation
$x = t^2 - 4$, $y = \frac{t}{2}$	$t = 2y$	$x = (2y)^2 - 4$	$x = 4y^2 - 4$

Now you can recognize that the equation $x = 4y^2 - 4$ represents a parabola with a horizontal axis and vertex at $(-4, 0)$.

When converting equations from parametric to rectangular form, you may need to alter the domain of the rectangular equation so that its graph matches the graph of the parametric equations. Such a situation is demonstrated in Example 2.

EXAMPLE 2 Eliminating the Parameter

Sketch the curve represented by the equations

$$x = \frac{1}{\sqrt{t + 1}} \quad \text{and} \quad y = \frac{t}{t + 1}$$

by eliminating the parameter and adjusting the domain of the resulting rectangular equation.

Solution Solving for t in the equation for x produces

$$x = \frac{1}{\sqrt{t + 1}} \quad \Longrightarrow \quad x^2 = \frac{1}{t + 1}$$

which implies that

$$t = \frac{1 - x^2}{x^2}.$$

Now, substituting in the equation for y, you obtain the rectangular equation

$$y = \frac{t}{t + 1} = \frac{\frac{1 - x^2}{x^2}}{\frac{1 - x^2}{x^2} + 1} = \frac{\frac{1 - x^2}{x^2}}{\frac{1 - x^2}{x^2} + 1} \cdot \frac{x^2}{x^2} = 1 - x^2.$$

From this rectangular equation, you can recognize that the curve is a parabola that opens downward and has its vertex at $(0, 1)$. Also, this rectangular equation is defined for all values of x. The parametric equation for x, however, is defined only when $t > -1$. This implies that you should restrict the domain of x to positive values, as shown in Figure 10.42.

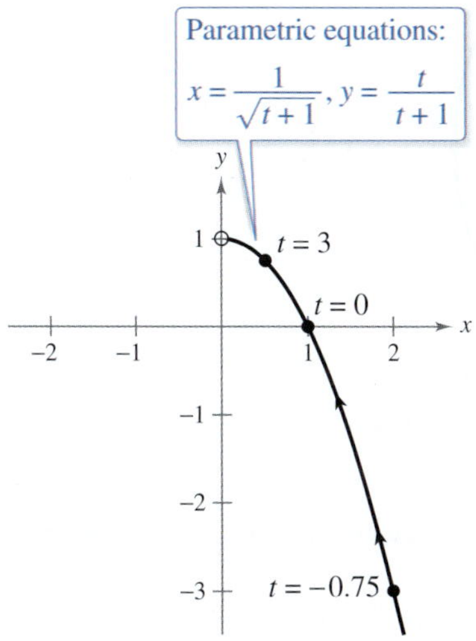

Figure 10.42

Checkpoint Audio-video solution in English & Spanish at LarsonPrecalculus.com.

Sketch the curve represented by the equations

$$x = \frac{1}{\sqrt{t - 1}} \quad \text{and} \quad y = \frac{t + 1}{t - 1}$$

by eliminating the parameter and adjusting the domain of the resulting rectangular equation.

It is not necessary for the parameter in a set of parametric equations to represent time. The next example uses an *angle* as the parameter.

REMARK To eliminate the parameter in equations involving trigonometric functions, try using identities such as

$$\sin^2 \theta + \cos^2 \theta = 1$$

or

$$\sec^2 \theta - \tan^2 \theta = 1$$

as shown in Example 3.

EXAMPLE 3 Eliminating an Angle Parameter

Sketch the curve represented by

$$x = 3 \cos \theta \quad \text{and} \quad y = 4 \sin \theta, \quad 0 \le \theta < 2\pi$$

by eliminating the parameter.

Solution Begin by solving for $\cos \theta$ and $\sin \theta$ in the equations.

$$\cos \theta = \frac{x}{3} \quad \text{and} \quad \sin \theta = \frac{y}{4}$$ Solve for $\cos \theta$ and $\sin \theta$.

Use the identity $\sin^2 \theta + \cos^2 \theta = 1$ to form an equation involving only x and y.

$$\cos^2 \theta + \sin^2 \theta = 1$$ Pythagorean identity

$$\left(\frac{x}{3}\right)^2 + \left(\frac{y}{4}\right)^2 = 1$$ Substitute $\frac{x}{3}$ for $\cos \theta$ and $\frac{y}{4}$ for $\sin \theta$.

$$\frac{x^2}{9} + \frac{y^2}{16} = 1$$ Rectangular equation

From this rectangular equation, you can see that the graph is an ellipse centered at $(0, 0)$, with vertices $(0, 4)$ and $(0, -4)$ and minor axis of length $2b = 6$, as shown below. Note that the elliptic curve is traced out *counterclockwise* as θ increases on the interval $[0, 2\pi)$.

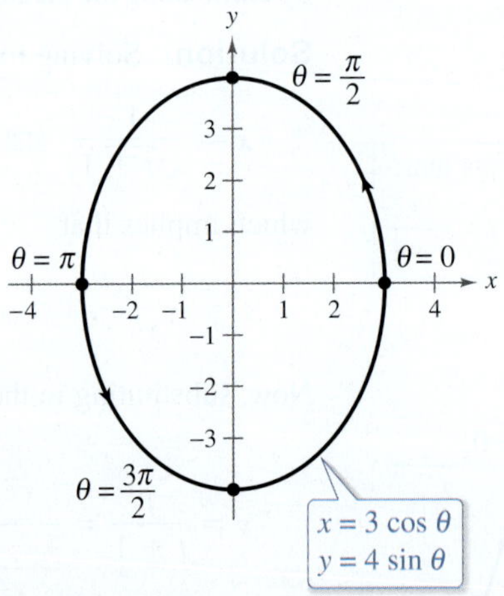

✓ **Checkpoint** *Audio-video solution in English & Spanish at LarsonPrecalculus.com.*

Sketch the curve represented by

$$x = 5 \cos \theta \quad \text{and} \quad y = 3 \sin \theta, \quad 0 \le \theta < 2\pi$$

by eliminating the parameter.

In Examples 2 and 3, it is important to realize that eliminating the parameter is primarily an *aid to curve sketching*. When the parametric equations represent the path of a moving object, the graph alone is not sufficient to describe the object's motion. You still need the parametric equations to tell you the *position, direction,* and *speed* at a given time.

Finding Parametric Equations for a Graph

You have been studying techniques for sketching the graph represented by a set of parametric equations. Now consider the *reverse* problem—that is, how can you find a set of parametric equations for a given graph or a given physical description? From the discussion following Example 1, you know that such a representation is not unique. That is, the equations

$$x = 4t^2 - 4 \quad \text{and} \quad y = t, \quad -1 \le t \le \frac{3}{2}$$

produced the same graph as the equations

$$x = t^2 - 4 \quad \text{and} \quad y = \frac{t}{2}, \quad -2 \le t \le 3.$$

This is further demonstrated in Example 4.

EXAMPLE 4 Finding Parametric Equations for a Graph

Find a set of parametric equations to represent the graph of $y = 1 - x^2$, using the following parameters.

a. $t = x$ **b.** $t = 1 - x$

Solution

a. Letting $t = x$, you obtain the parametric equations

$$x = t \quad \text{and} \quad y = 1 - x^2 = 1 - t^2.$$

The curve represented by the parametric equations is shown in Figure 10.43.

b. Letting $t = 1 - x$, you obtain the parametric equations

$$x = 1 - t \quad \text{and} \quad y = 1 - x^2 = 1 - (1 - t)^2 = 2t - t^2.$$

The curve represented by the parametric equations is shown in Figure 10.44. Note that the graphs in Figures 10.43 and 10.44 have opposite orientations.

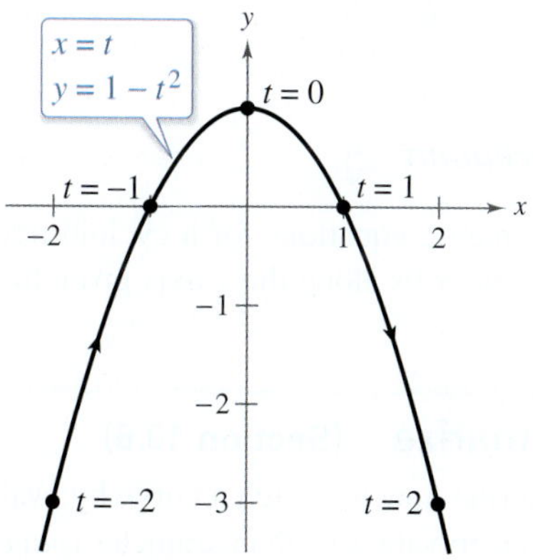

Figure 10.43

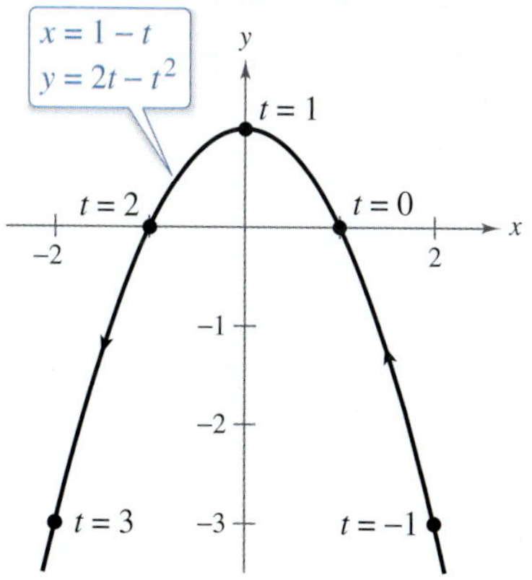

Figure 10.44

Checkpoint Audio-video solution in English & Spanish at LarsonPrecalculus.com.

Find a set of parametric equations to represent the graph of $y = x^2 + 2$, using the following parameters.

a. $t = x$ **b.** $t = 2 - x$

A **cycloid** is a curve traced out by a point P on a circle as the circle rolls along a straight line in a plane.

EXAMPLE 5 Parametric Equations for a Cycloid

Write parametric equations for a cycloid traced out by a point P on a circle of radius a as the circle rolls along the x-axis given that P is at a minimum when $x = 0$.

Solution As the parameter, let θ be the measure of the circle's rotation, and let the point $P(x, y)$ begin at the origin. When $\theta = 0$, P is at the origin; when $\theta = \pi$, P is at a maximum point $(\pi a, 2a)$; and when $\theta = 2\pi$, P is back on the x-axis at $(2\pi a, 0)$. From the figure below, you can see that $\angle APC = \pi - \theta$. So, you have

$$\sin\theta = \sin(\pi - \theta) = \sin(\angle APC) = \frac{AC}{a} = \frac{BD}{a}$$

$$\cos\theta = -\cos(\pi - \theta) = -\cos(\angle APC) = -\frac{AP}{a}$$

REMARK In Example 5, $\overset{\frown}{PD}$ represents the arc of the circle between points P and D.

which implies that $BD = a\sin\theta$ and $AP = -a\cos\theta$. Because the circle rolls along the x-axis, you know that $OD = \overset{\frown}{PD} = a\theta$. Furthermore, because $BA = DC = a$, you have

$$x = OD - BD = a\theta - a\sin\theta$$

and

$$y = BA + AP = a - a\cos\theta.$$

So, the parametric equations are $x = a(\theta - \sin\theta)$ and $y = a(1 - \cos\theta)$.

TECHNOLOGY You can use a graphing utility in *parametric* mode to obtain a graph similar to the one in Example 5 by graphing the following equations.

$X_{1T} = T - \sin T$

$Y_{1T} = 1 - \cos T$

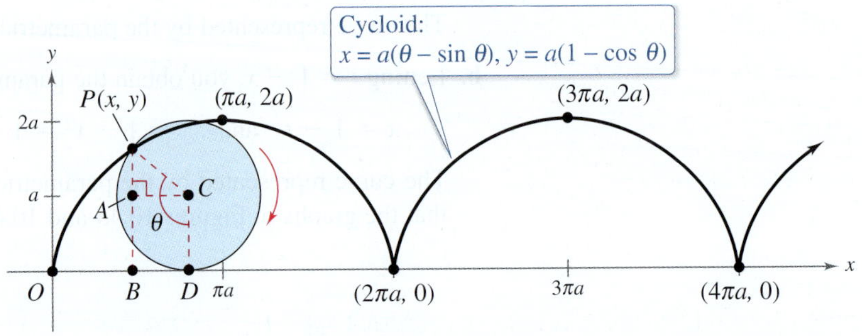

Checkpoint Audio-video solution in English & Spanish at LarsonPrecalculus.com.

Write parametric equations for a cycloid traced out by a point P on a circle of radius a as the circle rolls along the x-axis given that P is at a maximum when $x = 0$.

Summarize (Section 10.6)

1. Describe how to sketch a curve by evaluating a set of parametric equations for given values of the parameter *(page 732)*. For an example of sketching a curve given by parametric equations, see Example 1.
2. Describe the process of eliminating the parameter *(page 733)*. For examples of sketching curves by eliminating the parameter, see Examples 2 and 3.
3. Describe how to find a set of parametric equations for a graph *(page 735)*. For examples of finding sets of parametric equations for graphs, see Examples 4 and 5.

10.6 Exercises

See CalcChat.com for tutorial help and worked-out solutions to odd-numbered exercises.

Vocabulary: Fill in the blanks.

1. If f and g are continuous functions of t on an interval I, then the set of ordered pairs $(f(t), g(t))$ is a ________ ________ C.
2. The ________ of a curve is the direction in which the curve is traced out for increasing values of the parameter.
3. The process of converting a set of parametric equations to a corresponding rectangular equation is called ________ the ________.
4. A curve traced out by a point on the circumference of a circle as the circle rolls along a straight line in a plane is called a ________.

Skills and Applications

5. **Sketching a Curve** Consider the parametric equations $x = \sqrt{t}$ and $y = 3 - t$.
   (a) Create a table of x- and y-values using $t = 0, 1, 2, 3$, and 4.
   (b) Plot the points (x, y) generated in part (a), and sketch a graph of the parametric equations.
   (c) Find the rectangular equation by eliminating the parameter. Sketch its graph. How do the graphs differ?
6. **Sketching a Curve** Consider the parametric equations $x = 4\cos^2\theta$ and $y = 2\sin\theta$.
   (a) Create a table of x- and y-values using $\theta = -\pi/2, -\pi/4, 0, \pi/4$, and $\pi/2$.
   (b) Plot the points (x, y) generated in part (a), and sketch a graph of the parametric equations.
   (c) Find the rectangular equation by eliminating the parameter. Sketch its graph. How do the graphs differ?

Sketching a Curve In Exercises 7–34, (a) sketch the curve represented by the parametric equations (indicate the orientation of the curve) and (b) eliminate the parameter and write the resulting rectangular equation whose graph represents the curve. Adjust the domain of the rectangular equation, if necessary.

7. $x = t$, $y = -4t$
8. $x = t$, $y = \frac{1}{2}t$
9. $x = t - 1$, $y = 3t + 1$
10. $x = 3 - 2t$, $y = 2 + 3t$
11. $x = \frac{1}{4}t$, $y = t^2$
12. $x = t$, $y = t^3$
13. $x = t + 2$, $y = t^2$
14. $x = t - 3$, $y = t^2$
15. $x = \sqrt{t}$, $y = 1 - t$
16. $x = \sqrt{t + 2}$, $y = t - 1$
17. $x = t^3$, $y = t^2$
18. $x = t^3$, $y = t^4$
19. $x = t + 1$, $y = \dfrac{t}{t + 1}$
20. $x = t - 1$, $y = \dfrac{t}{t - 1}$
21. $x = 2(t + 1)$, $y = |t - 2|$
22. $x = |t - 1|$, $y = t + 2$
23. $x = 4\cos\theta$, $y = 2\sin\theta$
24. $x = 2\cos\theta$, $y = 3\sin\theta$
25. $x = 6\sin 2\theta$, $y = 6\cos 2\theta$
26. $x = \cos\theta$, $y = 2\sin 2\theta$
27. $x = 1 + \cos\theta$, $y = 1 + 2\sin\theta$
28. $x = 2 + 5\cos\theta$, $y = -6 + 4\sin\theta$
29. $x = e^t$, $y = e^{2t}$
30. $x = e^t$, $y = e^{3t}$
31. $x = e^{-t}$, $y = e^{3t}$
32. $x = e^{2t}$, $y = e^t$
33. $x = t^3$, $y = 3\ln t$
34. $x = \ln 2t$, $y = 2t^2$

Graphing a Curve In Exercises 35–46, use a graphing utility to graph the curve represented by the parametric equations.

35. $x = t$, $y = t^2$
36. $x = t + 4$, $y = t^2$
37. $x = t$, $y = \sqrt{t}$
38. $x = t + 1$, $y = \sqrt{2 - t}$
39. $x = 2t$, $y = |t + 1|$
40. $x = |t + 2|$, $y = 3 - t$
41. $x = 4 + 3\cos\theta$, $y = -2 + \sin\theta$
42. $x = 4 + 3\cos\theta$, $y = -2 + 2\sin\theta$
43. $x = 4\sec\theta$, $y = 2\tan\theta$
44. $x = \sec\theta$, $y = \tan\theta$
45. $x = \frac{1}{2}t$, $y = \ln(t^2 + 1)$
46. $x = 10 - 0.01e^t$, $y = 0.4t^2$

Comparing Plane Curves **In Exercises 47 and 48, determine how the plane curves differ from each other.**

47. (a) $x = t$, $y = 2t + 1$ (b) $x = \cos\theta$, $y = 2\cos\theta + 1$
(c) $x = e^{-t}$, $y = 2e^{-t} + 1$ (d) $x = e^{t}$, $y = 2e^{t} + 1$

48. (a) $x = t$, $y = t^2 - 1$ (b) $x = t^2$, $y = t^4 - 1$
(c) $x = \sin t$, $y = \sin^2 t - 1$ (d) $x = e^{t}$, $y = e^{2t} - 1$

Eliminating the Parameter **In Exercises 49–52, eliminate the parameter and obtain the standard form of the rectangular equation.**

49. Line passing through (x_1, y_1) and (x_2, y_2):
$$x = x_1 + t(x_2 - x_1),\ y = y_1 + t(y_2 - y_1)$$

50. Circle: $x = h + r\cos\theta,\ y = k + r\sin\theta$

51. Ellipse: $x = h + a\cos\theta,\ y = k + b\sin\theta$

52. Hyperbola: $x = h + a\sec\theta,\ y = k + b\tan\theta$

Finding Parametric Equations for a Graph **In Exercises 53–60, use the results of Exercises 49–52 to find a set of parametric equations to represent the graph of the line or conic.**

53. Line: passes through $(0, 0)$ and $(3, 6)$

54. Line: passes through $(3, 2)$ and $(-6, 3)$

55. Circle: center: $(3, 2)$; radius: 4

56. Circle: center: $(5, -3)$; radius: 4

57. Ellipse: vertices: $(\pm 5, 0)$; foci: $(\pm 4, 0)$

58. Ellipse: vertices: $(3, 7), (3, -1)$; foci: $(3, 5), (3, 1)$

59. Hyperbola: vertices: $(\pm 4, 0)$; foci: $(\pm 5, 0)$

60. Hyperbola: vertices: $(\pm 2, 0)$; foci: $(\pm 4, 0)$

Finding Parametric Equations for a Graph **In Exercises 61–76, find a set of parametric equations to represent the graph of the rectangular equation using (a) $t = x$ and (b) $t = 2 - x$.**

61. $y = 3x - 2$ 62. $y = 5x - 3$

63. $y = 2 - x$ 64. $y = 4 - 7x$

65. $x = 2y + 1$ 66. $x = 3y - 2$

67. $y = x^2 + 1$ 68. $y = x^2 - 3$

69. $y = 3x^2 + 1$ 70. $y = 6x^2 - 5$

71. $y = 1 - 2x^2$ 72. $y = 2 - 5x^2$

73. $y = \dfrac{1}{x}$ 74. $y = \dfrac{1}{2x}$

75. $y = e^x$ 76. $y = e^{2x}$

Graphing a Curve **In Exercises 77–86, use a graphing utility to graph the curve represented by the parametric equations.**

77. Cycloid: $x = 4(\theta - \sin\theta),\ y = 4(1 - \cos\theta)$

78. Cycloid: $x = \theta + \sin\theta,\ y = 1 - \cos\theta$

79. Prolate cycloid: $x = \theta - \frac{3}{2}\sin\theta,\ y = 1 - \frac{3}{2}\cos\theta$

80. Prolate cycloid: $x = 2\theta - 4\sin\theta,\ y = 2 - 4\cos\theta$

81. Epicycloid: $x = 8\cos\theta - 2\cos 4\theta$
$y = 8\sin\theta - 2\sin 4\theta$

82. Epicycloid: $x = 15\cos\theta - 3\cos 5\theta$
$y = 15\sin\theta - 3\sin 5\theta$

83. Hypocycloid: $x = 3\cos^3\theta,\ y = 3\sin^3\theta$

84. Curtate cycloid: $x = 8\theta - 4\sin\theta,\ y = 8 - 4\cos\theta$

85. Witch of Agnesi: $x = 2\cot\theta,\ y = 2\sin^2\theta$

86. Folium of Descartes: $x = \dfrac{3t}{1 + t^3},\ y = \dfrac{3t^2}{1 + t^3}$

Matching **In Exercises 87–92, match the parametric equations with the correct graph and describe the domain and range. [The graphs are labeled (a)–(f).]**

(a)
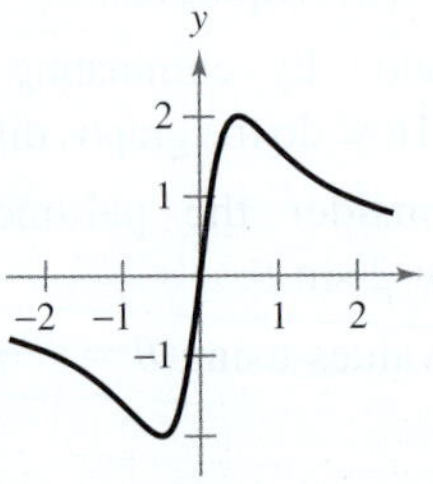

(b)
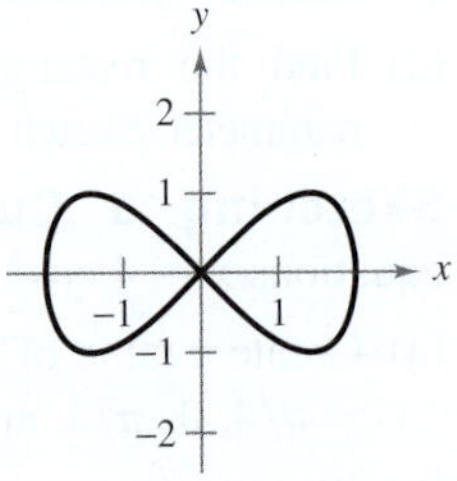

(c)
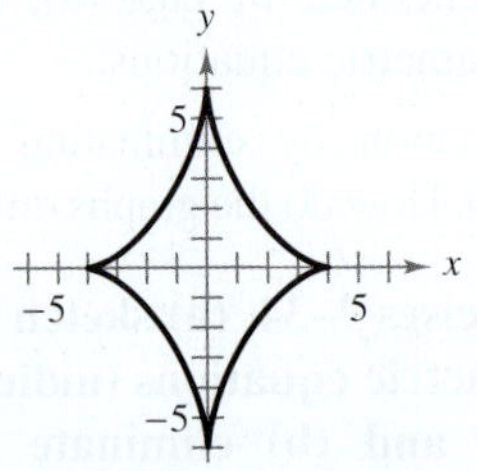

(d)
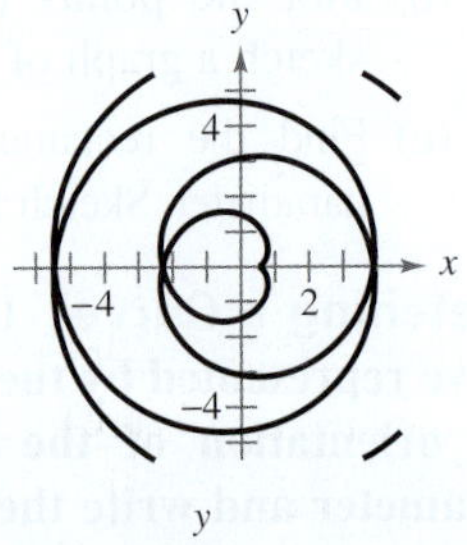

(e)
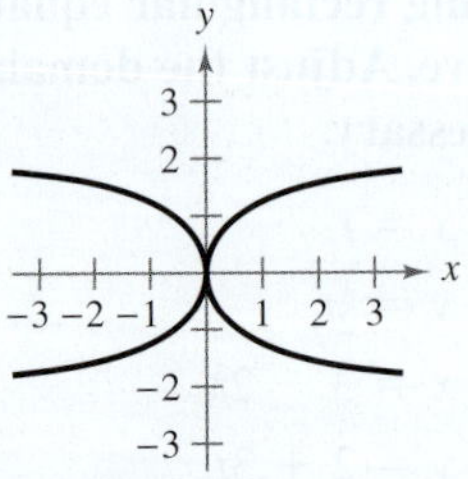

(f)
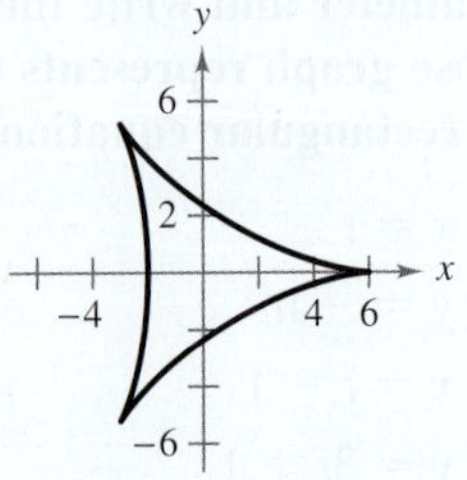

87. Lissajous curve: $x = 2\cos\theta,\ y = \sin 2\theta$

88. Evolute of ellipse: $x = 4\cos^3\theta,\ y = 6\sin^3\theta$

89. Involute of circle: $x = \frac{1}{2}(\cos\theta + \theta\sin\theta)$
$y = \frac{1}{2}(\sin\theta - \theta\cos\theta)$

90. Serpentine curve: $x = \frac{1}{2}\cot\theta,\ y = 4\sin\theta\cos\theta$

91. Tricuspoid: $x = 4\cos\theta + 2\cos 2\theta$
$y = 4\sin\theta - 2\sin 2\theta$

92. Kappa curve: $x = 2\cos\theta\cot\theta,\ y = 2\cos\theta$

Projectile Motion **A projectile is launched at a height of h feet above the ground at an angle of θ with the horizontal. The initial velocity is v_0 feet per second, and the path of the projectile is modeled by the parametric equations**

$$x = (v_0 \cos \theta)t$$

and

$$y = h + (v_0 \sin \theta)t - 16t^2.$$

In Exercises 93 and 94, use a graphing utility to graph the paths of a projectile launched from ground level at each value of θ and v_0. For each case, use the graph to approximate the maximum height and the range of the projectile.

93. (a) $\theta = 60°$, $v_0 = 88$ feet per second

(b) $\theta = 60°$, $v_0 = 132$ feet per second

(c) $\theta = 45°$, $v_0 = 88$ feet per second

(d) $\theta = 45°$, $v_0 = 132$ feet per second

94. (a) $\theta = 15°$, $v_0 = 50$ feet per second

(b) $\theta = 15°$, $v_0 = 120$ feet per second

(c) $\theta = 10°$, $v_0 = 50$ feet per second

(d) $\theta = 10°$, $v_0 = 120$ feet per second

95. Path of a Baseball The center field fence in Yankee Stadium is 7 feet high and 408 feet from home plate. A baseball is hit at a point 3 feet above the ground. It leaves the bat at an angle of θ degrees with the horizontal at a speed of 100 miles per hour (see figure).

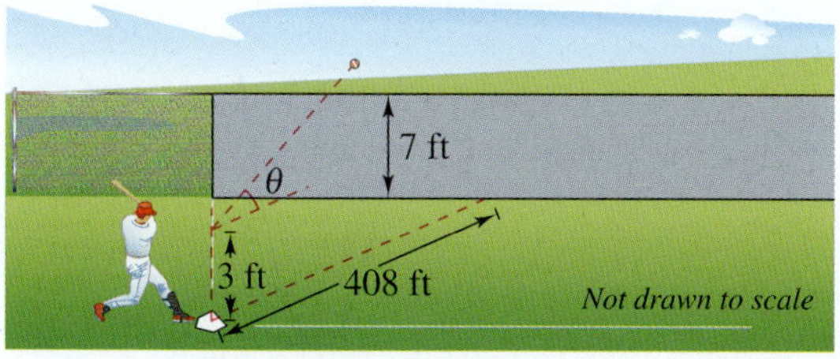

(a) Write a set of parametric equations that model the path of the baseball. (See Exercises 93 and 94.)

(b) Use a graphing utility to graph the path of the baseball when $\theta = 15°$. Is the hit a home run?

(c) Use the graphing utility to graph the path of the baseball when $\theta = 23°$. Is the hit a home run?

(d) Find the minimum angle required for the hit to be a home run.

96. Path of an Arrow An archer releases an arrow from a bow at a point 5 feet above the ground. The arrow leaves the bow at an angle of 15° with the horizontal and at an initial speed of 225 feet per second.

(a) Write a set of parametric equations that model the path of the arrow. (See Exercises 93 and 94.)

(b) Assuming the ground is level, find the distance the arrow travels before it hits the ground. (Ignore air resistance.)

(c) Use a graphing utility to graph the path of the arrow and approximate its maximum height.

(d) Find the total time the arrow is in the air.

97. Path of a Football A quarterback releases a pass at a height of 7 feet above the playing field, and a receiver catches the football at a height of 4 feet, 30 yards directly downfield. The pass is released at an angle of 35° with the horizontal.

(a) Write a set of parametric equations for the path of the football. (See Exercises 93 and 94.)

(b) Find the speed of the football when it is released.

(c) Use a graphing utility to graph the path of the football and approximate its maximum height.

(d) Find the time the receiver has to position himself after the quarterback releases the football.

98. Projectile Motion Eliminate the parameter t from the parametric equations

$$x = (v_0 \cos \theta)t$$

and

$$y = h + (v_0 \sin \theta)t - 16t^2$$

for the motion of a projectile to show that the rectangular equation is

$$y = -\frac{16 \sec^2 \theta}{v_0^2}x^2 + (\tan \theta)x + h.$$

99. Path of a Projectile The path of a projectile is given by the rectangular equation

$$y = 7 + x - 0.02x^2.$$

(a) Use the result of Exercise 98 to find h, v_0, and θ. Find the parametric equations of the path.

(b) Use a graphing utility to graph the rectangular equation for the path of the projectile. Confirm your answer in part (a) by sketching the curve represented by the parametric equations.

(c) Use the graphing utility to approximate the maximum height of the projectile and its range.

100. Path of a Projectile Repeat Exercise 99 for a projectile with a path given by the rectangular equation

$$y = 6 + x - 0.08x^2.$$

Randy Faris/Corbis Super RF/Alamy

101. Curtate Cycloid A wheel of radius a units rolls along a straight line without slipping. The curve traced by a point P that is b units from the center $(b < a)$ is called a **curtate cycloid** (see figure). Use the angle θ shown in the figure to find a set of parametric equations for the curve.

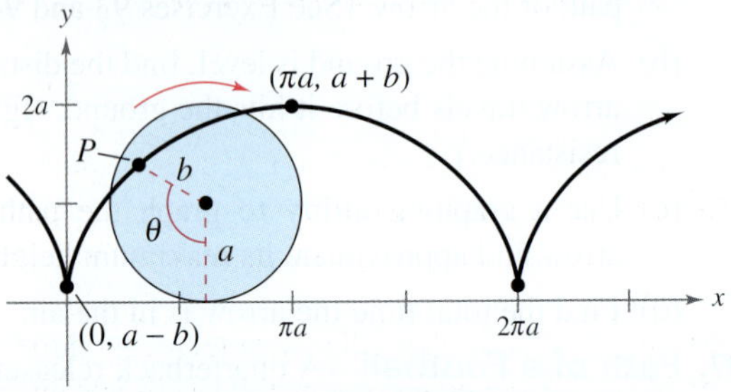

102. Epicycloid A circle of radius one unit rolls around the outside of a circle of radius two units without slipping. The curve traced by a point on the circumference of the smaller circle is called an **epicycloid** (see figure). Use the angle θ shown in the figure to find a set of parametric equations for the curve.

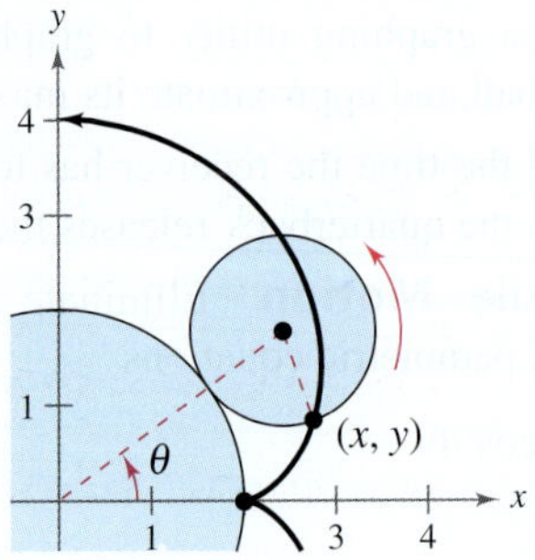

Exploration

True or False? **In Exercises 103–106, determine whether the statement is true or false. Justify your answer.**

103. The two sets of parametric equations

$$x = t,\ y = t^2 + 1 \quad \text{and} \quad x = 3t,\ y = 9t^2 + 1$$

correspond to the same rectangular equation.

104. Because the graphs of the parametric equations

$$x = t^2,\ y = t^2 \quad \text{and} \quad x = t,\ y = t$$

both represent the line $y = x$, they are the same plane curve.

105. If y is a function of t and x is a function of t, then y must be a function of x.

106. The parametric equations

$$x = at + h \quad \text{and} \quad y = bt + k$$

where $a \neq 0$ and $b \neq 0$, represent a circle centered at (h, k) when $a = b$.

107. Writing Write a short paragraph explaining why parametric equations are useful.

108. Writing Explain the process of sketching a plane curve given by parametric equations. What is meant by the orientation of the curve?

109. Think About It The graph of the parametric equations $x = t^3$ and $y = t - 1$ is shown below. Would the graph change for the parametric equations $x = (-t)^3$ and $y = -t - 1$? If so, how would it change?

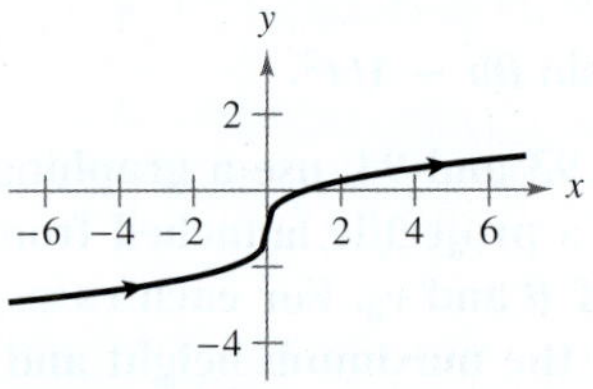

110. Think About It The graph of the parametric equations $x = t^2$ and $y = t + 1$ is shown below. Would the graph change for the parametric equations $x = (t + 1)^2$ and $y = t + 2$? If so, how would it change?

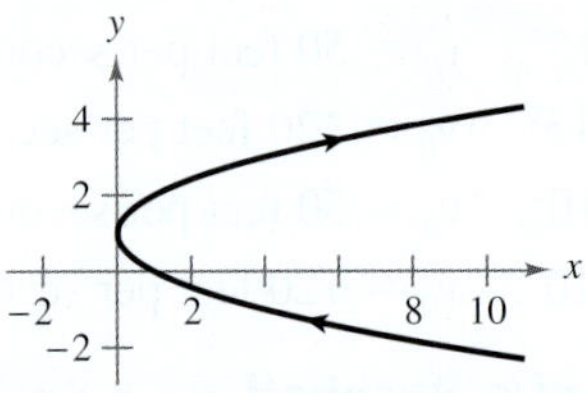

111. Think About It Use a graphing utility set in *parametric* mode to enter the parametric equations from Example 2. Over what values should you let t vary to obtain the graph shown in Figure 10.42?

112. HOW DO YOU SEE IT? The graph of the parametric equations $x = t$ and $y = t^2$ is shown below. Determine whether the graph would change for each set of parametric equations. If so, how would it change?

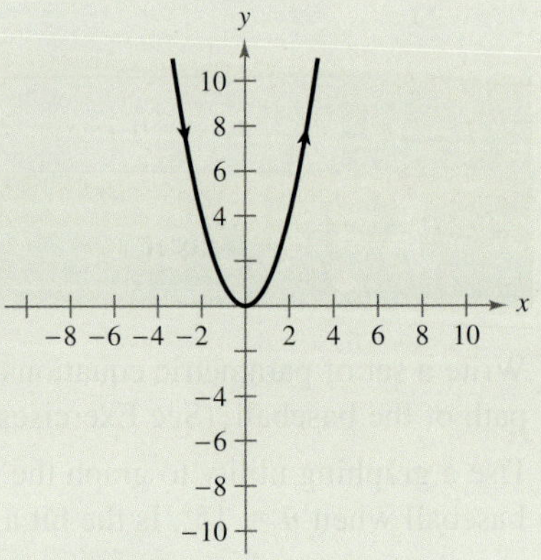

(a) $x = -t,\ y = t^2$

(b) $x = t + 1,\ y = t^2$

(c) $x = t,\ y = t^2 + 1$

10.7 Polar Coordinates

You can use polar coordinates in mathematical modeling. For instance, in Exercise 127 on page 746, you will use polar coordinates to model the path of a passenger car on a Ferris wheel.

- **Plot points in the polar coordinate system.**
- **Convert points from rectangular to polar form and vice versa.**
- **Convert equations from rectangular to polar form and vice versa.**

Introduction

So far, you have been representing graphs of equations as collections of points (x, y) in the rectangular coordinate system, where x and y represent the directed distances from the coordinate axes to the point (x, y). In this section, you will study a different system called the **polar coordinate system.**

To form the polar coordinate system in the plane, fix a point O, called the **pole** (or **origin**), and construct from O an initial ray called the **polar axis,** as shown below. Then each point P in the plane can be assigned **polar coordinates** (r, θ) as follows.

1. $r =$ *directed distance* from O to P

2. $\theta =$ *directed angle,* counterclockwise from polar axis to segment $\overline{OP}$

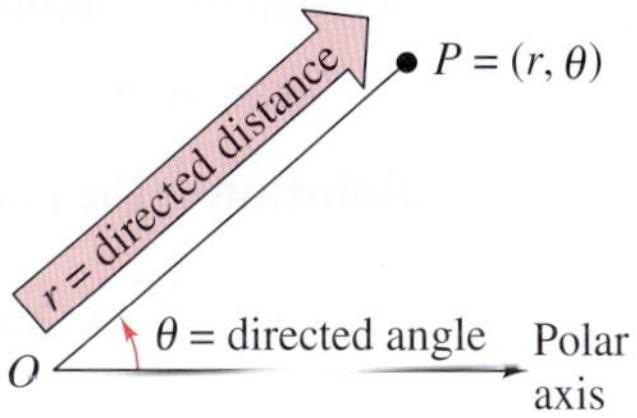

EXAMPLE 1 Plotting Points in the Polar Coordinate System

a. The point $(r, \theta) = (2, \pi/3)$ lies two units from the pole on the terminal side of the angle $\theta = \pi/3$, as shown in Figure 10.45.

b. The point $(r, \theta) = (3, -\pi/6)$ lies three units from the pole on the terminal side of the angle $\theta = -\pi/6$, as shown in Figure 10.46.

c. The point $(r, \theta) = (3, 11\pi/6)$ coincides with the point $(3, -\pi/6)$, as shown in Figure 10.47.

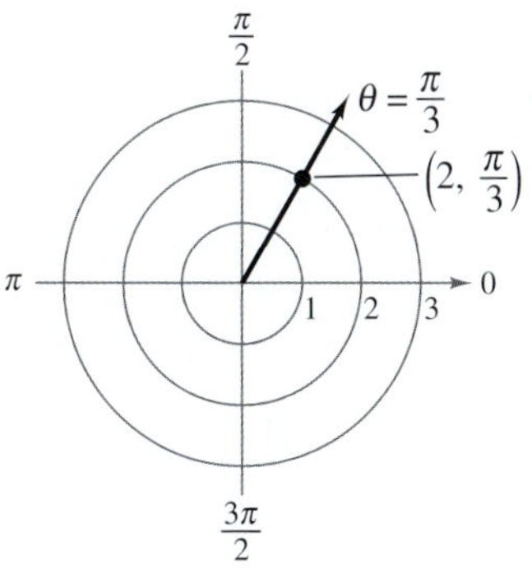

Figure 10.45

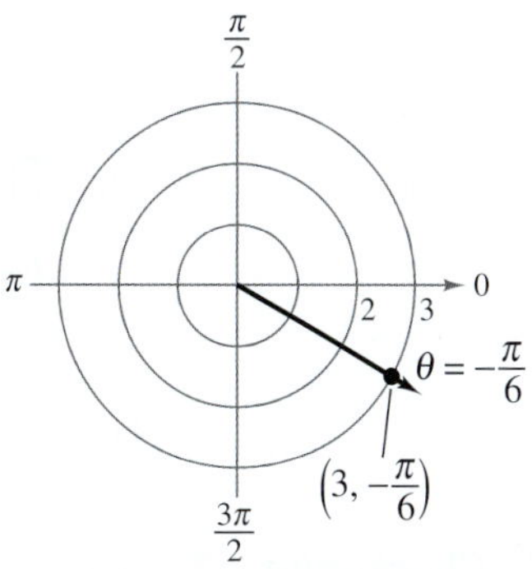

Figure 10.46

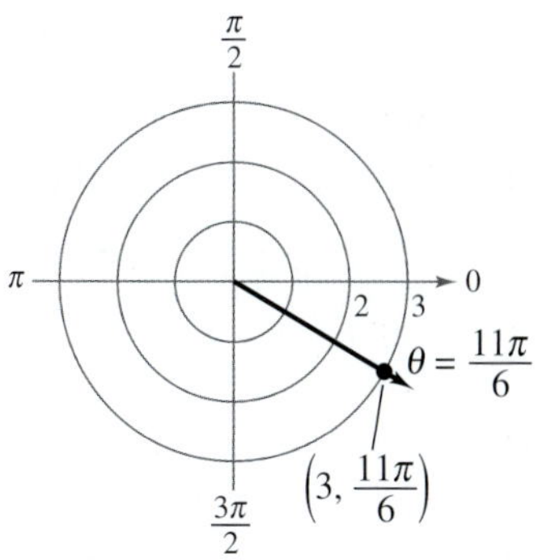

Figure 10.47

✓ ***Checkpoint*** *Audio-video solution in English & Spanish at LarsonPrecalculus.com.*

Plot each point given in polar coordinates.

a. $(3, \pi/4)$ **b.** $(2, -\pi/3)$ **c.** $(2, 5\pi/3)$

Pavzyuk Svitlana/Shutterstock.com

In rectangular coordinates, each point (x, y) has a unique representation. This is not true for polar coordinates. For instance, the coordinates

$$(r, \theta) \quad \text{and} \quad (r, \theta + 2\pi)$$

represent the same point, as illustrated in Example 1. Another way to obtain multiple representations of a point is to use negative values for r. Because r is a *directed distance,* the coordinates

$$(r, \theta) \quad \text{and} \quad (-r, \theta + \pi)$$

represent the same point. In general, the point (r, θ) can be represented as

$$(r, \theta) = (r, \theta \pm 2n\pi) \quad \text{or} \quad (r, \theta) = (-r, \theta \pm (2n + 1)\pi)$$

where n is any integer. Moreover, the pole is represented by $(0, \theta)$, where θ is any angle.

EXAMPLE 2 Multiple Representations of Points

Plot the point

$$\left(3, -\frac{3\pi}{4}\right)$$

and find three additional polar representations of this point, using

$$-2\pi < \theta < 2\pi.$$

Solution The point is shown below. Three other representations are as follows.

$$\left(3, -\frac{3\pi}{4} + 2\pi\right) = \left(3, \frac{5\pi}{4}\right) \qquad \text{Add } 2\pi \text{ to } \theta.$$

$$\left(-3, -\frac{3\pi}{4} - \pi\right) = \left(-3, -\frac{7\pi}{4}\right) \qquad \text{Replace } r \text{ by } -r\text{; subtract } \pi \text{ from } \theta.$$

$$\left(-3, -\frac{3\pi}{4} + \pi\right) = \left(-3, \frac{\pi}{4}\right) \qquad \text{Replace } r \text{ by } -r\text{; add } \pi \text{ to } \theta.$$

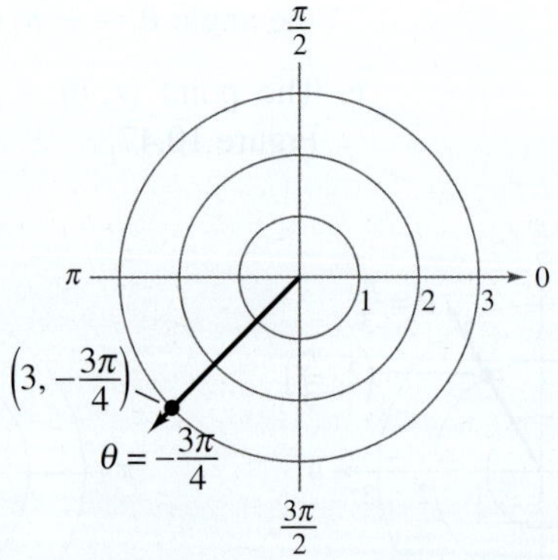

$$\left(3, -\frac{3\pi}{4}\right) = \left(3, \frac{5\pi}{4}\right) = \left(-3, -\frac{7\pi}{4}\right) = \left(-3, \frac{\pi}{4}\right) = \ldots$$

✓ ***Checkpoint*** *Audio-video solution in English & Spanish at LarsonPrecalculus.com.*

Plot each point and find three additional polar representations of the point, using $-2\pi < \theta < 2\pi$.

a. $\left(3, \frac{4\pi}{3}\right)$ **b.** $\left(2, -\frac{5\pi}{6}\right)$ **c.** $\left(-1, \frac{3\pi}{4}\right)$

Coordinate Conversion

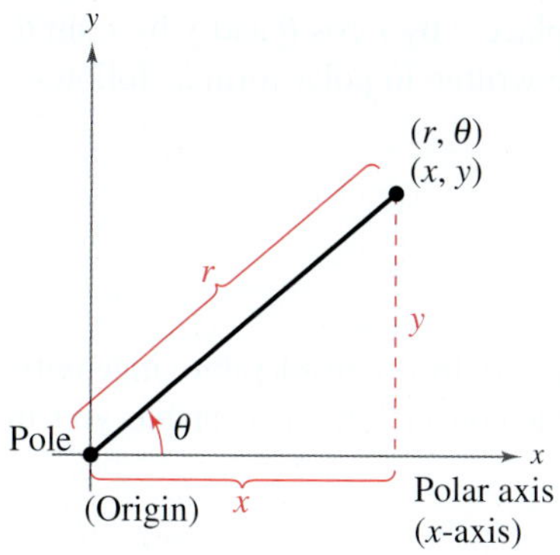

Figure 10.48

To establish the relationship between polar and rectangular coordinates, let the polar axis coincide with the positive x-axis and the pole with the origin, as shown in Figure 10.48. Because (x, y) lies on a circle of radius r, it follows that $r^2 = x^2 + y^2$. Moreover, for $r > 0$, the definitions of the trigonometric functions imply that

$$\tan\theta = \frac{y}{x}, \quad \cos\theta = \frac{x}{r}, \quad \text{and} \quad \sin\theta = \frac{y}{r}.$$

You can show that the same relationships hold for $r < 0$.

> **Coordinate Conversion**
>
> The polar coordinates (r, θ) are related to the rectangular coordinates (x, y) as follows.
>
> | **Polar-to-Rectangular** | **Rectangular-to-Polar** |
> |---|---|
> | $x = r\cos\theta$ | $\tan\theta = \dfrac{y}{x}$ |
> | $y = r\sin\theta$ | $r^2 = x^2 + y^2$ |

EXAMPLE 3 Polar-to-Rectangular Conversion

Convert $\left(\sqrt{3}, \dfrac{\pi}{6}\right)$ to rectangular coordinates.

Solution For the point $(r, \theta) = (\sqrt{3}, \pi/6)$, you have the following.

$$x = r\cos\theta = \sqrt{3}\cos\frac{\pi}{6} = \sqrt{3}\left(\frac{\sqrt{3}}{2}\right) = \frac{3}{2}$$

$$y = r\sin\theta = \sqrt{3}\sin\frac{\pi}{6} = \sqrt{3}\left(\frac{1}{2}\right) = \frac{\sqrt{3}}{2}$$

The rectangular coordinates are $(x, y) = \left(\dfrac{3}{2}, \dfrac{\sqrt{3}}{2}\right)$. (See Figure 10.49.)

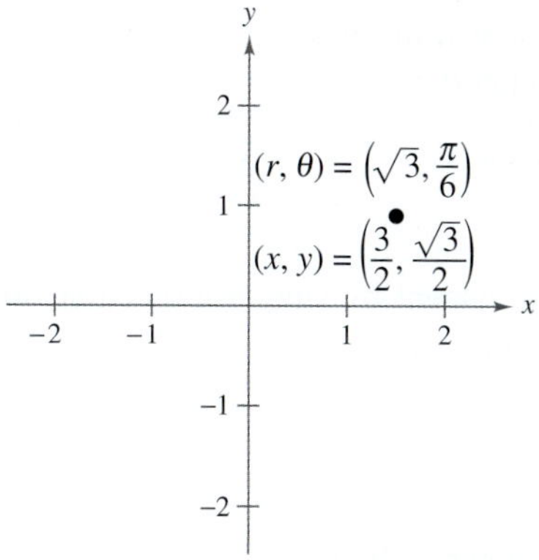

Figure 10.49

✓ Checkpoint Audio-video solution in English & Spanish at LarsonPrecalculus.com.

Convert $(2, \pi)$ to rectangular coordinates.

EXAMPLE 4 Rectangular-to-Polar Conversion

Convert $(-1, 1)$ to polar coordinates.

Solution For the second-quadrant point $(x, y) = (-1, 1)$, you have

$$\tan\theta = \frac{y}{x} = \frac{1}{-1} = -1 \quad \Rightarrow \quad \theta = \pi + \arctan(-1) = \frac{3\pi}{4}.$$

Because θ lies in the same quadrant as (x, y), use positive r.

$$r = \sqrt{x^2 + y^2} = \sqrt{(-1)^2 + (1)^2} = \sqrt{2}$$

So, *one* set of polar coordinates is $(r, \theta) = (\sqrt{2}, 3\pi/4)$, as shown in Figure 10.50.

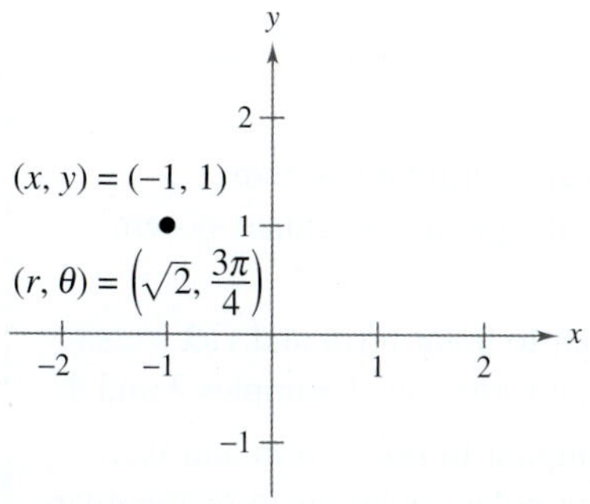

Figure 10.50

✓ Checkpoint Audio-video solution in English & Spanish at LarsonPrecalculus.com.

Convert $(0, 2)$ to polar coordinates.

Equation Conversion

To convert a rectangular equation to polar form, replace x by $r\cos\theta$ and y by $r\sin\theta$. For instance, the rectangular equation $y = x^2$ can be written in polar form as follows.

$$y = x^2 \qquad \text{Rectangular equation}$$

$$r\sin\theta = (r\cos\theta)^2 \qquad \text{Polar equation}$$

$$r = \sec\theta\tan\theta \qquad \text{Simplest form}$$

Converting a polar equation to rectangular form requires considerable ingenuity. Example 5 demonstrates several polar-to-rectangular conversions that enable you to sketch the graphs of some polar equations.

EXAMPLE 5 Converting Polar Equations to Rectangular Form

Convert each polar equation to a rectangular equation.

a. $r = 2$ **b.** $\theta = \pi/3$ **c.** $r = \sec\theta$

Solution

a. The graph of the polar equation $r = 2$ consists of all points that are two units from the pole. In other words, this graph is a circle centered at the origin with a radius of 2, as shown in Figure 10.51. You can confirm this by converting to rectangular form, using the relationship $r^2 = x^2 + y^2$.

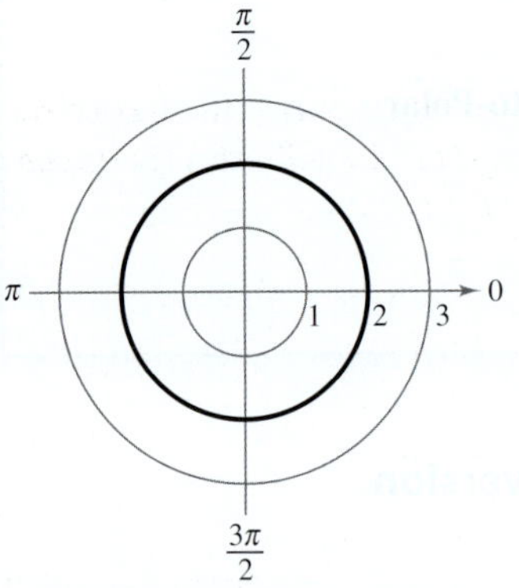

Figure 10.51

$$\underbrace{r = 2}_{\text{Polar equation}} \Rightarrow r^2 = 2^2 \Rightarrow \underbrace{x^2 + y^2 = 2^2}_{\text{Rectangular equation}}$$

b. The graph of the polar equation $\theta = \pi/3$ consists of all points on the line that makes an angle of $\pi/3$ with the positive polar axis, as shown in Figure 10.52. To convert to rectangular form, make use of the relationship $\tan\theta = y/x$.

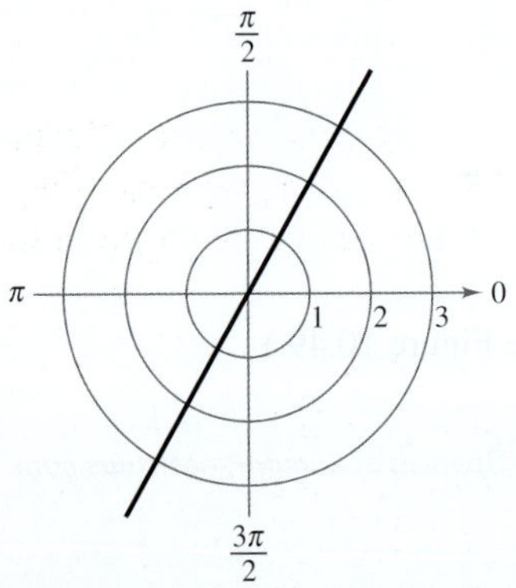

Figure 10.52

$$\underbrace{\theta = \pi/3}_{\text{Polar equation}} \Rightarrow \tan\theta = \sqrt{3} \Rightarrow \underbrace{y = \sqrt{3}x}_{\text{Rectangular equation}}$$

c. The graph of the polar equation $r = \sec\theta$ is not evident by simple inspection, so convert to rectangular form by using the relationship $r\cos\theta = x$.

$$\underbrace{r = \sec\theta}_{\text{Polar equation}} \Rightarrow r\cos\theta = 1 \Rightarrow \underbrace{x = 1}_{\text{Rectangular equation}}$$

Now you see that the graph is a vertical line, as shown in Figure 10.53.

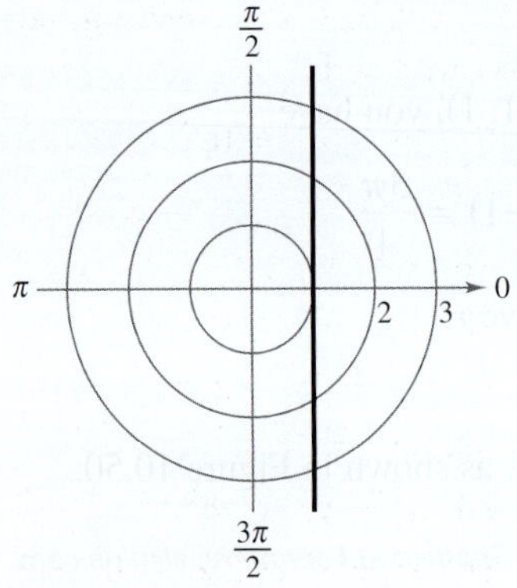

Figure 10.53

✓ **Checkpoint** *Audio-video solution in English & Spanish at LarsonPrecalculus.com.*

Convert $r = 6\sin\theta$ to a rectangular equation.

Summarize (Section 10.7)

1. Describe how to plot the point (r, θ) in the polar coordinate system *(page 741)*. For examples of plotting points in the polar coordinate system, see Examples 1 and 2.
2. Describe how to convert points from rectangular to polar form and vice versa *(page 743)*. For examples of converting between forms, see Examples 3 and 4.
3. Describe how to convert equations from rectangular to polar form and vice versa *(page 744)*. For an example of converting polar equations to rectangular form, see Example 5.

10.7 Exercises

See CalcChat.com for tutorial help and worked-out solutions to odd-numbered exercises.

Vocabulary: Fill in the blanks.

1. The origin of the polar coordinate system is called the ________.
2. For the point (r, θ), r is the ________ ________ from O to P and θ is the ________ ________, counterclockwise from the polar axis to the line segment $\overline{OP}$.
3. To plot the point (r, θ), use the ________ coordinate system.
4. The polar coordinates (r, θ) are related to the rectangular coordinates (x, y) as follows:

   $x =$ ________ $y =$ ________ $\tan\theta =$ ________ $r^2 =$ ________

Skills and Applications

Plotting Points in the Polar Coordinate System In Exercises 5–18, plot the point given in polar coordinates and find two additional polar representations of the point, using $-2\pi < \theta < 2\pi$.

5. $(2, 5\pi/6)$
6. $(3, 5\pi/4)$
7. $(4, -\pi/3)$
8. $(-1, -3\pi/4)$
9. $(2, 3\pi)$
10. $(4, 5\pi/2)$
11. $(-2, 2\pi/3)$
12. $(-3, 11\pi/6)$
13. $(0, -7\pi/6)$
14. $(0, -7\pi/2)$
15. $(\sqrt{2}, 2.36)$
16. $(2\sqrt{2}, 4.71)$
17. $(-3, -1.57)$
18. $(-5, -2.36)$

Polar-to-Rectangular Conversion In Exercises 19–34, a point in polar coordinates is given. Convert the point to rectangular coordinates.

19. $(0, \pi)$
20. $(0, -\pi)$
21. $(3, \pi/2)$
22. $(3, 3\pi/2)$
23. $(2, 3\pi/4)$
24. $(1, 5\pi/4)$
25. $(-1, 5\pi/4)$
26. $(-2, 7\pi/4)$
27. $(-2, 7\pi/6)$
28. $(-3, 5\pi/6)$
29. $(-3, -\pi/3)$
30. $(-2, -4\pi/3)$
31. $(2, 2.74)$
32. $(1.5, 3.67)$
33. $(-2.5, 1.1)$
34. $(-2, 5.76)$

Using a Graphing Utility to Find Rectangular Coordinates In Exercises 35–42, use a graphing utility to find the rectangular coordinates of the point given in polar coordinates. Round your results to two decimal places.

35. $(2, 2\pi/9)$
36. $(4, 11\pi/9)$
37. $(-4.5, 1.3)$
38. $(8.25, 3.5)$
39. $(2.5, 1.58)$
40. $(5.4, 2.85)$
41. $(-4.1, -0.5)$
42. $(8.2, -3.2)$

Rectangular-to-Polar Conversion In Exercises 43–60, a point in rectangular coordinates is given. Convert the point to polar coordinates.

43. $(1, 1)$
44. $(2, 2)$
45. $(-3, -3)$
46. $(-4, -4)$
47. $(-6, 0)$
48. $(3, 0)$
49. $(0, -5)$
50. $(0, 5)$
51. $(-3, 4)$
52. $(-4, -3)$
53. $(-\sqrt{3}, -\sqrt{3})$
54. $(-\sqrt{3}, \sqrt{3})$
55. $(\sqrt{3}, -1)$
56. $(-1, \sqrt{3})$
57. $(6, 9)$
58. $(6, 2)$
59. $(5, 12)$
60. $(7, 15)$

Using a Graphing Utility to Find Polar Coordinates In Exercises 61–70, use a graphing utility to find one set of polar coordinates of the point given in rectangular coordinates.

61. $(3, -2)$
62. $(-4, -2)$
63. $(-5, 2)$
64. $(7, -2)$
65. $(\sqrt{3}, 2)$
66. $(5, -\sqrt{2})$
67. $\left(\frac{5}{2}, \frac{4}{3}\right)$
68. $\left(\frac{9}{5}, \frac{11}{2}\right)$
69. $\left(\frac{7}{4}, \frac{3}{2}\right)$
70. $\left(-\frac{7}{9}, -\frac{3}{4}\right)$

Converting a Rectangular Equation to Polar Form In Exercises 71–90, convert the rectangular equation to polar form. Assume $a > 0$.

71. $x^2 + y^2 = 9$
72. $x^2 + y^2 = 16$
73. $y = x$
74. $y = -x$
75. $x = 10$
76. $x = a$
77. $y = 1$
78. $y = -2$
79. $3x - y + 2 = 0$
80. $3x + 5y - 2 = 0$
81. $xy = 16$
82. $2xy = 1$
83. $x^2 + y^2 = a^2$
84. $x^2 + y^2 = 9a^2$
85. $x^2 + y^2 - 2ax = 0$
86. $x^2 + y^2 - 2ay = 0$
87. $(x^2 + y^2)^2 = x^2 - y^2$
88. $(x^2 + y^2)^2 = 9(x^2 - y^2)$
89. $y^3 = x^2$
90. $y^2 = x^3$

Converting a Polar Equation to Rectangular Form In Exercises 91–116, convert the polar equation to rectangular form.

91. $r = 4 \sin \theta$
92. $r = 2 \cos \theta$
93. $r = -2 \cos \theta$
94. $r = -5 \sin \theta$
95. $\theta = 2\pi/3$
96. $\theta = 5\pi/3$
97. $\theta = 11\pi/6$
98. $\theta = 5\pi/6$
99. $r = 4$
100. $r = 10$
101. $r = 4 \csc \theta$
102. $r = 2 \csc \theta$
103. $r = -3 \sec \theta$
104. $r = -\sec \theta$
105. $r^2 = \cos \theta$
106. $r^2 = 2 \sin \theta$
107. $r^2 = \sin 2\theta$
108. $r^2 = \cos 2\theta$
109. $r = 2 \sin 3\theta$
110. $r = 3 \cos 2\theta$
111. $r = \dfrac{2}{1 + \sin \theta}$
112. $r = \dfrac{1}{1 - \cos \theta}$
113. $r = \dfrac{6}{2 - 3 \sin \theta}$
114. $r = \dfrac{5}{1 - 4 \cos \theta}$
115. $r = \dfrac{6}{2 \cos \theta - 3 \sin \theta}$
116. $r = \dfrac{5}{\sin \theta - 4 \cos \theta}$

Converting a Polar Equation to Rectangular Form In Exercises 117–126, convert the polar equation to rectangular form. Then sketch its graph.

117. $r = 6$
118. $r = 8$
119. $\theta = \pi/6$
120. $\theta = 3\pi/4$
121. $r = 2 \sin \theta$
122. $r = 4 \cos \theta$
123. $r = -6 \cos \theta$
124. $r = -3 \sin \theta$
125. $r = 3 \sec \theta$
126. $r = 2 \csc \theta$

127. **Ferris Wheel**
The center of a Ferris wheel lies at the pole of the polar coordinate system, where the distances are in feet. Passengers enter a car at $(30, -\pi/2)$. It takes 45 seconds for the wheel to complete one clockwise revolution.

(a) Write a polar equation that models the possible positions of a passenger car.

(b) Passengers enter a car. Find and interpret their coordinates after 15 seconds of rotation.

(c) Convert the point in part (b) to rectangular coordinates. Interpret the coordinates.

128. **Ferris Wheel** Repeat Exercise 127 when the distance from a passenger car to the center is 35 feet and it takes 60 seconds to complete one clockwise revolution.

Exploration

True or False? In Exercises 129–132, determine whether the statement is true or false. Justify your answer.

129. If $\theta_1 = \theta_2 + 2\pi n$ for some integer n, then (r, θ_1) and (r, θ_2) represent the same point in the polar coordinate system.

130. If (r_1, θ_1) and (r_2, θ_2) represent the same point in the polar coordinate system, then $\theta_1 = \theta_2 + 2\pi n$ for some integer n.

131. If $|r_1| = |r_2|$, then (r_1, θ) and (r_2, θ) represent the same point in the polar coordinate system.

132. If (r_1, θ_1) and (r_2, θ_2) represent the same point in the polar coordinate system, then $|r_1| = |r_2|$.

133. **Converting a Polar Equation to Rectangular Form** Convert the polar equation

$$r = 2(h \cos \theta + k \sin \theta)$$

to rectangular form and verify that it is the equation of a circle. Find the radius of the circle and the rectangular coordinates of the center of the circle.

134. **Converting a Polar Equation to Rectangular Form** Convert the polar equation $r = \cos \theta + 3 \sin \theta$ to rectangular form and identify the graph.

135. **Think About It**

(a) Show that the distance between the points (r_1, θ_1) and (r_2, θ_2) is $\sqrt{r_1^2 + r_2^2 - 2r_1 r_2 \cos(\theta_1 - \theta_2)}$.

(b) Simplify the Distance Formula for $\theta_1 = \theta_2$. Is the simplification what you expected? Explain.

(c) Simplify the Distance Formula for $\theta_1 - \theta_2 = 90°$. Is the simplification what you expected? Explain.

136. **HOW DO YOU SEE IT?** Use the polar coordinate system shown below.

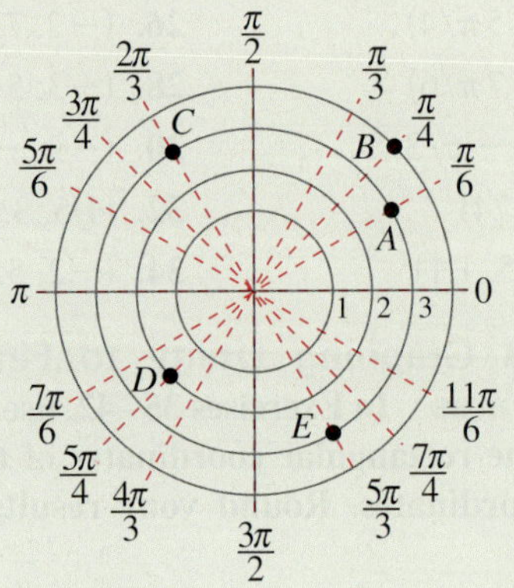

(a) Identify the polar coordinates of the points.

(b) Which points lie on the graph of $r = 3$?

(c) Which points lie on the graph of $\theta = \pi/4$?

Pavzyuk Svitlana/Shutterstock.com

10.8 Graphs of Polar Equations

You can use graphs of polar equations in mathematical modeling. For instance, in Exercise 69 on page 754, you will graph the pickup pattern of a microphone in a polar coordinate system.

- **Graph polar equations by point plotting.**
- **Use symmetry, zeros, and maximum *r*-values to sketch graphs of polar equations.**
- **Recognize special polar graphs.**

Introduction

In previous chapters, you learned how to sketch graphs in rectangular coordinate systems. You began with the basic point-plotting method. Then you used sketching aids such as symmetry, intercepts, asymptotes, periods, and shifts to further investigate the natures of graphs. This section approaches curve sketching in the polar coordinate system similarly, beginning with a demonstration of point plotting.

EXAMPLE 1 Graphing a Polar Equation by Point Plotting

Sketch the graph of the polar equation $r = 4 \sin \theta$.

Solution The sine function is periodic, so you can get a full range of r-values by considering values of θ in the interval $0 \le \theta \le 2\pi$, as shown in the following table.

θ	0	$\frac{\pi}{6}$	$\frac{\pi}{3}$	$\frac{\pi}{2}$	$\frac{2\pi}{3}$	$\frac{5\pi}{6}$	π	$\frac{7\pi}{6}$	$\frac{3\pi}{2}$	$\frac{11\pi}{6}$	2π
r	0	2	$2\sqrt{3}$	4	$2\sqrt{3}$	2	0	-2	-4	-2	0

By plotting these points, as shown in Figure 10.54, it appears that the graph is a circle of radius 2 whose center is at the point $(x, y) = (0, 2)$.

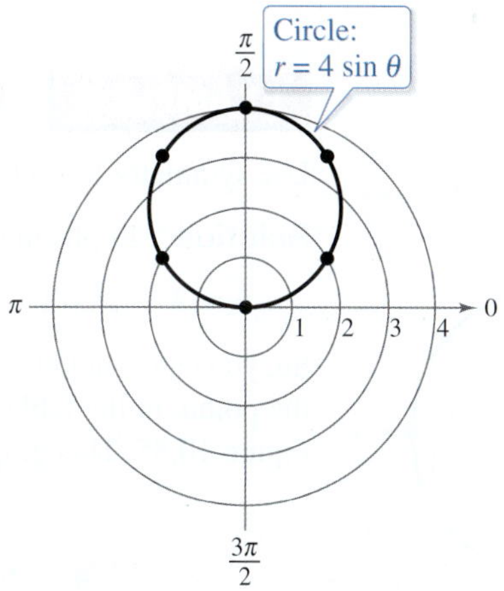

Figure 10.54

✓ ***Checkpoint*** *Audio-video solution in English & Spanish at LarsonPrecalculus.com.*

Sketch the graph of the polar equation $r = 6 \cos \theta$.

You can confirm the graph in Figure 10.54 by converting the polar equation to rectangular form and then sketching the graph of the rectangular equation. You can also use a graphing utility set to *polar* mode and graph the polar equation or set the graphing utility to *parametric* mode and graph a parametric representation.

nikkytok/Shutterstock.com

Symmetry, Zeros, and Maximum *r*-Values

In Figure 10.54 on the preceding page, note that as θ increases from 0 to 2π the graph is traced out twice. Moreover, note that the graph is *symmetric with respect to the line* $\theta = \pi/2$. Had you known about this symmetry and retracing ahead of time, you could have used fewer points. The three important types of symmetry to consider in polar curve sketching are shown below.

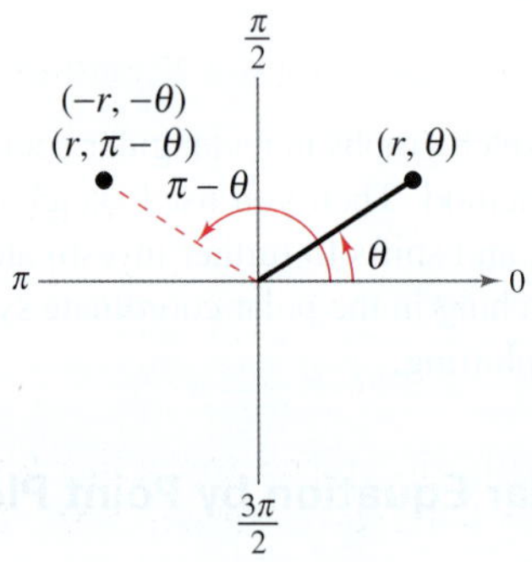

Symmetry with Respect to the Line $\theta = \dfrac{\pi}{2}$

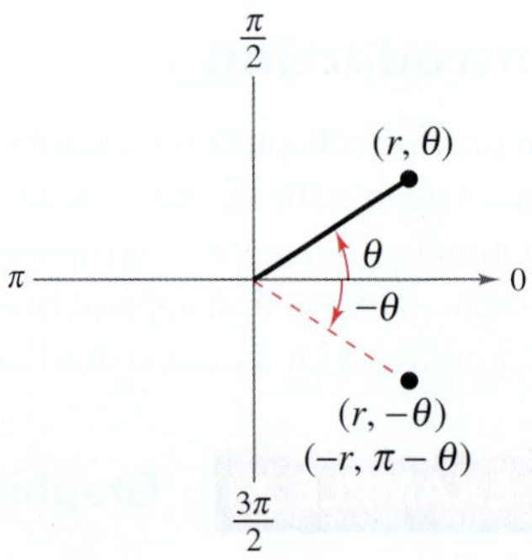

Symmetry with Respect to the Polar Axis

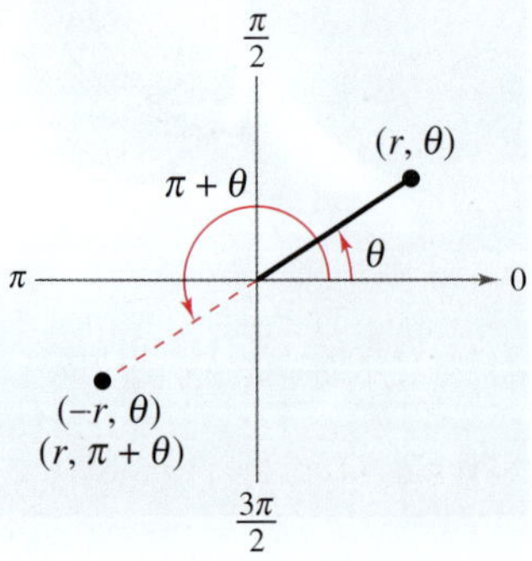

Symmetry with Respect to the Pole

Tests for Symmetry in Polar Coordinates

The graph of a polar equation is symmetric with respect to the following when the given substitution yields an equivalent equation.

1. The line $\theta = \pi/2$: Replace (r, θ) by $(r, \pi - \theta)$ or $(-r, -\theta)$.
2. The polar axis: Replace (r, θ) by $(r, -\theta)$ or $(-r, \pi - \theta)$.
3. The pole: Replace (r, θ) by $(r, \pi + \theta)$ or $(-r, \theta)$.

EXAMPLE 2 Using Symmetry to Sketch a Polar Graph

Use symmetry to sketch the graph of $r = 3 + 2\cos\theta$.

Solution Replacing (r, θ) by $(r, -\theta)$ produces

$$r = 3 + 2\cos(-\theta) = 3 + 2\cos\theta. \qquad \cos(-\theta) = \cos\theta$$

So, you can conclude that the curve is symmetric with respect to the polar axis. Plotting the points in the table and using polar axis symmetry, you obtain the graph shown in Figure 10.55. This graph is called a **limaçon.**

θ	0	$\frac{\pi}{6}$	$\frac{\pi}{3}$	$\frac{\pi}{2}$	$\frac{2\pi}{3}$	$\frac{5\pi}{6}$	π
r	5	$3 + \sqrt{3}$	4	3	2	$3 - \sqrt{3}$	1

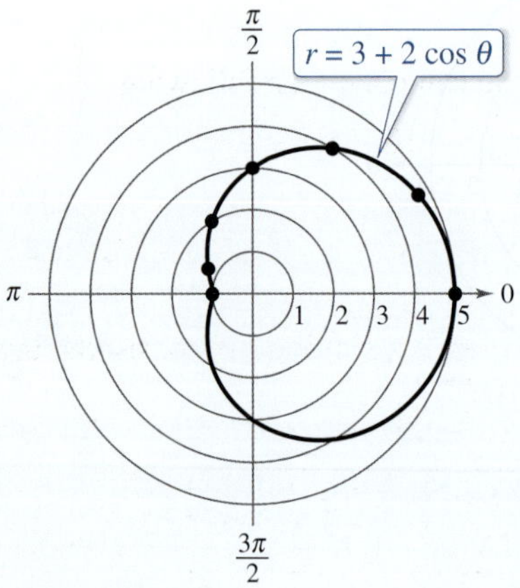

Figure 10.55

Checkpoint Audio-video solution in English & Spanish at LarsonPrecalculus.com.

Use symmetry to sketch the graph of $r = 3 + 2\sin\theta$.

Note in Example 2 that $\cos(-\theta) = \cos\theta$. This is because the cosine function is *even*. Recall from Section 4.2 that the cosine function is even and the sine function is odd. That is, $\sin(-\theta) = -\sin\theta$.

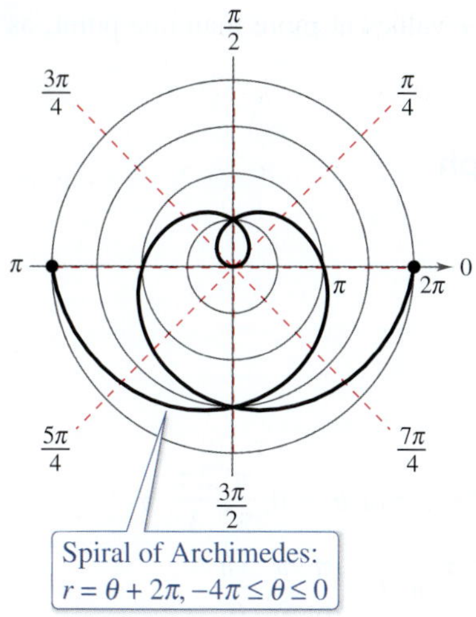

Figure 10.56

The three tests for symmetry in polar coordinates listed on page 748 are sufficient to guarantee symmetry, but they are not necessary. For instance, Figure 10.56 shows the graph of

$$r = \theta + 2\pi \quad \text{Spiral of Archimedes}$$

to be symmetric with respect to the line $\theta = \pi/2$, and yet the tests on page 748 fail to indicate symmetry because neither of the following replacements yields an equivalent equation.

Original Equation	Replacement	New Equation
$r = \theta + 2\pi$	(r, θ) by $(-r, -\theta)$	$-r = -\theta + 2\pi$
$r = \theta + 2\pi$	(r, θ) by $(r, \pi - \theta)$	$r = -\theta + 3\pi$

The equations discussed in Examples 1 and 2 are of the form

$$r = 4 \sin \theta = f(\sin \theta) \quad \text{and} \quad r = 3 + 2 \cos \theta = g(\cos \theta).$$

The graph of the first equation is symmetric with respect to the line $\theta = \pi/2$, and the graph of the second equation is symmetric with respect to the polar axis. This observation can be generalized to yield the following tests.

Quick Tests for Symmetry in Polar Coordinates

1. The graph of $r = f(\sin \theta)$ is symmetric with respect to the line $\theta = \dfrac{\pi}{2}$.
2. The graph of $r = g(\cos \theta)$ is symmetric with respect to the polar axis.

Two additional aids to sketching graphs of polar equations involve knowing the θ-values for which $|r|$ is maximum and knowing the θ-values for which $r = 0$. For instance, in Example 1, the maximum value of $|r|$ for $r = 4 \sin \theta$ is $|r| = 4$, and this occurs when $\theta = \pi/2$, as shown in Figure 10.54. Moreover, $r = 0$ when $\theta = 0$.

EXAMPLE 3 Sketching a Polar Graph

Sketch the graph of $r = 1 - 2 \cos \theta$.

Solution From the equation $r = 1 - 2 \cos \theta$, you can obtain the following.

Symmetry: With respect to the polar axis

Maximum value of $|r|$: $r = 3$ when $\theta = \pi$

Zero of r: $r = 0$ when $\theta = \pi/3$

The table shows several θ-values in the interval $[0, \pi]$. By plotting the corresponding points, you can sketch the graph shown in Figure 10.57.

θ	0	$\frac{\pi}{6}$	$\frac{\pi}{3}$	$\frac{\pi}{2}$	$\frac{2\pi}{3}$	$\frac{5\pi}{6}$	π
r	-1	$1 - \sqrt{3}$	0	1	2	$1 + \sqrt{3}$	3

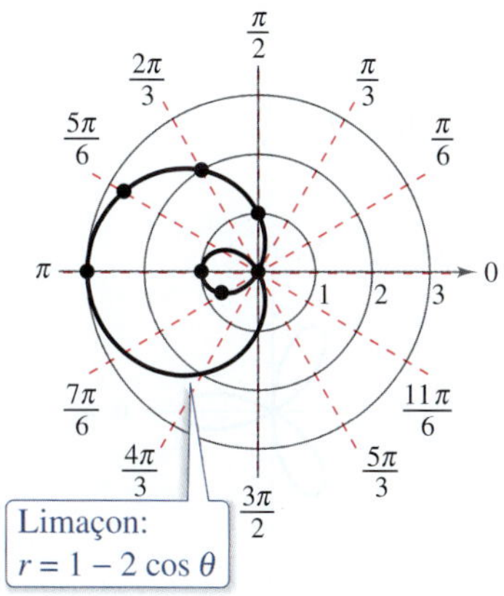

Figure 10.57

Note how the negative r-values determine the *inner loop* of the graph in Figure 10.57. This graph, like the one in Figure 10.55, is a limaçon.

✓ ***Checkpoint*** *Audio-video solution in English & Spanish at LarsonPrecalculus.com.*

Sketch the graph of $r = 1 + 2 \sin \theta$.

Some curves reach their zeros and maximum r-values at more than one point, as shown in Example 4.

EXAMPLE 4 Sketching a Polar Graph

Sketch the graph of

$$r = 2\cos 3\theta.$$

Solution

Symmetry: With respect to the polar axis

Maximum value of $|r|$: $|r| = 2$ when $3\theta = 0, \pi, 2\pi, 3\pi$ or $\theta = 0, \frac{\pi}{3}, \frac{2\pi}{3}, \pi$

Zeros of r: $r = 0$ when $3\theta = \frac{\pi}{2}, \frac{3\pi}{2}, \frac{5\pi}{2}$ or $\theta = \frac{\pi}{6}, \frac{\pi}{2}, \frac{5\pi}{6}$

θ	0	$\frac{\pi}{12}$	$\frac{\pi}{6}$	$\frac{\pi}{4}$	$\frac{\pi}{3}$	$\frac{5\pi}{12}$	$\frac{\pi}{2}$
r	2	$\sqrt{2}$	0	$-\sqrt{2}$	-2	$-\sqrt{2}$	0

By plotting these points and using the specified symmetry, zeros, and maximum values, you can obtain the graph, as shown below. This graph is called a **rose curve,** and each loop on the graph is called a *petal*. Note how the entire curve is generated as θ increases from 0 to π.

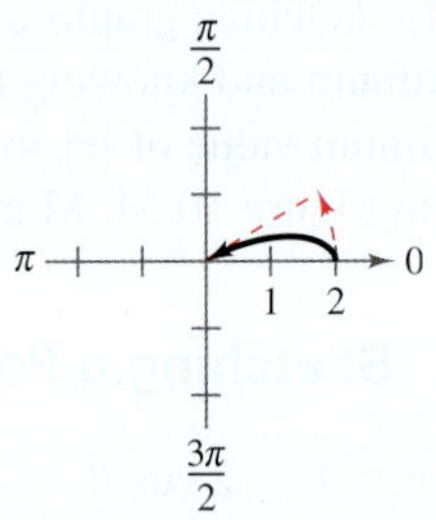

$0 \le \theta \le \frac{\pi}{6}$

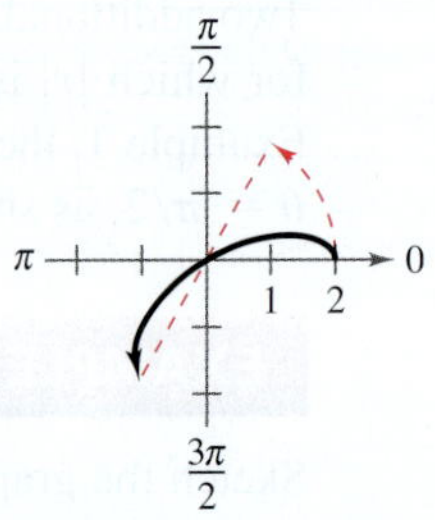

$0 \le \theta \le \frac{\pi}{3}$

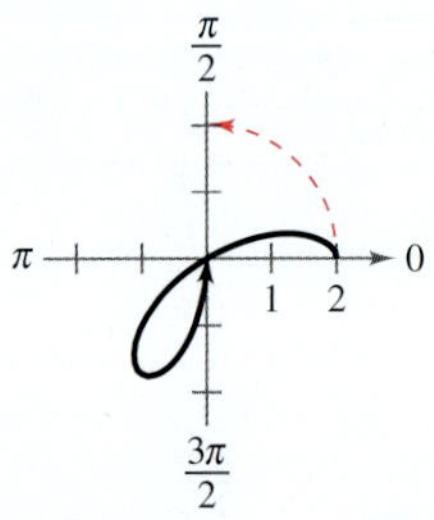

$0 \le \theta \le \frac{\pi}{2}$

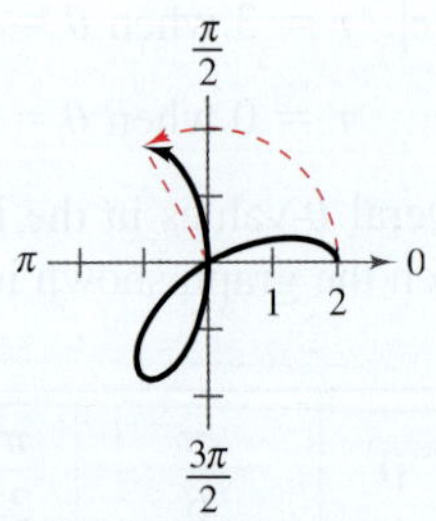

$0 \le \theta \le \frac{2\pi}{3}$

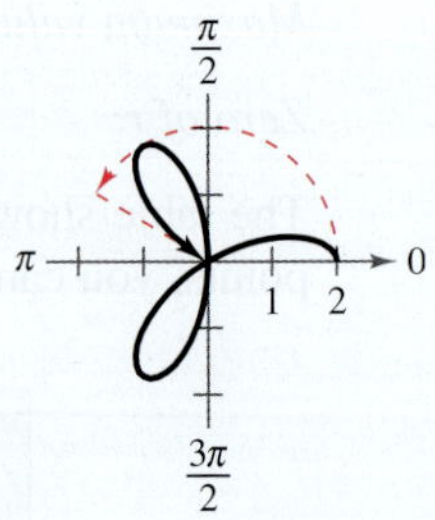

$0 \le \theta \le \frac{5\pi}{6}$

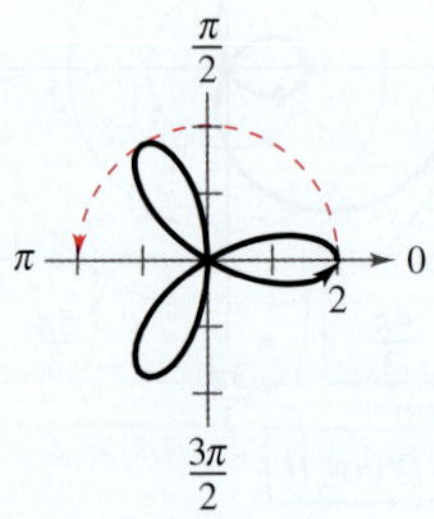

$0 \le \theta \le \pi$

TECHNOLOGY Use a graphing utility in *polar* mode to verify the graph of $r = 2\cos 3\theta$ shown in Example 4.

Checkpoint *Audio-video solution in English & Spanish at LarsonPrecalculus.com.*

Sketch the graph of

$$r = 2\sin 3\theta.$$

Special Polar Graphs

Several important types of graphs have equations that are simpler in polar form than in rectangular form. For example, the circle

$$r = 4 \sin \theta$$

in Example 1 has the more complicated rectangular equation

$$x^2 + (y - 2)^2 = 4.$$

Several other types of graphs that have simple polar equations are shown below.

Limaçons

$r = a \pm b \cos \theta,\ r = a \pm b \sin \theta \quad (a > 0, b > 0)$

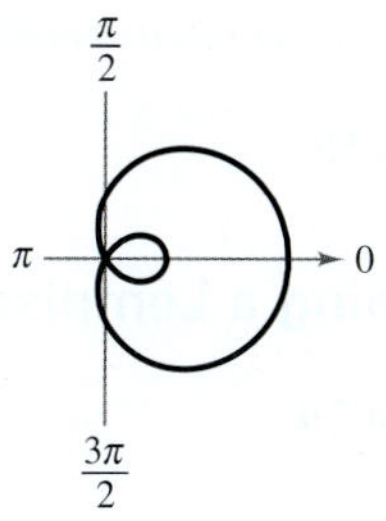

$\dfrac{a}{b} < 1$

Limaçon with inner loop

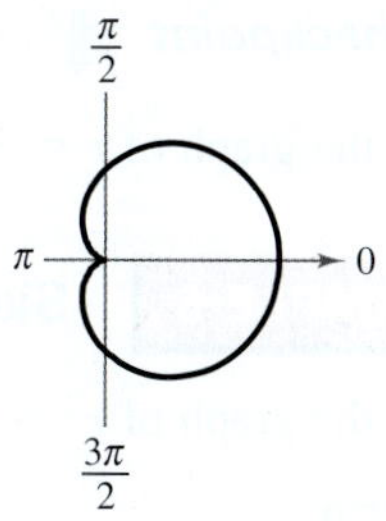

$\dfrac{a}{b} = 1$

Cardioid (heart-shaped)

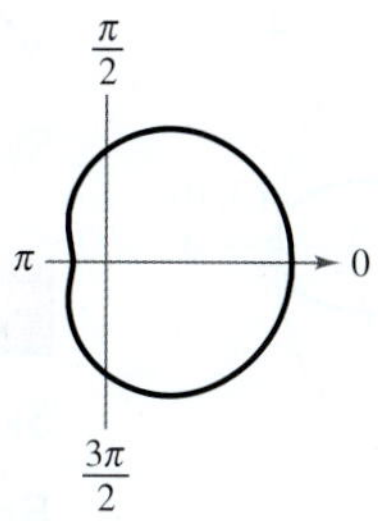

$1 < \dfrac{a}{b} < 2$

Dimpled limaçon

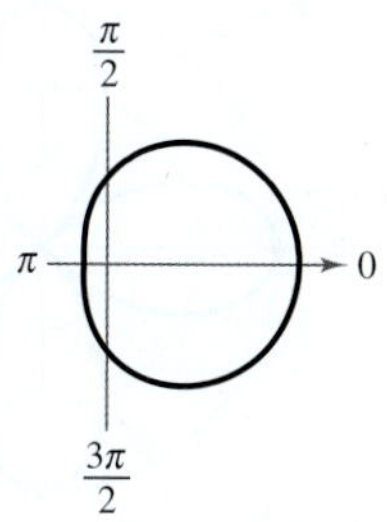

$\dfrac{a}{b} \geq 2$

Convex limaçon

Rose Curves

n petals when n is odd, $2n$ petals when n is even $\quad (n \geq 2)$

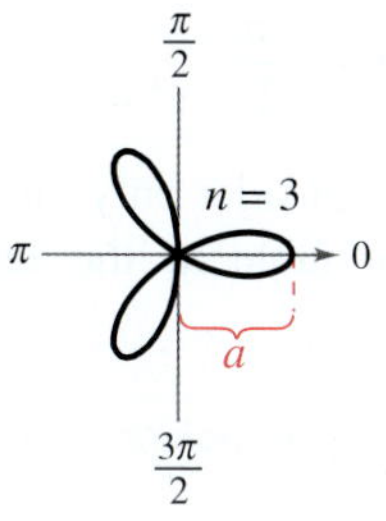

$r = a \cos n\theta$

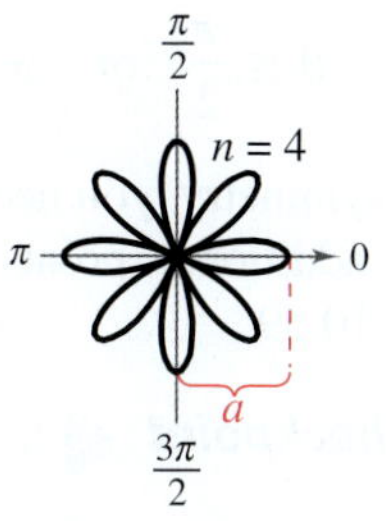

$r = a \cos n\theta$

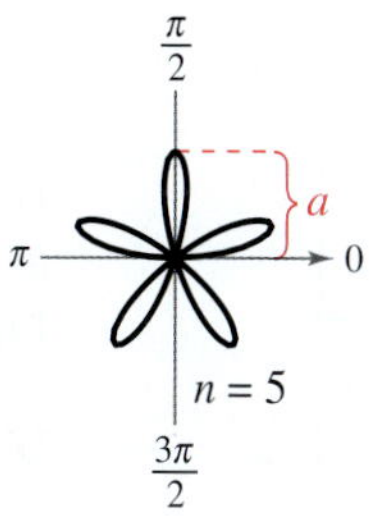

$r = a \sin n\theta$

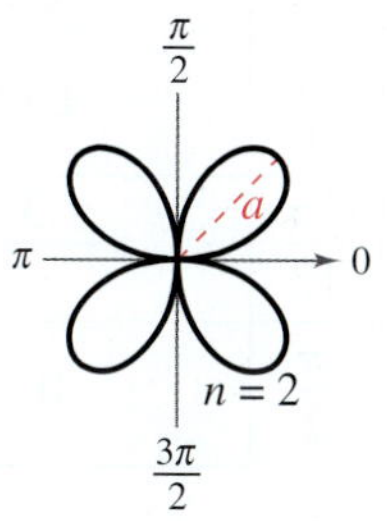

$r = a \sin n\theta$

Circles and Lemniscates

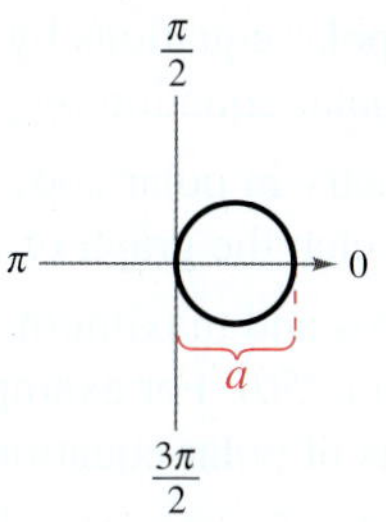

$r = a \cos \theta$

Circle

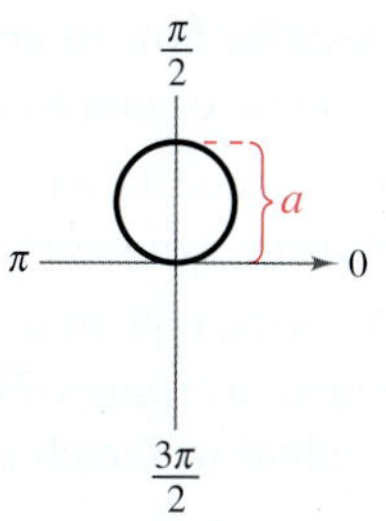

$r = a \sin \theta$

Circle

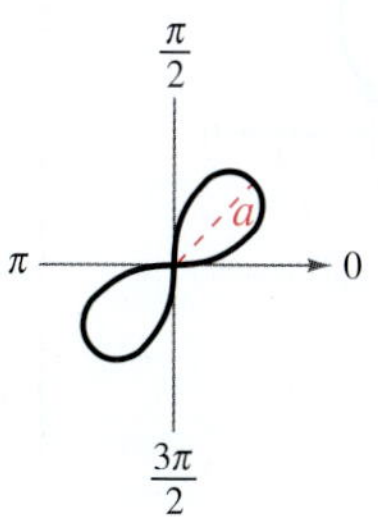

$r^2 = a^2 \sin 2\theta$

Lemniscate

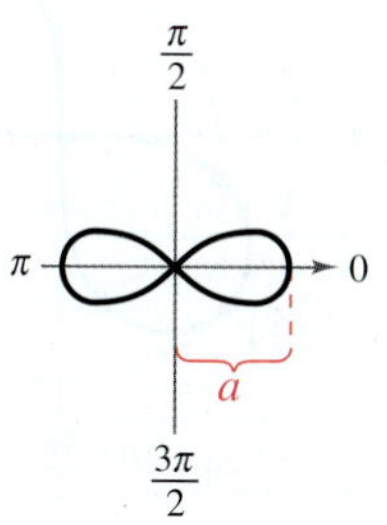

$r^2 = a^2 \cos 2\theta$

Lemniscate

θ	r
0	3
$\frac{\pi}{6}$	$\frac{3}{2}$
$\frac{\pi}{4}$	0
$\frac{\pi}{3}$	$-\frac{3}{2}$

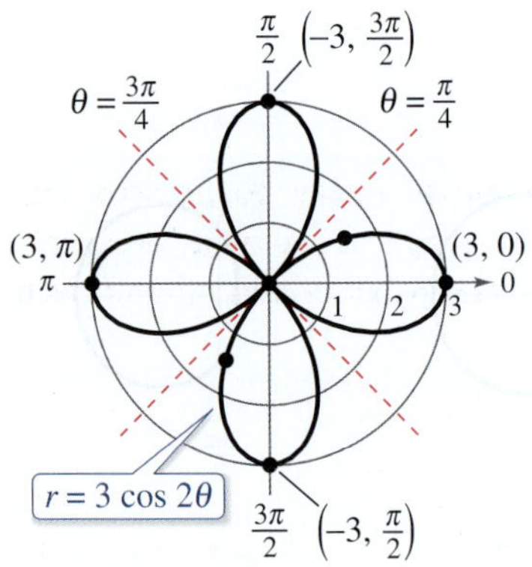

Figure 10.58

EXAMPLE 5 Sketching a Rose Curve

Sketch the graph of $r = 3 \cos 2\theta$.

Solution

Type of curve: Rose curve with $2n = 4$ petals

Symmetry: With respect to the polar axis, the line $\theta = \pi/2$, and the pole

Maximum value of $|r|$: $|r| = 3$ when $\theta = 0, \pi/2, \pi, 3\pi/2$

Zeros of r: $r = 0$ when $\theta = \pi/4, 3\pi/4$

Using this information together with the additional points shown in the table at the left, you obtain the graph shown in Figure 10.58.

✓ **Checkpoint** Audio-video solution in English & Spanish at LarsonPrecalculus.com.

Sketch the graph of $r = 3 \cos 3\theta$.

EXAMPLE 6 Sketching a Lemniscate

Sketch the graph of $r^2 = 9 \sin 2\theta$.

Solution

Type of curve: Lemniscate

Symmetry: With respect to the pole

Maximum value of $|r|$: $|r| = 3$ when $\theta = \pi/4$

Zeros of r: $r = 0$ when $\theta = 0, \pi/2$

When $\sin 2\theta < 0$, this equation has no solution points. So, you restrict the values of θ to those for which $\sin 2\theta \ge 0$.

$$0 \le \theta \le \frac{\pi}{2} \quad \text{or} \quad \pi \le \theta \le \frac{3\pi}{2}$$

Using symmetry, you need to consider only the first of these two intervals. By finding a few additional points (see table at the left), you obtain the graph shown in Figure 10.59.

θ	$r = \pm 3\sqrt{\sin 2\theta}$
0	0
$\frac{\pi}{12}$	$\pm\frac{3}{\sqrt{2}}$
$\frac{\pi}{4}$	± 3
$\frac{5\pi}{12}$	$\pm\frac{3}{\sqrt{2}}$
$\frac{\pi}{2}$	0

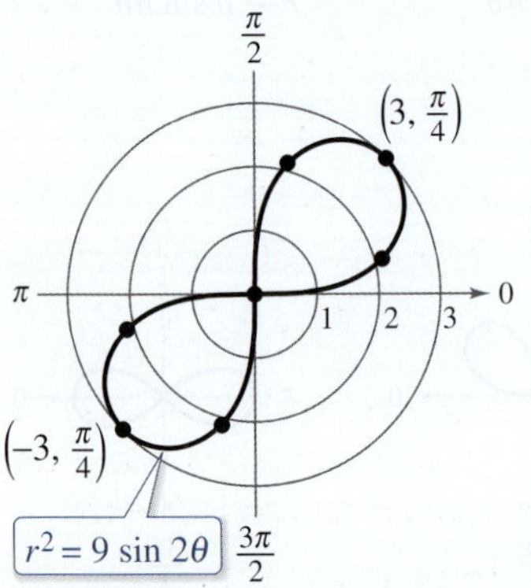

Figure 10.59

✓ **Checkpoint** Audio-video solution in English & Spanish at LarsonPrecalculus.com.

Sketch the graph of $r^2 = 4 \cos 2\theta$.

Summarize (Section 10.8)

1. Describe how to graph polar equations by point plotting *(page 747)*. For an example of graphing a polar equation by point plotting, see Example 1.
2. State the tests for symmetry in polar coordinates *(page 748)*. For an example of using symmetry to sketch the graph of a polar equation, see Example 2.
3. Describe how to use zeros and maximum r-values to sketch graphs of polar equations *(pages 749 and 750)*. For examples of using zeros and maximum r-values to sketch graphs of polar equations, see Examples 3 and 4.
4. State and give an example of a special polar graph covered in this lesson *(page 751)*. For examples of sketching special polar graphs, see Examples 5 and 6.

10.8 Exercises

See CalcChat.com for tutorial help and worked-out solutions to odd-numbered exercises.

Vocabulary: Fill in the blanks.

1. The graph of $r = f(\sin \theta)$ is symmetric with respect to the line ________.
2. The graph of $r = g(\cos \theta)$ is symmetric with respect to the ________ ________.
3. The equation $r = 2 + \cos \theta$ represents a ________ ________.
4. The equation $r = 2 \cos \theta$ represents a ________.
5. The equation $r^2 = 4 \sin 2\theta$ represents a ________.
6. The equation $r = 1 + \sin \theta$ represents a ________.

Skills and Applications

Identifying Types of Polar Graphs In Exercises 7–12, identify the type of polar graph.

7. 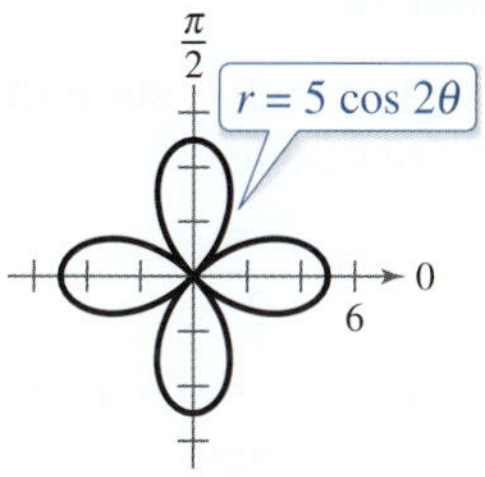

8. 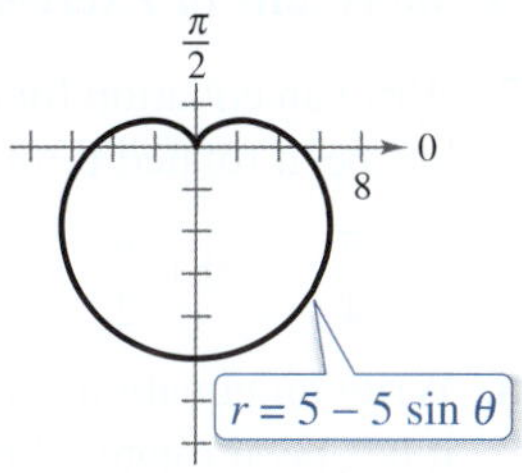

9. 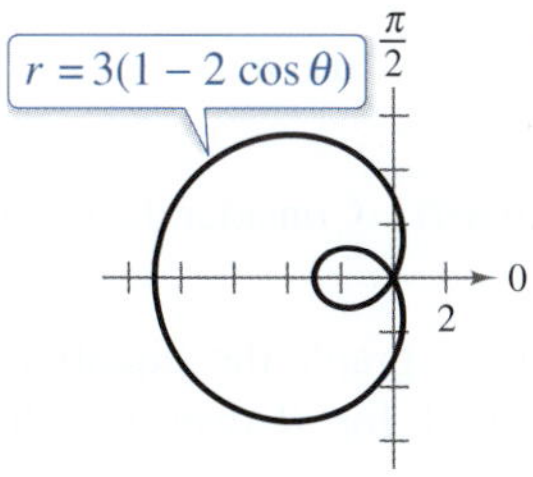

10. 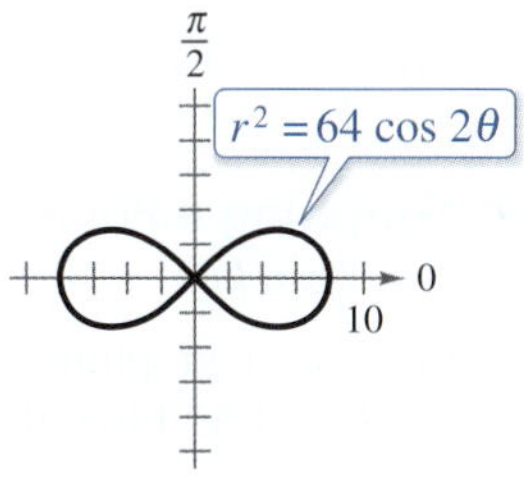

11. 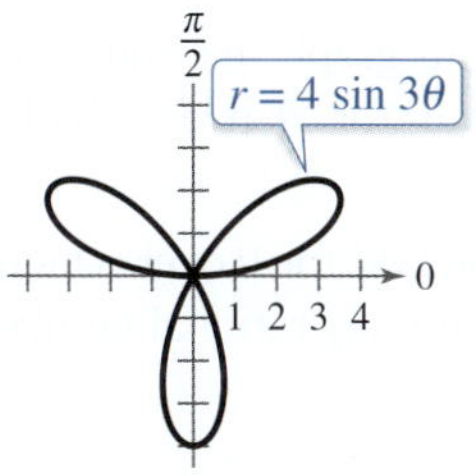

12. 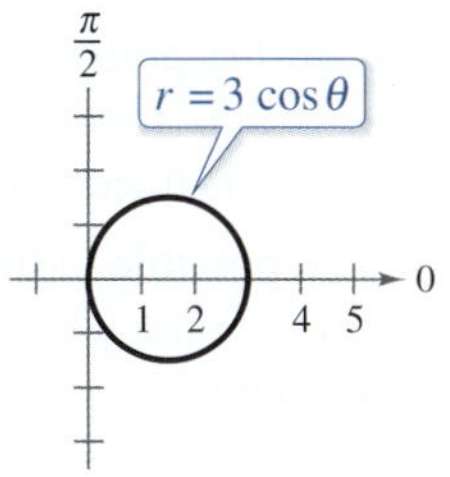

Testing for Symmetry In Exercises 13–18, test for symmetry with respect to the line $\theta = \pi/2$, the polar axis, and the pole.

13. $r = 5 + 4 \cos \theta$
14. $r = 9 \cos 3\theta$
15. $r = \dfrac{2}{1 + \sin \theta}$
16. $r = \dfrac{3}{2 + \cos \theta}$
17. $r^2 = 36 \cos 2\theta$
18. $r^2 = 25 \sin 2\theta$

Finding the Maximum Value of $|r|$ and Zeros of r In Exercises 19–22, find the maximum value of $|r|$ and any zeros of r.

19. $r = 10 - 10 \sin \theta$
20. $r = 6 + 12 \cos \theta$
21. $r = 4 \cos 3\theta$
22. $r = 3 \sin 2\theta$

Sketching the Graph of a Polar Equation In Exercises 23–48, sketch the graph of the polar equation using symmetry, zeros, maximum r-values, and any other additional points.

23. $r = 4$
24. $r = -7$
25. $r = \pi/3$
26. $r = -3\pi/4$
27. $r = \sin \theta$
28. $r = 4 \cos \theta$
29. $r = 3(1 - \cos \theta)$
30. $r = 4(1 - \sin \theta)$
31. $r = 4(1 + \sin \theta)$
32. $r = 2(1 + \cos \theta)$
33. $r = 4 + 3 \cos \theta$
34. $r = 4 - 3 \sin \theta$
35. $r = 1 - 2 \sin \theta$
36. $r = 2 - 4 \cos \theta$
37. $r = 3 - 4 \cos \theta$
38. $r = 3 + 6 \sin \theta$
39. $r = 5 \sin 2\theta$
40. $r = 2 \cos 2\theta$
41. $r = 6 \cos 3\theta$
42. $r = 3 \sin 3\theta$
43. $r = 2 \sec \theta$
44. $r = 5 \csc \theta$
45. $r = \dfrac{3}{\sin \theta - 2 \cos \theta}$
46. $r = \dfrac{6}{2 \sin \theta - 3 \cos \theta}$
47. $r^2 = 9 \cos 2\theta$
48. $r^2 = 4 \sin \theta$

Graphing a Polar Equation In Exercises 49–58, use a graphing utility to graph the polar equation. Describe your viewing window.

49. $r = 9/4$
50. $r = -5/2$
51. $r = 5\pi/8$
52. $r = -\pi/10$
53. $r = 8 \cos \theta$
54. $r = \cos 2\theta$
55. $r = 3(2 - \sin \theta)$
56. $r = 2 \cos(3\theta - 2)$
57. $r = 8 \sin \theta \cos^2 \theta$
58. $r = 2 \csc \theta + 5$

Finding an Interval In Exercises 59–64, use a graphing utility to graph the polar equation. Find an interval for θ for which the graph is traced *only once*.

59. $r = 3 - 8 \cos \theta$
60. $r = 5 + 4 \cos \theta$
61. $r = 2 \cos\left(\dfrac{3\theta}{2}\right)$
62. $r = 3 \sin\left(\dfrac{5\theta}{2}\right)$
63. $r^2 = 16 \sin 2\theta$
64. $r^2 = 1/\theta$

Asymptote of a Graph of a Polar Equation **In Exercises 65–68, use a graphing utility to graph the polar equation and show that the indicated line is an asymptote of the graph.**

	Name of Graph	Polar Equation	Asymptote
65.	Conchoid	$r = 2 - \sec\theta$	$x = -1$
66.	Conchoid	$r = 2 + \csc\theta$	$y = 1$
67.	Hyperbolic spiral	$r = \dfrac{3}{\theta}$	$y = 3$
68.	Strophoid	$r = 2\cos 2\theta\sec\theta$	$x = -2$

69. Microphone

The sound pickup pattern of a microphone is modeled by the polar equation

$$r = 5 + 5\cos\theta$$

where $|r|$ measures how sensitive the microphone is to sounds coming from the angle θ.

(a) Sketch the graph of the model and identify the type of polar graph.

(b) At what angle is the microphone most sensitive to sound?

Exploration

70. Area The area of the lemniscate $r^2 = a^2\cos 2\theta$ is a^2.

(a) Sketch the graph of $r^2 = 16\cos 2\theta$.

(b) Find the area of one loop of the graph from part (a).

True or False? **In Exercises 71 and 72, determine whether the statement is true or false. Justify your answer.**

71. The graph of $r = 10\sin 5\theta$ is a rose curve with five petals.

72. A rose curve will always have symmetry with respect to the line $\theta = \pi/2$.

73. Sketching the Graph of a Polar Equation Sketch the graph of $r = 6\cos\theta$ over each interval. Describe the part of the graph obtained in each case.

(a) $0 \le \theta \le \dfrac{\pi}{2}$ (b) $\dfrac{\pi}{2} \le \theta \le \pi$

(c) $-\dfrac{\pi}{2} \le \theta \le \dfrac{\pi}{2}$ (d) $\dfrac{\pi}{4} \le \theta \le \dfrac{3\pi}{4}$

74. Graphical Reasoning Use a graphing utility to graph the polar equation $r = 6[1 + \cos(\theta - \phi)]$ for (a) $\phi = 0$, (b) $\phi = \pi/4$, and (c) $\phi = \pi/2$. Use the graphs to describe the effect of the angle ϕ. Write the equation as a function of $\sin\theta$ for part (c).

75. Rotation Through an Angle The graph of $r = f(\theta)$ is rotated about the pole through an angle ϕ. Show that the equation of the rotated graph is $r = f(\theta - \phi)$.

76. Rotation Through an Angle Consider the graph of $r = f(\sin\theta)$.

(a) Show that when the graph is rotated counterclockwise $\pi/2$ radians about the pole, the equation of the rotated graph is $r = f(-\cos\theta)$.

(b) Show that when the graph is rotated counterclockwise π radians about the pole, the equation of the rotated graph is $r = f(-\sin\theta)$.

(c) Show that when the graph is rotated counterclockwise $3\pi/2$ radians about the pole, the equation of the rotated graph is $r = f(\cos\theta)$.

Rotation Through an Angle **In Exercises 77 and 78, use the results of Exercises 75 and 76.**

77. Write an equation for the limaçon $r = 2 - \sin\theta$ after it has been rotated through the given angle.

(a) $\dfrac{\pi}{4}$ (b) $\dfrac{\pi}{2}$ (c) π (d) $\dfrac{3\pi}{2}$

78. Write an equation for the rose curve $r = 2\sin 2\theta$ after it has been rotated through the given angle.

(a) $\dfrac{\pi}{6}$ (b) $\dfrac{\pi}{2}$ (c) $\dfrac{2\pi}{3}$ (d) π

79. Graphing a Polar Equation Consider the equation $r = 3\sin k\theta$.

(a) Use a graphing utility to graph the equation for $k = 1.5$. Find the interval for θ over which the graph is traced only once.

(b) Use the graphing utility to graph the equation for $k = 2.5$. Find the interval for θ over which the graph is traced only once.

(c) Is it possible to find an interval for θ over which the graph is traced only once for any rational number k? Explain.

80. HOW DO YOU SEE IT? Determine which graph matches each polar equation.

(a) $r = 5\sin\theta$

(b) $r = 2 + 5\sin\theta$

(c) $r = 5\cos 2\theta$

(i)
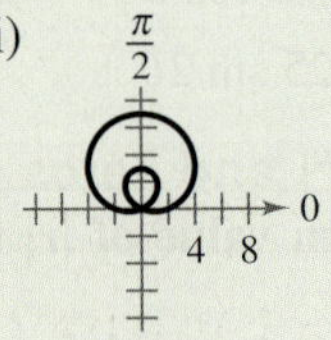

(ii)
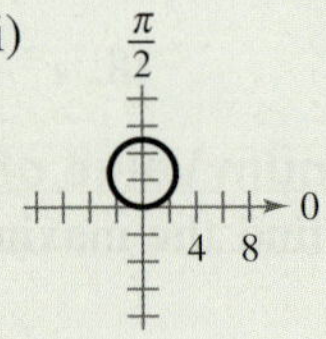

(iii)
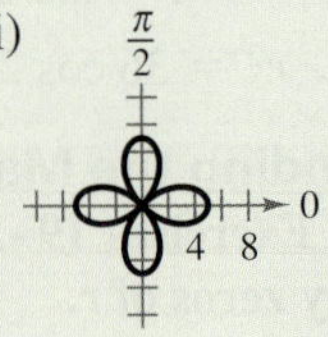

nikkytok/Shutterstock.com

10.9 Polar Equations of Conics

You can model the orbits of planets and satellites with polar equations. For instance, in Exercise 64 on page 760, you will use a polar equation to model the orbit of a satellite.

- **Define conics in terms of eccentricity, and write and graph equations of conics in polar form.**
- **Use equations of conics in polar form to model real-life problems.**

Alternative Definition and Polar Equations of Conics

In Sections 10.3 and 10.4, you learned that the rectangular equations of ellipses and hyperbolas take simple forms when the origin lies at their *centers*. As it happens, there are many important applications of conics in which it is more convenient to use one of the *foci* as the origin. In this section, you will learn that polar equations of conics take simple forms when one of the foci lies at the pole.

To begin, consider the following alternative definition of a conic that uses the concept of eccentricity.

> **Alternative Definition of a Conic**
>
> The locus of a point in the plane that moves such that its distance from a fixed point (focus) is in a constant ratio to its distance from a fixed line (directrix) is a **conic.** The constant ratio is the eccentricity of the conic and is denoted by e. Moreover, the conic is an **ellipse** when $0 < e < 1$, a **parabola** when $e = 1$, and a **hyperbola** when $e > 1$. (See the figures below.)

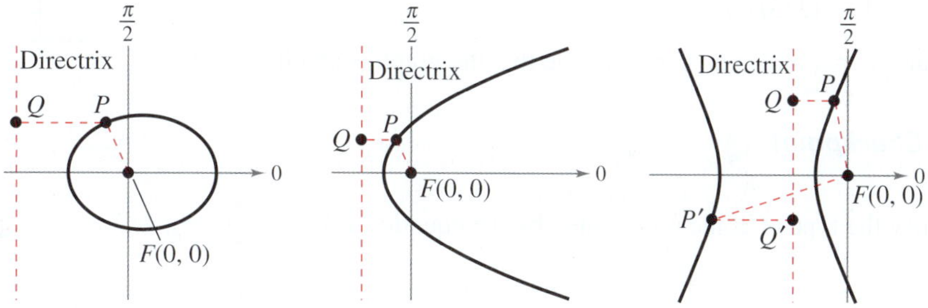

Ellipse: $0 < e < 1$ $\quad \frac{PF}{PQ} < 1$

Parabola: $e = 1$ $\quad \frac{PF}{PQ} = 1$

Hyperbola: $e > 1$ $\quad \frac{PF}{PQ} = \frac{P'F}{P'Q'} > 1$

In the figures, note that for each type of conic, the focus is at the pole. The benefit of locating a focus of a conic at the pole is that the equation of the conic takes on a simpler form. For a proof of the polar equations of conics, see Proofs in Mathematics on page 770.

> **Polar Equations of Conics**
>
> The graph of a polar equation of the form
>
> **1.** $r = \dfrac{ep}{1 \pm e \cos \theta}$ or **2.** $r = \dfrac{ep}{1 \pm e \sin \theta}$
>
> is a conic, where $e > 0$ is the eccentricity and $|p|$ is the distance between the focus (pole) and the directrix.

Cristi Matei/Shutterstock.com

An equation of the form

$$r = \frac{ep}{1 \pm e \cos \theta}$$ Vertical directrix

corresponds to a conic with a vertical directrix and symmetry with respect to the polar axis. An equation of the form

$$r = \frac{ep}{1 \pm e \sin \theta}$$ Horizontal directrix

corresponds to a conic with a horizontal directrix and symmetry with respect to the line $\theta = \pi/2$. Moreover, the converse is also true—that is, any conic with a focus at the pole and having a horizontal or vertical directrix can be represented by one of these equations.

EXAMPLE 1 Identifying a Conic from Its Equation

Identify the type of conic represented by the equation $r = \dfrac{15}{3 - 2 \cos \theta}$.

Algebraic Solution

To identify the type of conic, rewrite the equation in the form

$$r = \frac{ep}{1 \pm e \cos \theta}.$$

$$r = \frac{15}{3 - 2 \cos \theta}$$ Write original equation.

$$= \frac{5}{1 - (2/3) \cos \theta}$$ Divide numerator and denominator by 3.

Because $e = \frac{2}{3} < 1$, you can conclude that the graph is an ellipse.

Graphical Solution

Use a graphing utility in *polar* mode and be sure to use a square setting, as shown below.

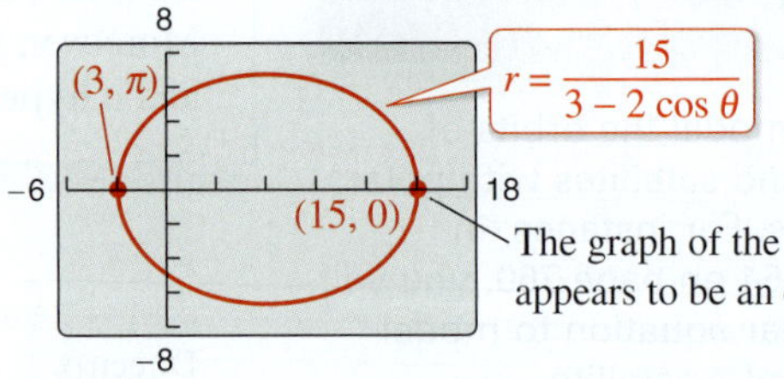

The graph of the conic appears to be an ellipse.

Checkpoint Audio-video solution in English & Spanish at LarsonPrecalculus.com.

Identify the type of conic represented by the equation $r = \dfrac{8}{2 - 3 \sin \theta}$.

For the ellipse in Example 1, the major axis is horizontal and the vertices lie at $(15, 0)$ and $(3, \pi)$. So, the length of the *major* axis is $2a = 18$. To find the length of the *minor* axis, you can use the equations $e = c/a$ and $b^2 = a^2 - c^2$ to conclude that

$$b^2 = a^2 - c^2$$
$$= a^2 - (ea)^2$$
$$= a^2(1 - e^2).$$ Ellipse

Because $e = \frac{2}{3}$, you have

$$b^2 = 9^2\left[1 - \left(\frac{2}{3}\right)^2\right] = 45$$

which implies that $b = \sqrt{45} = 3\sqrt{5}$. So, the length of the minor axis is $2b = 6\sqrt{5}$. A similar analysis for hyperbolas yields

$$b^2 = c^2 - a^2$$
$$= (ea)^2 - a^2$$
$$= a^2(e^2 - 1).$$ Hyperbola

EXAMPLE 2 Sketching a Conic from Its Polar Equation

Identify the conic $r = \dfrac{32}{3 + 5 \sin \theta}$ and sketch its graph.

Solution Dividing the numerator and denominator by 3, you have

$$r = \frac{32/3}{1 + (5/3) \sin \theta}.$$

Because $e = \frac{5}{3} > 1$, the graph is a hyperbola. The transverse axis of the hyperbola lies on the line $\theta = \pi/2$, and the vertices occur at $(4, \pi/2)$ and $(-16, 3\pi/2)$. Because the length of the transverse axis is 12, you can see that $a = 6$. To find b, write

$$b^2 = a^2(e^2 - 1) = 6^2\left[\left(\tfrac{5}{3}\right)^2 - 1\right] = 64.$$

So, $b = 8$. You can use a and b to determine that the asymptotes of the hyperbola are $y = 10 \pm \frac{3}{4}x$. The graph is shown in Figure 10.60.

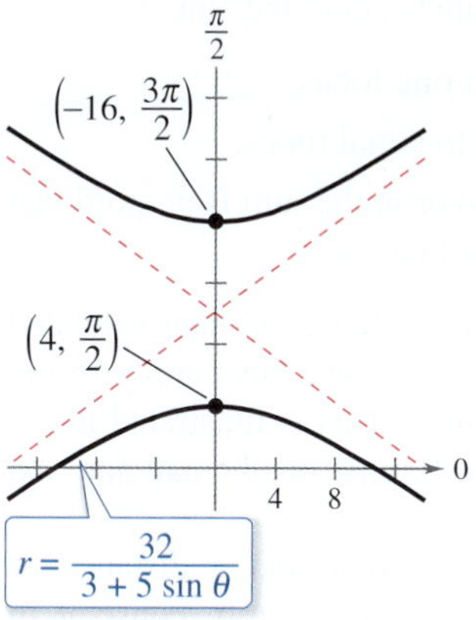

Figure 10.60

Checkpoint Audio-video solution in English & Spanish at LarsonPrecalculus.com.

Identify the conic $r = \dfrac{3}{2 - 4 \sin \theta}$ and sketch its graph.

In the next example, you are asked to find a polar equation of a specified conic. To do this, let p be the distance between the pole and the directrix.

1. Horizontal directrix above the pole: $r = \dfrac{ep}{1 + e \sin \theta}$
2. Horizontal directrix below the pole: $r = \dfrac{ep}{1 - e \sin \theta}$
3. Vertical directrix to the right of the pole: $r = \dfrac{ep}{1 + e \cos \theta}$
4. Vertical directrix to the left of the pole: $r = \dfrac{ep}{1 - e \cos \theta}$

TECHNOLOGY Use a graphing utility set in *polar* mode to verify the four orientations listed at the right. Remember that e must be positive, but p can be positive or negative.

EXAMPLE 3 Finding the Polar Equation of a Conic

Find a polar equation of the parabola whose focus is the pole and whose directrix is the line $y = 3$.

Solution Because the directrix is horizontal and above the pole, use an equation of the form

$$r = \frac{ep}{1 + e \sin \theta}.$$

Moreover, because the eccentricity of a parabola is $e = 1$ and the distance between the pole and the directrix is $p = 3$, you have the equation

$$r = \frac{3}{1 + \sin \theta}.$$

The parabola is shown in Figure 10.61.

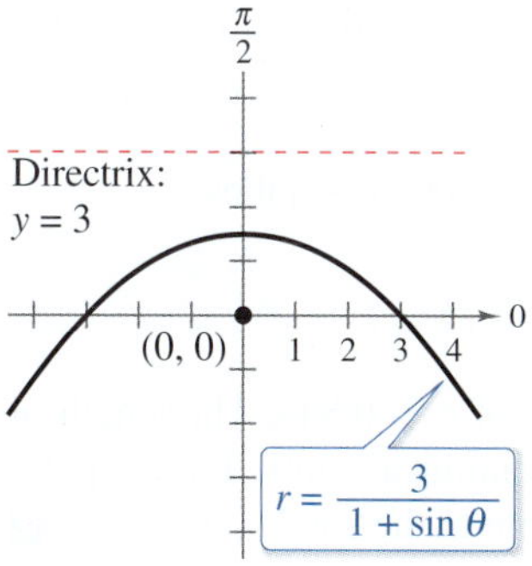

Figure 10.61

Checkpoint Audio-video solution in English & Spanish at LarsonPrecalculus.com.

Find a polar equation of the parabola whose focus is the pole and whose directrix is the line $x = -2$.

Application

Kepler's Laws (listed below), named after the German astronomer Johannes Kepler (1571–1630), can be used to describe the orbits of the planets about the sun.

1. Each planet moves in an elliptical orbit with the sun at one focus.
2. A ray from the sun to a planet sweeps out equal areas in equal times.
3. The square of the period (the time it takes for a planet to orbit the sun) is proportional to the cube of the mean distance between the planet and the sun.

Although Kepler stated these laws on the basis of observation, they were later validated by Isaac Newton (1642–1727). In fact, Newton showed that these laws apply to the orbits of all heavenly bodies, including comets and satellites. This is illustrated in the next example, which involves the comet named after the English mathematician and physicist Edmund Halley (1656–1742).

If you use Earth as a reference with a period of 1 year and a distance of 1 astronomical unit (about 93 million miles), then the proportionality constant in Kepler's third law is 1. For example, because Mars has a mean distance to the sun of $d \approx 1.524$ astronomical units, its period P is given by $d^3 = P^2$. So, the period of Mars is $P \approx 1.88$ years.

EXAMPLE 4 Halley's Comet

Halley's comet has an elliptical orbit with an eccentricity of $e \approx 0.967$. The length of the major axis of the orbit is approximately 35.88 astronomical units. Find a polar equation for the orbit. How close does Halley's comet come to the sun?

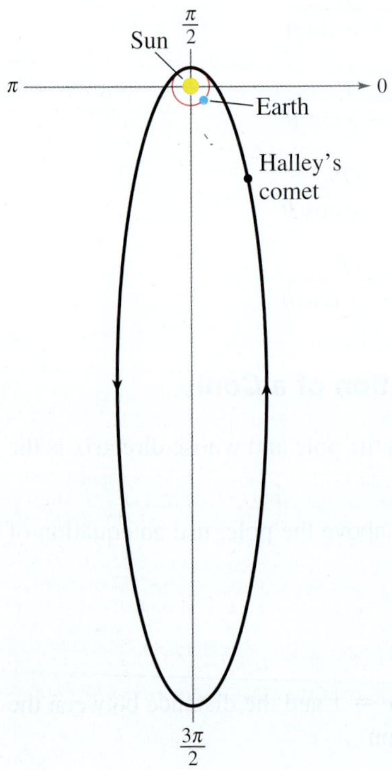

Figure 10.62

Solution Using a vertical major axis, as shown in Figure 10.62, choose an equation of the form $r = ep/(1 + e \sin \theta)$. Because the vertices of the ellipse occur when $\theta = \pi/2$ and $\theta = 3\pi/2$, you can determine the length of the major axis to be the sum of the r-values of the vertices. That is,

$$2a = \frac{0.967p}{1 + 0.967} + \frac{0.967p}{1 - 0.967} \approx 29.79p \approx 35.88.$$

So, $p \approx 1.204$ and $ep \approx (0.967)(1.204) \approx 1.164$. Using this value of ep in the equation, you have

$$r = \frac{1.164}{1 + 0.967 \sin \theta}$$

where r is measured in astronomical units. To find the closest point to the sun (the focus), substitute $\theta = \pi/2$ into this equation to obtain

$$r = \frac{1.164}{1 + 0.967 \sin(\pi/2)} \approx 0.59 \text{ astronomical unit} \approx 55{,}000{,}000 \text{ miles.}$$

✓ ***Checkpoint*** *Audio-video solution in English & Spanish at LarsonPrecalculus.com.*

Encke's comet has an elliptical orbit with an eccentricity of $e \approx 0.848$. The length of the major axis of the orbit is approximately 4.429 astronomical units. Find a polar equation for the orbit. How close does Encke's comet come to the sun?

Summarize (Section 10.9)

1. State the definition of a conic in terms of eccentricity *(page 755)*. For examples of identifying and sketching conics in polar form, see Examples 1 and 2.
2. Describe a real-life problem that can be modeled by an equation of a conic in polar form *(page 758, Example 4)*.

10.9 Exercises

See CalcChat.com for tutorial help and worked-out solutions to odd-numbered exercises.

Vocabulary

In Exercises 1–3, fill in the blanks.

1. The locus of a point in the plane that moves such that its distance from a fixed point (focus) is in a constant ratio to its distance from a fixed line (directrix) is a ________.
2. The constant ratio is the ________ of the conic and is denoted by ________.
3. An equation of the form $r = \dfrac{ep}{1 + e\cos\theta}$ has a ________ directrix to the ________ of the pole.
4. Match the conic with its eccentricity.

(a) $0 < e < 1$ (b) $e = 1$ (c) $e > 1$

(i) parabola (ii) hyperbola (iii) ellipse

Skills and Applications

Identifying a Conic **In Exercises 5–8, write the polar equation of the conic for $e = 1$, $e = 0.5$, and $e = 1.5$. Identify the conic for each equation. Verify your answers with a graphing utility.**

5. $r = \dfrac{2e}{1 + e\cos\theta}$
6. $r = \dfrac{2e}{1 - e\cos\theta}$
7. $r = \dfrac{2e}{1 - e\sin\theta}$
8. $r = \dfrac{2e}{1 + e\sin\theta}$

Matching **In Exercises 9–14, match the polar equation with its graph. [The graphs are labeled (a), (b), (c), (d), (e), and (f).]**

(a)

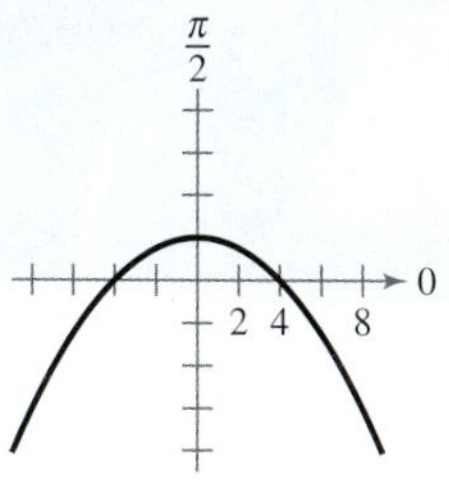

(b)

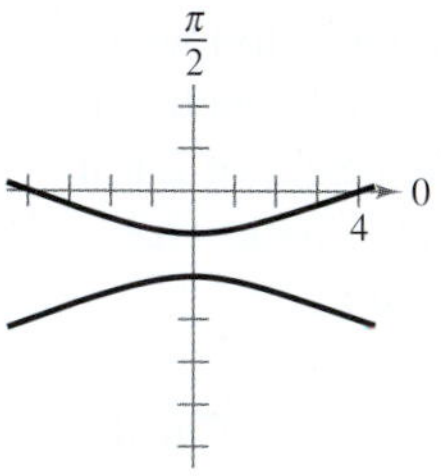

(c)

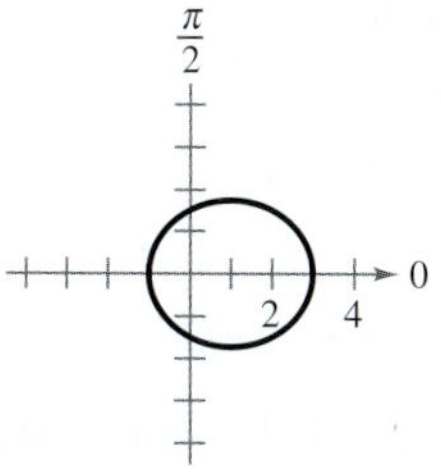

(d)

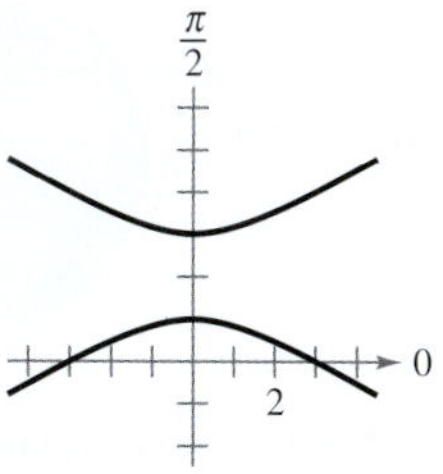

(e)

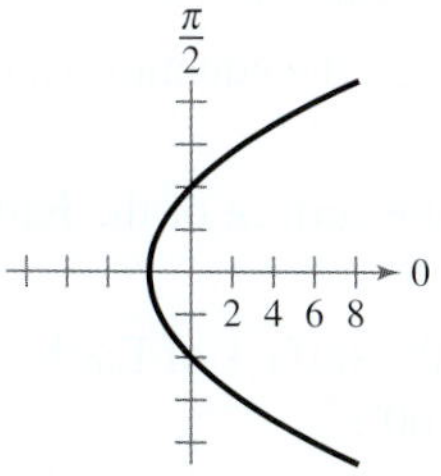

(f)

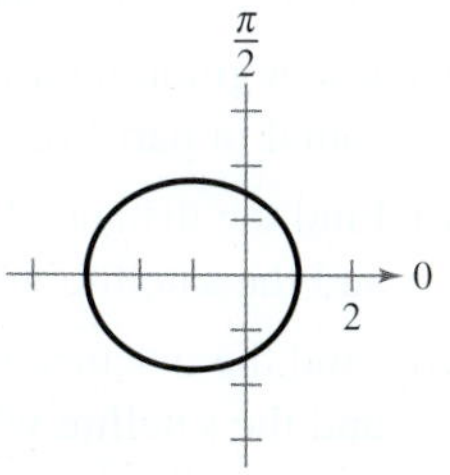

9. $r = \dfrac{4}{1 - \cos\theta}$
10. $r = \dfrac{3}{2 - \cos\theta}$
11. $r = \dfrac{3}{1 + 2\sin\theta}$
12. $r = \dfrac{3}{2 + \cos\theta}$
13. $r = \dfrac{4}{1 + \sin\theta}$
14. $r = \dfrac{4}{1 - 3\sin\theta}$

Sketching a Conic **In Exercises 15–26, identify the conic and sketch its graph.**

15. $r = \dfrac{3}{1 - \cos\theta}$
16. $r = \dfrac{7}{1 + \sin\theta}$
17. $r = \dfrac{5}{1 + \sin\theta}$
18. $r = \dfrac{6}{1 + \cos\theta}$
19. $r = \dfrac{2}{2 - \cos\theta}$
20. $r = \dfrac{4}{4 + \sin\theta}$
21. $r = \dfrac{6}{2 + \sin\theta}$
22. $r = \dfrac{9}{3 - 2\cos\theta}$
23. $r = \dfrac{3}{2 + 4\sin\theta}$
24. $r = \dfrac{5}{-1 + 2\cos\theta}$
25. $r = \dfrac{3}{2 - 6\cos\theta}$
26. $r = \dfrac{3}{2 + 6\sin\theta}$

Graphing a Polar Equation **In Exercises 27–34, use a graphing utility to graph the polar equation. Identify the graph.**

27. $r = \dfrac{-1}{1 - \sin\theta}$
28. $r = \dfrac{-5}{2 + 4\sin\theta}$
29. $r = \dfrac{3}{-4 + 2\cos\theta}$
30. $r = \dfrac{4}{1 - 2\cos\theta}$
31. $r = \dfrac{4}{3 - \cos\theta}$
32. $r = \dfrac{2}{2 + 3\sin\theta}$
33. $r = \dfrac{14}{14 + 17\sin\theta}$
34. $r = \dfrac{12}{2 - \cos\theta}$

Graphing a Rotated Conic **In Exercises 35–38, use a graphing utility to graph the rotated conic.**

35. $r = \dfrac{3}{1 - \cos(\theta - \pi/4)}$ (See Exercise 15.)

36. $r = \dfrac{4}{4 + \sin(\theta - \pi/3)}$ (See Exercise 20.)

37. $r = \dfrac{6}{2 + \sin(\theta + \pi/6)}$ (See Exercise 21.)

38. $r = \dfrac{5}{-1 + 2\cos(\theta + 2\pi/3)}$ (See Exercise 24.)

Finding the Polar Equation of a Conic **In Exercises 39–54, find a polar equation of the conic with its focus at the pole.**

	Conic	Eccentricity	Directrix
39.	Parabola	$e = 1$	$x = -1$
40.	Parabola	$e = 1$	$y = -4$
41.	Ellipse	$e = \frac{1}{2}$	$y = 1$
42.	Ellipse	$e = \frac{3}{4}$	$y = -2$
43.	Hyperbola	$e = 2$	$x = 1$
44.	Hyperbola	$e = \frac{3}{2}$	$x = -1$

	Conic	Vertex or Vertices
45.	Parabola	$(1, -\pi/2)$
46.	Parabola	$(8, 0)$
47.	Parabola	$(5, \pi)$
48.	Parabola	$(10, \pi/2)$
49.	Ellipse	$(2, 0), (10, \pi)$
50.	Ellipse	$(2, \pi/2), (4, 3\pi/2)$
51.	Ellipse	$(20, 0), (4, \pi)$
52.	Hyperbola	$(2, 0), (8, 0)$
53.	Hyperbola	$(1, 3\pi/2), (9, 3\pi/2)$
54.	Hyperbola	$(4, \pi/2), (1, \pi/2)$

55. Planetary Motion The planets travel in elliptical orbits with the sun at one focus. Assume that the focus is at the pole, the major axis lies on the polar axis, and the length of the major axis is $2a$ (see figure). Show that the polar equation of the orbit is $r = a(1 - e^2)/(1 - e\cos\theta)$, where e is the eccentricity.

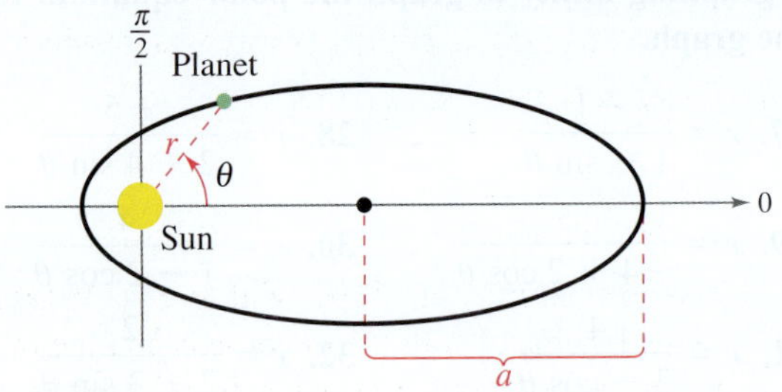

56. Planetary Motion Use the result of Exercise 55 to show that the minimum distance (*perihelion*) from the sun to the planet is

$r = a(1 - e)$

and the maximum distance (*aphelion*) is

$r = a(1 + e)$.

Planetary Motion **In Exercises 57–62, use the results of Exercises 55 and 56 to find (a) the polar equation of the planet's orbit and (b) the perihelion and aphelion.**

57.	Earth	$a \approx 9.2956 \times 10^7$ miles, $e \approx 0.0167$
58.	Saturn	$a \approx 1.4267 \times 10^9$ kilometers, $e \approx 0.0542$
59.	Venus	$a \approx 1.0821 \times 10^8$ kilometers, $e \approx 0.0068$
60.	Mercury	$a \approx 3.5983 \times 10^7$ miles, $e \approx 0.2056$
61.	Mars	$a \approx 1.4163 \times 10^8$ miles, $e \approx 0.0934$
62.	Jupiter	$a \approx 7.7841 \times 10^8$ kilometers, $e \approx 0.0484$

63. Astronomy The comet Hale-Bopp has an elliptical orbit with an eccentricity of $e \approx 0.995$. The length of the major axis of the orbit is approximately 500 astronomical units. Find a polar equation for the orbit. How close does the comet come to the sun?

64. Satellite Orbit

A satellite in a 100-mile-high circular orbit around Earth has a velocity of approximately 17,500 miles per hour. If this velocity is multiplied by $\sqrt{2}$, then the satellite will have the minimum velocity necessary to escape Earth's gravity and will follow a parabolic path with the center of Earth as the focus (see figure).

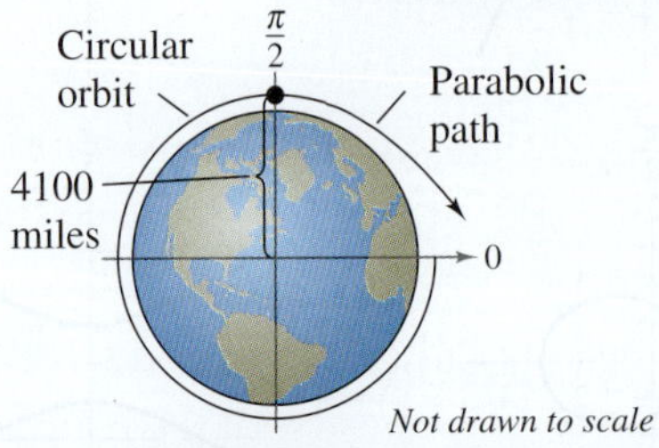

(a) Find a polar equation of the parabolic path of the satellite (assume the radius of Earth is 4000 miles).

(b) Use a graphing utility to graph the equation you found in part (a).

(c) Find the distance between the surface of the Earth and the satellite when $\theta = 30°$.

(d) Find the distance between the surface of Earth and the satellite when $\theta = 60°$.

Cristi Matei/Shutterstock.com

Exploration

True or False? **In Exercises 65–68, determine whether the statement is true or false. Justify your answer.**

65. For values of $e > 1$ and $0 \le \theta \le 2\pi$, the graphs of the following equations are the same.

$$r = \frac{ex}{1 - e\cos\theta} \quad \text{and} \quad r = \frac{e(-x)}{1 + e\cos\theta}$$

66. The graph of $r = 4/(-3 - 3\sin\theta)$ has a horizontal directrix above the pole.

67. The conic represented by the following equation is an ellipse.

$$r^2 = \frac{16}{9 - 4\cos\left(\theta + \frac{\pi}{4}\right)}$$

68. The conic represented by the following equation is a parabola.

$$r = \frac{6}{3 - 2\cos\theta}$$

69. Writing Explain how the graph of each conic differs from the graph of $r = \dfrac{5}{1 + \sin\theta}$. (See Exercise 17.)

(a) $r = \dfrac{5}{1 - \cos\theta}$ (b) $r = \dfrac{5}{1 - \sin\theta}$

(c) $r = \dfrac{5}{1 + \cos\theta}$ (d) $r = \dfrac{5}{1 - \sin[\theta - (\pi/4)]}$

70. HOW DO YOU SEE IT? The graph of

$$r = \frac{e}{1 - e\sin\theta}$$

is shown for different values of e. Determine which graph matches each value of e.

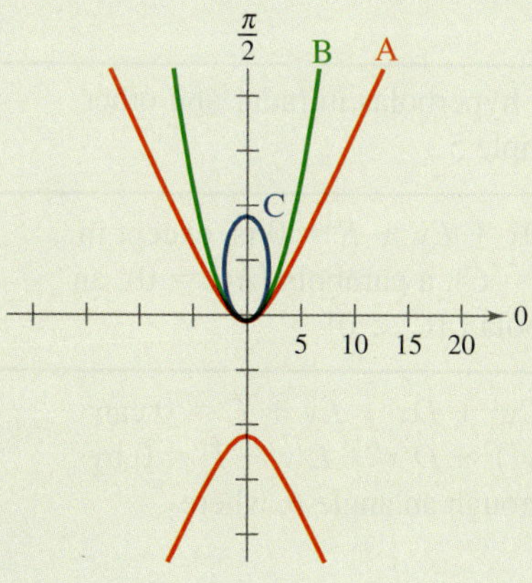

(a) $e = 0.9$ (b) $e = 1.0$ (c) $e = 1.1$

71. Verifying a Polar Equation Show that the polar equation of the ellipse

$$\frac{x^2}{a^2} + \frac{y^2}{b^2} = 1 \quad \text{is} \quad r^2 = \frac{b^2}{1 - e^2\cos^2\theta}.$$

72. Verifying a Polar Equation Show that the polar equation of the hyperbola

$$\frac{x^2}{a^2} - \frac{y^2}{b^2} = 1 \quad \text{is} \quad r^2 = \frac{-b^2}{1 - e^2\cos^2\theta}.$$

Writing a Polar Equation **In Exercises 73–78, use the results of Exercises 71 and 72 to write the polar form of the equation of the conic.**

73. $\dfrac{x^2}{169} + \dfrac{y^2}{144} = 1$

74. $\dfrac{x^2}{25} + \dfrac{y^2}{16} = 1$

75. $\dfrac{x^2}{9} - \dfrac{y^2}{16} = 1$

76. $\dfrac{x^2}{36} - \dfrac{y^2}{4} = 1$

77. Hyperbola
One focus: $(5, 0)$
Vertices: $(4, 0)$, $(4, \pi)$

78. Ellipse
One focus: $(4, 0)$
Vertices: $(5, 0)$, $(5, \pi)$

79. Reasoning Consider the polar equation

$$r = \frac{4}{1 - 0.4\cos\theta}.$$

(a) Identify the conic without graphing the equation.

(b) Without graphing the following polar equations, describe how each differs from the given polar equation.

$$r_1 = \frac{4}{1 + 0.4\cos\theta}$$

$$r_2 = \frac{4}{1 - 0.4\sin\theta}$$

(c) Use a graphing utility to verify your results in part (b).

80. Reasoning The equation

$$r = \frac{ep}{1 \pm e\sin\theta}$$

is the equation of an ellipse with $e < 1$. What happens to the lengths of both the major axis and the minor axis when the value of e remains fixed and the value of p changes? Use an example to explain your reasoning.

Chapter Summary

Section	What Did You Learn?	Explanation/Examples	Review Exercises
Section 10.1	Find the inclination of a line *(p. 688).*	If a nonvertical line has inclination θ and slope m, then $m = \tan \theta$.	1–4
	Find the angle between two lines *(p. 689).*	If two nonperpendicular lines have slopes m_1 and m_2, then the tangent of the angle between the lines is $\tan \theta = \lvert(m_2 - m_1)/(1 + m_1 m_2)\rvert$.	5–8
	Find the distance between a point and a line *(p. 690).*	The distance between the point (x_1, y_1) and the line $Ax + By + C = 0$ is $d = \lvert Ax_1 + By_1 + C\rvert / \sqrt{A^2 + B^2}$.	9, 10
Section 10.2	Recognize a conic as the intersection of a plane and a double-napped cone *(p. 695).*	In the formation of the four basic conics, the intersecting plane does not pass through the vertex of the cone. (See Figure 10.7.)	11, 12
	Write equations of parabolas in standard form *(p. 696).*	**Horizontal Axis** $(y - k)^2 = 4p(x - h),\ p \neq 0$ **Vertical Axis** $(x - h)^2 = 4p(y - k),\ p \neq 0$	13–16
	Use the reflective property of parabolas to solve real-life problems *(p. 698).*	The tangent line to a parabola at a point P makes equal angles with (1) the line passing through P and the focus and (2) the axis of the parabola.	17–20
Section 10.3	Write equations of ellipses in standard form and graph ellipses *(p. 705).*	**Horizontal Major Axis** $\dfrac{(x - h)^2}{a^2} + \dfrac{(y - k)^2}{b^2} = 1$ **Vertical Major Axis** $\dfrac{(x - h)^2}{b^2} + \dfrac{(y - k)^2}{a^2} = 1$	21–24, 27–30
	Use properties of ellipses to model and solve real-life problems *(p. 708).*	You can use the properties of ellipses to find distances from Earth's center to the moon's center in the moon's orbit. (See Example 4.)	25, 26
	Find eccentricities *(p. 709).*	The eccentricity e of an ellipse is given by $e = c/a$.	27–30
Section 10.4	Write equations of hyperbolas in standard form *(p. 714)*, and find asymptotes of and graph hyperbolas *(p. 715).*	**Horizontal Transverse Axis** $\dfrac{(x - h)^2}{a^2} - \dfrac{(y - k)^2}{b^2} = 1$ **Vertical Transverse Axis** $\dfrac{(y - k)^2}{a^2} - \dfrac{(x - h)^2}{b^2} = 1$	31–38
	Use properties of hyperbolas to solve real-life problems *(p. 718).*	You can use the properties of hyperbolas in radar and other detection systems. (See Example 5.)	39, 40
	Classify conics from their general equations *(p. 719).*	The graph of $Ax^2 + Cy^2 + Dx + Ey + F = 0$ is, except in degenerate cases, a circle $(A = C)$, a parabola $(AC = 0)$, an ellipse $(AC > 0)$, or a hyperbola $(AC < 0)$.	41–44
Section 10.5	Rotate the coordinate axes to eliminate the *xy*-term in equations of conics *(p. 723).*	The equation $Ax^2 + Bxy + Cy^2 + Dx + Ey + F = 0$ can be rewritten as $A'(x')^2 + C'(y')^2 + D'x' + E'y' + F' = 0$ by rotating the coordinate axes through an angle θ, where $\cot 2\theta = (A - C)/B$.	45–48
	Use the discriminant to classify conics *(p. 727).*	The graph of $Ax^2 + Bxy + Cy^2 + Dx + Ey + F = 0$ is, except in degenerate cases, an ellipse or a circle $(B^2 - 4AC < 0)$, a parabola $(B^2 - 4AC = 0)$, or a hyperbola $(B^2 - 4AC > 0)$.	49–52

	What Did You Learn?	Explanation/Examples	Review Exercises
Section 10.6	Evaluate sets of parametric equations for given values of the parameter *(p. 731)*.	If f and g are continuous functions of t on an interval I, then the set of ordered pairs $(f(t), g(t))$ is a plane curve C. The equations $x = f(t)$ and $y = g(t)$ are parametric equations for C, and t is the parameter.	53, 54
	Sketch curves that are represented by sets of parametric equations *(p. 732)*.	Sketching a curve represented by parametric equations requires plotting points in the xy-plane. Each set of coordinates (x, y) is determined from a value chosen for t.	55–60
	Rewrite sets of parametric equations as single rectangular equations by eliminating the parameter *(p. 733)*.	To eliminate the parameter in a pair of parametric equations, solve for t in one equation and substitute the value of t into the other equation. The result is the corresponding rectangular equation.	55–60
	Find sets of parametric equations for graphs *(p. 735)*.	When finding a set of parametric equations for a given graph, remember that the parametric equations are not unique.	61–66
Section 10.7	Plot points in the polar coordinate system *(p. 741)*.	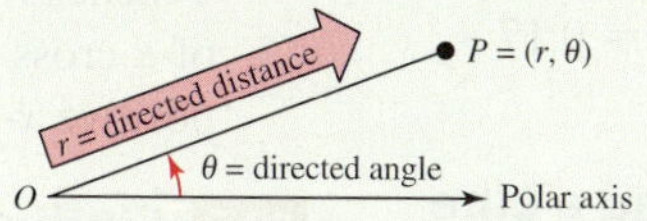	67–70
	Convert points *(p. 743)* and equations *(p. 744)* from rectangular to polar form and vice versa.	**Polar Coordinates (r, θ) and Rectangular Coordinates (x, y)** Polar-to-Rectangular: $x = r\cos\theta,\ y = r\sin\theta$ Rectangular-to-Polar: $\tan\theta = y/x,\ r^2 = x^2 + y^2$	71–90
Section 10.8	Graph polar equations by point plotting *(p. 747)*.	Graphing a polar equation by point plotting is similar to graphing a rectangular equation.	91–100
	Use symmetry, zeros, and maximum r-values to sketch graphs of polar equations *(p. 748)*.	The graph of a polar equation is symmetric with respect to the following when the given substitution yields an equivalent equation. **1.** Line $\theta = \pi/2$: Replace (r, θ) by $(r, \pi - \theta)$ or $(-r, -\theta)$. **2.** Polar axis: Replace (r, θ) by $(r, -\theta)$ or $(-r, \pi - \theta)$. **3.** Pole: Replace (r, θ) by $(r, \pi + \theta)$ or $(-r, \theta)$. Other aids to graphing polar equations are the θ-values for which $\lvert r\rvert$ is maximum and the θ-values for which $r = 0$.	91–100
	Recognize special polar graphs *(p. 751)*.	Several types of graphs, such as limaçons, rose curves, circles, and lemniscates, have equations that are simpler in polar form than in rectangular form. (See page 751.)	101–104
Section 10.9	Define conics in terms of eccentricity, and write and graph equations of conics in polar form *(p. 755)*.	The eccentricity of a conic is denoted by e. **ellipse:** $0 < e < 1$ **parabola:** $e = 1$ **hyperbola:** $e > 1$ The graph of a polar equation of the form (1) $r = (ep)/(1 \pm e\cos\theta)$ or (2) $r = (ep)/(1 \pm e\sin\theta)$ is a conic, where $e > 0$ is the eccentricity and $\lvert p\rvert$ is the distance between the focus (pole) and the directrix.	105–112
	Use equations of conics in polar form to model real-life problems *(p. 758)*.	You can use the equation of a conic in polar form to model the orbit of Halley's comet. (See Example 4.)	113, 114

Review Exercises

See CalcChat.com for tutorial help and worked-out solutions to odd-numbered exercises.

10.1 Finding the Inclination of a Line **In Exercises 1–4, find the inclination θ (in radians and degrees) of the line with the given characteristics.**

1. Passes through the points $(-1, 2)$ and $(2, 5)$
2. Passes through the points $(3, 4)$ and $(-2, 7)$
3. Equation: $y = 2x + 4$
4. Equation: $x - 5y = 7$

Finding the Angle Between Two Lines **In Exercises 5–8, find the angle θ (in radians and degrees) between the lines.**

5. $4x + y = 2$
   $-5x + y = -1$
6. $-5x + 3y = 3$
   $-2x + 3y = 1$
7. $2x - 7y = 8$
   $\frac{2}{5}x + y = 0$
8. $0.02x + 0.07y = 0.18$
   $0.09x - 0.04y = 0.17$

Finding the Distance Between a Point and a Line **In Exercises 9 and 10, find the distance between the point and the line.**

	Point	Line
9.	$(5, 3)$	$x - y = 10$
10.	$(0, 4)$	$x + 2y = 2$

10.2 Forming a Conic Section **In Exercises 11 and 12, state what type of conic is formed by the intersection of the plane and the double-napped cone.**

11. 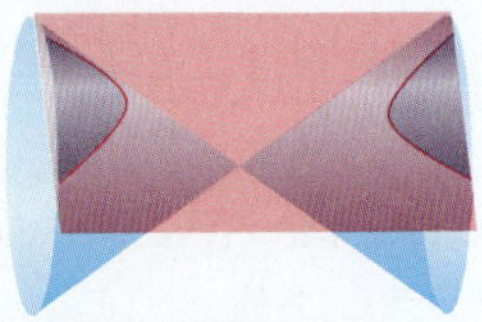

12. 

Finding the Standard Equation of a Parabola **In Exercises 13–16, find the standard form of the equation of the parabola with the given characteristics. Then sketch the parabola.**

13. Vertex: $(0, 0)$
    Focus: $(4, 0)$
14. Vertex: $(2, 0)$
    Focus: $(0, 0)$
15. Vertex: $(0, 2)$
    Directrix: $x = -3$
16. Vertex: $(-3, -3)$
    Directrix: $y = 0$

Finding the Tangent Line at a Point on a Parabola **In Exercises 17 and 18, find the equation of the tangent line to the parabola at the given point.**

17. $y = 2x^2$, $(-1, 2)$
18. $x^2 = -2y$, $(-4, -8)$

19. **Architecture** A parabolic archway is 10 meters high at the vertex. At a height of 8 meters, the width of the archway is 6 meters (see figure). How wide is the archway at ground level?

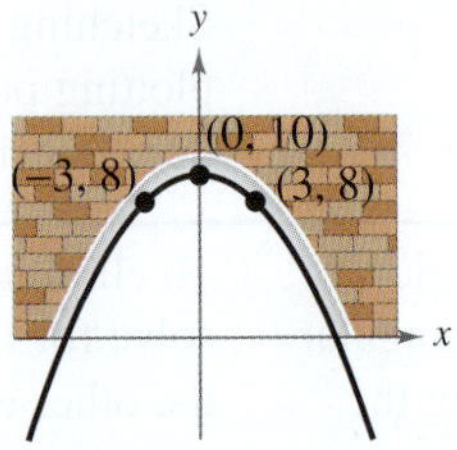

Figure for 19

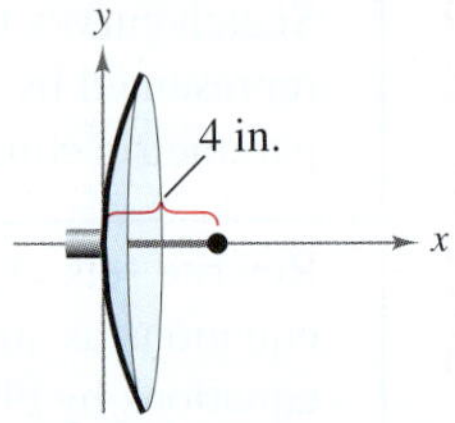

Figure for 20

20. **Parabolic Microphone** The receiver of a parabolic microphone is at the focus of its parabolic reflector, 4 inches from the vertex (see figure). Write an equation of a cross section of the reflector with its focus on the positive x-axis and its vertex at the origin.

10.3 Finding the Standard Equation of an Ellipse **In Exercises 21–24, find the standard form of the equation of the ellipse with the given characteristics.**

21. Vertices: $(-2, 0)$, $(8, 0)$; foci: $(0, 0)$, $(6, 0)$
22. Vertices: $(4, 3)$, $(4, 7)$; foci: $(4, 4)$, $(4, 6)$
23. Vertices: $(0, 1)$, $(4, 1)$; endpoints of the minor axis: $(2, 0)$, $(2, 2)$
24. Vertices: $(-4, -1)$, $(-4, 11)$; endpoints of the minor axis: $(-6, 5)$, $(-2, 5)$

25. **Architecture** A contractor plans to construct a semielliptical arch 10 feet wide and 4 feet high. Where should the foci be placed in order to sketch the arch?

26. **Wading Pool** You are building a wading pool that is in the shape of an ellipse. Your plans give an equation for the elliptical shape of the pool measured in feet as

$$\frac{x^2}{324} + \frac{y^2}{196} = 1.$$

Find the longest distance across the pool, the shortest distance, and the distance between the foci.

Sketching an Ellipse **In Exercises 27–30, find the center, vertices, foci, and eccentricity of the ellipse. Then sketch the ellipse.**

27. $\frac{(x + 1)^2}{25} + \frac{(y - 2)^2}{49} = 1$
28. $\frac{(x - 5)^2}{1} + \frac{(y + 3)^2}{36} = 1$
29. $16x^2 + 9y^2 - 32x + 72y + 16 = 0$
30. $4x^2 + 25y^2 + 16x - 150y + 141 = 0$

10.4 Finding the Standard Equation of a Hyperbola **In Exercises 31–34, find the standard form of the equation of the hyperbola with the given characteristics.**

31. Vertices: $(0, \pm 1)$; foci: $(0, \pm 2)$
32. Vertices: $(3, 3), (-3, 3)$; foci: $(4, 3), (-4, 3)$
33. Foci: $(\pm 5, 0)$; asymptotes: $y = \pm \frac{3}{4}x$
34. Foci: $(0, \pm 13)$; asymptotes: $y = \pm \frac{5}{12}x$

Sketching a Hyperbola **In Exercises 35–38, find the center, vertices, foci, and the equations of the asymptotes of the hyperbola. Then sketch the hyperbola using the asymptotes as an aid.**

35. $\dfrac{(x-5)^2}{36} - \dfrac{(y+3)^2}{16} = 1$
36. $\dfrac{(y-1)^2}{4} - x^2 = 1$
37. $9x^2 - 16y^2 - 18x - 32y - 151 = 0$
38. $-4x^2 + 25y^2 - 8x + 150y + 121 = 0$

39. **Navigation** Radio transmitting station A is located 200 miles east of transmitting station B. A ship is in an area to the north and 40 miles west of station A. Synchronized radio pulses transmitted at 186,000 miles per second by the two stations are received 0.0005 second sooner from station A than from station B. How far north is the ship?

40. **Locating an Explosion** Two of your friends live 4 miles apart and on the same "east-west" street, and you live halfway between them. You are having a three-way phone conversation when you hear an explosion. Six seconds later, your friend to the east hears the explosion, and your friend to the west hears it 8 seconds after you do. Find equations of two hyperbolas that would locate the explosion. (Assume that the coordinate system is measured in feet and that sound travels at 1100 feet per second.)

Classifying a Conic from a General Equation **In Exercises 41–44, classify the graph of the equation as a circle, a parabola, an ellipse, or a hyperbola.**

41. $5x^2 - 2y^2 + 10x - 4y + 17 = 0$
42. $-4y^2 + 5x + 3y + 7 = 0$
43. $3x^2 + 2y^2 - 12x + 12y + 29 = 0$
44. $4x^2 + 4y^2 - 4x + 8y - 11 = 0$

10.5 Rotation of Axes **In Exercises 45–48, rotate the axes to eliminate the xy-term in the equation. Then write the equation in standard form. Sketch the graph of the resulting equation, showing both sets of axes.**

45. $xy + 3 = 0$
46. $x^2 - 4xy + y^2 + 9 = 0$
47. $5x^2 - 2xy + 5y^2 - 12 = 0$
48. $4x^2 + 8xy + 4y^2 + 7\sqrt{2}x + 9\sqrt{2}y = 0$

Rotation and Graphing Utilities **In Exercises 49–52, (a) use the discriminant to classify the graph of the equation, (b) use the Quadratic Formula to solve for y, and (c) use a graphing utility to graph the equation.**

49. $16x^2 - 24xy + 9y^2 - 30x - 40y = 0$
50. $13x^2 - 8xy + 7y^2 - 45 = 0$
51. $x^2 + y^2 + 2xy + 2\sqrt{2}x - 2\sqrt{2}y + 2 = 0$
52. $x^2 - 10xy + y^2 + 1 = 0$

10.6 Sketching a Curve **In Exercises 53 and 54, (a) create a table of x- and y-values for the parametric equations using $t = -2, -1, 0, 1$, and 2, and (b) plot the points (x, y) generated in part (a) and sketch a graph of the parametric equations.**

53. $x = 3t - 2$ and $y = 7 - 4t$
54. $x = \dfrac{1}{4}t$ and $y = \dfrac{6}{t+3}$

Sketching a Curve **In Exercises 55–60, (a) sketch the curve represented by the parametric equations (indicate the orientation of the curve) and (b) eliminate the parameter and write the resulting rectangular equation whose graph represents the curve. Adjust the domain of the rectangular equation, if necessary. (c) Verify your result with a graphing utility.**

55. $x = 2t$, $y = 4t$
56. $x = 1 + 4t$, $y = 2 - 3t$
57. $x = t^2$, $y = \sqrt{t}$
58. $x = t + 4$, $y = t^2$
59. $x = 3\cos\theta$, $y = 3\sin\theta$
60. $x = 3 + 3\cos\theta$, $y = 2 + 5\sin\theta$

Finding Parametric Equations for a Graph **In Exercises 61–64, find a set of parametric equations to represent the graph of the rectangular equation using (a) $t = x$, (b) $t = x + 1$, and (c) $t = 3 - x$.**

61. $y = 2x + 3$
62. $y = 4 - 3x$
63. $y = x^2 + 3$
64. $y = 2 - x^2$
65. $y = 2x^2 + 2$
66. $y = 1 - 4x^2$

10.7 Plotting Points in the Polar Coordinate System **In Exercises 67–70, plot the point given in polar coordinates and find two additional polar representations of the point, using $-2\pi < \theta < 2\pi$.**

67. $\left(2, \dfrac{\pi}{4}\right)$
68. $\left(-5, -\dfrac{\pi}{3}\right)$
69. $(-7, 4.19)$
70. $(\sqrt{3}, 2.62)$

Polar-to-Rectangular Conversion In Exercises 71–74, a point in polar coordinates is given. Convert the point to rectangular coordinates.

71. $\left(-1, \frac{\pi}{3}\right)$ **72.** $\left(2, \frac{5\pi}{4}\right)$

73. $\left(3, \frac{3\pi}{4}\right)$ **74.** $\left(0, \frac{\pi}{2}\right)$

Rectangular-to-Polar Conversion In Exercises 75–78, a point in rectangular coordinates is given. Convert the point to polar coordinates.

75. $(0, 1)$ **76.** $(-\sqrt{5}, \sqrt{5})$

77. $(4, 6)$ **78.** $(3, -4)$

Converting a Rectangular Equation to Polar Form In Exercises 79–84, convert the rectangular equation to polar form.

79. $x^2 + y^2 = 81$ **80.** $x^2 + y^2 = 48$

81. $x^2 + y^2 - 6y = 0$ **82.** $x^2 + y^2 - 4x = 0$

83. $xy = 5$ **84.** $xy = -2$

Converting a Polar Equation to Rectangular Form In Exercises 85–90, convert the polar equation to rectangular form.

85. $r = 5$ **86.** $r = 12$

87. $r = 3 \cos \theta$ **88.** $r = 8 \sin \theta$

89. $r^2 = \sin \theta$ **90.** $r^2 = 4 \cos 2\theta$

10.8 Sketching the Graph of a Polar Equation In Exercises 91–100, sketch the graph of the polar equation using symmetry, zeros, maximum r-values, and any other additional points.

91. $r = 6$ **92.** $r = 11$

93. $r = 4 \sin 2\theta$ **94.** $r = \cos 5\theta$

95. $r = -2(1 + \cos \theta)$ **96.** $r = 1 - 4 \cos \theta$

97. $r = 2 + 6 \sin \theta$ **98.** $r = 5 - 5 \cos \theta$

99. $r = -3 \cos 2\theta$ **100.** $r^2 = \cos 2\theta$

Identifying Types of Polar Graphs In Exercises 101–104, identify the type of polar graph and use a graphing utility to graph the equation.

101. $r = 3(2 - \cos \theta)$ **102.** $r = 5(1 - 2 \cos \theta)$

103. $r = 8 \cos 3\theta$ **104.** $r^2 = 2 \sin 2\theta$

10.9 Sketching a Conic In Exercises 105–108, identify the conic and sketch its graph.

105. $r = \dfrac{1}{1 + 2 \sin \theta}$ **106.** $r = \dfrac{6}{1 + \sin \theta}$

107. $r = \dfrac{4}{5 - 3 \cos \theta}$ **108.** $r = \dfrac{16}{4 + 5 \cos \theta}$

Finding the Polar Equation of a Conic In Exercises 109–112, find a polar equation of the conic with its focus at the pole.

109. Parabola Vertex: $(2, \pi)$

110. Parabola Vertex: $(2, \pi/2)$

111. Ellipse Vertices: $(5, 0), (1, \pi)$

112. Hyperbola Vertices: $(1, 0), (7, 0)$

113. Explorer 18 On November 27, 1963, the United States launched Explorer 18. Its low and high points above the surface of Earth were 110 miles and 122,800 miles, respectively. The center of Earth was at one focus of the orbit (see figure). Find the polar equation of the orbit and find the distance between the surface of Earth (assume Earth has a radius of 4000 miles) and the satellite when $\theta = \pi/3$.

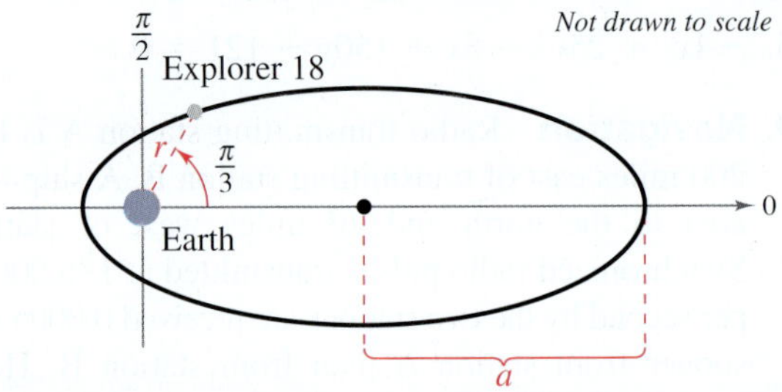

114. Asteroid An asteroid takes a parabolic path with Earth as its focus. It is about 6,000,000 miles from Earth at its closest approach. Write the polar equation of the path of the asteroid with its vertex at $\theta = \pi/2$. Find the distance between the asteroid and Earth when $\theta = -\pi/3$.

Exploration

True or False? In Exercises 115–117, determine whether the statement is true or false. Justify your answer.

115. The graph of $\frac{1}{4}x^2 - y^4 = 1$ is a hyperbola.

116. Only one set of parametric equations can represent the line $y = 3 - 2x$.

117. There is a unique polar coordinate representation of each point in the plane.

118. Think About It Consider an ellipse with the major axis horizontal and 10 units in length. The number b in the standard form of the equation of the ellipse must be less than what real number? Explain the change in the shape of the ellipse as b approaches this number.

119. Think About It What is the relationship between the graphs of the rectangular and polar equations?

(a) $x^2 + y^2 = 25, \quad r = 5$

(b) $x - y = 0, \quad \theta = \frac{\pi}{4}$

Chapter Test

See CalcChat.com for tutorial help and worked-out solutions to odd-numbered exercises.

Take this test as you would take a test in class. When you are finished, check your work against the answers given in the back of the book.

1. Find the inclination of the line $2x - 5y + 5 = 0$.
2. Find the angle between the lines $3x + 2y = 4$ and $4x - y = -6$.
3. Find the distance between the point $(7, 5)$ and the line $y = 5 - x$.

In Exercises 4–7, identify the conic and write the equation in standard form. Find the center, vertices, foci, and the equations of the asymptotes (if applicable). Then sketch the conic.

4. $y^2 - 2x + 2 = 0$
5. $x^2 - 4y^2 - 4x = 0$
6. $9x^2 + 16y^2 + 54x - 32y - 47 = 0$
7. $2x^2 + 2y^2 - 8x - 4y + 9 = 0$

8. Find the standard form of the equation of the parabola with vertex $(2, -3)$ and a vertical axis that passes through the point $(4, 0)$.
9. Find the standard form of the equation of the hyperbola with foci $(0, \pm 2)$ and asymptotes $y = \pm \frac{1}{9}x$.
10. (a) Determine the number of degrees the axes must be rotated to eliminate the xy-term of the conic $x^2 + 6xy + y^2 - 6 = 0$.
    (b) Sketch the conic from part (a) and use a graphing utility to confirm your result.
11. Sketch the curve represented by the parametric equations $x = 2 + 3\cos\theta$ and $y = 2\sin\theta$. Eliminate the parameter and write the resulting rectangular equation.
12. Find a set of parametric equations to represent the graph of the rectangular equation $y = 3 - x^2$ using (a) $t = x$ and (b) $t = x + 2$.
13. Convert the polar coordinates $\left(-2, \frac{5\pi}{6}\right)$ to rectangular form.
14. Convert the rectangular coordinates $(2, -2)$ to polar form and find two additional polar representations of the point, using $-2\pi < \theta < 2\pi$.
15. Convert the rectangular equation $x^2 + y^2 - 3x = 0$ to polar form.

In Exercises 16–19, sketch the graph of the polar equation. Identify the type of graph.

16. $r = \dfrac{4}{1 + \cos\theta}$
17. $r = \dfrac{4}{2 + \sin\theta}$
18. $r = 2 + 3\sin\theta$
19. $r = 2\sin 4\theta$

20. Find a polar equation of the ellipse with focus at the pole, eccentricity $e = \frac{1}{4}$, and directrix $y = 4$.
21. A straight road rises with an inclination of 0.15 radian from the horizontal. Find the slope of the road and the change in elevation over a one-mile stretch of the road.
22. A baseball is hit at a point 3 feet above the ground toward the left field fence. The fence is 10 feet high and 375 feet from home plate. The path of the baseball can be modeled by the parametric equations $x = (115\cos\theta)t$ and $y = 3 + (115\sin\theta)t - 16t^2$. Will the baseball go over the fence when it is hit at an angle of $\theta = 30°$? Will the baseball go over the fence when $\theta = 35°$?

Proofs in Mathematics

Inclination and Slope ***(p. 688)***

If a nonvertical line has inclination θ and slope m, then $m = \tan \theta$.

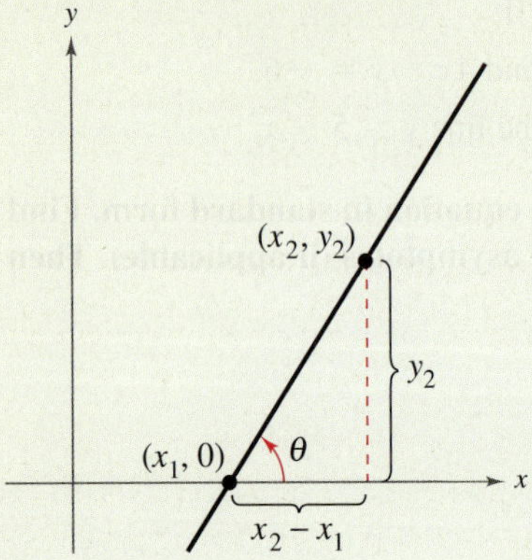

Proof

If $m = 0$, then the line is horizontal and $\theta = 0$. So, the result is true for horizontal lines because $m = 0 = \tan 0$.

If the line has a positive slope, then it will intersect the x-axis. Label this point $(x_1, 0)$, as shown in the figure. If (x_2, y_2) is a second point on the line, then the slope is

$$m = \frac{y_2 - 0}{x_2 - x_1} = \frac{y_2}{x_2 - x_1} = \tan \theta.$$

The case in which the line has a negative slope can be proved in a similar manner. ■

Distance Between a Point and a Line ***(p. 690)***

The distance between the point (x_1, y_1) and the line $Ax + By + C = 0$ is

$$d = \frac{|Ax_1 + By_1 + C|}{\sqrt{A^2 + B^2}}.$$

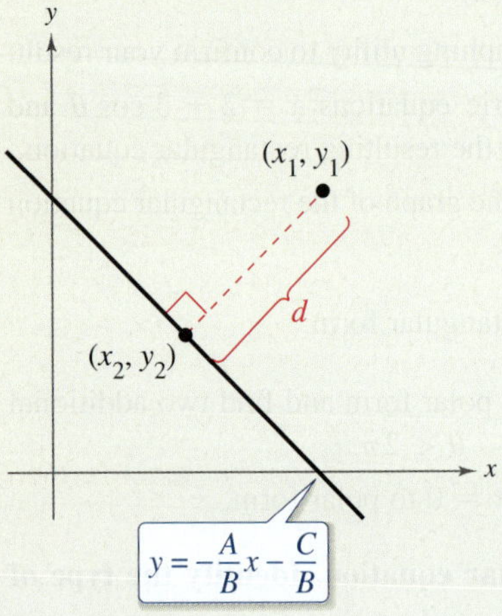

Proof

For simplicity, assume that the given line is neither horizontal nor vertical (see figure). By writing the equation $Ax + By + C = 0$ in slope-intercept form

$$y = -\frac{A}{B}x - \frac{C}{B}$$

you can see that the line has a slope of $m = -A/B$. So, the slope of the line passing through (x_1, y_1) and perpendicular to the given line is B/A, and its equation is $y - y_1 = (B/A)(x - x_1)$. These two lines intersect at the point (x_2, y_2), where

$$x_2 = \frac{B(Bx_1 - Ay_1) - AC}{A^2 + B^2} \quad \text{and} \quad y_2 = \frac{A(-Bx_1 + Ay_1) - BC}{A^2 + B^2}.$$

Finally, the distance between (x_1, y_1) and (x_2, y_2) is

$$\begin{aligned} d &= \sqrt{(x_2 - x_1)^2 + (y_2 - y_1)^2} \\ &= \sqrt{\left(\frac{B^2x_1 - ABy_1 - AC}{A^2 + B^2} - x_1\right)^2 + \left(\frac{-ABx_1 + A^2y_1 - BC}{A^2 + B^2} - y_1\right)^2} \\ &= \sqrt{\frac{A^2(Ax_1 + By_1 + C)^2 + B^2(Ax_1 + By_1 + C)^2}{(A^2 + B^2)^2}} \\ &= \frac{|Ax_1 + By_1 + C|}{\sqrt{A^2 + B^2}} \end{aligned}$$

■

PARABOLIC PATHS

There are many natural occurrences of parabolas in real life. For instance, the famous astronomer Galileo discovered in the 17th century that an object that is projected upward and obliquely to the pull of gravity travels in a parabolic path. Examples of this are the center of gravity of a jumping dolphin and the path of water molecules in a drinking fountain.

Standard Equation of a Parabola *(p. 696)*

The standard form of the equation of a parabola with vertex at (h, k) is as follows.

$(x - h)^2 = 4p(y - k), \quad p \neq 0$ — Vertical axis; directrix: $y = k - p$

$(y - k)^2 = 4p(x - h), \quad p \neq 0$ — Horizontal axis; directrix: $x = h - p$

The focus lies on the axis p units (*directed distance*) from the vertex. If the vertex is at the origin, then the equation takes one of the following forms.

$x^2 = 4py$ — Vertical axis

$y^2 = 4px$ — Horizontal axis

Proof

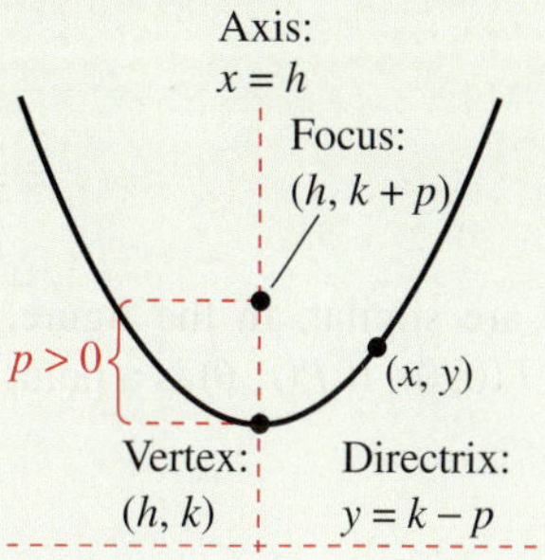

Parabola with vertical axis

For the case in which the directrix is parallel to the x-axis and the focus lies above the vertex, as shown in the top figure, if (x, y) is any point on the parabola, then, by definition, it is equidistant from the focus

$$(h, k + p)$$

and the directrix

$$y = k - p.$$

So, you have

$$\begin{aligned}
\sqrt{(x-h)^2 + [y-(k+p)]^2} &= y - (k - p) \\
(x-h)^2 + [y-(k+p)]^2 &= [y-(k-p)]^2 \\
(x-h)^2 + y^2 - 2y(k+p) + (k+p)^2 &= y^2 - 2y(k-p) + (k-p)^2 \\
(x-h)^2 + y^2 - 2ky - 2py + k^2 + 2pk + p^2 &= y^2 - 2ky + 2py + k^2 - 2pk + p^2 \\
(x-h)^2 - 2py + 2pk &= 2py - 2pk \\
(x-h)^2 &= 4p(y-k).
\end{aligned}$$

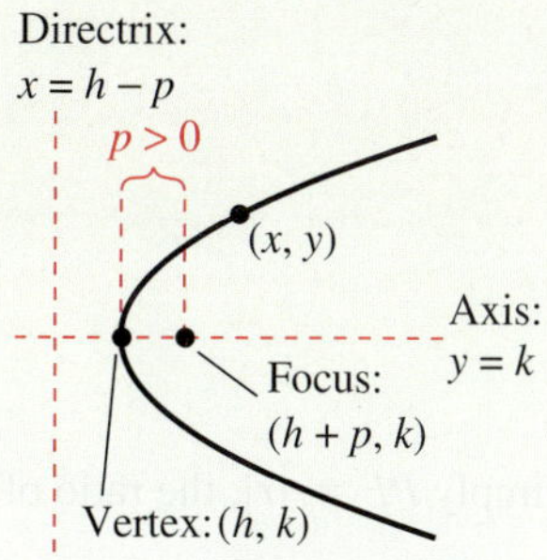

Parabola with horizontal axis

For the case in which the directrix is parallel to the y-axis and the focus lies to the right of the vertex, as shown in the bottom figure, if (x, y) is any point on the parabola, then, by definition, it is equidistant from the focus

$$(h + p, k)$$

and the directrix

$$x = h - p.$$

So, you have

$$\begin{aligned}
\sqrt{[x-(h+p)]^2 + (y-k)^2} &= x - (h - p) \\
[x-(h+p)]^2 + (y-k)^2 &= [x-(h-p)]^2 \\
x^2 - 2x(h+p) + (h+p)^2 + (y-k)^2 &= x^2 - 2x(h-p) + (h-p)^2 \\
x^2 - 2hx - 2px + h^2 + 2ph + p^2 + (y-k)^2 &= x^2 - 2hx + 2px + h^2 - 2ph + p^2 \\
-2px + 2ph + (y-k)^2 &= 2px - 2ph \\
(y-k)^2 &= 4p(x-h).
\end{aligned}$$

Note that if a parabola is centered at the origin, then the two equations above would simplify to $x^2 = 4py$ and $y^2 = 4px$, respectively. ■

Polar Equations of Conics *(p. 755)*

The graph of a polar equation of the form

1. $r = \dfrac{ep}{1 \pm e \cos \theta}$

or

2. $r = \dfrac{ep}{1 \pm e \sin \theta}$

is a conic, where $e > 0$ is the eccentricity and $|p|$ is the distance between the focus (pole) and the directrix.

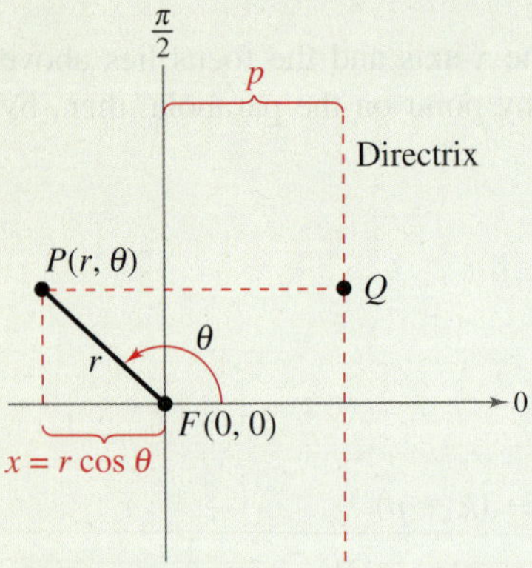

Proof

A proof for

$$r = \frac{ep}{1 + e \cos \theta}$$

with $p > 0$ is shown here. The proofs of the other cases are similar. In the figure, consider a vertical directrix, p units to the right of the focus $F(0, 0)$. If $P(r, \theta)$ is a point on the graph of

$$r = \frac{ep}{1 + e \cos \theta}$$

then the distance between P and the directrix is

$$\begin{aligned} PQ &= |p - x| \\ &= |p - r \cos \theta| \\ &= \left| p - \left(\frac{ep}{1 + e \cos \theta} \right) \cos \theta \right| \\ &= \left| p \left(1 - \frac{e \cos \theta}{1 + e \cos \theta} \right) \right| \\ &= \left| \frac{p}{1 + e \cos \theta} \right| \\ &= \left| \frac{r}{e} \right|. \end{aligned}$$

Moreover, because the distance between P and the pole is simply $PF = |r|$, the ratio of PF to PQ is

$$\begin{aligned} \frac{PF}{PQ} &= \frac{|r|}{\left| \frac{r}{e} \right|} \\ &= |e| \\ &= e \end{aligned}$$

and, by definition, the graph of the equation must be a conic. ■

P.S. Problem Solving

1. **Mountain Climbing** Several mountain climbers are located in a mountain pass between two peaks. The angles of elevation to the two peaks are 0.84 radian and 1.10 radians. A range finder shows that the distances to the peaks are 3250 feet and 6700 feet, respectively (see figure).

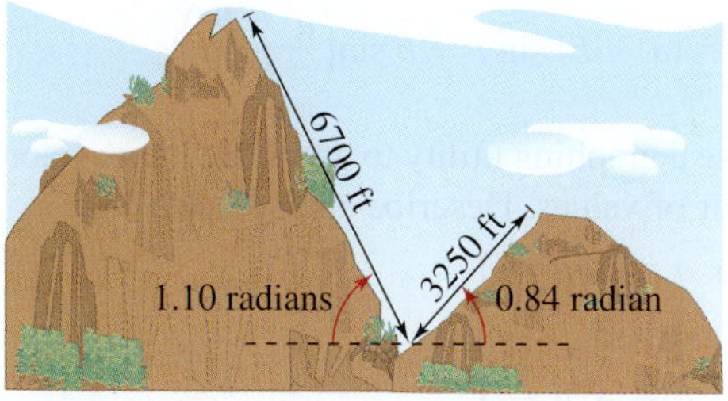

   (a) Find the angle between the two lines.

   (b) Approximate the amount of vertical climb that is necessary to reach the summit of each peak.

2. **Finding the Equation of a Parabola** Find the general equation of a parabola that has the x-axis as the axis of symmetry and the focus at the origin.

3. **Area** Find the area of the square inscribed in the ellipse, as shown below.

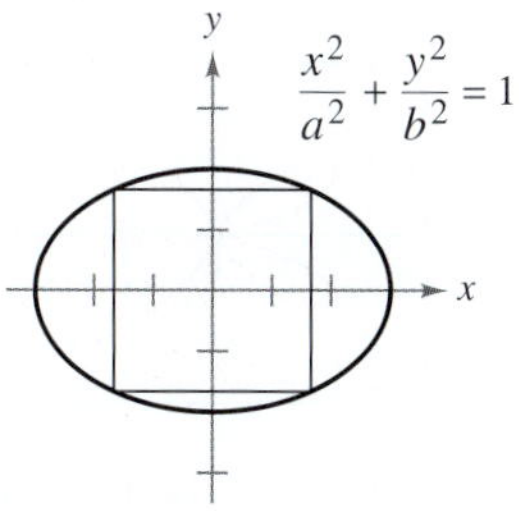

4. **Involute** The *involute* of a circle is described by the endpoint P of a string that is held taut as it is unwound from a spool (see figure). The spool does not rotate. Show that

   $x = r(\cos\theta + \theta\sin\theta)$

   and

   $y = r(\sin\theta - \theta\cos\theta)$

   is a parametric representation of the involute of a circle.

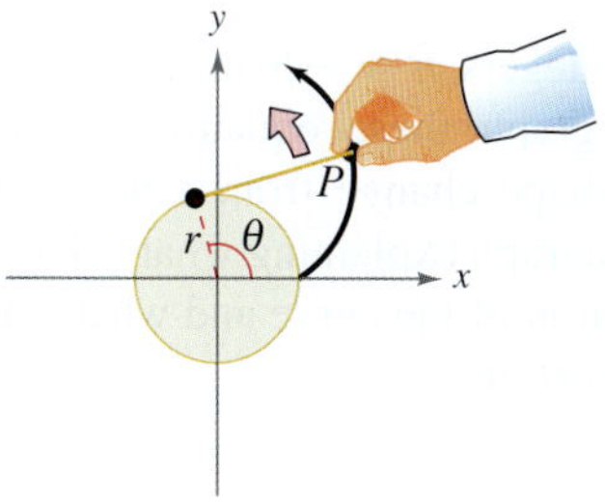

5. **Tour Boat** A tour boat travels between two islands that are 12 miles apart (see figure). There is enough fuel for a 20-mile trip.

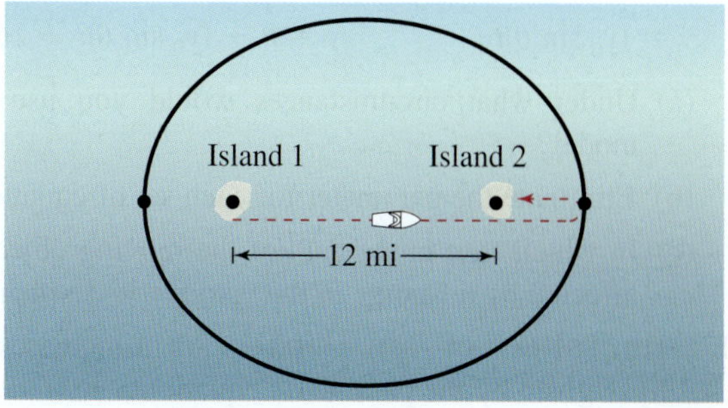

   (a) Explain why the region in which the boat can travel is bounded by an ellipse.

   (b) Let $(0, 0)$ represent the center of the ellipse. Find the coordinates of each island.

   (c) The boat travels from Island 1, past Island 2 to a vertex of the ellipse, and then to Island 2. How many miles does the boat travel? Use your answer to find the coordinates of the vertex.

   (d) Use the results from parts (b) and (c) to write an equation of the ellipse that bounds the region in which the boat can travel.

6. **Finding the Equation of a Hyperbola** Find an equation of the hyperbola such that for any point on the hyperbola, the absolute value of the difference of its distances from the points $(2, 2)$ and $(10, 2)$ is 6.

7. **Proof** Prove that the graph of the equation

   $Ax^2 + Cy^2 + Dx + Ey + F = 0$

   is one of the following (except in degenerate cases).

Conic	Condition
(a) Circle	$A = C$
(b) Parabola	$A = 0$ or $C = 0$ (but not both)
(c) Ellipse	$AC > 0$
(d) Hyperbola	$AC < 0$

8. **Proof** Prove that

   $c^2 = a^2 + b^2$

   for the equation of the hyperbola

   $$\frac{x^2}{a^2} - \frac{y^2}{b^2} = 1$$

   where the distance from the center of the hyperbola $(0, 0)$ to a focus is c.

9. **Projectile Motion** The following sets of parametric equations model projectile motion.

$x = (v_0 \cos \theta)t$ $\quad\quad$ $x = (v_0 \cos \theta)t$

$y = (v_0 \sin \theta)t$ $\quad\quad$ $y = h + (v_0 \sin \theta)t - 16t^2$

(a) Under what circumstances would you use each model?

(b) Eliminate the parameter for each set of equations.

(c) In which case is the path of the moving object not affected by a change in the velocity v? Explain.

10. **Orientation of an Ellipse** As t increases, the ellipse given by the parametric equations

$$x = \cos t$$

and

$$y = 2 \sin t$$

is traced out *counterclockwise*. Find a parametric representation for which the same ellipse is traced out *clockwise*.

11. **Rose Curves** The rose curves described in this chapter are of the form

$$r = a \cos n\theta$$

or

$$r = a \sin n\theta$$

where n is a positive integer that is greater than or equal to 2. Use a graphing utility to graph $r = a \cos n\theta$ and $r = a \sin n\theta$ for some noninteger values of n. Describe the graphs.

12. **Strophoid** The curve given by the parametric equations

$$x = \frac{1 - t^2}{1 + t^2}$$

and

$$y = \frac{t(1 - t^2)}{1 + t^2}$$

is called a **strophoid.**

(a) Find a rectangular equation of the strophoid.

(b) Find a polar equation of the strophoid.

(c) Use a graphing utility to graph the strophoid.

13. **Hypocycloid** A **hypocycloid** has the parametric equations

$$x = (a - b) \cos t + b \cos\left(\frac{a - b}{b}t\right)$$

and

$$y = (a - b) \sin t - b \sin\left(\frac{a - b}{b}t\right).$$

Use a graphing utility to graph the hypocycloid for each pair of values. Describe each graph.

(a) $a = 2, b = 1$

(b) $a = 3, b = 1$

(c) $a = 4, b = 1$

(d) $a = 10, b = 1$

(e) $a = 3, b = 2$

(f) $a = 4, b = 3$

14. **Butterfly Curve** The graph of the polar equation

$$r = e^{\cos \theta} - 2 \cos 4\theta + \sin^5\left(\frac{\theta}{12}\right)$$

is called the *butterfly curve,* as shown in the figure.

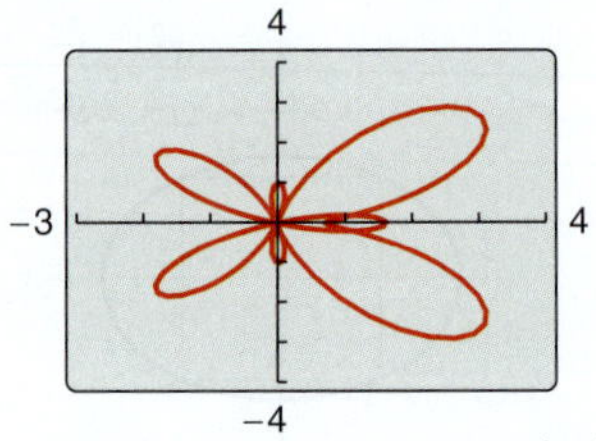

$r = e^{\cos \theta} - 2 \cos 4\theta + \sin^5\left(\frac{\theta}{12}\right)$

(a) The graph shown was produced using $0 \le \theta \le 2\pi$. Does this show the entire graph? Explain your reasoning.

(b) Approximate the maximum r-value of the graph. Does this value change when you use $0 \le \theta \le 4\pi$ instead of $0 \le \theta \le 2\pi$? Explain.

15. **Writing** Use a graphing utility to graph the polar equation

$$r = \cos 5\theta + n \cos \theta$$

for the integers $n = -5$ to $n = 5$ using $0 \le \theta \le \pi$. As you graph these equations, you should see the graph's shape change from a heart to a bell. Write a short paragraph explaining what values of n produce the heart portion of the curve and what values of n produce the bell portion.

A.1 Real Numbers and Their Properties

Real numbers can represent many real-life quantities. For example, in Exercises 55–58 on page A12, you will use real numbers to represent the federal deficit.

- **Represent and classify real numbers.**
- **Order real numbers and use inequalities.**
- **Find the absolute values of real numbers and find the distance between two real numbers.**
- **Evaluate algebraic expressions.**
- **Use the basic rules and properties of algebra.**

Real Numbers

Real numbers can describe quantities in everyday life such as age, miles per gallon, and population. Symbols such as

$$-5,\ 9,\ 0,\ \tfrac{4}{3},\ 0.666\ .\ .\ .\ ,\ 28.21,\ \sqrt{2},\ \pi,\ \text{and}\ \sqrt[3]{-32}$$

represent real numbers. Here are some important **subsets** (each member of a subset B is also a member of a set A) of the real numbers. The three dots, called *ellipsis points,* indicate that the pattern continues indefinitely.

$\{1, 2, 3, 4, . . .\}$ Set of natural numbers

$\{0, 1, 2, 3, 4, . . .\}$ Set of whole numbers

$\{. . . , -3, -2, -1, 0, 1, 2, 3, . . .\}$ Set of integers

A real number is **rational** when it can be written as the ratio p/q of two integers, where $q \neq 0$. For instance, the numbers

$$\tfrac{1}{3} = 0.3333\ .\ .\ . = 0.\overline{3},\ \tfrac{1}{8} = 0.125,\ \text{and}\ \tfrac{125}{111} = 1.126126\ .\ .\ . = 1.\overline{126}$$

are rational. The decimal representation of a rational number either repeats $\left(\text{as in } \tfrac{173}{55} = 3.1\overline{45}\right)$ or terminates $\left(\text{as in } \tfrac{1}{2} = 0.5\right)$. A real number that cannot be written as the ratio of two integers is called **irrational.** Irrational numbers have infinite nonrepeating decimal representations. For instance, the numbers

$$\sqrt{2} = 1.4142135 . . . \approx 1.41 \quad \text{and} \quad \pi = 3.1415926 . . . \approx 3.14$$

are irrational. (The symbol $\approx$ means "is approximately equal to.") Figure A.1 shows subsets of real numbers and their relationships to each other.

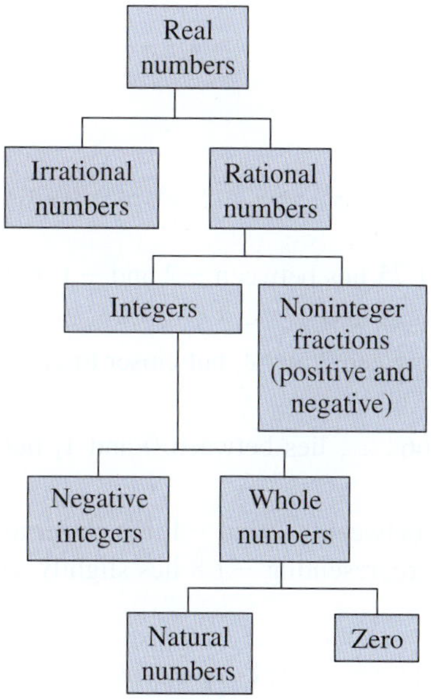

Subsets of real numbers

Figure A.1

EXAMPLE 1 Classifying Real Numbers

Determine which numbers in the set $\left\{-13, -\sqrt{5}, -1, -\tfrac{1}{3}, 0, \tfrac{5}{8}, \sqrt{2}, \pi, 7\right\}$ are (a) natural numbers, (b) whole numbers, (c) integers, (d) rational numbers, and (e) irrational numbers.

Solution

a. Natural numbers: $\{7\}$

b. Whole numbers: $\{0, 7\}$

c. Integers: $\{-13, -1, 0, 7\}$

d. Rational numbers: $\left\{-13, -1, -\dfrac{1}{3}, 0, \dfrac{5}{8}, 7\right\}$

e. Irrational numbers: $\left\{-\sqrt{5}, \sqrt{2}, \pi\right\}$

✓ **Checkpoint** *Audio-video solution in English & Spanish at LarsonPrecalculus.com.*

Repeat Example 1 for the set $\left\{-\pi, -\tfrac{1}{4}, \tfrac{6}{3}, \tfrac{1}{2}\sqrt{2}, -7.5, -1, 8, -22\right\}$.

Michael G Smith/Shutterstock.com

Real numbers are represented graphically on the **real number line.** When you draw a point on the real number line that corresponds to a real number, you are **plotting** the real number. The point 0 on the real number line is the **origin.** Numbers to the right of 0 are positive, and numbers to the left of 0 are negative, as shown below. The term **nonnegative** describes a number that is either positive or zero.

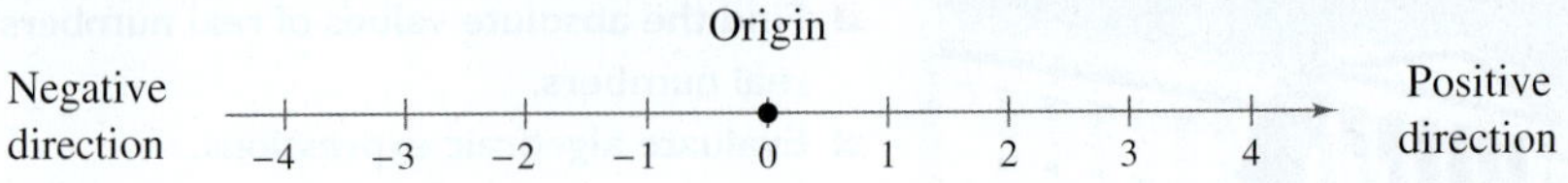

As illustrated below, there is a *one-to-one correspondence* between real numbers and points on the real number line.

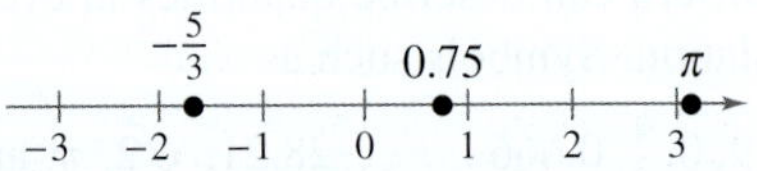

Every real number corresponds to exactly one point on the real number line.

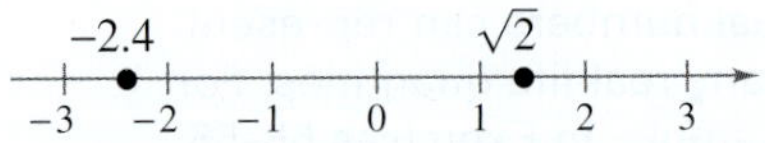

Every point on the real number line corresponds to exactly one real number.

EXAMPLE 2 Plotting Points on the Real Number Line

Plot the real numbers on the real number line.

a. $-\dfrac{7}{4}$

b. 2.3

c. $\dfrac{2}{3}$

d. -1.8

Solution The following figure shows all four points.

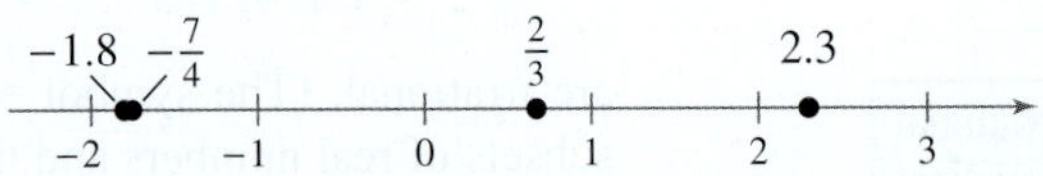

a. The point representing the real number $-\frac{7}{4} = -1.75$ lies between -2 and -1, but closer to -2, on the real number line.

b. The point representing the real number 2.3 lies between 2 and 3, but closer to 2, on the real number line.

c. The point representing the real number $\frac{2}{3} = 0.666\ldots$ lies between 0 and 1, but closer to 1, on the real number line.

d. The point representing the real number -1.8 lies between -2 and -1, but closer to -2, on the real number line. Note that the point representing -1.8 lies slightly to the left of the point representing $-\frac{7}{4}$.

✓ ***Checkpoint*** *Audio-video solution in English & Spanish at LarsonPrecalculus.com.*

Plot the real numbers on the real number line.

a. $\dfrac{5}{2}$ **b.** -1.6

c. $-\dfrac{3}{4}$ **d.** 0.7

Ordering Real Numbers

One important property of real numbers is that they are *ordered.*

> **Definition of Order on the Real Number Line**
>
> If a and b are real numbers, then a is less than b when $b - a$ is positive. The **inequality** $a < b$ denotes the **order** of a and b. This relationship can also be described by saying that b is *greater than* a and writing $b > a$. The inequality $a \le b$ means that a is *less than or equal to* b, and the inequality $b \ge a$ means that b is *greater than or equal to* a. The symbols $<$, $>$, $\le$, and $\ge$ are *inequality symbols.*

$a < b$ if and only if a lies to the left of b.

Figure A.2

Geometrically, this definition implies that $a < b$ if and only if a lies to the *left* of b on the real number line, as shown in Figure A.2.

EXAMPLE 3 Ordering Real Numbers

Place the appropriate inequality symbol ($<$ or $>$) between the pair of real numbers.

a. $-3, 0$ **b.** $-2, -4$ **c.** $\frac{1}{4}, \frac{1}{3}$ **d.** $-\frac{1}{5}, -\frac{1}{2}$

Solution

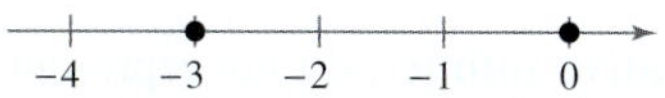

Figure A.3

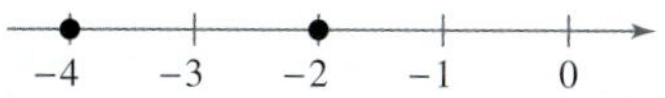

Figure A.4

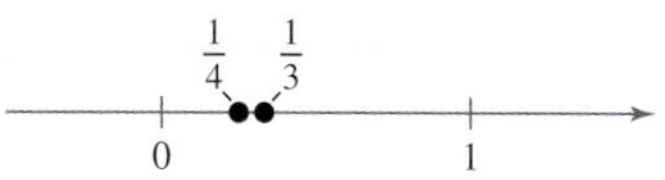

Figure A.5

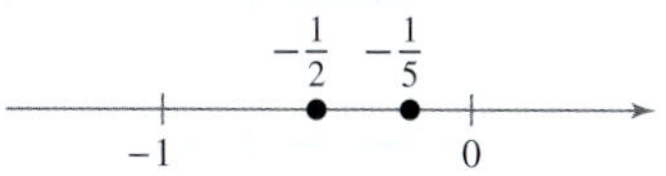

Figure A.6

a. Because -3 lies to the left of 0 on the real number line, as shown in Figure A.3, you can say that -3 is *less than* 0, and write $-3 < 0$.

b. Because -2 lies to the right of -4 on the real number line, as shown in Figure A.4, you can say that -2 is *greater than* -4, and write $-2 > -4$.

c. Because $\frac{1}{4}$ lies to the left of $\frac{1}{3}$ on the real number line, as shown in Figure A.5, you can say that $\frac{1}{4}$ is *less than* $\frac{1}{3}$, and write $\frac{1}{4} < \frac{1}{3}$.

d. Because $-\frac{1}{5}$ lies to the right of $-\frac{1}{2}$ on the real number line, as shown in Figure A.6, you can say that $-\frac{1}{5}$ is *greater than* $-\frac{1}{2}$, and write $-\frac{1}{5} > -\frac{1}{2}$.

✓ ***Checkpoint*** *Audio-video solution in English & Spanish at LarsonPrecalculus.com.*

Place the appropriate inequality symbol ($<$ or $>$) between the pair of real numbers.

a. $1, -5$ **b.** $\frac{3}{2}, 7$ **c.** $-\frac{2}{3}, -\frac{3}{4}$ **d.** $-3.5, 1$

EXAMPLE 4 Interpreting Inequalities

Describe the subset of real numbers that the inequality represents.

a. $x \le 2$ **b.** $-2 \le x < 3$

Solution

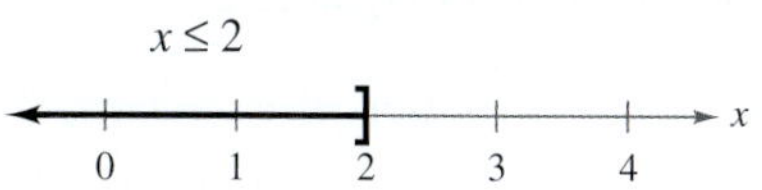

Figure A.7

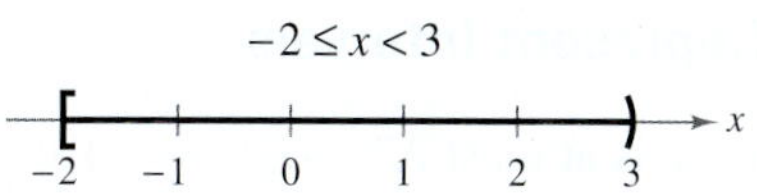

Figure A.8

a. The inequality $x \le 2$ denotes all real numbers less than or equal to 2, as shown in Figure A.7.

b. The inequality $-2 \le x < 3$ means that $x \ge -2$ *and* $x < 3$. This "double inequality" denotes all real numbers between -2 and 3, including -2 but not including 3, as shown in Figure A.8.

✓ ***Checkpoint*** *Audio-video solution in English & Spanish at LarsonPrecalculus.com.*

Describe the subset of real numbers that the inequality represents.

a. $x > -3$ **b.** $0 < x \le 4$

Inequalities can describe subsets of real numbers called **intervals.** In the bounded intervals below, the real numbers a and b are the **endpoints** of each interval. The endpoints of a closed interval are included in the interval, whereas the endpoints of an open interval are not included in the interval.

REMARK The reason that the four types of intervals at the right are called *bounded* is that each has a finite length. An interval that does not have a finite length is *unbounded* (see below).

Bounded Intervals on the Real Number Line

Notation	Interval Type	Inequality	Graph
$[a, b]$	Closed	$a \le x \le b$	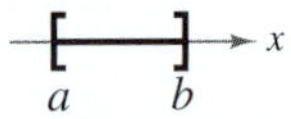
(a, b)	Open	$a < x < b$	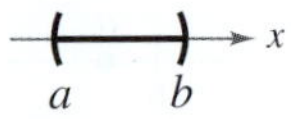
$[a, b)$		$a \le x < b$	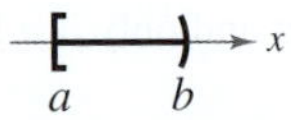
$(a, b]$		$a < x \le b$	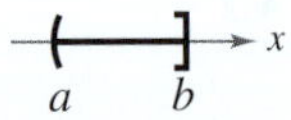

The symbols ∞, **positive infinity,** and $-\infty$, **negative infinity,** do not represent real numbers. They are simply convenient symbols used to describe the unboundedness of an interval such as $(1, \infty)$ or $(-\infty, 3]$.

REMARK Whenever you write an interval containing ∞ or $-\infty$, always use a parenthesis and never a bracket next to these symbols. This is because ∞ and $-\infty$ are never an endpoint of an interval and therefore are not included in the interval.

Unbounded Intervals on the Real Number Line

Notation	Interval Type	Inequality	Graph
$[a, \infty)$		$x \ge a$	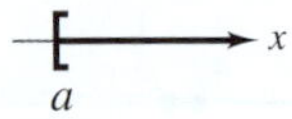
(a, ∞)	Open	$x > a$	
$(-\infty, b]$		$x \le b$	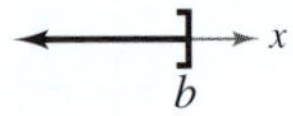
$(-\infty, b)$	Open	$x < b$	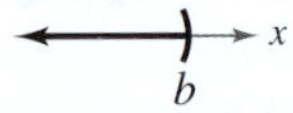
$(-\infty, \infty)$	Entire real line	$-\infty < x < \infty$	

EXAMPLE 5 **Interpreting Intervals**

a. The interval $(-1, 0)$ consists of all real numbers greater than -1 and less than 0.

b. The interval $[2, \infty)$ consists of all real numbers greater than or equal to 2.

✓ ***Checkpoint*** *Audio-video solution in English & Spanish at LarsonPrecalculus.com.*

Give a verbal description of the interval $[-2, 5)$.

EXAMPLE 6 **Using Inequalities to Represent Intervals**

a. The inequality $c \le 2$ can represent the statement "c is at most 2."

b. The inequality $-3 < x \le 5$ can represent "all x in the interval $(-3, 5]$."

✓ ***Checkpoint*** *Audio-video solution in English & Spanish at LarsonPrecalculus.com.*

Use inequality notation to represent the statement "x is greater than -2 and at most 4."

Absolute Value and Distance

The **absolute value** of a real number is its *magnitude*, or the distance between the origin and the point representing the real number on the real number line.

> **Definition of Absolute Value**
>
> If a is a real number, then the absolute value of a is
>
> $$|a| = \begin{cases} a, & \text{if } a \geq 0 \\ -a, & \text{if } a < 0 \end{cases}.$$

Notice in this definition that the absolute value of a real number is never negative. For instance, if $a = -5$, then $|-5| = -(-5) = 5$. The absolute value of a real number is either positive or zero. Moreover, 0 is the only real number whose absolute value is 0. So, $|0| = 0$.

EXAMPLE 7 **Finding Absolute Values**

a. $|-15| = 15$ **b.** $\left|\frac{2}{3}\right| = \frac{2}{3}$

c. $|-4.3| = 4.3$ **d.** $-|-6| = -(6) = -6$

Checkpoint *Audio-video solution in English & Spanish at LarsonPrecalculus.com.*

Evaluate each expression.

a. $|1|$ **b.** $-\left|\frac{3}{4}\right|$

c. $\frac{2}{|-3|}$ **d.** $-|0.7|$

EXAMPLE 8 **Evaluating the Absolute Value of a Number**

Evaluate $\frac{|x|}{x}$ for (a) $x > 0$ and (b) $x < 0$.

Solution

a. If $x > 0$, then $|x| = x$ and $\frac{|x|}{x} = \frac{x}{x} = 1$.

b. If $x < 0$, then $|x| = -x$ and $\frac{|x|}{x} = \frac{-x}{x} = -1$.

Checkpoint *Audio-video solution in English & Spanish at LarsonPrecalculus.com.*

Evaluate $\frac{|x + 3|}{x + 3}$ for (a) $x > -3$ and (b) $x < -3$.

The **Law of Trichotomy** states that for any two real numbers a and b, *precisely* one of three relationships is possible:

$$a = b, \quad a < b, \quad \text{or} \quad a > b. \qquad \text{Law of Trichotomy}$$

EXAMPLE 9 Comparing Real Numbers

Place the appropriate symbol ($<$, $>$, or $=$) between the pair of real numbers.

a. $|-4| \;\square\; |3|$ **b.** $|-10| \;\square\; |10|$ **c.** $-|-7| \;\square\; |-7|$

Solution

a. $|-4| > |3|$ because $|-4| = 4$ and $|3| = 3$, and 4 is greater than 3.

b. $|-10| = |10|$ because $|-10| = 10$ and $|10| = 10$.

c. $-|-7| < |-7|$ because $-|-7| = -7$ and $|-7| = 7$, and -7 is less than 7.

✓ ***Checkpoint*** *Audio-video solution in English & Spanish at LarsonPrecalculus.com.*

Place the appropriate symbol ($<$, $>$, or $=$) between the pair of real numbers.

a. $|-3| \;\square\; |4|$ **b.** $-|-4| \;\square\; -|4|$ **c.** $|-3| \;\square\; -|-3|$

Properties of Absolute Values

1. $|a| \geq 0$

2. $|-a| = |a|$

3. $|ab| = |a||b|$

4. $\left|\dfrac{a}{b}\right| = \dfrac{|a|}{|b|}, \quad b \neq 0$

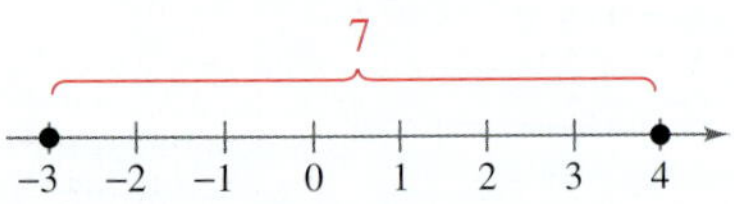

The distance between -3 and 4 is 7.
Figure A.9

Absolute value can be used to define the distance between two points on the real number line. For instance, the distance between -3 and 4 is

$$\begin{aligned} |-3 - 4| &= |-7| \\ &= 7 \end{aligned}$$

as shown in Figure A.9.

Distance Between Two Points on the Real Number Line

Let a and b be real numbers. The **distance between a and b** is

$$d(a, b) = |b - a| = |a - b|.$$

EXAMPLE 10 Finding a Distance

Find the distance between -25 and 13.

Solution

The distance between -25 and 13 is

$$|-25 - 13| = |-38| = 38. \qquad \text{Distance between } -25 \text{ and } 13$$

The distance can also be found as follows.

$$|13 - (-25)| = |38| = 38 \qquad \text{Distance between } -25 \text{ and } 13$$

✓ ***Checkpoint*** *Audio-video solution in English & Spanish at LarsonPrecalculus.com.*

a. Find the distance between 35 and -23.

b. Find the distance between -35 and -23.

c. Find the distance between 35 and 23.

Algebraic Expressions

One characteristic of algebra is the use of letters to represent numbers. The letters are **variables,** and combinations of letters and numbers are **algebraic expressions.** Here are a few examples of algebraic expressions.

$$5x, \qquad 2x - 3, \qquad \frac{4}{x^2 + 2}, \qquad 7x + y$$

> **Definition of an Algebraic Expression**
>
> An **algebraic expression** is a collection of letters **(variables)** and real numbers **(constants)** combined using the operations of addition, subtraction, multiplication, division, and exponentiation.

The **terms** of an algebraic expression are those parts that are separated by *addition.* For example, $x^2 - 5x + 8 = x^2 + (-5x) + 8$ has three terms: x^2 and $-5x$ are the **variable terms** and 8 is the **constant term.** The numerical factor of a term is called the **coefficient.** For instance, the coefficient of $-5x$ is -5, and the coefficient of x^2 is 1.

EXAMPLE 11 Identifying Terms and Coefficients

	Algebraic Expression	Terms	Coefficients
a.	$5x - \frac{1}{7}$	$5x, -\frac{1}{7}$	$5, -\frac{1}{7}$
b.	$2x^2 - 6x + 9$	$2x^2, -6x, 9$	$2, -6, 9$
c.	$\frac{3}{x} + \frac{1}{2}x^4 - y$	$\frac{3}{x}, \frac{1}{2}x^4, -y$	$3, \frac{1}{2}, -1$

✓ ***Checkpoint*** *Audio-video solution in English & Spanish at LarsonPrecalculus.com.*

Identify the terms and coefficients of $-2x + 4$.

To **evaluate** an algebraic expression, substitute numerical values for each of the variables in the expression, as shown in the next example.

EXAMPLE 12 Evaluating Algebraic Expressions

	Expression	Value of Variable	Substitute.	Value of Expression
a.	$-3x + 5$	$x = 3$	$-3(3) + 5$	$-9 + 5 = -4$
b.	$3x^2 + 2x - 1$	$x = -1$	$3(-1)^2 + 2(-1) - 1$	$3 - 2 - 1 = 0$
c.	$\frac{2x}{x + 1}$	$x = -3$	$\frac{2(-3)}{-3 + 1}$	$\frac{-6}{-2} = 3$

Note that you must substitute the value for *each* occurrence of the variable.

✓ ***Checkpoint*** *Audio-video solution in English & Spanish at LarsonPrecalculus.com.*

Evaluate $4x - 5$ when $x = 0$.

Use the **Substitution Principle** to evaluate algebraic expressions. It states that "If $a = b$, then b can replace a in any expression involving a." In Example 12(a), for instance, 3 is *substituted* for x in the expression $-3x + 5$.

Basic Rules of Algebra

There are four arithmetic operations with real numbers: *addition, multiplication, subtraction,* and *division,* denoted by the symbols $+$, $\times$ or $\cdot$, $-$, and $\div$ or $/$, respectively. Of these, addition and multiplication are the two primary operations. Subtraction and division are the inverse operations of addition and multiplication, respectively.

Definitions of Subtraction and Division

Subtraction: Add the opposite.

$$a - b = a + (-b)$$

Division: Multiply by the reciprocal.

If $b \neq 0$, then $a/b = a\left(\frac{1}{b}\right) = \frac{a}{b}$.

In these definitions, $-b$ is the **additive inverse** (or opposite) of b, and $1/b$ is the **multiplicative inverse** (or reciprocal) of b. In the fractional form a/b, a is the **numerator** of the fraction and b is the **denominator.**

Because the properties of real numbers below are true for variables and algebraic expressions as well as for real numbers, they are often called the **Basic Rules of Algebra.** Try to formulate a verbal description of each property. For instance, the first property states that *the order in which two real numbers are added does not affect their sum.*

Basic Rules of Algebra

Let a, b, and c be real numbers, variables, or algebraic expressions.

Property		Example
Commutative Property of Addition:	$a + b = b + a$	$4x + x^2 = x^2 + 4x$
Commutative Property of Multiplication:	$ab = ba$	$(4 - x)x^2 = x^2(4 - x)$
Associative Property of Addition:	$(a + b) + c = a + (b + c)$	$(x + 5) + x^2 = x + (5 + x^2)$
Associative Property of Multiplication:	$(ab)c = a(bc)$	$(2x \cdot 3y)(8) = (2x)(3y \cdot 8)$
Distributive Properties:	$a(b + c) = ab + ac$	$3x(5 + 2x) = 3x \cdot 5 + 3x \cdot 2x$
	$(a + b)c = ac + bc$	$(y + 8)y = y \cdot y + 8 \cdot y$
Additive Identity Property:	$a + 0 = a$	$5y^2 + 0 = 5y^2$
Multiplicative Identity Property:	$a \cdot 1 = a$	$(4x^2)(1) = 4x^2$
Additive Inverse Property:	$a + (-a) = 0$	$5x^3 + (-5x^3) = 0$
Multiplicative Inverse Property:	$a \cdot \frac{1}{a} = 1, \quad a \neq 0$	$(x^2 + 4)\left(\frac{1}{x^2 + 4}\right) = 1$

Because subtraction is defined as "adding the opposite," the Distributive Properties are also true for subtraction. For instance, the "subtraction form" of $a(b + c) = ab + ac$ is $a(b - c) = ab - ac$. Note that the operations of subtraction and division are neither commutative nor associative. The examples

$$7 - 3 \neq 3 - 7 \quad \text{and} \quad 20 \div 4 \neq 4 \div 20$$

show that subtraction and division are not commutative. Similarly

$$5 - (3 - 2) \neq (5 - 3) - 2 \quad \text{and} \quad 16 \div (4 \div 2) \neq (16 \div 4) \div 2$$

demonstrate that subtraction and division are not associative.

EXAMPLE 13 Identifying Rules of Algebra

Identify the rule of algebra illustrated by the statement.

a. $(5x^3)2 = 2(5x^3)$ **b.** $(4x + 3) - (4x + 3) = 0$

c. $7x \cdot \frac{1}{7x} = 1, \quad x \neq 0$ **d.** $(2 + 5x^2) + x^2 = 2 + (5x^2 + x^2)$

Solution

a. This statement illustrates the Commutative Property of Multiplication. In other words, you obtain the same result whether you multiply $5x^3$ by 2, or 2 by $5x^3$.

b. This statement illustrates the Additive Inverse Property. In terms of subtraction, this property states that when any expression is subtracted from itself the result is 0.

c. This statement illustrates the Multiplicative Inverse Property. Note that x must be a nonzero number. The reciprocal of x is undefined when x is 0.

d. This statement illustrates the Associative Property of Addition. In other words, to form the sum $2 + 5x^2 + x^2$, it does not matter whether 2 and $5x^2$, or $5x^2$ and x^2 are added first.

✓ ***Checkpoint*** *Audio-video solution in English & Spanish at LarsonPrecalculus.com.*

Identify the rule of algebra illustrated by the statement.

a. $x + 9 = 9 + x$ **b.** $5(x^3 \cdot 2) = (5x^3)2$ **c.** $(2 + 5x^2)y^2 = 2 \cdot y^2 + 5x^2 \cdot y^2$

REMARK Notice the difference between the *opposite of a number* and a *negative number.* If a is negative, then its opposite, $-a$, is positive. For instance, if $a = -5$, then

$$-a = -(-5) = 5.$$

Properties of Negation and Equality

Let a, b, and c be real numbers, variables, or algebraic expressions.

Property	Example
1. $(-1)a = -a$	$(-1)7 = -7$
2. $-(-a) = a$	$-(-6) = 6$
3. $(-a)b = -(ab) = a(-b)$	$(-5)3 = -(5 \cdot 3) = 5(-3)$
4. $(-a)(-b) = ab$	$(-2)(-x) = 2x$
5. $-(a + b) = (-a) + (-b)$	$-(x + 8) = (-x) + (-8)$ $= -x - 8$
6. If $a = b$, then $a \pm c = b \pm c$.	$\frac{1}{2} + 3 = 0.5 + 3$
7. If $a = b$, then $ac = bc$.	$4^2 \cdot 2 = 16 \cdot 2$
8. If $a \pm c = b \pm c$, then $a = b$.	$1.4 - 1 = \frac{7}{5} - 1 \Rightarrow 1.4 = \frac{7}{5}$
9. If $ac = bc$ and $c \neq 0$, then $a = b$.	$3x = 3 \cdot 4 \Rightarrow x = 4$

REMARK The "or" in the Zero-Factor Property includes the possibility that either or both factors may be zero. This is an *inclusive or,* and it is generally the way the word "or" is used in mathematics.

Properties of Zero

Let a and b be real numbers, variables, or algebraic expressions.

1. $a + 0 = a$ and $a - 0 = a$ **2.** $a \cdot 0 = 0$

3. $\frac{0}{a} = 0, \quad a \neq 0$ **4.** $\frac{a}{0}$ is undefined.

5. Zero-Factor Property: If $ab = 0$, then $a = 0$ or $b = 0$.

Properties and Operations of Fractions

Let a, b, c, and d be real numbers, variables, or algebraic expressions such that $b \neq 0$ and $d \neq 0$.

1. **Equivalent Fractions:** $\frac{a}{b} = \frac{c}{d}$ if and only if $ad = bc$.
2. **Rules of Signs:** $-\frac{a}{b} = \frac{-a}{b} = \frac{a}{-b}$ and $\frac{-a}{-b} = \frac{a}{b}$
3. **Generate Equivalent Fractions:** $\frac{a}{b} = \frac{ac}{bc}, \quad c \neq 0$
4. **Add or Subtract with Like Denominators:** $\frac{a}{b} \pm \frac{c}{b} = \frac{a \pm c}{b}$
5. **Add or Subtract with Unlike Denominators:** $\frac{a}{b} \pm \frac{c}{d} = \frac{ad \pm bc}{bd}$
6. **Multiply Fractions:** $\frac{a}{b} \cdot \frac{c}{d} = \frac{ac}{bd}$
7. **Divide Fractions:** $\frac{a}{b} \div \frac{c}{d} = \frac{a}{b} \cdot \frac{d}{c} = \frac{ad}{bc}, \quad c \neq 0$

REMARK In Property 1 of fractions, the phrase "if and only if" implies two statements. One statement is: If $a/b = c/d$, then $ad = bc$. The other statement is: If $ad = bc$, where $b \neq 0$ and $d \neq 0$, then $a/b = c/d$.

EXAMPLE 14 **Properties and Operations of Fractions**

a. Equivalent fractions: $\frac{x}{5} = \frac{3 \cdot x}{3 \cdot 5} = \frac{3x}{15}$ **b.** Divide fractions: $\frac{7}{x} \div \frac{3}{2} = \frac{7}{x} \cdot \frac{2}{3} = \frac{14}{3x}$

✓ **Checkpoint** Audio-video solution in English & Spanish at LarsonPrecalculus.com.

a. Multiply fractions: $\frac{3}{5} \cdot \frac{x}{6}$ **b.** Add fractions: $\frac{x}{10} + \frac{2x}{5}$

If a, b, and c are integers such that $ab = c$, then a and b are **factors** or **divisors** of c. A **prime number** is an integer that has exactly two positive factors—itself and 1—such as 2, 3, 5, 7, and 11. The numbers 4, 6, 8, 9, and 10 are **composite** because each can be written as the product of two or more prime numbers. The **Fundamental Theorem of Arithmetic** states that every positive integer greater than 1 is a prime number or can be written as the product of prime numbers in precisely one way (disregarding order). For instance, the *prime factorization* of 24 is $24 = 2 \cdot 2 \cdot 2 \cdot 3$.

REMARK The number 1 is neither prime nor composite.

Summarize (Appendix A.1)

1. Describe how to represent and classify real numbers *(pages A1 and A2)*. For examples of representing and classifying real numbers, see Examples 1 and 2.
2. Describe how to order real numbers and use inequalities *(pages A3 and A4)*. For examples of ordering real numbers and using inequalities, see Examples 3–6.
3. State the absolute value of a real number *(page A5)*. For examples of using absolute value, see Examples 7–10.
4. Explain how to evaluate an algebraic expression *(page A7)*. For examples involving algebraic expressions, see Examples 11 and 12.
5. State the basic rules and properties of algebra *(pages A8–A10)*. For examples involving the basic rules and properties of algebra, see Examples 13 and 14.

A.1 Exercises

See **CalcChat.com** for tutorial help and worked-out solutions to odd-numbered exercises.

Vocabulary: Fill in the blanks.

1. ________ numbers have infinite nonrepeating decimal representations.
2. The point 0 on the real number line is called the ________.
3. The distance between the origin and a point representing a real number on the real number line is the ________ ________ of the real number.
4. A number that can be written as the product of two or more prime numbers is called a ________ number.
5. The ________ of an algebraic expression are those parts separated by addition.
6. The ________ ________ states that if $ab = 0$, then $a = 0$ or $b = 0$.

Skills and Applications

Classifying Real Numbers **In Exercises 7–10, determine which numbers in the set are (a) natural numbers, (b) whole numbers, (c) integers, (d) rational numbers, and (e) irrational numbers.**

7. $\left\{-9, -\frac{7}{2}, 5, \frac{2}{3}, \sqrt{2}, 0, 1, -4, 2, -11\right\}$
8. $\left\{\sqrt{5}, -7, -\frac{7}{3}, 0, 3.12, \frac{5}{4}, -3, 12, 5\right\}$
9. $\{2.01, 0.666\ldots, -13, 0.010110111\ldots, 1, -6\}$
10. $\left\{25, -17, -\frac{12}{5}, \sqrt{9}, 3.12, \frac{1}{2}\pi, 7, -11.1, 13\right\}$

Plotting Points on the Real Number Line **In Exercises 11 and 12, plot the real numbers on the real number line.**

11. (a) 3 (b) $\frac{7}{2}$ (c) $-\frac{5}{2}$ (d) -5.2
12. (a) 8.5 (b) $\frac{4}{3}$ (c) -4.75 (d) $-\frac{8}{3}$

Plotting and Ordering Real Numbers **In Exercises 13–16, plot the two real numbers on the real number line. Then place the appropriate inequality symbol (< or >) between them.**

13. $-4, -8$
14. $1, \frac{16}{3}$
15. $\frac{5}{6}, \frac{2}{3}$
16. $-\frac{8}{7}, -\frac{3}{7}$

Interpreting an Inequality or an Interval **In Exercises 17–24, (a) give a verbal description of the subset of real numbers represented by the inequality or the interval, (b) sketch the subset on the real number line, and (c) state whether the interval is bounded or unbounded.**

17. $x \le 5$
18. $x < 0$
19. $[4, \infty)$
20. $(-\infty, 2)$
21. $-2 < x < 2$
22. $0 < x \le 6$
23. $[-5, 2)$
24. $(-1, 2]$

Using Inequality and Interval Notation **In Exercises 25–30, use inequality notation and interval notation to describe the set.**

25. y is nonnegative.
26. y is no more than 25.
27. t is at least 10 and at most 22.
28. k is less than 5 but no less than -3.
29. The dog's weight W is more than 65 pounds.
30. The annual rate of inflation r is expected to be at least 2.5% but no more than 5%.

Evaluating an Absolute Value Expression **In Exercises 31–40, evaluate the expression.**

31. $|-10|$
32. $|0|$
33. $|3 - 8|$
34. $|4 - 1|$
35. $|-1| - |-2|$
36. $-3 - |-3|$
37. $\dfrac{-5}{|-5|}$
38. $-3|-3|$
39. $\dfrac{|x + 2|}{x + 2}, \quad x < -2$
40. $\dfrac{|x - 1|}{x - 1}, \quad x > 1$

Comparing Real Numbers **In Exercises 41–44, place the appropriate symbol (<, >, or =) between the two real numbers.**

41. $|-4| \;\square\; |4|$
42. $-5 \;\square\; -|5|$
43. $-|-6| \;\square\; |-6|$
44. $-|-2| \;\square\; -|2|$

Finding a Distance **In Exercises 45–50, find the distance between a and b.**

45. $a = 126, b = 75$
46. $a = -126, b = -75$
47. $a = -\frac{5}{2}, b = 0$
48. $a = \frac{1}{4}, b = \frac{11}{4}$
49. $a = \frac{16}{5}, b = \frac{112}{75}$
50. $a = 9.34, b = -5.65$

Using Absolute Value Notation **In Exercises 51–54, use absolute value notation to describe the situation.**

51. The distance between x and 5 is no more than 3.
52. The distance between x and -10 is at least 6.
53. y is at most two units from a.
54. The temperature in Bismarck, North Dakota, was 60°F at noon, then 23°F at midnight. What was the change in temperature over the 12-hour period?

Federal Deficit In Exercises 55–58, use the bar graph, which shows the receipts of the federal government (in billions of dollars) for selected years from 2004 through 2010. In each exercise you are given the expenditures of the federal government. Find the magnitude of the surplus or deficit for the year. *(Source: U.S. Office of Management and Budget)*

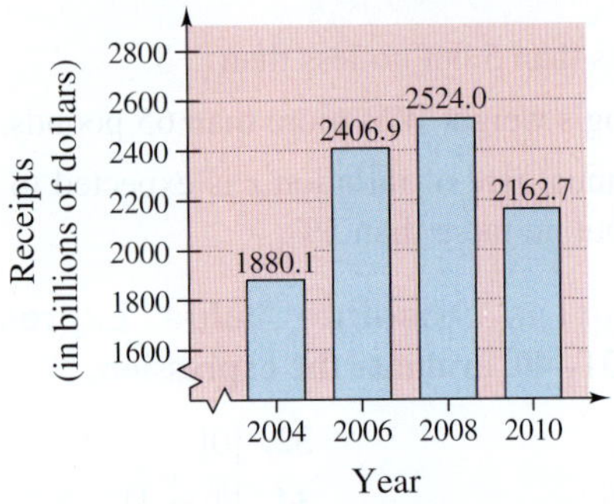

	Year	Receipts, R	Expenditures, E	$\|R - E\|$
55.	2004		\$2292.8 billion	
56.	2006		\$2655.1 billion	
57.	2008		\$2982.5 billion	
58.	2010		\$3456.2 billion	

Identifying Terms and Coefficients In Exercises 59–62, identify the terms. Then identify the coefficients of the variable terms of the expression.

59. $7x + 4$ **60.** $6x^3 - 5x$

61. $4x^3 + 0.5x - 5$ **62.** $3\sqrt{3}x^2 + 1$

Evaluating an Algebraic Expression In Exercises 63–66, evaluate the expression for each value of x. (If not possible, then state the reason.)

	Expression	Values	
63.	$4x - 6$	(a) $x = -1$	(b) $x = 0$
64.	$9 - 7x$	(a) $x = -3$	(b) $x = 3$
65.	$-x^2 + 5x - 4$	(a) $x = -1$	(b) $x = 1$
66.	$(x + 1)/(x - 1)$	(a) $x = 1$	(b) $x = -1$

Identifying Rules of Algebra In Exercises 67–72, identify the rule(s) of algebra illustrated by the statement.

67. $\dfrac{1}{h+6}(h + 6) = 1, \quad h \neq -6$

68. $(x + 3) - (x + 3) = 0$

69. $2(x + 3) = 2 \cdot x + 2 \cdot 3$

70. $(z - 2) + 0 = z - 2$

71. $x(3y) = (x \cdot 3)y = (3x)y$

72. $\frac{1}{7}(7 \cdot 12) = \left(\frac{1}{7} \cdot 7\right)12 = 1 \cdot 12 = 12$

Operations with Fractions In Exercises 73–76, perform the operation(s). (Write fractional answers in simplest form.)

73. $\frac{5}{8} - \frac{5}{12} + \frac{1}{6}$ **74.** $-\left(6 \cdot \frac{4}{8}\right)$

75. $\dfrac{2x}{3} - \dfrac{x}{4}$ **76.** $\dfrac{5x}{6} \cdot \dfrac{2}{9}$

Exploration

77. Determining the Sign of an Expression Use the real numbers A, B, and C shown on the number line to determine the sign of (a) $-A$, (b) $B - A$, (c) $-C$, and (d) $A - C$.

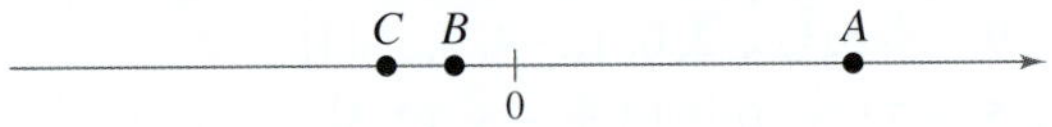

78. HOW DO YOU SEE IT? Match each description with its graph. Which types of real numbers shown in Figure A.1 on page A1 may be included in a range of prices? a range of lengths? Explain.

(i)

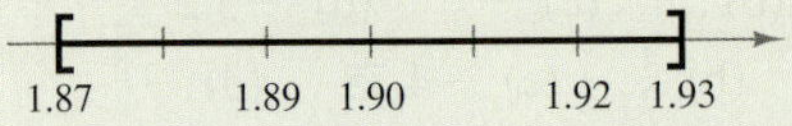

(ii)

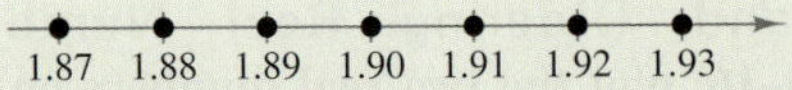

(a) The price of an item is within \$0.03 of \$1.90.

(b) The distance between the prongs of an electric plug may not differ from 1.9 centimeters by more than 0.03 centimeter.

True or False? In Exercises 79 and 80, determine whether the statement is true or false. Justify your answer.

79. Every nonnegative number is positive.

80. If $a > 0$ and $b < 0$, then $ab > 0$.

81. Conjecture

(a) Use a calculator to complete the table.

n	0.0001	0.01	1	100	10,000
$5/n$					

(b) Use the result from part (a) to make a conjecture about the value of $5/n$ as n (i) approaches 0, and (ii) increases without bound.

Michael G Smith/Shutterstock.com

A.2 Exponents and Radicals

Real numbers and algebraic expressions are often written with exponents and radicals. For instance, in Exercise 79 on page A24, you will use an expression involving rational exponents to find the times required for a funnel to empty for different water heights.

- **Use properties of exponents.**
- **Use scientific notation to represent real numbers.**
- **Use properties of radicals.**
- **Simplify and combine radicals.**
- **Rationalize denominators and numerators.**
- **Use properties of rational exponents.**

Integer Exponents

Repeated *multiplication* can be written in **exponential form.**

Repeated Multiplication	Exponential Form
$a \cdot a \cdot a \cdot a \cdot a$	a^5
$(-4)(-4)(-4)$	$(-4)^3$
$(2x)(2x)(2x)(2x)$	$(2x)^4$

Exponential Notation

If a is a real number and n is a positive integer, then

$$a^n = \underbrace{a \cdot a \cdot a \cdots a}_{n \text{ factors}}$$

where n is the **exponent** and a is the **base.** You read a^n as "a to the nth **power.**"

An exponent can also be negative. In Property 3 below, be sure you see how to use a negative exponent.

Properties of Exponents

Let a and b be real numbers, variables, or algebraic expressions, and let m and n be integers. (All denominators and bases are nonzero.)

Property	Example
1. $a^m a^n = a^{m+n}$	$3^2 \cdot 3^4 = 3^{2+4} = 3^6 = 729$
2. $\dfrac{a^m}{a^n} = a^{m-n}$	$\dfrac{x^7}{x^4} = x^{7-4} = x^3$
3. $a^{-n} = \dfrac{1}{a^n} = \left(\dfrac{1}{a}\right)^n$	$y^{-4} = \dfrac{1}{y^4} = \left(\dfrac{1}{y}\right)^4$
4. $a^0 = 1, \quad a \neq 0$	$(x^2 + 1)^0 = 1$
5. $(ab)^m = a^m b^m$	$(5x)^3 = 5^3 x^3 = 125x^3$
6. $(a^m)^n = a^{mn}$	$(y^3)^{-4} = y^{3(-4)} = y^{-12} = \dfrac{1}{y^{12}}$
7. $\left(\dfrac{a}{b}\right)^m = \dfrac{a^m}{b^m}$	$\left(\dfrac{2}{x}\right)^3 = \dfrac{2^3}{x^3} = \dfrac{8}{x^3}$
8. $\lvert a^2 \rvert = \lvert a \rvert^2 = a^2$	$\lvert(-2)^2\rvert = \lvert -2 \rvert^2 = (2)^2 = 4$

micropic/iStockphoto.com

TECHNOLOGY You can use a calculator to evaluate exponential expressions. When doing so, it is important to know when to use parentheses because the calculator follows the order of operations. For instance, you would evaluate $(-2)^4$ on a graphing calculator as follows.

((−) 2) ^ 4 ENTER

The display will be 16. If you omit the parentheses, then the display will be -16.

It is important to recognize the difference between expressions such as $(-2)^4$ and -2^4. In $(-2)^4$, the parentheses indicate that the exponent applies to the negative sign as well as to the 2, but in $-2^4 = -(2^4)$, the exponent applies only to the 2. So, $(-2)^4 = 16$ and $-2^4 = -16$.

The properties of exponents listed on the preceding page apply to *all* integers m and n, not just to positive integers, as shown in the examples below.

EXAMPLE 1 Evaluating Exponential Expressions

a. $(-5)^2 = (-5)(-5) = 25$ Negative sign is part of the base.

b. $-5^2 = -(5)(5) = -25$ Negative sign is *not* part of the base.

c. $2 \cdot 2^4 = 2^{1+4} = 2^5 = 32$ Property 1

d. $\dfrac{4^4}{4^6} = 4^{4-6} = 4^{-2} = \dfrac{1}{4^2} = \dfrac{1}{16}$ Properties 2 and 3

✓ **Checkpoint** *Audio-video solution in English & Spanish at LarsonPrecalculus.com.*

Evaluate each expression.

a. -3^4 **b.** $(-3)^4$

c. $3^2 \cdot 3$ **d.** $\dfrac{3^5}{3^8}$

EXAMPLE 2 Evaluating Algebraic Expressions

Evaluate each algebraic expression when $x = 3$.

a. $5x^{-2}$ **b.** $\frac{1}{3}(-x)^3$

Solution

a. When $x = 3$, the expression $5x^{-2}$ has a value of

$$5x^{-2} = 5(3)^{-2} = \frac{5}{3^2} = \frac{5}{9}.$$

b. When $x = 3$, the expression $\frac{1}{3}(-x)^3$ has a value of

$$\frac{1}{3}(-x)^3 = \frac{1}{3}(-3)^3 = \frac{1}{3}(-27) = -9.$$

✓ **Checkpoint** *Audio-video solution in English & Spanish at LarsonPrecalculus.com.*

Evaluate each algebraic expression when $x = 4$.

a. $-x^{-2}$ **b.** $\frac{1}{4}(-x)^4$

EXAMPLE 3 Using Properties of Exponents

Use the properties of exponents to simplify each expression.

a. $(-3ab^4)(4ab^{-3})$ **b.** $(2xy^2)^3$ **c.** $3a(-4a^2)^0$ **d.** $\left(\frac{5x^3}{y}\right)^2$

Solution

a. $(-3ab^4)(4ab^{-3}) = (-3)(4)(a)(a)(b^4)(b^{-3}) = -12a^2b$

b. $(2xy^2)^3 = 2^3(x)^3(y^2)^3 = 8x^3y^6$

c. $3a(-4a^2)^0 = 3a(1) = 3a, \quad a \neq 0$

d. $\left(\frac{5x^3}{y}\right)^2 = \frac{5^2(x^3)^2}{y^2} = \frac{25x^6}{y^2}$

✓ Checkpoint *Audio-video solution in English & Spanish at LarsonPrecalculus.com.*

Use the properties of exponents to simplify each expression.

a. $(2x^{-2}y^3)(-x^4y)$ **b.** $(4a^2b^3)^0$ **c.** $(-5z)^3(z^2)$ **d.** $\left(\frac{3x^4}{x^2y^2}\right)^2$

REMARK Rarely in algebra is there only one way to solve a problem. Do not be concerned when the steps you use to solve a problem are not exactly the same as the steps presented in this text. It is important to use steps that you understand and, of course, steps that are justified by the rules of algebra. For instance, you might prefer the following steps for Example 4(d).

$$\left(\frac{3x^2}{y}\right)^{-2} = \left(\frac{y}{3x^2}\right)^2 = \frac{y^2}{9x^4}$$

Note how the first step of this solution uses Property 3. The fractional form of this property is

$$\left(\frac{a}{b}\right)^{-m} = \left(\frac{b}{a}\right)^m.$$

EXAMPLE 4 Rewriting with Positive Exponents

a. $x^{-1} = \frac{1}{x}$ — Property 3

b. $\frac{1}{3x^{-2}} = \frac{1(x^2)}{3}$ — The exponent -2 does not apply to 3.

$= \frac{x^2}{3}$ — Simplify.

c. $\frac{12a^3b^{-4}}{4a^{-2}b} = \frac{12a^3 \cdot a^2}{4b \cdot b^4}$ — Property 3

$= \frac{3a^5}{b^5}$ — Property 1

d. $\left(\frac{3x^2}{y}\right)^{-2} = \frac{3^{-2}(x^2)^{-2}}{y^{-2}}$ — Properties 5 and 7

$= \frac{3^{-2}x^{-4}}{y^{-2}}$ — Property 6

$= \frac{y^2}{3^2x^4}$ — Property 3

$= \frac{y^2}{9x^4}$ — Simplify.

✓ Checkpoint *Audio-video solution in English & Spanish at LarsonPrecalculus.com.*

Rewrite each expression with positive exponents.

a. $2a^{-2}$ **b.** $\frac{3a^{-3}b^4}{15ab^{-1}}$

c. $\left(\frac{x}{10}\right)^{-1}$ **d.** $(-2x^2)^3(4x^3)^{-1}$

Scientific Notation

Exponents provide an efficient way of writing and computing with very large (or very small) numbers. For instance, there are about 359 billion billion gallons of water on Earth—that is, 359 followed by 18 zeros.

359,000,000,000,000,000,000

It is convenient to write such numbers in **scientific notation.** This notation has the form $\pm c \times 10^n$, where $1 \le c < 10$ and n is an integer. So, the number of gallons of water on Earth, written in scientific notation, is

$$3.59 \times 100{,}000{,}000{,}000{,}000{,}000{,}000 = 3.59 \times 10^{20}.$$

The *positive* exponent 20 indicates that the number is *large* (10 or more) and that the decimal point has been moved 20 places. A *negative* exponent indicates that the number is *small* (less than 1). For instance, the mass (in grams) of one electron is approximately

$$9.0 \times 10^{-28} = 0.0000000000000000000000000009.$$

28 decimal places

EXAMPLE 5 Scientific Notation

a. $0.0000782 = 7.82 \times 10^{-5}$

b. $836{,}100{,}000 = 8.361 \times 10^{8}$

Checkpoint *Audio-video solution in English & Spanish at LarsonPrecalculus.com.*

Write 45,850 in scientific notation.

EXAMPLE 6 Decimal Notation

a. $-9.36 \times 10^{-6} = -0.00000936$

b. $1.345 \times 10^{2} = 134.5$

Checkpoint *Audio-video solution in English & Spanish at LarsonPrecalculus.com.*

Write -2.718×10^{-3} in decimal notation.

TECHNOLOGY Most calculators automatically switch to scientific notation when they are showing large (or small) numbers that exceed the display range.

To enter numbers in scientific notation, your calculator should have an exponential entry key labeled

EE or EXP.

Consult the user's guide for instructions on keystrokes and how your calculator displays numbers in scientific notation.

EXAMPLE 7 Using Scientific Notation

Evaluate $\dfrac{(2{,}400{,}000{,}000)(0.0000045)}{(0.00003)(1500)}$.

Solution Begin by rewriting each number in scientific notation and simplifying.

$$\begin{aligned}\frac{(2{,}400{,}000{,}000)(0.0000045)}{(0.00003)(1500)} &= \frac{(2.4 \times 10^{9})(4.5 \times 10^{-6})}{(3.0 \times 10^{-5})(1.5 \times 10^{3})}\\ &= \frac{(2.4)(4.5)(10^{3})}{(4.5)(10^{-2})}\\ &= (2.4)(10^{5})\\ &= 240{,}000\end{aligned}$$

Checkpoint *Audio-video solution in English & Spanish at LarsonPrecalculus.com.*

Evaluate (24,000,000,000)(0.00000012)(300,000).

Radicals and Their Properties

A **square root** of a number is one of its two equal factors. For example, 5 is a square root of 25 because 5 is one of the two equal factors of 25. In a similar way, a **cube root** of a number is one of its three equal factors, as in $125 = 5^3$.

Definition of *n*th Root of a Number

Let a and b be real numbers and let $n \geq 2$ be a positive integer. If

$$a = b^n$$

then b is an ***n*th root of *a*.** When $n = 2$, the root is a **square root.** When $n = 3$, the root is a **cube root.**

Some numbers have more than one nth root. For example, both 5 and -5 are square roots of 25. The *principal square* root of 25, written as $\sqrt{25}$, is the positive root, 5. The **principal *n*th root** of a number is defined as follows.

Principal *n*th Root of a Number

Let a be a real number that has at least one nth root. The **principal *n*th root of *a*** is the nth root that has the same sign as a. It is denoted by a **radical symbol**

$\sqrt[n]{a}$. Principal nth root

The positive integer $n \geq 2$ is the **index** of the radical, and the number a is the **radicand.** When $n = 2$, omit the index and write $\sqrt{a}$ rather than $\sqrt[2]{a}$. (The plural of index is *indices.*)

A common misunderstanding is that the square root sign implies both negative and positive roots. This is not correct. The square root sign implies only a positive root. When a negative root is needed, you must use the negative sign with the square root sign.

Incorrect: ~~$\sqrt{4} = \pm 2$~~ *Correct:* $-\sqrt{4} = -2$ *and* $\sqrt{4} = 2$

EXAMPLE 8 Evaluating Expressions Involving Radicals

a. $\sqrt{36} = 6$ because $6^2 = 36$.

b. $-\sqrt{36} = -6$ because $-(\sqrt{36}) = -(\sqrt{6^2}) = -(6) = -6$.

c. $\sqrt[3]{\frac{125}{64}} = \frac{5}{4}$ because $\left(\frac{5}{4}\right)^3 = \frac{5^3}{4^3} = \frac{125}{64}$.

d. $\sqrt[5]{-32} = -2$ because $(-2)^5 = -32$.

e. $\sqrt[4]{-81}$ is not a real number because no real number raised to the fourth power produces -81.

✓ ***Checkpoint*** *Audio-video solution in English & Spanish at LarsonPrecalculus.com.*

Evaluate each expression (if possible).

a. $-\sqrt{144}$ **b.** $\sqrt{-144}$

c. $\sqrt{\frac{25}{64}}$ **d.** $-\sqrt[3]{\frac{8}{27}}$

Here are some generalizations about the nth roots of real numbers.

Generalizations About nth Roots of Real Numbers			
Real Number a	Integer $n > 0$	Root(s) of a	Example
$a > 0$	n is even.	$\sqrt[n]{a},\ -\sqrt[n]{a}$	$\sqrt[4]{81} = 3,\ -\sqrt[4]{81} = -3$
$a > 0$ or $a < 0$	n is odd.	$\sqrt[n]{a}$	$\sqrt[3]{-8} = -2$
$a < 0$	n is even.	No real roots	$\sqrt{-4}$ is not a real number.
$a = 0$	n is even or odd.	$\sqrt[n]{0} = 0$	$\sqrt[5]{0} = 0$

Integers such as 1, 4, 9, 16, 25, and 36 are called **perfect squares** because they have integer square roots. Similarly, integers such as 1, 8, 27, 64, and 125 are called **perfect cubes** because they have integer cube roots.

Properties of Radicals

Let a and b be real numbers, variables, or algebraic expressions such that the indicated roots are real numbers, and let m and n be positive integers.

Property	Example
1. $\sqrt[n]{a^m} = \left(\sqrt[n]{a}\right)^m$	$\sqrt[3]{8^2} = \left(\sqrt[3]{8}\right)^2 = (2)^2 = 4$
2. $\sqrt[n]{a} \cdot \sqrt[n]{b} = \sqrt[n]{ab}$	$\sqrt{5} \cdot \sqrt{7} = \sqrt{5 \cdot 7} = \sqrt{35}$
3. $\dfrac{\sqrt[n]{a}}{\sqrt[n]{b}} = \sqrt[n]{\dfrac{a}{b}},\quad b \neq 0$	$\dfrac{\sqrt[4]{27}}{\sqrt[4]{9}} = \sqrt[4]{\dfrac{27}{9}} = \sqrt[4]{3}$
4. $\sqrt[m]{\sqrt[n]{a}} = \sqrt[mn]{a}$	$\sqrt[3]{\sqrt{10}} = \sqrt[6]{10}$
5. $\left(\sqrt[n]{a}\right)^n = a$	$\left(\sqrt{3}\right)^2 = 3$
6. For n even, $\sqrt[n]{a^n} = \lvert a \rvert$.	$\sqrt{(-12)^2} = \lvert -12 \rvert = 12$
For n odd, $\sqrt[n]{a^n} = a$.	$\sqrt[3]{(-12)^3} = -12$

A common use of Property 6 is $\sqrt{a^2} = \lvert a \rvert$.

EXAMPLE 9 Using Properties of Radicals

Use the properties of radicals to simplify each expression.

a. $\sqrt{8} \cdot \sqrt{2}$ **b.** $\left(\sqrt[3]{5}\right)^3$

c. $\sqrt[3]{x^3}$ **d.** $\sqrt[6]{y^6}$

Solution

a. $\sqrt{8} \cdot \sqrt{2} = \sqrt{8 \cdot 2} = \sqrt{16} = 4$ **b.** $\left(\sqrt[3]{5}\right)^3 = 5$

c. $\sqrt[3]{x^3} = x$ **d.** $\sqrt[6]{y^6} = \lvert y \rvert$

✓ ***Checkpoint*** *Audio-video solution in English & Spanish at LarsonPrecalculus.com.*

Use the properties of radicals to simplify each expression.

a. $\dfrac{\sqrt{125}}{\sqrt{5}}$ **b.** $\sqrt[3]{125^2}$ **c.** $\sqrt[3]{x^2} \cdot \sqrt[3]{x}$ **d.** $\sqrt{\sqrt{x}}$

Simplifying Radicals

An expression involving radicals is in **simplest form** when the following conditions are satisfied.

1. All possible factors have been removed from the radical.
2. All fractions have radical-free denominators (a process called *rationalizing the denominator* accomplishes this).
3. The index of the radical is reduced.

To simplify a radical, factor the radicand into factors whose exponents are multiples of the index. Write the roots of these factors outside the radical. The "leftover" factors make up the new radicand.

REMARK When you simplify a radical, it is important that both expressions are defined for the same values of the variable. For instance, in Example 10(c), $\sqrt{75x^3}$ and $5x\sqrt{3x}$ are both defined only for nonnegative values of x. Similarly, in Example 10(e), $\sqrt[4]{(5x)^4}$ and $5|x|$ are both defined for all real values of x.

EXAMPLE 10 Simplifying Radicals

Perfect cube, Leftover factor

a. $\sqrt[3]{24} = \sqrt[3]{8 \cdot 3} = \sqrt[3]{2^3 \cdot 3} = 2\sqrt[3]{3}$

Perfect 4th power, Leftover factor

b. $\sqrt[4]{48} = \sqrt[4]{16 \cdot 3} = \sqrt[4]{2^4 \cdot 3} = 2\sqrt[4]{3}$

c. $\sqrt{75x^3} = \sqrt{25x^2 \cdot 3x} = \sqrt{(5x)^2 \cdot 3x} = 5x\sqrt{3x}$

d. $\sqrt[3]{24a^4} = \sqrt[3]{8a^3 \cdot 3a} = \sqrt[3]{(2a)^3 \cdot 3a} = 2a\sqrt[3]{3a}$

e. $\sqrt[4]{(5x)^4} = |5x| = 5|x|$

Checkpoint *Audio-video solution in English & Spanish at LarsonPrecalculus.com.*

Simplify each radical expression.

a. $\sqrt{32}$ **b.** $\sqrt[3]{250}$ **c.** $\sqrt{24a^5}$ **d.** $\sqrt[3]{-135x^3}$

Radical expressions can be combined (added or subtracted) when they are **like radicals**—that is, when they have the same index and radicand. For instance, $\sqrt{2}$, $3\sqrt{2}$, and $\frac{1}{2}\sqrt{2}$ are like radicals, but $\sqrt{3}$ and $\sqrt{2}$ are unlike radicals. To determine whether two radicals can be combined, you should first simplify each radical.

EXAMPLE 11 Combining Radicals

a.
$$\begin{aligned} 2\sqrt{48} - 3\sqrt{27} &= 2\sqrt{16 \cdot 3} - 3\sqrt{9 \cdot 3} && \text{Find square factors.} \\ &= 8\sqrt{3} - 9\sqrt{3} && \text{Find square roots and multiply by coefficients.} \\ &= (8 - 9)\sqrt{3} && \text{Combine like radicals.} \\ &= -\sqrt{3} && \text{Simplify.} \end{aligned}$$

b.
$$\begin{aligned} \sqrt[3]{16x} - \sqrt[3]{54x^4} &= \sqrt[3]{8 \cdot 2x} - \sqrt[3]{27 \cdot x^3 \cdot 2x} && \text{Find cube factors.} \\ &= 2\sqrt[3]{2x} - 3x\sqrt[3]{2x} && \text{Find cube roots.} \\ &= (2 - 3x)\sqrt[3]{2x} && \text{Combine like radicals.} \end{aligned}$$

Checkpoint *Audio-video solution in English & Spanish at LarsonPrecalculus.com.*

Simplify each radical expression.

a. $3\sqrt{8} + \sqrt{18}$ **b.** $\sqrt[3]{81x^5} - \sqrt[3]{24x^2}$

Rationalizing Denominators and Numerators

To rationalize a denominator or numerator of the form $a - b\sqrt{m}$ or $a + b\sqrt{m}$, multiply both numerator and denominator by a **conjugate:** $a + b\sqrt{m}$ and $a - b\sqrt{m}$ are conjugates of each other. If $a = 0$, then the rationalizing factor for $\sqrt{m}$ is itself, $\sqrt{m}$. For cube roots, choose a rationalizing factor that generates a perfect cube.

EXAMPLE 12 Rationalizing Single-Term Denominators

Rationalize the denominator of each expression.

a. $\dfrac{5}{2\sqrt{3}}$

b. $\dfrac{2}{\sqrt[3]{5}}$

Solution

a. $\dfrac{5}{2\sqrt{3}} = \dfrac{5}{2\sqrt{3}} \cdot \dfrac{\sqrt{3}}{\sqrt{3}}$ $\quad$ $\sqrt{3}$ is rationalizing factor.

$= \dfrac{5\sqrt{3}}{2(3)}$ $\quad$ Multiply.

$= \dfrac{5\sqrt{3}}{6}$ $\quad$ Simplify.

b. $\dfrac{2}{\sqrt[3]{5}} = \dfrac{2}{\sqrt[3]{5}} \cdot \dfrac{\sqrt[3]{5^2}}{\sqrt[3]{5^2}}$ $\quad$ $\sqrt[3]{5^2}$ is rationalizing factor.

$= \dfrac{2\sqrt[3]{5^2}}{\sqrt[3]{5^3}}$ $\quad$ Multiply.

$= \dfrac{2\sqrt[3]{25}}{5}$ $\quad$ Simplify.

✓ ***Checkpoint*** *Audio-video solution in English & Spanish at LarsonPrecalculus.com.*

Rationalize the denominator of each expression.

a. $\dfrac{5}{3\sqrt{2}}$ $\quad$ **b.** $\dfrac{1}{\sqrt[3]{25}}$

EXAMPLE 13 Rationalizing a Denominator with Two Terms

$$\frac{2}{3+\sqrt{7}} = \frac{2}{3+\sqrt{7}} \cdot \frac{3-\sqrt{7}}{3-\sqrt{7}}$$ Multiply numerator and denominator by conjugate of denominator.

$$= \frac{2(3-\sqrt{7})}{3(3) + 3(-\sqrt{7}) + \sqrt{7}(3) - (\sqrt{7})(\sqrt{7})}$$ Use Distributive Property.

$$= \frac{2(3-\sqrt{7})}{(3)^2 - (\sqrt{7})^2}$$ Simplify.

$$= \frac{2(3-\sqrt{7})}{2} = 3 - \sqrt{7}$$ Simplify.

✓ ***Checkpoint*** *Audio-video solution in English & Spanish at LarsonPrecalculus.com.*

Rationalize the denominator: $\dfrac{8}{\sqrt{6} - \sqrt{2}}$.

Sometimes it is necessary to rationalize the numerator of an expression. For instance, in Appendix A.4 you will use the technique shown in the next example to rationalize the numerator of an expression from calculus.

EXAMPLE 14 Rationalizing a Numerator

$$\frac{\sqrt{5}-\sqrt{7}}{2} = \frac{\sqrt{5}-\sqrt{7}}{2}\cdot\frac{\sqrt{5}+\sqrt{7}}{\sqrt{5}+\sqrt{7}}$$ Multiply numerator and denominator by conjugate of denominator.

$$= \frac{(\sqrt{5})^2-(\sqrt{7})^2}{2(\sqrt{5}+\sqrt{7})}$$ Simplify.

$$= \frac{5-7}{2(\sqrt{5}+\sqrt{7})}$$ Square terms of numerator.

$$= \frac{-2}{2(\sqrt{5}+\sqrt{7})} = \frac{-1}{\sqrt{5}+\sqrt{7}}$$ Simplify.

REMARK Do not confuse the expression $\sqrt{5}+\sqrt{7}$ with the expression $\sqrt{5+7}$. In general, $\sqrt{x+y}$ does not equal $\sqrt{x}+\sqrt{y}$. Similarly, $\sqrt{x^2+y^2}$ does not equal $x+y$.

Checkpoint *Audio-video solution in English & Spanish at LarsonPrecalculus.com.*

Rationalize the numerator: $\frac{2-\sqrt{2}}{3}$.

Rational Exponents

Definition of Rational Exponents

If a is a real number and n is a positive integer such that the principal nth root of a exists, then $a^{1/n}$ is defined as

$a^{1/n} = \sqrt[n]{a}$, where $1/n$ is the **rational exponent** of a.

Moreover, if m is a positive integer that has no common factor with n, then

$$a^{m/n} = (a^{1/n})^m = (\sqrt[n]{a})^m \quad \text{and} \quad a^{m/n} = (a^m)^{1/n} = \sqrt[n]{a^m}.$$

The numerator of a rational exponent denotes the *power* to which the base is raised, and the denominator denotes the *index* or the *root* to be taken.

Power, Index

$$b^{m/n} = (\sqrt[n]{b})^m = \sqrt[n]{b^m}$$

REMARK You must remember that the expression $b^{m/n}$ is not defined unless $\sqrt[n]{b}$ is a real number. This restriction produces some unusual results. For instance, the number $(-8)^{1/3}$ is defined because $\sqrt[3]{-8} = -2$, but the number $(-8)^{2/6}$ is undefined because $\sqrt[6]{-8}$ is not a real number.

When you are working with rational exponents, the properties of integer exponents still apply. For instance, $2^{1/2}2^{1/3} = 2^{(1/2)+(1/3)} = 2^{5/6}$.

EXAMPLE 15 Changing From Radical to Exponential Form

a. $\sqrt{3} = 3^{1/2}$ **b.** $\sqrt{(3xy)^5} = \sqrt[2]{(3xy)^5} = (3xy)^{5/2}$

c. $2x\sqrt[4]{x^3} = (2x)(x^{3/4}) = 2x^{1+(3/4)} = 2x^{7/4}$

Checkpoint *Audio-video solution in English & Spanish at LarsonPrecalculus.com.*

Write each expression in exponential form.

a. $\sqrt[3]{27}$ **b.** $\sqrt{x^3y^5z}$ **c.** $3x\sqrt[3]{x^2}$

> **TECHNOLOGY** There are four methods of evaluating radicals on most graphing calculators. For square roots, you can use the *square root key* $\boxed{\sqrt{\ }}$. For cube roots, you can use the *cube root key* $\boxed{\sqrt[3]{\ }}$. For other roots, you can first convert the radical to exponential form and then use the *exponential key* $\boxed{\wedge}$, or you can use the xth root key $\boxed{\sqrt[x]{\ }}$ (or menu choice). Consult the user's guide for your calculator for specific keystrokes.

EXAMPLE 16 Changing From Exponential to Radical Form

a. $(x^2 + y^2)^{3/2} = \left(\sqrt{x^2 + y^2}\right)^3 = \sqrt{(x^2 + y^2)^3}$

b. $2y^{3/4}z^{1/4} = 2(y^3z)^{1/4} = 2\sqrt[4]{y^3z}$

c. $a^{-3/2} = \dfrac{1}{a^{3/2}} = \dfrac{1}{\sqrt{a^3}}$

d. $x^{0.2} = x^{1/5} = \sqrt[5]{x}$

✓ **Checkpoint** Audio-video solution in English & Spanish at LarsonPrecalculus.com.

Write each expression in radical form.

a. $(x^2 - 7)^{-1/2}$ **b.** $-3b^{1/3}c^{2/3}$

c. $a^{0.75}$ **d.** $(x^2)^{2/5}$

Rational exponents are useful for evaluating roots of numbers on a calculator, for reducing the index of a radical, and for simplifying expressions in calculus.

EXAMPLE 17 Simplifying with Rational Exponents

a. $(-32)^{-4/5} = \left(\sqrt[5]{-32}\right)^{-4} = (-2)^{-4} = \dfrac{1}{(-2)^4} = \dfrac{1}{16}$

b. $(-5x^{5/3})(3x^{-3/4}) = -15x^{(5/3)-(3/4)} = -15x^{11/12}, \quad x \neq 0$

c. $\sqrt[9]{a^3} = a^{3/9} = a^{1/3} = \sqrt[3]{a}$ Reduce index.

d. $\sqrt[3]{\sqrt{125}} = \sqrt[6]{125} = \sqrt[6]{(5)^3} = 5^{3/6} = 5^{1/2} = \sqrt{5}$

e. $(2x - 1)^{4/3}(2x - 1)^{-1/3} = (2x - 1)^{(4/3)-(1/3)} = 2x - 1, \quad x \neq \frac{1}{2}$

> **REMARK** The expression in Example 17(e) is not defined when $x = \frac{1}{2}$ because
> $$\left(2 \cdot \frac{1}{2} - 1\right)^{-1/3} = (0)^{-1/3}$$
> is not a real number.

✓ **Checkpoint** Audio-video solution in English & Spanish at LarsonPrecalculus.com.

Simplify each expression.

a. $(-125)^{-2/3}$ **b.** $(4x^2y^{3/2})(-3x^{-1/3}y^{-3/5})$

c. $\sqrt[3]{\sqrt[4]{27}}$ **d.** $(3x + 2)^{5/2}(3x + 2)^{-1/2}$

Summarize (Appendix A.2)

1. Make a list of the properties of exponents *(page A13)*. For examples that use these properties, see Examples 1–4.
2. State the definition of scientific notation *(page A16)*. For examples of scientific notation, see Examples 5–7.
3. Make a list of the properties of radicals *(page A18)*. For an example that uses these properties, see Example 9.
4. Explain how to simplify a radical expression *(page A19)*. For examples of simplifying radical expressions, see Examples 10 and 11.
5. Explain how to rationalize a denominator or a numerator *(page A20)*. For examples of rationalizing denominators and numerators, see Examples 12–14.
6. State the definition of a rational exponent *(page A21)*. For an example of simplifying expressions with rational exponents, see Example 17.

A.2 Exercises

See CalcChat.com for tutorial help and worked-out solutions to odd-numbered exercises.

Vocabulary: Fill in the blanks.

1. In the exponential form a^n, n is the ________ and a is the ________.
2. A convenient way of writing very large or very small numbers is called ________ ________.
3. One of the two equal factors of a number is called a __________ __________ of the number.
4. In the radical form $\sqrt[n]{a}$, the positive integer n is the ________ of the radical and the number a is the ________.
5. Radical expressions can be combined (added or subtracted) when they are ________ ________.
6. The expressions $a + b\sqrt{m}$ and $a - b\sqrt{m}$ are ________ of each other.
7. The process used to create a radical-free denominator is known as ________ the denominator.
8. In the expression $b^{m/n}$, m denotes the ________ to which the base is raised and n denotes the ________ or root to be taken.

Skills and Applications

Evaluating Exponential Expressions **In Exercises 9–14, evaluate each expression.**

9. (a) $3 \cdot 3^3$ (b) $\dfrac{3^2}{3^4}$
10. (a) $(3^3)^0$ (b) -3^2
11. (a) $(2^3 \cdot 3^2)^2$ (b) $\left(-\frac{3}{5}\right)^3\left(\frac{5}{3}\right)^2$
12. (a) $\dfrac{3}{3^{-4}}$ (b) $48(-4)^{-3}$
13. (a) $\dfrac{4 \cdot 3^{-2}}{2^{-2} \cdot 3^{-1}}$ (b) $(-2)^0$
14. (a) $3^{-1} + 2^{-2}$ (b) $(3^{-2})^2$

Evaluating an Algebraic Expression **In Exercises 15–20, evaluate the expression for the given value of x.**

15. $-3x^3$, $x = 2$
16. $7x^{-2}$, $x = 4$
17. $6x^0$, $x = 10$
18. $2x^3$, $x = -3$
19. $-3x^4$, $x = -2$
20. $12(-x)^3$, $x = -\frac{1}{3}$

Using Properties of Exponents **In Exercises 21–26, simplify each expression.**

21. (a) $(-5z)^3$ (b) $5x^4(x^2)$
22. (a) $(3x)^2$ (b) $(4x^3)^0$, $x \neq 0$
23. (a) $6y^2(2y^0)^2$ (b) $(-z)^3(3z^4)$
24. (a) $\dfrac{7x^2}{x^3}$ (b) $\dfrac{12(x + y)^3}{9(x + y)}$
25. (a) $\left(\dfrac{4}{y}\right)^3\left(\dfrac{3}{y}\right)^4$ (b) $\left(\dfrac{b^{-2}}{a^{-2}}\right)\left(\dfrac{b}{a}\right)^2$
26. (a) $[(x^2y^{-2})^{-1}]^{-1}$ (b) $(5x^2z^6)^3(5x^2z^6)^{-3}$

Rewriting with Positive Exponents **In Exercises 27–30, rewrite each expression with positive exponents and simplify.**

27. (a) $(x + 5)^0$, $x \neq -5$ (b) $(2x^2)^{-2}$
28. (a) $(4y^{-2})(8y^4)$ (b) $(z + 2)^{-3}(z + 2)^{-1}$
29. (a) $\left(\dfrac{x^{-3}y^4}{5}\right)^{-3}$ (b) $\left(\dfrac{a^{-2}}{b^{-2}}\right)\left(\dfrac{b}{a}\right)^3$
30. (a) $3^n \cdot 3^{2n}$ (b) $\dfrac{x^2 \cdot x^n}{x^3 \cdot x^n}$

Scientific Notation **In Exercises 31–34, write the number in scientific notation.**

31. 10,250.4
32. -0.000125
33. One micron (millionth of a meter): 0.00003937 inch
34. Land area of Earth: 57,300,000 square miles

Decimal Notation **In Exercises 35–38, write the number in decimal notation.**

35. 3.14×10^{-4}
36. -1.801×10^5
37. Light year: 9.46×10^{12} kilometers
38. Width of a human hair: 9.0×10^{-5} meter

Using Scientific Notation **In Exercises 39 and 40, evaluate each expression without using a calculator.**

39. (a) $(2.0 \times 10^9)(3.4 \times 10^{-4})$
(b) $(1.2 \times 10^7)(5.0 \times 10^{-3})$
40. (a) $\dfrac{6.0 \times 10^8}{3.0 \times 10^{-3}}$ (b) $\dfrac{2.5 \times 10^{-3}}{5.0 \times 10^2}$

Evaluating Expressions Involving Radicals **In Exercises 41 and 42, evaluate each expression without using a calculator.**

41. (a) $\sqrt{9}$ (b) $\sqrt[3]{\frac{27}{8}}$
42. (a) $\sqrt[3]{27}$ (b) $\left(\sqrt{36}\right)^3$

Using Properties of Radicals **In Exercises 43 and 44, use the properties of radicals to simplify each expression.**

43. (a) $\left(\sqrt[5]{2}\right)^5$ (b) $\sqrt[5]{32x^5}$
44. (a) $\sqrt{12} \cdot \sqrt{3}$ (b) $\sqrt[4]{(3x^2)^4}$

Simplifying Radical Expressions In Exercises 45–54, simplify each radical expression.

45. (a) $\sqrt{20}$ (b) $\sqrt[3]{128}$

46. (a) $\sqrt[3]{\frac{16}{27}}$ (b) $\sqrt{\frac{75}{4}}$

47. (a) $\sqrt{72x^3}$ (b) $\sqrt{54xy^4}$

48. (a) $\sqrt{\frac{18^2}{z^3}}$ (b) $\sqrt{\frac{32a^4}{b^2}}$

49. (a) $\sqrt[3]{16x^5}$ (b) $\sqrt{75x^2y^{-4}}$

50. (a) $\sqrt[4]{3x^4y^2}$ (b) $\sqrt[5]{160x^8z^4}$

51. (a) $10\sqrt{32} - 6\sqrt{18}$ (b) $\sqrt[3]{16} + 3\sqrt[3]{54}$

52. (a) $5\sqrt{x} - 3\sqrt{x}$ (b) $-2\sqrt{9y} + 10\sqrt{y}$

53. (a) $-3\sqrt{48x^2} + 7\sqrt{75x^2}$
 (b) $7\sqrt{80x} - 2\sqrt{125x}$

54. (a) $-\sqrt{x^3 - 7} + 5\sqrt{x^3 - 7}$
 (b) $11\sqrt{245x^3} - 9\sqrt{45x^3}$

Rationalizing a Denominator In Exercises 55–58, rationalize the denominator of the expression. Then simplify your answer.

55. $\frac{1}{\sqrt{3}}$ 56. $\frac{8}{\sqrt[3]{2}}$

57. $\frac{5}{\sqrt{14} - 2}$ 58. $\frac{3}{\sqrt{5} + \sqrt{6}}$

Rationalizing a Numerator In Exercises 59–62, rationalize the numerator of the expression. Then simplify your answer.

59. $\frac{\sqrt{8}}{2}$ 60. $\frac{\sqrt{2}}{3}$

61. $\frac{\sqrt{5} + \sqrt{3}}{3}$ 62. $\frac{\sqrt{7} - 3}{4}$

Writing Exponential and Radical Forms In Exercises 63–68, fill in the missing form of the expression.

	Radical Form	Rational Exponent Form
63.	$\sqrt[3]{64}$	
64.		$(2ab)^{3/4}$
65.		$3x^{-2/3}$
66.	$x^2\sqrt{x}$	
67.	$x\sqrt{3xy}$	
68.		$a^{0.4}$

Simplifying Expressions In Exercises 69–78, simplify each expression.

69. (a) $32^{-3/5}$ (b) $\left(\frac{16}{81}\right)^{-3/4}$

70. (a) $100^{-3/2}$ (b) $\left(\frac{9}{4}\right)^{-1/2}$

71. (a) $(2x^2)^{3/2} \cdot 2^{-1/2} \cdot x^{-4}$ (b) $(x^4y^2)^{1/3}(xy)^{-1/3}$

72. (a) $x^{-3} \cdot x^{1/2} \cdot x^{-3/2} \cdot x$ (b) $5^{-1/2} \cdot 5x^{5/2}(5x)^{-3/2}$

73. (a) $\sqrt[4]{3^2}$ (b) $\sqrt[6]{(x + 1)^4}$

74. (a) $\sqrt[6]{x^3}$ (b) $\sqrt[4]{(3x^2)^4}$

75. (a) $\sqrt{\sqrt{32}}$ (b) $\sqrt{\sqrt[4]{2x}}$

76. (a) $\sqrt{\sqrt{243(x + 1)}}$ (b) $\sqrt{\sqrt[3]{10a^7b}}$

77. (a) $(x - 1)^{1/3}(x - 1)^{2/3}$
 (b) $(x - 1)^{1/3}(x - 1)^{-4/3}$

78. (a) $(4x + 3)^{5/2}(4x + 3)^{-5/3}$
 (b) $(4x + 3)^{-5/2}(4x + 3)^{1/2}$

79. **Mathematical Modeling**
A funnel is filled with water to a height of h centimeters. The formula

$$t = 0.03[12^{5/2} - (12 - h)^{5/2}], \quad 0 \le h \le 12$$

represents the amount of time t (in seconds) that it will take for the funnel to empty.

(a) Use the *table* feature of a graphing utility to find the times required for the funnel to empty for water heights of $h = 0, h = 1, h = 2, \ldots, h = 12$ centimeters.

(b) What value does t appear to be approaching as the height of the water becomes closer and closer to 12 centimeters?

80. **HOW DO YOU SEE IT?** Package A is a cube with a volume of 500 cubic inches. Package B is a cube with a volume of 250 cubic inches. Is the length x of a side of package A greater than, less than, or equal to twice the length of a side of package B? Explain.

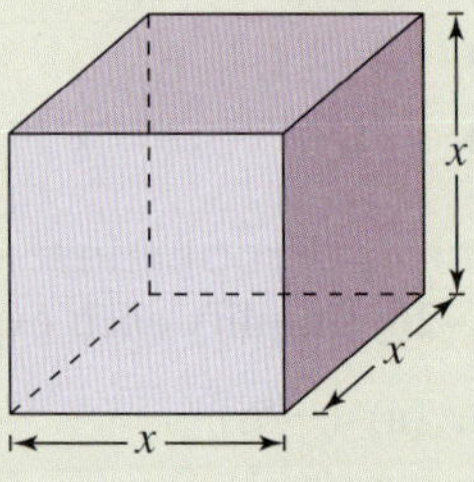

Exploration

True or False? In Exercises 81–84, determine whether the statement is true or false. Justify your answer.

81. $\frac{x^{k+1}}{x} = x^k$ 82. $(a^n)^k = a^{n^k}$

83. $(a + b)^2 = a^2 + b^2$ 84. $\frac{a}{\sqrt{b}} = \frac{a^2}{(\sqrt{b})^2} = \frac{a^2}{b}$

micropic/iStockphoto.com

A.3 Polynomials and Factoring

Polynomial factoring can help you solve real-life problems. For instance, in Exercise 96 on page A34, you will use factoring to develop an alternative model for the rate of change of an autocatalytic chemical reaction.

- **Write polynomials in standard form.**
- **Add, subtract, and multiply polynomials.**
- **Use special products to multiply polynomials.**
- **Remove common factors from polynomials.**
- **Factor special polynomial forms.**
- **Factor trinomials as the product of two binomials.**
- **Factor polynomials by grouping.**

Polynomials

One of the most common types of algebraic expressions is the **polynomial.** Some examples are $2x + 5$, $3x^4 - 7x^2 + 2x + 4$, and $5x^2y^2 - xy + 3$. The first two are *polynomials in x* and the third is a *polynomial in x and y*. The terms of a polynomial in x have the form ax^k, where a is the **coefficient** and k is the **degree** of the term. For instance, the polynomial $2x^3 - 5x^2 + 1 = 2x^3 + (-5)x^2 + (0)x + 1$ has coefficients $2, -5, 0$, and 1.

REMARK Expressions are not polynomials when a variable is underneath a radical or when a polynomial expression (with degree greater than 0) is in the denominator of a term. For example, the expressions $x^3 - \sqrt{3x} = x^3 - (3x)^{1/2}$ and $x^2 + (5/x) = x^2 + 5x^{-1}$ are not polynomials.

Definition of a Polynomial in x

Let $a_0, a_1, a_2, . . . , a_n$ be real numbers and let n be a nonnegative integer. A polynomial in x is an expression of the form

$$a_nx^n + a_{n-1}x^{n-1} + \cdot \cdot \cdot + a_1x + a_0$$

where $a_n \neq 0$. The polynomial is of **degree** n, a_n is the **leading coefficient,** and a_0 is the **constant term.**

Polynomials with one, two, and three terms are called **monomials, binomials,** and **trinomials,** respectively. A polynomial written with descending powers of x is in **standard form.**

EXAMPLE 1 **Writing Polynomials in Standard Form**

Polynomial	**Standard Form**	**Degree**	**Leading Coefficient**
a. $4x^2 - 5x^7 - 2 + 3x$	$-5x^7 + 4x^2 + 3x - 2$	7	-5
b. $4 - 9x^2$	$-9x^2 + 4$	2	-9
c. 8	$8 \ (8 = 8x^0)$	0	8

Checkpoint *Audio-video solution in English & Spanish at LarsonPrecalculus.com.*

Write the polynomial $6 - 7x^3 + 2x$ in standard form. Then identify the degree and leading coefficient of the polynomial.

A polynomial that has all zero coefficients is called the **zero polynomial,** denoted by 0. No degree is assigned to the zero polynomial. For polynomials in more than one variable, the degree of a *term* is the sum of the exponents of the variables in the term. The degree of the *polynomial* is the highest degree of its terms. For instance, the degree of the polynomial $-2x^3y^6 + 4xy - x^7y^4$ is 11 because the sum of the exponents in the last term is the greatest. The leading coefficient of the polynomial is the coefficient of the highest-degree term.

mikeledray/Shutterstock.com

Operations with Polynomials

You can add and subtract polynomials in much the same way you add and subtract real numbers. Add or subtract the *like terms* (terms having the same variables to the same powers) by adding or subtracting their coefficients. For instance, $-3xy^2$ and $5xy^2$ are like terms and their sum is

$$-3xy^2 + 5xy^2 = (-3 + 5)xy^2 = 2xy^2.$$

EXAMPLE 2 Sums and Differences of Polynomials

a. $(5x^3 - 7x^2 - 3) + (x^3 + 2x^2 - x + 8)$

$= (5x^3 + x^3) + (-7x^2 + 2x^2) - x + (-3 + 8)$ — Group like terms.

$= 6x^3 - 5x^2 - x + 5$ — Combine like terms.

b. $(7x^4 - x^2 - 4x + 2) - (3x^4 - 4x^2 + 3x)$

$= 7x^4 - x^2 - 4x + 2 - 3x^4 + 4x^2 - 3x$ — Distributive Property

$= (7x^4 - 3x^4) + (-x^2 + 4x^2) + (-4x - 3x) + 2$ — Group like terms.

$= 4x^4 + 3x^2 - 7x + 2$ — Combine like terms.

REMARK When a negative sign precedes an expression inside parentheses, remember to distribute the negative sign to each term inside the parentheses, as shown.

$$-(3x^4 - 4x^2 + 3x)$$
$$= -3x^4 + 4x^2 - 3x$$

✓ Checkpoint *Audio-video solution in English & Spanish at LarsonPrecalculus.com.*

Find the difference $(2x^3 - x + 3) - (x^2 - 2x - 3)$ and write the resulting polynomial in standard form.

To find the *product* of two polynomials, use the left and right Distributive Properties. For example, if you treat $5x + 7$ as a single quantity, then you can multiply $3x - 2$ by $5x + 7$ as follows.

$$(3x - 2)(5x + 7) = 3x(5x + 7) - 2(5x + 7)$$
$$= (3x)(5x) + (3x)(7) - (2)(5x) - (2)(7)$$
$$= 15x^2 + 21x - 10x - 14$$

Product of First terms — Product of Outer terms — Product of Inner terms — Product of Last terms

$$= 15x^2 + 11x - 14$$

Note in this **FOIL Method** (which can only be used to multiply two binomials) that the outer (O) and inner (I) terms are like terms and can be combined.

EXAMPLE 3 Finding a Product by the FOIL Method

Use the FOIL Method to find the product of $2x - 4$ and $x + 5$.

Solution

F O I L

$$(2x - 4)(x + 5) = 2x^2 + 10x - 4x - 20$$
$$= 2x^2 + 6x - 20$$

✓ Checkpoint *Audio-video solution in English & Spanish at LarsonPrecalculus.com.*

Use the FOIL Method to find the product of $3x - 1$ and $x - 5$.

Special Products

Some binomial products have special forms that occur frequently in algebra. You do not need to memorize these formulas because you can use the Distributive Property to multiply. However, becoming familiar with these formulas will enable you to manipulate the algebra more quickly.

Special Products

Let u and v be real numbers, variables, or algebraic expressions.

Special Product	Example
Sum and Difference of Same Terms	
$(u + v)(u - v) = u^2 - v^2$	$(x + 4)(x - 4) = x^2 - 4^2$ $= x^2 - 16$
Square of a Binomial	
$(u + v)^2 = u^2 + 2uv + v^2$	$(x + 3)^2 = x^2 + 2(x)(3) + 3^2$ $= x^2 + 6x + 9$
$(u - v)^2 = u^2 - 2uv + v^2$	$(3x - 2)^2 = (3x)^2 - 2(3x)(2) + 2^2$ $= 9x^2 - 12x + 4$
Cube of a Binomial	
$(u + v)^3 = u^3 + 3u^2v + 3uv^2 + v^3$	$(x + 2)^3 = x^3 + 3x^2(2) + 3x(2^2) + 2^3$ $= x^3 + 6x^2 + 12x + 8$
$(u - v)^3 = u^3 - 3u^2v + 3uv^2 - v^3$	$(x - 1)^3 = x^3 - 3x^2(1) + 3x(1^2) - 1^3$ $= x^3 - 3x^2 + 3x - 1$

EXAMPLE 4 Sum and Difference of Same Terms

Find each product.

a. $(5x + 9)(5x - 9)$ **b.** $(x + y - 2)(x + y + 2)$

Solution

a. The product of a sum and a difference of the *same* two terms has no middle term and takes the form $(u + v)(u - v) = u^2 - v^2$.

$$(5x + 9)(5x - 9) = (5x)^2 - 9^2 = 25x^2 - 81$$

b. By grouping $x + y$ in parentheses, you can write the product of the trinomials as a special product.

$$\begin{aligned}(x + y - 2)(x + y + 2) &= [\underset{\text{Difference}}{(x + y) - 2}][\underset{\text{Sum}}{(x + y) + 2}] \\ &= (x + y)^2 - 2^2 \quad \text{Sum and difference of same terms} \\ &= x^2 + 2xy + y^2 - 4\end{aligned}$$

✔ **Checkpoint** *Audio-video solution in English & Spanish at LarsonPrecalculus.com.*

Find the product: $(x - 2 + 3y)(x - 2 - 3y)$.

Polynomials with Common Factors

The process of writing a polynomial as a product is called **factoring.** It is an important tool for solving equations and for simplifying rational expressions.

Unless noted otherwise, when you are asked to factor a polynomial, assume that you are looking for factors that have integer coefficients. If a polynomial does not factor using integer coefficients, then it is **prime** or **irreducible over the integers.** For instance, the polynomial

$$x^2 - 3$$

is irreducible over the integers. Over the *real numbers,* this polynomial factors as

$$x^2 - 3 = (x + \sqrt{3})(x - \sqrt{3}).$$

A polynomial is **completely factored** when each of its factors is prime. For instance,

$$x^3 - x^2 + 4x - 4 = (x - 1)(x^2 + 4) \qquad \text{Completely factored}$$

is completely factored, but

$$x^3 - x^2 - 4x + 4 = (x - 1)(x^2 - 4) \qquad \text{Not completely factored}$$

is not completely factored. Its complete factorization is

$$x^3 - x^2 - 4x + 4 = (x - 1)(x + 2)(x - 2).$$

The simplest type of factoring involves a polynomial that can be written as the product of a monomial and another polynomial. The technique used here is the Distributive Property, $a(b + c) = ab + ac$, in the *reverse* direction.

$$ab + ac = a(b + c) \qquad a \text{ is a common factor.}$$

Removing (factoring out) any common factors is the first step in completely factoring a polynomial.

EXAMPLE 5 Removing Common Factors

Factor each expression.

a. $6x^3 - 4x$

b. $-4x^2 + 12x - 16$

c. $(x - 2)(2x) + (x - 2)(3)$

Solution

a. $6x^3 - 4x = 2x(3x^2) - 2x(2)$ — $2x$ is a common factor.

$= 2x(3x^2 - 2)$

b. $-4x^2 + 12x - 16 = -4(x^2) + (-4)(-3x) + (-4)4$ — -4 is a common factor.

$= -4(x^2 - 3x + 4)$

c. $(x - 2)(2x) + (x - 2)(3) = (x - 2)(2x + 3)$ — $(x - 2)$ is a common factor.

Checkpoint Audio-video solution in English & Spanish at LarsonPrecalculus.com.

Factor each expression.

a. $5x^3 - 15x^2$

b. $-3 + 6x - 12x^3$

c. $(x + 1)(x^2) - (x + 1)(2)$

Factoring Special Polynomial Forms

Some polynomials have special forms that arise from the special product forms on page A27. You should learn to recognize these forms.

Factoring Special Polynomial Forms

Factored Form	Example
Difference of Two Squares	
$u^2 - v^2 = (u + v)(u - v)$	$9x^2 - 4 = (3x)^2 - 2^2 = (3x + 2)(3x - 2)$
Perfect Square Trinomial	
$u^2 + 2uv + v^2 = (u + v)^2$	$x^2 + 6x + 9 = x^2 + 2(x)(3) + 3^2 = (x + 3)^2$
$u^2 - 2uv + v^2 = (u - v)^2$	$x^2 - 6x + 9 = x^2 - 2(x)(3) + 3^2 = (x - 3)^2$
Sum or Difference of Two Cubes	
$u^3 + v^3 = (u + v)(u^2 - uv + v^2)$	$x^3 + 8 = x^3 + 2^3 = (x + 2)(x^2 - 2x + 4)$
$u^3 - v^3 = (u - v)(u^2 + uv + v^2)$	$27x^3 - 1 = (3x)^3 - 1^3 = (3x - 1)(9x^2 + 3x + 1)$

The factored form of a difference of two squares is always a set of *conjugate pairs.*

$$u^2 - v^2 = (u + v)(u - v)$$ Conjugate pairs

Difference (the minus sign in $u^2 - v^2$); Opposite signs (the signs in $(u + v)$ and $(u - v)$)

To recognize perfect square terms, look for coefficients that are squares of integers and variables raised to *even powers.*

REMARK In Example 6, note that the first step in factoring a polynomial is to check for any common factors. Once you have removed any common factors, it is often possible to recognize patterns that were not immediately obvious.

EXAMPLE 6 Removing a Common Factor First

$$3 - 12x^2 = 3(1 - 4x^2)$$ 3 is a common factor.

$$= 3[1^2 - (2x)^2]$$

$$= 3(1 + 2x)(1 - 2x)$$ Difference of two squares

✓ Checkpoint *Audio-video solution in English & Spanish at LarsonPrecalculus.com.*

Factor $100 - 4y^2$.

EXAMPLE 7 Factoring the Difference of Two Squares

a. $(x + 2)^2 - y^2 = [(x + 2) + y][(x + 2) - y]$

$$= (x + 2 + y)(x + 2 - y)$$

b. $16x^4 - 81 = (4x^2)^2 - 9^2$

$$= (4x^2 + 9)(4x^2 - 9)$$ Difference of two squares

$$= (4x^2 + 9)[(2x)^2 - 3^2]$$

$$= (4x^2 + 9)(2x + 3)(2x - 3)$$ Difference of two squares

✓ Checkpoint *Audio-video solution in English & Spanish at LarsonPrecalculus.com.*

Factor $(x - 1)^2 - 9y^4$.

A **perfect square trinomial** is the square of a binomial, and it has the form

$$u^2 + 2uv + v^2 = (u + v)^2 \quad \text{or} \quad u^2 - 2uv + v^2 = (u - v)^2.$$

Like signs   Like signs

Note that the first and last terms are squares and the middle term is twice the product of u and v.

EXAMPLE 8 Factoring Perfect Square Trinomials

Factor each trinomial.

a. $x^2 - 10x + 25$ **b.** $16x^2 + 24x + 9$

Solution

a. $x^2 - 10x + 25 = x^2 - 2(x)(5) + 5^2 = (x - 5)^2$

b. $16x^2 + 24x + 9 = (4x)^2 + 2(4x)(3) + 3^2 = (4x + 3)^2$

✓ ***Checkpoint*** *Audio-video solution in English & Spanish at LarsonPrecalculus.com.*

Factor $9x^2 - 30x + 25$.

The next two formulas show the sums and differences of cubes. Pay special attention to the signs of the terms.

Like signs   Like signs

$$u^3 + v^3 = (u + v)(u^2 - uv + v^2) \qquad u^3 - v^3 = (u - v)(u^2 + uv + v^2)$$

Unlike signs   Unlike signs

EXAMPLE 9 Factoring the Difference of Cubes

$x^3 - 27 = x^3 - 3^3$ Rewrite 27 as 3^3.

$= (x - 3)(x^2 + 3x + 9)$ Factor.

✓ ***Checkpoint*** *Audio-video solution in English & Spanish at LarsonPrecalculus.com.*

Factor $64x^3 - 1$.

EXAMPLE 10 Factoring the Sum of Cubes

a. $y^3 + 8 = y^3 + 2^3$ Rewrite 8 as 2^3.

$= (y + 2)(y^2 - 2y + 4)$ Factor.

b. $3x^3 + 192 = 3(x^3 + 64)$ 3 is a common factor.

$= 3(x^3 + 4^3)$ Rewrite 64 as 4^3.

$= 3(x + 4)(x^2 - 4x + 16)$ Factor.

✓ ***Checkpoint*** *Audio-video solution in English & Spanish at LarsonPrecalculus.com.*

Factor each expression.

a. $x^3 + 216$ **b.** $5y^3 + 135$

Trinomials with Binomial Factors

To factor a trinomial of the form $ax^2 + bx + c$, use the following pattern.

Factors of a

$$ax^2 + bx + c = (\square x + \square)(\square x + \square)$$

Factors of c

The goal is to find a combination of factors of a and c such that the outer and inner products add up to the middle term bx. For instance, for the trinomial $6x^2 + 17x + 5$, you can write all possible factorizations and determine which one has outer and inner products that add up to $17x$.

$$(6x + 5)(x + 1),\ (6x + 1)(x + 5),\ (2x + 1)(3x + 5),\ (2x + 5)(3x + 1)$$

You can see that $(2x + 5)(3x + 1)$ is the correct factorization because the outer (O) and inner (I) products add up to $17x$.

$$\overset{\text{F}}{} \quad \overset{\text{O}}{} \quad \overset{\text{I}}{} \quad \overset{\text{L}}{} \qquad \overset{\text{O + I}}{}$$

$$(2x + 5)(3x + 1) = \underset{\text{F}}{6x^2} + \underset{\text{O}}{2x} + \underset{\text{I}}{15x} + \underset{\text{L}}{5} = 6x^2 + \underset{\text{O + I}}{17x} + 5$$

EXAMPLE 11 **Factoring a Trinomial: Leading Coefficient Is 1**

Factor $x^2 - 7x + 12$.

Solution For this trinomial, you have $a = 1$, $b = -7$, and $c = 12$. Because b is negative and c is positive, both factors of 12 must be negative. So, the possible factorizations of $x^2 - 7x + 12$ are

$$(x - 2)(x - 6), \quad (x - 1)(x - 12), \quad \text{and} \quad (x - 3)(x - 4).$$

Testing the middle term, you will find the correct factorization to be

$$x^2 - 7x + 12 = (x - 3)(x - 4).$$

REMARK Factoring a trinomial can involve trial and error. However, once you have produced the factored form, it is relatively easy to check your answer. For instance, verify the factorization in Example 11 by multiplying $(x - 3)$ by $(x - 4)$ to see that you obtain the original trinomial $x^2 - 7x + 12$.

✓ ***Checkpoint*** *Audio-video solution in English & Spanish at LarsonPrecalculus.com.*

Factor $x^2 + x - 6$.

EXAMPLE 12 **Factoring a Trinomial: Leading Coefficient Is Not 1**

Factor $2x^2 + x - 15$.

Solution For this trinomial, you have $a = 2$ and $c = -15$, which means that the factors of -15 must have unlike signs. The eight possible factorizations are as follows.

$$(2x - 1)(x + 15) \quad (2x + 1)(x - 15) \quad (2x - 3)(x + 5) \quad (2x + 3)(x - 5)$$

$$(2x - 5)(x + 3) \quad (2x + 5)(x - 3) \quad (2x - 15)(x + 1) \quad (2x + 15)(x - 1)$$

Testing the middle term, you will find the correct factorization to be

$$2x^2 + x - 15 = (2x - 5)(x + 3). \qquad \text{O + I} = 6x - 5x = x$$

✓ ***Checkpoint*** *Audio-video solution in English & Spanish at LarsonPrecalculus.com.*

Factor each trinomial.

a. $2x^2 - 5x + 3$ **b.** $12x^2 + 7x + 1$

Factoring by Grouping

Sometimes, polynomials with more than three terms can be **factored by grouping.**

REMARK Sometimes, more than one grouping will work. For instance, another way to factor the polynomial in Example 13 is as follows.

$$\begin{aligned} &x^3 - 2x^2 - 3x + 6 \\ &= (x^3 - 3x) - (2x^2 - 6) \\ &= x(x^2 - 3) - 2(x^2 - 3) \\ &= (x^2 - 3)(x - 2) \end{aligned}$$

As you can see, you obtain the same result as in Example 13.

EXAMPLE 13 Factoring by Grouping

$$\begin{aligned} x^3 - 2x^2 - 3x + 6 &= (x^3 - 2x^2) - (3x - 6) && \text{Group terms.} \\ &= x^2(x - 2) - 3(x - 2) && \text{Factor each group.} \\ &= (x - 2)(x^2 - 3) && \text{Distributive Property} \end{aligned}$$

✓ ***Checkpoint*** *Audio-video solution in English & Spanish at LarsonPrecalculus.com.*

Factor $x^3 + x^2 - 5x - 5$.

Factoring by grouping can save you some of the trial and error involved in factoring a trinomial. To factor a trinomial of the form $ax^2 + bx + c$ by grouping, rewrite the middle term using the sum of two factors of the product ac that add up to b. Example 14 illustrates this technique.

EXAMPLE 14 Factoring a Trinomial by Grouping

In the trinomial $2x^2 + 5x - 3$, $a = 2$ and $c = -3$, so the product ac is -6. Now, -6 factors as $(6)(-1)$ and $6 - 1 = 5 = b$. So, rewrite the middle term as $5x = 6x - x$.

$$\begin{aligned} 2x^2 + 5x - 3 &= 2x^2 + 6x - x - 3 && \text{Rewrite middle term.} \\ &= (2x^2 + 6x) - (x + 3) && \text{Group terms.} \\ &= 2x(x + 3) - (x + 3) && \text{Factor groups.} \\ &= (x + 3)(2x - 1) && \text{Distributive Property} \end{aligned}$$

✓ ***Checkpoint*** *Audio-video solution in English & Spanish at LarsonPrecalculus.com.*

Use factoring by grouping to factor $2x^2 + 5x - 12$.

Summarize (Appendix A.3)

1. State the definition of a polynomial in x and explain what is meant by the standard form of a polynomial *(page A25)*. For an example of writing polynomials in standard form, see Example 1.
2. Describe how to add and subtract polynomials *(page A26)*. For an example of adding and subtracting polynomials, see Example 2.
3. Describe the FOIL Method *(page A26)*. For an example of finding a product using the FOIL Method, see Example 3.
4. Explain how to find binomial products that have special forms *(page A27)*. For an example of binomial products that have special forms, see Example 4.
5. Describe what it means to completely factor a polynomial *(page A28)*. For an example of removing common factors, see Example 5.
6. Make a list of the special polynomial forms of factoring *(page A29)*. For examples of factoring these special forms, see Examples 6–10.
7. Describe how to factor a trinomial of the form $ax^2 + bx + c$ *(page A31)*. For examples of factoring trinomials of this form, see Examples 11 and 12.
8. Explain how to factor a polynomial by grouping *(page A32)*. For examples of factoring by grouping, see Examples 13 and 14.

A.3 Exercises

See **CalcChat.com** for tutorial help and worked-out solutions to odd-numbered exercises.

Vocabulary: Fill in the blanks.

1. For the polynomial $a_nx^n + a_{n-1}x^{n-1} + \cdot \cdot \cdot + a_1x + a_0$, $a_n \neq 0$, the degree is ________, the leading coefficient is ________, and the constant term is ________.

2. A polynomial with one term is called a ________, while a polynomial with two terms is called a ________ and a polynomial with three terms is called a ________.

3. To add or subtract polynomials, add or subtract the ________ ________ by adding their coefficients.

4. The letters in "FOIL" stand for the following. F ________ O ________ I ________ L ________

5. The process of writing a polynomial as a product is called ________.

6. A polynomial is ________ ________ when each of its factors is prime.

7. A ________ ________ ________ is the square of a binomial, and it has the form $u^2 + 2uv + v^2$ or $u^2 - 2uv + v^2$.

8. When a polynomial has more than three terms, a method of factoring called ________ ________ ________ may be used.

Skills and Applications

Polynomials **In Exercises 9–18, (a) write the polynomial in standard form, (b) identify the degree and leading coefficient of the polynomial, and (c) state whether the polynomial is a monomial, a binomial, or a trinomial.**

9. $14x - \frac{1}{2}x^5$ **10.** $7x$

11. $3 - x^6$ **12.** $-y + 25y^2 + 1$

13. 3 **14.** $-8 + t^2$

15. $1 + 6x^4 - 4x^5$ **16.** $3 + 2x$

17. $4x^3y$ **18.** $-x^5y + 2x^2y^2 + xy^4$

Operations with Polynomials **In Exercises 19–26, perform the operation and write the result in standard form.**

19. $(6x + 5) - (8x + 15)$

20. $(2x^2 + 1) - (x^2 - 2x + 1)$

21. $(15x^2 - 6) - (-8.3x^3 - 14.7x^2 - 17)$

22. $(15.6w^4 - 14w - 17.4) - (16.9w^4 - 9.2w + 13)$

23. $3x(x^2 - 2x + 1)$ **24.** $y^2(4y^2 + 2y - 3)$

25. $-5z(3z - 1)$ **26.** $(-3x)(5x + 2)$

Multiplying Polynomials **In Exercises 27–40, multiply or find the special product.**

27. $(x + 3)(x + 4)$ **28.** $(x - 5)(x + 10)$

29. $(x^2 - x + 1)(x^2 + x + 1)$

30. $(2x^2 - x + 4)(x^2 + 3x + 2)$

31. $(x + 10)(x - 10)$ **32.** $(4a + 5b)(4a - 5b)$

33. $(2x + 3)^2$ **34.** $(8x + 3)^2$

35. $(x + 1)^3$ **36.** $(3x + 2y)^3$

37. $[(m - 3) + n][(m - 3) - n]$

38. $[(x - 3y) + z][(x - 3y) - z]$

39. $[(x - 3) + y]^2$ **40.** $[(x + 1) - y]^2$

Factoring Out a Common Factor **In Exercises 41–44, factor out the common factor.**

41. $2x^3 - 6x$ **42.** $3z^3 - 6z^2 + 9z$

43. $3x(x - 5) + 8(x - 5)$ **44.** $(x + 3)^2 - 4(x + 3)$

Greatest Common Factor **In Exercises 45–48, find the greatest common factor such that the remaining factors have only integer coefficients.**

45. $\frac{1}{2}x^3 + 2x^2 - 5x$ **46.** $\frac{1}{3}y^4 - 5y^2 + 2y$

47. $\frac{2}{3}x(x - 3) - 4(x - 3)$ **48.** $\frac{4}{5}y(y + 1) - 2(y + 1)$

Factoring the Difference of Two Squares **In Exercises 49–52, completely factor the difference of two squares.**

49. $x^2 - 81$ **50.** $x^2 - 64$

51. $(x - 1)^2 - 4$ **52.** $25 - (z + 5)^2$

Factoring a Perfect Square Trinomial **In Exercises 53–58, factor the perfect square trinomial.**

53. $x^2 - 4x + 4$ **54.** $4t^2 + 4t + 1$

55. $9u^2 + 24uv + 16v^2$ **56.** $36y^2 - 108y + 81$

57. $z^2 + z + \frac{1}{4}$ **58.** $9y^2 - \frac{3}{2}y + \frac{1}{16}$

Factoring the Sum or Difference of Cubes **In Exercises 59–62, factor the sum or difference of cubes.**

59. $x^3 - 8$ **60.** $27 - x^3$

61. $27x^3 + 8$ **62.** $u^3 + 27v^3$

Factoring a Trinomial **In Exercises 63–70, factor the trinomial.**

63. $x^2 + x - 2$ **64.** $s^2 - 5s + 6$

65. $20 - y - y^2$ **66.** $24 + 5z - z^2$

67. $3x^2 - 5x + 2$ **68.** $2x^2 - x - 1$

69. $5x^2 + 26x + 5$ **70.** $-9z^2 + 3z + 2$

Factoring by Grouping **In Exercises 71–76, factor by grouping.**

71. $x^3 - x^2 + 2x - 2$ **72.** $x^3 + 5x^2 - 5x - 25$

73. $2x^3 - x^2 - 6x + 3$ **74.** $6 + 2x - 3x^3 - x^4$

75. $x^5 + 2x^3 + x^2 + 2$ **76.** $8x^5 - 6x^2 + 12x^3 - 9$

Factoring a Trinomial by Grouping **In Exercises 77–80, factor the trinomial by grouping.**

77. $2x^2 + 9x + 9$ **78.** $6x^2 + x - 2$

79. $6x^2 - x - 15$ **80.** $12x^2 - 13x + 1$

Factoring Completely **In Exercises 81–94, completely factor the expression.**

81. $6x^2 - 54$ **82.** $12x^2 - 48$

83. $x^3 - x^2$ **84.** $x^3 - 16x$

85. $x^2 - 2x + 1$ **86.** $16 + 6x - x^2$

87. $2x^2 + 4x - 2x^3$ **88.** $13x + 6 + 5x^2$

89. $5 - x + 5x^2 - x^3$ **90.** $3u - 2u^2 + 6 - u^3$

91. $5(3 - 4x)^2 - 8(3 - 4x)(5x - 1)$

92. $2(x + 1)(x - 3)^2 - 3(x + 1)^2(x - 3)$

93. $x^4(4)(2x + 1)^3(2x) + (2x + 1)^4(4x^3)$

94. $x^3(3)(x^2 + 1)^2(2x) + (x^2 + 1)^3(3x^2)$

95. Geometry The cylindrical shell shown in the figure has a volume of

$$V = \pi R^2 h - \pi r^2 h.$$

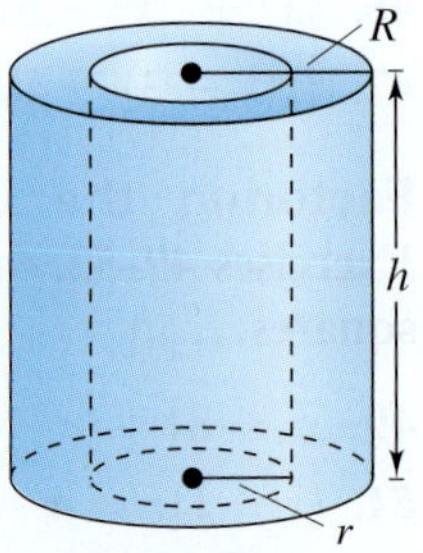

(a) Factor the expression for the volume.

(b) From the result of part (a), show that the volume is 2π(average radius)(thickness of the shell)h.

96. Chemistry The rate of change of an autocatalytic chemical reaction is $kQx - kx^2$, where Q is the amount of the original substance, x is the amount of substance formed, and k is a constant of proportionality. Factor the expression.

Exploration

True or False? **In Exercises 97–100, determine whether the statement is true or false. Justify your answer.**

97. The product of two binomials is always a second-degree polynomial.

98. The sum of two binomials is always a binomial.

99. The difference of two perfect squares can be factored as the product of conjugate pairs.

100. The sum of two perfect squares can be factored as the binomial sum squared.

101. Degree of a Product Find the degree of the product of two polynomials of degrees m and n.

102. Degree of a Sum Find the degree of the sum of two polynomials of degrees m and n, where $m < n$.

103. Think About It When the polynomial

$$-x^3 + 3x^2 + 2x - 1$$

is subtracted from an unknown polynomial, the difference is $5x^2 + 8$. If it is possible, then find the unknown polynomial.

104. Logical Reasoning Verify that $(x + y)^2$ is not equal to $x^2 + y^2$ by letting $x = 3$ and $y = 4$ and evaluating both expressions. Are there any values of x and y for which $(x + y)^2 = x^2 + y^2$? Explain.

105. Think About It Give an example of a polynomial that is prime with respect to the integers.

106. HOW DO YOU SEE IT? The figure shows a large square with an area of a^2 that contains a smaller square with an area of b^2.

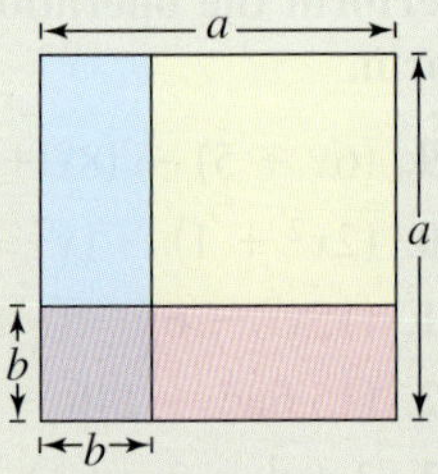

(a) Describe the regions that represent $a^2 - b^2$. How can you rearrange these regions to show that $a^2 - b^2 = (a - b)(a + b)$?

(b) How can you use the figure to show that $(a - b)^2 = a^2 - 2ab + b^2$?

(c) Draw another figure to show that $(a + b)^2 = a^2 + 2ab + b^2$. Explain how the figure shows this.

Factoring with Variables in the Exponents **In Exercises 107 and 108, factor the expression as completely as possible.**

107. $x^{2n} - y^{2n}$ **108.** $x^{3n} + y^{3n}$

mikeledray/Shutterstock.com

A.4 Rational Expressions

Rational expressions can help you solve real-life problems. For instance, in Exercise 73 on page A43, you will use a rational expression to model the temperature of food in a refrigerator.

- **Find domains of algebraic expressions.**
- **Simplify rational expressions.**
- **Add, subtract, multiply, and divide rational expressions.**
- **Simplify complex fractions and rewrite difference quotients.**

Domain of an Algebraic Expression

The set of real numbers for which an algebraic expression is defined is the **domain** of the expression. Two algebraic expressions are **equivalent** when they have the same domain and yield the same values for all numbers in their domain. For instance,

$$(x + 1) + (x + 2) \quad \text{and} \quad 2x + 3$$

are equivalent because

$$\begin{aligned}(x + 1) + (x + 2) &= x + 1 + x + 2 \\ &= x + x + 1 + 2 \\ &= 2x + 3.\end{aligned}$$

EXAMPLE 1 Finding the Domain of an Algebraic Expression

a. The domain of the polynomial

$$2x^3 + 3x + 4$$

is the set of all real numbers. In fact, the domain of any polynomial is the set of all real numbers, unless the domain is specifically restricted.

b. The domain of the radical expression

$$\sqrt{x - 2}$$

is the set of real numbers greater than or equal to 2, because the square root of a negative number is not a real number.

c. The domain of the expression

$$\frac{x + 2}{x - 3}$$

is the set of all real numbers except $x = 3$, which would result in division by zero, which is undefined.

✓ ***Checkpoint*** *Audio-video solution in English & Spanish at LarsonPrecalculus.com.*

Find the domain of each expression.

a. $4x^3 + 3, \quad x \geq 0$ **b.** $\sqrt{x + 7}$ **c.** $\dfrac{1 - x}{x}$

The quotient of two algebraic expressions is a *fractional expression.* Moreover, the quotient of two *polynomials* such as

$$\frac{1}{x}, \quad \frac{2x - 1}{x + 1}, \quad \text{or} \quad \frac{x^2 - 1}{x^2 + 1}$$

is a **rational expression.**

Filipe B. Varela/Shutterstock.com

Simplifying Rational Expressions

Recall that a fraction is in simplest form when its numerator and denominator have no factors in common aside from ± 1. To write a fraction in simplest form, divide out common factors.

$$\frac{a \cdot \cancel{c}}{b \cdot \cancel{c}} = \frac{a}{b}, \quad c \neq 0$$

The key to success in simplifying rational expressions lies in your ability to *factor* polynomials. When simplifying rational expressions, factor each polynomial completely to determine whether the numerator and denominator have factors in common.

EXAMPLE 2 Simplifying a Rational Expression

$$\frac{x^2 + 4x - 12}{3x - 6} = \frac{(x + 6)\cancel{(x - 2)}}{3\cancel{(x - 2)}} \quad \text{Factor completely.}$$

$$= \frac{x + 6}{3}, \quad x \neq 2 \quad \text{Divide out common factor.}$$

Note that the original expression is undefined when $x = 2$ (because division by zero is undefined). To make the simplified expression *equivalent* to the original expression, you must list the domain restriction $x \neq 2$ with the simplified expression.

REMARK In Example 2, do not make the mistake of trying to simplify further by dividing out terms.

$$\cancel{\frac{x + 6}{3} = \frac{x + \cancel{6}}{\cancel{3}} = x + 2}$$

Remember that to simplify fractions, divide out common *factors,* not terms.

Checkpoint Audio-video solution in English & Spanish at LarsonPrecalculus.com.

Write $\dfrac{4x + 12}{x^2 - 3x - 18}$ in simplest form.

Sometimes it may be necessary to change the sign of a factor by factoring out (-1) to simplify a rational expression, as shown in Example 3.

EXAMPLE 3 Simplifying a Rational Expression

$$\frac{12 + x - x^2}{2x^2 - 9x + 4} = \frac{(4 - x)(3 + x)}{(2x - 1)(x - 4)} \quad \text{Factor completely.}$$

$$= \frac{-\cancel{(x - 4)}(3 + x)}{(2x - 1)\cancel{(x - 4)}} \quad (4 - x) = -(x - 4)$$

$$= -\frac{3 + x}{2x - 1}, \quad x \neq 4 \quad \text{Divide out common factor.}$$

Checkpoint Audio-video solution in English & Spanish at LarsonPrecalculus.com.

Write $\dfrac{3x^2 - x - 2}{5 - 4x - x^2}$ in simplest form.

In this text, when writing a rational expression, the domain is usually not listed with the expression. It is *implied* that the real numbers that make the denominator zero are excluded from the expression. Also, when performing operations with rational expressions, this text follows the convention of listing *by the simplified expression* all values of x that must be specifically excluded from the domain in order to make the domains of the simplified and original expressions agree. Example 3, for instance, lists the restriction $x \neq 4$ with the simplified expression to make the two domains agree. Note that the value $x = \frac{1}{2}$ is excluded from *both* domains, so it is not necessary to list this value.

Operations with Rational Expressions

To multiply or divide rational expressions, use the properties of fractions discussed in Appendix A.1. Recall that to divide fractions, you invert the divisor and multiply.

EXAMPLE 4 **Multiplying Rational Expressions**

$$\frac{2x^2+x-6}{x^2+4x-5}\cdot\frac{x^3-3x^2+2x}{4x^2-6x}=\frac{(2x-3)(x+2)}{(x+5)(x-1)}\cdot\frac{x(x-2)(x-1)}{2x(2x-3)}$$

$$=\frac{(x+2)(x-2)}{2(x+5)},\quad x\neq 0, x\neq 1, x\neq \tfrac{3}{2}$$

REMARK Note that Example 4 lists the restrictions $x \neq 0, x \neq 1$, and $x \neq \frac{3}{2}$ with the simplified expression in order to make the two domains agree. Also note that the value $x = -5$ is excluded from both domains, so it is not necessary to list this value.

Checkpoint Audio-video solution in English & Spanish at LarsonPrecalculus.com.

Multiply and simplify: $\dfrac{15x^2+5x}{x^3-3x^2-18x}\cdot\dfrac{x^2-2x-15}{3x^2-8x-3}$.

EXAMPLE 5 **Dividing Rational Expressions**

$$\frac{x^3-8}{x^2-4}\div\frac{x^2+2x+4}{x^3+8}=\frac{x^3-8}{x^2-4}\cdot\frac{x^3+8}{x^2+2x+4}\qquad\text{Invert and multiply.}$$

$$=\frac{(x-2)(x^2+2x+4)}{(x+2)(x-2)}\cdot\frac{(x+2)(x^2-2x+4)}{(x^2+2x+4)}$$

$$=x^2-2x+4,\quad x\neq\pm 2\qquad\text{Divide out common factors.}$$

Checkpoint Audio-video solution in English & Spanish at LarsonPrecalculus.com.

Divide and simplify: $\dfrac{x^3-1}{x^2-1}\div\dfrac{x^2+x+1}{x^2+2x+1}$.

To add or subtract rational expressions, use the LCD (least common denominator) method or the *basic definition*

$$\frac{a}{b}\pm\frac{c}{d}=\frac{ad\pm bc}{bd},\quad b\neq 0, d\neq 0.\qquad\text{Basic definition}$$

This definition provides an efficient way of adding or subtracting *two* fractions that have no common factors in their denominators.

EXAMPLE 6 **Subtracting Rational Expressions**

$$\frac{x}{x-3}-\frac{2}{3x+4}=\frac{x(3x+4)-2(x-3)}{(x-3)(3x+4)}\qquad\text{Basic definition}$$

$$=\frac{3x^2+4x-2x+6}{(x-3)(3x+4)}\qquad\text{Distributive Property}$$

$$=\frac{3x^2+2x+6}{(x-3)(3x+4)}\qquad\text{Combine like terms.}$$

REMARK When subtracting rational expressions, remember to distribute the negative sign to all the terms in the quantity that is being subtracted.

Checkpoint Audio-video solution in English & Spanish at LarsonPrecalculus.com.

Subtract: $\dfrac{x}{2x-1}-\dfrac{1}{x+2}$.

For three or more fractions, or for fractions with a repeated factor in the denominators, the LCD method works well. Recall that the least common denominator of several fractions consists of the product of all prime factors in the denominators, with each factor given the highest power of its occurrence in any denominator. Here is a numerical example.

$$\frac{1}{6}+\frac{3}{4}-\frac{2}{3}=\frac{1\cdot 2}{6\cdot 2}+\frac{3\cdot 3}{4\cdot 3}-\frac{2\cdot 4}{3\cdot 4}$$

The LCD is 12.

$$=\frac{2}{12}+\frac{9}{12}-\frac{8}{12}$$

$$=\frac{3}{12}$$

$$=\frac{1}{4}$$

Sometimes, the numerator of the answer has a factor in common with the denominator. In such cases, simplify the answer. For instance, in the example above, $\frac{3}{12}$ simplifies to $\frac{1}{4}$.

EXAMPLE 7 **Combining Rational Expressions: The LCD Method**

Perform the operations and simplify.

$$\frac{3}{x-1}-\frac{2}{x}+\frac{x+3}{x^2-1}$$

Solution Using the factored denominators

$$(x-1),\quad x,\quad \text{and}\quad (x+1)(x-1)$$

you can see that the LCD is $x(x+1)(x-1)$.

$$\frac{3}{x-1}-\frac{2}{x}+\frac{x+3}{(x+1)(x-1)}$$

$$=\frac{3(x)(x+1)}{x(x+1)(x-1)}-\frac{2(x+1)(x-1)}{x(x+1)(x-1)}+\frac{(x+3)(x)}{x(x+1)(x-1)}$$

$$=\frac{3(x)(x+1)-2(x+1)(x-1)+(x+3)(x)}{x(x+1)(x-1)}$$

$$=\frac{3x^2+3x-2x^2+2+x^2+3x}{x(x+1)(x-1)}$$

Distributive Property

$$=\frac{3x^2-2x^2+x^2+3x+3x+2}{x(x+1)(x-1)}$$

Group like terms.

$$=\frac{2x^2+6x+2}{x(x+1)(x-1)}$$

Combine like terms.

$$=\frac{2(x^2+3x+1)}{x(x+1)(x-1)}$$

Factor.

✓ **Checkpoint** *Audio-video solution in English & Spanish at LarsonPrecalculus.com.*

Perform the operations and simplify.

$$\frac{4}{x}-\frac{x+5}{x^2-4}+\frac{4}{x+2}$$

Complex Fractions and the Difference Quotient

Fractional expressions with separate fractions in the numerator, denominator, or both are called **complex fractions.** Here are two examples.

$$\frac{\left(\frac{1}{x}\right)}{x^2+1} \quad \text{and} \quad \frac{\left(\frac{1}{x}\right)}{\left(\frac{1}{x^2+1}\right)}$$

To simplify a complex fraction, combine the fractions in the numerator into a single fraction and then combine the fractions in the denominator into a single fraction. Then invert the denominator and multiply.

EXAMPLE 8 Simplifying a Complex Fraction

$$\frac{\left(\frac{2}{x}-3\right)}{\left(1-\frac{1}{x-1}\right)} = \frac{\left[\frac{2-3(x)}{x}\right]}{\left[\frac{1(x-1)-1}{x-1}\right]}$$ Combine fractions.

$$= \frac{\left(\frac{2-3x}{x}\right)}{\left(\frac{x-2}{x-1}\right)}$$ Simplify.

$$= \frac{2-3x}{x} \cdot \frac{x-1}{x-2}$$ Invert and multiply.

$$= \frac{(2-3x)(x-1)}{x(x-2)}, \quad x \neq 1$$

Checkpoint Audio-video solution in English & Spanish at LarsonPrecalculus.com.

Simplify the complex fraction $\frac{\left(\frac{1}{x+2}+1\right)}{\left(\frac{x}{3}-1\right)}$.

Another way to simplify a complex fraction is to multiply its numerator and denominator by the LCD of all fractions in its numerator and denominator. This method is applied to the fraction in Example 8 as follows.

$$\frac{\left(\frac{2}{x}-3\right)}{\left(1-\frac{1}{x-1}\right)} = \frac{\left(\frac{2}{x}-3\right)}{\left(1-\frac{1}{x-1}\right)} \cdot \frac{x(x-1)}{x(x-1)}$$ LCD is $x(x-1)$.

$$= \frac{\left(\frac{2-3x}{x}\right) \cdot x(x-1)}{\left(\frac{x-2}{x-1}\right) \cdot x(x-1)}$$

$$= \frac{(2-3x)(x-1)}{x(x-2)}, \quad x \neq 1$$

The next three examples illustrate some methods for simplifying rational expressions involving negative exponents and radicals. These types of expressions occur frequently in calculus.

To simplify an expression with negative exponents, one method is to begin by factoring out the common factor with the *lesser* exponent. Remember that when factoring, you *subtract* exponents. For instance, in $3x^{-5/2} + 2x^{-3/2}$, the lesser exponent is $-\frac{5}{2}$ and the common factor is $x^{-5/2}$.

$$\begin{aligned} 3x^{-5/2} + 2x^{-3/2} &= x^{-5/2}[3(1) + 2x^{-3/2-(-5/2)}] \\ &= x^{-5/2}(3 + 2x^1) \\ &= \frac{3 + 2x}{x^{5/2}} \end{aligned}$$

EXAMPLE 9 **Simplifying an Expression**

Simplify the following expression containing negative exponents.

$$x(1 - 2x)^{-3/2} + (1 - 2x)^{-1/2}$$

Solution Begin by factoring out the common factor with the *lesser exponent.*

$$\begin{aligned} x(1 - 2x)^{-3/2} + (1 - 2x)^{-1/2} &= (1 - 2x)^{-3/2}[x + (1 - 2x)^{(-1/2)-(-3/2)}] \\ &= (1 - 2x)^{-3/2}[x + (1 - 2x)^1] \\ &= \frac{1 - x}{(1 - 2x)^{3/2}} \end{aligned}$$

✓ ***Checkpoint*** *Audio-video solution in English & Spanish at LarsonPrecalculus.com.*

Simplify $(x - 1)^{-1/3} - x(x - 1)^{-4/3}$.

The next example shows a second method for simplifying an expression with negative exponents.

EXAMPLE 10 **Simplifying an Expression**

$$\begin{aligned} \frac{(4 - x^2)^{1/2} + x^2(4 - x^2)^{-1/2}}{4 - x^2} &= \frac{(4 - x^2)^{1/2} + x^2(4 - x^2)^{-1/2}}{4 - x^2} \cdot \frac{(4 - x^2)^{1/2}}{(4 - x^2)^{1/2}} \\ &= \frac{(4 - x^2)^1 + x^2(4 - x^2)^0}{(4 - x^2)^{3/2}} \\ &= \frac{4 - x^2 + x^2}{(4 - x^2)^{3/2}} \\ &= \frac{4}{(4 - x^2)^{3/2}} \end{aligned}$$

✓ ***Checkpoint*** *Audio-video solution in English & Spanish at LarsonPrecalculus.com.*

Simplify

$$\frac{x^2(x^2 - 2)^{-1/2} + (x^2 - 2)^{1/2}}{x^2 - 2}.$$

EXAMPLE 11 Rewriting a Difference Quotient

The following expression from calculus is an example of a *difference quotient.*

$$\frac{\sqrt{x+h}-\sqrt{x}}{h}$$

Rewrite this expression by rationalizing its numerator.

Solution

$$\begin{aligned}\frac{\sqrt{x+h}-\sqrt{x}}{h} &= \frac{\sqrt{x+h}-\sqrt{x}}{h} \cdot \frac{\sqrt{x+h}+\sqrt{x}}{\sqrt{x+h}+\sqrt{x}} \\ &= \frac{\left(\sqrt{x+h}\right)^2-\left(\sqrt{x}\right)^2}{h\left(\sqrt{x+h}+\sqrt{x}\right)} \\ &= \frac{x+h-x}{h\left(\sqrt{x+h}+\sqrt{x}\right)} \\ &= \frac{h}{h\left(\sqrt{x+h}+\sqrt{x}\right)} \\ &= \frac{1}{\sqrt{x+h}+\sqrt{x}}, \quad h \neq 0\end{aligned}$$

ALGEBRA HELP You can review the techniques for rationalizing a numerator in Appendix A.2.

Notice that the original expression is undefined when $h = 0$. So, you must exclude $h = 0$ from the domain of the simplified expression so that the expressions are equivalent.

Checkpoint *Audio-video solution in English & Spanish at LarsonPrecalculus.com.*

Rewrite the difference quotient

$$\frac{\sqrt{9+h}-3}{h}$$

by rationalizing its numerator.

Difference quotients, such as that in Example 11, occur frequently in calculus. Often, they need to be rewritten in an equivalent form that can be evaluated when $h = 0$.

Summarize (Appendix A.4)

1. State the definition of the domain of an algebraic expression *(page A35)*. For an example of finding the domains of algebraic expressions, see Example 1.
2. State the definition of a rational expression and describe how to simplify a rational expression *(pages A35 and A36)*. For examples of simplifying rational expressions, see Examples 2 and 3.
3. Describe how to multiply, divide, add, and subtract rational expressions *(page A37)*. For examples of operations with rational expressions, see Examples 4–7.
4. State the definition of a complex fraction *(page A39)*. For an example of simplifying a complex fraction, see Example 8.
5. Describe how to rewrite a difference quotient and why you would want to do so *(page A41)*. For an example of rewriting a difference quotient, see Example 11.

A.4 Exercises

See **CalcChat.com** for tutorial help and worked-out solutions to odd-numbered exercises.

Vocabulary: Fill in the blanks.

1. The set of real numbers for which an algebraic expression is defined is the ________ of the expression.
2. The quotient of two algebraic expressions is a fractional expression, and the quotient of two polynomials is a ________ ________.
3. Fractional expressions with separate fractions in the numerator, denominator, or both are called ________ fractions.
4. Two algebraic expressions that have the same domain and yield the same values for all numbers in their domains are called ________.

Skills and Applications

Finding the Domain of an Algebraic Expression

In Exercises 5–16, find the domain of the expression.

5. $3x^2 - 4x + 7$
6. $6x^2 - 9, \quad x > 0$
7. $\dfrac{1}{3 - x}$
8. $\dfrac{x + 6}{3x + 2}$
9. $\dfrac{x^2 - 1}{x^2 - 2x + 1}$
10. $\dfrac{x^2 - 5x + 6}{x^2 - 4}$
11. $\dfrac{x^2 - 2x - 3}{x^2 - 6x + 9}$
12. $\dfrac{x^2 - x - 12}{x^2 - 8x + 16}$
13. $\sqrt{4 - x}$
14. $\sqrt{2x - 5}$
15. $\dfrac{1}{\sqrt{x - 3}}$
16. $\dfrac{1}{\sqrt{x + 2}}$

Simplifying a Rational Expression

In Exercises 17–30, write the rational expression in simplest form.

17. $\dfrac{15x^2}{10x}$
18. $\dfrac{18y^2}{60y^5}$
19. $\dfrac{3xy}{xy + x}$
20. $\dfrac{4y - 8y^2}{10y - 5}$
21. $\dfrac{x - 5}{10 - 2x}$
22. $\dfrac{12 - 4x}{x - 3}$
23. $\dfrac{y^2 - 16}{y + 4}$
24. $\dfrac{x^2 - 25}{5 - x}$
25. $\dfrac{x^3 + 5x^2 + 6x}{x^2 - 4}$
26. $\dfrac{x^2 + 8x - 20}{x^2 + 11x + 10}$
27. $\dfrac{2 - x + 2x^2 - x^3}{x^2 - 4}$
28. $\dfrac{x^2 - 9}{x^3 + x^2 - 9x - 9}$
29. $\dfrac{z^3 - 8}{z^2 + 2z + 4}$
30. $\dfrac{y^3 - 2y^2 - 3y}{y^3 + 1}$

31. **Error Analysis** Describe the error.

$$\frac{5x^3}{2x^3 + 4} = \frac{5x^3}{2x^3 + 4} = \frac{5}{2 + 4} = \frac{5}{6}$$

32. **Evaluating a Rational Expression** Complete the table. What can you conclude?

x	0	1	2	3	4	5	6
$\dfrac{x - 3}{x^2 - x - 6}$							
$\dfrac{1}{x + 2}$							

Multiplying or Dividing Rational Expressions

In Exercises 33–38, perform the multiplication or division and simplify.

33. $\dfrac{5}{x - 1} \cdot \dfrac{x - 1}{25(x - 2)}$
34. $\dfrac{r}{r - 1} \div \dfrac{r^2}{r^2 - 1}$
35. $\dfrac{4y - 16}{5y + 15} \div \dfrac{4 - y}{2y + 6}$
36. $\dfrac{t^2 - t - 6}{t^2 + 6t + 9} \cdot \dfrac{t + 3}{t^2 - 4}$
37. $\dfrac{x^2 + xy - 2y^2}{x^3 + x^2y} \cdot \dfrac{x}{x^2 + 3xy + 2y^2}$
38. $\dfrac{x^2 - 14x + 49}{x^2 - 49} \div \dfrac{3x - 21}{x + 7}$

Adding or Subtracting Rational Expressions

In Exercises 39–46, perform the addition or subtraction and simplify.

39. $6 - \dfrac{5}{x + 3}$
40. $\dfrac{2x - 1}{x + 3} + \dfrac{1 - x}{x + 3}$
41. $\dfrac{3}{x - 2} + \dfrac{5}{2 - x}$
42. $\dfrac{2x}{x - 5} - \dfrac{5}{5 - x}$
43. $\dfrac{4}{2x + 1} - \dfrac{x}{x + 2}$
44. $\dfrac{1}{x^2 - x - 2} - \dfrac{x}{x^2 - 5x + 6}$
45. $-\dfrac{1}{x} + \dfrac{2}{x^2 + 1} + \dfrac{1}{x^3 + x}$
46. $\dfrac{2}{x + 1} + \dfrac{2}{x - 1} + \dfrac{1}{x^2 - 1}$

Error Analysis In Exercises 47 and 48, describe the error.

47. $$\frac{x+4}{x+2} - \frac{3x-8}{x+2} = \frac{x+4-3x-8}{x+2} = \frac{-2x-4}{x+2} = \frac{-2(x+2)}{x+2} = -2$$

48. $$\frac{6-x}{x(x+2)} + \frac{x+2}{x^2} + \frac{8}{x^2(x+2)} = \frac{x(6-x)+(x+2)^2+8}{x^2(x+2)} = \frac{6x-x^2+x^2+4+8}{x^2(x+2)} = \frac{6(x+2)}{x^2(x+2)} = \frac{6}{x^2}$$

Simplifying a Complex Fraction In Exercises 49–54, simplify the complex fraction.

49. $\dfrac{\left(\dfrac{x}{2}-1\right)}{(x-2)}$

50. $\dfrac{(x-4)}{\left(\dfrac{x}{4}-\dfrac{4}{x}\right)}$

51. $\dfrac{\left[\dfrac{x^2}{(x+1)^2}\right]}{\left[\dfrac{x}{(x+1)^3}\right]}$

52. $\dfrac{\left(\dfrac{x^2-1}{x}\right)}{\left[\dfrac{(x-1)^2}{x}\right]}$

53. $\dfrac{\left(\sqrt{x}-\dfrac{1}{2\sqrt{x}}\right)}{\sqrt{x}}$

54. $\dfrac{\left(\dfrac{t^2}{\sqrt{t^2+1}}-\sqrt{t^2+1}\right)}{t^2}$

Factoring an Expression In Exercises 55–60, factor the expression by removing the common factor with the lesser exponent.

55. $x^5 - 2x^{-2}$

56. $5x^5 + x^{-3}$

57. $x^2(x^2+1)^{-5} - (x^2+1)^{-4}$

58. $2x(x-5)^{-3} - 4x^2(x-5)^{-4}$

59. $2x^2(x-1)^{1/2} - 5(x-1)^{-1/2}$

60. $4x^3(2x-1)^{3/2} - 2x(2x-1)^{-1/2}$

Simplifying an Expression In Exercises 61 and 62, simplify the expression.

61. $\dfrac{3x^{1/3} - x^{-2/3}}{3x^{-2/3}}$

62. $\dfrac{-x^3(1-x^2)^{-1/2} - 2x(1-x^2)^{1/2}}{x^4}$

Mateusz Wolski/Shutterstock.com

Simplifying a Difference Quotient In Exercises 63–66, simplify the difference quotient.

63. $\dfrac{\left(\dfrac{1}{x+h}-\dfrac{1}{x}\right)}{h}$

64. $\dfrac{\left[\dfrac{1}{(x+h)^2}-\dfrac{1}{x^2}\right]}{h}$

65. $\dfrac{\left(\dfrac{1}{x+h-4}-\dfrac{1}{x-4}\right)}{h}$

66. $\dfrac{\left(\dfrac{x+h}{x+h+1}-\dfrac{x}{x+1}\right)}{h}$

Rewriting a Difference Quotient In Exercises 67–72, rewrite the difference quotient by rationalizing the numerator.

67. $\dfrac{\sqrt{x+2}-\sqrt{x}}{2}$

68. $\dfrac{\sqrt{z-3}-\sqrt{z}}{3}$

69. $\dfrac{\sqrt{t+3}-\sqrt{3}}{t}$

70. $\dfrac{\sqrt{x+5}-\sqrt{5}}{x}$

71. $\dfrac{\sqrt{x+h+1}-\sqrt{x+1}}{h}$

72. $\dfrac{\sqrt{x+h-2}-\sqrt{x-2}}{h}$

73. Refrigeration

After placing food (at room temperature) in a refrigerator, the time required for the food to cool depends on the amount of food, the air circulation in the refrigerator, the original temperature of the food, and the temperature of the refrigerator. The model that gives the temperature of food that has an original temperature of 75°F and is placed in a 40°F refrigerator is

$$T = 10\left(\frac{4t^2+16t+75}{t^2+4t+10}\right)$$

where T is the temperature (in degrees Fahrenheit) and t is the time (in hours).

(a) Complete the table.

t	0	2	4	6	8	10	12
T							

t	14	16	18	20	22
T					

(b) What value of T does the mathematical model appear to be approaching?

74. Rate A digital copier copies in color at a rate of 50 pages per minute.

(a) Find the time required to copy one page.

(b) Find the time required to copy x pages.

(c) Find the time required to copy 120 pages.

Probability In Exercises 75 and 76, consider an experiment in which a marble is tossed into a box whose base is shown in the figure. The probability that the marble will come to rest in the shaded portion of the box is equal to the ratio of the shaded area to the total area of the figure. Find the probability.

75.

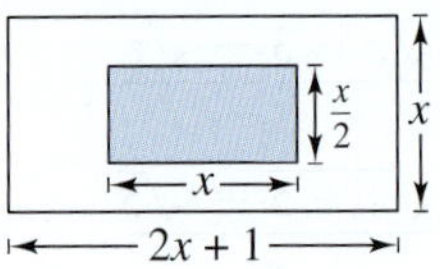

76.

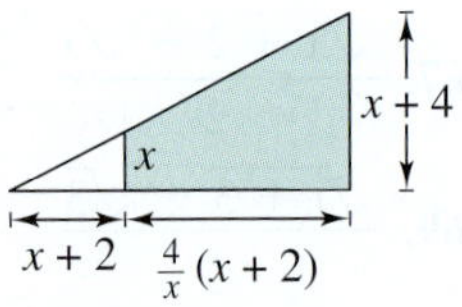

77. Interactive Money Management The table shows the numbers of U.S. households (in millions) banking online and paying bills online from 2005 through 2010. *(Source: Fiserv, Inc.)*

DATA

Year	Banking	Paying Bills
2005	46.7	17.9
2006	58.6	26.3
2007	62.8	28.2
2008	67.0	30.0
2009	69.7	32.6
2010	72.5	36.4

Spreadsheet at LarsonPrecalculus.com

Mathematical models for the data are

$$\text{Number banking online} = \frac{-33.74t + 121.8}{-0.40t + 1.0}$$

and

$$\text{Number paying bills online} = \frac{0.307t^2 - 6.54t + 24.6}{0.015t^2 - 0.28t + 1.0}$$

where t represents the year, with $t = 5$ corresponding to 2005.

(a) Using the models, create a table showing the numbers of households banking online and the numbers of households paying bills online for the given years.

(b) Compare the values given by the models with the actual data.

(c) Determine a model for the ratio of the number of households paying bills online to the number of households banking online.

(d) Use the model from part (c) to find the ratios for the given years. Interpret your results.

78. Finance The formula that approximates the annual interest rate r of a monthly installment loan is

$$r = \frac{24(NM - P)}{N} \div \left(P + \frac{NM}{12}\right)$$

where N is the total number of payments, M is the monthly payment, and P is the amount financed.

(a) Approximate the annual interest rate for a five-year car loan of \$28,000 that has monthly payments of \$525.

(b) Simplify the expression for the annual interest rate r, and then rework part (a).

79. Electrical Engineering The total resistance R_T (in ohms) of two resistors connected in parallel is given by

$$R_T = \frac{1}{\dfrac{1}{R_1} + \dfrac{1}{R_2}}$$

where R_1 and R_2 are the resistance values of the first and second resistors, respectively. Simplify the expression for the total resistance R_T.

80. HOW DO YOU SEE IT? The mathematical model

$$P = 100\left(\frac{t^2 - t + 1}{t^2 + 1}\right), \quad t \ge 0$$

gives the percent P of the normal level of oxygen in a pond, where t is the time (in weeks) after organic waste is dumped into the pond. The bar graph shows the situation. What conclusions can you draw from the bar graph?

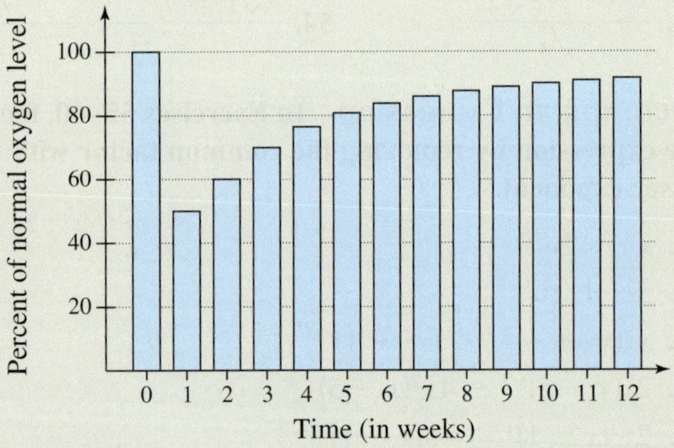

Exploration

True or False? In Exercises 81 and 82, determine whether the statement is true or false. Justify your answer.

81. $\dfrac{x^{2n} - 1^{2n}}{x^n - 1^n} = x^n + 1^n$

82. $\dfrac{x^2 - 3x + 2}{x - 1} = x - 2$, for all values of x

A.5 Solving Equations

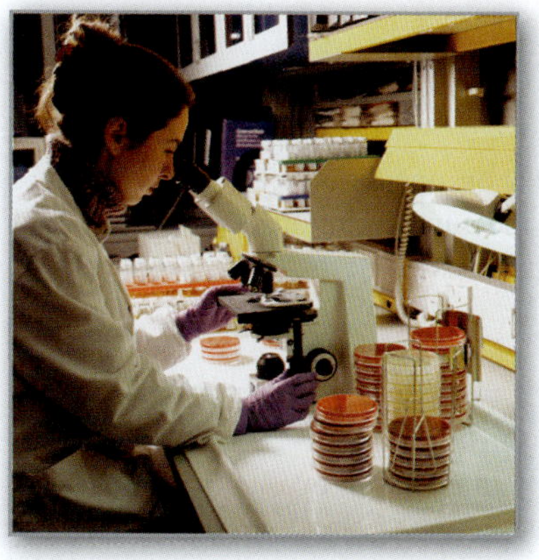

You can use linear equations in many real-life applications. For example, you can use linear equations in forensics to determine height from femur length. See Exercises 95 and 96 on page A57.

- Identify different types of equations.
- Solve linear equations in one variable and rational equations that lead to linear equations.
- Solve quadratic equations by factoring, extracting square roots, completing the square, and using the Quadratic Formula.
- Solve polynomial equations of degree three or greater.
- Solve radical equations.
- Solve absolute value equations.
- Use common formulas to solve real-life problems.

Equations and Solutions of Equations

An **equation** in x is a statement that two algebraic expressions are equal. For example,

$$3x - 5 = 7, \quad x^2 - x - 6 = 0, \quad \text{and} \quad \sqrt{2x} = 4$$

are equations. To **solve** an equation in x means to find all values of x for which the equation is true. Such values are **solutions.** For instance, $x = 4$ is a solution of the equation $3x - 5 = 7$ because $3(4) - 5 = 7$ is a true statement.

The solutions of an equation depend on the kinds of numbers being considered. For instance, in the set of rational numbers, $x^2 = 10$ has no solution because there is no rational number whose square is 10. However, in the set of real numbers, the equation has the two solutions $x = \sqrt{10}$ and $x = -\sqrt{10}$.

An equation that is true for *every* real number in the domain of the variable is called an **identity.** For example,

$$x^2 - 9 = (x + 3)(x - 3) \qquad \text{Identity}$$

is an identity because it is a true statement for any real value of x. The equation

$$\frac{x}{3x^2} = \frac{1}{3x} \qquad \text{Identity}$$

where $x \neq 0$, is an identity because it is true for any nonzero real value of x.

An equation that is true for just *some* (but not all) of the real numbers in the domain of the variable is called a **conditional equation.** For example, the equation

$$x^2 - 9 = 0 \qquad \text{Conditional equation}$$

is conditional because $x = 3$ and $x = -3$ are the only values in the domain that satisfy the equation.

A **contradiction** is an equation that is false for *every* real number in the domain of the variable. For example, the equation

$$2x - 4 = 2x + 1 \qquad \text{Contradiction}$$

is a contradiction because there are no real values of x for which the equation is true.

Linear and Rational Equations

Definition of Linear Equation in One Variable

A **linear equation in one variable** x is an equation that can be written in the standard form

$$ax + b = 0$$

where a and b are real numbers with $a \neq 0$.

Andrew Douglas/Masterfile

A linear equation has exactly one solution. To see this, consider the following steps. (Remember that $a \neq 0$.)

$$ax + b = 0 \quad \text{Write original equation.}$$

$$ax = -b \quad \text{Subtract } b \text{ from each side.}$$

$$x = -\frac{b}{a} \quad \text{Divide each side by } a.$$

To solve a conditional equation in x, isolate x on one side of the equation by a sequence of **equivalent equations,** each having the same solution(s) as the original equation. The operations that yield equivalent equations come from the properties of equality reviewed in Appendix A.1.

Generating Equivalent Equations

An equation can be transformed into an *equivalent equation* by one or more of the following steps.

	Given Equation	Equivalent Equation
1. Remove symbols of grouping, combine like terms, or simplify fractions on one or both sides of the equation.	$2x - x = 4$	$x = 4$
2. Add (or subtract) the same quantity to (from) *each* side of the equation.	$x + 1 = 6$	$x = 5$
3. Multiply (or divide) *each* side of the equation by the same *nonzero* quantity.	$2x = 6$	$x = 3$
4. Interchange the two sides of the equation.	$2 = x$	$x = 2$

The following example shows the steps for solving a linear equation in one variable x.

REMARK After solving an equation, you should check each solution in the original equation. For instance, you can check the solution of Example 1(a) as follows.

$3x - 6 = 0$ Write original equation.

$3(2) - 6 \stackrel{?}{=} 0$ Substitute 2 for x.

$0 = 0$ Solution checks. ✓

Try checking the solution of Example 1(b).

EXAMPLE 1 Solving a Linear Equation

a. $3x - 6 = 0$ Original equation

$3x = 6$ Add 6 to each side.

$x = 2$ Divide each side by 3.

b. $5x + 4 = 3x - 8$ Original equation

$2x + 4 = -8$ Subtract $3x$ from each side.

$2x = -12$ Subtract 4 from each side.

$x = -6$ Divide each side by 2.

✓ **Checkpoint** *Audio-video solution in English & Spanish at LarsonPrecalculus.com.*

Solve each equation.

a. $7 - 2x = 15$

b. $7x - 9 = 5x + 7$

REMARK An equation with a *single fraction* on each side can be cleared of denominators by **cross multiplying**. To do this, multiply the left numerator by the right denominator and the right numerator by the left denominator as follows.

$$\frac{a}{b} = \frac{c}{d} \quad \text{Original equation}$$

$$ad = cb \quad \text{Cross multiply.}$$

A **rational equation** is an equation that involves one or more fractional expressions. To solve a rational equation, find the least common denominator (LCD) of all terms and multiply every term by the LCD. This process will clear the original equation of fractions and produce a simpler equation to work with.

EXAMPLE 2 Solving a Rational Equation

Solve $\frac{x}{3} + \frac{3x}{4} = 2$.

Solution

$$\frac{x}{3} + \frac{3x}{4} = 2 \quad \text{Original equation}$$

$$(12)\frac{x}{3} + (12)\frac{3x}{4} = (12)2 \quad \text{Multiply each term by the LCD.}$$

$$4x + 9x = 24 \quad \text{Simplify.}$$

$$13x = 24 \quad \text{Combine like terms.}$$

$$x = \frac{24}{13} \quad \text{Divide each side by 13.}$$

The solution is $x = \frac{24}{13}$. Check this in the original equation.

✓ **Checkpoint** Audio-video solution in English & Spanish at LarsonPrecalculus.com.

Solve $\frac{4x}{9} - \frac{1}{3} = x + \frac{5}{3}$.

When multiplying or dividing an equation by a *variable expression,* it is possible to introduce an **extraneous solution** that does not satisfy the original equation.

EXAMPLE 3 An Equation with an Extraneous Solution

Solve $\frac{1}{x-2} = \frac{3}{x+2} - \frac{6x}{x^2-4}$.

REMARK Recall that the least common denominator of two or more fractions consists of the product of all prime factors in the denominators, with each factor given the highest power of its occurrence in any denominator. For instance, in Example 3, by factoring each denominator you can determine that the LCD is $(x + 2)(x - 2)$.

Solution The LCD is $x^2 - 4 = (x + 2)(x - 2)$. Multiply each term by this LCD.

$$\frac{1}{x-2}(x+2)(x-2) = \frac{3}{x+2}(x+2)(x-2) - \frac{6x}{x^2-4}(x+2)(x-2)$$

$$x + 2 = 3(x - 2) - 6x, \quad x \neq \pm 2$$

$$x + 2 = 3x - 6 - 6x$$

$$x + 2 = -3x - 6$$

$$4x = -8$$

$$x = -2 \quad \text{Extraneous solution}$$

In the original equation, $x = -2$ yields a denominator of zero. So, $x = -2$ is an extraneous solution, and the original equation has *no solution.*

✓ **Checkpoint** Audio-video solution in English & Spanish at LarsonPrecalculus.com.

Solve $\frac{3x}{x-4} = 5 + \frac{12}{x-4}$.

Quadratic Equations

A **quadratic equation** in x is an equation that can be written in the general form

$$ax^2 + bx + c = 0$$

where a, b, and c are real numbers with $a \neq 0$. A quadratic equation in x is also called a **second-degree polynomial equation** in x.

You should be familiar with the following four methods of solving quadratic equations.

Solving a Quadratic Equation

Factoring

If $ab = 0$, then $a = 0$ or $b = 0$.

Example:

$$x^2 - x - 6 = 0$$
$$(x - 3)(x + 2) = 0$$
$$x - 3 = 0 \Rightarrow x = 3$$
$$x + 2 = 0 \Rightarrow x = -2$$

Square Root Principle

If $u^2 = c$, where $c > 0$, then $u = \pm\sqrt{c}$.

Example:

$$(x + 3)^2 = 16$$
$$x + 3 = \pm 4$$
$$x = -3 \pm 4$$
$$x = 1 \quad \text{or} \quad x = -7$$

REMARK The Square Root Principle is also referred to as *extracting square roots.*

Completing the Square

If $x^2 + bx = c$, then

$$x^2 + bx + \left(\frac{b}{2}\right)^2 = c + \left(\frac{b}{2}\right)^2 \qquad \text{Add } \left(\frac{b}{2}\right)^2 \text{ to each side.}$$
$$\left(x + \frac{b}{2}\right)^2 = c + \frac{b^2}{4}.$$

Example:

$$x^2 + 6x = 5$$
$$x^2 + 6x + 3^2 = 5 + 3^2 \qquad \text{Add } \left(\frac{6}{2}\right)^2 \text{ to each side.}$$
$$(x + 3)^2 = 14$$
$$x + 3 = \pm\sqrt{14}$$
$$x = -3 \pm \sqrt{14}$$

Quadratic Formula

If $ax^2 + bx + c = 0$, then $x = \dfrac{-b \pm \sqrt{b^2 - 4ac}}{2a}$.

Example:

$$2x^2 + 3x - 1 = 0$$
$$x = \frac{-3 \pm \sqrt{3^2 - 4(2)(-1)}}{2(2)}$$
$$= \frac{-3 \pm \sqrt{17}}{4}$$

REMARK You can solve every quadratic equation by completing the square or using the Quadratic Formula.

EXAMPLE 4 Solving a Quadratic Equation by Factoring

a. $2x^2 + 9x + 7 = 3$ Original equation

$2x^2 + 9x + 4 = 0$ Write in general form.

$(2x + 1)(x + 4) = 0$ Factor.

$2x + 1 = 0 \implies x = -\frac{1}{2}$ Set 1st factor equal to 0.

$x + 4 = 0 \implies x = -4$ Set 2nd factor equal to 0.

The solutions are $x = -\frac{1}{2}$ and $x = -4$. Check these in the original equation.

b. $6x^2 - 3x = 0$ Original equation

$3x(2x - 1) = 0$ Factor.

$3x = 0 \implies x = 0$ Set 1st factor equal to 0.

$2x - 1 = 0 \implies x = \frac{1}{2}$ Set 2nd factor equal to 0.

The solutions are $x = 0$ and $x = \frac{1}{2}$. Check these in the original equation.

Checkpoint *Audio-video solution in English & Spanish at LarsonPrecalculus.com.*

Solve $2x^2 - 3x + 1 = 6$ by factoring.

Note that the method of solution in Example 4 is based on the Zero-Factor Property from Appendix A.1. This property applies *only* to equations written in general form (in which the right side of the equation is zero). So, all terms must be collected on one side *before* factoring. For instance, in the equation $(x - 5)(x + 2) = 8$, it is *incorrect* to set each factor equal to 8. Try to solve this equation correctly.

EXAMPLE 5 Extracting Square Roots

Solve each equation by extracting square roots.

a. $4x^2 = 12$

b. $(x - 3)^2 = 7$

Solution

a. $4x^2 = 12$ Write original equation.

$x^2 = 3$ Divide each side by 4.

$x = \pm\sqrt{3}$ Extract square roots.

The solutions are $x = \sqrt{3}$ and $x = -\sqrt{3}$. Check these in the original equation.

b. $(x - 3)^2 = 7$ Write original equation.

$x - 3 = \pm\sqrt{7}$ Extract square roots.

$x = 3 \pm \sqrt{7}$ Add 3 to each side.

The solutions are $x = 3 \pm \sqrt{7}$. Check these in the original equation.

Checkpoint *Audio-video solution in English & Spanish at LarsonPrecalculus.com.*

Solve each equation by extracting square roots.

a. $3x^2 = 36$

b. $(x - 1)^2 = 10$

When solving quadratic equations by completing the square, you must add $(b/2)^2$ to *each side* in order to maintain equality. When the leading coefficient is *not* 1, you must divide each side of the equation by the leading coefficient *before* completing the square, as shown in Example 7.

EXAMPLE 6 Completing the Square: Leading Coefficient Is 1

Solve $x^2 + 2x - 6 = 0$ by completing the square.

Solution

$$x^2 + 2x - 6 = 0 \qquad \text{Write original equation.}$$

$$x^2 + 2x = 6 \qquad \text{Add 6 to each side.}$$

$$x^2 + 2x + 1^2 = 6 + 1^2 \qquad \text{Add } 1^2 \text{ to each side.}$$

(half of 2)^2

$$(x + 1)^2 = 7 \qquad \text{Simplify.}$$

$$x + 1 = \pm\sqrt{7} \qquad \text{Extract square roots.}$$

$$x = -1 \pm \sqrt{7} \qquad \text{Subtract 1 from each side.}$$

The solutions are

$$x = -1 \pm \sqrt{7}.$$

Check these in the original equation.

✓ **Checkpoint** *Audio-video solution in English & Spanish at LarsonPrecalculus.com.*

Solve $x^2 - 4x - 1 = 0$ by completing the square.

EXAMPLE 7 Completing the Square: Leading Coefficient Is Not 1

Solve $3x^2 - 4x - 5 = 0$ by completing the square.

Solution

$$3x^2 - 4x - 5 = 0 \qquad \text{Write original equation.}$$

$$3x^2 - 4x = 5 \qquad \text{Add 5 to each side.}$$

$$x^2 - \frac{4}{3}x = \frac{5}{3} \qquad \text{Divide each side by 3.}$$

$$x^2 - \frac{4}{3}x + \left(-\frac{2}{3}\right)^2 = \frac{5}{3} + \left(-\frac{2}{3}\right)^2 \qquad \text{Add } \left(-\tfrac{2}{3}\right)^2 \text{ to each side.}$$

$\left(\text{half of } -\frac{4}{3}\right)^2$

$$\left(x - \frac{2}{3}\right)^2 = \frac{19}{9} \qquad \text{Simplify.}$$

$$x - \frac{2}{3} = \pm\frac{\sqrt{19}}{3} \qquad \text{Extract square roots.}$$

$$x = \frac{2}{3} \pm \frac{\sqrt{19}}{3} \qquad \text{Add } \tfrac{2}{3} \text{ to each side.}$$

✓ **Checkpoint** *Audio-video solution in English & Spanish at LarsonPrecalculus.com.*

Solve $3x^2 - 10x - 2 = 0$ by completing the square.

REMARK When using the Quadratic Formula, remember that *before* applying the formula, you must first write the quadratic equation in general form.

EXAMPLE 8 The Quadratic Formula: Two Distinct Solutions

Use the Quadratic Formula to solve $x^2 + 3x = 9$.

Solution

$$x^2 + 3x = 9 \quad \text{Write original equation.}$$

$$x^2 + 3x - 9 = 0 \quad \text{Write in general form.}$$

$$x = \frac{-b \pm \sqrt{b^2 - 4ac}}{2a} \quad \text{Quadratic Formula}$$

$$x = \frac{-3 \pm \sqrt{(3)^2 - 4(1)(-9)}}{2(1)} \quad \text{Substitute } a = 1, b = 3, \text{ and } c = -9.$$

$$x = \frac{-3 \pm \sqrt{45}}{2} \quad \text{Simplify.}$$

$$x = \frac{-3 \pm 3\sqrt{5}}{2} \quad \text{Simplify.}$$

The two solutions are

$$x = \frac{-3 + 3\sqrt{5}}{2} \quad \text{and} \quad x = \frac{-3 - 3\sqrt{5}}{2}.$$

Check these in the original equation.

Checkpoint *Audio-video solution in English & Spanish at LarsonPrecalculus.com.*

Use the Quadratic Formula to solve $3x^2 + 2x - 10 = 0$.

EXAMPLE 9 The Quadratic Formula: One Solution

Use the Quadratic Formula to solve $8x^2 - 24x + 18 = 0$.

Solution

$$8x^2 - 24x + 18 = 0 \quad \text{Write original equation.}$$

$$4x^2 - 12x + 9 = 0 \quad \text{Divide out common factor of 2.}$$

$$x = \frac{-b \pm \sqrt{b^2 - 4ac}}{2a} \quad \text{Quadratic Formula}$$

$$x = \frac{-(-12) \pm \sqrt{(-12)^2 - 4(4)(9)}}{2(4)} \quad \text{Substitute } a = 4, b = -12, \text{ and } c = 9.$$

$$x = \frac{12 \pm \sqrt{0}}{8} = \frac{3}{2} \quad \text{Simplify.}$$

This quadratic equation has only one solution:

$$x = \tfrac{3}{2}.$$

Check this in the original equation.

Checkpoint *Audio-video solution in English & Spanish at LarsonPrecalculus.com.*

Use the Quadratic Formula to solve $18x^2 - 48x + 32 = 0$.

Note that you could have solved Example 9 without first dividing out a common factor of 2. Substituting $a = 8$, $b = -24$, and $c = 18$ into the Quadratic Formula produces the same result.

Polynomial Equations of Higher Degree

The methods used to solve quadratic equations can sometimes be extended to solve polynomial equations of higher degrees.

REMARK A common mistake in solving an equation such as that in Example 10 is to divide each side of the equation by the variable factor x^2. This loses the solution $x = 0$. When solving an equation, always write the equation in general form, then factor the equation and set each factor equal to zero. Do not divide each side of an equation by a variable factor in an attempt to simplify the equation.

EXAMPLE 10 Solving a Polynomial Equation by Factoring

Solve $3x^4 = 48x^2$.

Solution First write the polynomial equation in general form with zero on one side. Then factor the other side, set each factor equal to zero, and solve.

$$3x^4 = 48x^2 \quad \text{Write original equation.}$$

$$3x^4 - 48x^2 = 0 \quad \text{Write in general form.}$$

$$3x^2(x^2 - 16) = 0 \quad \text{Factor out common factor.}$$

$$3x^2(x + 4)(x - 4) = 0 \quad \text{Write in factored form.}$$

$$3x^2 = 0 \Rightarrow x = 0 \quad \text{Set 1st factor equal to 0.}$$

$$x + 4 = 0 \Rightarrow x = -4 \quad \text{Set 2nd factor equal to 0.}$$

$$x - 4 = 0 \Rightarrow x = 4 \quad \text{Set 3rd factor equal to 0.}$$

You can check these solutions by substituting in the original equation, as follows.

Check

$$3(0)^4 = 48(0)^2 \quad \text{0 checks. ✓}$$

$$3(-4)^4 = 48(-4)^2 \quad \text{−4 checks. ✓}$$

$$3(4)^4 = 48(4)^2 \quad \text{4 checks. ✓}$$

So, the solutions are

$$x = 0, \quad x = -4, \quad \text{and} \quad x = 4.$$

✓ ***Checkpoint*** *Audio-video solution in English & Spanish at LarsonPrecalculus.com.*

Solve $9x^4 - 12x^2 = 0$.

EXAMPLE 11 Solving a Polynomial Equation by Factoring

Solve $x^3 - 3x^2 - 3x + 9 = 0$.

Solution

$$x^3 - 3x^2 - 3x + 9 = 0 \quad \text{Write original equation.}$$

$$x^2(x - 3) - 3(x - 3) = 0 \quad \text{Factor by grouping.}$$

$$(x - 3)(x^2 - 3) = 0 \quad \text{Distributive Property}$$

$$x - 3 = 0 \Rightarrow x = 3 \quad \text{Set 1st factor equal to 0.}$$

$$x^2 - 3 = 0 \Rightarrow x = \pm\sqrt{3} \quad \text{Set 2nd factor equal to 0.}$$

The solutions are $x = 3$, $x = \sqrt{3}$, and $x = -\sqrt{3}$. Check these in the original equation.

✓ ***Checkpoint*** *Audio-video solution in English & Spanish at LarsonPrecalculus.com.*

Solve each equation.

a. $x^3 - 5x^2 - 2x + 10 = 0$

b. $6x^3 - 27x^2 - 54x = 0$

Radical Equations

REMARK When squaring each side of an equation or raising each side of an equation to a rational power, it is possible to introduce extraneous solutions. In such cases, checking your solutions is crucial.

A **radical equation** is an equation that involves one or more radical expressions.

EXAMPLE 12 Solving Radical Equations

a.

$\sqrt{2x+7} - x = 2$	Original equation
$\sqrt{2x+7} = x + 2$	Isolate radical.
$2x + 7 = x^2 + 4x + 4$	Square each side.
$0 = x^2 + 2x - 3$	Write in general form.
$0 = (x+3)(x-1)$	Factor.
$x + 3 = 0 \Rightarrow x = -3$	Set 1st factor equal to 0.
$x - 1 = 0 \Rightarrow x = 1$	Set 2nd factor equal to 0.

By checking these values, you can determine that the only solution is $x = 1$.

b.

$\sqrt{2x-5} - \sqrt{x-3} = 1$	Original equation
$\sqrt{2x-5} = \sqrt{x-3} + 1$	Isolate $\sqrt{2x-5}$.
$2x - 5 = x - 3 + 2\sqrt{x-3} + 1$	Square each side.
$2x - 5 = x - 2 + 2\sqrt{x-3}$	Combine like terms.
$x - 3 = 2\sqrt{x-3}$	Isolate $2\sqrt{x-3}$.
$x^2 - 6x + 9 = 4(x-3)$	Square each side.
$x^2 - 10x + 21 = 0$	Write in general form.
$(x-3)(x-7) = 0$	Factor.
$x - 3 = 0 \Rightarrow x = 3$	Set 1st factor equal to 0.
$x - 7 = 0 \Rightarrow x = 7$	Set 2nd factor equal to 0.

The solutions are $x = 3$ and $x = 7$. Check these in the original equation.

REMARK When an equation contains two radicals, it may not be possible to isolate both. In such cases, you may have to raise each side of the equation to a power at *two* different stages in the solution, as shown in Example 12(b).

✓ ***Checkpoint*** *Audio-video solution in English & Spanish at LarsonPrecalculus.com.*

Solve $-\sqrt{40 - 9x} + 2 = x$.

EXAMPLE 13 Solving an Equation Involving a Rational Exponent

Solve $(x - 4)^{2/3} = 25$.

Solution

$(x-4)^{2/3} = 25$	Write original equation.
$\sqrt[3]{(x-4)^2} = 25$	Rewrite in radical form.
$(x-4)^2 = 15{,}625$	Cube each side.
$x - 4 = \pm 125$	Extract square roots.
$x = 129, \quad x = -121$	Add 4 to each side.

✓ ***Checkpoint*** *Audio-video solution in English & Spanish at LarsonPrecalculus.com.*

Solve $(x - 5)^{2/3} = 16$.

Absolute Value Equations

An **absolute value equation** is an equation that involves one or more absolute value expressions. To solve an absolute value equation, remember that the expression inside the absolute value bars can be positive or negative. This results in *two* separate equations, each of which must be solved. For instance, the equation

$$|x - 2| = 3$$

results in the two equations

$$x - 2 = 3 \quad \text{and} \quad -(x - 2) = 3$$

which implies that the equation has two solutions: $x = 5$ and $x = -1$.

EXAMPLE 14 Solving an Absolute Value Equation

Solve $|x^2 - 3x| = -4x + 6$.

Solution Because the variable expression inside the absolute value bars can be positive or negative, you must solve the following two equations.

First Equation

$x^2 - 3x = -4x + 6$ Use positive expression.

$x^2 + x - 6 = 0$ Write in general form.

$(x + 3)(x - 2) = 0$ Factor.

$x + 3 = 0 \Rightarrow x = -3$ Set 1st factor equal to 0.

$x - 2 = 0 \Rightarrow x = 2$ Set 2nd factor equal to 0.

Second Equation

$-(x^2 - 3x) = -4x + 6$ Use negative expression.

$x^2 - 7x + 6 = 0$ Write in general form.

$(x - 1)(x - 6) = 0$ Factor.

$x - 1 = 0 \Rightarrow x = 1$ Set 1st factor equal to 0.

$x - 6 = 0 \Rightarrow x = 6$ Set 2nd factor equal to 0.

Check

$|(-3)^2 - 3(-3)| \stackrel{?}{=} -4(-3) + 6$ Substitute -3 for x.

$18 = 18$ -3 checks. ✓

$|(2)^2 - 3(2)| \stackrel{?}{=} -4(2) + 6$ Substitute 2 for x.

$2 \neq -2$ 2 does not check.

$|(1)^2 - 3(1)| \stackrel{?}{=} -4(1) + 6$ Substitute 1 for x.

$2 = 2$ 1 checks. ✓

$|(6)^2 - 3(6)| \stackrel{?}{=} -4(6) + 6$ Substitute 6 for x.

$18 \neq -18$ 6 does not check.

The solutions are $x = -3$ and $x = 1$.

✓ ***Checkpoint*** *Audio-video solution in English & Spanish at LarsonPrecalculus.com.*

Solve $|x^2 + 4x| = 5x + 12$.

Common Formulas

The following geometric formulas are used at various times throughout this course. For your convenience, some of these formulas along with several others are also provided on the inside cover of this text.

Common Formulas for Area *A*, Perimeter *P*, Circumference *C*, and Volume *V*

Rectangle
$A = lw$
$P = 2l + 2w$

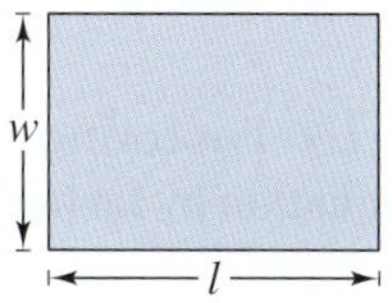

Circle
$A = \pi r^2$
$C = 2\pi r$

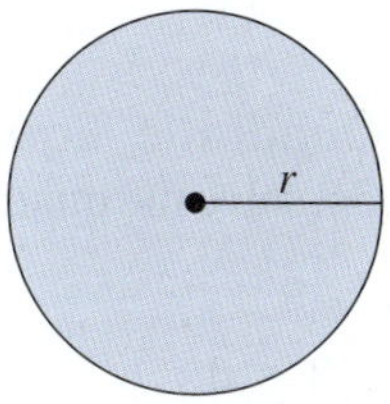

Rectangular Solid
$V = lwh$

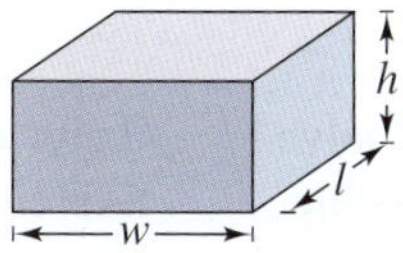

Circular Cylinder
$V = \pi r^2 h$

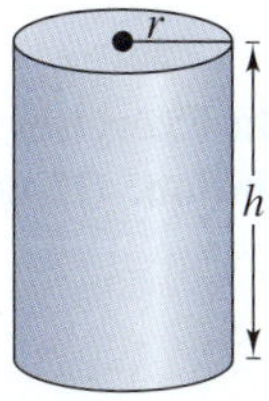

Sphere
$V = \frac{4}{3}\pi r^3$

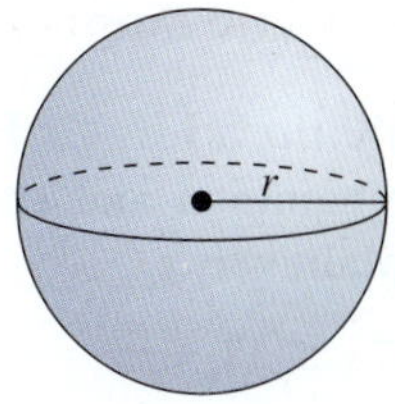

Figure A.10

EXAMPLE 15 **Using a Geometric Formula**

The cylindrical can shown in Figure A.10 has a volume of 200 cubic centimeters (cm^3). Find the height of the can.

Solution The formula for the *volume of a cylinder* is $V = \pi r^2 h$. To find the height of the can, solve for h. Then, using $V = 200$ and $r = 4$, find the height.

$$V = \pi r^2 h \quad \Longrightarrow \quad h = \frac{V}{\pi r^2} = \frac{200}{\pi(4)^2} = \frac{200}{16\pi} \approx 3.98$$

So, the height of the can is about 3.98 centimeters.

REMARK You can use unit analysis to check that the answer in Example 15 is reasonable.

$$\frac{200\ \text{cm}^3}{16\pi\ \text{cm}^2} = \frac{200\ \cancel{\text{cm}} \cdot \cancel{\text{cm}} \cdot \text{cm}}{16\pi\ \cancel{\text{cm}} \cdot \cancel{\text{cm}}}$$
$$= \frac{200}{16\pi}\ \text{cm}$$
$$\approx 3.98\ \text{cm}$$

✓ ***Checkpoint*** *Audio-video solution in English & Spanish at LarsonPrecalculus.com.*

A cylindrical container has a volume of 84 cubic inches and a radius of 3 inches. Find the height of the container.

Summarize (Appendix A.5)

1. State the definition of an identity, a conditional equation, and a contradiction *(page A45)*.
2. State the definition of a linear equation *(page A45)*. For examples of solving linear equations and rational equations that lead to linear equations, see Examples 1–3.
3. List the four methods of solving quadratic equations discussed in this section *(page A48)*. For examples of solving quadratic equations, see Examples 4–9.
4. Describe how to solve a polynomial equation by factoring *(page A52)*. For examples of solving polynomial equations by factoring, see Examples 10 and 11.
5. Describe how to solve a radical equation *(page A53)*. For an example of solving radical equations, see Example 12.
6. Describe how to solve an absolute value equation *(page A54)*. For an example of solving an absolute value equation, see Example 14.
7. Describe real-life problems that can be solved using common geometric formulas *(page A55)*. For an example that uses a volume formula, see Example 15.

A.5 Exercises

See CalcChat.com for tutorial help and worked-out solutions to odd-numbered exercises.

Vocabulary: Fill in the blanks.

1. An ________ is a statement that equates two algebraic expressions.
2. A linear equation in one variable x is an equation that can be written in the standard form ________.
3. An ________ solution is a solution that does not satisfy the original equation.
4. Four methods that can be used to solve a quadratic equation are ________, extracting ________ ________, ________ the ________, and the ________ ________.

Skills and Applications

Solving a Linear Equation In Exercises 5–12, solve the equation and check your solution. (If not possible, explain why.)

5. $x + 11 = 15$
6. $7 - x = 19$
7. $7 - 2x = 25$
8. $3x - 5 = 2x + 7$
9. $4y + 2 - 5y = 7 - 6y$
10. $0.25x + 0.75(10 - x) = 3$
11. $x - 3(2x + 3) = 8 - 5x$
12. $9x - 10 = 5x + 2(2x - 5)$

Solving a Rational Equation In Exercises 13–24, solve the equation and check your solution. (If not possible, explain why.)

13. $\dfrac{3x}{8} - \dfrac{4x}{3} = 4$
14. $\dfrac{5x}{4} + \dfrac{1}{2} = x - \dfrac{1}{2}$
15. $\dfrac{5x - 4}{5x + 4} = \dfrac{2}{3}$
16. $\dfrac{10x + 3}{5x + 6} = \dfrac{1}{2}$
17. $10 - \dfrac{13}{x} = 4 + \dfrac{5}{x}$
18. $\dfrac{1}{x} + \dfrac{2}{x - 5} = 0$
19. $\dfrac{x}{x + 4} + \dfrac{4}{x + 4} + 2 = 0$
20. $\dfrac{7}{2x + 1} - \dfrac{8x}{2x - 1} = -4$
21. $\dfrac{2}{(x - 4)(x - 2)} = \dfrac{1}{x - 4} + \dfrac{2}{x - 2}$
22. $\dfrac{4}{x - 1} + \dfrac{6}{3x + 1} = \dfrac{15}{3x + 1}$
23. $\dfrac{1}{x - 3} + \dfrac{1}{x + 3} = \dfrac{10}{x^2 - 9}$
24. $\dfrac{1}{x - 2} + \dfrac{3}{x + 3} = \dfrac{4}{x^2 + x - 6}$

Solving a Quadratic Equation by Factoring In Exercises 25–34, solve the quadratic equation by factoring.

25. $6x^2 + 3x = 0$
26. $9x^2 - 1 = 0$
27. $x^2 - 2x - 8 = 0$
28. $x^2 - 10x + 9 = 0$
29. $x^2 + 10x + 25 = 0$
30. $4x^2 + 12x + 9 = 0$
31. $x^2 + 4x = 12$
32. $-x^2 + 8x = 12$
33. $\frac{3}{4}x^2 + 8x + 20 = 0$
34. $\frac{1}{8}x^2 - x - 16 = 0$

Extracting Square Roots In Exercises 35–42, solve the equation by extracting square roots. When a solution is irrational, list both the exact solution *and* its approximation rounded to two decimal places.

35. $x^2 = 49$
36. $x^2 = 32$
37. $3x^2 = 81$
38. $9x^2 = 36$
39. $(x - 12)^2 = 16$
40. $(x + 9)^2 = 24$
41. $(2x - 1)^2 = 18$
42. $(x - 7)^2 = (x + 3)^2$

Completing the Square In Exercises 43–50, solve the quadratic equation by completing the square.

43. $x^2 + 4x - 32 = 0$
44. $x^2 - 2x - 3 = 0$
45. $x^2 + 6x + 2 = 0$
46. $x^2 + 8x + 14 = 0$
47. $9x^2 - 18x = -3$
48. $7 + 2x - x^2 = 0$
49. $2x^2 + 5x - 8 = 0$
50. $3x^2 - 4x - 7 = 0$

Using the Quadratic Formula In Exercises 51–64, use the Quadratic Formula to solve the equation.

51. $2x^2 + x - 1 = 0$
52. $2x^2 - x - 1 = 0$
53. $2 + 2x - x^2 = 0$
54. $x^2 - 10x + 22 = 0$
55. $2x^2 - 3x - 4 = 0$
56. $3x + x^2 - 1 = 0$
57. $12x - 9x^2 = -3$
58. $9x^2 - 37 = 6x$
59. $9x^2 + 30x + 25 = 0$
60. $28x - 49x^2 = 4$
61. $8t = 5 + 2t^2$
62. $25h^2 + 80h + 61 = 0$
63. $(y - 5)^2 = 2y$
64. $(z + 6)^2 = -2z$

Choosing a Method **In Exercises 65–72, solve the equation using any convenient method.**

65. $x^2 - 2x - 1 = 0$

66. $11x^2 + 33x = 0$

67. $(x + 3)^2 = 81$

68. $x^2 - 14x + 49 = 0$

69. $x^2 - x - \frac{11}{4} = 0$

70. $x^2 + 3x - \frac{3}{4} = 0$

71. $(x + 1)^2 = x^2$

72. $3x + 4 = 2x^2 - 7$

Solving a Polynomial Equation **In Exercises 73–76, solve the equation. Check your solutions.**

73. $6x^4 - 14x^2 = 0$

74. $36x^3 - 100x = 0$

75. $5x^3 + 30x^2 + 45x = 0$

76. $x^3 - 3x^2 - x = -3$

Solving a Radical Equation **In Exercises 77–84, solve the equation. Check your solutions.**

77. $\sqrt{3x} - 12 = 0$

78. $\sqrt{x - 10} - 4 = 0$

79. $\sqrt[3]{2x + 5} + 3 = 0$

80. $\sqrt[3]{3x + 1} - 5 = 0$

81. $-\sqrt{26 - 11x} + 4 = x$

82. $x + \sqrt{31 - 9x} = 5$

83. $\sqrt{x} - \sqrt{x - 5} = 1$

84. $2\sqrt{x + 1} - \sqrt{2x + 3} = 1$

Solving an Equation Involving a Rational Exponent **In Exercises 85–88, solve the equation. Check your solutions.**

85. $(x - 5)^{3/2} = 8$

86. $(x + 2)^{2/3} = 9$

87. $(x^2 - 5)^{3/2} = 27$

88. $(x^2 - x - 22)^{3/2} = 27$

Solving an Absolute Value Equation **In Exercises 89–92, solve the equation. Check your solutions.**

89. $|2x - 5| = 11$

90. $|3x + 2| = 7$

91. $|x^2 + 6x| = 3x + 18$

92. $|x - 15| = x^2 - 15x$

93. Volume of a Billiard Ball A billiard ball has a volume of 5.96 cubic inches. Find the radius of a billiard ball.

94. Length of a Tank The diameter of a cylindrical propane gas tank is 4 feet. The total volume of the tank is 603.2 cubic feet. Find the length of the tank.

18percentgrey/Shutterstock.com

Forensics

In Exercises 95 and 96, use the following information. The relationship between the length of an adult's femur (thigh bone) and the height of the adult can be approximated by the linear equations

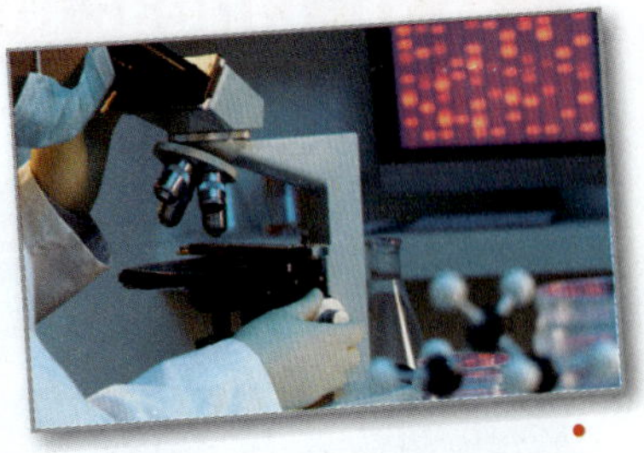

$y = 0.432x - 10.44$ Female

$y = 0.449x - 12.15$ Male

where y is the length of the femur in inches and x is the height of the adult in inches (see figure).

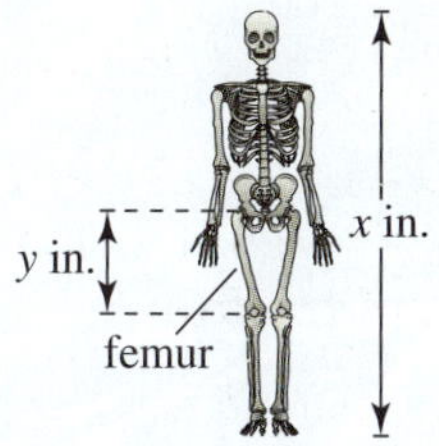

95. A crime scene investigator discovers a femur belonging to an adult human female. The bone is 18 inches long. Estimate the height of the female.

96. Officials search a forest for a missing man who is 6 feet 2 inches tall. They find an adult male femur that is 21 inches long. Is it possible that the femur belongs to the missing man?

Exploration

True or False? **In Exercises 97–99, determine whether the statement is true or false. Justify your answer.**

97. An equation can never have more than one extraneous solution.

98. The equation $2(x - 3) + 1 = 2x - 5$ has no solution.

99. The equation $\sqrt{x + 10} - \sqrt{x - 10} = 0$ has no solution.

100. **HOW DO YOU SEE IT?** The figure shows a glass cube partially filled with water.

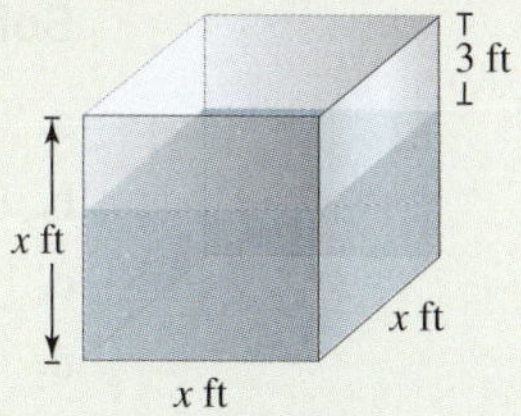

(a) What does the expression $x^2(x - 3)$ represent?

(b) Given $x^2(x - 3) = 320$, explain how you can find the capacity of the cube.

A.6 Linear Inequalities in One Variable

You can use inequalities to model and solve real-life problems. For instance, in Exercise 114 on page A66, you will use an absolute value inequality to analyze the cover layer thickness of a Blu-ray Disc™.

- **Represent solutions of linear inequalities in one variable.**
- **Use properties of inequalities to create equivalent inequalities.**
- **Solve linear inequalities in one variable.**
- **Solve absolute value inequalities.**
- **Use inequalities to model and solve real-life problems.**

Introduction

Simple inequalities were discussed in Appendix A.1. There, you used the inequality symbols $<$, $\leq$, $>$, and $\geq$ to compare two numbers and to denote subsets of real numbers. For instance, the simple inequality

$$x \geq 3$$

denotes all real numbers x that are greater than or equal to 3.

Now, you will expand your work with inequalities to include more involved statements such as

$$5x - 7 < 3x + 9 \quad \text{and} \quad -3 \leq 6x - 1 < 3.$$

As with an equation, you **solve an inequality** in the variable x by finding all values of x for which the inequality is true. Such values are **solutions** and are said to **satisfy** the inequality. The set of all real numbers that are solutions of an inequality is the **solution set** of the inequality. For instance, the solution set of

$$x + 1 < 4$$

is all real numbers that are less than 3.

The set of all points on the real number line that represents the solution set is the **graph of the inequality.** Graphs of many types of inequalities consist of intervals on the real number line. See Appendix A.1 to review the nine basic types of intervals on the real number line. Note that each type of interval can be classified as *bounded* or *unbounded.*

EXAMPLE 1 Intervals and Inequalities

Write an inequality to represent each interval. Then state whether the interval is bounded or unbounded.

a. $(-3, 5]$ **b.** $(-3, \infty)$

c. $[0, 2]$ **d.** $(-\infty, \infty)$

Solution

a. $(-3, 5]$ corresponds to $-3 < x \leq 5$. Bounded

b. $(-3, \infty)$ corresponds to $x > -3$. Unbounded

c. $[0, 2]$ corresponds to $0 \leq x \leq 2$. Bounded

d. $(-\infty, \infty)$ corresponds to $-\infty < x < \infty$. Unbounded

✓ **Checkpoint** *Audio-video solution in English & Spanish at LarsonPrecalculus.com.*

Write an inequality to represent each interval. Then state whether the interval is bounded or unbounded.

a. $[-1, 3]$ **b.** $(-1, 6)$

c. $(-\infty, 4)$ **d.** $[0, \infty)$

Jaroslav74/Shutterstock.com

Properties of Inequalities

The procedures for solving linear inequalities in one variable are much like those for solving linear equations. To isolate the variable, you can make use of the **properties of inequalities.** These properties are similar to the properties of equality, but there are two important exceptions. When each side of an inequality is multiplied or divided by a negative number, the direction of the inequality symbol must be reversed. Here is an example.

$-2 < 5$ Original inequality

$(-3)(-2) > (-3)(5)$ Multiply each side by -3 and reverse inequality symbol.

$6 > -15$ Simplify.

Notice that when you do not reverse the inequality symbol in the example above, you obtain the false statement

$6 < -15.$ False statement

Two inequalities that have the same solution set are **equivalent.** For instance, the inequalities

$$x + 2 < 5$$

and

$$x < 3$$

are equivalent. To obtain the second inequality from the first, you can subtract 2 from each side of the inequality. The following list describes the operations that can be used to create equivalent inequalities.

Properties of Inequalities

Let a, b, c, and d be real numbers.

1. Transitive Property

$a < b$ and $b < c \Rightarrow a < c$

2. Addition of Inequalities

$a < b$ and $c < d \Rightarrow a + c < b + d$

3. Addition of a Constant

$a < b \Rightarrow a + c < b + c$

4. Multiplication by a Constant

For $c > 0$, $a < b \Rightarrow ac < bc$

For $c < 0$, $a < b \Rightarrow ac > bc$ Reverse the inequality symbol.

Each of the properties above is true when the symbol $<$ is replaced by $\le$ and the symbol $>$ is replaced by $\ge$. For instance, another form of the multiplication property is shown below.

For $c > 0$, $a \le b \Rightarrow ac \le bc$

For $c < 0$, $a \le b \Rightarrow ac \ge bc$

On your own, try to verify each of the properties of inequalities by using several examples with real numbers.

Solving a Linear Inequality in One Variable

The simplest type of inequality is a **linear inequality** in one variable. For instance,

$$2x + 3 > 4$$

is a linear inequality in x.

EXAMPLE 2 Solving a Linear Inequality

Solve $5x - 7 > 3x + 9$. Then graph the solution set.

REMARK Checking the solution set of an inequality is not as simple as checking the solutions of an equation. You can, however, get an indication of the validity of a solution set by substituting a few convenient values of x. For instance, in Example 2, try substituting $x = 5$ and $x = 10$ into the original inequality.

Solution

$5x - 7 > 3x + 9$ Write original inequality.

$2x - 7 > 9$ Subtract $3x$ from each side.

$2x > 16$ Add 7 to each side.

$x > 8$ Divide each side by 2.

The solution set is all real numbers that are greater than 8, which is denoted by $(8, \infty)$. The graph of this solution set is shown below. Note that a parenthesis at 8 on the real number line indicates that 8 *is not* part of the solution set.

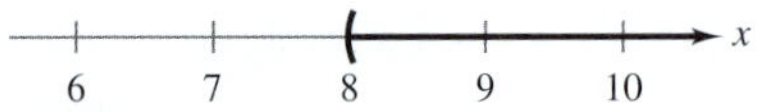

Solution interval: $(8, \infty)$

✓ Checkpoint *Audio-video solution in English & Spanish at LarsonPrecalculus.com.*

Solve $7x - 3 \le 2x + 7$. Then graph the solution set.

EXAMPLE 3 Solving a Linear Inequality

Solve $1 - \frac{3}{2}x \ge x - 4$.

Algebraic Solution

$1 - \dfrac{3x}{2} \ge x - 4$ Write original inequality.

$2 - 3x \ge 2x - 8$ Multiply each side by 2.

$2 - 5x \ge -8$ Subtract $2x$ from each side.

$-5x \ge -10$ Subtract 2 from each side.

$x \le 2$ Divide each side by -5 and reverse the inequality symbol.

The solution set is all real numbers that are less than or equal to 2, which is denoted by $(-\infty, 2]$. The graph of this solution set is shown below. Note that a bracket at 2 on the real number line indicates that 2 *is* part of the solution set.

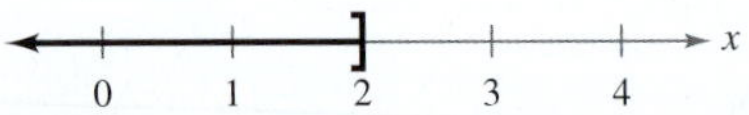

Solution interval: $(-\infty, 2]$

Graphical Solution

Use a graphing utility to graph $y_1 = 1 - \frac{3}{2}x$ and $y_2 = x - 4$ in the same viewing window, as shown below.

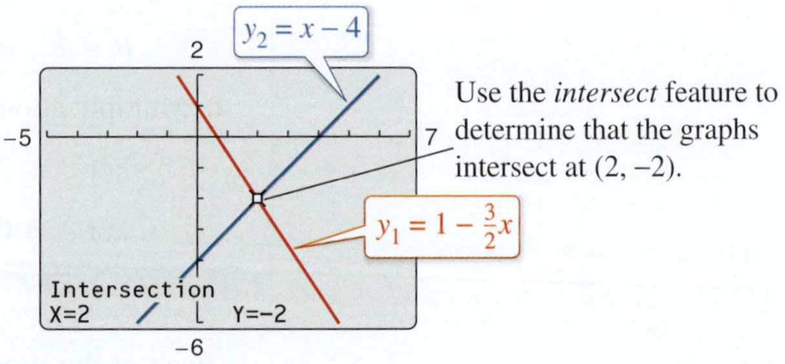

The graph of y_1 lies above the graph of y_2 to the left of their point of intersection, which implies that $y_1 \ge y_2$ for all $x \le 2$.

✓ Checkpoint *Audio-video solution in English & Spanish at LarsonPrecalculus.com.*

Solve $2 - \frac{5}{3}x > x - 6$ (a) algebraically and (b) graphically.

Sometimes it is possible to write two inequalities as a **double inequality.** For instance, you can write the two inequalities

$$-4 \le 5x - 2$$

and

$$5x - 2 < 7$$

more simply as

$$-4 \le 5x - 2 < 7. \quad \text{Double inequality}$$

This form allows you to solve the two inequalities together, as demonstrated in Example 4.

EXAMPLE 4 Solving a Double Inequality

Solve $-3 \le 6x - 1 < 3$. Then graph the solution set.

Solution To solve a double inequality, you can isolate x as the middle term.

$$-3 \le 6x - 1 < 3 \quad \text{Write original inequality.}$$

$$-3 + 1 \le 6x - 1 + 1 < 3 + 1 \quad \text{Add 1 to each part.}$$

$$-2 \le 6x < 4 \quad \text{Simplify.}$$

$$\frac{-2}{6} \le \frac{6x}{6} < \frac{4}{6} \quad \text{Divide each part by 6.}$$

$$-\frac{1}{3} \le x < \frac{2}{3} \quad \text{Simplify.}$$

The solution set is all real numbers that are greater than or equal to $-\frac{1}{3}$ and less than $\frac{2}{3}$, which is denoted by $\left[-\frac{1}{3}, \frac{2}{3}\right)$. The graph of this solution set is shown below.

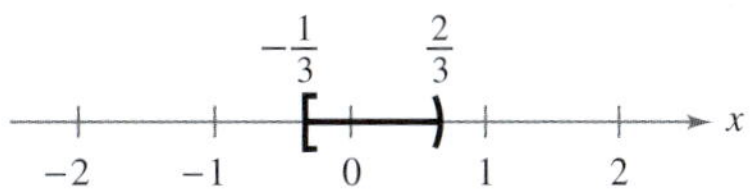

Solution interval: $\left[-\frac{1}{3}, \frac{2}{3}\right)$

✓ Checkpoint *Audio-video solution in English & Spanish at LarsonPrecalculus.com.*

Solve $1 < 2x + 7 < 11$. Then graph the solution set.

You can solve the double inequality in Example 4 in two parts, as follows.

$$-3 \le 6x - 1 \quad \text{and} \quad 6x - 1 < 3$$

$$-2 \le 6x \qquad 6x < 4$$

$$-\frac{1}{3} \le x \qquad x < \frac{2}{3}$$

The solution set consists of all real numbers that satisfy *both* inequalities. In other words, the solution set is the set of all values of x for which

$$-\frac{1}{3} \le x < \frac{2}{3}.$$

When combining two inequalities to form a double inequality, be sure that the inequalities satisfy the Transitive Property. For instance, it is *incorrect* to combine the inequalities $3 < x$ and $x \le -1$ as $3 < x \le -1$. This "inequality" is wrong because 3 is not less than -1.

Absolute Value Inequalities

Solving an Absolute Value Inequality

Let u be an algebraic expression and let a be a real number such that $a > 0$.

1. $|u| < a$ if and only if $-a < u < a$.
2. $|u| \le a$ if and only if $-a \le u \le a$.
3. $|u| > a$ if and only if $u < -a$ or $u > a$.
4. $|u| \ge a$ if and only if $u \le -a$ or $u \ge a$.

TECHNOLOGY A graphing utility can be used to identify the solution set of an inequality. For instance, to find the solution set of $|x - 5| < 2$ (see Example 5a), rewrite the inequality as $|x - 5| - 2 < 0$, enter

Y1 = abs (X − 5) − 2

and press the *graph* key. The graph should look like the one shown below.

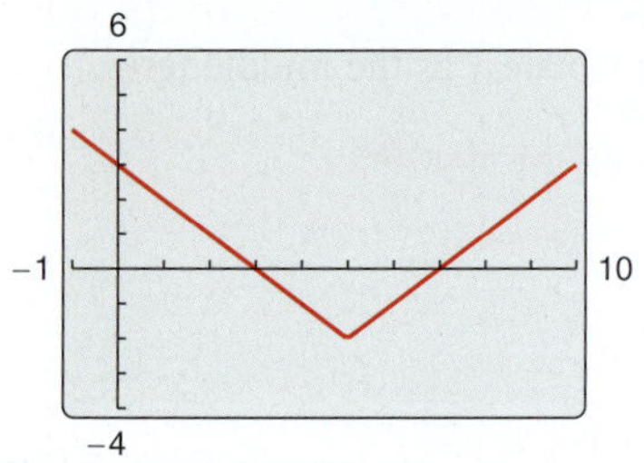

Notice that the graph is below the x-axis on the interval $(3, 7)$.

EXAMPLE 5 Solving an Absolute Value Inequality

Solve each inequality. Then graph the solution set.

a. $|x - 5| < 2$ **b.** $|x + 3| \ge 7$

Solution

a.

$$|x - 5| < 2 \qquad \text{Write original inequality.}$$

$$-2 < x - 5 < 2 \qquad \text{Write equivalent inequalities.}$$

$$-2 + 5 < x - 5 + 5 < 2 + 5 \qquad \text{Add 5 to each part.}$$

$$3 < x < 7 \qquad \text{Simplify.}$$

The solution set is all real numbers that are greater than 3 and less than 7, which is denoted by $(3, 7)$. The graph of this solution set is shown below. Note that the graph of the inequality can be described as all real numbers less than two units from 5.

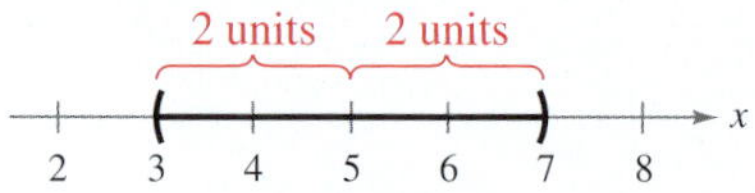

$|x - 5| < 2$: Solutions lie inside $(3, 7)$.

b.

$$|x + 3| \ge 7 \qquad \text{Write original inequality.}$$

$$x + 3 \le -7 \quad \text{or} \quad x + 3 \ge 7 \qquad \text{Write equivalent inequalities.}$$

$$x + 3 - 3 \le -7 - 3 \qquad x + 3 - 3 \ge 7 - 3 \qquad \text{Subtract 3 from each side.}$$

$$x \le -10 \qquad x \ge 4 \qquad \text{Simplify.}$$

The solution set is all real numbers that are less than or equal to -10 *or* greater than or equal to 4, which is denoted by $(-\infty, -10] \cup [4, \infty)$. The symbol $\cup$ is the *union* symbol, which denotes the combining of two sets. The graph of this solution set is shown below. Note that the graph of the inequality can be described as all real numbers at least seven units from -3.

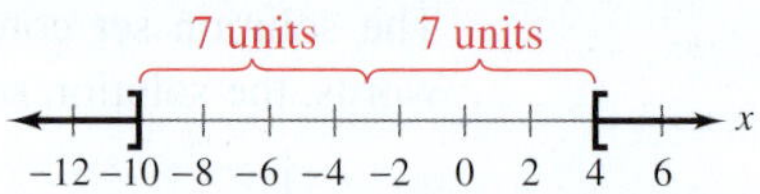

$|x + 3| \ge 7$: Solutions lie outside $(-10, 4)$.

Checkpoint *Audio-video solution in English & Spanish at LarsonPrecalculus.com.*

Solve $|x - 20| \le 4$. Then graph the solution set.

Applications

EXAMPLE 6 Comparative Shopping

Cell Phone Plans
Plan A:
$49.99 per month for 500 minutes plus $0.40 for each additional minute
Plan B:
$45.99 per month for 500 minutes plus $0.45 for each additional minute

Figure A.11

Consider the two cell phone plans shown in Figure A.11. How many *additional* minutes must you use in one month for plan B to cost more than plan A?

Solution Let m represent your additional minutes in one month. Write and solve an inequality.

$$0.45m + 45.99 > 0.40m + 49.99$$

$$0.05m > 4$$

$$m > 80$$

Plan B costs more when you use more than 80 additional minutes in one month.

✓ **Checkpoint** Audio-video solution in English & Spanish at LarsonPrecalculus.com.

Rework Example 6 when plan A costs $54.99 per month for 500 minutes plus $0.35 for each additional minute.

EXAMPLE 7 Accuracy of a Measurement

You buy chocolates that cost $9.89 per pound. The scale used to weigh your bag is accurate to within $\frac{1}{32}$ pound. According to the scale, your bag weighs $\frac{1}{2}$ pound and costs $4.95. How much might you have been undercharged or overcharged?

Solution Let x represent the actual weight of your bag. The difference of the actual weight and the weight shown on the scale is at most $\frac{1}{32}$ pound. That is, $\left|x - \frac{1}{2}\right| \le \frac{1}{32}$. You can solve this inequality as follows.

$$-\frac{1}{32} \le x - \frac{1}{2} \le \frac{1}{32}$$

$$\frac{15}{32} \le x \le \frac{17}{32}$$

The least your bag can weigh is $\frac{15}{32}$ pound, which would have cost $4.64. The most the bag can weigh is $\frac{17}{32}$ pound, which would have cost $5.25. So, you might have been overcharged by as much as $0.31 or undercharged by as much as $0.30.

✓ **Checkpoint** Audio-video solution in English & Spanish at LarsonPrecalculus.com.

Rework Example 7 when the scale is accurate to within $\frac{1}{64}$ pound.

Summarize (Appendix A.6)

1. Describe how to use inequalities to represent intervals *(page A58, Example 1).*
2. State the properties of inequalities *(page A59).*
3. Describe how to solve a linear inequality *(pages A60 and A61, Examples 2–4).*
4. Describe how to solve an absolute value inequality *(page A62, Example 5).*
5. Describe a real-life example that uses an inequality *(page A63, Examples 6 and 7).*

A.6 Exercises

See CalcChat.com for tutorial help and worked-out solutions to odd-numbered exercises.

Vocabulary: Fill in the blanks.

1. The set of all real numbers that are solutions of an inequality is the ________ ________ of the inequality.
2. The set of all points on the real number line that represents the solution set of an inequality is the ________ of the inequality.
3. It is sometimes possible to write two inequalities as one inequality, called a ________ inequality.
4. The symbol $\cup$ is the ________ symbol, which denotes the combining of two sets.

Skills and Applications

Intervals and Inequalities In Exercises 5–12, write an inequality that represents the interval. Then state whether the interval is bounded or unbounded.

5. $[0, 9)$
6. $(-7, 4)$
7. $[-1, 5]$
8. $(2, 10]$
9. $(11, \infty)$
10. $[-5, \infty)$
11. $(-\infty, -2)$
12. $(-\infty, 7]$

Solving a Linear Inequality In Exercises 13–42, solve the inequality. Then graph the solution set.

13. $4x < 12$
14. $10x < -40$
15. $-2x > -3$
16. $-6x > 15$
17. $x - 5 \geq 7$
18. $x + 7 \leq 12$
19. $2x + 7 < 3 + 4x$
20. $3x + 1 \geq 2 + x$
21. $2x - 1 \geq 1 - 5x$
22. $6x - 4 \leq 2 + 8x$
23. $4 - 2x < 3(3 - x)$
24. $4(x + 1) < 2x + 3$
25. $\frac{3}{4}x - 6 \leq x - 7$
26. $3 + \frac{2}{7}x > x - 2$
27. $\frac{1}{2}(8x + 1) \geq 3x + \frac{5}{2}$
28. $9x - 1 < \frac{3}{4}(16x - 2)$
29. $3.6x + 11 \geq -3.4$
30. $15.6 - 1.3x < -5.2$
31. $1 < 2x + 3 < 9$
32. $-9 \leq -2x - 7 < 5$
33. $0 < 3(x + 7) \leq 20$
34. $-1 \leq -(x - 4) < 7$
35. $-4 < \dfrac{2x - 3}{3} < 4$
36. $0 \leq \dfrac{x + 3}{2} < 5$
37. $-1 < \dfrac{-x - 2}{3} \leq 1$
38. $-1 \leq \dfrac{-3x + 5}{7} \leq 2$
39. $\dfrac{3}{4} > x + 1 > \dfrac{1}{4}$
40. $-1 < 2 - \dfrac{x}{3} < 1$
41. $3.2 \leq 0.4x - 1 \leq 4.4$
42. $1.6 < 0.3x + 1 < 2.8$

Solving an Absolute Value Inequality In Exercises 43–58, solve the inequality. Then graph the solution set. (Some inequalities have no solution.)

43. $|x| < 5$
44. $|x| \geq 8$
45. $\left|\dfrac{x}{2}\right| > 1$
46. $\left|\dfrac{x}{5}\right| > 3$
47. $|x - 5| < -1$
48. $|x - 7| < -5$
49. $|x - 20| \leq 6$
50. $|x - 8| \geq 0$
51. $|3 - 4x| \geq 9$
52. $|1 - 2x| < 5$
53. $\left|\dfrac{x - 3}{2}\right| \geq 4$
54. $\left|1 - \dfrac{2x}{3}\right| < 1$
55. $|9 - 2x| - 2 < -1$
56. $|x + 14| + 3 > 17$
57. $2|x + 10| \geq 9$
58. $3|4 - 5x| \leq 9$

Graphical Analysis In Exercises 59–68, use a graphing utility to graph the inequality and identify the solution set.

59. $6x > 12$
60. $3x - 1 \leq 5$
61. $5 - 2x \geq 1$
62. $20 < 6x - 1$
63. $4(x - 3) \leq 8 - x$
64. $3(x + 1) < x + 7$
65. $|x - 8| \leq 14$
66. $|2x + 9| > 13$
67. $2|x + 7| \geq 13$
68. $\frac{1}{2}|x + 1| \leq 3$

Graphical Analysis In Exercises 69–74, use a graphing utility to graph the equation. Use the graph to approximate the values of x that satisfy each inequality.

	Equation	Inequalities	
69.	$y = 2x - 3$	(a) $y \geq 1$	(b) $y \leq 0$
70.	$y = \frac{2}{3}x + 1$	(a) $y \leq 5$	(b) $y \geq 0$
71.	$y = -\frac{1}{2}x + 2$	(a) $0 \leq y \leq 3$	(b) $y \geq 0$
72.	$y = -3x + 8$	(a) $-1 \leq y \leq 3$	(b) $y \leq 0$
73.	$y = |x - 3|$	(a) $y \leq 2$	(b) $y \geq 4$
74.	$y = \left|\frac{1}{2}x + 1\right|$	(a) $y \leq 4$	(b) $y \geq 1$

Finding an Interval In Exercises 75–80, find the interval on the real number line for which the radicand is nonnegative.

75. $\sqrt{x - 5}$
76. $\sqrt{x - 10}$
77. $\sqrt{x + 3}$
78. $\sqrt{3 - x}$
79. $\sqrt[4]{7 - 2x}$
80. $\sqrt[4]{6x + 15}$

81. **Think About It** The graph of $|x - 5| < 3$ can be described as all real numbers less than three units from 5. Give a similar description of $|x - 10| < 8$.

82. Think About It The graph of $|x - 2| > 5$ can be described as all real numbers more than five units from 2. Give a similar description of $|x - 8| > 4$.

Using Absolute Value **In Exercises 83–90, use absolute value notation to define the interval (or pair of intervals) on the real number line.**

83. 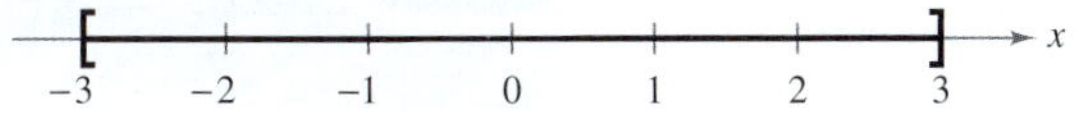

84. 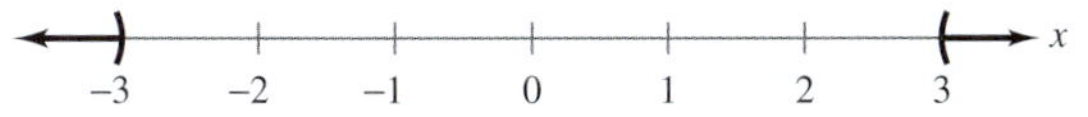

85. 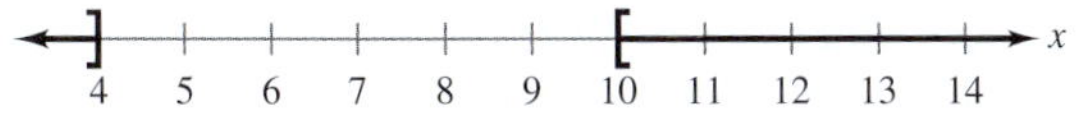

86. 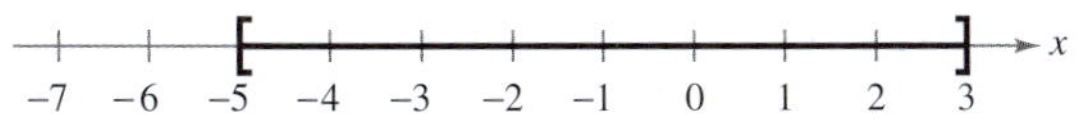

87. All real numbers at least 10 units from 12

88. All real numbers at least five units from 8

89. All real numbers more than four units from -3

90. All real numbers no more than seven units from -6

Writing an Inequality **In Exercises 91–94, write an inequality to describe the situation.**

91. A company expects its earnings per share E for the next quarter to be no less than \$4.10 and no more than \$4.25.

92. The estimated daily oil production p at a refinery is greater than 2 million barrels but less than 2.4 million barrels.

93. The return r on an investment is expected to be no more than 8%.

94. The net income I of a company is expected to be no less than \$239 million.

Physiology **The maximum heart rate r (in beats per minute) of a person in normal health is related to the person's age A (in years) by the equation**

$$r = 220 - A.$$

In Exercises 95 and 96, determine the interval in which the person's heart rate is from 50% to 85% of the maximum heart rate. *(Source: American Heart Association)*

95. a 20-year-old **96.** a 40-year-old

97. Job Offers You are considering two job offers. The first job pays \$13.50 per hour. The second job pays \$9.00 per hour plus \$0.75 per unit produced per hour. How many units must you produce per hour for the second job to pay more per hour than the first job?

98. Job Offers You are considering two job offers. The first job pays \$3000 per month. The second job pays \$1000 per month plus a commission of 4% of your gross sales. How much must you earn in gross sales for the second job to pay more per month than the first job?

99. Investment For what annual interest rates will an investment of \$1000 grow to more than \$1062.50 in 2 years? $[A = P(1 + rt)]$

100. Investment For what annual interest rates will an investment of \$750 grow to more than \$825 in 2 years? $[A = P(1 + rt)]$

101. Cost, Revenue, and Profit The revenue from selling x units of a product is $R = 115.95x$. The cost of producing x units is $C = 95x + 750$. To obtain a profit, the revenue must be greater than the cost. For what values of x will this product return a profit?

102. Cost, Revenue, and Profit The revenue from selling x units of a product is $R = 24.55x$. The cost of producing x units is $C = 15.4x + 150{,}000$. To obtain a profit, the revenue must be greater than the cost. For what values of x will this product return a profit?

103. Daily Sales A doughnut shop sells a dozen doughnuts for \$7.95. Beyond the fixed costs (rent, utilities, and insurance) of \$165 per day, it costs \$1.45 for enough materials (flour, sugar, and so on) and labor to produce a dozen doughnuts. The daily profit from doughnut sales varies between \$400 and \$1200. Between what levels (in dozens of doughnuts) do the daily sales vary?

104. Weight Loss Program A person enrolls in a diet and exercise program that guarantees a loss of at least $1\frac{1}{2}$ pounds per week. The person's weight at the beginning of the program is 164 pounds. Find the maximum number of weeks before the person attains a goal weight of 128 pounds.

105. Data Analysis: IQ Scores and GPA The admissions office of a college wants to determine whether there is a relationship between IQ scores x and grade-point averages y after the first year of school. An equation that models the data the admissions office obtained is $y = 0.067x - 5.638$.

(a) Use a graphing utility to graph the model.

(b) Use the graph to estimate the values of x that predict a grade-point average of at least 3.0.

(c) Verify your estimate from part (b) algebraically.

(d) List other factors that may influence GPA.

106. Data Analysis: Weightlifting You want to determine whether there is a relationship between an athlete's weight x (in pounds) and the athlete's maximum bench-press weight y (in pounds). An equation that models the data you obtained is $y = 1.3x - 36$.

(a) Use a graphing utility to graph the model.

(b) Use the graph to estimate the values of x that predict a maximum bench-press weight of at least 200 pounds.

(c) Verify your estimate from part (b) algebraically.

(d) List other factors that might influence an individual's maximum bench-press weight.

107. Teachers' Salaries The average salaries S (in thousands of dollars) for public elementary school teachers in the United States from 2001 through 2011 can be modeled by

$$S = 1.36t + 41.1, \quad 1 \le t \le 11$$

where t represents the year, with $t = 1$ corresponding to 2001. *(Source: National Education Association)*

(a) According to the model, when was the average salary at least \$45,000, but no more than \$50,000?

(b) Use the model to predict when the average salary will exceed \$62,000.

108. Milk Production Milk production M (in billions of pounds) in the United States from 2002 through 2010 can be modeled by

$$M = 3.24t + 161.5, \quad 2 \le t \le 10$$

where t represents the year, with $t = 2$ corresponding to 2002. *(Source: U.S. Department of Agriculture)*

(a) According to the model, when was the annual milk production greater than 178 billion pounds, but no more than 190 billion pounds?

(b) Use the model to predict when milk production will exceed 225 billion pounds.

109. Geometry The side of a square is measured as 10.4 inches with a possible error of $\frac{1}{16}$ inch. Using these measurements, determine the interval containing the possible areas of the square.

110. Geometry The side of a square is measured as 24.2 centimeters with a possible error of 0.25 centimeter. Using these measurements, determine the interval containing the possible areas of the square.

111. Accuracy of Measurement You stop at a self-service gas station to buy 15 gallons of 87-octane gasoline at \$3.61 per gallon. The gas pump is accurate to within $\frac{1}{10}$ gallon. How much might you be undercharged or overcharged?

112. Accuracy of Measurement You buy six T-bone steaks that cost \$8.99 per pound. The weight that is listed on the package is 5.72 pounds. The scale that weighed the package is accurate to within $\frac{1}{32}$ pound. How much might you be undercharged or overcharged?

113. Time Study A time study was conducted to determine the length of time required to perform a particular task in a manufacturing process. The times required by approximately two-thirds of the workers in the study satisfied the inequality

$$|t - 15.6| \le 1.9$$

where t is time in minutes. Determine the interval in which these times lie.

Jaroslav74/Shutterstock.com

114. Error Tolerance

The protective cover layer of a Blu-ray Disc™ is 100 micrometers thick with an error tolerance of 3 micrometers. Write an absolute value inequality for the possible thicknesses of the cover layer. Then graph the solution set. *(Source: Blu-ray Disc™ Association)*

Exploration

True or False? **In Exercises 115–117, determine whether the statement is true or false. Justify your answer.**

115. If a, b, and c are real numbers, and $a < b$, then $a + c < b + c$.

116. If a, b, and c are real numbers, and $a \le b$, then $ac \le bc$.

117. If $-10 \le x \le 8$, then $-10 \ge -x$ and $-x \ge -8$.

118. HOW DO YOU SEE IT? Michael Kasha of Florida State University used physics and mathematics to design a new classical guitar. The model he used for the frequency of the vibrations on a circular plate was $v = (2.6t/d^2)\sqrt{E/\rho}$, where v is the frequency (in vibrations per second), t is the plate thickness (in millimeters), d is the diameter of the plate, E is the elasticity of the plate material, and ρ is the density of the plate material. For fixed values of d, E, and ρ, the graph of the equation is a line (see figure).

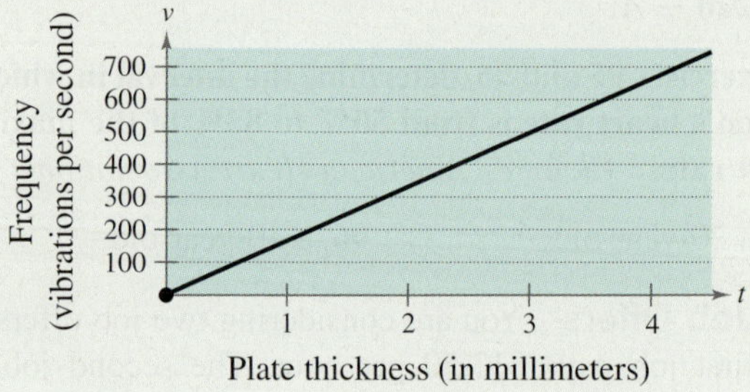

(a) Estimate the frequency when the plate thickness is 2 millimeters.

(b) Approximate the interval for the frequency when the plate thickness is greater than or equal to 0 millimeters and less than 3 millimeters.

A.7 Errors and the Algebra of Calculus

- **Avoid common algebraic errors.**
- **Recognize and use algebraic techniques that are common in calculus.**

Algebraic Errors to Avoid

This section contains five lists of common algebraic errors: errors involving parentheses, errors involving fractions, errors involving exponents, errors involving radicals, and errors involving dividing out. Many of these errors are made because they seem to be the *easiest* things to do. For instance, the operations of subtraction and division are often believed to be commutative and associative. The following examples illustrate the fact that subtraction and division are neither commutative nor associative.

Not commutative	**Not associative**
$4 - 3 \neq 3 - 4$	$8 - (6 - 2) \neq (8 - 6) - 2$
$15 \div 5 \neq 5 \div 15$	$20 \div (4 \div 2) \neq (20 \div 4) \div 2$

Errors Involving Parentheses

Potential Error	Correct Form	Comment
$a - (x - b) = a - x - b$	$a - (x - b) = a - x + b$	Distribute to each term in parentheses.
$(a + b)^2 = a^2 + b^2$	$(a + b)^2 = a^2 + 2ab + b^2$	Remember the middle term when squaring binomials.
$\left(\frac{1}{2}a\right)\left(\frac{1}{2}b\right) = \frac{1}{2}(ab)$	$\left(\frac{1}{2}a\right)\left(\frac{1}{2}b\right) = \frac{1}{4}(ab) = \frac{ab}{4}$	$\frac{1}{2}$ occurs twice as a factor.
$(3x + 6)^2 = 3(x + 2)^2$	$(3x + 6)^2 = [3(x + 2)]^2 = 3^2(x + 2)^2$	When factoring, raise all factors to the power.

Errors Involving Fractions

Potential Error	Correct Form	Comment
$\frac{2}{x + 4} = \frac{2}{x} + \frac{2}{4}$	Leave as $\frac{2}{x + 4}$.	The fraction is already in simplest form.
$\frac{\left(\frac{x}{a}\right)}{b} = \frac{bx}{a}$	$\frac{\left(\frac{x}{a}\right)}{b} = \left(\frac{x}{a}\right)\left(\frac{1}{b}\right) = \frac{x}{ab}$	Multiply by the reciprocal when dividing fractions.
$\frac{1}{a} + \frac{1}{b} = \frac{1}{a + b}$	$\frac{1}{a} + \frac{1}{b} = \frac{b + a}{ab}$	Use the property for adding fractions.
$\frac{1}{3x} = \frac{1}{3}x$	$\frac{1}{3x} = \frac{1}{3} \cdot \frac{1}{x}$	Use the property for multiplying fractions.
$(1/3)x = \frac{1}{3x}$	$(1/3)x = \frac{1}{3} \cdot x = \frac{x}{3}$	Be careful when expressing fractions in the form $1/a$.
$(1/x) + 2 = \frac{1}{x + 2}$	$(1/x) + 2 = \frac{1}{x} + 2 = \frac{1 + 2x}{x}$	Be careful when expressing fractions in the form $1/a$. Be sure to find a common denominator before adding fractions.

Errors Involving Exponents

Potential Error	Correct Form	Comment
$(x^2)^3 = x^5$	$(x^2)^3 = x^{2 \cdot 3} = x^6$	Multiply exponents when raising a power to a power.
$x^2 \cdot x^3 = x^6$	$x^2 \cdot x^3 = x^{2+3} = x^5$	Add exponents when multiplying powers with like bases.
$(2x)^3 = 2x^3$	$(2x)^3 = 2^3x^3 = 8x^3$	Raise each factor to the power.
$\dfrac{1}{x^2 - x^3} = x^{-2} - x^{-3}$	Leave as $\dfrac{1}{x^2 - x^3}$.	Do not move term-by-term from denominator to numerator.

Errors Involving Radicals

Potential Error	Correct Form	Comment
$\sqrt{5x} = 5\sqrt{x}$	$\sqrt{5x} = \sqrt{5}\sqrt{x}$	Radicals apply to every factor inside the radical.
$\sqrt{x^2 + a^2} = x + a$	Leave as $\sqrt{x^2 + a^2}$.	Do not apply radicals term-by-term when adding or subtracting terms.
$\sqrt{-x + a} = -\sqrt{x - a}$	Leave as $\sqrt{-x + a}$.	Do not factor minus signs out of square roots.

Errors Involving Dividing Out

Potential Error	Correct Form	Comment
$\dfrac{a + bx}{a} = 1 + bx$	$\dfrac{a + bx}{a} = \dfrac{a}{a} + \dfrac{bx}{a} = 1 + \dfrac{b}{a}x$	Divide out common factors, not common terms.
$\dfrac{a + ax}{a} = a + x$	$\dfrac{a + ax}{a} = \dfrac{a(1 + x)}{a} = 1 + x$	Factor before dividing out common factors.
$1 + \dfrac{x}{2x} = 1 + \dfrac{1}{x}$	$1 + \dfrac{x}{2x} = 1 + \dfrac{1}{2} = \dfrac{3}{2}$	Divide out common factors.

A good way to avoid errors is to *work slowly, write neatly,* and *talk to yourself.* Each time you write a step, ask yourself why the step is algebraically legitimate. You can justify the step below because *dividing the numerator and denominator by the same nonzero number produces an equivalent fraction.*

$$\frac{2x}{6} = \frac{2 \cdot x}{2 \cdot 3} = \frac{x}{3}$$

EXAMPLE 1 Describing and Correcting an Error

Describe and correct the error. $\dfrac{1}{2x} + \dfrac{1}{3x} = \dfrac{1}{5x}$

Solution Use the property for adding fractions: $\dfrac{1}{a} + \dfrac{1}{b} = \dfrac{b + a}{ab}$.

$$\frac{1}{2x} + \frac{1}{3x} = \frac{3x + 2x}{6x^2} = \frac{5x}{6x^2} = \frac{5}{6x}$$

✓ ***Checkpoint*** *Audio-video solution in English & Spanish at LarsonPrecalculus.com.*

Describe and correct the error. $\sqrt{x^2 + 4} = x + 2$

Some Algebra of Calculus

In calculus it is often necessary to take a simplified algebraic expression and rewrite it. See the following lists, taken from a standard calculus text.

Unusual Factoring

Expression	Useful Calculus Form	Comment
$\frac{5x^4}{8}$	$\frac{5}{8}x^4$	Write with fractional coefficient.
$\frac{x^2 + 3x}{-6}$	$-\frac{1}{6}(x^2 + 3x)$	Write with fractional coefficient.
$2x^2 - x - 3$	$2\left(x^2 - \frac{x}{2} - \frac{3}{2}\right)$	Factor out the leading coefficient.
$\frac{x}{2}(x + 1)^{-1/2} + (x + 1)^{1/2}$	$\frac{(x + 1)^{-1/2}}{2}[x + 2(x + 1)]$	Factor out the variable expression with the lesser exponent.

Writing with Negative Exponents

Expression	Useful Calculus Form	Comment
$\frac{9}{5x^3}$	$\frac{9}{5}x^{-3}$	Move the factor to the numerator and change the sign of the exponent.
$\frac{7}{\sqrt{2x - 3}}$	$7(2x - 3)^{-1/2}$	Move the factor to the numerator and change the sign of the exponent.

Writing a Fraction as a Sum

Expression	Useful Calculus Form	Comment
$\frac{x + 2x^2 + 1}{\sqrt{x}}$	$x^{1/2} + 2x^{3/2} + x^{-1/2}$	Divide each term of the numerator by $x^{1/2}$.
$\frac{1 + x}{x^2 + 1}$	$\frac{1}{x^2 + 1} + \frac{x}{x^2 + 1}$	Rewrite the fraction as a sum of fractions.
$\frac{2x}{x^2 + 2x + 1}$	$\frac{2x + 2 - 2}{x^2 + 2x + 1}$	Add and subtract the same term.
	$= \frac{2x + 2}{x^2 + 2x + 1} - \frac{2}{(x + 1)^2}$	Rewrite the fraction as a difference of fractions.
$\frac{x^2 - 2}{x + 1}$	$x - 1 - \frac{1}{x + 1}$	Use long division. (See Section 2.3.)
$\frac{x + 7}{x^2 - x - 6}$	$\frac{2}{x - 3} - \frac{1}{x + 2}$	Use the method of partial fractions. (See Section 7.4.)

Inserting Factors and Terms

Expression	Useful Calculus Form	Comment
$(2x - 1)^3$	$\frac{1}{2}(2x - 1)^3(2)$	Multiply and divide by 2.
$7x^2(4x^3 - 5)^{1/2}$	$\frac{7}{12}(4x^3 - 5)^{1/2}(12x^2)$	Multiply and divide by 12.
$\frac{4x^2}{9} - 4y^2 = 1$	$\frac{x^2}{9/4} - \frac{y^2}{1/4} = 1$	Write with fractional denominators.
$\frac{x}{x+1}$	$\frac{x + 1 - 1}{x + 1} = 1 - \frac{1}{x + 1}$	Add and subtract the same term.

The next five examples demonstrate many of the steps in the preceding lists.

EXAMPLE 2 Factors Involving Negative Exponents

Factor $x(x + 1)^{-1/2} + (x + 1)^{1/2}$.

Solution When multiplying powers with like bases, you add exponents. When factoring, you are undoing multiplication, and so you *subtract* exponents.

$$\begin{aligned} x(x + 1)^{-1/2} + (x + 1)^{1/2} &= (x + 1)^{-1/2}[x(x + 1)^0 + (x + 1)^1] \\ &= (x + 1)^{-1/2}[x + (x + 1)] \\ &= (x + 1)^{-1/2}(2x + 1) \end{aligned}$$

Checkpoint Audio-video solution in English & Spanish at LarsonPrecalculus.com.

Factor $x(x - 2)^{-1/2} + 3(x - 2)^{1/2}$.

Another way to simplify the expression in Example 2 is to multiply the expression by a fractional form of 1 and then use the Distributive Property.

$$\begin{aligned} x(x + 1)^{-1/2} + (x + 1)^{1/2} &= [x(x + 1)^{-1/2} + (x + 1)^{1/2}] \cdot \frac{(x + 1)^{1/2}}{(x + 1)^{1/2}} \\ &= \frac{x(x + 1)^0 + (x + 1)^1}{(x + 1)^{1/2}} = \frac{2x + 1}{\sqrt{x + 1}} \end{aligned}$$

EXAMPLE 3 Inserting Factors in an Expression

Insert the required factor: $\frac{x + 2}{(x^2 + 4x - 3)^2} = (\quad)\frac{1}{(x^2 + 4x - 3)^2}(2x + 4)$.

Solution The expression on the right side of the equation is twice the expression on the left side. To make both sides equal, insert a factor of $\frac{1}{2}$.

$$\frac{x + 2}{(x^2 + 4x - 3)^2} = \left(\frac{1}{2}\right)\frac{1}{(x^2 + 4x - 3)^2}(2x + 4)$$

Right side is multiplied and divided by 2.

Checkpoint Audio-video solution in English & Spanish at LarsonPrecalculus.com.

Insert the required factor: $\frac{6x - 3}{(x^2 - x + 4)^2} = (\quad)\frac{1}{(x^2 - x + 4)^2}(2x - 1)$.

EXAMPLE 4 Rewriting Fractions

Explain why the two expressions are equivalent.

$$\frac{4x^2}{9} - 4y^2 = \frac{x^2}{9/4} - \frac{y^2}{1/4}$$

Solution To write the expression on the left side of the equation in the form given on the right side, multiply the numerator and denominator of each term by $\frac{1}{4}$.

$$\frac{4x^2}{9} - 4y^2 = \frac{4x^2}{9}\left(\frac{1/4}{1/4}\right) - 4y^2\left(\frac{1/4}{1/4}\right) = \frac{x^2}{9/4} - \frac{y^2}{1/4}$$

✓ **Checkpoint** *Audio-video solution in English & Spanish at LarsonPrecalculus.com.*

Explain why the two expressions are equivalent.

$$\frac{9x^2}{16} + 25y^2 = \frac{x^2}{16/9} + \frac{y^2}{1/25}$$

EXAMPLE 5 Rewriting with Negative Exponents

Rewrite each expression using negative exponents.

a. $\dfrac{-4x}{(1-2x^2)^2}$ **b.** $\dfrac{2}{5x^3} - \dfrac{1}{\sqrt{x}} + \dfrac{3}{5(4x)^2}$

Solution

a. $\dfrac{-4x}{(1-2x^2)^2} = -4x(1-2x^2)^{-2}$

b. $$\frac{2}{5x^3} - \frac{1}{\sqrt{x}} + \frac{3}{5(4x)^2} = \frac{2}{5x^3} - \frac{1}{x^{1/2}} + \frac{3}{5(4x)^2}$$
$$= \frac{2}{5}x^{-3} - x^{-1/2} + \frac{3}{5}(4x)^{-2}$$

✓ **Checkpoint** *Audio-video solution in English & Spanish at LarsonPrecalculus.com.*

Rewrite $\dfrac{-6x}{(1-3x^2)^2} + \dfrac{1}{\sqrt[3]{x}}$ using negative exponents.

EXAMPLE 6 Rewriting a Fraction as a Sum of Terms

Rewrite each fraction as the sum of three terms.

a. $\dfrac{x^2 - 4x + 8}{2x}$ **b.** $\dfrac{x + 2x^2 + 1}{\sqrt{x}}$

Solution

a. $$\frac{x^2 - 4x + 8}{2x} = \frac{x^2}{2x} - \frac{4x}{2x} + \frac{8}{2x}$$
$$= \frac{x}{2} - 2 + \frac{4}{x}$$

b. $$\frac{x + 2x^2 + 1}{\sqrt{x}} = \frac{x}{x^{1/2}} + \frac{2x^2}{x^{1/2}} + \frac{1}{x^{1/2}}$$
$$= x^{1/2} + 2x^{3/2} + x^{-1/2}$$

✓ **Checkpoint** *Audio-video solution in English & Spanish at LarsonPrecalculus.com.*

Rewrite $\dfrac{x^4 - 2x^3 + 5}{x^3}$ as the sum of three terms.

A.7 Exercises

See **CalcChat.com** for tutorial help and worked-out solutions to odd-numbered exercises.

Vocabulary: Fill in the blanks.

1. To rewrite the expression $\frac{3}{x^5}$ using negative exponents, move x^5 to the ________ and change the sign of the exponent.
2. When dividing fractions, multiply by the ________.

Skills and Applications

Describing and Correcting an Error In Exercises 3–22, describe and correct the error.

3. $2x - (3y + 4) = 2x - 3y + 4$
4. $5z + 3(x - 2) = 5z + 3x - 2$
5. $\frac{4}{16x - (2x + 1)} = \frac{4}{14x + 1}$
6. $\frac{1 - x}{(5 - x)(-x)} = \frac{x - 1}{x(x - 5)}$
7. $(5z)(6z) = 30z$
8. $x(yz) = (xy)(xz)$
9. $a\left(\frac{x}{y}\right) = \frac{ax}{ay}$
10. $(4x)^2 = 4x^2$
11. $\sqrt{x + 9} = \sqrt{x} + 3$
12. $\sqrt{25 - x^2} = 5 - x$
13. $\frac{2x^2 + 1}{5x} = \frac{2x + 1}{5}$
14. $\frac{6x + y}{6x - y} = \frac{x + y}{x - y}$
15. $\frac{1}{a^{-1} + b^{-1}} = \left(\frac{1}{a + b}\right)^{-1}$
16. $\frac{1}{x + y^{-1}} = \frac{y}{x + 1}$
17. $(x^2 + 5x)^{1/2} = x(x + 5)^{1/2}$
18. $x(2x - 1)^2 = (2x^2 - x)^2$
19. $\frac{3}{x} + \frac{4}{y} = \frac{7}{x + y}$
20. $\frac{1}{2y} = (1/2)y$
21. $\frac{x}{2y} + \frac{y}{3} = \frac{x + y}{2y + 3}$
22. $5 + (1/y) = \frac{1}{5 + y}$

Inserting Factors in an Expression In Exercises 23–44, insert the required factor in the parentheses.

23. $\frac{5x + 3}{4} = \frac{1}{4}(\square)$
24. $\frac{7x^2}{10} = \frac{7}{10}(\square)$
25. $\frac{2}{3}x^2 + \frac{1}{3}x + 5 = \frac{1}{3}(\square)$
26. $\frac{3}{4}x + \frac{1}{2} = \frac{1}{4}(\square)$
27. $x^2(x^3 - 1)^4 = (\square)(x^3 - 1)^4(3x^2)$
28. $x(1 - 2x^2)^3 = (\square)(1 - 2x^2)^3(-4x)$
29. $2(y - 5)^{1/2} + y(y - 5)^{-1/2} = (y - 5)^{-1/2}(\square)$
30. $3t(6t + 1)^{-1/2} + (6t + 1)^{1/2} = (6t + 1)^{-1/2}(\square)$
31. $\frac{4x + 6}{(x^2 + 3x + 7)^3} = (\square)\frac{1}{(x^2 + 3x + 7)^3}(2x + 3)$
32. $\frac{x + 1}{(x^2 + 2x - 3)^2} = (\square)\frac{1}{(x^2 + 2x - 3)^2}(2x + 2)$
33. $\frac{3}{x} + \frac{5}{2x^2} - \frac{3}{2}x = (\square)(6x + 5 - 3x^3)$
34. $\frac{(x - 1)^2}{169} + (y + 5)^2 = \frac{(x - 1)^3}{169(\square)} + (y + 5)^2$
35. $\frac{25x^2}{36} + \frac{4y^2}{9} = \frac{x^2}{(\square)} + \frac{y^2}{(\square)}$
36. $\frac{5x^2}{9} - \frac{16y^2}{49} = \frac{x^2}{(\square)} - \frac{y^2}{(\square)}$
37. $\frac{x^2}{3/10} - \frac{y^2}{4/5} = \frac{10x^2}{(\square)} - \frac{5y^2}{(\square)}$
38. $\frac{x^2}{5/8} + \frac{y^2}{6/11} = \frac{8x^2}{(\square)} + \frac{11y^2}{(\square)}$
39. $x^{1/3} - 5x^{4/3} = x^{1/3}(\square)$
40. $3(2x + 1)x^{1/2} + 4x^{3/2} = x^{1/2}(\square)$
41. $(1 - 3x)^{4/3} - 4x(1 - 3x)^{1/3} = (1 - 3x)^{1/3}(\square)$
42. $\frac{1}{2\sqrt{x}} + 5x^{3/2} - 10x^{5/2} = \frac{1}{2\sqrt{x}}(\square)$
43. $\frac{1}{10}(2x + 1)^{5/2} - \frac{1}{6}(2x + 1)^{3/2} = \frac{(2x + 1)^{3/2}}{15}(\square)$
44. $\frac{3}{7}(t + 1)^{7/3} - \frac{3}{4}(t + 1)^{4/3} = \frac{3(t + 1)^{4/3}}{28}(\square)$

Rewriting with Negative Exponents In Exercises 45–50, rewrite the expression using negative exponents.

45. $\frac{7}{(x + 3)^5}$
46. $\frac{2 - x}{(x + 1)^{3/2}}$
47. $\frac{2x^5}{(3x + 5)^4}$
48. $\frac{x + 1}{x(6 - x)^{1/2}}$
49. $\frac{4}{3x} + \frac{4}{x^4} - \frac{7x}{\sqrt[3]{2x}}$
50. $\frac{x}{x - 2} + \frac{1}{x^2} + \frac{8}{3(9x)^3}$

Rewriting a Fraction as a Sum of Terms In Exercises 51–56, rewrite the fraction as a sum of two or more terms.

51. $\frac{x^2 + 6x + 12}{3x}$
52. $\frac{x^3 - 5x^2 + 4}{x^2}$
53. $\frac{4x^3 - 7x^2 + 1}{x^{1/3}}$
54. $\frac{2x^5 - 3x^3 + 5x - 1}{x^{3/2}}$
55. $\frac{3 - 5x^2 - x^4}{\sqrt{x}}$
56. $\frac{x^3 - 5x^4}{3x^2}$

Simplifying an Expression **In Exercises 57–68, simplify the expression.**

57. $\dfrac{-2(x^2-3)^{-3}(2x)(x+1)^3-3(x+1)^2(x^2-3)^{-2}}{[(x+1)^3]^2}$

58. $\dfrac{x^5(-3)(x^2+1)^{-4}(2x)-(x^2+1)^{-3}(5)x^4}{(x^5)^2}$

59. $\dfrac{(6x+1)^3(27x^2+2)-(9x^3+2x)(3)(6x+1)^2(6)}{[(6x+1)^3]^2}$

60. $\dfrac{(4x^2+9)^{1/2}(2)-(2x+3)\left(\frac{1}{2}\right)(4x^2+9)^{-1/2}(8x)}{[(4x^2+9)^{1/2}]^2}$

61. $\dfrac{(x+2)^{3/4}(x+3)^{-2/3}-(x+3)^{1/3}(x+2)^{-1/4}}{[(x+2)^{3/4}]^2}$

62. $(2x-1)^{1/2}-(x+2)(2x-1)^{-1/2}$

63. $\dfrac{2(3x-1)^{1/3}-(2x+1)\left(\frac{1}{3}\right)(3x-1)^{-2/3}(3)}{(3x-1)^{2/3}}$

64. $\dfrac{(x+1)\left(\frac{1}{2}\right)(2x-3x^2)^{-1/2}(2-6x)-(2x-3x^2)^{1/2}}{(x+1)^2}$

65. $\dfrac{1}{(x^2+4)^{1/2}}\cdot\dfrac{1}{2}(x^2+4)^{-1/2}(2x)$

66. $\dfrac{1}{x^2-6}(2x)+\dfrac{1}{2x+5}(2)$

67. $(x^2+5)^{1/2}\left(\frac{3}{2}\right)(3x-2)^{1/2}(3)$
$+ (3x-2)^{3/2}\left(\frac{1}{2}\right)(x^2+5)^{-1/2}(2x)$

68. $(3x+2)^{-1/2}(3)(x-6)^{1/2}(1)$
$+ (x-6)^3\left(-\frac{1}{2}\right)(3x+2)^{-3/2}(3)$

69. Athletics An athlete has set up a course for training as part of her regimen in preparation for an upcoming triathlon. She is dropped off by a boat 2 miles from the nearest point on shore. The finish line is 4 miles down the coast and 2 miles inland (see figure). She can swim 2 miles per hour and run 6 miles per hour. The time t (in hours) required for her to reach the finish line can be approximated by the model

$$t=\frac{\sqrt{x^2+4}}{2}+\frac{\sqrt{(4-x)^2+4}}{6}$$

where x is the distance down the coast (in miles) to the point at which she swims and then leaves the water to start her run.

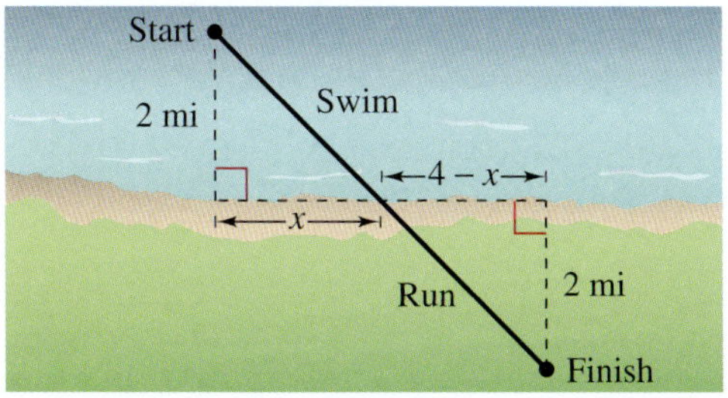

(a) Find the times required for the triathlete to finish when she swims to the points $x = 0.5$, $x = 1.0, \ldots, x = 3.5$, and $x = 4.0$ miles down the coast.

(b) Use your results from part (a) to determine the distance down the coast that will yield the minimum amount of time required for the triathlete to reach the finish line.

(c) The expression below was obtained using calculus. It can be used to find the minimum amount of time required for the triathlete to reach the finish line. Simplify the expression.

$\frac{1}{2}x(x^2+4)^{-1/2}+\frac{1}{6}(x-4)(x^2-8x+20)^{-1/2}$

70. Verifying an Equation

(a) Verify that $y_1 = y_2$ analytically.

$$y_1 = x^2\left(\frac{1}{3}\right)(x^2+1)^{-2/3}(2x)+(x^2+1)^{1/3}(2x)$$

$$y_2=\frac{2x(4x^2+3)}{3(x^2+1)^{2/3}}$$

(b) Complete the table and demonstrate the equality in part (a) numerically.

x	-2	-1	$-\frac{1}{2}$	0	1	2	$\frac{5}{2}$
y_1							
y_2							

Exploration

71. Writing Write a paragraph explaining to a classmate why $\dfrac{1}{(x-2)^{1/2}+x^4} \neq (x-2)^{-1/2}+x^{-4}$.

72. Think About It You are taking a course in calculus, and for one of the homework problems you obtain the following answer.

$$\frac{1}{10}(2x-1)^{5/2}+\frac{1}{6}(2x-1)^{3/2}$$

The answer in the back of the book is

$$\frac{1}{15}(2x-1)^{3/2}(3x+1).$$

Show how the second answer can be obtained from the first. Then use the same technique to simplify each of the following expressions.

(a) $\frac{2}{3}x(2x-3)^{3/2}-\frac{2}{15}(2x-3)^{5/2}$

(b) $\frac{2}{3}x(4+x)^{3/2}-\frac{2}{15}(4+x)^{5/2}$

Answers to Odd-Numbered Exercises and Tests

Chapter 1

Section 1.1 (page 8)

1. Cartesian **3.** Distance Formula

5.

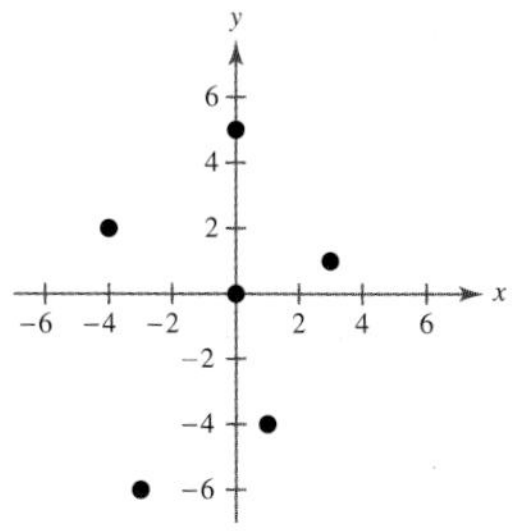

7. $(-3, 4)$
9. Quadrant IV
11. Quadrant II
13. Quadrant III

15.

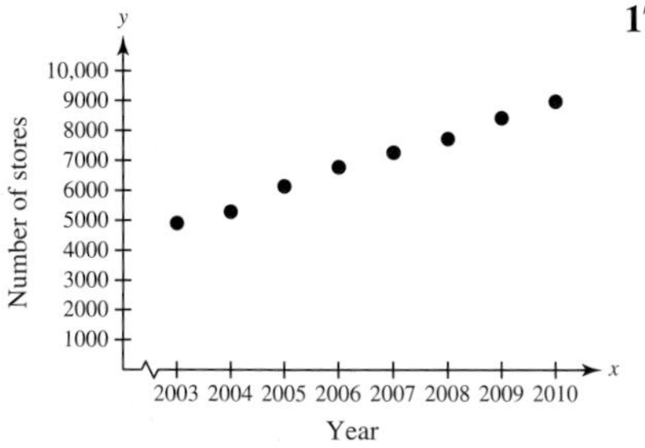

17. 13 **19.** $\sqrt{61}$

21. $\dfrac{\sqrt{277}}{6}$ **23.** (a) 5, 12, 13 (b) $5^2 + 12^2 = 13^2$

25. $(\sqrt{5})^2 + (\sqrt{45})^2 = (\sqrt{50})^2$

27. Distances between the points: $\sqrt{29}, \sqrt{58}, \sqrt{29}$

29. (a)

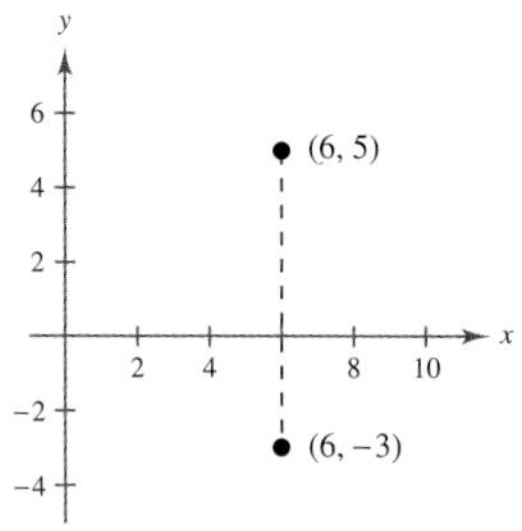

(b) 8 (c) $(6, 1)$

31. (a)

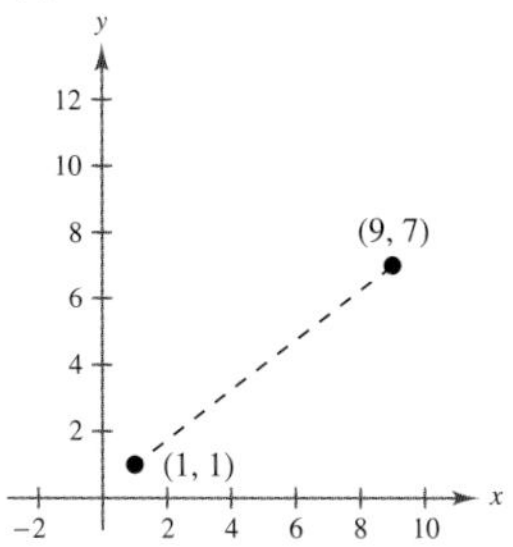

(b) 10 (c) $(5, 4)$

33. (a)

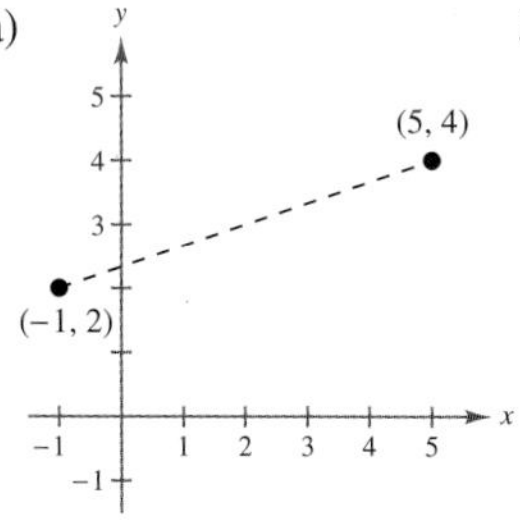

(b) $2\sqrt{10}$ (c) $(2, 3)$

35. (a)

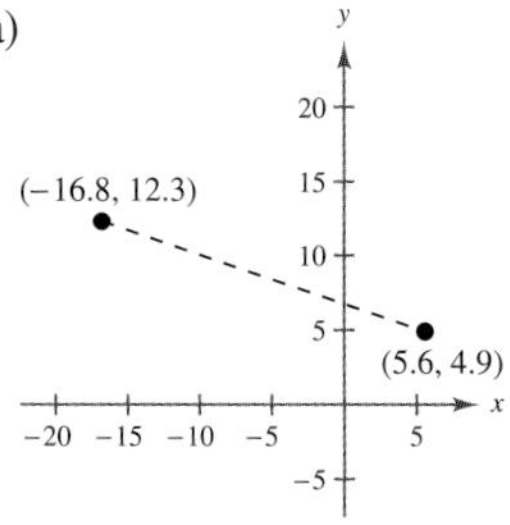

(b) $\sqrt{556.52}$
(c) $(-5.6, 8.6)$

37. $30\sqrt{41} \approx 192$ km **39.** \$27,343.5 million

41. $(0, 1), (4, 2), (1, 4)$ **43.** $(-3, 6), (2, 10), (2, 4), (-3, 4)$

45. (a) 2000s (b) 11.8%; 70.6% (c) \$12.37
(d) Answers will vary. *Sample answer:* No, the prediction is too high because it is likely that the percent increase over a 7-year period (2009–2016) will be less than the percent increase over a 16-year period (1995–2011).

47. $(2x_m - x_1, 2y_m - y_1)$

49. $\left(\dfrac{3x_1 + x_2}{4}, \dfrac{3y_1 + y_2}{4}\right), \left(\dfrac{x_1 + x_2}{2}, \dfrac{y_1 + y_2}{2}\right),$
$\left(\dfrac{x_1 + 3x_2}{4}, \dfrac{y_1 + 3y_2}{4}\right)$

51. (a)

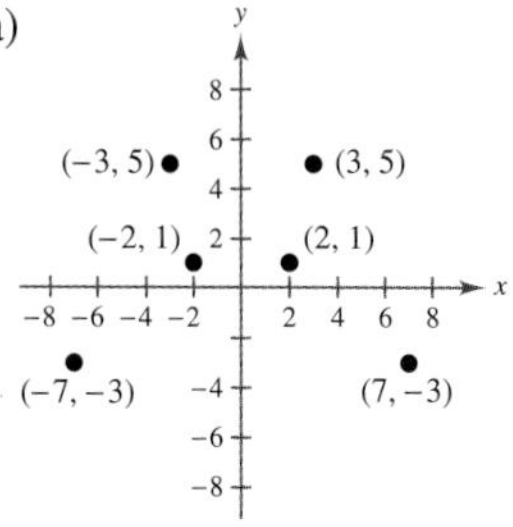

The point is reflected through the y-axis.

(b)

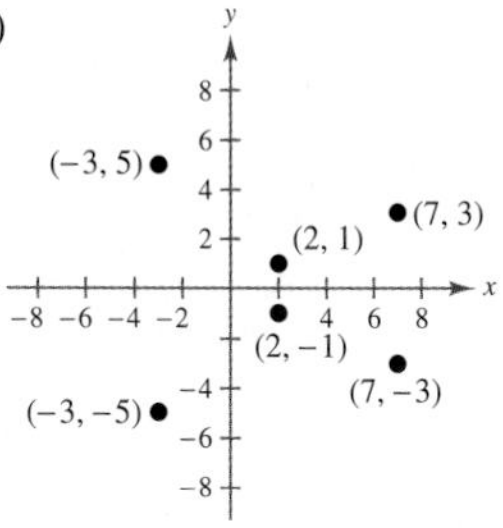

The point is reflected through the x-axis.

(c)

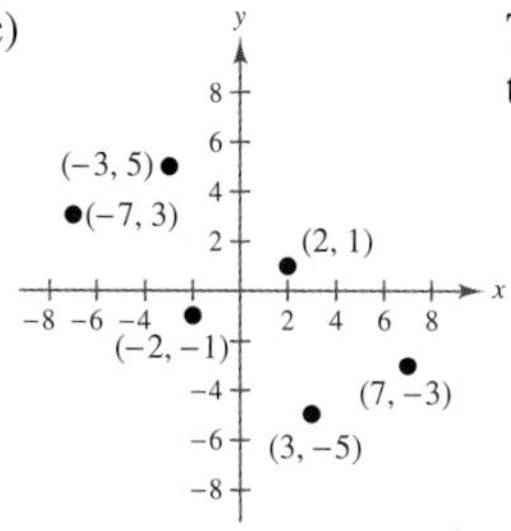

The point is reflected through the origin.

53. No. It depends on the magnitudes of the quantities measured.

55. False. The Midpoint Formula would be used 15 times.

57. False. The polygon could be a rhombus.

59. Use the Midpoint Formula to prove that the diagonals of the parallelogram bisect each other.

$$\left(\frac{b + a}{2}, \frac{c + 0}{2}\right) = \left(\frac{a + b}{2}, \frac{c}{2}\right)$$

$$\left(\frac{a + b + 0}{2}, \frac{c + 0}{2}\right) = \left(\frac{a + b}{2}, \frac{c}{2}\right)$$

Section 1.2 (page 19)

1. solution or solution point **3.** intercepts

5. circle; (h, k); r **7.** (a) Yes (b) Yes

9. (a) Yes (b) No **11.** (a) Yes (b) No

13. (a) No (b) Yes

15.

x	-1	0	1	2	$\frac{5}{2}$
y	7	5	3	1	0
(x, y)	$(-1, 7)$	$(0, 5)$	$(1, 3)$	$(2, 1)$	$(\frac{5}{2}, 0)$

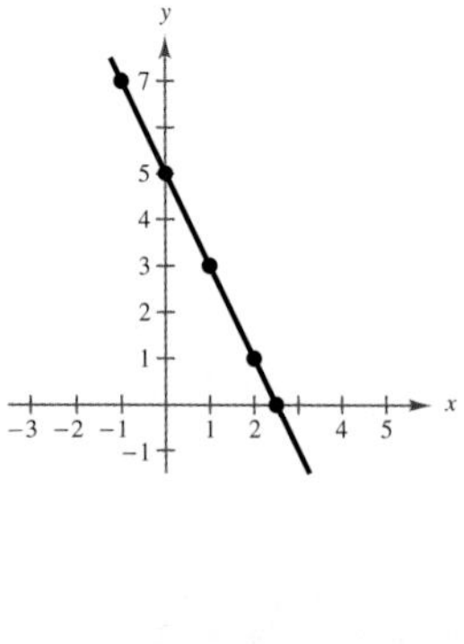

17.

x	-1	0	1	2	3
y	4	0	-2	-2	0
(x, y)	$(-1, 4)$	$(0, 0)$	$(1, -2)$	$(2, -2)$	$(3, 0)$

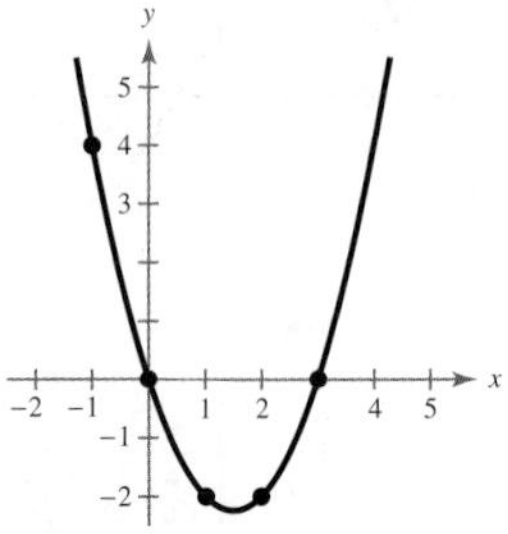

19. x-intercept: $(3, 0)$
y-intercept: $(0, 9)$

21. x-intercept: $(-2, 0)$
y-intercept: $(0, 2)$

23. x-intercept: $\left(\frac{6}{5}, 0\right)$
y-intercept: $(0, -6)$

25. x-intercept: $(-4, 0)$
y-intercept: $(0, 2)$

27. x-intercept: $\left(\frac{7}{3}, 0\right)$
y-intercept: $(0, 7)$

29. x-intercepts: $(0, 0), (2, 0)$
y-intercept: $(0, 0)$

31. x-intercept: $(6, 0)$
y-intercepts: $\left(0, \pm\sqrt{6}\right)$

33. y-axis symmetry

35. Origin symmetry

37. Origin symmetry

39. x-axis symmetry

41.

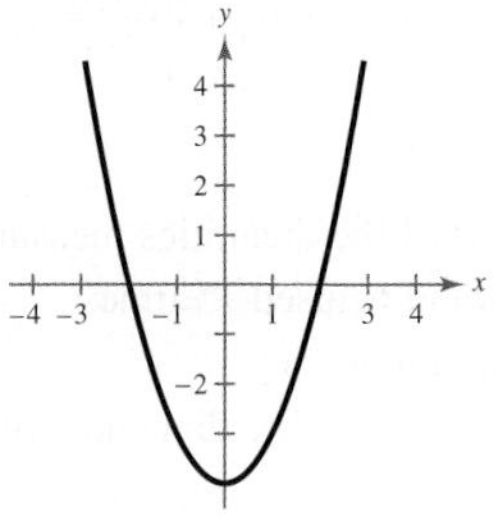

43.

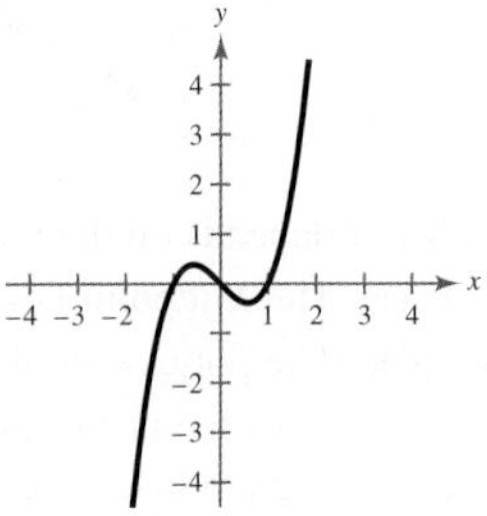

45. x-intercept: $\left(\frac{1}{3}, 0\right)$
y-intercept: $(0, 1)$
No symmetry

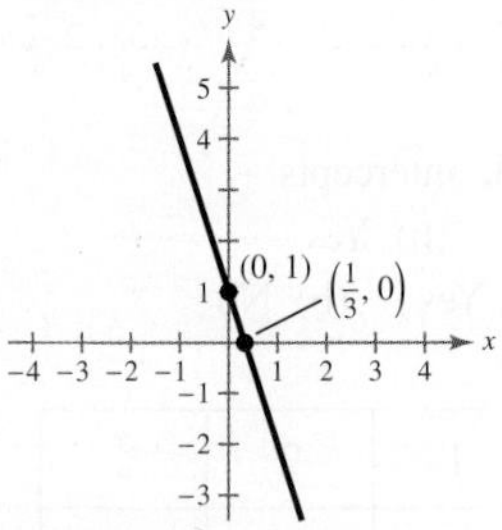

47. x-intercepts: $(0, 0), (2, 0)$
y-intercept: $(0, 0)$
No symmetry

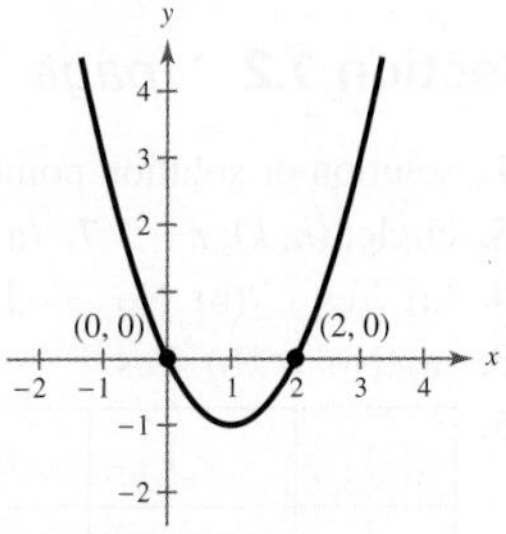

49. x-intercept: $\left(\sqrt[3]{-3}, 0\right)$
y-intercept: $(0, 3)$
No symmetry

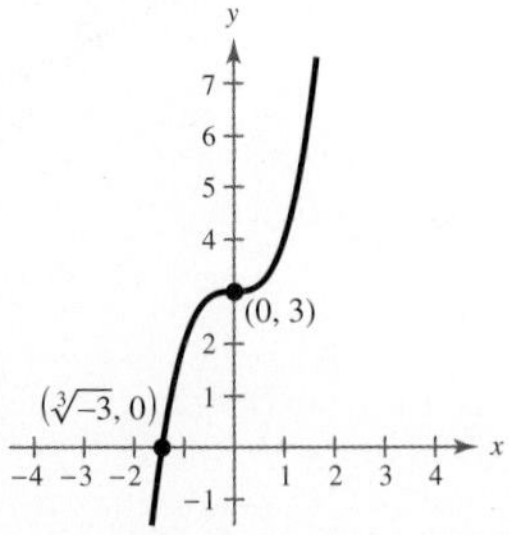

51. x-intercept: $(3, 0)$
y-intercept: None
No symmetry

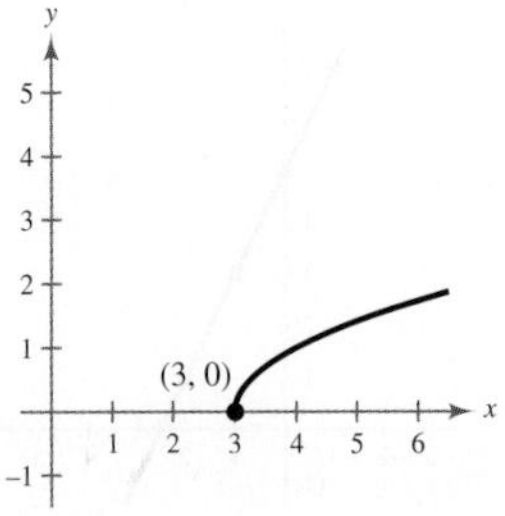

53. x-intercept: $(6, 0)$
y-intercept: $(0, 6)$
No symmetry

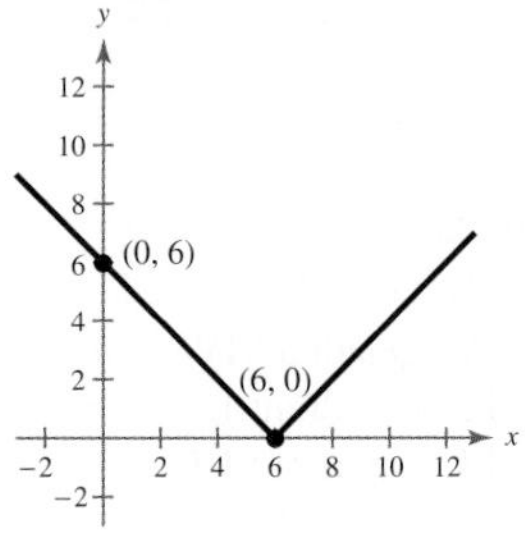

55. x-intercept: $(-1, 0)$
y-intercepts: $(0, \pm 1)$
x-axis symmetry

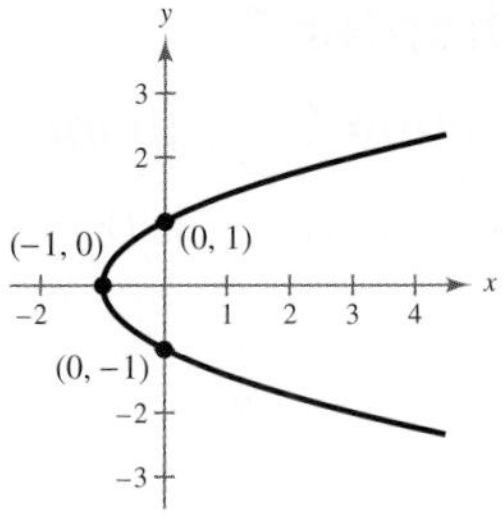

57.

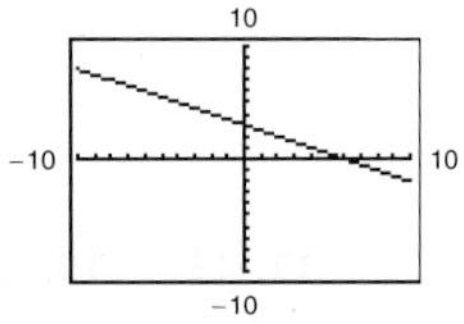

Intercepts: $(6, 0), (0, 3)$

59.

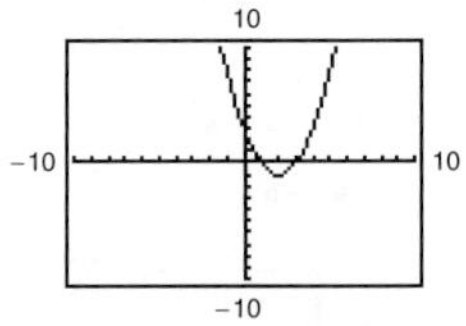

Intercepts:
$(3, 0), (1, 0), (0, 3)$

61.

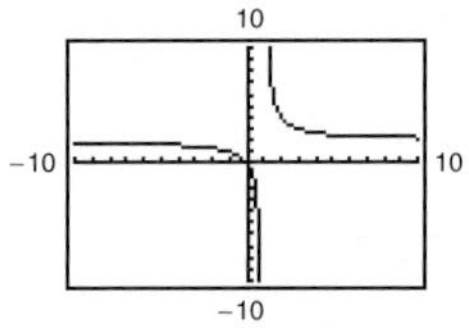

Intercept: $(0, 0)$

63.

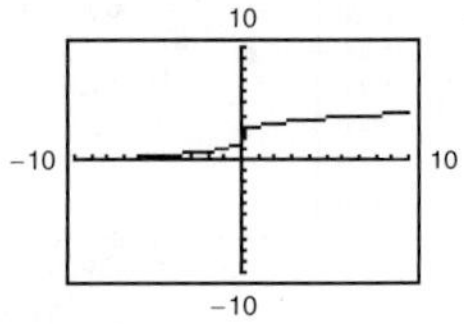

Intercepts: $(-8, 0), (0, 2)$

65.

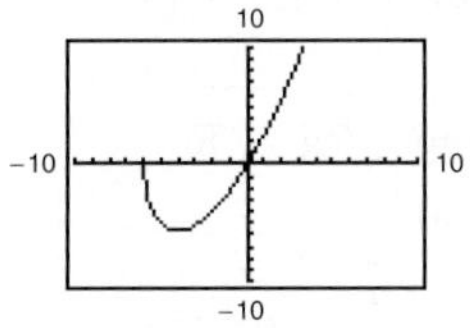

Intercepts: $(0, 0), (-6, 0)$

67.

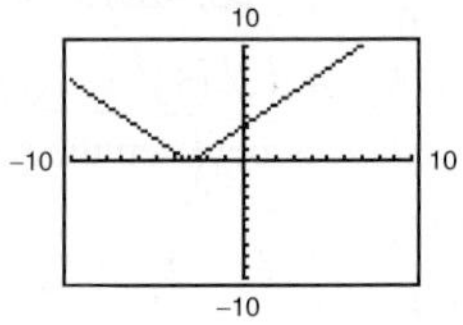

Intercepts: $(-3, 0), (0, 3)$

69. $x^2 + y^2 = 16$

71. $(x - 2)^2 + (y + 1)^2 = 16$

73. $(x + 1)^2 + (y - 2)^2 = 5$

75. $(x - 3)^2 + (y - 4)^2 = 25$

77. Center: $(0, 0)$; Radius: 5

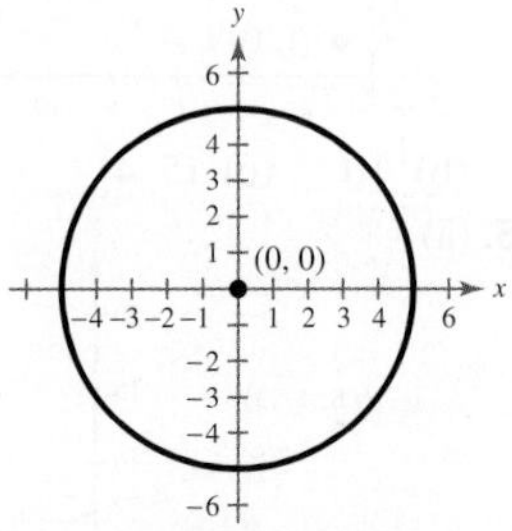

79. Center: $(1, -3)$; Radius: 3

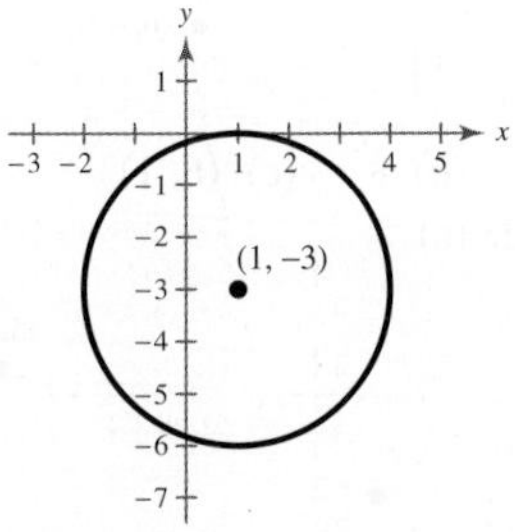

81. Center: $\left(\frac{1}{2}, \frac{1}{2}\right)$; Radius: $\frac{3}{2}$

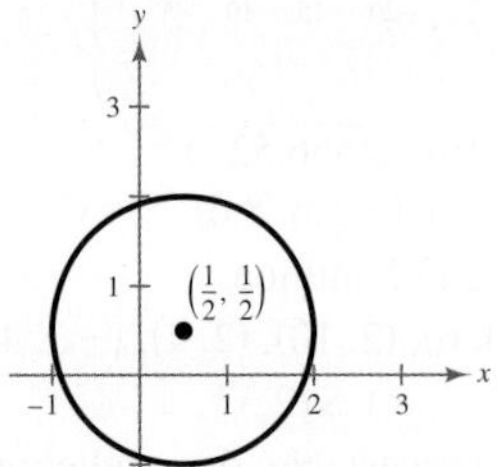

83.

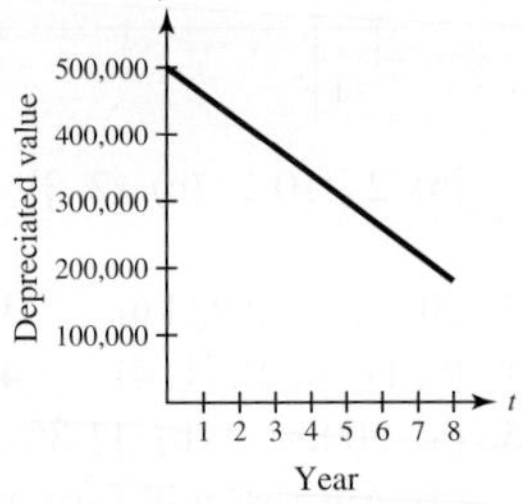

85. (a)

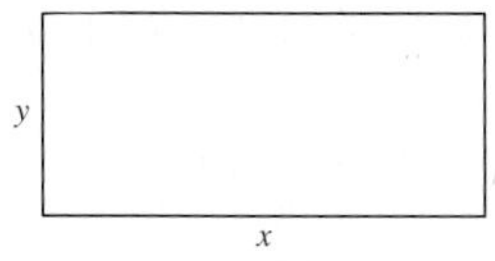

(b) Answers will vary.

(c)

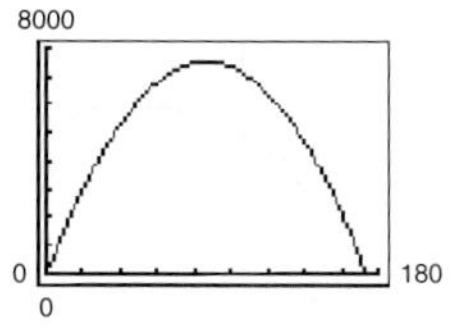

(d) $x = 86\frac{2}{3}$, $y = 86\frac{2}{3}$

(e) A regulation NFL playing field is 120 yards long and $53\frac{1}{3}$ yards wide. The actual area is 6400 square yards.

87. (a)

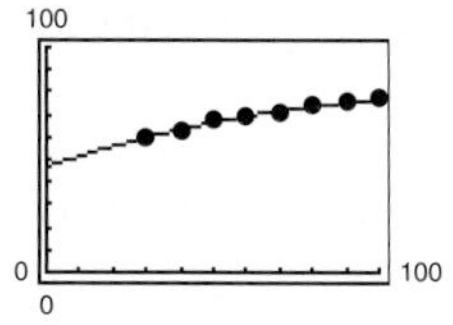

The model fits the data very well.

(b) 75.4 yr (c) 1994

(d) 77.65 yr; The projection given by the model is slightly less.

(e) Answers will vary.

89. (a) $a = 1, b = 0$ (b) $a = 0, b = 1$

Section 1.3 *(page 31)*

1. linear **3.** point-slope **5.** perpendicular

7. linear extrapolation **9.** (a) L_2 (b) L_3 (c) L_1

11.

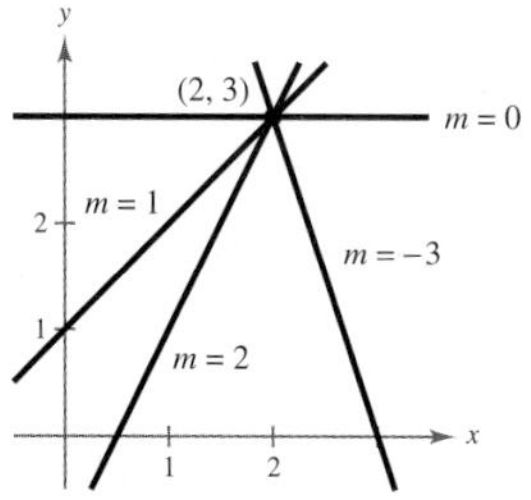

13. $\frac{3}{2}$

15. $m = 5$
y-intercept: $(0, 3)$

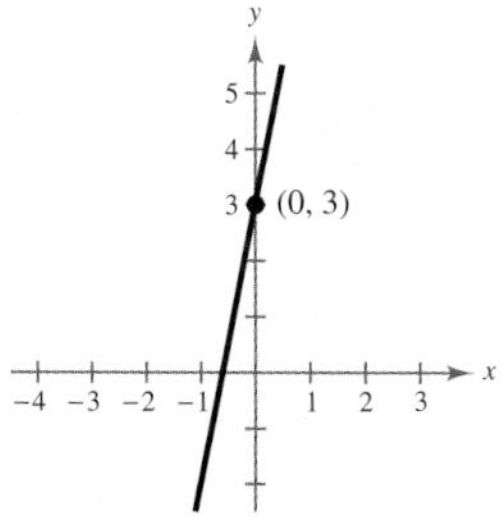

17. $m = -\frac{1}{2}$
y-intercept: $(0, 4)$

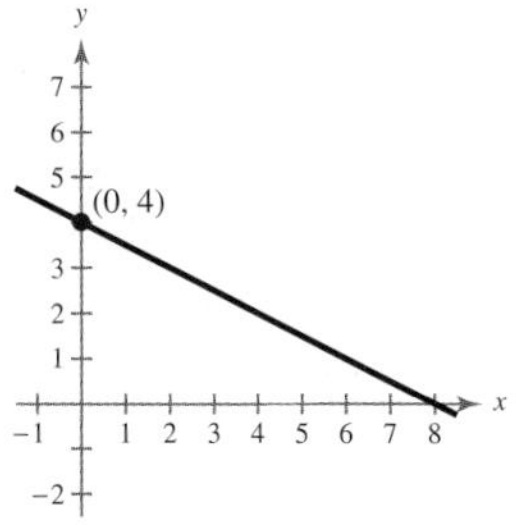

19. $m = 0$
y-intercept: $(0, 3)$

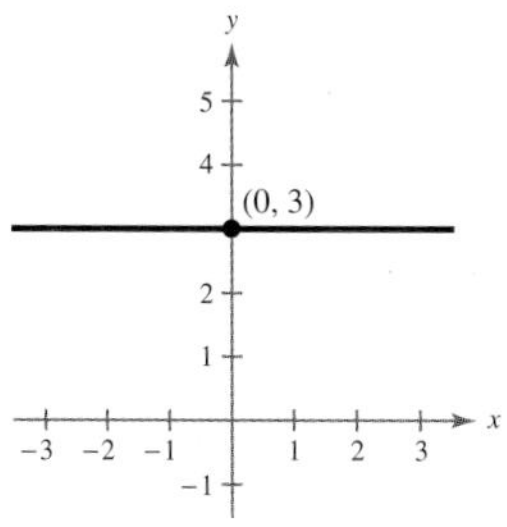

21. m is undefined.
There is no y-intercept.

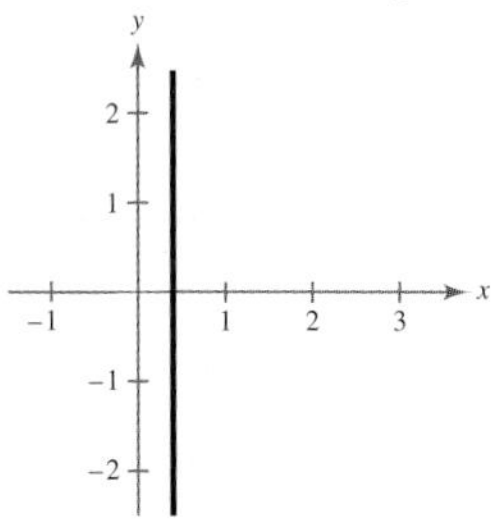

23. $m = \frac{7}{6}$; y-intercept: $(0, -5)$

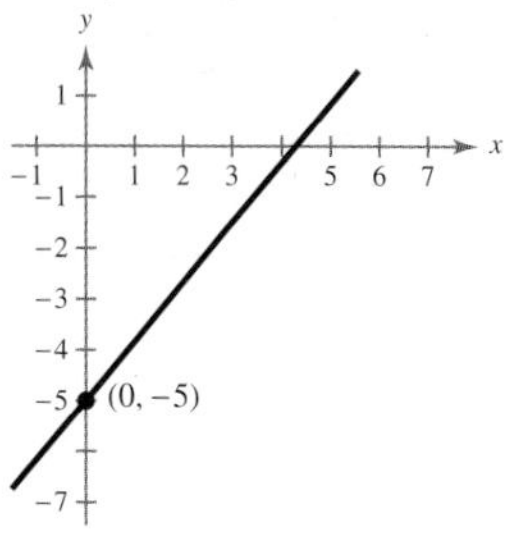

25.

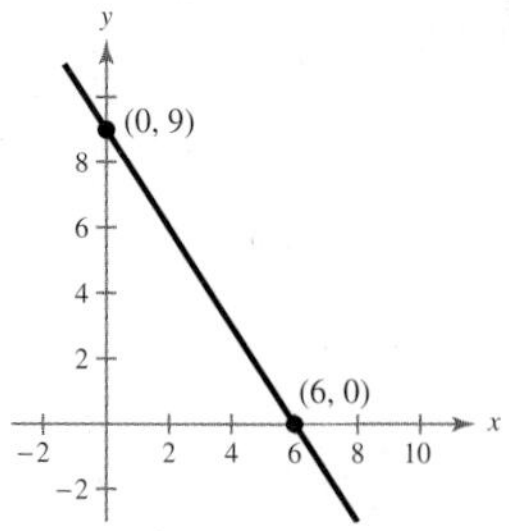

$m = -\frac{3}{2}$

27.

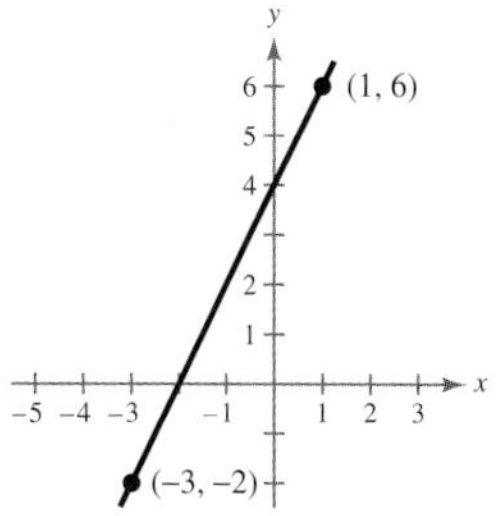

$m = 2$

29.

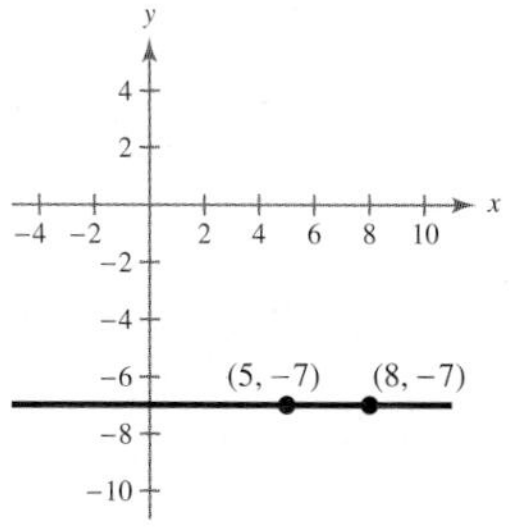

$m = 0$

31.

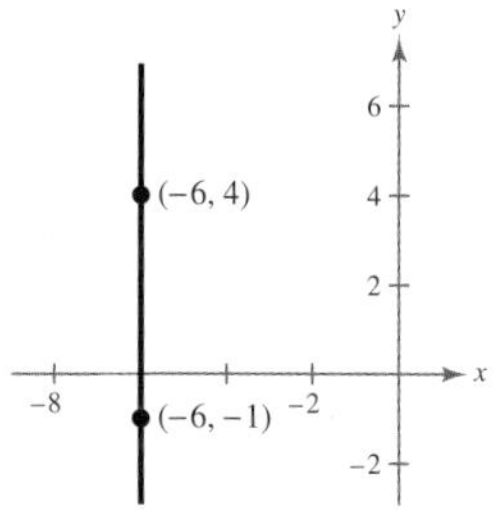

m is undefined.

33.

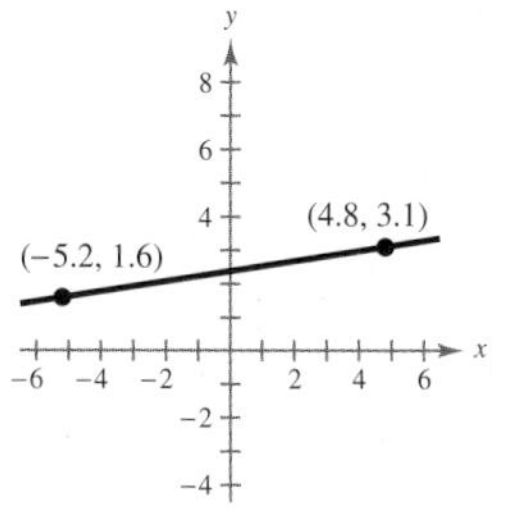

$m = 0.15$

35. $(0, 1), (3, 1), (-1, 1)$ **37.** $(-8, 0), (-8, 2), (-8, 3)$

39. $(-4, 6), (-3, 8), (-2, 10)$

41. $(-3, -5), (1, -7), (5, -9)$

43. $y = 3x - 2$

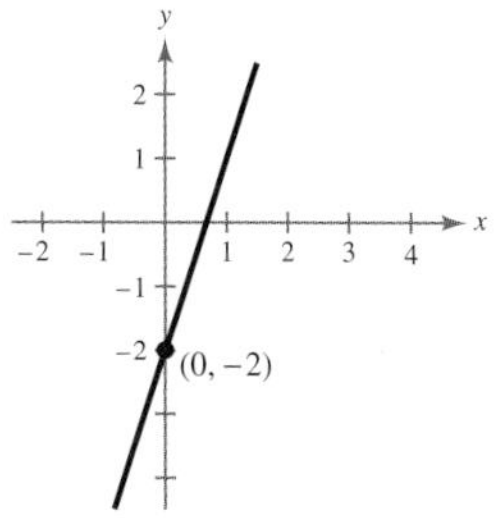

45. $y = -2x$

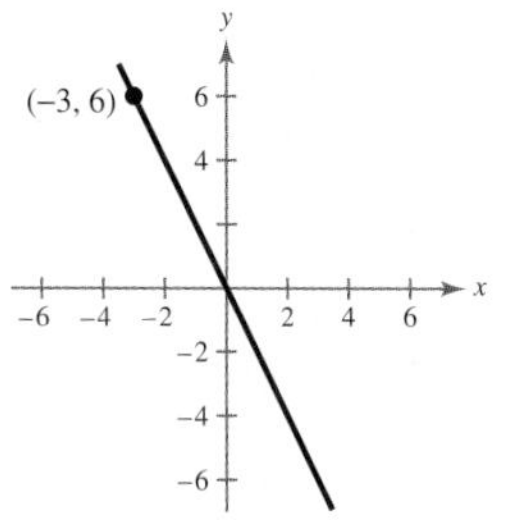

47. $y = -\frac{1}{3}x + \frac{4}{3}$

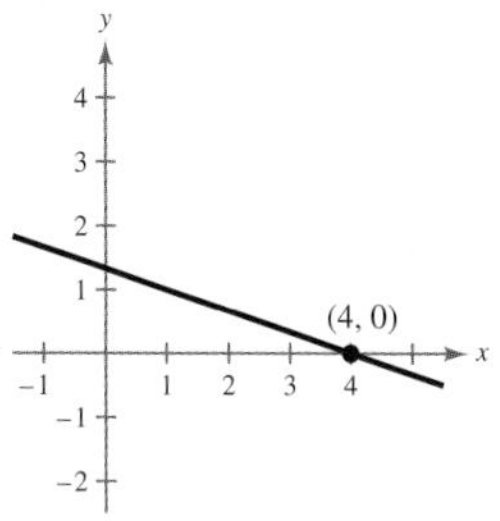

49. $y = -\frac{1}{2}x - 2$

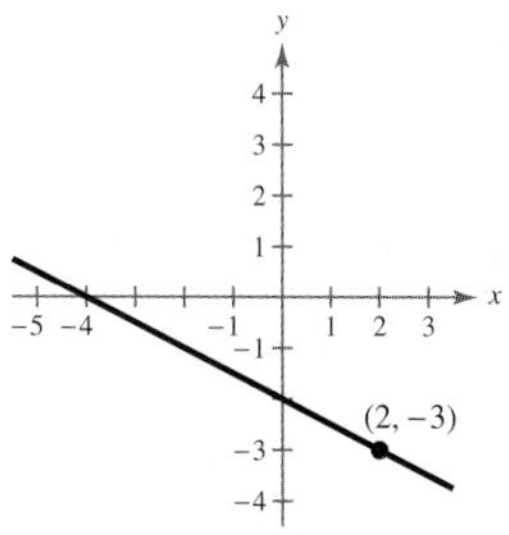

51. $x = 6$

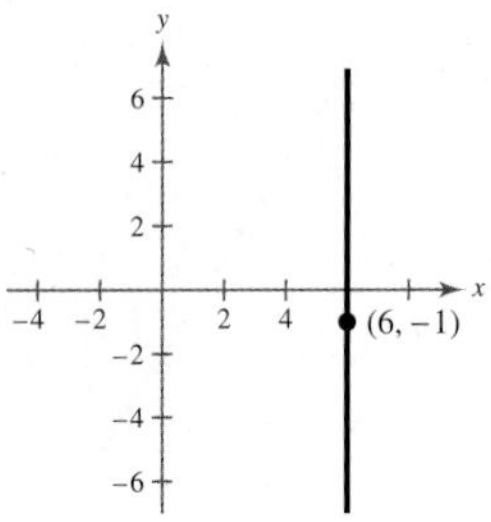

53. $y = \frac{5}{2}$

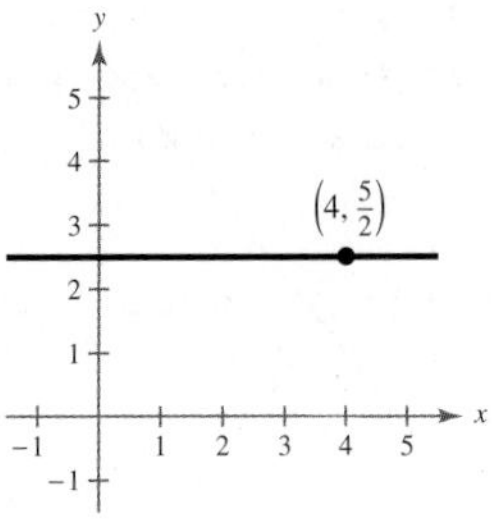

55. $y = -\frac{3}{5}x + 2$

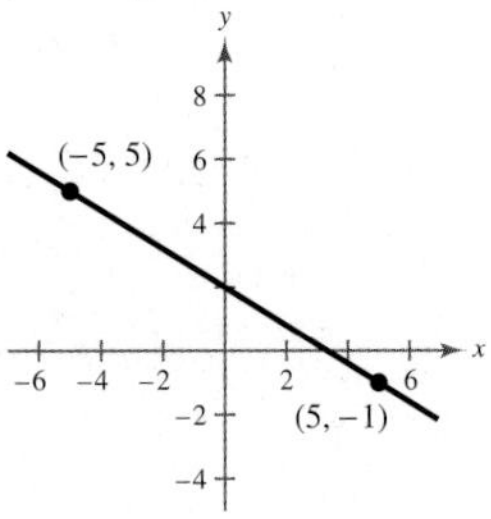

57. $x = -8$

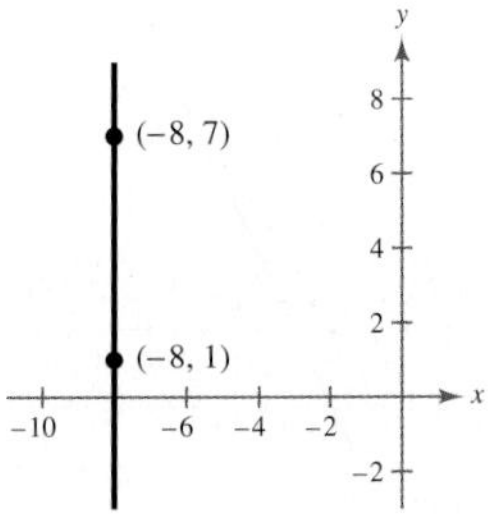

59. $y = -\frac{1}{2}x + \frac{3}{2}$

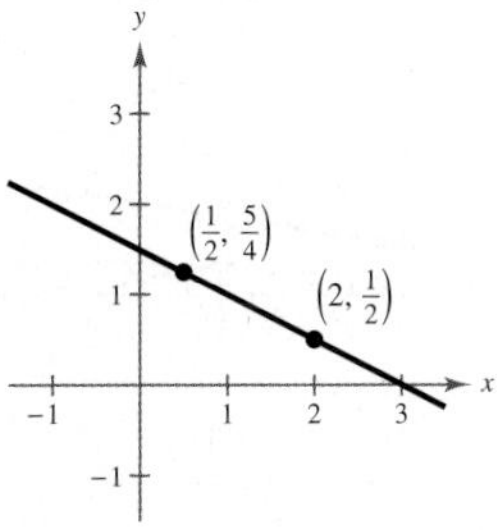

61. $y = 0.4x + 0.2$

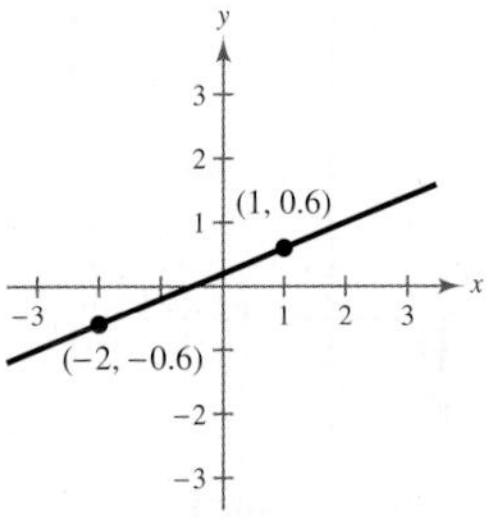

63. $y = -1$

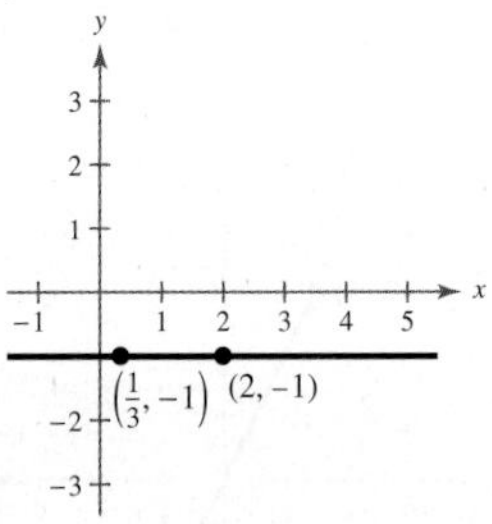

65. Parallel
67. Neither
69. Perpendicular
71. Parallel

73. (a) $y = 2x - 3$ (b) $y = -\frac{1}{2}x + 2$
75. (a) $y = -\frac{3}{4}x + \frac{3}{8}$ (b) $y = \frac{4}{3}x + \frac{127}{72}$
77. (a) $y = 0$ (b) $x = -1$
79. (a) $y = x + 4.3$ (b) $y = -x + 9.3$
81. $3x + 2y - 6 = 0$ **83.** $12x + 3y + 2 = 0$
85. $x + y - 3 = 0$
87. (a) Sales increasing 135 units/yr (b) No change in sales (c) Sales decreasing 40 units/yr
89. 12 ft **91.** $V(t) = -125t + 4165,\ 13 \le t \le 18$
93. C-intercept: fixed initial cost; Slope: cost of producing an additional laptop bag
95. $V(t) = -175t + 875,\ 0 \le t \le 5$
97. $F = 1.8C + 32$ or $C = \frac{5}{9}F - \frac{160}{9}$
99. (a) $C = 21t + 42{,}000$ (b) $R = 45t$ (c) $P = 24t - 42{,}000$ (d) 1750 h
101. False. The slope with the greatest magnitude corresponds to the steepest line.
103. Find the slopes of the lines containing each two points and use the relationship $m_1 = -1/m_2$.
105. No. The slope cannot be determined without knowing the scale on the y-axis. The slopes could be the same.
107. No. The slopes of two perpendicular lines have opposite signs (assume that neither line is vertical or horizontal).
109. The line $y = 4x$ rises most quickly, and the line $y = -4x$ falls most quickly. The greater the magnitude of the slope (the absolute value of the slope), the faster the line rises or falls.
111. $3x - 2y - 1 = 0$ **113.** $80x + 12y + 139 = 0$

Section 1.4 *(page 44)*

1. domain; range; function **3.** implied domain
5. Yes, each input value has exactly one output value.
7. No, the input values 7 and 10 each have two different output values.
9. (a) Function
(b) Not a function, because the element 1 in A corresponds to two elements, -2 and 1, in B.
(c) Function
(d) Not a function, because not every element in A is matched with an element in B.
11. Not a function **13.** Function **15.** Function
17. Function **19.** Function
21. (a) -1 (b) -9 (c) $2x - 5$
23. (a) 15 (b) $4t^2 - 19t + 27$ (c) $4t^2 - 3t - 10$
25. (a) 1 (b) 2.5 (c) $3 - 2|x|$
27. (a) $-\dfrac{1}{9}$ (b) Undefined (c) $\dfrac{1}{y^2 + 6y}$
29. (a) 1 (b) -1 (c) $\dfrac{|x - 1|}{x - 1}$
31. (a) -1 (b) 2 (c) 6
33.

x	-2	-1	0	1	2
$f(x)$	1	-2	-3	-2	1

35.

x	-2	-1	0	1	2
$f(x)$	5	$\frac{9}{2}$	4	1	0

37. 5 **39.** $\frac{4}{3}$
41. ± 3 **43.** $0, \pm 1$ **45.** $-1, 2$ **47.** $0, \pm 2$
49. All real numbers x **51.** All real numbers t except $t = 0$
53. All real numbers y such that $y \ge 10$
55. All real numbers x except $x = 0, -2$
57. All real numbers s such that $s \ge 1$ except $s = 4$
59. All real numbers x such that $x > 0$
61. (a) The maximum volume is 1024 cubic centimeters.
(b)

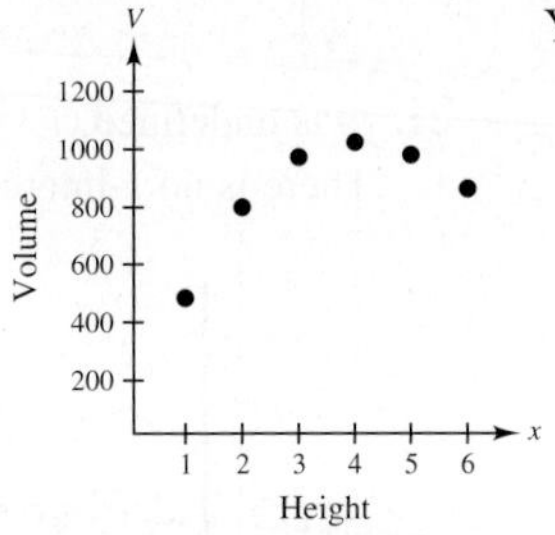

Yes, V is a function of x.
(c) $V = x(24 - 2x)^2,\ 0 < x < 12$
63. $A = \dfrac{P^2}{16}$ **65.** Yes, the ball will be at a height of 6 feet.

67. $A = \dfrac{x^2}{2(x-2)}, \quad x > 2$

69. 2004: 45.58%
2005: 50.15%
2006: 54.72%
2007: 59.29%
2008: 64.40%
2009: 67.75%
2010: 71.10%

71. (a) $C = 12.30x + 98{,}000$ (b) $R = 17.98x$
(c) $P = 5.68x - 98{,}000$

73. (a)

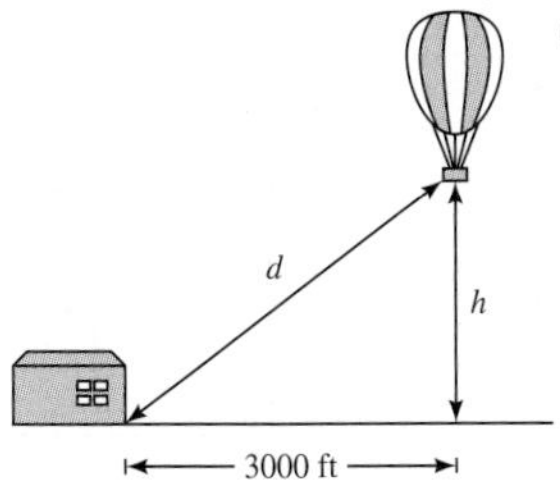

(b) $h = \sqrt{d^2 - 3000^2}$,
$d \geq 3000$

75. (a) $R = \dfrac{240n - n^2}{20}, \quad n \geq 80$

(b)

n	90	100	110	120	130	140	150
$R(n)$	\$675	\$700	\$715	\$720	\$715	\$700	\$675

The revenue is maximum when 120 people take the trip.

77. $3 + h, \ h \neq 0$ **79.** $3x^2 + 3xh + h^2 + 3, \ h \neq 0$

81. $-\dfrac{x+3}{9x^2}, \ x \neq 3$ **83.** $\dfrac{\sqrt{5x} - 5}{x - 5}$

85. $g(x) = cx^2; c = -2$ **87.** $r(x) = \dfrac{c}{x}; c = 32$

89. False. A function is a special type of relation.

91. False. The range is $[-1, \infty)$.

93. Domain of $f(x)$: all real numbers $x \geq 1$
Domain of $g(x)$: all real numbers $x > 1$
Notice that the domain of $f(x)$ includes $x = 1$ and the domain of $g(x)$ does not because you cannot divide by 0.

95. No; x is the independent variable, f is the name of the function.

97. (a) Yes. The amount you pay in sales tax will increase as the price of the item purchased increases.
(b) No. The length of time that you study will not necessarily determine how well you do on an exam.

Section 1.5 *(page 56)*

1. Vertical Line Test **3.** decreasing

5. average rate of change; secant

7. Domain: $(-\infty, \infty)$; Range: $[-4, \infty)$
(a) 0 (b) -1 (c) 0 (d) -2

9. Domain: $(-\infty, \infty)$; Range: $(-2, \infty)$
(a) 0 (b) 1 (c) 2 (d) 3

11. Function **13.** Not a function **15.** $-\frac{5}{2}, 6$ **17.** 0

19. $0, \pm\sqrt{2}$ **21.** $\pm 3, 4$ **23.** $\frac{1}{2}$

25.

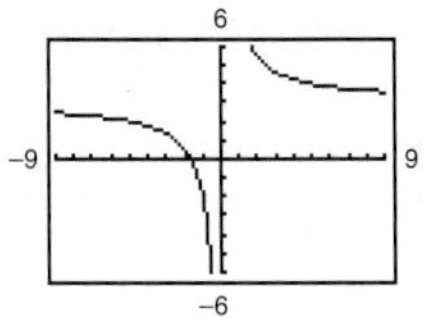

$-\frac{5}{3}$

27.

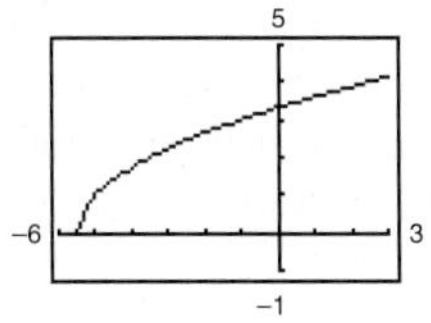

$-\frac{11}{2}$

29.

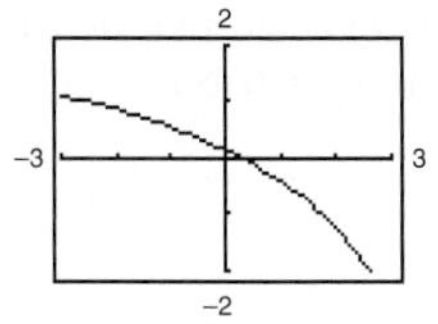

$\frac{1}{3}$

31. Increasing on $(-\infty, \infty)$

33. Increasing on $(-\infty, 0)$ and $(2, \infty)$; Decreasing on $(0, 2)$

35. Increasing on $(1, \infty)$; Decreasing on $(-\infty, -1)$
Constant on $(-1, 1)$

37. Increasing on $(-\infty, 0)$ and $(2, \infty)$; Constant on $(0, 2)$

39.

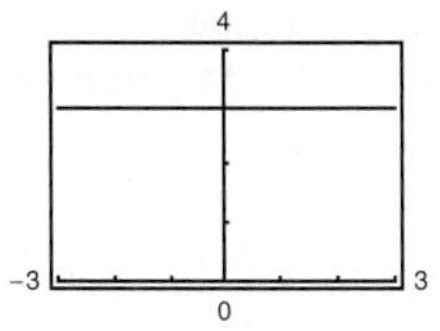

Constant on $(-\infty, \infty)$

41.

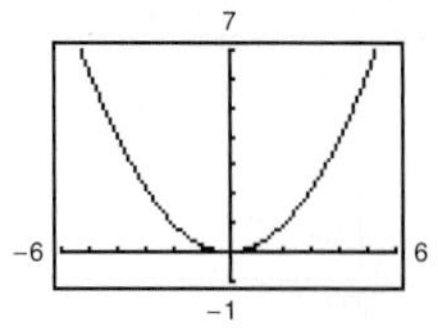

Decreasing on $(-\infty, 0)$
Increasing on $(0, \infty)$

43.

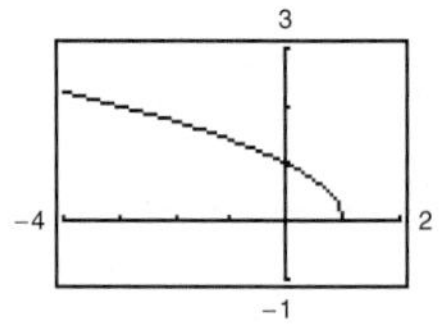

Decreasing on $(-\infty, 1)$

45.

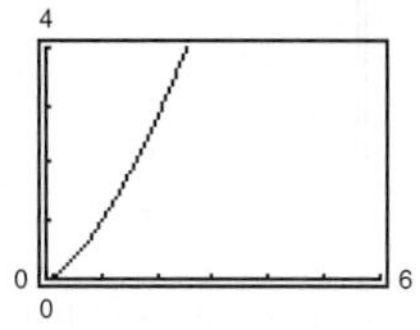

Increasing on $(0, \infty)$

47.

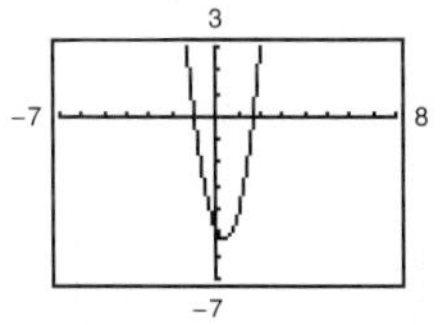

Relative minimum:
$\left(\frac{1}{3}, -\frac{16}{3}\right)$

49.

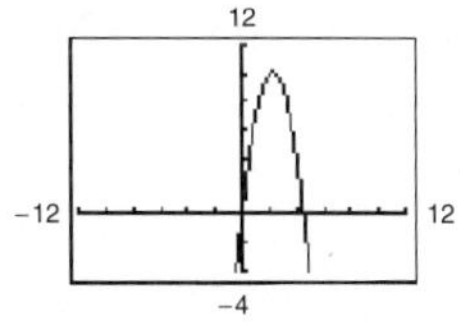

Relative maximum:
$(2.25, 10.125)$

51.

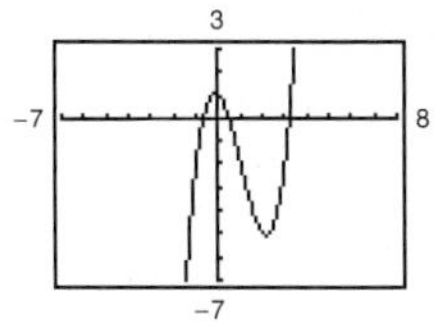

Relative maximum:
$(-0.15, 1.08)$
Relative minimum:
$(2.15, -5.08)$

53.

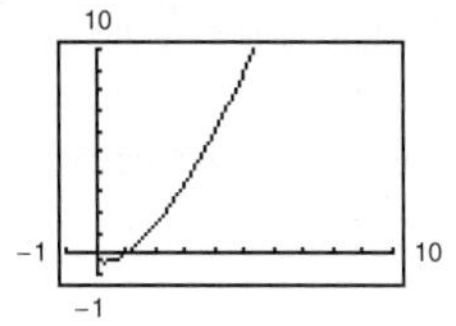

Relative minimum:
$(0.33, -0.38)$

55.

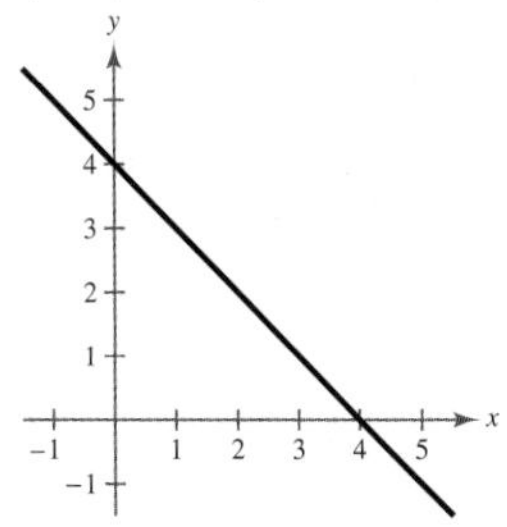

$(-\infty, 4]$

57.

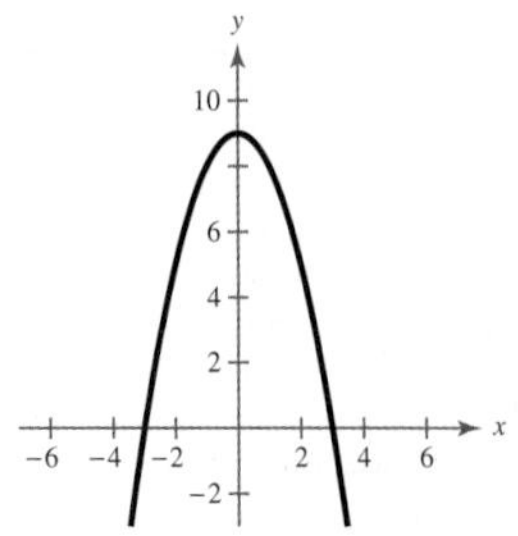

$[-3, 3]$

CHAPTER 1

59.

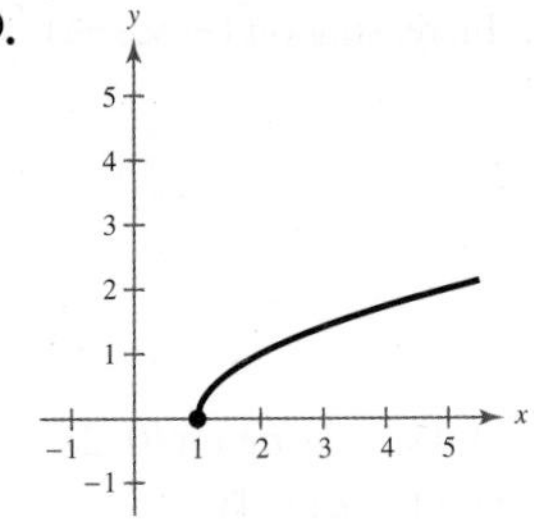

$[1, \infty)$

61. The average rate of change from $x_1 = 0$ to $x_2 = 3$ is -2.

63. The average rate of change from $x_1 = 1$ to $x_2 = 3$ is 0.

65. (a)

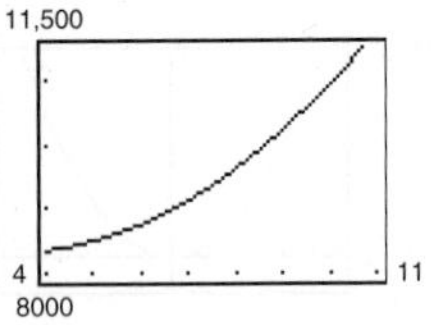

(b) 484.75 million; The amount the U.S. Department of Energy spent for research and development increased by about \$484.75 million each year from 2005 to 2010.

67. (a) $s = -16t^2 + 64t + 6$

(b)

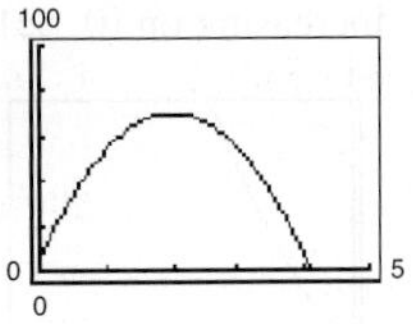

(c) Average rate of change $= 16$

(d) The slope of the secant line is positive.

(e) Secant line: $16t + 6$

(f)

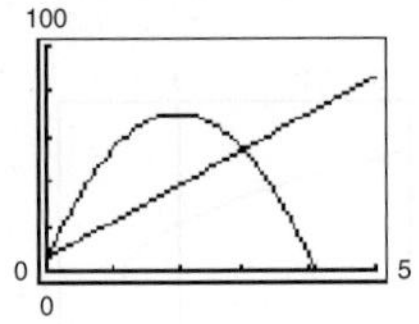

69. (a) $s = -16t^2 + 120t$

(b)

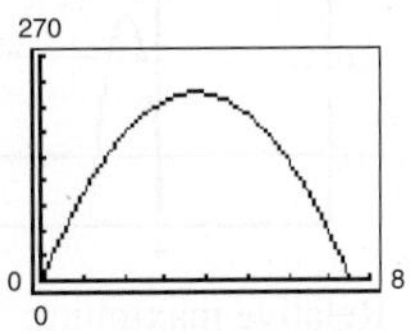

(c) Average rate of change $= -8$

(d) The slope of the secant line is negative.

(e) Secant line: $-8t + 240$

(f)

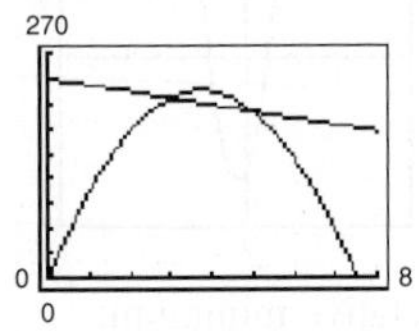

71. Even; y-axis symmetry

73. Odd; origin symmetry

75. Neither; no symmetry

77.

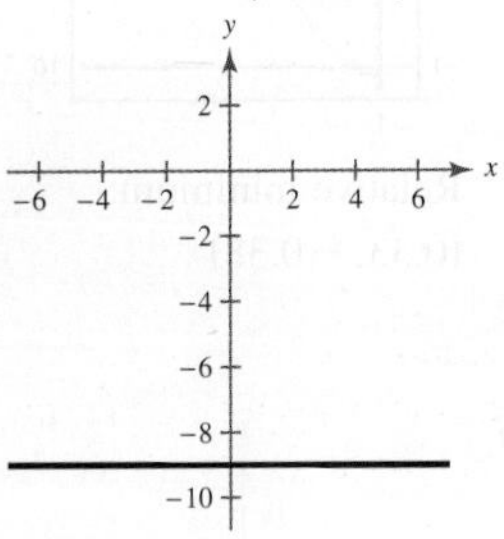

Even

79.

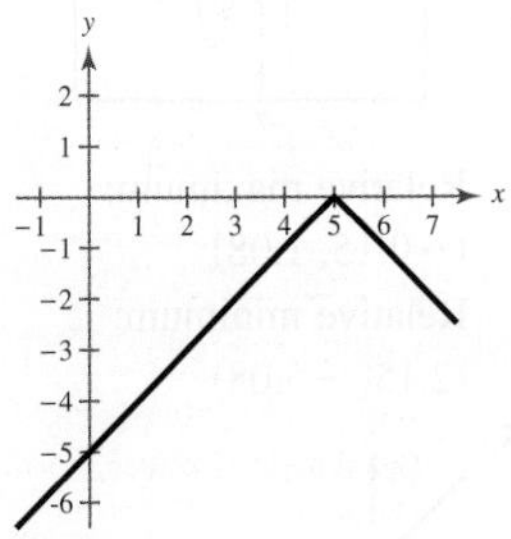

Neither

81.

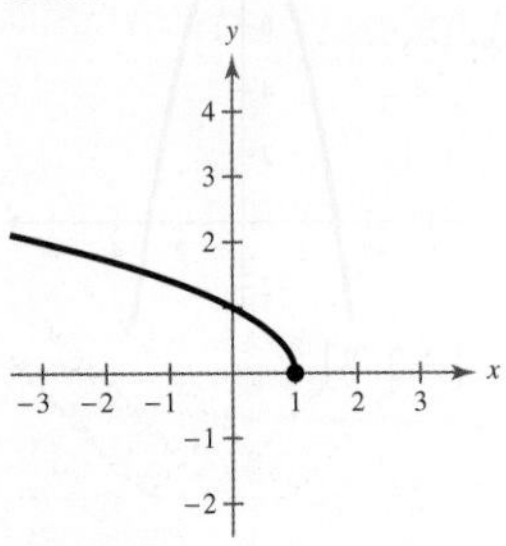

Neither

83. $h = 3 - 4x + x^2$

85. $L = 2 - \sqrt[3]{2y}$

87. (a)

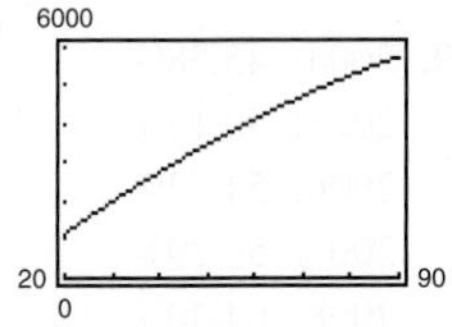

(b) 30 W

89. (a) Ten thousands (b) Ten millions (c) Percents

91. (a)

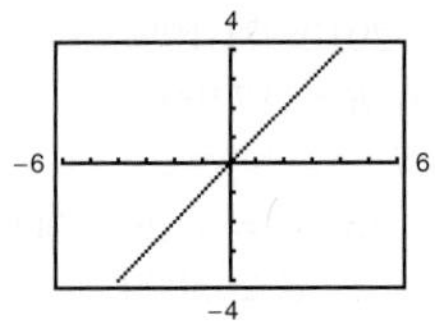

(b)

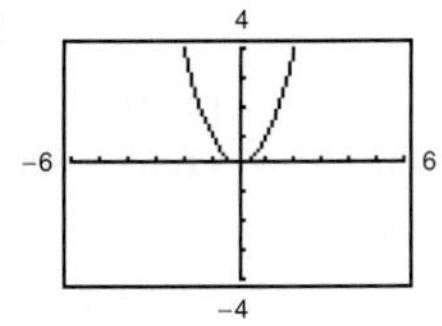

(c)

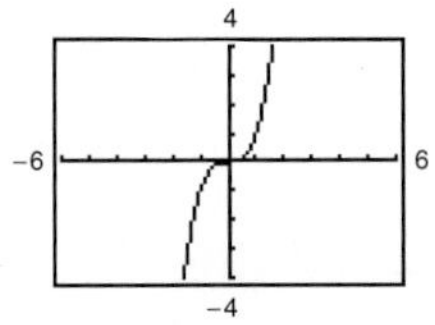

(d)

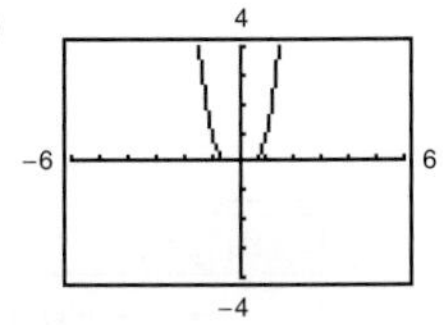

(e)

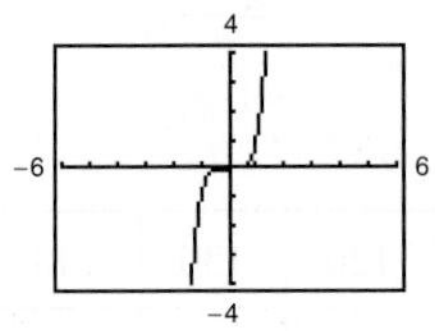

(f)

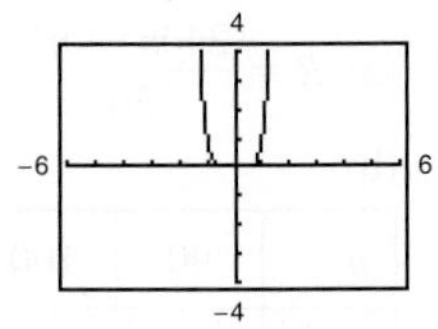

All the graphs pass through the origin. The graphs of the odd powers of x are symmetric with respect to the origin, and the graphs of the even powers are symmetric with respect to the y-axis. As the powers increase, the graphs become flatter in the interval $-1 < x < 1$.

93. False. The function $f(x) = \sqrt{x^2 + 1}$ has a domain of all real numbers.

95. (a) $\left(\frac{5}{3}, -7\right)$ (b) $\left(\frac{5}{3}, 7\right)$

97.

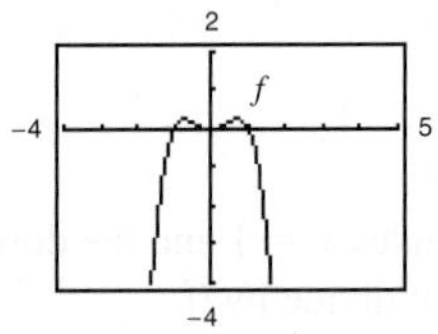

Even

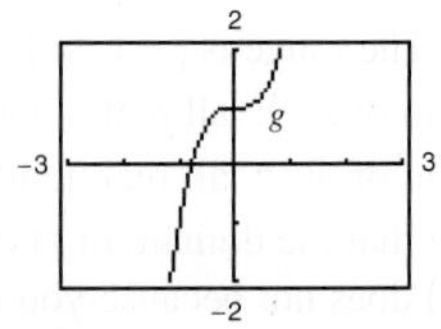

Neither

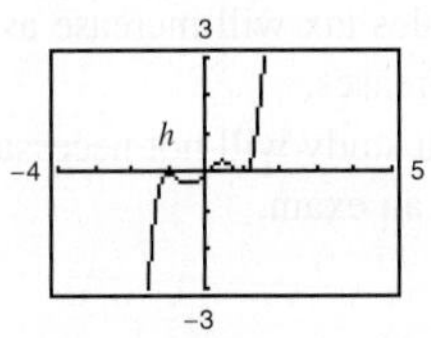

Odd

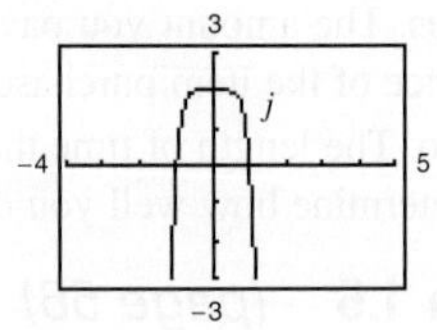

Even

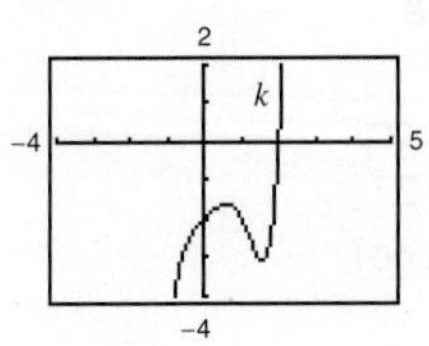

Neither

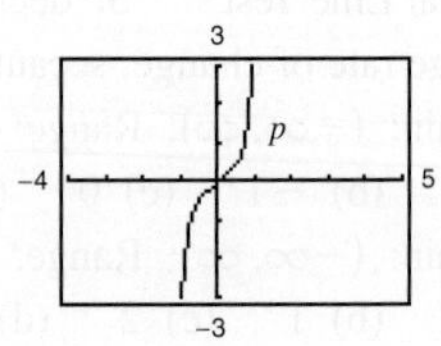

Odd

Equations of odd functions contain only odd powers of x. Equations of even functions contain only even powers of x. Odd functions have all variables raised to odd powers and even functions have all variables raised to even powers. A function that has variables raised to even and odd powers is neither odd nor even.

Section 1.6 (page 65)

1. g **2.** i **3.** h **4.** a **5.** b **6.** e **7.** f
8. c **9.** d

11. (a) $f(x) = -2x + 6$
(b)

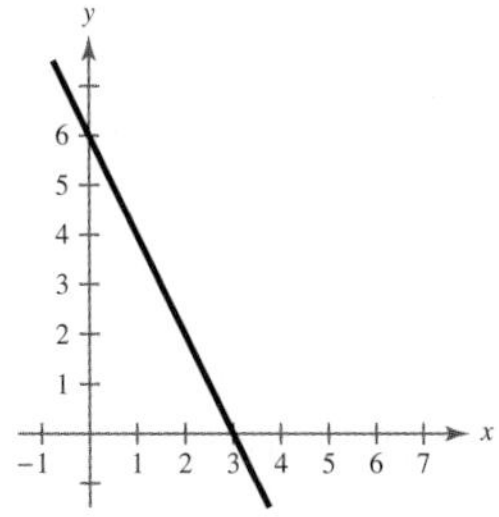

13. (a) $f(x) = -1$
(b)

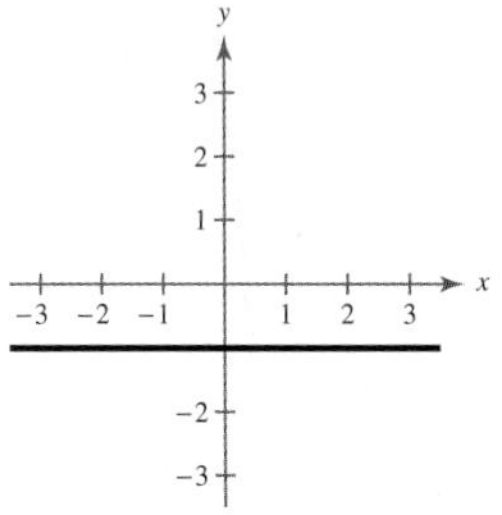

15.

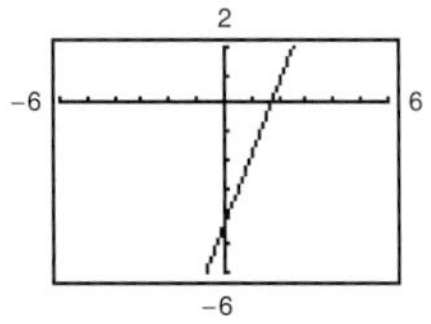

17.

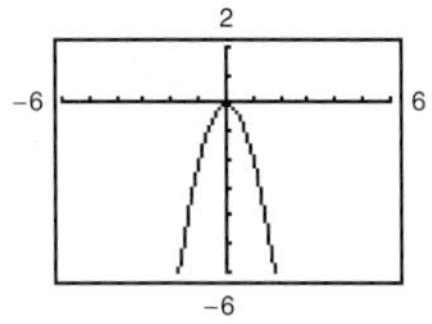

19.

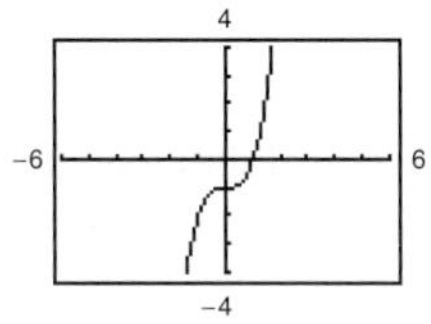

21.

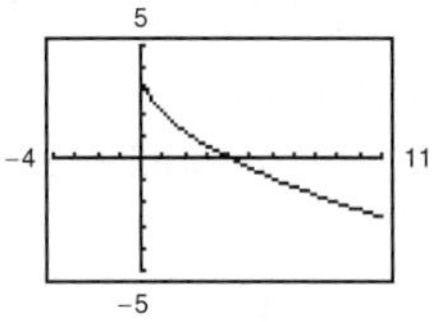

23.

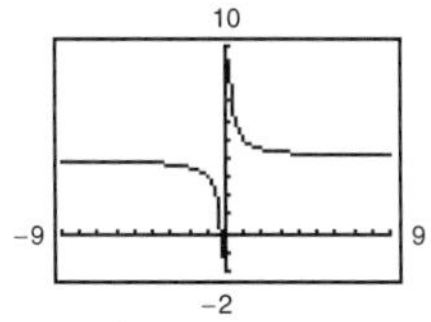

25.

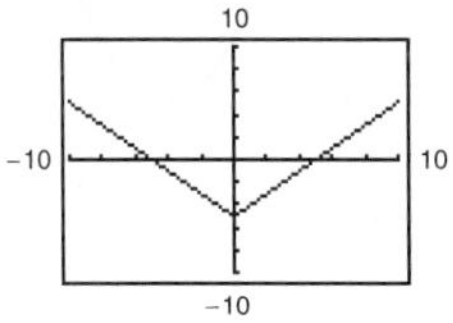

27. (a) 2 (b) 2 (c) -4 (d) 3

29. (a) 8 (b) 2 (c) 6 (d) 13

31.

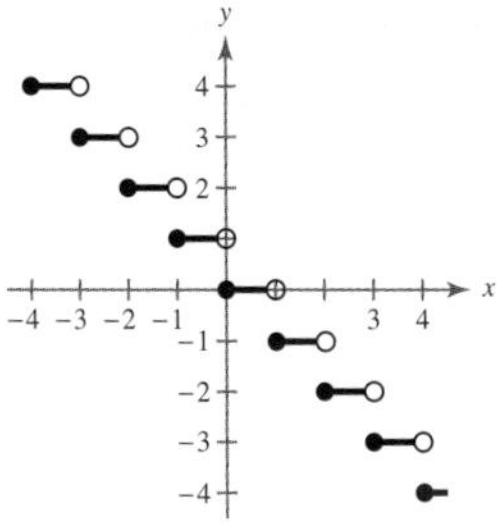

33.

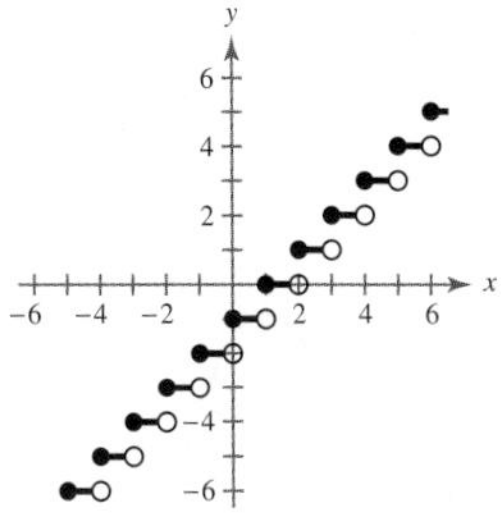

35.

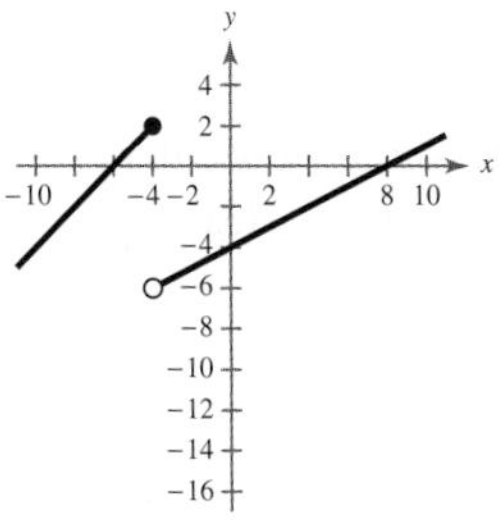

37.

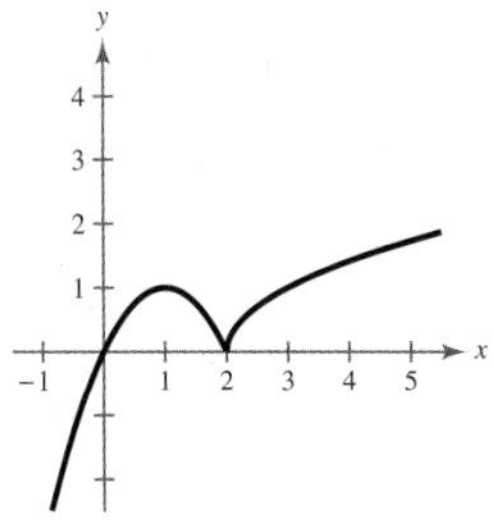

39.

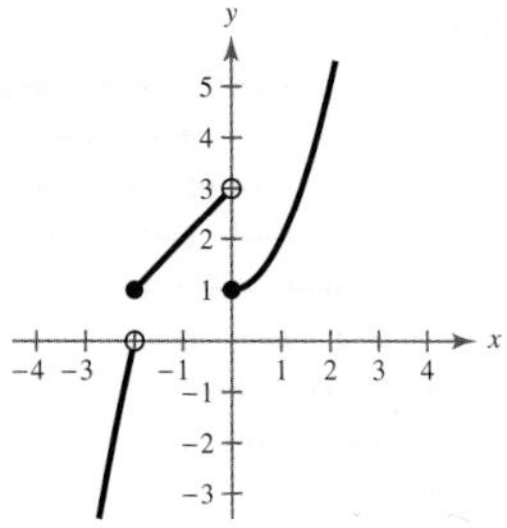

41. (a)

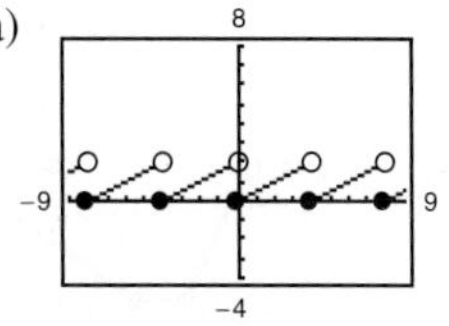

(b) Domain: $(-\infty, \infty)$
Range: $[0, 2)$

43. (a) $W(30) = 420$; $W(40) = 560$;
$W(45) = 665$; $W(50) = 770$

(b) $W(h) = \begin{cases} 14h, & 0 < h \le 45 \\ 21(h - 45) + 630, & h > 45 \end{cases}$

45.

Interval	Input Pipe	Drain Pipe 1	Drain Pipe 2
[0, 5]	Open	Closed	Closed
[5, 10]	Open	Open	Closed
[10, 20]	Closed	Closed	Closed
[20, 30]	Closed	Closed	Open
[30, 40]	Open	Open	Open
[40, 45]	Open	Closed	Open
[45, 50]	Open	Open	Open
[50, 60]	Open	Open	Closed

47. $f(t) = \begin{cases} t, & 0 \le t \le 2 \\ 2t - 2, & 2 < t \le 8 \\ \frac{1}{2}t + 10, & 8 < t \le 9 \end{cases}$

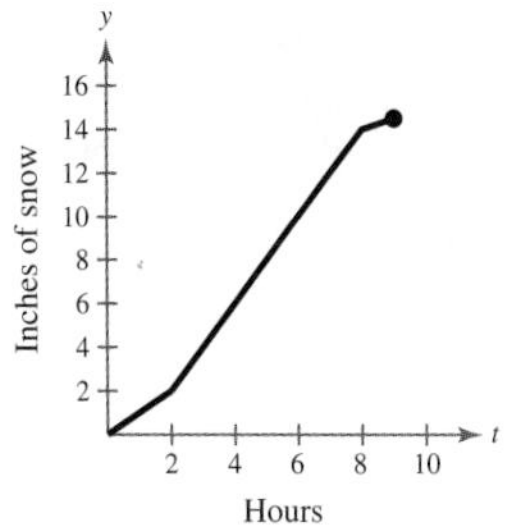

Total accumulation = 14.5 in.

49. False. A piecewise-defined function is a function that is defined by two or more equations over a specified domain. That domain may or may not include x- and y-intercepts.

Section 1.7 (page 72)

1. rigid **3.** vertical stretch; vertical shrink

5. (a)

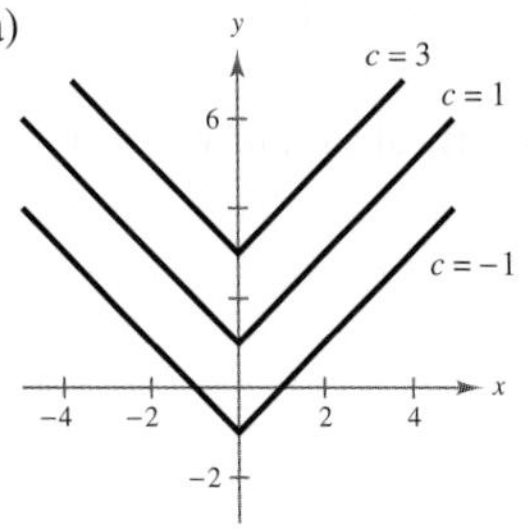

(b)

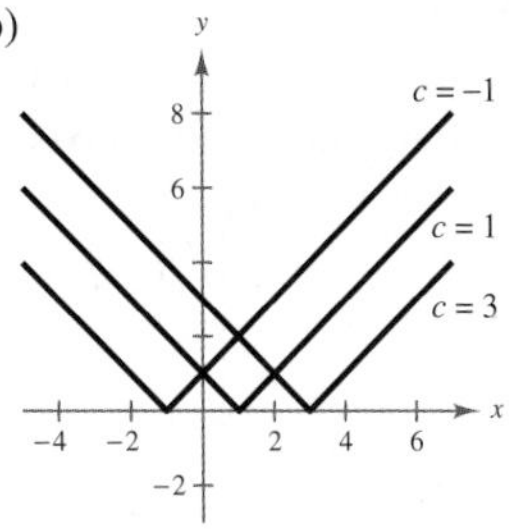

7. (a)

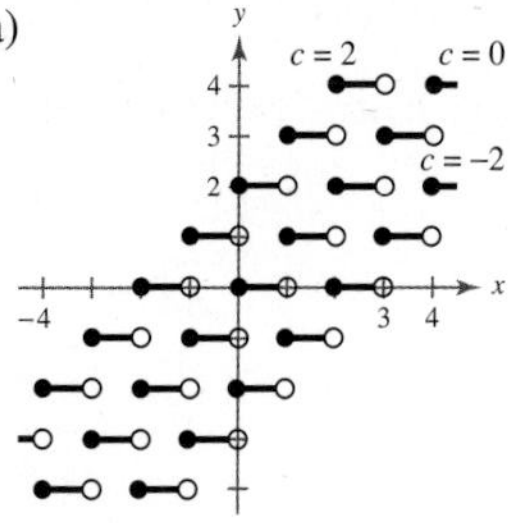

(b)

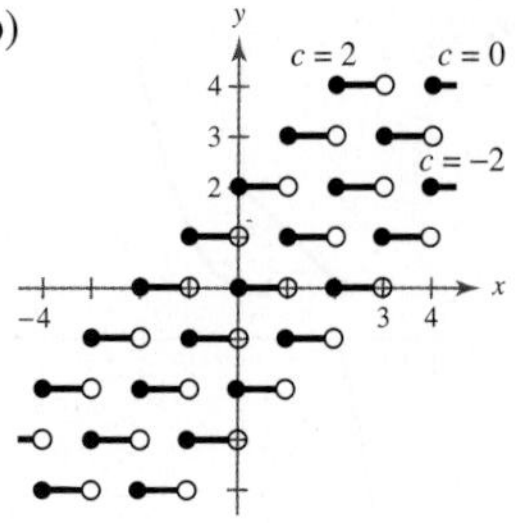

9. (a)

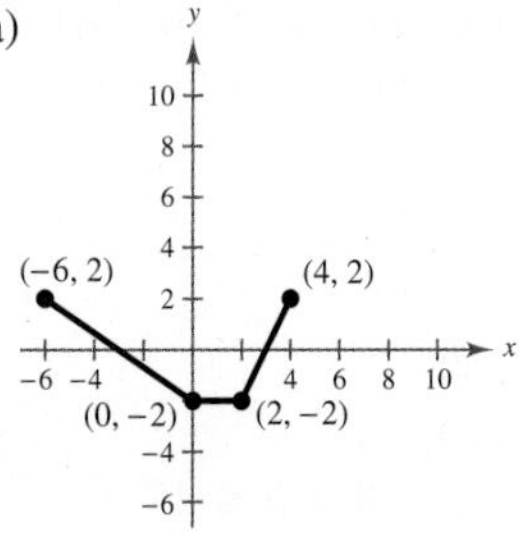

(b)

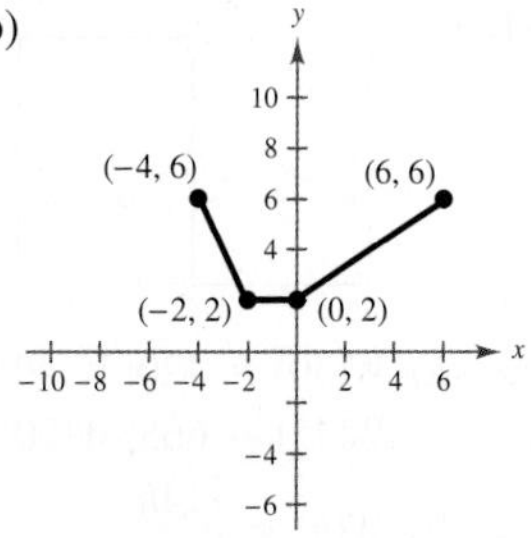

(c)

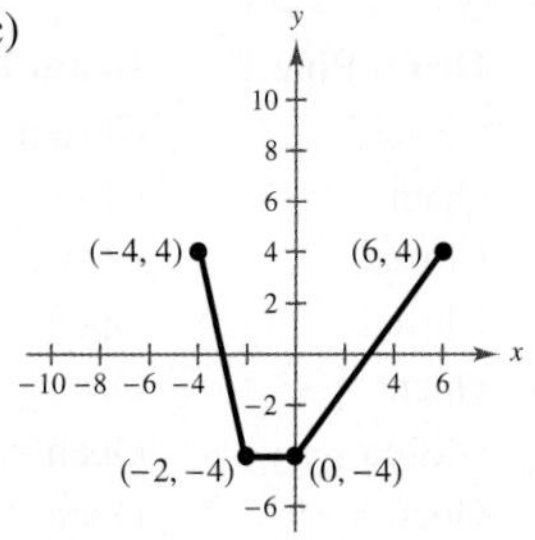

(d)

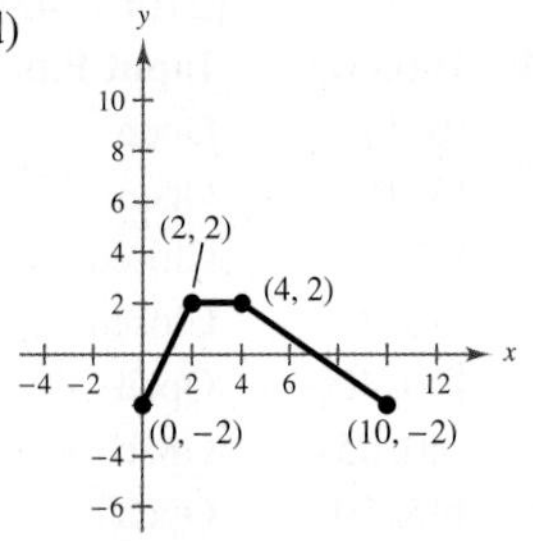

(e)

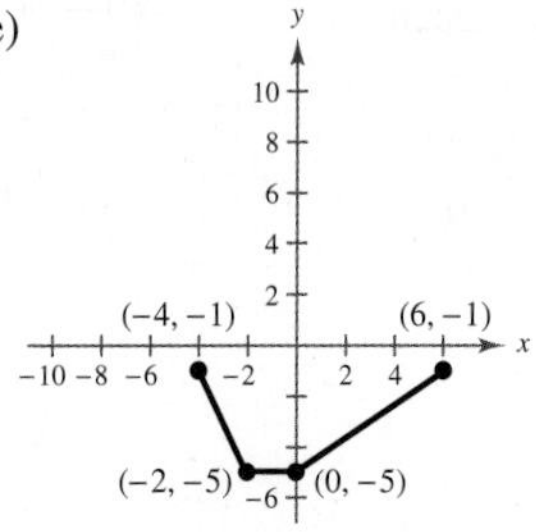

(f)

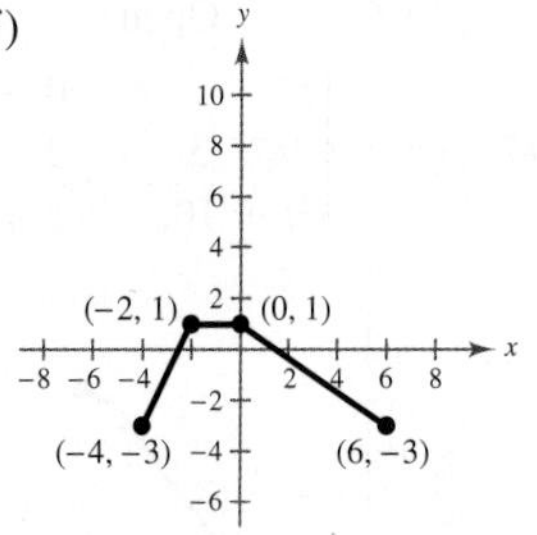

(g)

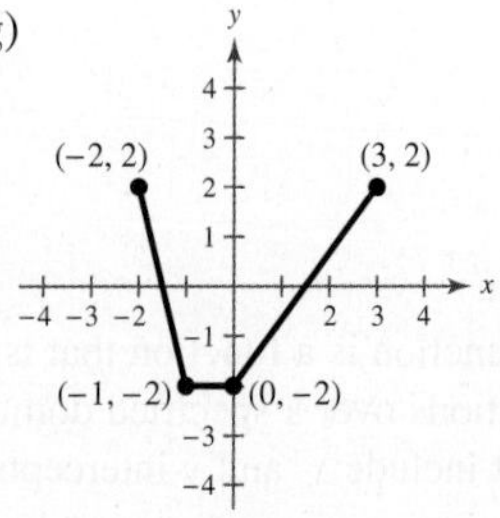

11. (a) $y = x^2 - 1$ (b) $y = 1 - (x + 1)^2$

13. (a) $y = -|x + 3|$ (b) $y = |x - 2| - 4$

15. Horizontal shift of $y = x^3$; $y = (x - 2)^3$

17. Reflection in the x-axis of $y = x^2$; $y = -x^2$

19. Reflection in the x-axis and vertical shift of $y = \sqrt{x}$; $y = 1 - \sqrt{x}$

21. (a) $f(x) = x^2$

(b) Reflection in the x-axis and vertical shift 12 units up

(c)

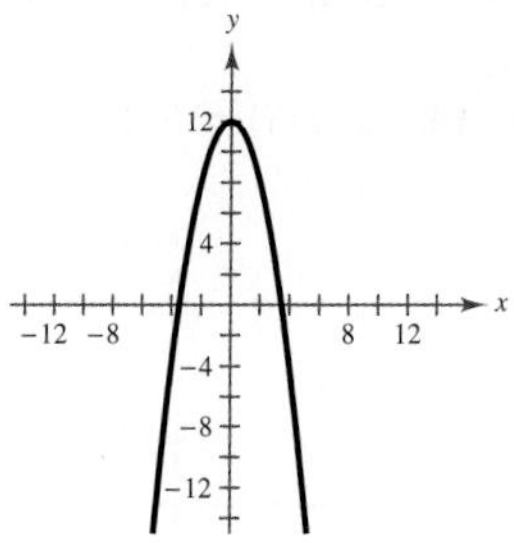

(d) $g(x) = 12 - f(x)$

23. (a) $f(x) = x^3$ (b) Vertical shift seven units up

(c)

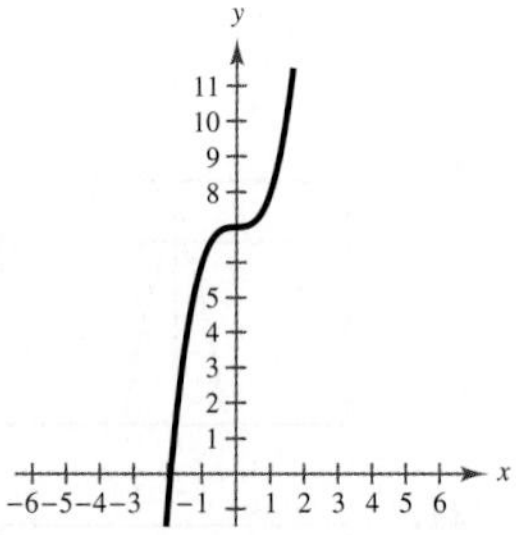

(d) $g(x) = f(x) + 7$

25. (a) $f(x) = x^2$

(b) Vertical shrink of two-thirds and vertical shift four units up

(c)

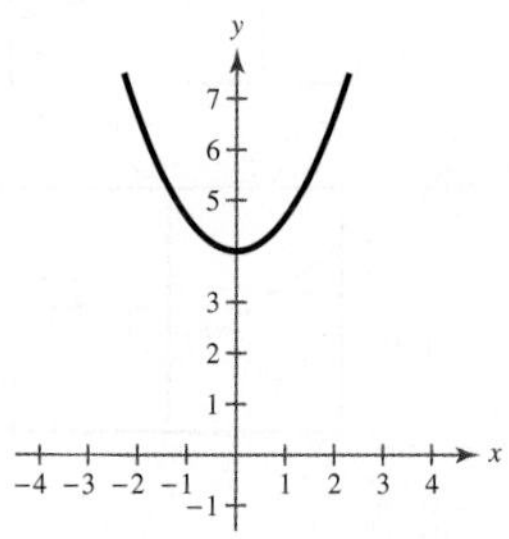

(d) $g(x) = \frac{2}{3}f(x) + 4$

27. (a) $f(x) = x^2$

(b) Reflection in the x-axis, horizontal shift five units to the left, and vertical shift two units up

(c)

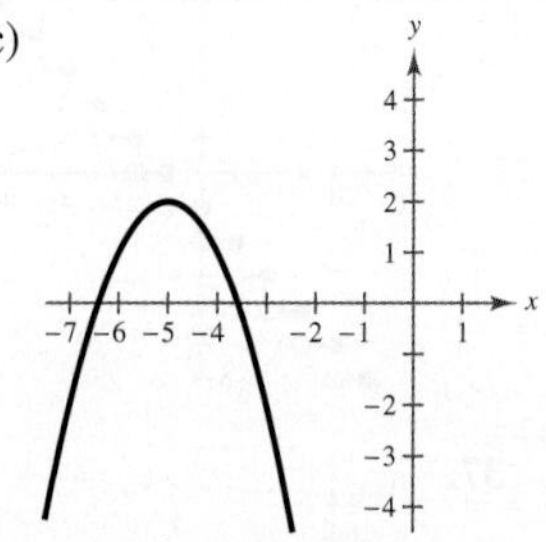

(d) $g(x) = 2 - f(x + 5)$

29. (a) $f(x) = \sqrt{x}$ (b) Horizontal shrink of one-third

(c)

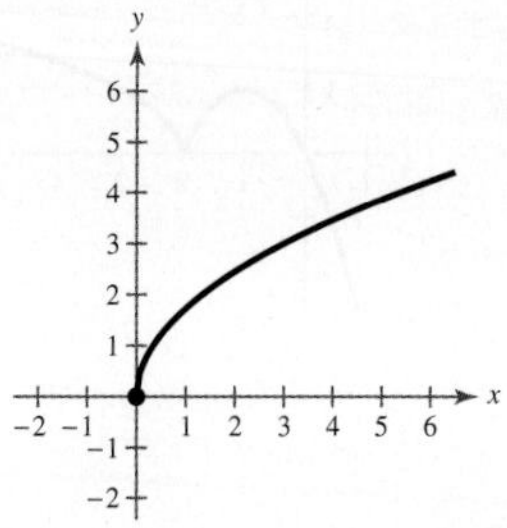

(d) $g(x) = f(3x)$

31. (a) $f(x) = x^3$

(b) Vertical shift two units up and horizontal shift one unit to the right

(c)

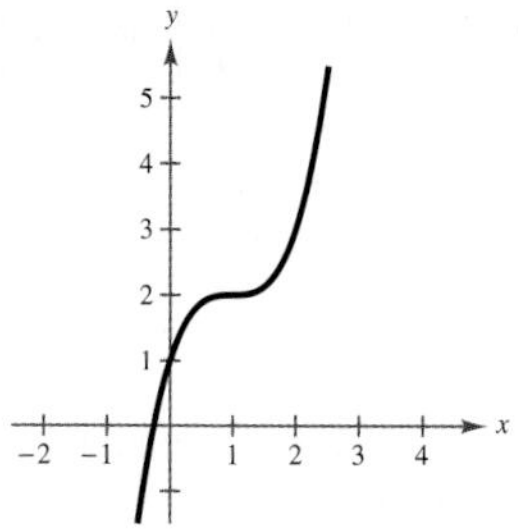

(d) $g(x) = f(x - 1) + 2$

33. (a) $f(x) = x^3$

(b) Vertical stretch of three and horizontal shift two units to the right

(c)

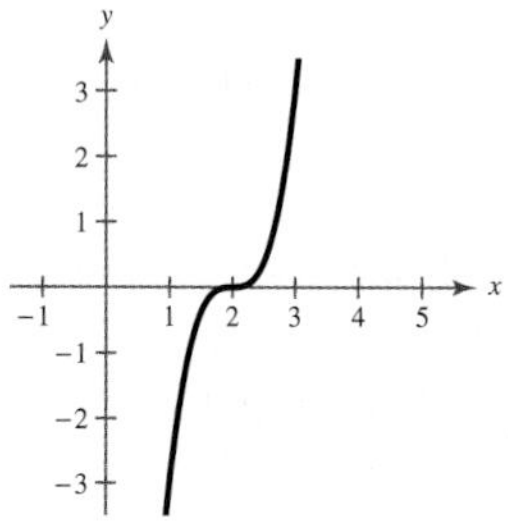

(d) $g(x) = 3f(x - 2)$

35. (a) $f(x) = |x|$

(b) Reflection in the x-axis and vertical shift two units down

(c)

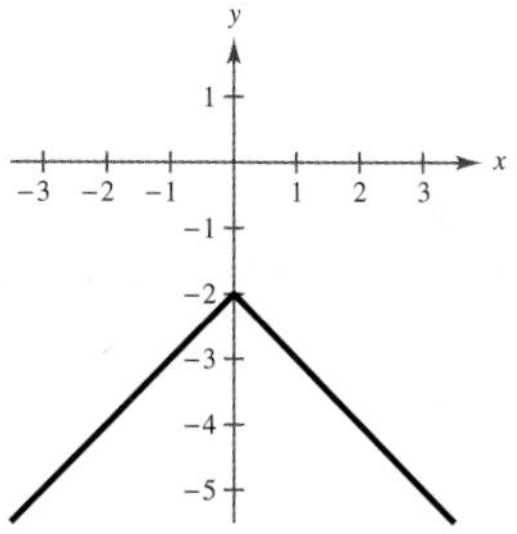

(d) $g(x) = -f(x) - 2$

37. (a) $f(x) = |x|$

(b) Reflection in the x-axis, horizontal shift four units to the left, and vertical shift eight units up

(c)

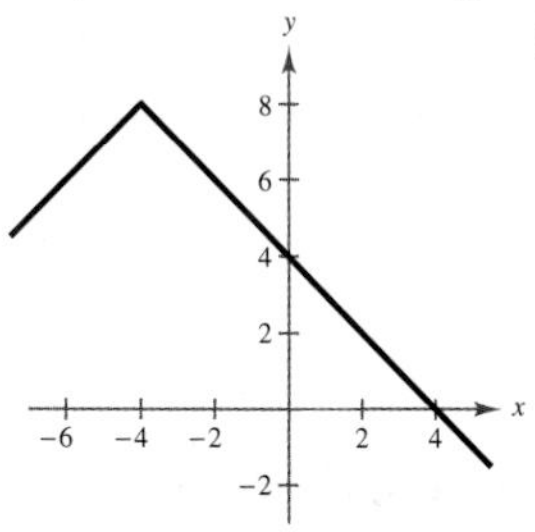

(d) $g(x) = -f(x + 4) + 8$

39. (a) $f(x) = |x|$

(b) Reflection in the x-axis, vertical stretch of two, horizontal shift one unit to the right, and vertical shift four units down

(c)

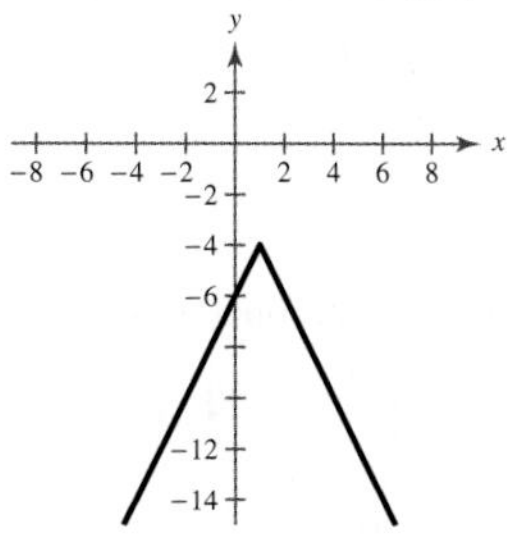

(d) $g(x) = -2f(x - 1) - 4$

41. (a) $f(x) = [\![x]\!]$

(b) Reflection in the x-axis and vertical shift three units up

(c)

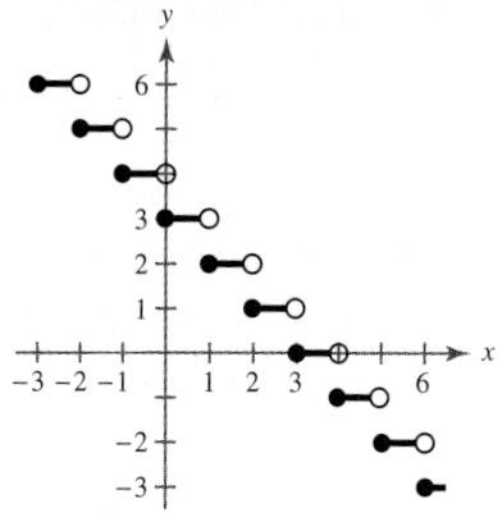

(d) $g(x) = 3 - f(x)$

43. (a) $f(x) = \sqrt{x}$

(b) Horizontal shift nine units to the right

(c)

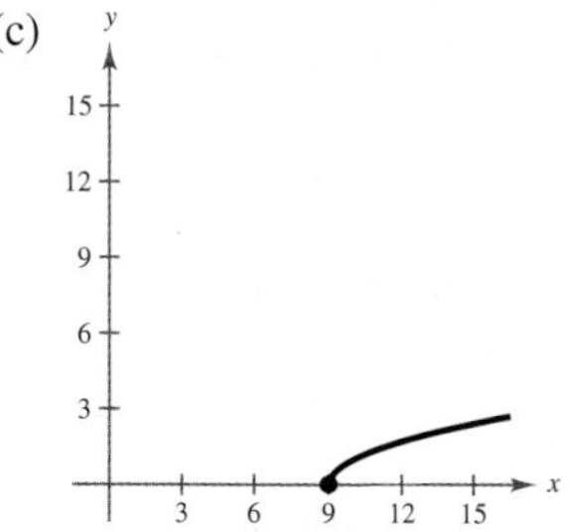

(d) $g(x) = f(x - 9)$

45. (a) $f(x) = \sqrt{x}$

(b) Reflection in the y-axis, horizontal shift seven units to the right, and vertical shift two units down

(c)

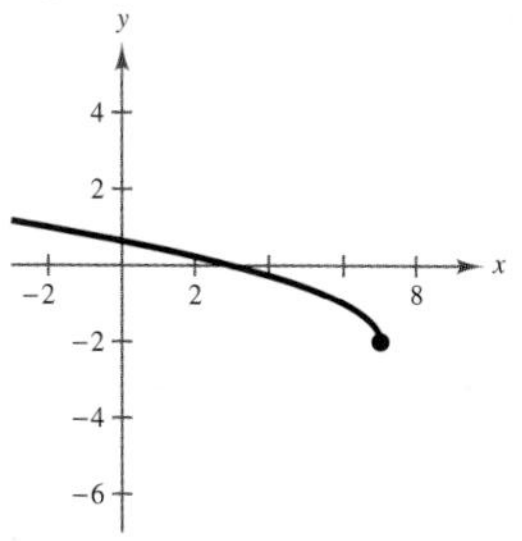

(d) $g(x) = f(7 - x) - 2$

47. $g(x) = (x - 3)^2 - 7$ **49.** $g(x) = (x - 13)^3$

51. $g(x) = -|x| + 12$ **53.** $g(x) = -\sqrt{-x + 6}$

55. (a) $y = -3x^2$ (b) $y = 4x^2 + 3$

57. (a) $y = -\frac{1}{2}|x|$ (b) $y = 3|x| - 3$

59. Vertical stretch of $y = x^3$; $y = 2x^3$

61. Reflection in the x-axis and vertical shrink of $y = x^2$; $y = -\frac{1}{2}x^2$

63. Reflection in the y-axis and vertical shrink of $y = \sqrt{x}$; $y = \frac{1}{2}\sqrt{-x}$

65. $y = -(x - 2)^3 + 2$ **67.** $y = -\sqrt{x} - 3$

69. (a)

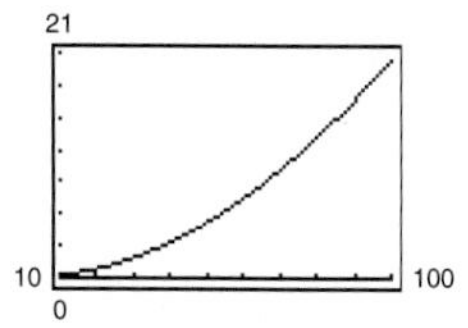

(b) $H\left(\dfrac{x}{1.6}\right) = 0.00078x^2 + 0.003x - 0.029,\ 16 \le x \le 160$;

Horizontal stretch

71. False. The graph of $y = f(-x)$ is a reflection of the graph of $f(x)$ in the y-axis.

73. True. $|-x| = |x|$ **75.** $(-2, 0), (-1, 1), (0, 2)$

77. (a)

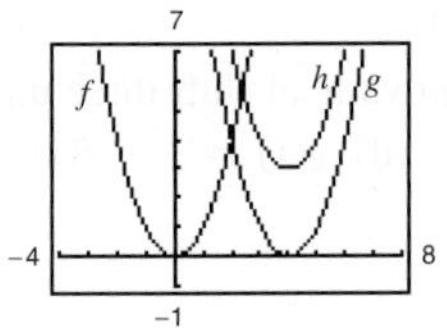

g is a right shift of four units. h is a right shift of four units and a shift of three units up.

(b)

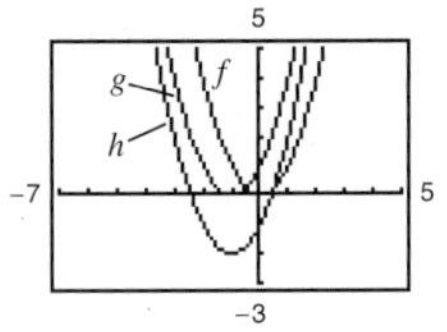

g is a left shift of one unit. h is a left shift of one unit and a shift of two units down.

(c)

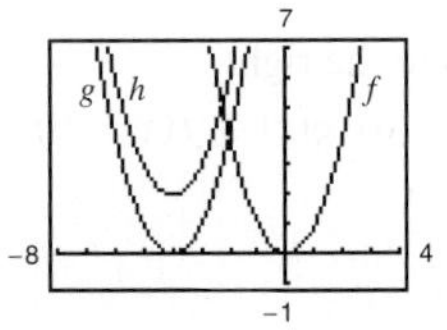

g is a left shift of four units. h is a left shift of four units and a shift of two units up.

79. (a) $g(t) = \frac{3}{4}f(t)$ (b) $g(t) = f(t) + 10{,}000$
(c) $g(t) = f(t - 2)$

Section 1.8 *(page 81)*

1. addition; subtraction; multiplication; division

3.

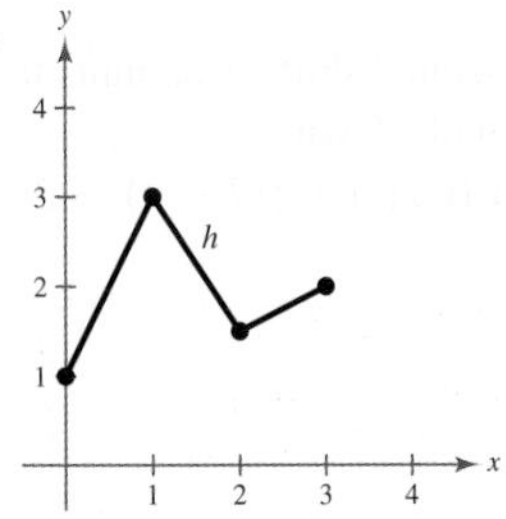

5. (a) $2x$ (b) 4 (c) $x^2 - 4$
(d) $\dfrac{x + 2}{x - 2}$; all real numbers x except $x = 2$

7. (a) $x^2 + 4x - 5$ (b) $x^2 - 4x + 5$ (c) $4x^3 - 5x^2$
(d) $\dfrac{x^2}{4x - 5}$; all real numbers x except $x = \dfrac{5}{4}$

9. (a) $x^2 + 6 + \sqrt{1 - x}$ (b) $x^2 + 6 - \sqrt{1 - x}$
(c) $(x^2 + 6)\sqrt{1 - x}$
(d) $\dfrac{(x^2 + 6)\sqrt{1 - x}}{1 - x}$; all real numbers x such that $x < 1$

11. (a) $\dfrac{x + 1}{x^2}$ (b) $\dfrac{x - 1}{x^2}$ (c) $\dfrac{1}{x^3}$
(d) x; all real numbers x except $x = 0$

13. 3 **15.** 5 **17.** $9t^2 - 3t + 5$

19. 74 **21.** 26 **23.** $\frac{3}{5}$

25.

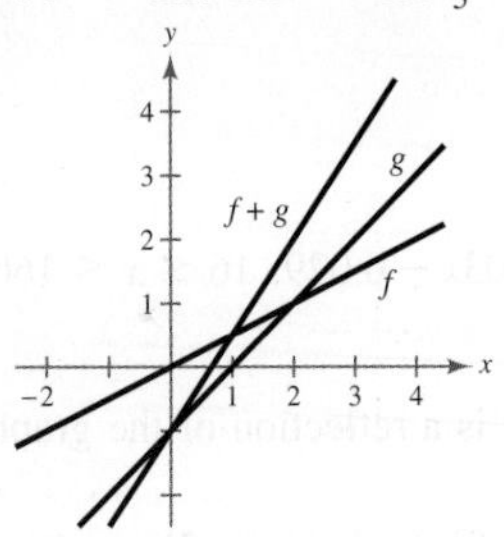

27.

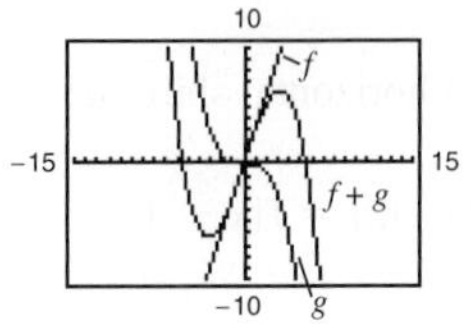

$f(x), g(x)$

29.

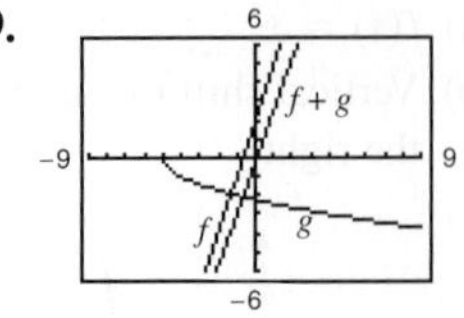

$f(x), f(x)$

31. (a) $(x - 1)^2$ (b) $x^2 - 1$ (c) $x - 2$

33. (a) x (b) x (c) $x^9 + 3x^6 + 3x^3 + 2$

35. (a) $\sqrt{x^2 + 4}$ (b) $x + 4$
Domains of f and $g \circ f$: all real numbers x such that $x \geq -4$
Domains of g and $f \circ g$: all real numbers x

37. (a) $x + 1$ (b) $\sqrt{x^2 + 1}$
Domains of f and $g \circ f$: all real numbers x
Domains of g and $f \circ g$: all real numbers x such that $x \geq 0$

39. (a) $|x + 6|$ (b) $|x| + 6$
Domains of $f, g, f \circ g$, and $g \circ f$: all real numbers x

41. (a) $\dfrac{1}{x + 3}$ (b) $\dfrac{1}{x} + 3$
Domains of f and $g \circ f$: all real numbers x except $x = 0$
Domain of g: all real numbers x
Domain of $f \circ g$: all real numbers x except $x = -3$

43. (a) 3 (b) 0 **45.** (a) 0 (b) 4

47. $f(x) = x^2, g(x) = 2x + 1$ **49.** $f(x) = \sqrt[3]{x}, g(x) = x^2 - 4$

51. $f(x) = \dfrac{1}{x}, g(x) = x + 2$ **53.** $f(x) = \dfrac{x + 3}{4 + x}, g(x) = -x^2$

55. (a) $T = \frac{3}{4}x + \frac{1}{15}x^2$
(b)

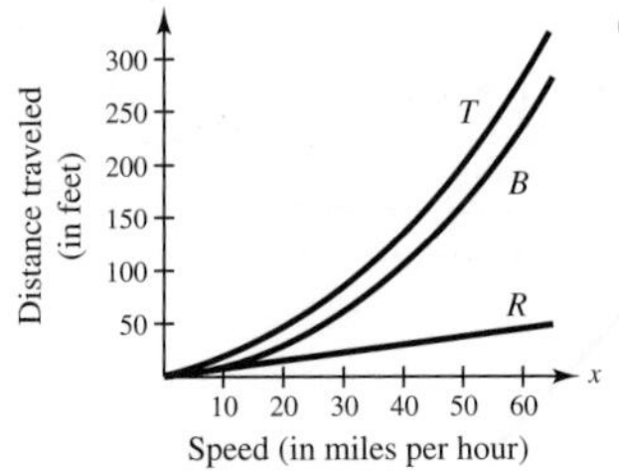

(c) The braking function $B(x)$. As x increases, $B(x)$ increases at a faster rate than $R(x)$.

57. (a) $p(t) = d(t) + c(t)$
(b) $p(13)$ is the number of dogs and cats in the year 2013.
(c) $h(t) = \dfrac{d(t) + c(t)}{n(t)}$;
The function $h(t)$ represents the number of dogs and cats per capita.

59. (a) $r(x) = \dfrac{x}{2}$ (b) $A(r) = \pi r^2$
(c) $(A \circ r)(x) = \pi\left(\dfrac{x}{2}\right)^2$;
$(A \circ r)(x)$ represents the area of the circular base of the tank on the square foundation with side length x.

61. $g(f(x))$ represents 3 percent of an amount over \$500,000.

63. (a) $O(M(Y)) = 2\left(6 + \frac{1}{2}Y\right) = 12 + Y$
(b) Middle child is 8 years old; youngest child is 4 years old.

65. False. $(f \circ g)(x) = 6x + 1$ and $(g \circ f)(x) = 6x + 6$

67–69. Proofs

Section 1.9 *(page 90)*

1. inverse **3.** range; domain **5.** one-to-one

7. $f^{-1}(x) = \dfrac{1}{6}x$ **9.** $f^{-1}(x) = \dfrac{x - 1}{3}$ **11.** $f^{-1}(x) = x^3$

13. $f(g(x)) = f\left(-\frac{2x+6}{7}\right) = -\frac{7}{2}\left(-\frac{2x+6}{7}\right) - 3 = x$

$g(f(x)) = g\left(-\frac{7}{2}x - 3\right) = -\frac{2\left(-\frac{7}{2}x - 3\right) + 6}{7} = x$

15. $f(g(x)) = f\left(\sqrt[3]{x-5}\right) = \left(\sqrt[3]{x-5}\right)^3 + 5 = x$

$g(f(x)) = g(x^3 + 5) = \sqrt[3]{(x^3+5) - 5} = x$

17.

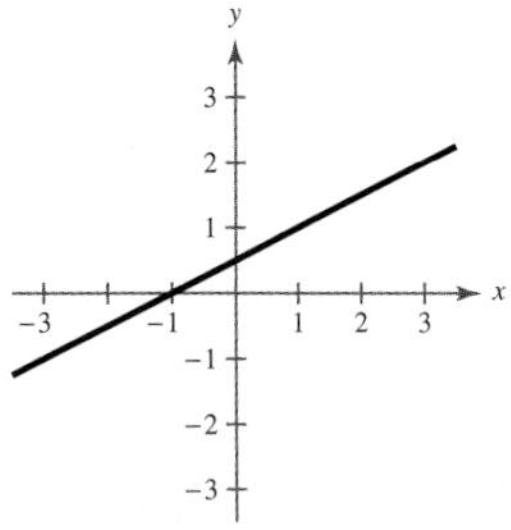

19.

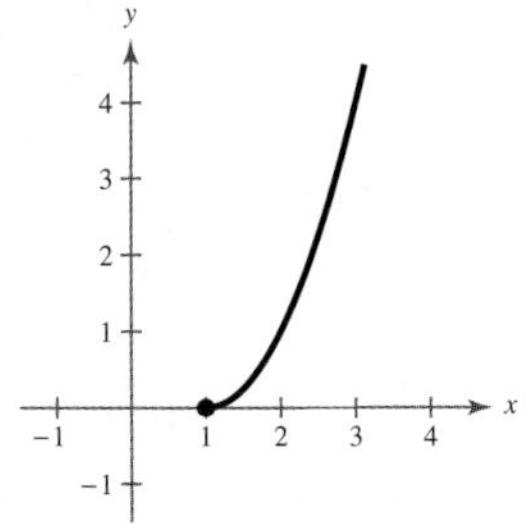

21. (a) $f(g(x)) = f\left(\frac{x}{2}\right) = 2\left(\frac{x}{2}\right) = x$

$g(f(x)) = g(2x) = \frac{(2x)}{2} = x$

(b)

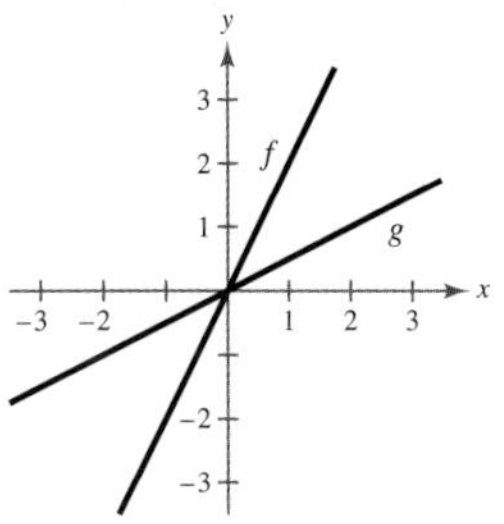

23. (a) $f(g(x)) = f\left(\frac{x-1}{7}\right) = 7\left(\frac{x-1}{7}\right) + 1 = x$

$g(f(x)) = g(7x + 1) = \frac{(7x+1) - 1}{7} = x$

(b)

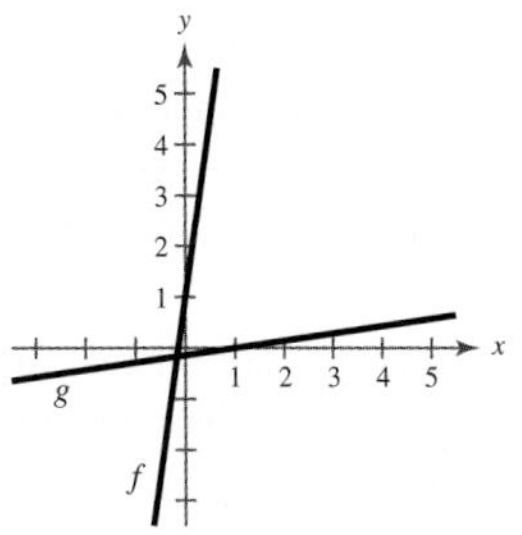

25. (a) $f(g(x)) = f\left(\sqrt[3]{8x}\right) = \frac{\left(\sqrt[3]{8x}\right)^3}{8} = x$

$g(f(x)) = g\left(\frac{x^3}{8}\right) = \sqrt[3]{8\left(\frac{x^3}{8}\right)} = x$

(b)

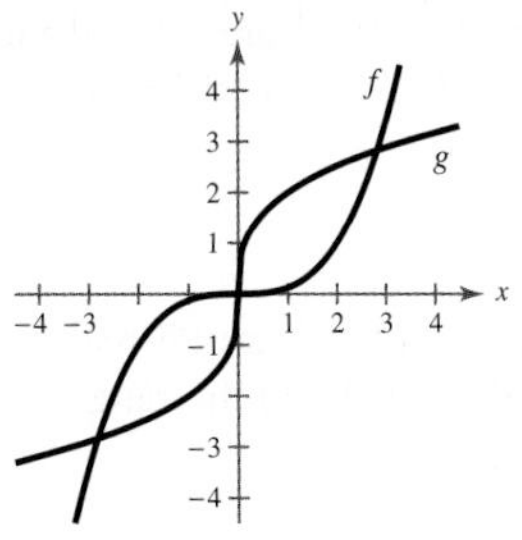

27. (a) $f(g(x)) = f(x^2 + 4),\ x \geq 0$

$= \sqrt{(x^2+4) - 4} = x$

$g(f(x)) = g\left(\sqrt{x-4}\right)$

$= \left(\sqrt{x-4}\right)^2 + 4 = x$

(b)

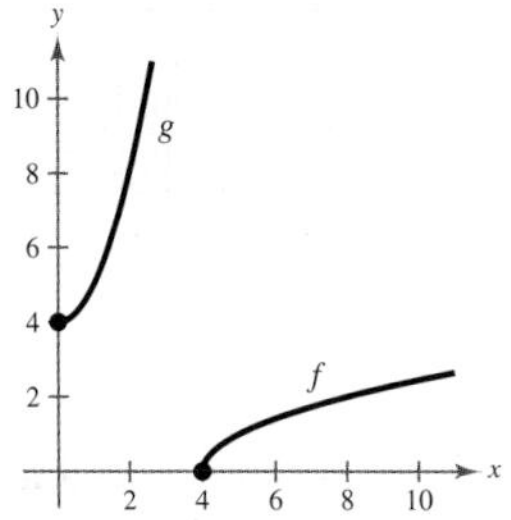

29. (a) $f(g(x)) = f\left(\sqrt{9-x}\right),\ x \leq 9$

$= 9 - \left(\sqrt{9-x}\right)^2 = x$

$g(f(x)) = g(9 - x^2),\ x \geq 0$

$= \sqrt{9 - (9 - x^2)} = x$

(b)

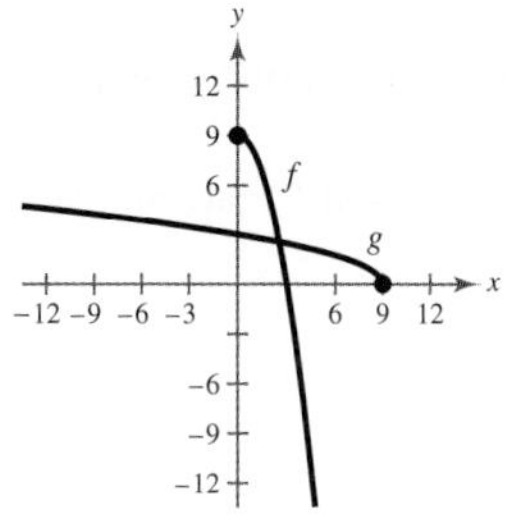

31. (a) $f(g(x)) = f\left(-\frac{5x+1}{x-1}\right) = \frac{-\left(\frac{5x+1}{x-1}\right) - 1}{-\left(\frac{5x+1}{x-1}\right) + 5}$

$= \frac{-5x - 1 - x + 1}{-5x - 1 + 5x - 5} = x$

$g(f(x)) = g\left(\frac{x-1}{x+5}\right) = \frac{-5\left(\frac{x-1}{x+5}\right) - 1}{\frac{x-1}{x+5} - 1}$

$= \frac{-5x + 5 - x - 5}{x - 1 - x - 5} = x$

(b)

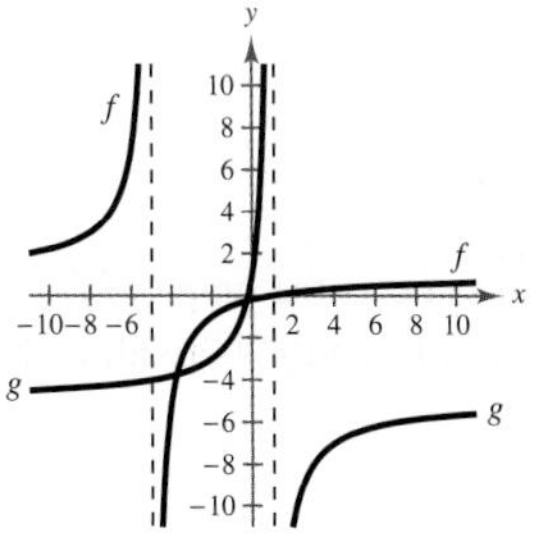

33. No

35.

x	-2	0	2	4	6	8
$f^{-1}(x)$	-2	-1	0	1	2	3

37. Yes **39.** No

41. 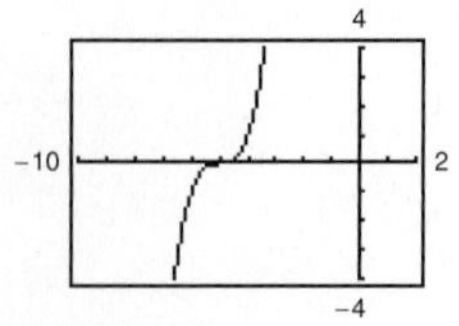

The function has an inverse function.

43. 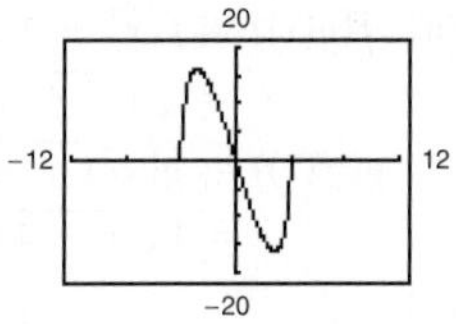

The function does not have an inverse function.

45. (a) $f^{-1}(x) = \dfrac{x+3}{2}$

(b) 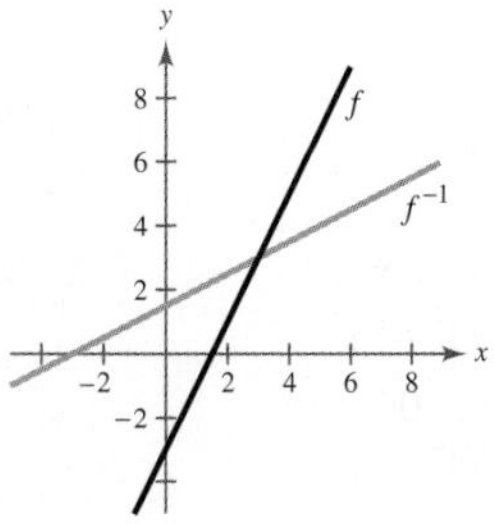

(c) The graph of f^{-1} is the reflection of the graph of f in the line $y = x$.

(d) The domains and ranges of f and f^{-1} are all real numbers.

47. (a) $f^{-1}(x) = \sqrt[5]{x+2}$

(b) 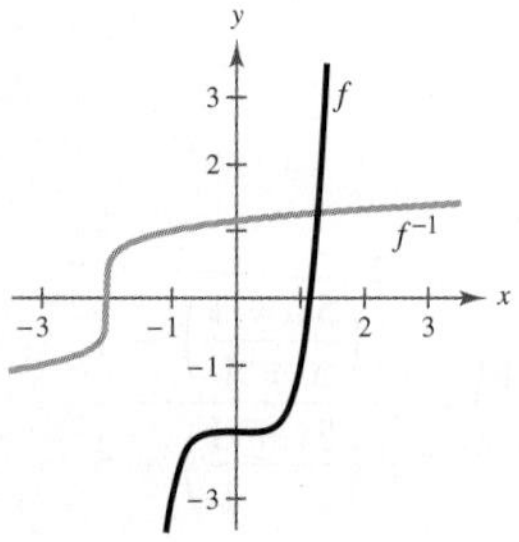

(c) The graph of f^{-1} is the reflection of the graph of f in the line $y = x$.

(d) The domains and ranges of f and f^{-1} are all real numbers.

49. (a) $f^{-1}(x) = \sqrt{4 - x^2},\ 0 \le x \le 2$

(b) 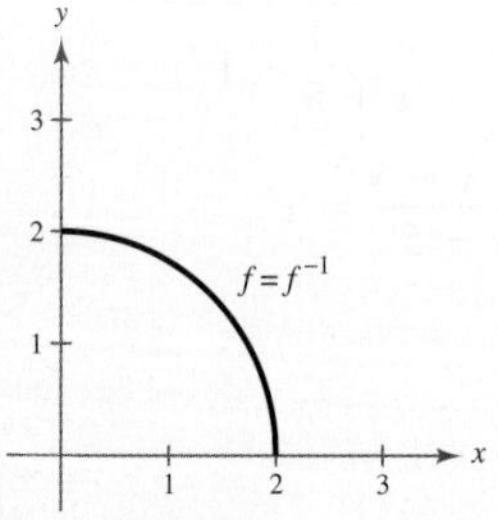

(c) The graph of f^{-1} is the same as the graph of f.

(d) The domains and ranges of f and f^{-1} are all real numbers x such that $0 \le x \le 2$.

51. (a) $f^{-1}(x) = \dfrac{4}{x}$

(b) 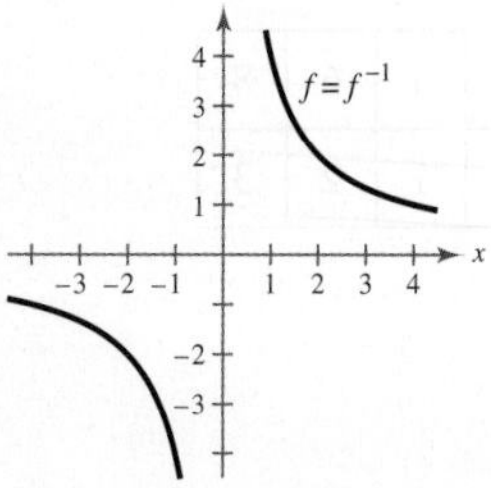

(c) The graph of f^{-1} is the same as the graph of f.

(d) The domains and ranges of f and f^{-1} are all real numbers x except $x = 0$.

53. (a) $f^{-1}(x) = \dfrac{2x+1}{x-1}$

(b) 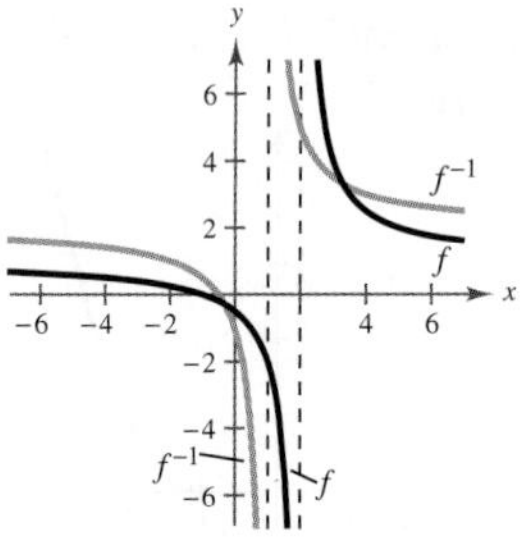

(c) The graph of f^{-1} is the reflection of the graph of f in the line $y = x$.

(d) The domain of f and the range of f^{-1} are all real numbers x except $x = 2$. The domain of f^{-1} and the range of f are all real numbers x except $x = 1$.

55. (a) $f^{-1}(x) = x^3 + 1$

(b) 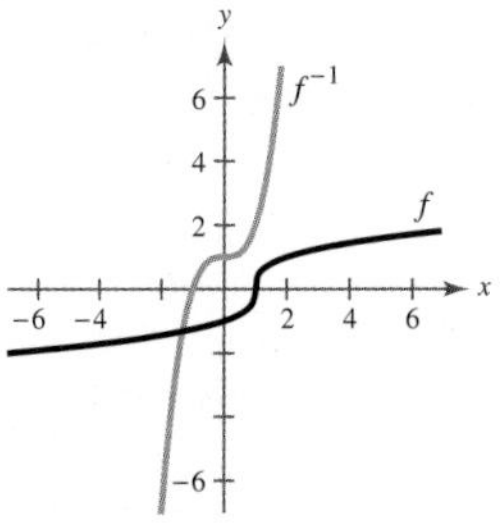

(c) The graph of f^{-1} is the reflection of the graph of f in the line $y = x$.

(d) The domains and ranges of f and f^{-1} are all real numbers.

57. No inverse function **59.** $g^{-1}(x) = 8x$

61. No inverse function **63.** $f^{-1}(x) = \sqrt{x} - 3$

65. No inverse function **67.** No inverse function

69. $f^{-1}(x) = \dfrac{x^2 - 3}{2},\ x \ge 0$ **71.** $f^{-1}(x) = \dfrac{5x - 4}{6 - 4x}$

73. $f^{-1}(x) = \sqrt{x} + 2$

The domain of f and the range of f^{-1} are all real numbers x such that $x \ge 2$. The domain of f^{-1} and the range of f are all real numbers x such that $x \ge 0$.

75. $f^{-1}(x) = x - 2$

The domain of f and the range of f^{-1} are all real numbers x such that $x \ge -2$. The domain of f^{-1} and the range of f are all real numbers x such that $x \ge 0$.

77. $f^{-1}(x) = \sqrt{x} - 6$

The domain of f and the range of f^{-1} are all real numbers x such that $x \ge -6$. The domain of f^{-1} and the range of f are all real numbers x such that $x \ge 0$.

79. $f^{-1}(x) = \dfrac{\sqrt{-2(x-5)}}{2}$

The domain of f and the range of f^{-1} are all real numbers x such that $x \ge 0$. The domain of f^{-1} and the range of f are all real numbers x such that $x \le 5$.

81. $f^{-1}(x) = x + 3$

The domain of f and the range of f^{-1} are all real numbers x such that $x \ge 4$. The domain of f^{-1} and the range of f are all real numbers x such that $x \ge 1$.

83. 32 **85.** 600 **87.** $2\sqrt[3]{x+3}$

89. $\dfrac{x+1}{2}$ **91.** $\dfrac{x+1}{2}$

93. (a) $y = \dfrac{x-10}{0.75}$

$x =$ hourly wage; $y =$ number of units produced

(b) 19 units

95. False. $f(x) = x^2$ has no inverse function.

97.

x	1	3	4	6
y	1	2	6	7

x	1	2	6	7
$f^{-1}(x)$	1	3	4	6

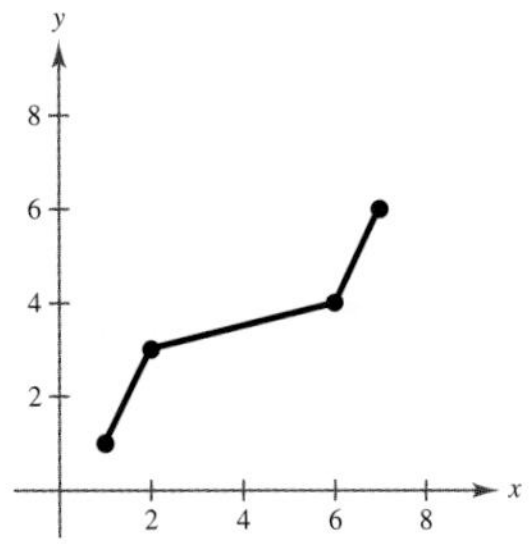

99. Proof **101.** $k = \frac{1}{4}$

103.

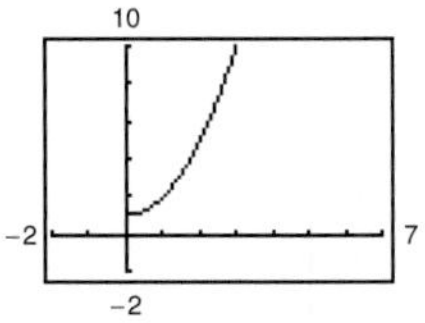

There is an inverse function $f^{-1}(x) = \sqrt{x-1}$ because the domain of f is equal to the range of f^{-1} and the range of f is equal to the domain of f^{-1}.

105. This situation could be represented by a one-to-one function if the runner does not stop to rest. The inverse function would represent the time in hours for a given number of miles completed.

Section 1.10 *(page 100)*

1. variation; regression **3.** least squares regression

5. directly proportional

7. directly proportional **9.** combined

11.

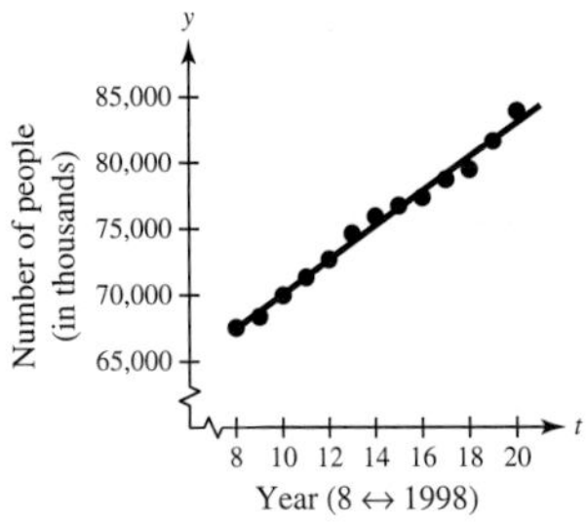

The model fits the data well.

13.

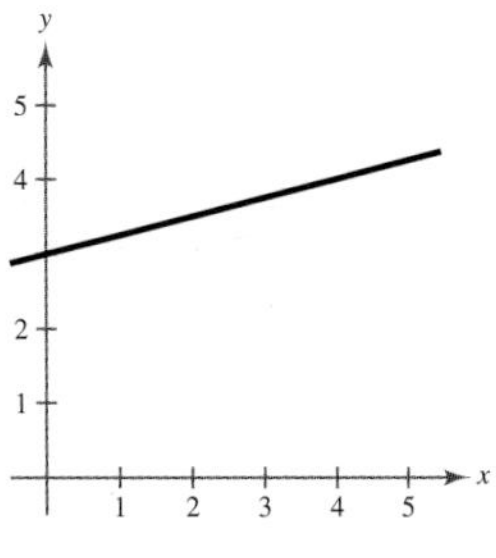

$y = \frac{1}{4}x + 3$

15.

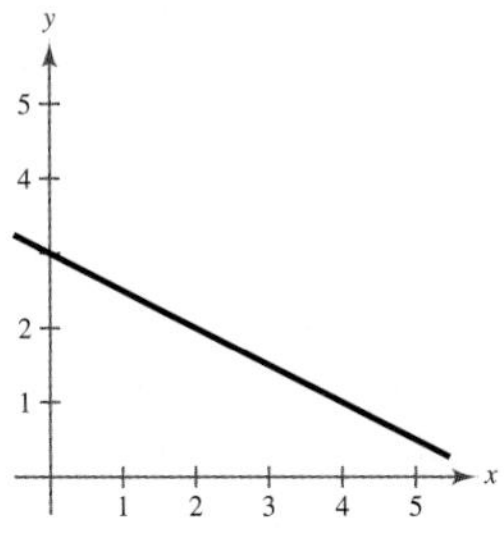

$y = -\frac{1}{2}x + 3$

17. (a) and (b)

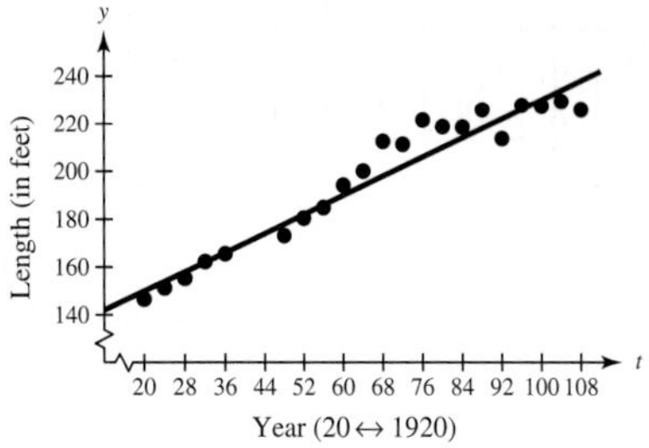

$y \approx t + 130$

(c) $y = 1.01t + 130.82$ (d) The models are similar.

19. $y = 7x$ **21.** $y = 205x$ **23.** $y = \frac{1}{5}x$ **25.** $y = 2\pi x$

27.

x	2	4	6	8	10
$y = x^2$	4	16	36	84	100

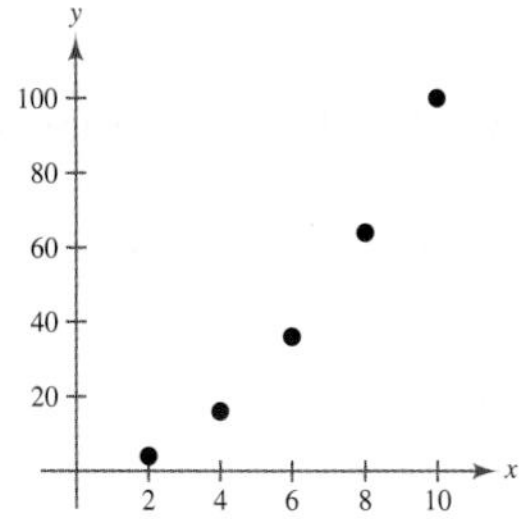

29.

x	2	4	6	8	10
$y = \frac{1}{2}x^3$	4	32	108	256	500

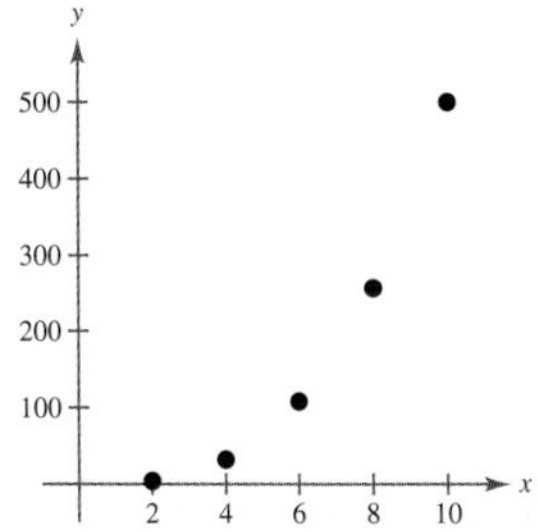

31.

x	2	4	6	8	10
$y = \dfrac{2}{x}$	1	$\frac{1}{2}$	$\frac{1}{3}$	$\frac{1}{4}$	$\frac{1}{5}$

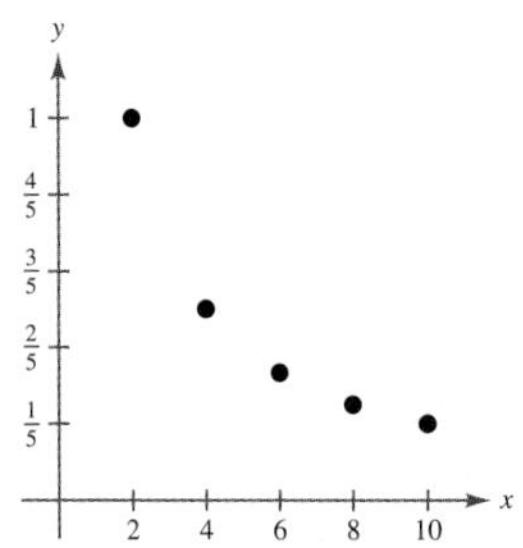

33.

x	2	4	6	8	10
$y = \dfrac{10}{x^2}$	$\frac{5}{2}$	$\frac{5}{8}$	$\frac{5}{18}$	$\frac{5}{32}$	$\frac{1}{10}$

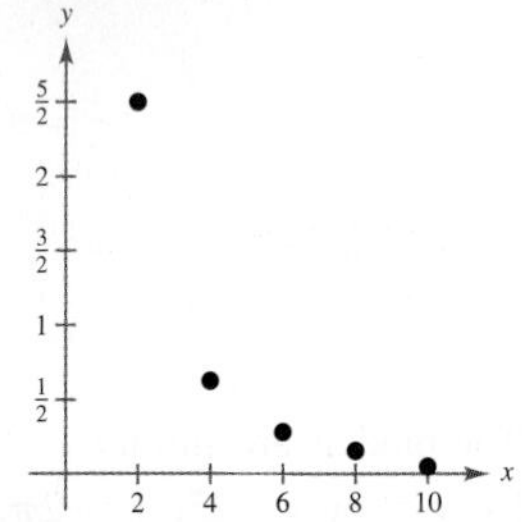

35. Inversely **37.** $y = \dfrac{5}{x}$ **39.** $y = -\dfrac{7}{10}x$ **41.** $A = kr^2$

43. $y = \dfrac{k}{x^2}$ **45.** $F = \dfrac{kg}{r^2}$ **47.** $R = k(T - T_e)$

49. $R = kS(L - S)$

51. The surface area of a sphere varies directly as the square of its radius.

53. The area of a triangle is jointly proportional to its base and height.

55. $A = \pi r^2$ **57.** $y = \dfrac{28}{x}$ **59.** $F = 14rs^3$

61. $z = \dfrac{2x^2}{3y}$ **63.** $I = 0.035P$

65. Model: $y = \frac{33}{13}x$; 25.4 cm, 50.8 cm **67.** $293\frac{1}{3}$ N

69. 39.47 lb **71.** About 0.61 mi/h **73.** 1470 J

75. (a)

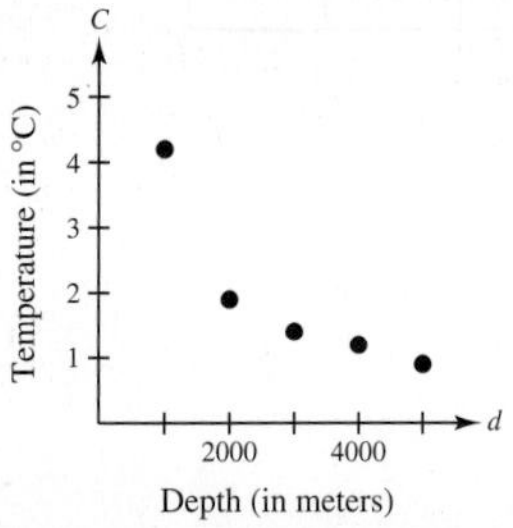

(b) Yes. $k_1 = 4200, k_2 = 3800, k_3 = 4200,$ $k_4 = 4800, k_5 = 4500$

(c) $C = \dfrac{4300}{d}$

(d)

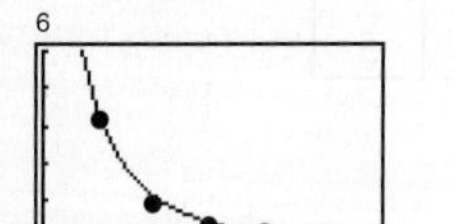

(e) About 1433 m

77. False. π is a constant, not a variable.

79. (a) y will change by a factor of four.

(b) y will change by a factor of one-fourth.

Review Exercises *(page 106)*

1.

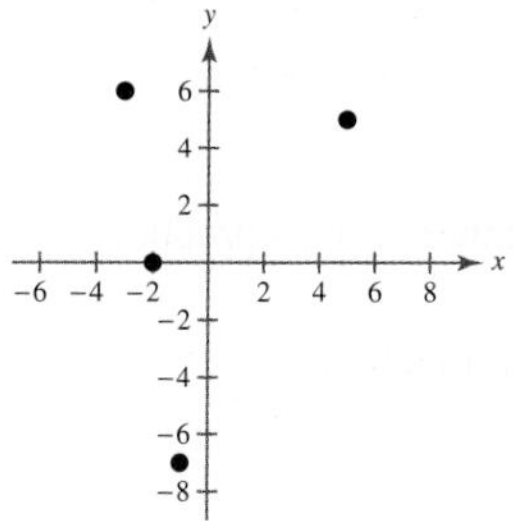

3. Quadrant IV

5.

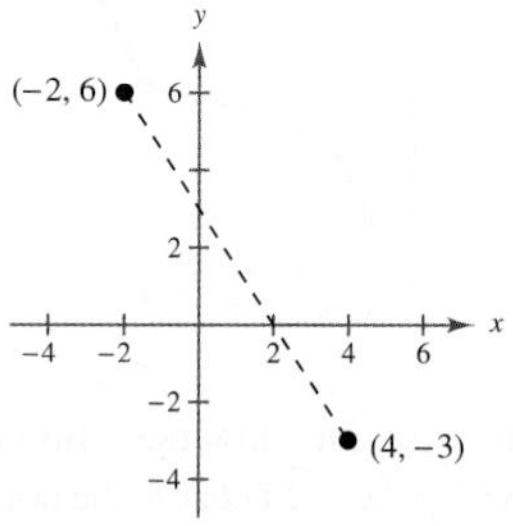

Distance: $3\sqrt{13}$
Midpoint: $\left(1, \frac{3}{2}\right)$

7.

x	−2	−1	0	1	2
y	−11	−8	−5	−2	1

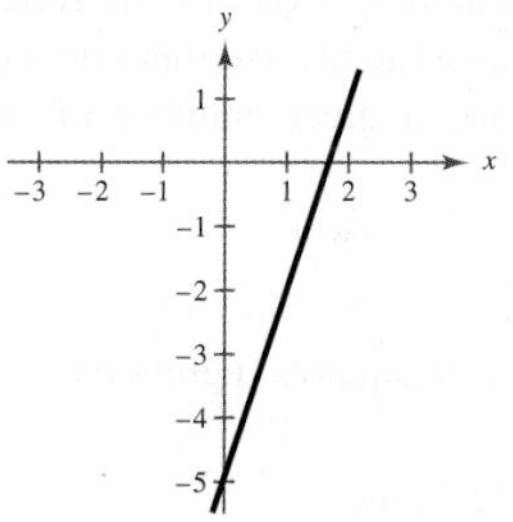

9.

x	−1	0	1	2	3	4
y	4	0	−2	−2	0	4

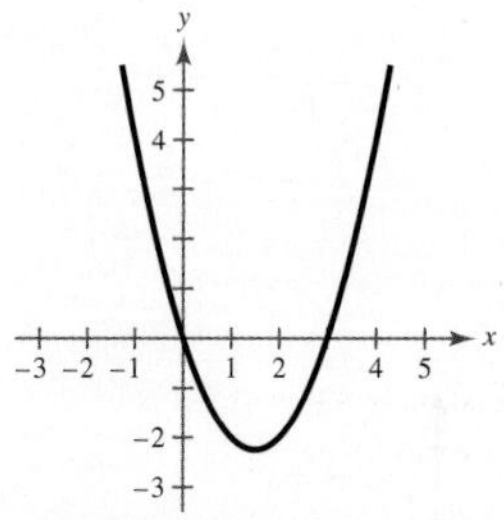

11. x-intercept: $\left(-\frac{7}{2}, 0\right)$
y-intercept: $(0, 7)$

13. x-intercepts: $(1, 0), (5, 0)$
y-intercept: $(0, 5)$

15. x-intercept: $\left(\frac{1}{4}, 0\right)$
y-intercept: $(0, 1)$
No symmetry

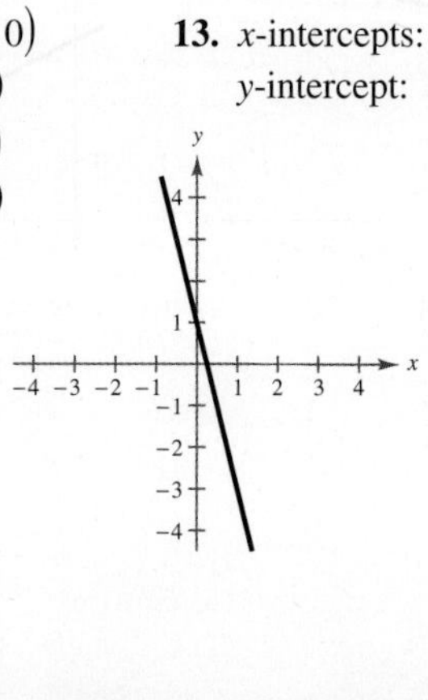

17. x-intercepts: $(\pm\sqrt{5}, 0)$
y-intercept: $(0, 5)$
y-axis symmetry

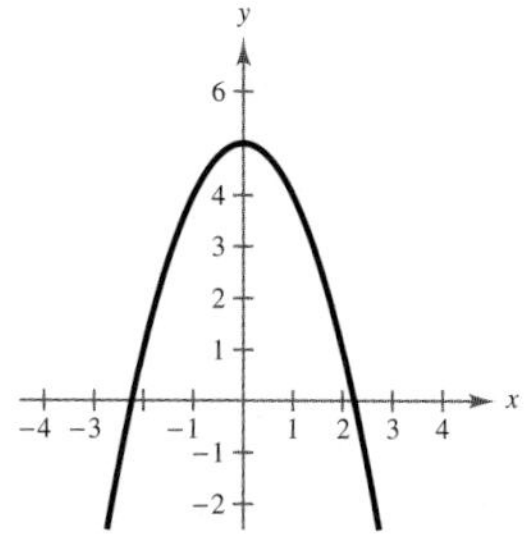

19. x-intercept: $(\sqrt[3]{-3}, 0)$
y-intercept: $(0, 3)$
No symmetry

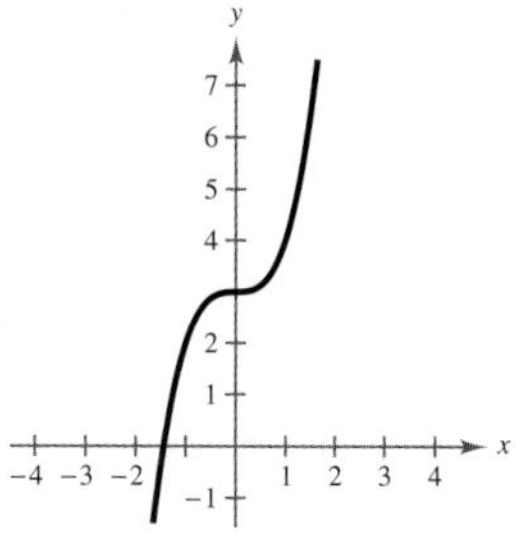

21. x-intercept: $(-5, 0)$
y-intercept: $(0, \sqrt{5})$
No symmetry

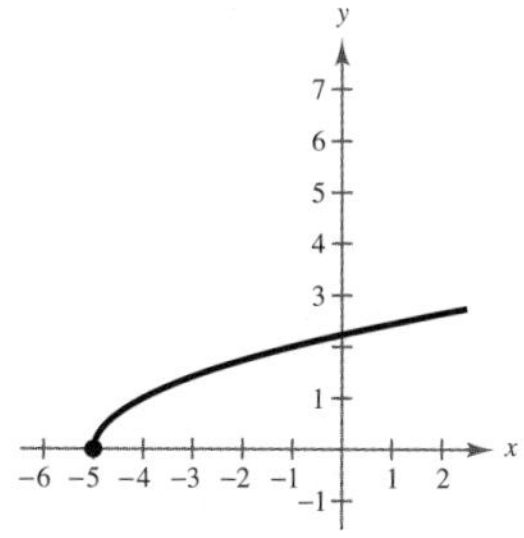

23. Center: $(0, 0)$
Radius: 3

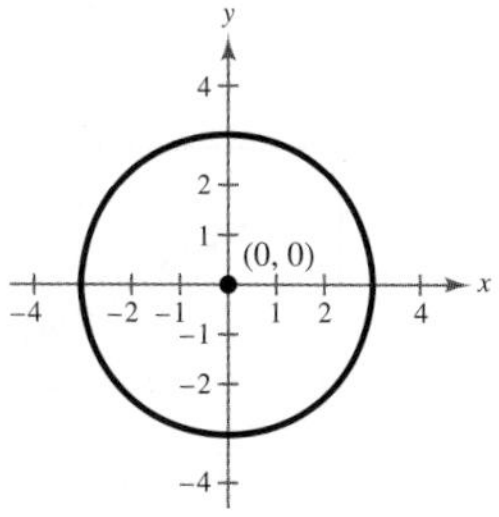

25. Center: $(-2, 0)$
Radius: 4

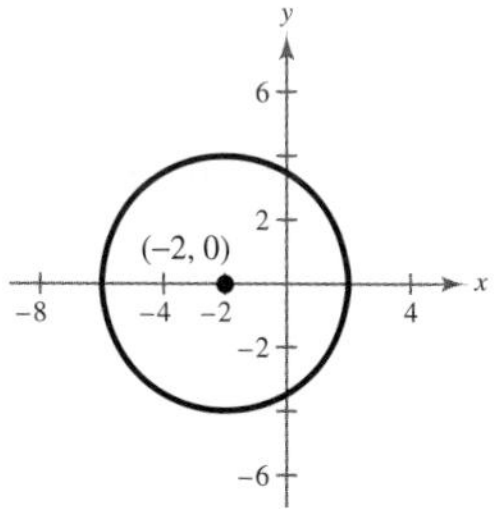

27. $(x - 2)^2 + (y + 3)^2 = 13$

29. slope: 3
y-intercept: 13

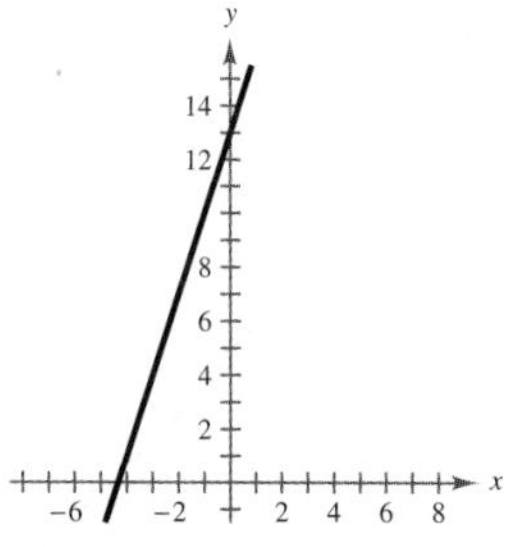

31. slope: 0
y-intercept: 6

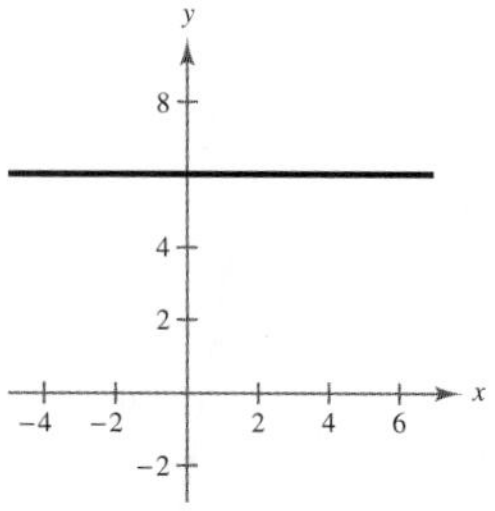

33.

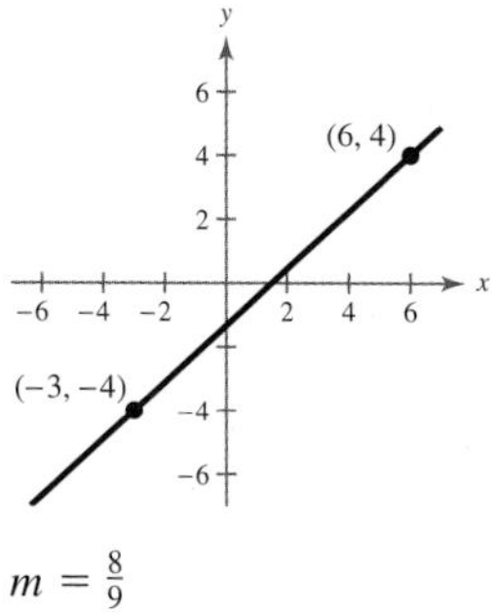

$m = \frac{8}{9}$

35. $y = -\frac{1}{2}x + 2$

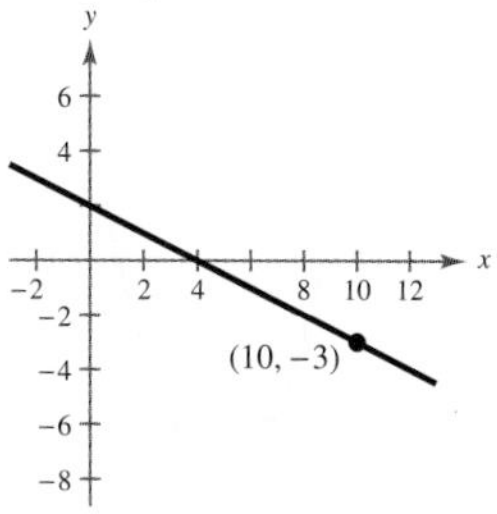

37. $y = \frac{2}{7}x + \frac{2}{7}$ **39.** (a) $y = \frac{5}{4}x - \frac{23}{4}$ (b) $y = -\frac{4}{5}x + \frac{2}{5}$
41. $S = 0.80L$ **43.** No **45.** Yes
47. (a) 5 (b) 17 (c) $t^4 + 1$ (d) $t^2 + 2t + 2$
49. All real numbers x such that $-5 \le x \le 5$

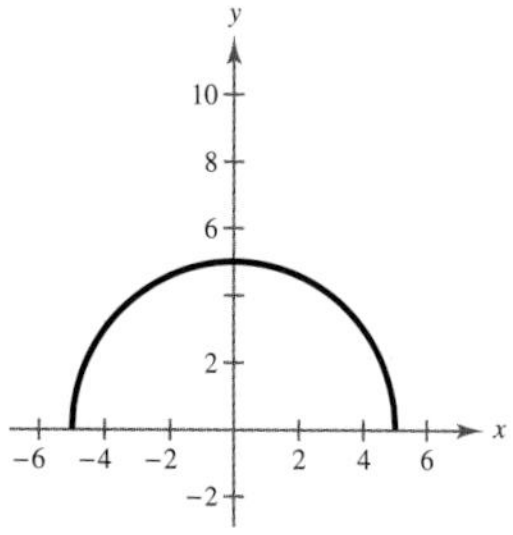

51. 16 ft/sec **53.** $4x + 2h + 3, \quad h \neq 0$
55. Function **57.** $\frac{7}{3}, 3$ **59.** $-\frac{3}{8}$
61.

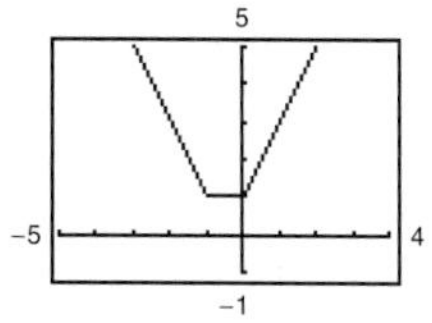

Increasing on $(0, \infty)$
Decreasing on $(-\infty, -1)$
Constant on $(-1, 0)$

63.

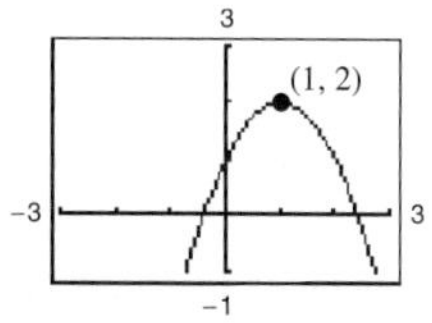

65. 4 **67.** Even; y-axis symmetry

69. (a) $f(x) = -3x$
(b)

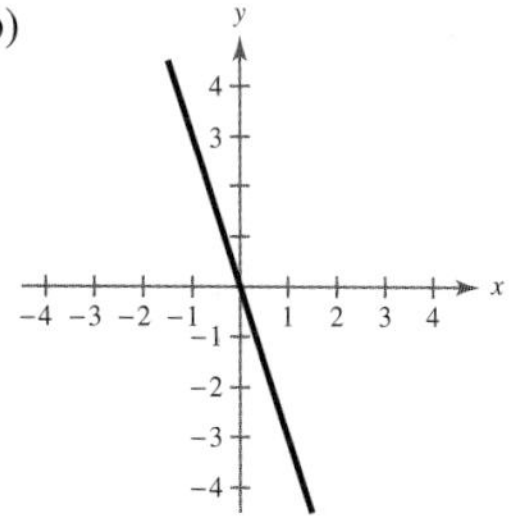

71.

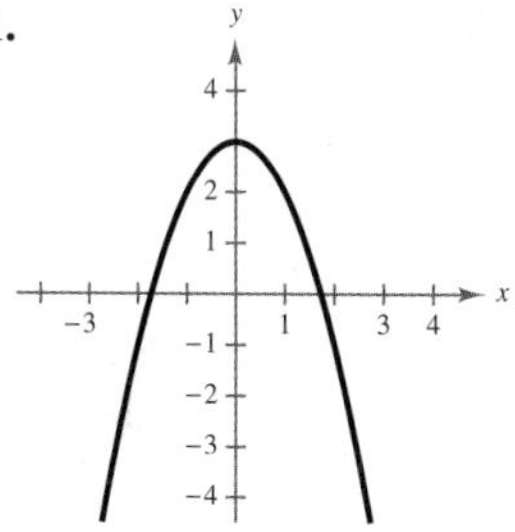

73.

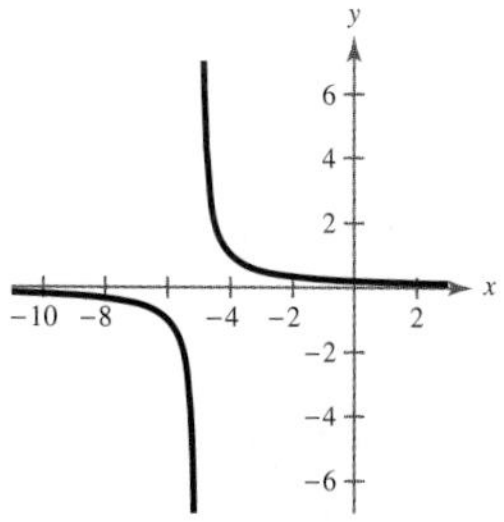

75. (a) $f(x) = x^2$
(b) Vertical shift nine units down
(c)

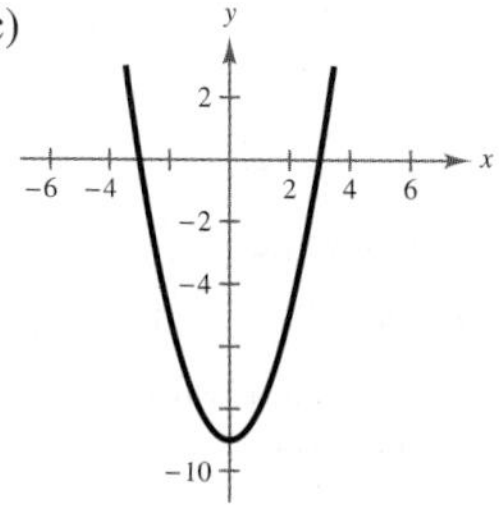

(d) $h(x) = f(x) - 9$

77. (a) $f(x) = \sqrt{x}$
(b) Reflection in the x-axis and vertical shift four units up
(c)

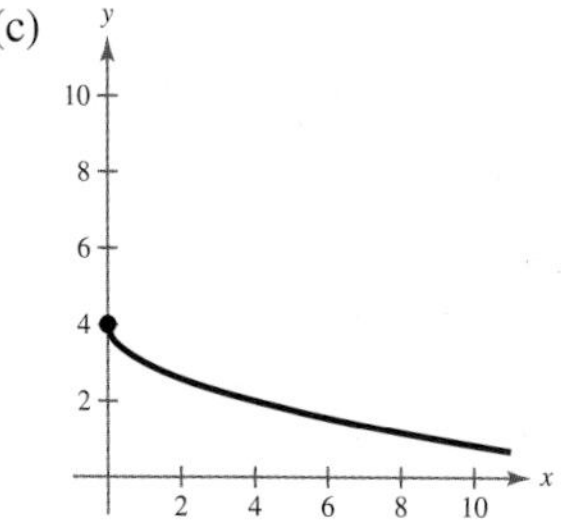

(d) $h(x) = -f(x) + 4$

79. (a) $f(x) = x^2$

(b) Reflection in the x-axis, horizontal shift two units to the left, and vertical shift three units up

(c)

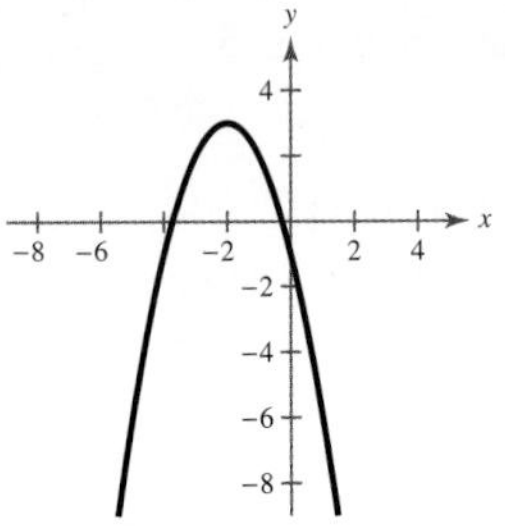

(d) $h(x) = -f(x + 2) + 3$

81. (a) $f(x) = [\![x]\!]$

(b) Reflection in the x-axis and vertical shift six units up

(c)

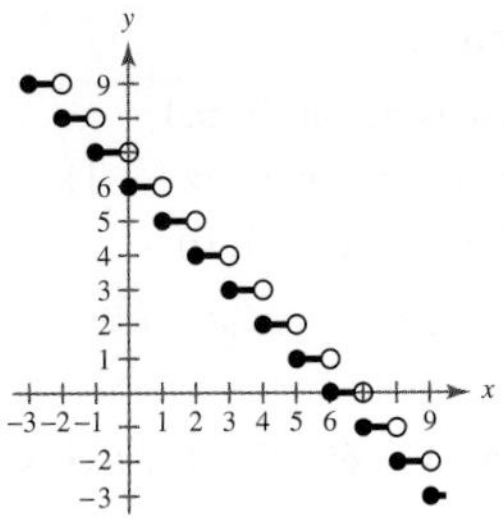

(d) $h(x) = -f(x) + 6$

83. (a) $f(x) = [\![x]\!]$

(b) Horizontal shift nine units to the right and vertical stretch

(c)

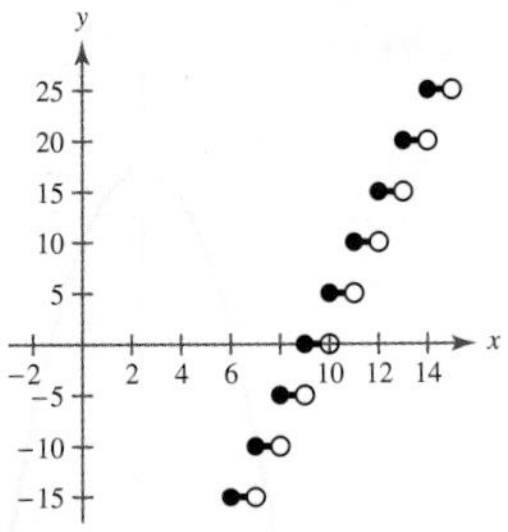

(d) $h(x) = 5\,f(x - 9)$

85. (a) $x^2 + 2x + 2$

(b) $x^2 - 2x + 4$

(c) $2x^3 - x^2 + 6x - 3$

(d) $\dfrac{x^2 + 3}{2x - 1}$; all real numbers x except $x = \dfrac{1}{2}$

87. (a) $x - \frac{8}{3}$

(b) $x - 8$

Domains of $f, g, f \circ g$, and $g \circ f$: all real numbers x

89. $N(T(t)) = 100t^2 + 275$

The composition function $N(T(t))$ represents the number of bacteria in the food as a function of time.

91. $f^{-1}(x) = \dfrac{x - 8}{3}$

$f(f^{-1}(x)) = 3\left(\dfrac{x - 8}{3}\right) + 8 = x$

$f^{-1}(f(x)) = \dfrac{3x + 8 - 8}{3} = x$

93.

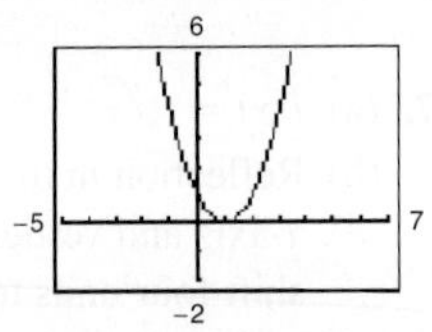

The function does not have an inverse function.

95. (a) $f^{-1}(x) = 2x + 6$

(b)

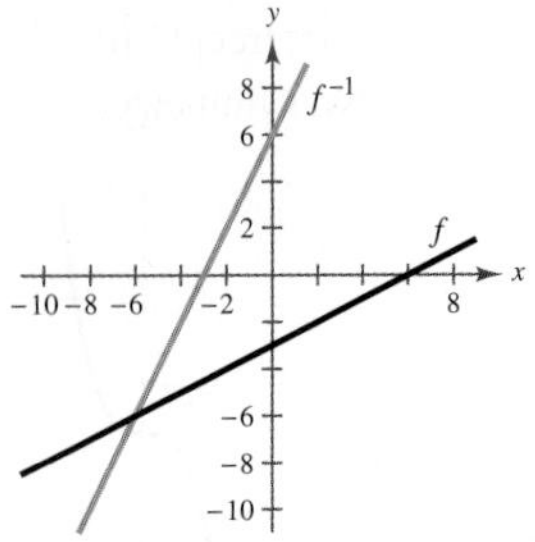

(c) The graphs are reflections of each other in the line $y = x$.

(d) Both f and f^{-1} have domains and ranges that are all real numbers.

97. $x > 4; f^{-1}(x) = \sqrt{\dfrac{x}{2}} + 4, x \neq 0$

99. (a) and (b)

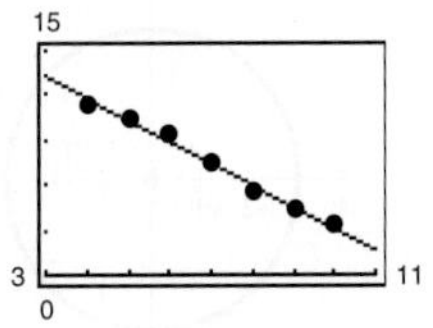

$V = -1.453t + 17.58$

The model is a good fit for the actual data.

101. \$44.80

103. True. If $f(x) = x^3$ and $g(x) = \sqrt[3]{x}$, then the domain of g is all real numbers, which is equal to the range of f, and vice versa.

Chapter Test *(page 109)*

1.

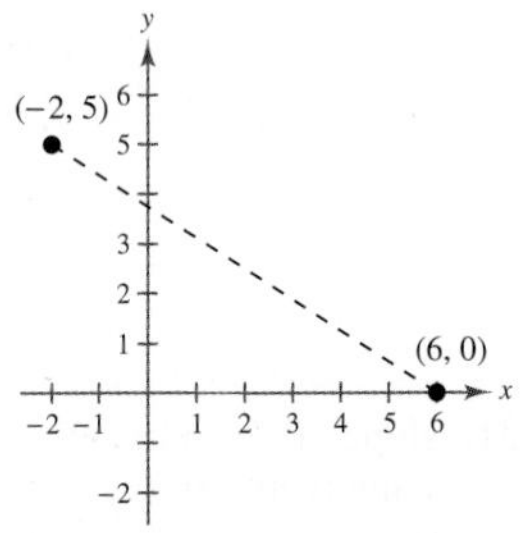

Midpoint: $\left(2, \frac{5}{2}\right)$; Distance: $\sqrt{89}$

2. $V(h) = 16\pi h$

3. No symmetry

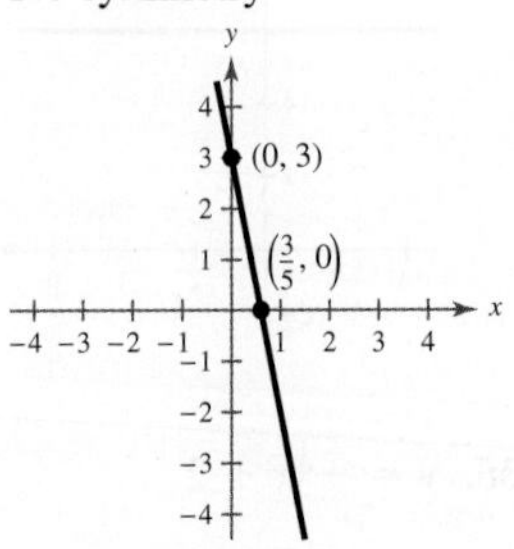

4. y-axis symmetry

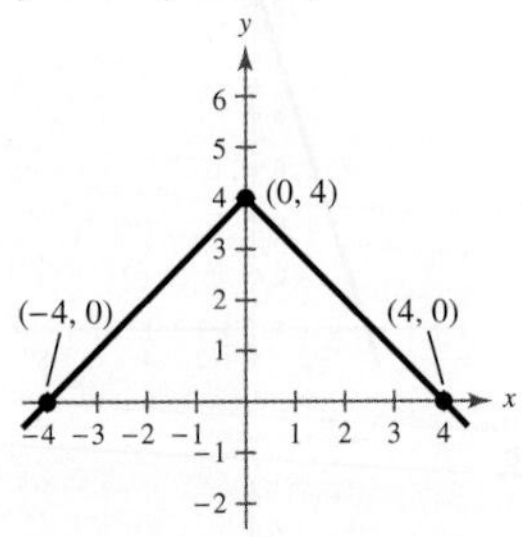

5. y-axis symmetry

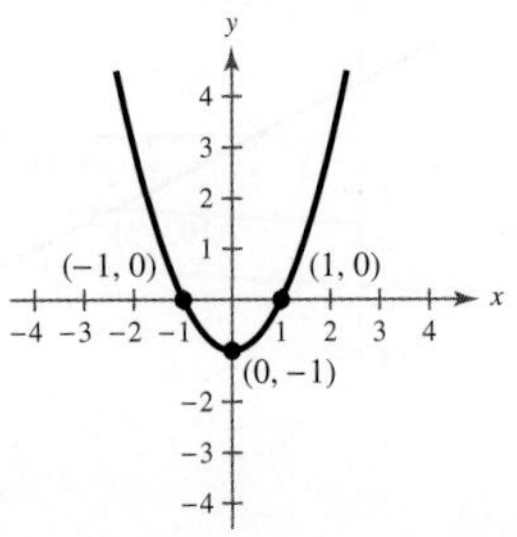

6. $(x - 1)^2 + (y - 3)^2 = 16$

7. $y = -2x + 1$ **8.** $y = -1.7x + 5.9$

9. (a) $5x + 2y - 8 = 0$ (b) $-2x + 5y - 20 = 0$

10. (a) $-\frac{1}{8}$ (b) $-\frac{1}{28}$ (c) $\frac{\sqrt{x}}{x^2 - 18x}$ **11.** $x \le 3$

12. (a) $0, \pm 0.4314$

(b)

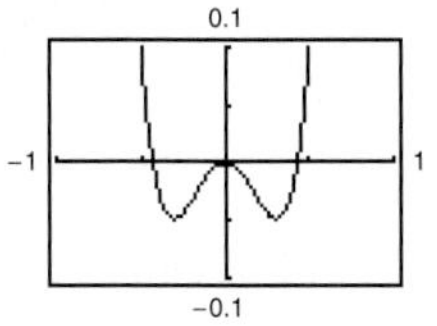

(c) Increasing on $(-0.31, 0), (0.31, \infty)$
Decreasing on $(-\infty, -0.31), (0, 0.31)$

(d) Even

13. (a) 0, 3

(b)

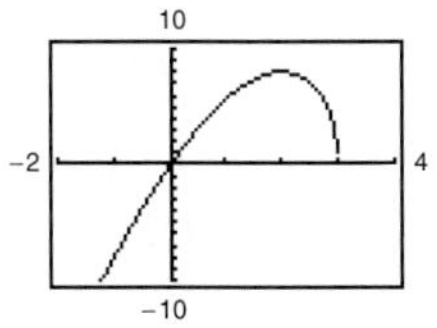

(c) Increasing on $(-\infty, 2)$
Decreasing on $(2, 3)$

(d) Neither

14. (a) -5

(b)

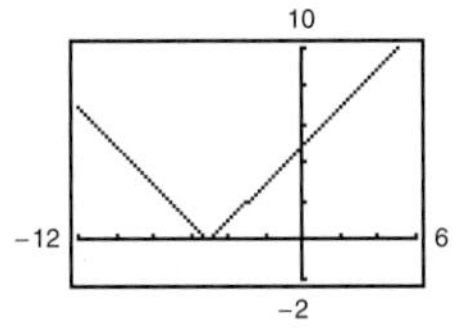

(c) Increasing on $(-5, \infty)$
Decreasing on $(-\infty, -5)$

(d) Neither

15.

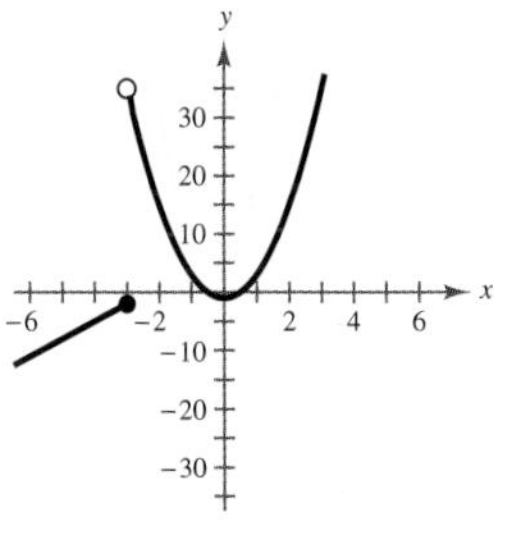

16. Reflection in the x-axis of $y = [\![x]\!]$

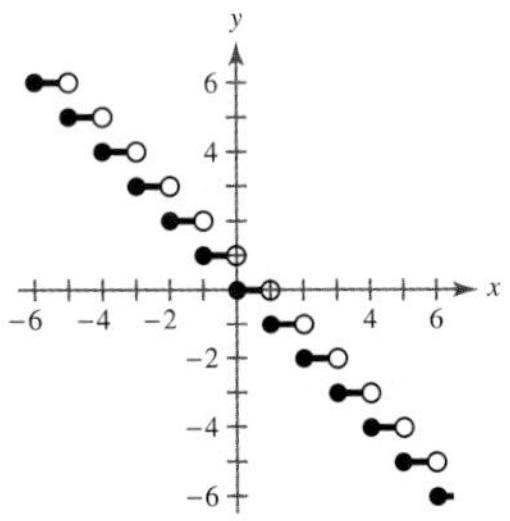

17. Reflection in the x-axis, horizontal shift, and vertical shift of $y = \sqrt{x}$

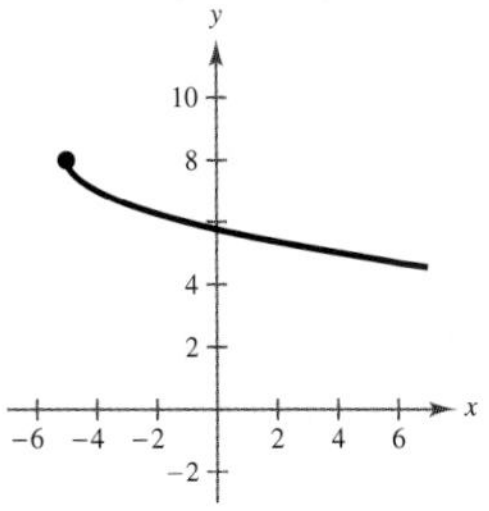

18. Reflection in the x-axis, vertical stretch, horizontal shift, and vertical shift of $y = x^3$

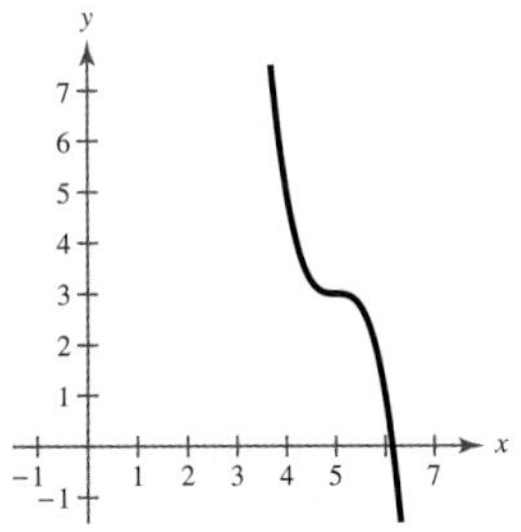

19. (a) $2x^2 - 4x - 2$ (b) $4x^2 + 4x - 12$

(c) $-3x^4 - 12x^3 + 22x^2 + 28x - 35$

(d) $\frac{3x^2 - 7}{-x^2 - 4x + 5}, \quad x \ne -5, 1$

(e) $3x^4 + 24x^3 + 18x^2 - 120x + 68$

(f) $-9x^4 + 30x^2 - 16$

20. (a) $\frac{1 + 2x^{3/2}}{x}, \quad x > 0$ (b) $\frac{1 - 2x^{3/2}}{x}, \quad x > 0$

(c) $\frac{2\sqrt{x}}{x}, \quad x > 0$ (d) $\frac{1}{2x^{3/2}}, \quad x > 0$

(e) $\frac{\sqrt{x}}{2x}, \quad x > 0$ (f) $\frac{2\sqrt{x}}{x}, \quad x > 0$

21. $f^{-1}(x) = \sqrt[3]{x - 8}$ **22.** No inverse

23. $f^{-1}(x) = \left(\frac{1}{3}x\right)^{2/3}, \ x \ge 0$ **24.** $v = 6\sqrt{s}$

25. $A = \frac{25}{6}xy$ **26.** $b = \frac{48}{a}$

Problem Solving *(page 111)*

1. (a) $W_1 = 2000 + 0.07S$ (b) $W_2 = 2300 + 0.05S$

(c)

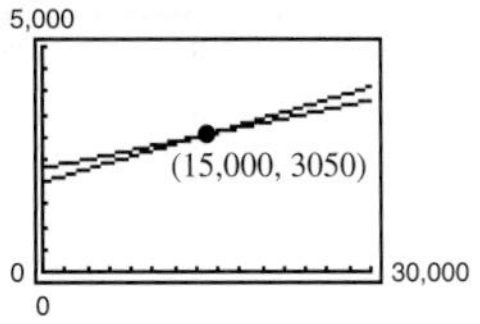

Both jobs pay the same monthly salary when sales equal \$15,000.

(d) No. Job 1 would pay \$3400 and job 2 would pay \$3300.

3. (a) The function will be even.

(b) The function will be odd.

(c) The function will be neither even nor odd.

5. $f(x) = a_{2n}x^{2n} + a_{2n-2}x^{2n-2} + \cdots + a_2x^2 + a_0$

$f(-x) = a_{2n}(-x)^{2n} + a_{2n-2}(-x)^{2n-2} + \cdots + a_2(-x)^2 + a_0$

$= f(x)$

7. (a) $81\frac{2}{3}$ h

(b) $25\frac{5}{7}$ mi/h

(c) $y = \frac{-180}{7}x + 3400$

Domain: $0 \le x \le \frac{1190}{9}$

Range: $0 \le y \le 3400$

(d)

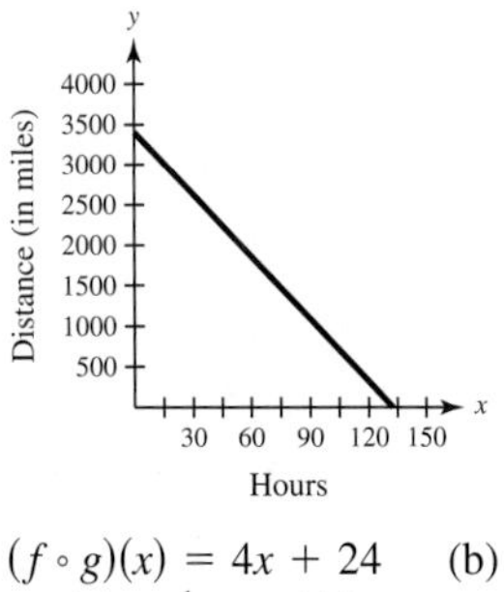

9. (a) $(f \circ g)(x) = 4x + 24$ (b) $(f \circ g)^{-1}(x) = \frac{1}{4}x - 6$

(c) $f^{-1}(x) = \frac{1}{4}x;\ g^{-1}(x) = x - 6$

(d) $(g^{-1} \circ f^{-1})(x) = \frac{1}{4}x - 6$; They are the same.

(e) $(f \circ g)(x) = 8x^3 + 1;\ (f \circ g)^{-1}(x) = \frac{1}{2}\sqrt[3]{x - 1};$
$f^{-1}(x) = \sqrt[3]{x - 1};\ g^{-1}(x) = \frac{1}{2}x;$
$(g^{-1} \circ f^{-1})(x) = \frac{1}{2}\sqrt[3]{x - 1}$

(f) Answers will vary.

(g) $(f \circ g)^{-1}(x) = (g^{-1} \circ f^{-1})(x)$

CHAPTER 1

11. (a)
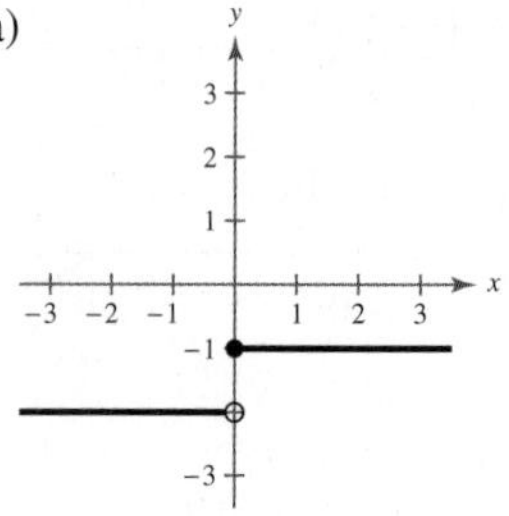

(b)
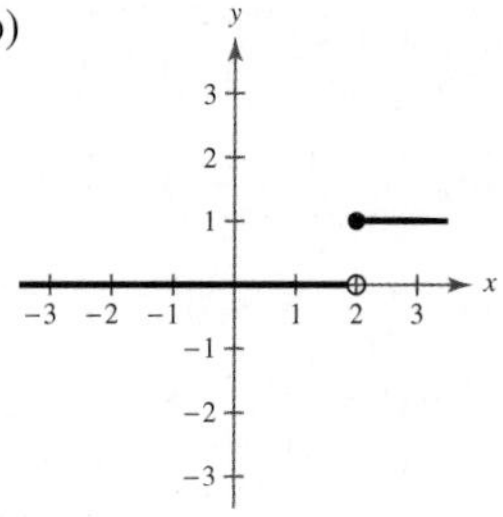

(c)
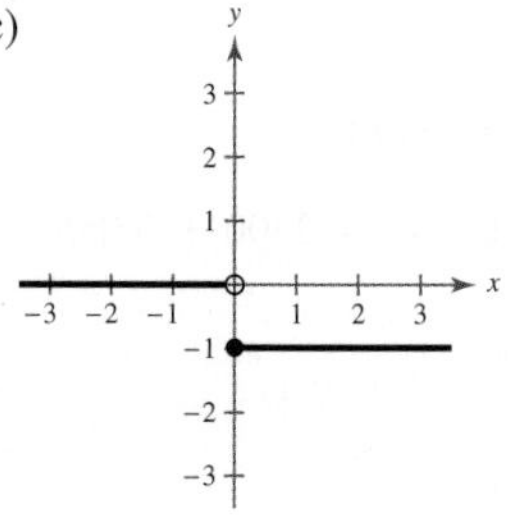

(d)
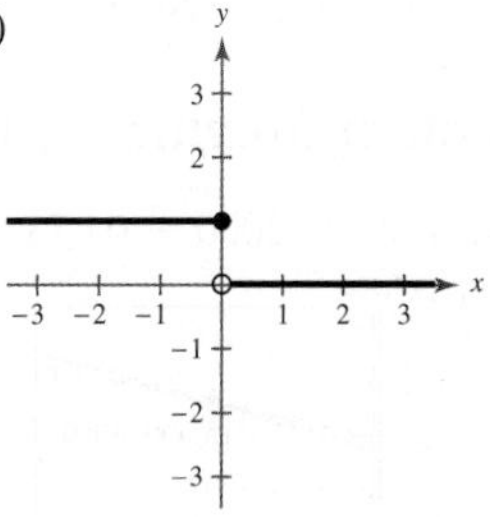

(e)
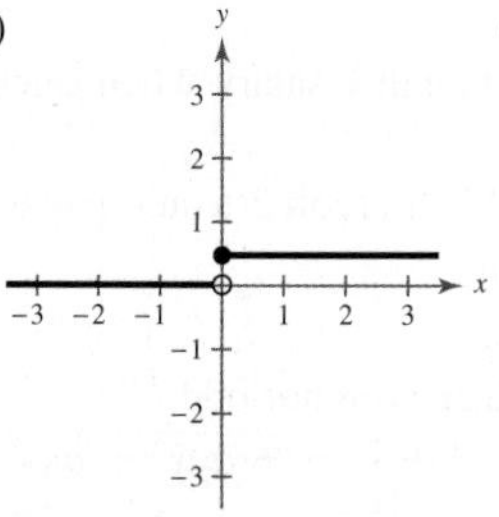

(f)
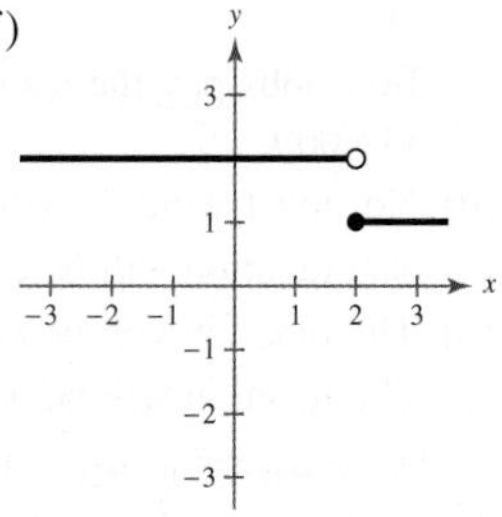

13. Proof

15. (a)

x	-4	-2	0	4
$f(f^{-1}(x))$	-4	-2	0	4

(b)

x	-3	-2	0	1
$(f + f^{-1})(x)$	5	1	-3	-5

(c)

x	-3	-2	0	1
$(f \cdot f^{-1})(x)$	4	0	2	6

(d)

x	-4	-3	0	4
$\lvert f^{-1}(x)\rvert$	2	1	1	3

Chapter 2

Section 2.1 *(page 120)*

1. polynomial **3.** quadratic; parabola
5. positive; minimum **7.** e **8.** c **9.** b
10. a **11.** f **12.** d

13. (a)
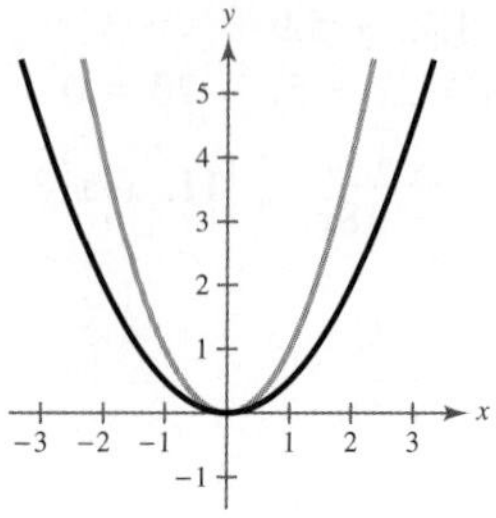

Vertical shrink

(b)
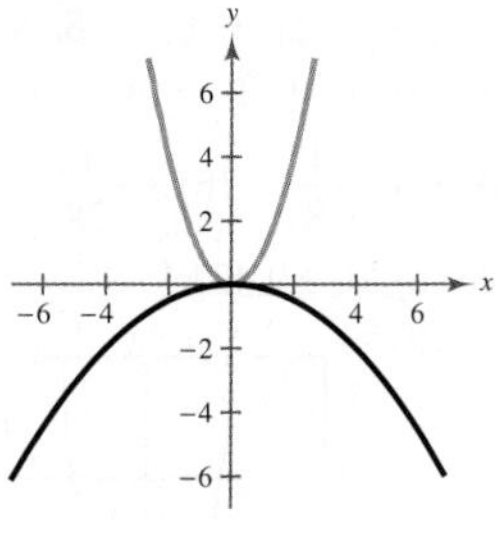

Vertical shrink and reflection in the x-axis

(c)
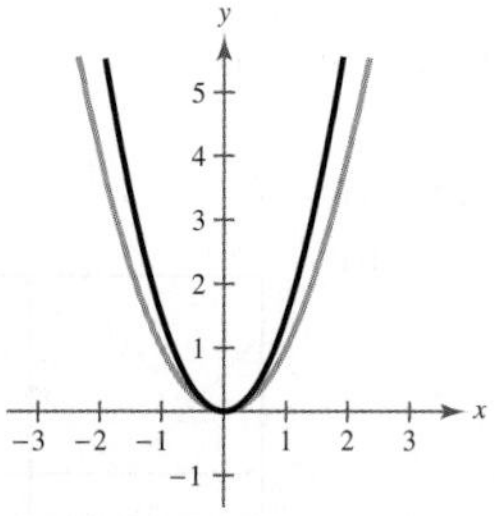

Vertical stretch

(d)
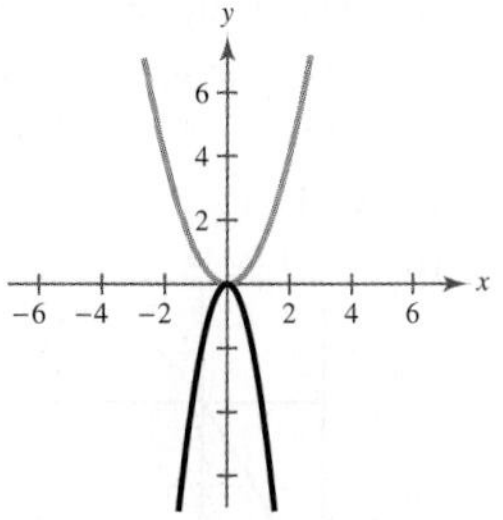

Vertical stretch and reflection in the x-axis

15. (a)
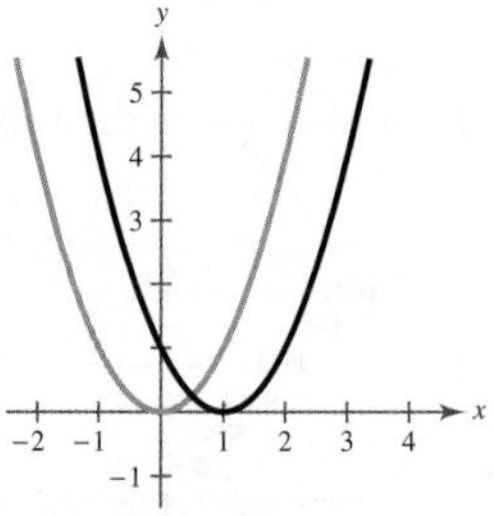

Horizontal shift one unit to the right

(b)
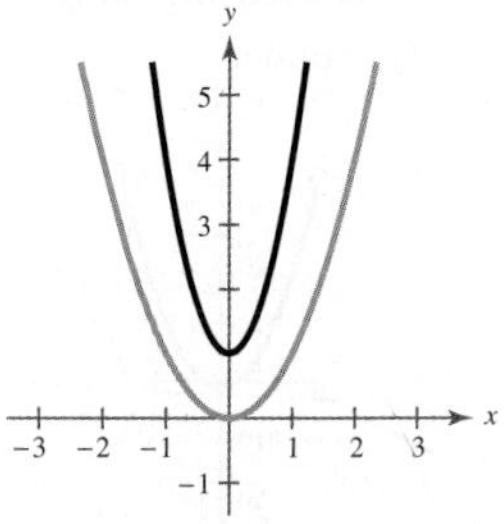

Horizontal shrink and vertical shift one unit up

(c)
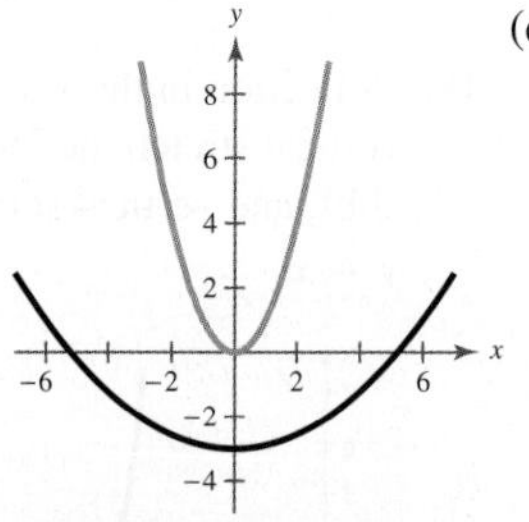

Horizontal stretch and vertical shift three units down

(d)
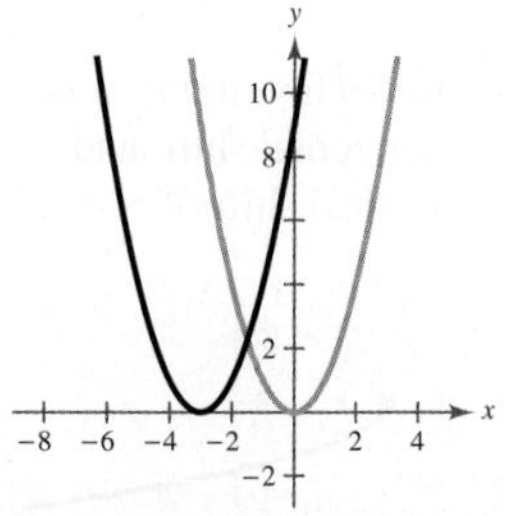

Horizontal shift three units to the left

17. $f(x) = (x - 3)^2 - 9$
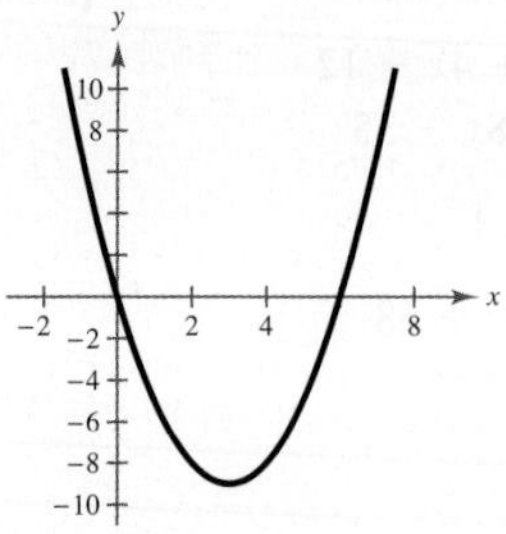

Vertex: $(3, -9)$
Axis of symmetry: $x = 3$
x-intercepts: $(0, 0)$, $(6, 0)$

19. $h(x) = (x - 4)^2$
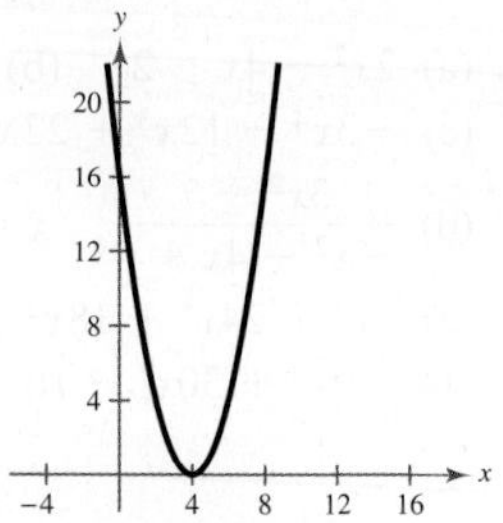

Vertex: $(4, 0)$
Axis of symmetry: $x = 4$
x-intercept: $(4, 0)$

21. $f(x) = (x + 4)^2 - 3$

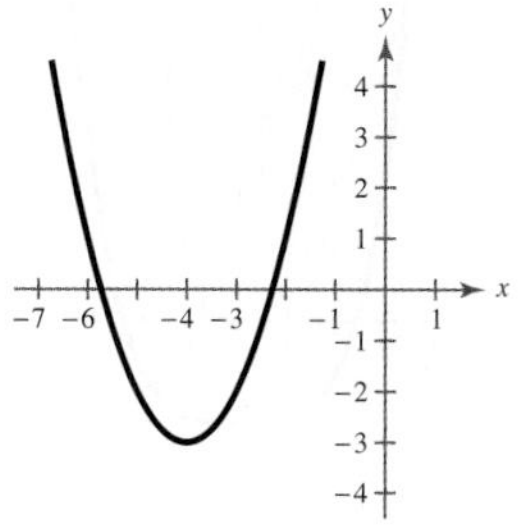

Vertex: $(-4, -3)$
Axis of symmetry: $x = -4$
x-intercepts: $(-4 \pm \sqrt{3}, 0)$

23. $f(x) = (x - 7)^2 + 5$

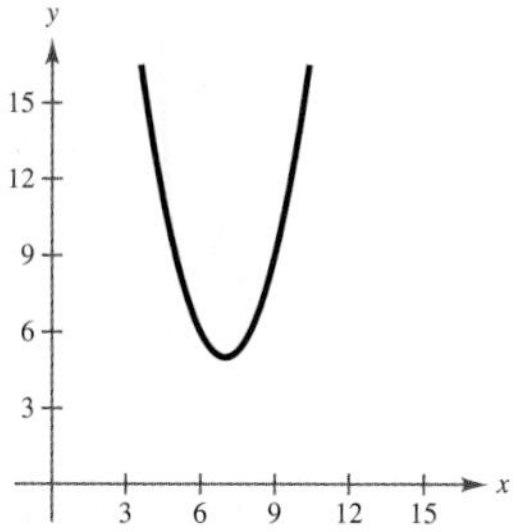

Vertex: $(7, 5)$
Axis of symmetry: $x = 7$
No x-intercept

25. $f(x) = (x + 17)^2$

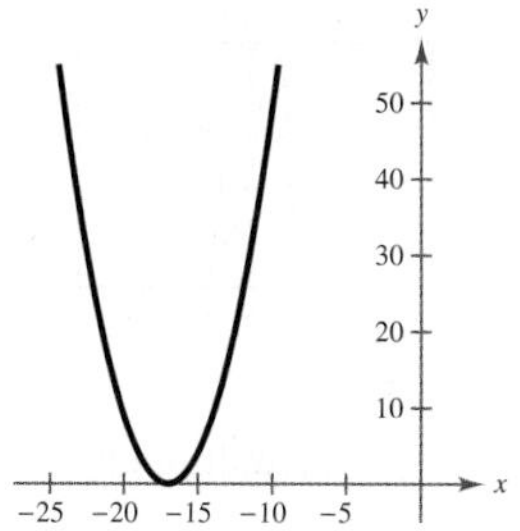

Vertex: $(-17, 0)$
Axis of symmetry: $x = -17$
x-intercept: $(-17, 0)$

27. $f(x) = \left(x - \frac{1}{2}\right)^2 + 1$

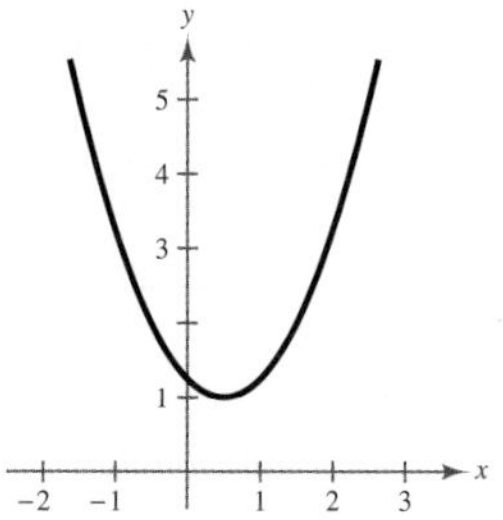

Vertex: $\left(\frac{1}{2}, 1\right)$
Axis of symmetry: $x = \frac{1}{2}$
No x-intercept

29. $f(x) = -(x - 1)^2 + 6$

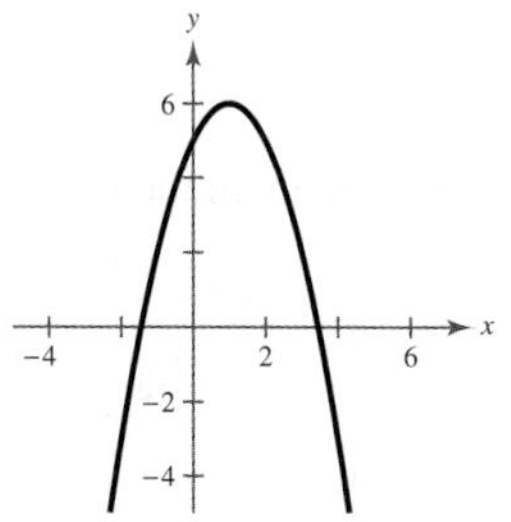

Vertex: $(1, 6)$
Axis of symmetry: $x = 1$
x-intercepts: $(1 \pm \sqrt{6}, 0)$

31. $h(x) = 4\left(x - \frac{1}{2}\right)^2 + 20$

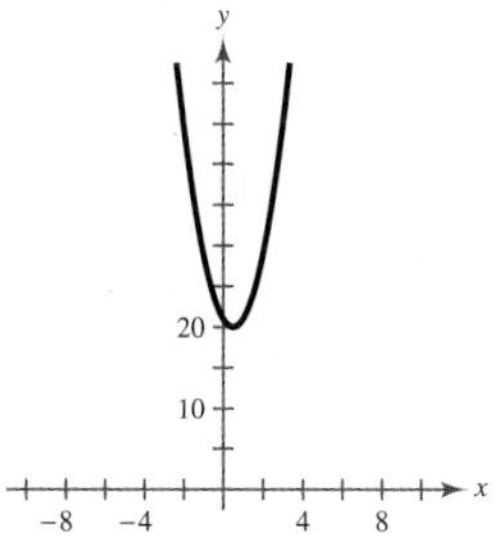

Vertex: $\left(\frac{1}{2}, 20\right)$
Axis of symmetry: $x = \frac{1}{2}$
No x-intercept

33. $f(x) = \frac{1}{4}(x - 4)^2 - 16$

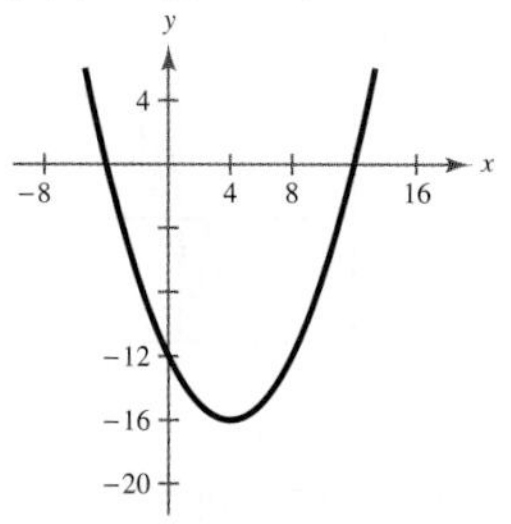

Vertex: $(4, -16)$
Axis of symmetry: $x = 4$
x-intercepts: $(-4, 0), (12, 0)$

35.

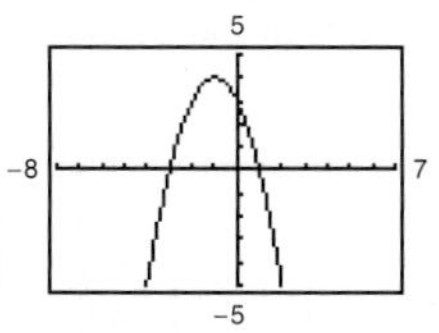

Vertex: $(-1, 4)$
Axis of symmetry: $x = -1$
x-intercepts: $(1, 0), (-3, 0)$

37.

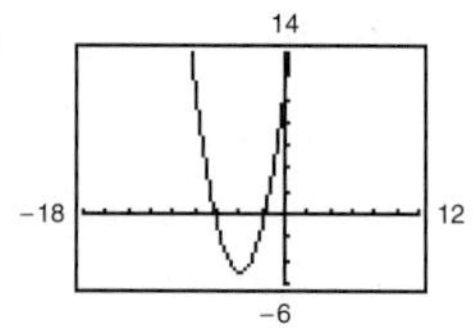

Vertex: $(-4, -5)$
Axis of symmetry: $x = -4$
x-intercepts: $(-4 \pm \sqrt{5}, 0)$

39.

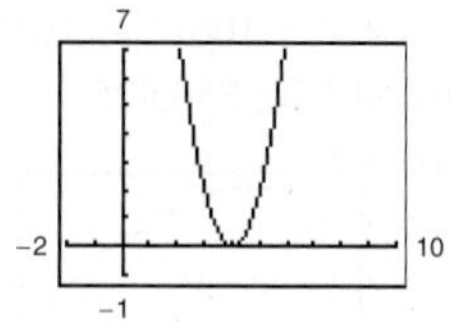

Vertex: $(4, 0)$
Axis of symmetry: $x = 4$
x-intercept: $(4, 0)$

41.

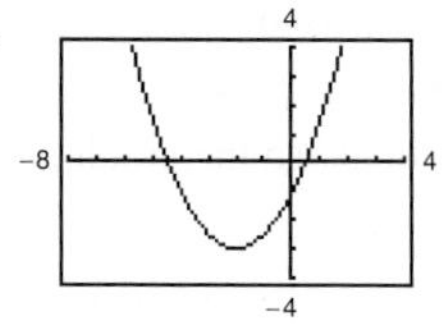

Vertex: $(-2, -3)$
Axis of symmetry: $x = -2$
x-intercepts: $(-2 \pm \sqrt{6}, 0)$

43. $y = -(x + 1)^2 + 4$ **45.** $y = -2(x + 2)^2 + 2$
47. $f(x) = (x + 2)^2 + 5$ **49.** $f(x) = 4(x - 1)^2 - 2$
51. $f(x) = \frac{3}{4}(x - 5)^2 + 12$ **53.** $f(x) = -\frac{24}{49}\left(x + \frac{1}{4}\right)^2 + \frac{3}{2}$
55. $f(x) = -\frac{16}{3}\left(x + \frac{5}{2}\right)^2$ **57.** $(5, 0), (-1, 0)$

59.

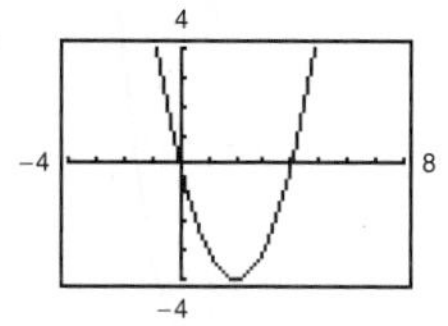

$(0, 0), (4, 0)$

61.

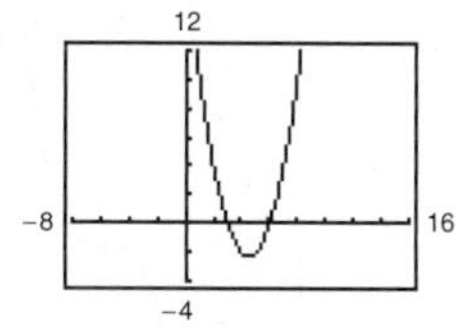

$(3, 0), (6, 0)$

63.

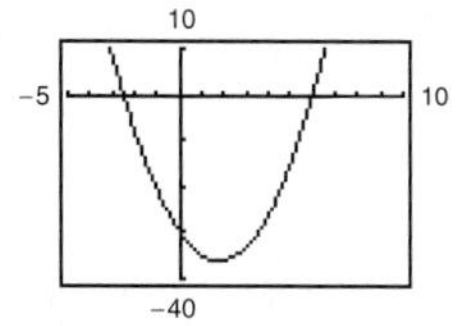

$\left(-\frac{5}{2}, 0\right), (6, 0)$

65. $f(x) = x^2 - 2x - 3$
$g(x) = -x^2 + 2x + 3$

67. $f(x) = x^2 - 10x$
$g(x) = -x^2 + 10x$

69. $f(x) = 2x^2 + 7x + 3$
$g(x) = -2x^2 - 7x - 3$

71. 55, 55 **73.** 12, 6 **75.** 16 ft **77.** 20 fixtures

79. (a) \$14,000,000; \$14,375,000; \$13,500,000
(b) \$24; \$14,400,000; Answers will vary.

81. (a) $A = \dfrac{8x(50 - x)}{3}$

(b)

x	5	10	15	20	25	30
a	600	$1066\frac{2}{3}$	1400	1600	$1666\frac{2}{3}$	1600

$x = 25$ ft, $y = 33\frac{1}{3}$ ft

(c)

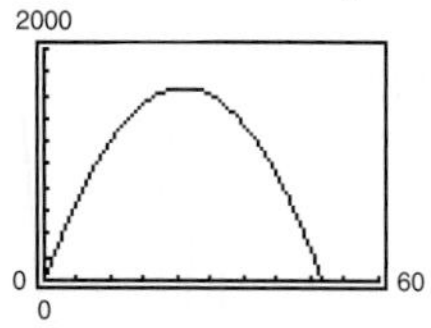

$x = 25$ ft, $y = 33\frac{1}{3}$ ft

(d) $A = -\frac{8}{3}(x - 25)^2 + \frac{5000}{3}$
(e) They are identical.

CHAPTER 2

83. (a) $R = -100x^2 + 3500x, \quad 15 \le x \le 20$
(b) \$17.50; \$30,625

85. (a)
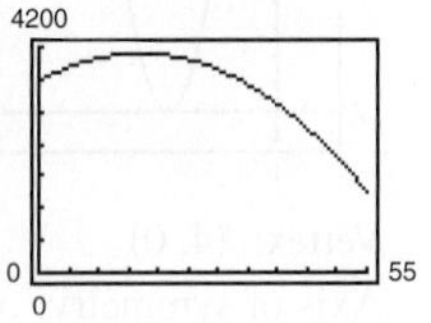

(b) 4075 cigarettes; Yes, the warning had an effect because the maximum consumption occurred in 1966.
(c) 7366 cigarettes per year; 20 cigarettes per day

87. True. The equation has no real solutions, so the graph has no x-intercepts.

89. $b = \pm 20$ **91.** $b = \pm 8$

93. $f(x) = a\left(x + \dfrac{b}{2a}\right)^2 + \dfrac{4ac - b^2}{4a}$ **95.** Proof

Section 2.2 *(page 133)*

1. continuous **3.** n; $n - 1$ **5.** touches; crosses
7. standard **9.** c **10.** f **11.** a **12.** e
13. d **14.** b

15. (a)
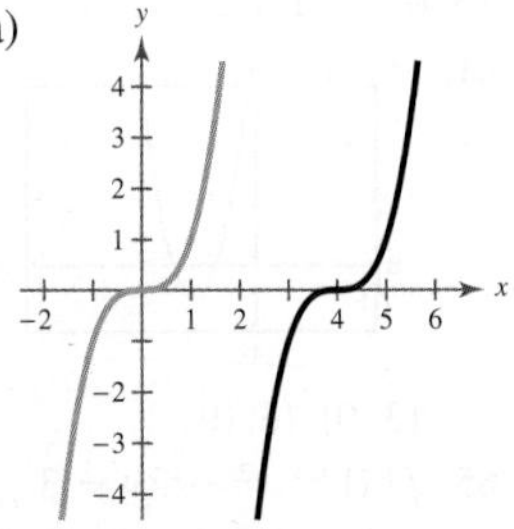

(b)
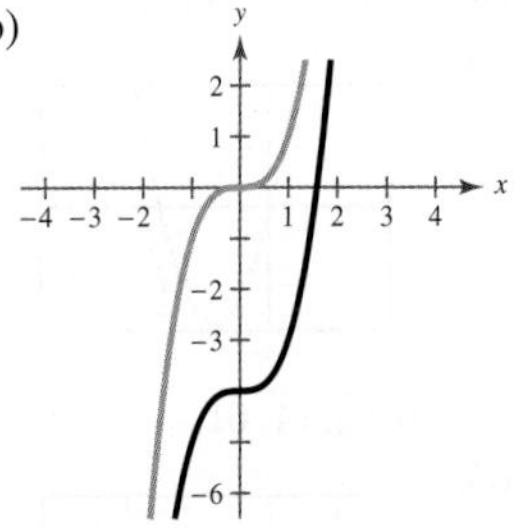

(c)
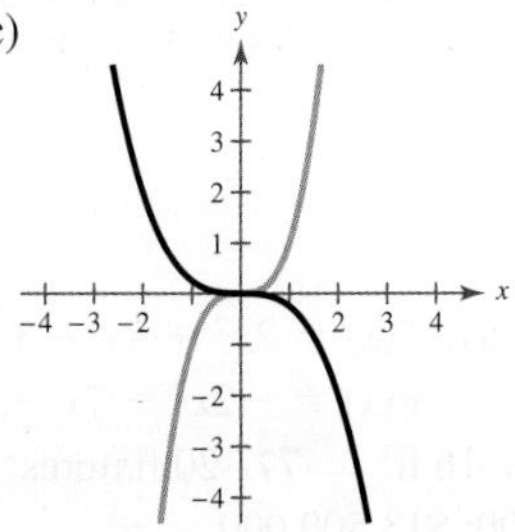

(d)
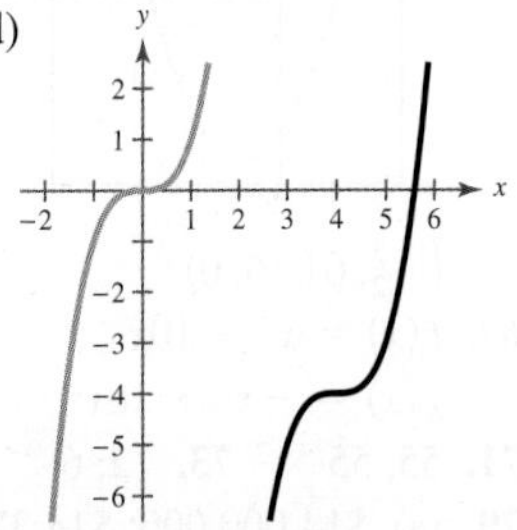

17. (a)
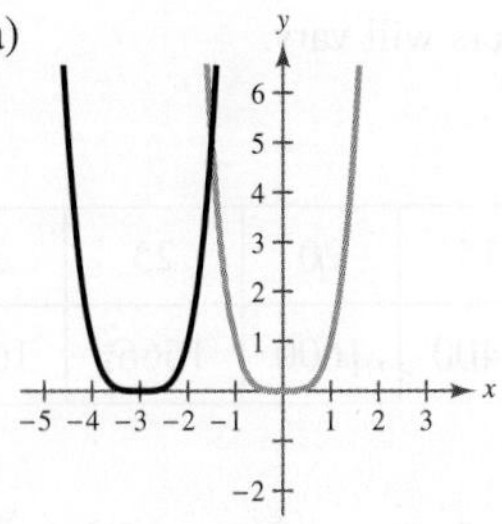

(b)
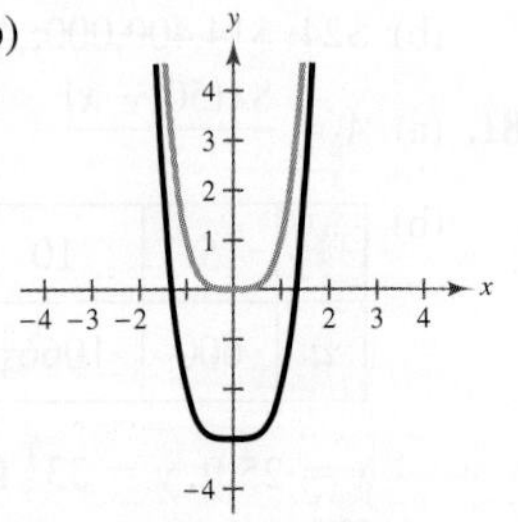

(c)
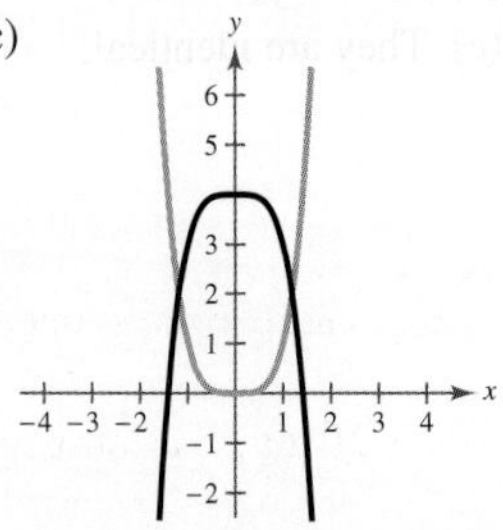

(d)
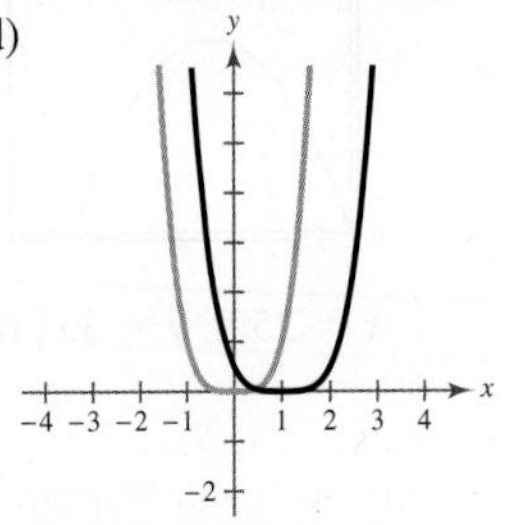

(e)
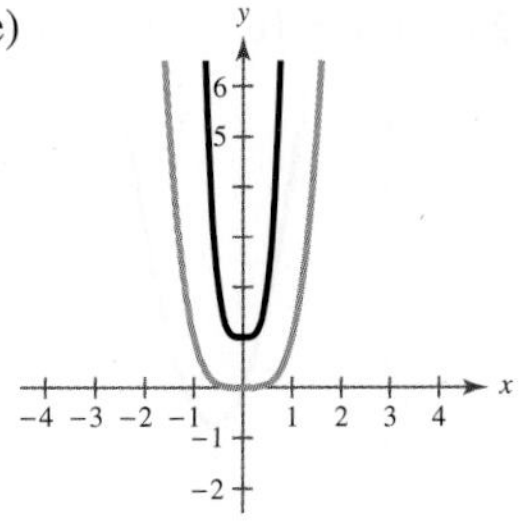

(f)
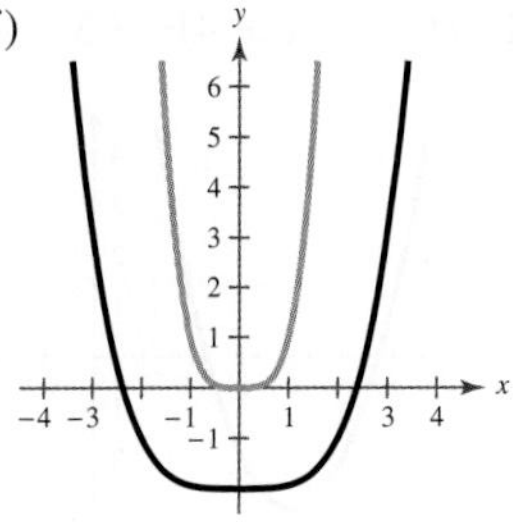

19. Falls to the left, rises to the right
21. Falls to the left, falls to the right
23. Rises to the left, falls to the right
25. Rises to the left, falls to the right
27. Rises to the left, falls to the right
29. Falls to the left, falls to the right

31.
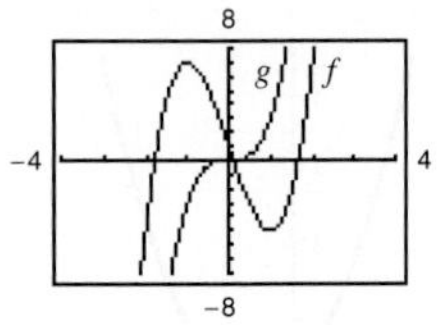

33.
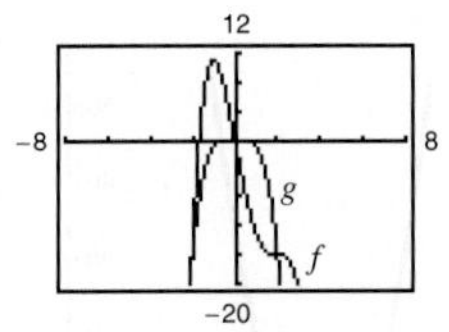

35. (a) ± 6
(b) Odd multiplicity
(c) 1
(d)
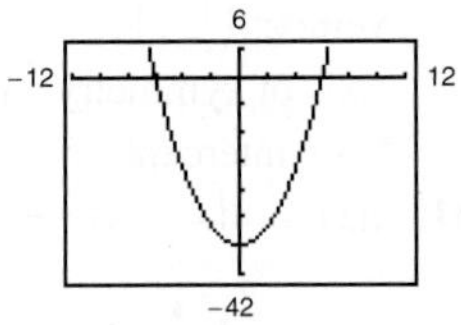

37. (a) 3
(b) Even multiplicity
(c) 1
(d)
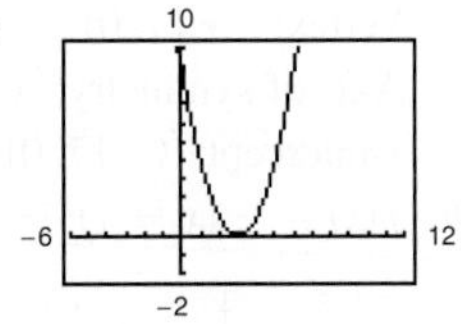

39. (a) $-2, 1$
(b) Odd multiplicity
(c) 1
(d)
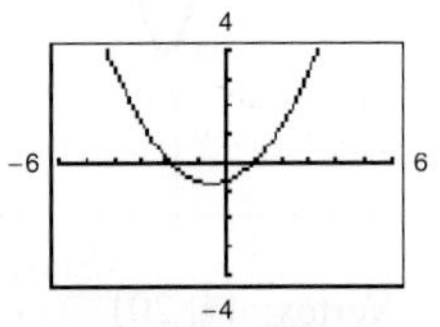

41. (a) $0, 2 \pm \sqrt{3}$
(b) Odd multiplicity
(c) 2
(d)
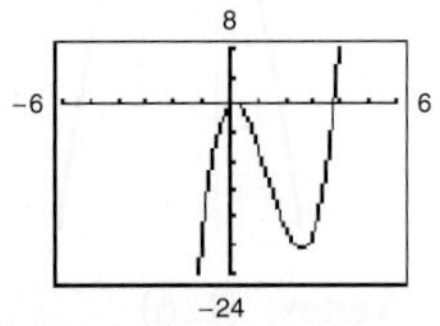

43. (a) 0, 4
(b) 0, odd multiplicity;
4, even multiplicity
(c) 2
(d)
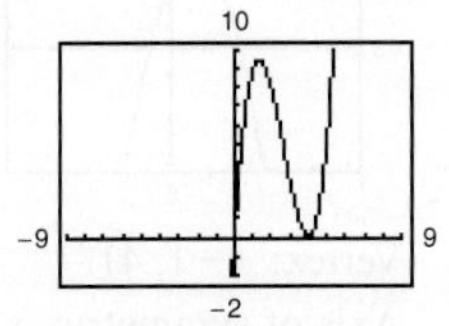

45. (a) $0, \pm\sqrt{3}$
(b) 0, odd multiplicity;
$\pm\sqrt{3}$, even multiplicity
(c) 4
(d)
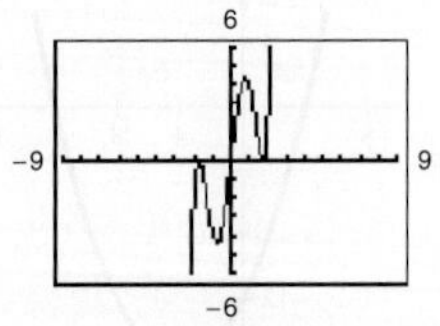

47. (a) No real zero
(b) No multiplicity
(c) 1
(d)
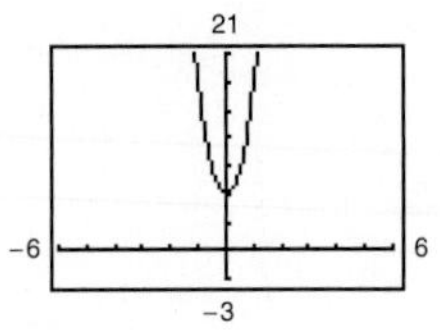

49. (a) $\pm 2, -3$
(b) Odd multiplicity
(c) 2
(d)
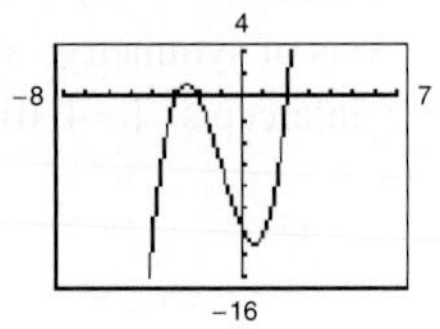

51. (a)
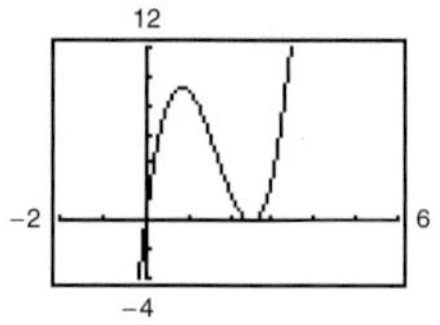

(b) x-intercepts: $(0, 0), \left(\frac{5}{2}, 0\right)$
(c) $x = 0, \frac{5}{2}$
(d) The answers in part (c) match the x-intercepts.

53. (a)
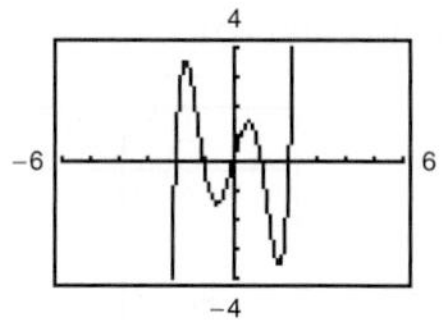

(b) x-intercepts: $(0, 0), (\pm 1, 0), (\pm 2, 0)$
(c) $x = 0, 1, -1, 2, -2$
(d) The answers in part (c) match the x-intercepts.

55. $f(x) = x^2 - 8x$ **57.** $f(x) = x^2 + 4x - 12$
59. $f(x) = x^3 + 9x^2 + 20x$
61. $f(x) = x^4 - 4x^3 - 9x^2 + 36x$ **63.** $f(x) = x^2 - 2x - 2$
65. $f(x) = x^2 + 6x + 9$ **67.** $f(x) = x^3 + 4x^2 - 5x$
69. $f(x) = x^3 - 3x$ **71.** $f(x) = x^4 + x^3 - 15x^2 + 23x - 10$
73. $f(x) = x^5 + 16x^4 + 96x^3 + 256x^2 + 256x$

75. (a) Falls to the left, rises to the right
(b) $0, 5, -5$
(c) Answers will vary.
(d)
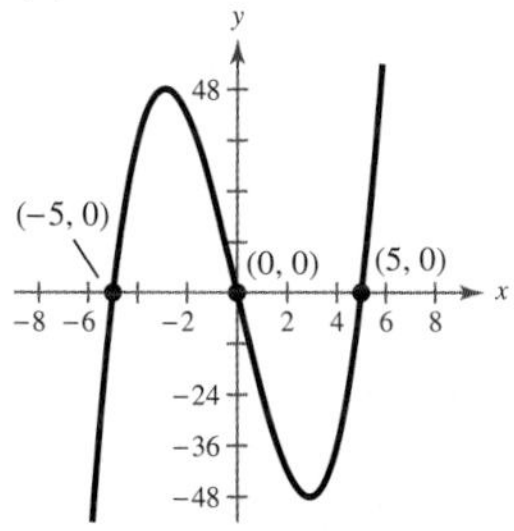

77. (a) Rises to the left, rises to the right
(b) No zeros
(c) Answers will vary.
(d)
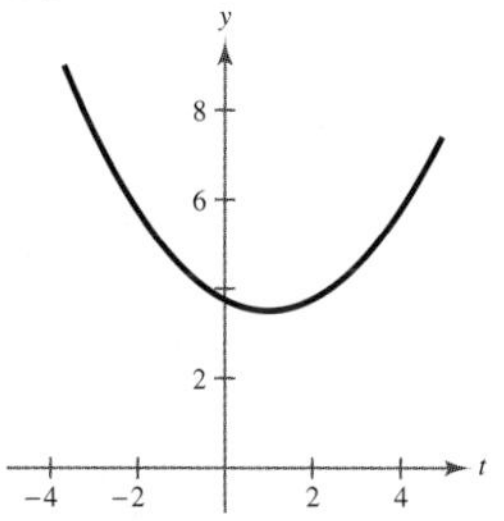

79. (a) Falls to the left, rises to the right
(b) 0, 2
(c) Answers will vary.
(d)
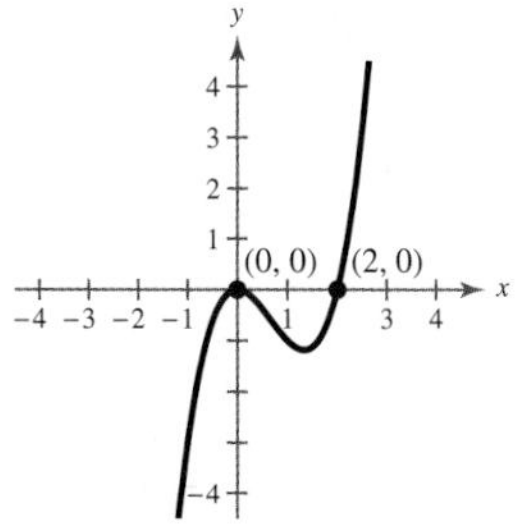

81. (a) Falls to the left, rises to the right
(b) 0, 2, 3
(c) Answers will vary.
(d)
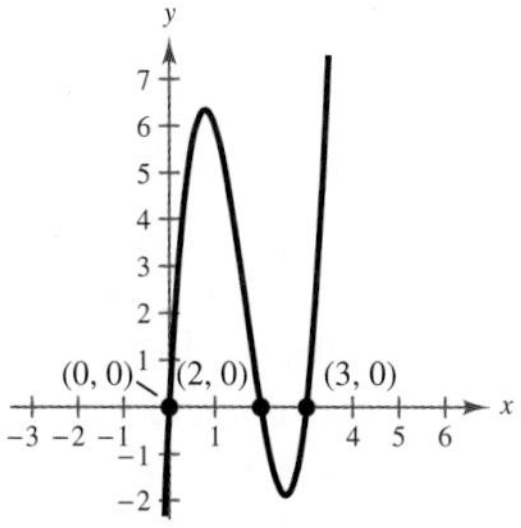

83. (a) Rises to the left, falls to the right
(b) $-5, 0$
(c) Answers will vary.
(d)
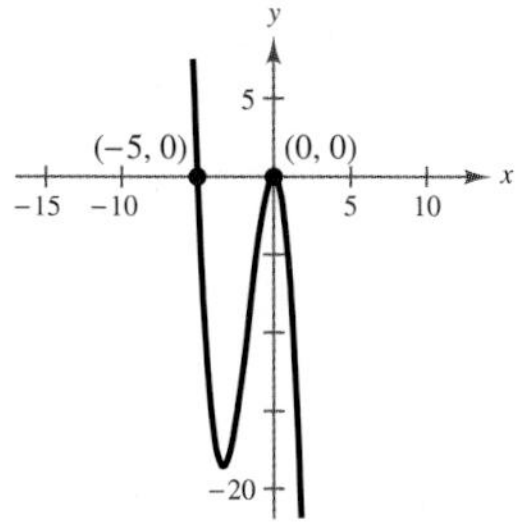

85. (a) Falls to the left, rises to the right
(b) 0, 4
(c) Answers will vary.
(d)
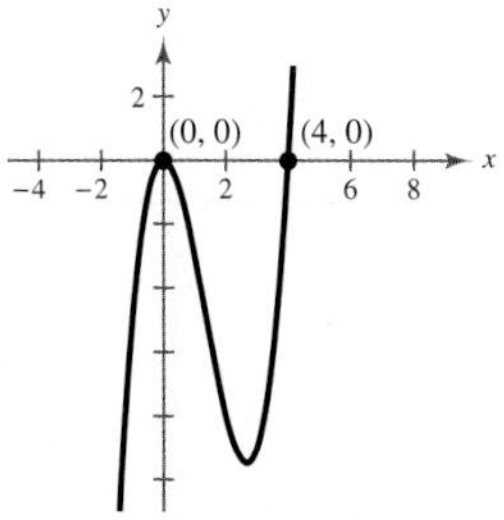

87. (a) Falls to the left, falls to the right
(b) ± 2 (c) Answers will vary.
(d)
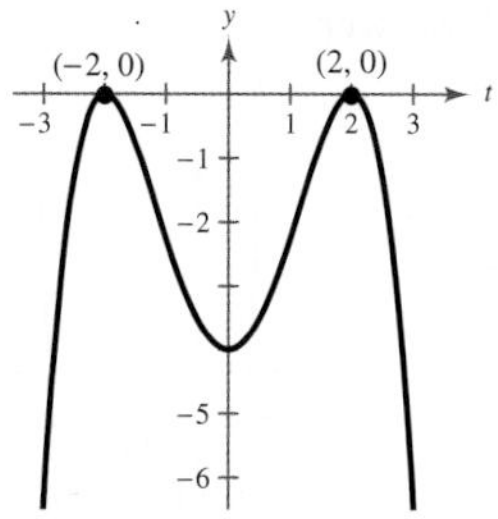

89.
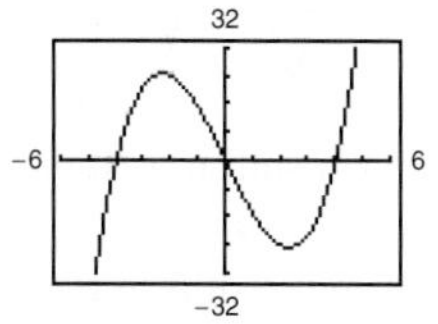

Zeros: $0, \pm 4$, odd multiplicity

91.
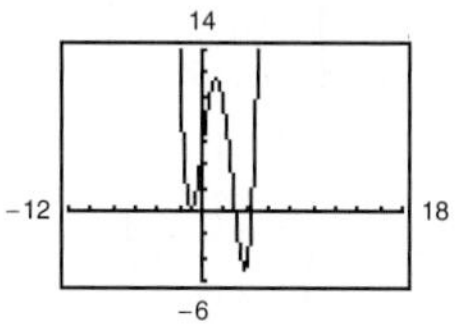

Zeros: -1, even multiplicity
$3, \frac{9}{2}$, odd multiplicity

93. (a) $[-1, 0], [1, 2], [2, 3]$
(b) About $-0.879, 1.347, 2.532$

95. (a) $[-2, -1], [0, 1]$ (b) About $-1.585, 0.779$

97. (a) $V(x) = x(36 - 2x)^2$ (b) Domain: $0 < x < 18$
(c)
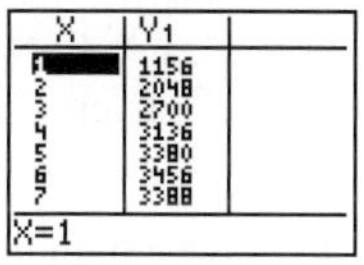

X	Y1
1	1156
2	2048
3	2700
4	3136
5	3380
6	3456
7	3388

X=1

6 in. × 24 in. × 24 in.

(d)
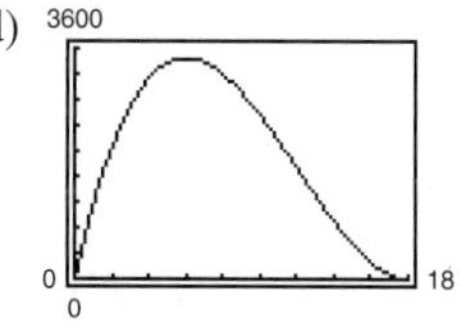

$x = 6$; The results are the same.

99. (a) $A = -2x^2 + 12x$ (b) $V = -384x^2 + 2304x$
(c) 0 in. $< x <$ 6 in.
(d)
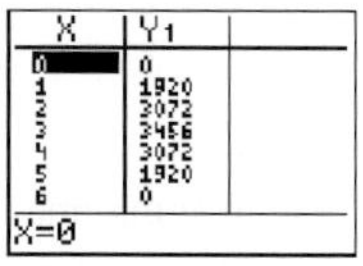

X	Y1
0	0
1	1920
2	3072
3	3456
4	3072
5	1920
6	0

X=0

When $x = 3$, the volume is maximum at $V = 3456$; dimensions of gutter are 3 in. × 6 in. × 3 in.

(e)
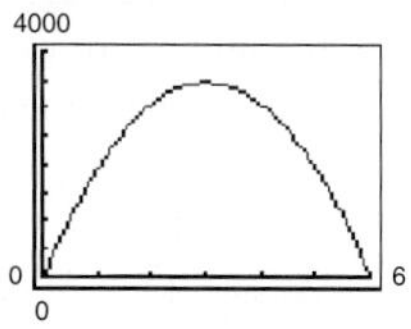

The maximum value is the same.
(f) No. Answers will vary.

101. (a) Relative maximum: (5.01, 655.75)
Relative minimum: (9.25, 417.42)
(b) Increasing: (3, 5.01), (9.25, 10)
Decreasing: (5.01, 9.25)
(c) Answers will vary.

103. $x \approx 200$

105. False. A fifth-degree polynomial can have at most four turning points.

107. False. The function $f(x) = (x - 2)^2$ has one turning point and two real (repeated) zeros.

109. False. $f(x) = -x^3$ rises to the left.

CHAPTER 2

111. True. The leading coefficient is negative and the degree is odd.

113. Answers will vary. *Sample answers:*

$a_4 < 0$

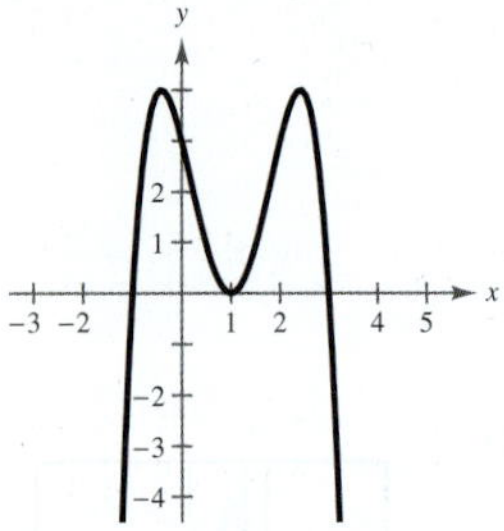

$a_4 > 0$

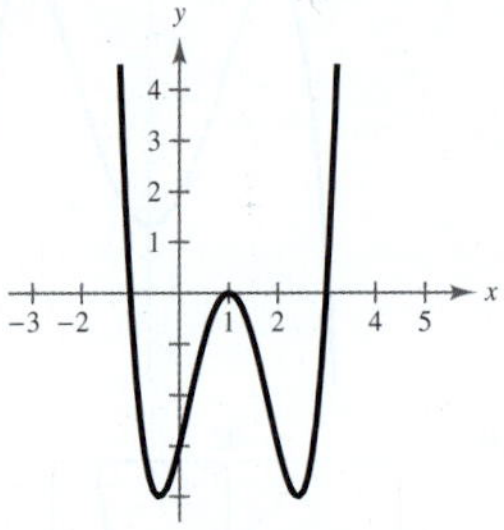

115.

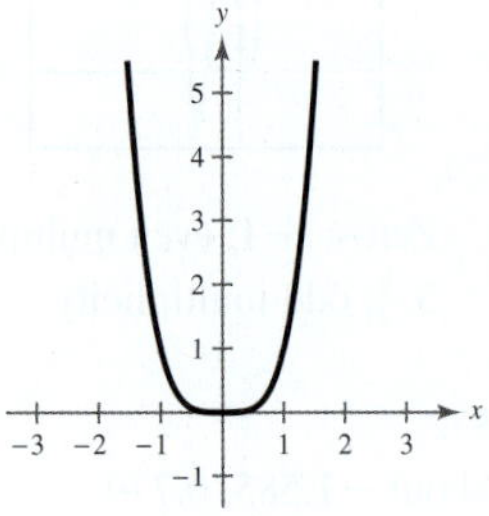

(a) Vertical shift of two units; Even
(b) Horizontal shift of two units; Neither
(c) Reflection in the y-axis; Even
(d) Reflection in the x-axis; Even
(e) Horizontal stretch; Even
(f) Vertical shrink; Even
(g) $g(x) = x^3, x \geq 0$; Neither
(h) $g(x) = x^{16}$; Even

117. (a)

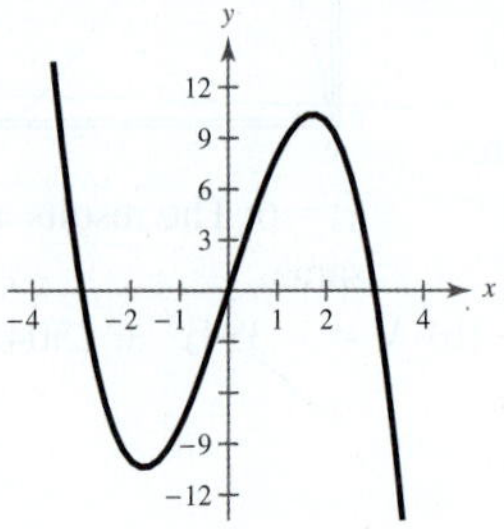

Zeros: 3
Relative minimum: 1
Relative maximum: 1
The number of zeros is the same as the degree, and the number of extrema is one less than the degree.

(b)

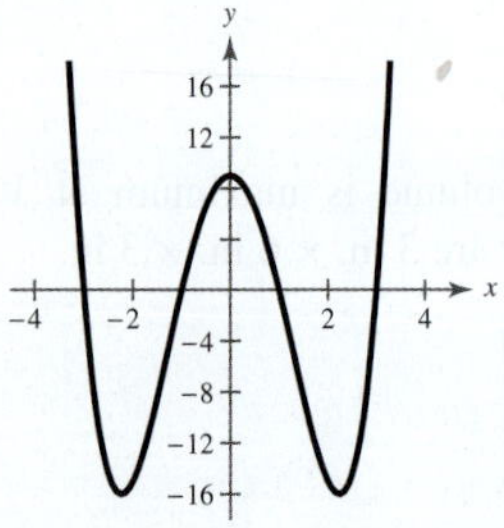

Zeros: 4
Relative minima: 2
Relative maximum: 1
The number of zeros is the same as the degree, and the number of extrema is one less than the degree.

(c)

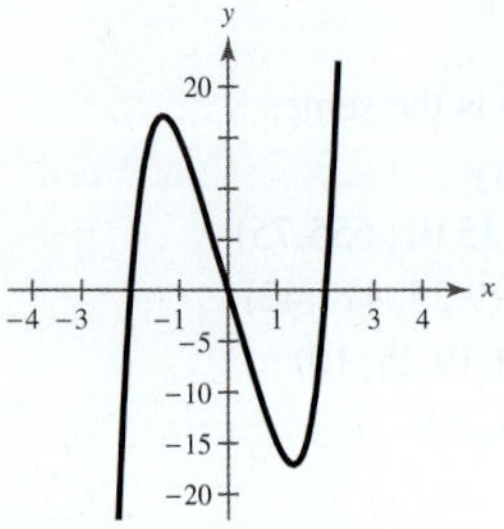

Zeros: 3
Relative minimum: 1
Relative maximum: 1
The number of zeros and the number of extrema are both less than the degree.

Section 2.3 *(page 144)*

1. $f(x)$: dividend; $d(x)$: divisor; $q(x)$: quotient; $r(x)$: remainder

3. improper **5.** Factor **7.** Answers will vary.

9. (a) and (b)

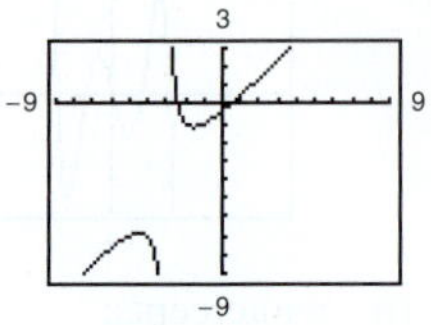

(c) Answers will vary.

11. $2x + 4, \quad x \neq -3$

13. $x^2 - 3x + 1, \quad x \neq -\frac{5}{4}$ **15.** $x^3 + 3x^2 - 1, \quad x \neq -2$

17. $x^2 + 3x + 9, \quad x \neq 3$ **19.** $7 - \dfrac{11}{x + 2}$

21. $x - \dfrac{x + 9}{x^2 + 1}$ **23.** $2x - 8 + \dfrac{x - 1}{x^2 + 1}$

25. $x + 3 + \dfrac{6x^2 - 8x + 3}{(x - 1)^3}$ **27.** $3x^2 - 2x + 5, \ x \neq 5$

29. $6x^2 + 25x + 74 + \dfrac{248}{x - 3}$ **31.** $4x^2 - 9, \quad x \neq -2$

33. $-x^2 + 10x - 25, \ x \neq -10$

35. $5x^2 + 14x + 56 + \dfrac{232}{x - 4}$

37. $10x^3 + 10x^2 + 60x + 360 + \dfrac{1360}{x - 6}$

39. $x^2 - 8x + 64, \quad x \neq -8$

41. $-3x^3 - 6x^2 - 12x - 24 - \dfrac{48}{x - 2}$

43. $-x^3 - 6x^2 - 36x - 36 - \dfrac{216}{x - 6}$

45. $4x^2 + 14x - 30, \ x \neq -\frac{1}{2}$

47. $f(x) = (x - 4)(x^2 + 3x - 2) + 3, \quad f(4) = 3$

49. $f(x) = \left(x + \frac{2}{3}\right)(15x^3 - 6x + 4) + \frac{34}{3}, \quad f\left(-\frac{2}{3}\right) = \frac{34}{3}$

51. $f(x) = \left(x - \sqrt{2}\right)\left[x^2 + \left(3 + \sqrt{2}\right)x + 3\sqrt{2}\right] - 8,$
$f\left(\sqrt{2}\right) = -8$

53. $f(x) = \left(x - 1 + \sqrt{3}\right)\left[-4x^2 + \left(2 + 4\sqrt{3}\right)x + \left(2 + 2\sqrt{3}\right)\right],$
$f\left(1 - \sqrt{3}\right) = 0$

55. (a) -2 (b) 1 (c) $-\frac{1}{4}$ (d) 5

57. (a) -35 (b) -22 (c) -10 (d) -211

59. $(x - 2)(x + 3)(x - 1)$; Solutions: $2, -3, 1$

61. $(2x - 1)(x - 5)(x - 2)$; Solutions: $\frac{1}{2}, 5, 2$

63. $\left(x + \sqrt{3}\right)\left(x - \sqrt{3}\right)(x + 2)$; Solutions: $-\sqrt{3}, \sqrt{3}, -2$

65. $(x - 1)\left(x - 1 - \sqrt{3}\right)\left(x - 1 + \sqrt{3}\right)$;
Solutions: $1, 1 + \sqrt{3}, 1 - \sqrt{3}$

67. (a) Answers will vary. (b) $2x - 1$
(c) $f(x) = (2x - 1)(x + 2)(x - 1)$ (d) $\frac{1}{2}, -2, 1$
(e)

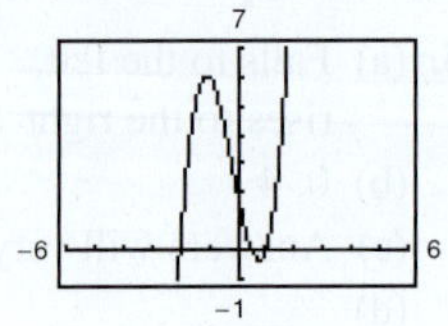

69. (a) Answers will vary. (b) $(x - 1), (x - 2)$
(c) $f(x) = (x - 1)(x - 2)(x - 5)(x + 4)$ (d) $1, 2, 5, -4$
(e)

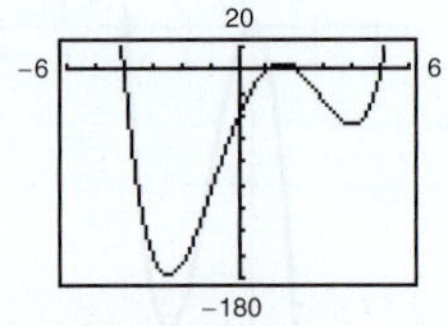

71. (a) Answers will vary. (b) $x + 7$

(c) $f(x) = (x + 7)(2x + 1)(3x - 2)$ (d) $-7, -\frac{1}{2}, \frac{2}{3}$
(e)

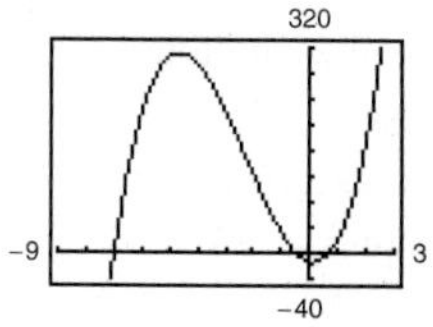

73. (a) Answers will vary. (b) $x - \sqrt{5}$
(c) $f(x) = (x - \sqrt{5})(x + \sqrt{5})(2x - 1)$ (d) $\pm\sqrt{5}, \frac{1}{2}$
(e)

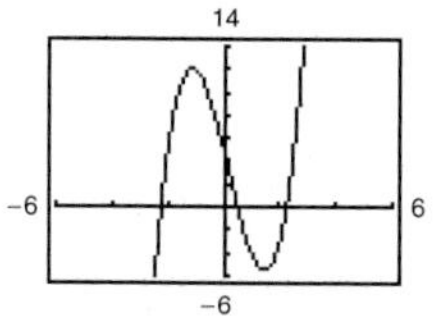

75. (a) Zeros are 2 and about ± 2.236.
(b) $x = 2$ (c) $f(x) = (x - 2)(x - \sqrt{5})(x + \sqrt{5})$

77. (a) Zeros are -2, about 0.268, and about 3.732.
(b) $t = -2$
(c) $h(t) = (t + 2)[t - (2 + \sqrt{3})][t - (2 - \sqrt{3})]$

79. (a) Zeros are 0, 3, 4, and about ± 1.414. (b) $x = 0$
(c) $h(x) = x(x - 4)(x - 3)(x + \sqrt{2})(x - \sqrt{2})$

81. $2x^2 - x - 1,\ x \neq \frac{3}{2}$ **83.** $x^2 + 3x,\ x \neq -2, -1$

85. (a)

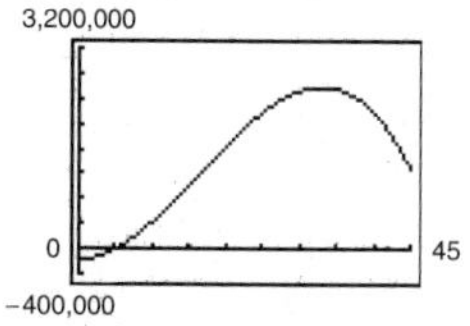

(b) \$250,366
(c) Answers will vary.

87. False. $-\frac{4}{7}$ is a zero of f.

89. True. The degree of the numerator is greater than the degree of the denominator.

91. $x^{2n} + 6x^n + 9,\ x^n \neq -3$ **93.** The remainder is 0.

95. $c = -210$ **97.** $k = 7$

Section 2.4 *(page 152)*

1. real **3.** pure imaginary **5.** principal square

7. $a = -12,\ b = 7$ **9.** $a = 6,\ b = 5$ **11.** $8 + 5i$

13. $2 - 3\sqrt{3}i$ **15.** $4\sqrt{5}i$ **17.** 14 **19.** $-1 - 10i$

21. $0.3i$ **23.** $10 - 3i$ **25.** 1 **27.** $3 - 3\sqrt{2}i$

29. $-14 + 20i$ **31.** $\frac{1}{6} + \frac{7}{6}i$ **33.** $5 + i$

35. $108 + 12i$ **37.** 24 **39.** $-13 + 84i$ **41.** -10

43. $9 - 2i, 85$ **45.** $-1 + \sqrt{5}i, 6$ **47.** $-2\sqrt{5}i, 20$

49. $\sqrt{6}, 6$ **51.** $-3i$ **53.** $\frac{8}{41} + \frac{10}{41}i$ **55.** $\frac{12}{13} + \frac{5}{13}i$

57. $-4 - 9i$ **59.** $-\frac{120}{1681} - \frac{27}{1681}i$ **61.** $-\frac{1}{2} - \frac{5}{2}i$

63. $\frac{62}{949} + \frac{297}{949}i$ **65.** $-2\sqrt{3}$ **67.** -15

69. $(21 + 5\sqrt{2}) + (7\sqrt{5} - 3\sqrt{10})i$ **71.** $1 \pm i$

73. $-2 \pm \frac{1}{2}i$ **75.** $-\frac{5}{2}, -\frac{3}{2}$ **77.** $2 \pm \sqrt{2}i$

79. $\dfrac{5}{7} \pm \dfrac{5\sqrt{15}}{7}$ **81.** $-1 + 6i$ **83.** $-14i$

85. $-432\sqrt{2}i$ **87.** i **89.** 81

91. (a) $z_1 = 9 + 16i,\ z_2 = 20 - 10i$
(b) $z = \dfrac{11{,}240}{877} + \dfrac{4630}{877}i$

93. False. When the complex number is real, the number equals its conjugate.

95. False.
$i^{44} + i^{150} - i^{74} - i^{109} + i^{61} = 1 - 1 + 1 - i + i = 1$

97. $i, -1, -i, 1, i, -1, -i, 1$; The pattern repeats the first four results. Divide the exponent by 4.
When the remainder is 1, the result is i.
When the remainder is 2, the result is -1.
When the remainder is 3, the result is $-i$.
When the remainder is 0, the result is 1.

99. $\sqrt{-6}\sqrt{-6} = \sqrt{6}i\sqrt{6}i = 6i^2 = -6$

101. Proof

Section 2.5 *(page 164)*

1. Fundamental Theorem of Algebra **3.** Rational Zero

5. linear; quadratic; quadratic **7.** Descartes's Rule of Signs

9. 1 **11.** 3 **13.** 2 **15.** $\pm 1, \pm 2$

17. $\pm 1, \pm 3, \pm 5, \pm 9, \pm 15, \pm 45, \pm\frac{1}{2}, \pm\frac{3}{2}, \pm\frac{5}{2}, \pm\frac{9}{2}, \pm\frac{15}{2}, \pm\frac{45}{2}$

19. $-2, -1, 3$ **21.** $1, -1, 4$ **23.** $-6, -1$

25. $\frac{1}{2}, -1$ **27.** $-2, 3, \pm\frac{2}{3}$ **29.** $-2, 1$ **31.** $-4, \frac{1}{2}, 1, 1$

33. (a) $\pm 1, \pm 2, \pm 4$
(b)

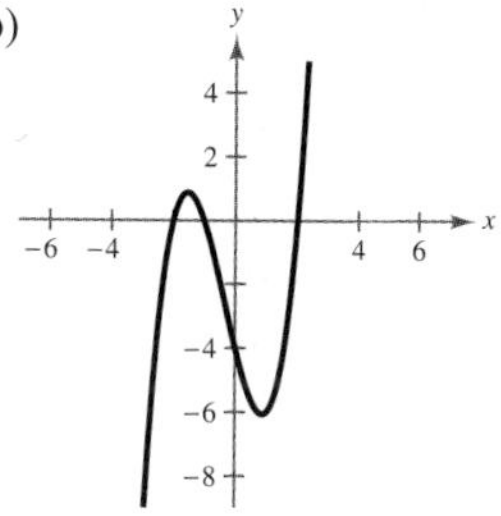

(c) $-2, -1, 2$

35. (a) $\pm 1, \pm 3, \pm\frac{1}{2}, \pm\frac{3}{2}, \pm\frac{1}{4}, \pm\frac{3}{4}$
(b)

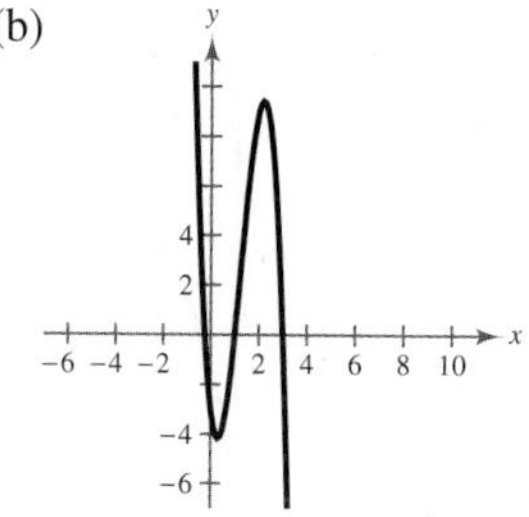

(c) $-\frac{1}{4}, 1, 3$

37. (a) $\pm 1, \pm 2, \pm 4, \pm 8, \pm\frac{1}{2}$
(b)

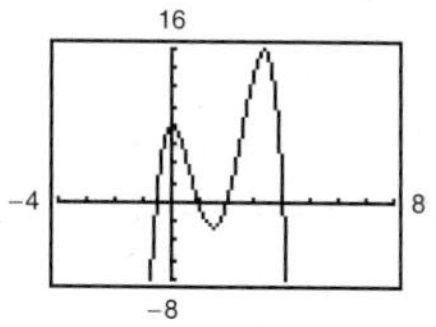

(c) $-\frac{1}{2}, 1, 2, 4$

39. (a) $\pm 1, \pm 3, \pm\frac{1}{2}, \pm\frac{3}{2}, \pm\frac{1}{4}, \pm\frac{3}{4}, \pm\frac{1}{8}, \pm\frac{3}{8}, \pm\frac{1}{16}, \pm\frac{3}{16}, \pm\frac{1}{32}, \pm\frac{3}{32}$
(b)

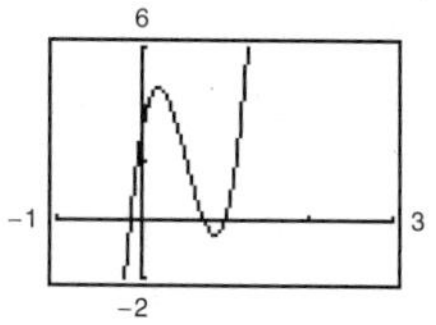

(c) $1, \frac{3}{4}, -\frac{1}{8}$

41. (a) ± 1, about ± 1.414 (b) $\pm 1, \pm\sqrt{2}$
(c) $f(x) = (x + 1)(x - 1)(x + \sqrt{2})(x - \sqrt{2})$

43. (a) 0, 3, 4, about ± 1.414 (b) $0, 3, 4, \pm\sqrt{2}$
(c) $h(x) = x(x - 3)(x - 4)(x + \sqrt{2})(x - \sqrt{2})$

45. $x^3 - x^2 + 25x - 25$ **47.** $x^3 - 12x^2 + 46x - 52$

49. $3x^4 - 17x^3 + 25x^2 + 23x - 22$

51. (a) $(x^2 + 9)(x^2 - 3)$ (b) $(x^2 + 9)(x + \sqrt{3})(x - \sqrt{3})$
(c) $(x + 3i)(x - 3i)(x + \sqrt{3})(x - \sqrt{3})$

53. (a) $(x^2 - 2x - 2)(x^2 - 2x + 3)$
(b) $(x - 1 + \sqrt{3})(x - 1 - \sqrt{3})(x^2 - 2x + 3)$
(c) $(x - 1 + \sqrt{3})(x - 1 - \sqrt{3})(x - 1 + \sqrt{2}i)$
$(x - 1 - \sqrt{2}i)$

55. $\pm 2i, 1$ **57.** $\pm 5i, -\frac{1}{2}, 1$ **59.** $-3 \pm i, \frac{1}{4}$

61. $2, -3 \pm \sqrt{2}i, 1$ **63.** $(x + 6i)(x - 6i); \pm 6i$

65. $(x - 1 - 4i)(x - 1 + 4i); 1 \pm 4i$

67. $(x - 2)(x + 2)(x - 2i)(x + 2i); \pm 2, \pm 2i$

69. $(z - 1 + i)(z - 1 - i)$; $1 \pm i$
71. $(x + 1)(x - 2 + i)(x - 2 - i)$; $-1, 2 \pm i$
73. $(x + 2)(x - 1 + \sqrt{2}i)(x - 1 - \sqrt{2}i)$; $-2, 1 \pm \sqrt{2}i$
75. $(5x + 1)(x - 1 + \sqrt{5}i)(x - 1 - \sqrt{5}i)$; $-\frac{1}{5}, 1 \pm \sqrt{5}i$
77. $(x - 2)^2(x + 2i)(x - 2i)$; $2, \pm 2i$
79. $(x + i)(x - i)(x + 3i)(x - 3i)$; $\pm i, \pm 3i$
81. $-10, -7 \pm 5i$ **83.** $-\frac{3}{4}, 1 \pm \frac{1}{2}i$ **85.** $-2, -\frac{1}{2}, \pm i$
87. One positive zero **89.** One negative zero
91. One positive zero, one negative zero
93. One or three positive zeros **95–97.** Answers will vary.
99. $1, -\frac{1}{2}$ **101.** $-\frac{3}{4}$ **103.** $\pm 2, \pm\frac{3}{2}$ **105.** $\pm 1, \frac{1}{4}$
107. d **108.** a **109.** b **110.** c
111. (a)

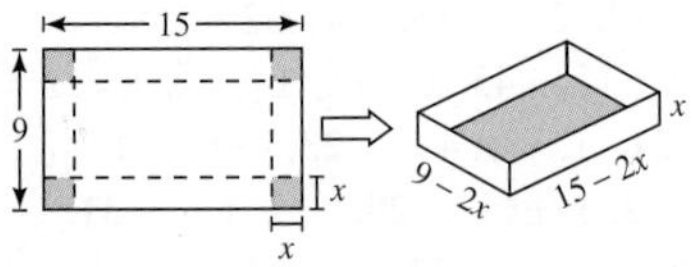

(b) $V(x) = x(9 - 2x)(15 - 2x)$
Domain: $0 < x < \frac{9}{2}$
(c)

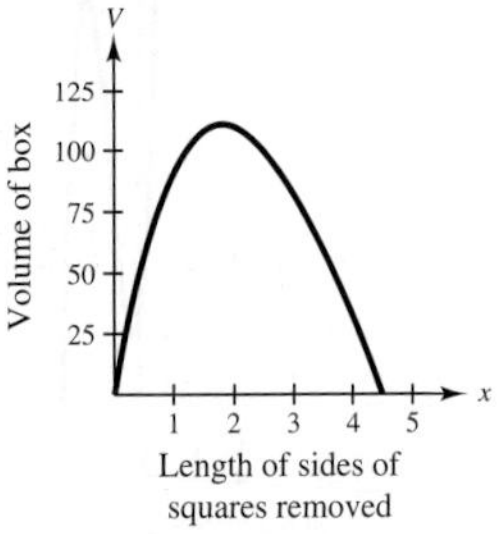

$1.82 \text{ cm} \times 5.36 \text{ cm} \times 11.36 \text{ cm}$
(d) $\frac{1}{2}, \frac{7}{2}, 8$; 8 is not in the domain of V.
113. (a) $V(x) = x^3 + 9x^2 + 26x + 24 = 120$ (b) 4 ft × 5 ft × 6 ft
115. False. The most complex zeros it can have is two, and the Linear Factorization Theorem guarantees that there are three linear factors, so one zero must be real.
117. r_1, r_2, r_3 **119.** $5 + r_1, 5 + r_2, 5 + r_3$
121. The zeros cannot be determined.
123. Answers will vary. There are infinitely many possible functions for f. *Sample equation and graph:*
$f(x) = -2x^3 + 3x^2 + 11x - 6$

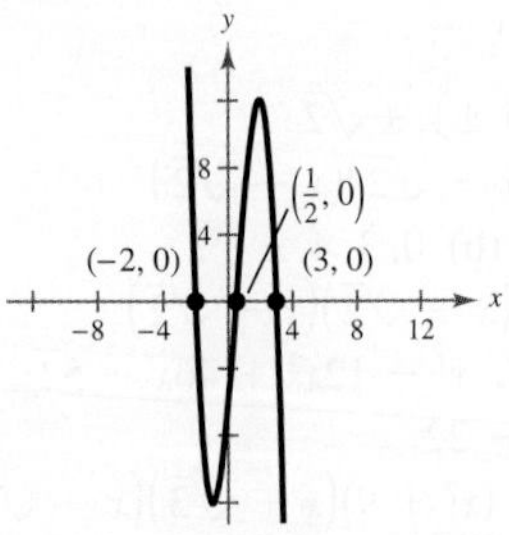

125. $f(x) = x^3 - 3x^2 + 4x - 2$ **127.** $f(x) = x^4 + 5x^2 + 4$
129. (a) $x^2 + b$ (b) $x^2 - 2ax + a^2 + b^2$
131. (a) Not correct because f has $(0, 0)$ as an intercept.
(b) Not correct because the function must be at least a fourth-degree polynomial.
(c) Correct function
(d) Not correct because k has $(-1, 0)$ as an intercept.

Section 2.6 *(page 177)*

1. rational functions **3.** horizontal asymptote
5. Domain: all real numbers x except $x = 1$
$f(x) \to -\infty$ as $x \to 1^-$, $f(x) \to \infty$ as $x \to 1^+$
7. Domain: all real numbers x except $x = \pm 1$
$f(x) \to \infty$ as $x \to -1^-$ and as $x \to 1^+$,
$f(x) \to -\infty$ as $x \to -1^+$ and as $x \to 1^-$
9. Vertical asymptote: $x = 0$
Horizontal asymptote: $y = 0$
11. Vertical asymptote: $x = 5$
Horizontal asymptote: $y = -1$
13. Vertical asymptotes: $x = \pm 1$
15. Horizontal asymptote: $y = 3$
17. (a) Domain: all real numbers x except $x = -2$
(b) y-intercept: $\left(0, \frac{1}{2}\right)$
(c) Vertical asymptote: $x = -2$
Horizontal asymptote: $y = 0$
(d)

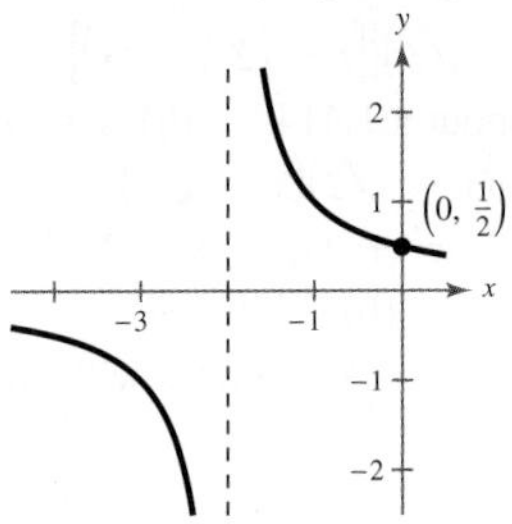

19. (a) Domain: all real numbers x except $x = -4$
(b) y-intercept: $\left(0, -\frac{1}{4}\right)$
(c) Vertical asymptote: $x = -4$
Horizontal asymptote: $y = 0$
(d)

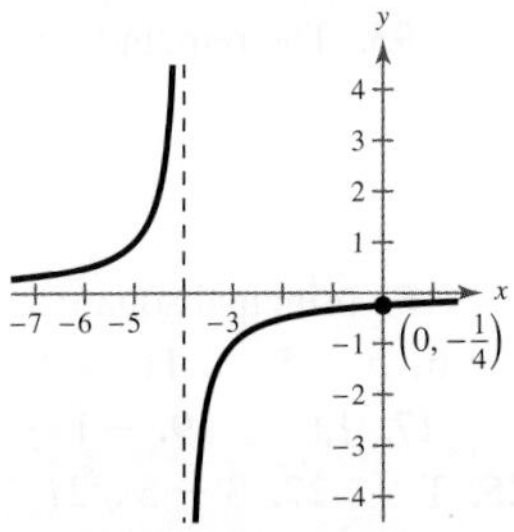

21. (a) Domain: all real numbers x except $x = -2$
(b) x-intercept: $\left(-\frac{7}{2}, 0\right)$
y-intercept: $\left(0, \frac{7}{2}\right)$
(c) Vertical asymptote: $x = -2$
Horizontal asymptote: $y = 2$
(d)

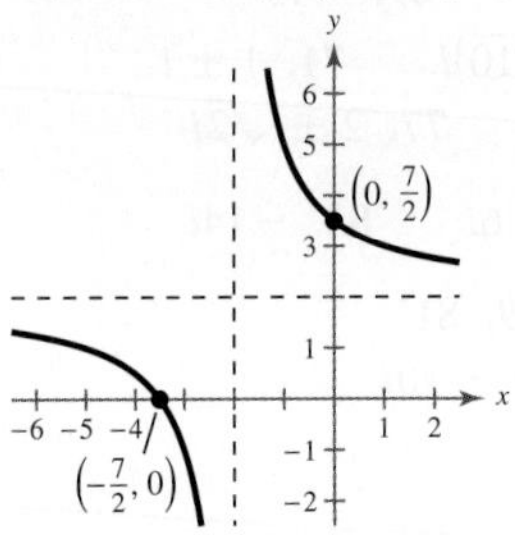

23. (a) Domain: all real numbers x (b) Intercept: $(0, 0)$
(c) Horizontal asymptote: $y = 1$
(d)

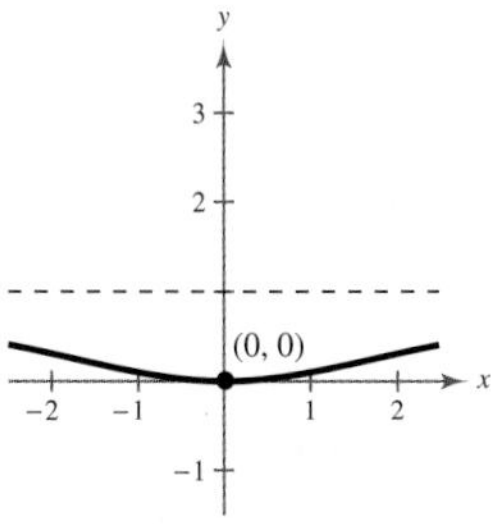

25. (a) Domain: all real numbers t except $t = 0$
(b) t-intercept: $\left(\frac{1}{2}, 0\right)$
(c) Vertical asymptote: $t = 0$
Horizontal asymptote: $y = -2$
(d)

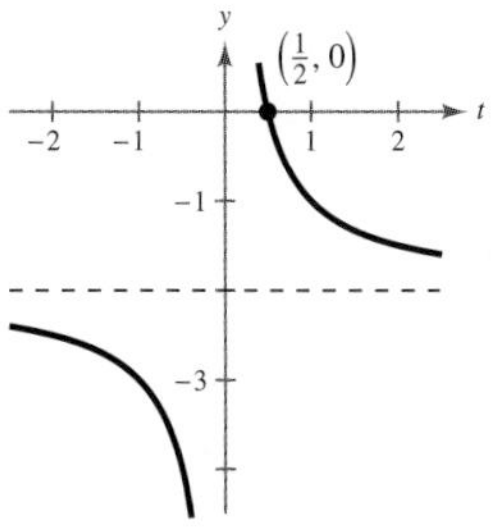

27. (a) Domain: all real numbers x except $x = \pm 2$
(b) x-intercepts: $(1, 0)$ and $(4, 0)$
y-intercept: $(0, -1)$
(c) Vertical asymptotes: $x = \pm 2$
Horizontal asymptote: $y = 1$
(d)

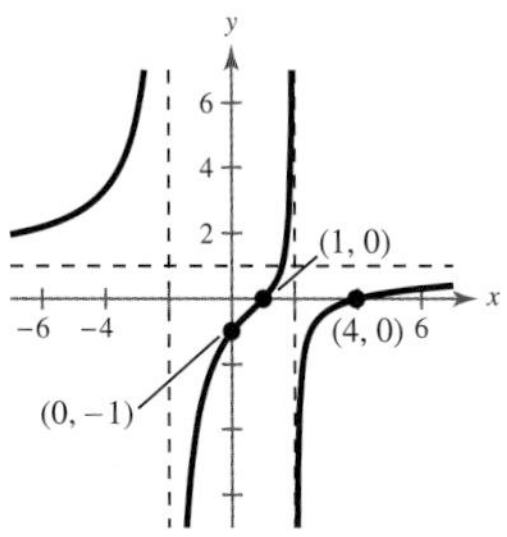

29. (a) Domain: all real numbers x except $x = \pm 4$
(b) y-intercept: $\left(0, \frac{1}{4}\right)$
(c) Vertical asymptote: $x = -4$
Horizontal asymptote: $y = 0$
(d)

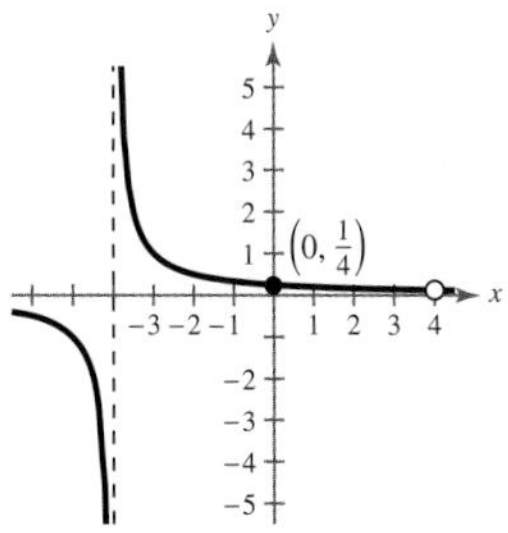

31. (a) Domain: all real numbers t except $t = 1$
(b) t-intercept: $(-1, 0)$
y-intercept: $(0, 1)$
(c) Vertical asymptote: None
Horizontal asymptote: None
(d)

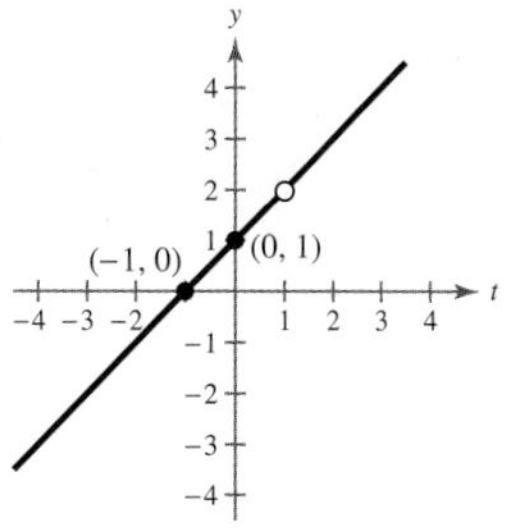

33. (a) Domain: all real numbers x except $x = -1, 5$
(b) x-intercept: $(-5, 0)$
y-intercept: $(0, 5)$
(c) Vertical asymptote: $x = -1$
Horizontal asymptote: $y = 1$
(d)

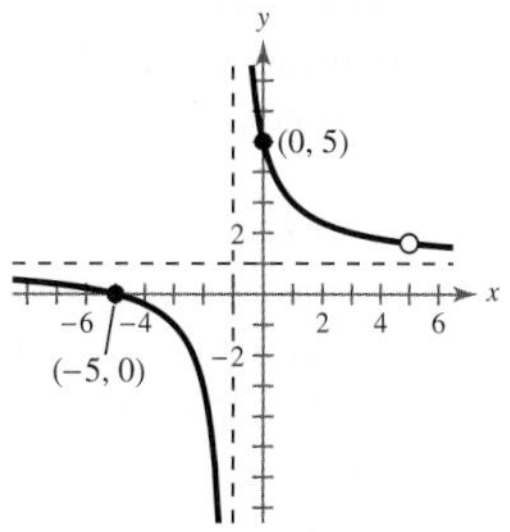

35. (a) Domain: all real numbers x except $x = 2, -3$
(b) Intercept: $(0, 0)$
(c) Vertical asymptote: $x = 2$
Horizontal asymptote: $y = 1$
(d)

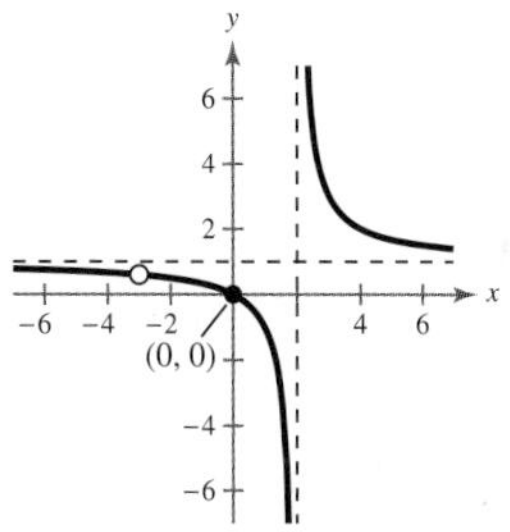

37. (a) Domain: all real numbers x except $x = -1, \frac{1}{2}$
(b) x-intercept: $(4, 0)$
y-intercept: $(0, 4)$
(c) Vertical asymptote: $x = \frac{1}{2}$
Horizontal asymptote: $y = \frac{1}{2}$
(d)

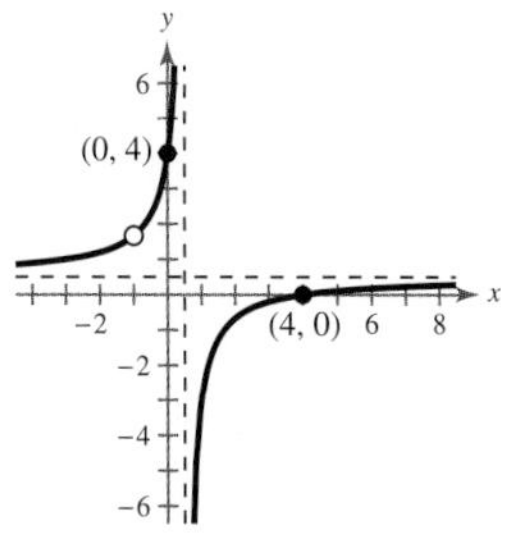

39. (a) Domain: all real numbers x except $x = \pm 1, 2$
(b) x-intercepts: $(3, 0), \left(-\frac{1}{2}, 0\right)$
y-intercept: $\left(0, -\frac{3}{2}\right)$
(c) Vertical asymptotes: $x = 2, x = \pm 1$
Horizontal asymptote: $y = 0$

CHAPTER 2

(d)

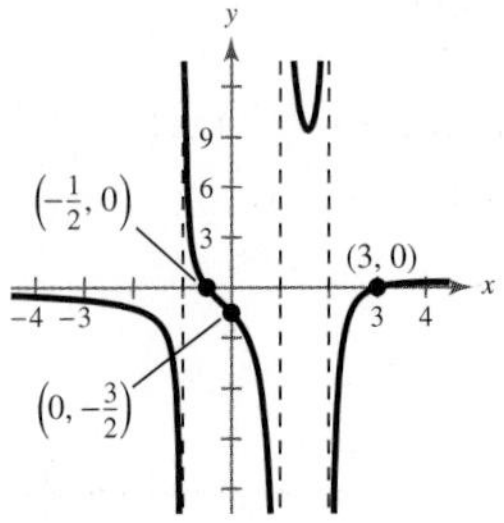

41. d **42.** a **43.** c **44.** b

45. (a) Domain of f: all real numbers x except $x = -1$
Domain of g: all real numbers x

(b)

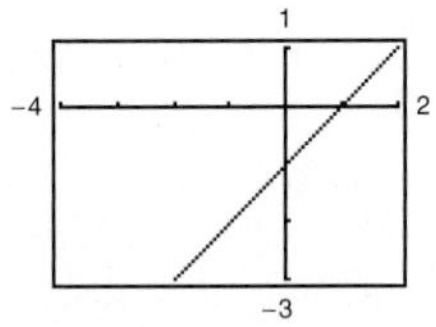

(c) Because there are only a finite number of pixels, the graphing utility may not attempt to evaluate the function where it does not exist.

47. (a) Domain of f: all real numbers x except $x = 0, 2$
Domain of g: all real numbers x except $x = 0$

(b)

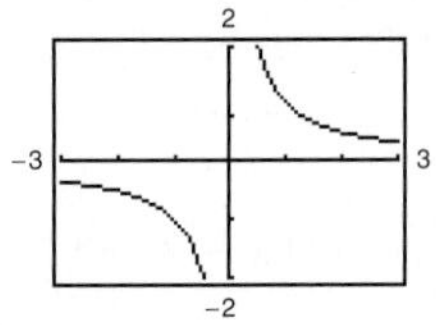

(c) Because there are only a finite number of pixels, the graphing utility may not attempt to evaluate the function where it does not exist.

49. (a) Domain: all real numbers x except $x = 0$
(b) x-intercepts: $(-3, 0)$, $(3, 0)$
(c) Vertical asymptote: $x = 0$
Slant asymptote: $y = x$
(d)

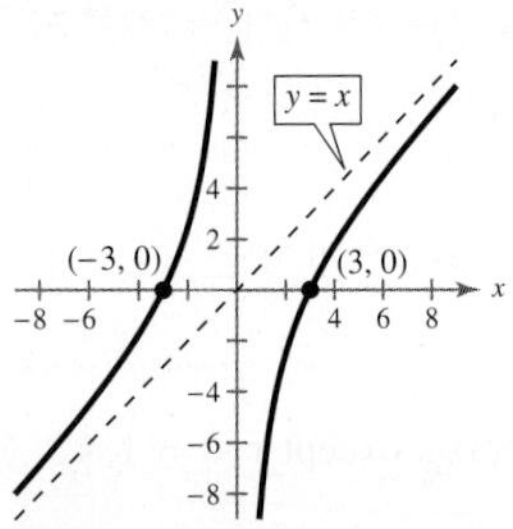

51. (a) Domain: all real numbers x except $x = 0$
(b) No intercepts
(c) Vertical asymptote: $x = 0$
Slant asymptote: $y = 2x$
(d)

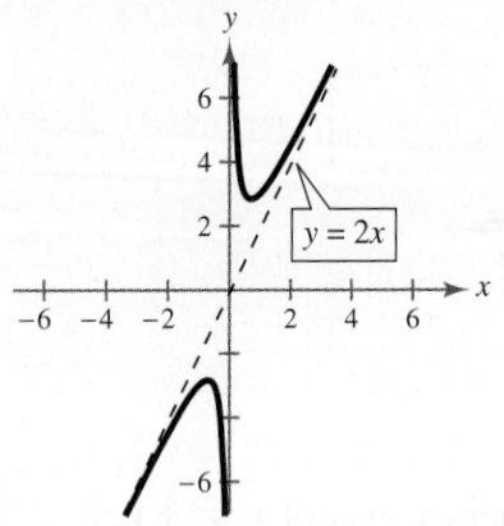

53. (a) Domain: all real numbers x except $x = 0$
(b) No intercepts
(c) Vertical asymptote: $x = 0$
Slant asymptote: $y = x$

(d)

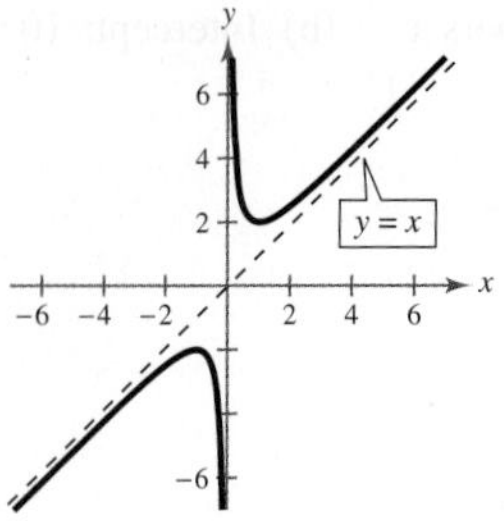

55. (a) Domain: all real numbers t except $t = -5$
(b) y-intercept: $\left(0, -\frac{1}{5}\right)$
(c) Vertical asymptote: $t = -5$
Slant asymptote: $y = -t + 5$
(d)

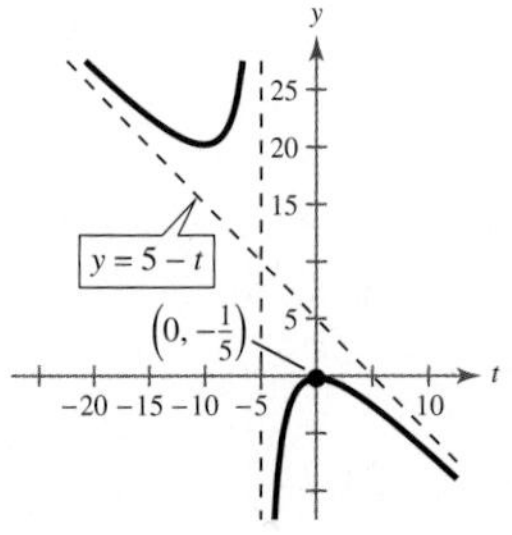

57. (a) Domain: all real numbers x except $x = \pm 2$
(b) Intercept: $(0, 0)$
(c) Vertical asymptotes: $x = \pm 2$
Slant asymptote: $y = x$
(d)

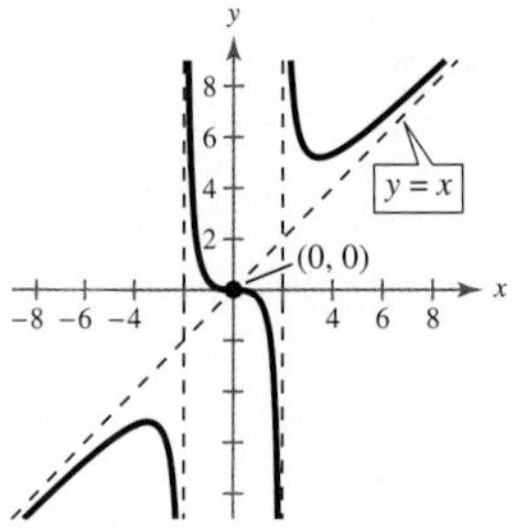

59. (a) Domain: all real numbers x except $x = 1$
(b) y-intercept: $(0, -1)$
(c) Vertical asymptote: $x = 1$
Slant asymptote: $y = x$
(d)

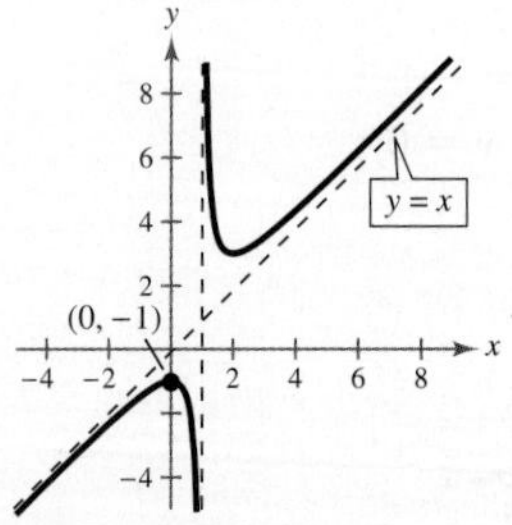

61. (a) Domain: all real numbers x except $x = -1, -2$
(b) y-intercept: $\left(0, \frac{1}{2}\right)$
x-intercepts: $\left(\frac{1}{2}, 0\right)$, $(1, 0)$
(c) Vertical asymptote: $x = -2$
Slant asymptote: $y = 2x - 7$

(d)

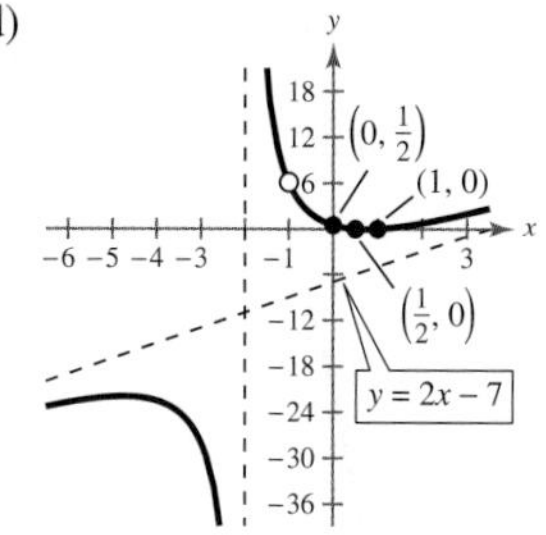

63.

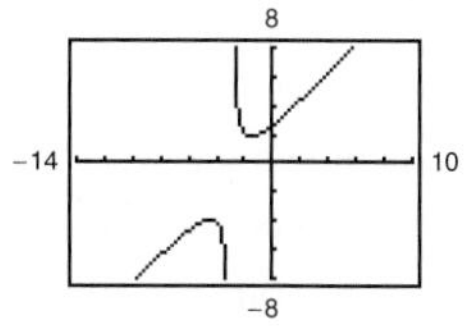

Domain: all real numbers x except $x = -3$
Vertical asymptote: $x = -3$
Slant asymptote: $y = x + 2$
$y = x + 2$

65.

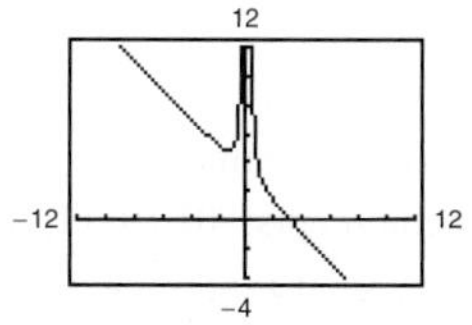

Domain: all real numbers x except $x = 0$
Vertical asymptote: $x = 0$
Slant asymptote: $y = -x + 3$
$y = -x + 3$

67. (a) $(-1, 0)$ (b) -1

69. (a) $(1, 0), (-1, 0)$ (b) ± 1

71. (a)

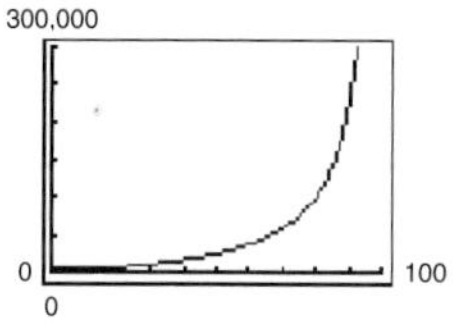

(b) \$4411.76; \$25,000; \$225,000
(c) No. The function is undefined at $p = 100$.

73. (a) $A = \dfrac{2x(x + 11)}{x - 4}$ (b) $(4, \infty)$

(c)

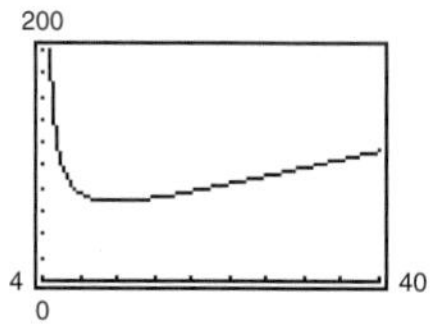

11.75 in. × 5.87 in.

75. (a) Answers will vary.
(b) Vertical asymptote: $x = 25$
Horizontal asymptote: $y = 25$
(c)

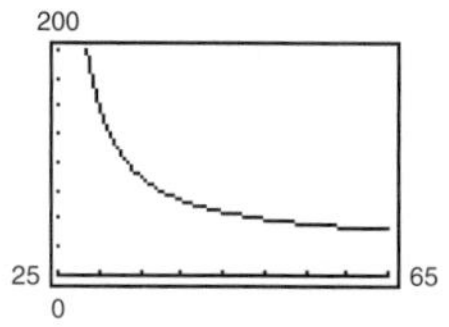

(d)

x	30	35	40	45	50	55	60
y	150	87.5	66.7	56.3	50	45.8	42.9

(e) *Sample answer:* No. You might expect the average speed for the round trip to be the average of the average speeds for the two parts of the trip.
(f) No. At 20 miles per hour you would use more time in one direction than is required for the round trip at an average speed of 50 miles per hour.

77. False. Polynomials do not have vertical asymptotes.

79. False. If the degree of the numerator is greater than the degree of the denominator, then no horizontal asymptote exists. However, a slant asymptote exists only if the degree of the numerator is one greater than the degree of the denominator.

81. b

Section 2.7 *(page 187)*

1. positive; negative **3.** zeros; undefined values

5. (a) No (b) Yes (c) Yes (d) No

7. (a) Yes (b) No (c) No (d) Yes

9. $-\frac{2}{3}, 1$ **11.** 4, 5

13. $(-3, 3)$

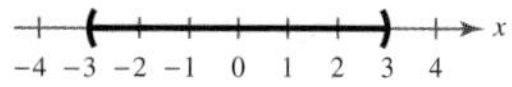

15. $[-7, 3]$

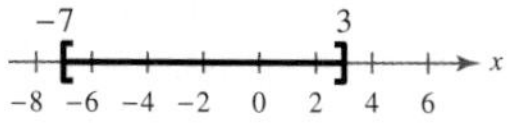

17. $(-\infty, -5] \cup [1, \infty)$

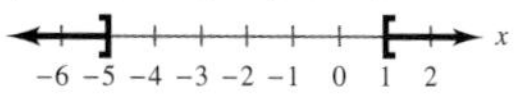

19. $(-3, 2)$

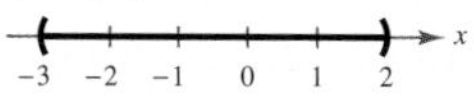

21. $(-3, 1)$

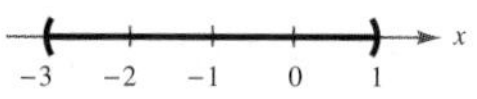

23. $\left(-\infty, -\frac{4}{3}\right) \cup (5, \infty)$

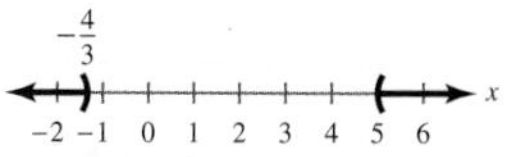

25. $(-\infty, -3) \cup (6, \infty)$

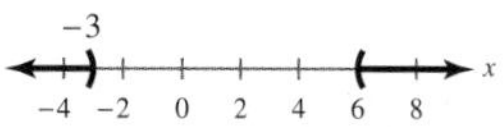

27. $(-1, 1) \cup (3, \infty)$

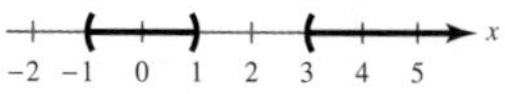

29. $(-\infty, 0) \cup \left(0, \frac{3}{2}\right)$

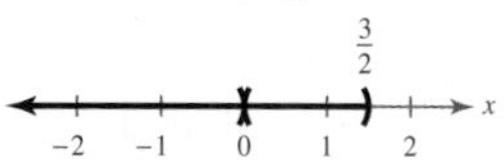

31. $[-2, 0] \cup [2, \infty)$

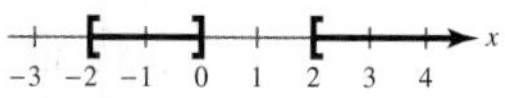

33. $[-2, \infty)$

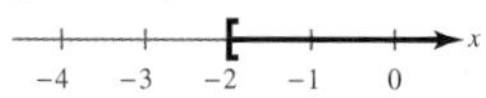

35. The solution set consists of the single real number $\frac{1}{2}$.

37. The solution set is empty.

39. $(-\infty, 0) \cup \left(\frac{1}{4}, \infty\right)$

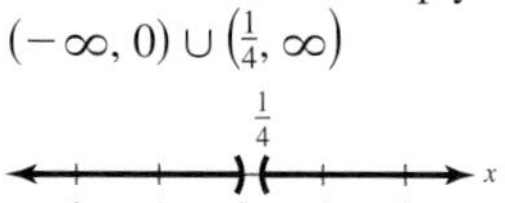

41. $\left(-\infty, \frac{5}{3}\right] \cup (5, \infty)$

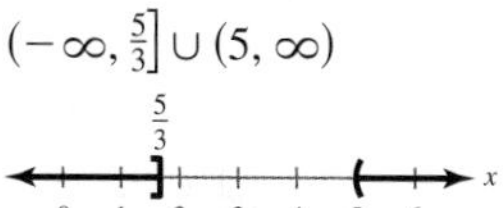

43. $(-\infty, -1) \cup (4, \infty)$

45. $(-5, 3) \cup (11, \infty)$

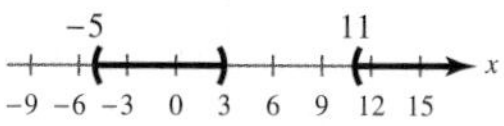

47. $\left(-\frac{3}{4}, 3\right) \cup [6, \infty)$

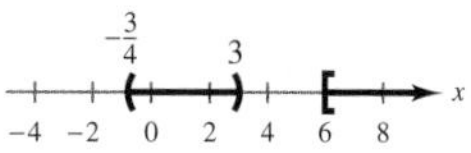

49. $(-3, -2] \cup [0, 3)$

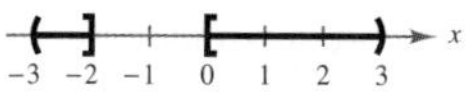

51. $(-\infty, -1) \cup (1, \infty)$

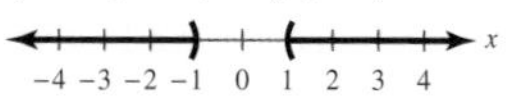

53.

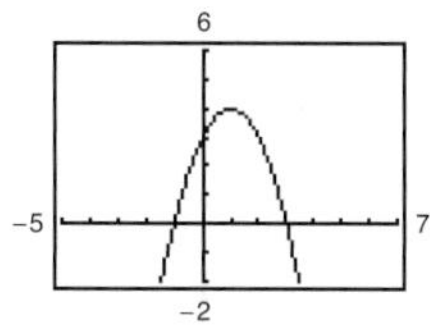

(a) $x \le -1, x \ge 3$
(b) $0 \le x \le 2$

55.

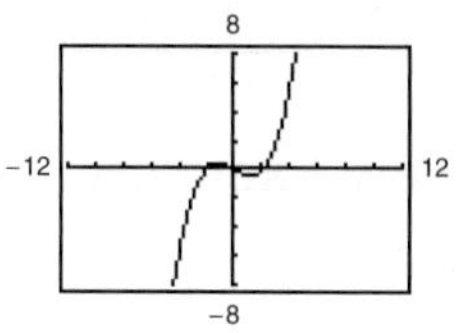

(a) $-2 \le x \le 0$, $2 \le x < \infty$
(b) $x \le 4$

57.

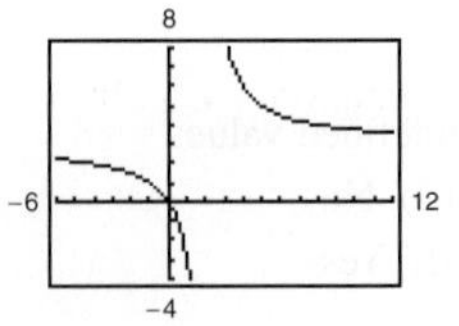

(a) $0 \le x < 2$
(b) $2 < x \le 4$

59.

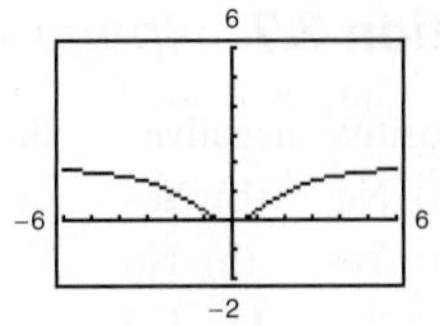

(a) $|x| \ge 2$
(b) $-\infty < x < \infty$

61. $[-2, 2]$ **63.** $(-\infty, 4] \cup [5, \infty)$ **65.** $(-5, 0] \cup (7, \infty)$
67. $(-3.51, 3.51)$ **69.** $(-0.13, 25.13)$ **71.** $(2.26, 2.39)$
73. (a) $t = 10$ sec (b) 4 sec $< t <$ 6 sec
75. 13.8 m $\le L \le$ 36.2 m
77. $40{,}000 \le x \le 50{,}000$; $\$50.00 \le p \le \$55.00$
79. (a) and (c)

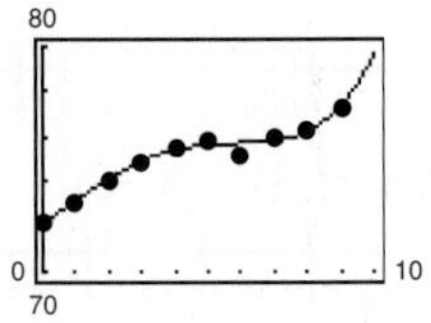

The model fits the data well.

(b) $N = 0.00406t^4 - 0.0564t^3 + 0.147t^2 + 0.86t + 72.2$
(d) 2001
(e) No. The model can be used to predict enrollments for years close to those in its domain but when you project too far into the future, the numbers predicted by the model increase too rapidly to be considered reasonable.

81. $R_1 \ge 2$ ohms
83. True. The test intervals are $(-\infty, -3)$, $(-3, 1)$, $(1, 4)$, and $(4, \infty)$.
85. (a) $(-\infty, -4] \cup [4, \infty)$
(b) When $a > 0$ and $c > 0$, $b \le -2\sqrt{ac}$ or $b \ge 2\sqrt{ac}$.
87. (a) $\left(-\infty, -2\sqrt{30}\right] \cup \left[2\sqrt{30}, \infty\right)$
(b) When $a > 0$ and $c > 0$, $b \le -2\sqrt{ac}$ or $b \ge 2\sqrt{ac}$.

89.

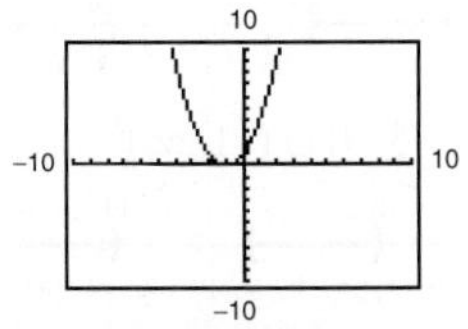

For part (b), the y-values that are less than or equal to 0 occur only at $x = -1$.

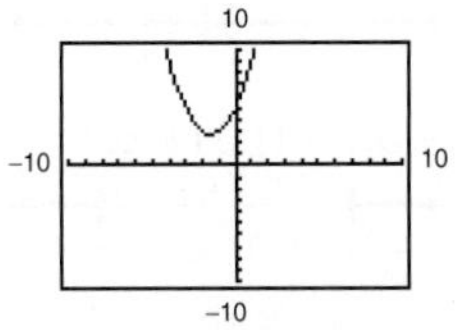

For part (c), there are no y-values that are less than 0.

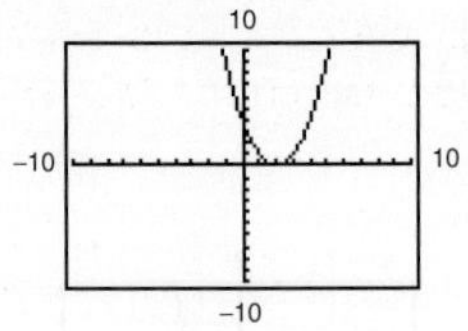

For part (d), the y-values that are greater than 0 occur for all values of x except 2.

Review Exercises *(page 192)*

1. (a)

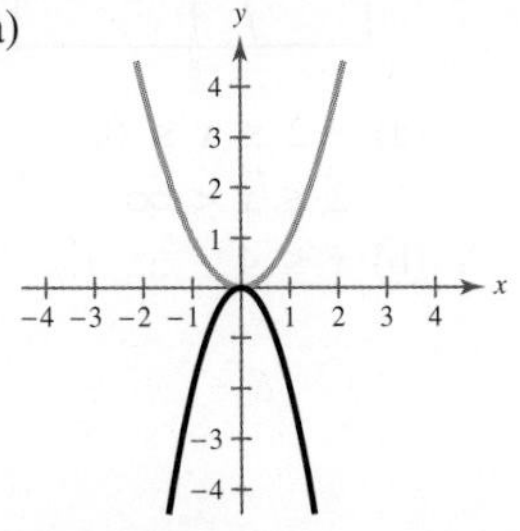

Vertical stretch and reflection in the x-axis

(b)

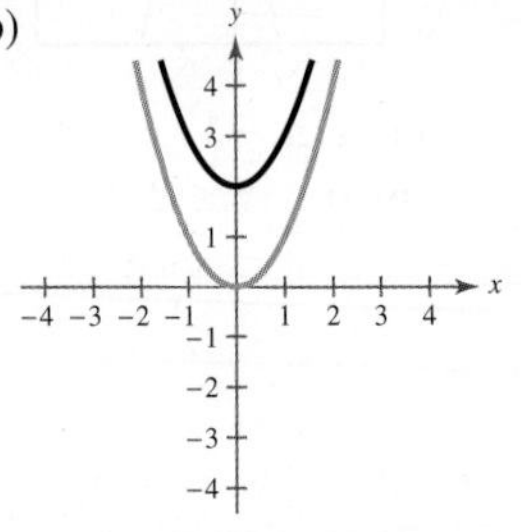

Vertical shift two units up

3. $g(x) = (x - 1)^2 - 1$

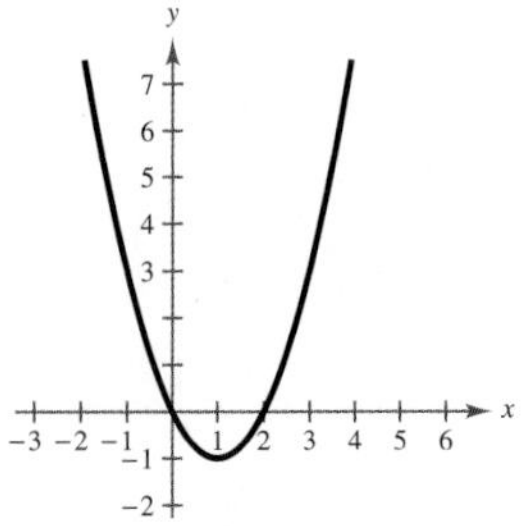

Vertex: $(1, -1)$
Axis of symmetry: $x = 1$
x-intercepts: $(0, 0)$, $(2, 0)$

5. $h(x) = -(x - 2)^2 + 7$

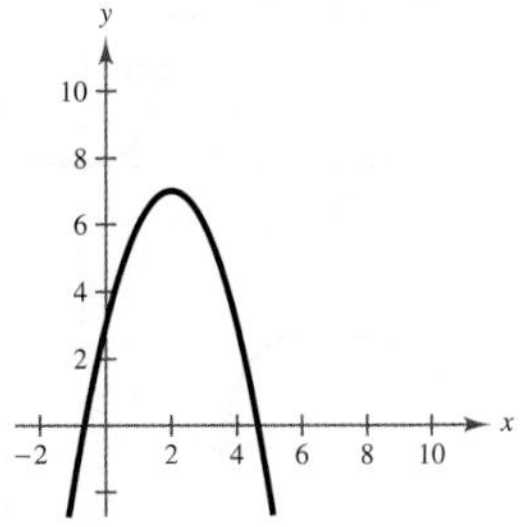

Vertex: $(2, 7)$
Axis of symmetry: $x = 2$
x-intercepts: $\left(2 \pm \sqrt{7}, 0\right)$

7. $h(x) = 4\left(x + \frac{1}{2}\right)^2 + 12$

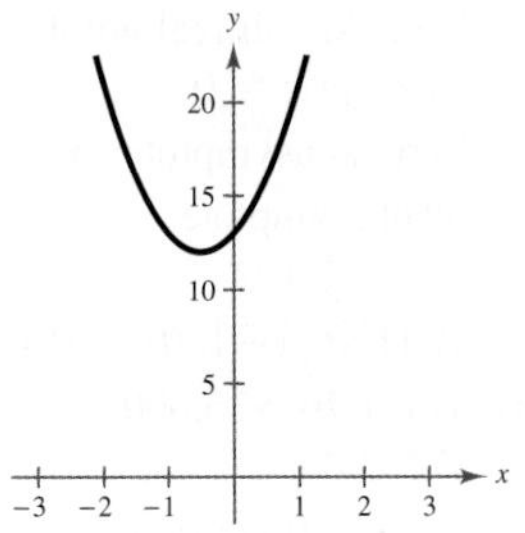

Vertex: $\left(-\frac{1}{2}, 12\right)$
Axis of symmetry: $x = -\frac{1}{2}$
No x-intercept

9. (a) $y = 500 - x$
$A(x) = 500x - x^2$
(b) $x = 250$, $y = 250$

11.

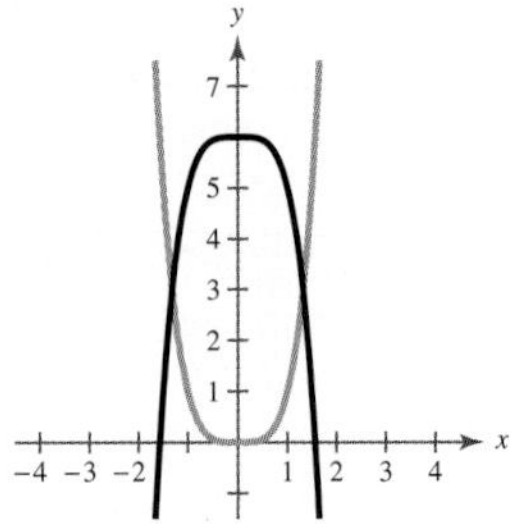

13. Falls to the left, falls to the right
15. Rises to the left, rises to the right

17. (a) Falls to the left, rises to the right
(b) $-2, 0$ (c) Answers will vary.
(d)

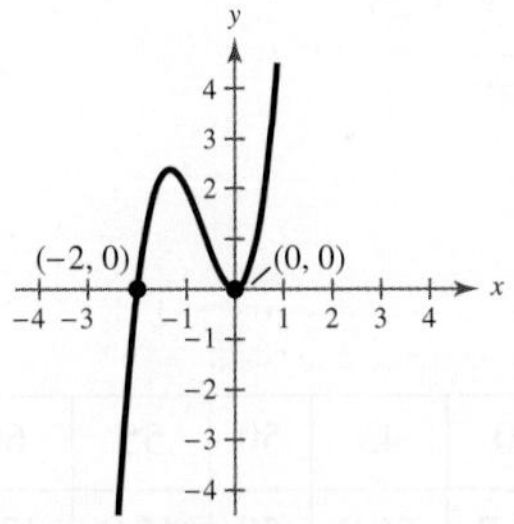

19. (a) Rises to the left, falls to the right
(b) -1 (c) Answers will vary.
(d)

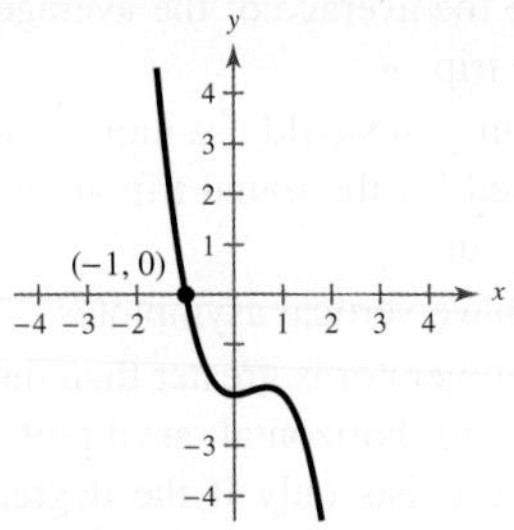

21. (a) $[-1, 0]$ (b) About -0.900

23. $6x + 3 + \dfrac{17}{5x - 3}$

25. (a) Yes (b) Yes (c) Yes (d) No

27. (a) Answers will vary. (b) $(2x + 5), (x - 3)$
(c) $f(x) = (2x + 5)(x - 3)(x + 6)$ (d) $-\frac{5}{2}, 3, -6$
(e)

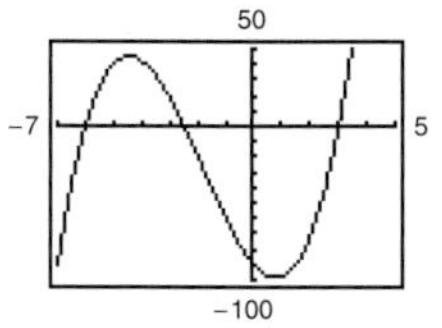

29. $8 + 10i$ **31.** $3 + 7i$ **33.** $63 + 77i$ **35.** $\frac{23}{17} + \frac{10}{17}i$

37. $1 \pm 3i$ **39.** 2 **41.** $-6, -2, 5$

43. $g(x) = (x + 2)(x - 3)(x - 6)$; $-2, 3, 6$

45. One or three positive zeros, two or no negative zeros

47. Domain: all real numbers x except $x = -10$
Vertical asymptote: $x = -10$
Horizontal asymptote: $y = 3$

49. (a) Domain: all real numbers x except $x = 0$
(b) No intercepts
(c) Vertical asymptote: $x = 0$
Horizontal asymptote: $y = 0$
(d)

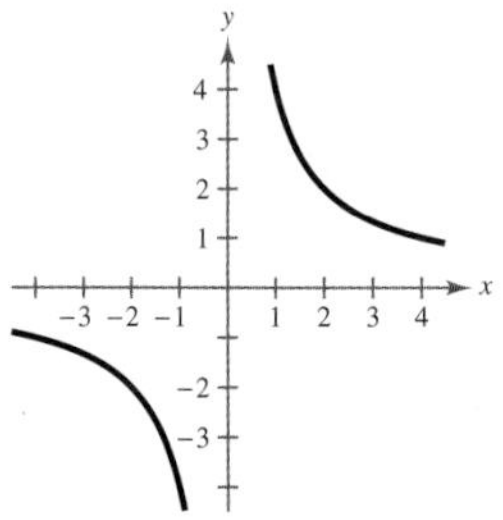

51. (a) Domain: all real numbers x (b) Intercept: $(0, 0)$
(c) Horizontal asymptote: $y = 0$
(d)

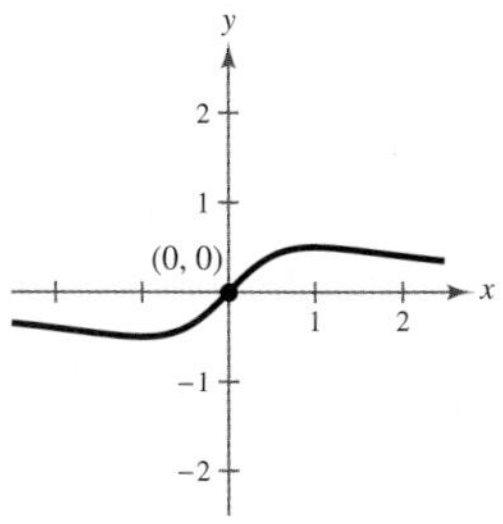

53. (a) Domain: all real numbers x except $x = 0, \frac{1}{3}$
(b) x-intercept: $\left(\frac{3}{2}, 0\right)$
(c) Vertical asymptote: $x = 0$
Horizontal asymptote: $y = 2$
(d)

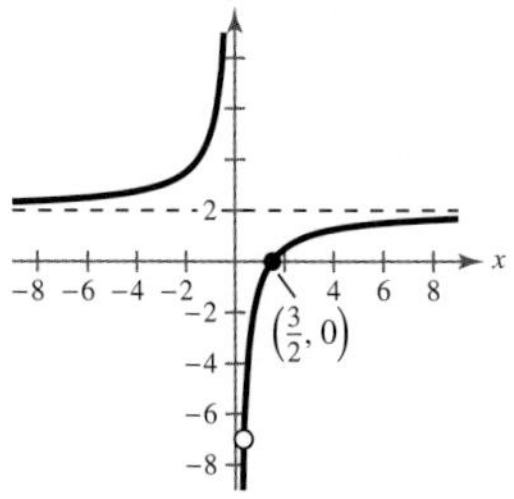

55. (a) Domain: all real numbers x
(b) Intercept: $(0, 0)$ (c) Slant asymptote: $y = 2x$
(d)

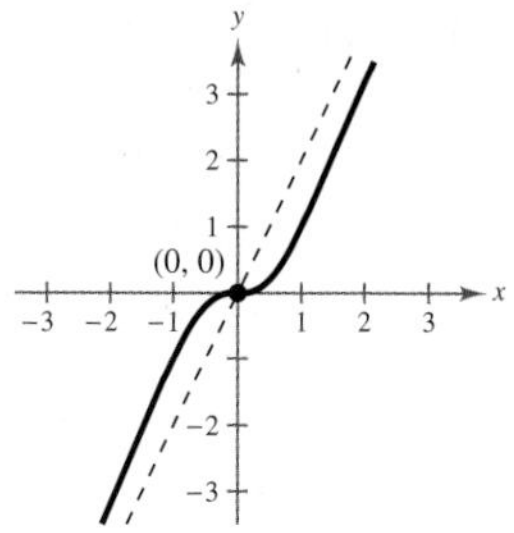

57. (a)

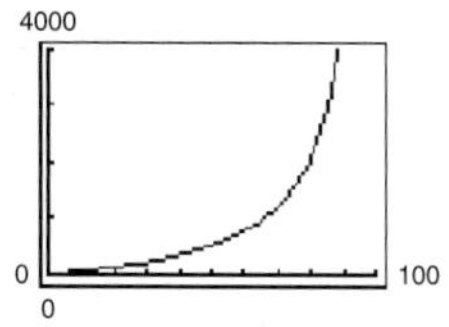

(b) \$176 million; \$528 million; \$1584 million (or \$1.584 billion)
(c) No

59. $\left(-\frac{2}{3}, \frac{1}{4}\right)$

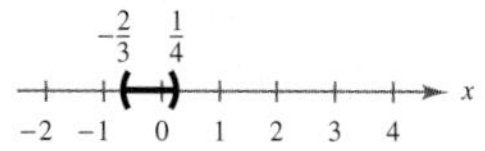

61. $[-5, -1) \cup (1, \infty)$

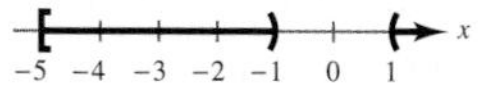

63. 9 days

65. False. A fourth-degree polynomial can have at most four zeros, and complex zeros occur in conjugate pairs.

Chapter Test *(page 194)*

1. (a) Reflection in the x-axis followed by a vertical shift two units up

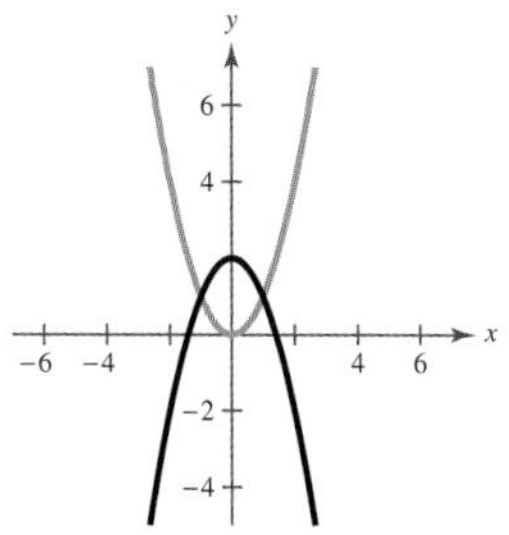

(b) Horizontal shift $\frac{3}{2}$ units to the right

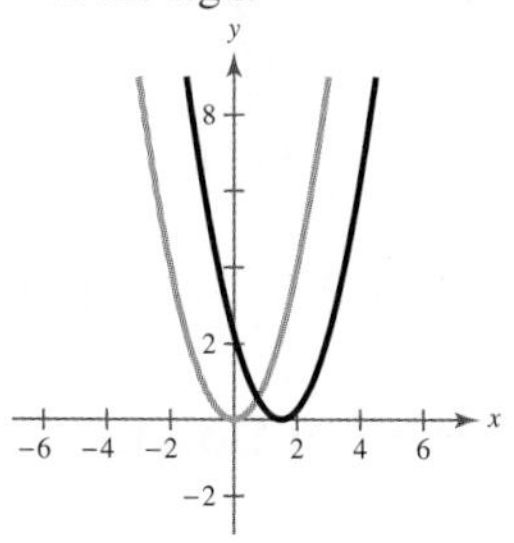

2. $y = (x - 3)^2 - 6$

3. (a) 50 ft
(b) 5. Yes, changing the constant term results in a vertical translation of the graph and therefore changes the maximum height.

4. Rises to the left, falls to the right

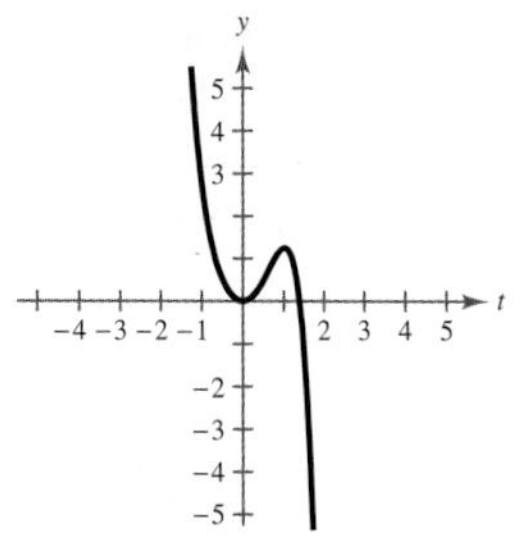

5. $3x + \dfrac{x - 1}{x^2 + 1}$ **6.** $2x^3 + 4x^2 + 3x + 6 + \dfrac{9}{x - 2}$

7. $(2x - 5)(x + \sqrt{3})(x - \sqrt{3})$;
Zeros: $\frac{5}{2}, \pm\sqrt{3}$

8. (a) $-3 + 5i$ (b) 7

9. $2 - i$ **10.** $f(x) = x^4 - 7x^3 + 17x^2 - 15x$

11. $f(x) = 4x^2 - 16x + 16$ **12.** $-5, -\frac{2}{3}, 1$

13. $-2, 4, -1 \pm \sqrt{2}i$

14. x-intercepts: $(-2, 0), (2, 0)$
Vertical asymptote: $x = 0$
Horizontal asymptote: $y = -1$

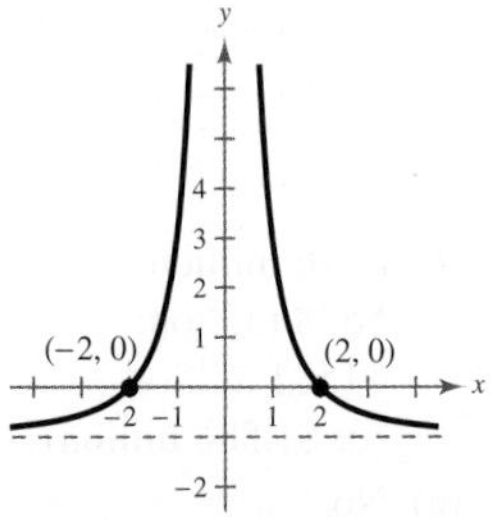

15. x-intercept: $\left(-\frac{3}{2}, 0\right)$
y-intercept: $\left(0, \frac{3}{4}\right)$
Vertical asymptote: $x = -4$
Horizontal asymptote: $y = 2$

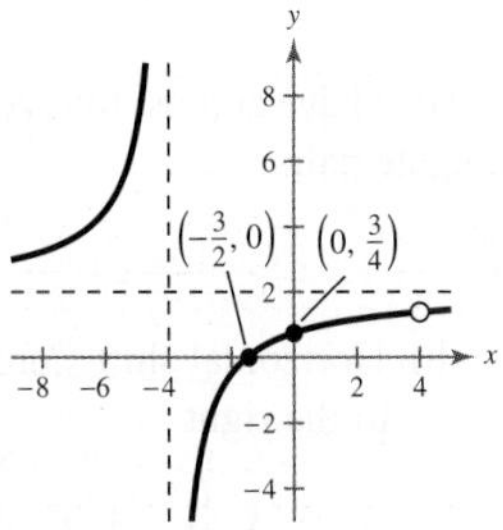

16. y-intercept: $(0, -2)$
Vertical asymptote: $x = 1$
Slant asymptote: $y = x + 1$

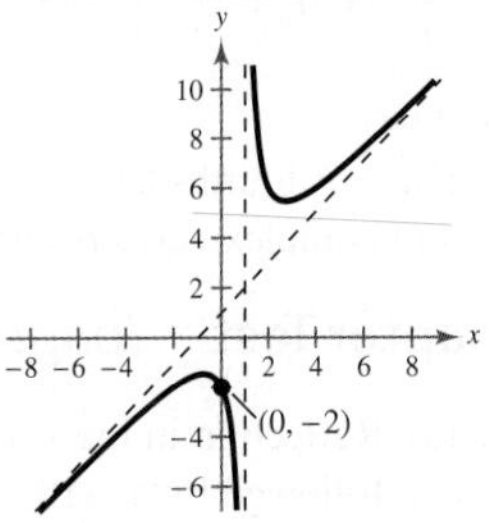

17. $x < -4$ or $x > \frac{3}{2}$

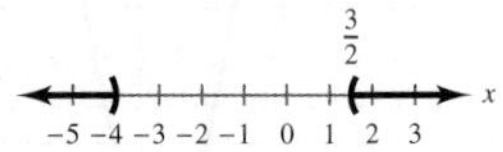

18. $x \leq -12$ or $-6 < x < 0$

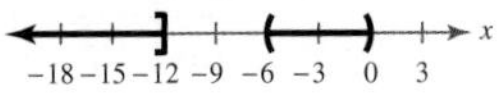

Problem Solving *(page 197)*

1. Answers will vary. **3.** 2 in. × 2 in. × 5 in.

5. (a) and (b) $y = -x^2 + 5x - 4$

7. (a) $f(x) = (x - 2)x^2 + 5 = x^3 - 2x^2 + 5$
(b) $f(x) = -(x + 3)x^2 + 1 = -x^3 - 3x^2 + 1$

9. $(a + bi)(a - bi) = a^2 + abi - abi - b^2i^2$
$= a^2 + b^2$

11. (a) As $|a|$ increases, the graph stretches vertically. For $a < 0$, the graph is reflected in the x-axis.
(b) As $|b|$ increases, the vertical asymptote is translated. For $b > 0$, the graph is translated to the right. For $b < 0$, the graph is reflected in the x-axis and is translated to the left.

13. No. Complex zeros always occur in conjugate pairs.

Chapter 3

Section 3.1 *(page 208)*

1. algebraic **3.** One-to-One **5.** $A = P\left(1 + \frac{r}{n}\right)^{nt}$

7. 0.863 **9.** 0.006 **11.** 1767.767

13. d **14.** c **15.** a **16.** b

17.

x	-2	-1	0	1	2
$f(x)$	4	2	1	0.5	0.25

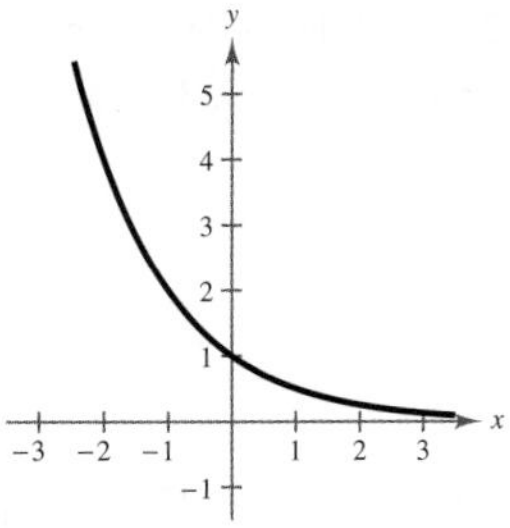

19.

x	-2	-1	0	1	2
$f(x)$	36	6	1	0.167	0.028

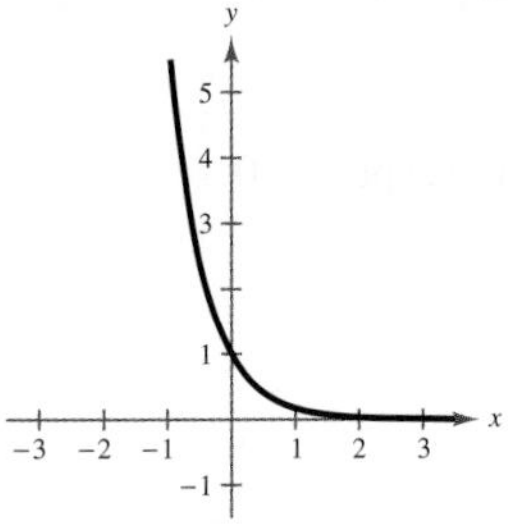

21.

x	-2	-1	0	1	2
$f(x)$	0.125	0.25	0.5	1	2

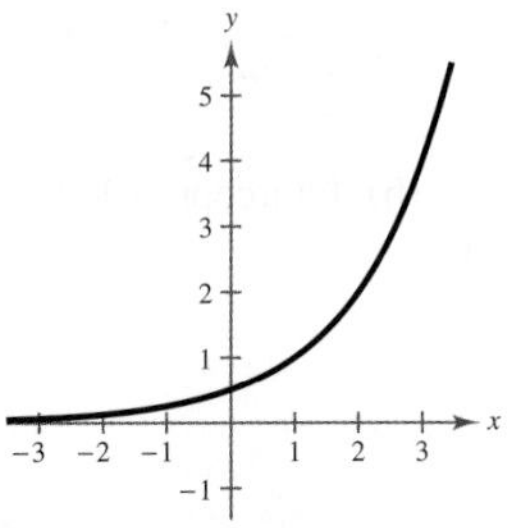

23. $x = 2$ **25.** $x = -5$ **27.** Shift the graph of f one unit up.

29. Reflect the graph of f in the origin.

31.

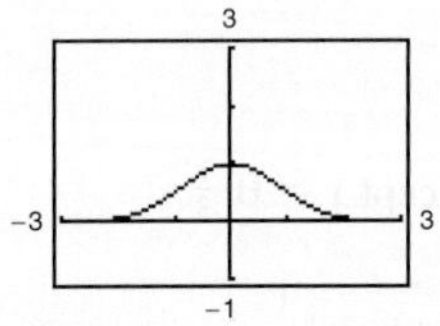

33.

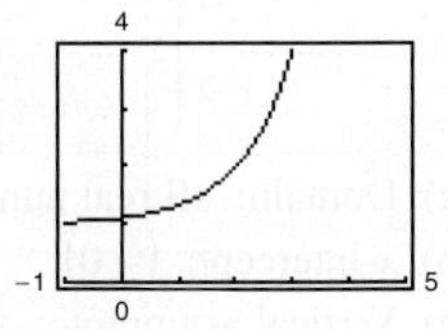

35. 24.533 **37.** 7166.647

39.

x	-2	-1	0	1	2
$f(x)$	0.135	0.368	1	2.718	7.389

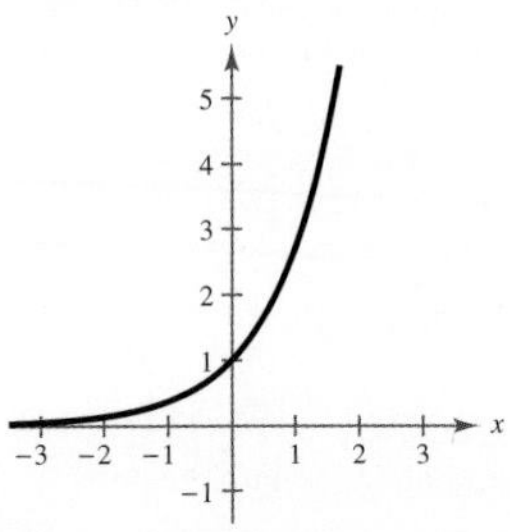

41.

x	-8	-7	-6	-5	-4
$f(x)$	0.055	0.149	0.406	1.104	3

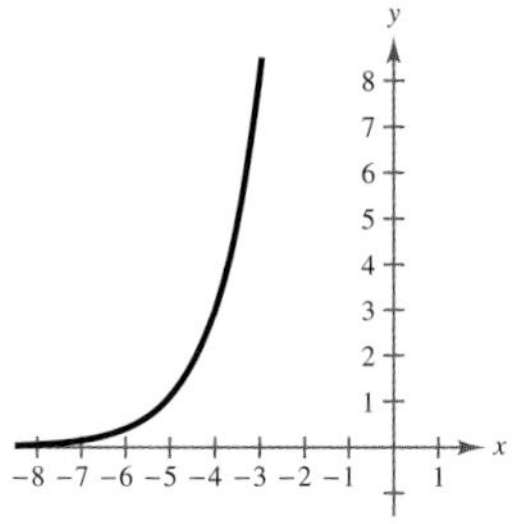

43.

x	-2	-1	0	1	2
$f(x)$	4.037	4.100	4.271	4.736	6

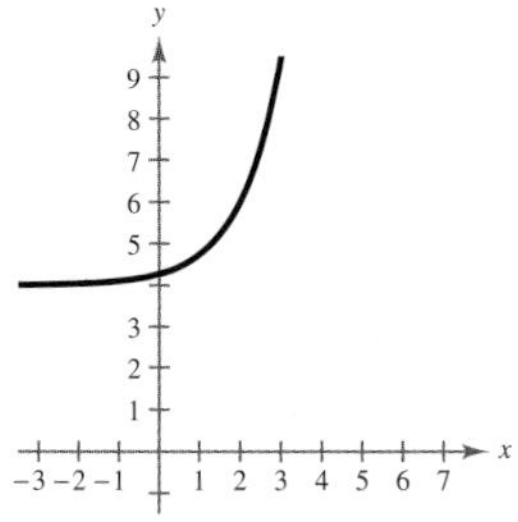

45. 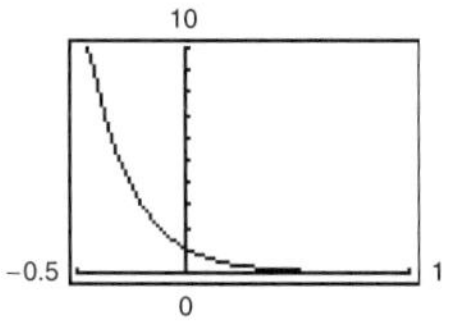

47. 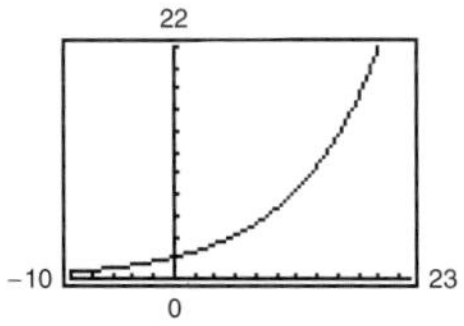

49. 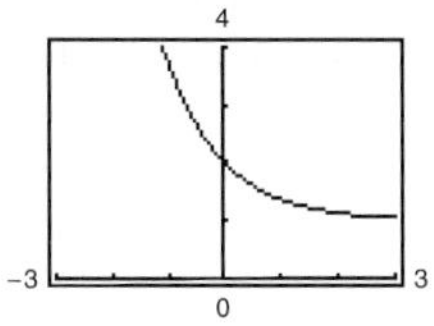

51. $x = \frac{1}{3}$ **53.** $x = 3, -1$

55.

n	1	2	4	12
A	\$1828.49	\$1830.29	\$1831.19	\$1831.80

n	365	Continuous
A	\$1832.09	\$1832.10

57.

n	1	2	4	12
A	\$5477.81	\$5520.10	\$5541.79	\$5556.46

n	365	Continuous
A	\$5563.61	\$5563.85

59.

t	10	20	30
A	\$17,901.90	\$26,706.49	\$39,841.40

t	40	50
A	\$59,436.39	\$88,668.67

61.

t	10	20	30
A	\$22,986.49	\$44,031.56	\$84,344.25

t	40	50
A	\$161,564.86	\$309,484.08

63. \$104,710.29 **65.** \$35.45

67. (a) 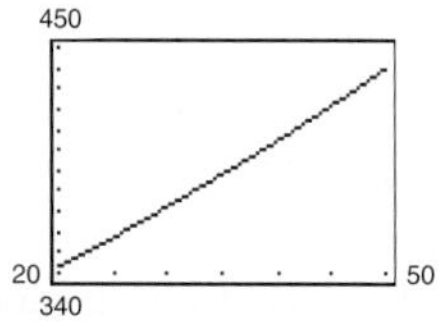

(b)

t	20	21	22	23
P (in millions)	342.748	345.604	348.485	351.389

t	24	25	26	27
P (in millions)	354.318	357.271	360.249	363.251

t	28	29	30	31
P (in millions)	366.279	369.331	372.410	375.513

t	32	33	34	35
P (in millions)	378.643	381.799	384.981	388.190

t	36	37	38	39
P (in millions)	391.425	394.687	397.977	401.294

t	40	41	42	43
P (in millions)	404.639	408.011	411.412	414.840

t	44	45	46	47
P (in millions)	418.298	421.784	425.300	428.844

t	48	49	50
P (in millions)	432.419	436.023	439.657

(c) 2038

69. (a) 16 g (b) 1.85 g (c) 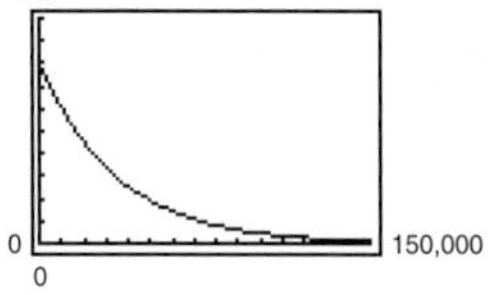

CHAPTER 3

71. (a) $V(t) = 49{,}810\left(\frac{7}{8}\right)^t$ (b) \$29,197.71

73. True. As $x \to -\infty$, $f(x) \to -2$ but never reaches -2.

75. $f(x) = h(x)$ **77.** $f(x) = g(x) = h(x)$

79.

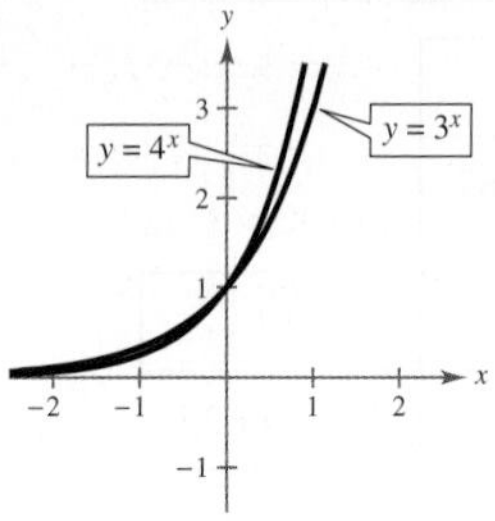

(a) $x < 0$ (b) $x > 0$

81.

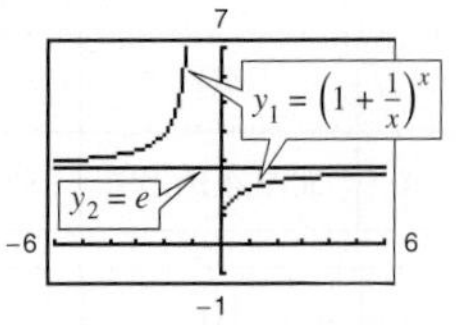

As the x-value increases, y_1 approaches the value of e.

83. (a)

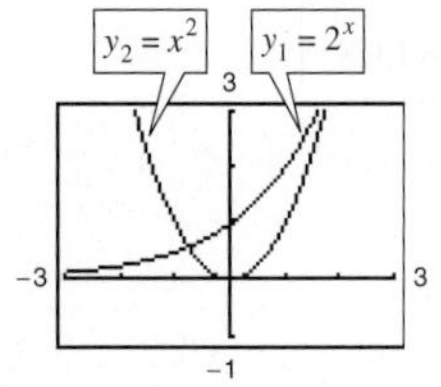

(b)

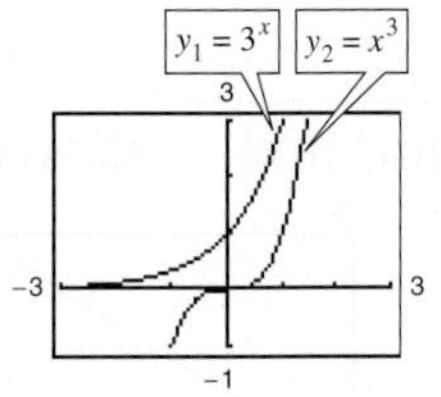

In both viewing windows, the constant raised to a variable power increases more rapidly than the variable raised to a constant power.

85. c, d

Section 3.2 *(page 218)*

1. logarithmic **3.** natural; e **5.** $x = y$ **7.** $4^2 = 16$

9. $32^{2/5} = 4$ **11.** $\log_5 125 = 3$ **13.** $\log_4 \frac{1}{64} = -3$

15. 6 **17.** 0 **19.** 2 **21.** -0.058 **23.** 1.097

25. 7 **27.** 1 **29.** $x = 5$ **31.** $x = 7$

33.

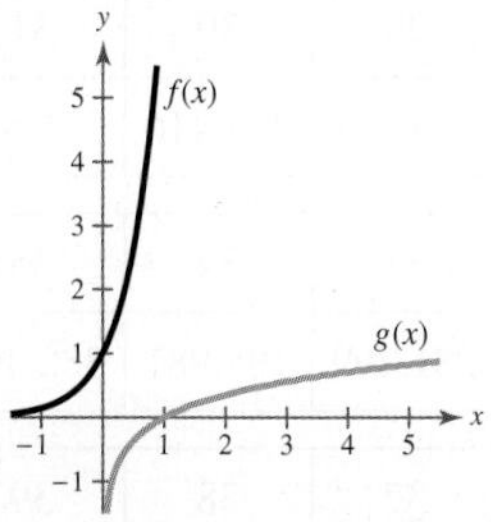

35.

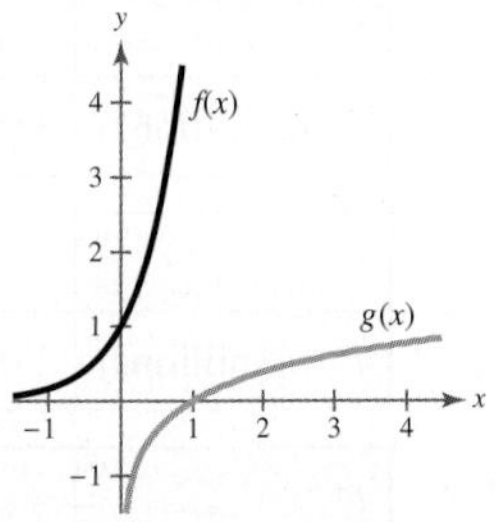

37. c **38.** d **39.** b **40.** a

41.

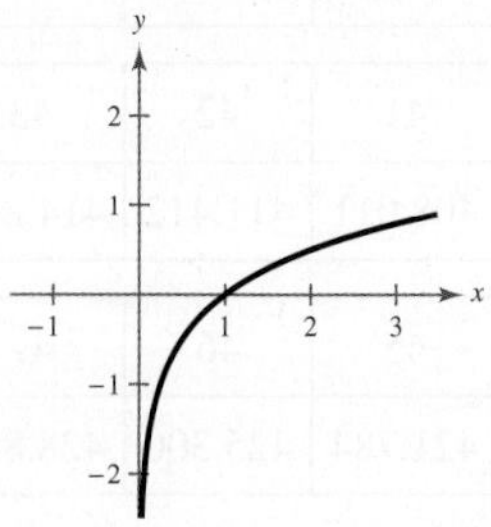

Domain: $(0, \infty)$
x-intercept: $(1, 0)$
Vertical asymptote: $x = 0$

43.

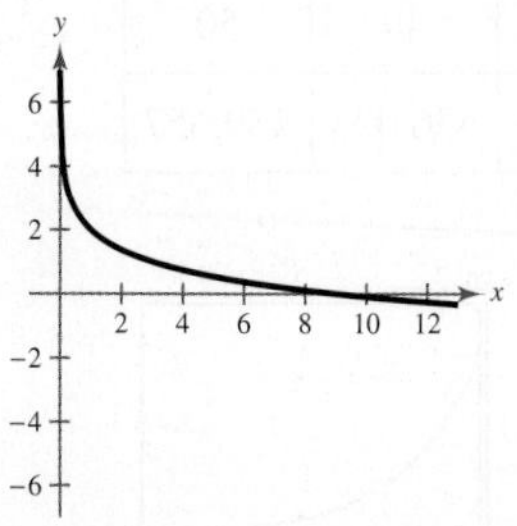

Domain: $(0, \infty)$
x-intercept: $(9, 0)$
Vertical asymptote: $x = 0$

45.

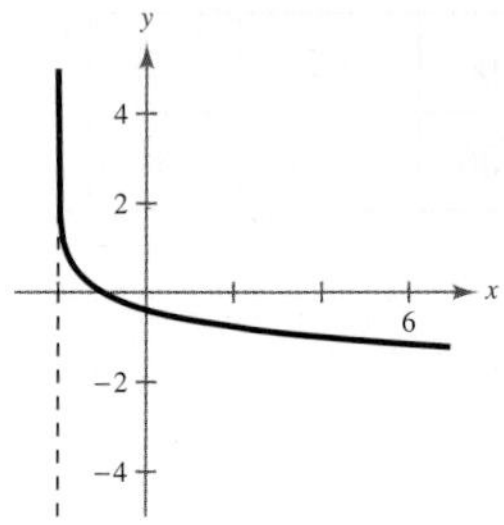

Domain: $(-2, \infty)$
x-intercept: $(-1, 0)$
Vertical asymptote: $x = -2$

47.

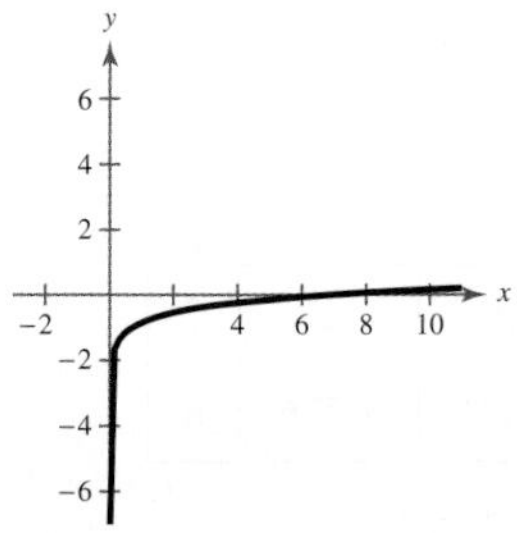

Domain: $(0, \infty)$
x-intercept: $(7, 0)$
Vertical asymptote: $x = 0$

49. $e^{-0.693\ldots} = \frac{1}{2}$ **51.** $e^{5.521\ldots} = 250$

53. $\ln 7.3890\ldots = 2$ **55.** $\ln 0.406\ldots = -0.9$

57. 2.913 **59.** -23.966 **61.** 5 **63.** $-\frac{5}{6}$

65.

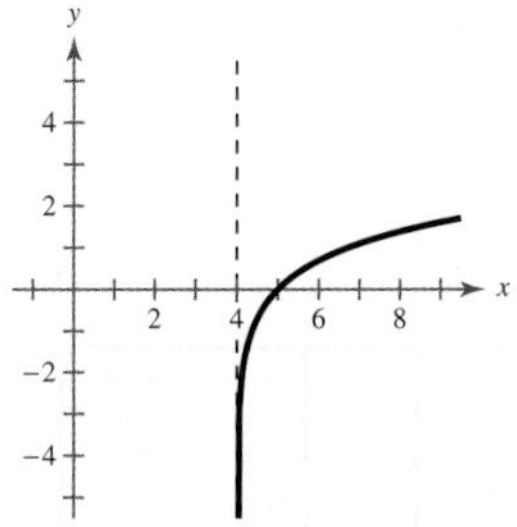

Domain: $(4, \infty)$
x-intercept: $(5, 0)$
Vertical asymptote: $x = 4$

67.

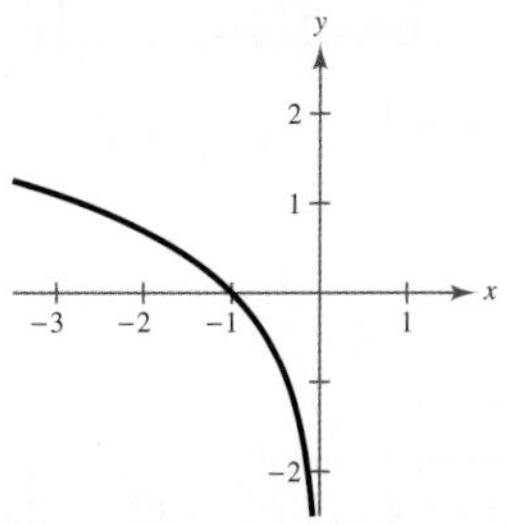

Domain: $(-\infty, 0)$
x-intercept: $(-1, 0)$
Vertical asymptote: $x = 0$

69.

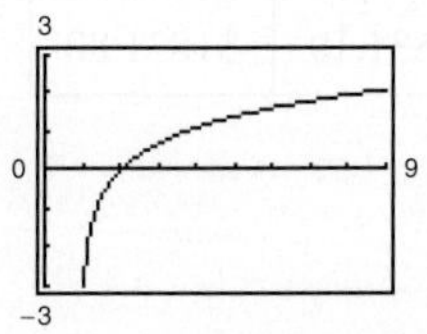

71.

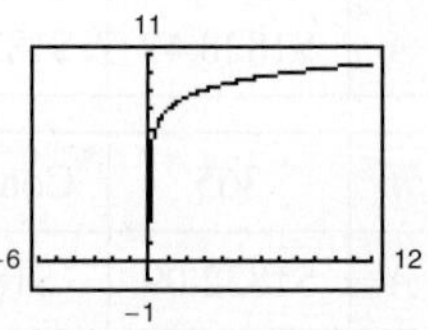

73. $x = 8$ **75.** $x = -5, 5$

77. (a) 30 yr; 10 yr (b) \$323,179; \$199,109
(c) \$173,179; \$49,109
(d) $x = 750$; The monthly payment must be greater than \$750.

79. (a)

r	0.005	0.010	0.015	0.020	0.025	0.030
t	138.6	69.3	46.2	34.7	27.7	23.1

As the rate of increase r increases, the time t in years for the population to double decreases.

(b) 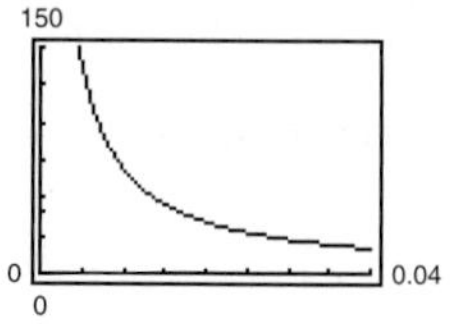

81. (a) 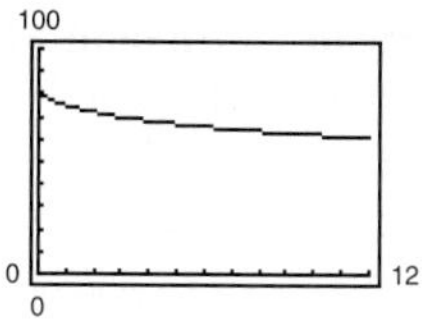 (b) 80 (c) 68.1 (d) 62.3

83. False. Reflecting $g(x)$ in the line $y = x$ will determine the graph of $f(x)$.

85. (a) 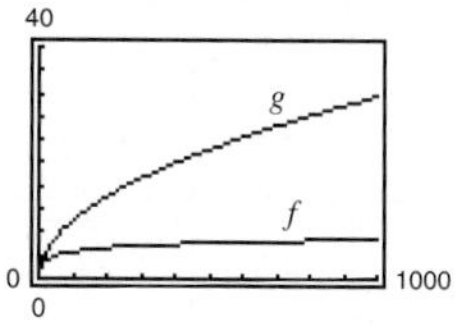

$g(x)$; The natural log function grows at a slower rate than the square root function.

(b) 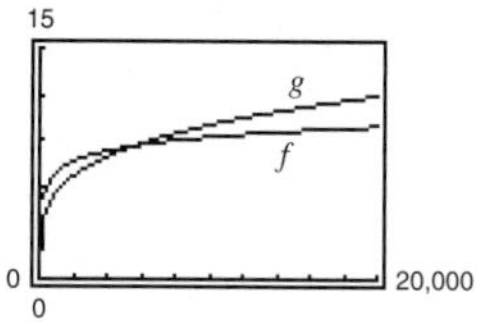

$g(x)$; The natural log function grows at a slower rate than the fourth root function.

87. (a) False (b) True (c) True (d) False

89. Answers will vary.

Section 3.3 *(page 225)*

1. change-of-base **3.** $\dfrac{1}{\log_b a}$ **4.** c

5. a **6.** b **7.** (a) $\dfrac{\log 16}{\log 5}$ (b) $\dfrac{\ln 16}{\ln 5}$

9. (a) $\dfrac{\log \frac{3}{10}}{\log x}$ (b) $\dfrac{\ln \frac{3}{10}}{\ln x}$ **11.** 1.771 **13.** -1.048

15. $\frac{3}{2}$ **17.** $-3 - \log_5 2$ **19.** $6 + \ln 5$ **21.** 2 **23.** $\frac{3}{4}$

25. 4 **27.** -2 is not in the domain of $\log_2 x$. **29.** 4.5

31. $-\frac{1}{2}$ **33.** 7 **35.** 2 **37.** $\ln 4 + \ln x$ **39.** $4 \log_8 x$

41. $1 - \log_5 x$ **43.** $\frac{1}{2} \ln z$ **45.** $\ln x + \ln y + 2 \ln z$

47. $\ln z + 2 \ln(z - 1)$ **49.** $\frac{1}{2} \log_2(a - 1) - 2 \log_2 3$

51. $\frac{1}{3} \ln x - \frac{1}{3} \ln y$ **53.** $2 \ln x + \frac{1}{2} \ln y - \frac{1}{2} \ln z$

55. $2 \log_5 x - 2 \log_5 y - 3 \log_5 z$ **57.** $\frac{3}{4} \ln x + \frac{1}{4} \ln(x^2 + 3)$

59. 1.1833 **61.** 1.0686 **63.** 1.9563 **65.** 2.5646

67. $\ln 2x$ **69.** $\log_2 x^2y^4$ **71.** $\log_3 \sqrt[4]{5x}$

73. $\log \dfrac{x}{(x + 1)^2}$ **75.** $\log \dfrac{xz^3}{y^2}$ **77.** $\ln \dfrac{x}{(x + 1)(x - 1)}$

79. $\ln \sqrt[3]{\dfrac{x(x + 3)^2}{x^2 - 1}}$ **81.** $\log_8 \dfrac{\sqrt[3]{y(y + 4)^2}}{y - 1}$

83. $\log_2 \frac{32}{4} = \log_2 32 - \log_2 4$; Property 2

85. $\beta = 10(\log I + 12)$; 60 dB **87.** 70 dB

89. $\ln y = \frac{1}{4} \ln x$ **91.** $\ln y = -\frac{1}{4} \ln x + \ln \frac{5}{2}$

93. $y = 256.24 - 20.8 \ln x$

95. (a) and (b) 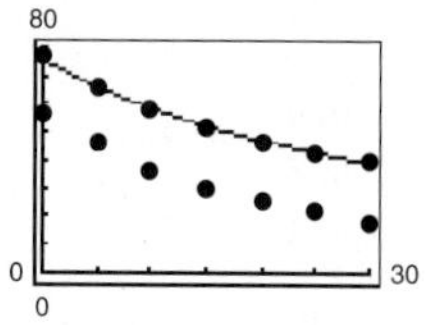

(c) 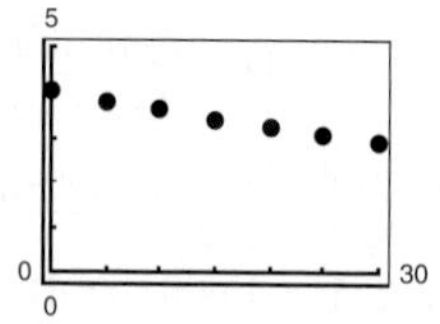

$T = 21 + e^{-0.037t + 3.997}$

The results are similar.

(d) 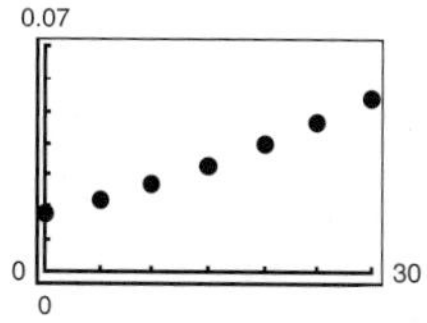

$$T = 21 + \frac{1}{0.001t + 0.016}$$

(e) Answers will vary.

97. False; $\ln 1 = 0$ **99.** False; $\ln(x - 2) \neq \ln x - \ln 2$

101. False; $u = v^2$

103. $f(x) = \dfrac{\log x}{\log 2} = \dfrac{\ln x}{\ln 2}$ 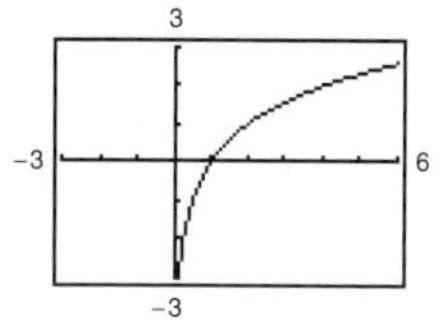

105. $f(x) = \dfrac{\log x}{\log \frac{1}{4}} = \dfrac{\ln x}{\ln \frac{1}{4}}$ 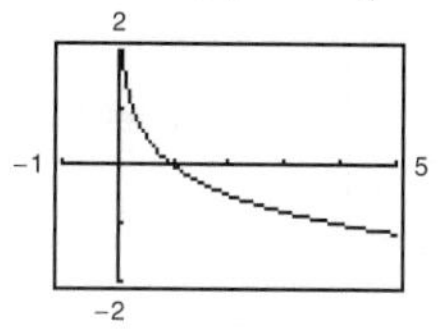

107. *Sample answers:*

(a) $\ln(1 + 3) \stackrel{?}{=} \ln(1) + \ln(3)$

$1.39 \neq 0 + 1.10$

$\ln u + \ln v = \ln(uv)$, but $\ln(u + v) \neq \ln u + \ln v$

(b) $\ln(3 - 1) \stackrel{?}{=} \ln(3) - \ln(1)$

$0.69 \neq 1.10 - 0$

$\ln u - \ln v = \ln \dfrac{u}{v}$, but $\ln(u - v) \neq \ln u - \ln v$

(c) $(\ln 2)^3 \stackrel{?}{=} 3 \ln 2$

$0.33 \neq 2.08$

$n(\ln u) = \ln u^n$, but $(\ln u)^n \neq n(\ln u)$

109.

$\ln 1 = 0$	$\ln 9 \approx 2.1972$
$\ln 2 \approx 0.6931$	$\ln 10 \approx 2.3025$
$\ln 3 \approx 1.0986$	$\ln 12 \approx 2.4848$
$\ln 4 \approx 1.3862$	$\ln 15 \approx 2.7080$
$\ln 5 \approx 1.6094$	$\ln 16 \approx 2.7724$
$\ln 6 \approx 1.7917$	$\ln 18 \approx 2.8903$
$\ln 8 \approx 2.0793$	$\ln 20 \approx 2.9956$

Section 3.4 *(page 235)*

1. (a) $x = y$ (b) $x = y$ (c) x (d) x

3. (a) Yes (b) No **5.** (a) Yes (b) No (c) No

7. 2 **9.** 2 **11.** $e^{-1} \approx 0.368$ **13.** 64 **15.** (3, 8)

17. 2, -1 **19.** $\dfrac{\ln 5}{\ln 3} \approx 1.465$ **21.** $\ln 28 \approx 3.332$

23. $\dfrac{\ln 80}{2 \ln 3} \approx 1.994$ **25.** $3 - \dfrac{\ln 565}{\ln 2} \approx -6.142$

27. $\dfrac{1}{3} \log\left(\dfrac{3}{2}\right) \approx 0.059$ **29.** $\dfrac{\ln 12}{3} \approx 0.828$

31. 0 **33.** $\frac{\ln \frac{8}{3}}{3 \ln 2} + \frac{1}{3} \approx 0.805$

35. $\frac{\ln 3}{\ln 2 - \ln 3} \approx -2.710$ **37.** $0, \frac{\ln 4}{\ln 5} \approx 0.861$

39. $\ln 5 \approx 1.609$ **41.** $2 \ln 75 \approx 8.635$

43. $\frac{\ln 4}{365 \ln\left(1 + \frac{0.065}{365}\right)} \approx 21.330$ **45.** $e^{-3} \approx 0.050$

47. $\frac{e^{2.1}}{6} \approx 1.361$ **49.** $\frac{e^{10/3}}{5} \approx 5.606$ **51.** $e^{-4/3} \approx 0.264$

53. $2(3^{11/6}) \approx 14.988$ **55.** No solution

57. No solution **59.** No solution **61.** 2

63.

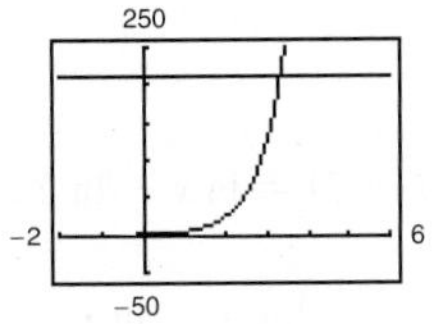

3.328

65.

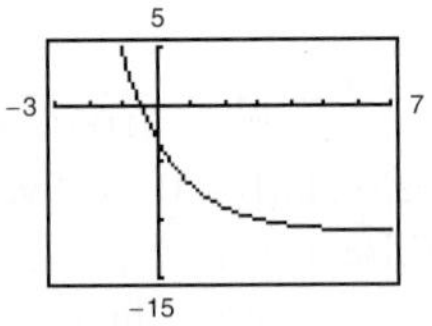

-0.478

67.

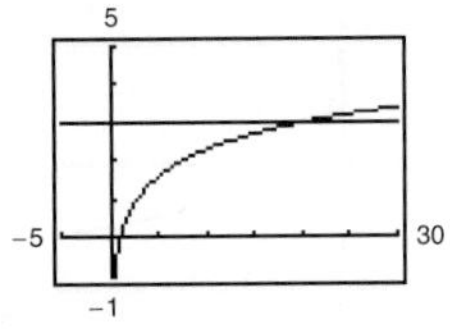

20.086

69.

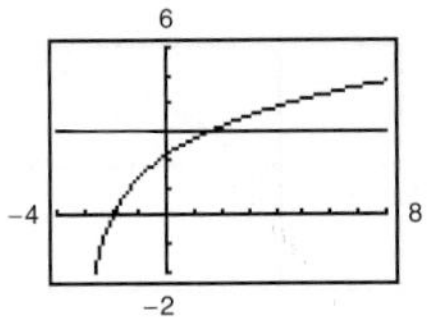

1.482

71. (a) 27.73 yr (b) 43.94 yr **73.** $-1, 0$ **75.** 1

77. $e^{-1/2} \approx 0.607$ **79.** $e^{-1} \approx 0.368$

81. (a) $y = 100$ and $y = 0$; The range falls between 0% and 100%.
(b) Males: 69.51 in. Females: 64.49 in.

83. 12.76 in.

85. (a)

x	0.2	0.4	0.6	0.8	1.0
y	162.6	78.5	52.5	40.5	33.9

(b)

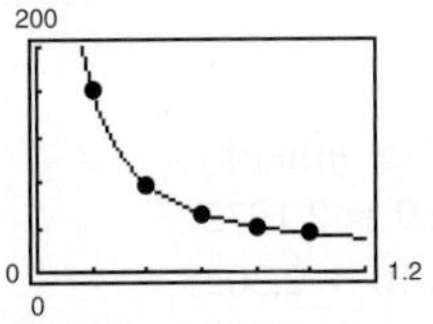

The model appears to fit the data well.

(c) 1.2 m

(d) No. According to the model, when the number of g's is less than 23, x is between 2.276 meters and 4.404 meters, which isn't realistic in most vehicles.

87. $\log_b uv = \log_b u + \log_b v$
True by Property 1 in Section 3.3.

89. $\log_b(u - v) = \log_b u - \log_b v$
False.
$1.95 \approx \log(100 - 10) \neq \log 100 - \log 10 = 1$

91. Yes. See Exercise 57.

93. For $rt < \ln 2$ years, double the amount you invest. For $rt > \ln 2$ years, double your interest rate or double the number of years, because either of these will double the exponent in the exponential function.

95. (a) 7% (b) 7.25% (c) 7.19% (d) 7.45%
The investment plan with the greatest effective yield and the highest balance after 5 years is plan (d).

Section 3.5 *(page 245)*

1. $y = ae^{bx}$; $y = ae^{-bx}$ **3.** normally distributed

5. (a) $P = \frac{A}{e^{rt}}$ (b) $t = \frac{\ln\left(\frac{A}{P}\right)}{r}$

	Initial Investment	Annual % Rate	Time to Double	Amount After 10 years
7.	$1000	3.5%	19.8 yr	$1419.07
9.	$750	8.9438%	7.75 yr	$1834.37
11.	$6376.28	4.5%	15.4 yr	$10,000.00

13. $303,580.52

15. (a) 7.27 yr (b) 6.96 yr (c) 6.93 yr (d) 6.93 yr

17. (a)

r	2%	4%	6%	8%	10%	12%
t	54.93	27.47	18.31	13.73	10.99	9.16

(b)

r	2%	4%	6%	8%	10%	12%
t	55.48	28.01	18.85	14.27	11.53	9.69

19.

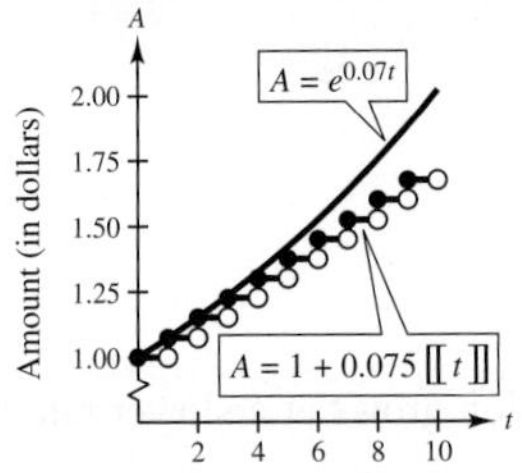

Continuous compounding

	Half-life (years)	Initial Quantity	Amount After 1000 Years
21.	1599	10 g	6.48 g
23.	5715	2.26 g	2 g

25. $y = e^{0.7675x}$ **27.** $y = 5e^{-0.4024x}$

29. (a)

Year	1980	1990	2000	2010
Population	106.1	143.15	196.25	272.37

(b) 2017

(c) No; The population will not continue to grow at such a quick rate.

31. $k = 0.2988$; About 5,309,734 hits **33.** About 800 bacteria

35. (a) $V = -300t + 1150$ (b) $V = 1150e^{-0.368799t}$

(c)

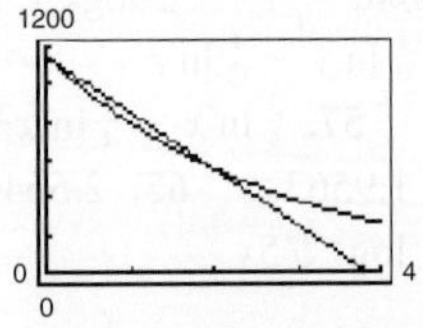

The exponential model depreciates faster.

(d)

t	1 yr	3 yr
$V = -300t + 1150$	850	250
$V = 1150e^{-0.368799t}$	795	380

(e) Answers will vary.

37. (a) About 12,180 yr old (b) About 4797 yr old

39. (a)

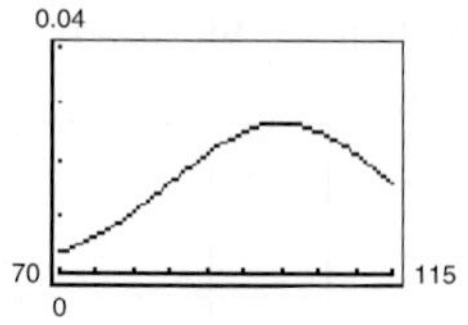

(b) 100

41. (a) 1998: 55,557 sites
2003: 147,644 sites
2006: 203,023 sites
(c) and (d) 2010

(b)

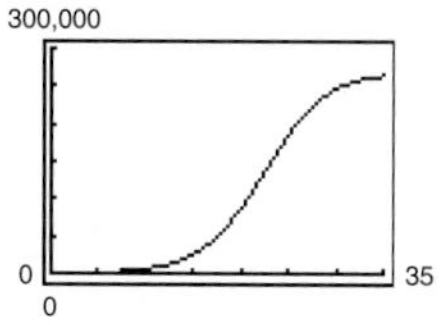

43. (a) 203 animals
(b) 13 mo
(c)

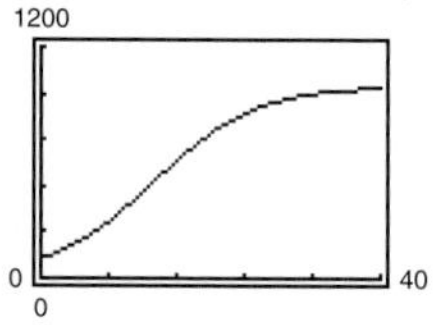

Horizontal asymptotes: $p = 0, p = 1000$. The population size will approach 1000 as time increases.

45. (a) $10^{6.6} \approx 3{,}981{,}072$ (b) $10^{5.6} \approx 398{,}107$
(c) $10^{7.1} \approx 12{,}589{,}254$

47. (a) 20 dB (b) 70 dB (c) 40 dB (d) 120 dB

49. 95% **51.** 4.64 **53.** 1.58×10^{-6} moles/L

55. $10^{5.1}$ **57.** 3:00 A.M.

59. (a)

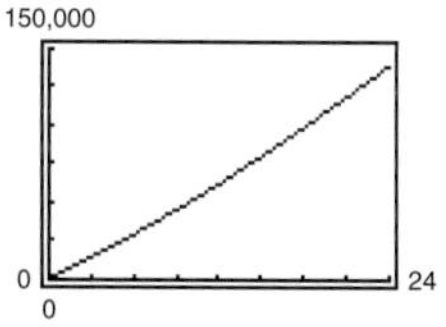

(b) $t \approx 21$ yr; Yes

61. False. The domain can be the set of real numbers for a logistic growth function.

63. False. The graph of $f(x)$ is the graph of $g(x)$ shifted five units up.

65. Answers will vary.

Review Exercises *(page 252)*

1. 0.164 **3.** 0.337 **5.** 1456.529

7. Shift the graph of f one unit up.

9. Reflect f in the x-axis and shift one unit up.

11.

x	-1	0	1	2	3
$f(x)$	8	5	4.25	4.063	4.016

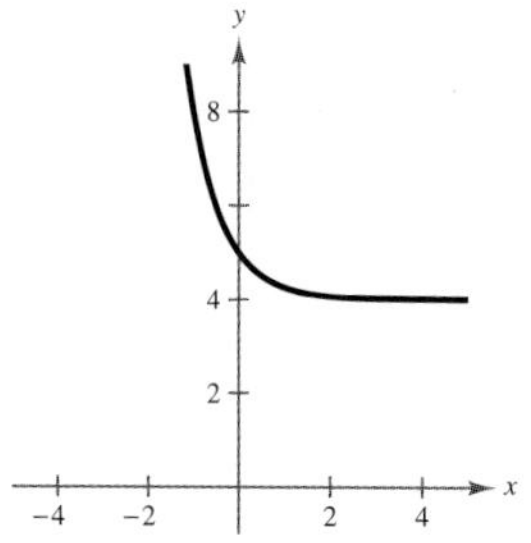

13.

x	-1	0	1	2	3
$f(x)$	4.008	4.04	4.2	5	9

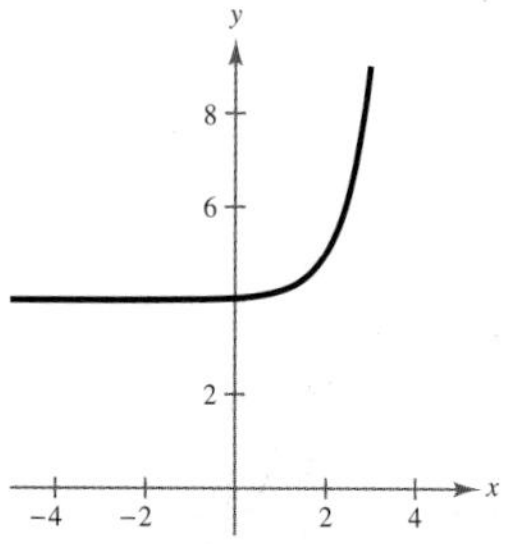

15.

x	-2	-1	0	1	2
$f(x)$	3.25	3.5	4	5	7

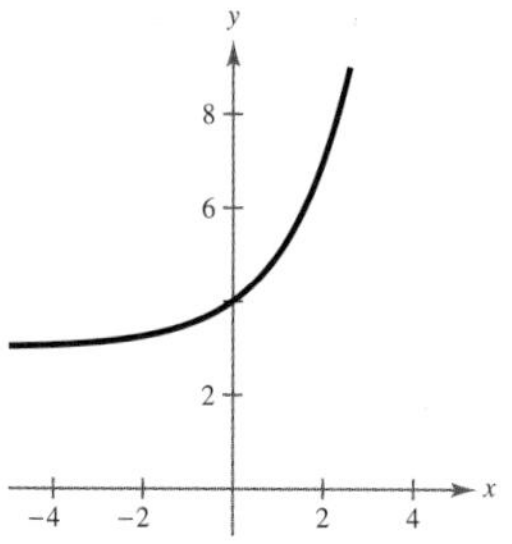

17. $x = 1$ **19.** $x = 4$ **21.** 2980.958 **23.** 0.183

25.

x	-2	-1	0	1	2
$h(x)$	2.72	1.65	1	0.61	0.37

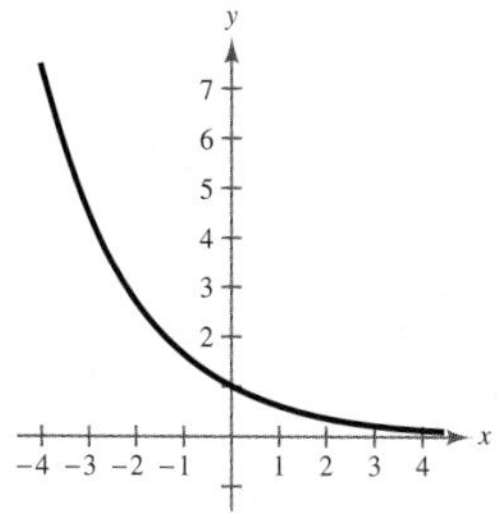

27.

x	-3	-2	-1	0	1
$f(x)$	0.37	1	2.72	7.39	20.09

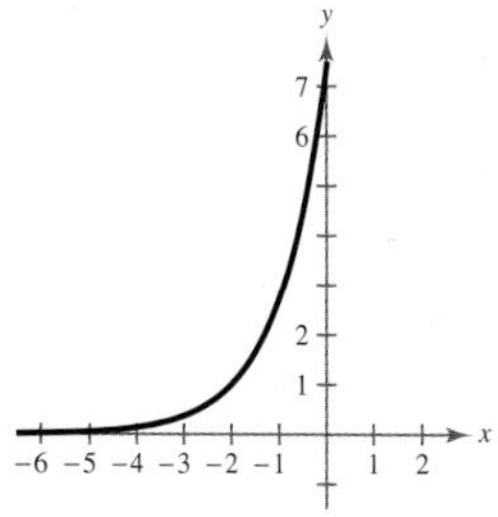

29. (a) 0.154 (b) 0.487 (c) 0.811

CHAPTER 3

31.

n	1	2	4	12
A	\$6719.58	\$6734.28	\$6741.74	\$6746.77

n	365	Continuous
A	\$6749.21	\$6749.29

33. $\log_3 27 = 3$ **35.** $\ln 2.2255\ldots = 0.8$ **37.** 3

39. -2 **41.** $x = 7$ **43.** $x = -5$

45. Domain: $(0, \infty)$
x-intercept: $(1, 0)$
Vertical asymptote: $x = 0$

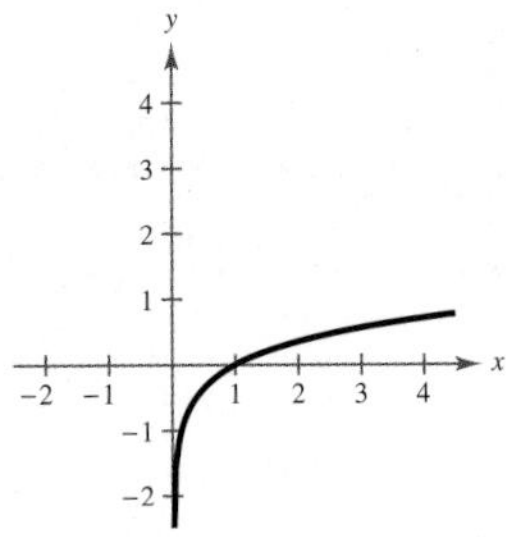

47. Domain: $(-5, \infty)$
x-intercept: $(9995, 0)$
Vertical asymptote: $x = -5$

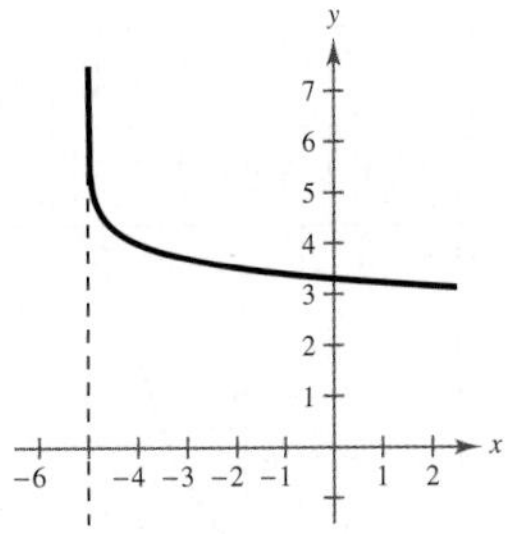

49. 3.118 **51.** 0.25

53. Domain: $(0, \infty)$
x-intercept: $(e^{-3}, 0)$
Vertical asymptote: $x = 0$

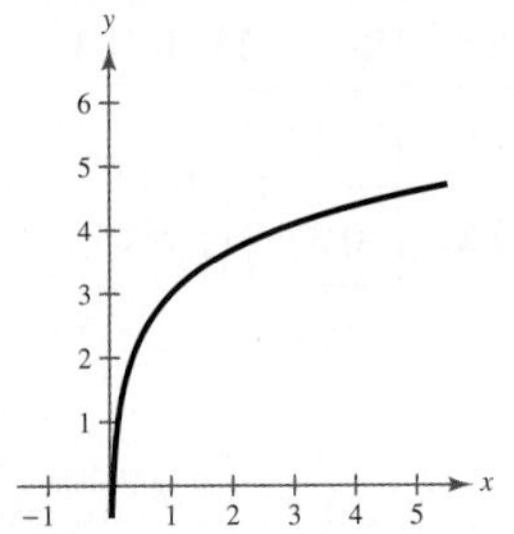

55. Domain: $(-\infty, 0), (0, \infty)$
x-intercepts: $(\pm 1, 0)$
Vertical asymptote: $x = 0$

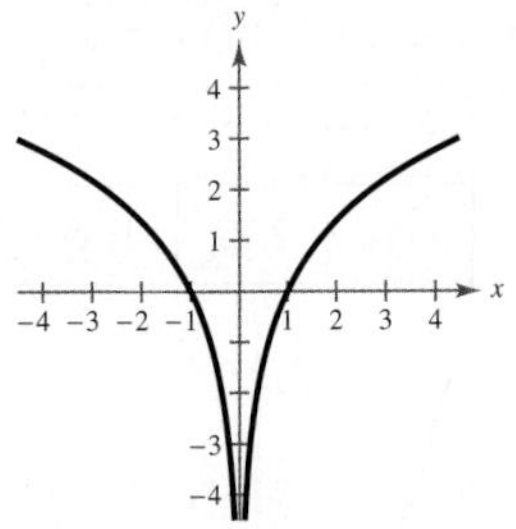

57. 53.4 in. **59.** (a) and (b) 2.585 **61.** (a) and (b) -2.322

63. $\log 2 + 2 \log 3 \approx 1.255$ **65.** $2 \ln 2 + \ln 5 \approx 2.996$

67. $1 + 2 \log_5 x$ **69.** $2 - \frac{1}{2}\log_3 x$ **71.** $2 \ln x + 2 \ln y + \ln z$

73. $\log_2 5x$ **75.** $\ln \dfrac{x}{\sqrt[4]{y}}$ **77.** $\log_3 \dfrac{\sqrt{x}}{(y + 8)^2}$

79. (a) $0 \leq h < 18{,}000$

(b)

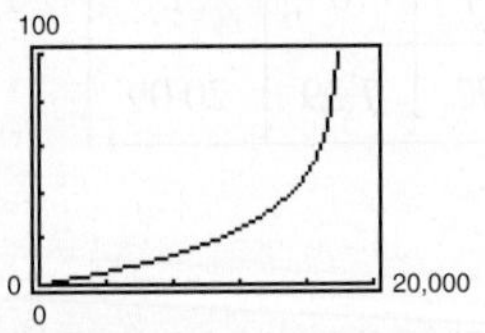

Vertical asymptote: $h = 18{,}000$

(c) The plane is climbing at a slower rate, so the time required increases.

(d) 5.46 min

81. 3 **83.** $\ln 3 \approx 1.099$ **85.** $e^4 \approx 54.598$ **87.** 1, 3

89. $\dfrac{\ln 32}{\ln 2} = 5$

91.

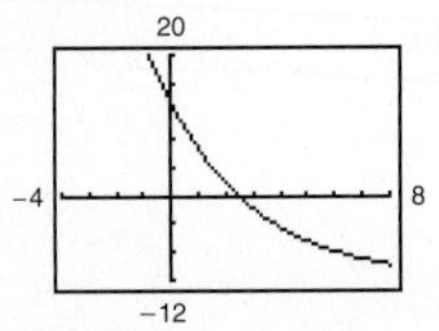

2.447

93. $\frac{1}{3}e^{8.2} \approx 1213.650$

95. $3e^2 \approx 22.167$ **97.** No solution **99.** 0.900

101.

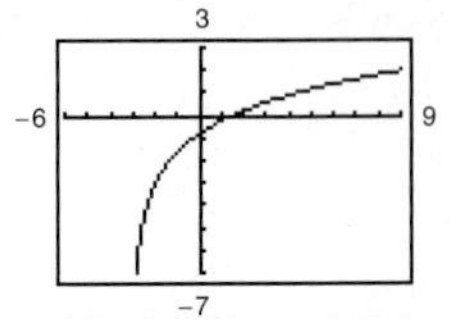

1.482

103.

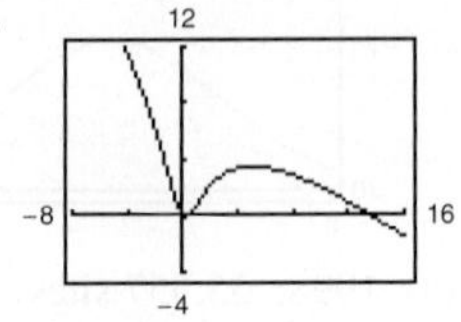

0, 0.416, 13.627

105. 73.2 yr **107.** e **108.** b **109.** f

110. d **111.** a **112.** c **113.** $y = 2e^{0.1014x}$

115.

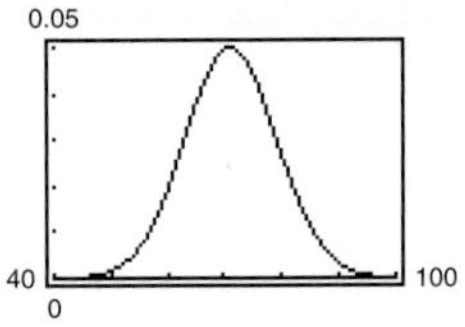

71

117. (a) 10^{-6} W/m^2
(b) $10\sqrt{10}$ W/m^2
(c) 1.259×10^{-12} W/m^2

119. True by the inverse properties.

Chapter Test *(page 255)*

1. 2.366 **2.** 687.291 **3.** 0.497 **4.** 22.198

5.

x	-1	$-\frac{1}{2}$	0	$\frac{1}{2}$	1
$f(x)$	10	3.162	1	0.316	0.1

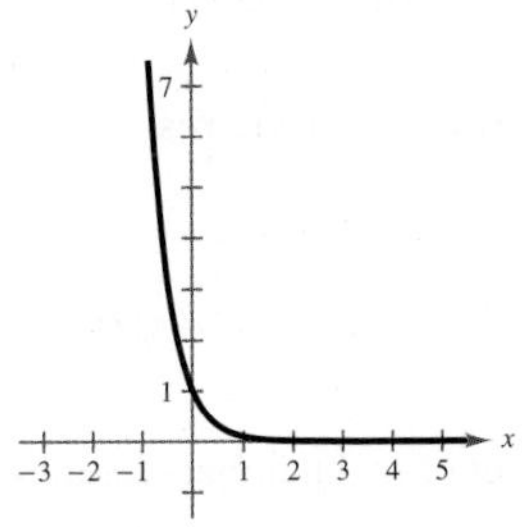

6.

x	-1	0	1	2	3
$f(x)$	-0.005	-0.028	-0.167	-1	-6

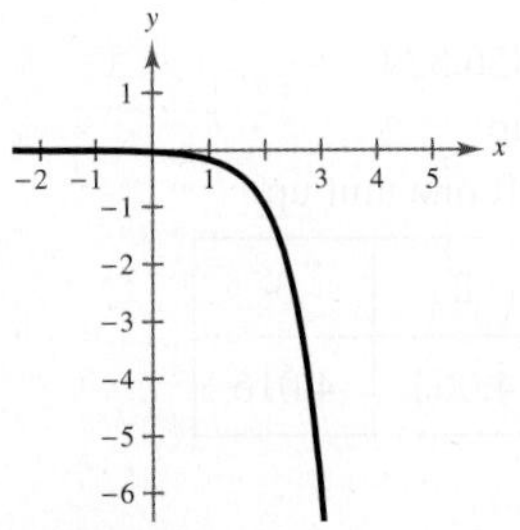

7.

x	-1	$-\frac{1}{2}$	0	$\frac{1}{2}$	1
$f(x)$	0.865	0.632	0	-1.718	-6.389

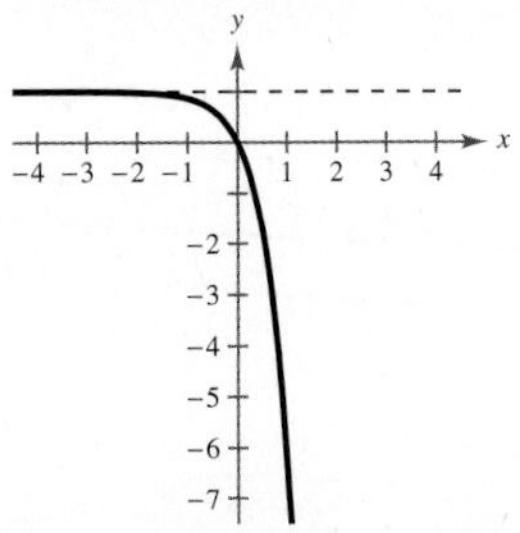

8. (a) -0.89 (b) 9.2

9. Domain: $(0, \infty)$
x-intercept: $(10^{-6}, 0)$
Vertical asymptote: $x = 0$

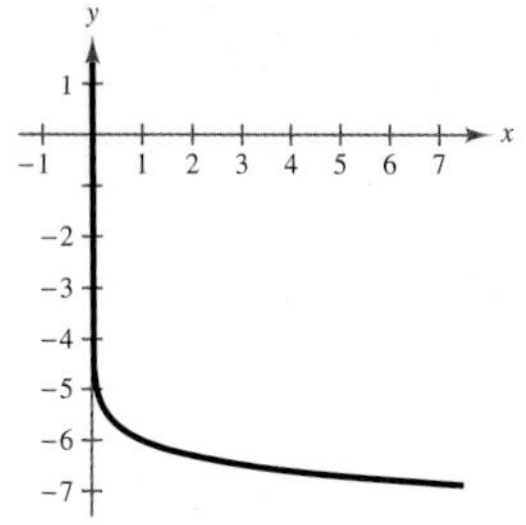

10. Domain: $(4, \infty)$
x-intercept: $(5, 0)$
Vertical asymptote: $x = 4$

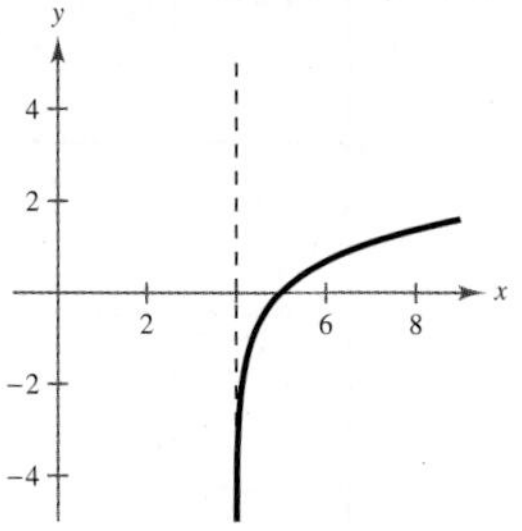

11. Domain: $(-6, \infty)$
x-intercept: $(e^{-1} - 6, 0)$
Vertical asymptote: $x = -6$

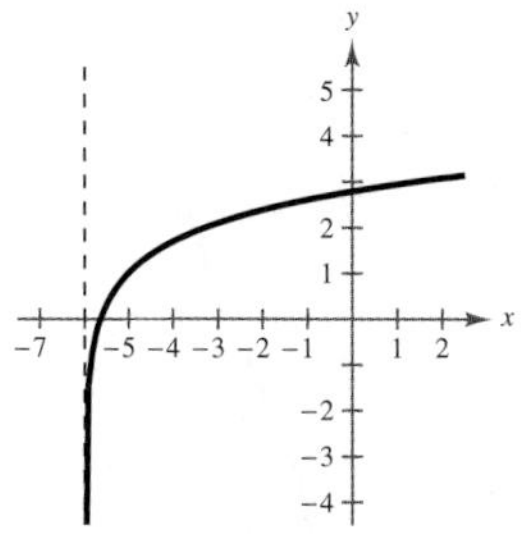

12. 1.945 **13.** -0.167 **14.** -11.047

15. $\log_2 3 + 4 \log_2 a$ **16.** $\ln 5 + \frac{1}{2}\ln x - \ln 6$

17. $3 \log(x - 1) - 2 \log y - \log z$ **18.** $\log_3 13y$

19. $\ln \dfrac{x^4}{y^4}$ **20.** $\ln \dfrac{x^3y^2}{x + 3}$ **21.** -2 **22.** $\dfrac{\ln 44}{-5} \approx -0.757$

23. $\dfrac{\ln 197}{4} \approx 1.321$ **24.** $e^{1/2} \approx 1.649$ **25.** $e^{-11/4} \approx 0.0639$

26. 20 **27.** $y = 2745e^{0.1570t}$ **28.** 55%

29. (a)

x	$\frac{1}{4}$	1	2	4	5	6
H	58.720	75.332	86.828	103.43	110.59	117.38

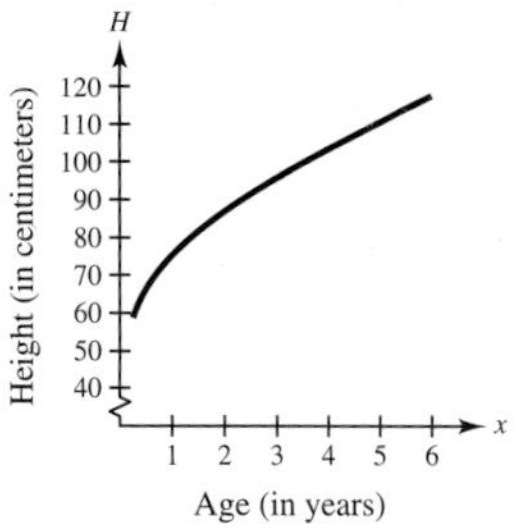

(b) 103 cm; 103.43 cm

Cumulative Test for Chapters 1–3 *(page 256)*

1.

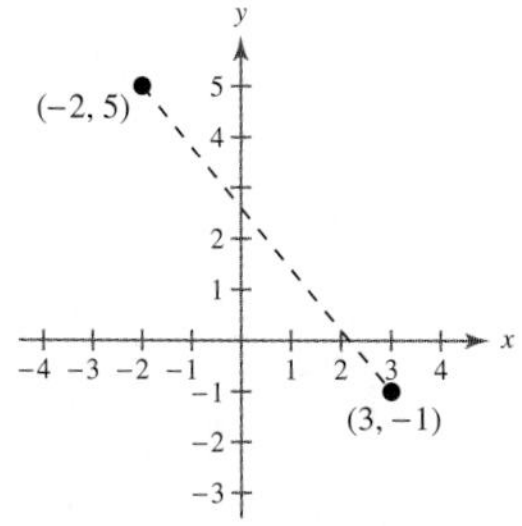

Midpoint: $\left(\frac{1}{2}, 2\right)$
Distance: $\sqrt{61}$

2.

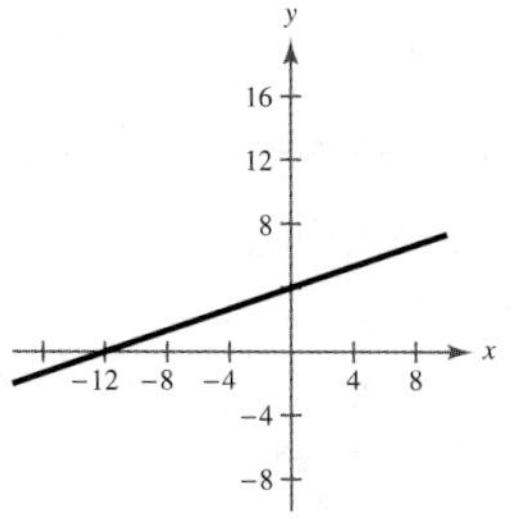

3.

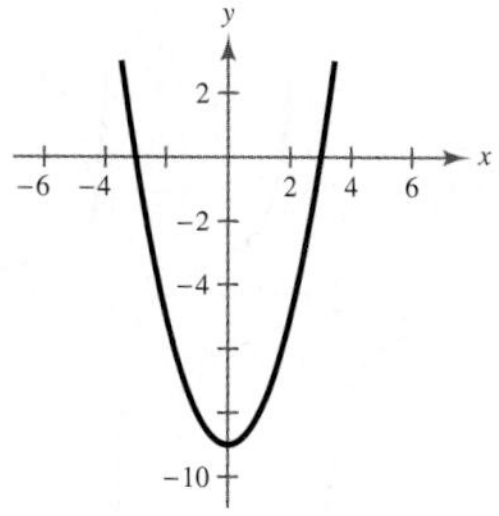

4.

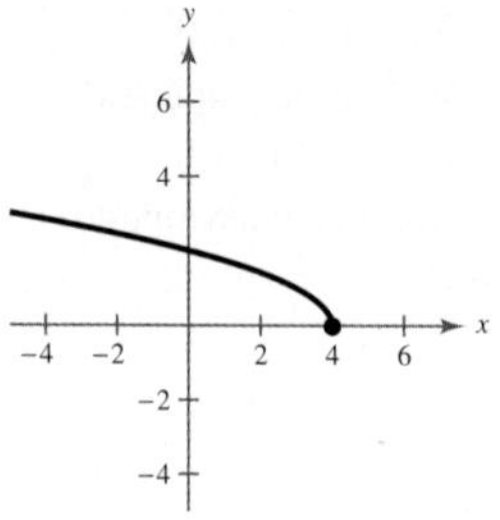

5. $y = 2x + 2$

6. For some values of x there correspond two values of y.

7. (a) $\dfrac{3}{2}$ (b) Division by 0 is undefined. (c) $\dfrac{s + 2}{s}$

8. (a) Vertical shrink by $\frac{1}{2}$ (b) Vertical shift two units up
(c) Horizontal shift two units to the left

9. (a) $5x - 2$ (b) $-3x - 4$ (c) $4x^2 - 11x - 3$
(d) $\dfrac{x - 3}{4x + 1}$; Domain: all real numbers x except $x = -\dfrac{1}{4}$

10. (a) $\sqrt{x - 1} + x^2 + 1$ (b) $\sqrt{x - 1} - x^2 - 1$
(c) $x^2\sqrt{x - 1} + \sqrt{x - 1}$
(d) $\dfrac{\sqrt{x - 1}}{x^2 + 1}$; Domain: all real numbers x such that $x \geq 1$

11. (a) $2x + 12$ (b) $\sqrt{2x^2 + 6}$
Domain of $f \circ g$: all real numbers x such that $x \geq -6$
Domain of $g \circ f$: all real numbers

12. (a) $|x| - 2$ (b) $|x - 2|$
Domain of $f \circ g$ and $g \circ f$: all real numbers

13. Yes; $h^{-1}(x) = -\frac{1}{5}(x - 3)$ **14.** 2438.65 kW

15. $y = -\frac{3}{4}(x + 8)^2 + 5$

16.

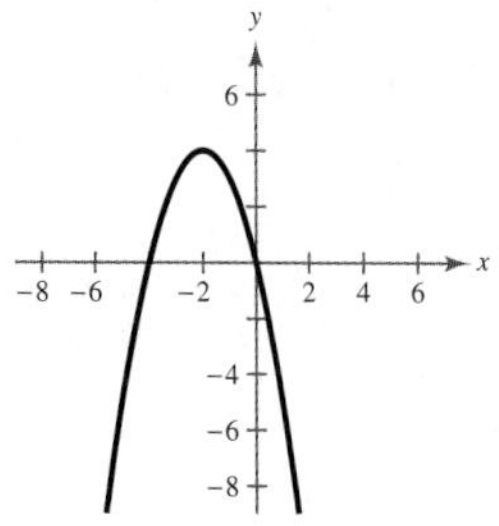

17.

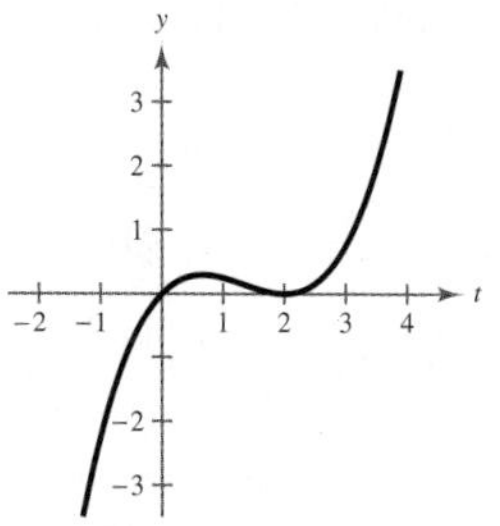

18.

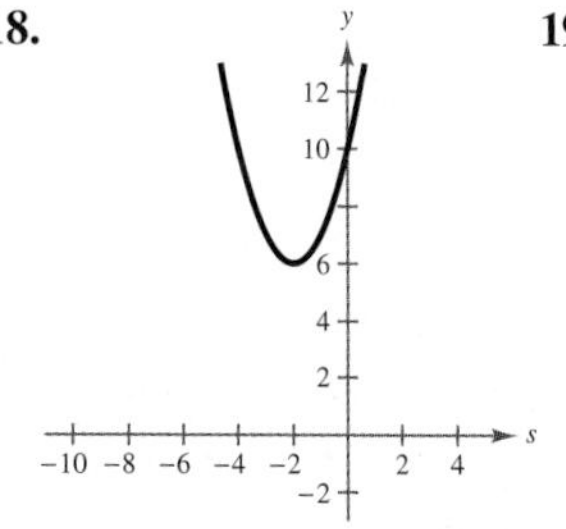

19. $-2, \pm 2i$; $(x + 2)(x + 2i)(x - 2i)$

20. $-7, 0, 3$; $x(x)(x - 3)(x + 7)$

21. $4, -\frac{1}{2}, 1 \pm 3i$; $(x - 4)(2x + 1)(x - 1 + 3i)(x - 1 - 3i)$

22. $3x - 2 - \dfrac{3x - 2}{2x^2 + 1}$ **23.** $3x^3 + 6x^2 + 14x + 23 + \dfrac{49}{x - 2}$

24.

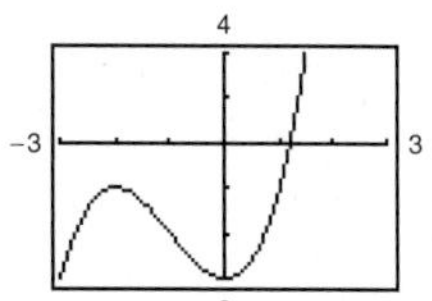

Interval: $[1, 2]$; 1.20

25. Intercept: $(0, 0)$
Vertical asymptotes:
$x = -3, x = 1$
Horizontal asymptote: $y = 0$

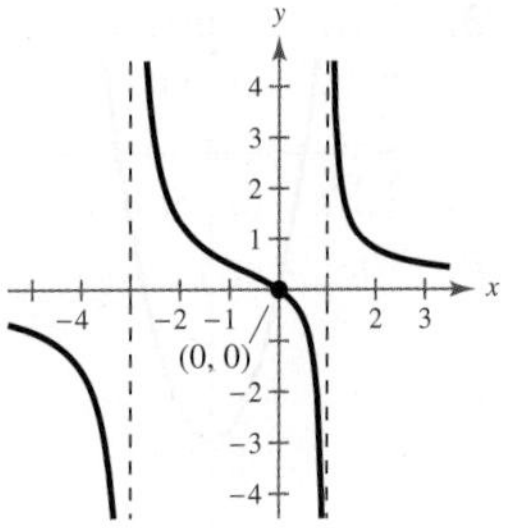

26. y-intercept: $(0, 2)$
x-intercept: $(2, 0)$
Vertical asymptote: $x = 1$
Horizontal asymptote:
$y = 1$

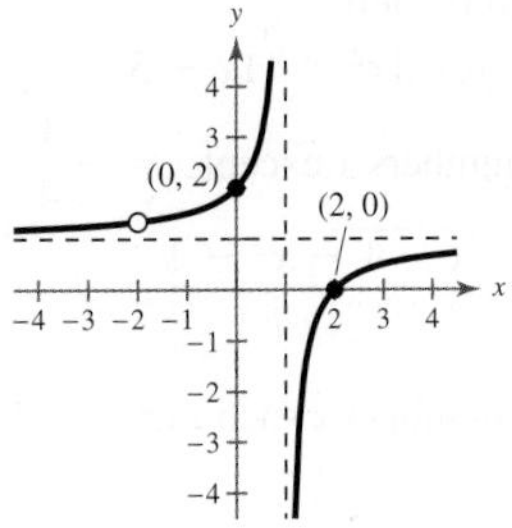

27. y-intercept: $(0, 6)$
x-intercepts: $(2, 0), (3, 0)$
Vertical asymptote: $x = -1$
Slant asymptote:
$y = x - 6$

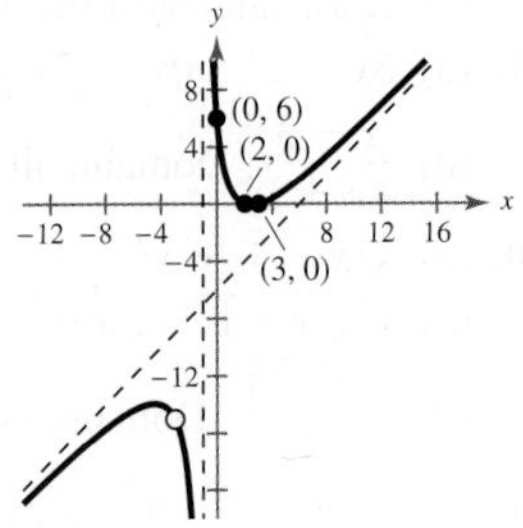

28. $x \le -3$ or $0 \le x \le 3$

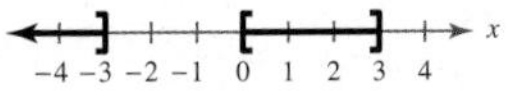

29. All real numbers x such that $x < -5$ or $x > -1$

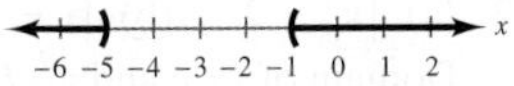

30. Reflect f in the x-axis and y-axis, and shift three units to the right.

31. Reflect f in the x-axis, and shift four units up.

32. 1.991 **33.** -0.067 **34.** 1.717 **35.** 0.281

36. $\ln(x + 4) + \ln(x - 4) - 4 \ln x, x > 4$

37. $\ln \dfrac{x^2}{\sqrt{x + 5}}, x > 0$ **38.** $x = \dfrac{\ln 12}{2} \approx 1.242$

39. $\ln 6 \approx 1.792$ or $\ln 7 \approx 1.946$ **40.** $e^6 - 2 \approx 401.429$

41. (a) and (c)

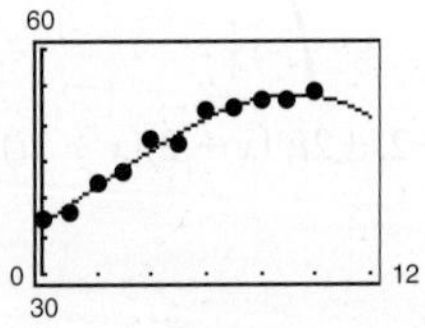

The model is a good fit for the data.

(b) $S = -0.0172t^3 + 0.119t^2 + 2.22t + 36.8$

(d) \$15.0 billion; No; The trend appears to show sales increasing.

42. 6.3 h

Problem Solving (page 259)

1.

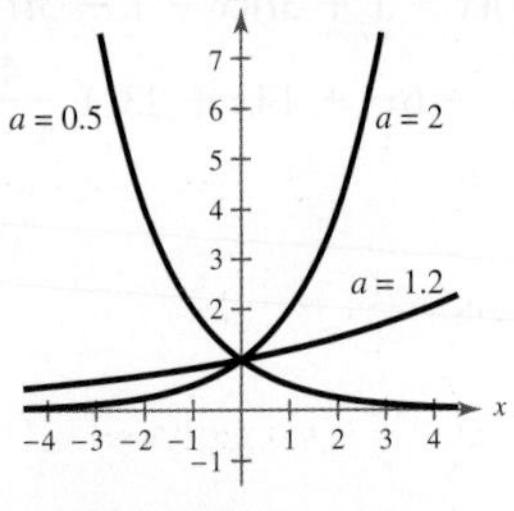

$y = 0.5^x$ and $y = 1.2^x$
$0 < a \le e^{1/e}$

3. As $x \to \infty$, the graph of e^x increases at a greater rate than the graph of x^n.

5. Answers will vary.

7. (a)

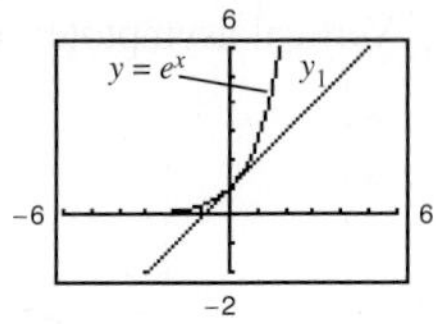

(b)

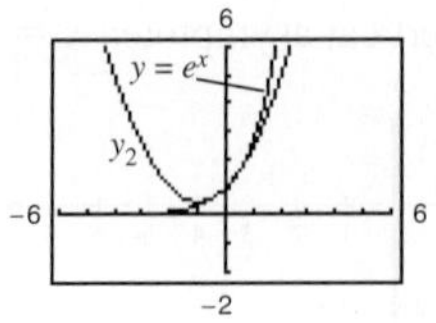

(c)

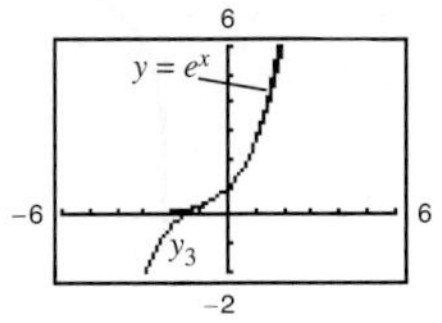

9.

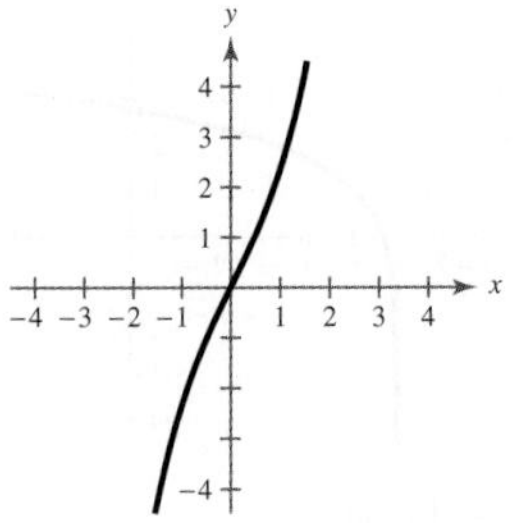

$$f^{-1}(x) = \ln\left(\frac{x + \sqrt{x^2 + 4}}{2}\right)$$

11. c **13.** $t = \dfrac{\ln c_1 - \ln c_2}{\left(\dfrac{1}{k_2} - \dfrac{1}{k_1}\right)\ln \dfrac{1}{2}}$

15. (a) $y_1 = 252{,}606(1.0310)^t$
(b) $y_2 = 400.88t^2 - 1464.6t + 291{,}782$
(c)

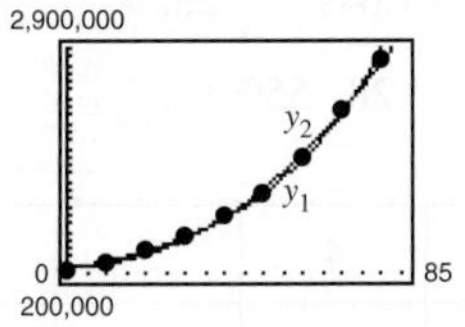

(d) The exponential model is a better fit. No, because the model is rapidly approaching infinity.

17. $1, e^2$

19. $y_4 = (x - 1) - \frac{1}{2}(x - 1)^2 + \frac{1}{3}(x - 1)^3 - \frac{1}{4}(x - 1)^4$

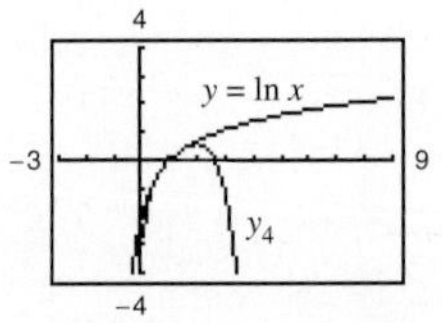

The pattern implies that
$\ln x = (x - 1) - \frac{1}{2}(x - 1)^2 + \frac{1}{3}(x - 1)^3 - \cdots$.

21.

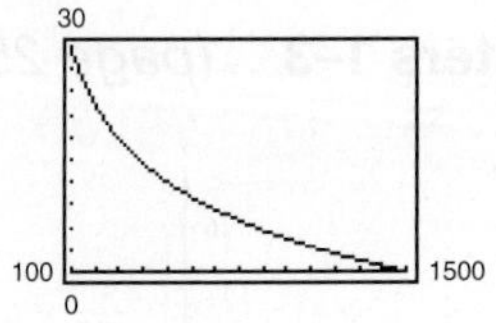

17.7 ft^3/min

23. (a)

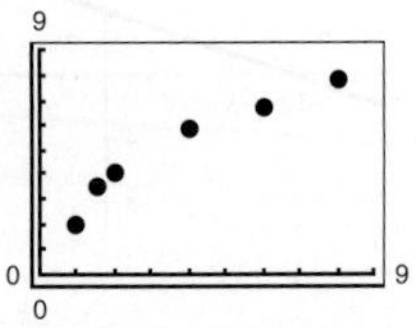

(b)–(e) Answers will vary.

25. (a)

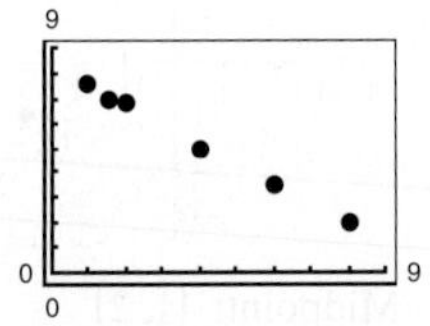

(b)–(e) Answers will vary.

Chapter 4

Section 4.1 *(page 269)*

1. coterminal **3.** complementary; supplementary
5. linear; angular **7.** 1 rad **9.** -3 rad
11. (a) Quadrant I (b) Quadrant III
13. (a)

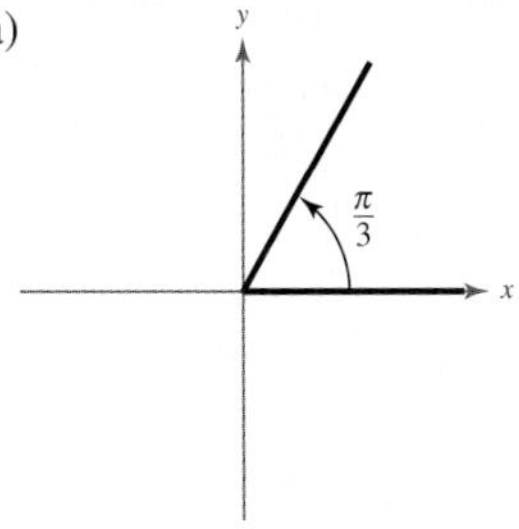

(b)

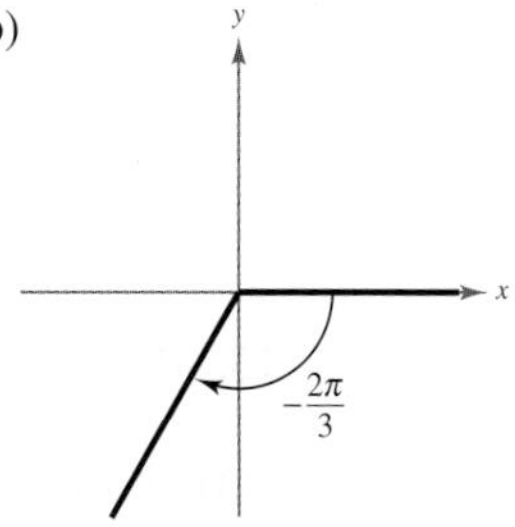

15. *Sample answers:* (a) $\frac{13\pi}{6}, -\frac{11\pi}{6}$ (b) $\frac{19\pi}{6}, -\frac{5\pi}{6}$

17. (a) Complement: $\frac{\pi}{6}$; Supplement: $\frac{2\pi}{3}$
(b) Complement: $\frac{\pi}{4}$; Supplement: $\frac{3\pi}{4}$

19. (a) Complement: $\frac{\pi}{2} - 1 \approx 0.57$;
Supplement: $\pi - 1 \approx 2.14$
(b) Complement: none; Supplement: $\pi - 2 \approx 1.14$

21. 210° **23.** $-60°$ **25.** (a) Quadrant II (b) Quadrant I

27. (a)

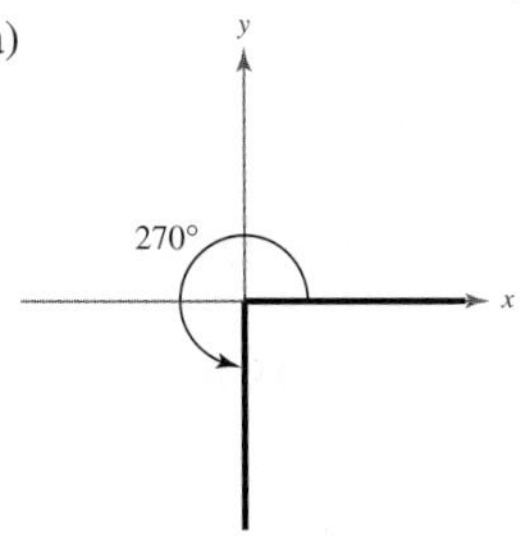

(b)

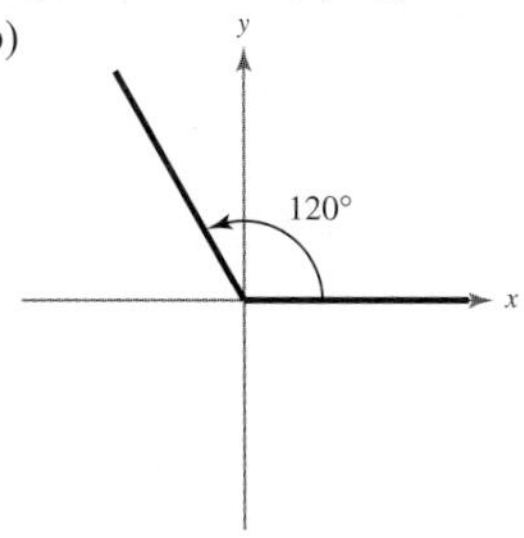

29. *Sample answers:* (a) 405°, $-315°$ (b) 324°, $-396°$
31. (a) Complement: 72°; Supplement: 162°
(b) Complement: 5°; Supplement: 95°
33. (a) Complement: none; Supplement: 30°
(b) Complement: 11°; Supplement: 101°
35. (a) $\frac{2\pi}{3}$ (b) $-\frac{\pi}{9}$ **37.** (a) 270° (b) 210°
39. 0.785 **41.** 0.009 **43.** 81.818° **45.** $-756.000°$
47. (a) 54.75° (b) $-128.5°$
49. (a) 240° 36′ (b) $-145°48′$
51. 10π in. ≈ 31.42 in. **53.** $\frac{15}{8}$ rad **55.** 4 rad
57. 18π mm² ≈ 56.55 mm² **59.** 591.3 mi **61.** 23.87°
63. (a) $10{,}000\pi$ rad/min $\approx 31{,}415.93$ rad/min
(b) 9490.23 ft/min
65. (a) $[400\pi, 1000\pi]$ rad/min (b) $[2400\pi, 6000\pi]$ cm/min
67. (a) 35.70 mi/h (b) 739.50 revolutions/min
69.

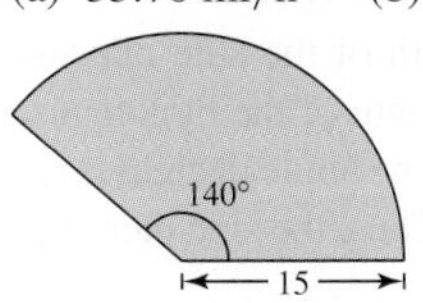

$A = 87.5\pi\ \text{m}^2 \approx 274.89\ \text{m}^2$

71. False. A measurement of 4π radians corresponds to two complete revolutions from the initial side to the terminal side of an angle.
73. False. The terminal side of the angle lies on the x-axis.
75. The speed increases. The linear velocity is proportional to the radius.
77. The arc length is increasing. If θ is constant, the length of the arc is proportional to the radius ($s = r\theta$).

Section 4.2 *(page 277)*

1. unit circle **3.** period

5. $\sin t = \frac{5}{13}$ $\csc t = \frac{13}{5}$
$\cos t = \frac{12}{13}$ $\sec t = \frac{13}{12}$
$\tan t = \frac{5}{12}$ $\cot t = \frac{12}{5}$

7. $\sin t = -\frac{3}{5}$ $\csc t = -\frac{5}{3}$
$\cos t = -\frac{4}{5}$ $\sec t = -\frac{5}{4}$
$\tan t = \frac{3}{4}$ $\cot t = \frac{4}{3}$

9. (0, 1) **11.** $\left(-\frac{\sqrt{3}}{2}, \frac{1}{2}\right)$

13. $\sin\frac{\pi}{4} = \frac{\sqrt{2}}{2}$
$\cos\frac{\pi}{4} = \frac{\sqrt{2}}{2}$
$\tan\frac{\pi}{4} = 1$

15. $\sin\left(-\frac{\pi}{6}\right) = -\frac{1}{2}$
$\cos\left(-\frac{\pi}{6}\right) = \frac{\sqrt{3}}{2}$
$\tan\left(-\frac{\pi}{6}\right) = -\frac{\sqrt{3}}{3}$

17. $\sin\left(-\frac{7\pi}{4}\right) = \frac{\sqrt{2}}{2}$
$\cos\left(-\frac{7\pi}{4}\right) = \frac{\sqrt{2}}{2}$
$\tan\left(-\frac{7\pi}{4}\right) = 1$

19. $\sin\frac{11\pi}{6} = -\frac{1}{2}$
$\cos\frac{11\pi}{6} = \frac{\sqrt{3}}{2}$
$\tan\frac{11\pi}{6} = -\frac{\sqrt{3}}{3}$

21. $\sin\left(-\frac{3\pi}{2}\right) = 1$
$\cos\left(-\frac{3\pi}{2}\right) = 0$
$\tan\left(-\frac{3\pi}{2}\right)$ is undefined.

23. $\sin\frac{2\pi}{3} = \frac{\sqrt{3}}{2}$ $\csc\frac{2\pi}{3} = \frac{2\sqrt{3}}{3}$
$\cos\frac{2\pi}{3} = -\frac{1}{2}$ $\sec\frac{2\pi}{3} = -2$
$\tan\frac{2\pi}{3} = -\sqrt{3}$ $\cot\frac{2\pi}{3} = -\frac{\sqrt{3}}{3}$

25. $\sin\frac{4\pi}{3} = -\frac{\sqrt{3}}{2}$ $\csc\frac{4\pi}{3} = -\frac{2\sqrt{3}}{3}$
$\cos\frac{4\pi}{3} = -\frac{1}{2}$ $\sec\frac{4\pi}{3} = -2$
$\tan\frac{4\pi}{3} = \sqrt{3}$ $\cot\frac{4\pi}{3} = \frac{\sqrt{3}}{3}$

27. $\sin\left(-\frac{5\pi}{3}\right) = \frac{\sqrt{3}}{2}$ $\csc\left(-\frac{5\pi}{3}\right) = \frac{2\sqrt{3}}{3}$
$\cos\left(-\frac{5\pi}{3}\right) = \frac{1}{2}$ $\sec\left(-\frac{5\pi}{3}\right) = 2$
$\tan\left(-\frac{5\pi}{3}\right) = \sqrt{3}$ $\cot\left(-\frac{5\pi}{3}\right) = \frac{\sqrt{3}}{3}$

29. $\sin\left(-\frac{\pi}{2}\right) = -1$ $\csc\left(-\frac{\pi}{2}\right) = -1$
$\cos\left(-\frac{\pi}{2}\right) = 0$ $\sec\left(-\frac{\pi}{2}\right)$ is undefined.
$\tan\left(-\frac{\pi}{2}\right)$ is undefined. $\cot\left(-\frac{\pi}{2}\right) = 0$

31. $\sin 4\pi = \sin 0 = 0$ **33.** $\cos \frac{7\pi}{3} = \cos \frac{\pi}{3} = \frac{1}{2}$

35. $\sin \frac{19\pi}{6} = \sin \frac{7\pi}{6} = -\frac{1}{2}$ **37.** (a) $-\frac{1}{2}$ (b) -2

39. (a) $-\frac{1}{5}$ (b) -5 **41.** (a) $\frac{4}{5}$ (b) $-\frac{4}{5}$ **43.** 1.7321

45. 1.3940 **47.** -4.4014

49. (a) 0.25 ft (b) 0.02 ft (c) -0.25 ft

51. False. $\sin(-t) = -\sin(t)$ means that the function is odd, not that the sine of a negative angle is a negative number.

53. True. The tangent function has a period of π.

55. (a) y-axis symmetry (b) $\sin t_1 = \sin(\pi - t_1)$
(c) $\cos(\pi - t_1) = -\cos t_1$

57. Answers will vary.

59. (a)

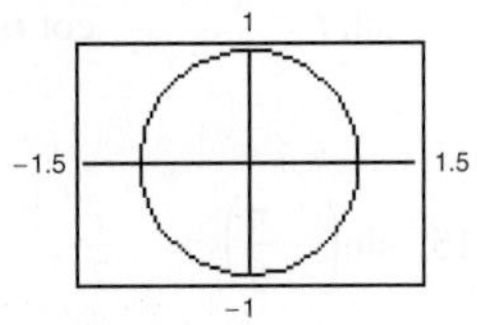

Circle of radius 1 centered at (0, 0)

(b) The t-values represent the central angle in radians. The x- and y-values represent the location in the coordinate plane.
(c) $-1 \le x \le 1, -1 \le y \le 1$

61. It is an odd function.

Section 4.3 *(page 286)*

1. (a) v (b) iv (c) vi (d) iii (e) i (f) ii

3. complementary

5. $\sin \theta = \frac{3}{5}$ $\csc \theta = \frac{5}{3}$
$\cos \theta = \frac{4}{5}$ $\sec \theta = \frac{5}{4}$
$\tan \theta = \frac{3}{4}$ $\cot \theta = \frac{4}{3}$

7. $\sin \theta = \frac{9}{41}$ $\csc \theta = \frac{41}{9}$
$\cos \theta = \frac{40}{41}$ $\sec \theta = \frac{41}{40}$
$\tan \theta = \frac{9}{40}$ $\cot \theta = \frac{40}{9}$

9. $\sin \theta = \frac{8}{17}$ $\csc \theta = \frac{17}{8}$
$\cos \theta = \frac{15}{17}$ $\sec \theta = \frac{17}{15}$
$\tan \theta = \frac{8}{15}$ $\cot \theta = \frac{15}{8}$
The triangles are similar, and corresponding sides are proportional.

11. $\sin \theta = \frac{1}{3}$ $\csc \theta = 3$
$\cos \theta = \frac{2\sqrt{2}}{3}$ $\sec \theta = \frac{3\sqrt{2}}{4}$
$\tan \theta = \frac{\sqrt{2}}{4}$ $\cot \theta = 2\sqrt{2}$
The triangles are similar, and corresponding sides are proportional.

13.

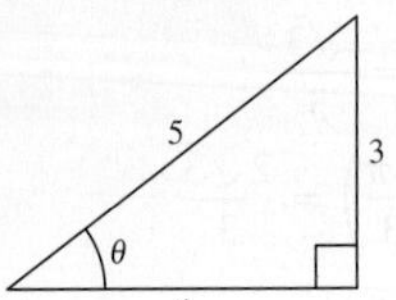

$\sin \theta = \frac{3}{5}$ $\csc \theta = \frac{5}{3}$
$\cos \theta = \frac{4}{5}$ $\sec \theta = \frac{5}{4}$
$\cot \theta = \frac{4}{3}$

15.

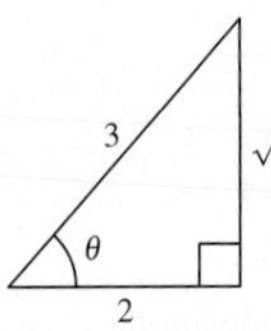

$\sin \theta = \frac{\sqrt{5}}{3}$ $\csc \theta = \frac{3\sqrt{5}}{5}$
$\cos \theta = \frac{2}{3}$
$\tan \theta = \frac{\sqrt{5}}{2}$ $\cot \theta = \frac{2\sqrt{5}}{5}$

17.

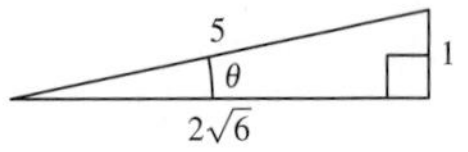

$\csc \theta = 5$
$\cos \theta = \frac{2\sqrt{6}}{5}$ $\sec \theta = \frac{5\sqrt{6}}{12}$
$\tan \theta = \frac{\sqrt{6}}{12}$ $\cot \theta = 2\sqrt{6}$

19.

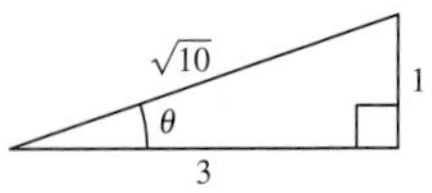

$\sin \theta = \frac{\sqrt{10}}{10}$ $\csc \theta = \sqrt{10}$
$\cos \theta = \frac{3\sqrt{10}}{10}$ $\sec \theta = \frac{\sqrt{10}}{3}$
$\tan \theta = \frac{1}{3}$

21. $\frac{\pi}{6}; \frac{1}{2}$ **23.** $45°; \sqrt{2}$ **25.** $60°; \frac{\pi}{3}$ **27.** 30°; 2

29. $45°; \frac{\pi}{4}$ **31.** (a) 0.1736 (b) 0.1736

33. (a) 0.2815 (b) 3.5523 **35.** (a) 0.9964 (b) 1.0036

37. (a) 5.0273 (b) 0.1989 **39.** (a) 1.8527 (b) 0.9817

41. (a) $\frac{1}{2}$ (b) $\frac{\sqrt{3}}{2}$ (c) $\sqrt{3}$ (d) $\frac{\sqrt{3}}{3}$

43. (a) $\frac{2\sqrt{2}}{3}$ (b) $2\sqrt{2}$ (c) 3 (d) 3

45. (a) $\frac{1}{5}$ (b) $\sqrt{26}$ (c) $\frac{1}{5}$ (d) $\frac{5\sqrt{26}}{26}$

47–55. Answers will vary. **57.** (a) $30° = \frac{\pi}{6}$ (b) $30° = \frac{\pi}{6}$

59. (a) $60° = \frac{\pi}{3}$ (b) $45° = \frac{\pi}{4}$

61. (a) $60° = \frac{\pi}{3}$ (b) $45° = \frac{\pi}{4}$ **63.** $x = 9, y = 9\sqrt{3}$

65. $x = \frac{32\sqrt{3}}{3}, r = \frac{64\sqrt{3}}{3}$ **67.** 443.2 m; 323.3 m

69. $30° = \frac{\pi}{6}$ **71.** (a) 219.9 ft (b) 160.9 ft

73. $(x_1, y_1) = (28\sqrt{3}, 28)$
$(x_2, y_2) = (28, 28\sqrt{3})$

75. $\sin 20° \approx 0.34, \cos 20° \approx 0.94, \tan 20° \approx 0.36,$
$\csc 20° \approx 2.92, \sec 20° \approx 1.06, \cot 20° \approx 2.75$

77. (a) 519.33 ft (b) 1174.17 ft (c) 173.11 ft/min

79. True, $\csc x = \frac{1}{\sin x}$. **81.** False, $\frac{\sqrt{2}}{2} + \frac{\sqrt{2}}{2} \neq 1$.

83. False, $1.7321 \neq 0.0349$.

85. Yes, $\tan \theta$ is equal to opp/adj. You can find the value of the hypotenuse by the Pythagorean Theorem. Then you can find $\sec \theta$, which is equal to hyp/adj.

87. (a)

θ	0°	18°	36°	54°	72°	90°
$\sin \theta$	0	0.3090	0.5878	0.8090	0.9511	1
$\cos \theta$	1	0.9511	0.8090	0.5878	0.3090	0

(b) Increasing function (c) Decreasing function
(d) As the angle increases, the length of the side opposite the angle increases relative to the length of the hypotenuse and the length of the side adjacent to the angle decreases relative to the length of the hypotenuse. Thus, the sine increases and the cosine decreases.

Section 4.4 *(page 296)*

1. $\frac{y}{r}$ **3.** $\frac{y}{x}$ **5.** $\cos\theta$ **7.** zero; defined

9. (a) $\sin\theta = \frac{3}{5}$ $\csc\theta = \frac{5}{3}$
$\cos\theta = \frac{4}{5}$ $\sec\theta = \frac{5}{4}$
$\tan\theta = \frac{3}{4}$ $\cot\theta = \frac{4}{3}$
(b) $\sin\theta = \frac{15}{17}$ $\csc\theta = \frac{17}{15}$
$\cos\theta = -\frac{8}{17}$ $\sec\theta = -\frac{17}{8}$
$\tan\theta = -\frac{15}{8}$ $\cot\theta = -\frac{8}{15}$

11. (a) $\sin\theta = -\frac{1}{2}$ $\csc\theta = -2$
$\cos\theta = -\frac{\sqrt{3}}{2}$ $\sec\theta = -\frac{2\sqrt{3}}{3}$
$\tan\theta = \frac{\sqrt{3}}{3}$ $\cot\theta = \sqrt{3}$
(b) $\sin\theta = -\frac{\sqrt{17}}{17}$ $\csc\theta = -\sqrt{17}$
$\cos\theta = \frac{4\sqrt{17}}{17}$ $\sec\theta = \frac{\sqrt{17}}{4}$
$\tan\theta = -\frac{1}{4}$ $\cot\theta = -4$

13. $\sin\theta = \frac{12}{13}$ $\csc\theta = \frac{13}{12}$
$\cos\theta = \frac{5}{13}$ $\sec\theta = \frac{13}{5}$
$\tan\theta = \frac{12}{5}$ $\cot\theta = \frac{5}{12}$

15. $\sin\theta = -\frac{2\sqrt{29}}{29}$ $\csc\theta = -\frac{\sqrt{29}}{2}$
$\cos\theta = -\frac{5\sqrt{29}}{29}$ $\sec\theta = -\frac{\sqrt{29}}{5}$
$\tan\theta = \frac{2}{5}$ $\cot\theta = \frac{5}{2}$

17. $\sin\theta = \frac{4}{5}$ $\csc\theta = \frac{5}{4}$
$\cos\theta = -\frac{3}{5}$ $\sec\theta = -\frac{5}{3}$
$\tan\theta = -\frac{4}{3}$ $\cot\theta = -\frac{3}{4}$

19. Quadrant I **21.** Quadrant II

23. $\sin\theta = \frac{15}{17}$ $\csc\theta = \frac{17}{15}$
$\cos\theta = -\frac{8}{17}$ $\sec\theta = -\frac{17}{8}$
$\tan\theta = -\frac{15}{8}$ $\cot\theta = -\frac{8}{15}$

25. $\sin\theta = \frac{3}{5}$ $\csc\theta = \frac{5}{3}$
$\cos\theta = -\frac{4}{5}$ $\sec\theta = -\frac{5}{4}$
$\tan\theta = -\frac{3}{4}$ $\cot\theta = -\frac{4}{3}$

27. $\sin\theta = -\frac{\sqrt{10}}{10}$ $\csc\theta = -\sqrt{10}$
$\cos\theta = \frac{3\sqrt{10}}{10}$ $\sec\theta = \frac{\sqrt{10}}{3}$
$\tan\theta = -\frac{1}{3}$ $\cot\theta = -3$

29. $\sin\theta = -\frac{\sqrt{3}}{2}$ $\csc\theta = -\frac{2\sqrt{3}}{3}$
$\cos\theta = -\frac{1}{2}$ $\sec\theta = -2$
$\tan\theta = \sqrt{3}$ $\cot\theta = \frac{\sqrt{3}}{3}$

31. $\sin\theta = 0$ $\csc\theta$ is undefined.
$\cos\theta = -1$ $\sec\theta = -1$
$\tan\theta = 0$ $\cot\theta$ is undefined.

33. $\sin\theta = \frac{\sqrt{2}}{2}$ $\csc\theta = \sqrt{2}$
$\cos\theta = -\frac{\sqrt{2}}{2}$ $\sec\theta = -\sqrt{2}$
$\tan\theta = -1$ $\cot\theta = -1$

35. $\sin\theta = -\frac{2\sqrt{5}}{5}$ $\csc\theta = -\frac{\sqrt{5}}{2}$
$\cos\theta = -\frac{\sqrt{5}}{5}$ $\sec\theta = -\sqrt{5}$
$\tan\theta = 2$ $\cot\theta = \frac{1}{2}$

37. 0 **39.** Undefined **41.** 1 **43.** Undefined

45. $\theta' = 20°$

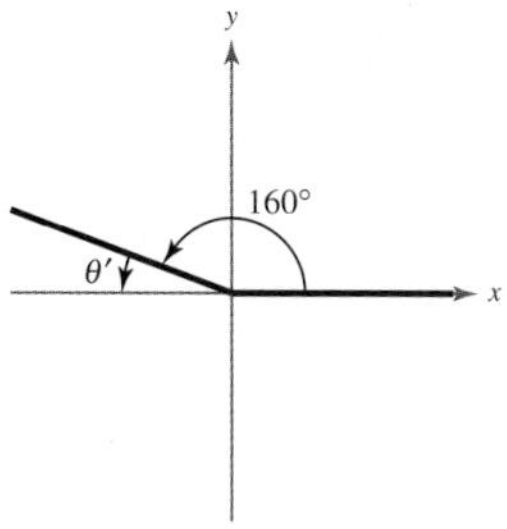

47. $\theta' = 55°$

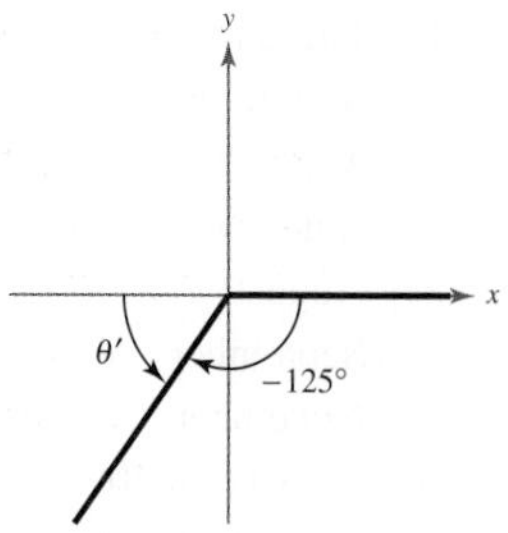

49. $\theta' = \frac{\pi}{3}$

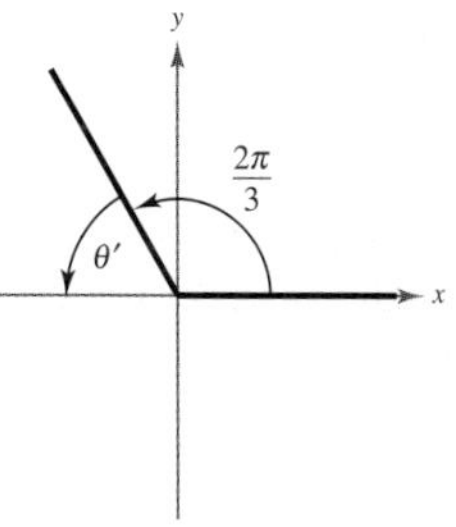

51. $\theta' = 2\pi - 4.8$

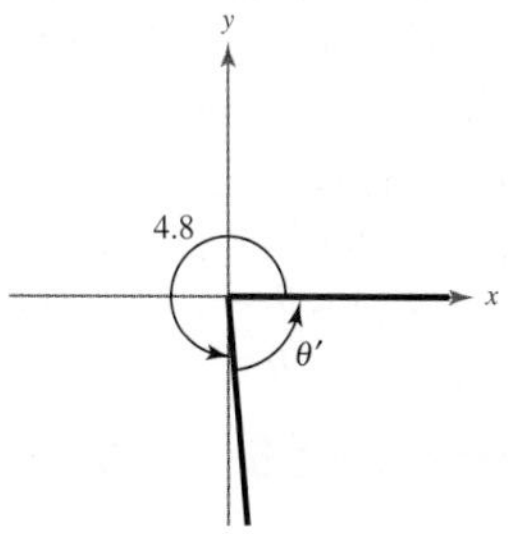

53. $\sin 225° = -\frac{\sqrt{2}}{2}$
$\cos 225° = -\frac{\sqrt{2}}{2}$
$\tan 225° = 1$

55. $\sin 750° = \frac{1}{2}$
$\cos 750° = \frac{\sqrt{3}}{2}$
$\tan 750° = \frac{\sqrt{3}}{3}$

57. $\sin(-840°) = -\frac{\sqrt{3}}{2}$
$\cos(-840°) = -\frac{1}{2}$
$\tan(-840°) = \sqrt{3}$

59. $\sin\frac{2\pi}{3} = \frac{\sqrt{3}}{2}$
$\cos\frac{2\pi}{3} = -\frac{1}{2}$
$\tan\frac{2\pi}{3} = -\sqrt{3}$

61. $\sin\frac{5\pi}{4} = -\frac{\sqrt{2}}{2}$
$\cos\frac{5\pi}{4} = -\frac{\sqrt{2}}{2}$
$\tan\frac{5\pi}{4} = 1$

63. $\sin\left(-\frac{\pi}{6}\right) = -\frac{1}{2}$
$\cos\left(-\frac{\pi}{6}\right) = \frac{\sqrt{3}}{2}$
$\tan\left(-\frac{\pi}{6}\right) = -\frac{\sqrt{3}}{3}$

65. $\sin\frac{9\pi}{4} = \frac{\sqrt{2}}{2}$
$\cos\frac{9\pi}{4} = \frac{\sqrt{2}}{2}$
$\tan\frac{9\pi}{4} = 1$

67. $\sin\left(-\frac{3\pi}{2}\right) = 1$
$\cos\left(-\frac{3\pi}{2}\right) = 0$
$\tan\left(-\frac{3\pi}{2}\right)$ is undefined.

69. $\frac{4}{5}$ **71.** $-\frac{\sqrt{13}}{2}$ **73.** $\frac{8}{5}$ **75.** 0.1736

77. -0.3420 **79.** -1.4826 **81.** 3.2361 **83.** 4.6373

85. 0.3640 **87.** -0.6052 **89.** -0.4142

91. (a) $30° = \frac{\pi}{6}, 150° = \frac{5\pi}{6}$ (b) $210° = \frac{7\pi}{6}, 330° = \frac{11\pi}{6}$

93. (a) $60° = \frac{\pi}{3}, 120° = \frac{2\pi}{3}$ (b) $135° = \frac{3\pi}{4}, 315° = \frac{7\pi}{4}$

95. (a) $45° = \frac{\pi}{4}, 225° = \frac{5\pi}{4}$ (b) $150° = \frac{5\pi}{6}, 330° = \frac{11\pi}{6}$

97. (a) 12 mi (b) 6 mi (c) 6.9 mi

99. (a) $N = 22.099 \sin(0.522t - 2.219) + 55.008$
$F = 36.641 \sin(0.502t - 1.831) + 25.610$
(b) February: $N = 34.6°$, $F = -1.4°$
March: $N = 41.6°$, $F = 13.9°$
May: $N = 63.4°$, $F = 48.6°$
June: $N = 72.5°$, $F = 59.5°$
August: $N = 75.5°$, $F = 55.6°$
September: $N = 68.6°$, $F = 41.7°$
November: $N = 46.8°$, $F = 6.5°$
(c) Answers will vary.

101. (a) 2 cm (b) 0.11 cm (c) -1.2 cm

103. False. In each of the four quadrants, the signs of the secant function and the cosine function are the same because these functions are reciprocals of each other.

105. Answers will vary.

107. (a) $\sin t = y$, $\cos t = x$ (b) $r = 1$ because it is a unit circle.
(c) $\sin \theta = y$, $\cos \theta = x$ (d) $\sin t = \sin \theta$ and $\cos t = \cos \theta$

Section 4.5 *(page 306)*

1. cycle **3.** phase shift **5.** Period: $\frac{2\pi}{5}$; Amplitude: 2

7. Period: 4π; Amplitude: $\frac{3}{4}$ **9.** Period: 6; Amplitude: $\frac{1}{2}$

11. Period: 2π; Amplitude: 4 **13.** Period: $\frac{\pi}{5}$; Amplitude: 3

15. Period: $\frac{5\pi}{2}$; Amplitude: $\frac{5}{3}$ **17.** Period: 1; Amplitude: $\frac{1}{4}$

19. g is a shift of f π units to the right.

21. g is a reflection of f in the x-axis.

23. The period of f is twice the period of g.

25. g is a shift of f three units up.

27. The graph of g has twice the amplitude of the graph of f.

29. The graph of g is a horizontal shift of the graph of f π units to the right.

31.
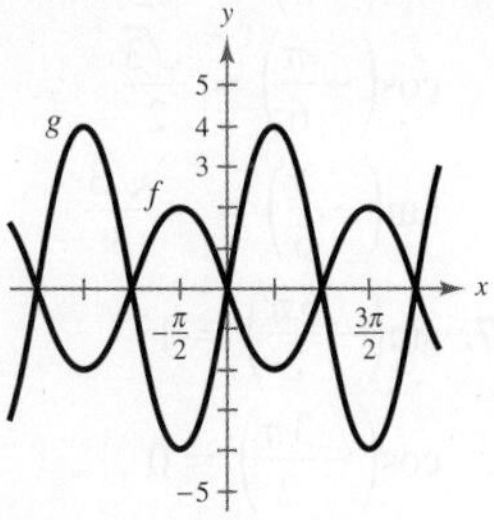

33.
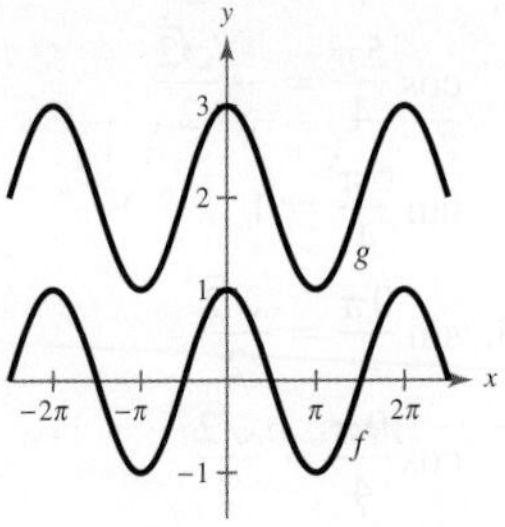

35.
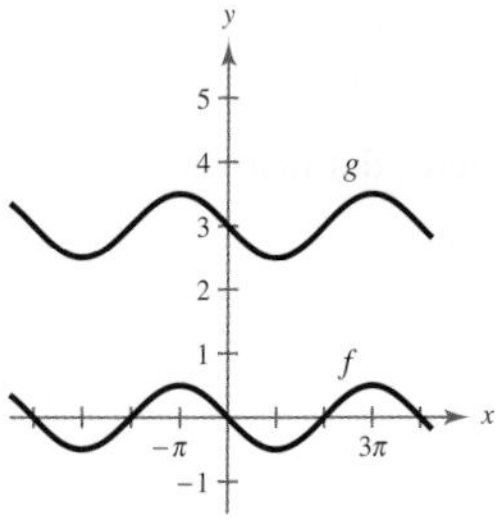

37.
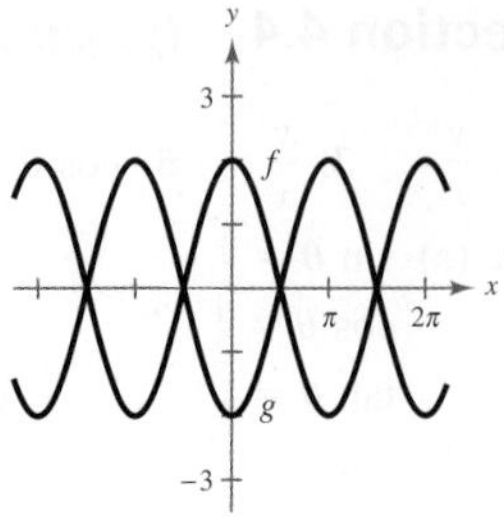

39.
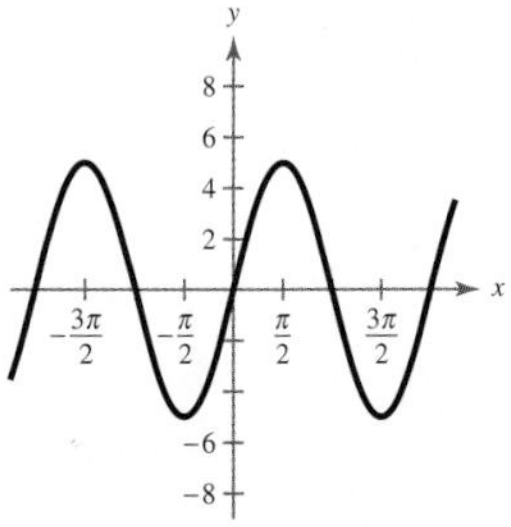

41.
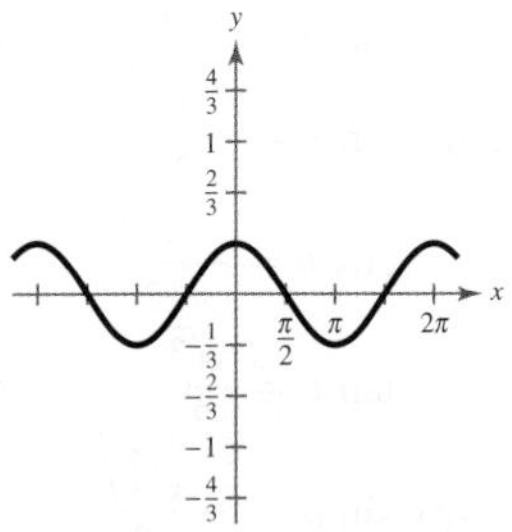

43.
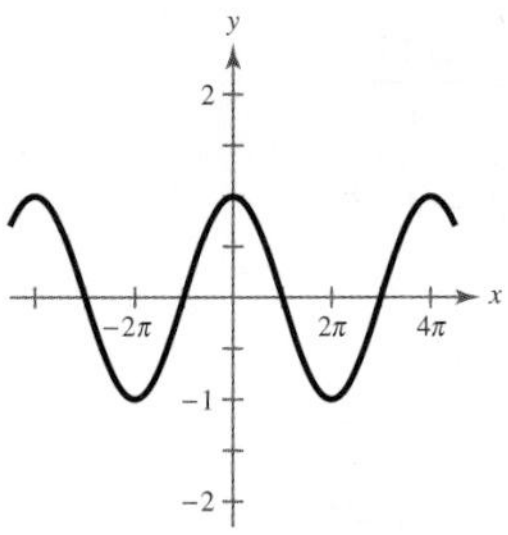

45.
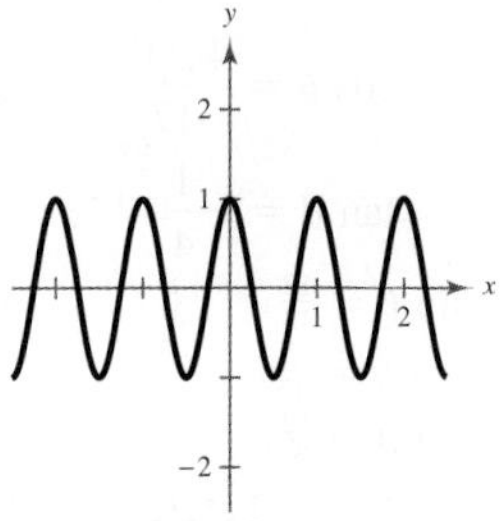

47.
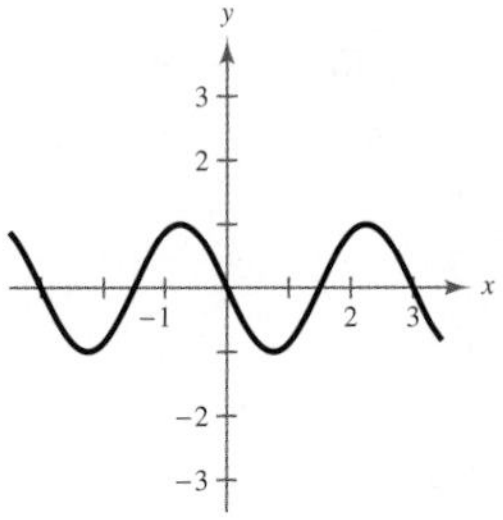

49.
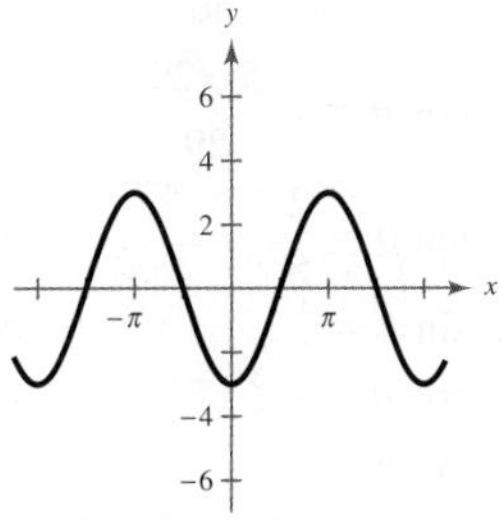

51.
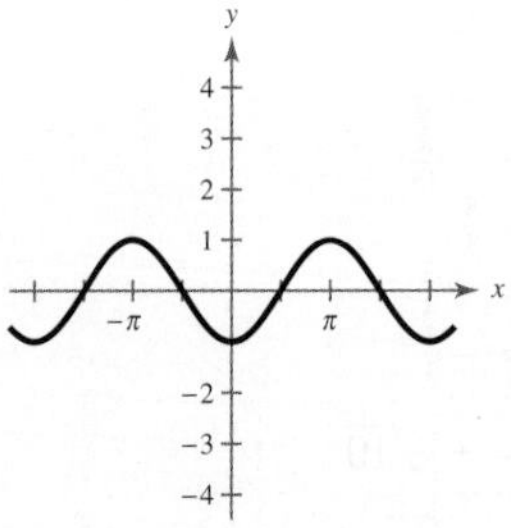

53.
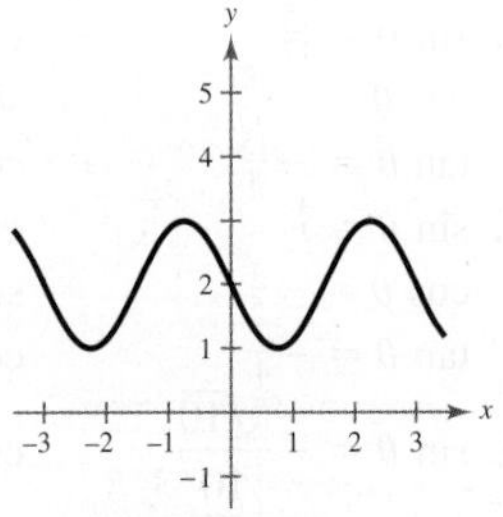

55.
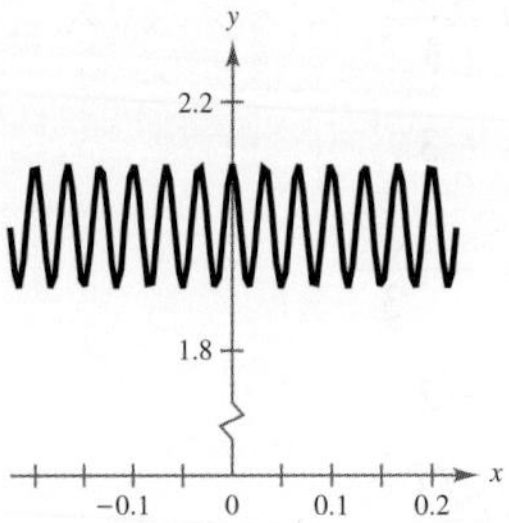

57.
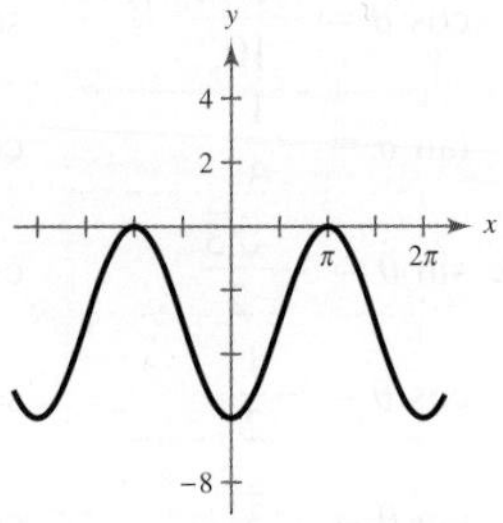

59.

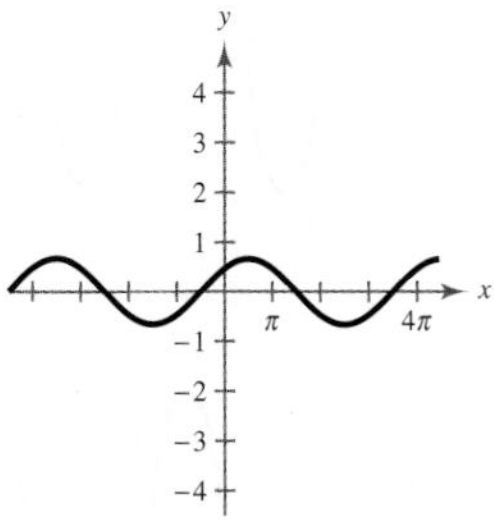

61. (a) $g(x)$ is obtained by a horizontal shrink of four, and one cycle of $g(x)$ corresponds to the interval $[\pi/4, 3\pi/4]$.

(b)

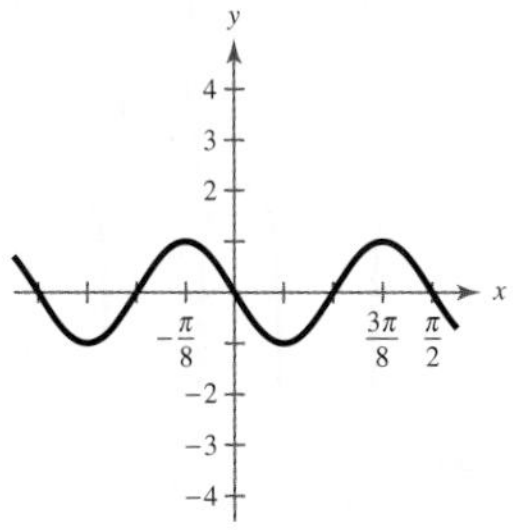

(c) $g(x) = f(4x - \pi)$

63. (a) One cycle of $g(x)$ corresponds to the interval $[\pi, 3\pi]$, and $g(x)$ is obtained by shifting $f(x)$ up two units.

(b)

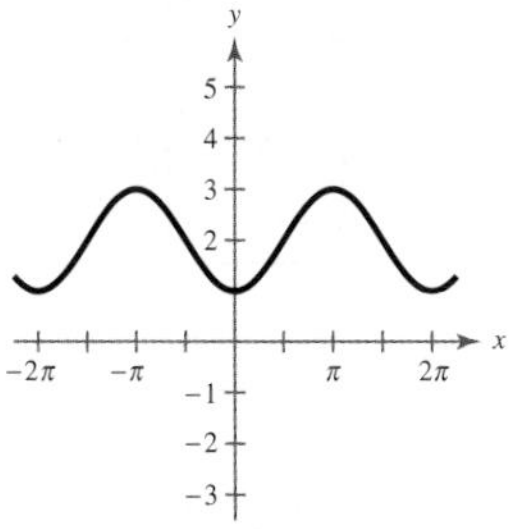

(c) $g(x) = f(x - \pi) + 2$

65. (a) One cycle of $g(x)$ is $[\pi/4, 3\pi/4]$. $g(x)$ is also shifted down three units and has an amplitude of two.

(b)

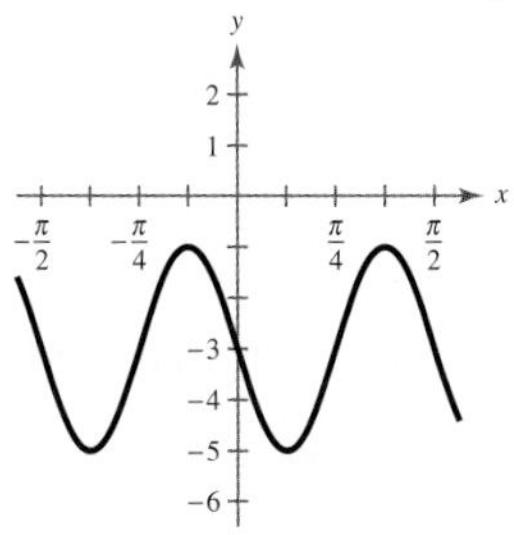

(c) $g(x) = 2f(4x - \pi) - 3$

67.

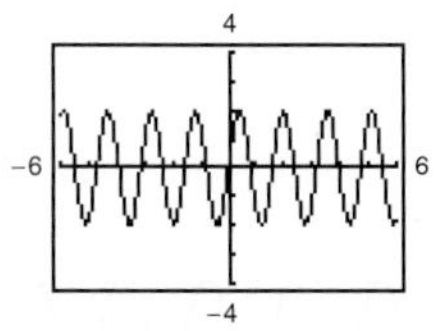

69.

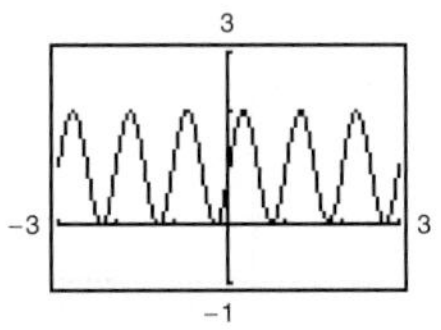

71.

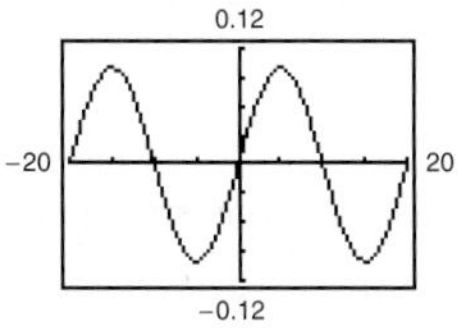

73. $a = 2, d = 1$

75. $a = -4, d = 4$ **77.** $a = -3, b = 2, c = 0$

79. $a = 2, b = 1, c = -\frac{\pi}{4}$

81.

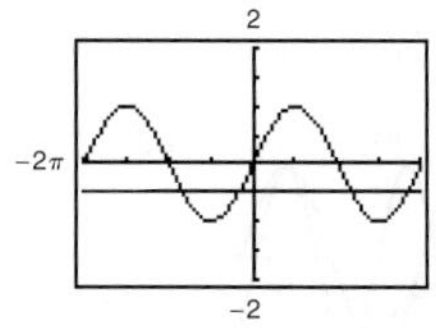

$x = -\frac{\pi}{6}, -\frac{5\pi}{6}, \frac{7\pi}{6}, \frac{11\pi}{6}$

83. $y = 1 + 2\sin(2x - \pi)$ **85.** $y = \cos(2x + 2\pi) - \frac{3}{2}$

87. (a) 4 sec (b) 15 cycles/min

(c)

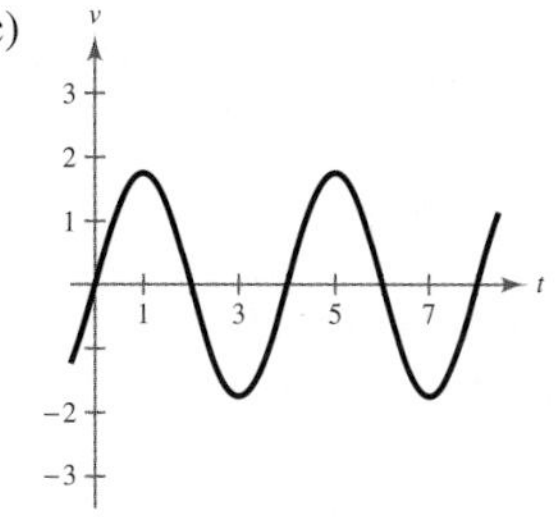

89. (a) $I(t) = 46.2 + 32.4\cos\left(\frac{\pi t}{6} - 3.67\right)$

(b)

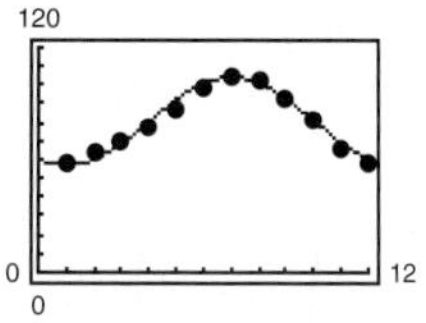

The model fits the data well.

(c)

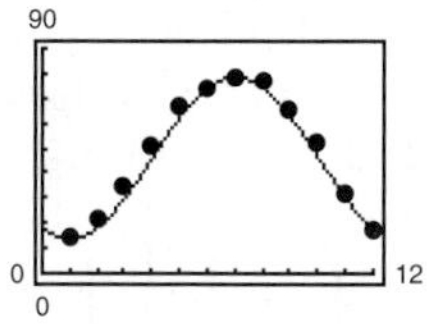

The model fits the data well.

(d) Las Vegas: 80.6°; International Falls: 46.2°
The constant term gives the annual average temperature.

(e) 12; yes; One full period is one year.

(f) International Falls; amplitude; The greater the amplitude, the greater the variability in temperature.

91. (a) $\frac{1}{440}$ sec (b) 440 cycles/sec

93. (a) 20 sec; It takes 20 seconds to complete one revolution on the Ferris wheel.

(b) 50 ft; The diameter of the Ferris wheel is 100 feet.

(c)

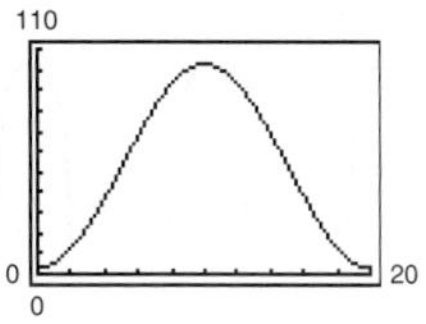

95. False. The graph of $f(x) = \sin(x + 2\pi)$ translates the graph of $f(x) = \sin x$ exactly one period to the left so that the two graphs look identical.

97.

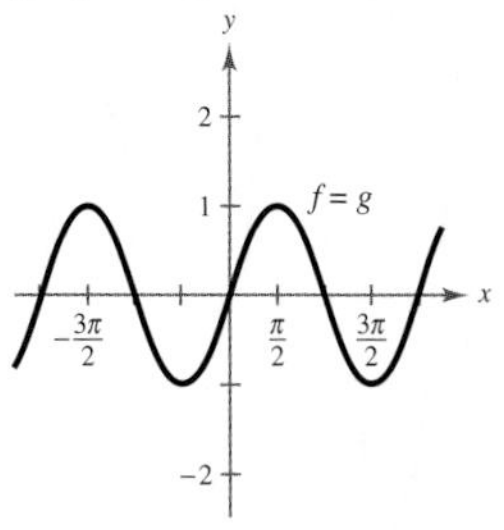

Conjecture: $\sin x = \cos\left(x - \frac{\pi}{2}\right)$

99.

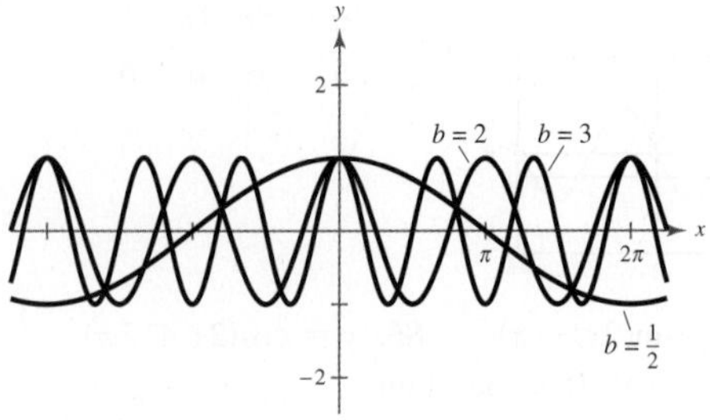

The value of b affects the period of the graph.

$b = \frac{1}{2} \rightarrow \frac{1}{2}$ cycle

$b = 2 \rightarrow 2$ cycles

$b = 3 \rightarrow 3$ cycles

101. (a) 0.4794, 0.4794 (b) 0.8417, 0.8415 (c) 0.5, 0.5 (d) 0.8776, 0.8776 (e) 0.5417, 0.5403 (f) 0.7074, 0.7071

The error increases as x moves farther away from 0.

Section 4.6 *(page 317)*

1. odd; origin **3.** reciprocal **5.** π
7. $(-\infty, -1] \cup [1, \infty)$ **9.** e, π **10.** c, 2π **11.** a, 1
12. d, 2π **13.** f, 4 **14.** b, 4

15.

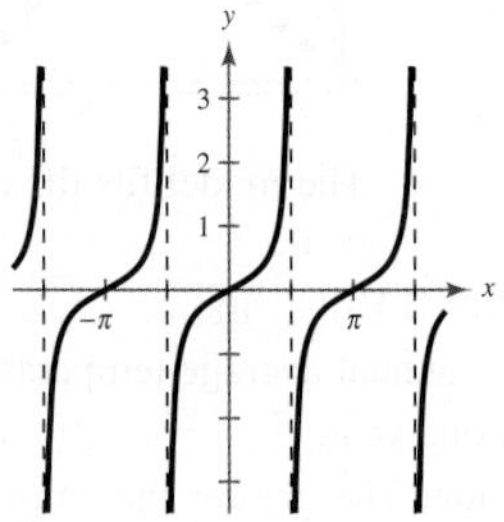

17.

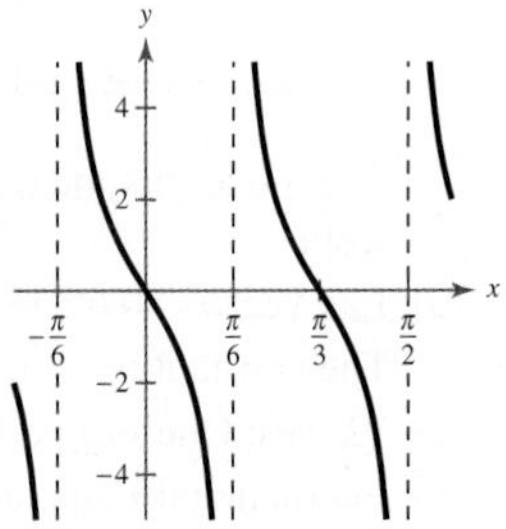

19.

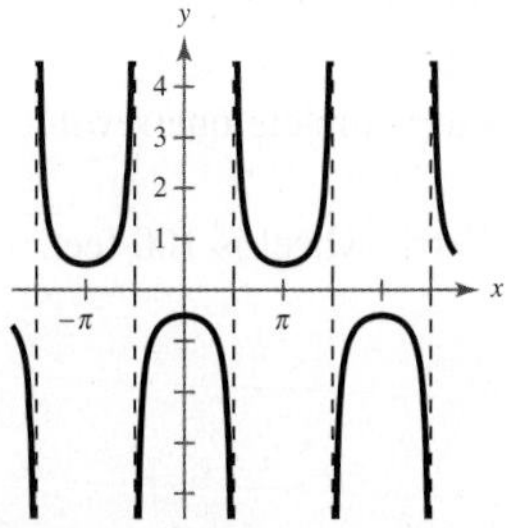

21.

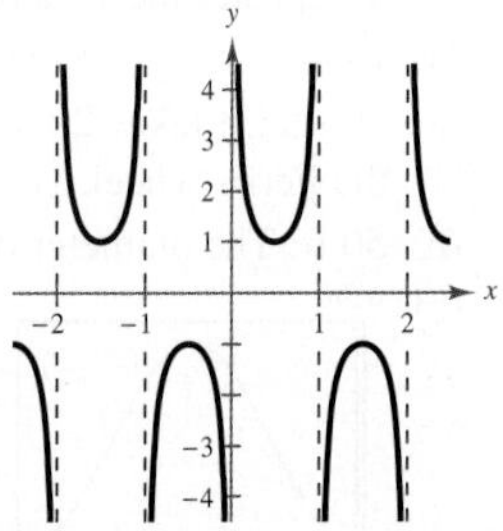

23.

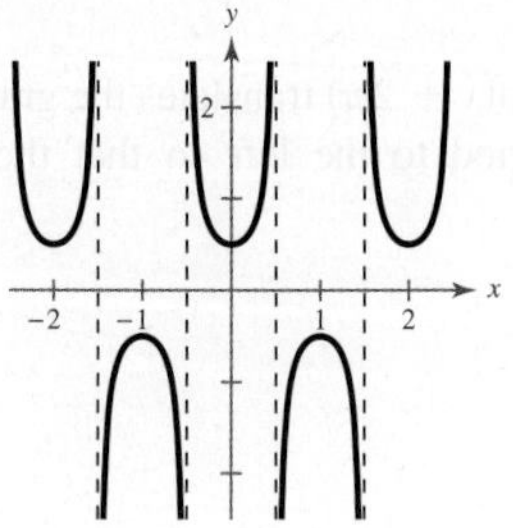

25.

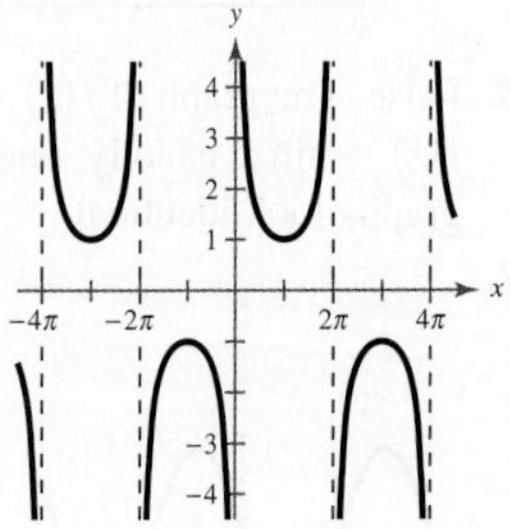

27.

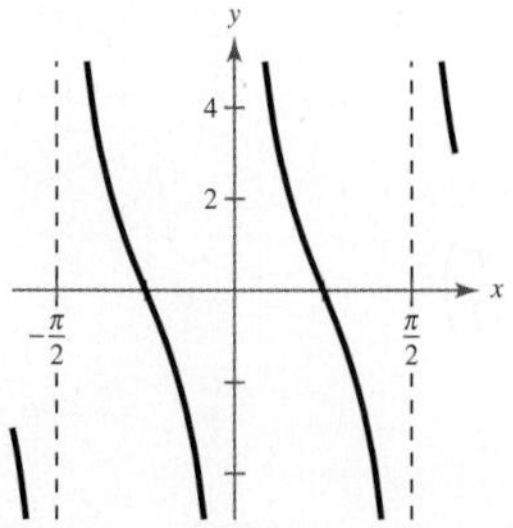

29.

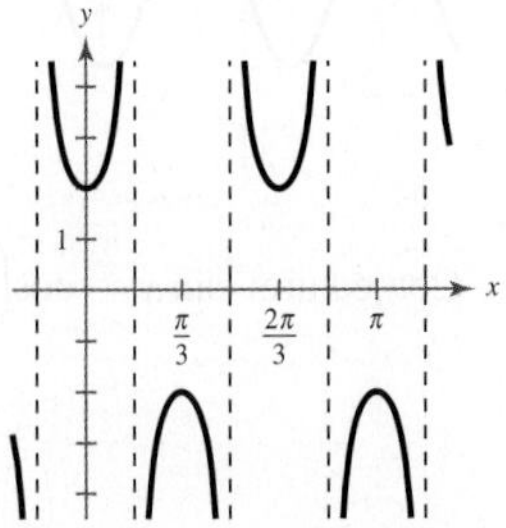

31.

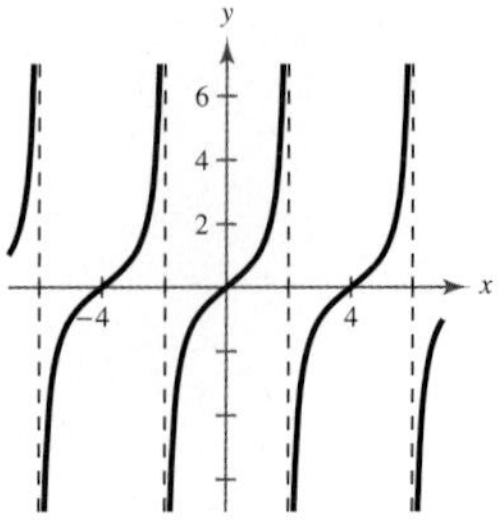

33.

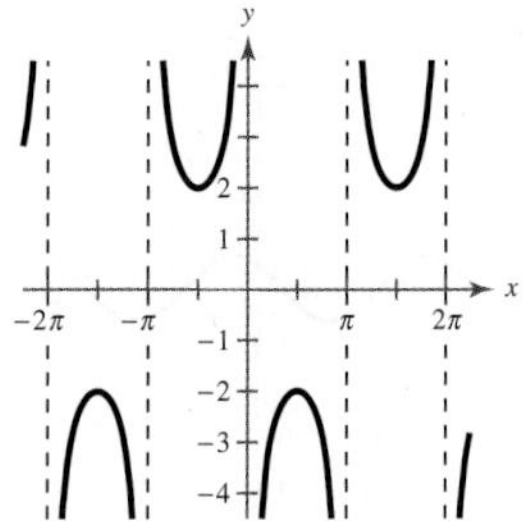

35.

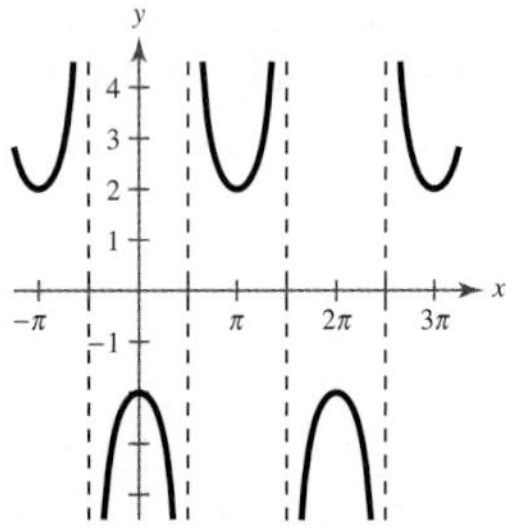

37.

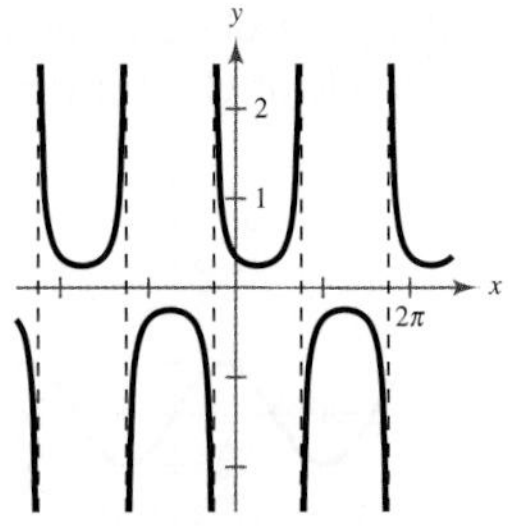

39.

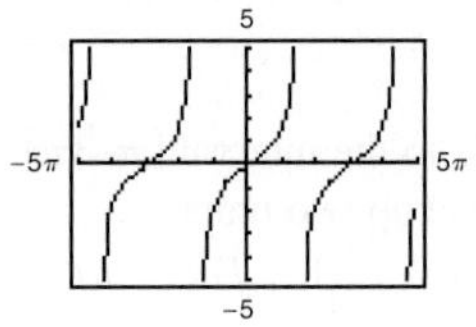

41.

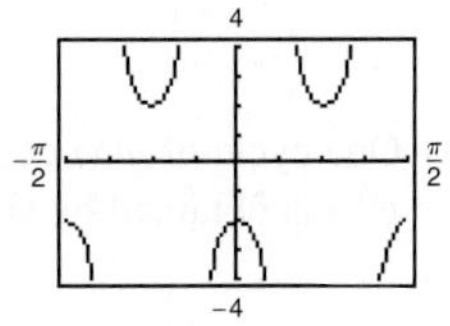

43.

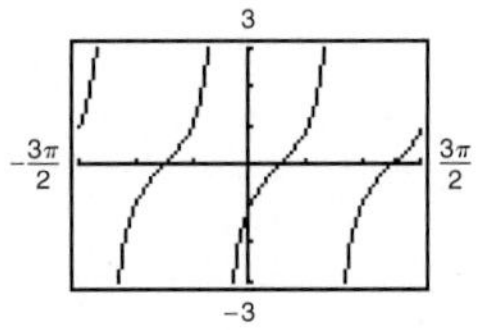

45.

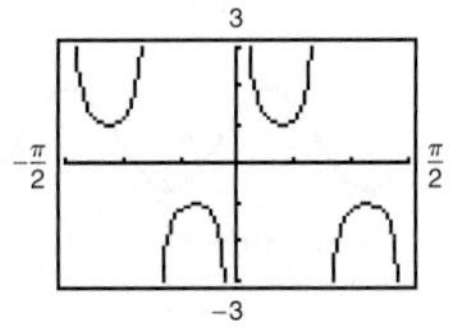

47.

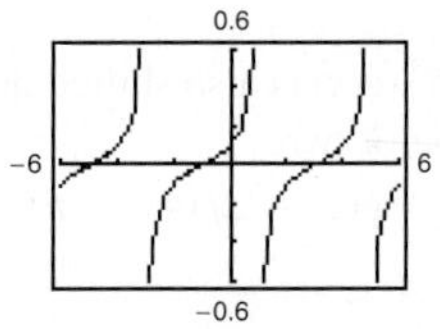

49. $-\frac{7\pi}{4}, -\frac{3\pi}{4}, \frac{\pi}{4}, \frac{5\pi}{4}$

51. $-\frac{4\pi}{3}, -\frac{\pi}{3}, \frac{2\pi}{3}, \frac{5\pi}{3}$ **53.** $-\frac{4\pi}{3}, -\frac{2\pi}{3}, \frac{2\pi}{3}, \frac{4\pi}{3}$

55. $-\frac{7\pi}{4}, -\frac{5\pi}{4}, \frac{\pi}{4}, \frac{3\pi}{4}$ **57.** Even **59.** Odd **61.** Odd

63. Even **65.** d, $f \rightarrow 0$ as $x \rightarrow 0$. **66.** a, $f \rightarrow 0$ as $x \rightarrow 0$.

67. b, $g \rightarrow 0$ as $x \rightarrow 0$. **68.** c, $g \rightarrow 0$ as $x \rightarrow 0$.

69.

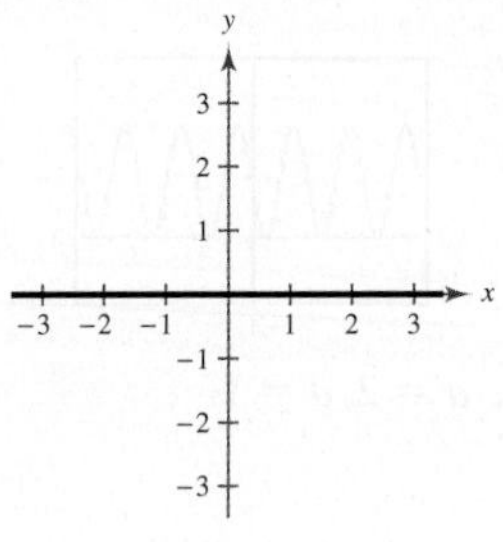

The functions are equal.

71.

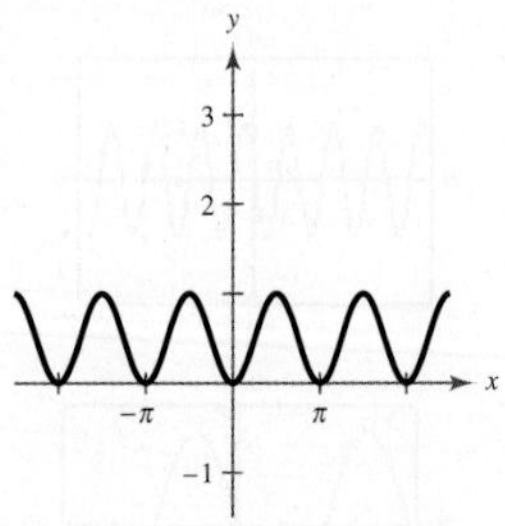

The functions are equal.

73.

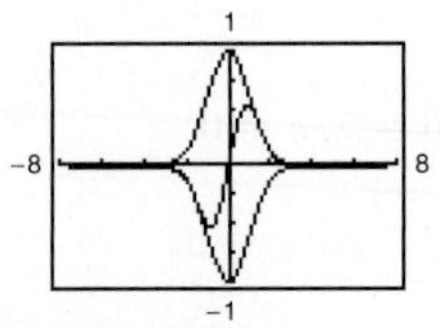

As $x \rightarrow \infty$, $g(x) \rightarrow 0$.

75.

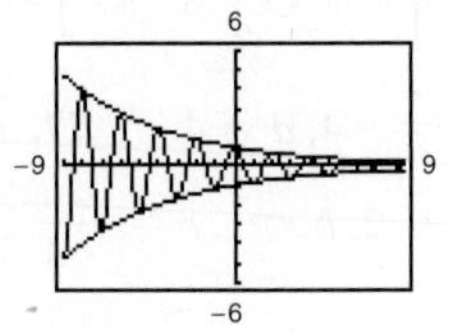

As $x \rightarrow \infty$, $f(x) \rightarrow 0$.

77. 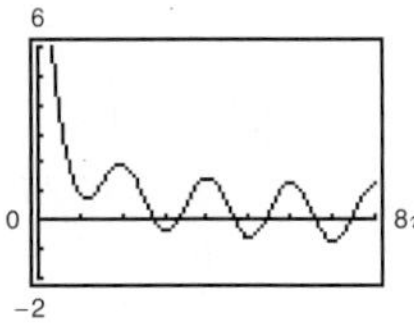

As $x \to 0, y \to \infty$.

79. 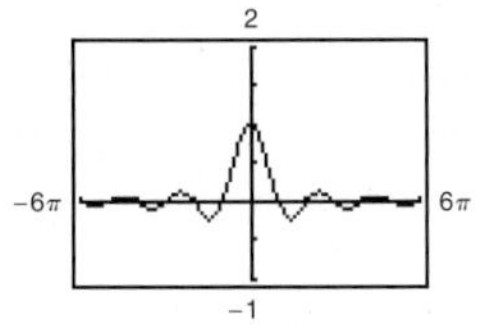

As $x \to 0, g(x) \to 1$.

81. 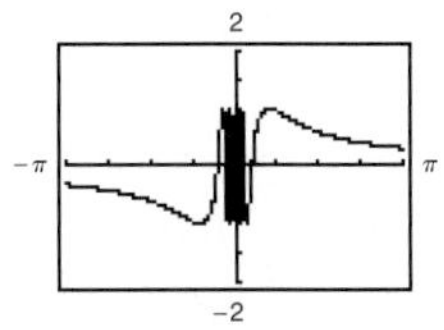

As $x \to 0$, $f(x)$ oscillates between 1 and -1.

83. (a) Period of $H(t)$: 12 mo
Period of $L(t)$: 12 mo
(b) Summer; winter
(c) About 0.5 mo

85. $d = 7 \cot x$

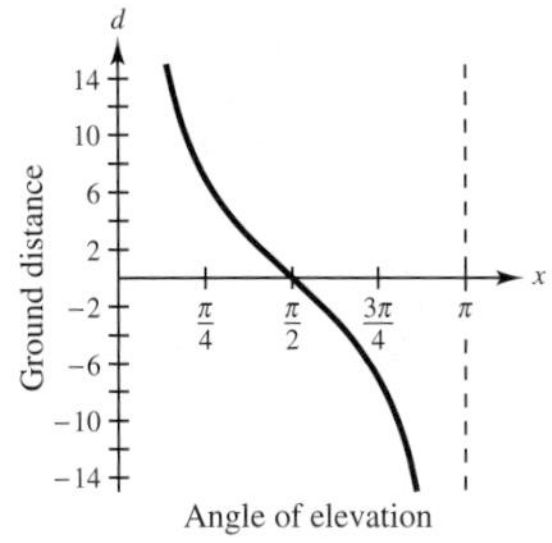

87. True. $y = \sec x$ is equal to $y = 1/\cos x$, and if the reciprocal of $y = \sin x$ is translated $\pi/2$ units to the left, then

$$\frac{1}{\sin\left(x + \frac{\pi}{2}\right)} = \frac{1}{\cos x} = \sec x.$$

89. (a) As $x \to 0^+$, $f(x) \to \infty$. (b) As $x \to 0^-$, $f(x) \to -\infty$.
(c) As $x \to \pi^+$, $f(x) \to -\infty$. (d) As $x \to \pi^-$, $f(x) \to \infty$.

91. (a) As $x \to \left(\frac{\pi}{2}\right)^+$, $f(x) \to -\infty$.

(b) As $x \to \left(\frac{\pi}{2}\right)^-$, $f(x) \to \infty$.

(c) As $x \to \left(-\frac{\pi}{2}\right)^+$, $f(x) \to \infty$.

(d) As $x \to \left(-\frac{\pi}{2}\right)^-$, $f(x) \to -\infty$.

93. (a) 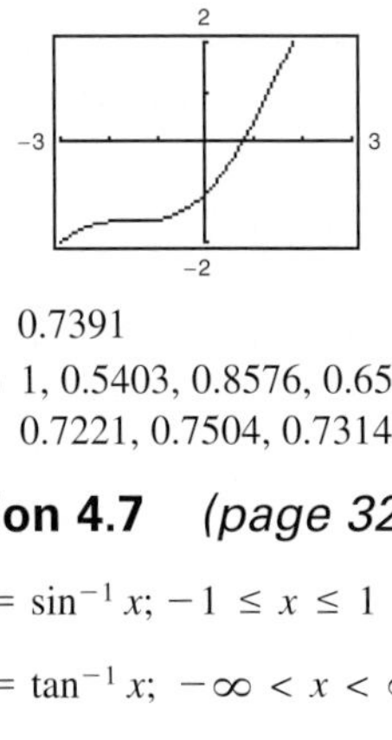

0.7391

(b) 1, 0.5403, 0.8576, 0.6543, 0.7935, 0.7014, 0.7640, 0.7221, 0.7504, 0.7314, . . . ; 0.7391

Section 4.7 *(page 326)*

1. $y = \sin^{-1} x;\ -1 \le x \le 1$

3. $y = \tan^{-1} x;\ -\infty < x < \infty;\ -\frac{\pi}{2} < y < \frac{\pi}{2}$

5. $\frac{\pi}{6}$ **7.** $\frac{\pi}{3}$ **9.** $\frac{\pi}{6}$ **11.** $\frac{5\pi}{6}$ **13.** $-\frac{\pi}{3}$

15. $\frac{2\pi}{3}$ **17.** $-\frac{\pi}{3}$

19. 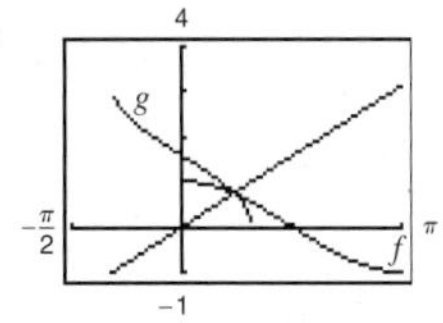

21. 1.19 **23.** -0.85 **25.** -1.25 **27.** 0.32
29. 1.99 **31.** 0.74 **33.** 1.07 **35.** 1.36 **37.** -1.52

39. $-\frac{\pi}{3}, -\frac{\sqrt{3}}{3}, 1$ **41.** $\theta = \arctan \frac{x}{4}$ **43.** $\theta = \arcsin \frac{x+2}{5}$

45. $\theta = \arccos \frac{x+3}{2x}$ **47.** 0.3 **49.** -0.1 **51.** 0

53. $\frac{3}{5}$ **55.** $\frac{\sqrt{5}}{5}$ **57.** $\frac{12}{13}$ **59.** $\frac{\sqrt{34}}{5}$ **61.** $\frac{\sqrt{5}}{3}$

63. 2 **65.** $\frac{1}{x}$ **67.** $\sqrt{1 - 4x^2}$ **69.** $\sqrt{1 - x^2}$

71. $\frac{\sqrt{9 - x^2}}{x}$ **73.** $\frac{\sqrt{x^2 + 2}}{x}$

75. 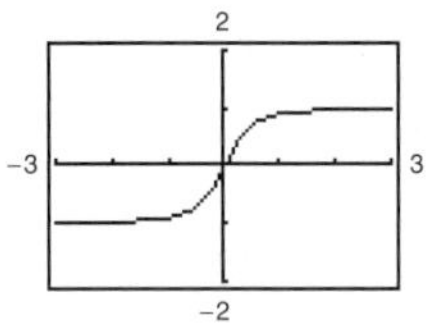

Asymptotes: $y = \pm 1$

77. $\frac{9}{\sqrt{x^2 + 81}}$

79. $\frac{|x - 1|}{\sqrt{x^2 - 2x + 10}}$

81. 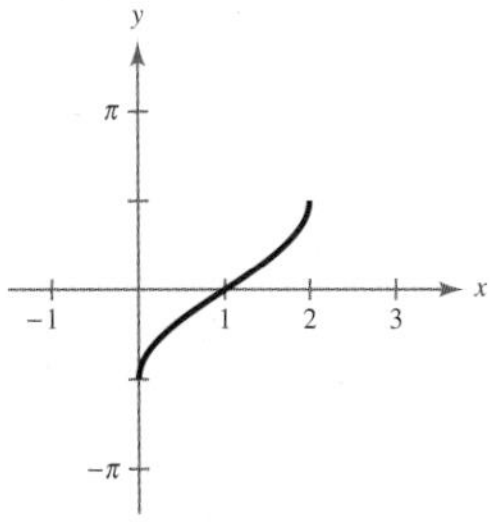

The graph of g is a horizontal shift one unit to the right of f.

83. 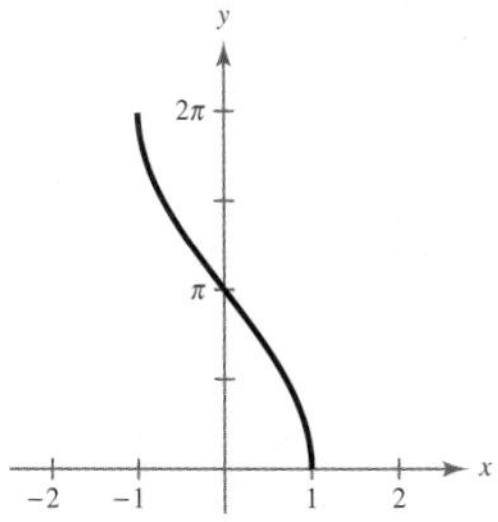

85. 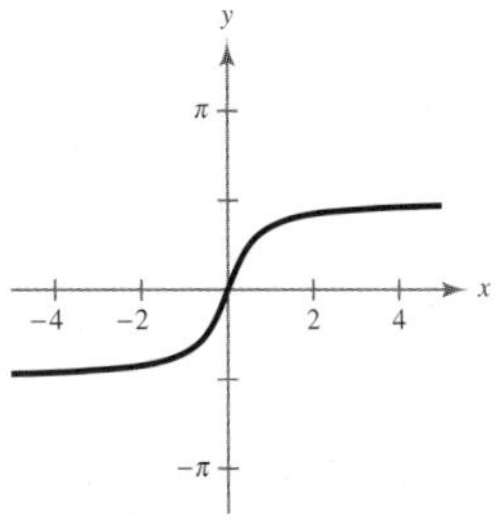

87. 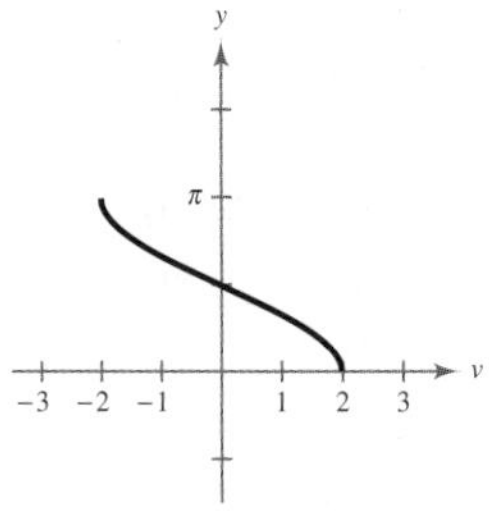

89. 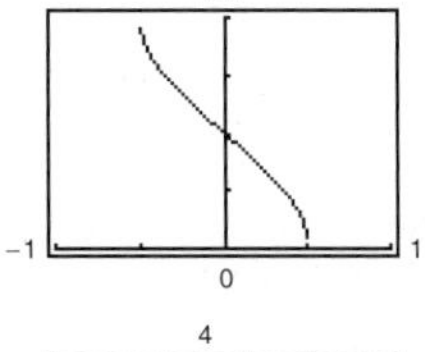

91. 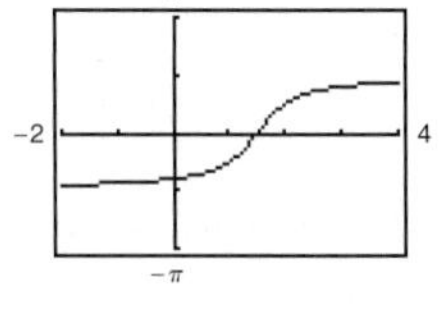

93. 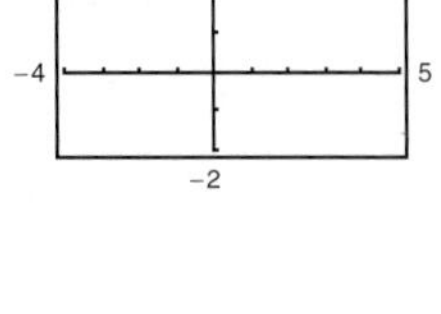

CHAPTER 4

95. $3\sqrt{2}\sin\left(2t + \frac{\pi}{4}\right)$

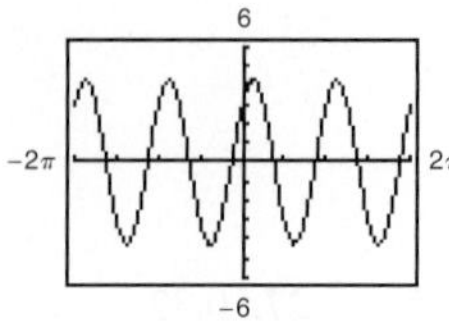

The graph implies that the identity is true.

97. $\frac{\pi}{2}$ **99.** $\frac{\pi}{2}$ **101.** π

103. (a) $\theta = \arcsin \frac{5}{s}$ (b) 0.13, 0.25

105. (a)

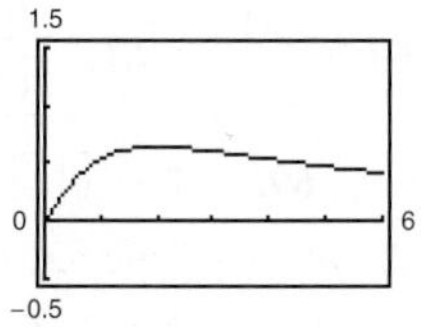

(b) 2 ft (c) $\beta = 0$; As x increases, β approaches 0.

107. (a) $\theta \approx 26.0°$ (b) 24.4 ft

109. (a) $\theta = \arctan \frac{x}{20}$ (b) 14.0°, 31.0°

111. False. $\frac{5\pi}{4}$ is not in the range of the arctangent.

113. False. $\sin^{-1} x$ is the inverse of $\sin x$, not the reciprocal.

115. Domain: $(-\infty, \infty)$ Range: $(0, \pi)$

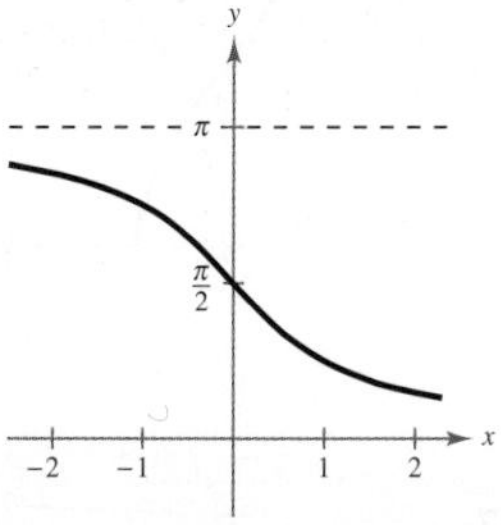

117. Domain: $(-\infty, -1] \cup [1, \infty)$ Range: $[-\pi/2, 0) \cup (0, \pi/2]$

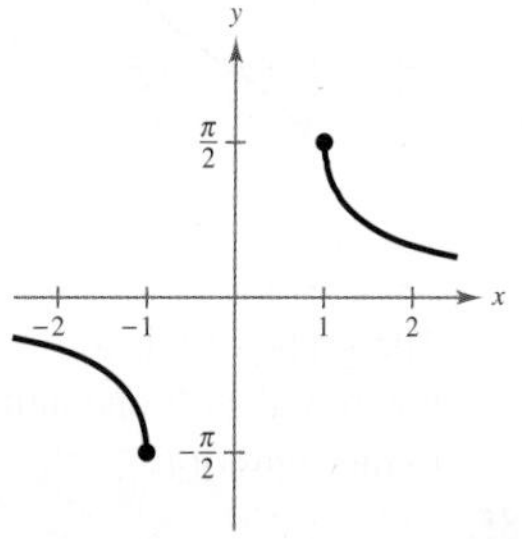

119. $\frac{\pi}{4}$ **121.** $\frac{3\pi}{4}$ **123.** $\frac{\pi}{6}$ **125.** $\frac{\pi}{3}$ **127.** 1.17

129. 0.19 **131.** 0.54 **133.** −0.12

135. (a) $\frac{\pi}{4}$ (b) $\frac{\pi}{2}$ (c) 1.25 (d) 2.03

137. (a) $f \circ f^{-1}$

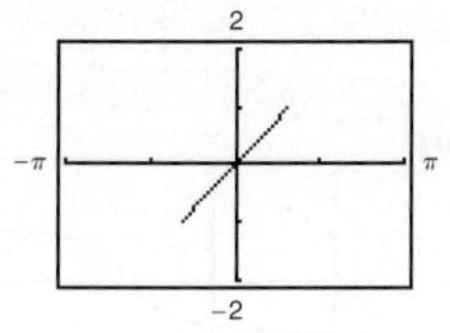

$f^{-1} \circ f$

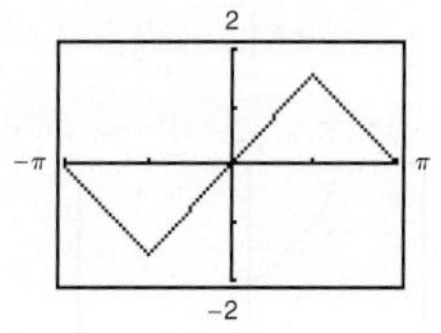

(b) The domains and ranges of the functions are restricted. The graphs of $f \circ f^{-1}$ and $f^{-1} \circ f$ differ because of the domains and ranges of f and f^{-1}.

Section 4.8 *(page 336)*

1. bearing **3.** period

5. $a \approx 1.73$, $c \approx 3.46$, $B = 60°$

7. $a \approx 8.26$, $c \approx 25.38$, $A = 19°$

9. $c = 5$, $A \approx 36.87°$, $B \approx 53.13°$

11. $a \approx 49.48$, $A \approx 72.08°$, $B \approx 17.92°$

13. $a \approx 91.34$, $b \approx 420.70$, $B = 77°45'$

15. 3.00

17. 2.50 **19.** 214.45 ft **21.** 19.7 ft

23. 19.9 ft **25.** 11.8 km **27.** 56.3° **29.** 2.06°

31. (a) $\sqrt{h^2 + 34h + 10{,}289}$ (b) $\theta = \arccos \frac{100}{l}$ (c) 53.02 ft

33. (a) $l = 250$ ft, $A \approx 36.87°$, $B \approx 53.13°$ (b) 4.87 sec

35. 554 mi north; 709 mi east

37. (a) 104.95 nm south; 58.18 nm west (b) S 36.7° W; distance ≈ 130.9 nm

39. N 56.31° W **41.** (a) N 58° E (b) 68.82 m

43. 35.3° **45.** 29.4 in. **47.** $d = 4 \sin \pi t$

49. $d = 3 \cos \frac{4\pi t}{3}$ **51.** $\omega = 528\pi$

53. (a) 9 (b) $\frac{3}{5}$ (c) 9 (d) $\frac{5}{12}$

55. (a) $\frac{1}{4}$ (b) 3 (c) 0 (d) $\frac{1}{6}$

57. (a)

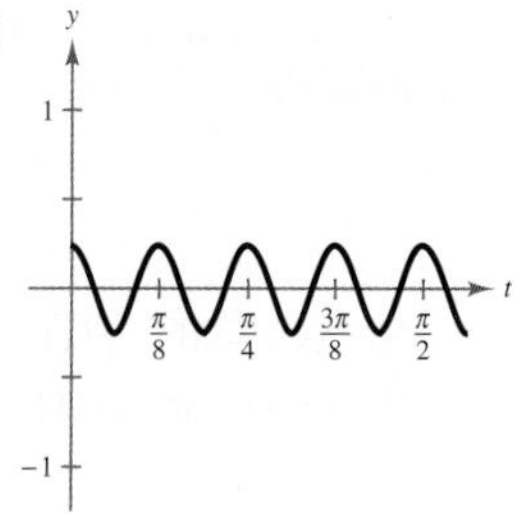

(b) $\frac{\pi}{8}$ (c) $\frac{\pi}{32}$

59. (a)

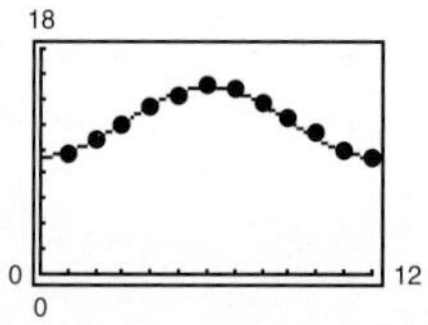

(b) 12; Yes, there are 12 months in a year.

(c) 2.77; The maximum change in the number of hours of daylight

61. False. The scenario does not create a right triangle because the tower is not vertical.

Review Exercises *(page 342)*

1. (a)

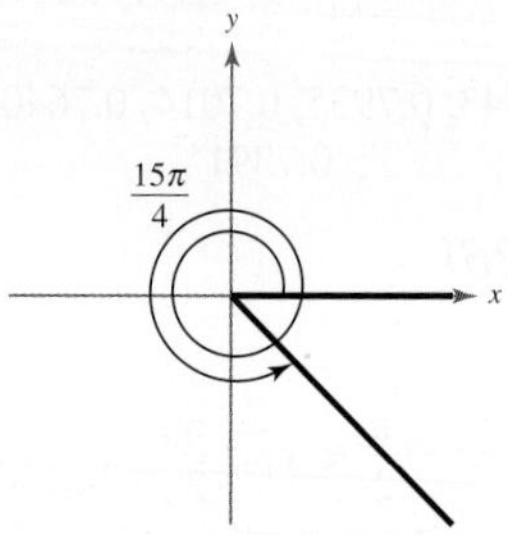

(b) Quadrant IV (c) $\frac{23\pi}{4}, -\frac{\pi}{4}$

3. (a)

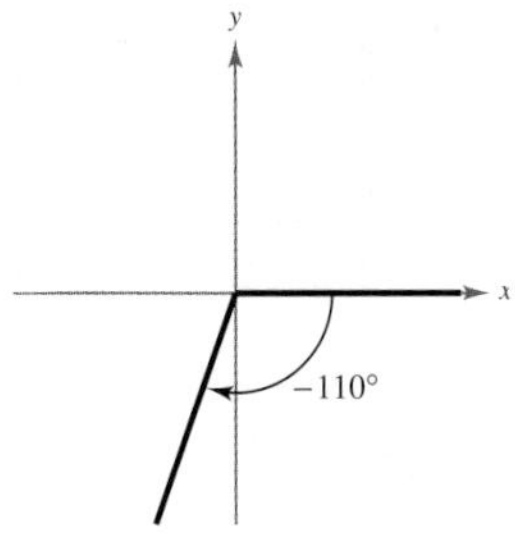

(b) Quadrant III (c) 250°, −470°

5. 7.854 **7.** −0.589 **9.** 54.000° **11.** −200.535°

13. 198° 24′ **15.** 48.17 in. **17.** Area ≈ 339.29 in.^{2}

19. $\left(-\frac{1}{2}, \frac{\sqrt{3}}{2}\right)$ **21.** $\left(-\frac{\sqrt{3}}{2}, -\frac{1}{2}\right)$

23. $\sin\frac{3\pi}{4} = \frac{\sqrt{2}}{2}$ $\csc\frac{3\pi}{4} = \sqrt{2}$

$\cos\frac{3\pi}{4} = -\frac{\sqrt{2}}{2}$ $\sec\frac{3\pi}{4} = -\sqrt{2}$

$\tan\frac{3\pi}{4} = -1$ $\cot\frac{3\pi}{4} = -1$

25. $\sin\frac{11\pi}{4} = \sin\frac{3\pi}{4} = \frac{\sqrt{2}}{2}$

27. $\sin\left(-\frac{17\pi}{6}\right) = \sin\frac{7\pi}{6} = -\frac{1}{2}$ **29.** −75.3130 **31.** 3.2361

33. $\sin\theta = \frac{4\sqrt{41}}{41}$ $\csc\theta = \frac{\sqrt{41}}{4}$

$\cos\theta = \frac{5\sqrt{41}}{41}$ $\sec\theta = \frac{\sqrt{41}}{5}$

$\tan\theta = \frac{4}{5}$ $\cot\theta = \frac{5}{4}$

35. 0.6494 **37.** 3.6722

39. (a) 3 (b) $\frac{2\sqrt{2}}{3}$ (c) $\frac{3\sqrt{2}}{4}$ (d) $\frac{\sqrt{2}}{4}$ **41.** 71.3 m

43. $\sin\theta = \frac{4}{5}$ $\csc\theta = \frac{5}{4}$
$\cos\theta = \frac{3}{5}$ $\sec\theta = \frac{5}{3}$
$\tan\theta = \frac{4}{3}$ $\cot\theta = \frac{3}{4}$

45. $\sin\theta = \frac{4}{5}$ $\csc\theta = \frac{5}{4}$
$\cos\theta = \frac{3}{5}$ $\sec\theta = \frac{5}{3}$
$\tan\theta = \frac{4}{3}$ $\cot\theta = \frac{3}{4}$

47. $\sin\theta = -\frac{\sqrt{11}}{6}$

$\cos\theta = \frac{5}{6}$

$\tan\theta = -\frac{\sqrt{11}}{5}$

$\csc\theta = -\frac{6\sqrt{11}}{11}$

$\cot\theta = -\frac{5\sqrt{11}}{11}$

49. $\sin\theta = \frac{\sqrt{21}}{5}$

$\tan\theta = -\frac{\sqrt{21}}{2}$

$\csc\theta = \frac{5\sqrt{21}}{21}$

$\sec\theta = -\frac{5}{2}$

$\cot\theta = -\frac{2\sqrt{21}}{21}$

51. $\theta' = 84°$

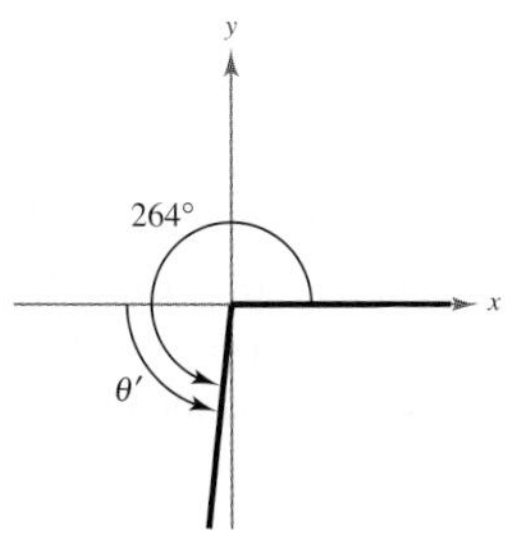

53. $\theta' = \frac{\pi}{5}$

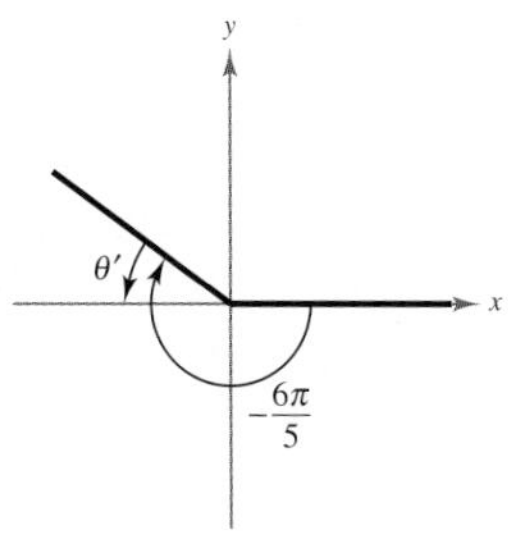

55. $\sin\frac{\pi}{3} = \frac{\sqrt{3}}{2}$; $\cos\frac{\pi}{3} = \frac{1}{2}$; $\tan\frac{\pi}{3} = \sqrt{3}$

57. $\sin(-150°) = -\frac{1}{2}$; $\cos(-150°) = -\frac{\sqrt{3}}{2}$;

$\tan(-150°) = \frac{\sqrt{3}}{3}$

59. −0.7568 **61.** 0.9511

63.

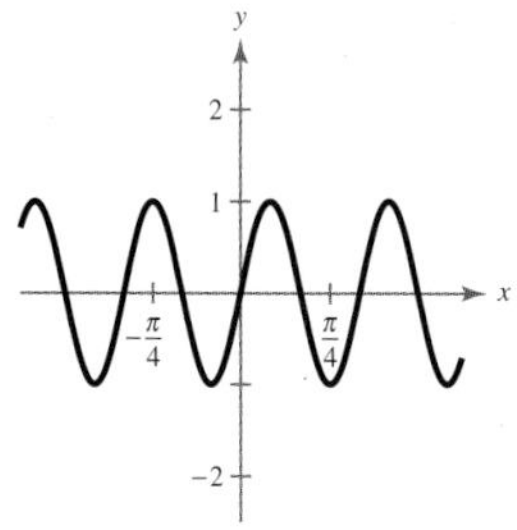

65.

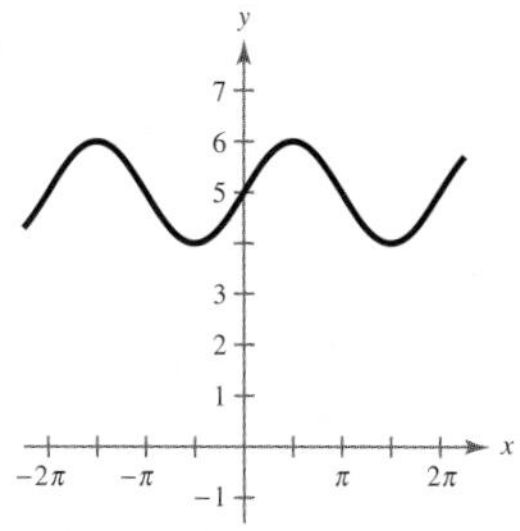

67.

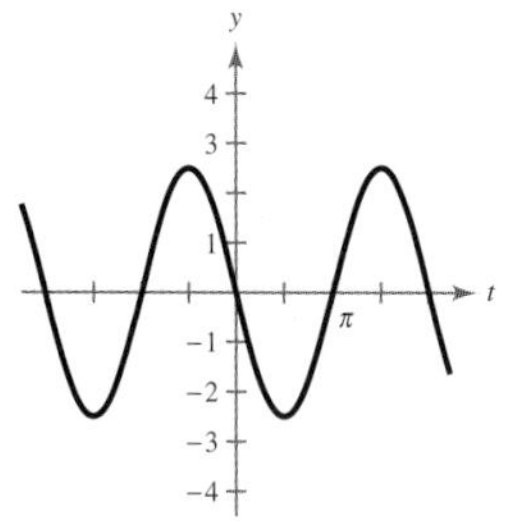

69. (a) $y = 2\sin 528\pi x$
(b) 264 cycles/sec

71.

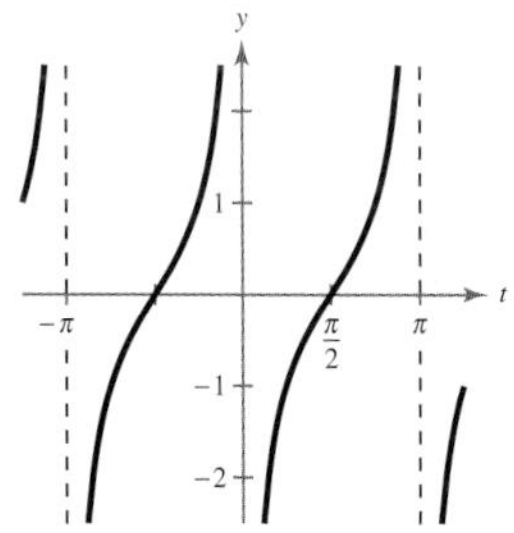

73.

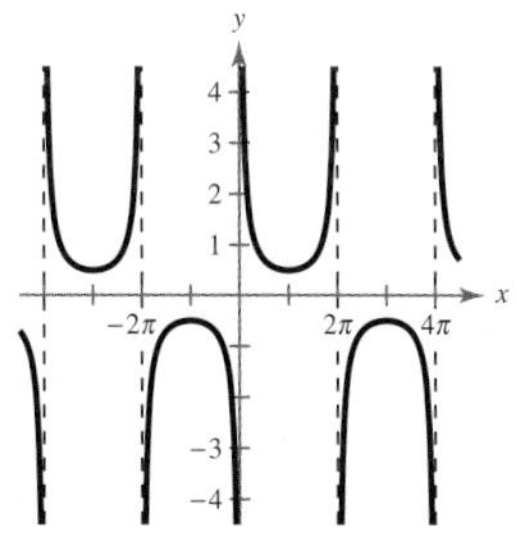

75.

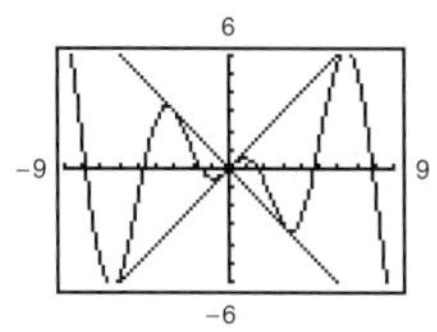

As $x \to +\infty$, $f(x)$ oscillates.

77. $-\frac{\pi}{2}$ **79.** $\frac{\pi}{6}$ **81.** −0.98 **83.** 0.09

85.

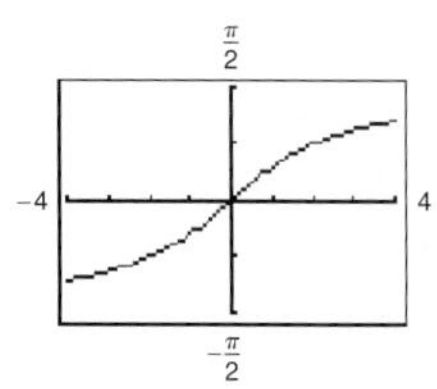

87. $\frac{4}{5}$ **89.** $\frac{4\sqrt{15}}{15}$

91. $\frac{\sqrt{4 - x^2}}{x}$ **93.**

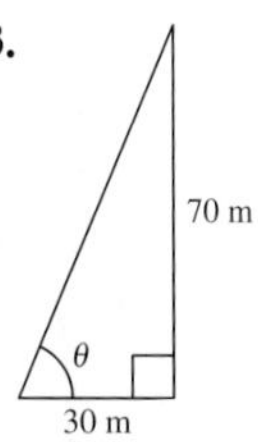

$\theta \approx 66.8°$

95. 1221 mi, 85.6°

97. False. For each θ there corresponds exactly one value of y.

99. The function is undefined because $\sec\theta = 1/\cos\theta$.

101. The ranges of the other four trigonometric functions are $(-\infty, \infty)$ or $(-\infty, -1] \cup [1, \infty)$.

Chapter Test *(page 345)*

1. (a)

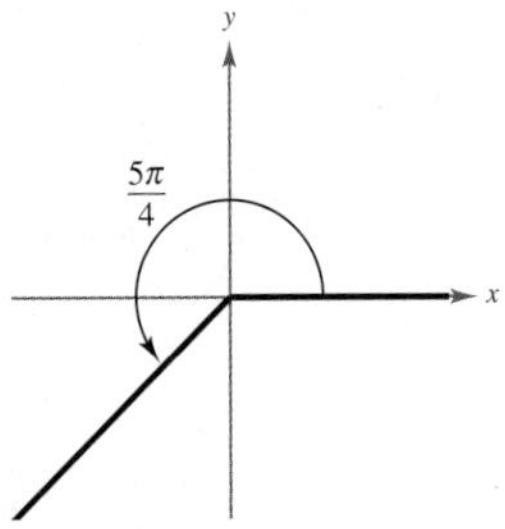

(b) $\frac{13\pi}{4}, -\frac{3\pi}{4}$ (c) $225°$

2. 3500 rad/min **3.** About 709.04 ft^2

4. $\sin\theta = \frac{3\sqrt{10}}{10}$ $\csc\theta = \frac{\sqrt{10}}{3}$

$\cos\theta = -\frac{\sqrt{10}}{10}$ $\sec\theta = -\sqrt{10}$

$\tan\theta = -3$ $\cot\theta = -\frac{1}{3}$

5. For $0 \le \theta < \frac{\pi}{2}$: For $\pi \le \theta < \frac{3\pi}{2}$:

$\sin\theta = \frac{3\sqrt{13}}{13}$ $\sin\theta = -\frac{3\sqrt{13}}{13}$

$\cos\theta = \frac{2\sqrt{13}}{13}$ $\cos\theta = -\frac{2\sqrt{13}}{13}$

$\csc\theta = \frac{\sqrt{13}}{3}$ $\csc\theta = -\frac{\sqrt{13}}{3}$

$\sec\theta = \frac{\sqrt{13}}{2}$ $\sec\theta = -\frac{\sqrt{13}}{2}$

$\cot\theta = \frac{2}{3}$ $\cot\theta = \frac{2}{3}$

6. $\theta' = 25°$

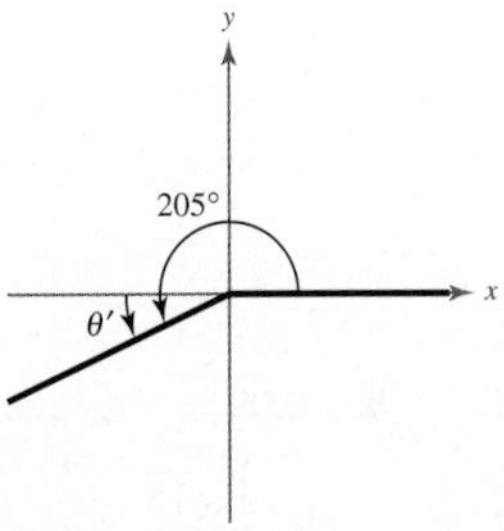

7. Quadrant III

8. 150°, 210°

9. 1.33, 1.81

10. $\sin\theta = -\frac{4}{5}$
$\tan\theta = -\frac{4}{3}$
$\csc\theta = -\frac{5}{4}$
$\sec\theta = \frac{5}{3}$
$\cot\theta = -\frac{3}{4}$

11. $\sin\theta = \frac{21}{29}$
$\cos\theta = -\frac{20}{29}$
$\tan\theta = -\frac{21}{20}$
$\csc\theta = \frac{29}{21}$
$\cot\theta = -\frac{20}{21}$

12.

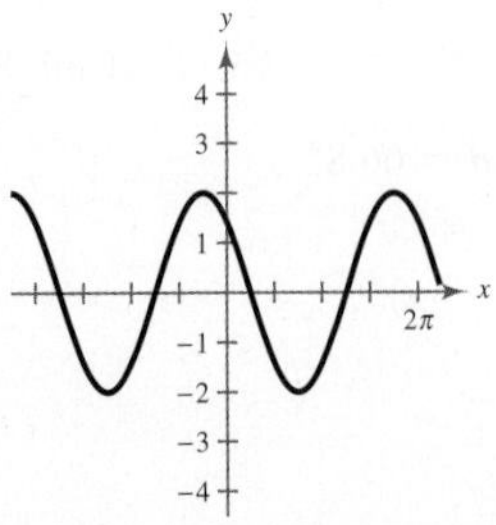

13.

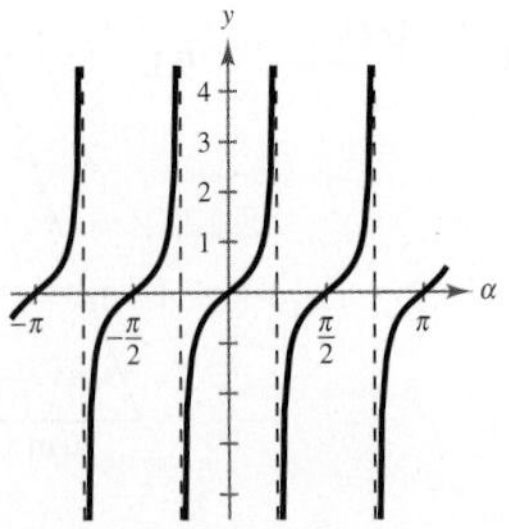

14.

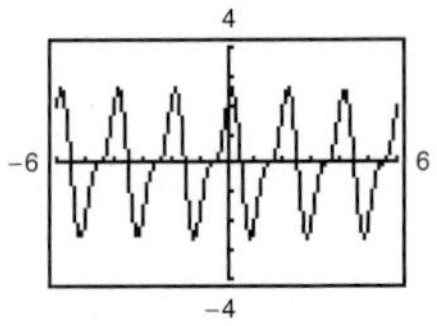

Period: 2

15.

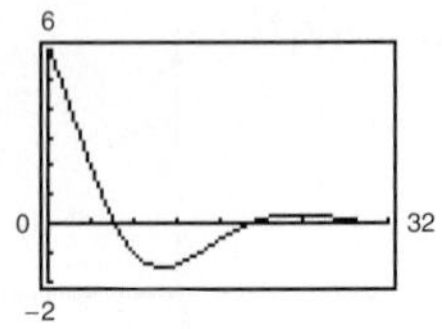

Not periodic

16. $a = -2, b = \frac{1}{2}, c = -\frac{\pi}{4}$ **17.** $\frac{\sqrt{55}}{3}$

18.

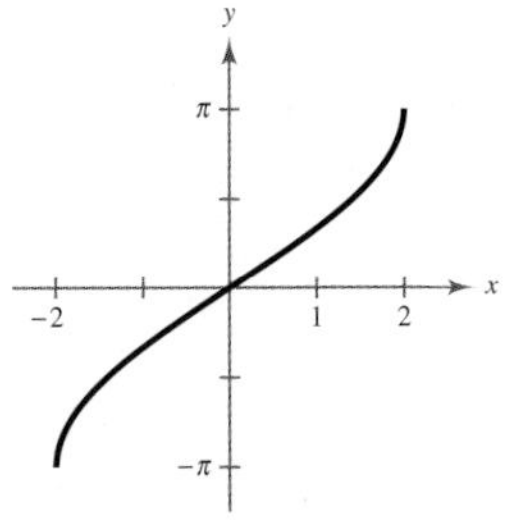

19. 309.3° **20.** $d = -6\cos\pi t$

Problem Solving *(page 347)*

1. (a) $\frac{11\pi}{2}$ rad or 990° (b) About 816.42 ft

3. (a) 4767 ft (b) 3705 ft

(c) $w \approx 2183$ ft, $\tan 63° = \frac{w + 3705}{3000}$

5. (a)

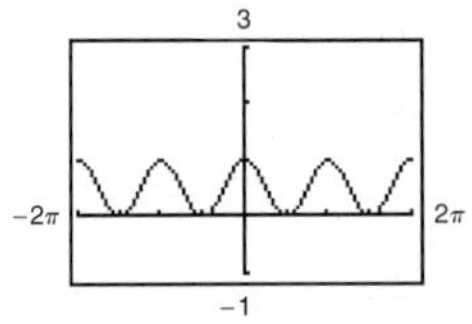

Even

(b)

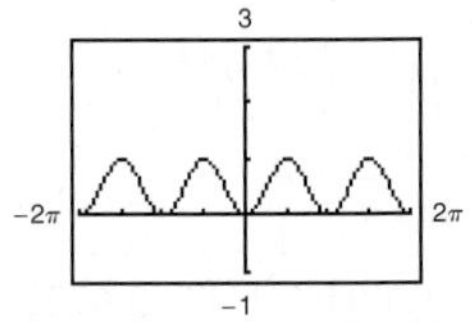

Even

7. $h = 51 - 50\sin\left(8\pi t + \frac{\pi}{2}\right)$

9. (a)

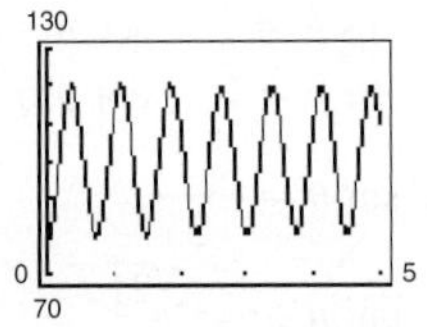

(b) Period $= \frac{3}{4}$ sec; Answers will vary.

(c) 20 mm; Answers will vary. (d) 80 beats/min

(e) Period $= \frac{15}{16}$ sec; $\frac{32\pi}{15}$

11. (a)

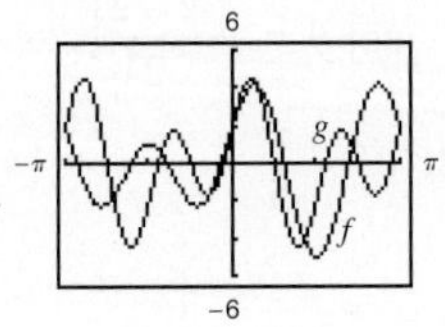

(b) Period of f: 2π; Period of g: π

(c) Yes, because the sine and cosine functions are periodic.

13. (a) 40.5° (b) $x \approx 1.71$ ft; $y \approx 3.46$ ft (c) About 1.75 ft

(d) As you move closer to the rock, d must get smaller and smaller. The angles θ_1 and θ_2 will decrease along with the distance y, so d will decrease.

Chapter 5

Section 5.1 (page 355)

1. $\tan u$ **3.** $\cot u$ **5.** $\cot^2 u$

7. $\sin x = \frac{1}{2}$

$\cos x = \frac{\sqrt{3}}{2}$

$\tan x = \frac{\sqrt{3}}{3}$

$\csc x = 2$

$\sec x = \frac{2\sqrt{3}}{3}$

$\cot x = \sqrt{3}$

9. $\sin x = \frac{3}{5}$

$\cos x = \frac{4}{5}$

$\tan x = \frac{3}{4}$

$\csc x = \frac{5}{3}$

$\sec x = \frac{5}{4}$

$\cot x = \frac{4}{3}$

11. $\sin x = \frac{\sqrt{15}}{4}$

$\cos x = \frac{1}{4}$

$\tan x = \sqrt{15}$

$\csc x = \frac{4\sqrt{15}}{15}$

$\sec x = 4$

$\cot x = \frac{\sqrt{15}}{15}$

13. $\sin \theta = -1$

$\cos \theta = 0$

$\tan \theta$ is undefined.

$\csc \theta = -1$

$\sec \theta$ is undefined.

$\cot \theta = 0$

15. c **16.** b **17.** f **18.** a **19.** e **20.** d

21. $\sin^2 x$ **23.** $\sec x + 1$ **25.** $\sin^4 x$

27. $\csc^2 x(\cot x + 1)$ **29.** $(3 \sin x + 1)(\sin x - 2)$

31. $(\csc x - 1)(\csc x + 2)$ **33.** $1 + 2 \sin x \cos x$

35. $\csc \theta$ **37.** $\cos^2 \phi$ **39.** $\sin^2 x$ **41.** $\tan x$

43. $\sec \beta$ **45.** $2 \csc^2 x$ **47.** $-\cot x$ **49.** $1 + \cos y$

51. $\csc x$ **53.** $3 \sin \theta$ **55.** $2 \tan \theta$

57. $3 \cos \theta = 3$; $\sin \theta = 0$; $\cos \theta = 1$ **59.** $0 \le \theta \le \pi$

61. $\ln|\cos x|$ **63.** $\ln|\csc t \sec t|$ **65.** $\mu = \tan \theta$

67. True. For example, $\sin(-x) = -\sin x$. **69.** $\infty, 0$

71. Not an identity because $\frac{\sin k\theta}{\cos k\theta} = \tan k\theta$ **73.** $a \sec \theta$

75. $\cos \theta = \pm\sqrt{1 - \sin^2 \theta}$

$\tan \theta = \pm\frac{\sin \theta}{\sqrt{1 - \sin^2 \theta}}$

$\cot \theta = \pm\frac{\sqrt{1 - \sin^2 \theta}}{\sin \theta}$

$\sec \theta = \pm\frac{1}{\sqrt{1 - \sin^2 \theta}}$

$\csc \theta = \frac{1}{\sin \theta}$

Section 5.2 (page 362)

1. identity **3.** $\tan u$ **5.** $\cos^2 u$ **7.** $-\csc u$

9–49. Answers will vary.

51. In the first line, $\cot(x)$ is substituted for $\cot(-x)$, which is incorrect; $\cot(-x) = -\cot(x)$.

53. (a)

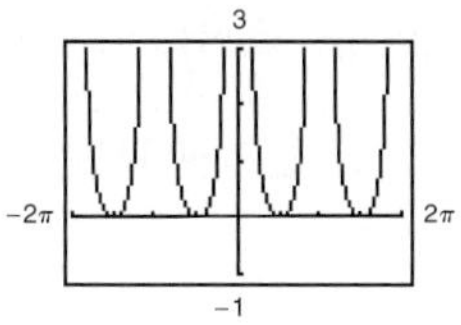

(b)

X	Y1	Y2
-3	49.214	49.214
-2	.20945	.20945
-1	.41228	.41228
0	ERROR	ERROR
1	.41228	.41228
2	.20945	.20945
3	49.214	49.214

X= -3

Identity

(c) Answers will vary.

55. (a)

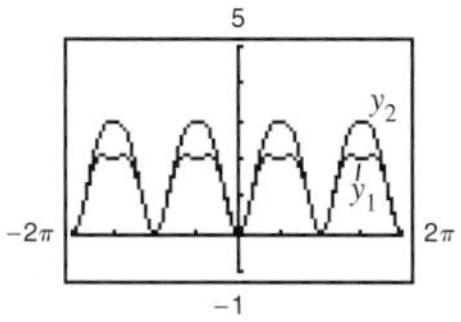

(b)

X	Y1	Y2
-4.712	2	3
-3.142	0	0
-1.571	2	3
0	0	0
1.5708	2	3
3.1416	0	0
4.7124	2	3

X= -4.71238898038

Not an identity

(c) Answers will vary.

57. (a)

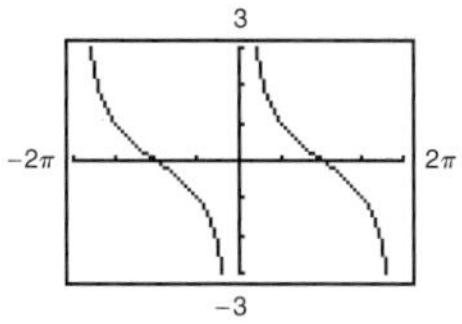

(b)

X	Y1	Y2
-3	-.0709	-.0709
-2	-.6421	-.6421
-1	-1.83	-1.83
0	ERROR	ERROR
1	1.8305	1.8305
2	.64209	.64209
3	.07091	.07091

X= -3

Identity

(c) Answers will vary.

59–61. Answers will vary. **63.** 1

65. Answers will vary.

67. True. Many different techniques can be used to verify identities.

69. False. $\sin x^2 = \sin(x \cdot x) \neq \sin^2 x = (\sin x)(\sin x)$

71. The equation is not an identity because $\sin \theta = \pm\sqrt{1 - \cos^2 \theta}$.

Possible answer: $\frac{7\pi}{4}$

73. The equation is not an identity because $1 - \cos^2 \theta = \sin^2 \theta$.

Possible answer: $-\frac{\pi}{2}$

Section 5.3 (page 371)

1. isolate **3.** quadratic **5–9.** Answers will vary.

11. $\frac{\pi}{3} + 2n\pi, \frac{2\pi}{3} + 2n\pi$ **13.** $\frac{2\pi}{3} + 2n\pi, \frac{4\pi}{3} + 2n\pi$

15. $\frac{\pi}{6} + n\pi, \frac{5\pi}{6} + n\pi$ **17.** $\frac{\pi}{3} + n\pi, \frac{2\pi}{3} + n\pi$

19. $\frac{\pi}{8} + \frac{n\pi}{2}, \frac{3\pi}{8} + \frac{n\pi}{2}$ **21.** $\frac{n\pi}{3}, \frac{\pi}{4} + n\pi$

23. $n\pi, \frac{3\pi}{2} + 2n\pi$ **25.** $0, \frac{\pi}{2}, \pi, \frac{3\pi}{2}$

27. $0, \pi, \frac{\pi}{6}, \frac{5\pi}{6}, \frac{7\pi}{6}, \frac{11\pi}{6}$ **29.** $\frac{\pi}{3}, \frac{5\pi}{3}, \pi$ **31.** No solution

33. $\pi, \frac{\pi}{3}, \frac{5\pi}{3}$ **35.** $\frac{\pi}{6}, \frac{5\pi}{6}, \frac{7\pi}{6}, \frac{11\pi}{6}$ **37.** $\frac{\pi}{2}$

39. $\frac{\pi}{6} + n\pi, \frac{5\pi}{6} + n\pi$ **41.** $\frac{\pi}{12} + \frac{n\pi}{3}$

43. $\frac{\pi}{2} + 4n\pi, \frac{7\pi}{2} + 4n\pi$ **45.** $3 + 4n$

47. $-2 + 6n, 2 + 6n$ **49.** 2.678, 5.820 **51.** 1.047, 5.236

53. 0.860, 3.426 **55.** 0, 2.678, 3.142, 5.820

57. 0.983, 1.768, 4.124, 4.910

59. 0.3398, 0.8481, 2.2935, 2.8018

61. 1.9357, 2.7767, 5.0773, 5.9183

63. $\arctan(-4) + \pi, \arctan(-4) + 2\pi, \arctan 3, \arctan 3 + \pi$

65. $\frac{\pi}{4}, \frac{5\pi}{4}$, arctan 5, arctan 5 + π **67.** $\frac{\pi}{3}, \frac{5\pi}{3}$

69. $\arctan\left(\frac{1}{3}\right), \arctan\left(\frac{1}{3}\right) + \pi, \arctan\left(-\frac{1}{3}\right) + \pi, \arctan\left(-\frac{1}{3}\right) + 2\pi$

71. $\arccos\left(\frac{1}{4}\right), 2\pi - \arccos\left(\frac{1}{4}\right)$

73. $\frac{\pi}{2}, \arcsin\left(-\frac{1}{4}\right) + 2\pi, \arcsin\left(\frac{1}{4}\right) + \pi$

75. $-1.154, 0.534$ **77.** 1.110

79. (a)

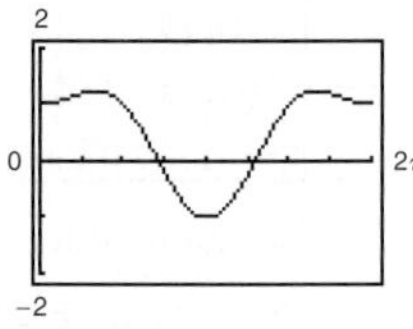

(b) $\frac{\pi}{3} \approx 1.0472$
$\frac{5\pi}{3} \approx 5.2360$
0
$\pi \approx 3.1416$

Maximum: (1.0472, 1.25)
Maximum: (5.2360, 1.25)
Minimum: (0, 1)
Minimum: (3.1416, −1)

81. (a)

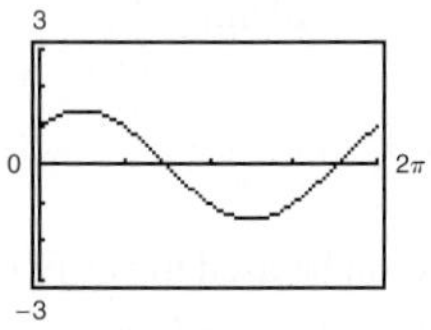

(b) $\frac{\pi}{4} \approx 0.7854$
$\frac{5\pi}{4} \approx 3.9270$

Maximum: (0.7854, 1.4142)
Minimum: (3.9270, −1.4142)

83. (a)

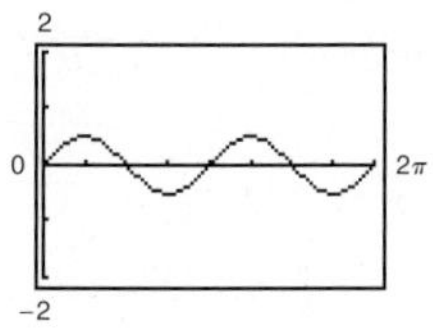

(b) $\frac{\pi}{4} \approx 0.7854$
$\frac{5\pi}{4} \approx 3.9270$
$\frac{3\pi}{4} \approx 2.3562$
$\frac{7\pi}{4} \approx 5.4978$

Maximum: (0.7854, 0.5)
Maximum: (3.9270, 0.5)
Minimum: (2.3562, −0.5)
Minimum: (5.4978, −0.5)

85. 1

87. (a) All real numbers x except $x = 0$
(b) y-axis symmetry (c) y approaches 1.
(d) Four solutions: $\pm\pi, \pm 2\pi$

89. 0.04 sec, 0.43 sec, 0.83 sec

91. January, November, December

93. (a) and (c)

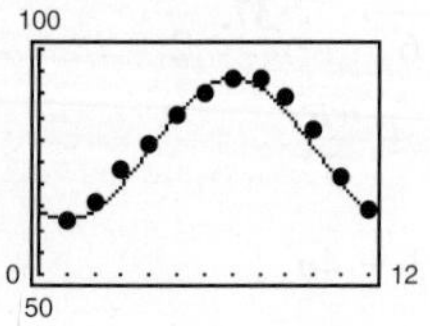

The model fits the data well.

(b) $H = 15.65\cos\left(\frac{\pi}{6}t - \frac{7\pi}{6}\right) + 77.95$ (d) 77.95°F

(e) Above 86°: June through September
Below 86°: October through May

95. (a)

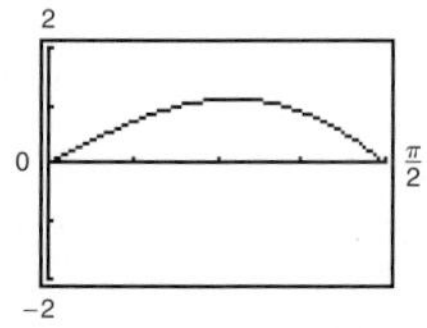

$A \approx 1.12$

(b) $0.6 < x < 1.1$

97. 1

99. True. The first equation has a smaller period than the second equation, so it will have more solutions in the interval $[0, 2\pi)$.

101. The equation would become $\cos^2 x = 2$; this is not the correct method to use when solving equations.

103. (a)

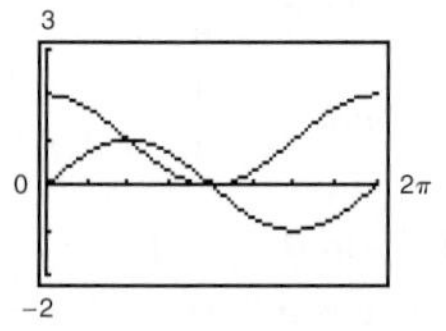

Graphs intersect when $x = \frac{\pi}{2}$ and $x = \pi$.

(b)

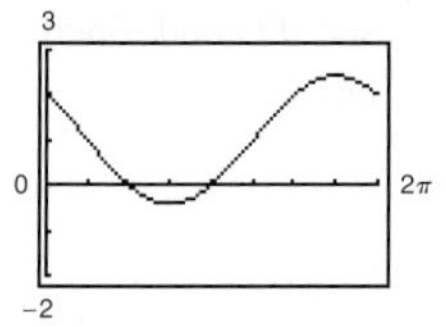

x-intercepts: $\left(\frac{\pi}{2}, 0\right), (\pi, 0)$

Both methods produce the same x-values.

Section 5.4 *(page 379)*

1. $\sin u \cos v - \cos u \sin v$ **3.** $\frac{\tan u + \tan v}{1 - \tan u \tan v}$

5. $\cos u \cos v + \sin u \sin v$

7. (a) $\frac{\sqrt{2} - \sqrt{6}}{4}$ (b) $\frac{\sqrt{2} + 1}{2}$

9. (a) $\frac{\sqrt{6} + \sqrt{2}}{4}$ (b) $\frac{\sqrt{2} - \sqrt{3}}{2}$

11. $\sin\frac{11\pi}{12} = \frac{\sqrt{2}}{4}(\sqrt{3} - 1)$
$\cos\frac{11\pi}{12} = -\frac{\sqrt{2}}{4}(\sqrt{3} + 1)$
$\tan\frac{11\pi}{12} = -2 + \sqrt{3}$

13. $\sin\frac{17\pi}{12} = -\frac{\sqrt{2}}{4}(\sqrt{3} + 1)$
$\cos\frac{17\pi}{12} = \frac{\sqrt{2}}{4}(1 - \sqrt{3})$
$\tan\frac{17\pi}{12} = 2 + \sqrt{3}$

15. $\sin 105° = \frac{\sqrt{2}}{4}(\sqrt{3} + 1)$
$\cos 105° = \frac{\sqrt{2}}{4}(1 - \sqrt{3})$
$\tan 105° = -2 - \sqrt{3}$

17. $\sin 195° = \frac{\sqrt{2}}{4}(1 - \sqrt{3})$
$\cos 195° = -\frac{\sqrt{2}}{4}(\sqrt{3} + 1)$
$\tan 195° = 2 - \sqrt{3}$

19. $\sin\frac{13\pi}{12} = \frac{\sqrt{2}}{4}(1 - \sqrt{3})$
$\cos\frac{13\pi}{12} = -\frac{\sqrt{2}}{4}(1 + \sqrt{3})$
$\tan\frac{13\pi}{12} = 2 - \sqrt{3}$

21. $\sin\left(-\frac{13\pi}{12}\right) = \frac{\sqrt{2}}{4}(\sqrt{3} - 1)$

$\cos\left(-\frac{13\pi}{12}\right) = -\frac{\sqrt{2}}{4}(\sqrt{3} + 1)$

$\tan\left(-\frac{13\pi}{12}\right) = -2 + \sqrt{3}$

23. $\sin 285° = -\frac{\sqrt{2}}{4}(\sqrt{3} + 1)$

$\cos 285° = \frac{\sqrt{2}}{4}(\sqrt{3} - 1)$

$\tan 285° = -(2 + \sqrt{3})$

25. $\sin(-165°) = -\frac{\sqrt{2}}{4}(\sqrt{3} - 1)$

$\cos(-165°) = -\frac{\sqrt{2}}{4}(1 + \sqrt{3})$

$\tan(-165°) = 2 - \sqrt{3}$

27. $\sin 1.8$ **29.** $\sin 75°$ **31.** $\tan 15°$ **33.** $\cos(3x - 2y)$

35. $\frac{\sqrt{3}}{2}$ **37.** $\frac{\sqrt{3}}{2}$ **39.** $-\sqrt{3}$ **41.** $-\frac{63}{65}$ **43.** $-\frac{63}{16}$

45. $\frac{65}{56}$ **47.** $\frac{3}{5}$ **49.** $-\frac{44}{117}$ **51.** $-\frac{125}{44}$ **53.** 1 **55.** 0

57–63. Proofs **65.** $-\sin x$ **67.** $-\cos\theta$ **69.** $\frac{\pi}{6}, \frac{5\pi}{6}$

71. $\frac{5\pi}{4}, \frac{7\pi}{4}$ **73.** $0, \frac{\pi}{3}, \pi, \frac{5\pi}{3}$ **75.** $\frac{\pi}{4}, \frac{7\pi}{4}$ **77.** $\frac{\pi}{2}, \pi, \frac{3\pi}{2}$

79. (a) $y = \frac{5}{12}\sin(2t + 0.6435)$ (b) $\frac{5}{12}$ ft (c) $\frac{1}{\pi}$ cycle/sec

81. True. $\sin(u \pm v) = \sin u \cos v \pm \cos u \sin v$

83. False. $\tan\left(x - \frac{\pi}{4}\right) = \frac{\tan x - 1}{1 + \tan x}$

85. (a) All real numbers h except $h = 0$

(b)

h	0.5	0.2	0.1
$f(h)$	0.267	0.410	0.456
$g(h)$	0.267	0.410	0.456

h	0.05	0.02	0.01
$f(h)$	0.478	0.491	0.496
$g(h)$	0.478	0.491	0.496

(c)

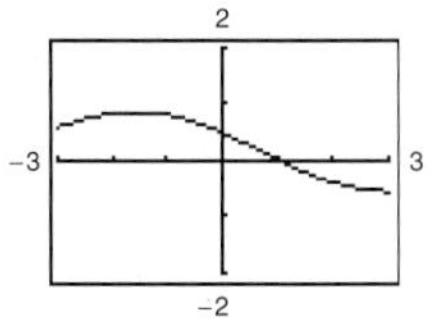

(d) As $h \to 0^+$, $f \to 0.5$ and $g \to 0.5$.

87–89. Answers will vary.

91. (a) $\sqrt{2}\sin\left(\theta + \frac{\pi}{4}\right)$ (b) $\sqrt{2}\cos\left(\theta - \frac{\pi}{4}\right)$

93. (a) $13\sin(3\theta + 0.3948)$ (b) $13\cos(3\theta - 1.1760)$

95. $\sqrt{2}\sin\theta + \sqrt{2}\cos\theta$ **97.** 15°

99.

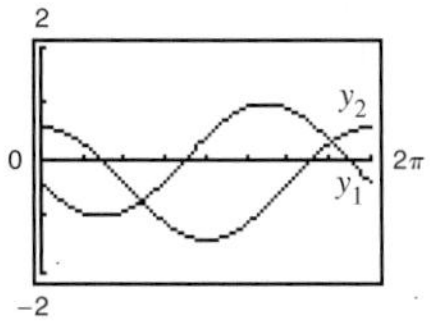

No, $y_1 \neq y_2$ because their graphs are different.

101. (a) and (b) Proofs

Section 5.5 *(page 389)*

1. $2 \sin u \cos u$ **3.** $\tan^2 u$

5. $\frac{1}{2}[\sin(u + v) + \sin(u - v)]$ **7.** $0, \frac{\pi}{3}, \pi, \frac{5\pi}{3}$

9. $0, \frac{2\pi}{3}, \frac{4\pi}{3}$ **11.** $0, \frac{\pi}{2}, \pi, \frac{3\pi}{2}$

13. $\frac{\pi}{2}, \frac{\pi}{6}, \frac{5\pi}{6}, \frac{7\pi}{6}, \frac{3\pi}{2}, \frac{11\pi}{6}$ **15.** $3\sin 2x$ **17.** $3\cos 2x$

19. $4\cos 2x$ **21.** $\sin 2u = -\frac{24}{25}$, $\cos 2u = \frac{7}{25}$, $\tan 2u = -\frac{24}{7}$

23. $\sin 2u = \frac{15}{17}$, $\cos 2u = \frac{8}{17}$, $\tan 2u = \frac{15}{8}$

25. $8\cos^4 x - 8\cos^2 x + 1$ **27.** $\frac{1}{8}(3 + 4\cos 2x + \cos 4x)$

29. $\frac{(3 - 4\cos 4x + \cos 8x)}{(3 + 4\cos 4x + \cos 8x)}$ **31.** $\frac{1}{8}(1 - \cos 8x)$

33. $\sin 75° = \frac{1}{2}\sqrt{2 + \sqrt{3}}$

$\cos 75° = \frac{1}{2}\sqrt{2 - \sqrt{3}}$

$\tan 75° = 2 + \sqrt{3}$

35. $\sin\frac{\pi}{8} = \frac{1}{2}\sqrt{2 - \sqrt{2}}$

$\cos\frac{\pi}{8} = \frac{1}{2}\sqrt{2 + \sqrt{2}}$

$\tan\frac{\pi}{8} = \sqrt{2} - 1$

37. (a) Quadrant I

(b) $\sin\frac{u}{2} = \frac{3}{5}$, $\cos\frac{u}{2} = \frac{4}{5}$, $\tan\frac{u}{2} = \frac{3}{4}$

39. (a) Quadrant II

(b) $\sin\frac{u}{2} = \frac{\sqrt{26}}{26}$, $\cos\frac{u}{2} = -\frac{5\sqrt{26}}{26}$, $\tan\frac{u}{2} = -\frac{1}{5}$

41. $|\sin 3x|$ **43.** $-|\tan 4x|$

45. π

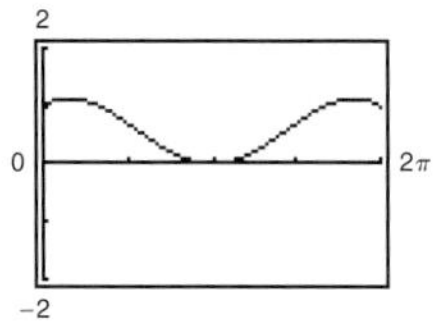

47. $\frac{\pi}{3}, \pi, \frac{5\pi}{3}$

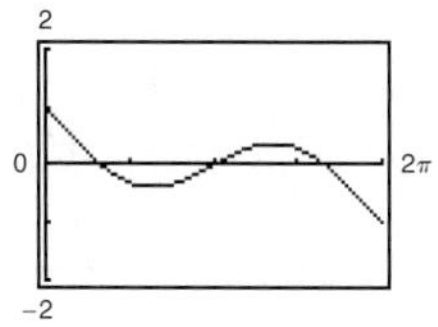

49. $\frac{1}{2}(\cos 2\theta - \cos 8\theta)$ **51.** $\frac{1}{2}(\cos(-2\theta) + \cos 6\theta)$

53. $2\cos 4\theta \sin\theta$ **55.** $2\cos 4x \cos 2x$ **57.** $\frac{\sqrt{6}}{2}$

59. $-\sqrt{2}$

61. $0, \frac{\pi}{4}, \frac{\pi}{2}, \frac{3\pi}{4}, \pi, \frac{5\pi}{4}, \frac{3\pi}{2}, \frac{7\pi}{4}$

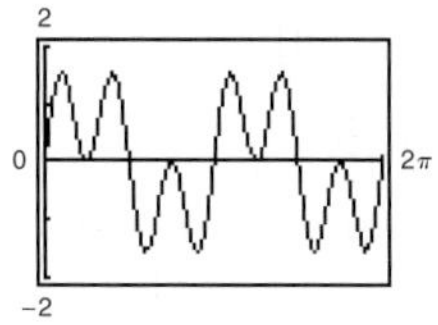

63. $\frac{\pi}{6}, \frac{5\pi}{6}$

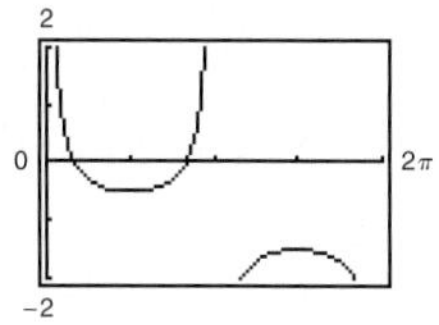

65–71. Answers will vary.

73. (a) $\cos\theta = \frac{M^2 - 2}{M^2}$ (b) π (c) 0.4482

(d) 760 mi/h; 3420 mi/h

75. $x = 2r(1 - \cos\theta)$

77. False. For $u < 0$,

$\sin 2u = -\sin(-2u)$

$= -2\sin(-u)\cos(-u)$

$= -2(-\sin u)\cos u$

$= 2\sin u \cos u.$

79. Answers will vary.

Review Exercises *(page 393)*

1. $\tan x$ **3.** $\cot x$

5. $\sin\theta = \frac{2\sqrt{13}}{13}$ $\cos\theta = \frac{3\sqrt{13}}{13}$

$\csc\theta = \frac{\sqrt{13}}{2}$ $\cot\theta = \frac{3}{2}$

7. $\sin^2 x$ **9.** 1 **11.** $\tan u \sec u$ **13.** $\cot^2 x$

15. $-2\tan^2\theta$ **17.** $5\cos\theta$ **19–25.** Answers will vary.

27. $\frac{\pi}{3} + 2n\pi, \frac{2\pi}{3} + 2n\pi$ **29.** $\frac{\pi}{6} + n\pi$

31. $\frac{\pi}{3} + n\pi, \frac{2\pi}{3} + n\pi$ **33.** $0, \frac{2\pi}{3}, \frac{4\pi}{3}$ **35.** $0, \frac{\pi}{2}, \pi$

37. $\frac{\pi}{8}, \frac{3\pi}{8}, \frac{9\pi}{8}, \frac{11\pi}{8}$ **39.** $\frac{\pi}{2}$

41. $0, \frac{\pi}{8}, \frac{3\pi}{8}, \frac{5\pi}{8}, \frac{7\pi}{8}, \frac{9\pi}{8}, \frac{11\pi}{8}, \frac{13\pi}{8}, \frac{15\pi}{8}$

43. $0, \pi, \arctan 2, \arctan 2 + \pi$

45. $\arctan(-3) + \pi, \arctan(-3) + 2\pi, \arctan 2, \arctan 2 + \pi$

47. $\sin 285° = -\frac{\sqrt{2}}{4}(\sqrt{3} + 1)$ **49.** $\sin\frac{25\pi}{12} = \frac{\sqrt{2}}{4}(\sqrt{3} - 1)$

$\cos 285° = \frac{\sqrt{2}}{4}(\sqrt{3} - 1)$ $\cos\frac{25\pi}{12} = \frac{\sqrt{2}}{4}(\sqrt{3} + 1)$

$\tan 285° = -2 - \sqrt{3}$ $\tan\frac{25\pi}{12} = 2 - \sqrt{3}$

51. $\sin 15°$ **53.** $-\frac{24}{25}$ **55.** -1

57–59. Answers will vary. **61.** $\frac{\pi}{4}, \frac{7\pi}{4}$ **63.** $\sin 2u = \frac{24}{25}$

$\cos 2u = -\frac{7}{25}$

$\tan 2u = -\frac{24}{7}$

65.

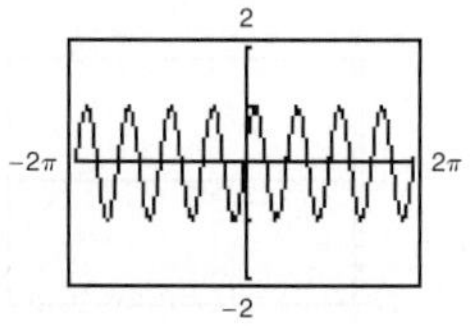

67. $\frac{1 - \cos 4x}{1 + \cos 4x}$

69. $\sin(-75°) = -\frac{1}{2}\sqrt{2 + \sqrt{3}}$

$\cos(-75°) = \frac{1}{2}\sqrt{2 - \sqrt{3}}$

$\tan(-75°) = -2 - \sqrt{3}$

71. (a) Quadrant II

(b) $\sin\frac{u}{2} = \frac{2\sqrt{5}}{5}, \cos\frac{u}{2} = -\frac{\sqrt{5}}{5}, \tan\frac{u}{2} = -2$

73. $-|\cos 5x|$ **75.** $\frac{1}{2}[\sin 10\theta - \sin(-2\theta)]$

77. $2\cos\frac{11\theta}{2}\cos\frac{\theta}{2}$ **79.** $\theta = 15°$ or $\frac{\pi}{12}$

81. False. If $(\pi/2) < \theta < \pi$, then $\cos(\theta/2) > 0$. The sign of $\cos(\theta/2)$ depends on the quadrant in which $\theta/2$ lies.

83. True. $4\sin(-x)\cos(-x) = 4(-\sin x)\cos x$

$= -4\sin x\cos x$

$= -2(2\sin x\cos x)$

$= -2\sin 2x$

85. No. For an equation to be an identity, the equation must be true for all real numbers x. $\sin\theta = \frac{1}{2}$ has an infinite number of solutions but is not an identity.

Chapter Test *(page 395)*

1. $\sin\theta = -\frac{6\sqrt{61}}{61}$ $\csc\theta = -\frac{\sqrt{61}}{6}$

$\cos\theta = -\frac{5\sqrt{61}}{61}$ $\sec\theta = -\frac{\sqrt{61}}{5}$

$\tan\theta = \frac{6}{5}$ $\cot\theta = \frac{5}{6}$

2. 1 **3.** 1 **4.** $\csc\theta\sec\theta$

5. $\theta = 0, \pi/2 < \theta \le \pi, 3\pi/2 < \theta < 2\pi$

6.

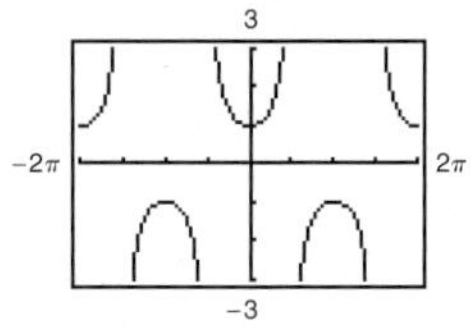

$y_1 = y_2$

7–12. Answers will vary. **13.** $\frac{1}{8}(3 - 4\cos x + \cos 2x)$

14. $\tan 2\theta$ **15.** $2(\sin 5\theta + \sin\theta)$ **16.** $-2\sin 2\theta\sin\theta$

17. $0, \frac{3\pi}{4}, \pi, \frac{7\pi}{4}$ **18.** $\frac{\pi}{6}, \frac{\pi}{2}, \frac{5\pi}{6}, \frac{3\pi}{2}$ **19.** $\frac{\pi}{6}, \frac{5\pi}{6}, \frac{7\pi}{6}, \frac{11\pi}{6}$

20. $\frac{\pi}{6}, \frac{5\pi}{6}, \frac{3\pi}{2}$ **21.** 0, 2.596 **22.** $\frac{\sqrt{2} - \sqrt{6}}{4}$

23. $\sin 2u = -\frac{20}{29}, \cos 2u = -\frac{21}{29}, \tan 2u = \frac{20}{21}$

24. Day 123 to day 223

25. $t = 0.26, 0.58, 0.89, 1.20, 1.52, 1.83$

Problem Solving *(page 399)*

1. $\sin\theta = \pm\sqrt{1 - \cos^2\theta}$

$\csc\theta = \pm\frac{1}{\sqrt{1 - \cos^2\theta}}$ $\sec\theta = \frac{1}{\cos\theta}$

$\tan\theta = \pm\frac{\sqrt{1 - \cos^2\theta}}{\cos\theta}$ $\cot\theta = \pm\frac{\cos\theta}{\sqrt{1 - \cos^2\theta}}$

3. Answers will vary. **5.** $u + v = w$

7. (a) $A = 100\sin\frac{\theta}{2}\cos\frac{\theta}{2}$ (b) $A = 50\sin\theta; \theta = \frac{\pi}{2}$

9. (a) $F = \frac{0.6W\cos\theta}{\sin 12°}$ (b)

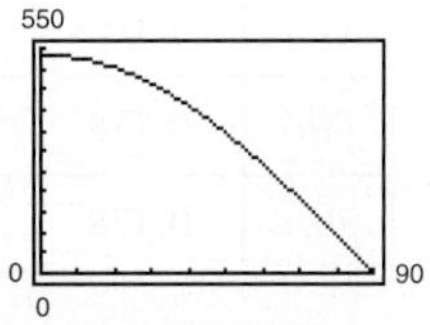

(c) Maximum: $\theta = 0°$

Minimum: $\theta = 90°$

11. (a) High tides: 6:12 A.M., 6:36 P.M.

Low tides: 12:00 A.M., 12:24 P.M.

(b) The water depth never falls below 7 feet. (c)

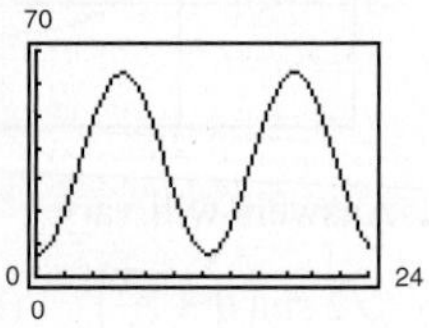

13. (a) $n = \frac{1}{2}\left(\cot\frac{\theta}{2} + \sqrt{3}\right)$ (b) $\theta \approx 76.5°$

15. (a) $\frac{\pi}{6} \le x \le \frac{5\pi}{6}$ (b) $\frac{2\pi}{3} \le x \le \frac{4\pi}{3}$

(c) $\frac{\pi}{2} < x < \pi, \frac{3\pi}{2} < x < 2\pi$

(d) $0 \le x \le \frac{\pi}{4}, \frac{5\pi}{4} \le x < 2\pi$

Chapter 6

Section 6.1 *(page 408)*

1. oblique **3.** angles; side
5. $A = 30°, a \approx 14.14, c \approx 27.32$
7. $C = 120°, b \approx 4.75, c \approx 7.17$
9. $B = 60.9°, b \approx 19.32, c \approx 6.36$
11. $B = 42°4', a \approx 22.05, b \approx 14.88$
13. $C = 80°, a \approx 5.82, b \approx 9.20$
15. $C = 83°, a \approx 0.62, b \approx 0.51$
17. $B \approx 21.55°, C \approx 122.45°, c \approx 11.49$
19. $A \approx 10°11', C \approx 154°19', c \approx 11.03$
21. $B \approx 9.43°, C = 25.57°, c \approx 10.53$
23. $B \approx 18°13', C \approx 51°32', c \approx 40.06$
25. $B \approx 48.74°, C \approx 21.26°, c \approx 48.23$ **27.** No solution
29. Two solutions:
$B \approx 72.21°, C \approx 49.79°, c \approx 10.27$
$B \approx 107.79°, C \approx 14.21°, c \approx 3.30$
31. No solution **33.** $B = 45°, C = 90°, c \approx 1.41$
35. (a) $b \le 5,\ b = \dfrac{5}{\sin 36°}$ (b) $5 < b < \dfrac{5}{\sin 36°}$
(c) $b > \dfrac{5}{\sin 36°}$
37. (a) $b \le 10.8,\ b = \dfrac{10.8}{\sin 10°}$ (b) $10.8 < b < \dfrac{10.8}{\sin 10°}$
(c) $b > \dfrac{10.8}{\sin 10°}$
39. 10.4 **41.** 20 **43.** 1675.2 **45.** 3204.5
47. 24.1 m **49.** 16.1° **51.** 240°
53. 3.2 mi **55.** $d = \dfrac{2 \sin \theta}{\sin(\phi - \theta)}$
57. True. If an angle of a triangle is obtuse (greater than 90°), then the other two angles must be acute and therefore less than 90°. The triangle is oblique.
59. False. When just three angles are known, the triangle cannot be solved.
61. (a) $A = 20\left(15 \sin \dfrac{3\theta}{2} - 4 \sin \dfrac{\theta}{2} - 6 \sin \theta\right)$
(b)
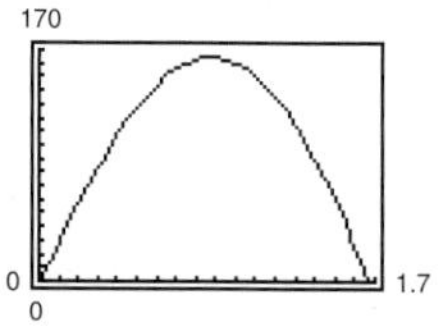

(c) Domain: $0 \le \theta \le 1.6690$
The domain would increase in length and the area would have a greater maximum value.

Section 6.2 *(page 415)*

1. Cosines **3.** $b^2 = a^2 + c^2 - 2ac \cos B$
5. $A \approx 38.62°, B \approx 48.51°, C \approx 92.87°$
7. $A \approx 26.38°, B \approx 36.34°, C \approx 117.28°$
9. $B \approx 23.79°, C \approx 126.21°, a \approx 18.59$
11. $B \approx 29.44°, C \approx 100.56°, a \approx 23.38$
13. $A \approx 30.11°, B \approx 43.16°, C \approx 106.73°$
15. $A \approx 92.94°, B \approx 43.53°, C \approx 43.53°$
17. $B \approx 27.46°, C \approx 32.54°, a \approx 11.27$
19. $A \approx 141°45', C \approx 27°40', b \approx 11.87$
21. $A = 27°10', C = 27°10', b \approx 65.84$
23. $A \approx 33.80°, B \approx 103.20°, c \approx 0.54$

	a	b	c	d	θ	ϕ
25.	5	8	12.07	5.69	45°	135°
27.	10	14	20	13.86	68.2°	111.8°
29.	15	16.96	25	20	77.2°	102.8°

31. Law of Cosines; $A \approx 102.44°,\ C \approx 37.56°,\ b \approx 5.26$
33. Law of Sines; No solution
35. Law of Sines; $C = 103°,\ a \approx 0.82,\ b \approx 0.71$
37. 43.52 **39.** 10.4 **41.** 52.11 **43.** 0.18
45.
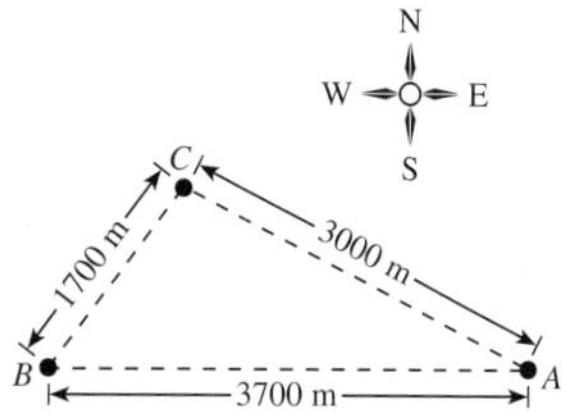

N 37.1° E, S 63.1° E
47. 373.3 m **49.** 43.3 mi
51. (a) N 59.7° E (b) N 72.8° E **53.** 63.7 ft **55.** 72.3°
57. 46,837.5 ft^2 **59.** $83,336.37
61. False. For s to be the average of the lengths of the three sides of the triangle, s would be equal to $(a + b + c)/3$.
63. $c^2 = a^2 + b^2$; The Pythagorean Theorem is a special case of the Law of Cosines.
65. The Law of Cosines can be used to solve the single-solution case of SSA. There is no method that can solve the no-solution case of SSA.
67. Proof

Section 6.3 *(page 427)*

1. directed line segment **3.** magnitude
5. magnitude; direction **7.** unit vector **9.** resultant
11. $\|\mathbf{u}\| = \|\mathbf{v}\| = \sqrt{17}$, $\text{slope}_{\mathbf{u}} = \text{slope}_{\mathbf{v}} = \frac{1}{4}$
u and **v** have the same magnitude and direction, so they are equal.
13. $\mathbf{v} = \langle 1, 3 \rangle, \|\mathbf{v}\| = \sqrt{10}$ **15.** $\mathbf{v} = \langle 4, 6 \rangle; \|\mathbf{v}\| = 2\sqrt{13}$
17. $\mathbf{v} = \langle 0, 5 \rangle; \|\mathbf{v}\| = 5$ **19.** $\mathbf{v} = \langle 8, 6 \rangle; \|\mathbf{v}\| = 10$
21. $\mathbf{v} = \langle -9, -12 \rangle; \|\mathbf{v}\| = 15$ **23.** $\mathbf{v} = \langle 16, 7 \rangle; \|\mathbf{v}\| = \sqrt{305}$
25.
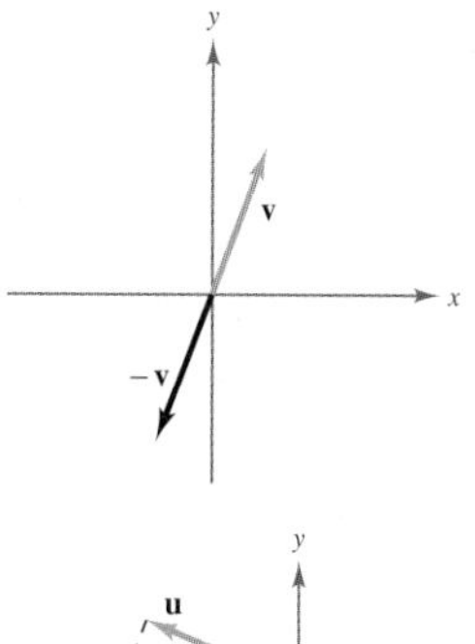

27.
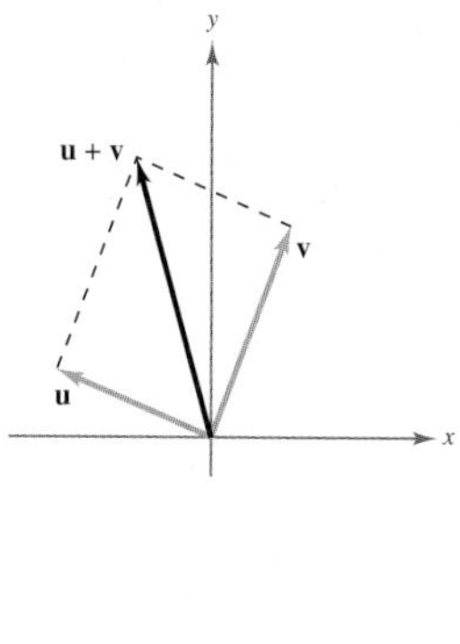

29.
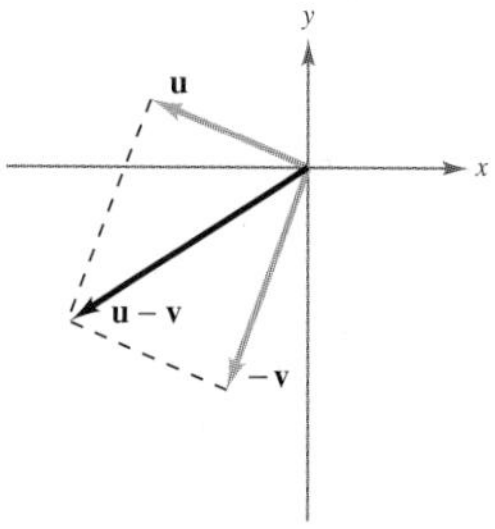

31. (a) $\langle 3, 4\rangle$

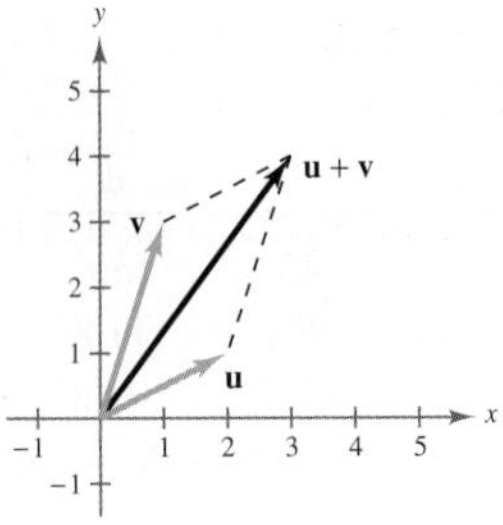

(b) $\langle 1, -2\rangle$

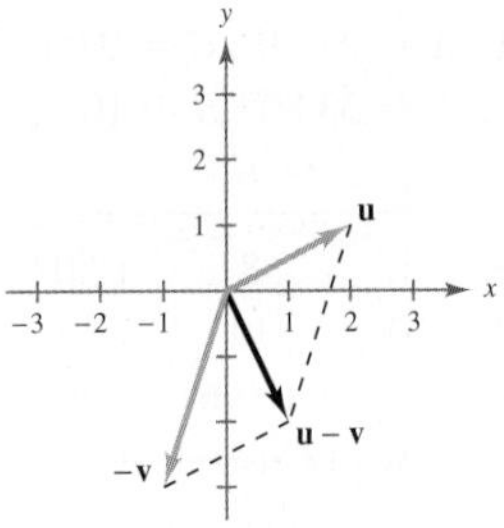

(c) $\langle 1, -7\rangle$

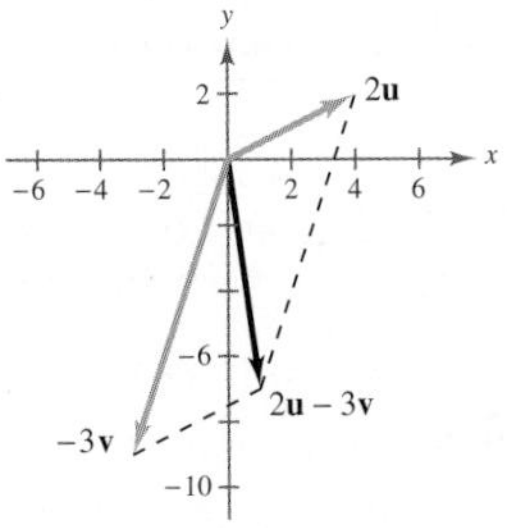

33. (a) $\langle -5, 3\rangle$

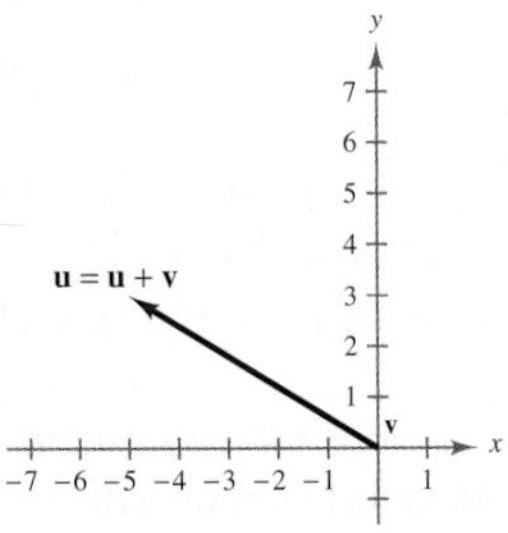

(b) $\langle -5, 3\rangle$

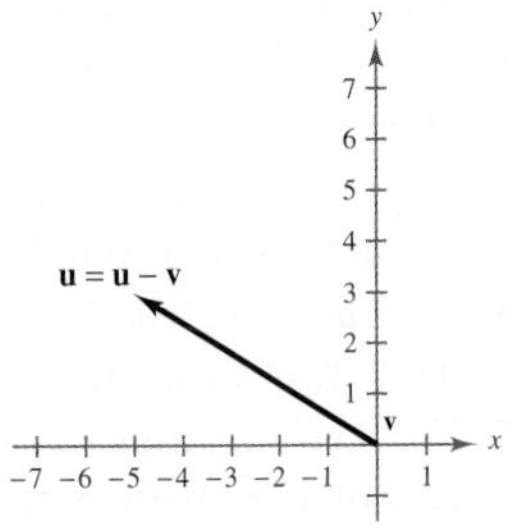

(c) $\langle -10, 6\rangle$

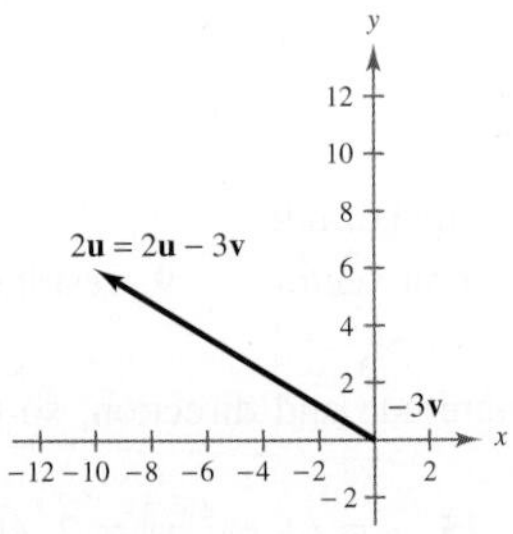

35. (a) $3\mathbf{i} - 2\mathbf{j}$

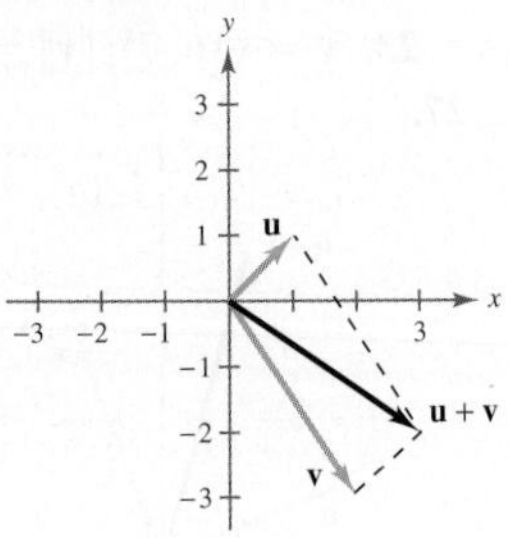

(b) $-\mathbf{i} + 4\mathbf{j}$

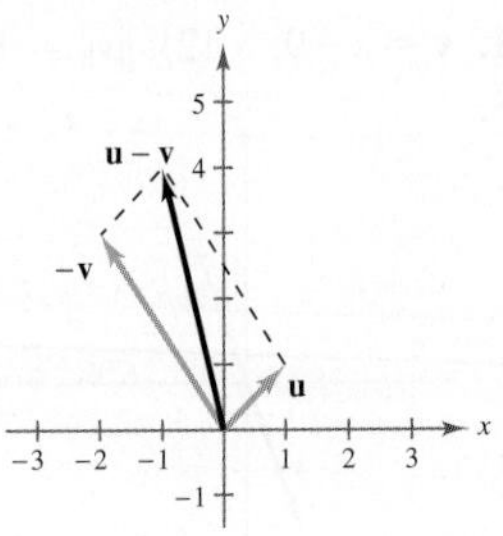

(c) $-4\mathbf{i} + 11\mathbf{j}$

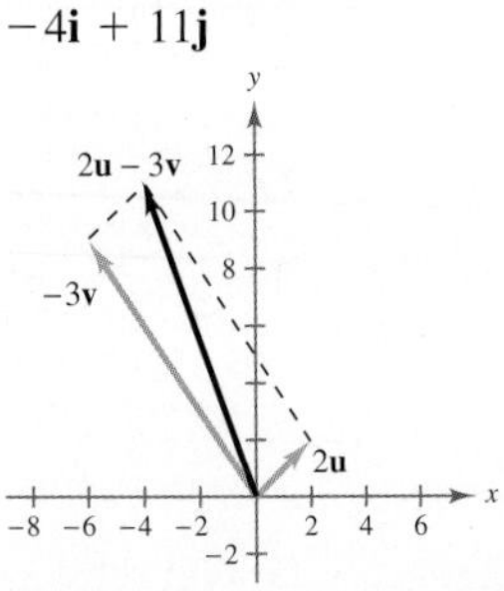

37. (a) $2\mathbf{i} + \mathbf{j}$

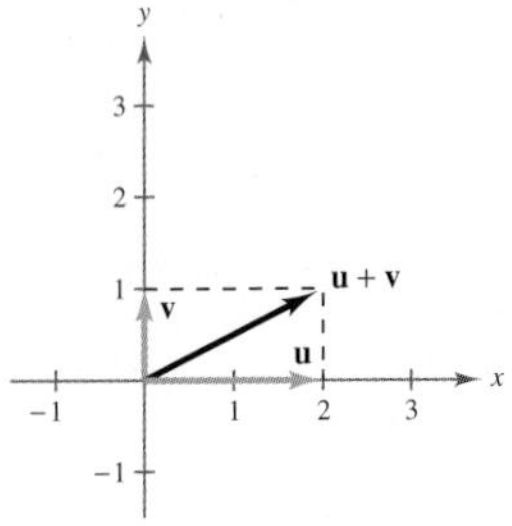

(b) $2\mathbf{i} - \mathbf{j}$

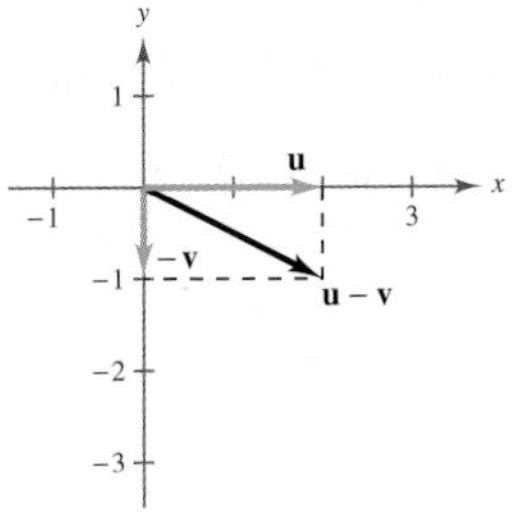

(c) $4\mathbf{i} - 3\mathbf{j}$

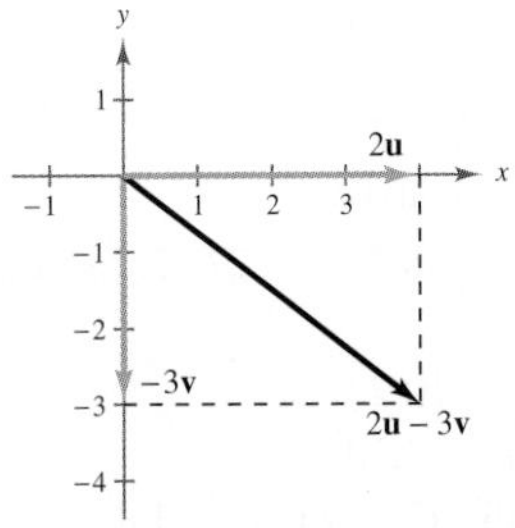

39. $\langle 1, 0\rangle$ **41.** $\left\langle -\frac{\sqrt{2}}{2}, \frac{\sqrt{2}}{2}\right\rangle$ **43.** $\frac{\sqrt{2}}{2}\mathbf{i} + \frac{\sqrt{2}}{2}\mathbf{j}$

45. $\mathbf{j}$ **47.** $\frac{\sqrt{5}}{5}\mathbf{i} - \frac{2\sqrt{5}}{5}\mathbf{j}$ **49.** $\mathbf{v} = \langle -6, 8\rangle$

51. $\mathbf{v} = \left\langle \frac{18\sqrt{29}}{29}, \frac{45\sqrt{29}}{29}\right\rangle$ **53.** $5\mathbf{i} - 3\mathbf{j}$ **55.** $6\mathbf{i} - 3\mathbf{j}$

57. $\mathbf{v} = \left\langle 3, -\frac{3}{2}\right\rangle$

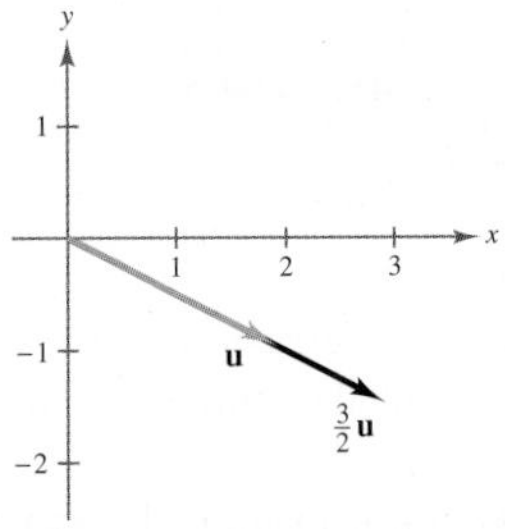

59. $\mathbf{v} = \langle 4, 3\rangle$

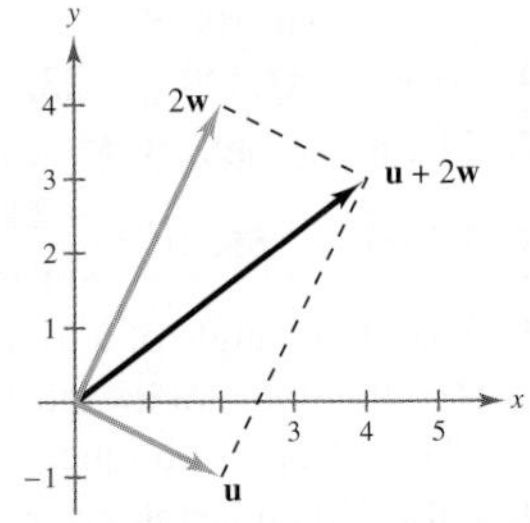

61. $\mathbf{v} = \left\langle \frac{7}{2}, -\frac{1}{2}\right\rangle$

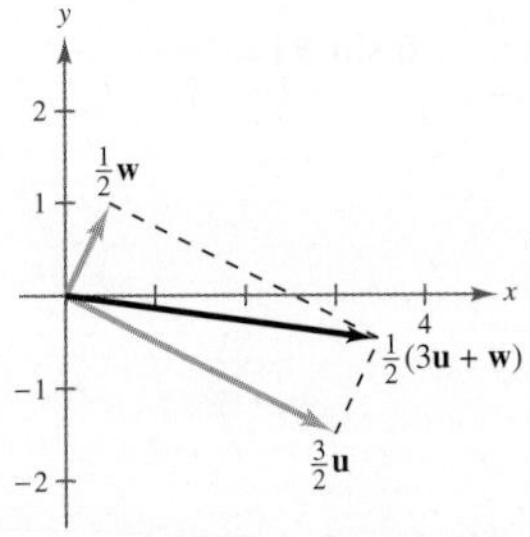

63. $\|\mathbf{v}\| = 6\sqrt{2}$; $\theta = 315°$ **65.** $\|\mathbf{v}\| = 3$; $\theta = 60°$

67. $\mathbf{v} = \langle 3, 0\rangle$

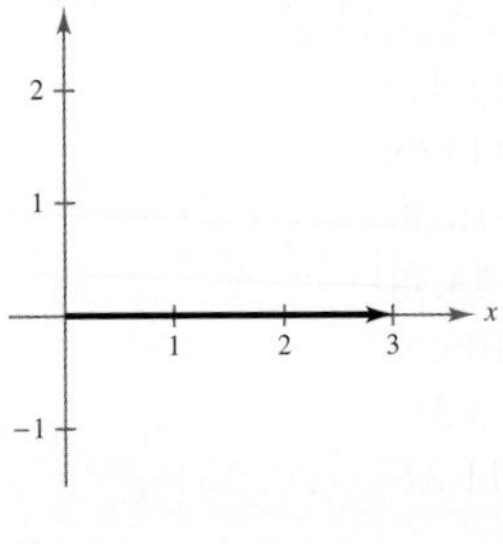

69. $\mathbf{v} = \left\langle -\frac{7\sqrt{3}}{4}, \frac{7}{4}\right\rangle$

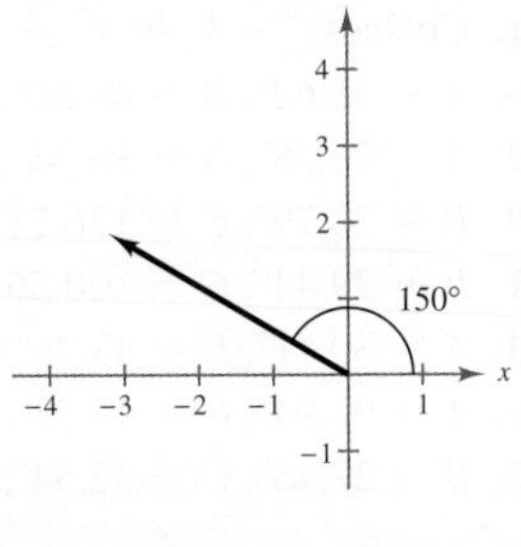

71. $\mathbf{v} = \langle \sqrt{6}, \sqrt{6} \rangle$

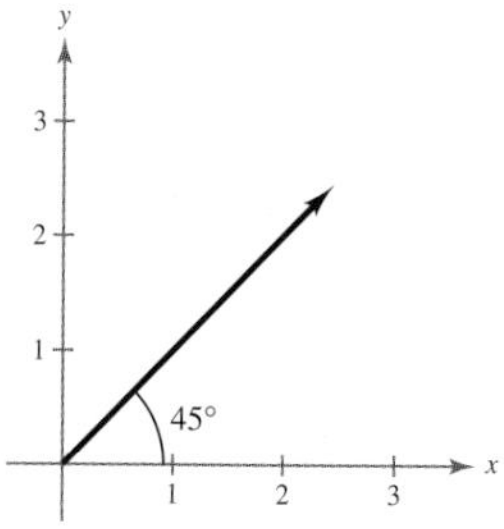

73. $\mathbf{v} = \left\langle \frac{9}{5}, \frac{12}{5} \right\rangle$

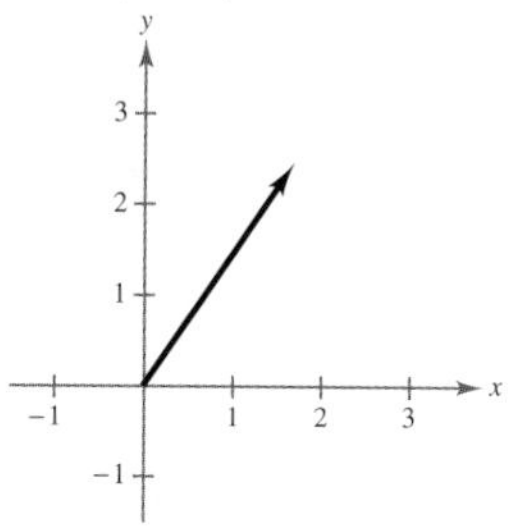

75. $\langle 5, 5 \rangle$ **77.** $\langle 10\sqrt{2} - 50, 10\sqrt{2} \rangle$ **79.** 90° **81.** 62.7°
83. Vertical ≈ 125.4 ft/sec, horizontal ≈ 1193.4 ft/sec
85. 12.8°; 398.32 N **87.** 71.3°; 228.5 lb
89. $T_L \approx 15{,}484$ lb
$T_R \approx 19{,}786$ lb
91. $T_{AC} \approx 1758.8$ lb
$T_{BC} \approx 1305.4$ lb
93. 3154.4 lb **95.** 20.8 lb **97.** 19.5°
99. 1928.4 ft-lb **101.** N 21.4° E; 138.7 km/h
103. True. See Example 1. **105.** True. $a = b = 0$
107. Proof **109.** $\langle 1, 3 \rangle$ or $\langle -1, -3 \rangle$
111. (a) $5\sqrt{5 + 4\cos\theta}$
(b)

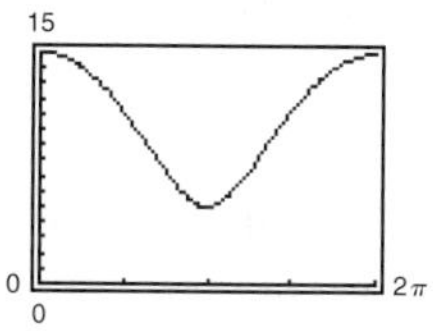

(c) Range: $[5, 15]$
Maximum is 15 when $\theta = 0$.
Minimum is 5 when $\theta = \pi$.
(d) The magnitudes of $\mathbf{F}_1$ and $\mathbf{F}_2$ are not the same.
113. Answers will vary.

Section 6.4 *(page 437)*

1. dot product **3.** $\frac{\mathbf{u} \cdot \mathbf{v}}{\|\mathbf{u}\| \|\mathbf{v}\|}$ **5.** $\left(\frac{\mathbf{u} \cdot \mathbf{v}}{\|\mathbf{v}\|^2}\right)\mathbf{v}$ **7.** -19
9. -11 **11.** 6 **13.** -12 **15.** 18; scalar
17. $\langle 24, -12 \rangle$; vector **19.** $\langle -126, -126 \rangle$; vector
21. $\sqrt{10} - 1$; scalar **23.** -12; scalar **25.** 17
27. $5\sqrt{41}$ **29.** 6 **31.** 90° **33.** 143.13°
35. 60.26° **37.** 90° **39.** $\frac{5\pi}{12}$

41.

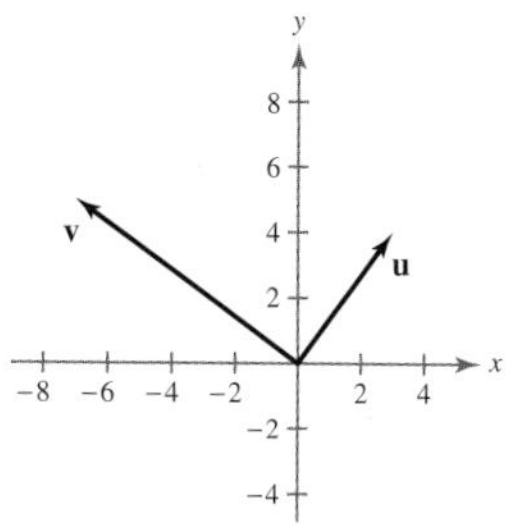

About 91.33°

43.

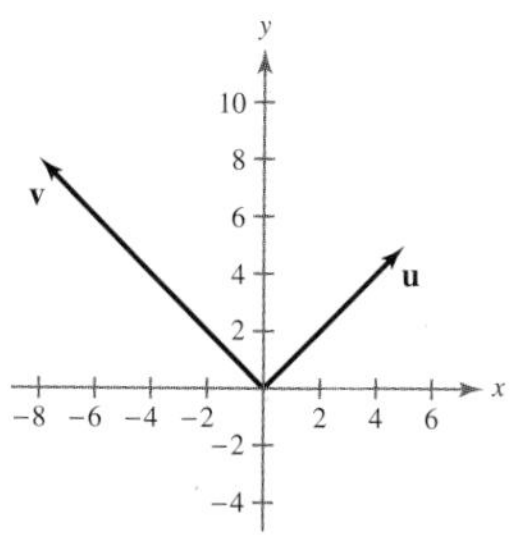

90°

45. 26.57°, 63.43°, 90° **47.** 41.63°, 53.13°, 85.24°
49. -20 **51.** -229.1 **53.** Not orthogonal
55. Not orthogonal **57.** Orthogonal
59. $\frac{1}{37}\langle 84, 14 \rangle, \frac{1}{37}\langle -10, 60 \rangle$ **61.** $\frac{45}{229}\langle 2, 15 \rangle, \frac{6}{229}\langle -15, 2 \rangle$
63. $\langle 3, 2 \rangle$ **65.** $\langle 0, 0 \rangle$ **67.** $\langle -5, 3 \rangle, \langle 5, -3 \rangle$
69. $\frac{2}{3}\mathbf{i} + \frac{1}{2}\mathbf{j}, -\frac{2}{3}\mathbf{i} - \frac{1}{2}\mathbf{j}$ **71.** 32
73. (a) \$35,727.50
This value gives the total amount paid to the employees.
(b) Multiply **v** by 1.02.
75. (a) Force = 30,000 sin d
(b)

d	0°	1°	2°	3°	4°	5°
Force	0	523.6	1047.0	1570.1	2092.7	2614.7

d	6°	7°	8°	9°	10°
Force	3135.9	3656.1	4175.2	4693.0	5209.4

(c) 29,885.8 lb
77. 735 N-m **79.** 779.4 ft-lb **81.** 10,282,651.78 N-m
83. 1174.62 ft-lb **85–87.** Answers will vary.
89. False. Work is represented by a scalar. **91.** Proof
93. (a) **u** and **v** are parallel. (b) **u** and **v** are orthogonal.

Section 6.5 *(page 448)*

1. absolute value **3.** DeMoivre's

5.

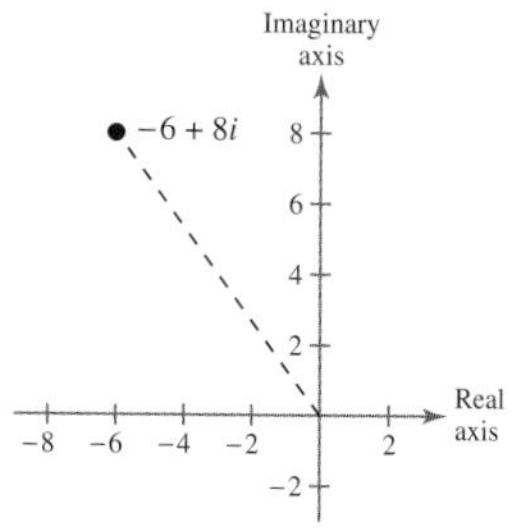

10

7.

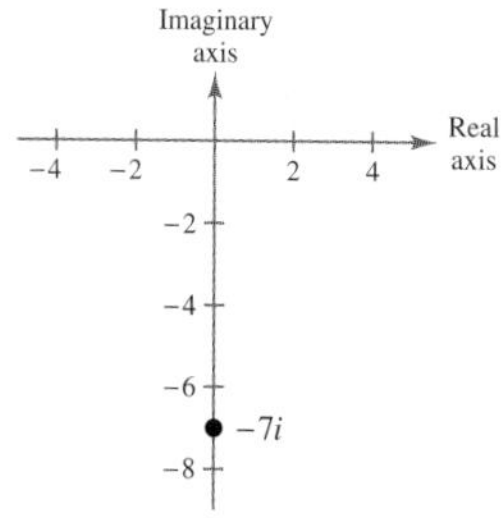

7

9.

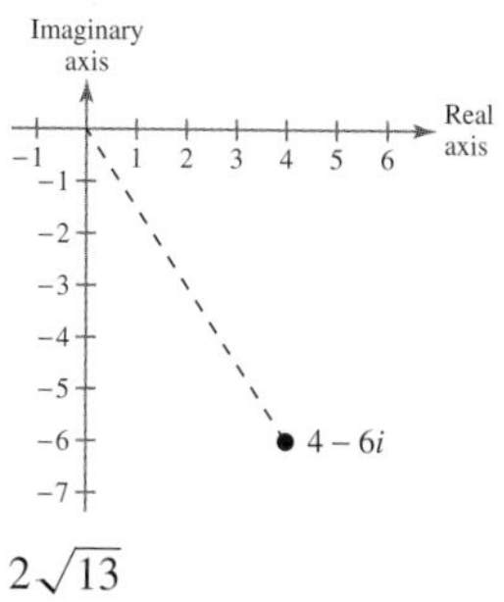

$2\sqrt{13}$

11.

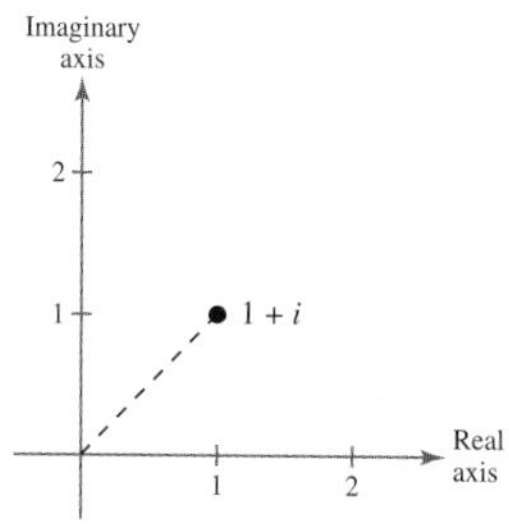

$\sqrt{2}\left(\cos\frac{\pi}{4} + i\sin\frac{\pi}{4}\right)$

13.

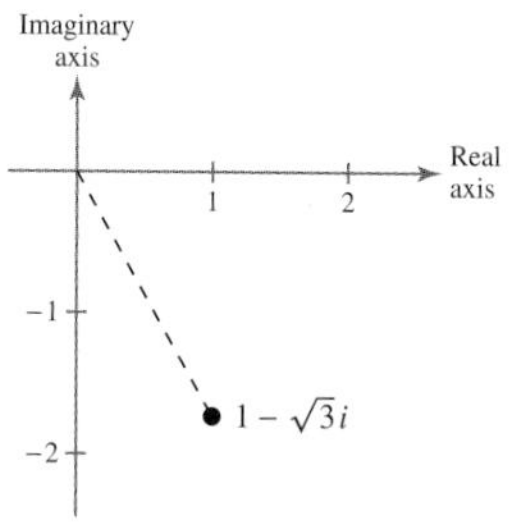

$2\left(\cos\frac{5\pi}{3} + i\sin\frac{5\pi}{3}\right)$

15.

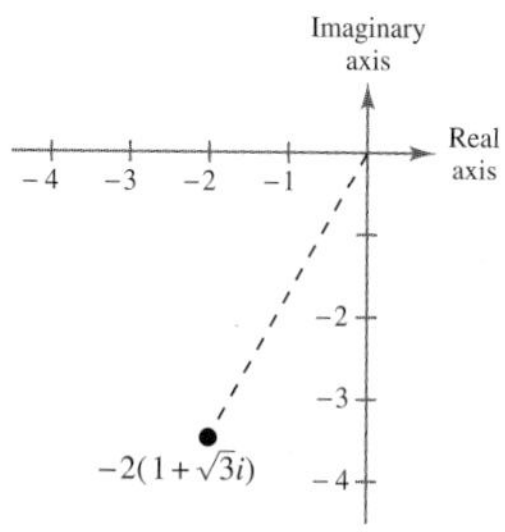

$4\left(\cos\frac{4\pi}{3} + i\sin\frac{4\pi}{3}\right)$

CHAPTER 6

17.

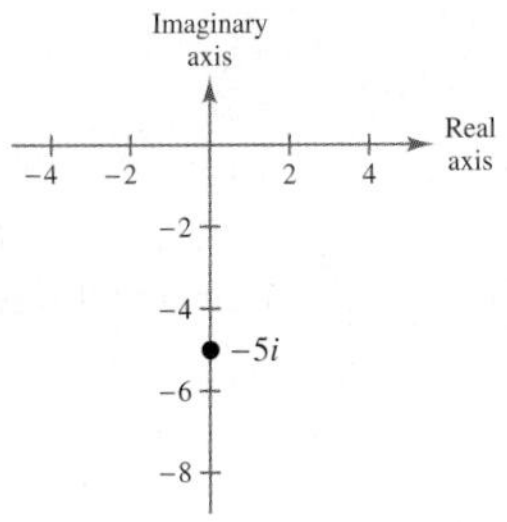

$5\left(\cos \frac{3\pi}{2} + i \sin \frac{3\pi}{2}\right)$

19.

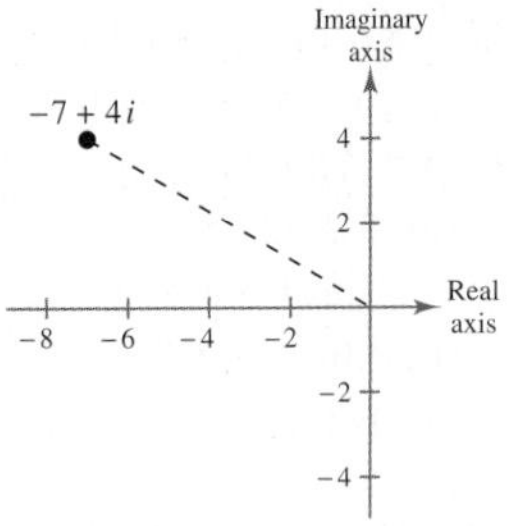

$\sqrt{65}(\cos 2.62 + i \sin 2.62)$

21.

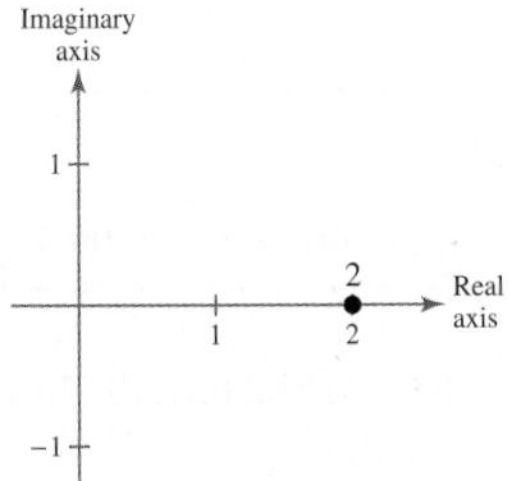

$2(\cos 0 + i \sin 0)$

23.

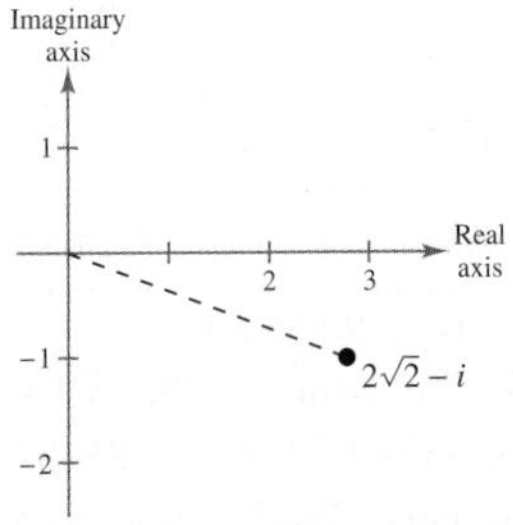

$3(\cos 5.94 + i \sin 5.94)$

25.

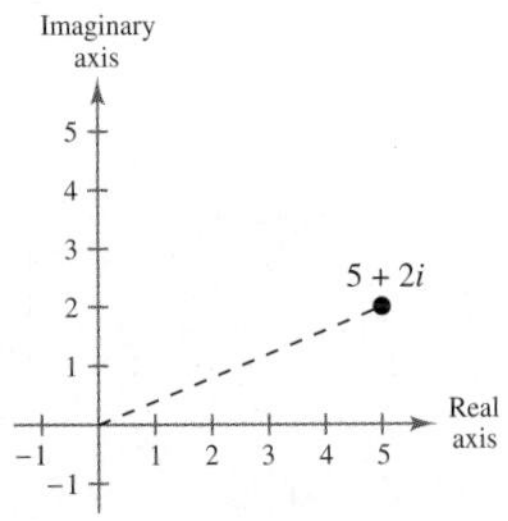

$\sqrt{29}(\cos 0.38 + i \sin 0.38)$

27.

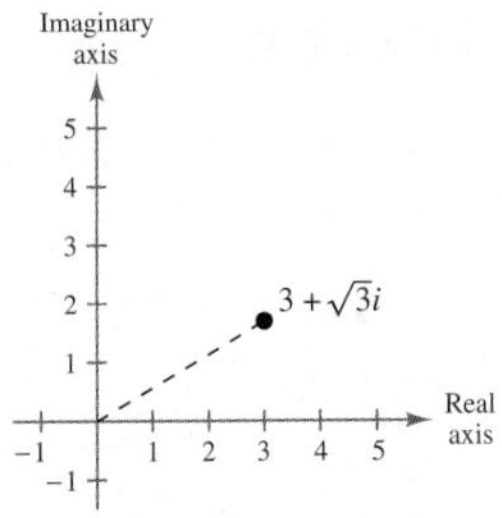

$2\sqrt{3}\left(\cos \frac{\pi}{6} + i \sin \frac{\pi}{6}\right)$

29.

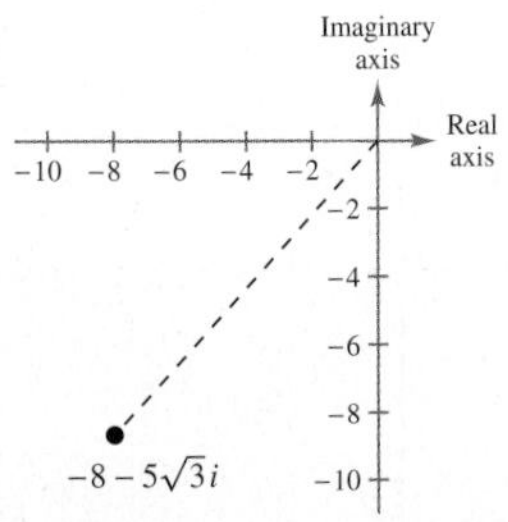

$\sqrt{139}(\cos 3.97 + i \sin 3.97)$

31.

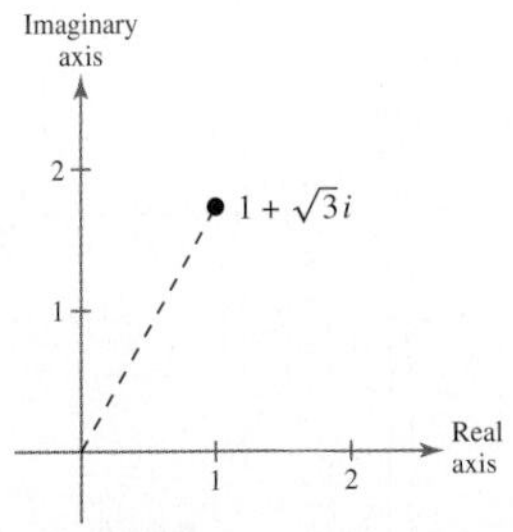

$1 + \sqrt{3}i$

33.

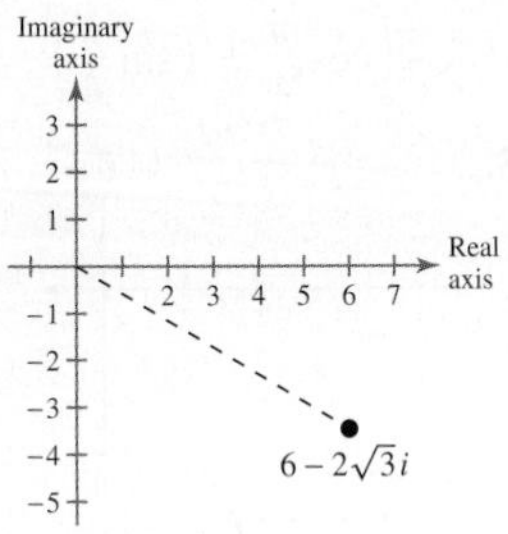

$6 - 2\sqrt{3}i$

35.

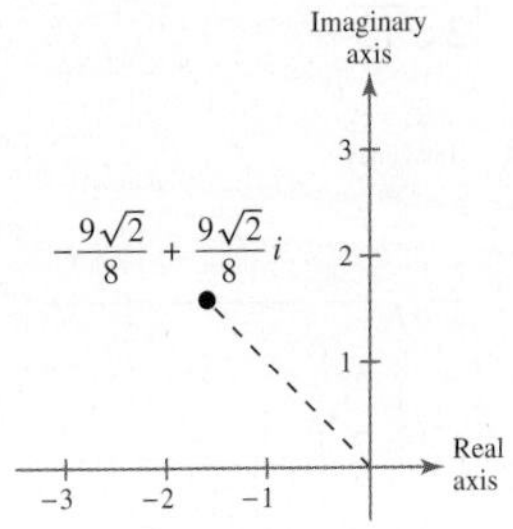

$-\frac{9\sqrt{2}}{8} + \frac{9\sqrt{2}}{8}i$

37.

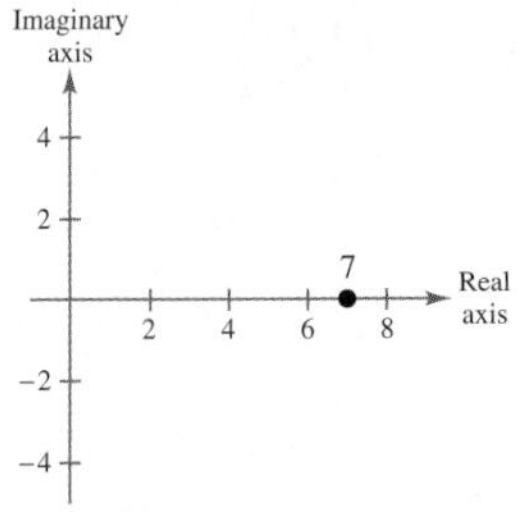

7

39.

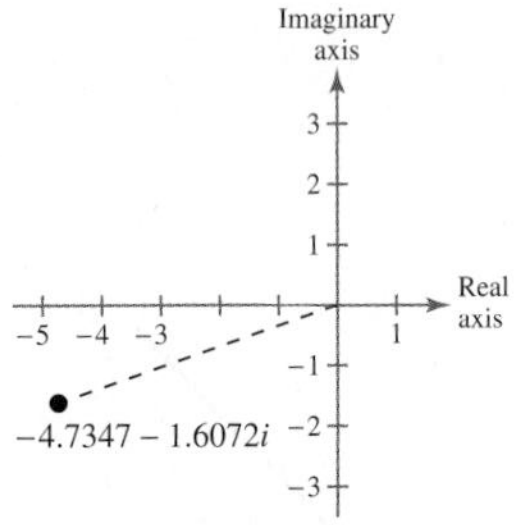

$-4.7347 - 1.6072i$

41. $4.6985 + 1.7101i$ **43.** $-1.8126 + 0.8452i$

45. $12\left(\cos \frac{\pi}{3} + i \sin \frac{\pi}{3}\right)$ **47.** $\frac{10}{9}(\cos 150^\circ + i \sin 150^\circ)$

49. $\cos 50^\circ + i \sin 50^\circ$ **51.** $\frac{1}{3}(\cos 30^\circ + i \sin 30^\circ)$

53. $\cos \frac{2\pi}{3} + i \sin \frac{2\pi}{3}$ **55.** $6(\cos 330^\circ + i \sin 330^\circ)$

57. (a) $\left[2\sqrt{2}\left(\cos \frac{\pi}{4} + i \sin \frac{\pi}{4}\right)\right]\left[\sqrt{2}\left(\cos \frac{7\pi}{4} + i \sin \frac{7\pi}{4}\right)\right]$
(b) $4(\cos 0 + i \sin 0) = 4$ (c) 4

59. (a) $\left[2\left(\cos \frac{3\pi}{2} + i \sin \frac{3\pi}{2}\right)\right]\left[\sqrt{2}\left(\cos \frac{\pi}{4} + i \sin \frac{\pi}{4}\right)\right]$
(b) $2\sqrt{2}\left(\cos \frac{7\pi}{4} + i \sin \frac{7\pi}{4}\right) = 2 - 2i$
(c) $-2i - 2i^2 = -2i + 2 = 2 - 2i$

61. (a) $[5(\cos 0.93 + i \sin 0.93)] \div \left[2\left(\cos \frac{5\pi}{3} + i \sin \frac{5\pi}{3}\right)\right]$
(b) $\frac{5}{2}(\cos 1.97 + i \sin 1.97) \approx -0.982 + 2.299i$
(c) About $-0.982 + 2.299i$

63.

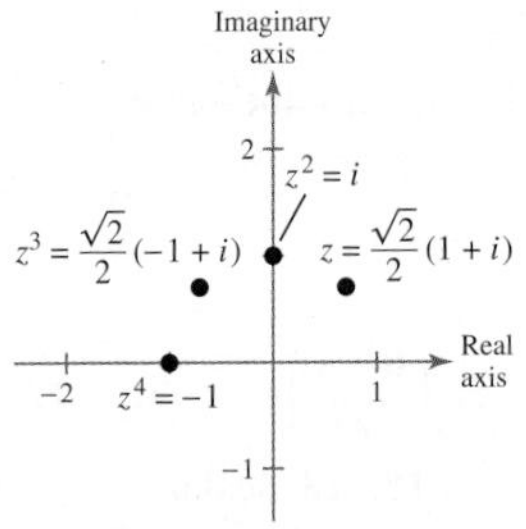

The absolute value of each is 1, and the consecutive powers of z are each 45° apart.

65. $-4 - 4i$ **67.** $8i$ **69.** $1024 - 1024\sqrt{3}i$

71. $\frac{125}{2} + \frac{125\sqrt{3}}{2}i$ **73.** -1 **75.** $608.0 + 144.7i$

77. $-597 - 122i$ **79.** $\frac{81}{2} + \frac{81\sqrt{3}}{2}i$

81. (a) $\sqrt{5}(\cos 60^\circ + i \sin 60^\circ)$
$\sqrt{5}(\cos 240^\circ + i \sin 240^\circ)$

(b)

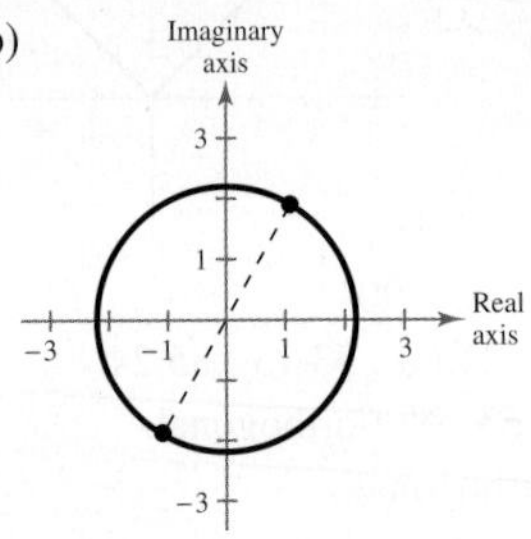

(c) $\frac{\sqrt{5}}{2} + \frac{\sqrt{15}}{2}i,$
$-\frac{\sqrt{5}}{2} - \frac{\sqrt{15}}{2}i$

83. (a) $2\left(\cos\frac{2\pi}{9} + i\sin\frac{2\pi}{9}\right)$
$2\left(\cos\frac{8\pi}{9} + i\sin\frac{8\pi}{9}\right)$
$2\left(\cos\frac{14\pi}{9} + i\sin\frac{14\pi}{9}\right)$
(b)

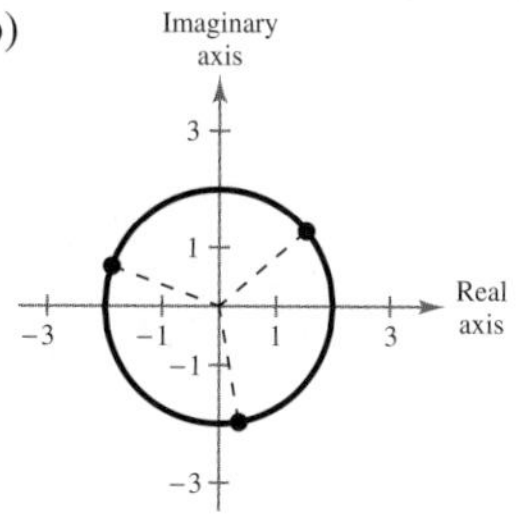

(c) $1.5321 + 1.2856i$, $-1.8794 + 0.6840i$, $0.3473 - 1.9696i$

85. (a) $5\left(\cos\frac{4\pi}{9} + i\sin\frac{4\pi}{9}\right)$
$5\left(\cos\frac{10\pi}{9} + i\sin\frac{10\pi}{9}\right)$
$5\left(\cos\frac{16\pi}{9} + i\sin\frac{16\pi}{9}\right)$
(b)

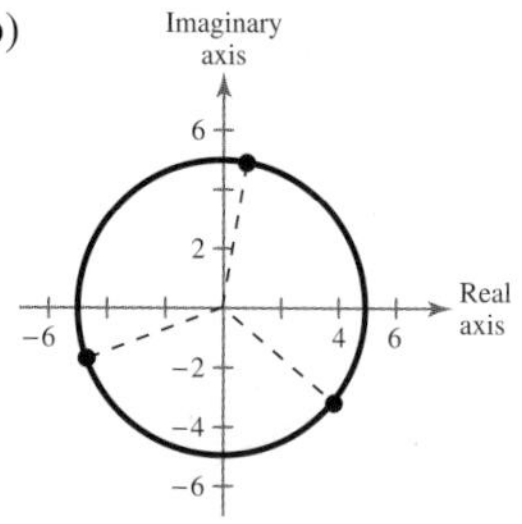

(c) $0.8682 + 4.9240i$, $-4.6985 - 1.7101i$, $3.8302 - 3.2140i$

87. (a) $5\left(\cos\frac{3\pi}{4} + i\sin\frac{3\pi}{4}\right)$
$5\left(\cos\frac{7\pi}{4} + i\sin\frac{7\pi}{4}\right)$
(b)

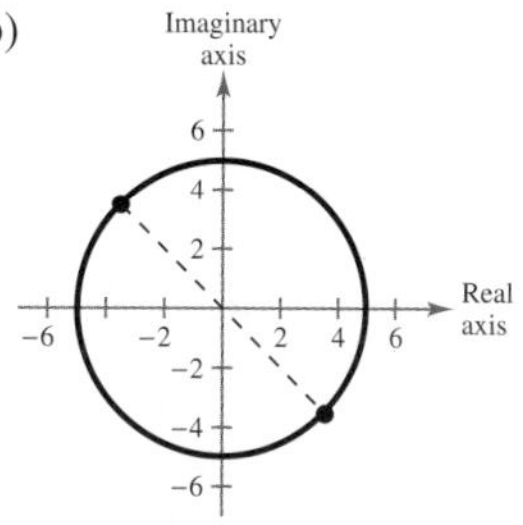

(c) $-\frac{5\sqrt{2}}{2} + \frac{5\sqrt{2}}{2}i$
$\frac{5\sqrt{2}}{2} - \frac{5\sqrt{2}}{2}i$

89. (a) $2(\cos 0 + i\sin 0)$
$2\left(\cos\frac{\pi}{2} + i\sin\frac{\pi}{2}\right)$
$2(\cos\pi + i\sin\pi)$
$2\left(\cos\frac{3\pi}{2} + i\sin\frac{3\pi}{2}\right)$
(b)

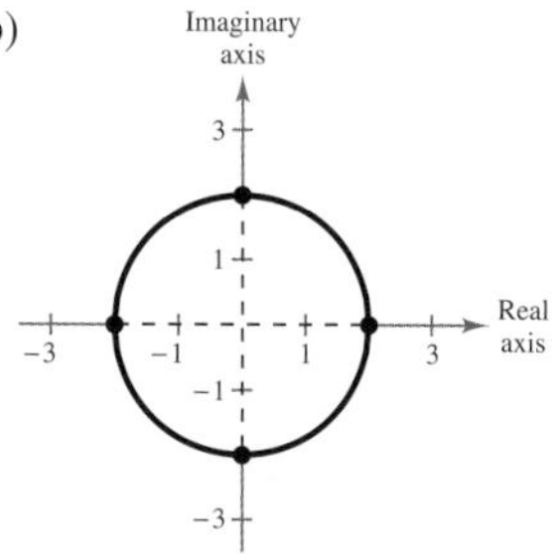

(c) $2, 2i, -2, -2i$

91. (a) $\cos 0 + i\sin 0$
$\cos\frac{2\pi}{5} + i\sin\frac{2\pi}{5}$
$\cos\frac{4\pi}{5} + i\sin\frac{4\pi}{5}$
$\cos\frac{6\pi}{5} + i\sin\frac{6\pi}{5}$
$\cos\frac{8\pi}{5} + i\sin\frac{8\pi}{5}$
(b)

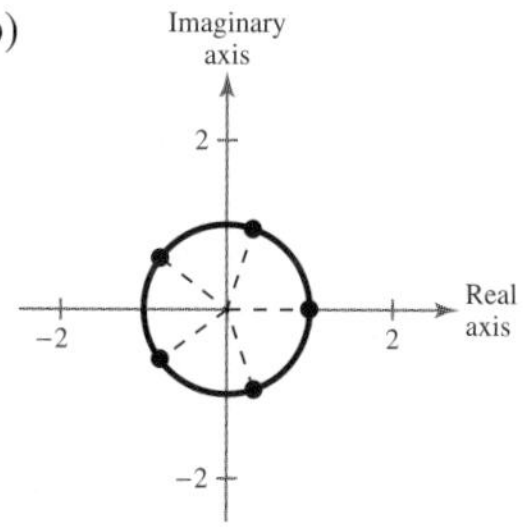

(c) $1, 0.3090 + 0.9511i$, $-0.8090 + 0.5878i$, $-0.8090 - 0.5878i$, $0.3090 - 0.9511i$

93. (a) $5\left(\cos\frac{\pi}{3} + i\sin\frac{\pi}{3}\right)$
$5(\cos\pi + i\sin\pi)$
$5\left(\cos\frac{5\pi}{3} + i\sin\frac{5\pi}{3}\right)$
(b)

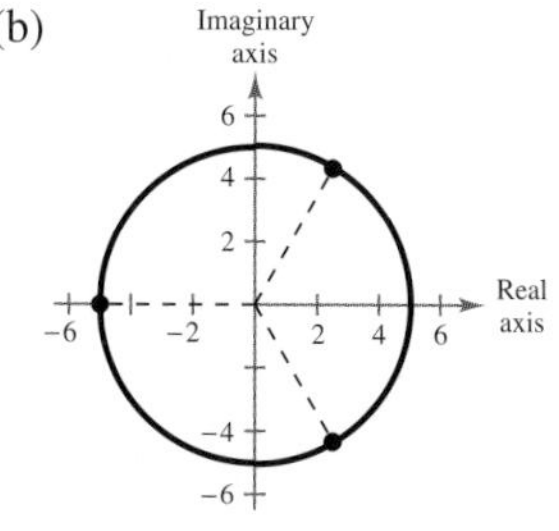

(c) $\frac{5}{2} + \frac{5\sqrt{3}}{2}i, -5,$
$\frac{5}{2} - \frac{5\sqrt{3}}{2}i$

95. (a) $\sqrt{2}\left(\cos\frac{7\pi}{20} + i\sin\frac{7\pi}{20}\right)$
$\sqrt{2}\left(\cos\frac{3\pi}{4} + i\sin\frac{3\pi}{4}\right)$
$\sqrt{2}\left(\cos\frac{23\pi}{20} + i\sin\frac{23\pi}{20}\right)$
$\sqrt{2}\left(\cos\frac{31\pi}{20} + i\sin\frac{31\pi}{20}\right)$
$\sqrt{2}\left(\cos\frac{39\pi}{20} + i\sin\frac{39\pi}{20}\right)$
(b)

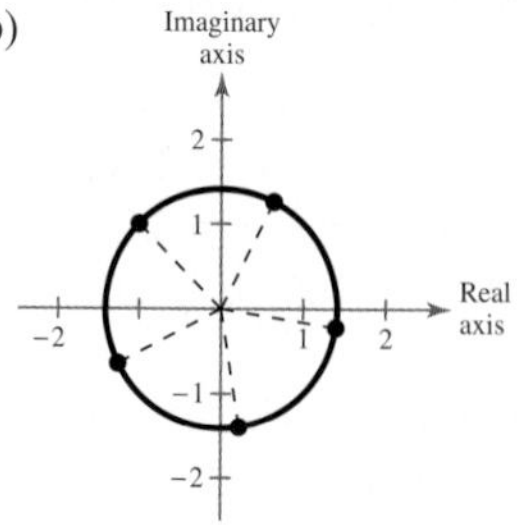

(c) $0.6420 + 1.2601i$, $-1 + i$, $-1.2601 - 0.6420i$, $0.2212 - 1.3968i$, $1.3968 - 0.2212i$

97. $\cos\frac{3\pi}{8} + i\sin\frac{3\pi}{8}$
$\cos\frac{7\pi}{8} + i\sin\frac{7\pi}{8}$
$\cos\frac{11\pi}{8} + i\sin\frac{11\pi}{8}$
$\cos\frac{15\pi}{8} + i\sin\frac{15\pi}{8}$

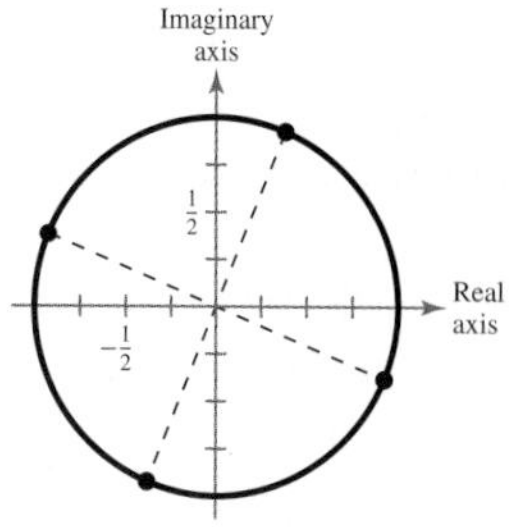

99. $3\left(\cos\frac{\pi}{5} + i\sin\frac{\pi}{5}\right)$
$3\left(\cos\frac{3\pi}{5} + i\sin\frac{3\pi}{5}\right)$
$3(\cos\pi + i\sin\pi)$
$3\left(\cos\frac{7\pi}{5} + i\sin\frac{7\pi}{5}\right)$
$3\left(\cos\frac{9\pi}{5} + i\sin\frac{9\pi}{5}\right)$

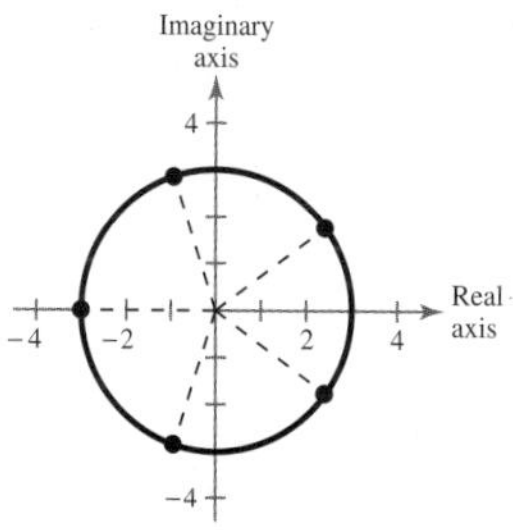

101. $2\left(\cos\frac{3\pi}{8} + i\sin\frac{3\pi}{8}\right)$
$2\left(\cos\frac{7\pi}{8} + i\sin\frac{7\pi}{8}\right)$
$2\left(\cos\frac{11\pi}{8} + i\sin\frac{11\pi}{8}\right)$
$2\left(\cos\frac{15\pi}{8} + i\sin\frac{15\pi}{8}\right)$

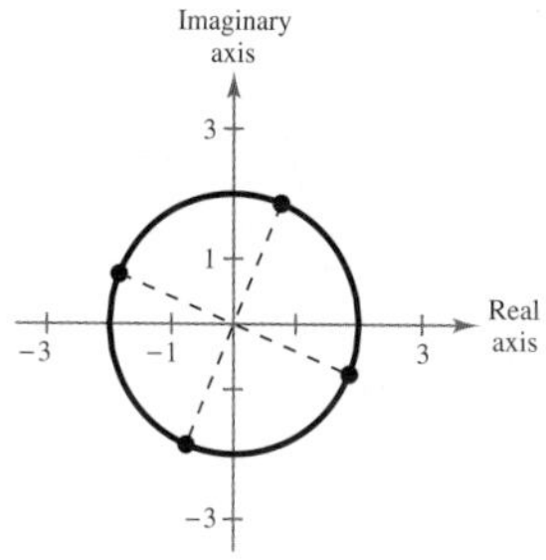

103. $\sqrt[6]{2}\left(\cos\frac{7\pi}{12} + i\sin\frac{7\pi}{12}\right)$
$\sqrt[6]{2}\left(\cos\frac{5\pi}{4} + i\sin\frac{5\pi}{4}\right)$
$\sqrt[6]{2}\left(\cos\frac{23\pi}{12} + i\sin\frac{23\pi}{12}\right)$

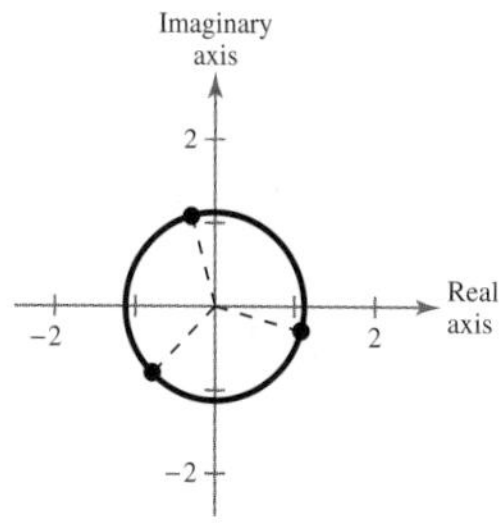

105. (a) $E = 24(\cos 30° + i\sin 30°)$ volts
(b) $E = 12\sqrt{3} + 12i$ volts (c) $|E| = 24$ volts

107. True. $z_1z_2 = r_1r_2[\cos(\theta_1 + \theta_2) + i\sin(\theta_1 + \theta_2)] = 0$ if and only if $r_1 = 0$ and/or $r_2 = 0$.

109–111. Answers will vary.

Review Exercises (page 452)

1. $C = 72°, b \approx 12.21, c \approx 12.36$
3. $A = 26°, a \approx 24.89, c \approx 56.23$
5. $C = 66°, a \approx 2.53, b \approx 9.11$
7. $B = 108°, a \approx 11.76, c \approx 21.49$
9. $A \approx 20.41°, C \approx 9.59°, a \approx 20.92$
11. $B \approx 39.48°, C \approx 65.52°, c \approx 48.24$
13. 19.06 **15.** 47.23 **17.** 31.1 m **19.** 31.01 ft
21. $A \approx 27.81°, B \approx 54.75°, C \approx 97.44°$
23. $A \approx 16.99°, B \approx 26.00°, C \approx 137.01°$
25. $A \approx 29.92°, B \approx 86.18°, C \approx 63.90°$
27. $A = 36°, C = 36°, b \approx 17.80$
29. $A \approx 45.76°, B \approx 91.24°, c \approx 21.42$
31. Law of Sines; $A \approx 77.52°, B \approx 38.48°, a \approx 14.12$
33. Law of Cosines; $A \approx 28.62°, B \approx 33.56°, C \approx 117.82°$
35. About 4.3 ft, about 12.6 ft **37.** 615.1 m **39.** 7.64
41. 8.36 **43.** $\|\mathbf{u}\| = \|\mathbf{v}\| = \sqrt{61}$, $\text{slope}_\mathbf{u} = \text{slope}_\mathbf{v} = \frac{5}{6}$
45. $\langle 7, -5\rangle$ **47.** $\langle 7, -7\rangle$ **49.** $\langle -4, 4\sqrt{3}\rangle$

51. (a) $\langle -4, 3\rangle$

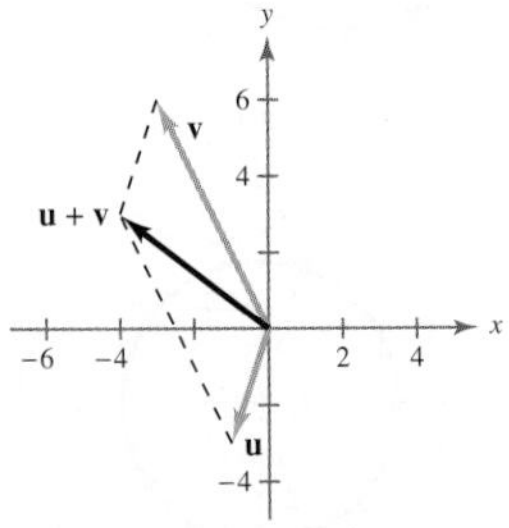

(b) $\langle 2, -9\rangle$

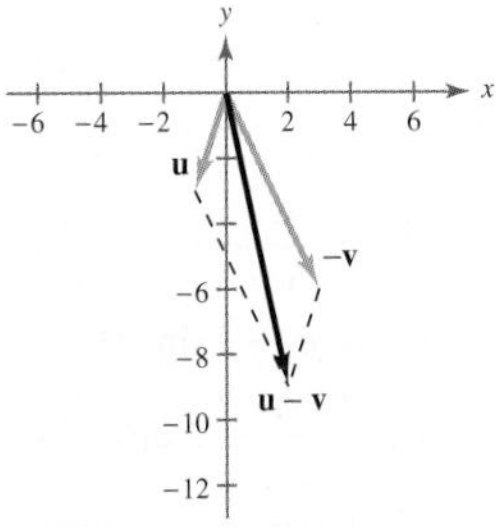

(c) $\langle -4, -12\rangle$

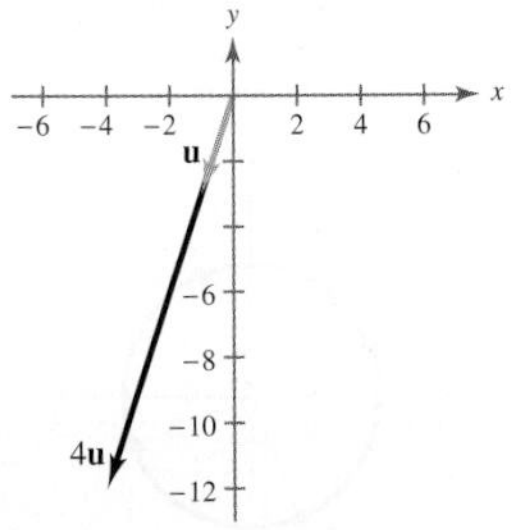

(d) $\langle -14, 3\rangle$

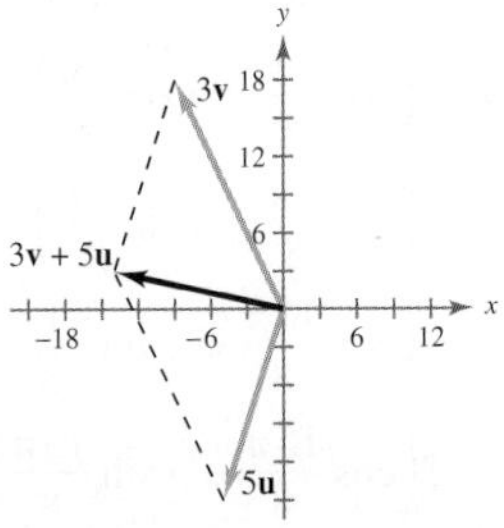

53. (a) $\langle -1, 6\rangle$

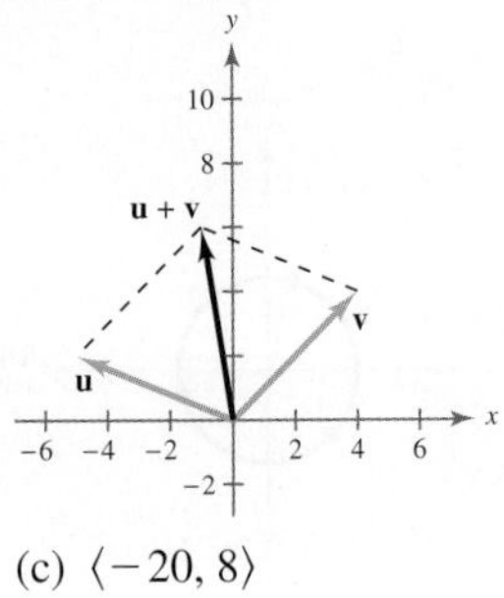

(b) $\langle -9, -2\rangle$

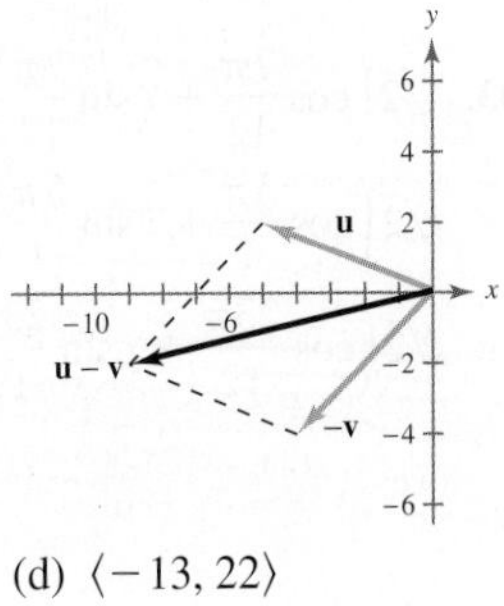

(c) $\langle -20, 8\rangle$

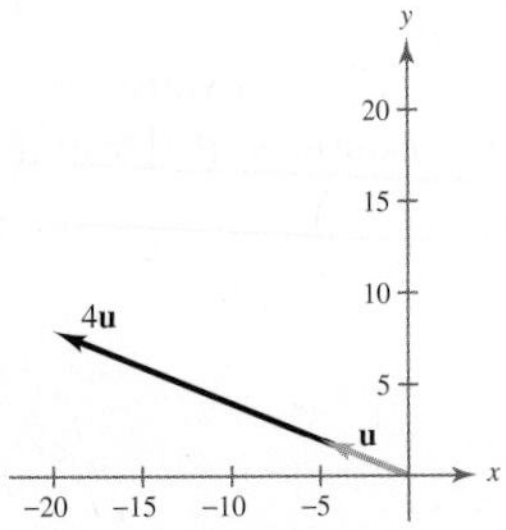

(d) $\langle -13, 22\rangle$

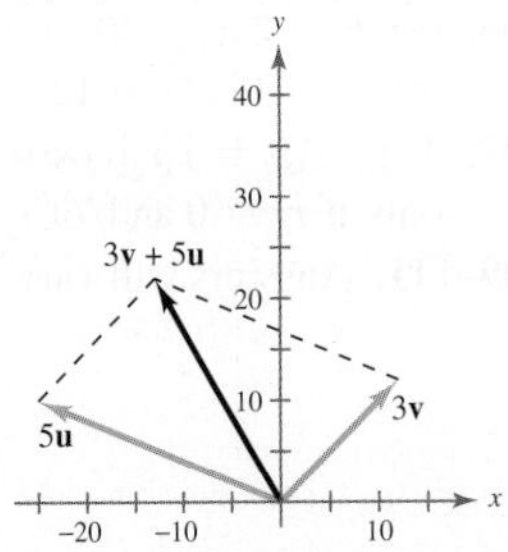

55. (a) $7\mathbf{i} + 2\mathbf{j}$

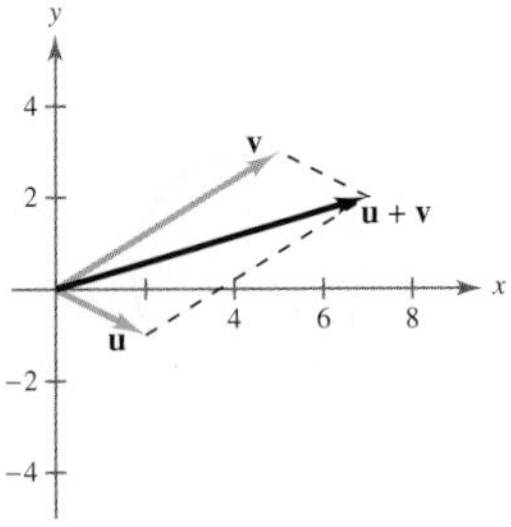

(b) $-3\mathbf{i} - 4\mathbf{j}$

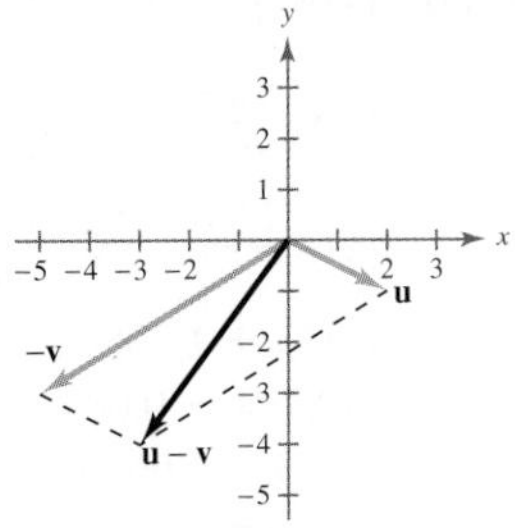

(c) $8\mathbf{i} - 4\mathbf{j}$

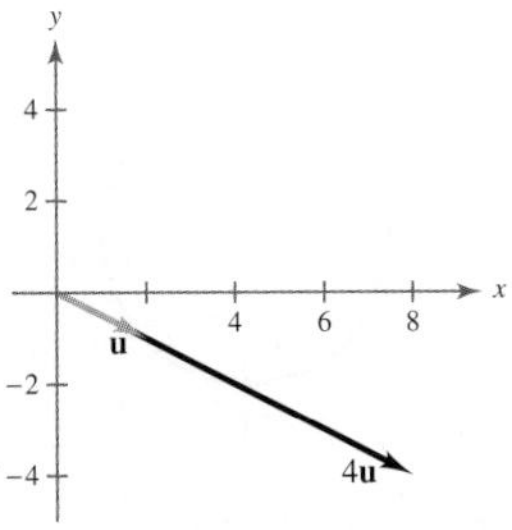

(d) $25\mathbf{i} + 4\mathbf{j}$

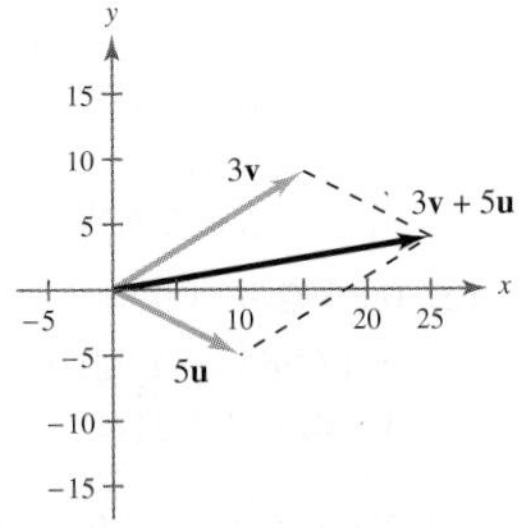

57. (a) $3\mathbf{i} + 6\mathbf{j}$

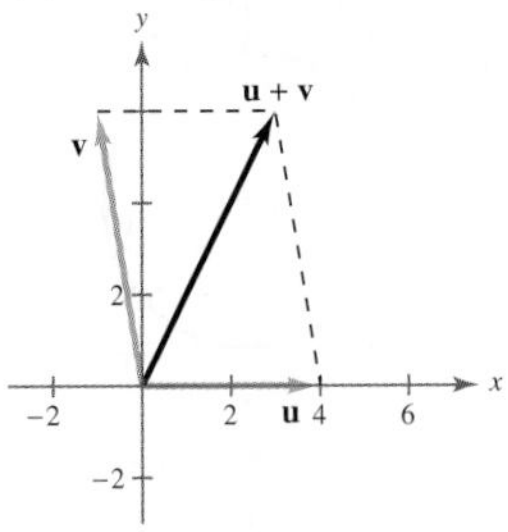

(b) $5\mathbf{i} - 6\mathbf{j}$

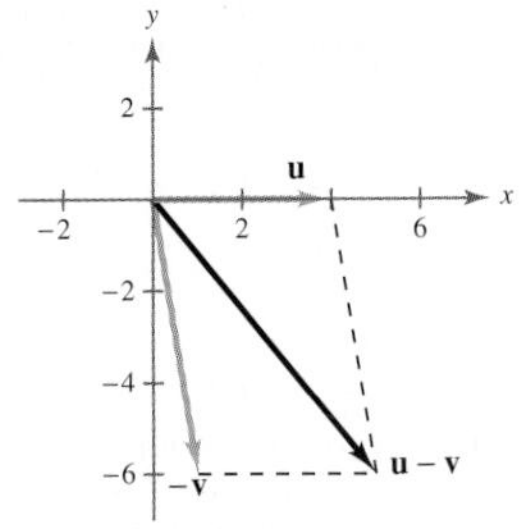

(c) $16\mathbf{i}$

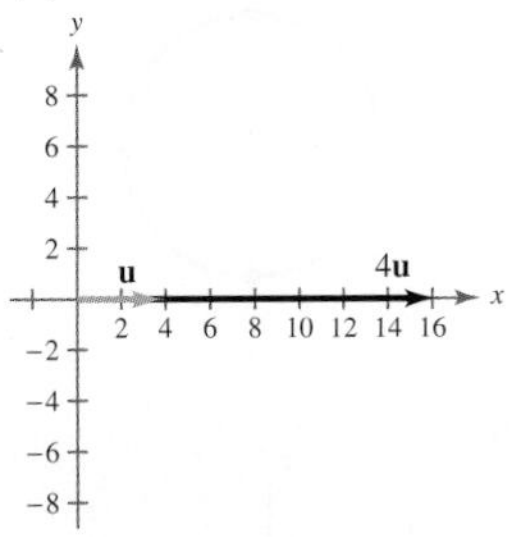

(d) $17\mathbf{i} + 18\mathbf{j}$

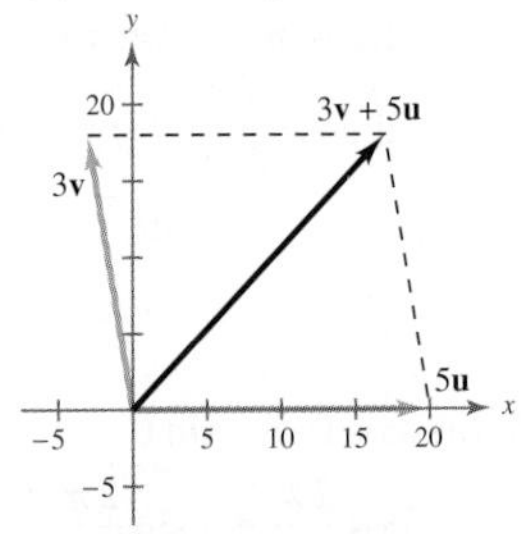

59. $\langle 30, 9\rangle$

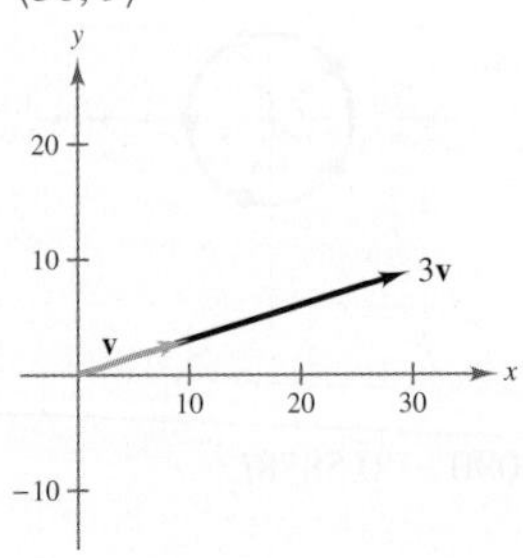

61. $\langle 22, -7\rangle$

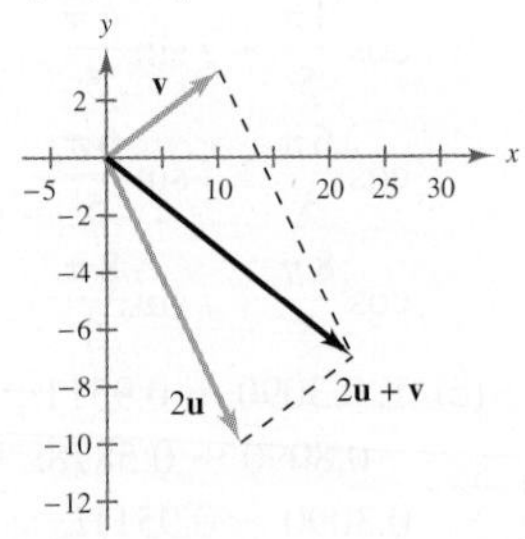

63. $\langle -10, -37 \rangle$

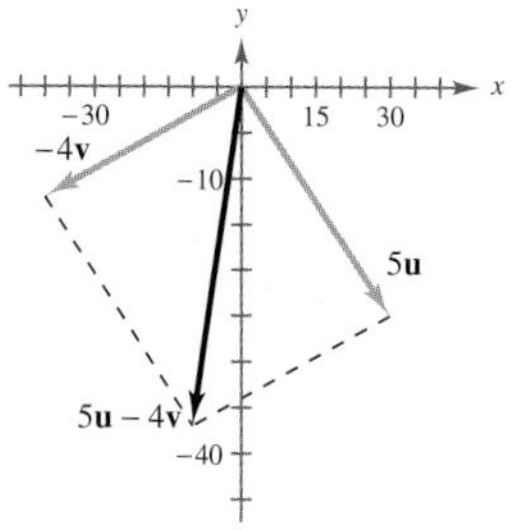

65. $-\mathbf{i} + 5\mathbf{j}$ **67.** $6\mathbf{i} + 4\mathbf{j}$ **69.** $\|\mathbf{v}\| = 7$; $\theta = 60°$

71. $\|\mathbf{v}\| = \sqrt{41}$; $\theta = 38.7°$ **73.** $\|\mathbf{v}\| = 3\sqrt{2}$; $\theta = 225°$

75. The resultant force is 133.92 pounds and 5.6° from the 85-pound force.

77. 422.30 mi/h; 130.4° **79.** 45 **81.** -2

83. 40; scalar **85.** $4 - 2\sqrt{5}$; scalar

87. $\langle 72, -36 \rangle$; vector **89.** 38; scalar **91.** $\frac{11\pi}{12}$

93. 160.5° **95.** Orthogonal **97.** Not orthogonal

99. $-\frac{13}{17}\langle 4, 1\rangle, \frac{16}{17}\langle -1, 4\rangle$ **101.** $\frac{5}{2}\langle -1, 1\rangle, \frac{9}{2}\langle 1, 1\rangle$ **103.** 48

105. 72,000 ft-lb

107.

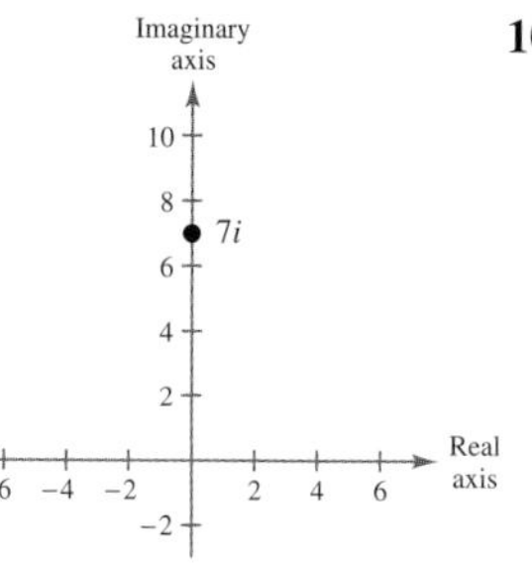

7

109.

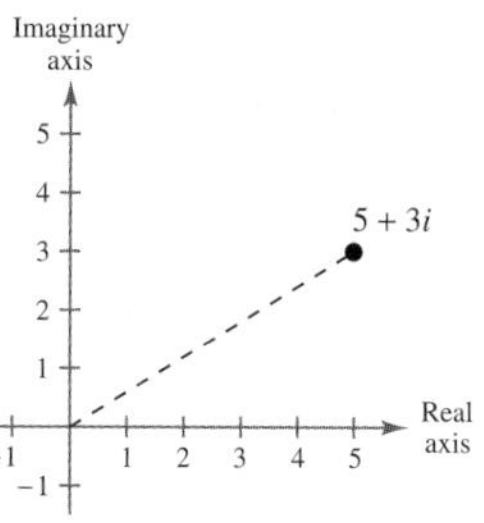

$\sqrt{34}$

111.

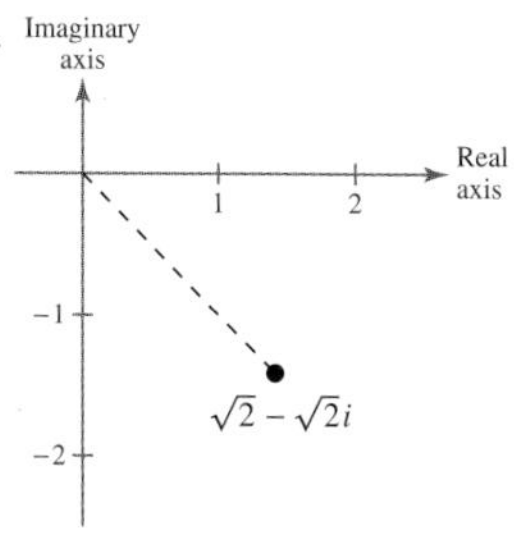

2

113. $4\left(\cos\frac{\pi}{2} + i\sin\frac{\pi}{2}\right)$ **115.** $7\sqrt{2}\left(\cos\frac{7\pi}{4} + i\sin\frac{7\pi}{4}\right)$

117. $13(\cos 4.32 + i\sin 4.32)$

119. (a) $z_1 = 4\left(\cos\frac{11\pi}{6} + i\sin\frac{11\pi}{6}\right)$

$z_2 = 10\left(\cos\frac{3\pi}{2} + i\sin\frac{3\pi}{2}\right)$

(b) $z_1z_2 = 40\left(\cos\frac{10\pi}{3} + i\sin\frac{10\pi}{3}\right)$

$\frac{z_1}{z_2} = \frac{2}{5}\left(\cos\frac{\pi}{3} + i\sin\frac{\pi}{3}\right)$

121. $\frac{625}{2} + \frac{625\sqrt{3}}{2}i$ **123.** $2035 - 828i$

125. (a) $3\left(\cos\frac{\pi}{4} + i\sin\frac{\pi}{4}\right)$

$3\left(\cos\frac{7\pi}{12} + i\sin\frac{7\pi}{12}\right)$

$3\left(\cos\frac{11\pi}{12} + i\sin\frac{11\pi}{12}\right)$

$3\left(\cos\frac{5\pi}{4} + i\sin\frac{5\pi}{4}\right)$

$3\left(\cos\frac{19\pi}{12} + i\sin\frac{19\pi}{12}\right)$

$3\left(\cos\frac{23\pi}{12} + i\sin\frac{23\pi}{12}\right)$

(b)

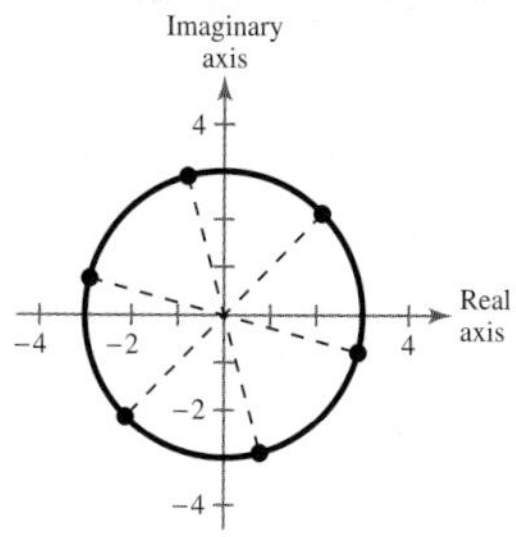

(c) $\frac{3\sqrt{2}}{2} + \frac{3\sqrt{2}}{2}i, -0.776 + 2.898i,$

$-2.898 + 0.776i, -\frac{3\sqrt{2}}{2} - \frac{3\sqrt{2}}{2}i,$

$0.776 - 2.898i, 2.898 - 0.776i$

127. (a) $2(\cos 0 + i\sin 0)$

$2\left(\cos\frac{2\pi}{3} + i\sin\frac{2\pi}{3}\right)$

$2\left(\cos\frac{4\pi}{3} + i\sin\frac{4\pi}{3}\right)$

(b)

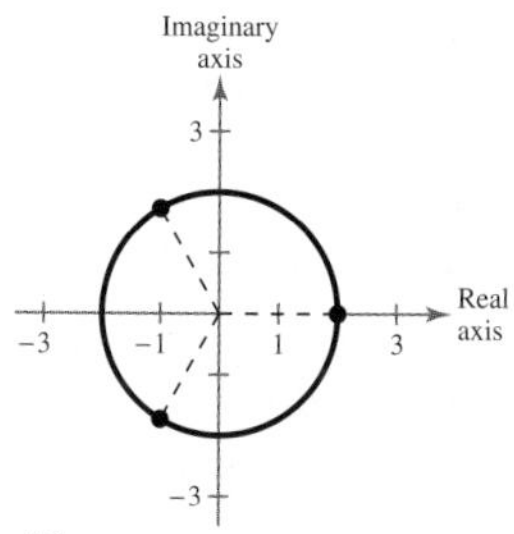

(c) $2, -1 + \sqrt{3}i,$ $-1 - \sqrt{3}i$

129. $3\left(\cos\frac{\pi}{4} + i\sin\frac{\pi}{4}\right) = \frac{3\sqrt{2}}{2} + \frac{3\sqrt{2}}{2}i$

$3\left(\cos\frac{3\pi}{4} + i\sin\frac{3\pi}{4}\right) = -\frac{3\sqrt{2}}{2} + \frac{3\sqrt{2}}{2}i$

$3\left(\cos\frac{5\pi}{4} + i\sin\frac{5\pi}{4}\right) = -\frac{3\sqrt{2}}{2} - \frac{3\sqrt{2}}{2}i$

$3\left(\cos\frac{7\pi}{4} + i\sin\frac{7\pi}{4}\right) = \frac{3\sqrt{2}}{2} - \frac{3\sqrt{2}}{2}i$

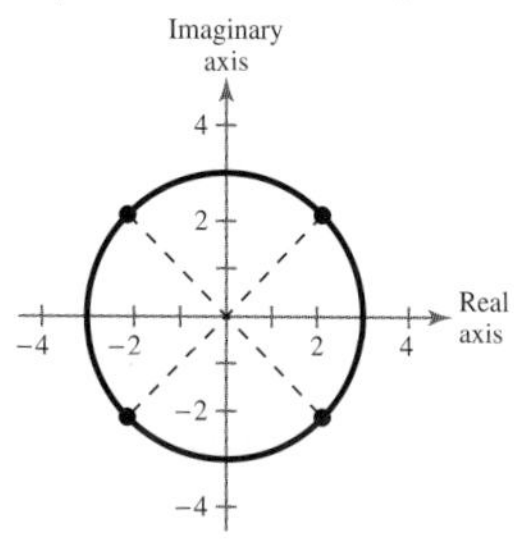

131. $2\left(\cos\frac{\pi}{2} + i\sin\frac{\pi}{2}\right) = 2i$

$2\left(\cos\frac{7\pi}{6} + i\sin\frac{7\pi}{6}\right) = -\sqrt{3} - i$

$2\left(\cos\frac{11\pi}{6} + i\sin\frac{11\pi}{6}\right) = \sqrt{3} - i$

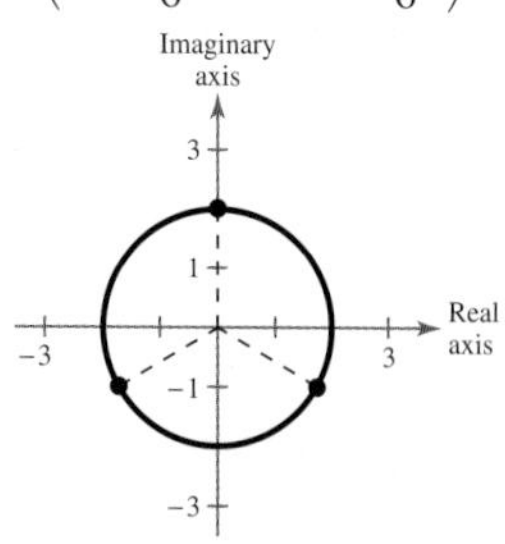

133. $\cos 0 + i \sin 0 = 1$

$\cos \dfrac{\pi}{2} + i \sin \dfrac{\pi}{2} = i$

$\cos \dfrac{2\pi}{3} + i \sin \dfrac{2\pi}{3} = -\dfrac{1}{2} + \dfrac{\sqrt{3}}{2}i$

$\cos \dfrac{4\pi}{3} + i \sin \dfrac{4\pi}{3} = -\dfrac{1}{2} - \dfrac{\sqrt{3}}{2}i$

$\cos \dfrac{3\pi}{2} + i \sin \dfrac{3\pi}{2} = -i$

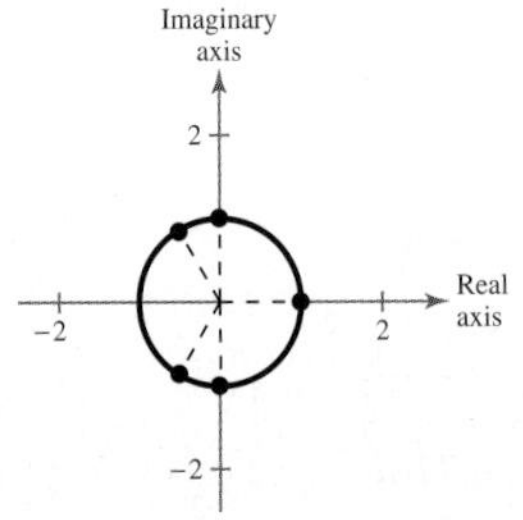

135. True. Sin 90° is defined in the Law of Sines.

137. False. The solutions to $x^2 - 8i = 0$ are $x = 2 + 2i$ and $x = -2 - 2i$.

139. $a^2 = b^2 + c^2 - 2bc \cos A$, $b^2 = a^2 + c^2 - 2ac \cos B$, $c^2 = a^2 + b^2 - 2ab \cos C$

141. A and C

143. For $k > 0$, the direction is the same and the magnitude is k times as great.

For $k < 0$, the result is a vector in the opposite direction and the magnitude is $|k|$ times as great.

145. (a) $4(\cos 60° + i \sin 60°)$ (b) -64
$4(\cos 180° + i \sin 180°)$
$4(\cos 300° + i \sin 300°)$

147. (a) $2(\cos 30° + i \sin 30°)$ (b) $8i$
$2(\cos 150° + i \sin 150°)$
$2(\cos 270° + i \sin 270°)$

149. $z_1 z_2 = -4$

$\dfrac{z_1}{z_2} = \cos(2\theta - \pi) + i \sin(2\theta - \pi)$

$= -\cos 2\theta - i \sin 2\theta$

Chapter Test *(page 456)*

1. Law of Sines; $C = 88°$, $b \approx 27.81$, $c \approx 29.98$

2. Law of Sines; $A = 42°$, $b \approx 21.91$, $c \approx 10.95$

3. Law of Sines

Two solutions: $B \approx 29.12°$, $C \approx 126.88°$, $c \approx 22.03$
$B \approx 150.88°$, $C \approx 5.12°$, $c \approx 2.46$

4. Law of Cosines; No solution

5. Law of Sines; $A \approx 39.96°$, $C \approx 40.04°$, $c \approx 15.02$

6. Law of Cosines; $A \approx 21.90°$, $B \approx 37.10°$, $c \approx 78.15$

7. 2052.5 m^2 **8.** 606.3 mi; 29.1° **9.** $\langle 14, -23 \rangle$

10. $\left\langle \dfrac{18\sqrt{34}}{17}, -\dfrac{30\sqrt{34}}{17} \right\rangle$

11. $\langle -4, 12 \rangle$

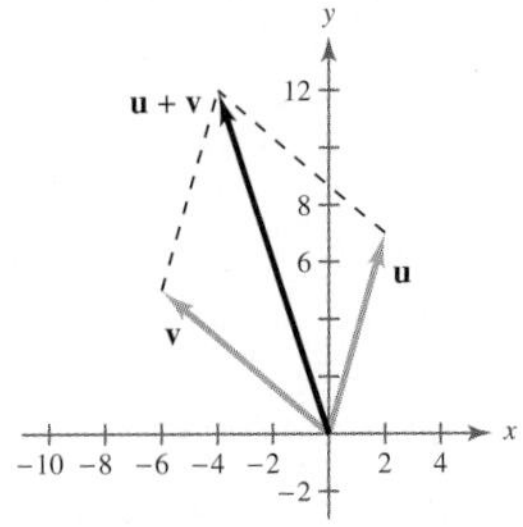

12. $\langle 8, 2 \rangle$

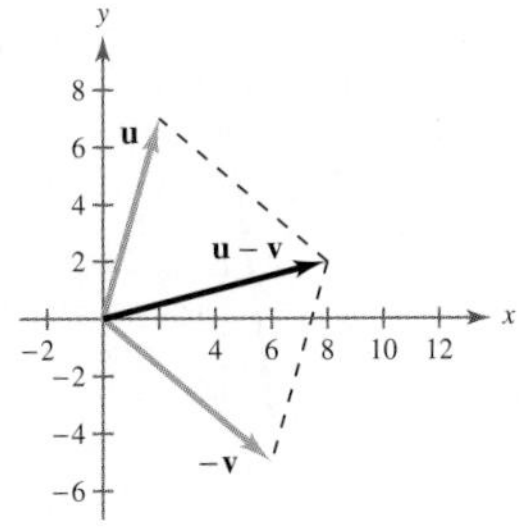

13. $\langle 28, 20 \rangle$

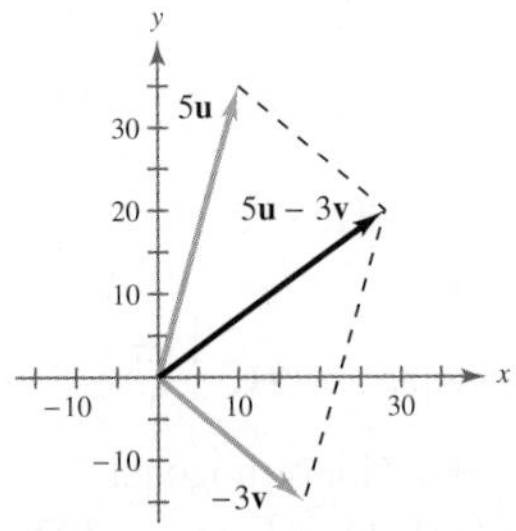

14. $\langle -4, 38 \rangle$

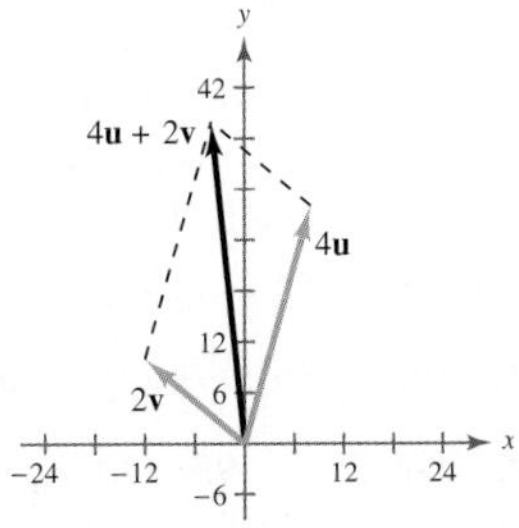

15. $\left\langle \frac{24}{25}, -\frac{7}{25} \right\rangle$ **16.** 14.9°; 250.15 lb **17.** 135°

18. Yes **19.** $\frac{37}{26}\langle 5, 1 \rangle$; $\frac{29}{26}\langle -1, 5 \rangle$ **20.** About 104 lb

21. $4\sqrt{2}\left(\cos \dfrac{7\pi}{4} + i \sin \dfrac{7\pi}{4}\right)$ **22.** $-3 + 3\sqrt{3}i$

23. $-\dfrac{6561}{2} - \dfrac{6561\sqrt{3}}{2}i$ **24.** $5832i$

25. $4\sqrt[4]{2}\left(\cos \dfrac{\pi}{12} + i \sin \dfrac{\pi}{12}\right)$

$4\sqrt[4]{2}\left(\cos \dfrac{7\pi}{12} + i \sin \dfrac{7\pi}{12}\right)$

$4\sqrt[4]{2}\left(\cos \dfrac{13\pi}{12} + i \sin \dfrac{13\pi}{12}\right)$

$4\sqrt[4]{2}\left(\cos \dfrac{19\pi}{12} + i \sin \dfrac{19\pi}{12}\right)$

26. $3\left(\cos \dfrac{\pi}{6} + i \sin \dfrac{\pi}{6}\right)$

$3\left(\cos \dfrac{5\pi}{6} + i \sin \dfrac{5\pi}{6}\right)$

$3\left(\cos \dfrac{3\pi}{2} + i \sin \dfrac{3\pi}{2}\right)$

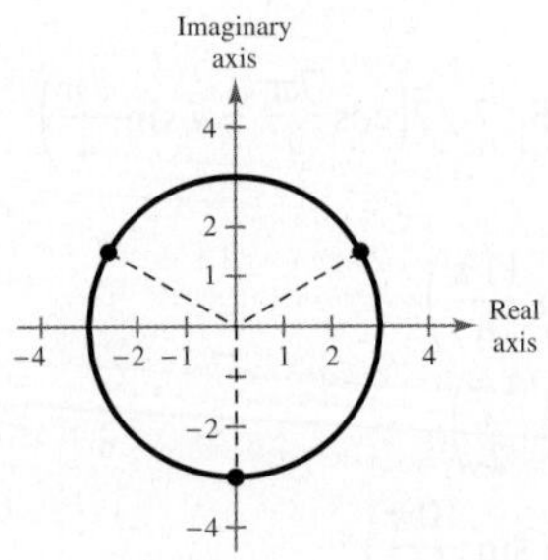

Cumulative Test for Chapters 4–6 *(page 457)*

1. (a)

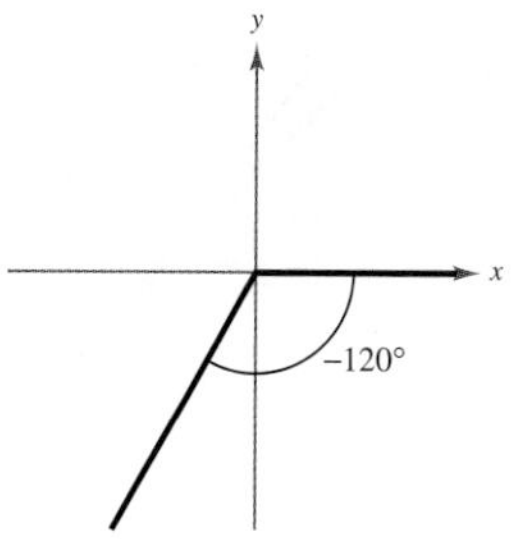

(b) $240°$
(c) $-\frac{2\pi}{3}$
(d) $60°$

(e) $\sin(-120°) = -\frac{\sqrt{3}}{2}$ $\quad \csc(-120°) = -\frac{2\sqrt{3}}{3}$
$\cos(-120°) = -\frac{1}{2}$ $\quad \sec(-120°) = -2$
$\tan(-120°) = \sqrt{3}$ $\quad \cot(-120°) = \frac{\sqrt{3}}{3}$

2. $-83.1°$ **3.** $\frac{20}{29}$

4.

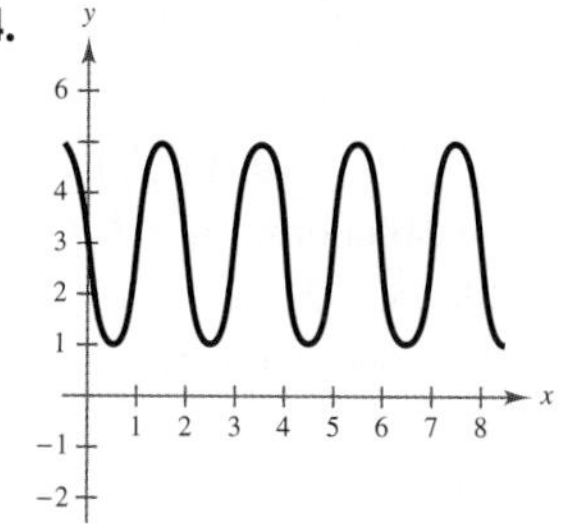

5.

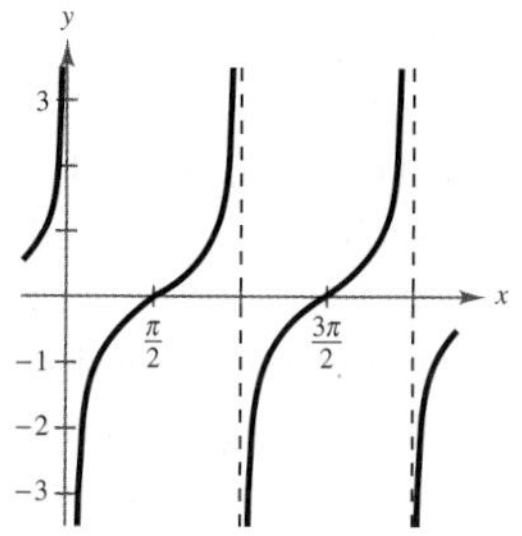

6.

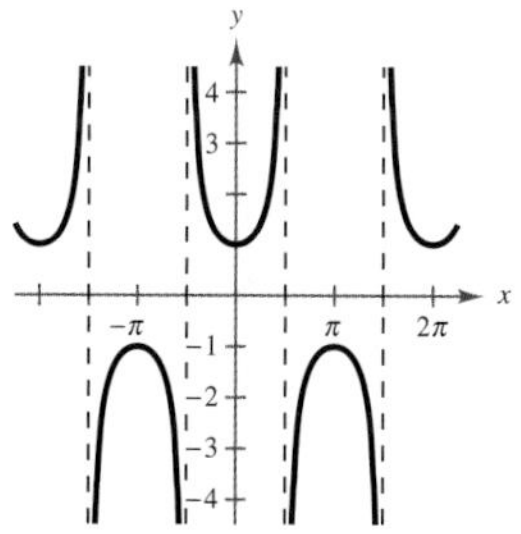

7. $a = -3, b = \pi, c = 0$

8.

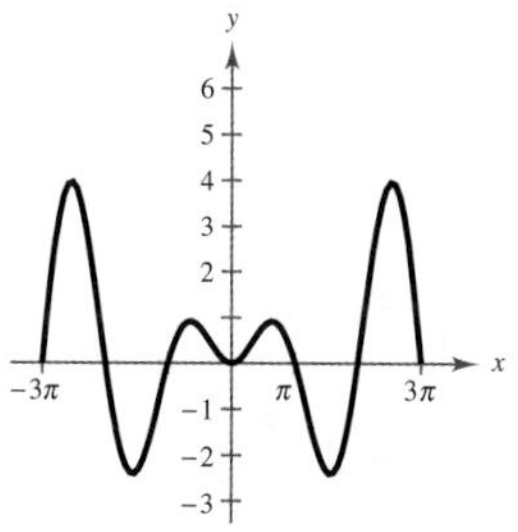

9. 4.9
10. $\frac{3}{4}$
11. $\sqrt{1 - 4x^2}$
12. 1
13. $2 \tan \theta$
14–16. Answers will vary.
17. $\frac{\pi}{3}, \frac{\pi}{2}, \frac{3\pi}{2}, \frac{5\pi}{3}$

18. $\frac{\pi}{6}, \frac{5\pi}{6}, \frac{7\pi}{6}, \frac{11\pi}{6}$ **19.** $\frac{3\pi}{2}$ **20.** $\frac{16}{63}$ **21.** $\frac{4}{3}$

22. $\frac{\sqrt{5}}{5}, \frac{2\sqrt{5}}{5}$ **23.** $\frac{5}{2}\left(\sin \frac{5\pi}{2} - \sin \pi\right)$ **24.** $-2 \sin 8x \sin x$

25. Law of Sines; $B \approx 26.39°, C \approx 123.61°, c \approx 14.99$
26. Law of Cosines; $B \approx 52.48°, C \approx 97.52°, a \approx 5.04$
27. Law of Sines; $B = 60°, a \approx 5.77, c \approx 11.55$
28. Law of Cosines; $A \approx 26.28°, B \approx 49.74°, C \approx 103.98°$
29. Law of Sines; $C = 109°, a \approx 14.96, b \approx 9.27$
30. Law of Cosines; $A \approx 6.75°, B \approx 93.25°, c \approx 9.86$
31. 41.48 in.^{2} **32.** 599.09 m^2 **33.** $7\mathbf{i} + 8\mathbf{j}$

34. $\left\langle \frac{\sqrt{2}}{2}, \frac{\sqrt{2}}{2} \right\rangle$ **35.** -5 **36.** $-\frac{1}{13}\langle 1, 5\rangle; \frac{21}{13}\langle 5, -1\rangle$

37. $2\sqrt{2}\left(\cos \frac{3\pi}{4} + i \sin \frac{3\pi}{4}\right)$ **38.** $-12\sqrt{3} + 12i$

39. $\cos 0 + i \sin 0 = 1$
$\cos \frac{2\pi}{3} + i \sin \frac{2\pi}{3} = -\frac{1}{2} + \frac{\sqrt{3}}{2}i$
$\cos \frac{4\pi}{3} + i \sin \frac{4\pi}{3} = -\frac{1}{2} - \frac{\sqrt{3}}{2}i$

40. $3\left(\cos \frac{\pi}{5} + i \sin \frac{\pi}{5}\right)$
$3\left(\cos \frac{3\pi}{5} + i \sin \frac{3\pi}{5}\right)$
$3(\cos \pi + i \sin \pi)$
$3\left(\cos \frac{7\pi}{5} + i \sin \frac{7\pi}{5}\right)$
$3\left(\cos \frac{9\pi}{5} + i \sin \frac{9\pi}{5}\right)$

41. About 395.8 rad/min; about 8312.7 in./min

42. 42π yd$^2 \approx 131.95$ yd^2 **43.** 5 ft **44.** 22.6°

45. $d = 4 \cos \frac{\pi}{4}t$ **46.** 32.6°; 543.9 km/h **47.** 425 ft-lb

Problem Solving *(page 463)*

1. 2.01 ft

3. (a)

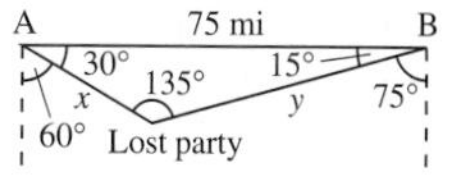

(b) Station A: 27.45 mi; Station B: 53.03 mi
(c) 11.03 mi; S 21.7° E

5. (a) (i) $\sqrt{2}$ (ii) $\sqrt{5}$ (iii) 1
(iv) 1 (v) 1 (vi) 1
(b) (i) 1 (ii) $3\sqrt{2}$ (iii) $\sqrt{13}$
(iv) 1 (v) 1 (vi) 1
(c) (i) $\frac{\sqrt{5}}{2}$ (ii) $\sqrt{13}$ (iii) $\frac{\sqrt{85}}{2}$
(iv) 1 (v) 1 (vi) 1
(d) (i) $2\sqrt{5}$ (ii) $5\sqrt{2}$ (iii) $5\sqrt{2}$
(iv) 1 (v) 1 (vi) 1

7. $\mathbf{u} \cdot \mathbf{v} = 0$ and $\mathbf{u} \cdot \mathbf{w} = 0$
$\mathbf{u} \cdot (c\mathbf{v} + d\mathbf{w}) = \mathbf{u} \cdot c\mathbf{v} + \mathbf{u} \cdot d\mathbf{w}$
$= c(\mathbf{u} \cdot \mathbf{v}) + d(\mathbf{u} \cdot \mathbf{w})$
$= 0$

9. (a) $\mathbf{u} = \langle 0, -120\rangle, \mathbf{v} = \langle 40, 0\rangle$

(b)

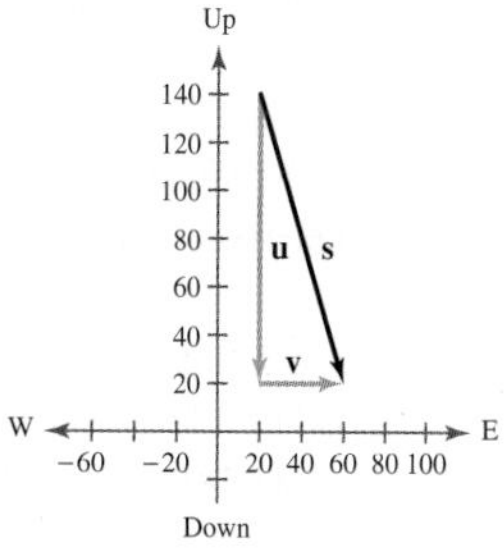

(c) 126.5 miles per hour; The magnitude gives the actual rate of the skydiver's fall.
(d) 71.57°

(e)

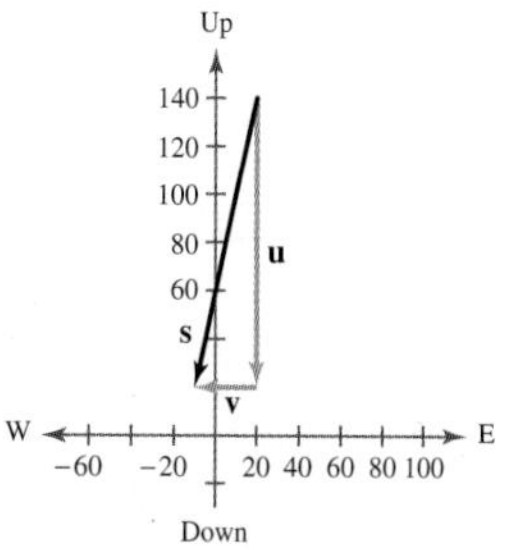

123.7 mi/h

Chapter 7

Section 7.1 (page 473)

1. solution **3.** point; intersection
5. (a) No (b) No (c) No (d) Yes **7.** (2, 2)
9. (2, 6), (−1, 3) **11.** (−3, −4), (5, 0)
13. (0, 0), (2, −4) **15.** (6, 4) **17.** $\left(\frac{1}{2}, 3\right)$ **19.** (1, 1)
21. $\left(\frac{20}{3}, \frac{40}{3}\right)$ **23.** No solution **25.** \$5000; \$7000
27. \$6000; \$6000 **29.** (−2, 4), (0, 0) **31.** No solution
33. (6, 2) **35.** $\left(-\frac{3}{2}, \frac{1}{2}\right)$ **37.** (2, 2), (4, 0)
39. (1, 4), (4, 7) **41.** No solution **43.** (4, 3), (−4, 3)

45.

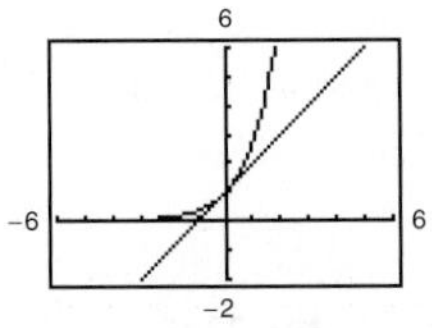

(0, 1)

47.

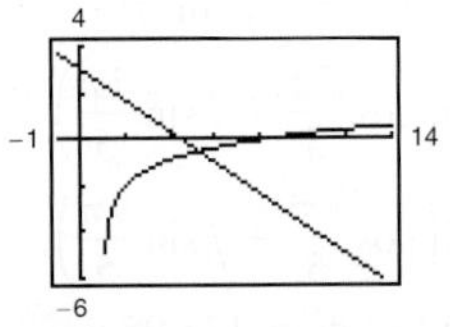

(5.31, −0.54)

49. (1, 2) **51.** No solution **53.** (0.287, 1.751)
55. $\left(\frac{1}{2}, 2\right), \left(-4, -\frac{1}{4}\right)$ **57.** 192 units
59. (a) 344 units (b) 2495 units
61. (a) 8 weeks
(b)

	1	2	3	4
360 − 24x	336	312	288	264
24 + 18x	42	60	78	96

	5	6	7	8
360 − 24x	240	216	192	168
24 + 18x	114	132	150	168

63. (a)

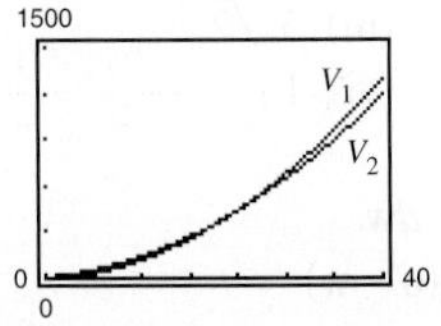

(b) 24.7 in.
(c) Doyle Log Rule; For large logs, the Doyle Log Rule gives a greater volume for a given diameter.

65. 12 m × 16 m **67.** 10 km × 12 km
69. False. To solve a system of equations by substitution, you can solve for either variable in one of the two equations and then back-substitute.
71. For a linear system, the result will be a contradictory equation such as 0 = N, where N is a nonzero real number. For a nonlinear system, there may be an equation with imaginary solutions.
73. (a)–(c) Answers will vary.

Section 7.2 (page 484)

1. elimination **3.** consistent; inconsistent

5. (2, 1)

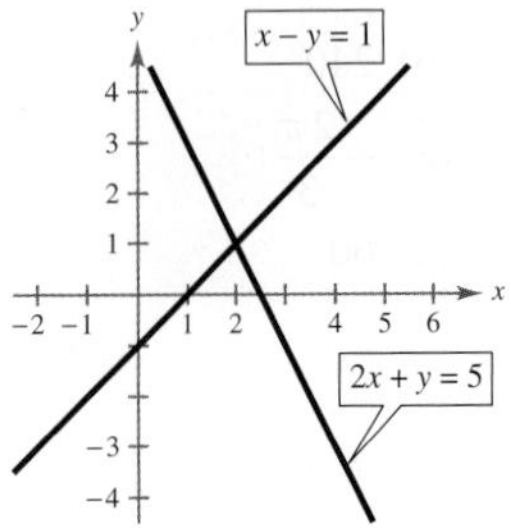

7. (1, −1)

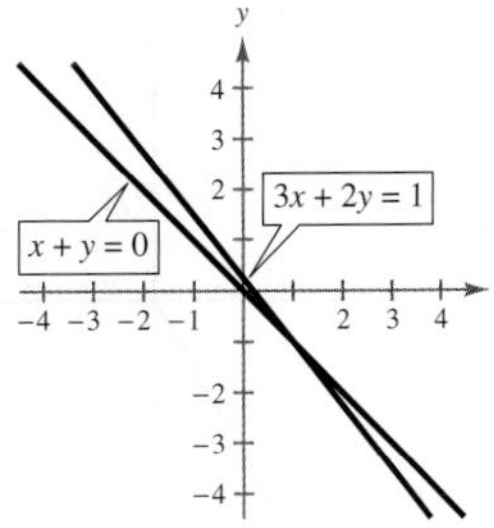

9. No solution

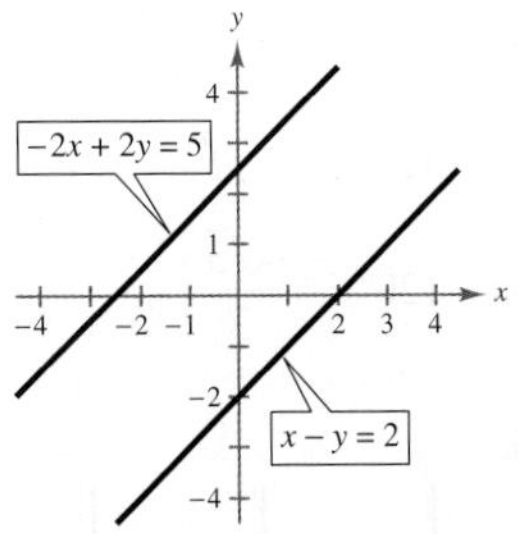

11. $\left(a, \frac{3}{2}a - \frac{5}{2}\right)$

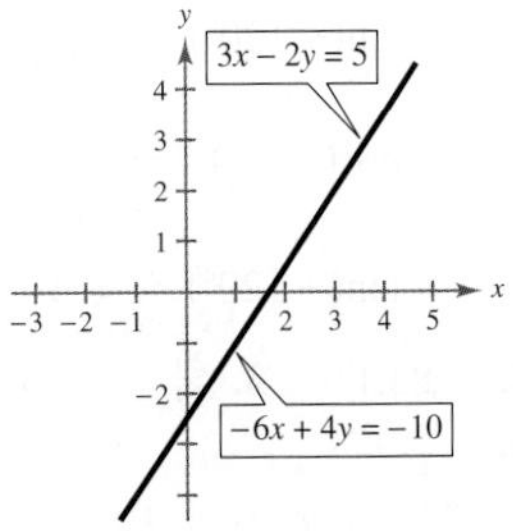

13. (4, 1) **15.** $\left(\frac{3}{2}, -\frac{1}{2}\right)$ **17.** (4, −1) **19.** $\left(\frac{12}{7}, \frac{18}{7}\right)$
21. No solution **23.** Infinitely many solutions: $\left(a, -\frac{1}{2} + \frac{5}{6}a\right)$
25. (101, 96) **27.** $\left(-\frac{6}{35}, \frac{43}{35}\right)$ **29.** (5, −2)
31. b; one solution; consistent **32.** c; one solution; consistent
33. a; infinitely many solutions; consistent
34. d; no solutions; inconsistent **35.** (4, 1) **37.** (2, −1)
39. (6, −3) **41.** 550 mi/h, 50 mi/h
43. Cheeseburger: 300 calories; French fries: 230 calories
45. (240, 404) **47.** (2,000,000, 100)
49. (a) $\begin{cases} x + y = 30 \\ 0.25x + 0.5y = 12 \end{cases}$
(b)

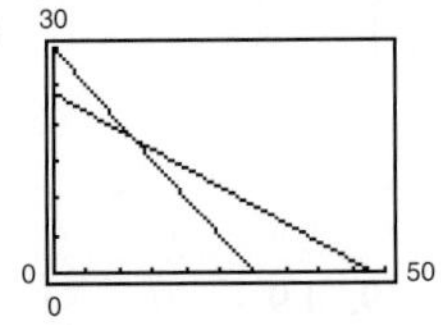

Decreases
(c) 25% solution: 12 L; 50% solution: 18 L

51. \$18,000
53. (a)

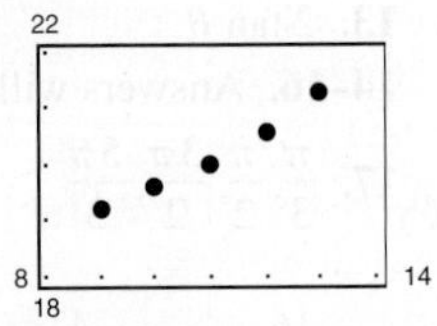

Pharmacy A: $P = 0.52t + 14.4$

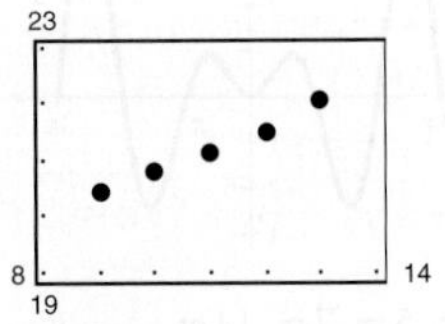

Pharmacy B: $P = 0.39t + 16.9$

(b) Yes. 2019

55. $y = 0.97x + 2.1$
57. (a) $y = 14x + 19$ (b) 41.4 bushels/acre
59. False. Two lines that coincide have infinitely many points of intersection.
61. $k = -4$
63. No. Two lines will intersect only once or will coincide, and if they coincide the system will have infinitely many solutions.
65. The method of elimination is much easier.

67. (39,600, 398). It is necessary to change the scale on the axes to see the point of intersection.

69. $u = 1, v = -\tan x$

Section 7.3 *(page 496)*

1. row-echelon **3.** Gaussian **5.** nonsquare

7. (a) No (b) No (c) No (d) Yes

9. (a) No (b) No (c) Yes (d) No

11. $(-13, -10, 8)$ **13.** $(3, 10, 2)$ **15.** $\left(\frac{11}{4}, 7, 11\right)$

17. $\begin{cases} x - 2y + 3z = 5 \\ y - 2z = 9 \\ 2x - 3z = 0 \end{cases}$

First step in putting the system in row-echelon form.

19. $(-2, 2)$ **21.** $(4, 3)$ **23.** $(1, -2)$ **25.** $(4, 1, 2)$

27. $(-4, 8, 5)$ **29.** No solution **31.** $\left(-\frac{1}{2}, 1, \frac{3}{2}\right)$

33. $(0, 0, 0)$ **35.** No solution **37.** $(-a + 3, a + 1, a)$

39. $(2a, 21a - 1, 8a)$ **41.** $(-3a + 10, 5a - 7, a)$

43. $\left(-\frac{3}{2}a + \frac{1}{2}, -\frac{2}{3}a + 1, a\right)$ **45.** $(1, 1, 1, 1)$

47. $s = -16t^2 + 144$

49. $y = \frac{1}{2}x^2 - 2x$

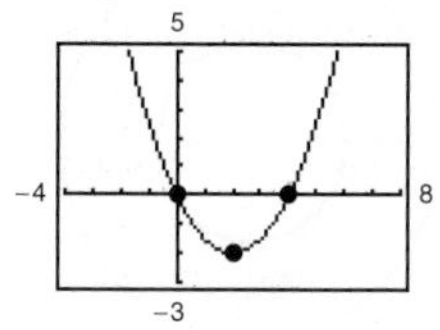

51. $y = x^2 - 6x + 8$

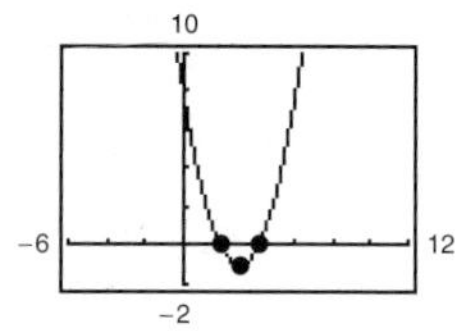

53. $y = 4x^2 - 2x + 1$

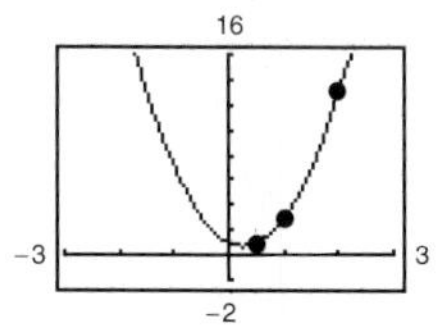

55. $x^2 + y^2 - 10x = 0$

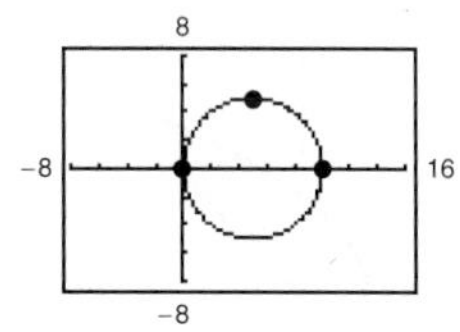

57. $x^2 + y^2 + 6x - 8y = 0$

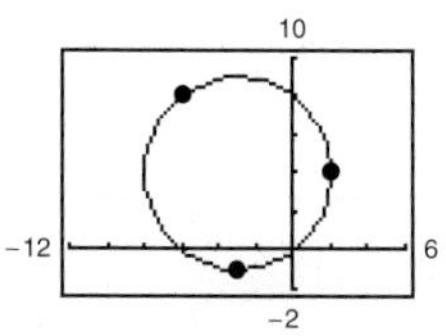

59. 6 touchdowns
6 extra-point kicks
1 field goal

61. \$300,000 at 8%
\$400,000 at 9%
\$75,000 at 10%

63. $x = 60°, y = 67°, z = 53°$ **65.** 75 ft, 63 ft, 42 ft

67. $I_1 = 1, I_2 = 2, I_3 = 1$ **69.** $y = x^2 - x$

71. (a) $y = 0.165x^2 - 6.55x + 103$

(b)

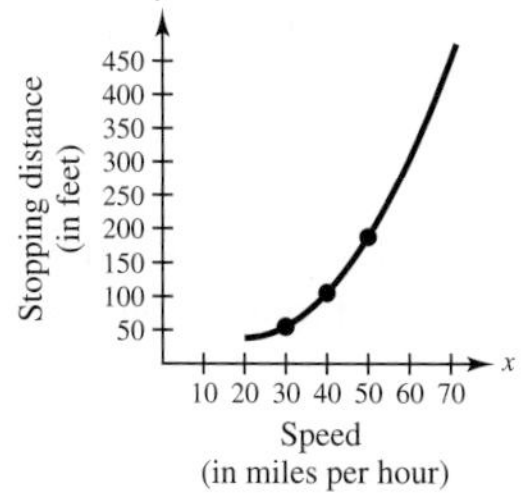

(c) 453 ft

73. $x = \pm\frac{\sqrt{2}}{2}, y = \frac{1}{2}, \lambda = 1$ or $x = 0, y = 0, \lambda = 0$

75. False. Equation 2 does not have a leading coefficient of 1.

77. No. Answers will vary.

79. *Sample answers:*

$\begin{cases} 2x + y - z = 0 \\ y + 2z = 0 \\ -x + 2y + z = -9 \end{cases}$ $\begin{cases} x + y + z = 1 \\ 2x - z = 4 \\ 4y + 8z = 0 \end{cases}$

81. *Sample answers:*

$\begin{cases} x + 2y + 4z = -14 \\ x - 12y = 0 \\ x - 8z = 8 \end{cases}$ $\begin{cases} 4x - 2y - 8z = -9 \\ -x + 4z = -1 \\ -7y + 2z = 0 \end{cases}$

Section 7.4 *(page 506)*

1. partial fraction decomposition **3.** partial fraction

5. b **6.** c **7.** d **8.** a **9.** $\frac{A}{x} + \frac{B}{x - 2}$

11. $\frac{A}{x} + \frac{B}{x^2} + \frac{C}{x - 7}$ **13.** $\frac{A}{x - 5} + \frac{B}{(x - 5)^2} + \frac{C}{(x - 5)^3}$

15. $\frac{A}{x} + \frac{Bx + C}{x^2 + 1} + \frac{Dx + E}{(x^2 + 1)^2}$ **17.** $\frac{1}{x} - \frac{1}{x + 1}$

19. $\frac{1}{x} - \frac{2}{2x + 1}$ **21.** $\frac{1}{x - 1} - \frac{1}{x + 2}$

23. $\frac{1}{2}\left(\frac{1}{x - 1} - \frac{1}{x + 1}\right)$ **25.** $-\frac{3}{x} - \frac{1}{x + 2} + \frac{5}{x - 2}$

27. $\frac{3}{x - 3} + \frac{9}{(x - 3)^2}$ **29.** $\frac{3}{x} - \frac{1}{x^2} + \frac{1}{x + 1}$

31. $\frac{3}{x} - \frac{2x - 2}{x^2 + 1}$ **33.** $-\frac{1}{x - 1} + \frac{x + 2}{x^2 - 2}$

35. $\frac{2}{x^2 + 4} + \frac{x}{(x^2 + 4)^2}$

37. $\frac{1}{8}\left(\frac{1}{2x + 1} + \frac{1}{2x - 1} - \frac{4x}{4x^2 + 1}\right)$

39. $\frac{1}{x + 1} + \frac{2}{x^2 - 2x + 3}$

41. $\frac{2}{x} - \frac{3}{x^2} - \frac{2x - 3}{x^2 + 2} - \frac{4x - 6}{(x^2 + 2)^2}$ **43.** $1 - \frac{2x + 1}{x^2 + x + 1}$

45. $2x - 7 + \frac{17}{x + 2} + \frac{1}{x + 1}$

47. $x + 3 + \frac{6}{x - 1} + \frac{4}{(x - 1)^2} + \frac{1}{(x - 1)^3}$

49. $x + \frac{2}{x} + \frac{1}{x + 1} + \frac{3}{(x + 1)^2}$ **51.** $\frac{3}{2x - 1} - \frac{2}{x + 1}$

53. $\frac{1}{2}\left[-\frac{1}{x} + \frac{5}{x + 1} - \frac{3}{(x + 1)^2}\right]$ **55.** $\frac{1}{x^2 + 2} + \frac{x}{(x^2 + 2)^2}$

57. $2x + \frac{1}{2}\left(\frac{3}{x - 4} - \frac{1}{x + 2}\right)$ **59.** $\frac{60}{100 - p} - \frac{60}{100 + p}$

61. False. The partial fraction decomposition is

$\frac{A}{x + 10} + \frac{B}{x - 10} + \frac{C}{(x - 10)^2}.$

63. True. The expression is an improper rational expression.

65. $\frac{1}{2a}\left(\frac{1}{a + x} + \frac{1}{a - x}\right)$ **67.** $\frac{1}{a}\left(\frac{1}{y} + \frac{1}{a - y}\right)$

69. Answers will vary. *Sample answer:* You can substitute any convenient values of x that will help determine the constants. You can also find the basic equation, expand it, then equate coefficients of like terms.

Section 7.5 *(page 515)*

1. solution **3.** solution

5.

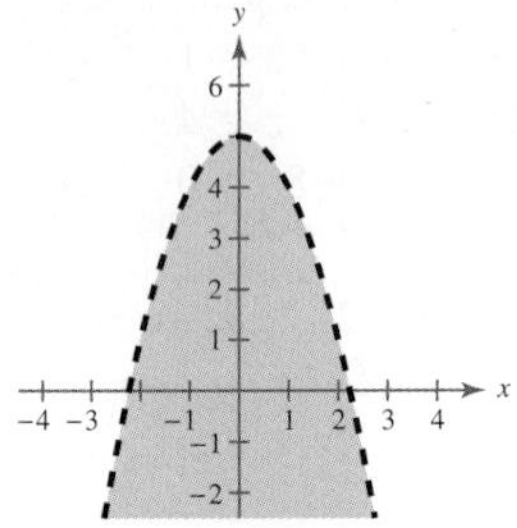

7.

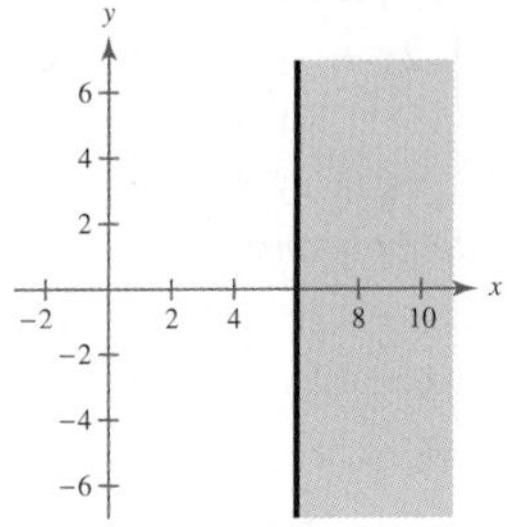

9.

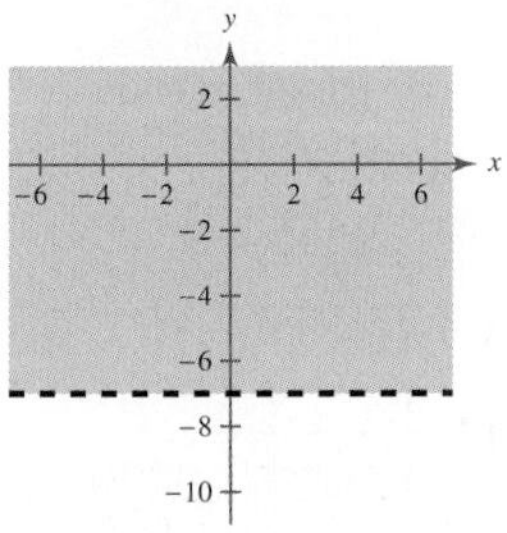

11.

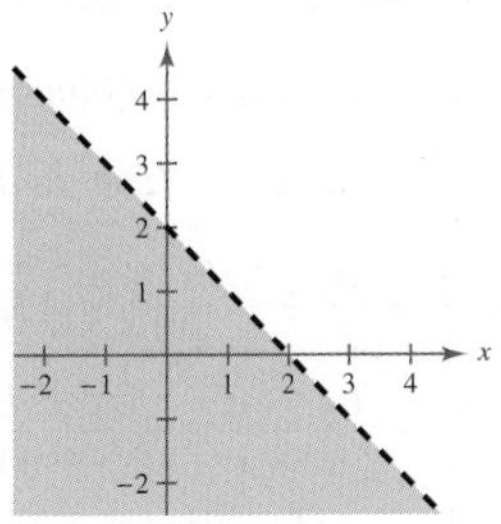

13.

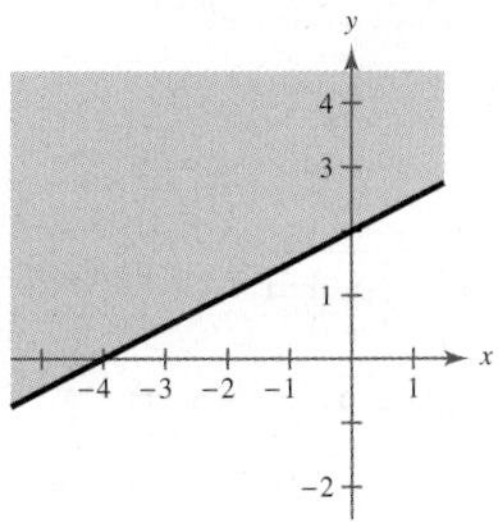

15.

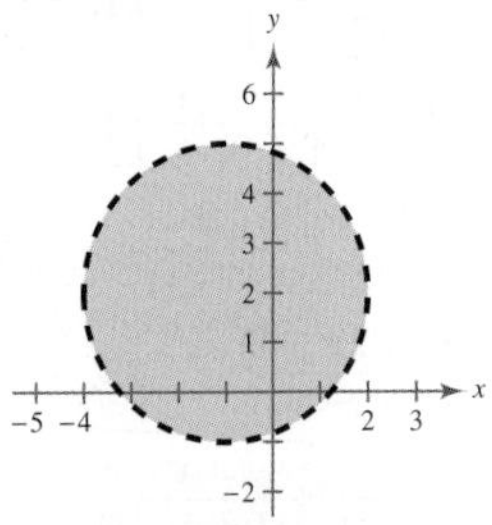

17.

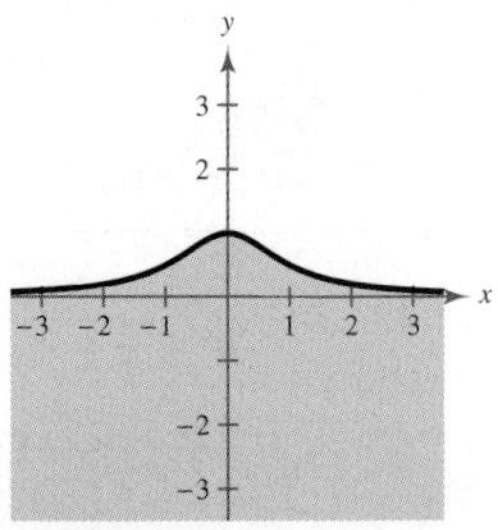

19.

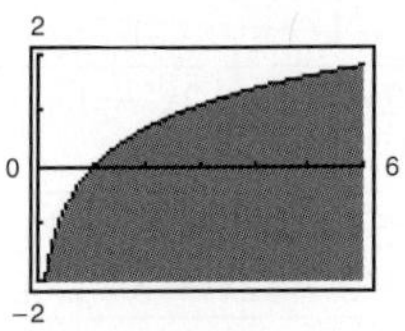

21.

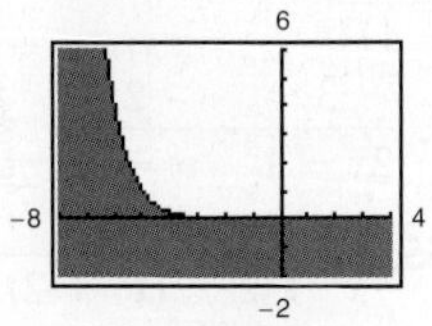

23.

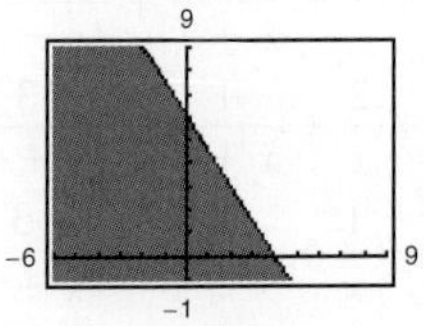

25.

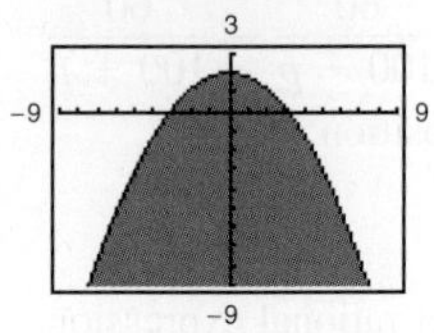

27.

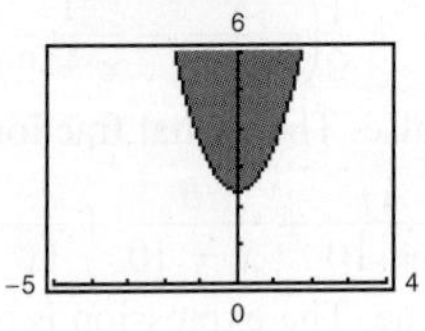

29. $y < 5x + 5$ **31.** $y \geq -\frac{2}{3}x + 2$

33.

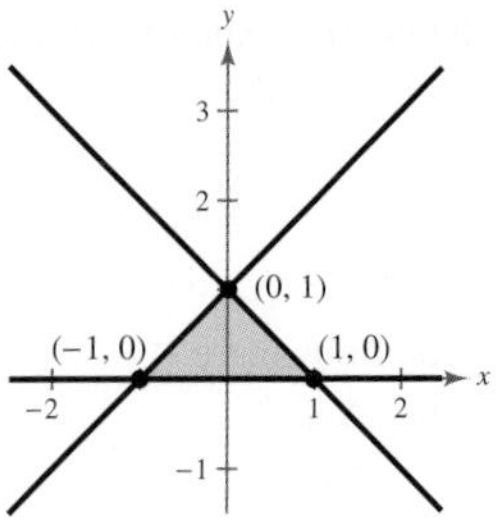

35.

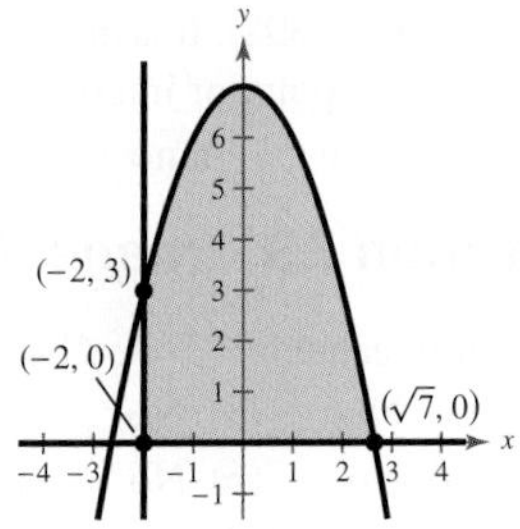

37.

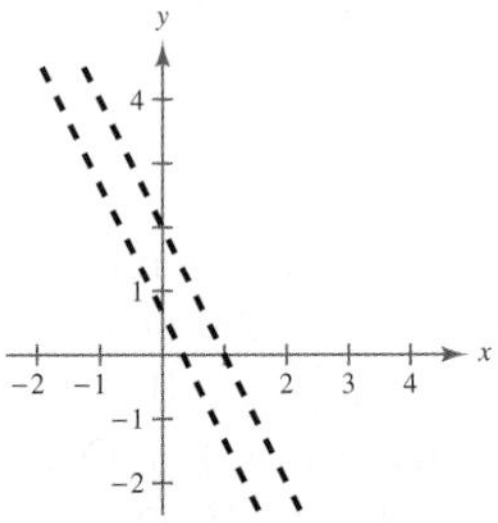

No solution

39.

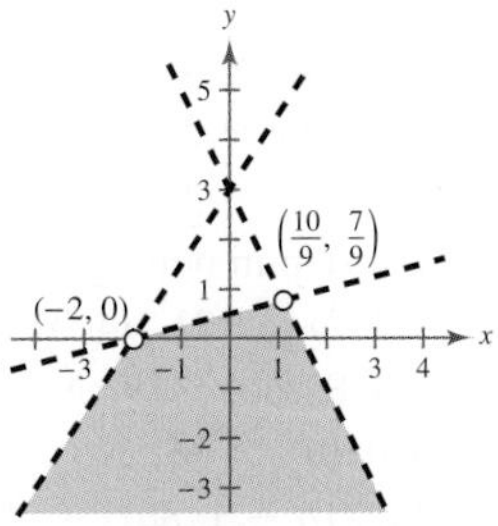

41.

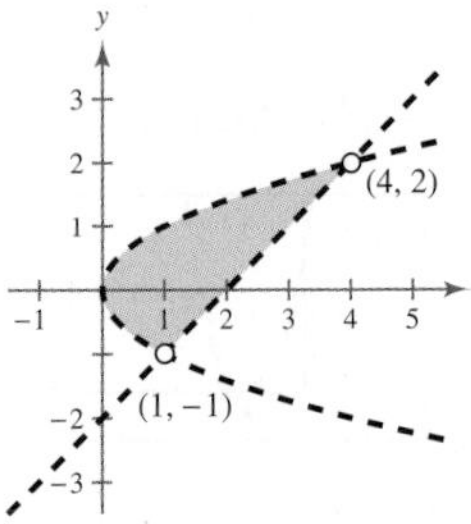

43.

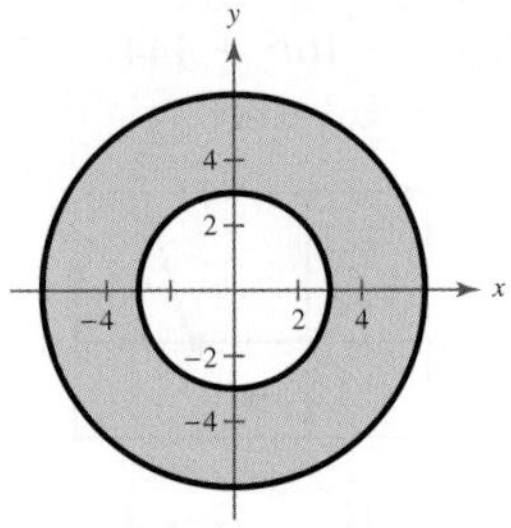

45.

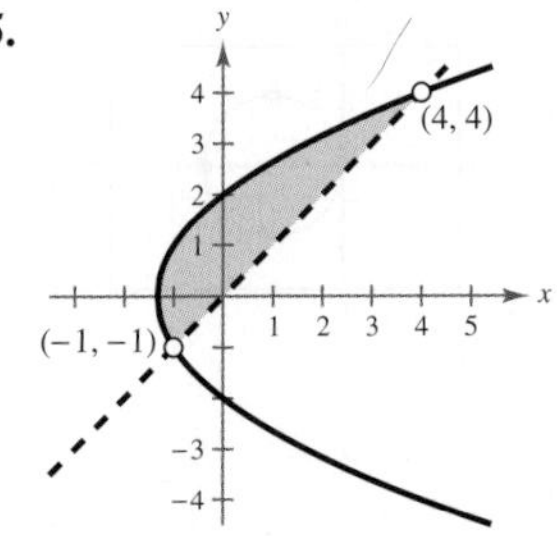

47.

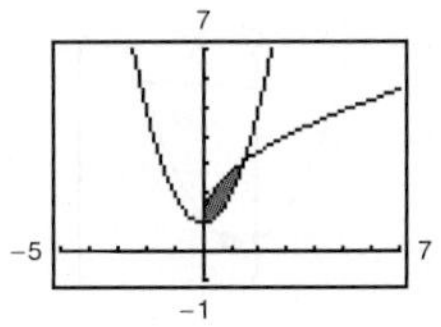

49.

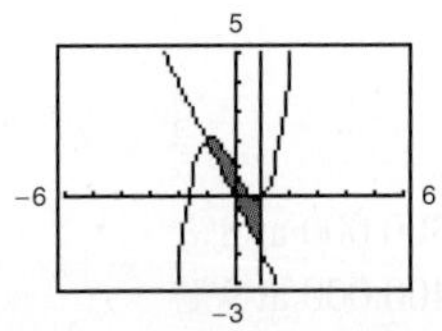

51.

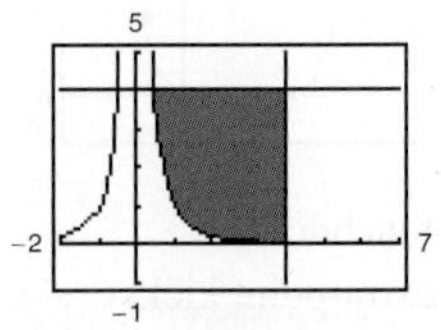

53. $\begin{cases} x \geq 0 \\ y \geq 0 \\ y \leq 6 - x \end{cases}$ **55.** $\begin{cases} x \geq 0 \\ y \geq 0 \\ x^2 + y^2 < 64 \end{cases}$ **57.** $\begin{cases} x \geq 4 \\ x \leq 9 \\ y \geq 3 \\ y \leq 9 \end{cases}$

59. $\begin{cases} y \geq 0 \\ y \leq 5x \\ y \leq -x + 6 \end{cases}$

61. (a)

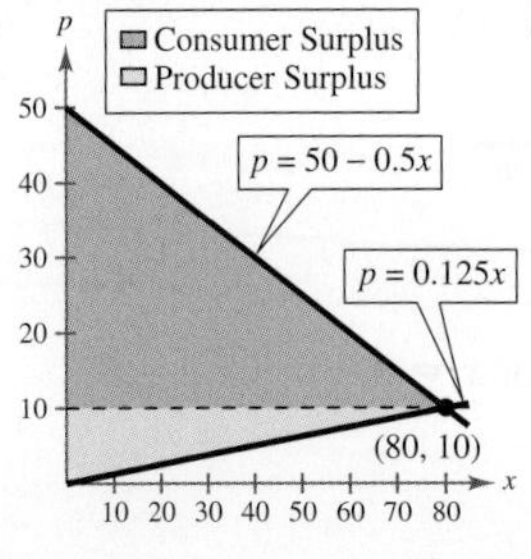

(b) Consumer surplus: \$1600
Producer surplus: \$400

63. (a)

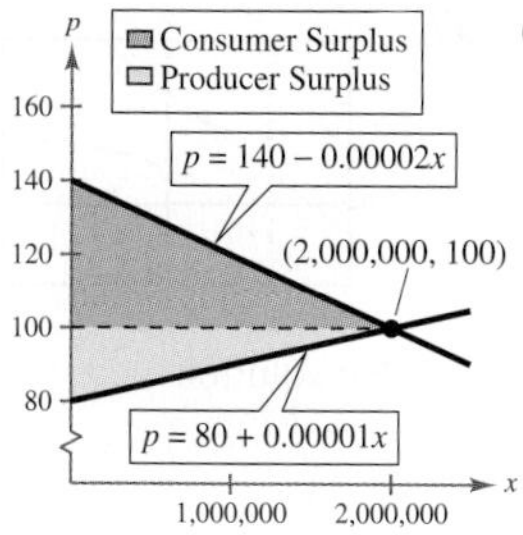

(b) Consumer surplus: \$40,000,000
Producer surplus: \$20,000,000

65. $\begin{cases} x + \frac{3}{2}y \le 12 \\ \frac{4}{3}x + \frac{3}{2}y \le 15 \\ x \ge 0 \\ y \ge 0 \end{cases}$

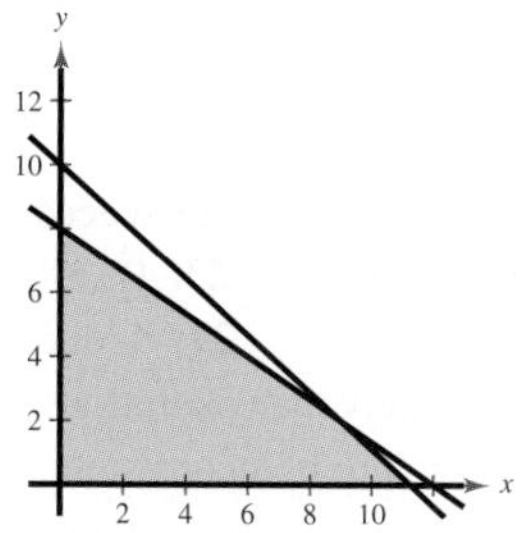

67. $\begin{cases} x + y \le 20{,}000 \\ y \ge 2x \\ x \ge 5{,}000 \\ y \ge 5{,}000 \end{cases}$

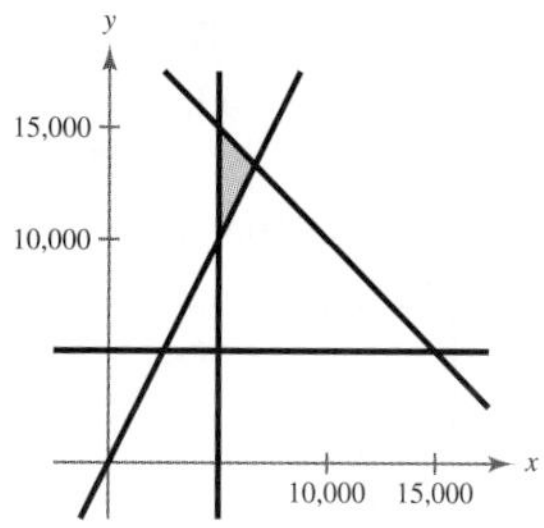

69. $\begin{cases} 6x + 4y \ge 15 \\ 3x + 6y \ge 16 \\ x \ge 0 \\ y \ge 0 \end{cases}$

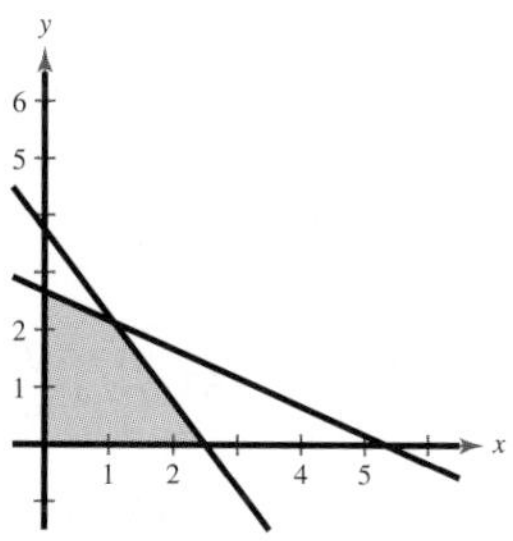

71. (a) $\begin{cases} y \ge 0.5(220 - x) \\ y \le 0.85(220 - x) \\ x \ge 20 \\ x \le 70 \end{cases}$

(b)

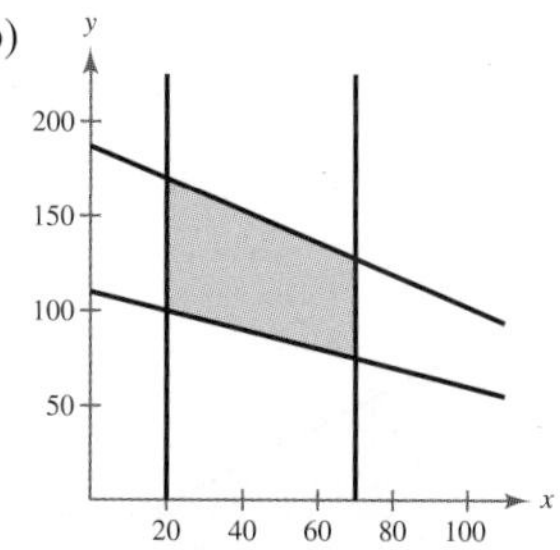

(c) Answers will vary.

73. True. The figure is a rectangle with a length of 9 units and a width of 11 units.

75. Test a point on each side of the line. Because the origin (0, 0) satisfies the inequality, the solution set of the inequality lies below the line.

77. (a) $\begin{cases} \pi y^2 - \pi x^2 \ge 10 \\ y > x \\ x > 0 \end{cases}$

(b)

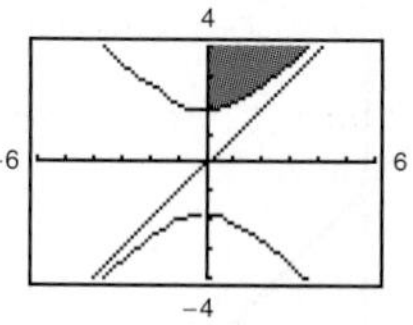

(c) The line is an asymptote to the boundary. The larger the circles, the closer the radii can be while still satisfying the constraint.

79. The graph is a half-line on the real number line; on the rectangular coordinate system, the graph is a half-plane.

Section 7.6 *(page 524)*

1. optimization **3.** objective **5.** inside; on

7. Minimum at (0, 0): 0
Maximum at (5, 0): 20

9. Minimum at (0, 0): 0
Maximum at (3, 4): 26

11. Minimum at (0, 0): 0
Maximum at (60, 20): 740

13.

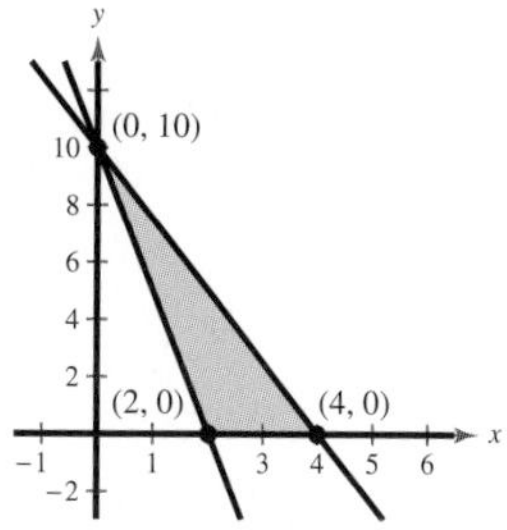

Minimum at (2, 0): 6
Maximum at (0, 10): 20

15.

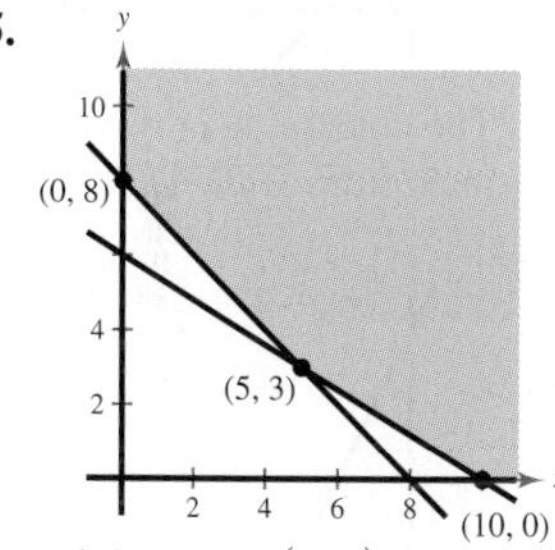

Minimum at (5, 3): 35
No maximum

17.

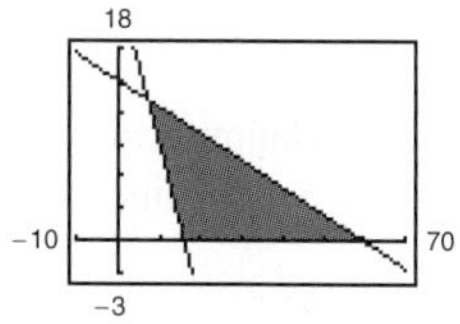

Minimum at (7.2, 13.2): 34.8
Maximum at (60, 0): 180

19.

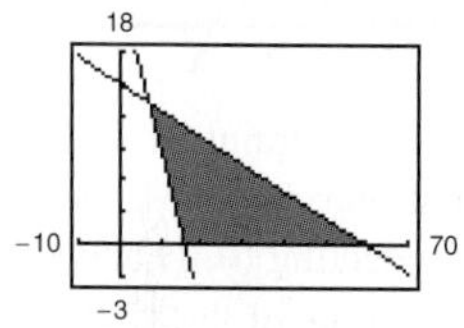

Minimum at (16, 0): 16
Maximum at any point on the line segment connecting (7.2, 13.2) and (60, 0): 60

21. Minimum at (0, 0): 0
Maximum at (3, 6): 12

23. Minimum at (0, 0): 0
Maximum at (0, 10): 10

25. Minimum at (0, 0): 0
Maximum at (0, 5): 25

27. Minimum at (0, 0): 0
Maximum at $\left(\frac{22}{3}, \frac{19}{6}\right)$: $\frac{271}{6}$

29.

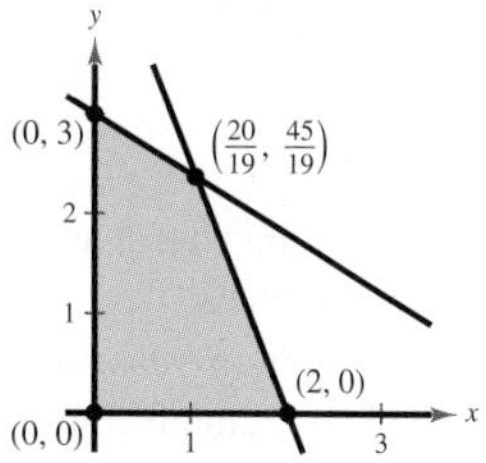

The maximum, 5, occurs at any point on the line segment connecting (2, 0) and $\left(\frac{20}{19}, \frac{45}{19}\right)$. Minimum at (0, 0): 0

31.

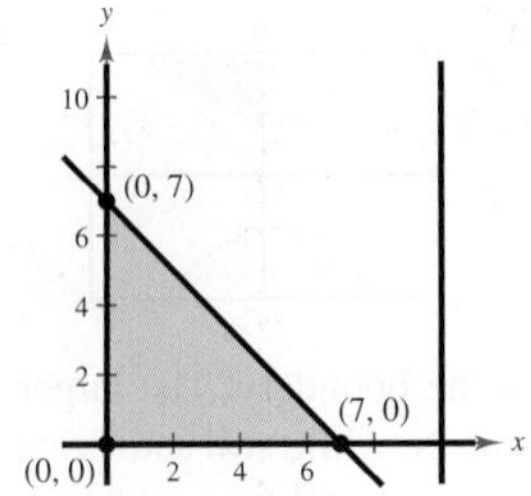

The constraint $x \leq 10$ is extraneous. Minimum at $(7, 0)$: -7; maximum at $(0, 7)$: 14

33.

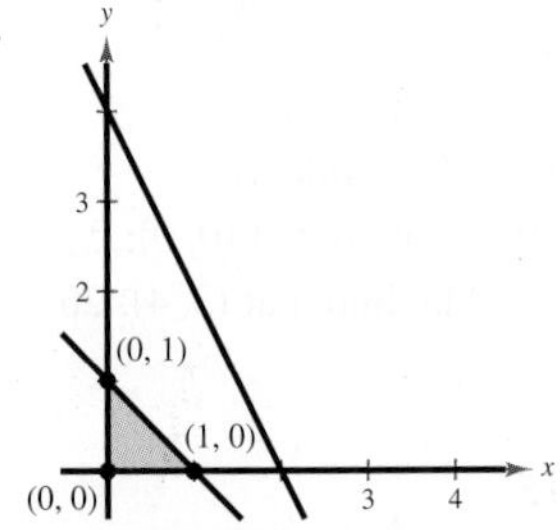

The constraint $2x + y \leq 4$ is extraneous. Minimum at $(0, 0)$: 0; maximum at $(0, 1)$: 4

35.

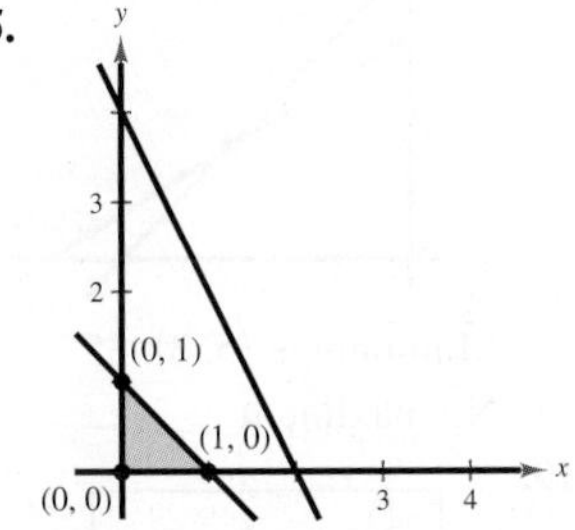

The constraint $2x + y \leq 4$ is extraneous. Minimum at $(0, 0)$: 0; The maximum, 1, occurs at any point on the line segment connecting $(0, 1)$ and $(1, 0)$.

37. 230 units of the \$225 model
45 units of the \$250 model
Optimal profit: \$8295

39. 3 bags of brand X
6 bags of brand Y
Optimal cost: \$195

41. 13 audits
0 tax returns
Optimal revenue: \$20,800

43. 60 acres for crop A
90 acres for crop B
Optimal yield: 63,000 bushels

45. \$0 on TV ads
\$1,000,000 on newspaper ads
Optimal audience: 250 million people

47. True. The objective function has a maximum value at any point on the line segment connecting the two vertices.

49. True. If an objective function has a maximum value at more than one vertex, then any point on the line segment connecting the points will produce the maximum value.

Review Exercises *(page 529)*

1. $(1, 1)$ **3.** $\left(\frac{3}{2}, 5\right)$ **5.** $(0.25, 0.625)$ **7.** $(5, 4)$
9. $(0, 0), (2, 8), (-2, 8)$ **11.** $(4, -2)$
13. $(1.41, -0.66), (-1.41, 10.66)$

15.

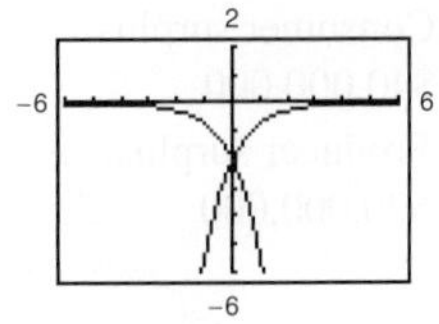

$(0, -2)$

17.

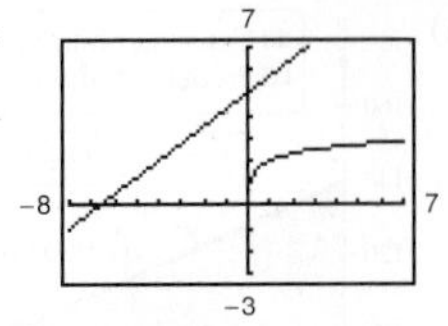

No solution

19.

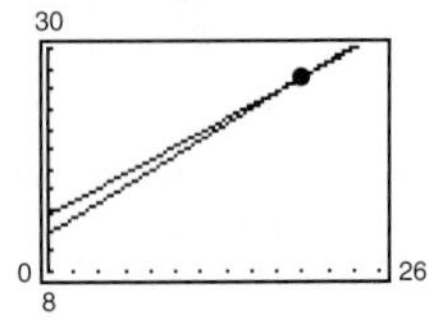

The BMI for males exceeds the BMI for females after age 20.

21. 16 ft × 18 ft **23.** $\left(\frac{5}{2}, 3\right)$ **25.** $(0, 0)$
27. $\left(\frac{8}{5}a + \frac{14}{5}, a\right)$ **29.** d, one solution, consistent
30. c, infinitely many solutions, consistent
31. b, no solution, inconsistent
32. a, one solution, consistent **33.** $(100{,}000, 23)$
35. $(2, -4, -5)$ **37.** $(-6, 7, 10)$ **39.** $\left(\frac{24}{5}, \frac{22}{5}, -\frac{8}{5}\right)$
41. $\left(-\frac{3}{4}, 0, -\frac{5}{4}\right)$ **43.** $(a - 4, a - 3, a)$
45. $y = 2x^2 + x - 5$ **47.** $x^2 + y^2 - 4x + 4y - 1 = 0$
49. (a) $y = -5.950t^2 + 104.45t - 312.9$
(b)

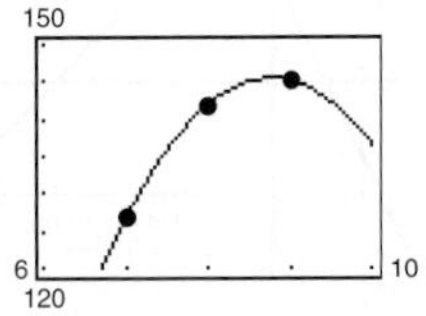

The model is a good fit for the data.

(c) $-\$84.9$ billion; no

51. \$16,000 at 7%
\$13,000 at 9%
\$11,000 at 11%

53. $s = -16t^2 + 150$

55. $\dfrac{A}{x} + \dfrac{B}{x + 20}$ **57.** $\dfrac{A}{x} + \dfrac{B}{x^2} + \dfrac{C}{x - 5}$

59. $\dfrac{3}{x + 2} - \dfrac{4}{x + 4}$ **61.** $1 - \dfrac{25}{8(x + 5)} + \dfrac{9}{8(x - 3)}$

63. $\dfrac{1}{2}\left(\dfrac{3}{x - 1} - \dfrac{x - 3}{x^2 + 1}\right)$ **65.** $\dfrac{3}{x^2 + 1} + \dfrac{4x - 3}{(x^2 + 1)^2}$

67.

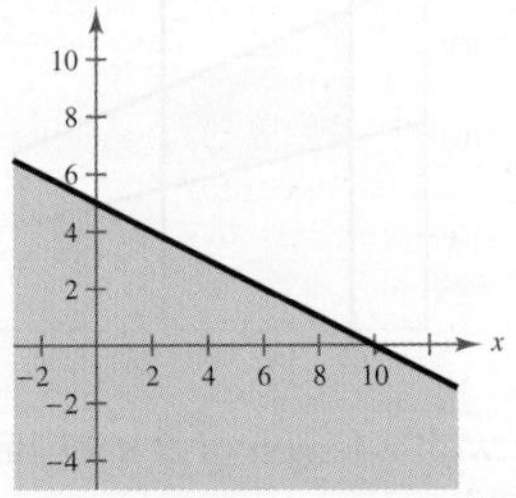

69.

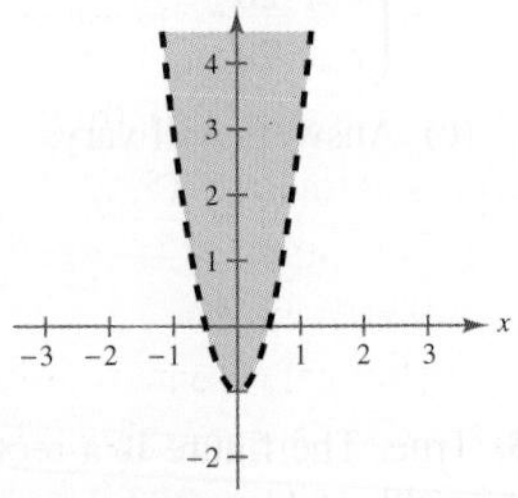

71.

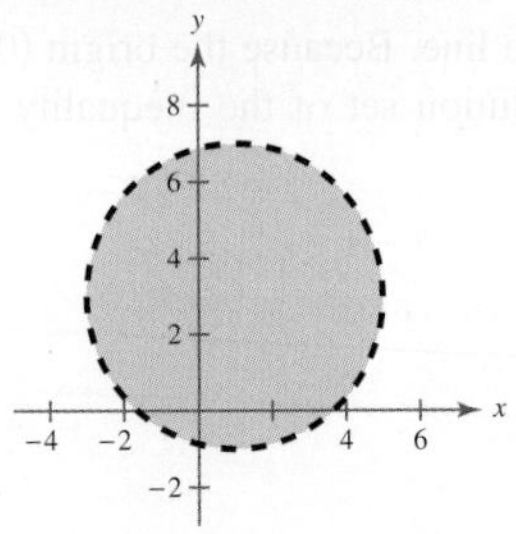

73.

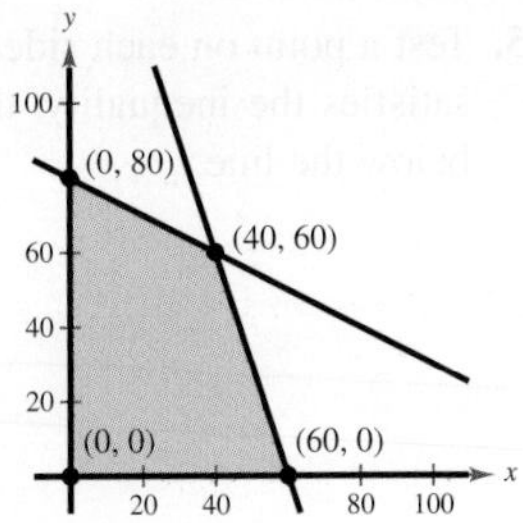

75.

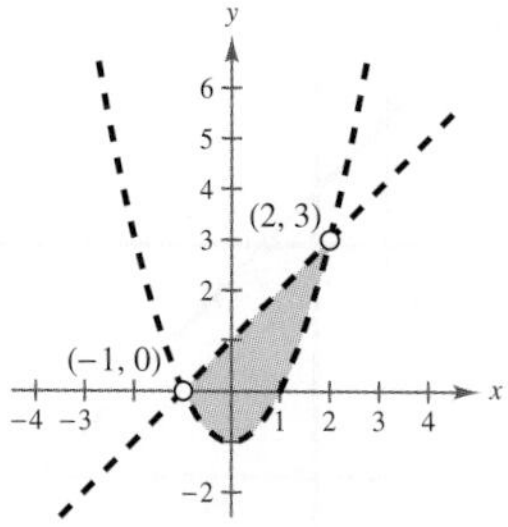

77.

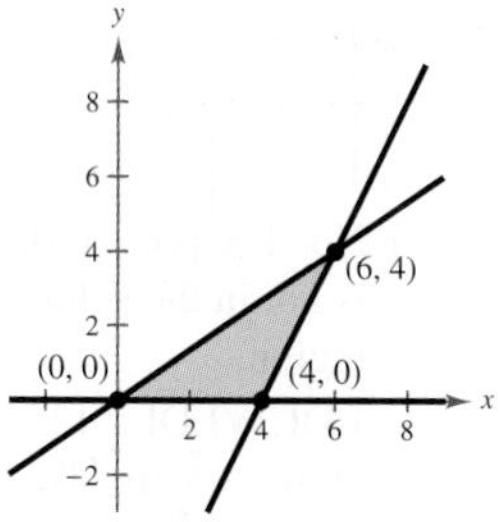

79. (a)

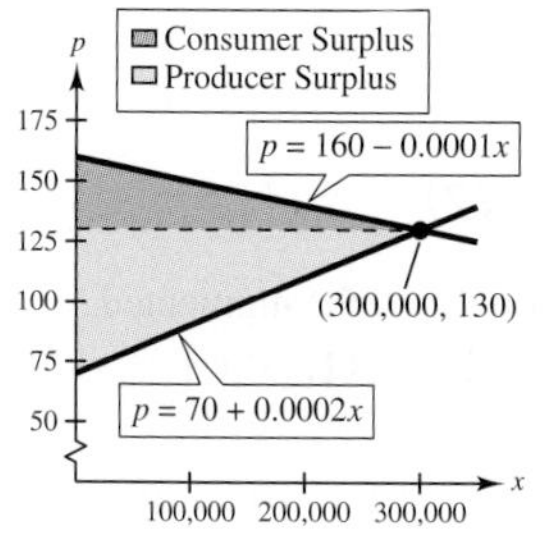

(b) Consumer surplus: \$4,500,000
Producer surplus: \$9,000,000

81. $\begin{cases} x \ge 3 \\ x \le 7 \\ y \ge 1 \\ y \le 10 \end{cases}$

83. $\begin{cases} 20x + 30y \le 24{,}000 \\ 12x + 8y \le 12{,}400 \\ x \ge 0 \\ y \ge 0 \end{cases}$

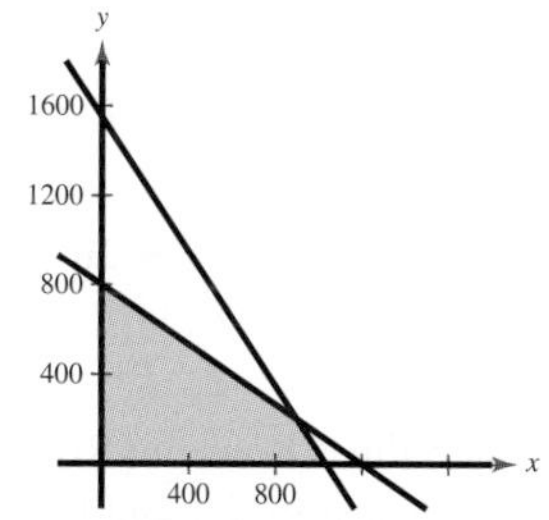

85.

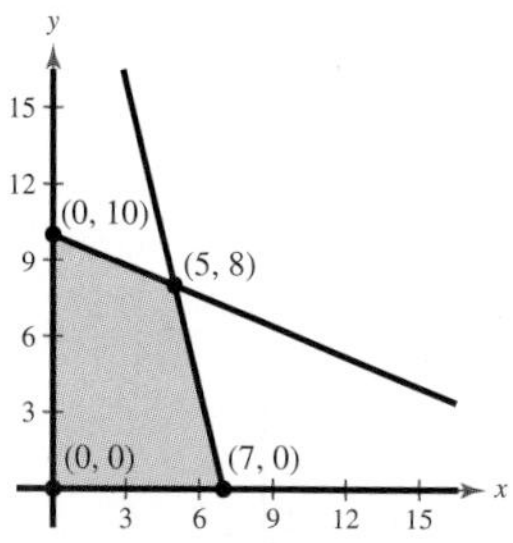

Minimum at $(0, 0)$: 0
Maximum at $(5, 8)$: 47

87.

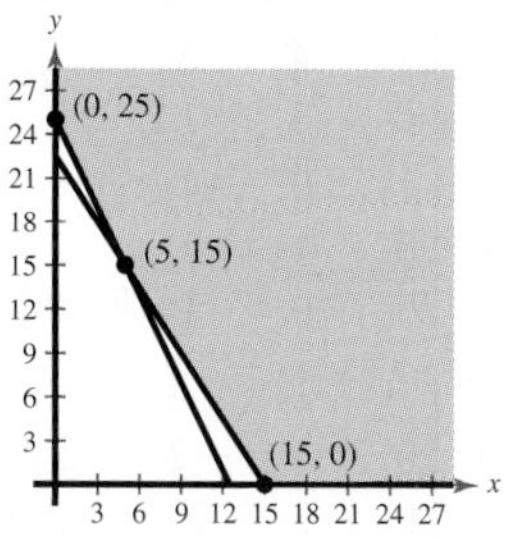

Minimum at $(15, 0)$: 26.25
No maximum

89. 72 haircuts
0 permanents
Optimal revenue: \$1800

91. True. The nonparallel sides of the trapezoid are equal in length.

93. $\begin{cases} 4x + y = -22 \\ \frac{1}{2}x + y = 6 \end{cases}$

95. $\begin{cases} 3x + y = 7 \\ -6x + 3y = 1 \end{cases}$

97. $\begin{cases} x + y + z = 6 \\ x + y - z = 0 \\ x - y - z = 2 \end{cases}$

99. $\begin{cases} 2x + 2y - 3z = 7 \\ x - 2y + z = 4 \\ -x + 4y - z = -1 \end{cases}$

101. An inconsistent system of linear equations has no solution.

Chapter Test *(page 533)*

1. $(-4, -5)$ **2.** $(0, -1), (1, 0), (2, 1)$ **3.** $(8, 4), (2, -2)$

4.

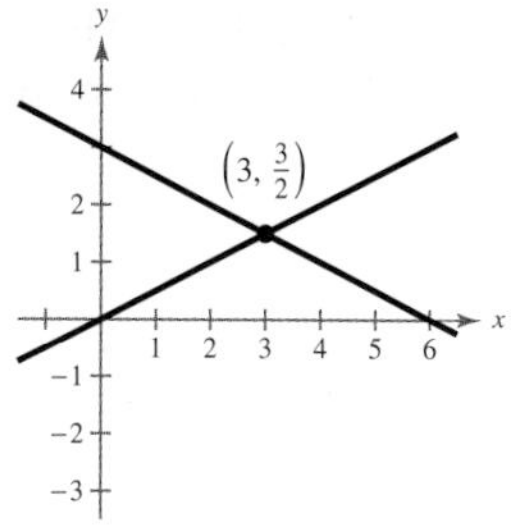

$\left(3, \frac{3}{2}\right)$

5.

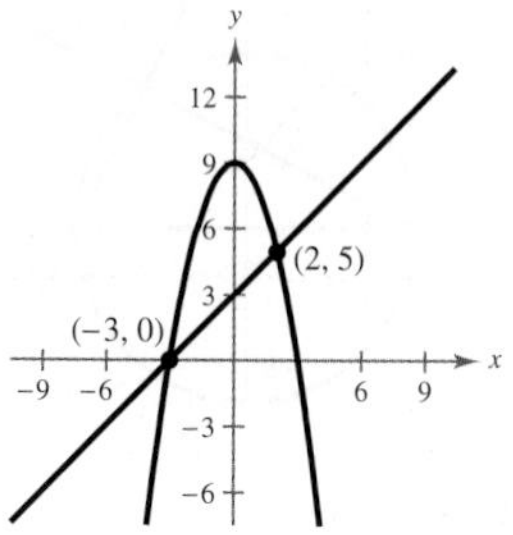

$(-3, 0), (2, 5)$

6.

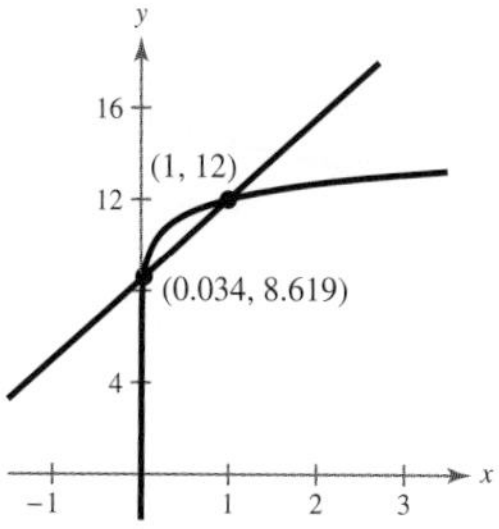

$(1, 12), (0.034, 8.619)$

7. $(-2, -5)$ **8.** $(10, -3)$ **9.** $(2, -3, 1)$

10. No solution

11. $-\dfrac{1}{x+1} + \dfrac{3}{x-2}$ **12.** $\dfrac{2}{x^2} + \dfrac{3}{2-x}$

13. $-\dfrac{5}{x} + \dfrac{3}{x+1} + \dfrac{3}{x-1}$ **14.** $-\dfrac{2}{x} + \dfrac{3x}{x^2+2}$

15.

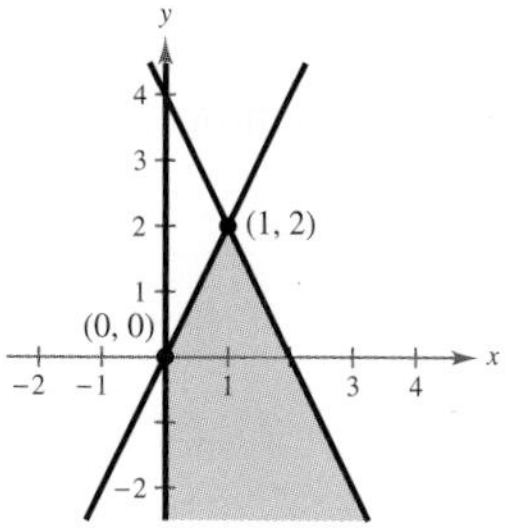

16.

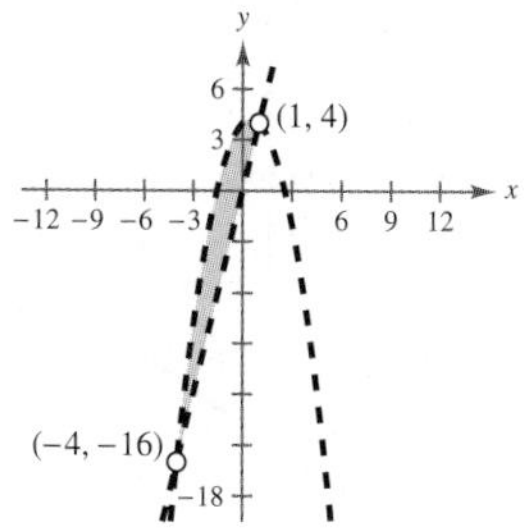

17.

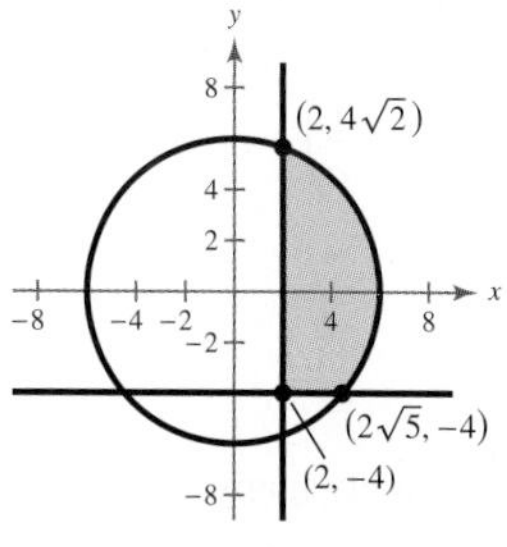

18. Minimum at $(0, 0)$: 0
Maximum at $(12, 0)$: 240

19. \$24,000 in 4% fund
\$26,000 in 5.5% fund

20. $y = -\frac{1}{2}x^2 + x + 6$

21. 0 units of model I
5300 units of model II
Optimal profit: \$212,000

Problem Solving *(page 535)*

1. 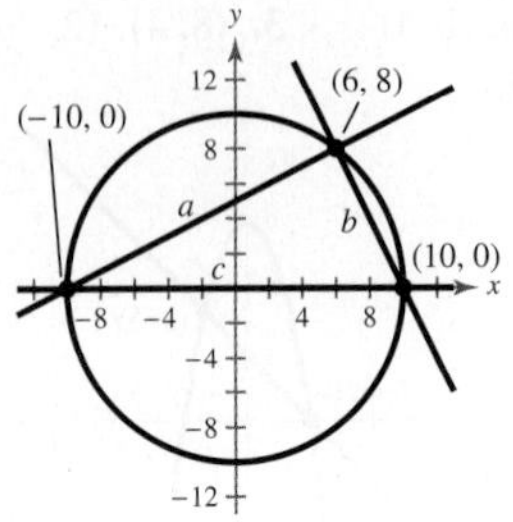

$a = 8\sqrt{5}, b = 4\sqrt{5}, c = 20$
$(8\sqrt{5})^2 + (4\sqrt{5})^2 = 20^2$
Therefore, the triangle is a right triangle.

3. $ad \neq bc$

5. (a) 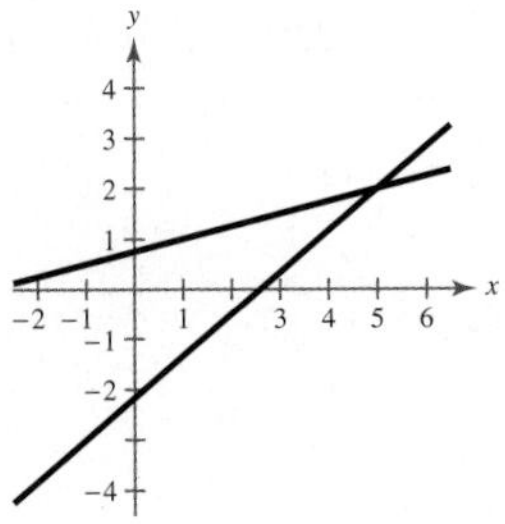 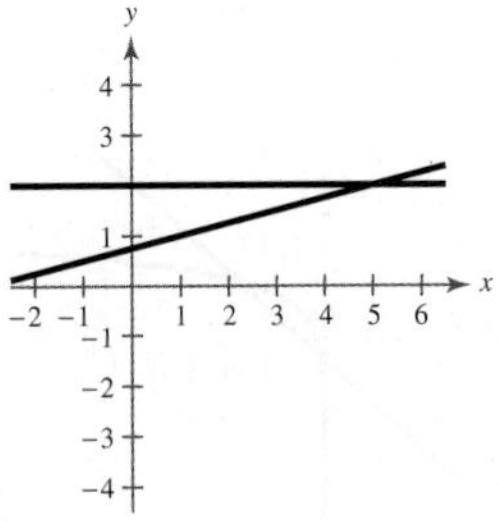

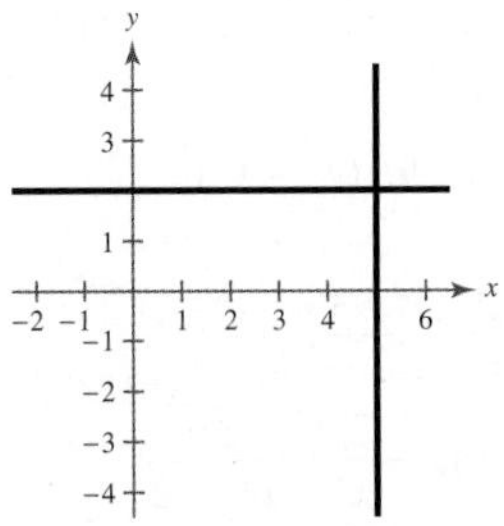

(5, 2)
Answers will vary.

(b) 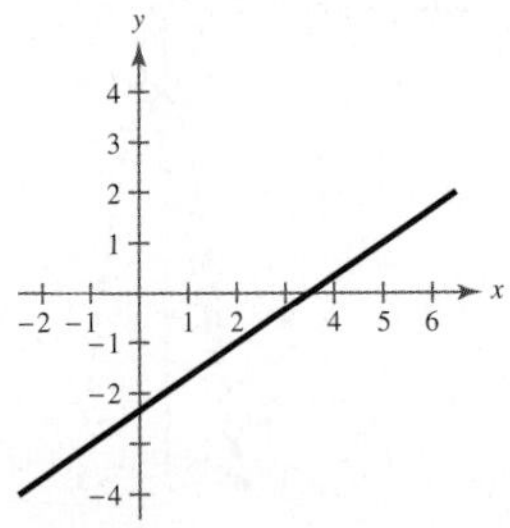

$\left(\frac{3}{2}a + \frac{7}{2}, a\right)$
Answers will vary.

7. 10.1 ft; About 252.7 ft **9.** \$12.00

11. (a) $(3, -4)$ (b) $\left(\dfrac{2}{-a+5}, \dfrac{1}{4a-1}, \dfrac{1}{a}\right)$

13. (a) $\left(\dfrac{-5a+16}{6}, \dfrac{5a-16}{6}, a\right)$

(b) $\left(\dfrac{-11a+36}{14}, \dfrac{13a-40}{14}, a\right)$

(c) $(-a+3, a-3, a)$ (d) Infinitely many

15. $\begin{cases} a + t \leq 32 \\ 0.15a \geq 1.9 \\ 193a + 772t \geq 11{,}000 \end{cases}$

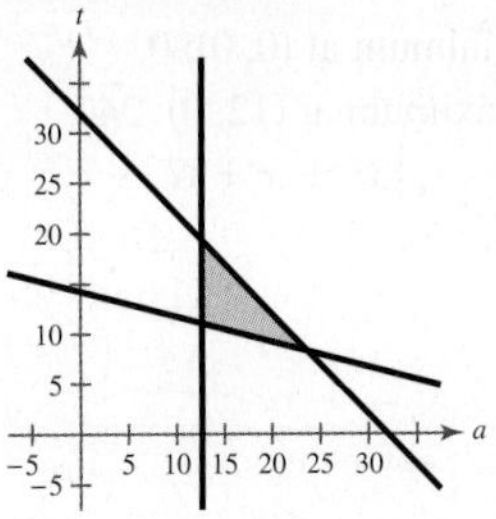

17. (a) $\begin{cases} 0 < y \leq 130 \\ x \geq 40 \\ x + y \leq 200 \end{cases}$

(b) 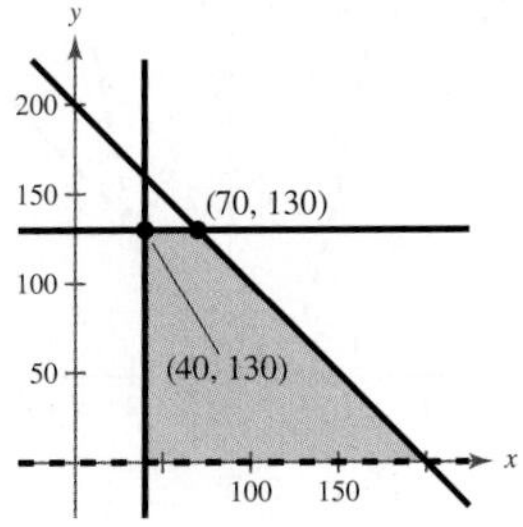

(c) No. The point (90, 120) is not in the solution region.

(d) LDL/VLDL: 135 mg/dL
HDL: 65 mg/dL

(e) $(75, 90); \frac{165}{75} = 2.2 < 4$

Chapter 8

Section 8.1 *(page 547)*

1. matrix **3.** main diagonal **5.** augmented

7. row-equivalent **9.** 1×2 **11.** 3×1

13. 2×2 **15.** 3×3

17. $\left[\begin{array}{rr:r} 4 & -3 & -5 \\ -1 & 3 & 12 \end{array}\right]$ **19.** $\left[\begin{array}{rrr:r} 1 & 10 & -2 & 2 \\ 5 & -3 & 4 & 0 \\ 2 & 1 & 0 & 6 \end{array}\right]$

21. $\left[\begin{array}{rrr:r} 7 & -5 & 1 & 13 \\ 19 & 0 & -8 & 10 \end{array}\right]$ **23.** $\begin{cases} x + 2y = 7 \\ 2x - 3y = 4 \end{cases}$

25. $\begin{cases} 2x + 5z = -12 \\ y - 2z = 7 \\ 6x + 3y = 2 \end{cases}$

27. $\begin{cases} 9x + 12y + 3z = 0 \\ -2x + 18y + 5z + 2w = 10 \\ x + 7y - 8z = -4 \\ 3x + 2z = -10 \end{cases}$

29. Add 5 times Row 2 to Row 1.

31. Interchange Row 1 and Row 2.
Add 4 times new Row 1 to Row 3.

33. $\begin{bmatrix} 1 & 4 & 3 \\ 0 & 2 & -1 \end{bmatrix}$ **35.** $\begin{bmatrix} 1 & 1 & 1 \\ 0 & -7 & -1 \end{bmatrix}$

37. $\begin{bmatrix} 1 & 0 & 14 & -11 \\ 0 & 1 & -2 & 2 \\ 0 & 0 & 1 & -7 \end{bmatrix}$

39. $\begin{bmatrix} 1 & 1 & 4 & -1 \\ 0 & 5 & -2 & 6 \\ 0 & 3 & 20 & 4 \end{bmatrix}\begin{bmatrix} 1 & 1 & 4 & -1 \\ 0 & 1 & -\frac{2}{5} & \frac{6}{5} \\ 0 & 3 & 20 & 4 \end{bmatrix}$

41. (a) (i) $\left[\begin{array}{rr:r} 3 & 0 & -6 \\ 6 & -4 & -28 \end{array}\right]$

(ii) $\left[\begin{array}{rr:r} 3 & 0 & -6 \\ 0 & -4 & -16 \end{array}\right]$

(iii) $\left[\begin{array}{rr:r} 3 & 0 & -6 \\ 0 & 1 & 4 \end{array}\right]$

(iv) $\left[\begin{array}{rr:r} 1 & 0 & -2 \\ 0 & 1 & 4 \end{array}\right]$

(b) $\begin{cases} -3x + 4y = 22 \\ 6x - 4y = -28 \end{cases}$

Solution: $(-2, 4)$

(c) Answers will vary.

43. Reduced row-echelon form **45.** Not in row-echelon form

47. $\begin{bmatrix} 1 & 1 & 0 & 5 \\ 0 & 1 & 2 & 0 \\ 0 & 0 & 1 & -1 \end{bmatrix}$ **49.** $\begin{bmatrix} 1 & -1 & -1 & 1 \\ 0 & 1 & 6 & 3 \\ 0 & 0 & 0 & 0 \end{bmatrix}$

51. $\begin{bmatrix} 1 & 0 & 0 \\ 0 & 1 & 0 \\ 0 & 0 & 1 \end{bmatrix}$ **53.** $\begin{bmatrix} 1 & 2 & 0 & 0 \\ 0 & 0 & 1 & 0 \\ 0 & 0 & 0 & 1 \\ 0 & 0 & 0 & 0 \end{bmatrix}$ **55.** $\begin{bmatrix} 1 & 0 & 3 & 16 \\ 0 & 1 & 2 & 12 \end{bmatrix}$

57. $\begin{cases} x - 2y = 4 \\ y = -3 \end{cases}$ **59.** $\begin{cases} x - y + 2z = 4 \\ y - z = 2 \\ z = -2 \end{cases}$

$(-2, -3)$ $(8, 0, -2)$

61. $(3, -4)$ **63.** $(-4, -10, 4)$ **65.** $(3, 2)$

67. $(-5, 6)$ **69.** $(-4, -3, 6)$ **71.** Inconsistent

73. $(3, -2, 5, 0)$ **75.** $(-1, -4)$ **77.** $\left(\frac{1}{2}, -\frac{3}{4}\right)$

79. $(5a + 4, -3a + 2, a)$ **81.** $(4, -3, 2)$

83. $(7, -3, 4)$ **85.** $(0, 2 - 4a, a)$ **87.** $(1, 0, 4, -2)$

89. $(-2a, a, a, 0)$ **91.** Yes; $(-1, 1, -3)$ **93.** No

95. $f(x) = -x^2 + x + 1$ **97.** $f(x) = -9x^2 - 5x + 11$

99. $f(x) = x^2 + 2x + 5$

101. $y = 7.5t + 28$; 133 cases; Yes, because the data values increase in a linear pattern.

103. $250,000 at 7%
$1,400,000 at 8.5%
$350,000 at 9.5%

105. False. It is a 2×4 matrix. **107.** They are the same.

Section 8.2 *(page 561)*

1. equal **3.** zero; O **5.** $x = -4, y = 22$

7. $x = 2, y = 3$

9. (a) $\begin{bmatrix} 3 & -2 \\ 1 & 7 \end{bmatrix}$ (b) $\begin{bmatrix} -1 & 0 \\ 3 & -9 \end{bmatrix}$ (c) $\begin{bmatrix} 3 & -3 \\ 6 & -3 \end{bmatrix}$

(d) $\begin{bmatrix} -1 & -1 \\ 8 & -19 \end{bmatrix}$

11. (a) $\begin{bmatrix} 9 & 5 \\ 1 & -2 \\ -3 & 15 \end{bmatrix}$ (b) $\begin{bmatrix} 7 & -7 \\ 3 & 8 \\ -5 & -5 \end{bmatrix}$ (c) $\begin{bmatrix} 24 & -3 \\ 6 & 9 \\ -12 & 15 \end{bmatrix}$

(d) $\begin{bmatrix} 22 & -15 \\ 8 & 19 \\ -14 & -5 \end{bmatrix}$

13. (a) $\begin{bmatrix} 5 & 5 & -2 & 4 & 4 \\ -5 & 10 & 0 & -4 & -7 \end{bmatrix}$

(b) $\begin{bmatrix} 3 & 5 & 0 & 2 & 4 \\ 7 & -6 & -4 & 2 & 7 \end{bmatrix}$

(c) $\begin{bmatrix} 12 & 15 & -3 & 9 & 12 \\ 3 & 6 & -6 & -3 & 0 \end{bmatrix}$

(d) $\begin{bmatrix} 10 & 15 & -1 & 7 & 12 \\ 15 & -10 & -10 & 3 & 14 \end{bmatrix}$

15. (a), (b), and (d) not possible (c) $\begin{bmatrix} 18 & 0 & 9 \\ -3 & -12 & 0 \end{bmatrix}$

17. $\begin{bmatrix} -8 & -7 \\ 15 & -1 \end{bmatrix}$ **19.** $\begin{bmatrix} -24 & -4 & 12 \\ -12 & 32 & 12 \end{bmatrix}$ **21.** $\begin{bmatrix} 10 & 8 \\ -59 & 9 \end{bmatrix}$

23. $\begin{bmatrix} -17.12 & 2.2 \\ 11.56 & 10.24 \end{bmatrix}$ **25.** $\begin{bmatrix} -1.581 & -3.739 \\ -4.252 & -13.249 \\ 9.713 & -0.362 \end{bmatrix}$

27. $\begin{bmatrix} -4 & 4 \\ 6 & 0 \\ -2 & -10 \end{bmatrix}$ **29.** $\begin{bmatrix} -1 & 1 \\ \frac{3}{2} & 0 \\ -\frac{1}{2} & -\frac{5}{2} \end{bmatrix}$ **31.** $\begin{bmatrix} 3 & 3 \\ -\frac{1}{2} & 0 \\ -\frac{13}{2} & \frac{11}{2} \end{bmatrix}$

33. $\begin{bmatrix} 2 & -5 \\ -5 & 0 \\ 5 & 6 \end{bmatrix}$ **35.** Not possible

37. $\begin{bmatrix} -2 & 51 \\ -8 & 33 \\ 0 & 27 \end{bmatrix}$ Order: 3×2 **39.** $\begin{bmatrix} 1 & 0 & 0 \\ 0 & 1 & 0 \\ 0 & 0 & \frac{7}{2} \end{bmatrix}$ Order: 3×3 **41.** $\begin{bmatrix} 60 & -20 & 10 & 60 \\ 72 & -24 & 12 & 72 \end{bmatrix}$ Order: 2×4

43. $\begin{bmatrix} 70 & -17 & 73 \\ 32 & 11 & 6 \\ 16 & -38 & 70 \end{bmatrix}$ **45.** $\begin{bmatrix} 151 & 25 & 48 \\ 516 & 279 & 387 \\ 47 & -20 & 87 \end{bmatrix}$

47. (a) $\begin{bmatrix} 0 & 15 \\ 6 & 12 \end{bmatrix}$ (b) $\begin{bmatrix} -2 & 2 \\ 31 & 14 \end{bmatrix}$ (c) $\begin{bmatrix} 9 & 6 \\ 12 & 12 \end{bmatrix}$

49. (a) $\begin{bmatrix} 0 & -10 \\ 10 & 0 \end{bmatrix}$ (b) $\begin{bmatrix} 0 & -10 \\ 10 & 0 \end{bmatrix}$ (c) $\begin{bmatrix} 8 & -6 \\ 6 & 8 \end{bmatrix}$

51. (a) $\begin{bmatrix} 7 & 7 & 14 \\ 8 & 8 & 16 \\ -1 & -1 & -2 \end{bmatrix}$ (b) $[13]$ (c) Not possible

53. $\begin{bmatrix} 5 & 8 \\ -4 & -16 \end{bmatrix}$ **55.** $\begin{bmatrix} -4 & 10 \\ 3 & 14 \end{bmatrix}$

57. (a) $\begin{bmatrix} -1 & 1 \\ -2 & 1 \end{bmatrix}\begin{bmatrix} x_1 \\ x_2 \end{bmatrix} = \begin{bmatrix} 4 \\ 0 \end{bmatrix}$ (b) $\begin{bmatrix} 4 \\ 8 \end{bmatrix}$

59. (a) $\begin{bmatrix} -2 & -3 \\ 6 & 1 \end{bmatrix}\begin{bmatrix} x_1 \\ x_2 \end{bmatrix} = \begin{bmatrix} -4 \\ -36 \end{bmatrix}$ (b) $\begin{bmatrix} -7 \\ 6 \end{bmatrix}$

61. (a) $\begin{bmatrix} 1 & -2 & 3 \\ -1 & 3 & -1 \\ 2 & -5 & 5 \end{bmatrix}\begin{bmatrix} x_1 \\ x_2 \\ x_3 \end{bmatrix} = \begin{bmatrix} 9 \\ -6 \\ 17 \end{bmatrix}$ (b) $\begin{bmatrix} 1 \\ -1 \\ 2 \end{bmatrix}$

63. (a) $\begin{bmatrix} 1 & -5 & 2 \\ -3 & 1 & -1 \\ 0 & -2 & 5 \end{bmatrix}\begin{bmatrix} x_1 \\ x_2 \\ x_3 \end{bmatrix} = \begin{bmatrix} -20 \\ 8 \\ -16 \end{bmatrix}$ (b) $\begin{bmatrix} -1 \\ 3 \\ -2 \end{bmatrix}$

65. $\begin{bmatrix} 110 & 99 & 77 & 33 \\ 44 & 22 & 66 & 66 \end{bmatrix}$

67. [$1037.50 $1400 $1012.50]
The entries represent the profits from both crops at each of the three outlets.

69. [$17,699,050 $17,299,050]
The entries represent the costs of the three models of the LCD televisions at the two warehouses.

71. (a)

Sales $	Profit
571.8	206.6
798.9	288.8
936	337.8

The entries represent the total sales and profits for milk on Friday, Saturday, and Sunday.

(b) $833.20

73. $P^3 = \begin{bmatrix} 0.300 & 0.175 & 0.175 \\ 0.308 & 0.433 & 0.217 \\ 0.392 & 0.392 & 0.608 \end{bmatrix}$

$P^4 = \begin{bmatrix} 0.250 & 0.188 & 0.188 \\ 0.315 & 0.377 & 0.248 \\ 0.435 & 0.435 & 0.565 \end{bmatrix}$

$P^5 = \begin{bmatrix} 0.225 & 0.194 & 0.194 \\ 0.314 & 0.345 & 0.267 \\ 0.461 & 0.461 & 0.539 \end{bmatrix}$

$P^6 = \begin{bmatrix} 0.213 & 0.197 & 0.197 \\ 0.311 & 0.326 & 0.280 \\ 0.477 & 0.477 & 0.523 \end{bmatrix}$

$P^7 = \begin{bmatrix} 0.206 & 0.198 & 0.198 \\ 0.308 & 0.316 & 0.288 \\ 0.486 & 0.486 & 0.514 \end{bmatrix}$

$P^8 = \begin{bmatrix} 0.203 & 0.199 & 0.199 \\ 0.305 & 0.309 & 0.292 \\ 0.492 & 0.492 & 0.508 \end{bmatrix}$

Approaches the matrix $\begin{bmatrix} 0.2 & 0.2 & 0.2 \\ 0.3 & 0.3 & 0.3 \\ 0.5 & 0.5 & 0.5 \end{bmatrix}$.

75. True. The sum of two matrices of different orders is undefined.

77. $\begin{bmatrix} 1 & 0 \\ 2 & 1 \end{bmatrix} \neq \begin{bmatrix} 0 & 0 \\ 3 & 2 \end{bmatrix}$ **79.** $\begin{bmatrix} 3 & -2 \\ 8 & 3 \end{bmatrix} \neq \begin{bmatrix} 2 & -2 \\ 5 & 4 \end{bmatrix}$

81. $AC = BC = \begin{bmatrix} 2 & 3 \\ 2 & 3 \end{bmatrix}$

83. AB is a diagonal matrix whose entries are the products of the corresponding entries of A and B.

85. Answers will vary.

Section 8.3 *(page 571)*

1. square **3.** nonsingular; singular

5–11. $AB = I$ and $BA = I$

13. $\begin{bmatrix} \frac{1}{2} & 0 \\ 0 & \frac{1}{3} \end{bmatrix}$ **15.** $\begin{bmatrix} -3 & 2 \\ -2 & 1 \end{bmatrix}$ **17.** $\begin{bmatrix} 1 & -\frac{1}{2} \\ -2 & \frac{3}{2} \end{bmatrix}$

19. $\begin{bmatrix} 1 & 1 & -1 \\ -3 & 2 & -1 \\ 3 & -3 & 2 \end{bmatrix}$ **21.** Does not exist

23. $\begin{bmatrix} -\frac{1}{8} & 0 & 0 & 0 \\ 0 & 1 & 0 & 0 \\ 0 & 0 & \frac{1}{4} & 0 \\ 0 & 0 & 0 & -\frac{1}{5} \end{bmatrix}$ **25.** $\begin{bmatrix} -175 & 37 & -13 \\ 95 & -20 & 7 \\ 14 & -3 & 1 \end{bmatrix}$

27. $\begin{bmatrix} -1.5 & 1.5 & 1 \\ 4.5 & -3.5 & -3 \\ -1 & 1 & 1 \end{bmatrix}$ **29.** $\begin{bmatrix} -12 & -5 & -9 \\ -4 & -2 & -4 \\ -8 & -4 & -6 \end{bmatrix}$

31. $\begin{bmatrix} 0 & -1.\overline{81} & 0.\overline{90} \\ -10 & 5 & 5 \\ 10 & -2.\overline{72} & -3.\overline{63} \end{bmatrix}$ **33.** $\begin{bmatrix} 1 & 0 & 1 & 0 \\ 0 & 1 & 0 & 1 \\ 2 & 0 & 1 & 0 \\ 0 & 1 & 0 & 2 \end{bmatrix}$

35. $\begin{bmatrix} \frac{5}{13} & -\frac{3}{13} \\ \frac{1}{13} & \frac{2}{13} \end{bmatrix}$ **37.** Does not exist **39.** $\begin{bmatrix} \frac{16}{59} & \frac{15}{59} \\ -\frac{4}{59} & \frac{70}{59} \end{bmatrix}$

41. $(5, 0)$ **43.** $(-8, -6)$ **45.** $(3, 8, -11)$

47. $(2, 1, 0, 0)$ **49.** $(2, -2)$ **51.** No solution

53. $(-4, -8)$ **55.** $(-1, 3, 2)$ **57.** $\left(\frac{5}{16}a + \frac{13}{16}, \frac{19}{16}a + \frac{11}{16}, a\right)$

59. $(-7, 3, -2)$

61. \$7000 in AAA-rated bonds
\$1000 in A-rated bonds
\$2000 in B-rated bonds

63. \$9000 in AAA-rated bonds
\$1000 in A-rated bonds
\$2000 in B-rated bonds

65. $I_1 = 0.5$ ampere
$I_2 = 3$ amperes
$I_3 = 3.5$ amperes

67. $I_1 = 4$ amperes
$I_2 = 1$ ampere
$I_3 = 5$ amperes

69. 100 bags of potting soil for seedlings
100 bags of potting soil for general potting
100 bags of potting soil for hardwood plants

71. (a) $\begin{cases} 2.5r + 4l + 2i = 300 \\ -r + 2l + 2i = 0 \\ r + l + i = 120 \end{cases}$

(b) $\begin{bmatrix} 2.5 & 4 & 2 \\ -1 & 2 & 2 \\ 1 & 1 & 1 \end{bmatrix} \begin{bmatrix} r \\ l \\ i \end{bmatrix} = \begin{bmatrix} 300 \\ 0 \\ 120 \end{bmatrix}$

(c) 80 roses, 10 lilies, 30 irises

73. True. If B is the inverse of A, then $AB = I = BA$.

75. Answers will vary.

77. (a) Answers will vary.

(b) $A^{-1} = \begin{bmatrix} \frac{1}{a_{11}} & 0 & 0 & \cdots & 0 \\ 0 & \frac{1}{a_{22}} & 0 & \cdots & 0 \\ 0 & 0 & \frac{1}{a_{33}} & \cdots & 0 \\ \vdots & \vdots & \vdots & \cdots & \vdots \\ 0 & 0 & 0 & \cdots & \frac{1}{a_{nn}} \end{bmatrix}$

Section 8.4 *(page 579)*

1. determinant **3.** cofactor **5.** 4 **7.** 16 **9.** 28

11. 0 **13.** 6 **15.** -9 **17.** -23 **19.** -24

21. $\frac{11}{6}$ **23.** 11 **25.** 1924 **27.** 0.02 **29.** 0.34

31. (a) $M_{11} = -6, M_{12} = 3, M_{21} = 5, M_{22} = 4$
(b) $C_{11} = -6, C_{12} = -3, C_{21} = -5, C_{22} = 4$

33. (a) $M_{11} = -4, M_{12} = -2, M_{21} = 1, M_{22} = 3$
(b) $C_{11} = -4, C_{12} = 2, C_{21} = -1, C_{22} = 3$

35. (a) $M_{11} = 3, M_{12} = -4, M_{13} = 1, M_{21} = 2, M_{22} = 2,$
$M_{23} = -4, M_{31} = -4, M_{32} = 10, M_{33} = 8$
(b) $C_{11} = 3, C_{12} = 4, C_{13} = 1, C_{21} = -2, C_{22} = 2,$
$C_{23} = 4, C_{31} = -4, C_{32} = -10, C_{33} = 8$

37. (a) $M_{11} = 10, M_{12} = -43, M_{13} = 2, M_{21} = -30,$
$M_{22} = 17, M_{23} = -6, M_{31} = 54, M_{32} = -53,$
$M_{33} = -34$
(b) $C_{11} = 10, C_{12} = 43, C_{13} = 2, C_{21} = 30, C_{22} = 17,$
$C_{23} = 6, C_{31} = 54, C_{32} = 53, C_{33} = -34$

39. (a) -75 (b) -75 **41.** (a) 96 (b) 96

43. (a) 225 (b) 225 **45.** (a) 170 (b) 170 **47.** 0

49. 0 **51.** -9 **53.** -58 **55.** -30 **57.** -168

59. 0 **61.** 412 **63.** -126 **65.** -336

67. (a) -3 (b) -2 (c) $\begin{bmatrix} -2 & 0 \\ 0 & -3 \end{bmatrix}$ (d) 6

69. (a) -8 (b) 0 (c) $\begin{bmatrix} -4 & 4 \\ 1 & -1 \end{bmatrix}$ (d) 0

71. (a) -21 (b) -19 (c) $\begin{bmatrix} 7 & 1 & 4 \\ -8 & 9 & -3 \\ 7 & -3 & 9 \end{bmatrix}$ (d) 399

73. (a) 2 (b) -6 (c) $\begin{bmatrix} 1 & 4 & 3 \\ -1 & 0 & 3 \\ 0 & 2 & 0 \end{bmatrix}$ (d) -12

75–79. Answers will vary. **81.** ± 2 **83.** $1 \pm \sqrt{2}$

85. $-1, 4$ **87.** $-1, -4$ **89.** $8uv - 1$ **91.** e^{5x}

93. $1 - \ln x$

95. True. If an entire row is zero, then each cofactor in the expansion is multiplied by zero.

97. Answers will vary.

99. A square matrix is a square array of numbers. The determinant of a square matrix is a real number.

101. (a) Columns 2 and 3 of A were interchanged.
$|A| = -115 = -|B|$
(b) Rows 1 and 3 of A were interchanged.
$|A| = -40 = -|B|$

103. (a) Multiply Row 1 by 5.
(b) Multiply Column 2 by 4 and Column 3 by 3.

105. (a) 28 (b) -10 (c) -12
The determinant of a diagonal matrix is the product of the entries on the main diagonal.

Section 8.5 *(page 591)*

1. Cramer's Rule **3.** $A = \pm\frac{1}{2}\begin{vmatrix} x_1 & y_1 & 1 \\ x_2 & y_2 & 1 \\ x_3 & y_3 & 1 \end{vmatrix}$

5. uncoded; coded **7.** $(-3, -2)$ **9.** Not possible

11. $\left(\frac{32}{7}, \frac{30}{7}\right)$ **13.** $(-1, 3, 2)$ **15.** $(-2, 1, -1)$ **17.** 7

19. 14 **21.** $\frac{33}{8}$ **23.** $\frac{5}{2}$ **25.** 28 **27.** $\frac{41}{4}$

29. $y = \frac{16}{5}$ or $y = 0$ **31.** $y = -3$ or $y = -11$

33. 250 mi^2 **35.** Collinear **37.** Not collinear

39. Collinear **41.** $y = -3$ **43.** $3x - 5y = 0$

45. $x + 3y - 5 = 0$ **47.** $2x + 3y - 8 = 0$

49. (a) Uncoded: [3 15], [13 5], [0 8], [15 13], [5 0], [19 15], [15 14]
(b) Encoded: 48 81 28 51 24 40 54 95 5 10 64 113 57 100

51. (a) Uncoded: [3 1 12], [12 0 13], [5 0 20], [15 13 15], [18 18 15], [23 0 0]
(b) Encoded: -68 21 35 -66 14 39 -115 35 60 -62 15 32 -54 12 27 23 -23 0

53. 1 -25 -65 17 15 -9 -12 -62 -119 27 51 48 43 67 48 57 111 117

55. -5 -41 -87 91 207 257 11 -5 -41 40 80 84 76 177 227

57. HAPPY NEW YEAR **59.** CLASS IS CANCELED

61. SEND PLANES **63.** MEET ME TONIGHT RON

65. $I_1 = -0.5$ ampere, $I_2 = 1$ ampere, $I_3 = 0.5$ ampere

67. False. The denominator is the determinant of the coefficient matrix.

69. False. When the determinant of the coefficient matrix is zero, the system has either no solution or infinitely many solutions.

71. Answers will vary. **73.** 12

Review Exercises *(page 596)*

1. 3×1 **3.** 1×1 **5.** $\left[\begin{array}{cc|c} 3 & -10 & 15 \\ 5 & 4 & 22 \end{array}\right]$

7. $\begin{cases} 5x + y + 7z = -9 \\ 4x + 2y = 10 \\ 9x + 4y + 2z = 3 \end{cases}$ **9.** $\begin{bmatrix} 1 & 2 & 3 \\ 0 & 1 & 1 \\ 0 & 0 & 1 \end{bmatrix}$

11. $\begin{cases} x + 2y + 3z = 9 \\ y - 2z = 2 \\ z = 0 \end{cases}$
$(5, 2, 0)$

13. $\begin{cases} x - 5y + 4z = 1 \\ y + 2z = 3 \\ z = 4 \end{cases}$
$(-40, -5, 4)$

15. $(10, -12)$ **17.** $\left(-\frac{1}{5}, \frac{7}{10}\right)$ **19.** Inconsistent

21. $(1, -2, 2)$ **23.** $\left(-2a + \frac{3}{2}, 2a + 1, a\right)$ **25.** $(5, 2, -6)$

27. $(1, 0, 4, 3)$ **29.** $(1, 2, 2)$ **31.** $(2, -3, 3)$

33. $(2, 3, -1)$ **35.** $(2, 6, -10, -3)$ **37.** $x = 12, y = -7$

39. $x = 1, y = 11$

41. (a) $\begin{bmatrix} -1 & 8 \\ 15 & 13 \end{bmatrix}$ (b) $\begin{bmatrix} 5 & -12 \\ -9 & -3 \end{bmatrix}$
(c) $\begin{bmatrix} 8 & -8 \\ 12 & 20 \end{bmatrix}$ (d) $\begin{bmatrix} -7 & 28 \\ 39 & 29 \end{bmatrix}$

43. (a) $\begin{bmatrix} 5 & 7 \\ -3 & 14 \\ 31 & 42 \end{bmatrix}$ (b) $\begin{bmatrix} 5 & 1 \\ -11 & -10 \\ -9 & -38 \end{bmatrix}$
(c) $\begin{bmatrix} 20 & 16 \\ -28 & 8 \\ 44 & 8 \end{bmatrix}$ (d) $\begin{bmatrix} 5 & 13 \\ 5 & 38 \\ 71 & 122 \end{bmatrix}$

45. $\begin{bmatrix} 17 & -17 \\ 13 & 2 \end{bmatrix}$ **47.** $\begin{bmatrix} 54 & 4 \\ -2 & 24 \\ -4 & 32 \end{bmatrix}$ **49.** $\begin{bmatrix} 48 & -18 & -3 \\ 15 & 51 & 33 \end{bmatrix}$

51. $\begin{bmatrix} -11 & -6 \\ 8 & -13 \\ -18 & -8 \end{bmatrix}$ **53.** $\begin{bmatrix} 3 & \frac{2}{3} \\ -\frac{4}{3} & \frac{11}{3} \\ \frac{10}{3} & 0 \end{bmatrix}$ **55.** $\begin{bmatrix} -30 & 4 \\ 51 & 70 \end{bmatrix}$

57. $\begin{bmatrix} 100 & 220 \\ 12 & -4 \\ 84 & 212 \end{bmatrix}$ **59.** $\begin{bmatrix} 14 & -2 & 8 \\ 14 & -10 & 40 \\ 36 & -12 & 48 \end{bmatrix}$

61. $\begin{bmatrix} 44 & 4 \\ 20 & 8 \end{bmatrix}$ **63.** $\begin{bmatrix} 14 & -22 & 22 \\ 19 & -41 & 80 \\ 42 & -66 & 66 \end{bmatrix}$

65. Not possible. The number of columns of the first matrix does not equal the number of rows of the second matrix.

67. $\begin{bmatrix} 76 & 114 & 133 \\ 38 & 95 & 76 \end{bmatrix}$

69. [\$2,396,539 \$2,581,388]
The merchandise shipped to warehouse 1 is worth \$2,396,539 and the merchandise shipped to warehouse 2 is worth \$2,581,388.

71–73. $AB = I$ and $BA = I$ **75.** $\begin{bmatrix} 4 & -5 \\ 5 & -6 \end{bmatrix}$

77. $\begin{bmatrix} \frac{1}{2} & -1 & -\frac{1}{2} \\ \frac{1}{2} & -\frac{2}{3} & -\frac{5}{6} \\ 0 & \frac{2}{3} & \frac{1}{3} \end{bmatrix}$ **79.** $\begin{bmatrix} 13 & 6 & -4 \\ -12 & -5 & 3 \\ 5 & 2 & -1 \end{bmatrix}$

81. $\begin{bmatrix} -3 & 6 & -5.5 & 3.5 \\ 1 & -2 & 2 & -1 \\ 7 & -15 & 14.5 & -9.5 \\ -1 & 2.5 & -2.5 & 1.5 \end{bmatrix}$ **83.** $\begin{bmatrix} 1 & -1 \\ 4 & -\frac{7}{2} \end{bmatrix}$

85. Does not exist **87.** $\begin{bmatrix} 2 & \frac{20}{3} \\ \frac{1}{10} & \frac{1}{6} \end{bmatrix}$ **89.** $\begin{bmatrix} \frac{20}{9} & \frac{5}{9} \\ -\frac{10}{9} & -\frac{25}{9} \end{bmatrix}$

91. $(36, 11)$ **93.** $(-6, -1)$ **95.** $(2, 3)$ **97.** $(-8, 18)$
99. $(2, -1, -2)$ **101.** $(6, 1, -1)$ **103.** $(-3, 1)$
105. $\left(\frac{1}{6}, -\frac{7}{4}\right)$ **107.** $(1, 1, -2)$ **109.** -42 **111.** 550
113. (a) $M_{11} = 4, M_{12} = 7, M_{21} = -1, M_{22} = 2$
(b) $C_{11} = 4, C_{12} = -7, C_{21} = 1, C_{22} = 2$
115. (a) $M_{11} = 30, M_{12} = -12, M_{13} = -21, M_{21} = 20,$
$M_{22} = 19, M_{23} = 22, M_{31} = 5, M_{32} = -2, M_{33} = 19$
(b) $C_{11} = 30, C_{12} = 12, C_{13} = -21, C_{21} = -20,$
$C_{22} = 19, C_{23} = -22, C_{31} = 5, C_{32} = 2, C_{33} = 19$
117. -6 **119.** 15 **121.** 130 **123.** -8 **125.** 279
127. $(4, 7)$ **129.** $(-1, 4, 5)$ **131.** 16 **133.** 10
135. Collinear **137.** $x - 2y + 4 = 0$
139. $2x + 6y - 13 = 0$
141. (a) Uncoded: [12 15 15], [11 0 15], [21 20 0], [2 5 12], [15 23 0]
(b) Encoded: -21 6 0 -68 8 45 102 -42 -60 -53 20 21 99 -30 -69
143. SEE YOU FRIDAY
145. False. The matrix must be square.
147. An error message appears because $1(6) - (-2)(-3) = 0$.
149. If A is a square matrix, then the cofactor C_{ij} of the entry a_{ij} is $(-1)^{i+j}M_{ij}$, where M_{ij} is the determinant obtained by deleting the ith row and jth column of A. The determinant of A is the sum of the entries of any row or column of A multiplied by their respective cofactors.
151. The part of the matrix corresponding to the coefficients of the system reduces to a matrix in which the number of rows with nonzero entries is the same as the number of variables.

Chapter Test *(page 601)*

1. $\begin{bmatrix} 1 & 0 & 0 \\ 0 & 1 & 0 \\ 0 & 0 & 1 \end{bmatrix}$

2. $\begin{bmatrix} 1 & 0 & -1 & 2 \\ 0 & 1 & 0 & -1 \\ 0 & 0 & 0 & 0 \\ 0 & 0 & 0 & 0 \end{bmatrix}$

3. $\left[\begin{array}{rrr|r} 4 & 3 & -2 & 14 \\ -1 & -1 & 2 & -5 \\ 3 & 1 & -4 & 8 \end{array}\right]$, $\left(1, 3, -\frac{1}{2}\right)$

4. (a) $\begin{bmatrix} 1 & 5 \\ 0 & -4 \end{bmatrix}$ (b) $\begin{bmatrix} 18 & 15 \\ -15 & -15 \end{bmatrix}$
(c) $\begin{bmatrix} 8 & 15 \\ -5 & -13 \end{bmatrix}$ (d) $\begin{bmatrix} 5 & -5 \\ 0 & 5 \end{bmatrix}$

5. $\begin{bmatrix} \frac{2}{7} & \frac{3}{7} \\ \frac{5}{7} & \frac{4}{7} \end{bmatrix}$ **6.** $\begin{bmatrix} -\frac{5}{2} & 4 & -3 \\ 5 & -7 & 6 \\ 4 & -6 & 5 \end{bmatrix}$ **7.** $(12, 18)$ **8.** -112

9. 29 **10.** 43 **11.** $(-3, 5)$ **12.** $(-2, 4, 6)$ **13.** 7
14. Uncoded: [11 14 15], [3 11 0], [15 14 0], [23 15 15], [4 0 0]
Encoded: 115 -41 -59 14 -3 -11 29 -15 -14 128 -53 -60 4 -4 0
15. 75 L of 60% solution, 25 L of 20% solution

Problem Solving *(page 603)*

1. (a) $AT = \begin{bmatrix} -1 & -4 & -2 \\ 1 & 2 & 3 \end{bmatrix}$, $AAT = \begin{bmatrix} -1 & -2 & -3 \\ -1 & -4 & -2 \end{bmatrix}$

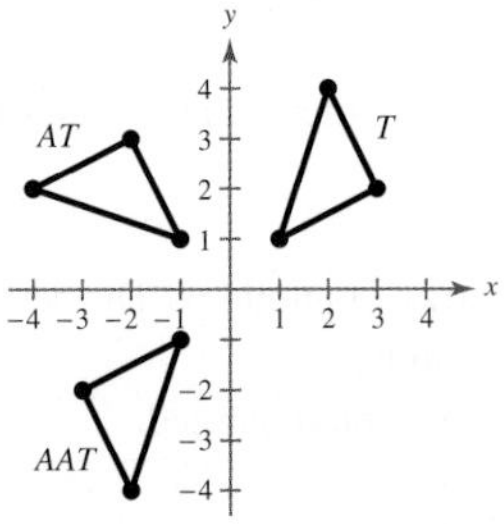

A represents a counterclockwise rotation.
(b) AAT is rotated clockwise 90° to obtain AT. AT is then rotated clockwise 90° to obtain T.
3. (a) Yes (b) No (c) No (d) No
5. (a) Gold Satellite System: 28,750 subscribers
Galaxy Satellite Network: 35,750 subscribers
Nonsubscribers: 35,500
Answers will vary.
(b) Gold Satellite System: 30,813 subscribers
Galaxy Satellite Network: 39,675 subscribers
Nonsubscribers: 29,513
Answers will vary.
(c) Gold Satellite System: 31,947 subscribers
Galaxy Satellite Network: 42,329 subscribers
Nonsubscribers: 25,724
Answers will vary.
(d) Satellite companies are increasing the number of subscribers, while the nonsubscribers are decreasing.
7. $x = 4$
9. Answers will vary. For example: $A = \begin{bmatrix} 1 & 1 \\ 0 & 0 \end{bmatrix}$
11. Answers will vary.
13. $\begin{vmatrix} x & 0 & 0 & d \\ -1 & x & 0 & c \\ 0 & -1 & x & b \\ 0 & 0 & -1 & a \end{vmatrix}$
15. Transformer: \$10.00
Foot of wire: \$0.20
Light: \$1.00
17. (a) $A^{-1} = \begin{bmatrix} 1 & -2 \\ 1 & -3 \end{bmatrix}$ (b) JOHN RETURN TO BASE
19. $|A| = 0$

Chapter 9

Section 9.1 (page 613)

1. infinite sequence **3.** recursively
5. index; upper; lower **7.** $-3, 1, 5, 9, 13$
9. $-2, 4, -8, 16, -32$ **11.** $\frac{1}{3}, \frac{1}{2}, \frac{3}{5}, \frac{2}{3}, \frac{5}{7}$ **13.** $0, 1, 0, \frac{1}{2}, 0$
15. $\frac{2}{3}, \frac{4}{9}, \frac{8}{27}, \frac{16}{81}, \frac{32}{243}$ **17.** $\frac{2}{3}, \frac{2}{3}, \frac{2}{3}, \frac{2}{3}, \frac{2}{3}$ **19.** $0, 0, 6, 24, 60$
21. $-\frac{1}{2}, \frac{2}{3}, -\frac{3}{4}, \frac{4}{5}, -\frac{5}{6}$ **23.** -73 **25.** $\frac{44}{239}$

27.

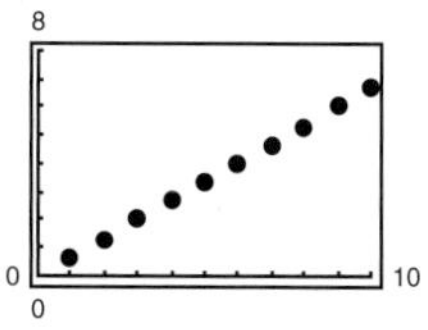

29.

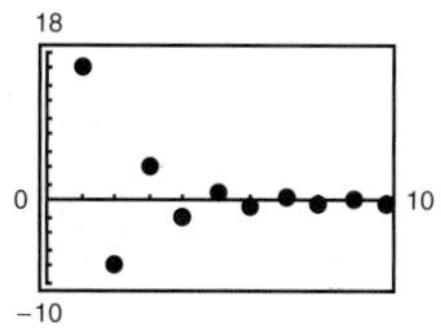

31.

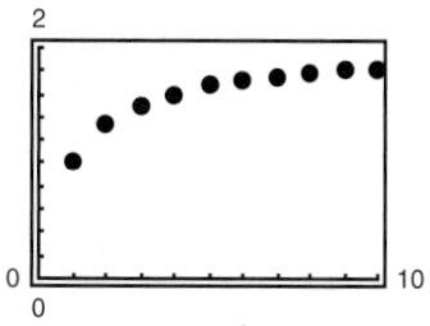

33. c **34.** b **35.** d
36. a **37.** $a_n = 4n - 1$ **39.** $a_n = \dfrac{(-1)^n(n+1)}{n+2}$
41. $a_n = \dfrac{n+1}{2n-1}$ **43.** $a_n = \dfrac{1}{n^2}$ **45.** $a_n = (-1)^{n+1}$
47. $a_n = \dfrac{3^{n-1}}{(n-1)!}$ **49.** $28, 24, 20, 16, 12$ **51.** $1, 2, 2, 3, \frac{7}{2}$
53. $6, 8, 10, 12, 14$; $a_n = 2n + 4$ **55.** $81, 27, 9, 3, 1$; $a_n = \dfrac{243}{3^n}$
57. $1, 1, 2, 3, 5, 8, 13, 21, 34, 55, 89, 144$
$1, 2, \frac{3}{2}, \frac{5}{3}, \frac{8}{5}, \frac{13}{8}, \frac{21}{13}, \frac{34}{21}, \frac{55}{34}, \frac{89}{55}$
59. $5, 5, \frac{5}{2}, \frac{5}{6}, \frac{5}{24}$ **61.** $1, \frac{1}{2}, \frac{1}{6}, \frac{1}{24}, \frac{1}{120}$ **63.** $\frac{1}{30}$
65. $n + 1$ **67.** 35 **69.** 40 **71.** 88 **73.** 30
75. $\dfrac{6508}{3465}$ **77.** $\dfrac{3}{8}$ **79.** $\displaystyle\sum_{i=1}^{9} \frac{1}{3i}$ **81.** $\displaystyle\sum_{i=1}^{8} \left[2\left(\frac{i}{8}\right) + 3\right]$
83. $\displaystyle\sum_{i=1}^{6} (-1)^{i+1}3^i$ **85.** $\displaystyle\sum_{i=1}^{20} \frac{(-1)^{i+1}}{i^2}$ **87.** $\displaystyle\sum_{i=1}^{5} \frac{2^i - 1}{2^{i+1}}$
89. $\frac{75}{16}$ **91.** $-\frac{3}{2}$ **93.** $\frac{2}{3}$ **95.** $\frac{7}{9}$
97. (a) $A_1 = \$10{,}087.50, A_2 \approx \$10{,}175.77, A_3 \approx \$10{,}264.80,$
$A_4 \approx \$10{,}354.62, A_5 \approx \$10{,}445.22, A_6 \approx \$10{,}536.62,$
$A_7 \approx \$10{,}628.81, A_8 \approx \$10{,}721.82$
(b) \$14,169.09
(c) No. $A_{80} \approx \$20{,}076.31 \neq 2A_{40} \approx \$28{,}338.18$
99. True by the Properties of Sums.
101. \$500.95 **103.** Proof **105.** $x, \dfrac{x^2}{2}, \dfrac{x^3}{6}, \dfrac{x^4}{24}, \dfrac{x^5}{120}$
107. (a)

Number of blue cube faces	0	1	2	3
3 × 3 × 3	1	6	12	8

(b)

Number of blue cube faces	0	1	2	3
4 × 4 × 4	8	24	24	8
5 × 5 × 5	27	54	36	8
6 × 6 × 6	64	96	48	8

(c) The different columns change at different rates.
(d)

Number of blue cube faces	0	1	2	3
$n \times n \times n$	$(n-2)^3$	$6(n-2)^2$	$12(n-2)$	8

Section 9.2 (page 622)

1. arithmetic; common **3.** recursion
5. Arithmetic sequence, $d = -2$
7. Not an arithmetic sequence
9. Arithmetic sequence, $d = -\frac{1}{4}$
11. Not an arithmetic sequence
13. $8, 11, 14, 17, 20$; Arithmetic sequence, $d = 3$ **15.** $7, 3, -1, -5, -9$; Arithmetic sequence, $d = -4$
17. $-1, 1, -1, 1, -1$; Not an arithmetic sequence **19.** $-3, \frac{3}{2}, -1, \frac{3}{4}, -\frac{3}{5}$; Not an arithmetic sequence
21. $a_n = 3n - 2$ **23.** $a_n = -8n + 108$
25. $a_n = -\frac{5}{2}n + \frac{13}{2}$ **27.** $a_n = \frac{10}{3}n + \frac{5}{3}$
29. $a_n = -3n + 103$ **31.** $5, 11, 17, 23, 29$
33. $-\frac{13}{5}, -3, -\frac{17}{5}, -\frac{19}{5}, -\frac{21}{5}$ **35.** $2, 6, 10, 14, 18$
37. $-2, 2, 6, 10, 14$ **39.** $15, 19, 23, 27, 31$ **41.** $\frac{5}{8}, \frac{1}{2}, \frac{3}{8}, \frac{1}{4}, \frac{1}{8}$
43. 59 **45.** 18.6 **47.** 110 **49.** -25 **51.** 10,000
53. 620 **55.** 17.4 **57.** 4000 **59.** 1275 **61.** 355
63. 129,250 **65.** b **66.** d **67.** c **68.** a

69.

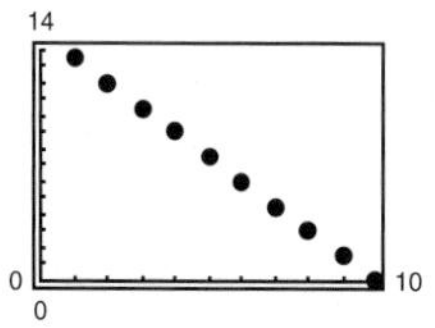

71.

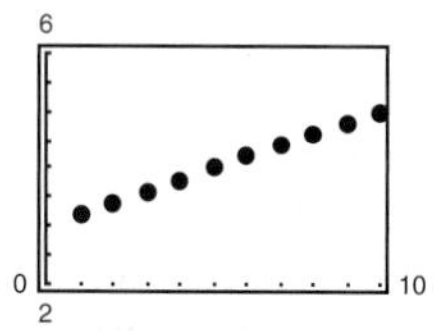

73. 0 **75.** 14,268 **77.** 2430 seats
79. (a) \$40,000 (b) \$217,500 **81.** 784 ft
83. \$375,000; Answers will vary.
85. (a)

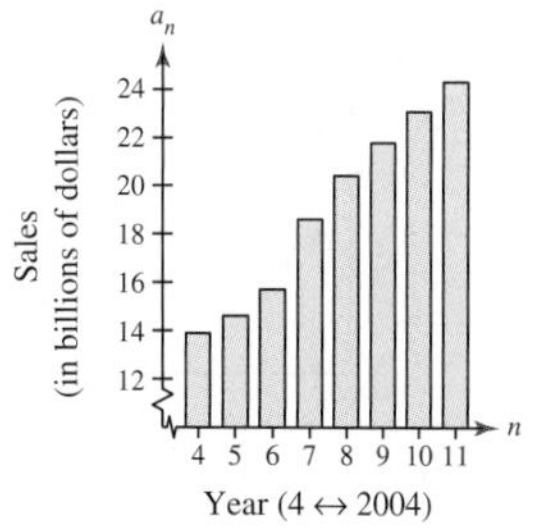

(b) $a_n = 7.0 + 1.61n$

(c)

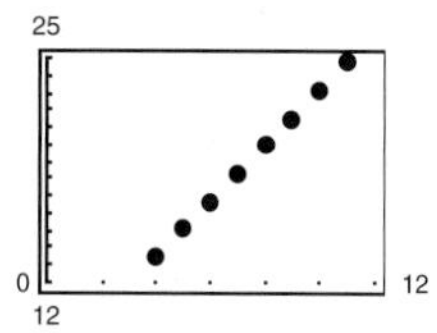

(d) $\displaystyle\sum_{n=4}^{11} a_n = \$152.4$ billion

87. True. Given a_1 and a_2, $d = a_2 - a_1$ and $a_n = a_1 + (n-1)d$.
89. $x, 3x, 5x, 7x, 9x, 11x, 13x, 15x, 17x, 19x$

91. (a)

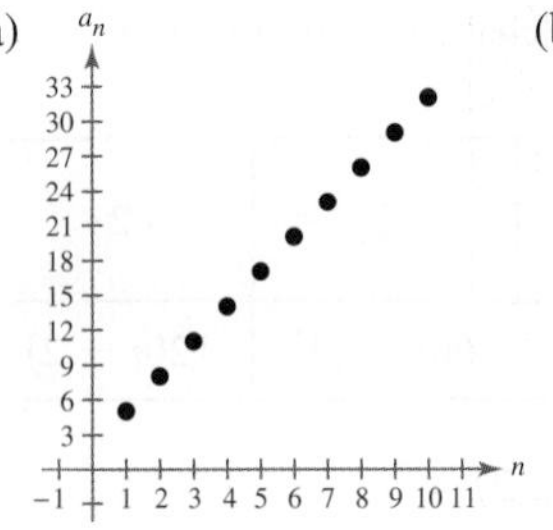

(b)

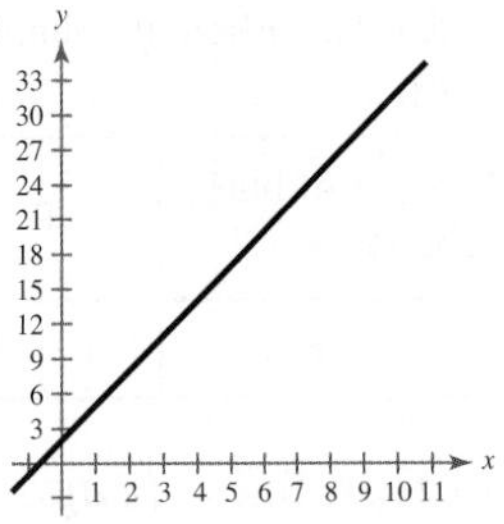

(c) The graph of $y = 3x + 2$ contains all points on the line. The graph of $a_n = 2 + 3n$ contains only points at the positive integers.

(d) The slope of the line and the common difference of the arithmetic sequence are equal.

93. (a) 4, 9, 16, 25, 36 (b) $S_n = n^2$; $S_7 = 49 = 7^2$

(c) $\frac{n}{2}[1 + (2n - 1)] = n^2$

Section 9.3 *(page 631)*

1. geometric; common **3.** $S_n = a_1\left(\frac{1 - r^n}{1 - r}\right)$

5. Geometric sequence, $r = 5$ **7.** Geometric sequence, $r = 2$

9. Not a geometric sequence

11. Geometric sequence, $r = -\sqrt{7}$ **13.** 4, 12, 36, 108, 324

15. $1, \frac{1}{2}, \frac{1}{4}, \frac{1}{8}, \frac{1}{16}$ **17.** $1, e, e^2, e^3, e^4$

19. $3, 3\sqrt{5}, 15, 15\sqrt{5}, 75$ **21.** $2, \frac{x}{2}, \frac{x^2}{8}, \frac{x^3}{32}, \frac{x^4}{128}$

23. $64, 32, 16, 8, 4$; $r = \frac{1}{2}$; $a_n = 128\left(\frac{1}{2}\right)^n$

25. $9, 18, 36, 72, 144$; $r = 2$; $a_n = \frac{9}{2}(2)^n$

27. $6, -9, \frac{27}{2}, -\frac{81}{4}, \frac{243}{8}$; $r = -\frac{3}{2}$; $a_n = -4\left(-\frac{3}{2}\right)^n$

29. $a_n = 4\left(\frac{1}{2}\right)^{n-1}$; $\frac{1}{128}$ **31.** $a_n = 6\left(-\frac{1}{3}\right)^{n-1}$; $-\frac{2}{59{,}049}$

33. $a_n = 100e^{x(n-1)}$; $100e^{8x}$ **35.** $a_n = \left(\sqrt{2}\right)^{n-1}$; $32\sqrt{2}$

37. $a_n = 500(1.02)^{n-1}$; About 1082.372 **39.** $a_9 = 72{,}171$

41. $a_8 = -\frac{1}{32{,}768}$ **43.** $a_3 = 9$ **45.** $a_6 = -2$

47. a **48.** c **49.** b **50.** d

51.

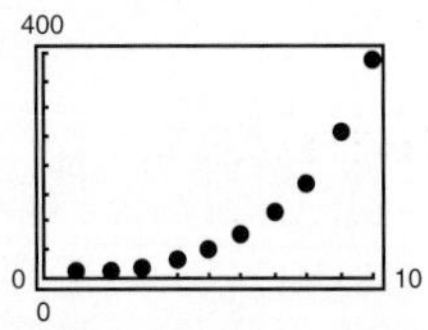

53.

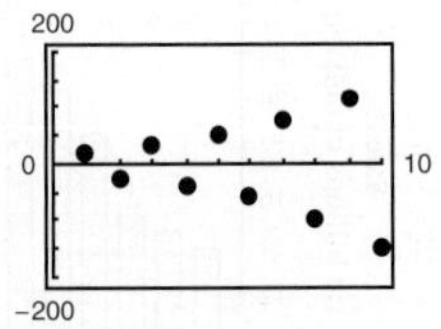

55. 5461 **57.** $-14{,}706$ **59.** 43 **61.** 29,921.311

63. 12.500 **65.** 1.600 **67.** 45.000 **69.** $\sum_{n=1}^{7} 10(3)^{n-1}$

71. $\sum_{n=1}^{5} 32\left(\frac{3}{4}\right)^{n-1}$ **73.** 2 **75.** $\frac{6}{5}$ **77.** 32

79. Undefined **81.** $\frac{4}{11}$

83.

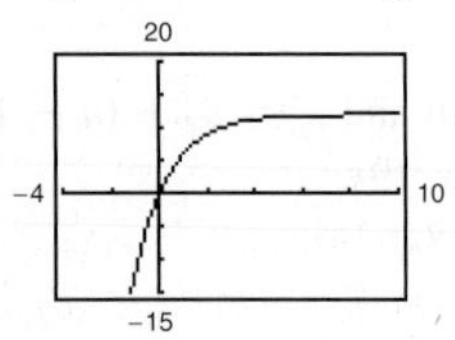

Horizontal asymptote: $y = 12$
Corresponds to the sum of the series

85. \$29,412.25 **87.** Answers will vary. **89.** \$1600

91. $273\frac{8}{9}$ in.^{2} **93.** \$5,435,989.84

95. False. A sequence is geometric when the ratios of consecutive terms are the same.

97. (a)

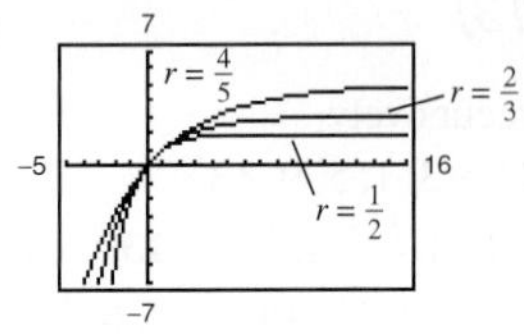

As $x \to \infty$, $y \to \frac{1}{1 - r}$.

(b)

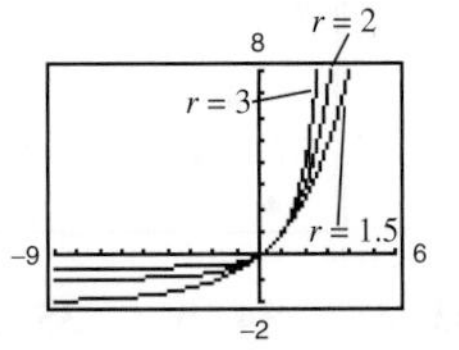

As $x \to \infty$, $y \to \infty$.

Section 9.4 *(page 642)*

1. mathematical induction **3.** arithmetic

5. $\frac{5}{(k + 1)(k + 2)}$ **7.** $\frac{(k + 1)^2(k + 4)^2}{6}$

9. $\frac{3}{(k + 3)(k + 4)}$ **11–39.** Proofs **41.** $S_n = n(2n - 1)$

43. $S_n = \frac{n}{2(n + 1)}$ **45.** 120 **47.** 91 **49.** 979

51. 70 **53.** -3402 **55.** Linear; $a_n = 8n - 3$

57. Quadratic; $a_n = 3n^2 + 3$ **59.** Quadratic; $a_n = n^2 - 3$

61. 0, 3, 6, 9, 12, 15
First differences: 3, 3, 3, 3, 3
Second differences: 0, 0, 0, 0
Linear

63. $3, 1, -2, -6, -11, -17$
First differences: $-2, -3, -4, -5, -6$
Second differences: $-1, -1, -1, -1$
Quadratic

65. 2, 4, 16, 256, 65,536, 4,294,967,296
First differences: 2, 12, 240, 65,280, 4,294,901,760
Second differences: 10, 228, 65,040, 4,294,836,480
Neither

67. 2, 0, 3, 1, 4, 2
First differences: $-2, 3, -2, 3, -2$
Second differences: $5, -5, 5, -5$
Neither

69. $a_n = n^2 - n + 3$ **71.** $a_n = \frac{1}{2}n^2 + n - 3$

73. $a_n = n^2 + 5n - 6$

75. (a) 7, 5, 6, 9, 14; *Sample answer:* $a_n = 7n + 629$
(b) $a_n = 7.7n + 623$; The models are similar.
(c) Part (b): 741,000; Part (c): 746,200
The values are similar.

77. False. P_1 must be proven to be true.

Section 9.5 *(page 649)*

1. binomial coefficients **3.** $\binom{n}{r}$; ${}_nC_r$ **5.** 10 **7.** 1

9. 15,504 **11.** 210 **13.** 4950 **15.** 6 **17.** 35

19. $x^4 + 4x^3 + 6x^2 + 4x + 1$

21. $a^4 + 24a^3 + 216a^2 + 864a + 1296$
23. $y^3 - 12y^2 + 48y - 64$
25. $x^5 + 5x^4y + 10x^3y^2 + 10x^2y^3 + 5xy^4 + y^5$
27. $8x^3 + 12x^2y + 6xy^2 + y^3$
29. $r^6 + 18r^5s + 135r^4s^2 + 540r^3s^3 + 1215r^2s^4 + 1458rs^5 + 729s^6$
31. $243a^5 - 1620a^4b + 4320a^3b^2 - 5760a^2b^3 + 3840ab^4 - 1024b^5$
33. $x^8 + 4x^6y^2 + 6x^4y^4 + 4x^2y^6 + y^8$
35. $\dfrac{1}{x^5} + \dfrac{5y}{x^4} + \dfrac{10y^2}{x^3} + \dfrac{10y^3}{x^2} + \dfrac{5y^4}{x} + y^5$
37. $\dfrac{16}{x^4} - \dfrac{32y}{x^3} + \dfrac{24y^2}{x^2} - \dfrac{8y^3}{x} + y^4$
39. $2x^4 - 24x^3 + 113x^2 - 246x + 207$
41. $32t^5 - 80t^4s + 80t^3s^2 - 40t^2s^3 + 10ts^4 - s^5$
43. $x^5 + 10x^4y + 40x^3y^2 + 80x^2y^3 + 80xy^4 + 32y^5$
45. $120x^7y^3$ **47.** $360x^3y^2$ **49.** $1{,}259{,}712x^2y^7$
51. $-4{,}330{,}260{,}000y^9x^3$ **53.** 1,732,104 **55.** 720
57. $-6{,}300{,}000$ **59.** 210 **61.** $x^{3/2} + 15x + 75x^{1/2} + 125$
63. $x^2 - 3x^{4/3}y^{1/3} + 3x^{2/3}y^{2/3} - y$
65. $81t^2 + 108t^{7/4} + 54t^{3/2} + 12t^{5/4} + t$
67. $3x^2 + 3xh + h^2,\ h \neq 0$
69. $6x^5 + 15x^4h + 20x^3h^2 + 15x^2h^3 + 6xh^4 + h^5,\ h \neq 0$
71. $\dfrac{1}{\sqrt{x+h} + \sqrt{x}},\ h \neq 0$ **73.** -4 **75.** $2035 + 828i$
77. 1 **79.** 1.172 **81.** 510,568.785
83.

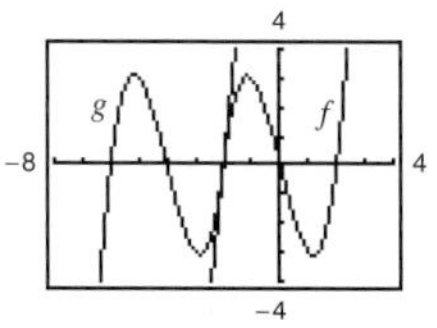

g is shifted four units to the left of f.
$g(x) = x^3 + 12x^2 + 44x + 48$

85. 0.273 **87.** 0.171 **89.** Fibonacci sequence
91. (a) $g(t) = -0.009t^2 + 0.96t + 23.025$
(b)

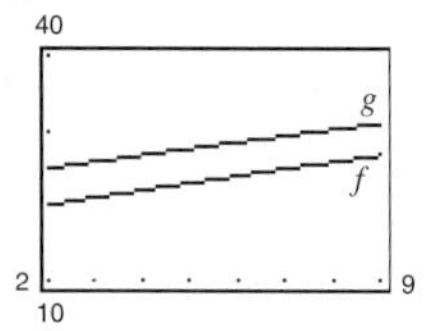

(c) 2007
93. True. The coefficients from the Binomial Theorem can be used to find the numbers in Pascal's Triangle.
95. The first and last numbers in each row are 1. Every other number in each row is formed by adding the two numbers immediately above the number.
97.

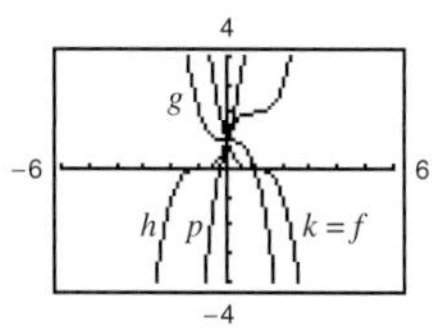

$k(x)$ is the expansion of $f(x)$.
99–101. Proofs

103.

n	r	${}_nC_r$	${}_nC_{n-r}$
9	5	126	126
7	1	7	7
12	4	495	495
6	0	1	1
10	7	120	120

${}_nC_r = {}_nC_{n-r}$
This illustrates the symmetry of Pascal's Triangle.

Section 9.6 *(page 659)*

1. Fundamental Counting Principle **3.** ${}_nP_r = \dfrac{n!}{(n-r)!}$
5. combinations **7.** 6 **9.** 5 **11.** 3 **13.** 8
15. 30 **17.** 30 **19.** 64 **21.** 175,760,000
23. (a) 900 (b) 648 (c) 180 (d) 600
25. 64,000 **27.** (a) 40,320 (b) 384 **29.** 120
31. 24 **33.** 380 **35.** 1,860,480 **37.** 970,200
39. 504 **41.** 1,816,214,400 **43.** 420 **45.** 2520
47. ABCD, ABDC, ACBD, ACDB, ADBC, ADCB, BACD, BADC, CABD, CADB, DABC, DACB, BCAD, BDAC, CBAD, CDAB, DBAC, DCAB, BCDA, BDCA, CBDA, CDBA, DBCA, DCBA
49. 10 **51.** 4 **53.** 4845 **55.** 850,668
57. AB, AC, AD, AE, AF, BC, BD, BE, BF, CD, CE, CF, DE, DF, EF
59. 5,586,853,480 **61.** 324,632
63. (a) 7315 (b) 693 (c) 12,628
65. (a) 3744 (b) 24 **67.** 292,600 **69.** 5 **71.** 20
73. 36 **75.** $n = 2$ **77.** $n = 3$ **79.** $n = 5$ or $n = 6$
81. $n = 10$ **83.** False. It is an example of a combination.
85. ${}_{10}P_6 > {}_{10}C_6$. Changing the order of any of the six elements selected results in a different permutation but the same combination.
87–89. Proofs
91. No. For some calculators the number is too great.

Section 9.7 *(page 670)*

1. experiment; outcomes **3.** probability
5. mutually exclusive **7.** complement
9. $\{(H, 1), (H, 2), (H, 3), (H, 4), (H, 5), (H, 6), (T, 1), (T, 2), (T, 3), (T, 4), (T, 5), (T, 6)\}$
11. {ABC, ACB, BAC, BCA, CAB, CBA}
13. {AB, AC, AD, AE, BC, BD, BE, CD, CE, DE}
15. $\frac{3}{8}$ **17.** $\frac{1}{2}$ **19.** $\frac{7}{8}$ **21.** $\frac{3}{13}$ **23.** $\frac{3}{26}$ **25.** $\frac{5}{36}$
27. $\frac{11}{12}$ **29.** $\frac{1}{3}$ **31.** $\frac{1}{5}$ **33.** $\frac{2}{5}$
35. (a) 1.375 million (b) $\frac{41}{100}$ (c) $\frac{3}{10}$ (d) $\frac{33}{100}$
37. (a) $\frac{13}{16}$ (b) $\frac{3}{16}$ (c) $\frac{1}{32}$ **39.** 19%
41. (a) $\frac{21}{1292}$ (b) $\frac{225}{646}$ (c) $\frac{49}{323}$ **43.** (a) $\frac{1}{120}$ (b) $\frac{1}{24}$
45. (a) $\frac{5}{13}$ (b) $\frac{1}{2}$ (c) $\frac{4}{13}$ **47.** (a) $\frac{14}{55}$ (b) $\frac{12}{55}$ (c) $\frac{54}{55}$
49. (a) $\frac{1}{4}$ (b) $\frac{1}{2}$ (c) $\frac{841}{1600}$ (d) $\frac{1}{40}$ **51.** 0.13
53. $\frac{3}{4}$ **55.** 0.77 **57.** $\frac{18}{35}$

59. (a) 0.9702 (b) 0.9998 (c) 0.0002

61. (a) $\frac{1}{38}$ (b) $\frac{9}{19}$ (c) $\frac{10}{19}$ (d) $\frac{1}{1444}$ (e) $\frac{729}{6859}$ **63.** $\frac{7}{16}$

65. True. Two events are independent when the occurrence of one has no effect on the occurrence of the other.

67. (a) As you consider successive people with distinct birthdays, the probabilities must decrease to take into account the birth dates already used. Because the birth dates of people are independent events, multiply the respective probabilities of distinct birthdays.

(b) $\frac{365}{365} \cdot \frac{364}{365} \cdot \frac{363}{365} \cdot \frac{362}{365}$ (c) Answers will vary.

(d) Q_n is the probability that the birthdays are *not* distinct, which is equivalent to at least two people having the same birthday.

(e)

n	10	15	20	23	30	40	50
P_n	0.88	0.75	0.59	0.49	0.29	0.11	0.03
Q_n	0.12	0.25	0.41	0.51	0.71	0.89	0.97

(f) 23; $Q_n > 0.5$ for $n \geq 23$.

Review Exercises *(page 676)*

1. 8, 5, 4, $\frac{7}{2}$, $\frac{16}{5}$ **3.** 72, 36, 12, 3, $\frac{3}{5}$ **5.** $a_n = 2(-1)^n$

7. $a_n = \dfrac{4}{n}$ **9.** 362,880 **11.** 1 **13.** $\dfrac{205}{24}$

15. $\displaystyle\sum_{k=1}^{20} \frac{1}{2k}$ **17.** $\dfrac{4}{9}$

19. (a) $A_1 = \$10{,}018.75$
$A_2 \approx \$10{,}037.54$
$A_3 \approx \$10{,}056.36$
$A_4 \approx \$10{,}075.21$
$A_5 \approx \$10{,}094.10$
$A_6 \approx \$10{,}113.03$
$A_7 \approx \$10{,}131.99$
$A_8 \approx \$10{,}150.99$
$A_9 \approx \$10{,}170.02$
$A_{10} \approx \$10{,}189.09$

(b) \$12,520.59

21. Arithmetic sequence, $d = -7$

23. Arithmetic sequence, $d = \frac{1}{2}$ **25.** $a_n = 12n - 5$

27. $a_n = -7n + 107$ **29.** 3, 14, 25, 36, 47 **31.** 35,350

33. 80 **35.** 88 **37.** (a) \$51,600 (b) \$238,500

39. Geometric sequence, $r = 2$

41. Geometric sequence, $r = -3$

43. 2, 30, 450, 6750, 101,250

45. 9, 6, 4, $\frac{8}{3}$, $\frac{16}{9}$ or 9, -6, 4, $-\frac{8}{3}$, $\frac{16}{9}$ **47.** $a_n = 18\left(-\frac{1}{2}\right)^{n-1}$; $\frac{9}{512}$

49. $a_n = 100(1.05)^{n-1}$; About 155.133 **51.** 127 **53.** $\frac{15}{16}$

55. 31 **57.** 24.85 **59.** 8 **61.** 12

63. (a) $a_n = 120{,}000(0.7)^n$ (b) \$20,168.40

65–67. Proofs **69.** $S_n = n(2n + 7)$

71. $S_n = \frac{5}{2}\left[1 - \left(\frac{3}{5}\right)^n\right]$ **73.** 2850

75. 5, 10, 15, 20, 25
First differences: 5, 5, 5, 5
Second differences: 0, 0, 0
Linear

77. 15 **79.** 21 **81.** $x^4 + 16x^3 + 96x^2 + 256x + 256$

83. $41 + 840i$ **85.** 11 **87.** 10,000 **89.** 120

91. 225,792,840 **93.** $\frac{1}{9}$ **95.** (a) 43% (b) 82%

97. $\frac{1}{1296}$ **99.** $\frac{3}{4}$

101. False. $\dfrac{(n+2)!}{n!} = \dfrac{(n+2)(n+1)n!}{n!} = (n+2)(n+1)$

103. True by the Properties of Sums.

105. The set of natural numbers

107. Each term of the sequence is defined in terms of preceding terms.

Chapter Test *(page 679)*

1. $-\dfrac{1}{5}, \dfrac{1}{8}, -\dfrac{1}{11}, \dfrac{1}{14}, -\dfrac{1}{17}$ **2.** $a_n = \dfrac{n+2}{n!}$

3. 60, 73, 86; 243 **4.** $a_n = 0.8n + 1.4$

5. $a_n = 7(4)^{n-1}$ **6.** 5, 10, 20, 40, 80 **7.** 86,100

8. 477 **9.** 4 **10.** Proof

11. (a) $x^4 + 24x^3y + 216x^2y^2 + 864xy^3 + 1296y^4$
(b) $3x^5 - 30x^4 + 124x^3 - 264x^2 + 288x - 128$

12. $-22{,}680$ **13.** (a) 72 (b) 328,440

14. (a) 330 (b) 720,720 **15.** 26,000 **16.** 720

17. $\dfrac{1}{15}$ **18.** $\dfrac{1}{27{,}405}$ **19.** 10%

Cumulative Test for Chapters 7–9 *(page 680)*

1. $(1, 2), \left(-\frac{3}{2}, \frac{3}{4}\right)$ **2.** $(-3, -1)$ **3.** $(5, -2, -2)$

4. $(1, -2, 1)$

5. \$0.75 mixture: 120 lb; \$1.25 mixture: 80 lb

6. $y = \frac{1}{4}x^2 - 2x + 6$

7.

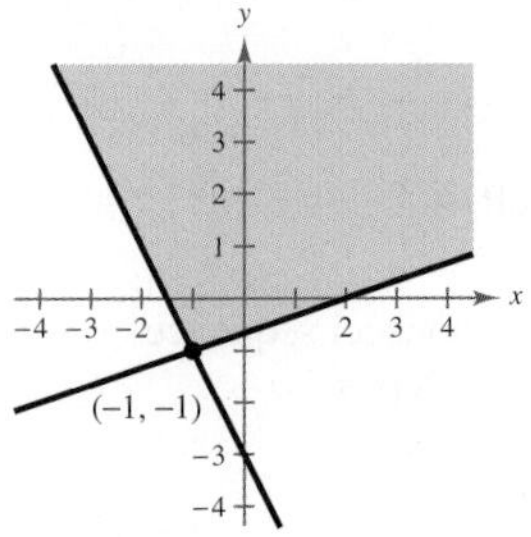

8.

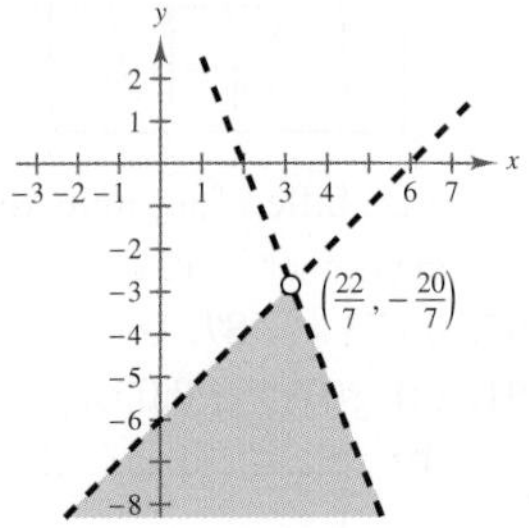

9.

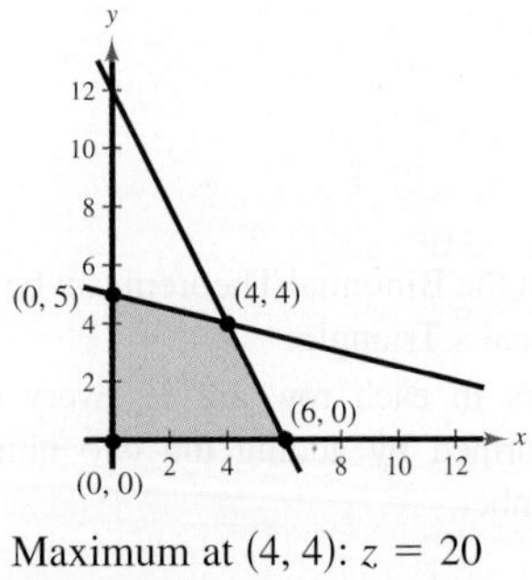

Maximum at (4, 4): $z = 20$
Minimum at (0, 0): $z = 0$

10. $\begin{bmatrix} -1 & 2 & -1 & \vdots & 9 \\ 2 & -1 & 2 & \vdots & -9 \\ 3 & 3 & -4 & \vdots & 7 \end{bmatrix}$

11. $(-2, 3, -1)$ **12.** $\begin{bmatrix} 1 & 5 \\ -1 & 3 \end{bmatrix}$ **13.** $\begin{bmatrix} 16 & -40 \\ 0 & 8 \end{bmatrix}$

14. $\begin{bmatrix} 16 & -25 \\ -2 & 13 \end{bmatrix}$ **15.** $\begin{bmatrix} -6 & 15 \\ 2 & -9 \end{bmatrix}$ **16.** $\begin{bmatrix} 9 & 0 \\ -7 & 16 \end{bmatrix}$

17. $\begin{bmatrix} -15 & 35 \\ 1 & -5 \end{bmatrix}$ **18.** $\begin{bmatrix} -175 & 37 & -13 \\ 95 & -20 & 7 \\ 14 & -3 & 1 \end{bmatrix}$ **19.** 203

20. Gym shoes: \$2539 million
Jogging shoes: \$2362 million
Walking shoes: \$4418 million

21. $(-5, 4)$ **22.** $(-3, 4, 2)$ **23.** 9

24. $\frac{1}{5}, -\frac{1}{7}, \frac{1}{9}, -\frac{1}{11}, \frac{1}{13}$

25. $a_n = \dfrac{(n+1)!}{n+3}$ **26.** 1536

27. (a) 65.4 (b) $a_n = 3.2n + 1.4$

28. 3, 6, 12, 24, 48 **29.** $\frac{130}{9}$ **30.** Proof

31. $w^4 - 36w^3 + 486w^2 - 2916w + 6561$

32. 2184 **33.** 600 **34.** 70

35. 462 **36.** 453,600

37. 151,200 **38.** 720 **39.** $\frac{1}{4}$

Problem Solving *(page 685)*

1. (a)

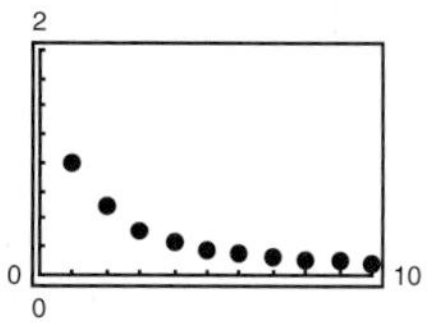

(b) 0

(c)

n	1	10	100	1000	10,000
a_n	1	0.1089	0.0101	0.0010	0.0001

(d) 0

3. $s_d = \dfrac{a_1}{1-r} = \dfrac{20}{1-\frac{1}{2}} = 40$

This represents the total distance Achilles ran.

$s_t = \dfrac{a_1}{1-r} = \dfrac{1}{1-\frac{1}{2}} = 2$

This represents the total amount of time Achilles ran.

5. (a) Arithmetic sequence, difference $= d$
(b) Arithmetic sequence, difference $= dC$
(c) Not an arithmetic sequence

7. (a) 7, 22, 11, 34, 17, 52, 26, 13, 40, 20, 10, 5, 16, 8, 4, 2, 1, 4, 2, 1
(b) $a_1 = 4$: 4, 2, 1, 4, 2, 1, 4, 2, 1, 4
$a_1 = 5$: 5, 16, 8, 4, 2, 1, 4, 2, 1, 4
$a_1 = 12$: 12, 6, 3, 10, 5, 16, 8, 4, 2, 1
Eventually, the terms repeat: 4, 2, 1

9. Proof

11. $S_n = \left(\frac{1}{2}\right)^{n-1}$; $A_n = \dfrac{\sqrt{3}}{4}S_n^2$ **13.** $\dfrac{1}{3}$

15. (a) 30 marbles
(b) 3 to 7; 7 to 3
(c) $P(E) = \dfrac{\text{odds in favor of } E}{\text{odds in favor of } E + 1}$
(d) Odds in favor of event $E = \dfrac{P(E)}{P(E')}$

Chapter 10

Section 10.1 *(page 692)*

1. inclination **3.** $\left|\dfrac{m_2 - m_1}{1 + m_1 m_2}\right|$ **5.** $\dfrac{\sqrt{3}}{3}$ **7.** -1

9. $\sqrt{3}$ **11.** 0.2660 **13.** 3.2236 **15.** -4.1005

17. $\dfrac{3\pi}{4}$ rad, 135° **19.** $\dfrac{\pi}{4}$ rad, 45° **21.** 0.6435 rad, 36.9°

23. 1.9513 rad, 111.8° **25.** $\dfrac{\pi}{6}$ rad, 30° **27.** $\dfrac{5\pi}{6}$ rad, 150°

29. 1.0517 rad, 60.3° **31.** 2.1112 rad, 121.0°

33. 1.6539 rad, 94.8° **35.** $\dfrac{3\pi}{4}$ rad, 135° **37.** $\dfrac{\pi}{4}$ rad, 45°

39. $\dfrac{5\pi}{6}$ rad, 150° **41.** 1.2490 rad, 71.6°

43. 2.4669 rad, 141.3° **45.** 1.1071 rad, 63.4°

47. 0.1974 rad, 11.3° **49.** 1.4289 rad, 81.9°

51. 0.9273 rad, 53.1° **53.** 0.8187 rad, 46.9°

55. $(1, 5) \leftrightarrow (4, 5)$: slope $= 0$
$(4, 5) \leftrightarrow (3, 8)$: slope $= -3$
$(3, 8) \leftrightarrow (1, 5)$: slope $= \frac{3}{2}$
(1, 5): 56.3°; (4, 5): 71.6°; (3, 8): 52.1°

57. $(-4, -1) \leftrightarrow (3, 2)$: slope $= \frac{3}{7}$
$(3, 2) \leftrightarrow (1, 0)$: slope $= 1$
$(1, 0) \leftrightarrow (-4, -1)$: slope $= \frac{1}{5}$
$(-4, -1)$: 11.9°; (3, 2): 21.8°; (1, 0): 146.3°

59. $\dfrac{\sqrt{2}}{2} \approx 0.7071$ **61.** $\dfrac{3\sqrt{5}}{5} \approx 1.3416$ **63.** $\dfrac{\sqrt{2}}{2} \approx 0.7071$

65. 0 **67.** $\dfrac{4\sqrt{10}}{5} \approx 2.5298$ **69.** 1

71. $\dfrac{5\sqrt{34}}{34} \approx 0.8575$ **73.** $\dfrac{8\sqrt{5}}{5} \approx 3.5777$

75. (a)

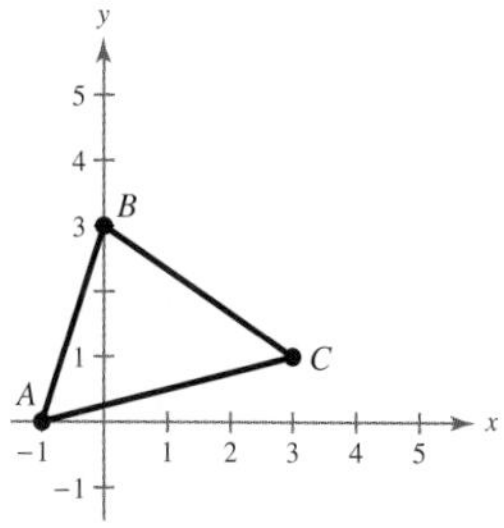

(b) $\dfrac{11\sqrt{17}}{17}$ (c) $\dfrac{11}{2}$

77. (a)

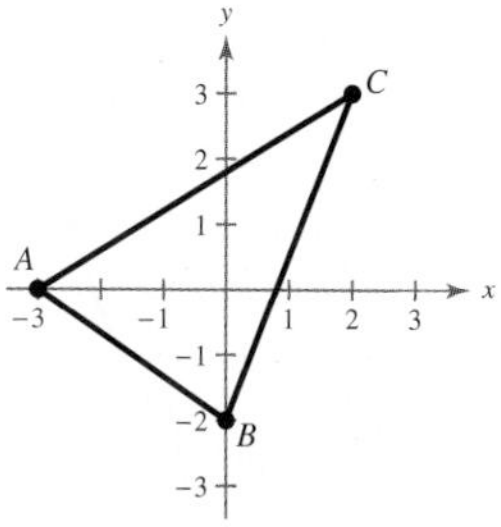

(b) $\dfrac{19\sqrt{34}}{34}$ (c) $\dfrac{19}{2}$

79. (a)

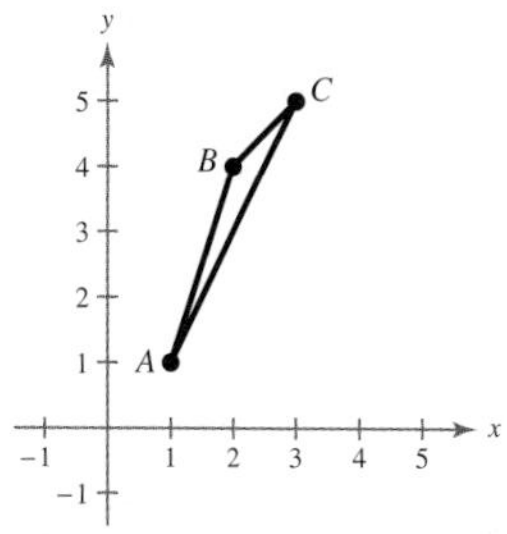

(b) $\dfrac{\sqrt{5}}{5}$ (c) 1

81. $2\sqrt{2}$ **83.** 0.1003, 1054 ft **85.** 31.0°

87. $\alpha \approx 33.69°$; $\beta \approx 56.31°$

89. True. The inclination of a line is related to its slope by $m = \tan \theta$. If the angle is greater than $\pi/2$ but less than π, then the angle is in the second quadrant, where the tangent function is negative.

91. False. The inclination is the positive angle measured counterclockwise from the x-axis.

93. (a) $d = \dfrac{4}{\sqrt{m^2 + 1}}$

(b)

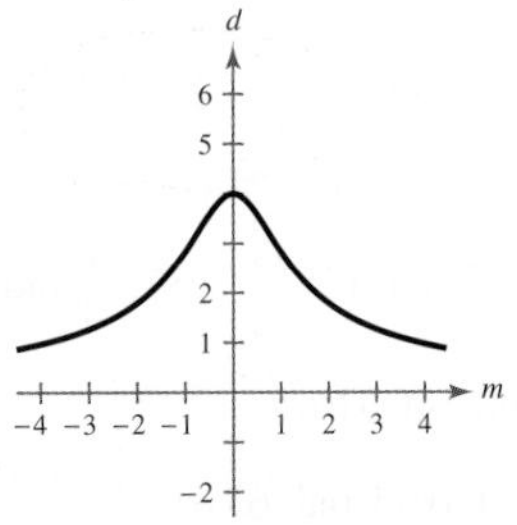

(c) $m = 0$

(d) The graph has a horizontal asymptote of $d = 0$. As the slope becomes larger, the distance between the origin and the line, $y = mx + 4$, becomes smaller and approaches 0.

95. The inclination of a line measures the angle of intersection (measured counterclockwise) of a line and the x-axis. The angle between two lines is the acute angle of their intersection, which must be less than $\pi/2$.

Section 10.2 *(page 700)*

1. conic **3.** locus **5.** axis **7.** focal chord

9. e **10.** b **11.** d **12.** f **13.** a **14.** c

15. $x^2 = \frac{3}{2}y$ **17.** $x^2 = 2y$ **19.** $y^2 = -8x$

21. $x^2 = -4y$ **23.** $y^2 = 4x$ **25.** $x^2 = \frac{8}{3}y$

27. $y^2 = -\frac{25}{2}x$

29. Vertex: $(0, 0)$
Focus: $\left(0, \frac{1}{2}\right)$
Directrix: $y = -\frac{1}{2}$

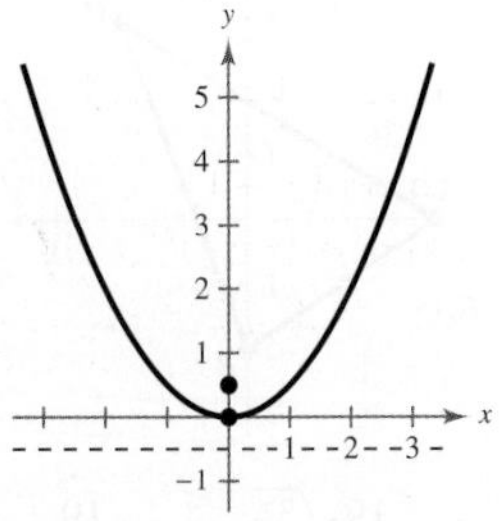

31. Vertex: $(0, 0)$
Focus: $\left(-\frac{3}{2}, 0\right)$
Directrix: $x = \frac{3}{2}$

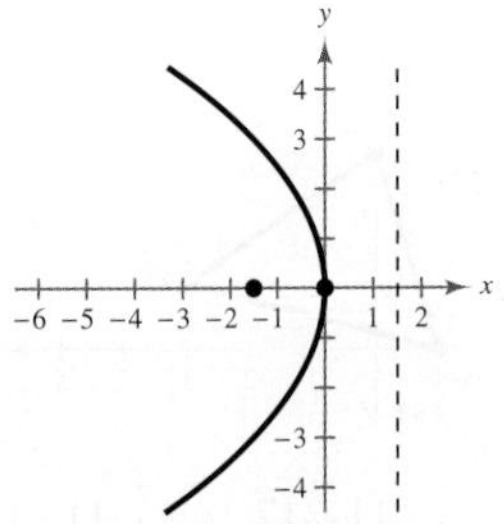

33. Vertex: $(0, 0)$
Focus: $\left(0, -\frac{3}{2}\right)$
Directrix: $y = \frac{3}{2}$

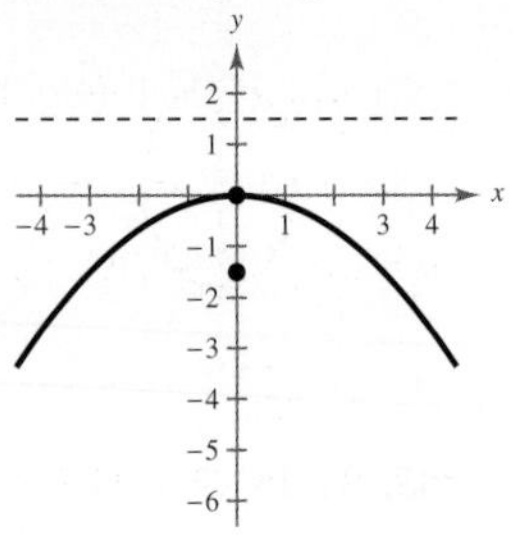

35. Vertex: $(1, -2)$
Focus: $(1, -4)$
Directrix: $y = 0$

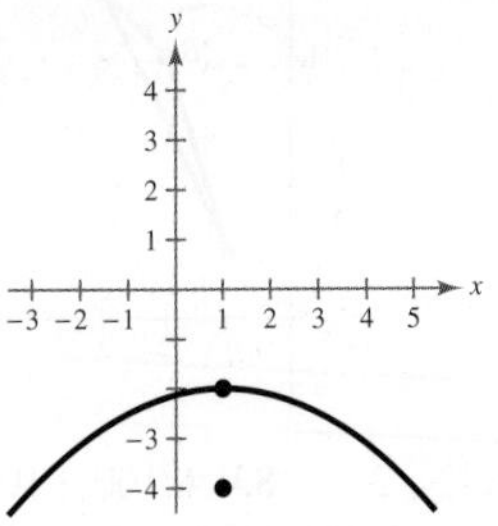

37. Vertex: $\left(-3, \frac{3}{2}\right)$
Focus: $\left(-3, \frac{5}{2}\right)$
Directrix: $y = \frac{1}{2}$

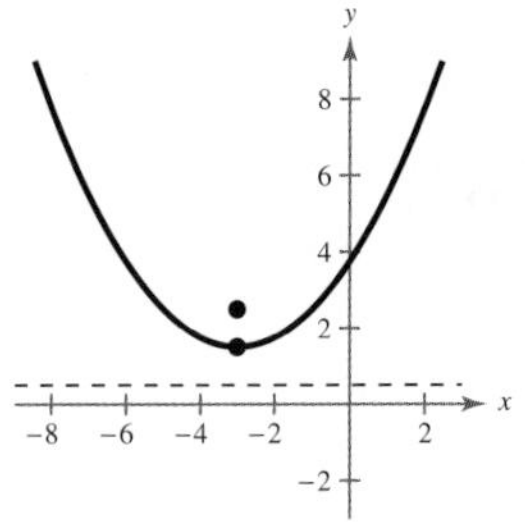

39. Vertex: $(1, 1)$
Focus: $(1, 2)$
Directrix: $y = 0$

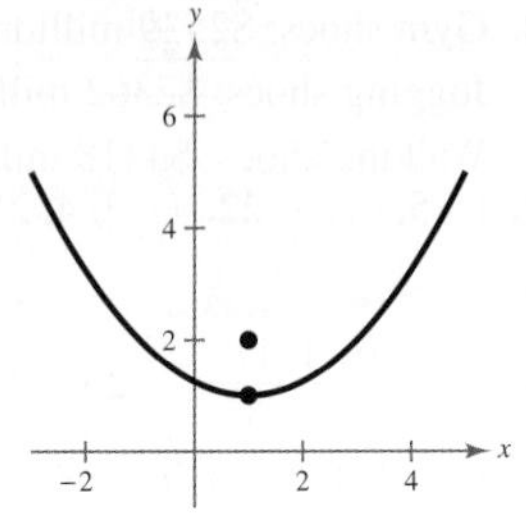

41. Vertex: $(-2, -3)$
Focus: $(-4, -3)$
Directrix: $x = 0$

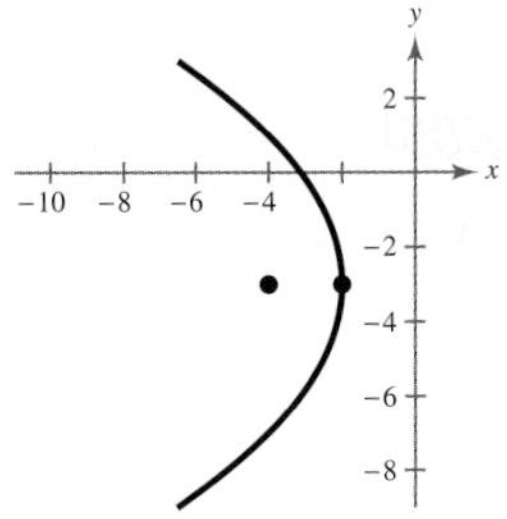

43. Vertex: $(-2, 1)$
Focus: $\left(-2, -\frac{1}{2}\right)$
Directrix: $y = \frac{5}{2}$

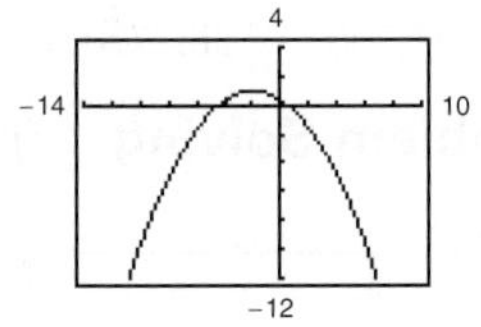

45. Vertex: $\left(\frac{1}{4}, -\frac{1}{2}\right)$
Focus: $\left(0, -\frac{1}{2}\right)$
Directrix: $x = \frac{1}{2}$

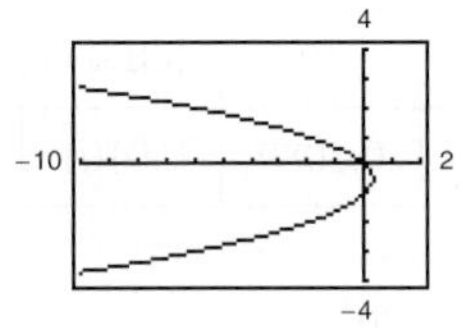

47. $(x - 3)^2 = -(y - 1)$ **49.** $y^2 = 4(x + 4)$

51. $(y - 3)^2 = 8(x - 4)$ **53.** $x^2 = -8(y - 2)$

55. $(y - 2)^2 = 8x$ **57.** $4x - y - 8 = 0$

59. $4x - y + 2 = 0$ **61.** $y^2 = 640x$ **63.** $y^2 = 6x$

65. (a) $x^2 = 12{,}288y$ (in feet) (b) 22.6 ft

67. $x^2 = -\frac{25}{4}(y - 48)$ **69.** About 19.6 m

71. (a) $(0, 45)$ (b) $y = \frac{1}{180}x^2$

73. (a) $17{,}500\sqrt{2}$ mi/h $\approx 24{,}750$ mi/h
(b) $x^2 = -16{,}400(y - 4100)$

75. (a) $x^2 = -49(y - 100)$ (b) 70 ft

77. False. If the graph crossed the directrix, then there would exist points closer to the directrix than the focus.

79. $m = \dfrac{x_1}{2p}$

81.

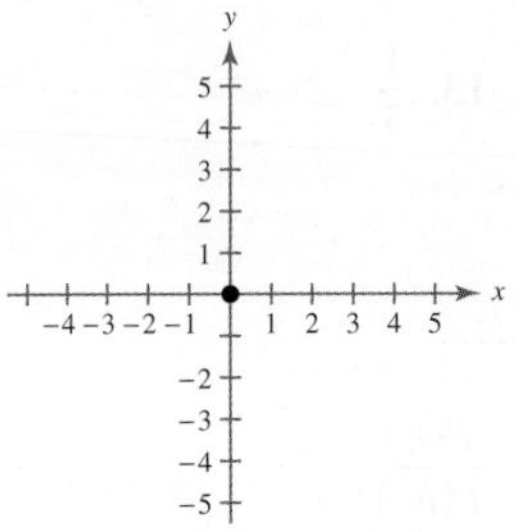

Single point $(0, 0)$; A single point is formed when a plane intersects only the vertex of the cone.

83. (a)
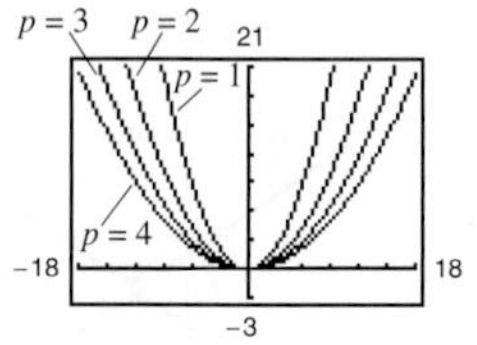

As p increases, the graph becomes wider.

(b) (0, 1), (0, 2), (0, 3), (0, 4) (c) 4, 8, 12, 16; $4|p|$

(d) It is an easy way to determine two additional points on the graph.

Section 10.3 *(page 710)*

1. ellipse; foci **3.** minor axis **5.** b **6.** c **7.** a

8. d **9.** $\frac{x^2}{4} + \frac{y^2}{16} = 1$ **11.** $\frac{x^2}{49} + \frac{y^2}{45} = 1$

13. $\frac{x^2}{49} + \frac{y^2}{24} = 1$ **15.** $\frac{x^2}{9} + \frac{y^2}{36} = 1$ **17.** $\frac{x^2}{36} + \frac{5y^2}{9} = 1$

19. $\frac{(x-2)^2}{1} + \frac{(y-3)^2}{9} = 1$ **21.** $\frac{(x-4)^2}{16} + \frac{(y-2)^2}{1} = 1$

23. $\frac{(x-2)^2}{9} + \frac{y^2}{5} = 1$ **25.** $\frac{(x-1)^2}{9} + \frac{(y-3)^2}{4} = 1$

27. $\frac{x^2}{60} + \frac{(y-2)^2}{64} = 1$ **29.** $\frac{(x-3)^2}{36} + \frac{(y-2)^2}{32} = 1$

31. $\frac{(x-2)^2}{4} + \frac{(y-2)^2}{1} = 1$

33. Center: (0, 0)
Vertices: $(\pm 5, 0)$
Foci: $(\pm 3, 0)$
Eccentricity: $\frac{3}{5}$

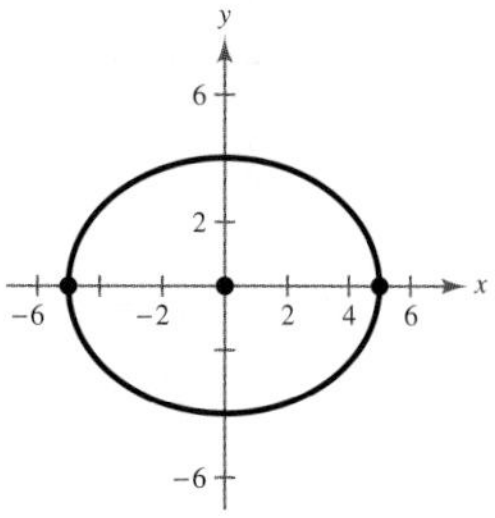

35. Center: (0, 0)
Vertices: $(0, \pm 3)$
Foci: $(0, \pm 2)$
Eccentricity: $\frac{2}{3}$

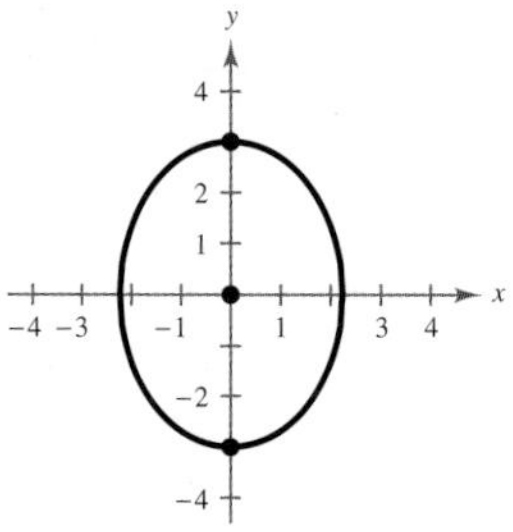

37. Center: $(4, -1)$
Vertices: $(4, -6), (4, 4)$
Foci: $(4, 2), (4, -4)$
Eccentricity: $\frac{3}{5}$

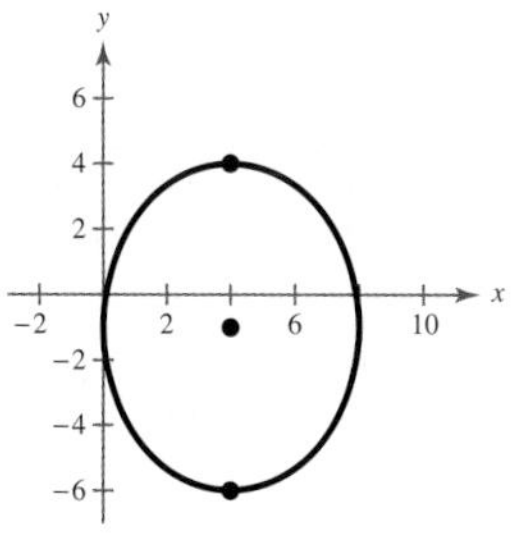

39. Center: $(-5, 1)$
Vertices: $\left(-\frac{7}{2}, 1\right), \left(-\frac{13}{2}, 1\right)$
Foci: $\left(-5 \pm \frac{\sqrt{5}}{2}, 1\right)$
Eccentricity: $\frac{\sqrt{5}}{3}$

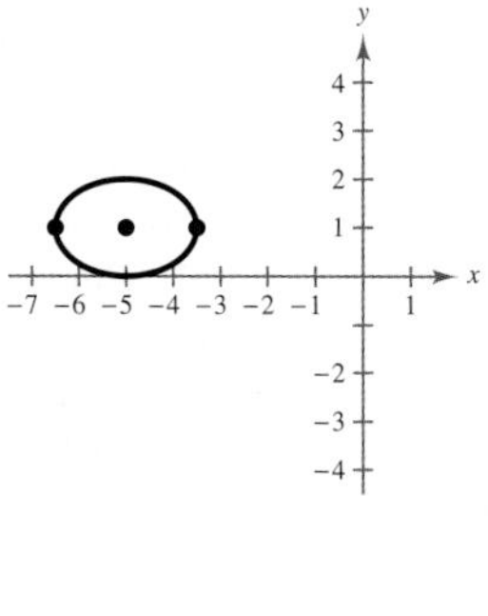

41. Center: $(-2, 3)$
Vertices: $(-2, 6), (-2, 0)$
Foci: $(-2, 3 \pm \sqrt{5})$
Eccentricity: $\frac{\sqrt{5}}{3}$

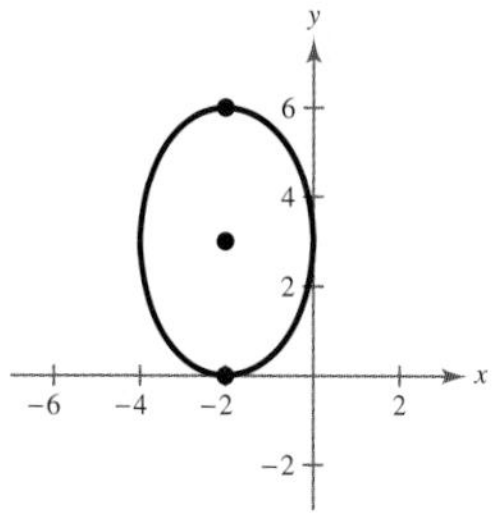

43. Center: (4, 3)
Vertices: $(14, 3), (-6, 3)$
Foci: $(4 \pm 4\sqrt{5}, 3)$
Eccentricity: $\frac{2\sqrt{5}}{5}$

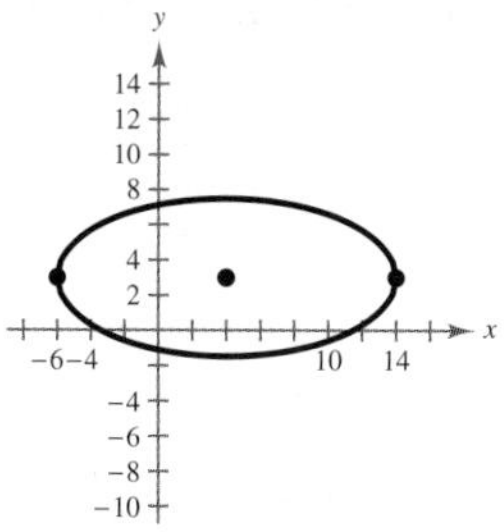

45. Center: $\left(-\frac{3}{2}, \frac{5}{2}\right)$
Vertices: $\left(-\frac{3}{2}, \frac{5}{2} \pm 2\sqrt{3}\right)$
Foci: $\left(-\frac{3}{2}, \frac{5}{2} \pm 2\sqrt{2}\right)$
Eccentricity: $\frac{\sqrt{6}}{3}$

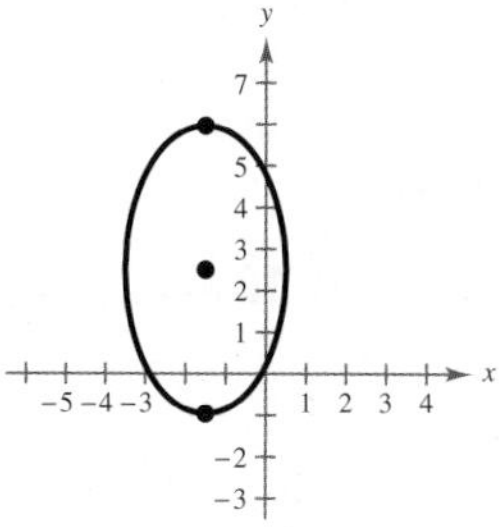

47. Center: $(1, -1)$
Vertices: $\left(\frac{9}{4}, -1\right), \left(-\frac{1}{4}, -1\right)$
Foci: $\left(\frac{7}{4}, -1\right), \left(\frac{1}{4}, -1\right)$
Eccentricity: $\frac{3}{5}$

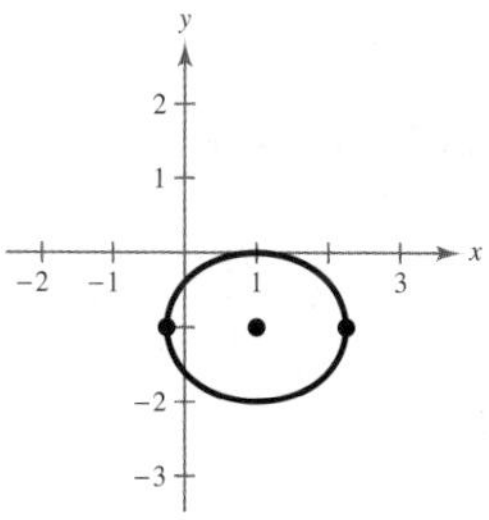

49.
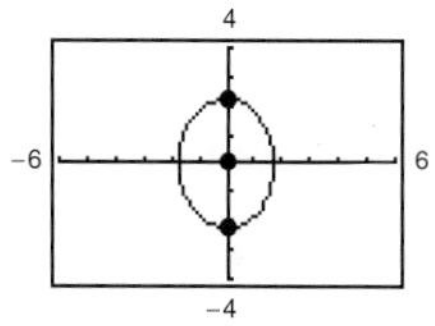

Center: (0, 0)
Vertices: $(0, \pm\sqrt{5})$
Foci: $(0, \pm\sqrt{2})$

51.
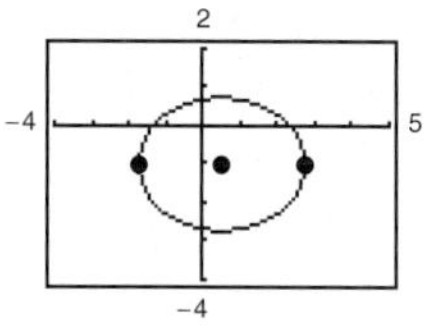

Center: $\left(\frac{1}{2}, -1\right)$
Vertices: $\left(\frac{1}{2} \pm \sqrt{5}, -1\right)$
Foci: $\left(\frac{1}{2} \pm \sqrt{2}, -1\right)$

53. $\frac{x^2}{25} + \frac{y^2}{16} = 1$

55. (a) $\frac{x^2}{2352.25} + \frac{y^2}{529} = 1$ (b) About 85.4 ft

57. About 229.8 mm **59.** $e \approx 0.052$

61.
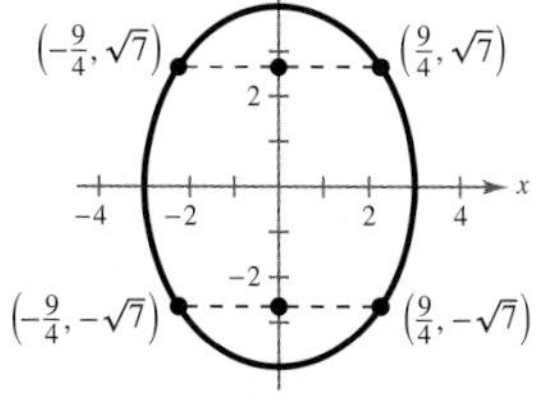

63.
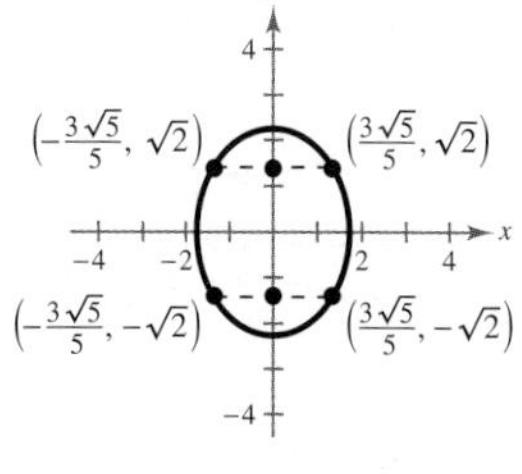

65. False. The graph of $(x^2/4) + y^4 = 1$ is not an ellipse. The degree of y is 4, not 2.

67. $\frac{(x-6)^2}{324} + \frac{(y-2)^2}{308} = 1$

69. (a) $A = \pi a(20 - a)$ (b) $\frac{x^2}{196} + \frac{y^2}{36} = 1$

(c)

a	8	9	10	11	12	13
A	301.6	311.0	314.2	311.0	301.6	285.9

$a = 10$, circle

(d)

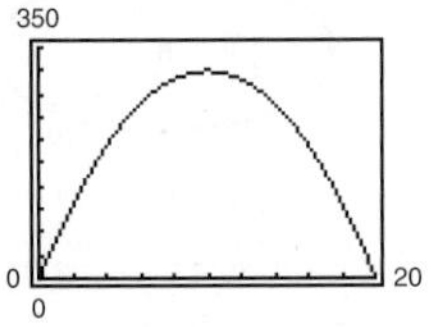

The shape of an ellipse with a maximum area is a circle. The maximum area is found when $a = 10$ (verified in part c) and therefore $b = 10$, so the equation produces a circle.

71. Proof

Section 10.4 *(page 720)*

1. hyperbola; foci **3.** transverse axis; center **5.** b

6. c **7.** a **8.** d **9.** $\frac{y^2}{4} - \frac{x^2}{12} = 1$

11. $\frac{(x-4)^2}{4} - \frac{y^2}{12} = 1$ **13.** $\frac{(y-5)^2}{16} - \frac{(x-4)^2}{9} = 1$

15. $\frac{y^2}{9} - \frac{4(x-2)^2}{9} = 1$ **17.** $\frac{(y-2)^2}{4} - \frac{x^2}{4} = 1$

19. Center: $(0, 0)$
Vertices: $(\pm 1, 0)$
Foci: $(\pm\sqrt{2}, 0)$
Asymptotes: $y = \pm x$

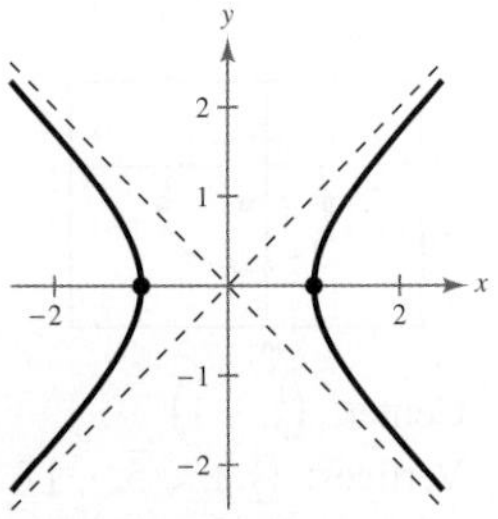

21. Center: $(0, 0)$
Vertices: $(0, \pm 5)$
Foci: $(0, \pm\sqrt{106})$
Asymptotes: $y = \pm\frac{5}{9}x$

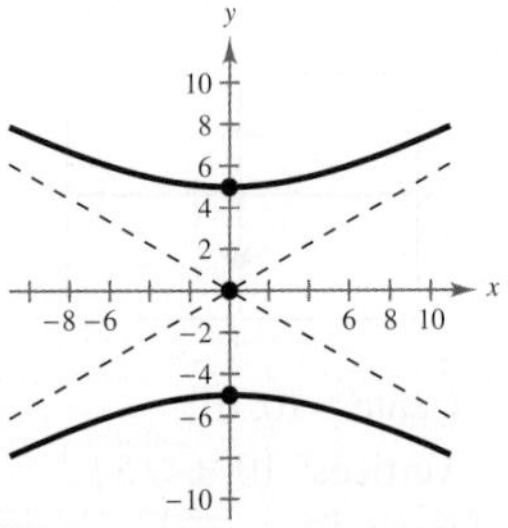

23. Center: $(0, 0)$
Vertices: $(0, \pm 1)$
Foci: $(0, \pm\sqrt{5})$
Asymptotes: $y = \pm\frac{1}{2}x$

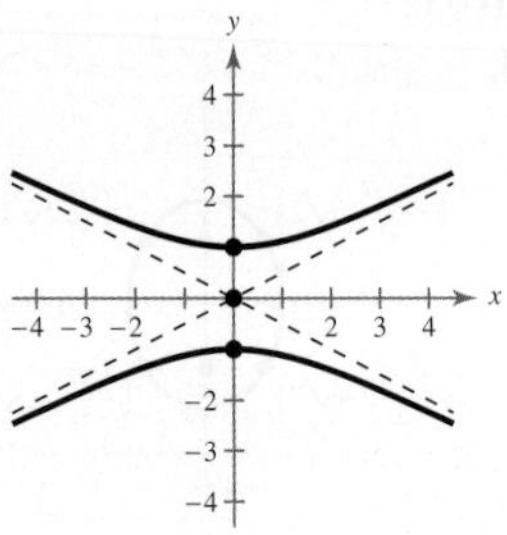

25. Center: $(1, -2)$
Vertices: $(3, -2), (-1, -2)$
Foci: $(1 \pm \sqrt{5}, -2)$
Asymptotes:
$y = -2 \pm \frac{1}{2}(x - 1)$

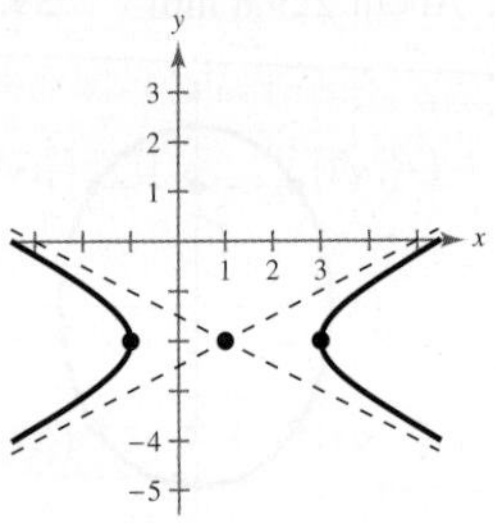

27. Center: $(2, -6)$
Vertices: $\left(2, -\frac{17}{3}\right), \left(2, -\frac{19}{3}\right)$
Foci: $\left(2, -6 \pm \frac{\sqrt{13}}{6}\right)$
Asymptotes: $y = -6 \pm \frac{2}{3}(x - 2)$

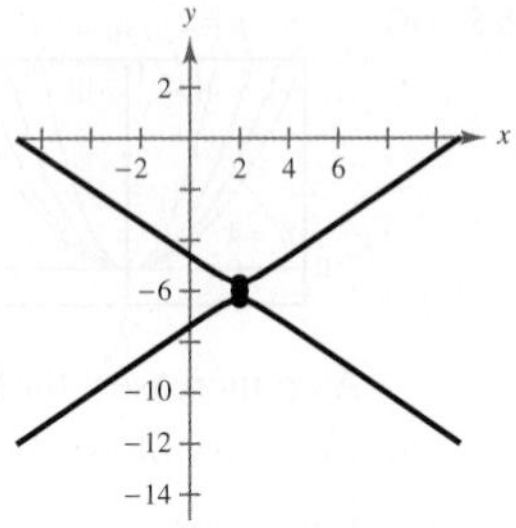

29. Center: $(2, -3)$
Vertices: $(3, -3), (1, -3)$
Foci: $(2 \pm \sqrt{10}, -3)$
Asymptotes:
$y = -3 \pm 3(x - 2)$

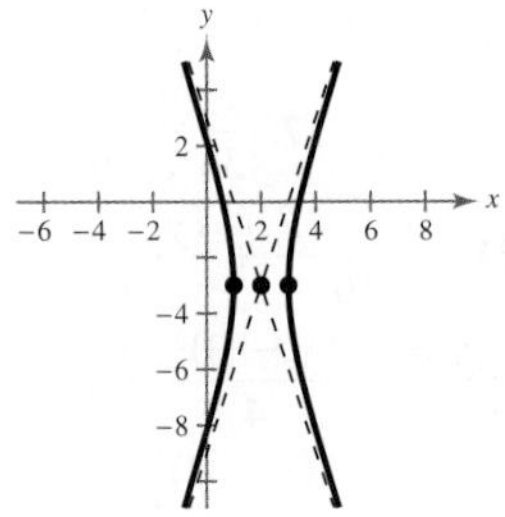

31. The graph of this equation is two lines intersecting at $(-1, -3)$.

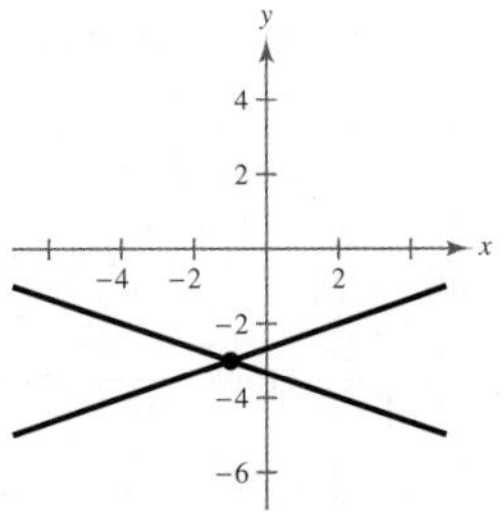

33. Center: $(0, 0)$
Vertices: $(\pm\sqrt{3}, 0)$
Foci: $(\pm\sqrt{5}, 0)$
Asymptotes: $y = \pm\frac{\sqrt{6}}{3}x$

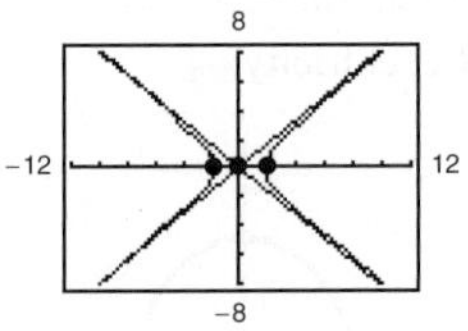

35. Center: $(0, 0)$
Vertices: $(\pm 3, 0)$
Foci: $(\pm\sqrt{13}, 0)$
Asymptotes: $y = \pm\frac{2}{3}x$

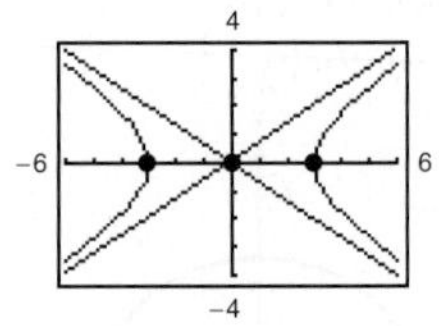

37. Center: $(1, -3)$
Vertices: $(1, -3 \pm \sqrt{2})$
Foci: $(1, -3 \pm 2\sqrt{5})$
Asymptotes: $y = -3 \pm \frac{1}{3}(x - 1)$

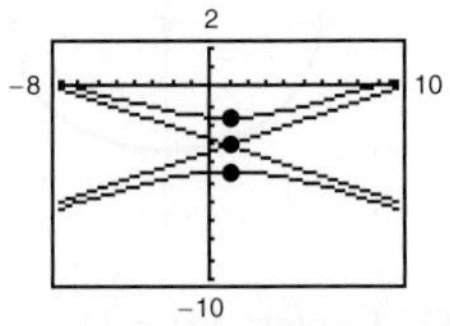

39. $\frac{x^2}{1} - \frac{y^2}{25} = 1$ **41.** $\frac{17y^2}{1024} - \frac{17x^2}{64} = 1$

43. $\frac{(x-2)^2}{1} - \frac{(y-2)^2}{1} = 1$ **45.** $\frac{(x-3)^2}{9} - \frac{(y-2)^2}{4} = 1$

47. (a) $\frac{x^2}{1} - \frac{y^2}{169/3} = 1$ (b) About 2.403 ft

49. $\frac{x^2}{98{,}010{,}000} - \frac{y^2}{13{,}503{,}600} = 1$

51. (a) $x \approx 110.3$ mi (b) 57.0 mi (c) 0.00129 sec
(d) The ship is at the position (144.2, 60).

53. Ellipse **55.** Hyperbola **57.** Hyperbola

59. Parabola **61.** Ellipse **63.** Parabola

65. Parabola **67.** Circle

69. True. For a hyperbola, $c^2 = a^2 + b^2$. The larger the ratio of b to a, the larger the eccentricity of the hyperbola, $e = c/a$.

71. False. When $D = -E$, the graph is two intersecting lines.

73. Answers will vary.

75. $y = 1 - 3\sqrt{\dfrac{(x - 3)^2}{4} - 1}$

77.

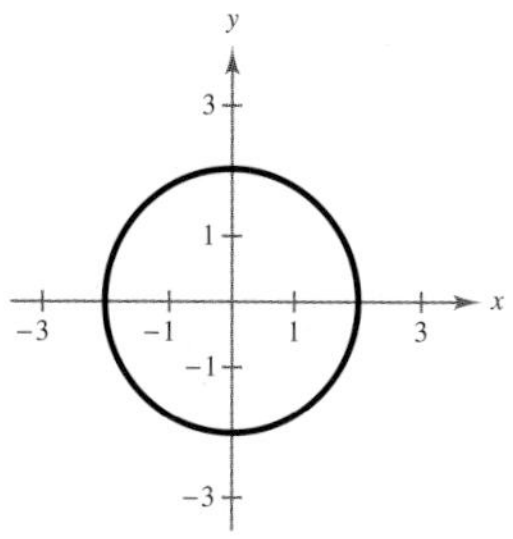

The equation $y = x^2 + C$ is a parabola that could intersect the circle in zero, one, two, three, or four places depending on its location on the y-axis.

(a) $C > 2$ and $C < -\frac{17}{4}$ (b) $C = 2$

(c) $-2 < C < 2$, $C = -\frac{17}{4}$ (d) $C = -2$

(e) $-\frac{17}{4} < C < -2$

Section 10.5 *(page 729)*

1. rotation; axes **3.** invariant under rotation **5.** $(3, 0)$

7. $\left(\dfrac{3 + \sqrt{3}}{2}, \dfrac{3\sqrt{3} - 1}{2}\right)$ **9.** $\left(\dfrac{3\sqrt{2}}{2}, -\dfrac{\sqrt{2}}{2}\right)$

11. $\left(\dfrac{2\sqrt{3} + 1}{2}, \dfrac{2 - \sqrt{3}}{2}\right)$

13. $\dfrac{(y')^2}{2} - \dfrac{(x')^2}{2} = 1$

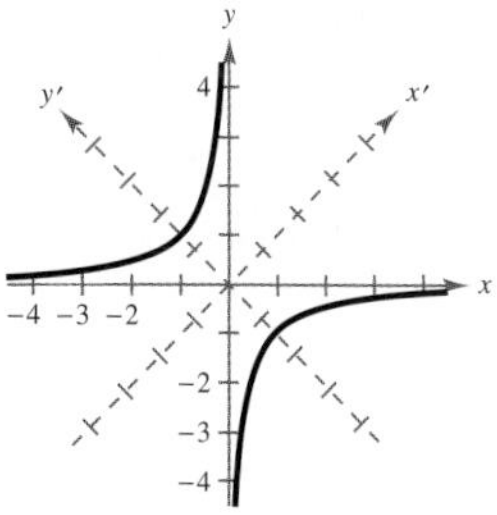

15. $\dfrac{\left(y' + \dfrac{3\sqrt{2}}{2}\right)^2}{12} - \dfrac{\left(x' + \dfrac{\sqrt{2}}{2}\right)^2}{12} = 1$

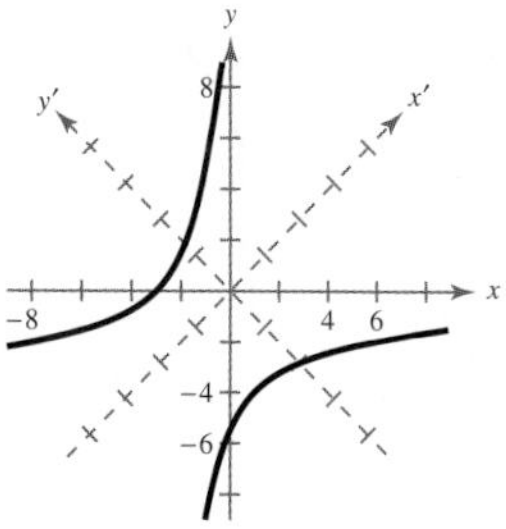

17. $\dfrac{(x')^2}{6} + \dfrac{(y')^2}{3/2} = 1$

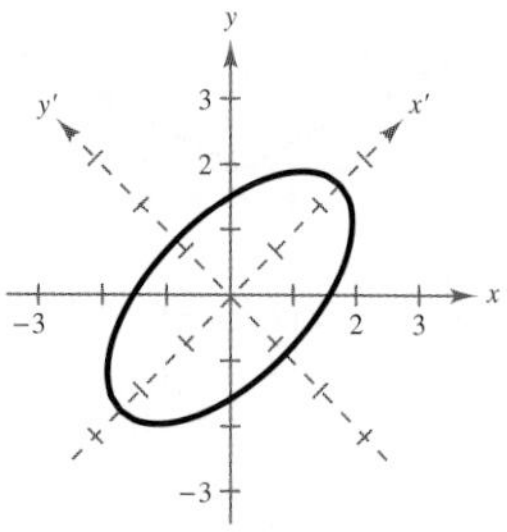

19. $\dfrac{(x')^2}{1} + \dfrac{(y')^2}{4} = 1$

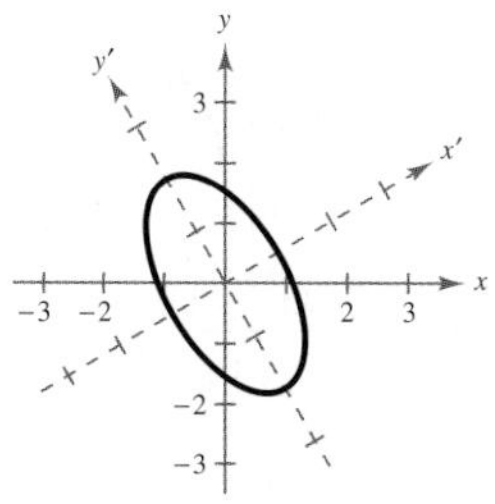

21. $(x')^2 = y'$

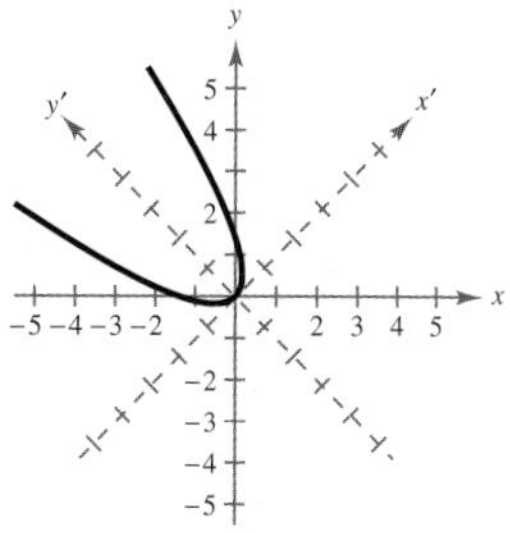

23. $(x' - 1)^2 = 6\left(y' + \frac{1}{6}\right)$

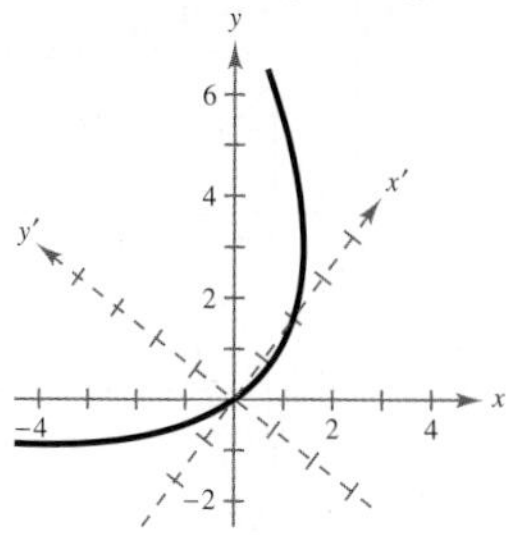

25.

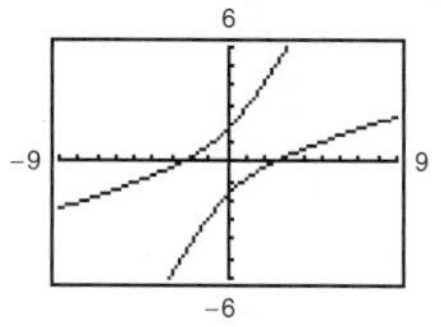

$\theta \approx 37.98°$

27.

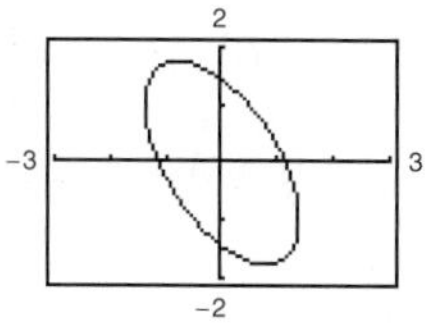

$\theta \approx 33.69°$

29.

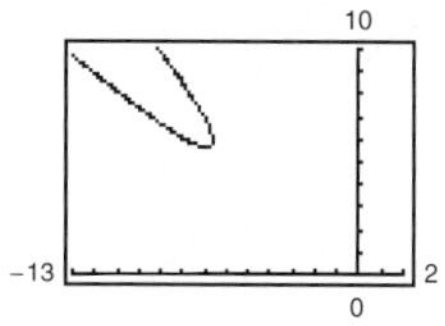

$\theta = 45°$

31. e **32.** f **33.** b **34.** a **35.** d **36.** c

37. (a) Parabola

(b) $y = \dfrac{(8x - 5) \pm \sqrt{(8x - 5)^2 - 4(16x^2 - 10x)}}{2}$

(c)

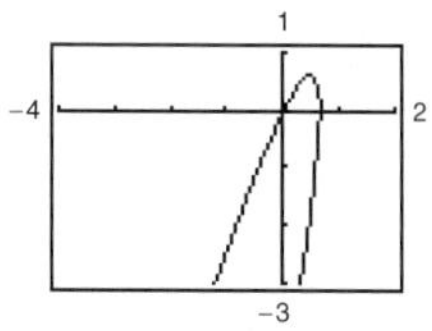

39. (a) Ellipse

(b) $y = \dfrac{6x \pm \sqrt{36x^2 - 28(12x^2 - 45)}}{14}$

(c)

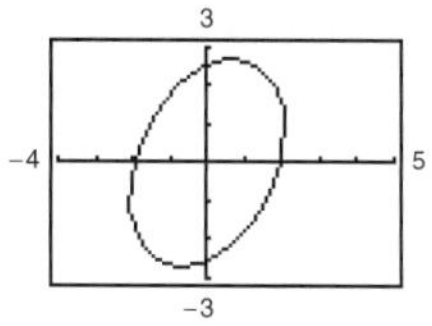

41. (a) Hyperbola

(b) $y = \dfrac{6x \pm \sqrt{36x^2 + 20(x^2 + 4x - 22)}}{-10}$

(c)

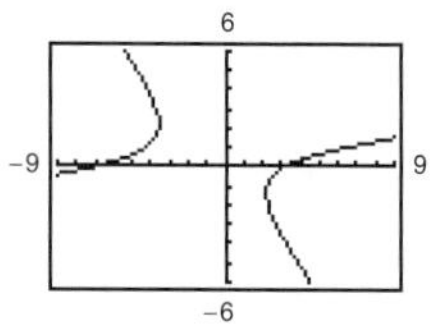

43. (a) Parabola

(b) $y = \dfrac{-(4x - 1) \pm \sqrt{(4x - 1)^2 - 16(x^2 - 5x - 3)}}{8}$

(c)

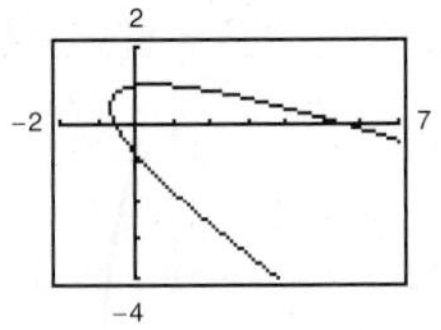

45.

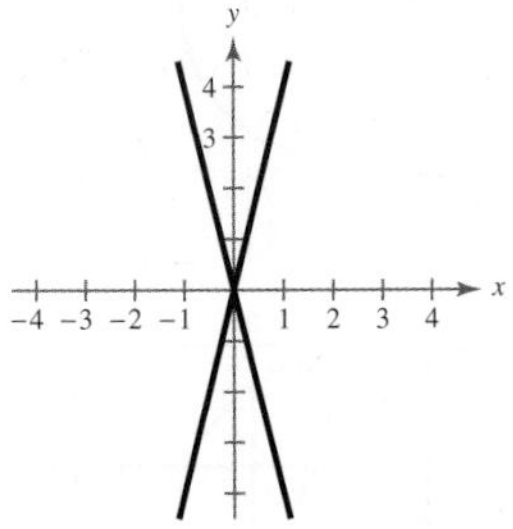

47.

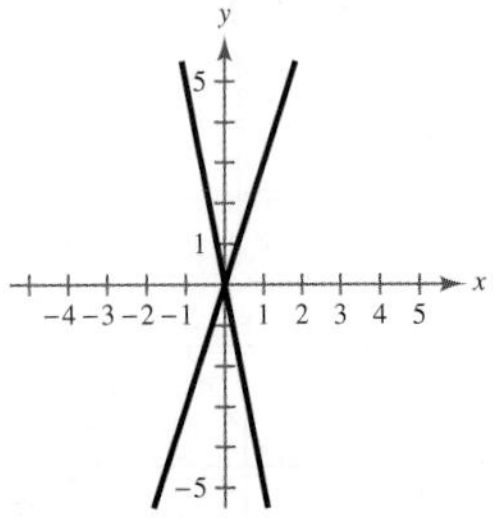

49.

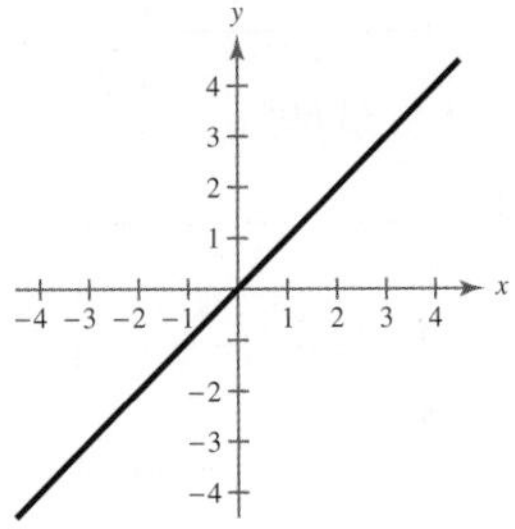

51.

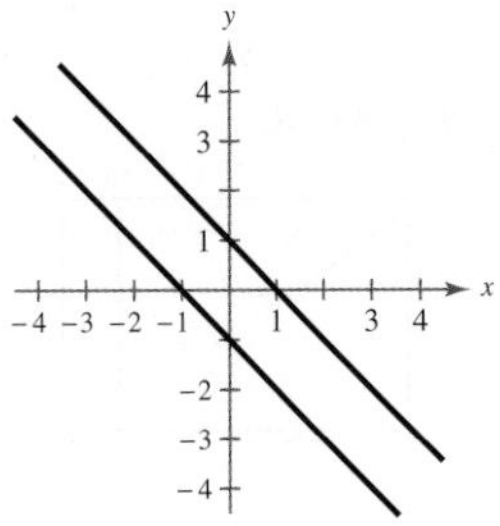

53.

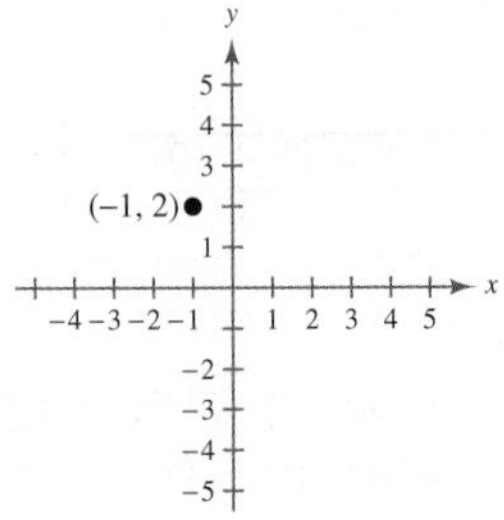

55. $(14, -8), (6, -8)$

57. $(1, 0)$ **59.** $(1, \sqrt{3}), (1, -\sqrt{3})$ **61.** $(2, 2), (2, 4)$

63. (a) $(y' + 9)^2 = 9(x' - 12)$ (b) 2.25 ft

65. True. The graph of the equation can be classified by finding the discriminant. For a graph to be a hyperbola, the discriminant must be greater than zero. If $k \geq \frac{1}{4}$, then the discriminant would be less than or equal to zero.

67. Major axis: 4; Minor axis: 2

Section 10.6 *(page 737)*

1. plane curve **3.** eliminating; parameter

5. (a)

t	0	1	2	3	4
x	0	1	$\sqrt{2}$	$\sqrt{3}$	2
y	3	2	1	0	-1

(b)

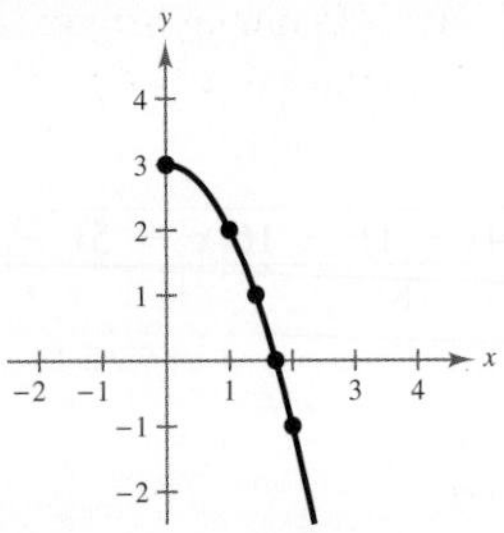

(c) $y = 3 - x^2$

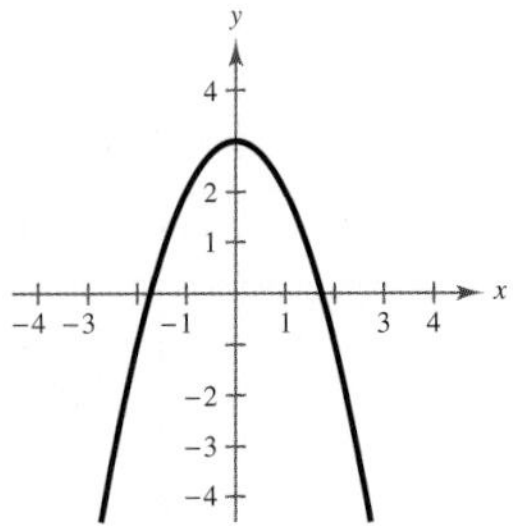

The graph of the rectangular equation shows the entire parabola rather than just the right half.

7. (a)

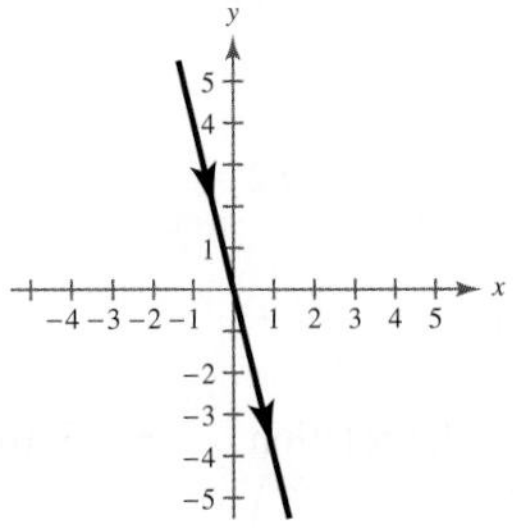

(b) $y = -4x$

9. (a)

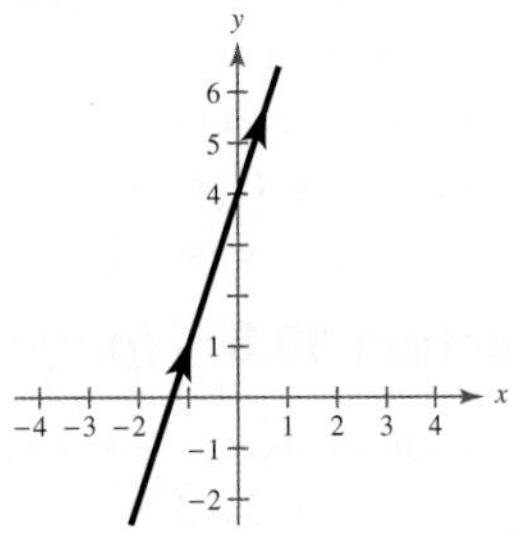

(b) $y = 3x + 4$

11. (a)

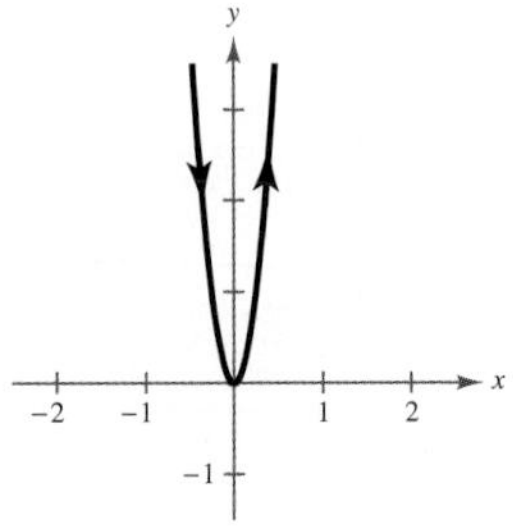

(b) $y = 16x^2$

13. (a)

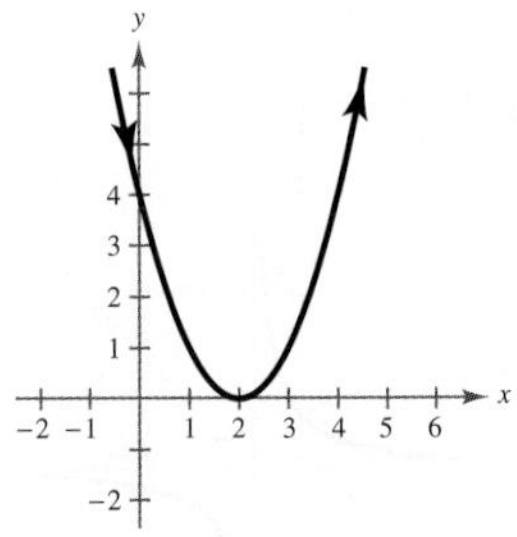

(b) $y = x^2 - 4x + 4$

15. (a)

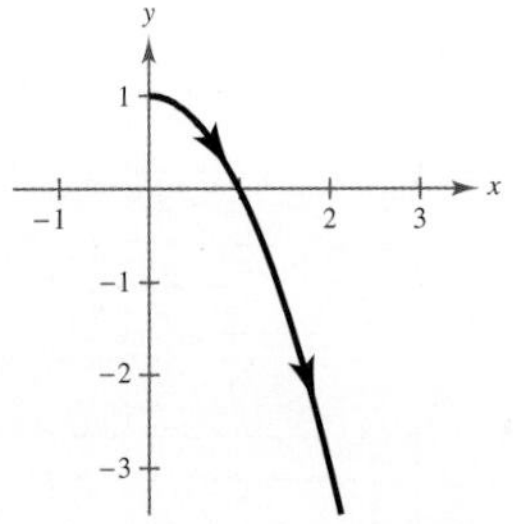

(b) $y = 1 - x^2, \quad x \geq 0$

17. (a)

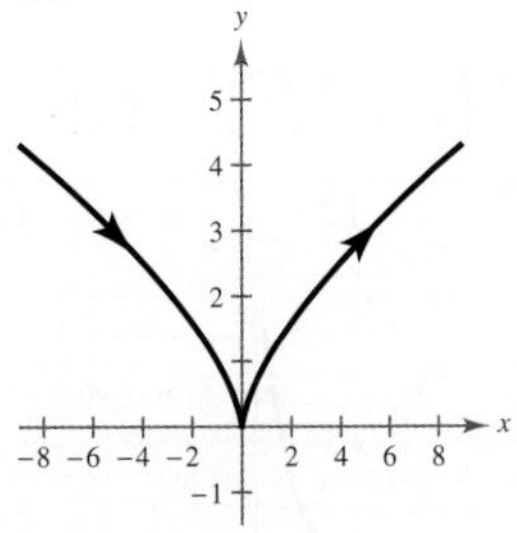

(b) $y = x^{2/3}$

19. (a)

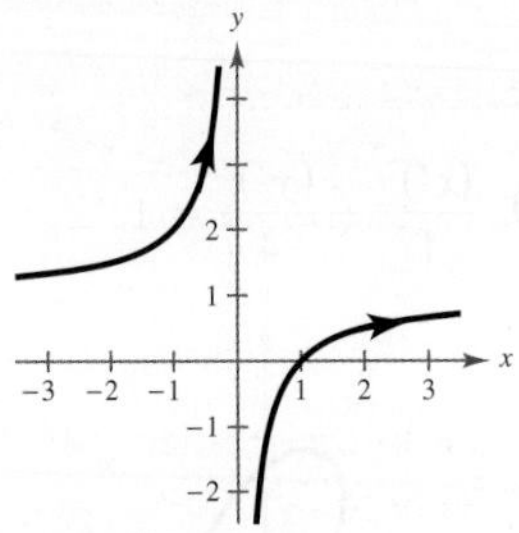

(b) $y = \dfrac{(x - 1)}{x}$

21. (a)

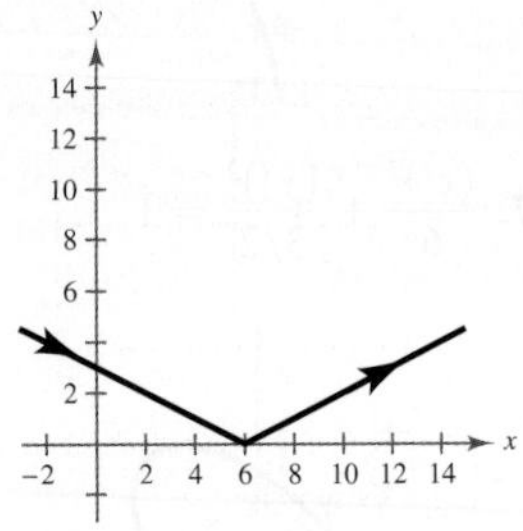

(b) $y = \left|\dfrac{x}{2} - 3\right|$

23. (a)

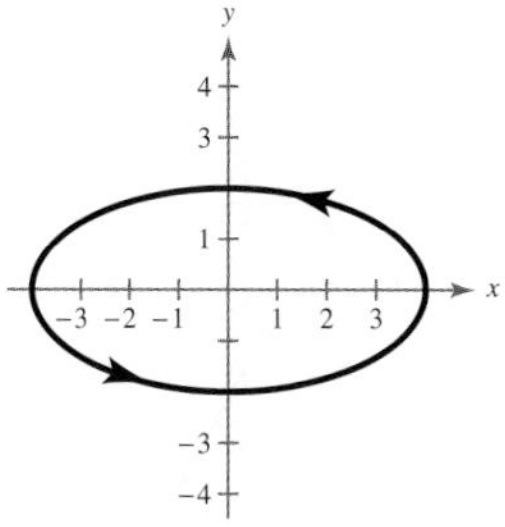

(b) $\dfrac{x^2}{16} + \dfrac{y^2}{4} = 1$

25. (a)

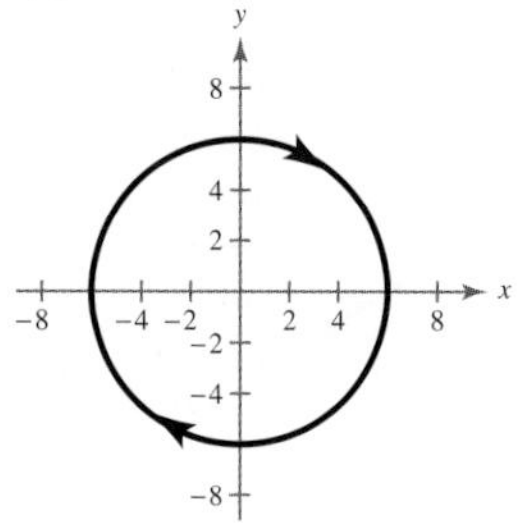

(b) $\dfrac{x^2}{36} + \dfrac{y^2}{36} = 1$

27. (a)

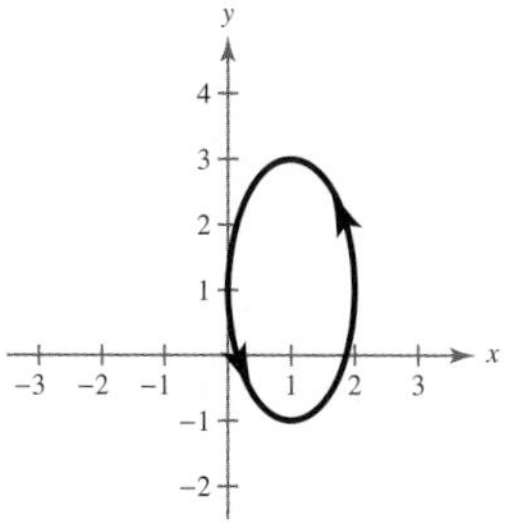

(b) $\dfrac{(x-1)^2}{1} + \dfrac{(y-1)^2}{4} = 1$

29. (a)

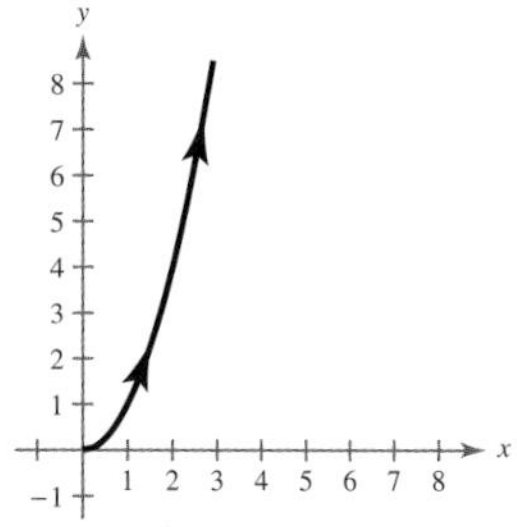

(b) $y = x^2,\ x > 0$

31. (a)

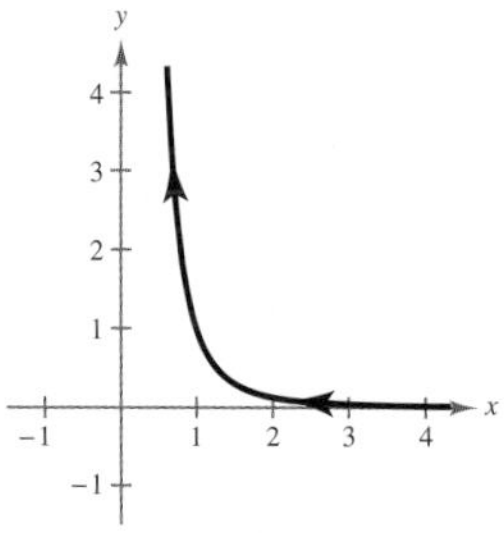

(b) $y = \dfrac{1}{x^3},\quad x > 0$

33. (a)

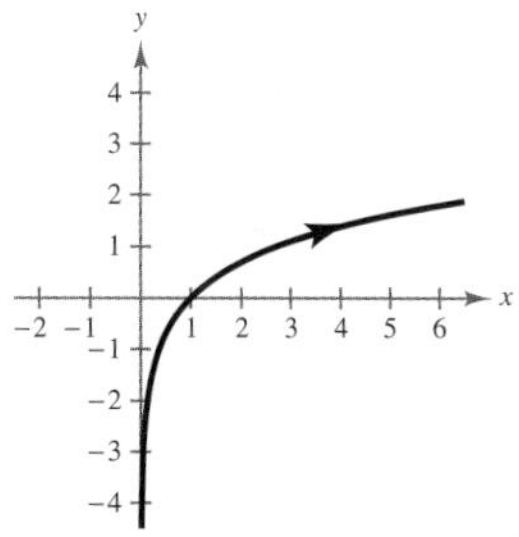

(b) $y = \ln x$

35.

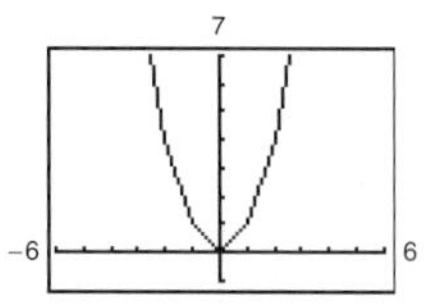

37.

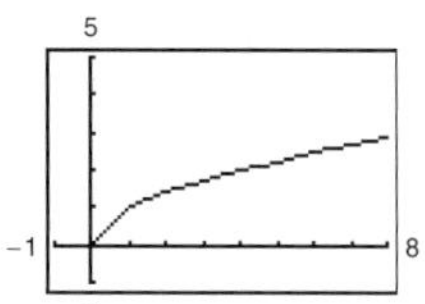

39.

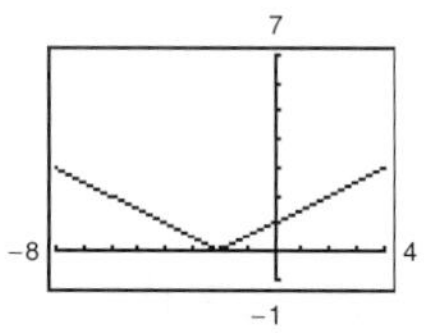

41.

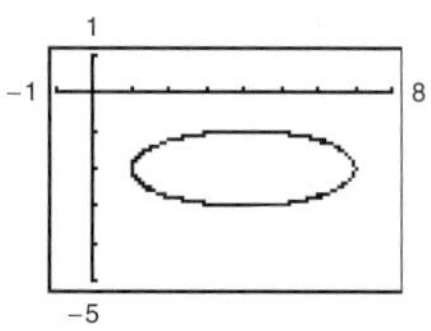

43.

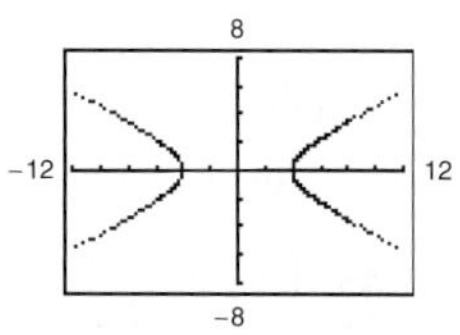

45.

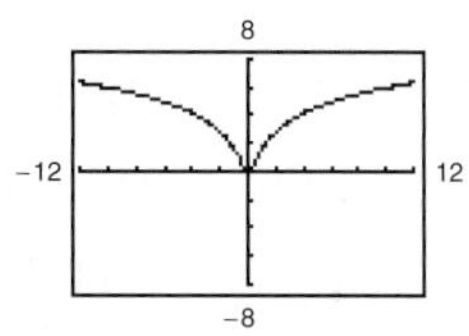

47. Each curve represents a portion of the line $y = 2x + 1$.

	Domain	**Orientation**
(a)	$(-\infty, \infty)$	Left to right
(b)	$[-1, 1]$	Depends on θ
(c)	$(0, \infty)$	Right to left
(d)	$(0, \infty)$	Left to right

49. $y - y_1 = m(x - x_1)$

51. $\dfrac{(x-h)^2}{a^2} + \dfrac{(y-k)^2}{b^2} = 1$

53. $x = 3t$
$y = 6t$

55. $x = 3 + 4\cos\theta$
$y = 2 + 4\sin\theta$

57. $x = 5\cos\theta$
$y = 3\sin\theta$

59. $x = 4\sec\theta$
$y = 3\tan\theta$

61. (a) $x = t,\ y = 3t - 2$ (b) $x = -t + 2,\ y = -3t + 4$

63. (a) $x = t, y = 2 - t$ (b) $x = -t + 2, y = t$

65. (a) $x = t, y = \frac{1}{2}(t - 1)$ (b) $x = -t + 2, y = -\frac{1}{2}(t - 1)$

67. (a) $x = t, y = t^2 + 1$ (b) $x = -t + 2, y = t^2 - 4t + 5$

69. (a) $x = t, y = 3t^2 + 1$
(b) $x = -t + 2, y = 3t^2 - 12t + 13$

71. (a) $x = t, y = 1 - 2t^2$
(b) $x = 2 - t, y = -2t^2 + 8t - 7$

73. (a) $x = t, y = \dfrac{1}{t}$ (b) $x = -t + 2, y = -\dfrac{1}{t-2}$

75. (a) $x = t, y = e^t$ (b) $x = -t + 2, y = e^{-t+2}$

77.

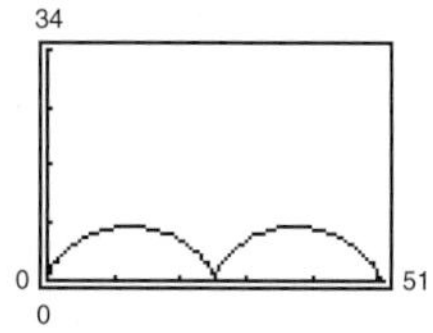

79.

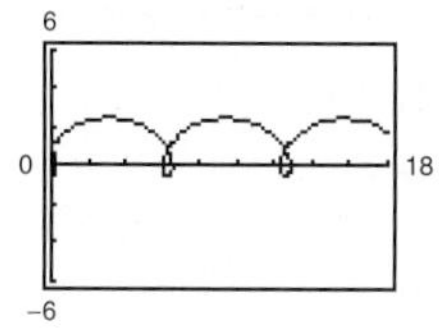

81.

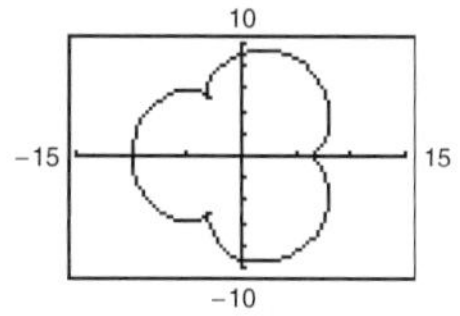

83.

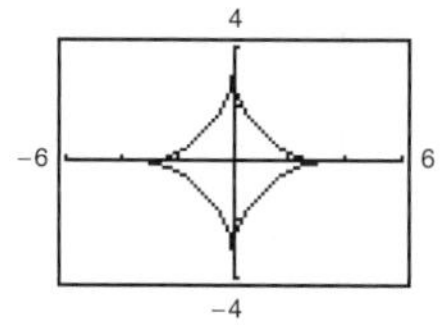

85.

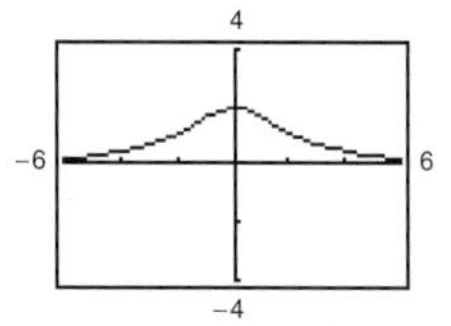

87. b
Domain: $[-2, 2]$
Range: $[-1, 1]$

88. c
Domain: $[-4, 4]$
Range: $[-6, 6]$

89. d
Domain: $(-\infty, \infty)$
Range: $(-\infty, \infty)$

90. a
Domain: $(-\infty, \infty)$
Range: $[-2, 2]$

91. f
Domain: $[-3, 6]$
Range: $[-3\sqrt{3}, 3\sqrt{3}]$

92. e
Domain: $(-\infty, \infty)$
Range: $(-2, 2)$

93. (a)

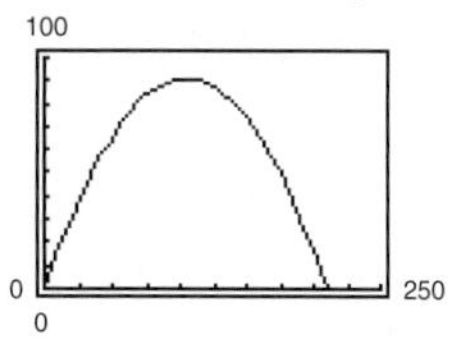

Maximum height: 90.7 ft
Range: 209.6 ft

(b)

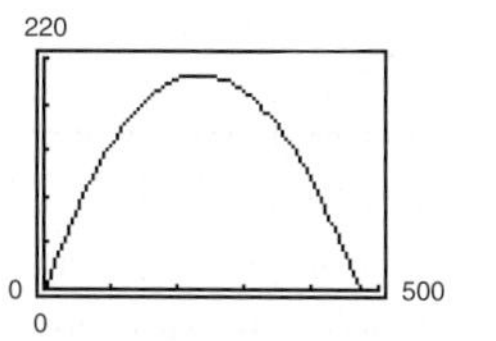

Maximum height: 204.2 ft
Range: 471.6 ft

(c)

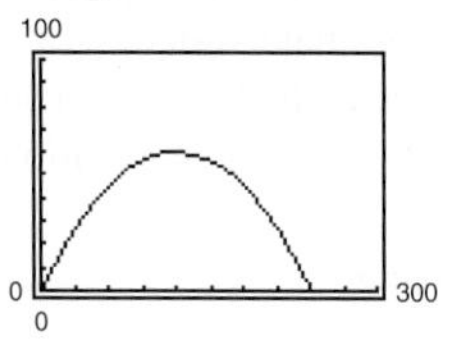

Maximum height: 60.5 ft
Range: 242.0 ft

(d) 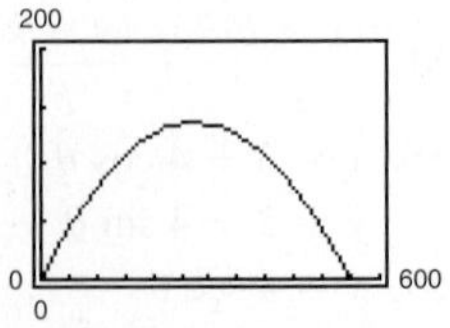

Maximum height: 136.1 ft
Range: 544.5 ft

95. (a) $x = (146.67 \cos \theta)t$
$y = 3 + (146.67 \sin \theta)t - 16t^2$

(b) 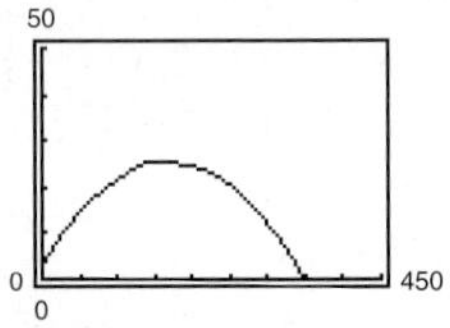

No

(c) 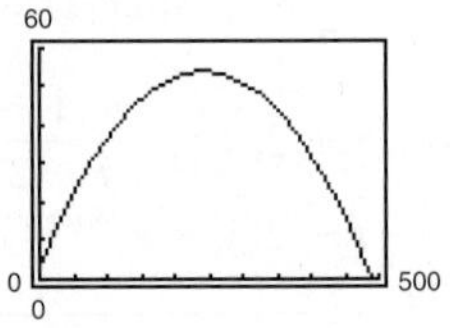

Yes

(d) 19.3°

97. (a) $x = (\cos 35°)v_0 t$
$y = 7 + (\sin 35°)v_0 t - 16t^2$

(b) About 54.09 ft/sec

(c) 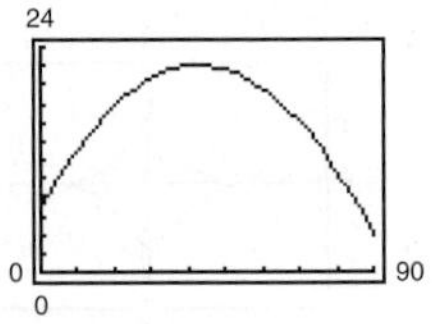

22.04 ft

(d) About 2.03 sec

99. (a) $h = 7, v_0 = 40, \theta = 45°$
$x = (40 \cos 45°)t$
$y = 7 + (40 \sin 45°)t - 16t^2$

(b) 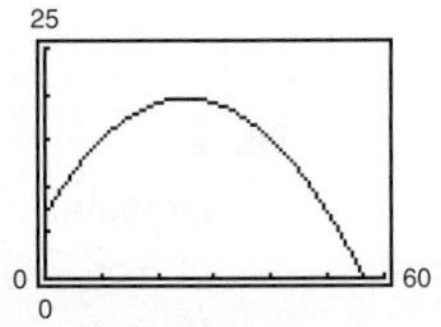

(c) Maximum height: 19.5 ft
Range: 56.2 ft

101. $x = a\theta - b \sin \theta$
$y = a - b \cos \theta$

103. True
$x = t$
$y = t^2 + 1 \Rightarrow y = x^2 + 1$
$x = 3t$
$y = 9t^2 + 1 \Rightarrow y = x^2 + 1$

105. False. The parametric equations $x = t^2$ and $y = t$ give the rectangular equation $x = y^2$, so y is not a function of x.

107. Parametric equations are useful when graphing two functions simultaneously on the same coordinate system. For example, they are useful when tracking the path of an object so that the position and the time associated with that position can be determined.

109. Yes. The orientation would change.

111. $-1 < t < \infty$

Section 10.7 *(page 745)*

1. pole **3.** polar

5. 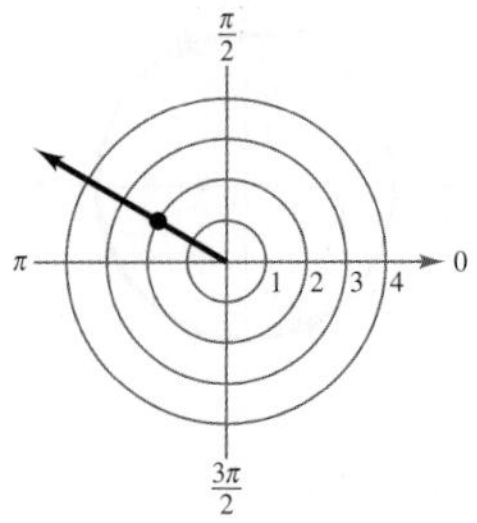

$\left(2, -\frac{7\pi}{6}\right), \left(-2, -\frac{\pi}{6}\right)$

7. 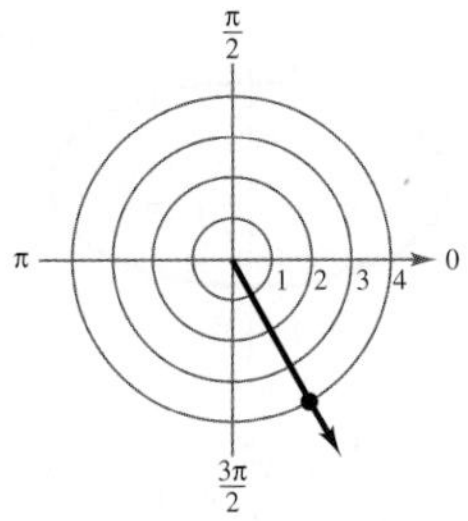

$\left(4, \frac{5\pi}{3}\right), \left(-4, -\frac{4\pi}{3}\right)$

9. 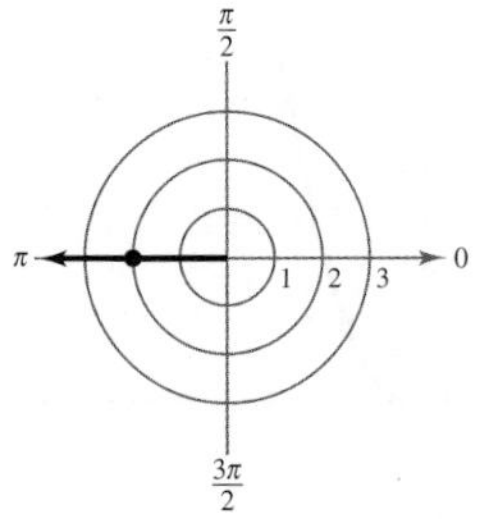

$(2, \pi), (-2, 0)$

11. 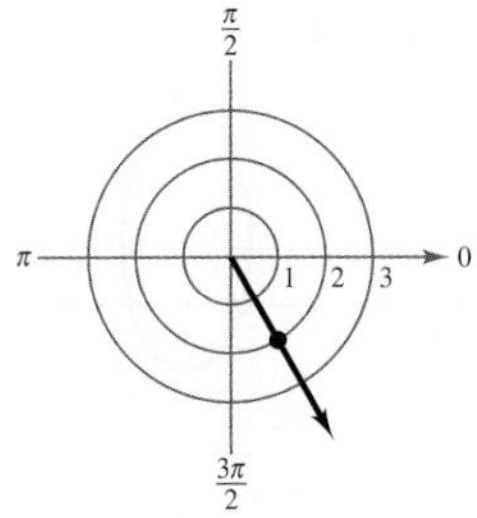

$\left(-2, -\frac{4\pi}{3}\right), \left(2, \frac{5\pi}{3}\right)$

13. 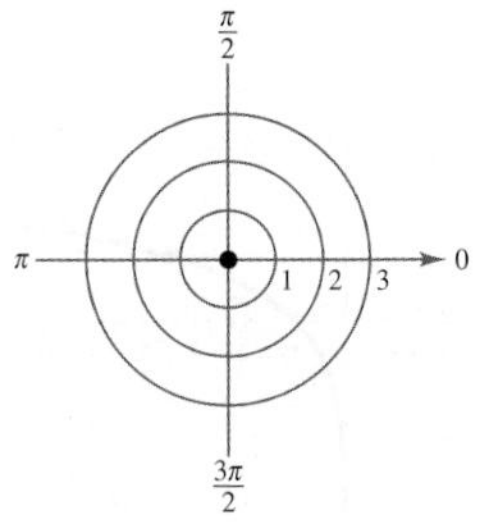

$\left(0, \frac{5\pi}{6}\right), \left(0, -\frac{\pi}{6}\right)$

15. 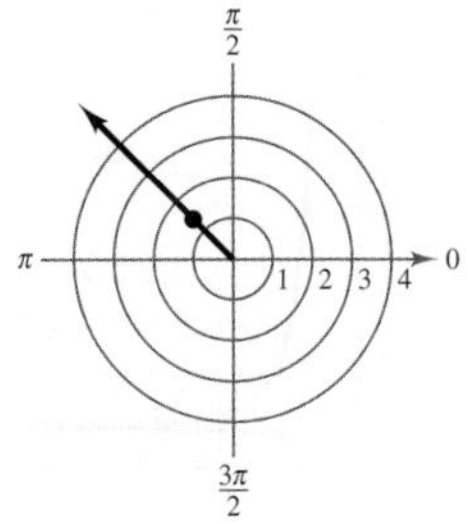

$(\sqrt{2}, -3.92), (-\sqrt{2}, -0.78)$

17. 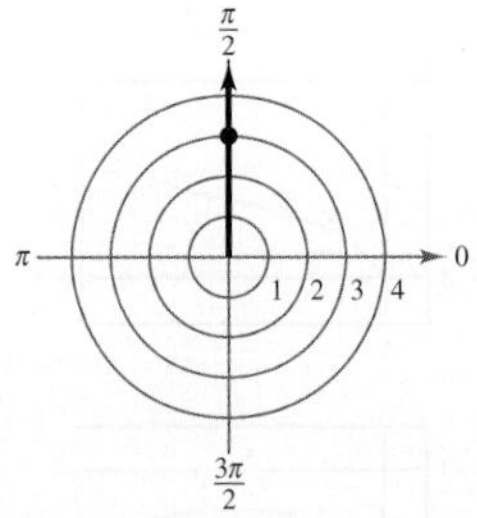

$(-3, 4.71), (3, 1.57)$

19. $(0, 0)$ **21.** $(0, 3)$

23. $(-\sqrt{2}, \sqrt{2})$ **25.** $\left(\frac{\sqrt{2}}{2}, \frac{\sqrt{2}}{2}\right)$ **27.** $(\sqrt{3}, 1)$

29. $\left(-\frac{3}{2}, \frac{3\sqrt{3}}{2}\right)$ **31.** $(-1.84, 0.78)$ **33.** $(-1.1, -2.2)$

35. $(1.53, 1.29)$ **37.** $(-1.20, -4.34)$ **39.** $(-0.02, 2.50)$

41. $(-3.60, 1.97)$ **43.** $\left(\sqrt{2}, \frac{\pi}{4}\right)$ **45.** $\left(3\sqrt{2}, \frac{5\pi}{4}\right)$

47. $(6, \pi)$ **49.** $\left(5, \frac{3\pi}{2}\right)$ **51.** $(5, 2.21)$ **53.** $\left(\sqrt{6}, \frac{5\pi}{4}\right)$

55. $\left(2, \frac{11\pi}{6}\right)$ **57.** $(3\sqrt{13}, 0.98)$ **59.** $(13, 1.18)$

61. $(\sqrt{13}, 5.70)$ **63.** $(\sqrt{29}. 2.76)$ **65.** $(\sqrt{7}, 0.86)$

67. $\left(\frac{17}{6}, 0.49\right)$ **69.** $\left(\frac{\sqrt{85}}{4}, 0.71\right)$ **71.** $r = 3$

73. $\theta = \frac{\pi}{4}$ **75.** $r = 10 \sec \theta$ **77.** $r = \csc \theta$

79. $r = \frac{-2}{3 \cos \theta - \sin \theta}$

81. $r^2 = 16 \sec \theta \csc \theta = 32 \csc 2\theta$ **83.** $r = a$

85. $r = 2a \cos \theta$ **87.** $r^2 = \cos 2\theta$ **89.** $r = \cot^2 \theta \csc \theta$

91. $x^2 + y^2 - 4y = 0$ **93.** $x^2 + y^2 + 2x = 0$

95. $\sqrt{3}x + y = 0$ **97.** $\frac{\sqrt{3}}{3}x + y = 0$

99. $x^2 + y^2 = 16$ **101.** $y = 4$

103. $x = -3$ **105.** $x^2 + y^2 - x^{2/3} = 0$

107. $(x^2 + y^2)^2 = 2xy$ **109.** $(x^2 + y^2)^2 = 6x^2y - 2y^3$

111. $x^2 + 4y - 4 = 0$ **113.** $4x^2 - 5y^2 - 36y - 36 = 0$

115. $2x - 3y = 6$

117. $x^2 + y^2 = 36$

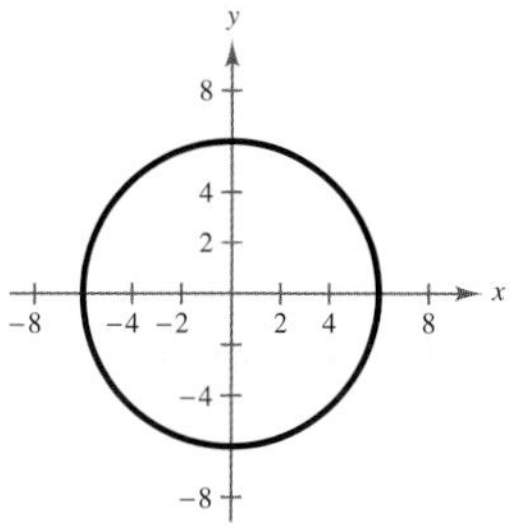

119. $-\sqrt{3}x + 3y = 0$

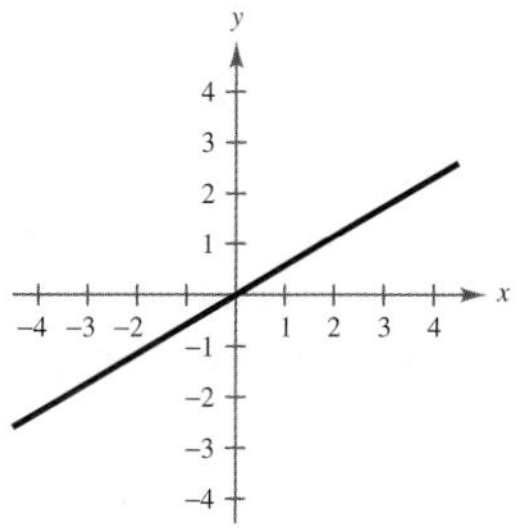

121. $x^2 + (y - 1)^2 = 1$

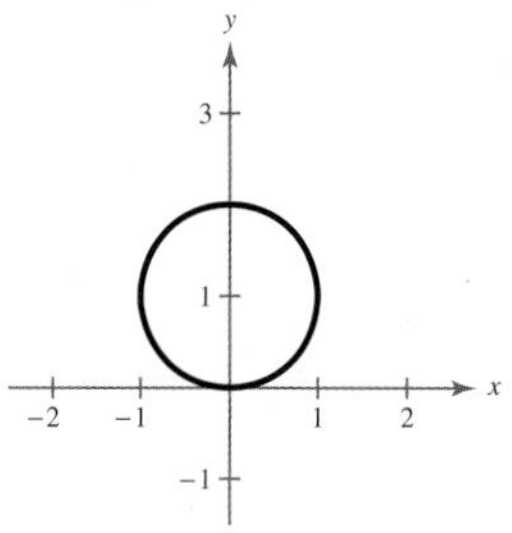

123. $(x + 3)^2 + y^2 = 9$

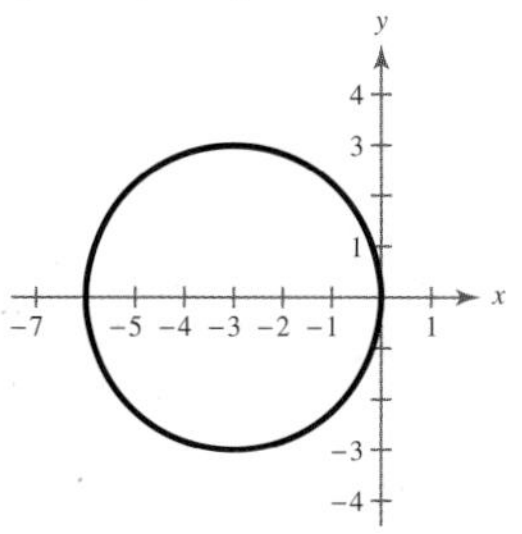

125. $x - 3 = 0$

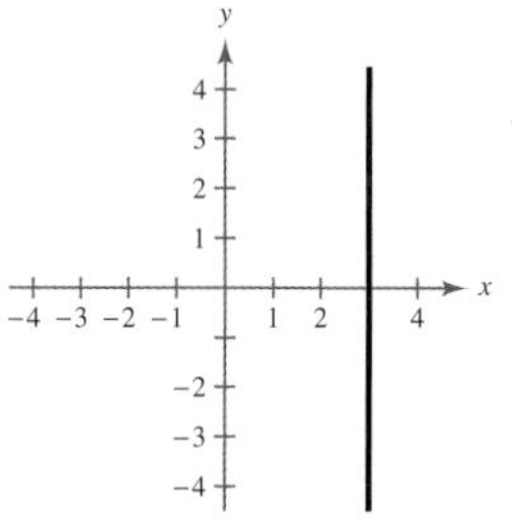

127. (a) $r = 30$

(b) $(30, 5\pi/6)$; 30 represents the distance of the passenger car from the center, and $5\pi/6 = 150°$ represents the angle to which the car has rotated.

(c) $(-25.98, 15)$; The car is about 25.98 feet to the left of the center and 15 feet above the center.

129. True. Because r is a directed distance, the point (r, θ) can be represented as $(r, \theta \pm 2\pi n)$.

131. False. If $r_1 = -r_2$, then (r_1, θ) and (r_2, θ) are different points.

133. $(x - h)^2 + (y - k)^2 = h^2 + k^2$
Radius: $\sqrt{h^2 + k^2}$
Center: (h, k)

135. (a) Answers will vary.

(b) (r_1, θ_1), (r_2, θ_2) and the pole are collinear.
$d = \sqrt{r_1^2 + r_2^2 - 2r_1r_2} = |r_1 - r_2|$
This represents the distance between two points on the line $\theta = \theta_1 = \theta_2$.

(c) $d = \sqrt{r_1^2 + r_2^2}$
This is the result of the Pythagorean Theorem.

Section 10.8 *(page 753)*

1. $\theta = \frac{\pi}{2}$ **3.** convex limaçon **5.** lemniscate

7. Rose curve with 4 petals **9.** Limaçon with inner loop

11. Rose curve with 3 petals **13.** Polar axis

15. $\theta = \frac{\pi}{2}$ **17.** $\theta = \frac{\pi}{2}$, polar axis, pole

19. Maximum: $|r| = 20$ when $\theta = \frac{3\pi}{2}$

Zero: $r = 0$ when $\theta = \frac{\pi}{2}$

21. Maximum: $|r| = 4$ when $\theta = 0, \frac{\pi}{3}, \frac{2\pi}{3}$

Zeros: $r = 0$ when $\theta = \frac{\pi}{6}, \frac{\pi}{2}, \frac{5\pi}{6}$

23.

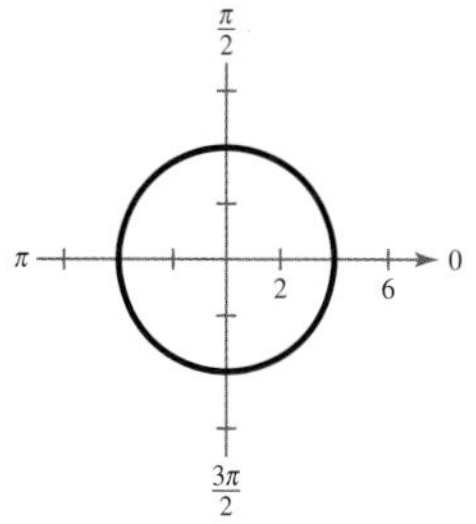

25.

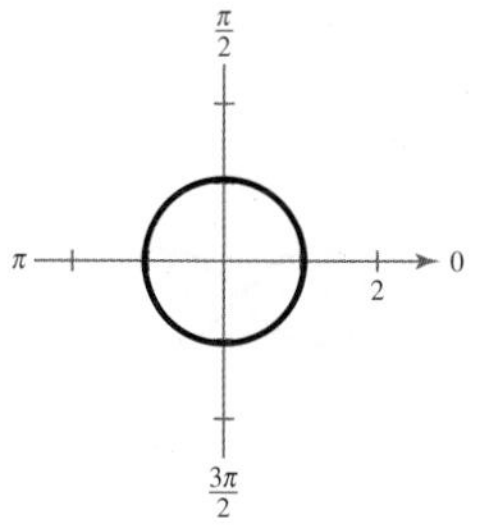

27.

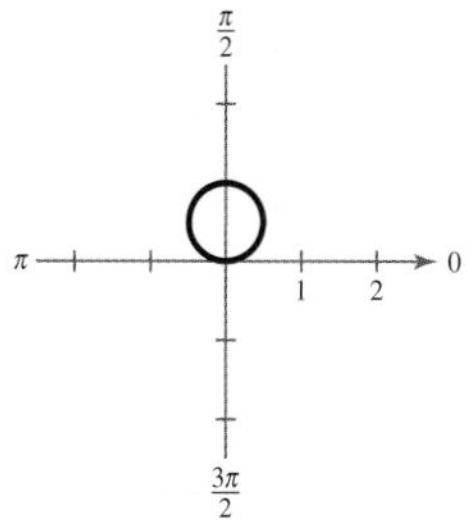

29.

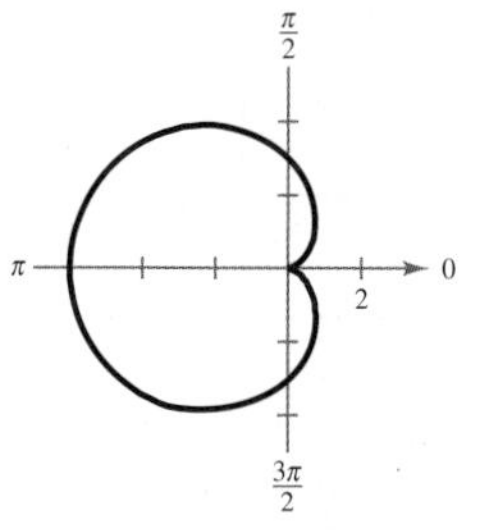

31.

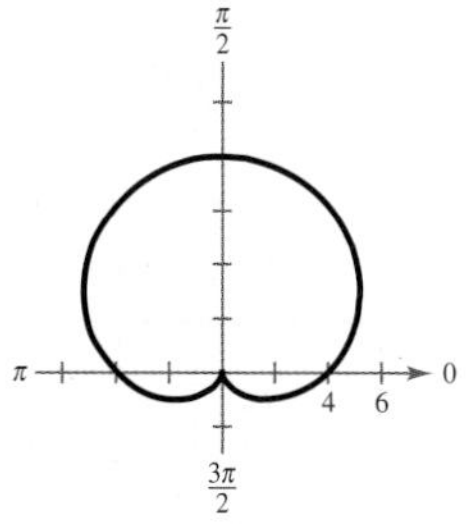

33.

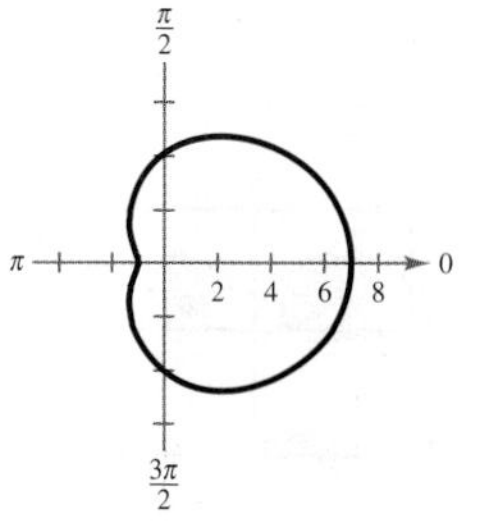

35.

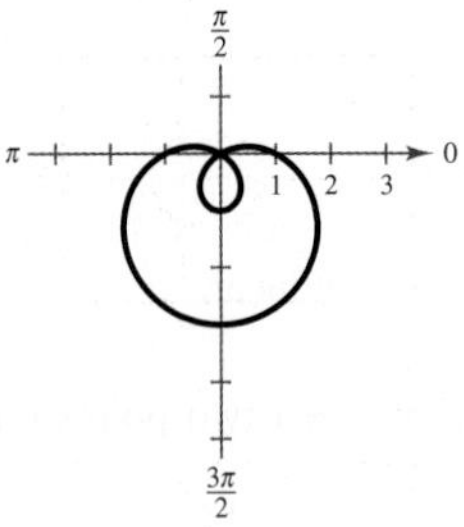

37.

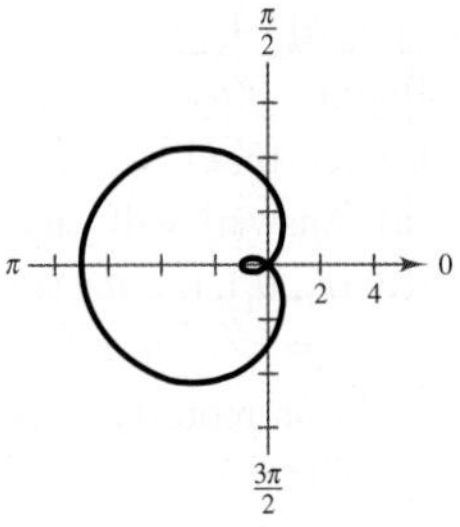

39.

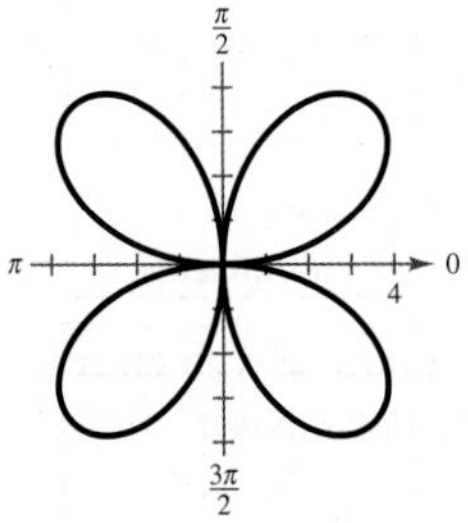

41.

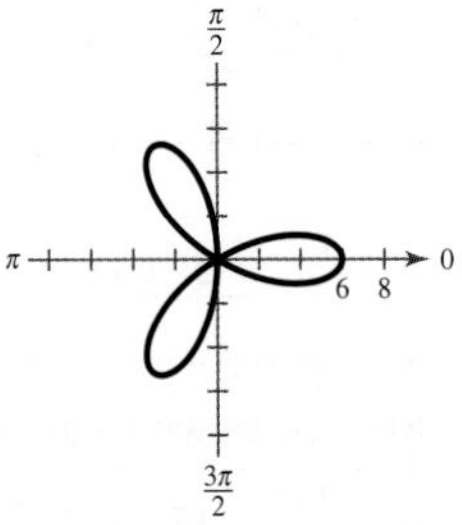

43.

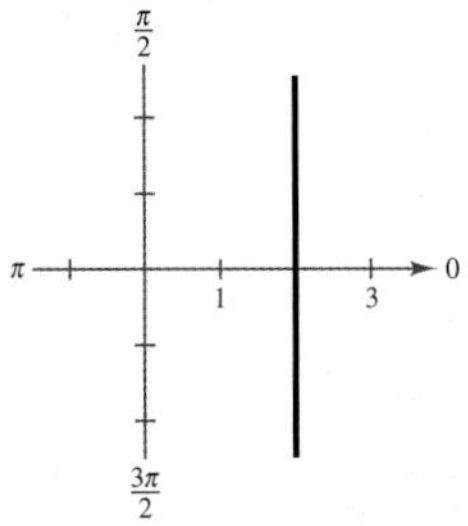

45.

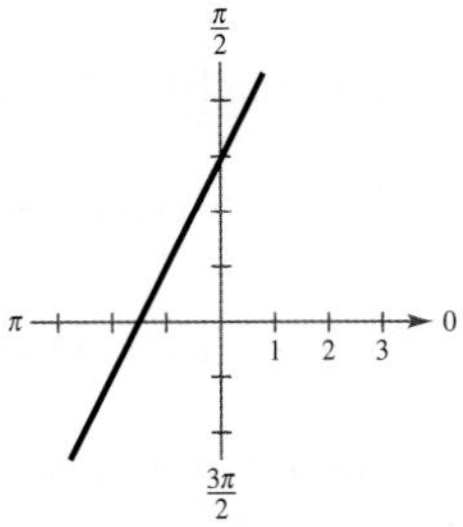

47.

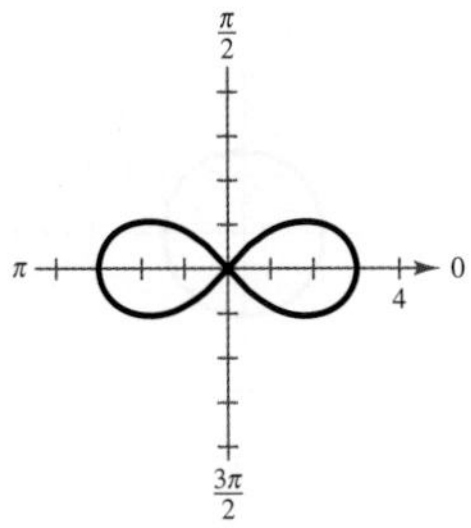

49.

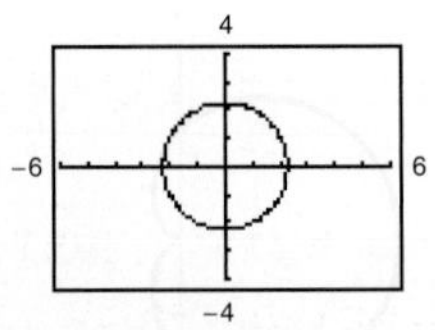

51.

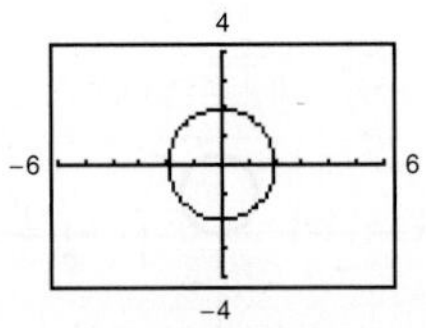

53.

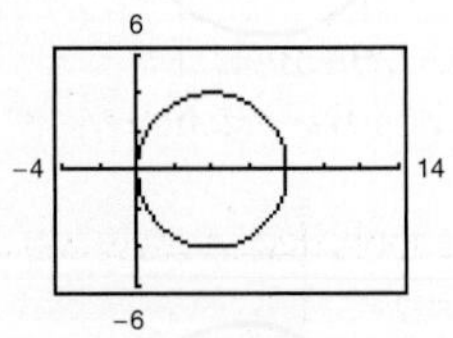

55.

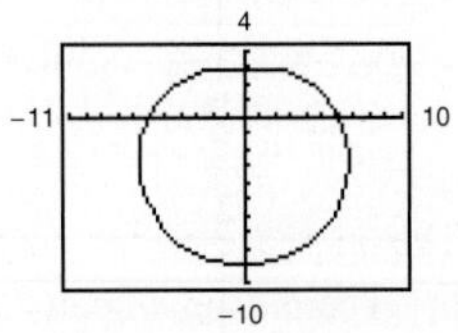

57.

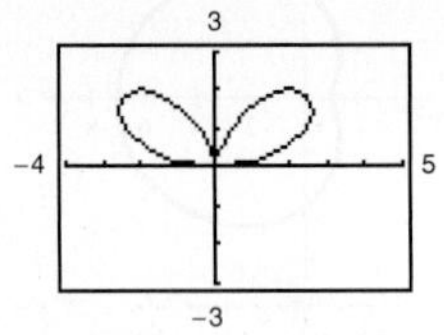

59.

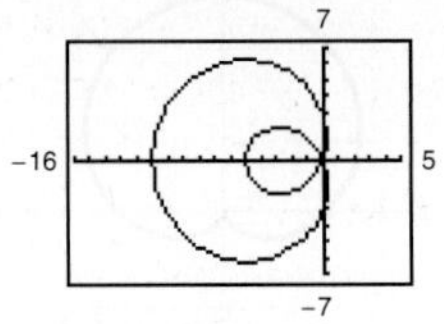

$0 \le \theta < 2\pi$

61.

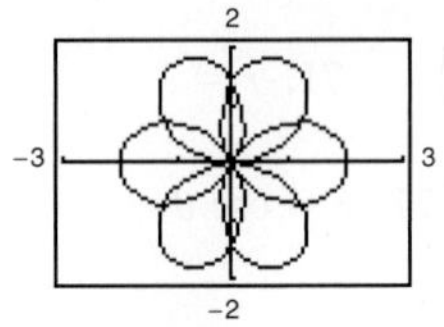

$0 \le \theta < 4\pi$

63.

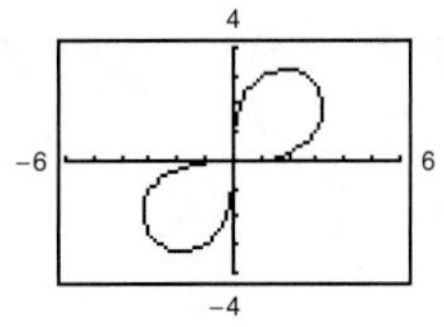

$0 \le \theta < \pi/2$

65.

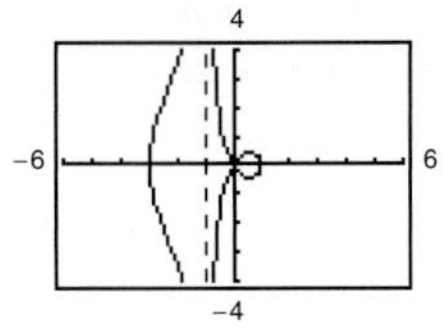

67.

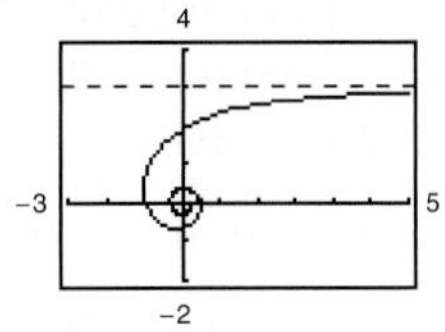

69. (a)

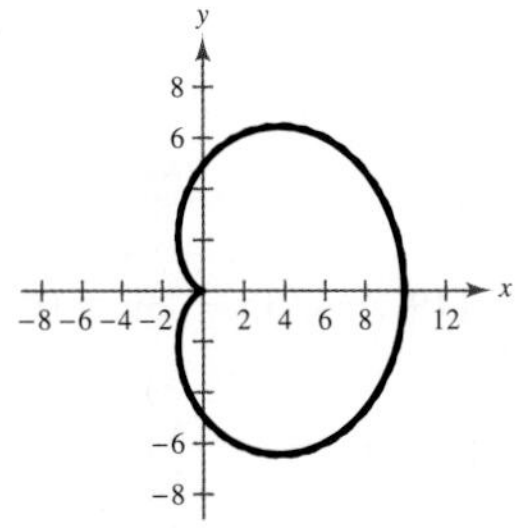

Cardioid

(b) 0 radians

71. True. The equation is of the form $r = a \sin n\theta$, where n is odd.

73. (a)

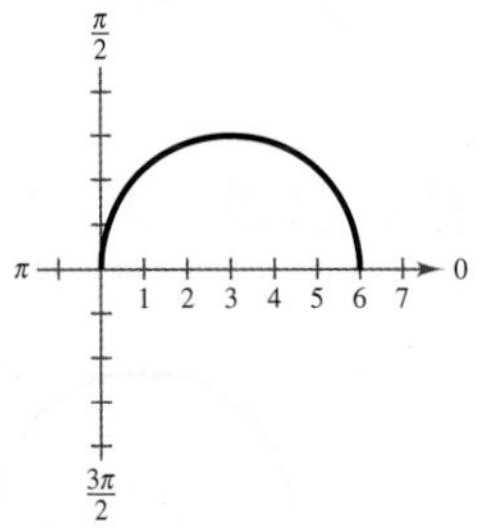

Upper half of circle

(b)

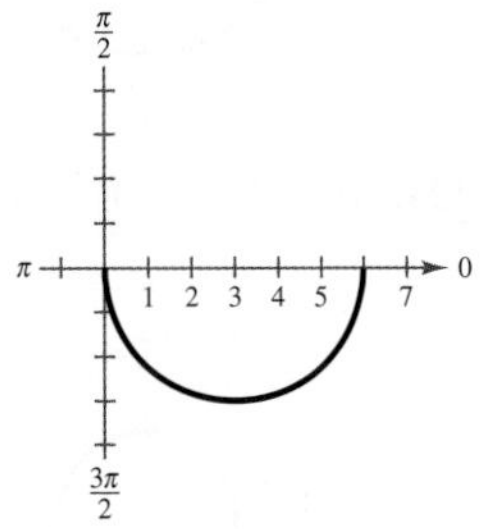

Lower half of circle

(c)

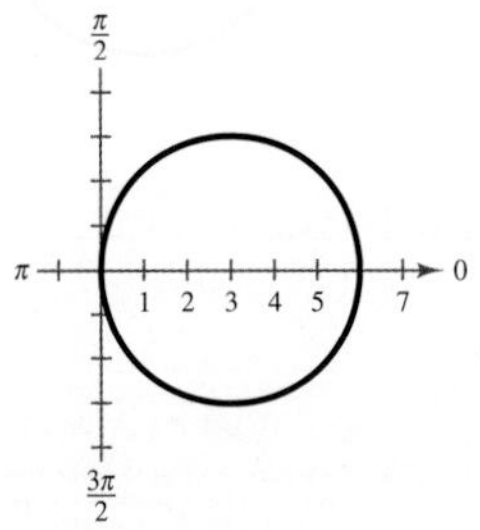

Full circle

(d)

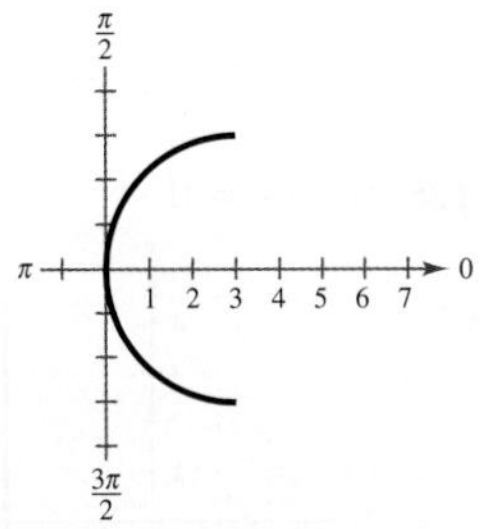

Left half of circle

75. Answers will vary.

77. (a) $r = 2 - \dfrac{\sqrt{2}}{2}(\sin\theta - \cos\theta)$ (b) $r = 2 + \cos\theta$

(c) $r = 2 + \sin\theta$ (d) $r = 2 - \cos\theta$

79. (a)

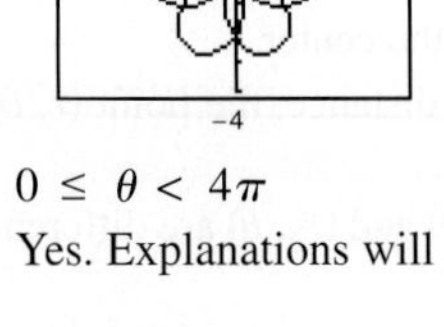

$0 \le \theta < 4\pi$

(b)

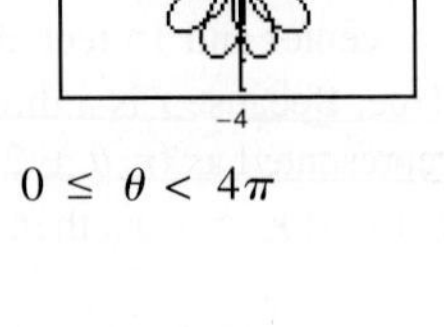

$0 \le \theta < 4\pi$

(c) Yes. Explanations will vary.

Section 10.9 (page 759)

1. conic **3.** vertical; right

5. $e = 1$: $r = \dfrac{2}{1 + \cos\theta}$, parabola

$e = 0.5$: $r = \dfrac{1}{1 + 0.5\cos\theta}$, ellipse

$e = 1.5$: $r = \dfrac{3}{1 + 1.5\cos\theta}$, hyperbola

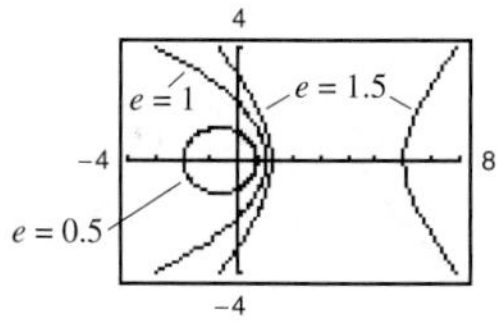

7. $e = 1$: $r = \dfrac{2}{1 - \sin\theta}$, parabola

$e = 0.5$: $r = \dfrac{1}{1 - 0.5\sin\theta}$, ellipse

$e = 1.5$: $r = \dfrac{3}{1 - 1.5\sin\theta}$, hyperbola

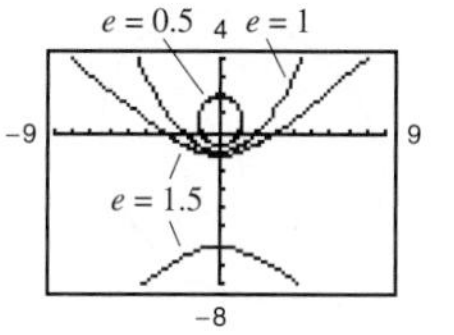

9. e **10.** c **11.** d **12.** f **13.** a **14.** b

15. Parabola

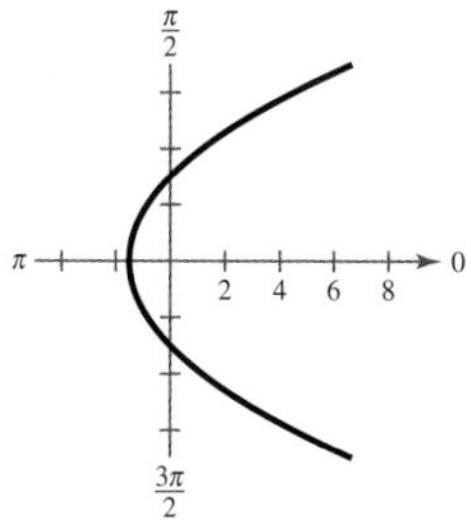

17. Parabola

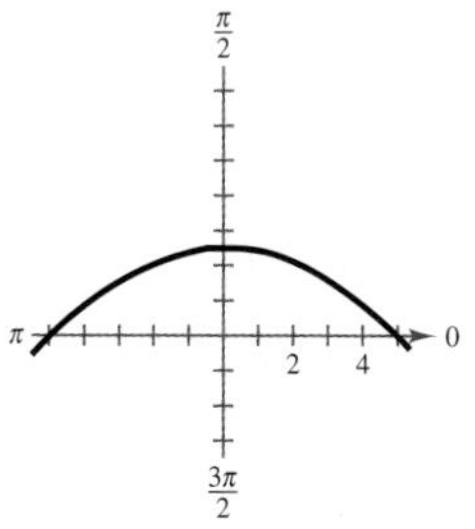

19. Ellipse

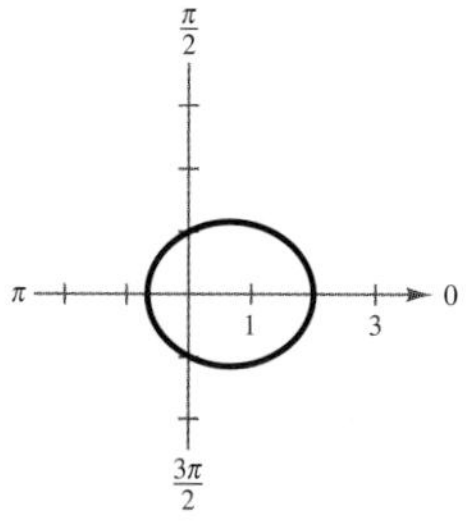

21. Ellipse

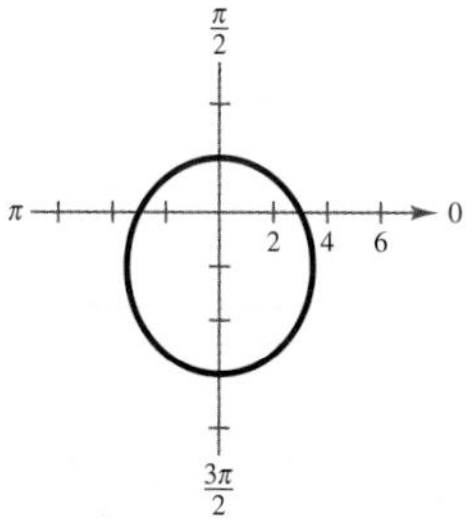

23. Hyperbola

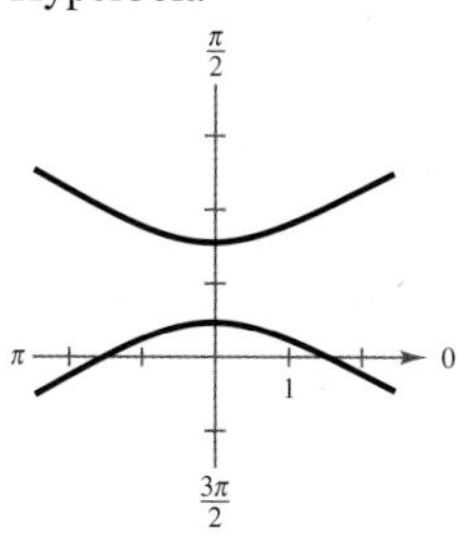

25. Hyperbola

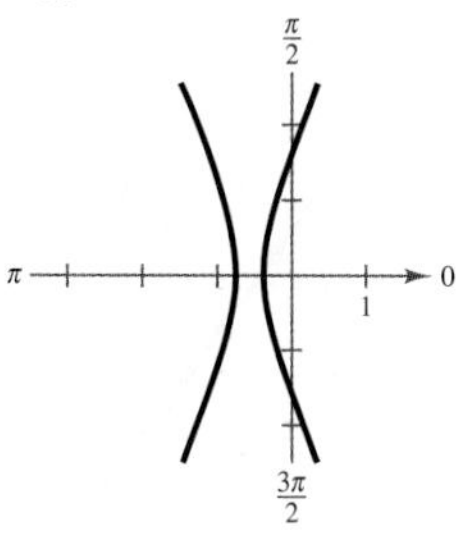

27.

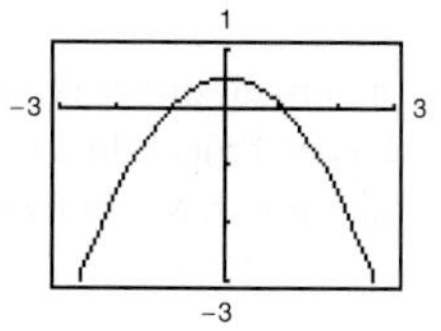

Parabola

29.

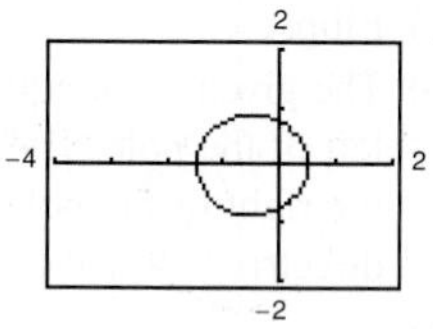

Ellipse

31.

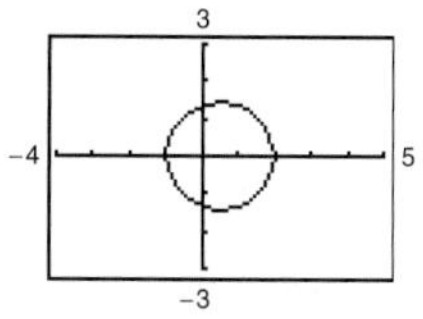

Ellipse

33.

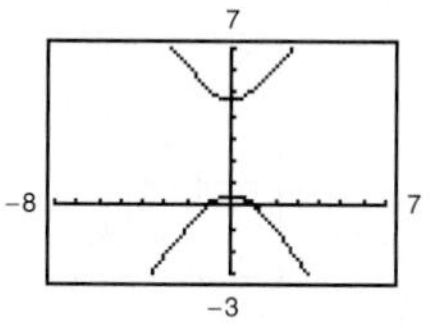

Hyperbola

35.

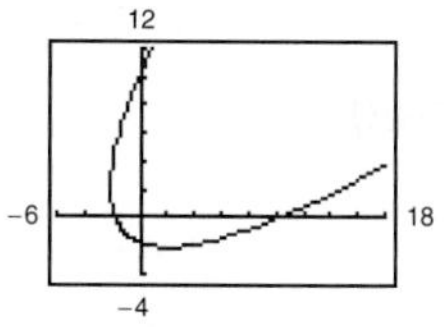

37.

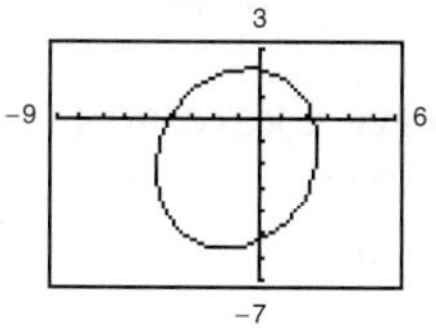

39. $r = \dfrac{1}{1 - \cos\theta}$ **41.** $r = \dfrac{1}{2 + \sin\theta}$

43. $r = \dfrac{2}{1 + 2\cos\theta}$ **45.** $r = \dfrac{2}{1 - \sin\theta}$

47. $r = \dfrac{10}{1 - \cos\theta}$ **49.** $r = \dfrac{10}{3 + 2\cos\theta}$

51. $r = \dfrac{20}{3 - 2\cos\theta}$ **53.** $r = \dfrac{9}{4 - 5\sin\theta}$

55. Answers will vary.

57. $r = \dfrac{9.2930 \times 10^7}{1 - 0.0167\cos\theta}$

Perihelion: 9.1404×10^7 mi

Aphelion: 9.4508×10^7 mi

59. $r = \dfrac{1.0820 \times 10^8}{1 - 0.0068\cos\theta}$

Perihelion: 1.0747×10^8 km

Aphelion: 1.0895×10^8 km

61. $r = \dfrac{1.4039 \times 10^8}{1 - 0.0934\cos\theta}$

Perihelion: 1.2840×10^8 mi

Aphelion: 1.5486×10^8 mi

63. $r = \dfrac{2.494}{1 + 0.995\sin\theta}$; $r \approx 1.25$ astronomical units

65. True. The graphs represent the same hyperbola.

67. True. The conic is an ellipse because the eccentricity is less than 1.

69. The original equation graphs as a parabola that opens downward.

(a) The parabola opens to the right.
(b) The parabola opens up.
(c) The parabola opens to the left.
(d) The parabola has been rotated.

71. Answers will vary.

73. $r^2 = \dfrac{24{,}336}{169 - 25\cos^2\theta}$ **75.** $r^2 = \dfrac{144}{25\cos^2\theta - 9}$

77. $r^2 = \dfrac{144}{25\cos^2\theta - 16}$

79. (a) Ellipse

(b) The given polar equation, r, has a vertical directrix to the left of the pole. The equation r_1 has a vertical directrix to the right of the pole, and the equation r_2 has a horizontal directrix below the pole.

(c)

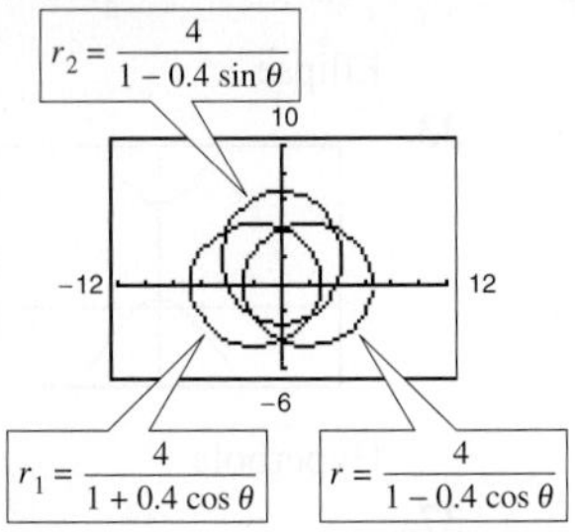

Review Exercises (page 764)

1. $\frac{\pi}{4}$ rad, 45° **3.** 1.1071 rad, 63.43° **5.** 0.4424 rad, 25.35°

7. 0.6588 rad, 37.75° **9.** $4\sqrt{2}$ **11.** Hyperbola

13. $y^2 = 16x$ **15.** $(y - 2)^2 = 12x$

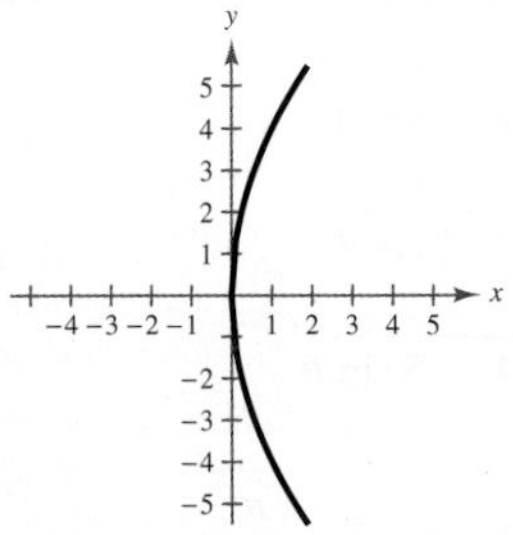

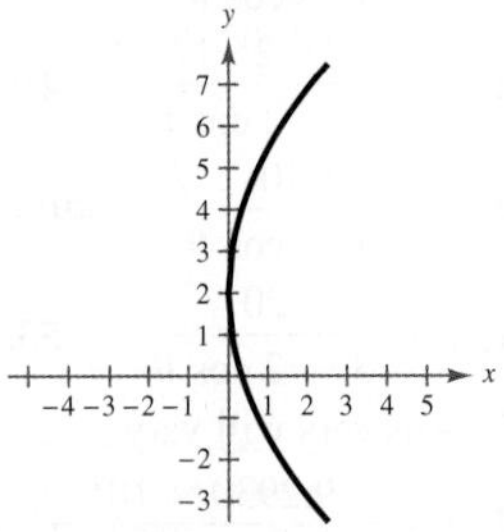

17. $y = -4x - 2$ **19.** $6\sqrt{5}$ m

21. $\frac{(x-3)^2}{25} + \frac{y^2}{16} = 1$ **23.** $\frac{(x-2)^2}{4} + \frac{(y-1)^2}{1} = 1$

25. The foci occur 3 feet from the center of the arch.

27. Center: $(-1, 2)$
Vertices: $(-1, 9), (-1, -5)$
Foci: $\left(-1, 2 \pm 2\sqrt{6}\right)$
Eccentricity: $\frac{2\sqrt{6}}{7}$

29. Center: $(1, -4)$
Vertices: $(1, 0), (1, -8)$
Foci: $\left(1, -4 \pm \sqrt{7}\right)$
Eccentricity: $\frac{\sqrt{7}}{4}$

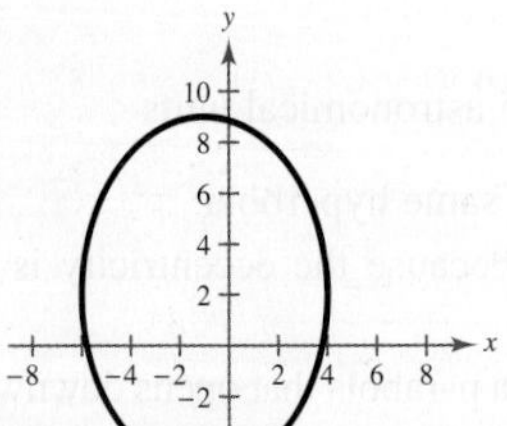

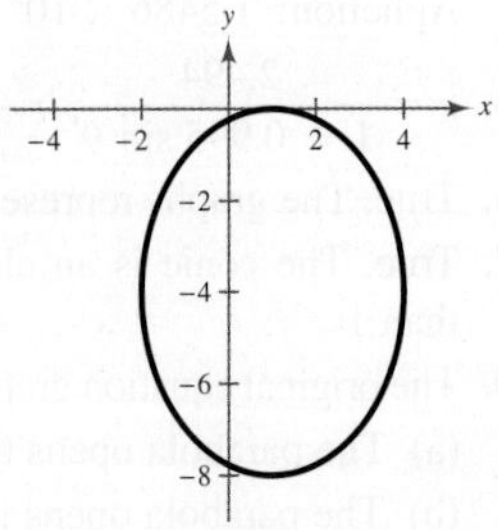

31. $\frac{y^2}{1} - \frac{x^2}{3} = 1$ **33.** $\frac{x^2}{16} - \frac{y^2}{9} = 1$

35. Center: $(5, -3)$
Vertices: $(11, -3), (-1, -3)$
Foci: $\left(5 \pm 2\sqrt{13}, -3\right)$
Asymptotes: $y = -3 \pm \frac{2}{3}(x - 5)$

37. Center: $(1, -1)$
Vertices: $(5, -1), (-3, -1)$
Foci: $(6, -1), (-4, -1)$
Asymptotes: $y = -1 \pm \frac{3}{4}(x - 1)$

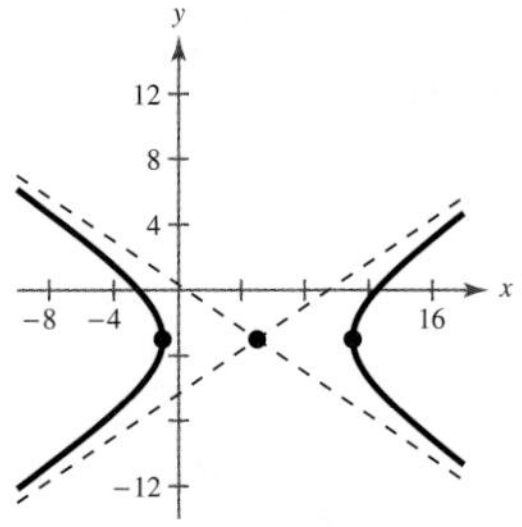

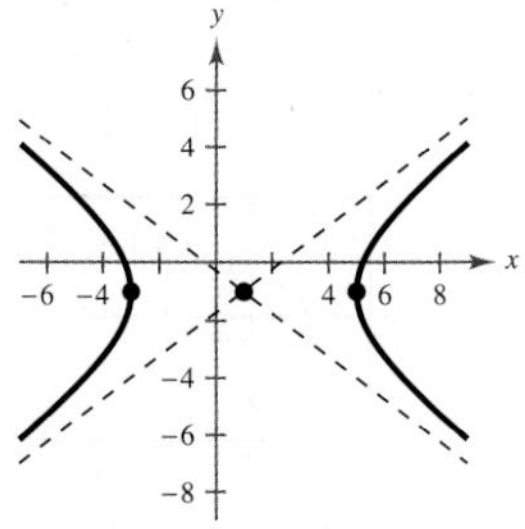

39. 72 mi **41.** Hyperbola **43.** Ellipse

45. $\frac{(y')^2}{6} - \frac{(x')^2}{6} = 1$ **47.** $\frac{(x')^2}{3} + \frac{(y')^2}{2} = 1$

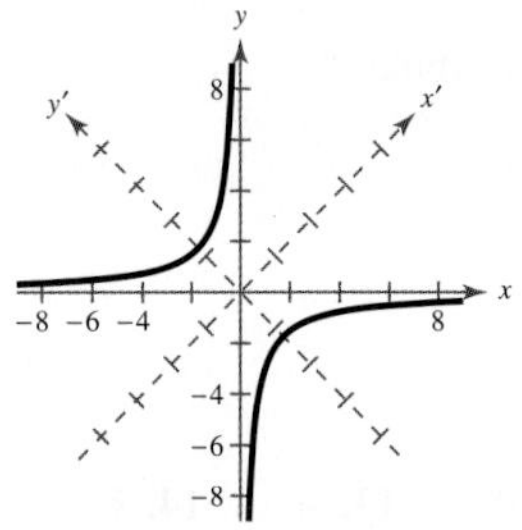

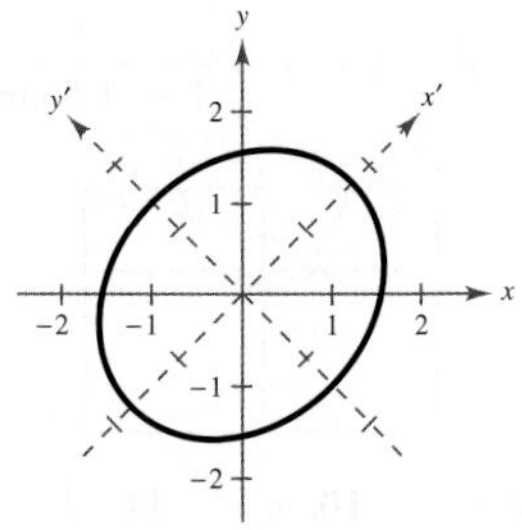

49. (a) Parabola

(b) $y = \frac{24x + 40 \pm \sqrt{(24x + 40)^2 - 36(16x^2 - 30x)}}{18}$

(c)

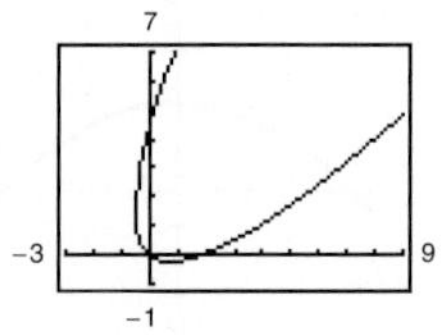

51. (a) Parabola

(b) $y = \frac{-\left(2x - 2\sqrt{2}\right) \pm \sqrt{\left(2x - 2\sqrt{2}\right)^2 - 4\left(x^2 + 2\sqrt{2}x + 2\right)}}{2}$

(c)

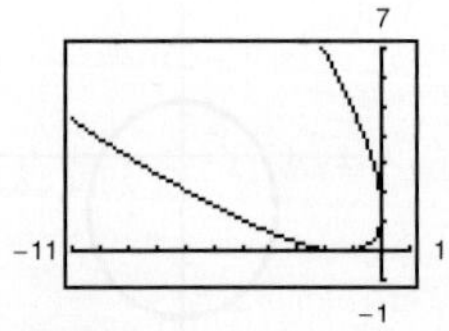

53. (a)

t	-2	-1	0	1	2
x	-8	-5	-2	1	4
y	15	11	7	3	-1

(b)

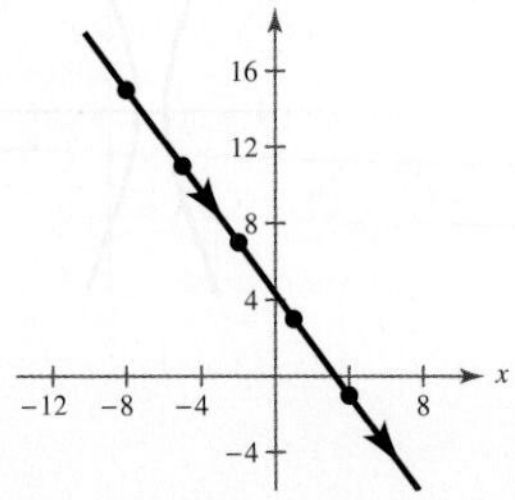

55. (a)

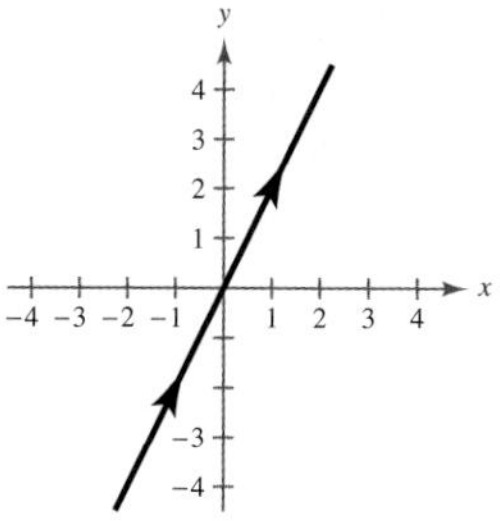

(b) $y = 2x$

57. (a)

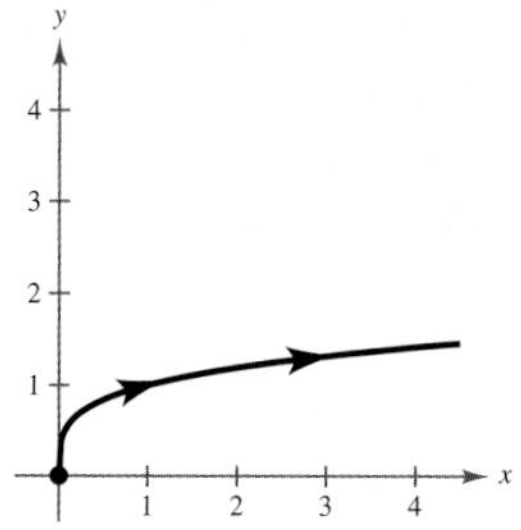

(b) $y = \sqrt[4]{x}$

59. (a)

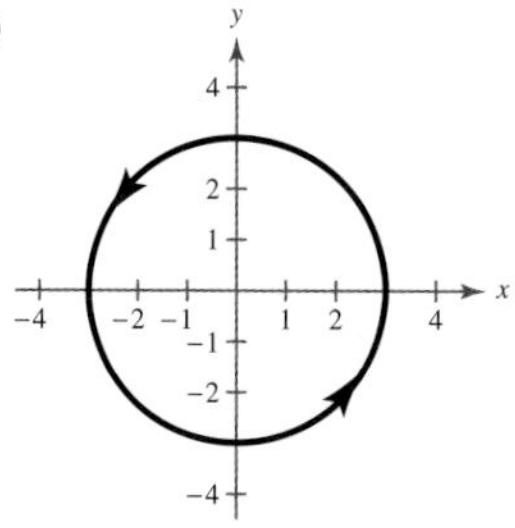

(b) $x^2 + y^2 = 9$

61. (a) $x = t, y = 2t + 3$ (b) $x = t - 1, y = 2t + 1$
(c) $x = 3 - t, y = 9 - 2t$

63. (a) $x = t, y = t^2 + 3$ (b) $x = t - 1, y = t^2 - 2t + 4$
(c) $x = 3 - t, y = t^2 - 6t + 12$

65. (a) $x = t, y = 2t^2 + 2$ (b) $x = t - 1, y = 2t^2 - 4t + 4$
(c) $x = 3 - t, y = 2t^2 - 12t + 20$

67.

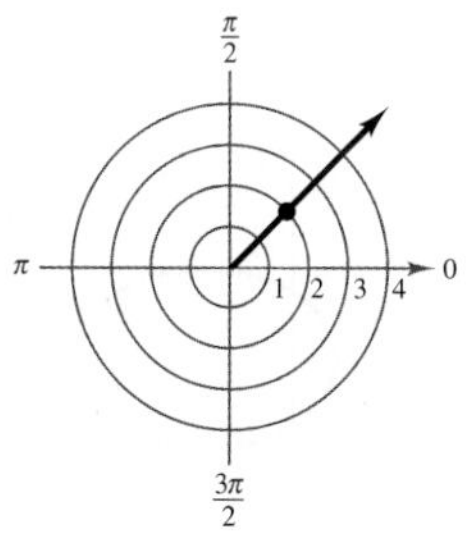

$\left(2, -\frac{7\pi}{4}\right), \left(-2, \frac{5\pi}{4}\right)$

69.

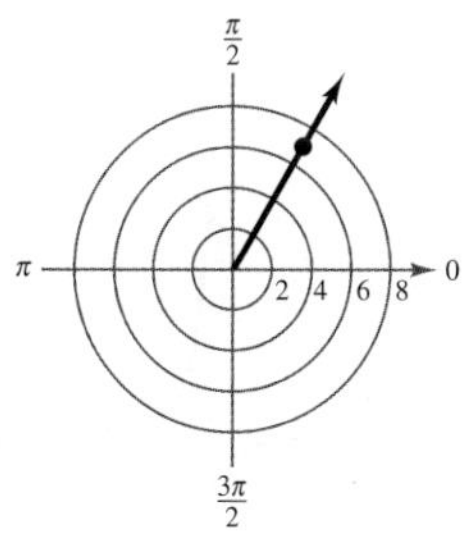

$(7, 1.05), (-7, -2.09)$

71. $\left(-\frac{1}{2}, -\frac{\sqrt{3}}{2}\right)$ **73.** $\left(-\frac{3\sqrt{2}}{2}, \frac{3\sqrt{2}}{2}\right)$ **75.** $\left(1, \frac{\pi}{2}\right)$

77. $(2\sqrt{13}, 0.9828)$ **79.** $r = 9$ **81.** $r = 6 \sin \theta$

83. $r^2 = 10 \csc 2\theta$ **85.** $x^2 + y^2 = 25$

87. $x^2 + y^2 = 3x$ **89.** $x^2 + y^2 = y^{2/3}$

91. Symmetry: $\theta = \frac{\pi}{2}$, polar axis, pole

Maximum value of $|r|$: $|r| = 6$ for all values of θ

No zeros of r

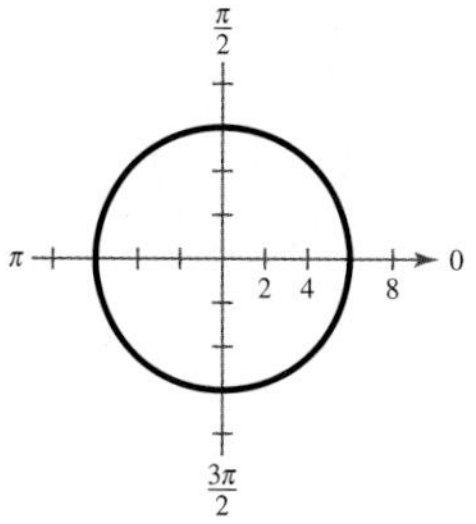

93. Symmetry: $\theta = \frac{\pi}{2}$, polar axis, pole

Maximum value of $|r|$: $|r| = 4$ when $\theta = \frac{\pi}{4}, \frac{3\pi}{4}, \frac{5\pi}{4}, \frac{7\pi}{4}$

Zeros of r: $r = 0$ when $\theta = 0, \frac{\pi}{2}, \pi, \frac{3\pi}{2}$

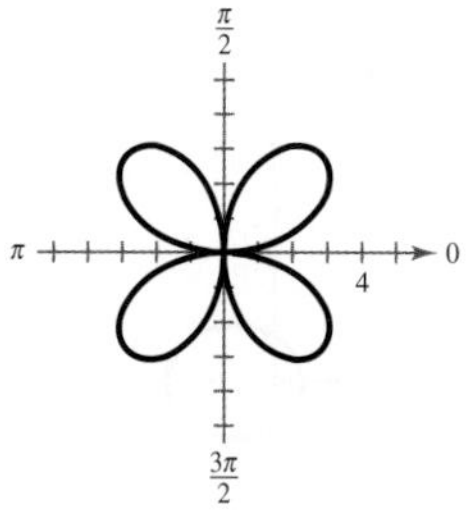

95. Symmetry: polar axis

Maximum value of $|r|$: $|r| = 4$ when $\theta = 0$

Zeros of r: $r = 0$ when $\theta = \pi$

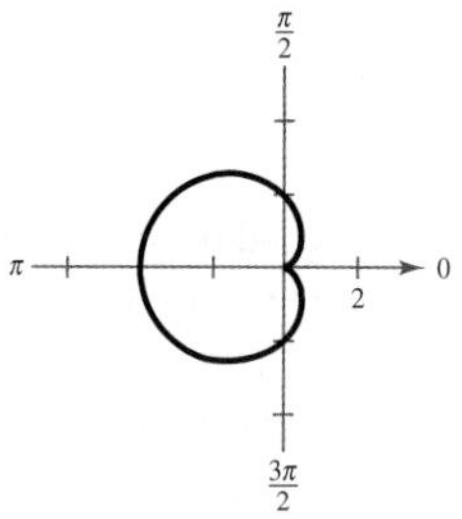

97. Symmetry: $\theta = \frac{\pi}{2}$

Maximum value of $|r|$: $|r| = 8$ when $\theta = \frac{\pi}{2}$

Zeros of r: $r = 0$ when $\theta = 3.4814, 5.9433$

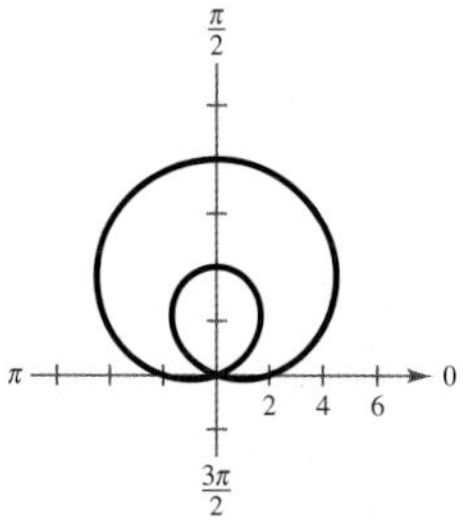

99. Symmetry: $\theta = \frac{\pi}{2}$, polar axis, pole

Maximum value of $|r|$: $|r| = 3$ when $\theta = 0, \frac{\pi}{2}, \pi, \frac{3\pi}{2}$

Zeros of r: $r = 0$ when $\theta = \frac{\pi}{4}, \frac{3\pi}{4}, \frac{5\pi}{4}, \frac{7\pi}{4}$

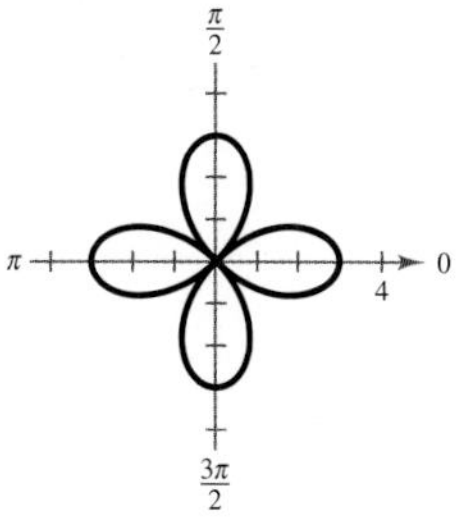

101. Limaçon

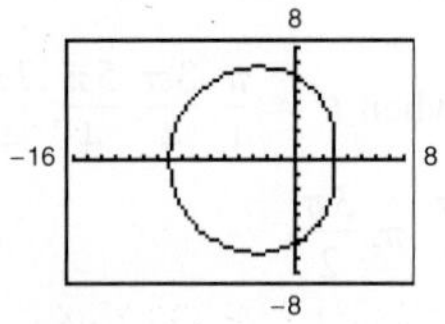

103. Rose curve

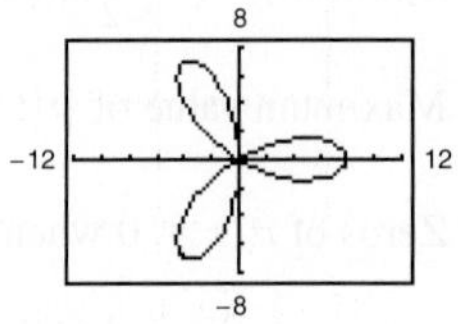

105. Hyperbola

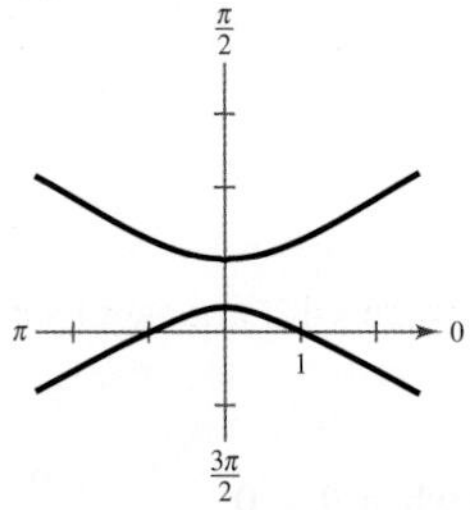

107. Ellipse

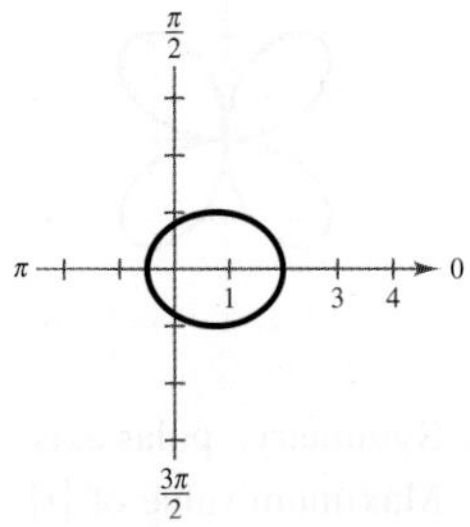

109. $r = \dfrac{4}{1 - \cos\theta}$ **111.** $r = \dfrac{5}{3 - 2\cos\theta}$

113. $r = \dfrac{7961.93}{1 - 0.937\cos\theta}$; 10,980.11 mi

115. False. The equation of a hyperbola is a second-degree equation.

117. False. $(2, \pi/4)$, $(-2, 5\pi/4)$, and $(2, 9\pi/4)$ all represent the same point.

119. (a) The graphs are the same. (b) The graphs are the same.

Chapter Test *(page 767)*

1. 0.3805 rad, 21.8° **2.** 0.8330 rad, 47.7° **3.** $\dfrac{7\sqrt{2}}{2}$

4. Parabola: $y^2 = 2(x - 1)$
Vertex: $(1, 0)$
Focus: $\left(\frac{3}{2}, 0\right)$

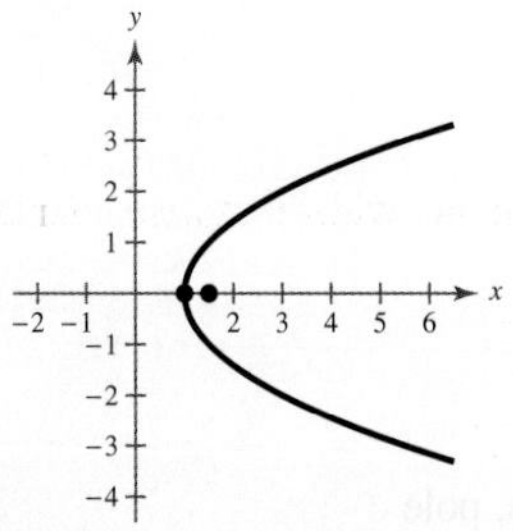

5. Hyperbola: $\dfrac{(x-2)^2}{4} - y^2 = 1$
Center: $(2, 0)$
Vertices: $(0, 0)$, $(4, 0)$
Foci: $\left(2 \pm \sqrt{5}, 0\right)$
Asymptotes: $y = \pm\frac{1}{2}(x - 2)$

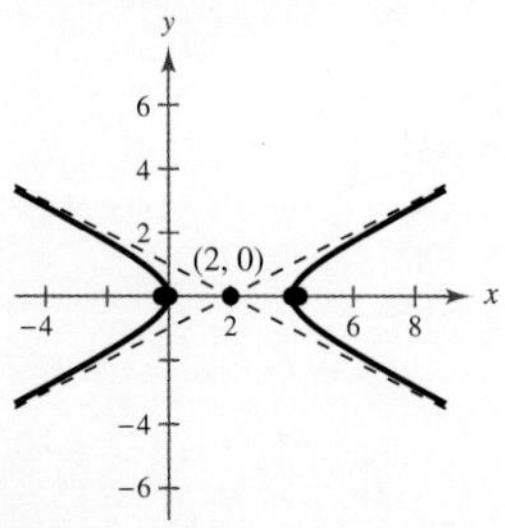

6. Ellipse: $\dfrac{(x+3)^2}{16} + \dfrac{(y-1)^2}{9} = 1$
Center: $(-3, 1)$
Vertices: $(1, 1)$, $(-7, 1)$
Foci: $\left(-3 \pm \sqrt{7}, 1\right)$

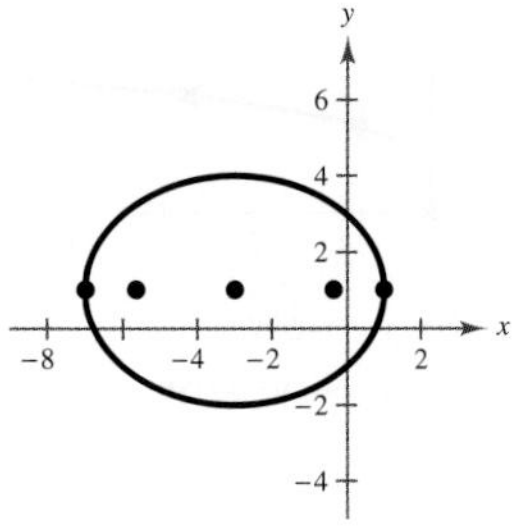

7. Circle: $(x - 2)^2 + (y - 1)^2 = \frac{1}{2}$
Center: $(2, 1)$

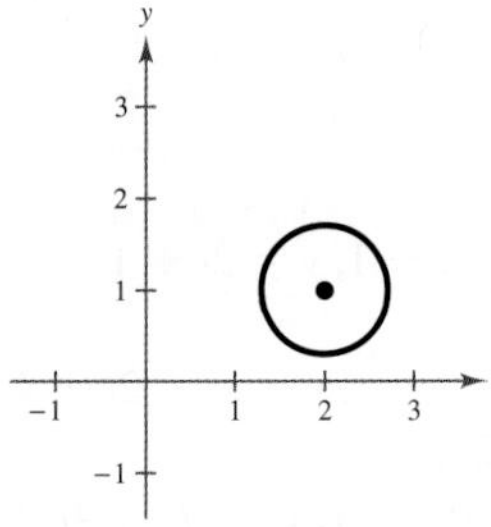

8. $(x - 2)^2 = \frac{4}{3}(y + 3)$ **9.** $\dfrac{y^2}{2/41} - \dfrac{x^2}{162/41} = 1$

10. (a) 45°
(b)

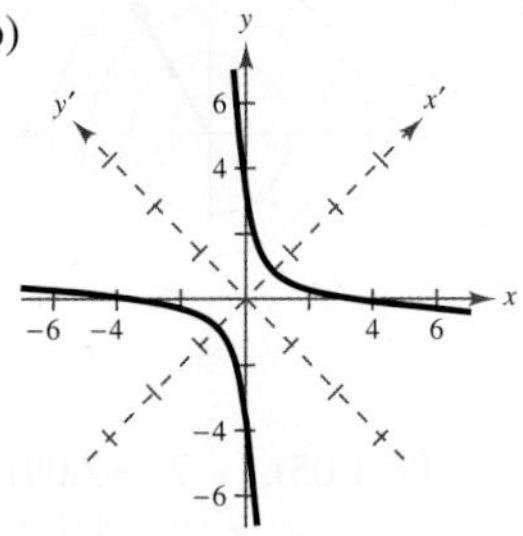

11.

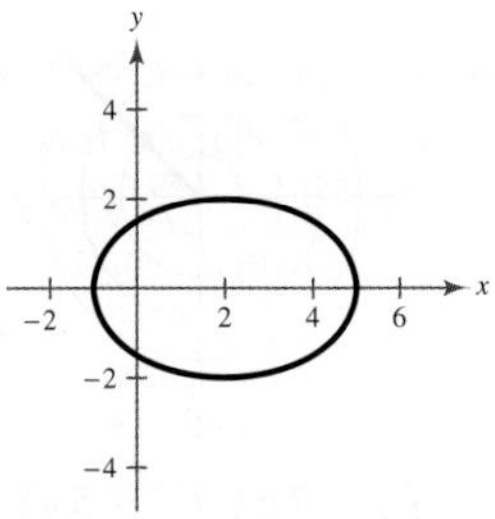

$\dfrac{(x-2)^2}{9} + \dfrac{y^2}{4} = 1$

12. (a) $x = t$, $y = 3 - t^2$
(b) $x = t - 2$, $y = -t^2 + 4t - 1$

13. $\left(\sqrt{3}, -1\right)$

14. $\left(2\sqrt{2}, \frac{7\pi}{4}\right), \left(-2\sqrt{2}, \frac{3\pi}{4}\right), \left(2\sqrt{2}, -\frac{\pi}{4}\right)$

15. $r = 3\cos\theta$

16.

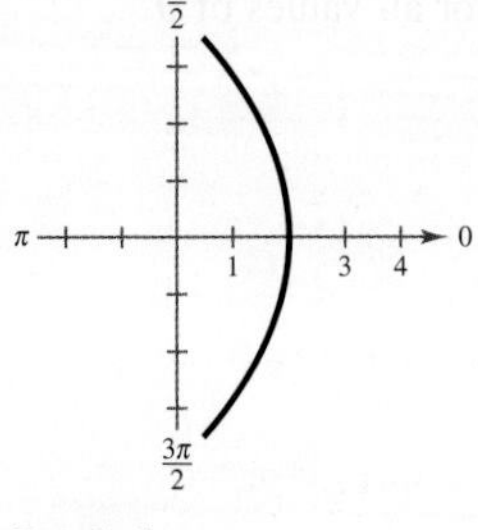

Parabola

17.

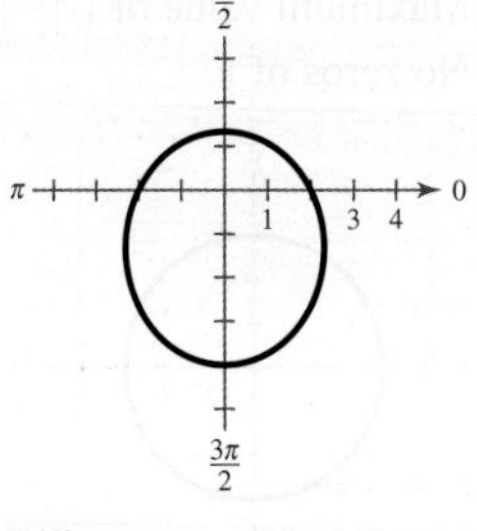

Ellipse

18.

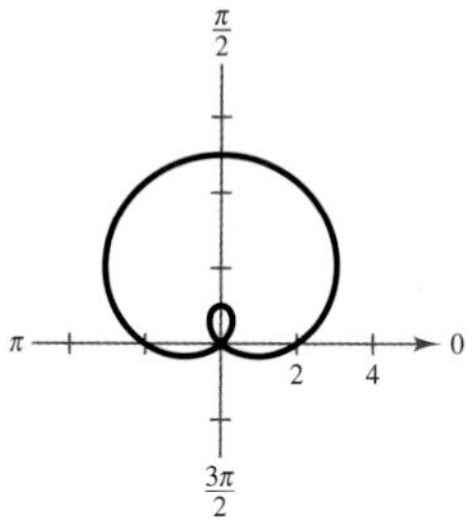

Limaçon with inner loop

19.

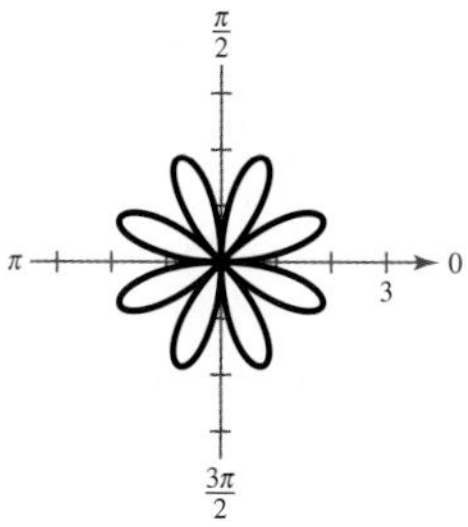

Rose curve

20. Answers will vary. For example: $r = \dfrac{1}{1 + 0.25 \sin \theta}$

21. Slope: 0.1511; Change in elevation: 789 ft

22. No; Yes

Problem Solving *(page 771)*

1. (a) 1.2016 rad (b) 2420 ft, 5971 ft **3.** $A = \dfrac{4a^2 b^2}{a^2 + b^2}$

5. (a) Because $d_1 + d_2 \le 20$, by definition, the outer bound that the boat can travel is an ellipse. The islands are the foci.

(b) Island 1: $(-6, 0)$;
Island 2: $(6, 0)$

(c) 20 mi; Vertex: $(10, 0)$

(d) $\dfrac{x^2}{100} + \dfrac{y^2}{64} = 1$

7. Proof

9. (a) The first set of parametric equations models projectile motion along a straight line. The second set of parametric equations models projectile motion of an object launched at a height of h units above the ground that will eventually fall back to the ground.

(b) $y = (\tan \theta)x$; $y = h + x \tan \theta - \dfrac{16x^2 \sec^2 \theta}{v_0^{\,2}}$

(c) In the first case, the path of the moving object is not affected by a change in the velocity because eliminating the parameter removes v_0.

11.

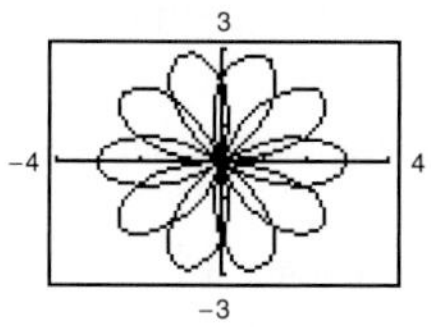

$r = 3 \sin\left(\dfrac{5\theta}{2}\right)$

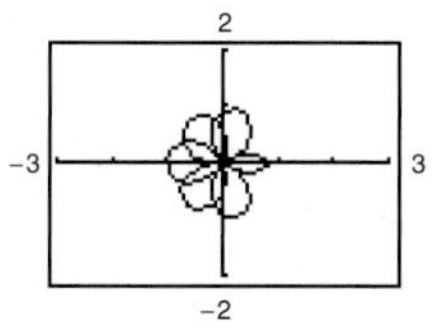

$r = -\cos(\sqrt{2}\theta)$,
$-2\pi \le \theta \le 2\pi$

Sample answer: If n is a rational number, then the curve has a finite number of petals. If n is an irrational number, then the curve has an infinite number of petals.

13. (a)

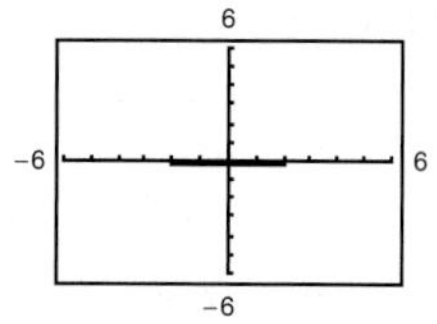

The graph is a line between -2 and 2 on the x-axis.

(b)

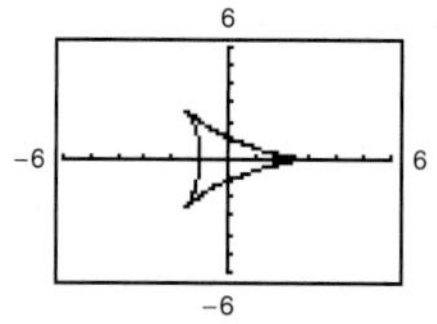

The graph is a three-sided figure with counterclock-wise orientation.

(c)

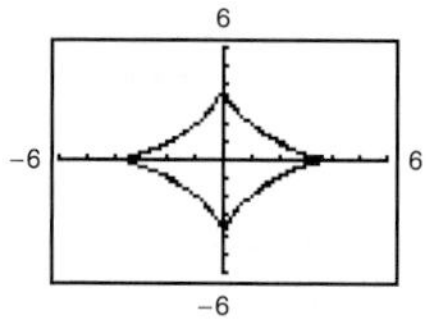

The graph is a four-sided figure with counterclock-wise orientation.

(d)

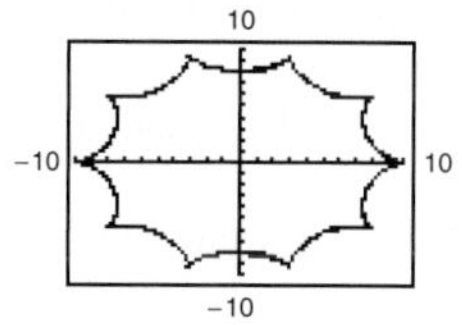

The graph is a 10-sided figure with counterclock-wise orientation.

(e)

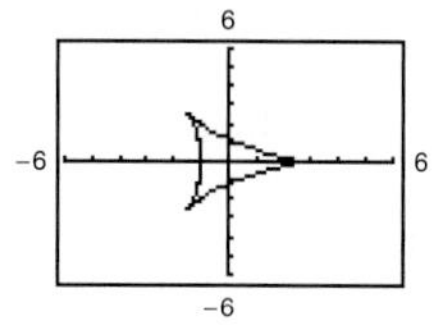

The graph is a three-sided figure with clockwise orientation.

(f)

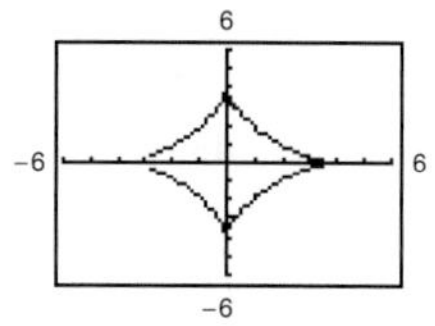

The graph is a four-sided figure with clockwise orientation.

15.

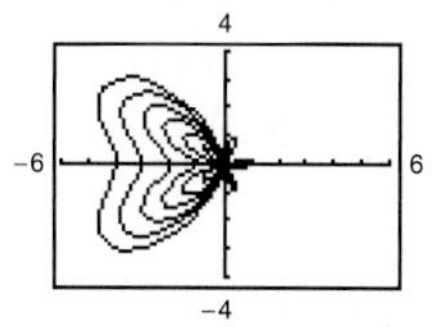

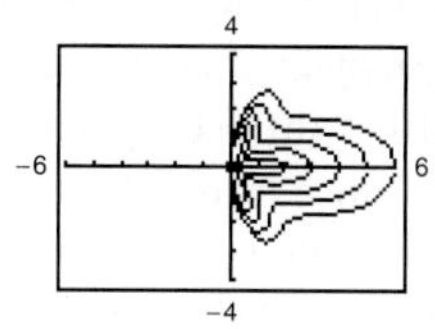

For $n \ge 1$, a bell is produced.
For $n \le -1$, a heart is produced.
For $n = 0$, a rose curve is produced.

Appendix A

Appendix A.1 *(page A11)*

1. irrational **3.** absolute value **5.** terms

7. (a) 5, 1, 2 (b) 0, 5, 1, 2 (c) $-9, 5, 0, 1, -4, 2, -11$
(d) $-\frac{7}{2}, \frac{2}{3}, -9, 5, 0, 1, -4, 2, -11$ (e) $\sqrt{2}$

9. (a) 1 (b) 1 (c) $-13, 1, -6$
(d) $2.01, -13, 1, -6, 0.666\ldots$ (e) $0.010110111\ldots$

11. (a)  (b) 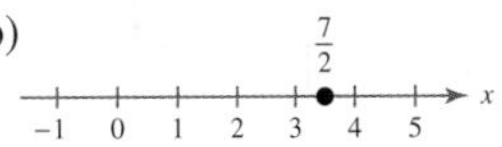

(c) 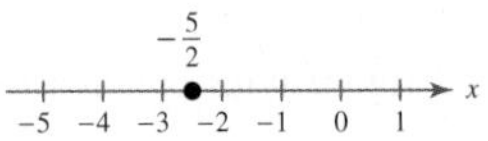 (d) 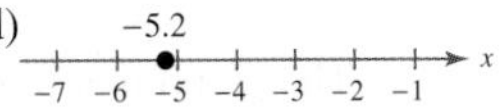

13. 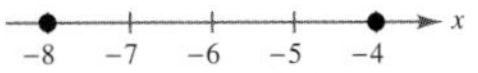
$-4 > -8$

15. 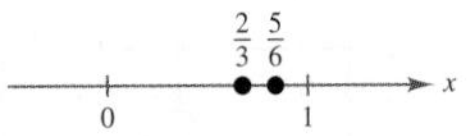
$\frac{5}{6} > \frac{2}{3}$

17. (a) $x \le 5$ denotes the set of all real numbers less than or equal to 5.
(b) 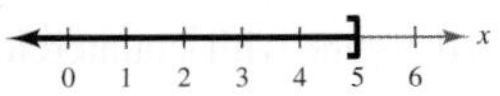 (c) Unbounded

19. (a) $[4, \infty)$ denotes the set of all real numbers greater than or equal to 4.
(b) 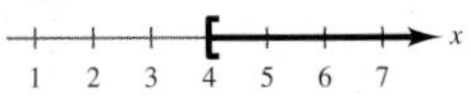 (c) Unbounded

21. (a) $-2 < x < 2$ denotes the set of all real numbers greater than -2 and less than 2.
(b) 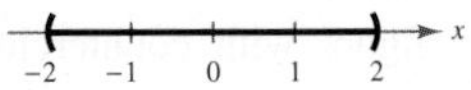 (c) Bounded

23. (a) $[-5, 2)$ denotes the set of all real numbers greater than or equal to -5 and less than 2.
(b) 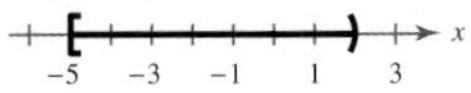 (c) Bounded

	Inequality	**Interval**
25.	$y \ge 0$	$[0, \infty)$
27.	$10 \le t \le 22$	$[10, 22]$
29.	$W > 65$	$(65, \infty)$

31. 10 **33.** 5 **35.** -1 **37.** -1 **39.** -1

41. $|-4| = |4|$ **43.** $-|-6| < |-6|$ **45.** 51 **47.** $\frac{5}{2}$

49. $\frac{128}{75}$ **51.** $|x - 5| \le 3$ **53.** $|y - a| \le 2$

55. \$1880.1 billion; \$412.7 billion

57. \$2524.0 billion; \$458.5 billion

59. $7x$ and 4 are the terms; 7 is the coefficient.

61. $4x^3$, $0.5x$, and -5 are the terms; 4 and 0.5 are the coefficients.

63. (a) -10 (b) -6 **65.** (a) -10 (b) 0

67. Multiplicative Inverse Property

69. Distributive Property

71. Associative and Commutative Properties of Multiplication

73. $\dfrac{3}{8}$ **75.** $\dfrac{5x}{12}$

77. (a) Negative (b) Negative (c) Positive (d) Positive

79. False. Zero is nonnegative, but not positive.

81. (a)

n	0.0001	0.01	1	100	10,000
$\frac{5}{n}$	50,000	500	5	0.05	0.0005

(b) (i) The value of $5/n$ approaches infinity as n approaches 0.
(ii) The value of $5/n$ approaches 0 as n increases without bound.

Appendix A.2 *(page A23)*

1. exponent; base **3.** square root **5.** like radicals

7. rationalizing **9.** (a) 81 (b) $\frac{1}{9}$

11. (a) 5184 (b) $-\frac{3}{5}$ **13.** (a) $\frac{16}{3}$ (b) 1

15. -24 **17.** 6 **19.** -48 **21.** (a) $-125z^3$ (b) $5x^6$

23. (a) $24y^2$ (b) $-3z^7$ **25.** (a) $\dfrac{5184}{y^7}$ (b) 1

27. (a) 1 (b) $\dfrac{1}{4x^4}$ **29.** (a) $\dfrac{125x^9}{y^{12}}$ (b) $\dfrac{b^5}{a^5}$

31. 1.02504×10^4 **33.** 3.937×10^{-5} in. **35.** 0.000314

37. 9,460,000,000,000 km

39. (a) 6.8×10^5 (b) 6.0×10^4 **41.** (a) 3 (b) $\frac{3}{2}$

43. (a) 2 (b) $2x$ **45.** (a) $2\sqrt{5}$ (b) $4\sqrt[3]{2}$

47. (a) $6x\sqrt{2x}$ (b) $3y^2\sqrt{6x}$

49. (a) $2x\sqrt[3]{2x^2}$ (b) $\dfrac{5|x|\sqrt{3}}{y^2}$ **51.** (a) $22\sqrt{2}$ (b) $11\sqrt[3]{2}$

53. (a) $23|x|\sqrt{3}$ (b) $18\sqrt{5x}$ **55.** $\dfrac{\sqrt{3}}{3}$

57. $\dfrac{\sqrt{14} + 2}{2}$ **59.** $\dfrac{2}{\sqrt{2}}$ **61.** $\dfrac{2}{3(\sqrt{5} - \sqrt{3})}$ **63.** $64^{1/3}$

65. $\dfrac{3}{\sqrt[3]{x^2}}$ **67.** $(3x^3y)^{1/2}$ **69.** (a) $\frac{1}{8}$ (b) $\frac{27}{8}$

71. (a) $\dfrac{2}{|x|}$ (b) $xy^{1/3}$ **73.** (a) $\sqrt{3}$ (b) $\sqrt[3]{(x + 1)^2}$

75. (a) $2\sqrt[4]{2}$ (b) $\sqrt[8]{2x}$ **77.** (a) $x - 1$ (b) $\dfrac{1}{x - 1}$

79. (a)

h	0	1	2	3	4	5	6
t	0	2.93	5.48	7.67	9.53	11.08	12.32

h	7	8	9	10	11	12
t	13.29	14.00	14.50	14.80	14.93	14.96

(b) $t \to 8.64\sqrt{3} \approx 14.96$

81. False. When $x = 0$, the expressions are not equal.

83. False. For instance, $(3 + 5)^2 = 8^2 = 64 \ne 34 = 3^2 + 5^2$.

Appendix A.3 *(page A33)*

1. n; a_n; a_0 **3.** like terms **5.** factoring

7. perfect square trinomial

9. (a) $-\frac{1}{2}x^5 + 14x$ (b) Degree: 5; Leading coefficient: $-\frac{1}{2}$
(c) Binomial

11. (a) $-x^6 + 3$ (b) Degree: 6; Leading coefficient: -1
(c) Binomial

13. (a) 3 (b) Degree: 0; Leading coefficient: 3
(c) Monomial

15. (a) $-4x^5 + 6x^4 + 1$
(b) Degree: 5; Leading coefficient: -4 (c) Trinomial

17. (a) $4x^3y$ (b) Degree: 4; Leading coefficient: 4
(c) Monomial

19. $-2x - 10$ **21.** $8.3x^3 + 29.7x^2 + 11$

23. $3x^3 - 6x^2 + 3x$ **25.** $-15z^2 + 5z$

27. $x^2 + 7x + 12$ **29.** $x^4 + x^2 + 1$ **31.** $x^2 - 100$
33. $4x^2 + 12x + 9$ **35.** $x^3 + 3x^2 + 3x + 1$
37. $m^2 - n^2 - 6m + 9$ **39.** $x^2 + 2xy + y^2 - 6x - 6y + 9$
41. $2x(x^2 - 3)$ **43.** $(x - 5)(3x + 8)$
45. $\frac{1}{2}x(x^2 + 4x - 10)$ **47.** $\frac{2}{3}(x - 6)(x - 3)$
49. $(x + 9)(x - 9)$ **51.** $(x + 1)(x - 3)$ **53.** $(x - 2)^2$
55. $(3u + 4v)^2$ **57.** $\left(z + \frac{1}{2}\right)^2$ **59.** $(x - 2)(x^2 + 2x + 4)$
61. $(3x + 2)(9x^2 - 6x + 4)$ **63.** $(x + 2)(x - 1)$
65. $-(y + 5)(y - 4)$ **67.** $(3x - 2)(x - 1)$
69. $(5x + 1)(x + 5)$ **71.** $(x - 1)(x^2 + 2)$
73. $(2x - 1)(x^2 - 3)$ **75.** $(x^2 + 2)(x + 1)(x^2 - x + 1)$
77. $(x + 3)(2x + 3)$ **79.** $(2x + 3)(3x - 5)$
81. $6(x + 3)(x - 3)$ **83.** $x^2(x - 1)$ **85.** $(x - 1)^2$
87. $-2x(x + 1)(x - 2)$ **89.** $(5 - x)(1 + x^2)$
91. $(3 - 4x)(23 - 60x)$ **93.** $4x^3(2x + 1)^3(2x^2 + 2x + 1)$
95. (a) $\pi h(R + r)(R - r)$ (b) $V = 2\pi\left[\left(\frac{R + r}{2}\right)(R - r)\right]h$
97. False. $(4x^2 + 1)(3x + 1) = 12x^3 + 4x^2 + 3x + 1$
99. True. $a^2 - b^2 = (a + b)(a - b)$
101. $m + n$ **103.** $-x^3 + 8x^2 + 2x + 7$
105. Answers will vary. *Sample answer:* $x^2 - 3$
107. $(x^n + y^n)(x^n - y^n)$

Appendix A.4 *(page A42)*

1. domain **3.** complex **5.** All real numbers x
7. All real numbers x such that $x \neq 3$
9. All real numbers x such that $x \neq 1$
11. All real numbers x such that $x \neq 3$
13. All real numbers x such that $x \leq 4$
15. All real numbers x such that $x > 3$
17. $\frac{3x}{2}$, $x \neq 0$ **19.** $\frac{3y}{y + 1}$, $x \neq 0$ **21.** $-\frac{1}{2}$, $x \neq 5$
23. $y - 4$, $y \neq -4$ **25.** $\frac{x(x + 3)}{x - 2}$, $x \neq -2$
27. $\frac{-(x^2 + 1)}{x + 2}$, $x \neq 2$ **29.** $z - 2$
31. When simplifying fractions, you can only divide out common factors, not terms.
33. $\frac{1}{5(x - 2)}$, $x \neq 1$ **35.** $-\frac{8}{5}$, $y \neq -3, 4$
37. $\frac{x - y}{x(x + y)^2}$, $x \neq -2y$ **39.** $\frac{6x + 13}{x + 3}$ **41.** $-\frac{2}{x - 2}$
43. $\frac{-2x^2 + 3x + 8}{(2x + 1)(x + 2)}$ **45.** $\frac{2 - x}{x^2 + 1}$, $x \neq 0$
47. The error is incorrect subtraction in the numerator.
49. $\frac{1}{2}$, $x \neq 2$ **51.** $x(x + 1)$, $x \neq -1, 0$
53. $\frac{2x - 1}{2x}$, $x > 0$ **55.** $\frac{x^7 - 2}{x^2}$ **57.** $\frac{-1}{(x^2 + 1)^5}$
59. $\frac{2x^3 - 2x^2 - 5}{(x - 1)^{1/2}}$ **61.** $\frac{3x - 1}{3}$, $x \neq 0$
63. $\frac{-1}{x(x + h)}$, $h \neq 0$ **65.** $\frac{-1}{(x - 4)(x + h - 4)}$, $h \neq 0$
67. $\frac{1}{\sqrt{x + 2} + \sqrt{x}}$ **69.** $\frac{1}{\sqrt{t + 3} + \sqrt{3}}$, $t \neq 0$
71. $\frac{1}{\sqrt{x + h + 1} + \sqrt{x + 1}}$, $h \neq 0$

73. (a)

t	0	2	4	6	8	10	12
T	75	55.9	48.3	45	43.3	42.3	41.7

t	14	16	18	20	22
T	41.3	41.1	40.9	40.7	40.6

(b) The model is approaching a T-value of 40.

75. $\frac{x}{2(2x + 1)}$, $x \neq 0$

77. (a)

Year, t	Banking (millions)	Paying Bills (millions)
5	46.9	17
6	57.6	25.6
7	63.5	27.3
8	67.3	28.8
9	69.9	30.8
10	71.9	33.7

(b) The estimates are fairly close to the actual numbers of households.

(c) $\text{Ratio} = \frac{-0.1228t^3 + 2.923t^2 - 16.38t + 24.6}{-0.5061t^3 + 11.2742t^2 - 67.844t + 121.8}$

(d)

Year	Ratio
2005	0.3625
2006	0.4449
2007	0.4292
2008	0.4282
2009	0.4403
2010	0.4685

Answers will vary.

79. $\frac{R_1R_2}{R_2 + R_1}$
81. False. In order for the simplified expression to be equivalent to the original expression, the domain of the simplified expression needs to be restricted. If n is even, $x \neq -1, 1$. If n is odd, $x \neq 1$.

Appendix A.5 *(page A56)*

1. equation **3.** extraneous **5.** 4 **7.** -9 **9.** 1
11. No solution **13.** $-\frac{96}{23}$ **15.** 4 **17.** 3
19. No solution. The variable is divided out.
21. No solution. The solution is extraneous.
23. 5 **25.** $0, -\frac{1}{2}$ **27.** $4, -2$ **29.** -5 **31.** $2, -6$
33. $-\frac{20}{3}, -4$ **35.** ± 7 **37.** $\pm 3\sqrt{3} = \pm 5.20$
39. 8, 16 **41.** $\frac{1 \pm 3\sqrt{2}}{2} \approx 2.62, -1.62$ **43.** $4, -8$
45. $-3 \pm \sqrt{7}$ **47.** $1 \pm \frac{\sqrt{6}}{3}$ **49.** $\frac{-5 \pm \sqrt{89}}{4}$

51. $\frac{1}{2}, -1$ **53.** $1 \pm \sqrt{3}$ **55.** $\frac{3}{4} \pm \frac{\sqrt{41}}{4}$ **57.** $\frac{2}{3} \pm \frac{\sqrt{7}}{3}$

59. $-\frac{5}{3}$ **61.** $2 \pm \frac{\sqrt{6}}{2}$ **63.** $6 \pm \sqrt{11}$ **65.** $1 \pm \sqrt{2}$

67. $6, -12$ **69.** $\frac{1}{2} \pm \sqrt{3}$ **71.** $-\frac{1}{2}$ **73.** $0, \pm\frac{\sqrt{21}}{3}$

75. $-3, 0$ **77.** 48 **79.** -16 **81.** $2, -5$ **83.** 9

85. 9 **87.** $\pm\sqrt{14}$ **89.** $8, -3$ **91.** $-6, -3, 3$

93. $\sqrt[3]{\frac{4.47}{\pi}} \approx 1.12$ in. **95.** 65.8 in.

97. False. See Example 14 on page A54.

99. True. There is no value that satisfies this equation.

Appendix A.6 *(page A64)*

1. solution set **3.** double

5. $0 \le x < 9$; Bounded **7.** $-1 \le x \le 5$; Bounded

9. $x > 11$; Unbounded **11.** $x < -2$; Unbounded

13. $x < 3$

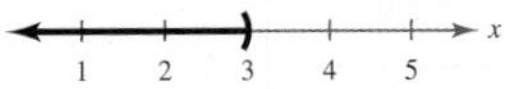

15. $x < \frac{3}{2}$

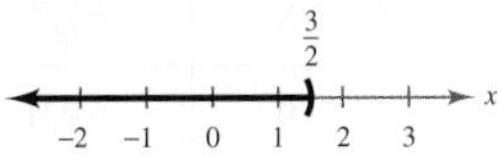

17. $x \ge 12$

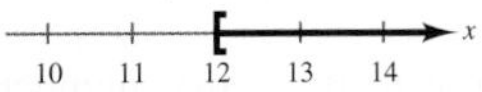

19. $x > 2$

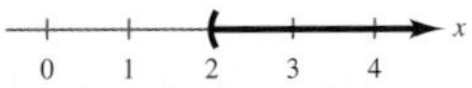

21. $x \ge \frac{2}{7}$

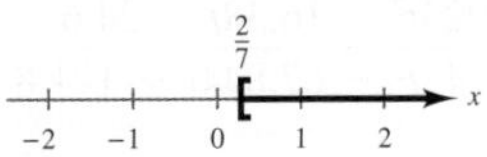

23. $x < 5$

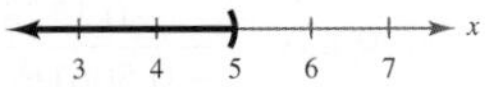

25. $x \ge 4$

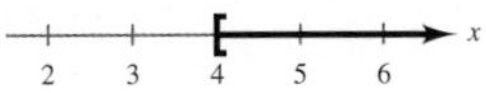

27. $x \ge 2$

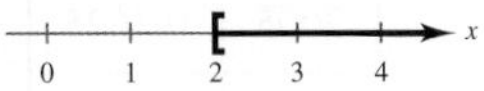

29. $x \ge -4$

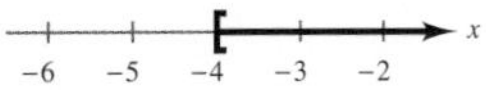

31. $-1 < x < 3$

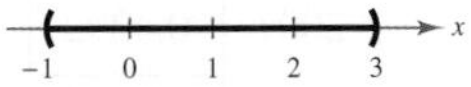

33. $-7 < x \le -\frac{1}{3}$

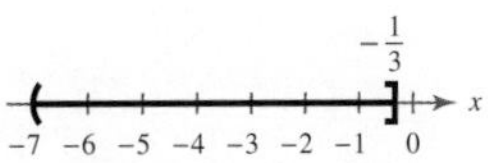

35. $-\frac{9}{2} < x < \frac{15}{2}$

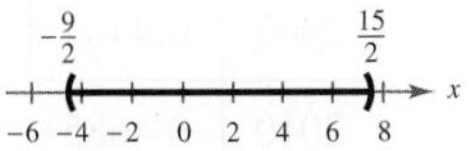

37. $-5 \le x < 1$

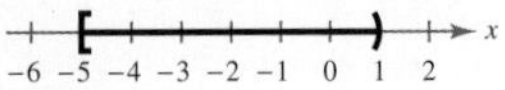

39. $-\frac{3}{4} < x < -\frac{1}{4}$

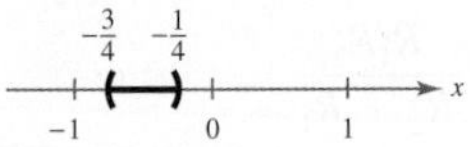

41. $10.5 \le x \le 13.5$

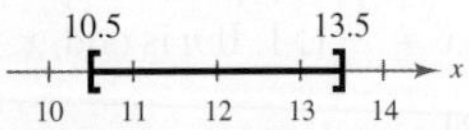

43. $-5 < x < 5$

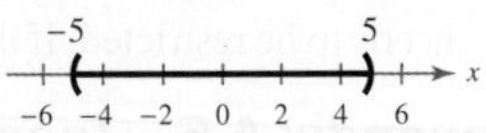

45. $x < -2, x > 2$

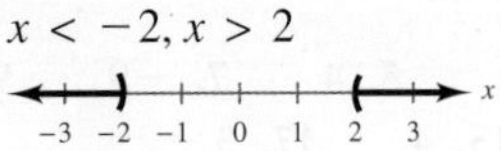

47. No solution

49. $14 \le x \le 26$

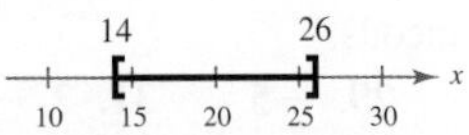

51. $x \le -\frac{3}{2}, x \ge 3$

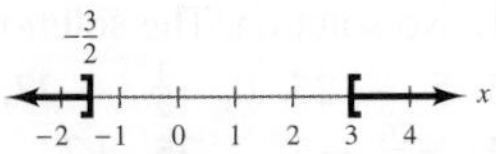

53. $x \le -5, x \ge 11$

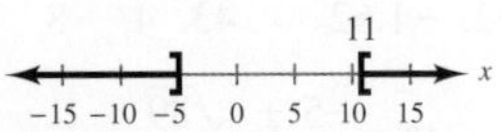

55. $4 < x < 5$

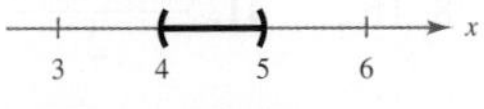

57. $x \le -\frac{29}{2}, x \ge -\frac{11}{2}$

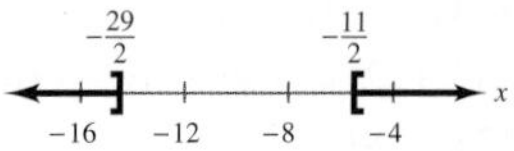

59.

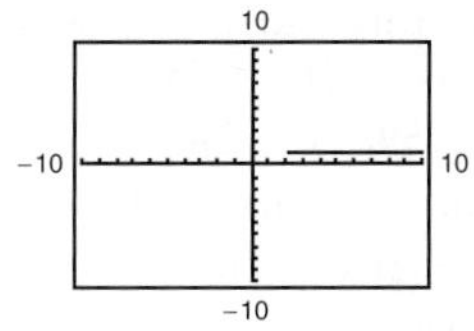

$x > 2$

61.

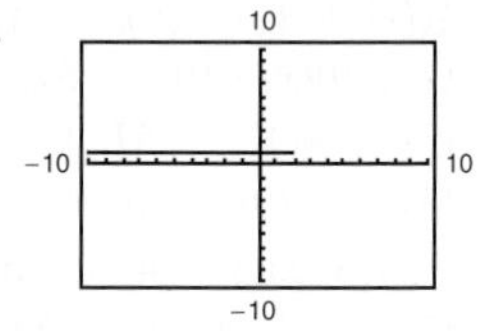

$x \le 2$

63.

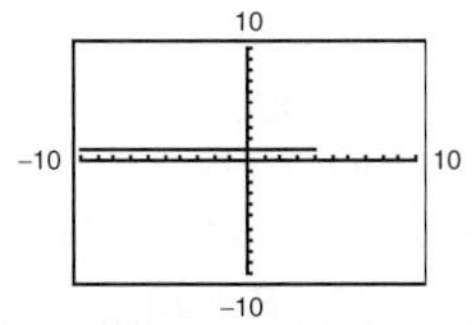

$x \le 4$

65.

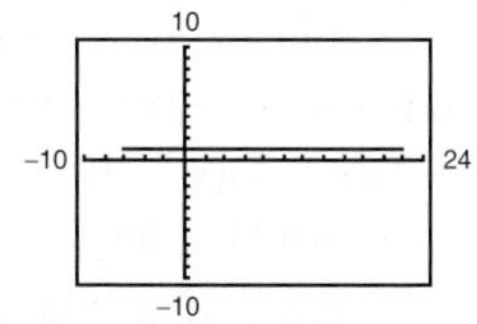

$-6 \le x \le 22$

67.

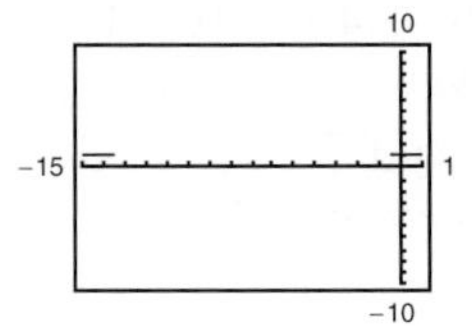

$x \le -\frac{27}{2}, x \ge -\frac{1}{2}$

69.

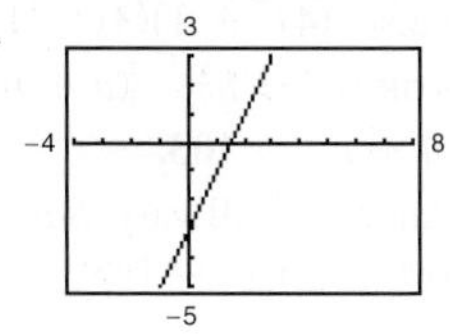

(a) $x \ge 2$ (b) $x \le \frac{3}{2}$

71.

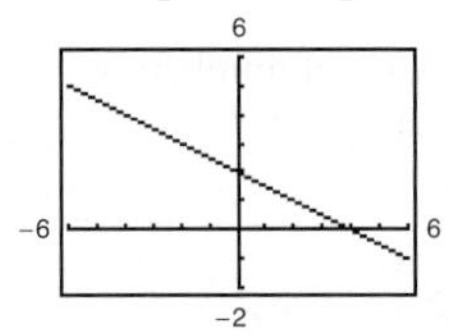

(a) $-2 \le x \le 4$
(b) $x \le 4$

73.

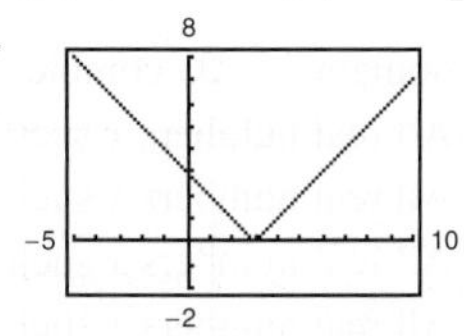

(a) $1 \le x \le 5$
(b) $x \le -1, x \ge 7$

75. $[5, \infty)$ **77.** $[-3, \infty)$ **79.** $\left(-\infty, \frac{7}{2}\right]$

81. All real numbers less than eight units from 10

83. $|x| \le 3$ **85.** $|x - 7| \ge 3$ **87.** $|x - 12| \ge 10$

89. $|x + 3| > 4$ **91.** $4.10 \le E \le 4.25$ **93.** $r < 0.08$

95. $100 \le r \le 170$ **97.** More than 6 units per hour

99. Greater than 3.125% **101.** $x \ge 36$

103. $87 \le x \le 210$

105. (a)

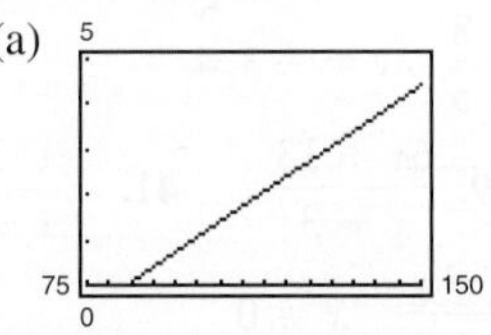

(b) $x \ge 129$ (c) $x \ge 128.93$ (d) Answers will vary.

107. (a) $2.87 \le t \le 6.54$ (Between 2002 and 2006)
(b) $t > 15.37$ (2015)

109. 106.864 in.^{2} $\le$ area $\le$ 109.464 in.^{2}

111. \$0.36 **113.** $13.7 < t < 17.5$

115. True by the Addition of a Constant Property of Inequalities.

117. False. If $-10 \le x \le 8$, then $10 \ge -x$ and $-x \ge -8$.

Appendix A.7 *(page A72)*

1. numerator

3. Change all signs when distributing the minus sign.
$2x - (3y + 4) = 2x - 3y - 4$

5. Change all signs when distributing the minus sign.
$$\frac{4}{16x-(2x+1)} = \frac{4}{14x-1}$$

7. z occurs twice as a factor. $(5z)(6z) = 30z^2$

9. The fraction as a whole is multiplied by a, not the numerator and denominator separately.
$$a\left(\frac{x}{y}\right) = \frac{ax}{y}$$

11. $\sqrt{x+9}$ cannot be simplified.

13. Divide out common factors, not common terms.
$\frac{2x^2+1}{5x}$ cannot be simplified.

15. To get rid of negative exponents:
$$\frac{1}{a^{-1}+b^{-1}} = \frac{1}{a^{-1}+b^{-1}} \cdot \frac{ab}{ab} = \frac{ab}{b+a}.$$

17. Factor within grouping symbols before applying exponent to each factor.
$$(x^2+5x)^{1/2} = [x(x+5)]^{1/2} = x^{1/2}(x+5)^{1/2}$$

19. To add fractions, first find a common denominator.
$$\frac{3}{x} + \frac{4}{y} = \frac{3y+4x}{xy}$$

21. To add fractions, first find a common denominator.
$$\frac{x}{2y} + \frac{y}{3} = \frac{3x+2y^2}{6y}$$

23. $5x+3$ **25.** $2x^2+x+15$ **27.** $\frac{1}{3}$ **29.** $3y-10$

31. 2 **33.** $\frac{1}{2x^2}$ **35.** $\frac{36}{25}, \frac{9}{4}$ **37.** 3, 4 **39.** $1-5x$

41. $1-7x$ **43.** $3x-1$ **45.** $7(x+3)^{-5}$

47. $2x^5(3x+5)^{-4}$ **49.** $\frac{4}{3}x^{-1}+4x^{-4}-7x(2x)^{-1/3}$

51. $\frac{x}{3}+2+\frac{4}{x}$ **53.** $4x^{8/3}-7x^{5/3}+\frac{1}{x^{1/3}}$

55. $\frac{3}{x^{1/2}}-5x^{3/2}-x^{7/2}$ **57.** $\frac{-7x^2-4x+9}{(x^2-3)^3(x+1)^4}$

59. $\frac{27x^2-24x+2}{(6x+1)^4}$ **61.** $\frac{-1}{(x+3)^{2/3}(x+2)^{7/4}}$

63. $\frac{4x-3}{(3x-1)^{4/3}}$ **65.** $\frac{x}{x^2+4}$

67. $\frac{(3x-2)^{1/2}(15x^2-4x+45)}{2(x^2+5)^{1/2}}$

69. (a)

x	0.50	1.0	1.5	2.0
t	1.70	1.72	1.78	1.89

x	2.5	3.0	3.5	4.0
t	2.02	2.18	2.36	2.57

(b) $x = 0.5$ mi

(c) $\frac{3x\sqrt{x^2-8x+20}+(x-4)\sqrt{x^2+4}}{6\sqrt{x^2+4}\sqrt{x^2-8x+20}}$

71. You cannot move term-by-term from the denominator to the numerator.

Index

O

P

Q

R

T

Definition of the Six Trigonometric Functions

Right triangle definitions, where $0 < \theta < \pi/2$

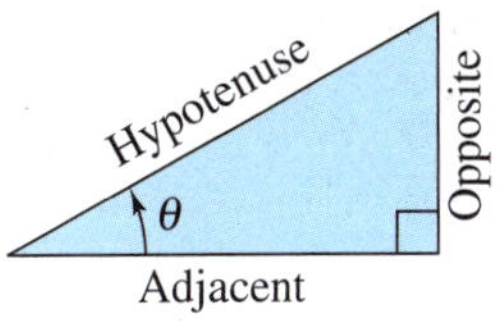

$$\sin\theta = \frac{\text{opp}}{\text{hyp}} \qquad \csc\theta = \frac{\text{hyp}}{\text{opp}}$$

$$\cos\theta = \frac{\text{adj}}{\text{hyp}} \qquad \sec\theta = \frac{\text{hyp}}{\text{adj}}$$

$$\tan\theta = \frac{\text{opp}}{\text{adj}} \qquad \cot\theta = \frac{\text{adj}}{\text{opp}}$$

Circular function definitions, where θ *is any angle*

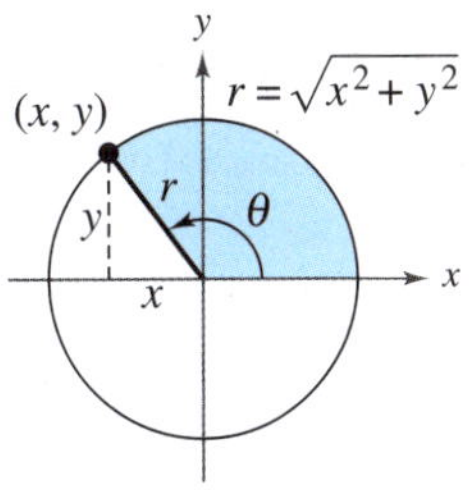

$$\sin\theta = \frac{y}{r} \qquad \csc\theta = \frac{r}{y}$$

$$\cos\theta = \frac{x}{r} \qquad \sec\theta = \frac{r}{x}$$

$$\tan\theta = \frac{y}{x} \qquad \cot\theta = \frac{x}{y}$$

Reciprocal Identities

$$\sin u = \frac{1}{\csc u} \qquad \cos u = \frac{1}{\sec u} \qquad \tan u = \frac{1}{\cot u}$$

$$\csc u = \frac{1}{\sin u} \qquad \sec u = \frac{1}{\cos u} \qquad \cot u = \frac{1}{\tan u}$$

Quotient Identities

$$\tan u = \frac{\sin u}{\cos u} \qquad \cot u = \frac{\cos u}{\sin u}$$

Pythagorean Identities

$$\sin^2 u + \cos^2 u = 1$$

$$1 + \tan^2 u = \sec^2 u \qquad 1 + \cot^2 u = \csc^2 u$$

Cofunction Identities

$$\sin\left(\frac{\pi}{2} - u\right) = \cos u \qquad \cot\left(\frac{\pi}{2} - u\right) = \tan u$$

$$\cos\left(\frac{\pi}{2} - u\right) = \sin u \qquad \sec\left(\frac{\pi}{2} - u\right) = \csc u$$

$$\tan\left(\frac{\pi}{2} - u\right) = \cot u \qquad \csc\left(\frac{\pi}{2} - u\right) = \sec u$$

Even/Odd Identities

$$\sin(-u) = -\sin u \qquad \cot(-u) = -\cot u$$

$$\cos(-u) = \cos u \qquad \sec(-u) = \sec u$$

$$\tan(-u) = -\tan u \qquad \csc(-u) = -\csc u$$

Sum and Difference Formulas

$$\sin(u \pm v) = \sin u \cos v \pm \cos u \sin v$$

$$\cos(u \pm v) = \cos u \cos v \mp \sin u \sin v$$

$$\tan(u \pm v) = \frac{\tan u \pm \tan v}{1 \mp \tan u \tan v}$$

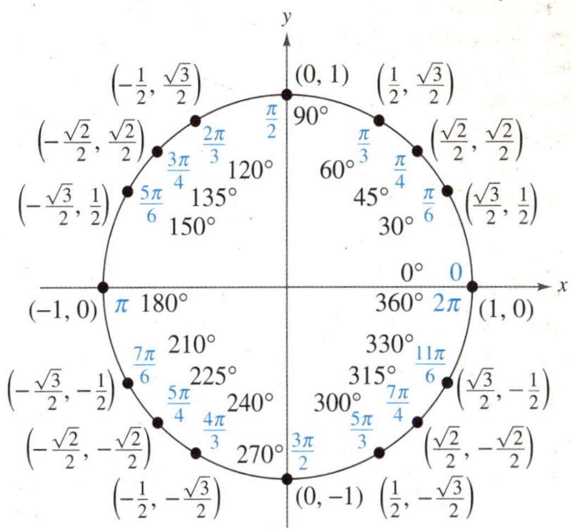

Double-Angle Formulas

$$\sin 2u = 2 \sin u \cos u$$

$$\cos 2u = \cos^2 u - \sin^2 u = 2\cos^2 u - 1 = 1 - 2\sin^2 u$$

$$\tan 2u = \frac{2\tan u}{1 - \tan^2 u}$$

Power-Reducing Formulas

$$\sin^2 u = \frac{1 - \cos 2u}{2}$$

$$\cos^2 u = \frac{1 + \cos 2u}{2}$$

$$\tan^2 u = \frac{1 - \cos 2u}{1 + \cos 2u}$$

Sum-to-Product Formulas

$$\sin u + \sin v = 2\sin\left(\frac{u+v}{2}\right)\cos\left(\frac{u-v}{2}\right)$$

$$\sin u - \sin v = 2\cos\left(\frac{u+v}{2}\right)\sin\left(\frac{u-v}{2}\right)$$

$$\cos u + \cos v = 2\cos\left(\frac{u+v}{2}\right)\cos\left(\frac{u-v}{2}\right)$$

$$\cos u - \cos v = -2\sin\left(\frac{u+v}{2}\right)\sin\left(\frac{u-v}{2}\right)$$

Product-to-Sum Formulas

$$\sin u \sin v = \frac{1}{2}[\cos(u - v) - \cos(u + v)]$$

$$\cos u \cos v = \frac{1}{2}[\cos(u - v) + \cos(u + v)]$$

$$\sin u \cos v = \frac{1}{2}[\sin(u + v) + \sin(u - v)]$$

$$\cos u \sin v = \frac{1}{2}[\sin(u + v) - \sin(u - v)]$$

FORMULAS FROM GEOMETRY

Triangle:

$h = a \sin \theta$

$\text{Area} = \frac{1}{2}bh$

$c^2 = a^2 + b^2 - 2ab \cos \theta$ (Law of Cosines)

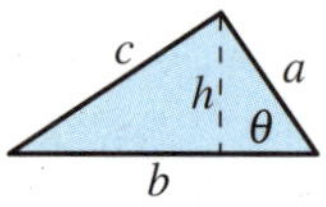

Right Triangle:

Pythagorean Theorem

$c^2 = a^2 + b^2$

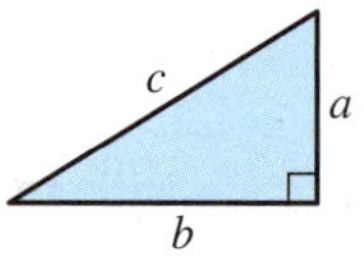

Equilateral Triangle:

$h = \frac{\sqrt{3}s}{2}$

$\text{Area} = \frac{\sqrt{3}s^2}{4}$

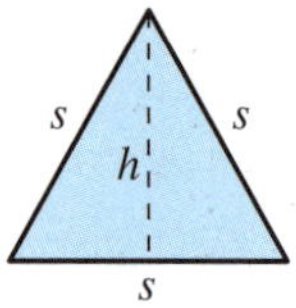

Parallelogram:

$\text{Area} = bh$

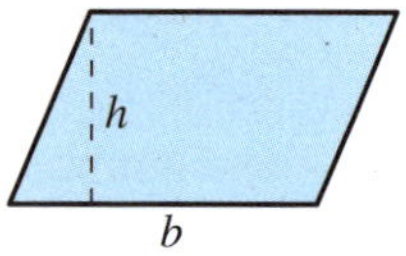

Trapezoid:

$\text{Area} = \frac{h}{2}(a + b)$

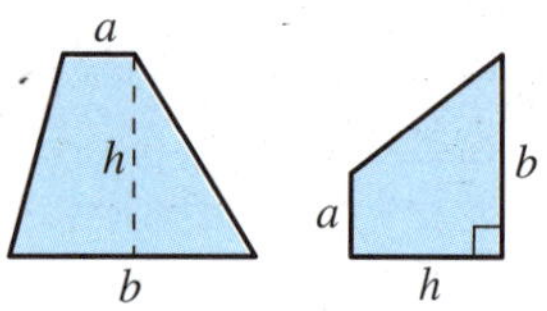

Circle:

$\text{Area} = \pi r^2$

$\text{Circumference} = 2\pi r$

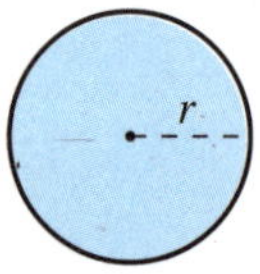

Sector of Circle:

$\text{Area} = \frac{\theta r^2}{2}$

$s = r\theta$

θ in radians

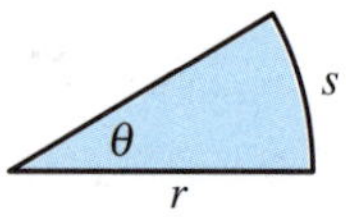

Circular Ring:

$\text{Area} = \pi(R^2 - r^2)$

$= 2\pi pw$

p = average radius,

w = width of ring

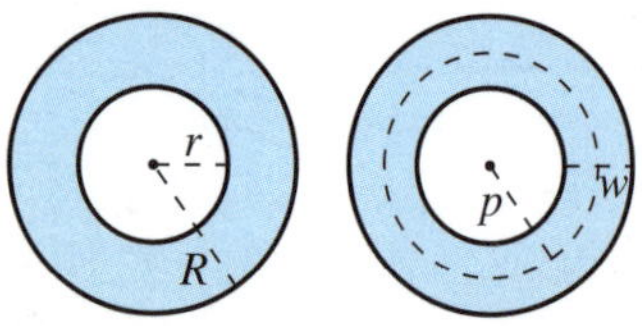

Sector of Circular Ring:

$\text{Area} = \theta pw$

p = average radius,

w = width of ring,

θ in radians

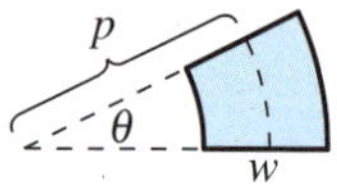

Ellipse:

$\text{Area} = \pi ab$

$\text{Circumference} \approx 2\pi\sqrt{\frac{a^2 + b^2}{2}}$

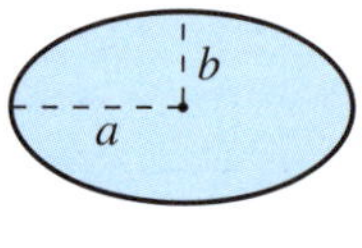

Cone:

$\text{Volume} = \frac{Ah}{3}$

A = area of base

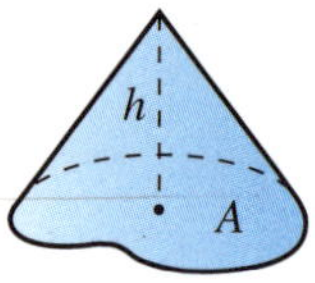

Right Circular Cone:

$\text{Volume} = \frac{\pi r^2 h}{3}$

$\text{Lateral Surface Area} = \pi r\sqrt{r^2 + h^2}$

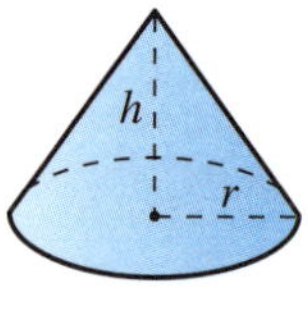

Frustum of Right Circular Cone:

$\text{Volume} = \frac{\pi(r^2 + rR + R^2)h}{3}$

$\text{Lateral Surface Area} = \pi s(R + r)$

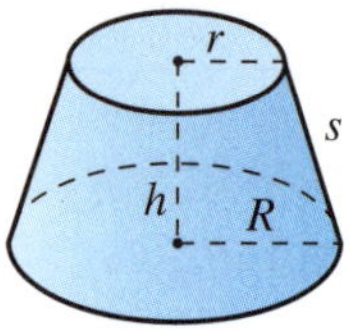

Right Circular Cylinder:

$\text{Volume} = \pi r^2 h$

$\text{Lateral Surface Area} = 2\pi rh$

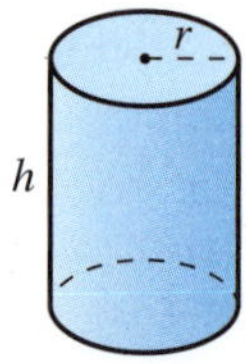

Sphere:

$\text{Volume} = \frac{4}{3}\pi r^3$

$\text{Surface Area} = 4\pi r^2$

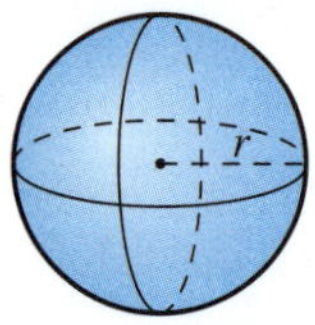

Wedge:

$A = B \sec \theta$

A = area of upper face,

B = area of base

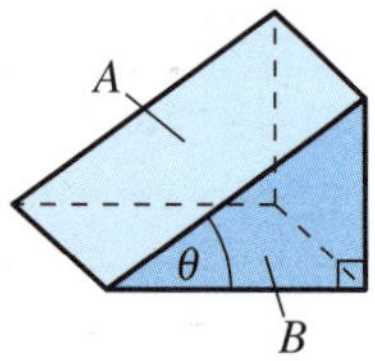